DIRECTORY OF GEOSCIENCE DEPARTMENTS

58th Edition

Christopher M. Keane, Editor

American Geosciences Institute
Alexandria, Virginia

american geosciences institute
connecting earth, science, and people

Directory of Geoscience Departments 2023, 58th Edition
Edited by Christopher M. Keane
ISBN-13: 978-0-922152-67-4
ISBN-10: 0-922152-67-5
ISSN: 0364-7811

Typeset in Times New Roman and Arial using Adobe InDesign CC.

Layout and programming: Christopher M. Keane

Advertising: John P. Rasanen

For more information on the American Geosciences Institute and its publications check us out at https://www.americangeosciences.org

Cover: The protostar within the dark cloud L1527, shown in this image from NASA's James Webb Space Telescope Near-Infrared Camera (NIRCam), is embedded within a cloud of material feeding its growth. Ejections from the star have cleared out cavities above and below it, whose boundaries glow orange and blue in this infrared view. The upper central region displays bubble-like shapes due to stellar "burps," or sporadic ejections. Webb also detects filaments made of molecular hydrogen that has been shocked by past stellar ejections. The edges of the cavities at upper left and lower right appear straight, while the boundaries at upper right and lower left are curved. The region at lower right appears blue, as there's less dust between it and Webb than the orange regions above it. Photo courtesy NASA, ESA, CSA, STScI.

INTRODUCTION

Thank you for purchasing the 58th edition of the Directory of Geoscience Departments. The American Geosciences Institute (AGI) is grateful to all the departments who updated their information and provided additional historical data of enrollments and awarded degrees for their department.

We have continued the enhancements introduced last year, including the expanded list of research specialties to better represent the various areas of research currently conducted in the geosciences, along with information about specific programs and certificates offered by individual programs, and indications of whether programs offered online or blended courses and degree or certificate programs.

This edition includes a listing of U.S. and Canadian student theses and dissertations from 2020 that have been reported to GeoRef Information Services. We would like to thank GeoRef for their diligent work compiling this list for use in this edition.

The data used to compile the directory is provided by the individuals and departments listed. AGI only edits the information for consistency and format. Any errors or omissions most likely reflect the entries of the individuals or the primary department contact. If you identify any issues, please email keane@americangoesciences.org with information about a current responsible contact for the department in question, so that we may further encourage updates to the next edition.

The American Geosciences Institute does not warrant the accuracy of any of the self-reported information, but we do encourage individuals and departments to update the material they believe is out-of-date.

Some basic statistics for the 58th edition of the Directory of Geoscience Departments: 1911 academic departments and programs globally, with 969 of those departments in the United States. Three hundred twenty-three departments in the United States are two-year institutions. There are 18,163 individuals globally identified in the Directory, with 13,304 in the United States and 1427 in Canada.

Christopher M. Keane

Editor

DIRECTORY OF GEOSCIENCE DEPARTMENTS
58th Edition 2023

TABLE OF CONTENTS

Directory of Geoscience Departments

U.S. Section 1

Non-U.S. Section 173

Theses and Dissertations 272

Geological Surveys 282

U.S. Federal Agencies and International Organizations 297

List of Faculty Specialty Codes 307

Faculty by Specialty Index 308

Faculty Index 416

USAGE KEY

Degrees Offered

A - Associate's or 2-year degree
B - Bachelor's or equivalent undergraduate degree
M - Master's Degree
D - Doctorate

Basic enrollment and degrees-granted data for the most recent year reported are shown below department information: degree level, enrollment, and degrees granted (in parentheses).

🗍 indicates thesis and dissertations were provided.
● indicates department offers a field camp only for their majors.
○ indicates department offers a field camp with open enrollment.
☒ indicates department does not offer any online courses
▣ indicates department offers blended learning courses
🖱 indicates department offers fully online courses
☑ indicates department offers fully online programs or certificates

The letters following individual faculty listings indicates their research specialties. Capital letters indicate general area, lowercase letters indicate subspecialty. Two or more consecutive lowercase letters refer to multiple subspecialties in the same major focus area referred to by the preceding capital letter.

DEPARTMENTS AND FACULTIES

This section contains a global listing of academic geoscience departments. Universities and colleges in the United States are arranged alphabetically by state. Non-U.S. institutions are listed alphabetically by country following the U.S. section.

All data is as reported by the departments and faculty themselves and current as of April 4, 2023. If you are aware of updates, corrections, or other means to ensure continued improvement of this listing, please email dgd@americangeosciences.org with the appropriate information so that it can be addressed during the next revision.

Alabama

Alabama A&M University

Dept of Biological and Environmental Sciences (A,B,M,D)
P.O. Box 1208
Normal, AL 35762
p. (256) 372-4214
wubishet.tadesse@aamu.edu
http://www.aamu.edu

Chair:
Anthony Overton

Professor:
Tommy L. Coleman, (D), Iowa State, 1980, So
Florence A. Okafor, (D), Nigeria, 1995, ZnEn
Govind Sharma, (D), Kansas State, 1970, Zn
James W. Shuford, (D), Penn State, 1975, So

Associate Professor:
Monday O. Mbila, (D), Iowa State, 2000, Sd

Secretary:
Martha Palmer

Auburn University

Dept of Crop, Soil & Environmental Sciences (B,M,D) ☒
201 Funchess Hall
Auburn University, AL 36849
p. (334) 844-4100
jpb0035@auburn.edu
Programs: Crop and Soil Sciences (B.S.)
Crop, Soil & Environmental Sciences (M.S., M.Ag., Ph.D.)
Enrollment (2006): M: 8 (2) D: 6 (3)

Professor:
Yucheng Feng, (D), Penn State, 1995, SbCb
Elizabeth A. Guertal, (D), Oklahoma State, 1993, Sc
Joey N. Shaw, (D), Georgia, 1997, Sd

Associate Professor:
Gobena E. Huluka, (D), Auburn, 1990, Sc

Assistant Professor:
Audrey V. Gamble, (D), Delaware, 2017, Sc
Thorsten J. Knappenberger, (D), Hohenheim (Germany), 2009, Sp
Rishi Prasad, (D), Florida, 2014, Sb
Di Tian, (D), Florida, 2014, At
Matthew N. Waters, (D), N Carolina, 2007, Gn

Dept of Geosciences (B,M) ●⊘
2050 Beard Eaves Coliseum
Auburn, AL 36849
p. (334) 844-4282
apr0022@auburn.edu
http://www.auburn.edu/cosam/departments/geosciences
Enrollment (2016): B: 62 (17) M: 30 (8)

Chair:
Ming-Kuo Lee, (D), Illinois, 1993, HwGe

Professor:
Willis E. Hames, (D), Virginia Tech, 1990, CcGpt
David T. King, Jr., (D), Missouri, 1980, GrZn
Charles E. Savrda, (D), S California, 1986, GdPeGr
Ashraf Uddin, (D), Florida State, 1996, GdoGt
Lorraine W. Wolf, (D), Alaska (Fairbanks), 1989, Ysg
Haibo Zou, (D), Florida State, 1999, CcGiv

Associate Professor:
Phil L. Chaney, (D), Louisiana State, 1999, ZnnZn
Ronald D. Lewis, (D), Texas, 1982, Pi
Luke J. Marzen, (D), Kansas State, 2001, Zir
Karen S. McNeal, (D), Texas A&M, 2007, ZeGeCm
Martin A. Medina Elizalde, (D), California (Santa Barbara), 2007, PeHyCl

Assistant Professor:
Christopher G. Burton, (D), South Carolina, 2012, Zi
Chandana Mitra, (D), Georgia, 2011, ZgAsZi
Stephanie L. Shepherd, (D), Arkansas, 2010, GmZi

Lecturer:
Carmen P. Brysch, (D), Texas State, 2014
John F. Hawkins, (M), Auburn, 2013, GcZeGt

Emeritus:
Robert B. Cook, (D), Georgia, 1971, Eg
Cyrus B. Dawsey, (D), Florida, 1975, Zy

On Leave:
Mark G. Steltenpohl, (D), N Carolina, 1985, Gtc

Dauphin Island Sea Lab

Marine Science Program (M,D)
P.O. Box 369
101 Bienville Boulevard
Dauphin Island, AL 36528
p. (334) 861-7528
langelo@disl.org
http://www.disl.org/aboutus.html
Department Secretary: Carolyn F. Wood

Director:
William W. Schroeder, (D), Texas A&M, 1971, Og

Professor:
Thomas S. Hopkins, (D), California (San Diego), 1967, Ob

Associate Professor:
Jonathan R. Pennock, (D), Delaware, 1983, Oc

Librarian:
Dennis Patronas, Zn

University of Alabama

Dept of Geological Sciences (B,M,D) O☒
Box 870338
201 7th Avenue
Room 2003 Bevill Bldg.
Tuscaloosa, AL 35487-0338
p. (205) 348-5095
geology@geo.ua.edu
http://www.geo.ua.edu
t: @UA_GEOLOGY
Programs: Geology
Enrollment (2022): B: 70 (11) M: 40 (11) D: 33 (2)

Chair:
Delores Robinson, (D), Arizona, 2001, Gtc

Professor:
C. Fred T. Andrus, (D), Georgia, 2000, PcGaCs
Ibrahim Çemen, (D), Penn State, 1983, Gct
Rona J. Donahoe, (D), Stanford, 1984, Cl
Samantha E. Hansen, (D), California (Santa Cruz), 2007, YsGt
Alberto Perez-Huerta, (D), Oregon, 2004, PgClGz
Harold H. Stowell, (D), Princeton, 1987, GpCcGt
Geoffrey Tick, (D), Arizona, 2003, Hw
Yong Zhang, (D), Nanjing (China), 1998, Hqw

Associate Dean of the Graduate School:
Andrew Goodliffe, (D), Hawaii, 1998, YgeGt

Associate Professor:
Natasha T Dimova, (D), Florida State, 2010, HwCcm
Kimberly Genareau, (D), Arizona State, 2009, Gvi
Yuehan Lu, (D), Michigan, 2008, GgOuGe
Tom S. Tobin, (D), Washington, 2014, PsCsPe
Bo Zhang, (D), Oklahoma, 2014, YeGoEo

Assistant Professor:
Julia Cartwright, (D), Manchester (UK), 2010, XmgCc
Marcello Minzoni, Kansas, 2007, GosEo
Alain Plattner, (D), ETH (Switzerland), 2011, YmuYg
Rebecca Totten, (D), Rice, 2015, GsPme

ALASKA

University of South Alabama

Dept of Earth Sciences (B) ○

5871 USA Dr. N.
Room 136
Mobile, AL 36688-0002
p. (251) 460-6381
dbeebe@southalabama.edu
http://www.usouthal.edu/earthsci
Programs: Geology (B); Geography (B); Meteorology (B); GIS (m)
Certificates: GIS
Enrollment (2017): B: 20 (20)

Chair:
Sytske K. Kimball, (D), Penn State, 2000, Am

Professor:
Roy Ryder, (D), Florida, 1989, SdZry

Associate Professor:
David T. Allison, (D), Florida State, 1992, Gc
Murlene W. Clark, (D), Florida State, 1983, PiGrg
Carol F. Sawyer, (D), Texas State, 2007, GmZye

Assistant Professor:
Alex Beebe, (D), Clemson, 2013, Ge
William T. Jackson, (D), Alabama, 2017, GsCc
Steven R. Schultze, (D), Michigan State, 2015, At
Wes Terwey, (D), Colorado State, 2007, Am

Instructor:
Karen J. Jordan, (M), Alabama, 2005, Zyr
Andrew Murray, (M), Florida State, 2010, Am
Diana Sturm, (D), Alabama, 2000, GeZge

Emeritus:
Keith G. Blackwell, (D), Texas A&M, 1990, Am
Douglas W. Haywick, (D), James Cook, 1990, Gs
Aaron Williams, (D), Oklahoma, 1971, Am

Dept of Marine Sciences (M,D)

307 University Drive
Mobile, AL 36688-0002
p. (334) 460-7136
keyser@southalabama.edu
Administrative Assistant: Jan Keyser

Chair:
Sean Powers, (D), Ob

Professor:
Ronald P. Kiene, (D), SUNY (Stony Brook), 1986, Oc

Adjunct Professor:
Douglas W. Haywick, (D), James Cook, 1990, Gs
Erich M. Mueller, (D), Miami, 1983, Ob

University of West Alabama

Dept of Biological & Environmental Sciences (B)

UWA Station 7
Livingston, AL 35470
p. (205) 652-3416
arindsberg@uwa.edu
http://www.uwa.edu/Biological_and_Environmental_Sciences.aspx
Programs: Environmental Science

Director, Black Belt Museum:
James P. Lamb, (B), Alabama, PvePg

Professor:
Andrew K. Rindsberg, (D), Colorado Mines, 1986, PeGePi

Alaska

University of Alaska Fairbanks

Dept of Civil, Geological, and Environmental Engineering (B,M,D) ● ☒

P.O. Box 755900
Fairbanks, AK 99775-5900
p. (907) 474-7241
mmdarrow@alaska.edu
https://www.uaf.edu/cem/programs/geological-engineering/
Programs: Geological Engineering
Enrollment (2021): B: 20 (7) M: 4 (1)

Professor:
Margaret M. Darrow, (D), Alaska, 2008, NxZnn

Assistant Professor:
Shishay Kidanu, (D), Yu

University of Alaska, Anchorage

Dept of Geological Sciences (B) ●

3211 Providence Drive
CPISB 101
Anchorage, AK 99508-4670
p. (907) 786-1298
jaschoff@alaska.edu
https://www.uaa.alaska.edu/academics/college-of-arts-and-sciences/departments/geology/
f: www.facebook.com/UAAGeologicalSciences
t: @UAAGeology
Enrollment (2015): B: 60 (8)

Chair:
Kristine J. Crossen, (D), Washington, 1997, Gl

Professor:
LeeAnn Munk, (D), Ohio State, 2001, Cle

Associate Professor:
Jennifer Aschoff, (D), Colorado Mines, 2008, GrdGo
Donald Matt Reeves, (D), Nevada, 2006, GeHgZn

Term Assistant Professor:
Terry R. Naumann, (D), Idaho, 1998, Gi

Assistant Professor:
Erin Shea, (D), MIT, 2014, CcGcCg

Term Instructor:
Peter J. Oswald, (M), Idaho, Gg
Mark Rivera, (M), New Mexico State, 2000, Gg

University of Alaska, Fairbanks

Alaska Quaternary Center

907 Yukon Drive
Fairbanks, AK 99775-1200
p. (907) 474-7758
nhbigelow@alaska.edu
https://www.uaf.edu/aqc/
Administrative Assistant: Leicha Welton

Professor:
W. Scott Ambruster, (D), California (Davis), 1981, Zn
Bruce P. Finney, (D), Oregon State, 1987, Ou
David M. Hopkins, (D), Harvard, 1955, Gg
Roger W. Powers, (D), Wisconsin, 1973, Zn
Dan L. Wetzel, (B), Alaska (Fairbanks), 1975, Zn

Research Associate:
Wendy H. Arundale, (D), Michigan State, 1976, Zn
Peter Bowers, (M), Washington State, 1980, Ga
Owen Mason, (D), Alaska, 1990, Ga

Adjunct Professor:
Daniel H. Mann, (D), Washington, 1983, Pe

Geology Librarian:
Judy Triplehorn, Zn

Dept of Civil & Environmental Engineering

Fairbanks, AK 99775-5900
jlhulsey@alaska.edu

Dept of Geosciences (B,M,D) ○

1930 Yukon Drive
P.O. Box 755780
Fairbanks, AK 99775-5780
p. (907) 474-7565
pjmccarthy@alaska.edu
http://www.uaf.edu/geology/
Programs: Geoscience; Geology; Geophysics; Earth Science; Geography
Enrollment (2019): B: 94 (10) M: 44 (10) D: 34 (5)

Chair:
Paul McCarthy, (D), Guelph, 1995, GsSaPc

Professor:
Bernard Coakley, (D), Columbia, 1991, Yg
Patrick S. Druckenmiller, (D), Calgary, 2006, Pv
Sarah J. Fowell, (D), Columbia, 1994, Pl
Regine M. Hock, (D), Swiss Fed Inst Tech, 1997, Gl
Jessica F. Larsen, (D), California (Santa Cruz), 1996, Gv
Franz J. Meyer, (D), Tech (Munich), 2004, Zr
Rainer J. Newberry, (D), Stanford, 1980, Em
Vladimir Romanovsky, (D), Alaska (Fairbanks), 1996, Yg
Michael T. Whalen, (D), Syracuse, 1993, Gs

Associate Professor:
Cary de Wit, (D), Kansas, 1997, Zg

Ronni Grapenthin, (D), Yd
Chris Maio, (D), Massachusetts (Boston), 2014, Zy
Dan Mann, (D), Washington, 1983, Zy
Elisabeth S. Nadin, (D), Caltech, 2007, Gtc
Carl Tape, (D), Caltech, 2009, Ys

Assistant Professor:
Sean Regan, (D), Gzx

Instructor:
Jochen E. Mezger, (D), Alberta, 1997, GcpCc

Emeritus:
James E. Beget, (D), Washington, 1981, Gm
Douglas Christensen, (D), Michigan, 1987, Ys
Catherine L. Hanks, (D), Alaska (Fairbanks), 1991, Go
Mary J. Keskinen, (D), Stanford, 1979, Gxz
Kenneth P. Severin, (D), California (Davis), 1987
David B. Stone, (D), Newcastle, 1963, Ym
Donald M. Triplehorn, (D), Illinois, 1961, Gs

Dept of Petroleum Engineering
Fairbanks, AK 99775-5880
adandekar@alaska.edu

University of Alaska Museum Earth Science (B,M,D) O☒
Box 756960
Fairbanks, AK 99775-6960
p. (907) 474-7505
psdruckenmiller@alaska.edu
http://www.uaf.edu/museum/collections/earth/
Programs: Geology
Enrollment (2020): M: 2 (1) D: 0 (1)

Curator of Earth Science:
Patrick S. Druckenmiller, (D), Calgary, 2006, Pv

University of Alaska, Southeast

Dept of Natural Sciences ⊘
11066 Auke Lake Way
Juneau, AK 99801
p. (907) 796-6580
sanagorski@alaska.edu
https://www.uas.alaska.edu/arts_sciences/naturalsciences/
Programs: Environmental Science (B), Environmental Resources (B), Environmental Studies (B)

Assistant Professor:
Sonia Nagorski, (D), Montana, 2001, GgZcCt

Arizona

Arizona State University

School of Earth and Space Exploration (B,M,D) ●
Box 871404
Tempe, AZ 85287-1404
p. (480) 965-5081
seseinfo@asu.edu
https://sese.asu.edu
f: https://www.facebook.com/SESE.at.ASU
t: @SESEASU
Programs: Earth and Environmental Studies,Earth and Space Exploration, Astrobiology and Biogeosciences, Astrophysics, Exploration Systems Design, Geological Sciences
Enrollment (2021): B: 72 (38) M: 12 (2) D: 66 (0)

Research Professor:
Laurence Garvie, (D), Bristol, 1991, GzXc
Amy Jurewicz, (D), Stanford, 1986, Xm
Lynda Williams, (D), Calgary, 2000, Gb
Mikhail Y. Zolotov, (D), Vernadsky Inst (Russia), 1990, Cg

Professor of Practice:
Catherine Coleman, (D), Massachusetts, 1991, Ze

Director:
Meenakshi Wadhwa, (D), Washington (St Louis), 1994, Xc

Professor:
Ariel D. Anbar, (D), Caltech, 1996, CbsXb
Ramon Arrowsmith, (D), Stanford, 1995, GcmGt
Gregory Asner, (D), Colorado, 1997, ZrGmSb
James Bell, (D), Hawaii, 1992, Xg
Judd Bowman, (D), MIT, 2007, Xa
Donald M. Burt, (D), Harvard, 1972, EgCgGz
Peter R. Buseck, (D), Columbia, 1962, Cg
Philip R. Christensen, (D), California (Los Angeles), 1981, Xg
Amanda Clarke, (D), Penn State, 2002, Gv
Steven Desch, (D), Illinois, 1998, Xa
Linda Elkins-Tanton, (D), MIT, 2002, Xg
Jack D. Farmer, (D), California (Davis), 1978, Pg
Edward Garnero, (D), Caltech, 1994, Ys
Christopher Groppi, (D), Arizona, 2003, XaZrn
Hilairy E. Hartnett, (D), Washington, 1998, ComOc
Arjun Heimsath, (D), California (Berkeley), 1999, Gc
Richard Hervig, (D), Chicago, 1979, Cg
Kip V. Hodges, (D), MIT, 1982, CcGtXg
Philip Mauskopf, (D), California (Berkeley), 1997, Xa
Stephen J. Reynolds, (D), Arizona, 1982, GctZe
Mark Robinson, (D), Hawaii, 1993, Xg
Evan Scannapieco, (D), California (Berkeley), 2001, Xa
Steven Semken, (D), MIT, 1989, ZeGgZg
Thomas G. Sharp, (D), Arizona State, 1990, Gz
Sang-Heon Shim, (D), Princeton, 2011, Gz
Everett Shock, (D), California (Berkeley), 1987, Cg
Sumner Starrfield, (D), California (Los Angeles), 1969, XaZnn
Francis Timmes, (D), California (Santa Cruz), 1992, Xa
Enrique Vivoni, (D), MIT, 2003, Hg
Kelin Whipple, (D), Washington, 1994, Gmt
Rogier Windhorst, (D), Leiden (Neth), 1984, Xa
Patrick Young, (D), Arizona, 2004, Xa

Research Professor:
Cassandra Bowman, (D), Harvard, 2008, Ze
Sean A. Bryan, (D), Case Western, 2014, Xa
Steven Ruff, (D), Arizona State, 1998, Xg
David Williams, (D), Alabama, 1998, Xg

Associate Professor:
Nathaniel Butler, (D), MIT, 2003, Xa
Jennifer Patience, (D), California (Los Angeles), 2000, Xa
Evgenya Shkolnik, (D), British Columbia, 2004, Xa
Heather Throop, (D), SUNY (Stony Brook), 2002, Cb
Christy Till, (D), MIT, 2011, Gi
Sara Walker, (D), Dartmouth, 2010, Xb

Research Professor:
Jnaneshwar Das, (D), S California, 2014, Zr
Duane DeVecchio, (D), California (Santa Barbara), 2009, GtmGc
Jordan Okie, (D), New Mexico, 2011, Cb
Devin Schrader, (D), Arizona, 2012, Xm

Assistant Professor:
Melanie Barboni, (D), Lusanne, 2011, Gi
Sanchayeeta Borthakur, (D), Massachusetts, 2010, Xa
Maitrayee Bose, (D), Washington (St Louis), 2011, Xc
Katrina Bossert, (D), Colorado, Xp
Craig Hardgrove, (D), Tennessee, 2011, Xg
Daniel Jacobs, (D), Pennsylvania, 2011, Xa
Mingming Li, (D), Arizona State, 2015, Yg
Michael Line, (D), Caltech, 2013, Xy
Allison Noble, (D), McGill, 2014, Xa
Joseph O'Rourke, (D), Caltech, Xy
Darryl Reano, (D), Purdue, 2019, Ze
Molly N. Simon, (D), Arizona, 2019, Xa
Elizabeth Trembath-Reichert, (D), Caltech, 2016, Cb
Alexander Van Engelen, (D), McGill, Xa

Research Professor:
Paul Scowen, (D), Rice, 1993, Xa

Emeritus:
L. Paul Knauth, (D), Caltech, 1979, Cs
Robert F. Lundin, (D), Illinois, 1962, Pm
Edmund Stump, (D), Ohio State, 1976, Gt
James A. Tyburczy, (D), Oregon, 1983, GyzYx
Stanley N. Williams, (D), Dartmouth, 1983, Gv

Arizona Western College

Dept of Geosciences (A)
2020 S. Avenue 8E
Yuma, AZ 85365
p. (928) 317-6000
fred.croxen@azwestern.edu
https://www.azwestern.edu/instruction/science
Programs: Geology (A); Geography (A); Applied Geology (A)
Certificates: GIS Technician, UAS Operator
Enrollment (2017): A: 1 (0)

Professor:
Fred W. Croxen III, (M), N Arizona, 1977, HwPvZi
Catherine R. Hill, (M), GgOgHw

Adjunct Professor:
Heather Casares, (M), Ball State, 2019, Gvi
Kelly L. Esslinger, (M), AsOpGg
Maureen Garrett, (M), Mississippi State, 2004, GgeOg
Todd Pinnt, (M), N Arizona, 2017, ZyiZg

Central Arizona College

Div of Science
Coolidge, AZ 85228
p. (602) 426-4444
diane.beecroft@centralaz.edu
Instructor:
Allan E. Morton, (M), Brigham Young, 1975, Ze

Cochise College

Geology Dept (A)
901 North Colombo Avenue
Sierra Vista, AZ 85635
p. (520) 515-5425
deakinj@cochise.edu
http://skywalker.cochise.edu/wellerr/aawellerweb.htm
Instructor:
Joann Deakin, (M), Ggg

Dine' College

Science Dept (A,B)
#1 Circle Drive
Tsaile, Navajo Nation, AZ 86556
p. (928) 724-6721
cacate@dinecollege.edu
http://www.dinecollege.edu
Chair:
Don Robinson, (D), Maharishi Mgmt
Instructor:
Margaret Mayer, (M), Rhode Island, GeZn

Mesa Community College

Dept of Physical Science (A)
1833 W. Southern Avenue
Mesa, AZ 85202
p. (480) 461-7015
kelli.santistevan@mesacc.edu
http://www.mesacc.edu/dept/d43/glg/
Professor:
Donna M. Benson, (M), Arizona State, 1996, Gg
Kaatje Kraft, (M), Arizona State, 1999, Gg
Robert S. Leighty, (D), Arizona State, 1997, Zg
Adjunct Professor:
Zack Bowles, (M), Arizona State, Gg
Chloe Branciforte, (M), SD Mines, Gg
Mike Grivois, (M), Texas Christian
Steve Guggino
Jack Kepper
Jill Lockard
Tony Occhiuzzi, (M), Arizona State, Zn
Donna Pollard, (M)
Joanna Scheffler, (M), Washington State, Gg
Melinda Shimizu, (M), Arizona State, 2008, Gg
Carolyn Taylor, (M), Georgia State, Gg
Kelli Wakefield, (M), Arizona State, Gg
Emeritus:
Armand J. Lombard, (M), Arizona, 1977, Gg

Northern Arizona University

Dept of Geography, Planning, and Recreation (B,M)
P.O. Box 15015
Flagstaff, AZ 86011
p. (928) 523-2650
geog@nau.edu
https://nau.edu/gpr/
f: https://www.facebook.com/NAUGPR
Programs: Parks and Recreation Mgmt, Planning, Geography
Certificates: GIS, Planning
Enrollment (2013): B: 39 (15) M: 44 (17)
Park Ranger Training Program Director:
Mark J. Maciha, (M), N Arizona, 2009, ZnnZn
Coordinator, Parks and Recreation Management Program:
Charles Hammersley, (D), New Mexico, 1988, Zn
Associate Professor:
Rebecca D. Hawley, (D), Arizona State, 1994, Zn
Ruihong Huang, (D), Wisconsin (Milwaukee), 2003
Associate Scientist:
Erik Schiefer, (D), British Columbia, 2004, GmZyi
Lecturer - Outdoor Leadership and Education:
Aaron Divine, (M), N Arizona, 2004, ZnnZn
Distance Learning Lecturer:
R. Marieke Taney, (M), N Arizona, 2002, ZnnZn
Senior Lecturer:
Judith Montoya, (M), New Mexico, 1985, Zy
Professor and Chair:
Pamela Foti, (D), Wisconsin, 1988, Zn
Emeritus:
Graydon L. Berlin, (D), Tennessee, 1970, Zr
Robert O. Clark, (D), Denver, 1970, Zn
Carolyn M. Daugherty, (D), Arizona State, 1987, Zn
Leland R. Dexter, (D), Colorado, 1986, Zy
Christina B. Kennedy, (D), Arizona, 1989, Zn
Alan A. Lew, (D), Oregon, 1986, Zgn
Stanley W. Swarts, (D), California (Los Angeles), 1975, Zy
George A. Van Otten, (D), Oregon State, 1977, Zn

Div of Geosciences, School of Earth and Sustainability (B,M,D) O
P O Box 4099
Building 12, Room 100
625 S Knoles Dr
Flagstaff, AZ 86011-4099
p. 928-523-9333
Clare.Aslan@nau.edu
https://nau.edu/ses/
Enrollment (2018): B: 120 (32) M: 26 (9)
Professor:
David K. Elliott, (D), Bristol, 1979, Pv
Thomas D. Hoisch, (D), S California, 1985, Gp
Darrell S. Kaufman, (D), Colorado, 1991, Zc
Michael H. Ort, (D), California (Santa Barbara), 1991, Gv
Roderic A. Parnell, (D), Dartmouth, 1982, Cl
Mary R. Reid, (D), MIT, 1987, CucGi
Nancy Riggs, (D), California (Santa Barbara), 1991, Gv
James C. Sample, (D), California (Santa Cruz), 1986, GtClOu
Abraham E. Springer, (D), Ohio State, 1994, Hw
Paul J. Umhoefer, (D), Washington, 1989, Gt
Associate Professor:
Laura Wasylenki, (D), Caltech, 1999, ClOoCs
Assistant Professor:
Nicholas McKay, (D), Arizona, 2012, PcZo
Ryan Porter, (D), Arizona, 2011, YgsGt
Michael E. Smith, (D), Wisconsin, 2007, Gsr
Research Associate:
Joseph E. Hazel, Jr., (M), N Arizona, 1991, Gs
Matthew A. Kaplinski, (M), N Arizona, 1990, Yg
Lecturer:
Lisa Skinner, (M), ZeGvZi
Kelsey Winsor, (D), Wisconsin, 2014, PcGm
Adjunct Professor:
Wendell A. Duffield, (D), Stanford, 1967, Gv
David D. Gillette, (D), S Methodist, 1974, Pv
James Wittke, (D), Texas, 1984, Gi
Emeritus:
Charles W. Barnes, (D), Wisconsin, 1965, GcXg
Ronald C. Blakey, (D), Iowa, 1973, Gr
David S. Brumbaugh, (D), Indiana, 1972, Ys
Ernest M. Duebendorfer, (D), Wyoming, 1986, Gct
Richard F. Holm, (D), Washington, 1969, Gv
James I. Mead, (D), Arizona, 1983, Pg
Larry T. Middleton, (D), Wyoming, 1979, Gs
Paul Morgan, (D), London (UK), 1973, Yh
Business Manager:
Sarah Colby, Zn

Phoenix College

Dept of Physical Sciences (A)
1202 West Thomas Road
Phoenix, AZ 85013

p. (602) 285-7244
abeer.hamdan@phoenixcollege.edu
Chair:
James J. White, (D), Arizona, 1986
Professor:
Richard Cups, (M), Arizona State, 1983, Gg
Don Speed, (M), Arizona, 1980, Gg

Pima Community College - Community Campus

Geology Dept
401 North Bonita Ave.
Tucson, AZ 85709-5000
p. (520) 206-4500
wrcavanaugh@pima.edu
http://www.pima.edu

Pima Community College - West Campus

Geology Dept
2202 West Anklam Rd.
Tucson, AZ 85709-0001
p. (520) 206-4500
jrasmussen@pima.edu
http://www.pima.edu

Prescott College

Dept of Environmental Studies (B)
220 Grove Avenue
Prescott, AZ 86301
p. (928) 350-2256
eboyer@prescott.edu
http://www.prescott.edu
Professor:
Kurt Refsnider, (D), Colorado, 2012, GmlPe

Scottsdale Community College

Dept of Mathematics and Sciences
9000 East Chaparral Road
Scottsdale, AZ 85256-2626
p. (480) 423-6111
merry.wilson@scottsdalecc.edu
http://www.scottsdalecc.edu/academics/departments/math-sciences

South Mountain Community College

Geology Program
7050 South 24th Street
Phoenix, AZ 85042-5806
p. (602) 243-8290
latricia.maddox@southmountaincc.edu
http://www.southmountaincc.edu/math-science-engineering/program/geology/

University of Arizona

Dept of Geosciences (B,M,D) O
Gould-Simpson Building
1040 E. Fourth Street
Tucson, AZ 85721-0077
p. (520) 621-6000
bcarrapa@email.arizona.edu
http://www.geo.arizona.edu
Administrative Assistant: Anne Chase, Sylvia Quintero
Enrollment (2006): B: 70 (0) M: 23 (0) D: 52 (0)
Head:
Peter Reiners, (D), Washington, Cg
Professor:
Victor R. Baker, (D), Colorado, 1971, GmXgRh
Mark D. Barton, (D), Chicago, 1981, Eg
Susan L. Beck, (D), Michigan, 1987, YsGt
Richard Bennett, (D), MIT, 1995, YdGt
Jon Chorover, California (Berkeley), 1993, Sc
Andrew S. Cohen, (D), California (Davis), 1982, Ps
Julia E. Cole, (D), Columbia, 1992, PeCsAs
Owen K. Davis, (D), Minnesota, 1981, Pl
Peter G. DeCelles, (D), Indiana, 1984, GtsGc
Robert T. Downs, (D), Virginia Tech, 1992, Gi
Mihai N. Ducea, (D), Caltech, 1998, Gt
Karl W. Flessa, (D), Brown, 1973, PgeGe
Jiba Ganguly, (D), Chicago, 1967, Gx
George E. Gehrels, (D), Caltech, 1986, Gt
Vance T. Holliday, (D), Colorado, 1982, GamSa
Roy A. Johnson, (D), Wyoming, 1984, Ys
A. J. Timothy Jull, (D), Bristol, 1976, CclCg
Paul A. Kapp, (D), California (Los Angeles), 2001, GtcGm
Alfred S. McEwen, (D), Arizona State, 1988, XgGmv
Jonathan T. Overpeck, (D), Brown, 1985, As
Jon D. Pelletier, (D), Cornell, 1997, Gm
Mary Poulton, (D), Arizona, 1990, Ng
Jay Quade, (D), Utah, 1990, Sc
Randall M. Richardson, (D), MIT, 1978, Yg
Joaquin Ruiz, (D), Michigan, 1983, CgcEz
Timothy Swindle, (D), Washington, 1986, Xc
Connie Woodhouse, (D), Arizona, 1996, Pe
Associate Specialist:
Michael A. Crimmins, (D), Arizona, 2004, Zc
Associate Professor:
Barbara Carrapa, (D), Vrije (Amsterdam), 2002, Gs
Joellen Russell, (D), Scripps, 1999, Oc
Eric Seedorff, (D), Stanford, 1987, Eg
Jessica Tierney, (D), Brown, 2010, Co
Marek Zreda, (D), New Mexico Tech, 1994, Cc
Assistant Professor:
Jennifer McIntosh, (D), Michigan, 2004, Hy
Jianjun Yin, (D), Illinois, Og
Lecturer:
Paul Goodman, Washington, 2000, As
Jessica Kapp, (D), California (Los Angeles), Gg
Adjunct Professor:
Robert J. Kamilli, (D), Harvard, 1976, EgmCg
Charles Prewitt, (D), MIT, 1969, Zn
Marc Sbar, (D), Columbia, 1972, Yes
Emeritus:
William B. Bull, (D), Stanford, 1960, Gm
Clement G. Chase, (D), California (San Diego), 1970, Yg
George H. Davis, (D), Michigan, 1971, Gc
William R. Dickinson, (D), Stanford, 1958, Gt
John M. Guilbert, (D), Wisconsin, 1962, Em
DeVerle P. Harris, (D), Penn State, 1965, Eg
C. Vance Haynes, Jr., (D), Arizona, 1965, Ga
Everett H. Lindsay, (D), California (Berkeley), 1967, Pv
Edgar J. McCullough, Jr., (D), Arizona, 1963, Ge
Jonathan Patchett, (D), Edinburgh, 1976
Joseph Schreiber, (D)
Spencer R. Titley, (D), Arizona, 1958, Eg
George Zandt, (D), MIT, 1978, Ys
Researcher:
David Dettman, (D), Michigan, 1994, Cs
Chris J. Eastoe, (D), Tasmania, 1979, CsGg

Dept of Hydrology & Atmospheric Sciences (B,M,D) O
122 Harshbarger Building #11
PO Box 210011
1133 E James E Rogers Way
Tucson, AZ 85721-0011
p. (520) 621-7120
tmeixner@email.arizona.edu
https://has.arizona.edu
f: https://www.facebook.com/UAzWaterWeatherClimate/
t: https://twitter.com/HAS_UArizona
Programs: Hydrology and Water Resources, Atmospheric Sciences
Certificates: Hydrology and Water Resources Grad Certificate
Administrative Assistant: Erma Santander
Enrollment (2014): B: 0 (2) M: 23 (2) D: 27 (1)
University Distinguished Scholar and Professor :
Ali Behrangi, (D), California (Irvine), 2009, As
Regents Professor:
Victor R. Baker, (D), Colorado, 1971, GmXgRh
Hoshin V. Gupta, (D), Case Western, 1984, HqZgAs
Professor:
Christopher Castro, (D), Colorado State, 2005, As
Xiquan Dong, (D), Penn State, 1996, As
Paul A. Ferre, (D), Waterloo, 1997, Hw
Jennifer C. McIntosh, (D), Michigan, 2004, Cg
Thomas Meixner, (D), Arizona, 1999, CgRw
Steven L. Mullen, (D), Washington, 1985, As

ARIZONA

Peter A. Troch, (D), Ghent, 1993, Hs
Tian-Chyi Jim Yeh, (D), New Mexico Tech, 1983, Hq
Marek G. Zreda, (D), New Mexico Tech, 1994, Hw
Associate Professor:
Arellano M. Avelino, (D), Duke, 2005
Guo-Yue Niu, (D), Chinese Acad Sci, 1996, Hg
Assistant Professor:
Laura Condon, (D), Colorado Mines, 2015, Hq
Bo Guo, (D), Princeton, 2016, Hw
Yang C. Song, (D), Illinois, 2015
Adjunct Associate Professor:
David C. Goodrich, (D), Arizona, 1990, Hs
Adjunct Professor:
Leo S. Leonhart, (D), Arizona, 1978, Hw
Regents Professor Emeritus:
Shlomo P. Neuman, (D), California (Berkeley), 1968, HwqGq
Professor Emeritus:
Thomas Maddock, III, (D), Harvard, 1972, Hq
Juan B. Valdes, (D), MIT, 1976, Hs
Distinguished Professor Emeritus:
Eric Betterton, (D), Witwatersrand, 1983, As
Cooperating Faculty:
Mark L. Brusseau, (D), Florida, 1989, Hw
Cooperating Faculty:
Jonathan D. Chorover, (D), California (Berkeley), 1993, Cg
Cooperating Faculty:
Bonnie C. Colby, (D), Wisconsin, 1983, Zn
Cooperating Faculty:
Kevin E. Lansey, (D), Texas, 1987, Hs
Cooperating Faculty:
Sharon B. Megdal, (D), Princeton, 1981, Hg
Cooperating Faculty:
Robert G. Varady, (D), Arizona, 1981, Zn

Dept of Mining & Geological Engineering (B,M,D)
1235 E. James E. Rogers Way
Mines Building, Room 229
PO Box 210012
Tucson, AZ 85721-0012
p. (520) 621-6063
ENGR-Mining@email.arizona.edu
http://www.mge.engineering.arizona.edu
f: https://www.facebook.com/UAMining/
Programs: Mining Engineering, Mining & Geological Engineering
Certificates: Mineral Processing and Extractive Metallurgy, Geomechanics and Rock Mechanics, Mine Production & Information Technology, Mining Occupational Health & Safety
Enrollment (2017): B: 71 (26) M: 24 (8) D: 11 (4)
Professor of Practice:
Victor O. Tenorio, (D), Arizona, 2012, Nmx
Professor:
Ben K. Sternberg, (D), Wisconsin, 1977, Ye
John M. Kemeny, (D), California (Berkeley), 1986, Nr
Interim Head:
Moe Momayez, (D), McGill, NmYxNr
Associate Professor:
Jaeheon Lee, (D), Arizona, 2004, Nx
Jinhong Zhang, (D), Virginia Tech, 2006, Nm
Assistant Professor:
Isabel Barton, (D), Arizona
Emeritus:
J. Brent Hiskey, (D), Utah, 1973, Nx

Dept of Soil, Water & Environmental Science (B,M,D) ☒
PO Box 210038
Shantz 429
Tucson, AZ 85721-0038
p. (520) 621-1646
wrhaley@email.arizona.edu
https://swes.cals.arizona.edu
Programs: Environmental Science (undergrad and grad), Sustainable Plant Systems (undergrad), Subsurface science (grad), Soil, plant, atmosphere systems (grad)
Head:
Jonathan Chorover, (D), California (Berkeley), 1993, ScClo
Associate Director, SAHRA Center:
Katharine L. Jacobs, (M), California (Berkeley), 1981, Hg
Professor:
Mark L. Brusseau, (D), Florida, 1989, Hw
Kevin Fitzsimmons, (D), Arizona, 1999, Zn
Charles P. Gerba, (D), Miami, 1973, Sb
Martha C. Hawes, (D)
Raina M. Maier, (D), Rutgers, 1988, Zn
Ian L. Pepper, (D), Ohio State, 1975, Zn
Craig Rasmussen, (D), SzGmCl
Charles Sanchez, (D), Iowa State, 1986, Zn
Jeffrey C. Silvertooth, (D), Oklahoma State, 1986, So
Markus Tuller, (D), Zn
James Walworth, (D), Georgia, 1985, Soc
Arthur W. Warrick, (D), Iowa State, 1967, Sp
Associate Professor:
Joan E. Curry, (D), California (Davis), 1992, Sc
Marcel Schaap, (D), So
Assistant Professor:
Karletta Chief, (D), Arizona, 2007, Zn
Associate Research Scientist:
Janick F. Artiola, (D), Arizona, 1980, ScHsCa
Associate Professor of Practice:
Thomas B. Wilson, (D), Zn
Adjunct Professor:
Floyd Adamsen, (D), Colorado State, 1983, So
Paul (Ty) Ferre, (D), So
Jim Yeh, (D), Sp
Emeritus:
Ed Glenn, (D), Hawaii, 1978, Zn
Donald F. Post, (D), Purdue, 1967, Sd
James Riley, (D), Arizona, 1968, Zn
Peter J. Wierenga, (D), California, 1968, Sp
Extension Specialist:
Paul W. Brown, (D), Wisconsin, 1981, Am

Laboratory of Tree Ring Research O
Bryant Bannister Tree-Ring Bldg
1215 E. Lowell St.
Tucson, AZ 85721
p. (520) 621-1608
webmaster@ltrr.arizona.edu
http://ltrr.arizona.edu/
f: https://www.facebook.com/pages/Laboratory-of-Tree-Ring-Research/112628272090047
t: @TreeRingLabUA
Certificates: Graduate Certificate in Dendrochronology
Professor:
David C. Frank, (D), Bern, 2005
Professor:
Malcolm K. Hughes, (D), Durham (UK), 1970, Pe
Steven W. Leavitt, (D), Arizona, 1982, Csc
Russ Monson, (D), Washington State, 1982
Associate Professor:
Bryan A. Black, (D), 2003
Paul Sheppard, (D), 1995
Ronald Towner, (D), Arizona, 1997, Gae
Valerie Trouet, (D), KU Leuven, 2004, PeZyAs
Assistant Professor:
Charlotte Pearson, (D), Reading (UK), 2003
Research Associate:
Matthew W. Salzer, (D), Arizona, 2000, PeGav
Christopher H. Guiterman, (D), Arizona, 2016
Kiyomi A. Morino, (D), Arizona, 2008
Emeritus:
Jeffrey S. Dean, (D), Arizona, 1967, Zn
Katherine K. Hirschboeck, (D), Arizona, 1985, AsHsZc
Charles W. Stockton, (D), Arizona, 1971, Hq
Thomas W. Swetnam, (D), Arizona, 1987, Pe
Research Professor:
David M. Meko, (D), Arizona, 1981, Hq
Ramzi Touchan, (D), Arizona, 1991, PcHsAs
Curator, Collections:
Peter Brewer, (D), Reading (UK), 2004
Associate Professor:
Pearce Paul Creasman, (D), Texas A&M, 2008
Assoc. Research Professor:
Irina Panyushkina, (D), Sukachev Inst Forest, 1997
Assistant Research Professor:
Margaret Evans, (D), Arizona, 2003

Lunar and Planetary Laboratory (M,D)
1629 E. University Boulevard

Tucson, AZ 85721-0092
p. (520) 621-6963
marley@lpl.arizona.edu
http://www.lpl.arizona.edu

Head:
Michael J. Drake, (D), Oregon, 1972, Xc

Research Professor:
Martin G. Tomasko, (D), Princeton, 1969, As

Professor:
Erik I. Asphaug, (D), Arizona, 1993, XygXa
Victor R. Baker, (D), Colorado, 1971, XgGmRh
William V. Boynton, (D), Carnegie Mellon, 1971, XcZr
Robert H. Brown, (D), Hawaii, 1982, Xc
Uwe Fink, (D), Penn State, 1965, As
Tom Gehrels, (D), Chicago, 1956, Zn
Richard J. Greenberg, (D), MIT, 1972, Xy
Donald M. Hunten, (D), McGill, 1950, As
Jack R. Jokipii, (D), Caltech, 1965, Xy
Harold P. Larson, (D), Purdue, 1967, Zn
Dante Lauretta, (D), Washington (St Louis), 1997, XcmXg
Jonathan I. Lunine, (D), Caltech, 1985, Xc
Renu Malhotra, (D), Cornell, 1988, Xs
Alfred McEwen, (D), Arizona State, 1988, Xg
Henry J. Melosh, (D), Caltech, 1972, Gt
George H. Rieke, (D), Harvard, 1969, Zn
Timothy D. Swindle, (D), Washington (St Louis), 1986, XmCc
Roger Yelle, (D), Wisconsin, 1984, Zn

Senior Research Scientist:
Lonnie L. Hood, (D), California (Los Angeles), 1979, XyAs
Larry A. Lebofsky, (D), MIT, 1974, ZngXm

Senior Scientist:
Jozsef Kota, (D), Roland Eotvos, 1980, As
David Kring, (D), Harvard, 1989, XcGiXg

Associate Professor:
Caitlin Griffith, (D), SUNY (Stony Brook), 1991, Zn
Robert Kursinski, (D), Caltech, 1997, As

Associate Research Scientist:
Robert S. McMillan, (D), Texas, 1977, ZnnZn

Assistant Professor:
Joe Giacalone, (D), Kansas, 1991, Xy
Adam Showman, (D), Caltech, 1998, Zn

Senior Research Scientist:
Lyle A. Broadfoot, (D), Saskatchewan, 1963, Zr
Jay B. Holberg, (D), California (Berkeley), 1974, As
Bill R. Sandel, (D), Rice, 1972, As

Research Associate:
Alexander Dessler, (D), Duke, 1956, Zn
Ralph Lorenz, (D), Kent (UK), 1994, Xg
Elizabeth Turtle, (D), Arizona, 1998, Xg

Associate Research Scientist:
Peter Smith, (M), Arizona, 1977, Zr

Emeritus:
William B. Hubbard, (D), California (Berkeley), 1967, XyYvAs
John S. Lewis, (D), California (San Diego), 1968, Xcm
Elizabeth Roemer, (D), California (Berkeley), 1955, Zn
Charles P. Sonett, (D), California (Los Angeles), 1954, Yg
Robert G. Strom, (M), Stanford, 1957, Xg

Graduate Secretary:
Mary Guerrieri, Zn

School of Geography and Development (B,M,D)
Harvill Building
PO Box 210076
Tucson, AZ 85721
p. (520) 621-1652
liverman@email.arizona.edu
http://geography.arizona.edu/
Enrollment (2010): B: 350 (0) M: 20 (0) D: 54 (0)

Director:
Paul Robbins, (D), Clark, 1996

Dean:
John Paul Jones, III, (D), Ohio State, 1984, Zn

Co-Director of the University of Arizona Institute of the Environment :
Diana Liverman, (D), California (Los Angeles), 1984

Professor:
Michael E. Bonine, (D), Texas, 1975, Zu
Andrew Comrie, (D), Pennsylvania, 1992, Zy
Sallie A. Marston, (D), Colorado, 1986, Zu
Beth A. Mitchneck, (D), Columbia, 1990, Zn
David A. Plane, (D), Pennsylvania, 1981, Zn

Lecturer:
Dereka Rushbrook, (D), Arizona, 2005

Associate Professor:
Keiron D. Bailey, (D), Kentucky, 2002, Zi
Carl J. Bauer, (D), California (Berkeley), 1995, Hs
Sandy Dall'Erba, (D), Pau, 2004
Elizabeth Oglesby, (D), California (Berkeley), 2002, Zn
Willem van Leeuwen, (D), Arizona, 1995, Zy
Marvin Waterstone, (D), Rutgers, 1983, Zn
Margaret Wilder, (D), Arizona, 2002, Zn

Director of GIST:
Christopher Lukinbeal, (D), San Diego State, 2000, Zi

Assistant Professor:
Jeffrey Banister, (D), Arizona, 2010
Gary Christopherson, (D), Arizona, 2000, Zi
Sarah Moore, (D), Kentucky, 2006
Tracey Osborne, (D), California (Berkeley), 2010
Daoqin Tong, (D), Ohio State, 2007, Zi

Adjunct Professor:
Julio Betancourt, (D), Arizona, 1989, Zn
Vance Holliday, (D), Colorado, 1982, Ga
Laura Huntoon, (D), Pennsylvania, 1991, Zn
Charles F. Hutchinson, (D), California (Riverside), 1978, Zr
Miranda Joseph, (D), Stanford, 1995, Zn
Barbara Morehouse, (D), Arizona, 1993, Zn
Thomas W. Swetnam, (D), Arizona, 1987, Zn

Emeritus:
Stephen R. Yool, (D), California (Santa Barbara), 1985, Zri

Director, International Studies-External Affairs:
Wayne Decker, (D), Johns Hopkins, 1979

Arkansas

Arkansas Tech University

Dept of Physical Sciences - Geology (B) ⊘
1701 N. Boulder Ave
McEver Science Building
Russellville, AR 72801
p. (479) 968-0293
jpatton@atu.edu
http://www.atu.edu/geology
Programs: Professional Geology, Environmental Geology, Petroleum Geology
Enrollment (2021): B: 25 (6)

Professor:
Jason A. Patton, (D), Arkansas, 2008, GgeGc

Associate Professor:
Michael Davis

Assistant Professor:
Jacob Grosskopf, (D)
Jacob Siebach, (D)

Northwest Arkansas Community College

Northwest Arkansas Community College
One College Drive
Bentonville, AR 72712
p. (479) 631-8661
dandroes@nwacc.edu
http://www.nwacc.edu

University of Arkansas at Little Rock

School of Physical Sciences - Geology Program (B)
2801 South University Avenue
Little Rock, AR 72204-1099
p. (501) 916-3070
mtdeangelis@ualr.edu
http://www.ualr.edu/geology/
f: https://www.facebook.com/ualrearthsciences
Programs: Geology (B); Environmental Geology (B)
Enrollment (2021): B: 28 (5)

Professor:
Jeffrey B. Connelly, (D), Tennessee, 1993, GcNg

Associate Professor:
Michael T. DeAngelis, (D), Tennessee, 2011, GpCp

Assistant Professor:
Rene A. Shroat-Lewis, (D), Tennessee, 2011, PeZe

Instructor:
Joshua C. Spinler, (D), Arizona, 2014, YdGt

University of Arkansas, Fayetteville

Arkansas Water Resources Center
113 Ozark Hall
Fayetteville, AR 72701
p. (501) 575-4403
haggard@uark.edu
http://www.uark.edu/depts/awrc
Administrative Assistant: Theresa J. Russell

Director:
Kenneth F. Steele, (D), North Carolina, 1971, Cg
Research Associate:
Terry E. Nichols, (D), Chicago, 1973, Zn
Laboratory Director:
Marc A. Nelson, (D), Arkansas, 1992, Cl

Dept of Geosciences (B,M,D) O⊘
340 N. Campus Drive
216 Gearhart Hall
Fayetteville, AR 72701
p. (479) 575-3355
geosdept@uark.edu
http://fulbright.uark.edu/departments/geosciences/
f: https://www.facebook.com/uageosciences/0396/
Programs: Earth Science (B); Geology (B,M); Geography (B,M); Geosciences (D)
Certificates: Geospatial Technologies (UG,G)
Enrollment (2019): B: 121 (39) M: 51 (29) D: 20 (3)

Maurice Storm Endowed Chair:
Christopher Liner, (D), Colorado Mines, 1989, YseGg
University Professor:
William (. Limp, (D), Indiana, 1983, Zfi
Thomas R. Paradise, (D), Arizona State, 1993, GmZyRn
Distinguished Professor:
David W. Stahle, (D), Arizona State, 1990, GeZon
Daniel Sui, (D), Georgia, 1993, Zf
Director of the Center for Advanced Spatial Technologies:
Jackson D. Cothren, (D), Ohio State, 2004, Zir
Professor:
Stephen K. Boss, (D), North Carolina, 1994, RcYgGg
Jason A. Tullis, (D), South Carolina, 2003, ZriZf
Associate Professor:
Mohamed H. Aly, (D), Texas A&M, 2006, Gg
Matthew Covington, (D), California, 2008, HqGmg
Fiona M. Davidson, (D), Nebraska, 1991, Zy
Gregory Dumond, (D), Massachusetts, 2008, GtcGg
Song Feng, (D), Chinese Acad of Sci, 1999, AsZyAm
Adriana Potra, (D), Florida Intl, 2011, EgCcGg
John B. Shaw, (D), Texas, 2013, GsrGg
Celina Suarez, (D), Kansas, 2010, ClPqGg
Assistant Professor:
Linyin Cheng, (D), California (Irvine), Zo
Edward C. Holland, (D), Colorado, Zy
Andrew P. Lamb, (D), Boise State, 2013, YumYv
Jill A. Marshall, (D), Oregon, 2015, Gm
Glenn R. Sharman, (D), Stanford, 2014, Go
Research Professor:
Phillip D. Hays, (D), Texas A&M, 1996, HwCsGg
Research Associate:
Barry J. Shaulis, (D), Houston, 2013
Instructor:
Paula Anderson, (M), Arkansas, 2012, Gg
Rashauna Hintz, (M), Arkansas, Zu
Henry Turner, III, (D), Arkansas, 2010, GgtGc
Adjunct Professor:
Fred Paillet, (D), Rochester, 1974, YgHw
Emeritus:
John Van Brahana, (D), Missouri, 1973, HwGe
Malcolm K. Cleaveland, (D), Arizona, 1983, PcZc
John C. Dixon, (D), Colorado, 1983, GmXgSo
Margaret J. Guccione, (D), Colorado, 1982, GmaGg
Walter Manger, (D), Iowa, Gr
Richard Smith, N Colorado, Gm
Doy L. Zachry, (D), Texas, 1969, Gr

California

American River College

Dept of Earth Science (A)
4700 College Oak Drive
Sacramento, CA 95841
p. (916) 484-8107
aubertj@arc.losrios.edu
http://arc.losrios.edu/~earthsci/

Chair:
Melissa H. Levy, (M), E Tennessee State, 1994, GgZgy
Professor:
John E. Aubert, (M), California (Davis), 1994, Zy
Charles E. Thomsen, (M), Cal State (Chico), 1994, Zy
GIS Coordinator:
Hugh H. Howard, (D), Kansas, 2003, Zi
Assistant Professor:
Glenn Jaecks, (D), California (Davis), 2002, GgOgPg
Adjunct:
Paul M. Veisze, (M), California (Berkeley), 1985, Zi
Adjunct Professor:
Terry J. Boroughs, (M), Ohio State, 1992, GgXgCg
Beth Dushman, (M), California (Davis), 2007, Gg
Robert Earle, (M), Zi
Nathan Jennings, (M), Zi
Tom Lupo, (M), Zi
Richard L. Oldham, (M), Nevada, 1972, Gg
Kimberly Olson, (M), Cal State (Chico), 2008, Zye
Steven C. Smith, (M), Cal State (Chico), 1994, ZynZn

Antelope Valley College

Div-Geosciences Prog
3041 West Avenue K
Lancaster, CA 93536
mpesses@avc.edu
https://www.avc.edu/sites/default/files/catalog/geosciences.pdf

Associate Professor:
Michael Pesses, Zyi
Assistant Professor:
Aurora Burd, (D), Washington, 2013, GgZgYm

Bakersfield College

Physical Science Dept
1801 Panorama Drive
Bakersfield, CA 93305
p. (661) 395-4391
kvaughan@bakersfieldcollege.edu

Acting Chair:
Robert A. Schiffman, (M), California (Santa Barbara), 1971, Ga
Professor:
John C. Lyman, (B), Cal State (San Diego), 1962, Gg
Instructor:
Robert Lewy, (M), Cal State (Bakersfield), 1992, Gx
Michael Oldershaw, (M), California (Davis), 1987, Eo

Cabrillo College

Dept of Geology, Oceanography and Environmental Sci (A) ⊘
6500 Soquel Drive
Aptos, CA 95003
p. (408) 479-6495
daschwar@cabrillo.edu
http://www.cabrillo.edu/academics/earthscience/
Programs: Geology (A); Environmental Science (A)
Enrollment (2015): A: 35 (0)

Chair:
David Schwartz, (M), San Jose State, 1983, Gu

California Institute of Technology

Div of Geological & Planetary Sciences (B,M,D) ● ☒
1200 East California Boulevard
MC 170-25
Pasadena, CA 91125
p. (626) 395-6111
grotz@gps.caltech.edu
http://www.gps.caltech.edu
Enrollment (2018): B: 17 (0) D: 112 (0)

Chair:
John P. Grotzinger, (D), Virginia Tech, 1985, GsPeXg
Professor:
Jess F. Adkins, (D), MIT, 1997, Cm
Jean-Paul Ampuero, (D), Paris VII, 2002, Ys
Paul D. Asimow, (D), Caltech, 1997, CpGzy
Jean-Philippe Avouac, (D), Inst Physique du Globe de Paris, 1991, GtYs
Geoffrey A. Blake, (D), Caltech, 1985, Xc
Simona Bordoni, (D), California (Los Angeles), 2007, Op
Michael E. Brown, (D), California (Berkeley), 1994, Xy
Robert W. Clayton, (D), Stanford, 1981, Ys
Bethany L. Ehlmann, (D), Brown, 2010, XgZrGz
John M. Eiler, (D), Wisconsin, 1994, Cg
Kenneth A. Farley, (D), California (San Diego), 1991, Cg
Woodward W. Fischer, (D), Harvard, 2007, Pg
Christian Frankenberg, (D), Ruprecht-Karls, 2005, Zr
Michael C. Gurnis, (D), Australian Nat, 1987, Ys
Thomas H. Heaton, (D), Caltech, 1979, Ne
Andrew P. Ingersoll, (D), Harvard, 1966, As
Jennifer M. Jackson, (D), Illinois, 2005, Gy
Joseph L. Kirschvink, (D), Princeton, 1979, Pg
Heather Knutson, (D), Harvard, 2009, Xg
Shrinivas R. Kulkarni, (D), California (Berkeley), 1983, Xa
Michael P. Lamb, (D), California (Berkeley), 2008, Gm
Nadia Lapusta, (D), Harvard, 2001, GcYs
Jared R. Leadbetter, (D), Michigan State, 1997, Pm
Dianne K. Newman, (D), MIT, 1997, Pg
Victoria Orphan, (D), California (Santa Barbara), 2001, Pg
George R. Rossman, (D), Caltech, 1971, GzCa
Tapio Schneider, (D), Princeton, 2001, AsmZg
Alex L. Sessions, (D), Indiana, 2001, CosCb
Mark Simons, (D), MIT, 1995, Ys
David J. Stevenson, (D), Cornell, 1976, Xy
Joann M. Stock, (D), MIT, 1988, GtcYr
Edward M. Stolper, (D), Harvard, 1979, Cp
Andrew F. Thompson, (D), Scripps, 2006, OpgZc
Victor Tsai, (D), Harvard, 2009, Yg
Paul O. Wennberg, (D), Harvard, 1994, As
Brian P. Wernicke, (D), MIT, 1982, Gtc
Yuk L. Yung, (D), Harvard, 1974, As
Assistant Professor:
Konstantin Batygin, (D), Caltech, 2012, Xg
Zhongwen Zhan, (D), Caltech, 2013, Ys
Research Professor:
Egill Hauksson, (D), Columbia, 1981, YsGtYg
Emeritus:
Arden L. Albee, (D), Harvard, 1957, GizXg
Clarence R. Allen, (D), Caltech, 1954, Ys
Donald S. Burnett, (D), California (Berkeley), 1963, Xc
Charles Elachi, (D), Caltech, 1971, Xg
Peter M. Goldreich, (D), Cornell, 1963, Xy
Donald V. Helmberger, (D), California (San Diego), 1967, Ys
Hiroo Kanamori, (D), Tokyo, 1964, Ysg
Duane O. Muhleman, (D), Harvard, 1963, Xy
Jason B. Saleeby, (D), California (Santa Barbara), 1975, Gc
Leon T. Silver, (D), Caltech, 1955, Cc
Hugh P. Taylor, (D), Caltech, 1959, Cs
Peter J. Wyllie, (D), St Andrews, 1958, Cp

California Lutheran University

Dept of Earth and Environmental Science (B) ☒
60 West Olsen Road #3700
Thousand Oaks, CA 91360-2787
p. (805) 493-3264
bilodeau@callutheran.edu
http://www.callutheran.edu/admission/undergraduate/majors/geology/
Programs: Geology (B)
Environmental Science (B)
Enrollment (2018): B: 4 (5)
Professor:
William L. Bilodeau, (D), Stanford, 1979, GcsGg
Chair:
Robert Dull, (D)
Assistant Professor:
Megan Fung, (D)

Emeritus:
Linda A. Ritterbush, (D), California (Santa Barbara), 1990, Pi

California Polytechnic State University

Department of Natural Resources Management and Environmental Sciences (B,M)
One Grand Avenue
San Luis Obispo, CA 93407
p. (805) 756-2261
jstevens@calpoly.edu
http://nres.calpoly.edu
Professor:
Delmar D. Dingus, (D), Oregon State, 1975, So
Brent G. Hallock, (D), California (Davis), 1979, Sf
William L. Preston, (D), Oregon, 1979, Ze
Thomas A. Ruehr, (D), Colorado State, 1975, Sb
Terry L. Smith, (D), Nebraska, 1980, So
Assistant Professor:
Christopher (Chip) S. Appel, (D), Florida, 2001, Sc
Emeritus:
Ronald D. Taskey, (D), Oregon State, 1977, Sf
Administrative Coordinator:
Joan M. Stevens, Zn
Instructional Tech:
Craig Stubler, (B), Cal Poly, 1996, So

Department of Physics (B) ☒
1 Grand Ave
180-204
San Luis Obispo, CA 93407
p. (805) 756-2448
geology@calpoly.edu
https://physics.calpoly.edu/geology
Administrative Assistant: Michell Maksoudian
Professor:
Antonio F. Garcia, (D), California (Santa Barbara), 2001, Gmg
John J. Jasbinsek, (D), Wyoming, 2008, YsuHw
Scott Johnston, (D), California (Santa Barbara), 2006, GcpCc
Emeritus:
Lawrence Balthaser, Indiana, 1969
David H. Chipping, (D), Stanford, 1970, Grt
Ken Hoffman, California (Berkeley)

California State Polytechnic University, Pomona

Dept of Geological Sciences (B,M)
3801 West Temple Avenue
Pomona, CA 91768
p. (909) 869-3454
janourse@cpp.edu
https://www.cpp.edu/sci/geological-sciences/
Programs: Geology (B,M)
Administrative Assistant: Monica Baez
Enrollment (2020): B: 115 (21) M: 30 (5)
Chair:
Jonathan A. Nourse, (D), Caltech, 1989, GctEm
Professor:
Jeffrey S. Marshall, (D), Penn State, 2000, GmtHs
Stephen G. Osborn, (D), Arizona, 2010, HwClSp
Jascha Polet, (D), Caltech, 1999, Ysg
Associate Professor:
Nicholas J. Van Buer, (D), Stanford, 2012, GxtGz
Assistant Professor:
Bryan P. Murray, (D), California (Los Angeles), 2013, Gsg
Emeritus:
John A. Klasik, (D), Louisiana State, 1976, Gu
Equipment Technician:
Frank Wille, (B), Cal State (Fullerton), 2009, Zn

California State University, Bakersfield

Dept of Geological Sciences (B,M)
9001 Stockdale Hwy, 66 SCI
Bakersfield, CA 93311
p. (661) 654-3027
arathburn@csub.edu
http://www.csub.edu/Geology/
f: https://www.facebook.com/groups/103146842589/
Programs: Geology; Natural Science

Administrative Assistant: Sue Holt
Enrollment (2020): B: 42 (25) M: 14 (9)
Professor:
Anthony E. Rathburn, (D), Duke, PmOoCb
Associate Professor:
Junhua Guo, (D), Missouri, Gs
W Chris Krugh, (D), Swiss Fed Inst Tech, 2008, GcmCc
Assistant Professor:
Mathew Herman, (D), Penn State, 2017, GtYsRn
Katie O'Sullivan, (D), Notre Dame, 2013, GviXg
Anna Paula . Soares Cruz, (D), Federal Fluminense, 2016, CmOoPc
Liaosha Song, (D), West Virginia, 2018, GoYeEo
Lecturer:
Robert Crewdson, (D), Colorado Mines, HwYgCg
Emeritus:
Dirk Baron, (D), Oregon, Hg
Janice Gillespie, (D), Wyoming, GorZi
Robert A. Horton, Jr., (D), Colorado Mines, 1985, GdsGo
Robert M. Negrini, (D), California (Davis), 1986, YmgGn
Department IT Support:
Elizabeth Powers, (B), N Arizona, 1998, Gg

California State University, Chico

Dept of Geological and Environmental Sciences (B,M) ● ☒
400 W First Street
Chico, CA 95929-0205
p. (530) 898-5262
geos@csuchico.edu
http://www.csuchico.edu/geos/
f: https://www.facebook.com/CSUChicoGEOS/
Programs: Geology, Environmental Science
Certificates: Geology (Minor)
Enrollment (2020): B: 200 (34) M: 14 (6)
Chair:
Todd Greene, (D), Stanford, 2000, Gsr
Professor:
Ann Bykerk-Kauffman, (D), Arizona, 1990, GcZe
Julie Monet, (D), Rutgers, 2006, GgNg
Russell S. Shapiro, (D), California (Santa Barbara), 1998, PgGdXb
Rachel Teasdale, (D), Idaho, 2001, Gv
Associate Professor:
Hannah M. Aird, (D), Duke, 2014, GxEg
Kristen M. Kaczynski, (D), Colorado State, 2015, ZnHg
Sandrine Matiasek, (D), California (Davis), 2015, CqRw
Shane D. Mayor, (D), Wisconsin, 2001, AsZrAp
Lecturer:
Carrie Monohan, (D), Washington, 2004, Hg
Jochen Nuester, (D), Max Planck, 2005, Cm
Emeritus:
Terence T. Kato, (D), California (Los Angeles), 1976, GtcGp

California State University, Dominguez Hills

Earth Sciences (B) ☒
1000 E. Victoria Street
NSM B202
Carson, CA 90747
p. (310) 243-3377
https://www.csudh.edu/earth/
https://www.csudh.edu/earth/
Programs: Bachelor of Science in Earth Science, Minor in Earth Sciences, Bachelor of Arts in Geography, Minor in Geography
Certificates: Certificate Program in Geotechniques
Enrollment (2014): B: 55 (9)
Professor:
Brendan A. McNulty, (D), California (Santa Cruz), 1996, Gc
Ashish Sinha, (D), S California, 1997, PeCs
Chair, Earth Sciences and Geography:
John Keyantash, (D), California (Los Angeles), 2001, HgAsZe
Associate Professor:
Ralph Saunders, (D), Arizona, 1996, Zg
Assistant Professor:
Parveen Chhetri, (D), Texas A&M, 2017, Geg
Tianjun Lu, (D), Virginia Tech, 2020, Zi
Emeritus:
Rodrick Hay, (D), Arizona State, 1996, Zr
David Sigurdson, (D), California (Riverside), 1974, GizEg
Related Staff:
Michael H. Ferris, (M), Cal State (Long Beach), 2012, Zr

California State University, East Bay

Dept of Anthropology, Geography, and Environmental Studies (B) ●
25800 Carlos Bee Blvd
Hayward, CA 94542
p. (510) 885-3193
michael.lee@csueastbay.edu
http://class.csueastbay.edu/geography
Programs: Anthropology BA, Environmental Studies BA, Geography BS, Global Studies BA
Certificates: Cartography & GIS, Sustainability
Enrollment (2022): B: 48 (20)
Chair:
Michael D. Lee, (D), London Schl of Econ, 1990, Hg
Assistant Professor:
Tony Marks-Block, (D), Stanford, Zc
Antonio Roman-Alcalá, (D), Intl Inst of Social Studies (Nether), 2021, Zn
Emeritus:
Karina Garbesi, (D), California (Berkeley), 1994, Cg
David J. Larson, (D), California (Berkeley), 1994, Zu

Dept of Earth & Environmental Sciences (B,M) ●
25800 Carlos Bee Blvd.
Hayward, CA 94542
p. (510) 885-3486
mitchell.craig@csueastbay.edu
http://www.csueastbay.edu/earth
Programs: Geology (B); Geoscience Education (B); Environmental Science (B); Environmental Geoscience (M); Environmental Health (B)
Certificates: Geology minor
Enrollment (2019): B: 75 (0) M: 12 (0)
Associate Professor and Chair:
Michael S. Massey, (D), Stanford, 2013, ScGeCg
Professor:
Mitchell S. Craig, (D), Georgia Tech, 1990, Yg
Jean E. Moran, (D), Rochester, 1994, HwCl
Jeffery C. Seitz, (D), Virginia Tech, 1994, CgZeGx
Associate Professor:
Luther M. Strayer, (D), Minnesota, 1998, Gc
Assistant Professor:
Patricia Oikawa, (D), Virginia, 2011, SbCoSf
Jose Rosario, (D), Cornell, 2015, PsSaGr

California State University, Fullerton

Dept of Geological Sciences (B,M) ● ⊘
800 N. State College Boulevard
Geological Sciences, MH 254
Fullerton, CA 92831
p. (657) 278-3882
awoods@fullerton.edu
http://www.fullerton.edu/geology/
f: https://www.facebook.com/CalStateFullerton.Geology/
t: @csufgeology
Programs: Geology (B), Earth Science (B), Geology (M)
Enrollment (2021): B: 80 (12) M: 9 (1)
Chair:
Adam D. Woods, (D), S California, 1998, GsPe
Professor:
Matthew E. Kirby, (D), Syracuse, 2001, GnPeGg
Associate Professor:
Nicole Bonuso, (D), S California, 2005, Pie
Joe Carlin, (D), Texas A&M, 2013, GuOuGs
W. Richard Laton, (D), W Michigan, 1997, HwGeHy
Sean Loyd, (D), S California, 2010, CgGe
Valbone Memeti, (D), S California, 2009, GivCa
James Parham, (D), California (Berkeley), 2003, PvgGg
Assistant Professor:
Sinan Akciz, (D), MIT, 2002, GcRn
Virginia Isava, (D), Stanford, 2021, ZeRc
Kathryn Metcalf, (D), Arizona, 2018, Gt
Lecturer:
Angela Aranda, (M), Cal State (Fullerton), 2016, Gg
Freddi Jo Bruschke, (D), Gg
Emily A. Hamecher, (D), Caltech, 2013, GgzGi
Wayne G. Henderson, (M), George Washington, 1997, PgGg

Joanna Hynes, (M), Cal State (Fullerton), Gg
Carolyn Rath, (M), Cal State (Fullerton), Gg
Kelly R. Ruppert, (M), California (Riverside), 2001, Gg
Emeritus:
Phillip A. Armstrong, (D), Utah, 1996, GcYg
Galen R. Carlson, (D), Zg
Diane Clemens-Knott, (D), Caltech, 1992, Gig
John H. Foster, (D), Ng
Jeffrey R. Knott, (D), California (Riverside), 1998, Gmg
Prem K. Saint, (D), Minnesota, 1973, Hw

California State University, Long Beach

Dept of Geological Sciences (B,M)
1250 Bellflower Boulevard
Long Beach, CA 90840-3902
p. (562) 985-4809
rfrancis@csulb.edu
http://www.csulb.edu/colleges/cnsm/departments/geology/
Administrative Assistant: Margaret Costello
Enrollment (2009): B: 67 (11) M: 16 (3)
Chair:
Richard J. Behl, (D), California (Santa Cruz), 1992, GsPeGd
Conrey Endowed Chair:
Matthew Becker, (D), Texas, 1996, Hwg
Professor:
Stanley C. Finney, (D), Ohio State, 1977, Psi
Robert D. Francis, (D), California (San Diego), 1980, Co
Roswitha B. Grannell, (D), California (Riverside), 1969, Yv
Jack Green, (D), Columbia, 1953, Gv
Gregory J. Holk, (D), Caltech, 1997, CsGxEm
Nate Onderdonk, (D), California (Santa Barbara), 2003, Gtm
Associate Professor:
Thomas Kelty, (D), California (Los Angeles), 1998, Gc
Assistant Professor:
Lora R. Stevens (Landon), (D), Minnesota, 1997, Gn
Lecturer:
Bruce Perry, (M), Cal State (Long Beach), 1993, Gc
Emeritus:
Kwan M. Chan, (D), Liverpool (UK), 1966, Oc
Paul J. Fritts, (D), Colorado, 1969, Pm
Charles T. Walker, (D), Leeds (UK), 1952, Cl
Robert F. Winchell, (D), Ohio State, 1963, Gz

California State University, Los Angeles

Department of Geography, Geology and Environment (B,M) ○⊘
5151 State University Drive
Los Angeles, CA 90032
p. (323) 343-2400
smulher@calstatela.edu
http://www.calstatela.edu/academic/geos
Programs: BA Geography, BS Geology, MS Geology, MS Environmental Science
Certificates: GIS certificate, undergraduate and post baccalaureate
Enrollment (2021): B: 30 (27) M: 15 (6)
Dept Chair:
Stephen Mulherin, (D), Ohio State, 1999
Professor:
Jennifer M. Garrison, (D), California (Los Angeles), 2004, GiCcGv
Barry Hibbs, (D), Texas, 1993, Hw
Associate Professor:
Kris Bezdecny, (D), S Florida, 2011, ZniZn
Jingjing Li, (D), California (Irvine), 2013
Assistant Professor:
Michael Beland, (D), California (Santa Barbara), 2018
Joshua Kelly, (D), San Diego State, 2020
Adjunct Professor:
Mohammad Hassan Rezaie-Boroon, (D), Erlangen (Germany), 1997, GeCg
Emeritus:
Qiu Hong-lie, (D), Louisiana State, 1994
Steve LaDochy, (D), Manitoba, 1985, Zn

California State University, Northridge

Dept of Geography and Environmental Studies (B,M) ⊘
18111 Nordhoff Street
Northridge, CA 91330-8249
p. (818) 677-3532
geography@csun.edu
http://www.csun.edu/social-behavioral-sciences/geography
Programs: Undergraduate: Geography: Standard and GIS (BA/MA);
Graduate (MA) GIS (MSc)
Enrollment (2020): B: 127 (33) M: 37 (18)
Professor:
Julie E. Laity, (D), California (Los Angeles), 1982, Zy
Amalie Orme, (D), California (Los Angeles), 1983, Gm
Yifei Sun, (D), SUNY (Buffalo), 2000, Zi
Associate Professor:
Soheil Boroushaki, (D), W Ontario, 2010, Zi
Mario A. Giraldo, (D), Georgia, 2007, ZriZg
Regan Maas, (D), California (Los Angeles), 2010, Zi
Assistant Professor:
Sanchayeeta Adhikari, (D), Florida, 2011, Zi
Luke Drake, (D), Rutgers, 2015, Zi

Dept of Geological Sciences (B,M) ●☒
18111 Nordhoff Street
MD: 8266
Northridge, CA 91330-8266
p. (818) 677-3541
geology@csun.edu
http://www.csun.edu/geology/
Programs: Geology, Geophysics
Administrative Assistant: Mari C. Flores-Garcia
Enrollment (2012): B: 72 (0) M: 23 (0)
Retired:
J. Douglas Yule, (D), Caltech, 1996, Gt
Professor:
Kathleen M. Marsaglia, (D), California (Los Angeles), 1989, Gs
Elena A. Miranda, (D), Wyoming, 2006, Gc
Dayanthie Weeraratne, (D), Carnegie Inst, 2005, Yg
Associate Professor:
M. Robinson Cecil, (D), Arizona, 2009, GiCcGz
Matthew d'Alessio, (D), California (Berkeley), 2004, ZeGt
Richard V. Heermance, (D), California (Santa Barbara), 2007, GrCc
Joshua J. Schwartz, (D), Wyoming, 2007, GiCc
Assistant Professor:
Jennifer Cotton, (D), Michigan, 2013, Pc
Eileen L. Evans, (D), Harvard, 2014, YdGtYg
Priya Ganguli, (D), California (Santa Cruz), 2013
Scott Hauswirth, (D), N Carolina, 2014, HgCg
Julian Lozos, (D), California (Riverside), 2013, YgGt
Emeritus:
Herbert G. Adams, (D), California (Los Angeles), 1971, Ng
Lorence G. Collins, (D), Illinois, 1959, Gzx
George C. Dunne, (D), Rice, 1972, Gc
Vicki A. Pedone, (D), SUNY (Stony Brook), 1990, GdCs
Gerald W. Simila, (D), California (Berkeley), 1979, Yg
Jon R. Sloan, (D), California (Davis), 1980, Pm
Richard L. Squires, (D), Caltech, 1973, Pi

California State University, San Bernardino

Dept of Geological Sciences (B,M) ☒
5500 University Parkway
San Bernardino, CA 92407-2397
p. (909) 537-5336
CPalmer@csusb.edu
https://cns.csusb.edu/geology
Administrative Assistant: Christina Palmer
Enrollment (2018): B: 47 (7) M: 5 (0)
Professor:
Sally F. McGill, (D), Caltech, 1992, GtYdGm
Professor:
Joan E. Fryxell, (D), North Carolina, 1984, Gct
Erik Melchiorre, (D), Washington (St Louis), 1997, EmHwCl
Associate Professor:
W. Britt Leatham, (D), Ohio State, 1987, PsmOg
Assistant Professor:
Kerry Cato, (D), Texas A&M, 1991, Ng
Codi Lazar, (D), California (Los Angeles), 2010, Cp
Emeritus:
Louis A. Fernandez, (D), Syracuse, 1969, Gi
Alan L. Smith, (D), California (Berkeley), 1969, GviGz

Cooperating Faculty:
David Maynard, (D), California (Riverside), 1992, Zn
Cooperating Faculty:
Stuart S. Sumida, (D), California (Los Angeles), 1987, Pv

California State University, Stanislaus

Dept of Physics and Geology (B) ●☒
One University Circle
Turlock, CA 95382
p. (209) 667-3466
HFerriz@csustan.edu
https://www.csustan.edu/geology
Programs: Geology (B); Applied Geology (B); Geology (minor)
Enrollment (2021): B: 9 (12)
Program coordinator:
Robert D. Rogers, (D), Texas, 2003, GcmGt
Professor:
Horacio Ferriz, (D), Stanford, 1984, NgHgYe
Mario J. Giaramita, (D), California (Davis), 1989, Gpz
Julia Sankey, (D), Louisiana State, 1998, Pve
Lecturer:
Garry F. Hayes, (M), Nevada (Reno), 1988, Gg
Ryan Hollister, (M), Mississippi State, 1995, ZeGg
Nari Miller, (D), Arizona State, 2021, CgYg
Maira Rainwater, (M), CSU Sacramento, 1997, GeHw

Cerritos College

Earth Science Dept (A) O
11110 E. Alondra Boulevard
Norwalk, CA 90650
p. (562) 860-2451
ddekraker@cerritos.edu
http://www.cerritos.edu/earth-science/
Programs: AS - Geology
AA - Geography
Enrollment (2012): A: 9 (0)
Chair:
Dan DeKraker, (M), Cal State (Fullerton), 2004, ZeOpGg
Professor:
Tor Lacy, (M), Cal State (Long Beach), 2005, GgcGc
Crystal LoVetere, (D), USC, 2012, Zy
Assistant Professor:
Ryan Goode, (D), Cal State (Fullerton), Zc
Nicole Iwaz, (M), California (San Diego), 2012, GgeOg

Chabot College

Div of Mathematics & Science (A)
25555 Hesperian Boulevard
Hayward, CA 94545
p. (415) 786-6865
kbononcini@chabotcollege.edu
Instructor:
Adolph A. Oliver, (M), Stanford, 1974, Ys
David J. Perry, (M), San Jose State, 1967, Op

Chaffey College

Earth Science/Geology Dept
5885 Haven Avenue
Rancho Cucamonga, CA 91737-3002
p. (909) 652-6402
henry.shannon@chaffey.edu
http://www.chaffey.edu/mathandscience/
Professor:
Jane Warger, (M), Columbia, 1990, ZgGg

City College of San Francisco

Earth Sciences (A)
Box S99
50 Frida Kahlo Way
San Francisco, CA 94112
p. (415) 452-7423
rduncan@ccsf.edu
http://www.ccsf.edu/Earth
Programs: Associate degrees in Geography, Geology and Oceanography
Certificates: GIS
Enrollment (2018): A: 10 (1)
Chair:
Ian Duncan, (D), Washington, 2013, Zyi
Professor:
Darrel E. Hess, (M), California (Los Angeles), 1990, Zy
Chris Lewis, (M), California (Berkeley), 1993, GgEmSo
Katryn Wiese, (M), Oregon State, 1992, GiOg
Adjunct Professor:
James Kuwabara, (D), Caltech, 1980, RwOgCg
Elizabeth Proctor, (M), San Francisco State, Zi
Gordon Ye, (M), California (Berkeley), 1993, Zi

College of Marin

Geology Dept (A)
835 College Ave
SC 192, Phone Ext: 7523
Kentfield, CA 94904
p. (415) 457-8811
dfoss@marin.edu
http://marin.edu/~geology
Professor:
Donald J. Foss, (M), Boise State, 1980, Zg
Emeritus:
James L. Locke, (M), San Jose State, 1971, Zg

College of San Mateo

Geological Sciences (A) ☒
1700 W. Hillsdale Blvd.
San Mateo, CA 94402
p. (650) 574-6633
hand@smccd.edu
http://www.collegeofsanmateo.edu/geologicalsciences
Programs: Geological Sciences, Geology
Enrollment (2018): A: 9 (1)
Professor:
Linda M. Hand, (M), Texas A&M, 1988, GgPgOg

College of the Canyons

Dept of Earth, Space, and Environmental Sciences (A)
26455 Rockwell Canyon Road
Santa Clarita, CA 91355
p. (661) 362-3658
Vincent.Devlahovich@canyons.edu
http://www.canyons.edu/departments/ESES/
Enrollment (2012): A: 4 (0)
Chair:
Vincent A. Devlahovich, (D), Cal State (Northridge), 2012, GgZye
Professor:
Mary Bates, (M), Cal State (Northridge), Zyg

College of the Desert

Dept of Science (A)
43-500 Monterey Avenue
Palm Desert, CA 92260
p. (760) 776-7272
nmoll@collegeofthedesert.edu
http://www.collegeofthedesert.edu/students/ap/masc/sci/Pages/sciencesphysical.aspx
Enrollment (2013): A: 3 (1)
Professor:
Nancy E. Moll, (D), Washington, 1981, Gge
Adjunct Professor:
Brian Koenig, (M), Arizona, 1978, Gge
Robert Pellenbarg, (D), Delaware, 1976, OcGu

College of the Sequoias

Science Div (A)
Visalia, CA 93277
p. (209) 730-3812
erich@cos.edu
Chair:
Eric D. Hetherington, (D), Minnesota, 1991, Gc
Emeritus:
John R. Crain, (M), Nevada, 1961, Gg

College of the Siskiyous

Dept of Biological & Physical Sciences (A)
800 College Avenue
Weed, CA 96094
p. (530) 938-5255
hirt@siskiyous.edu
http://www.siskiyous.edu/instruction/programs/geology/
Enrollment (2015): A: 1 (0)

Instructor:
William H. Hirt, (D), California (Santa Barbara), 1989, Gi

Contra Costa College

Astronomy/Engineering/Geology/Physics Dept (A)
2600 Misson Bell Drive
San Pablo, CA 94806
p. (510) 235-7800
jsmithson@contracosta.edu
http://www.contracosta.edu/home/programs-departments/engineering/

Professor:
Mary Lewis, Gg
Jayne Smithson, Gg

Cosumnes River College

Dept of Science, Mathematics & Engineering (A)
8401 Center Parkway
Sacramento, CA 95823
p. (916) 691-7210
MuranaB@crc.losrios.edu
Department Secretary: Sue McCoy

Chair:
Debra Sharkey, (M), California (Davis), 1994, Zy

Assistant Professor:
Hiram Jackson, (M), California (Davis), 1992, Gg

Adjunct Professor:
Gerry Drobny, (M), Washington State, 1981, Gg

Crafton Hills College

Geology Dept (A) ⃠
11711 Sand Canyon Road
Yucaipa, CA 92399
p. (909) 794-2161
rihughes@craftonhills.edu
http://www.craftonhills.edu/courses_&_programs/Physical_Science/Geology/
Programs: Geology (A), Geography, and Environmental Science
Enrollment (2020): A: 20 (8)

Professor:
Richard O. Hughes III, (M), Ohio, 1994, GlZe

Cuesta College

Physical Sciences Div (A)
P. O. Box 8106
San Luis Obispo, CA 93403
p. (805) 546-3230
jgrover@cuesta.edu
http://academic.cuesta.edu/physci/Geology
Enrollment (2014): A: 6 (0)

Professor:
Jeffrey A. Grover, (M), Arizona, 1982, Gg
Debra Stakes, (D), Gg

Cuyamaca College

Dept of Science and Engineering (A)
900 Rancho San Diego Parkway
El Cajon, CA 92019
p. (619) 660-4345
Glenn.Thurman@gcccd.edu
http://www.cuyamaca.edu/

Instructor:
Lisa Chaddock, Zy
Michael Farrell, Gg
Bryan Miller-Hicks, Gg
Agatha Wein, Gg
Ray Wolcott, Og

Emeritus:
Waverly Ray, Gg

Cypress College

Geology Dept (A)
9200 Valley View Street
Cypress, CA 90630
p. (714) 484-7153
RArmale@CypressCollege.edu

Professor:
Russell L. Flynn, (M), San Diego State, 1971, Op

Adjunct Professor:
Hank Wadleigh, (B), Cal State (Long Beach), 1957, Gg

Visiting Professor:
Curtis J. Williams, (M), Cal State (Los Angeles), 1996, Gg

Emeritus:
Keith E. Green, (M), S California, 1958, Pg
Altus Simpson, (M), S California, 1958, Go

Diablo Valley College

Div of Physical Sciences (A)
321 Golf Club Road
Pleasant Hill, CA 94523
p. (925) 685-1230 x46
JHetherington@dvc.edu
http://www.dvc.edu/org/departments/physics/

Chair:
Jean Hetherington, (M), Washington, 1983, Gg

East Los Angeles College

Anthropology, Geography, and Geology
1301 Avenida Cesar Chavez
Monterey Park, CA 91754
p. (323) 265-8843
adsitrj@elac.edu
http://www.elac.edu/academics/departments/anthrogeo

Chair:
John Grimmer

Gold Creek Coordinator:
Robert West, (D), California (Santa Barbara), 2004, GgmSd

El Camino College

Dept of Earth Sciences (A)
16007 Crenshaw Blvd.
Torrance, CA 90506
p. (310) 660-3593
jholliday@elcamino.edu
http://www.elcamino.edu/academics/naturalsciences/earth/

Professor:
Jerry Brothen, (M), California (Los Angeles), Zy
Sara Di Fiori, (M), California (Los Angeles), GgOg
Matt Ebiner, (M), California (Los Angeles), Zy
Joseph W. Holliday, (M), Oregon State, 1982, Og

Associate Professor:
Chuck Herzig, (D), California (Riverside), GgOg

Instructor:
Gary Booher, Zg
Robin Bouse, Gg
Charles Dong, Og
Lynn Fielding, Zg
Patricia Neumann, Zg
Douglas Neves, Zg
Jim Noyes, (D), Scripps, Og
Ebenezer Peprah, Zg

Feather River College

Feather River College (A)
570 Golden Eagle Ave.
Quincy, CA 95971
dlerch@frc.edu
http://www.frc.edu/

Instructor:
Derek Lerch, Zg

Folsom Lake College

Geosciences (A) ●
10 College Parkway
Folsom, CA 95630
p. (916) 608-6668
pittmaj@flc.losrios.edu
http://www.flc.losrios.edu/Academics/Geology.htm
Enrollment (2015): A: 7 (1)
Professor:
Jason Pittman, (M), Oregon State, 1999, ZgiZi

Fresno City College

Earth & Physical Science Dept (A) ⊘
Math, Science & Engineering Div
1101 E. University Avenue
Fresno, CA 93741
p. (559) 442-4600
alexandra.priewisch@fresnocitycollege.edu
Programs: Geology (A)
Enrollment (2018): A: 40 (1)
Instructor:
T. Craig Poole, (M), Cal State (Fresno), 1987, Gg
Alexandra Priewisch, (D), New Mexico, 2014, Gg

Fresno State University

Dept of Earth & Environmental Sciences (B,M)
2345 E. San Ramon Avenue
MH24
Fresno, CA 93740-8031
p. (559) 278-3086
rdundas@csufresno.edu
http://www.csufresno.edu/ees/
Enrollment (2009): B: 51 (13) M: 30 (5)
Chair:
Stephen D. Lewis, (D), Columbia, 1982, Yg
Professor:
Keith D. Putirka, (D), Columbia, 1997, GviGp
Zhi (Luke) Wang, (D), Leuven (Belgium), 1997, SpHwZi
Associate Professor:
Christopher J. Pluhar, (D), California (Santa Cruz), 2003, GtYmNg
John Wakabayashi, (D), California (Berkeley), GmcGt
Assistant Professor:
Robert G. Dundas, (D), California (Berkeley), 1994, PvGr
Mathieu Richaud, (D), N Illinois, 2006, GusCs
Peter Van de Water, (D), Arizona, GrPbGf
Lecturer:
Jeff Anglen, (M), Texas Tech, 2001, GsdPg
Susan Bratcher, (M), Cal State (Fresno), 2008, Hw
Kerry Workman-Ford, (M), Cal State (Fresno), 2003, Gc
Adjunct Professor:
Jerry DeGraff, (M), Utah State, 1976, Zn
Dong Wang, (D), HwSpZr
Emeritus:
Jon C. Avent
Bruce A. Blackerby, (D), California (Los Angeles), Giz
Roland H. Brady, (D), California (Davis), 1986, NgZuGa
Seymour Mack, YrHgOn
Robert D. Merrill, (D), Texas, 1974, GsmPe
C. John Suen, (D), MIT, 1978, HwCsGe

Fullerton College

Div of Natural Sciences (Geology) (A) 🖱
321 E. Chapman Ave
Fullerton, CA 92832
p. (714) 992-7445
mwillis@fullcoll.edu
http://natsci.fullcoll.edu/
Programs: AS, AS-T Geology
AS Earth Science
Enrollment (2021): A: 2 (1)
Professor:
William S. Chamberlin, (D), S California, 1989, OgbZe
Marc Willis, (M), Gg
Associate Professor:
Roman DeJesus, (D), Og

Gavilan College

Dept of Physical Sciences
5055 Santa Theresa Boulevard
Gilroy, CA 95020
p. (408) 848-4701
dwillahan@gavilan.edu
http://www.gavilan.edu/natural_sciences/
Instructor:
Duane Willahan, (M), Santa Fe State, 1992, Gg

Golden West College

Physical Science Program (A)
15744 Golden West Street
Huntington Beach, CA 92647
p. (714) 892-7711 x51116
msouto@gwc.cccd.edu
http://www.goldenwestcollege.edu/campus/physicalscience.html
Professor:
Ronald C. Gibson, (M), California (Riverside), 1964, Gc
Bernard J. Gilpin, (M), California (Riverside), 1976, Ys

Grossmont College

Dept of Earth Sciences (A) ○🖱
8800 Grossmont College Drive
El Cajon, CA 92020
p. (619) 644-7887
gary.jacobson@gcccd.edu
https://www.grossmont.edu/academics/programs-departments/earth-sciences/
f: https://www.facebook.com/grossmontearthsciences/
Programs: Geology (A); Geography (A); Oceanography (A)
Professor:
Chris Hill, (D), S California, 2000, GgOg
Instructor:
Gary L. Jacobson, (M), San Diego State, 1982, GucEm

Hartnell College

Div of Math and Science (A)
411 Central Avenue
Salinas, CA 93901
myahdi@hartnell.edu
http://www.hartnell.edu/academics/math.html
Instructor:
Robert Barminski, (M), Moss Landing Marine Lab, GgOg

Humboldt State University

Dept of Geology (B,M) ●⊘
1 Harpst Street
Arcata, CA 95521-8299
p. (707) 826-3931
geology@humboldt.edu
http://www.humboldt.edu/geology/
Programs: Geology (B); Geoscience (B)
Enrollment (2022): B: 60 (15) M: 6 (3)
Chair:
Brandon L. Browne, (D), Alaska (Fairbanks), 2005, GivZe
Professor:
Mark Hemphill-Haley, (D), Oregon, 2000, Gt
Associate Professor:
Melanie Michalak, (D), California (Santa Cruz), 2013, GtCcGc
Assistant Professor:
Jaclyn S. Baughman, (D), Colorado, 2018, GtCgGc
Laura B. Levy, (D), Dartmouth, 2014, GslPc
Research Associate:
Eileen Hemphill-Haley, (D), California (Santa Cruz), 1991, Pm
Harvey M. Kelsey, (D), California (Santa Cruz), 1977, Gmt
Robert C. McPherson, (M), Humboldt State, 1989, Gt
Lecturer:
Amanda R. Admire, (M), Humboldt State, 2013, Zg
Adjunct Professor:
Jason R. Patton, (D), Oregon State, 2014, GuOuGg
Emeritus:
Raymond M. Burke, (D), Colorado, 1976, Gm
Gary A. Carver, (D), Washington, 1972, Gmt
Susan M. Cashman, (D), Washington, 1977, Gc
Lorinda Dengler, (D), California (Berkeley), 1978, Yg
Andre K. Lehre, (D), California (Berkeley), 1982, GmHs

Alistair W. McCrone, (D), Kansas, 1961, Gs
William C. Miller, (D), Tulane, 1984, Pi

Irvine Valley College

Dept of Geology (A) ☒
5500 Irvine Center Drive
Irvine, CA 92618
p. (949) 451-5561
astinson@ivc.edu
http://www.ivc.edu
Programs: Geology (A)
Enrollment (2007): A: 20 (5)

Professor:
Amy L. Stinson, (M), San Diego State, 1990, Gc

Adjunct Professor:
Mark Bordelon, (M)
Scott Mata, (D), S California
Doug Neves, (D)
Jim Schneider, (M)

Laney College

Dept of Geography/Geology (A)
900 Fallon Street
Oakland, CA 94607
p. (510) 464-3233
dwoodrow@peralta.edu
http://www.laney.peralta.edu/apps/comm.asp?Q=30123

Las Positas College

Dept of Geology (A)
3000 Campus Hill Drive
Livermore, CA 94551
p. (925) 424-1319
ochristen@laspositascollege.edu
http://www.laspositascollege.edu/geology/
Programs: Geology (AS and AS-T)
Certificates: Geology Certificate being approved now.
Enrollment (2018): A: 10 (2)

Instructor:
Ruth L. Hanna, (M), California (Davis), 1988, Gg

Geology Lab Technician:
Carol Edson, Gg

Loma Linda University

Dept of Earth and Biological Sciences (B,M,D) ●
Loma Linda University
Loma Linda, CA 92350
p. (909) 824-4530
pbuchheim@llu.edu
https://medicine.llu.edu/research/department-earth-and-biological-sciences
f: https://www.facebook.com/lomalindageology
Enrollment (2016): B: 1 (0) M: 1 (4) D: 11 (2)

Professor:
Leonard R. Brand, (D), Cornell, 1970, Pv
Paul Buchheim, (D), Wyoming, 1978, GnsPe
Leroy Leggitt, (D), Loma Linda, 2005, Pi

Senior Scientist:
Benjamin L. Clausen, (D), Colorado, 1987, GiCtYg

Associate Professor:
Kevin Nick, (D), Oklahoma, 1990, GsbGr

Assistant Professor:
Ronald Nalin, (D), Padua, 2006, Gsr

Adjunct Professor:
Raul Esperante, (D), Loma Linda, 2002, Pv

Los Angeles County Museum of Natural History

Research and Collections Branch
900 Exposition Boulevard
Los Angeles, CA 90007
p. (213) 763-3360
acelesti@nhm.org
https://www.nhm.org/site/research-collections/mineral-sciences

Associate Curator:
Luis M. Chiappe, (D), Buenos Aires, 1992, Pv

Curator:
Kenneth E. Campbell, (D), Florida, 1973, PvGgr
Anthony R. Kampf, (D), Chicago, 1976, Gz

Associate Curator:
Aaron Celestian, (D)
Nathan Smith, (D)

Collections Manager:
Samuel A. McLeod, (D), California (Berkeley), 1981, Pv

Collections Manager:
Christopher A. Shaw, (M), Cal State (Long Beach), 1981, Pv

Los Angeles Harbor College

Dept of Earth Science (A)
1111 Figueroa Place
Wilmington, CA 90744
p. (310) 233-4000
munasit@lahc.edu
http://www.lahc.cc.ca.us/

Instructor:
Patricia Kellner, (D), Zg
John Mack, Zg
Tissa Munasinghe, (D), Zg
Melanie Renfrew, (D), Zg
Susan White, Zg

Los Angeles Pierce College

Physics and Planetary Sciences (A)
6201 Winnetka Ave
Woodland Hills, CA 91371
p. (818) 710-2218
zayacjm@piercecollege.edu
Enrollment (2012): A: 11 (0)

Professor:
John M. Zayac, (M), California (Santa Barbara), 2006, Ggv

Professor:
Jason P. Finley, (M), California (Los Angeles), Am
Stephen C. Lee, (B), Illinois, 1971, Og
W. Craig Meyer, (M), S California, 1973, GeuPm

Adjunct Professor:
Harry Filkorn, (D), Kent State, GgPi
James P. Krohn, (M), S California, 1974, NgGg
Donald Prothero, (D), Columbia, Gg

Emeritus:
Ruth Y. Lebow, (M), Chicago, 1941, Op
Mark L. Powell, (M), Cal State (Northridge), 1967, As
William H. Russell, (M), Cal State (Northridge), 1970, Am
James Y. Vernon, (M), California (Los Angeles), 1951, Am

Los Angeles Southwest College

Geology Dept (A)
1600 West Imperial Highway
Los Angeles, CA 90047
p. (323) 241-5297
RobertTJ@lasc.edu
http://www.lasc.edu

Chair:
Glenn Yoshida, Gg

Professor:
Paul R. Doose, (D), California (Los Angeles), 1980

Los Angeles Valley College

Earth Science (A)
5800 Fulton Ave.
Valley Glen, CA 91401
p. (818) 778-5566
hamsje@lavc.edu
https://lavc.edu/earthscience
Programs: AS-Transfer Geology, AS Earth Science, AA-Transfer Geography
Enrollment (2020): A: 3 (3)

Chair:
Jacquelyn E. Hams, (M), Cal State (Los Angeles), 1987, GgOuZe

Professor:
Meredith L. Leonard, (M), Cal State (Northridge), AmZei

Academic Senate President:
Donald J. Gauthier, (M), California (Los Angeles), 1993, ZiyAm

CALIFORNIA

Mendocino College

Earth Science (A)
1000 Hensley Creek Road
Ukiah, CA 95482
p. (707) 468-3002
scardimo@mendocino.edu
Programs: Earth Science (A); Natural Resources (A); Geology (A); Geography (A)
Enrollment (2018): A: 74 (3)
Professor:
Steve Cardimona, (D), Texas, 1992, Ys

Merced College

Science, Technology, Engineering & Mathematics (A) ☒
3600 M Street
Merced, CA 95348
p. (209) 384-6293
kain.d@mccd.edu
http://www.mccd.edu/academics/sme
Programs: Geology
Professor:
Robert Davies

MiraCosta College

Dept of Geography (A)
1 Barnard Drive
Oceanside, CA 92056
lmiller@miracosta.edu
https://www.miracosta.edu/Instruction/Geography/
Chair:
Herschel I. Stern, (D), Oregon, 1988, Zy

Dept of Physical Sciences (A)
1 Barnard Drive
Oceanside, CA 92056
p. (760) 944-4449 x7738
lmiller@miracosta.edu
https://www.miracosta.edu/Instruction/Geology/
Professor:
Keith H. Meldahl, (D), Arizona, 1990, Gg
Christopher V. Metzler, (D), California (San Diego), 1987, Gg
John Turbeville, GgOg
Adjunct Professor:
Phil Farquharson, Gg
Laboratory Director:
Larry Hernandez, Gg

Modesto Junior College

Dept of Science, Mathematics & Engineering (A) ⃠
435 College Avenue
Modesto, CA 95350
p. (209) 575-6172
hayesg@yosemite.edu
https://www.mjc.edu/instruction/sme/earthscience.php
Programs: Geology
Enrollment (2022): A: 9 (2)
Professor:
Noah C. Hughes, (M), Montana, 2001, ZgAs
Instructor:
Garry F. Hayes, (M), Nevada (Reno), 1985, Gg

Monterey Peninsula College

Earth Sciences Dept (A)
980 Fremont St.
Monterey, CA 93940
p. (831) 646-4149
ahochstaedter@mpc.edu
http://www.mpc.edu/academics/academic-divisions/physical-science/earth-science-eart
Programs: Geology
Enrollment (2018): A: 1 (1)
Chair:
Alfred Hochstaedter, (D), California (Santa Cruz), 1991, GgOgZg

Moorpark College

Geology Program (A)
7075 Campus Road
Moorpark, CA 93021
p. (805) 553-4161
rharma@vcccd.edu
http://www.moorparkcollege.edu/geology
Enrollment (2016): A: 24 (2)
Instructor:
Roberta L. Harma, (M), Hawaii, 1982, Gg

Moss Landing Marine Laboratories

CSU Consortium for Marine Sciences (M) O☒
8272 Moss Landing Road
Moss Landing, CA 95039
p. (831) 771-4400
jharvey@mlml.calstate.edu
http://www.mlml.calstate.edu
Enrollment (2014): M: 7 (25)
Director:
James T. Harvey, (D), Oregon, 1987, Ob
Librarian:
Joan Parker, (M), California (Los Angeles), 1986, Zn
Professor:
Ivano Aiello, (D), Bologna, 1997, OuGmu
Kenneth H. Coale, (D), California (Santa Cruz), 1988, CtcOc
Jonathan Geller, (D), California (Berkeley), 1988, Zn
Michael Graham, (D), Scripps, 2000, Ob
Nicholas A. Welschmeyer, (D), Washington, 1982, Ob
Research Faculty:
John S. Oliver, (D), Scripps, 1980, Ob
Assistant Professor:
Scott Hamilton, (D), California (Santa Barbara), 2008
Birgitte McDonald, (D)
Research Faculty:
Simona Bartl, (D), California (San Diego), 1989, Zn
Laurence Breaker, (D), Naval Postgrad Sch, 1983, Op
David Ebert, (D), Rhodes, 1990, Zn
Stacy Kim, (D), MIT/WHOI, 1996, Zn
Valerie Loeb, (D), Scripps, 1979, Zn
Richard Starr, (D), Auto de Baja California, 2002, Zn
Emeritus:
Gregor M. Cailliet, (D), California (Santa Barbara), 1972, Zn
Michael Foster, (D), California (Santa Barbara), 1971, Zn
H. Gary Greene, (D), Stanford, 1977, Gu
Diving Safety Officer:
Diana Steller, (D), California (Santa Cruz), 2003, Zn

Mt. San Antonio College

Dept of Earth Sciences & Astronomy (A) ☒
1100 North Grand Avenue
Walnut, CA 91789
p. (909) 594-5611
cwebb@mtsac.edu
Enrollment (2010): A: 16 (0)
Acting Chair:
Julie Ali-Bray, (M), S California, 1999, Zn
Professor:
Micol Christopher, (D), Caltech, 2007, Zn
Craig A. Webb, (M), Duke, 1996, ZgAmOg
Instructor:
Mark Boryta, (D), New Mexico Tech, 1997
Barbara Grubb, (M), Cal State (Long Beach), 1991, Ps
Larry Mendenhall, (D), Oregon State, 1961, Am
Charles Roberts, (M), Ohio, 1972, Pm
Emeritus:
Hallock J. Bender, (D), San Gabriel, 1959, Gg
John W. Burns, (M), Pittsburgh, 1960, Zn
Damon P. Day, (M), Michigan Tech, 1965, Gg
Ron N. Hartman, (M), Cal State (Los Angeles), 1973, Xm
Kazimierz M. Pohopien, (M), McGill, 1951, Ge
Harold V. Thurman, (M), Cal State (Los Angeles), 1966, Og
Geotechnician:
Mark Koestel, (B), Arizona, 1978, Em

Naval Postgraduate School

Dept of Meteorology (M,D)
589 Dyer Road
Root Hall, Room 254
Monterey, CA 93943-5114
p. (831) 656-2516
nuss@nps.edu
http://www.nps.edu/Academics/Schools/GSEAS/Departments/Meteorology/
Enrollment (2012): M: 45 (36) D: 9 (2)
Chair:
Wendell A. Nuss, (D), Washington, 1986, Am
Distinguished Professor:
Michael T. Montgomery, (D), Harvard, 1990, AmsZg
Dean:
Philip A. Durkee, (D), Colorado State, 1984, AmZr
Professor:
Patrick A. Harr, (D), Naval Postgrad Sch, 1993, Am
Qing Wang, (D), Penn State, 1993, Am
Associate Professor:
Joshua P. Hacker, (D), British Columbia, 2001, Am
Assistant Professor:
Richard W. Moore, (D), Colorado State, 2004, Am
Barbara V. Scarnato, (D), ETH (Switzerland), 2008, Am
Research Associate:
Hway-Jen Chen, (M), California (Los Angeles), 1993, Am
Paul A. Frederickson, (M), Maryland, 1989, Am
Mary S. Jordan, (M), Naval Postgrad Sch, 1985, Am
Kurt E. Nielsen, (M), Oklahoma, 1988, Am
Andrew Penny, (M), Arizona, 2009, Am
Emeritus:
Robert Haney, (D), California (Los Angeles), 1971, Am
Robert J. Renard, (D), Am
Carlyle H. Wash, (D), Wisconsin, 1978, Am
Forrest Williams, (M), MIT, 1972, Am
Roger T. Williams, (D), California (Los Angeles), 1963, Am
Research Professor:
Kenneth L. Davidson, (D), Michigan, 1970, Am
Peter S. Guest, (D), Naval Postgrad Sch, 1992, Am
Research Associate Professor:
James Thomas Murphree, (D), California (Davis), 1989, Am
NRC Postdoctoral Fellow:
Myung-Sook Park, (D), Am
Distinguished Research Professor:
Russell L. Elsberry, (D), Colorado State, 1968, Am
Distinguished Professor:
Chih-Pei Chang, (D), Washington, 1972, Am
Meteorologist:
Robert L. Creasey, (M), Am

Dept of Oceanography (M,D) ☒
833 Dyer Road, Room 328
Monterey, CA 93943-5122
p. (831) 656-2673
pcchu@nps.edu
http://www.nps.edu/Academics/GSEAS/Oceanography/
Enrollment (2014): M: 14 (14) D: 2 (2)
Chair:
Peter C. Chu, (D), Chicago, 1985, OpZr
Professor:
John A. Closi, California (Santa Cruz), 1993
Jamie MacMahan, (D), Florida, 2003, Onp
Jeffrey D. Paduan, (D), Oregon State, 1987, Op
Timour Radko, Florida State, 1997, Op
Assistant Professor:
Derek Olson, (D), Penn State, 2017, Zn
Emeritus:
Robert H. Bourke, (D), Oregon State, 1972, Op
Ching-Sang Chiu, (D), MIT/WHOI, 1985, Op
Curtis A. Collins, (D), Oregon State, 1967, Op
Roland W. Garwood, (D), Washington, 1976, Op
Thomas H. Herbers, (D), California (San Diego), 1990, Op
Timothy P. Stanton, (M), Auckland, 1978, Op
Edward B. Thornton, (D), Florida, 1970, On
Robin T. Tokmakian, (D), Naval Postgrad Sch, 1997, Op
Joseph J. Von Schwind, (D), Texas A&M, 1968, Op
Research Professor:
Wieslaw Maslowski, (D), Alaska (Fairbanks), 1994, Op

Occidental College

Dept of Geology (B) ☒
1600 Campus Road
Los Angeles, CA 90041
p. (323) 259-2823
coze@oxy.edu
http://www.oxy.edu/Geology.xml
Administrative Assistant: Elisa Ruiz
Enrollment (2018): B: 33 (0)
Chair:
Christopher Oze, (D), Stanford
Professor:
Scott W. Bogue, (D), California (Santa Cruz), 1982, Ym
Darren J. Larsen, (D), Colorado
Margaret E. Rusmore, (D), Washington, 1985, Gc
James L. Sadd, (D), South Carolina, 1987, ZiGeZr
Related Staff:
Jan Garcia, Zn

Ohlone College

Dept of Geology (A) ⦸
43600 Mission Boulevard
Fremont, CA 94539
p. (510) 979-7938
pbelasky@ohlone.edu
http://www.ohlone.edu/instr/geology/
Programs: Associate in Geology
Certificates: Geology, Paleobiology, GIS (through Geography Department)
Professor:
Paul Belasky, (D), California (Los Angeles), 1994, PgqGs

Orange Coast College

Div of Mathematics & Science (A) ☒
2701 Fairview Road
Box 5005
Costa Mesa, CA 92626-5005
p. (714) 432-5647
ebender@occ.cccd.edu
http://www.orangecoastcollege.edu/academics/divisions/math_science/geology/Pages/
Programs: Geology; Earth Science
Enrollment (2020): A: 17 (3)
Chair:
E. Erik Bender, (D), S California, 1994, GitGz
Assistant Professor:
Christopher A. Berg, (D), Texas, 2007, GpZeGt
Lecturer:
Michael Van Ry, (M), Cal State (Fullerton), 2010, Gv

Palomar College

Earth, Space, and Environmental Sciences (A)
1140 West Mission Road
San Marcos, CA 92069
p. (760) 744-1150
sfigg@palomar.edu
http://www.palomar.edu/earthscience/
Programs: Geology. Palomar College offers two-degree options. Geology A.S. (Associates) and A.S.-T (geology for transfer). A.A. Drone Technology.
Certificates: GIS; Astronomy; Drone Technology
Enrollment (2014): A: 35 (0)
Professor:
Wing H. Cheung, (D), California (Irvine), 2017, ZiRnZr
Doug Key, (M), San Diego State, Zy
Lisa Yon, (D), Brown, 1994, OgZg
Associate Professor:
Sean Figg, (M), N Colorado, 2012, Gg
Mark Lane, (M), San Diego State, 1996, Zn
Assistant Professor:
Steven Crook, (D), Zi

Pasadena City College

Dept of Geology ●
NAtural Sciences Division
1570 E. Colorado Boulevard
Pasadena, CA 91106

p. (626) 585-7138
naturalsciences@pasadena.edu
http://pasadena.edu/academics/divisions/natural-sciences/areas-of-study/geology.php
Dean, Natural Sciences:
David N. Douglass, (D), Dartmouth, 1987, Ze
Acting Chair:
Martha House, (D), MIT, Gg
Professor:
Elizabeth Nagy-Shadman, (D), Caltech, Gg
Yuet-Ling O'Connor, (D), S California, Ge
Bryan Wilbur, (D), California (Los Angeles), Gg
Instructor:
Michael Vendrasco, (D), California (Los Angeles)
Emeritus:
Gerald L. Lewis, (M), Cal State (Long Beach), 1965, Ps
Geology Lab Tech:
Debra A. Cantarero, (A), Pasadena City Coll, 2000, Zn

Pomona College

Geology Dept (B) ☒
185 East Sixth Street
Claremont, CA 91711-6339
p. (909) 621-8675
LKeala@pomona.edu
http://www.geology.pomona.edu
f: https://www.facebook.com/pomonacollegegeology?mibextid=ZbWKwL
Programs: Geology
Administrative Assistant: Lori Keala
Enrollment (2022): B: 22 (0)
Professor:
Robert R. Gaines, (D), California (Riverside), 2003, GsClPi
Eric B. Grosfils, (D), Brown, 1996, XgGvq
Linda A. Reinen, (D), Brown, 1993, Gct
Chair:
Jade Star Lackey, (D), Wisconsin, 2005, GipCs
Department Technician:
Jonathan Harris

Riverside City College

Dept of Physical Science: Geology (A)
4800 Magnolia Avenue
Riverside, CA 92506-1299
p. (951) 222-8350
william.phelps@rcc.edu
Enrollment (2016): A: 11 (0)
Assistant Professor:
William Phelps, (D), California (Riverside), 2007, PeOpPg

Sacramento City College

Dept of Physics, Astronomy, & Geology (A)
3835 Freeport Blvd.
Sacramento, CA 95822
p. (916) 558-2343
stantok@scc.losrios.edu
https://scc.losrios.edu/academics/programs-and-majors/geology
Programs: AS-T (Associate of Science for Transfer) in Geology
Enrollment (2021): A: 20 (1)
Professor:
Kathryn Stanton, (D), California (Davis), 2006, GgPg

Sacramento State University

Dept of Geology (B,M) ○
6000 J Street
Placer Hall
Sacramento, CA 95819-6043
p. (916) 278-6337
geology@csus.edu
http://www.csus.edu/geology/
Enrollment (2010): B: 77 (16) M: 18 (3)
Chair:
Tim C. Horner, (D), Ohio State, 1992, HwGs
Professor:
Kevin J. Cornwell, (D), Nebraska, 1994, GmeHs
David G. Evans, (D), Louisiana State, 1989, HqYg
Lisa Hammersley, (D), California (Berkeley), 2003, Gig
Brian Hausback, (D), California (Berkeley), 1984, Gv
Judith E. Kusnick, (D), California (Davis), 1996, Ze
Lecturer:
Barbara J. Munn, (D), Virginia Tech, 1997, Ggp
Adjunct Professor:
Brian Bergamaschi, (D), Washington, 1995, HwOc
Emeritus:
Diane H. Carlson, (D), Washington State, 1984, Gc
Charles C. Plummer, (D), Washington, 1969, Gp
Greg Wheeler, (D), Washington, Em
Secretary:
Stacy Lindley, (B), California (Sacramento), 2010, Zn
Instructional Support Tech:
Steven W. Rounds, (B), Cal State (Sacramento), 1990, Gg

Saddleback Community College

Dept of Earth and Ocean Sciences (A) ☒
28000 Marguerite Parkway
Mission Viejo, CA 92692
p. (949) 582-4820
jrepka@saddleback.edu
http://www.saddleback.edu/mse/geo/
f: https://www.facebook.com/Saddleback-College-Geology-731491460197935/
Enrollment (2015): A: 7 (1)
Chair:
James Repka, (D), California (Santa Cruz), 1998, GgmZe

San Bernardino County Museum

Geological Sciences Div
2024 Orange Tree Lane
Redlands, CA 92374
p. (909) 307-2669
melissa.russo@sbcm.sbcounty.gov
http://www.co.san-bernardino.ca.us/museum/
Senior Curator:
Kathleen B. Springer, (M), California (Riverside), Zg
Curator:
J. Chris Sagebiel, (M), Texas, 1998, Pv

San Bernardino Valley College

Earth and Spatial Sciences Program (A)
701 South Mount Vernon Ave.
San Bernardino, CA 92410
p. (909) 384-8638
theibel@valleycollege.edu
Head:
Todd Heibel, Zyi
Professor:
Stephen H. Sandlin, Zy
Instructor:
Gary M. Croft, Zy
Vanessa Engstrom, Zy
Walter Grossman, Og
Jeffrey Krizek, Zi
William Muir, Og
Solomon Nana Kwaku Nimako, Zi
Edmund Jekwu Ogbuchiekwe, Zy
Lisa Schmidt, Zy
Adjunct Professor:
Donald G. Buchanan, (M), Naval Postgrad Sch, 1975, Og

San Diego State University

Dept of Geological Sciences (B,M,D) ●☒
5500 Campanile Drive
San Diego, CA 92182-1020
p. (619) 594-5586
dkimbrough@mail.sdsu.edu
http://www.geology.sdsu.edu
f: https://www.facebook.com/SDSU.Geology
t: @sdsugeology
Administrative Assistant: Irene Occhiello, Pia Parrish
Enrollment (2015): B: 72 (9) M: 25 (12) D: 9 (0)
Chair:
David L. Kimbrough, (D), California (Santa Barbara), 1982, Cc
Associate Dean, Division of Undergraduate Studies:
Stephen A. Schellenberg, (D), S California, 2000, Pe

Professor:
Eric G. Frost, (D), S California, 1983, Zr
Gary H. Girty, (D), Columbia, 1983, Gc
Kim B. Olsen, (D), Utah, 1994, Ye
Thomas K. Rockwell, (D), California (Santa Barbara), 1983, Gm
Senior Scientist:
Barry B. Hanan, (D), Virginia Tech, 1980, Cc
Associate Professor:
Shuo Ma, (D), California (Santa Barbara), 2006, Ys
Kathryn W. Thorbjarnarson, (D), California (Los Angeles), 1990, Hq
Lecturer:
Victor E. Camp, (D), Washington State, 1976, Gv
Kevin Robinson, (M), San Diego State, 1996, Gc
Isabelle Sacramentogrilo, (M), San Diego State, 1999, Gg
Adjunct Professor:
Mario V. Caputo, (D), Cincinnati, 1988, GsdOn
Emeritus:
Patrick L. Abbott, (D), Texas, 1973, Gs
Kathe K. Bertine, (D), Yale, 1970, Cl
Steven M. Day, (D), California (San Diego), 1977, Ys
Clive E. Dorman, (D), Oregon State, 1974, OpAmt
George R. Jiracek, (D), California (Berkeley), 1972, Ye
J. Philip Kern, (D), California (Los Angeles), 1968, Pg
Daniel Krummenacher, (D), Geneva, 1959, Cc
Monte Marshall, (D), Stanford, 1971, YmGtc
Richard H. Miller, (D), California (Los Angeles), 1975, Ps
Gary L. Peterson, (D), Washington, 1963, Gr
Anton D. Ptacek, (D), Washington (St Louis), 1965, Gx

San Francisco State University

Dept of Earth & Climate Sciences (B,M)
1600 Holloway Avenue TH 509
San Francisco, CA 94132
p. (415) 338-2061
geosci@sfsu.edu
http://tornado.sfsu.edu
Enrollment (2013): B: 56 (0) M: 22 (0)
Chair:
Karen Grove, (D), Stanford, 1989, Gs
Professor:
David P. Dempsey, (D), Washington, 1985, As
Oswaldo Garcia, (D), SUNY (Albany), 1976, As
Mary L. Leech, (D), Stanford, 1999, GptGz
John P. Monteverdi, (D), California (Berkeley), 1977, As
David A. Mustart, (D), Stanford, 1972, GizGv
Raymond Pestrong, (D), Stanford, 1965, Ng
Associate Professor:
John Caskey, (D), Nevada (Reno), 1996, Gc
Petra Dekens, (D), California (Santa Cruz), 2007, Pe
Newell (Toby) Garfield, (D), Rhode Island, 1990, Op
Jason Gurdak, (D), Colorado Mines, 2006, HwCqAt
Leonard Sklar, (D), California (Berkeley), 2003, Gm
Assistant Professor:
Alexander Stine, (D), California (Berkeley), At
Adjunct Professor:
E. Jan Null, (D), California (Davis), 1974, Ge
Emeritus:
Charles E. Bickel, (D), Harvard, 1971, Gx
York T. Mandra, (D), Stanford, 1958, Pg
Erwin Seibel, (D), Michigan, 1972, On
Raymond Sullivan, (D), Glasgow, 1960, Go
Lisa D. White, (D), California (Santa Cruz), 1989, Pg
Geosciences Tech:
Russell McArthur, Zn

San Joaquin Delta College

Dept of Geology (A)
Stockton, CA 95204
p. (209) 954-5354
gfrost@deltacollege.edu
http://www.deltacollege.edu/div/scimath/geology.html
Professor:
Gina Marie Frost, (D), California (Santa Cruz), Gg

San Jose City College

Dept of Earth and Space Sciences
2100 Moorpark Ave
San Jose, CA 95128
p. (408) 288-3716
jessica.smay@sjcc.edu
http://www.sjcc.edu/academics/departments-divisions/physical-sciences
Programs: Earth Science
Oceanography
Introduction to Astronomy
Associate Professor:
Jessica J. Smay, (M), California (Santa Barbara), 2002, ZeGm

San Jose State University

Dept of Geology (B,M) O☒
One Washington Square
San Jose, CA 95192-0102
p. (408) 924-5050
leslie.blum@sjsu.edu
http://www.sjsu.edu/geology/
Enrollment (2009): B: 26 (7) M: 24 (6)
Chair:
Jonathan S. Miller, (D), North Carolina, 1994, GiCc
Professor:
Emmanuel Gabet, (D), California (Santa Barbara), 2002, Gm
Paula Messina, (D), CUNY, 1998, ZeGm
Ellen P. Metzger, (D), Syracuse, 1984, GpZe
Robert B. Miller, (D), Washington, 1980, Gct
June A. Oberdorfer, (D), Hawaii, 1983, HwGeHy
Donald L. Reed, (D), California (San Diego), 1985, GuYg
Assistant Professor:
Kimberly Blisniuk, (D), California (Davis), 2011, Gmt
Carlie Pietsch, (D), S California, 2015, PiePc
Ryan Portner, (D), Macquarie, 2010, Gdv
Emeritus Professor:
David W. Andersen, (D), Utah, 1973, GsCl
Calvin H. Stevens, (D), S California, 1963, PsGrs
John W. Williams, (D), Stanford, 1970, Ng

Santa Barbara City College

Dept of Earth & Planetary Sciences (A)
721 Cliff Drive
Santa Barbara, CA 93109
p. (805) 965-0581
ebkitao@sbcc.edu
Chair:
Michael A. Robinson, (D), California (Santa Barbara), 2009, ZyiAm
Professor:
Jeffrey W. Meyer, (D), California (Santa Barbara), 1992, GgzGx
Jan Schultz, (M), California (Santa Barbara), 1991, GgePg
Assistant Professor:
William Dinklage, (D), California (Santa Barbara), Gge
Emeritus:
Robert S. Gray, (D), Arizona, 1965, PvGzx

Santa Monica College

Earth Science Dept (A) O
1900 Pico Blvd.
Santa Monica, CA 90405
p. (310) 434-8652
drake_vicki@smc.edu
http://www.smc.edu/AcademicPrograms/EarthScience/Pages/
Programs: Geography, Geospatial Technologies
Certificates: Geospatial Technologies
Professor:
Pete Morris, (M), Wisconsin, 1999, Zyc
Associate Professor:
Jing Liu, (D), Wisconsin, 2017, ZyiZr
Adjunct Professor:
Alessandro Grippo, (D), Gs

Santa Rosa Junior College

Earth and Space Sciences ●
1501 Mendocino Avenue
Santa Rosa, CA 95401-4395
p. (707) 527-4365
aschott@santarosa.edu
http://online.santarosa.edu/presentation/?2992

Santiago Canyon College

Dept of Earth, Space, & Physical Sciences
8045 East Chapman Avenue
Orange, CA 92869-4512
p. (714) 564-4788
hannes_susan@sccollege.edu
Chair:
Debra A. Brooks, (M), Texas A&M, 1989, Yg
Adjunct Professor:
Gail F. Montwill, (M), Cal State (Los Angeles), 1988, Gg
Lizanne V. Simmons, (B), Brigham Young, 1979, Gr
Amy L. Stinson, (M), San Diego State, 1990, Gc
Adam D. Woods, (D), S California, 1998, Gs

Solano Community College

School of Mathematics and Science
4000 Suisun Valley Road
Fairfield, CA 94534
p. (707) 864-7211
danielle.widemann@solano.edu
http://www.solano.edu/

Sonoma State University

Dept of Geology (B) ☒
1801 East Cotati Avenue
Rohnert Park, CA 94928
p. (707) 664-2334
geology@sonoma.edu
http://geology.sonoma.edu/
Programs: B.S. in Geology, B.A. in Earth Science
Enrollment (2017): B: 80 (30)
Department Chair:
Matty Mookerjee, (D), Rochester, 2005, GcYgZi
Professor:
Matthew J. James, (D), California (Berkeley), 1987, PiRhPg
Associate Professor:
Owen Anfinson, (D), Calgary, 2012, GsdGg
Assistant Professor:
Marissa Mnich, (D), Massachusetts, 2019, GvpGg
Emeritus:
Thomas B. Anderson, (D), Colorado, 1969, GsrGd
Rolfe C. Erickson, (D), Arizona, 1968, Gx
Retired:
Daniel B. Karner, (D), California (Berkeley), 1997, Cc
Instructional Support Technician:
Phillip Mooney, (M), California (Davis), 2010, ZnGc

Stanford University ❐

Dept of Geological and Environmental Sciences (B,M,D) ☒
450 Jane Stanford Way Bldg 320
Stanford, CA 94305-2215
p. (650) 723-5002
cbaroni@stanford.edu
http://pangea.stanford.edu/GES/
Enrollment (2010): B: 21 (9) M: 11 (5) D: 45 (8)
Chair:
Kevin Boyce, (D)
Professor:
Jef Caers, (D)
C. Page Chamberlain, (D), Harvard, 1985, Yr
Robert B. Dunbar, (D), California (San Diego), 1981, On
W. Gary Ernst, (D), Johns Hopkins, 1959, Cp
Rodney C. Ewing, (D), Stanford, 1974, GzeCl
Stephan A. Graham, (D), Stanford, 1976, Go
Martin Grove, (D)
George Hilley, (D)
Donald R. Lowe, (D), Illinois, 1967, Gd
Wendy Mao, (D)
Elizabeth L. Miller, (D), Rice, 1977, Gc
Jonathan Payne, (D)
Associate Professor:
Jane Willenbring, (D)
Assistant Professor:
Mathieu Lapôtre, (D)
Andrew Leslie, (D)
Ayla Pamukcu, (D)
Laura Schaefer, (D)
Erik Sperling, (D)
Courtesy Professor:
James O. Leckie, (D), Harvard, 1970, Cl
Emeritus:
Atilla Aydin, (D), Stanford, 1978, Gc
Dennis K. Bird, (D), California (Berkeley), 1978, Cg
Gordon E. Brown, Jr., (D), Virginia Tech, 1970, Gz
Robert G. Coleman, (D), Stanford, 1957, Gi
Robert R. Compton, (D), Stanford, 1949, Gx
John W. Harbaugh, (D), Wisconsin, 1955, Gq
James C. Ingle, Jr., (D), S California, 1966, Pm
Andre G. Journel, (D), Nancy (France), 1977, Gq
Juhn G. Liou, (D), California (Los Angeles), 1970, Gp
Ronald J. P. Lyon, (D), California (Berkeley), 1954, Zr
Gail A. Mahood, (D), California (Berkeley), 1979, Gi
David D. Pollard, (D), Stanford, 1969, Gc
Irwin Remson, (D), Columbia, 1954, Hw
Jonathan F. Stebbins, (D), California (Berkeley), 1983, Cg

Dept of Geophysics (B,M,D) ☒
397 Panama Mall
Stanford, CA 94305-2215
p. (650) 497-3498
biondo@stanford.edu
https://earth.stanford.edu/geophysics/
Programs: Geophysics
Enrollment (2018): B: 2 (2) M: 7 (5) D: 76 (19)
Chair:
Howard A. Zebker, (D), Stanford, 1984, Yx
Research Professor:
William L. Ellsworth, (D), MIT, 1978, Ys
Courtesy Associate Professor:
Simon L. Klemperer, (D), Cornell, 1985, GtYes
Professor:
Gregory C. Beroza, (D), MIT, 1989, Ys
Biondo L. Biondi, (D), Stanford, 1990, Ye
Jerry M. Harris, (D), Caltech, 1980, Ys
Rosemary J. Knight, (D), Stanford, 1985, Hy
Paul Segall, (D), Stanford, 1981, Yg
Norman H. Sleep, (D), MIT, 1972, Yg
Mark D. Zoback, (D), Stanford, 1975, Yg
Associate Professor:
Eric M. Dunham, (D), California (Santa Barbara), 2005, Ys
Assistant Professor:
Dustin M. Schroeder, (D), Texas, 2014, Gl
Jenny Suckale, (D), Yg
Tiziana Vanorio, (D), Yx
Adjunct Professor:
Steven Gorelick, Yr
Emeritus:
Jon F. Claerbout, (D), MIT, 1967, Ye
Antony C. Fraser-Smith, (D), Auckland, 1966, Ym
Robert L. Kovach, (D), Caltech, 1962, Ys
Gerald M. Mavko, (D), Stanford, 1977, Yg
Amos M. Nur, (D), MIT, 1969, Yg
Joan Roughgarden, (D), Harvard, 1971, Ob
Department Manager:
Jen Kidwell, Zn

Taft College

Math/Science Div
29 Emmons Park Drive
Taft, CA 93268
p. (661) 763-7932
ggolling@taftcollege.edu

University of California, Berkeley

Dept of Earth & Planetary Science (B,M,D)
307 McCone
Berkeley, CA 94720-4767
p. (510) 642-3993
bbuffett@berkeley.edu
http://eps.berkeley.edu
Chair:
Hans-Rudolf Wenk, (D), Zurich, 1965, Gz
Professor:
Walter Alvarez, (D), Princeton, 1967, GrRh
Jillian Banfield, (D), Johns Hopkins, 1990, CoGz

James K. Bishop, (D), MIT/WHOI, 1977, Oc
George H. Brimhall, (D), California (Berkeley), 1972, Eg
Roland Burgmann, (D), Stanford, 1993, GtYd
Donald J. DePaolo, (D), Caltech, 1978, Cc
William E. Dietrich, (D), Washington, 1982, Gm
Inez Fung, (D), MIT, 1977, As
B. Lynn Ingram, (D), Stanford, 1992, Cs
Raymond Jeanloz, (D), Caltech, 1979, Gy
James W. Kirchner, (D), California (Berkeley), 1990, Ge
Michael Manga, (D), Harvard, 1994, GvXyHw
Mark A. Richards, (D), Caltech, 1986, Yg
Barbara A. Romanowicz, (D), Paris, 1979, Ys
Associate Professor:
Kristie Boering, (D), Stanford, 1991, As
Douglas S. Dreger, (D), Caltech, 1992, Ys
Assistant Professor:
Richard M. Allen, (D), Princeton, 2001, Ys
Burkhard Militzer, (D), Illinois (Urbana), 2000, Gy
Adjunct Associate Professor:
David L. Alumbaugh, (D), California (Berkeley), 1993, Ye
Paul Renne, (D), California (Berkeley), 1987, Cc
Adjunct Professor:
Steven Pride, (D), Texas A&M, 1991, Ys
Visiting Professor:
William D. Collins, (D), Chicago, 1988, As
Emeritus:
Mark S. Bukowinski, (D), California (Los Angeles), 1975, Gy
Ian S. Carmichael, (D), London (UK), 1958, Gi
Garniss H. Curtis, (D), California (Berkeley), 1951, Cc
Lane R. Johnson, (D), Caltech, 1966, Ys
Doris Sloan, (D), California (Berkeley), 1981, Pme
Chi-Yuen Wang, (D), Harvard, 1964, Yg

Dept of Environmental Science, Policy and Management (B,M,D) O☒
140 Mulford Hall
Berkeley, CA 94720-3110
p. (510) 643-3788
earthy@berkeley.edu
http://espm.berkeley.edu/
Enrollment (2018): B: 200 (200) D: 30 (10)

Dept of Integrative Biology (B,M,D)
3040 Valley Life Sciences Building #3140
Berkeley, CA 94720-3140
p. (510) 502-5887
johnh@berkeley.edu
http://ib.berkeley.edu/
Curator:
Carole S. Hickman, (D), Stanford, 1975, PgePg
Professor:
F. Stuart Chapin, (D), Stanford, 1973, Pe
William A. Clemens, (D), California (Berkeley), 1960, Pv
Robert J. Full, (D), SUNY (Buffalo), 1984, Pi
Harry W. Green, (D), Tennessee, 1977, Pv
Ned K. Johnson, (D), California (Berkeley), 1961, Pv
Mimi A. R. Koehl, (D), Duke, 1976, Pg
Brent Mishler, (D), Harvard, 1984, Pe
Kevin Padian, (D), Yale, 1980, Pv
Wayne P. Sousa, (D), California (Santa Barbara), 1977, Ob
Associate Professor:
Mary E. Power, (D), Washington, 1981, Pe
Emeritus:
Roy L. Caldwell, (D), Iowa, 1969, Pi
Carole S. Hickman, (D), Stanford, 1975, Pg
William Z. Lidicker, Jr., (D), Illinois, 1957, Pv
David R. Lindberg, (D), California (Santa Cruz), Pi
James L. Patton, (D), Arizona, 1968, Pv
Thomas M. Powell, (D), California (Berkeley), 1970, Op
Montgomery Slatkin, (D), Harvard, 1970, Pg
Glennis Thompson, (D), Melbourne (Australia), 1974, Zn
James W. Valentine, (D), California (Los Angeles), 1958, Pg
David B. Wake, (D), S California, 1964, Pg
Marvalee H. Wake, (D), S California, 1968, Pg

Dept of Materials Science and Engineering (B,M,D)
577 Evans Hall
Berkeley, CA 94720-1760
p. (510) 642-3801
hfmorrison@berkeley.edu
http://www.mse.berkeley.edu/
Chair:
Robert O. Ritchie, (D), Cambridge, 1973, Zm
Professor:
Alex Becker, (D), McGill, 1963, Yg
George A. Cooper, (D), Cambridge, 1967, Np
Didier deFontaine, (D), Northwestern, 1967, Zm
Lutgard DeJonghe, (D), California (Berkeley), 1970, Zm
Thomas M. Devine, (D), MIT, 1974, Zm
Fiona M. Doyle, (D), Imperial Coll (UK), 1983, Nx
James W. Evans, (D), SUNY (Buffalo), 1970, Nx
Douglas W. Fuerstenau, (D), MIT, 1953, Nx
Andreas Glaeser, (D), MIT, 1981, Zm
Ronald Gronsky, (D), California (Berkeley), 1977, Zm
Eugene Haller, (D), Basel (Switzerland), 1970, Zm
J. W. Morris, Jr., (D), MIT, 1969, Zm
H. Frank Morrison, (D), California (Berkeley), 1967, Yg
T. N. Narasimhan, (D), California (Berkeley), 1975, Hw
Timothy Sands, (D), California (Berkeley), 1984, Zm
Kalanadh V. S. Sastry, (D), California (Berkeley), 1970, Nx
Eicke Weber, (D), Cologne, 1976, Zm
Associate Professor:
Tad W. Patzek, (D), Silesian Tech, 1979, Np
James W. Rector, III, (D), Stanford, 1990, Ys
Assistant Professor:
Daryl Chrzan, (D), California (Berkeley), 1989, Zm
Mauro Ferrari, (D), California (Berkeley), 1989, Zm

GeoEngineering Program (B,M,D)
Berkeley, CA 94720
p. (510) 642-3157
pestana@ce.berkeley.edu
http://www.ce.berkeley.edu/geo/
Chair:
Lisa Alvarez-Cohen
Professor:
Alex Becker, (D), McGill, 1964, Ye
Huntly Frank Morrison, (D), California (Berkeley), 1967, Ye
Senior Scientist:
Ki Ha Lee, (D), California (Berkeley), 1978, Ye
Assistant Professor:
James W. Rector, (D), Stanford, 1990, Ye

Museum of Paleontology (D)
1101 Valley Life Sciences Bldg #4780
Berkeley, CA 94720-4780
p. (510) 642-1821
crmarshall@berkeley.edu
http://www.ucmp.berkeley.edu
Enrollment (2016): D: 25 (3)
Professor:
Charles R. Marshall, (D), Chicago, 1989, PqgPi
Curator:
Kevin Padian, (D), Yale, 1980, Pv
Principal Museum Scientist:
Mark B. Goodwin, (D), California (Davis), 2008, Pv
Museum Scientist:
Diane Erwin, (D), Alberta, 1990, Pb
Pat Holroyd, (D), Duke, 1994, Pv
Curator:
Leslea Hlusko, (D), Penn State, 2000, PgZnn
Associate Professor:
Cynthia Looy, (D), Utrecht, 1995, Pb
Assistant Professor:
Seth Finnegan, (D), California (Riverside), 1994, Pig
Curatorial Associate:
Walter Alvarez, (D), Princeton, 1967, Gr
Curator:
Anthony D. Barnosky, (D), Washington, 1983, Pv
William A. Clemens, (D), California (Berkeley), 1960, Pv
David R. Lindberg, (D), California (Santa Cruz), 1983, Pg
Emeritus:
Roy Caldwell, (D), Iowa, 1969, Pg
James W. Valentine, (D), California (Los Angeles), 1958, Pg
Assistant Director:
Lisa D. White, California (Santa Cruz), 1989, ZePim

University of California, Davis

Dept of Earth and Planetary Sciences (B,M,D) ●⊘
2119 Earth & Physical Sciences Building
One Shields Ave.
Davis, CA 95616-5270
p. (530) 752-0350
geology@ucdavis.edu
http://eps.ucdavis.edu
f: https://www.facebook.com/UCDavisEarthPlanetarySci/
Programs: Geology (B,M,D); Marine and Coastal Science (B)
Enrollment (2021): B: 172 (34) M: 6 (5) D: 39 (5)

Chair:
Kari M. Cooper, (D), California (Los Angeles), 2001, Gi

Professor:
Eliot A. Atekwana, (D), W Michigan, 1996, CsHgGe
Magali I. Billen, (D), Caltech, 2001, Yg
Sandra J. Carlson, (D), Michigan, 1986, Pg
Eric S. Cowgill, (D), California (Los Angeles), 2001, Gtc
Tessa M. Hill, (D), California (Santa Barbara), 2004, Pe
Isabel P. Montañez, (D), Virginia Tech, 1990, Gd
Ryosuke Motani, (D), Toronto, 1997, Pv
Sujoy Mukhopadhyay, (D), Caltech, 2002, Cg
Michael Oskin, (D), Caltech, 2002, GmtGc
David A. Osleger, (D), Virginia Tech, 1990, Pc
Nicholas Pinter, (D), California (Santa Barbara), 1992, Gm
John Rundle, (D), California (Los Angeles), 1976, Yd
Howard J. Spero, (D), California (Santa Barbara), 1986, Cm
Sarah T. Stewart, (D), Caltech, 2002
Dawn Y. Sumner, (D), MIT, 1995, Gs
Geerat J. Vermeij, (D), Yale, 1971, Pe
Qing-zhu Yin, (D), Max Planck, 1995, Cc

Assistant Professor:
David Gold, (D), California (Los Angeles), 2014, Pg
Barbara C. Ratschbacher, (D), S California, 2017, GgCg
Maxwell L. Rudolph, (D), California (Berkeley), 2012, Yg

Research Associate:
Oliver Kreylos, (D), California (Davis), 2003, Ng
Ann D. Russell, (D), Washington, 1994, Cm
Peter Thy, (D), Aarhus (Denmark), 1982, Gx
Burak Yikilmaz, California (Davis), 2010, Gc

Emeritus:
Cathy J. Busby, (D), Princeton, 1983, Gt
Richard Cowen, (D), Cambridge, 1966, Pg
Howard W. Day, (D), Brown, 1971, Gx
John Dewey, (D), London (UK), 1960, Gt
James A. Doyle, (D), Harvard, 1971, Pb
Graham E. Fogg, (D), Texas, 1986, Hw
Charles G. Higgins, (D), California (Berkeley), 1950, Gm
Charles E. Lesher, (D), Harvard, 1985, Gi
James S. McClain, (D), Washington, 1979, Yr
Jeffrey F. Mount, (D), California (Santa Cruz), 1980, Gm
Alexandra Navrotsky, (D), Chicago, 1967, Zm
Sarah M. Roeske, (D), California (Santa Cruz), 1988, GtcGp
James R. Rustad, (D), Minnesota, 1992, Cl
Peter Schiffman, (D), Stanford, 1978, Gp
Donald L. Turcotte, (D), Caltech, 1958, Yg
Robert J. Twiss, (D), Princeton, 1970, Gc
Kenneth L. Verosub, (D), Stanford, 1973, Ym
Robert A. Zierenberg, (D), Wisconsin, 1983, Cs

Chief Administrative Officer:
Corinne Esser, (M), California (Davis), Zn

Cooperating Faculty:
William H. Casey, (D), Penn State, 1985, Cl

Dept of Land, Air & Water Resources (B,M,D) O☒
1110 Plant and Environmental Sciences Building
One Shields Avenue
Davis, CA 95616
p. (530) 752-1130
rjsouthard@ucdavis.edu
http://lawr.ucdavis.edu/
Programs: Atmospheric Science; Hydrology/Hydrologic Science; Environmental Science and Management; Soils and Biogeochemistry
Enrollment (2014): B: 5 (5) M: 54 (19) D: 55 (8)

Specialist in Cooperative Extension:
Thomas Harter, (D), Arizona, 1994, Hg
Toby O'Geen, (D), Idaho, 2002, So

Professor:
Cort Anastasio, (D), Duke, 1994, As
Shu-Hua Chen, (D), Purdue, 1999, ZrAsm
Randy A. Dahlgren, (D), Washington, 1987, So
Graham E. Fogg, (D), Texas, 1986, Hw
Richard Grotjahn, (D), Florida State, 1979, As
Peter J. Hernes, (D), Washington, 1999, Hg
Jan W. Hopmans, (D), Auburn, So
William R. Horwath, (D), Michigan State, 1993, Sob
Louise E. Jackson, (D), Washington, 1982, So
Terrence R. Nathan, (D), SUNY (Albany), 1985, Asm
Kate Scow, (D), Cornell, 1988, So
Randal J. Southard, (D), North Carolina State, 1983, So
Susan L. Ustin, (D), California (Davis), 1983, Zr

Associate Professor:
Ian C. Faloona, (D), Penn State, 2001, AscAp
Ben Houlton, (D), Princeton, So
Sanjai Parikh, (D), Arizona, So

20% Cooperative Extension:
Samuel Sandoval Solis, (D), Texas, 2011, Hg

Assistant Professor:
Yufang Jin, (D), Boston, 2002, Zr

Adjunct Professor:
Travis A. O'Brien, (D), California (Santa Cruz), 2011, AstAp
Minghua Zhang, (D), California (Davis), 1993, Hg

Emeritus:
James H. Richards, (D), Alberta, 1981, So

Asst Proj Scientist:
Michael L. Grieneisen, (D), North Carolina, 1992, Hg

Asst Professional Researcher:
Lucas CR Silva, (D), Guelph, 2011, Sb

Cooperating Faculty:
Daniel Geisseler, (D), California (Davis), 2009, So

Cooperating Faculty:
Stephen R. Grattan, (D), California (Riverside), 1984, Sp

Cooperating Faculty:
Richard L. Snyder, (D), Iowa State, 1980, As

Cooperating Faculty:
Daniele Zaccaria, (D), Utah State, 2011, Hg

Dept of Land, Air & Water Resources - Hydrology Program (B,M,D) ☒
1110A Plant Environmental Sciences
One Shields Avenue
Davis, CA 95616-8628
p. (530) 752-3060
thharter@ucdavis.edu
http://lawr.ucdavis.edu/
Programs: B.S. in Hydrology, Soil Sciences, Environmental Science and Management, Atmospheric Sciences. M.S. and Ph.D. in Hydrologic Sciences, Soils and Biogeochemistry, Atmospheric Sciences.
Enrollment (2012): B: 25 (0) M: 16 (0) D: 21 (0)

Groundwater Hydrology Specialist:
Thomas L. Harter, (D), Arizona, 1993, Hw

Professor:
Cort Anastasio, (D), Duke, Acs
Shu-Hua Chen, (D), Purdue, Apm
Mark E. Grismer, (D), Colorado State, 1984, Hw
Peter J. Hernes, (D), Washington, CoHg
William R. Horwath, (D), Michigan State, ScCo
Terry Nathan, SUNY (Albany), AsZc
Toby A. O'Geen, Idaho, Sod
Gregory B. Pasternack, (D), Johns Hopkins, 1998, GmHg
Kyaw Tha Paw U, (D), Yale, Asm
Carlos E. Puente, (D), MIT, 1984, Hqg
Jorge Rodrigues, Michigan State, SbPo
Kate M. Scow, Cornell, Sb
Anthony S. Wexler, (D), Caltech, ZcAsZn
Minghua Zhang, (D), California (Davis), HgEg

Associate Professor:
Majdi Abou Najm, (D), CbSb
Helen E. Dahlke, (D), Cornell, 2011, HgsZi
Yufang Jin, Boston, ZrHg
Isaya Kisekka, Florida, ZuRw
Sanjai Parikh, Arizona, ScbCb
Kerri Steenwerth, Cb
Paul Ullrich, Waterloo, As

Associate Scientist:
Mallika Nocco, Wisconsin, Sp
Kosana Suvocarev, (D), Zaragoza, Asm
Daniele Zaccaria, Utah State (Logan), HgRw
Assistant Professor:
Rebecca Hernandez, (D), Stanford, Zcn
Adele Igel, (D), Colorado State, Asp
Rebecca A. Lybrand, SbCb
Erwan Monier, California (Davis), Atm
Samuel Sandoval, (D), Texas, 2011, Hg
Da Yang, Caltech, Ast
Research Associate:
Victor P. Claassen, (D), California (Davis), So
Daniel Geisseler, (D), California (Davis), Sb
Adjunct Professor:
Laura Foglia, (D), ETH (Switzerland), HwZc
Matthew Igel, (D), Colorado State, Asm

Graduate Group in Hydrologic Sciences (M,D)
1152 PES
One Shields Ave
Davis, CA 95616
p. (530) 752-1669
lawrgradadvising@ucdavis.edu
http://hydscigrad.ucdavis.edu/
Administrative Assistant: Diane Swindall
Enrollment (2010): M: 14 (1) D: 14 (0)
Chair:
Graham E. Fogg, (D), Texas, 1986, Hy
Professor:
William H. Casey, (D), Pennsylvania, 1985, Cl
Jeannie Darby, (D), Texas, 1988, Hg
Mark E. Grismer, (D), Colorado State, 1984, Hw
Jan W. Hopmans, (D), Auburn, 1985, So
Theodore C. Hsiao, (D), Illinois, 1964, Zn
B. E. Larock, (D), Stanford, 1966, Zn
Jay R. Lund, (D), Washington, 1986, Hs
M. A. Marino, (D), California (Los Angeles), 1972, Hw
James S. McClain, (D), Washington, 1979, Yr
Eldridge M. Moores, (D), Princeton, 1963, Gt
Jeffrey F. Mount, (D), California (Santa Cruz), 1980, Gs
Dennis E. Rolston, (D), California (Davis), 1971, So
K. K. Tanji, (M), California (Davis), 1961, Zn
Wes W. Wallender, (D), Utah State, 1982, Hg
B. C. Weare, (D), SUNY (Buffalo), 1974, As
Associate Professor:
Bruce Kutter, (D), Cambridge, 1983, Ne
Carlos E. Puente, (D), MIT, 1984, Hg
Emeritus:
Robert A. Matthews, (B), California (Berkeley), 1953, Ge

University of California, Irvine

Dept of Earth System Science (B,D)
School of Physical Sciences
Irvine, CA 92697-3100
p. (949) 824-8794
fprimeau@uci.edu
http://www.ess.uci.edu
Enrollment (2013): B: 166 (64) D: 49 (5)
Chair:
Michael L. Goulden, (D), Stanford, 1992, Zg
Advance Professor:
Ellen R. M. Druffel, (D), California (San Diego), 1980, Oc
Professor:
James Famiglietti, (D), Princeton, Hw
Gudrun Magnusdottir, (D), Colorado State, 1989, As
Michael J. Prather, (D), Yale, 1975, As
Eric Rignot, (D), S California, 1991, GlOpZr
Eric S. Saltzman, (D), Rosenstiel, 1986, As
Soroosh Sorooshian, (D), California (Los Angeles), 1978, Hg
Susan E. Trumbore, (D), Columbia, 1989, Sc
Jin-Yi Yu, (D), Washington, AsOpZc
Charles Zender, (D), Colorado, 1996, As
Associate Professor:
Steven J. Davis, (D), Stanford, 2008, GeCsZu
Jefferson Keith Moore, (D), Oregon State, 1999, Op
Francois Primeau, (D), MIT/WHOI, 1998, Op

Assistant Professor:
Claudia Czimczik, (D)
Todd Dupont, (D), Penn State, 2004, Gl
Julie Ferguson, (D), Oxford, 2008, Ze
Kathleen Johnson, (D), California (Berkeley), 2004, Zg
Saewung Kim, (D), Georgia Tech, 2007, As
Adam Martiny, (D), Tech (Denmark), 2003, Ob
Michael Pritchard, (D)
Isabella Velicogna, (D), Trieste, 1999, Zr
Laboratory Director:
John Southon, (D), Auckland, 1976, CcOcPe

University of California, Los Angeles

Dept of Atmospheric and Oceanic Sciences (B,M,D) ☒
7127 Math Sciences
Box 951565
Los Angeles, CA 90095
p. (310) 825-1217
paulson@atmos.ucla.edu
http://www.atmos.ucla.edu
f: https://www.facebook.com/AOS.UCLA
Chair:
Suzanne Paulson, (D), Caltech, 1991, As
Professor:
Jacob Bortnik, (D)
Marcelo Chamecki
Gang Chen
Rong Fu
Alex Hall, (D)
Qinbin Li
Kuo Nan Liou, (D), New York, 1970, As
James McWilliams, (D), Harvard, 1971, OgAs
J. David Neelin, (D), Princeton, 1987, As
Jochen P. Stutz, (D), Heidelberg, 1996, As
Tina Treude
Yongkang Xue, (D)
Associate Professor:
Jasper Kok , (D)
Ulli Seibt , (D)
Arahdna Tripati, (D)
Assistant Professor:
Daniele Bianchi , (D)
Rob Eagle
Andrew Stewart, (D)
Lecturer:
Jeffrey Lew, (D), California (Los Angeles), 1985, As
Adjunct Professor:
Yi Chao
Emeritus:
Akio Arakawa, (D), Tokyo, 1961, As
Robert Fovell, (D), Illinois, 1988, As
Michael Ghil, (D), Courant Inst, 1975, As
Lawrence Lyons, (D), California (Los Angeles), 1972, As
Carlos R. Mechoso, (D), Princeton, 1978, As
Richard M. Thorne, (D), MIT, 1968, As
Richard Turco, (D), Illinois, 1971, As

Dept. of Earth, Planetary, and Space Sciences (B,M,D) O☒
595 Charles E Young Drive East
3806 Geology Building, Box 951567
Los Angeles, CA 90095-1567
p. 310.825.3880
info@epss.ucla.edu
https://epss.ucla.edu
f: https://www.facebook.com/uclaepss
t: @uclaepss
Programs: Geology (BS); Geology/Engineering Geology (BS); Geophysics (BS); Earth & Environmental Science (BA)
Enrollment (2022): B: 89 (28) M: 3 (15) D: 84 (11)
Chair:
Carolina Lithgow-Bertelloni, (D), California (Berkeley), 1994, Yg
Associate Dean for Research:
Hilke Schlichting, (D), Caltech, 2009, Xy
Professor:
Vassilis Angelopoulos, (D), California (Los Angeles), 1993, XyAsYm
Jonathan M. Aurnou, (D), Johns Hopkins, 1999, YgmYx
T. Mark Harrison, (D), Australian Nat, 1981, CcGt
David Jewitt, (D), Caltech, 1983, XyZnn

Abby Kavner, (D), California (Berkeley), 1997, GyYg
Craig E. Manning, (D), Stanford, 1989, GxCp
Jean-Luc Margot, (D), Cornell, 1999, XyYdv
Kevin D. McKeegan, (D), Washington (St Louis), 1987, XcCsc
Jonathan Mitchell, (D), Chicago, 2007, AsXg
William I. Newman, (D), Cornell, 1979, Xy
David A. Paige, (D), Caltech, 1985, Xy
Gilles Peltzer, (D), Paris VII, 1987, ZrGt
Edwin A. Schauble, (D), Caltech, 2002, Csg
J. William Schopf, (D), Harvard, 1968, Pg
Lars Stixrude, (D), California (Berkeley), 1991, YgXy
Tina Treude, (D), Max Planck, 2004, ObPgCm
Aradhna Tripati, (D), California (Santa Cruz), 2002, CmPeCs
Marco Velli, (D), Pisa, 1985, Yg
An Yin, (D), S California, 1988, Gt
Edward D. Young, (D), S California, 1990, CsXc

Associate Professor:
Caroline Beghein, (D), Utrecht, 2003, Ysg
Lingsen Meng, (D), Caltech, 2012, Ys
Seulgi Moon, (D), Stanford, 2013, Gm
Ulrike Seibt, (D), Hamburg, 2003, PgCg

Assistant Professor:
Hao Cao, (D), California (Los Angeles), 2014, Xy
Mackenzie Day, (D), Caltech, 2017, Ggs

Adjunct Professor:
Robert C. Newton, (D), California (Los Angeles), 1963, CpGp
Edward J. Rhodes, (D), Oxford, 1990, CcGma

Emeritus:
G. Peter Bird, (D), MIT, 1976, GtYsRn
Paul M. Davis, (D), Queensland, 1974, YsmYg
Wayne A. Dollase, (D), MIT, 1966, Gz
Clarence A. Hall, (D), Stanford, 1956, GgtGe
Raymond V. Ingersoll, (D), Stanford, 1976, Gst
David D. Jackson, (D), MIT, 1969, YsdYg
Margaret G. Kivelson, (D), Radcliffe, 1957, Xy
Robert L. McPherron, (D), California (Berkeley), 1968, Xy
Paul M. Merifield, (D), Colorado, 1963, Ng
Bruce Runnegar, (D), Queensland, 1967, Pg
Christopher T. Russell, (D), California (Los Angeles), 1968, Xy
Gerald Schubert, (D), California (Berkeley), 1964, YgXy
Laurence C. Smith, (D), Cornell, 1996, Hs
Raymond J. Walker, (D), California (Los Angeles), 1973, Xy

Inst of Geophysics & Planetary Physics
3845 Slichter Hall
Los Angeles, CA 90095-1567
p. (310) 825-1664
jnakatsu@igpp.ucla.edu
http://www.igpp.ucla.edu

Professor:
Vassilis Angelopoulos, (D), California (Los Angeles), 1993, Yg
Maha Ashour-Abdalla, (D), Imperial Coll (UK), 1971, Zn
Friedrich H. Busse, (D), Munich, 1967, Zn
Richard E. Dickerson, (D), Minnesota, 1957, Zn
Michael Ghil, (D), New York, 1975, AsYdOp
T. Mark Harrison, (D), Australian Nat, 1980, Cc
Charles F. Kennel, (D), Princeton, 1964, Zn
Margaret G. Kivelson, (D), Radcliffe, 1957, Xy
Robert L. McPherron, (D), California (Berkeley), 1968, Xy
James C. McWilliams, (D), Harvard, 1971, As
Paul H. Roberts, (D), Cambridge, 1967, Ym
Bruce Runnegar, (D), Queensland, 1967, Pg
Christopher T. Russell, (D), California (Los Angeles), 1968, Yg
J. William Schopf, (D), Harvard, 1968, Pg
Gerald Schubert, (D), California (Berkeley), 1964, Yg
Karl O. Stetter, (D), Tech (Munich), 1973, Pg
Richard Turco, (D), Illinois, 1971, As
Raymond J. Walker, (D), California (Los Angeles), 1973, Zn
John T. Wasson, (D), MIT, 1958, Xm
Edward Young, Cg

Senior Scientist:
Stanislav I. Braginsky, (D), Moscow Inst, 1948, Ym
Robert J. Strangeway, (D), London (UK), 1978, Zn
M I. Venkatesan, (D), Madras (India), 1973, Co
Paul Warren, (D), California (Los Angeles), 1979, Xm

Associate Professor:
Abby Kavner, (D), California (Berkeley), 1997, Yg
J. David Neelin, (D), Princeton, 1987, As

Associate Research Scientist:
Jean Berchem, (D), California (Los Angeles), 1986, Zn
Gregory Kallemeyn, (D), California (Los Angeles), 1982, Cg
Frank T. Kyte, (D), California (Los Angeles), 1983, Ct
Alan E. Rubin, (D), New Mexico, 1982, Xm
David Schriver, (D), California (Los Angeles), 1988, Yg
Fred Schwab, (B), California (Los Angeles), 1960, Ys

Assistant Research Scientist:
Kayo Ide, (D), Caltech, 1990, Zg
Krishan Khurana, (D), Durham (UK), 1984, Zn
Robert Richard, (D), California (Los Angeles), 1988, Yg
William Smythe, (D), California (Los Angeles), 1979, Ys
Ferenc D. Varadi, (D), California (Los Angeles), 1989, Zn

Emeritus:
Orson L. Anderson, (D), Utah, 1951, Yx
Paul J. Coleman, Jr., (D), California (Los Angeles), 1966, Gc
Isaac R. Kaplan, (D), S California, 1962, Cg
Leon Knopoff, (D), Caltech, 1949, Ys
Ronald L. Shreve, (D), Caltech, 1958, Gm

Department Manager:
James Nakatsuka, Zn

University of California, Riverside

Department of Earth Sciences (B,M,D) ●

Department of Earth Sciences
University of California, Riverside
Riverside, CA 92521
p. (951) 827-3434
david.oglesby@ucr.edu
http://earthsciences.ucr.edu
f: https://www.facebook.com/ucrearthsciences/
Programs: Geology, Earth Sciences, Geophysics
Enrollment (2013): B: 51 (13) M: 15 (8) D: 38 (4)

Chair:
David D. Oglesby, (D), California (Santa Barbara), 1999, Ys

Distinguished Professor:
Timothy W. Lyons, (D), Yale, 1992, CgOcCl

Professor:
Mary L. Droser, (D), S California, 1987, Pg
Nigel C. Hughes, (D), Bristol, 1990, Pg
Gordon D. Love, (D), Strathclyde, 1995, Cog
Richard A. Minnich, (D), California (Los Angeles), 1978, Zy
Andy Ridgwell, (D), E Anglia (UK), 2001, Zn
Peter M. Sadler, (D), Bristol, 1973, GrPs

Associate Professor:
Robert J. Allen, (D), Yale, 2009, As
Andrey Bekker, (D), Virginia Tech, 2001, GsrCg
Gareth J. Funning, (D), Oxford, 2005, YdgYs
Abhijit Ghosh, (D), Washington, 2011, YsGt
Michael A. McKibben, (D), Penn State, 1984, Cg

Assistant Professor:
Nicolas Barth, (D), Otago (NZ), 2013, Gt
Maryjo Brounce, (D), Rhode Island, 2014, Cg
Roby Douilly, (D), Purdue, 2016, YsGtYg
Heather Ford, (D), Brown, 2013, Ys
Sandra Kirtland Turner, (D), Scripps, Cs

Adjunct Assistant Professor:
Katherine J. Kendrick, (D), California (Riverside), 1999, Gm
Thomas A. Scott, (D), California (Berkeley), 1987, Zy

Adjunct Professor:
Elizabeth Cochran, California (Los Angeles), 2005, Ys
Larissa F. Dobrzhinetskaya, (D), Inst of Physics of Earth (Moscow), 1978, Gx
Douglas M. Morton, (D), California (Los Angeles), 1966, Gp

Professor of the Graduate Division:
James H. Dieterich, (D), Yale, 1968, Yg

Emeritus:
Wilfred A. Elders, (D), Durham (UK), 1961, Gi
Michael A. Murphy, (D), California (Los Angeles), 1954, Ps
Stephen K. Park, (D), MIT, 1984, Ym
Michael O. Woodburne, (D), California (Berkeley), 1966, Pv

University of California, San Diego

Scripps Institution of Oceanography (B,M,D) ☒

Graduate & Undergraduate Office
9500 Gilman Dr.
Mail Code 0208
La Jolla, CA 92093-0208

p. (858) 534-3206
siodept@sio.ucsd.edu
https://scripps.ucsd.edu/education/
Programs: Geoscience, Earth Sciences, Environmental Systems, Oceanic and Atmospheric Sciences, Oceanography, Marine Biology
Enrollment (2022): B: 114 (26) M: 10 (11) D: 73 (7)

Vice Chancellor:
Margaret Leinen, (D), Rhode Island, 1980, OgPe

Professor:
Brian Palenik, (D), MIT/WHOI, 1989, Ob

Teaching Professor:
Janet Becker, (D), California (San Diego), Op
Geoffrey Cook, (D), Washington State, Gv

Department Vice Chair:
Lihini I. Aluwihare, (D), MIT/WHOI, 1999, Gu

Department Chair:
Sarah T. Gille, (D), MIT/WHOI, 1995, Op

Professor:
Eric E. Allen, (D), California (San Diego), 2002, Ob
Andreas Andersson, (D), Hawaii, 2006, Oc
Laurence Armi, (D), California (Berkeley), 1975, Op
Katherine A. Barbeau, (D), MIT/WHOI, 1998, Oc
Douglas H. Bartlett, (D), Illinois, 1985, Ob
Adrian Borsa, (D), Scripps, 2005, Hg
Michael J. Buckingham, (D), Reading (UK), 1971, Op
Ronald S. Burton, (D), Stanford, 1981, Ob
Paterno R. Castillo, (D), Washington (St Louis), 1987, Giu
Paola Cessi, (D), MIT, 1987, Op
Christopher D. Charles, (D), Columbia, 1991, Pe
Catherine G. Constable, (D), California (San Diego), 1987, Ymg
Steven C. Constable, (D), Australian Nat, 1983, Yr
James Day, (D), Durham (UK), Cg
Andrew G. Dickson, (D), Liverpool (UK), 1978, Cm
Neal W. Driscoll, (D), Columbia, 1992, Gu
Yuri A. Fialko, (D), Princeton, 1998, YgdGt
Peter J. S. Franks, (D), MIT/WHOI, 1990, Ob
Helen A. Fricker, (D), Tasmania, 1999, Yr
Terry Gaasterland, (D), Maryland, 1992, Ob
Jeffrey S. Gee, (D), California (San Diego), 1991, Gu
William H. Gerwick, (D), California (San Diego), 1981, Cm
Carl H. Gibson, (D), Stanford, 1962, No
Vicki Grassian, (D), California (Berkeley), 1987, Cm
Philip A. Hastings, (D), Arizona, 1987, Ob
John A. Hildebrand, (D), Stanford, 1983, Yr
Ralph F. Keeling, (D), Harvard, 1988, As
Gabi Laske, (D), Karlsruhe, 1993, Ys
James J. Leichter, (D), Stanford, 1997, Ob
Todd Martz, (D), Montana, 2005, Gg
J. Bernard H. Minster, (D), Caltech, 1974, Ys
Mario J. Molina, (D), California (Berkeley), 1972, As
Bradley S. Moore, (D), Zurich, 1995, Cm
Joel R. Norris, (D), Washington, 1997, As
Richard D. Norris, (D), Harvard, 1990, Gu
Mark D. Ohman, (D), Washington, 1983, Ob
Kimberly A. Prather, (D), California (Davis), 1990, AsCmHg
V. Ramanathan, (D), SUNY (Stony Brook), 1973, As
Dean H. Roemmich, (D), MIT/WHOI, 1980, Op
Gregory W. Rouse, (D), Sydney, 1991, Ob
Daniel L. Rudnick, (D), California (San Diego), 1987, Op
Lynn M. Russell, (D), Caltech, 1995, As
David T. Sandwell, (D), California (Los Angeles), 1981, Yr
Uwe Send, (D), California (San Diego), 1988, Op
Jeffrey P. Severinghaus, (D), Columbia, 1995, Pe
Peter M. Shearer, (D), California (San Diego), 1986, Ysg
Jennifer E. Smith, (D), Hawaii, 2003, Obn
Len Srnka, (D), Newcastle upon Tyne, 1974, Ym
Dave Stegman, (D), California (Berkeley), Yd
Dariusz Stramski, (D), Gdansk (Poland), 1985, Op
George Sugihara, (D), Princeton, 1983, Ob
Lynne D. Talley, (D), MIT/WHOI, 1982, OpZc
Lisa Tauxe, (D), Columbia, 1983, Ym
Bradley T. Werner, (D), Caltech, 1987, Gm
William R. Young, (D), MIT/WHOI, 1981, Op

Associate Professor:
Jade D'Alpoim Guedes, (D), Harvard, Ga
Alice Gabriel, (D), ETH Zurich, 2013, Ys
Sarah Giddings, (D), Stanford, 2010, Op
Dave May, (D), Monash
Isabel Rivera-Collazo, (D), Univ (London), Ga

Assistant Professor:
Sarah Aarons, (D), Michigan
Emily Chin, (D), Rice, CuGx
Julia Diaz, (D), Georgia Tech, 2011
Wenyuan Fan, (D), California (San Diego), 2017, Ys
Morgan Levy, (D), California (Berkeley), Hg
Andrew Lucas, (D), California (San Diego), 2009, Op
Nicholas Lutsko, (D), Princeton, As
Dierdre Lyons, (D), California (Berkeley), Ob
Florian Meyer, (D), TU Wein, Austria, Op
Ross Parnell-Turner, (D), Cambridge
Colleen Petrik, (D), MIT-WHOI, 2011, Ob
Sarah Purkey, (D), Washington, 2014, Op
Katherine Ricke, (D), Carnegie Mellon, Zo
Amina Schartup, (D), Connecticut, Ct
Vashan Wright, (D), S Methodist, Gt

Adjunct Professor:
Jay P. Barlow, (D), California (San Diego), 1982, Ob

Emeritus:
Duncan C. Agnew, (D), California (San Diego), 1979, Ys
Farooq Azam, (D), Czech Acad of Sci, 1968, Ob
George E. Backus, (D), Chicago, 1956, YmsOp
Jeffrey L. Bada, (D), California (San Diego), 1968, Co
Kevin M. Brown, (D), Durham (UK), 1987, Gu
David M. Checkley, (D), California (San Diego), 1978, Ob
Paul Crutzen, (D), Oc
Joseph R. Curray, (D), Scripps, 1959, Gu
Paul K. Dayton, (D), Washington, 1970, Ob
Leroy M. Dorman, (D), Wisconsin, 1970, Yg
Horst Felbeck, (D), Muenster (Germany), 1979, Ob
William H. Fenical, (D), California (Riverside), 1968, Oc
Joris M. Gieskes, (D), Manitoba, 1965, Oc
Robert T. Guza, (D), California (San Diego), 1974, On
James W. Hawkins, (D), Washington, 1963, OuGip
Anthony DJ Haymet, (D), Chicago, 1981, OcZgn
Myrl C. Hendershott, (D), Harvard, 1966, Op
William S. Hodgkiss, Jr., (D), Duke, 1975, No
Nicholas D. Holland, (D), Stanford, 1965, Ob
Jeremy B. C. Jackson, (D), Yale, 1971, PgePi
Miriam Kastner, (D), Harvard, 1970, CmlCm
Charles F. Kennel, (D), Princeton, 1964, As
Gerald L. Kooyman, (D), Arizona, 1966, Ob
William A. Kuperman, (D), Maryland, 1972, Op
Michael R. Landry, (D), Washington, 1976, Ob
Lisa A. Levin, (D), California (San Diego), 1982, Ob
Peter F. Lonsdale, (D), California (San Diego), 1974, Gu
J. Douglas Macdougall, (D), California (San Diego), 1972, Cg
T. Guy Masters, (D), Cambridge, 1979, YsGy
John A. McGowan, (D), California (San Diego), 1960, Ob
John A. Orcutt, (D), California (San Diego), 1976, YrsYd
Robert L. Parker, (D), Cambridge, 1966, Ym
Robert Pinkel, (D), California (San Diego), 1974, Op
Richard L. Salmon, (D), California (San Diego), 1976, Op
John G. Sclater, (D), Cambridge, 1966, Yh
Richard C. J. Somerville, (D), New York, 1966, As
Victor D. Vacquier, (D), California (Berkeley), 1968, Ob
Ray F. Weiss, (D), California (San Diego), 1970, Cm
Clinton D. Winant, (D), S California, 1972, On

University of California, Santa Barbara

Dept of Earth Science (B,M,D) ☒
Room 1006
Webb Hall
Santa Barbara, CA 93106-9630
p. (805) 893-4688
chair@geol.ucsb.edu
http://www.geol.ucsb.edu
Enrollment (2012): B: 81 (22) M: 12 (2) D: 36 (3)

Professor:
Stanley M. Awramik, (D), Harvard, 1973, Pg
Jordan F. Clark, (D), Columbia, 1995, ClHg
John Cottle, (D), Oxford, 2007, Gi
Phillip B. Gans, (D), Stanford, 1987, Gt
Matt Jackson, (D), Cu
Chen Ji, (D), Caltech, 2002, Ys
Edward A. Keller, (D), Purdue, 1973, Gm
David W. Lea, (D), MIT, 1990, PeCm

Lorraine Lisiecki, (D), Brown, 2005, Gn
Francis Macdonald, (D), Harvard, Gg
Susannah Porter, (D), Harvard, 2002, Gn
Roberta Rudnick, (D)
Alexander Simms, (D), Rice, 2006, GsuGm
Toshiro Tanimoto, (D), California (Berkeley), 1982, Ys
David Valentine, (D), Cm
Andre R. Wyss, (D), Columbia, 1989, Pv

Associate Professor:
Robin Matoza, (D), Gv
Kristen Morell, (D)
Syee Weldeab, (D), Tubingen, 2002, PeCm

Assistant Professor:
Zach Eilon, (D), Colombia, Ys
Gen Li, (D), S California, 2017, Gms
Morgan Raven, (D), Caltech, Cl

Lecturer:
Matt Rioux, (D), California (Santa Barbara), Cc

Emeritus:
Ralph J. Archuleta, (D), California (San Diego), 1976, Ys
Doug Burbank, (D), Dartmouth, 1982, GmtGs
Bradley R. Hacker, (D), California (Los Angeles), 1998, Gp
Frank J. Spera, (D), California (Berkeley), 1977, Gi
Bruce H. Tiffney, (D), Harvard, 1977, Pb

Earth Research Institute ☒
Mail Code 1100
Santa Barbara, CA 93106-1100
p. (805) 893-8231
director@eri.ucsb.edu
http://eri.ucsb.edu
Administrative Assistant: Kathy J. Scheidemen

Director:
Kelly Caylor, (D), Virginia, 2003, HgZcy

Associate Director:
Susannah Porter, (D), Harvard, 2002, Pg

University of California, Santa Cruz

Center for the Dynamics & Evolution of the Land-Sea Interface
1156 High Street
Room A234, Earth & Marine Sciences Building
Santa Cruz, CA 95064
p. (831) 459-4089
acr@es.ucsc.edu

Professor:
Robert S. Anderson, (D), Washington, 1986, Gm
Kenneth W. Bruland, (D), California (San Diego), 1974, OcCtm
Margaret L. Delaney, (D), MIT/WHOI, 1983, Oc
Russell Flegal, (D), Oregon State, Cg
Laurel R. Fox, (D), California (Santa Barbara), Zn
Dianne Gifford-Gonzalez, (D), California (Berkeley), Zn
Mark S. Mangel, (D), British Columbia, 1978, Zn
Donald C. Potts, (D), California (Santa Barbara), Ob
Mary W. Silver, (D), California (San Diego), 1971, Ob
Jonathan P. Zehr, (D), California (Davis), 1985, Ob

Associate Professor:
Mark Carr, (D), California (Santa Barbara), Ob
Brent Haddad, (D), California (Berkeley), 1996, Zn
Karen D. Holl, (D), Virginia Tech, 1994, Zu
Christina Ravelo, (D), Columbia, 1991, Cs
Donald R. Smith, (D), California (Santa Cruz), Zn

Assistant Professor:
Don Croll, (D), California (Santa Cruz), Ob
Raphael M. Kudela, (D), S California, 1995, Zr
Margaret A. McManus, (D), Old Dominion, 1996, On

Center for the Origin, Dynamics, & Evolution of Planets
1156 High Street
Room A234, Earth & Marine Sciences Building
Santa Cruz, CA 95064
p. (831) 459-4089
dkorycan@ucsc.edu

Professor:
Robert S. Anderson, (D), Washington, 1986, Gm
Peter Bodenheimer, (D), California (Berkeley), 1965, Xc
Frank Bridges, (D), California (San Diego), 1968, Yx
Douglas Lin, (D), Cambridge, 1976, Xc
Steven Vogt, (D), Texas, 1978, Xc

Other:
Don Korycansky, (D), California (Santa Cruz), Xc

Earth & Planetary Sciences Dept (B,M,D) ●
1156 High Street
Earth & Marine Sciences Bldg, Rm A232
Santa Cruz, CA 95064
p. (831) 459-4089
lms@ucsc.edu
http://www.eps.ucsc.edu
f: https://www.facebook.com/UcscEPS
t: @EpsUcsc
Programs: Earth Sciences B.S., Earth Sciences B.S.with Geology, Earth Sciences B.S. with Ocean Science, Earth Sciences B.S. with Planetary Sciences, Earth Sciences B.S. with Geophysics, Earth Sciences/Anthropology B.A., Earth Sciences Minor, Environmental Sciences B.S.
Enrollment (2022): B: 384 (60) M: 5 (5) D: 52 (7)

Distinguished Professor, Department Chair:
Quentin Williams, (D), California (Berkeley), 1988, Gy

Distinguished Professor:
Thorne Lay, (D), Caltech, 1983, Ys

Professor:
Emily Brodsky, (D), Caltech, 2001, YsgHw
Patrick Y. Chuang, (D), Caltech, 1999, As
Matthew Clapham, (D), S California, 2006
Andrew T. Fisher, (D), Miami, 1989, Hw
Gary Griggs, (D), Oregon State, 1968, GuOnp
Mathis Hain, (D), Princeton, 2013, Cb
Elise Knittle, (D), California (Berkeley), 1988, Gy
Paul L. Koch, (D), Michigan, 1989, Pv
Francis Nimmo, (D), Cambridge, 1996, Xy
Tamara Pico, (D), Harvard, 2019, Gl
Heather Savage, (D), Penn State, 2007
Susan Schwartz, (D), Michigan, 1988, Ys
Slawek Tulaczyk, (D), Caltech, 1998, GlmNg
James Zachos, Rhode Island, 1988, Pc

Associate Professor:
Noah J. Finnegan, (D), Washington, 2007, Gm
Ian Garrick-Bethell, (D), MIT, 2009, Xy
Jeremy Hourigan, (D), Stanford, 2002, Cc

Assistant Professor:
Terrence Blackburn, (D), MIT, 2012, Cc
Nicole Feldl, (D), Washington, 2013, At
Myriam Telus, (D), Hawaii, 2015, Xc
Xi Zhang, (D), Caltech, 2013, AsXy
Margaret Zimmer, (D), Duke, 2017, Hw

Senior Lecturer:
Hilde Schwartz, (D), California (Santa Cruz), 1983, Pv

Lecturer:
Peter Weiss, (D), Washington, 1995, Cb

Graduate Program Advisor:
Jennifer M. Fish, (B), California (Santa Cruz), Zn

Inst of Geophysics & Planetary Physics
1156 High Street
Earth & Marine Sciences Building, Room A234
Santa Cruz, CA 95064
p. (831) 459-4089
shalevst@gmail.com

http://igpp.ucsc.edu

Director:
Ana Christina Ravelo, (D), OcGe

University of San Diego

Dept of Environmental and Ocean Sciences (B,M) ☒
Alcala Park
San Diego, CA 92110
p. (619) 260-4795
eosc@sandiego.edu
https://www.sandiego.edu/cas/environmental-ocean-sciences/
Programs: B.A. Marine Ecology, B.A. Environmental Science and B.A. Environmental Studies, M.S. Environmental and Ocean Science

Chair:
Sarah Gray, (D), California (Santa Cruz), Zn

Professor:
Zhi-Yong Yin, (D), Georgia, Zir
Graduate Director:
Ronald S. Kaufmann, (D), California (San Diego), 1992, Ob
Associate Professor:
Michel A. Boudrias, (D), California (San Diego), 1992, Ob
Bethany O'Shea, (D), New South Wales, 2006, CqGeCt
Nathalie Reyns, (D), North Carolina State
Drew Talley, (D), California (San Diego), 2000, ObZe
Assistant Professor:
Ruth Coffey, (D), Stony Brook, 2014, Gug
Jennifer Prairie, (D), Scripps, Ob
Suzanne Walther, (D), Oregon, GmZi
Adjunct Assistant Professor and Field Trip Coordinator:
Elizabeth Baker-Treloar, (M), Scripps, 1995, Geu
Adjunct Assistant Professor :
Eric Cathcart, (M), California (San Diego), 1996, Geg
Andrew Nosal, (D), Scripps, Ob
Steven Searcy, (D), North Carolina State, Ob

University of Southern California

Dept of Earth Sciences (B,M,D) ●☒
3651 Trousdale Parkway
ZHS 117
Los Angeles, CA 90089-0740
p. (213) 740-6106
earthsci@usc.edu
https://dornsife.usc.edu/earth/
t: https://twitter.com/USC_earth
Programs: Geological Sciences (BS), Earth Sciences (BA), Geological Sciences (MS), Geological Sciences (PhD)
Administrative Assistant: Darlene Garza, Vardui Ter-Simonian
Enrollment (2022): B: 19 (5) M: 1 (0) D: 59 (8)
Chair:
Frank A. Corsetti, (D), California (Santa Barbara), 1998, Gs
Research Professor:
David A. Okaya, (D), Stanford, 1985, Ys
Professor:
Jan Amend, (D), California (Berkeley), 1995, ClmCo
Yehuda Ben-Zion, (D), S California, 1990, Ys
William M. Berelson, (D), S California, 1986, Cm
David J. Bottjer, (D), Indiana, 1978, Pe
James F. Dolan, (D), California (Santa Cruz), 1988, Gt
Douglas E. Hammond, (D), Columbia, 1975, Cm
Heidi B. Houston, (D), Caltech, 1987, YsGt
Thomas H. Jordan, (D), Caltech, 1972, Ys
Steven P. Lund, (D), Minnesota, 1981, Ym
James W. Moffett, (D), Miami, 1986, OcCba
Scott R. Paterson, (D), California (Santa Cruz), 1986, Gc
John P. Platt, (D), California (Santa Barbara), 1973, GctGp
Sergio Sanudo, (D), California (Santa Cruz), 1993, Pg
Lowell D. Stott, (D), Rhode Island, 1989, Pm
Ta-liang Teng, (D), Caltech, 1966, YsGtYe
John E. Vidale, (D), Caltech, 1986, YsGtYg
Associate Professor:
Julien Emile-Geay, (D), Columbia, 2006, Oc
Sarah Feakins, (D), Columbia, 2006, CoGuCs
A. Joshua (Josh) West, (D), Cambridge, 2007, ClGmHs
Research Associate Professor:
Yong-Gang Li, (D), S California, 1988, Ys
Ellen S. Platzman, (D), ETH (Switzerland), 1990
Assistant Professor:
Sam Silva, (D), MIT, 2019, Ac
Emeritus:
Gregory A. Davis, (D), California (Berkeley), 1961, GtcGe
Alfred G. Fischer, (D), Columbia, 1950, Gs
Thomas L. Henyey, (D), Caltech, 1968, Yg
Teh-Lung Ku, (D), Columbia, 1966, CgcGn
Kenneth Nealson, (D), Chicago, 1969, Pg
Bernard W. Pipkin, (D), Arizona, 1964, Ng
Charles G. Sammis, (D), Caltech, 1971, Yg

University of the Pacific

Department of Geological & Environmental Sciences (B) ☒
3601 Pacific Avenue
Stockton, CA 95211
p. (209) 946-2482
lkfox@pacific.edu
https://liberalarts.pacific.edu/liberalarts/academics/departments-and-programs/geological-and-environmental-sciences/progr
f: https://liberalarts.pacific.edu/liberalarts/academics/departments-and-programs/geological-and-environmental-sciences
t: @PacificGESC
Programs: BS in Geological and Environmental Sciences with a concentration in Geology or Environmental Science; BA in Geological and Environmental Sciences; 3+3 (BA+JD) in Geological and Environmental Sciences / Environmental Law
Certificates: Wilderness first aid
Enrollment (2020): B: 35 (16)
Professor:
Thomas Naehr, (D)
Eugene F. Pearson, (D), Wyoming, 1972, GsPgZg
Associate Professor:
Lydia K. Fox, (D), California (Santa Barbara), 1989, GizZe
Laura K. Rademacher, (D), California (Santa Barbara), 2002, ClHySo
Emeritus:
Roger T. Barnett, (D), California (Berkeley), 1973, Zy
J. Curtis Kramer, (D), California (Davis), 1976, Gg

Ventura College

Dept of Geology
4667 Telegraph Road
Ventura, CA 93003
spalladino@vcccd.edu
Chair:
Luke D. Hall, (M), W Kentucky, 1975, Zy
Assistant Professor:
Steve D. Palladino, (M), California (Santa Barbara), 1994, Zi
Instructor:
William Budke, (M), Cal Poly, 2001, So

Victor Valley College

Victor Valley College
18422 Bear Valley Road
Victorville, CA 92395
p. (415) 506-0234
carol.delong@vvc.edu
http://www.vvc.edu/

West Valley College

Dept of Geology (A) ●⏚
14000 Fruitvale Avenue
Saratoga, CA 95070-5698
p. (408) 741-2437
robert_lopez@westvalley.edu
Programs: Geology, Oceanography, Earth Science
Emeritus:
Theodore C. Herman, (M), Michigan, 1961, Ze

Whittier College

Environmental Science Program (B)
Whittier, CA 90608
p. (562) 907-4220
cswift@whittier.edu
http://www.whittier.edu/Academics/EnvironmentalSciences/
Chair:
Cheryl Swift
Visiting Professor:
Andrew H. Wulff, (D), Massachusetts, 1998, Gi
Emeritus:
William B. Wadsworth, (D), Northwestern, 1966, Gi

Colorado

Adams State University

Department of Geosciences (B) ☒
208 Edgemont Blvd
Suite 3060
208 Edgemont Blvd.
Alamosa, CO 81101
p. (719) 587-7256
matt.nehring@adams.edu
https://www.adams.edu/academics/undergraduate/earth-science/
Programs: Geology, Physical Geography and Conservation,

Science Education (Secondary Teacher Licensure)
Enrollment (2020): B: 14 (2)
Assistant Professor:
Chayan Lahiri, (D), Mississippi, 2018, HwZr

Aims Community College

Dept of Sciences (A) ☒
5401 West 20th Street
Greeley, CO 80634
p. (970) 339-6637
jim.stone@aims.edu
http://www.aims.edu/academics/sciences
Instructor:
Jim Stone, Zg

Arapahoe Community College

Geology Program (A)
5900 South Santa Fe Drive
Littleton, CO 80120
p. (303) 797-5831
henry.weigel@arapahoe.edu
http://www.arapahoe.edu/departments-and-programs/a-z-offerings/geology
Program Chair:
Henry Weigel

Colorado College

Geology Dept (B) ☒
14 E Cache La Poudre
Colorado Springs, CO 80903
p. (719) 389-6621
geology@coloradocollege.edu
http://www.coloradocollege.edu/academics/dept/geology
f: https://www.facebook.com/geodept.coloradocollege
Programs: Geology
Enrollment (2018): B: 24 (17)
Professor:
Eric M. Leonard, (D), Colorado, 1981, Gl
Paul M. Myrow, (D), Memorial, 1987, Gsr
Jeffrey B. Noblett, (D), Stanford, 1980, Gxv
Christine S. Siddoway, (D), California (Santa Barbara), 1995, GctGg
Department Chair:
Henry C. Fricke, (D), Michigan, 1997, Cs
Technical Director:
Stephen G. Weaver, (D), Colorado Mines, 1988, Gx

Colorado Mesa University

Dept of Physical & Environmental Sciences (A,B) ○✓
1100 North Avenue
Grand Junction, CO 81501-3122
p. (970) 248-1993
gbaker@coloradomesa.edu
http://www.coloradomesa.edu/geosciences/
Programs: Geosciences (B); Geology (B); Environmental Geology (B); Secondary Education (B); Geology (A); Geology (minor); Watershed Science (minor); GIST (minor)
Certificates: GIS Technology
Enrollment (2017): B: 75 (15)
Professor:
Andres Aslan, (D), Colorado, 1994, Gm
Rex D. Cole, (D), Utah, 1975, GsrGo
Verner C. Johnson, (D), Tennessee, 1975, Yg
Richard F. Livaccari, (D), New Mexico, 1994, Gc
Gigi A. Richard, (D), Colorado State, 2001, Hsg
Instructor:
Cassandra Fenton, (D), Utah, 2002, CcGmCl
Lecturer:
Lawrence S. Jones, (D), Wyoming, 1996, GsmGr
Adjunct Professor:
William C. Hood, (D), Montana, 1964, Gg
Julia McHugh, (D), Iowa, 2012, Pgv
Dave Wolny, (B), Mesa State, 1992, Ys
Emeritus:
James B. Johnson, (D), Colorado, 1979, Gl
Jack E. Roadifer, (D), Arizona, 1966, Gx

Colorado Mountain College

Geology Div (A)
3000 County Road 114
Glenwood Springs, CO 81601
p. (303) 945-7481 x263
mhaselhorst2@coloradomtn.edu
Professor:
Garrett E. Zabel, (M), Houston, 1977, Gg

Colorado School of Mines

Department of Chemistry (B,M,D) ☒
1012 14th Street, CO 204
Golden, CO 80401
p. (303) 273-3610
chemistry@mines.edu
https://chemistry.mines.edu/
f: https://www.facebook.com/csmchemistry/
t: @MinesChemistry
Enrollment (2016): B: 78 (27) M: 14 (5) D: 52 (10)
Department Head:
Thomas Gennett, (D), Vermont
Professor:
Mark E. Eberhart, (D), MIT, 1983, Zm
Mark Jensen, (D), Florida State, 1994
Daniel M. Knauss, (D), Virginia Tech, 1994
James Ranville, (D), Colorado Mines, Ca
Ryan M. Richards, (D), Michigan State, 2000
Bettina Voelker, (D), Swiss Fed Inst Tech, 1994, Ca
Kim R. Williams, (D), Michigan State, 1986, Ca
David T. Wu, (D), California (Berkeley), 1991, Zn
Associate Professor:
Renee L. Falconer, (D), South Carolina, 1994
Matthew C. Posewitz, (D), Dartmouth, 1995
Alan Sellinger, (D), Michigan, 1997
Angela C. Sower, (D), New Mexico
Assistant Professor:
Dylan Domaille, (D), California (Berkeley)
Amanda Jameer, (M)
Svitlana Pylypenko, (D), New Mexico, Zmn
Jenifer C. Shafer, (D), Washington State, 2010
Brian G. Trewyn, (D), Iowa State, 2006
Shubham Vyas, (D), Ohio State
Emeritus:
Dean W. Dickerhoof, (D), Illinois, 1961
Donald L. Macalady, (D), Wisconsin, 1969, Cl
Patrick MacCarthy, (D), Cincinnati, 1975, Co
Craig Simmons, (D), SUNY (Stony Brook), 1976, Cp
Kent J. Voorhees, (D), Utah State, 1970, Co
Thomas R. Wildeman, (D), Wisconsin, 1967, Ct

Dept of Geology & Geological Engineering (B,M,D) ●⦸
1516 Illinois Street
Golden, CO 80401-1887
p. (303) 273-3800
bohrson@mines.edu
https://geology.mines.edu/
Programs: BS Geology & Geological Engineering, Master of Science (Geology), Master of Science (Geological Engineering), Doctor of Philosophy (Geology), Doctor of Philosophy (Geological Engineering), Master of Engineering (Geological Engineer) (Non-Thesis), Professional Master Degree (Petroleum Reservoir Systems), (Non- Thesis), Professional Master Degree (Mineral Exploration) (Non-Thesis)
Certificates: GIS, Economic Geology, Exploration Methods, Earth Resource Data Science
Enrollment (2022): B: 112 (44) M: 152 (22) D: 75 (13)
Professor:
Wendy Bohrson, (D), California, 1993, Gv
Professor:
David A. Benson, (D), Nevada (Reno), 1998, Hq
Zhaoshan Chang, (D), Peking, 1997, Eg
Brendan Hanger, (D), Australian Nat, 2014, GiCg
Yvette D. Kuiper, (D), New Brunswick, 2003, Gc
Thomas Monecke, (D), Germany, 2003, Em
Piret Plink-Bjorklund, (D), Goteborg (Sweden), 1998, Gs
Paul M. Santi, (D), Colorado Mines, 1995, Ng
Kamini Singha, (D), Stanford, 2005, Hw
Stephen A. Sonnenberg, (D), Colorado Mines, 1981, Go

Lesli Wood, (D), Colorado State, 1992, GsoGm
Associate Professor:
Mathias Burisch, (D), Tuebingen, 2016, Gz
Alexis Navarre-Sitchler, (D), Penn State, 2008, Ca
Bruce D. Trudgill, (D), Imperial Coll (UK), 1989, Gc
Wendy W. Zhou, (D), Missouri S&T, 2001, NgZir
Assistant Professor:
Kevin Cannon, (D), Brown, 2017, ZnnZr
Adrienne Marshall, (D), Idaho, HgAt
Danica Roth, (D), California, 2016, GmYs
Ryan Venturelli, (D), S Florida, 2021, CgcZo
Gabriel Walton, (D), Queen's, 2014, NgrYg
Lecturer:
Christian V. Shorey, (D), Iowa, 2002, Gg
Emeritus:
L. Graham Closs, (D), Queen's, 1973, Ce
John B. Curtis, (D), Ohio State, 1989, GoCo
Wendy J. Harrison, (D), Manchester (UK), 1979, Cl
Jerry D. Higgins, (D), Missouri (Rolla), 1980, NxgNt
Keenan Lee, (D), Stanford, 1969, Zr
Eileen P. Poeter, (D), Washington State, 1980, Ng
Samuel B. Romberger, (D), Penn State, 1968, Em
John E. Warme, (D), California (Los Angeles), 1966, GsPeGu
Richard F. Wendlandt, (D), Penn State, 1978, GizCp

Dept of Geophysics (B,M,D) ●

924 16th Street
Room 283
Golden, CO 80401
p. (303) 273-3451
geophysics@mines.edu
https://geophysics.mines.edu/
f: https://www.facebook.com/MinesGeophysics/
t: @MinesGeophysics
Programs: Geophysical Engineering (B,M,D); Geophysics (M,D); Humanitarian Geophysics (M); Hydrogeophysics (M)
Certificates: 12-credit Certificate Program in Petroleum Geophysics
Enrollment (2022): B: 63 (24) M: 20 (14) D: 33 (8)
W. M. Keck Distinguished Chair in Professional Development Education:
Roel Snieder, (D), Utrecht, 1987, Ye
Vice Provost for External Initiatives and Dean of Earth Resources and Environmental Programs:
John Bradford, (D), Rice, 1999, Yue
Director, Reservoir Characterization Project:
Ali Tura, (D), California (Berkeley), YsNr
Director, Center for Wave Phenomena:
Ilya D. Tsvankin, (D), Moscow State (Russia), 1982, Ye
Director, Center for Gravity, Electrical, and Magnetic Studies:
Yaoguo Li, (D), British Columbia, 1992, Yev
C.H. Green Professor of Exploration Geophysics and Department Head:
Paul Sava, (D), Stanford, 2005, Ye
Baker Hughes Chair of Petrophysics and Borehole Geophysics and Associate Dept Head:
Brandon Dugan, (D), Penn State, 2003, GuYrHw
Professor:
Manika Prasad, (D), Kiel, 1990, NrYx
Associate Professor:
Ebru Bozdag, (D), Utrecht, 2009, Ys
Jeffrey Shragge, (D), Stanford, 2009, Ys
Assistant Professor:
Ge Jin, (D), Columbia, 2014, Ye
Eileen Martin, (D), Stanford, 2018, YuZu
Matthew Siegfried, (D), Scripps, 2015, ZrGl
Bia Villas Boas, (D), California (San Diego), 2020, OpZrHs
Affiliate Faculty:
Andrei Swidinsky, (D), Toronto, 2011, Ye

Dept of Mining Engineering (B,M,D) ●

1600 Illinois
Golden, CO 80401
p. (303) 273-3700
mining@mines.edu
https://mining.mines.edu/
Enrollment (2015): B: 100 (41) M: 18 (18) D: 14 (6)
Head:
M. Stephen Enders, Gg
Haddon Endowed Chair:
Jamal Rostami, Nr
Associate DH & Fred Banfield Chair:
Sebnem Duzgun, (D), Nm
Professor:
Corby Anderson, Nx
Kadri Dagdelen, (D), Colorado Mines, 1985, Nm
Priscilla P. Nelson, (D), Cornell, 1983, NrgNm
Social Anthropologist:
Nicole Smith, Zn
Associate Professor:
Elizabeth Holley, (D), Colorado Mines, 2012, Eg
Rennie Kaunda, (D), W Michigan, 2007, Nm
Jaeheon Lee, Nx
Hugh Miller, (D), Colorado Mines, 1996, Nmg
Research Associate:
Erik Spiller, Nx
Director of Research Edgar Mine:
Jürgen F. Brune, (D), Tech (Clausthal), 1994, Nm
Director f Professional Masters Program:
Paul Zink, Zn

Colorado State University

Dept of Atmospheric Science (M,D) ☒

1371 Campus Delivery
Fort Collins, CO 80523-1371
p. (970) 491-8360
info@atmos.colostate.edu
http://www.atmos.colostate.edu
Programs: Master of Science in Atmospheric Science, Doctorate in Atmospheric Science
Enrollment (2018): M: 34 (12) D: 43 (6)
Head:
Jeffrey L. Collett, Jr., (D), Caltech, 1989, As
Professor:
A. Scott Denning, (D), Colorado State, 1994, AsZg
James Hurrell, (D), Purdue, 1990, As
Sonia M. Kreidenweis, (D), Caltech, 1989, As
Christian D. Kummerow, (D), Minnesota, 1987, Yr
Eric Maloney, (D), Washington, 2000, As
David A. Randall, (D), California (Los Angeles), 1976, As
A.R. Ravishankara, (D), Florida, 1975, AsZg
Steven A. Rutledge, (D), Washington, 1983, As
David W. J. Thompson, (D), Washington, 2000, Yr
Sue van den Heever, (D), Colorado State, 2001, As
Peter Jan van Leeuwen, (D), Delft (Neth), 1992, As
Associate Professor:
Elizabeth A. Barnes, (D), Washington, 2012, As
Michael M. Bell, (D), Naval Postgrad Sch, 2010, As
Christine Chiu, (D), Purdue, 2003, As
Jeff Pierce, (D), Carnegie Mellon, 2008, As
Russ Schumacher, (D), Colorado State, 2008, As
Assistant Professor:
Emily Fischer, (D), Washington, 2010, As
Kristen L. Rasmussen, (D), Washington, 2014, As

Dept of Geosciences (B,M,D) ●

1482 Campus Delivery
Fort Collins, CO 80523-1482
p. (970) 491-5661
WCNR_GEO_Info@mail.colostate.edu
https://warnercnr.colostate.edu/geosciences/
Programs: Geology (B); Geology, Environmental Geology Concentration (B); Geology, Geophysics Concentration (B); Geology, Hydrogeology Concentration (B); Geosciences (M,D)
Administrative Assistant: Kailarae Lilly, Angela Sharpe, Patti Uman
Enrollment (2022): B: 104 (34) M: 35 (11) D: 20 (2)
Head:
Richard C. Aster, (D), California (San Diego), 1991, YsGv
Professor:
Judith L. Hannah, (D), California (Davis), 1980, GiCc
Dennis L. Harry, (D), Texas (Dallas), 1989, Yg
Sara L. Rathburn, (D), Colorado State, 2001, Ggm
Derek L. Schutt, (D), Oregon, 2000, Ysg
Ellen E. Wohl, (D), Arizona, 1988, GmHs
Associate Professor:
Jerry F. Magloughlin, (D), Minnesota, 1993, Gpz
John R. Ridley, (D), Edinburgh, 1982, Eg
Michael J. Ronayne, (D), Stanford, 2008, Hwq
William E. Sanford, (D), Cornell, 1992, Hw

COLORADO

John Singleton, (D), Texas, 2011, Gct
Lisa Stright, (D), Stanford, 2011, GoNpYe
Assistant Professor:
Sean Gallen, (D), North Carolina State, 2013, GemGt
Daniel McGrath, (D), Colorado, 2013, GlZrYu
Jeremy Caves Rugenstein, (D), Stanford, 2016, ClsPc
Research Associate:
James R. Chappell, (B), Colorado State, 2000, Ge
Ronald J. Karpilo, Jr, (M), Denver, 2004, Zy
Stephanie A. O'Meara, (M), Colorado State, 1997, Gc
Trista L. Thornberry-Ehrlich, (M), Colorado State, 2001, GgcGx
Senior Instructor:
Sean Bryan, (D), Colorado, 2010, GgOoPc
Emeritus:
Eric A. Erslev, (D), Harvard, 1981, GctGg
Frank G. Ethridge, (D), Texas A&M, 1970, Gs
Sally Sutton, (D), Cincinnati, 1987, CgGd
Tommy Thompson, (D), New Mexico, 1966, Egm

Community College of Aurora

Dept of Science
16000 East Centretech Parkway
Aurora, CO 80011
p. (303) 361-7398
Admissions@CCAurora.edu
http://www.ccaurora.edu

Denver Museum of Nature & Science

Dept of Earth Sciences ☒
2001 Colorado Boulevard
Denver, CO 80205-5798
p. (303) 370-6000
libby.couch@dmns.org
http://www.dmns.org
Director of Earth and Space Sciences Branch, Department Chair of Earth Sciences, Curator of Paleobotany:
Ian Miller, (D), Yale, 2007, PbGg
Tim and Kathryn Ryan Curator of Geology:
James W. Hagadorn, (D), S California, 1998, GgsPg
Preparator:
Natalie Toth, (M), SD Mines, 2010, Pg
Curator of Vertebrate Paleontology:
Tyler Lyson, (D), Yale, 2012, Pv
Joseph Sertich, (D), SUNY (Stony Brook), 2011, Pv
Collections Manager:
Kristen Mackenzie, (M), Oregon, 2013, Zg
Business Specialist:
Libby Couch, Zn

Fort Lewis College

Dept of Geosciences (B) O⃠
1000 Rim Drive
Durango, CO 81301
p. (970) 247-7278
gonzales_d@fortlewis.edu
http://geo.fortlewis.edu
Programs: Geology
Certificates: GIS certificate
Enrollment (2021): B: 80 (0)
Chair:
David A. Gonzales, (D), Kansas, 1997, Gx
Professor:
Gary Gianniny, (D), Wisconsin, 1995, Gs
Associate Professor:
Melissa Clutter, (D), 2019, Ge
Carolyn Tewksbury-Christle, (D), 2021, Gc
Assistant Professor:
Jon Harvey, 2015, Gm
Adjunct Professor:
Mary L. Gillam, (D), Colorado, 1998, Gm

Front Range Community College - Larimer

Natural Sciences (A) ☒
4616 S. Shields Street
Fort Collins, CO 80526
p. (970) 204-8607
stephanie.irwin@frontrange.edu
http://www.frontrange.edu/Academics/Academic-Departments/Larimer-Campus/Natural-Applied-Environment-Science/
Programs: Geology (A)
Instructor:
Andy Caldwell, (M), N Colorado, 1999, Gg
Mike Smith, (M), N Colorado, 1998, GgZe

Front Range Community College - Westminster

Science and Technology (A)
3645 West 112th Avenue
Westminster, CO 80031
p. (303) 404-5279
Angela.Greengarcia@frontrange.edu
https://www.frontrange.edu/programs-and-courses/a-z-program-list/geology
Programs: A.S. Geology; A.S.
Enrollment (2018): A: 2 (2)

Metropolitan State College of Denver

Earth & Atmospheric Sciences Dept (B)
P.O. Box 173362, Campus Box 22
Denver, CO 80217-3362
p. (303) 556-3143
eas@msudenver.edu
Administrative Assistant: Diane Hollenbeck
Chair:
James M. Cronoble, (D), Colorado Mines, 1977, Gg
Ken Engelbrecht
Professor:
John R. Kilcoyne, (D), Washington, 1973, Zy
Anthony A. Rockwood, (M), Colorado State, 1976, Am
Roberta A. Smilnak, (D), Clark, 1973, Zg
Associate Professor:
Thomas J. Corona, (M), Colorado State, 1978, Am
Assistant Professor:
Robert E. Leitz, (M), California (Berkeley), 1974, Gz
Rafael Moreno, (D), Colorado State, 1992, Zg
Emeritus:
James MacLachlan, (D), Princeton, Gg
H. Dixon Smith, (D), Minnesota, 1960, Zg

Northeastern Junior College

Math, Science, and Health (A) ☒
100 College Avenue
Sterling, CO 80751
p. (970) 521-6753
david.coles@njc.edu
http://njc.edu
Programs: Associate of Science with Geology Designation

Red Rocks Community College

Geology Program, Science Dept (A) ●☒
13300 West Sixth Avenue
Campus Box 20
Lakewood, CO 80228
p. (303) 914-6290
eleanor.camann@rrcc.edu
https://www.rrcc.edu/geology
Programs: Geology
Enrollment (2018): A: 5 (6)
Professor:
Eleanor J. Camann, (D), N Carolina, 2005, OnZeGs

United States Air Force Academy

Dept of Economics & Geosciences (B)
HQ USAFA/DFEG
2354 Fairchild Drive, Suite 6K110
USAF Academy, CO 80840-5701
p. (719) 333-3080
ian.irmischer@usafa.edu
http://www.usafa.edu/df/dfeg/?catname=dfeg
Enrollment (2013): B: 90 (27)
Professor:
Terry W. Haverluk, (D), Minnesota, 1993, Zg
Associate Professor:
Steven J. Gordon, (D), Arizona State, 1999, Gm

Thomas Koehler, (D), Wisconsin, As
Lt Col :
Matthew Tracy, (D), Arizona State, 2008, Zn
Assistant Professor:
Glen Gibson, (D), Virginia Tech, 2012, Zri
Brett Machovina, (D), Denver, 2010, Zir
Evan Palmer, (D), Arizona State, 2014, Zig
Sarah Robinson, (D), Arizona State, GgZi
Instructor:
Scott Dubsky, (M), North Dakota, 2006, ZyGm
Related Staff:
Danny Portillo, (B), Zi

University of Colorado

Dept of Geography (B,M,D) ☒
Campus Box 260
Boulder, CO 80309-0260
p. (303) 492-8312
darla.shatto@colorado.edu
http://www.colorado.edu/geography
Programs: Geography
Certificates: Certificate in GIS and Computational Science
Enrollment (2020): B: 168 (58) M: 21 (10) D: 41 (12)
Professor:
Waleed Abdalati, (D), Colorado, 1996, Zr
Peter D. Blanken, (D), British Columbia, 1997, HsAsm
Mark Serreze, (D), Colorado, 1989, As
Associate Professor:
Jennifer K. Balch, (D), Yale, 2008, Zg
Holly R. Barnard, (D), Oregon State, 2009, Hg
Noah Molotch, (D), Arizona, 2004, Hg
Assistant Professor:
Katherine Lininger, (D), Colorado State, 2018, Gm
Emeritus:
John Pitlick, (D), Colorado State, 1988, Gm
Thomas T. Veblen, (D), California (Berkeley), 1975, Zg
Mark W. Williams, (D), California (Santa Barbara), 1990, Zg

Dept of Geological Sciences (B,M,D)
Campus Box 399
Boulder, CO 80309-0399
p. (303) 492-8141
robert.s.anderson@Colorado.edu
https://www.colorado.edu/geologicalsciences/
Administrative Assistant: Carmen Juszczyk
Enrollment (2014): B: 248 (29) M: 13 (7) D: 45 (10)
Associate Dean:
Mary J. Kraus, (D), Colorado, 1983, GsSa
Professor:
Robert Anderson, (D), Washington, 1986, Gm
David A. Budd, (D), Texas, 1984, GdsEo
Jaelyn J. Eberle, (D), Wyoming, 1996, Pv
G. Lang Farmer, (D), California (Los Angeles), 1983, Cg
Shemin Ge, (D), Johns Hopkins, 1990, HwEoYg
Bruce M. Jakosky, (D), Caltech, 1982, Xg
Craig H. Jones, (D), MIT, 1987, YsGtYg
Gifford H. Miller, (D), Colorado, 1975, GlCc
Stephen J. Mojzsis, (D), California (San Diego), 1997, XcCcGp
Peter Molnar, (D), Columbia, 1970, Gt
Karl J. Mueller, (D), Wyoming, 1992, GtcGm
Anne F. Sheehan, (D), MIT, 1991, YszYe
Joseph R. Smyth, (D), Chicago, 1970, Gz
Charles R. Stern, (D), Chicago, 1973, Gi
Eric E. Tilton, (D), California (Santa Cruz), 1998, Hw
Gregory E. Tucker, (D), Penn State, 1996, GmtGs
Paul Weimer, (D), Texas, 1989, Gor
James W. C. White, (D), Columbia, 1983, CsAt
Associate Professor:
Karen Chin, (D), California (Santa Barbara), 1996, Pg
Rebecca M. Flowers, (D), MIT, 2005, CgcGt
Brian M. Hynek, (D), Washington, 2003, Xg
Thomas M. Marchitto, (D), MIT, 1999, Oc
Dena M. Smith, (D), Arizona, 2000, Pg
Alexis Templeton, (D), Stanford, 2002, ClZaCq
Assistant Professor:
Kevin H. Mahan, (D), Massachusetts, 2005, Gpc
Julio C. Sepúlveda, (D), Bremen, 2008, Co
Instructor:
Lon Abbott, (D), California (Santa Cruz), 1993, GgtGm
Emeritus:
John T. Andrews, (D), Nottingham (UK), 1965, GluGs
William W. Atkinson, Jr., (D), Harvard, 1973, EgmCg
Peter W. Birkeland, (D), Stanford, 1961, Sd
William C. Bradley, (D), Stanford, 1956, Gm
Don L. Eicher, (D), Yale, 1958, Pm
Alexander Goetz, (D), Caltech, 1967, Zr
Edwin E. Larson, (D), Colorado, 1965, Ym
James L. Munoz, (D), Johns Hopkins, 1966, Cp
Peter Robinson, (D), Yale, 1960, Pv
Donald D. Runnells, (D), Harvard, 1964, Cl
Hartmut A. Spetzler, (D), Caltech, 1969, Ys
Theodore R. Walker, (D), Wisconsin, 1952, Gd

Univ of Colorado Museum
Campus Box 265
Boulder, CO 80309-265
p. (303) 492-6165
karen.chin@colorado.edu
http://cumuseum.colorado.edu/
Associate Professor:
Jaelyn J. Eberle, (D), Wyoming, 1996, Pve
Assistant Professor:
Karen Chin, (D), California (Santa Barbara), 1996, Pg
Research Associate:
Dena M. Smith, (D), Arizona, 2000, PeiPb
Adjunct Curator:
Kenneth Carpenter, Pv
Mary Dawson, Pv
Trihn Dzanh, Pv
Emmett Evanoff, Pi
Jeff Indeck, Pv
Jonathan Marcot, Pv
Greg McDonald, Pv
Karen Sears, Pv
Emeritus:
Judith A. Harris, (D), Cambridge, 1972, Pv
Peter Robinson, (D), Yale, 1960, Pv
Museum Associate:
Emily Bray, Pg
Frank Fisher, Pg
Pat Monaco, Pg
Steve Wallace, Pg
Collection Manager:
Tonia Superchi-Culver, (M), SD Mines, 2001, Pg
Associate Curator, Micropaleontology:
Donald Eicher, (D), Yale, 1958, Pm
Associate Curator, Fossil Primates:
Herbert Covert, Pv

University of Colorado, Denver

Dept of Geography & Environmental Science
CB 172
P.O. Box 173364
Denver, CO 80217-3364
p. (303) 556-2276
sue.eddleman@ucdenver.edu
Associate Professor:
John W. Wyckoff, (D), Utah, 1980, Zy
Emeritus:
Wesley E. LeMasurier, (D), Stanford, 1965, Gv
Martin G. Lockley, (D), Birmingham (UK), 1977, Pi

University of Denver

Dept of Geography and the Environment (B,M,D)
2050 E. Iliff Avenue
Boettcher Center West, Room 120
Denver, CO 80208
p. (303) 871-2654
Michael.Keables@du.edu
http://www.du.edu/geography
f: https://www.facebook.com/DUGeography/
Programs: Environmental Science (B); Geography (B,M); Geographic Information Science (M); Geography (D)
Certificates: Geographic Information Systems (G)
Enrollment (2018): B: 198 (34) M: 46 (23) D: 10 (1)

Professor:
Andrew R. Goetz, (D), Ohio State, 1987, Zgu
Paul C. Sutton, (D), California (Santa Barbara), 1999, Zc
Matthew Taylor, (D), Arizona State, 2003, Zc
Teaching Professor:
Hillary Hamann, (D), Colorado, 2002, HgZy
Teaching Associate Professor:
Helen Hazen, (D), Minnesota, 2006, Zy
Erika Trigoso, (D), Oxford, 2010, Zc
Associate Professor:
E. Eric Boschmann, (D), Ohio State, 2008, Zc
J. Michael Daniels, (D), Wisconsin, 2002, GmSp
Michael J. Keables, (D), Wisconsin, 1986, AstAm
Michael W. Kerwin, (D), Colorado, 2000, GgeZc
Jing Li, (D), George Mason, 2012, Zi
Rebecca Powell, (D), California (Santa Barbara), 2006, Zr
Donald G. Sullivan, (D), California (Berkeley), 1988, Pc
Teaching Assistant Professor:
Kristopher Kuzera, (D), San Diego State, 2011, ZyHg
Assistant Professor:
Hanson Nyantakyi-Frimpong, (D), W Ontario, 2014, Zc
Guiming Zhang, (D), Wisconsin, 2018, Zi
Adjunct Professor:
G. Thomas Lavanchy, (D), Denver, 2015, Hg
Michelle Moran-Taylor, (D), Arizona State, 2003, Zn
Martha A. Narey, (D), Denver, 1999, Zg
Sean Tierney, (D), Denver, 2009, Zc
Emeritus:
David Longbrake, (D), Iowa, 1972, Zi
Terrence J. Toy, (D), Denver, 1973, Gm

University of Northern Colorado

Earth and Atmospheric Sciences (B,M)
Campus Box 100
Greeley, CO 80639
p. (970) 351-2647
timothy.grover@unco.edu
http://esci.unco.edu
Programs: Environmental Earth Sciences (B); Geology (B); Meteorology (B); Secondary Teaching (B); Earth Sciences (M); Environmental Geosciences (PSM)
Certificates: Safety Science Certificate (12 credits), including Industrial Safety course and HAZWOPER certificate (OSHA Hazardous Waste Operations and Emergency Response)
Enrollment (2020): B: 67 (17) M: 7 (4)
Chair:
Timothy W. Grover, (D), Oregon, 1988, GpzGt
Professor:
Steven W. Anderson, (D), Arizona, 1990, GvZeXg
Graham Baird, (D), Minnesota, 2006, GcgGz
Lucinda Shellito, (D), California (Santa Cruz), 2004, AsPeAm
Associate Professor:
Joe T. Elkins, (D), Georgia, 2002, ZeGgCl
Emmett Evanoff, (D), Colorado, 1990, GrPgGs
Wendilyn Flynn, (D), Illinois, 2012, Ams
David G. Lerach, (D), Colorado State, 2012, Ams
Assistant Professor:
Sharon Bywater-Reyes, (D), Montana, 2015, HqsHt
Senior Lecturer:
Byron Straw, (M), N Colorado, 2010, ZeGgl
Adjunct Professor:
Todd A. Dallegge, (D), Alaska (Fairbanks), 2002, GosGr
Emeritus:
Richard D. Dietz, (D), Colorado, 1965, Xg
Kenneth D. Hopkins, (D), Washington, 1976, Gml
William H. Hoyt, (D), Delaware, 1982, OgGsZe
William D. Nesse, (D), Colorado, 1977, Gxz
K. Lee Shropshire, (D), Colorado, 1974, Pg
Other:
Carolyn D. Lambert, (M), Arizona, 2003, Hw
Walter A. Lyons, (D), Chicago, 1970, Asm

Western Colorado University

Geology Program in the Natural and Environmental Sciences Dept (B) O☒
1 Western Way
Gunnison, CO 81231
p. (970) 943-2015
dmarchetti@western.edu
http://www.western.edu/geology
Programs: Geology; Petroleum Geology; Geoarchaeology; Environmental Geology, Earth Science Education
Enrollment (2018): B: 46 (14)
Professor:
Robert P. Fillmore, (D), Kansas, 1994, Gs
David W. Marchetti, (D), Utah, 2006, GmClc
Allen L. Stork, (D), California (Santa Cruz), 1984, Giv
Rady Chair:
Bradford R. Burton, (D), Wyoming, 1997, GocGt
Moncrief Chair:
Elizabeth S. Petrie, (D), Utah State, 2014, GcoYe
Lecturer:
Holly Brunkal, (D), Colorado Mines, 2015, NgGm

Connecticut

Central Connecticut State University

Dept of Geography
New Britain, CT 06050-4010
p. (860) 832-2785
cannatadij@ccsu.edu
http://www.ccsu.edu/geography

Dept of Physics & Earth Sciences (B,M)
1615 Stanley Street
New Britain, CT 06050-4010
p. (860) 832-2930
antar@ccsu.edu
http://www.physics.ccsu.edu/
Department Secretary: Sandra O'Day
Chair:
Ali A. Antar, (D), Connecticut, Zn
Professor:
Sandra Burns, (D), Connecticut, 1972, Ze
Steven B. Newman, (D), SUNY (Albany), As
Michael Wizevich, (D), Virginia Tech, GsdGm
Associate Professor:
Marsha Bednarski, (D), Connecticut, 1997, Ze
Mark Evans, (D), Pittsburgh, 1989, GciGz
Kristine Larsen, (D), Connecticut, 1988, Xy
Jennifer L. Piatex, (D), Pittsburgh, XgZr
Emeritus:
Charles W. Dimmick, (D), Tulane, 1969, Ge
Related Staff:
R. Craig Robinson, (B), Millersville, 1971, Xy

Eastern Connecticut State University

Environmental Earth Science (B) ●
83 Windham Street
Willimantic, CT 06226
p. (860) 465-4317
cunninghamw@easternct.edu
https://www.easternct.edu/environmental-earth-science/
f: https://www.facebook.com/groups/181927638678890/
Programs: Environmental Earth Science; Environmental Science, General Earth Science; Sustainable Energy Science
Enrollment (2020): B: 75 (19)
Chair:
Stephen Nathan , (D), Massachusetts, 2005, PmEo
Professor:
Catherine A. Carlson, (D), Michigan State, 1994, Hwg
Dickson Cunningham, (D), Texas, 1993, GtcGx
Peter A. Drzewiecki, (D), Wisconsin, 1996, GrsGd
James A. Hyatt, (D), Queens, 1993, Gml
Associate Professor:
Meredith Metcalf, (D), Connecticut, 2013, ZirHw
Bryan Oakley, (D), Rhode Island, 2012, OnGml
Adjunct Professor:
Heath Carlson, (M), E Connecticut State, 2013, Zg
Lynn-Ann DeLima, (M), S Connecticut State, 2000, Zg
Vishnu R. Khade, (D), Cincinnati, 1987, Zg
Emile Levasseur, (M), Sacred Heart, 1992, Zg
Bruce Morton, (M), Connecticut, 1983, Zg
James Motyka, (M), E Connecticut State, 1987, Zg
Julie Sandeen, (M), Connecticut, 1991
Wesley Winterbottom, (M), Connecticut, 1988, Zg

Emeritus:
Sherman M. Clebnik, (D), Massachusetts, 1975, Gl
Fred Loxsom, (D), Dartmouth, 1969, Zn
Henry I. Snider, (D), New Mexico, 1966, Ge
Roy R. Wilson, (D), Oregon State, 1984, Zi

Middlesex Community College

Div of Science, Allied Health and Engineering (A) ⃠
100 Training Hill Road
Middletown, CT 06457
p. (860) 343-5779
MBusa@mxcc.commnet.edu
http://www.mxcc.commnet.edu/Content/Environmental_Science_1.asp
Programs: Physics studies, Chemistry studies, Biology studies
Certificates: Manufacturing Engineering Tech, Computer Engineering Tech
Professor:
Mark Busa, (D), Connecticut, ZgEgYg

Naugatuck Valley Community College

Science, Technology, Engineering, and Mathematics (STEM)
750 Chase Parkway
Waterbury, CT 06708
p. (203) 596-8690
cdonaldson@nv.edu
http://www.nvcc.commnet.edu

Norwalk Community College

Sciences Deptartment
188 Richards Avenue
Norwalk, CT 06854
p. (203) 857-7275
mbarber@ncc.commnet.edu
http://www.ncc.commnet.edu/dept/science/

Quinebaug Valley Community College

Dept of Environmental Science
742 Upper Maple Street
Danielson, CT 06239
p. (860) 412-7230
JLewis@qvcc.edu
http://www.qvcc.commnet.edu

Southern Connecticut State University

Department of Earth Science (B) O⃠
501 Crescent Street
New Haven, CT 06515
p. (203) 392-5835
flemingt1@southernct.edu
https://inside.southernct.edu/earth-science
Programs: Geology (BS); Environmental Earth Science (BS); General Earth Science (BS); Earth Science Education (BS); Natural Resources (BS); Natural Hazards; Earth Science (BA)
Enrollment (2014): B: 46 (7)
Professor:
Thomas H. Fleming, (D), Ohio State, 1995, GiCgc
Assistant Professor:
Jennifer Cooper-Boemmels, (D), Connecticut, 2020, GcgOg
Nicholas D. Fedorchuk, (D), Wisconsin (Milwaukee), 2019, GslGg
Dushmantha Jayawickreme, (D), Michigan State, 2008, YgSpNg
Instructor:
Bryan Adinolfi, (M), SUNY (Environ), 2006, GgZg
Christopher Balsley, (M), Wesleyan, 1972, Gg
Daniel Coburn, (M), C Connecticut, 2003, Ze
Yolanda Lee-Gorishti, (M), Connecticut, 2006, GaZe
Julie Rumrill, (M), Vermont, 2009, GgeGl
Emeritus:
Cynthia R. Coron, (D), Toronto, 1982, Eg
John W. Drobnyk, (D), Rutgers, 1962, RhGs
James W. Fullmer, (D), MIT, 1979, AmXa
Robert Radulski, (D), Rhode Island, Og
William Tolley, (M), Syracuse, Gg

University of Connecticut

Department of Marine Sciences (B,M,D) ⃠
1080 Shennecossett Road
Groton, CT 06340
p. (860) 405-9152
marinesciences@uconn.edu
http://www.marinesciences.uconn.edu/
Programs: Marine Sciences(B), Oceanography (M,D)
Enrollment (2018): B: 50 (10) M: 9 (0) D: 33 (0)
Head:
J. Evan Ward, (D), Delaware, 1989, Ob
Professor:
Ann Bucklin, (D), California (Berkeley), 1980, Ob
Hans G. Dam, (D), SUNY (Stony Brook), 1989, Ob
Heidi Dierssen, (D), California, 2000, Op
Senjie Lin, (D), SUNY (Stony Brook), 1995, Ob
Robert P. Mason, (D), Connecticut, 1991, OcCtm
George B. McManus, (D), SUNY (Stony Brook), 1986, Ob
James O'Donnell, (D), Delaware, 1986, Op
Craig Tobias, (D), William & Mary, 1999, Oc
Pieter T. Visscher, (D), Groningen (Neth), 1991, Co
Associate Professor:
Hannes Baumann, (D), Hamburg, 2006, Ob
Julie Granger, (D), British Columbia, 2006, Oc
David Lund, (D), MIT/WHOI, 2006, Oc
Annelie Skoog, (D), Goteborg (Sweden), Oc
Penny Vlahos, (D), Massachusetts, 2001, OcGeu
Michael Whitney, (D), Delaware, 2003, Op
Assistant Professor:
Cara Manning, (D), MIT/WHOI, 2016, Oc
Catherine Matassa, (D), Northeastern, 2014, Ob
Cesar Rocha, (D), Scripps, 2018, Op
Leonel Romero, (D), Scripps, 2008, Op
Samantha Siedlecki , (D), Chicago, 2010, On
Research Associate:
Paola Batta-Lona, (D), Connecticut, 2014, Ob
Zofia Baumann, (D), SUNY (Stony Brook), 2011, ObcCm
Jennifer O'Donnell, (D), Cambridge, 1989
Jamie Vaudrey, (D), Connecticut, 2007, Ob
Huan Zhang, (D), Tokyo Fisheries, 1995, Ob
Emeritus:
Peter Auster, (D), National (Ireland), 2000, Ob
Walter F. Bohlen, (D), MIT/WHOI, 1969, Op
James Edson, (D), Penn State, Op
William F. Fitzgerald, (D), MIT/WHOI, 1970, Oc
Edward C. Monahan, (D), MIT, 1966, AsOp
Sandra Shumway, (D), Coll of North Wales, 1976, Ob

Dept of Geography
Storrs, CT 06269-4148
p. (860) 486-2610
cindy.zhang@uconn.edu
https://geography.uconn.edu/

Dept of Geosciences (B,M,D)
354 Mansfield Road
U-1045
Storrs, CT 06269-1045
p. (860) 486-4432
hren@uconn.edu
http://geosciences.uconn.edu
Programs: Geology, Geophysics, Climate Modeling
Enrollment (2021): B: 48 (15) M: 5 (2) D: 16 (2)
Department Head:
Tracy Frank, (D), Michigan, 1996, CgGs
Professor:
Vernon F. Cormier, (D), Columbia, 1976, YsGt
Chris Fielding, (D), Durham, 1982, Gs
Lanbo Liu, (D), Stanford, 1993, Yg
Gary A. Robbins, (D), Texas A&M, 1983, Hw
Robert M. Thorson, (D), Washington, 1970, Gm
Pieter Visscher, (D), Groningen (Neth), 1991, Co
Associate Department Head:
Michael Hren, (D), Stanford, 2007, CslCo
Associate Professor:
Andrew M. Bush, (D), Harvard, 2005, PgqPi
Timothy Byrne, (D), California (Santa Cruz), 1981, Gc
Jean M. Crespi, (D), Colorado, 1985, Gc
William Ouimet, (D), MIT, 2007, GmZy

Assistant Professor:
Ran Feng, (D), Michigan, 2015, PcAt
Julie Fosdick, (D), Stanford, 2011, CcGt
Clay Tabor, (D), Michigan, 2015, PcAt
Lecturer:
Benjamin Chilson-Parks, (D), Brown, 2019, Gi
Tammo Reichgelt, (D), Otago, 2015, Pbc
Emeritus:
Raymond Joesten, (D), Caltech, 1974, GzZmGp

University of New Haven

Dept of Environmental Sciences (B,M) ○☒
300 Boston Post Rd
West Haven, CT 06516
p. (203) 932-7101
rldavis@newhaven.edu
http://www.newhaven.edu
Programs: Environmental Science (BS, MS), Marine Biology (BS), Marine Affairs (BS), MS in Environmental Science offers concentrations in Environmental Geoscience, Environmental Ecology, GIS, Environmental Health and Management
Certificates: GIS
Enrollment (2016): B: 28 (5) M: 29 (8)
Provost and Senior Vice President of Academic Affairs:
Daniel J. May, (D), California (Santa Barbara), 1986, GtxGe
Coordinator-Graduate Environmental Science:
Roman N. Zajac, (D), Connecticut, 1985, Zni
Professor:
Carmela Cuomo, (D), Yale, CmmGu
R. Laurence Davis, (D), Rochester, 1980, GemHg
Coordinator-Undergraduate Environmental Science:
Kristen Przyborski, (D)
Associate Professor:
Amy L. Carlile, (D), Washington, ObZn
John Kelly, (D), California (Davis), ObZn
Lecturer:
Jean-Paul Simjouw, (D), Old Dominion, 2004, OcCma
Practitioner-in-Residence:
Paul Bartholemew, (D), British Columbia, GeZiGz

Wesleyan University

Dept of Earth & Environmental Sciences (B,M) ☒
265 Church Street
Room 455
Middletown, CT 06459-0139
p. (860) 685-2244
vharris@wesleyan.edu
http://www.wesleyan.edu/ees
Programs: Earth & Environmental Sciences
Administrative Assistant: Virginia M. Harris
Enrollment (2021): B: 29 (12) M: 5 (4)
Professor:
Barry Chernoff, (D), Michigan, 1983, Ge
Martha S. Gilmore, (D), Brown, 1997, XgGmZr
Suzanne B. OConnell, (D), Columbia, 1986, GsuGe
Phillip G. Resor, (D), Stanford, 2003, Gct
Dana Royer, (D), Yale, 2002, PegPb
Chair:
Timothy C.W. Ku, (D), Michigan, 2001, Cl
Associate Professor:
James P. Greenwood, (D), Brown, 1997, Xc
Associate Professor of the Practice:
Kim Diver, (D), Syracuse, 2004, ZiyZg
Emeritus:
James T. Gutmann, (D), Stanford, 1972, GviGz
Peter C. Patton, (D), Texas, 1976, Gm
Ellen Thomas, (D), Utrecht, 1979, Pm
Johan C. Varekamp, (D), Utrecht, 1979, CgGzv
Facilities Manager:
Joel LaBella, (B), S Connecticut, 1987, Zn

Yale University

Dept of Earth and Planetary Sciences (B,D) ☒
210 Whitney Avenue
P.O. Box 208109
New Haven, CT 06520-8109
p. (203) 432-3114
rebecca.pocock@yale.edu
http://earth.yale.edu
Enrollment (2019): B: 18 (7) D: 54 (13)
Curator:
Derek E. G Briggs, (D), Cambridge, 1976, PigPg
Professor:
Jay J. Ague, (D), California (Berkeley), 1987, Gx
David Bercovici, (D), California (Los Angeles), 1989, Yg
Ruth E. Blake, (D), Michigan, 1997, Cg
Mark T. Brandon, (D), Washington, 1984, Gc
Derek E G Briggs, (D), Cambridge, 1976, Yg
David A D Evans, (D), Caltech, 1998, YmGt
Alexey V. Fedorov, (D), Scripps, 1997, Op
Jacques Gauthier, (D), California (Berkeley), 1984, Pv
Shun-ichiro Karato, (D), Tokyo, 1977, Gy
Jun Korenaga, (D), MIT, 2000, YgsCg
Maureen D. Long, (D), MIT, 2006, Ysg
Jeffrey J. Park, (D), California (San Diego), 1985, YsZgc
Danny M. Rye, (D), Minnesota, 1972, Cs
Mary-Louise Timmermans, (D), Cambridge, 2000, Oc
John S. Wettlaufer, (D), Washington, 1991, Yg
Senior Research Scientist:
Edward W. Bolton, (D), California (Los Angeles), 1985, GqHqCq
Ellen Thomas, (D), Utrecht, 1979, Pm
Associate Professor:
Noah Planavsky, (D), California (Riverside), 2012, CsbCm
Assistant Professor:
Bhart-Anjan Bhullar, (D), Harvard, 2014
Pincelli Hull, (D), California (San Diego), 2010
Juan Lora, (D), Arizona, 2009, As
Juan Lora, (D), Arizona, 2014, ZoAs
Alan Rooney, (D), Durham (UK), 2011, Cc
Lidya Tarhan, (D), California (Riverside), 2013, PiGsPe
Emeritus:
Robert B. Gordon, (D), Yale, 1955, Nr
Ronald B. Smith, (D), Johns Hopkins, 1975, As
Elisabeth S. Vrba, (D), Cape Town, 1974, Pv
Research Affiliate:
William C. Graustein, (D), Yale, 1981, Cl
Related Staff:
James O. Eckert, (D), Texas A&M, 1988, Gp

Peabody Museum of Natural History ☒
PO Box 208118
170 Whitney Avenue
New Haven, CT 06520-8118
p. (203) 432-3752
peabody.director@yale.edu
http://www.peabody.yale.edu/
Emeritus Curator:
Leo W. Buss, (D), Johns Hopkins, 1979, Pi
Elisabeth S. Vrba, (D), Cape Town, 1974, Pvg
Curator:
Jay J. Ague, (D), California (Berkeley), 1987, Gzx
Michael J. Donoghue, (D), Harvard, 1982, Pbg
Jacques A. Gauthier, (D), California (Berkeley), 1984, Pv
Assistant Curator:
Bhart-Anjan Bhullar, (D), Pv
Senior Collections Manager:
Susan H. Butts, (D), Idaho, 2003, PiGsPe
Collections Manager:
Shusheng Hu, (D), Florida, 2006, PblGs
Stefan Nicolescu, (D), Gothenburg, 1998, CtGpz

Delaware

University of Delaware

Dept of Earth Sciences (B,M,D) ●☒
103 Penny Hall
Newark, DE 19716
p. (302) 831-2569
jmadsen@udel.edu
https://www.udel.edu/academics/colleges/ceoe/departments/es/
f: https://www.facebook.com/UDCEOE
t: @udceoe
Programs: BS Geological Sciences, BA Geological Sciences, BS Earth Science Education, MS Geological Sciences, PhD Geological Sciences
Enrollment (2020): B: 43 (19) M: 20 (7) D: 14 (2)

Director, DGS, and State Geologist:
David R. Wunsch, (D), Kentucky, 1992, Ge
Acting Chair:
John A. Madsen, (D), Rhode Island, 1987, YrOuZe
Professor:
Eliot A. Atekwana, (D), W Michigan, 1996, CgsCq
Ronald E. Martin, (D), California (Berkeley), 1981, PmePg
Holly A. Michael, (D), MIT, 2005, Hwq
Michael ONeal, (D), Washington, 2005, Gm
James E. Pizzuto, (D), Minnesota, 1982, Gm
Neil C. Sturchio, (D), Washington (St Louis), 1983, ClcCa
Associate Professor:
Clara S. Chan, (D), California (Berkeley), 2006, PoObGz
Susan McGeary, (D), Stanford, 1984, Ze
Adam F. Wallace, (D), Virginia Tech, 2008, ClGzZm
Jessica Warren, (D), MIT/WHOI, 2007, GxCpGt
Assistant Professor:
Chandranath Basak, (D), Florida, 2011, ClOcCs
Colton Lynner, (D), Yale, 2015, GtYsGc
Visiting Professor:
Claire J. O'Neal, (D), Washington, 2005, Ze
Director Emeritus, DGS:
Robert R. Jordan, (D), Bryn Mawr, 1964, Gr
John H. Talley, (M), Franklin and Marshall, 1974, Eg
Emeritus:
Billy P. Glass, (D), Columbia, 1968, XmcGz
Peter B. Leavens, (D), Harvard, 1967, Gz
John F. Wehmiller, (D), Columbia, 1971, Cl
Laboratory Director:
Bill Parnella, (B), Buffalo, 1984, Gg
Cooperating Faculty:
A. Scott Andres, (M), Lehigh, 1984, Hw
Katharina Billups, (D), California (Santa Cruz), 1998, Ou
Laurent E. Cartier, (D), Basel, 2014, Gz
Shreeram Inamdar, (D), Virginia Tech, 1996, Hg
Deb Jaisi, (D), Miami, 2007, Cb
Kendra J. Lynn, (D), Hawaii, 2017, Gi
Thomas E. McKenna, (D), Texas, 1997, H
Peter P. McLaughlin, Jr., (D), Louisiana State, 1989, Gr
Jack Puleo, (D), Florida, 2004, Hw
Kelvin W. Ramsey, (D), Delaware, 1988, Gs
William S. Schenck, (M), Delaware, 1997, Gg
Angelia L. Seyfferth, (D), California (Riverside), 2008, Sc
Art Trembanis, (D), Virginia Inst of Marine Sci, 2004, Ou
William J. Ullman, (D), Chicago, 1982, Oc

Dept of Geography ●☒
125 Academy Street
216 Pearson Hall
Newark, DE 19716-2541
p. (302) 831-3218
dlevia@udel.edu
http://www.ceoe.udel.edu/schools-departments/department-of-geography

Oceanography Program (M,D)
700 Pilottown Road
Lewes, DE 19958
p. (302) 645-4279
kbillups@udel.edu
http://www.ocean.udel.edu
Professor:
Thomas M. Church, (D), California (San Diego), 1970, Cm
Victor Klemas, (D), Braunschweig (Germany), 1965, Zr
George W. Luther, III, (D), Pittsburgh, 1972, Cm
Jonathan H. Sharp, (D), Dalhousie, 1972, Oc
Christopher K. Sommerfield, (D), SUNY (Stony Brook), 1997, OuGsCc
William J. Ullman, (D), Chicago, 1982, Cm
Ferris Webster, (D), MIT, 1961, Op
Xiao-Hai Yan, (D), SUNY (Stony Brook), 1989, Zr
Associate Professor:
Katharina Billups, (D), California (Santa Cruz), 1998, Pe
Douglas C. Miller, (D), Washington, 1985, Ob
Associate Scientist:
Charles H. Culberson, (D), Oregon State, 1972, Oc
Richard T. Field, (D), Delaware, 1994, Og
Assistant Professor:
Matthew J. Oliver, (D), Rutgers, 2006, ObZrOg

Adjunct Professor:
Richard B. Coffin, (D), Delaware, 1986, Zn
James Crease, (D), Cambridge, 1951, Op
Marilyn L. Fogel, (D), Texas, 1977, Oc
Norden E. Huang, (D), Johns Hopkins, 1967, Zn
David E. Krantz, (D), South Carolina, Cs
Kamlesh Lulla, (D), Indiana State, 1963, Zr
Donald B. Nuzzio, (D), Rutgers, 1982, Oc
Manmohan Sarin, (D), Gujarat (India), 1984, Zn
Alain J. Veron, (D), Paris, 1988, Oc
John F. Wehmiller, (D), Columbia, Gu
Emeritus:
Jin Wu, (D), Iowa, 1964, Op

Physical Ocean Science and Engineering
Lewes, DE 19958
p. (302) 831-6640
carcher@udel.edu
http://www.ceoe.udel.edu/schools-departments/school-of-marine-science-and-policy/pose-program

District of Columbia

Carnegie Institution for Science

Earth and Planets Laboratory ☒
5241 Broad Branch Road, N.W.
Washington, DC 20015-1305
p. (202) 478-8900
mwalter@carnegiescience.edu
https://epl.carnegiescience.edu
Emeritus Staff Scientist:
Alan T. Linde, (D), Queensland, 1972, Ys
Ho-kwang Mao, (D), Rochester, 1968, Yx
Douglas Rumble, III, (D), Harvard, 1969, Gp
I. Selwyn Sacks, (D), Witwatersrand, 1961, Ys
Fouad Tera, (D), Vienna, 1962, Xm
Staff Scientist:
Conel M.O D. Alexander, (D), Essex, 1987, Xc
Alan Boss, (D), California (Santa Barbara), 1979, Xa
Paul Butler, (D), Maryland, 1993, Xa
John Chambers, (D), Manchester, 1994, Xs
George D. Cody, (D), Penn State, 1992, Co
Ronald E. Cohen, (D), Harvard, 1985, Gy
Peter E. Driscoll, (D), Johns Hopkins, 2010, Ym
Yingwei Fei, (D), CUNY, 1989, Cp
Peter Gao, (D), Caltech, 2017, Xg
Alexander F. Goncharov, (D), Russian Acad of Sci, 1983, GyYxh
Robert M. Hazen, (D), Harvard, 1975, Gz
Helene Le Mevel, (D), Wisconsin, 2016, Gv
Bjorn O. Mysen, (D), Penn State, 1974, Cp
Anne Pommier, (D), Orléans (France), 2009, Gy
Diana C. Roman, (D), Oregon, 2004, Gv
Anat Shahar, (D), California (Los Angeles), 2008, CgYxCs
Scott S. Sheppard, (D), Hawaii, 2004, Xs
Steven B. Shirey, (D), SUNY - Stony Brook, 1984, Gi
Andrew Steele, (D), Portsmouth, 1996, Xy
Timothy A. Strobel, (D), Colorado Mines, 2008, Zm
Johanna K. Teske, (D), Arizona, 2014, Xg
Sally June Tracy, (D), CalTech, 2016, Yg
Peter E. van Keken, (D), Arizona, 2005, Ys
Lara Wagner, (D), Arizona, 2005, Ys
Alycia J. Weinberger, (D), CalTech, 1998, Xa
Research Scientist:
Dionysis Foustoukos, (D), Minnesota, 2005, Cg
Zachary Geballe, (D), California (Berkeley), 2014, Gy
Amol Karandikar, (D), Johannes Wolfgang Goethe, 2016, Gy
Shaunna Morrison, (D), Arizona, 2017, Zn
Librarian:
Shaun J. Hardy, (M), SUNY (Buffalo), 1987, Zn
Geo-Informatics Scientist:
Anirudh Prabhu, (D), Rensselaer Polytech, 2021, Zn
Emeritus GL Director:
Wesley T. Huntress, Jr., (D), Stanford, 1968, Xc
Director:
Michael Walter, (D), Texas (Dallas), 1991, Cg

George Washington University

Dept of Geography (B,M)
2036 H St NW
Washington, DC 20052
p. (202) 994-6185
geog@gwu.edu
https://geography.columbian.gwu.edu/
f: https://www.facebook.com/GWGeography/
Enrollment (2018): B: 60 (0) M: 21 (0)
Chair:
Lisa Benton-Short, (D), Syracuse
Professor:
Elizabeth Chacko, (D), California (Los Angeles), 1997, Zy
Marie D. Price, (D), Syracuse, 1990, Zy
Associate Professor:
Mona Atia, (D), Washington
Ryan Engstrom, (D), San Diego State, Zr
David Rain, (D), Penn State
Nikolay Shiklomanov, (D)
Dmitry Streletskiy, (D)
Assistant Professor:
Ginger Allington, (D), Saint Louis
Melissa Keeley, Tech (Berlin)
Michael Mann, (D)

National Academies of Sciences, Engineering, and Medicine

Board on Earth Sciences and Resources ☒
Keck Center-6th Floor
500 5th Street, NW
Washington, DC 20001
p. (202) 334-2744
besr@nas.edu
http://dels.nas.edu/besr
f: https://www.facebook.com/Board-on-Earth-Sciences-and-Resources-260128822801/
t: @NASEM_Earth
Director:
Elizabeth A. Eide, (D), Stanford, 1993, GgoGt
Senior Program Officer:
Sammantha L. Magsino, (M), Florida Intl, 1993, GvNg
Scholar:
Anne M. Linn, (D), California (Los Angeles), 1991, Gs
Senior Program Officer:
Deborah Glickson, (D), Washington, 2007, GutGg

Smithsonian Institution / National Museum of Natural History

Dept of Mineral Sciences ☒
NHB MRC 119
10th & Constitution Avenue, NW
PO Box 37012
Washington, DC 20013-7012
p. (202) 633-1860
NMNH-MineralSciences@si.edu
https://naturalhistory.si.edu/research/mineral-sciences
Administrative Assistant: Phyllis McKenzie Claire Puig
Research Geologist:
Elizabeth Cottrell, (D), Columbia, 2004, Cp
Research Geologist:
Michael Ackerson, (D), Rensselaer Polytech, 2015, Gip
Benjamin Andrews, (D), Texas, 2009, GvCpGi
Catherine Corrigan, (D), Case Western, 2004, XmcXg
Gabriela Farfan, (D), MIT/WHOI, 2019, GzCbGu
Ioan Lascu, (D), Minnesota, 2011, GzYm
Glenn J. MacPherson, (D), Princeton, 1981, XcGi
Timothy J. McCoy, (D), Hawaii, 1994, Xmg
Jeffrey E. Post, (D), Arizona State, 1981, Gz
Museum Technician (GVP):
Kadie Bennis, (M), Missouri (Kansas City), 2019, Gv
Museum Specialist (GVP):
Sally K. Sennert, (M), Pittsburgh, 2003, GvZr
Edward Venzke, (M), Minnesota (Duluth), 1993, Gv
Analytical Laboratories - Manager:
Timothy Rose, (M), Maryland, 1991, CaGv
Education and Outreach:
Marion Le Voyer, (D), Blaise Pascal, 2009, Cu
Museum Specialist (Labs):
Timothy Gooding, (B), Hampshire Coll, 1990
Rob Wardell, (B), N Carolina (Pembroke), 2014, Zn
Mineralogist:
Michael A. Wise, (D), Manitoba, 1987, Gz
IT Specialist:
Adam Mansur, (M), Maryland, 2008, Zf
Collection Technician:
Kelsey Falquero, (M), George Washington, 2011, Xm
Greg Polley, (B), Maryland, 2012, Gxz
Collection Specialist:
Cathleen Brown, (M), Maryland, 1996, GiCgGz
Collection Manager:
Russell Feather, (B), George Mason, 1982, Gz
Leslie J. Hale, (B), Maryland, 1989, GgxZf
Julie Hoskin, (B), Appalachian State, 1997, Xm
Paul Pohwat, 1982, Gz

Dept of Paleobiology
Dept. of Paleobiology, MRC121
NMNH, Smithsonian Institution
P.O. Box 37012
Washington, DC 20013-7012
p. (202) 633-1320
huntg@si.edu
https://naturalhistory.si.edu/research/paleobiology
Curator:
Scott L. Wing, (D), Yale, 1981, Pb
Curator:
Anna K. Behrensmeyer, (D), Harvard, 1973, Pv
William A. DiMichele, (D), Illinois, 1979, Pb
Brian T. Huber, (D), Ohio State, 1988, Pm
Conrad C. Labandeira, (D), Chicago, 1991, Pi
Ian G. Macintyre, (D), McGill, 1967, Gs
Curator:
Martin A. Buzas, (D), Yale, 1963, Pm
Douglas H. Erwin, (D), California (Santa Barbara), 1985, Pi
Daniel J. Stanley, (D), Grenoble, 1964, Ou
Curator:
Matthew T. Carrano, (D), Chicago, 1998, Pv
Curator:
Gene Hunt, (D), Chicago, 2003, Pm
Geologist:
Thomas Dutro, (D), Pi
Bevan French, (D), Xm
John Pojeta, (D), Cincinnati, 1963, Pi
Curator:
Alan H. Cheetham, (D), Columbia, 1959, Pg
Robert Emry, (D), Columbia, 1970, Pv
Clayton E. Ray, (D), Harvard, 1962, Pv
Thomas R. Waller, (D), Columbia, 1966, Pi
Associate Chair & Collections Manager:
Jann W. M. Thompson, (B), George Washington, 1970, Pg

Smithsonian Institution, National Air & Space Museum

Center for Earth & Planetary Studies ☒
PO Box 37012, MRC 315
Washington, DC 20013-7012
p. (202) 633-3632
irwinr@si.edu
https://airandspace.si.edu/about/organization/staff?field_department_type_target_id=52
Supervisory Geologist:
Rossman P. Irwin, (D), Virginia, 2005, XgGm
Senior Geologist:
Bruce A. Campbell, (D), Hawaii, 1991, Zr
John A. Grant, (D), Brown, 1990, XgGm
Thomas R. Watters, (D), George Washington, 1985, Xg
Senior Geologist:
Ted A. Maxwell, (D), Utah, 1977, Xg
James R. Zimbelman, (D), Arizona State, 1984, Xg
Geologist:
Robert A. Craddock, (D), Virginia, 1999, XgGm
Emily S. Martin, (D), Idaho, 2014, XgGt
Sharon A. Wilson Purdy, (D), Virginia, 2017, XgGm
Photo Librarian:
Rosemary Aiello

Florida

Broward College

Environmental Science Department (A,B)
3501 SW Davie Road
Davie, FL 33314
p. (954) 201-6771
jmuza@broward.edu
http://www.broward.edu/academics/programs/environmental/Pages/
Administrative Assistant: Nicki Pickett
Chair:
Valerio Bartolucci, (M), Bologna, Ge
Professor:
Lewis Fox, (D), Delaware, 1981, Oc
Jay P. Muza, (D), Florida State, 1996, OuPmOo
Laura Precedo, (D), Emory, 1993, Oc
Lab Manager:
Lynn A. Curtis, (B), Florida Atlantic, 1997, GgPg

Natural Science Dept (A)
111 East Las Olas Boulevard
Fort Lauderdale, FL 33301
p. (954) 201-7650
mdugan@broward.edu
Instructor:
Xenia Conquy, (M), Florida Atlantic, Gg
Henri L. Liauw, (M), S Florida, Gg
William Opperman, (M), Florida, Gg

Chipola College

Dept of Natural Science (A)
3094 Indian Circle
Mariana, FL 32446
p. (850) 526-2761
hiltond@chipola.edu
http://www.chipola.edu/instruct/science
Professor:
Allan Tidwell, (M), Troy State, Zg

Daytona State College

School of Biological and Physical Sciences (A)
1200 W. International Speedway Blvd.
Daytona Beach, FL 32114
horikas@daytonastate.edu
http://www.daytonastate.edu/CampusDirectory/deptInfo.jsp?dept=SCI
Professor:
Debra W. Woodall, (D), Florida Inst of Tech, GgOg

Eckerd College

Dept of Geosciences (B) ☒
Galbraith Marine Science Laboratory
4200 54th Avenue South
St. Petersburg, FL 33711
p. (727) 864-8200
wetzellr@eckerd.edu
https://www.eckerd.edu/marinescience/
Programs: Geosciences and Marine Science
Enrollment (2018): B: 5 (5)
Professor:
Gregg R. Brooks, (D), S Florida, 1986, Gu
Joel B. Thompson, (D), Syracuse, 1989, Pg
Laura R. Wetzel, (D), Washington (St Louis), 1997, YrOuGg

Florida Atlantic University

Dept of Geosciences (B,M,D) O
777 Glades Road
Boca Raton, FL 33431
p. (561) 297-3250
warburto@fau.edu
http://www.geosciences.fau.edu
Programs: BS Geoscience, MS Geoscience, PhD Geoscience
Certificates: GIS certificate, Advanced GIS certificate
Enrollment (2018): B: 103 (25) M: 27 (8) D: 34 (1)
Chair:
Zhixiao Xie, (D), SUNY (Buffalo), 2002, ZirZy
Director, Florida Center for Environmental Studies:
Colin Polsky, (D), Penn State, 2002, Znc
Associate Provost for Programs and Assessment :
Russell L. Ivy, (D), Florida, 1992, Zn
Professor:
Leonard Berry, (D), Bristol, 1969, Zg
Assistant Chair:
David L. Warburton, (D), Chicago, 1978, CgGe
Associate Professor:
Xavier Comas, (D), Rutgers, 2005, Yg
Maria Fadiman, (D), Texas, 2003, Zn
Scott H. Markwith, (D), Georgia, 2007, Zy
Anton Oleinik, (D), Purdue, 1998, Pi
Tara L. Root, (D), Wisconsin, 2005, Hg
Caiyun Zhang, (D), Texas (Dallas), 2010, Zr
Associate Scientist:
Tobin Hindle, (D), Florida Atlantic, 2006, ZyiZc
Assistant Professor:
Tiffany Roberts M. Briggs, (D), S Florida, 2012, On
Erik Johanson, (D), Tennessee, 2016, RnZy
Weibo Liu, (D), Kansas, 2016, Zi
Instructor:
James Gammack-Clark, (M), Florida Atlantic, 2001, Zri
Emeritus:
Howard Hanson, (D), Miami, 1979, AsOg
Edward J. Petuch, (D), Miami, 1980, Pg
Jorge I. Restrepo, (D), Colorado State, 1987, Hqw
Charles E. Roberts, (D), Penn State, 1991, Zri

Florida Gateway College

Mathematics/Science Div (A,B)
149 SE College Place
Lake City, FL 32025
p. (386) 752-1822
mustapha.kane@fgc.edu
https://www.fgc.edu/academics/liberal-arts--sciences/math-and-science.aspx
Programs: Geology, Earth Science Education
Professor:
Mustapha Kane, (D), Zg
Instructor:
Avo Oymayan, Zg

Florida Gulf Coast University

Dept of Marine and Earth Sciences ☒
10501 FGCU Blvd. S.
Fort Myers, FL 33965
p. (239) 590-1000
jmacdona@fgcu.edu
https://www.fgcu.edu/cas/departments/mes/envirogeobs/
Programs: Environmental Geology BS
Professor:
James H. MacDonald, (D), SUNY (Albany), 2006, GizGt
Michael Savarese, (D), California (Davis)
Department Chair:
Joanne Muller, (D), James Cook
Assistant Professor:
Rachel Rotz, (D), Georgia, 2020, HwGg
Instructor:
Mary Abercrombie, (D), S Florida
Alayde Barbosa, (D)

Florida Institute of Technology

Ocean Engineering and Marine Science (B,M,D) O☒
150 West University Boulevard
Melbourne, FL 32901
p. (321) 674-8034
oems@fit.edu
https://www.fit.edu/engineering-and-science/academics-and-learning/ocean-engineering-and-marine-sciences/
f: https://www.facebook.com/pg/FLTechOEMS
Programs: Biological Sciences, D, Conservation Technology, M, Earth Remote Sensing, M, Ecology, M, Environmental Resource Management, M, Environmental Science, B, Environmental Science, M, Environmental Science, D , Fisheries and Aquaculture,

B, General Biology, B, Marine Biology, B, Marine Biology, M, Marine Conservation, B, Meteorology, B, Meteorology, M, Ocean Engineering, B, Ocean Engineering, M, Ocean Engineering, D, Oceanography, B, Oceanography, M, Oceanography, D, Sustainability Studies, B
Enrollment (2014): B: 100 (25) M: 60 (20)
Head:
George A. Maul, (D), Miami, 1974, Op
Professor:
Kevin Johnson, (D), Oregon, 1998, Ob
Steven M. Lazarus, (D), Oklahoma, 1996, As
Geoffrey Swain, (D), Southampton (UK), 1981, Ob
John G. Windsor, Jr., (D), William & Mary, Oc
Gary Zarillo, (D), Georgia, OupOn
Associate Professor:
Charles Bostater, (D), Delaware, 1990, Op
Assistant Professor:
Austin Fox, (D), Florida Inst of Tech, 2015, Oc
Kelli Hunsucker, (D), Florida Inst of Tech, 2013, Ob
Pallav Ray, (D), Miami, 2008, AstZo

Florida International University

Center for the Study of Matter at Extreme Conditions (M,D)
11200 SW 8th ST
VH 150, CeSMEC
Miami, FL 33199
p. (305) 348-3030
chenj@fiu.edu
http://cesmec.fiu.edu
Enrollment (2012): D: 2 (2)
Director:
Jiuhua Chen, (D), Japan Grad Univ Adv Studies, 1994, Gyz
Emeritus:
Surendra K. Saxena, (D), Uppsala, 1967, Gy

Dept of Earth & Environment (B,M,D)
Miami, FL 33199
p. (305) 348-2365
collinsl@fiu.edu
http://www.fiu.edu/orgs/geology
Enrollment (2007): B: 30 (7) M: 11 (5) D: 20 (5)
Chair:
Rene Price, (D), Miami, 2001, HwClRw
Professor:
William T. Anderson, (D), ETH (Switzerland), 2000, CsPcCl
Grenville Draper, (D), West Indies, 1979, GctRh
Rosemary Hickey-Vargas, (D), MIT, 1983, Cg
Jose F. Longoria, (D), Texas (Dallas), 1972, ZeGoe
Florentin J-M.R Maurrasse, (D), Columbia, 1973, PsGsCm
Gautam Sen, (D), Texas (Dallas), 1981, Gi
Michael C. Sukop, (D), Kentucky, 2001, HwqSp
Dean Whitman, (D), Cornell, 1993, YgGet
Associate Professor:
Laurel S. Collins, (D), Yale, 1988, Pg
Michael Gross, (D), Penn State, 1993, Gc
Andrew W. Macfarlane, (D), Harvard, 1989, Em
Shimon Wdowinski, (D), Harvard, 1990, YdGt
Assistant Professor:
Ping Zhu, (D), Miami, 2002, As
Distinguished Research Professor:
Stephen E. Haggerty, (D), London (UK), 1968, GiEgGz
Hugh E. Willoughby, (D), Miami, 1977, As
Lecturer:
Neptune Srimal, (D), Rochester, 1986, Gt
Adjunct Professor:
Jose Antonio Barros, (D), Miami, 1995, Zr
Michael Wacker, (M), Florida Intl, Gg
Research Scientist:
Edward Robinson, (D), London (UK), 1969, Pm
Research Lab Manager:
Diane H. Pirie, (B), Florida Intl, 1980, Gg
Cooperating Faculty:
Gabriel Guitierrez-Alonso, (D), Oviedo, 1992, Gc

Florida State University

Dept of Earth, Ocean, and Atmospheric Science (B,M,D)
108 Carraway Building
Tallahassee, FL 32306-4100
p. (850) 644-5861
rhart@fsu.edu
https://www.eoas.fsu.edu/
Program Assistant: Tami S. Karl
Chair:
LeRoy A. Odom, (D), N Carolina, 1971, Cg
James F. Tull, (D), Rice, 1973, Gct
Professor:
James B. Cowart, (D), Florida State, 1974, Cc
Lynn M. Dudley, (D), Washington State, 1983, Sc
Philip Froelich, (D), Rhode Island, 1979, Og
Bill X. Hu, (D), Purdue, 1996, Hw
Munir Humayun, (D), Chicago, 1994, XcCgGi
Yang Wang, (D), Utah, 1992, CgsPc
Associate Professor:
Anthony J. Arnold, (D), Harvard, 1983, Pm
William C. Parker, (D), Chicago, 1983, PqeGq
Vincent J. Salters, (D), MIT, 1989, Cg
Assistant Professor:
Jennifer Georgen, (D), MIT/WHOI, 2001, Gu
Ming Ye, (D), Arizona, 2002, Hg
Emeritus:
George W. Devore, (D), Chicago, 1952, Gp
John K. Osmond, (D), Wisconsin, 1954, Cc
Paul C. Ragland, (D), Rice, 1962, Ca
Sherwood W. Wise, Jr, (D), Illinois, 1970, PmGu

Hillsborough Community College

Earth Science (A)
2112 North 15th St.
Tampa, FL 33605
p. (813) 253-7647
jolney2@hccfl.edu
http://www.hccfl.edu
Instructor:
Marianne O. Caldwell, (D), S Florida, 2012, Ggs
James W. Fatherree, Zg
Thomas M. Klee, (M), S Illinois, 1986, Gg
Jessica L. Olney, (D), N Illinois, 2006, ZgGg
Matthew J. Werhner, (M), Adelphi, 1974, Gg
James F. Wysong, Jr., (M), S Florida, 1989, Am
Adjunct Professor:
Joseph D. Brod, Zg
Kyle M. Champion, Zg
James R. Douthat, Jr., Zg
Van E. Hayes, Zg
James E. MacNeil, Zg
Brian W. Marlowe, Zg
Matthew P. Olney, (D), N Illinois, 2006, ZgGg
Leonard T. Roth, Zg
Norman E. Soash, Zg
Poetchanaporn Tongdee, Zg
Kevan A. Van Cleave, Zg
William V. Wills, Zg

Jacksonville University

Dept of Biology & Marine Science (B,M) ☒
2800 University Boulevard North
Jacksonville, FL 32211
p. (904) 256-7302
ngoldbe@ju.edu
Programs: Biology; Marine Science
Enrollment (2018): B: 10 (10)
Chair:
Nisse Goldberg, (D)

Miami Dade College (Kendall Campus)

Chemistry-Physics & Earth Science Dept (A)
11011 S.W. 104 St.
Room 3291
Miami, FL 33176
p. (305) 237-2492
sorbon@mdc.edu
http://www.mdc.edu/kendall/chmphy/
Enrollment (2010): A: 100 (0)
Assistant Professor:
Michael G. McGauley, (M), Miami, 2003, AmOp

Emeritus:
John M. Steger, (D), Naval Postgrad Sch, 1997, OgAmZg

Miami-Dade College (Wolfson Campus)

Dept of Natural Sciences, Health & Wellness (A)
300 NE 2nd Avenue
Miami, FL 33132
p. (305) 237-3658
fgrataco@mdc.edu
Administrative Assistant: Ileana Baldizon
Professor:
Tony Barros, (D), Miami, 1995, Og
Michael Kaldor, (M), SUNY (Buffalo), 1969, Gg

Palm Beach StateCollege

Environmental Science (A)
MS #56 4200 Congress Avenue
Lake Worth, FL 33467
p. (561) 868-3475
milesj@palmbeachstate.edu
https://www.palmbeachstate.edu/areasofstudy/AreasofStudy-Environment.aspx
Programs: Environmental Science Technology A.S.
Certificates: Hazardous Materials Specialist College Credit Certificate, Environmental Science Technician College Credit Certificate, HAZWOPER 40-hr Certificate

Pensacola State College

Natural Sciences Department
1000 Collge Blvd
Pensacola, FL 32504-8998
p. (850) 484-1189
dgrasso@pensacolastate.edu
https://www.pensacolastate.edu/departments/natural-sciences/
Administrative Assistant: Kimberly LaFlamme
District Academic Department Head:
Edwin Stout
Professor:
Lois Dixon
Brooke L. Towery, (M), Ball State, 1969, Gg
Wayne Wooten
Joseph Zayas
Assistant Professor:
Thor Garber
Timothy Hathaway
Instructor:
Bobby Roberson
Michael Stumpe
Emeritus:
Thomas Gee
Science Lab Specialist:
Darrell Kelly

Saint Petersburg College, Clearwater

Dept of Natural Sciences (A)
2465 Drew Street
Clearwater, FL 33765
p. (727) 791-2534
Andrasik.Stephen@spcollege.edu
http://www.spcollege.edu/clw/science/
Professor:
Carl Opper, (M), Florida, 1982, Hw
Adjunct Professor:
Neva Duncan Tabb, Am
Hilary Flower, (M), California (Santa Barbara), Gg
Joseph C. Gould, (D), Nova, Gg
Heather L. Judkins, (D), S Florida, Og

Sante Fe Community College

Physical Science Dept
3000 NW 83rd Street
Gainesville, FL 32606
p. (505) 428-1307
david.johnson@sfcc.edu
http://www.sfcollege.edu

University of Florida

Dept of Geological Sciences (B,M,D)
P.O. Box 112120
241 Williamson Hall
1843 Stadium Road
Gainesville, FL 32611-2120
p. (352) 392-2231
info@geology.ufl.edu
geology.ufl.edu/
Programs: Geology
Certificates: Geology
Enrollment (2022): M: 2 (5) D: 40 (0)
Chair:
Ellen E. Martin, (D), California (San Diego), 1993, Cl
Professor:
Thomas S. Bianchi, (D), Maryland, 1987, Co
Mark Brenner, (D), Florida, 1983, Gn
James E T Channell, (D), Newcastle upon Tyne, 1975, YmGt
David A. Foster, (D), SUNY (Albany), 1989, GtCc
Douglas S. Jones, (D), Princeton, 1980, PeiPg
Jonathan B. Martin, (D), California (San Diego), 1993, ClHg
Joseph G. Meert, (D), 1993, YgGt
Elizabeth J. Screaton, (D), Lehigh, 1995, Hg
Associate Professor:
Peter N. Adams, (D), California (Santa Cruz), 2004, Gm
Rene Gassmoeller, (D)
John M. Jaeger, (D), SUNY (Stony Brook), 1998, Gs
Raymond Russo, (D), Northwestern, 1990, Yg
Courtney Sprain, (D), California (Berkeley), 2017, CgYg
Amy Williams, (D), California (Davis), 2014, Co
Andrew Zimmerman, (D), William & Mary, 2000, ComSb
Associate Scientist:
Ann L. Heatherington, (D), Washington (St Louis), 1988, CcGti
Kyoungwon Min, (D), California (Berkeley), 2002, Cc
Assistant Professor:
Juliane Dannberg, (D)
Stephen Elardo, (D), New Mexico, 2014, GiXgCp
Robert G. Hatfield, (D), Lancaster, 2008
Lecturer:
Anita Marshall, (D)
Matthew C. Smith, (D), Gi
Jim Vogl, (D), California (Santa Barbara), 2000, Gct
Laboratory Director:
Jason H. Curtis, (D), Florida, 1997, Pe
George D. Kamenov, (D), Florida, 2004, CgaEm
Assistant Curator:
Jonathan Bloch, (D), Michigan, 2001, Pv
Cooperating Faculty:
Bruce J. McFadden, (D), Columbia, 1976, Pv

Florida Museum of Natural History ☒
PO Box 117800
Gainesville, FL 32611-7800
p. (352) 846-2000
bmacfadd@flmnh.ufl.edu
http://www.flmnh.ufl.edu
Programs: We are not a degree-granting unit, but our students get their degree from the Dept of Geological Sciences at University of Florida
Administrative Assistant: Brittany Snipes
Director:
Douglas S. Jones, (D), Princeton, 1980, Pg
Distinguished Professor:
Bruce J. MacFadden, (D), Columbia, 1976, Pv
Curator:
Jonathan Bloch, Michigan, 2003, Pv
Professor:
Michal Kowalewski, (D), Arizona, 2000, Pi
Steven R. Manchester, (D), Indiana, 1981, PbGs

Soil & Water Science Dept (B,M,D)
PO Box 110290
2181 McCarty Hall A
Gainesville, FL 32611-0290
p. (352) 294-3151
mwhiles@ufl.edu
http://soils.ifas.ufl.edu
Chair:
K Ramesh Reddy, (D), Louisiana State, 1976, CbSbc

Professor and Center Director:
Nicholas B. Comerford, (D), SUNY (Syracuse), 1980, Sf
John E. Rechcigl, (D), Virginia Tech, 1986, So
Charley Wesley Wood, (D), So
Professor:
Teri Balser, (D), Soo
Mary E. Collins, (D), Iowa State, 1980, Sd
James H. Graham, Jr., (D), Oregon State, 1980, Sb
Sabine -. Grunwald, (D), Giessen (Germany), 1996, ZiSoZr
Willie G. Harris, Jr., (D), Virginia Tech, 1984, Scd
George J. Hochmuth, (D), Wisconsin, 1980, So
Yuncong Li, (D), Maryland, 1993, Sco
Lena Q. Ma, (D), Colorado State, 1991, Sc
Peter Nkedi-Kizza, (D), California, 1979, Sp
George A. O'Connor, (D), Colorado State, 1970, Sc
Thomas A. Obreza, (D), Florida, 1983, So
Andrew V. Ogram, (D), Tennessee, 1988, Sb
Jerry B. Sartain, (D), North Carolina, 1973, So
Associate Professor:
Zhenli He, (D), Zhejang Ag, 1988, Sc
James W. Jawitz, (D), Florida, 1999, SpHsw
Marc Kramer, (D), So
S Rao Mylavarapu, (D), Clemson, 1996, Sc
Arnold W. Schumann, (D), Georgia, 1997, So
P. Christopher Wilson, (D), Clemson, 1999, So
Assistant Professor:
Mark W. Clark, (D), Florida, 2000, Sf
Samira H. Daroub, (D), Michigan State, 1994, So
Stefan Gerber, (D), So
Patrick Inglett, (D), Florida, 2005, Sf
Cheryl Mackowiak, (D), Utah State, 2001, So
Kelly Morgan, (D), So
Maria Silveira, (D), Sao Paulo, 2003, So
Max Teplitski, (D), Ohio State, 2002, Zn
Gurpal Toor, (D), Lincoln (NZ), 2002, Sc
Alan Wright, (D), Sc
Research Associate Professor:
Vimala D. Nair, (D), Gottingen, 1978, Sc
Ann C. Wilkie, (D), Univ Coll (Ireland), 1984, Sb
Lecturer:
James Bonczek, (D), SoHsg
Susan Curry, (M), ZiSo
Research Assistant Professor:
Todd Osborne, (D), Sf
John Thomas, (D), Sp
Assistant In:
Mengsheng Gao, (D), Zn

University of Miami

Dept of Geography and Regional Studies
Coral Gables , FL 33124
p. (305) 284-6695
dofuller@miami.edu
http://www.as.miami.edu/geography/

Dept of Geological Sciences (B,M,D)
1301 Memorial Drive
43 Cox Science Building
Coral Gables, FL 33124
p. (305) 284-4253
ruthgoodin@miami.edu
http://www.as.miami.edu/geology/
Administrative Assistant: Ruth Goodin
Enrollment (2016): B: 37 (7)
Professor:
Larry C. Peterson, (D), Brown, 1984, Gr
Peter Swart
Harold R. Wanless, (D), Johns Hopkins, 1971, GseOn
Senior Lecturer:
Teresa A. Hood, (D), Miami, 1991, Gg
Lecturer:
Peter J. Leech, (D), Georgia Tech, 2013
Ta-Shana A. Taylor, (M), Arizona, 2006, ZePvGe
Emeritus:
David E. Fisher, (D), Florida, 1958, Cs
John R. Southam, (D), Illinois, 1974, Op

Dept of Marine Geosciences (M,D) ☒
RSMAS
4600 Rickenbacker Causeway
Miami, FL 33149-1098
p. (305) 421-4662
spurkis@rsmas.miami.edu
https://marine-geosciences.rsmas.miami.edu/
Programs: PhD and MSc in Marine Geosciences, BS Dual major in Marine Science/Geological Sciences, BS in Geological Sciences, BS/MS Marine Geology, BA in Geological Sciences
Certificates: Applied Carbonate Geology
Enrollment (2021): M: 17 (2) D: 5 (10)
Chair:
Sam J. Purkis, (D), Vrije (Amsterdam), 2004, Gu
Professor:
Falk Amelung, (D), Lous Pasteur, 1996, Zn
Keir Becker, (D), Scripps, 1981, YrGu
Gregor P. Eberli, (D), ETH (Switzerland), 1985, GusYx
Guoqing Lin, (D), Scripps, 2005, Yg
Larry C. Peterson, (D), Brown, 1984, GuPcm
Pam Reid, (D), Miami, 1979, Gs
Peter K. Swart, (D), London (UK), 1980, ClPsCs
Scientist:
Donald F. McNeill, (D), Miami, 1989, GsrGu
Associate Professor:
Adam Francis Holt, (D), Los Angeles, 2016
James S. Klaus, (D), Illinois, 2005, Gg
Ali Pourmand, (D), Tulane, 2005, Cga
Assistant Professor:
Amanda M. Oehlert, (D), Miami, 2014, Gu
Cooperating Faculty:
Patricia L. Blackwelder, (D), South Carolina, 1976, Pg

University of South Florida

College of Marine Science (M,D)
140 7th Avenue South
St. Petersburg, FL 33701
p. (727) 553-1130
Mitchum@usf.edu
http://www.marine.usf.edu
Enrollment (2009): M: 30 (13) D: 50 (6)
Distinguished Research Professor:
Robert H. Byrne, (D), Rhode Island, 1974, Cm
John H. Paul, (D), Miami, 1980, Ob
John J. Walsh, (D), Miami, 1969, Ob
Robert H. Weisberg, (D), Rhode Island, 1975, Op
Professor:
Luis H. Garcia-Rubio, (D), McMaster, 1981, Oc
Pamela Hallock-Muller, (D), Hawaii, 1977, OuGsPm
Gary T. Mitchum, (D), Florida State, 1984, Op
Frank E. Muller-Karger, (D), Maryland, 1988, Zr
Joseph J. Torres, (D), California, 1980, Ob
Associate Professor:
Paula G. Coble, (D), MIT/WHOI, 1992, Cm
Kendra Lee Daly, (D), Tennessee, 1995, Ob
Boris Galperin, (D), Israel Inst of Tech, 1982, Op
David J. Hollander, (D), Swiss Fed Inst Tech, 1989, Oc
Mark E. Luther, (D), N Carolina, 1982, Op
David F. Naar, (D), Scripps, 1990, Ou
Ernst B. Peebles, (D), S Florida, 1996, ObCs
Assistant Professor:
Mya Breitbart, (D), California (San Diego), 2006, Ob
Ashanti J. Pyrtle, (D), Texas A&M, 1999, Oc
Research Associate:
Rick Cole, (B), Florida Inst of Tech, 1983, Op
Dwight A. Dieterle, (M), California (San Jose), 1979, Ob
Jeff C. Donovan, (B), Florida, 1985, Op
Debra E. Huffman, (D), S Florida, 1994, Zn
Stanley D. Locker, (D), Rhode Island, 1989, Ou
Wensheng Yao, (D), Miami, 1995, Oc
Research Assistant:
David C. English, (M), Washington, 1983, Op
Adjunct Professor:
Serge Andrefouet, (D), Polynesie Francaise, 1998, OpZr
Bruce Barber, (D), Ob
Leonard Ciaccio, (D)
Thomas Cuba, (D), S Florida, Ob
Christopher D'Elia, (D), Georgia, 1974, Ob

George Denton, (D), Og
Cynthia Heil, (D), S Florida, Ob
Brian Keller, (D)
John Lisle, (D)
Anne Meylan, (D), Florida, 1984, Ob
Terrence Quinn, (D), Ou
Harunur Rashid, (D)
Eugene Shinn, (B), Miami, 1957, Gg
Randy Wells, (D), California (Santa Cruz), 1986, Ob

Emeritus:
Peter R. Betzer, (D), Rhode Island, 1970, Cm
Norman J. Blake, (D), Rhode Island, 1972, Ob
John C. Briggs, (D), Stanford, 1951, Ob
Kendall L. Carder, (D), Oregon State, 1970, Op
Kent A. Fanning, (D), Rhode Island, 1973, OcGu
Albert C. Hine, (D), South Carolina, 1975, Gus
Thomas L. Hopkins, (D), Florida State, 1964, Ob
Harold J. Humm, (D), Duke, 1945, Ob
Edward S. Van Vleet, (D), Rhode Island, 1978, Co
Gabriel A. Vargo, (D), Rhode Island, 1976, Ob

School of Geosciences (B,M,D)
4202 East Fowler Avenue
NES 107
Tampa, FL 33620-5201
p. (813) 974-2236
mrains@usf.edu
http://hennarot.forest.usf.edu/main/depts/geosci/
f: https://www.facebook.com/USFGeosciences
Programs: Geology, Geography, and Environmental Science & Policy
Certificates: GIS, Environmental Science & Policy

Director:
Mark C. Rains, (D), California (Davis), 2002, Hw

Professor:
Fenda Akiwumi, (D), Texas State, 2006, Zy
Jennifer Collins, (D), Univ Coll (London), 2002, As
Charles B. Connor, (D), Dartmouth, 1987, Gv
Barnali Dixon, (D), Arkansas, 2001, Zi
Tim Dixon, Scripps, 1979, Yd
Sarah E. Kruse, (D), MIT, 1989, Yg
Steve McNutt, (D), Columbia, 1985, Ys
Ambe njoh, (D), London, 1990, Zn
Petroniu (Bogdan) Onac, (D), Babes-Bolyai, Pc
Matthew Pasek, (D), Arizona, 2006, Cq
Ruiliang Pu, (D), Chinese Acad of Sci, 2000, Zr
Jeffrey G. Ryan, (D), Columbia, 1989, Ct
Joseph (Donny) Smoak, (D), S Carolina, Sf
Ping Wang, (D), S Florida, 1995, Gs

Associate Professor:
Kamal Alsharif, (D), Minnesota, 2004, Rw
Martin Bosman, (D), Kentucky, 1999, Zn
Sylvain Charbonnier, (D), Keele, 2008, Gv
Joni Downs, (D), Florida State, 2008, Zi
Jason Gulley, Florida, 2010, Hw
Gregory Herbert, (D), California (Davis), 2005, Pe
Rebecca Johns, (D), Wisconsin, Zn
Shawn Landry, (D), S Florida, 2013, Zi
Rocco Malservisi, (D), Penn State, 2002, Yd
Richard Mbatu, (D), Oklahoma State, 2006, Zy
Christopher Meindl, (D), Florida, 1996, Hg
Kai Rains, (D), California (Davis), 2006, Sf
Steven Reader, (D), Bristol, 1989, Zi
Phil van Beynen, (D), McMaster, 1998, Pc
Paul Wetmore, (D), S California, 2003, Gc

Assistant Professor:
Zachary Atlas, (D), Miami, 2008, Gi
Jochen Braunmiller, (D), Ys
Yasin Elshorbany, Wuppertal, Ac
Aurelie Germa, (D), Paris-Sud, 2008, Gv
Joeseph Panzik, (D), Yale, 2015, Ym
Yi Qiang, (D), Ghent, 2012, Zi
Mel Rodgers, (D), S Florida, 2013, Gv
Sarah Sheffield, (D), Tennessee, 2017, Pe
Patricia Spellman, (D), Michigan Tech, 2016, Hq
Ran Tao, (D), N Carolina (Charlotte), 2017, Zi
zglenn Thompson, (D), Leeds, 1999, Ys

Research Associate:
Laura Connor, (M), San Antonio, 2000, Zf
Anthony Menicucci, (D), California (Davis), Cs

Instructor:
Seth Cavello, (D), SUNY Buffalo, 2019, Zn
James Ivey, (D), S Florida, 2009, Zy
He (Hannah) Jin, (D), Texas State, 2019, Zi
Thomas C. Juster, (D), S Florida, 1995, Hw
Judy McIlrath, (M), S Florida, 2004, Gg
Connie Mizak, (D), S Florida, 2004, Zy

University of West Florida

Dept of Earth and Environmental Sciences (B,M)
11000 University Parkway
Bldg 13
Pensacola, FL 32514
p. (850) 474-3377
environmental@uwf.edu
http://uwf.edu/hmcse/departments/earth-and-environmental-sciences/
f: https://www.facebook.com/pages/UWF-Department-of-Environmental-Studies/205988446096964
t: @UWF_EES
Programs: Environmental Science (B,M); GIS Administration (M)
Certificates: GIS (UG,G)
Enrollment (2016): B: 133 (39) M: 40 (6)

Chair:
Johan Liebens, (D), Michigan State, 1995, So

Professor:
Zhiyong Hu, (D), Georgia, 2003, Zr
Jason Ortegren, (D), North Carolina (Greensboro), 2008, PeZu
Matthew C. Schwartz, (D), Delaware, 2002, OcCl

Associate Professor:
John D. Morgan, (D), Florida State, 2010, Zi
Phillip P. Schmutz, (D), Louisiana State, 2014, Gm

Instructor:
Chasidy Hobbs, (M), W Florida, 2005, ZgeZy

Adjunct Professor:
Wilbur G. Hugli, (D), W Florida, 2001, As
Taylor Kirschenfeld, (M), W Florida, 1988, Cg
Hilde Snoeckx, (D), Michigan, 1995, OuZg

Online GIS Program Coordinator:
Amber Bloechle, (M), W Florida, 2007, Zi

Valencia Community College

Valencia Community College
P.O. Box 3028
Orlando, FL 32802
p. (407) 299-5000
abosley@valenciacollege.edu
http://www.valenciacc.edu

Georgia

Columbus State University

Earth & Space Sciences (B,M) ☒
4225 University Avenue
Columbus, GA 31907-5645
p. (706) 507-8091
barineau_clinton@columbusstate.edu
http://ess.columbusstate.edu/
f: https://www.facebook.com/Earth-and-Space-Sciences-at-Columbus-State-University-GA-347767372234/
Programs: AS Engineering Studies, BS Robotics Engineering, BS Earth and Space Sciences with 4 concentrations: Astrophysics and Planetary Geology, Environmental Science, Geology, Secondary Education, MS Natural Sciences with 2 tracks: Environmental Science, Geosciences
Certificates: Robotics
Enrollment (2020): B: 108 (18) M: 12 (0)

Professor:
Clinton I. Barineau, (D), Florida State, 2009, GctGg
Warren Church, (D), Yale, 1996, ZnPcZc
Shawn Cruzen, (D), Nevada, 1997, Xa
Zdeslav Hrepic, (D), Kansas State, 2004, Zn
Troy Keller, (D), Michigan, 1997, RwZc
Shaw Kimberly, (D), Florida State, 1997, Zn
Mahmut Reyhanoglu, (D), Zn
David R. Schwimmer, (D), SUNY (Stony Brook), 1973, Pv

GEORGIA

Abiye Seifu, (D), Rensselaer, 1991, Zn
Rosa Williams, (D), Illinois, 1999, Xa
Associate Professor:
Andrew Puckett, (D), Chicago, 2007, Xa
Assistant Professor:
Stacey Sloan Blersch, (D), SUNY (Buffalo), 2016, HsgRw
Mohammad Jafari, (D), Zn
Stephen Jessup, (D), Am
Emeritus:
Thomas B. Hanley, (D), Indiana, 1975, Gc

Dalton State College

Dept of Physical Sciences ☒
650 College Drive
Dalton, GA 30720
p. (706) 272-4440
jmjohnson@daltonstate.edu
https://www.daltonstate.edu/academics/physical-sciences-overview.cms
Professor:
Jean M. Johnson, (D), Michigan, 1995, Zg

East Georgia State College

Science & Mathematics (A,B) ☒
131 College Circle
Swainsboro, GA 30401
p. (478) 289-2073
wedincamp@ega.edu
http://faculty.ega.edu/facweb/stracher/stracher.html
Enrollment (2015): A: 3 (1)
Professor Emeritus:
Glenn B. Stracher, (D), Nebraska, 1989, GycEc

Emory University

Dept of Environmental Sciences (B)
Mathematics & Science Center
400 Dowman Drive
Atlanta, GA 30322
p. (404) 727-4216
jwegner@emory.edu
http://www.envs.emory.edu/
Head:
Joy Budensiek, (D)
Chair:
Lance Gunderson, (D), Florida, 1992, Sf
Professor:
William B. Size, (D), Illinois, 1971, GivGa
Associate Professor:
Thomas Gillespie, (D), Florida, 2004, Zn
Assistant Professor:
Tracy Yandle, (D), Indiana, 2001, Ob
Instructor:
Anne M. Hall, (M), Georgia Tech, 1985, Gz
Campus Env Officer:
John Wegner, (D), Carleton, 1995, Sf
Lecturer:
Anthony J. Martin, (D), Georgia, 1991, Pe
Adjunct Professor:
Pamela J. W. Gore, (D), George Washington, 1983, Gs
Senior Lecturer:
Charles W. Hickcox, (D), Rice, 1971, Gg
Emeritus:
Howard R. Cramer, (D), Northwestern, 1954, Pg
Willard H. Grant, (D), Johns Hopkins, 1955, Cg
Lore Ruttan, (D), California (Davis), 1999, Ob

Fort Valley State University

Cooperative Developmental Energy Program ☒
Box 5800 FVSU
1005 State University Drive
Fort Valley, GA 31030
p. (912) 825-6454
crumblyi@fvsu.edu
http://www.fvsu.edu/academics/cdep
Director:
Isaac J. Crumbly, (D), North Dakota State, 1970, Zn
Assistant Director:
Jackie Hodges, (B), Georgia Southern, 1983, Zn
Associate Professor:
Aditya Kar, (D), Oklahoma, 1997, Ct

Georgia Highlands College

Div of Science and Physical Education (A,B)
Main Campus 3175 Hwy 27 South
P.O. Box 1864
Rome, GA 30162
p. (706) 368-7528
bmorris@highlands.edu
http://www.highlands.edu/site/geology
Associate Professor:
Billy Morris, Gg
Instructor:
Tracy Hall, Gg

Georgia Institute of Technology

School of Earth & Atmospheric Sciences (B,M,D)
311 Ferst Drive
Atlanta, GA 30332-0340
p. (404) 894-3893
greg.huey@eas.gatech.edu
http://www.eas.gatech.edu
Enrollment (2011): B: 63 (21) M: 34 (19) D: 76 (13)
Chair:
Judith A. Curry, (D), Chicago, 1992, AsZrn
Professor:
Annalisa Bracco, (D), Genova (Italy), 2000, OpZnn
Kim M. Cobb, (D), Scripps, 2002, PeCsOg
Felix J. Herrmann, (D), Delft (Neth), 1997, YesZn
Gregory L. Huey, (D), Wisconsin, 1992, As
Ellery D. Ingall, (D), Yale, 1991, CmlOc
Jean Lynch-Stieglitz, (D), Columbia, 1995, OoPcCs
Andrew V. Newman, (D), Northwestern, 2000, YdsYg
Edward Michael Perdue, (D), Georgia Tech, 1973, Co
Irina Sokolik, (D), Russian Acad of Sci, 1989, AsZr
Martial Taillefert, (D), Northwestern, 1997, ClmCa
Rodney J. Weber, (D), Minnesota, 1995, As
Peter J. Webster, (D), MIT, 1972, AsOp
Senior Scientist:
Hai-ru Chang, (D), As
Robert Stickel, (D), Rice, 1979, As
Viatcheslav Tatarskii, (D), As
Hsiang-Jui Wang, (D), Georgia Tech, 1995, As
Associate Professor:
Michael H. Bergin, (D), Carnegie Mellon, 1995, As
Robert X. Black, (D), MIT, 1990, As
Yi Deng, (D), Illinois, 2005, AsZoAm
Emanuele Di Lorenzo, (D), Scripps, 2003, OpZnn
Takamitsu Ito, (D), MIT, 2005
Athanasios Nenes, (D), Caltech, 2002, As
Marc Stieglitz, (D), Columbia, 1995, Hsg
Yuhang Wang, (D), Harvard, 1997, As
James Wray, (D), Cornell, 2010
Assistant Professor:
Josef Dufek, (D), Washington, 2006, GvZnn
Carol M. Paty, (D), Washington, 2006, YmgZn
Zhigang Peng, (D), S California, 2004, YsgZn
Andrew Stack, (D), Wyoming, 2002, Cg
Research Scientist:
Carlos Hoyos, (D), Georgia Tech, 2008, As
Hyemi Kim, (D), As
Jiping Liu, (D), Columbia, As
Chao Luo, (D), Peking, 1990, As
James C. St. John, (D), Georgia Tech, 1997, As
Henian Zhang, (D), As
Adjunct Associate Professor:
Jay Brandes, (D), Washington, Og
Carmen Nappo, (D), Georgia Tech, 1989, As
Valerie Thomas, (D), Cornell, 1987, Zu
Adjunct Assistant Professor:
Karim Sabra, (D), Michigan, 2003, Yg
Adjunct Professor:
Clark R. Alexander, (D), North Carolina State, 1990, Cl
Dominic Assimaki, (D)
Jackson O. Blanton, (D), Oregon State, 1968, Cl

Thomas D. Christina, (D), Caltech, 1989, Cl
James Crawford, (D), Georgia Tech, As
Heidi Cullen, (D), Columbia, Am
Rong Fu, (D)
Leonid Germanovich, (D), Moscow State (Russia), 1982, NrYg
Gary G. Gimmestad, (D), Colorado, 1978, ZrAs
Richard Jahnke, (D), Washington, 1976, Cl
Yongqiang Liu, (D), Am
Joseph Montoya, (D), Harvard, 1990, Cm
Armistead G. Russell, (D), Caltech, 1985, As
Stuart G. Wakeham, (D), Washington, 1976, Co
Herbert L. Windom, (D), California (San Diego), 1968, Cm

Emeritus:
Paul H. Wine, (D), Florida State, 1974, As

Georgia Southern University

Applied Coastal Research Laboratory
10 Ocean Science Circle
Savannah, GA 31411
p. (912) 598-2329
icps@georgiasouthern.edu
http://cosm.georgiasouthern.edu/icps/acrl/

Director:
Clark R. Alexander, (D), North Carolina State, 1990, OuGsOn

Assistant Professor:
Chester M. Jackson, (D), Georgia, 2010, OnGs

Research Associate:
Michael Robinson, (B), Georgia Southern, 2002, ZiOn

Dept of Geology and Geography (B,M)
68 Georgia Avenue
Building 201
PO Box 8149
Statesboro, GA 30460-8149
p. (912) 478-5361
cneville@georgiasouthern.edu
https://cosm.georgiasouthern.edu/geo/
Programs: BA Geology, BS Geology, BA Geography, BS Geography, MS Applied Geography
Enrollment (2019): B: 71 (23) M: 6 (0)

Chair:
James S. Reichard, (D), Purdue, 1995, HwGe

Professor:
Wei Tu, (D), Texas A&M, 2004, Zi
Robert Kelly Vance, (D), New Mexico Tech, 1989, GiEgGg

Associate Professor:
Christine Hladik, (D), Georgia, 2012, SbZr
Chester W. Jackson, (D), Georgia, 2010, GsmGu
Jacque L. Kelly, (D), Hawaii, 2012, Cl
Amy E. Potter, (D), Louisiana State, 2011, Zn
Kathlyn M. Smith, (D), Michigan, 2010, Pv
John Van Stan, (D), Delaware, 2012, HgClSb
Robert A. Yarbrough, (D), Georgia, 2006, Zn

Assistant Professor:
Steve N. Guggino, (D), Arizona State, 2012, GgCg
Meimei Lin, (D), Miami, 2015, Zi
Travis Swanson, (D), Texas, 2015, Gsm

Lecturer:
Nicholas Radko, (M), Georgia, 2011, Geg

Emeritus:
Gale A. Bishop, (D), Texas, 1971, Pi
James H. Darrell, (D), Louisiana State, 1973, Pl
Daniel B. Good, (D), Tennessee, 1973, Zn
Dallas D. Rhodes, (D), Syracuse, 1973, Gm
Fredrick J. Rich, (D), Penn State, 1979, Pl

Georgia Southwestern State University

Dept of Geology & Physics (B)
800 GSW State University Drive
Americus, GA 31709
p. (229) 931-2353
Deborah.Standridge@gsw.edu
http://www.gsw.edu/%7Egeology/
Administrative Assistant: Debbie Standridge

Professor:
Burchard D. Carter, (D), West Virginia, 1981, Pg
Thomas J. Weiland, (D), North Carolina, 1988, Gi

Associate Professor:
Samuel T. Peavy, (D), Virginia Tech, 1997, Yg

Assistant Professor:
Svilen Kostov, (D), CUNY, 1992, Zn

Emeritus:
Daniel D. Arden, (D), California (Berkeley), 1961, Ps
Harland E. Cofer, (D), Illinois, 1957, Gz
John P. Manker, (D), Rice, 1975, Gs

Georgia State University

Dept of Geosciences (B,M,D)
730 Langdale Hall
PO Box 3965
38 Peachtree Center Avenue
Atlanta, GA 30302-3965
p. (404) 413-5750
khankins@gsu.edu
http://geosciences.gsu.edu
f: https://www.facebook.com/gsugeosciences
t: @gsugeos
Programs: B.S. in Geosciences with concentrations in Geography, Urban Studies, Environmental Geosciences, and Geology M.S. in Geosciences with concentrations in Geography, Geology, or Water Science, Certificates: undergraduate certificates: Water Science, GIS, Sustainability, graduate certificates: GIS
Administrative Assistant: Basirat Lawal
Enrollment (2019): B: 113 (29) M: 48 (22) D: 3 (0)

Chair:
Katherine B. Hankins, (D), Georgia, 2004, Zn

Associate Dean:
Daniel M. Deocampo, (D), Rutgers, 2001, Gsn

Professor:
Jeremy E. Diem, (D), Arizona, 2000, At

Associate Professor:
Hassan A. Babaie, (D), Northwestern, 1984, GcZf
Dajun Dai, (D), S Illinois, 2007, Zi
W. Crawford Elliott, (D), Case Western, 1988, Clc
Nadine Kabengi, (D), Florida, ScClGe
Lawrence M. Kiage, (D), Louisiana State, 2007, ZyPlZr

Assistant Professor:
Sarah Ledford, (D), Syracuse, 2016, HsRw
Richard Milligan, (D), Georgia, 2016, Zn
Luke Panglo, (D), Oregon State, 2013, HqCs

Lecturer:
Armita Davarpanah, (D), Georgia State, 2015, Ggc
Paulo J. Hidalgo, (D), Michigan State, 2011, GivGz
Brian Meyer, (D), Georgia State, 2013, GesHw
Ricardo Nogueira, (D), Louisiana State, 2009, Asm
Christy Visaggi, (D), North Carolina (Wilmington), 2011, PiZe

Adjunct Professor:
J. Marion Wampler, (D), Columbia, 1963, Cc

Emeritus:
Sanford H. Bederman, (D), Minnesota, 1973, Zn
W. Robert Power, (D), Johns Hopkins, 1959, En
Seth E. Rose, (D), Arizona, 1987, Hw

Related Staff:
Atieh Tajik, (M), Georgia State, 2005, Gg

Georgia State University, Perimeter College, Alpharetta Campus

Dept of Life & Earth Sciences (A)
3705 Brookside Parkway
Alpharetta, GA 30022
p. (678) 240-6227
dstewart29@gsu.edu
Programs: Geology
Enrollment (2020): A: 2 (0)

Professor:
Dion C. Stewart, (D), Penn State, 1980, Gzi

Georgia State University, Perimeter College, Clarkston Campus

Department of Life & Earth Sciences (A)
555 North Indian Creek Drive
Building CC
Clarkston, GA 30021
p. (678) 891-3750

les@gsu.edu
https://perimeter.gsu.edu/life-and-earth-sciences/
Programs: Geology
Enrollment (2020): A: 4 (0)

Professor:
Pamela J. W. Gore, (D), George Washington, 1983, GsZeGn

Associate Professor:
E. Lynn Zeigler, (M), Emory, 1989, Gdg

Assistant Professor:
Victor J. Ricchezza, (D), S Florida, 2019, Gg

Senior Lecturer:
Stephan D. Fitzpatrick, (M), Georgia, 2011, HwCgSo

Georgia State University, Perimeter College, Decatur Campus

Dept of Life & Earth Sciences (A)
3251 Panthersville Road
Decatur, GA 30034-3897
p. 678 891-2641
les@gsu.edu
https://perimeter.gsu.edu/life-and-earth-sciences/
Programs: Geology
Enrollment (2020): A: 1 (0)

Assistant Professor:
Mengist Teklay, (D), Johannes-Gutenberg (Germany), 1996, Gg

Georgia State University, Perimeter College, Dunwoody Campus

Dept of Life & Earth Sciences (A)
2101 Womack Road
Dunwoody, GA 30338-4497
p. (770) 274-5050
Programs: Geology
Enrollment (2020): A: 2 (0)

Assistant Professor:
Robin John McDowell, (D), Kentucky, 1992, Gge

Instructor:
Alexander David Ullrich, (M), Florida, 2010, GgZnGu

Georgia State University, Perimeter College, Newton Campus

Dept of Geology (A)
239 Cedar Lane
Covington, GA 30014
p. (770) 278-1263
Polly.bouker@gpc.edu
http://www.gpc.edu/~newsci/

Associate Professor:
Polly A. Bouker, (M), Georgia, 2006, Gg

Georgia State University, Perimeter College, Online

Dept of Life & Earth Sciences (A)
555 North Indian Creek Drive
Clarkston, GA 30021
p. (678) 891-3750
les@gsu.edu
Programs: Geology

Interim Chair:
Deniz Z. A. Ballero, (D), Georgia, 2013, GgPm

Adjunct Professor:
Ed Albin, (D), GgXg
Jefferson B. Chaumba
Kimberly D. Schulte, (M), Gg
Christine D. Valenti, Gg

Mercer University

Dept of Environmental Engineering (B,M)
1400 Coleman Avenue
Macon, GA 31207
p. (912) 744-2597
lackey_l@mercer.edu
Department Secretary: Brenda Walraven

Professor:
Bruce D. Dod, (D), S Mississippi, 1973, Zg
Dan R. Quisenberry, (D), World Open, 1980, Zr

Assistant Professor:
Geoffrey W. Hayden, (D), Pennsylvania, 1987, Zm
Robert L. Huffman, (D), Massachusetts, 1985, Zn

Research Associate:
Paul P. Sipiera, (D), Otago (NZ), 1985, Xm

Instructor:
Barbara D. Henley, (M), Mercer, 1978, Gg

Middle Georgia College

Geology Dept (A)
1100 Second Street
Cochran, GA 31014
dawn.sherry@mga.edu
http://www.mgc.edu

Associate Professor:
Tina Mahaffee, (M), Georgia State, Gg

Assistant Professor:
Daniel Snyder, (D), Iowa, Gg

University of Georgia

Dept of Crop & Soil Science (B,M,D)
311 Plant Sciences Building
Athens, GA 30602-7272
p. (706) 542-2461
dgs@uga.edu
http://www.cropsoil.uga.edu/

Chair:
Don Shilling

Professor:
Domy C. Adriano, (D), Kansas State, 1970, Sc
Paul M. Bertsch, (D), Kentucky, 1983, Sc
Gary J. Gascho, (D), Michigan State, 1968, Sc
James E. Hook, (D), Penn State, 1975, Sp
Edward T. Kanemasu, (D), Wisconsin, 1969, Sp
David E. Kissel, (D), Kentucky, 1969, Sc
William P. Miller, (D), Virginia Tech, 1981, Sc
David E. Radcliffe, (D), Kentucky, 1984, Sp
William P. Segars, (D), Clemson, 1972, Sc
Larry M. Shuman, (D), Penn State, 1970, Sc

Associate Professor:
Miguel L. Cabrera, (D), Kansas State, 1986, So
Peter G. Hartel, (D), Oregon State, 1984, Sb
Owen C. Plank, (D), Virginia Tech, 1973, Sc
Larry T. West, (D), Texas A&M, 1986, Sd

Assistant Professor:
Glendon H. Harris, (D), Michigan State, 1993, Sb

Dept of Geology (B,M,D) ●⊘
210 Field Street
Room 308
Athens, GA 30602
p. (706) 542-2652
geology@uga.edu
http://geology.uga.edu/
Enrollment (2011): B: 69 (9) M: 32 (11) D: 6 (0)

Head:
Paul A. Schroeder, (D), Yale, 1992, GzEnCl

Graduate Coordinator :
Steven M. Holland, (D), Chicago, 1990, Gr

Professor:
Douglas E. Crowe, (D), Wisconsin, 1990, Em
Ervan G. Garrison, (D), Missouri, 1979, GamGn
Robert B. Hawman, (D), Princeton, 1988, YseYu
Valentine A. Nzengung, (D), Georgia Tech, 1993, ClGeCq
L. Bruce Railsback, (D), Illinois, 1989, PcClGd
Sally E. Walker, (D), California (Berkeley), 1988, Pm

Associate Head:
Adam Milewski, (D), W Michigan, 2008, Hw

Assistant Professor:
Charlotte Garing, (D), Montpellier, 2011, Hw
Christian Klimczak, (D), Nevada, 2011, GcmGt
Mattia Pistone , (D), ETH (Switzerland), 2012, Gv

Senior Lecturer:
Marta Patino-Douce, (D), Buenos Aires, 1990, Gi

Lecturer:
Andy Darling, (D), Arizona State, 2016

Adjunct Professor:
Omer Isik Ece

Elizabeth J. Reitz, (D), Florida, 1979, Ga
John Washington, (D), Penn State, 1991
Sandra Whitney , (D), Georgia, Ga

Emeritus:
Gilles O. Allard, (D), Johns Hopkins, 1956, Eg
R. David Dallmeyer, (D), SUNY (Stony Brook), 1972, Gt
John F. Dowd, (D), Yale, 1984, Hy
Raymond Freeman-Lynde, Columbia, 1980, Gu
Susan T. Goldstein, (D), California (Berkeley), 1984, Pm
John Noakes, (D)
Alberto E. Patino-Douce, (D), Oregon, 1990, ZnnZn
Michael F. Roden, (D), MIT, 1982, Gi
Sam Swanson , (D), Stanford, 1974, Gi
David B. Wenner, (D), Caltech, 1971, Cs
James A. Whitney, (D), Stanford, 1972, GivCg
James (Jim) E. Wright, (D), California (Santa Barbara), 1981, Gt

University of West Georgia

Dept of Geosciences (B)
1601 Maple Street
Callaway Building, Room 148
Carrollton, GA 30118
p. (678) 839-6479
jmayer@westga.edu
http://www.westga.edu/geosci/
Administrative Assistant: Anita M. Bryant
Enrollment (2016): B: 122 (22)

Chair:
James R. Mayer, (D), Texas, 1995, Hw

Professor:
David M. Bush, (D), Duke, 1991, Ou
Curtis L. Hollabaugh, (D), Washington State, 1980, GzCg
Randal L. Kath, (D), SD Mines, 1990, GcNgHg
Jeong C. Seong, (D), Georgia, 1999, Zir

Associate Professor:
Brad Deline, (D), Cincinnati, 2009, Pi
Georgina G. DeWeese, (D), Tennessee, 2007, Zy
Hannes Gerhardt, (D), Arizona, 2007, Zn
Leanna Shea Rose, (D), Florida State, 2008, ZyAsm
Karen S. Tefend, (D), Michigan State, 2005, CgGe
Nathan A. Walter, (D), Florida State, 2005, Zg

Assistant Professor:
Christopher A. Berg, (D), Texas, 2007, Gpi
Jessie Hong, (D), Colorado, 2012, ZieZy

Research Associate:
Randa R. Harris, (M), Tennessee, 2001, Hs

Emeritus:
Timothy M. Chowns, (D), Newcastle, 1968, GsrGg
Thomas J. Crawford, (M), Emory, 1957, Eg

Lab Coordinator:
John D. Congleton, (M), S Methodist, 1990, GeZiHs

Valdosta State University

Dept of Physics, Astronomy & Geosciences (B)
1500 North Patterson Street
Valdosta, GA 31698-0055
p. (229) 333-5752
echatela@valdosta.edu
http://www.valdosta.edu/phy/
Enrollment (2010): B: 77 (10)

Head:
Edward E. Chatelain, (D), Iowa, 1984, PiGlPv

Professor:
Cecilia S. Barnbaum, (D), California (Los Angeles), 1992, Zn
Frank A. Flaherty, (D), Fordham, 1983, Zn
Martha A. Leake, (D), Arizona, 1982, Xg
Kenneth S. Rumstay, (D), Ohio State, 1984, Znn

Associate Professor:
Can Denizman, (D), Florida, 1998, Hw
Mary A. Fares, (D), Tennessee Tech, 1992, Zn
Judy Grable, (D), Tennessee, 2001, Hs
Mark S. Groszos, (D), Florida State, 1996, GctEg
Michael G. Noll, (D), Kansas, 2000, Zn
Paul C. Vincent, (D), Texas A&M, 2004, Zig

Pre-Engineering Director:
Barry Hojjatie, (D), Florida, 1990, Zm

Assistant Professor:
Jason Allard, (D), Penn State, 2006, AsZg

Instructor:
Perry A. Baskin, (M), Valdosta State, 1972, Zn
Donald Thieme, (D), Georgia, 2003, GaSdGm

Emeritus:
Dennis W. Marks, (D), Michigan, 1970, Zn
Arnold E. Somers, Jr., (D), Virginia Tech, 1975, Zn

Guam

University of Guam

College of Natural and Applied Sciences (B,M)
Mohammad H. Golabi, PhD
Mangilao, GU 96923
p. (671) 734-9305
mgolabi@uguam.uog.edu

Professor:
Mohammad H. Golabi, (D), Georgia, 1991, SzpSc

Water & Environmental Research Institute of the Western Pacific (M) ☒
303 University Drive
UOG Station
Mangilao, GU 96923
p. (671) 735-2685
jjenson@triton.uog.edu
https://weri.uog.edu/
Programs: Environmental Science (M)
Certificates: Groundwater Hydrology
Enrollment (2017): M: 7 (2)

Director:
John W. Jenson, (D), Oregon State, 1993, HwGel

Associate Professor:
Nathan C. Habana, (D), Guam, 2014, Hw
Yuming Wen, (D), Rhode Island, 2004, Zir

Assistant Professor:
Yongsang (Barry) Kim, (D), Purdue, 2008, Hw
Mark A. Lander, (D), Hawaii, 1990, Am
MYEONG-HO (CHRIS) YEO, (D), McGill, 2014, Hwy

Professor of Hydrology:
Leroy F. Heitz, (D), Idaho, Hg

Professor of Geology:
Henry Galt Siegrist, jr, (D), Penn State, 1961, GdzHy

Emeritus:
Gary R. W. Denton, (D), London (UK), 1974, Ob

Hawaii

Honolulu Community College

Natural Sciences (A)
874 Dillingham Blvd.
Honolulu, HI 96817
p. (808) 845-9488
brill@hawaii.edu
http://libart.honolulu.hawaii.edu/natsci/geology.php

Professor:
Richard C. Brill, Jr., (M), Hawaii, Gg

University of Hawai'i at Manoa

S.O.E.S.T. Atmospheric Sciences (B,M,D) ☒
2525 Correa Road, HIG 350
Honolulu, HI 96822
p. (808) 956-8775
smalljen@hawaii.edu
http://www.soest.hawaii.edu/atmo
Programs: Atmospheric Sciences BS
- with NWS Forecasting or US Government Track available
- ATMO Minor possible with other Geo-science, Engineering, or Math or Computer Programming degrees.

Atmospheric Sciences Masters Plan A & B
Plan A is Thesis track
Plan B is non-thesis track. Requires a general capstone research project.

Atmospheric Sciences Ph.D.
at this time

Enrollment (2021): M: 12 (7) D: 16 (1)
Chair:
Steven Businger, (D), Washington, 1986, Am
Professor:
Yi-Leng Chen, (D), Illinois, 1980, Asm
Pao-Shin Chu, (D), Wisconsin, 1981, As
Fei Fei Jin, (D), Acad Sinica (Beijing), 1985, AmsZo
Tim Li, (D), Hawaii, 1993, Asm
Bin Wang, (D), Florida State, 1984, Am
Yuqing Wang, (D), Monash, 1995, As
Associate Professor:
Jennifer Small Griswold, (D), California (Santa Cruz), AmZr
Assistant Professor:
Christina Karamperidou, (D), Columbia, 2012, AsZoc
Alison D. Nugent, (D), Yale, 2014, AsZnn
Giuseppe Torri , (D), Imperial Coll (UK), 2012, ApsAm
Jingxia Zhao, (D), California (Los Angeles), 1993, AmmAc

University of Hawai'i, Hilo

Dept of Geology (B)
200 W. Kawili Street
Hilo, HI 96720-4091
p. (808) 933-3383
jene@hawaii.edu
https://hilo.hawaii.edu/depts/geology/
Chair:
Ken Hon, (D), Colorado, 1987, Gvi
Professor:
Steven P. Lundblad, (D), North Carolina, 1994, GraCz
Jene D. Michaud, (D), Arizona, 1992, HqwGm
Associate Professor:
James L. Anderson, (D), S California, 1987, GctYd
Affiliate Faculty:
Cheryl A. Gansecki, (D), Stanford, 1998, GviZe
John P. Lockwood, (D), Princeton, 1966, Gvg
Emeritus:
Joseph B. Halbig, (D), Penn State, 1969, Ca
Cooperating Faculty:
Christina C. Heliker, (M), W Washington, 1984, Gv

University of Hawai'i, Manoa

Dept of Earth Sciences (B,M,D) ⦸
1680 East-West Road, POST 701
Honolulu, HI 96822
p. (808) 956-7640
earth-chair@soest.hawaii.edu
http://www.soest.hawaii.edu/earthsciences
f: https://www.facebook.com/UHEarthSciences
t: https://twitter.com/uh_earthscience
Programs: Earth Sciences, Environmental Earth Science, Earth and Planetary Sciences
Enrollment (2020): B: 46 (16) M: 26 (14) D: 25 (6)
Undergraduate Chair:
Scott K. Rowland, (D), Hawaii, 1987, GvZn
Graduate Chair:
Henrietta Dulai, (D), Florida State, 2005, CcHw
Department Chair:
Garrett T. Ito, (D), MIT/WHOI, 1996, YgrGt
Dean:
Brian Taylor, (D), Columbia, 1982, Gtu
Assoc. Dean:
Charles H. Fletcher, (D), Delaware, 1986, GsOnZu
Professor:
Robert A. Dunn, (D), Oregon, 1999, Yrs
Eric J. Gaidos, (D), MIT, 1996, Xsb
Craig R. Glenn, (D), Rhode Island Grad Sch Ocean, 1987, GeCmGs
Julia E. Hammer, (D), Oregon, 1998, GiCpGv
Bruce F. Houghton, (D), Otago (NZ), 1977, Gv
Jasper G. Konter, (D), California (San Diego), 2007, CcGiCt
Brian N. Popp, (D), Illinois, 1986, Csb
Kenneth H. Rubin, (D), California (San Diego), 1991, CcGvCg
Kathleen Ruttenberg, (D), Yale, 1990, Cg
Bridget R. Smith-Konter, (D), California (San Diego), 2005, YdXy
Paul Wessel, (D), Columbia, 1989, YrGt
Associate Professor:
Gregory Ravizza, (D), Yale, 1991, Cm
Associate Scientist:
Aaron Pietruszka, (D), Hawaii, 2008, Ca

Assistant Professor:
Sloan Coats, (D), Columbia, 2015, Zo
Helen Janiszewski, (D), Columbia, 2017, Ys
Thomas Shea, (D), Hawaii, 2010, Gv
Emeritus:
Michael O. Garcia, (D), California (Los Angeles), 1976, Gi
Kevin T. M. Johnson, (D), MIT, 1990, Gi
Stephen J. Martel, (D), Stanford, 1987, Ng
Ralph Moberly, (D), Princeton, 1956, GutGs
Gregory F. Moore, (D), Cornell, 1977, Gt
John M. Sinton, (D), Otago (NZ), 1976, Gi
Director of Student Services:
Heather Saito, Zn
Cooperating Faculty:
Margo H. Edwards, (D), Columbia, 1992, Gu
Sarah Fagents, (D), Lancaster (UK), 1994, Gv
Luke P. Flynn, (D), Hawaii, 1992, Zr
Patricia B. Fryer, (D), Hawaii, 1981, Gu
Milton A. Garces, (D), California (San Diego), 1995, Ys
Alexander N. Krot, (D), Moscow State (Russia), 1989, Xm
Paul G. Lucey, (D), Hawaii, 1986, Xy
Murli H. Manghnani, (D), Montana State, 1962, Yx
Fernando Martinez, (D), Columbia, 1988, Yr
Floyd W. McCoy, Jr., (D), Harvard, 1974, Gu
Peter J. Mouginis-Mark, (D), Lancaster (UK), 1977, Xg
Shiv K. Sharma, (D), Indian Inst of Tech, 1973, Gx
G. Jeffrey Taylor, (D), Rice, 1970, Xm
Donald M. Thomas, (D), Hawaii, 1977, Ca
Affiliate Graduate Faculty:
Donald A. Swanson, (D), Johns Hopkins, 1964, Gv

Dept of Oceanography (B,M,D)
1000 Pope Road
Honolulu, HI 96822
p. (808) 956-7633
ocean@soest.hawaii.edu
http://www.soest.hawaii.edu/oceanography/
Administrative Assistant: Kristin Momohara
Professor:
Barbara Bruno, (D), Hawaii, 1994, Ob
Eric H. De Carlo, (D), Hawaii, 1982, Cm
Jeffrey C. Drazen, (D), California (San Diego), 2000, Ob
Eric Firing, (D), MIT, 1978, Op
Pierre J. Flament, (D), California (San Diego), 1986, Op
David Ho, (D), Columbia, 2001, OcCb
David M. Karl, (D), California (San Diego), 1978, Ob
Paul Kemp, (D), Oregon State, 1985, Ob
Rudolf C. Kloosterziel, (D), Utrecht, 1990, Op
Douglas S. Luther, (D), MIT, 1980, Op
Julian P. McCreary, (D), California (San Diego), 1977, Op
Margaret Anne McManus, (D), Old Dominion, 1996, Op
Christopher Measures, (D), Southampton (UK), 1978, Oc
Mark A. Merrifield, (D), California (San Diego), 1989, Op
Michael J. Mottl, (D), Harvard, 1976, OuCgm
Bo Qiu, (D), Kyoto, 1990, Op
Kelvin J. Richards, (D), Southampton (UK), 1978, Op
Kathleen C. Ruttenberg, (D), Yale, Cg
Francis J. Sansone, (D), N Carolina, 1980, Cm
Niklas Schneider, (D), Hawaii, 1992, Og
Craig R. Smith, (D), California (San Diego), 1983, Ob
Grieg F. Steward, (D), California (San Diego), 1996, Ob
Axel Timmermann, (D), Hamburg, 1999, Op
Richard E. Zeebe, (D), Bremen, 1998, Ou
Associate Professor:
Glenn Carter, (D), Washington, 2005, Op
Matthew Church, (D), William & Mary, 2003, Ob
Brian Glazer, (D), Cm
Erica Goetze, (D), California (San Diego), 2004, Ob
Christopher Kelley, (D), Hawaii, 1995, Ob
Gary M. McMurtry, (D), Hawaii, 1979, CaGvu
Brian Powell, (D), Colorado, 2005, Op
Karen E. Selph, (D), Hawaii, 1999, Ob
Affiliate Graduate Faculty:
Allen H. Andrews, (D), Rhodes, 2009, Oc
Emeritus:
Paul K. Bienfang, (D), Hawaii, 1977, Ob
Antony D. Clarke, (D), Washington, 1983, As
Richard W. Grigg, (D), California (San Diego), 1969, Ob
Barry J. Huebert, (D), Northwestern, 1970, OcCg

Yuan-Hui Li, (D), Columbia, 1967, Cm
Roger Lukas, (D), Hawaii, 1981, Op
Fred T. Mackenzie, (D), Lehigh, 1962, CmGes
Lorenz Magaard, (D), Kiel, 1963, Op
Alexander Malahoff, (D), Hawaii, 1965, Cm
Peter Muller, (D), Hamburg, 1974, Op
Jane E. Schoonmaker, (D), Northwestern, 1981, Cm
Stephen V. Smith, (D), Hawaii, 1970, Ou
Richard E. Young, (D), Miami, 1968, Ob

Affiliate Graduate Faculty:
Russell E. Brainard, (D), Naval Postgrad Sch, 1994
Paul G. Falkowski, (D), British Columbia, 1975, Og
Carolyn Jones, (D), Georgia Tech, 2004
Dennis Moore, (D), Harvard, 1968, Og
Jaromir Ruzicka, (D), Tech (Czech), 1963, Zn
John R. Sibert, (D), Columbia, 1968, Ob
Kevin Weng, (D), Stanford, 2007, Ob

Cooperating Faculty:
Marlin J. Atkinson, (D), Hawaii, 1981, Ob
Whitlow W L. Au, (D), Washington State, 1970, Zn
Janet M. Becker, (D), California (San Diego), 1989, Og
Michael Cooney, (D), California (Davis), 1992, Og
Walter Dudley, (D), Hawaii, 1976, Ou
Eric J. Gaidos, (D), MIT, 1996, Yg
Ruth D. Gates, (D), Newcastle (UK), 1990, Ob
Petra H. Lenz, (D), California (Santa Barbara), 1983, Og
Jeffrey J. Polovina, (D), California (Berkeley), 1974, Zn
Brian N. Popp, (D), Illinois, 1986, Cm
Michael S. Rappe, (D), Oregon State, 1997, Og
Florence Thomas, (D), California (Berkeley), 1992, Ob
Robert Toonen, (D), California (Davis), 2001, Ob
John C. Wiltshire, (D), Hawaii, 1983, Og

Windward Community College

Natural Sciences (A) O⊘
45-720 Kea'ahala Rd.
Kane'ohe, HI 96744
p. (808) 236-9115
fmccoy@hawaii.edu
http://windward.hawaii.edu
Programs: AA in Natural Sciences
Certificates: certificate programs in Natural Science; Sustainability; Marine Science [Marine Option Program])
Enrollment (2021): A: 2 (2)

Professor:
Floyd W. McCoy, (D), Harvard, 1974, GgaGu

Idaho

Boise State University

Dept of Geosciences (B,M,D) ●
1910 University Drive
Mail Stop 1535
Boise, ID 83725
p. (208) 426-2902
geoadmin-group@boisestate.edu
http://earth.boisestate.edu/
Programs: Geology, Geophysics, Hydrology, Earth Science Education
Certificates: Geographic Information Systems
Administrative Assistant: Susan Hamblin
Enrollment (2022): B: 86 (27) M: 23 (13) D: 19 (3)

Acting Chair and Associate Dean of College of Arts & Sciences:
Clyde J. Northrup, (D), MIT, 1996, Gc

Vice President for Research:
Nancy Glenn, (D), Nevada (Reno), 2000, ZrNgZi

University Distinguished Professor:
Matthew Kohn, (D), Rensselaer, 1991, CsGpPv
James P. McNamara, (D), Alaska (Fairbanks), 1997, Hg
Mark D. Schmitz, (D), MIT, 2001, CcaCg

Interim Dean of College of Innovation and Design:
Shawn Benner, (D), Waterloo, 2000, Clt

Professor:
Jeffrey Johnson, (D), Washington, 2000, YsGv
Jennifer L. Pierce, (D), New Mexico, 2004, GmRcZc

Senior Scientist:
James Crowley, (D), Carleton, 1997, Ccg

Associate Professor:
Brittany Brand, (D), Arizona State, 2008, GsvRn
Alejandro Flores, (D), MIT, 2009, HqAs
Hans-Peter Marshall, (D), Colorado, YxGl
Dorsey Wanless, (D), Florida, 2010, Gi

Associate Scientist:
Darin Schwartz, (D), Boise State, 2019, CuGi

Assistant Professor:
Jake Anderson, (D), Boise State, 2018, YxGv
Anna Bergstrom, (D), Colorado, 2020, HgGlCb
Ellyn Enderlin, (D), Ohio State, 2013, GlYg
Qifei Niu, (D), Hong Kong, 2014, Ygu
Karen Viskupic, (D), MIT, 2002, ZeGc

Research Professor:
Lee M. Liberty, (M), Wyoming, 1992, Ys

Assistant Research Professor:
Kendra Kaiser, (D), Duke, 2017, Hsq
Linda Reynard, (D), Oxford, 2007, Cs

Assistant Clinical Professor:
Daniel Luna, (D), California (Santa Barbara), 2015, GgPv

Lecturer:
Kerrie Weppner, (M), Boise State, 2012, GeZc

Idaho Geological Survey:
Virginia S. Gillerman, (D), California (Berkeley), 1982, Eg

Emeritus:
Warren Barrash, (D), Idaho, 1986, HwYxGe
Vladimir Davydov, (D), All-Russian Geo Res Inst, 1982, PmGr
Paul R. Donaldson, (D), Colorado Mines, 1974, Ye
Paul Michaels, (D), Utah, 1993, Ne
John R. Pelton, (D), Utah, 1979, YgeYs
Walter S. Snyder, (D), Stanford, 1977, GrcGt
Craig M. White, (D), Oregon, 1980, Giv
David E. Wilkins, (D), Utah, 1997, ZyGm
Spencer H. Wood, (D), Caltech, 1975, Gmm

Brigham Young University - Idaho

Dept of Geology (B) ●⊘
525 South Center Street
Rexburg, ID 83460-0510
p. (208) 496-7670
geologysec@byui.edu
http://www.byui.edu/Geology
Programs: Geology; Earth Science Education; Environmental Geoscience
Certificates: GIS
Enrollment (2022): B: 215 (27)

Chair:
Gregory T. Roselle, (D), Wisconsin, 1997, GpCaHg

Professor:
Robert W. Clayton, (D), S California, 1993, GcYgNg
Forest J. Gahn, (D), Michigan, 2004, PggGs
Megan Sjoblom, (D), Penn State, 2015, GiZeGe
Julie B. Willis, (D), Utah, 2009, GtcZi

College of Southern Idaho

Dept of Physical Science (A)
Box 1238
315 Falls Avenue
Twin Falls, ID 83301
p. (208) 732-6400
swillsey@csi.edu
http://physsci.csi.edu/geology/
Enrollment (2013): A: 8 (0)

Professor:
Shawn P. Willsey, (M), N Arizona, 2000, GgcGv

College of Western Idaho

Physical & Agricultural Sciences (A)
5500 E Opportunity Drive
Nampa, ID 83687
andreaschumaker@cwidaho.cc
http://cwidaho.cc
Programs: Geology; Geography; Biology-Natural Resources
Certificates: GIS
Enrollment (2018): A: 38 (3)

Assistant Professor:
Ander Sundell, (M), Boise State, Ggc

IDAHO

Idaho State University

Department of Geosciences (B,M,D) O☒
921 S. 8th Ave STOP 8072
Pocatello, ID 83209-8072
p. (208) 282-3365
geology@isu.edu
http://geology.isu.edu/
f: https://www.facebook.com/idahostategeosciences/
t: @ISUGeoscience
Programs: BA/BS Geology, BA/BS Earth and Environmental Science, BS Earth and Environmental Science - Geospatial Systems, Minor Geology or Geotechnology, MS Geology, MS Geographic Information Science, PhD Geosciences
Certificates: GIS
Enrollment (2022): B: 54 (0) M: 12 (0) D: 5 (0)

GIS Training and Research Center:
Keith Weber, (M), Montana, 2000, Zi
Director, Idaho Museum of Natural History:
Leif Tapanila, (D), Utah, 2005, Pig
Department Chair:
Benjamin T. Crosby, (D), MIT, 2006, GmHsRn
Associate Vice President for Research:
David W. Rodgers, (D), Stanford, 1987, Gct
Professor:
Glenn D. Thackray, (D), Oregon State, 1989, Glm
Associate Professor:
Donna M. Delparte, (D), Calgary, 2008, Zir
Sarah E. Godsey, (D), California (Berkeley), 2009, HgsHq
Shannon E. Kobs-Nawotniak, (D), Buffalo, 2007, Gv
David M. Pearson, (D), Arizona, 2012, Gct
Field Camp Director:
Ryan Anderson, (D), Washington State, 2019, GcCaZe
Assistant Professor & Director of Geotechnologies:
H. Carrie Bottenberg, (D), Missouri S&T, 2012, ZiGtc
Assistant Professor:
Kendra Murray, (D), Arizona, 2016, ClGtm
Emeritus:
Scott S. Hughes, (D), Oregon State, 1983, GvXgGi
Paul K. Link, (D), California (Santa Barbara), 1982, GsgGr
Michael O. McCurry, (D), California (Los Angeles), 1985, Gzv

Lewis-Clark State College

Earth Sciences (B) ☒
Div of Natural Science & Mathematics
500 8th Ave
Lewiston, ID 83501
p. (208) 792-2283
klschmidt@lcsc.edu
http://www.lcsc.edu/science/degree-programs/earth-science/
Programs: Earth Science (B)
Geochemistry emphasis in Chemistry (B)
Earth Science Secondary Education
Certificates: GIS
Geology
Enrollment (2018): B: 13 (3)

Professor:
Keegan L. Schmidt, (D), S California, 2000, Gc

North Idaho College

Dept of Geosciences (A)
1000 West Garden Avenue
Coeur d'Alene, ID 83814
p. (208) 769-3477
Bill_Richards@nic.edu
https://www.nic.edu/
Programs: Geology
Geography
Enrollment (2021): A: 4 (1)

Associate Professor:
Bill Richards, (M), Kansas State, 1984, GgzZe

University of Idaho

College of Natural Resources
Moscow, ID 83844-3019
drbecker@uidaho.edu

Dept of Civil & Environmental Engineering
875 Perimeter Drive - MS 1022
Moscow, ID 83844-1022
p. 208-885-6782
sjung@uidaho.edu
https://www.uidaho.edu/engr/departments/cee
Programs: Civil : M. Eng., MS, PhD
Geological Engineering: M Eng, MS
Certificates: N/A

Professor:
SJ Jung, (D), West Virginia, 1989, NrxNr

Dept of Geography
875 Perimeter Drive MS 3021
Moscow, ID 83844-3021
p. (208) 885-6216
geography@uidaho.edu
https://www.uidaho.edu/sci/geography
f: https://www.facebook.com/groups/502335576777518
t: @GeogUidaho
Programs: B.S. Geography
M.S. Geography
Ph.D. Geography
Certificates: GIS certificate
Climate Change certificate

Professor:
Raymond Dezzani, (D), California, 1996, RhZnn
Jeffrey Hicke, (D), Colorado, 2000, ZcyZn
Karen Humes, (D), Arizona, 1992, ZriZn
Associate Professor:
John Abatzoglou, (D), California, 2006, ZoyZc
Assistant Professor:
Chao Fan, Arizona State, 2016, Zir
Grant Harley, (D), Tennessee, 2012, ZynZn
Haifeng (Felix) Liao, (D), Utah, 2014, ZuiZn
Tom Ptak, (D), Oregon, 2016, Zu
Steven Radil, (D), Illinois, 2011, Zn
Research Scientist:
Katherine Hegewisch, (D), Washington State, 2010, Zo

Dept of Soil and Water Systems (B,M,D)
Moscow, ID 83844-2339
p. (208) 885-7012
jmaynard@uidaho.edu
https://www.uidaho.edu/cals/soil-and-water-systems

Professor:
Guy Knudsen, (D), Cornell, 1984, Sb
Robert L. Mahler, (D), North Carolina State, 1980, Zn
Daniel G. Strawn, (D), Delaware, 1999, Sc
Associate Professor:
Bradford D. Brown, (D), Utah State, 1985, Zn
Matthew J. Morra, (D), Ohio State, 1986, Sb
Assistant Professor:
Jodi Johnson-Maynard, (D), California (Riverside), 1999, Sd
Paul A. McDaniel, (D), North Carolina State, 1989, Sd

Geological Sciences (B,M,D) O⊘
875 Perimeter Drive
MS 3022
Moscow, ID 83844-3022
p. (208) 885-6216
Earthsciences@uidaho.edu
http://www.uidaho.edu/sci/ess
Programs: Geography, Geology, Environmental Hydrogeology,Geographical information Systems, Energy Resiliency, Climate
Certificates: GIS, Climate Change, Geography minor, Geology minor, Groundwater Hydrology minor
Enrollment (2022): B: 63 (0) M: 24 (0) D: 13 (0)

Professor:
Alistair Smith, (D), Kings (London), 2004, SfZyAm
Distinguished Professor:
Robert Smith, (D), New Mexico Tech, 1984, Cb
Professor:
Jerry P. Fairley, (D), California (Berkeley), 2000, HwGv
Jeff Hicke, (D), Colorado, 2000, ZcyZr
Karen Humes, (D), Arizona, 1992, ZrHgZy
Kenneth F. Sprenke, (D), Alberta, 1983, Ye

Clinical Associate Professor:
Thomas Williams, (D), Maryland, 1994, GfzEm
Associate Professor:
Grant Harley, (D), Tennessee, 2012, ZyAtRn
Haifeng (Felix) Liao, (D), Utah, 2014, ZugZi
Assistant Professor:
Timothy Bartholomaus, (D), Alaska, 2013, Gl
Elizabeth J. Cassel, (D), Stanford, 2010, Gs
Jeffrey Langman, (D), Texas, 2008, ClHws
Eric L. Mittelstaedt, (D), Hawaii, 2008, Yr
Erika Rader, (D), Idaho, 2014, GviXg
Jessica Stanley, (D), Colorado, 2015, Gt
Instructor:
Malinda Ritts, (D), Stanford, 2009, GsCg
Emeritus University Distinguished Professor:
Mickey E. Gunter, (D), Virginia Tech, 1987, GzEnGb
Emeritus:
Leslie L. Baker, (D), Brown, 1996, ClXgGe
Dennis J. Geist, (D), Oregon, 1985, Giv
Peter E. Isaacson, (D), Oregon State, 1974, Psi
Clinical Faculty:
Renee Love, (D), Idaho, 2011, PgEoGo

Illinois

Augustana College

Department of Geology (B) ☒
639 38th St.
Rock Island, IL 61201
p. (309) 794-7318
jeffreystrasser@augustana.edu
https://augustana.edu/academics/areas-of-study/geology
f: https://www.facebook.com/AugustanaGeology/
Programs: B.A. major in geology
Administrative Assistant: Jennifer Milner
Enrollment (2020): B: 26 (4)
Professor:
Jeffrey C. Strasser, (D), Lehigh, 1996, GmlGe
Michael B. Wolf, (D), Caltech, 1992, Gi
Professional Faculty:
Jenny C. Arkle, (D), Cincinnati, 2019, GtmCc
Assistant Professor:
Kelsey M. Arkle, (D), Cincinnati, 2015, PeGsg
Emeritus:
William R. Hammer, (D), Wayne State, 1979, Pv
Assistant Curator:
John Oostenryk
Assistant Curator and Educational Programs Coordinator:
Susan Kornreich Wolf, (B), Hamilton, 1987, Ze

College of Lake County

Earth Science Dept (A)
19351 W. Washington Steet
Grayslake, IL 60030-1198
p. (847) 543-2504
rcumpston@clcillinois.edu
http://www.clcillinois.edu/programs/esc/
Professor:
Eric Priest, (M), Creighton, 1986, AmZe
Xiaoming Zhai, (D), California (Davis), 1997, Gp

Eastern Illinois University

Geology and Geography Department (B,M)
600 Lincoln Avenue
Charleston, IL 61920-3099
p. (217) 581-2626
geoscience@eiu.edu
http://www.eiu.edu/~geoscience/
Programs: Geology (B.S); accelerated B.S Geology/law degree with NIU; Geography (B.S.); Science Teacher Certification (B.S); Geographic Information Systems (MBA, PSM)
Certificates: GISci Graduate Certificate, co-offer Certificate in Public Planning (with Political Science)
Administrative Assistant: Michelle Rhine
Enrollment (2020): B: 30 (13) M: 11 (1)
Associate Professor:
Diane M. Burns, (D), Wyoming, 2006, GsrGe
Vice President of Academic Affairs:
Jay Gatrell, (D), West Virginia, 1997, Zi
Professor:
Michael W. Cornebise, (D), Tennessee, 2003, Zg
Christopher R. Laingen, (D), Kansas State, 2009, SfZy
Katherine Lewandowski, (D), Ohio State, 2008, PmcZe
Associate Professor:
James A. Davis, (D), Kansas State, 2001, Zg
Barry J. Kronenfeld, (D), SUNY (Buffalo), 2004, Zi
James D. Riley, (D), Illinois, 2013, ZyHsq
David C. Viertel, (D), Texas State, 2008, Zru
Assistant Professor:
Jake R. Crandall, (D), S Illinois (Carbondale), 2021, GzXgEd
Instructor:
Cameron D. Craig, (M), Indiana State, 2003, ZgAm
Emeritus:
Craig A. Chesner, (D), Michigan Tech, 1988, GivGz
Belayet H. Khan, (D), Pittsburgh, 1985, Zy
Betty E. Smith, (D), SUNY (Buffalo), 1994, Zu

Elgin Community College

Dept of Geology
1700 Spartan Drive
Elgin, IL 60123
p. (847) 214-7359
agoyal@elgin.edu
http://www.elgin.edu/employeelisting/faculty/
Adjunct Professor:
Mark R. Kuntz, Gg
Timothy M. Millen, (M), N Illinois, 1982, Gg
Joseph E. Peterson, Gg

Field Museum of Natural History

Dept of Geology
1400 S. Lake Shore Drive
Chicago, IL 60605-2496
p. (312) 665-7621
klawson@fieldmuseum.org
http://www.fieldmuseum.org
Administrative Assistant: Karsten L.. Lawson, Elaine Zeiger
Chair:
Olivier C. Rieppel, (D), Basel (Switzerland), 1978, Pv
Curator, Meteoritics:
Meenakshi Wadhwa, (D), Washington (St Louis), 1994, Xm
Curator, Fossil Fishes:
Lance Grande, (D), CUNY, 1983, Pv
Associate Curator, Fossil Invertebrates:
Scott H. Lidgard, (D), Johns Hopkins, 1984, Pi
Curator, Fossil Amphibians & Reptiles:
John R. Bolt, (D), Chicago, 1968, Pv
Emeritus:
Matthew H. Nitecki, (D), Chicago, 1968, Pi
Bertram G. Woodland, (D), Chicago, 1962, Gp
Associate Curator, Fossil Invertebrates:
Peter J. Wagner, (D), Chicago, 1995, Pi

Harold Washington College

Physcial Science Dept (A)
30 E. Lake Street
Room 903
Chicago, IL 60601
p. (312) 553-5791
aescuadro@ccc.edu
http://www.ccc.edu/colleges/washington/departments/Pages/Physical-Sciences.aspx
Enrollment (2012): A: 17 (0)

Heartland Community College

Math And Science Div (A)
2400 Instructional Commons Building
1500 W Raab Road
Normal, IL 61761
p. (309) 268-8640
mark.finley@heartland.edu
http://www.heartland.edu/ms/easc/
Professor:
Robert Dennison, (M), S Mississippi, Zgn

Mark Finley, (M), Iowa State, Gg
Adjunct Professor:
Janet Beach Davis, (B), Illinois (Springfield), Zg
Paul Ritter, (M), E Illinois, Zg
Steven Travers, (M), Illinois State, Zg
Mark Yacucci, (B), Youngstown State, Zg

Illinois Central College

Math, Science, and Engineering Dept (A)
One College Drive
East Peoria, IL 61611
p. (309) 694-5364
info@icc.edu
Administrative Assistant: Diane Weber
Professor:
Martin A. Petit, (M), Illinois State, 1967, Ze
Associate Professor:
Cheryl R. Emerson, (M), N Arizona, 1989, Ze
Assistant Professor:
Ed Stermer, (M), Iowa, Zg
Adjunct Professor:
Linda Aylward, (B), Bradley, 1980, Ze
Sean Mulkey, (M), Iowa, 1994, Ze

Illinois State University

Dept of Geography, Geology, and the Environment (B,M)
○⊘
101 South School Street
Campus Box 4400
206 Felmley Hall of Science
Normal, IL 61790-4400
p. (309) 438-7649
geo@illinoisstate.edu
https://geo.illinoisstate.edu/
f: https://www.facebook.com/groups/21051078663/
Programs: Geography (BA or BS); Geography Social Science Education (BS); Earth and Space Science Education (BS); Environmental Systems and Sustainability (BS) Geology (BS); Hydrogeology (M) Geography Minor, Geology Minor, Environmental Studies Minor, Urban Studies Minor
Minors in Geography, Geology, and Environmental Studies
Certificates: Geographic Information Systems, Hydrogeology Geographic Information Systems
Administrative Assistant: Karen K. Dunton
Enrollment (2022): B: 62 (14) M: 10 (8)
University Professor:
Eric W. Peterson, (D), Missouri, 2002, HwsHt
Distinguished University Professor:
David H. Malone, (D), Wisconsin, 1994, GcsEg
Chair:
Dagmar Budikova, (D), Calgary, 2001, AsZyi
Professor:
James E. Day, (D), Iowa, 1988, PisPm
John Kostelnick, (D), Kansas, 2006, Ziy
Catherine M. O'Reilly, (D), Arizona, 2001, HgCgGg
Rex J. Rowley, (D), Kansas, 2009, Zi
Jonathan B. Thayn, (D), Kansas, 2009, ZirZf
Associate Professor:
Tenley Banik, (D), Vanderbilt, 2015, GviCc
Wondwosen M. Seyoum, (D), Georgia, 2016, HwqZr
Lisa M. Tranel, (D), Virginia Tech, 2010, GmtZi
Assistant Professor:
Melissa Heil, (D), Illinois, 2020, Zy
Daniel Kpienbaareh, (D), W Ontario, 2021, ZuyZr
Adjunct Professor:
Toby J. Dogwiler, (D), Missouri, 2002, Gm
Emeritus:
Robert S. Nelson, (D), Iowa, 1970, Gm
Coordinator of Academic Services :
Paul A. Meister, (M), Illinois State, 2016, GgHgZn

Illinois Valley Community College

Dept of Geology (A)
Natural Sciences, Math, & Business Division
815 North Orlando Smith Road
Oglesby, IL 61348-9692
p. (815) 224-0394
Mike_Phillips@ivcc.edu
https://facultyweb.ivcc.edu/mphillips/
Programs: Geology
Enrollment (2021): A: 2 (1)
Professor:
Michael Phillips, (M), S Illinois, 1990, GgeGm

Kaskaskia College

Life and Physical Sciences Dept (A)
27210 College Road
Centralia, IL 62801
pvig@kaskaskia.edu
http://www.kaskaskia.edu/LPDept/
Instructor:
Pradeep K. Vig, Gg

Lincoln Land Community College

Math and Sciences (A)
Springfield, IL 62794-9256
p. (217) 786-4923
dean.butzow@llcc.edu
http://www.llcc.edu/mtsc/Sciences/tabid/3280/
Enrollment (2006): A: 7 (0)
Professor:
Dean G. Butzow, (M), W Michigan, Zy
Instructor:
Samantha Reif, (M), Michigan Tech, Gg

McHenry County College

Earth Science and Geography (A)
8900 US Hwy 14
Crystal Lake, IL 60012
p. (815) 455-3700
kkramer@mchenry.edu
http://www.mchenry.edu/EarthScience
Instructor:
Theodore Erski, (M), Akron, 1995, ZgEg
Paul Hamill, (M), Wisconsin, AmZg
Kate Kramer, (M), Indiana-Purdue, 2008, GgZg

Moraine Valley Community College

Physical Sciences Dept (A) ☒
9000 W. College Pwky
Palos Hills, IL 60465-2478
p. 708-974-5615
syrup@morainevalley.edu
Professor:
Krista A. Syrup, (M), W Michigan, 2002, Cs

Northeastern Illinois University

Dept of Earth Science (B) ●⊘
5500 N. St. Louis Avenue
Chicago, IL 60625
p. (773) 442-6050
E-Head@neiu.edu
https://www.neiu.edu/academics/college-of-arts-and-sciences/departments/earth-science
f: https://www.facebook.com/neiuearthscience/
Programs: Earth Science (B)
Administrative Assistant: Lidia Costea
Enrollment (2018): B: 28 (6)
Professor:
Laura L. Sanders, (D), Kent State, 1986, HwGe
Associate Professor:
Elisabet M. Head, (D), GvZrCg
Kenneth M. Voglesonger, (D), Arizona State, 2004, ClGe
Assistant Professor:
Nadja Insel, (D), Michigan, 2011, Gt
Instructor:
Mohammad Fariduddin, (D), N Illinois, 1999, Pe
Rebekah Fitchett, (M), Illinois (Chicago), 2006, GgCg
Jean M. Hemzacek Laukant, (M), N Illinois, 1986, Sc

Northern Illinois University

Dept of Geology and Environmental Geosciences (B,M,D)

O✓🖱
1425 W. Lincoln Hwy
De Kalb, IL 60115-2854
p. (815) 753-1943
askgeology@niu.edu
http://www.niu.edu/geology/
f: https://www.facebook.com/NIU-Geology-and-Environmental-Geosciences-346654145320/?ref=aymt_homepage_panel&eid=ARAaKxUV4veRGbvNDUVXDD7c6lp2k-JCEuQQs-CNdgA7YzlE4mAD9EaJKJTiNmFLcJa019c_6zO2seUZ
Programs: B.S. Geology and Environmental Geosciences High School Teacher Licensure in Earth and Space Science Education, M.S. Geology and Environmental Geosciences Ph.D. Geology and Environmental Geosciences
Administrative Assistant: Lesslie Erickson, Patricia Liberty-Baczek
Enrollment (2020): B: 33 (11) M: 15 (9) D: 6 (0)

Graduate Program Director:
Mark R. Frank, (D), Maryland, 2001, Cp
Chair:
Mark P. Fischer, (D), Penn State, 1994, GcoHy
Professor:
Philip J. Carpenter, (D), New Mexico Tech, 1984, YgHwNg
Melissa E. Lenczewski, (D), Tennessee, 2001, HwGeCo
Ross D. Powell, (D), Ohio State, 1980, GslZc
Reed P. Scherer, (D), Ohio State, 1991, Pm
Associate Professor:
Justin P. Dodd, (D), New Mexico, 2011, Cs
Nicole D. LaDue, (D), Michigan State, 2013, Ze
Nathan D. Stansell, (D), Pittsburg State, 2009, Gl
Assistant Professor:
Megan R.M. Brown, (D), Colorado, 2020, HwRw
Research Associate:
Anna Buczynska, (D), Antwerp (Belgium), 2014, Csa
Guillaume Girard, (D), McGill, 2009, GviCu
Adjunct Professor:
B. Brandon Curry, (D), Illinois, 1995, GelPs
Virginia Naples, (D), Massachusetts, 1980, Pv
Karen Samonds, (D), SUNY (Stony Brook), 2006, Pv
Laboratory Manager:
Joshua Schwartz, (B)

Northwestern University

Earth and Planetary Sciences (B,D) ●☒
2145 Sheridan Rd - TECH-F374
Evanston, IL 60208-3130
p. (847) 491-3238
earth@northwestern.edu
http://www.earth.northwestern.edu/
t: @NU_EARTHSCI
Programs: Earth and Planetary Sciences
Administrative Assistant: Lisa J. Collins, Murielle M. Harris, Tia Ng, Negatwa Tewodros
Enrollment (2021): B: 17 (14) D: 32 (5)

Department Chair, Professor:
Matthew T. Hurtgen, (D), Penn State, 2003, GsPgGr
Director of Graduate Studies, Professor:
Suzan van der Lee, (D), Princeton, 1996, YsxYg
Professor:
Craig R. Bina, (D), Northwestern, 1987, YgZmGy
Neal E. Blair, (D), Stanford, 1980, CoSbOc
Steven D. Jacobsen, (D), Colorado, 2001, GyzZm
Andrew D. Jacobson, (D), Michigan, 2001, ClPeCa
Bradley B. Sageman, (D), Colorado, 1991, PsCbPc
Seth Stein, (D), Caltech, 1978, YsdYg
Director of Graduate Admissions & Recruitment, Associate Professor:
Yarrow L. Axford, (D), Colorado, 2007, GnPmi
Associate Professor:
Magdalena R. Osburn, (D), Caltech, 2013, PoCbs
Assistant Professor:
Daniel E. Horton, (D), Michigan, 2011, AsPeZn
Elvira Mulyukova, (D), GFZ German Res Ctr Geosciences, 2015
IRMS Lab Manager:
Andrew Masterson, (D), CslCa
Research Associate:
Mitchell Barklage, (D), YgeYs
Director of Environmental Sciences Program, Assistant Chair, Associate Prof. of Instruction:
Patricia A. Beddows, (D), Bristol, 2004, HyGma
Assistant Professor of Instruction, Weinberg College Adviser:
Rosemary Bush, (D), Northwestern, 2014, Pc
Adjunct Professor:
Gilbert Klapper, (D), Iowa, 1962, Pm
Emeritus:
Donna M. Jurdy, (D), Michigan, 1974, YgXyg
Abraham Lerman, (D), Harvard, 1964, CgGnCm
Emile A. Okal, (D), Caltech, 1978, YsgYr
Research Lab Manager:
Grace C. Schellinger, (B), Wisconsin, 2012

Oakton Community College

Earth Science (A) ☒
1600 E. Golf Road
Des Plaines, IL 60016
p. (847) 376-7042
clandrie@oakton.edu
http://www.oakton.edu/academics/academic_departments/earth_science/

Chair:
Chad Landrie, (D), Illinois (Chicago)
Lecturer:
Richard DiMaio, (M), N Illinois
David LoBue, (M), Kansas
William Tong, (M), NE Illinois
Raymond Wiggers, (B), Purdue
Julie Wulff, (D), Ohio State

Olivet Nazarene University

Dept of Chemistry and Geosciences (B) ☒
One University Avenue
Bourbonnais, IL 60914
p. (815) 939-5394
ccarriga@olivet.edu
http://geology.olivet.edu/
f: https://www.facebook.com/pages/Olivet-Nazarene-University-Geosciences/50458842809
Programs: Geology (B.S.), Earth and Space Science (B.A.), Science Education (B.S.), Geography (B.A.), Environmental Science (B.S.).
Enrollment (2018): B: 14 (4)

Acting Director:
Charles W. Carrigan, (D), Michigan, 2005, GzCgCc
Director, Strickler Planetarium:
Stephen Case, (D), Notre Dame, 2014, RhXg
Emeritus:
Max W. Reams, (D), Washington (St Louis), 1968, GsmPg

Prarie State College

Dept of Physical Science, Earth Science Discipline (A) ☒
202 S. Halsted St
Chicago Heights, IL 60411
p. (708) 709-3674
lburrough@prairiestate.edu
https://prairiestate.edu/academics/academic-programs/natural-sciences/physical-scienceearth-sciencegeology.aspx
Programs: Associates in Science

Principia College

Biology and Natural Resources Department (B) ☒
1 Maybeck Place
Elsah, IL 62028
scott.eckert@principia.edu
http://www.principiacollege.edu/biology-env-studies

Chair:
Scott Eckert, (D), Georgia
Professor:
Greg Bruland, (D), Duke
Christine McAllister, (D), St Louis
Instructor:
John Lovseth, (D), S Illinois, 2018, Zni

Richard J. Daley Community College

Physical Sciences
7500 South Pulaski Rd
Chciago, IL 60652
p. (773) 838-7636

efuoco@ccc.edu
http://daley.ccc.edu/

Richland Community College

Richland Community College
One College Park
Decatur, IL 62521
p. (217) 875-7211
jsmith@richland.edu
http://www.richland.edu/

Rock Valley College

Rock Valley College
3301 North Mulford Road
Rockford, IL 61114
E.Adame@RockValleyCollege.edu
http://ednet.rockvalleycollege.edu/

Sauk Valley Community College

Earth Science
173 IL Rt. 2
Dixon, IL 61021
james.r.chisholm@svcc.edu
http://www.svcc.edu

South Suburban College

Dept of Physical Science
15800 South State Street
South Holland, IL 60473
p. (708) 596-2000 x2417
ahelwig@ssc.edu
http://www.ssc.edu

Southern Illinois University Carbondale

Geology Programs (B,M,D) O☒
1259 Lincoln Drive, Mailcode 4324
Parkinson 102
Carbondale, IL 62901
p. (618) 453-3351
geology@geo.siu.edu
http://www.geology.siu.edu/
f: https://www.facebook.com/Geology-at-Southern-Illinois-University-191268227586004/
Programs: BA Geology, BS Geology, MA Geology, MS Geology, PhD Geosciences
Administrative Assistant: Dana Wise
Enrollment (2018): B: 30 (14) M: 22 (9) D: 8 (1)

Director:
Justin Schoof, (D), Indiana, 2004

Professor:
Ken B. Anderson, (D), Melbourne (Australia), 1989, Co
James A. Conder, (D), Brown, 2001, Ysr
Liliana Lefticariu, (D), N Illinois, 2004, CsbGe
John L. Sexton, (D), Indiana, 1974, Ye

Associate Professor:
Harvey Henson, (M), S Illinois, 1989, Yg
Sally Potter-McIntyre, (D), Utah, 2013, Gs

Assistant Professor:
Daniel R. Hummer, (D), Penn State, 2010, GzyCu

Research Associate:
William Huggett, (M), S Illinois, 1981, Ec

Emeritus:
Steven P. Esling, (D), Iowa, 1983, HwGl
Richard Fifarek, (D), Oregon State, 1985, Em
Charles O. Frank, (D), Syracuse, 1973, Gx
Scott E. Ishman, (D), Ohio State, 1990, Pmc
Susan M. Rimmer, (D), Penn State, 1985, EcCgGo
Jay Zimmerman, Jr., (D), Princeton, 1968, Gc

Southwestern Illinois College - Belleville Campus

Dept of Earth Science
2500 Carlyle Avenue
Belleville, IL 62221
p. (618) 235-2700
http://www.swic.edu

Southwestern Illinois College - Sam Wolf Granite City Campus

Physical Science (A)
4950 Maryville Road
Granite City, IL 62040
p. (618) 235-2700
http://www.swic.edu

Professor:
Joy Branlund, (D), Washington (St Louis), 2008, Zg

Triton College

Triton College
2000 Fifth Ave.
River Grove, IL 60171
p. (708) 456-0300
austinweinstock@triton.edu
http://www.triton.edu

University of Chicago

Dept of the Geophysical Sciences (B,M,D) ☒
5734 S. Ellis Avenue
Chicago, IL 60637
p. (773) 702-8101
info@geosci.uchicago.edu
http://geosci.uchicago.edu
Enrollment (2009): B: 20 (9) D: 31 (4)

Professor:
David Archer, (D), Washington, 1990, Yr
Andrew Davis, (D), Yale, 1977, Xc
Michael J. Foote, (D), Chicago, 1989, Pg
John E. Frederick, (D), Colorado, 1975, As
David Jablonski, (D), Yale, 1979, Pi
Susan M. Kidwell, (D), Yale, 1982, GsPe
Douglas R. MacAyeal, (D), Princeton, 1983, Yr
Frank M. Richter, (D), Chicago, 1972, Gt
David B. Rowley, (D), SUNY (Albany), 1983, GtgGc

Associate Professor:
Dion L. Heinz, (D), California (Berkeley), 1986, GyzYx
Noboru Nakamura, (D), Princeton, 1989, As

Assistant Professor:
Fred Ciesla, (D), Arizona, 2003, Xcg
Nicolas Dauphas, (D), Ctr Pétro et Géochimiques, 2002, Xc
Elisabeth Moyer, (D), Caltech, 2001, As
Mark Webster, (D), California (Riverside), 2003, Pi

Emeritus:
Alfred T. Anderson, Jr., (D), Princeton, 1963, Gv
Robert N. Clayton, (D), Caltech, 1955, Cs
Ramesh C. Srivastava, (D), McGill, 1964, AsmAm

Student Services Administrator:
Sarah Lippert, Zn

University of Illinois Chicago

Dept of Earth and Environmental Sciences (B,M,D) ⃠
845 W. Taylor Street
2440 SES
MC 186
Chicago, IL 60607-7059
p. (312) 996-3155
eaes@uic.edu
http://eaes.uic.edu/
Programs: Earth and Environmental Sciences
Enrollment (2022): B: 112 (16) M: 6 (2) D: 9 (0)

Head:
Kathryn L. Nagy, (D), Texas A&M, 1988, ClbGe

Professor:
Andrew J. Dombard, (D), Washington (St Louis), 2000, Xy
Fabien Kenig, (D), Orleans (France), 1991, CoXbGe
Carol A. Stein, (D), Columbia, 1984, Yg

Associate Professor:
Max Berkelhammer, (D), S California, 2010, At
DArcy Meyer Dombard, (D), Washington (St Louis), 2004, PoXb

Assistant Professor:
Akintomide Akinsanola, (D), Hong Kong, 2019, Zo
Kerry Callaghan, (D), Minnesota, 2020, Hw
Gavin McNicol, (D), California (Berkeley), 2016, Sbf

B2F Postdoctoral Research Associate:
David Hernández Uribe, (D), Colorado, 2020, Gpt
Adjunct Assistant Professor:
Andrew King, (D), California (San Diego), 2008, Og
Adjunct Professor:
Peter T. Doran, (D), Nevada (Reno), 1996, ZcHwZy
Paul Fenter, (D), Pennsylvania, 1990, Gy
Barry Lesht, (D), Chicago, 1977, Op
Eugene Yan, (D), Ohio State, 1998, Hg
Visiting Clinical Assistant Professor:
Andrew Malone, (D), Illinois (Chicago), 2017, GlAtZc
Emeritus:
Jean E. Bogner, (D), N Illinois, 1996, ClSbHy
Martin F.J Flower, (D), Manchester (UK), 1971, GiCuGt
Stephen J. Guggenheim, (D), Wisconsin, 1976, Gz
August F. Koster Van Groos, (D), Leiden (Neth), 1966, Cp
Roy E. Plotnick, (D), Illinois (Chicago), 1983, PgiGq
Kelvin S. Rodolfo, (D), S California, 1967, GueGs
Visiting Research Specialist:
Jack Breiter, (B), Illinois (Urbana), 2021, Sf
Clinical Associate Professor:
Stefany Sit, (D), Miami, 2013, YgsZe
Office Support Specialist:
Minnie O. Jones, Zn
Coordinator of Undergrad Labs:
Lee Falkena, (M), Iowa, 2012, ZeGig
Asst to the Head/Business Manager:
Edna L. Fuentes, (B), Illinois (Chicago), 2004, Zn

University of Illinois, Urbana-Champaign

Dept of Atmospheric Sciences (B,M,D) ●⊘
Nat Hist Bldg, Rm 3081, MC-104
1301 W Green Street
Urbana, IL 61801
p. (217) 333-2046
atmos-sci@illinois.edu
http://www.atmos.illinois.edu
f: https://www.facebook.com/dasuiuc/
Programs: Atmospheric Sciences
Administrative Assistant: Tammy R. Warf
Enrollment (2018): B: 71 (17) M: 12 (10) D: 34 (8)
Director:
Robert M. Rauber, (D), Colorado State, 1985, As
Head:
Robert J. Trapp, (D), Oklahoma, 1994, As
Professor:
Larry Di Girolamo, (D), McGill, 1996, Zr
Atul K. Jain, (D), India, 1988, As
Sonia Lasher-Trapp, (D), Oklahoma, 1998, As
Donald J. Wuebbles, (D), California (Davis), 1983, As
Associate Professor:
Francina Dominguez, (D), Illinois, 2007, AsHg
Stephen W. Nesbitt, (D), Utah, 2003, As
Nicole Riemer, (D), Karlsruhe, 2002, As
Zhuo Wang, (D), Hawaii, 2004, As
Assistant Professor:
Deanna A. Hence, (D), Washington, 2011, AsmAp
Cristian (Cristi) Proistosescu, (D), Harvard, 2017, AsGgAp
Ryan Sriver, (D), Purdue, 2008, As
Research Scientist:
Brian Jewett, (D), Illinois, 1996, As
Advisor:
Jessica (Jessie) Choate, (M), Illinois, 2016, Asm
Lecturer:
Donna J. Charlevoix, (M), California (Davis), 1996, As
Emeritus:
Kenneth V. Beard, (D), California (Los Angeles), 1970, As
Mankin Mak, (D), MIT, 1968, As
Walter A. Robinson, (D), Columbia, 1985, As
Robert B. Wilhelmson, (D), Illinois, 1972, As
Teaching Asst Professor:
Alicia Klees, (D), Penn State, 2020, As
Office Manager:
Joe Jeffries, (B)
Clinical Assistant Professor:
Jeffrey Frame, (D), Penn State, 2008, Asm

Dept of Geology (B,M,D) ●
1301 W Green Street
3081 Natural History Building
Urbana, IL 61801
p. (217) 333-3540
geology@illinois.edu
http://www.geology.illinois.edu
Programs: Geology (B, M, D); Geology and Geophysics (B); Earth Science Teaching (B); Environmental Geology (fully online M)
Certificates: Certificate in Environmental Geology
Enrollment (2022): B: 43 (13) M: 7 (4) D: 29 (5)
Threet Professor:
James L. Best, (D), London (UK), 1985, Gs
Head:
Craig C. Lundstrom, (D), California (Santa Cruz), 1996, Cpg
Grim Professor:
Bruce W. Fouke, (D), SUNY (Stony Brook), 1993, GsdGb
Affiliate Faculty:
Stanley Ambrose, (D), California (Santa Barbara)
Kenneth T. Christensen, (D), Illinois
Marcelo H. Garcia, (D), Minnesota, 1989, Gmu
Bruce Rhoads, (D), Arizona, Gm
Charles J. Werth, (D), Stanford
Professor:
Thomas M. Johnson, (D), California (Berkeley), 1995, HwCls
Lijun Liu, (D), Caltech, 2010, YgGtm
Gillen D. Wood, (D), Columbia, 2000, Zc
Associate Head:
Patricia Gregg, (D), MIT/WHOI, 2008, YgCg
Affiliate Faculty:
Scott Olson, (D), Illinois
Associate Professor:
Alison M. Anders, (D), Washington, 2005, Gm
Jessica Conroy, (D), Arizona, 2011, Cs
Jennifer Druhan, (D), California (Berkeley), 2012, HwCs
Wendy Yang, (D), California (Berkeley), 2010
Affiliate Faculty:
Surangi W. Punyasena, (D), Chicago, 2007, PblPe
Assistant Professor:
Willy Guenthner, (D), Arizona, 2013, GtzCg
Cristian Proistosescu, (D), Harvard, 2017
Lecturer:
Max L. Christie, (D), Penn State, 2017, PgGr
Adjunct Professor:
Ercan Alp, (D), S Illinois, 1984
Kurtis Burmeister, (D), Illinois, 2005
Brandon Curry, Illinois, 1995
Przemyslaw Dera, (D), Mickiewicz (Poland), 2000
Robert J. Finley, (D), South Carolina, 1975, Go
Leon R. Follmer, (D), Illinois, 1970, SaGlm
Hannes E. Leetaru, (D), Illinois, 1997, GoYe
Morris W. Leighton, (D), Chicago, 1951, Eo
M. Scott Wilkerson, (D), Illinois, 1991, Gc
Walgreen Chair Emeritus Professor:
Susan W. Kieffer, (D), Caltech, 1971, Gv
Johnson Professor:
Gary Parker, (D), Minnesota, 1974, Gm
Grim Professor:
Jay D. Bass, (D), SUNY (Stony Brook), 1982, Gy
Feng-Sheng Hu, (D), Washington, 1994, Pe
Grim Emeritus Professor:
Craig M. Bethke, (D), Illinois, 1984, Hw
Emeritus:
Stephen P. Altaner, (D), Illinois, 1985, Gz
Thomas F. Anderson, (D), Columbia, 1967, CslCm
Daniel B. Blake, (D), California (Berkeley), 1966, Pi
Chu-Yung Chen, (D), MIT, 1983, Cg
Wang-Ping Chen, (D), MIT, 1979, YsGt
Donald L. Graf, (D), Columbia, 1950, Cl
Stephen Marshak, (D), Columbia, 1983, Gct
Philip A. Sandberg, (D), Stockholm, 1965, Gd
Xiaodong Song, (D), Caltech, 1994, Ys
Teaching Specialist:
Ann D. Long, (M), Leeds (UK), 1981, Gm
Teaching Assistant Professor:
J. Cory Pettijohn, (D), Boston, 2008, Hg
Research Scientist:
Jonathan H. Tomkin, (D), Australian Nat, 2002, Gm

Research Associate Professor:
Robert A. Sanford, (D), Michigan State, 1996, CbSbPo
Clinical Associate Professor:
Michael A. Stewart, (D), Duke, 2000, Gi

Dept of Natural Resources & Environmental Sciences (B,M,D)
W-503 Turner Hall
1102 South Goodwin Avenue
Urbana, IL 61801-4798
p. (217) 333-2770
g-rolfe@uiuc.edu
http://nres.illinois.edu/
Chair:
Jeffrey Brawn
Professor:
Charles W. Boast, (D), Iowa State, 1970, Sp
Mark B. David, (D), New York, 1983, Sf
John J. Hassett, (D), Utah State, 1970, Sc
Robert L. Jones, (D), Illinois, 1962, Sc
Richard L. Mulvaney, (D), Illinois, 1983, Sc
Theodore R. Peck, (D), Wisconsin, 1962, Sc
Joseph W. Stucki, (D), Purdue, 1975, Sc
Associate Professor:
Robert G. Darmody, (D), Maryland, 1980, Sd
Timothy R. Ellsworth, (D), California, 1989, Sp
Kenneth R. Olson, (D), Cornell, 1983, Sc
F. William Simmons, (D), North Carolina State, 1987, Sp
Assistant Professor:
Robert J. M. Hudson, (D), MIT, 1989, Cg
Gregory F. McIsaac, (D), Illinois, 1994, Hs

Waubonsee Community College

Earth Sciences (A)
Rt. 47 at Waubonsee Dr.
Sugar Grove, IL 60554
p. (630) 466-2783
dvoorhees@waubonsee.edu
https://www.waubonsee.edu/programs-courses/programs-subject/science-technology-engineering-and-mathematics/earth-science
Programs: Geology (A)
Certificates: GIS
Enrollment (2022): A: 5 (2)
Professor:
David H. Voorhees, (M), Rensselaer, 1982, ZgGg
Associate Professor:
Karl Schulze, (M), Texas A&M, 2003, ZgAms
Assistant Professor:
Alfred W. Weiss, (M), S Illinois, 2000, Ziy

Western Illinois University

Department of Geography, Geographic Information Science and Meteorology
Macomb, IL 61455
p. (309) 298-1648
s-thompson@wiu.edu
http://www.wiu.edu/cas/geography/

Dept of Geology (B) ○
1 University Circle
Macomb, IL 61455
p. (309) 298-1151
geology@wiu.edu
http://www.wiu.edu/cas/geology/
Programs: Geology; Paleontology
Administrative Assistant: Cerese Wright
Enrollment (2017): B: 16 (0)
Professor:
Kyle R. Mayborn, (D), California (Davis), 2000, Gic
Leslie A. Melim, (D), S Methodist, 1991, Gd
Associate Professor:
Steven W. Bennett, (D), Indiana, 1994, Hw
Assistant Professor:
Thomas A. Hegna, (D), Yale, 2012, PigPg
Instructor:
Sara Bennett, (M), Indiana, Gg

Emeritus:
Jack B. Bailey, (D), Illinois, 1975, Pg
Peter L. Calengas, (D), Indiana, 1977, En

Wheaton College

Dept of Earth & Environmental Science (B) ○☒
501 College Ave.
Wheaton, IL 60187
p. (630) 752-5063
geology@wheaton.edu
http://www.wheaton.edu/geology/
Programs: Geology, Environmental Science
Enrollment (2021): B: 13 (4)
Chair:
Stephen O. Moshier, (D), Louisiana State, 1986, Gd
Professor:
Chris Keil, (D), Illinois (Chicago), 1994, Rm
Assistant Professor:
Andrew J. Luhmann, (D), Minnesota, 2011, HwGmCl
Kathryn A. Maneiro, (D), Boston, 2016, CcGic
Lab and Office Coordinator:
Katy Foltz, (M), Oklahoma State Uni, 2015
Instructor:
Lisa Heidlauf, (M), Illinois (Urbana), 1986, Gg

Indiana

Ball State University

Dept of Geography
Muncie, IN 47306
p. 765-285-1776
turk@bsu.edu
https://www.bsu.edu/academics/collegesanddepartments/geography

Ball State University

Dept of Geology (B,M,D) ○
2000 University Avenue
AR117
Muncie, IN 47306
p. (765) 285-8270
egnr@bsu.edu
https://www.bsu.edu/academics/collegesanddepartments/environment-geology-natural-resources
Administrative Assistant: Brenda J. Rathel
Enrollment (2016): B: 47 (8) M: 14 (4) D: 4 (0)
Chair:
Richard H. Fluegeman, (D), Cincinnati, 1987, PmGro
Associate Dean:
Jeffry D. Grigsby, (D), Cincinnati, 1989, GdClGo
Professor:
Kirsten N. Nicholson, (D), Joseph Fourier, 1999, Gi
R. Scott Rice-Snow, (D), Penn State, 1983, Gm
Associate Professor:
Carolyn B. Dowling, (D), Rochester, 2002, HgGeCg
Klaus Neumann, (D), Alabama, 1999, Cl
Assistant Professor:
Shawn J. Malone, (D), Iowa, 2012, GtxGm
Emeritus:
Alan C. Samuelson, (D), Penn State, 1972, Hw
Technician, Environmental Science:
Eric Lange, (M), Ball State, 2014, Cg
Geological Tech:
Michael Kutis, (B), Wisconsin (Platteville), 1989, Gg

College of the Holy Cross

Dept of Geosciences (A,B) ☒
P.O. Box 308
Notre Dame, IN 46556
p. (219) 239-8417
smitchel@holycross.edu

DePauw University

Dept of Geology and Environmental Geoscience (B) ●☒
2 East Hanna Street
Julian Science Center
Greencastle, IN 46135

p. (765) 658-4654
tcope@depauw.edu
https://www.depauw.edu/academics/departments-programs/geosciences/
f: https://www.facebook.com/DePauwGeosciences/
t: https://twitter.com/depauwgeo
Programs: Geology, Environmental Geoscience, Earth Science
Administrative Assistant: Rachel Curtis
Enrollment (2022): B: 23 (8)

Professor:
Tim D. Cope, (D), Stanford, 2003, GsZiGr
M. Scott Wilkerson, (D), Illinois, 1991, GctZi

Assistant Professor:
Kenneth L. Brown, (D), Miami (Ohio), 2015, GizGe

Emeritus:
James G. Mills, (D), Michigan State, 1991, Gi
Frederick M. Soster, (D), Case Western, 1984, Gs

Cooperating Faculty:
Jeanette K. Pope, (D), Virginia Tech, 2002, Ge

Earlham College

Geology Dept (B) ☒
801 National Road West
Richmond, IN 47374-4095
p. (765) 983-1429
streeme@earlham.edu
http://www.earlham.edu/geology/
Programs: Geology
Enrollment (2017): B: 14 (4)

Associate Professor:
Cynthia M. Fadem, (D), Washington (St Louis), 2009, GaSoCs
Andrew Moore, (D), Washington, 1999, Gm
Meg Streepey Smith, (D), Michigan, 2001, Gt

Emeritus:
Jon W. Branstrator, (D), Cincinnati, 1975, Pg

Hanover College

Dept of Geology (B)
PO Box 890
Hanover, IN 47243-0890
p. (812) 866-7306
worcestr@hanover.edu
Enrollment (2007): B: 12 (5)

Chair:
Peter A. Worcester, (D), Miami (Ohio), 1976, Gi

Professor:
Kenneth A. Bevis, (D), Oregon State, 1995, GlmGe
Heyo Van Iten, (D), Michigan, 1989, PgHgRn

Emeritus:
Stanley M. Totten, (D), Illinois, 1962, Gl

Indiana State University

Dept of Earth and Environmental Systems (B,M,D) ☒
159 Science Building
Terre Haute, IN 47809
p. (812) 237-2444
isu-ees@mail.indstate.edu
https://www.indstate.edu/cas/ees
f: www.facebook.com/isu.ees
Certificates: GIS
Enrollment (2009): B: 87 (13) M: 16 (7) D: 6 (2)

Professor:
Stephen Aldrich, (D), Ziy
Gregory Bierly, (D), Michigan State, Asm
Jennifer C. Latimer, (D), Indiana, 2004, ClGeb
Nancy J. Obermeyer, (D), Chicago, 1987, Zi
Shawn Phillips, (D), New York, 2001
James Speer, (D), Tennessee, 2001, PeZyGe
Jeffery Stone, (D), PceGs
Qihao Weng, (D), Georgia, Zru

Associate Professor:
Susan Berta, (D), Oklahoma, 1986, GmZr
Kathleen M. Heath, (D), Utah, 2001, Pb

Assistant Professor:
Elvis Alex Badillo, (D), ZiGa

Emeritus:
Sandra S. Brake, (D), Colorado Mines, 1989, GeEm

Prodip K. Dutta, (D), Indiana, 1983, Gs
C. Russell Stafford, (D), Arizona State, 1981, Gam

Indiana University - Purdue University Indianapolis

Dept of Earth Sciences (B,M,D) ●🖱
723 West Michigan Street
Indianapolis, IN 46202-5132
p. (317) 274-7484
geology@iupui.edu
https://science.iupui.edu/earthsciences
Programs: Geology, Environmental Science

Professor:
Andrew P. Barth, (D), S California, 1989, Gi
Gabriel M. Filippelli, (D), California (Santa Cruz), 1994, CmGb
Pierre-Andre Jacinthe, (D), Ohio State, 1995, Cl
Lin Li, (D), Brown, 2002, ZrXgZg
Kathy J. Licht, (D), Colorado, 1999, Gl

Associate Professor:
Broxton Bird, (D), Pittsburgh, 2009, Gmn
Greg Druschel, (D), Wisconsin, 2002, Cl
William Gilhooly, (D), Virginia, 2006, ClOc
Lixin Wang, (D), Virginia, 2008, HyCl

Assistant Professor:
Catherine Macris, (D), California, 2012, GzCu

Lecturer:
Anna Jessee, (D), Indiana, 2017, GgYg
Jennifer Nelson, (M), Indiana, 2006, Gg
Thomas Rossbach, (D), N Carolina, 1992, GgPg

Indiana University Northwest

Dept of Geosciences (A,B) 🖱
3400 Broadway
Gary, IN 46408
p. (219) 980-6738
zkilibar@iun.edu
https://www.iun.edu/geosciences
Programs: Geology (B); Environmental Science (B)
Enrollment (2021): B: 20 (2)

Chair:
Zoran Kilibarda, (D), Nebraska, 1994, GsmGd

Professor:
Erin Argyilan, (D), Illinois (Chicago), 2004, GeHwAm

Associate Professor:
Kristin Huysken, (D), Michigan State, 1996, GizGc

Adjunct Professor:
Aldo Durazzo, (D), Italy, 1967, Ggi

Emeritus:
Robert Votaw, (D), PiGrs

Indiana University, Bloomington

Dept of Earth and Atmospheric Sciences (B,M,D) ○🖱
1001 E. Tenth Street
Bloomington, IN 47405
p. (812) 855-5582
geoinfo@indiana.edu
https://earth.indiana.edu/
Programs: Earth Sciences, Atmospheric Sciences, Environmental Geoscience, Environmental Sciences
Certificates: Atmospheric Sciences
Enrollment (2020): B: 51 (10) M: 15 (5) D: 27 (6)

Chair:
P. David Polly, (D), California (Berkeley), 1993, PvgPq

Professor:
Simon C. Brassell, (D), Bristol, 1980, Cob
Michael W. Hamburger, (D), Cornell, 1986, YsGtv
Claudia C. Johnson, (D), Colorado, 1993, Peg
Kaj Johnson, (D), Stanford, 2004, Yg
Juergen Schieber, (D), Oregon, 1985, Gs
Chen Zhu, (D), Johns Hopkins, 1992, CgHg

Assistant Research Scientist:
Shelby Rader, (D), Arizona, 2018, Ct

Senior Scientist:
Chusi Li, (D), Toronto, 1993, EgGiEm
Arndt Schimmelmann, (D), California (Los Angeles), 1985, CsGsPe

Adjunct Associate Professor:
Michael C. Rygel, (D), Dalhousie, 2005, Gg

Associate Professor:
Douglas A. Edmonds, (D), Penn State, 2009, Gsm
Chanh Q. Kieu, (D), Maryland, 2008, As
Jackson K. Njau, (D), Rutgers, 2006, Pv
Brian Yanites, (D), Colorado, 2009, Gm
Associate Scientist:
Edward W. Herrmann, (D), Indiana, 2013, Sa
Jessica Miller-Camp, (D), Iowa, 2015, Pv
Peter Sauer, (D), Colorado, 1997, Pe
Adjunct Associate Professor:
Adam V. Maltese, (D), Virginia, 2008, Ze
Adjunct Assistant Professor:
Page Quinton, (D), Missouri, 2016, Pc
Adam Ward, (D), Penn State, 2011, Hs
Assistant Professor:
Ben Kravitz, (D), Rutgers, 2011, At
Travis O'Brien, (D), California (Santa Cruz), 2011, As
Paul W. Staten, (D), Utah, 2013, As
Andrea L. Stevens-Goddard, (D), Arizona, 2017, Cc
Adjunct Research Scientist:
Jose Luis Antinao-Rojas, (D), Dalhousie, 2009
Lee Florea, (D), S Florida, Hg
Maria D. Mastalerz, (D), Silesian Tech, 1988, Ec
Patrick McLaughlin, (D), Cincinnati, 2006, Ps
Tara M. Smiley, (D), Michigan, 2016, Pe
Todd Thompson, (D), Indiana, 1987, Gg
Senior Lecturer:
Bruce Douglas, (D), Princeton, 1983, Gcg
Erika R. Elswick, (D), Cincinnati, 1998, Cl
Cody Kirkpatrick, (D), Alabama (Huntsville), 2010, As
Emeritus:
Abhijit Basu, (D), Indiana, 1975, Gd
David L. Bish, (D), Penn State, 1977, Gz
James G. Brophy, (D), Johns Hopkins, 1984, Gi
Jeremy D. Dunning, (D), N Carolina, 1978, Yg
Enrique Merino, (D), California (Berkeley), 1973, Cl
Greg A. Olyphant, (D), Iowa, 1979, Hw
Gary L. Pavlis, (D), Washington, 1982, YsGtYe
Lisa M. Pratt, (D), Princeton, 1981, Co
Lee J. Suttner, (D), Wisconsin, 1966, Gd
Robert P. Wintsch, (D), Brown, 1975, GpcGt

Indiana University, Indianapolis

Dept of Geography (A,B)
213 Cavanaugh Hall
Indianapolis, IN 46202
p. (317) 274-8877
tbrother@iupui.edu
http://www.iupui.edu/~geogdept/
Chair:
Jeffey S. Wilson, (D), Indiana State, 1998, Zy
Professor:
F. L. Bein, (D), Florida, 1974, Zy
Associate Professor:
Timothy S. Brothers, (D), California (Los Angeles), 1985, Zy
Catherine J. Souch, (D), British Columbia, 1990, Gm

Purdue University

Department of Earth, Atmospheric, and Planetary Sciences (B,M,D)
550 Stadium Mall Drive
West Lafayette, IN 47907-2051
p. (765) 494-3258
eas-info@purdue.edu
http://www.eaps.purdue.edu
f: https://www.facebook.com/EAPSPurdue/
t: @PurdueEAPS
Programs: Atmospheric Science/Meteorology, Environmental Geosciences, Geodata Science, Geology & Geophysics, Planetary Science
Certificates: Geospatial Information Science (GIS)
Enrollment (2022): B: 224 (56) M: 19 (11) D: 112 (4)
Department Head:
Daniel Cziczo, (D), Chicago, 1999, As
Senior Associate Dean for Faculty Affairs:
Lucy M. Flesch, (D), SUNY (Stony Brook), 2002, Yg
Professor:
Andrew M. Freed, (D), Arizona, 1998, Gg
Darryl E. Granger, (D), California (Berkeley), 1996, Gg
Matthew Huber, (D), California (Santa Cruz), 2001, As
Cliff Johnston, (D), California (Riverside), 1983, Sc
Greg Michalski, (D), California (San Diego), 2003, CsGeAs
Robert L. Nowack, (D), MIT, 1985, Ys
Kenneth D. Ridgway, (D), Rochester, 1992, Gs
Douglas R. Schmitt, (D), Caltech, 1987, YxNrYe
Daniel P. Shepardson, (D), Iowa, 1990, Ze
Qianlai Zhuang, (D), Alaska (Fairbanks), As
Associate Professor:
Chris Andronicos, (D), Princeton, 1999, Gg
Michael Baldwin, (D), Oklahoma, 2003, As
Daniel Chavas, (D), MIT, 2013, As
Marty Frisbee, (D), New Mexico Tech, 2010, Hws
Briony Horgan, (D), Cornell, 2010, XgZr
Brandon Johnson, (D), Purdue, 2013, As
Yunyue Elita Li, (D), Stanford, 2014, Yg
Nathaniel A. Lifton, (D), Arizona, 1997, GmCc
David Minton, (D), Arizona, 2009, Xg
Robin Tanamachi, (D), Oklahoma, 2011, Asm
Wen-wen Tung, (D), California (Los Angeles), 2002, As
Yuan Wang, (D), Texas A&M, 2013, As
Lisa Welp, (D), Caltech, 2006, Cs
Assistant Professor:
Ali Bramson, (D), Arizona, 2018, Xg
Daniel Dawson, (D), Oklahoma, 2009, As
Jonathan Delph, (D), Arizona, 2016, Yg
Michael Eddy, (D), MIT, 2017, Gt
Saad Haq, (D), SUNY (Stony Brook), 2004, Gt
Alexandria Johnson, (D), Purdue, 2014, As
Stephanie Olson, (D), California (Riverside), 2018, CgGg
Gouri Prabhakar, (D), Arizona, 2014, As
Michael Sori, (D), MIT, 2014, Gg
Michelle Thompson, (D), Arizona, 2016, Xg
Marissa Tremblay, (D), California (Berkeley), 2017, Cc
Lei Wang, (D), Chicago, 2016, As
Xiaotao Yang, (D), Indiana, 2016, Yg
Emeritus:
Ernest M. Agee, (D), Missouri, 1968, As
Lawrence W. Braile, (D), Utah, 1973, Ys
John Cushman, (D), Iowa State, 1978, Sp
Alexander Gluhovsky, (D), USSR Acad of Sci, 1973, As
Jon M. Harbor, (D), Washington, 1990, GmlGe
Harshvardhan, (D), SUNY (Stony Brook), 1976, As
William J. Hinze, (D), Wisconsin, 1957, YevGt
Paul B. Shepson, (D), Penn State, 1982, As
Yuch-Ning Shieh, (D), Caltech, 1968, CsGpv
Phillip J. Smith, (D), Wisconsin, 1967, As
Wen-Yih Sun, (D), Chicago, 1975, AsZoAm
Terry R. West, (D), Purdue, 1966, Ng
William J. Zinsmeister, (D), California (Riverside), 1974, Ps

University of Notre Dame

Dept of Civil & Environmental Engineering & Earth Sciences (B,M,D) ☒
156 Fitzpatrick Hall
Notre Dame, IN 46556
p. (219) 631-5380
ceees@nd.edu
http://www.nd.edu/~ceees
Programs: Environmental Earth Sciences; Environmental Science with a Concentration in Earth Sciences
Certificates: Earth Sciences (minor)
Chair:
Joannes J. Westerink, (D), MIT, 1984, On
Professor:
Peter C. Burns, (D), Manitoba, 1994, Gz
Jeremy B. Fein, (D), Northwestern, 1989, Cl
Joe Fernando, (D), Johns Hopkins, 1983, As
Clive R. Neal, (D), Leeds (UK), 1985, Gi
Associate Professor:
Diogo Bolster, (D), California (San Diego), 2007, Hw
Andrew Kennedy, (D), Monash, 1998, On
Tony Simonetti, (D), Carleton, 1994, Ct
Assistant Professor:
Melissa Berke, (D), Minnesota, 2011, Cs
Alan Hamlet, (D), Washington, 2006, Hs
Amy E. Hixon, (D), Clemson, 2013, ClqCc

Marc Muller, (D), Rw
David Richter, (D), Stanford, 2011, As
Instructor:
Stephanie Simonetti, (D), McGill, 2002, Zn

University of Southern Indiana

Dept of Geology, Physics, and Environmental Science (B)
8600 University Boulevard
Evansville, IN 47712
p. (812) 464-1701
ahmaria@usi.edu
http://www.usi.edu/science/geology-and-physics/
Programs: Environmental Science; Geology
Administrative Assistant: Hannah Woodring
Enrollment (2022): B: 56 (9)
Chair:
Tony Maria, (D), Rhode Island, 2000, Gv
Professor:
Joseph A. DiPietro, (D), Oregon State, 1990, GcpGm
Paul K. Doss, (D), N Illinois, 1991, HwGeZu
William S. Elliott, Jr., (D), Indiana, 2002, GsrCl
Associate Professor:
James Durbin, (D), Nebraska, 1999, Gm
Instructor:
Carrie L. Wright, (M), Wright State, 2006, Ze
Emeritus:
Norman R. King, (D), Indiana, 1973, Gr

Iowa

Cornell College

Dept of Geology (B) ●☒
600 First Street SW
Mount Vernon, IA 52314
p. (319) 895-4306
rdenniston@cornellcollege.edu
http://www.cornellcollege.edu/geology
Programs: Geology
Enrollment (2021): B: 17 (5)
Professor:
Rhawn F. Denniston, (D), Iowa, 2000, GeCs
Emily O. Walsh, (D), California (Santa Barbara), 2003, GptCc
Assistant Professor:
Drew Muscente, (D), Virginia Tech, 2017, PgGs

Drake University

Dept of Environmental Science and Policy (B)
Des Moines, IA 50311
p. (515) 271-2803
thomas.rosburg@drake.edu

Iowa State University of Science & Technology

Dept of Agronomy (B,M,D)
2101 Agronomy Hall
Ames, IA 50011-1010
p. (515) 294-1360
agron@iastate.edu
http://www.agron.iastate.edu/
Department Secretary: Pam Hinderaker
Director:
Dennis R. Keeney, (D), Iowa, 1965, So
Chair:
Kendall Lamkey, (D), Iowa State, 1985
Professor:
Raymond W. Arritt, (D), Colorado State, 1985, As
Richard E. Carlson, (D), Iowa State, 1971, Am
Richard M. Cruse, (D), Minnesota, 1978, So
V. P. (Bill) Evangelou, (D), California (Davis), 1981, Sc
William J. Gutowski, (D), MIT, 1984, Am
Robert Horton, (D), New Mexico State, 1981, Sp
Douglas L. Karlen, (D), Kansas State, 1978, So
Randy J. Killorn, (D), Idaho, 1983, So
Thomas B. Moorman, (D), Washington State, 1983, So
Jonathan A. Sandor, (D), California (Berkeley), 1983, Sd
Eugene S. Takle, (D), Iowa State, 1971, Asm
Elwynn Taylor, (D), Washington (St Louis), 1970, As
Michael L. Thompson, (D), Ohio State, 1980, Sc
Regis D. Voss, (D), Iowa State, 1962, So
Douglas N. Yarger, (D), Arizona, 1967, As
Associate Professor:
Cynthia Cambardella, (D), Colorado, 1991, So
Thomas A. Kaspar, (D), Iowa State, 1982, So
David A. Laird, (D), Iowa State, 1987, Sc
Antonio W. Mallarino, (D), Iowa State, 1991, So
Assistant Professor:
Lee Burras, (D), Ohio State, 1992, Sd
Larry Halverson, (D), Wisconsin, 1991, So
Stanley J. Henning, (D), Oregon State, 1975, So
Thomas A. Polito, (D), Iowa State, 1987, So
Associate Dean, College of Agriculture:
Gerald A. Miller, (D), Iowa State, 1974, SdoGm
Emeritus:
Thomas E. Loynachan, (D), North Carolina State, 1975, So
Geology Librarian:
Peter A. Peterson, (D), Illinois, 1953, Zn

Dept of Geological & Atmospheric Sciences (B,M,D) ○⊘
253 Science I
2237 Osborn Drive
Ames, IA 50011-3212
p. (515) 294-4477
geology@iastate.edu
http://www.ge-at.iastate.edu/
f: https://www.facebook.com/ISUgeology
Programs: Geology; Earth Science, Meteorology
Administrative Assistant: DeAnn M. Frisk
Enrollment (2018): B: 136 (43) M: 26 (8) D: 22 (4)
Chair:
Sven S. Morgan, (D), Virginia Tech, 1998, Gc
Professor:
Igor A. Beresnev, (D), USSR Acad of Sci, 1986, YgsYe
Cinzia C. Cervato, (D), ETH (Switzerland), 1990, GgZe
William A. Gallus, (D), Colorado State, 1993, As
William J. Gutowski, (D), MIT, 1984, ZoAtZc
Neal R. Iverson, (D), Minnesota, 1989, Gl
William W. Simpkins, (D), Wisconsin, 1989, HwGlCs
Paul G. Spry, (D), Toronto, 1984, Em
Xiaoqing Wu, (D), California (Los Angeles), 1992, As
Associate Professor:
Kristie Franz, (D), California (Irvine), 2006, Hs
Chris Harding, (D), Houston, 2001, Gq
Alan D. Wanamaker, (D), Maine, 2007, CsPeOg
Assistant Professor:
Beth E. Caissie, (D), Massachusetts, 2012, Pe
Alex Gonzalez, (D), Colorado State, 2015, As
Jacqueline Reber, (D), Oslo, 2012, Gct
Yuyu Zhou, (D), Rhode Island, 2008, Zyi
Senior Lecturer:
James V. Aanstoos, (D), Purdue, 1996, ZrAs
Jane P. Dawson, (D), New Mexico, 1995, Gp
David M. Flory, (M), Iowa State, 2003, AmsZn
Lecturer:
Rachindra Mawalagedara, (D), Nebraska, 2013, As
Aaron R. Wood, (D), Michigan, 2009, PvGrs
Adjunct Professor:
Michael R. Burkart, (D), Iowa, 1976, Hy
Emeritus:
Tsing-Chang Chen, (D), Michigan, 1975, As
Robert Cody, (D), Colorado, 1968, Gz
Carl E. Jacobson, (D), California (Los Angeles), 1980, GtcGp
Karl E. Seifert, (D), Wisconsin, 1963, Ct
Eugene S. Takle, (D), Iowa State, 1971, As
Carl F. Vondra, (D), Nebraska, 1963, Gr
Kenneth E. Windom, (D), Penn State, 1976, Cp
Douglas N. Yarger, (D), Arizona, 1967, As
Lab Equipment Specialist:
Alison Whale, (M), Iowa State, 2018

Scott Community College

Environmental Science
500 Belmont Road
Bettendorf, IA 52722
p. 563-441-4001
eiccinfo@eicc.edu
http://www.eicc.edu

University of Iowa

Earth & Environmental Sciences (B,M,D) ●⦸
115 Trowbridge Hall
123 N. Capitol Street
Iowa City, IA 52242
p. (319) 335-1818
geology@uiowa.edu
http://clas.uiowa.edu/ees/
f: https://www.facebook.com/UIowaGeoscience
Programs: Geoscience, BA, BS, MS, PhD, Environmental Sciences, BA, BS
Administrative Assistant: Angela Bellew
Enrollment (2022): B: 46 (15) M: 14 (8) D: 13 (2)

Chair:
David W. Peate, (D), Open (UK), 1989, GiCa

Professor:
Jonathan M. Adrain, (D), Alberta, 1993, Pi
Christopher A. Brochu, (D), Texas, 1997, Pv
Charles T. Foster Jr., (D), Johns Hopkins, 1975, GptGg
Jane A. Gilotti, (D), Johns Hopkins, 1987, GctGp
William C. McClelland, (D), Arizona, 1990, GtCcGc

Associate Professor:
Bradley D. Cramer, (D), Ohio State, 2009, GrPsGs
Jeffrey A. Dorale, (D), Minnesota, 2001, CsPc
Emily Finzel, (D), Purdue, 2010, GstEo
Frank H. Weirich, (D), Toronto, 1982, Gm

Assistant Professor:
Shamar C. Chin, (D), Nebraska-Lincoln, 2020, PmOoPe
Jessica R. Meyer, (D), Guelph, 2013, Hw
Valerie Payré, Lorraine, 2017, XgGiz

Associate Professor of Instruction:
Kate Tierney, (D), Ohio State, 2010, GrCsGg

Lecturer:
Benjamin Swanson, (D), New Mexico, 2012, Gme

Adjunct Professor:
Keith E. Schilling, (D), Iowa, 2009, Hgs

Adjunct Associate Professor:
Ingrid Ukstins, (D), Royal Holloway (UK), 2003, GviXg
Brian J. Witzke, (D), Iowa, 1981, GrPsg
Kathleen Woida, (D), Iowa, 1991, So

Adjunct Assistant Professor:
Raymond R. Anderson, (D), Iowa, 1992, Gr
Alyssa M. Bancroft, (D), Ohio State, 2014, Gr
Lihan Li, Houston, 2022, Cg
Drew Muscente, (D), Virginia Tech, 2016, Peq

Adjunct Professor:
Leon Aden, (M), Iowa, 1982, Go
Neil P. Bernstein, (D), Minnesota, 1982, Ge
Rhawn F. Denniston, (D), Iowa, 2000, Cs
Emily O. Walsh, (D), California (Santa Barbara), 2003, GptCc

Associate:
Robert Brenner, Missouri-Columbia, 1973, Gds

Emeritus:
Richard G. Baker, (D), Colorado, 1969, PebPc
E. Arthur Bettis III, (D), Iowa, 1995, SdGmSa
Ann F. Budd, (D), Johns Hopkins, 1978, Pgi
Robert S. Carmichael, (D), Pittsburgh, 1967, YexGg
Lon D. Drake, (D), Ohio State, 1968, Hy
Philip H. Heckel, (D), Rice, 1966, GrPsGd
George R. McCormick, (D), Ohio State, 1964, GzEmCa
Mark K. Reagan, (D), California-Santa Cruz, 1987, Gi
Holmes A. Semken, Jr., (D), Michigan, 1965, Pv
You-Kuan Zhang, (D), Arizona, 1990, Hw

Rock Lab Manager:
Matthew J. Wortel, (M), Iowa, 2007, Gx

Collections Manager, Paleontology:
Tiffany S. Adrain, (M), Iowa, 2003, Pg

Office Coordinator:
Alexandra Geraets

University of Northern Iowa

Dept of Earth and Environmental Sciences (B) ⦸
121 Latham Hall
Cedar Falls, IA 50614-0335
p. (319) 273-2759
siobahn.morgan@uni.edu
http://www.earth.uni.edu/
f: https://www.facebook.com/pages/University-of-Northern-Iowa-Earth-and-Environmental-Science/56860203682
Programs: Earth Science; Earth Science Teaching; Environmental Sciences, Environmental Resource Management
Administrative Assistant: Noel Graff
Enrollment (2022): B: 92 (18)

Head:
Siobahn M. Morgan, (D), Washington, 1991, Xa

Professor:
Alan C. Czarnetzki, (D), Wisconsin, 1992, Am
Chad E. Heinzel, (D), N Illinois, 2005, GmaSa
Thomas A. Hockey, (D), New Mexico State, 1988, Zn
Mohammad Z. Iqbal, (D), Indiana, 1994, Hw

Associate Professor:
Kyle R. Gray, (D), Akron, 2009, ZeGg
Alexa Sedlacek, (D), Ohio State, 2013, CsGr
Xinhua Shen, (D), Colorado State, 2011, Asm

Instructor:
Aaron Spurr, (M), N Iowa, 1997, Zeg

Emeritus:
Wayne I. Anderson, (D), Iowa, 1964, Pg
Lynn A. Brant, (D), Penn State, 1980, Gn
Timothy M. Cooney, (D), N Colorado, 1976, Ze
Walter E. De Kock, (D), Ohio State, 1972, Ze
Kenneth J. De Nault, (D), Stanford, 1974, Gz
James C. Walters, (D), Rutgers, 1975, Gm

Lab Tech:
Steven J. Smith, (M), N Iowa, 2000, Zg

Kansas

Emporia State University

Earth Science Program (A,B,M)
1 Kellogg Circle
Emporia, KS 66801-5087
p. (620) 341-5330
mmorales@emporia.edu
http://www.emporia.edu/earthsci/
Programs: BS undergraduate degree in Earth Science with concentrations in Physical Geology, Environmental Geology, Soil Science, and Atmospheric Science, and minors in Paleontology and Geospatial Analysis/GIS., Master's degree with an Earth Science concentration and BS and BSE Undergraduate degrees in Earth Science
Certificates: Undergraduate and graduate certificates in Geospatial Analysis/GIS.
Enrollment (2018): B: 19 (9) M: 15 (5)

Professor:
Richard O. Sleezer, (D), Kansas, 2001, ZiSdHs

Associate Professor:
Michael A. Morales, (D), California (Berkeley), 1987, PvGrg

Assistant Professor:
Alivia J. Allison, (D), Missouri (Kansas City), 2013, GgaGm
Paul Zunkel, (D), Texas State, 2017, AmZig

Emeritus:
James S. Aber, (D), Kansas, 1978, Gl
Paul L. Johnston, (M), Kansas, 1959, Gg
Kenneth W. Thompson, (D), Iowa State, 1991, Ze

Fort Hays State University

Department of Geosciences (B,M) O
600 Park Street
Tomanek Hall
Hays, KS 67601-4099
p. (785) 628-5389
geosciences@fhsu.edu
http://www.fhsu.edu/geo
f: https://www.facebook.com/GeoFHSU/
t: @GeoFHSU
Programs: Geology, Geography, Geographic Information Systems, Environmental Geosciences
Certificates: Geographic Information Systems, Museum Studies, Crime Mapping and Analysis
Administrative Assistant: Patricia Duffey
Enrollment (2019): B: 152 (27) M: 22 (6)

Emeritus:
Kenneth R. Neuhauser, (D), South Carolina, 1973, GceGs

Paul Phillips, (D), Kansas, 1977, Zen
Richard J. Zakrzewski, (D), Michigan, 1968, PvsPe
Professor:
Richard Lisichenko, (D), Kansas State, 2000, Zie
Interim Chair, Geosciences & Curator, Sternberg Museum:
Laura E. Wilson, (D), Colorado, 2012, PvePg
Associate Professor:
Hendratta N. Ali, (D), Oklahoma State, 2010, GogYs
Tom Schafer, (D), Kansas State, 2000, ZyiRn
Assistant Professor:
Henry Agbogun, (D), New Brunswick, 2012, GgYg
Keith A. Bremer, (D), Texas State, 2011, ZuiZn
Jonathan B. Sumrall, (D), Mississippi State, 2013, GgxZn
Instructor:
Eamonn Coveney, (M), Fort Hays State, 2006, Zyn
William H. Heimann, (M), Fort Hays State, 1987, HgwGm
Kara Kuntz, (M), Kansas State, 1990, Zn
Jami J. Seirer, (M), Montana, 2015, ZrIZn
Jeanne L. Sumrall, (D), Mississippi State, 2015, Zgn
Cooperating Faculty (Emeritus) Paleobiology:
Joseph R. Thomasson, (D), Iowa State, 1976, Pb
Dean, College of Science Technology and Mathematics:
P. Grady Dixon, (D), Arizona State, 2005, AmZyAs

Johnson County Community College

Science Division (A)
12345 College Blvd
Overland Park, KS 66210
p. (913) 469-3826
mwisgird@jccc.edu
http://www.jccc.edu/academics/math-science
Enrollment (2015): A: 7 (2)
Professor:
Lynne Beatty, (M), S Illinois, 1985, GgZy
Assistant Professor:
John P. Harty, (D), Zy
Adjunct Professor:
Bruce Frederick, (D), Gg
John Maher, (D), Maryland, 1999, Zy

Kansas State University

Dept of Agronomy (B,M,D)
2004 Throckmorton
Manhattan, KS 66506
p. (785) 532-6101
agronomy@ksu.edu
http://www.agronomy.k-state.edu/
f: https://www.facebook.com/kstate.agronomy
t: @KStateAgron
Enrollment (2014): B: 70 (33) M: 41 (9) D: 43 (10)
Deparment Head:
Gary M. Pierzynski, (D), Ohio State, 1989, Sc
Professor:
Stewart Duncan, (D), Kansas State, 1991, So
Walter H. Fick, (D), Texas Tech, 1978, Sf
Dale Fjell, (D), Kansas State, 1982, So
Allan Fritz, (D), Kansas State, 1994, So
Mary Beth Kirkham, (D), Wisconsin, Sp
Gerard J. Kluitenberg, (D), Iowa State, 1989, Sp
David Mengel, (D), Purdue, 1975, So
Clenton E. Owensby, (D), Kansas State, 1969, Sf
Dallas Peterson, (D), North Dakota State, 1987, So
Vara Prasad, (D), Reading (UK), 1999, So
Michel D. Ransom, (D), Ohio State, 1984, SdcGm
Chuck W. Rice, (D), Kentucky, 1983, Sb
Bill T. Schapaugh, (D), Purdue, 1979, So
Alan Schlegel, (D), Purdue, 1985, So
Phillip Stahlman, (D), Wyoming, 1989, So
Curtis Thompson, (D), Idaho, 1993, So
Steve M. Welch, (D), Michigan State, 1977, Sp
Associate Professor:
Ignacio Ciampitti, (D), Purdue, 2012, So
Gary Cramer, (D), Nebraska, 1998, So
Ganga Hettiarachchi, (D), Kansas State, 2000, Sc
John Holman, (D), Idaho, 2005, So
Doo-Hong Min, (D), Maryland, 1998, So
Nathan Nelson, (D), North Carolina State, 2004, So
DeAnn Presley, (D), Kansas State, 2007, So
Kraig Roozeboom, (D), Kansas State, 2006, So
Dorivar Ruiz-Diaz, (D), Iowa State, 2007, So
Gretchen Sassenrath, (D), Illinois, 1988, So
Tesfaye Tesso, (D), Kansas State, 2002, So
Assistant Professor:
Eric Adee, (D), Wisconsin, 1993, So
Lucas Haag, (D), Kansas State, 2013, So
Mithila Jugulam, (D), Guelph, 2004, So
Xiaomao Lin, (D), Nebraska, 1999, Zn
Colby Moorberg, (D), North Carolina, 2014, So
Geoffrey Morris, (D), Chicago, 2007, So
Augustine Obour, (D), Florida, 2010, So
Ram Perumal, (D), Tamil Nadu Ag, 1993, So
Eduardo Santos, (D), Guelph, 2011, So
Peter Tomlinson, (D), Arkansas, 2006, So
Guorong Zhang, (D), North Dakota State, 2007, So
Emeritus:
Mark Claassen, (D), Iowa State, 1971, So
Bill Eberle, (D), Illinois, 1973, SoZu
Stan Ehler, (D), Missouri, 1975, So
Barney Gordon, (D), South Dakota, 1990, So
Keith Janssen, (D), Michigan State, 1973, So
George Liang, (D), Wisconsin, 1965, So
Gerry L. Posler, (D), Iowa State, 1969, Sf
Kevin Price, (D), Utah, 1987, Zr
David Regehr, (D), Illinois, 1975, So
Jim Shroyer, (D), Iowa State, 1980, So
Loyd Stone, (D), South Dakota, 1973, Sp
Steve J. Thien, (D), Purdue, 1971, Sc
Richard Vanderlip, (D), So
D.A. Whitney, (D), Iowa State, 1966, So
Agronomist:
Doug Shoup, (D), Kansas State, 2006, So
Librarian:
Nancy William, (N), Zn

Dept of Geology (B,M)
108 Thompson Hall
Manhattan, KS 66506-3201
p. (785) 532-6724
rocknrat@ksu.edu
http://www.k-state.edu/geology/
Administrative Assistant: Lori Page-Willyard
Enrollment (2013): B: 69 (15) M: 27 (5)
Department Head:
Pamela D. Kempton, (D), S Methodist, 1984, GiCtc
Professor:
Sambhudas Chaudhuri, (D), Ohio State, 1966, GeoCc
Associate Professor:
Allen W. Archer, (D), Indiana, 1983, Gr
Matthew E. Brueseke, (D), Miami (Ohio), 2006, GitGv
Abdelmoneam E. Raef, (D), AGH (Poland), 2001, YesZr
Matthew W. Totten, (D), Oklahoma, 1992, Gso
Assistant Professor:
Saugata Datta, (D), W Ontario, 2001, ClHwGe
Keith B. Miller, (D), Rochester, 1988, Pe
Joel Q.G. Spencer, (D), Glasgow, 1996, CcGs
Emeritus:
Robert L. Cullers, (D), Wisconsin, 1971, Ct
Charles G. Oviatt, (D), Utah, 1984, Gm
Ronald R. West, (D), Oklahoma, 1970, PgeGs

University of Kansas

Dept of Geology (B,M,D) O
1414 Naismith Dr
Room 254
Ritchie Hall
Lawrence, KS 66045-7613
p. (785) 864-4974
fowle@ku.edu
http://geo.ku.edu/
f: https://www.facebook.com/KUGeology
Programs: BA, BS, MS, PhD in Geology, Professional Science MS in Environmental Geology
Certificates: Environmental Geology
Enrollment (2013): B: 72 (14) M: 50 (17) D: 34 (9)
Union Pacific Distinguished Professor:
J. Douglas Walker, (D), MIT, 1985, Gc

Ritchie Distinguished Professor:
Michael D. Blum, (D), Texas, 1997, GsrGm
Professor and Chair:
Jennifer A. Roberts, (D), Texas, 2000, Cbl
Gulf-Hedberg Distinguished Professor:
Paul A. Selden, (D), Cambridge, 1979, Pig
Distinguished Professor:
Robert H. Goldstein, (D), Wisconsin, 1986, Gs
Courtesy Professor:
Rolfe Mandel, (D), Kansas, 1991, Gam
Professor:
J. F. Devlin, (D), Waterloo, 1994, Hw
David A. Fowle, (D), Notre Dame, 2000, Cb
Evan K. Franseen, (D), Wisconsin, 1989, Gd
Luis A. Gonzalez, (D), Michigan, 1989, CsGd
Stephen T. Hasiotis, (D), Colorado, 1997, Pe
Mary C. Hill, (D), Princeton, 1985, HgqHw
Gwendolyn L. Macpherson, (D), Texas, 1989, Hw
Gene Rankey, (D), Kansas, 1996, GsrGu
Michael H. Taylor, (D), California (Los Angeles), 2004, GtmZr
George P. Tsoflias, (D), Texas, 1999, YgeGe
Associate Professor:
Craig Marshall, (D), Tech (Australia), 2001, GzZmXb
Andreas Möller, (D), Christian-Albrechts (Germany), 1996, Gt
Alison Olcott Marshall, (D), S California, 2006, PgCoPo
Leigh Stearns, (D), Maine, 2008, GlZrf
Anthony W. Walton, (D), Texas, 1972, GsEoGv
Associate Scientist:
Diane Kamola, (D), Georgia, 1989, Gs
Randy Stotler, (D), Waterloo, 2008, Hw
Assistant Professor:
Noah McLean, (D), MIT, 2012, CcGq
Chi Zhang, (D), Rutgers, 2012, YgxYu
Courtesy Professor:
John H. Doveton, (D), Edinburgh, 1969, Gq
William C. Johnson, (D), Wisconsin, 1976, Grm
Leonard Krishtalka, (D), Kansas, 1976, Pv
Bruce S. Lieberman, (D), Columbia, 1994, Pgi
Gregory A. Ludvigson, (D), Iowa, 1988, Gs
Richard D. Miller, (M), Kansas, 1983, Ye
Edith Taylor, (D), Ohio State, 1983, Pb
W. Lynn Watney, (D), Kansas, 1985, Gr
Donald O. Whittemore, (D), Penn State, 1973, Cl
Courtesy Associate Professor:
Geoff Bohling, (D), Kansas, 1999, Hw
Gaisheng Liu, (D), Alabama, 2004, Hg
Courtesy Assistant Professor:
Andrea Brookfield, (D), Waterloo, 2009, Hw
Jon Smith, (D), Kansas, 2008, GsPe
Adjunct Professor:
Timothy R. Carr, (D), Wisconsin, 1981, Go
John Gosse, (D), Lehigh, 1994, Gm
Emeritus:
Ernest E. Angino, (D), Kansas, 1961, Cl
Ross A. Black, (D), Wyoming, 1989, Ye
Wakefield Dort, Jr., (D), Stanford, 1955, Gm
Gisela Dreschoff, (D), Tech (Braunschweig), 1972, Ce
Paul Enos, (D), Yale, 1965, GsdGu
Lee C. Gerhard, (D), Kansas, 1964, GsoGd
Carl D. McElwee, (D), Kansas, 1970, HgYg
Richard A. Robison, (D), Texas, 1962, PiGr
Albert J. Rowell, (D), Leeds (UK), 1953, Pi
Don W. Steeples, (D), Stanford, 1975, YseGe
W. Randall Van Schmus, (D), California (Los Angeles), 1964, Cc

Wichita State University

Dept of Geology (B,M)
1845 Fairmount
Wichita, KS 67260-0027
p. (316) 978-3140
bill.bischoff@wichita.edu
https://www.wichita.edu/academics/fairmount_college_of_liberal_arts_and_sciences/geology/
Administrative Assistant: K. L. Smith
Chair:
William C. Parcell, (D), Alabama, 2000, Gr
Professor:
William D. Bischoff, (D), Northwestern, 1985, Cl
John C. Gries, (D), Texas, 1970, Gt
Salvatore J. Mazzullo, (D), Rensselaer, 1974, Go
Associate Professor:
Collette D. Burke, (D), Wisconsin (Milwaukee), 1983, Pm
Assistant Professor:
Hongsheng Cao, (D), Florida State, 2001, Cl
Wan Yang, (D), Texas, 1999, Gs
Lecturer:
Toni K. Jackman, (M), Wichita State, 1984, Ge
David L. Schaffer, (B), Utah, 1982, Am
Emeritus:
James N. Gundersen, (D), Minnesota, 1958, Ga
Daniel F. Merriam, (D), Kansas, 1961, Gr
Peter G. Sutterlin, (D), Northwestern, 1958, Gd

Kentucky

Alice Lloyd College

Div of Natural Sciences & Mathematics
Pippa Passes, KY 41844-9701
p. (606) 368-2101 x5405
paulyeary@alc.edu
https://www.alc.edu/academics/natural-science-and-mathematics/

Bluegrass Community and Technical College

Environmental Science Technology Program
470 Cooper Dr
Lexington, KY 40506
p. (859) 246-6448
tracy.knowles@kctcs.edu
https://bluegrass.kctcs.edu/education-training/program-finder/environmental-science-technology.aspx

Eastern Kentucky University

Dept of Physics, Geosciences & Astronomy (B)
521 Lancaster Avenue
Science 3140
Richmond, KY 40475-3102
p. (859) 622-1521
alice.jones@eku.edu
https://www.eku.edu/phygeosast/
f:https://www.facebook.com/EKU-Geosciences-447266678713162
Programs: B.S. Environmental and Applied Geology
Certificates: GIS Certificate
Enrollment (2018): B: 61 (20)
Acting Chair:
Alice Jones, (D), Ohio State, 1998, Zyn
Professor:
Walter S. Borowski, (D), North Carolina, 1998, GsrOu
Melissa S. Dieckmann, (D), Notre Dame, 1995, ZeCl
John C. White, (D), Baylor, 2002, GiCgt
Donald M. Yow, (D), South Carolina, 2003, AtZye
Associate Professor:
F. Tyler Huffman, (D), Connecticut, 2006, Zi
Kelly Watson, (D), Florida State, 2012, ZriZn
Assistant Professor:
Haley Cabaniss, (D), Illinois, 2020, GvuYr
Estifanos Haile, (D), Kentucky, 2011, Hw
Instructor:
Cory BlackEagle, (M), E Kentucky, 1989, HwZg
Ann Harris, (D), Kentucky, 2018, GgZy
Sung Bae Jeon, (D), Zy
Craig Webb, (M), W Kentucky, 1994, Zyn
Peter Worcester, (D), Miami, 1977, GgOuZe
Senior Lecturer:
Glenn A. Campbell, (M), Marshall, 1995, Zy
Emeritus:
Gary L. Kuhnhenn, (D), Illinois, 1976, GgZe
David Zurick, (D), Hawaii, 1986, Zn

Morehead State University

Dept of Earth and Space Sciences (B)
235 Martindale Drive
Morehead, KY 40351
p. (606) 783-2381
c.mason@moreheadstate.edu

http://www.moreheadstate.edu/physsci/
Administrative Assistant: Amanda Holbrook
Enrollment (2014): B: 40 (4)
Department Chair:
Benjamin K. Malphrus, (D), West Virginia, 1990
Professor:
Charles E. Mason, (M), George Washington, 1981, PiGrs
Associate Professor:
Marshall Chapman, (D), Massachusetts, 1996, Gv
Eric Jerde, (D), California (Los Angeles), 1991, Cg
Jennifer O'Keefe, (D), Kentucky, 2008, PlEcZe
Steven K. Reid, (D), Texas A&M, 1991, GsoGe

Murray State University

Dept of Earth and Environmental Sciences (B,M)
334 Blackburn Science Building
Murray, KY 42071
p. (270) 809-2591
qzhang@murraystate.edu
http://murraystate.edu/ees
f: https://www.facebook.com/murraystateees/
t: @MurrayStateEES
Programs: BS Tracks: Geology, Earth Science Education, Environmental Science, Geography & GIS, Archaeology, MS Concentrations: Environmental Geology, Geoinformatics, Watershed Science, Archaeology
Certificates: GIS (undergraduate), Geospatial Data Science (graduate)
Enrollment (2022): B: 68 (11) M: 10 (4)
Chair:
Qiaofeng (Robin) Zhang, (D), W Ontario, 2002, ZirZu
Professor:
Haluk Cetin, (D), Purdue, 1993, ZrGe
Associate Professor:
Bassil El Masri, (D), Indiana, 2011, Zc
Anthony L. Ortmann, (D), Tulane, 2007, GaYg
Assistant Professor:
Sung-ho Hong, (D), New Mexico Tech, 2008, HyGe
Marcie Venter, (D), Kentucky, Ga
Lecturer:
Jane L. Benson, (M), Murray State, 1986, Zi
Michael R. Busby, (M), Murray State, 1996, Zi

Northern Kentucky University

Dept of Geology (B)
204H Natural Sciences Center
Highland Heights, KY 41099
p. (859) 572-5309
rockawayj@nku.edu
http://nku.edu/
Programs: Geology; Earth Science Education
Certificates: GIS
Enrollment (2017): B: 30 (60)
Associate Professor:
Janet Bertog, (D), Cincinnati, 2002, Gd
John D. Rockaway, (D), Purdue, 1968, Ng
Assistant Professor:
Trent Garrison, (D), Kentucky, 2015, GeHwEc
Lecturer:
Sarah E. Johnson, (M), Purdue, 1997, GmNgZi

Owensboro Community and Technical College

Owensboro Community and Technical College
4800 New Hartford Road
Owensboro, KY 42303
http://www.owensboro.kctcs.edu/

University of Kentucky

Dept of Earth and Environmental Sciences (B,M,D)
101 Slone Research Building
121 Washington St.
Lexington, KY 40506-0053
p. (859) 257-3758
ewoolery@uky.edu
http://ees.as.uky.edu
f:https://www.facebook.com/UKEarthandEnvironmentalSciences/
Programs: Geological Sciences
Enrollment (2022): B: 28 (7) M: 15 (2) D: 12 (1)
Chair:
Edward W. Woolery, (D), Kentucky, 1998, YusNg
Professor:
Frank R. Ettensohn, (D), Illinois, 1975, PsGdPi
Alan E. Fryar, (D), Alberta, 1992, HwGeCl
William C. Haneberg, (D), Cincinnati, 1989, NgZi
James C. Hower, (D), Penn State, 1978, Ec
David P. Moecher, (D), Michigan, 1988, GxtCg
Dhananjay Ravat, (D), Purdue, 1989, YgmYv
Kevin M. Yeager, (D), Texas A&M, 2002, GsCmc
Associate Professor:
Andrea M. Erhardt, (D), Stanford, 2013, CgsCm
Rebecca Freeman, (D), Tulane, 2011, GrPi
Michael M. McGlue, (D), Arizona, 2011, GnrGm
Ryan Thigpen, (D), Virginia Tech, 2009, Gtc
Assistant Professor:
Keely A. O'Farrell, (D), Toronto, 2013, Ygd
Laboratory Manager:
Kimberly Schindler, (M), Kentucky, Cc
Lecturer:
Summer Brown, (M), Virginia Tech, 2010, GgZei
Kent Ratajeski, (D), North Carolina, 1999, Gi
Adjunct Professor:
William Andrews, (D), Kentucky, 2004, Gg
John R. Bowersox, (D), S Florida, 2006, God
N. Seth Carpenter, (D), Kentucky, 2019, Ys
Matthew Crawford, (D), Kentucky, 2018, Ng
Jason Dortch, (D), Cincinnati, 2010, GmZr
Cortland F. Eble, (D), West Virginia, 1988, Pl
Stephen F. Greb, (D), Kentucky, 1992, GsEcGr
Hickman B. John, (D), Kentucky, 2011, Gto
Georgina Lukoczki, (D), Oklahoma State, 2019, GdsCg
Matthew Massey, (D), Kentucky, 2010, Ggc
Thomas M. Parris, (D), California (Santa Barbara), 1998, Cg
Benjamin Tobin, (D), Texas State (San Marcos), 2013, Hy
Zhenming Wang, (D), Kentucky, 1998, YsNeRn
Junfeng Zhu, (D), Arizona, 2005, Hw
Emeritus:
William H. Blackburn, (D), MIT, 1967, GxCg
Bruce R. Moore, (D), Melbourne (Australia), 1967, Gd
Kieran D. O'Hara, (D), Brown, 1984, GcCg
Lyle V. A. Sendlein, (D), Iowa State, 1964, Hw
Ronald L. Street, (D), St Louis, 1975, Ys
William A. Thomas, (D), Virginia Tech, 1959, Gt
Laboratory Director:
Aaron Shultis, (D), Nebraska, 2019, Csg

Dept of Mining Engineering (B,M,D)
230 Mining & Mineral Resources Building
Lexington, KY 40506-0107
p. (859) 257-8026
rick.honaker@uky.edu
http://www.engr.uky.edu/mng/
t: @UK_Mining
Enrollment (2015): B: 125 (36) M: 14 (7) D: 17 (3)
Chair:
Rick Honkaer, (D), Virginia Tech, 1992, Nm
Professor:
Zach Agioutantis, (D), Virginia Tech, 1987, NrmEc
Braden Lusk, (D), Missouri (Rolla), 2006, Nm
Thomas Novak, (D), Penn State, 1984, Nm
Joseph Sottile, Jr., (D), Penn State, 1991, Nm
Assistant Professor:
Kyle Perry , (D), Kentucky, 2010, Nr
Jhon Silva-Castro, (D), Kentucky, 2012, Nm
William Chad Wedding, (D), Kentucky, 2014, Nm
Emeritus:
Kot F. Unrug, (D), Acad Mining-Metallurgy, 1966, Nr
Andrew M. Wala, (D), Acad Mining-Metallurgy, 1972, Nm

Western Kentucky University

Dept of Geography & Geology (B,M)
1906 College Heights Blvd
#31066
Bowling Green, KY 42101-1066

p. (270) 745-4555
fred.siewers@wku.edu
http://www.wku.edu/geoweb
Programs: Geology, Geography, Meteorology, GIS
Certificates: GIS, grad and undergrad
Enrollment (2016): B: 110 (33) M: 25 (9)

Chair:
Fredrick D. Siewers, (D), Illinois, 1995, GsPiZc

Professor:
Catherine Algeo, (D), Louisiana State, 1997, Zi
Margaret Gripshover, (D), Tennessee, Zy
Christopher Groves, (D), Virginia, 1993, Hg
David J. Keeling, (D), Oregon, 1992, Zg
Michael May, (D), Indiana, 1993, Ge
Jun Yan, (D), Buffalo, 2004, Zi

Associate Professor:
Josh Durkee, (D), Georgia, 2008, AsZy
Xingang Fan, (D), Lanzhou, As
M. Royhan Gani, (D), Texas (Dallas), 2005, GsrGg
Nahid Gani, (D), Texas (Dallas), Gc
Greg Goodrich, (D), Arizona State, 2005, Ast
Jason Polk, (D), S Florida, Gm
Andrew Wulff, (D), Massachusetts, 1993, Gv

Associate Scientist:
Leslie North, (D), S Florida, Zg

Instructor:
William Blackburn, (M), W Kentucky, 2002, Zy
Kevin Cary, (M), W Kentucky, 2000, Zi
Margaret Crowder, (D), W Kentucky, 2012, Ze
Scott Dobler, (M), Bowling Green, 1996, As
Patricia Kambesis, (D), Mississippi State, 2014, ZigHw

Instructor:
Amy Nemon, (M), W Kentucky, 2005, Zig

Emeritus:
Nicholas Crawford, (D), Clark, 1977, Hg

Louisiana

Centenary College of Louisiana

Dept of Geology (B) ☒
2911 Centenary Boulevard
Shreveport, LA 71104
p. (318) 869-5234
dbieler@centenary.edu
Programs: Geology (B)
Enrollment (2020): B: 7 (6)

Professor:
David B. Bieler, (D), Illinois, 1983, GtrYe
Scott K. Vetter, (D), South Carolina, 1989, Gi

Delgado Community College

Science & Math Division (A)
615 City Park Avenue
New Orleans, LA 70119
p. (504) 671-6480
registrar@dcc.edu
https://www.dcc.edu/academics/science-math/

Acting Head:
Raymond Duplessis, (M), New Orleans

Louisiana State University

Dept of Geography & Anthropology (B,M,D)
227 Howe-Russell Geoscience Complex
Baton Rouge, LA 70803
p. (225) 578-5942
gachair@lsu.edu
http://www.ga.lsu.edu
Enrollment (2006): B: 54 (23) M: 12 (6) D: 22 (7)

Chair:
Kevin Robbins

Professor:
Patrick A. Hesp, (D), Sydney, 1981, Gm
Richard H. Kesel, (D), Maryland, 1971, Hg

Associate Professor:
Steven Namikas, (D), S California, 1999, Gms

Dept of Geology & Geophysics (B,M,D) O☒
E235 Howe Russell Kniffen Geoscience Complex
Baton Rouge, LA 70803-4101
p. (225) 578-3353
geology@lsu.edu
http://www.lsu.edu/science/geology/
f: http://www.facebook.com/LSUGeology
t: @LSUGeology
Enrollment (2017): B: 86 (31) M: 28 (16) D: 28 (7)

Chair:
Carol M. Wicks, (D), Virginia, 1992, HwClHy

Professor:
Huiming Bao, (D), Princeton, 1998, CsAcCu
Samuel J. Bentley, (D), SUNY (Stony Brook), 1998, Gu
Peter Clift, (D), Edinburgh, 1993, GsuGo
Peter Doran, (D), Nevada (Reno), 1996, Hws
Barbara L. Dutrow, (D), S Methodist, 1985, Gz
Brooks B. Ellwood, (D), Rhode Island, 1977, Ym
Darrell J. Henry, (D), Wisconsin, 1981, Gp

Associate Professor:
Philip J. Bart, (D), Rice, 1998, Gr
Juan M. Lorenzo, (D), Columbia, 1991, Ys
Sophie Warny, (D), Catholic (Belgium), 1999, Pl

Assistant Professor of Research:
Xiaobin Cao, (D), Chinese Acad of Sci, 2012, Cs
Yongbo Peng, (D), Louisiana State, Cg

Assistant Professor - Professional Practice:
Amy Luther, (D), New Mexico Tech, GcZe

Assistant Professor:
Adam Forte, (D), California (Davis), 2012, Gcm
Achim Herrmann, (D), Penn State, 2005, PeGd
Suniti Karuntillake, (D), Cornell, 2008, Xg
Karen Luttrell, (D), Scripps, Yg
Patricia Persaud, (D), Caltech, Ys
Jianwei Wang, (D), Illinois (Urbana), 2004, ClZmGy
Carol A. Wilson, (D), Boston, 2013, GsOnCc
Guangsheng Zhuang, (D), California (Santa Cruz), 2011, CocGs

Instructor:
Yanxia Ma, (D), Og

Associate Curator, LSU Museum of Natural Science:
Judith A. Schiebout, (D), Texas, 1973, Pv

Emeritus:
Ajoy K. Baksi, (D), Toronto, 1970, Cc
Gary R. Byerly, (D), Michigan State, 1974, Gi
Ray E. Ferrell, Jr., (D), Illinois, 1966, Cl
Jeffrey S. Hanor, (D), Harvard, 1967, Cl
George Hart, (D), Sheffield, 1961, Gg
Clyde Moore, Zn
Jeffrey A. Nunn, (D), Northwestern, 1981, Yg
James E. Roche, (D), Illinois, 1969, Gg
Barun K. Sen Gupta, (D), Indian Inst of Tech, 1963, Pm

Accountant Technician:
Jeanne L. Johnson, Zn

Dept of Oceanography & Coastal Sciences (M,D)
1002 Energy Coast and Environment Building
Baton Rouge, LA 70803
p. (225) 578-6308
ocean@lsu.edu
http://www.ocean.lsu.edu
Administrative Assistant: Gaynell Gibbs
Enrollment (2010): M: 63 (11) D: 92 (11)

Professor:
Robert P. Gambrell, (D), North Carolina State, 1974, Cg
Paul A. LaRock, (D), Rensselaer, 1968, Ob
Irving A. Mendelssohn, (D), North Carolina State, 1974, Ob
Richard F. Shaw, (D), Maine, 1981, Ob
Robert E. Turner, (D), Georgia, 1974, Ob

Associate Professor:
Donald M. Baltz, (D), California (Davis), 1980, Ob
Robert S. Carney, (D), Oregon State, 1977, Ob
Lawrence J. Rouse, Jr., (D), Louisiana State, 1969, Op

Assistant Professor:
Mark C. Benfield, (D), Texas A&M, 1991, Ob

Adjunct Professor:
Dubravko Justic, (D), Zagreb, 1989, Op
Nancy N. Rabalais, (D), Texas, 1983, Ob
Harry H. Roberts, (D), Louisiana State, 1969, Gs
Paul W. Sammarco, (D), SUNY, 1978, Zn
Nan D. Walker, (D), Cape Town, 1989, Og

School of Plant, Environmental and Soil Sciences (B,M,D) ☒
104 M. B. Sturgis Hall
Baton Rouge, LA 70803
p. (225) 578-2110
spess@lsu.edu
http://www.spess.lsu.edu
f: https://www.facebook.com/LsuSchoolOfPlantEnvironmental-SoilSciences/
Programs: Plant, Environ Management and Soil Sciences (M,D)
Enrollment (2017): B: 3 (0) M: 3 (3) D: 13 (0)

Professor:
Lewis A. Gaston, (D), Florida, 1987, Sc
Hussein M. Selim, (D), Iowa State, 1971, Sp
Maud M. Walsh, (D), Louisiana State, 1989, GeRcZe
Jim Wang, (D), Iowa State, 1990, Szc

Associate Professor:
Brenda S. Tubana, (D), Oklahoma State, 2007, So

Assistant Professor:
Lisa M. Fultz, Texas Tech, 2012, Sb

Louisiana Tech University

Environmental Science Program (B,M,D) ☒
600 W. Arizona Street
Ruston, LA 71272
p. (318) 257-3972
campbell@latech.edu
Programs: B.S. in Biology, B.S. in Environmental Science, B.S. in Medical Laboratory Science, M.S. in Biology, M.S. and Ph.D. in Molecular Sciences and Nanotechnology

Northwestern State University

Dept of Chemistry and Physics (B)
Natchitoches, LA 71497
p. (318) 357-5501
graham@nsula.edu
http://chemphys.nsula.edu/

Director:
Paul Withey

Professor:
Carol S. Chin, (D)
Kelly Knowlton, (D), Texas A&M, 1991, Gg

South Louisiana Community College

South Louisiana Community College
320 Devalcourt
Lafayette, LA 70506
p. (337) 521-8983
http://www.slcc.cc.la.us

Tulane University

Dept of Earth and Environmental Sciences (B,M,D) ☒
6823 St. Charles Ave.
101 Blessey Hall
New Orleans, LA 70118
p. (504) 865-5198
ees@tulane.edu
http://www.tulane.edu/sse/eens
f: https://www.facebook.com/EENS6823/
Programs: Geology BS, Environmental Earth Science BS, Earth & Env Sciences MS, Earth & Env Sciences PhD
Certificates: GIS Certificates (undergraduate and graduate certificates available)
Enrollment (2015): B: 6 (0) M: 0 (2) D: 0 (2)

Chair:
Torbjörn E. Törnqvist, (D), Utrecht, 1993, Gs

Professor:
Mead A. Allison, (D), SUNY (Stony Brook), 1993, Gs
Karen Haley Johannesson, (D), Nevada, 1993, HgCg

Associate Professor:
Nancye H. Dawers, (D), Columbia, 1997, GcoGm
George C. Flowers, (D), California (Berkeley), 1979, Hw

Associate Scientist:
Nicole Gasparini, (D), MIT, 2003, GmHs

Assistant Professor:
Brent M. Goehring, (D), Columbia, 2010, CcGlm
Kyle Martin Straub, (D), MIT, 2007, GgCg

Lecturer:
Jeffrey G. Agnew, (D), Louisiana State, 2008, Pge

Visiting Professor:
Reda Amer, (D), St Louis, 2011, ZirGe

Emeritus:
Stephen A. Nelson, (D), California (Berkeley), 1979, GivRn

University of Louisiana at Lafayette

School of Geosciences (B,M,D) O☒
611 McKinley St
Lafayette, LA 70504
p. (337) 482-6468
geosciences@louisiana.edu
https://geos.louisiana.edu/
f: UL Lafayette School of Geosciences
t: @ULGeosciences
Programs: Bachelors of Science in Environmental Geology and Petroleum Geology, Master's in Geology, Ph.D. in Earth and Energy Sciences, Bachelors of Science in Environmental Sciences, Master's in Environmental Resource Science
Certificates: GIS Certificate
Administrative Assistant: Nadean S.. Bienvenu
Administrative Assistant: Rachel M. Eastin Hughes
Enrollment (2022): B: 80 (20) M: 30 (16) D: 16 (1)

Director:
Eric C. Ferre', (D), Toulouse, 1989, YmGc

Assistant Director:
Durga Poudel, (D), Georgia, 1998, Ge

Professor:
Gary L. Kinsland, (D), Rochester, 1974, YgeGt
Carl Richter, (D), Tubingen, 1990, Ym

Senior Scientist:
James E. Martin, (D), Washington, 1979, PvGg

Resource Facilitator:
Jim Foret, (M), Iowa State, 1971, Ge

Associate Professor:
Timothy W. Duex, (D), Texas, 1983, HgGex

Geology Program Director:
Brian Schubert, (D), SUNY (Binghamton), 2008, Cg

Field Camp Director:
Raphael Gottardi, (D), Minnesota, 2012, Gcp

Assistant Professor:
Davide Oppo, (D), Bologna, 2012, Gs
Anna Paltseva, (D), CUNY, 2019, Ge
Jorge Villa, (D), Ohio State, 2014, Ge
Rui Zhang, (D), Houston, 2010, Ye

Instructor:
Kristie Cornell, (M), Louisiana (Lafayette), 2003, Gg
Jennifer E. Hargrave, (D), Oklahoma, 2009, PvGrZe

Emeritus:
Brian E. Lock, (D), Cambridge, 1969, Gs

Laboratory Director:
Yingfeng XU, (D), Tulane, 2004, ZgGe

Adjunct Instructor:
William R. Finley, (M), SW Louisiana, 1975, YsZi

University of Louisiana, Monroe

School of Science, Atmospheric Science Program (B)
700 University Avenue
Monroe, LA 71209-0550
p. (318) 342-1822
casehanks@ulm.edu
http://www.ulm.edu/atmos/
Enrollment (2012): B: 43 (9)

Associate Dean of Arts and Sciences:
Michael A. Camille, (D), Texas A&M, 1991, Zy

Professor:
Eric A. Pani, (D), Texas Tech, 1987, As

Associate Professor:
M. Sean Chenoweth, (D), Wisconsin (Milwaukee), 2003, GmZir

Department Head:
Anne T. Case Hanks, (D), Georgia Tech, 2008, AsZe

Assistant Professor:
Ken Leppert, (D), Alabama (Huntsville), As
Todd Murphy, (D), Alabama (Huntsville), As

Station Archeologist/Poverty Point:
Diana M. Greenlee, (D), Washington, 2002, GaCo

University of New Orleans

Dept of Earth and Environmental Sciences (B,M,D) ☒
2000 Lakeshore Drive
New Orleans, LA 70148
p. (504) 280-6325
igeorgio@uno.edu
http://www.uno.edu/geology/

Professor:
William H. Busch, (D), Oregon State, 1981, Ou
Terry L. Pavlis, (D), Utah, 1982, Gt
Denise J. Reed, (D), Cambridge, 1986, GmOn
A. K. Mostofa Sarwar, (D), Indiana, 1983, Ye
Laura F. Serpa, (D), Cornell, 1986, Ys
William B. Simmons, (D), Michigan, 1973, Gz
Ronald K. Stoessell, (D), California (Berkeley), 1977, Cl

Associate Professor:
Kraig L. Derstler, (D), California (Davis), 1985, PgiPv
Mark A. Kulp, (D), Kentucky, 2000, GrmGs

Assistant Professor:
Christopher D. Parkinson, (D), London (UK), 1991, Gp

Adjunct Professor:
Miles O. Hayes, (D), Texas, 1965, Gs
Karen L. Webber, (D), Rice, 1988, Gv
Michael A. Wise, (D), Manitoba, 1987, Gz

Emeritus:
Gary C. Allen, (D), N Carolina, 1968, Gp
Jacqueline Michel, (D), South Carolina, 1980, Cg

Maine

Bates College

Dept of Earth and Climate Science (B) ☒
Carnegie Science Center
44 Campus Avenue
Lewiston, ME 04240-6084
p. (207) 786-6490
bjohnso3@bates.edu
http://www.bates.edu/geology/
f: https://www.facebook.com/groups/243958415633596/
Programs: Geology
Administrative Assistant: Sylvia Deschaine
Enrollment (2018): B: 15 (12)

Chair:
J. Dykstra Eusden, (D), Dartmouth, 1988, Gc

Professor:
Beverly J. Johnson, (D), Colorado, 1995, CsbCl
Michael J. Retelle, (D), Massachusetts, 1985, Gl

Assistant Professor:
Genevieve Robert, (D), Missouri, 2014, CpGiz

Assistant Instructor:
Marita Bryant, (M), Free (Berlin), 1984, Gg

Lecturer:
Raj Saha, (D), N Carolina, 2011

Visiting Professor:
Alice Doughty, Victoria (NZ), 2013, Gl

Emeritus:
Gene A. Clough, (D), Caltech, 1978, Ym

Bowdoin College

Dept of Earth and Oceanographic Science (B) ☒
6800 College Station
Brunswick, ME 04011
p. (207) 725-3628
epeterma@bowdoin.edu
http://www.bowdoin.edu/earth-oceanographic-science/
Programs: Earth & Oceanographic Science
Administrative Assistant: Bridget Spaeth
Enrollment (2022): B: 38 (13)

William R. Kenan Professor of Earth and Oceanographic Science:
Collin Roesler, (D), Washington, 1992, OgpZr

Rusack Professor of Environmental Studies and Earth and Oceanographic Science:
Philip Camill III, (D), Duke, 1999, PebSb

Anne T. and Robert M. Bass Professor of Natural Sciences:
Rachel J. Beane, (D), Stanford, 1997, GzxZe

Chair:
Emily M. Peterman, (D), California (Santa Barbara), 2009, Gtp

Associate Professor:
Michèle LaVigne, (D), Rutgers, 2010, OcPe

Laboratory Instructor:
Cathryn K. Field, (M), Smith, 2000, Ge

Lab Instructor:
Joanne Urquhart, (M), Dartmouth, 1987, Gge

Associate Professor:
Edward P. Laine, (D), MIT, 1977, Gu

Colby College

Department of Geology (B) ☒
5800 Mayflower Hill
Waterville, ME 04901-8858
p. (207) 859-5800
alice.ridky@colby.edu
https://www.colby.edu/geologydept/
t: @colbygeology
Programs: Geology
Administrative Assistant: Alice M. Ridky
Enrollment (2022): B: 17 (2)

Chair:
Tasha L. Dunn, (D), Tennessee, 2008, XmGiz

Associate Professor:
Walter Sullivan, (D), Wyoming, 2007, Gct

Assistant Professor:
Bess Koffman, (D), Maine, 2013, PcClZc
Alejandra C. Ortiz, (D), MIT/WHOI, 2015, Gm

Visiting Professor:
Alena Giesche, (D), Cambridge (UK), 2020, ZcGa
Bruce F. Rueger, (D), Colorado, 2002, PlGg

University of Maine

School of Earth and Climate Sciences (B,M,D) ☒
5790 Bryand Global Sciences Center
Orono, ME 04469-5790
p. (207) 581-2152
Karl.Kreutz@maine.edu
http://www.umaine.edu/earthclimate/
f: https://www.facebook.com/pages/UMaine-School-of-Earth-and-Climate-Sciences/238244500701
Enrollment (2015): B: 42 (6) M: 13 (4) D: 14 (1)

Research Professor:
Edward S. Grew, (D), Harvard, 1971, Gpz

Director:
Karl J. Kreutz, (D), New Hampshire, 1998, Cs

Professor:
George H. Denton, (D), Yale, 1965, Gl
Brenda L. Hall, (D), Maine, 1997, Gl
Scott E. Johnson, (D), James Cook, 1989, Gc
Kirk A. Maasch, (D), Yale, 1989, As
Paul A. Mayewski, (D), Ohio State, 1973, Pe
Aaron E. Putnam, (D), Maine, 2011, Gll
Andrew S. Reeve, (D), Syracuse, 1996, Hw

Associate Professor:
Christopher C. Gerbi, (D), Maine, 2005, Gt

Assistant Professor:
Katherine A. Allen, (D), Columbia, 2013, Gu
Alicia M. Cruz-Uribe, (D), Penn State, 2014, GpCpt
Amanda A. Olsen, (D), Virginia Tech, 2007, Cl
Sean Smith, (D), Johns Hopkins, 2010, Hs

Research Assistant Professor:
Seth W. Campbell, (D), Maine, 2010, Gl

Instructor:
Alice R. Kelley, (D), Maine, 2006, GalGm
Martin G. Yates, (D), Indiana, 1988, EgGzx

Emeritus:
Daniel F. Belknap, (D), Delaware, 1979, GusGa
Joseph T. Kelley, (D), Lehigh, 1980, Gua
Peter O. Koons, (D), ETH (Switzerland), 1982, Gt
Stephen A. Norton, (D), Harvard, 1967, Cl

Research Assistant Professor:
Sean Birkel, (D), Maine, 2010, Gl

School of Marine Sciences (B,M,D) O
5706 Aubert Hall, Rm 360
Orono, ME 04469-5706
p. (207) 581-4381
fchai@maine.edu

http://www.umaine.edu/marine/
Enrollment (2010): B: 145 (38) M: 56 (0) D: 9 (0)

University of Maine, Farmington

Dept of Geology (B) ☒
173 High Street
Farmington, ME 04938
p. (207) 778-7402
reusch@maine.edu
http://sciences.umf.maine.edu
Programs: Geology (B); Earth and Environmental Science (B)
Certificates: GIS Certificate
Enrollment (2022): B: 12 (4)
Professor:
Douglas N. Reusch, (D), Maine, 1998, GtCgOu
Associate Professor:
Julia F. Daly, (D), Maine, 2002, GmHsGu
Assistant Professor:
Marlon Jean, (D), Utah State, 2012, GizGp

University of Maine, Presque Isle

Div of Mathematics & Science (B)
181 Main Street
Presque Isle, ME 04769
p. (207) 768-9482
kevin.mccartney@umpi.edu
http://www.umpi.edu/
Department Secretary: Connie Leveque
Chair:
Michael Knopp
Professor:
Kevin McCartney, (D), Florida State, 1988, Pm

Maryland

College of Southern Maryland

Biological and Physical Sciences (A)
8730 Mitchell Rd
La Plata, MD 20646
p. (301) 934-7841
askme@csmd.edu
http://www.csmd.edu/bio
Professor:
Jean M. Russ, (M), Kutztown, Zey

Community College of Baltimore County, Catonsville

School of Mathematics & Science (A) ☒
800 S. Rolling Road
MASH 015B
Catonsville, MD 21228
p. (443) 840-5935
DLudwikoski@ccbcmd.edu
http://www.ccbcmd.edu/math_science/geology.html
Programs: Earth Science; Geology
Certificates: GIS
Administrative Assistant: Annjeannette Black
Enrollment (2015): A: 3 (0)
Associate Professor:
David J. Ludwikoski, (M), Toledo, 1993, ZegZn

Frederick Community College

Science Dept (A)
7932 Opossumtown Pike
Frederick, MD 21702
p. (301) 846-2510
shsmith@frederick.edu
http://www.frederick.edu/courses_and_programs/dept_science.aspx
Professor:
Natasha Cleveland, (M), Utah, 2001, Zge

Frostburg State University

Dept of Geography (B)
101 Braddock Rd.
Frostburg, MD 21532
p. (301) 687-4369
jsaku@frostburg.edu
http://www.frostburg.edu/dept/geog/
Administrative Assistant: Gale Yutzy
Enrollment (2010): B: 65 (16)
Chair:
Craig L. Caupp, (D), Utah State, 1986, ZnHs
Professor:
Henry W. Bullamore, (D), Iowa, 1978, Zn
Francis L. Precht, (D), Georgia, 1989, Zy
James C. Saku, (D), Saskatchewan, 1995, Zn
Associate Professor:
Phillip Allen, (D), Coventry (UK), 2005, GmePe
Fritz Kessler, (D), Kansas, 1999, Zr
George W. White, (D), Oregon, 1994, Zn
Assistant Professor:
David L. Arnold, (D), Indiana, 1994, As
Matthew E. Ramspott, (D), Kansas, 2006, Zr

Howard Community College

Science, Engineering, and Technology Division
10901 Little Patuxent Pkwy
Columbia, MD 21044
p. (443) 518-1000
pturner@howardcc.edu
http://www.howardcc.edu/programs-courses/academics/academic-divisions/science-engineering-technology/sciences/

Johns Hopkins University

Dept of Environmental Health & Engineering (B,M,D)
313 Ames Hall
34th & Charles Streets
Baltimore, MD 21218-2681
p. (410) 516-7092
mwkarp@jhu.edu
http://www.jhu.edu/dogee
Professor:
William P. Ball, (D), Stanford, 1989, Zn
Edward Bouwer, (D), Stanford, 1982, RwHws
Grace S. Brush, (D), Harvard, 1956, PleGn
Hugh Ellis, (D), Waterloo, 1984, Zn
Steve H. Hanke, (D), Colorado, 1969, Zn
Benjamin F. Hobbs, (D), Cornell, 1983, Zn
A. Lynn Roberts, (D), MIT, 1991, Zn
Erica Schoenberger, (D), California (Berkeley), 1984, Zn
Alan T. Stone, (D), Caltech, 1983, ClSc
Peter W. Wilcock, (D), MIT, 1987, Gm
Associate Professor:
Markus Hilpert, (D), Zn
Assistant Professor:
Kai Loon Chen, (D), Yale, 2008, Zn
Seth Guikema, (D), Stanford, 2003, Zn
Catherine Norman, (D), California (Santa Barbara), 2005, Zn
Lecturer:
Hedy Alavi, (D), Ohio State, 1983, Zn
Emeritus:
John J. Boland, (D), Johns Hopkins, 1973, Zn
Charles R. O'Melia, (D), Michigan, 1963, Zn
Eugene D. Shchukin, (D), Moscow State (Russia), 1958, Zn
Senior Academic Program Coordinator:
Adena Rojas, (M), Zn

The Morton K. Blaustein Dept of Earth & Planetary Sciences (B,M,D)
3400 N Charles Street
301 Olin Hall
Baltimore, MD 21218
p. (410) 516-7135
sabine@jhu.edu
http://eps.jhu.edu/
Enrollment (2013): D: 33 (5)
Chair:
Thomas W.N. Haine, (D), Southampton (UK), 1993, OpgAs
Professor:
Anand Gnanadesikan, (D), MIT/WHOI, 1994, OpcAs
Peter L. Olson, (D), California (Berkeley), 1977, Yg
Darrell F. Strobel, (D), Harvard, 1969, AsZnn
Dimitri A. Sverjensky, (D), Yale, 1980, Cl
Darryn W. Waugh, (D), Cambridge, 1991, As

Assistant Professor:
Sarah Horst, (D), Arizona, 2011, XaAcXb
Naomi Levin, (D), Utah, 2008, Gs
Kevin Lewis, (D), Caltech, 2009
Benjamin Zaitchik, (D), Yale, 2006, As
Research Professor:
Katalin Szlavecz, (D), Eotvos Lorand, 1981, Pi

Montgomery College

Dept of Physics, Engineering & Geosciences (A)
51 Mannakee Street
Rockville, MD 20850
p. (301) 279-5230
alan.cutler@montgomerycollege.edu
https://www.montgomerycollege.edu/Departments/phengrv/
Administrative Assistant: Mary (Deep) McGregor
Professor:
Alan Cutler, (D), Geology, PgRh
Instructional Laboratory Coordinator for Geosciences:
Kimberly Kelly, (B), Pittsburg State, Zn

St. Charles Community College

Geology
4601 Mid Rivers Mall Drive
Cottleville, MD 63376
p. (636) 922-8000
jmatheney-rood@stchas.edu
https://www.stchas.edu/academics/departments/Geology

Towson University

Dept of Physics, Astronomy & Geosciences (B) ☒
8000 York Road
Towson, MD 21252-0001
p. (410) 704-3020
dschaefer@towson.edu
https://www.towson.edu/fcsm/departments/physics/
Programs: Geology, Earth Space Science (for secondary education majors as well as for science majors), Environmental Science, Physics
Enrollment (2020): B: 66 (12)
Professor:
Rachel J. Burks, (D), Texas, 1985, Gc
Joel Moore, (D), Penn State, 2008, ClsSc
David A. Vanko, (D), Northwestern, 1982, Gxz
Associate Professor:
Wendy Nelson, (D), Penn State, 2009, GiCc
Assistant Professor:
Michelle Casey, (D), Yale, 2011, Pe
Lecturer:
Andrew Hawkins, (D), Virginia Tech, 2017, Pg
Eriks Perkons, (M), Penn State, 2016, Ge

United States Naval Academy

Dept of Oceanography (B) ☒
572C Holloway Road
Annapolis, MD 21402-5026
p. (410) 293-6550
natunewi@usna.edu
https://www.usna.edu/Oceanography
f: https://www.facebook.com/usnaoceanography
Administrative Assistant: Ira Ostrowski
Enrollment (2020): B: 133 (32)
Chair:
Cecily N. Steppe, (D), Delaware, 2001, Obp
Professor:
Bradford S. Barrett, (D), Oklahoma, 2007, As
Peter L. Guth, (D), MIT, 1980, GcZiy
CAPT:
Elizabeth R. Sanabia, (D), Naval Postgrad Sch, 2010, Am
Associate Professor:
Gina R. Henderson, (D), Delaware, 2010, As
Andrew C. Muller, (D), Old Dominion, 1999, OpnOu
Joseph P. Smith, (D), Massachusetts (Boston), 2007, CmOg
CDR:
Shawn Gallaher, (D), Naval Postgrad Sch
LCDR:
Matthew Burich, (M)
CDR:
Francis Carmody, (M)
Michael Cornelius, (M)
Allon Turek, (M)
Instructor:
Brianna Tracy, (M)
Related Staff:
Benjamin Hickman, (M)

University of Maryland

Dept of Geographical Studies
College Park, MD 20742
p. (301) 405-1600
dubayah@umd.edu
https://geog.umd.edu/

Dept of Geology (B,M,D) ☒
Geology Building (#237)
8000 Regents Drive
College Park, MD 20742
p. (301) 405-4082
geology@umd.edu
http://www.geol.umd.edu
f: https://www.facebook.com/UMDGeology
Administrative Assistant: Dorothy Brown
Enrollment (2018): B: 54 (12) M: 0 (3) D: 33 (2)
Chair:
Richard J. Walker, (D), SUNY (Stony Brook), 1984, CcXcCg
Affiliate Professor:
Fernando Mirales-Wilhelm, (D)
Raghuram G. Murtugudde, (D), Columbia, 1994, ObZr
Jessica Sunshine, (D), Brown, 1993
Ning Zeng, (D), Arizona, 1994, AsCgGl
Professor:
Michael Brown, (D), Keele, 1975, GxpGt
James Farquhar, (D), Alberta, 1995, Cs
Alan J. Kaufman, (D), Indiana, 1990, Csg
Daniel Lathrop, (D), Texas, 1991, Yg
William F. McDonough, (D), Australian Nat, 1988, Cg
Laurent G.J. Montesi, (D), MIT, 2002, YgXyGt
Wenlu Zhu, (D), SUNY (Stony Brook), 1996, YrNrHg
Senior Research Scientist:
Philip M. Piccoli, (D), Maryland, 1992, Cg
Igor Puchtel, (D), Russian Acad of Sci, 1992, CcgGi
Associate Professor:
Ricardo Arevalo, Jr, (D), Maryland, 2010
Michael N. Evans, (D), Columbia, 1999, PeAsCs
Sujay Kaushal, (D), Colorado, 2003, ZuGeHs
Sarah Penniston-Dorland, (D), Johns Hopkins, 2005, Gp
Karen L. Prestegaard, (D), California (Berkeley), 1982, Hgw
Associate Research Scientist:
Richard Ash, (D), Open, 1990, Ca
Affiliate Assistant Professor:
Derrick Lampkin, (D), GlAsZr
Assistant Professor:
Mong-Han Huang, (D), California (Berkeley), 2017
Vedran Lekic, (D), California (Berkeley), 2009, Ysg
Megan E. Newcombe, (D), Caltech, 2017
Nicholas C. Schmerr, (D), Arizona State, 2008, XyYs
Research Associate:
Jabrane Labidi, (D), Inst Physique du Globe de Paris, 2012, Cg
Research Assistant Scientist:
Katherine R. Bermingham, (D), Muenster (Germany), 2011, XcCcg
Shuiwang Duan, (D), Tulane, 2005, HsZu
Principal Lecturer:
Thomas R. Holtz, Jr., (D), Yale, 1992, PvgPg
John W. Merck, Jr., (D), Texas, 1997, Pv
Lecturer:
Tracey Centorbi, (B), Maryland, 2003, Cg
Adjunct Professor:
Anat Shahar, (D), California (Los Angeles), 2008, CspCa
Adjunct Associate Professor:
Elizabeth Cottrell, (D), Columbia, 2004, CpgGx
Adjunct Professor:
Yingwei Fei, (D), CUNY, 1989, CpGxCg
Jeffrey Plesia, (D)
Deborah Smith, (D), California (San Diego), 1985

Professor Emeritus:
Philip A. Candela, (D), Harvard, 1982, CpgEg
Galt Siegrist, (D)
Emeritus Professor:
Ann G. Wylie, (D), Columbia, 1972, Gz
Emeritus Affiliate Research Professor:
George Helz, (D), Penn State, 1970, HsGeCg
Associate Professor Emeritus:
Peter B. Stifel, (D), Utah, 1964, Pg
Adjunct Professor Emeritus:
Roberta L. Rudnick, (D), Australian Nat, 1988, CgsCt
Faculty Research Assistant:
Todd Karwoski, (B), Maryland, 2005, Gg
Valentina Puchtel, (M), Moscow Geol Prospect Acad, 1983

Dept of Plant Science & Landscape Architecture (B,M,D)
2102 Plant Sciences Building
College Park, MD 20742-4432
p. (301) 405-4356
asmurphy@umd.edu
http://www.psla.umd.edu/
Professor:
Christopher Walsh, (D), Cornell, 1980, Zn
Associate Professor:
Gary D. Coleman, (D), Nebraska, 1989, Zn
Jack B. Sullivan, (M), Virginia, 1980, Zu
Program Management Specialist:
Kathy Hunt, Zn
Coordinator:
Sue Burk, Zn

Marine-Estuarine-Environmental Sciences Graduate Program (M,D) ☒
1213 HJ Patterson Hall
University of Maryland
College Park, MD 20742
p. (301) 405-6938
mees@umd.edu
http://www.mees.umd.edu
Enrollment (2013): M: 6 (0) D: 12 (2)
Professor:
Shenn-Yu Chao, (D), North Carolina State, 1979, Op
Keith N. Eshleman, (D), MIT, 1985, Hg
Thomas R. Fisher, Jr., (D), Duke, 1975, ObZiCl
Patricia M. Gilbert, (D), Harvard, 1982, Ob
Lawrence P. Sanford, (D), MIT, 1984, Op
Diane Stoecker, (D), SUNY (Stony Brook), 1979, Ob
Associate Professor:
William Boicourt, (D), Johns Hopkins, 1973, Op
James Carton, (D), Princeton, 1983, Op
Micheal S. Kearney, (D), Ontario, 1981, Gu
Karen L. Prestegaard, (D), California (Berkeley), 1982, Hw
Research Associate Professor:
Jeffery C. Cornwell, (D), Alaska, 1983, Oc
Assistant Professor:
Mark S. Castro, (D), Virginia, 1991, As
Raleigh Hood, (D), California (San Diego), 1990, Ob
Alba Torrents, (D), Johns Hopkins, 1992, Sb
Research Associate Professor:
Todd M. Kana, (D), Harvard, 1982, Ob

Massachusetts

Amherst College

Dept of Geology (B) ☒
P.O. Box 2238
Amherst, MA 01002-5000
p. (413) 542-2233
dbhutton@amherst.edu
https://www.amherst.edu/academiclife/departments/geology
Enrollment (2022): B: 5 (5)
Professor:
John T. Cheney, (D), Wisconsin, 1975, Gi
Tekla A. Harms, (D), Arizona, 1986, Gt
Anna M. Martini, (D), Michigan, 1997, Cl
Associate Professor:
David S. Jones, (D), Harvard, 2009, GsClGr
Assistant Professor:
Rachael E. Bernard, (D), Texas, 2018, Gc
Victor E. Guevara, (D), Virginia Tech, 2017, GpiGg
Nicholas D. Holschuh, (D), Penn State, 2016, GlYg
Emeritus:
Margery C. Coombs, (D), Columbia, 1971, Pv
Peter D. Crowley, (D), MIT, 1985, Gc

Bard College at Simon's Rock

Bard College at Simon's Rock
84 Alford Rd.
Great Barrington, MA 01230
p. (413) 644-4400
admin@simons-rock.edu
http://www.simons-rock.edu/

Bentley University

Dept of Natural and Applied Sciences (B) ☒
175 Forest Street
Waltham, MA 02452-4705
p. (781) 891-2980
dszymanski@bentley.edu
https://www.bentley.edu/academics/departments/natural-and-applied-sciences
Programs: BA in Sustainability Science
Enrollment (2018): B: 1 (0)
Dean of Arts and Sciences:
Rick Oches, (D), Massachusetts, 1994, ZcePe
Professor:
P. Thompson Davis, (D), Colorado, 1980, GlmZc
Associate Professor:
David Szymanski, (D), Michigan State, 2007, GifCg
Lecturer:
George Fishman, (M), Boston, 1993, XgbXc
Nicole Hill, (D), Cornell, 2016, Ge
Betsy Stoner, (D), Florida Intl, 2014, ObPe
Adjunct Professor:
Janette Gartner, (M), Massachusetts, 2000, Hy

Berkshire Community College

Environmental and Life Science Dept (A)
Pittsfield, MA 01201
p. (413) 236-4601
tflanagan@berkshirecc.edu
Chair:
Clifford D. Myers, (D), Maine, Zn
Professor:
Timothy Flanagan, (M), Antioch, 1983, Zei
Thomas F. Tyning, (M), Zn
Charles E. Weinstein, (M), Wisconsin, Zn
Emeritus:
Richard L. Ferren, (M), Louisiana State, Zn
George Hamilton, (M), North Adams State Coll, Zn
Mary R. Mercuri, (M), Catholic

Boston College

Dept of Earth & Environmental Sciences (B,M,D) ☒
140 Commonwealth Avenue
213 Devlin Hall
Chestnut Hill, MA 02467-3809
p. (617) 552-3640
earthenvsci@bc.edu
https://www.bc.edu/bc-web/schools/mcas/departments/eesc.html
Programs: Geological sciences, Environmental geoscience, Environmental studies
Enrollment (2020): B: 43 (15) M: 14 (8) D: 3 (0)
chair:
Noah Snyder, (D), MIT, 2001, GmHg
Professor:
Ethan Baxter, (D), California (Berkeley), 2000, Ccg
Mark Behn, (D), MIT/WHOI, 2002, YgrGt
John E. Ebel, (D), Caltech, 1981, Ys
Yi Ming, (D), Princeton, 2003, ApZo
Laura Steinberg, (D), Ng
Hanqin Tian, (D), SUNY ESF, 1996, Zo
Associate Professor:
Alan L. Kafka, (D), SUNY (Stony Brook), 1980, Yg
Seth C. Kruckenberg, (D), Minnesota, 2009, Gct

Tara Pisani-Gareau, (D), California (Santa Cruz), 2008, Ge
Jeremy D. Shakun, (D), Oregon State, 2010, Gg
Assistant Professor:
Hilary I. Palevsky, (D), Washington, 2016, Oc
Xingchen Wang, (D), Princeton, 2016, Cb
Kat Wilson, (D), Texas (Austin), 2022, GsmOn
Lab Coordinator:
Kenneth G. Galli, (D), Massachusetts, 2003, GsdGg
Adjunct Professor:
Paul K. Strother, (D), Harvard, 1980, PlbPg
Emeritus:
J. Christopher Hepburn, (D), Harvard, 1972, Gpt
Gail C. Kineke, (D), Washington, 1993, On

Weston Observatory
381 Concord Road
Weston, MA 02493
p. (617) 552-8300
weston.observatory@bc.edu
http://www.bc.edu/westonobservatory
Director:
Alan L. Kafka, (D), SUNY (Stony Brook), 1980, Ys
Science Education:
Michael Barnett, (D), Indiana, 2003, Ze
Senior Scientist:
John E. Ebel, (D), Caltech, 1981, Yg
Seismology, Seismic Network Development:
Michael Hagerty, (D), California (Santa Cruz), 1998, Ys
Seismic Analyst and Educational Seismologist:
Anastasia Moulis, (M), Boston, 2003, Ys
Research Scientist:
Seth Kruckenberg, (D), Minnesota, 2009, Gc
Research Scientist:
John J. Cipar, (D), Caltech, 1981, Ys
Research Associate:
John H. Beck, (D), Boston, 1998, Pb
Adjunct Professor:
Paul L. Strother, (D), Harvard, 1980, PlbPm
Alfredo Urzua, (D), MIT, 1981, Ng
Visiting Professor:
Vincent Murphy, (M), Boston, 1957, Yg
Emeritus:
J. Christopher Hepburn, (D), Harvard, 1972, Gg
James W. Skehan, (D), Harvard, 1953, Gc

Boston University

Center for Remote Sensing
685 Commonwealth Avenue
Room 433
Boston, MA 02215
p. (617) 353-9709
crsadmin@bu.edu
https://www.bu.edu/remotesensing/
Administrative Assistant: Emily P. Johnson
Director:
Farouk El-Baz, (D), Missouri (Rolla), 1964, Zr
Professor:
Sucharita Gopal, (D), California (Santa Barbara), 1988, Zi
Alan Strahler, (D), Johns Hopkins, 1969, Zr
Curtis E. Woodcock, (D), California (Santa Barbara), 1986, Zr
Research Associate Professor:
Magaly Koch, (D), Boston, 1993, HyZri
Associate Professor:
Mark A. Friedl, (D), California (Santa Barbara), 1994, Zr
Kenneth L. Kvamme, (D), California (Santa Barbara), 1983, Ga
Guido D. Salvucci, (D), MIT, 1994, Hq
Research Associate Professor:
Cordula Robinson, (D), Univ Coll (UK), 1991, Gm
Crystal Schaaf, (D), Boston, 1994, Zr
Research Associate:
Eman Ghoneim, (D), Southampton (UK), 2002, Gm
Geology Librarian:
Nasim Momen, Zn

Dept of Earth & Environment (B,M,D)
675 Commonwealth Avenue
Boston, MA 02215
p. (617) 353-2525
gdsalvuc@bu.edu
http://www.bu.edu/earth/
Enrollment (2015): B: 70 (15) M: 15 (10) D: 50 (8)
Chair:
David R. Marchant, (D), Edinburgh, 1994, Gm
Professor:
Bruce Anderson, (D), Scripps, 1998, AsOp
James Lawford Anderson, (D), Gi
Duncan M. FitzGerald, (D), South Carolina, 1977, OnGsu
Mark Friedl, (D), California (Santa Barbara), 1993, Zr
Sucharita Gopal, California (Santa Barbara), 1988, Zi
Tony Janetos, (D), 1980, Zn
Richard Murray, (D), California (Berkeley), 1991, Cm
Ranga Myneni, (D), Antwerp (Belgium), 1985, Zr
Nathan Phillips, (D), Duke, 1997, Zy
Guido D. Salvucci, (D), MIT, 1994, Hq
Curtis Woodcock, (D), California (Santa Barbara), 1986, Zr
Associate Professor:
Rachel Abercrombie, (D), Reading (UK), 1991, Gt
Michael Dietze, (D), Duke, 2006, SfZr
Sergio Fagherazzi, (D), Padua, 1999, GmOnGs
Robinson Fulweiler, (D), Rhode Island, 2007, Cm
Lucy Hutyra, (D), Harvard, 2007, AsZr
Andrew Kurtz, (D), Cornell, 2000, Clg
Assistant Professor:
Dan Li, (D), Princeton, 2013, HqAs
Christine Regalla, (D), Penn State, 2013, GtcGm
Diane Thompson, Arizona, 2013, ObPe
Research Associate:
Farouk El-Baz, (D), Missouri, 1964, Zr

Bridgewater State University

Dept of Geological Sciences (B)
Conant Science Building
Room 131
Bridgewater, MA 02325
p. (508) 531-1390
k1lawlor@bridgew.edu
https://www.bridgew.edu/department/geological-sciences
Programs: Geological Sciences; Environmental Geosciences; Earth Science
Administrative Assistant: Kiera Lawlor
Enrollment (2022): B: 41 (9)
Chairperson:
Michael A. Krol, (D), Lehigh, 1996, GzxGt
Professor:
Robert D. Cicerone, (D), MIT, 1991, YsgXy
Richard L. Enright, (D), Rutgers, 1969, HgZrEg
Peter J. Saccocia, (D), Minnesota, 1991, CqsCm
Associate Professor:
Christine M. Brandon, (D), Massachusetts, 2015, Gs

Bristol Community College

Div of Mathematics, Science and Engineering (A)
777 Elsbree Street
Fall River, MA 02720
p. (508) 678-2811
John.Ahola@bristolcc.edu
Instructor:
John Ahola, Gg

Cape Cod Community College

Environmental Technology Program
2240 Iyannough Rd
West Barnstable, MA 02668
p. (508) 362-2131 x4468
admiss@capecod.edu
http://www.capecod.edu/web/natsci/env

Fitchburg State University

Earth and Geographic Sciences (B)
160 Pearl Street
Fitchburg, MA 01420-2697
p. (978) 665-4636
egordon3@fitchburgstate.edu
http://www.fitchburgstate.edu/academics/academic-departments/earth-and-geographic-sciences-dept/

f: https://www.facebook.com/FitchburgStateUniversityEGS
Programs: Environmental and Earth Science; Geographic Science and Technology
Enrollment (2018): B: 43 (5)
Professor:
Jane Huang, (D), Zi
Associate Professor:
Elizabeth S. Gordon, (D), OgAsZg
Assistant Professor:
Elyse Clark, (D), Cq
Reid A. Parsons, (D), XgHs

Hampshire College

School of Natural Science (B) ☒
Amherst, MA 01002
p. (413) 582-5373
sroof@hampshire.edu
Professor:
Steven Roof, (D), Massachusetts, 1995, ZiPe
Associate Professor:
Christina Cianfrani, (D), Vermont, Hg

Harvard University

Dept of Earth and Planetary Sciences (B,D) ● ☒
Hoffman Laboratory
20 Oxford Street
Cambridge, MA 02138-2902
p. (617) 495-2351
shaw@eps.harvard.edu
http://www.eps.harvard.edu
f: https://www.facebook.com/Harvard-University-Earth-and-Planetary-Sciences-6322073679/
Programs: Earth and Planetary Sciences
Enrollment (2018): B: 34 (6) D: 58 (10)
Chair:
John Shaw, (D), Princeton, 1993, Gt
Affiliated Faculty with EPS:
Naomi Oreskes, (D), Stanford, 1990, Rh
Professor:
James G. Anderson, (D), Colorado, 1970, As
Jeremy Bloxham, (D), Cambridge, 1985, Yg
Brian F. Farrell, (D), Harvard, 1981, Am
John P. Holdren, (D), Stanford, 1970, Zn
Peter Huybers, (D), MIT, 2004, Pe
Miaki Ishii, (D), Harvard, 2003, Yg
Daniel J. Jacob, (D), Caltech, 1985, As
Stein B. Jacobsen, (D), Caltech, 1980, Cc
David T. Johnston, (D), Maryland, 2007, CgPg
Andrew H. Knoll, (D), Harvard, 1977, Pb
Zhiming Kuang, (D), Caltech, 2003, As
Charles H. Langmuir, (D), SUNY (Stony Brook), 1980, Cg
Scot T. Martin, (D), Caltech, 1995, Cg
James T. McCarthy, (D), Scripps, 1971, Ob
Michael B. McElroy, (D), Queen's, 1962, As
Brendan Meade, (D), MIT, 2004, Yg
Jerry X. Mitrovica, (D), Toronto, 1991, Yg
Ann Pearson, (D), MIT/WHOI, 2000, Cb
James R. Rice, (D), Lehigh, 1964, Ygu
Daniel P. Schrag, (D), California (Berkeley), 1993, Cg
Eli Tziperman, (D), MIT/WHOI, 1987, Op
Steven C. Wofsy, (D), Harvard, 1971, As
Affiliated Faculty with EPS:
Robin Wordsworth, (D), Oxford, 2008, Xg
Assistant Professor:
Marine Denolle, (D), Stanford, 2014, Ys
Rebecca A. Fischer, (D), Chicago, 2015, Gy
Roger Fu, (D), MIT, 2015, YmXym
Kaighin McColl, (D), MIT, 2016, Hg
Visiting Professor:
Carl Wunsch, (D), MIT, 1966, Op
Emeritus:
Charles W. Burnham, (D), MIT, 1961, GzyGe
Paul F. Hoffman, (D), Johns Hopkins, 1970, Gc

Massachusetts Institute of Technology

Dept of Earth, Atmospheric, & Planetary Sciences (B,M,D) ● ☒
77 Massachusetts Avenue, 54-918
Cambridge, MA 02139
p. (617) 253-2127
eapsinfo@mit.edu
http://eapsweb.mit.edu/
f: https://www.facebook.com/EAPS.MIT
t: @eapsmit
Programs: Earth, Atmospheric and Planetary Sciences (B, M, D); Atmospheric Science (M,D); Climate Science (M,D); Geology (M, D); Geochemistry (M, D); Geobiology (M, D); Geophysics (M, D); Planetary Sciences (M, D)

Enrollment (2020): B: 21 (5) M: 0 (5) D: 166 (20)
Vice-President for Research:
Maria T. Zuber, (D), Brown, 1986, Xy
Department Head:
Robert van der Hilst, (D), Utrecht, 1990, Ys
Professor:
Edward A. Boyle, (D), MIT, 1976, Oc
Dara Entekhabi, (D), MIT, 1990, Hg
Raffaele Ferrari, (D), Scripps, 2000, OpAsOb
Glenn R. Flierl, (D), Harvard, 1975, Op
Michael Follows, (D), E Angola (UK), 1991, Op
Timothy L. Grove, (D), Harvard, 1976, Gi
Bradford H. Hager, (D), Harvard, 1978, Ys
Thomas A. Herring, (D), MIT, 1983, Yd
Paola M. Malanotte-Rizzoli, (D), California (San Diego), 1978, Op
John C. Marshall, (D), Imperial Coll (UK), 1980, Op
Taylor Perron, (D), California (Berkeley), 2006, Gm
Ronald G. Prinn, (D), MIT, 1971, As
Daniel H. Rothman, (D), Stanford, 1986, Yg
Leigh H. Royden, (D), MIT, 1982, Gt
Sara Seager, (D), Harvard, 1999, Xy
Susan Solomon, (D), California (Berkeley), 1981, As
Roger Summons, (D), New South Wales, 1972, CobCs
Jack Wisdom, (D), Caltech, 1981, Zn
Senior Research Scientist:
William Durham, (D), MIT, 1975, Cg
Michael Fehler, (D), MIT, 1979, Yg
C. Adam Schlosser, (D), Maryland, 1995, As
Principal Research Scientist:
Nilanjan Chatterjee, (D), CUNY, 1989, Cg
Stephanie Dutkiewicz, (D), Rhode Island, 1997, Cm
Eduardo Andrade Lima, (D), Catholic (Brazil), Ym
Nori Nakata, (D), Colorado Mines, 2013
Sai Ravela, (D), Massachusetts, 2002, Zn
Ryan Woosley, (D), Miami, 2012
Associate Professor:
Tanja Bosak, (D), Caltech, 2004, Pg
Kerri Cahoy, (D), Stanford, 2008, Zn
Laurent Demanet, (D), Caltech, 2006, Zn
Gregory Fournier, (D), Connecticut, 2009, Pg
Colette Heald, (D), Harvard, 2005, As
Oliver Jagoutz, (D), ETH (Switzerland), 2004, GitGc
Ruben Juanes, (D), California (Berkeley), 2003, Zn
David McGee, (D), Columbia, 2009, Cl
Paul O'Gorman, (D), Caltech, 2004, As
Shuhei Ono, (D), Penn State, 2001, PgCs
Noelle Selin, (D), Harvard, 2007, As
Assistant Professor:
Andrew Babbin, (D), Princeton, 2014, Cb
Kristin Bergmann, (D), Caltech, 2013, GsCl
Camilla Cattania, (D), GFZ German Res Ctr Geosci, 2015, Ys
Timothy W. Cronin, (D), MIT, 2014, As
Julien de Wit, (D), MIT, 2014, Zn
William Frank, (D), Institut de Physique du Globe de Paris, 2014
Wanying Kang, (D), Harvard, AsZo
Brent Minchew, (D), Caltech, 2016, Yg
Matej Pec, (D), Basel (Switzerland), 2012, Nr
Gaia Stucky de Quay, (D), Imperial (London), ZgGms
Richard Teague, (D), Max-Planck Astronomy, XsZn
Senior Lecturer:
Lodovica Illari, (D), Imperial Coll (UK), 1982, Am
Emeritus:
Richard P. Binzel, (D), Texas, 1986, Xm
B. Clark Burchfiel, (D), Yale, 1961, Gc
Charles C. Counselman, III, (D), 1969
Kerry A. Emanuel, (D), MIT, 1978, AmRn

J. Brian Evans, (D), MIT, 1978, Yx
Frederick A. Frey, (D), Wisconsin, 1967, CtGvi
Richard S. Lindzen, (D), Harvard, 1964, Am
F Dale Morgan, (D), MIT, 1981, Yg
Gordon H. Pettengill, (D), California (Berkeley), 1955, Zg
Raymond A. Plumb, (D), Manchester (UK), 1972, As
M. Gene Simmons, (D), Harvard, 1962, Yg
John B. Southard, (D), Harvard, 1966, Gs
Peter H. Stone, (D), Harvard, 1964, As
M Nafi Toksoz, (D), Caltech, 1963, Ys
Carl I. Wunsch, (D), MIT, 1967, Opo

Principal Research Engineer:
Christopher Hill, (D), Zn

Mount Holyoke College

Dept of Geology (B)
50 College Street
Clapp Laboratory #304
South Hadley, MA 01075-6419
p. (413) 538-2278
rforjwuo@mtholyoke.edu
http://www.mtholyoke.edu/acad/geology
f: https://www.facebook.com/groups/mhcgeoalums/
Administrative Assistant: Rhodaline Forjwuor
Enrollment (2016): B: 22 (14)

Professor:
Steven R. Dunn, (D), Wisconsin, 1989, Gp
Girma Kebbede, (D), Syracuse, 1981, Zy
Mark McMenamin, (D), California (Santa Barbara), 1984, PgGst
Thomas L. Millette, (D), Clark, 1989, ZriGm
Alan Werner, (D), Colorado, 1988, GleGm

Associate Professor:
Michelle J. Markley, (D), Minnesota, 1998, GcgGt

Assistant Professor:
Serin D. Houston, (D), Syracuse, 2012, Zyn

Geoprocessing Lab Manager:
Eugenio J. Marcano, (D), Cornell, 1994, ZiSoZn

Visiting Assistant Professor:
Samuel Tuttle, (D), Boston, 2015, Gq

Emeritus:
Martha M. Godchaux, (D), Oregon, 1969, Gv

Laboratory Director:
Penny M. Taylor, (M), SUNY (Oneonta), 2000, Gg

Geology Technician:
Gerard Marchand, (B), Westfield State, 1990, Gga

Northeastern University

Dept of Marine and Environmental Sciences (B,M,D) O☒
14 Holmes Hall
360 Huntington Ave
Boston, MA 02115
p. (617) 373-3176
environment@neu.edu
http://www.northeastern.edu/mes
Programs: Environmental Science (B); Environmental Studies (B); Ecology, Evolution, and Marine Biology (B), Marine Biology (B,M,D); Ecology (D); Evolution (D)
Enrollment (2018): B: 591 (154) M: 18 (1) D: 65 (13)

Professor:
Joseph Ayers, (D), California (Santa Cruz), 1975, Zn
Richard H. Bailey, (D), N Carolina, 1973, Pi
William Detrich, (D), Yale, Zn
Brian Helmuth, (D), Washington, 1997, OnZcRc
Mark Patterson, (D), Harvard, 1995, Zn
Hanumant Singh, (D), MIT, 1995, Zn

Chair:
Geoffrey Trussell, (D), William & Mary, 1998, Zn

Associate Professor:
Jonathan Grabowski, (D), North Carolina, 2012, GuEg
Malcolm D. Hill, (D), California (Santa Cruz), 1979, ZiGze
Justin Ries, (D), Johns Hopkins, 2005, Cm
Rebeca Rosengaus, (D), Boston, 1993, Zn
Steven Scyphers, (D), S Alabama, 2012, Zn
Aron Stubbins, (D), Newcastle upon Tyne, 2001, CgbZc

Associate Chair:
Jennifer Bowen, (D), Boston, Zn

Assistant Professor:
Loretta Fernandez, (D), MIT, 2010, Co
Tarik Gouhier, (D), McGill, Zn
Randall Hughes, (D), California (Davis), 2006, Zn
David Kimbro, (D), California, Zn
Kathleen Lotterhos, (D), Florida State, 2011, Zn
Amy Mueller, (D), MIT, 2012, Zn
Samuel Munoz, (D), Wisconsin, 2015, GsPcRn
Samuel Scarpino, (D), Texas, Zn
Steve Vollmer, (D), Harvard, Zn

Lecturer:
Daniel C. Douglass, (D), Wisconsin, 2005, GlSdPc
Tara Duffy, (D), SUNY (Stony Brook), Ob
Stephanie Eby, (D), Syracuse, 2010, Zn

Emeritus:
Donald Cheney, (D), S Florida, Zn
Gwilym Jones, (D), Indiana State, Zn
Peter S. Rosen, (D), William & Mary, 1976, OnZu
Martin E. Ross, (D), Idaho, 1978, Gie

Business and Operations Manager:
Elizabeth Magee, Zn

Academic Program Assistant:
Jessica Smith-Japhet

Salem State University

Geological Sciences Dept (B) O⊘
352 Lafayette Street
Salem, MA 01970
p. (978) 542-6282
bhubeny@salemstate.edu
https://www.salemstate.edu/academics/college-arts-and-sciences/geological-science
Programs: BS Geology, BS Applied Geosciences: Sustainability, BS Applied Geosciences: Forensic, BS Earth Science Education, BS Earth Studies
Enrollment (2018): B: 50 (17)

Chair:
J Bradford Hubeny, (D), Rhode Island, 2006, GeOnCs

Professor:
Douglas Allen, (D), Minnesota, 2003, CgHsCt
James L. Cullen, (D), Brown, 1984, GsPmOu

Assistant Professor:
Sara Mana, (D), Rutgers, 2013, GciGv
Erkan Toraman, (D), Minnesota, 2014, GpiGt

Emeritus:
Lindley S. Hanson, (D), Boston, 1988, GmgOn

Laboratory Director:
Renee Knudstrup, (B), Ca

Smith College

Dept of Geosciences (B) ☒
Clark Science Center
44 College Lane
Northampton, MA 01063
p. (413) 585-3805
arhodes@smith.edu
https://www.smith.edu/academics/geosciences
Enrollment (2010): B: 27 (6)

Chair:
Sara B. Pruss, (D), S California, 2004, Pg

Professor:
Bosiljka Glumac, (D), Tennessee, 1997, Gs
Robert M. Newton, (D), Massachusetts, 1978, Gm
Amy L. Rhodes, (D), Dartmouth, 1996, ClGe

Associate Professor:
Jack Loveless, (D), Cornell, 2008, Gtc

Assistant Professor:
Sarah Mazza, (D), Virginia Tech, 2016, Gx

Lecturer:
Mark E. Brandriss, (D), Stanford, 1994, Gx

Emeritus:
John B. Brady, (D), Harvard, 1975, GxzZe
H. Robert Burger, (D), Indiana, 1966, GcYeZi
H. Allen Curran, (D), N Carolina, 1968, Pg

Tufts University

Dept of Earth and Climate Sciences (B) ☒
Lane Hall
2 N Hill Road

Medford, MA 02155
p. (617) 627-3494
jack.ridge@tufts.edu
https://as.tufts.edu/ecs/
Programs: Geological Sciences (B); Environmental Geology (B)
Enrollment (2022): B: 27 (6)
Professor:
John C. Ridge, (D), Syracuse, 1985, GlnGg
Associate Professor:
Andrew Kemp, (D), Pennsylvania, 2009, OnPmOg
Jill VanTongeren, (D), Columbia, 2010, GiCu
Lecturer:
Noel A. Heim, (D), Georgia, 2008, PgiPs
Emeritus:
Robert L. Reuss, (D), Michigan, 1970, Gzi

University of Massachusetts, Amherst

Dept of Geosciences (B,M,D) ●☒
233 Morrill Science Center BLDG2
627 North Pleasant St.
Amherst, MA 01003-9297
p. (413) 545-2286
sburns@geo.umass.edu
http://www.geo.umass.edu/
f: https://www.facebook.com/umassgeo
t: https://twitter.com/umassgeo
Programs: Geology (BS, BA), Earth Systems (BS), Geosciences (MS & PhD), Geography (BA, BS & MS)
Certificates: GIST
Enrollment (2019): B: 77 (27) M: 26 (13) D: 33 (2)
Department Head:
Stephen J. Burns, (D), Duke, 1987, Cs
Distinguished University Professor:
Raymond S. Bradley, (D), Colorado, 1974, As
Professor:
Julie Brigham-Grette, (D), Colorado, 1985, GlPcGu
Michele L. Cooke, (D), Stanford, 1996, Gc
Robert DeConto, (D), Colorado, 1996, As
Piper Gaubatz, (D), California (Berkeley), 1968, Zy
R. Mark Leckie, (D), Colorado, 1984, PmGru
J. Michael Rhodes, (D), Australian Nat, 1970, GviCa
Sheila J. Seaman, (D), New Mexico, 1988, Gi
Michael L. Williams, (D), New Mexico, 1987, Gc
Extension Associate Professor:
Michael Rawlins, (D), New Hampshire, 2006, As
Extension Assistant Professor:
Christine Hatch, (D), California (Santa Cruz), 2007, HwgSp
Associate Extension Professor:
William P. Clement, (D), Wyoming, 1995, Yue
Associate Professor:
David Boutt, (D), New Mexico Tech, 2004, HwqHy
Steven Petsch, (D), Yale, 2000, Col
Stan Stevens, (D), California (Berkeley), 1983, Zy
Eve Vogel, (D), Oregon, 2007, Zy
Jonathan D. Woodruff, (D), MIT, 2008, Gs
Qian Yu, (D), California (Berkeley), 2005, Zri
Research Assistant Professor:
Ambarish Karmalkar, (D), Massachusetts, 2010, As
Brian Yellen, (D), Gs
Assistant Professor:
Isla Castañeda, (D), Minnesota, 2007, Ca
Haiying Gao, (D), Rhode Island, 2012, Ys
Isaac J. Larsen, (D), Washington, 2013, Gm
Justin Richardson, (D), Dartmouth, 2015, CbSc
Stephen Turner, (D), Harvard, 2015, Giv
Matthew Winnick, Stanford, 2015, Cbq
Research Associate:
Tim Cook, (D), Massachusetts, 2009, Pc
Lab Manager:
Jeff Salacup, (D), Brown, 2014, PcCb
Assistant Professor :
Forrest J. Bowlick, (D), Texas A&M, 2016, ZieZg
Lecturer:
Michael J. Jercinovic, (D), New Mexico, 1988, Gx
Seda lalap-Ayça, (D), San Diego State, 2018, Zi
Adjunct Professor:
Eileen McGowan, (D), Massachusetts, 2010, Xg
Peter T. Panish, (D), Massachusetts, 1989, Gp

Emeritus:
Laurie L. Brown, (D), Oregon State, 1974, Ymg
Christopher D. Condit, (D), New Mexico, 1984, Gi
James A. Hafner, (D), Michigan, 1970, Zn
William D. McCoy, (D), Colorado, 1981, Zy
George E. McGill, (D), Princeton, 1958, Gc
Stearns A. Morse, (D), McGill, 1962, Gi
Rutherford H. Platt, (D), Chicago, 1971, Zu
Peter Robinson, (D), Harvard, 1963, Gc
Richard W. Wilkie, (D), Washington, 1968, ZynZn
Richard F. Yuretich, (D), Princeton, 1976, Cl
Massachusetts State Geologist:
Stephen B. Mabee, (D), Massachusetts, 1992, GgHw

University of Massachusetts, Boston

School for the Environment (B,M,D) ☒
100 Morrissey Boulevard
Boston, MA 02125
p. (617) 287-7440
sfe@umb.edu
http://www.umb.edu/environment
Enrollment (2010): B: 153 (27) M: 22 (12) D: 18 (4)
Dean:
Robyn Hannigan, (D), Rochester, ClmCt
Professor:
Bob Chen, (D), California (San Diego), 1992, CoOc
Zhongping Lee, (D), S Florida, 1994, ZrOg
William Robinson, (D), Northeastern, 1981, Zn
Crystal Schaaf, (D), Boston, 1994, ZrAm
Director for Research, New England Aquarium:
Michael Tlusty, (D), Syracuse, 1996, Ob
Director, Environmental Studies Program:
Alan D. Christian, (D), Miami, 2002, HsGmCs
Associate Professor:
Robert Bowen, (D), S California, 1981, ZnnZn
Ellen Douglas, (D), Tufts, 2002, HwqHy
John Duff, (D), Washington, 1995, ZnnZn
Eugene Gallagher, (D), Washington, 1983, Ob
Juanita Urban-Rich, (D), Memorial, Ob
Assistant Professor:
Jennifer Bowen, (D), Boston, 2005, ObPgZn
Helen Poynton, (D), California (Berkeley), Zn
Research Engineer:
Francesco Peri, (M), Massachusetts (Boston), Zn
Manager, GIS Lab:
Helenmary Hotz, (M), Massachusetts (Boston), Zi
Director, Green Harbors Project:
Anamarija Frankic, (D), Virginia Inst of Marine Sci, Ze

University of Massachusetts, Lowell

Dept of Environmental, Earth, & Atmospheric Sciences (B,M,D) ⊘
1 University Avenue
University of Massachusetts
Lowell, MA 01854
p. (978) 934-3903
Erica_Gavin@uml.edu
https://www.uml.edu/sciences/eeas/
f: https://www.facebook.com/Earth.Sciences.UMass.Lowell/
Programs: Geology, Environmental Science, Atmospheric Science, Earth Systems Science
Enrollment (2020): B: 28 (6) M: 18 (1)
Acting Chair:
Mathew Barlow, (D), Maryland, 1999, As
Professor:
Frank P. Colby, (D), MIT, 1983, Am
G. Nelson Eby, (D), Boston, 1971, CgGiCt
Daniel Obrist, (D), Nevada, 2002, As
Juliette Rooney-Varga, (D), ZcRc
Associate Professor:
Richard M. Gaschnig, (D), Washington State, 2010, CctCa
Kate Swanger, (D), Boston, 2009, Gl
Assistant Professor:
James Heiss, (D), Delaware, 2017, Hw
Christopher Skinner, (D), Stanford, 2014, At
Joy Winbourne, (D), California, 2015, CbZc
Visiting Faculty Lecturer:
Kimberly Howell, (D), Gso

Associate Chair of Undergraduate Programs:
Lori Weeden, (M), Boston Coll, 2002, GgeGs
Emeritus:
Arnold L. O'Brien, (D), Boston, 1973, Hw

Wellesley College

Dept of Geosciences (B) ☒
106 Central Street
Wellesley, MA 02481-8203
p. (781) 283-3151
kmonecke@wellesley.edu
http://www.wellesley.edu/Geosciences/
Programs: Geosciences
Administrative Assistant: Carol Gagosian
Enrollment (2021): B: 10 (4)
Chair:
Daniel J. Brabander, (D), Brown, 1997, Cl
Associate Scientist:
Katrin Monecke, (D), ETH (Switzerland), Gsn
Assistant Professor:
Adrian Castro, (D), Rensselaer, 2018, Gp
Emeritus:
James Besancon, (D), MIT, 1975, Hww
David Hawkins, (D), MIT, 1996, Gzi
Margaret D. Thompson, (D), Harvard, 1976, Gc
Instructor:
Kathleen W. Gilbert, (M), Miami, 1995, Cs

Williams College

Dept of Geosciences (B) ☒
18 Hoxsey Street
Wachenheim Science Center
Williamstown, MA 01267
p. (413) 597-2221
pac3@williams.edu
http://www.williams.edu/Geoscience
f: https://www.facebook.com/groups/williamsgeosciences
Programs: Geosciences
Administrative Assistant: Patricia E. Acosta
Enrollment (2021): B: 24 (11)
Chair:
Phoebe A. Cohen, (D), Harvard, 2010, Pgg
Professor:
Mea S. Cook, (D), MIT/WHOI, 2006, OuGe
Ronadh Cox, (D), Stanford, 1993, GsXg
Paul Karabinos, (D), Johns Hopkins, 1981, Gc
Assistant Professor:
Alice Bradley, (D), Colorado, 2016, AtOn
Jose A. Constantine, (D), California (Santa Barbara), 2008, GmZi
Research Associate:
B. Gudveig Baarli, (D), Oslo, 1988, PssPi
Mark E. Brandriss, (D), Stanford, 1993, Gi
Lecturer:
Alex Apotsos, (D), MIT, 2007, Ge
Emeritus:
David P. Dethier, (D), Washington, 1977, Gm
Markes E. Johnson, (D), Chicago, 1977, PseGm
Reinhard A. Wobus, (D), Stanford, 1966, GivGz
Science Librarian:
Abigail Cahill, (M), Wisconsin, 2017, Zn

Woods Hole Oceanographic Institution

Dept of Geology & Geophysics (D)
Woods Hole, MA 02543-1541
p. (508) 289-2388
mburke@whoi.edu
http://www.whoi.edu/page.do?pid=7145
Administrative Assistant: Maryanne F. Ferreira
Enrollment (2014): D: 26 (0)
Chair:
Robert L. Evans, (D), Cambridge, 1991, Yrr
Senior Scientist:
Mark D. Behn, (D), MIT/WHOI, 2002, Yr
Joan M. Bernhard, (D), California (San Diego), 1990, Cg
Henry J B. Dick, (D), Yale, 1975, GitGc
Jeffrey Donnelly, (D), Brown, 2000, Gu
Daniel J. Fornari, (D), Columbia, 1978, Gu
Chris German, (D), Cambridge, 1988, Gt
Susan E. Humphris, (D), MIT/WHOI, 1977, GuCm
Lloyd D. Keigwin, (D), Rhode Island, 1979, CsGul
Jian Lin, (D), Brown, 1988, Gt
Olivier Marchal, (D), Paris, 1996, Pe
Daniel C. McCorkle, (D), Washington, 1987, Cm
Delia W. Oppo, (D), Columbia, 1989, Pm
Ralph A. Stephen, (D), Cambridge, 1978, Ys
Maurice A. Tivey, (D), Washington, 1988, Yrm
Tenured Associate Scientist:
Andrew Ashton, (D), Duke, 2005, On
Juan Pablo Canales Cisneros, (D), Barcelona, 1997, Yr
Anne L. Cohen, (D), Cape Town, 1993, PecOo
Sarah B. Das, (D), Penn State, 2003, GlAtZg
Virginia Edgcomb, (D), Delaware, 1997, ObZc
Glenn A. Gaetani, (D), MIT, 1996, Gi
Liviu Giosan, (D), SUNY (Stony Brook), 2001, GusPc
Daniel Lizarralde, (D), MIT/WHOI, 1997, Ys
Jeffrey J. McGuire, (D), MIT, 2000, Ys
Robert A. Sohn, (D), California (San Diego), 1996, Yr
S. Adam Soule, (D), Oregon, 2003, Gvu
Associate Scientist:
Weifu Guo, (D), Caltech, 2008, CslPe
Sune G. Nielsen, (D), ETH (Switzerland), 2005, Cs
Assistant Scientist:
Veronique Le Roux, (D), Macquarie, 2008, Gi
Senior Research Specialist:
James E. Broda, (B), Penn State, 1970, Gs
John A. Collins, (D), MIT/WHOI, 1989, Ys
Ann P. McNichol, (D), MIT, 1986, Oc
Mark L. Roberts, (D), Duke, 1988, Yg
Research Specialist:
Jurek Blusztajn, (D), Polish Acad of Sci, 1985, CcGi
Alan R. Gagnon, (B), New Hampshire, 1983, Cs
Li Xu, (D), Xiamen, 1992, Yr
Engineer:
Peter B. Landry, (B), NEIT, 1990, Ca
Research Associate:
Kathryn L. Elder, (B), Massachusetts, 1985, Cs
Kalina D. Gospodinova, (M), MIT/WHOI, 2012, Cm
Joshua D. Hlavenka, (B), N Texas
Peter C. Lemmond, (B), Lehigh, 1978, Zn
Brett Longworth, (M), Massachusetts, 2005, Ou
Brian D. Monteleone, (D), Syracuse, 2000, Cc
Kathryn R. Pietro, (M), California (Davis), 2007, Gu
Adjunct Scientist:
Peter D. Bromirski, (D), Hawaii, 1993
Johnson R. Cann, (D), Cambridge, 1963, Ys
Colin Devey, (D), Oxford, 1986
Javier Escartin, (D), MIT, 1996, Gg
Andrea D. Hawkes, (D), Pennsylvania, 2008
Gregory Hirth, (D), Brown, 1991, Gc
Kuo-Fang (Denner) Huang, (D), Natl Cheng Kung, 2007
Peter B. Kelemen, (D), Washington, 1987, Gi
Yajing Liu, (D), Harvard, 2007
John Maclennan, (D), Cambridge, 2000
Larry Mayer, (D), Scripps, 1979, Gg
Andrew M. McCaig, (D), Cambridge, 1983
Jerry F. McManus, (D), Columbia, 1997, Pe
Uri S. ten Brink, (D), Columbia, 1986, Yg
David Thornalley, (D), Churchill, 2008
Masako Tominaga, (D), Texas A&M, 2009
Oceanographer Emeritus:
Graham S. Giese, (D), Chicago, 1966, On
Steven J. Manganini, (B), Nasson, 1974, Ou
Robert J. Schneider, (D), Tufts, 1968, Cc
Stephen A. Swift, (D), MIT/WHOI, 1986, Yr
Karl F. Von Reden, (D), Hamburg, 1983, Ct
Emeritus:
William A. Berggren, (D), Stockholm, 1962, Ps
Carl O. Bowin, (D), Princeton, 1960, YrGtg
William B. Curry, (D), Brown, 1980, Pm
Stanley R. Hart, (D), MIT, 1960, CgGvCp
Susumu Honjo, (D), Hokkaido, 1961, Ou
George P. Lohmann, (D), Brown, 1972, Pe
David A. Ross, (D), California (San Diego), 1965, GuOg
Hans Schouten, (D), Utrecht, 1970, Yr
Nobumichi Shimizu, (D), Tokyo, 1968, Ca
William G. Thompson, (D), Columbia, 2005, PeCc

Brian E. Tucholke, (D), MIT/WHOI, 1973, Gut
Elazar Uchupi, (D), S California, 1962, Ou
Frank B. Wooding, (B), Harvard, 1965, Yr

Dept of Marine Chemistry & Geochemistry (M,D) ☒
360 Woods Hole Road
MS 25
Woods Hole, MA 02543-1541
p. (508) 289-3696
mcg@whoi.edu
http://www.whoi.edu/page.do?pid=7146
Programs: Chemical Oceanography
Administrative Assistant: Linda Cannata, Sheila A. Clifford, Donna Mortimer, Mary Zawoysky
Enrollment (2015): D: 28 (5)

Dept. Chair:
Bernhard Peucker-Ehrenbrink, (D), Max Planck, 1994, CmlHs
Associate Dean:
Margaret K. Tivey, (D), Washington, 1989, Cm
Senior Scientist:
Ken O. Buesseler, (D), MIT/WHOI, 1986, Oc
Matthew A. Charette, (D), Rhode Island, 1998, Cm
Colleen Hansel, (D), Stanford, 2004, Cm
Julie Huber, (D), Washington, 2004, Cm
Konrad A. Hughen, (D), Colorado, 1997, Cc
Elizabeth B. Kujawinski, (D), MIT/WHOI, 2000, Oc
Mark D. Kurz, (D), MIT/WHOI, 1982, Cc
Christopher M. Reddy, (D), Rhode Island, 1997, Co
Daniel J. Repeta, (D), MIT/WHOI, 1982, Co
Mak A. Saito, (D), MIT/WHOI, 2001, Oc
Jeffrey S. Seewald, (D), Minnesota, 1990, Cp
Benjamin Van Mooy, (D), Washington, 2003, Oc
Associate Scientist w/Tenure:
Amy Apprill, (D), Hawaii, 2009, Oc
Valier Galy, (D), Inst Nat Polytechnique de Lorrain, 2007, Cm
Frieder Klein, (D), Bremen, 2009, Cm
Z. Aleck Wang, (D), Georgia, 2003, Cm
Scott Wankel, (D), Stanford, 2007, Cm
Associate Scientist:
Tristan Horner, (D), Oxford, 2012, Cm
Matthew Long, (D), Virginia, 2013, Cm
David Nicholson, (D), Washington, 2009, Gu
Assistant Scientist:
Peter Barry, (D), California (San Diego), 2012, Cm
Ann Dunlea, (D), Boston, 2016, Cm
Hyewon (Heather) Kim, (D), Columbia, 2017, Cm
Adam Subhas
Collin Ward, (D), Michigan, 2015, Cm
Sr. Research Specialist:
Robert K. Nelson, (B), C Connecticut, 1980, Co
Senior Research Specialist (ret.):
Dempsey E. Lott, (M), Florida State, 1973, Oc
Research Specialist:
Heather Benway, (D), Oregon State, 2005, OcCmOo
Helen Fredricks, (D), Plymouth (UK), 2000, Oc
Paul Henderson, (B), Bowling Green, 1997, Cm
Carl G. Johnson, (M), Maine, 1983, Ca
Ivan D. Lima, (D), Miami, 1999, Oc
Krista Longnecker, (D), Oregon State, 2004, Oc
Matt McIlvin, (D), Massachusetts, 2004, Oc
Steven M. Pike, (M), Rhode Island, 1998, Ca
Jennie Rheuban, (M), Virginia, 2013, Oc
Melissa Soule, (D), Oregon, 2007
Research Assistant:
Samual Bowman
Kevin Cahill, (B), Massachusetts, 1999, Cm
Jessica Drysdale, (M), Queens (Ontario), 2011, Oc
Kate Morkeski, (M), Virginia Tech, 2007
Justin Ossolinski
Information Systems Associate:
Dana Gerlach, (M), MIT/WHOI, 1997, Og
Adam Shepherd, (B), Northeastern, 2001
Communications Specialist for WHOI SeaGrant:
Stephanie Murphy, (B), California (Santa Cruz), 1985
Research Associate:
Maureen Auro, (M), San Francisco State, 2007
Joshua M. Curtice, (B), Massachusetts, 1992, Cc
Carolyn Miller, (D), Boston, 2006
Dawn Moran, (B), Bridgewater State, 2003, Ob
Margrethe Serres, (D), Wisconsin (Madison), 1995
Gretchen Swarr, (B), Colgate, 2007, Oc
Sean Sylva, (B), Rhode Island, 1997, Oc
Adjunct Scientist:
Tim Eglinton, (D), Newcastle (UK), 1988, Oc
Margaret Estapa, (D), Maine, 2011, Oc
Adjunct Professor:
Minhan Dai, (D), Pierre & Marie Curie, 1995, Cm
Rachel Stanley
Scientist Emeritus:
Michael P. Bacon, (D), MIT/WHOI, 1976, Oc
Werner G. Deuser, (D), Penn State, 1963, CsGsu
John W. Farrington, (D), Rhode Island, 1972, ComOc
William R. Martin, (D), MIT/WHOI, 1985, Oc
Frederick L. Sayles, (D), Manchester (UK), 1968, Oc
Edward R. Sholkovitz, (D), California (San Diego), 1972, Oc
Geoffrey Thompson, (D), Manchester (UK), 1965, Cm
Oliver C. Zafiriou, (D), Johns Hopkins, 1966, Oc
Oceanographer Emeritus:
Cyndy Chandler, (B), SUNY (Geneseo), 1975, Og
Nelson M. Frew, (D), Washington, 1971, Ca
David M. Glover, (D), Alaska, 1985, Oc
Jean K. Whelan, (D), MIT, 1965, Co
Marine Education Specialist:
Grace Simpkins, (M), Alaska, 2000
Manager, Community Engagement - OCB:
Mairead Mahegan, (M), Cal State (Northridge), Ob
Department Administrator:
Jennifer Carter
Communications and Outreach Specialist:
Jeffrey Brodeur

Worcester State University

Earth, Environment and Physics (B)
486 Chandler Street
Worcester, MA 01602-2597
p. (508) 929-8583
adunn@worcester.edu
http://www.worcester.edu/Earth-Environment-and-Physics/
f: https://www.facebook.com/worcesterstatedeep
Programs: Geography, Environmental Science
Enrollment (2022): B: 81 (19)

Professor:
Allison L. Dunn, (D), Harvard, 2006, AsZyHg
William J. Hansen, (D), CUNY, 2002, ZiyZr
Assistant Professor:
Anne Armstrong, (D), Cornell, 2022, Zec
Nabin Malakar, (D), SUNY (Albany), 2012, Ap
Laura Reynolds, (D), California (Santa Barbara), 2018, Gs
Alexander Tarr, (D), California (Berkeley), 2014, Zg

Michigan

Adrian College

Geology Dept (A,B) ☒
110 S. Madison St.
Adrian, MI 49221
p. (517) 265-5161
tmuntean@adrian.edu
http://adrian.edu/academics/academic-departments/geology/
f: https://www.facebook.com/groups/55153346895/
Programs: Geology, Environmental Geology
Enrollment (2018): B: 14 (4)

Chair:
Thomas Muntean, (D), Nevada, 2012, Gde
Professor:
Sarah L. Hanson, (D), Utah, 1995, Giz

Albion College

Dept of Earth & Environment (B) O⊘
611 E. Porter St.
Albion, MI 49224
p. (517) 629-0312
cmenold@albion.edu
https://web.albion.edu/academics/departments/geological-sciences
Programs: Geology; Earth Science; Earth Science with

Secondary Education Certification; Geology with Secondary Education Certification; Environmental Science; Environmental Studies
Certificates: Geology; GIS; Paleontology; Environmental Geology; Earth Science with Secondary Education Certification (minors); Environmental Science Concentration; Environmental Studies Concentration
Enrollment (2021): B: 34 (10)

Chair:
Carrie A. Menold, (D), California (Los Angeles), 2006, GpCsGt
Professor:
Thomas I. Wilch, (D), New Mexico Tech, 1997, GlvGm
Associate Professor:
Michael McRivette, (D), California (Los Angeles), 2011, GtZir
Assistant Professor:
Joe Lee-Cullin, (D), Michigan State, 2020, Ge
Madeline Marshall, (D), Chicago, 2018, GsPi
Emeritus:
William S. Bartels, (D), Michigan, 1986, Pv
Russell G. Clark, (D), Dartmouth, 1972, GiZi
Beth Z. Lincoln, (D), California (Los Angeles), 1985, Gc
Lawrence D. Taylor, (D), Ohio State, 1962, Gl

Calvin College

Dept of Geology, Geography & Environmental Studies (B)
3201 Burton SE
Grand Rapids, MI 49546
p. (616) 526-8415
jbascom@calvin.edu
http://www.calvin.edu/academic/geology/
Enrollment (2012): B: 60 (14)

Chair:
Jason E. VanHorn, (D), Ohio State, 2007, ZirRh
Chair:
Johnathan Bascom, (D), Iowa, 1989, Zy
Professor:
Henry Aay, (D), Clark, 1978, Zyu
Janel M. Curry, (D), Minnesota, 1985, Zy
Ralph F. Stearley, (D), Michigan, 1990, PvOgRh
Associate Professor:
Deanna van Dijk, (D), Waterloo, 1998, GmOnZy
Assistant Professor:
Kenneth A. Bergwerff, (M), Grand Valley State, 1988, Ze
James Skillen, (D), Cornell, 2006, Zu
Laboratory Manager/Instructor:
Margene Brewer, (M), W Michigan, 1991, ZnGgZn
Emeritus:
Clarence Menninga, (D), Purdue, 1966, CcGg
Gerald K. Van Kooten, (D), California (Santa Barbara), 1980, GoCePi
Davis A. Young, (D), Brown, 1969, GiRhGz

Central Michigan University

Dept of Earth and Atmospheric Sciences (B)
314 Brooks Hall
Mount Pleasant, MI 48859
p. (989) 774-3179
l.d.lemke@cmich.edu
https://www.cmich.edu/colleges/se/earth_atmos/pages/
Programs: Geology; Meteorology; Environmental Science
Enrollment (2021): B: 29 (3)

Professor:
Martin Baxter, (D), St Louis, 2006, Ams
Anthony Chappaz, (D), Quebec, 2008, Cat
Lawrence D. Lemke, (D), Michigan, 2003, HwGer
Mona Sirbescu, (D), Missouri, 2002, GiCup
Associate Professor:
John Allen, (D), Australian Nat, 2013, Ams
Daria Kluver, (D), Delaware, 2011, Asm
Wendy Robertson, (D), Texas, 2014, HwCg
Assistant Professor:
Jason Keeler, (D), Illinois, 2015, Am
Natalia Zakharova, (D), Columbia, 2014, YgGe
Lecturer:
Rachael Agardy, (M), Texas A&M, 2005, OuGrs
Maria Mercedes Gonzalez, (D), Nac del Sur (Argentina), 1997, Gig
Adjunct Professor:
Megan Rohrssen, (D), California, 2013, Co
Nicole West, (D), Penn State, 2014, Gm

Emeritus:
Richard Mower, (D), Wisconsin, 1981, Am
Related Staff:
James J. Student, (D), Virginia Tech, 2002, CaGiz

Charles Stewart Mott Community College

Div of Science and Mathematics (A)
1401 East Court St.
Flint, MI 48503
p. (810) 762-0279
sheila.swyrtek@mcc.edu
https://www.mcc.edu/science-math/index.shtml

Professor:
Sheila Swyrtek, (M), Minnesota (Duluth), 1996, Ga

Concordia University

Div of Natural Sciences (A,B,M)
4090 Geddes Road
Ann Arbor, MI 48105-2750
p. (734) 995-7300
James.Refenes@cuaa.edu
Programs: Earth Science Education

Assoc. Prof. Biology:
James L. Refenes, (D), E Michigan, 2016, Ze

Delta College

Dept of Geology (A)
1961 Delta Road
University Center, MI 48710-0002
p. (989) 686-9252
lochoa-sanchez@deltacollege.edu
Department Secretary: Barb Jurmanovich
Enrollment (2011): A: 8 (4)

Department Chair:
Timothy L. Clarey, (D), W Michigan, 1996, HwGcPv
Lecturer:
Mary C. Gorte, (M), Rensselaer, 1983, Gs
Emeritus:
Barry A. Carlson, (D), Michigan State, 1974, Yg
Paul A. Catacosinos, (D), Michigan State, 1972, Gr
Cooperating Faculty:
Kevin T. Dehne, (M), E Michigan, 1992, Ze

Eastern Michigan University

Dept of Geography & Geology (B)
Ypsilanti, MI 48197
p. (743) 487-8589
rsambroo@emich.edu

Head:
Michael Kasenow, (D), W Michigan, 1994, Hg
Richard A. Sambrook, (D)
Professor:
Eugene Jaworski, (D), Louisiana State, 1971, Zr
Carl F. Ojala, (D), Georgia, 1972, As
Constantine N. Raphael, (D), Louisiana State, 1967, Gu
Associate Professor:
Steven T. LoDuca, (D), Rochester, 1990, Pi
Assistant Professor:
Kevin Blake, (D), Wisconsin, 1998, Gl
Michael Bradley, (D), Utah, 1988, Gc
Maria-Serena Poli, (D), Padua, 1995, Pe

Ferris State University

Dept of Physical Sciences ☒
ASC-3021
Big Rapids, MI 49307
p. (231) 591-2588
heckf@ferris.edu
https://ferris.edu/arts-sciences/departments/physical-sciences
Programs: None

Professor:
Frederick R. Heck, (D), Northwestern, 1987, GgZg

Gogebic Community College

Math-Science Div (A)
E4946 Jackson Rd.

Ironwood, MI 49938
p. (906) 932-4231
admissions@gogebic.edu
http://www.gogebic.edu/academics/Math_Science/
Professor:
Bill Perkis, Gg

Grand Rapids Community College

Physical Sciences Dept (A)
143 Bostwick Avenue, NE
Grand Rapids, MI 49503
p. (616) 234-4248
jqualls@grcc.edu
http://www.grcc.edu/physicalscience/geology
f: https://www.facebook.com/GRCCphysicalsciences
Programs: Geology, Environmental Studies and Sustainablility
Enrollment (2021): A: 5 (0)
Assistant Professor:
Tari Mattox, (M), N Illinois, 1989, GgvCg

Grand Valley State University

Dept of Geology (B)
1 Campus Drive
118 Padnos Hall
Allendale, MI 49401
p. (616) 331-3728
geodept@gvsu.edu
http://www.gvsu.edu/geology
Programs: Geology; Geology with Environmental Emphasis; Geology-Chemistry
Administrative Assistant: Janet H. Potgeter
Enrollment (2018): B: 100 (23)
Head:
Figen A. Mekik, (D), N Illinois, 1999, GuCm
Professor:
Patrick M. Colgan, (D), Wisconsin, 1996, GmlPe
Stephen R. Mattox, (D), N Illinois, 1992, ZeGv
Virginia L. Peterson, (D), Massachusetts, 1992, GxcGt
Peter J. Wampler, (D), Oregon State, 2004, HsZiGm
John C. Weber, (D), Northwestern, 1995, GctYd
Associate Professor:
Kevin C. Cole, (D), Arizona, 1990, Gz
Peter E. Riemersma, (D), Wisconsin, 1997, Hw
Assistant Professor:
Caitlin N. Callahan, (D), W Michigan, 2013, GgZg
Tara A. Kneeshaw, (D), Texas A&M, 2008, ClGe
Instructor:
Kelly L. Heid, (M), Mississippi State, 2011, Ze
John M. VanRegenmorter, (M), Colorado, 2011, PvGsr
Ian Z. Winkelstern, (D), Michigan, 2018, GsOo
Visiting Professor:
Jeremy C. Gouldey, (D), Northwestern, 2015, CgPeCg
Emeritus:
Thomas E. Hendrix, (D), Wisconsin, 1960, Gc
William J. Neal, (D), Missouri, 1968, GsdGr
Norman W. Ten Brink, (D), Washington, 1971, Gm
Patricia E. Videtich, (D), Brown, 1982, Gd

Henry Ford Community College

Science Div
Dearborn, MI 48128
p. (313) 845-9632
cjacobs@hfcc.edu

Hope College

Dept of Geological & Environmental Sciences (B) ☒
P.O. Box 9000
Holland, MI 49422-9000
p. (616) 395-7540
bodenbender@hope.edu
https://hope.edu/academics/geological-environmental-sciences/
Programs: BA in Geology, BS in Geology, Minor in Geology, Minor in Environmental Science, BS in Environmental Science with Concentration in Biology, BS in Environmental Science with Concentration in Chemistry, BS in Environmental Science with Concentration in Geology
Enrollment (2020): B: 16 (5)
Chair:
Brian E. Bodenbender, (D), Michigan, 1994, GsPiGe
Professor:
Jon W. Peterson, (D), Chicago, 1989, Ge
Assistant Professor:
Uzonna O. Anyiam, (D), Sci & Tech China, 2022, Yge
Sarah L. Dean, (D), Rice, 2014, Gc
Suzanne DeVries-Zimmerman, (M), Princeton, 1989, Gge
Michael Philben, (D), S Carolina, 2014, CboSb
Carolyn R. Sparks, (D), Kansas, 2001, GzHg

Jackson College

Dept of Geology & Geography (A,B)
2111 Emmons Road
Jackson, MI 49201
p. ((51) 7) -787- (Ext. 00 x157)
AlbeeScSteven@jccmi.edu
http://www.jccmi.edu/academics/science/geo
Enrollment (2012): A: 200 (20)
Wilbur L. Dungy Endowed Chair in the Sciences:
Steven R. Albee-Scott, (D), Michigan, 2005, GePeSf

Lake Michigan College

Lake Michigan College (A)
Napier Avenue Campus
2755 E. Napier Avenue
Benton Harbor, MI 49022
lovett@lakemichigancollege.edu
http://www.lakemichigancollege.edu/index.php?option=com_content&task=view&id=509&Itemid=157
Instructor:
Cole Lovett, (D), W Michigan, 1995, Gg

Lake Superior State University

Geology (B) O☒
650 W. Easterday Avenue
Sault Ste. Marie, MI 49783
p. (906) 635-2267
pkelso@lssu.edu
https://www.lssu.edu/geology/
Programs: B.S. Geology
B.S. Geology - Environmental option
Certificates: GIS certificate
GIS associates degree
Administrative Assistant: Donna White
Enrollment (2018): B: 38 (12)
Professor:
Paul R. Kelso, (D), Minnesota, 1993, YmGct
Assistant Professor:
William (Bill) Houston, (D), Michigan Tech, 2002, Gsg
Hari Kandel, (D), Florida Intl, 2015, HgZi
MaryKathryn (Kat) Rocheford, (D), Iowa, 2014, GeZiGa
Matt Spencer, (D), Penn State, 2005, Gl
Emeritus:
Lewis M. Brown, (D), New Mexico, 1973, PiZe
Science Lab Manager:
Benjamin Southwell, (M), C Michigan, 2012, Zn

Macomb Community College, Center Campus

Dept of Science (Geology) (A)
44575 Garfield Road
Clinton Township, MI 48038-1139
p. (586) 286-2154
schaferc@macomb.edu
Enrollment (2014): A: 9 (0)
Professor:
Carl M. Schafer, (M), Montana, 1998, Gg

Michigan State University

Dept of Earth and Environmental Sciences (B,M,D)
288 Farm Lane
East Lansing, MI 48824-1115
p. (517) 355-4626
geosci@msu.edu
https://glg.natsci.msu.edu/
Enrollment (2014): M: 4 (0) D: 20 (0)

Chair:
Ralph E. Taggart, (D), Michigan State, 1971, Pb
Professor:
jiquan Chen, (D), Washington, 1991, Zg
Kazuya Fujita, (D), Northwestern, 1979, Ys
Julie C. Libarkin, (D), Arizona, 1999, Zen
David T. Long, (D), Kansas, 1977, ClGeb
Michael A. Velbel, (D), Yale, 1984, ClGdSc
Associate Professor:
Bruno Basso, (D), Michigan State, 2002, Zn
Danita S. Brandt, (D), Yale, 1985, PiGr
Michael D. Gottfried, (D), Kansas, 1991, Pv
David W. Hyndman, (D), Stanford, 1995, Hy
Assistant Professor:
Tyrone Rooney, (D), Penn State, 2006, Gi
Matt Schrenk, (D), Washington, 2005, Pg
Jay Zarnetske, (D), Oregon State, 2011, Hwg

Michigan Technological University

A. E. Seaman Mineral Museum ☒
1404 E. Sharon Avenue
Houghton, MI 49931
p. (906) 487-2572
tjb@mtu.edu
http://www.museum.mtu.edu
Director:
Theodore J. Bornhorst, (D), New Mexico, 1980, EgCgGx
Associate Curator:
Christopher J. Stefano, (D), Michigan, 2010, GzxCg

Dept of Geological & Mining Engineering & Sciences
(B,M,D) O⃠
1400 Townsend Drive
Dow Building, Room 630
Houghton, MI 49931-1295
p. (906) 487-2531
asmirnov@mtu.edu
http://www.mtu.edu/geo
f: https://www.facebook.com/mtugeomining/
t: https://twitter.com/mtugeomining
Programs: Geology, Applied Geophysics, Geological Engineering, Mining Engineering
Certificates: Graduate Certificate in Geoinformatics, Graduate Certificate in Natural Hazards and Disaster Risk Reduction
Enrollment (2018): B: 78 (30) M: 39 (5) D: 7 (5)
Chair and Professor:
Aleksey V. Smirnov, (D), Rochester, 2002, YmgGt
Professor:
Simon Carn, (D), Cambridge, 1999, ZrGv
John S. Gierke, (D), Michigan Tech, 1990, Hy
Thomas Oommen, (D), Tufts, 2009, NeZiNg
Gregory P. Waite, (D), Utah, 2004, YsGv
Shiliang Wu, (D), Harvard, 2007, AsCg
Associate Professor:
Snehamoy Chatterjee, (D), Indian Inst of Tech, 2007, NmGqNx
Chad Deering, (D), Canterbury (NZ), 2009, Ggi
Assistant Professor:
Roohollah Askari, (D), Calgary, 2013, YxxNp
Xin Xi, (D), Georgia Tech, 2014, AsZor
Senior Lecturer:
Nathan Manser, (D), S Florida, 2015, GeNm
Principal Lecturer:
Jeremy Shannon, (D), Michigan Tech, 2006, Gg
Adjunct Professor:
Guy A. Meadows, (D), Purdue, 1977, Gu
Research Professor:
James DeGraff, (D), Purdue, Gc
Research Scientist:
Carol Asiala, (B), Michigan Tech, 1985
Cooperating Faculty:
William I. Rose, (D), Dartmouth, 1970, Gv
Roger M. Turpening, (D), Michigan, 1966, Ye

Mott Community College

Science and Mathematics (A) ⌚
1401 East Court St.
Flint, MI 48503
p. (810) 232-9312
sheila.swyrtek@mcc.edu
http://www.mcc.edu
Enrollment (2015): A: 4 (0)
Professor:
Sheila M. Swyrtek, (M), Minnesota (Duluth), 1997, ZeGa

Muskegon Community College

Dept of Mathematics & Physical Sciences (A)
221 S. Quarterline Road
Muskegon, MI 49442
p. (231) 777-0289
amber.kumpf@muskegoncc.edu
http://www.muskegoncc.edu/pages/652.asp
Administrative Assistant: Tamera Owens
Instructor:
Amber C. Kumpf, (M), Rhode Island, 2010, GgYrOu

Northwestern Michigan College

Northwestern Michigan College
1701 East Front Street
Traverse City, MI 49686
p. (231) 995-1000
admissions@nmc.edu
http://www.nmc.edu

Oakland Community College

Natural Sciences Dept (A) ☒
2900 Featherstone Road
Auburn Hills, MI 48326
p. (248) 232-4538
jmbaker@oaklandcc.edu
Programs: Science (A)

Schoolcraft College

Dept of Geology (A)
18600 Haggerty Road
Livonia, MI 48151
p. (734) 462-4400
jrexius@schoolcraft.edu
Instructor:
James E. Rexius, (M), E Michigan, 1978, Gl

University of Michigan

Dept of Climate and Space Sciences and Engineering
(B,M,D) ☒
2455 Hayward
Ann Arbor, MI 48109-2143
p. (734) 615-3583
clasp--um@umich.edu
http://clasp.engin.umich.edu/
f: www.facebook.com/umclasp
t: @umclasp
Programs: Climate & Space Science & Engineering PhD; Applied Climate MEng; Space Engineering MEng; Atmospheric & Space Sciences MS; Climate & Meteorology BSE; Space Sciences & Engineering BSE
Enrollment (2019): B: 30 (7) M: 49 (26) D: 122 (5)
Chair:
Tuija Pulkkinen, (D), Helsinki, 1992, Zn
Professor:
Sushil Atreya, (D), Michigan, ApZy
John Boyd, (D), Harvard, 1976, OpAs
R. Paul Drake, (D), Johns Hopkins
Lennard Fisk, (D), California (San Diego)
Brian Gilchrist, (D), Stanford
Tamas Gombosi, (D), Lóránd Eötvös
Enrico Landi, (D), Florence
Michael Liemohn, (D), Michigan
Mark B. Moldwin, (D), Boston, 1993, XpYz
Joyce Penner, (D), Harvard
Nilton Renno, (D), MIT
Aaron Ridley, (D), Michigan
Richard B. Rood, (D), Florida State, 1982, ZcAsm
Christopher Ruf, (D), Massachusetts
Perry Samson, (D), Wisconsin, 1979, AsZeAm

James Slavin, (D), California (Los Angeles)
Associate Professor:
Jeremy Bassis, (D), Scripps, 2007, GlOgZg
Mark Flanner, (D), California (Irvine)
Xianglei Huang, (D), Caltech
Christiane Jablonowski, (D), Michigan, 2004, AsmZo
Xianzhe Jia, (D), California (Los Angeles)
Justin C. Kasper, (D), MIT, 2003
Susan Lepri, (D), Michigan
Allison Steiner, (D), Georgia Tech
Shasha Zou, (D), California (Los Angeles)
Assistant Professor:
Angel F. Adames-Corraliza, (D), Washington, Ast
Gretchen Keppel-Aleks, (D), Caltech
Eric A. Kort, (D), Harvard
Ashley E. Payne, California (Irvine), 2016, At

Dept of Earth and Environmental Sciences (B,M,D)
2534 C.C. Little Building
1100 North University Avenue
Ann Arbor, MI 48109-1005
p. (734) 764-1435
mukasa@umich.edu
http://www.lsa.umich.edu/geo/
Administrative Assistant: Robert J. Patterer
Professor:
Tomasz K. Baumiller, (D), Chicago, 1990, Pg
Joel D. Blum, (D), Caltech, 1990, CsaCl
Maria Clara Castro, (D), Paris, 1995, Hy
Daniel C. Fisher, (D), Harvard, 1975, Pi
Stephen E. Kesler, (D), Stanford, 1966, Eg
Rebecca A. Lange, (D), California (Berkeley), 1989, Gi
Kyger C. Lohmann, (D), SUNY (Stony Brook), 1977, CsGsCl
Samuel B. Mukasa, (D), California (Santa Barbara), 1984, Cc
Henry N. Pollack, (D), Michigan, 1963, Yg
Jeroen Ritsema, (D), California (Santa Cruz), 1995, Ysg
Larry J. Ruff, (D), Caltech, 1981, Ys
Gerald R. Smith, (D), Michigan, 1965, Ps
Ben van der Pluijm, (D), New Brunswick, 1984, GcZce
Lynn M. Walter, (D), Miami, 1983, Cl
Youxue Zhang, (D), Columbia, 1989, Cu
Associate Professor:
Udo Becker, (D), Virginia Tech, 1995, Zn
Robyn J. Burnham, (D), Washington, 1987, Pb
Todd A. Ehlers, (D), Utah, 2001, Yg
Nathan A. Niemi, (D), Caltech, 2001, Gtc
Benjamin Passey, (D), Utah, 2007, CsaCl
Christopher J. Poulsen, (D), Penn State, 1999, Pe
Peter J. van Keken, (D), Utrecht, 1989, Yg
Associate Research Scientist:
Shaopeng Huang, (D), Acad Sinica (Beijing), 1990, YhZcGt
Assistant Professor:
Marin Clark, (D), MIT, 2003, Gm
Ingrid Hendy, (D), California (Santa Barbara), 2000, Pg
Jeffrey A. Wilson, (D), Chicago, 1999, Gg
Research Scientist:
Jeffrey C. Alt, (D), Miami, 1984, Ou
Associate Research Scientist:
Chris Hall, (D), Toronto, 1982, Cc
Josep M. Pares, (D), Barcelona, 1988, Ym
Assistant Research Scientist:
James D. Gleason, (D), Arizona, 1994, Cg
Jie Lian, (D), Michigan, 2003, Gz
Mirjam Schaller, (D), Bern, 2001, Gm
Adjunct Assistant Research Scientist:
Roland C. Rouse, (D), Michigan, 1972, Gz
Adjunct Assistant Professor:
Karen L. Webber, (D), Rice, 1988, Gv
Adjunct Professor:
John W. Geissman, (D), Michigan, 1980, Ym
William B. Simmons, (D), Michigan, 1973, Gz
Emeritus:
Charles Beck, (D), Cornell, 1955, Pg
William Farrand, (D), Michigan, 1960, Ga
William C. Kelly, (D), Columbia, 1954, Eg
Philip A. Meyers, (D), Rhode Island, 1972, CoGnu
Theodore C. Moore, (D), California (San Diego), 1968, OuPmc
James O'Neil, (D), Chicago, 1963, Cs
Robert M. Owen, (D), Wisconsin, 1975, CmGu
Donald R. Peacor, (D), MIT, 1962, Gz
David K. Rea, (D), Oregon State, 1974, Gu
Rob Van der Voo, (D), Utrecht, 1969, YmGtc
Bruce H. Wilkinson, (D), Texas, 1973, GseGt

University of Michigan, Dearborn

Dept of Natural Sciences (B,M)
4901 Evergreen Road
Dearborn, MI 48128
p. (313) 593-5277
kmurray@umd.umich.edu
http://www.umd.umich.edu/?id=570101
Enrollment (2013): B: 30 (4) M: 14 (3)
Acting Chair:
Kent S. Murray, (D), California (Davis), 1981, Hw
Professor:
Don Bord, (D), Dartmouth, 1976, Xy
Jacob A. Napieralski, (D), Purdue, 2005, GmlZi
Associate Professor:
John Riebesell, (D), Chicago, 1975, Zu
Lecturer:
David Matzke, (M), Michigan, 1975, Xy
Adjunct Professor:
Michael Favor, (B), Michigan, 1985, Zn

Washtenaw Community College

Dept of Geology (A)
4800 E. Huron River Drive
Ann Arbor, MI 48105
p. (734) 677-5111
salbach@wccnet.edu
Head:
Suzanne M. Albach, (M), Mississippi State, 2005, ZegGe

Wayne State University

Environmental Science and Geology (B,M) ⊘
0224 Old Main Building
4831 Cass Avenue
Detroit, MI 48201
p. (313) 577-2506
baskaran@wayne.edu
https://clas.wayne.edu/esg
t: @WSUgeology
Programs: Geology (B.S., B.A.); Environmental Science (B.S.); M.A. (Geology), M.S. (Geology), M.A. (Environmental Science, Online)
Certificates: Graduate Certificate on Emerging Contaminants (being developed)
Enrollment (2021): B: 139 (26) M: 8 (2)
Director of Environmental Science Program:
Donna R. Kashian, (D), Zoology/Ecotoxicology, 2002
Chair:
Mark Baskaran, (D), Isotope Geochemistry, 1985, Ccl
Associate Professor:
Sarah J. BrownLee, (D), California (Berkeley), 2009, GxcGy
Bratton F. John, (D), California (Berkeley), 1997, Cg
Shirley Papuga, (D), Colorado, 2006, Hgs
Assistant Professor:
Scott Burdick, (D), MIT, 2014, Ysg
Zhijiang Lu, (D), Environmental Science, 2014
Instructor:
Walter Pociask, (M), Zn
Lecturer:
Felice Sperone, (M), Calabria - Rende (Italy), 2003, GgZr

Western Michigan University

Dept of Geosciences (B,M,D)
1903 W Michigan Ave
Kalamazoo, MI 49008-5241
p. (269) 387-5486
kathryn.wright@wmich.edu
http://www.wmich.edu/geology/
f: https://www.facebook.com/wmugeosciences
Programs: Earth Science (B); Earth Science Education (B); Geochemistry (B); Geology (B); Geophysics (B); Hydrogeology (B); Accelerated MA in Earth Sciences (M), Earth Sciences (M), Geosciences (M, D)

Certificates: Applied Hydrogeology (UG, G)
Enrollment (2015): B: 108 (18) M: 51 (17) D: 12 (4)

Chair:
Mohamed Sultan, (D), Washington, 1984, ZriGe

Dean of the College of Arts and Sciences:
Carla Koretsky, (D), Johns Hopkins, 1998, CglHs

Professor:
Alan E. Kehew, (D), Idaho, 1977, GmlHw
Michelle A. Kominz, (D), Columbia, 1986, OuYgGu
R. V. Krishnamurthy, (D), Phy Res Lab (India), 1984, Cs

Geosciences Specialist and Co-Director of the Hydrogeology Field Course:
Thomas R. Howe III, (M), W Michigan, 2017, HwGem

Co-Director of the Hydrogeology Field Course:
Donald Matthew M. Reeves, (D), Nevada (Reno), 2006, HwqNr

Associate Professor:
Daniel P. Cassidy, (D), Notre Dame, 1995, Hw
Johnson R. Haas, (D), Washington, 1993, Cl
Duane R. Hampton, (D), Colorado State, 1989, HwSpHg
William A. Sauck, (D), Arizona, 1972, YgeYv

Assistant Professor:
Robb Gillespie, (D), SUNY, GsoGm
Stephen E. Kaczmarek, (D), Michigan State, 2005, GdCgGz
Joyashish Thakurta, (D), Indiana, 2008, GiEg

Senior Research Associate and Part-time Instructor:
Andrew Caruthers, (D), British Columbia, 2013, GsCsGr

Director of Michigan Geological Repository for Research and Education and Professor Emeritus:
William B. Harrison, III, (D), Cincinnati, 1974, GsrPg

Teaching Faculty Specialist and Director of CoreKids:
Peter J. Voice, (D), Virginia Tech, 2010, GsZn

Minnesota

Bemidji State University

Center for Environmental, Economic, Earth, & Space Studies (A,B,M)
#27, 1500 Birchmont Drive NE
Bemidji, MN 56601
p. (218) 755-2783
carl.isaacson@bemidjistate.edu
https://www.bemidjistate.edu/academics/departments/css/
Enrollment (2012): B: 109 (17) M: 23 (3)

Professor:
Dragoljub D. Bilanovic, (D), Tech Israel Inst Tech, 1990, GePgCa

Assistant Professor:
Carl Isaacson, (D), Oregon State, 2007, Ge
Miriam Rios-Sanchez, (D), Michigan Tech, 2012, HwZrGg

Carleton College

Dept of Geology (B) ☒
One North College Street
Northfield, MN 55057
p. (507) 222-5769
tlittle@carleton.edu
https://apps.carleton.edu/curricular/geol/
Programs: Geology
Administrative Assistant: Tamara Little
Enrollment (2022): B: 16 (18)

Professor:
Clinton A. Cowan, (D), Queen's, 1992, GsPe
Cameron Davidson, (D), Princeton, 1991, GptCc
Bereket Haileab, (D), Utah, 1994, GxzGi
Sarah J. Titus, (D), Wisconsin, 2006, Gc

Assistant Professor:
Chloé Fandel, (D), Arizona, 2021, Hw
Dan Maxbauer, (D), Minnesota, 2017, PcYm

Director, Science Ed Resource Center:
Cathryn A. Manduca, (D), Caltech, 1988, Gi

Charles L. Denison Professor of Geology, Emerita:
Mary E. Savina, (D), California (Berkeley), 1982, Gm

Emeritus:
Caryl E. Buchwald, (D), Kansas, 1966, Gg

Technical Director:
Jonathon L. Cooper, (B), W Washington, Gg

Century College

Earth Science (A) 🖱
3300 Century Avenue North
White Bear Lake, MN 55110
p. (651) 779-3242
joe.osborn@century.edu
http://www.century.edu/futurestudents/programs/pnd.aspx?id=66
Programs: AS in Earth Science
Enrollment (2020): A: 2 (2)

Head:
Joe Osborn, (M), Zg

Instructor:
Jill Bries-Korpik, (M), Zg

Fond du Lac Tribal and Community College

Fond du Lac Tribal and Community College (A) ☒
2101 14th Street
Cloquet, MN 55720
http://www.fdltcc.edu/

Gustavus Adolphus College

Environment, Geography, and Earth Sciences (B) ☒
800 West College Avenue
St Peter, MN 56082
p. (507) 933-7333
jbartley@gustavus.edu
http://www.gustavus.edu/ege
Programs: Geology, Geography, Environmental Studies
Administrative Assistant: Jennifer Kruse
Enrollment (2022): B: 84 (37)

Co-chair:
Julie K. Bartley, (D), California (Los Angeles), 1994, GsPgo
Joaquin Villanueva, (D)

Chair:
Laura Triplett, (D), Minnesota, 2008, GemCl

Associate Professor:
Jeff La Frenierre, (D), Ohio State, 2014, GlHg
Anna Versluis, (D)

Assistant Professor:
Tiffany Grobelski, (D)
Erik Gulbranson, (D), PcGsSc
Rory McFadden, (D), Minnesota, 2009, GcxZe

Visiting Assistant Professor:
John E. Bailey, (D), Hawaii, 2005, GvmZr

Emeritus:
James L. Welsh, (D), Wisconsin, 1982, Gxc

Cooperating Faculty:
Daniel Mollner, Zn

Inver Hills Community College

Geology Dept (A)
2500 East 80th Street
Inver Grove Heights, MN 55076
admissions@inverhills.edu
http://www.inverhills.edu/Departments/Geology/

Instructor:
Jill Bries Korpik, Gg

Itasca Community College

Geography/Geographic Info Systems (GIS) (A) 🖱
1851 East Highway 169
Grand Rapids, MN 55744
p. (218) 322-2364
timothy.fox@itascacc.edu
https://www.itascacc.edu/gis
Programs: Geography/GIS A.S. Degree, Environmental Studies A.S. Degree, Natural Resources- Geospatial A.A.S. Degree
Certificates: GIS Professional Certificate (fully online)

Head:
Tim Fox, (M), Zi

Lake Superior College

Liberal Arts & Sciences Dept (A)
2101 Trinity Rd.
Duluth, MN 55811
m.whitehill@lsc.edu
http://www.lsc.edu

Instructor:
Matthew Whitehill, Gg

Macalester College

Geology Department (B) ☒
1600 Grand Avenue
St Paul, MN 55105
p. (651) 696-6000
rogers@macalester.edu
http://www.macalester.edu/geology/
Programs: Geology
Enrollment (2021): B: 23 (12)
Chair:
Raymond R. Rogers, (D), Chicago, 1995, GsrPv
Professor:
Kristina A. Curry Rogers, (D), SUNY (Stony Brook), 2001, PveGg
Kelly MacGregor, (D), California (Santa Cruz), 2002, GmlGg
Associate Professor:
Alan Chapman, (D), Caltech, 2011, GctCc
Assistant Professor:
Emily First, (D), Hawaii (Manoa), 2017, GivXg
Geology Lab Supervisor:
Jeffrey T. Thole, (M), Washington State, 1991, GgCaHw

Minneapolis Community and Technical College

Div of Arts and Sciences
1501 Hennepin Ave S
Minneapolis, MN 55403
admissions.office@minneapolis.edu
http://www.minneapolis.edu/academics/artsandsciences.cfm

Minnesota State University

Chemistry and Geology (B)
Department of Chemistry and Geology
241 Ford Hall
Mankato, MN 56001
p. (507) 389-1963
christine.cords@mnsu.edu
http://cset.mnsu.edu/chemgeol/programs/geol/
Programs: Geology (B); Earth Science Education (B); Earth Science (B)
Certificates: Environmental Geology
Professor:
Bryce W. Hoppie, (D), California (Santa Cruz), 1996, GeHw
Steven Losh, (D), Yale, 1985, GzxGo
Associate Professor:
Chad Wittkop, (D), Minnesota, GsCgGl

Normandale Community College

Dept of Geography and Geology (A)
9700 France Avenue South
Bloomington, MN 55431
p. (952) 358-8668
lindsay.iredale@normandale.edu
http://www.normandale.edu/departments/stem-and-education/geography-and-geology
Instructor:
David J. Berner, (M), Colorado
Douglas J. Claycomb, (D), Texas A&M
Richard P. Dunning, (D), Wisconsin
Carolyn Dykoski, (M), Minnesota
Annia Fayon, (D), Arizona State
Lindsay Iredale, (M), Minnesota
Paul D. Sabourin, (D), Minnesota
Ronald D. Ward, (D), Georgia
Other:
Cary Komoto, (D), Minnesota, 1994, ZnnZn

North Hennepin Community College

Earth & Environmental Science Dept (A) ⃠
7411 85th Avenue N.
Brooklyn Park, MN 55445
p. (763) 424-0869
megan.jones@nhcc.edu
http://www.nhcc.edu/academic-programs/academic-departments/geology
Programs: Environmental Science (AS)
Enrollment (2017): A: 1 (0)

Northland Community & Technical College

Liberal Arts Program
2022 Central Avenue NE
East Grand Forks, MN 56721
p. (218) 683-8694
http://www.northlandcollege.edu

Rochester Community & Technical College

Department of Earth Sciences
851 30th Ave SE
Rochester, MN 55904
p. (507) 285-7220
Sadredin.Moosavi@rctc.edu
http://www.rctc.edu
Instructor:
John C. Tacinelli, (D), Minnesota, 2000, Gig

Saint Cloud State University

Dept of Atmospheric and Hydrologic Sciences (B)
720 4th Avenue South
St Cloud, MN 56301-4498
p. (320) 308-3260
arhansen@stcloudstate.edu
http://www.stcloudstate.edu/eas/
Administrative Assistant: Debbie Schlumpberger
Enrollment (2012): B: 119 (31)
Professor:
Anthony R. Hansen, (D), Iowa State, 1981, Am
Professor:
Kate S. Pound, (D), Otago (NZ), 1993, GgZeGd
Robert A. Weisman, (D), SUNY (Albany), 1988, Am
Associate Professor:
Juan J. Fedele, (D), Illinois, 2003, Hg
Jean L. Hoff, (D), North Dakota, 1989, HwZeGg
Rodney Kubesh, (D), Illinois, 1991, Am
Assistant Professor:
Brian J. Billings, (D), Nevada (Reno), 2009, Am

St. Cloud State University

Department of Geography & Planning
720 4th Ave S
Stewart Hall 359
St. Cloud, MN 56301-4498
p. (320) 308-3160
geog@stcloudstate.edu
https://www.stcloudstate.edu/geogplan/
Programs: B.A. Geography
M.S. Geography with GIS emphasis
Certificates: Land Surveying

University of Minnesota Duluth

Earth & Environmental Science (B,M) O☒
230 Heller Hall
1114 Kirby Drive
Duluth, MN 55812
p. (218) 726-8385
dees@d.umn.edu
http://www.d.umn.edu/dees/
f: https://www.facebook.com/umdees/
Programs: Geological Sciences; Environmental Science; Earth Sci
Administrative Assistant: April A. Finkenhoefer
Enrollment (2021): B: 161 (30) M: 10 (10)
Head:
Karen B. Gran, (D), Washington, 2005, Gms
Emeritus:
John C. Green, (D), Harvard, 1960, GieGv
Timothy Holst, (D), Minnesota, GcYnGt
Thomas C. Johnson, (D), California (San Diego), 1975, PcGn
Ronald L. Morton, (D), Carleton, 1976, GvEg
Professor:
Erik T. Brown, (D), MIT, 1990, Og
John W. Goodge, (D), California (Los Angeles), 1987, Gp
Vicki L. Hansen, (D), California (Los Angeles), 1987, GtXgGc
Howard D. Mooers, (D), Minnesota, 1988, GleHw

Emeritus:
James D. Miller, (D), Minnesota, 1986, GiEd
Penelope C. Morton, (D), Carleton, 1982, EgGpCg
Director of Graduate Studies:
Byron A. Steinman, (D), Pittsburgh, 2011, Gn
Associate Professor:
Fred A. Davis, (D), Minnesota, 2012, CpGi
Christina D. Gallup, (D), Minnesota, 1997, Cct
John B. Swenson, (D), Minnesota, 2000, GrsGm
Nigel J. Wattrus, (D), Minnesota, 1984, Yr
Assistant Professor:
Latisha A. Brengman, (D), Tennessee, 2015, Gd
Salli F. Dymond, (D), Minnesota, 2014, Hg
Wendy F. Smythe, (D), Oregon Health & Sci, 2015, Ou

University of Minnesota, Morris

Div of Science & Mathematics (B)
Geology Discipline
Science Building
600 East 4th Street
Morris, MN 56267
p. (320) 589-6300
geol@mrs.umn.edu
http://www.mrs.umn.edu
Coordinator:
James F. Cotter, (D), Lehigh, 1984, Gl
Professor:
Keith A. Brugger, (D), Minnesota, 1992, GlGm
James B. Van Alstine, (D), North Dakota, 1980, Pg
Emeritus:
Peter M. Whelan, (D), California (Santa Cruz), 1988, Gx

University of Minnesota, Twin Cities

Dept of Civil, Environmental and Geo-Engineering (B,M,D)
500 Pillsbury Drive SE
Minneapolis, MN 55455
p. (612) 625-5522
cege@umn.edu
http://www.ce.umn.edu/
Enrollment (2010): B: 26 (8) M: 11 (4) D: 49 (2)
Head:
John S. Gulliver, (D), Minnesota, 1980, Hs
Professor:
Emmanuel M. Detournay, (D), Minnesota, 1983, NrpNm
Andrew Drescher, (D), Inst of Fund Tech Res (Poland), 1968, So
Efi Foufoula-Georgiou, (D), Florida, 1985, Hg
Joseph F. Labuz, (D), Northwestern, 1985, Nr
Otto D. Strack, (D), Delft (Neth), 1973, Hw
Vaughan R. Voller, (D), Sunderland, 1980, Zm
Associate Professor:
Randal J. Barnes, (D), Colorado Mines, 1985, Zn
Bojan B. Guzina, (D), Colorado, 1996, Nr
Karl A. Smith, (D), Minnesota, 1980, Nx
Research Associate:
Sonia Mogilevskaya, (D), Russian Acad of Sci, 1987, Ng
Adjunct Professor:
Peter A. Cundall, (D), Nr

Dept of Earth & Environmental Sciences (B,M,D) O☒
116 Church Street SE
150 John T. Tate Hall
Minneapolis, MN 55455
p. (612) 624-1333
esci@umn.edu
http://www.esci.umn.edu
t: @UMNEarthScience
Programs: Earth Sciences; Geology; Geophysics; Geobiology; Environmental Sciences; Hydrogeology
Enrollment (2016): B: 48 (15) M: 15 (2) D: 41 (7)
Director, PGC:
Paul Morin, Zn
Head:
Donna L. Whitney, (D), Washington, 1991, Gpt
Teaching Professor:
Kent C. Kirkby, (D), Wisconsin, 1994, Gg
Regents Professor:
R. Lawrence Edwards, (D), Caltech, 1988, Cc
Director, Minnesota Geological Survey:
Harvey Thorleifson, (D), Colorado, 1989, Gg
Director of IRM:
Bruce M. Moskowitz, (D), Minnesota, 1980, Ym
Assoc Director of IRM:
Joshua Feinberg, (D), California (Berkeley), 2005, YmGze
Professor:
David L. Fox, (D), Michigan, 1999, Pg
Marc M. Hirschmann, (D), Washington, 1992, Gi
Emi Ito, (D), Chicago, 1979, ClGnCs
Sally G. Kohlstedt, (D), Illinois, 1972, Rh
Peter Makovicky, (D), Columbia, 2002, Pv
Katsumi Matsumoto, (D), Columbia, 2000, Oc
Samuel Mukasa, (D), Cg
Christopher Paola, (D), MIT/WHOI, 1983, Gs
Vera Pospelova, (D), McGill, 2003, GePl
Justin Revenaugh, (D), MIT, 1989, Ys
William E. Seyfried, Jr., (D), S California, 1977, Cm
Christian P. Teyssier, (D), Monash, 1986, GctGg
Senior Scientist:
Amanda Dillman, (D), Yx
Associate Professor:
Jake Bailey, (D), S California, 2008, Pg
Cara M. Santelli, (D), MIT/WHOI, 2007, Pg
Assistant Professor:
Max Bezada, (D), Rice, 2010, Ys
Anna Graber, (D), Yale, 2016, Rh
Lars N. Hansen, (D), Minnesota, 2012, Yx
Peter Kang, (D), Hg
Crystal Ng, (D), MIT, 2008, Hy
Ikuko Wada, (D), Victoria, 2009, ZnYgGt
Andrew D. Wickert, (D), Colorado, 2014, Gml
Xinyuan Zheng, (D), Oxford, 2014, CsOc
XRCT Lab Manager:
Brian Bagley, (D), Minnesota, 2011, YsZn
Research Associate Professor:
Michael Jackson, (D), Michigan, 1986, Ym
Microprobe Manager:
Anette von der Handt, (D), GiCt
Research Associate:
Dario Bilardello, (D), Lehigh, 2009, Ym
Randy Calcote, (D), Minnesota, 2000, Ple
Hai Cheng, (D), Nanjing (China), 1988, CcPe
Kang Ding, (D), Acad Sinica (Beijing), 1987, Cg
Beverly Flood, (D), Pg
Jed Mosenfelder, (D), Cp
Mark Shapley, Minnesota, 2005, Gn
Ivanka Stefanova, (D), Sofia, 1991, Pl
Alexander Stone, (M), Gn
Chunyang Tan, (D), Zhejiang, 2011, Cm
Mark Zimmerman, (D), Minnesota, 1999, Yx
Adjunct Professor:
James E. Almendinger, (D), Minnesota, 1988, Pe
Val W. Chandler, (D), Purdue, 1977, Ye
Mark B. Edlund, (D), Michigan, 1998, Pe
Daniel R. Engstrom, (D), Minnesota, 1983, Pe
Annia K. Fayon, (D), Arizona State, 1997, Gc
Robert G. Johnson, (D), Iowa State, 1952, Pe
Shenghua Mei, (D), Yx
Kristina Curry Rogers, (D), Pg
Raymond Rogers, (D), Pg
Anthony Runkel, (D), Texas, 1988, Gs
Emeritus:
E. Calvin Alexander, Jr., (D), Missouri (Rolla), 1970, HwCcl
Subir K. Banerjee, (D), Cambridge, 1963, Ym
Roger LeB Hooke, (D), Gl
Peter J. Hudleston, (D), Imperial Coll (UK), 1969, Gc
Karen L. Kleinspehn, (D), Princeton, 1982, Gt
David L. Kohlstedt, (D), Illinois, 1970, YxGzi
Hans Olaf Pfannkuch, (D), Paris, 1962, HwGeRh
James H. Stout, (D), Harvard, 1970, GpzGt
Paul W. Weiblen, (D), Minnesota, 1965, Gi
Student Services:
Jennifer Petrie
Researcher:
Michelle LaRue, (D), Zi
Remote Sensing Specialist:
Cathleen Torres Parisian, (M), Zr

Postdoc:
Joseph Byrnes, (D), Oregon, 2016, Ys
Director, CSDCO:
Anders Noren, Gn
Dept Administrator:
Sharon J. Kressler, (B), Zn
Department Safety Officer:
Scott Alexander, (B), Hw
Curator, LacCore:
Kristina Brady, (M), Minnesota, 2006, Gn
Research Fellow:
Cole Kelleher, (M), Zi
Claire Porter, (M), Zi
Geology Librarian:
Carolyn Rauber, (M), Zn
Coordinator - PGC:
Jonathan Pundsack, Zn
Cooperating Faculty:
Martin O. Saar, (D), California (Berkeley), 2003, Hg

Dept of Soil, Water & Climate (B,M,D) ●

1991 Upper Buford Circle
St. Paul, MN 55108-6028
p. (612) 625-8114
crosen@umn.edu
http://www.swac.umn.edu
Programs: Environmental Science, Policy and Management (BS); Land and Atmospheric Science (MS, PhD)
Administrative Assistant: Nicole Smiley
Enrollment (2022): B: 323 (90) M: 17 (7) D: 30 (4)
Head:
Carl J. Rosen, (D), California (Davis), 1983, Sc
Professor:
James C. Bell, (D), Penn State, 1990, SdfZe
Timothy J. Griffis, (D), McMaster, 2000, As
Dylan B. Millet, (D), California (Berkeley), 2003, As
David J. Mulla, (D), Purdue, 1983, Sp
Michael A. Schmitt, (D), Illinois, 1985, Sc
Jeffrey S. Strock, (D), North Carolina State, 1999, So
Brandy M. Toner, (D), California (Berkeley), 2004, ClmSc
Kyungsoo Yoo, (D), California (Berkeley), 2003, Sd
Associate Professor:
Melinda L. Erickson, Minnesota, 2005, Hw
Fabian G. Fernandez, (D), Purdue, 2006, ScbSo
Jessica L. M. Gutknecht, (D), Wisconsin, 2007, Sb
Satoshi Ishii, (D), Minnesota, 2007, SbPg
Nicolas Jelinski, (D), Minnesota, 2014, SdCs
Daniel E. Kaiser, (D), Iowa State, 2007, Sc
Yuxin Miao, (D), Minnesota, 2005, ScZf
Paulo H. Pagliari, (D), Wisconsin, 2012, Sc
Albert L. Sims, (D), North Carolina State, 1992, Sc
Peter K. Snyder, (D), Wisconsin, 2003, As
Tracy E. Twine, (D), Wisconsin, 2003, As
Melissa L. Wilson, (D), Minnesota, 2012, Sco
Assistant Professor:
Anna Cates, (D), Wisconsin, 2018, Sb
Peter Neff, (D), Victoria (Wellington), 2015, Pc
Linday Pease, (D), Ohio State, 2016
Heidi Roop, Victoria (Wellington), 2015, At
Vasudha Sharma, (D), Nebraska, 2018, SpHg
Adjunct Professor:
John M. Baker, (D), Texas A&M, 1987, Sp
Gary Feyereisen, (D), Minnesota, 2005, Sp
Jane M F Johnson, (D), Minnesota, 1995, So
Randall K. Kolka, (D), Minnesota, 1996, So
Pamela J. Rice, (D), Iowa State, 1996, So
Kurt Spokas, (D), Minnesota, 2005, Spo
Rodney T. Venterea, (D), California (Davis), 2000, Sp
Emeritus:
Deborah L. Allan, (D), California (Riverside), 1987, Sb
James L. Anderson, (D), Wisconsin, 1976, Sd
Paul R. Bloom, (D), Cornell, 1978, Sc
Terence H. Cooper, (D), Michigan State, 1975, Sd
Robert H. Dowdy, (D), Michigan State, 1966, Sc
Satish C. Gupta, (D), Utah State, 1972, Sp
William C. Koskinen, (D), Washington State, 1980, Sc
John A. Lamb, (D), Nebraska, 1984, Sc
Gary L. Malzer, (D), Purdue, 1973, Sc
Jean-Alex E. Molina, (D), Cornell, 1967, SbCgSo
John F. Moncrief, (D), Wisconsin, 1981, Sp
Edward A. Nater, (D), California (Davis), 1987, Sd
Gyles W. Randall, (D), Wisconsin, 1972, So
George W. Rehm, (D), Minnesota, 1969, Sc
Michael P. Russelle, (D), Nebraska, 1982, Sc
Michael J. Sadowsky, (D), Hawaii, 1983, Sb
Mark W. Seeley, (D), Nebraska, 1977, As
Executive Secretary:
Kari A. Jarcho, (B), Zn

University of Saint Thomas

Earth, Environment and Society, Geology Program (B) ●☒

OWS 153
2115 Summit Avenue
St Paul, MN 55105
p. (651) 962-5241
tahickson@stthomas.edu
http://www.stthomas.edu/geology/
f: https://www.facebook.com/ustgeology
Programs: Geology BA, BS and Minor
Environmental Science with Geology, Biology, or Chemistry Concentrations
Geography BA, Minor
GIS Minor
Sustainability Minor
Certificates: GIS
Enrollment (2020): B: 14 (7)
Director:
Thomas A. Hickson, (D), Stanford, 1999, GsmGe
Professor:
Melissa A. Lamb, (D), Stanford, 1998, GtcGs
Kevin Theissen, (D), Stanford, GnOg
Associate Professor:
Jeni McDermott, (D), Arizona State, GmHw

Vermilion Community College

Vermilion Community College

1900 East Camp Street
Ely, MN 55731
p. (218) 235-2173
http://www.vcc.edu/

Winona State University

Dept of Geoscience (B) ☒

P.O. Box 5838
Winona, MN 55987
p. (507) 457-5260
geoscience@winona.edu
https://www.winona.edu/geoscience/
f: https://www.facebook.com/GeoscienceWSU
Programs: Geoscience (B); Geology (B); Environmental Science (B); Earth Science Teaching (B)
Enrollment (2020): B: 50 (16)
Professor:
Stephen T. Allard, (D), Wyoming, 2003, GctGx
Jennifer LB Anderson, (D), Brown, 2004, XgYgZe
William L. Beatty, (D), Pittsburgh, 2003, PgiGs
Candace L. Kairies-Beatty, (D), Pittsburgh, 2003, GeClRw
Associate Professor:
Dylan Blumentritt, (D), Minnesota, 2013, HsGm
Emeritus:
John F. Donovan, (D), Cornell, 1963, Eg
Nancy O. Jannik, (D), New Mexico Tech, 1989, Hw
Jamie Ann Meyers, (D), Indiana, 1971, Gsd
Dennis N. Nielsen, (D), North Dakota, 1973, Gm
College Laboratory Services Specialist:
Luke Zwiefelhofer, (B), Wyoming, 2004, Zn

Mississippi

Jackson State University

Chemistry, Physics, and Atmospheric Sciences (B,M,D) ☒

1400 J. R. Lynch St.
JSU Box 17660
Jackson, MS 39217
p. (601) 979-7012
mfadavi@jsums.edu

http://www.jsums.edu/cset/phyat.htm
Programs: Physics (B); Chemistry (B); Meteorology (B); Earth System Science (B); Chemistry (M,D)
Enrollment (2014): B: 9 (4)

Millsaps College

Dept of Geosciences (B) O☒
Box 150648
1701 North State Street
Jackson, MS 39210
p. (601) 974-1344
galics@millsaps.edu
http://www.millsaps.edu/geology/
Enrollment (2018): B: 8 (5)

Chair:
Zachary A. Musselman, (D), Kentucky, 2006, Gm

Professor:
Stan Galicki, (D), Mississippi, 2002, Ged
James B. Harris, (D), Kentucky, 1992, Ye

Emeritus:
Delbert E. Gann, (D), Missouri Sch of Mines, 1976, Gz

Mississippi State University

Dept of Geosciences (B,M,D)
P. O. Box 5448
108 Hilbun Hall
East Lee Blvd.
Mississippi State, MS 39762
p. (662) 325-3915
rclary@geosci.msstate.edu
http://www.geosciences.msstate.edu/
f: https://www.facebook.com/MSUGeosciences/
Enrollment (2014): B: 505 (86) M: 224 (103) D: 24 (4)

Professor:
Michael E. Brown, (D), North Carolina, 1999, Am
Brenda L. Kirkland, (D), Louisiana State, 1992, GdEoPo
Darrel W. Schmitz, (D), Texas A&M, 1991, Hw

Associate Professor:
Shrinidhi Ambinakudige, (D), Florida State, 2006, ZiyZr
Renee M. Clary, (D), Louisiana State, 2003, ZePgGe
Jamie Dyer, (D), Georgia, 2005, Am
Andrew Mercer, (D), Oklahoma, 2008, Ams
John C. Rodgers, (D), Georgia, 1999, Zy
Kathleen M. Sherman-Morris, (D), Florida State, 2006, As

Assistant Professor:
Padmanava Dash, (D), Louisiana State, 2011, ZrHg
Christopher Fuhrmann, (D), 2011, As
Rinat Gabitov, (D), Rensselaer, 2005, Cas
Qingmin Meng, (D), Georgia, 2006, Zig
Adam Skarke, (D), Delaware, 2013, GuYrOn
Kim Wood, (D), Arizona, 2012, AsmZr

Research Associate:
Katarzynz Grala, (M), Iowa State, 2004, Zi

Instructor:
Christa M. Haney, (M), Mississippi State, 1999, Am
Amy P. Moe-Hoffman, (M), Colorado, 2002, Zn
John A. Morris, (M), Mississippi State, 2007, ZirAm
Lindsey Morschauser, (M), Mississippi State, Am
Athena Nagel, (D), Mississippi State, 2014, ZgiZr
Greg Nordstrom, (M), Mississippi State, 2007, Am
Tim Wallace, (M), Mississippi State, 1994, Am

Adjunct Professor:
Paul J. Croft, (D), Rutgers, 1991, Am
Patrick J. Fitzpatrick, (D), Colorado State, 1995, Am
James May, (D), Texas A&M, 1988, Hw
Jack C. Pashin, (D), Kentucky, 1990, GoEcGs
Janet E. Simms, (D), Texas A&M, 1991, Yg
Jayaram Veeramony, (D), Delaware, 1999, Og

Emeritus:
John M. Kaye, (D), Louisiana State, 1974, Gg
John E. Mylroie, (D), Rensselaer, 1977, GmHyGs
Charles L. Wax, (D), Louisiana State, 1977, As

Distance Academic Coordinator:
Mary A. Dean

Collection Manager:
Amy P. Moe-Hoffman, (M), Colorado, 2002, PgZe

Business Manager:
Jerri Wright, (B), Mississippi Women, 1996

Academic Coordinator:
Tina Davis, (M), Mississippi Women

Related Staff:
Cynthia Bell

University of Mississippi

Dept of Geology & Geological Engineering (B,M,D) ●
School of Engineering
P.O. Box 1848
120A Carrier Hall
University, MS 38677-1848
p. (662) 915-7498
geology@olemiss.edu
http://www.engineering.olemiss.edu/gge/
Enrollment (2016): B: 301 (45) M: 21 (6) D: 5 (0)

Professor:
Adnan Aydin, (D), Memorial, 1994, NgrYe
Gregg R. Davidson, (D), Arizona, 1995, Hw
Gregory L. Easson, (D), Missouri (Rolla), 1996, Zr
Robert M. Holt, (D), New Mexico Tech, 2000, Hq

Associate Professor:
Louis G. Zachos, (D), Texas, 2008, PiGeNx

Assistant Professor:
Jennifer N. Gifford, (D), Florida, 2013, GtxCg
Andrew M. O'Reilly, (D), Florida, 2012, Hwq
Brian F. Platt, (D), Kansas, 2009, GsPe
Lance D. Yarbrough, (D), Mississippi, 2006, Ng

Instructor:
Inoka Widanagamage, (D), Kent State, 2015, ClGep

Lecturer:
Cathy A. Grace, (M), Mississippi, 1996, Gg

University of Southern Mississippi

Dept of Geography and Geology (B,M,D) O
118 College Drive, Box 5051
Walker Science Building, Room 127
Hattiesburg, MS 39406
p. (601) 266-4729
david.cochran@usm.edu
http://www.usm.edu/geography-geology
f: https://www.facebook.com/groups/Geography.GeologyUSM/
t: @geousm
Programs: Geology (B, M); Geography (B, M, D)
Certificates: Geographic Information Technologies
Enrollment (2014): B: 114 (19) M: 30 (5) D: 4 (1)

Professor:
Greg Carter, (D), Wyoming, 1985, Zyi
Clifton Dixon, (D), Texas A&M, 1988, Zn
Maurice A. Meylan, (D), Hawaii, 1978, Gu
Mark Miller, (D), Arizona, 1988, Zn
David M. Patrick, (D), Oklahoma, 1972, Ng
Mark Puckett, (D), Alabama, PmiGg
Andy Reese, (D), Louisiana State, 2003, Zy

Associate Professor:
Jerry Bass, (D), Texas, 2003, Zn
David Cochran, (D), Kansas, 2005, Zn
Franklin Heitmuller, (D), Texas, 2009, GmsGe
David Holt, (D), Arkansas, 2002, Zyi
Bandana Kar, (D), South Carolina, 2008, Zi
George Raber, (D), South Carolina, 2003, Zi

Assistant Professor:
Jeremy Deans, (D), Texas Tech, 2016, YrGc

Instructor:
Evan Bagley, (M), GgEo
Alyson Brink, (D), Texas Tech, 2016, PgvGg

Gulf Coast Research Laboratory (M,D) O⊘
703 E. Beach Drive
Ocean Springs, MS 39564
p. (228) 818-8804
read.hendon@usm.edu
http://gcrl.usm.edu

Director:
Joe Griffitt, (D)

Chair:
Leila Hamdan, (D)

Marine Science (B,M,D) ☒
1020 Balch Boulevard
National Aeronautics and Space Administration
Stennis Space Center, MS 39529
p. (228) 688-3177
Carlton.P.Anderson@usm.edu
http://www.usm.edu/marine
f: https://www.facebook.com/SouthernMissDivisionOfMarine-Science/
t: @USMMarineSci
Programs: Marine Science; Hydrographic Science; Ocean Eng
Certificates: UMS Certificate Program - Tier 1
UMS Operator Certificate Program - Tier 2
Enrollment (2021): B: 8 (1) M: 26 (15) D: 22 (2)

Associate Director, SOSE, & Professor:
Jerry Wiggert, (D), S California, 1995, Op

Professor:
Stephan Howden, (D), Rhode Island, 1996, Op
Alan M. Shiller, (D), California (San Diego), 1982, OcCtHs
Denis A. Wiesenburg, (D), 1980, Og
Xiaodong Zhang, (D), Dalhousie, 2002, Op

Associate Professor:
Maarten Buijsman, (D), Utrecht, 2007, Op
Christopher T. Hayes, (D), 2013, Oc
Scott P. Milroy, (D), S Florida (St. Petersburg), 2007, Ob
Dmitri Nechaev, (D), Shirshov Inst, 1986, Op
Davin Wallace, (D), Rice, 2010, Ou

Assistant Professor:
Mustafa Kemal Cambazoglu, (D), Georgia Tech, 2009, No
Anand Devappa Hiroji, (D), 2016, Yd
Kristina Mojica, (D), Amsterdam, 2015, Ob
Gero Nootz, (D), C Florida, 2010, No
Johnson Olusegun Oguntuase, (D), S Mississippi, 2020
Mahdi Razaz, (D), Hiroshima, 2010

Associate Research Professor:
Arne R. Diercks, (D), S Mississippi, 1995, Ou
Leonardo Macelloni, (D), Rome, 2005, Yr

Assistant Research Faculty:
Diana N Bernstein, (D), Hebrew, 2014

Program Coordinator, Hyd Sci Deg Programs:
Alberto Pedrassani Costa Neves, (M), Naval War Brazil, 1996

Assistant Teaching Professor:
Joris L. Van Der Ham, (D), Louisiana, 2007

Professor:
Vernon L. Asper, (D), MIT, 1986
Jerald W. Caruthers, (D), Texas A&M, 1968
Vladimir M. Kamenkovich, (D), Oceanology, USSR Acad Sci, 1957
Steven E. Lohrenz, (D), MIT/WHOI, 1985

Missouri

Metropolitan Community College-Blue River

Geology and Geography Program (A)
20301 E. 78 Highway
Independence, MO 64057-2053
p. (816) 220-6622
benjamin.wolfe@mcckc.edu
http://www.mcckc.edu/progs/geol/geology/overview.asp

Metropolitan Community College-Kansas City

Geology Dept (A)
3200 Broadway
Kansas City, MO 64111
p. (816) 604-3335
william.young@mcckc.edu
http://www.mcckc.edu/programs/geology/

Instructor:
Alice Fuerst, Gg
John Horn, (D), Nebraska, GgZy
Carl Priesendorf, (M), Missouri, 1987, GgZyGe
Laura Veverka, (M), Missouri (Kansas City), Zy
Ben Wolfe, (M), Alaska (Fairbanks), 2001, GgZyg

Adjunct Professor:
Janet Raymer, (M), Missouri S&T, GgEo

Mineral Area College

Science Dept (A)
5270 Flat River Road
P.O. Box 1000
Park Hills, MO 63601
p. (573) 518-2314
bscheidt@MineralArea.edu
http://www.mineralarea.edu/faculty/academicDepartments/science.aspx
Programs: Associates Degree of Arts
Enrollment (2016): A: 2 (2)

Professor:
Brian Scheidt, (M), S Illinois, Hw

Missouri State University

Dept of Geography, Geology & Planning (B,M)
901 S. National
Springfield, MO 65897
p. (417) 836-5800
ggp@missouristate.edu
https://geosciences.missouristate.edu
f: https://www.facebook.com/MSUggp/
t: @MSUggp
Programs: https://geosciences.missouristate.edu
Certificates: https://geosciences.missouristate.edu
Administrative Assistant: Tracy Carroll, Deana Gibson
Enrollment (2016): B: 185 (56) M: 38 (8)

Distinguished Professor:
Kevin L. Mickus, (D), Texas (El Paso), 1989, GtYv
Robert T. Pavlowsky, (D), Wisconsin, 1995, Gm

Department Head:
Toby Dogwiler, (D), Missouri, 2002, ZiGmZy

Professor:
Kevin R. Evans, (D), Kansas, 1997, GrsGt
Douglas R. Gouzie, (D), Kentucky, 1986, HwGeCl
Melida Gutierrez, (D), Texas (El Paso), 1992, CgHy
Rajinder S. Jutla, (D), Virginia Tech, 1995, Zn
Jun Luo, (D), Wisconsin (Milwaukee), Zi
Judith Meyer, (D), Wisconsin, 1994, Zn
Xin Miao, (D), California (Berkeley), 2005, Zri
Charles W. Rovey, (D), Wisconsin (Milwaukee), 1990, HyGle

Associate Professor:
Alice (Jill) Black, (D), Missouri, 2003, ZeAsHw

Assistant Professor:
Mario Daoust, (D), McGill, 1992, AtZy
Krista M. Evans, (D), Clemson, 2017, Zn
Ron Malega, (D), Georgia, Zn
Matthew P. McKay, (D), West Virginia, 2015, Gct
Gary Michelfelder, (D), Montana State, 2015, GviCg
David R. Perkins, (D)

Senior Instructor:
Damon Bassett, (M), Missouri, 2003, Gg
Deborah Corcoran, (M), Michigan State, 1980, Zn
Linnea Iantria, (M), George Washington, Zn

Instructor:
Melanie E. Carden-Jessen, (M), Missouri State, 1998, Ze

Emeritus:
David A. Castillon, (D), Michigan State, 1972, Gm
John C. Catau, (D), Michigan State, 1973, Zn
William H. Cheek, (D), Michigan State, 1976, Zn
William Corcoran, (D), Michigan State, 1981, As
Dimitri Ioannides, (D), Rutgers, 1994, Zn
Elias Johnson, (D), Oklahoma, 1977, Zr
Vincent E. Kurtz, (D), Oklahoma, 1960, Ps
Erwin J. Mantei, (D), Missouri (Rolla), 1965, EdGez
Diane M. May, (M), S Illinois (Edwardsville), 1974, Zn
James F. Miller, (D), Wisconsin, 1970, Ps
Thomas D. Moeglin, (D), Nebraska, 1978, Ng
Thomas G. Plymate, (D), Minnesota, 1986, Gx
Paul A. Rollinson, (D), Illinois, 1988, Zn

Missouri University of Science and Technology

Geosciences and Geological and Petroleum Engineering (B,M,D) O
129 McNutt Hall
Rolla, MO 65409-0410
p. (573) 341-4616
rocks@mst.edu
https://ggpe.mst.edu/academic-programs/geology-and-geophysics/
f: https://www.facebook.com/MissouriSandTGGPE

t: https://twitter.com/MissouriSandT
Programs: Geology and Geophysics and Geological Engineering
Certificates: Petroleum Systems, Geoenvironmental Science and Engineering, Geophysics, Geoanalytics and Geointelligence, Space Resources, Natural Hazards, Subsurface Water Resources, Geotechnics, Military Geology
Enrollment (2021): B: 41 (10) M: 14 (5) D: 31 (6)

Professor:
David M. Borrok, (D), Notre Dame, 2005, ClqCs
Stephen Gao, (D), California (Los Angeles), 1995, Yg
Kelly Liu, (D), California (Los Angeles), 1998, Ye
Francisca Oboh-Ikuneobe, (D), Cambridge, 1991, Pg
Wan Yang, (D), Texas, 1995, Gs

Associate Professor:
Andreas Eckert, (D), Yg
John Hogan, (D), Gc
David Wronkiewicz, (D), Cl

Assistant Professor:
Marek Locmelis, (D), Macquarie, 2011, Eg
Jeremy Maurer, Stanford, 1997, YdRn
Jonathan Obrist Farner, (D), Missouri S&T, 2015, Gr
Ryan Smith, (D), Stanford, 2017, HwYg

Emeritus:
Richard D. Hagni, (D), Missouri, 1962, Em
Robert Laudon

Moberly Area Community College - Columbia Campus

Moberly Area Community College - Columbia Campus
601 Business Loop 70 West
Columbia, MO 65203
p. (573) 234-1067
TinaHying@macc.edu
http://www.macc.edu/

Northwest Missouri State University

Dept of Geology-Geography (B,M)
800 University Drive
Maryville, MO 64468
p. (660) 562-1723
crater@nwmissouri.edu

Chair:
C. Renee Rohs, (D), Kansas, 2000, Cg

Professor:
Gregory D. Haddock, (D), Idaho, 1996, Zi

Associate Professor:
Theodore L. Goudge, (D), Oklahoma State, 1984, Zy

Assistant Professor:
James Hickey, (D), Dartmouth, 2006, Ge
Ming-Chih Hung, (D), Utah, 2003, Zr
Yanfen Le, (D), Georgia, 2005, Zi
Leah D. Manos, (M), Tennessee, 1997, Zy

Instructor:
Jeffrey Bradley, (M), Oklahoma State, 1991, Zg

Emeritus:
Richard M. Felton, (M), Missouri, 1979, Pg

Ozarks Technical Community College

Physical Sciences
1001 E. Chestnut Expressway
Springfield, MO 65802
p. (417) 447-8238
ehrichp@otc.edu

Saint Louis University

Earth & Atmospheric Sciences (B,M,D) ☒
3642 Lindell Blvd
O'Neil Hall
Room 205
St. Louis, MO 63108
p. (314) 977-3116
linda.warren@slu.edu
http://www.slu.edu/department-of-earth-and-atmospheric-sciences-home
Programs: Geoscience with Concentrations in Geology, Geophysics and Environmental Geoscience; Meteorology; GIS; Geoinformatics
Certificates: GIS
Enrollment (2018): B: 69 (10) M: 7 (4) D: 5 (0)

Chair:
Charles E. Graves, (D), Iowa State, 1988, As

Professor:
Benjamin de Foy, (D), Cambridge, 1998, As
John Encarnacion, (D), Michigan, 1994, Gx
Jack Fishman, (D), Saint Louis, As
Daniel M. Hanes, (D), Scripps, 1983, GesGm
Zaitao Pan, (D), Iowa State, 1996, As
Lupei Zhu, (D), Caltech, 1998, YsGt

Associate Professor:
Elizabeth Hasenmueller, (D), Washington (St Louis), 2011, ClHs
Robert W. Pasken, (D), St Louis, 1981, Am
Vasit Sagan, (D), Zi
Linda M. Warren, (D), California (San Diego), 2003, YgsZn

Assistant Professor:
Zachary Phillips, (D), North Dakota State, 2020, Zi

Emeritus:
David J. Crossley, (D), British Columbia, 1973, YdsXy
Robert B. Herrmann, (D), St Louis, 1974, Ys

Systems Engineer:
Eric J. Haug, (M)

Computer/Electronics Tech:
Robert Wurth

University of Missouri

Dept of Geological Sciences (B,M,D) O☒
101 Geology Building
Columbia, MO 65211
p. (573) 882-6785
HuckabeyM@missouri.edu
http://geology.missouri.edu/
Administrative Assistant: Marsha Huckabey
Enrollment (2013): B: 60 (7) M: 13 (11) D: 21 (0)

Director of Geology Field Studies:
Miriam Barquero-Molina, (D), Texas, 2009, Gct

Chair:
Kevin L. Shelton, (D), Yale, 1982, EmCs
Alan Whittington, (D), Open (UK), 1997, GivCp

Professor:
Mian Liu, (D), Arizona, 1989, Yg
Kenneth A. MacLeod, (D), Washington, 1992, Pg
Peter I. Nabelek, (D), SUNY (Stony Brook), 1983, GipCu
Eric A. Sandvol, (D), New Mexico State, 1995, Ys

Associate Professor:
Martin S. Appold, (D), Johns Hopkins, 1998, HwEm
Francisco Gomez, (D), Cornell, 1999, Gt

Assistant Professor:
John Huntley, (D), Virginia Tech, 2007, PqgPe
James Schiffbauer, (D), Virginia Tech, 2009, Pgi

Emeritus:
Michael B. Underwood, (D), Cornell, 1984, GsuGt

School of Natural Resources (B,M,D) ●✓🖱
203 Natural Resources Building
Columbia, MO 65211
p. 15738827717
heh@missouri.edu
http://www.snr.missouri.edu/
Programs: Natural Resources
Certificates: Graduate and undergraduate GIS certificate
Enrollment (2021): B: 850 (0) M: 75 (0) D: 50 (0)

Professor:
Stephen H. Anderson, (D), North Carolina State, 1985, Sp
Hong S. He, (D), Chinese Acad Sci, 1989, ZicZi
Anthony R. Lupo, (D), Purdue, 1995, As
Patrick S. Market, (D), St Louis, 1999, As

Extension:
Patrick E. Guinan, (D), Missouri, 2004, As

Associate Professor:
Neil I. Fox, (D), Salford (UK), 1998, As

Instructor:
Eric A. Aldrich, (M), Missouri, 2011, As

Adjunct Professor:
Newell R. Kitchen, (D), Colorado State, 1990, So
Robert J. Kremer, (D), Mississippi State, 1981, Sb
Robert N. Lerch, (D), Colorado State, 1990, So

Emeritus:
Robert W. Blanchar, (D), Minnesota, 1964, Sc
Clark J. Gantzer, (D), Minnesota, 1980, So
Randall J. Miles, (D), Texas A&M, 1981, Sd

University of Missouri, Kansas City

Dept of Geosciences (B,M,D) ●⃠
5100 Rockhill Road
Room 420, Robert H. Flarsheim Hall
Kansas City, MO 64110-2499
p. (816) 235-1334
geosciences@umkc.edu
http://cas.umkc.edu/Geosciences/
t: @UMKC_geosci
Programs: Geology (B); Geography (B); Environmental Science (B)
Certificates: GIS
Enrollment (2018): B: 108 (56) M: 0 (6) D: 26 (2)

Chair:
Wei Ji, (D), Connecticut, 1991, ZriZy

Professor:
Jimmy Adegoke, (D), Penn State, 2000, AsZyr
James B. Murowchick, (D), Penn State, 1984, CgGz
Tina M. Niemi, (D), Stanford, 1992, Gt

Associate Professor:
Caroline P. Davies, (D), Arizona State, 2000, GnZce
Jejung Lee, (D), Northwestern, 2001, GqHw

Assistant Professor:
Alison Graettinger, (D), Pittsburg State, 2012, Gv
Fengpeng Sun, (D), California (Irvine), 2008, ZoiZc

Emeritus:
Raymond M. Coveney Jr, (D), Michigan, 1972, EgmCe
Steven L. Driever, (D), Georgia, 1977, ZnnZn
Richard J. Gentile, (D), Missouri, 1965, GrsNx
Syed E. Hasan, (D), Purdue, 1978, Ng

Washington University in St. Louis

Dept of Earth & Planetary Sciences (B,M,D) ☒
Mail Stop 1169 204 110
Rudolph Hall
1 Brookings Drive
St Louis, MO 63130-4899
p. (314) 935-5610
slava@wustl.edu
http://eps.wustl.edu
f: https://www.facebook.com/WashU.EPSci/
t: @WUSTL_EPS
Programs: Geology; Geochemistry; Geophysics; Environmental Earth Sciences
Administrative Assistant: Katherine M.. Totty
Enrollment (2017): B: 56 (15) M: 0 (3) D: 31 (5)

Chair:
Viatcheslav S. Solomatov, (D), Moscow Inst, 1990, Yg

Rudolph Professor of Earth & Planeary Sciences:
Bradley L. Jolliff, (D), SD Mines, 1987, GxzGi

Robert S. Brookings Distinguished Professor:
Douglas A. Wiens, (D), Northwestern, 1985, YsrYg

James S. McDonnell Distinguished University Professor:
Raymond E. Arvidson, (D), Brown, 1974, Xg

Professor:
Jeffrey G. Catalano, (D), Stanford, 2004, Cl
Robert F. Dymek, (D), Caltech, 1977, GpiGa
M. Bruce Fegley, (D), MIT, 1980, Xc
David A. Fike, (D), MIT, 2007, CsmPg
William B. McKinnon, (D), Caltech, 1981, Xg
Philip Skemer, (D), Yale, 2007, GcCp
Jennifer R. Smith, (D), Pennsylvania, 2001, Ga
William H. Smith, (D), Princeton, 1966, Xc
Michael E. Wysession, (D), Northwestern, 1991, Ys

Research Professor:
Katharina Lodders-Fegley, (D), Max Planck, 1991, Xc
Alian Wang, (D), Sci/Tech (France), 1987, Ca

Associate Professor:
Alexander S. Bradley, (D), MIT, 2008, CboGg
Paul Byrne, (D), Trinity (Ireland), 2010, Xg

Assistant Professor:
Bronwen L. Konecky, (D), Brown, 2014, PcGn
Michael J. Krawczynski, (D), MIT, 2011, Cg
Claire C. Masteller, (D), California (Santa Cruz), 2017, Hg
Rita Parai, (D), Harvard, 2014, Cs
Kun Wang, (D), Washington (St Louis), 2013, XcCsXm

Emeritus:
Robert E. Criss, (D), Caltech, 1981, Cs
Jill D. Pasteris, (D), Yale, 1980, GbzGe

Research Professor:
Anne M. Hofmeister, (D), Caltech, 1984, YghXa

Laboratory Aministrator:
Glenn (Bob) Osburn, (M), New Mexico Tech, 1978, Zn

Administrative Officer:
Robert Gemignani, (M), Notre Dame de Namure, 1998

Environmental Studies Program (B)
Box 1169
One Brookings Drive
St. Louis, MO 63130-4899
p. (314) 935-7047
enstadmin@levee.wustl.edu
http://enst.wustl.edu
Administrative Assistant: Barbara Winston
Enrollment (2010): B: 161 (59)

Director:
Jan P. Amend, (D), California (Berkeley), 1995, Co

Professor:
Raymond E. Arvidson, (D), Brown, 1974, Zn
Richard Axelbaum, (D), California (Davis), 1988, Ng
Pratim Biswas, (D), Caltech, 1985, Ng
Robert Blankenship, (D), California (Berkeley), 1975, Zn
Robert E. Criss, (D), Caltech, 1981, CsHqXa
Willem H. Dickhoff, (D), Free (Amsterdam), Zn
Mike Dudukovic, (D), Illinois Inst Tech, 1971, Zn
Robert F. Dymek, (D), Caltech, 2006, Gi
Claude Evans, (D), SUNY (Stony Brook), Zn
Bruce Fegley, (D), MIT, 1980, Xc
T.R. Kidder, (D), Harvard, 1988, Zn
Maxine I. Lipeles, (D), Harvard, 1979, Zn
William R. Lowry, (D), Stanford, 1988, Zn
Jill D. Pasteris, (D), Yale, 1980, Zn
Bruce Petersen, (D), Harvard, Zn
Robert Pollak, (D), MIT, 1964, Zn
Tab Rasmussen, (D), Duke, Zn
Barbara Schaal, (D), Yale, 1974, Zn
Glenn D. Stone, (D), Arizona, 1988, Zn
Robert W. Sussman, (D), Duke, 1972, Zn
Alan R. Templeton, (D), Michigan, 1972, ZicZr

Associate Professor:
Jon M. Chase, (D), Chicago, Zn
Clare Palmer, (D), Zn
Jen R. Smith, (D), Pennsylvania, 2001, Ga
Jay Turner, (D), Washington (St Louis), 1993, Zn

Assistant Professor:
Jeff Catalano, (D), Stanford, 2004, Ca
Geoff Childs, (D), Indiana, Zn
Ellen Damschen, (D), North Carolina State, Zn
Daniel Giammar, (D), Caltech, 2001, Zn
Young-Shin Jun, Harvard, 2005, Zn
Tiffany Knight, (D), Pittsburgh, 2003, Pe
John Orrock, (D), Iowa State, 2004, Zn

Engineering & Science Director:
Beth Martin, (M), Washington (St Louis), 1996, Ng

Geology Librarian:
Clara McLeod, Zn

William Jewell College

Dept of Biology
Liberty, MO 64068
p. (816) 781-3806 x230
allent@william.jewell.edu
http://www.jewell.edu

Chair:
Tara Allen

Associate Professor:
Charles F. J. Newlon, (M), Missouri, 1962, Zg

Montana

Flathead Valley Community College

Geology and Geography (A)
777 Grandview Drive
Kalispell, MT 59901
p. (406) 756-3873
aho@fvcc.edu
f: https://www.facebook.com/fvccmt/
t: @fvccmt
Programs: Transfer programs in geology/earth science, geography, environmental science and environmental studies. Also forestry, parks, tourism and recreation management, environmental biology and wildlife biology.
Certificates: AAS or certificates in natural resources conservation and management, geospatial technology and surveying.

Associate Professor:
Anita Ho, (D), Oregon, Gg

Montana State University

Dept of Earth Sciences (B,M,D) O☒
P.O. Box 173480
226 Traphagen Hall
Bozeman, MT 59717-3480
p. (406) 994-3331
earth@montana.edu
http://www.montana.edu/wwwes/
Enrollment (2018): B: 228 (45) M: 21 (8) D: 22 (1)

Regents Professor:
Cathy Whitlock, (D), Washington, 1983, Peg

Professor:
Mary Hubbard, (D), MIT, 1988
David R. Lageson, (D), Wyoming, 1980, GctGs
David W. Mogk, (D), Washington, 1984, GpZeGz
Mark L. Skidmore, (D), Alberta, 2001, ClPgGl
David J. Varricchio, (D), Montana State, 1995, Pv
William K. Wyckoff, (D), Syracuse, 1982, Zn

Associate Professor:
Jordy Hendrikx, (D), Canterbury (NZ), 2005, ZnHsAs

Assistant Professor:
Jean Dixon, (D), Dartmouth, 2009, GmScZg
Julia H. Haggerty, (D), Colorado, 2004, Zu
Andrew Laskowski, (D), Arizona, 2016
Jamie McEvoy, (D), Arizona, 2013, Ze
David B. McWethy, (D), Montana State, 2007, PeSb
Madison Myers, (D), Oregon, 2017
Devon Orme, (D), Arizona, 2015
Eric Sproles, (D), Oregon, 2012

Research Assistant Professor:
Frankie Jackson, (D), Montana State, 2007, Pv
Colin Shaw, (D), New Mexico, 2001, GcEmGp

Assistant Teaching Professor:
Christopher Organ, (D), Montana State, 2003

Research Assistant Professor:
David W. Bowen, (D), Colorado, 2001, GroGs

Regents Professor:
John R. Horner, Penn State, 2006, Pv

Dept of Land Resources & Environmental Sciences (B,M,D)
334 Leon Johnson Hall
P.O. Box 173120
Bozeman, MT 59717-3120
p. (406) 994-7060
lresfrontdesk@montana.edu
http://landresources.montana.edu/
Programs: Environmental Science
Land Rehabilitation
Land Resources & Environmental Sciences
Ecology & Environmental Sciences
Administrative Assistant: Dorie Seymour

Department Head:
Tracy M. Sterling, (D), Wisconsin, 1988, Zn

Professor:
William P. Inskeep, (D), Minnesota, 1985, Sc
Clain Jones, (D), So
Timothy McDermott, (D), Po
Geoff Poole, (D), Hg
John Priscu, (D), Po

Senior Scientist:
John Dore, (D), Hawaii, 1995, Ca

Associate Professor:
Jack Brookshire, (D), Zc
Stephanie Ewing, (D), Sd
Tony Hartshorn, (D), So
Robert Payn, (D), Hg
Scott Powell, (D), Zr

Assistant Professor:
Adam Sigler, (D), Rw

Research Associate:
William Kleindl, (D), Sf
Luke McKay, (D), Po
Ann Marie Reinhold, (D), Montana State, 2014, Hq

Instructor:
Nicholas Fox, (M), Zi
Paul B. Hook, (D), Colorado State, 1992, Sf

Emeritus:
James W. Bauder, (D), Utah State, 1974, Sf
Richard E. Engel, (D), Minnesota, 1983, Sc
Jeffrey Jacobsen, So
Rick L. Lawrence, (D), Oregon State, 1998, Zr
Cliff Montagne, (D), Montana State, 1976, Sd
Gerald A. Nielsen, (D), Wisconsin, 1963, Sd
David M. Ward, (D), Wisconsin, 1975, Po

Montana State University, Billings

Dept of Biological & Physical Sciences
1500 N. 30th Street
Billings, MT 59101
p. (406) 657-2341 x2028
ccastles@msubillings.edu
http://www.msubillings.edu/cas/sciences

Head:
Stanley Wiatr

Dean:
Tasneem Khaleel, (D), Bangalore, 1970, Pg

Professor:
Matt Benacquista, (D), Montana State, 1989, Xy
Thomas T. Zwick, (D), N Colorado, 1977, Ze

Montana Tech of the University of Montana

Dept of Chemistry & Geochemistry (B,M)
1300 West Park Street
Butte, MT 59701-8997
p. (406) 496-4207
acox@mtech.edu
https://mtech.edu/gradschool/degreeprograms/geochemistry.html
Department Secretary: Wilma Immonen

Professor:
Douglas A. Coe, (D), Oregon State, 1974, Cg
Douglas A. Drew, (D), Wyoming, 1971, Cg
Donald Stierle, (D), California (Riverside), 1979, Co

Assistant Professor:
Alysia Cox, (D), MIT/WHOI, 2011, CqlCm
John D. Hobbs, (D), New Mexico, 1991, Zn

Research Associate:
Wayne Olmsted, (B), Montana State, 1962, Ca
Andrea Stierle, (D), Montana State, Zn

Instructor:
Stephen R. Parker, (M), Indiana, 1972, Zn

Emeritus:
Frank E. Diebold, (D), Colorado Mines, 1967, Cl
Alexis Volborth, (D), Helsinki, 1954, Ca

Dept of Geological Engineering (B,M) ●☒
1300 West Park Street
Butte, MT 59701
p. (406) 496-4262
dconrad@mtech.edu
https://www.mtech.edu/mines-engineering/geological
Programs: Geological Engineering (B, M); Geotechnical (B); Mining Geology (B); Hydrogeology (B, M); Petroleum Geology (B); Geology (M); Hydrogeological Engineering (M)
Enrollment (2018): B: 37 (9) M: 20 (7)

Chair:
Larry N. Smith, (D), New Mexico, 1988, GsmEo

Professor:
Christopher H. Gammons, (D), Penn State, 1988, CgEmCs
Mary M. MacLaughlin, (D), California (Berkeley), 1997, NrgNt
Diane Wolfgram, (D), California (Berkeley), 1977, EgoNm
Associate Professor:
Glenn D. Shaw, (D), California (Merced), 2009, HwCsHs

Geophysical Engineering Department (B,M) ○☒
1300 West Park Street
Butte, MT 59701
p. (406) 496-4401
mspeece@mtech.edu
https://www.mtech.edu/mines-engineering/geophysical
f: https://www.facebook.com/GeophysicalEngineeringMontanaTech/
Programs: Geophysical Engineering (B,M); Geoscience (M)
Enrollment (2018): B: 15 (5) M: 6 (4)
Head:
Marvin A. Speece, (D), Wyoming, 1992, YeuYs
Associate Professor:
Xiaobing Zhou, (D), Alaska (Fairbanks), 2002, Zr
Assistant Professor:
Trevor Irons, (D), Colorado Mines, Yue
Mohamed Khalil, (D), Giessen, 2002, YeuGe
Emeritus:
Curtis A. Link, (D), Houston, 1993, Ye

Rocky Mountain College
Dept of Geology (B) ☒
1511 Poly Drive
Billings, MT 59102
p. (406) 657-1101
kalakayt@rocky.edu
http://www.rocky.edu/academics/academic-programs/undergraduate-majors/geology/
f: https://www.facebook.com/rmcgeology/
Programs: Geology, Geology-Petroleum Systems Concentration
Enrollment (2021): B: 8 (2)
Professor:
Thomas J. Kalakay, (D), Wyoming, 2001, GcpGi
Associate Professor:
Emily Geraghty Ward, (D), Montana, 2007, ZeGc
Derek Sjostrom, (D), Dartmouth, 2002, ClGsHs

Salish Kootenai College
Dept of environmental Science
58138 US Hwy 93
Ronan, MT 59855
chris_frissell@skc.edu
http://www.skc.edu

University of Montana
Dept. of Geosciences (B,M,D)
32 Campus Drive #1296
Missoula, MT 59812-1296
p. (406) 243-2341
bendick@mso.umt.edu
http://www.umt.edu/geosciences
Programs: Geosciences; Geosciences Earth Education; International Field Geosciences
Administrative Assistant: Loreene Skeel
Enrollment (2017): B: 68 (0) M: 16 (0) D: 10 (0)
Chair:
James R. Staub, (D), South Carolina, 1985, GsrEo
Professor:
Joel Harper, (D), Wyoming, 1998, GlZc
Marc S. Hendrix, (D), Stanford, 1992, Gs
Nancy W. Hinman, (D), California (San Diego), 1987, Cgl
James W. Sears, (D), Queens, 1979, GctEg
George D. Stanley, Jr., (D), Kansas, 1977, Pi
Associate Professor:
Julia Baldwin, (D), MIT, 2003, Gp
Rebecca Bendick, (D), Colorado, 2000, Yg
Marco Maneta, (D), Extremadura (Spain), 2006, Hsg
Andrew Wilcox, (D), Colorado State, 2005, Gm
Assistant Professor:
Payton Gardner, (D), Utah, 2009, Hw
Hilary R. Martens, (D), Caltech, 2016, YgsYd
Research Associate:
Michael Hofmann, (D), Montana, 2005, GosGr

University of Montana Western
Environmental Sciences Dept (B) ●⊘
710 South Atlantic Street
Dillon, MT 59725-3598
p. (406) 683-7615
rebekah.levine@umwestern.edu
https://w.umwestern.edu/department/environmental-sciences/
f: https://www.facebook.com/UMWenvirosciences/
Programs: Environmental Science (B); Environmental Sustainability (B); Natural Resource Science (B); Earth Science Ed (B)
Enrollment (2022): B: 53 (12)
Chair:
Rebekah Levine, (D), New Mexico, 2016, GmHsSd
Professor:
Spruce W. Schoenemann, (D), Washington, 2015, GeCgs
Robert C. Thomas, (D), Washington, 1993, GseZe
Associate Professor:
Arica Crootof, (D), Arizona, 2019, RwcZi
Assistant Professor:
Foley T. Neil, (D), California (Santa Cruz), 2020, YuGl
Adjunct Professor:
Heidi Anderson-Folnagy, (D), Idaho, 2011, Gs
Emeritus:
R. Stephen Mock, (D), Montana State, 1989, Ca
Sheila M. Roberts, (D), Calgary, 1996, GeSo
Craig E. Zaspel, (D), Montana State, 1975, Yg

Nebraska

Central Community College
Physical Science (A)
4500 63rd Street
PO Box 1027
Columbus, NE 68602
p. (402) 562-1216
wadlonhilker@cccneb.edu
https://www.cccneb.edu/physicalscience
Instructor:
Denise Condreay, Zg

Chadron State College
Dept of Geosciences (B,M) ○
1000 Main Street
Chadron, NE 69337
p. (308) 432-6377
mleite@csc.edu
http://www.csc.edu/geoscience
f: https://www.facebook.com/CSCgeoscience/
Programs: Geoscience
Environmental Geoscience
Certificates: Water Resources Management
Administrative Assistant: Stacy Mittleider
Enrollment (2022): B: 12 (2) M: 0 (1)
Professor:
Michael B. Leite, (D), Wyoming, 1992, Gg
Assistant Professor:
Tawny Tibbits, (D), Ggz
Adjunct Professor:
Matthew Tibbits, (D), Gg

Creighton University
Dept of Atmospheric Sciences (B,M)
2500 California Plaza
Omaha, NE 68178
p. (402) 280-2641
zehnder@creighton.edu
Enrollment (2012): B: 10 (0) M: 3 (0)
Professor:
Joseph A. Zehnder, (D), Chicago, 1986, As
Associate Professor:
Jon M. Schrage, (D), Purdue, 1998, Am
Assistant Professor:
Timothy J. Wagner, (D), Wisconsin, 2011, As

Adjunct Professor:
Richard Ritz, (M), Texas A&M, Am
Emeritus:
Arthur V. Douglas, (D), Arizona, 1976, Zg

University of Nebraska, Kearney

Dept of Geography (B)
203 Copeland Hall
Kearney, NE 68849
p. (308) 865-8355
combshj@unk.edu
Chair:
Jason Combs, (D), Nebraska, 2000
Professor:
Vijay Boken, (D), Manitoba, 1999
Paul Burger, (D), Oklahoma State, 1997
Jeremy Dillon, (D), Kansas, 2001
Associate Professor:
John Bauer, (D), Kansas, 2006
Instructor:
Nate Eidem, (D), Oregon State, 2011
Matt Engel, (D), Nebraska, 2007

University of Nebraska, Lincoln

Dept of Earth & Atmospheric Sciences (B,M,D) ● ☒
126 Bessey Hall
Lincoln, NE 68588-0340
p. (402) 472-2663
crowe1@unl.edu
http://eas.unl.edu/
f: https://www.facebook.com/UNLEarthAtmosSci
Programs: Geology (B); Meteorology-Climatology (B); Earth and Atmospheric Sciences (M,D)
Administrative Assistant: Janelle Gerry, Tina M. Schinstock
Enrollment (2018): B: 97 (19) M: 38 (18) D: 20 (2)
Chair:
Clinton M. Rowe, (D), Delaware, 1988, As
Professor:
Mark R. Anderson, (D), Colorado, 1985, As
Christopher R. Fielding, (D), Durham (UK), 1982, Gs
Tracy D. Frank, (D), Michigan, 1996, GssCs
Sherilyn C. Fritz, (D), Minnesota, 1985, Pem
David M. Harwood, (D), Ohio State, 1986, PmGul
Adam L. Houston, (D), Illinois, 2004, Asm
Qi S. Hu, (D), Colorado State, 1992, As
R. M. Joeckel, (D), Iowa, 1993, GrsGm
David K. Watkins, (D), Florida State, 1984, PmGu
Vitaly A. Zlotnik, (D), Natl Inst Hydro Engi Geo (Russia), 1979, HwyHq
Associate Professor:
Caroline M. Burberry, (D), Imperial Coll (UK), 2008, Gct
Richard M. Kettler, (D), Michigan, 1990, Cl
Ross Secord, (D), Michigan, 2004, PvCs
Matthew S. Van Den Broeke, (D), Oklahoma, 2011, AsZr
Peter J. Wagner, (D), Chicago, 1995, PigPq
Karrie A. Weber, (D), Alabama, 2002, PgCl
Assistant Professor:
Lynne J. Elkins, (D), MIT/WHOI, 2009, GivCt
Irina Filina, (D), Texas, 2007, YgGtYe
Erin Haacker, (D), Hw
Research Assistant Professor:
Mindi L. Searls, (D), Washington, 2007, YgGg
Emeritus:
Ronald G. Goble, (D), Toronto, GxEg
Priscilla C. Grew, (D), California (Berkeley), 1967, GpeZe
Mary Anne Holmes, (D), Florida State, 1989, GsOuZn
Robert M. Hunt, (D), Columbia, 1971, Pv
Merlin P. Lawson, (D), Clark, 1973, AmtZr
David B. Loope, (D), Wyoming, 1981, Gsg
Darryll T. Pederson, (D), North Dakota, 1971, Hw
Norman D. Smith, (D), Brown, 1967, Gsm
Michael R. Voorhies, (D), Wyoming, 1966, Pv
Assistant Professor of Practice:
Dawn Kopacz, (D), Ast

School of Natural Resource Sciences (B,M,D) ● ☒
Hardin Hall
3310 Holdrege Street
Lincoln, NE 68583-0961
p. (402) 472-3471
jcarroll2@unl.edu
http://snr.unl.edu/
Programs: Applied Climate Science, Water Science, Environmental Restoration Science, Environmental Science.
Certificates: GIS
Enrollment (2019): B: 48 (21)
Research Hydrogologist:
Jozsef Szilagyi, (D), California (Davis), 1997, Hw
Quaternary Geologist:
Paul Hanson, (D), Nebraska, 2005, Gm
Nebraska State Geologist/Research Geologist:
R.M. Joeckel, (D), Iowa, 1993, Grs
Hydrogeochemist/Lab Services Director:
Daniel Snow, (D), Nebraska, 1996, Ge
Environmental Biophysicist:
Elizabeth A. Walter-Shea, (D), Nebraska, 1987, As
Earth Systems Scientist:
David C. Gosselin, (D), SD Mines, 1987, Ze
Climatologist:
Michael J. Hayes, (D), Missouri, 1994, As
Agricultural Climatologist:
Qi (Steve) Hu, (D), Colorado State, 1992, As
Nebraska State Climatologist/Applied Climatologist:
Martha D. Shulski, (D), Minnesota, 2002, As
Micrometeorologist:
Andrew E. Suyker, (D), Nebraska, 2000, As
Climatologist:
Tsegaye Tadesse, (D), Nebraska, 2002, ZirAs
Climate Scientist:
Deborah J. Bathke, (D), Ohio State, 2004, AsZe
Pedologist:
Judith Turk, (D), California (Riverside), 2012, Sd
Groundwater Hydrologist:
Troy Gilmore, (D), N Carolina State, 2015, Hw
Groundwater Geologist:
Jesse Korus, (D), Nebraska, 2015, Gs

University of Nebraska, Omaha

Dept of Geography and Geology (B,M)
6001 Dodge Street
DSC 260
Omaha, NE 68182-0199
p. (402) 554-2662
rshuster@unomaha.edu
http://www.unomaha.edu/college-of-arts-and-sciences/geology/
Programs: Geology, Environmental Science-Earth Science Option, Geography
Certificates: GIS
Administrative Assistant: Brenda Todd
Enrollment (2019): B: 64 (18) M: 20 (8)
Professor:
George F. Engelmann, (D), Columbia, 1978, PvGdr
Harmon D. Maher, Jr., (D), Wisconsin, 1984, GcsGt
Associate Professor:
Bradley Bereitschaft, (D), North Carolina (Greensboro), 2011, Zuy
Ashlee LD Dere, (D), Penn State, 2014, SdGmCl
Robert D. Shuster, (D), Kansas, 1985, GiaZe
Assistant Professor:
Zac Suriano, (D), Delaware, 2018, Atm
Emeritus:
Jeffrey S. Peake, (D), Louisiana State, 1977, ZyAtRm
John F. Shroder, Jr., (D), Utah, 1967, GmlZy

Nevada

College of Southern Nevada - West Charleston Campus

Dept of Physical Sciences (A)
6375 W. Charleston Blvd.
Las Vegas, NV 89146
p. (702) 651-7475
physic@csn.edu
http://www.csn.edu/pages/2497.asp
Lead Faculty:
Barbara Graham, (M), ZyAm
John E. Keller, (D), S Illinois, 2009, GeHw
Cynthia S. Shroba, (D), Illinois, Gg

Professor:
Patrick D. Clennan, (M), ZyGe
Gale D. Martin, (M), Gg
Instructor:
Douglas Sims, (D), Kingston, 2011, GeSp

Desert Research Institute

Earth & Ecosystems Sciences
2215 Raggio Parkway
Reno, NV 89512-1095
p. (775) 673-7300
ken.adams@dri.edu
Assistant Research Professor:
Thomas F. Bullard, (D), New Mexico, 1995, Gma
Professor:
John Arnone, (D), Yale, 1988, Pg
Colleen M. Beck, (D), California (Berkeley), 1979, Sa
Christian H. Fritsen, (D), S California, 1996, Pg
Eric McDonald, (D), New Mexico, 1994, Sd
Alison E. Murray, (D), California (Santa Barbara), 1998, Pg
David E. Rhode, (D), Washington, 1987, Pe
Staff Geomorphologist:
Sophie Baker, (M), Dalhousie, 2005, Gm
Associate Research Geomorphologist:
Steven N. Bacon, (M), Humboldt State, 2003, Gm
Associate Professor:
Kenneth D. Adams, (D), Nevada (Reno), 1997, Gm
Thomas F. Bullard, (D), New Mexico, 1995, Gm
Mary Cablk, (D), Oregon State, 1997, Zr
Lynn Fenstermaker, (D), Nevada, 2003, Zr
Giles Marion, (D), California (Berkeley), 1974, Sc
Kenneth C. McGwire, (D), California (Santa Barbara), 1992, Zy
David A. Mouat, (D), Oregon State, 1974, Ge
Assistant Professor:
JoseLuis Antinao, (D), Dalhousie, 2009, Gm
Donald E. Sabol, Jr, (D), Washington, 1991, Zr
GIS/Remote Sensing Scientist:
Timothy B. Minor, (M), California (Santa Barbara), 1982, Zr
Assistant Research Ecologist:
Richard Jasoni, (D), Texas A&M, 1998, So
Archaeological Technician:
David Page, (M), Nevada (Reno), 2008, Ga

Great Basin College

Science Dept (A,B) ⦸
1500 College Parkway
Elko, NV 89801
p. (775) 753-2120
caroline.meisner@gbcnv.edu
http://www2.gbcnv.edu/departments/SCI.html
Programs: Associates of Science - Pattern of Study Geosciences
Associates of Science - Natural Resources
Bachelor's of Arts - Natural Resources
Professor:
Caroline Bruno Meisner, (M), Oregon State, 2003, ZgSo

Truckee Meadows Community College

Dept of Physical Sciences
7000 Dandini Boulevard
Reno, NV 89512
p. (775) 673-7183
loanderson@tmcc.edu
http://www.tmcc.edu/physicalsci/

University of Nevada, Las Vegas

Geoscience Department (B,M,D) ●
4505 S. Maryland Parkway
Box 454010
Las Vegas, NV 89154-4010
p. (702) 895-3262
geodept@unlv.edu
http://geoscience.unlv.edu
f: https://www.facebook.com/UNLVGeoscience/
t: @UNLVGeoscience
Programs: Geology BS
Earth & Environmental Science BS
PhD & MS in Geoscience
Administrative Assistant: Maria I. Rojas
Enrollment (2022): B: 0 (29) M: 15 (1) D: 25 (2)
Chair:
Matthew S. Lachniet, (D), Syracuse, 2001, Css
Professor:
Brenda J. Buck, (D), New Mexico State, 1996, GbSd
Pamela C. Burnley, (D), California (Davis), 1990, GpyGg
Elisabeth M. Hausrath, (D), Penn State, 2007, CgScCl
Ganqing Jiang, (D), Columbia, 2002, GsClGr
David K. Kreamer, (D), Arizona, 1982, Hw
Michael L. Wells, (D), Cornell, 1991, Gc
Water Resourse Director/ Co-Chair:
Michael J. Nicholl, (D), Nevada (Reno), 1993, Hq
Associate Professor:
Andrew D. Hanson, (D), Stanford, 1998, Co
Jowitt Simon, (D), Leicester (UK), 2009, Eg
Associate Scientist:
Kathleen Zanetti, (M), Idaho, 1997, Gg
Assistant Professor:
Tomas Capaldi, (D), Texas, 2019, Gst
Kevin Konrad, (D), Oregon State, 2017, Cc
Andrew S. Marting, (D), Cardiff, 2019, EgCg
Margo Odlum , (D), Texas, 2019, Gt
Arya Udry, (D), Tennessee, 2014, GiXmGg
Research Associate:
Christopher Adcock, (D), Nevada (Las Vegas), 2014, Xg
Minghua Ren, (D), Baylor, 2003, Gig
Lecturer:
Gabriel Judkins, (D), Arizona State, 2009, Gg
Jeremy Koonce, (D), Nevada (Las Vegas), 2016, Hw
Visiting Professor:
Aubrey Bonde, (D), Nevada (Las Vegas), 2013, Pg
Emeritus:
Jean S. Cline, (D), Virginia Tech, 1990, EmCe
Rodney Metcalf, New Mexico, 1990, Gp
Margaret Rees, (D), Kansas, 1984, Gs
Stephen Rowland, (D), California (Santa Cruz), 1978, Pi
Eugene I. Smith, (D), New Mexico, 1970, Gi

University of Nevada, Reno

Center for Neotectonic Studies
MS 0169
Reno, NV 89557-0169
p. (775) 784-6067
wesnousky@unr.edu
http://neotectonics.seismo.unr.edu/CNSHome.html
Director:
Steven Wesnousky, (D), Columbia, 1982, Ys

Center for Research in Economic Geology (M,D) ☒
Mail Stop 1169
Reno, NV 89557-1169
p. (775) 784-1382
dawnsnell@unr.edu
Programs: Graduate program
Administrative Assistant: Dawn Lee. Snell
Enrollment (2021): M: 8 (8) D: 4 (4)
Director:
John Muntean, (D), Stanford, Eg

Dept of Geography (B,M,D) ☒
Mail Stop 0154
Reno, NV 89557-0154
p. (775) 784-6995
geography@unr.edu
http://www.unr.edu/geography/
Enrollment (2020): B: 53 (16) M: 16 (6) D: 17 (3)
Chair:
Douglas Boyle, (D), Arizona, 2000, Hg
Professor:
Kate Berry, (D), Colorado, 1993, Zy
Jill Heaton, (D), Oregon State, 2001, ZniZy
Scott A. Mensing, (D), California (Berkeley), 1993, Zy
Anne Nolin, (D), California (Santa Barbara), 1993, Hg
Associate Professor:
Thomas Albright, (D), Wisconsin, 2007, Zy
Scott Bassett, (D), Harvard, 2001, Zu
Stephanie McAfee, (D), Arizona, 2009, Zy

Paul White, (D), Brown, 2008, Zn
Assistant Professor:
Jessie Clark, (D), Arizona, 2012, Zu
Adam Csank, (D), Arizona, 2011, Zy
Jia Feng, (D), Michigan State, 2016, Zi
Scott Kelley, (D), Arizona State, 2015, Zi
Casey Lynch, (D), Arizona, 2019, Zn
Kenneth Nussear, (D), Nevada, 2004, Zn
Kerri Jean Ormerod, (D), Arizona, 2015, Zu

Dept of Geological Sciences and Engineering (B,M,D) ○⦸
1664 N. Virginia St., MS 0172
Reno, NV 89557-0172
p. (775) 784-6050
geology@unr.edu
http://www.unr.edu/geology
Programs: Geology, Hydrogeology, Geophysics Geological Engineering, Dual degree Geology and Secondary Education via NevadaTeach
Enrollment (2021): B: 108 (33) M: 27 (0) D: 18 (0)
Chair:
Paula J. Noble, (D), Texas, 1993, PmGnPs
Professor:
Wendy M. Calvin, (D), Colorado, 1991, XgZrYg
Stacia M. Gordon, (D), Minnesota, 2009, Gpz
Scott W. Tyler, (D), Nevada (Reno), 1990, Hw
Robert J. Watters, (D), Imperial Coll (UK), 1972, Nrg
Research Professor:
Simon R. Poulson, (D), Penn State, 1990, Cs
Associate Professor:
Scott W. McCoy, (D), Colorado, 2012, GmNg
Assistant Professor:
Wenrong Cao, (D), S California, 2015, Gtc
Michael H. Gardner, (D), California (Berkeley), 2018, NxRn
Kaitlin M. Keegan, (D), Dartmouth, 2014, PcNg
Philipp P. Ruprecht, (D), Washington, 2009, Giv
Joel S. Scheingross, (D), Caltech, 2016, Gsm
Lecturer:
John K. McCormack, (D), Nevada (Reno), 1997, GzEm
Laboratory Director:
Joel DesOrmeau, (D), Nevada (Reno), 2016, Gp
CTEMPs Laboratory Coordinator:
Chris Kratt, (M), YvZrYe
Cooperating Faculty:
John G. Anderson, (D), Columbia, 1976, Ys
Geoffrey Blewitt, (D), Caltech, 1986, Yd
Mike Darin, (D), N Arizona, 2019, Gc
James E. Faulds, (D), New Mexico, 1989, Gc
William C. Hammond, (D), Oregon, 2000, Yd
Graham Kent, (D), Scripps, 1992, Ys
Rich D. Koehler, (D), Nevada (Reno), 2009, Rn
Corne Kreemer, (D), SUNY (Stony Brook), 2001, Yd
John N. Louie, (D), Caltech, 1987, Ys
Carolina Muñoz-Saez, (D), California (Berkeley), 2016, Cu
John L. Muntean, (D), Stanford, 1998, Em
Seth Saltiel, (D), California (Berkeley), 2017, Yx
Lisa L. Stillings, (D), Penn State, 1994, Cl
Steven G. Wesnousky, (D), Columbia, 1982, Ys
Andrew Zuza, (D), California (Los Angeles), 2016, Gc

Dept of Mining Engineering (B,M,D)
Mail Stop 0173
Reno, NV 89557-0173
p. (775) 784-6961
jameshendrix@unr.edu
http://www.unr.edu/cos/mining/
Administrative Assistant: Carla Scott
Enrollment (2011): B: 66 (5) M: 8 (0) D: 5 (0)
Chair:
Danny L. Taylor, (D), Colorado Mines, 1980, Nm
Professor:
Jaak Daemen, (D), Minnesota, 1975, Nr
George Danko, (D), Budapest, 1985, Nm
Associate Professor:
Carl Nesbitt, (D), Nevada, 1990, Nx
Thom Seal, (D), Idaho, 2004, Nx
Emeritus:
Maurice Feunstenau, (D), Nx
Pierre Mousset-Jones, (D), London (UK), 1988, Nm

Development Technician:
John D. Leland, (M), Stanford, 1983

Graduate Program of Hydrologic Sciences (M,D)
1664 N. Virginia Street
MS 0186
Reno, NV 89557-0175
p. (775) 784-6221
hydro@unr.edu
http://www.hydro.unr.edu
Enrollment (2015): M: 32 (10) D: 8 (1)
Research Professor:
Kumud Acharya, (D), Saitama, HsGeHq
Kenneth Adams, (D), Nevada (Reno), 1997, Gam
Braimah Apambire, (D)
John J. Arnone, (D), Yale, 1988, Sp
Gayle L. Dana, (D), Nevada (Reno), 1997, Zr
Joseph Grzymski, (D)
Roger Jacobson, (D), Penn State, 1973, Cg
Nick Lancaster, (D), Cambridge, 1977, Gm
Joseph McConnell, (D), Arizona, 1997, Zn
Eric McDonald, (D), New Mexico, 1994, Sp
Alison Murray, (D), California (Santa Barbara), 1998, ObCg
Daniel Obrist, (D), Nevada (Reno), 2002
Simon Poulson, (D), Penn State, 1990, Cls
Associate Director:
Rina Schumer, (D), Nevada (Reno), 2002, HqwGm
Research Professor:
Christian H. Fritsen, (D), S California, 1995, Og
Professor:
Franco Biondi, (D), Arizona, 1994
Wendy Calvin, (D), Colorado, 1991, YgZr
George Danko, (D), Hungarian Acad of Sci, 1985, Hq
Mae Gustin, (D), Arizona, 1988, Cg
David Kreamer, (D), Arizona, 1982, HwGn
John Louie, (D), Caltech, 1987, Yg
Maureen McCarthy, (D)
Paula Noble, (D), Texas, 1993, Gn
Anna Panorska, (D), California (Santa Barbara), 1992
Mark Pinsky, (D)
Greg Pohll, (D), Nevada (Reno), 1996, Hwq
Robert G. Qualls, (D), Georgia, 1989, SoHsSf
Loretta Singletary, (D)
Scott Tyler, (D), Nevada (Reno), 1990, Hg
Mark Walker, (D), Cornell, 1998, Zn
Research Hydrologist:
Brian Andraski, (D), Nevada (Reno)
Senior Scientist:
Richard Niswonger, (D)
Research Hydrogeologist and Civil Engineer:
Dave Decker, (D), Nevada (Reno), Hw
Associate Research Professor:
Marcus Berli, (D), Swiss Fed Inst Tech, Sp
Clay A. Cooper, (D), Nevada (Reno), 1999, HySpHq
Alan Heyvaert, (D), California (Davis), 1998, GnHs
Justin Huntington, (D), Nevada (Reno), 2011, HgZr
Richard Jasoni, (D), Texas A&M
Alexandra Lutz, (D), Nevada (Reno), Hw
Kenneth McGwire, (D), California (Santa Barbara), 1992, Zr
Don Sada, (D)
Rick Susfalk, (D)
Julian Zhu, (D), Dalhousie
Associate Professor:
Sudeep Chandra, (D), California (Davis), 2003, Hs
Keith E. Dennett, (D), Georgia Tech, 1995, Hs
Eric Marchand, (D), Colorado, 2000
Sherman Swanson, (D), Oregon State, 1983, Hs
Aleksey S. Telyakovskiy, (D), Wyoming, 2002, Hqy
Associate Research Professor:
Ronald L. Hershey, (D), Nevada (Reno), 2010, HwCls
Associate Scientist:
Lisa Shevenell, (D), Nevada (Reno), 1990, Hw
Assistant Research Professor:
Rishi Parashar, (D), Purdue, 2008, Hq
Seshadri Rajagopal, (D)
Casey Schmidt, (D)
Assistant Research Hydrogeologist:
Rosemary Carroll, (D), Nevada (Reno), 2010, Hq

Assistant Professor:
Ronald Breitmeyer, (D), Wisconsin, 2011, HwGe
Adrian Harpold, (D), Cornell, 2010, HqsZr
Scott McCoy, (D)
Ben Sullivan, (D)
Paul Verburg, (D), Wageningen (Neth), 1998, SdcSb
Steve G. Wells, (D), Cincinnati, 1976, Zn
Yu (Frank) Yang, (D)
Research Associate:
Ramon Naranjo, (D), Nevada (Reno), 2012, CgHw
Lisa Stillings, (D), Penn State, 1994, Cm
Assistant Research Professor:
Tom Bullard, (D), New Mexico, 1995, Gm
Adjunct Professor:
Jonathan Price, (D)
Jim Thomas, (D), Nevada (Reno), 1996, Cg
Research Professor:
Ken Taylor, (D)
Emeritus:
Dale Johnson, (D)
Wally Miller, (D)
Steve Wheatcraft, (D), Hawaii, 1979, Hw
Cooperating Faculty:
Chris Benedict
Jeanne Chambers, (D), Sf
David Prudic, (D), Nevada (Reno)
Michael R. Rosen, (D), Texas, 1989, Gn
Keirith Snyder, (D)
Mark Weltz, (D)

Great Basin Center for Geothermal Energy ☒
Mail Stop 0172
1664 N. Virginia St
Reno, NV 89557-0172
p. (775) 784-7018
geothermal@unr.edu
http://www.gbcge.org
Director:
Wendy Calvin, (D), Colorado, 1991, ZrYgXg
Professor:
John Louie, (D), Caltech, 1987, Yse
Research Professor:
James Faulds, (D), New Mexico, 1989, Gc
Research Associate:
Nick Hinz, (M), Nevada (Reno), 2004, Gcg

Mackay School of Earth Sciences and Engineering (Director's Office) ☒
Mail Stop 0168
Reno, NV 89557-0168
p. (775) 784-6987
juliehill@unr.edu
http://www.mines.unr.edu/Mackay/
Administrative Assistant: Julie Hill
Director:
Thom Seal, (D), Idaho, 2004, NxEmNm

Nevada Seismological Lab ☒
Mail Stop 0174
Reno, NV 89557-0174
p. (775) 784-4975
mainofc@seismo.unr.edu
http://www.seismo.unr.edu
Administrative Assistant: Lori McClelland
Administrative Assistant: Erik Williams
Director:
Graham Kent, (D), California (San Diego), 1992, Ys
Professor:
John G. Anderson, (D), Columbia, 1976, Ys
John N. Louie, (D), Caltech, 1987, YesRn
Steve Wesnousky, (D), Columbia, 1982, Ys
Research Associate Professor:
Glenn P. Biasi, (D), Oregon, 1994, Ys
Ileana Tibuleac, (D), S Methodist, 1999, Ys
Assistant Director, Seismic Network Manager & Development Director:
Ken D. Smith, (D), Nevada (Reno), 1991, YsGt
Adjunct Research Associate:
Bill Honjas, (M), Nevada (Reno), Ys
Emeritus Seismic Network Manager:
David H. von Seggern, (D), Penn State, 1982, YsgEo
Emeritus:
James N. Brune, (D), Columbia, 1961, Ys
Development Technician:
Ryan Presser, (A)
Volunteer Adjunct Faculty:
Aasha Pancha, (D), Nevada (Reno), 2007, Ys
Seismic Systems Analyst:
David Slater, (B), Calgary, 1990, Ys
Seismic Records Technician:
Tom Rennie, (D), Nevada (Reno), 2007, Ys
Programmer/Analyst, Seismic Network:
Gabriel Plank, (B), Cornell, 1994, Ys
Network Seismologist:
Diane dePolo, (M), Nevada (Reno), 1989, Ys
Development Technician:
Kent Straley
Associate Engineer:
John Torrisi, (A)

Western Nevada College

Western Nevada College
2201 West College Parkway
Carson City, NV 89703
p. (775) 445-4442
Winnie.Kortemeier@wnc.edu
http://www.wnc.edu/academics/division/sme/

New Hampshire

Dartmouth College

Dept of Earth Sciences (B,M,D) ☒
227 Fairchild Hall, HB 6105
Hanover, NH 03755
p. (603) 646-9037
phyllis.ford@dartmouth.edu
http://www.dartmouth.edu/~earthsci
f: www.facebook.com/DartmouthEarthSciences
Programs: Earth Sciences (B,M,D)
Enrollment (2018): B: 22 (0) M: 10 (0) D: 15 (0)
Professor:
Xiahong Feng, (D), Case Western, 1991, Cs
Carl E. Renshaw, (D), Stanford, 1993, HgGc
Associate Professor:
Robert L. Hawley, (D), Washington, 2005, Gl
Meredith Kelly, (D), Bern, 2003, Gl
Mukul Sharma, (D), Rochester, Ge
Leslie J. Sonder, (D), Harvard, 1986, Gq
Assistant Professor:
William D. Leavitt, (D), Harvard
Erich C. Osterberg, (D), Maine, 2007, AsCl
Marisa Palucis, (D)
Justin V. Strauss, (D), Harvard, 2015, GsCgGt
Emeritus:
William Brian Dade, (D), Washington, GsHgEo
Gary D. Johnson, (D), Iowa State, 1971, GrdYm
Research Professor:
Brian P. Jackson, (D), Georgia, 1998, CaGeCt

Keene State College

Dept of Geology (B) ☒
Mail Stop 2001
229 Main Street
Keene, NH 03435-2001
p. (603) 358-2553
pnielsen@keene.edu
http://www.keene.edu/academics/programs/geol/
Enrollment (2014): B: 17 (4)
Adjunct Professor:
Charles M. Kerwin, (D), New Hampshire, 2006, Gg
Emeritus:
Steven D. Bill, (D), Case Western, 1982, Pg
Peter A. Nielsen, (D), Alberta, 1977, GxzGc

Plymouth State University

Environmental Science and Policy Program (B)
17 High Street
Campus Box 48
Plymouth, NH 03264
p. (603) 536-2573
ladoner@plymouth.edu
https://campus.plymouth.edu/esp/
Programs: Environmental Science & Policy
Certificates: GIS (offered by Plymouth State's Geography program)
Enrollment (2021): B: 73 (19)

Professor:
Eric G. Hoffman, (D), SUNY (Albany), 2000, Am
Associate Professor:
Lisa A. Doner, (D), Colorado, 2001, PcGes
Assistant Professor:
Simon Pendleton, (D), Colorado, 2019, GlCcGe
Research Associate:
Eric Kelsey, (D), New Hampshire, 2014, AtsGl

University of New Hampshire

Dept of Earth Sciences (B,M,D)
214 James Hall
56 College Road
Durham, NH 03824
p. (603) 862-1718
earth.sciences@unh.edu
http://ceps.unh.edu/earth-sciences
f: https://www.facebook.com/UNHEarthSciences
Programs: Earth Sciences; Geochemistry; Geology; Hydrology; Ocean Mapping; Oceanography
Administrative Assistant: Susan E. Clark
Enrollment (2022): B: 49 (13) M: 39 (5) D: 18 (4)

Chair:
J. Matthew Davis, (D), New Mexico Tech, 1994, HwYh
Research Professor:
Stephen E. Frolking, (D), New Hampshire, 1993, As
Cameron P. Wake, (D), New Hampshire, 1993, Gl
Professor:
Julia G. Bryce, (D), California (Santa Barbara), 1998, CgcCb
William C. Clyde, (D), Michigan, 1997, PgYmGs
John E. Hughes-Clarke, (D), Dalhousie, 1988, Og
Joel E. Johnson, (D), Oregon State, 2004, Gus
Joseph M. Licciardi, (D), Oregon State, 2000, Gl
Joseph M. Licciardi, (D), Oregon State, 2000, Gl
Thomas C. Lippmann, (D), Oregon State, 1992, Onp
Larry A. Mayer, (D), California (San Diego), 1979, Gu
David C. Mosher, (D), Dalhousie, 1993, Og
James M. Pringle, (D), MIT/WHOI, 1998, Op
Ruth K. Varner, (D), New Hampshire, 2000, Cl
Research Associate Professor:
Jack E. Dibb, (D), SUNY (Binghamton), 1988, AcClAs
Larry G. Ward, (D), South Carolina, 1978, Gu
Affiliate Associate Professor:
Mary D. Stampone, (D), Delaware, 2009, Zy
Associate Professor:
Rosemarie E. Came, (D), MIT/WHOI, 2005, OoZnn
Jo Laird, (D), Caltech, 1977, Gp
Anne F. Lightbody, (D), MIT, 2007, HgsZn
Michael W. Palace, (D), New Hampshire, Zn
Assistant Professor:
Robert T. Letscher, (D), Rosenstiel, 2012, OcCbZo
Emeritus:
Wallace A. Bothner, (D), Wyoming, 1967, Gc
S. Lawrence Dingman, (D), Harvard, 1970, Hq
Henri E. Gaudette, (D), Illinois, 1963, Cc
Theodore C. Loder, (D), Alaska, 1971, Oc
Affiliate Research Professor:
Brian Calder, (D), Heriot-Watt, 1997, Og
Erik Hobbie
Joseph Salisbury, (D), New Hampshire, Og
Douglas C. Vandemark, (M), New Hampshire, 2005, Zr
Affiliate Research Associate Professor:
Mark A. Fahnestock, (D), Caltech, 1991, Gl
Kai Ziervogel, Rostock, 2004, Gu
Affiliate Research Assistant Professor:
Elizabeth Burakowski, (D), New Hampshire, At
Alexandra Contosta, (D), New Hampshire, 2011, So
Christos Kastrisios, (D), Natl Tech Athens, 2017, Zn
Affiliate Professor:
Andrew Armstrong, (M), Johns Hopkins, 1991, Zn
Christopher E. Parrish, (D), Wisconsin, Zr
Affiliate Clinical Professor:
Semme Dijkstra, (D), New Brunswick, Og
Research Scientist II:
Florencia Fahnestock, (D), New Hampshire, 2022
Affiliate Faculty:
Rochelle Wigley, (D), Cape Town, 2005, CgGu

Dept of Natural Resources and the Environment (B,M,D)
114 James Hall
56 College Rd
Durham, NH 03824-3589
p. 603-862-1022
wendy.rose@unh.edu
https://colsa.unh.edu/natural-resources-environment
Programs: Environmental Science
Certificates: GIS certificate
Enrollment (2018): B: 70 (25) M: 20 (0) D: 20 (0)

Chair:
Mark Ducey, (D), Zc
Professor:
Russell Congalton, (D), Zi
Serita Frey, (D), Sb
William H. McDowell, (D), HgRw
Scott Ollinger, (D), Zc
Associate Professor:
Stuart Grandy, (D), Sc
Wil Wollheim, (D), Zu
Assistant Professor:
Jessica Ernakovich, (D), Zc
Research Associate:
Adam Wymore, (D), Hg

New Jersey

Bergen Community College

Department of Physical Sciences
400 Paramus Rd
Paramus, NJ 07652-1508
p. (201) 447-7100
mnotholt@bergen.edu
http://bergen.edu/academics/academic-divisions-departments/physical-science/
Programs: AS in Physics, Chemistry, and Engineering Science.

College of New Jersey

Physics Dept
2000 Pennington Rd.
Ewing , NJ 08628
p. 609-771-2569
physics@tcnj.edu

Associate Professor:
Margaret Benoit, Penn State, 2005, Ys
Nathan Magee, (D), Penn State, 2006, As
Adjunct Professor:
Thomas Gillespie, (M), Rutgers, 1986, Gc

Kean University

School of Environmental and Sustainability Sciences (B)
1000 Morris Avenue
Union, NJ 07083-0411
p. (908) 737-3737
fqi@kean.edu
http://www.kean.edu/KU/College-of-Natural-Applied-Health-Sciences
Enrollment (2006): B: 160 (0)

Professor:
Robert Metz, (D), Rensselaer, 1967, Gr
Shing Yoh, (D), Drexel, 1989, Am
Constantine S. Zois, (D), Rutgers, 1980, Am
Executive Director:
Paul J. Croft, (D), Rutgers, 1991, AmsZe
Associate Professor:
Carrie M. Manfrino, (D), Miami, 1995, GuOb

Feng Qi, (D), Wisconsin, 2005, ZiyZg
Assistant Professor:
Kikombo Ngoy, (D), Oregon, 1996, Zy
Lecturer:
William C. Heyniger, (B), Montclair State, 2011, Am
Secretary:
Christina Pacia, (N), Zn

Montclair State University

Dept of Earth & Environmental Studies (B,M)
1 Normal Avenue
Upper Montclair, NJ 07043
p. (973) 655-4448
ophorid@montclair.edu
http://www.csam.montclair.edu/earth/eesweb
Chair:
Jonathan M. Lincoln, (D), Northwestern, 1990, Gr
Associate Dean, Science & Mathematics:
Michael A. Kruge, (D), California (Berkeley), 1985, Co
Professor:
Mark J. Chopping, (D), Nottingham (UK), 1998, ZriZe
Huan E. Feng, (D), SUNY (Stony Brook), 1997, Cm
Gregory A. Pope, (D), Arizona State, 1994, GmaZy
Harbans Singh, (D), Rutgers, 1973, Zn
William Solecki, (D), Rutgers, 1990, Zu
Rolf Sternberg, (D), Syracuse, 1971, Zn
Robert W. Taylor, (D), St Louis, 1971, Zn
John V. Thiruvathukal, (D), Oregon State, 1968, Yg
Associate Dean:
Duke U. Ophori, (D), Alberta, 1986, Hw
Associate Professor:
Matthew L. Gorring, (D), Cornell, 1997, Gi
Adjunct Professor:
Kathryn Black, (M), Oklahoma, 1966, Zy
Matthew S. Tomaso, (M), Texas, 1995, Ga
Christine Valenti, (M), Montclair State, 1997, Gg
Emeritus:
Barbara De Beus, Zn
Laboratory Director:
Yoko Sato, (M), Montclair State, 2000, Gg

New Jersey City University

Dept of Earth and Environmental Sciences (B)
Rossey Hall - Room 608
2039 Kennedy Boulevard
Jersey City, NJ 07305-1597
p. (201) 200-3161
lengland@njcu.edu
http://www.njcu.edu/dept/geoscience%5Fgeography/
Enrollment (2006): B: 37 (19)
Chair:
Deborah Freile, (D), Boston, 1992, GseGg
Professor:
Martin Abend, (D), Syracuse, 1955, Zy
Research Associate:
John M. O'Brien, (D), California (Santa Barbara), 1973, Gs
Lecturer:
William W. Montgomery, (D), W Michigan, 1998, Hw
Adjunct Professor:
George Papcun, (M), Ze
Howard Zlotkin, (M), Ze
Emeritus:
John Marchisin, (M), Montclair State, 1965, Gg

Princeton University

Dept of Geosciences (B,D) ☒
113 Guyot Hall
Princeton, NJ 08544-1003
p. (609) 258-4101
mrusso@princeton.edu
http://www.geoweb.princeton.edu
f: https://www.facebook.com/GeosciencesPU/
Programs: Environmental Biogeochemistry; Geophysical & Geology; Ocean, Atmosphere & Climate
Administrative Assistant: Mary Rose Russo
Enrollment (2022): B: 11 (7) D: 36 (4)
Chair:
Thomas S. Duffy, (D), Caltech, 1992, GyYgZm
Associate Chair:
Frederik J. Simons, (D), MIT, 2002, YsdYm
Professor:
Curtis Deutsch, (D), Princeton, 2003, As
Stephan A. Fueglistaler, (D), ETH (Switzerland), 2002, As
Adam C. Maloof, (D), Harvard, 2004, Ym
Satish C B. Myneni, (D), Ohio State, 1995, Sc
Michael Oppenheimer, (D), Chicago, 1970, AsGe
Allan M. Rubin, (D), Stanford, 1988, Yg
Blair Schoene, (D), MIT, 2006, Cc
Daniel M. Sigman, (D), MIT, 1997, ClsPc
Jeroen Tromp, (D), Princeton, 1992, Yss
Gabriel Vecchi, (D), Washington, 2000, ZoOp
Bess B. Ward, (D), Washington, 1982, ObCb
Professional Specialist:
Amal Jayakumar, (D), Goa, 1999, Cm
Sergey Oleynik, (D), Moscow State (Russia), 1999, Cs
Associate Professor:
John A. Higgins, (D), Harvard, 2009, ClPcCs
Research Scholar:
Anne Morel-Kraepiel, (D), Princeton, 2001, Em
Assistant Professor:
Jie Deng, (D), Yale, 2019, Gy
Ching-Yao Lai, (D), Princeton, 2018, YgZoGl
Elizabeth Niespolo, (D), California (Berkeley), 2019, CgGg
Laure Resplandy, (D), Sorbonne (France), 2010, Ob
Xinning Zhang, (D), Caltech, 2010, Cb
Academic Lab Manager:
Laurel P. Goodell, (M), Princeton, 1983, ZeGgc
Danielle M. Schmitt, (M), W Michigan, 1999, Ze
Emeritus:
Michael L. Bender, (D), Columbia, 1970, Cg
Lincoln S. Hollister, (D), Caltech, 1966, GptGz
Gerta Keller, (D), Stanford, 1978, Pm
Francois M M. Morel, (D), Caltech, 1971, Cl
S. George H. Philander, (D), Harvard, 1970, Op
Robert A. Phinney, (D), Caltech, 1961, Ys
Jorge L. Sarmiento, (D), Columbia, 1978, Oc
Post Doctoral Research Fellow:
Alliya Akhtar, (D), Princeton, 2021, CsOoPc
Lauren N. Pincus, (D), Yale, 2020, CqlCg
Post Doctoral Research Associate:
Francisco Apen, (D), California (Santa Barbara), 2022, CcGpt
Thomas Boag, (D), Stanford, 2020, As
Riley Culberg, (D), Stanford, 2022, GlYu
Gregory Davies, (D), Princeton, 2018, Yg
Mathurin Dongmo Wamba, (D), Paris, 2020, YsnYg
Samantha Fortin, (D), Virginia Inst Marine Science, 2021, ObCbOc
Cedric Hagen, (D), Oregon State, 2021, GrsPc
Huai-Hsuan Huang, (D), Hong Kong, 2019, OoPm
Haiyang Luo, (D), Louisiana State, 2021, CusGy
Dario Marconi, (D), Princeton, 2017, Cms
Andrew Margolskee, (D), Washington, 2021, As
Yue (Olivia) Meng, (D), MIT, 2022, As
Justin Penn, (D), Washington, 2020, ZcObCb
Linta Reji, (D), Stanford, 2020, CbPoOb
Tommy Rock, (D), N Arizona, 2017, ZgGg
Sarah Shackelton, (D), California (San Diego), 2016, Pc
Linhan Shen, (D), Caltech, 2017, Cas
Joel D. Simon, (D), Princeton, 2020, YsgYr
Weiyi Tang, (D), Duke, 2019, CbOcb
Mathurin D. Wamba, (D), Paris, 2020, Ys
Xianhui S. Wan, (D), Xiamen, 2017, Cm
Xiaoyu Wang, Rice, 2021, Yg
Yongji Wang, (D), MIT, 2021, As
Mebrahtu Weldeghebriel, (D), Binghamton, 2021, Cl
Julia Wilcots, (D), MIT, 2022, GgPc
Associate Research Scholar:
Etienne Bachmann, (D), Toulouse, 2016, Ys
Pierre Cadeau, (D), Institut de Physique du Globe de Paris, 2017, Cs
Evan Howard, (D), MIT, 2017, As
Renxing Liang, (D), Oklahoma, 2015, Po
Wenchang Yang, Columbia, 2014, As
Associate Professional Specialist:
Sean Gaynor, (D), North Carolina, 2018, GiEmCc
Stefania Gili, (D), Nacl de Córdoba (Argentina), 2014, CgGgCo
Shannon Haynes, (D), West Virginia, 2019, CsOoGg

Valens Hishamunda, (D), Indian Inst Sci, 2021, ClsCq
Joseph Vineis, (D), Northeastern, 2022, Cb
Undergraduate/Graduate Coordinator:
Sheryl A. Robas, (A), 1975, Zn

Environmental Engineering & Water Resources Program (B,M,D) ☒
E-220 Engineering Quad
Princeton, NJ 08544
p. (609) 258-4655
cee@princeton.edu
https://www.princeton.edu/cee/graduate/programs/eewr/
Programs: Civil and Environmental Engineering (B,M,D); Geosciences (B,M,D)
Enrollment (2018): B: 30 (0) M: 4 (0) D: 40 (0)
Director:
Elie Bou-Zeid, (D), Johns Hopkins, 2005, As
Professor:
Michael A. Celia, (D), Princeton, 1983, Hw
Peter R. Jaffe, (D), Vanderbilt, 1981, SbHgw
James A. Smith, (D), Johns Hopkins, 1981, Hg
Eric F. Wood, (D), MIT, 1974, HgZrHq
Assistant Professor:
Catherine A. Peters, (D), Carnegie Mellon, 1992, Hw

Program in Atmospheric & Oceanic Sciences (D) ☒
300 Forrestal Road, Sayre Hall
Princeton, NJ 08540-6654
p. (609) 258-6677
apval@princeton.edu
http://www.princeton.edu/aos/
Enrollment (2018): D: 15 (6)
Professor:
Denise L. Mauzerall, (D), Harvard, 1996, As
Michael Oppenheimer, (D), Chicago
Stephen Pacala, (D), Stanford
James Smith, (D), Johns Hopkins
Gabriel Vecchi, (D), Washington
Senior Scientist:
Isaac Held, (D), Princeton
Syukuro Manabe, (D), Tokyo, 1958, AstPc
Associate Professor:
Stephan Fueglistaler, (D), ETH (Switzerland)
Mark Zondlo, (D), Colorado
Assistant Professor:
Laure Resplandy, 2010
Lecturer:
Thomas Delworth, (D), Wisconsin
Leo Donner, (D), Chicago
Stephen Garner, (D), MIT
Stephen Griffies, (D), Pennsylvania
Robert Hallberg, (D), Washington
Larry Horowitz, (D), Harvard
Sonya Legg, (D), Imperial Coll (UK)
Yi Ming, (D), Princeton
V. Ramaswamy, (D), SUNY (Albany)
Rong Zhang, (D), MIT, Op
Emeritus:
Michael Bender, (D), Columbia
George Mellor, (D), MIT, 1957, Op
Isidoro Orlanski, (D), MIT, 1967, AsZoc
George Philander, (D), Harvard
Jorge L. Sarmiento, (D), Columbia, 1978, Zn

Raritan Valley Community College

Dept of Biology
118 Lamington Road
Branchburg, NJ 08878
p. (908) 526-1200
marianne.baricevic@raritanval.edu

Rider University

Earth and Chemical Sciences (B)
2083 Lawrenceville Road
Lawrenceville, NJ 08648-3099
p. (609) 895-5422
grushow@rider.edu
www.rider.edu/gems
f: www.facebook.com/riderclas/
t: @RiderCLAS
Programs: Geology (B.S.); Environmental Sciences (B.S.); Environmental Studies (B.A.); Marine Sciences (B.S.); Earth Sciences (B.A.), Integrated Sciences and Math (B.A.)
Enrollment (2019): B: 62 (7)
Professor:
Alex Grushow, (D), Zn
Professor:
Daniel L. Druckenbrod, (D), Virginia, 2003, AtZu
Hongbing Sun, (D), Florida State, 1995, HwCgSc
Associate Professor:
Kathleen M. Browne, (D), Miami, 1993, GuZe
Reed A. Schwimmer, (D), Delaware, 1999, GsOgn
Gabriela W. Smalley, (D), Maryland, 2002, ObcOp
Assistant Professor:
Randy Kertes, (M), Ge
Adjunct Professor:
William B. Gallagher, (D), Pennsylvania, 1990, PvgGr
Emeritus:
Mary Jo Hall, (D), Lehigh, 1981, Gs
Jonathan M. Husch, (D), Princeton, 1982, GieXg

Rutgers, The State University of New Jersey

Earth and Planetary Sciences (B,M,D)
Wright Lab
610 Taylor Road
Piscataway, NJ 08854-8066
p. (848) 445-2044
gmtn@eps.rutgers.edu
http://geology.rutgers.edu/
f: https://www.facebook.com/Earth-and-Planetary-Sciences-at-Rutgers-University-187118004101/
Programs: Geological Sciences (B.S.), Geology (B.A.), Environmental Geology (B.S.), Planetary Sciences (B.S.)
Administrative Assistant: Michael Flak, Katanya Myers, Wendy Rodriguez, Tonya Rufus
Enrollment (2018): B: 30 (10) M: 6 (4) D: 23 (0)
Chair:
Gregory S. Mountain, (D), Columbia, 1981, GuYrGr
Associate Chair:
Ying Fan Reinfelder, (D), Utah State, 1992, HgqGe
Undergraduate Director:
Roy W. Schlische, (D), Columbia, 1990, GctEo
Graduate Program Director:
James D. Wright, (D), Columbia, 1991, CsOuPm
Distinguished Professor:
Gail M. Ashley, (D), British Columbia, 1977, GsmSa
Marie-Pierre Aubry, (D), Marie Curie, Pms
Paul G. Falkowski, (D), British Columbia, 1975, ObCmPe
Dennis V. Kent, (D), Columbia, 1974, YmZcGr
George R. McGhee, Jr., (D), Rochester, 1978, PgqPi
Kenneth G. Miller, (D), MIT/WHOI, 1982, GurPm
Yair Rosenthal, (D), MIT/WHOI, CmlOc
Professor:
Craig S. Feibel, (D), Utah, 1988, GrsGa
Mark D. Feigenson, (D), Princeton, 1982, CgGiv
Claude T. Herzberg, (D), Edinburgh, 1975, CpGi
Robert E. Kopp, (D), Caltech, 2007, ZcOoZg
Vadim Levin, (D), Columbia, 1996, Ysg
Robert M. Sherrell, (D), MIT/WHOI, 1991, CmOu
Carl C. Swisher III, (D), California (Berkeley), 1992, Cc
Martha O. Withjack, (D), Brown, 1977, GctEo
Nathan Yee, (D), Notre Dame, 2001, PgCo
Associate Professor:
Juliane Gross, (D), Ruhr (Germany), 2009, Xcm
Assistant Professor:
Sonia M. Tikoo, (D), MIT, 2014, YmXy
Jill A. Van Tongeren, (D), Columbia, 2010, GiCgGt
Research Professor:
Brent D. Turrin, (D), California (Berkeley), 1996, CcGvYm
Assistant Research Professor:
James V. Browning, (D), Rutgers, 1996, GgsGr
Linda Godfrey, (D), Cambridge, 1990, CmlCq
Richard Mortlock, (D), Rutgers, 2017, CgmCs
Co-Director, Rutgers Geology Museum:
Lauren Neitzke Adamo, (D), Rutgers, 2016, GuPmZe

Research Associate:
Don Monteverde, (D), Rutgers, 2008, GruGg
Peter P. Sugarman, (D), Rutgers, 1995, Gr
Lecturer:
Christopher H. Lepre, (D), Rutgers, 2009, YmSaPv
Distinguished Visiting Professor:
William A. Berggren, (D), Stockholm, 1962, PmsPe
Emeritus:
Michael J. Carr, (D), Dartmouth, 1974, Gv
Richard K. Olsson, (D), Princeton, 1958, PmGr
Robert E. Sheridan, (D), Columbia, 1968, YrGu
IT Specialist:
Jason Pappas
Cooperating Faculty:
Jeremy S. Delaney, (D), Belfast, 1978, Xg

Rutgers, The State University of New Jersey, Newark

Dept of Earth & Environmental Sciences (B,M,D) ●⦸
101 Warren Street
Smith Hall, room 135
Newark, NJ 07102
p. (973) 353-5100
nicky.agate@rutgers.edu
http://www.ncas.rutgers.edu/ees
Programs: Geology; Geoscience Engineering; Environmental Sciences
Certificates: Environmental Geology (G)
Enrollment (2017): B: 71 (13) M: 3 (0) D: 13 (2)
Chair:
Lee S. Slater, (D), Lancaster (UK), 1997, YeHw
Professor:
Yuan Gao, (D), Rhode Island, 1994, As
Alexander E. Gates, (D), Virginia Tech, 1986, Gc
Associate Research Professor:
Francisco Artigas, Ohio State, Zi
Associate Professor:
Evert J. Elzinga, (D), Delaware, 2000, Sc
Kristina M. Keating, (D), Stanford, 2009, Yug
Adam B. Kustka, (D), SUNY (Stony Brook), 2002, OgCmHs
Dimitrios Ntarlagiannis, (D), Rutgers, 2006, Yu
Ashaki Rouff, (D), SUNY (Stony Brook), 2004, Cl
Assistant Professor:
Mihaela Glamoclija, (D), D'Annunzio (Italy), 2005, PgXgGe
Research Associate:
Judith Robinson, (D), Rutgers, 2015, Yg
Emeritus:
Warren Manspeizer, (D), Rutgers, 1964, Gr
John H. Puffer, (D), Stanford, 1969, GivGe
Andreas H. Vassiliou, (D), Columbia, 1969, Gz

Stockton University

Dept of Environmental Sciences (B,M) ☒
Division of Natural Science and Mathematics
101 Vera King Farris Drive
Galloway, NJ 08205
p. (609) 652-4620
jeffrey.webber@stockton.edu
http://intraweb.stockton.edu/eyos/page.cfm?siteID=183&pageID=23
Professor:
Tait Chirenje, (D), Florida, Sc
Weihong Fan, (D), Colorado State, 1993, Zi
Associate Professor:
Daniel Moscovici, (D), Pennsylvania, 2009, Zu
Matthew Severs, (D), Virginia Tech, 2009, Gg
Assistant Professor:
Jessica Favorito, (D), Virginia Tech, 2017, So
Jeffrey Webber, (D), Massachusetts, 2015, Gg
Emma Witt, (D), Kentucky, 2012, Hg

Geology (B) ●☒
101 Vera King Farris Dr
Galloway, NJ 08205
p. 609-626-6857
matthew.severs@stockton.edu
http://intraweb.stockton.edu/eyos/page.cfm?siteID=183&pageID=33
Programs: Geology BS and BA
Enrollment (2021): B: 20 (5)
Associate Professor:
Matthew Rocky Severs, (D), Virginia Tech, 2007, GxCgEg
Assistant Professor:
Jessica Favorito, (D), Virginia Tech
Susanne Moskalski, (D), Delaware, GsOn
Jeffrey R. Webber, (D), Massachusetts, 2016, GcpGt
Emma Witt, (D), Hgs
Emeritus:
Stewart C. Farrell, (D), Massachusetts, 1972, On
Michael J. Hozik, (D), Massachusetts, 1976, GcYmg

William Paterson University

Dept of Environmental Science (B)
Science Hall
Wayne, NJ 07470
p. (201) 595-2721
beckerm2@wpunj.edu
http://www.wpunj.edu/cosh/departments/environmental-science/
Enrollment (2011): B: 90 (3)
Chairman:
Martin A. Becker, (D), Brooklyn Coll, 1997, PgAm
Professor:
Richard R. Pardi, (D), Pennsylvania, 1983, Cc
Assistant Professor:
Jennifer R. Callanan, (D), Montclair State, 2008, ScGmSz
Karen Swanson, (D), Penn State, 1989, Cc

New Mexico

Eastern New Mexico University

Dept of Physical Sciences (B) ⦸
1500 S Ave K
STA 33
Portales, NM 88130
p. (575) 562-2174
jim.constantopoulos@enmu.edu
https://www.enmu.edu/academics/degrees-programs/undergraduate-degree/bachelor/environmental-science-degree
f: https://www.facebook.com/ENMUENVS
Programs: Environmental Science with Geology emphasis
Enrollment (2022): B: 9 (3)
Professor:
James T. Constantopoulos, (D), Idaho, 1989, GezGx

Mesalands Community College

Mesalands Dinosaur Museum and Natural Sciences Laboratory (A) ○☒
911 South Tenth Street
Tucumcari, NM 88401
p. (575) 461-4413
axelh@mesalands.edu
http://www.mesalands.edu/
Programs: Associate of Arts Degree in Natural Sciences, option Paleontology
Associate of Arts Degree in Natural Sciences, option Geology
Enrollment (2018): A: 5 (2)
Instructor:
Axel Hungerbuehler, (D), Bristol, 1998, PvgGg

New Mexico Community College

New Mexico Community College
525 Buena Vista Dr. SE
Albuquerque, NM 87106
p. (505) 224-3000
pcarman@cnm
http://www.cnm.edu

New Mexico Highlands University

Natural Resources Management Dept (B,M)
P.O. Box 9000
Las Vegas, NM 87701
p. (505) 454-3000
mspetro@nmhu.edu
http://www.nmhu.edu/academics/undergraduate/arts_science/natural_resources/

Enrollment (2012): B: 16 (4) M: 6 (1)
Professor:
Jennifer Lindline, (D), Bryn Mawr, 1997, GizZe
Associate Professor:
Michael S. Petronis, (D), New Mexico, 2005, YmGcv

New Mexico Institute of Mining and Technology

Dept of Earth & Environmental Science (B,M,D) O✓
801 Leroy Place
Socorro, NM 87801
p. (575) 835-5634
geos.dept@npe.nmt.edu
https://www.nmt.edu/academics/ees/
Programs: B.S. in Earth Science, B.S. in Environmental Science, M.S. in Geology, M.S. in Geochemistry, M.S. in Geophysics, M.S. in Hydrology, Ph.D. in Earth and Environmental Science, Professional Master of Hydrology
Certificates: Graduate Certificate Program in Hydrology
Enrollment (2018): B: 33 (10) M: 36 (16) D: 18 (2)
Professor:
Susan L. Bilek, (D), California (Santa Cruz), 2001, Ys
Peter S. Mozley, (D), California (Santa Barbara), 1988, Gs
Mark A. Person, (D), Johns Hopkins, 1990, HyYhg
Glenn Spinelli, (D), California (Santa Cruz), 2002, Hw
Senior Volcanologist, NMBG:
William C. McIntosh, (D), New Mexico Tech, 1990, Cc
Associate Professor:
Gary Axen, (D), Harvard, 1991, Gct
Daniel Cadol, (D), Colorado State, 2010, HsGe
Bruce I. Harrison, (D), New Mexico, 1992, GmSof
Jolante Van Wijk, (D), Vrije (Amsterdam), 2002, GtoYe
Assistant Professor:
Ronni Grapenthin, (D), Alaska (Fairbanks), 2012, Gv
Daniel Jones, (D), Cb
Ryan Leary, Gs
Kierran Maher, Washington State, Eg
Associate Research Professor:
Mark Murray, (D), MIT, Yds
David B. Reusch, (D), Penn State, 2003, Zn
Dana S. Ulmer-Scholle, (D), S Methodist, 1992, GsdCl
Map Production Coordinator, NMBG:
Phillip Miller
Technical Staff Member, Seimologist:
Charlotte A. Rowe, (D), New Mexico Tech, 2000, Ys
Sr. Geochronologist/Co-Director NM Geochronology Research:
Matthew T. Heizler, (D), California (Los Angeles), 1993, Cc
Senior Scientist:
Robert S. Balch, (D), New Mexico Tech, 1997, Yse
Senior Mineralogist/Economic Geologist/ Director XRD Lab/Curator Mineral Museum:
Virgil L. Lueth, (D), Texas (El Paso), 1988, GzEg
Senior Field Geologist:
Steven M. Cather, (D), Texas, 1986, Gc
Daniel Koning, (M), New Mexico, 1999
Senior Economic Geologist:
Virginia T. McLemore, (D), Texas (El Paso), 1993, Eg
Research Associate, NMBG:
Matthew Zimmerer, New Mexico Tech
Principal Senior Petroleum Geologist:
Ronald F. Broadhead, (M), Cincinnati, 1979, GorGs
Principal Senior Environmental Geologist:
David W. Love, (D), New Mexico, 1980, Ge
Principal Geologist:
Paul W. Bauer, (D), New Mexico Tech, 1988, Gc
Postdoc Fellow, USGS:
Jesus Gomez, New Mexico Tech, 2014
Planetary Protection Officer, NASA:
Catharine A. Conley, (D), Cornell, 1994, Pg
Geophysicist/Field Geologist/Web Information Specialist:
Shari A. Kelley, (D), S Methodist, 1984, Gt
Geologic Mapping Program Manager:
J Michael Timmons, (D), New Mexico, 2004, GctGs
Geochemist & Deputy Director Manger of Electron Microprobe Lab:
Nelia W. Dunbar, (D), New Mexico Tech, 1989, Cg
Emeritus Senior Principal Geophysicist:
Marshall A. Reiter, (D), Virginia Tech, 1969, Yh
Emeritus Senior Field Geologist:
Richard Chamberlin, (D), Colorado Mines, 1980, GcvGg
Emeritus Director and State Geologist:
Charles E. Chapin, (D), Colorado Mines, 1965, Gv
Distinguished Member of Technical Staff:
Vincent C. Tidwell, (D), New Mexico Tech, 1999, Hwq
Cave & Karst Hydrologist:
Lewis Land, (D), North Carolina, 1999, Hy
Assistant Professor of Hydrogeology and Applied Geology, Purdue:
Marty Frisbee, New Mexico Tech, 2010
Assistant Professor:
Nigel J.F. Blamey, (D), New Mexico Tech, 2000, CaeCl
Adjunct Professor:
Denis Cohen
Charles (Jack) Oviatt
Michael Underwood
Emeritus Senior Environmental Geologist:
John W. Hawley, (D), Illinois (Urbana), 1962, GeHwGr
Emeritus:
Antonius J. Budding, (D), Amsterdam
Andrew R. Campbell, (D), Harvard, 1984, CsEmGz
Kent C. Condie, (D), California (San Diego), 1965, Ct
Gerardo W. Gross, (D), Penn State, 1959, Yx
Jan M. H. Hendrickx, (D), New Mexico State, 1984, Hw
David B. Johnson, (D), Iowa, 1978, Ps
Philip R. Kyle, (D), Victoria (NZ), 1976, Giv
Fred M. Phillips, (D), Arizona, 1981, Hw
John W. Schlue, (D), California (Los Angeles), 1975, YsgZg
John L. Wilson, (D), MIT, 1974, Hw

Dept of Mineral Engineering (B,M,D)
Campus Station
Socorro, NM 87801-9990
p. (505) 835-5345
navid.mojtabai@nmt.edu
http://nmt.edu/academics/mining
Department Secretary: Lucero Joanna
Chair:
Navid Mojtabai, (D), Arizona, 1990, Nr
Professor:
William X. Chavez, Jr., (D), California (Berkeley), 1984, Em
Associate Professor:
Cathrine T. Aimone-Martin, (D), Northwestern, 1982, Ng
Assistant Professor:
Baolin Deng, (D), Johns Hopkins, 1995, Cg
Randal S. Martin, (D), Washington State, 1992, As
Adjunct Professor:
William Haneberg, (D), Cincinnati, 1989, Ng
Per-Anders Persson, (D), Cambridge, 1960, Nr
Ingar F. Walder, (D), New Mexico Tech, 1991, Cl
Emeritus:
George B. Griswold, (D), Arizona, Nx
Kalman I. Oravecz, (D), Witwatersrand, 1967, Nr

New Mexico State University, Alamogordo

New Mexico State University, Alamogordo
2400 N. Scenic Drive
Alamogordo, NM 88310
hrnmsua@nmsu.edu
http://www.nmsua.edu

New Mexico State University, Grants

Dept of Natural Sciences ☒
1500 N Third St
Grants, NM 87020
p. (505) 287-6678
ssgrants@nmsu.edu
http://www.grants.nmsu.edu

New Mexico State University, Las Cruces

Department of Physics (M,D) ☒
MSC 3D
P.O. Box 30001
Las Cruces, NM 88003-8001
p. (505) 646-3831
physics@nmsu.edu
http://physics.nmsu.edu
Programs: Physics, Geophysics
Enrollment (2020): D: 1 (0)

Head:
Heinz Nakote, (D), Arizona State, 1991, Zm
Associate Professor:
Thomas M. Hearn, (D), Caltech, 1985, Ys
Assistant Professor:
Lauren Waszek, (D), Cambridge, 2012, Ys
Emeritus:
James F. Ni, (D), Cornell, 1984, YsGtYg

Dept of Geological Sciences (B,M) ●☒
MSC 3AB, Box 30001
1255 N. Horseshoe
Gardiner Hall, Room 171
Las Cruces, NM 88003
p. (575) 646-2708
geology@nmsu.edu
http://geology.nmsu.edu/
Programs: Geology (B,M); Geological Sciences (B); Earth and Environmental Systems (B); Earth Science Education (B)
Administrative Assistant: Lee Hubbard
Enrollment (2019): B: 40 (7) M: 17 (5)
Head:
Nancy J. McMillan, (D), S Methodist, 1986, Gi
Professor:
Jeffrey M. Amato, (D), Stanford, 1995, GcCc
Frank C. Ramos, (D), California (Los Angeles), 2000, CcuGi
Assistant Professor:
Reed J. Burgette, (D), Oregon, 2008, Gtc
Brian A. Hampton, (D), Purdue, 2006, Gst
Emily R. Johnson, (D), Oregon, 2008, Gvi

Dept of Plant & Environmental Sciences (A,B,M,D) ⦸
Box 30003
Dept. 3Q
Las Cruces, NM 88003-0003
p. (505) 646-3405
mhammitt@nmsu.edu
http://aces.nmsu.edu/academics/pes/
Programs: Agronomy, Environmental Science, Genetics, Horticulture, Soil Science, Turfgrass Science and Management, and interdisciplinary degree program in Water Science and Management
Department Secretary: Paula Ross
Head:
LeRoy A. Daugherty, (D), Cornell, 1975, Sd
Richard Pratt, (D)
Professor:
William C. Lindemann, (D), Minnesota, 1978, Sb
Bobby D. McCaslin, (D), Minnesota, 1974, Sc
Theodore W. Sammis, (D), Arizona, 1974, Sp
Assistant Professor:
Dean Heil, (D), California (Berkeley), 1991, Sc
Tim L. Jones, (D), Washington State, 1989, Sp
H. C. Monger, (D), New Mexico State, 1990, Sa

San Juan College

San Juan College (A)
4601 College Blvd.
Farmington, NM 87402
p. (505) 566-3325
burrisj@sanjuancollege.edu
http://www.sanjuancollege.edu/geology
Programs: Geology (A)
Enrollment (2021): A: 10 (5)
Professor:
John H. Burris, (D), Michigan State, 2004, GgRhGz

University of New Mexico

Dept of Earth & Planetary Sciences (B,M,D) O☒
221 Yale Blvd NE
Northrop Hall, Room 141
MSC03 2040
Albuquerque, NM 87131-0001
p. (505) 277-4204
epsdept@unm.edu
http://eps.unm.edu/
Enrollment (2019): B: 68 (33) M: 12 (17) D: 33 (1)
Chair:
Peter J. Fawcett, (D), Penn State, 1994, Pe
Professor:
Carl A. Agee, (D), Columbia, 1988, Cp
Yemane Asmerom, (D), Arizona, 1988, CcPeCg
Adrian J. Brearley, (D), Manchester (UK), 1984, Gz
Laura J. Crossey, (D), Wyoming, 1985, Cl
Tobias Fischer, (D), Arizona State, 1999, Gv
Joseph Galewsky, (D), California (Santa Cruz), 1996, Am
David J. Gutzler, (D), MIT, 1986, As
Karl E. Karlstrom, (D), Wyoming, 1980, Gt
James J. Papike, (D), Minnesota, 1964, Ca
Louis A. Scuderi, (D), California (Los Angeles), 1984, GmPcGs
Zachary D. Sharp, (D), Michigan, 1987, Cs
Gary S. Weissmann, (D), California (Davis), 1999, HwGsr
Senior Scientist:
Nieu-Viorel Atudorei, (D), Lausanne, 1998, Cs
Victor J. Polyak, (D), Texas Tech, 1998, CcGzm
Associate Professor:
Brandon Schmandt, (D), Oregon, 2011, Yg
Assistant Professor:
Tyler Mackey, (D), California (Davis), 2016, GdPo
Corinne E. Myers, (D), Kansas, 2013, PeqPi
Lindsay Lowe Worthington, (D), Texas, 2010, Yg
Jin Zhang, (D), Illinois, 2014, Gyp
Research Associate:
Frans J.M. Rietmeijer, (D), Utrecht, 1979, Gp
Lecturer:
Aurora Pun, (D), New Mexico, 1996, Ca
Adjunct Professor:
Fraser Goff, (D), California (Santa Cruz), 1977, Cg
Duane M. Moore, (D), Illinois (Urbana), 1963, Sc
Chester J. Weiss, (D), Texas A&M, 1998, Ye
Thomas E. Williamson, (D), New Mexico, 1993, Pv
Emeritus:
Roger Y. Anderson, (D), Stanford, 1960, Gn
Maya Elrick, (D), Virginia Tech, 1990, Gs
Rodney E. Ewing, (D), Stanford, 1974, Gz
John W. Geissman, (D), Michigan, 1980, Ym
Rhian Jones, (D), Manchester (UK), 1986, Gz
Barry S. Kues, (D), Indiana, 1974, Pi
Leslie M. McFadden, (D), Arizona, 1982, Sd
Grant A. Meyer, (D), New Mexico, 1993, Gm
Jane E. Selverstone, (D), MIT, 1985, Gpt
Lee A. Woodward, (D), Washington, 1962, Gc

Institute of Meteoritics (B,M,D) ☒
MSC03 2050
1 University of New Mexico
Albuquerque, NM 87131
p. (505) 277-1644
iom@unm.edu
http://meteorite.unm.edu
Administrative Assistant: Beth Ha
Enrollment (2017): M: 1 (1) D: 3 (1)
Professor:
Carl B. Agee, (D), Columbia, 1988, XcGyCp
Senior Scientist III:
Horton E. Newsom, (D), Arizona, 1981, XgcXm
Senior Scientist III:
Charles K. Shearer, Jr., (D), Massachusetts, 1983, Gi
Senior Research Scientist III:
Steven B. Simon, (D), SD Mines, 1988, XmcGi
Assistant Professor:
Jin Zhang, (D), Illinois, 2014, Gy
Senior Scientist:
Karen Ziegler, (D), Reading (UK), 1993, CsXc
Research Scientist III:
Michael N. Spilde, (M), SD Mines, 1987, Gz
Program Manager:
Shannon Clark, Zn

Water Resources Program
1915 Roma NE, Room 1044
Albuquerque, NM 87131-1217
p. (505) 277-5249
fleckj@unm.edu
http://www.unm.edu/~wrp/

Director:
Michael E. Campana, Zn

University of New Mexico, Gallup

Div of Arts and Sciences
200 College Road
Gallup, NM 87301
p. (505) 863-7500
jzimm58@unm.edu
http://www.gallup.unm.edu/

University of New Mexico, Taos

University of New Mexico - Taos (A,B) ●
1157 County Road 110
Ranchos de Taos, NM 87557
colnic@unm.edu
http://www.taos.unm.edu
Adjunct Professor:
Deborah Ragland, (D), Gg

Western New Mexico University

Dept of Natural Sciences ☒
P.O. Box 680
1000 West College Avenue
Silver City, NM 88062
p. (575) 538-6352
corrie.neighbors@wnmu.edu
http://natsci.wnmu.edu/
Assistant Professor:
Corrie Neighbors, (D), California (Riverside), 2015, Ysr

New York

Adelphi University

Environmental Studies Program (B,M) ○☒
South Avenue
Garden City, NY 11530
p. (516) 877-4170
bwygal@adelphi.edu
http://environmental-studies.adelphi.edu
t: @mammothunter
Enrollment (2018): B: 30 (7) M: 8 (3)
Associate Professor:
Brian Wygal, (D), Nevada (Reno), 2009, GaRmGe

Adirondack Community College

Science Div (A)
640 Bay Road
Queensbury, NY 12804
p. (518) 743-2325
minkeld@sunyacc.edu
http://www.sunyacc.edu

Alfred University

Dept of Geology (B) ☒
Saxon Drive
Alfred, NY 14802
p. (607) 871-2208
beaudry@alfred.edu
https://www.alfred.edu/academics/undergrad-majors-minors/geology.cfm
Programs: Geology
Enrollment (2018): B: 9 (2)
Professor:
Michele M. Hluchy, (D), Dartmouth, 1988, Cl
Otto H. Muller, (D), Rochester, 1974, Gc

American Museum of Natural History

Dept of Earth & Planetary Sciences ☒
Central Park West at 79th Street
New York, NY 10024-5192
p. (212)769-5100
debel@amnh.org
https://www.amnh.org/research/physical-sciences/earth-and-planetary-sciences
Administrative Assistant: Cheryl Mayefsky
Chair and Curator:
Denton S. Ebel, (D), Purdue, 1993, XmCgZe
Emeritus Curator:
George E. Harlow, (D), Princeton, 1977, Gz
Associate Curator:
Nathalie Goodkin, (D), MIT/WHOI, 2007
Research Scientist:
Céline Martin, (D), Henri Poincaré, 2009
Nicholas Tailby, (D), Australian Nat, 2014
Senior Museum Specialist:
Saebyul Choe, (B), Bates, 2014
Jamie Newman, (M), Brooklyn Coll, Gg
Museum Specialist:
Sam Alpert, (B), Case Western, 2012, Xm
Keiji Hammond, (B), Northeastern, 2016
Curator Emeritus:
Edmond A. Mathez, (D), Washington, 1981, Gx
Postdoctoral Fellow:
Steven Jaret, (D), SUNY (Stony Brook), 2017
Leanne Melbourne, (D), Bristol, 2018
Natalie Umling, (D), South Carolina, 2017

Div of Paleontology (D)
Central Park West at 79th Street
New York, NY 10024
p. (212) 769-5815
jmeng@amnh.org
https://www.amnh.org/research/paleontology
Administrative Assistant: Judy Galkin
Enrollment (2009): D: 8 (0)
Chair, Professor and Curator:
Mark A. Norell, (D), Yale, 1989, Pv
Provost, Professor and Curator:
Michael J. Novacek, (D), California (Berkeley), 1978, Pv
Professor and Curator:
Niles Eldredge, (D), Columbia, 1969, Pi
Neil H. Landman, (D), Yale, 1982, Pi
John G. Maisey, (D), London (UK), 1974, Pv
Jin Meng, (D), Columbia, 1991, Pv
Dean of the Richard Gilder Graduate School, Professor, and Frick Curator:
John J. Flynn, (D), Columbia, 1983, Pv
Frick Curator Emeritus:
Richard H. Tedford, (D), California (Berkeley), 1960, Pv
Curator Emeritus:
Roger L. Batten, (D), Columbia, 1956, Pi
Eugene S. Gaffney, (D), Columbia, 1969, Pv

Binghamton University

Dept of Geological Sciences and Environmental Studies (B,M,D) ☒
PO Box 6000
Binghamton, NY 13902-6000
p. (607) 777-2264
tdesmet@binghamton.edu
http://geology.binghamton.edu
Administrative Assistant: Carol Slavetskas
Enrollment (2012): B: 30 (0) M: 4 (0) D: 12 (0)
Chair:
Tim K. Lowenstein, (D), Johns Hopkins, 1982, Cl
Professor:
Joseph R. Graney, (D), Michigan, 1994, ClGeHs
David M. Jenkins, (D), Chicago, 1980, CpGzp
H. Richard Naslund, (D), Oregon, 1980, Gi
Associate Professor:
Peter L. K. Knuepfer, (D), Arizona, 1984, Gt
Thomas Kulp, (D), Indiana, 2002, Pg
Assistant Professor:
Alex Nikulin, (D), Rutgers, 2011, Ys
Molly Patterson, (D), Victoria (NZ), 2015, CmGl
Jeff Pietras, (D), Wisconsin, 2003, GsEo
Research Associate:
Alan Jones, (D), Purdue, 1964, Ys
Emeritus:
Richard E. Andrus, (D), SUNY (Syracuse), 1974, Sf
Jeffrey S. Barker, (D), Penn State, 1984, Ys
Donald R. Coates, (D), Columbia, 1956, Gm
Robert V. Demicco, (D), Johns Hopkins, 1981, Gs
Steven R. Dickman, (D), California (Berkeley), 1977, Yg

Thomas W. Donnelly, (D), Princeton, 1959, Gg
William D. MacDonald, (D), Princeton, 1965, Gc
Karen M. Salvage, (D), Penn State, 1998, Hw
James E. Sorauf, (D), Kansas, 1962, Pi
Francis T. Wu, (D), Caltech, 1966, Ys
Related Staff:
Michael Hubenthal, (M), SUNY (Binghamton), 2010, ZeYg

Brooklyn College (CUNY)

Dept of Earth and Environmental Science (B,M) ●☒
2900 Bedford Avenue
Brooklyn, NY 11210
p. (718) 951-5416
wpowell@brooklyn.cuny.edu
http://depthome.brooklyn.cuny.edu/geology/
Enrollment (2009): B: 24 (5) M: 10 (4) D: 3 (1)
Chair:
Wayne G. Powell, (D), Queen's, 1994, GaCsEm
Jennifer Cherrier, OcCmRw
Professor:
John A. Chamberlain, (D), Rochester, 1971, Pg
Zhongqi Cheng, (D), Ohio State, 2001, CaScGe
Constantin Cranganu, (D), Oklahoma, 1997, GoYhHw
Peter Groffman, SbCl
John Marra, (D), Dalhousie, 1977, Ob
David E. Seidemann, (D), Yale, 1975, Cc
Associate Professor:
Stephen U. Aja, (D), Washington State, 1989, Cl
Rebecca Boger, (D), William & Mary, 2002, ZiHs
Assistant Professor:
Brett Branco, (D), Connecticut, 2007, OnHsOg
Kennet Flores, (D), Gpc
Brianne Smith, (D), Hs
Lecturer:
Matt Garb, (M), Brooklyn Coll, Pg
Laboratory Director:
Guillermo Rocha, (M), CUNY, 1994, GgeCg

Broome Community College

Dept of Physical Sciences (A)
Upper Front Street
Box 1017
Binghamton, NY 13902
p. (607) 778-5000
smithjj@sunybroome.edu
Professor:
Bruce K. Oldfield, (M), SUNY (Binghamton), 1988, Gg
Assistant Professor:
Jason J. Smith, (M), SUNY (Binghamton), 2009, Ggs

Buffalo State College

Earth Sciences (B) ●⦸
SAMC 160
1300 Elmwood Avenue
Buffalo, NY 14222
p. (716) 878-6731
solargs@buffalostate.edu
https://earthsciences.buffalostate.edu
Programs: Geology; Earth Sciences
Administrative Assistant: Cindy Wong
Enrollment (2021): B: 36 (16)
Planetarium Director:
Kevin K. Williams, (D), Johns Hopkins, 2002, Xg
Chair:
Gary S. Solar, (D), Maryland, 1999, GcpGt
Associate Professor:
Elisa T. Bergslien, (D), SUNY (Buffalo), 2002, ClHw
Kevin K. Williams, (D), Johns Hopkins, 2002, GmXg
Adjunct Professor:
Daniel E. MacIssac, (D), XaZe
Emeritus:
John E. Mack, (D), Fordham, 1971, Xg
Jill K. Singer, (D), Rice, 1986, GsOp
Irving Tesmer, (D), Syracuse, Pg

Cayuga Community College

Math and Science (A)
197 Franklin Street
Auburn, NY 13021
p. (315) 255-1743
jennifer.nichols@cayuga-cc.edu
http://www.cayuga-cc.edu/academics/programs_of_study/math_and_science.php
Professor:
Abu Z. Badruddin, (D), SUNY (Environ), Zi
Raymond F. Leszczynski, (M), SUNY (Albany), 1965, GglGm

City College (CUNY)

Dept of Earth & Atmospheric Sciences (B,M)
New York, NY 10031
p. (212) 650-6984
zluo@ccny.cuny.edu
Chair:
Jeffrey Steiner, (D), Stanford, 1970, Cp
Professor:
Stanley Gedzelman, (D), MIT, 1970, As
Edward E. Hindman, (D), Washington, 1975, As
Margaret A. Winslow, (D), Columbia, 1979, Gc
Associate Professor:
Patricia M. Kenyon, (D), Cornell, 1986, Yg
Federica Raia, (D), Naples, 1997, Giv
Pengfei Zhang, (D), Utah, 2000, Hw

Colgate University

Department of Earth and Environmental Geosciences (B) ●☒
13 Oak Drive
Hamilton, NY 13346
p. (315) 228-7201
http://www.colgate.edu/academics/departments-and-programs/geology
Programs: Geology, Environmental Geology
Enrollment (2022): B: 28 (11)
Chair:
William H. Peck, (D), Wisconsin, 2000, GpiCs
Professor:
Karen Harpp, (D), Cornell, 1994, GvCgGi
Amy Leventer, (D), Rice, 1988, Ou
Associate Professor:
Martin Wong, (D), California (Santa Barbara), 2005, Gtc
Aubreya Adams, (D), Penn State, 2010, Yg
Assistant Professor:
Paul Harnik, (D), Chicago, 2009, PieOb
Joseph Levy, (D), Brown, 2009, GlsGm
Senior Lecturer:
Dianne M. Keller, (M), Colgate, 1988, Gz
Alison Koleszar, (D), Oregon State, 2011, GviCu
Emeritus:
Richard April, (D), Massachusetts, 1978, CgGze
Paul Pinet, (D), Rhode Island, 1972, Ou
Constance M. Soja, (D), Oregon, 1985, PiGs

College of Staten Island/CUNY

Engineering and Environmental Science (A,B) ☒
2800 Victory Boulevard
Staten Island, NY 10314-6609
p. 718-982-2800
alan.benimoff@csi.cuny.edu
https://csivc.csi.cuny.edu/geology/files/geo.html
Professor:
Alan I. Benimoff, (D), Lehigh, 1984, GizGp
William J. Fritz, (D), Montana, 1980, Grv
Athanasios Koutavas, (D), Columbia, 2003, PeOg
Assistant Professor:
Jane L. Alexander, (D), Univ Coll (UK), 1998, GsClt
Adjunct Lecturer:
Imad Harone, (M), CUNY (Staten Island), Gg
Edward Johnson, (M), CUNY (Staten Island), Gg
Vladimir Jovanovic, (M), CUNY (Staten Island), Ge
Adjunct Associate Professor:
Mosbah Kolkas, (D), CUNY, Gg
Adjunct Assistant Professor:
Noureddin Amaach, (D), CUNY, Gg

Rosemary McCall, (D), Gg

Columbia University

Dept of Earth & Environmental Engineering
Henry Krumb School of Mines
500 West 120 Street
918 Mudd Bldg
New York, NY 10027
p. (212) 894-2905
schlosser@ldeo.columbia.edu
http://www.eee.columbia.edu/
Administrative Assistant: Co'Quesie Gilbert
Department Administrator: Barbara Algin

Acting Chair:
Nickolas J. Themelis, (D), McGill, 1961, Nx

Professor:
Paul F. Duby, (D), Columbia, 1962, Nx
Peter Schlosser, (D), Heidelberg, 1985, Cg
Ponisseril Somasundaran, (D), California (Berkeley), 1964, Nx
Tuncel M. Yegulalp, (D), Columbia, 1968, Nm

Associate Professor:
Ross Bagtzoglou, (D), California (Berkeley), 1990, Hw

Senior Research Scientist:
Roelof Versteeg, (D), Paris VII, 1991, Yg

Adjunct Professor:
Vasilis M. Fthenakis, (D), New York, 1991, As

Emeritus:
Stefan H. Boshkov, (M), Columbia, 1942, Nm
John T. Kuo, (D), Stanford, 1958, Yg
Malcolm T. Wane, (M), Columbia, 1954, Nm

Dept of Earth & Environmental Sciences (B,M,D) ☒
P.O. Box 1000
61 Route 9W
Palisades, NY 10964
p. (845) 365-8550
kaleighm@ldeo.columbia.edu
http://eesc.columbia.edu
Programs: Earth and Environmental Sciences

Chair:
Sidney R. Hemming, (D), SUNY (Stony Brook), 1994, CscPe

Vice Chair:
Peter B. de Menocal, (D), Columbia, 1991, PeOuCg

Director, Lamont Doherty Earth Observatory:
Sean Solomon, (D), MIT, 1971, Xy

Dir. Graduate Studies:
Goran Ekstrom, (D), Harvard, 1987, Ys

Professor:
Wallace S. Broecker, (D), Columbia, 1958, CmOcPe
Nicholas Christie-Blick, (D), California (Santa Barbara), 1979, Gst
Joel E. Cohen, (D), Harvard, 1970, Zn
Hugh Ducklow, (D), Harvard, 1977, ObCb
Peter M. Eisenberger, (D), Harvard, 1967, Zn
Steven L. Goldstein, (D), Columbia, 1986, Csg
Arnold L. Gordon, (D), Columbia, 1965, Op
Kevin L. Griffin, (D), Duke, 1994, PbeZn
Peter B. Kelemen, (D), Washington, 1988, GiCg
Jerry F. McManus, (D), Columbia, 1989, PeOu
William H. Menke, (D), Columbia, 1981, YsGvq
John C. Mutter, (D), Columbia, 1982, YrGtYs
Paul E. Olsen, (D), Yale, 1984, PvgGr
Terry A. Plank, (D), Columbia, 1993, GxCg
Lorenzo M. Polvani, (D), MIT, 1988, AmsGq
G. Michael Purdy, (D), Cambridge, 1974, Yr
Peter Schlosser, (D), Heidelberg, 1985, Hw
Adam H. Sobel, (D), MIT, 1998, As
Marc W. Spiegelman, (D), Cambridge, 1989, GqxCg
Martin Stute, (D), Heidelberg, 1989, CsHgGe
Felix Waldhauser, (D), ETH (Switzerland), 1996, YsgGt

Associate Professor:
Sonya Dyhrman, Scripps, 1999, Ob
Arlene M. Fiore, (D), Harvard, 2003, As
Baerbel Hoenisch, (D), Bremen, 2002, PcClm
Meredith Nettles, (D), Harvard, 2005, YsGl
Maria Tolstoy, (D), California (San Diego), 1994, Yr

Assistant Professor:
Ryan P. Abernathey, (D), MIT, 2012, Op

Lecturer:
Alberto Malinverno, (D), Columbia, 1989, Gug
Benjamin S. Orlove, (D), California (Berkeley), 1975, As
Andreas M. Thurnherr, (D), Southampton (UK), 2000, Op
Christopher J. Zappa, (D), Washington, 1999, OpAsZr

Adjunct Professor:
Robert F. Anderson, (D), MIT, 1981, OcCmPe
W. Roger Buck, IV, (D), MIT, 1984, YgGt
John J. Flynn, (D), Columbia, 1983, PvgYm
Alessandra Giannini, (D), Columbia, 2001, AsOpZc
Lisa M. Goddard, (D), Princeton, 1995, AstZc
Arthur L. Lerner-Lam, (D), California (San Diego), 1982, Ys
Douglas G. Martinson, (D), Columbia, 1982, OpGq
Ronald L. Miller, (D), MIT, 1990, As
Mark A. Norell, (D), Yale, 1988, Pv
Dorothy M. Peteet, (D), New York, 1983, PelGn
Andrew W. Robertson, (D), Reading (UK), 1984, As
Joerg Schaefer, (D), ETH (Switzerland), 2000, Cg
Christopher Small, (D), California (San Diego), 1993, ZrYr
Taro Takahashi, (D), Columbia, 1957, OcCg
Mingfang Ting, (D), Princeton, 1990, As
Spahr C. Webb, (D), California (San Diego), 1984, Yrg
Gisela Winckler, (D), Heidelberg, 1998, Cm

Emeritus:
James D. Hays, (D), Columbia, 1964, PemOu
Paul G. Richards, (D), Caltech, 1970, YsZn
Christopher H. Scholz, (D), MIT, 1967, YxNr
H. James Simpson, Jr., (D), Columbia, 1970, Cm
Lynn R. Sykes, (D), Columbia, 1965, YsGtZn
David Walker, (D), Harvard, 1972, CpGzCg

Dir. Academic Admin & Finance:
Sarah K. Odland, (M), Colorado, 1981, Zn

Asst. Director Climate and Society Program:
Cynthia Thomson, (M), Columbia, 2009, Zn

Lamont-Doherty Earth Observatory (M,D) ☒
P.O. Box 1000
61 Route 9W
Palisades, NY 10964
p. (845) 359-2900
director@ldeo.columbia.edu
http://www.ldeo.columbia.edu
f: https://www.facebook.com/Lamont.Doherty/
t: @LamontEarth
Programs: M.S. in Sustainability Science; M.A. in Climate and Society
Enrollment (2006): M: 40 (0) D: 83 (6)

Professor:
Sonya Dyhrman, (D), California (San Diego), 1999, Zg

Lamont Research Professor:
Andrew Juhl, (D), Scripps, 2000, Ob

Higgins Professor of Earth and Environmental Sciences:
Steven Goldstein, (D), Columbia, 1986, Cg

Arthur D. Storke Memorial Professor:
Peter Kelemen, (D), Washington, 1987, Gi
Paul E. Olsen, (D), Yale, 1983, Gm
Marc Spiegelman, (D), Cambridge, 1989, Ys
Sidney Hemming, (D), SUNY (Stony Brook), 1994, Cg

Professor:
Nicholas Christie-Blick, (D), California (Santa Barbara), 1979, Gs
Peter B. deMenocal, (D), Columbia, 1991, Pe
Hugh W. Ducklow, (D), Harvard, 1977
Peter Eisenberger, Zg
Goran Ekstrom, (D), Harvard, 1987, Yg
Arlene Fiore, (D), Harvard, 2003, As
Arnold L. Gordon, (D), Columbia, 1965, Op
Kevin Griffin, (D), Duke, 1994, Zn
Baerbel Hoenisch, (D), Alfred Wegener Inst (Germany), 2002, Cm
Galen McKinley, (D), MIT, 2002, Cg
Jerry McManus, (D), Cm
William H. Menke, (D), Columbia, 1981, Ys
John C. Mutter, (D), Columbia, 1982, Yr
Meredith Nettles, (D), Harvard, 2005, YsGl
Adam Sobel, (D), MIT, 1998, As
Renata Wentzcovitch, California (Berkeley), 1988, Ys

Sr. PGI Research Scientist:
Richard Seager, (D), Columbia, 1990, Am

Senior Research Scientist:
Robert Newton, (D), Columbia, 2001, Ct

Professor:
Maria Tolstoy, (D), California (San Diego), 1994, Yr
Palisades Geophysical Institute/Lamont Research Professor:
Robin E. Bell, (D), Columbia, 1989, Yr
Marie Tharp Lamont Research Professor:
Suzana Camargo, (D), Tech (Munich), 1992, As
Lamont Research Professor, Sr.:
Radley Horton, (D), Columbia, 2007, Og
Lamont Research Professor:
Roger W. Buck, (D), MIT, 1984, Yr
Brendon Buckley, (D), Tasmania, 1997, Pe
Steven Chillrud, (D), Columbia, 1995, Cg
Rosanne D'Arrigo, (D), Columbia, 1989, Zn
James Davis, (D), MIT, 1986, Yd
Joaquim Goes, (D), Nagoya, 1996, Gu
Won-Young Kim, (D), Uppsala, 1986, Ys
Braddock Linsley, (D), New Mexico, 1990, Pe
Alberto Malinverno, (D), Columbia, 1989, Go
Robert Pincus, Og
Joerg Schaefer, (D), Swiss Fed Inst Tech, 2000, Pe
Bruce Shaw, (D), Chicago, 1989, Ys
Christopher Small, (D), California (San Diego), 1993, Zr
Jason Smerdon, (D), Michigan, 2004, Pe
Michael Steckler, (D), Columbia, 1980, Yv
Ajit Subramaniam, (D), SUNY (Stony Brook), 1995, ObZr
Marco Tedesco, (D), Italian Nat Res Cncl, 2003, GlZr
Andreas Thurnherr, (D), Southampton (UK), 1999, Op
Mingfang Ting, (D), Princeton, 1990, As
Alexander Van Geen, (D), MIT/WHOI, 1989, Cg
Felix Waldhauser, (D), ETH (Switzerland), 1996, Ys
Gisela Winckler, (D), Heidelberg, 1998, Cg
Xiaojun Yuan, (D), California (San Diego), 1994, Op
Christopher Zappa, (D), Washington, 1999, Op
Jerome M. Paros/Lamont Research Professor of Observational Physics :
Spahr Webb, (D), California (San Diego), 1984, Yr
Ewing Lamont Research Professor:
Robert F. Anderson, (D), MIT, 1981, Cg
Edward R. Cook, (D), Arizona, 1985, Hw
Taro Takahashi, (D), Columbia, 1957, Cm
Doherty Senior Research Scientist:
Kerstin Lehnert, (D), Albert-Ludwigs Freiburg, 1989, Gi
Director, Marine Operations; Senior Research Scientist:
Sean M. Higgins, (D), Columbia, 2002, Gs
Director, Lamont-Doherty Earth Observatory, G. Unger Vetlesen Prof:
Maureen Raymo, (D), Columbia, 1989, Pe
Deputy Director, Lamont Research Professor:
David S. Goldberg, (D), Columbia, 1985, Yr
Deputy Director:
Arthur L. Lerner-Lam, (D), California (San Diego), 1982, Ys
Bruce Heezen Lamont Research Scientist:
Suzanne Carbotte, (D), California (Santa Barbara), 1992, Yr
Arthur D. Storke Memorial Professor:
Terry Plank, (D), Columbia, 1993, Cg
Lamont Associate Research Professor:
Christine McCarthy, (D), Brown, 2009, Ys
Associate Professor:
Kerry Key, (D), California (San Diego)
Ryan Abernathy, (D), MIT, 2012, Og
Lamont Research Professor:
Benjamin Holtzman, (D), Minnesota, 2003, Ys
Michael Kaplan, (D), Colorado, 1999, Gl
Lamont Associate Research Scientist:
Kirsteen Tinto, (D), Otago (NZ), 2010, Gu
Lamont Associate Research Professor:
Laia Andreu-Hayles, (D), Barcelona, 2007, Pe
Anne Becel, (D), Inst de Physique du Globe de Paris, 2006, Gu
Michela Biasutti, (D), Washington, 2003, Am
Benjamin C. Bostick, (D), Stanford, 2002, Sc
Connie Class, (D), Karlsruhe, 1994, Cg
Indrani Das, Gujarat (India), 2007, Gu
Alexey Kaplan, (D), Gubkin Inst of Tech (Moscow), 1990, Op
Jennifer L. Middleton, Cg
Raymond N. Sambrotto, (D), Alaska, 1983, Ob
David Schaff, (D), Stanford, 2001, Ys
Susanne Straub, (D), Kiel, 1991, Gv
Kevin Uno, (D), Utah, 2012, Pe
Yutian Wu, (D), Columbia, 2011, Og
Associate Research Scientist:
Jacob A. Tielke, Ys
Patrick M. Alexander, (D), Graduate Center, CUNY, 2015
Dhruv Balwada
Valerie Bennington
Louise Bolge, (D), Rutgers, 2005, Cg
Brian B. Boston, Gu
Yue Cai, (D), Columbia, 2009, Cg
Heng Chen, Cg
Renata Regina Constantino, Gu
Carol J. Cotterill, Zg
Shuo Ding, Cg
Will Frances Jayaram Fortin, (D), Wyoming, Ys
Andrew Goodwillie, Gu
Peng Ji, Shandong, 2009, Yr
Yael Kiro Feldman, Hebrew, 2013, Cg
Jennifer L. Lamp, Cg
Chiara Lepore, Og
David Porter, Gu
Lamont Assistant Research Professor:
Sarah J. Hurley, Pg
Assistant Professor:
Jacqueline Austermann, Harvard, 2016, Ys
Jonathan Kingslake, Sheffield, 2012, Yr
Yves Moussallam, Cg
Roisin Commane, (D), Leeds (UK), 2009, Yr
Lamont Associate Research Professor:
Einat Lev, (D), MIT, 2009, Gv
Special Research Scientist:
Pierre E. Biscaye, (D), Yale, 1964, Cm
Enrico Bonatti, (D), Pisa, 1967, Yr
Dake Chen, (D), SUNY (Stony Brook), 1989, Op
James R. Cochran, (D), Columbia, 1977, Yr
James D. Hays, Pg
Klaus H. Jacob, (D), Goethe (Frankfurt), 1968, Ys
Stanley Jacobs, (B), MIT, 1962, Op
Kim Anne Kastens, Gu
Mikhail Kogan, (D), Inst of Physics (Moscow), 1977, Yd
Yochanan Kushnir, (D), Oregon State, 1985, As
John Marra, Pg
Douglas G. Martinson, (D), Columbia, 1982, Op
Paul G. Richards, (D), Caltech, 1970, Ys
William B. F. Ryan, (D), Columbia, 1961, Yr
Christopher H. Scholz, (D), MIT, 1967, Ys
Leonardo Seeber, (B), Columbia, 1964, Ys
William Smethie, (D), Washington, 1979, Cg
Lamont Assistant Research Professor:
Reinhard Kozdon, (D), Christian-Albrechts (Germany), 2007, PgGg
G. Unger Vetlesen Professor of Earth and Climate Sciences:
Mark A. Cane, (D), MIT, 1975, Op
Emeritus:
David Walker, (D), Harvard, 1972, Gx
Senior Research Scientist:
Andrew Barclay, (D), Oregon, 1998, Ys
Research Scientist:
Victoria Ferrini, (D), SUNY (Stony Brook), 2004, Ou
Helga Gomes, (D), Bombay, 1985, Ob
Gilles Guerin, (D), Columbia, 2000, Gu
Naomi Henderson, (D), Wisconsin, 1987, Yr
Timothy Kenna, (D), MIT/WHOI, 2002, Cg
Frank Nitsche, (D), Alfred Wegener Inst (Germany), 1997, Gu
Angela Slagle, (D), Columbia, 2007, Gu
Lamont Research Professor:
Natalie Boelman, (D), Columbia, 2004, Zn
Lamont Associate Research Professor, Jr.:
Nicolas Young, (D), Buffalo, 2012, Cg
Lamont Associate Research Professor:
Timothy Crone, Washington, 2007, Gu
William Joseph D'Andrea, (D), Brown, 2008, Pe
Solange Duhamel, (D), Aix-Maraseille II, Pe
Jonathan E. Nichols, (D), Brown, 2009, Pe
Michael Previdi, (M), Rutgers, 2006, As
Heather M. Savage, (D), Penn State, 2007, Ys
Beizhan Yan, (D), Rensselaer, 2004, Cg
Lamont Assistant Research Scientist:
Daniel Westervelt, Carnegie Mellon, 2013, Og
Lamont Associate Research Professor:
Timothy T. Creyts, (D), British Columbia, 2007, Gl
Associate Research Professor:
Kathrin Schillling, Cg

NEW YORK

Cornell University

Dept of Earth & Atmospheric Sciences (B,M,D) ● ☒
2160 Snee Hall
Ithaca, NY 14853-1504
p. (607) 255-3474
easinfo@cornell.edu
http://www.eas.cornell.edu/
Programs: Earth and Atmospheric Sciences (B); Atmospheric Science (B,M,PhD); Geological Sciences (M,PhD); Geological Sciences (MEng)
Certificates: Cornell Satellite Remote Sensing Program (Professional Training Certificate Program)
Enrollment (2018): B: 43 (13) M: 3 (1) D: 24 (8)

William and Katherine Snee Professor of Geological Sciences:
Geoffrey A. Abers, (D), MIT, 1989, YsGt
Wold Family Professor in Environmental Balance and Human Sustainability :
Karin Elizabeth Olson Hoal, (D), Massachusetts, Gz
Thomas R Briggs Professor in Engineering:
David Hysell, (D), Cornell, 1992, Znr
Sidney Kaufman Professor in Geophysics:
Larry D. Brown, (D), Cornell, 1976, Ye
Irving Porter Church Professor of Engineering:
Natalie M. Mahowald, (D), MIT, 1996, As
Hunter R. Rawlings III Professor of Paleontology:
Warren D. Allmon, (D), Harvard, 1988, PgePi
Charles L. Pack Professor:
Susan Riha, (D), Washington, 1980, Sf
Professor:
Arthur DeGaetano, (D), Rutgers, 1989, As
Louis A. Derry, (D), Harvard, 1989, Cl
Rowena B. Lohman, (D), Caltech, 2004, YsZrGt
Matthew E. Pritchard, (D), Caltech, 2003, YdGvl
Sara C. Pryor, (D), E Anglia (UK), 1992, AsZn
Charles N. Mellowes Professor in Engineering:
Esteban Gazel, (D), Rutgers, 2009, CeGiv
Associate Professor:
Toby R. Ault, (D), Arizona, 2011, Am
Kade M. Keranen, (D), Stanford, 2008, Ys
Britney Elyce Schmidt, (D), California (Los Angeles), 2010, Ye
Assistant Professor:
Grace Barcheck, (D), California (Santa Cruz), 2018, Gg
Nicole Fernandez, (D), Illinois, 2020, Gg
Patrick Fulton, (D), Penn State, 2008
Peter Hitchcock, (D), Toronto, 2012, As
Megan Holycross, (D), Rensselaer Polytech, 2017, CmGv
Flavio Lehner, (D), Bern, 2013, Atm
Jonathan Lin, (D), MIT, 2021, As
Angeline G. Pendergrass, (D), Washington, 2013, As
Stephen H. Weiss Provost's Teaching Fellow & Senior Lecturer:
Bruce Monger, (D), Hawaii, 1993, Ob
Senior Lecturer:
Mark Wysocki, (M), Cornell, 1988, As
David C. Duncan Professor in the Physical Sciences:
Jonathan I. Lunine, (D), Xc
Adjunct Associate Professor:
Gregory P. Dietl, (D), North Carolina State, 2002, Peg
Adjunct Professor:
Jonathan Robert Hendricks, (D), Cornell, 2005, Zg
Jason Phipps Morgan, (D), Brown, 1985, GvCmPe
Robert M. Ross, (D), Harvard, 1990, PgZeGs
John F.H Thompson, (D), Toronto, 1982, EmgNx
Martyn Unsworth, (D), Cambridge, 1991, Ye
Emeritus:
Richard W. Allmendinger, (D), Stanford, 1979, Gc
William A. Bassett, (D), Columbia, 1959, Gz
Lawrence M. Cathles, (D), Princeton, 1971, GqYgEg
John L. Cisne, (D), Chicago, 1973, PgsPe
Stephen J. Colucci, (D), SUNY (Albany), 1982, As
Charles H. Greene, (D), Washington, 1985, Ob
Bryan L. Isacks, (D), Columbia, 1965, GtYsGm
Teresa E. Jordan, (D), Stanford, 1979, GrtPc
Daniel E. Karig, (D), California (San Diego), 1970, YrGl
Robert W. Kay, (D), Columbia, 1970, Giz
Suzanne M. Kay, (D), Brown, 1975, GiCuGz
Warren Knapp, (D), Wisconsin, 1968, As
Daniel Wilks, (D), Oregon, 1986, As
Cooperating Faculty:
Rebecca J. Barthelmie, (D), As
Maria A. Gandolfo-Nixon, (D), Buenos Aires, 1994, Pb
Oliver H. Gao, (D), California (Davis), 2004, Zn
Alexander G. Hayes, (D), Caltech, 2011, Xg
Peter G. Hess, (D), Ac
Sturt W. Manning, (D), Cambridge, 1995, Ga
Greg C. McLaskey, (D), California (Berkeley), 2011, Ys
Tammo S. Steenhuis, (D), Wisconsin, 1977, Hg
Scott Steinschneider, (D), Rw
Jefferson W. Tester, (D), MIT, 1971, Ng
Max Zhang, (D), California (Davis), 2004, Zn

Institute for the Study of the Continents ☒
2122 Snee Hall
Ithaca, NY 14853-1504
p. (607) 255-3474
easinfo@cornell.edu
http://www.eas.cornell.edu/
Enrollment (2010): M: 9 (0) D: 29 (4)

Professor:
Geoffrey A. Abers, (D), MIT, 1989, Ys
Richard W. Allmendinger, (D), Stanford, 1979, Gct
Larry D. Brown, (D), Cornell, 1976, YesGt
Louis A. Derry, (D), Harvard, 1989, Cg
David L. Hysell, (D), Cornell, 1992, Znr
Teresa E. Jordan, (D), Stanford, 1979, GrtZg
Suzanne M. Kay, (D), Brown, 1975, GiCuGt
William M. White, (D), Rhode Island, 1977, Ce
Associate Professor:
Katie M. Keranen, (D), Stanford, 2008, Ys
Rowena B. Lohman, (D), Caltech, 2004, Yg
Matthew E. Pritchard, (D), Caltech, 2003, YdGvl
Assistant Professor:
Greg C. McLaskey, (D), California (Berkeley), 2011, YsNre
Visiting Professor:
Franklin G. Horowitz, (D), Cornell, 1989, YgZnYe
Emeritus:
Muawia Barazangi, (D), Columbia, 1971, Ys
Bryan L. Isacks, (D), Gmt
Robert W. Kay, (D), Columbia, 1970, Gi

Dowling College

Dept of Earth & Marine Sciences
Oakdale, NY
asmirnov@dowling.edu

Dutchess Community College

Physical Science (A)
53 Pendell Road
Poughkeepsie, NY 12601
p. (845) 431-8550
rambo@sunydutchess.edu
http://www.sunydutchess.edu/academics/departments/mathematicsphysicalandcomputersciences/
Enrollment (2016): A: 5 (2)

Chair:
Tim Welling, (M), GeSo
Professor:
Mark McConnaughhay, ZgAmEo
Associate Professor:
Susan H. Conrad, GsmHs

Graduate School of the City University of New York

PhD Program in Earth & Environmental Sciences (D) ☒
365 Fifth Avenue, Room 4306
New York, NY 10016
p. (212) 817-8240
ees@gc.cuny.edu
http://www.gc.cuny.edu/Page-Elements/Academics-Research-Centers-Initiatives/Doctoral-Programs/Earth-and-Environmental-Sciences
f: https://www.facebook.com/CunyGCEES/
t: @cunygcees
Programs: Environmental and Geological Sciences; Geography
Certificates: GIS
Enrollment (2020): D: 24 (24)

President, CSI:
William Fritz, (D), Montana, Zg
Herbert Kayser Professor:
Samir Ahmed, (D), Univ Coll (UK), Gg
Distinguished Professor:
George Hendrey, (D), Washington, Gg
Discipline Cooordinator of Geology:
Nazul Khandaker, (D), Penn State, Gg
Professor:
Teresa Bandosz, (D), Tech (Cracow), Gge
Homar S. Barcena, (D), New York, Gg
Sunil Bhaskaran, (D), New South Wales, 2003, ZfrZi
Jeffrey Bird, (D), California (Davis), 2001, Sbf
Anthony Carpi, (D), Cornell, Gg
John A. Chamberlain, (D), Rochester, 1971, Pg
Zhongqi (Joshua) Cheng, (D), Ohio State, Zg
Constantin Cranganu, (D), Oklahoma, Zg
Eric Delson, (D), Columbia, 1973, PvqPs
Timothy Eaton, (D), Wisconsin, Gg
Allan Frei, (D), Rutgers, Gg
Alexander Gilerson, (D), Tech, Kazan, Gu
Yuri Gorokhovich, (D), CUNY (Grad Ctr), Gg
Peter Groffman, (D), Georgia, Zg
Carsten Kessler, (D), Muenster (Germany), Gg
Reza M. Khanbilvardi, (D), Penn State, 1983, Hy
Athanasios Koutavas, (D), Columbia, Gg
Nir Krakauer, (D), Caltech, Gg
Tammy Lewis, (D), California (Davis), Gg
Allan Ludman, (D), Pennsylvania, 1969, Gg
Johnny Luo, (D), Columbia, Zg
John Marra, (D), Dalhousie, Zg
Kyle McDonald, (D), Michigan, As
Cecilia M. McHugh, (D), Columbia, 1993, Ou
Fred Moshary, (D), Columbia, Gg
Wenge Ni-Meister, (D), Boston, Gg
Stephen Pekar, (D), Rutgers, Zg
Wayne G. Powell, (D), Queen's, 1994, EmGaCs
David E. Seidemann, (D), Yale, 1976, Cc
Gillian Stewart, (D), SUNY (Stony Brook), Oc
Maria Tzortziou, (D), Maryland, As
Charles Vorosmarty, (D), New Hampshire, Ng
John Waldman, (D), CUNY (Grad Ctr), ObHsZc
William G. Wallace, (D), SUNY (Stony Brook), 1996, Og
Zhengrong Wang, (D), Caltech, Gge
Michael Weisberg, (D), CUNY (Grad Ctr), Zy
Chuixiang Yi, (D), Nanjing (China), Zg
Pengfei Zhang, (D), Utah, Zg
Chairperson, Queens College:
Gregory O'Mullan, (D), Princeton, Zg
Chairperson:
Jennifer Cherrier, (D), Florida State, Zg
Associate Professor:
Stephen U. Aja, (D), Washington State, 1989, Cl
Karin Block, (D), CUNY (Grad Ctr), Ze
Rebecca Boger, (D), Virginia Inst of Marine Sci, Zg
James Booth, (D), Washington, Gge
Brett Branco, (D), Connecticut, Gg
Frank S. Buonaiuto Jr., (D), SUNY (Stony Brook), Ze
Nathalie Goodkin, (D), MIT/WHOI, Gg
Jean Grassman, (D), California (Berkeley), Gg
Dianne I. Greenfield, (D), SUNY (Stony Brook), Gu
Kieren Howard, (D), Tasmania, Ggz
Urs Jans, (D), Swiss Fed Inst Tech, Gg
Patricia M. Kenyon, (D), Cornell, 1986, Yg
Jacob Mey, (D), CUNY (Grad Ctr), Gg
Hamidreza Norouzi, (D), CUNY (Grad Ctr), Gg
Michael Piasecki, (D), Michigan, Gg
Andrew Reinmann, (D), Boston, Gg
Randye Rutberg, (D), Columbia, Gg
Haydee Salmun, (D), Johns Hopkins, Gg
Heather Sloan, (D), Paris, Gg
Assistant Professor:
Benjamin Black, (D), MIT, As
William Blanford, (D), Arizona, Ze
Cheila Cullen, (D), CUNY (Grad Ctr), Rn
Kennet Flores, (D), Lausann, Zg
Steven Kidder, (D), Caltech, As
David Lindo Atachati, (D), Georgia, Og
Marc-Antoine Longpre, (D), Trinity (Dublin), Zg
Hari Pant, (D), Dalhousie, Zg
Jisun Park, (D), Tokyo Grad Sch Sci, Xm
Adjunct Professor:
Denton Ebel, (D), Purdue, Gg
John T. Flynn, (D), Columbia, Gg
George E. Harlow, (D), Princeton, 1977, Gz
Neil H. Landman, (D), Yale, 1982, Ps
Jin Meng, (D), Columbia, 1991, Pv
Robert P. Nolan, (D), CUNY, 1986, Co

Hamilton College

Geosciences Department (B) ☒
198 College Hill Road
Clinton, NY 13323
p. (315) 859-4142
ccbeck@hamilton.edu
https://my.hamilton.edu/academics/departments?dept=Geosciences
Programs: Geoscience; Geoarchaeology
Enrollment (2020): B: 20 (13)
Chair:
David G. Bailey, (D), Washington State, 1990, GizGa
Professor:
Cynthia R. Domack, (D), Rice, 1985, Pg
Todd W. Rayne, (D), Wisconsin, 1993, Hy
Barbara J. Tewksbury, (D), Colorado, 1981, Gc
Associate Professor:
Michael L. McCormick, (D), Michigan, 2002, Cb
Assistant Professor:
Catherine C. Beck, (D), Rutgers, 2015, GsnGr

Hartwick College

Dept of Geological and Environmental Sciences (B)
Johnstone Science Center
1 Hartwick Drive
Oneonta, NY 13820
p. (607) 431-4658
griffingd@hartwick.edu
https://www.hartwick.edu/academics/academic-departments/geology-environmental-sciences-department/
Administrative Assistant: Nancy Heffernan
Enrollment (2011): B: 28 (9)
Chair:
David H. Griffing, (D), SUNY (Binghamton), 1994, GduGm
Professor Emeritus:
David Hutchison, (D), West Virginia, 1968, GxgGi
Professor:
Eric L. Johnson, (D), SUNY (Binghamton), 1990, GpcGi
Robert C. Titus, (D), Boston, 1974, Ps
Associate Professor:
Zsuzsanna Balogh-Brunstad, (D), Washington State, 2006, ClHgSo

Hobart & William Smith Colleges

Dept of Geoscience (B) ●☒
300 Pulteney Street
Geneva, NY 14456
p. (315) 781-3586
geoscience@hws.edu
http://www.hws.edu/academics/geoscience/
f: https://www.facebook.com/geoscience.hws
Programs: Geoscience (B)
Enrollment (2021): B: 49 (18)
Professor:
Nan Crystal Arens, (D), Harvard, 1993, Pg
John D. Halfman, (D), Duke, 1987, GeHsGn
Neil Laird, (D), Illinois, 2001, As
Associate Professor:
Tara M. Curtin, (D), Arizona, 2001, PeGs
David Finkelstein, (D), Illinois (Urbana), 1997, CgbCs
David C. Kendrick, (D), Harvard, 1997, PgGg
Nicholas Metz, (D), SUNY (Albany), 2011, As
Technician:
Barbara Halfman, Zn

Hofstra University

Dept of Geology, Environment and Sustainability (B,M) ☒
114 Hofstra University

Hempstead, NY 11549
p. (516) 463-5564
j.b.bennington@hofstra.edu
http://www.hofstra.edu/Academics/Colleges/HCLAS/GEOL/
f: https://www.facebook.com/GESatHU/
Programs: BA / BS Geology, BS Environmental Resources, BA /BS Sustainability Studies, MA Sustainability
Enrollment (2018): B: 50 (10) M: 25 (7)

Professor and Chair:
J Bret Bennington, (D), Virginia Tech, 1994, PeGs
Professor:
Emma Christa Farmer, (D), Columbia, 2005, Ou
Associate Professor:
Antonios Marsellos, (D), SUNY (Albany), 2008, GtqZf
Assistant Professor:
Jase Bernhardt, Penn State, 2016, Ams
Instructor:
Annetta Centrella-Vitale, (M), SUNY (Stony Brook), 1998
Adina Hakimian, (M), CUNY (Queens), 2012, Gg
Steven C. Okulewicz, (M), CUNY (Brooklyn), 1979, CgGue
Adjunct Professor:
Nehru Cherukupalli, (D), Madras (India), 1963, Gg
Sandra J. Garren, (D), S Florida, 2014, Zi
Richard Liebling, (D), Columbia, 1963, Gg
Emeritus:
Charles M. Merguerian, (D), Columbia, 1985, Gc
Dennis Radcliffe, Queens, 1966, Gzx

Hudson Valley Community College

Biology, Chemistry, Physics Dept
80 Vandenburgh Ave.
Troy, NY 12180
p. (518) 629-7453
p.schaefer@hvcc.edu

Assistant Professor:
Ruth H. Major, (M), Syracuse, 1989, GgRh

Hunter College (CUNY)

Dept of Geography (B,M)
695 Park Avenue
Room 1006 North Building
New York, NY 10021
p. (212) 772-5265
imiyares@hunter.cuny.edu
http://www.geo.hunter.cuny.edu
Programs: Geography; Environmental Studies; Earth Science Education; Geoinformatics (M); Geography (M)
Certificates: GIS
Administrative Assistant: Dana G.. Reimer
Enrollment (2006): B: 135 (30) M: 40 (8)

Chair:
Allan Frei, (D), Rutgers, 1997, AtHgs
Professor:
Sean C. Ahearn, (D), Wisconsin, 1986, Zi
Jochen Albrecht, (D), Vechta (Germany), 1995, ZiyZu
Hongmian Gong, (D), Georgia, 1997, Zn
Ines Miyares, (D), Arizona State, 1994, Zen
Wenge Ni-Meister, (D), Boston, 1997, Zr
Marianna Pavlovskaya, (D), Clark, 1998, Zi
William Solecki, (D), Rutgers, 1990, Zu
Associate Professor:
Frank Buonaiuto, (D), SUNY (Stony Brook), 2003, On
Mohamed Ibrahim, (D), Alberta, 1985, ZgRwZc
Rupal Oza, (D), Rutgers, 1999, Zn
Haydee Salmun, (D), Johns Hopkins, 1989, OpZgAs
Assistant Professor:
Randye L. Rutberg, (D), Columbia, 2000, CmOcZc
Director, SPARs Lab:
Thomas Walter, (M), Miami, 1984, ZyeAm
Lecturer:
Anthony Grande, (M), CUNY (Baruch), 1999, Zn
Adjunct Professor:
Jack Eichenbaum, (D), Michigan, 1972, Zie
Edward Linky, (D), Duquesne Law, 1973, Zn
Teodosia Manecan, (D), Bucharest, 1985, Gp
Faye Melas, (D), CUNY, 1989, Gs
Douglas Williamson, (D), CUNY, 2003, Zi

Emeritus:
Charles A. Heatwole, (D), Michigan State, 1974, Zn
Karl H. Szekielda, (D), Marseille, 1967, Zrg
GeoScience Lab Tech:
Amy Jeu, (M), Minnesota, 2002, Zi

Lehman College (CUNY)

Earth, Environmental and Geospatial Sciences (B) ☒
250 Bedford Park Boulevard West
Bronx, NY 10468-1589
p. (718) 960-8660
heather.sloan@lehman.cuny.edu
http://www.lehman.edu/academics/eggs/
Programs: BA in Earth Science, BA in Geography, BS in Environmental Science, MS in GISc
Certificates: Certificate in Earth Science, GIS
Enrollment (2014): B: 24 (5)

Chair:
Hari Pant, (D), Dalhousie, Cg
Professor:
Irene S. Leung, (D), California (Berkeley), 1969, Gz
Juliana Maantay, (D), Rutgers, Zi
Associate Professor:
Elia Merchado
Heather Sloan, (D), Paris VI, 1993, Yr
Emeritus:
Frederick C. Shaw, (D), Harvard, 1965, Ps

Long Island University, Brooklyn Campus

Dept of Physics
1 University Plaza
Brooklyn, NY 11201-8423
p. (718) 488-1011
bkln-admissions@liu.edu
http://www.liu.edu/Home/Brooklyn

Professor:
Richard Macomber, (D), Iowa, 1963, Pg
Adjunct Professor:
Richard A. Jackson, (D), Massachusetts, 1980, Gc
Alan Siegelberg, (M), Brooklyn Coll, 1977, Gg
Emeritus:
Samuel R. Kamhi, (D), Columbia, 1963, Gz

Long Island University, C.W. Post Campus

Dept of Biological and Environmental Sciences (B,M) ⦸
720 Northern Boulevard
Brookville, NY 11548-1300
p. (516) 299-2318
margaret.boorstein@liu.edu
Programs: Biology BS. Biology Adolescent-ed BS
Our Geography BA, Biology BA, Geology BA and BS, Earth System Sciences BA, Earth Science Adol Ed BS were frozen [no new majors are accepted] by the Administration
Enrollment (2020): B: 3 (0) M: 10 (11)

Chair:
Margaret F. Boorstein, (D), Columbia, 1977, Zg
Associate Professor:
Scott Carlin, (D), Zn
Lillian Hess-Tanguay, (D), CUNY, 1993, Gs
Emeritus:
Robert S. Harrison, (D), Cambridge, 1965, Zy
Heinrich Toots, (D), Wyoming, 1965, Pg

Monroe Community College

Geoscience Dept (A)
1000 E. Henrietta Road
Rochester, NY 14623
p. (716) 292-2425
amcbride14@monroecc.edu
http://www.monroecc.edu/depts/geochem/
Administrative Assistant: Judy Miller
Enrollment (2013): A: 7 (0)

Associate Professor:
Jessica Barone, (M), Ball State, Ge
Michael Boester, (M), Zy
Amanda Colosimo, (M), North Carolina, 2004, Gg
Daniel E. Robertson, (M), Arizona State, 1986, Eg

Assistant Professor:
Jonathan Little, (M), ZyGlZi
Jason Szymanski, (M), GlPe
Instructor:
Heather Pierce, (M), Connecticut, Zyi

Orange County Community College

Dept of Science, Engineering, and Architecture (A)
115 South Street
Middletown, NY 10940
p. (845) 341-4570
lawrenceobrien@sunyorange.edu
Enrollment (2015): A: 3 (1)
Professor:
Lawrence E. O'Brien, (M), Michigan, 1972, Gg

Pace University, New York Campus

Dept of Chemistry & Physical Sciences
1 Pace Plaza
New York, NY 10038
p. (212) 346-1502
mshirigarakani@pace.edu
http://www.pace.edu/dyson/academic-departments-and-programs/chemistry-and-physical-sciences---nyc
Department Secretary: Pat Calegari
Chair:
Nigel Yartlett, (D)
Assistant Professor:
Stephen T. Lofthouse, (M), Hunter (CUNY), 1974, Zg
Adjunct Professor:
Anatole Dolgoff, (M), Miami (Ohio), 1960, Zg
William Hansen, (M), Hunter, 1991, As
John Marchisin, (M), Rutgers, 1965, Zg
Nathan Reiss, (D), New York, 1973, As

Paleontological Research Institution

Paleontological Research Institution ☒
1259 Trumansburg Road
Ithaca, NY 14850
p. (607) 273-6623
info@priweb.org
http://www.priweb.org
f: https://www.facebook.com/museumoftheearth/
t: @PRInstitution
Director:
Warren D. Allmon, (D), Harvard, 1988, Pg
Associate Director for Outreach:
Robert M. Ross, (D), Harvard, 1990, PgZePc
Director of Collections:
Gregory Dietl, (D), North Carolina State, 2002, Pg
Associate Director of Science Communication:
Jonathan R. Hendricks, (D), Cornell, 2005, Pig

Queens College (CUNY)

School of Earth & Environmental Sciences (B,M,D)
65-30 Kissena Boulevard
Flushing, NY 11367
p. (718) 997-3300
gregory.omullan@qc.cuny.edu
http://www.qc.edu/EES
Administrative Assistant: Gladys Sapigao
Enrollment (2012): B: 150 (19) M: 19 (6) D: 11 (2)
Distinguished Professor, Director & Chair:
George Hendrey, (D), Washington, 1973, Zg
Professor:
Nicholas K. Coch, (D), Yale, 1965, GseOu
N. Gary Hemming, (D), SUNY (Stony Brook), 1993, CaOcGg
Allan Ludman, (D), Pennsylvania, 1969, GgtGr
Steven Markowitz, (D), Columbia, 1981, GbZn
Cecilia McHugh, (D), Columbia, 1993, Ou
Alfredo Morabia, (D), Johns Hopkins, 1989, GbZn
Stephen Pekar, (D), Rutgers, 1999, PeGrOu
Gillian Stewart, (D), SUNY (Stony Brook), 2005, OcCmOz
Yan Zheng, (D), Columbia, 1998, CgHwCm
Associate Professor:
Jeffrey Bird, (D), California (Davis), 2001, Sbc
Timothy Eaton, (D), Wisconsin, 2002, HgwGg
Assistant Professor:
Gregory O'Mullan, (D), Princeton, 2005, PgOb
Ashaki Rouff, (D), SUNY (Stony Brook), 2004, Cac
Chuixiang Yi, (D), Nanjing (China), 1991, Asm
Emeritus:
Eugene A. Alexandrov, (D), Columbia, 1959, Eg
Patrick W. G. Brock, (D), Leeds (UK), 1963, Gg
Hannes K. Brueckner, (D), Yale, 1968, GgtCg
Robert M. Finks, (D), Columbia, 1959, Pi
Daniel Habib, (D), Penn State, 1965, Pl
Peter H. Mattson, (D), Princeton, 1957, Gc
Andrew McIntyre, (D), Columbia, 1967, Pe
B. Charlotte Schreiber, (D), Rensselaer, 1974, Gd
David H. Speidel, (D), Penn State, 1964, Cgp
David L. Thurber, (D), Columbia, 1964, Cl

Queensborough Community College

Dept of Biological Sciences and Geology
222-05 56th Avenue
Bayside, NY 11364
p. (718) 631-6335
MGorelick@qcc.cuny.edu
http://www.qcc.cuny.edu/biologicalsciences/advisors.asp

Rensselaer Polytechnic Institute

Dept of Earth & Environmental Sciences (B,M,D) ☒
Science Center 1W19
110 8th Street
Troy, NY 12180-3590
p. (518) 276-6474
ees@rpi.edu
https://science.rpi.edu/earth
Programs: Geology (B,M,D); Hydrogeology (B); Environmental Science (B)
Enrollment (2021): B: 0 (7) M: 1 (2) D: 11 (0)
Head:
Frank S. Spear, (D), California (Los Angeles), 1976, GpCpGc
Professor:
Steven W. Roecker, (D), MIT, 1981, Yg
Associate Professor:
Miriam E. Katz, (D), Rutgers, 2001, PmeGu
Karyn L. Rogers, (D), Washington, 2006, PgCl
Morgan F. Schaller, (D), Rutgers, 2011, CsPe
Assistant Professor:
Sasha Wagner, (D), Florida Intl, 2015, Cob
Lecturer:
Sarah Cadieux, (D), Indiana, 2015, ZeCb
Emeritus:
E. Bruce Watson, (D), MIT, 1976, CpcCt

Skidmore College

Dept of Geosciences (B) ☒
815 North Broadway
Saratoga Springs, NY 12866
p. (518) 580-5190
afrappie@skidmore.edu
https://www.skidmore.edu/geosciences/
Programs: Geosciences
Enrollment (2018): B: 14 (3)
Associate Professor:
Amy Frappier, (D), New Hampshire, 2006, GePe
Kyle K. Nichols, (D), Vermont, 2002, Gm
Visiting Assistant Professor:
Margaret Estapa, (D), Maine, 2011, Oc
Assistant Professor:
Greg Gerbi, (D), MIT/WHOI, OpYg
Victor Guevara, (D), Virginia Tech, 2017, Gp
Instructor:
Jennifer Cholnoky, (M), Rensselaer, 2013, Gg

St. Lawrence University

Department of Geology (B) ●☒
23 Romoda Dr.
Brown Hall
Canton, NY 13617-1475
p. (315) 229-5851
astewart@stlawu.edu

https://www.stlawu.edu/geology
f: https://www.facebook.com/SLUGeology
Programs: Geology
Administrative Assistant: Sherrie Kelly
Enrollment (2022): B: 20 (14)

Chair:
Judith Nagel-Myers, (D), Muenster (Germany), 2006, PisPe
Professor:
Jeffrey R. Chiarenzelli, (D), Kansas, 1989, GzCg
Antun Husinec, (D), Zagreb, 2002, GsCsGo
Associate Professor:
Alexander K. Stewart, USA (ret), (D), Cincinnati, 2007, GmlHw
Research Associate:
George W. Robinson, (D), Queens, 1978, Gz
Visiting Professor:
Sarah Brehm, (D), Michigan, 2021, Gi
Emeritus:
J. Mark Erickson, (D), North Dakota, 1971, Pgi
Technician:
Matthew F. Van Brocklin, (M), Akron, 1996, GgZgn

Stony Brook University

Dept of Geosciences (B,M,D)
Nicolls Road
Stony Brook, NY 11794-2100
p. (631) 632-8200
brian.phillips@stonybrook.edu
https://www.stonybrook.edu/geosciences
Programs: Geology
Earth and Space Sciences
Master of Arts in Teaching

Professor:
Timothy Glotch, (D), Arizona State, 2004, XgGz
William E. Holt, (D), Arizona, 1989, Ys
Baosheng Li, (D), Stony Brook, 1996, Yx
Scott M. McLennan, (D), Australian Nat, 1981, Cg
Hanna Nekvasil, (D), Penn State, 1986, Cp
John B. Parise, (D), James Cook, 1980, Gz
Brian L. Phillips, (D), Illinois, 1990, GzCl
E. Troy Rasbury, (D), SUNY (Stony Brook), 1998, Cc
Richard J. Reeder, (D), California (Berkeley), 1980, ClGz
Martin A. Schoonen, (D), Penn State, 1989, Cl
Donald J. Weidner, (D), MIT, 1972, Gy
Lianxing Wen, (D), Caltech, 1998, Ys
Associate Professor:
Lars Ehm, (D), Christian-Albrechts (Germany), 2003, Gz
Joel A. Hurowitz, (D), Stony Brook, 2006, Xg
A. Deanne Rogers, (D), Arizona State, 2005, XgZr
Assistant Professor:
Marine Frouin, (D), Bordeaux, 2014, Cc
Gregory Henkes, (D), Johns Hopkins, 2014, Cs
Qingyun Li, (D), Washington (St. Louis), 2016, Cl
Weisen Shen, (D), Colorado, 2014, Ys
Research Associate:
Michael Sperazza, (D), Montana, 2006, Zn
Lecturer:
Christiane W. Stidham, (D), California (Berkeley), 1999, YgGe
Adjunct Professor:
Robert C. Aller, (D), Yale, 1977, Cm
Henry J. Bokuniewicz, (D), Yale, 1976, Yr
J. Kirk Cochran, (D), Yale, 1979, Oc
Maureen O'Leary, (D), Johns Hopkins, 1997, Pv
Emeritus:
Daniel M. Davis, (D), MIT, 1983, Yg
Gilbert N. Hanson, (D), Minnesota, 1964, GgZe
Robert C. Lieberman, (D), Columbia, 1969, GyYsx
Donald H. Lindsley, (D), Johns Hopkins, 1961, Cp
Teng-fong Wong, (D), MIT, 1981, Yx
Laboratory Director:
Owen C. Evans, (D), SUNY (Stony Brook), 1994, Cg

Suffolk County Community College, Ammerman Campus

Dept of Physical Science (A)
533 College Road
Selden, NY 11784
p. (631) 451-4338
butkosd@sunysuffolk.edu
http://depthome.sunysuffolk.edu/Selden/PhysicalScience/
Enrollment (2016): A: 27 (0)

Professor:
Darryl J. Butkos, (M), GgHw
Michael Inglis, (D), Zg
Scott A. Mandia, (M), Penn State, 1990, Amt
Associate Professor:
Matthew Pappas, Zg
Assistant Professor:
Sean Tvelia, (M), SUNY (Stony Brook), Ggl
Adjunct Professor:
Jessica Dutton, Zg
Michael Flanagan, Zg
Philip Harrington, Zg
Margaret Lomaga, Zg
Brian Vorwald, Zg

Sullivan County Community College

Mathematics and Natural Sciences
112 College Road
Lock Sheldrake, NY 12759
dlewkiewicz@sunysullivan.edu

SUNY Jefferson

Science
1220 Coffeen Street
Watertown, NY 13601
p. (315) 786-2200
cebeyhoneycutt@sunyjefferson.edu
http://www.sunyjefferson.edu/academics/programs-study/liberal-arts-sciences-mathematics-science

SUNY Potsdam

Department of Earth and Environmental Sciences (B) ☒
220 Timerman Hall
44 Pierrepont Avenue
Potsdam, NY 13676
p. (315) 267-2286
earth@potsdam.edu
http://www.potsdam.edu/academics/AAS/Geology/
f: https://www.facebook.com/potsdamgeology/
Programs: Geology, Geographic Information Science
Administrative Assistant: Beth A. Fayette
Enrollment (2020): B: 34 (0)

Professor:
Michael C. Rygel, (D), Dalhousie, 2005, GsrZi
Assistant Professor:
Sara E. Bier, (D), Penn State, 2010, GctGg
Kamal Humagain, (D), Texas Tech, 2016, Zir
Adam Pearson, (D), Delaware, 2015, Gm
Page C. Quinton, (D), Missouri, 2016, PcCsPm
Christian M. Schrader, (D), Georgia, 2009, GiEmGv
Emeritus:
Robert L. Badger, (D), Virginia Tech, 1989, GipGc
Frank A. Revetta, (D), Rochester, 1970, YgGtg

SUNY, Albany

Dept of Atmospheric and Environmental Sciences (B,M,D)
1400 Washington Avenue
Albany, NY 12222
p. (518) 442-4466
daeschair@albany.edu
http://www.atmos.albany.edu

Chair:
Vincent P. Idone, (D), SUNY (Albany), 1982, As
Professor:
Lance F. Bosart, (D), MIT, 1969, As
Daniel Keyser, (D), Penn State, 1981, As
John E. Molinari, (D), Florida State, 1979, As
Senior Research Professor:
David R. Fitzjarrald, (D), Virginia, 1980, Asm
Richard R. Perez, (D), SUNY (Albany), 1983, As
James J. Schwab, (D), Harvard, 1983, Acs
Christopher J. Walcek, (D), California (Los Angeles), 1983, As
Wei-Chyung Wang, (D), Columbia, 1973, As
Associate Professor:
Robert G. Keesee, (D), Colorado, 1979, As

Christopher D. Thorncroft, (D), Reading (UK), 1988, As
Research Associate:
Stephen S. Howe, (M), Penn State, 1981, Cs
David Knight, (D), Washington, 1987, As

SUNY, Buffalo

Dept of Geology (B,M,D) ○⊘
126 Cooke Hall
Buffalo, NY 14260
p. (716) 645-3489
geology@buffalo.edu
http://www.geology.buffalo.edu
t: @UBGeology
Programs: BA, BS Geological Sciences
MS, PhD Geological Sciences
MS Computation Earth Science
Enrollment (2022): B: 34 (16) M: 24 (14) D: 17 (1)
Chair:
Beata M. Csatho, (D), Miskolc (Hungary), 1993, GlZrYg
UB Distinguished Professor:
Gregory Valentine, (D), California (Santa Barbara), 1988, Gv
Empire Innovation Professor/Director of the Center for Geological and Climate Hazards:
Sophie Nowicki, (D), Univ Coll (London), 2007, Gl
Director of Graduate Studies:
Richelle Allen-King, (D), Waterloo, 1991, Cg
Professor:
Jason P. Briner, (D), Colorado, 2003, Gl
Tracy K. P. Gregg, (D), Arizona State, 1995, Gv
Associate Professor:
Margarete Jadamec, (D), California (Davis), 2009, GtYgZf
Christopher S. Lowry, (D), Wisconsin, 2008, Hw
Elizabeth K. Thomas, (D), Brown, 2014, Cos
Clinical Assistant Professor:
James Boyle, (D), Buffalo, 2018, Pq
Nicholas DiFrancesco, (D), SUNY at Stony Brook, 2018, Gg
Assistant Professor:
Stephan Kolzenburg, (D), Studi di Torino, 2016, Gv
Richard Marinos, (D), Duke, 2018, Sc
Kristin Poinar, (D), Washington, 2017, Gl
Adjunct Assistant Professor:
Julio Sepulveda, (D), Bremen, 2008
Professor Emeritus:
Marcus I. Bursik, (D), Caltech, 1989, Gv
Mary Alice Coffroth, (D), Miami, 1988, Gu
Dennis S. Hodge, (D), Wyoming, 1966
Robert D. Jacobi, (D), Columbia, 1980, GctGs
Chester C. Langway, (D), Michigan, 1965, Gl
Howard R. Lasker, (D), Chicago, 1978, Gu
Michael F. Sheridan, (D), Stanford, 1965, Gv
SUNY Distinguished Teaching Professor Emeritus:
Charles E. Mitchell, (D), Harvard, 1983, Pg
Research Support Specialist:
Ivan Parmuzin, (M), Moscow State
Research Professor:
Galina L. Rogova, (D), Moscow State (Russia), 1988, Gq
Anton Schenk, (D), Switzerland (Berne), 1972, Zr
Research Assistant Professor:
M. Oryaelle Chevrel, (D), Ludwig Maximilians, 2013, Gv
Mattia de' Michieli Vitturi, (D), Pisa, 2004, Gv
Kirstie Haynie, (D), Buffalo, 2019, Gv
Angela Martinez Quintana, (D), Buffalo, 2022
Kimberly Meehan, (D), CUNY, 2014, Pe
Susan Sakimoto, (D), Johns Hopkins, 1995
Ingo Sonder, (D), Wurzburg, 2010, Gv
Neil Terry, (D), Rutgers (Newark), 2016
Laboratory Technician:
Owen Cowling, (M), Buffalo, 2018, Cs
Department Administrator, Senior Staff:
Alison A. Lagowski, (M), Buffalo, 1996, Zn
Adjunct Assistant Professor:
Michael McHale, (D), SUNY Env. Science & Forestry, 1999, Hg

SUNY, Cortland

Geology Dept (B,M)
342 Bowers Hall
PO Box 2000
Cortland, NY 13045
p. (607) 753-2815
gleasong@cortland.edu
http://www.cortland.edu/geology/
Administrative Assistant: Susan K. Nevins
Enrollment (2007): B: 48 (15) M: 8 (6)
Professor:
David J. Barclay, (D), SUNY (Buffalo), 1998, Glm
Christopher P. Cirmo, (D), Syracuse, 1994, Hg
Robert S. Darling, (D), Syracuse, 1992, GzpCp
Associate Professor:
Christopher A. McRoberts, (D), Syracuse, 1994, Pg
Associate Scientist:
Gayle C. Gleason, (D), Brown, 1993, Gct
Lecturer:
Julie L. Barclay, (M), SUNY (Buffalo), 1997, Ggm
Laboratory Director:
John R. Driscoll, (A), Northwest Electronic, 1974, Zn

SUNY, Fredonia

Department of Geology and Environmental Sciences (B)
280 Central Avenue
Fredonia, NY 14063-1020
p. (716) 673-3303
earth@fredonia.edu
http://www.fredonia.edu/earth
Enrollment (2016): B: 40 (7)
Chair:
Gordon C. Baird, (D), Rochester, 1975, PseGs
Professor:
Gordon C. Baird, (D), Rochester, 1975, GrPgGs
Gary G. Lash, (D), Lehigh, 1980, Gr
Sherri A. Mason, (D), Montana, 2001, ZgGe
Associate Professor:
Ann K. Deakin, (D), SUNY (Buffalo), 1996, ZiGgZr
Instructor:
Kimberly Weborg-Benson, (M), Illinois, 1991, GgAsPg

SUNY, Geneseo

Dept of Geological Sciences (B) ☒
1 College Circle
ISC 235
Geneseo, NY 14454
p. (585) 245-5291
gsci@geneseo.edu
http://www.geneseo.edu/geology
Programs: Geological Sciences, Geological Sciences w/ adol. Edu. Cert., Geophysics, Geochemistry
Administrative Assistant: Laurie A. Montgomery
Enrollment (2022): B: 69 (16)
Chair:
Scott D. Giorgis, (D), Wisconsin, 2003, GctYm
Professor:
D. Jeffrey Over, (D), Texas Tech, 1990, Ps
Associate Professor:
Dori J. Farthing, (D), Johns Hopkins, 2001, Gz
Amy L. Sheldon, (D), Utah, 2002, Hw
Nicholas H. Warner, (D), Arizona State, 2008, XgGrHg
Assistant Professor:
Jacalyn M. (Wittmer) Malinowski, (D), Virginia Tech, 2014, PiGsPe
Emeritus:
Phillip D. Boger, (D), Ohio State, 1976, Cg
Richard B. Hatheway, (D), Cornell, 1969, GxzGp
Richard A. Young, (D), Washington (St Louis), 1966, GmXgZr
Instructional Support Specialist:
Sarah J. Gaudio, (D), California (Davis), 2013, GzgGg

SUNY, Maritime College

Science Dept (B)
6 Pennyfield Avenue
Bronx, NY 10465
p. (718) 409-7380
dan@sunymaritime.edu
Enrollment (2009): B: 68 (22)
Chair:
Kathy Olszewski, (D), SUNY (Stony Brook), 1994, Cg
Associate Professor:
Marie deAngelis, (D), Washington, 1989, Oc

Assistant Professor:
Anthony Manzi, (M), Montclair State, Am

SUNY, New Paltz

Geology (B) ☒
1 Hawk Drive
New Paltz, NY 12561
p. (845) 257-3760
vollmerf@newpaltz.edu
http://www.newpaltz.edu/geology/
Programs: Geology, Environmental Geoscience, Earth Science Education
Enrollment (2019): B: 45 (11)

Professor:
Frederick W. Vollmer, (D), Minnesota, 1985, GctGx

Professor:
John A. Rayburn, (D), SUNY Binghamton, 2004, Gml

Associate Professor:
Alexander J. Bartholomew, (D), Cincinnati, 2006, GrPi
Shafiul H. Chowdhury, (D), W Michigan, 1999, HwGe

Assistant Professor:
Gordana Garapic, (D), Boston, 2013, Gz

Lecturer:
Kaustubh Patwardhan, (D), Johns Hopkins, 2009, Ggi

Adjunct Professor:
Laurel Mutti, (M), Johns Hopkins, 2004, Gg

Emeritus:
Gilbert J. Brenner, (D), Penn State, 1962, Pl
Alvin S. Konigsberg, (D), Syracuse, 1969, As
Constantine Manos, (D), Illinois, 1963, Gs
Martin S. Rutstein, (D), Brown, 1969, Gz

Related Staff:
Donald R. Hodder, (B), SUNY (Geneseo), 1985, Gg

SUNY, Oneonta

Dept of Earth and Atmospheric Sciences (B) ☒
108 Ravine Parkway
209 Sci. #1
Oneonta, NY 13820-4015
p. (607) 436-3707
Lisa.Hoffman@oneonta.edu
http://www.oneonta.edu/academics/earths/
Programs: Geology; Earth Science; Meteorology; Earth Science Education
Enrollment (2017): B: 91 (0)

Chair:
Jerome B. Blechman, (D), Wisconsin, 1979, As

Distinguished Teaching Professor:
James R. Ebert, (D), SUNY (Binghamton), 1984, GrsZe

Associate Professor:
Keith A. Brunstad, (D), Washington State, 2013, GviEg
Leigh M. Fall, (D), Texas A&M, 2010, Pgq
Melissa Godek, (D), Delaware, 2009, AmsZg
Leslie E. Hasbargen, (D), Minnesota, 2003, GmHgZr

Assistant Professor:
Christopher Karmosky, (D), Penn State, 2013, AmsZy

Dr.:
Marta L. Clepper, (D), Kentucky, 2011, ZeGgs

Lecturer:
Kathryn Metcalf, (M), Arizona, 2012, GctGg

Emeritus:
P. Jay Fleisher, (D), Washington State, 1967, GlmGe
Arthur N. Palmer, (D), Indiana, 1969, HgqGm

SUNY, Oswego

Dept of Atmospheric and Geological Sciences (B)
394 Shineman Science Center
Oswego, NY 13126
p. (315) 312-3065
tomascak@oswego.edu
http://www.oswego.edu/ags
Programs: Geology; Meteorology
Enrollment (2015): B: 160 (25)

Professor:
Alfred J. Stamm, (D), Wisconsin, 1976, Am
Paul B. Tomascak, (D), Maryland, 1995, CgGzi
David W. Valentino, (D), Virginia Tech, 1993, GtcYu

Associate Professor:
Scott Steiger, (D), Texas A&M, 2005, Am

Assistant Professor:
Rachel J. Lee, (D), Pittsburg State, 2013, GvZr
Steven T. Skubis, (D), SUNY (Albany), 1994, Am
Justin Stroup, (D), Dartmouth, 2015, GmHwGe
Michael Veres, (D), Nebraska, 2014, As

Lab/Field Technician:
Richard Frieman, (M), Buffalo, 2016, Gg

SUNY, Plattsburgh

Center for Earth & Environmental Science (B) ☒
101 Broad Street
132 Hudson Hall
Plattsburgh, NY 12901
p. (518) 564-2028
cees@plattsburgh.edu
http://www.plattsburgh.edu/cees
f: https://www.facebook.com/SUNYPlattsburghCEES/
Programs: Geology, Environmental Geology, and Earth Science (with a 5 year Ba/MST)
Enrollment (2019): B: 28 (4)

Director:
Edwin A. Romanowicz, (D), Syracuse, 1993, HwGcYu

Professor:
David A. Franzi, (D), Syracuse, 1984, Gl

Associate Professor:
Eric Leibensperger, (D), Harvard, 2011, AsOg

Assistant Professor:
Nancy A. Price, (D), Maine, 2012, GzCgGc
James R. Thomka, (D), Cincinnati, 2015, GsPg

Lecturer:
Patrick Korths, (M), Syracuse, 2006, RwGg

SUNY, Purchase

Environmental Studies (B) ⦸
735 Anderson Hill Road
Purchase, NY 10577
p. (914) 251-6646
naturalsciences@purchase.edu
https://www.purchase.edu/academics/environmental-studies/
Programs: Environmental Studies: Ecology and Policy Concentrations
Enrollment (2021): B: 43 (10)

Director of Teaching and Learning Technologies:
Keith Landa, (D), Michigan, 1989, XgbGg

Professor:
George P. Kraemer, (D), California (Los Angeles), 1989, ObnCm

Department Chair:
Ryan W. Taylor, (D), Oregon State, 2006, ZiyZg

Assistant Professor:
Allyson K. Jackson, (D), Oregon State, 2017, ZgcZe

SUNY, Stony Brook

School of Marine and Atmospheric Sciences (B,M,D)
145 Endeavour Hall
Stony Brook, NY 11794-5000
p. (516) 632-8700
minghua.zhang@stonybrook.edu
http://www.somas.stonybrook.edu

Professor:
Josephine Y. Aller, (D), S California, 1975, ObAs
Edmund K. M. Chang, (D), Princeton, 1993, As
J. Kirk Cochran, (D), Yale, 1979, Oc
Brian Colle, (D), Washington, 1997, As
David O. Conover, (D), Massachusetts, 1981, Ob
Roger D. Flood, (D), MIT/WHOI, 1978, Ou
Marvin A. Geller, (D), MIT, 1969, As
Christopher Gobler, (D), SUNY (Stony Brook), 1999, Ob
Sultan Hameed, (D), Manchester (UK), 1968, As
Darcy J. Lonsdale, (D), Maryland, 1979, Ob
Glenn R. Lopez, (D), SUNY (Stony Brook), 1976, Ob
John E. Mak, (D), California (San Diego), 1992, AsCas
Anne E. McElroy, (D), MIT/WHOI, 1985, ObCb
Ellen K. Pikitch, Indiana, 1983, Ob
Mary I. Scranton, (D), MIT/WHOI, 1977, Oc
R. Lawrence Swanson, (D), Oregon State, 1971, Og

Gordon T. Taylor, (D), S California, 1983, Obc
Minghua Zhang, (D), Inst of Atm Physics, 1987, As
Associate Professor:
Bassem Allam, (D), W Brittany (France), 1998, Ob
Robert A. Armstrong, (D), Minnesota, 1975, Ocb
David Black, (D), Ou
Bruce J. Brownawell, (D), MIT/WHOI, 1986, Oc
Robert M. Cerrato, (D), Yale, 1980, Ob
Jackie Collier, (D), Stanford, 1994, Ob
Michael Frisk, (D), Maryland, 2004
Marat Khairoutdinov, (D), Oklahoma, 1997, As
Daniel A. Knopf, (D), Swiss Fed Inst Tech, 2003, As
Kamazima M. Lwiza, (D), Wales, 1991, Oc
Bradley Peterson, (D), S Alabama, 1998, Ob
Joseph Warren, (D), MIT, 2001, Ob
Robert E. Wilson, (D), Johns Hopkins, 1973, Op
Assistant Science Director, IOCS:
Demian Chapman, (D), Nova Southeastern, 2007
Assistant Professor:
Anthony Dvarskas, (D), Maryland, 2007
Hyemi Kim, (D), Seoul Nat, 2008
Janet Nye, (D), Maryland, 2008, Ob
Christopher Wolfe, (D), Oregon State, 2006, Op
Qingzhi Zhu, (D), Xiamen, 1997, Oc
Faculty Director Semester By The Sea:
Kurt Bretsch, (D), South Carolina, 2005
Lecturer:
Lesley Thorne, (D), Duke, 2010
Research Scientist:
Wuyin Lin, (D), SUNY (Stony Brook), 2002, As
Research Professor:
Charles Flagg, (D), MIT, 1977, Op
Engineer:
Douglas Hill, (D), Columbia, 1977
Ecologist, Author:
Carl Safina, (D), Rutgers, 1987
Director Riverhead Foundation:
Robert A. DiGiovanni, Jr., (M), SUNY (Stony Brook), 2002
Adjunct Professor:
James Ammerman, (D), Scripps, 1983
Paul Bowser, (D), Auburn, 1978, Obb
Carl Brenninkmeijer, (D), Groningen (Neth), 1983
Michael J. Cahill, (D), DePaul, 1978, Zn
Alistair Dove, Queensland, 1999
Anga Engel, Oc
Emmanuelle pales Espinosa, (D), Nante (France), 1999
Mark Fast, (D), Dalhousie, 2005, Ob
Scott Ferson, (D), SUNY (Stony Brook), 1988
Scott Fowler, (D), 1969
Roxanne Karimi, (D), Dartmouth, 2007
Kathryn Kavanagh, (D), James Cook, 1998
Jeffrey Levinton, (D), Yale, 1971, Ob
Yangang Liu, (D), Nevada (Reno), 1998
Stephan Munch, (D), SUNY (Stony Brook), 2002, Ob
John Rapaglia, (D), SUNY (Stony Brook), 2007
Frank J. Roethel, (D), SUNY (Stony Brook), 1981, Oc
Jeffrey Tongue, As
Andrew Vogelmann, (D), Penn State, 1994, As
Duane E. Waliser, (D), As
Douglas W. R. Wallace, (D), Dalhousie, 1985, Oc
Jian Wang, (D), Caltech, 2002, As
Distinguished Professor:
Cindy Lee, (D), California (San Diego), 1975, OcCo
Emeritus:
Dong-Ping Wang, (D), Miami, 1975
Affiliated and Joint Faculty:
Heather L. Lynch, (D), Harvard, 2006
Cooperating Faculty:
Resit Akcakaya, (D), SUNY (Stony Brook), 1989
Cooperating Faculty:
Stephen Baines, (D), Yale, 1993
Lee K. Koppelman, (D), New York, 1968, On
Dianna K. Padilla, (D), Alberta, 1987, Ob
Sheldon Reaven, (D), California (Berkeley), 1975, Zn

SUNY, The College at Brockport

Dept of the Earth Sciences (B)
350 New Campus Drive
Brockport, NY 14420-2936
p. (585) 395-2636
earthsci@esc.brockport.edu
http://www.brockport.edu/esc
Enrollment (2012): B: 117 (16)
Associate Professor:
James A. Zollweg, (D), Cornell, 1994, Hs
Professor:
Mark R. Noll, (D), Delaware, 1989, Cl
Associate Professor:
Paul L. Richards, (D), Penn State, 1999, Hg
Scott M. Rochette, (D), St Louis, 1998, Am
Adjunct Professor:
David A. Boehm, (M), SUNY (Buffalo), 2003, Gg
Christine Crafts, (B), SUNY (Brockport), 2000, Am
Jutta S. Dudley, (D), SUNY (Buffalo), 1998, Gg
William G. Glynn, (M), Texas A&M, 1984, Gg
Linda J. Schaffer, (M), SUNY (Brockport), 1988, Ze
Department Secretary:
Lauri A. Kifer, (A), Monroe Comm Coll, 1997, Zn
Emeritus:
Robert W. Adams, (D), Johns Hopkins, 1964, Gs
Whitney J. Autin, (D), Louisiana State, 1989, Gs
John E. Hubbard, (D), Colorado State, 1968, Hg
Richard M. Liebe, (D), Iowa, 1962, Ps
Judy A. Massare, (D), Johns Hopkins, 1983, Pv
John M. Williams, (B), Goshen, As
On Leave:
Robert S. Weinbeck, (D), Iowa State, 1980, As
Systems Administrator:
Thomas M. McDermott, Zn

SUNY, Ulster County Community College

STEM (A)
Burroughs 105
491 Cottekill Road
Stone Ridge, NY 12484
p. (845) 687-5230
schimmrs@sunyulster.edu
http://people.sunyulster.edu/esc
Programs: Liberal Arts & Sciences (A); Mathematics & Science (A)
Professor:
Steven Schimmrich, (M), SUNY (Albany), 1991, GgZgn
Assistant Professor:
Karen Helgers, (M), SUNY, 1987, GeZgy

Syracuse University

Department of Earth & Environmental Sciences (B,M,D)
204 Heroy Geology Laboratory
Syracuse, NY 13244-1070
p. (315) 443-2672
earthenvsciences@syr.edu
http://earthsciences.syr.edu
f: https://www.facebook.com/SUEarthEnvSci
t: @SUEarthEnvSci
Programs: B.A. in Earth Sciences, B.S. in Earth Sciences, B.A. in Environment Sustainability and Policy, B.S. in Environment Sustainability and Policy, B.A. in Energy and Its Impacts, B.S. in Energy and Its Impacts, M.A. in Earth Sciences, M.S. in Earth Sciences, Ph.D. in Earth Sciences
Enrollment (2021): B: 35 (8) M: 13 (3) D: 25 (4)
Chair:
Gregory Hoke, (D), Cornell, 2006, ZgGmt
Professor:
Jeffrey Karson, (D), SUNY (Albany), 1977, GtcGv
Suzanne L. Baldwin, (D), SUNY (Albany), 1988, Cc
Paul G. Fitzgerald, (D), Melbourne (Australia), 1988, GtCc
Linda C. Ivany, (D), Harvard, 1997, PegCs
Zunli Lu, (D), Rochester, 2008, Cg
Cathryn R. Newton, (D), California (Santa Cruz), 1983, Pg
Scott D. Samson, (D), Arizona, 1990, Cc
Christopher A. Scholz, (D), Duke, 1989, Gs
Associate Professor:
Christopher Junium, (D), Penn State, 2010, CsPeCo
Robert Moucha, (D), Toronto, 2003, GtYeg
Jay Thomas, (D), Virginia Tech, 2003, Gi
Assistant Professor:
Tripti Bhattacharya, (D), California (Berkeley), 2016, Ge

Melissa Chipman, (D), Illinois (Urbana), 2017, GelGn
Samuel Tuttle, (D), Boston, 2015, Ge
Tao Wen, (D), Michigan, 2017, Ge

Instructor:
Daniel Curewitz, (D), Duke, 1999, OgGct

Emeritus:
M. E. Bickford, (D), Illinois, 1960, CcGiz
Donald I. Siegel, (D), Minnesota, 1981, Hw

Other:
Bruce Wilkinson, (D), Texas, 1974, Gs

Union College

Geology Dept (B)
807 Union Street
Schenectady, NY 12308-3107
p. (518) 388-6770
geology@union.edu
http://www.union.edu/academic_depts/geology/
Administrative Assistant: Deborah A. Klein
Enrollment (2015): B: 40 (15)

Chair:
Donald T. Rodbell, (D), Colorado, 1991, GlmGe

Professor:
John I. Garver, (D), Washington, 1989, GtCcGr
Kurt T. Hollocher, (D), Massachusetts, 1985, GxCga

Associate Professor:
Holli M. Frey, (D), Michigan, 2005, GvCag
David P. Gillikin, (D), Vrije (Brussel), 2005, CsmCl

Lecturer:
Matthew R. Manon, (D), Michigan, 2008, Gp
Anouk Verheyden-Gillikin, (D), Vrije (Brussel), 2004

Emeritus:
George H. Shaw, (D), Washington, 1971, Yx

Related Staff:
William S. Neubeck, (M), SUNY (Binghamton), 1980, Gm

United States Military Academy

Dept of Geography & Environmental Engineering (B) ●☒
745 Brewerton Rd
West Point, NY 10996
p. (914) 938-2300
william.wright@westpoint.edu
https://westpoint.edu/academics/academic-departments/geography-and-environmental-engineering
f: https://www.facebook.com/WPGENE
t: usma_gene
Programs: Geospatial Information Science; Geography; Environmental Science; Environmental Engineering
Enrollment (2021): B: 110 (44)

COL:
William C. Wright, (D), Florida, 2017, ZriYd

Assistant Professor:
Matthew S. O'Banion, (D), Oregon State, 2017, ZrGg

University of Rochester

Department of Earth & Environmental Sciences (B,M,D) ☒
120 Trustee Road
227 Hutchison Hall
Box 270221
Rochester, NY 14627
p. (585) 275-5713
ees@ur.rochester.edu
https://www.sas.rochester.edu/ees/
Programs: Geological sciences (BS), Geological sciences (BA), Environmental science (BS), Environmental studies (BA), Geomechanics (BS)
Enrollment (2022): B: 49 (22) M: 1 (5) D: 28 (3)

Chair:
John Kessler, (D), California (Irvine), 2005, Oc

Professor:
John A. Tarduno, (D), Stanford, 1987, YmGtXm

Associate Professor:
Vasilii Petrenko, (D), California (San Diego), 2008, Act
Dustin Trail, (D), Rensselaer Polytech, 2010, Cp

Associate Scientist:
Rory D. Cottrell, (D), Rochester, 2000, Ym

Assistant Professor:
Chiara Borrelli, (D), Rensselaer Polytech, 2014, PmcCb
Erin Black, (D), MIT, 2018, Cbt
Rachel Glade, (D), Colorado, 2019, SpHt
Lee Murray, (D), Harvard, 2013, Ac
Miki Nakajima, (D), Caltech, 2016, Xyg
Tolulope Olugboji, (D), Yale, 2014, Ys
Thomas Weber, (D), Californa (Los Angeles), 2013, Obu

Associate Professor of Instruction:
Karen Berger, (D), MIT, 2000, Rw

Utica University

Dept of Geology (B)
Gordon Science Center
1600 Burrstone Road
Utica, NY 13502
p. (315) 792-3134
skanfoush@utica.edu
https://www.utica.edu/academic/as/geoscience/new/bachelors.cfm
Enrollment (2015): B: 10 (6)

Chair:
Adam Schoonmaker, (D), SUNY (Albany), 2005, GczGx

Associate Professor:
Sharon L. Kanfoush, (D), Florida, 2002, GsOuGn

Adjunct Professor:
Lindsey Geary, (M), Florida State, 2008, Geg
Tiffany McGivern, (M), Utica Coll, 2014, Geg

Emeritus:
Herman Muskatt, (D), Syracuse, 1963, GgrPg

Vassar College

Dept of Earth Science & Geography (B)
Box 735
124 Raymond Avenue
Poughkeepsie, NY 12604-0735
p. (845) 437-5540
geo@vassar.edu
http://earthscienceandgeography.vassar.edu/
Administrative Assistant: Lois Horst
Enrollment (2015): B: 12 (3)

Chair:
Mary A. Cunningham, (D), Minnesota, 2001, ZiyZu

Professor:
Brian J. Godfrey, (D), California (Berkeley), 1984, Zu
Kirsten M. Menking, (D), California (Santa Cruz), 1995, GmPeGc
Joseph Nevins, (D), California (Los Angeles), Zu
Jill S. Schneiderman, (D), Harvard, 1987, Gs
Jeffrey R. Walker, (D), Dartmouth, 1987, GzScGv
Yu Zhou, (D), Minnesota, 1995, Zun

Lab Technician and Collections Manager:
Richard Jones, (M), California (Santa Cruz), 1996

GIS specialist:
Neil Curri, (B), Zi

York College (CUNY)

Dept of Earth and Physical Sciences (B) ○⦸
94-20 Guy R. Brewer Blvd
Jamaica, NY 11451
p. (718)262- 2654
nkhandaker@york.cuny.edu
http://www.york.cuny.edu/academics/departments/earth-and-physical-sciences/
f: https://www.facebook.com/groups/787724658666512
Programs: BS degree in Geology, Environmental Health Science, K7-12 Earth Science Teacher Certification Program and minor in Environmental Science.
Enrollment (2021): B: 9 (8)

Chair:
Timothy Paglione, (D), Boston, Xa

Geology Discipline Coordinator & CUNY Doctoral Faculty in Earth and Environmental Sciences:
Nazrul I. Khandaker, (D), Iowa State, 1991, GdeZe

Associate Professor:
Ratan K. Dhar, (D), CUNY (Grad Ctr), 2006, Hw

Assistant Professor:
Dawn Roberts-Semple, (D), Rutgers, 2012, Gem

Instructor:
Malek Shami, (M), Brooklyn Coll, 2019, Gcg
Emeritus:
Stephen Lakatos, (D), Rensselaer, 1971, Cc
Arthur P. Loring, (D), New York, 1966, Gm
Stanley Schleifer, (D), CUNY, 1996, Ge

North Carolina

Appalachian State University

Dept of Geography & Planning (B,M) ☒
323 Rankin Science West
ASU Box 32066
Boone, NC 28608-2066
p. (828) 262-3000
schroederk@appstate.edu
http://www.geo.appstate.edu
f: Appalachian State Geography & Planning
t: AppstateGhyPln
Professor:
Jeff Colby, (D), Colorado, 1995, Zi
Michael W. Mayfield, (D), Tennessee, 1984, Hs
Baker Perry, (D), North Carolina, 2006, Zg
Kathleen Schroeder, (D), Minnesota, 1995, Zn
Peter T. Soule, (D), Georgia, 1989, As
Saskia van de Gevel, (D), Tennessee, 2008, Zg
Associate Professor:
Rob Brown, (D), Louisiana State, 2001, Zn
Richard J. Crepeau, (D), California (Irvine), 1995, Zn
Derek Martin , (D), Tennessee
Assistant Professor:
Kara Dempsey, (D)
Elizabeth Shay, (D)
Song Shu, (D), 2019
Johnathan Sugg, (D)
Maggie Sugg, (D)

Dept of Geological and Environmental Sciences (B) O☒
PO Box 32067
033 Rankin Science West
Boone, NC 28608-2067
p. (828) 262-3049
earth_env_sci@appstate.edu
http://www.earth.appstate.edu/
Programs: Geology, Environmental Science, Earth/Environmental Science Secondary Education
Enrollment (2022): B: 292 (0)
Chair:
Cynthia M. Liutkus-Pierce, (D), Rutgers, 2005, GsCsPs
Professor:
William P. Anderson, (D), North Carolina State, 1999, HwqHs
Sarah Carmichael, (D), Johns Hopkins, 2006, GxCgGd
Ellen A. Cowan, (D), N Illinois, 1988, GsmGu
Steven J. Hageman, (D), Illinois, 1992, PieGg
Andrew B. Heckert, (D), New Mexico, 2001, PvgZe
Scott T. Marshall, (D), Massachusetts, 2008, YxGc
Associate Professor:
Gabriele M. Casale, (D), Gct
Cole T. Edwards, (D), Ohio State, 2015, GsCs
Jamie SF Levine, (D), Texas, 2010, Gzp
Assistant Professor:
William H Armstrong, (D), Colorado, 2017, Glm
Sarah G. Evans, (D), Colorado, 2017, Hwq
Senior Lecturer:
Laura D. Mallard, (M), Vermont, 2000, GtZe
Marta Toran, (M), Montana State, 2011, Ge
Lauren H. Waterworth, (M), Texas A&M, 2002, Gcg
Brian W. Zimmer, (M), N Arizona, 2007, Gg
Lecturer:
Joey D. Mosteller, (M), Appalachian State, 2009, Gg
Visiting Professor:
Hannah B. Riegel, (D), Camerino (Italy), 2019, Ggc

Asheville-Buncombe Technical Community College

Dept of Chemistry and Physics (A)
340 Victoria Road
Asheville, NC 28801
p. (828) 254-1921
mfender@abtech.edu
http://www1.abtech.edu/content/arts-and-sciences/chemistry-physics/chemistry-and-physics-overview
Instructor:
John Bultman, Gg
Adjunct Professor:
Dan Murphy, Gg

Cape Fear Community College

Dept of Science
411 N. Front Street
Wilmington, NC 28401
lerotanz59@cfcc.edu
https://cfcc.edu/associate-in-science/
Chair:
Joy Smoots, (D)

Science Dept (A)
411 N Front St
Wilmington, NC 28401
p. (910) 362-7674
baparnell251@cfcc.edu
http://cfcc.edu/programs/science/
Enrollment (2016): A: 5 (0)
Instructor:
Alvin L. Coleman, (M), Tennessee, 2008, Gg
Phil Garwood, (D), Edith Cowans (Australia), 1978, Gg

Central Piedmont Community College

Sciences Div (A)
PO Box 35009
Charlotte, NC 28235
p. (704) 330-6750
David.Privette@cpcc.edu
http://www.cpcc.edu/
Chair:
Steppen Murphy, (M), S Illinois, Gue
Division Director:
David Privette, (M), Georgia, 1978, Zy
Instructor:
Alisa Hylton, (M), Wichita State, Gg

Coastal Carolina University

Dept of Marine Science
Conway, NC 29528
jguentze@coastal.edu

Duke University

Div of Earth & Ocean Sciences (B,M,D)
Nicholas School of the Environment and Earth Sciences
Box 90227
Durham, NC 27708-0227
p. (919) 684-5847
bill.chameides@duke.edu
http://www.nicholas.duke.edu/eos
Enrollment (2010): B: 37 (15) D: 21 (0)
Chair:
M. Susan Lozier, (D), Washington, 1989, Op
Professor:
Paul A. Baker, (D), California (San Diego), 1981, PcCgGg
Alan E. Boudreau, (D), Washington, 1986, Gi
Bruce H. Corliss, (D), Rhode Island, 1978, Pm
Peter K. Haff, (D), Virginia, 1970, Zn
Robert B. Jackson, (D), Utah State, 1992, Zn
Emily M. Klein, (D), Columbia, 1988, Gi
A. Bradshaw Murray, (D), Minnesota, 1995, Gm
Lincoln F. Pratson, (D), Columbia, 1993, Gs
Avner Vengosh, (D), Australian Nat, 1990, Hw
Associate Professor:
Nicolas Cassar, (D), Hawaii, 2003, CbOcb
Associate Scientist:
Gary S. Dwyer, (D), Duke, 1996, Gs
Assistant Professor:
Wenhong Li, (D), Georgia Tech, Zn
Lecturer:
Alexander Glass, (D), Illinois (Urbana), 2006, PiGgZe

Adjunct Professor:
David J. Erickson, (D), Rhode Island, 1987
Peter E. Malin, (D), Princeton, 1978, Ys
Bruce F. Molnia, (D), South Carolina, 1972, ZrGlu
Daniel D. Richter, (D), Duke, 1980, So
William H. Schlesinger, (D), Cornell, 1976, Zn
Emeritus:
Richard T. Barber, (D), Stanford, 1967, Ob
Duncan Heron, (D), North Carolina, 1958, Grg
Daniel A. Livingstone, (D), Yale, 1953, Pe
Ronald D. Perkins, (D), Indiana, 1962, Gdo
Orrin H. Pilkey, Jr., (D), Florida State, 1962, Ou
Other:
James S. Clark, (D), Minnesota, 1988, Zn
Mark N. Feinglos, (D), McGill, 1973, Gz
Richard F. Kay, (D), Yale, 1973, Zn

East Carolina University

Dept of Geological Sciences (B,M)
101 Graham Building
Greenville, NC 27858-4353
p. (252) 328-6360
morrisonl14@ecu.edu
http://www.geology.ecu.edu
f: https://www.facebook.com/ecugeology/
t: @Ecugeology
Programs: Geology (B,M)
Certificates: Hydrogeology and Environmental Geology (G)
Administrative Assistant: Emily Kancharla, Lauren Morrison
Enrollment (2022): B: 37 (9) M: 22 (7)
Professor:
Stephen J. Culver, (D), Wales, 1976, PmGu
David Mallinson, (D), S Florida, 1995, Gu
Siddhartha Mitra, (D), William & Mary, 1997, Co
Stephen Moysey, (D), Stanford, 2005, Hg
Associate Professor:
Adriana Heimann, (D), Iowa State, 2006, GzpGi
Eric M. Horsman, (D), Wisconsin, 2006, GcYgGt
Eduardo Leorri, (D), Basque (Spain), 2003, GsPm
Alex K. Manda, (D), Massachusetts, 2009, HwgHq
Assistant Professor:
Margaret Blome, (D), Arizona, 2012, Ggn
David Farris, (D), S California, 2006, Ggc
Distinguished Research Professor:
Stanley R. Riggs, (D), Montana, 1967, GusGe
Emeritus:
Catherine A. Rigsby, (D), California (Santa Cruz), 1989, Gs

Guilford College

Dept of Geology & Earth Science (B)
5800 West Friendly Avenue
Greensboro, NC 27410-4173
p. (336) 316-2263
ddobson@guilford.edu
http://www.guilford.edu/academics/departments-and-programs/geology/
f: https://www.facebook.com/guilfordgeo/
Enrollment (2016): B: 22 (7)
Professor:
David M. Dobson, (D), Michigan, 1997, GusPc
Marlene McCauley, (D), California (Los Angeles), 1986, Cp
Assistant Professor:
Holly Peterson, (D), British Columbia, 2014, HwGe
Emeritus:
Cyril H. Harvey, (D), Nebraska, 1960, Gr

North Carolina Agricultural & Tech State University

Dept of Natural Resources and Environmental Design (B,M)
1601 E. Market St
Carver Hall Room 238
Greensboro, NC 27411
p. (336) 334-7543
uzo@ncat.edu
Programs: Environmental Studies
Certificates: Certificate In Waste Management
Enrollment (2018): B: 79 (22) M: 13 (7)
Professor:
Godfrey A. Uzochukwu, (D), Nebraska, 1983, Gz
Assistant Professor:
Niroj Aryal, (D), Michigan State, 2015, HsgZi

North Carolina Central University

Department of Environmental, Earth and Geospatial Sciences (B,M)
1801 Fayetteville Street
Durham, NC 27707-19765
p. (919) 530-5296
gvlahovic@nccu.edu
http://www.nccu.edu/academics/sc/artsandsciences/geospatialscience
f: https://www.facebook.com/groups/142873172423480/
t: @DEEGS_NCCU
Programs: Environmental and Geographic Sciences (B); Earth Science (M)
Certificates: USGIF Accredited Geospatial Intelligence Certificate (undergraduate and graduate)
Enrollment (2020): B: 37 (9) M: 14 (7)
Professor :
Gordana Vlahovic, (D), North Carolina, 1999, YsZeRn
John Bang, (D), Texas (El Paso), GeZa
Associate Professor:
Rakesh Malhotra, (D), Georgia, Zri
Chris McGinn, (D), North Carolina (Greensboro), Zgi
Timothy Mulrooney, (D), North Carolina (Greensboro), Zir
Zhiming Yang, (D), Oklahoma State, Zry
Assistant Professor:
Carresse Gerald, (D), North Carolina A&T, GeZg
Christopher Zarzar, (D), Mississippi State, 2017, At

North Carolina State University

Dept of Marine, Earth & Atmospheric Sciences (B,M,D)
P.O. Box 8208
Raleigh, NC 27695-8208
p. (919) 515-3711
meas-grad-program@ncsu.edu
http://www.meas.ncsu.edu
f: https://www.facebook.com/measncsu
t: @ncsumeas
Programs: Geology
Marine Sciences
Meteorology
Natural Resources
Certificates: Climate Change and Society (Graduate Program)
Enrollment (2020): B: 205 (44) M: 27 (7) D: 51 (16)
Head:
Lewis A. Owen, (D), Leicester, 1988, GmtGl
Director of Graduate Programs:
Elana L. Leithold, (D), Washington, 1987, Gs
Professor:
Viney P. Aneja, (D), North Carolina State, 1977, Ac
DelWayne Bohnenstiehl, (D), Columbia, 2002, YrsGt
David B. Eggleston, (D), William & Mary, 1991, Ob
Ruoying He, (D), S Florida, 2002, Op
Gary M. Lackmann, (D), SUNY (Albany), 1995, As
Jay F. Levine, (D)
Jingpu P. Liu, (D), William & Mary, 2001, Gu
David A. McConnell, (D), Texas A&M, 1987, ZeRcGc
Nicholas Meskhidze, (D), Georgia Tech, 2003, As
Helena Mitasova, (D), Slovak Tech, 1987, Zi
Matthew Parker, (D), Colorado State, 2002, Am
Walter Robinson, (D), Columbia, 1985, As
Lian Xie, (D), Miami, 1992, As
Sandra Yuter, (D), Washington, 1996, As
Associate Professor:
Anantha Aiyyer, (D), SUNY (Albany), 2003, As
Chris Osburn, (D), Lehigh, 2000, Cms
Astrid Schnetzer, (D), Vienna, 2001, Ob
Karl Wegmann, (D), Lehigh, 2008, Gm
Assistant Professor:
Stuart P. Bishop, (D), Rhode Island, 2012, OpGq
Magdalena E. Curry, (D), North Carolina, 2015, Gt
Catherine Davis, (D), California (Davis), 2016, Cm
Lisa Falk, (D), Columbia, 2014, Zg
Ethan Hyland, (D), Michigan, 2014, GrCsPe

Markus Petters, (D), Wyoming, 2004, As
Camilo Rey-Sanchez, (D), Ohio State, 2018, As
Arianna Soldati, (D), Missouri, 2018, Gv
Arianna Soldati, (D), Missouri, 2018, Gv
Emeritus:
Satyapal S. Arya, (D), Colorado State, 1968, As
David J. DeMaster, (D), Yale, 1979, OcCbc
Ronald V. Fodor, (D), New Mexico, 1972, Gi
John C. Fountain, (D), California (Santa Barbara), 1975, Hw
James P. Hibbard, (D), Cornell, 1988, GctZn
Daniel Kamykowski, (D), California (San Diego), 1973, ObpOc
Charles E. Knowles, (D), Texas A&M, 1970, Op
Sethu S. Raman, (D), Colorado, 1972, As
Allen J. Riordan, (D), Wisconsin, 1977, As
Dale A. Russell, (D), Columbia, 1964, Pv
Fred H. M. Semazzi, (D), Nairobi, 1983, As
Ping-Tung Shaw, (D), MIT/WHOI, 1982, Op
William J. Showers, (D), Hawaii, 1982, Cs
Edward F. Stoddard, (D), California (Los Angeles), 1976, Gp
Charles W. Welby, (D), MIT, 1952, Hw
Donna L. Wolcott, (D), California (Berkeley), 1972, Ob
Thomas G. Wolcott, (D), California (Berkeley), 1971, Ob

Dept of Soil Science (B,M,D)
Box 7619
Raleigh, NC 27695-7619
p. (919) 515-2655
jeff_mullahey@ncsu.edu
Administrative Assistant: Ashru Shah
Professor:
Aziz Amoozegar, (D), Arizona, 1977, Sp
Stephen W. Broome, (D), North Carolina, 1973, Sf
Donald K. Cassel, (D), California (Davis), 1968, Sp
John L. Havlin, (D), Colorado State, 1983, Sc
Dean L. Hesterberg, (D), California (Riverside), 1988, Sc
Michael T. Hoover, (D), Penn State, 1983, Sd
Greg D. Hoyt, (D), Georgia, 1981, Sb
Daniel W. Israel, (D), Oregon State, 1973, Sb
Harold J. Kleiss, Illinois, 1972, Sd
Deanna L. Osmond, (D), Cornell, 1991, Sb
Wayne P. Robarge, (D), Wisconsin, 1975, Sc
Thomas J. Smyth, (D), North Carolina State, 1981, Sc
Michael J. Vepraskas, (D), Texas A&M, 1980, Sd
Michael G. Wagger, (D), Kansas State, 1983, Sb
Associate Professor:
David A. Crouse, (D), North Carolina State, 1996, Sc
Carl Crozier, (D), North Carolina State, 1992, Sb
David Lindbo, (D), Massachusetts, 1990, Sd
Richard A. McLaughlin, (D), Purdue, 1985, Sc
Jeffrey G. White, (D), Cornell, 1988, SoZri
Assistant Professor:
Alexandria Graves, (D), Virginia Tech, 2003, Sb
Wei Shi, (D), Purdue, Sb
Emeritus:
James W. Gilliam, (D), Mississippi State, 1965, Sb

University of North Carolina, Asheville

Dept of Atmospheric Sciences (B) ●☒
CPO 2450
One University Heights
Asheville, NC 28804-3299
p. (828) 251-6149
chennon@unca.edu
https://atms.unca.edu
f: https://www.facebook.com/uncaweather/
t: @uncaweather
Programs: Atmospheric Sciences
Enrollment (2018): B: 36 (11)
Chair:
Chris Hennon, (D), Ohio State, 2003, AstZn
Professor:
Christopher Godfrey, (D), Oklahoma, 2006, AsZnn
Alex Huang, (D), Purdue, 1984, As
Doug Miller, (D), Purdue, 1996, AsZnn

Environmental Studies (B) ☒
1 University Heights, CPO 2330
Asheville, NC 28804-8511
p. (828) 251-6441
bmcnamee@unca.edu
http://envr.unca.edu
f: https://www.facebook.com/uncaenvr
Programs: B.S. Environmental Studies, including concentrations in Earth Science, Ecology and Environmental Biology, and Environmental Management and Policy.
Certificates: Sustainability
Enrollment (2020): B: 39 (9)
Professor:
Irene M. Rossell, (D), SUNY (Syracuse), 1995, Zn
Associate Professor:
Delores M. Eggers, (D), N Carolina, 1999, Zn
David P. Gillette, (D), Oklahoma, 2008, Zn
Jackie M. Langille, (D), Tennessee, 2012, GctGe
Brittani D. McNamee, (D), Idaho, 2013, GzEgGe
Jeffrey D. Wilcox, (D), Wisconsin, 2007, HwClGg
Assistant Professor:
Andrew Laughlin, (D), Tulane, Zn
Lecturer:
Jake Hagedorn, (D), Maryland, SbZun
Alison Ormsby, (D), Antioch New England Grad, Zn
Landon Ward, (M), Virginia Inst of Marine Sci, ZnnZn

University of North Carolina, Chapel Hill

Dept of Earth, Marine and Environmental Sciences (B,M,D)
CB 3315, Mitchell Hall
Chapel Hill, NC 27599-3315
p. (919) 966-4516
ekirby@unc.edu
http://www.emes.unc.edu
t: @geosciUNC
Programs: BA Earth Science, BS Earth Science, BS Environmental Geoscience, MS Geological Sciences, PhD Geological Sciences, MS Marine Sciences, PhD Marine Sciences
Enrollment (2016): B: 66 (21) M: 4 (5) D: 20 (2)
Chair:
Eric Kirby, (D), MIT, 2001, GtmGc
Professor:
Drew S. Coleman, (D), Kansas, 1991, CcGit
Jonathan M. Lees, (D), Washington, 1989, YsGv
Laura J. Moore, (D), California (Santa Cruz), 1998, Gme
Donna M. Surge, (D), Michigan, 2001, Pe
Associate Professor:
Xiaoming Liu, (D), Maryland, 2013, Cgl
Tamlin M. Pavelsky, (D), California (Los Angeles), 2008, Hg
Kevin G. Stewart, (D), California (Berkeley), 1987, Gc
Assistant Professor:
Kennet Flores, (D), Lausanne, 2010, Gpt
Emeritus:
Larry Benninger, (D), Yale
Paul D. Fullagar, (D), Illinois, 1963, CcGaZe
Allen F. Glazner, (D), California (Los Angeles), 1981, GitZf
Daniel A. Textoris, (D), Illinois, 1963, Gd

University of North Carolina, Charlotte

Dept of Geography & Earth Sciences (B,M,D)
9201 University City Boulevard
Charlotte, NC 28223-0001
p. (704) 687-5973
wjgarcia@uncc.edu
http://www.geoearth.uncc.edu/
Enrollment (2010): B: 280 (70) M: 86 (16) D: 32 (8)
Professor:
John F. Bender, (D), SUNY (Stony Brook), 1980, GiCtOu
John A. Diemer, (D), SUNY (Binghamton), 1985, Gs
Associate Professor:
Craig J. Allan, (D), York, 1992, Hg
Andy R. Bobyarchick, (D), SUNY (Albany), 1983, Gc
Scott P. Hippensteel, (D), Delaware, 2000, Gr
Walter Martin, (D), Tennessee, 1984, As
Ross Meentemeyer, (D), North Carolina, 2000, Zi
Assistant Professor:
Manda S. Adams, (D), Wisconsin, 2005, Am
Matt Eastin, (D), Colorado State, 2003, Am
Martha Cary Eppes, (D), New Mexico, 2002, So

Lecturer:
Jake Armour, (M), New Mexico, 2002, GmlGg
Terry Shirley, (M), Penn State, 2003, Asm
Emeritus:
Anne Jefferson, (D), Oregon State, 2006, HgGm
Laboratory Director:
William Garcia, (M), Cincinnati, Pv

University of North Carolina, Pembroke

Geology & Geography Dept (B)
PO Box 1510
Pembroke, NC 28372-1510
p. (910) 775-4024
geo@uncp.edu
https://www.uncp.edu/departments/geology-and-geography
Programs: Geo-Environmental Studies
Certificates: Geospatial Technologies
Enrollment (2021): B: 27 (7)
Chair:
Martin B. Farley, (D), Penn State, 1987, PlZe
Associate Professor:
Jeff B. Chaumba, (D), Georgia, 2009, GxEgGz
Dennis J. Edgell, (D), Kent State, 1992, As
Assistant Professor:
Madan Maharjan, (D), West Virginia, 2017, Hw
Jesse Rouse, (D), West Virginia, 2018, Zni
Lecturer:
Amy Gross, (M), North Carolina (Wilmington), 2006, Gg
Nathan E. Phillippi, (M), South Dakota State, 2004, Zgi
Emeritus:
Thomas E. Ross, (D), Tennessee, 1977, Zy

University of North Carolina, Wilmington

Dept of Earth and Ocean Sciences (B,M) O
601 South College Road
Wilmington, NC 28403-5944
p. (910) 962-3490
lynnl@uncw.edu
http://www.uncw.edu/earsci/
Administrative Assistant: Alexis Lee
Enrollment (2007): B: 78 (21) M: 32 (5)
Chair:
Lynn A. Leonard, (D), S Florida, 1993, Gu
Professor:
Michael M. Benedetti, (D), Wisconsin, 2000, GmSaZy
Nancy R. Grindlay, (D), Rhode Island, 1991, Yg
Michael S. Smith, (D), Washington (St Louis), 1990, GzpGa
Associate Professor:
Lewis J. Abrams, (D), Rhode Island, 1992, Gu
David E. Blake, (D), Washington State, 1991, Gp
Douglas W. Gamble, (D), Georgia, 2000, Zy
Joanne N. Halls, (D), South Carolina, 1996, Zi
Eric J. Henry, (D), Arizona, 2001, Hw
Mary E. Hines, (D), Louisiana State, 1992, Zu
Lecturer:
Roger D. Shew, (M), North Carolina, 1979, Eo
Emeritus:
Robert T. Argenbright, (D), California (Berkeley), 1990, Zn
William J. Cleary, (D), South Carolina, 1972, Gu
James A. Dockal, (D), Iowa, 1980, GdcEm
W. Burleigh Harris, (D), North Carolina, 1975, GrdGs
Patricia H. Kelley, (D), Harvard, 1979, Pgi
Paul A. Thayer, (D), North Carolina, 1967, GdoHg
Laboratory Director:
Yvonne Marsan, (B), North Carolina (Wilmington), Zi

Wake Technical Community College

Natural Sciences Dept
9101 Fayetteville Rd
Raleigh, NC 27603
p. (919) 866-5000
mnbalachander@waketech.edu
http://www.waketech.edu/programs-courses/credit/natural-sciences

Western Carolina University

Dept of Geosciences & Natural Resources (B) ☒
Stillwell Building Room 331
Cullowhee, NC 28723-9047
p. (828) 227-7367
mlord@wcu.edu
http://geology.wcu.edu
Programs: Geology,
Science Education: Earth Science
Enrollment (2021): B: 37 (10)
Whitmire Prof Env Sci:
Jerry R. Miller, (D), S Illinois, 1990, GmeGf
Director Prog Study Dev Shorelines:
Robert S. Young, (D), Duke, 1995, GsOn
Associate Provost:
Brandon E. Schwab, (D), Oregon, 2000, GizGv
Professor:
Mark L. Lord, (D), North Dakota, 1988, HwGmZe
Science Education Program Dir.:
Frank Forcino, (D), Alberta, 2013, ZePgGr
Dept. Head:
Diane M. Styers, Auburn, 2008, Zrn
Dean, Arts & Sciences:
David A. Kinner, (D), Colorado, 2003, HgZe
Associate Professor:
Cheryl Waters-Tormey, (D), Wisconsin, 2004, Gct
Assistant Professor:
Amy Fagan, (D), Notre Dame, 2013, GiXg
Xizhen (Jenny) Schenk, (D), Florida State, 2012, HgRwZi
Instructor:
Emily Stafford, (D), Alberta, 2014, PgGg
Emeritus:
Steven P. Yurkovich, (D), Brown, 1972, Gx

North Dakota

Dickinson State University

Dept of Natural Science (B)
Dickinson, ND 58601
p. (701) 227-2114
Michael.Hastings@dickinsonstate.edu
http://www.dsu.nodak.edu/
Chair:
Michael Hastings
Associate Professor:
Larry D. League, (M), Kansas, 1971, Zy

Minot State University

Dept of Geoscience (B) ☒
500 University Avenue West
Minot, ND 58707
p. (701) 858-3873
john.webster@minotstateu.edu
http://www.minotstateu.edu/geology/
Programs: Geology; Earth Science Education
Enrollment (2021): B: 16 (1)
Head:
John R. Webster, (D), Indiana, 1992, Giz
Associate Professor:
Joseph Collette, (D), California (Riverside), 2014, PgGsr
Assistant Professor:
Kathyrn Kilroy, (D), Nevada (Reno), 1992, Hg

North Dakota State University

Dept of Soil Science (B,M,D)
Walster Hall
Fargo, ND 58105
p. (701) 231-8690
Thomas.Desutter@ndsu.edu
Administrative Assistant: Jacinda Wollan
Enrollment (2014): B: 18 (0) M: 10 (0) D: 4 (0)
Professor:
Francis Casey, (D), Iowa State, 2000, SpHq
Dave Franzen
Robert J. Goos, (D), Colorado, 1980, Sc
Associate Professor:
Larry J. Cihacek, (D), Iowa State, 1976, Zg

Assistant Professor:
Amitava Chatterjee, Wyoming
Aaron Daigh, (D), Iowa State
Tom DeSutter, (D), Kansas State, So
Ann-Marie Fortuna
David G. Hopkins, (D), North Dakota State, 1997, Sd
Abbey Wick, Wyoming

Earth, Environmental, and Geospatial Sciences (B)
NDSU Dept. 2745
P.O. Box 6050
Fargo, ND 58108-6050
p. (701) 231-8455
stephanie.day@ndsu.edu
https://www.ndsu.edu/eegs/
f: https://www.facebook.com/NDSUGeosciences
Programs: Geology (B); Geography (minor); Geology (minor); Environmental Geology (minor)
Certificates: GIS and Remote Sensing
Enrollment (2022): B: 30 (2)
Chair:
Stephanie S. Day, (D), Minnesota, 2012, GmZiu
Professor:
Kenneth Lepper, (D), Oklahoma State, 2001, CcGml
Peter Oduor, (D), Missouri (Rolla), 2004, ZiRwZr
Bernhardt Saini-Eidukat, (D), Minnesota, 1991, GxCqEg
Scott A. Wood, (D), Princeton, 1985, Cq
Associate Professor:
Benjamin J. C. Laabs, (D), Wisconsin, 2004, GlPcGm
Lydia S. Tackett, (D), S California, 2014, PeGsPi
Lecturer:
Jessie L. Rock, (M), North Dakota State, 2009, ZgPcg
Emeritus:
Allan C. Ashworth, (D), Birmingham (UK), 1969, Pe
Donald P. Schwert, (D), Waterloo, 1978, PeGe

University of North Dakota

Dept of Geography and Geographic Information Science (B,M)
221 Centennial Drive
Stop 9020
Grand Forks, ND 58202
p. (701) 777-4246
und.geography@und.edu
http://arts-sciences.und.edu/geography/
f: https://www.facebook.com/groups/91074880028/
Programs: Geography (B,M); Geospatial Technologies (minor); Environmental Studies (BA, BS)
Certificates: GISc (G)
Enrollment (2022): B: 12 (2) M: 3 (3)
Chair:
Douglas C. Munski, (D), Illinois, 1978, ZeRhZu
Professor:
Bradley C. Rundquist, (D), Kansas State, 2000, ZriZy
Gregory S. Vandeberg, (D), Kansas State, 2005, GmZiGl
Associate Professor:
Enru Wang, (D), Washington, 2005, Zgi
Instructor:
Mbongowo J. Mbuh, (D), George Mason, 2015, ZrHs

Harold Hamm School of Geology & Geological Engineering (B,M,D)
81 Cornell Street
Stop 8358
Grand Forks, ND 58202
p. (701) 777-2248
kathie.johnke@und.edu
http://engineering.und.edu/geology-and-geological-engineering/
Programs: BS, MS, and PhD Geology, BS Earth Science, BS Environmental Geoscience, BS, MS, and PhD Geological Engineering
Certificates: Certificate for Petroleum Geology
Enrollment (2014): B: 78 (0) M: 17 (0) D: 10 (0)
Director:
Jaakko Putkonen, (D), Washington, 1997, Gml
Professor:
William D. Gosnold, (D), S Methodist, 1976, YghCt
Joseph H. Hartman, (D), Minnesota, 1984, Pi
Dexter Perkins, III, (D), Michigan, 1979, Gp
Associate Professor:
I-Hsuan Ho, (D), Iowa, Ng
Ronald K. Matheney, (D), Arizona State, 1989, Cs
Stephan Nordeng, (D), Michigan, Go
Dongmei Wang, (D), China, NgGo
Emeritus:
Nels F. Forsman, (D), North Dakota, 1985, GgXgRh
Philip J. Gerla, (D), Arizona, 1983, Hw
Richard D. LeFever, (D), California (Los Angeles), 1979, GsrGo

Ohio

Bowling Green State University

Dept of Geology (B,M)
190 Overman Hall
Bowling Green, OH 43403
p. (419) 372-2886
jasnyd@bgsu.edu
https://www.bgsu.edu/arts-and-sciences/earth-environment-and-society/geology.html
Administrative Assistant: Pat A. Wilhelm
Enrollment (2010): B: 29 (13) M: 23 (2)
Professor:
James E. Evans, (D), Washington, 1988, GsHsGe
Charles M. Onasch, (D), Penn State, 1977, Gc
Robert K. Vincent, (D), Michigan, 1973, Yg
Peg M. Yacobucci, (D), Harvard, 1999, Pgi
Associate Professor:
John R. Farver, (D), Brown, 1988, Gy
Enrique Gomezdelcampo, (D), Tennessee, 2003, Ge
Peter Gorsevski, (D), Idaho, 2002, ZirGm
Kurt S. Panter, (D), New Mexico Tech, 1995, Giv
Jeffrey A. Snyder, (D), Ohio State, 1996, Gm
Lecturer:
Nichole Elkins, (M), Georgia, 2002, GaZe
Christopher Pepple, (M), Ze
Paula J. Steinker, (D), Bowling Green, 1982, Pg
Emeritus:
Don C. Steinker, (D), California (Berkeley), 1969, Pg
Department IT:
William Butcher, (B), Rochester, 1972, Gg

Case Western Reserve University

Dept of Earth, Environmental and Planetary Sciences (B,M,D) ☒
10900 Euclid Avenue
A.W. Smith #112
Cleveland, OH 44106-7216
p. (216) 368-3690
steven.hauck@case.edu
http://eeps.case.edu/
Programs: Geological Sciences (BA, BS, MS, PhD), Environmental Geology (BA)
Enrollment (2022): B: 17 (3) D: 4 (0)
Professor and Chair:
Steven A. Hauck, II, (D), Washington (St. Louis), 2001, XyYgXg
Professor:
Ralph P. Harvey, (D), Pittsburgh, 1990, Xmg
Beverly Z. Saylor, (D), MIT, 1996, Gs
James A. Van Orman, (D), MIT, 2000, CgGyCp
Peter J. Whiting, (D), California (Berkeley), 1990, Gm
Visiting Assistant Professor:
Onema Adojoh, (D), Liverpool, 2017, Pm
Anne W. Omta, (D), Vrije, 2009, Op
Adjunct Associate Professor:
Mulugeta Alene Araya, (D), Turin
Andrew Dombard, (D), Washington (St. Louis), 2000, Xy
Mark B. Green, (D), Minnesota, 2007, Hg
Adjunct Assistant Professor:
Joseph T. Hannibal, (D), Kent State, 1990, Pi
Scott Hardy, (D), Ohio State, 2007, Rw
David Saja, (D), Pennsylvania, 1999, Gc
Emeritus:
Gerald Matisoff, (D), Johns Hopkins, 1978, Cbc
Peter L. McCall, (D), Yale, 1975, Pg
Samuel M. Savin, (D), Caltech, 1967, Cs

OHIO

Cedarville University

Dept of Science and Mathematics (B)
251 North Main Street
Cedarville, OH 45314
p. (937) 766-7940
trice@cedarville.edu
https://www.cedarville.edu/Academic-Programs/Geology.aspx
f: https://www.facebook.com/cedarvillegeology
Programs: Geology (BS) with two tracks to choose from - Physical Geology (BS) and Geoscience (BS)
Enrollment (2022): B: 27 (2)

Professor:
Mark Gathany, (D), Colorado State, ZiuSb
Steven Gollmer, (D), Purdue, GzAmOg
John H. Whitmore, (D), Loma Linda, 2003, PgGsg

Assistant Professor:
Thomas L. Rice, (M), Colorado Mines, 1987, GemEo

Adjunct Professor:
Steve A. Austin, (D), Penn State, 1979, GsEcGd

Central State University

Intl Center for Water Resources Management (B) ☒
C.J. McLin Bldg
Wilberforce, OH 45384
p. (937) 376-6212
knedunuri@centralstate.edu
Enrollment (2017): B: 35 (4)

Chairperson and Director:
Krishna Kumar Nedunuri, (D), Purdue, 1999, HwCaSo

Associate Ditector for Land Grant Research:
Subramania I. Sritharan, (D), Colorado State, 1984, Hg

Professor:
Ramanitharan Kandiah, (D), Tulane, 2004, HqSoZi
Sam Laki, (D), Michigan State, 1992, EgSoHs
Xiaofang Wei, (D), Indiana State, 2008, ZrOgZe

Associate Professor:
De Bonne N. Wishart, (D), Rutgers (Newark), 2008, YgCa
Ning Zhang, (D), West Virginia, 2012, RwEo

Emeritus:
Samuel Okunade, (D), Kent State, 1986, Gm

Cincinnati Museum Center

Geier Collections and Research Center ☒
1301 Western Avenue
Cincinnati, OH 45203
p. (513) 287-7000
information@cincymuseum.org
http://www.cincymuseum.org

Associate Vice President:
Glenn W. Storrs, (D), Yale, 1986, PvGgr

Collection Manager:
Cameron E. Schwalbach, (M), Cincinnati, 2017, Pi

Curator:
Brenda R. Hunda, (D), California (Riverside), 2004, Pi

Research Associate:
Carlton E. Brett, (D), Michigan, 1978, Ps
Brooke E. Crowley, (D), California (Santa Cruz), 2009, Pe
Nigel C. Hughes, (D), Bristol, 1990, Pi
Takuya Konishi, (D), Alberta, 2009, Pv
David L. Meyer, (D), Yale, 1971, Pi
Arnold I. Miller, (D), Chicago, 1986, Pq
Joshua H. Miller, (D), Chicago, 2009, Pe
Ryan C. Shell, (D), Wright State, 2020, Pv
Andrew J. Webber, (D), Cincinnati, 2003, Pi

Cleveland Museum of Natural History

Department of Paleobotany & Paleoecology ☒
1 Wade Oval Drive
Cleveland, OH 44106-1767
p. (216) 231-4600 (Ext. 3529)
mdonovan@cmnh.org
https://www.cmnh.org/research-collections/paleobotany-paleoecology

Dept of Paleontology ☒
1 Wade Oval Drive
University Circle
Cleveland, OH 44106-1767
p. (216) 231-4600
https://www.cmnh.org/research-collections/vertebrate-paleontology

Emeritus:
Joseph T. Hannibal, (D), Kent State, 1990, PiEn

Senior Collections Manager:
Michael Donovan, (D), Penn State, 2017, Pb

Collections Manager:
Amanda R. McGee, (M), Calgary, 2011, Pvg

Assistant Curator:
Caitlin Colleary, (D), Virginia Tech, 2018, Pv

Mineralogy ☒
1 Wade Oval Drive
University Circle
Cleveland, OH 44106-1767
p. (216) 231-4600
mineralogy@cmnh.org
http://www.cmnh.org/mineralogy

Emeritus:
David Saja, (D), Pennsylvania, 1999, GdcZe

College of Wooster

Dept of Earth Sciences (B) ☒
Scovel Hall
944 College Mall
Wooster, OH 44691-2363
p. (330) 263-2380
barmstrong@wooster.edu
https://www.wooster.edu/departments/earth-sciences/
f: https://www.facebook.com/pages/College-of-Wooster-Geology-Department/143144126437
Programs: Environmental Geoscience or Geology Major and Minor; Earth Science Minor

Enrollment (2020): B: 28 (7)

Chair:
Shelley Judge, (D), Ohio State, 2007, GtsGc

Professor:
Gregory C. Wiles, (D), SUNY (Buffalo), 1992, Gl
Mark A. Wilson, (D), California (Berkeley), 1982, Pi

Associate Professor:
Meagen Pollock, (D), Duke, 2007, CuGiv

Cuyahoga Community College - Western Campus

Earth Science (A)
11000 Pleasant Valley Road
Parma, OH 44130
p. (216) 987-5278
enroll@tri-c.edu

Assistant Professor:
Robert Zaleha, Zg

Instructor:
Carol Fondran, Zg
Joseph M. Lane, Zg
Abby N. Norton-Krane, Zg
Kathryn Sasowsky, Zg

Adjunct Professor:
Gloria CC Britton, (D), ZgGmZn
Jennifer Deka, Zg
John L. Ezerskis, (M), Toledo, 1988, ZgGcHw

Denison University

Dept of Geosciences (B)
F.W. Olin Science Hall
100 W. College Street
Granville, OH 43023
p. (740) 587-6217
hall@denison.edu
http://www.denison.edu/academics/departments/geosciences/
Administrative Assistant: Jude Hall
Enrollment (2013): B: 9 (9)

Professor:
Tod A. Frolking, (D), Wisconsin, 1985, Zy

David C. Greene, (D), Nevada (Reno), 1995, GctGe
Associate Professor:
David H. Goodwin, (D), Arizona, 2003, PiCs
Assistant Professor:
Erik Klemetti, (D), Oregon State, 2005, GvxGg
Kate E. Tierney, (D), Ohio State, 2010, CgGrZg
Emeritus:
Kennard B. Bork, (D), Indiana, 1967, PiRhGs
Robert J. Malcuit, (D), Michigan State, 1973, Gx

Hocking College

Natural Resources ●☒
3301 Hocking Parkway
Nelsonville, OH 45764-9704
p. (740) 753-6277
admissions@hocking.edu
https://www.hocking.edu/majors
Programs: Environmental Science, Wildlife Management, Forest Management, Parks and Museum Education
Certificates: GIS/GPS, Forestry

Kent State University

Dept of Geology (B,M,D) ○☒
221 McGilvrey Hall
325 South Lincoln Street
Kent, OH 44242
p. (330) 672-2680
geology@kent.edu
http://www.kent.edu/geology/
Programs: BS Geology, BS Environmental Geology Concentration, BA Geology, BA Geology Earth Science, MS Geology, PhD Applied Geology
Enrollment (2015): B: 154 (23) M: 22 (10) D: 9 (3)
Chair:
Daniel K. Holm, (D), Harvard, 1992, Gc
Professor:
David B. Hacker, (D), Kent State, 1998, GcHw
Joseph D. Ortiz, (D), Oregon State, 1995, Gs
Carrie E. Schweitzer, (D), Kent State, 2000, Pi
Alison J. Smith, (D), Brown, 1991, GnPce
Neil A. Wells, (D), Michigan, 1984, Gs
Associate Professor:
Anne J. Jefferson, (D), Oregon State, 2006, HgGmCs
David M. Singer, (D), Stanford, 2008, Gze
Assistant Professor:
Timothy M. Gallagher, (D), Michigan, 2016
Christopher J. Rowan, (D), Southampton (UK), 2006, Gt
Kuldeep Singh, (D), Texas, 2013
Allyson Tessin, (D), Michigan, 2016
Jeremy C. Williams, (D), Massachusetts (Boston), 2014, Cg
Emeritus:
Rodney M. Feldmann, (D), North Dakota, 1967, Pi
Donald F. Palmer, (D), Princeton, 1968, Yg

Kent State University at Stark

Dept of Geology (B) ○🖱
6000 Frank Avenue NW
North Canton, OH 44720
p. 330-244-3303
cschweit@kent.edu
https://www.kent.edu/geology
Programs: Geology, Earth Science
Administrative Assistant: Debra Stimer
Enrollment (2019): B: 22 (0)
Chair:
Carrie E. Schweitzer, (D), Kent State, 2000, PiGg
Associate Professor:
Eric Taylor, (D), Ohio State, GezZe

Lakeland Community College

Geoscience Dept (A)
7700 Clocktower Drive
Kirtland, OH 44094
p. (440) 525-7341
dpierce@lakelandcc.edu
http://www.lakelandcc.edu/academic/sh/geol
Professor:
David Pierce, (D), GgHsAm

Marietta College

Dept of Petroleum Engineering & Geology (B) ☒
215 Fifth Street
Marietta, OH 45750
p. (740) 376-4775
sch030@marietta.edu
http://w3.marietta.edu/departments/Petroleum_Engineering/
Programs: Geology; Petroleum Engineering; Environmental Engineering
Enrollment (2017): B: 36 (13)
Professor:
David L. Jeffery, (D), Texas A&M, 1996, PsGo
Associate Professor:
Tej P. Gautam, (D), Kent State, 2012, NgGeZi
Instructor:
Wendy Bartlett, (M), Texas A&M, 1980, Goe
Veronica Freeman, (M), Texas (Arlington), 1993, Pg
Visiting Professor:
Paul A. Washington, (D), Connecticut, 1987, Gcs
Administrative Coordinator:
Susan Hiser, (M), Springfield Coll, 1992, Zn

Miami University

Dept of Geology and Environmental Earth Sciences (B,M,D) ○⊘
250 S,. Patterson Avenue
118 Shideler Hall
Oxford, OH 45056
p. (513) 529-3216
burgerge@MiamiOh.edu
http://www.miamioh.edu/geology
f: https://www.facebook.com/MiamiGeology
t: @geologymiamioh
Programs: Geology (B.S.); Environmental Earth Science (B.S.)
Enrollment (2022): B: 144 (41) M: 15 (8) D: 13 (1)
Janet & Elliot Baines Bicentenntial Professor & Chair:
Elisabeth Widom, (D), California (Santa Cruz), 1991, CcGve
Professor:
Michael Brudzinski, (D), Illinois, 2002, Ys
Brian S. Currie, (D), Arizona, 1998, GstGc
Yildirm Dilek, (D), California (Davis), 1989, Gt
Hailiang Dong, (D), Michigan, 1997, Cg
Jason Rech, (D), Arizona, 2001, Gm
Associate Professor (Hamilton Campus:
Mark Krekeler, (D), Illinois (Chicago), 2003, Ge
Associate Professor & Director of IES:
Jonathan Levy, (D), Wisconsin, 1993, Hw
Associate Professor:
Claire McLeod, (D), Durham (UK), 2012, Gx
Instructor:
Jill Mignery, (M), Miami, 2004, Zge
Lecturer (Middletown Campus:
Tammie Gerke, (D), Cincinnati, 1995, ClGea
Adjunct Professor:
Patri Larrea, (D), Zaragoza, 2014
Emeritus:
A. Dwight Baldwin, Jr., (D), Stanford, 1966, Hw
Mark R. Boardman, (D), North Carolina, 1978, On
William K. Hart, (D), Case Western, 1982, GivCc
John M. Hughes, (D), Dartmouth, 1981, Gz
Robert G. McWilliams, (D), Washington, 1968, Ps
John K. Pope, (D), Cincinnati, 1966, Pi
Isotope Geochemistry & Mass Spec Lab Mgr:
David C. Kuentz, (M), Texas (Arlington), 1986, Ca
Geochemistry Lab Manager:
Marion Lytle, (D), Rhode Island, 2013
Associate Teaching Professor:
Todd Dupont, (D), Penn State, 2004, Gl
Assistant Teaching Professor:
Maija Sipola, (D), Iowa, 2018, Gg
Accounting Technician:
Gail Burger
Director Limper Geology Museum:
Kendall L. Hauer, (D), Miami (Ohio), 1995, GgZeCg

Cooperating Faculty:
Jerome Bellian, (D), Texas, 2009, Go
Carina Colombi, (D), National (San Juan), 2007, Gr
R. Hays Cummins, (D), Texas A&M, 1984, Ge
J Christophe Haley, (D), Johns Hopkins, 1986, Gt
Hassan Mirnejad, (M), Carleton, 2001, Gx
Monica Rakovan, (D), Miami (Ohio), 2011, Hg
Fara Rasoazanamparany, (D), Miami, 2015

Mount Union College

Dept of Geology (B)
Alliance, OH 44601
p. (330) 823-3672
graylm@mountunion.edu
http://www.mountunion.edu/gy
Enrollment (2010): B: 8 (2)
Professor:
Lee M. Gray, (D), Rochester, 1985, PgGg
Department Chair:
Mark A. McNaught, (D), Rochester, 1991, Gcg
Adjunct Professor:
Leonard G. Epp, (D), Penn State, 1970, Ob

Muskingum University

Dept of Geology (B)
163 Stormont Street
New Concord, OH 43762
p. (740) 826-8306
svanhorn@muskingum.edu
http://muskingum.edu/dept/geology
Enrollment (2016): B: 33 (6)
Associate Professor:
Stephen R. Van Horn, (D), Connecticut, 1996, GeZiEo
Associate Professor:
Eric W. Law, (D), Case Western, 1982, Gp
David L. Rodland, (D), Virginia Tech, 2003, PieGr

Northwest State Community College

Div of Arts and Sciences
22600 State Route 34
Archbold, OH 43502
levans@northweststate.edu
http://www.northweststate.edu

Oberlin College

Dept of Geosciences (B) ☒
52 West Lorain Street
Oberlin, OH 44074-1044
p. (440) 775-8350
geosciences@oberlin.edu
http://new.oberlin.edu/arts-and-sciences/departments/geology/
Enrollment (2022): B: 29 (0)
Chair:
Amanda H. Schmidt, (D), Washington, 2010, Gm
Professor:
F Zeb Page, (D), Michigan, 2005, GpCsg
Associate Professor:
Rachel Eveleth, (D)
Assistant Professor:
Veronica Vriesman, (D)
Emeritus:
Karla M. Parsons-Hubbard, (D), Rochester, 1993, PgGu
Bruce M. Simonson, (D), Johns Hopkins, 1982, GsXmHw
Steven F. Wojtal, (D), Johns Hopkins, 1982, Gc
Chair:
Dennis K. Hubbard, (D), South Carolina, 1977, Gsu
Cooperating Faculty:
Alison Ricker, (M), Rhode Island, 1977, Zn

Ohio State University

Dept of Civil, Environmental & Geodetic Engineering (B,M,D) ☒
470 Hitchcock Hall
2070 Neil Avenue
Columbus, OH 43210
p. (612) 292-2771
mackay.49@osu.edu
http://ceg.osu.edu
Programs: Civil Engineering
minor in Survey Engineering
Enrollment (2016): M: 11 (4) D: 8 (2)
Professor and Chair:
Allison A. MacKay, (D), MIT, 1998, Cq
Research Professor:
Charles K. Toth, (D), ZrYdZi
Professor:
Harvey J. Miller, (D), Ohio State, 1991, Zi
Alper Yilmaz, (D), C Florida, 2004, ZfrZu
Assistant Professor:
Rongjun Qin, (D), Zrf
Lei Wang, (D), Ohio State, Zri

Dept of Geography (B,M,D)
1036 Derby Hall
154 North Oval Mall
Columbus, OH 43210-1361
p. (614) 292-2514
geog_webmaster@osu.edu
http://www.geography.osu.edu/
Chair:
Daniel Sui
Professor:
David H. Bromwich, (D), Wisconsin, 1979, As
Ellen E. Mosley-Thompson, (D), Ohio State, 1979, As
Jeffrey C. Rogers, (D), Colorado, 1979, As
Associate Professor:
Jay S. Holgood, (D), Ohio State, 1984, As
Jialin Lin, (D), SUNY (Stony Brook), 2001, As
Bryan G. Mark, (D), Syracuse, 2001, As
Assistant Professor:
Alvaro Montenegro, (D), Florida State, 2004, OgAsZy
Emeritus:
A. John Arnfield, (D), McMaster, 1973, As
John N. Rayner, (D), Canterbury (NZ), 1965, As

Geography (B,M,D) ☒
1036 Derby Hall
154 North Oval Mall
Columbus, OH 43210-1361
p. (614) 292-2514
geog_webmaster@osu.edu
https://geography.osu.edu/
Programs: BA in Geography, BS in Geography, BS in Atmospheric Sciences, MA in Geography, MS in Atmospheric Sciences, PhD in Geography, PhD in Atmospheric Sciences
Associate Professor:
Alvaro Montenegro, (D), Florida State, 2003, AsOg

School of Earth Sciences (B,M,D) O☒
275 Mendenhall Lab
125 South Oval Mall
Columbus, OH 43210-1398
p. (614) 292-2721
earthsciences@osu.edu
http://www.earthsciences.osu.edu
f: https://www.facebook.com/OSUGeology/
t: @osuearthscience
Programs: Earth History, Geodetic Science, Solid Earth Dynamics, and Water, Climate, and the Environment
Administrative Assistant: Jill Bryant, Theresa Mooney, Angeletha M. Rogers
Enrollment (2011): M: 36 (16) D: 40 (2)
Undergraduate Coordinator:
Anne E. Carey, (D), Nevada (Reno), 1995, HwCg
School Director:
Matthew R. Saltzman, (D), California (Los Angeles), 1996, GsClg
Ohio Research Scholar:
David R. Cole, (D), Penn State, 1980, CgsZa
Ohio Eminent Scholar:
Michael G. Bevis, (D), Cornell, 1982, Yx
Frank W. Schwartz, (D), Illinois, 1972, Hw
Distinguished University Scholar:
William Berry Lyons, (D), Connecticut, 1979, HwCgZg
CK Shum, (D), Texas, 1982, YdHg
Director, Byrd Polar Research Center:
Ian M. Howat, (D), California (Santa Cruz), 2006, GlZr

Professor:
Douglas E. Alsdorf, (D), Cornell, 1996, HgYg
Loren E. Babcock, (D), Kansas, 1990, Pg
Michael Barton, (D), Manchester (UK), 1975, GiCpt
Yu-Ping Chin, (D), Michigan, 1988, CqHw
Andrea G. Grottoli, (D), Houston, 1998, ObCbOo
Christopher Jekeli, (D), Ohio State, 1981, Ydv
Mark A. Kleffner, (D), Ohio State, 1988, PgsGr
Wendy R. Panero, (D), California (Berkeley), 2001, Yg
Burkhard A. Schaffrin, (D), Bonn, 1983, Yd
Lonnie G. Thompson, (D), Ohio State, 1976, GlZgPc
Design Engineer:
Dana Caccamise, (M)
Senior Scientist:
Junyi Guo
John W. Olesik, (D), Wisconsin, 1982, Ca
Yuchan Yi, (D), Ohio State, 1995, Ydv
Associate Professor:
Ann Cook, (D), Columbia, 2010, GuYg
Ozeas S. Costa, Jr, (D), Plymouth (UK), 2002, CblZu
Thomas Darrah, (D), Rochester, 2009, Cg
Michael T. Durand, (D), California, 2007, HsZr
William Ashley Griffith, (D), Stanford, 2008, GcNr
Motomu Ibaraki, (D), Waterloo, 1994, Hg
Steven K. Lower, (D), Virginia Tech, 2001, PoZa
Alan J. Saalfeld, (D), Maryland, 1993, Zi
Senior Research Associate:
Christopher Gardner, (M)
Research Scientist:
Paolo Gabrielli, (D), LGGE Grenoble, 2004, CtPeGl
Susan A. Welch, (D), Delaware, 1997, ClGeCs
Instructional Lab Supervisor:
Christena Cox, (D), Ohio State, 1994, GeoGc
Assistant Professor:
Joel D. Barker, (D), Alberta, 2007, CblPe
Elizabeth Griffith, (D), Stanford, 2008, CsOoCb
Jill Leonard-Pingel, (D), Scripps, 2012, Pi
Joachim Moortgat, (D), Radboud, 2006, HwNp
Audrey Sawyer, (D), Texas, 2011, HwSpHs
Derek Sawyer, (D), Texas, 2010, YgGoRn
Research Associate:
Gerald Allen, (D), Ohio State, 2011
Chunli Dai, (D), Ohio State, 2015, ZrYg
Renato Frasson, (D), Iowa, 2011, Hg
Siddharth Gautam, (D), Mumbai, 2009
Eric Kendrick, (D), Hawaii
How-wai (Peter) Luk
Anthony Lutton, (D)
Myoung-Jong Noh, (D), Inha, 2011, Zr
Julie Sheets, (D), Ohio State, 1994, Gz
Stephanie Sherman, (D), Ohio State
Museum Curator :
Dale M. Gnidovec, (M), Fort Hays State, 1978, PvGr
Adjunct Professor:
Thomas G. Naymik, (D), Ohio State, 1978, Hw
Visiting Professor:
David Young, (D), California, 2005, GcpCc
Academy Professor:
William I. Ausich, (D), Indiana, 1978, PgGs
Stig M. Bergstrom, (D), Lund (Sweden), 1961, PgGg
David H. Elliot, (D), Birmingham (UK), 1965, GxtGv
Lawrence A. Krissek, (D), Oregon State, 1982, Gsu
Teresa Mensing, (D), Ohio State, 1987, Cg
Ralph R. B. von Frese, (D), Purdue, 1980, Ye
Emeritus:
E. Scott Bair, (D), Penn State, 1980, HwGe
James Bradley
James W. Collinson, (D), Stanford, 1966, PgGsg
Charles E. Corbato, (D), California (Los Angeles), 1960, Ye
Jeffrey J. Daniels, (D), Colorado Mines, 1974, Yg
James W. Downs, (D), Virginia Tech, 1983, Gz
Gunter Faure, (D), MIT, 1961, CgHw
Kenneth A. Foland, (D), Brown, 1972, CcgCs
Charles E. Herdendorf, (D), Ohio State, 1970, OgGuOn
Kenneth Jezek, (D), Wisconsin, 1980, YgPcGl
Garry D. McKenzie, (D), Ohio State, 1968, GmeZe
Ivan Mueller, (D), Ohio State, 1960, YddYv
David Nickey, (D), Penn State, 1966
Hallan C. Noltimier, (D), Newcastle upon Tyne, 1965, YmgGt
Douglas E. Pride, (D), Illinois, 1969, Eg
Richard H. Rapp, (D), Ohio State, 1964, Yd
Rodney T. Tettenhorst, (D), Illinois, 1960, Gz
Russell O. Utgard, (D), Indiana, 1969, YgEnZe
Peter N. Webb, (D), Utrecht, 1966, PmRhGr
Terry J. Wilson, (D), Columbia, 1983, GctGl
Program Coordinator:
Mike Kositzke
Courtesy Appt.:
Bryan Mark
Undergraduate Advisor:
Karen Royce, (D), Ohio State
Space Management Coordinator:
Dan Dunlap
Related Staff:
Christina Millan, (D), Ggc

School of Environment and Natural Resources (B,M,D)
2021 Coffey Road
Room 210 Koffman Hall
Columbus, OH 43210
p. (614) 292-2265
sullivan.191@osu.edu
http://senr.osu.edu/
Administrative Assistant: Mary Capoccia
Eminent Scholar:
Richard P. Dick, (D), Iowa State, 1986, Sb
Associate Director:
Donald J. Eckert, (D), Ohio State, 1978, Sc
Professor:
Jerry M. Bigham, (D), North Carolina State, 1977, Sd
Frank G. Calhoun, (D), Florida, 1971, Sd
Rattan Lal, (D), Ohio State, 1968, Sp
Associate Professor:
Nicholas T. Basta, (D), Iowa State, 1989, Sc
Edward L. McCoy, (D), Oregon State, 1984, Sp
Brian K. Slater, (D), Wisconsin, 1994, SdZi
Assistant Professor:
Dawn Ferris, (D), Minnesota, 1997, Sf
Emeritus:
Warren A. Dick, (D), Iowa State, 1980, SbCbHs

Ohio University

Geological Sciences (B,M) ●

316 Clippinger Lab
Athens, OH 45701
p. (740) 593-1101
geological.sciences@ohio.edu
http://www.ohio.edu/geology/
Programs: BS Geological Sciences, BS Environmental Water Resources, BA Geological Sciences, Minor Geological Science, Minor Paleontology, MS with Thesi, MS with Portfolio (online program)
Enrollment (2021): B: 71 (71) M: 56 (56)
Professor:
Daniel Hembree, (D), Kansas, 2005, Pi
Dina L. Lopez, (D), Louisiana State, 1992, CgHwGv
R. Damian Nance, (D), Cambridge, 1978, Gtc
Alycia L. Stigall, (D), Kansas, 2004, Pg
Chair:
Gregory S. Springer, (D), Colorado State, 2002, GmPcGa
Associate Professor:
Eung Seok Lee, (D), Indiana, 1999, Hw
Keith A. Milam, (D), Tennessee, 2006, Xg
Assistant Professor:
Katherine Fornash, (D), Minnesota, 2016, GpCga

Shawnee State University

Dept of Natural Sciences (B) ● ☒

940 Second Street
Portsmouth, OH 45662
p. (740) 351-3395
kshoemaker@shawnee.edu
https://www.shawnee.edu/areas-study/college-arts-sciences/natural-sciences/bachelor-degrees/geology
f: https://www.facebook.com/DeptNaturalSciencesShawneeState University/
t: @SSU_Science

Programs: Geology
Administrative Assistant: Heather Thacker
Enrollment (2021): B: 9 (6)
Professor:
Kurt A. Shoemaker, (D), Miami (Ohio), 2004, GxzGm
Associate Professor:
Erik B. Larson, (D), Mississippi State, 2014, GmsHg
Other:
Jeffrey A. Bauer, (D), Ohio State, 1987, Ps

University of Akron

Dept of Geosciences (B,M) O

325 Buchtel Ave
Akron, OH 44325-4101
p. (330) 972-7630
steer@uakron.edu
http://www.uakron.edu/geology/
t: @UAkron_Geosci
Programs: Geology (B,M); Earth Science (B); Environmental Science (B); Environmental Geology (M)
Certificates: Environmental Studies
Administrative Assistant: Lynne Suponcic
Enrollment (2021): B: 82 (10) M: 7 (3)
Professor of Instruction:
John F. Beltz, (M), Akron, 1992, GgRh
Meera Chatterjee, (D), Zy
Professor:
John A. Peck, (D), Rhode Island, 1995, GseGn
Ira D. Sasowsky, (D), Penn State, 1992, HwGmo
David N. Steer, (D), Cornell, 1996, Yes
Associate Professor:
John M. Senko, (D), Oklahoma, 2004, Cg
Assistant Professor of Instruction:
Thomas J. Quick, (M), Akron, 1983, GgCaYe
Assistant Professor:
Caleb Holyoke, (D), Brown, 2005, GcNrYx

University of Akron - Wayne College

University of Akron - Wayne College
1901 Smucker Road
Orrville, OH 44667
p. (330) 972-8934
WayneCommunityRelations@uakron.edu
http://www.wayne.uakron.edu/

University of Cincinnati

Dept of Geology (B,M,D) ☒
500 Geology/Physics Building
P. O. Box 210013
Cincinnati, OH 45221-0013
p. (513) 556-3732
krista.smilek@uc.edu
http://www.artsci.uc.edu/departments/geology.html
Programs: Geology
Enrollment (2018): B: 83 (10) M: 13 (8) D: 17 (2)
Head:
Lewis Owen, (D), Leicester (UK), 1988, GmlGt
Professor:
Thomas J. Algeo, (D), Michigan, 1989, Gs
Carlton E. Brett, (D), Michigan, 1978, Pe
Attila I. Kilinc, (D), Penn State, 1969, CpGvCu
Thomas V. Lowell, (D), SUNY (Buffalo), 1986, Gl
Associate Professor:
Christopher L. Atchison, (D), Ohio State, 2011, Ze
Brooke E. Crowley, (D), California (Santa Cruz), 2009, CsPev
Andrew D. Czaja, (D), California (Los Angeles), 2006, PmCl
Craig Dietsch, (D), Yale, 1985, Gp
Amy Townsend-Small, (D), Texas, 2006, Co
Dylan Ward, (D), Colorado, 2010, GmqZi
Assistant Professor:
Aaron Diefendorf, (D), Penn State, 2010, Co
Joshua H. Miller, (D), Chicago, 2009, PgvPe
Reza Soltanian, (D), Wright State, 2015, Hw
Daniel M. Sturmer, (D), Cal State (Fullerton), 2003, YeGdo
Yurena Yanes, (D), La Laguna (Spain), 2005, PeCsPg
Adjunct Professor:
Brenda R. Hunda, (D), California (Riverside), 2004, Pi
Glenn W. Storrs, (D), Yale, 1986, Pv
Emeritus:
Madeleine Briskin, (D), Brown, 1973, Pe
John E. Grover, (D), Yale, 1972, Gz
Warren D. Huff, (D), Cincinnati, 1963, GzvGr
J. Barry Maynard, (D), Harvard, 1972, Cl
David L. Meyer, (D), Yale, 1971, Pi
Arnold I. Miller, (D), Chicago, 1986, Pi
David Nash, (D), Michigan, 1977, Gm
Paul E. Potter, (D), Chicago, 1952, Gs

University of Dayton

Dept of Geology (B) ●
300 College Park
SC 179
Dayton, OH 45469-2364
p. (937) 229-3432
dgoldman1@udayton.edu
http://www.udayton.edu/artssciences/geology/
Administrative Assistant: Darla Titus
Enrollment (2016): B: 40 (7)
Chair:
Daniel Goldman, (D), SUNY (Buffalo), 1993, PgqGr
Professor:
Donald Pair, (D), Syracuse, 1991, Gl
Michael R. Sandy, (D), London (UK), 1984, Pi
Associate Professor:
Umesh Haritashya, (D), Indian Inst of Tech, 2005, ZrGm
Andrea M. Koziol, (D), Chicago, 1988, Cp
Allen J. McGrew, (D), Wyoming, 1992, GtcGp
Shuang-Ye Wu, (D), Cambridge, 2000, Zy
Visiting Assistant Professor:
Zelalem Bedaso, (D), S Florida, 2011, CslGs
Adjunct Professor:
Sue Klosterman, (M), Wright State, 2005, Gg
Andrew Rettig, (D), Cincinnati, 2014, Zy
Emeritus:
Charles J. Ritter, (D), Michigan, 1971, Ct

University of Toledo

Dept of Environmental Sciences (B,M,D) ☒
2801 W. Bancroft Street
Toledo, OH 43606-3390
p. (419) 530-2009
jonathan.bossenbroek@utoledo.edu
http://www.utoledo.edu/nsm/envsciences/
Administrative Assistant: Shirley Michel
Enrollment (2019): B: 33 (13) M: 4 (2)
Chair:
Timothy G. Fisher, (D), Calgary, 1993, GlmGn
Director, Lake Erie Center :
Thomas Bridgeman, (D), Michigan, 2001, Zn
Dean, Jesup Scott Honors College:
Heidi Appel, (D), Michigan, 1990, Zn
Professor:
Jonathon Bossenbroek, (D), Colorado State, 2004, Zn
Scott Heckathorn, (D), Illinois, 1995, Zn
Christine M. Mayer, (D), Illinois, 1998, Zn
Daryl L. Moorhead, (D), Tennessee, 1985, Zn
William V. Sigler, (D), Purdue, 1999, Zn
Alison L. Spongberg, (D), Texas A&M, 1994, Co
Michael N. Weintraub, California, 2004, Zn
Director, Stranahan Arboretum:
Daryl F. Dwyer, (D), Michigan State, 1986, Zn
Associate Professor:
Richard H. Becker, (D), W Michigan, 2008, ZrHzGe
Mark J. Camp, (D), Ohio State, 1974, PiRhEs
David E. Krantz, (D), South Carolina, 1988, Gs
James Martin-Hayden, (D), Connecticut, 1994, Hq
Song Qian, (D), Duke, 1995, Zn
Assistant Professor:
Kennedy Doro, (D), Tuebingen, 2015, Zn
William Hintz, (D), S Illinois, 2014, Zn
Jeanine Refsnider-Streby, (D), Iowa, 2012, Zn
Trisha Spanbauer, (D), Nebraska, 2015, Zn
Henry Streby, (D), Minnesota, 2010, Zn
Lecturer:
Todd Crail, (D), Toledo, 2012, Zn

Wittenberg University

Dept of Geology (B) ☒
P.O. Box 720
Springfield, OH 45501-0720
p. (937) 327-6475
mzaleha@wittenberg.edu
http://www.wittenberg.edu/academics/geology
f: https://www.facebook.com/WittenbergGeology
Programs: B.S. & B.A. in Geology, B.S. & B.A. in Environmental Science, B.A. in Earth Science Education
Enrollment (2018): B: 11 (4)

Professor:
John B. Ritter, (D), Penn State, 1990, Gm

Associate Professor:
Sarah K. Fortner, (D), Ohio State, 2008, ClZeHg
Michael J. Zaleha, (D), SUNY (Binghamton), 1994, Gs

Emeritus:
Katherine L. Bladh, (D), Arizona, 1976, Gi
Kenneth W. Bladh, (D), Arizona, 1978, GzEm
Robert W. Morris, (D), Columbia, 1969, PiGg

Wright State University

Dept of Earth and Environmental Science (B,M)
3640 Colonel Glenn Highway
260 Brehm Lab
Dayton, OH 45435
p. (937) 775-2201
david.schmidt@wright.edu
http://www.wright.edu/ees/
Enrollment (2010): B: 66 (8) M: 48 (15)

Chair:
David F. Dominic, (D), West Virginia, 1988, GsrRc

Director of Undergraduate Programs:
David Schmidt, (D), Ohio State, PiGd

Professor:
Christopher C. Barton, (D), Yale, 1983, GqYgHq
C. B. Gregor, (D), Utrecht, 1967, Gs
Chad Hammerschmidt, (D), Connecticut, 2005, CmOcCb
Allen G. Hunt, (D), California (Riverside), 1983, GqSpCl
Robert W. Ritzi, Jr., (D), Arizona, 1989, Hw

Associate Professor:
Abinash Agrawal, (D), North Carolina, 1990, ClHwPg
Songlin Cheng, (D), Arizona, 1984, Hw
Ernest C. Hauser, (D), Wisconsin, 1982, YeGtYu
William Slattery, (D), CUNY, 1994, ZeGr
Rebecca Teed, (D), Minnesota, 1999, ZePeGn
Doyle Watts, (D), Michigan, 1979, YeZr

Director of Sustainability:
Huntting (Hunt) Brown, (D), Ge

Emeritus:
Byron Kulander, (D), West Virginia, 1966, Gc
Benjamin H. Richard, (D), Indiana, 1966, Ye
Paul J. Wolfe, (D), Case Western, 1966, Ye

Youngstown State University

Dept of Geological & Environmental Sciences (B,M)
One University Plaza
2120 Moser Hall
Youngstown, OH 44555
p. (330) 941-3612
amjacobs@ysu.edu
http://www.as.ysu.edu/~geology/
Enrollment (2007): B: 48 (4) M: 21 (1)

Chair:
Jeffrey C. C., (D), Kent State, 1992, EoNgHy

Director, Env Studies Program:
Isam E. Amin, (D), Nevada (Reno), 1987, Hw

Professor:
Raymond E. Beiersdorfer, (D), California (Davis), 1992, Cg
Alan M. Jacobs, (D), Indiana, 1967, Ge

Assistant Professor:
Joseph E. Andrew, (D), Kansas, 2002, Gc
Felicia P. Armstrong, (D), Oklahoma State, 2003, Sb
Shane V. Smith, (D), Washington State, 2005, Gs

Instructor:
Harry Bircher, (M), Youngstown State, 1995, Ge
Brian M. Greene, (D), Kent State, 2001, Ng
Lawrence P. Gurlea, (M), Pennsylvania, 1971, Cg

Engineering Adjunct:
Scott C. Martin, (D), Clarkson, 1984, Ge
Douglas M. Price, (D), Notre Dame, 1988, Cg

Emeritus:
Ann G. Harris, (M), Miami (Ohio), 1958, Ge
Ikram U. Khawaja, (D), Indiana, 1969, Ec
Charles R. Singler, (D), Nebraska, 1969, Gs

Cooperating Faculty:
Thomas P. Diggins, (D), SUNY, 1997, Pg

Cooperating Faculty:
Carl G. Johnston, (D), Cincinnati, 1992, Pg

Zane State College

Oil and Gas Engineering Technology (A)
1555 Newark Rd
Zanesville, OH 43701
p. (740) 588-1282
admissions@zanestate.edu
http://www.zanestate.edu
f: https://www.facebook.com/ZaneStateCollege
t: @ZaneStateC

Oklahoma

Oklahoma City Community College

Physical Sciences
7777 South May Avenue
Oklahoma City, OK 73159
p. (405) 682-1611
physicalscience@occc.edu
http://www.occc.edu

Oklahoma State University

Boone Pickens School of Geology (B,M,D) ○⊘
105 Noble Research Center
Stillwater, OK 74078-3031
p. (405) 744-6358
camelia.knapp@okstate.edu
http://geology.okstate.edu
Programs: BS in Geology
BS in Geophysics
MS in Geology
PhD in Geology
Administrative Assistant: Sandy Earls
Enrollment (2021): B: 46 (12) M: 38 (14) D: 32 (3)

Head:
Camelia Knapp, (D), Cornell, 2000, GtYgs

Professor:
Mohamed G. Abdelsalam, (D), Texas (Dallas), 1993, GctYg
Michael Grammer, (D), Miami, 1991, GsEoGr
Todd Halihan, (D), Texas, 2000, HwYuRc
James Knapp, (D), MIT, 1989, GcYgs
Jack C. Pashin, (D), Kentucky, 1990, GrcGo

Associate Professor:
Priyank Jaiswal, (D), Yg
Daniel A. Lao, (D), Pittsburgh, 2008, GctGg
James Puckette, (D), Oklahoma State, 1996, Go
Tracy Quan, (D), MIT/WHOI, 2005, Og

Assistant Professor:
Ashley Burkett, (D), Indiana State, 2015, Pme
Ahmed Ismail, (D), Missouri S&T, 2006, Ysu
Natascha Riedinger, (D), Bremen, 2005, CgGu
Javier Vilcaez, (D), Tohoku, 2009, XzNpCl
Tingying Xu, (D), Washington, 2018, Cbq

Visiting Professor:
Mary E. Hileman, (D), Michigan, 1973, GorGd

Emeritus:
Jay M. Gregg, (D), Michigan State, 1982, Gs
Arthur Hounslow, (D), Carleton, 1968, Cl
Douglas Kent, (D), Iowa State, 1969, Hw
Vernon Scott, (D), Utah, 1975, Ze
Gary Stewart, (D), Kansas, 1973, Go
John D. Vitek, (D), Iowa, 1973, Gm

Teaching Assistant Professor:
Brandon Spencer, (M)

Rogers State University

Dept of Mathematics and Physical Sciences
1701 W. Will Rogers Blvd.
Claremore, OK 74107
p. (918) 343-6812
sparker@rsu.edu
http://www.rsu.edu

Tulsa Community College

Science and Math
909 S. Boston Avenue
Tulsa, OK 74119
p. (918) 595-7085
Kelly.E.Allen@tulsacc.edu
http://www.tulsacc.edu
Certificates: GIS
Enrollment (2018): A: 19 (5)
Professor:
Kelly E. Allen, (D), Texas Tech, 2000, ZyiZe
Adjunct Professor:
Douglas Ashe, (M), Oklahoma State, 2018, GgZe
Joshua Florie, (M), Tulsa, 2018, OgGg

University of Oklahoma

Mewbourne School of Petroleum & Geological Engineering (B,M,D)
100 East Boyd Street
Sarkeys Energy Center 1210
Norman, OK 73019-1001
p. (405) 325-2921
mpge@ou.edu
http://mpge.ou.edu/
f: https://www.facebook.com/OUMPGE
t: @ou_mpge
Enrollment (2015): B: 975 (133) M: 101 (25) D: 37 (4)
Director:
Chandra S. Rai, (D), Hawaii, 1977, Np
Graduate Liaison:
Deepak Devegowda, (D), Texas A&M, 2008, Np
Director Natural Gas Engineering & Management:
Suresh Sharma, (D), Oklahoma, 1968, Np
Professor:
Younane Abousleiman, (D), Delaware, 1991, Nr
Ramadan Ahmed, (D), Norwegian Inst of Tech, 2001, Np
Jeff Callard, (D), Louisiana State, 1994, Np
Ahmed Ghassemi, (D), Oklahoma, 1996, Nr
Ben Shiau, (D), Oklahoma, 1995
Carl H. Sondergeld, (D), Cornell, 1977, NpYex
Musharraf Zaman, (D), Bangladesh Eng and Tech, 1975, Np
Associate Professor:
Mashhad Fahes, (D), Imperial Coll, 2006, Np
Ahmad Jamili, (D), Kansas, 2004, Np
Rouzbeh Moghanloo, (D), Texas, 2012, Np
Maysam Pournik, (D), Texas A&M, 2008, Np
Catalin Teodoriu, (D), Ploiesti, Np
Xingru Wu, (D), Texas, 2006, Np
Assistant Professor:
Siddharth Misra, (D), Texas, 2015, Np
Ahmad Sakhaee-Pour, (D), Texas, 2012, Np
Instructor:
Ilham El-Monier, (D), Texas A&M, 2012, Np
Emeritus:
Faruk Civan, (D), Oklahoma, 1978, Np
Roy M. Knapp, (D), Kansas, 1973, Np
Jean-Claude Roegiers, (D), Minnesota, 1974, Nr
Subhash N. Shah, (D), New Mexico, 1974, Np

School of Geosciences (B,M,D) O☒
100 East Boyd
710 Energy Center
Norman, OK 73019-0628
p. (405) 325-3253
geosciences@ou.edu
https://geosciences.ou.edu
Programs: Geology and Geophysics
Enrollment (2020): B: 90 (10) M: 38 (16) D: 34 (10)
Director:
Gerilyn S. Soreghan, (D), Arizona, 1992, GsPcZg
Professor:
Younane N. Abousleiman, (D), Delaware, 1991, Gg
R. Douglas Elmore, (D), Michigan, 1981, GdYm
Andrew S. Elwood Madden, (D), Virginia Tech, 2005, ClZaCb
Megan E. Elwood Madden, (D), Virginia Tech, 2005, ClXgCg
Michael H. Engel, (D), Arizona, 1980, Co
Kurt J. Marfurt, (D), Columbia, 1978, Ys
Matthew J. Pranter, (D), Colorado Mines, Go
Stephen R. Westrop, (D), Toronto, 1984, Pi
Associate Professor:
Xiaowei Chen, (D), California (San Diego), 2013, YsGt
Richard Lupia, (D), Chicago, 1997, Pm
John D. Pigott, (D), Northwestern, 1981, GoYeGd
Michael J. Soreghan, (D), Arizona, 1994, Gs
Barry L. Weaver, (D), Birmingham (UK), 1980, Ct
Assistant Professor:
Heather Bedle, (D), Northwestern, 2008, Yg
Brett Carpenter, (D), Penn State, 2012, Gc
Kato Dee, (D), Colorado Mines, HgCg
Shannon Dulin, (D), Oklahoma, 2014, YmGs
Junle Jiang, (D), Caltech, 2016, Yg
Xiaolei Liu, (D), Bremen (Germany), 2011, Cg
Emeritus:
Judson L. Ahern, (D), Cornell, 1980, Yg
M. Charles Gilbert, (D), California (Los Angeles), 1965, Cp
Charles W. Harper, Jr., (D), Caltech, 1964, Pi
G. Randy Keller, (D), Texas Tech, 1973, YgGtEo
David London, (D), Arizona State, 1981, CpGiz
Shankar Mitra, (D), Johns Hopkins, 1977, Gc
R. Paul Philp, (D), Sydney, 1972, Co
David W. Stearns, (D), Texas A&M, 1969, Gc

University of Tulsa

Dept of Geosciences (B,M) ☒
800 S. Tucker Drive
Tulsa, OK 74104-9700
p. (918) 631-3014
jingyi-chen@utulsa.edu
https://engineering.utulsa.edu/geosciences/
Programs: Geosciences with Geology, Environmental, Geophysics and Petroleum Engineering Options (B); Geosciences (M)
Enrollment (2021): B: 12 (6) M: 7 (5)
Decker Dawson Professor of Geophysics:
Jingyi Chen, (D), Chinese Acad of Sci, 2005, Ygs
Associate Professor:
Junran Li, (D), Virginia, 2008, Gme
Assistant Professor:
Christine J. Ruhl, (D), Nevada (Reno), 2016, YsGtc
Professor:
Peter J. Michael, (D), Columbia, 1983, GivGz
Associate Professor:
Dennis R. Kerr, (D), Wisconsin, 1989, GsrGt
Steven L. Roche, (D), Colorado Mines, 1997, YesYg
Applied Associate Professor:
Winton Cornell, (D), Rhode Island, 1987, GivCa
Research Associate:
Bethany P. Theiling, (D), New Mexico, 2012, Cs
Adjunct Professor:
Aaron Ball
Amy Richardson
Emeritus:
J. Bryan Tapp, (D), Oklahoma, 1983, GceZi

Oregon

Central Oregon Community College

Dept of Science (A)
2600 NW College Way
Bend, OR 97701
p. (541) 383-7557
chigginbotham@cocc.edu
http://science.cocc.edu/Programs_Classes/Geology/
Associate Professor:
Robert W. Reynolds, (D), Idaho, 1994, Gv

Oregon State University

College of Earth, Ocean, and Atmospheric Sciences (B,M,D)

●⊘
104 CEOAS Administration Building
Corvallis, OR 97331-5503
p. (541) 737-1201
contact@coas.oregonstate.edu
http://ceoas.oregonstate.edu
Programs: Earth Science (B); Environmental Sciences (B); Geography and Geospatial Science (B); Geography (M,D); Geology (M,D); Marine Resource Management (M,D); Ocean, Earth, and Atmospheric Sciences (M,D)
Certificates: GIS (UG); GIS (G), Marine Resource Management (G), Water Conflict Management and Transformation (G)
Enrollment (2018): B: 706 (98) M: 64 (21) D: 67 (14)

Senior Research:
Mitchell Lyle, (D), Oregon State, 1978, GgYg
David Mellinger, Cb

R.S. Yeats Professor of Earthquake Geology and Active Tectonics and Associate Dean for Academic Programs:
Eric Kirby, (D), MIT, 2001, GtmGc

Professor, Associate Dean for Strategic Initiatives, and director of the Oregon Climate Change Research Institute (OCCRI):
Philip Mote, (D), Washington, 1994, AsZg

Exec Dir-Marine Studies Initiative:
Jack A. Barth, (D), MIT, 1987, On

Director-Budget & Fiscal Planning:
Sherman H. Bloomer, (D), California (San Diego), 1982, Gi

Director of Ocean Science Program:
Rob Wheatcroft, (D), Washington, 1990, OgGg

Director of Marine Resource Management Program:
Flaxen D. Conway, (M), Oregon State, 1986, Gu

Director of Geology Program:
Adam J. R Kent, (D), Australian Nat, 1995, GiCga

Director of Geography Program:
Julia A. Jones, (D), Johns Hopkins, 1983, So

Director of Environmental Science Program:
Laurence Becker, (D), California (Berkeley), 1989, Zn

Professor:
Jeffrey R. Barnes, (D), Washington, 1983, As
Kelly Benoit-Bird, (D), Hawaii, 2003, Gu
Edward J. Brook, (D), MIT/WHOI, 1993, ClPeGl
Michael E. Campana, (D), Arizona, 1975, Hg
Lorenzo Ciannelli, (D), Washington, 2002, Og
Frederick (Rick) Colwell, (D), Virginia Tech, 1986, Zn
Byron Crump, (D), Washington, 1999
Shanika de Silva, (D), Open (UK), 1987, Gv
John H. Dilles, (D), Stanford, 1984, Em
Gary D. Egbert, (D), Washington, 1987, Yr
Chris Goldfinger, (D), Oregon State, 1994, Yr
Miguel A. Goni, (D), Washington, 1992, Gu
David W. Graham, (D), MIT, 1987, Cm
Burke R. Hales, Washington, 1995, Oc
Merrick C. Haller, (D), Delaware, 1999, OnZrOp
Robert N. Harris, (D), Utah, 1996, Yhr
Michael Harte, (D), Victoria, 1994, Gu
Anthony Koppers, (D), Free (Amsterdam), 1988, CgGv
Michael Kosro, (D), Scripps, 1985, OpZrOn
Jim Lerczak, (D), Scripps, 2000, Op
Ricardo Letelier, (D), Hawaii, 1994, ObZr
Ricardo P. Matano, (D), Princeton, 1991, OpAs
Andrew J. Meigs, (D), S California, 1995, Gc
James N. Moum, (D), British Columbia, 1984, Op
John L. Nabelek, (D), MIT/WHOI, 1984, GtYs
Jonathan Nash, (D), Oregon State, 2000, Op
Roger L. Nielsen, (D), S Methodist, 1983, GizGv
David Noone, (D), Melbourne (Australia), 2001, AmZg
Clare Reimers, (D), Oregon State, 1982, Oc
Peter Ruggiero, (D), Oregon State, 1997, On
Roger Samelson, (D), Oregon State, 1987, Op
Andreas Schmittner, (D), Bern, 1999, AsYr
Adam Schultz, (D), Washington, 1986, YgmRn
Eric Skyllingstad, (D), Wisconsin, 1986, Am
William D. Smyth, (D), Toronto, 1990, Op
Yvette H. Spitz, (D), Old Dominion, 1995, Yr
Frank J. Tepley, III, (D), California (Los Angeles), 1999, GiCs
Marta E. Torres, (D), Oregon State, 1988, Cm
Anne M. Trehu, (D), MIT/WHOI, 1982, YrsGt
Aaron T. Wolf, (D), Wisconsin, 1992, Hg

Senior Research:
Theodore Durland, (D), Hawaii, 2006, Yg
Brian Haley, (D), Oregon State, 2004, Cs
Haruyoshi Matsumoto
Nicholas Tufillaro, (D), Bryn Mawr, 1990, Cb

Director, Water Resources Graduate Program:
Mary V. Santelmann, (D), Minnesota, 1988, Zy

Director of Atmospheric Science Program:
Karen M. Shell, (D), Scripps, 2004, As

Associate Professor:
Anders Carlson, (D), Oregon State, 2006, GlCt
Patrick Corcoran, 1989
Simon P. de Szoeke, (D), Washington, 2004, OpAs
Hannah Gosnell, (D), Colorado, 2000, Zu
Alexander Kurapov, (D), St Petersburg, 1994, Yr
Stephen Lancaster, (D), MIT, 1999, Hs
Kipp Shearman, (D), Oregon State, 1999, Opn
Emily L. Shroyer, (D), Oregon State, 2009, On
Joseph Stoner, (D), Québec (Montréal), 1995, Gsr
George Waldbusser, (D), Maryland (Baltimore), 2008, Ob

Senior Research:
Christo Buizert, (D), Copenhagen, 2012, Yr
Louise A. Copeman, (D), Memorial, 2011, Zn
Jonathan Fram, (D), California (Berkeley), 2005, Op
Joseph Haxel
W. Todd Jarvis, (D), Oregon State, 2006, Hw
Maria Kavanaugh, Cb
Jennifer L. McKay, (D), British Columbia, 2004, Cm
David Rupp, (D), Oregon State, 2005, AsOp
Jenna Tilt, (D), Washington, 2007, Zu

Assistant Professor:
Kim S. Bernard, (D), Rhodes, 2007
Jessica Creveling, (D), Harvard, 2012, GgsGr
Jennifer Fehrenbacher, (D), N Illinois, 1997, Cm
Jennifer Hutchings, Gl
Lauren W. Juranek, Washington, 2007, Cm
Robert Kennedy, (D), Oregon State, 2004, Zri
Larry O'Neill, (D), Oregon State, 2007, AmOg
Alyssa E. Shiel, (D), British Columbia, 2010, ClsCt
Nick Siler
Andrew Thurber, (D), California (San Diego), 2010, Cb
Jamon Van Den Hoek, (D), Wisconsin, 2012, Zur
James R. Watson, Zy
Justin Wettstein, (D), Washington, 2007, As
Greg Wilson, (D), Oregon State, 2013, Onp
Bo Zhao, (D), Ohio State, 2015, Zi

Senior Instructor:
Lorene Yokoyama Becker, (M), Wisconsin, 1999, Zi

Instructor:
Andrea Allan, (D), Oregon State, 2012, As
Lynette de Silva, (M), Indiana, 2000
Randall A. Keller, (D), Oregon State, 1996, Gu
Randall Milstein, (D), Oregon State, 1994
Rebecca Yalcin, (M), Maine, 2001, Gg

Emeritus:
A. Jon Kimerling, (D), Wisconsin, 1976, Zir
Gary P. Klinkhammer, (D), Rhode Island, 1979, CaHsRw
Robert Lloyd Smith, (D), Oregon State, 1964, OpgZn
Robert S. Yeats, (D), Washington, 1958, GtYsNg

Dean of College of Earth, Ocean, and Atmospheric Sciences and Professor:
Roberta Marinelli, (D), South Carolina, 1991, GuOcGe

Portland Community College - Sylvania Campus

Physical Science Dept (A) O
12000 SW 49th Ave.
Portland, OR 97219
p. 971-722-8209
mnarayan@pcc.edu
http://www.pcc.edu/programs/geology/

Instructor:
Talal Abdulkareem, (D)
Kali Abel, (M)
Sharon Delcambre, (D), AsOg
Gretchen Gebhardt, (M)
Melinda Hutson, (D), Xm
Hollie Oakes-Miller, (M), Gg
Kristy Schepker, (M)
Steve Todd, Am
Jonathan Weatherford, (M), Gg

Portland State University

Dept of Geology (B,M,D)
P.O. Box 751
Portland, OR 97207
p. (503) 725-3022
geology@pdx.edu
http://www.pdx.edu/geology
t: @PSUGeology
Programs: BA,BS,MS in Geology and Earth Science, PhD in Earth and Environmental Science
Certificates: Environmental Geology, Engineering Geology, Hydrogeology
Enrollment (2022): B: 70 (0) M: 23 (0) D: 2 (0)

Associate Professor:
John T. Bershaw, (D), Rochester, 2011, GsCsHy
Professor:
Alexander (Alex) M. Ruzicka, (D), Arizona, 1996, Xm
Martin J. Streck, (D), Oregon State, 1994, GivCt
Associate Professor:
Adam M. Booth, (D), Oregon, 2012, GmZr
Robert Benjamin Perkins, (D), Portland State, 2000, Hq
Ashley Streig, (D), Oregon, 2014, YsGt
Research Associate:
Richard Hugo, (D), Washington State, 1998, Gy
David Percy, (B), Portland State, 1999, Zi
Adjunct Professor:
Sheila Alfsen, (D), GgPgOu
Matthew Brunengo, (D), Portland State, 2012, GgNgGm
Megan Faust, (D), ZeGr
Frank D. Granshaw, (D), Portland State, 2011, ZnGl
Melinda Hutson, (D), Arizona, 1996, Xmc
William Orr, (D), Michigan State, 1968, GgOg
Erik Shafer, Portland State, 2015, Gg
Barry Walker, (D), Oregon State, 2011, GiCgGv
Emeritus:
Scott F. Burns, (D), Colorado, 1980, Ng
Kenneth M. Cruikshank, (D), Purdue, 1991, Gc
Michael L. Cummings, (D), Wisconsin, 1978, GgHg
Andrew G. Fountain, (D), Washington, 1992, Gl
Paul E. Hammond, (D), Washington, 1963, Gi
Ansel G. Johnson, (D), Stanford, 1973, Ye
Richard E. Thoms, (D), California, 1965, Ps
Cooperating Faculty:
Christina L. Hulbe, (D), Chicago, 1998, Gl

Rogue Community College

Science Dept-Physical Sciences-Geology
3345 Redwood Highway
Grants Pass, OR 97527
p. (541) 245-7527
jvanbrunt@roguecc.edu
http://learn.roguecc.edu/science/physical.htm

Southern Oregon University

Dept of Geology (B)
1250 Siskiyou Boulevard
Ashland, OR 97520
p. (541) 552-6479
lane@sou.edu
Administrative Assistant: Susan Koralek
Enrollment (2006): B: 45 (5)

Chair:
Charles L. Lane, (D), California (Los Angeles), 1987, Hg
Dean, Science:
Joseph L. Graf, Jr., (D), Yale, 1975, Em
Professor:
Jad A. D'Allura, (D), California (Davis), 1977, Gc
Associate Professor:
Eric Dittmer, (M), San Jose State, 1972, Gg
Adjunct Professor:
Vernon J. Crawford, (M), Oregon, 1970, Gg
Harry W. Smedes, (D), Washington, 1959, Gi
Richard Ugland, (M), Utah, 1974, Gg
Emeritus:
Monty A. Elliott, (D), Oregon State, 1971, Gr
William B. Purdom, (D), Arizona, 1960, Gx

Southwestern Oregon Community College

Dept of Geology (A)
1988 Newmark Ave
Coos Bay, OR 97420-2912
p. (541) 888-7216
rmetzger@socc.edu
https://www.socc.edu/geology
Enrollment (2013): A: 2 (0)

Professor:
Ronald A. Metzger, (D), Iowa, 1991, PmZePs

Tillamook Bay Community College

Associate of Science
2510 First Street
Tillamook, OR 97141
robertpietruszka@tillamookbaycc.edu

Treasure Valley Community College

Career and Technical Education ● ☒
650 College Blvd.
Ontario, OR 97914
p. (541) 881-8866
dtinkler@tvcc.cc
http://www.tvcc.cc.or.us/science
f: https://www.facebook.com/profile.php?id=100005312898088

Instructor:
Dorothy Tinkler, (D), Texas Tech, 2003, Zi

University of Oregon

Dept of Earth Sciences (B,M,D)
1272 University of Oregon
Eugene, OR 97403-1272
p. (541) 346-4573
shopkins@uoregon.edu
https://earthsciences.uoregon.edu
Programs: Geology; Earth Science; Geophysics; Paleontology; Environmental Geoscience;
Enrollment (2021): B: 100 (0) M: 1 (0) D: 54 (0)

Lillis Professor of Volcanology:
Josef Dufek, (D), Washington, 2006, Gv
Department Head:
Samantha Hopkins, (D), California (Berkeley), 2005, Pg
Professor:
Ilya N. Bindeman, (D), Chicago, 1998, CsGvCl
Rebecca J. Dorsey, (D), Princeton, 1989, Gr
Eugene D. Humphreys, (D), Caltech, 1985, YsGt
Qusheng Jin, (D), Illinois (Urbana), 2003, Cg
Mark H. Reed, (D), California (Berkeley), 1977, CgEm
Alan W. Rempel, (D), Cambridge, 2001, YgNrGl
Gregory J. Retallack, (D), New England, 1978, PbSoGa
Joshua J. Roering, (D), California (Berkeley), 2000, Gm
Douglas R. Toomey, (D), MIT/WHOI, 1987, YsGt
Paul Wallace, (D), California (Berkeley), 1991, GviCg
Ray J. Weldon, (D), Caltech, 1986, Gc
Associate Professor:
Edward B. Davis, (D), California (Berkeley), 2005, PvePq
Thomas Giachetti, (D), Clermont-Ferrand, 2010, Gvu
Emilie E. Hooft, (D), MIT, 1996, Yr
Leif Karlstrom, (D), California (Berkeley), 2011, Yx
Diego Melgar, (D), Scripps, 2014, Yd
Carol Paty, (D), Washington, 2006
Matthew Polizzotto, (D), Stanford, 2007, CqSc
Dave Sutherland, (D), MIT, 2008, Op
Amanda Thomas, (D), California (Berkeley), 2012, Ys
James Watkins, (D), California (Berkeley), 2010, Cg
Assistant Professor:
Brittany Erickson, (D), California, 2010, Ygs
Valerie J. Sahakian, (D), Scripps, 2015, Yg
Meredith Townsend, (D), Stanford, 2017, Gv
Instructor:
Marli G. Miller, (D), Washington, 1997, Gc
Emeritus:
A. Dana Johnston, (D), Minnesota, 1983, Cp
Marvin A. Kays, (D), Washington (St Louis), 1960, GpcGt
William N. Orr, (D), Michigan State, 1967, Pm
Norman M. Savage, (D), Sydney, 1968, Pi

Office & Business Manager:
Sandy K. Thoms, (B), Portland State, 1993

Dept of Geography (B,M,D) ⊘
107 Condon Hall
1251 University of Oregon
Eugene, OR 97403-1251
p. (541) 346-4555
uogeog@uoregon.edu
http://geography.uoregon.edu
f: https://www.facebook.com/universityoforegongeography/
t: @uogeog
Enrollment (2009): B: 167 (55) M: 22 (9) D: 19 (5)
Department Head:
Xiaobo Su, (D), Singapore, 2007, Zn
Professor:
Patrick Bartlein, (D), Wisconsin, 1978, Pe
Dan Gavin, (D), Washington, 2000, ZyPle
Amy Lobben, (D), Michigan State, 1999, ZinZn
Patricia McDowell, (D), Wisconsin, 1980, Gm
Alexander Murphy, (D), Chicago, 1987, Zn
Laura Pulido, (D), California (Los Angeles), 1991
Peter Walker, (M), Oregon, 1966, Zn
Associate Professor:
Daniel Buck, (D), California (Berkeley), 2002, Zn
Shaul Cohen, (D), Chicago, 1991, Zn
Mark Fonstad, (D), Arizona State, 2000, Zy
Lucas Silva, (D), Guelph, 2011
Assistant Research Professor:
Melissa Lucash, (D), SUNY, 2005
Assistant Professor:
Sarah Cooley, (D), Brown, 2020
Carolyn Fish, (D), Penn State, 2018
Leigh Johnson, (D), California (Berkeley), 2011
Henry Hui Luan, (D), Waterloo, 2016
Johnny Ryan, (D), Aberystwyth, 2018
Sr. Instructor & Undergrad Coordinator:
Leslie McLees, (D), Oregon, 2012
Sr. Instructor:
Nicholas Kohler, (D), Oregon, 2004, Zi
Instructor:
Donald Holtgrieve, (D), Oregon, 1973
Emeritus:
Stanton Cook, (D), California (Berkeley), 1960, Zn
W. Andrew Marcus, (D), Colorado, 1987
Alvin Urquhart, California (Berkeley), 1962, Zn
Ronald Wixman, (D), Columbia, 1978, Zn

Western Oregon University

Earth and Environmental Science Department (B) ⊘
345 N. Monmouth Ave.
Monmouth, OR 97361
p. (503) 838-8398
taylors@mail.wou.edu
http://www.wou.edu/earthscience
Programs: B.S. / B.A. Earth and Environmental Science
Certificates: Minor / Certificate in Geographic Information Science
Enrollment (2022): B: 40 (5)
Professor:
Jeffrey A. Myers, (D), California (Santa Barbara), 1998, Gs
Stephen B. Taylor, (D), West Virginia, 1999, Gm
Jeffrey H. Templeton, (D), Oregon State, 1998, GivZe
Assistant Professor:
David Szpakowski, (D), Texas State, 2019, ZirRn
Instructor:
Jeremiah Oxford, (M), Oregon State, 2006, Zg
Grant Smith, (D), Oregon State, 2012, Zg

Willamette University

Department of Environmental Science (B) ☒
900 State Street
Salem, OR 97301
p. (503) 370-6587
spike@willamette.edu
http://willamette.edu/cla/envs
f: https://www.facebook.com/groups/186102921735448/
Programs: Environmental Science
Enrollment (2018): B: 99 (0)
Chair:
Scott Pike, (D), Georgia, 2000, Gag
Professor:
Karen Arabas, (D), Penn State, 1997, Zn
Endowed Dempsey Chair:
Joe Bowersox, (D), Wisconsin, 1995, Zn
Assistant Professor:
Melinda Butterworth, (D), Arizona, 2015
Katja Meyer, (D), Penn State, 2008

Pennsylvania

Bloomsburg University

Dept of Environmental, Geographical, and Geological Sciences (B) ☒
400 East Second Street
Bloomsburg, PA 17815-1301
p. (570) 389-4108
mshepard@bloomu.edu
https://www.bloomu.edu/offices-directory/environmental-geographical-and-geological-sciences
Programs: Professional Geology, Environmental Geoscience, Geography and Planning. Minors in Soil Science, Hydrology, GIS & Spatial Analysis, Geoscience, Geography.
Administrative Assistant: Cheryl Smith
Enrollment (2022): B: 107 (0)
Professor:
John E. Bodenman, (D), Penn State, 1995, Zu
Brett T. McLaurin, (D), Wyoming, 2000, GrsGd
Michael K. Shepard, (D), Washington (St Louis), 1994, XgYgZr
Associate Professor:
Patricia J. Beyer, (D), Arizona State, 1997, HsZyGm
Jeffrey C. Brunskill, (D), SUNY (Buffalo), 2005, ZiAmZy
Jennifer B. Whisner, (D), Tennessee, 2010, GmHw
S. Christopher Whisner, (D), Tennessee, 2005, GciGz
Assistant Professor:
Rebecca Bourgault, (D), Vermont, 2014, SdfSb
Tina Delahunty, (D), Florida, 2002, Zri
Benjamin Franek, (D), Connecticut, 2013, HsZy
Alan D. Gishlick, (D), Yale, 2002, GgPgOg
John G. Hintz, (D), Kentucky, 2005, Zin

Bryn Mawr College

Dept of Geology (B) ☒
101 North Merion Avenue
Bryn Mawr, PA 19010-2899
p. (610) 526-5115
aweil@brynmawr.edu
http://www.brynmawr.edu/geology/
Programs: Geology
Enrollment (2021): B: 32 (7)
Professor:
Arlo B. Weil, (D), Michigan, 2001, GctYm
Associate Professor:
Donald C. Barber, (D), Colorado, 2001, GesOn
Selby Cull-Hearth, (D), Washington, 2011, XgGzx
Pedro J. Marenco, (D), S California, 2007, PgCsPg
Instructor:
Katherine N. Marenco, (D), S California, 2008, PiePg
Emeritus:
Maria Luisa B. Crawford, (D), California (Berkeley), 1965, GxzGp

Bucknell University

Geology and Environmental Geosciencs (B) ☒
231 O'Leary Center
Lewisburg, PA 17837
p. (570) 577-1382
mbgray@bucknell.edu
http://www.bucknell.edu/Geology
f: http://www.facebook.com/BucknellGeology
Administrative Assistant: Carilee Dill
Enrollment (2021): B: 32 (9)
(Chair):
Mary Beth Gray, (D), Rochester, 1991, Gc
Professor:
Christopher G. Daniel, (D), Rensselaer, 1998, Gp
Robert W. Jacob, (D), Brown, 2006, Yg

Jeffrey M. Trop, (D), Purdue, 2000, Gs
Associate Professor:
Ellen K. Herman, (D), Penn State, 2006, Hw
Assistant Professor:
Lorelei Curtin, (D), Columbia, 2021, Cl
Emeritus:
Carl S. Kirby, (D), Virginia Tech, 1993, Cl
R. Craig Kochel, (D), Texas, 1980, Gm
Laboratory Director:
Gavin Davidson, (M), SD Mines, 2021, Gg

California University of Pennsylvania

Dept of Earth Sciences (B)
250 University Avenue
California, PA 15419
p. (724) 938-4180
wickham@calu.edu
http://www.cup.edu/eberly/earthscience
Administrative Assistant: Pamela Higinbotham
Enrollment (2015): B: 195 (25)
Chair:
Thomas Wickham, (D), Penn State, 2000, Zu
Professor:
Kyle Fredrick, (D), SUNY (Buffalo), 2008, GgHgGm
Associate Professor:
Thomas Mueller, (D), Illinois, 1999, Zi
Assistant Professor:
John Confer, (D), Penn State, 1997, Zn
Swarndeep S. Gill, (D), Wyoming, 2002, As
Chad Kauffman, (D), Nebraska, 2000, As
Susan Ryan, (D), Calgary, 2005, Zg

Carnegie Museum of Natural History

Section of Vertebrate Paleontology
4400 Forbes Avenue
Pittsburgh, PA 15213
p. (412) 622-5782
beardc@carnegiemnh.org
http://www.carnegiemnh.org/vp/
Curator:
K. Christopher Beard, (D), Johns Hopkins, 1989, Pv
Assistant Curator:
Matthew C. Lamanna, (D), Pennsylvania, 2004, PvgPe
Curator:
David S. Berman, (D), California (Los Angeles), 1969, PggPv
Mary R. Dawson, (D), Kansas, 1957, Pv
Preparator:
Dan Pickering, (B), Carnegie Mellon, 1983, Zn
Alan R. Tabrum, (M), SD Mines, 1981, Pv
Norman Wuerthele, (B), Pittsburgh, 1966, Pv
Curator & Associate Director:
Zhexi Luo, (D), California (Berkeley), 1987, Pv
Collections Manager:
Amy C. Henrici, (M), Pittsburgh, 1990, Pv

Clarion University

Dept of Biology and Geosciences (B) O☝
840 Wood Street
Clarion, PA 16214
p. (814) 393-2317
cking@clarion.edu
http://www.clarion.edu/BIGS
Programs: B.S. Geology, B.S. Environmental Geoscience, Minors in Geology and Environmental Geoscience
Certificates: GIS
Enrollment (2019): B: 60 (17)
Professor:
Yasser M. Ayad, (D), Montreal, 2000, Zi
Valentine U. James, (D), Texas A&M, ZuRwZn
Anthony J. Vega, (D), Louisiana State, 1994, AtmOp
Craig E. Zamzow, (D), Texas, 1983, Gi
Adjunct Professor:
Sheila Kasar, (D), ZreGe
Shane Smith, (D), Washington State, 2005

Delaware County Community College

STEM (A)
901 Media Line Road
Media, PA 19063
p. (610) 359-5082
wecare@dccc.edu
http://www.dccc.edu/
Professor:
Daniel Childers, (D), Delaware, 2014, ZgOuZi
Associate Professor:
Jennifer L. Snyder, (D), W Michigan, 1998, Zg

Dickinson College

Dept of Earth Sciences (B) ☒
P.O. Box 1773
Carlisle, PA 17013-2896
p. (717) 245-1355
edwardsb@dickinson.edu
https://www.dickinson.edu/earthsciences
f: https://www.facebook.com/pages/category/Product-Service/Dickinson-College-Department-of-Earth-Sciences-108884862477555/
t: @dsonearthsci
Programs: Earth Sciences (geoscience and environmental geoscience tracks)
Administrative Assistant: Debra Peters
Enrollment (2020): B: 16 (6)
Chair:
Benjamin R. Edwards, (D), British Columbia, 1997, GivCg
Professor:
Marcus M. Key, Jr., (D), Yale, 1988, PiGsa
Associate Professor:
Peter B. Sak, (D), Penn State, 2002, Gcm
Assistant Professor:
Jorden L. Hayes, (D), Wyoming, 2016, Ygs
Alyson M. Thibodeau, (D), Arizona, 2012, CcGaCs
Emeritus:
Jeffery W. Niemitz, (D), S California, 1978, ClOuGn
Noel Potter, Jr., (D), Minnesota, 1969, Gmc
Technician:
Robert Dean, (M), Texas (El Paso), 2004, Gi

Environmental Studies & Environmental Science
P.O. Box 1773
Carlisle, PA 17013-2896
p. (717) 245-1355
arnoldt@dickinson.edu
Administrative Assistant: Patricia Braught
Associate Professor:
Candie Wilderman, (D), Johns Hopkins, 1984, Hs
Visiting Professor:
Kirsten Hural, (D), Cornell, 1997, Gg

Drexel University

Dept of Biodiversity, Earth & Environmental Science (B,M,D) ☒
2024 MacAlister Hall
3250 Chestnut Street
Philadelphia, PA 19104
p. (215) 571-4651
bees@drexel.edu
http://drexel.edu/bees/
f: https://www.facebook.com/groups/DrexelBEES/
Programs: Geoscience; Environmental Science; Environmental Studies
Enrollment (2019): B: 15 (5) D: 5 (0)
Pilsbry Chair of Malacology:
Gary Rosenberg, (D), Harvard, 1989, Pi
Professor:
David Velinsky, (D), Old Dominion, 1987, OcCms
Associate Professor:
Ted Daeschler, (D), Pennsylvania, 1998, PvGgPg
Loyc Vanderkluysen, (D), Hawaii, 2008, GviGz
Assistant Professor:
Marie J. Kurz, (D), Florida, 2013, CqbHw
Amanda Lough, (D), Washington (St Louis), 2014, Ysg
Jocelyn A. Sessa, (D), Penn State, 2009, PeCsPq
Elizabeth Watson, (D), California (Berkeley), Zc

Adjunct Professor:
Mitch Cron, (M), Pennsylvania, 2013, Gge
Stephen Huxta, (M), Pennsylvania, 2017, Gg
Carl Mastropaolo, (M), Drexel, Hw

Edinboro University of Pennsylvania

Dept of Geosciences (B) ☒
126 Cooper Hall
230 Scotland Road
Edinboro, PA 16444-0001
p. (814) 732-2529
geosciences@edinboro.edu
http://www.edinboro.edu/academics/schools-and-departments/cshp/departments/geosciences/
f: https://www.facebook.com/pages/Edinboro-University-Geosciences-Department/426407214146067
Programs: Geology (B); Environmental Studies (B); GIS (B)
Enrollment (2017): B: 108 (24)

Chair:
Brian S. Zimmerman, (D), Washington State, 1991, Gzx

Professor:
Baher A. Ghosheh, (D), SUNY (Buffalo), 1988, Zn
David Hurd, (D), Cleveland State, 1997, Zg
Kerry A. Moyer, (D), Penn State, 1993, As
Laurie A. Parendes, (D), Oregon State, 1997, Zn
Joseph F. Reese, (D), Texas, 1995, GctZe
Eric Straffin, (D), Nebraska, 2000, GmsSo
Dale Tshudy, (D), Kent State, 1993, Pi

Associate Professor:
Karen Eisenhart, (D), Colorado, 2004, Zy
Wook Lee, (D), Ohio State, Zi

Assistant Professor:
Richard Deal, (D), South Carolina, 2000, Zi
Tamara Misner, (D), Pittsburg State, 2013, GmHs

Elizabethtown College

Dept of Engineering and Physics (B) ☒
One Alpha Drive
Esbenshade Room 160
Elizabethtown, PA 17022-2298
p. (717) 361-1392
mcfaddenj@etown.edu
http://www.etown.edu/PhysicsEngineering.aspx
Administrative Assistant: Jennifer McFadden

Associate Professor:
Michael A. Scanlin, (D), Penn State, Ye

Emeritus:
David Ferruzza, (M), MIT, 1967, AsZn

Franklin and Marshall College

Dept of Earth and Environment (B) ⊘
PO Box 3003
Lancaster, PA 17604-3003
p. (717) 358-4133
mbetrone@fandm.edu
http://www.fandm.edu/earth-environment
f: https://www.facebook.com/FM-Department-of-Earth-Environment
t: @FandMENE
Programs: Geosciences, Environmental Science, Environmental Studies majors; minor in Science, Technology and Society
Enrollment (2020): B: 11 (9)

Director, Center for Sustainable Environment:
Sarah Dawson, (D), Utah State, 2008, Zn

Chair of Environmental Science Program:
Andrew P. deWet, (D), Cambridge, 1989, Ge

Chair Department of Earth and Environment:
Christopher J. Williams, (D), Pennsylvania, 2002, PeSb

Professor:
Carol B. de Wet, (D), Cambridge, 1989, GsdGn
Dorothy J. Merritts, (D), Arizona, 1987, Gm
Stanley A. Mertzman, (D), Case Western, 1971, GizZm
James E. Strick, (D), Princeton, 1997, RhGe
Robert C. Walter, (D), Case Western, 1981, Cc

Associate Professor:
Eilzabeth De Santo, (D), Univ Coll (UK), Zn
Zeshan Ismat, (D), Rochester, 2002, Gc

Assistant Professor:
Eve Bratman, (D), American, 2009, Zn
Eric Hirsch, (D), Chicago, Zn

Emeritus:
Robert S. Sternberg, (D), Arizona, 1982, YmGa
Roger D. K. Thomas, (D), Harvard, 1970, PgRh

Laboratory Director:
Emily Wilson, (M), Idaho, 2013, CgGi

Director of F&M Science Outreach & Teaching Professor of Geosciences:
Timothy D. Bechtel, (D), Brown, 1989, GeYg

Gannon University

Environmental Science & Engineering (B,M) ☒
109 University Square
Erie, PA 16541
p. (814) 871-7000
kim008@gannon.edu
http://www.gannon.edu/enviro
f: www.facebook.com/GUENVIRONMENTAL
t: GUENVIRONMENTAL
Programs: Environmental Science, Environmental Engineering, GIS Minor, M.S. in Environmental Science & Management

Harrisburg Area Community College

Science (A)
One HACC Drive
Harrisburg, PA 17110
p. 800-222-4222
rkdennis@hacc.edu
http://www.hacc.edu
f: https://www.facebook.com/HACC64
t: @hacc_info
Enrollment (2016): A: 2 (0)

Professor:
James E. Baxter, P.G., (M), Penn State, 1983, GgHwGm

Indiana University of Pennsylvania

Dept of Geography, Geology, Environment and Planning (B) O⁄🖱
56 Weyandt Hall
Kopchick College Natural Sciences
Indiana, PA 15705
p. (724) 357-2379
geoscience-info@iup.edu
https://www.iup.edu/geoplanning/
Programs: Geology, Geology + Earth and Space Science Education certificate
Enrollment (2019): B: 40 (16)

Chair:
Nicholas Deardorff, (D), Oregon, Gv

Professor:
Karen Rose Cercone, (D), Michigan, 1984, Zeg
Kenneth S. Coles, (D), Columbia, 1988, XgZeYg
Steven A. Hovan, (D), Michigan, 1993, Ou
Jon C. Lewis, (D), Connecticut, 1998, Gc

Associate Professor:
Katie Farnsworth, (D), Virginia Inst of Marine Sci, 2003, Ong

Assistant Professor:
Jonathan P. Warnock, (D), N Illinois, 2013, PemPv

Emeritus:
Joseph C. Clark, (D), Stanford, 1966, Gr
Frank W. Hall, (D), Montana, 1969, Gc
Darlene S. Richardson, (D), Columbia, 1974, Gd
Connie J. Sutton, (M), Indiana (Penn), 1968, Ze
John F. Taylor, (D), Missouri, 1984, PsiGs

Dept of Geography, Geology, Environment, and Planning ●☒
Indiana, PA 15705
p. (724) 357-2379
n.deardorff@iup.edu
https://www.iup.edu/geoplanning/
Programs: Geology, BS; Geography, BA; Regional Planning, BS; Geography, MS;
Geology, BS + Secondary Science Education Teaching certificate
Certificates: Unmanned Aerial Systems Science and Applications Certificate
Geospatial Intelligence Analysis Certificate

Juniata College

Dept of Environmental Science & Studies
1700 Moore Street
Huntingdon, PA 16652
yohn@juniata.edu
http://www.juniata.edu/departments/environmental/

Dept of Geology (B) ●☒
1700 Moore St
Huntingdon, PA 16652
p. (814) 641-3601
johanesen@juniata.edu
http://www.juniata.edu/academics/departments/geology/
Programs: Geology; Environmental Geology; Earth and Space Science Education
Enrollment (2017): B: 25 (10)

Dept. Chair:
Ryan Mathur, (D), Arizona, 2000, CeYgHw

Associate Professor:
Matthew G. Powell, (D), Johns Hopkins, 2005, PiqPe

Assistant Professor:
Katharine Johanesen, (D), S California, 2011, GzxGc

Research Associate:
Adam J. Ianno, (D), S California, 2015, GgiCu

Emeritus:
Laurence J. Mutti, (D), Harvard, 1978, GxzSc
J. Peter Trexler, (D), Michigan, 1964, Ps
Robert H. Washburn, (D), Columbia, 1966, Gs

Kutztown University of Pennsylvania

Dept of Physical Science (B) ☒
Boehm Science Building Room 135
Kutztown, PA 19530
p. (610) 683-4447
simpson@kutztown.edu
http://www.kutztown.edu/acad/geology
Administrative Assistant: Caecilia Holt
Enrollment (2018): B: 35 (11)

Chair:
Edward L. Simpson, (D), Virginia Tech, 1987, Gs

Professor:
Kurt Friehauf, (D), Stanford, 1998, EmCgGx
Laura Sherrod, (D), W Michigan, 2007, YgHwGa
Sarah E. Tindall, (D), Arizona, 2000, Gc

Assistant Professor:
Erin Kraal, (D), California (Santa Cruz), XgGm
Adrienne Oakley, (D), Hawaii, 2009, YrOun
Jacob Sewall, (D), California (Santa Cruz), 2004, GeAs

Lafayette College

Dept of Geology & Environmental Geosciences (B) ☒
116 Van Wickle Hall
4 South College Drive
Easton, PA 18042
p. (610) 330-5193
geology@lafayette.edu
http://geology.lafayette.edu
f: @LafayetteGeology
Administrative Assistant: Rohana Meyerson
Enrollment (2022): B: 43 (16)

Professor:
Dru Germanoski, (D), Colorado State, 1989, Gm
Kira Lawrence, (D), Brown, 2006, OoPc

Associate Professor:
Tamara Carley, (D), Vanderbilt, 2014, GiCgc
Lawrence L. Malinconico, (D), Dartmouth, 1982, GctYu
David Sunderlin, (D), Chicago, 2005, GgPg

Research Associate:
Mary Ann Malinconico, (D), Columbia, 2002, Gt

Emeritus:
Guy L. Hovis, (D), Harvard, 1971, GzxCg

Museum Curator:
Lejeune Brian, (D), Northeastern, 2019, GzXmZm

Laboratory Coordinator:
John R. Wilson, (M), Virginia Tech, 2001, ZirGe

Instrument Support Specialist:
Jennie Pinho, (M), Lehigh, 2013, GeHwZu

Lehigh University

Dept of Earth & Environmental Sciences (B,M,D) ○☒
1 W. Packer Ave.
Bethlehem, PA 18015-3001
p. (610) 758-3660
jij422@lehigh.edu
http://www.ees.lehigh.edu
f: https://www.facebook.com/LehighEES/
Enrollment (2018): B: 61 (15) M: 8 (3) D: 12 (1)

Department Chair:
Gray E. Bebout, (D), California (Los Angeles), 1989, Gp

Professor:
David J. Anastasio, (D), Johns Hopkins, 1988, Gca
Robert K. Booth, (D), Wyoming, 2003, ZnnPc
Edward B. Evenson, (D), Michigan, 1972, Gl
Kenneth P. Kodama, (D), Stanford, 1978, Ym
Anne S. Meltzer, (D), Rice, 1988, Ys
Frank J. Pazzaglia, (D), Penn State, 1993, Gm
Dork Sahagian, (D), Chicago, 1987, PeGvr
Zicheng Yu, (D), Toronto, 1997, Pe
Peter K. Zeitler, (D), Dartmouth, 1983, Cc

Associate Professor:
Benjamin S. Felzer, (D), Brown, 1995, ZcoZn
Donald P. Morris, (D), Colorado, 1990, Zn
Stephen C. Peters, (D), Michigan, 2001, Cl
Joan Ramage, (D), Cornell, 2001, ZrGlm

Assistant Professor:
Jill I. McDermott, (D), MIT/WHOI, 2015, Cm

Research Associate:
Bruce D. Idleman, (D), SUNY (Albany), 1990, Cc
Joshua Stachnik, (D), Wyoming, 2010, Ys

Emeritus:
Bobb Carson, (D), Washington, 1971, GusGt
Bruce R. Hargreaves, (D), California (Berkeley), 1977, Ob

Lock Haven University

Dept of Geology & Physics (B) ☒
301 West Church Street
Lock Haven, PA 17745-2390
p. (570) 484-2048
mkhalequ@lockhaven.edu
Programs: Geology
Administrative Assistant: Denise Rupert
Enrollment (2018): B: 30 (6)

Professor:
Loretta D. Dickson, (D), Connecticut, 2006, GizGe
Md. Khalequzzaman, (D), 1998, HwOnZi

Associate Scientist:
Thomas C. Wynn, (D), Virginia Tech, 2004, GsPiEo

Mansfield University

Dept of Geosciences (A,B) ○☒
Belknap Hall
Mansfield, PA 16933
p. (570) 662-4613
jdemchak@mansfield.edu
http://geoggeol.mansfield.edu/
Enrollment (2016): A: 10 (6) B: 124 (42)

Chair:
Jennifer Demchak, (D), West Virginia, 2005, Hs

Associate Professor:
Christopher F. Kopf, (D), Massachusetts, 1999, Gcp
Lee Stocks, (D), Kent State, 2010, ZyGgm

Assistant Professor:
Linda Kennedy, (D), North Carolina (Greensboro), 2012, Zy

Mercyhurst University

Dept of Geology (B,M,D)
501 East 38th Street
Erie, PA 16546
p. (814) 824-2581
rbuyce@mercyhurst.edu
http://mai.mercyhurst.edu
Enrollment (2012): B: 16 (3)

Director:
James M. Adovasio, (D), Utah, 1970, GafGs
Professor:
M. Raymond Buyce, (D), Rensselaer, 1975, GsaOn
Assistant Professor:
Nicholas Lang, (D), Minnesota, 2006, GvXgGc
Scott C. McKenzie, (B), Edinboro, 1976, PgXmZe
Lyman Perscio, (D), New Mexico, 2012, GsSpHg
Adjunct Professor:
Frank Vento, (D), Pittsburg State, 1985, GaSoGm

Millersville University

Dept of Earth Sciences (B,M)
PO Box 1002
Millersville, PA 17551
p. (717) 871-4359
esci@millersville.edu
http://www.millersville.edu/esci
f: Millersville University Department of Earth Science
t: @mu_earthscience
Programs: BS Environmental Earth and Ocean Sciences with concentrations in Environmental Geology, Environmental Ocean Sciences, and Environmental Earth Sciences, BS Meteorology, BS Emergency Management
Certificates: Space Weather, Emergency Management
Enrollment (2021): B: 174 (99) M: 15 (6)
Chair:
Richard D. Clark, (D), Wyoming, 1987, As
Professor:
Alex J. DeCaria, (D), Maryland, 2000, As
Ajoy Kumar, (D), Old Dominion, 1996, OpZr
L. Lynn Marquez, (D), Northwestern, 1998, Cg
Todd D. Sikora, (D), Penn State, 1996, As
Sepideh Yalda, (D), St Louis, 1997, As
Associate Professor:
Sam Earman, (D), New Mexico Tech, 2004, HwCsl
Duane E. Hagelgans, Rnm
Robert Vaillancourt, (D), Rhode Island, 1996, Obc
Talor B. Walsh, (D), Rochester, 2015, GcmGe
Instructor:
Joseph Calhoun, (B), Penn State, As
Mary Ann Schlegel, (M), MIT/WHOI, 1998, Og
Emeritus:
William M. Jordan, (D), Wisconsin, 1965, Gs
Bernard L. Oostdam, (D), Delaware, 1971, Ou
Robert S. Ross, (D), Florida State, 1977, As
Charles K. Scharnberger, (D), Washington (St Louis), 1971, Gc

Montgomery County Community College

Dept of Science, Technology, Engineering, and Math (A)
Blue Bell, PA 19422
p. (215) 641-6446
rkuhlman@mc3.edu
Professor:
Robert Kuhlman, (M), Bryn Mawr, 1975, Gg
Instructor:
George Buchanan, (M), Drexel, 1992, Ng
Adjunct Professor:
Laurie Martin-Vermilyea, (D), South Carolina, 1992, Ze
Frank Roberts, (D), Bryn Mawr, 1969, Gp
Kelly C. Spangler, (M), Drexel, 2003
Anthony Stevens, (M), Florida, 1981, Ge

Moravian College

Dept of Physics & Earth Science (B) ☒
1200 Main Street
Bethlehem, PA 18018-6650
p. (610) 861-1437
krieblek@moravian.edu
http://www.physics.moravian.edu
Programs: Geology; General Science; Physics
Administrative Assistant: Ann Sywensky
Enrollment (2017): B: 1 (1)
Chair:
Kelly Krieble, (D), Lehigh, 1993, ZnnZn

Pennsylvania State University, Erie

Geoscience Dept
Erie, PA 16510
p. (814) 898-6277
amf11@psu.edu
http://www.personal.psu.edu/faculty/a/m/amf11/
Chair:
Anthony M. Foyle
Assistant Professor:
Eva Tucker, (M), Cincinnati, 1962, Gg

Pennsylvania State University, Monaca

Dept of Geosciences (B)
100 University Drive
Monaca, PA 15061
p. (412) 773-3867
jac7@psu.edu
http://www.br.psu.edu/
Assistant Professor:
John A. Ciciarelli, (D), Penn State, 1971, Gg

Pennsylvania State University, University Park

Dept of Geosciences (B,M,D) ● ☒
503 Deike Building
University Park, PA 16802-2714
p. (814) 865-6711
demian@psu.edu
http://www.geosc.psu.edu/
Programs: Geosciences BS, Earth Science & Policy BS, Earth Sciences BS, Geosciences MS, PhD
Certificates: Sustainability, MEd
Administrative Assistant: Amy L.. Homan
Enrollment (2018): B: 133 (0) M: 20 (0) D: 59 (0)
Head:
Demian M. Saffer, (D), California (Santa Cruz), 1999, Hw
Evan Pugh Professor:
Richard B. Alley, (D), Wisconsin, 1987, Gl
James F. Kasting, (D), Michigan, 1979, As
Distinguished Professor:
Katherine H. Freeman, (D), Indiana, 1991, Cos
Director, Earth & Mineral Science Museum:
Russell W. Graham, (D), Texas, 1976, Pv
Director, Earth & Env Systems Inst:
Susan L. Brantley, (D), Princeton, 1987, Cg
Director, Astrobiology Research Center:
Christopher H. House, (D), California (Los Angeles), 1999, Pg
Associate Head, Graduate Programs:
Mark E. Patzkowsky, (D), Chicago, 1992, Pg
Professor:
Charles J. Ammon, (D), Penn State, 1991, Ys
Sridhar Anandakrishnan, (D), Wisconsin, 1990, Ys
David M. Bice, (D), California (Berkeley), 1989, Gg
Timothy J. Bralower, (D), California (San Diego), 1986, Pe
Donald M. Fisher, (D), Brown, 1988, GctZm
Kevin P. Furlong, (D), Utah, 1981, Gt
Tanya Furman, (D), MIT, 1989, GiCu
Peter J. Heaney, (D), Johns Hopkins, 1989, GzClGe
Klaus Keller, (D), Princeton, 2000, Og
Lee R. Kump, (D), S Florida, 1986, ClPeCm
Michael E. Mann, (D), Yale, 1998, As
Chris Marone, (D), Columbia, 1988, Yx
Andrew A. Nyblade, (D), Michigan, 1992, Yg
Peter D. Wilf, (D), Pennsylvania, 1998, Pg
Senior Scientist:
Todd Sowers, (D), Rhode Island, 1991, Cs
(Associate Head for Undergraduate Programs):
Maureen D. Feineman, (D), California (Berkeley), 2004, Cp
Associate Professor:
Matthew S. Fantle, (D), California (Berkeley), 2005, Cs
Elizabeth Hajek, (D), Wyoming, 2009, Gs
Peter C. LaFemina, (D), Miami, 2005, Yd
Jennifer L. Macalady, (D), California (Davis), 2000, PgClSb
Eliza Richardson, (D), MIT, 2002, YsZe
Assistant Professor:
Roman DiBiase, (D), Arizona State, 2011, Gmt
Bradford Foley, (D), Yale, 2014, YgXyGt
Sarah Ivory, (D), Arizona, 2013, Pl
Tess A. Russo, (D), California (Santa Cruz), 2012, Hw

Christelle Wauthier, (D), Liege, 2011, Zn

Emeritus:
Shelton S. Alexander, (D), Caltech, 1963, Ys
Michael A. Arthur, (D), Princeton, 1979, OuClGs
Hubert L. Barnes, (D), Columbia, 1958, CgEmCe
Roger J. Cuffey, (D), Indiana, 1966, PiePv
David H. Eggler, (D), Colorado, 1967, CpGiv
Terry Engelder, (D), Texas A&M, 1973, Nr
David (Duff) P. Gold, (D), McGill, 1963, GcgGx
Earl K. Graham, Jr., (D), Penn State, 1969, Yx
Roy J. Greenfield, (D), MIT, 1965, YgsYu
Albert L. Guber, (D), Illinois, 1962, Pe
Benjamin F. Howell, Jr., (D), Caltech, 1949, Ys
Derrill M. Kerrick, (D), California (Berkeley), 1968, Gp
Hiroshi Ohmoto, (D), Princeton, 1969, Cs
Richard R. Parizek, (D), Illinois, 1961, Hw
Arthur W. Rose, (D), Caltech, 1958, Cge
Rudy L. Slingerland, (D), Penn State, 1977, Gs
Barry Voight, (D), Columbia, 1965, GveGg
William B. White, (D), Penn State, 1962, CgGzy

Dept of Meteorology and Atmospheric Science (B,M,D) ○✓
503 Walker Building
University Park, PA 16802-5013
p. (814) 865-0478
meteodept@meteo.psu.edu
http://www.met.psu.edu
Programs: Meteorology and Atmospheric Science, Dual Title PhD in Climate Science
Certificates: Certificate of Achievement in Weather Forecasting, Graduate Certificate in Weather and Climate Analytics

Professor and Head:
David J. Stensrud, (D), Penn State, 1992, Am

Evan Pugh Professor:
James F. Kasting, (D), Michigan, 1979, AsPc

Distinguished Professor:
William H. Brune, (D), Johns Hopkins, 1978, Am
Michael E. Mann, (D), Yale, 1998, AsZoPc

Director, Institute for Computational and Data Sciences:
Jenni L. Evans, (D), Monash, 1990, Am

Associate Head, Undergraduate Program:
Jon Nese, (D), Penn State, 1989, Am

Associate Head, Graduate Program:
Paul M. Markowski, (D), Oklahoma, 2000

Professor:
Eugene E. Clothiaux, (D), Brown, Am
Kenneth J. Davis, (D), Colorado, 1992, Am
Jose D. Fuentes, (D), Guelph, 1992, As
John Harlim, (D), Maryland, 2006, Am
Jerry Y. Harrington, (D), Colorado State, 1997, Am
Gregory S. Jenkins, (D), Am
Sukyoung Lee, (D), Princeton, 1991, Am
Raymond G. Najjar, (D), Princeton, 1990, OcpAt
Johannes Verlinde, (D), Colorado State, 1992, Am
George S. Young, (D), Colorado State, Am

Associate Professor:
Steven J. Greybush, (D), Maryland, 2011, Am
Matthew R. Kumjian, (D), Oklahoma, 2012, Am
Kelly A. Lombardo, (D), Stoney Brook, 2011, As

Assistant Professor:
Anthony Didlake, Jr., (D), Washington, As
Melissa Gervais, (D), McGill, As
Laifang Li, (D), Duke, 2014, AsZo
Ying Pan, (D), Penn State, 2014, As
Colin Zarzycki, (D), Michigan, 2014, As

Research Assistant:
William F. Ryan, (M), Maryland, 1990, Am

Director of Meteorological Computing:
Chad Bahrmann, (M), North Carolina State, 1997, Am

Research Associate:
Arthur Person, (M), Penn State, 1983, Am
William Syrett, (M), Penn State, 1987, Am

Emeritus:
Peter R. Bannon, (D), Colorado, 1979, Am
Craig F. Bohren, (D), Arizona, 1975, Am
John J. Cahir, (D), Penn State, 1971, Am
Toby N. Carlson, (D), Imperial Coll (UK), 1965, Am
John H. E. Clark, (D), Florida State, 1969, Am
John A. Dutton, (D), Wisconsin, Am
William M. Frank, (D), Colorado State, 1976, Am
Alistair B. Fraser, (D), Imperial Coll (UK), 1968, Am
J. Michael Fritsch, (D), Colorado State, 1978, Am
Charles L. Hosler, (D), Penn State, 1951, Am
Dennis Lamb, (D), Washington, 1970, Am
Nelson L. Seaman, (D), Penn State, 1977, Am
Hampton N. Shirer, (D), Penn State, 1978, Am
David R. Stauffer, (D), Penn State, 1990, Am
Dennis W. Thomson, (D), Wisconsin, 1968, Am
John C. Wyngaard, (D), Penn State, 1967, Am

Associate Dean of Undergraduate Education:
Yvette P. Richardson, (D), Oklahoma, 1999, Ams

Dept of Plant Science (A,B,M,D)
119 Tyson Building
University Park, PA 16802
p. (814) 865-6541
rpm12@psu.edu
http://plantscience.psu.edu/

Professor:
Jean-Marc Bollag, (D), Basel (Switzerland), 1959, Sb
Edward J. Ciolkosz, (D), Wisconsin, 1967, Sd
Daniel D. Fritton, (D), Iowa State, 1968, Sp
Sridhar Komarneni, (D), Wisconsin, 1973, Sc
Gary W. Petersen, (D), Wisconsin, 1965, Sd

Associate Professor:
Peter J. Landschoot, (D), Rhode Island, 1988, So
Gregory W. Roth, (D), Penn State, 1987, So

Assistant Professor:
Rick L. Day, (D), Penn State, 1991, Zu

Research Associate:
Barry M. Evans, (M), Penn State, 1977, Sd

Adjunct Professor:
Andrew S. Rogowski, (D), Iowa State, 1964, Sp
Lawrence A. Schardt, (D), Penn State, 2000, SoHw

Earth and Mineral Sciences Museum & Art Gallery ☒
116 Deike Building
University Park, PA 16802
p. (814) 863-8554
museum@ems.psu.edu
https://museum.ems.psu.edu/
f: https://www.facebook.com/EMSMAAG
t: @emsmaag

Director:
Jane Cook, (D), Wisconsin (Madison), 1998, ZmeZn

John and Willie Leone Family Dept of Energy and Mineral Engineering (B,M,D) ☒
110 Hosler Building
University Park, PA 16802
p. (814) 865-3437
eme@ems.psu.edu
http://www.eme.psu.edu/mnge

Program Chair, and Deike Endowed Chair in Mining Engineering:
Jeffery L. Kohler, (D), Penn State, 1982, Nm

Assistant Professor:
Shimin Liu, (D), S Illinois, 2013, Nm

Point Park University

Dept of Environmental Studies (A,B)
201 Wood Street
Pittsburgh, PA 15222-1994
p. (412) 392-3900
jkudlac@pointpark.edu
http://www.pointpark.edu/Academics/Schools/SchoolofArtsandSciences/Departments/NaturalSciencesandEngineeringTechnology
Administrative Assistant: Roberta T. Gallick

Head:
Mark O. Farrell, (D), Carnegie Mellon, 1978, Zn

Professor:
John J. Kudlac, (D), Pittsburgh, Ng

Shippensburg University

Geography-Earth Science Department (B,M) ●✓
1871 Old Main Drive
Shearer Hall 104

Shippensburg, PA 17257
p. (717) 477-1685
tlmyers@ship.edu
https://www.ship.edu/geo-ess/
f: https://www.facebook.com/ShipGeoESS/
Programs: B.Sc. in Geoenvironmental Studies, B.Sc. in Environmental Sustainability, B.Sc. in Geography - GIS, M.Sc. Geoenvironmental Studies
Certificates: GIS Certificate, GIS Minor, Geography Minor, Marine Science Minor
Administrative Assistant: Tammy Myers
Enrollment (2022): B: 94 (41) M: 14 (11)

Professor:
Scott A. Drzyzga, (D), Michigan State, 2007, ZiuGl
Alison E. Feeney, (D), Michigan State, 1998, ZinZn
Thomas P. Feeney, (D), Georgia, 1997, ZyGeHw
Timothy W. Hawkins, (D), Arizona State, 2004, AsmHg
Claire A. Jantz, (D), Maryland, 2005, ZuiZc
Paul G. Marr, (D), Denver, 1996, ZfyZi
George M. Pomeroy, (D), Akron, 1999, ZuyZn

Associate Professor:
Michael T. Applegarth, (D), Arizona State, 2001, GmZrSo
Sean R. Cornell, (D), Cincinnati, 2008, GgsOn
Joseph T. Zume, (D), Oklahoma, 2007, HgYug

Assistant Professor:
Russell Hedberg, (D), Penn State, 2018, ZiSoZc

Slippery Rock University

Dept of Geography, Geology, and the Environment (B) ●☒
1 Morrow Way
Slippery Rock, PA 16057
p. (724) 738-2048
cynthia.schnur@sru.edu
https://www.sru.edu/academics/colleges-and-departments/ches/departments/geography-geology-and-the-environment
f: facebook.com/GGESRU
t: twitter.com/ggesru
Programs: BS Environmental Geoscience: Geology, BS Environmental Geoscience: Environmental Science, BA Environmental Geoscience, BS Geography: Environmental Studies and Sustainability, BS Geography: Applied Geographic Technology, BS Geography: Global Studies
Certificates: Cert. in Sustainability, Cert. in Geographic Information Technology
Enrollment (2022): B: 120 (49)

Professor:
Patrick A. Burkhart, (D), Lehigh, 1994, HgGl
Heike Hartmann, (D), Justus Leibig, 2007, Amt
Jack Livingston, (D), Kansas, 2001, ZiSo
Tamra A. Schiappa, (D), Idaho, 1999, PiGrZe
Langdon Smith, (D), Kansas, 2002, Zn
Julie A. Snow, (D), Rhode Island, 2002, As

Associate Professor:
Patricia A. Campbell, (D), Pittsburgh, 1994, Gc
Xianfeng Chen, (D), West Virginia, 2005, Zr
Stentor B. Danielson, (D), Clark, 2007, Rn
Wang Jialing, (D), Florida State, 2005, Zi

State Museum of Pennsylvania

Section of Paleontology & Geology
300 North Street
Harrisburg, PA 17120-0024
p. (717) 783-9897
c-sjasinsk@pa.gov
http://www.statemuseumpa.org/geologyc.html
f: https://www.facebook.com/StateMuseumofPA

Acting Curator:
Steven E. Jasinski, (D), Pennsylvania, pend, PvgPg

Susquehanna University

Dept of Earth & Environmental Sciences (B)
514 University Ave
Selinsgrove, PA 17870
p. (570) 372-4216
straubk@susqu.edu
http://www.susqu.edu/ees
Enrollment (2009): B: 29 (7)

Chair:
Jennifer M. Elick, (D), Tennessee, 1999, Pe
Katherine H. Straub, (D), Colorado State, 2002, Am

Associate Professor:
Daniel E. Ressler, (D), Iowa State, 1998, Sp
Derek J. Straub, (D), As

Assistant Professor:
Ahmed Lachhab, (D), Iowa, 2006, Hw

Temple University

Earth & Environmental Science (B,M,D)
1901 N. 13th Street
Beury Hall, Rm. 326
Philadelphia, PA 19122-6081
p. (215) 204-8227
scox@temple.edu
http://www.temple.edu/geology
f: https://www.facebook.com/tu.geology
Programs: BS, Geology, BA, Geology, Minor, Geology, BS, Environmental Science, Masters, Geology, PhD, Geoscience
Certificates: Environmental Professional Training Undergraduate Certificate
Administrative Assistant: Shelah Cox

Chair:
Nicholas Davatzes, (D), Stanford, 2003, GcNrZn

Professor:
David E. Grandstaff, (D), Princeton, 1974, Cl
Jonathan Nyquist, (D), Wisconsin, 1986, Ygu
Laura Toran, (D), Wisconsin, 1986, HwZu

Associate Professor:
Ilya Buynevich, (D), Boston, 2001, GmPeOn
Alexandra K. Davatzes, (D), Stanford, 2007, GsXgZe
Dennis O. Terry, (D), Nebraska, 1998, SaPcGr
Allison Tumarkin-Deratzian, (D), Pennsylvania, 2003, Pv

Assistant Professor:
Steven Chemtob, (D), 2012
Natalie Flynn, (D), Temple, 2015, Gx
Bojeong Kim, (D), GzZa
Atsuhiro Muto, (D), 2010, GlYg
Sujith Ravi, (D), HqZuRw
Jesse Thornburg, (D), Rutgers, 2016, GsPs

Instructor:
Timothy Davis, (M), Temple, 2011, Gg

Emeritus:
George Myers, (D), Yale, 1965, Gz

Asst. Lab Manager/Bldg. Coordinator:
James Ladd, (M)

Thiel College

Dept of Environmental Science (B)
Greenville, PA 16125
p. (412) 589-2821
areinsel@thiel.edu

Professor:
James H. Barton, (D), N Colorado, 1977, Zy

University of Pennsylvania

Dept of Earth & Environmental Science (B,M,D) ⃠
240 S. 33rd Street
Hayden Hall
Philadelphia, PA 19104-6316
p. (215) 898-5724
earth@sas.upenn.edu
http://www.sas.upenn.edu/earth/
Programs: Undergraduate majors in Earth Science (EASC) and Environmental Studies (ENVS), as well as minors in Environmental Science, Environmental Studies, Geology, and Sustainability and Environmental Management.
Certificates: Various certificates within our professional masters programs; see MES and MSAG websites at: https://www.sas.upenn.edu/earth/graduate/professional-masters-programs
Administrative Assistant: Sally A.. Cardy
Enrollment (2020): B: 30 (19) M: 29 (11) D: 17 (12)

Chair:
Brenda B. Casper, (D), Utah, 1982

Undergraduate Chair:
Alain F. Plante, (D), Alberta, 2001, SboCo

Professor of Practice:
Howard Neukrug, (B), Pennsylvania, 1978, Hs
Distinguished Professor:
Joseph Francisco, (D), MIT, 1983, Ac
Professor:
Reto Gieré, (D), ETH (Switzerland), 1990, GzCgGe
Douglas Jerolmack, (D), MIT, 2006, GmYxHq
Graduate Chair:
David Goldsby, (D), Minnesota, 1997, Yg
Associate Professor:
Irina Marinov, (D), Princeton, 2005, ZoOg
Assistant Professor:
Ileana Pérez-Rodríguez, (D), Rutgers, 2012, Po
Lauren Sallan, (D), Chicago, 2012, PqvPe
Lecturer:
Willig B. Sarah, (D), Pennsylvania, 1988, Ge
Emeritus:
Robert F. Giegengack, Jr., (D), Yale, 1968, Gg
Arthur H. Johnson, (D), Cornell, 1975, So
Hermann W. Pfefferkorn, (D), Muenster (Germany), 1968, PbePs
Stephen P. Phipps, (D), Princeton, 1984, Gc
Laboratory Manager:
David R. Vann, (D), Pennsylvania, 1993, GeSfXm
Teaching Faculty:
Jane Dmochowski, (D), Caltech, 2004, Yg
Gomaa I. Omar, (D), Pennsylvania, 1985, Cc
Education Administrator:
Nicholas Crivaro
Director, Professional Masters Programs:
Yvette Bordeaux, (D), Pennsylvania, 2000, Pg
Department Administrator:
Joan Buccilli
Associate Director:
Maria Andrews, (M)

University of Pittsburgh

Dept of Geology & Environmental Science (B,M,D) ●☒
200 SRCC Building
4107 O'Hara Street
Pittsburgh, PA 15260-3332
p. (412) 624-8780
gpsgrad@pitt.edu
http://www.geology.pitt.edu
Enrollment (2018): B: 218 (79) M: 9 (2) D: 24 (7)
Chair:
Josef Werne, (D), Northwestern, 2000, CosGn
Professor:
Mark B. Abbott, (D), Minnesota, 1995, Gs
Emily M. Elliott, (D), Johns Hopkins, 2003, Cbs
William P. Harbert, (D), Stanford, 1987, Ym
Nadine McQuarrie, (D), Arizona, 2001, Gct
Michael S. Ramsey, (D), Arizona State, 1996, Zr
Associate Professor:
Daniel J. Bain, (D), Johns Hopkins, 2004, HgGmZu
Rosemary C. Capo, (D), California (Los Angeles), 1990, Cl
Brian W. Stewart, (D), California (Los Angeles), 1990, CclCq
Assistant Professor:
Eitan Shelef
Brian Thomas
Environmental Reporter, Pittsburgh Post-Gazette:
S. Don Hopey
Instructor:
Emily Collins
Lecturer:
R. Ward Allebach
Danielle Andrews-Brown, (D)
Charles E. Jones, (D), Oxford, 1992, GgsPg
Adjunct Professor:
Matthew C. Lamanna, (D), Pennsylvania, 2004, Pv
Steven C. Latta
John S. Pallister, (D), California (Santa Barbara), 1980, GviGg
Serge Shapiro
Matthew Watson
Emeritus:
Thomas Anderson
Bruce W. Hapke, (D), Cornell, 1962, Xy
Edward G. Lidiak, (D), Rice, 1963, Gi

University of Pittsburgh, Bradford

Dept of Petroleum Technology (A)
300 Campus Drive
Bradford, PA 16701-2898
p. (814) 362-7569
aap@pitt.edu
http://www.upb.pitt.edu/academics/petroleumtechnology.aspx
Administrative Assistant: Janet Shade
Enrollment (2010): A: 28 (8)
Program Director:
Assad I. Panah, (D), Oklahoma, 1966, GorGc

West Chester University

Dept of Earth and Space Sciences (B,M) ●⃠
750 South Church Street
Merion Science Center
West Chester, PA 19383
p. (610) 436-2727
hbosbyshell@wcupa.edu
https://www.wcupa.edu/sciences-mathematics/earthSpace-Sciences/
f: https://www.facebook.com/WCU-Earth-Space-Sciences-143447267219/
Programs: Geoscience (B,M); Earth Systems (B); Geology (B); Earth and Space Science Education (B)
Certificates: GIS (B,M)
Enrollment (2019): B: 85 (16) M: 25 (12)
Chair:
Joby Hilliker, (D), Penn State, 2002, Am
Professor:
Richard M. Busch, (D), Pittsburgh, 1984, GrPgZe
Marc R. Gagne, (D), Georgia, 1994, Xa
Martin F. Helmke, (D), Iowa State, 2003, HwSdZg
Associate Professor:
Howell Bosbyshell, (D), Bryn Mawr, 2001, GcpGt
Karen Schwarz, (D), Arizona State, XaZe
Assistant Professor:
Yong Hoon Kim, (D), South Carolina, 2005, OgZr
Christopher Roemmele, (D), Purdue, 2017, ZeGgZg
Adjunct Professor:
Cynthia V. Hall, (D), Georgia Tech, 2008, Cg
Vicky Helmke, (M), Iowa State, Gg
James "Sandy" Maxwell, (B), Ze
Jamie Vann, (M), CgGg
Emeritus:
LeeAnn Srogi, (D), Pennsylvania, 1988, Gi

Wilkes University

Dept of Environmental Engineering & Earth Sciences (B) ○☒
84 West South Street
Wilkes-Barre, PA 18766
p. (570) 408-4610
sid.halsor@wilkes.edu
http://wilkes.edu/academics/colleges/science-and-engineering/environmental-engineering-earth-sciences/
Programs: Geology; Earth & Environmental Sciences; Environmental Engineering
Certificates: Sustainability
Enrollment (2018): B: 7 (0)
Chair:
Marleen Troy, (D), Drexel, 1989, Ht
Professor:
Dale A. Bruns, (D), Idaho State, 1981, ZirSf
Sid P. Halsor, (D), Michigan Tech, 1989, Giv
Kenneth M. Klemow, (D), SUNY (Syracuse), 1982, Pg
Prahlad N. Murthy, (D), Texas A&M, 1993, AsZnn
Michael A. Steele, (D), Wake Forest, 1988, Pg
Brian E. Whitman, (D), Michigan Tech, 1998, HqwSb
Associate Professor:
Holly Frederick, (D), Penn State, 1999, Sdb
Assistant Professor:
Matthew S. Finkenbinder, (D), Pittsburg State, 2015, GsmGl
Bobak Karimi, (D), Pittsburg State, 2014, GctYg
Lecturer:
Mark A. Kaster, (M), Saint Louis, 1993, Am
Julie McMonagle, (M), Lehigh, 1991, Gge
Emeritus:
James M. Case, (D), Dalhousie, 1979, Ob

Brian T. Redmond, (D), Rensselaer, 1982, GsHwZg

York College of Pennsylvania

Dept of Physical Science
York, PA 17405
p. (717) 846-7788 x333
jforesma@ycp.edu
Assistant Professor:
William (Bill) Kreiger, (D), Penn State, 1976, GisZg
Adjunct Professor:
Ralph Eisenhart, (M), Penn State, 1994, Gg
Jeri L. Jones, (B), Catawba, 1977, Ga

Puerto Rico

University of Puerto Rico

Dept of Geology (B,M) O
PO Box 9000
Mayaguez, PR 00681-9000
p. (787) 265-3845
geol@uprm.edu
https://www.uprm.edu/geology/
f: https://www.facebook.com/geologyuprm/
t: @GeologyUPRM
Programs: Geology
Enrollment (2022): B: 108 (12) M: 18 (1)
Director:
Lizzette A. Rodriguez, (D), Michigan Tech, 2007, GviZr
Professor:
Fernando Gilbes, (D), S Florida, 1996, ZrGuZi
Kenneth Stephen Hughes, (D), N Carolina State, 2014, Gcm
James Joyce, (D), Northwestern, 1985, NgGcp
Alberto Lopez, (D), Northwestern, 2006, YdGtYs
Wilson R. Ramirez, (D), Tulane, 2000, GudOu
Hernan Santos, (D), Colorado, 1999, PiGdr
Associate Professor:
Thomas Hudgins, (D), Michigan, 2014, GiCpGv
Elizabeth A. Vanacore, (D), Rice, 2008, YsuYx
Director of Puerto Rico Seismic Network:
Victor Huerfano, (D), Puerto Rico, 2004, YsgGt

Rhode Island

Brown University

Dept of Earth, Environmental and Planetary Sciences (B,M,D) ☒
Box 1846, 324 Brook Street
Providence, RI 02912
p. (401) 863-3339
DEEPS@brown.edu
https://deeps.brown.edu/
f: https://www.facebook.com/BrownGeologicalSciences
t: @BrownGeoSci
Enrollment (2018): B: 0 (12) M: 0 (10) D: 55 (8)
Chair:
James M. Russell, (D), Minnesota, 2004, Gn
Director of the Institute at Brown for Environment and Society:
Kim Cobb, (D), GeOp
Professor:
Reid F. Cooper, (D), Cornell, 1983, Gy
Emanuele Di Lorenzo, (D), GeOp
Karen M. Fischer, (D), MIT, 1988, Ys
Baylor Fox-Kemper, (D), MIT/WHOI, 2003, OpYnAt
Meredith Hastings, (D), Princeton, 2004, As
Timothy D. Herbert, (D), Princeton, 1987, Pe
Greg Hirth, (D), Brown, 1991, GcyYg
Yongsong Huang, (D), Bristol, 1997, Co
Christian Huber, (D), California (Berkeley), 2009, Gv
Yan Liang, (D), Chicago, 1994, Cp
Amanda H. Lynch, (D), Melbourne (Australia), 1993, AsGe
John F. Mustard, (D), Brown, 1990, Zr
Alberto E. Saal, (D), MIT/WHOI, 2000, CgGi
Professor of Research:
Steven C. Clemens, (D), Brown, 1990, Ou
Associate Professor:
Colleen Dalton, (D), Harvard, 2007, Ys
Jung-Eun Lee, (D), California (Berkeley), 2005
Ralph E. Milliken, (D), Brown, 2006
Stephen Parman, (D), MIT, 2001, CpgCt
Victor Tsai, (D), Ys
Assistant Professor (Research):
Ingrid Dauber, (D)
Assistant Professor:
Gerrit Budde, (D), Cs
Alexander J. Evans, (D)
Daniel Ibarra, (D), GeCm
Emeritus:
Donald W. Forsyth, (D), MIT/WHOI, 1974, YrsGt
L. Peter Gromet, (D), Caltech, 1979, Cc
James W. Head, III, (D), Brown, 1969, XgGvt
John F. Hermance, (D), Toronto, 1967, HqZrHw
Paul C. Hess, (D), Harvard, 1968, Gi
E. Marc Parmentier, (D), Cornell, 1975, Yg
Carle M. Pieters, (D), MIT, 1977, Zr
Warren L. Prell, (D), Columbia, 1974, Ou
Malcolm J. Rutherford, (D), Johns Hopkins, 1968, Cp
Peter H. Schultz, (D), Texas, 1972, Xg
Jan A. Tullis, (D), California (Los Angeles), 1971, Gcy
Terry E. Tullis, (D), California (Los Angeles), 1971, Yx
Thompson Webb, III, (D), Wisconsin, 1971, PeIAs
Academic Program Manager:
Patricia M. Davey, (B), Rhode Island Coll, 1986, Zn
Academic Department Manager:
Dina Egge, (M), 2015

Community College of Rhode Island

Dept of Physics and Engineering (Geology & Oceanography Div) (A)
400 East Avenue
Warwick, RI 02886
p. (401) 825-2156
kkortz@ccri.edu
http://www.ccri.edu/physics/
Professor:
Emily Burns, (D), Rhode Island, GgOgZi
Karen M. Kortz, (D), Rhode Island, 2009, ZeGg
Paul White, (D), Gg
Associate Professor:
Duayne Rieger, (D), Ys
Assistant Professor:
Roger M. Hart, (M), 2018, Gg

Roger Williams University

College of Arts & Sciences (B)
Bristol, RI 02809
p. (401) 254-3087
jborden@rwu.edu
Department Secretary: Valerie Catalano
Head:
Mark D. Gould, (D), Rhode Island, 1973, Ob
Chair:
Paul Webb
Professor:
Thomas Doty, (D), Rhode Island, 1977, Ob
Richard Heavers, (D), Rhode Island, 1977, Op
Thomas J. Holstein, (D), Brown, 1969, Zn
Martine Villalard-Bohnsack, (D), Rhode Island, 1971, Ob
Assistant Professor:
Tim Scott, (D), SUNY (Stony Brook), 1993, Ob

University of Rhode Island

Dept of Geosciences (B,M,D) ☒
9 East Alumni Ave.,
Kingston, RI 02881
p. (401) 874-2265
http://web.uri.edu/geo/
t: @URI_GEO
Programs: Geology; Geological Oceanography
Certificates: Hydrology
Enrollment (2018): B: 42 (8) M: 6 (0) D: 6 (0)
Associate Dean:
Anne I. Veeger, (D), Arizona, 1991, HwCl
Professor:
Thomas B. Boving, (D), Arizona, 1999, Hw
David E. Fastovsky, (D), Wisconsin, 1986, PvGsPs

Chair:
Brian K. Savage, (D), Caltech, 2004, Ys
Associate Professor:
Dawn Cardace, (D), Washington, GpPg
Simon E. Engelhart, (D), Pennsylvania, OnGmYs
Assistant Professor:
Ananya Mallik, (D), Rice, 2014, Gx
Soni M. Pradhanang, (D), HsqHq
Emeritus:
J. Allan Cain, (D), Northwestern, 1962, Gx
O Don Hermes, (D), North Carolina, 1967, Gi
Daniel P. Murray, (D), Brown, 1976, ZeGtp

Graduate School of Oceanography (M,D)
215 South Ferry Road
Narragansett, RI 02882
p. (401) 874-6222
bcorliss@uri.edu
http://www.gso.uri.edu
Enrollment (2010): M: 41 (12) D: 31 (8)
Research Professor:
Theodore J. Smayda, (D), Oslo, 1967, Ob
Associate Dean:
Mark Wimbush, (D), California (San Diego), 1969, Op
Professor:
Robert D. Ballard, (D), Rhode Island, 1974, Ga
Steven N. Carey, (D), Rhode Island, 1982, Ou
Jeremy S. Collie, (D), MIT/WHOI, 1985, Ob
Peter Cornillon, (D), Cornell, 1973, Op
Steven L. D'Hondt, (D), Princeton, 1989, Ou
Edward G. Durbin, (D), Rhode Island, 1976, Ob
Isaac Ginis, (D), Inst Exp Meteor, 1986, Op
Tetsu Hara, (D), MIT, 1990, Op
Paul E. Hargraves, (D), William & Mary, 1968, Ob
David L. Hebert, (D), Dalhousie, 1988, Op
Christopher Kincaid, (D), Johns Hopkins, 1990, Ou
John King, (D), Minnesota, 1983, Ou
Roger Larson, (D), California (San Diego), 1970, Ou
Margaret Leinen, (D), Rhode Island, 1980, Ou
John T. Merrill, (D), Colorado, 1976, Oc
S. Bradley Moran, (D), Dalhousie, 1991, Oc
Scott W. Nixon, (D), North Carolina, 1970, Ob
Candace Oviatt, (D), Rhode Island, 1967, Ob
Hans Thomas Rossby, (D), MIT, 1966, Op
Lewis Rothstein, (D), Hawaii, 1983, Op
Haraldur Sigurdsson, (D), Durham (UK), 1970, Ou
Jennifer Specker, (D), Oregon State, 1980, Ob
Robert Tyce, (D), California (San Diego), 1976, Ou
D. Randolph Watts, (D), Cornell, 1973, Op
Karen Wishner, (D), California (San Diego), 1979, Ob
Associate Professor:
Brian G. Heikes, (D), Michigan, 1984, As
Yang Shen, (D), Brown, 1994, Ou
David C. Smith, (D), California (San Diego), 1994, Ob
Marine Research Scientist:
Percy Donaghay, (D), Oregon State, 1980, Ob
Kathleen Donohue, (D), Rhode Island, 1996, Op
Alfred K. Hanson, Jr., (D), Rhode Island, 1981, Oc
Barbara K. Sullivan-Watts, (D), Oregon State, 1977, Og
Associate Dean:
John Farrell, (D), Brown, 1991, Ou
Adjunct Professor:
Lawrence J. Buckley, (D), New Hampshire, 1975, Oc
Richard J. Pruell, (D), Rhode Island, 1984, Co
Charles T. Roman, (D), Delaware, 1981, Ob
Marine Research Scientist:
Dian J. Gifford, (D), Dalhousie, 1986, Ob
Robert D. Kenney, (D), Rhode Island, 1984, Ob
Emeritus:
H. Perry Jeffries, (D), Rutgers, 1959, Ob
John A. Knauss, (D), California, 1959, Op
Theodore A. Napora, (D), Yale, 1964, Ob
Michael E. Pilson, (D), California (San Diego), 1964, Oc
James G. Quinn, (D), Connecticut, 1967, Oc
Kenneth A. Rahn, (D), Michigan, 1971, Oc
Saul B. Saila, (D), Cornell, 1952, Ob
Jean-Guy E. Schilling, (D), MIT, 1966, OuCuc
John M. Sieburth, (D), Minnesota, 1954, Ob
Elijah V. Swift, (D), Johns Hopkins, 1967, Ob

South Carolina

Clemson University

Bob Campbell Geology Museum (A,B,M,D)
140 Discovery Lane
Clemson, SC 29634-0130
p. (864) 656-4602
tsteadm@clemson.edu
http://www.clemson.edu/geomuseum
Director:
Todd A. Steadman, (M), Louisiana State, 1987, ZnnZn
Curator:
David J. Cicimurri, (M), SD Mines, 1998, Pv
Christian M. Cicimurri, (M), SD Mines, 1999, Pv

Environmental Engineering and Earth Sciences (B,M)
321 Calhoun Drive
Room 445 Brackett Hall
Clemson, SC 29634-0919
p. (864) 656-3438
dfreedm@clemson.edu
http://www.clemson.edu/ces/departments/eees/
Administrative Assistant: Cynthia Rae Gravely
Enrollment (2013): B: 40 (9) M: 8 (6)
Chair:
Tanju Karanfil, (D), Michigan, 1995, NgZnn
Professor:
James W. Castle, (D), Illinois, 1978, GseHw
Ronald W. Falta, (D), California (Berkeley), 1990, Hq
Cindy M. Lee, (D), Colorado Mines, 1990, GeClZe
Lawrence C. Murdoch, (D), Cincinnati, 1991, Hw
Mark Schlautman, (D), Caltech, 1992, ClHgSc
Assistant Professor:
Stephen M.J Moysey, (D), Stanford, 2005, HwYuZe
Brian A. Powell, (D), Clemson, 2004, Cg
Lindsay C. Shuller-Nickles, (D), Michigan, 2010, Gz
Research Associate:
Scott E. Brame, (M), Clemson, 1993, Hw
Lecturer:
Alan B. Coulson, (D), South Carolina, 2009, PgGgCs
Adjunct Professor:
C. Brannon Andersen, (D), Syracuse, 1994, Cl
Christian M. Cicimurri, (M), South Dakota, 1999, Pv
Brian Looney, (D), Minnesota, 1984, Hq
Vaneaton Price, (D), North Carolina, 1969, Ce
Tommy Temples, (D), South Carolina, 1996, GoYg
Emeritus:
Lois B. Krause, (D), Clemson, 1996, Ze
Fred Molz, (D), Stanford, 1970, Sp
John R. Wagner, (D), South Carolina, 1993, ZeGm
Richard D. Warner, (D), Stanford, 1971, Gz

College of Charleston

Dept of Geology & Environmental Geosciences (B,M) O⊘
66 George Street
Charleston, SC 29424
p. (843) 953-5589
callahant@cofc.edu
http://geology.cofc.edu/
f: https://www.facebook.com/Geology.CofC
t: @cofcgeology
Programs: Geology, Environmental Geosciences
Enrollment (2022): B: 126 (28)
Chair:
Timothy J. Callahan, (D), New Mexico Tech, 2001, Hwq
Professor:
Norman S. Levine, (D), Purdue, 1995, ZiGeNg
Vijay M. Vulava, (D), Swiss Fed Inst Tech, 1998, CgHwSc
Director, Graduate Program in Environmental and Sustainability Studies:
K. Adem Ali, (D), Kent State, 2011, ZrHwNg
Associate Professor:
Barbara Beckingham, (D), Maryland, 2011, ClHwGe
Erin K. Beutel, (D), Northwestern, 2000, Gct
John Chadwick, (D), Florida, 2002, GivZr
Mitchell W. Colgan, (D), California (Santa Cruz), 1990, GrsZc
M. Scott Harris, (D), Delaware, 2000, OnGam

Steven C. Jaumé, (D), Columbia, 1994, YsRnYg
Cassandra R. Runyon, (D), Hawaii, 1988, XgGmZe
Curator, Mace Brown Museum of Natural History:
W. Scott Persons, (D), Alberta, 2016, PegGe
Assistant Professor:
Theodore R. Them, II, (D), Virginia Tech, 2016, CmOcZc
Emeritus:
James L. Carew, (D), Texas, 1978, PgGde
Michael P. Katuna, (D), N Carolina, 1974, Gus
Robert L. Nusbaum, (D), Missouri (Rolla), 1984, Gz
Leslie R. Sautter, (D), South Carolina, 1990, OuGuOn
Laboratory Director:
Robin Humphreys, (M), Charleston, 2000, Ge

Furman University

Earth, Environmental, and Sustainability Sciences (B) ☒
3300 Poinsett Highway
Greenville, SC 29613
p. (864) 294-2052
nina.anthony@furman.edu
https://www.furman.edu/academics/earth-environmental-sciences/
f: https://www.facebook.com/EESdepartment/
Programs: Earth and Environmental Sciences, and Sustainability Science
Administrative Assistant: Nina Anthony
Enrollment (2021): B: 25 (11)
Chair:
Suresh Muthukrishnan, (D), Purdue, 2002, GmZir
Professor:
C. Brannon Andersen, (D), Syracuse, 1994, ClGe
Geoffrey Habron, (D), Oregon State, 1999, Zn
Associate Professor:
Matt Cohen, (D), Arizona State, 2015, Zn
Assistant Professor:
Karen Allen, (D), Georgia, 2016, Zn
Ruth F. Aronoff, (D), Purdue, 2016, Gc
Emma Hunt, (D), St. Andrews, 2015, GpzEg
Lecturer:
Melissa Ranhofer, (D), South Carolina, 2009, Gg
Adjunct Professor:
Valeska Minkowski, (D), Florida, 2015, Zn
Courtney Quinn, (D), Nebraska, 2012, Zn
Emeritus:
John M. Garihan, (D), Penn State, 1973, Gct
William A. Ranson, (D), Massachusetts, 1979, GxzGp
Kenneth A. Sargent, (D), Oklahoma, 1973, Hy
Laboratory Manager:
Christopher Romanek, (D), Texas A&M, 1991, Cls

University of South Carolina

School of the Earth, Ocean & Environment (B,M,D) ●⦸
701 Sumter St
EWS 617
Columbia, SC 29208
p. (803) 777-4535
gyogodzin@geol.sc.edu
http://sc.edu/seoe
f: https://www.facebook.com/MarineScienceSC/
Programs: Geological Sciences (B,M,D); Marine Science (B,M,D); Environmental Science (B); Environmental Studies (B); Earth & Environmental Resources Management (M,JD);
Enrollment (2022): B: 580 (125) M: 33 (12) D: 29 (0)
Director of Graduate Studies - School of Earth, Ocean, and Environment:
Michael Bizimis, (D), Florida State, 2001, GiCg
Associate Director - School of Earth, Ocean, and Environment:
Gene Yogodzinkski, (D), Cornell, 1993, GiCuc
Associate Dean for Instruction, Community Engagement and Research:
Claudia Benitez-Nelson, (D), MIT/WHOI, 1999, Oc
Acting Director - School of Earth, Ocean, and Environment:
Peter Chametzky, (D), CUNY, 1991, Zn
Professor:
Carol Boggs, (D), Texas, 1979, Zn
Subrahmanyam Bulusu, (D), Southampton (UK), 1998, Op
James Pinckney, (D), South Carolina, 1992, Obn
Joseph Quattro, (D), Rutgers, 1991, Ob
Tammi Richardson, (D), Dalhousie, 1996, Ob
Raymond Torres, (D), California (Berkeley), 1997, Hy
George Voulgaris, (D), Southampton (UK), 1992, OnpOg
Scott M. White, (D), California (Santa Barbara), 2001, YrZr
Alicia M. Wilson, (D), Johns Hopkins, 1999, Hw
Alexander E. Yankovsky, (D), Marine Hydrophysical Inst (Ukraine), 1991, OpnAm
Associate Professor:
David Barbeau, Jr., (D), Arizona, 2003, Gs
Jessica Barnes, (D), Columbia, 2010, Zn
Conor Harrison, (D), North Carolina, 2014, Zn
David Kneas, (D), Yale, 2014, Zn
Andrew L. Leier, (D), Arizona, 2005, Gst
Katherine Ryker, (D), North Carolina State, 2014, Ze
Howie Scher, (D), Florida, 2005, GuCl
Lori A. Ziolkowski, (D), California (Irvine), 2009, CoOcCs
Assistant Professor:
Monica Barra, (D), New York, 2019, Zn
Annie Bourbonnais, (D), Victoria, 2012, Oc
Besim Dragovic, (D), Boston, 2013
Daniel Frost, (D), California (Berkeley), 2014, Ys
David E. Fuente, (D), North Carolina, 2017, Zn
R. Dean Hardy, (D), Georgia, 2016, Zn
Erin Meyer-Gutbrod, (D), Cornell, 2016, Ob
Research Associate Professor:
Matthew Kimball, (D), Rutgers, 2008, Ob
Erik Smith, (D), Maryland, 2000, HsOb
William Strosnider, (D), Oklahoma, 2010, On
Research Assistant Professor:
Robert Dunn, (D), California (Davis), 2019, Ob
Research Associate:
Philip Crotwell, (D), S Carolina, 2007, Ys
Eric Tappa, (M), S Carolina, 1984, Oo
Director of Undergraduate Studies:
William Joseph Jones, (D), California (Santa Cruz), 2001, Ob
Emeritus:
Philip Barnes, (D), Zn
Ron Benner, (D), Georgia, 1984, Ob
Bruce Coull, (D), Lehigh, 1968, Ob
John Mark Dean, (D), Purdue, 1962, Ob
Robert Feller, (D), Washington, 1977, Ob
Madilyn Fletcher, (D), Univ Coll (Wales), 1975, Ob
Gwendelyn Geidel, (D), South Carolina, 1982, HwGeCg
James N. Kellogg, (D), Princeton, 1981, Yg
Christopher G. Kendall, (D), Imperial Coll (UK), 1966, Gs
Willard S. Moore, (D), SUNY (Stony Brook), 1969, Cc
Thomas J. Owens, (D), Utah, 1984, Ys
Jennifer R. Pournelle, (D), California (San Diego), 2003, GaZuSa
W. Edwin Sharp, (D), California (Los Angeles), 1964, Gz
Stephen Stancyk, (D)
Pradeep Talwani, (D), Stanford, 1973, Ys
Sarah A. Woodin, (D), Washington, 1972, Ob
Richard Zingmark, (D)

University of South Carolina, Lancaster

University of South Carolina, Lancaster (A) ⦸
P.O. Box 889
Lancaster, SC 29721
p. (803) 313-7129
martek@mailbox.sc.edu
http://usclancaster.sc.edu
Instructor:
Lynnette Flann Martek, (M), Emporia State, 1994, ZgAm

Winthrop University

Dept of Chemistry, Physics, & Geology
Sims Science Building
Winthrop University
Rock Hill, SC 29733
p. (803) 323-4949
boyerd@winthrop.edu
https://www.winthrop.edu/cas/chemistry/
Chair, Environmental Sciences and Studies Program:
Marsha S. Bollinger, (D), South Carolina, 1986, CmOcCc
Professor:
Irene B. Boland, (D), South Carolina, 1996, GtgZe
Associate Professor:
Gwen M. Daley, (D), Virginia Tech, 1999, PqgGs
Scott P. Werts, (D), Johns Hopkins, 2006, ScbPg

Wofford College

Dept of Geology
Wofford College
429 North Church Street
Spartanburg, SC 29303-3663
p. (864) 597-4527
fergusonta@wofford.edu
http://www.wofford.edu/geology/
Director:
Terry A. Ferguson, (D), Tennessee, 1988, Ga

South Dakota

Black Hills State University

School of Natural Sciences (B) ●☒
1200 University Street, Unit 9008
Spearfish, SD 57799-9008
p. (605) 642-6506
abigail.domagall@bhsu.edu
https://www.bhsu.edu/Academics/Natural-Sciences/Environmental-Physical-Science
f: https://www.facebook.com/EnvPhysSciBHSU/
t: Eps@BHSU
Programs: Environmental Physical Science, Earth Science minor, Composite Science Education
Enrollment (2021): B: 30 (9)
Associate Professor:
Abigail M S Domagall, (D), SUNY (Buffalo), 2008, GveZe

Oglala Lakota College

Dept of Math, Science & Technology
P.O. Box 490
Kyle, SD 57755
p. (605) 455-6124
hlagarry@olc.edu
http://www.olc.edu/local_links/smet/

South Dakota School of Mines & Technology

Dept of Atmospheric and Environmental Sciences (B,M,D)
501 E. St. Joseph Street
Rapid City, SD 57701-3995
p. (605) 394-2291
william.capehart@sdsmt.edu
http://www.ias.sdsmt.edu/
Enrollment (2011): B: 15 (0) M: 14 (3)
Associate Professor:
William J. Capehart, (D), Penn State, 1997, AsZoHq
Donna V. Kliche, (D), SD Mines, 2007, AssAs
P. V. Sundareshwar, (D), South Carolina, 2002, ZeCgSb
Assistant Professor:
Adam French, (D), North Carolina State, 2011, Asm
Lisa Kunza, (D), Wyoming, 2012, HsZe
Instructor:
Darren R. Clabo, (M), SD Mines, 2009, AssAs
Emeritus:
Andrew G. Detwiler, (D), SUNY (Albany), 1980, AspAm
John H. Helsdon, (D), SUNY (Albany), 1979, As
Mark R. Hjelmfelt, (D), Chicago, 1980, As
Paul L. Smith, (D), Carnegie Inst, 1960, As

Dept of Geology & Geological Engineering (B,M,D) ○⊘
501 E. Saint Joseph St.
Rapid City, SD 57701-3901
p. (605) 394-2461
laurie.anderson@sdsmt.edu
https://www.sdsmt.edu/GGE/
f: https://www.facebook.com/SDSMTGeologyGeologicalEngineering
Programs: Geology (B); Geological Engineering (B); Geology and Geological Engineering (M,D); Paleontology (M)
Certificates: Geospatial Technology (UG,G); Petroleum Systems (G)
Administrative Assistant: Cleo J. Heenan
Enrollment (2022): B: 112 (16) M: 23 (8) D: 3 (2)
Field Station Director:
Nuri Uzunlar, (D), SD Mines, 1993, EmCeGc
Department Head; Director Museum of Geology:
Laurie C. Anderson, (D), Wisconsin, 1991, PiePq
Dean of Graduate Education:
Maribeth H. Price, (D), Princeton, 1995, Zir
Professor:
Edward F. Duke, (D), Dartmouth, 1984, GpiZr
Timothy L. Masterlark, (D), Wisconsin, 2000, YdhNr
Larry D. Stetler, (D), Washington State, 1993, NxGsm
Research Scientist:
Roger L. Nielsen, (D), Southern Methodist, 1983, CptGi
Associate Professor:
Kurt W. Katzenstein, (D), Nevada (Reno), 2008, NxtZr
Liangping Li, (D), Polytech Valencia (Spain), 2011, HwqHt
Darrin C. Pagnac, (D), California (Riverside), 2005, PvGrZe
Gokce K. Ustunisik, (D), Cincinnati, 2009, Cp
Assistant Professor:
Sarah W. Keenan, (D), Tennessee, 2014, PvClb
Trevor S. Waldien, (D), California (Davis), 2021, GctGp
Kevin M. Ward, (D), Arizona, 2016, YsGtYg
Coordinator and Instructor:
Christopher J. Pellowski, (D), SD Mines, 2012, Gg
Associate Director, Museum of Geology:
Nathaniel Fox, (D), California (Merced), 2020, Pgv
Instructor:
Victoria M. Karnes, (M), Arkansas (Fayetteville), 2022, ZifZr
Emeritus:
Phillip R. Bjork, (D), Michigan, 1968, Pg
Arden D. Davis, (D), SD Mines, 1983, HwNx
James E. Fox, (D), Wyoming, 1972, Gs
James E. Martin, (D), Washington, 1979, Ps
Colin J. Paterson, (D), Otago (NZ), 1978, Eg
Perry H. Rahn, (D), Penn State, 1965, Ng
William M. Roggenthen, (D), Princeton, 1980, NgYg
J. Foster Sawyer, (D), SD Mines, 2006, GsoHw
Preparator and Lab Manager:
Kayleigh Johnson, (M), SD Mines, 2018, Pgv

University of South Dakota

Dept of Earth Sciences & Physics (B)
414 East Clark Street
Vermillion, SD 57069-2390
p. (605) 677-5649
esci@usd.edu
http://www.usd.edu/earthsciences/
Enrollment (2011): B: 25 (9)
Chair:
Timothy H. Heaton, (D), Harvard, 1988, PvOg
Associate Professor:
Brennan T. Jordan, (D), Oregon State, 2002, GiAm
Mark R. Sweeney, (D), Washington State, 2004, Gms
Instructor:
Jeanne M. Fromm, (M), Idaho State, 1995, GgHs

Tennessee

Austin Peay State University

Geosciences Dept (B)
601 College St
Clarksville, TN 37044
p. (931) 221-7454
deibertj@apsu.edu
http://www.apsu.edu/geosciences
Programs: Geosciences
Enrollment (2019): B: 85 (10)
Chair:
Jack Deibert, (D), Wyoming, GsrGo
Professor:
Phyllis A. Camilleri, (D), Wyoming, 1994, Gct
Daniel L. Frederick, (D), Tennessee, 1994, PgGsr
Gregory D. Ridenour, (D), Texas A&M, 1993, HsOgZn
Associate Professor:
Kallina M. Dunkle, (D), Wisconsin, 2012, HwGle
Christopher Gentry, (D), Indiana State, 2008, Ziy
Christine Mathenge, (D), Indiana, 2008, Zcy
Assistant Professor:
Madeline M. Giefer, (D), N Carolina, 2020, ZcrZi
Erik L. Haroldson, (D), Wisconsin, 2016, EdGzCg

Emeritus:
Phillip R. Kemmerly, (D), Oklahoma State, 1973, GmeHg
Laboratory Director:
Randal P. Roberson, (M), Murray State, 2018, Gg
Other:
Richard F. Wheeler, (M), Brigham Young, 1980, Ggo

Middle Tennessee State University

Dept of Geosciences (B,M)
Box 9
Davis Science Building
Room 241
Murfreesboro, TN 37132
p. (615) 898-2726
karen.wolfe@mtsu.edu
https://www.mtsu.edu/geosciences/
f: https://www.facebook.com/groups/88868852075/
t: @MTSUGeosciences
Programs: Geology; Earth Science; Earth Science for Teachers; Physical Geography; Geospatial Analysis; Environmental Science
Administrative Assistant: Karen M. Wolfe
Enrollment (2022): B: 154 (38) M: 22 (8)
Chair:
Henrique G. Momm, (D), Mississippi, 2008, SpZfRw
Director:
Zada Law, (M), Wisconsin, 1980, Zin
Professor:
Mark J. Abolins, (D), Caltech, 1999, Gc
Warner Cribb, (D), Ohio State, 1993, GipGz
Associate Professor:
Jeremy Aber, (D), Kansas State, 2011, Zyi
Racha El Kadiri, (D), W Michigan, 2014, HgGge
Clay D. Harris, (D), Indiana, 1992, Gse
Melissa Lobegeier, (D), James Cook, 2001, Pmg
Assistant Professor:
Alisa L. Hass, (D), Tennessee, 2019, AtZy
Instructor:
Laura R. Collins, (M), Mississippi State, 2005, Gg
Leslie Valentine, (D), Louisiana State, 2022, Gms
Lecturer:
Alan Brown, (M), Illinois State, 2005, Gg

Motlow State Community College

Dept of Natural Sciences (A)
PO Box 8500
Lynchburg, TN 37352-8500
p. (931) 393-1810
lmayo@mscc.edu
http://www.mscc.edu/natural_science/
Instructor:
Lisa L Herring Mayo, (M), Mississippi State, 2000, GgZge

Pellissippi State Community College

Natural and Behavioral Sciences
10915 Hardin Valley Road
P.O. Box 22990
Knoxville, TN 37801
p. (865) 694-6685
kmbarker1@pstcc.edu
http://www.pstcc.edu
Adjunct Professor:
Peter J. Lemiszki, (D), Tennessee, 1992, GcgZi

Roane State Community College - Oak Ridge

Mathematics and Sciences (Geology) (A)
276 Patton Lane
Harriman , TN 37748
p. (865) 481-2000
leea@roanestate.edu
http://aclee1234.fortunecity.com
Professor:
Arthur C. Lee, (D), S California, 1994, Ges

Sewanee: University of the South

Dept of Earth and Environmental Systems (B)
735 University Avenue
Sewanee, TN 37383-1000
p. (931) 598-1271
Sherwood@sewanee.edu
http://www.sewanee.edu/EnvStudies
Chair:
Scott Torreano, (D), Georgia, 1991, Sf
Professor:
Martin A. Knoll, (D), Texas (El Paso), 1988, Hw
Donald B. Potter, Jr., (D), Massachusetts, 1985, Gc
Stephen A. Shaver, (D), Stanford, 1984, Eg
Associate Professor:
C. Ken Smith, (D), Florida, 1996, Sf
Adjunct Professor:
Glendon W. Smalley, (D), Tennessee, 1975, Sf

Tennessee Tech University

Dept of Earth Sciences (B)
PO Box 5062
Cookeville, TN 38505
p. (931) 372-3121
MHarrison@tntech.edu
https://www.tntech.edu/cas/earth/index.php
Programs: Geosciences
Administrative Assistant: Peggy Medlin
Enrollment (2022): B: 40 (12)
Chair:
Jeannette Wolak, (D), Montana State, 2011, Gsr
Professor:
Michael J. Harrison, (D), Illinois (Urbana), 2002, Gc
Evan A. Hart, (D), Tennessee, 2000, Zy
Ping-Chi Li, (D), Iowa, 1992, Zi
Associate Professor:
Joseph Asante, (D), Nevada, 2012, HwZrGe
Assistant Professor:
Lauren Michel, (D), Baylor, 2014, PcSaCl
Emeritus:
Larry W. Knox, (D), Indiana, 1974, Pmg
H. Wayne Leimer, (D), Missouri, 1969, Gz
Laboratory Director:
Philip Roberson, (M), Murray State, 2018, GgPg

University of Memphis

Center for Earthquake Research & Information (CERI) (M,D)
3876 Central Avenue, Suite 1
Memphis, TN 38152-3050
p. (901) 678-2007
clangstn@memphis.edu
http://www.memphis.edu/ceri
Programs: M.S. and PhD concentration in Geophysics
Enrollment (2018): M: 9 (2) D: 17 (1)
Director:
Charles A. Langston, (D), Caltech, 1976, Ys
Graduate Coordinator:
Christine A. Powell, (D), Princeton, 1976, Ys
Professor:
Randel Tom Cox, (D), Missouri, 1995, GtcGm
Associate Professor:
Eunseo Choi, (D), Caltech, 2008, Gtq
Assistant Professor:
Thomas H. Goebel, (D), S California, 2013, Yxs
Christodoulos Kyriakopoulos, (D), Bologna, 2011, Yd
Research Professor :
James Dorman, (D), Columbia, 1961, Yse
CERI Founding Director:
Archibald C. Johnston, (D), Colorado, 1979, Ys
Emeritus:
Jer-Ming Chiu, (D), Cornell, 1982, Ys
Research Scientist:
Stephen P. Horton, (D), Nevada (Reno), 1992, Ys
Research Professor :
Chris Cramer, (D), Stanford, 1976, Ys
Robert Smalley, Jr., (D), Cornell, 1988, Yd
Assoc. Research Professor:
Mitchell M. Withers, (D), New Mexico Tech, 1997, Ys
Seismic Network Engineer:
James Bollwerk, (M), Memphis
Research Equipment Technician II:
Chris McGoldrick

David Steiner
Research Equipment Tech II:
Patrick Shivers
Research Associate Technician:
John Parker
Research Associate II:
Kent Moran, (D), Memphis
Holly Withers
Local Technical Support Provider II:
Robert Debula
Deshone Marshall
Local Technical Support Provider I:
James Davis, (D), Memphis, 2013
Director Education & Outreach:
Gary Patterson, (M), Memphis, Gg
Digital Seismic Systems Supervisor:
Steve Brewer, (M), Memphis
Assistant Director Administration & Finance:
Michelle Smith, (B)

Dept of Earth Sciences (B,M,D)
109 Johnson Hall
488 Patterson Street
Memphis, TN 38152-3550
p. (901) 678-4571 or 678-2177
aahill@memphis.edu
http://memphis.edu/earthsciences/
Programs: Earth Sciences (B, M, D); Geoarchaeology (B); Geography (B, M); Geology (B, M); Archaeology (M); Interdisciplinary Studies (M)
Certificates: GIS (G)
Enrollment (2019): B: 60 (7) M: 23 (13) D: 33 (5)
Chair:
Daniel Larsen, (D), New Mexico, 1994, ClGsHw
Director, Confucius Institute:
Hsiang-Te Kung, (D), Tennessee, 1980, Zy
Professor:
Mervin J. Bartholomew, (D), Virginia Tech, 1971, Gtc
Randel T. Cox, (D), Missouri, 1995, GcmGt
David H. Dye, (D), Washington, 1980, Ga
Arleen A. Hill, (D), South Carolina, 2002, Rn
Esra Ozdenerol, (D), Louisiana State, 2000, Zi
Jose Pujol, (D), Wyoming, 1985, Ye
Roy B. Van Arsdale, (D), Utah, 1979, Gcm
Associate Professor:
Anzhelika Antipova, (D), Louisiana State, 2010, Zg
Dorian Burnette, (D), Arkansas, 2009, Atm
Andrew M. Mickelson, (D), Ohio State, 2002, Ga
Ryan M. Parish, (D), Memphis, 2013, Ga
Assistant Professor:
Youngsang Kwon, (D), SUNY (Buffalo), 2012, Zi
Deborah Leslie, (D), Ohio State, 2013, HwCl
Visiting Professor:
Elizabeth Rhenberg, (D), West Virginia, 2015, Pg
Emeritus:
Phili B. Deboo, (D), Louisiana State, 1963, Pg
Robert W. Deininger, (D), Rice, 1964, Gx
James Dorman, (D), Columbia, 1961, Ys
Archibald C. Johnston, (D), Colorado, 1979, Ys
David N. Lumsden, (D), Illinois (Urbana), 1965, Gd
Instructor/Coordinator:
Julie Johnson, (D), Florida Intl, 2012, Gx

University of Tennessee, Chattanooga

Biology, Geology, and Environmental Science (B) ☒
615 McCallie Ave., Dept. 2653
Chattanooga, TN 37403
p. (423) 425-4341
http://www.utc.edu/biology-geology-environmental-science/division-geology/
f: https://www.facebook.com/GeologyatUTC
Programs: Geology:Geology,
Geology: Environmental Geology, Geology: STEM Education
Enrollment (2021): B: 33 (13)
Associate Dept. Head:
Amy Brock-Hon, (D), Nevada, 2007, GmSc
Professor:
Jonathan W. Mies, (D), North Carolina, 1990, GctHg
Assistant Professor:
Stephanie DeVries, (D), HgCq
A K M Azad Hossain, (D), Mississippi, 2008, ZriZg
Ashley Manning-Berg, (D), GsPo
Laboratory Coordinator:
Jason Muhlbauer, (D), Tennessee, 2021, Gg
Adjunct Professor:
Gregory Brodie, (M), Purdue, 1979, Ge

University of Tennessee, Knoxville

Dept. of Earth & Planetary Sciences (B,M,D)
602 Strong Hall
1621 Cumberland Ave.
Knoxville, TN 37996-1526
p. (865) 974-2366
eps@utk.edu
https://eps.utk.edu/
f: https://www.facebook.com/UTEPS
Programs: Geology; Environmental Studies
Enrollment (2021): B: 45 (30) M: 20 (6) D: 30 (1)
Acting Head:
Harry Y. McSween, Jr., (D), Harvard, 1977, XcGi
Governors Chair:
Terry C. Hazen, (D), Michigan State, 1973, Po
Professor:
William M. Dunne, (D), Bristol, 1980, Gc
Annette S. Engel, (D), Texas, 2004, Cl
Christopher Fedo, (D), Virginia Tech, 1994, GsXgCg
Linda Kah, (D), Harvard, 1997, Gs
Larry D. McKay, (D), Waterloo, 1991, HyGe
Michael L. McKinney, (D), Yale, 1985, GePg
Jeffery E. Moersch, (D), Cornell, 1997, XgZr
Associate Professor:
Micah Jessup, (D), Virginia Tech, 2007, Gc
Molly McCanta, (D), Brown, 2004, GiXg
Colin Sumrall, (D), Texas, 1997, Pi
Anna Szynkiewicz, (D), Wroclaw, 2004, CslXg
ORNL Joint Faculty:
Elizabeth M. Herndon, (D), Penn State, 2012, Cg
Melanie A. Mayes, (D), Tennessee, 2006, ClGeHw
Assistant Professor:
Nick Dygert, (D), Brown, 2015, Giz
Andrew Steen, (D), North Carolina, 2009, CoOb
Bradley J. Thomson, (D), Brown, 2006, Xg
Lecturer:
Md Iftekhar Alam, (D), Oklahoma State, 2016, Yu
Jake Benner, (M), Utah, 2002, ZePg
Stephanie K. Drumheller-Horton, (D), Iowa, 2012, Pg
Estifanos Haile, (D), Kentucky, 2011, Hy
Adjunct Professor:
Gary G. Bible, (D), Iowa State, 1978, Go
Emily J. Chin, (D), Rice, 2013, Cu
John M. Cottle, (D), Oxford, 2008, Gt
William E. Doll, (D), Wisconsin, 1983, Ys
Steven G. Driese, (D), Wisconsin, 1982, SaClGs
Craig J. Hardgrove, (D), Tennessee, 2011, Xg
Christian Klimczak, (D), Nevada, 2011, Xg
Jasper F. Kok, (D), Michigan, 2009, Ap
Peter J. Lemiszki, (D), Tennessee, 1992, Gc
Claudia I. Mora, (D), Wisconsin, 1988, Cs
Matthew L. Niemiller, (D), Tennessee, 2011, Zn
Tommy J. Phelps, (D), Wisconsin, 1985, Cb
Louise M. Prockter, (D), Brown, 2000, Xg
Ganapathy Shanmugam, (D), Tennessee, 1978, Gs
Ben Thuy, (D), Gottingen, Germany, 2012, Pg
Christina E. Viviano-Beck, (D), Tennessee, 2012, Xg
Steven W. Wilhelm, (D), W Ontario, 1994, Po
Interim Head:
Edmund Perfect, (D), Cornell, 1986, SpHwGq
Emeritus:
Thomas W. Broadhead, (D), Iowa, 1978, Pi
Robert D. Hatcher Jr., (D), Tennessee, 1965, Gct
Kula C. Misra, (D), W Ontario, 1973, Eg
Research Professor:
Liyuan Liang, (D), Caltech, 1988, Cb
Robert Riding, (D), Sheffield, UK, Cb

University of Tennessee, Martin

Dept of Agriculture, Geosciences, and Natural Resources (B)
256 Brehm Hall
Martin, TN 38238
p. (731) 881-7260
mehlhorn@utm.edu
http://www.utm.edu/departments/caas/agnr/geosciences/
Enrollment (2013): B: 29 (13)

Professor:
Paula M. Gale, (D), Arkansas, 1988, SodSc
Michael A. Gibson, (D), Tennessee, 1988, PgiZe
Jefferson S. Rogers, (D), Illinois, 1995, Zn
Robert M. Simpson, (D), Indiana State, 2000, AsmZi

Associate Professor:
Stan P. Dunagan, (D), Tennessee, 1998, GsSa

Assistant Professor:
Thomas A. DePriest, (D), Union (Jackson), 2009, Ze
Benjamin P. Hooks, (D), Maine, 2009, GciNr

Instructor:
Eleanor E. Gardner, (M), Georgia, GgPg

Emeritus:
William T. McCutchen, (M), Berea, 1967, Gg
Robert P. Self, (D), Rice, 1971, Gs
Helmut C. Wenz, (M), W Michigan, 1968, Zn

Vanderbilt University

Earth & Environmental Sciences (B,M,D) ☒
2301 Vanderbilt Place
VU Station B 351805
Nashville, TN 37235-1805
p. (615) 322-2976
gale.newton@vanderbilt.edu
http://www.vanderbilt.edu/ees/
f: https://www.facebook.com/groups/393646734010753/
t: @vanderbiltEES
Enrollment (2020): B: 45 (10) D: 17 (2)

Chair:
Steven L. Goodbred, Jr., (D), William & Mary, 1999, GsOn

Professor:
John C. Ayers, (D), Rensselaer, 1991, CgGx
Ralf Bennartz, (D), Free (Berlin), 1997, AsZr
David J. Furbish, (D), Colorado, 1985, GmIg
Guilherme Gualda, (D), Chicago, 2010, GivGz
George Hornberger, (D), Stanford, 1970, HgwHs

Associate Professor:
Larisa R.G DeSantis, (D), Florida, 2009, Pve
Jonathan M. Gilligan, (D), Yale, 1991, ZcRwZu
Jessica L. Oster, (D), California (Davis), 2010, ClsPc

Assistant Professor:
Simon A.F. Darroch, (D), Yale, 2015, PgiPe
Kristen Fauria, (D), California (Berkeley), 2017, GvqYu
Maria Luisa Jorge, (D), Illinois (Chicago), 2007, Pg
Neil P. Kelley, (D), California (Davis), 2012, GgOgPg

Principal Senior Lecturer:
Lily L. Claiborne, (D), Vanderbilt, 2011, GivCg
Daniel J. Morgan, (D), Washington, 2009, GmCc

Emeritus:
Leonard P. Alberstadt, (D), Oklahoma, 1967, Gd
Calvin F. Miller, (D), California (Los Angeles), 1977, Gi
Molly F. Miller, (D), California (Los Angeles), 1977, PeGs
Arthur L. Reesman, (D), Missouri, 1966, Gg
William G. Siesser, (D), Cape Town, 1971, PmGsu

Volunteer State Community College

Volunteer State Community College
1800 Nashville Pike
Gallatin, TN 37006
p. (615) 230-3294
Clark.Cropper@volstate.edu
http://www.volstate.edu

Walters State Community College

Walters State Community College
500 South Davy Crockett Parkway
Morristown, TN 37813
p. (423) 585-6764
http://www.ws.edu

Texas

Alamo Colleges, Palo Alto College

Dept of Geology (A)
1400 W. Villaret Blvd
San Antonio, TX 78224
p. (210) 486-3000
swilkins6@alamo.edu
http://alamo.edu/pac/geology/

Alamo Colleges, San Antonio College

Natural Sciences
1819 N Main Ave
San Antonio, TX 78212
p. (210) 486-0840
dlambert@alamo.edu
http://alamo.edu/sac/earthsci/

Professor:
Dean Lambert, (D), Texas, Zy

Adjunct Professor:
Dwight Jurena, (M), Renssalear, GgZg
Ryan E. Rudnicki, (D), Penn State, Zy
Charles K. Smith, (M), Texas State, Zy

Alvin Community College

Physcial Science Program (A)
3110 Mustang Rd
Alvin, TX 77511
p. (281) 756-5670
pobrien@alvincollege.edu
https://www.alvincollege.edu/physical-science/faculty.html
Programs: Physical Sciences
Enrollment (2012): A: 2 (1)

Geology Instructor:
Philip O'Brien, (M), Texas Tech, 2014, Gg

Amarillo College

Dept of Physical and Biological Sciences
P.O. Box 447
Amarillo, TX 79178
p. (806) 371-5030
rdhobbs@actx.edu
https://www.actx.edu/pscience/
Enrollment (2009): A: 3 (0)

Professor:
Richard D. Hobbs, (D), Wyoming, 1998, ZrGc

Adjunct Professor:
David Pertl, (M), W Texas A&M, 1984, Go

Angelo State University

Dept of Physics and Geosciences (B) O
ASU Station #10904
San Angelo, TX 76909
p. (325) 942-2242
joseph.satterfield@angelo.edu
https://www.angelo.edu/departments/physics-geosciences/Geosciences/
Programs: Geosciences (B)
Enrollment (2022): B: 42 (8)

Professor:
Joseph I. Satterfield, (D), Rice, 1995, Gc

Associate Professor:
Heather L. Lehto, (D), S Florida, 2012, GvYsZe

Assistant Professor:
Elizabeth Koeman-Shields, (D), Notre Dame, 2015, CgXg

Instructor:
Jessica Garza, (M), Hawaii, 2011, AmGg
Stephen Shields, (M), Missouri State, 2013, Ggm

Adjunct Professor:
Cary D. Carman, (B), Angelo State, 1999, Hg
Steven Lyons, (D), Hawaii, 1981, Am
Fred L. Wilson, (D), Kansas, 1964, Gm

TEXAS

Austin Community College District

Dept of Earth and Environmental Sciences (A)

6101 Highland Campus Drive
Austin, TX 78752
p. (512) 223-7157
mark.shepherd@austincc.edu
https://sites.austincc.edu/ees/
Programs: Environmental Studies (A); Environmental Technology (A); Geology (A)
Certificates: Environmental Technology
Administrative Assistant: Oralia Guerra
Enrollment (2020): A: 227 (12)

Chair:
Mark A. Shepherd, (D), Nebraska (Medical Center), 2015, AsObZn
Associate Dean:
Ronald A. Johns, (D), Texas, 1993, Pi
Professor:
Robert H. Blodgett, (D), Texas, 1990, RnGsZe
Associate Professor:
Heather R. Miller, (D), Texas A&M, 2012, Zeg
Peter J. Wehner, (M), Vanderbilt, 1992, GieXs
Adjunct Professor:
Brittany N. Blomberg, (D), Texas A&M (Corpus Christi), 2015, OnbRn
Peter A. Boone, (D), Texas A&M, 1972, Gro
Thomas W. Brown, (M), Indiana, 1987, Ges
Laura E. Chapa, (M), Texas State, 2014, ZgeZr
M. Jennifer Cooke, (D), Texas, 2005, GmCl
Leslie M. Davis, (M), Florida State, 1986, Op
Meredith Y. Denton-Hedrick, (M), Texas A&M, 1992, ZeYeEo
Maedeh Faraji, (D), Texas, 2007, AsZge
Kusali R. Gamage, (D), Florida, 2005, HyOuPm
Khaled W. Hasan, (D), Texas A&M, 1995, HsZri
Elena K. Keen, (D), Texas A&M (Corpus Christi), 2017
Zackary E. Martin, (M), Texas State, 2006, RwZn
Amy L. Moreland, (D), Texas, 2011, Zge
Ata U. Rahman, (D), Texas Tech, 1987, GgdGe
Fabienne M. Rambaud, (M), Texas, 2005, Eg
Carolyn M. Riess, (M), Texas (El Paso), 1984, GoeGg
Alina A. Satkoski, (M), Syracuse, 2011, GemGg
Raymond M. Slade, Jr., (B), SW Texas, 1971, HgsHq
Paepin K. Starr, (D), Texas State, 2019, GlZrRc
Jason H. Stephens, (D), Texas, 2014, YrGuYe
Anne Turner, (M), Texas, 1986, Hg
Holly Zafian, (M), Texas State, 2011, Ze
Science Laboratory Technician:
Shannon M. Grace, (M), Illinois, 2017, Zn
Amie C. Hammond, (B), Texas, 2014, ZeGg
Sean C. Murphy, (M), Georgia, 1984, EgGpRc
Cooperating Faculty:
David J. Froehlich, (D), Texas, 1996, Pv

Baylor University

Dept of Environmental science

Waco, TX 76706
p. (254) 710-3406
George_Cobb@baylor.edu
https://www.baylor.edu/environmentalscience/index.php?id=55209

Dept of Geosciences (B,M,D) ● ⃠

One Bear Place #97354
101 Bagby Ave.
BSB, 4th Floor, Rm. D409
Waco, TX 76798-7354
p. (254) 710-2361
paulette_penney@baylor.edu
http://www.baylor.edu/geosciences
f: https://www.facebook.com/baylorgeosciences
t: @BaylorGeo
Programs: Geology; Geophysics; Earth Science
Administrative Assistant: Janelle Atchley
Administrative Assistant: Jamie J.. Ruth
Enrollment (2022): B: 36 (4) M: 10 (5) D: 31 (2)

W.M. Keck Foundation Professor of Geophysics:
Robert Jay Pulliam, (D), California (Berkeley), 1991, YsgYe
Keck Foundation Prof. of Geophysics:
Jay Pulliam, (D), California (Berkeley), 1991, YsgYe
Chair:
Joe C. Yelderman, Jr., (D), Wisconsin, 1983, HwRwHw
Professor:
Stacy C. Atchley, (D), Nebraska, 1990, GroGo
Stephen I. Dworkin, (D), Texas, 1991, ClGd
Stephen Forman, (D), Colorado, CcPc
Lee C. Nordt, (D), Texas A&M, 1996, SdGa
Joseph D. White, (D), Montana, 1998, Zr
Associate Dean of Research Graduate School:
William C. Hockaday, (D), Ohio State, 2006, CoaSb
Associate Professor:
Kenneth S. Befus, (D), Texas, 2014, GviGv
Daniel J. Peppe, (D), Yale, 2009, PbcYm
Assistant Professor:
James Fulton, (D), Penn State, 2010, Cbo
Peter B. James, (D), MIT, Ygg
Elizabeth Petsios, (D), S California, 2016, Pie
Emeritus:
Peter M. Allen, (D), S Methodist, 1977, HgNg
Rena M. Bonem, (D), Oklahoma, 1975, PieOb
Vincent S. Cronin, (D), Texas A&M, 1988, GctNg
Steven G. Driese, (D), Wisconsin, 1982, SaClGs
John A. Dunbar, (D), Texas, 1989, YgHg
Thomas T. Goforth, (D), S Methodist, 1973, Yg
Don M. Greene, (D), Oklahoma, 1980, ZyAm
Don F. Parker, (D), Texas, 1976, Giv
Kenneth Wilkins, (D), Florida, 1982, Pv
Luminescence Geochronology Research Specialist:
Liliana Marin, (M), Illinois (Chicago), 2005, ZiGgCc
Instrumentation Specialist:
Timothy Meredith
Ren Zhang, (D), McMaster, 2007, Cs
Laboratory Director:
Sharon Browning, (M), Gg
Office Manager:
Paulette Penney, (A), Zn

Blinn College

Agricultural and Natural Science Programs (A)

902 College Avenue
Brenham, TX 77833
p. (979) 830-4200
cl.metz@blinn.edu
http://www.blinn.edu/natscience

Professor:
Michael Dalman, (M), W Michigan, 1999, Ze
Cheyl L. Metz, (D), Texas A&M, GgOg

Brookhaven College

Science/Math Div - Geology Dept

3939 Valley View Lane
Farmers Branch, TX 75244
p. (972) 860-4758
LannaBradshaw@dcccd.edu
https://www.brookhavencollege.edu/cd/credit/geology/pages/

Chair:
Lanna K. Bradshaw, (M), Texas A&M, 2000, GgZg

Coastal Bend College

Science Div (A)

3800 Charco Road
Beeville, TX 78102
p. (361) 354-2423
ewilliams4@coastalbend.edu
http://www.coastalbend.edu/acdem/science/
Administrative Assistant:
Enrollment (2011): A: 3 (0)

Instructor:
Danny Burns, (M), Ball State, 1984, Zgn
Richard Cowart, (D), Texas A&M (Corpus Christi), 2008, GeEo

Collin College

District Department of Geology (A) ⃠

Wylie Campus
Student Center 331
391 Country Club Road
Wylie , TX 75098

p. (972) 378-8835
dbabcock@collin.edu
http://www.collin.edu/academics/programs/geology.html
Enrollment (2012): A: 10 (0)
Professor:
Neal Alexandrowicz, (D)
Instructor:
Stacey Bilich, (B)

Collin College - Central Park Campus

Deptartment of Geology (A)
2200 W. University Drive
McKinney, TX 75071
p. (972) 548-6790
nalexandrowicz@collin.edu
http://www.collin.edu/geology/
Enrollment (2016): A: 30 (0)
Professor:
Brett Burkett, (M), SUNY (Buffalo), 2008, Gvg
Related Staff:
Shannon Burkett, (M), SUNY (Buffalo), 2005, Gv

Collin College - Preston Ridge Campus

Dept of Geology and Environmental Science (A)
9700 Wade Boulevard
Frisco, TX 75035
p. (972) 377-1635
smay@collin.edu
Professor:
Heinrich Goetz, (M), Texas A&M, 1997, Ge
Paul Manganelli, (M), Boston Coll, 1998
S Judson May, (D), New Mexico, 1980, GgoGc

Dallas College

Engineering, Technology, Mathematics, and Sciences (A)
O
1601 S. Lamar St.
Dallas, TX 75215
p. 214-378-1500
dallascollegeacademics@dcccd.edu
https://www.dcccd.edu/cd/schools/stem/pages/
Programs: See link for a full list of course offerings.
https://www1.dcccd.edu/catalog/coursedescriptions/detail.cfm?loc=DCCCD&heading=Geology

Del Mar College

Dept of Natural Sciences (A)
Corpus Christi, TX 78404
p. (512) 886-1240
jhalcomb@delmar.edu
Professor:
Roger T. Steinberg, (M), Tennessee, 1981, GgoPg
Associate Professor:
Walter V. Kramer, (M), Texas (El Paso), 1970, GgoGx
Emeritus:
Mary S. Thorpe, (M), Baylor, 1966, Ge

El Centro College - Dallas Community College District

Geology Program (A)
801 Main Street
Dallas, TX 75202
p. (214) 860-2429
nfields@dcccd.edu
http://elcentrocollege.edu/programs/geology
Coordinator:
Nancy Fields, (M), Baylor, Zg
Professor:
Steven McCauley, (M), Mississippi State, GgAm
Bethan Salle, (M), N Colorado, 1999, GgeZg
Adjunct Professor:
Anna F. Banda, (M), Baylor, 2007, GgeGv
David Coffman, (M), Baylor, Gg
Stephanie Coffman, (M), Baylor, Gg
Alice Ruffel, (M), Oklahoma, Gg

El Paso Community College

Dept of Geological Sciences (A)
P.O. Box 20500
El Paso, TX 79998
p. (915) 831-5161
jvillal6@epcc.edu
https://www.epcc.edu/Academics/Geology
Programs: Associates of Science in Geology
Enrollment (2021): A: 37 (3)
Chair:
Joshua Villalobos, (M), Texas (El Paso), 2001, ZeYgu
Coordinator - Transmountain Campus:
Kathleen Devaney, (D), California (Los Angeles), 1992, Gg
Professor:
Sulaiman Abushagur, (D), Texas (El Paso), 1991, Gg
Assistant Professor:
Tina L. Carrick, (D), Texas (El Paso), 2014, ZeGgYg
Adriana Perez, (M), Texas (El Paso), Gg
Robert Rohbaugh, (M), Texas (El Paso), Ggt
Adjunct Professor:
Brenda Barnes, (D), Texas (El Paso), Gg
Sabrina Canalda, (M), Texas (El Paso), Gg
Emile Couroux, Gg
Kirk Rothemund, Gg

Hardin-Simmons University

Dept of Geology & Environmental Science (B)
Box 16164
2200 Hickory Street
Abilene, TX 79698-6164
p. (325) 670-1383
srosscoe@hsutx.edu
http://www.hsutx.edu/academics/undergraduate/holland/geology
Enrollment (2015): B: 20 (5)
Head:
Mark A. Ouimette, (D), Texas (El Paso), 1994, GieGt
Associate Professor:
Marla Potess, (D), Texas Tech, 2011, Zgu
Steven Rosscoe, (D), Texas Tech, 2008, PmGsPs

Hill College

Div of Mathematics and Sciences (Environmental Science)
112 Lamar
Hillsboro, TX 76645
p. 254-659-7500 or 817 295-7392
rroberts@hillcollege.edu
http://www.hillcollege.edu/academics/Traditional/Math_Science_EdServices/Geology.html
Programs: Associate of Science

Houston Community College System

Geology Dept (A)
1010 W. Sam Houston Pkwy.N.
Houston, TX 77043
p. (713) 718-5641
dwight.kranz@hccs.edu
http://learningwebsys.hccs.edu/discipline/geology/
Professor:
Dwight S. Kranz, (M), Texas A&M, 1980, Gg

Kilgore College

Dept of Chemistry and Geology (A)
1100 Broadway
Kilgore, TX 75662
p. (903) 983-8253
pbuchanan@kilgore.edu
Programs: Geology
Enrollment (2016): A: 5 (2)
Instructor:
Paul C. Buchanan, (D), Houston, 1995, GgiXm

Lamar University

Dept of Earth and Space Sciences (B) ☒
P.O. Box 10031
Beaumont, TX 77710

p. (409) 880-8236
jim.jordan@lamar.edu
http://ess.lamar.edu

Professor:
Jim L. Jordan, (D), Rice, 1975, XcgXm

Professor:
Roger W. Cooper, (D), Minnesota, 1978, Gi
Donald E. Owen, (D), Kansas, 1963, Gr
James W. Westgate, (D), Texas, 1988, Pev

Associate Professor:
Joseph M. Kruger, (D), Arizona, 1991, YgGgZi

Instructor:
Bennetta Schmidt, (D), Gg

Adjunct Professor:
Mark Adams, (M), Houston (Clear Lake), Xg
Cynthia L. Parish, (M), Lamar, 2004, Gg
Carla M. Tucker, (M), Texas, 1990, Hw

Laboratory Coordinator:
Karen M. Woods, (B), Lamar, 1991, Zg

Laredo College

Dept of Natural Sciences
West End Washington Street
Laredo, TX 78040
p. (956) 721-5195
glenn.blaylock@laredo.edu
http://www.laredo.edu/cms/LCC/Instruction/Divisions/Sciences/Natural_Sciences/Science/

Professor:
Glenn W. Blaylock, (M), Brigham Young, 1998, Ggv

Lee College

Dept of Physical Sciences (A)
P. O.Box 818
Baytown, TX 77522
p. (281) 425-6552
erichards@lee.edu
Enrollment (2016): A: 2 (0)

Professor:
Sharon Gabel, (D), SUNY, 1991, GseGg

Lonestar College - CyFair

Geology Dept
9191 Barker Cypress Road
Cypress, TX 77433
p. (281) 290-3919
michael.r.konvicka@lonestar.edu
http://www.lonestar.edu/geology-dept-cyfair.htm

Lonestar College - Kingwood

Geology Dept
20000 Kingwood Drive
Kingwood, TX 77339
p. (281) 312-1629
Jean.Whileyman@lonestar.edu
http://www.lonestar.edu/geology-dept-kingwood.htm

Lonestar College - Montgomery

Geology Dept
3200 College Park Drive
Conroe, TX 77384
p. (936) 273-7077
Michael.J.Sundermann@lonestar.edu
http://www.lonestar.edu/geology-dept-montgomery.htm

Professor:
Nathalie N. Brandes, (M), New Mexico Tech
Cynthia Lawry-Berkins, (D), Texas A&M, 2009, Ges

Lonestar College - North Harris

Geology Dept
2700 W.W. Thorne Drive
Houston, TX 77073-3499
p. (281) 618-5685
richard.a.owen@lonestar.edu
Enrollment (2007): A: 20 (0)

Head:
Thomas M. C. Hobbs, (M), Texas (El Paso), 1979, Gg

Professor:
Peter E. Price, (M), Kentucky, 1979, Zi

Adjunct Professor:
Penni Major, (M), Gg
Michelle Mc Mahon, (D), Aberdeen, 1993, Go
Victor S. Resnic, (D), Nat Pet Inst (Russia), 1971, Go
Linda C. Tran, (B), Texas A&M, 1999, Zi

Lonestar College - Tomball

Geology Dept
30555 Tomball Parkway
Tomball, TX 77375
p. (281) 351-3324
David.O.Bary@lonestar.edu
http://www.lonestar.edu/geology-dept-tomball.htm

McLennan Community College

Geology Dept (A)
1400 College Drive
Waco, TX 76708
p. (254) 299-8442
efagner@mclennan.edu
http://www.mclennan.edu/departments/geol/

Instructor:
Elaine Alexander, Gg

Midland College

Geology (A)
3600 N. Garfield
Midland, TX 79705
p. (432) 685-4612
agiles@midland.edu
https://www.midland.edu/academics/degrees/stem/geology.php
Enrollment (2011): A: 14 (6)

Associate Professor:
Joan Gawloski, (M), Baylor, GgzGe
Antony Giles, (M), Sul Ross State, 2006, GgvGi

Assistant Professor:
Keonho Kim, (D), GgAmPi

Adjunct Professor:
Karen Waggoner, (D), Texas Tech, GgRhGc

Midwestern State University

Kimbell School of Geosciences (B,M)
3410 Taft
Wichita Falls, TX 76308
p. (940) 397-4250
geology.program@msutexas.edu
http://www.msutexas.edu/academics/scienceandmath/geosciences
f: https://www.facebook.com/MSUGeosProgram/
t: @KSG_msutexas
Programs: Geology; Environmental Sciences
Enrollment (2021): B: 53 (13) M: 19 (1)

Chair & Prothro Distinguished Associate Professor of Geological Science:
Jonathan D. Price, (D), Oklahoma, 1998, GiCuGz

Robert L. Bolin Distinguished Professor of Petroleum Geology:
W. Scott Meddaugh, (D), Harvard, 1982, GoEgCg

Assistant Professor and Graduate Coordinator:
Andrew Katumwehe, (D), Oklahoma State, 2016, YuvYe

Assistant Professor:
Anna M. Weiss, (D), Texas, 2019, PcgGs

Emeritus:
Rebecca L. Dodge, (D), Colorado Mines, 1982, Zre
John Kocurko, (D), Texas Tech, 1972, Gs

Odessa College

Dept of Geology, Anthropology & Geography (A)
201 W. University
Odessa, TX 79762
p. (915) 335-6558
jbolton@odessa.edu
https://catalog.odessa.edu/preview_program.php?catoid=9&poid=1102&returnto=379
Enrollment (2010): A: 1 (1)

Associate Professor:
Gerald B. McAfee, (M), Sul Ross State, 1966, Pi

Paris Junior College

Dept of Science
2400 Clarksville Street
Paris, TX 75460
p. 903-782-0481
mbarnett@parisjc.edu
http://www.parisjc.edu/index.php/pjc2/directory-index/C208

Rice University

Center for Computational Earth Science (M,D)
MS-126
PO Box 1892
Houston, TX 77251-1892
p. (713) 348-3574
dmberry@rice.edu
http://earthscience.rice.edu/centers/ccg/
Enrollment (2016): M: 7 (2) D: 335 (4)

Associate Director:
Alan Levander, (D), Stanford, 1984, Ys

Professor:
Richard G. Gordon, (D), Stanford, 1979, YmGtYd
Adrian Lenardic, (D), California (Los Angeles), 1995, Gc
Julia K. Morgan, (D), Cornell, 1993, Gc
Fenglin Niu, (D), Tokyo, 1997, Ysg
Dale S. Sawyer, (D), MIT, 1982, Ys
Colin A. Zelt, (D), British Columbia, 1989, Ys

Assistant Professor:
Helge Gonnermann, (D), California (Berkeley), 2004, Gv

Dept of Civil and Environmental Engineering (M,D) ☒
P.O. Box 1892
MS 519
Houston, TX 77251-1892
p. (713) 348-4949
bedient@rice.edu
http://ceve.rice.edu/
Certificates: UG - Bachelor of Science in Civil Engineering, UG - B.A. in Civil and Environmental Engineering, UG - Minor in energy and water sustainability (EWSU), GR - Master of Science (M.S.), Doctor of Philosophy (Ph.D.) , Master of Civil and Environmental Engineering (M.C.E.E.)

Dept of Earth, Environmental and Planetary Sciences (B,M,D) ☒
MS 126
PO Box 1892
Houston, TX 77251-1892
p. (713) 348-4880
eeps@rice.edu
https://eeps.rice.edu/
f: https://www.facebook.com/RiceEarthScience
t: https://twitter.com/RiceEarthSci
Enrollment (2015): B: 28 (8) M: 45 (18) D: 40 (5)

Chair:
Julia K. Morgan, (D), Cornell, 1993, GctGv

Professor:
Jonathan Ajo-Franklin, (D), Stanford, 2005, Yg
Rajdeep Dasgupta, (D), Minnesota, 2006, GxCg
Helge Gonnermann, (D), California (Berkeley), 2004, Gv
Richard G. Gordon, (D), Stanford, 1979, YmGtYd
Cin-Ty A. Lee, (D), Harvard, 2001, CgGiv
Adrian Lenardic, (D), California, 1995, Yg
Alan R. Levander, (D), Stanford, 1984, Ye
Caroline A. Masiello, (D), California (Irvine), 1999, Zn
Fenglin Niu, (D), Tokyo, 1997, YsGt
Colin A. Zelt, (D), British Columbia, 1989, Yg

Assistant Professor:
Sylvia G. Dee, (D), S California, 2015, Zo
Melodie E. French, (D), Texas A&M, 2014, Gc
Kirsten L. Siebach, Caltech, 2016, Xg
Mark Torres, (D), S California, 2015, Cb

Lecturer:
Zhuoya He, (D), California (Irvine), 2021, Ou

Adjunct Professor:
K. K. Bissada, (D), Washington (St Louis), 1967, Cg
Jeffrey J. Dravis, (D), Rice, 1980, Gs
Patrick J. McGovern, (D), MIT, 1996, Xy

Emeritus:
John B. Anderson, (D), Florida State, 1972, Gu
H. C. Clark, (D), Stanford, 1966, YgGeYu
Andre W. Droxler, (D), Miami, 1984, Gs
Dieter Heymann, (D), Amsterdam, 1958, Xm
William P. Leeman, (D), Oregon, 1974, Gi
Andreas Luttge, (D), Tubingen, 1990, Cg
John C. Stormer, Jr., (D), California (Berkeley), 1971, Gi
Manik Talwani, (D), Columbia, 1959, Yr
Peter R. Vail, (D), Northwestern, 1959, Gr

On Leave:
Laurence Y. Yeung, (D), Caltech, 2010, CsAsCl

Department Administrator:
Lee Willson, Zn

Saint Mary's University

Dept of Physics and Earth Sciences
One Camino Santa Maria
San Antonio, TX 78228-8569
p. (210) 436-3235
rcardenas@stmarytx.edu
http://www.stmarytx.edu/acad/physicsandearthscience

Chair:
Paul Nienaber

Professor:
David Fitzgerald, (M), Iowa, 1977, Gd

Associate Professor:
Gene W. Lene, (D), Texas, 1981, Ge

Sam Houston State University

Department of Environmental and Geosciences
1900 Ave. I
Huntsville, TX 77341-2148
p. (936) 294-1451
pat-harris@shsu.edu
https://www.shsu.edu/academics/environmental-and-geosciences/
t: @SHSUgeo
Programs: Geography, Geology, GIS, Environmental Science
Certificates: GIS

Chair:
Pat Harris, (D), Texas A&M, 1997, ClGdSc

San Antonio Community College

Dept of Chemistry/Earth Sciences/Astronomy (A)
1300 San Pedro Avenue
CG Rm 207
San Antonio, TX 78212
p. (210) 486-0045
tstaggs@alamo.edu
http://www.alamo.edu/sac/earthsci

Professor:
Dean P. Lambert, (D), Texas, 1992, ZyiZz

Associate Professor:
Anne D. Dietz, (M), Texas A&M (Kingsville), 1989, PgZg
George R. Stanley, (M), Texas (San Antonio), 2006, XzZgRm
David A. Wood, (D), Arizona, 2000, XsZnn

Adjunct Professor:
Thomas Adams, (D), 2011, PvGg
T Scott Girhard, (M), SW Texas State, ZyAm
Robert Janusz, (M), Texas (San Antonio), Gg
Dwight J. Jurena, (M), Rensselaer, 2002, XgYvGz
Ryan E. Rudnicki, (D), Penn State, 1979, Zyr
C. Keith Smith, (M), Texas State, 2005, ZyzZz

Cooperating Faculty:
Steve Dingman, Zn

San Jacinto Community College-Central

Geology Dept
8060 Spencer Hwy.
Pasadena, TX 77089
p. (281) 998-6150 x1882
Karen.Purpera@sjcd.edu
http://www.sanjac.edu/

TEXAS

San Jacinto Community College-North

Geology Dept
5800 Uvalde
Houston, TX 77049
p. (281) 998-6150 x7210
information@sjcd.edu
http://www.sanjac.edu/

San Jacinto Community College-South

Geology Dept
13735 Beamer Rd.
Houston, TX 77089
p. (281) 998-6150 x4662
Tyler.Olivier@sjcd.edu
http://www.sanjac.edu/

Southern Methodist University

Roy M. Huffington Dept of Earth Sciences (B,M,D) ☒
Post Office Box 750395
Dallas, TX 75275-0395
p. (214) 768-2750
geol@smu.edu
https://www.smu.edu/Dedman/Academics/Departments/Earth-Sciences
Programs: Geology, Geophysics
Administrative Assistant: Stephanie L. Schwob
Enrollment (2021): B: 23 (9) M: 10 (1) D: 20 (4)

Chair:
Heather R. DeShon, (D), California (Santa Cruz), 2004, Ysr
Shuler-Foscue Chair in Earth Sciences:
Zhong Lu, (D), Alaska (Fairbanks), 1996, Zr
Albritton Chair in Earth Sciences:
Brian W. Stump, (D), California (Berkeley), 1979, Ys
Professor:
Robert T. Gregory, (D), Caltech, 1981, Cs
Matthew J. Hornbach, (D), Wyoming, 2005, YrhYe
M. Beatrice Magnani, (D), Perugia, 2000, YsGtYe
James E. Quick, (D), Caltech, 1981, Gv
Neil J. Tabor, (D), California (Davis), 2002, SdGs
Crayton J. Yapp, (D), Caltech, 1980, Csl
Hamilton Chair in Earth Sciences:
Stephen J. Arrowsmith, (D), Leeds (UK), 2004, Ys
Associate Scientist:
Christopher T. Hayward, (D), S Methodist, 1997, Ys
Ian J. Richards, (D), Tennessee, 1994, Cs
Assistant Professor:
Rita C. Economos, (D), California (Los Angeles), 2009, CsGc
Adjunct Professor:
Anthony R. Fiorillo, (D), Pennsylvania, 1989, Pv
Matthew Siegler, (D), California (Los Angeles), 2011, YgXy
Alisa Winkler, (D), S Methodist, 1990, Pv
Dale A. Winkler, (D), Texas, 1985, Pv
Emeritus:
David D. Blackwell, (D), Harvard, 1968, Yh
Michael J. Holdaway, (D), California (Berkeley), 1963, Gp
Bonnie F. Jacobs, (D), Arizona, 1983, Pl
Louis L. Jacobs, (D), Arizona, 1977, Pv
Robert L. Laury, (D), Wisconsin, 1966, Gd
John V. Walther, (D), California (Berkeley), 1978, Cg

Stephen F. Austin State University

Dept of Geology (B,M) O
PO Box 13011 SFA Station
Nacogdoches, TX 75962
p. (936) 468-3701
geology@sfasu.edu
https://www.sfasu.edu/academics/colleges/sciences-math/geology
f: https://www.facebook.com/SFAGeology
t: https://twitter.com/SFAGeology
Programs: B.S. Geology, M.S. Geology
Administrative Assistant: Shana Scott
Enrollment (2021): B: 48 (14) M: 37 (9)

Chair:
Wesley A. Brown, (D), Texas (El Paso), 2004, YgsGt
Professor:
R. LaRell Nielson, (D), Utah, 1981, Grs
Kevin W. Stafford, (D), New Mexico Tech, 2008, Hw
Assistant Professor:
Julie M. Bloxson, (D), Case Western, 2017, Go
Melinda S. Faulkner, (D), Stephen F. Austin, 2016, GeClEg
Liane M. Stevens, (D), Montana, 2015, Gpc
Lab Coordinator:
Wesley L. Turner, (M), Stephen F. Austin, 2016, Gg
Lecturer:
Michael T. Read, (D), Texas (Arlington), 2018, PgGg
Visiting Professor:
Jenny M. Rashall, (D), Texas (Arlington), 2020, GgPg

Sul Ross State University

Dept of Biology, Geology and Physical Sciences (B,M) O☒
Box C-64
Alpine, TX 79832
p. (432) 837-8112
measures@sulross.edu
http://www.sulross.edu/BGPS
Programs: Geology
Teacher Certification in Geology
Enrollment (2019): B: 28 (8) M: 21 (4)

Professor:
Elizabeth A. Measures, (D), Idaho, 1992, Gdq
Kevin Urbanczyk, (D), Washington State, 1993, GiZi
Assistant Professor:
Thomas Shiller, (D), Texas Tech, 2017, PvGrd
Lecturer:
Jesse Kelsch, (M), New Mexico, Gct
Emeritus:
David M. Rohr, (D), Oregon State, 1978, PiGd

Tarleton State University

Chemistry, Geosciences, and Physics (B,M) ☒
Box T-0540
Stephenville, TX 76402
p. (254) 968-9143
rmorgan@tarleton.edu
http://www.tarleton.edu/CHGP/
Programs: Geosciences Bachelor's of Science (concentrations in Earth Science, Environmental Science, Geology, Petroleum Geology, Hydrogeology), Master's of Science in Environmental Science
Certificates: GIS
Administrative Assistant: Eva Moody
Enrollment (2021): B: 37 (0) M: 17 (0)

Head:
Ryan Morgan, (D), Baylor, 2015, Pei
Professor:
Stephen W. Field, (D), Massachusetts, 1988, Gz
Associate Professor:
Catherine Ronck, (D), God
Instructor:
Joree Burnett, (M), Tarleton State, 2015, HwGe
Kristopher Juntunen, (M), Tarleton State, Pv

Tarrant County College, Northeast Campus

Natural Science Dept (A)
828 Harwood Road
Hurst, TX 76054
p. (817) 515-6565
marles.mccurdy@tccd.edu
https://www.tccd.edu/academics/tcc-catalog/courses-and-programs/geology/
Enrollment (2011): A: 650 (30)

Associate Professor:
Meena Balakrishnan, (D), GggGg
Kevin M. Barrett, (D), Texas State, 2012, AsZgr
Professor:
Hayden R. Chasteen, (M), NE Louisiana, 1981, GgeGg

Tarrant County College, Southeast Campus

Physical Sciences Dept ☒
2100 Southeast Pkwy
Arlington, TX 76018
p. (817) 515-3238
thomas.awtry@tccd.edu

https://www.tccd.edu/academics/
Programs: Associate of Science

Assistant Professor:
Stephanie Mitzman, (M), Texas A&M, 1999, Gg

Instructor:
Samantha Caputi, (M), 1997, Gg
Darren Greenaway, (M), 1997, Gg

Temple College

Temple College
2600 South First Street
Temple, TX 76504
p. (254) 298-8472
john.mcclain@templejc.edu
http://www.templejc.edu/

Texas A&M University

Andreas Kronenberg (M,D) ☒
MS 3115
Department Geology & Geophysics
College of Arts and Sciences
College Station, TX 77843
p. (979) 845-2400
a-kronenberg@tamu.edu
http://tectono.tamu.edu/
Enrollment (2022): M: 5 (0) D: 9 (4)

Director:
Andreas Kronenberg, (D), Brown, 1983, NrGyYx

Department Head:
Julie Newman, (D), Rochester, 1992, Gct

Professor:
Frederick Chester, (D), Texas A&M, 1988, GcYxGt
Judith Chester, (D), Texas A&M, 1992, GcNrGo
Marcelo J. Sanchez, (D), Politecnica de Catalunya, 2004, NtrNx
David Sparks, (D), Brown, 1992, GtYhn

Laboratory Director:
Hiroko Kitajima, (D), Texas A&M, 2010, YxSpGc

Assoc Director:
Benchuan Duan, (D), California (Riverside), 2006, YsgGq

Associate Professor:
xiaowei Chen, (D), California (San Diego), 2013, YsuGt
Kan Wu, (D), Texas (Austin), 2013, NpYeGo
Jin Zhang, (D), Illinois, 2014, GytYx

Dept of Atmospheric Sciences (B,M,D) ☒
3150 TAMU
College Station, TX 77843-3150
p. (979) 845-7671
sarava@tamu.edu
http://atmo.tamu.edu/
Programs: Meteorology (B); Atmospheric Sciences (M,D)
Enrollment (2018): B: 120 (0) M: 20 (0) D: 20 (0)

Instructional Professor:
Don T. Conlee, (D), Texas A&M, 1994

Distinguished Professor:
Gerald North, (D), 1966, As

Distinguished Professor :
Renyi Zhang, (D), MIT, 1993, As

Department Head:
Ping Yang, (D), Utah, 1995, As

Professor:
Kenneth P. Bowman, (D), Princeton, 1984, Am
Sarah D. Brooks, (D), Colorado, 2002, AsZc
Ping Chang, (D), Princeton, 1988, Op
Andrew E. Dessler, (D), Harvard, 1994, As
John Nielsen-Gammon, (D), MIT, 1990, As
Richard L. Panetta, (D), Wisconsin, 1978, Am
R. Saravanan, (D), Princeton, 1990, As
Courtney Schumacher, (D), Washington, 2003, As
Istvan Szunyogh, (D), Hungarian Acad of Sci, 1994, As

Associate Professor:
Craig Epifanio, (D), Washington, As
Robert Korty, (D), MIT, 2005, As
Gunnar Schade, (D), Johannes-Gutenberg (Germany), 1997, As

Instructional Assistant Professor:
Tim Logan

Assistant Professor:
Christopher J. Nowotarski, (D), Penn State, 2013, Ams
Anita Rapp, (D), Colorado State, 2008, Zr
Yangyang Xu

Adjunct Professor:
Alex Dessler, (D), Duke, 1956, As

Emeritus:
Richard E. Orville, (D), Arizona, 1966, As

Dept of Geography (B,M,D)
810 Eller O&M Building
3147 TAMU
College Station, TX 77843-3147
p. (979) 845-7141
cbruton@geog.tamu.edu
http://geog.tamu.edu/
Enrollment (2009): B: 156 (40) M: 16 (8) D: 18 (3)

Head:
David M. Cairns, (D), Iowa, 1995, Zy

Professor:
Sarah W. Bednarz, (D), Texas A&M, 1992, Ze
John R. Giardino, (D), Nebraska, 1979, Gm
Andrew G. Klein, (D), Cornell, 1997, ZriGl
Michael R. Waters, (D), Arizona, 1983, Ga

Associate Professor:
Daniel Z. Sui, (D), Goergia, 1993, Zr
Vatche P. Tchakerian, (D), California (Los Angeles), 1989, Gm

Assistant Professor:
Anne Chin, (D), Arizona State, 1994, Gm

Research Associate:
Jean A. Bowman, (M), Rutgers, 1984, Hg

Emeritus:
Robert S. Bednarz, (D), Chicago, 1975, Zeu
Clarissa T. Kimber, (D), Wisconsin, 1969, Zy

Dept of Geology & Geophysics (B,M,D) ●✓🖱
3115 TAMU
College Station, TX 77843-3115
p. (979) 845-6401
julie-newman@tamu.edu
http://geoweb.tamu.edu
Programs: BA Geology, BS Geology, BS Geophysics, MS Geology, MS Geophysics, PhD Geology, PhD Geophysics
Certificates: Environmental and Engineering Geology (undergraduate), Petroleum Geology (graduate)
Enrollment (2022): B: 135 (0) M: 35 (0) D: 50 (0)

Head:
Julie Newman, (D), Rochester, Gct

Professor:
Tom Blasingame, (D), Texas A&M, 1989, Np
Frederick M. Chester, (D), Texas A&M, 1988, GcNr
Judith Chester, (D), Texas A&M, 1992, GcNr
Benchun Duan, (D), California (Riverside), 2006, YsGtYg
Mark Everett, (D), Toronto, 1991, Ym
Ryan Ewing, (D), Texas, Gs
Ethan L. Grossman, (D), S California, 1982, Csl
Andreas Kronenberg, (D), Brown, 1983, GyNr
Will Lamb, (D), Wisconsin, 1987, GpCpGz
Franco Marcantonio, (D), Columbia, 1994, Ct
Brent Miller, (D), Dalhousie, 1997, Cc
Michael Pope, (D), Virginia Tech, 1995, GrCc
Anne Raymond, (D), Chicago, 1983, Pb
David Sparks, (D), Brown, 1992, YgGqZe
Yuefeng Sun, (D), Columbia, 1994, GoYe
Debbie Thomas, (D), North Carolina, 2002, Ou
Thomas E. Yancey, (D), California (Berkeley), 1971, Pg
Hongbin Zhan, (D), Nevada (Reno), 1996, Hw

Associate Professor:
Xiaowei Chen, (D), California (San Diego), 2013, Ys
Hiroko Kitajima, (D), Texas A&M, 2010, Nr
Peter Knappett, (D), Tennessee, 2010, Hw
Nicholas Perez, (D), Texas, 2015, Gt
Julia Reece, (D), Texas, 2011, Gs
Jin Zhang, (D), Illinois, 2014, GyYx

Instructional Assistant Professor:
David Bapst, (D), Chicago, 2013, Pq
Brandi Lenz, (D), Ohio State, 2021, Ys

Assistant Professor:
Christina Belanger, (D), Chicago, 2011, Pe
Juan Carlos Laya, (D), Durham, 2012, Gs
Lucien Nana Yobo, (D), Houston, 2021, Cc

Research Scientist:
Mike Tice, (D), Stanford, 2006, Pg
Professor of Practice:
Mauro Becker, (D), Texas, 1996, Eo
Art Donovan, (D), Colorado Mines, Gs
Lecturer:
Alfonso Benavides-Iglesias, (D), Texas A&M, 2007, Ys

Dept of Oceanography (B,M,D)
1204 Eller O&M Blg
MS 3146
College Station, TX 77843-3146
p. (979) 845-7211
syvon-lewis@tamu.edu
http://ocean.tamu.edu
Programs: Oceanography (B, M, D); Ocean Science and Technology (M)
Enrollment (2017): B: 8 (0) M: 22 (10) D: 42 (6)
Director, Texas Sea Grant:
Pamela Plotkin, (D), Texas A&M, 1994, Ob
Interim Dean:
Debbie Thomas, (D), North Carolina, 2002, Ou
Head:
Shari A. Yvon-Lewis, (D), Miami, 1994, Oc
Executive Associate Dean for Research:
Jack G. Baldauf, (D), California (Berkeley), 1984, Ou
Director, Geochemical and Environmental Research Group:
Anthony Knap, (D), Southampton (UK), Oc
Professor:
Rainier MW Amon, (D), Texas, 1995, CbOcb
David A. Brooks, (D), Miami, 1975, Op
Lisa Campbell, (D), SUNY (Stony Brook), 1985, Ob
Ping Chang, (D), Princeton, 1988, Op
Piers Chapman, (D), Univ Coll (N Wales), 1982, OcCm
Steven F. DiMarco, (D), Texas (Dallas), 1991, OpnOg
Wilford D. Gardner, (D), MIT/WHOI, 1978, OugZr
Benjamin S. Giese, (D), Washington, 1989, Op
Gerardo Gold-Bouchot, (D), Ctr Res Adv Stud (Mexico), 1991, OcCm
Norman L. Guinasso, (D), Texas A&M, 1984, Opc
Robert D. Hetland, (D), Florida State, 1999, Op
Alejandro H. Orsi, (D), Texas A&M, 1993, Op
Mary Jo Richardson, (D), MIT/WHOI, 1980, Gu
Gilbert T. Rowe, (D), Duke, 1968, Ob
Peter H. Santschi, (D), Switzerland (Berne), 1975, Oc
Niall C. Slowey, (D), MIT, 1991, Ou
Senior Scientist:
Troy L. Holcombe, (D), Columbia, 1972, OuGcg
Matthew K. Howard, (D), Texas A&M, 1992, Op
Ann E. Jochens, (D), Texas A&M, 1977, Op
Adam Klaus, (D), Hawaii, 1991, Gu
Associate Professor:
Ayal Anis, (D), Oregon State, 1993, Opn
Timothy M. Dellapenna, (D), William & Mary, 1999, Ou
Anja Schulze, (D), Victoria, 2001, Ob
Achim Stoessel, (D), Hamburg, 1990, OpAmGl
Daniel C.O Thornton, (D), Queen Mary (London), 1995, Ob
Associate Scientist:
Steven K. Baum, (D), Texas A&M, 1996, Op
Jose L. Sericano, (D), Texas A&M, 1993, Cm
Assistant Professor:
Jessica N. Fitzsimmons, (D), MIT/WHOI, 2013, OcCta
Henry Potter, (D), Miami, 2014, Op
Kathryn E. F. Shamberger, (D), Washington, 2011, OcZc
Jason B. Sylvan, (D), Rutgers, 2008, ObXb
Chrissy Wiederwohl, (D), Texas A&M, 2012, Op
Yige Zhang, (D), Yale, 2014, Co
Research Associate:
Shinichi Kobara, (D), Texas A&M, Zi
Marion Stoessel, (M), Hamburg, 1985, Op
Zhankun Wang, (D), Massachusetts (Dartmouth), 2009, Op
Distinguished Professor Emeritus:
Robert A. Duce, (D), MIT, 1964, Oc
Worth D. Nowlin, Jr., (D), Texas A&M, 1966, OpZn
Distinguished Professor:
Gerald R. North, (D), Wisconsin, 1966, Am
Director, Texas Sea Grant College Program:
Robert R. Stickney, (D), Florida State, 1971, Ob
Emeritus:
Douglas C. Biggs, (D), MIT/WHOI, 1976, Ob
George A. Jackson, (D), Caltech, 1976, Op
Bobby J. Presley, (D), California, 1969, Oc
Robert H. Stewart, (D), California (San Diego), 1969, Op
Laboratory Director:
Terry L. Wade, (D), Rhode Island, 1978, OcCao

Dept of Soil & Crop Sciences (B,M,D) ☒
TAMU 2474
College Station, TX 77843-2474
p. (979) 845-3603
cmorgan@tamu.edu
http://soilcrop.tamu.edu
Programs: Environmental Plant and Soil Science
Enrollment (2017): B: 32 (15) M: 19 (3) D: 21 (3)
Professor:
David D. Baltensperger, (D), New Mexico State, 1981, PvCbSb
Terry Gentry, (D), Arizona, Sb
Kevin J. McInnes, (D), Kansas State, 1985, Sp
Cristine L. S. Morgan, (D), Wisconsin, 2003, Spd
Tony L. Provin, (D), Purdue, 1995, Sc
Paul Schwab, (D), Colorado State, 1981, Sc
Associate Professor:
Jacqueline A. Aitkenhead-Peterson, (D), New Hampshire, 2000, HsSoGf
Paul DeLaune, (D), Arkansas, 2002, So
Youjun Deng, (D), Texas A&M, 2001, GzSc
Fugen Dou, (D), Texas A&M, 2005, So
Julie Howe, (D), Wisconsin, 2004, Sc
Donald McGahan, (D), California (Davis), 2007, Sdc
Assistant Professor:
Jourdan Bell, Texas A&M, 2014
Katie Lewis, (D), Texas A&M, 2014, Sco
Jake Mowrer, (D), Georgia, 2014, Sc
Haly L. Neely, (D), Texas A&M, 2014, Spd
Anil Somenhally, (D), Texas A&M, 2010, Sb

Texas A&M University, Commerce

Dept of Biological & Environmental Sciences (B,M)
Commerce, TX 75429
p. (903) 886-5378
DongWon.Choi@tamuc.edu
http://www.tamu-commerce.edu/biology/
Assistant Professor:
Haydn A "Chip" Fox, (D), South Carolina, 1994, HgGe

Texas A&M University, Corpus Christi

Dept of Physical and Environmental Sciences (B) O
6300 Ocean Drive
Corpus Christi, TX 78412
p. (361) 825-6000
Valeriu.Murgulet@tamucc.edu
http://geology.tamucc.edu/
Programs: General Geology, Environmental Geology, Geochemistry, Energy Resources
Enrollment (2016): B: 84 (13)
Chair, Dept, Physical and Environmental Sciences:
Richard Coffin, (D), Delaware, 1986, Ouc
Professor:
Jennifer M. Smith-Engle, (D), Georgia, 1983, Gs
Director, Center for Water Supply Studies:
Dorina Murgulet, (D), Alabama, 2009, HwsGe
Assistant Professor:
Mohamed Ahmed, (D), W Michigan, 2012, YguYe
Lindsay Prothro, (D), Rice, 2018, GsuGl
Endowed Associate Research Professor:
James Gibeaut, (D), S Florida, 1991, OnZi
Adjunct Professor:
Clinton Randall (R Bissell, (M), Oklahoma State, 1984, GosGr
Coordinator, Geology Program:
Valeriu Murgulet, (D), Alabama, 2010, Cg

Environmental Science Program (B,M)
6300 Ocean Drive
Corpus Christi, TX 78412
p. (361) 825-2814
jennifer.smith-engle@tamucc.edu
http://sci.tamucc.edu/PENS
Programs: Environmental Science

Enrollment (2020): B: 222 (52) M: 18 (9)
Director of National Spill Control School:
Howard Wood, (M), American Military, 2011, ZeHsZn
Program Coordinator:
Jennifer M. Smith-Engle, (D), Georgia, 1983, Gs
Endowed Chair for Socioeconomics:
David Yoskowitz, (D), Texas Tech, 1997, Zn
Endowed Chair for Fisheries and Ocean Health:
Greg Stunz, (D), Texas A&M, 1999, Zn
Endowed Chair for Ecosystems Studies and Modeling:
Paul Montagna, (D), South Carolina, 1983, ObCmHs
Director of Center for Coastal Studies:
Paul Zimba, (D), Mississippi State, 1990, Zn
Chair, Physical and Environmental Sciences Dept.:
Richard Coffin, (D), Delaware, 1986, Oc
Professor:
Fereshteh Billiot, (D), Louisiana State, 2000, Zn
Cherie McCollough, (D), Texas, 2005, Ze
Richard McLaughlin, (D), California (Berkeley), 1997, Zn
Toshiaki Shinoda, (D), Hawaii, 1993, As
Chair, Center for Water Supply Studies:
Dorina Murgulet, (D), Alabama, 2009, Hw
Associate Professor:
Hussain Abdulla, (D), Old Dominion, 2009, Co
Darek Bogucki, (D), S California, 1996, Op
Gregory Buck, (D), Georgia State, 1999, Zn
Kirk Cammerata, (D), Kentucky, 1987, Zn
Jeremy Conkle, (D), Louisiana State, 2010, Zn
Xinping Hu, (D), Old Dominion, 2007, Cms
Patrick Larkin, (D), Texas A&M, 1999, Zn
Riccardo Mozzachiodi, (D), Pisa, 1999, Zn
Jennifer Pollack, (D), South Carolina, 2006, Ob
James Silliman, (D), Michigan, 1998, Co
Michael Starek, (D), Florida, 2008, Zi
Michael Wetz, (D), Oregon State, 2006, Ob
Assistant Professor:
Mohamed Ahmed, (D), W Michigan, 2012, YgZri
Sharon Derrick, (D), Texas A&M, 2001, Zn
Joseph David Felix, (D), Pittsburg State, 2012, Zn
Chuntao Liu, (D), Wyoming, 2003, As
Lindsay O. Prothro, (D), Rice, 2018, GluGs
Kim Withers, (D), Texas A&M, 1994, Zn
Lin Zhang, (D), Rhode Island, 2012, Cs
Research Associate:
Philippe Tissot, (D), Texas A&M, 1994, Zn

Texas A&M University, Kingsville

Department of Physics & Geosciences (B,M) ○⊘
MSC 175
700 University Blvd.
Kingsville, TX 78363
p. (361)593-3310
brent.hedquist@tamuk.edu
https://www.tamuk.edu/artsci/departments/phge/geo/
Programs: Geology, Geoscience, Geophysics
Certificates: GIS, Geophysics
Enrollment (2022): B: 21 (6) M: 6 (3)
Chair:
Brent Hedquist, (D), Arizona State, 2010, ZyAtZi
Professor:
Haibin Su, (D), Cincinnati, 2011, Zir
Associate Professor:
Subbarao Yelisetti, (D), Victoria (Canada), 2014, Ygs
Assistant Professor:
Veronica I. Sanchez, (D), Houston, 2011, GctGm
Robert V. Schneider, (D), Texas (El Paso), 1990, GoYes
Emeritus:
Thomas L. McGehee, (D), Texas (Dallas), 1987, Cl

Texas Christian University

Department of Geological Sciences (B,M) ☒
TCU Box 298830
2950 West Bowie
Fort Worth, TX 76129
p. (817) 257-7270
geology@tcu.edu
https://geology.tcu.edu/
Programs: Geology; Applied Geoscience
Administrative Assistant: Krista Scapelli
Enrollment (2016): B: 76 (13) M: 30 (6)
TCU Provost and Academic Vice Chancellor:
R Nowell Donovan, (D), Newcastle upon Tyne, 1972, GdcGt
Professor of Professional Practice:
Richard A. Denne, (D), Louisiana State, 1990, GoPms
Professor:
Richard E. Hanson, (D), Columbia, 1983, GxvGt
John M. Holbrook, (D), Indiana, 1992, Gsr
Chair:
Helge Alsleben, (D), S California, 2005, GctGg
Associate Professor:
Arthur B. Busbey, (D), Chicago, 1982, PgvGr
Xiangyang Xie, (D), Wyoming, 2007, GoEog
Assistant Professor:
Omar R. Harvey, (D), Texas A&M, 2010, GeHwg
Adjunct Professor:
Floyd Henk, Jr., (M), Texas Christian, 1981, Go
Emeritus:
John Breyer, (D), Nebraska, 1977, GsdEo
Professor of Professional Practice:
Milton Enderlin, (M), Texas Christian, 2010, Nr
Tamie Morgan, (M), Texas Christian, 1984, Zi

Texas State University

Dept of Geography (B,M,D) ☒
Evans Liberal Arts, Room 139
601 University Drive
San Marcos, TX 78666
p. (512) 245-2170
rd11@txstate.edu
http://www.geo.txstate.edu
Chair:
Richard W. Dixon, (D), Zy

Geography and Environmental Studies
601 University Drive
Evans Liberal Arts Room 139
San Marcos, TX 78666
p. (512) 245-2170
rd11@txstate.edu
https://www.geo.txstate.edu
Programs: Physical Geography (BS), Resource and Environmental Studies (BA, BS, MAG), Geographic Information Science (BA, BS, MAG), Water Resources (BA, BS), Geology (minor), Geography (MS, PhD)
Professor of Practice:
Robert Mace, (D), Texas, 1998, HwgHy
Department Chair:
Yongmei Lu, (D), SUNY (Buffalo), 2001, Zi
Professor:
R. Denise Blanchard, (D), Colorado, 1992, RnmRc
Edwin Chow, (D), S Carolina, 2005, ZiRn
Richard Dixon, (D), Texas A&M, 1996, AsOpRn
Alberto Giordano, (D), Syracuse, 1999, Zi
Ronald Hagelman III, (D), Texas State, 2001, Rnm
Jennifer Jensen, (D), Idaho, 2009, Zru
Jason Julian, (D), N Carolina, 2007, HsRwZc
Oswaldo Muniz-Solari, (D), Tennessee, 1991, Ze
John Tiefenbacher, (D), Rutgers, 1992, Rnm
F Benjamin Zhan, (D), SUNY (Buffalo), 1994, Zi
Associate Professor:
Nathan Currit, (D), Penn State, 2003, ZruZc
Injeong Jo, (D), Texas A&M, 2011, Ze
Kimberly Meitzen, (D), S Carolina, 2011, HsGmZy
Yihong Yuan, (D), California (Santa Barbara), 2013, Ze
Assistant Professor:
Su Y. Han, (D), SDSU/UCSB, 2016, Zi
Samantha Krause, (D), Texas, 2018, ScHyZc
Yanan Li, (D), Tennessee, 2015, GmCcPc
Thomas Ptak, (D), Oregon, 2016, Zy
Rosalie Ray, (D), Columbia, 2020, Zu
Senior Lecturer:
Benjamin Price, (D), Texas State, 2016, Zu
Christi Townsend, (D), Texas State, 2012, Zy
Shelly J. Wernette, (D), California (Riverside), 2020, GgPge
Lecturer:
Dane Atkins, (D), Texas State, 2022, Zy

Brandon Radoman-Shaw, (D), Case Western, 2019, Cg

Texas Tech University

Dept of Geosciences (B,M,D) O
Box 41053
2500 Broadway
Science 125
Lubbock, TX 79409-1053
p. (806) 834-0497
callum.hetherington@ttu.edu
http://www.geosciences.ttu.edu
Enrollment (2015): B: 390 (33) M: 48 (21) D: 23 (0)

Chair:
Jeffrey A. Lee, (D), Arizona State, 1990, Zy
Horn Professor:
Sankar Chatterjee, (D), Calcutta, 1970, PvGct
Professor:
George B. Asquith, (D), Wisconsin, 1966, Go
Calvin G. Barnes, (D), Oregon, 1982, Gi
James E. Barrick, (D), Iowa, 1978, Psm
Gary S. Elbow, (D), Pittsburgh, 1972
Juske Horita, (D), Texas A&M, 1997
Thomas M. Lehman, (D), Texas, 1985, Gs
Moira K. Ridley, (D), Nebraska, 1997, CqlZm
John L. Schroeder, (D), Texas Tech, 1999, As
Paul J. Sylvester, (D)
Aaron S. Yoshinobu, (D), S California, 1999, GctGu
Senior Scientist:
Melanie A. Barnes, (D), Texas Tech, 2001, CaGie
Associate Professor:
Eric C. Bruning, (D), Oklahoma, 2008, As
Perry L. Carter, (D), Ohio State, 1998
Harold Gurrola, (D), California (San Diego), 1995, Ys
Callum J. Hetherington, (D), Basel (Switzerland), 2001, GxzCg
Haraldur R. Karlsson, (D), Chicago, 1988, Cl
David W. Leverington, (D), Manitoba, 2001, GmZrXg
Kevin R. Mulligan, (D), Texas A&M, 1997, ZiGmSo
Seiichi Nagihara, (D), Texas, 1992, Zi
Dustin E. Sweet, (D), Oklahoma, 2009, GsdGr
Christopher C. Weiss, (D), Oklahoma, 2004, AssAs
Assistant Professor:
Brian C. Ancell, (D), Washington, 2006, As
Guofeng Cao, (D), California (Santa Barbara), 2011, Zi
Johannes M L Dahl, (D), Ludwig-Maximilians (Germany), 2010, Ams
Song-Lak Kang, (D), Penn State, 2007, AsHg
Instructor:
Steven R. Cobb, As
Linda L. Jones, (M), California (Los Angeles), 1986
Justin E. Weaver, (M), Texas Tech, 1992, As
Professor:
Richard E. Peterson, (D), Missouri, 1971, As
Unit Coordinator:
Alisan C. Sweet, (M), Oklahoma, 2011, Gs
Debra J. Walker
Senior Technician:
James M. Browning, (B), Lamar, 1983, Ggx
Senior Business Assistant:
Alison Winton, (B), Texas Tech, 1992, ZnnZn
Computer Technician:
Darren W. Hedrick
Academic Advisor:
Celeste N. Yoshinobu, (M), San Diego State, 1994

Trinity University

Department of Geosciences (B) ☒
One Trinity Place
San Antonio, TX 78212-7200
p. (210) 999-7092
ksurples@trinity.edu
https://www.trinity.edu/academics/departments/geosciences
Programs: Geosciences (B); Earth Systems Science (B)
Enrollment (2022): B: 39 (10)

Chair:
Kathleen D. Surpless, (D), Stanford, 2001, Gst
Professor:
Daniel J. Lehrmann, (D), Kansas, 1993, PiGs
Diane R. Smith, (D), Rice, 1984, Gi
Benjamin E. Surpless, (D), Stanford, 1999, GctGi
Associate Professor:
Glenn C. Kroeger, (D), Stanford, 1987, Yg
Assistant Professor:
Kurt Knesel, (D), California (Los Angeles), 1998, Gvi
Brady A. Ziegler, (D), Virginia Tech, 2018, HwCqGe
Adjunct Professor:
Leslie F. Bleamaster III, (D), S Methodist, 2003, Xg
Emeritus:
Walter Coppinger, (D), Miami (Ohio), 1974, Gcg
Robert L. Freed, (D), Michigan, 1966, Gz
Thomas W. Gardner, (D), Cincinnati, 1978, GmtHg

Tyler Junior College

Dept of Geology
1327 South Baxter Avenue
Tyler, TX 75701
p. (903) 510-2232
thoo@tjc.edu
http://www.tjc.edu/

University of Houston

Allied Geophysical Laboratories (AGL) (B,M,D) O☒
SR1 131C
Houston, TX 77204-4231
p. (713) 743-9150
rrstewart@uh.edu
http://www.agl.uh.edu
Programs: Research and Development in Applied Geophysics especially exploration seismology

Professor:
Robert R. Stewart, (D), MIT, 1983, YesYu

Dept of Earth and Atmospheric Sciences (B,M,D) O☒
Science & Research Building 1
3507 Cullen Blvd, Rm. 312
Houston, TX 77204-5007
p. (713) 743-3399
tjlapen@uh.edu
http://www.eas.uh.edu
Administrative Assistant: Edwina Boateng Kumi, Jim Parker, Kirene Ramesar, Anja Wells
Enrollment (2016): B: 413 (69) M: 129 (41) D: 128 (10)

Chair:
Thomas Lapen, (D), Wisconsin, 2005, Gz
Professor:
Alan Brandon, (D), Alberta, 1992, GiCac
John F. Casey, (D), SUNY (Albany), 1980, GtiGu
John P. Castagna, (D), Texas, 1983, Ye
Henry S. Chafetz, (D), Texas, 1970, Gds
Evgeny Chesnokov, (D), Russian Acad of Sci, 1987, YseYx
Peter Copeland, (D), SUNY (Albany), 1990, CcGt
Stuart A. Hall, (D), Newcastle, 1976, Ym
Shuhab D. Khan, (D), Texas (Dallas), 2001, ZrGtYg
Aibing Li, (D), Brown, 2000, Ys
Rosalie F. Maddocks, (D), Kansas, 1965, Pm
Paul Mann, (D), SUNY (Albany), 1983
Michael Murphy, (D), California (Los Angeles), 2000, Gct
Bernhard Rappenglueck, (D), Munich, 1996, As
Arch M. Reid, (D), Pittsburgh, 1964, GizXm
William W. Sager, (D), Hawaii, 1983, GtYrm
Jonathan Snow, (D), MIT/WHOI, 1992, Ca
Robert Stewart, (D), MIT, 1983, Ye
John Suppe, (D), Yale, 1969, Gtc
Robert Talbot, (D), Wisconsin, 1981, ActAm
Arthur B. Weglein, (D), CUNY, 1980, Ye
Hua-Wei Zhou, (D), Caltech, 1989, YseYg
University Distinguished Research Professor:
Fred Hilterman, (D), Colorado Mines, 1970, Ye
Research Professor:
Adry Bissada, (D), Washington, 1967, Co
De-hua Han, (D), Stanford, 1987, Ye
Leon Thomsen, (D), Columbia, 1969, Ye
Research Associate Professor:
Donald Van Niewenhuise, (D), South Carolina, 1978, PsGoPe
Associate Professor:
Regina M. Capuano, (D), Arizona, 1988, HwCqs
William R. Dupre, (D), Stanford, 1975, On

Xun Jiang, (D), Caltech, 2006
Alexander Robinson, (D), California (Los Angeles), 2005, Gc
Guoquan Wang, (D), Inst of Geology (China), 2001

Research Associate Professor:
Yongjun Gao, (D), Goettingen (Germany), 2004, CstCa
Virginia Sisson, (D), Princeton, 1985, Gpt

Instructional Assistant Professor:
Daniel Hauptvogel, (D), CUNY, 2015
Jennifer N. Lytwyn, (D), Houston, 1993, Gi

Assistant Professor:
Emily Beverly, (D), Baylor
Yunsoo Choi, (D), Georgia Tech, 2007
Qi Fu, (D), Minnesota, 2006, CosCa
Joel Saylor, (D), Arizona, 2008
Yuxuan Wang, (D), Harvard, 2005, As
Julia Wellner, (D), Rice, 2001, GslOu
Jonny Wu, (D), Royal Holloway (UK), 2010, GtcGo
Yingcai Zheng, (D), California (Santa Cruz), 2007

Research Scientist:
Tom Bjorklund, (D), Houston, 2002, Gco
Martin Cassidy, (D), Houston, 2005, GoCo
Nikolay Dyaur, (D), Russian Acad of Sci, 1986, Yxs
Xiangshan Li, (D), Tulane, 2000, Asm

Research Professor:
James Lawrence, (D), Caltech, 1970, Cl
Peter Percell, (D), California (Berkeley), 1973, As

Research Associate Professor:
Dale Bird, (D), Houston, 2004
Robert Wiley, (D), Colorado Mines, 1980

Research Assistant Professor:
James Flynn, (M), Houston, 1991
Charlotte Sjunneskog, (D), Uppsala, 2002

Adjunct Professor:
Peter Bartok, (M), SUNY (Buffalo), 1972, GoYeNr

Senior Researcher:
Mike Darnell, (B), Texas A&M, 1974

Postdoctoral Fellow:
Hao Hu, (D), CAS (China), 2015, Ye

Adjunct:
Amy Kelly, (D), MIT, 2009
Gary Morris, (D), Rice, 1995

Researcher:
Min Sun
Ewa Szymczyk
Fuyong Yan

Laboratory Supervisor:
Minako Righter, (D), Grad Adv Studies, 2006

University of Houston Downtown

Dept of Natural Sciences
1 Main Street
Houston, TX 77002
p. (713) 221-8015
lyonsp@uhd.edu

Professor:
Glen K. Merrill, (D), Louisiana State, 1968, Pi
Penny A. Morris-Smith, (D), California (Berkeley), 1975

Associate Professor:
Kenneth Johnson, (D), Texas Tech, 1995, GiCac

Lecturer:
Donald S. Musselwhite, (D), 1995

University of North Texas

Dept of Geography (B,M)
1155 Union Circle #305279
Denton, TX 76203
p. (940) 565-2091
geog@unt.edu
http://www.geography.unt.edu
Certificates: GIS
Administrative Assistant: Tami Deaton

Professor:
Pinliang Dong, (D), New Brunswick, 2003, Zir
C. Reid Ferring, (D), Texas (Dallas), 1993, Ga
Paul F. Hudak, (D), California (Santa Barbara), 1991, Hw
Joseph R. Oppong, (D), Alberta, 1992, Zn
Murray Rice, (D), Saskatchewan, 1995, Zn
Harry F. L. Williams, (D), Simon Fraser, 1989, Gm
Steve Wolverton, (D), N Texas, 2001, Ga

Associate Professor:
Waquar Ahmed, (D), Clark, 2007, Zn
Ipsita Chatterjee, (D), Clark, 2007, Zn
Matthew Fry, (D), Texas, 2008, Zn
Kent M. McGregor, (D), Kansas, 1982, AsZg
Lisa A. Nagaoka, (D), Washington, 2000, Ga
Feifei Pan, (D), Georgia Tech, 2002, Hw
Alexandra Ponette-Gonzalez, (D), Yale, 2011, Zyc
Chetan Tiwari, (D), Iowa, 2008, Zi

Assistant Professor:
Lu Liang, (D), Zi

University of Texas Permian Basin

Geosciences (B,M) O☒
4901 E. University Boulevard
Geosciences
Odessa, TX 79762-0001
p. (432) 552-2243
zavada_m@utpb.edu
http://www.utpb.edu/geosciences
Programs: BS Geology with optional concentrations in Petroleum Geology, Environmental Geology and Paleontology. MS Geology with both Thesis and Non-Thesis options.
Certificates: Undergraduate and Graduate certificate in GIS
Enrollment (2022): B: 31 (8) M: 35 (16)

Chairman:
Michael Zavada, (D), Connecticut, Ge

Professor:
Robert C. Trentham, (D), Texas (El Paso), 1981, Go

Associate Professor:
Sumit Verma, (D), Oklahoma, 2015, YeGo

Assistant Professor:
Miles Henderson, (D), Tennessee, 2018, GsdGg
Joon Heo, (D), Texas A&M, 2012, Hg
Mohamed Zobaa, (D), Missouri S&T, 2011, PlEoCo

University of Texas Rio Grande Valley

School of Earth, Environmental, and Marine Sciences
Brownsville, TX 78520
p. (956) 761-2644
abdullah.rahman@utrgv.edu
https://www.utrgv.edu/graduate/for-future-students/graduate-programs/program-requirements/ocean-coastal-and-earth-sciences-ms/in

University of Texas, Arlington

Dept of Earth & Environmental Sciences (B,M,D) O
Box 19049
500 Yates Street
Arlington, TX 76019
p. (817) 272-2987
geology@uta.edu
http://www.uta.edu/ees/
Enrollment (2014): B: 189 (29) M: 70 (21) D: 28 (3)

Professor:
Asish Basu, (D), California (Davis), 1975, GxCcs
Glen Mattioli, (D), Northwestern, 1987, GitYd
Merlynd K. Nestell, (D), Oregon State, 1966, PmiPs

Associate Professor:
Qinhong (Max) Hu, (D), Arizona, 1995, GoHwNg
Andrew Hunt, (D), Liverpool (UK), 1988, GbCg
Arne M. Winguth, (D), Hamburg, 1997, OpAs

Assistant Professor:
Majie Fan, (D), Arizona, 2009, GstCs
Ashley Griffith, (D), Stanford, 2008, Gct
Liz Griffith, (D), Stanford, 2008, CslCm
Ashanti Johnson, (D), OgGe

Lecturer:
Cornelia Winguth, (D), Hamburg, 1998, GugZm

Adjunct Professor:
John E. Damuth, (D), Columbia, 1973, GusYr
Galina P. Nestell, (D), VSEGEI (Russia), 1990, PmsPi

Emeritus:
Brooks Ellwood, (D), Rhode Island, 1977, Yg
Christopher R. Scotese, (D), Chicago, 1985, Gt
John S. Wickham, (D), Johns Hopkins, 1969, Gco

TEXAS

University of Texas, Austin

Dept of Marine Science (M,D) ●☒
750 Channel View Drive
Port Aransas, TX 78373-5015
p. (361) 749-6711
ed.buskey@utexas.edu
http://www.utmsi.utexas.edu
f: www.facebook.com/utmsi
Programs: Marine Science
Enrollment (2015): M: 16 (6) D: 16 (1)

Associate Chair:
Edward J. Buskey, (D), Rhode Island, 1983, Ob

Professor:
Kenneth H. Dunton, (D), Alaska, 1985, Ob
Lee A. Fuiman, (D), Michigan, 1983, Ob
James W. McClelland, (D), Boston, 1998, CgHsCs
Peter Thomas, (D), Leicester (UK), 1977, Ob
Tracy A. Villareal, (D), Rhode Island, 1989, Ob

Associate Professor:
Deana L. Erdner, (D), MIT/WHOI, 1997, Ob
Andrew J. Esbaugh, (D), Queens, 2005, Ob
Zhanfei Liu, (D), SUNY (Stony Brook), 2006, Og

Assistant Professor:
Brett Baker, (D), Michigan, 2014, Ze
Brad Erisman, (D), California (San Diego), 2008, Ob
Amber K. Hardison, (D), William & Mary, 2010, Cm
Lauren A. Yeager, (D), Florida Intl, 2013, Og

Lecturer:
Dong-Ha Min, (D), Scripps, 1999, Og

Emeritus:
Wayne S. Gardner, (D), Wisconsin, 1971, Ob
Gloria J. Holt, (D), Texas A&M, 1976, Ob

Inst for Geophysics
JJ Pickle Research Campus
10100 Burnet Road, Bldg. 196 (ROC)
Austin, TX 78758
p. (512) 471-6156
utig@ig.utexas.edu
http://www.ig.utexas.edu/

Associate Director:
Ian W. D. Dalziel, (D), Edinburgh, 1963, Gt
Cliff Frohlich, (D), Cornell, 1976, Ys

Research Professor:
Stephen P. Grand, (D), Caltech, 1986, Ys
Yosio Nakamura, (D), Penn State, 1963, Ys
Mrinal K. Sen, (D), Hawaii, 1987, Ye

Professor:
Sean S. Gulick, (D), Lehigh, 1999, YrGtb
Paul L. Stoffa, (D), Columbia, 1974, Ye

Research Scientist:
Gail L. Christeson, (D), MIT/WHOI, 1994, YrsGt
Craig S. Fulthorpe, (D), Northwestern, 1988, Gu

Senior Scientist:
James A. Austin, Jr., (D), MIT/WHOI, 1979, Gut
Nathan L. Bangs, (D), Columbia, 1990, Gu
John A. Goff, (D), MIT/WHOI, 1990, Yr
Charles Jackson, (D), Chicago, 1998, ZoOoAt
Lawrence A. Lawver, (D), California (San Diego), 1976, YrGt
Paul Mann, (D), SUNY (Albany), 1983, Gt
Thomas H. Shipley, (D), Rice, 1975, Yr
Frederick W. Talyor, (D), Cornell, 1979, Pe

Research Scientist:
Donald D. Blankenship, (D), Wisconsin, 1989, Yg
Kirk D. McIntosh, (D), California (Santa Cruz), 1992, Yr
Robert J. Pulliam, (D), California (Berkeley), 1991, Ys

Associate Professor:
Luc L. Lavier, (D), Columbia, 1999, YgGt

Research Associate:
John W. Holt, (D), Caltech, 1997, Ye
David L. Morse, (D), Washington, 1997, Yg
Robert B. Scott, (D), McGill, 1999, Og
Roustam K. Seifoullaev, (D), Baku State, 1979, Ye
Harm Van Avendonk, (D), Scripps, 1998, Yg

Research Professor:
William E. Galloway, (D), Texas, 1971, GsoGr

Emeritus:
Milo M. Backus, (D), MIT, 1956, Ye
Arthur E. Maxwell, (D), Scripps, 1959, Og

Postdoc:
Christina Holland, (D), S Florida, 2003, Og
Matthew Hornbach, (D), Wyoming, 2004, Yg
Timothy Whiteaker, (D), Texas, 2004, Gs

Project Coordinator:
Patricia E. Ganey-Curry, (B), Texas A&M, 1978, Zn

Program Manager:
Katherine K. Ellins, (D), Columbia, 1988, Zn

Related Staff:
Mark Wiederspahn, (B), Bucknell, 1975, Zn

Jackson School of Geosciences (B,M,D) ●
Jackson School of Geosciences
2225 Speedway, Stop C1160
Austin, TX 78712-1692
p. (512) 471-5172
stockli@jsg.utexas.edu
http://www.jsg.utexas.edu
f: https://www.facebook.com/UTJSG
t: @txgeosciences
Programs: (1) General Geology (BS and BA), (2) Geophysics (BS), (3) Hydrogeology (BS), (4) Climate Sciences (BS), (5) UTeach Geoscience (BS)
Certificates: (1) Machine Learning and Data Analysis
Enrollment (2019): B: 190 (46) M: 22 (24) D: 140 (14)

Research Professor:
Sean S. Gulick, (D), Lehigh, 1999, Gc

Director, UTIG:
Demian Saffer, (D), California (Santa Cruz), 1999, Gu

Director, Bureau of Economic Geology:
Scott W. Tinker, (D), Colorado, 1996, Go

Dean, Jackson School of Geosciences:
Claudia Mora, (D), Wisconsin, Gg

Chair, Department of Geological Sciences:
Daniel Stockli, (D), Stanford, 1999, CcGct

Associate Dean for Research, Jackson School:
David Mohrig, (D), Washington, 1994, GsmGr

Professor:
Jay L. Banner, (D), SUNY (Stony Brook), 1986, Cl
Jaime D. Barnes, (D), New Mexico, 2006, CgsCc
Thorsten Becker, (D), Harvard, Yg
Christopher J. Bell, (D), California (Berkeley), 1997, Pv
Daniel O. Breecker, (D), New Mexico, 2008, ClsSc
Bayani Cardenas, (D), New Mexico Tech, 2006, Hg
Ginny A. Catania, (D), Washington, 2004, Gl
Julia A. Clarke, (D), Yale, 2002, PvgPq
Kerry H. Cook, (D), North Carolina State, 1984, AsPe
Ian W. D. Dalziel, (D), Edinburgh, 1963, Gt
Peter B. Flemings, (D), Cornell, 1990, Gr
Sergey B. Fomel, (D), Stanford, 2001, YesEo
James E. Gardner, (D), Rhode Island, 1993, Gv
Omar Ghattas, (D), Duke, 1988, ZnYg
Stephen P. Grand, (D), McGill, 1986, Yg
Patrick Heimbach, Max Planck, 1998, Og
Brian Horton, (D), Arizona, 1998, Gs
Charles Kerans, (D), Carleton, 1982, GsrGo
Richard A. Ketcham, (D), Texas, 1995, GgCc
John Lassiter, (D), California (Berkeley), 1995, Cc
Luc L. Lavier, (D), Columbia, 1999, Gt
Jung-Fu Lin, (D), Chicago, 2002, GyYm
Dev Niyogi, (D), North Carolina State, 2000, As
Timothy B. Rowe, (D), California (Berkeley), 1987, Pv
Mrinal K. Sen, (D), Hawaii, 1987, Ye
Zong-Liang Yang, (D), Macquarie, 1992, AmsHq

Senior Research Scientist:
James A. Austin, Jr., (D), MIT/WHOI, 1979, YrGr
Nathan L. Bangs, (D), Columbia, 1990, Gc
Donald D. Blankenship, (D), Wisconsin, 1989, Gl
John A. Goff, (D), MIT, 1990, Gu
Bob A. Hardage, (D), Oklahoma State, 1967, Yes
Susan D. Hovorka, (D), Texas, 1990, Gs
Michael R. Hudec, (D), Wyoming, 1990, Gc
Stephen E. Laubach, (D), Illinois, 1986, GcoNp
Robert G. Loucks, (D), Texas, 1976, Gso
Kitty L. Milliken, (D), Texas, 1985, Gd
Jeffrey G. Paine, (D), Texas, 1991, On

Stephen C. Ruppel, (D), Tennessee, 1979, GsrGo
Bridget R. Scanlon, (D), Kentucky, Hg
Thomas H. Shipley, (D), Rice, 1975, Ys
Michael H. Young, (D), Arizona, 1995, SpHwGe
Hongliu Zeng, (D), Texas, 1994, YsGs

Research Scientist:
Tim P. Dooley, (D), London (UK), 1994, Gt
Peter Eichhubl, (D), California (Santa Barbara), 1997, GcCgNr
Timothy A. Meckel, (D), Texas, 2003, Gr

Senior Scientist:
Jacob Covault, (D), Stanford, Gso
Peter Hennings, (D), Texas (Austin), 1991, GcoYs
Timothy Lawton, (D), Arizona, Gst
Lorena Moscardelli, (D), Texas (Austin), 2007, Gso
Jean-Philippe Nicot, (D), Texas (Austin), 1998, Hgt
Jud Partin, (D), Georgia Tech, CsAt
Alexandros Savvaidis, (D), Aristotle Thessaloniki, 1998, Ys
Hosseini Seyyed, (D), Tulsa, 2008, Go
Harm J. Van Avendonk, (D), Scripps, 1998, Ys

Associate Professor:
Elizabeth J. Catlos, (D), California (Los Angeles), 2000, GzCg
Marc A. Hesse, (D), Stanford, 2008, GqoYg
Joel P. Johnson, (D), MIT, 2007, Gs
Rowan Martindale, (D), S California, 2014, PgGs
Timothy M. Shanahan, (D), Arizona, 2007, PeGsCg
Kyle T. Spikes, (D), Stanford, 2008, Ye

Research Scientist:
Xavier Janson, (D), Miami, 2002, Gs
Laura Wallace, (D), California (Santa Cruz), 2002, GtYdZr

Associate Scientist:
Sheng Peng, (D), Arizona, 2004, GoHt
Katherine Romanak, (D), Texas (Austin), GxHt
Saad Saleh, (D), Wisconsin, 1991, Go
Krista Soderlund, (D), California (Los Angeles), XyYm
Duncan Young, (D), S Methodist, GlXy

Assistant Professor:
Timothy Goudge, (D), Brown, 2015, Gs
Ashley Matheny, (D), Ohio State, 2016, Hg
Geeta Persad, Princeton, 2016, As
Daniella Rempe, (D), California (Berkeley), Hg
Chenguang Sun, (D), Brown, 2014, Gx
Nicola Tisato, (D), ETH, 2013, Yg

Energy Economist:
Svetlana Ikonnikova, (D), Humboldt (Germany), 2007, Ego

Research Associate:
Eric Attias, (D), Southampton, 2017, GuYr
Robert Cunningham, (D), Texas (Dallas), Go
Andras Fall, (D), Virginia Tech, 2008, CgGcEg
Peter P. Flaig, (D), Alaska (Fairbanks), 2010, GsrEo
Qilong Fu, (D), Regina, 2005, Gso
Cyril Grima, (D), Grenoble, XyGlYg
Shuoshuo Han, (D), Columbia, YrGt
Seyyed Abolfazi Hosseini, (D), Tulsa, 2008, Eo
Benjamin Keisling, (D), Massachusetts, GlZc
Carey King, (D), Texas, 2004, Eog
Toti Larson, (D), New Mexico, 2003, GsoEo
Chris Lowery, (D), Massachusetts, OoPm
Maria-Aikaterini Nikolinakou, (D), MIT, 2008, Np
Yuko Okumura, (D), Hawaii, 2005, AsZoOp
Cornel Olariu, (D), Texas (Dallas), 2005, Gr
Mariana Olariu, (D), Texas (Dallas), 2007, GruGs
Alexey Portnov, (D), Arctic Norway, EgGx
Cornelia Rasmussen, (D), Utah, 2018, XbPgCc
Diana C. Sava, (D), Stanford, 2004, Yg
Timothy L. Whiteaker, (D), Texas, 2004, Zi
Brad Wolaver, (D), Texas, 2008, HwGe
Xiaohua (Eric) Xu, (D), California (San Diego), 2017, Yd
Changbing Yang, Hw
Kehua You, (D), Texas A&M, HtEo
Christopher K. Zahm, (D), Colorado Mines, 2002, Yg
Mehdi Zeidouni, (D), Calgary, 2011, Eo
Tongwei Zhang, (D), Chinese Acad of Sci, 1999, Cgs
Zeyu Zhao, (D), Ys

Lecturer:
Hillary C. Olson, (D), Stanford, 1988, Ps
Mary F. Poteet, (D), California (Berkeley), 2001, Pg

Adjunct Professor:
Marcus Gary, (D), Texas, 2009, Hgs

Senior Research Scientist:
Clifford A. Frohlich, (D), Cornell, 1976, YsGt
Frederick W. Taylor, (D), Cornell, 1978, GtmGe

Senior Lecturer:
Mark A. Helper, (D), Texas, 1985, GctXg

Emeritus:
Philip Bennett, (D), Syracuse, 1988, Hg
William D. Carlson, (D), California (Los Angeles), 1980, GpzCp
Mark P. Cloos, (D), California (Los Angeles), 1981, Gcx
William L. Fisher, (D), Kansas, 1961, GsrGo
William E. Galloway, (D), Texas, 1971, Gs
Gary A. Kocurek, (D), Wisconsin, 1980, Gs
J. Richard Kyle, (D), W Ontario, 1977, EmnCe
Leon E. Long, (D), Columbia, 1959, Cc
Earle F. McBride, (D), Johns Hopkins, 1960, Gd
Sharon Mosher, (D), Illinois, 1978, Gc
Yosio none Nakamura, (D), Penn State, 1963, YsGt
John M. Sharp, Jr., (D), Illinois, 1974, HqwGe
Douglas Smith, (D), Caltech, 1969, GipGz
James T. Sprinkle, (D), Harvard, 1971, PigGr
Ronald J. Steel, (D), Glasgow, 1970, Gs
Paul L. Stoffa, (D), Columbia, 1974, Ye
Robert H. Tatham, (D), Columbia, 1975, Yg
Clark R. Wilson, (D), California (San Diego), 1975, Yg

University of Texas, Dallas

Dept of Geosciences (B,M,D)
Mail Stop ROC21
800 W Campbell Rd
Richardson, TX 75083-3021
p. (972) 883-2401
david.lumley@utdallas.edu
https://geosciences.utdallas.edu/
Programs: No full degrees on line
Enrollment (2021): B: 38 (10) M: 18 (9) D: 12 (3)

Professor and Dept Head:
David Lumley, (D), Stanford, 1995, YseYu

Professor:
William I. Manton, (D), Witwatersrand, 1968, Cc
Robert J. Stern, (D), California (San Diego), 1979, Gt

Associate Professor:
Tom H. Brikowski, (D), Arizona, 1987, Hw
Hojun Zhu, (D), Princeton, 2014, Ysx

Research Professor:
Robert B. Finkelman, (D), Maryland, 1980, GbEcCa

Senior Lecturer:
William R. Griffin, (D), Texas (Dallas), 2008, Gg
Ignacio Pujana, (D), Texas (Dallas), 1997, Pm

Retired Professor:
Richard M. Mitterer, (D), Florida State, 1966, ColOc
Emile A. Pessagno, Jr., (D), Princeton, 1960, Pm
Dean C. Presnall, (D), Penn State, 1963, Cp

Retired President and Professor:
Robert H. Rutford, (D), Minnesota, 1969, Gl

Professor:
John F. Ferguson, (D), S Methodist, 1981, Yg
John Dr. Geissman, (D), Michigan, 1980, GtYm

In Memoriam:
Carlos L. V. Aiken, (D), Arizona, 1976, Zr

Emeritus:
George A. McMechan, (M), Toronto, 1972, Ys

Senior Lecturer:
Mortaza X. Pirouz, (D), Geneva, 2013, GcZiGc

Science/Mathematics Education Program (M)
P. O. Box 830688 FN33
Richardson, TX 75080-9688
p. (972) 883-2496
mont@utdallas.edu
http://www.utdallas.edu/dept/SciMathEd

Professor:
Thomas R. Butts, (D), Michigan State, 1973, Zn
Fred L. Fifer, (D), Vanderbilt, 1973, Zn

Associate Professor:
Cynthia E. Ledbetter, (D), Texas A&M, 1987, Zn

Assistant Professor:
Homer A. Montgomery, (D), Texas (Dallas), 1988, Pg
Mary Urquhart, (D), Colorado, 1999, Xy

Instructor:
Barbara Curry, (M), Texas (Dallas), 1998, Zn

University of Texas, El Paso

Dept of Earth, Environmental and Resource Sciences (B,M,D) O☒
500 W. University Avenue
101 Geological Sciences
El Paso, TX 79968-0555
p. (915) 747-5501
jdkubicki@utep.edu
http://science.utep.edu/geology/
Programs: BS Geology, BS Geology – Secondary Education Minor – 7-12 Science Certification, BS Environmental Science - Biology Concentration, BS Environmental Science - Chemistry Concentration, BS Environmental Science - Geoscience Concentration, BS Environmental Science - Hydroscience Concentration, BS Environmental Science – Secondary Education Minor – 7 – 12 Science Concentration, Fast Track Programs:BS Geology to MS Geology, BS Geology to MBA, Graduate Program: MS Geology, MS Geophysics, MS Environmental Science, PhD Geology
Certificates: Geospatial Information Science and Technology - GIST
Enrollment (2021): B: 187 (42) M: 36 (17) D: 29 (8)

National Seismic Source Facility:
Galen M. Kaip, (D), Texas (El Paso), 1998, Yx

Professor:
James D. Kubicki, (D), Yale, 1989, ClGe

Professor:
Antonio Arribas, (D), Michigan, 1992, EmCe
Mark Engle, (D), Nevada (Reno), 2005, HwGe
Katherine A. Giles, (D), Arizona, 1991, Go
Thomas E. Gill, (D), California (Davis), 1995, GmAsCl
Philip C. Goodell, (D), Harvard, 1970, Ce
Jose M. Hurtado, (D), MIT, 2002, Gt
Lixin Jin, (D), Michigan, 2007, Ge
Richard P. Langford, (D), Utah, 1989, Gs
Lin Ma, (D), Michigan, 2008, HgCg
Nicholas E. Pingitore, Jr., (D), Brown, 1973, Cl
Aaron A. Velasco, (D), California (Santa Cruz), 1993, Yx

Associate Professor:
Marianne Karplus, (D), Stanford, 2012, Ysg
Deana Pennington, (D), Oregon State, 2002, ZyfZe

Assistant Professor:
Laura Alvarez, (D), Arizona State, 2015, ZiGs
Benjamin Brunner, (D), ETH (Switzerland), 2003, Cs
Julien Chaput, (D), New Mexico Tech, 2012, GeYes
Hugo A. Gutierrez-Jurado, (D), New Mexico Tech, 2011, HgsHw
Hernan Moreno, (D), Arizona State, 2012, Zir
Jason Ricketts, (D), New Mexico, 2014, GctGi
Jie Xu, (D), Wisconsin, 2011, ClPoGz

Lecturer:
Vicki Harder, (D), Texas (El Paso), 1997, Gg
Musa Hussein, (D), Texas (El Paso), 2007, YgeYm
Adriana Perez, (D), Texas (El Paso), 2008, Gg

Associate Professor Emeritus:
William Cornell , (D), California (Los Angeles), Gg

Emeritus:
Kenneth F. Clark, (D), New Mexico, 1966, Eg
George Randy Keller , (D), Texas Tech, 1973, Yg

Research Assistant Professor :
Steven H. Harder, (D), Texas (El Paso), 1986, Yg

University of Texas, Rio Grande Valley

School of Earth, Environmental and Marine Sciences (B,M) ⊘
1201 W. University Drive
ESCNE 2.620
Edinburg, TX 78539
p. (956) 882-5040
juan.l.gonzalez@utrgv.edu
https://www.utrgv.edu/seems/
Programs: BS Environmental Science - Earth and Ocean Sciences Concentration., MS Ocean, Coastal and Earth Sciences MS Agricultural, Environmental, and Sustainability Sciences
Enrollment (2020): B: 140 (0) M: 32 (0)

Director:
David Hicks, (D), Texas (Arlington), 1999, Ob

Professor:
Carlos Cintra-Buenrostro, (D), Arizona, 2006, PiCsPe
Abdullah F. Rahman, (D), Arizona, 1996, Zr

Associate Professor:
Jude Benavides, (D), Rice, 2005, Hsq
John Breier, (D), Texas, 2006, Cma
Juan L. Gonzalez, (D), Tulane, 2008, GmsGa
Elizabeth Heise, (D), Texas A&M, 2001, OnZeGg
Ruben Mazariegos, (D), Texas A&M, 1993, ZiYe

Assistant Professor:
Chu-Lin Cheng, (D), Iowa State, 2009, HgwHq
Christopher Gabler, (D), Rice, 2012, PeSf
Cheryl Harrison, (D), California (Santa Cruz), 2012, Op
James Kang, (D), N Carolina State, 2007, SpcZu
Engil I. Pereira, (D), California (Davis), 2014, SbZcSo
Amit Raysoni, (D), Texas (El Paso), 2011, Asc

Research Associate:
Lorena Longoria, (B), Texas (Pan American), 2012, Ob

Lecturer:
Sarah M. Hardage, (M), New Orleans, 2004, GmZiGg
Saad Mohamed, (D), Dalhousie
Idua Olunwa, (M), Texas (Dallas), 2009

Adjunct Professor:
James R. Hinthorne, (D), California (Santa Barbara), 1974, GgzCa

University of Texas, San Antonio

Dept of Earth and Planetary Sciences (B,M) ●⊘
One UTSA Circle
San Antonio, TX 78249-0663
p. (210) 458-4455
geosciences@utsa.edu
https://www.utsa.edu/sciences/earth-planetary-sciences/
f: https://www.facebook.com/Dpt-of-Geological-Sciences-UTSA-881208175556941/
Programs: Geosciences (B,M); Geoinformatics (M)
Certificates: GIS (G) (U)
Enrollment (2017): B: 107 (0) M: 35 (0)

University of the Incarnate Word

Dept of Math, Science, and Engineering (B)
Bonilla Science Hall 105
4305 Broadway
San Antonio, TX 78209
recarde1@uiwtx.edu
https://www.uiw.edu/smse/academics/departments/atmosscien-cephysics/

Victoria College

Science, Mathematics, & Physical Education (A)
2200 E. Red River
Victoria, TX 77901
p. (361)573-3291 (Ext.3432)
Alisha.Stearman@VictoriaCollege.edu
https://www.victoriacollege.edu/sciencemathematicsphysical-education

Wayland Baptist University

Geology Department (B) ●✓
1900 W. 7th Street
Plainview, TX 79072
p. (806) 291-1115
walsht@wbu.edu
http://www.wbu.edu/academics/schools/math_and_science/geology/
f: https://www.facebook.com/Wayland-Baptist-University-Geology-Department-198757676854342/
Programs: Geology
Enrollment (2018): B: 4 (1)

Professor:
Don Parker, (D), GivCg
Tim R. Walsh, (D), Texas Tech, 2002, GsPmGo

Assistant Professor:
Mark Bryan, (M), Oklahoma State, 2001, GeoPs

Weatherford College

Weatherford College
225 College Park Drive

Weatherford, TX 76086
p. (817) 598-6277
http://www.wc.edu

West Texas A&M University

Dept of Life, Earth & Environmental Sciences (B,M)
P. O. Box 60808, WT Station
Canyon, TX 79016-0001
p. (806) 651-2570
dsissom@wtamu.edu
https://www.wtamu.edu/academics/college-agriculture-natural-sciences/department-life-earth-environmental-sciences/
Administrative Assistant: Debi Adams
Enrollment (2013): B: 13 (2)
Chair:
David Sissom, (D), Zn
Professor:
Joseph C. Cepeda, (D), Texas, 1977, GiHw
David B. Parker, (D), Nebraska, 1996, Sbf
William J. Rogers, (D), Texas A&M, 1999, Zn
Gerald E. Schultz, (D), Michigan, 1966, PvGzZg
Associate Professor:
Gary C. Barbee, (D), Texas A&M, 2004, HwSoZi
Instructor:
Cindy D. Meador, (M), W Texas A&M, 1989, Zg
Joe D. Rogers, (M), WTSU (WTAMU), 1987, Ga
William C. Rogers, (M), WTAMU, 2005, Ge
Lynn C. Rosa, (M), Oklahoma, 1984, Zge

Wharton County Junior College - Sugarland Campus

Wharton County Junior College - Sugarland Campus
14004 University Blvd.
Sugarland, TX 77479
p. (281) 239-1559
dannyg@wcjc.edu
http://www.wcjc.edu

Wharton County Junior College - Wharton Campus

Geology
911 Boling Highway
Wharton, TX 77488
p. (979) 532-6506
dannyg@wcjc.edu
https://www.wcjc.edu/Programs/math-and-science/geology/

Utah

Brigham Young University

Dept of Geography (B) ☒
690 SWKT
Provo, UT 84602
p. (801) 378-3851
dholsen@byu.edu
https://geography.byu.edu/
Programs: Geospatial Science & Technology, Geospatial Intelligence, Environmental Systems, Urban & Regional Planning, Global Studies, Travel and Tourism Studies
Chair:
Ryan R. Jensen, (D), Florida, 2000, Zii
Professor:
Perry J. Hardin, (D), Utah, 1989, Zr
Samuel M. Otterstrom, (D), Louisiana State, 1997, Zu
Matthew J. Shumway, (D), Indiana, 1991, Zn
Associate Professor:
Matthew F. Bekker, (D), Iowa, 2002, Zy
Michael J. Clay, (D), California (Davis), 2005
James A. Davis, (D), Arizona State, 1992, Zn
Jeffrey O. Durrant, (D), Hawaii, 2001, Ge
Chad Emmett, (D), Chicago, 1991, Zn
Clark S. Monson, (D), Hawaii, 2004
Daniel H. Olsen, (D), Waterloo, 2008
Brandon Plewe, (D), Buffalo, 1997

Dept of Geological Sciences (B,M) ●⦸
S 389 ESC
Provo, UT 84602
p. (801) 422-3918
brooks_britt@byu.edu
https://geology.byu.edu/
Programs: Geology, Earth Space and Science Education
Enrollment (2019): B: 89 (24) M: 32 (13)
Research Professor:
Michael J. Dorais, (D), Georgia, 1987, Gi
Teaching Professor:
Randall Skinner, (M), Brigham Young, 1996, Gg
Professor:
Barry R. Bickmore, (D), Virginia Tech, 2000, ClGz
Eric H. Christiansen, (D), Arizona State, 1981, GiXgGv
Ron Harris, (D), London (UK), 1989, Gc
Jeffrey D. Keith, (D), Wisconsin, 1982, Eg
John H. McBride, (D), Cornell, 1987, YeGoe
Stephen T. Nelson, (D), California (Los Angeles), 1991, Cs
Jani Radebaugh, (D), Arizona, 2005, XgGvm
Scott M. Ritter, (D), Wisconsin, 1986, Ps
Research Professor:
David G. Tingey, (M), Brigham Young, 1989, Zn
Associate Professor:
Brooks B. Britt, (D), Calgary, 1993, Pg
Gregory T. Carling, (D), Utah, 2012, HwCts
Assistant Professor:
Samuel M. Hudson, (D), Utah, 2008, GosGr
Visiting Professor:
R. William Keach II, (M), Cornell, 1986, YeEoGg
Emeritus:
Myron G. Best, (D), California (Berkeley), Gg
Dana T. Griffen, (D), Virginia Tech, 1976, Gz
Lehi F. Hintze, (D), Columbia, 1951, Pg
Bart J. Kowallis, (D), Wisconsin, 1981, GgCcGz
Alan L. Mayo, (D), Idaho, 1982, Hw
Wade E. Miller, (D), California (Berkeley), 1968, Pv
Thomas H. Morris, (D), Wisconsin, 1986, Gr
R. Paul Nixon, (D), Brigham Young, 1972, Go
Morris S. Petersen, (D), Iowa, 1962, Gso
William R. Phillips, (D), Utah, 1954, Gz

Dept of Plant and Wildlife Sciences ☒
Provo , UT 84602
p. 801-422-9389
pws-grad-secretary@byu.edu
http://pws.byu.edu/
Programs: Environmental Science & Sustainability, Wildlife and Wildlands Conservation, Genetics, Genomics & Biotechnology, and Plant & Landscape Systems

Salt Lake City Community College-Jordan Campus

Dept of Geosciences
4600 South Redwood Road
Salt Lake City, UT 84123
p. (801)957-4150
adam.dastrup@slcc.edu
http://www.slcc.edu

Snow College

Dept of Geology (A)
150 College Ave E
Ephraim, UT 84627
p. (435) 283-7519
renee.faatz@snow.edu
https://www.snow.edu/academics/science_math/geology
f: https://www.facebook.com/Snow-College-Geology-616452215106045/
Enrollment (2016): A: 7 (1)
Chair:
Renee M. Faatz, (M), Ohio State, 1985, Gg
Assistant Professor:
Ted L. Olson, (M), Utah, 1976, Ys

Southern Utah University

Dept of Geosciences (B) O
351 West University Blvd.
Cedar City, UT 84720

p. (435) 586-7900
jasonkaiser@suu.edu
https://www.suu.edu/cos/geosciences/
f: https://www.facebook.com/SUUGeologyClub
Programs: B.S. in Geology, B.A. in Geosciences, B.S. in Environmental Science, B.A. in Environmental Studies, B.S. in Geospatial Science
Certificates: GIS
Enrollment (2022): B: 65 (0)
Associate Professor:
Jacqueline Grant, (D), Zn
Assistant Professor:
Jason Kaiser, (D), Oregon State, 2014, GzvCg
Aspen Manning, (D), Zry
Dave Maxwell, (M), Zi
Erich Mueller, (D), HgGm
Grant Shimer, (D), Alaska, Gs
Jamie Spinney, (D), Zyu
Casey Webb, (M), GtZg

University of Utah

Dept of Atmospheric Sciences (B,M,D) ☒
135 S 1460 E, Rm 819
Salt Lake City, UT 84112-0110
p. (801) 581-6136
atmos-info@lists.utah.edu
https://www.atmos.utah.edu
f: www.facebook.com/UUAtmos
t: @UofUATMOS
Programs: Atmospheric Sciences
Certificates: N/A
Enrollment (2021): B: 62 (16) M: 27 (2) D: 13 (5)
Chair:
John Horel, (D), Washington, 1982, AsZf
Professor:
Timothy J. Garrett, (D), Washington, 2000, As
Steven Krueger, (D), California (Los Angeles), 1985, As
John Chun-Han Lin, (D), Harvard, 2003, As
Gerald Mace, (D), Penn State, 1994, As
Zhaoxia Pu, (D), Lanzhou, 1997, As
Jim Steenburgh, (D), Washington, 1995, As
Edward Zipser, (D), Florida State, 1965, As
Associate Professor:
Anna Gannet Hallar, (D), Colorado, 2003, AspAc
Kevin D. Perry, (D), Washington, 1995, As
Thomas Reichler, (D), California (San Diego), 2003, As
Courtenay Strong, (D), Virginia, 2005, As

Dept of Geography (B,M,D)
260 S. Central Campus Drive
Rm 270
Salt Lake City, UT 84112-9155
p. (801) 581-8218
dennison@geog.utah.edu
http://www.geog.utah.edu
Administrative Assistant: Susan Van Roosendaal
Chair:
George F. Hepner, (D), Arizona State, 1979, Zy
Professor:
Donald R. Currey, (D), Kansas, 1969, Gm
Thomas M. Kontuly, (D), Pennsylvania, 1978, Zn
Chung M. Lee, (D), Michigan, 1961, Zu
Assistant Professor:
Thomas J. Cova, (D), California (Santa Barbara), 1999, Zi
Richard R. Forster, (D), Cornell, 1997, Zr
Adjunct Professor:
Jeffrey R. Keaton, (D), Texas A&M, 1988, Zn
Elliott W. Lips, (M), Colorado State, 1990, Gm
Emeritus:
Philip C. Emmi, (D), North Carolina, 1979, Zu
Roger M. Mccoy, (D), Kansas, 1967, Zr
Merrill K. Ridd, (D), Northwestern, 1963, Zr
Professor-Lecturer:
Arthur Hampson, (D), Hawaii, 1980, Zy
Assistant Professor:
Trevor J. Davis, (D), British Columbia, 1999, Zi
Other:
Fred E. May, (D), Virginia Tech, 1976, Zn

Dept of Geology & Geophysics (B,M,D) ●☒
Frederick Albert Sutton Building, Rm 383
115 South 1460 East
Salt Lake City, UT 84112-0102
p. (801) 581-7062
william.johnson@utah.edu
http://www.earth.utah.edu/
f: https://www.facebook.com/University-of-Utah-Geology-and-Geophysics-361264640646
Programs: Geoscience (emphasis in Geology, Environmental Geoscience, or Geophysics), Geological Engineering, Earth Science Composite Teaching, Earth Science Minor
Certificates: Interdisciplinary Graduate Certificate in Hydrology and Water Resources (can be taken as a non-degree seeking student with a completed Bachelor's degree from an accredited college or university)
Enrollment (2020): B: 91 (45) M: 14 (6) D: 37 (10)
CMES Student Services Director:
Michelle Tuitupou, (M), 2004
Chair:
Thure E. Cerling, (D), California (Berkeley), 1977, ClsZc
Professor Lecturer:
David A. Dinter, (D), MIT, 1994, GtcGu
Director, Seismograph Stations:
Keith Koper, (D), Washington, 1998
Associate Director, Global Change and Sustainability Center:
Brenda Bowen, (D), Gse
Professor:
Gabriel Bowen, (D), California (Santa Cruz), 2003, Cs
Paul D. Brooks, (D), Colorado, 1995, HgCbZg
Marjorie A. Chan, (D), Wisconsin, 1982, Gs
Paul W. Jewell, (D), Princeton, 1989, GmHg
Cari Johnson, (D), Stanford, 2003, Gs
William P. Johnson, (D), Colorado, 1993, Ng
Douglas K. Solomon, (D), Waterloo, 1992, Hw
Tonie van Dam, (D), Yd
Michael S. Zhdanov, (D), Moscow State (Russia), 1968, Ye
Research Associate Professor:
Diego Fernandez, (D), Buenos Aires, 1991, Cg
Marie D. Jackson, (D), Gv
James M. Karner, (D), Xm
Huilian Ma, (D)
Kristine L. Pankow, (D), California, 1999, Ys
James C. Pechmann, (D), Caltech, 1983, YsGt
Chief Curator and Curator of Paleontology (NHMU):
Randall B. Irmis, (D), California (Berkeley), 2008, Sa
Associate Professor Lecturer:
Holly Godsey, (D), Ze
Mark Loewen, (D), Pvg
Associate Professor:
Lauren Birgenheier, (D), Nebraska, 2007, Gso
Fan-Chi Lin, (D), Colorado, 2009, Ygs
Peter C. Lippert, (D), California (Santa Cruz), 2010, YmGtPe
Lowell Miyagi, (D), California (Berkeley), 2009, Gz
Jeffrey Moore, (D), California (Berkeley), 2007, NgrGm
Michael S. Thorne, (D), Arizona State, 2005, YsxYv
Research Assistant Professor:
Amir Allam, (D)
Jamie Farrell, (D), Ys
Richard Fiorella, (D)
Alex Gribenko, (D), Yg
Scott R. Miller, (D)
Assistant Professor:
Kathleen Allabush, (D), Pe
Sarah Lambart, (D), Cpg

Dept of Mining Engineering (B,M,D)
135 South 1460 East
Room 313
Salt Lake City, UT 84112-0113
p. (801) 581-7198
charles.kocsis@utah.edu
http://www.mines.utah.edu/mining
Enrollment (2009): B: 48 (11) M: 6 (3) D: 1 (0)
Chair:
Michael G. Nelson, (D), West Virginia, 1989, Nm
Professor:
Michael K. McCarter, (D), Utah, 1972, Nm

William G. Pariseau, (D), Minnesota, 1966, Nr
Adjunct Associate Professor:
Stephen Bessinger, (D), Nm
Associate Professor:
Felipe Calizaya, (D), Colorado Mines, 1985, Nm
Adjunct Associate Professor:
Helmut H. Doelling, (D), Utah, 1964, Gg
Duane L. Whiting, (B), Utah, 1959, Hw
Adjunct Professor:
James Donovan, (D), Virginia Tech, 2003, NmrZr
Krishna P. Sinha, (D), Minnesota, 1979, Nr
Jeffrey Whyatt, (D), Nm
Zavis Zavodni, (D), Nm

Energy & Geoscience Institute (M) ●
423 Wakara Way
Suite 300
Salt Lake City, UT 84108
p. (801) 581-5126
egidirector@egi.utah.edu
http://www.egi.utah.edu
Director:
Raymond A. Levey, (D), South Carolina, 1981, Eo
Professor:
Richardson B. Allen, (D), Columbia, 1983, GctZi
Alastair Fraser, (D), Edinburgh, Gco
Brian McPherson, (D), Utah, 1996, Ye
Joseph N. Moore, (D), Penn State, 1975, Gv
Michal Nemcok, (D), Comenius (Slovakia), 1991, GctGo
Peter E. Rose, (D), Utah, 1993, Np
Stuart Simmons, (D), Minnesota, 1986, EmCqHg
Rasoul Sorkhabi, (D), Japan, 1991, GctGc
Phillip E. Wannamaker, (D), Utah, 1983, Ye
Scientific Staff :
Julia Kotulova, (D), Komenius (Slovakia), CoGo
Senior Scientist:
Bryony Richards-McClung, (D), Go
Lansing Taylor, (D), Stanford, 1999, Gc
David Thul, (D), Colorado Mines, 2014, Gg
Associate Professor:
Glenn W. Johnson, (D), South Carolina, 1997, Gq
Greg Nash, (D), Utah, Gt
Marylin Segall, (D), 1991, GeOu
Assistant Professor:
Shu Jiang , (D), China Geosciences, GoYsNp
Sudeep Kanungo, (D), Univ Coll (UK), Pms
Research Associate:
Kenneth L. Shaw, (B), British Columbia, 1965, Ye
Instructor:
William Keach, (M), Cornell, 1986, GoYes
Adjunct Professor:
Ian Walton, (D), Manchester (UK), 1972, Gq

Utah State University

Department of Geosciences (B,M,D) ●
4505 Old Main Hill
Logan, UT 84322-4505
p. (435) 797-1273
geo@usu.edu
http://geo.usu.edu/
Programs: Geology; Applied Environmental Geoscience; Geo-Workforce, Earth Science Composite Teaching
Enrollment (2022): B: 52 (0) M: 16 (0) D: 6 (0)
Head:
Joel L. Pederson, (D), New Mexico, 1999, Gm
Professor:
Carol M. Dehler, (D), New Mexico, 2001, Gs
Tammy M. Rittenour, (D), Nebraska, 2004, GmCcGa
Professional Practice:
Kelly K. Bradbury, (D), Utah State, 2012, Gc
Associate Professor:
Alexis K. Ault, (D), Colorado, 2012, GtCc
Benjamin J. Burger, Colorado, 2009, PvGs
Thomas E. Lachmar, (D), Idaho, 1989, Hw
Anthony R. Lowry, (D), Utah, 1994, YdsGt
Dennis L. Newell, (D), New Mexico, 2007, Cls
Professional Practice:
Katherine E. Potter, (D), Utah State, 2014, GiZeEg
Assistant Professor:
Evelyn Gannaway Dalton, (D), Texas (El Paso), 2019, GstZe
Donald E. Penman, (D), California (Santa Cruz), 2015, GsCbPc
Srisharan Shreedharan, (D), Penn State, 2021, NrYsGq
Emeritus:
James P. Evans, (D), Texas A&M, 1987, Gc
Susanne U. Janecke, (D), Utah, 1991, Gt
W. David Liddell, (D), Michigan, 1979, GsPg
Robert Q. Oaks, Jr., (D), Yale, 1965, GsHwYv
John W. Shervais, (D), California (Santa Barbara), 1979, Gi

Utah State University Eastern

Dept of Geology (A) ☒
451 East 400 North
Price, UT 84501
p. (435) 613-5232
michelle.fleck@usu.edu
Programs: Associate of Science
Enrollment (2017): A: 4 (0)
Associate Professor:
Michelle Cooper Fleck, (D), Wyoming, 2001, GgZy

Utah Valley University

Dept. of Earth Science (A,B) ○
800 West University Parkway
MS 179
Orem, UT 84058
p. (801) 863-8582
hornsda@uvu.edu
https://www.uvu.edu/earthscience/
f: https://www.facebook.com/groups/uvuearthscience/
Programs: Geology; Geography; Environmental Science; Earth Science Education
Certificates: GIS
Enrollment (2021): B: 140 (23)
Professor:
Michael Bunds, (D), GctRn
Daniel Horns, (D), NgGeRn
Associate Professor:
Joel Bradford, (M), Utah, RwGe
Eddy Cadet, (D), Ge
James Callison, (D), GeSfo
Hilary Hungerford, (D), ZuRw
Daniel Stephen, (D), PiGsPe
Nathan Toke', (D), Arizona State, 2011, GtmZi
Weihong Wang, (D), CmGeZi
Alessandro Zanazzi, (D), CgsPe
Assistant Professor:
Doug Czajka, (D), N Carolina State, 2018
Daren Nelson, (D)
Michael Stearns, (D), California (Santa Barbara), 2014, Gip
Justin White, ZyiZu

Weber State University

Department of Earth and Environmental Sciences (B) ●
1415 Edvalson St. Dept 2507
Ogden, UT 84408-2507
p. (801) 626-7139
mmatyjasik@weber.edu
https://www.weber.edu/ees/
f: https://www.facebook.com/WSUGeosciences
Programs: Geology (B); Applied Environmental Geoscience (B); Earth Science Teaching (B)
Certificates: Geospatial Analysis (GIS)
Administrative Assistant: Anna Cich
Enrollment (2022): B: 60 (12)
Chair:
Marek Matyjasik, (D), Kent State, 1997, HwGe
Professor:
Michael W. Hernandez, (D), Utah, 2004, Zir
W. Adolph Yonkee, (D), Utah, 1990, Gct
Associate Professor:
Elizabeth A. Balgord, (D), Arizona, 2015, Grt
Carie M. Frantz, (D), S California, 2013, PoCbZn
Assistant Professor:
Ryan J. Frazier, (D), British Columbia, 2016, ZriZc
Caitlin E. Tems, (D), S California, 2016, OoCs

Adjunct Professor:
Amanda L. Gentry, (M), Nevada (Las Vegas), 2016, GgsGg
Gregory B. Nielsen, (D), Utah, 2010, Gg
Emeritus:
Jeffrey G. Eaton, (D), Colorado, 1987, PsGs
Richard L. Ford, (D), California (Los Angeles), 1997, GmZeEo
David J. Matty, (D), Rice, 1984, Gig
James R. Wilson, (D), Utah, 1976, Gz
Laboratory Director:
Sara Summers, (M), Notre Dame, 2012, Gg

Vermont

Castleton University

Dept of Natural Sciences (Geology) (B)
Castleton University
233 South Street
Castleton, VT 05735
p. (802) 468-1238
helen.mango@castleton.edu
http://www.castleton.edu/academics/undergraduate-programs/geology/
Enrollment (2015): B: 12 (3)
Professor:
Timothy W. Grover, (D), Oregon, 1988, GptGc
Helen N. Mango, (D), Dartmouth, 1992, CgGe

Middlebury College

Geology Dept (B)
276 Bicentennial Hall
Middlebury, VT 05753
p. (802) 443-5029
geology_chair@middlebury.edu
http://www.middlebury.edu/academics/geol
f: https://www.facebook.com/MiddleburyGeology
Administrative Assistant: Eileen Brunetto
Enrollment (2016): B: 30 (15)
Chair:
David P. West, (D), Maine, 1993, Gc
Professor:
Patricia L. Manley, (D), Columbia, 1989, YrGu
Jeffrey S. Munroe, (D), Wisconsin, 2001, Gln
Peter C. Ryan, (D), Dartmouth, 1994, Cl
Assistant Professor:
Will Amidon, (D), Caltech, 2010, CcGmt
Kristina J. Walowski, (D), Oregon, 2015, Gi
Visiting Professor:
Thomas O. Manley, (D), Columbia, 1981, Op

Northern Vermont University-Lyndon

Dept of Atmospheric Sciences (B) ☒
1001 College Road
PO Box 919
Lyndonville, VT 05851
p. (802) 626-6225
jason.shafer@northernvermont.edu
http://meteorology.lyndonstate.edu
f: https://www.facebook.com/nvuatm/
t: @lyndonweather
Enrollment (2018): B: 73 (9)
Chair:
Janel Hanrahan, (D), Wisconsin, 2010, AsZoRc
Professor:
Jason Shafer, (D), Utah, 2005, As
Research Associate:
David Siuta, (D), British Columbia, 2017, AsZn
Adjunct Professor:
George Loriot, (D), Connecticut, AsZn
Visiting Professor:
Aaron Preston, (D), Florida State, AsZrAc
Atmospheric Sciences Data Systems Administrator:
Jason Kaiser, (M), Plymouth State, 2013, AsZn

Geology Minor (A,B)
1001 College Rd
Lyndonville, VT 05851
p. (802) 635-1325
brad.moskowitz@northernvermont.edu
Enrollment (2010): B: 37 (11)
Professor:
Tania S. Bacchus, (D), Maine, 1993, GuAm

Norwich University

Dept of Earth and Environmental Sciences (B) ☒
158 Harmon Drive
Northfield, VT 05663
p. (802) 485-2304
rdunn@norwich.edu
https://www.norwich.edu/programs/geology
f: https://www.facebook.com/groups/102347383133958/
Programs: Geology, and Environmental Science
Enrollment (2019): B: 40 (6)
Professor:
Richard K. Dunn, (D), Delaware, 1998, GalGs
Associate Professor:
G. Christopher Koteas, (D), Massachusetts, 2010, GitYh
Assistant Professor:
Laurie Grigg, (D), Oregon, 2000, PeGnZi
Research Associate:
George E. Springston, (M), Massachusetts, 1990, GmZiYg
Lecturer:
John Gartner, (D), Dartmouth, 2015, GmHy
Emeritus:
David S. Westerman, (D), Lehigh, 1972, Gxi

University of Vermont

Dept of Geology (B,M) ☒
Delehanty Hall
180 Colchester Avenue
Burlington, VT 05405-0122
p. (802) 656-3396
geology@uvm.edu
http://www.uvm.edu/geology
f: https://www.facebook.com/UVMGeology/
Programs: Geology
Administrative Assistant: Robin Hopps
Administrative Assistant: Srebrenka Mrsic
Enrollment (2015): B: 48 (0) M: 13 (0)
Professor:
Paul R. Bierman, (D), Washington, 1993, GmeCc
John M. Hughes, (D), Dartmouth, 1981, Gz
Keith A. Klepeis, (D), Texas, 1993, Gct
Research Associate Professor:
Andrew W. Schroth, (D), Dartmouth, 2007, ClqCb
Chair:
Andrea Lini, (D), ETH (Switzerland), 1994, CsGn
Associate Professor:
Laura E. Webb, (D), Stanford, 1999, GtCcGc
Assistant Professor:
Julia Perdrial, (D), Strasbourg (France), 2008, ClSc
Nicolas Perdrial, (D), Strasbourg (France), 2007, GezCl
Senior Lecturer:
Stephen F. Wright, (D), Minnesota, 1988, Glc
Emeritus:
David P. Bucke, (D), Oklahoma, 1969, GgdGs
Barry L. Doolan, (D), SUNY (Binghamton), 1970, Gx
John C. Drake, (D), Harvard, 1967, Cl
Charlotte J. Mehrtens, (D), Chicago, 1979, Gs
Senior Research Technician:
Gabriela Mora-Klepeis, (M), Texas, 1992, CacGi

Dept of Plant & Soil Science (B,M,D)
Jeffords Hall
63 Carrigan Drive
Burlington, VT 05405-1737
p. (802) 656-2630
pss@uvm.edu
http://www.uvm.edu/~pss/

Virginia

Central Virginia Community College

Science, Math, and Engineering (A)
3506 Wards Road
Lynchburg, VA 24502

p. (434) 832-7707
counseling@centralvirginia.edu
http://www.cvcc.vccs.edu/Academics/SME/
Instructor:
Mark Tinsley, Zg

George Mason University

Department of Geography and Geoinformation Science
Fairfax, VA 22030
p. (703) 993-1210
ggs@gmu.edu
https://cos.gmu.edu/ggs/

Dept of Atmospheric, Oceanic, and Earth Sciences (B,M,D)
Research Hall 280
Mail Stop 6C5
4400 University Drive
Fairfax, VA 22030-4444
p. (703) 993-5700
mdsouza2@gmu.edu
https://cos.gmu.edu/aoes/
t: @MasonAOES
Programs: Atmospheric Sciences, Geology, Climate Dynamics
Enrollment (2020): B: 85 (21) M: 9 (0) D: 16 (3)
Chair:
Mark D. Uhen, (D), Michigan, 1996, Pv
University Professor:
Jagadish Shukla, (D), MIT, 1976, As
Professor:
Timothy DelSole, (D), Harvard, 1993, As
Paul A. Dirmeyer, (D), Maryland, 1992, As
Linda Hinnov, (D), Johns Hopkins, 1994, GrYg
Bohua Huang, (D), Maryland, 1992, Op
Jim Kinter, (D), Princeton, 1984, As
David M. Straus, (D), Cornell, 1977, As
Stacey Verardo, (D), CUNY, 1995, Pei
Associate Professor:
Zafer Boybeyi, (D), North Carolina State, 1993, As
Long S. Chiu, (D), MIT, 1980, AsZrAm
Barry Klinger, (D), MIT/WHOI, 1992, Op
Randolph McBride, (D), Louisiana State, 1997, On
Julia Nord, (D), CUNY (Brooklyn), 1989, Gzg
Cristiana Stan, (D), Colorado State, 2005, As
Daniel Tong, (D), N Carolina State, 2003, Ac
Assistant Professor:
Paul Betka, (D), Texas, 2013, GctGg
Natalie Burls, (D), Cape Town, 2010, AsOp
Geoff Gilleaudeau, (D), Tennessee, 2013, Gs
Brittany Hupp, (D), Wisconsin, 2021, Cg
Giuseppina Kysar Mattietti, (D), George Washington, 2001, GiZeGa
Luis Ortiz, (D), CUNY, 2018, As
University Professor:
Edwin K. Schneider, (D), Harvard, 1976, As
Emeritus:
Richard J. Diecchio, (D), N Carolina, 1980, Gr
Paul S. Schopf, (D), Princeton, 1978, OpAs

Hampton University

Center for Marine & Coastal Environmental Studies (B)
100 Queen Street
Hampton, VA 23668
p. (757) 727-5783
cas@hamptonu.edu
Enrollment (2010): B: 32 (9)
Chair:
George P. Burbanck, (D), Delaware, 1981, On
Professor:
Benjamin E. Cuker, (D), North Carolina State, 1981, Ob
Associate Professor:
Robert A. Jordan, (D), Michigan, 1970, Ob
Assistant Professor:
Deidre M. Gibson, (D), Georgia, 2000, Ob
Adjunct Faculty:
Emory Morgan, (M), Ze
Related Staff:
Gary Morgan, (B), North Carolina State, 2007, ZnnZn

James Madison University

Dept of Geology & Environmental Science (B)
MSC 6903
Memorial Hall
Harrisonburg, VA 22807
p. (703) 568-6130
ulansksl@jmu.edu
http://www.jmu.edu/geology/
Department Secretary: Sandra Delawder
Head:
Stanley L. Ulanski, (D), Virginia, 1977, Og
Professor:
Roddy V. Amenta, (D), Bryn Mawr, 1971, Gc
Lynn S. Fichter, (D), Michigan, 1972, Gr
William C. Sherwood, (D), Lehigh, 1961, So
Steven J. Whitmeyer, (D), Boston, 2004, GctZi
Associate Professor:
Steven J. Baedke, (D), Indiana, 1998, Hy
Lewis S. Eaton, (D), Virginia, 1999, Gm
Eric J. Pyle, (D), Georgia, 1995, Ze
Kristen E. St. John, (D), Ohio State, 1998, Ou
Emeritus:
Lance E. Kearns, (D), Delaware, 1977, Gz

Mary Washington College

Dept of Geography (B)
1301 College Avenue
Fredericksburg, VA 22401-5358
p. (540) 654-1470
jhayob@umw.edu
http://www.mwc.edu/geog
Enrollment (2006): B: 80 (0)
Chair:
Joseph W. Nicholas, (D), Georgia, 1991, Zy
Associate Professor:
Donald N. Rallis, (D), Penn State, 1992, Zn
Assistant Professor:
Dawn S. Bowen, (D), Queen's, 1998, Zn
Stephen P. Hanna, (D), Kentucky, 1997, Zr
Farhang Rouhani, (D), Arizona, 2001, Zn

Mountain Empire Community College

Mountain Empire Community College
3441 Mountain Empire Road
Big Stone Gap, VA 24219
p. (276) 523-7460
sfritts@mecc.edu
http://www.me.vccs.edu/

Northern Virginia Community College - Alexandria

Geology Program (A)
5000 Dawes Avenue
Alexandria, VA 22311-5097
p. 703.845.6507
vzabielski@nvcc.edu
http://www.nvcc.edu/campuses-and-centers/alexandria/academic-divisions/science/geology.html
Assistant Dean of Geology:
Victor Zabielski, (D), Brown, 2001, GgOg

Northern Virginia Community College - Annandale

Geology Program (A)
8333 Little River Turnpike
Annandale, VA 22003
p. (703) 323-3276
bwang@nvcc.edu
http://www.nvcc.edu/campuses-and-centers/annandale/academic-divisions/math-science--engineering/gol.html
Programs: Science (B)
Enrollment (2011): A: 10 (10)
Professor:
Kenneth Rasmussen, (D), North Carolina, 1989, Gus
Assistant Professor:
Callan Bentley, (M), Maryland, 2004, Gg

Shelley Jaye, (M), Wayne State, 1984, Giy

Northern Virginia Community College - Loudoun Campus

Mathematics, Sciences, Technologies and Business Div - Dept of Geology (A)
21200 Campus Drive
Sterling, VA 20164
p. (703) 450-2612
wbour@nvcc.edu
https://www.nvcc.edu/academics/divisions/mstb
Programs: Associates degree in Science
Associate Professor:
William Bour, (M), George Washington, 1993, Gge
William Straight, (D), North Carolina State, GgPv
Assistant Professor:
Okia Ikwuazorm, (M), Gg

Northern Virginia Community College - Woodbridge

Geology Program (A) ☒
2645 College Drive
Woodbridge, VA 22191
p. (703) 878-5614
vkrishnan@nvcc.edu
http://www.nvcc.edu/woodbridge/divisions/natural.html
Enrollment (2014): A: 6 (1)
Instructor:
Erik Burtis, (M), Montana, GgcGi

Old Dominion University

Department of Ocean and Earth Sciences (B,M,D) ⊘
4600 Elkhorn Avenue
Norfolk, VA 23529
p. (757) 683-4285
OESadmin@odu.edu
http://www.odu.edu/oes
f: @odu.oes
t: @odu_oes
Programs: BS in Ocean and Earth Sciences; MS in Ocean and Earth Sciences; PhD in Oceanography
Certificates: GIS; Environmental Science; Marine Science Technology
Enrollment (2021): B: 49 (6) M: 4 (0) D: 6 (1)
Chair:
Fred C. Dobbs, (D), Florida State, 1987, Ob
Associate Dean:
H. Rodger Harvey, (D), 1985, Com
Professor:
Alexander Bochdansky, (D), Memorial, 1997, Ob
David J. Burdige, (D), California (San Diego), 1983, OcCbo
Gregory A. Cutter, (D), California (Santa Cruz), 1982, OcCtHs
Tal Ezer, (D), Florida State, 1989, Op
Eileen E. Hofmann, (D), North Carolina State, 1980, Op
John M. Klinck, (D), Iowa, 1980, Op
Margaret Mulholland, (D), Maryland, 1998, Ob
Hans-Peter Plag, (D), Free (Berlin), 1988, YgZc
Peter Sedwick, (D), Hawaii, 1991, CmOc
Richard Zimmerman, (D), S California, 1983, Ob
Associate Chair:
P. Dreux Chappell, (D), MIT/WHOI, 2009, Oc
Associate Professor:
Jennifer Georgen, (D), MIT/WHOI, 2001, Gu
Nora K. Noffke, (D), Oldenburg, 1997, GsPoGu
Matthew Schmidt, (D), California, 2005, Ou
Assistant Professor:
Sophie Clayton, MIT/WHOI, 2013, Cb
Rip Hale, (D), Washington, 2014, Ou
Victoria Hill, Nottingham Trent, 2002, Ob
Joseph Tamborski, Stony Brook, 2016, Hy
Emeritus:
Dennis A. Darby, (D), Wisconsin, 1971, Ou
G. Richard Whittecar, Jr., (D), Wisconsin, 1979, GmHwZe

Patrick Henry Community College

Art, Science, Business, and Technology Dept (A)
645 Patriot Avenue
Martinsville, VA 24112
abroome@patrickhenry.edu
http://www.ph.vccs.edu/
Assistant Professor:
Brett Dooley, (M), Virginia Tech, 2005, Gg

Piedmont Virginia Community College

Associate of Science
501 College Drive
Charlottesville, VA 22902
p. (434) 961-5446
khudson@pvcc.edu
http://www.pvcc.edu/

Radford University

Dept of Geology (B)
Box 6939
Radford University
Radford, VA 24142
p. (540) 831-5652
geology@radford.edu
http://www.radford.edu/~geol-web
Administrative Assistant: Theresa Gawthrop
Enrollment (2013): B: 62 (12)
Associate Professor:
Jonathan L. Tso, (D), Virginia Tech, 1987, Gcp
Professor:
Rhett B. Herman, (D), Montana State, 1996, YuXaGa
Parvinder S. Sethi, (D), North Carolina State, 1994, Gz
Chester F. Watts, (D), Purdue, 1983, NgrHw
Director, Museum of the Earth Sciences:
Stephen W. Lenhart, (D), Kentucky, 1985, Pgg
Associate Professor:
Elizabeth McClellan, (D), Tennessee, GipGd
Emeritus:
Robert C. Whisonant, (D), Florida State, 1967, GsaGg

Randolph-Macon College

Environmental Studies Program (B) ☒
P.O. Box 5005
Ashland, VA 23005
p. (804) 752-3745
mfenster@rmc.edu
http://www.rmc.edu/academics/environmental-studies.aspx
Programs: Environmental Studies major with an Area of Expertise in Geology
Enrollment (2018): B: 31 (15)
Watts Professor of Science:
Michael S. Fenster, (D), Boston, 1995, OnGue
Adjunct Professor:
Charles Saunders, (M), E Carolina, 1990, ZgGrPg

Rappa Hannock Community College

Rappa Hannock Community College
12745 College Drive
Glenns, VA 23149
p. (804) 435-8970
babdul-malik@rappahannock.edu
http://www.rappahannock.edu

Tidewater Community College

Geophysical Sciences Dept (A) ☒
1700 College Crescent
Virginia Beach, VA 23453
p. (757) 822-7089
RClayton@tcc.edu
Programs: Physical and Historical Geology, Oceanography I & II, Earth Science
Certificates: GIS and Remote Sensing
Professor:
Rodney L. Clayton, (D), NCU, 2016, GgOg
Instructor:
Mike Lyle, Gg
John Waugh, Gg

University of Lynchburg

Environmental Sciences, Studies, and Sustainability (B,M) O☒
1501 Lakeside Dr.
Lynchburg, VA 24501
p. (434) 544-8415
haiar@lynchburg.edu
https://www.lynchburg.edu/academics/majors-and-minors/environmental-science/
Programs: Environmental Science; Environmental Studies
Enrollment (2019): B: 44 (16)

Professor:
David R. Perault, (D), Oklahoma, 1998, Ge

Associate Professor:
Brooke Haiar, (D), Oklahoma, 2008, PgGeg

University of Mary Washington

Dept of Earth and Environmental Sciences (B) ●⃠
1301 College Avenue
Fredericksburg, VA 22401-5358
p. (540) 654-1016
jhayob@umw.edu
https://cas.umw.edu/ees/
f: https://www.facebook.com/UMW-Earth-Environmental-Sciences-259585170726418/?view_public_for=259585170726418
Programs: Geology; Environmental Geology; Environmental Science-Natural Track; Environmental Science-Social Track
Certificates: GIS (through the Geography department)
Enrollment (2021): B: 25 (10)

Chair:
Jodie Hayob, (D), Michigan, 1994, GxzGg

Professor:
Ben O. Kisila, (D), Arkansas, 2002, HsSfZu
Grant R. Woodwell, (D), Yale, 1985, Gc

Associate Professor:
Melanie Szulczewski, (D), Wisconsin, 1999, ScClZg

Assistant Professor:
Tyler E. Frankel, (D), Maryland, Rw
Pamela Grothe, (D), Georgia Tech, GsOgZc

Instructor:
Sarah A. Morealli, (M), Pittsburg State, 2010, Gg

Adjunct Professor:
John Tippett, (M), Duke, Zn

University of Virginia

Blandy Experimental Farm
400 Blandy Farm Lane
Boyce, VA 22620
p. (540) 837-1758
Blandy@virginia.edu
http://blandy.virginia.edu/

Director:
David E. Carr, (D), Maryland, 1990, Zn

Associate Director:
Kyle J. Haynes, (D), Louisiana State, 2004, Zn

Dept of Environmental Sciences (B,M,D) ☒
Clark Hall
291 McCormick Road
Box 400123
Charlottesville, VA 22904
p. (434) 924-7761
cba4a@virginia.edu
https://evsc.as.virginia.edu/
f: uvaevsc
t: uvaevsc
Programs: MA, MS and PhD in Environmental Sciences.
Enrollment (2018): B: 170 (91) M: 17 (6) D: 31 (9)

Dept Chair:
Howard E. Epstein, (D), Colorado State, 1997, Zn

Research Full Professor:
Peter Berg, (D), Tech (Denmark), 1988, So
David E. Carr, (D), Maryland, 1990, Zn
William Keene, (M), Virginia, 1981, As
G. Carleton Ray, (D), Columbia, 1960, Ob

Professor:
Lawrence Band, (D), California, 1983
Robert Davis, (D), Delaware, 1988, AsmZy
Stephan F. J DeWekker, (D), British Columbia, 2002, As
Scott Doney, (D), MIT, 1991, Oc
James N. Galloway, (D), California (San Diego), 1972, As
Janet S. Herman, (D), Penn State, 1982, ClHw
Alan D. Howard, (D), Johns Hopkins, 1970, GmXg
Deborah Lawrence, (D), Duke, 1998, ZcoZg
Manuel Lerdau, (D), Stanford, 1994, Zn
Stephen A. Macko, (D), Texas, 1981, CosOc
Karen McGlathery, (D), Cornell, 1992, Ob
Aaron L. Mills, (D), Cornell, 1975, So
Michael L. Pace, (D), Georgia, 1981, Zn
Matthew A. Reidenbach, (D), Stanford, 2004, Hg
Todd M. Scanlon, (D), Virginia, 2002, Hg
Herman H. Shugart, Jr., (D), Georgia, 1971, Zg
David E. Smith, (D), Texas A&M, 1982, Ob
Patricia L. Wiberg, (D), Washington, 1987, Og

Research Associate Professor:
Linda K. Blum, (D), Cornell, 1980, Sb
Kyle J. Haynes, (D), Louisiana State, 2004, Zn
Jennie L. Moody, (D), Michigan, 1986, As
John H. Porter, (D), Virginia, 1988, Zif
Tai Roulston

Associate Professor:
Kevin M. Grise, (D), Colorado State, 2011, As
Thomas M. Smith, (D), Tennessee, 1982, Zg

Research Assistant Professor:
Karen C. Rice, (D), Virginia, 2001, HgCg

Assistant Professor:
Max Castorani, (D), California, 2014
Ajay Limaye
Sally Pusede, (D), California, 2014, As
Kathleen Schiro
Lauren Simkins
Xi Yang, (D), Brown, 2014

Emeritus:
Bruce W. Nelson, (D), Illinois, 1955, Gs
Wallace E. Reed, (D), Chicago, 1967, Zr
William F. Ruddiman, (D), Columbia, 1969, Gu

Shenandoah Watershed Study (B,M,D)
Department of Environmental Sciences
Clark Hall
Charlottesville, VA 22904
p. (434) 924-3382
tms2v@virginia.edu
http://people.virginia.edu/~swas/POST/scripts/overview.php

Associate Professor:
Todd M. Scanlon, (D), Virginia, 2002, Hgs

Research Scientist:
Ami L. Riscassi, (D), Virginia, 2009

Virginia Coast Reserve Long Term Ecological Research ☒
291 McCormick Road
P.O. Box 400123
Charlottesville, VA 22904-4123
p. 434-924-0558
kjm4k@virginia.edu
https://www.vcrlter.virginia.edu
f: https://www.facebook.com/virginia.coast.reserve
t: https://twitter.com/vcrlter

Professor:
Karen J. McGlathery, (D), Cornell, 1992, Obn

University of Virginia's College at Wise

Dept of Natural Science (B) ☒
1 College Avenue
Wise, VA 24293
p. (276) 328-0203
maw4v@uvawise.edu
https://www.uvawise.edu/academics/department-natural-sciences/
f: https://www.facebook.com/UVaWiseNaturalSciences
t: @UVaWiseNatSci
Programs: BA in Earth Science
Environmental Science, geology track
Administrative Assistant: Alan West

Instructor:
Robert D. VanGundy, (M), North Carolina, 1983, GeHgZg

Virginia Highlands Community College

Virginia Highlands Community College
100 VHCC Drive
Abingdon, VA 24212
p. (276)739-2433
jsurber@vhcc.edu
http://www.vhcc.edu

Virginia Polytechnic Institute & State University

Department of Geosciences (B,M,D)
4044 Derring Hall
Blacksburg, VA 24061
p. (540) 231-6521
wstevenh@vt.edu
http://www.geos.vt.edu
f: https://www.facebook.com/vtgeosciences/
t: @vtgeosciences
Programs: Six options: Geology, Geophysics, Geochemistry, Environmental and Engineering Geoscience, Paleobiology, and Geoscience Education
Enrollment (2021): B: 74 (0) M: 10 (0) D: 38 (0)
Department Head:
W. Steven Holbrook, (D), Stanford, 1989, Yur
Research Professor:
Martin C. Chapman, (D), Virginia Tech, 1998, Ys
Professor:
Robert J. Bodnar, (D), Penn State, 1985, CgEmGv
Patricia M. Dove, (D), Princeton, 1991, Cl
Kenneth A. Eriksson, (D), Witwatersrand, 1977, Gs
John A. Hole, (D), British Columbia, 1993, Ye
Scott D. King, (D), Caltech, 1990, Yg
Richard D. Law, (D), London (UK), 1981, Gc
Nancy L. Ross, (D), Arizona State, 1985, Gz
Madeline E. Schreiber, (D), Wisconsin, 1999, Hw
James A. Spotila, (D), Caltech, 1998, Gtm
Robert Weiss, (D), Westfalische–Wilhelms, 2005, RnOn
Shuhai Xiao, (D), Harvard, 1998, Pi
Associate Collegiate Professor:
John A. Chermak, (D), Virginia Tech, 1989, ClGeEd
Associate Professor:
Mark Caddick, (D), Cambridge, 2005, GpCa
Benjamin Gill, (D), California (Riverside), 2009, CsPcGs
F. Marc Michel, (D), SUNY (Stony Brook), 2007, ZaGz
Sterling J. Nesbitt, (D), Columbia, 2009, Pv
Ryan M. Pollyea, (D), Idaho, 2012, Hq
Brian W. Romans, (D), Stanford, 2008, GsrEo
Manoochehr Shirzaei, (D), Potsdam, 2010, ZrYd
D. Sarah Stamps, (D), Purdue, 2013, YdGtRn
Michael J. Willis, (D), Ohio State, 2008, GlZr
Ying Zhou, (D), Princeton, 2004, Ys
Assistant Professor:
George H. Allen, (D), North Carolina, 2017, HsZrGm
Megan S. Duncan, (D), Rice, 2015, GiCp
Christina Dura, (D), Pennsylvania, 2014, OnRnPm
Michelle Stocker, (D), Texas, 2013, Pv
Susanna Werth, (D), Potsdam, 2010, ZrHgYv
Sr. Research Associate:
Jing Zhao, (D), Chinese Acad of Sci, 1997
Research Associate:
Nizhou Han, (D), Iowa State, 1996, Sc
Advanced Instructor:
Neil E. Johnson, (D), Virginia Tech, 1986, GzEg
Instructor:
Laura Neser, (D), North Carolina, 2014, Gsc
Adjunct Professor:
James S. Beard, (D), California (Davis), 1985, GiClg
Benedetto DeVivo, (D)
William S. Henika, (M), Virginia, 1969, Ng
William Schmachtenbert, (D), Chicago, 1983, Pi
Karen M. Weber, (M), Virginia Tech, 2019, Ge
Emeritus:
G. A. Bollinger, (D), St Louis, 1967, Ys
Thomas J. Burbey, (D), Nevada (Reno), 1994, HwyZr
Cahit Coruh, (D), Istanbul, 1970, YesGt
James R. Craig, (D), Lehigh, 1965, Eg
Jerry Gibbs, (D), Penn State, Gz
David A. Hewitt, (D), Yale, 1970, Cp
Michael F. Hochella, Jr., (D), Stanford, 1981, ClGze
Wallace D. Lowry, (D), Rochester, 1943, Gc
J. Fred Read, (D), W Australia, 1971, Gs
A. Krishna Sinha, (D), California (Santa Barbara), 1969, Gt
J. Arthur Snoke, (D), Yale, 1969, Ys
Research Scientist, Stable Isotope Lab Manager:
Rachel E.B Reid, (D), California (Santa Cruz), 2014, CsPcCb
Research Scientist, Field Operations Lab Manager:
Sean P. Bemis, Oregon, 2010, Gt
Research Associate, Paleontology Lab Manager:
Vicki Yarborough, (B), Texas, 1991, Pv
Research Associate, Microbeam Facility Manager:
Lowell P. Moore, (D), Virginia Tech, 2019, Ca
A/P Faculty:
Gary Glesener, (M), California (Los Angeles), 2016, Ze
April Newcomer, (M), Ze
Cooperating Faculty:
Edward Lener, (M), Virginia Tech, 1997, Zn

Dept of Civil & Environmental Engineering (B,M,D) ☒
750 Drillfield Drive
200 Patton Hall
Blacksburg, VA 24061-0105
p. (703) 231-6635
mwiddows@vt.edu
http://www.cee.vt.edu/
Programs: BS, MS and PhD in Civil Engineering; MS in Environmental Engineering; MS in Environmental Science and Engineering
Administrative Assistant: Beth Lucas
Department Head:
Mark A. Widdowson, (D), Auburn, HwSpHq
Professor:
Stanley Grant, (D), Caltech, HsRw
Jennifer Irish, (D), Delaware, NoOg
Associate Professor:
Erich Hester, (D), North Carolina, Hsw
Kyle Strom, (D), Iowa, HsGs
Assistant Professor:
Landon Marston, (D), Illinois, 2017, HqRw
Megan Rippy, (D), Scripps, Rw
Suddarth Saksena, (D), Purdue, 2019, HqZiRn
Emeritus:
Randel Dymond, (D), Penn State, 1987, Hq

Dept of Crop & Soil Environmental Sciences (B,M,D)
240 Smyth Hall
Blacksburg, VA 24061-0404
p. (540) 231-6305
tlthomps@vt.edu
http://www.cses.vt.edu
Department Secretary: Nancy Shields
Head:
John R. Hall, III, (D), Ohio State, 1971, So
Professor:
Marcus M. Alley, (D), Virginia Tech, 1975, Sc
James C. Baker, (D), Virginia Tech, 1978, Sd
Walter L. Daniels, (D), Virginia Tech, 1985, SdGeCg
Stephen J. Donohue, (D), Purdue, 1974, Sc
Gregory K. Evanylo, (D), Georgia, 1982, Sc
Charles Hagedorn, (D), Iowa, 1974, Sb
Gregory L. Mullins, (D), Purdue, 1985, Sc
Raymond B. Reneau, (D), Florida, 1969, Sc
Lucian W. Zelazny, (D), Virginia Tech, 1970, Sc
Associate Professor:
Duane F. Berry, (D), Michigan State, 1984, Sb
Matthew J. Eick, (D), Delaware, 1995, Sc
Naraine Persaud, (D), Florida, 1978, Sp
Assistant Professor:
John M. Galbraith, (D), Cornell, 1997, Sp
Carl E. Zipper, (D), Virginia Tech, 1986, Ze
Adjunct Professor:
Domy C. Adriano, (D), Kansas State, 1970, Sc
V. C. Baligar, (D), Mississippi State, 1975, Sc
Pamela J. Thomas, (D), Virginia Tech, 1998, Sd

Dept of Geography (B,M,D)
115 Major Williams Hall
Blacksburg, VA 24061-0115

p. (540) 231-7557
carstens@vt.edu
Enrollment (2010): B: 170 (49) M: 17 (9) D: 8 (0)

Head:
Laurence W. Carstensen, (D), North Carolina, 1981, Zi

Professor:
James B. Campbell, (D), Kansas, 1976, Zr

Associate Professor:
Lawrence S. Grossman, (D), Australian Nat, 1979, Zn
Lisa M. Kennedy, (D), Tennessee, Pe
Resler M. Lynn, (D), Texas State, Zy

Instructor:
David Carroll, (M), Mississippi State, Am

Virginia Wesleyan College

Dept of Earth and Environmental Sciences (B)
1584 Wesleyan Dr
Norfolk, VA 23502
p. 757.455.3200
emalcolm@vwu.edu
http://www.vwc.edu/earth-and-envirmental-sciences/
f: https://www.facebook.com/vwc.ees
Enrollment (2015): B: 22 (0)

Professor:
John C. Haley, (D), Johns Hopkins, 1986, GgeZi

Associate Professor:
Elizabeth Malcolm, (D), Michigan, 2002, AsCgOg
Garry Noe, (D), California (Riverside), 1982, HsZin

Virginia Western Community College

Geology Dept (A)
3095 Colonial Avenue
Roanoke, VA 24038
p. (540) 857-7273
hrapps@virginiawestern.edu

Professor:
Anna Balog-Szabo, (D), Virginia Tech, 1996, Gs

Washington & Lee University

Dept of Geology (B) ● ☒
204 West Washington Street
Lexington, VA 24450
p. (540) 458-8800
geology@wlu.edu
http://geology.wlu.edu
Programs: Geology
Administrative Assistant: Sarah Barrash. Wilson
Enrollment (2021): B: 13 (14)

Professor:
Christopher Connors, (D), Princeton, 1999, GcYeGo
Lisa Greer, (D), Miami, 2001, GsPe
David J. Harbor, (D), Colorado State, 1990, Gm
Elizabeth P. Knapp, (D), Virginia, 1997, Cl
Jeffrey Rahl, (D), Yale, 2005, Gt

Assistant Professor:
Margaret Anne Hinkle, (D), Washington (St. Louis), 2015, Cqb

Emeritus:
Frederick Lyon Schwab, (D), Harvard, 1968, Gdg

Laboratory Supervisor:
Emily Flowers Falls, (M), E Kentucky, 2009, Gg

William & Mary

Department of Geology (B) ⦸
PO Box 8795
Williamsburg, VA 23187-8795
p. (757) 221-2440
rxlock@wm.edu
http://www.wm.edu/as/geology
Programs: B.S. in Geology; B.S. in Environmental Geology
Certificates: GIS
Administrative Assistant: Carol Roe
Enrollment (2020): B: 63 (23)

Chair:
Rowan Lockwood, (D), Chicago, 2001, PgePi

Professor:
Christopher M. Bailey, (D), Johns Hopkins, 1994, Gc
Gregory S. Hancock, (D), California (Santa Cruz), 1998, Gm
R. Heather Macdonald, (D), Wisconsin, 1984, GsZe
Brent E. Owens, (D), Washington (St Louis), 1992, Gx

Associate Professor:
James Kaste, (D), Dartmouth, 2003, ClcSc

Assistant Professor:
Nicholas Balascio, (D), Massachusetts, 2011, ZyGnl

Research Associate:
Carl R. Berquist, Jr., (D), William & Mary, 1986, Ou

Lecturer:
Rebecca Jiron, (D), California (Santa Barbara), 2015, GgsZi
Linda D. Morse, (B), Virginia Tech, 1983, GeZg

Visiting Professor:
Jonathan Kay, (D), Illinois (Chicago), 2017, ZyiGc

Emeritus:
Stephen C. Clement, (D), Cornell, 1964, Gz
P. Geoffrey Feiss, (D), Harvard
Gerald H. Johnson, (D), Indiana, 1965, Pg

School of Marine Science (M,D) ☒
Virginia Institute of Marine Science
P. O. Box 1346
Gloucester Point, VA 23062-1246
p. (804) 684-7105
ad-as@vims.edu
http://www.vims.edu
Programs: Biological Sciences, Fisheries Science, Aquatic Health Sciences, Physical Sciences (Physical Oceanography, Chemical Oceanography, Marine Geology)
Enrollment (2021): M: 5 (3) D: 10 (3)

Research Professor:
Marjorie Friedrichs, (D), Old Dominion, 1999, Obp
Jian Shen, (D), William & Mary, 1996, Op
Y. Joseph Zhang, (D), Wollongong, 1996, Opn

Department Chair:
Courtney K. Harris, (D), Virginia, 1999, On

Associate Director, Chesapeake Bay National Estuarine Research Reserve in VA:
Carl T. Friedrichs, (D), MIT/WHOI, 1993, On

Professor:
Steven A. Kuehl, (D), NC State, 1985, Ou
Harry Wang, (D), Johns Hopkins, 1983, Op

Research Associate Professor:
William G. Reay, (D), Virginia Tech, 1992, Hg

Associate Professor:
Donglai Gong, (D), Rutgers, 2010, Op
Amber Hardison, (D), William & Mary, 2010, Cm
Christopher J. Hein, (D), Boston, 2012, Ou
Matthew L. Kirwan, (D), Duke, 2007, Gm

Research Assistant Professor:
Jon Derek Loftis, (D), William & Mary, 2014, OnpZi

Assistant Professor:
Pierro Mazzini, (D), Oregon State, 2014, Op

Washington

Bellevue College

Earth and Space Sciences Program (A)
3000 Landerholm Circle SE
Bellevue, WA 98007
p. (425) 564-3158
kshort@bellevuecollege.edu
http://scidiv.bellevuecollege.edu/

Instructor:
Cary Easterday, (M), GgPg
Gwyn Jones, (M), Gg
Deborah Minium, (M), Gg
Rob Viens, (D), Washington, GgZg

Central Washington University

Dept of Geological Sciences (B,M) ●
400 East University Way
MS 7418
Ellensburg, WA 98926-7418
p. (509) 963-2701
chair@geology.cwu.edu
http://www.geology.cwu.edu
Programs: Geology (B,M); Environmental Geology (B)
Administrative Assistant: Moriah Kauer

Enrollment (2022): B: 59 (23) M: 27 (5)
Director, PANGA Laboratory:
Timothy I. Melbourne, (D), Caltech, 1998, Yd
Chair:
Anne E. Egger, (D), Stanford, 2010, ZeGtRn
Professor:
Lisa L. Ely, (D), Arizona, 1992, GmHs
Carey A. Gazis, (D), Caltech, 1994, CsHw
Susan Kaspari, (D), Maine, 2007, PcAtGl
Chris Mattinson, (D), Stanford, 2006, GpCcGt
Co-chair:
Breanyn MacInnes, (D), Washington, 2010, GseGu
Associate Professor:
Audrey Huerta, (D), MIT, 1998, Ygd
Walter Szeliga, (D), Colorado, 2010, YgsYd
GPS Data Analyst:
Marcelo Santillan, (M), Memphis, 2003, Yd
Assistant Professor:
Hannah Shamloo, (D), Arizona State, 2020, Gvi
Geologist, Washington Geological Survey:
Andrew Sadowski, (M), Nevada (Reno), 2016, Gg
Research Associate:
Megan Anderson, (D), Arizona, 2005, YgvYm
Beth Pratt-Sitaula, (D), California (Santa Barbara), 2005, YdZe
Paul Winberry, Penn State, 2008, Yg
Science Outreach & Education Coordinator:
Nick Zentner, (M), Idaho State, 1989, Gg
Lecturer:
Jordan Carey, (M), California (Davis), 2017, Geg
Keegan Fengler, (M), C Washington, 2007, Gg
Adrienne Sorenson, (M), Washington, 2019, Gg
John Sugden, (M), Montana State, 2015, Geg
Emeritus:
Wendy A. Bohrson, (D), California (Los Angeles), 1993, GivCc
Steven Farkas, (D), New Mexico, 1965, Gs
James R. Hinthorne, (D), California (Santa Barbara), 1974, Ca
Jeffrey Lee, (D), Stanford, 1990, Gct
M. Meghan Miller, (D), Stanford, 1987, Yd
Winston Norrish, (D), Cincinnati, 1989, Gog
Systems Administrator:
Craig Scrivner, (D), Caltech, 1998
Field Engineer:
Rex Flake, (M), C Washington, 2002, Yd

Centralia College

Geosciences (A) ⦸
600 Centralia College Blvd
Centralia, WA 98531
p. (360) 623-8472
michelle.harris@centralia.edu
http://www.centralia.edu/academics/earthscience
Programs: Geology and Earth Science
Enrollment (2019): A: 6 (1)
Associate Professor:
Michelle Harris, (M), Gg
Emeritus:
Patrick T. Pringle, (M), Akron, 1982, ZgGvZn

Eastern Washington University

Department of Geosciences (B) O⦸
129 Patterson Hall
Cheney, WA 99004-2439
p. (509) 359-2286
geology@ewu.edu
https://www.ewu.edu/cstem/geosciences/
Programs: Geosciences BA/BS; Professional Geology BS; Environmental Science; Earth and Space Science Education; Certificates: GIS; Remote Sensing; Digital Geoarchaeology; Wetland Science and Management; Water Resources; Environmental Justice; Climate Change; River Studies and Leadership; Wildfire Science and Management; Applied Geology
Enrollment (2022): B: 80 (7)
Chair:
Chad Pritchard, (D), Washington State, GcHg
Professor:
Carmen A. Nezat, (D), Michigan, 2006, ClGeSc
Richard L. Orndorff, (D), Kent State, 1994, Ge
Jennifer A. Thomson, (D), Massachusetts, 1992, Gpz
Stacy Warren, (D), Ziu
Associate Professor:
Lauren Stachowiak, (D), ZiAt
Stephen Tsikalas, (D), AsGlOp
Assistant Professor:
Brian Buchanan, (D), ZrGaZi
Erin Dascher, (D), ZySf
Lindsay MacKenzie, (D), Pg
Lecturer:
Sharen Keattch, (M), Kent State, 1993, Gg
Emeritus:
Ernest H. Gilmour, (D), Montana, 1967, Pi
Eugene P. Kiver, (D), Wyoming, 1968, Gl
Linda B. McCollum, (D), SUNY (Binghamton), 1980, Pg

Edmonds College

Geology (A)
20000 68th Ave W
Lynnwood, WA 98036
p. (425) 640-1918
mkelly@edmonds.edu
https://www.edmonds.edu/programs/stem/geology/default.html
Programs: Associate in Science (AS-T) degree designed to prepare geology and earth science majors for transfer to a four-year institution with junior standing.
Enrollment (2013): A: 4 (0)
Instructor:
Maria Kelly, (M), California (Davis), Zg
Adjunct Professor:
Dylan Ahearn, (D), California (Davis)

Everett Community College

Dept of Physical Sciences (A)
2000 Tower Street
Everett, WA 98201
p. (425) 388-9429
sdamp@everettcc.edu
http://www.everettcc.edu/programs/mathsci/physical/
Instructor:
Steve Grupp, (M), Colorado Mines, GgOg
Adjunct Professor:
Alecia Spooner, Zg

Grays Harbor College

Dept of Geology & Oceanography (A)
Aberdeen, WA 98520
p. (206) 538-4299
john.hillier@ghc.edu
Professor:
John Hillier, (D), Cornell, ZgCa
Emeritus:
James B. Phipps, (D), Oregon State, 1974, Gu

Green River Community College

Dept of Geology (A)
12401 S.E. 320th
Auburn, WA 98092-3699
p. (253) 833-9111 (Ext. 4248)
khoppe@greenriver.edu
https://www.greenriver.edu/grc/course-descriptions/course-description-detail.aspx?desc=GEOL&deptname=Geology
Programs: AA, AS; & BAS in Natural Resources in Forest Resource Management
Instructor:
Kathryn A. Hoppe, (D), Princeton, 1999, ZgOgPg
Katy Shaw, (M), Washington, ZgOgGg

Highline College

Physical Sciences Dept/Geology Program (A)
MS 29-3
PO Box 98000
2400 S. 240th St.
Des Moines, WA 98198-9800
p. (206) 878-3710
geology@highline.edu
http://geology.highline.edu/
Programs: A.S. Geology, A.A.

Enrollment (2022): A: 6 (0)
Instructor:
Eric M. Baer, (D), California (Santa Barbara), 1995, GgvZg
Stephaney Puchalski, (D), Indiana, 2011, PgGg
Jacob Selander, (D), California (Davis)
Carla Whittington, (M), Indiana, GgvZg

Lower Columbia College

Earth Sciences (Natural Sciences Dept.) (A)
P.O. Box 3010
Longview, WA 98632
p. (360) 442-2883
msalisbury@lowercolumbia.edu
https://lowercolumbia.edu/programs/natural-science.php
Professor:
David I. Cordero, (M), Portland State, 1997, Zg

North Seattle Community College, North Campus

Earth Science
9600 College Way North
Seattle, WA 98103
p. (206) 934-4509
john.figge@seattlecolleges.edu
http://www.northseattle.edu/

Northwest Indian College

Native Environmental Studies
2522 Kwina Road
Bellingham, WA 98226
p. (360) 392-4256
sarques@nwic.edu
http://www.nwic.edu/degrees-and-certificates/bsnes-bachelors-degree

Olympic College

Mathematics, Engineering, Sciences, and Health Div (A)
Bremerton, WA 98310
p. (360) 475-7777
SMacias@olympic.edu
http://www.olympic.edu/Students/AcadDivDept/MESH/Sciences/Geology/
Instructor:
Steve E. Macias, (M), Washington, 1996, Gg
Adjunct Professor:
Katie Howard, Gg

Pacific Lutheran University

Dept of Geosciences (B)
12180 S. Park
Tacoma, WA 98447
p. (253) 535-7378
geos@plu.edu
http://www.plu.edu/geosciences/
Enrollment (2013): B: 52 (11)
Chair:
Jill M. Whitman, (D), California (San Diego), 1989, OuGuZe
Professor:
Duncan Foley, (D), Ohio State, 1978, GeHyZe
Associate Professor:
Rosemary McKenney, (D), Penn State, 1997, Gm
Claire E. Todd, (D), Washington, 2007, Glm
Assistant Professor:
Peter B. Davis, (D), Minnesota, 2008, GcpGz

Peninsula College

Environmental Science
1502 East Lauridsen Boulevard
Port Angeles, WA 98362
p. (360) 452-9277
pchr@pencol.edu
http://www.pc.ctc.edu

Pierce College

Pierce College
9401 Farwest Drive SW
Lakewood, WA 98498
p. (253) 964-6676
zayacjm@piercecollege.edu
http://www.piercecollege.edu/departments/physics_planetary_sciences/geology.asp

Seattle Central College

Div of Science & Mathematics
1701 Broadway
Seattle, WA 98122
p. (206) 587-3858
Nancy.Sola-Llonch2@seattlecolleges.edu
http://seattlecentral.edu/sci-math/
Instructor:
Katie Gagnon, (D), California (San Diego), 2007, OuZg
Joseph M. Hull, (D), Rochester, 1988, Gc
Adjunct Professor:
Michael Harrell, Gg

Shoreline Community College

Geology & Earth Sciences (A)
16101 Greenwood Avenue North
Seattle, WA 98133
p. (206) 546-4659
eagosta@shoreline.edu
http://www.shoreline.edu/science/geology.htm

Spokane Community College

Dept of Earth Science
1810 N. Greene Street
Spokane, WA 99217
Andy.Buddington@scc.spokane.edu
http://www.scc.spokane.edu/ArtsSciences/Science/

Tacoma Community College

Dept of Earth Sciences (A)
6501 South 19th Street
Tacoma, WA 98466
p. (253) 566-5060
rhitz@tacomacc.edu
https://www.tacomacc.edu/academics-programs/programs/geology
Programs: 1) Associate of Science with an Earth Sciences Specialization
2) Associate of Arts
Enrollment (2018): A: 3 (0)
Professor:
Ralph B. Hitz, (D), California (Santa Barbara), 1997, PvSaZi
Adjunct Professor:
Jim McDougall, (D), Zg
James Peet, (D), Washington
Michael Valentine, (D), Zg

University of Puget Sound

Geology Dept (B) ☒
1500 N. Warner Street
Tacoma, WA 98416-1048
p. (253) 879-3814
kena@pugetsound.edu
http://www.pugetsound.edu/academics/departments-and-programs/undergraduate/geology/
f: https://www.facebook.com/upsgeology
Programs: B.S in Geology; B.S.in Natural Science
Administrative Assistant: Leslie Levenson
Enrollment (2019): B: 20 (12)
Chair:
Michael J. Valentine, (D), Massachusetts, 1990, YmGc
Professor:
Barry Goldstein, (D), Minnesota, 1985, GlmGs
Jeffrey H. Tepper, (D), Washington, 1991, GiCgGv
Associate Professor:
Kena L. Fox-Dobbs, (D), California (Santa Cruz), 2006, PgCsGe
Instructor:
Ken P. Clark, (M), W Washington, 1989, GgcGi
Emeritus:
Albert A. Eggers, (D), Dartmouth, 1972, Gv

WASHINGTON

University of Washington

Dept of Atmospheric Sciences (B,M,D) ☒
Box 351640
Seattle, WA 98195-1640
p. (206) 543-4250
chair@atmos.washington.edu
http://www.atmos.washington.edu
f: www.facebook.com/UWAtmosSci
t: @UWAtmosSci
Programs: Atmospheric Sciences
Enrollment (2017): B: 73 (18) M: 34 (16) D: 29 (7)

Affiliate Associate Professor:
Philip W. Mote, (D), Washington, 1994, As

Professor:
Cecilia M. Bitz, (D), Washington, 1997, As

Research Associate Professor:
Roger T. Marchand, (D), Virginia Tech, 1997, Ass

Professor:
Thomas P. Ackerman, (D), Washington, 1976, As
Becky Alexander, (D), California (San Diego), 2002, AsCsPe
David S. Battisti, (D), Washington, 1988, As
Christopher S. Bretherton, (D), MIT, 1984, As
Shuyi S. Chen, (D), Penn State, 1990, As
Dale R. Durran, (D), MIT, 1981, As
Qiang Fu, (D), Utah, 1991, As
Gregory J. Hakim, (D), SUNY (Albany), 1997, As
Dennis L. Hartmann, (D), Princeton, 1975, As
Lyatt Jaegle, (D), Caltech, 1996, As
Daniel A. Jaffe, (D), Washington, 1987, As
Clifford F. Mass, (D), Washington, 1978, As
Joel A. Thornton, (D), California (Berkeley), 2002, As
Robert Wood, (D), Manchester (UK), 1997, As

Associate Professor:
Dargan M. Frierson, (D), Princeton, 2005, As
Abigail L. S. Swann, (D), California (Berkeley), 2010, AsZn

Assistant Professor:
Daehyun Kim, (D), Seoul Nat, 2010, As

Adjunct Associate Professor:
Jessica D. Lundquist, (D), Scripps, 2004, As

Adjunct Professor:
Gerard H. Roe, (D), MIT, 1999, As
Eric J. Steig, (D), Washington, 1996, CsGIAs
LuAnne Thompson, (D), MIT, 1990, Og
Ka-Kit Tung, (D), Harvard, 1977, As

Research Professor Emeritus:
James E. Tillman, (M), MIT, As

Emeritus:
Marcia B. Baker, (D), Washington, As
Robert A. Brown, (D), Washington, 1969, As
Joost A. Businger, (D), State (Utrecht), As
Robert J. Charlson, (D), Washington, As
David S. Covert, (D), Washington, 1974, As
Thomas C. Grenfell, (D), Washington, 1972, As
Halstead Harrison, (D), Stanford, 1960, As
Dean A. Hegg, (D), Washington, 1979, As
Robert A. Houze, (D), MIT, 1972, As
Gary A. Maykut, (D), Washington, As
Peter B. Rhines, (D), Trinity Coll (Cambridge), 1967, Op
Edward S. Sarachik, (D), Brandeis, As
John M. Wallace, (D), MIT, 1966, As
Stephen G. Warren, (D), Harvard, 1973, ApGl

Research Associate Professor:
Lynn A. McMurdie, (D), Washington, 1989, As

Research Assistant Professor:
Edward Blanchard, (D), Washington, 2013, As

Affiliate Professor:
James E. Overland, (D), New York, 1973, As
Lawrence F. Radke, (D), Washington, 1968, As
Chidong Zhang, (D), Penn State, 1989, AsmOp

Affiliate Associate Professor:
Timothy S. Bates, (D), Washington, 1992, As
Nicholas A. Bond, (D), Washington, 1986, As
Bradley R. Colman, (D), MIT, 1984, As
Mark T. Stoelinga, (D), Washington, 1993, As

Affiliate Assistant Professor:
Bonnie Light, (D), Washington, 2000, As

Dept of Earth & Space Sciences (B,M,D) ●☒
070 Johnson Hall
Box 351310
Seattle, WA 98195-1310
p. (206) 543-1190
essadv@uw.edu
http://www.ess.washington.edu
f: https://www.facebook.com/pages/University-of-Washington-Earth-and-Space-Sciences/132505186787510
t: @UW_ESS
Programs: Earth and Space Sciences, Earth and Space Sciences: Geology, Earth and Space Sciences: Biology, Earth and Space Sciences: Environmental, Earth and Space Sciences: Physics, Applied Geosciences, Geological Sciences, Geophysics
Administrative Assistant: Scott Dakins
Enrollment (2019): B: 172 (81) M: 20 (10) D: 22 (12)

Chair:
Kenneth C. Creager, (D), California (San Diego), 1984, Ys

Academic Services:
Noell Bernard-Kingsley, Ze

Research Professor:
Howard B. Conway, (D), Canterbury (NZ), 1986, Gl
Dale P. Winebrenner, (D), Washington, 1985, GlZr

Pacific Northwest Seismic Network:
Harold Tobin

Professor:
George W. Bergantz, (D), Johns Hopkins, 1988, Gi
J. Michael Brown, (D), Minnesota, 1980, Gy
Roger Buick, (D), W Australia, 1986, CbXbPm
David C. Catling, (D), Oxford, 1994, CgAsXb
Bernard Hallet, (D), California (Los Angeles), 1975, Gml
Robert H. Holzworth, (D), California (Berkeley), 1977, XyAsZg
David R. Montgomery, (D), California (Berkeley), 1991, Gm
Bruce K. Nelson, (D), California (Los Angeles), 1985, CcGie
Charles A. Nittrouer, (D), Washington, 1978, GuYr
Mark Richards
Eric J. Steig, (D), Washington, 1996, CsGlAs
Edwin D. Waddington, (D), British Columbia, 1981, Gl
Peter D. Ward, (D), McMaster, 1976, Pg
Stephen G. Warren, (D), Harvard, 1973, AsGl
Robert M. Winglee, (D), Sydney, 1984, XyYx

Research Associate Professor:
Evan H. Abramson, (D), MIT, 1985, Gy
Paul A. Bodin, (D), Colorado, 1992, Ys
Michael P. McCarthy, (D), Washington, 1988, XyYx
Ronald S. Sletten, (D), Washington, 1995, Sob

Research Assistant Professor:
Erika M. Harnett, (D), Washington, 2003, XyGev

Affiliate Associate Professor:
Amit Mushkin, (D), ZrCcGm

Associate Professor:
Rolf E. Aalto, Gms
Derek Booth, Gm
Juliet G. Crider, (D), Stanford, 1998, GcmNr
Drew J. Gorman-Lewis, (D), Notre Dame, 2006, PgCl
Bob Hawley, Gl
Katharine W. Huntington, (D), MIT, 2006, Gt
Seth Moran, Gv
Gerard H. Roe, (D), MIT, 1999, AsGml
David Schmidt, (D), California (Berkeley), 2002, Gt
Ben Smith, GlZr
John O.H Stone, (D), Cambridge, 1986, Cca
Caroline Stromberg, Pbe
Fangzhen Teng, (D), Maryland, 2005, Cg
Jeremy Thomas, ZrXa
Gregory Wilson, (D), California (Berkeley), 2004, Pvg

Research Assistant Professor:
Michelle Koutnik, (D), Washington, 2009, Gll

Adjunct Assistant Professor:
Christian A. Sidor, (D), Chicago, 2000, Pv

Assistant Professor:
Scott Bennett, (D), California (Davis), 2013, YsGtm
Tom Brown, CcAt
Knut Christianson, (D), GlZr
Trenton Cladouhos, Gcg
Cailey Condit
Janice DeCosmo, AtZe
Alison Duvall, (D), Michigan, 2011, GmcGt
T.J. Fudge, (D), Gl
Alex Grant

Alexis Licht, (D), Gsn
Kenichi Matsuoka, Gl
Tom Neumann, Gl
Gregg Petrie, Zr
David Reusch
Erin Wirth, (D), Yale, 2014, YsRn
Tim Ziemba, Xa

Research Associate:
Marc Biundo, Ys
Doug Gibbons, ZeYs
Renate Hartog, Ys
Scott Henderson
Alex Hutko
Abram Jacobsen, Xg
Scott Kuehner, Ggz

Scientific Instructional Designer:
Michael Harrell, (D), Washington, Ze

Senior Lecturer:
Terry W. Swanson, (D), Washington, 1994, CcGe
Kathy G. Troost, (D), Zun

Lecturer:
Brian Collins, Gm
Ruth Martin
Steven Walters, Zir

Adjunct Professor:
David S. Battisti, (D), Washington, 1988, As
John R. Delaney, (D), Arizona, 1977, GuYr
Harlan Paul Johnson, (D), Washington, 1972, Gu
Randy LeVeque, (D), Stanford, 1982, GvYs
Andrea Ogston, Og
John D. Sahr, (D), Cornell, 1990, ZrXym
William S.D Wilcock, (D), MIT, 1992, Gu

Research Professor Emeritus:
Stephen D. Malone, (D), Nevada (Reno), 1972, YsGt
Gary A. Maykut, (D), Washington, 1969, AsGlYr

Research Professor:
Alan R. Gillespie, (D), Caltech, 1982, GmlZr

Research Associate Professor:
Robert I. Odom, (D), Washington, 1980, YrGu

Professor:
Joost Businger, Am

Affiliate Professor:
Arthur Frankel, (D), Ys
Joan Gomberg, (D), YsGt
Frank I. Gonzalez, (D), Hawaii, 1977, OpRn
Tony Irving, (D), CgXmGi
Richard Sack, (D), CgZn
George Shaw
Brian Atwater, (D), Delaware, 1980, RnGr
Charlotte Schreiber, (D), Gsd

Affiliate Assistant Professor:
Ralph Haugerud, (D), Zi
Brian L. Sherrod, (D), YsGt

Emeritus:
John B. Adams, (D), Washington, 1961, GcXgZr
Patricia M. Anderson, (D), Brown, 1982, Ple
Marcia B. Baker, (D), Washington, 1971, AsYg
John Booker, (D), California (San Diego), 1968, YmGct
Joanne (Jody) Bourgeois, (D), Wisconsin, 1980, GsrRh
Bob Charlson, As
Eric S. Cheney, (D), Yale, 1964, Eg
Darrel S. Cowan, (D), Stanford, 1972, Gct
Robert S. Crosson, (D), Stanford, 1966, YsGct
Bernard W. Evans, (D), Oxford, 1959, GpzGi
Heidi Houston
Estella Leopold
Ronald T. Merrill, (D), California (Berkeley), 1967, Ygm
George K. Parks, (D), California (Berkeley), 1966, YxXy
Charles F. Raymond, (D), Caltech, 1969, Gl
John M. Rensberger, (D), California (Berkeley), 1967, Pvg
Stewart W. Smith, (D), Caltech, 1961, Ys
Minze Stuiver, (D), Groningen (Neth), 1958, Cc
Brian Swanson, Asp
Joseph A. Vance, (D), Washington, 1957, Gi

Seismology Lab:
Bill Steele, Ys

Senior Computer Specialist:
Harvey Greenberg, (M), Cal State (Chico), 1977, Gml

Postdoctoral Researcher:
Jon Toner, (D)

Other:
Nick Lancaster

School of Oceanography (B,M,D)
Box 357940
Seattle, WA 98195-7940
p. (206) 543-5060
admin@ocean.washington.edu
http://www.ocean.washington.edu/

Research Professor:
Mark L. Holmes, (D), Washington, 1975, OuGl
Ronald L. Shreve, (D), Caltech, 1959, Yr

Professor:
Knut Aagaard, (D), Washington, 1966, Op
John A. Baross, (D), Washington, 1972, Ob
Roy Carpenter, (D), California (San Diego), 1968, No
William O. Criminale, Jr., (D), Johns Hopkins, 1960, Op
Eric A. D'Asaro, (D), MIT/WHOI, 1980, Op
John R. Delaney, (D), Arizona, 1977, Yr
Jody W. Deming, (D), Maryland, 1981, Ob
Allan H. Devol, (D), Washington, 1975, Oc
Steven R. Emerson, (D), Columbia, 1974, OcCmOg
Charles C. Eriksen, (D), MIT, 1977, Op
Bruce W. Frost, (D), California (San Diego), 1969, Ob
Michael C. Gregg, (D), California (San Diego), 1971, Op
G. Ross Heath, (D), Scripps, 1968, Cm
Barbara M. Hickey, (D), California (San Diego), 1975, Op
H. Paul Johnson, (D), Washington, 1972, Yr
Deborah S. Kelley, (D), Dalhousie, 1990, OuGvp
Marvin D. Lilley, (D), Oregon State, 1983, Ob
Seelye Martin, (D), Johns Hopkins, 1967, Op
James W. Murray, (D), MIT/WHOI, 1973, Oc
Charles A. Nittrouer, (D), Washington, 1978, Ou
Arthur R. M. Nowell, (D), British Columbia, 1975, Ou
Paul D. Quay, (D), Columbia, 1977, Oc
Peter B. Rhines, (D), Cambridge, 1967, Op
Jeffrey E. Richey, (D), California (Davis), 1973, CbHs
Stephen C. Riser, (D), Rhode Island, 1981, Op

Associate Professor:
Virginia E. Armbrust, (D), MIT/WHOI, 1990, Ob
Susan L. Hautala, (D), Washington, 1992, Op
Bruce M. Howe, (D), California (San Diego), 1986, Op
Mitsuhiro Kawase, (D), Princeton, 1986, Op
Richard G. Keil, (D), Delaware, 1991, Oc
Evelyn J. Lessard, (D), Rhode Island, 1984, Ob
Parker MacCready, (D), Washington, Op
LuAnne Thompson, (D), MIT/WHOI, 1990, Op
Mark J. Warner, (D), California (San Diego), 1988, Op
William S. D. Wilcock, (D), MIT/WHOI, 1992, Ou
Kevin L. Williams, (D), Washington State, 1985, Op

Research Assistant Professor:
Miles G. Logsdon, (D), Washington, 1997, Zi
Andrea S. Ogston, (D), Washington, 1997, Ou

Assistant Professor:
Andrew Barclay, (D), Oregon, 1998, Gu
Daniel Grunbaum, (D), Cornell, 1992, Ob
Anitra E. Ingalls, (D), SUNY (Stony Brook), 2002, Oc
Jeffrey D. Parsons, (D), Illinois, 1998, Ou
Gabrielle L. Rocap, (D), MIT/WHOI, 2000, Ob

Senior Lecturer:
Christina M. Emerick, (D), Oregon State, 1985, Ou

Lecturer:
Richard M. Strickland, (M), Washington, 1975, ObgZg

Adjunct Professor:
Rose Ann Cattolico, (D), SUNY (Stony Brook), 1973, Zn
Robert Francis, (D), Washington, 1970, Ob
Barbara B. Krieger-Brockett, (D), Wayne State, 1976, Ob
Ronald T. Merrill, (D), California (Berkeley), 1967, Ym
Edward Sarachik, (D), Brandeis, 1966, As
Robert C. Spindel, (D), Yale, 1971, Op

Emeritus:
George C. Anderson, (D), Washington, 1954, Ob
Karl Banse, (D), Kiel, 1955, Ob
Joe S. Creager, (D), Texas A&M, 1958, Ou
Alyn C. Duxbury, (D), Texas A&M, 1963, On
Terry E. Ewart, (D), Washington, 1965, Op
Richard H. Gammon, (D), Harvard, 1970, Oc

Joyce C. Lewin, (D), Yale, 1953, Ob
Brian T. R. Lewis, (D), Wisconsin, 1970, Yr
Russell E. McDuff, (D), California (San Diego), 1978, OcGu
Dean A. McManus, (D), Kansas, 1959, Ou
Gunnar I. Roden, (M), California (Los Angeles), 1956, Op
David A. Rothrock, (D), Cambridge, 1968, Op
Thomas B. Sanford, (D), MIT, 1967, Op
Richard W. Sternberg, (D), Washington, 1965, Ou

Cooperating Faculty:
Matthew H. Alford, (D), California (San Diego), 1998, Op
Edward T. Baker, (D), Washington, 1973, Ou
Laurie Balistrieri, (M), Washington, 1977, Oc
John Bullister, (D), California (San Diego), 1984, Oc
David A. Butterfield, (D), Washington, 1990, Yr
Glenn A. Cannon, (D), Johns Hopkins, 1969, Op
Meghan F. Cronin, (D), Rhode Island, 1993, Op
Brian D. Dushaw, (D), California (San Diego), 1992, Op
Richard A. Feely, (D), Texas A&M, 1974, Oc
Don E. Harrison, (D), Harvard, 1977, Op
Albert Hermann, (D), Washington, 1988, Op
Robin T. Holcomb, (D), Stanford, 1981, Ou
Gregory C. Johnson, (D), MIT/WHOI, 1990, Op
Peter A. Jumars, (D), California (San Diego), 1974, Ob
Kathryn A. Kelly, (D), California (San Diego), 1983, Op
William S. Kessler, (D), Washington, 1989, Op
Craig M. Lee, (D), Washington, 1995, Op
Michael J. McPhaden, (D), California (San Diego), 1980, Og
Curtis D. Mobley, (D), Maryland, 1977, Ob
Harold O. Mofjeld, (D), Washington, 1970, Op
Dennis W. Moore, (D), Harvard, 1968, Op
James H. Morison, (D), Washington, 1980, Op
Jeffrey Napp, (D), California (San Diego), 1986, Ob
Jan Newton, (D), Washington, 1989, Ob
Jeffrey A. Nystuen, (D), California (San Diego), 1985, Op
Joan M. Oltman-Shay, (D), California, 1986, Op
Mary Jane Perry, (D), California (San Diego), 1974, Ob
Thomas Pratt, (N), Gu
Joseph A. Resing, (D), Hawaii, 1997, Oc
Christopher L. Sabine, (D), Hawaii, 1992, Oc
Randy Shuman, (D), Washington, 1978, Ob
Laurenz A. Thomsen, (D), Kiel, 1992, Ob
Cynthia T. Tynan, (D), California (San Diego), 1993, Ob
Rebecca A. Woodgate, (D), Oxford, 1994, Op

Walla Walla Community College

Walla Walla Community College
500 Tausick Way
Walla Walla, WA 99362
p. (509) 527-4278
steven.may@wwcc.edu
http://www.wwcc.edu

Washington State University

Dept of Crop & Soil Sciences (B,M,D)
201 Johnson Hall
P.O. Box 646420
Pullman, WA 99164-6420
p. (509) 335-3475
bc.johnson@wsu.edu
http://css.wsu.edu/

Chair:
Richard Koenig, (D)

Professor:
David F. Bezdicek, (D), Minnesota, 1967, Sb
James B. Harsh, (D), California (Berkeley), 1983, Sc
Shiou Kuo, (D), Maine, 1973, Sc
Thomas A. Lumpkin, (D), Hawaii, 1978, Zn
William L. Pan, (D), North Carolina, 1983, Sb
John P. Reganold, (D), California (Davis), 1980, Soo

Senior Scientist:
Robert G. Stevens, (D), Colorado State, 1971, Sb

Associate Professor:
Joan R. Davenport, (D), Guelph, 1985, Sb
Frank J. Peryea, (D), California, 1983, Sc

Assistant Professor:
Markus Flury, (D), Swiss Fed Inst Tech, 1994, Sp

Adjunct Professor:
Alan J. Busacca, (D), California (Davis), 1982, Sa
Ann C. Kennedy, (D), North Carolina State, 1985, Sb
Robert I. Papendick, (D), South Dakota State, 1962, Sp
Jeffery L. Smith, (D), Washington State, 1983, Sb

Peter Hooper GeoAnalytical Lab (B,M,D)
School of the Environment
1228 Webster Physical Sciences Building
Washington State University
Pullman, WA 99164-2812
p. (509) 335-1626
jawolff@wsu.edu
https://environment.wsu.edu/facilities/geoanalytical-lab/

Professor:
John A. Wolff, (D), London (UK), 1983, Gi

Research Associate:
Owen K. Neill, (D), Alaska (Fairbanks), 2012, Ca

Emeritus:
Franklin F. Foit, Jr., (D), Michigan, 1968, Gz

Laboratory Manager:
Scott Boroughs, (D), Washington State, 2010, GiCa

Research Technologist:
Charles Knaack, (M), Washington State, 1991, Ca

School of the Environment (B,M,D) ●✓🖱
Webster 1228
P.O. Box 642812
Pullman, WA 99164-2812
p. (509) 335-3009
soe@wsu.edu
http://environment.wsu.edu/
Programs: Bachelors of Science degree in Earth and Environmental Sciences:with a major in:
Earth Science, Environmental & Ecosystem Sciences; Forest Ecology and Management; Wildlife Ecology and Conservation (Pre-Vet Option)
Enrollment (2009): B: 34 (11) M: 30 (9) D: 11 (5)

Director:
C. Kent Keller, (D), Waterloo, 1987, Hw

Professor:
Stephen Bollens, (D), Washington, 1990, Ob
Matt Carroll, (D), Washington, 1984, Zn
Peter B. Larson, (D), Caltech, 1983, Cs
Jeffrey D. Vervoort, (D), Cornell, 1994, CcaGt
John A. Wolff, (D), Imperial Coll (UK), 1983, Giv

Clinical Associate Professor:
Allyson Beall King, (D), Washington State

Associate Professor:
Sean P. Long , (D), Princeton, 2010, Gc

Clinical Assistant Professor :
Mike Berger , (D), Oregon, 2004

Assistant Professor:
Catherine Cooper, (D), Rice, 2005, Yg

Research Associate:
Scott Boroughs, (D), Washington State, 2010, Gv
Victor A. Valencia, (D), Arizona, 2005, ZgCcEg

Instructor:
Kurt Wilkie, (D)

Adjunct Professor:
Regan L. Patton, (D), Washington State, 1997, Gc
Stephen P. Reidel, (D), Washington State, 1978, GivGt
Dirk Schulze-Makuch, (D), Wisconsin (Milwaukee), 1996, Hw

Emeritus:
Keith Blatner , (D), 1983, Zn
Franklin F. Foit, Jr., (D), Michigan, 1968, Gza
Andrew Ford, (D), Dartmouth, 1975, Zu
Eldon H. Franz, (D), Illinois, 1971, Sf
David R. Gaylord, (D), Wyoming, 1983, Gs
George Hinman
Philip E. Rosenberg, (D), Penn State, 1960, GzCg
A. John Watkinson, (D), Imperial Coll (UK), 1972, Gc
Gary D. Webster, (D), California (Los Angeles), 1966, Pis

Wenatchee Valley College

Earth Sciences (A)
1300 Fifth Street
Wenatchee, WA 98801
p. (509) 682-6754
rdawes@wvc.edu

https://www.wvc.edu/directory/departments/geology/
Professor:
Ralph Dawes, (D), Washington, 1993, GgeAm
Adjunct Professor:
Kelsay Stanton, (M), W Washington, 2005, GgZe

Western Washington University

Geology Department (B,M) O⦸
516 High Street
Bellingham, WA 98225-9080
p. (360) 650-3582
Geology@wwu.edu
https://geology.wwu.edu/
Programs: Geology; Geophysics; Earth Science Education;
Enrollment (2022): B: 148 (55) M: 48 (13)
Chair:
Bernard A. Housen, (D), Michigan, 1994, Ym
Director Honors Program:
Scott R. Linneman, (D), Wyoming, 1990, GmZe
Professor:
Colin B. Amos, (D), California (Santa Barbara), 2007, Gc
Jackie Caplan-Auerbach, (D), Hawaii, Ys
Susan M. DeBari, (D), Stanford, 1991, GxZe
Robert J. Mitchell, (D), Michigan Tech, 1996, Hw
Melissa S. Rice, (D), Cornell, 2012, XgGsm
Elizabeth R. Schermer, (D), MIT, 1989, Gt
Associate Professor:
Douglas H. Clark, (D), Washington, 1995, Gl
Brady Z. Foreman, (D), Wyoming, 2012, Grs
Sean R. Mulcahy, (D), California (Davis), 2009, GpzGp
Assistant Professor:
Asmaa Boujibar, (D), Blaise Pascal (France), 2014, CpXcy
Robyn Dahl, (D), California (Riverside), 2015, PiZe
Allison Pfeiffer, (D), California (Santa Cruz), 2017, Gm
Camilo Ponton, (D), MIT, 2012, CsoPc
Emily Roland, (D), MIT, 2011, GuYrOn
Casey Saenger, (D), MIT, 2009, Pc
Kristina Walowski, (D), Oregon, 2015, GxvCg
Paleomagnetism Lab Manager:
Cristina Garcia Lasanta, (D), Zaragoza, 2016, YmGtc
Geochemistry Research Associate:
Mai Sas, (D), Auckland, 2020, GiCaGv
Research Associate:
Russell F. Burmester, (D), Princeton, 1974, Ym
Eric Grossman, (D), Hawaii, 2001, GusGm
George Mustoe, (M), W Washington, 1972, PbCaPo
Charles A. Ross, (D), Yale, 1964, Pm
Brian Rusk, (D), Oregon, 2003
Pete Stelling, (D), Alaska (Fairbanks), 2003, GxEgGv
Senior Instructor:
Paul Thomas, (M), W Washington, 1997, Gm
Emeritus:
R. Scott Babcock, (D), Washington, 1970, Cg
Edwin H. Brown, (D), California (Berkeley), 1966, Gp
Don J. Easterbrook, (D), Washington, 1962, GlmGe
David C. Engebretson, (D), Stanford, 1983, Gt
Thor A. Hansen, (D), Yale, 1978, Pg
James L. Talbot, (D), Adelaide, 1962, Gc
Undergraduate Program Advisor:
Hye In Park, (B), Washington, 2016
Instruct/Clsrm Support Tech 4 and Safety Coordinator:
Ben Paulson, (M), W Washington, 2010
Academic Department Manager:
Kate Blizzard, (B), Western Governors, 2019

Whatcom Community College

Sciences & Engineering Dept (A)
237 W. Kellogg Road
Bellingham, WA 98226
kkraft@whatcom.edu
Programs: We offer an associate in science transfer degree in geology and Earth science
Enrollment (2022): A: 6 (1)
Professor:
Kaatje Kraft, (D), Arizona State, 2014, Gg
Adjunct Professor:
Bernie Dougan, (M), W Washington, 1990, Gg

Whitman College

Dept of Geology (B) ⊠
345 Boyer
Walla Walla, WA 99362
p. (509) 527-5225
baderne@whitman.edu
http://www.whitman.edu/geology
Programs: Geology
Administrative Assistant: Patti Moss
Enrollment (2018): B: 40 (19)
Chair:
Kevin R. Pogue, (D), Oregon State, 1993, Gc
Professor:
Patrick K. Spencer, (D), Washington, 1984, Pg
Associate Professor:
Nicholas E. Bader, (D), California (Santa Cruz), 2006, SoHgZi
Kirsten P. Nicolaysen, (D), MIT, 2001, GiCgGa
Assistant Professor:
Lyman P. Persico, (D), New Mexico, 2012, Gm
Emeritus:
Robert J. Carson, (D), Washington, 1970, GmeGl
Geology Technician:
Elliot Broze, (M), Tromsø, 2017, Gg

Yakima Valley College

Dept of Geology (A)
P. O. Box 22520
16th Ave. & Nob Hill Blvd.
Yakima, WA 98907
p. (509) 575-2350 x2366
dhuycke@yvcc.edu
http://www.yvcc.edu/FutureStudents/AcademicOptions/Programs/PhysicalSciences/Geology/Pages/
Instructor:
David Huycke, (M), Wyoming, 1979, Gg

West Virginia

Concord University

Dept of Physical & Environmental Sciences (B) ●
1000 Vermillion St.
Athens, WV 24712-1000
p. (304) 384-5327
allenj@concord.edu
https://www.concord.edu/academics/college-of-science-mathematics-and-health/department-of-physical-and-environmental-sciences
f: www.facebook.com/ConcordGeoscience/
t: @CUGeology
Programs: Environmental Geoscience
Enrollment (2022): B: 15 (4)
Chair:
Joseph L. Allen, (D), Kentucky, 1994, Gct
Professor:
Alice M. Hawthorne Allen, (D), Indiana, 1998, Xa
Associate Professor:
Stephen C. Kuehn, (D), Washington State, 2002, CaGv
Thomas Saladyga, (D), West Virginia, 2011, Zyi
Adjunct Professor:
Alyce Lee, (D), Texas A&M, 2009, Og

Fairmont State University

Natural Sciences (B)
1201 Locust Avenue
Fairmont, WV 26554
p. (304) 367-4393
dhemler@fairmontstate.edu
Programs: Earth & Space Science Education
Certificates: Surveying
Enrollment (2017): B: 3 (0)
Coordinator of Geoscience Program:
Deb Hemler, (D), West Virginia, 1997, Ze
Instructor:
Todd I. Ensign, (D), West Virginia, 2017, Ze
Josh Revels, (M), Fairmont State, 2012, Ze
Adjunct Professor:
Ronald McDowell, (D), Gge

Barnes Nugent, (M), Gg

Marshall University

Dept of Geology (B,M)
One John Marshall Drive
Huntington, WV 25755
p. (304) 696-6720/(304) 696-6756
geology@marshall.edu
http://marshall.edu/geology
Programs: B.S. Geology
B.A. Geology
M.S. Physical Science, Geology Concentration
Enrollment (2022): B: 23 (11)
Chair:
Aley El Shazly, (D), Stanford, 1991, GpiCg
Professor:
Michael Caudill, (D), Tennessee (Knoxville), 1996, GmNgHw
Ronald L. Martino, (D), Rutgers, 1981, GsrPe
Adjunct Professor:
Ann Harris, (D), Kentucky, 2018, PgZei
Emeritus:
William L. Niemann, (D), Missouri (Rolla), 1999, Ng

West Virginia University

Dept of Geology & Geography (B,M,D) ●
330 Brooks Hall
P.O. Box 6300
98 Beechurst Ave.
Morgantown, WV 26506-6300
p. (304) 293-5603
jtoro@wvu.edu
http://geo.wvu.edu
Programs: Geology B.S., Geography B.A., Environmental Sciences B.A.
Certificates: GIS
Enrollment (2021): B: 148 (30) M: 37 (6) D: 13 (2)
Associate Chair:
Jaime Toro, (D), Stanford, 1998, GctGo
Professor:
Timothy R. Carr, (D), Wisconsin, 1981, GoYeGg
Dengliang Gao, (D), Duke, 1997, Ye
Shikha Sharma, (D), Lucknow (India), 1998, Csa
Dorothy Vesper, (D), Penn State, 2002, Hw
Associate Professor:
Kathleen Benison, (D), GrsCm
Amy L. Weislogel, (D), Stanford, 2006, Gs
Teaching Professor:
Joe Lebold, (D), West Virginia, 2005, PeGrg
Assistant Professor:
Christopher Russoniello, (D), Delaware, 2018, Hw
Charles Shobe, (D), Colorado, 2019, Gm
Research Assistant Professor:
Payam Kavousi Ghahfarohki, (D), West Virginia, 2016, Gog
Adjunct Professor:
Katherine Lee Avary, (M), North Carolina, 1978, Go
Bascombe (Mitch) Blake, (D), West Virginia, 2009, GsEcPb
Alan Brown, (D), Louisiana State, 2002, EoYe
Katherine R. Bruner, (D), West Virginia, 1991, Gd
Phillip Dinterman
Jason A. Hubbart, (D), Idaho, 2007, HqgHs
Ronald McDowell, (D), Colorado Mines, 1987, GsPiCe
Paul F. Ziemkiewicz, , (D), Hg
Emeritus:
Joseph J. Donovan, (D), Penn State, 1992, Hq
Thomas W. Kammer, (D), Indiana, 1982, Pi
J. Steven Kite, (D), Wisconsin, Gm
Henry W. Rauch, (D), Penn State, 1972, HwGe
Robert C. Shumaker, (D), Cornell, 1960, Go
Timothy A. Warner, (D), Purdue, 1992, Zr
Thomas H. Wilson, (D), West Virginia, 1980, YeGcYg

Wisconsin

Beloit College

R.D. Salisbury Dept of Geology (B) ☒
700 College Street
Beloit, WI 53511
p. (608) 363-2132
swansons@beloit.edu
http://www.beloit.edu/geology/
f: https://www.facebook.com/groups/47867916990/
Programs: Geology
Environmental Geology
Enrollment (2021): B: 21 (9)
Professor:
Susan K. Swanson, (D), Wisconsin, 2001, HwGmZi
Associate Professor:
James R. Rougvie, (D), Texas, 1999, GxpGz
James J. Zambito, (D), Cincinnati, 2011, PcGsPg
Emeritus:
Carol Mankiewicz, (D), Wisconsin, 1987, Gs
Carl V. Mendelson, (D), California (Los Angeles), 1981, PmGs

Lawrence University

Department of Geosciences (B) ☒
711 E Boldt Way
Appleton, WI 54911
p. (920) 832-6731
knudsena@lawrence.edu
http://www.lawrence.edu/dept/geology/
Administrative Assistant: Ellen c. Walsh
Enrollment (2018): B: 16 (5)
Professor:
Marcia Bjornerud, (D), Wisconsin, 1987, GctRc
Associate Professor:
Jeffrey J. Clark, (D), Johns Hopkins, 1997, Gm
Andrew Knudsen, (D), Idaho, 2002, CgGez
Assistant Professor:
Relena R. Ribbons, (D), Copenhagen, 2017, SbGe
Emeritus:
John C. Palmquist, (D), Iowa, 1961, GcpGg
Theodore W. Ross, (D), Washington State, 1969, Gg
Ronald W. Tank, (D), Indiana, 1962, Ge

Milwaukee Public Museum

Dept of Geology
800 W. Wells Street
Milwaukee, WI 53233
p. (414) 278-2741
sheehan@mpm.edu
http://www.mpm.edu/collect/geology/geosec-noframes.html
Chair:
Peter M. Sheehan, (D), California (Berkeley), 1971, Pi
Adjunct Professor:
Gary D. Rosenberg, (D), California (Los Angeles), 1972, RhPg

Northland College

Dept of Geosciences (B) ●☒
1411 Ellis Avenue
Ashland, WI 54806
p. (715) 682-1852
tfitz@northland.edu
Programs: Geology and Water science

Enrollment (2022): B: 15 (5)
Professor:
Thomas J. Fitz, (D), Delaware, 1999, GgxGc
Assistant Professor:
Cynthia May, (M), Penn State, 2007, ZiyZn
David J. Ullman, (D), Wisconsin, 2013, GlPeGm

Saint Norbert College

Geology Dept (B) ☒
100 Grant Street
De Pere, WI 54115
p. (920) 403-3987
tim.flood@snc.edu
http://www.snc.edu
Programs: Geology
Enrollment (2020): B: 12 (12)
Professor:
Tim P. Flood, (D), Michigan State, 1987, GvZe
Nelson R. Ham, (D), Wisconsin, 1994, Gl
Associate Professor:
Rebecca McKean , (D), Nebraska, 2009, GdPgRh

University of Wisconsin Oshkosh

Geography Department (B)
800 Algoma Blvd
Oshkosh, WI 54901
p. (920) 424-4105
geography@uwosh.edu
https://geography.uwosh.edu/
f: https://www.facebook.com/UWOgeography/
Programs: Geography
Certificates: GIS
Enrollment (2018): B: 32 (14)

Chair:
Angela G. Subulwa, (D), Kansas, 2009, Zn

Associate Dean:
Colin Long, (D), Oregon, 2003, PeZyc

Professor:
Heike C. Alberts, (D), Minnesota, 2003, Zn
John A. Cross, (D), Illinois, 1979, ZyAst
Kazimierz J. Zaniewski, (D), Wisconsin (Milwaukee), 1987, Zn

Associate Professor:
Mamadou Coulibaly, (D), S Illinois, 2004, ZiyHg

Assistant Professor:
Elizabeth Barron, (D), Rutgers, 2010, Zn

University of Wisconsin, Eau Claire

Dept of Geology (B) ● ☒
157 Phillips Hall
Eau Claire, WI 54702-4004
p. (715) 836-3732
steinklm@uwec.edu
http://www.uwec.edu/academic/geology
Administrative Assistant: Lorilie M. Steinke
Enrollment (2011): B: 88 (0)

Chair:
Scott K. Clark, (D), Illinois (Urbana), 2007, Ze

Professor:
Robert L. Hooper, (D), Washington State, 1983, Gz
Phillip D. Ihinger, (D), Caltech, 1991, Gx
J. Brian Mahoney, (D), British Columbia, 1994, GstEg
Kent M. Syverson, (D), Wisconsin, 1992, GlEnGe

Assistant Professor:
Robert Lodge, (D), EmCuGv
Sarah A. Vitale, (D), Connecticut, 2017, Hq

University of Wisconsin, Extension

Dept of Environmental Sciences
3817 Mineral Point Road
Madison, WI 53705-5100
p. (608) 262-1705
tricia.gorby@wisc.edu

Professor:
John W. Attig, (D), Wisconsin, 1984, Gl
Thomas J. Evans, (D), Wisconsin, 1994, Gg
James M. Robertson, (D), Michigan, 1972, Eg

Associate Professor:
Madeline B. Gotkowitz, (M), New Mexico, 1993, Hw
David J. Hart, (D), Wisconsin, 2000, Hw

Assistant Professor:
Eric C. Carson, (D), Wisconsin, 2003, Gl
Patrick I. McLaughlin, (D), Cincinnati, 2006, Gr

Emeritus:
Lee Clayton, (D), Illinois, 1965, Gl

University of Wisconsin, Fox Valley

Dept of Geography and Geology
1478 Midway Rd
Menasha, WI 54952
p. (920) 832-2600
michael.jurmu@uwc.edu
http://uwc.edu/depts/geography-geology/geography-geology

University of Wisconsin, Green Bay

Dept. of Natural & Applied Sciences - Geoscience Program (B,M) ⦸
LS-465
2420 Nicolet Drive
Green Bay, WI 54311
p. (920) 465-2371
luczajj@uwgb.edu
http://www.uwgb.edu/geoscience/
Programs: Geoscience; Earth Science Teaching; Accelerated Emphasis merged with Environmental Science & Policy Masters Program
Enrollment (2022): B: 19 (5) M: 3 (0)

Chair of Geoscience:
John A. Luczaj, (D), Johns Hopkins, 2000, HwGsCl

Associate Professor:
Steven J. Meyer, (D), Nebraska, 1990, AsmZg

Assistant Professor:
Erin Berns-Herrboldt, (D), Texas (Austin), 2020, CbSp
Shawn Malone, (D), Iowa, 2012, GgpGz

Assistant Teaching Professor:
William Jacobson, (D), Wisconsin (Milwaukee), 2015, GgZg

Professor:
Kevin J. Fermanich, (D), Wisconsin, 1995, SpRwHt

Emeritus:
Steven I. Dutch, (D), Columbia, 1976, GcgGe
Ronald D. Stieglitz, (D), Illinois, 1972, GrlHy
Keith West, Zgn

University of Wisconsin, Madison

Center for Limnology (M,D) ☒
Hasler Lab
680 N. Park Street
Madison, WI 53706
p. (608)262-3014
mjvanderzand@wisc.edu
https://limnology.wisc.edu/
f: https://www.facebook.com/centerforlimnology/
t: @WiscLimnology

Director:
Jake Vander Zanden, (D), McGill, 1999, On

Chair:
Kenneth Potter, (D), Johns Hopkins, Ng

Professor:
Calvin B. DeWitt, (D), Michigan, 1963, Ob
Linda K. Graham, (D), Michigan, 1975, Ob
John A. Hoopes, (D), MIT, 1965, No
John E. Kutzbach, (D), Wisconsin, 1966, Pe
Joy Zedler, (D), Wisconsin, 1968, Ob

Associate Professor:
Emily Stanley, (D), Arizona State, 1993, Ob

Assistant Professor:
Sarah Hotchkiss, (D), Minnesota, 1998, Ob
Carol Lee, (D), Washington, 1998, Ob
Zheng-yu Liu, (D), MIT, 1991, Op
Katherine McMahon, (D), California (Berkeley), 2002, Ng
Chin Wu, (D), MIT, Ng

Emeritus:
Michael S. Adams, (D), California (Riverside), 1968, Ob
Anders W. Andren, (D), Florida State, 1972, Oc
David E. Armstrong, (D), Wisconsin, 1966, Oc
Steve Carpenter, (D), Wisconsin, 1979, Ob
James F. Kitchell, (D), Colorado, 1970, Ob
William C. Sonzogni, (D), Wisconsin, 1974, Oc

Department of Geography (B,M,D)
550 N. Park St.
Room 160 Science Hall
Madison, WI 53706
p. (608) 262-2138
mason@geography.wisc.edu
http://www.geography.wisc.edu
f: https://www.facebook.com/UWMadisonGeography/
t: @UWMadisonGeog
Programs: GIS Development (PSM), Geography - MS (incl Physical Geography & Earth System Science focus), Cartography / GIS - MS, Geography - PhD (incl Physical Geography & Earth System, Science focus; Cartography/GIS focus)
Certificates: GIS Fundamentals; Advanced GIS
Enrollment (2020): B: 112 (36) M: 16 (3) D: 46 (15)

Professor:
Erika Marin-Spiotta, Cb
Joseph Mason, GmSd

John W. Williams, (D), Brown, 1999, PeZyPl
Assistant Professor:
Christian Andresen, Zy
Ken Keefover-Ring, Zg

Department of Geoscience (B,M,D) ●⃠
1215 West Dayton St
Madison, WI 53706
p. (608) 262-8960
geodept@geology.wisc.edu
http://geoscience.wisc.edu/geoscience/
Programs: Geoscience
Enrollment (2009): B: 58 (17) M: 20 (6) D: 42 (4)
Chair:
D. Charles DeMets, (D), Northwestern, 1988, Ym
Professor:
Alan R. Carroll, (D), Stanford, 1991, Gs
Kurt Feigl, (D), MIT, 1991, Yd
Dante Fratta, (D), Georgia Tech, 1999, Ng
Laurel B. Goodwin, (D), California (Berkeley), 1988, Gc
Clark M. Johnson, (D), Stanford, 1986, CgcCs
D. Clay Kelly, (D), North Carolina, 1999, Pm
Stephen R. Meyers, (D), Northwestern, 2003, PcOoGs
Shanan Peters, (D), Chicago, 2003, Gs
Eric E. Roden, (D), Maryland, 1990, PgSbCl
Bradley S. Singer, (D), Wyoming, 1990, Cc
Clifford H. Thurber, (D), MIT, 1981, YsGv
Basil Tikoff, (D), Minnesota, 1994, Gc
Harold Tobin, (D), California (Santa Cruz), 1995, Yg
John W. Valley, (D), Michigan, 1980, CsGxz
Huifang Xu, (D), Johns Hopkins, 1993, Gz
Senior Scientist:
Brian L. Beard, (D), Wisconsin, 1992, Cc
John Fournelle, (D), Johns Hopkins, 1989, GiZn
Associate Professor:
Michael A. Cardiff, (D), Stanford, 2010, HwYg
Assistant Professor:
Annie Bauer, (D), MIT, 2017, Cca
Chloe Bonamici, (D), Wisconsin, 2013, GxCa
Ken Ferrier, (D), California (Berkeley), 2009, GmZo
Shaun Marcott, (D), Oregon State, 2011, GlPe
Lucas Zoet, (D), Penn State, 2012, GllGm
Emeritus:
Mary P. Anderson, (D), Stanford, 1973, Hw
Jean M. Bahr, (D), Stanford, 1987, Hw
Carl J. Bowser, (D), California (Los Angeles), 1965, Cl
Philip E. Brown, (D), Michigan, 1980, EmGzp
Charles W. Byers, (D), Yale, 1973, Grs
Nikolas I. Christensen, (D), Wisconsin, 1963, Yx
David L. Clark, (D), Iowa, 1957, Pm
Dana H. Geary, (D), Harvard, 1986, Pg
Louis J. Maher, Jr., (D), Minnesota, 1961, Pl
Levi Gordon Medaris, Jr., (D), California, 1966, GipSa
David M. Mickelson, (D), Ohio State, 1971, Glm
Herbert F. Wang, (D), MIT, 1971, YgHwYx
Museum Director:
Richard Slaughter, (D), Iowa, 2001, Pg
Geology Librarian:
Marie Dvorzak, (M), Zn
Related Staff:
David J. Hart, (D), Wisconsin, 2000, Hg
Thomas S. Hooyer, (D), Iowa State, 1999, Hg
David Krabbenhoft, (D), Wisconsin, 1988, ClGeHw
James M. Robertson, (D), Michigan, 1972, Eg
Cooperating Faculty:
Kenneth R. Bradbury, (D), Wisconsin, 1982, Hw

Dept of Atmospheric & Oceanic Sciences (B,M,D) ☒
1225 W. Dayton Street
Atmospheric, Oceanic & Space Science Building
Madison, WI 53706-1695
p. (608)262-2828
aos@aos.wisc.edu
http://www.aos.wisc.edu
f: https://www.facebook.com/uwaos
t: @uw_aos
Enrollment (2021): M: 17 (0) D: 26 (0)
Chair:
Ankur R. Desai, (D), Penn State, 2006, Asm
Affiliate :
Chris Kucharik, (D), Wisconsin, 1997, As
Affiliate:
Galen McKinley, (D), MIT, 2002, OcpOg
Samuel Stechmann, (D), Courant Inst, 2008, ZnAs
Professor:
Steven A. Ackerman, (D), Colorado State, 1987, Zr
Matthew H. Hitchman, (D), Washington, 1985, As
Tracey Holloway, (D), Princeton, 2001, AscZn
Tristan S. L'Ecuyer, (D), Colorado State, 2001, As
Jonathan E. Martin, (D), Washington, 1992, Am
Michael C. Morgan, (D), MIT, 1994, Am
Grant W. Petty, (D), Washington, 1990, AsZrAm
Gregory J. Tripoli, (D), Colorado State, 1986, Am
Daniel J. Vimont, (D), Washington, 2001, As
Senior Scientist:
Edwin W. Eloranta, (D), Wisconsin, 1972, Yr
Associate Professor:
Larissa E. Back, (D), Washington, 2007, As
Adjunct Associate Professor:
Jeffrey R. Key, (D), Colorado, 1988, ZrAs
Adjunct :
Andrew Heidinger, (D), Colorado State, 1998, Zr
Adjunct Professor:
James Kossin, (D), Colorado State, 2000, As
Steven Platnick, (D), Arizona, Zr
Emeritus:
Stefan L. Hastenrath, (D), Bonn, 1959, As
Linda M. Keller, (M), Wisconsin, 1971, As
Zhengyu Liu, (D), MIT, 1991, Op
John M. Norman, (D), Wisconsin, 1971, So
Robert A. Ragotzkie, (D), Wisconsin, 1953, Ob
Pao-Kuan Wang, (D), California (Los Angeles), 1978, As
John A. Young, (D), MIT, 1966, AstOp
Research Specialist:
Dierk T. Polzin, (B), Wisconsin, As

Dept of Soil Science (B,M,D)
1525 Observatory Drive
Madison, WI 53706-1299
p. (608) 262-2633
slspeth@wisc.edu
http://www.soils.wisc.edu
Enrollment (2011): B: 34 (4) M: 16 (2) D: 8 (2)
Chair:
William L. Bland, (D), Wisconsin, 1984, Sp
Professor:
Phillip W. Barak, (D), Hebrew, 1988, ScGzRw
William F. Bleam, (D), Cornell, 1984, Sc
James G. Bockheim, (D), Washington, 1972, Sf
William J. Hickey, (D), California (Riverside), 1990, Sb
King-Jau S. Kung, (D), Cornell, 1984, Sp
Birl Lowery, (D), Oregon, 1980, Sp
Frederick W. Madison, (D), Wisconsin, 1972, Sd
J. Mark Powell, (D), Texas A&M, 1989, So
Stephen J. Ventura, (D), Wisconsin, 1989, ZrSo
Associate Professor:
Teri C. Balser, (D), California (Berkeley), 2000, Sb
Nick J. Balster, (D), Idaho, 1999, Sf
Carrie A.M Laboski, (D), Minnesota, 2001, So
Joel A. Pedersen, (D), California (Los Angeles), 2001, So
Assistant Professor:
Matthew D. Ruark, (D), Purdue, 2006, So
Douglas J. Soldat, (D), Cornell, 2007, So
Emeritus:
Larry G. Bundy, (D), Iowa, 1973, So
Robin F. Harris, (D), Wisconsin, 1972, Sb
Philip A. Helmke, (D), Wisconsin, 1971, Sc
Keith A. Kelling, (D), Wisconsin, 1974, So
Wayne R. Kusssow, (D), Wisconsin, 1966, So
Kevin McSweeney, (D), Illinois, 1984, SdaSf
John M. Norman, (D), Wisconsin, 1971, Spo
E. Jerry Tyler, (D), North Carolina, 1975, Sd

Geological Engineering (B,M,D) ●☒
1415 Engineering Drive
2205 Engineering Hall
Madison, WI 53706-1691
p. (608) 890-2662

jmtinjum@wisc.edu
http://www.gle.wisc.edu
Programs: Geological Engineering
Enrollment (2018): B: 92 (0) M: 9 (0) D: 5 (0)

Chair:
William J. Likos, (D), Colorado Mines, 2000, NgSp

Professor:
Jean M. Bahr, (D), Stanford, 1987, Hw
Kurt L. Feigl, (D), Gt
Laurel Goodwin, California (Berkeley), 1988
Tracey Holloway, (D), Princeton, As
Steven P. Loheide, (D), Stanford, 2006, HwSp
Clifford Thurber, (D), MIT, 1981, Ysx
Basil Tikoff, (D), Minnesota, 1997, Gc
Chin Wu, (D), MIT, Onp

Associate Professor:
Mike Cardiff, Stanford, 2010, Hw
Dante Fratta, (D), Georgia Tech, 1999, Yx
James M. Tinjum, (D), Wisconsin, 2003, Ng

Assistant Professor:
Greeshma Gadikota, (D), Columbia, 2014, GzCqu
Matt Ginder-Vogel, (D), Stanford, 2006, Sc
Andrea Hicks, (D), Illinois (Chicago), Zn
Hiroki Sone, (D), Stanford, 2012, NrYxGt
Lucas Zoet, (D), Penn State, Gl

Emeritus:
Tuncer B. Edil, (D), Northwestern, 1973, Ng
Bezalel C. Haimson, (D), Minnesota, 1968, Nr
Herbert F. Wang, (D), MIT, 1971, YgHwYx

University of Wisconsin, Milwaukee

Dept of Geosciences (B,M,D)
P.O. Box 413
Milwaukee, WI 53201-0413
p. (414) 229-4561
geosci-office@uwm.edu
https://uwm.edu/geosciences/
f: https://www.facebook.com/UWMgeosci
Enrollment (2020): B: 91 (20) M: 25 (3) D: 5 (2)

Chair:
John L. Isbell, (D), Ohio State, 1990, Gs

Professor:
Dyanna M. Czeck, (D), Minnesota, 2001, Gc
Timothy J. Grundl, (D), Colorado Mines, 1988, HwCl
Mark T. Harris, (D), Johns Hopkins, 1988, GsrGd
Lindsay J. McHenry, (D), Rutgers, 2004, GzaXg
Keith A. Sverdrup, (D), Scripps, 1981, YsGt

Associate Professor:
Barry I. Cameron, (D), N Illinois, 1998, Gvi
Shangping Xu, (D), Princeton, 2005, Hg

Assistant Professor:
Julie Bowles, (D), California (San Diego), 2005, Ym
Charles Paradis, (D), Tennessee, 2017, Hw

Adjunct Professor:
Stephen Q. Dornbos, (D), S California, 2003, PeiPo
Daniel T. Feinstein, (M), Wisconsin, 1986, Hw
Margaret L. Fraiser, (D), S California, 2005, PiePg

Visiting Assistant Professor:
Victoria McCoy, (D), Yale, 2015, Pb

Emeritus:
Douglas S. Cherkauer, (D), Princeton, 1972, HwGe
William F. Kean, Jr., (D), Pittsburgh, 1972, Ym
Norman P. Lasca, (D), Michigan, 1965, GmlSd
Peter M. Sheehan, (D), California (Berkeley), 1971, Pie

University of Wisconsin, Oshkosh

Geology Dept (B) O☒
645 Dempsey Trail
Oshkosh, WI 54901-8649
p. (920) 424-4460
mode@uwosh.edu
http://www.uwosh.edu/geology/
f: http://www.facebook.com/UWOgeology
Programs: Geology; Earth Science Secondary Education
Administrative Assistant: Courtney Maron
Enrollment (2018): B: 40 (15)

Chair:
William N. Mode, (D), Colorado, 1980, Gl

Professor:
Eric E. Hiatt, (D), Colorado, 1997, GsOcGu
Maureen A. Muldoon, (D), Wisconsin, 1999, Hw
Timothy S. Paulsen, (D), Illinois, 1997, Gc
Jennifer M. Wenner, (D), Boston, 2001, Gi

Assistant Professor:
Benjamin W. Hallett, (D), Rensselaer, 2012, Gpx
Joseph E. Peterson, (D), N Illinois, 2010, PgGr

Emeritus:
Norris W. Jones, (D), Virginia Tech, 1968, Gi
Gene L. La Berge, (D), Wisconsin, 1963, Eg
James W. McKee, (D), Louisiana State, 1967, Gr
Brian K. McKnight, (D), Oregon State, 1970, Gdu
Thomas J. Suszek, (M), Minnesota (Duluth), 1991, Gg

Instrumentation Specialist:
James A. Amato, (M), Wisconsin (Milwaukee), 2017, GgZi

University of Wisconsin, Parkside

Dept of Geosciences (B)
Box 2000
900 Wood Road
Kenosha, WI 53141
p. (262) 595-2744
li@uwp.edu
Programs: Earth Sciences; Environmental Geosciences
Enrollment (2020): B: 24 (8)

Professor:
John D. Skalbeck, (D), Nevada (Reno), HwYgGg
Li Zhaohui, (D), SUNY (Buffalo), 1994, ClGzHw

Research Associate:
Julie Kindelman, (D), Hs

Emeritus:
Gerald A. Fowler, PgGg
Allan F. Schneider, (D), Gl
James H. Shea, (D), Gs

University of Wisconsin, Platteville

Dept of Geography & Geology (B)
Platteville, WI 53818
p. (608) 342-1791
underwoodc@uwplatt.edu
http://www.uwplatt.edu/geography/

Professor:
Charles W. Collins, (D), Nova, 1985, Gm

Associate Professor:
Richard A. Waugh, (D), Wisconsin, 1995, Gc

Assistant Professor:
Mari A. Vice, (D), S Illinois, 1993, Gd

Emeritus:
William A. Broughton, (B), Wisconsin, 1938, Em
Kenneth A. Shubak, (M), Michigan, 1966, Pi

University of Wisconsin, River Falls

Dept of Plant & Earth Science (B) ☒
410 S. Third Street
River Falls, WI 54022
p. (715) 425-3345
holly.dolliver@uwrf.edu
http://www.uwrf.edu/pes/geol/
Program Assistant: Sue Freiermuth
Enrollment (2016): B: 50 (0)

Chair:
Holly A.S. Dolliver, (D), Minnesota, 2007, GmSd

Professor:
Kerry L. Keen, (D), Minnesota, 1992, HwGse
Ian S. Williams, (D), California (Santa Barbara), 1981, Yg

Lecturer:
Kevin Thaisen, (D)

University of Wisconsin, Stevens Point

Dept of Geography and Geology (B,M)
2001 Fourth Avenue, SCI D332
Stevens Point, WI 54481
p. (715) 346-2629
David.Ozsvath@uwsp.edu
http://www.uwsp.edu/geo/
Programs: Geospatial Sciences (B)

Certificates: GIS
Enrollment (2017): B: 63 (27)
Professor:
Neil C. Heywood, (D), Colorado, 1988, Zy
Eric J. Larson, Oregon State, 2001, Zr
David L. Ozsvath, (D), SUNY (Binghamton), 1985, Hw
Keith W. Rice, (D), Kansas, 1989, Zir
Associate Professor:
Samantha W. Kaplan, (D), Wisconsin, 2003, GnPlGd
Ismaila Odogba, (D), Louisville, 2009, Zui

University of Wisconsin, Superior

Dept of Natural Sciences (B,M)
PO Box 2000
Belknap & Catlin
Superior, WI 54880
p. (715) 394-8322
wbajjali@uwsuper.edu
Professor:
William Bajjali, (D), Ottawa, 1994, HwCsg
Associate Professor:
Andy Breckenridge, (D), Minnesota, 2005, GlnGs
Assistant Professor:
Kristin E. Riker-Coleman, (D), Minnesota, 2008, Gg

University of Wisconsin, Whitewater

Geography, Geology, and Environmental Science (B)
800 W Main Street
Whitewater, WI 53190
p. (262) 472-1071
filipiap@uww.edu
http://www.uww.edu/cls/departments/geography-geology-env-sci
Programs: Geography, Geology, and Environmental Science majors; Geography, Geology, GIS, and Environmental Studies minor.
Certificates: GIS
Enrollment (2020): B: 150 (45)
Professor:
Prajukti Bhattacharyya, (D), Minnesota, 2000, Gce
Eric Compas, (D), Wisconsin, 2008, ZiRw
Rex Hanger, (D), California (Berkeley), 1992, PigGs
Peter Jacobs, (D), Wisconsin, 1994, SdaGm
Dale Splinter, (D), Oklahoma State, 2006, Hs
Associate Professor:
Rocio Duchesne, (D), Montclair State, 2015, Zri
John Frye, (D), Georgia, 2011, AtZr
Jeff Olson, (D), Ohio State, 2013, Ziu
Assistant Professor:
Stephen J. Levas, (D), Ohio State, 2015, CmGe

Wyoming

Casper College

Department of Earth and Environmental Sciences (A)
School of Science
125 College Drive
Casper, WY 82601
p. (307) 268-2017
mconnely@caspercollege.edu
http://www.caspercollege.edu/geology/
Programs: Geology (A); Environmental Science (A); GIS (A), Certificates: GIS (A)
Enrollment (2019): A: 5 (3)
Klaenhammer Earth Science Chair:
Melissa Connely, (M), Utah State, 2000, GsPev
Instructor:
Jeff Sun, (M), Texas State, Zi
Kent A. Sundell, (D), California (Santa Barbara), 1985, GgvPv
Beth Wisely, (D), Oregon, 2005, YgHwGc

Central Wyoming College

Earth, Energy, Environment (A)
2660 Peck Avenue
Riverton, WY 82501
p. (307) 855-2000
jklanche@cwc.edu
http://www.cwc.edu/academics/programs/earthenviron
f: https://www.facebook.com/cwc.edu/
Enrollment (2015): A: 5 (0)
Professor:
Suzanne M. Smaglik, (M), Colorado Mines, 1987, GgZeCg
Assistant Professor:
Jacki Klancher, (M), Zi

Eastern Wyoming College

Eastern Wyoming College (A)
3200 West C. Street
Torrington, WY 82240
p. (307) 532-8330
bob.creagar@ewc.wy.edu
http://www.ewc.wy.edu/
Instructor:
Christopher Wenzel, (D), Wyoming, 2019, SbfHg

Northwest College

Northwest College
231 West 6th Street
Powell, WY 82435
p. 307.754.6405
Eric.Atkinson@nwc.edu
http://www.nwc.cc.wy.us/

University of Wyoming

Dept of Geology and Geophysics (B,M,D) ●⊘
Dept. 3006
1000 E. University Ave.
Laramie, WY 82071
p. (307) 766-3386
carrick@uwyo.edu
http://www.uwyo.edu/geolgeophys/
Programs: BS in Geology, BA in Earth Science, BS in Environmental Geology and Geohydrology, MS in Geology, MS in Geophysics, PhD in Geology
Enrollment (2015): B: 157 (26) M: 28 (7) D: 31 (2)
Head:
Carrick M. Eggleston, (D), Stanford, 1991, ClqCb
Professor:
Kevin R. Chamberlain, (D), Washington (St Louis), 1990, CcGt
Mark T. Clementz, (D), California (Santa Cruz), 2002, Pg
Carol D. Frost, (D), Cambridge, 1984, CcGit
Neil F. Humphrey, (D), Washington, 1987, GlmYh
Barbara E. John, (D), California (Santa Barbara), 1987, Gci
Subhashis Mallick, (D), Hawaii, 1987, YseYg
James D. Myers, (D), Johns Hopkins, 1979, Gi
Bryan N. Shuman, (D), Brown, 2001, PeGne
Kenneth W.W Sims, (D), California (Berkeley), 1995, CgGvCa
Ye Zhang, (D), Indiana, 2005, HyGeYu
Senior Scientist:
Janet Dewey, (M), Auburn, 1993, CaGg
Susan M. Swapp, (D), Yale, 1982, Gp
Associate Professor:
Michael J. Cheadle, (D), Cambridge, 1989, Yg
Po Chen, (D), S California, 2005, YseYg
Ellen Currano, (D), Penn State, 2008, Pbe
Kenneth G. Dueker, (D), Oregon, 1994, Ys
Dario Grana, (D), Stanford, 2013, Ye
John P. Kaszuba, (D), Colorado Mines, 1997, CgGze
Brandon McElroy, (D), Texas, 2009, GsmGr
Clifford S. Riebe, (D), California (Berkeley), 2000, ClGme
Associate Scientist:
Laura Vietti, (D), Minnesota, 2014, Pv
Assistant Professor:
James Chapman, (D), Arizona, 2018, Gct
Kimberly Lau, (D), Stanford, 2017, Cb
Andrew D. Parsekian, (D), Rutgers, 2011, Yg
Simone Runyon, (D), Arizona, 2017, EdGip
Adjunct Professor:
Vladimir Alvarado, (D), Minnesota, 1996, Np
Eric A. Erslev, (D), Harvard, 1981, GctGg
Peter H. Hennings, (D), Texas, 1991, Gc
Ranie Lynds, (D), Wyoming, 2005, Gsr
Emeritus:
B. Ronald Frost, (D), Washington, 1973, GpiEd

Robert R. Howell, (D), Arizona, 1980, XgZr
Department Secretary:
Deborah Prusia

Western Wyoming Community College

Western Wyoming Community College
2500 College Drive
Rock Springs, WY 82901
p. (307) 382-1662
mustangcentral@westernwyoming.edu
http://www.wwcc.wy.edu/academics/geology/

Algeria

Algerian Petroleum Institute

Algerian Petroleum Institute (B)
Avenue 1er Novembre
Boumerdes, Skikda, Oran 35000
p. +213 (0) 24 81 90 56
iap@iap.dz
http://www.iap.dz/

Centre Universitaire de Khemis Miliana

Institut de Sciences de la Nature et de la Terre (B)
Route de Theniet El-Had
Khemis Miliana, Khemis Miliana 44225
p. +213-27-66-42-32
kouben55@hotmail.fr
http://www.cukm.org/

Universite Abou Bekr Belkaid de Tlemcen

Faculté des sciences de la nature et de la vie et sciences de la terre et de l'univers (B)
BP 119
Tlemcen 13000
p. (+213) 040.91.59.09
webcri@mail.univ-tlemcen.dz
http://snv.univ-tlemcen.dz/

Universite Badji Mokhtar

Faculte des Sciences de la Terre: Amenagement du territoire, Geologie, Hydraulique, Mines (B)
BP 12
Annaba 23000
d.fst@univ-annaba.dz
http://www.univ-annaba.org/

Universite D'Oran

Faculté des Sciences de la Terre, de la Géographie et de l'Amenagement du Territoire (B)
Rue du Colonel Lofti
Es-Senia, El-Mnouar, Oran 31000
p. +213 (0) 41 58 19 47
labo.egeat@univ-oran2.dz
http://www.univ-oran.dz/facultes/F_Terre

Universite de Annaba

Inst of Natural Sciences
BP 12
El Hadjar
Annaba
http://www.univ-annaba.org/

Universite de Batna

Dept de Sciences de la Terre (B)
1, Rue Chahid Boukhlouf Mohamed
El Hadi, Batna 5000
p. (+213) 033.86.06.02
recteur@univ-batna.dz
http://www.univ-batna.dz/

Universite de Jijel

Dept de Geologie (B,M,D)
Ouled Aissa, Jijel
p. +213 (34) 498016
webmaster@univ-jijel.dz
http://www.univ-jijel.dz/

Universite de Mentouri

Dept des Sciences de la Terre/Geologie (B)
Campus Ahmed Zouaghi
B.P. 325 Route d Ain El Bey
, Constantine 25000
p. 213 031 90 38 52
univ-constantine@fr.fm
http://www.umc.edu.dz/fst/

Dept de l'Amenagement du territoire (B)
Campus Ahmed Zouaghi
B.P. 325 Route d Ain El Bey
, Constantine 25000
p. (+213) 31.90.02.07
fstgat@umc.edu.dz
http://www.umc.edu.dz/vf/index.php/recherche-scientifique/annuaire-des-laboratoires/110-faculte-des-sciences-de-la-terre-/398-de

Universite des Sciences et de la Technologie

Inst of Earth Sciences
Houari Boumediene
B.P. 139 Dar El Beida
Eldjazair

Faculte de Sciences de la Tere, Geographie et Amenagement du Territoire (B)
USTHB-IST BP 32 El Alia
Bab-Ezzouar Alger 16123
p. (+213) 21247904
aouabadi@usthb.dz
http://www.usthb.dz/fst/

Universite Djillali Liabes Sidi Bel Abbes

Faculte des Sciences de la Nature et de la vie (B)
BP 89
Sidi Bel Abbes 22000
p. +213 (48) 543018
mbouziani@univ-sba.dz
http://www.univ-sba.dz/

Universite Kasdi Merbah Ouargla

Dept de Geologie (B)
Route de Ghardaia
Ouargla
p. +213 (29) 712468
info@ouargla-univ.dz
http://193.194.92.30/spip/dept/filiere-hydrocarbures-chimie.htm

Dept des Hydrocarbures (B)
Route de Ghardaia
Ouargia
p. +213 (29) 712468
info@ouargla-univ.dz
http://www.ouargla-univ.dz/

Universite M'hamed Bouguerra de Boumerdes

Faculte des Hydrocarbures et de la chimie (B)
Avenue de Ildependance
Boumerdes 35000
p. +213 (24) 816420
doyen_fhc@umbb.dz
http://www.umbb.dz/

Universite Saad Dahlab, Blida

Dept des Sciences de l'Eau et de l'Environnement (B)
BP 270
Blida 9000
p. +213 (25) 433625
contact@univ-blida.dz
http://www.univ-blida.dz/fac_ingenieur

Angola

Universidade Agostinho Neto

Faculdade de Geologia (B)
Av. 4 de Fevereiro
No. 7
2 andar
Luanda C. P. 815
p. +244-222-333-816
info@geologia-uan.com
http://www.geologia-uan.com/

Universidade Independente de Angola

Engenharia dos Recursos Naturais e Ambiente (B)
Rua da Missao
Bairro Morro Bento II - Corimba
Luanda
p. (+244) 222 33 89 70
unia@unia.ao
http://www.unia.ao/curso.php?cr=6
Vice-Rector:
Nuno Nascimento Gomes, (M), Lisbon, 1992, GesGu

Universite de Angola

Dept of Geology
Avenida 4 de Fevereiro 7
Caixa postal 815-C
Luanda

Argentina

Servicio Geologico Minero Argentino

Servicio Geologico Minero Argentino
Av. Julio Argentino Roca 651. P.B
Capital Federal
mjanit@mecon.gov.ar
http://www.segemar.gov.ar/

Universidad de Buenos Aires

Departamento de Ciencias Geológicas - FCEN (B,D) O☒
Ciudad Universitaria-Pabellón 2
Intendente Güiraldes 2160
Ciudad Autónoma de Buenos Aires C1428EHA
p. (5411) 528-58249
geologia@gl.fcen.uba.ar
http://www.gl.fcen.uba.ar/
Programs: Geology
Paleontology

Universidad Nacional de Catamarca

Facultad de Tecnologia Y Ciencias Aplicadas
Maximio Victoria No. 55
San Fernando del Valle de Catamarca (470, Catamarca CP 4700
p. (54) 0383-4435 112 int. 112
daa@tecno.unca.edu.ar
http://tecno.unca.edu.ar

Universidad Nacional de Jujuy

Instituto de Geologia y Mineria
San Salvador de Jujuy , Jujuy
narce@idgym.unju.edu.ar
http://www.idgym.unju.edu.ar/

Universidad Nacional de la Pampa

Dept of Geologia
Uruguay 151, 6300 Santa Rosa
La Pampa
geologia@exactas.unlpam.edu.ar
http://www.exactas.unlpam.edu.ar/

Universidad Nacional de la Patagonia San Juan Bosco

Facultad de Ciencias Naturales
Comodoro Rivadavia , Chubut
secacademica@unp.edu.ar
http://www.unp.edu.ar/

Universidad Nacional de La Plata

School of Astronomy and Geophysics
Paseo del Bosque s/n
B1900FWA
p. (0221)-423-6593
academic@fcaglp.unlp.edu.ar
http://www.fcaglp.unlp.edu.ar/

Universidad Nacional de Rio Cuarto

Dept. of Geology (B,M,D) O
ocampanella@gmail.com
mvillegas@exa.unrc.edu.ar
Rio Cuarto , Cordoba C.P. X5804BYA
p. (0358) 467-6198
webgeo@exa.unrc.edu.ar
http://geo.exa.unrc.edu.ar/
Enrollment (2016): B: 32 (0) M: 5 (0) D: 2 (0)
Associate Professor:
Monica B. Villegas, (M), UNRC, 1995, Gsc
Associate Scientist:
Juan Enrique Otamendi, (D), Nac de Río Cuarto, 2000, CgaCp
Adjunct Professor:
Hector Daniel Origlia, (M), Arizona State, 1999, NgZn

Universidad Nacional de Rio Negro

Dept. of Geology
Building Perito Moreno - Pacheco 460
General Roca , Rio Negro
sdeyve@unrn.edu.ar
http://www.unrn.edu.ar

Universidad Nacional de Salta

Facultad de Ciencias Naturales
Salta
p. (0387) 425-5413
decnat@unsa.edu.ar
http://naturales.unsa.edu.ar/

Universidad Nacional de San Juan

Facultad de Ciencias Exactas, Físicas y Naturales
Parque de Mayo
5400 San Juan
comunicacionfcefyn1@gmail.com
http://exactas.unsj.edu.ar/

Universidad Nacional de San Luis

Facultad de Ciencias Exactas, Fisico, Matematicas Y Naturales
Ejército de los Andes 950
San Luis D5700HHW
p. +54 (2652) 424027
sacadfmn@unsl.edu.ar
http://webfmn.unsl.edu.ar

Universidad Nacional de Tucuman

Facultad de Ciencias Naturales E Instituto Miguel Lillo (D)
Miguel Lillo 205
S.M. de Tucuman
San Miguel de Tucuman, Tucuman CP4000
p. 0381-4239456
info@csnat.unt.edu.ar
http://www.csnat.unt.edu.ar/

Universidad Nacional del Comahue

Dept of Geology and Petroleum
School of Engineering
Buenos Aires 1400
Neuquen 18300 , Patagonia
p. +54-299-4490300
sacadfi@uncoma.edu.ar
http://www.uncoma.edu.ar/

Universidad Nacional del Sur

Departamento de Geologia
San Juan 670
Primer Piso
Bahia Blanca
Buenos Aires 8000
p. 54-(0291)-4595147
secgeo@uns.edu.ar
http://www.uns.edu.ar/

Australia

Australian National University

Research School of Earth Sciences (M,D) ☒
142 Mills Rd.
Canberra, ACT 0200
p. +61 2 6125 3406
Director.rses@anu.edu.au
http://rses.anu.edu.au/
Emeritus:
Patrick De Deckker, (D), Adelaide, 1981, GuPmGn

Dept of Earth & Marine Sciences (B,M,D) ☒
47 Daley Road
Canberra, ACT 0200
p. 02 6125 2056
director.rses@anu.edu.au
http://rses.anu.edu.au/

Curtin University

Dept of Mining Engineering (B,M,D)
P. O. Box 597
Kalgoorlie, WA
e.topal@curtin.edu.au

School of Earth and Planetary Sciences (B,M,D) O✓🖹
GPO Box U1987
Perth, WA 6845
p. +618 92667968
eps-admin@curtin.edu.au
https://www.curtin.edu.au/about/learning-teaching/science-engineering/school-of-earth-and-planetary-sciences/
f: https://www.facebook.com/CurtinEPS
t: @CurtinGeology
Programs: Geology, Geophysics, Earth Sciences, Surveying, GIS, Exploration and Mining Geology
Enrollment (2022): B: 525 (94) M: 125 (48) D: 100 (21)
Head of School:
Ian Fitzsimons, (D), GptCc
Professor:
David Antoine, (D), ObZrc
Joseph Awange, (D), ZfGqZr
Gretchen Benedix, (D), XgmXc
Phil Bland, (D), XcgXm
Chris Clark, (D), GpCcGt
Ahmed El-Mowafy, (D), YdZrf
Christine Erbe, (D), ObpOg
Katy Evans, (D), GpEmCp
Noreen Evans, (D), CacCl
Kliti Grice, (D), CosPc
Boris Gurevich, (D), Yse
Brett Harris, (D), YegHw
Tim Johnson, (D), Gpt
Fred Jourdan, (D), CcGiXm
Chris Kirkland, (D), Cca
Zheng-Xiang Li, (D), YmGtg
Rob McCauley, (D), ObYr
Alexander Nemchin, (D), CcXc
Roman Pevzner, (D), Ys
Steve Reddy, (D), ZaGzCc
Associate Professor:
Milo Barham, (D), Gsg
Marco Coolen, (D), PoCbPe
Ashraf Dewan, (D), RnZof
Alexander Gavrilov, (D), YrOb
Petra Helmholz, (D), Zfr
Jon Kirby, (D), Ydv
Michael Kuhn, (D), YdvOp
Katarina Miljkovic, (D), Xyg
Todd Robinson, (D), ZfrZi
Nick Timms, (D), GzcXg
Dom Wolff-Boenisch, (D), CqGe
Senior Lecturer:
Mehrooz Aspandiar, (D), CeZaSa
Dave Belton, (D), Zfr
Alison Blyth, (D), PeCsZc
Sten Claessens, (D), Yd
Mick Filmer, (D), YdZrOp
Ivana Ivanova, (D), Zfi
Senior Research Fellow:
Aaron Cavosie, (D), GzCuXm
Navdeep Dhami, (D), Po
Alec Duncan, (D), Yr
Paul Greenwood, (D), CaoCe
Vassily Mikhaltsevitch, (D), Yxs
Iain Parnum, (D), YrOup
Sergei Pisarevskiy, (D), Ym
Konstantin Tertyshnikov, (D), Ys
Stéphanie Vialle, (D), YxCq
Research Fellow:
Michael Carson, (D), Ye
Olivia Collet, (D), Yes
Luc Doucet, (D), GiCus
Lucy Forman, (D), Xmg
Denis Fougerouse, (D), ZaCcEm
Hamed Gamaleldien, (D), GtCuGx
Alex Holman, (D), CasCe
Chuan Huang, (D), YgGt
Eriita Jones, (D), ZrXaZi
Uwe Kirscher, (D), Ym
Anthony Lagain, (D), Xg
Juan Li, (D), ZrObZi
Yebo Liu, (D), Ym
Hoang Nguyen, (D), Ye
Hugo Olierook, (D), CceEm
Chandanlal Parida, (D), ObpZr
Ellie Sansom, (D), Xm
Pavel Shashkin, (M), Ys
Leonid Shumlyanskyy, (D), CcGiEm
Martin Towner, (D), Xgm
Daniel Wilkes, (D), Yr
Research Associate:
Amir Allahvirdizadeh, (M), Yd
SM Mahbubul Ameen, (D), CgcGx
Lyn Beahan, (D), Zr
Katelyn Boase, (B), Po
Hadrien Devillepoix, (D), Xa
Max Dröllner, (M), GsCcEn
Amy Elson, (D), CoGs
Michael Hartnady, (D), Cc
Joshua Hedgepath, (D), Xy
Robert Howie, (D), Xm
Janne Liebmann, (D), Ccs
Anh Ngo, (M), Zif
Richard Palmer, (D), Zr
Steve Poropat, (D), Pg
Alan Scarlett, (D), CbaEo
Bettina Schaefer, (D), Co
Tommaso Tacchetto, (M), ZaGp
Christopher Taylor, (D), Pe
Lewis Trotter, (M), Zi
Bruno Vieira Ribeiro, (D), GxCcGt
Alex Walker, (D), EmGt
Chong Wei, (D), Ob
Lecturer:
Sasha Ziramov, (M), Yse
Adjunct Senior Research Fellow:
John Simmonds, (D), Em
Chris Swain, (D), Yg
Adjunct Research Fellow:
Capri Beck, (D), Ob
Jason Bennett, (D), EmGz
Jane Cuneen, (D), GstGc
Bisun Datt, (D), Zr
Timmons Erickson, (D), GzCc
Paul Johnston, (D), Yd
Trudi Kennedy, (D), XgCc

Brian Luinstra, (D), Hw
Darryl McMahon, (D), Yr
Lynne Milne, (D), Plb
Hoa Thi Pham, (D), Zf
Vladimir Puzyrev, (D), Ye
Justy Siwabessy, (D), YrZr
Hugh Smithies, (D), CuGi
Brian Tattitch, (D), CpEm
Andy Wilson, (D), Grs

Adjunct Research Associate:
Paul Baker, (M), Ng
Dimitri Bulatov, (D), Zf
Gan Duan, (D), Cql
Charlotte Mack, (D), Pls
Shyam Madhusudhana, (D), Ob

Adjunct Associate Professor:
Lesley Arnold, (D), Zf
Marisa Ioppolo-Armanios, (D), CoAc
Jessica Whiteside, (D), Co
Jianhong (Cecilia) Xia, (D), Zi

Adjunct Professor:
Bill Collins, (D), GiCgGt
Chris Elders, (D), GcoGt
Pete Kinny, (D), Cca
Derek Lichti, (D), Zf
Bob Pidgeon, (D), Cc
Andrew Putnis, (D), Gz
Lorenz Schwark, (D), CoOo
Robert Seggie, (B), GsoGc
Greg Smith, (D), GoCoGq
Peter Teunissen, (D), YdZrf
Simon Wilde, (D), CcGit

Visiting Professor:
Michael Brown, (D), GpiGt

Emeritus Professor:
Merv Lynch, (D), Zr
John Penrose, (D), YrZrOg
Geoff West, (D), Zf

Systems Engineer:
Daniel-Cristian Busan, (B), Zn

Software Developer:
Martin Cupak, (M), Zn

Senior Technical Officer:
Peter Hopper, (B), CaEoRw

Senior Research Technician:
Malcolm Perry, Og

Senior Research Officer:
Josh Beardmore, (D), Ym

Research Assistant:
Jacob Cook, (B), Zn
Fergus Downey, (B), Zn
Chelsea Tay, (B), Zn

Project Manager:
Ben Hartig, (B), Zn

Embedded Software Engineer:
Stuart Buchan, (B), Zn
Tristan Ward, (M), Zn

Dept of Exploration Geophysics (B,M,D)
GPO Box U1987
Perth, WA 6845
p. +61 8 9266-3565
Geophysics-GeneralEnquiries@curtin.edu.au
http://www.geophysics.curtin.edu.au/
f: https://www.facebook.com/ExplorationGeophysicsCurtinUniversity
t: @CurtinGeophys
Enrollment (2014): B: 27 (13) M: 26 (4) D: 28 (6)

Professor:
Boris Gurevich, (D), Moscow (Russia), 1988, Ye
Anton W. Kepic, (D), British Columbia, 1990, YexYg
Maxim Lebedev, (D), Moscow State (Russia), 1990, YxNr

Senior Lecturer:
Vassily Mikhaltsevitch, (D), Kaliningrad State, 1997, Ye
Andrew Peter Squelch, (D), Nottingham (UK), 1998, NmYeZe

Research Fellow:
Michael Carson, (D), Univ Coll (Dublin), 2000, Yx
Aleksandar Dzunic, (M), Belgrade, 1989, YgsYe

Senior Scientist:
Stephanie Vialle, (D), Paris 7, 2008, YxCqNp

Head of Department:
Andrej Bona, (D), Calgary, 2002, Yx

Associate Professor:
Brett D. Harris, (D), Curtin, 2002, HwYve
Roman Pevzner, (D), Moscow State (Russia), 2004, YxxYz
Milovan Urosevic, (D), Curtin, 1998, YerYx

Lecturer:
Robert Galvin, (D), Curtin, 2006, Ye
Stanislav Glubokovskikh, (D), Lomosov (Moscow), 2012, Yxe
Mahyar Madadi, (D), Inst Adv Studies Basic Sci (Zanjan), 1997, Ye
Andrew Pethick, (D), Curtin, 2013, YevYg
Konstantin Tertyshnikov, (D), Curtin, 2014, Ye
Sasha Ziramov, (M), Belgrade, 2005, Ye

Senior Technical Officer:
Dominic J. Howman, (B), Curtin, 1994, Ye

Federation University Australia

Dept of Geology (B,M,D) ● ☒
Federation University
P. O. Box 663
Ballarat, VIC 3353
p. 613 53279354
k.dowling@federation.edu.au
Programs: Bachelor of Geosciences

Flinders University

School of the Environment (B,M,D) ●
GPO Box 2100
Adelaide, SA 5001
p. +61 8 8201 7577
jeanne.young@flinders.edu.au
http://www.flinders.edu.au/science_engineering/environment/
Enrollment (2012): B: 44 (14) M: 15 (5) D: 37 (4)

Professor of Hydrogeology:
Okke Batelaan, (D), Free (Brussels), Hw

Research Fellow:
Peter Cook, (D), Flinders, Hw

ARC Future Fellow:
Adrian Werner, (D), Queensland, Hw

Academic Status:
Nancy Cromar, (D), Napier (UK), Zn

Professor:
Howard Fallowfield, (D), Dundee (UK), Zn
Iain Hay, (D), Washington, 1989, Zyy
Patrick Hesp, (D), Sydney, GmZy
Andrew Millington, (D), Sussex (UK), ZuyZr
Craig Simmons, (D), Adelaide, Hw

Director ARA:
Jorg Hacker, (D), Bonn, ZrAs

Academic Status:
John Edwards, (D), Adelaide, Zn
Andew Love, (D), Hw

Associate Professor:
Erick Bestland, (D), Oregon, GsSa
Beverley Clarke, (D), Adelaide, Zyn
Huade Guan, (D), New Mexico Tech, 2005, HgsHg
Udoy Saikia, (D), Flinders, Zn

DECRA Research Fellow:
Margaret Shanafield, (D), Nevada, Hw

Sessional Lecturer:
Harpinder Sandhu, (D), Lincoln (NZ), Zn
Maria Zotti, Zn

Senior Lecturer:
David Bass, (D), New England, Zy
Kirstin Ross, (D), S Australia, Zn

Adjunct Senior Lecturer:
Simon Benger, (D), Australian Nat, Ziy
John Hutson, (D), Natal, Spc
Vincent Post, (D), Amsterdam, Hw

Adjunct Lecturer:
Stephen Fildes, (M), Adelaide, Zri
Ben van den Akker, (D), Flinders, Zn

Lecturer:
Dylan Irvine, (D), Flinders, Hw
Mark Lethbridge, (D), S Australia, Ziy
Graziela Miot da Silva, (D), Fed Rio Grande do Sul (Brazil), Gug

Michael Taylor, (D), Flinders, Zn
Ilka Wallis, (D), Flinders, HqCg
Harriet Whiley, (D), Flinders, Zn
Academic Status:
Gour Dasvarma, (D), Australian Nat, Zn
Adjunct Professor:
Jim Smith, Zn
Research Fellow:
Eddie Banks, (D), Flinders, Hw
Adjunct Academic Status:
Samantha de Ritter, (M), Flinders, Ze
Academic Status:
Glenn Harrington, (D), Flinders, Hw
Andrew McGrath, (D), ZrAs

Geoscience Australia

Geoscience Australia ☒
GPO Box 378
Canberra, ACT 2601
p. +61 2 6249 9111
ref.library@ga.gov.au
http://www.ga.gov.au/
t: @GeoAusLibrary

Government of South Australia - Department for Energy and Mining

Geological Survey of South Australia ☒
GPO Box 320
Adelaide, SA 5001
p. +61 8 8463 3000
DEM.customerservices@sa.gov.au
http://www.energymining.sa.gov.au
t: @DEM_sagov
Programs: not applicable.
Certificates: not applicable.

James Cook University

Geosciences (B,M,D) O☒
Building 034
College of Science & Engineering
James Cook University
Townsville, QLD 4811
p. 07 4781 6947
eric.roberts@jcu.edu.au
https://www.jcu.edu.au/college-of-science-and-engineering
Enrollment (2016): B: 85 (54) M: 6 (2) D: 25 (7)
Associate Professor:
Eric M. Roberts, (D), Utah, 2005, GsPvCc
Professor:
Michael I. Bird, (D), Australian Nat, 1988, ClsSa
Senior Scientist:
Karen E. Joyce, (D), Quebec, 2005
Associate Professor:
Zhaoshan Chang, (D), Washington State, 2003, EmCce
Carl Spandler, (D), 2005, GxEmCg
Postdoctoral Research Fellow:
Arianne Ford, (D), James Cook, 2008, GqZiEg
Lecturer:
James J. Daniell, (D), Sydney, 2010, GumYr
Jan Marten Huizenga, (D), Vrije (Amsterdam), 1995, GpCgGc
Christa Placzek, (D), Arizona, 2006, CclGe
Ioan Sanislav, (D), James Cook, 2009, GctGp
Peter W. Whitehead, (B), La Trobe, 1986, Ggv
Adjunct Professor:
Robert Henderson, (D), Victoria (NZ), 1967, GtPiGg
Postdoctoral Research Fellow & Lecturer:
Hannah L. Hilbert-Wolf, (D), James Cook, 2016, GsCc

Economic Geology Research Centre (A,B,M,D) O⦸
EGRU
James Cook University
Townsville, QLD 4811
p. +61 7 4781 4726
egru@jcu.edu.au
http://www.jcu.edu.au/economic-geology-research-centre-egru
Programs: Bachelor of Science, Bachelor of Advanced Science Bachelor of Geology, Bachelor of Science Honours, Bachelor Geology Honours, Diploma of Science, Master of Science, Master of Science (Professional), Doctorate (PhD)
Certificates: Professional Development Training relevant to the minerals industry

Macquarie University

Dept of Earth and Planetary Sciences (B,M,D) O☒
North Ryde
Sydney, NSW 2109
p. 61-2-98508426
fse.earthenv-admin@mq.edu.au
http://www.eps.mq.edu.au/
f: https://www.facebook.com/MQeps

Department of Environmental Sciences (B,M,D)
North Ryde, NSW 2109
neil.saintilan@mq.edu.au
http://www.mqu.edu.au/

Monash University

School of Earth, Atmosphere and Environment (B,M,D) ☒
PO Box 28E
Clayton, VIC 3800
p. +61 3 99054884
earth-atmosphere-environment@monash.edu
http://www.earth.monash.edu.au/
Programs: Generic, Earth's physical environment, Earths climate, Geoscience
Professor:
Peter A. Cawood, (D), Sydney, 1980, GtcGg

New South Wales Resources & Energy

New South Wales Resources & Energy
NSW Resource & Energy, PO Box 344
Hunter Regional Mail Centre
, NSW 2310
p. 1300 736 122
geologicalsurvey.info@trade.nsw.gov.au
http://www.resourcesandenergy.nsw.gov.au/

Queensland University of Technology

School of Earth & Atmospheric Sciences (B,M,D) ●⦸
GPO Box 2434
2 George Street
Brisbane, QLD 4001
p. 61 7 3138 2324
sef.easenquiry@qut.edu.au
https://www.qut.edu.au/about/faculty-of-science/school-of-earth-and-atmospheric-sciences
f: https://www.facebook.com/QUTEarthScienceUndergraduate-Page/
Programs: Bachelor of Science (3 yrs) with majors in Earth Science, Environmental Science; Geology Extension Minor Climate Science Minor (from 2022)
Enrollment (2020): B: 32 (16) M: 1 (0) D: 12 (2)
Professor:
Zoran Ristovski, (D), Belgrade, 1996, Asp
Distinguished Professor:
Lidia Morawska, (D), Jagiellonian, 1982, Asc
Professor:
David A. Gust, (D), Australian Nat, 1982, GiCpGt
Balz Kamber, (D), Bern, 1995, GpCgGz
Senior Research Fellow :
Charlotte M. Allen, (D), Virginia Tech, CcaGi
Research Fellow:
Henrietta Cathey, (D), Utah, 2006, Giv
Associate Professor:
Scott E. Bryan, (D), Monash, 1999, GitGs
Senior Lecturer:
Craig Sloss, (D), Wollongong (Australia), 2005, GsrGm
Jessica Trofimovs, (D), Monash, 2003, GsvGu
Research Associate:
David Flannery, (D), New South Wales, 2012, PoCaXg
Senior Lecturer:
Graham Johnson, (D), Queensland Tech, 2005, Asp
Branka Miljevic, (D), Queensland Tech, 2010, Acs

Lecturer:
Patrick C. Hayman, (D), Monash, 2009, EgGvi
David T. Murphy, (D), Queensland, 2002, CgEgCc
Luke Nothdurft, (D), Queensland Tech, 2009, GdCmGg
Christoph Schrank, (D), Toronto, 2009, GcqGt
Emeritus:
John Rigby, Pb
Analytical Laboratory Coordinator :
Wan-Ping (Sunny) Hu, (D), Canterbury (NZ), Ca
Senior Technician - Geology :
Will Stearman, (B), Queensland Tech, 2011, Geg

Southern Cross University

Faculty of Science & Engineering (A,B,M,D) O
P. O. Box 157
Lismore, NSW 2480
p. +61 2 6620 3766
fse@scu.edu.au
https://www.scu.edu.au/science-and-engineering
f: https://www.facebook.com/scu
Programs: BSc. with four specializations in Earth & Environmental Systems, Marine Systems, Forestry Systems and Regenerative Agriculture
Certificates: Diploma of Science
Biogeochemistry:
Bradley Eyre, (D), Queensland Tech, CoGeCs
Professor:
Nicholas J. Ashbolt, (D), Tasmaina, 1985, PoCb
Edward Burton, (D), Southern Cross, 2013, CaHy
Peter Harrison, (D), James Cook, Cm
Renaud Joannes-Boyau, (D), France, CcaCs
Scott Johnston, (D), CaGe
Damien Maher, (D), Sourthern Cross, 2001, GeCs
Amanda Reichelt-Brushett, (D), Southern Cross, Hs
Adjunct:
Graham Jones, (D), James Cook, Ca
Lecturer:
Sumith Pathirana, (D), Kent State, Zr
Emeritus:
Bill Boyd, (D), Glasgow, 1982, Pe

University of Adelaide

School of Earth and Environmental Sciences
Adelaide, SA 5000
p. +61 8 8303 3999
graham.heinson@adelaide.edu.au
http://www.ees.adelaide.edu.au/

Australian School of Petroleum and Energy Resources (B,M,D) O☒
Santos Petroleum Engineering Building
North Terrace
Adelaide, SA 5005
p. 61-8-8313-8000
admin@asp.adelaide.edu.au
https://ecms.adelaide.edu.au/petroleum-engineering/
Programs: Petroleum Geoscience, Petroleum Engineering - Bachelors, Masters and Doctorate
Certificates: Graduate Certificate in Petroleum Geoscience. Graduate Certificate and Graduate Diploma in Petroleum Engineering
Dr:
Kathryn Amos, (D), E Anglia (UK), Gs
Professor:
Pavel Bedrikovetski, (D), Gubkin Russian State Oil and Gas (Moscow), Np
Associate Professor:
Manouchehr Haghighi, (D), S California, Np
Simon Holford, (D), Birmingham, Go
Research Associate:
Alex Badalyan, (D), Grozny State Oil Tech, Np
Sara Borazjani, (D), Adelaide, Np
Themis Carageorgos, (D), Imperial (London), Np
Ulrike Schacht, (D), Kiel, Gd
Carmine Wainman, (D), Adelaide, Grs
Senior Lecturer:
Khalid Amrouch, (D), Pierre & Marie Curie, Gc
Abbas Zeinijahromi, (D), Adelaide, Np
Lecturer:
Mark Bunch, (D), Birmingham, Ye
Maria Gonzalez-Perdomo, (B), Zulia, Np
Alireza Salmachi, (D), Adelaide, Np
Mohammad Sayyafzadeh, (D), Adelaide, Np
Professor - Decision Making:
Stephen Begg, (D), Reading, Eo
Professor:
John Kaldi, (D), Cambridge, Go
Professor :
Peter McCabe, (D), Keele, Go

University of Canberra

Institute for Applied Ecology
Duanne.White@canberra.edu.au
https://www.canberra.edu.au/research/institutes/iae

Faculty of Science and Technology ☒
P. O. Box 1
Belconnen, ACT
SciTech-Dean@canberra.edu.au
http://scides.canberra.edu.au/rehs/

University of Melbourne

School of Earth Sciences (B,M,D)
Melbourne, VIC 3010
p. +61 3 8344 9866
head@earthsci.unimelb.edu.au
http://www.earthsci.unimelb.edu.au

University of New England

Earth Sciences (A,B,M,D) O
Armidale
NSW
2351
Armidale, NSW 2351
p. +61-2-67732101
geology@une.edu.au
http://www.une.edu.au/about-une/academic-schools/school-of-environmental-and-rural-science/research/life-earth-and-environment/e
Associate Professor:
John R. Paterson, (D), Macquarie, 2005, PigPs
Lecturer:
Phil R. Bell, (D), Alberta, 2011, PgvPg
Luke Milan, (D), GtcGp
Nancy Vickery, (D), New England, GgeYg
Adjunct Professor:
Ian Metcalfe, (D), Leeds (UK), 1976, PmGtg
Emeritus:
Paul Ashley, (D), Macquarie, EgGeEm
Peter Flood, (D), EcGt

University of New South Wales

School of Biological, Earth & Environmental Sciences (B,M,D) ☒
UNSW
Sydney, NSW 2052
p. 61 2 93852961
bees@unsw.edu.au
http://www.bees.unsw.edu.au
Programs: Geology
Enrollment (2022): B: 117 (34) D: 28 (6)
Associate Professor:
David R. Cohen, (D), New South Wales, 1990, Cet
Professor:
Michael Archer, (D), Pv
Andy Baker, (D), HwCs
Suzanne J. Hand, (D), Pv
Shawn Laffan, (D), Ziy
Martin van Kranendonk, (D), Queens (Canada), GtPeGd
Dr:
Ian T. Graham, (D), UTS, 1998, Gzx
Hamid Zekri, (D), Isfahan Tech, 2018, Gz
Associate Professor:
Bryce Kelly, (D), New South Wales, 1995, HwAcGe
Paul G. Lennox, (D), Monash, 1985, GcgZe

University of Newcastle

Discipline of Earth Sciences (B,M,D) O
School of Environmental & Life Sciences
University Drive
Callaghan, NSW 2308
p. +61 2 4921 8976
name.surname@newcastle.edu.au
http://www.newcastle.edu.au/school/environ-life-science
Enrollment (2014): B: 132 (121) D: 5 (3)

Associate Professor:
Silvia Frisia, (D), Milan (Italy), 1991, ClPeGd
Phil Geary, (D), UWS, HwSoHw
Gregory Hancock, (D), ZiGmm

Lecturer:
Judy Bailey, (D), Newcastle, 1993, Ec
David Boutelier, (D), Gtc
Alistair Hack, (D), GcCgEm
Anthony Kiem, (D), HqAsZi
Bill Landenberger, (D), Newcastle, 1997, GiCg
Danielle Verdon-Kidd, (D), HqAsZi

University of Queensland

School of Earth Sciences (B,M,D) O
Faculty of Science
Steele Building
Brisbane, QLD 4072
p. +61 7 3365 1180
sees.hos@uq.edu.au
https://sees.uq.edu.au/
Enrollment (2015): B: 106 (57) M: 38 (5) D: 60 (6)

University of South Australia

School of Natural and Built Environments (B) O
GPO Box 2471
Adelaide, SA 5001
Amie.Albrecht@unisa.edu.au
http://www.unisa.edu.au/it-engineering-and-the-environment/natural-and-built-environments/

Research Fellow:
Juliette Woods, (D), Adelaide, Hwq

University of Sydney

School of Geosciences (A,B,M,D) ⊘
Room 348
Madsen Building F09
Sydney, NSW 2006
geosciences.enquiries@sydney.edu.au
https://sydney.edu.au/science/schools/school-of-geosciences.html
f: https://www.facebook.com/sydneyunigeo/
t: @sydneyunigeo

Professor:
Jonathan Aitchison, (D), UNE, 1989, GrtPm

Professor:
Geoffrey L. Clarke, (D), Melbourne (Australia), 1987, Gp
Dietmar Muller, (D), Scripps, 1989, Yrg

University of Tasmania

ARC Center of Excellence in Ore Deposits (M,D) O
Private Bag 79
Hobart, TAS 7001
steve.calladine@utas.edu.au
http://www.utas.edu.au/codes

School of Earth Sciences (B,M,D)
Private Bag 79
Hobart, TAS 7001
p. 61 3 6226 2476
Rose.Pongratz@utas.edu.au
http://www.utas.edu.au/earth-sciences/home

Director:
J Bruce Gemmell, (D), Dartmouth, 1987, Eg
Ross R. Large, (D), New England, Eg

Professor:
David R. Cooke, (D), Monash, Eg
Anthony J. Crawford, (D), Melbourne (Australia), 1983, Git
Jocelyn McPhie, (D), New England, Gv

Associate Professor:
Ron F. Berry, (D), Flinders, Gc

Lecturer:
Garry J. Davidson, (D), Tasmania, Cs
Peter J. McGoldrick, (D), Melbourne (Australia), Ce
Anya Reading, (D), Ysg
Michael Roach, (N), Tasmania, Yg

University of Western Australia

School of Earth & Environment (B,M,D)
M004
35 Stirling Highway
Crawley, WA 6009
p. 6488 1921
enquiry-see@uwa.edu.au
http://www.see.uwa.edu.au/

University of Wollongong

School of Earth, Atmospheric and Life Sciences (B,M,D) ● ☒
Northfields Avenue
Wollongong, NSW 2522
p. 61 2 4221 4419
seals-admin@uow.edu.au
https://www.uow.edu.au/science-medicine-health/schools-entities/school-of-earth-atmospheric-and-life-sciences/
Programs: Earth Science, Geology, Environmental Sciences, Conservation Biology, Archaeoology

Austria

Geological Survey of Austria

Geologische Bundesanstalt von Osterreich ☒
Neulinggasse 38, A 1030
Vienna
p. 43-1-712 56 74
office@geologie.ac.at
https://www.geologie.ac.at/

Karl-Franzens-Universitaet Graz

Inst for Earth Science (A,B,M,D) O
Heinrichstraße 26
Graz 8010
p. +43 316 380-5587
erdwissenschaften@uni-graz.at
http://erdwissenschaften.uni-graz.at/index_en.php
Enrollment (2016): A: 10 (1) B: 40 (35) M: 20 (14) D: 12 (2)

Leopold-Franzens-Universitaet Innsbruck

Inst fuer Geologie
Innrain 52-f
6020 Innsbruck
p. +43 (0) 512 / 507 - 96125
dekanat-geowiss@uibk.ac.at
http://www.uibk.ac.at/fakultaeten/geo_und_atmosphaerenwissenschaften/

Montan Universitaet

Dept of Applied Geological Sciences and Geophysics
Inst. for Prospecting
A-8700
Leoben
geologie@mu-leoben.at

Technische Universitaet Graz

Earth Sciences
Rechbauerstrasse 12
A-8010 Graz
p. +43(0)316 873 6360
martin.dietzel@tugraz.at

University Leoben

Chair of Geology and Economic Geology (B,M,D) ☒
Montauniversitat Leoben
Peter-Tunnerstrasse 5
Leoben A-8700
p. +43 3842 402 6101
geologie@unileoben.ac.at
http://www.unileoben.ac.at/~buero62/geologie/geologie.html
Programs: Applied Geosciences and Geophysics
Enrollment (2018): B: 300 (0) M: 35 (0) D: 4 (0)
Professor:
Frank Melcher, (D), Leoben, 1993, GgEgm
Senior Scientist:
Heinrich Mali, (D), Leoben, GcEgNm
Associate Professor:
Gerd Rantitsch, (D), Graz, Ggq
Assistant Professor:
Phillip Gopon, (D)
Research Associate:
Viktor Erlandsson, (M), Gothenburg, 2018, Gg

University of Innsbruck

Institute of Geology (A,B,M,D) O☒
Innrain 52f
A-6020
Innsbruck 6020
p. +43 512 507-54300
regina.gratzl@uibk.ac.at
http://www.uibk.ac.at/geologie
Programs: Earth Science (Bachelor), Geology (MSc, PhD)

University of Salzburg

Inst fuer Geowissenschaften
Hellbrunner Strasse 34
A-5020 Salzburg
p. +43-662-8044-5200 oder 5400
office.geographie@sbg.ac.at
http://www.uni-salzburg.at/geo

University of Vienna

Dept of Meteorology and Geophysics (B,M,D) ●☒
Althanstrasse 14
A-1090 Vienna
p. 0043 1 4277 53701
img-wien@univie.ac.at
http://imgw.univie.ac.at/en/imgw/
Programs: Master programme "Physics of the Earth", in collaboration between the University of Vienna (Austria) and the Comenius University Bratislava (Slovakia).
Enrollment (2021): M: 6 (0) D: 16 (0)
Univ.-Prof.:
Goetz HR Bokelmann, (D), Princeton, 1992, Ygs

Dept of Geography
Althanstrasse 14
Vienna A-1090
p. 00431427753401
elisabeth.aufhauser@univie.ac.at
http://www.univie.ac.at/Geologie/
Head:
Thilo Hofmann, (D)

Azerbaijan

Baku State University

Faculty of Geology
23 Z. Khalilov Street
370145 Baku
m.mahluga@rambler.ru
http://www.ceebd.co.uk/ceeed/un/az/az003.htm

Bangladesh

Jahangirnagar University

Dept of Geological Sciences
University Campus, Savar, Dhaka
Dhaka
p. 0088 2 7791045-51 (Ext. 1402)
rabiulju@gmail.com
http://www.juniv.edu/home.php?pg=faculty_science

Rajshahi University

Dept of Geology & Mining
Motihar, Rajshahi
Rajshahi
p. +880 721-750041
chair.geology@ru.ac.bd
http://www.ru.ac.bd/geol

University of Dhaka

Department of Geology (B,M,D) ☒
Curzon Hall Campus
Dhaka 1000
p. (880) 2- 966 1920-73 Ext. 7301
geology@du.ac.bd
http://www.du.ac.bd/academic/department_item/GLG
Enrollment (2017): B: 190 (37) M: 85 (31) D: 2 (0)
Chair:
M Aziz Hasan, (D), KTH, Stockholm, 2008, CqHwGg
Professor:
Kazi Matin Ahmed, (D), UK, 1994, HwGgHg

Belgium

Faculte Polytechnique de Mons

Dept of Geology
Fue de Houdain 9
7000 Mons
yves.quinif@fpms.ac.be

Faculte Univ Catholique de Mons

Dept of Geology
Chaussee de Binche 151
B-7000 Mons

Facultes Univ Notre Dame de la Paix

Dept Geologie
Rue de Bruxelles 61
B-5000 Namur
vincent.hallet@fundp.ac.be

Geological Survey of Belgium

Service geologique de Belgique
Department VII
of the Royal Belgian Institute of Natural Sciences (RBINS)
13, Rue Jenner
1000 Brussels
p. +32 (0)2.788.76.61
earthlife@naturalsciences.be
http://www.naturalsciences.be/geology/

Ghent University

Dept of Geology (B,M,D) ●⃠
Krijgslaan 281-S8
Gent B-9000
p. +32 9 264 45 94
we13@ugent.be
https://www.ugent.be/we/geologie/nl/nl
f: https://www.facebook.com/GeologieUGent
t: @GeologyUGent
Programs: Geology; Marine and Lacustrine Science and Management; Physical Land Resources
Enrollment (2021): B: 57 (13) M: 33 (16) D: 58 (8)
Head:
Marc De Batist, (D), Ghent, 1989, GnuGs

Professor:
Veerle Cnudde, (D), Ghent, 2005, EsHt
Stephen Louwye, (D), Ghent, PmZgOo
Kristine Walraevens, (D), Ghent, 1987, HwCqYu
Associate Professor:
Johan De Grave, (D), Ghent, CcZgGx
Stijn Dewaele, (D), KU Leuven, 2004, EgGzCg
David Van Rooij, (D), Ghent, YrGu
Thijs Vandenbroucke, (D), Ghent, PsmPc
Assistant Professor:
Thomas Hermans, (D), Liege, HwYeu
Maarten Van Daele, (D), Ghent, 2013, GsRnZc

Katholieke Universiteit Leuven

Dept of Earth and Environmental Science (B,M,D)
Department of Earth & Environmental Sciences
Redingenstraat 16
B-3000 Leuven-Heverlee
p. (+32) 016 321450
erik.mathjis@ees.kuleaven.be
http://ees.kuleuven.be/
Professor:
Patrick Degryse, (D), GaCeGf
Jan Elsen, (D), Gz
Philippe Muchez, (D), 1988, EmCgGd
Manuel Sintubin, (D), GtcGa
Robert Speijer, (D), Utrecht, 1994, PmcOo
Rudy Swennen, Gso
Associate Professor:
Okke Batelaan, (D), Hw
Sarah Fowler, (D), Gip
Assistant Professor:
Marijke Huysmans, (D), Hw

Earth and Environmental Sciences (B,M,D) ☒
Celestijnenlaan 200E PO 2408
Leuven BE-3001
p. +32 16 327580
gert.verstraeten@kuleuven.be
http://ees.kuleuven.be
Programs: Geography; Geology

Universite Catholique de Louvain

Dept of Geology
Bâtiment Mercator
place Louis Pasteur 3
B-1348 Louvain-la-Neuve
p. (32) 10 47 32 97
sophie.vanwambeke@uclouvain.be
http://www.uclouvain.be/geo.html

Universite de Liege

Dept of Geology (B,M,D)
Batiment B18 (secretariat)
Sart Tilman
Liege B-4000
p. 32 4 366 22 51
TH.Billen@ulg.ac.be

http://www.ulg.ac.be/geolsed/geologie
Chair:
Frederic P. Boulvain, (D), Brussels, 1990, GdgGs
chargé de cours:
Nathalie Fagel, (D), GsuGe
Hans-Balder Havenith, (D), Liege, Ge
Professor:
Andre-Mathieu Fransolet, (D), Liege, Gz
Emmanuelle Javaux, (D), Dalhousie, 1999, Pl
Jacqueline Vander Auwera, (D), Louvain-la-Neuve, 1988, Giv
Emeritus:
Edouard Poty, (D), Liege, 1981, PiGsPe

Universite Libre de Bruxelles (ULB)

Département des Sciences de la Terre et de l'Environnement
50, Ave. F. Roosevelt
Brussels 1050
Julie.Paraire@ulb.ac.be

University of Mons

Dept of Geology and Applied Geology (M,D) ○
rue de Houdain, 9
Mons
gfa@umons.ac.be

Vrije University Brussel

Analytical-Environmental and Geo-Chemistry (D) ☒
AMGC-WE-VUB
Faculty of Sciences
Pleinlaan 2
Brussels 1050
p. +3226293394
phclaeys@vub.be
https://we.vub.ac.be/en/analytical-environmental-and-geo-chemistry
f: https://www.facebook.com/amgcvub/
Enrollment (2021): D: 26 (5)
Head:
Philippe Claeys, (D), California (Davis), 1993, CgaXc

Bolivia

Universidad Mayor de San Andres

Dept of Geology
P. O. Box 12198
Campus Universitario Cota Cota, Calle 27
La Paz
p. (591-2) 2441983
webmaster@umsa.bo
http://www.geologia.umsa.bo/

Botswana

Geological Survey of Botswana

Geological Survey of Botswana
Khama One Avenue
Plot 1734
Lobatse
p. +267 5330327
http://www.gov.bw/en/Ministries

University of Botswana

Dept of Geology (B)
4775 Notwane Rd.
Private Bag UB00704
Gaborone
p. (267)355 2529
geology@mopipi.ub.bw
http://www.ub.bw/home/ac/1/fac/1/dep/79/Geology/

Dept of Environmental Science (B)
4775 Notwane Road
Private Bag UB 0022
Gaborone
p. (267) 355-0000
Sedilamoya.Pansiri@mopipi.ub.bw
http://www.ub.bw/learning_faculties.cfm?pid=588

Brazil

Federal University of Bahia Geophysics

CPGG
Salvador
BA
geofisic@ufba.br
http://www/pppg.ufba.br/

Geological Survey of Brazil

Servico Geologico do Brasil
Av. SGAN- Quadra 603 - conjunto J
Parte A - 1º andar
Brasilia - DF 70830-030
seus@cprm.gov.br
http://www.cprm.gov.br/

Universidad Federal de Rio Grande do Sul

Institute de Geociencia
Avenida Bento Gonçalves, 9500
Porto Alegre, RS - 91.501-970
p. +55 51 3308-6337
igeo@ufrgs.br
http://www.ufrgs.br/english/the-university/institutes-faculties-and-schools/institute-of-geoscience

Universidade de Brasília

Instituto de Geociências
Campus Universitário Darcy Ribeiro ICC - Ala Central
CEP 70.910-900 - Brasilia DF
Caixa Postal 04465.
CEP 70919-970
p. +61 3307-2433
igd@unb.br
http://www.igd.unb.br/

Universidade de Sao Paulo

Inst de Geociencias
Rua do Lago
562 Cidade Universitaria
05508-080 São Paulo
p. 63.025.530/0007-08
gmgigc@usp.br
http://www.igc.usp.br/

Inst Oceanografico
Cidade Universitaria, Av. Prof. Luciano Gualberto, Travessa 3 no 380, 05508-900 Sao Paulo - SP

Universidade do Vale do Rio Dos Sinos

Inst de Geociencias
Av. Unisinos, 950 - Cristo Rei, 93 000 Sao Leopoldo - RS

Universidade Federal da Bahia

Inst de Geociencias
Rua Augusto Viana s/n - Canela, 40 410 Salvador - BA
olivia@ufba.br

Universidade Federal de Minas Gerais

Inst de Geociencias
Av. Antonio Carlos 6.627
Pampulha , 31 270 Belo Horizont
p. 55(31) 3409-5420
dir@igc.ufmg.br
http://www.igc.ufmg.br/

Universidade Federal de Ouro Preto

Dept de Geologia
Campus Morro do Cruzeiro
35 400 Ouro Preto
p. (31) 3559-1600
geologia.degeo@ufop.edu.br
http://www.degeo.ufop.br/

Universidade Federal de Pernambuco

Centro de Tecnologia - Geociencias
Av. Agamenon Magalhaes s/n - Santo Amaro, 50 000 Recife - PE
p. (81) 2126.8105
secretaria.proacad@ufpe.br
http://www.ufpe.br/proacad/index.php?option=com_content&view=article&id=150&Itemid=138

Universidade Federal do Ceara

Inst de Geociencias
Av. Da Universidade, 2853 - Benfica, 60 000 Fortaleza - CE

Universidade Federal do Para

Inst. de Geociencias (B,M,D) ○⊘
Augusto Correa Street, 01
Campus Guamá
Belem, Pará 66075-110
p. +55 91 3201 7107
dirig@ufpa.br
http://www.ig.ufpa.br/
Programs: We have at the Geosciences Institute (IG): bachelor's degree in Geophysics, Geology, Meteorology and Oceanography; Master's and Doctoral's programs in Geophysics, Geology and Geochemistry; Professional Masters in Environmental Sciences, Water Resources, Natural Hazards and Risk Management in the Amazon, Environmental Sciences Teaching; Specialization in Water and Environmental Management, Mine Geology and Open-pit mining, Sedimentary Basin Analysis.

Universidade Federal Fluminense

Dept de Geociencias
Av. Gal. Milton Tavares de Souza
p. +55 (21) 2629-5951
gge@vm.uff.br
http://www.uff.br/degeografia/

Universidade Federal Rural do Rio de Janeiro

Inst de Geociencias
Km 47 da Antigua Rodovia Rio/Sao Paulo - Seropedico, 23 460 Itaguai - RJ

University of Campinas Geoscience Institute

Inst de Geosciencias
Caixa Postal: 6152
13083-970
Campinas
secretaria.prpg@reitoria.unicamp.br
http://www.ige.unicamp.br/

Bulgaria

Bulgarian Academy of Sciences

Geological Institute (D) ☒
Acad.G.Bonchev st. bl.24
Sofia 1113
p. +0359 2 8723 563
geolinst@geology.bas.bg
http://www.geology.bas.bg/
Programs: Paleontology and stratigraphy, Geotectonics, Regional Geology, Mineralogy, Petrology, Geochemistry, Lithology, Hydrogeology, Engineering geology,
Professor:
Radoslav Alexandrov Nakov, (D), Geological Inst, 1989, GtEgm
Professor:
Iliana Boncheva, (D)
Professor Dr, DSc:
Kristalina Christova Stoykova, (D), Bulgaria Acad Sci, 2008, PemPi
Professor:
Thomas Noubar Kerestedjian, (D), Bulgaria Acad Sci, 1990, GzeEg

Mining and Geology University

Geology Dept
Studentski Grad
Sofia 1756
p. 8060221
dekangpf@mgu.bg

Ministry of Environment and Water

Ministry of Environment and Water
22 Maria Louiza Blvd.
Sofia, 1000
minister@moew.government.bg
http://www.moew.government.bg/

Sofia University St. Kliment Ohridski

Dept of Geology and Paleontology
1504 Sofia
15 Tsar Osvobodtel Blvd.
Geology of Fossil, Fuel, Organic Petrology, Organic Geochemistry room 276
gpff@gea.uni-sofia.bg
http://www.uni-sofia.bg/newweb/faculties/geo/departments/geo

Dept of Cartography & GIS (B,M,D)
1504 Sofia
15 Tsar Osvobodtel Blvd.
GIS, General Physical Geography, Landscape Ecology, Agroecology
Sofia, Sofia 1504
p. 9308261
popov@gea.uni-sofia.bg
http://www.gis.gea.uni-sofia.bg/

University of Mining and Geology, St. Ivan Rilski

Faculty of Geology
Studentski grad
1100 Sofia
dekangpf@mgu.bg
http://www.mgu.bg/frame/html

Burkina Faso

Universite de Ouagadougou

L'Unité de Formation et de Recherche en Sciences de la Vie et de la Terre (UFR/SVT) (B)
03 BP 7021, Ougadougou 03
p. +226 50-30-70-64/65
webmaster@univ-ouaga.bf
http://www.univ-ouaga.bf.html/formations/ufr_SVT/frFormtnsS-VTdip1.html

Burundi

Universite du Burundi

Dept of Earth Sciences (B) ○
B.P. 2700
Bujumbura, Burundi
p. 00257 22 22 55 56
webmaster@ub.edu.bi
http://www.ub.edu.bi/

Faculté des Sciences
B.P. 2700
Bjumbura
p. 0025722222059
http://www.ub.edu.bi/ub-fac6.php

Cameroon

Bamenda University of Science and Technology

Dept of Geology (B)
PO Box 277
Bamenda, NW Province
p. +237 7726 1789
http://www.bamendauniversity.com/

Universite de Buea

Dept of Geology and Environmental Science (B)
PO Box 63
Buea, South West Province
p. 237-332-2134
vpktitanji@yahoo.co.uk
http://ubuea.cm/

Universite de Douala

Faculty of Sciences (B,M,D)
BP 2701
Douala, Cameroun
p. (237) 33 40 75 69
daac@univ-douala.com
http://www.facsciences-univ-douala.cm/index.php/contact

Universite de Dschang

Faculte D'Agronomie et des Sciences Agricoles (B,M)
POB 96
Dschang
p. (237) 33 45 15 66
udsrectorate@univ-dschang.org

Universite de Ngaoundere

School of Geology and Mining (B)
BP 454
Ngaoundere
p. +(237) 225 2767
http://www.univ-ndere.cm/index.php?LANG=EN&ETS=0433A030F35FE61&RUB=E2C0BE24560D78C

Universite de Yaounde 1

Dept des Sciences de la Terre (B)
BP 812
Yanounde
p. (237) 222 56 60
doyen@facsciences.uy1.cm
http://www.facsciences.uninet.cm/act_aca_dst.html

Canada

Royal Tyrrell Museum of Palaeontology ☒
P.O. Box 7500
Drumheller, AB T0J 0Y0
p. (403) 823-7707
tyrrell.info@gov.ab.ca
http://www.tyrrellmuseum.com/
Director, Preservation and Research:
Craig S. Scott, (D), Alberta, 2008, Pv
Curator:
James D. Gardner, (D), Alberta, 2000, Pv
Donald M. Henderson, (D), Bristol, 2000, Pv
Curator:
Francois Therrien, (D), Johns Hopkins, 2004, Pv
Postdoctoral Fellow:
Femke M. Holwerda, (D), Ludwig-Maximilians, 2019, Pv
Curator:
Caleb M. Brown, (D), Toronto, 2013, Pv

Aboriginal Affairs and Northern Development Canada

Mineral Resources Division
Mineral Resources Division
PO Box 100
Iqaluit, NU X0A 0H0
p. (867) 975-4293
nrcan.nunavutgeoscience-geosciencenunavut.rncan@canada.ca
http://nunavutgeoscience.ca/

Acadia University

Dept of Earth and Environmental Science (B,M) ●☒
Huggins Science Hall
12 University Avenue HSH 327
Wolfville, NS B4P 2R6
p. (902) 585-1208
ees@acadiau.ca
http://ees.acadiau.ca/
f: https://www.facebook.com/Acadia-Department-of-Earth-and-Environmental-Science-335568173300748/
Programs: Geology; Environmental Geoscience; Environmental Science
Enrollment (2022): B: 0 (19) M: 4 (10)
Head:
Ian S. Spooner, (D), Calgary, 1994, Ge
Professor:
Sandra M. Barr, (D), British Columbia, 1973, GitGg
Nelson O'Driscoll, (D), Ottawa, 2003, CgGgZn
Clifford R. Stanley, (D), British Columbia, 1988, Cea
Deanne van Rooyen, (D), Gpt
Associate Professor:
Alice Cohen, (D), British Columbia, 2011, ZnnZn
Assistant Professor:
Morgan E. Snyder, (D), Calgary, 2019, Gc
Instructor:
David W.A. McMullin, (D), British Columbia, 1991, Ze
Emeritus:
Robert P. Raeside, (D), Calgary, 1982, GptGc

CANADA

Geology Tech:
Lucas Evans, Gg

Brandon University

Dept of Geology (B,M) ○☒
270 18th Street
Brandon, MB R7A 6A9
p. (204) 727-9677
mumin@brandonu.ca
http://www.brandonu.ca/geology
Programs: Geology
Mineral Resources
Energy Resources
Environmental Geoscience
Paleontology
Enrollment (2020): B: 26 (5)

Professor:
Rong-Yu Li, (D), Alberta, 2002, PgmGg
A. Hamid Mumin, (D), W Ontario, 1994, EgGzt
Simon A. J Pattison, (D), McMaster, 1992, GsoGd
Alireza Somarin, (D), New England, 1999, Gig

Emeritus:
Robert K. Springer, (D), California (Davis), 1971, Gi
Harvey R. Young, (D), Queen's, 1973, Gd

Laboratory Director:
Ayat Baig, (M), Lakehead, 2016, EdmEg

Other:
Paul Alexandre, (D), Loraine, 2000, EgCgGz

Brock University

Dept of Earth Sciences (B,M) ●
1812 Sir Isaac Brock Way
St. Catharines, ON L2S 3A1
p. 905 688-5550
earth@brocku.ca
http://www.brocku.ca/mathematics-science/departments-and-centres/earth-sciences
f: https://www.facebook.com/Brock-University-Department-of-Earth-Sciences-101358005388895
Programs: Earth Sciences; Environmental Geoscience
Enrollment (2021): B: 70 (14) M: 13 (0)

Professor:
Uwe Brand, (D), Ottawa, 1979, ClsAs
Richard J. Cheel, (D), McMaster, 1984, Gs
Frank Fueten, (D), Toronto, 1989, Gc
Martin J. Head, (D), Aberdeen, 1990, Pl
Francine G. McCarthy, (D), Dalhousie, 1992, Pm
John Menzies, (D), Edinburgh, 1976, GlmGs

Associate Professor:
Gregory C. Finn, (D), Memorial, 1989, Gx
Daniel P. McCarthy, (D), Saskatchewan, 1993, Gl
Mariek Schmidt, (D), Oregon State, 2005, GivXg

Adjunct Professor:
Paul Budkewitsch, (M), Toronto, 1990, GeZrHq
Phil McCausland, (D), Gz
Andrew W. Panko, (D), McMaster, 1985, Ge

Cape Breton University

Math, Physics, Geology (B)
P.O. Box 5300
Sydney, NS B1P 6L2
p. (902) 539-5300
fenton_isenor@cbu.ca

Instructor:
Fenton M. Isenor, (M), Acadia, 2000, GgNgGm

Emeritus:
Erwin L. Zodrow, (D), Pi

Capilano University

Geology Dept (A,B)
2055 Purcell Way
North Vancouver, BC V7J 3H5
p. (604) 986-1911
sciences@capilanou.ca
http://www.capilanou.ca/programs/geology.html

Head:
Dileep J A Athaide, (M), British Columbia, 1974, GgZge

Professor:
Jennifer Getsinger, (D), British Columbia

Carleton University

Dept of Earth Sciences (B,M,D) ●
1125 Colonel By Drive
Ottawa, ON K1S 5B6
p. (613) 520-5633
earth.sciences@carleton.ca
http://earthsci.carleton.ca/
f: https://www.facebook.com/pages/Department-of-Earth-Sciences-Carleton-University/510369329037382
t: @ErthSciCarleton

Professor:
Keith Bell, (D), Oxford, 1964, Cc
Brian L. Cousens, (D), California (Santa Barbara), 1990, GiCgc
George R. Dix, (D), Syracuse, 1988, Gsd
James E. Mungall, (D), McGill, 1993, GiEgCg
R. Timothy Patterson, (D), California (Los Angeles), 1986, Pm
Giorgio Ranalli, (D), Illinois, 1970, Yg
Claudia Schroder-Adams, (D), Dalhousie, 1986, Pm
George B. Skippen, (D), Johns Hopkins, 1966, Cg
Richard P. Taylor, (D), Leicester (UK), 1980, Cg
David H. Watkinson, (D), Penn State, 1965, Em

Associate Professor:
Gail M. Atkinson, (D), W Ontario, 1993, Ne
John Blenkinsop, (D), British Columbia, 1972, Cc
Sharon D. Carr, (D), Carleton, 1990, Gt
Fred A. Michel, (D), Waterloo, 1982, Hw

Lecturer:
Ildi Munro, (M), Waterloo, 1975, Pg

Adjunct Professor:
Robert Berman, (D), British Columbia, 1983, Gp
Steve L. Cumbaa, (D), Florida, 1975, Pv
J. Allan Donaldson, (D), Johns Hopkins, 1960, Gs
T. Scott Ercit, (D), Manitoba, 1986, Gz
Harold Gibson, (D), Carleton, 1990, Em
Simon Hanmer, (D), Chelsea (UK), 1977, Gc
Mark D. Hannington, (D), Toronto, 1989, Em
Jarmila Kukalova-Peck, (D), Charles (Prague), 1962, Pi
Dale A. Leckie, (D), McMaster, 1983, Gs
R. R. Rainbird, (D), Western, 1991, Gs

Emeritus:
F. K. North, (D), Oxford, 1951, Go

Dalhousie University

Dept of Earth Sciences (B,M,D) ●☒
Halifax, NS B3H 3J5
p. (902) 494-2358
earth.sciences@dal.ca
http://earthsciences.dal.ca
Programs: Earth Sciences (minor,B,M,D); Geography (minor)
Certificates: Geographic Information Science; Materials Science; Science Leadership and Communication
Enrollment (2017): B: 94 (39) M: 24 (9) D: 13 (2)

Chair:
James M. Brenan, (D), Rensselaer, 1990, CptEm

Professor:
John C. Gosse, (D), Lehigh, 1994, CcGmRn
Djordje Grujic, (D), ETH (Switzerland), 1992, Gt
Mladen Nedimovic, (D), Toronto, 2000, Ysg
Grant D. Wach, (D), Oxford, 1993, GorGs

Retired:
Nicholas Culshaw, (D), Ottawa, 1983, Gc

Associate Professor:
Isabelle Coutand, (D), Rennes, Gg
Yana Fedortchouk, (D), Victoria, 2006, CpEdGi
Shannon Sterling, (D), Duke, 2005, HgCbHs

Assistant Professor:
Lawrence Plug, (D), Alaska (Fairbanks), 2000, Gm
Owen Sherwood, (D), Dalhousie, 2006, CsOoGo

Research Associate:
Alan Ruffman, (M), Dalhousie, 1966, Ys

Senior Instructor:
Mike Young, (M), Queens, 2003, GgcGp

Retired:
Charles C. Walls, (M), Dalhousie, 1996, Zi

Instructor:
Richard Cox, (D), Memorial, 1999, GzEdGg
Chris Greene, (D), Ryerson, 2015, Zi
Honorary Research Associate:
Prasanta Mukhopadhyay, (D), Jadavpur, 1971, Ec
Adjunct Professor:
Sandra Barr, (D), British Columbia, 1973, Gx
D. Brown, (B), Dalhousie
John Calder, (D), Dalhousie, 1991, Ec
C. Campbell, (D), Dalhousie
T. Claire, (D), McMaster
D Barrie Clarke, (D), Edinburgh, Gi
M. Deptuck, (D), Dalhousie
Jarda Dostal, (D), McMaster, 1974, Cg
A. Dyke, (D), Colorado
T. J. Fedak, (D), Dalhousie
R. Fensome, (D), Nottingham (UK)
D. Forbes, (D), British Columbia
M. Fowler, (D), Newcastle upon Tyne
D.G. Froese, (D), Calgary
C. Gerbi, (D), Maine
D. Gibson, (D), Simon Fraser
J. Hanley, (D), Toronto
C. Keen, (D), Cambridge
Lisa M. Kellman, (D), Quebec (Montreal), 1998, Cg
Y. Kettanah, (D), Southampton (UK)
E. Kirby, (D), MIT
T. Lakeman, (D), Alberta
M. Lavigne, (D), Yale
K. E. Louden, (D), MIT
J. Marsh, (D), Maine
T. Martel, (D), Dalhousie
Michael Melchin, (D), Western, 1987, Pg
D. Mosher, (D), Dalhousie
Peta J. Mudie, (D), Dalhousie, 1980, Pl
J. E. Mungall, (D), McGill
J. B. Murphy, (D), McGill
J. Brendan Murphy, (D), McGill, 1982, Gx
Michael Parsons, (D), Stanford, Gg
David Piper, (D), Cantab, 1969, Gs
P. Pufhal, (D), British Columbia, 2002
F. W. Richards, (M), Imperial Coll (UK)
D. Risk, (D), Dalhousie
Cliff Shaw, (D), W Ontario
J. Shimeld, (M), Dalhousie
D. W. Simpson, (D), Australian Nat
S. Swinden, (D), Memorial
H. Vincent, (D), Dalhousie
J. Waldron, (D), Edinburgh
C. Warron, (D), Oxford
T. Webster, (D), Dalhousie
D. M. Whipp, (D), Michigan
B. Wilson, (D), Wales
Emeritus:
H.B. S. Cooke, (D), Witwatersrand, 1947, Pv
Martin R. Gibling, (D), Ottawa, 1978, Gs
G. Clinton Milligan, (D), Harvard, 1961, Gc
Peter H. Reynolds, (D), British Columbia
Patrick J.C. Ryall, (D), Dalhousie, 1974, Yg
David B. Scott, (D), Dalhousie, 1977, Pm
Marcos Zentilli, (D), Queen's, 1974, Eg
University Teaching Fellow:
Anne-Marie Ryan, (D), Dalhousie, 2006, GeZeCg
Administrator:
Ann Bannon, (B), Dalhousie, 2001, Zn
Cooperating Faculty:
Christopher Beaumont, (D), Dalhousie, 1973, Ys

Dept of Oceanography (B,M,D)
Life Sciences Centre
1355 Oxford Street
PO Box 15000
Halifax, NS B3H 4R2
p. (902) 494-3557
oceanography@dal.ca
http://oceanography.dal.ca/
Enrollment (2015): M: 22 (6) D: 31 (2)
Chair:
Paul S. Hill, (D), Washington, 1992, Gs
Professor:
Christopher Beaumont, (D), Dalhousie, 1973, Yg
Bernard P. Boudreau, (D), Yale, 1985, OcCml
Katja Fennel, (D), Rostock, 1998, Ob
Jonathan Grant, (D), South Carolina, 1981, Ob
Alex E. Hay, (D), British Columbia, 1981, OpnGu
Dan Kelley, (D), Dalhousie, 1986, Op
Markus Kienast, (D), British Columbia, 2002, Cs
Hugh MacIntyre, (D), Deleware, 1996, Ob
Anna Metaxas, (D), Dalhousie, 1994, Ob
Jinyu Sheng, (D), Memorial, 1991, Op
Christopher T. Taggart, (D), McGill, 1986, Ob
Helmuth Thomas, (D), Rostock, 1997, Oc
Keith R. Thompson, (D), Liverpool (UK), 1979, Op
Douglas Wallace, (D), Dalhousie, 1985, Oc
Associate Professor:
Tetjana Ross, (D), Manitoba, 2003, Op
Assistant Professor:
Christopher Algar, (D), Dalhousie, 2009, Obc
David R. Barclay, (D), California (San Diego), 2011, OpNo
Carolyn Buchwald, (D), MIT/WHOI, 2013, Oc
Stephanie Kienast, (D), British Columbia, 2002, Gu
Eric Oliver, (D), Dalhousie, 2011, Op
Lecturer:
Barry R. Ruddick, (D), MIT, 1977, Op
Adjunct Professor:
Mark Baumgartner, (D), Oregon, 2002, Ob
Susanne Craig, (D), Strathclyde, 2000, Op
Peter Cranford, (D), Dalhousie, 1998, Ob
Claudio DiBacco, (D), California (San Diego), 1999, Ob
Dale Ellis, (D), McMaster, 1976, Op
Kenneth Frank, (D), Toledo, 1978, Ob
Richard Greatbatch, (D), Cambridge, 1981, Op
David Greenberg, (D), Liverpool (UK), 1975, Op
David Hebert, (D), Dalhousie, 1988, Op
Paul Hines, (D), Bath, 1989, Op
Bruce D. Johnson, (D), Dalhousie, 1979, Oc
Sebastian Krastel, (D), Kiel, 1999, Yr
William K. W. Li, (D), Dalhousie, 1978, Ob
Keith E. Louden, (D), MIT, 1976, Yr
Youyu Lu, (D), Victoria, 1997, Op
Timothy Milligan, (M), Dalhousie, 1997, Gs
David C. Mosher, (D), Dalhousie, 1993, Ou
Andreas Oschlies, (D), Kiel, 1994, Ob
William Perrie, (D), MIT, 1979, OpAm
David J.W. Piper, (D), Cambridge, 1969, GusGo
Harold C. Ritchie, (D), McGill, 1982, Am
Peter C. Smith, (D), MIT/WHOI, 1973, Op
Ulrich Sommer, (D), Vienna, 1977, Ob
Toste Tanhua, (D), Goteborg (Sweden), 1997, Oc
Emeritus:
Anthony J. Bowen, (D), California (San Diego), 1967, On
John J. Cullen, (D), California (San Diego), 1980, Ob
Robert O. Fournier, (D), Rhode Island, 1967, Ob
Marlon R. Lewis, (D), Dalhousie, 1984, Ob
Eric L. Mills, (D), Yale, 1964, ObZnn
Robert M. Moore, (D), Southampton, 1977, Oc

Douglas College

Dept of Earth and Environmental Sciences, Faculty of Science and Technology (A) ●☒
P.O. Box 2503
New Westminster, BC V3L 5B2
p. (604) 527-5400
vigourouxcaillibotn@douglascollege.ca
http://www.douglascollege.ca/programs-courses/faculties/science-technology/earth-and-environmental-sciences, Programs: Earth and Environmental Science (A); Geological Resources Diploma;, Environmental Science (A)
Enrollment (2019): A: 39 (0)
Chair:
Nathalie Vigouroux-Caillibot, (D), Simon Fraser, 2011, GviCg
David C. Waddington, (M), Queen's (Canada), 1991, ZgGuEd
Lab Facilitator:
Denis Beausoleil, (M), Victoria, GgZg
Corinne Griffing, (D), Simon Fraser, 2018, GgeGm
Christine Shiels, (M), Saskatchewan, 2017, GgxZg

Instructor:
Reid Staples, (D), Simon Fraser, 2014, GpxGg
Selina Tribe, (D), Simon Fraser, 2004, EgGge
Derek Turner, (D), Simon Fraser, 2014, GemGl
Emeritus:
Michael C. Wilson, (D), Calgary, 1981, GaPgGs

Lakehead University

Geology (B,M) ● ☒
955 Oliver Road
Thunder Bay, ON P7B 5E1
p. (807) 343-8461
kristine.carey@lakeheadu.ca
https://www.lakeheadu.ca/academics/departments/geology
f: https://www.facebook.com/lakeheaduniversity
t: @mylakehead
Programs: The Department of Geology offers programs leading to Bachelor and Master of Science degrees. These undergraduate programs are offered in Geology and Earth Science majors. A Geography major combined with a Geology minor and an HBESc in Earth Science are available. Geology and Earth Science programs are also offered concurrently with a BEd from the Department of Undergraduate Studies in Education.
Administrative Assistant: Kristine M. Carey
Enrollment (2021): B: 27 (0) M: 16 (0)
Professor:
Philip W. Fralick, (D), Toronto, 1985, Gs
Mary Louise Hill, (D), Princeton, 1985, Gc
Peter N. Hollings, (D), Saskatchewan, 1998, Eg
Chair:
Andrew G. Conly, (D), Toronto, 2003, EgCgGz
Associate Professor:
Amanda Diochon, (D), Dalhousie, 2009, Ge
Shannon Zurevinski, (D), Alberta, 2009, Gz
Assistant Professor:
Noah J. Phillips, (D), McGill, 2020, GcNrZm
Emeritus:
Stephen A. Kissin, (D), Toronto, 1974, Em
Roger H. Mitchell, (D), McMaster, 1969, Gi
Geology Technician:
Kristi Tavener, Zn
Jonas K. Valiunas, (D), Lakehead, 2017, Zm

Laurentian University, Sudbury

Harquail School of Earth Sciences (B,M,D) ●
Ramsey Lake Road
Sudbury, ON P3E 2C6
p. (705) 675-1151 Ext. 6575
hes@laurentian.ca
http://hes.laurentian.ca/
Programs: Geology
Administrative Assistant: Roxane J. Mehes
Enrollment (2022): B: 57 (22) M: 56 (23) D: 29 (10)
Director:
Pedro J. Jugo, (D), Alberta, 2003, EgCpGi
Director, Metal Earth:
Ross Sherlock, (D), Waterloo, 1993, Eg
Professor:
Daniel J. Kontak, (D), Queens, 1985, Eg
Bruno Lafrance, (D), New Brunswick, 1990, GcEmGt
Andrew M. McDonald, (D), Carleton, 1992, Gz
Richard S. Smith, (D), Toronto, Ye
Elizabeth C. Turner, (D), Queen's (Canada), 1999, GrdEm
Associate Professor:
Stéphane Perrouty, (D), Toulouse, 2012, EgGcYg
Doug Tinkham, (D), Alabama, 2002, Gp
Emeritus:
Anthony E. Beswick, (D), London (UK), 1965, Gi
Paul Copper, (D), Imperial Coll (UK), 1965, PieZc
Harold L. Gibson, (D), Carleton, 1990, EmGv
Richard James, (D), Manchester (UK), 1967, GpiEm
Reid R. Keays, (D), McMaster, 1968, Em
Michael Lesher, (D), W Australia, 1984, Em
Darrel Long, (D), W Ontario, 1976, GsaGl
Don H. Rousell, (D), Manitoba, 1965, Gc
Robert E. Whitehead, (D), New Brunswick, 1973, Ce
On Leave:
Alessandro Ielpi, (D), Siena, 2013, GsmGr

Related Staff:
Tobias Roth, (M), Karlsruhe Inst, 2006
Cooperating Faculty:
Phillips C. Thurston, (D), Western, 1981, Zg

Manitoba Museum

Dept. of Geology & Paleontology ☒
190 Rupert Avenue
Winnipeg, MB R3B 0N2
p. (204) 956-2830
gyoung@manitobamuseum.ca
http://www.manitobamuseum.ca
Curator:
Graham A. Young, (D), New Brunswick, 1988, PieRc

McGill University

Dept of Atmospheric & Oceanic Sciences (B,M,D) ☒
805 Sherbrooke Street West
Room 945
Montreal, QC H3A 0B9
p. (514) 398-3764
john.gyakum@mcgill.ca
http://www.mcgill.ca/meteo
Administrative Assistant: Lucy Nunez
Enrollment (2017): B: 6 (10) M: 13 (10) D: 31 (5)
Professor:
Parisa A. Ariya, (D), York, 1996, As
Peter Bartello, (D), McGill, 1988, As
John R. Gyakum, (D), MIT, 1981, Am
Man Kong Yau, (D), MIT, 1977, As
Associate Professor:
Frederic Fabry, (D), McGill, 1994, As
Daniel Kirshbaum, (D), Washington, 2005, Asm
David Straub, (D), Washington, 1990, Op
Bruno Tremblay, (D), McGill, Op
Assistant Professor:
Yi Huang, (D), Princeton, 2007, ZrYgAs
Timothy Merlis, (D), Caltech, 2011, Asm
Thomas Colin Preston, (D), Duke, 2011, AsZnn
Andreas Zuend, (D), ETH (Switzerland), 2008, Asm
Adjunct Professor:
Gilbert Brunet, (D), McGill, 1989, Am
Ashu Dastoor, (D), IIT
Luc Fillion, (D), McGill, 1991, Ams
Pavlos Kollias, (D), Miami, 2000, ZrAs
Hai Lin, (D), McGill, 1995, As
Emeritus:
Jacques F. Derome, (D), Michigan, 1968, As
Henry G. Leighton, (D), Alberta, 1968, As
Lawrence A. Mysak, (D), Harvard, 1966, Op
Isztar Zawadzki, (D), McGill, 1972, AsZr

Dept of Mining & Materials Engineering (B,M,D) ☒
Rm 125 FDA Building
3450 University Street
Montreal, QC H3A 0E8
p. (514) 398-4986
roussos.dimitrakopoulos@mcgill.ca
http://www.mcgill.ca/minmat/mining
Programs: Mining Engineering
Chair:
Stephen Yue, (D), Leeds (UK), 1979, Nx
Professor:
George Demopoulos, (D), McGill, 1982, Nx
Roussos Dimitrakopoulos, (D), Ecole Polytechnique, 1989, Nm
Raynald Gauvin, (D)
Roderick I. Guthrie, (D), London (UK), 1967, Nx
Ferri Hassani, (D), Nottingham (UK), 1981, Nm
Hani Mitri, (D), Nottingham (UK), 1981, Nm
Associate Professor:
Mathieu Brochu, (D), McGill
Mainul Hasan, (D), McGill, 1987, Nx
Showan Nazhat, (D)
Mihriban Pekguleryuz, (D), McGill
Assistant Professor:
Kirk Bevan, (D), Purdue
Marta Cerruti, (D)
Richard Chromik, (D), SUNY

Nathaniel Quitoriano, (D), MIT
Jun Song, (D), Princeton
Kristian Waters, (D)

Lecturer:
Forence Paray, (D), McGill

Emeritus:
John J. Jonas, (D), Cambridge, 1960, Nx

Dept of Earth & Planetary Sciences (B,M,D) ●☒
3450 University Street
Room 238
Montreal, QC H3A 0E8
p. (514) 398-6767
kristy.thornton@mcgill.ca
http://www.mcgill.ca/eps
Programs: Geology (B); Planetary Sciences (B); Earth and Planetary Sciences (M,D)
Enrollment (2021): B: 12 (4) M: 34 (5) D: 40 (7)

Chair:
Jeffrey M. McKenzie, (D), Syracuse, 2005, Hw

Professor:
Don Baker, (D), Penn State, 1985, CuGvCp
Eric Galbraith, (D), British Columbia, 2006, Zo
Galen Halverson, (D), Harvard, 2003, GsrCs
Olivia G. Jensen, (D), British Columbia, 1971, Yg
John Stix, (D), Toronto, 1989, Gv
Anthony E. Williams-Jones, (D), Queens, 1973, Ce

Associate Professor:
Kim Berlo, (D), Bristol, 2006, CgGv
Nicolas B. Cowan, (D), Washington, 2009, AsXyg
Natalya Gomez, (D), Harvard, 2013, Og
James Kirkpatrick, (D), Glasgow, 2008, Gct
Yajing Liu, (D), Harvard, 2007, Ygs
Jeanne Paquette, (D), SUNY (Stony Brook), 1991, Gz
Christie Rowe, (D), California (Santa Cruz), 2007, GctEm
Vincent van Hinsberg, (D), Bristol, 2006, EgCg

Assistant Professor:
Peter Douglas, (D), Yale, 2014, Cs
Nagissa Mahmoudi, (D), McMaster, 2013, Cb

Earth System Science Faculty Lecturer:
William G. Minarik, (D), Rensselaer, 1993, Cp

Adjunct Professor:
Rebecca Harrington, (D), California (Los Angeles), 2008, Yg
Richard Léveillé, (D), W Ontario, 2000, CgXb
Bjorn Sundby, (D), Bergen (Norway), 1966, Cm

Emeritus:
Jafar Arkani-Hamed, (D), MIT, 1969, Xy
Don Francis, (D), MIT, 1974, Gi
Reinhard Hesse, (D), Tech (Munich), 1964, Gs
Andrew J. Hynes, (D), Cambridge, 1972, Gc
Robert F. Martin, (D), Stanford, 1969, GziEm
Alfonso Mucci, (D), Miami, 1981, CmlOc
Colin Stearn, (D), Yale, Ps

McMaster University

School of Earth, Environment & Society (B,M,D) ●☒
1280 Main Street West
General Science Building
Room 206
Hamilton, ON L8S 4K1
p. (905) 525-9140 (Ext. 24535)
geograd@mcmaster.ca
https://www.science.mcmaster.ca/ees/
f: @McMASTERSEES
Programs: Honours Earth & Environmental Sciences, Honours Environmental Sciences, Honours Biodiversity & Environmental Sciences, Environmental Sciences, Minor in Earth Sciences, Minor in Environmental Sciences, Minor in Geography and Earth Sciences, Minor in Environment & Society, Minor in Geography, Minor in GIS, Earth & Environmental Sciences (PhD), Earth & Environmental Sciences - Astrobiology (PhD), Geography (PhD), Earth & Environmental Sciences (MSc), Earth & Environmental Sciences -Astrobiology (MSc), Geography (MA/MSc), Certificates: Geographic Information Sciences (GIS), Urban Studies & Planning
Enrollment (2021): B: 249 (66) M: 24 (10) D: 33 (5)

Director:
K. Bruce Newbold, (D), McMaster, 1994, Ge

Professor:
M. Altaf Arain, (D), Arizona, 1997, Hg
Janok Bhattacharya, (D), McMaster, 1989, Gr
Joseph I. Boyce, (D), Toronto, 1997, Zg
Sean Carey, (D), McMaster, 2000, Hg
Vera Chouinard, (D), McMaster, 1986, Zn
Paulin Coulibaly, (D), Laval, 2000, Hg
Carolyn H. Eyles, (D), Toronto, 1986, GslGm
Karen Kidd, (D), Alberta, 1996, Gu
H. Antonio Paez, (D), Tohoku, 2000, Eg
Edward G. Reinhardt, (D), Carleton, 1996, PmGam
Darren M. Scott, (D), McMaster, 2000, Zi
Gregory F. Slater, (D), Toronto, 2001, Cg
James E. Smith, (D), Waterloo, 1995, Hw
J. Michael Waddington, (D), York, 1995, Pe
Allison M. Williams, (D), York, 1997, Zn
Robert D. Wilton, (D), S California, 1999, Zn

Associate Professor:
Sang-Tae Kim, (D), McGill, 2006, Cg
Gita Ljubicic, (D)
Suzanne Mills, (D), Saskatchewan, 2007, Zn
Maureen Padden, (D), ETH (Switzerland), 2001, CsPc
Niko Yiannakoulias, (D), Alberta, 2006, Ge

Assistant Professor:
Melanie Bedore, (D), Queens, 2010, Zu
Luc Bernier, (D), McMaster, 2007, Cg
Alemu Gonsamo, (D), Helsinki, 2009
Michael Mercier, (D), McMaster, 2004, Zu
Elli Papangelakis, (D), Waterloo, 2019, Gm
Alexander Peace, (D), Durham (UK), 2016

Lecturer:
John MacLachlan, (D), McMaster, 2011, Gs

Adjunct Professor:
Matthew Adams, (D), McMaster, 2015
Patricia Beddows, (D), Bristol, 2004
Jing Chen, (D), Reading (UK), 1986, AmZry
Susan J. Elliott, (D), McMaster, 1992, Ge
Beverly Goodman-Tchernov, (D), McMaster, 2006
Peter Kitchen, (D), Ottawa
Chris McLaughlin, (D), McMaster
Hamid Mehmood, (D), Asian Inst Tech, 2012
Nidhi Nagabhatla, (D), ISRO, 2005
Duminda Perera, (D), Kyushu, 2010
Matthew Peros, (D), Toronto, GaPg
Manzoor Qadir, (D), Ag, Faisalabad, 1995
Dominique Rissolo, (D), California (Riverside), 2001
James Roy, (D), Waterloo, 2004
Matthias Sweet, (D), Pennsylvania, 2012
Lina Taing, (D), Cape Town, 2015
S Martin Taylor, (D), Victoria (Canada), 1974
Shuzen Wang
Lesley A. Warren, (D), Toronto, 1994, CgHg
Amin Yazdani, (D), Waterloo, 2015

University Professor:
John D. Eyles, (D), London (UK), 1983, Ge
Henry P. Schwarcz, (D), Caltech, 1960, Cs

Emeritus:
Paul Clifford, (D), London (UK), 1956
James H. Crocket, (D), MIT
Alan P. Dickin, (D), Oxford, 1981, Cc
John J. Drake, (D), McMaster, 1973
Derek C. Ford, (D), Oxford, 1963, GmHwCc
Doug Grundy, (D), Manchester (UK), 1966
Fred L. Hall, (D), MIT, 1975
Richard S. Harris, (D), Queens, 1981, Zu
Leslie J. King, (D), Iowa, 1960
James R. Kramer, (D), Michigan, 1958, Cl
Kao Lee Liaw, (D), Clark, 1972
Gerry V. Middleton, (D), London (UK), 1954
William A. Morris, (D), 1974, Yg
Yorgos Papageorgiou, (D), Ohio State, 1970
W. Jack Rink, (D), Florida State, 1990, CcGam
Michael J. Risk, (D), S California, 1971, Pg
Wayne R. Rouse, (D), McGill, 1968
Roger G. Walker, (D), Oxford, 1964
Ming-Ko Woo, (D), British Columbia, 1972, HgZy

Memorial University of Newfoundland

Dept of Earth Sciences (B,M,D) ☒
Alexander Murray Building
St. John's, NL A1B 3X5
p. (709) 864-8142
earthsci@mun.ca
http://www.mun.ca/earthsciences

Professor:
Karem Azmy, (D), Cl
Farquharson Colin, (D), Yg
Gregory R. Dunning, (D), Memorial, 1984, CcGit
John Hanchar, (D), Cg
Aphrodite D. Indares, (D), Montreal, 1989, Gp
Graham Layne, (D), Cg
Duncan McIlroy, (D), Gg
Penny Morrill, Cg
Michael A. Slawinski, (D), Calgary, 1996, Ys

Associate Professor:
Luke Beranek, (D), GgsCc
Tao Cheng, (D), Hw
John Jamieson, (D), Gu
Alison Leitch, (D), Australian Nat, 1986, Yg
Alison Malcolm, (D), Yg

Assistant Professor:
Michael Babechuk, (D), Cu
Philippe Belley, (D), British Columbia, 2019, Gz
David Lowe, (D), Ottawa, 2016, Gs
Eric Thiessen, (D), Simon Fraser, 2019, Gc

First Year Lab Instructor:
Amanda Langille, (M), Gg

Emeritus:
Ali E. Aksu, (D), Dalhousie, 1980, Gu
Jeremy Hall, (D), Glasgow, 1971, YsGtYr
Richard N. Hiscott, (D), McMaster, 1977, Gs
Joseph P. Hodych, (D), Toronto, 1971, Ym
Henry Longerich, (D), Indiana, 1967, Cg
Toby C. J. S. Rivers, (D), Ottawa, 1976, Gp
Michael G. Rochester, (D), Utah, 1959, Yg

Memorial University of Newfoundland, Grenfell Campus

Environmental Science Program (B) ●⊘
School of Science and Environment
University Drive
Corner Brook, NL A2H 6P9
p. (709) 637-6215
jsircom@grenfell.mun.ca
https://grenfell.mun.ca/
Programs: Environmental Science
Enrollment (2018): B: 56 (12)

Mount Royal University

Dept of Earth and Environmental Sciences (A,B)
4825 Mount Royal Gate SW
Calgary, AB T3E 6K6
p. (403) 440-6165
ttaylor@mtroyal.ca
http://www.mtroyal.ab.ca/scitech/earth.shtml
Administrative Assistant: Leona Stadnyk

Chair:
Paul Johnston, (D), W Australia, 1986, Pi

Professor:
John Cox, (D), Aberdeen, 1994, EoGsHw
Barbara McNicol, (D), Calgary, 1997, Zg

Associate Professor:
Katherine Boggs, (D), Calgary, 2004, Gt
Pamela MacQuarrie, (M), Calgary, 1988, Zy

Related Staff:
Michael Clark, (B), Adams, 1976, Zg

Natural Resources Canada

Ressources Naturelles Canada
580 Booth Street , 21st Floor
Ottawa, ON K1A 0E4
debra.tompkinscaron@canada.ca
http://www.nrcan.gc.ca/

Senior Scientist:
Sergey V. Samsonov, (D), Western, 2007, YdZrYg

Other:
Esther Asselin, (M), Laval, 1988, PlGuRh

New Brunswick Department of Natural Resources and Energy Development

New Brunswick Department of Natural Resources and Energy Development ☒
Hugh John Flemming Forestry Centre
P.O. Box 6000
Fredericton, NB E3B 5H1
p. (506) 453-3826
geoscience@gnb.ca
http://www.gnb.ca/minerals

Queen's University

Dept of Geological Sciences and Geological Engineering (B,M,D) ●
#240, Bruce Wing, Miller Hall
36 Union St.
Kingston, ON K7L 3N6
p. (613) 533-2597
geolundergradassistant@queensu.ca
http://www.queensu.ca/geol
Enrollment (2011): B: 232 (47) M: 46 (20) D: 18 (5)

Head:
Jean Hutchinson, (D), Toronto, 1992

Professor:
Mark Diederichs, (D), Toronto, Nr
Georgia Fotopoulos, (D), Calgary, 2003, Yd
Laurent Godin, (D), Carleton, 1999, Gct
Noel P. James, (D), McGill, 1972, Gd
Guy M. Narbonne, (D), Ottawa, 1981, PigGs
Gema Olivo, (D), Québec (Montréal), 1995, Eg
Ronald C. Peterson, (D), Virginia Tech, 1980, Gz
Victoria H. Remenda, (D), Waterloo, 1993, Hw

Associate Professor:
Alexander Braun, (D), Goethe (Frankfurt), 1999, YgdYv
John A. Hanes, (D), Toronto, 1979, Cc
Heather E. Jamieson, (D), Queen's, 1982, Ge
Daniel Layton-Matthews, (D), Toronto, 2006, EgCte

Research Associate:
Doug A. Archibald, (D), Queen's, 1982, Cc

Adjunct Professor:
Rob Harrap, (M), Carleton, 1990, Gc

Emeritus:
Alan H. Clark, (D), Manchester (UK), 1964, Em
Robert W. Dalrymple, (D), McMaster, 1977, Gs
John M. Dixon, (D), Connecticut, 1974, Gct
Herwart Helmstaedt, (D), New Brunswick, 1968, Gc
Raymond A. Price, (D), Princeton, 1958, GtcGe

Royal Ontario Museum

Dept of Palaeobiology
100 Queen's Park
Toronto, ON M5S 2C6
p. (416) 586-5591
davidru@rom.on.ca

Curator Of Vertebrate Palaeontology:
David C. Evans, (D), Toronto, 2007, PvgPq

Department of Natural History ☒
100 Queen's Park
Toronto, ON M5S 2C6
p. (416) 586-5820
ktait@rom.on.ca

Emeritus:
Robert I. Gait, (D), Manitoba, 1967, Gz
Frederick J. Wicks, (D), Oxford, 1969, Gz

Curator:
Kimberly T. Tait, (D), Arizona, 2007, GzyXm

Saint Francis Xavier University

Dept of Earth Sciences (B,M) ○☒
P.O Box 5000
Antigonish, NS B2G 2W5

p. (902) 867-5109
lsampson@stfx.ca
https://www.stfx.ca/academics/science/earth-sciences
Programs: Geoscience, Earth Science, Environmental Earth Science
Enrollment (2015): B: 11 (15) M: 13 (3)

Professor:
Alan J. Anderson, (D), Queen's, 1990, Gx
Hugo Beltrami, (D), Quebec (Montreal), 1993, YhAs
Lisa M. Kellman, (D), Quebec (Montreal), 1997, CbSoCs
Michael J. Melchin, (D), Western, 1987, Pi
Dave A. Risk, (D), Dalhousie, 2006, SoZr

Associate Professor:
James Braid, (D), Dalhousie U, Gc

Instructor:
Donnelly Archibald, (D), Gg
Cindy Murphy, (M), McGill, 1986, Gg
Colette Rennie, (M), Queen's, 1987, Gg
Matthew Schumacher, (M), Waterloo, 2006, Zg

Emeritus:
J. Brendan Murphy, (D), McGill, 1982, Gt

Simon Fraser University

Department of Earth Sciences (B,M,D) ●⊘
8888 University Drive
Burnaby, BC V5A 1S6
p. 778-782-5387
http://www.sfu.ca/earth-sciences

Professor:
Diana M. Allen, (D), Carleton, 1996, Hw
Andrew J. Calvert, (D), Cambridge, 1985, Ys
Shahin Dashtgard, (D), Gs
Gwenn Flowers, (D), British Columbia, 2000, Gl
H. Daniel (Dan) Gibson, (D), Carleton, 2003, GcCcGp
James A. Mac Eachern, (D), Alberta, 1994, Gs
Derek J. Thorkelson, (D), Carleton, 1992, Gt
Eileen van der Flier-Keller, (D), Western Ontario, 1985, Ze
Brent C. Ward, (D), Alberta, 1992, GlmGe
Glyn Williams-Jones, (D), Open (UK), 2001, GvCeYe

FRBC Chair in Resource Geoscience & Geotechnics:
Sergio Sepulveda, (D), Leeds, UK, 2004, NgRn

Assistant Professor:
Jessica Pilarczyk, (D), McMaster, 2011, Rn

Lecturer:
Kevin Cameron, (M), Memorial, 1986, Ggz
Cindy Hansen, (M), Gg
Reid Staples, (D), GpcCg

Emeritus:
John J. Clague, (D), British Columbia, 1973, Rn
Dan D. Marshall, (D), Lausanne, 1995, Ca
Douglas Stead, (D), Nottingham (UK), 1984, Ng

Universite du Quebec

INRS, centre Eau Terre Environnement (Quebec Geoscience Center) (M,D) ☒
490 de la Couronne
Quebec, QC G1K 9A9
p. + (418) 654-4677
info@ete.inrs.ca
http://www.ete.inrs.ca
Programs: PhD in Earth sciences, Master's in Earth Sciences, Professional Master's in Earth Sciences

Professor:
Normand Bergeron, (D), SUNY (Buffalo), 1994, Gm
Monique Bernier, (D), Zr
Fateh Chebana, Hq
Karem Chokmani, Zi
Pierre Francus, (D), 1997, PeGse
Bernard Giroux, Yeg
Erwan Gloaguen, Ye
Lyal B. Harris, GcYg
Marc R. LaFleche, (D), Montpellier II, 1991, Ct
Rene Lefebvre, (D), Laval, 1994, Hw
Richard Martel, (D), Laval, 1996, Hw
Claudio Paniconi, (D), Hw
Pierre-Simon Ross, GvEg

Senior Scientist:
Jean H. Bedard, (D), Montreal, 1985, Gi
Christian Begin, (D), Laval, 1991, PeGmSo
Benoit Dube, (D), Quebec, 1990, Em
Denis Lavoie, (D), Laval, 1988, Gs
Yves Michaud, (D), Laval, 1991, Gm
Michel Parent, (D), W Ontario, 1987, Gl
Didier Henri Perret, (D), Laval, 1995, NgeZu

Research Associate:
Eric Boisvert, (M), Quebec (Montreal), 1994, Zn
Kathleen Lauziere, (M), Quebec, 1989, Gg
Marc R. Luzincourt, (B), Quebec (Montreal), 1982, Cs
Anna Smirnov, (M), Memorial, 1997, Cs

Adjunct Professor:
Geneviève Bordeleau, (D), INRS, Hw
Louise Corriveau, (D), McGill, 1989, EmGpg
Saied Homayouni, (D), Zrf
Damien Pham van Bang, (D), HtOn
Jasmin Raymond, (D), Hy
Renaud Soucy de la Roche, (D), Gc

Universite du Quebec a Chicoutimi

Sciences de la Terre (B,M,D) ●☒
555, Boulevard de l'Universite
Chicoutimi, QC G7H 2B1
p. (418) -545- 5011 (Ext. 5202)
sue_sc-terre@uqac.ca
Programs: B.Sc Geology, B.Sc. geological engineering, M.Sc. Earth science and geological engineering, M. Sc. Professional, exploration geology, Ph.D. Earth sciences
Enrollment (2018): B: 55 (13) M: 34 (8) D: 6 (1)

Professor:
Sarah- Jane Barnes, (D), Toronto, 1983, EgGiCt
Paul Bedard, (D), Quebec, 1992, CaGzEg
Romain Chesnaux, (D), École Polytechnique (Canada), 2005, Hw
Damien Gaboury, (D), Quebec, 1999, EmCe
Ali Saeidi, (D), Lorraine INP (France), 2010, Nrm
Edward W. Sawyer, (D), Toronto, 1983, GptGi
Julien Walter, (D), Quebec, 2018, HwqGm

Assistant Professor:
Lucie Mattieu, (D), Trinity (Dublin), 2010, EgmCe

Research Associate:
Silvain Rafini, (D), Quebec (Montreal), 2008, GcHwEg

Instructor:
Denis Cote, (M), Quebec, 1980, GiEgZi

Lecturer:
Dominique Genna, (D), Quebec, 2009, GvEn
Philippe Page, (D), INRS-ETE, 2006, GiCgEm

Adjunct Professor:
Sylvain Raffini, (D), Quebec (Montreal), 2008, GcHw

Emeritus:
Guy Archambault, (D), Nr
Pierre Cousineau, (D), Laval, 1986, Gd
Jayanta Guha, (D), Jadavpur, 1967, Eg
Michael D. Higgins, (D), McGill, 1980, Gi
Alain Rouleau, (D), Waterloo, 1984, Hw
Denis W. Roy, (D), Princeton, 1976, Gc

On Leave:
Real Daigneault, (D), Laval, 1991, GcEgZi

Laboratory Director:
Dany Savard, (M), Quebec, 2010, Ca

Universite du Quebec a Montreal

Département des sciences de la Terre et de l'atmosphere (B,M,D)
C.P. 8888, succursale Centre-ville
Montreal, QC H3C 3P8
p. +1 514-987-3000
dept.sct@uqam.ca
http://scta.uqam.ca
Administrative Assistant: France Beauchemin

Professor:
Florent Barbecot, (D), Hw
Jean-Pierre Blanchet, (D), Toronto, 1984, Pe
Gilles Couture, (D), British Columbia, 1987, Yg
Fiona Ann Darbyshire, (D), Cambridge, 2000, Ys
Anne de Vernal, (D), Montreal, 1986, Pm
Alessandro Marco Forte, (D), Toronto, 1989, Yg
Pierre Gauthier, (D), McGill, 1988, Am
Eric Girard, (D), McGill, 1999, Am

Normand Goulet, (D), Queen's, 1976, Gc
Cherif Hamzaoui, (D), Alziers, 1980, Yg
Alfred Jaouich, (D), Minnesota, 1975, Sc
Michel Jebrak, (D), Orleans (France), 1984, Eg
Michel Lamothe, (D), W Ontario, 1985, Gl
Rene Laprise, (D), Toronto, 1988, As
Marie Larocque, (D), de Poitiers (France), 1998, Hw
Marc Michel Lucotte, (D), McGill, 1987, Oc
Daniele Luigi Pinti, (D), France, 1993, Cs
Martin Roy, (D), Oregon, 1998, Gl
Ross Stevenson, (D), Arizona, 1989, Cg
Laxmi Sushama, (D), Melbourne (Australia), 1999, Hy
Julie Mireille Thériault, (D), McGill, 2009, Zn
Enrico Torlaschi, (D), Ecole Polytechnique (Milan), 1976, Am
Alain Tremblay, (D), Laval, 1989, Gct
David Widory, (D), IPGP, 1999, Cs

Associate Professor:
Sanda Balescu, (D), Free (Brussels), 1988, Pe
Jean Côté, As
Bernard Dugas, As
Stephane Faure, (D), INRS-Georessources, 1995, Gc
Philippe Gachon, (D), Quebec (Montreal), 1999, As
Michel Gauthier, (D), Polytechnique (Montreal), 1982, Eg
Jean-François Helie, (D), Quebec, 2004, Cs
Claude Hillaire-Marcel, (D), Paris VI, 1979, Cs
Jean-Claude Mareschal, (D), Texas A&M, 1975, Yg
Andre Poirier, (D), Quebec (Montreal), 2005, Cs
Gilbert P. Prichonnet, (D), Bordeaux, 1967, Pi
William W. Shilts
Gabriel-Constantin Voicu, (D), Quebec (Montreal), 1995, GcCg

Universite du Quebec a Rimouski

Inst des sciences de la mer de Rimouski (M,D)
310, Allee des Ursulines
Rimouski, QC G5L 3A1
p. (418) 723- 8617
ismer@uqar.ca
http://www.ismer.ca

Director:
Serge Demers, (D), Laval, 1981, Ob

Professor:
Celine Audet, (D), Laval, 1985, Ob
Jean-Claude Brethes, (D), Aix-Marseille, 1978, Ob
Jean-Pierre Gagne, (D), Montreal, 1993, OcCmo
Michel Gosselin, (D), Laval, 1990, Ob
Vladimir G. Koutitonsky, (D), SUNY (Stony Brook), 1985, Op
Jocelyne Pellerin, (D), Laval, 1982, Ob
Emilien Pelletier, (D), McGill, 1983, Oc
Suzanne Roy, (D), Dalhousie, 1986, Ob
Bjorn Sundby, (D), Bergen (Norway), 1966, Oc
Bruno Zakardjian, (D), Paris, 1994, Ob

Universite Laval

Dept de geologie (B,M,D)
Pavillon Pouliot
Faculte des Sciences et de genie
Ste-Foy, QC G1K 7P4
p. (418) 656-2193
marc.constantin@ggl.ulaval.ca
http://www.ggl.ulaval.ca

Head:
Marc Constantin, (D), Brest, 1995, Gi
Josee Duchesne, (D), Laval, 1993, Gz

Professor:
Georges Beaudoin, (D), Ottawa, 1991, EgGzCa
Richard Fortier, (D), Montreal, 1994, Yg
Paul W. Glover, (D), E Anglia (UK), 1989, Yx
Rejean J. Hebert, (D), Brest, 1985, Gi
Jacques E. Locat, (D), Sherbrooke, 1982, Ng
Fritz Neuweiler, (D), Berlin, 1995, GsdPg
Rene Therrien, (D), Waterloo, 1992, HwGe

Eletron Microprobe Specialist:
Marc Choquette, (D), Laval, 1988, Zn

Associate Scientist:
Pauline Dansereau, (M), Laval, 1988, Gs
Andre Levesque, (B), Laval, 1979, Gz
Pierre Therrien, (M), Laval, 1986, Gq

Research Associate:
Danielle Cloutier, (D), Laval, GeOg

Adjunct Professor:
Benoit Fournier, (D), Laval, 1993, NgZmGg

Technician:
Jean Frenette, (B), Laval, 1985, Gz
Martin Plante, Cg

University of Alberta

Dept of Earth & Atmospheric Sciences (B,M,D) ●⦸
1-26 Earth Sciences Building
Edmonton, AB T6G 2E3
p. (780) 492-3265
eas@ualberta.ca
http://www.ualberta.ca/EAS/
f: https://www.facebook.com/easualberta
t: @UofA_EAS
Programs: Geology, Environmental Earth Sciences, Human Geography, Urban and Regional Planning, Paleontology
Enrollment (2016): B: 449 (129) M: 0 (17) D: 0 (18)

Chair:
Stephen T. Johnston, (D), Alberta, 1993, GctGg

Professor & Inaugural Director, Planning Program:
Sandeep Agrawal, (D), Zu

Professor:
Robert W. Luth, (D), California (Los Angeles), 1985, Gi
Martin J. Sharp, (D), Aberdeen, 1982, Gl

Canada Research Chair:
Thomas Stachel, (D), Wurzburg, 1991, GiCs

Associate Dean (Research):
Larry M. Heaman, (D), McMaster, 1986, Cc

Associate Chair:
Murray Gingras, (D), North Carolina, 1987, Go
Kurt Konhauser, (D), W Ontario, 1993, Pg

Professor:
Andrew B.G Bush, (D), Toronto, 1995, AsOpPe
Michael W. Caldwell, (D), McGill, 1995, Pv
Octavian Catuneanu, (D), Toronto, 1996, Gs
Thomas Chacko, (D), N Carolina, 1987, Gp
Robert A. Creaser, (D), La Trobe, 1990, Cc
Philippe Erdmer, (D), Queen's, 1982, Gc
Duane Froese, (D), Calgary, 2001, GlmPe
John Gamon, (D), California (Davis), 1989, Zr
Christopher DK Herd, (D), New Mexico, 2001, XcGiCp
Brian Jones, (D), Ottawa, 1974, Ps
Lindsey Leighton, (D), Michigan, 1999, Pi
Hans G. Machel, (D), McGill, 1985, Go
Karlis Muehlenbachs, (D), Chicago, 1971, Cs
Graham Pearson, (D), Leeds (UK), Cec
David Potter, (D), Newcastle upon Tyne, Ye
Gerhard Reuter, (D), McGill, 1985, Am
Benoit Rivard, (D), Washington (St Louis), 1990, Zr
Benjamin J. Rostron, (D), Alberta, 1995, HwGoCg
G. Arturo Sanchez-Azofeifa, (D), New Hampshire, 1996, ZirZg
Bruce Sutherland, (D), Toronto, 1994, Cm
Martyn Unsworth, (D), Cambridge, GtvYe
John W.F Waldron, (D), Edinburgh, 1981, Gc
John D. Wilson, (D), Guelph, 1980, Am
John-Paul Zonneveld, (D), Alberta, 1999, Go

Associate Chair:
Tara McGee, (D), Australian Nat, 1996, Zn

Associate Professor:
Damian Collins, (D), Simon Fraser, 2004, Zn
Theresa D. Garvin, (D), McMaster, 1999, Zn
Nicholas Harris, (D), Stanford, Gso
Jeffrey Kavanaugh, (D), British Columbia, 2001, Ng
Paul Myers, (D), Victoria, 1996, Og

Assistant Professor:
Daniel S. Alessi, (D), Notre Dame, 2009, ClGeHw
Jeff Birchall, (D), Canterbury (NZ), 2013, ZuyZg
Monireh Faramarzi, (D), ETH (Switzerland), 2010, HgwHq
Long Li, (D), Lehigh, 2006, CsGpv
Manish Shirgaokar, (D), Zu
Amrita Singh, (D), California (Irvine), 2015, Zu

Distinguished University Professor:
Nathaniel W. Rutter, (D), Alberta, 1966, Ge

Emeritus:
Ronald A. Burwash, (D), Minnesota, 1955, Gp

Ian A. Campbell, (D), Colorado, 1968, Gm
Brian D. Chatterton, (D), Australian Nat, 1970, Pi
David M. Cruden, (D), London (UK), 1969, Nr
John England, (D), Colorado, 1974, Gml
Leszek A. Kosinski, (D), Warsaw, 1958, Zn
Arleigh H. Laycock, (D), Minnesota, 1957, Hg
Edward P. Lozowski, (D), Toronto, 1970, As
Carl A. Mendoza, (D), Waterloo, 1993, HwSpHq
O.F. George Sitwell, (D), Toronto, 1968, Zn
Peter J. Smith, (D), Edinburgh, 1964, Zn
József Tóth, (D), Utrecht, 1965, Hw

Related Staff:
Robert Summers, (D), Guelph, 2005, Zn

Inst of Geophysical Research (B,M,D)
Mailstop #615
CEB/Physics
Edmonton, AB T6G 2G7
p. (780) 492-3521
msacchi@ualberta.ca
http://www-geo.phys.ualberta.ca/institute/
Administrative Assistant: Lee Grimard

Professor:
Robert A. Creaser, (D), La Trobe, 1992, Cc
T. Bryant Moodie, (D), Toronto, 1972, Zn
Robert Rankin, (D), North Wales, 1984, Xy
Gerhard W. Reuter, (D), McGill, 1985, As
Wojciech Rozmus, (D), Inst Nuc Res (Poland), Zn
John C. Samson, (D), Alberta, 1971, Xy
Martin J. Sharp, (D), Aberdeen, Gl
John Shaw, (D), Reading (UK), 1969, Gl
Samuel S. Shen, (D), Wisconsin, As
Bruce R. Sutherland, (D), Toronto, 1994, Zn
Gordon E. Swaters, (D), British Columbia, 1985, Op
Richard D. Sydora, (D), Texas, 1985, Xy
Martyn Unsworth, (D), Cambridge, Zn
John D. Wilson, (D), Guelph, 1980, As

Associate Professor:
Andrew B. G. Bush, (D), Toronto, As
Carl Mendoza, (D), Waterloo, 1993, Hw
Benoit Rivard, (D), Washington, 1990, Zr
Ben Rostron, (D), Alberta, Zn
Mauricio D. Sacchi, (D), British Columbia, Ys

Assistant Professor:
Francis Fenrich, (D), Alberta, 1997, Xy
Jeff Gu, (D), Harvard, 2001, Ys
Moritz Heimpel, (D), Johns Hopkins, 1995, Zn
Vadim Kravchinsky, (D), Irkutsk, 1996, Ym
Paul Myers, (D), Victoria, 1992, Op

Emeritus:
Michael E. Evans, (D), Australian Nat, 1969, Ym
Keith D. Hage, (D), Chicago, 1957, As
F. Walter Jones, (D), McGill, 1968, Ym
Edward P. Lozowski, (D), Toronto, 1970, As
Roger D. Morton, (D), Nottingham (UK), 1959, Em
Edo Nyland, (D), California (Los Angeles), 1967, Ys
David Rankin, (D), Alberta, 1960, Ym
Gordon Rostoker, (D), British Columbia, 1966, Xy
T.J.T (Tim) Spanos, (D), Alberta, 1977, YgsEo

University of British Columbia

Dept of Earth, Ocean, and Atmospheric Sciences (B,M,D) O
2020-2207 Main Mall
Vancouver, BC V6T 1Z4
p. (604) 827-5284
acairns@eos.ubc.ca
http://www.eoas.ubc.ca
Enrollment (2010): B: 361 (105) M: 111 (20) D: 93 (15)

Director, Geological Engineering Program:
Roger D. Beckie, (D), Princeton, 1992, HwCl

Honorary Professor:
Mati Raudsepp, (D), Manitoba, 1984, GzeZm

Dean, Science:
Simon M. Peacock, (D), California (Los Angeles), 1985, GtpYs

Canada Research Chair:
Roger Francois, (D), British Columbia, 1987, Cm
Dominique A.M. Weis, (D), Free (Brussels), 1982, CaGeCu

Professor:
Susan E. Allen, (D), Cambridge, 1988, Op
Raymond J. Andersen, (D), California (San Diego), 1975, Oc
Neil Balmforth, (D), Cambridge, 1990, As
Michael G. Bostock, (D), Australian Nat, 1991, YsGt
R. Marc Bustin, (D), British Columbia, 1979, Ec
Gregory M. Dipple, (D), Johns Hopkins, 1991, Gp
Erik Eberhardt, (D), Saskatchewan, 1998, Ng
Lee A. Groat, (D), Manitoba, 1988, GzEg
Oldrich Hungr, (D), Alberta, 1981, Gm
Mark Jellinek, (D), Australian Nat, 1999, Gg
Catherine L. Johnson, (D), California (San Diego), 1994
Ulrich Mayer, (D), Waterloo, 1999, Nm
James K. Mortensen, (D), California (Santa Barbara), 1983, Cc
Douglas W. Oldenburg, (D), California (Santa Barbara), 1974, Yg
Evgeny Pakhomov, (D), Russian Acad of Sci, 1992, Ob
James Kelly Russell, (D), Calgary, 1984, GviCg
James S. Scoates, (D), Wyoming, 1994, GizGv
Paul L. Smith, (D), McMaster, 1981, Pg
Douw G. Steyn, (D), British Columbia, 1980, Asm
Roland B. Stull, (D), Washington, 1975, As
Curtis Suttle, (D), British Columbia, 1987, Ob

NSERC Industrial Research Chair in Computational Geoscience:
Eldad Haber, (D), British Columbia, 1997, YgZn

Director, MDRU:
Craig J.R. Hart, (D), W Australia, 2004, EmGiCc

Canada Research Chair:
Maria Maldonado, (D), McGill, 1999, Ob
Christian Schoof, (D), Oxford, 2002, Gl

Associate Professor:
Philip Austin, (D), Washington, 1987, As
Kurt A. Grimm, (D), California (Santa Cruz), 1992, Gs
Mark S. Johnson, (D), Cornell, 2005, SpHsq
Lori Kennedy, (D), Texas A&M, 1996, Gc
Maya G. Kopylova, (D), Moscow (Russia), 1990, Gi
Kristin J. Orians, (D), California (Santa Cruz), 1988, Cm
Richard A. Pawlowicz, (D), MIT/WHOI, 1994, Op
Philippe Tortell, (D), Princeton, 2001, Ob

Assistant Professor:
Kenneth A. Hickey, (D), James Cook, 1995, EgGcg
Valentina Radic, (D), Alaska (Fairbanks), 2008, GlAsZg

Honorary Research Associate:
Kevin Kingdon, (M), British Columbia, 1998, Yg
Michael Maxwell, (D), British Columbia, 1986, Yg

Research Associate:
Thomas Bissig, (D), Queens, 2001, EmCg
Farhad Bouzari, (D), Queens, 2003, Eg
Amanda Bustin, (D), Victoria, 2006, YgNg
Philip T. Hammer, (D), California (San Diego), 1991, Ys
Brian Hunt, (D), Tasmania, 2005, Ob
Bruno Kieffer, (D), Grenoble, 2002, Cs
Henryk Modzelewski, (D), British Columbia, 2004, As
Roger Pieters, (D), California (Santa Barbara), Op
Lin-Ping Song, (D), Sichuan (China), 1996, Ng
Dave E. Williams, (D), British Columbia, 1987, Oc

Instructor:
Brett H. Gilley, (M), Simon Fraser, 2003, GsZeRn
Sara Harris, (D), Oregon State, 1998, Ou
Tara Ivanochko, (D), British Columbia, 2004, OoZce
Stuart Sutherland, (D), Leicester (UK), 1992, Pg

Lecturer:
Francis H. M. Jones, (M), British Columbia, 1987, YeZe

Honorary Lecturer:
Dileep Athaide, (M), British Columbia, 1975, Gg

Adjunct Professor:
Robert G. Anderson, (D), Carleton, 1983, Gt
Stephen Billings, (D), Sydney, 1998, Yg
Alex Cannon, (D), British Columbia, 2009, As
Edward Carmack, (D), Washington, 1972, Op
Michael G. G. Foreman, (D), British Columbia, 1984, Op
James Haggart, (D), California (Davis), 1984, PgGr
Catherine J. Hickson, (D), British Columbia, 1987, EgGvZe
Mark Holzer, (D), Simon Fraser, 1990, As
R. Lynn Kirlin, (D), Utah State, 1968, As
Doug McCollor, (D), British Columbia, 2008, Asm
Barry Narod, (D), British Columbia, 1979, Yx
Michael Orchard, (D), Hull, 1975, PgGg
K. Wayne Savigny, (D), Alberta, 1980, Ng
Barbara H. Scott Smith, (D), Edinburgh, 1977, Gz

John F. H. Thompson, (D), Toronto, 1984, Em
Richard E. Thomson, (D), British Columbia, 1971, Op
Richard Tosdal, (D), California (Santa Barbara), 1988, Em
Knut von Salzen, (D), Hamburg, 1997, As
Chi S. Wong, (D), California (San Diego), 1968, Cm

Emeritus:
Mary Lou Bevier, (D), California (Santa Barbara), 1982
Peter M. Bradshaw, (D), Durham (UK), 1965, EgCeGe
Stephen E. Calvert, (D), California (San Diego), 1964, Cm
Richard L. Chase, (D), Princeton, 1963, Gu
Garry K. C. Clarke, (D), Toronto, 1967, Yg
Ronald M. Clowes, (D), Alberta, 1969, Ys
Robert M. Ellis, (D), Alberta, 1964, Ys
William K. Fletcher, (D), Imperial Coll (UK), 1968, Ce
Michael Healey, (D), Aberdeen, 1969, Ob
William W. Hsieh, (D), British Columbia, 1981, OpAs
Alan G. Lewis, (D), Hawaii, 1961, Ob
Stephen G. Pond, (D), British Columbia, 1965, Op
R. Doncaster Russell, (D), Toronto, 1954, Yg
Alastair J. Sinclair, (D), British Columbia, 1964, EmGg
J. Leslie Smith, (D), British Columbia, 1978, Hw
Frank J. R. Taylor, (D), Cape Town, 1965, Ob

Undergraduate Program Coordinator:
Teresa Woodley, Zn

Secretary to the Head:
Selene Chan, (B), British Columbia, 1995, Zn

Office Support:
Alicia Warkentin, Zn

Human Resources Manager:
Cary Thomson, Zn

Graduate Coordinator:
Audrey Van Slyck

Finance Clerk:
Anita Lam, Zn
Kathy Scott, Zn

Director of Resources and Operations:
Renee Haggart, Zn

Computer Department Manager:
John Amor, Zn

Soil Science (B,M,D) O

2357 Main Mall
Vancouver, BC V6T 1Z4
p. (604) 822-0252
maja.krzic@ubc.ca
http://www.landfood.ubc.ca/academics/graduate/soil-science-msc-phd/
Programs: Soil Science
Enrollment (2021): B: 19 (19) M: 8 (3) D: 3 (0)

Professor:
T. Andrew Black, (D), Wisconsin, 1969, Sp
Christopher Chanway, (D), British Columbia, 1987, Sb
Sue Grayston, (D), Sheffield, 1988, Sb
Cindy Prescott, (D), Calgary, 1988, Sb
Suzanne Simard, (D), Oregon State, 1995, Sb

Associate Professor:
Maja Krzic, (D), British Columbia, 1997, Sf
Sean Smukler, (D), California (Davis), 2008, So

Assistant Professor:
Sandra Brown, (D), British Columbia, 1997, Sp
Jean-Thomas Cornelis, (D), UCLouvain, 2010, Sd

Emeritus:
Arthur A. Bomke, (D), Illinois, 1972, Sc
Leslie M. Lavkulich, (D), Cornell, 1969, Sd
Hans D. Schreier, (D), British Columbia, 1976, Zg

University of British Columbia, Okanagan Campus

Earth, Environmental and Geographic Sciences ● ☒

Science 305
1177 Research Road
Kelowna, BC V1V 1V7
p. (250) 807-8757
eegs.okanagan@ubc.ca
https://eegs.ok.ubc.ca/
Programs: Earth and Environmental Sciences
Freshwater Science
Geography
Minor in Geospatial Information Science
Administrative Assistant: Janet Heisler
Administrative Assistant: Natalie Kirkwood

Head:
Edward RC Hornibrook, (D), W Ontario, 1997, CbsSc

Forest Renewal BC Chair of Watershed Research:
Xiaohua (Adam) Wei, (D), Northeast Forestry, 1990, HqgSf

Director, Fipke Laboratory for Trace Element Research:
Kyle Larson, (D), Queens, 2009, GtpCc

Associate Dean, Research, Graduate and Post-doctoral Studies; Director Okanagan Institute for Biodiversity, Resilience, and Ecosystem Services:
Lael Parrot, (D), McGill, 2000, ZgiZf

Professor:
John D. Greenough, (D), Memorial, 1984, GiCtGa
Ian R. Walker, (D), Simon Fraser, 1988, PeGnHs

Forest Renewal BC Chair of Watershed Research:
David F. Scott, (D), KwaZulu-Natal, 1994, SpHgs

Associate Head; Associate Professor of Teaching:
Craig F. Nichol, (D), British Columbia, 2002, HwSpZe

Associate Dean Teaching, Learning and Curriculum; Associate Professor of Teaching:
Trudy A. Kavanagh, (D), W Ontario, 2000, ZyRcZg

Associate Professor:
Yuan Chen, (D), W Ontario, 1993, EmCeGz
P. Jefferson Curtis, (D), Manitoba, 1991, CbHsRw
Kevin Hanna, (D), Toronto, 1998, ZnRcZu
Michael Pidwirny, (D), Simon Fraser, 1994, AtZyc
Ian Saunders, (D), Simon Fraser, 1990, AtGlZy
Robert R. Young, (D), Calgary, 1995, GlmZy

Assistant Professor:
Mathieu L. Bourbonnais, (D), Victoria, 2018, ZifRn
Brendan Dyck, (D), Oxford, 2016, Gpx

Emeritus:
Bernard O. Bauer, (D), Johns Hopkins, 1988, GmOnZy
Fes De Scally, (D), Waterloo, 1989, RnZyRm

Watershed Management Extension Facilitator:
Marni Turek, (B), Thomson Rivers, 1998, Hg

Laboratory Program Manager:
Stuart J. MacKinnon, (M), Victoria, 2016, ZeRcZg

University of Calgary

Dept of Geoscience (B,M,D) ● ⃠

2500 University Drive NW
ES118C
Calgary, AB T2N 1N4
p. (403) 220-5184
geoscience@ucalgary.ca
https://science.ucalgary.ca/geoscience
Programs: BSc Geology Major, BSc Geology Major (Petroleum Geology Concentration), BSc Geology Major Hydrogeology-Environmental Geoscience), BSc Geology Major (Solid Earth), BSc Geophysics Major
Enrollment (2020): M: 78 (7) D: 64 (3)

Head:
Stephen M. Hubbard, (D), Stanford, 2006, Gs

Professor:
Benoit Beauchamp, (D), Calgary, 1987, Pm
Christopher Clarkson, (D), British Columbia, 1998, Eo
David Eaton, (D), Calgary, 1992, Ys
Masaki Hayashi, (D), Waterloo, 1997, HwSp
Charles M. Henderson, (D), Calgary, 1988, Pm
Kristopher Innanen, (D), British Columbia, 2003, Ye
Stephen Larter, (D), Newcastle (UK), 1978
Donald C. Lawton, (D), Auckland, 1979, Ye
David R. Pattison, (D), Edinburgh, 1985, Gp
Cathy Ryan, (D), Waterloo, 1994, Hw

Associate Professor:
Edwin Cey, (D), Waterloo, 2008, Hw
Robert Ferguson, (D), Calgary, 2000, Ye
Alan R. Hildebrand, (D), Arizona, 1992, XmgXc
Per Pedersen, (D), Aarhus (Denmark), 1999, Go
Adam Pidlisecky, (D), Stanford, 2006
Dan Shugar, (D), Simon Fraser, 2011
Daniel o. Trad

Assistant Professor:
Sytle Antao, (D), Stoney Brook, 2006, Gz
Jan Dettmer, (D), Victoria (Australia), 2007, Ys
Eva Enkelmann, (D), 2005, CcGmt

Hersh Gilbert, (D), Colorado, 2001, Ys
Rachel Lauer, (D), Penn State, 2013, Hy
Benjamin Tutolo
Darla Zelenitsky, (D), Calgary, 2004, Pg
Instructor:
Brandon Karchewski, (D), McMaster U, 2015
Adjunct Professor:
Stephen E. Grasby, (D), Calgary, 1997, Hw
Thomas F. Moslow, (D), South Carolina, 1980, Co
Senior Instructor:
Jon W. Jones, (D), Calgary, 1972, Gp
Emeritus:
Laurence R. Bentley, (D), Princeton, 1990, Hw
Frederick A. Cook, (D), Cornell, 1980, YeGtYs
Edward D. Ghent, (D), California, 1964, Gp
Terence M. Gordon, (D), Princeton, 1969, Gq
Ian E. Hutcheon, (D), Carleton, 1977, Cl
Federico F. Krause, (D), Calgary, 1979, GsdGo
Edward S. Krebes, (D), Alberta, 1980, Ys
Gary F. Margrave, (D), Alberta, 1981, Ye
James W. Nicholls, (D), California, 1969, Gi
Gerald D. Osborn, (D), California, 1972, Gm
Philip S. Simony, (D), London (UK), 1963, Gc
Ronald J. Spencer, (D), Johns Hopkins, 1981, Cg
Deborah A. Spratt, (D), Johns Hopkins, 1980, Gc
Norman C. Wardlaw, (D), Glasgow, 1960, Go
Teaching Professor:
Jennifer Cuthbertson, (D), Carleton, 2006
Senior Instructor:
Glenn Dolphin
Alex Dutchak
Marius (Dan) Georgescu
Rudi Meyer, (D), Calgary, 1998, Gs
Rajeev S. Nair, (D), Alberta, 2008
Interim Dean, Faculty of Sciece:
Bernhard Mayer, (D), LMU Munich, 1993, CslGe
Instructor:
Ann Quinney

University of Guelph

School of Environmental Sciences (B,M,D)
Guelph, ON N1G 2W1
p. (519) 824-2052
earnaud@uoguelph.ca
http://www.ses.uoguelph.ca
Director:
Jonathan Newman, (D), Albany, 1990, Zn
Director, Controlled Environment Systems Research Facility:
Michael Dixon, (D), Edinburgh, Zn
Canada Research Chair in Recombinant Antibody Technology:
Christopher C. Hall, (D), Alberta, Zn
Canada Research Chair in Microbial Ecology:
Kari Dunfield, (D), Saskatchewan, 2002, SbZn
Associate Dean, Research, OAC:
Beverley A. Hale, (D), Guelph, 1989, Ct
Associate Dean, Academics, OAC:
Jonathan Schmidt, (D), Toronto, Zn
Professor:
Paul Goodwin, (D), California (Davis), Zn
Andrew Gordon, (D), Alaska (Fairbanks), 1984, Zn
Ernesto Guzman, (D), California (Davis), Zn
Tom Hsiang, (D), Washington, Zn
Hung Lee, (D), McGill, Zn
Stephen Marshall, (D), Guelph, Zn
Gard Otis, (D), Kansas, Zn
Cynthia Scott-Dupree, (D), Simon Fraser, 1986, Zn
Jack Trevors, (D), Waterloo, Zn
R. Paul Voroney, (D), Saskatchewan, 1983, Sb
Claudia Wagner Riddle, (D), Guelph, 1992, Am
Director of the Arboretum:
Shelley Hunt, (D), Guelph, Zn
Dean, OAC:
Rob J. Gordon, (D), Guelph, As
Associate Professor:
Madhur Anand, (D), W Ontario, Zn
Emmanuelle Arnaud, (D), McMaster, 2002, Gls
Susan Glasauer, (D), Tech (Munich), 1995, Pg
Marc Habash, (D), Guelph, Zn
Rebecca Hallett, (D), Simon Fraser, Zn
Richard Heck, (D), Saskatchewan, Sd
John Lauzon, (D), Guelph, So
Ivan O'Halloran, (D), Saskatchewan, 1986, So
Paul Sibley, (D), Waterloo, Zn
Jon Warland, (D), Guelph, 1999, As
Assistant Professor:
Tim Rennie, Zn
Laura Van Eerd, (D), Guelph, Zn
Alan Watson, (M), Guelph, Zn
Instructor:
Neil Rooney, (D), McGill, Zn
Adjunct Professor:
Gary Parkin, (D), Guelph, 1994, HwSp
Emeritus:
George Barron, (D), Iowa State, Zn
Greg J. Boland, (D), Guelph, Zn
Ward Chesworth, (D), McMaster, 1967, ClGe
Les J. Evans, (D), Wales, 1974, Cl
Austin Fletcher, (D), Alberta, Zn
Terry J. Gillespie, (D), Guelph, 1968, Am
Michael J. Goss, (D), Reading (UK), So
Pieter H. Groenevelt, (D), Wageningen (Neth), 1969, Sp
Robert Hall, (D), Melbourne (Australia), Zn
Stewart G. Hilts, (D), Toronto, 1981, Zn
Peter Kevan, (D), Alberta, Zn
Kenneth M. King, (D), Wisconsin, 1956, As
I. Peter Martini, (D), McMaster, 1966, Gsl
Raymond G. McBride, (D), Guelph, 1982, So
Murray H. Miller, (D), Purdue, 1957, So
Kaushik Narinder, Zn
Leonard Ritter, Zn
Keith Solomon, (D), Illinois, Zn
Gerry Stephenson, Zn
George W. Thurtell, (D), Wisconsin, 1965, As
H. Peter Van Straaten, (D), Goettingen (Germany), 1974, EnScCb

University of Lethbridge

Dept of Geography (B,M)
4401 University Drive
Lethbridge, AB T1K 3M4
p. (403) 329-2225
geography@uleth.ca
http://home.uleth.ca/geo
Administrative Assistant: Margaret Cook
Professor:
Walter E. Aufrecht, (D), Toronto, Zn
Rene W. Barendregt, (D), Queen's (Canada), 1977, Gm
Robert J. Rogerson, (D), Macquarie, 1979, Gm
Associate Professor:
James M. Byrne, (D), Alberta, 1990, Hg
Hester Jiskoot, (D), Leeds (UK), 2000, GlmZy
Thomas Johnston, (D), Waterloo, 1988, Ze
Derek R. Peddle, (D), Waterloo, 1997, Zr
Assistant Professor:
Craig Coburn, (D), Simon Fraser, 2002, Zy
Susan Dakin, (D), Waterloo, 2000, Ze
Stefan Kienzle, (D), Heidelberg, 1993, Zi
Ivan Townshend, (D), Calgary, 1997, Zn
Wei Xu, (D), Guelph, 1998, Zi
Adjunct Professor:
John Dormaar, (D), Alberta, 1961, So
Ron Hall, Zn
Larry Herr, (D), Zn
Daniel L. Johnson, (D), British Columbia, 1983, Zn
Pano George Karkanis, (D), Uppsala, 1966, So
Ross McKenzie, (D), Zn
Anne Smith, (D)
Derald Smith, (D), Zn
Emeritus:
Roy J. Fletcher, (D), Clark, 1968, As
Ian R. MacLachlan, (D), Toronto, 1990, Ze

University of Manitoba

Geological Sciences (B,M,D) O⊘
240 Wallace Building
125 Dysart Road
Winnipeg, MB R3T 2N2

p. (204) 474-9371
alfredo.camacho@umanitoba.ca
http://www.umanitoba.ca/geoscience
Administrative Assistant: Brenda Miller
Enrollment (2014): M: 12 (3) D: 8 (2)

Professor:
Anton Chakhmouradian, (D), St Petersburg State (Russia), 1997, GziCg
Nancy Chow, (D), Memorial, 1986, Gs
Robert J. Elias, (D), Cincinnati, 1979, Pi
Mostafa Fayek, (D), Saskatchewan, 1996, Cs
Ian J. Ferguson, (D), Australian Nat, 1988, YemYu
Andrew Frederiksen, (D), British Columbia, 2001, YsGt
Norman M. Halden, (D), Glasgow, 1983, Cg
Frank C. Hawthorne, (D), McMaster, 1973, Gz
William M. Last, (D), Manitoba, 1980, GsnGo
Soren Rysgaard, (D), Aarhus (Denmark), 1995, Gl
Elena Sokolova, (D), Moscow (Russia), 1980, Gz

Senior Scholar:
George S. Clark, (D), Columbia, 1967, Cc
Barbara L. Sherriff, (D), McMaster, 1988, Gz
Allan C. Turnock, (D), Johns Hopkins, 1960, Gp

Associate Professor:
Alfredo Camacho, (D), Australian Nat, 1998, Gt

Assistant Professor:
Genevieve Ali, (D), Montreal, 2010, Hw
Zou Zou Kuzyk, (D), Manitoba, 2009, Cg

Research Associate:
Yassir Abdu, (D), Uppsala, 2004, Gz

Instructor:
Karen Ferreira, (M)
William S. Mandziuk, (M), Manitoba, 1989, Gg
Jeffrey Young, (M), Manitoba, 1992, GcZe

Adjunct Professor:
Scott Anderson, (D), Dalhousie, 1998
Christian Bohm, (D), ETH (Switzerland), 1996, Cg
William Buhay
David Corrigan, (D), Carleton, Gt
Jody Deming, (D), Maryland, 1981, Zn
Michel Houle, (D), Laurentian, 2008, Gv
Brooke Milne, (D), Ga
Vince Palace, (D), Manitoba, 1996
James Reist, (D), Toronto, 1983
Graham A. Young, (D), New Brunswick, 1988, Pi

Emeritus:
William C. Brisbin, (D), California (Los Angeles), 1970, Gc
Robert B. Ferguson, (D), Toronto, 1948, Gz
Wooil M. Moon, (D), British Columbia, 1976, ZrYnOp
James T. Teller, (D), Cincinnati, 1970, Gs

Related Staff:
Neil Ball, (B), Manitoba, 1982, Gz
Laura Bergen, (M), Manitoba, 2013
Mark Cooper, (M), Manitoba, 1997, Gz
Mulu Serzu, (D), Manitoba, 1990, Yg
Ryan Sharpe, (M), Manitoba, 2012
Ravinder Sidhu
Panseok Yang, (D), Memorial, 2002, CaGp
Misuk Yun, (M), Manitoba, 1986, Cs

University of New Brunswick

Dept of Earth Sciences (B,M,D)
Fredericton, NB E3B 5A3
p. (506) 453-4803
geology@unb.ca
Administrative Secretary: Merrill Ann Beatty

Chair:
Cliff S.J. Shaw, (D), W Ontario, 1994, Giv
Joseph C. White, (D), W Ontario, 1979, Gc

Professor:
Bruce E. Broster, (D), W Ontario, 1982, Gl
John Todd Dunn, (D), Alberta, 1983, Gi
John G. Spray, (D), Cambridge, 1980, GpXgGz
Paul F. Williams, (D), Sydney, 1969, Gc

Associate Professor:
Nicholas J. Susak, (D), Princeton, 1981, CglEm

Assistant Professor:
Karl Butler, (D), British Columbia, 1996, Yg

Research Associate:
James Whitehead, (D), New Brunswick, 1998, Gg

Honorary Research Professor:
Henk W. Van De Poll, (D), Swansea, 1970, Gs

Adjunct Professor:
Richard A. F. Grieve, (D), Toronto, 1970, Xg
David R. Lentz, (D), Ottawa, 1992, Cg
Randall F. Miller, (D), Waterloo, 1984, Pg

Emeritus:
Arnold L. McAllister, (D), McGill, 1950, Em

Geology Librarian:
Eszter Schwenke, Zn

University of New Brunswick Saint John

Dept of Biological Sciences (B,M,D) ⊘
PO Box 5050
100 Tucker Park Road
Saint John, NB E2L 4L5
p. (506) 648-5607
lwilson@unbsj.ca
http://www.unbsj.ca/sase/biology/
Programs: Primarily a Biology department, with a Geology minor as an option.
Enrollment (2021): D: 1 (0)

Professor:
Lucy A. Wilson, (D), Paris VI, 1986, Ga

University of Ottawa

Department of Earth and Environmental Sciences (B,M,D) ●☒
STEM Complex
150 Louis Pasteur Pvt
Ottawa, ON K1N 6N5
p. (613) 562-5773
estchair@uottawa.ca
https://science.uottawa.ca/earth/
Programs: Geology, Geology-Physics, Environmental Geosciences, Environmental Sciences
Enrollment (2021): B: 236 (40) M: 27 (8) D: 30 (2)

Chair:
David Schneider, (D), Lehigh, 2000, GtCcGc

Professor:
Tom Al, (D), Waterloo, 1996, HwCgZg
R. William Arnott, (D), Alberta, 1987, Gsd
Pascal Audet, (D), British Columbia, 2008, Ygs
Danielle Fortin, (D), Quebec, 1992, CblGe
Mark Hannington, (D), Toronto, 1989, EgGuEm
Keiko Hattori, (D), Tokyo, 1977, CgEmGz
Glenn A. Milne, (D), Toronto, 1998, YgAsZg

Associate Professor:
Clement Bataille, (D), Utah, 2014, ClGeCs
Jonathan O'Neil, (D), McGill, 2009, Cgc

Assistant Professor:
Brett Walker, (D), California (Santa Cruz), 2010, OgcCc

Lecturer:
Simone Dumas, (D), Ottawa, 2004, GsOg

Adjunct Professor:
John Crowley
Eric de Kemp
Andre Desrochers, (D), Memorial, 1986, Gs
David A. Fisher, (D), Copenhagen, 1978, Gl
Dorothy Godfrey-Smith
Richard Goulet, (D), Ottawa, 2001, CgPgGe
Jeffrey Hedenquist, (D), Auckland, 1983, Eg
Matthew Herod
Christopher Lawley
Aaron Lussier
Dogan Paktunc, (D), Ottawa, 1983, Gz
Claire Perry
Pat E. Rasmussen, (D), Waterloo, 1993, GeAsGb

Emeritus:
Ian D. Clark, (D), Paris, 1988, HwGel
Jan Veizer, (D), Australian Nat, 1971, Cl

University of Regina

Dept of Geology (B,M,D) ●☒
3737 Wascana Parkway

Regina, SK S4S 0A2
p. (306) 585-4147
geology.office@uregina.ca
http://www.uregina.ca/science/geology/
Programs: Geology and Environmental Geoscience
Administrative Assistant: Van Tran
Enrollment (2021): B: 0 (10) M: 16 (1) D: 5 (2)

Professor:
Kathryn Bethune, (D), Queens, GcpGt
Guoxiang Chi, (D), Queens, Eg
Ian Coulson, (D), Birmingham (UK), 1996, Gvi
Hairuo Qing, (D), McGill, 1991, GsCg
Maria Velez, (D), Amsterdam

Associate Professor:
Osman Salad Hersi

Assistant Professor:
Joyce McBeth, (D)
Tsilavo Raharimahefa, (D), Gc
Leslie Robbins, (D)

Adjunct Professor:
Kenneth E. Ashton, (D), Gp
Pier L. Binda, (D), Alberta, 1970, Ge
Janis Dale, (D), Queens, Gml
Donald M. Kent, (D), Alberta, 1968, Gs

Emeritus:
Stephen L. Bend, (D), Newcastle, Gox

University of Saskatchewan

Dept of Geological Sciences (B,M,D) O

114 Science Place
Saskatoon, SK S7N 5E2
p. (306) 966-5683
sam.butler@usask.ca
http://artsandscience.usask.ca/geology/
Programs: Geology; Geophysics; Paleobiology; Environmental Geoscience
Enrollment (2021): B: 89 (19) M: 40 (7) D: 23 (0)

Head:
Sam Butler, (D), Toronto, 2000, Yg

Professor:
Kevin M. Ansdell, (D), Saskatchewan, 1992, EgCgGt
Luis Buatois, (D), Buenos Aires, 1992, Ps
Graham George, (D), Sussex (UK), 1983, Cg
Chris Holmden, (D), Alberta, 1995, Csc
Gabriela Mangano, (D), Buenos Aires, 1992, PsGs
Igor B. Morozov, (D), Moscow State (Russia), 1985, YsGy
Yuanming Pan, (D), W Ontario, 1990, Gx
William P. Patterson, (D), Michigan, 1995, CsPeGn
Ingrid J. Pickering, (D), Imperial Coll (UK), 1990, CgtZn
Brian R. Pratt, (D), Toronto, 1989, PgGsr

Associate Professor:
Matthew B. Lindsay, (D), Waterloo, 2009, ClGg
Camille Partin, (D), Gtc

Adjunct Professor:
Irvine Annesley, (D), Ottawa, 1989, Gi
Emily Bamforth, Pv
Ning Chen, (D), Saskatchewan, 2001, Gz
Meagan Gilbert, (D), Saskatchewan, Gs
David Greenwood, (D), Pg
Tom Kotzer, (D), Saskatchewan, 1993, CgEg
Joyce M. McBeth, (D), Manchester (UK), 2007, PgGeSb
Brett Moldovan, (D), Cg
Michael Moncur, Ge
Jenni Scott, (D), Saskatchewan, Gs
Len Wassenaar, (D), Waterloo, 1990, Hw
Derek A. Wyman, (D), Saskatchewan, 1990, Cg

Emeritus:
James F. Basinger, (D), Alberta, 1979, Pb
William G. E. Caldwell, (D), Glasgow, 1957, Ps
Donald J. Gendzwill, (D), Saskatchewan, 1969, Ye
Zoltan Hajnal, (D), Manitoba, 1970, YsGtc
Jim Hendry, (D), Waterloo, 1984, HwClHy
James B. Merriam, (D), York, 1976, Yg
Robin W. Renaut, (D), London (UK), 1982, Gsn
Mel R. Stauffer, (D), Australian Nat, 1964, Gc

University of Toronto

Department of Physics (B,M,D) ⦸

60 St. George Street
Toronto, ON M5S 1A7
p. (416) 978-5175
chair@physics.utoronto.ca
http://www.physics.utoronto.ca
f: https://www.facebook.com/PhysicsUofT/
t: https://twitter.com/uoftphysics
Programs: Physics (Minor, Major, Specialist BSc); Physics and Earth Sciences (Joint Specialist BSc); other joint undergraduate programs; Physics (MSc and PhD)

Professor:
Qinya Liu, (D), Caltech, 2006, Ys
Julian Lowman, (D), York, 1997, Gc

Emeritus:
Richard C. Bailey, (D), Cambridge, 1970, Yg
Richard N. Edwards, (D), Cambridge, 1970, Yr
Bernd Milkereit, (B), Kiel, 1984, Ye

Dept of Earth Sciences (B,M,D) ●☒

Earth Sciences Centre
22 Russell Street
Toronto, ON M5S 3B1
p. (416) 978-3022
welcome@es.utoronto.ca
http://www.es.utoronto.ca
Business Officer: Silvanna Papaleo

Chair:
Russell N. Pysklywec, (D), Toronto, 1998, Yg

Professor:
Nicholas Eyles, (D), Leicester (UK), 1978, Gl
Grant F. Ferris, (D), Guelph, 1985, Gn
Henry C. Halls, (D), Toronto, 1970, Ym
Kenneth W.F. Howard, (D), Birmingham (UK), 1979, Hw
Andrew D. Miall, (D), Ottawa, 1969, Gso
Barbara Sherwood Lollar, (D), Waterloo, 1990, Cs

Director, Jack Satterly Geochronology Lab:
Michael A. Hamilton, (D), Massachusetts, 1993, CcGiCu

Associate Professor:
Bridget Bergquist, (D), MIT, 2004, Cgb
Donald W. Davis, (D), Alberta, 1978, Cc
Grant S. Henderson, (D), W Ontario, 1983, Gz
Qinya Liu, (D), Caltech, 2006, Ygx
Daniel J. Schulze, (D), Texas, 1982, Gi

Assistant Professor:
Carl-Georg Bank, (D), British Columbia, 2002, Zg
Jörg Bollmann, (D), Swiss Fed Inst Tech, 1995, Pm
Xu Chu, (D), Yale, 2015, Gx
Rebecca Ghent, (D), S Methodist, 2002, Zr
Jochen Halfar, (D), Stanford, 1999, Pe
Ulrich B. Wortmann, (D), Tech (Munich), Gg
Zoltan Zajacz, (D), ETH (Switzerland), 2007, Cu

Research Associate:
Colin Bray, (D), Oxford, 1980, Cs

Adjunct Professor:
Jean-Bernard Caron, (D), Toronto, 2005, Pi
John H. McAndrews, (D), Minnesota, 1964, Pl

Emeritus:
G M. Anderson, (D), Toronto, 1961, Cp
Richard C. Bailey, (D), Cambridge, 1970, Yg
J. J. Fawcett, (D), Manchester (UK), 1961, Gp
John Gittins, (D), Cambridge, 1959, GipGz
Anthony J. Naldrett, (D), Queen's, 1964, EmGiCp
Geoffrey Norris, (D), Cambridge, 1964, Pl
Pierre-Yves F. Robin, (D), MIT, 1974, Gc
John C. Rucklidge, (D), Manchester (UK), 1962, Ge
Walfried M. Schwerdtner, (D), Free (Berlin), 1961, GctGq
Steven D. Scott, (D), Penn State, 1968, GuEmCm
Edward T. C. Spooner, (D), Manchester (UK), 1976, En
Peter H. von Bitter, (D), Kansas, 1971, Pm
John A. Westgate, (D), Alberta, 1964, Gl
Frederick J. Wicks, (D), Oxford, 1969, Gz

University of Victoria

School of Earth & Ocean Sciences (B,M,D) O☒

P.O. Box 1700
3800 Finnerty Road

Victoria, BC V8W 2Y2
p. (250) 721-6120
seos@uvic.ca
http://www.uvic.ca/science/seos/
Associate Director (SCIENCE) NEPTUNE Canada:
Kim Juniper, (D), Canterbury (NZ), 1982, Ob
Associate Dean of Science:
Kathryn Gillis, (D), Dalhousie, 1987, Gp
Professor:
Dante Canil, (D), Alberta, 1989, Gi
Laurence Coogan, (D), Leicester (UK), Gi
Stanley E. Dosso, (D), British Columbia, 1990, Op
Adam Monahan, (D), British Columbia, 2000, Zn
Thomas F. Pedersen, (D), Edinburgh, 1979, CmOcCs
Vera Pospelova, (D), McGill, 2003, PelOu
George D. Spence, (D), British Columbia, 1984, Ys
Verena Tunnicliffe, (D), Yale, 1980, Ob
Michael J. Whiticar, (D), Kiel, 1978, CosGo
Associate Professor:
Jay Cullen, (D), Rutgers, 2001, Oc
John Dower, (D), Victoria, 1994, Ob
Roberta C. Hamme, (D), Washington, 2003, Oc
Thomas S. James, (D), Princeton, 1991, YdvYg
Jody Klymak, (D), Washington, 2001, Op
Eileen Van Der Flier-Keller, (D), W Ontario, 1985, Cl
Diana Varela, (D), British Columbia, 1998, Ob
Assistant Professor:
Colin Goldblatt, (D), E Anglia (UK), 2008, As
Lucinda Leonard, (D), Victoria, Yg
Adjunct Professor:
Vaughn Barrie, (D), Wales, 1986, Gu
John Cassidy, (D), British Columbia, Ys
James R. Christian, (D), Hawaii, 1995, Obc
Kenneth L. Denman, (D), British Columbia, 1972, Op
Richard K. Dewey, (D), British Columbia, 1987, Opg
Gregory M. Flato, (D), Dartmouth, 1991, As
John C. Fyfe, (D), McGill, 1987, As
Richard J. Hebda, (D), British Columbia, 1977, Pl
Roy D. Hyndman, (D), Australian Nat, 1967, Ys
Stephen T. Johnston, (D), Alberta, 1993, Gc
Victor M. Levson, (D), Alberta, 1995, Gl
David L. Mackas, (D), Dalhousie, 1977, Ob
Norman McFarlane, (D), Michigan, 1974, As
Garry C. Rogers, (D), British Columbia, 1983, Ys
George J. Simandl, (D), Ecole Polytechnique, 1992, En
Richard Thomson, (D), British Columbia, 1971, Og
Kelin Wang, (D), W Ontario, 1989, Yg
Michael Wilmut, (D), Queen's, 1971, Zn
Emeritus:
Christopher R. Barnes, (D), Ottawa, 1964, PmOgPe
Ross N. Chapman, (D), British Columbia, 1975, Op
Christopher J. R. Garrett, (D), Cambridge, 1968, Op
John T. Weaver, (D), Saskatchewan, 1959, Ym
On Leave:
Andrew J. Weaver, (D), British Columbia, 1987, Op
Administrative Officer:
Terry P. Russell, (B), Victoria, 2000, Zn

University of Waterloo

Dept of Earth and Environmental Sciences (B,M,D) ●
Waterloo, ON N2L 3G1
p. (519) 888-4567, ext. 32069
klalbrec@uwaterloo.ca
https://uwaterloo.ca/earth-environmental-sciences/
Administrative Assistant: Lorraine Albrecht
Chair:
Barry G. Warner, (D)
Professor:
David W. Blowes, (D), Waterloo, 1990, Cg
Mario Coniglio, (D), Memorial, 1985, Gr
Maurice B. Dusseault, (D), Alberta, 1977, Ng
Thomas W. D. Edwards, (D), Waterloo, 1987, Cs
Stephen George Evans, (D), Alberta, 1983, Ng
Shaun K. Frape, (D), Queen's, 1979, Cl
Shoufa Lin, (D), New Brunswick, 1992, Zn
Carol J. Ptacek, (D), Waterloo, 1992, Cg
David L. Rudolph, (D), Waterloo, 1989, Hq
Sherry L. Schiff, (D), Columbia, 1986, Ng
Edward A. Sudicky, (D), Waterloo, 1983, Hw
Philippe Van Cappellen, (D)
Associate Professor:
Anthony E. Endres, (D), British Columbia, 1991, Pe
Walter Illman, (D)
Martin Ross, (D)
Andre Unger, (D), Waterloo, 1995, Hw
Assistant Professor:
Nandita Basu, (D)
Carl Guilmette, (D)
Brian Kendall, (D)
Lingling Wu, (D)
Lecturer:
Eric C. Grunsky, (D)
John Johnson, (D)
Isabelle McMartin, (D)
Brent Wolfe, (D)
Adjunct Professor:
Edward C. Appleyard, (D), Cambridge, 1963, Gp
Gail Atkinson, (D)
James F. Barker, (D), Waterloo, 1979, Co
Steven Berg, (D)
Alec Blyth, (D)
Thomas Bullen, (D)
Lauren Charlet, (D)
John A. Cherry, (D), Illinois, 1966, Hw
Peter Condon, (D)
Rick Devlin
Michael English, (D)
Nicholas Eyles, (D)
John F. Gartner, Zn
John J. Gibson, (D), Waterloo, 1996, Cs
Susan Glasuer, (D)
David Good, (D)
John Gosse, (D)
Douglas Gould, (D)
Norman Halden, (D)
Daniel Hammarlund, (D)
Jens Hartman, (D)
Lin Huang, (D)
Daniel Hunkeler
Hyoun-Tae Hwang, (D)
Richard Jackson, (D)
Sun-Wook Jeen, (D)
Michael Krom, (D)
David R. Lee, (D), Virginia Tech, 1976, Hw
Yuri Leonenko, (D)
John Lin, (D)
Robert Linnen, (D), McGill, 1992, Eg
Benoit Made, (D)
Uli Mayer, (D)
Hossein Memarian, (D)
John Molson
Alan V. Morgan, (D), Birmingham (UK), 1970, Gl
Christopher Neville, (D)
Dogan Paktunc, (D)
Sorab Panday, (D)
Young-Jin Park, (D), Hq
Gary Parkin, (N)
Peter Pehme, (D)
Jennifer Pell, (D)
Richard Peltier, (D)
Terry D. Prowse, (D), Canterbury (NZ), 1981, Cg
William Quinton, (D)
Eric J. Reardon, (D), Penn State, 1974, Cl
Rashid Rehan, (D)
Iain Samson, (D)
Houston C. Saunderson, (N)
James Sloan, (D)
James Smith
John Spoelstra, (D)
Andrew Stumpf, (D)
William Taylor, (D)
Rene Therrien, (D)
Harvey Thorleifson, (D)
Martin Thullner, (D)
Benoit Valley, (D)
Garth Van der Kamp, (D), Amsterdam, 1973, Hw
Jan van der Kruk, (D)

Cees van Staal
Andrea Vander Woude, (D)
Owen L. White, (D), Illinois, 1970, Ng
C. Wolkersdorfer, (D)
Wenjiao Xiao, (D)
Leiming Zhang, (D)

Emeritus:
Emil O. Frind, (D), Toronto, 1971, Hq
Robert W. Gillham, (D), Illinois, 1973, Hq
Paul Karrow, (D), Illinois, 1957, Gl

Research Professor:
Ramon Aravena, (D), Waterloo, 1993, Co

Research Associate Professor:
Will Robertson, (D), Waterloo, 1992, Hw

Other:
Brewster Conant, Hw

University of Windsor

Dept of Earth and Environmental Sciences (B,M,D) ⃠
401 Sunset Avenue
Windsor, ON N9B 3P4
p. (519)253-3000 (Ext. x2486)
earth@uwindsor.ca
http://www.uwindsor.ca/ees
Programs: Research-based, graduate degree program in Earth Sciences at the levels of M.Sc. and Ph.D. Research partnerships with the Great Lakes Institute for Environmental Studies (GLIER)
Administrative Assistant: Connie Iaquinta

Director:
Robert Mike McKay, (D), McGill, 1991

Dean and Acting Director:
Chris Houser, (D)

University President:
Robert Gordon

Tier 1 Canada Research Chair in Aquatic Invasive Species:
Hugh MacIssac, (D), Dartmouth, 1990

Professor:
Ihsan S. Al-Aasm, (D), Ottawa, 1985, GdClGo
Ken Drouillard, (D), Trent, 2001
Aaron Fisk, (D), Manitoba, 1998, ObCs
Joel E. Gagnon, (D), McGill, 2006, CaqEg
V. Chris Lakhan, (D), Toronto, 1982, ZriOn
Ali Polat, (D), Saskatchewan, 1998, GixCc
Iain M. Samson, (D), Strathclyde, 1983, CgEmGx
Frank -. Simpson, (D), Jagellonian (Krakow), 1968, GsrGe
Alan S. Trenhaile, (D), Wales, 1969, Gm

Associate Professor:
Maria T. Cioppa, (D), Lehigh, 1997, Ym
Phil A. Graniero, (D), Toronto, 2001, ZinZy
Cyril G. Rodrigues, (D), Carleton, 1980, Pm
Christopher Weisener, (D), South Australia, 2003, CoGePg
Jianwen Yang, (D), Toronto, 1997, HwYg

Cross Appointed:
Scott Mundle, (D), Toronto, 2010

Assistant Professor:
Cameron Proctor, (D), Toronto, 2017, Zi

Learning Specialist:
Jasra Pardeep

Academic Advisor:
Alice Grigcak-Mannion, (B), Waterloo, 1996, Zi
Denis Tetrault, (D), Western, 2002, Gg

Instructor:
Michelle Bondy

Lecturer:
Neil Porter

Emeritus:
Brian J. Fryer, (D), MIT, 1971, Cla
Peter P. Hudec, (D), Rensselaer, 1965, NgEmGe
Terence E. Smith, (D), Wales, 1963, Gi
David T. A. Symons, (D), Toronto, 1965, YmGtEm
Andrew Turek, (D), Australian Nat, 1966, Cc

Western University

Dept of Earth Sciences (B,M,D) O☒
Biology & Geology Building
Room 1026
1151 Richmond Street North
London, ON N6A 5B7
p. (519) 661-3187
earth-sc@uwo.ca
http://www.uwo.ca/earth/
t: @westernuEarth
Administrative Assistant: Erika Gongora
Enrollment (2016): B: 123 (15) M: 77 (16) D: 54 (8)

Chair:
Patricia Corcoran, (D), Dalhousie, 2001, Gd

Robert Hodder Chair:
Robert Linnen, (D), McGill, 1992, EmCe

NSERC Industrial Research Chair:
Gail M. Atkinson, (D), W Ontario, 1993, YssYs

Joint appointment with Geography, U.W.O.:
Desmond Moser, (D), Queens, 1993, GtCcXg

Industrial Research Chair, Joint Appt Physics & Astronomy, U.W.O:
Gordon R. Osinski, (D), New Brunswick, 2004, XgcGp

Cross-Appt.UWO: Physics&Astronomy:
Sean R. Shieh, (D), Hawaii, 1998, YxGzZm

Cross-Appt.-Home Dept: Physics & Astronomy, UWO:
Peter Brown, (D), W Ontario, 1999, Xym

Canada Research Chair, Cross Appt. Biology&Geography, UWO:
Brian Branfireun, (D), McGill, Ge

Canada Research Chair:
Frederick J. Longstaffe, (D), McMaster, 1978, Cs

(P.Eng.,P.Geo.):
Robert A. Schincariol, (D), Ohio State, 1993, HwsNg

Professor:
Dazhi Jiang, (D), New Brunswick, 1996, GcgGg
Jisuo Jin, (D), Saskatchewan, 1988, Pi
A Guy Plint, (D), Oxford, 1980, Gs
R. Gerhard Pratt, (D), Imperial Coll (UK), 1989, Yeg
Richard A. Secco, (D), W Ontario, 1988, Gy

Industrial Research Chair:
Neil Banerjee, (D), Victoria, 2001, EgCgGx

Cross-Appt.UWO: Physics&Astronomy:
Robert Shcherbakov, (D), Cornell, 2002, YgsYd

Canada Research Chair:
Audrey Bouvier, (D), École Normale Supérieure de Lyon, 2005, Xc

Associate Professor:
Burns A. Cheadle, (D), W Ontario, 1986, Go
Roberta L. Flemming, (D), Queen's (Canada), 1997, GzXmZm
Katsu Goda, (D), Rn
Elizabeth A. Webb, (D), W Ontario, 2000, CslSa

Assistant Professor:
Sheri Molnar, (D), Victoria, 2011, Ys
Catherine Neish, (D), Arizona, 2008, Xg
Cameron J. Tsujita, (D), McMaster, 1995, Pi
Tony Withers, (D), Bristol, 1997, Cp

Research Technician:
Peter Christoffersen, (M), W Ontario
Claire Montera

Research Scientist:
Li Huang, (M), Concordia, 1992, Yr

Research Engineer:
Matthew Bourassa, (D), Xg

Research Associate:
Hadi Ghofrani, (D)

W.S. Fyfe Visiting Scientist- in- Residence:
David Good, (D), McMaster, 1992, Em

Research Associate:
Brian R. Hart, (D), W Ontario, 1995, Ca

Research Adjunct Professor:
Natalie Pietrzak-Renaud, (D), W Ontario, 2011, Eg

Associate Curator of Mineralogy, Royal Ontario Museum/Professor, University of Toronto:
Kim Tait, (D), Arizona, 2007, Xm

Adjunct Professor:
Keith Baron, (D)
Melissa Battler
Rob Carpenter, (D)
Ed Cloutis, (D), Alberta, 1989, Zi
Claudia Cochrane, (M), W Ontario, Go
Yunpeng Dong, (D), Northwest (China), 1997, GtcCg
Richard Grieve, (D), Toronto, 1970, Xg
Matt Izawa, (D), W Ontario, 2012, Gz
Laura Jansen
Peter Lightfoot
Phil JA McCausland, (D), Western, 2002, YmXm
Gero Michel, (D), Harvard, Ys

Sobhi Nasir, (D), Wuerzburg, 1989
Sergey Samsonov
Brenden Smithyman, (D), Ye
Gordon Southam, (D), Guelph, 1990, Pg
Yves Thibault, (D), W Ontario, 1990, Gz
Kristy F. Tiampo, (D), Colorado, 2000, Yd
Livio Tornabene, (D), Tennessee, 2007, XgZr
Lisa Van Loon, (D), Ohio State, 2007, Ca

Emeritus:
Alan Beck, (D), Australian Nat, 1964, Zn
W. Glen E. Caldwell, (D), Glasgow, 1957, Pm
William R. Church, (D), Wales, 1961, Gt
Norman A. Duke, (D), Manitoba, 1983, Eg
Akio Hayatsu, (D), Toronto, 1965, Cc
Stephen R. Hicock, (D), W Ontario, 1980, Gl
Alfred C. Lenz, (D), Princeton, 1959, PiePi
Robert F. Mereu, (D), W Ontario, 1962, Ys
H Wayne Nesbitt, (D), Johns Hopkins, 1975, Cl
H. Currie Palmer, (D), Princeton, 1963, YmGv
Grant M. Young, (D), Glasgow, 1967, GrCgGt

Laboratory Director:
Kim Law, Cs
Wenjun Yong, (D), Yx

Technical Specialist:
Barry Price

Research Technologist:
Jon Jacobs, Gg

Research Technician:
Marc Beauchamp, Ca
Grace Yau

Lab Technician:
Steven Wood

Geoscience Collections Curator:
Alysha McNeil, (D), W Ontario, 2017

Financial Assistant:
Miyako Maekawa

Computer Technician:
Bernie Dunn, Ye

Administrative Officer:
Kristen Harris

Academic Program Coordinator:
Amy N. Wickham, (M), Connecticut, 2016

York University

Earth and Space Science and Engineering (B,M,D) ●
4700 Keele Street
Toronto, ON M3J 1P3
p. (416) 736-5245
essedept@yorku.ca
http://www.yorku.ca/esse
Administrative Assistant: Paola Panaro
Enrollment (2010): B: 100 (38)

Professor:
Qiuming Cheng, (D), Ottawa, 1994, Zi
Christian Haas, (D), Bremen, 1996, Yg
Gary T. Jarvis, (D), Cambridge, 1978, Yg
Ian C. McDade, (D), Belfast, 1979, Zr
Tom McElroy, (D), York, 1985, As
Spiros Pagiatakis, (D), New Brunswick, 1988, Yd
Jinjun Shan, (D)
Peter A. Taylor, (D), Bristol, 1967, As

Chair:
Regina Lee, (D), Toronto, 2000

Associate Professor:
Costas Armenakis, (D), New Brunswick, 1988, Zr
Sunil Bisnath, (D), New Brunswick, 2004
Yongsheng Chen, As
Michael Daly, (D)
Baoxin Hu, (D), Boston, 1998, Zr
Mary Ann Jenkins, (D), Toronto, 1986, As
Gary P. Klaassen, (D), Toronto, 1983, As
Brendan Quine, (D)
Gunho Sohn, (D)
George Vukovich, (D), As
James Whiteway, (D), York, 1994, As

Assistant Professor:
William Colgan, (D), Yg
Mark Gordon, (D), As
John E. Moores, (D), Arizona, 2008

Lecturer:
Hugh Chesser, (M), Toronto, 1987
Franz Newland, (D), Southampton (UK), 2000, Zr
Jian-Guo Wang, (D)

Emeritus:
Keith D. Aldridge, (D), MIT, 1967, YmdYh
John Miller, (D), Saskatchewan, 1969, Zr
Gordon G. Shepherd, (D), Toronto, 1956, AsZrAp
Douglas E. Smylie, (D), Toronto, 1963, Yg
Anthony M. K. Szeto, (D), Australian Nat, 1982, Yg

Cape Verde

Universidade de Cabo Verde

Dept de Ciencia e Tecnologia (M,D)
Campus de Palmarejo - CP 279
Palmarejo, Praia - Cabo Verde
Praia
Praca Antonio Lereno, Praia CP 379C
p. +238 261 99 01
joao.semedo@docente.unicv.edu.cv
http://www.unicv.edu.cv/dct

Faculdade de Engenharia e Ciencias do Mar (B,M,D) ☒
Ribeira de Julião
CP 163
Mindelo, São Vicente CP 163
p. +238 2326561/62
amdelgado@uta.cv
http://www.unicv.edu.cv

Universidade Jean Piaget de Cabo Verde

Ecologia e Desenvolvimento (B)
Campus Universitario da Cidade da Praia
Praia
p. +238 629085
info@unipiaget.cv
http://www.unipiaget.cv/pdf/cursos/ecdm.pdf

Central African Republic

Universite de Bangui

Dept de Chimie-Biologie-Geologie (B)
BP 1450
Avenue des Martyrs
Bangui
p. (236) 61 20 05
info@univ-bangui.org
http://www.univ-bangui.org/

Chad

Universite de N'Djamena

Dept de Geologie (B)
BP 1117
Avenue Mobutu
N'Djamena
abomnia50@gmail.com
http://www.undt.info/

Chile

Servicio National de Geologia y Mineria de Chile

Servicio National de Geologia y Mineria de Chile
Avda Sta Maria Nº 0104
Providence
p. 56-2-7375050
oirs@sernageomin.cl
http://www.sernageomin.cl/

Universidad Austral de Chile

Inst de Geociencias
Campus "Isla Teja", Casilla 567, Valdivia
p. +56 63 2293861
cienciasdelatierra@uach.cl

Universidad Catolica del Norte

Dept de Ciencias Geologicas
Avenida Angamos 610, ,
Casilla 1280, Antofagasta
p. (55) 2355968
mbembow@ucn.cl
http://www.ucn.cl/facultades/SitioDelnteres/?cod=2&codItem=110&codPrincipal=1124

Universidad de Chile

Dept de Geologia
Plaza Ercilla 803, Casilla 13148, Correo 21, Santiago
colegios@ing.uchile.cl

Dept de Geofisica
Blanco Encalada 2002
Casilla 2777
Santiago
p. (56 2)696 6563
contacto@dgf.uchile.cl
http://www.dgf.uchile.cl

Universidad de Concepcion

Inst de Geologia Economica Aplicada
Barrio Universitario - Victor Lamas 1290
Casilla 4107
Concepcion, Region del Bio-Bio
p. (56) 41 220 44 88
maravenah@udec.cl
http://www.udec.cl/postgrado/?q=node/39&codigo=4161

Dept de Geociencia
Cabina 13 - Barrio Universitario, Casilla 3-C, 4250-1 Concepcion, Region del Bio-Bio

China

Capital Normal University

College of Resources, Environment and Tourism
West 3rd Ring Road North 105#
Beijing 100048
p. 86 10 68903321
info@mail.cnu.edu.cn
http://www.cnu.edu.cn

China Geological Survey

China Geological Survey
24 Huangsi Dajie,
Xicheng District
Beijing 100011
p. +86 10 51632963 51632906
enwebmaster@mail.cgs.gov.cn
http://www.cgs.gov.cn/

China University of Geosciences, Beijing

Geosciences
No.29, Xueyuan Road,
Haidian District
Beijing 100083
http://www.cugb.edu.cn/EnglishWeb

China University of Geosciences, Wuhan

Faculty of Geosciences
No. 388, Lumo Road, Hongshan District
Wuhan 430074
p. 86-27-87481030
wudany@cug.edu.cn
http://en.cug.edu.cn/cug

China University of Mining and Technology, Beijing

College of Geoscience and Surveying Engineering
Beijing
dcxy@cumtb.edu.cn
http://dcxy.cumtb.edu.cn/

Nanjing University

Dept of Earth Sciences
22 Hankou Road
Nanjing , Jiangsu
wsj@nju.edu.cn
http://www.nju.edu.cn/cps/site/njueweb/fg

University of Hong Kong

Dept of Earth Sciences (B,M,D) ● ☒
James Lee Science Building
Pokfulam Road
Hong Kong
p. (852) 2859 1084
earthsci@hku.hk
https://www.earthsciences.hku.hk/
f: https://www.facebook.com/HKUEarthSciences/
Programs: Geology, Earth System Science

Colombia

EAFIT University

Dept of Geology
Carrera 49 No. 7 Sur - 50
Medellin
p. 01 8000515 900
contacto@eafit.edu.co

Escuela de Ingenieria de Antioquia

Dept of Geologic Engineering ☒
Calle 25 Sur
#42-63 Envigado
Medellin
maria.espinosa68@eia.edu.co
http://www.eia.edu.co/site/index.php/pregrados/programas/ing-geologica.html

Universidad de Santander

Environmental Engineering (B,M)
Facultad Ingenierias
Cll 70 No. 55-210 Campus Universitario Lagos del Cacique
Bucaramanga, Santander
p. +57 7 6516500
nmantilla@udes.edu.co
http://www.udes.edu.co/programas-profesionales/facultad-ingenierias/ingenieria-ambiental.html

Universidad Industrial de Santander

School of Geology (B,M) ○
Cra 27 Calle 9 Ciudad Universitaria
Bucaramanga, Santander 1
p. 57-7-6343457
escgeo@uis.edu.co
http://geologia.uis.edu.co/eisi/

Director:
Juan Diego Colegial Gutierrez, (D), Politecnica de Madrid (Spain), 2004, GeNgZr

Professor:
Luis Carlos Mantilla Figueroa, (D), EgCec

Universidad Nacional de Colombia

Dept de Geociencias (B,M,D) ☒
Carrera 30 No. 45-03, Edificio 224
Calle 45, Cr. 30
Bogota
depgeocie_fcbog@unal.edu.co
http://ciencias.bogota.unal.edu.co/departamentos/departamento-de-geociencias/
Programs: Geology, Msc. Geophysics, Msc. Geology, Msc. Meteorology, PhD. Geosciences
Enrollment (2022): B: 55 (0) M: 18 (0) D: 2 (0)

Costa Rica

Universidad de Costa Rica

Escuela Centroamericana de Geologia (B,M) ●☒
Apartado 35-2060 UCR, San Pedro de Montes de Oca, San José San José
p. 506 25118129
geologia@ucr.ac.cr
http://www.geologia.ucr.ac.cr/
f: https://www.facebook.com/ECG.UCR/
t: @geologiaucr
Programs: Geology (BS)
Certificates: NA

Director:
Mauricio Manuel Mora Fernández, (D), Savoie (France), 2003

Professor:
Mario Enrique Arias Salguero, (M)
Marco Barahona Palomo, (D)
Guaria Cárdenes Sandí, (D)
Percy Denyer Chavarría, (D)
Lepolt Linkimer Abarca, (D)
Rolando Mora Chinchilla, (M)

Associate Professor:
Elena Badilla Coto, (M)
Oscar Lucke Castro, (D)
María Isabel Sandoval Gutiérrez, (D)

Instructor:
María Cristina Araya , (D), 2020, YdGtYs
Patrick Durán Leiva, (B)
Maximiliano Garnier Villareal, (D), NxYgGg
María del Pilar Madrigal Quesada, (D)
Stephanie Murillo Maikut, (D)
Giovanni Marino Peraldo Huertas, (M)
Vanessa Rojas Herrera, (A)
Paulo Ruiz Cubillo, (D), GvmRn
Luis Guillermo Salazar Mondragón, (M)
César Sequeira Peraza, (D), Rutgers, 2019
Ingrid Vargas Azofeifa, (M)

Escuela Centroamericana de Geología (B,M) ●⃠
Apartado 214-2060 UCR
San Pedro de Montes de Oca
San Jose, San José
p. +506 2511 8128
geologia@ucr.ac.cr
http://www.geologia.ucr.ac.cr/
f: https://www.facebook.com/ECG.UCR/
Programs: Bachillerato en Geología, Licenciatura en Geología

Cote d'Ivoire

Universite d'Abobo-Adjame

UFR des Sciences de la Nature (B)
info@uabobo.ci

Universite de Cocody

Dept des Sciences de la Terre et des Resources Minieres (B,M,D)
BP V 34
Abidjan
p. +225 (22) 44 35 31
sinfuc@univ-cocody.ci
http://www.univ-cocody.ci/

Université Félix Houphouët-Boigny

UFR des Sciences de la Terre et de Ressources Minieres (B)
BP V 34
Abidjan 01
p. +225 22 44 35 31
ufrstrm@univ-fhb.edu.ci
http://www.univ-cocody.ci/struct_formation/strm

UFR des Sciences des Structures de la Matiere et Technologie (B)
BP V 34
, Abidjan 01
p. +225 22 44 35 31
secretariat.presidence@univ-fhb.edu.ci
http://www.univ-cocody.ci/struct_formation/ssmt

Croatia

Croatian Geological Survey

Croatian Geological Survey ☒
Sachsova 2
P.O.Box 268
Zagreb HR-10000
p. +38516160888
ured@hgi-cgs.hr
http://www.hgi-cgs.hr/
f: https://www.facebook.com/HGI.CGS/

University of Zagreb

Dept of Geology, Faculty of Science (B,M,D)
Horvatovac 102a
Zagreb HR-10000
p. +38514605960
godsjek@geol.pmf.hr
http://www.geol.pmf.hr
Enrollment (2010): B: 121 (27) M: 28 (11) D: 30 (6)

full professor:
Mladen Juraèiæ, (D), Zagreb, 1987, GueCm

associate professor:
Dražen Balen, (D), Zagreb, 1999, GxpGi

full professor:
Darko Tibljaš, (D), Zagreb, 1996, GzScGx
Vlasta Æosoviæ, (D), Zagreb, 1996, PemPs

emeritus:
Ivan Gušiæ, (D), Zagreb, 1974, GgPsm

Cyprus

Ministry of Agriculture, Rural Development and Environment

Cyprus Geological Survey Department ☒
1 Lefkonos Street
Strovolos 2064
p. +357 22409213
director@gsd.moa.gov.cy
http://www.moa.gov.cy/gsd
f: https://www.facebook.com/Cyprus-Geological-Survey-Department-894868840721111
t: @CY_earthquakes

Czech Republic

Charles University

Inst of Hydrogeology, Engineering Geology and Applied Geophysics
Albertov 6
128 43 Praha 2
fischer@natur.cuni.cz
http://prfdec.natur.cuni.cz/~geophys/ustav/katedra.htm

Institute of Petrology and Structural Geology (B,M,D)
Albertov 6
Praha 2 128 43
p. 00420-221951524
petrol@natur.cuni.cz
http://petrol.natur.cuni.cz
Enrollment (2012): B: 3 (5) M: 15 (4) D: 15 (0)

Assoc. Prof.:
David Dolejs, (D), 2004, GiCpg

Professor:
Shah Wali Faryad, (D), 1990, GpzGi

Dept of Geophysics (B,M,D) ☒
KG MFF UK
V Holesovickach 2
180 00 Prague 8, Czech Republic
Prague 180 00

geo@mff.cuni.cz
http://geo.mff.cuni.cz
Programs: Solid Earth Geophysics (B); Planetary Science (B)
Enrollment (2017): B: 5 (0) M: 3 (0) D: 2 (0)

Czech Geological Survey

Czech Geological Survey ☒
Klarov 3
Praha 1 118 21
p. +420 257 089 500
secretar@geology.cz
http://www.geology.cz

Geofond
Kostelni 26
Prague PSC 170 06
p. +420 234 742 111
jana.kubova@geologycz
http://www.geofond.cz/cz/domu

Masaryk University

Dept of Geological Sciences (B,M,D) O☒
Faculty of Science
Kotlarska 2
Brno 611 37
p. +420 549 49 4322
geologie@sci.muni.cz
http://ugv.cz/
f: https://www.facebook.com/chci.byt.geolog
Enrollment (2017): B: 134 (30) M: 80 (35) D: 81 (5)

Head:
Zdenek Losos, (D), Gzy

Professor:
Ondrej Babek, (D), Gds
Jiri Kalvoda, (A), PsePm
Milan Novak, (A), GzpGi
Antonín Prichystal, (A), GgaGv

Associate Professor:
Jan Cenpírek, Gz
Eva Geršlová, Ge
Martin Ivanov, (A), PgiPe
Jaromir Leichmann, (A), GxpGi
Rostislav Melichar, (A), GtcGr
Slavomír Nehyba, (A), GsdGg
Radek Skoda, (A), Masaryk (Czech Republic), 2017, GzZmGi
Marek Slobodnik, (A), GeEg
Josef Zeman, (D), Comenius (Slovakia), 1987, CgpCe

Assistant Professor:
Martin Knizek, (D), NgmNr
Tomas Kuchovsky, (D), Hgy
Tomas Kumpan, Ps
Pavel Pracny, Cg
Adam Ricka, (D), Hgy

Research Associate:
Renata Copjakova, (D), Gz

Lecturer:
Nela Dolakova, (A), Plg

Emeritus:
Rostislav Brzobohaty, (A), PgsPe
Rudolf Musil, (D), Charles (Prague), 1968, PgvGe

Dept of Geological Sciences
Kotlarska 2
611 37 Brno
Brno 611 37
pechmannova@sci.muni.cz
https://www.muni.cz/en/about-us/organizational-structure/faculty-of-science/315010-deptof-geological-sciences/teaching

Dept of Geography
Kotlarska 2
611 37 Brno
dobro@sci.muni.cz

http://www.muni.cz/sci/structure/315030.html

Technical University of Ostrava

Faculty of Mining & Geology
Vysoka Skola Banska v Ostrava
17.listopadu 15/2172
708-33 Ostrava
p. +420 597 325 456
dekan.hgf@vsb.cz
http://www.hgf.vsb.cz/cs

D.R. of Congo

Marien Ngouabi University

Faculty of Science and Technology (B,M,D) O
Face Square Général Charles De Gaulle
Bacongo
BP:69
Brazzaville, Congo
p. (+242) 06 623 61 22
jm_ouamba@yahoo.fr
http://www.univ-mngb.net/fs/

Universite de Kinshasa

Dept des Sciences de la Terre (B)
PO Box 190
Kinshasa XI
p. (+243) 82 333 96 93
contact@facsciences-unikin.ac.cd
http://www.unikin.cd

Départements et filières Sciences Agronomiques (B)
PO Box 190
Kinshas XI
p. +243 89 89 20 507
fbkapuku@hotmail.com
http://www.unikin.cd/ogec/

Universite de Lumumbashi

Faculté des sciences (B)
BP 1825
Lubumbashi
p. +243 263-22-5403
kaumbalufunda@unilu.ac.cd
http://www.unilu.ac.cd/En/Pages/

Denmark

Aarhus University

Department of Geoscience (B,M,D) ☒
Hoegh-Guldbergs Gade 2
DK-8000 Arhus C
geologi@au.dk
http://geo.au.dk
Programs: Geology
Geophysics

Geological Survey of Denmark and Greenland (GEUS)

Geological Survey of Denmark and Greenland (GEUS) ☒
Oester Voldgade 10
Copenhagen DK-1350
p. +45 38 20 00
geus@geus.dk
http://www.geus.dk/
Programs: As a research institute affiliated to the Danish Ministry of Climate, Energy and Utilities we do not offer degree programs but some of our researchers act as co-supervisors with university colleagues for both master- and PhD-students, and in some cases, PhD-students are employed at GEUS .
on a regular basis

Roskilde University

QuadLab, ENSPAC
Universitetsvej 1
Roskilde DK-4000
p. 45 46743097
ruc@ruc.dk

http://www.quadlab.dk/bcms-ui-base/

Technical University of Denmark

Dept of Earth Sciences
Anker Engelunsved
DK-2800 Lyngby
thho@env.dtu.dk
http://www.igg.dtu.dk/index_z.htm

University of Copenhagen

Inst for Geography and Geology
Oster Voldgage 5-7
DK-1350 Copenhagen K
p. +4535322400
pab@ign.ku.dk
http://geo.ku.dk/

Dept of Geophysics ☒
Juliane Maries Vej 30
Copenhagen 2100 Copenhagen
p. +45 353 20605
losos@sci.muni.cz
http://www.nbi.ku.dk/theinstitute/page52794.htm

Geological Museum ☒
Oster Voldgade 5-7
Copenhagen K DK-1350
p. +45 35322345
rcp@snm.ku.dk
http://geologi.snm.ku.dk/english/

Djibouti

Insitut de Physique du Globe de Paris, (IPGP)

Dept of Volcanic Systems (B)
Observatoire Géophysique d'Arta, BP 1888
p. (253) 42 21 92
komorow@ipgp.fr
http://volcano.ipgp.jussieu.fr/djibouti/stationdj.html

Universite de Djibouti

Dept de Geologie/Biologie (B)
BP 1904
p. +253-250459
webmaster@univ.edu.dj
http://www.univ.edu.dj/facultes.fsti.bg.html

Dominican Republic

Mineterio de Energie y Minas Republica Dominica

Servicio Geologico Nacional Republica Dominica
Ave Winston Churchill No 75
Edificio J.F. Martinez 3er Piso
Santa Domingo
p. (809) 732 0363
info@mem.gob.do
http://www.sgn.gov.do

Ecuador

Escuela Politecnica Nacional

Facultad de Geologia, Minas y Petroleos
Ladron de Guevara E11
253 Quito
p. (593-02) 2-507 - 126
info@epn.edu.ec
http://www.epn.edu.ec/index.php?option=com_content&view=article&id=1202%3Adepartamento-de-geologia-dg&catid=163&Itemid=342

Escuela Superior Politecnica Del Litoral

Facultad de Ciencias de la Tierra
Km 30.5, Via Perimentral
Guayaquil EC090150

Universidad de Guayaquil

Facultad de Ciencias Naturales
Av Raul Gomez Lynx s / n Av Juan Tanca Marengo
Guayaquil
p. 3080777 to 3080758
carmenbonifaz@hotmail.com
http://fccnnugye.com/

Egypt

Ain Shams University

Dept of Geology (B)
Abbassia 11566
Cairo 11566
p. +20(2)6830963
deans@asunet.shams.edu.eg
http://sci.shams.edu.eg/Departments_Geology_Index.ASP

Dept of Geophysics (B)
Abbassia 11566
Cairo 11566
p. +20(2)6830963
deans@asunet.shams.edu.eg
http://sci.shams.edu.eg/Departments_Geophysics_Index.ASP

Al Azhar University

Dept of Geology (B)
Yosief Abbas Street
Cairo 11787
p. +20 (2) 262 3274
k.ubeid@alazhar.edu.ps
http://www.uazhar.edu.eg/bfac/sci/jiolojy.htm

Alexandria University

Dept of Environmental Sciences (B)
22 Al-Guish Avenue
Alexandria
p. +20 (3) 591 1152
anwar_elfiky@hotmail.com
http://www.alex.edu.eg/dept.jsp?FC=4&CODE=09

Dept of Geology (A,B,M,D) O
Baghdad Street
Qism Moharram Bek
Alexandria 21511
p. (+203) 3921595
sc-dean@alexu.edu.eg
http://www.sci.alexu.edu.eg/en/Departments/?Dept=4
f: https://www.facebook.com/groups/GSAU.alex/
Enrollment (2014): B: 26 (0)

Professor:
Galal Mohamed I Galal, (D), Germany, 1994, GgPeGe
Khalil I. Khalil Ebeid, (D), Germany, 1995, GgEmm
Kadry Nasser Sediek, (D), USSR Acad of Sci, 1991, GgsGd
Mohamad Nasser Shaaban, (D), 1992, GgsGd

Assistant Lecturer:
Ahmed Ibrahim Dyab, (M), Egypt, 2013, Gg

Assistant Professor:
Hossam EL-Din Ahmed EL-S Helba, (D), Germany, 1994, GgZm
Ahmed Sadek Mansour, (D), Egypt, 1999, Ggs

Instructor:
Sara Akram M. Mahmoud, (B), Egypt, 2006, Gg

Emeritus Professor:
Mohamed Waguih El Dakkak, (D), Egypt, 1971, GgsGd
Galal Abd EL-Ham Ewas, (D), Norway, 1969, GgNr
Hanafy M. Holail, (D), 1988, Ggs
Rousine Tanios Toni, (D), Egypt, 1967, GgPg

Emeritus Lecturer:
Mohamed M. Tamish, (D), W Germany, 1988, GgsCg

Emeritus Professor:
Mohamed Ahmed Rashed, (D), Moscow (Russia), 1984, GgSo

Oceanography Dept (B)
22 Al-Guish Avenue
Alexandria
p. +20 (3) 591 1152
h_mitwally@sci.alex.edu.eg

http://www.alex.edu.eg/dept.jsp?FC=4&CODE=07

American University of Cairo

Dept of Petroleum and Energy Engineering (B)
New Cairo Campus: AUC Avenue
PO Box 74
New Cairo 11835
p. +20.2.2615.1000
amrserag@aucegypt.edu
http://www.aucegypt.edu/academics/dept/peng/Pages/

Assiut University

Dept of Geology (B)
Assiut Governorate
Assiut City
PO Box 71515
Assiut 71515
p. +20 (88) 235 7007
sci@aun.edu.eg
http://www.aun.edu.eg/fac_sci/depart/ge1.htm

Mining & Metallurgical Engineering Dept (B,M) ☒
Assiut Governorate
Assiut City
PO Box 71515
Assiut 71515
p. +20 (88) 235 7007
gamalshoeib@eng.aun.edu.eg
http://www.aun.edu.eg/fac_eng?Dpart/Mining/Mining.htm

Beni Suef University

Geology Dept (B)
Salah Salem Street
Beni Suef 62511
p. +(20) 822324879, +(20) 2082232
elsherif_zakaria@yahoo.com
http://193.227.1.224/sci/

British University in Egypt

Petroleum Engineering (B)
Misr Ismalia Road
El Sherouk City 11837
enas.sabry@bue.edu.eg
http://www.bue.edu.eg/

Cairo University

Dept of Geophysics (B)
University Avenue - Univeristy Square
Giza
p. +20 (2) 572 9584
http://www.freewebtown.com/geophysics2/

Mining, Petroleum, and Metallurgical Engineering (B)
Univerisity Avenue - University Square
Giza
p. +20 (2) 572 9584
nahed_ecae2003@yahoo.com
http://www.eng.cu.edu.eg/dept/en/mpm

Dept of Geology (B)
University Avenue - University Square
Giza, Cairo
p. +20 (2) 572 9584
portal@cu.edu.eg
http://www.cu.edu.eg/english/

Egyptian Geological Survey and Mining Authority

egov@ad.gov.eg
http://www.egsma.gov.eg/

El Mansoura University

Dept of Geology (B,M,D)
Faculty of Science
at Mansoura
Department of Geology
Mansoura, Dakahliya governorat 35516
p. +20 050 2242388 Ext. 582
ghazala@mans.edu.eg
http://scifac.mans.edu.eg/en/scientific-dept/geology-department
Enrollment (2012): B: 250 (230) M: 25 (13) D: 15 (7)
Head:
Hosni H. Ghazala, (D), Mansoura, 1990, Ygx

El Zagazig University

Dept of Geology (B,M,D) ●⊘
Zagazig
p. +20 (55) 324 577
Khaledgemail@zu.edu.eg
http://www.zu.edu.eg/science
Programs: Geophysics (B.sc.)
Geology (B.sc.)
Petroleum and water geology (B.sc.)
Geology and chemistry (B.sc.)
Enrollment (2020): B: 78 (78) M: 11 (9) D: 7 (3)
Professor:
Khaled Said Gemail, (D), Zagazig, 2003, YuRcYe

Fayoum University

Geology Dept (B,M,D)
fayoum-Egypt University Zone
ags00@fayoum.edu.eg
Fayoum 63514
p. +20 01005856505
ags00@fayoum.edu.eg
http://www.fayoum.edu.eg/English/Science/Geology/About-Board.aspx
Enrollment (2016): B: 20 (18) M: 4 (0) D: 2 (0)
Professor:
ahmad gaber shedied, (D), Cairo, 1995, GgHww

Minia University

Geology & Chemistry Dept (B)
Minia
p. +(208-6) 324 420 #321 443
minia@frcu.eun.rg
http://www.minia.edu.eg

Sohag University

Geology Dept (B)
EL Kawaser
PO Box 82524
Sohag 82524
p. +(20) 93 4605745
ahmed_aziz9791@science.sohag.edu.eg
http://www.sohag-univ.edu.eg/

South Valley University

Dept of Geology (Aswan Campus) (B)
Qena 83523
p. +20 (96) 339 756
mohammedsm2003@yahoo.com
http://www.svu.edu.eg/arabic/aswan/sci/en/depart/Geology/Geology.htm

Tanta University

Dept of Geology (B)
El-Geish Street
Tanta 31527
p. +20 (40) 337 7929
president@tainta.edu.eg
http://www.tanta.edu.eg/ar/Tanta/3elom/geology.html

Zagazig University

Dept of Geology (B,M,D)
Sharkia Governorate
Zagazig City
geology@zu.edu.eg
http://www.zu.edu.eg/

Eritrea

University of Asmara

Dept of Marine Sciences (B)
PO Box 1220
Asmara
p. 291 1 161926 (Ext. 259)
zekeria@marine.uoa.edu.er
http://www.uoa.edu.er/academics/dmarine

Estonia

Geological Survey of Estonia

Eesti Geoloogiakeskus OU
Juniper tee 82
Tallinn 12618
p. 672 0094
egk@egk.ee
http://www.egk.ee/

University of Tartu

Inst of Geology (B,M,D) ○⊘
Ravila 14a
Tartu 50411
p. +372 7 375 891
geol@ut.ee
http://www.geoloogia.ut.ee/et
Programs: Geology (B, M, D); Environmental Technology (B)
Enrollment (2021): B: 57 (9) M: 10 (10) D: 15 (3)

Ethiopia

Addis Ababa University

Dept of Planetary and Earth Sciences
PO Box 1176
Addis Ababa
p. 251-111239462
balem@geo.aau.edu.at
http://www.aau.edu.et/index.php/earth-sciences

School of Civil & Environmental Engineering (B)
PO Box 1176
Addis Ababa
youngkyunkim@aait.edu.et
http://www.aau.edu.et/index.php/earth-sciences

Arba Minch University

Dept of Geology (B)
PO Box 21
Arba Minch
p. 251-468814972
yoditayalew@fastmail.fm
http://www.arbaminch-univ.com/

College of Natural Sciences (B)
PO Box 21
Arba Minch
p. +251-46-8810070
alemayehu.hailemicael@amu.edu.et
http://www.arbamich-univ.com/WTIMeteorlogy.html#

Water & Environmental Engineering (B)
PO Box 21
Arba Minch
p. +251-46-8810070
nigussie_tg@yahoo.com
http://www.arbaminch-univ.com/WTIWEE.html

Mekelle University

Dept of Earth Science (B)
P.O.Box: 231
Management Building, Main Campus
Mekelle
p. (+251) 344 40 40 05
cciad.mu@gmail.com
http://www.mu.edu.et/index.php/department-of-earth-science

Institute of Geo-information and Earth Observation Sciences (B)
P.O.Box: 231
Management Building, Main Campus
Makelle
p. +251 914 720398
meseleagw@yahoo.com
http://www.mu.edu.et/index.php/programs/ethiopia-institute-of-technology-mekelle/institute-of-geo-information-and-earth-observat

Semera University

Dept of Geology (B)
PO Box 132
Semera
p. 251-336660603
semerauniversity@ethionet.et

Wollega University

Dept of Geology (B)
PO Box 395 Nekemte
Nekemte
p. 251-576615038
wu@ethionet.et
http://www.wuni.edu.et/

Fiji

Ministry of Lands and Mineral Resources

Mineral Resources Dept of Fiji
Private Mail Bag, GPO
Suva
p. (679) 338 1611
director@mrd.gov.fj
http://www.mrd.gov.fj/

University of the South Pacific

Div of Earth and Environment Science
Private Mail Bag
GPO Suva
Suva, Fiji Islands
helpdesk@student.usp.ac.fj
http://www.sidsnet.org/pacific/usp/earth/

Finland

Aalto University

Dept of civil engineering (B,M,D) ⊘
Rakentajanaukio 4
Espoo
SF-02150 Espoo 00076 AALTO
leena.korkiala-tanttu@aalto.fi
https://www.aalto.fi/en/department-of-civil-engineering
Programs: Geoengineering
European Mining and Mineral Education Programme
Certificates: No
Associate Professor:
Leena Korkiala-Tanttu, (D), Helsinki Tech, 2009
Leena Korkiala-Tanttu, (D), Helsinki Tech, 2009

Abo Akademi University

Geology and Mineralogy (B,M,D) ○
Domkyrkotorget 1
Turku Fi-20500
p. (+)358442956429
geologi@abo.fi
http://www.abo.fi/fakultet/geologi
Enrollment (2015): B: 26 (8) M: 50 (4) D: 9 (2)
Professor:
Olav Eklund, (D), Abo Akademi, 1993, GxeGz

Geological Survey of Finland

☒
P.O.Box 96
Espoo FI-02151
p. +358 29 503 0000
gtk@gtk.fi

https://www.gtk.fi/en/
f: https://www.facebook.com/GTK.FI
t: @GTK_FI
Director, Science and Innovation:
Saku Vuori, (D), Gg
Director, Operative Units:
Olli Breilin, (M), Oulu, 1988, GmlGs
Director, Human Resources, Talent Management and Working Environments:
Hannu Sivula, (M)
Director General:
Kimmo Tiilikainen, (M)

University of Helsinki

Dept of Geosciences and Geography (B,M,D)
Gustaf Hällströmink 2
P.O. BOX 68
Helsinki 00014
p. 358-294150827
Mia.Kotilainen@helsinki.fi
http://www.helsinki.fi/geology/
Enrollment (2007): B: 0 (6) M: 26 (15) D: 6 (0)
Head:
Juha A. Karhu, (D), Helsinki, 1993, Csg

University of Oulu

Dept of Geology (B,M,D) O
Linnanmaa
FIN-90014
Oulu
arja.alin@oulu.fi
http://cc.oulu.fi/~geolwww/Geology.htm

Dept of Geophysics
riitta.keiski@oulu.fi
http://www.gh.oulu.fi/

France

Bureau de Recherches Geologiques et Minieres

BRGM-Orleans
3 avenue Claude-Guillemin
BP 36009
Orleans 45060 Cedex 2
p. +33 (0)2 38 64 34 34
http://www.brgm.fr/

Catholic University of the West

Environmental Management
3, place Andre Leroy
BP 808
49008 Anger
sciences@uco.fr

Centre de Recheches Petrographiques et geochimiques

Center of Petrographic and Geochemical Research (CRPG du CNRS)
15, rue Notre Dame des Pauvres
BP 20
54501 Vandoeuvre Les Nancy Cedex
p. + 33 (0)3 83 59 42 02
barbey@crpg.cnrs-nancy.fr
http://www.crpg.cnrs-nancy.fr

Ecole des Mines de Paris

Ecole des Mines de Paris
60-62, Boulevard Saint Michel
75272 PARIS cedex 06
Paris, France 75272
xavier.caillard@mines-paristech.fr
http://www.ensmp.fr/Eng/ENSMP/aboutENSMP.html

Ecole Nationale Supérieure de Géologie (ENSG)

ENSG Géologie (M,D) O☒
2 Rue du Doyen Marcel Roubault
BP 10162
VANDOEUVRE-LES-NANCY, Lorraine F-54505
p. 33 (0)3 72 74 46 00
ensg-contact@univ-lorraine.fr
http://www.ensg.univ-lorraine.fr/
Programs: Geology, Earth Science
Director:
Sausse Judith, (D), GgeGq

Ecole Nationale Superieure des Mines de Nancy

Ecole des Mines de Nancy (M,D) ●☒
Campus ARTEM
92 rue du Sergent Blandan
CS14234
NANCY 54042
p. +33 0355662600
mines-nancy-scolarite-ficm@univ-lorraine.fr
http://www.mines-nancy.univ-lorraine.fr/content/g%C3%A9oing%C3%A9nierie
f: https://www.facebook.com/groups/geoingenierie/
t: @minesnancy
Professor:
Olivier DECK

Ecole Nationale Superieure des Mines de Saint Etiene

Earth and Universe Sciences
158 Cours Fauriel
CS 62362
Saint Etienne 42023
p. (+33) (0)4 77 420 278
bernard.guy@mines-stetienne.fr
https://www.mines-stetienne.fr/en/academics/doctoral-studies-phd/

Ecole Normale Superieure de Paris

Dept of Geosciences (M,D) ●
24 rue Lhomond
Paris 75005
p. (+33) (0)1 44 32 22 11
delescluse@geologie.ens.fr
http://www.geosciences.ens.fr
Enrollment (2015): M: 20 (10) D: 29 (6)

ENSPM - Institut Francais du Petrole

Centre Exploration
1-4 avenue de Bois Preau
BP 311
92506 Rueil Malmaison
vincent.richard@ifpen.fr

French Institute for Research Exploitation of the Sea (IFREMER)

Dept of Marine Geosciences
BP 70
29263 Plouzane

Higher National School of Mines in Paris (ENSMP)

Center of Geosciences and Geoengineering (M,D) O☒
35, rue Saint-Honore
Fontainebleau Cedex 77305
contact@geosciences.mines-paristech.fr
https://www.geosciences.minesparis.psl.eu/en/home/
Programs: Master's Degree in Science and Executive Engineering, Master in earth and planetary science, environment - PSL university, Master Recherche Sciences de l'Univers, Environnement, Écologie (SDUEE), Doctorat of PSL university at MINES ParisTech

Higher Natl School of Mines at Saint-Etienne

(ENSME)

Departement Geosciences et environnement (A,M,D) ☒
158, cours Fauriel
42023 Saint-Etienne Cedex 02
steve.peuble@emse.fr
https://www.mines-stetienne.fr/formation/master-geospheres/#new_tab
Programs: Geology, Earth Science Education

Inst of Physics of the Globe-Paris VII

Lab de Geochimie, Geomateriaux, Geomag, Sismo, Tech, Obser Volcanologiques ☒
Tour 14/24 - 2e etage, 4, place Jussieu
75252 Paris Cedex 05
cartigny@ipgp.fr

Institut de Physique Du Globe De Paris

Institut de Physique Du Globe De Paris (M,D)
1 rue Jussieu
Paris 75005
p. 01 83 95 74 00
accueil@ipgp.fr
http://www.ipgp.fr/

Institut Polytechnique LaSalle Beauvais (ex-IGAL)

Dept of Geosciences (B,M,D)
19 Rue Pierre WAGUET - BP 30313
Beauvais 60026 cedex
p. +33 (0) 3 44 06 89 91
yannick.vautier@lasalle-beauvais.fr
http://www.lasalle-beauvais.fr/
Administrative Assistant: Nathalie Lermurier

Director of the Geosciences Department:
Yannick Vautier, Inst Géologique Albert-de-Laparent, 1999, GodCo

Director of the Education Program in Geosciences:
Hervé Leyrit, (D), Gv

Director of School/Companies relations:
Pascal Barrier, (D), Inst Géologique Albert-de-Laparent, PmGsPs

Professor:
Olivier Pourret, (D), Rennes I (France), 2006, CgaGe
Lahcen Zouhri, (D), Lille (France), 2000, HwgGe

Geotechnics:
Bassam Barakat, (D), Ecole Centrale Paris (France), 1991, NrgGq

Engineer in Mining & Quarry:
Lucien Corbineau, Inst Géologique Albert-de-Laparent, 2007, EgGgNm

Engineer in Marine Geology:
Olivier Bain, Inst Géologique Albert-de-Laparent, 2000, OuGgZi

Engineer in Geotechnics:
Jean-David Vernhes, Polytech Paris UPMC (France), 1998, YgNgr

Engineer in Geology:
Benoit Proudhon, Inst Géologique Albert-de-Laparent, 1997, GgcGt

Associate Professor:
Jessica Bonhoure, (D), CREGU-Nancy, 2007, GziCc
Sadek Brahmi, (D), UPMC Paris VI (France), 1991, Nr
Claudia Cherubini, (D), Bari (Italia), 2007, HwGe
Cyril Gagnaison, (D), Sorbonne (France), 2006, PgGsa
Sébastien Laurent-Charvet, (D), Orléans (France), 2001, GctZe
Pascale Lutz, (D), Pau (France), 2002, YgxYd
Mohamed Nasraoui, (D), ENSMP, 1996, EgGzEm
Elsa Ottavi-Pupier, (D), CRPG-CNRS Nancy, 1996, GvzGi
Sébastien Potel, (D), Bâle, 2001, GpzGi
Elodie Saillet, (D), Glasgow, 2009, GctGo
Renaud Toullec, (D), Bordeaux, 2006, GsoEo
Ghislain Trullenque, (D), Basel (Switzerland), 2005, GcNrGg

National Institute of Applied Science

Lab de Mineralogie et Geotechnique
20, avenue des Buttes de Coesmes
35043 Rennes Cedex 7
p. (+33) (0)2 23 23 82 00
olivier.guillou@insa-rennes.fr
http://www.insa-rennes.fr/insa-rennes.html

National Polytechnic Inst. of Grenoble

Ecole Doctorale Terre, Univers, Environnement
Domaine Univ de Grnoble - BP 95
46 avenue Félix Viallet
Cedex 1
Grenoble 38031

Pytheas Institute - Earth Sciences and Astronomy Observatory (PYTHEAS)

Faculte des Sciences de Luminy
Laboratoires de Geologie Marine et Sedimentologie
70 avenue Leon Lachamp
13288 Marseille
secretariat-direction@osupytheas.fr

Toulouse University

Satellite Geophysics and Oceanography Laboratory (M,D)
14, Avenue Edouard Belin
31400 Toulouse
p. (+33) (0)5 61 33 29 02
directeur@legos.obs-mip.fr
http://www.legos.obs-mip.fr/Presentation-generale?set_language=en&cl=en

Universite Blaise Pascal (Clermont Ferrand II)

Dept des Sciences de la Terra
34 Avenue Carnot
63038 Clermont Ferrand Cedex
p. (+33) 04 73 34 67 22
cecile.sergere@univ-bpclermont.fr
http://wwwobs.univ-bpclermont.fr/lmv/cursus/

Université Claude Bernard Lyon 1

Laboratoire de Géologie de Lyon: Terre, Planètes, Environnement (B,M,D) ☒
Bd du 11 Novembre
Campus de La Doua
Bâtiment Géode
Villeurbanne 69622
dir-umr5276@osulistes.univ-lyon1.fr
http://lgltpe.ens-lyon.fr/
Programs: Geoscience
Administrative Assistant: Marie-Jeanne Barrier
Enrollment (2012): B: 36 (30) M: 16 (16) D: 10 (10)

Professor:
Pascal Allemand
Fabrice Cordey
Gilles Cuny
Isabelle Daniel
Gilles Dromart
Stephane Labrosse
Christophe Lécuyer
Emanuela Mattioli
Guillemette Ménot
Cathy Quantin
Pierre Thomas

Senior Scientist:
Thierry Alboussiere
Vincent Balter
janne Blichert toft
Bernard Bourdon
Eric Debayle
Vincent Grossi
Philippe Herve Leloup
Bruno Reynard
Yanick Ricard
Jean Vannier

Associate Professor:
Muriel Andréani
Anne-Marie Aucourt
Frederic Chambat
Regis Chirat
Claude Colombié
Véronique Daviéro-Gomez
Véronique Gardien
Bernard Gomez

Vincent Langlois
Gweltaz Mahéo
Matthew Makou
Jean-Emmanuel Martelat
Davide Olivero
Vincent Perrier
Jean-Philippe Perrillat
Sylvain Pichat
Frédéric Quillévéré
Stéphane Reboulet
Philippe Sorrel
Guillaume Suan
Benoît Tauzin

Associate Scientist:
Romain Amiot
Thomas Bodin
caroline Fitoussi
Bertrand Lefebvre
Laurence Lemelle
Jérémy Martin

Emeritus:
francis Albarede

Related Staff:
Emmanuelle Albalat
Ingrid Antheaume
Florent Arnaud-Godet
Ghislaine Broillet
Herve Cardon
Fabien Dubuffet
Philippe Grandjean
Aline Lamboux
Gilles Montagnac
Sophie Passot
Emmanuel Robert
Magali Seris
Philippe Telouk

Universite d'Orleans

Dept des Sciences de la Terre
BP. 6759
45067 Orleans
scolarite-osuc@univ-orleans.fr

Universite de Bordeaux I

Dept des Sciences de la Terra et de la Mer
341 Cours de la Liberation
Talence 33400
p. (+33) 05 40 00 88 79
termer@adm.u-bordeaux1.fr
http://www.u-bordeaux1.fr/universite/organisation/composantes-ufr-instituts/ufr-des-sciences-de-la-terre-et-de-la-mer.html

Universite de Bourgogne

UFR Sciences Vie, Terre & Environnement (B,M,D)
6 boulevard Gabriel
21100 Dijon
secretariat-ufrsvte@u-bourgogne.fr
http://ufr-svte.u-bourgogne.fr/

Universite de Bretagne Occidentale

Departement des Sciences de la Terre
3 Rue des Archives
Brest 29287
p. (+33) 02 98 01 61 88
muriel.vidal@univ-brest.fr
http://www.univ-brest.fr/ufr-sciences/menu/Les_departements/Sciences_de_la_Terre

Universite de Caen

Dept des Sciences de la Terre
Esplanade de la Paix
Caen 14000
p. (+33) 02 31 56 55 87
isabelle.villette@unicagen.fr
http://ufrsciences.unicaen.fr/departements/departement-des-sciences-de-la-terre/

Universite de Lorraine - Faculte des Sciences et Technologies

Dept Geosciences (B,M) ○☒
Boulevard des Aiguillettes
BP 239
Vandoeuvre Les Nancy, 54506
p. (+33) (0)3 83 68 47 18
cecile.fabre@univ-lorraine.fr
http://fst-en.univ-lorraine.fr/la-faculte/department-of-geosciences
Programs: Geology
Geophysics
Mineralogy
Metallogeny
Environment
Soils
Waters
Planetology
Enrollment (2022): B: 100 (100) M: 100 (100)

Head:
Cecile FABRE, (D), 2000, Gg

Universite de Montpellier

Geosciences Montpellier (M,D) ☒
Campus Triolet
CC 060
Montpellier cedex 5 34095
p. (+33) (0)4 67 14 36 43
dirgm@gm.univ-montp2.fr
http://www.gm.univ-montp2.fr/

Universite de Paris Saclay (Orsay)

Dept Sciences de la Terre (B,M,D) ●☒
15 Rue Georges Clemenceau
Orsay 91400
p. (+33) (0)1 69 15 49 09
alexandra.courtin@universite-paris-saclay.fr
http://geosciences.geol.u-psud.fr/
f: https://www.facebook.com/geosciences.paris.sud/
t: @GEOPS_Orsay
Programs: Hydro(geo)logy; Ressources; Sedimentary Basins; Planetology; Paleoclimate; Engineering Geology
Enrollment (2019): B: 31 (27) M: 68 (63) D: 56 (20)

Universite de Paris VI

Lab of Sub-Marine Geodynamics (CEROV)
Port de la Darse - BP 48
06230 Villefranche-Sur-Mer

Center of Geosynamics Research
4 Place Jussieu
Paris 75005
p. (+33) 01 44 27 46 98
licence.sciterre@upmc.fr
http://www.upmc.fr/en/education/diplomas/sciences_and_technologies/bachelor_s_degrees/department_of_earth_sciences.html

Universite de Paris VII

Dept des Sciences Physiques de la Terre
4 place Jussieu
75251 Paris
p. (+33) (0)1 57 27 57 27
zarie.rouas@univ-paris-diderot.fr

Universite de Pau

Dept of Geology (A,B,M,D)
I.P.R.A. - Geologie
Avenue de L'Universite
Pau 64230
anne-sophie.laloge@univ-pau.fr
http://www.univ-pau.fr/RECHERCHE/GEOPHY/

Universite de Pau et des Pays de l'Adour

Faculte des Sciences
Avenue de l'Universite

64000 Pau

Universite de Rennes I

Dept. de Geosciences
2 Rue du Thabot
35065 Rennes Cedex
p. (+33) (0)2 23 23 60 76
olivier.dauteuil@univ-rennes1.fr
http://www.geosciences.univ-rennes1.fr/?lang=en

Universite de Strasbourg

Ecole et Observatoire des Sciences de la Terre (B,M,D) ●
5 rue Descartes
67084 Strasbourg cedex F
p. (+33) (0)3 68 85 03 53
eost-contact@unistra.fr
http://eost.unistra.fr/nouveautes-du-site/
Director:
Frederic Masson, (D)
Laboratory Director:
Ulrich Achauer, (D), Karlsruhe, 1990, Ysg

Université Jean Monnet, Saint-Etienne

Département de Géologie - Faculte des Sciences et Techiques (B,M,D) ●☒
23 rue du Dr. Paul Michelon
Saint-Etienne Cedex F-42023
p. (+33) (0)4 77 48 51 23
gerbe@univ-st-etienne.fr
https://fac-sciences.univ-st-etienne.fr/en/
Programs: Bachelor of Earth Sciences, Professional bachelor for technical competences in environnement (mid time at the university, mid time in a company)
Head:
Marie-Christine GERBE, (D), GiCsg
Professor:
Jérôme BASCOU, (D), Gcy
Damien GUILLAUME, (D), Gz
Jean-François MOYEN, (D), 2000, GiCtGp
Associate Professor:
Adélie DELACOUR, (D), GiCsOu
Jean-Luc FROGER, (D), Zr
Antonin LAURENT, (D), Gp
Véronique LAVASTRE, (D), Cs

Universite Paul Sabatier (Toulouse III)

Dept Biologie & Geosciences (B,M) ●☒
118 Route de Narbonne
31062 Toulouse
p. (+33) (0)5 82 52 57 21
scolarite.inscriptions@univ-tlse3.fr
http://www.univ-tlse3.fr/04628859/0/fiche___pagelibre/&RH=ACCUEIL&RF=1237305837890

Université de Franche-Comté

Faculte des Sciences et Techniques
16, route de Gray
25030 Besancon Cedex
p. (+33) (0)3 81 66 62 09
accueil.ufr-st@univ-fcomte.fr
http://sciences.univ-fcomte.fr/pages/fr/menu3795/formations/licence--sciences-de-la-vie-16862-15340.html

University of Caen

Dept des Sciences de la Terre
Lab de Geologie structurals
Esplanadde de la Paix
14032 Caen Cedex

University of Franche-Comte

Lab de Geol Strucurale et appliquee
Faculte des Sci et Techniques
Place Leclerc
25030 Besancon
master.geologie@univ-fcomte.fr

Sciences Environnementales
Sciences et techniques
Service Scolarité
16, route de Gray
25030 Besançon cedex
p. 03 81 66 62 09 / 62 11
Scolarite.UFR-ST@univ-fcomte.fr
http://sciences.univ-fcomte.fr/formations/listedesformations.htm

University of Francois Rabelais

Laboratoire de Geologie
Parc de Grandmot
37200 Tours

University of Lille

Dept of Geology and Laboratory of Oceanology and Geosciences (LOG) (B,M,D) ●☒
Cite Scientifique
Batiment SN-5
Villeneuve D'Ascq 59655
bruno.vendeville@univ-lille.fr
http://log.cnrs.fr
Programs: Geology (B,M,D)

University of Louis Pasteur (Strasbourg 1)

School and Observatory of Earth Sciences
1, rue Blessing
67084 Strasbourg Cedex

University of Maine ❐

Faculte des Sci - Lab de Geologie ☒
Avenue Olivier Messiasen
72085 Le Mans
p. (+33) (0)2 43 83 30 00
http://sciences.univ-lemans.fr/Biologie-Geosciences

University of Montpellier II

Observatoire des sciences de l'univers OREME
2, place Eugene Bataillon
34095 Montpellier Cedex 05
eric.servat@msem.univ-montp2.fr

University of Nantes

Departement des Sci de la Terre
2, rue de la Houssiniere
44072 Nantes Cedex 03
secretariat.stu@univ-nantes.fr

University of Nice

Dept of Earth Science
Faculte des Sciences
28, avenue Valrose
06034 Nice Cedex
secretariat-master-PPA@unice.fr

University of Nice-Sopia Antipollis-1

Laboratoire des Sciences de la Terre
Rue A. Einstein
06560 Valbonne
dir.recherche@unice.fr
http://www-geoazur.unice.fr/

University of Orleans

Dept des Sciences de la Terre
Domaine de la Source
BP 6759
45067 Orleans Cedex 02

University of Perpignan

Centre de Sed et Geochimie marines
52 Avenue Paul Alduy
66100 Perpignan
p. (+33) (0)4 68 66 21 39
facscien@univ-perp.fr

University of Picardy

Dept de Geologie
Chemin du Thil
80000 Amiens
p. (+33) (0)4 22 82 76 65
mohamed.benlahsen@u-picardie.fr
http://www.u-picardie.fr/jsp/fiche_structure.jsp?STNAV=US&RUBNAV=&CODE=US&LANGUE=0

University of Pierre & Marie Curie (Paris VI)

Dept de Geotectonique
Tour 26 - ler etage (Boite 219)
4, place Jussieu
75252 Paris Cedex 05

Dept of Living Earth and Environment
Tour 15 - 4e etage
4, place Jussieu
75252 Paris Cedex 05
chrystele.sanloup@upmc.fr
http://www.ipgp.jussieu.fr/

University of Poitiers

Dept des Sciences de la Terre
40, ave du Recteur Pineau
86022 Poitiers Cedex

University of Provence (Aix-Marseille I)

Lab de Geol Structurale et appliquee
Centre Saint Charles
3, place Victor Hugo - Case 28
13331 Marseille Cedex 3
bertrand.martin-garin@univ-amu.fr

University of Reims-Champagne

Dept des Sciences de la Terre
Moulin de la Housse - BP 1039
51687 Reims Cedex
p. (+33) (0)3 26 91 34 19
scolarite.sceiences@univ-reims.fr
http://www.univ-reims.fr/formation/ufr-instituts-et-ecoles/ufr-sciences-exactes-et-naturelles/presentation,8370.html?

University of Rouen-Upper Normandy

Dept de Geologie
1 Rue Thomas Becket
76821 Mont-Saint-Aignan
p. (+33) (0)2 35 14 68 26
francoise.baillot@univ-rouen.fr

University of Savoy

Laborotoire de Geologie
Faculte des Sci et Techniques
Campus de Technolac, BP 1104
73011 Chambery Cedex
scolarite.centrale@univ-smb.fr

University of Science & Technology of Lille

Inst Des Sciences De La Terre
Flandres - Artois, Cite Scientifique
Bat SN5, BP 36
59655 Villeneuve D'Ascq Cedex
alain.trentesaux@univ-lille1.fr

University of West Brittany

Earth Sciences
Groupement de Recherche: GEDO
6, avenue Le Gorgeu
29287 Brest

Gabon

Universite des Sciences et Techniques de Masuku

Dept de Geologie (B)
BP 901
Franceville
p. (+241) 01 67 75 78
http://www.labogabon.net/ustm/facscience

Universite Omar Bongo

Dept de Geologie (B)
BP 13 131
Boulevard Leon Mba
Libreville
p. +241 73 20 45, (241-72) 69 10
uob@internetgabon.com
http://www.uob.ga/

Germany

Aachen University of Technology

Inst of Structural Geology, Tectonics and Geomechanics
Lochnerstr 4-20
D-52056
ged@ged.rwth-aachen.de
http://www.ged.rwth-aachen.de

Baden-Wuerttemberg - Landesamt fuer Geologie, Rohstoffe und Bergbau (LGRB)

Regional Government of Freiburg State Office of Geology, Raw Materials and Mining ☒
Albertstrasse 5
Freiburg, Baden-Wuerttemberg 79104
p. ++49 9761 208 3000
abteilung9@rpf.bwl.de
http://www.lgrb-bw.de/

Bavarian Environment Agency

Department Geological Survey ☒
Buergermeister-Ulrich-Strasse
160
Augsburg , Bavaria 86179
poststelle@lfu.bayern.de
https://www.lfu.bayern.de/geologie/

Brandenburg - Landesamt fuer Geowissenschaften und Rohstoffe (LGRB)

State Office for Mining , Geology and Minerals of Brandenburg
Inselstrabe 26
03046 Cottbus
lbgr@lbgr.brandenburg.de
http://www.lbgr.brandenburg.de

Bundesanstalt fur Geowissenschaften und Rohstoffe (BGR)

Federal Institute for Geosciences and Natural Resources ☒
Stilleweg 2
Hannover 30655
poststelle@bgr.de
http://www.bgr.bund.de

Christian-Albrechts-Universitaet

Geology Dept
Olshausenstrasse 40/60
2300 Kiel 1

Ernst-Moritz-Arndt Universitaet

Institut fur Geographie und Geologie
Domstrabe 11
17489 Griesswald
p. (+49) (0) 3834 86-4502
geogra@uni-greifswald.de
http://www.mnf.uni-greifswald.de/institute/geo.html

GERMANY

Freie Universitaet Berlin

Institute of Geological Sciences, Planetary Sciences and Remote Sensing ☒
Malteserstr. 74-100
12249 Berlin
p. (030) 838-70 575
plansec@zedat.fu-berlin.de
http://www.geo.fu-berlin.de/geol/fachrichtungen/planet

Inst fuer Palaontologie
Maltese-Strasse
74-100, Haus D
D-12249 Berlin
kim.kuehne@fu-berlin.de
http://userpage.fu-berlin.de/~palaeont/WELCOME.HTM

Friedrich-Schiller-University Jena

Institute for Geosciences (B,M,D) ●
Burgweg 11
Jena 07749
p. 0049(0)3641/948600
geowissenschaften@uni-jena.de
http://www.geo.uni-jena.de
Enrollment (2014): B: 17 (43) M: 16 (22) D: 69 (12)
Univ.-Prof. Dr.:
Sabine Attinger, (D), Hy
Georg Büchel, (D), Gg
Christoph Heubeck, (D), Stanford, 1994, GsdPg
Nina Kukowski, (D), Ygx
Falko H. Langenhorst, (D), Gz
Juraj Majzlan, (D), Gz
Kai U. Totsche, (D), Bayreuth, 1994, HwClSo
Lothar Viereck, (D)
Jun.-Prof. Dr.:
Anke Hildebrandt, (D), MIT, 2005, HgSpf
Univ.-Prof. Dr.:
Kamil M. Ustaszewski, (D), 2004, Gct

Garching Technical University

Inst fur Geologie und Mineralogie
Lichtenbergergstrasse 4
0-8046 Garching
rieder@tum.de

Geologischer Dienst Nordrhein-Westfalen

NRW Geological Survey
Postfach 10 07 63
De-Greiff-Strasse 195
Krefeld D-47707
p. +49-2151-897-0
geoinfo@gd.nrw.de
http://www.gd.nrw.de/home.php

Geologisches Institut der Universitaet

Dept of Geology
Plelcherwael 1
87 Wuerzburg

Georg-August University of Goettingen

Dept of Geobiology (B,M,D) ☒
Goldschmidstr. 3
Goettingen, Lower Saxony 37077
p. +49 551 397951
jreitne@gwdg.de
http://www.geobiologie.uni-goettingen.de/
Programs: Bachelor & Master Geosciences/Geobiology
Enrollment (2018): B: 2 (2) M: 4 (1) D: 10 (4)
Prof.Dr.:
Joachim Reitner, (D), PggPg
Professor:
Volker Thiel, Dr., (D), CooCo
Dr.:
Gernot Arp, Dr., (D), PggPg
Andreas Reimer, (D), OcCmm
Jan-Peter Duda, (D), 2014, PgCoGs

Goethe Universitaet

Fachbereich Geowissenschaften
Altenhoferallee 1
60438 Frankfurt
p. (+49) (0)69 798 40208
dekanat-geowiss@em.uni-frankfurt.de
http://www.uni-frankfurt.de/fb/fb11

Hamburg - Geological Survey

Ministry of Urban Development and Environmental Protection Agency for the Environment
Neuenfelder Straße 19
21109 Hamburg
gla@bue.hamburg.de
http://www.geologie.hamburg.de

Helmholtz-Zentrum Potsdam Deutsches GeoForschungsZentrum

GeoForschungs Zentrum GFZ
Telegrafenberg 14473
14473 Potsdam
p. (+49) 331 288-1045
presse@gfz-potsdam.de
http://www.gfz-potsdam.de/startseite/

Hessen - Landesamt fuer Umwelt und Geologie

65203 Wiesbaden, Rheingaustraße 186
p. 0611-6939-0
http://www.hlug.de/

Humboldt Universitaet zu Berlin

Palaontologisches Museum
Invaliden Strasse 43
0-4010 Berlin

Institut fur Palaontologie

Dept of Geology
O. Weidlich, Fohrenweg 19
W-8504 Stein
palsek@univie.ac.at

Karlsruhe Institute of Technology

Geophysical Institut (B,M,D)
Hertzstr. 16
Karlsruhe 76187
p. +49-721-6084431
geophysics@gpi.kit.edu
http://www-gpi.physik.uni-karlsruhe.de/
Enrollment (2010): B: 15 (0) D: 10 (0)
Professor:
Thomas Bohlen, (D), Yx
Senior Scientist:
Thomas Forbriger, (D), Ys
Associate Professor:
Joachim R.R. Ritter, (D), Universit, 1996, YsGtv
Research Associate:
Rebecca Harrington, (D), Ys

Dept of Geology
Kaiserstrasse 12
76131 Karlsruhe
p. (+49) 721 608 4219 2/43651
geophysik@gpi.kit.edu
http://www.bgu.kit.edu

Institute for Applied Geosciences
Hertzstr. 16
Karlsruhe
nevena.tomasevic@kit.edu
http://www.agw.kit.edu/

Institute for Mineralogy and Geochemistry
Adenauerring 20b
Karlsruhe D-76131

p. +49 721 608- 43323
img@img.uka.de
http://www.img.kit.edu/

Ludwig-Maximilians-Universitaet Muenchen

Dept of Earth & Enviromental Sciences (B,M,D)
Theresienstr. 41/III
Munich 80333
p. 0049/89/21804250
dingwell@lmu.de
http://min.geo.uni-muenchen.de/
Enrollment (2007): B: 173 (28) M: 12 (0) D: 24 (0)

Director:
Donald Bruce Dingwell, Giv

Professor:
Alexander Altenbach, Pg
Wladyslaw Altermann, Gg
Michael Amler, Pg
Valerian Bachtadse, Yg
Hans-Peter Bunge, Yg
Christina De Campos, Gz
Karl Thomas Fehr, Gz
Friedrich Frey, Gz
Anke Friedrich, (D)
Anke Friedrich, (D), MIT, 1998, GgtCc
Helmut Gebrande, (D), LMU Munich, 1975, YgsEo
Stuart Gilder, (D), Ym
Peter Gille, (D), Humboldt (Germany), 1984, GzZm
Wolfgang Heckel, Gz
Ernst Hegner, Gg
Soraya Heuss-Aßbichler, Gz
Heiner Igel, Yg
Harald Immel, Pg
Bernd L. Lammerer, (D), LMU Munich, 1972, GgtGc
Reinhold Leinfelder, Pg
Rocco Malservisi, Yg
Robert Marschik, (D), EgmCg
Ludwig Masch, Gz
Wolfgang Moritz, Gz
Bettina Reichenbacher, (D), Johann Wolfgang Goethe, 1992, PgvPe
Wolfgang Schmahl, Gz
Klaus Weber-Diefenbach, Gg
Christian Wolkersdorfer, Gg

Adjunct Professor:
Frank Trixler, (D), GzZmXc

Emeritus:
Nikolai Petersen, (D), 1964, Ygm

Other:
Kirill Aldushin, Gz
Greta Barbi, Yg
Robert Barsch, Yg
Johannes Birner, Gg
Florian Bleibinhaus, Yg
Hans Boysen, Gz
Gilbert Britzke, Yg
Benoit Cordonnier, Gz
Alexander Dorfman, Gz
Thomas Dorfner, Gz
Barbara Emmer, Yg
Werner Ertel-Ingrisch, Gz
Andreas Fichtner, Yg
Michaela Frei, Gg
Frantisek Gallovic, Yg
Alexander Gigler, Gz
Stefan Grießl, Gz
Hagen Göttlich, Gz
Marc Hennemeyer, Gz
Katja Henßel, Pg
Kai-Uwe Hess, Gz
Maria Linda Iaccheri, Gg
Giampiero Iaffaldano, Yg
Guntram Jordan, Gz
Ines Kaiser-Bischoff, Gz
Thorsten Kowalke, Pg
Thomas Kunzmann, Gz
Martin Käser, Yg
Markus Lackinger, Gz
Yan Lavallee, Yg
Maike Lübbe, Gz
Götz Meisterernst, Gz
Timo Casjen Merkel, Gz
Marcus Mohr, Yg
Lena Müller, Gz
Dieter Müller-Sohnius, Gz
Malik Naumann, Pg
Jens Oeser, (D), Yg
Sohyun Park, Gz
Karin Paschert, Gg
Rossitza Pentcheva, Gz
Helen Pfuhl, Yg
Antonio Sebastian Piazzoni, Yg
Christina Plattner, Yg
Josep Puente Alvarez de la
Oliver Riedel, Gz
Alexander Rocholl, Gz
Javier Rubio-Sierra, Gz
Gertrud Rößner, Pg
Dieter Schmid, Pg
Julius Schneider, Gz
Bernhard Schuberth, Yg
Oliver Spieler, Gz
Robert Stark, Gz
Katja Steffens, Gg
Stefan Strasser, Gz
Marco Stupazzini, Yg
Frank Söllner, Gg
Ferdinand Walther, Gz
Joachim Wassermann, Yg
Laura Wehrmann, Pg
Michael Winklhofer, Yg
Ayhan Yurtsever, Gz
Matthias Zeitlhöfler, Gg
Albert Zink, Gz

Palaeontology & Geobiology, Dept of Earth and Env Sci (B,M,D) ○☒
Richard-Wagner-Strasse 10
Muenchen, Bavaria 80333
geobiologie@geo.lmu.de
http://www.palmuc.de
Programs: Geobiology & Paleobiology (M)

Prof. Dr.:
Gert Woerheide, (D), Pgi

Prof. Dr.:
William Orsi, (D), PgZn
Bettina Reichenbacher, (D), PggPv

Marburg Chome University

Geology Dept
Deutschhausstrabe 10
35032 Marburg
p. (+49) 06421/28-24257
dekan19@staff.uni-marburg.de
https://www.uni-marburg.de/fb19_alt

Martin-Luther-Universitaet Halle-Wittenberg

Inst for Geosciences and Geography (A,B,M,D) ●⃠
Von-Seckendorff-Platz 3
D-06120 Halle
p. ++49-045-55 26010
direktor@geo.uni-halle.de
http://www.geo.uni-halle.de
Programs: Geography (B,M); Geosciences (B,M); Natural Resource Management (B,M); Intl Area Studies (M); Geography Education

Director:
Christine Fürst, (D), Bonn, 2013, ZucZi

Mecklenburg-Vorpommern - Landesamt fuer Umwelt, Naturschutz und Geologie

State office for the Environment, Nature Conservation and Geology ☒
Goldberger Strasse 12
18273 Gustrow
p. ++49-3843-777 0

poststelle@lung.mv-regierung.de
http://www.lung.mv-regierung.de/

Niedersachsen - Landesamt fuer Bodenforschung

p. +49 (0)511-643-0
poststelle-hannover@lbeg.niedersachsen.de
http://www.lbeg.niedersachsen.de

Rheinisch-Westfaelische Technische Hochschule Aachen

Fachgruppe fuer Geowissenschaften und Geographie
Lochnerstr. 4-20 (Haus B)
Aachen 52056
p. 0241 80 96219
geowiss@rwth-aachen.de
http://www.fgeo.rwth-aachen.de/

Rheinland-Pfalz - Landesamt fuer Geologie und Bergbau

State Office for Geology and Mining
Emy-Roeder-Strabe 5
PO Box 100 255
Mainz-Hechtsheim D-55129
office@lgb-rlp.de
http://www.lgb-rlp.de/

RWTH Aachen University

Institute of Geology and Palaeontology (B,M,D) ☒
Faculty Georessources and
Materials Engineering
Wuellnerstr. 2
52062 Aachen 52062
geo.sek@emr.rwth-aachen.de
Programs: Applied Geosciences, Applied Geophysics, Georessources Management

Sachsen - Landesamt fuer Umwelt und Geologie (LfUG)

Saxon State Agency for Environment, Agriculture and Geology
Postfach 54 01 37
Dresden 01 311
andreas.voellings@smul.sachsen.de
http://www.smul.sachsen.de/

State Survey for Geology and Mining of Sachsen-Anhalt

State Survey for Geology and Mining of Sachsen-Anhalt ☒
Halle, Saxony-Anhalt D-06035
p. +49 345 52 12 0
poststelle@lagb.mw.sachsen-anhalt.de
http://www.lagb.sachsen-anhalt.de

Technical University of Munich

Engineering Geology (B,M,D) ●⃠
Arcisstr. 21
Munich, Bavaria 80333
p. +49 89 28925850
geologie@tum.de
https://www.cee.ed.tum.de/eng/
Programs: Geosciences (B); Engineering Geology & Hydrogeology (M); Geothermal Energy (M)
Enrollment (2017): B: 0 (1) M: 0 (3) D: 11 (0)

Prof. Dr.:
Kurosch Thuro, (D), Munich (Germany), 1995, NgrGg
Dr.:
Gerhard Lehrberger, (D), Munich (Germany), 1992, EgGxg
Bernhard Lempe, (D), TUM, 2012, GlgNg
Prof. Dr.:
Hans Albert Gilg, (D), Munich (Germany), 1995, CsEgn
Dr.-Ing.:
Bettina Menschik, (D), Munich (Germany), 2014, NgrNx
Dr.:
Katja R. Lokau, (D), TUM, 1999, Ng
Marion Nickmann, (D), TUM, 2006, Ng
M.Sc.:
Matthias Brugger, (M), TUM, 2015, NgGg
Moritz Gamperl, (M), Munich (Germany), 2018, NgrYu
M.A.:
Margreta Sonnenwald, (M), TUM, 2016, Ng
Dr.-Ing.:
Martin Potten, (D), Munich (Germany), 2020, NgrNx
Georg Maximilian Stockinger, (D), TUM, 2021, NrgNx
Dr.:
Heiko Käsling, (D), Minich (Germany), 2009, NrmNg

Technische Hochschule

Institute of Applied Geoscience
Karolinenplatz 5
64289 Darmstadt
p. (+49) 6151 16-2171
herrmann@geo.tu-darmstadt.de
http://www.geo.tu-darmstadt.de/iag/index.de.jsp

Technische Universitaet Bergakademie Freiberg

Faculty for Geosciences, Geoengineering and Mining (A,B,M,D) ☒
Dekanat Fak. 3
Postfach 47/Bernhard-Cotta Strasse
Freiberg 09599
p. +49 (0)3731 / 39 - 3249
Andrea.Thuemmel@geort.tu-freiberg.de
https://tu-freiberg.de/fakultaet3
Programs: Geology, Mineralogy, Mining, Geotechnics, Petroleum Engineering, Geophysics, ...

Technische Universitaet Berlin

Applied Geosciences
Str. Des 17 Juni 135
W-1000 Berlin 12
p. +49 30314-7260 5
http://www.geo.tu-berlin.de/

Technische Universitaet C.W. Braunschweig

Institute of Environmental Geology
Postfach 3329
Braunschweig D-38023
geosecret@tu-bs.de
http://www.tu-braunschweig.de/iug

Inst fuer Geologie und Palaontologie
Fachbereich fur Physik und Geowissenschaften
Pockelstrasse 14
D-3300 Braunschweig DEN

Inst fuer Geophysik und Meteorologie
Mendelssohnstr 2-3
D-3300 Braunschweig

Technische Universitaet Clausthal

Institut fuer Geologie und Palaeontologie
LeibnizstraBe 10
D-38678
Clausthal-Zellerfeld 38678
Office@geologie.tu-clausthal.de
http://www.geologie.tu-clausthal.de/

Inst for Petroleum Engineering
Agricolastrasse 10
38678 Clausthal-Zellerfeld
p. +49-5323-72-2239
marion.bischof@tu-clausthal.de
http://www.ite.tu-clausthal.de/

Technische Universitaet Darmstadt

Institute of Applied Geosciences (B,M,D)
Schnittspahnstr. 9
Darmstadt D-64287
p. +49 6151 16 2171
henk@geo.tu-darmstadt.de

http://www.geo.tu-darmstadt.de/
Enrollment (2011): B: 280 (3) M: 54 (0) D: 29 (2)
Acting head:
Hans-Joachim Kleebe, GzZm
Professor:
Rafael Ferreiro Maehlmann, Gxp
Matthias Hinderer, Gs
Andreas Hoppe, GgZui
Stephan Kempe, GgCgHs
Ingo Sass, Ng
Christoph Schueth, Hw
Stephan Weinbruch, GzAmZm

Technische Universitat Dresden

Dept of Geosciences
Mommsenstr 13
0-8027 Dresden
p. +49 351 463-32863
doris.salomon@tu-dresden.de
http://tu-dresden.de/die_tu_dresden/fakultaeten/fakultaet_forst_geo_und_hydrowissenschaften/fachrichtung_geowissenschaften

Thuringen - Landesanstalt fuer Umwelt und Geologie

Thuringen State Institute for Environment and Geology
Goeschwitzer Str forty-one
07745 Jena
Poststelle@tlug.thueringen.de
http://www.tlug-jena.de/de/tlug/

Universitaet Bochum

Institut fuer Geologie, Mineralogie und Geophysik (B,M,D)
Universitaetsstr. 150
Bochum 44780
p. +49 234 32 23233
sabine.sitter@ruhr-uni-bochum.de
http://www.ruhr-uni-bochum.de/gmg/

Universitaet Bonn

Steinmann-Institut für Geologie, Mineralogie und Paläontologie (B,M,D) ●
Nussallee 8
Bonn 53115
p. +49 228 73 4803
tmartin@uni-bonn.de
http://www.steinmann.uni-bonn.de/

Universitaet Erlangen-Nuernberg

Geozentrum Nordbayern
Schlossgarten 5
91054 Erlangen
p. 09131/85-22615
veronika.kuehnert@fau.de
http://www.geol.uni-erlangen.de

Institut fur Geographie
Kochstr. 4/4
91054 Erlangen
p. 09131/85-22633
common@geographie.uni-erlangen.de
http://www.geographie.uni-erlangen.de/

Universitaet Frankfurt

Institut fur Geowissenschaften
Fachbereisch 17
Senckenberganlage 32-34
6000 Frankfurt
geowissenschaften@em.uni-frankfurt.de
http://www.geo.uni-frankfurt.de/ifg/

Universitaet Freiburg

Institut fur Geo- und Umweltnaturwissenschaften - Geologie (B,M,D) ☒
Albertstr. 23-B
Freiburg i. Br., Baden-Wuerttemberg 79104
ulmer@uni-freiburg.de
http://portal.uni-freiburg.de/geowissenschaften
Programs: Geowissenschaften, Geology, Sustainable Materials
Enrollment (2017): B: 140 (50) M: 52 (19) D: 2 (1)

Universitaet Giessen

Inst fuer Geowissenschaften und LithosphSrenforshung
Senckenbergstrasse 3
35390 Giessen
p. (+49) 641 990
auskunft@bibsys.uni-giessen.de
http://www.uni-giessen.de/fbr08/geolith/

Universitaet Goettingen

Geowissenschaftliches Zentrum
Goldschmidstr. 3
Lower Saxony
Gottingen D-37077
p. +49 551 397951
abeckma1@gwdg.de
http://www.uni-goettingen.de/de/125309.html

Universitaet Greifswald

Institute for Geography and Geology
F.L. Jahn Strasse 17
Greifswald
p. +49 (0)3834 86-4570
geologie@uni-greifswald.de
http://www.mnf.uni-greifswald.de/institute/geo.html

Universitaet Halle-Wittenberg

Institute for Geosciences and Geographie (A,B,M,D) ●☒
Von-Seckendorff-Platz 3/4
Halle D-06120
p. +49-0345-55 26055
direktor@geo.uni-halle.de
http://www.geo.uni-halle.de/
Programs: Geography, Geology, International Area Studies, Natural Resource Management, Teachers
Certificates: great variety

Universitaet Hamburg

Center for Earth System Research and Sustainability (M,D) ☒
Bundesstrasse 53
Hamburg 20146
anke.allner@uni-hamburg.de
http://www.cen.uni-hamburg.de/

Institute of Geophysics (B,M,D) ●☒
Bundesstrasse 55
Hamburg 20146
p. +4940428382973
dirk.gajewski@uni-hamburg.de
http://www.geo.uni-hamburg.de/de/geophysik.html
Programs: BSc Geophysics/Oceanography, MSc Geophysics
Enrollment (2016): B: 47 (23) M: 14 (11) D: 12 (8)
Professor:
Dirk J. Gajewski, (D), Karlsruhe, 1987, YesYg

Universitaet Hannover

Institut fur Geologie
Fachbeleich Erdwissenschaften
Callinstrasse 30
D-30167 Hannover
p. +49-(0)511-762 2343
sekretariat@geowi.uni-hannover.de
http://www.geologie.uni-hannover.de

Universitaet Heidelberg

Insitut fuer Geowissenschaften
Im Neuenheimer Feld 236
Heidelberg D-69120
p. 06221-54-8291
Elfriede.Hofmann@geow.uni-heidelberg.de
http://www.geow.uni-heidelberg.de/

Facultat fur Chemie und Geowissenschaften
Im Neuenheimer Feld 234
69120 Heidelberg
p. (+49) 06221/544 844
dcg@urz.uni-heidelberg.de
http://www.chemgeo.uni-hd.de/

Universitaet Leipzig

Institut fur Geophysik und Geologie
Talstrasse 35
04103 Leipzig
geologie@rz.uni-leipzig.de
http://www.uni-leipzig.de/~geo/

Universitaet Marburg

Inst fuer Geologie und Palaontologie
Biergenstrasse 12
Lahnberge
D-3550 Marburg-Lahn

Universitaet Muenster

Institut fuer Mineralogie ☒
Corrensstrasse 24
Muenster 48149
p. +49 251 83-33472
minsek@uni-muenster.de
http://www.uni-muenster.de/Mineralogie/

Institut fuer Geologie und Palaeontologie (B,M,D) ●☒
Corrensstr. 24
D-48149 Muenster
p. +49 251-83 33974
sklaus@uni-muenster.de
http://www.uni-muenster.de/GeoPalaeontologie/en
t: @GeoMuenster
Programs: BSc Geosciences, MSc Geosciences, MSc Water Sciences
Enrollment (2016): B: 270 (41) M: 46 (21) D: 25 (3)
Professor:
Christine Achten, (D), Frankfurt (Germany), 2002, CloEg
Ralph Thomas Becker, (D), Ruhr (Germany), 1990, PisPm
Benjamin Bomfleur, (D), PblGg
David De Vleeschouwer, (D), PcGg
Ralf Hetzel, (D), Gcm
Harald Strauss, (D), Goettingen (Germany), 1985, CsGg
Laura Stutenbecker, (D), GsgCc

Universitaet of Kiel

Institute of Geoscience
Christian-Albrechts-Platz 4
24118 Kiel
p. (+49) 431 880 3900
geschaeftsfuehrung@ifg.uni-kiel.de
http://www.ifg.uni-kiel.de/6+M52087573ab0.html

Universitaet Oldenburg

Institute for Chemistry and Biology of the Marine Environment
Carl-von-Ossietzky-Str. 9-11
Box 2503
Oldenburg 26111
p. +49-0441-798-5342
director@icbm.de
http://www.icbm.uni-oldenburg.de/

Universitaet Potsdam

Institute for Earth and Environmental Sciences (B,M,D)
Karl-Liebknecht-Str. 24-25
Potsdam 14476
p. +49 331 977 2116
sekretariat@geo.uni-potsdam.de
http://www.geo.uni-potsdam.de/
Enrollment (2013): B: 319 (14) M: 79 (11) D: 122 (16)

Universitaet Stuttgart

Inst fuer Geol und Palaontogisch
Fakultaet-7-Geo. und Biowissenschaften
Boblinger Str. 72
D-7000 Stuttgart 10

Universitaet Trier

Regional and Environmental Science
Behringstrabe 21
D-54296 Trier
p. (+49) (0) 651-201-4528
dekanatfb6@uni-trier.de
http://www.uni-trier.de/index.php?id=2220

Lehrstuhl fuer Geologie (B,M,D) ☒
Geowissenschaften (FB VI)
Trier 54286
p. 0651 201 4647
ensch@uni-trier.de
http://www.uni-trier.de/index.php?id=2632
Programs: Geography; Environmental Sciences; Geoarchaeology
Enrollment (2020): B: 14 (4) M: 12 (3) D: 3 (1)

Universitaet Tuebingen

Institute for Geoscience
72076 Tuebingen
Sigwartstr. 10
Christophe.Pascal@rub.de
http://www.ifg.uni-tuebingen.de/

Universitaet Würzburg

Institut für Paläontologie
Pleicherwall 1
D-97070 Würzburg
p. 49 0931- 31 25 97
baumhauer@uni-wuerzburg.de
http://www.palaeontologie.uni-wuerzburg.de

Institut für Geographie und Geologie
Am Hubland
97074 Würzburg
p. +49 (0) 931 / 31-85555
http://www.geographie.uni-wuerzburg.de

Universitaet zu Koeln

Institute for Geology and Mineralogy
Cologne
p. +49 221 470-5619
ellen.stefan@uni-koeln.de
http://www.geologie.uni-koeln.de/

University of Bayreuth

Bayerisches Geoinstitut (M,D)
Bayreuth 95440
p. +49(0)921 55-3700
bayerisches.geoinstitut@uni-bayreuth.de
http://www.bgi.uni-bayreuth.de/

University of Bremen

Geosciences (A,B,M,D) ●☒
Klagenfurter Strasse 2-4
Bremen, Bremen 28359
p. +49 421 218 - 65010
info@geo.uni-bremen.de
http://www.geo.uni-bremen.de
Programs: Geosciences (BSc,MSc); Marine Geosciences (MSc); Materials Chemistry and Mineralogy (MSc)
Enrollment (2019): B: 199 (0) M: 289 (0)

University of Cologne

Dept of Geology and Mineralogy (B,M,D) ●
Zuelpicherstr. 49a
50674 Koeln
p. +49-221-470-5619

geo-pruefungsamt@uni-koeln.de
http://www.uni-koeln.de/math-nat-fak/geologie

Institute of Geophysics and Meteorology (B,M,D)
Pohligstraße 3
D-50969 Köln
p. +49 (0)221 470 2552
tezkan@geo.uni-koeln.de
http://www.geomet.uni-koeln.de/en/general/home/

Universität Mainz

Institut für Geowissenschaften (B,M,D) ● ☒
Joh.-J.-Becher-Weg 21
55128 Mainz
p. +49 6131 39 24373
igw@uni-mainz.de
http://www.geowiss.uni-mainz.de/

Head:
Boris PJ Kaus, (D), Yg

Associate Professor:
Roman Botcharnikov, (D), Gx
Jonathan M. Castro, (D), Gv
Michael Kersten, (D), Ca
Cees W. Passchier, (D), Gt
Denis Scholz, (D), Cl
Bernd R. Schöne, (D), PegPi
Frank Sirocko, (D), Gs

Adjunct Professor:
Kirsten I. Grimm, (D), Pm
Dieter Mertz, (D), Cc
Thomas Tütken, (D), Pv

West Wilhelms Universitat

Dept of Applied Geology
Schlossplatz 2
4400 Muenster

Westfälische Wilhelms-Universität Münster

Fachbereich Geowissenschaften (B,M,D)
Heisenbergstrasse 2
Münster D-48149
p. +49 251 83-30002
dekangeo@uni-muenster.de
http://www.uni-muenster.de/Geowissenschaften/

Ghana

Kwame Nkrumah University of Science and Technology

Petroleum Engineering (B)
http://www.knust.edu.gh/ceng/faculties.php

Dept of Geomatic Engineering (B,M,D) ○
geomaticeng@knust.edu.gh
head.geomatic@knust.edu.gh
Kumasi PMB KNUST
p. +233 (0)3220 60227
geomaticeng@knust.edu.gh
http://www.knust.edu.gh/ceng/faculties.php

Geological Engineering (B,M,D) ●
Geological Engineering Department
College of Engineering
KNUST
Kumasi PMB
geologicaleng@knust.edu.gh
https://coe.knust.edu.gh/

Senior Scientist:
Emmanuel K. Appiah-Adjei, (D), Hohai, 2013, HwgGg

Associate Professor:
Simon K. Gawu, (D), Witwatersrand, 2004, GgEgZn

Lecturer:
Bukari Ali, (D), Birmingham (UK), 1996, HwNgYg
Godfrey C. Amedjoe, (M), Ghana, 2005, GidGc
Samuel Banash, (M)
Gordon Foli, (M), 2004
Solomon S. Gidigasu, (M), Sci & Tech (China), 2013, GeZrNr
Emmanuel Mensah, (M), Sci & Tech (China), 1998, EgGg

University of Development Studies

Earth and Environmental Science (B)
PO Box 1350
Tamale
p. +233 (71) 22422
edambayi@uds.edu.gh
http://www.uds.edu.gh/

University of Ghana

Dept of Earth Science (B,M,D) ● ☒
PO Box LG 58
Legon
Accra
p. +233-244-116-879
pmnude@ug.edu.gh
http://www.ug.edu.gh/index1.php?linkid=185&sublinkid=40&subsublinkid=36
Programs: Earth Science, Geology, Applied Geochemistry, Applied Geophysics, Petroleum Geoscience
Certificates: Not available
Enrollment (2013): B: 336 (104) M: 105 (33) D: 2 (1)

Professor:
Daniel K. Asiedu, (D), GsCgGd
David Atta-Peters, (D), Ghana, PlsPl
Bruce K. Banoeng-Yakubo, (D), Ghana, GcHw
Prosper M. NUDE, (D), Ghana, GgpGx
Frank K. Nyame, (D), Cg
Sandow M. Yidana, (D), Hwq

Senior Scientist:
Thomas K. Armah, (D), YgeYe
Yvonne A.S Loh, (M), Hw

Associate Professor:
Thomas M. Akabzaa, (D), Ghana, 2004, EgGeNm
Chris Y. Anani, (D), Gsd
Larry-Pax CHEGBELEH, (D), Ng
Johnson Manu, (D), GzEgGz
Patrick Asamoah Sakyi, (D), Cg

University of Mines and Technology

Dept of Geological Engineering (B)
PO Box 237
Tarkwa
p. +233 3123 20935
sps@umat.edu.gh
http://www2.umat.edu.gh/dept/gl/

Dept of Geomatic Engineering (B)
PO Box 237
Tarkwa
p. +233(0)362 20324
gm@umat.edu.gh
http://www.umat.edu.gh/UndergraduatePrograms/geomatic.html

Dept of Mineral Engineering (B)
PO Box 237
Tarkwa
p. +233(0)362 21136
mr@umat.edu.gh
http://www.umat.edu.gh/UndergraduatePrograms/mineral.html

Dept of Mining Engineering (B)
PO Box 237
Tarkwa
p. +233 3123 20935
mn@umat.edu.gh
http://www.umat.edu.gh/UndergraduatePrograms/Mining.html

Greece

Aristotle University of Thessaloniki

School of Geology
GR-541 24
Thessaloniki 54124
p. (+30) 2310 99.8450
kantira@geo.auth.gr

http://www.geo.auth.gr/index_en.htm

Greek Institute of Geology & Mineral Exploration

Greek Institute of Geology & Mineral Exploration ☒
1, Spirou Louis St.,
Olympic Village
Acharnae P.C. 13677
dirgen@igme.gr
http://www.igme.gr

National Technical University of Athens

Dept of Mining Engineering and Metallurgy
Zographou Campus
15 780 Athens
secretary@metal.ntua.gr
http://www.metal.ntua.gr/div-geology

University of Athens

Faculty of Geology and Geoenvironment
National & Capodistran Univ. of Athens
Panepistimioupolis, Ilistra
Athens
secr@geol.uoa.gr
http://www.geol.uoa.gr/engindex.htm

University of Patras

Dept of Geology
26110 Patras

Guatemala

Universidad sde San Carlos de Guatemala

Centro de Estudios Superiores de Energia y Minas
Ciudad Universitaria Zona 12, Guatemala City 01012
p. (502) 2418-9139 ex. 86211
usacesem@ing.usac.edu.gt
http://cesem.ingenieria.usac.edu.gt/

Guinea

Univerité Gamal Abdel Nasser de Conakry

Faculty of Geology and Mining (M,D) ●☒
BP 1147
Conakry, Guinée
p. +224 631 54 48 38
sekoumoussa@gmail.com
http://www.uganc.org/index.php/2013-01-12-06-22-44/cere
Enrollment (2018): M: 30 (0)

Guyana

Guyana Geology and Mines Commission

Guyana Geology and Mines Commission
Upper Brickdam
Georgetown
p. 226-5591, 225-2862
http://www.ggmc.gov.gy/

Haiti

Ecole Nacionale de Geologie Appliquee

Ecole Nacionale de Geologie Appliquee
B.P. 1560-Varreux, Port-au-Prince

Hungary

Eotvos Lorand University

Insitute of Geography and Earth Sciences (B,M,D) ●
Pazmany Peter setany 1/C
Budapest H-1117
p. (+36 1) 381 2191
altfoldtan@ttk.elte.hu
http://geosci.elte.hu/en_index.htm
Enrollment (2014): B: 900 (360) M: 300 (120) D: 40 (18)

Professor:
Judit Bartholy, (D), Eotvos Lorand, 1978, Am

Associate Professor:
Gábor Timár, (D), Eotvos Lorand, 2003, YgdZi

Research Professor:
János Lichtenberger, (D), Hungarian Acad of Sci, 1996, AsZrXy

Professor:
Kristof Petrovay, (D), Eotvos Lorand, 1992, Xy
Miklos Kazmer, (D), Eotvos Lorand, 1982, PggRh

Associate Professor:
Agnes Gorog, (D), Eotvos Lorand, Pmi
László Lenkey, (D), Vrije (Amsterdam), 1999, YhHwGa
Balázs Székely, (D), Tubingen, 2001, GmZri

Associate Scientist:
Orsolya Ferencz, (D), Budapest, 2001, XyZr
Anikó Kern, (D), Eotvos Lorand, 2012, ZrAms
Gábor Molnár, (D), Eotvos Lorand, 2004, ZrYgZi

Assistant Professor:
László Balázs, (D), Eotvos Lorand, 2009, EoYeg
Attila Galsa, (D), Eotvos Lorand, 2004, YgHwYe
Attila Osi, (D), Eotvos Lorand, PviPe
Emoke Toth, (D), Eotvos Lorand, PmsPb

Research Associate:
Istvan Szente, (D), Eotvos Lorand, Pig

Professor:
Ferenc Horváth, (D), Eotvos Lorand, 1971, YgRhGc
Péter Márton, (D), Eotvos Lorand, 1971, YmgYs

Hungarian Geological Survey

Hungarian Geological Institute
Stefánia út 14
Budapest HU-1143
p. 36 1267 1433
muzeum@mbfsz.gov.hu

Jozsef Attila University

Dept of Mineralogy, Geochem & Petrology
P O Box 651
H-6701 Szeged

Dept of Geology & Paleontology
Egyetum u. 2-6
H-6722 Szeged

Mining and Geological Survey of Hungary

Mining and Geological Survey of Hungary
PO Box 95
1590 Budapest 1145
p. (+36-1) 301-2900
info@mbfsz.gov.hu
https://mbfsz.gov.hu/en

Univesity of Szeged

Department of Geoinformatics, Physical and Environmental Geography (B,M,D) ☒
SZTE Geoinformatikai, Természet- és Környezetföldrajzi Tanszék
Szeged, Egyetem utca 2-6.
Szeged 6722
p. 0036 62544158
gysipos@geo.u-szeged.hu
http://www.geo.u-szeged.hu/
f: @foldrajz
Programs: Geography (MSc) in English
Geography (BSc, MSc) in Hungarian
Geoinformatics (MSc) in Hungarian
Environmental Engineer (BSc, MSc) in Hungarian
Enrollment (2021): B: 33 (0) M: 13 (0) D: 5 (0)

vice head of department:
Andrea Farsang, (D), Szeged, 1997, SdZe

vice head of department:
Károly Barta, Szeged, 2004, SpZy

head of department:
György Sipos, (D), Szeged, 2007, GmCcHs

Associate Professor:
Timea Kiss, (D), Debrecen, 2001, GmZyHs
Ferenc Kovács, (D), Szeged, 2007, ZriZy
László Mucsi, (D), Szeged, 1997, ZirZy
József Szatmári, (D), Szeged, 2007, ZriZy
Péter Szilassi, (D), Szeged, 2004, Zuy

Assistant Professor:
Viktória Blanka-Végi, (D), Szeged, 2010, HsZyRn
Zsuzsanna Ladányi, (D), Szeged, 2011, ZyeRn
Zalán Tobak, (D), Szeged, 2013, ZirZy
Boudewijn van Leeuwen, (D), Szeged, 2012, ZirZy
Emeritus:
Gábor Mezősi, (D), Szeged, 1983, ZyGmRn
János Rakonczai, (D), Szeged, 1995, ZeHwZy

Iceland

Landmælingar Íslands

National Land Survey of Iceland ☒
Stillholt 16-18 300 Akranes
Akranes 300
p. 4309000
lmi@lmi.is
http://www.lmi.is/

University of Iceland

Faculty of Earth Science
Öskju, Sturlugötu 7
101 Reykjavik
p. (+354) 5254600
nemvon@hi.is
http://english.hi.is/sens/faculty_of_earth_sciences/about_faculty

India

Aligarh Muslim University

Dept of Geology (B,M,D) ●
Aligarh - U.P.
Aligarh, Uttar Pradesh 202002
p. 0571-2700615
lakrao@yahoo.com
http://www.amu.ac.in/fsc/no4.html

Andhra University

Dept of Geology
Vishakha-patnam-530 003
College of Science and Technology
, Andhra Pradesh
p. 91-891-2844888
head.geology@andhrauniversity.edu.in
http://www.andhrauniversity.info/

Anna University

Dept of Geology
Guindy, Chennai-600 025,
Tamil Nadu
p. 044-22358442 / 8444 / 8452
geonag@gmail.com
http://www.annauniv.edu

Annamalai University

Faculty of Marine Sciences
Annamalainagar-608 025
Tamil Nadu
p. 91 - 4144 - 238248
deanfms1@gmail.com
http://annamalaiuniversity.ac.in/marinesciences.htm

Banaras Hindu University

HARI SRIVASTAVA (B,M,D) ☒
316/17
Ambrosia Appartment
Lanka
Varanasi, U.P. 221005
p. +91941535360
hbsrivastava@gmail.com
http://www.bhu.ac.in/Geology
Programs: Geology
Enrollment (2017): B: 150 (0) M: 56 (0) D: 50 (0)
Professor:
HARI BAHADUR SRIVASTAVA, (D), Banaras Hindu, 1980, GctGp

Bharathidasan University

Dept of Geology
School of Geosciences
Tiruchirapalli 620 024
p. +91 431 2407034
drmm_bdu@yahoo.co.uk
http://www.bdu.ac.in/schools/geo_sciences/geology/

Cochin University of Science & Technology

Dept of Geology
Kochi
p. +91(484)2577550
akv@cusat.ac.in
http://www.cusat.ac.in/

Geological Survey of India

Geological Survey of India ○☒
27, J.L.Nehru Road
Kolkata
Kolkata, West Bengal 700016
dg.gsi@gov.in
http://www.gsi.gov.in
f: https://www.facebook.com/gsipage/
t: @GeologyIndia
Certificates: Advanced GIS, Digital Cartography, Geological Mapping, Mineral Exploration, Geophysical Survey, etc

Goa University

School of Earth, Ocean and Atmospheric Sciences (M,D) ○☒
Taleigao Plateau
Goa University P.O.
Taleigao Plateau
Panaji, Goa 403 206
p. +91-8669609193
mkotha@unigoa.ac.in
http://www.unigoa.ac.in/department.php?adepid=11&mdepid=3
Programs: Applied Geology
Certificates: GIS
Enrollment (2018): M: 22 (22) D: 1 (1)
Professor:
Mahender Kotha, (D), IIT (Bombay), 1987, GsoZi
Associate Professor:
Anthony V. Viegas, (D), Goa, 1997, GiEmGx
On Contract:
Poornima Dhavaskar, (M), Geology, 2013, Gge
Purushottam Verlekar, (M), GrmGg
Assistant Professor:
Niyati Kalangutkar, (D), Goa, GugGe
Cooperating Faculty:
Sohini Ganguly, (D), Gi

Guru Nanak Dev University

Dept of Geology
Amritsar
http://www.gndu.ac.in/

Indian Institute of Science Education and Research, Kolkata

Dept of Earth Science (B,M,D) ●☒
Main Campus
Nadia
Mohanpur (Kolkata) , West Bengal 741246
des.chair@iiserkol.ac.in
http://earth.iiserkol.ac.in/
Programs: Geological Sciences

Indian Institute of Technology

Dept of Applied Geology
Bombay, Powai
Bombay
ayaz@iitb.ac.in
http://www.geos.iitb.ac.in/

Indian Institute of Technology, Kharagpur

Dept of Geology & Geophysics (B,M,D)
Kharagpur, Kharagpur
West Bengal 721302
Kharagpur, West Bengal 721302
p. +91-3222-282268
head@gg.iitkgp.ernet.in
http://www.iitkgp.ac.in/departments/home.php?deptcode=MG

Indian Institute of Technology, Roorkee

Dept of Earth Sciences (M,D)
Department of Earth Sciences
IITRoorkee
Roorkee, Uttrakhand 247667
p. +911332285532
dpkesfes@iitr.ac.in
http://members.tripod.com/rurkiu/acd-earth

Indian School of Mines

Dept of Applied Geology
Dhanbad
West Bengal 826 004
p. +91-326-2296616
agl@iitism.ac.in
http://www.ismdhanbad.ac.in/depart/geology

Institute of Science

Dept of Geology
Aurangabad 431004
Aurangabad , Maharashtra
p. +91 (0240) 2400586
director@inosca.org
http://www.inosca.org

Jadavpur University

Dept of Applied Geology (B,M,D)
Jadavpur
Calcutta
p. +913324572268
hod@geology.jdvu.ac.in
http://www.jaduniv.edu.in/view_department.php?deptid=76

Jmia Millia Islamia

Dept of Geography
Jamia Nagar
New Delhi-25
rocketibrahim@yahoo.com
http://jmi.ac.in/aboutjamia/departments/geography/introduction

Lucknow University

Dept of Geology
Lucknow 226007 U.P.
ajaimishra2007@yahoo.co.in
http://www.lkouniv.ac.in/dept_geology.htm

Maharaja Sayajirao University of Baroda

Dept of Geology
Baroda - Gujarat
p. (919) 426721547
aditya.joshi-geology@msubaroda.ac.in
http://www.msubaroda.ac.in/science

Nagpur University

Dept of Geology
Rao Bahadur D. Laxminarayan Educational Campus
Law College Square
Amravati Road
Nagpur 440 001
p. +91-712-253241
geodeptplacecell@gmail.com
http://www.nagpuruniversity.org/links/FacultyofScience.htm

Osmania University

Dept of Geology
Hyderabad 500007 A.P.
http://www.osmania.ac.in/

Centre of Exploration Geophysics (M,D) ○ ☒
Department of Geophysics
Osmania University Campus
Hyderabad, Telengana 500007
p. (+91) 40-27097116
head.geophysics@osmania.ac.in
http://www.osmania.ac.in/
Programs: Geophysics (M.Sc.)
Enrollment (2019): M: 32 (27) D: 11 (11)
Chairman, Board of Studies in Geophysics:
Veeraiah B, (D), Osmania, 2005, YexYg
Head, Department of Geophysics:
Ram Raj Mathur, (D), Osmania, 1989, YeuYm
Assistant Professor:
Manjula Dharmasoth, (M), Osmania (India), 2012, YmeYg
Vijay Kumar Dubba, (D), Osmania (India), 2015, YevYs
Udaya Laxm Gakka, (D), Osmania, 2009, HwYge

Panjab University

Dept of Geology (B,M,D) ● ☒
Sector-14
Chandigarh, Punjab 160014
p. (+91) 1722534250
chairperson_geology@pu.ac.in
http://geology.puchd.ac.in/
Programs: Geology

Centre for Petroleum and Applied Geology (U.I.E.A.S.T.)
Chandigarh 160014
rpatnaik@pu.ac.in
http://pu.ac.in/

Pondicherry University

Dept of Earth Science (B,M,D)
The Head
Department of Earth Sciences
Pondicherry University
Pondicherry 605 014
p. +914132656741
head.esc@pondiuni.edu.in
http://www.pondiuni.edu.in/department/department-earth-sciences
Professor:
Balakrishnan Srinivasan, (D), Jawaharlal Nehru (India), 1986, CcGiz

Presidency College

Dept of Geology
College Street
Calcutta
jm.geol@presiuniv.ac.in
http://www.concentric.net/~slahiri/pc/Department/Geology/geo

Pt. Ravishankar Shukla University

School of Studies in Geology and Water Resource Management (M,D)
SOS Geology
Pt. Ravishankar Shukla UNiversity
Raipur, Chhattisgarh 492010
vc_raipur@prsu.ac.in
http://www.prsu.ac.in/
Enrollment (2013): M: 15 (0) D: 4 (1)
Professor:
Srikant k. Pande, (D), Nagoya, 1986, GgHyg
Professor:
Ninad Bodhankar, (D), Raipur, 1992, GcNm
Kosiyath Rghavan Na Hari, (D), Vikram Ujjain, 1992, GiiCp
Mohammad Wahdat Yar Khan, (D), AMU, Aligarh, 1979, GdEgCl

Savitribai Phule Pune University,

Dept of Geology (M,D) ☒
Ganeshkhind Road

Pune, Maharashtra 411 007
p. +91-020-25601360
geology@unipune.ac.in
http://www.unipune.ac.in/dept/science/geology/
Programs: M.Sc. Geology, Ph.D Geology
Certificates: Nil
Professor:
Makarand Ganesh Kale, (D), 1992, Gso

Sri Sri University

Faculty of Science (B,M,D) O⃠
Faculty of Science
Sri Sri University
Cuttack
Odisha
p. (981) 117-4867
chandra.d@srisriuniversity.edu.in
http://www.srisriuniversity.edu.in
Programs: Environment Sciences
Enrollment (2014): B: 105 (1200) M: 60 (600) D: 30 (130)
Director COL:
C S. Dubey, (D), Delhi, 1993, GexGt

University of Baroda

Dept of Geology
Baroda , Vadodara 390 002
swanikhil@yahoo.co.in
http://www.msubaroda.ac.in/deptindex.php?ffac_code=2&fdept_code=6

University of Madras

Dept of Geology
Madras, 600 025
Maraimalai Adigalar Campus
spmohan50@hotmail.com
http://www.unom.ac.in/departments/geology/geology.html

Dept of Geology
Maraimalai Adigalar Campus
Chennai 600 005
p. 044 - 22202790
spmohan50@hotmail.com
http://www.unom.ac.in/departments/geology/geology.html

University of Mysore

Dept of Geology
Mysore 570 005
p. +91 821 2419724
bb@geology.uni-mysore.ac.in
http://www.uni-mysore.ac.in/geology/#A

Vijayanagara SriKrishnaDevaraya University

Dept of Applied Geology (M,D) O☒
Dr.C.Venkataiah
Professor of Geology
Vijayanagara SriKrishnaDevaraya University
Bellary, Karnataka 583 104
p. (937) 909-0588
venkataiah.c@gmail.com
http://www.vskub.ac.in
Programs: M.Sc two years Applied Geology

Indonesia

Direktorat Vulkanologi

Volcanological Survey of Indonesia
Jl. Diponegoro 57
Bandung , West Java
http://portal.vsi.esdm.go.id/

Gadjah Mada University

Dept of Geology
Jalan Bulaksmur
Yogzakarta 55281
p. (+62) 274 6492340
geografi@geo.ugm.ac.id
http://geo.ugm.ac.id/main/

Indonesian Directorate General of Geology and Mineral Resources

Ministry of Mines and Energy
contactcenter136@esdm.go.id
http://www.esdm.go.id

Institut Teknologi Bandung

Dept of Geology
Jalan Ganesa, No. 10
Bandung 40132, Jawa Barat
p. 022 2514990
humas@itb.ac.id
http://www.fitb.itb.ac.id/en/

Dept of Geological Engineering
Jl. Ganesha 10
Bandung, 40135
geologi@gc.itb.ac.id

Trisakti University

Faculty of Earth and Energy Technologies
Jl. Kiai Tapa, Grogel
Jakarta 11440
p. 5663232 Ext. 8510

Universitas Hasanudin

Dept of Geological Engineering
Jl. Perintis Kemerdekaan
Ujung Pandang 90245

Universitas Padjadjaran

Dept of Geological Engineering
Jl. Raya Bandung-Sumedang km.21,
Jatinangor

Universitas Pakuan

Dept of Geological Engineering
Jl. Pakuan, PO Box 452
Bogor 16143, Jawa Barat
p. 0251-8312206
rektorat@unpak.ac.id
http://www.unpak.ac.id/

Ireland

Geological Survey of Ireland

Beggars Bush
Haddington Road
Dublin 4
john.butler@gsi.ie
http://www.gsi.ie/

National University of Ireland Galway

Earth and Ocean Sciences (B,D)
Earth & Ocean Sciences
National University of Ireland
University Road
Galway
p. + 353 (0)91 492 126
lorna.larkin@nuigalway.ie
http://www.nuigalway.ie/eos/
Professor:
Peter Croot, (D), Otago (NZ), Ct
Lecturer:
Rachel R. Cave, (D), Southampton (UK), Cm
Eve Daly, (D), National (Ireland), Ygu
Tiernan Henry, (D), National (Ireland), 2014, HwyEg
John Murray, (D), Trinity Coll (Cambridge), PgGs
Robin Raine, (D), National (Ireland), Yg
Tyrrell Shane, (D), Univ Coll (Dublin), Gs
Martin White, (D), Southampton (UK), Op
Emeritus:
Martin Feely, (D), National (Ireland), 1982, GzxEg

Trinity College

Dept of Geology
Department of Geology
Museum Building
Trinity College-Dublin
Dublin 2
p. +353 01 896 1074
dickensg@tcd.ie
http://www.tcd.ie/Geology/
Chair:
Balz S. Kamber, (D), CgGgp
Associate Professor:
David Chew, (D), Univ Coll (Dublin), 2001, CcGgt
Catherine Coxon, GeOu
Robin Edwards, Ge
Patrick N. Wyse Jackson, (D), Dublin, 1992, PiRh
Assistant Professor:
Quentin G. Crowley, Cc
Seán h. mcClenaghan, Eg
Chris Nicholas, Eo
Juan Diego Rodriguez-Blanco, (D), 2006, GzClZm
Catherine V. Rose, Cg
Emma L. Tomlinson, Cc

University College Cork

Dept of Geology (B,M,D)
Donovans Road
Cork
p. +353 21 4902657
s.culloty@ucc.ie
http://www.ucc.ie/ucc/depts/geology/
Administrative Assistant: Patricia Hegarty
Head:
John Gamble, Gi
Chair:
Andy J. Wheeler, (D), Cambridge, 1994, GusGg
Professor:
Ken Higgs, Pl
Research Associate:
Tara Davis, Yg
Jim Smith, Pl
Lecturer:
Bettie M. Higgs, (D), Sheffield, 1977, YgGqRh
Ed Jarvis, Pl
Ivor MacCarthy, (D), 1974, Gs
Pat Meere, (D), National (Ireland), 1992, Gc
John Reavy, (D), GicGt
Other:
Mary Lehane, Zn
Mick O'Callaghan, Zn
Dan Rose, Zn

University College Dublin

School of Earth Sciences
Science Centre West
University College Dublin
Belfield, Dublin 4
p. +353 1 716 2331
geology@ucd.ie
http://www.ucd.ie/geology/
Head:
J. Stephen Daly
Professor:
Peter D. W. Haughton
Frank McDermott, (D), Open (UK), 1987, GgCgZg
Patrick M. Shannon
John J. Walsh
Associate Professor:
Christopher J. Bean
Julian F. Menuge, (D), Cambridge, 1983, EmCcs
Ian D. Somerville
Lecturer:
Conrad Childs
Aggeliki Georgiopopoulou, (D), Southampton (UK), 2006, GusOg
Ivan Lokmer
Patrick J. Orr
Adjunct Professor:
Tom Manzocchi, (D), Goc

Israel

Ben Gurion University of the Negev

Dept of Geological and Env Sciences (B,M,D) ● ☒
P.O. Box 653
84105 Beer Sheva
Beer-Sheva 84105
p. +972-8-6461290
katzir@bgu.ac.il
http://www.bgu.ac.il/geol/

Geological Survey of Israel

30 Malkhe Israel St.
Jerusalem, 95501
ask_gsi@gsi.gov.il
http://www.gsi.gov.il/

Hebrew University of Jerusalem

Faculty of Advanced Environmental Studies
Givat Ram
Jerusalem 91904
msfeitel@mscc.huji.ac.il

Institute of Earth Sciences (B,M,D) ● ☒
Jerusalem 91904
Jerusalem 91904
p. 97226584686
arimatmon@mail.huji.ac.il
http://earth.huji.ac.il/
f: https://www.facebook.com/earthhuji/
Programs: Geology (B,M); Atmosphere, Oceanography & Climate (B); Environmental Studies (B); Atmospheric Sciences (M); Oceanography (M); Hydrology (M); Environmental Studies (M)
Enrollment (2017): B: 70 (0)

Tel Aviv University

Dept of Geophysical, Atmospheric and Planetary Sciences (B,M,D) ● ☒
Ramat Aviv
P O Box 30940
Tel Aviv 69978 69978
p. 972-3-6408633
http://geophysics.tau.ac.il/
Full Professor:
Shmuel Marco, (D), Gg

Geosciences (B,M,D) ●
Levanon Road
Ramat Aviv
Tel Aviv 6997801
p. 972-3-6408633
shmulikm@tau.ac.il
http://geophysics.tau.ac.il/
Professor:
Pinhas Alpert, (D), Hebrew, 1980, AsZrAm
Zvi Ben-Avraham, (D), MIT/WHOI, 1973, YrGt
Shmulik Marco, (D), Hebrew, 1997, Ggc
Morris Podolak, (D), Yeshiva, 1974, ZnnZn
Colin G. Price, (D), Columbia, 1993, As
Moshe Reshef, (D), Tel Aviv, 1985, Yeg
Senior Scientist:
Pavel Kishcha, (D), Russian Acad of Sci, 1985, As
Dr:
Nili Harnik, (D), MIT, 2000, As
Lecturer:
Gilles Hillel Wust-Bloch, (D), 1990, Ys
Alon Ziv, (D), Ys
Principal Research Assoc. (Assoc. Professor):
Lev Eppelbaum, (D), Inst of Geophysics of Georgia, 1989, YvmGt

Italy

Alma Mater Studiorum Università di Bologna

Dipartimento di Scienze Biologiche, Geologiche e Ambientali (B,M,D) O⊘
Piazza di Porta San Donato 1
Bologna 40126
p. +39 051 2094238
bigea.direttore@unibo.it
http://www.bigea.unibo.it/en
f: https://www.facebook.com/bigea.unibo
Programs: BSc in Geology
MSc in Geology & Territory
PhD in Earth Sciences
Enrollment (2021): B: 184 (34) M: 76 (27) D: 21 (5)

Professor:
Alessandro Amorosi, (D), Gs
Roberto Barbieri, (D), Pg
Matteo Berti, (D), NgZuNt
William Cavazza, (D), Gd
Jo de Waele, (D), Gm
Enrico Dinelli, (D), Cl
Antonio Navarra, (D), As
Nicholas Pierce, (D), Cg
Giovanni Valdrè, (D), Gz
Giulio Viola, (D), Gct
Alberto Vitale Brovarone, Gp

Associate Professor:
Marco Antonellini, (D), Gc
Francesco Brardinoni, (D), Gm
Rossella Capozzi, (D), Gr
Barbara Cavalazzi, (D), Xb
Claudia Romagnoli, (D), Gu
Daniele Scarponi, (D), Pe
Alessandro Simoni, (D), Nrx

Associate Scientist:
Roberto Braga, (D), Gp
Giorgio Gasparotto, (D), Gz
Federico Lucchi, (D), Gv
Gianluca Vignaroli, (D), Gc

Assistant Professor:
Alessandro Buscaroli, (D), Sd
Bruno Campo, (D), Gsr
Luigi Cantelli, (D), ZriGc
Stefano Cremonini, (D), Gma
Maria Filippini, Hq
Paolo Stefano Garofalo, (D), Eg
Veronica Rossi, (D), Gs
Natalia Rovella, (D), Zm
Sonia Silvestri, (D), GmZr
Claudio Tranne, (D), Gv
Gianfranco Ulian, (D), Gy
Stefano Claudio Vaiani, (D), Pm

Adjunct Professor:
Federico Fanti, Pv

Servizio Geologico d'Italia

Via Vitaliano Brancati
Roma 48-00144
p. (+39) 0650071
portalesgi@isprambiente.it
http://www.isprambiente.gov.it/it/servizi-per-lambiente/il-servizio-geologico-ditalia

Università degli Studi di Padova

Department of Geosciences (B,M,D) ●☒
Via G. Gradenigo 6
Padova 35131
p. +390498279110
geoscienze.direzione@unipd.it
http://www.geoscienze.unipd.it/
Programs: Geological Sciences (Bachelor), Geology and Technical Geology (Master), Earth Sciences (Doctorate)
Enrollment (2017): B: 259 (49) M: 112 (38) D: 17 (7)

Full Professor:
Fabrizio Nestola, (D), Modena & Reggio Emilia, 2003, Gz

Full Professor:
Gilberto Artioli, (D), Chicago, 1985, GzyZm
Alessandro Caporali, (D), Ludwig-Maximilian (Germany), 1979, YdsZr
Alberto Carton, (M), Gm
Giorgio Cassiani, (D), YgHs
Bernardo Cesare, (D), Gp
Giulio Di Toro, (D), Gc
Silvana Martin, Gc
Giorgio Pennacchioni, Gc
Cristina Stefani, Gd
Nicola Surian, Gm
Massimiliano Zattin, (D), Gs

Professor:
Francesca Da Porto

Associate Professor:
Claudia Agnini, (D), Padua, 2007, PmeGu
Lapo Boschi, Yg
Luca Capraro, (D), Padua, 2002, PelGg
Andrea D'Alpaos, (D), Hs
Maria Chiara Dalconi, (D), Gz
Paolo Fabbri, Hw
Manuele Faccenda, (D), Gq
Mario Floris, (D), NgZr
Alessandro Fontana, (D), Gm
Eliana Fornaciari, (D), Pgm
Massimiliano Ghinassi, (D), Gs
Luca Giusberti, (D), Pm
Lara Maritan, (D), Gx
Andrea Marzoli, (D), Cg
Matteo Massironi, (D), XgZr
Claudio Mazzoli, Gx
Stefano Monari, Pg
Paolo Mozzi, (D), Gm
Paolo Nimis, Eg
Nereo Preto, (D), Gs
Manuel Rigo, (D), Gs
Gabriella Salviulo, Gz
Raffaele Sassi, Ggx
Paolo Scotton, Hg
Luciano Secco, Gz
Alberta Silvestri, (D), Gz
Richard Spiess, Gp
Annalisa Zaja, Yg
Dario Zampieri, Gc

Assistant Professor:
Omar Bartoli, Gp
Jacopo Boaga, Yg
Anna Breda, (D), Padua, 2003, GsrGg
Antonio Galgaro, Ng
Roberto Gatto, Pg
Christine Marie Meyzen, (D), Cg
Leonardo Piccinini, (D), Hg
Telemaco Tesei
Luca Valentini

Università degli Studi di Pavia

Dept of Earth and Environmental Sciences (A,B,M,D) O
via Ferrata, 1
Pavia 27100
p. +30 0382 985754
dvagnini@unipv.it
http://sciter.unipv.eu/site/home.html
Enrollment (2014): B: 45 (25) M: 25 (14) D: 15 (5)

Universita Degli Studi di Siena

Centro di GeoTecnologie
Via Vetri Vecchi 34
San Giovanni Valdarno 52027
p. +390559119400
bottacchi@unisi.it
http://www.geotecnologie.unisi.it

Universita di Bari

Department of Earth and Geo-environmental Sciences
Via E. Orabona
4-70125 Bari
scandale@geomin.uniba.it
http://www.geomin.uniba.it/

Universita di Cagliari

Dept di Geoingegneria e Tecnologie Ambientali
Via Marengo,3
09124 Cagliari
p. +39 70 675 52/29
mazzella@unica.it
http://geoing.unica.it/digita.htm
Professor:
Antonio AM MAZZELLA, Gq

Universita di Calabria

Dept of Biology Ecology and Earth Science (DIBEST) (D)
87036 Arcavacata di Rende
Calabria
crisci@unical.it
http://www.unical.it

Universita di Camerino

Dept Scienze della Terra (B,M,D) ○☒
Via Gentile III da Varano
Camerino, MC 62032
p. (+39) 737402126
protocollo@pec.unicam.it
http://www.sst.unicam.it/SST/en/course-of-degree
Programs: Geology (B); Natural Science (B); Geoenvironmental Resources and Risks (M)

Universita di Catania

Inst Scienze della Terra
Corso Italia 57
95129 Catania
p. (+39) 095-7195730
giovali@unict.it
http://www3.unict.it/idgeg/

Universita di Firenze

Dept Scienze della Terra (B,M,D) ●
via La Pira 4
50121 Firenze
direttore@geo.unifi.it
http://www.dst.unifi.it/

Universita di Genova

Dept Scienze della Terra Ambiente e Vita - DISTAV (B,D)
Corso Europa 26
16132 Genoa
p. +39 010 353 8311
direttore@dipteris.unige.it
http://www.distav.unige.it/drupalint
Professor:
Egidio Armadillo, (D), Ye
Research Associate:
Donato Belmonte, (D), Cg

Università di Modena e Reggio Emilia

Dept di Scienze Chimiche e Geologiche (B,M,D)
Largo S. Eufemia, 19
Modena, Italy 41121
p. 390592055885
direttore.chimgeo@unimore.it
http://www.dscg.unimore.it/
Enrollment (2013): B: 23 (13) M: 14 (4) D: 10 (9)

Universita di Napoli FEDERICO II

Dipartimento di Scienze della Terra, dell'Ambiente e delle Risorse (B,M,D) ●
L.go S. Marcellino, 10
Naples 80138
p. +390812538112
domenico.calcaterra@unina.it
http://www.distar.unina.it

Universita di Perrugia

Dept Scienze della Terra
Piazza Università, 1
06100 Perugia
cclgeol@unipg.it
http://cclgeol.unipg.it/cclgeol/

Università di Torino

Dept Scienze della Terra (B,M,D) ●☒
via Valperga Caluso 35
Torino 10125
p. 011.6705184
direzione.scienzeterra@unito.it
http://www.dst.unito.it/do/home.pl
Programs: Geological Sciences
Science in applied Geology
Environmental monitoring, protection and remediation
Enrollment (2020): B: 25 (25) M: 15 (15)
Professor:
Elena Belluso, (D), Gz
Giorgio Carnevale, (D), Pg
Rodolfo Carosi, (D), Gc
Daniele Castelli, (D), Gx
Anna Maria Ferrero, (D), Nr
Alessandro Pavese, Gz
Associate Professor:
Piera Benna, (M), Gz
Alessandro Borghi, (D), Gx
Marco Bruno, (D), Gz
Paola Cadoppi, (D), Gc
Cesare Comina, (D), Yx
Diego Coppola, Gv
Domenico Antonio De Luca, (D), Hy
Francesco dela Pierre, (D), Gd
Massimo Delfino, (D), Pg
Andrea Festa, (D), Gc
Maria Gabriella Forno, (M), Gg
Giandomenico Fubelli, (D), Gm
Marco Gattiglio, (D), Gc
Rocco Gennari, Gn
Marco Giardino, (D), Gm
Roberto Giustetto, (D), Gz
Chiara Teresa Groppo, (D), Gx
Luca Martire, (D), Gd
Michele Motta, (M), Zy
Pierluigi Pieruccini, Gm
Mauro Prencipe, (D), Gz
Franco Rolfo, (D), Gx
Sergio Carmelo Vinciguerra, (D), Nr
Assistant Professor:
Gianni Balestro, (D), Gc
Carlo Bertok, (D), Gd
Sabrina Maria Rita Bonetto, (D), Ng
Corrado Cigolini, (M), Gv
Emanuele Costa, (D), Gz
Anna d'Atri, (D), Gr
Simona Ferrando, (D), Gx
Franco Gianotti, (D), Gg
Daniele Giordano, (D), Gv
Salvatore Iaccarino, Gc
Francesca Lozar, (D), Pg
Edoardo Martinetto, (D), Pb
Luciano Masciocco, (M), Ge
Luigi Motta, (M), Zy
Marcello Natalicchio, Gs
Marco Davide Tonon, (M), Ze
Elena Zanella, (D), Ym
Emeritus:
Emiliano Bruno, (M), Gz
Ezio Callegari, (M), Gx
Roberto Compagnoni, (M), Gx
Giovanni Ferraris, (M), Gz
Giulio Pavia, (M), Pg

Università di Trieste

Dipartimento di Matematica e Geoscienze (B,M,D) ☒
via Weiss 2
Trieste , Italy I-34128

p. +39 040 5582055
dmg@pec.units.it
https://dmg.units.it/
Programs: bachelor in Geology, master in Geosciences, master in Geophysics and Geodata, PhD in Earth Sciences, Fluid Mechanics and Mathematics.
Certificates: GIS

Università di Udine

Dipartimento di Georisorse e Territorio
Dipartimento di Georisorse e Territorio
Via Cotonificio 114
33100 Udine
ta.tempoindeterminato@uniud.it
http://udgtls.dgt.uniud.it/

Universita di Urbino

Inst di Geologia Applicata
Via Muzio Oddi 14
61029 Urbino

Universita Pisa

Dept of Geosciences (D)
via S. Maria 53
56126 Pisa
p. +39050847260
martinelli@dst.unipi.it
http://www.dst.unipi.it/

University of Ferrara

Department of Physics and Earth Sciences ☒
Via G. Saragat 1
44122 Ferrara
p. (+39) 0532 974211
dip.fisicascienzeterra@unife.it
http://fst.unife.it/it

University of Milano

Dept of Geology
Via Festa Del Perdono 7
20126 Milano
p. (+39) 02 6448 1
welcome.desk@unimib.it
http://www.unimib.it/go/46204/Home/English/Academic-Programs/Mathematics-Physics-and-Natural-Sciences/Geological-Sciences-and-Te

University of Siena

Dept of Physical Science, Earth, and Environment (B,M,D)
●⃠
Strada Laterina, 8
Siena 53100
p. (+39) 0577 233938
pec.dsfta@pec.unisipec.it
http://www.dsfta.unisi.it/it
f: https://www.facebook.com/dsfta.siena
Programs: A) - Undergraduate Course on "Geological Sciences" (duration: 3 academic years); B) Graduate Advanced/Magistral Course on "Geological Sciences and Technologies" (duration: 2 academic years); C) Doctoral/PhD Programme in "Environmental, Geological and Polar Sciences and Technologies".
Certificates: 1st Level Master Courses in: a) Geomatics; 2nd Level Master Courses in: a) Engineering Geology; b) Geomatics; c) Environmental Geotechnologies; d) Geotechnologies for Archaeology;
Enrollment (2021): B: 14 (12) M: 12 (12) D: 4 (4)

Professor:
Dario Albarello, (D), Bologna (Italy), YgsYx
Enrico Tavarnelli, (D), Siena (Italy), 1993, GgcGt

Associate Professor:
Gianluca Cornamusini, (D), Siena (Italy), GgsGr
Pier Lorenzo Fantozzi, (D), Pisa (Italy), Gme
Luca Maria Foresi, (D), PmGs
Ivan Martini, (D), Siena (Italy), GsaGg
Massimo Salleolini, (M), Siena (Italy), HwyHs
Riccardo Salvini, (D), Siena (Italy), ZruNg
Cecilia Viti, Gz

Research Associate:
Paolo Conti, (D), Basel (Switzerland), GctGg
Leonardo Disperati, (D), Pisa (Italy), 1992, NgtZr
Marco Giamello, Siena (Italy), GxCgZm
Giuseppe Protano, (M), Siena (Italy), CgaSc
Marcello Viti, YgsYh

Jamaica

University of the West Indies Mona Campus

Dept of Geography and Geology (B,M,D)
University of the West Indies
Mona
Kingston KGN7
p. 876-927-2728
geoggeol@uwimona.edu.jm
http://myspot.mona.uwi.edu/dogg/

Head:
Simon F. Mitchell, (D), Liverpool (UK), 1993, GsgGs

Japan

Akita University

Deparment of Earth Science and Technology
1-1 Tagata Gakuen-cho, Akita-shi
Akita

Ehime University

Department of Earth Sciences (B,M) ○⃠
Bunkyo-cho
2-5
Matsuyama City, Ehime Prefecture 790-8577
p. +81-89-927-9623
hori.rie.mm@ehime-u.ac.jp
http://www.ehime-u.ac.jp/~cutie
Programs: Geoscience, Earth Science
Certificates: field surveying, Paleontology
Enrollment (2020): B: 30 (5) M: 18 (6)

Geological Survey of Japan

http://www.gsj.jp/

Hirosaki University

Dept of Earth Science
wata@cc.hirosaki-u.ac.jp
http://sci.hirosaki-u.ac.jp/~earth

Hiroshima University

Dept of Earth and Planetary Systems Science
Kagami-yama 1-3-1
Higashi-Hiroshima
Hiroshima 739
web-admin@office.hiroshima-u.ac.jp
http://www.geol.sci.hiroshima-u.ac.jp/index_e.html

Kagoshima University

Faculty of Science
portraits@niad.ac.jp
http://earth.sci.kagoshima-u.ac.jp

Kanazawa University

Dept of Earth Sciences
Kakuma-machi
Kanazawa 920-1192
fsci-pla-director@edu.kobe-u.ac.jp
http://earth.s.kanazawa-u.ac.jp/

Kobe University

Earth & Planetary Sciences
fsci-pla-director@edu.kobe-u.ac.jp
http://shidahara1.earth.s.kobe-u.ac.jp

Kumamoto University

Dept of Earth and Environment (B,M,D)
2-39-1, Kurokami
Kumamoto City 860-8555
p. 81-96-342-3411
tadao@sci.kumamoto-u.ac.jp
http://www.sci.kumamoto-u.ac.jp/earthsci

Professor:
Shiro Hasegawa, (D), Pm
Toshiaki Hasenaka, (D), Gv
Hiroki Matsuda, (D), Gs
Tadao Nishiyama, (D), Gp
Hidetoshi Shibuya, (D), Ymg
Jun Shimada, (D), Hw
Akira Yoshiasa, (D), Gz

Nagoya University

Dept of Earth and Planetary Sciences
env@post.jimu.nagoya-u.ac.jp
http://www.eps.nagoya-u.ac.jp/

Shizuoka University

Inst of Geosciences
Shizuoka 422-8529
kawamoto.tatsuhiko@shizuoka.ac.jp
http://www.sci.shizuoka.ac.jp/~geo/Welcome.html

Tohoku University

Dept of Mineralogy, Petrology and Economic Geology
Sendai 980-8578
sci-syom@grp.tohoku.ac.jp
http://www.ganko.tohoku.ac.jp/

Tsukuba University

College of Geoscience
kankyojoho@ynu.ac.jp
http://www.geo.tsukuba.ac.jp/

University of Tokyo

Deparment of earth and Planetary Science
3-1, Hongo 7chome, Bunkyo-ku
Tokyo
Hirata@eri.u-tokyo.ac.jp
http://www.eri.u-tokyo.ac.jp/

University of Toyama

Deptt of Earth Sciences (B,M,D)
3190 Gofuku
Toyama City, Toyama Prefecture 930-8555
p. 81-76-445-6654
matsuura@sci.u-toyama.ac.jp
http://www.sci.u-toyama.ac.jp/earth/index-en.html
Enrollment (2015): B: 208 (0) M: 36 (0) D: 7 (0)

Professor:
Akira Takeuchi, (D), Osaka, 1979, GtcYe
Tohru Watanabe, (D), Tokyo, 1991, YxsGv

Associate Professor:
Shigekazu Kusumoto, (D), Kyoto, 1999, YvdGt

Yokohama National University

Dept of Environment and Natural Sciences
kankyojoho@ynu.ac.jp
http://chigaku.ed.ynu.ac.jp/geology-e.html

Kenya

Jomo Kenyatta University of Agriculture & Technology

Geomatic Engineering and Geospatial Information Systems (B,M,D) ☒
PO Box 62000
-00200 Nairobi
Nairobi
p. +254-67-5352391
gegis@eng.jkuat.ac.ke
http://www.jkuat.ac.ke/departments/gegis/
t: @JKUAT_GEGIS
Programs: Geomatic Engineering; Geospatial Information Sciences
Enrollment (2014): B: 175 (35) M: 45 (19) D: 8 (0)

Chair:
Thomas G. Ngigi, (D), Chiba (Japan), 2007, ZriZr

Lecturer:
Nathan O. Agutu, Jomo Kenyatta, ZgiZr
Mark Boitt, (M), Stuttgart, 2010, ZgrZi
George W. Chege, (M), Jomo Kenyatta, 2014, ZgrZi
Charles Gaya, (M), Stuttgart, 2002, ZrgZi
Andrew Imwati, (D), Nairobi, ZggZi
Benson K. Kenduiywo, (D), Tech (Darmstadt), 2016, ZriYd
Fridah K. Kirimi, (M), Jomo Kenyatta, 2010, ZiiZr
Moffat G. Magondu, (M), Jomo Kenyatta, 2014, ZirZg
Felix N. Mutua, (D), Tokyo, 2012, ZigZr
Nancy Mwangi, (M), Jomo Kenyatta, 2008, ZiiZr
Mercy W. Mwaniki, (M), Jomo Kenyatta, 2010, ZiiZr
Eunice W. Nduati, (M), Jomo Kenyatta, 2014, ZrrZi
Patroba A. Odera, (D), Kyoto, 2012, YdZge
Hunja Waithaka, (D), Hokkaido, YdZgi
Charles B. Wasomi, (M), Stuttgart, ZrrZi

Dept of Environmental Studies (B)
P.O. Box 62000 00200
Nairobi
pro@jkuat.ac.ke
http://www.jkuat.ac.ke/

Kenyatta University

School of Environmental Sciences (B,M,D)
Maseno
p. +254-057-351620/2
dean-envs@ku.ac.ke
http://www.maseno.ac.ke/index3.php?section=schools&page=school-encironment

University of Nairobi

Dept of Geology (B)
PO Box 30197
Nairobi
p. +254 20 4449856
geology@uonbi.ac.ke
http://geology.uonbi.ac.ke/

Dept of Geography (B)
PO Box 30917
Nairobi
p. +254 (020) 318262 (28400)
samowuor@uonbi.ac.ke
http://uonbi.ac.ke/departments/?dept_code=HF&&face_code=32

Dept of Meteorology (B,M,D)
PO Box 30197
Nairobi 00100
p. 254 020-4449004 (2070)
dept-meteo@uonbi.ac.ke
http://www.uonbi.ac.ke

Korea, South

Korea University

Earth and Environmental Sciences (B,M,D) ☒
Anam-dong, Seongbuk-Gu
Seoul 02841
p. 82-2-3290-3170
sjchoh@korea.ac.kr
http://ees.korea.ac.kr/
Enrollment (2021): B: 36 (30) M: 20 (16) D: 10 (6)

Head:
Hyeong Soo Kim, (D), James Cook, 2003, GpxGt

Professor:
Suk-Joo Choh, (D), Texas, 2004, GdsPe
Ho Young Jo, (D), Wisconsin, 2003, NgCaGe
Meehye Lee, (D), Rhode Island, 1999, AsOgCm
Young Jae Lee, (D), SUNY (Stony Brook), 2005, ScGze

Jin-Han Ree, (D), SUNY (Albany), 1991, Gct
Seong-Taek Yun, (D), Korea, 1991, CqHyCl
Associate Professor:
Man Jae Kwon, (D), Illinois, 2009, CbSbGe
Soon Jae Lee, (D), Korea, 2013, HwyCq
Assistant Professor:
Seong Ryong Kim, (D), Seoul Nat, 2015, Ys

Latvia

Latvija Valst Geologijas Dienests

Latvia State Geological Survey
Moscow street 165
Riga, LV-1019
lvgma@lvgma.gov.lv
http://mapx.map.vgd.gov.lv/geo3/

University of Latvia

Faculty of Geographical and Earth Sciences
10 Alberta Str.
202-203 Room
Riga
zeme@lu.lv
http://www.lu.lv/e_strukt/fakult/geog/info/

Lebanon

American University of Beirut

Dept of Geology
Faculty of Arts and Sciences
P.O.Box 11-0236
Riad El-Solh
Beirut 1107 2020
p. (961)-1-340460/ext. 4160
arahman@aub.edu.lb

Lesotho

National University of Lesotho

Dept of Geology & Mines (B)
P.O. Box 750, Maseru
p. (266) 34 0601
registrar@nul.ls
http://www.nul.ls/

Dept of Geography and Environmental Science (B)
P.O. Roma 180
Maseru 100
p. +266 2234 0601
registrar@nul.ls
http://www.nul.ls/faculties/fost/geography

Liberia

University of Liberia

Dept of Geology (B)
Capitol Hill
PO Box 9020
Monrovia 9020
p. 231-6-422-304
weekso@fiu.edu
http://www.universityliberia.org/ul_course_master_list_biology.htm#geology

Dept of Mining Engineering (B)
Captiol Hill
PO Box 9020
Monrovia 9020
p. +231 6422304
weekso@fiu.edu
http://www.universityliberia.org/ul_course_master_list_mining.htm#mine

Libya

Oil Companies School

Oil Companies School (B)
info@ocslibya.com
http://www.geocites.com/chatalaine/OCS.html

Petroleum Training and Qualifying Institute (PTQI)

Gergarish Road 9KM - Asiahia
Tripoli
p. +218 21 4833771-5
info@ptqi.edu.ly
https://ptqi.edu.ly/online1/_ptqi_en_tdc.php

University of Garyounis

Dept of Geology (B)
PO Box 1308
Benghazi
p. +218-61-86304
uni.office@uob.edu.ly
http://www.garyounis.edu/

Lithuania

Lithuanian Geological Survey under the Ministry of Environment

Lithuanian Geological Survey
S. Konarskio St. 35
LT-03123 Vilnius
indre.virbickiene@lgt.lt
http://www.lgt.lt/

Vilniaus Pedagoginis Universitetas

Dept of Geology
Studentu 39, 2034 Vilnius
p. (8 5) 275 89 35
gmtf.dekanatas@leu.lt
http://www.leu.lt/lt/gmtf/gmtf_apie_mus/all.html

Vilniaus Universitetas

Dept of Hydrogeology & Engin Geol
Giurlionio 21/27, 2009, Vilnius
p. 239 8278
robert.mokrik@gf.vu.lt
http://www.vu.lt/en/scientific-report-2012/faculties-and-institutes/faculty-of-natural-sciences#DEPARTMENT_OF_HYDROGEOLOGY_AND_E

Dept of Geology and Mineralogy (B,M,D) ●
Ciurlionio str. 21/27
LT03101
Vilnius LT03101
p. +370 5 2398272
eugenija.rudnickaite@gf.vu.lt
http://www.geol.gf.vu.lt/lt
Enrollment (2015): B: 8 (8) M: 11 (11) D: 10 (3)

Luxembourg

Administration des ponts et chaussées

Service Geologique du Luxembourg ☒
23, rue du Chemin de Fer
L-8057 Bertrange
geologie@pch.etat.lu
http://www.geology.lu

Madagascar

Universite d' Antananarivo

Dept of Chemistry (B)
BP 566
Antananarivo 101
p. +261 20 22 326 39
rtianasoa.manoelson@gmail.com
http://www.univ-antananarivo.mg/

Dept of Biology (B)
BP 566
Antananarivo 101
p. +261 20 22 329 39
arovonjy@yahoo.fr
http://www.univ-antananarivo.mg/

Dept des Sciences de la Terre (A,B,M,D)
BP 906
Antananarivo 101 , Madagascar
p. 00261324205328
saphirzzf@gmail.com
http://www.univ-antananarivo.mg/
Enrollment (2016): A: 144 (138) B: 127 (111) M: 63 (14) D: 1 (0)

Malawi

Mzuzu University

Faculty of Environmental Sciences (B)
Private Bag 201
Mzuzu
p. +(265) 1 320 722/ 320 575
ur@mzuni.ac.mw
http://www.mzuni.ac.mw/index.php?option=com_content&view=article&id=21&Itemid=15

University of Malawi

Dept of Natural Resources Management (B)
PO Box 219
Lilongwe
p. (265) 01 277 260
david.mkwambisi@bunda.luanar.mw
http://www.bunda.unima.mw/nrm.thm

Chancellor College, Natural Resources and Environment Centre (B)
PO Box 280
Zomba
p. +(265) 1 524 685
geo@chanco.unima.mw
http://www.chanco.unima.mw/department/department.php?DepartmentID=4&Source=Department_of_Geography_and_Earth_Sciences

Malaysia

Minerals and Geoscience Department Malaysia

Minerals and Geoscience Dept Malaysia
20th Floor, Bangunan Tabung Haji
Jalan Tun Razak
50658 Kuala Lumpur
jmgkll@jmg.gov.my
http://www.jmg.gov.my/

National University of Malaysia

Geology Dept
43600 Bangi
Selangor
dftsm@ftsm.ukm.my

Universiti Putra Malaysia

Dept of Environmental Sciences
43400 UPM Serdang
zakh@upm.edu.my
http://fsas.upm.edu.my/~sas/envpage/Dept.html

University of Malaya

Dept of Geology (B,M,D)
Dept. of Geology
University of Malaya
50603 Kuala Lumpur, Malaysia.
Kuala Lumpur, Selangor 50603
p. 03-79674203
ketua_geologi@um.edu.my
http://www.um.edu.my

Head:
Wan Hasiah Abdullah, (D), Newcastle upon Tyne, 1994, Ec
Professor:
Teh Guan Hoe, (D), Heidelberg, Ca
Lee Chai Peng, (D), Liverpool (UK), Pg
John Kuna Raj, (D), Malaya, 1983, Ng
Denis N.K Tan, (M), Malaya, Go
Associate Professor:
Abd Rashid Ahmad, (D), Oxford, Sc
Azman Abdul Ghani, (D), Liverpool (UK), Gi
Mohamed Ali Hasan, (M), Ulster, Hg
Azhar Hj Hussin, (D), London (UK), Gs
Tajul Anuar Jamaluddin, (D), Wales, 1997, Ng
Mustaffa Kamal Shuib, (M), London (UK), 1986, Gc
Samsudin Hj. Taib, (D), Durham (UK), Yg
Lecturer:
Nuraiteng Tee Abdullah, (D), London (UK), Gr
Ahmad Tajuddin Hj Ibrahim, (D), Newcastle upon Tyne, Ng
Che Noorliza Lat, (M), Nevada (Reno), 1989, Yg
Mat Ruzlin Maulud, (B), Malaya, Zn
Nur Iskandar Taib, (D), Indiana, Gi
Ismail Yusoff, (D), Norwich, Hg

Mali

Universite de Bamako

Sciences de la Terre (B)
BP 3206
Bamako
p. (223) 222 32 44
phine.romagnoli@unige.ch
http://www.ml.refer.org/u-bamako/spip.php?article.137

Malta

University of Malta

Dept of Geosciences (B,M,D) ☒
Department of Geosciences
University of Malta
Msida Campus
Msida MSD2080
p. (+356) 2340 2362
geo.sci@um.edu.mt
http://www.um.edu.mt/science/geosciences
f: https://www.facebook.com/um.geosciences/
Programs: Geoscience and Physics (B); Geoscience and Mathematics (B); Geoscience and Statistics and Operations Research (B); Geoscience (M); Applied Oceanography (M); Geosciences(D)
Enrollment (2018): M: 5 (8) D: 8 (0)
Dr:
Pierre-Sandre Farrugia, (D), Malta
Professor:
Raymond Ellul, (D), As
Alfred Micallef, (D), Ap
Associate Professor:
Alan Deidun, (D)
Aldo Drago, (D), Southampton (UK), Opg
Pauline Galea, (D), Victoria (NZ), 1993, YsgZg
Aaron Micallef, (D), Southampton (UK), 2008, GumOu
Dr:
Noel Aquilina, (D), Birmingham (UK), Asm
Sebastiano D'Amico, (D), YsgZg
Anthony Galea, (D), Trieste, Opg
Lecturer:
Adam Gauci, (D), Malta, 2018, Op

Mauritania

Universite de Nouakchott

Dept de Geologie (B)
BP 5026
Nouakchott
p. (222) 25 13 82
awa@univ-nkc.mr
http://www.univ-nkc.mr/spip.php?rubrique34

Mauritius

University of Mauritius

Mauritius Radio Telescope (B)
Reduit
p. (230) 454 1041
nalini@uom.ac.mu
http://www.uom.ac.mu/mrt/mrt2.html

Dept of Chemical and Environmental Engineering (B)
Reduit
p. (230) 454 1041
d.surroop@uom.ac.mu
http://www.uom.ac.mu/Faculties/FOE/CEE

Mexico

Centro de Investigación Científica y de Educación Superior de Ensenada

Earth Sciences Division (M,D)
Carretera Tijuana-Ensenada # 3918
Zona Playitas
Ensenada, Baja California 22860
p. (01152-664)1750500
dir-ct@cicese.mx
http://www.cicese.mx
Enrollment (2010): M: 44 (8) D: 20 (2)

Titular researcher:
Edgardo Canon-Tapia, (D), Hawaii, 1996, GvYmGt

Research Associate:
Jose G. Acosta, (M), CICESE (Ensenada), 1980, Ne
Jesus M. Brassea, (M), Cinvestav, 1986, Ye
Juan M. Espinosa, (M), CICESE (Ensenada), 1983, Ye
Jose J. Gonzalez, (M), CICESE (Ensenada), 1986, Yd
Alejandro Hinojosa, (M), CICESE (Ensenada), 1988, Zr
Luis H. Mendoza, (M), CICESE (Ensenada), 1982, Ne
Alfonso Reyes, (B), UNAM, Ne

Titular researcher:
Raul Castro, (D), Nevada (Reno), 1991, Ys
Juan Contreras, (D), Columbus, 1999, Gt
Luis A. Delgado-Argote, (D), UNAM, 2000, GitGu
Francisco Esparza, (D), CICESE (Ensenada), 1991, Ye
John Fletcher, (D), Utah, 1994, Gpt
Carlos Flores, (D), Toronto, 1986, Ye
Jose Frez, (M), California (Los Angeles), 1980, Ys
Juan Garcia, (D), Oregon State, 1990, Yv
Ewa Glowacka, (D), Polish Acad of Sci, 1991, Ys
Enrique Gomez, (D), Toronto, 1981, Ye

Titular Researcher:
Mario Gonzalez-Escobar, (D), CICESE (Ensenada), 2002, YesYs

Titular researcher:
Antonio Gonzalez-Fernandez, (D), Complutense (Madrid), 1996, Gu
Javier Helenes, (D), Stanford, 1980, Gr
Thomas Kretzschmar, (D), Tubingen, 1995, Ge
Margarita Lopez, (D), Toronto, 1985, Cc
Arturo Martin, (D), Paris XI Orsay, 1988, Gd
Luis Munguia, (D), California (San Diego), 1982, Ys
Alejandro Nava, (D), California (San Diego), 1980, Ys
Marco A. Perez, (D), CICESE (Ensenada), 1995, Ye
Jose M. Romo, (D), CICESE (Ensenada), 2002, Ye
Pratap Sahay, (D), Alberta, 1986, Ys
Rogelio Vazquez, (D), CICESE (Ensenada), 2002, Ye
Antonio Vidal, (D), CICESE (Ensenada), 2001, Ys
Bodo Weber, (D), Gp

Ciudad Universitaria

Facultad de Ingenieria
C.P. 04510
p. 56 22 08 66
fainge@servidor.unam.mx
http://www.ingenieria.unam.mx/

SGM

Servicio Geológico Mexicano
Blvd. F. Angeles 93.50-4 km.
Hidalgo 42080, Pachuca 42083
gciadoctec@sgm.gob.mx
http://www.sgm.gob.mx/

Universidad Autonoma de Baja California Sur

Dept of Marine Geology (B)
Carretera al Sur km 5.5, Box 19-B
La Paz, BS 23000
p. 682/2-47-55/2-01-40/2-45-69
oceanologia.fcm@uabc.edu.mx

Chair:
Alejandro Alvarez-Arellano, (M), Nac Auton, 1984, Gs

Professor:
Rodolfo Cruz-Orozco, (D), Louisiana State, 1974, Gu
Javier Gaitan-Moran, (M), Intl Inst-Aerospace Sur & Earth Sci, 1986, Zr
Carlos A. Galli-Olivier, (D), Utah, 1968, Gd
Jose I. Peredo-Jaime, (B), Auto de Baja California, 1978, Og
Luis R. Segura-Vernis, (D), Nac Auton, 1977, Pi

Associate Professor:
Alejandro J. Carillo-Chavez, (M), Cincinnati, 1981, Gp
Efrain Cornejo-Luna, (B), Nac Auton, 1969, Gq
Genaro Martinez-Gutierrez, (B), Inst Politecnico Nac, 1982, Gr
Cesar Martinez-Noriega, (B), Nac Auton, Ou
Jaime I. Monroy-Sanchez, (M), 1987, Ge
Miriam Nunez-Velazco, (B), Tech (Madero), 1980, Go
Jose A. Perez-Venzor, (B), Autonoma (SLP), 1978, Gi
Ramon Pimentel-Hernandez, (B), Inst Politecnico Nac, 1980, Yg
Humberto Rojas-Soriano, (B), Nac Auton, 1983, Gz
Paulino Rojo-Garcia, (B), Inst Politecnico Nac, 1983, Gc

Assistant Professor:
Luis A. Herrera-Gil, (B), Nac Auton, 1980, Pg

Lecturer:
Cesar A. Lopez-Ferreira, (B), Nac Auton, 1979, Hw

Cooperating Faculty:
Oscar Rodriguez-Plasencia, (B), Escuela Militar de Met, 1959, Am

Universidad Autonoma de Chihuahua

Facultad de Ingenieria (B,M)
Circuito Universitario Campus 2
Chihuahua, CH 31170
p. (614) 442-9500
ancorral@uach.mx

Director:
Arturo Leal-Bejarano, (B), Auto de Chihuahua, 1976, Zn

Head:
Arturo Lujan-Lopez, (M), Essex (UK), 1975, Zn

Chair:
Socorro I. Aguirre-Moriel, (B), ITR (Switzerland), 1979, Nx
Hector M. Mendoza-Aguilar, (B), Auto de Chihuahua, 1980, Nx

Professor:
Rafael Chavez-Aguirre, (M), Auto de Chihuahua, 1993, Hw
Adolfo Chavez-Rodriguez, (D), Arizona, 1987, Hy
Miguel Franco-Rubio, (M), Nac Auton, 1978, Eg
Rafael Madrigal-Rubio, (B), Nac Auton, 1969, Ng
Teodulo Mena-Zambrano, (B), Inst Politecnico Nac, 1964, Ng
Hector Minor-Velazquez, (B), Auto de Chihuahua, 1978, Nx
Ignacio A. Reyes-Cortes, (B), Texas (El Paso), 1997, Gx
Manuel Reyes-Cortes, (B), Nac Auton, 1970, Gi
Miguel Royo-Ochoa, (M), Auto de Chihuahua, 1997, Hw
David H. Ruiz-Cisneros, (B), Auto de Chihuahua, 1974, Nx
Luis M. Trevizo-Cano, (B), Auto de Chihuahua, 1978, Nx

Universidad Autonoma de Nuevo Leon

Facultad de Ciencias de la Tiera
Hacienda Guadalupe Km. 8 Camino a Cerro Prieto. - Linares, N 67700
p. (81) 8329 4170 Ext. 4170
fmedina@fct.uanl.mx
http://www.fct.uanl.mx/portal

Universidad Autonoma de San Luis Potosi

Area de Ciencias de la Terra (B,M)
Facultad de Ingenieri#2.a
Av. Dr. Manuel Nava No. 8
San Luis Potosi#2., SL 78290
p. (48) 13-82-22
aaguillonr@uaslp.mx
Department Secretary: Juan Manuel Torres Aguilera

Director:
Hector David Atisha-Castillo, (M), Texas, 1978, Zn
Head:
Joel Milan-Navarro, (M), Nancy (France), 1979, Zr
Professor:
Jose Refugio Acevedo-Arroyo, (B), Nac Auton, 1957, Ng
Luis Garcia-Gutierrez, (M), Stanford, 1951, Eg
Panfilo R. Martinez-Macias, (M), Nac Auton, 1995, Gc
Francisco Javier Orozco-Villasenor, (M), Colorado State, 1983, Eg
Ramon Ortiz-Aguirre, (M), Madrid, 1979, Hg
Carlos Francisco Puente-Muniz, (B), Autonoma (SLP), 1979, Hw
Delfino C. Ruvalcaba-Ruiz, (D), Colorado State, 1982, Em
Juan Manuel Torres-Agulera, (B), Autonoma (SLP), 1980, Gx
Librarian:
Panfilo R. Martinez-Macias, Zn

Universidad Autonoma de Sonora

Ejecutivo de la Unidad Regional Norte
Av. Universidad e Irigoyen
Col. Ortiz
Hermosillo

Universidad de Guanajuato

Departamento de Ingeniería en Minas, Metalurgia y Geología (B) O
Sede San Matias
Ex Hda. de San Matías s/n
Col. San Javier
Guanajuato, Guanajuato C.P. 36025
p. ((+473) 73 2 2291 (Ext. 5304)
juancho@ugto.mx
http://www.di.ugto.mx/minas/index.php/antecedentes
Enrollment (2015): B: 7 (7)

Universidad Estatal de Sonora

Escuela Superior de Geociencias (B) O☒
Col. Apolo
Hermosillo, SO 83240
p. (662) 6890100
maricela.lopez@ues.mx
https://www.ues.mx/?p=especiales/ofertaeducativa/malla.aspx&cid=0&sid=13&smid=0&latder=0&params=pa=007_pe=14_tipopa=l
f: https://www.facebook.com/ues.mx/
Programs: minning, metallurgy and geology
Enrollment (2020): B: 450 (23)
Director:
Maricela López, (M), Sonora, 2015, NxGg
Associate Professor:
Gustavo E. Durazo-Tapia, (B), Sonora, 1981, Em
David García-Martínez, (D), 2017, GzEmNx
Gonzalo de Jesus Ibarra-Dessens, (M), Sonora, 2019, Nm
Rafael Jordan-Hernandez, (D), La Habana, 2005, Zn
Veronica Moreno-Rodriguez, (D), Madrid, Eg
Christian A. Murguía-Romero, (M), Sonora, 2018, Nx
Hector O. Murillo-Valenzuela, (B), CESUES, 1989, Nx
Paula C. Santos-Munguia, (D), CINVESTAV, 2019, Nx

Universidad Nacional Autonoma de Mexico

Ciencias de la Tierra (B) ☒
Ciudad Universitaria,CDMX
Facultad de Ciencias
Ciudad de Mexico 04510
tierra.coord@ciencias.unam.mx
Programs: Earth Sciences
Enrollment (2017): B: 700 (84)
Dr.:
Blanca Mendoza, (D), Oxford, 1985, Zn

Facultad de Ingeniería (B)
Av. Universidad 3000, Ciudad Universitaria
Coyoacán, CP 4510
p. 56 22 08 66
fainge@servidor.unam.mx
http://www.ingenieria.unam.mx/paginas/Carreras/ingenieriaMinas/ingMinas_Desc.php
Director:
Antonio Nieto Antunez, (M), Stanford, 1970, Nm
Professor:
Angelica Casillas, (B), Inst Politecnico Nac, 1976, Hw
Esteban Cedillo, (D), Heidelberg, 1988, Em
Juventino Martinez, (D), Paris VI, 1980, Gt
Edgardo Meave, (B), Guanajuato, 1947, Gg
Francisco Medina, (B), Guanajuato, 1982, Zg
Ricardo Navarro, (B), Inst Politecnico Nac, 1973, Nm
Salvador Ulloa, (B), Guanajuato, 1947, Eg
Fernando Vasallo, (D), Lomonosov (USSR), 1982, Em
Carlos Yanez, (B), Inst Politecnico Nac, 1976, Ce

Inst de Geologia (M,D)
Apdo. postal 70-296
Ciudad Universitaria
Delegacion Coyoacan, DF 04510
p. (52) 5622 4314
academicaigl@geologia.unam.mx
https://www.geologia.unam.mx/
Administrative Assistant: Ana María Rodriguez Simental
Director:
Dante J. Moran-Zenteno, (D), Nac Auton, 1992, Cc
Head:
Susana A. Alaniz Alvarez, (D), Nac Auton, 1995, Gc
Luca Ferrari Pedraglio, (D), Milan, 1992, Gt
Sergio Cevallos Ferriz, (D), Alberta, 1990, Pg
Carlos M. Gonzalez Leon, (D), Arizona, 1992, Gr
Klavdia Oleschko Loutkova, (D), Lomosov (Moscow), 1984, So
Maria S. Lozano Garcia, (D), d'Aix (Marseilles), 1979, Pe
Francisco J. Vega Vera, (D), Nac Auton, 1988, Pg
Chair:
Gustavo Tolson Jones, (D), Nac Auton, 1998, Gc
Senior Scientist:
Shelton Applegate-Pleasants, (D), Chicago, 1961, Pv
Jorge Aranda, (D), Oregon, 1982, Gi
Blanca E. Buitron-Sanchez, (D), Nac Auton, 1974, Pi
Oscar Carranza-Castaneda, (D), Nac Auton, 1989, Pv
Gerardo Carrasco-Nunez, (D), Michigan Tech, 1993, Gv
Ana L. Carreno, (D), d'Orsay (Paris), 1979, Pm
Miguel Carrillo Martinez, (D), Paris VI, 1976, Pb
Liberto De Pablo, (D), Ohio State, 1958, Gz
Rodolfo Del Arenal-Capetillo, (B), Nac Auton, 1960, Hw
Ismael Ferrusquia, (D), Houston, 1971, Gr
David Flores-Roman, (D), Nac Auton, 1981, So
Celestina Gonzalez-Arreola, (D), Nac Auton, 1989, Pi
Jose C. Guerrero-Garcia, (D), Texas (Dallas), 1975, Ym
John D. Keppie, (D), Glasgow, 1964, Gt
Victor M. Malpica-Cruz, (D), Bordeaux, 1980, Gs
Enrique Martinez Hernandez, (D), Michigan State, 1979, Pl
Juventino Martinez-Reyes, (D), Paris VI, 1980, Gt
Luis M. Mitre-Salazar, (D), Paris IV, 1978, Ge
Adrian Ortega, (D), Waterloo, 1993, Hw
Fernando Ortega-Gutierrez, (D), Leeds (UK), 1975, Gp
Sergio Palacios-Mayorga, (M), Nac Auton, 1972, So
Jerjes Pantoja-Alor, (M), Arizona, 1963, Gg
Maria del Carmen Perrilliat-Montoya, (D), Nac Auton, 1969, Pi
Angel Nieto Samaniego, (D), Nac Auton, 1994, Gc
Christina D. Siebe Grabach, (D), Hohenheim (Germany), 1993, Sc
Alicia Silva-Pineda, (D), Nac Auton, 1980, Pg
Max Suter-Cargnelutti, (D), Basel (Switzerland), 1978, Gt
Jordi Tritlla, (D), Barcelona, 1994, Eg
Ana B. Villasenor-Martinez, (D), Nac Auton, 1991, Pi
Reinhard Weber-Gobel, (D), Tubingen, 1967, Pg
Associate Scientist:
Gerardo J. Aguirre Diaz, (D), Texas, 1993, Gv
Thierry Calmus, (D), Paris VI, 1983, Gu
Antoni Camprubi, (D), Barcelona, 1998, Eg
Alejandro J. Carillo Chavez, (D), Wyoming, 1996, Hg
Elena Centeno Garcia, (D), Arizona, 1994, Gt
Rodolfo Corona-Esquivel, (M), Nac Auton, 1985, Em
Mariano Elias-Herrera, (M), Nac Auton, 1982, Gp
Maria L. Flores-Delgadillo, (M), Nac Auton, 1987, So
Gilberto Hernandez-Silva, (D), Nac Auton, 1983, So
Rafael Huizar-Alvarez, (D), Franche Comte, 1989, Hw
Cesar Jacques Ayala, (D), Cincinnati, 1983, Gr
Marisol Montellano-Ballesteros, (D), California (Berkeley), 1986, Pv
Amabel M. Oretega Rivera, (D), Queens, 1997, Cc
Odranoel Quintero, (D), Sorbonne (France), 1995, Gc

Jose L. Rodriguez-Castaneda, (M), Pittsburgh, 1984, Gg
Jaime Roldan Quintana, (M), Iowa, 1976, Gv
Gerardo Sanchez-Rubio, (M), Imperial Coll (UK), 1984, Gv
Jesu Sole, (D), Barcelona, 1996, Cc
Luis F. Vassallo-Morales, (D), Lomonosov (USSR), 1981, Em
Maria G. Villasenor-Cabral, (M), Leeds (UK), 1974, Ca
Research Associate:
Irma Aguilera-Ortiz, (B), Iberoameric, 1971, Ca
Victor M. Davila-Alcocer, (M), Texas (Dallas), 1986, Ps
Jose G. Solorio-Munguia, (B), Nac Auton, 1958, Gz
Adjunct Professor:
J. Duncan Keppie, (D), Glasgow, 1967, Gt
Luis Silva-Mora, (D), d'Aix (Marseilles), 1979, Gi
Emeritus:
Gloria Alencaster-Ybarra, (D), Nac Auton, 1969, Pi
Zoltan De Cserna, (D), Columbia, 1955, Gt
Geology Librarian:
Teresa Soledad Medina Malagon, (B), Zn

Inst de Geofisica
losos@sci.muni.cz
http://tlacaelel.igeofcu.unam.mx/index.eng.html

Mongolia

Gazarchin Institute

Gazarchin Institute
Bayanzurkh District
Ulaanbaatar 46
marketing.gazarchin@gmail.com
http://www.gazarchin.edu.mn/

Mineral Resources Authority of Mongolia

Mineral Resources Authority of Mongolia
info@mram.gov.mn
http://www.mram.gov.mn/

Mongolian University of Science and Technology

School of Geology and Petroleum Engineering
Bagatoiruu-46, P. O. Box-520
Ulaanbaatar 46
p. 976 11 312291
jepces@must.edu.mn
http://www.gs.edu.mn

National University of Mongolia

Geology and Geography Faculty
Ikh Surguuliin gudamj - 1, Baga Toiruu,
Sukhbaatar district,
Ulaanbaatar
p. 976-11-311890
int_rel@num.edu.mn
http://geo.num.edu.mn

Morocco

Ecole Nationale de l' Industrie Minerale

Dept de Sciences de la Terre (B)
Avenue Hadj Ahmed Cherkaoui
BP 753
Agdal, Rabat
p. (+212) 037 68 02 30
zaydi@enim.ac.ma
http://www.enim.ac.ma/formation/sciences_de_la_terre/

Dept de Mines (B)
Avenue Hadj Ahmed Cherkaoui
BP 753
Adgal, Rabat
p. (+212) 037 68 02 30
zaydi@enim.ac.ma
http://www.enim.ac.ma/formation/mines

Dept de Sciences des Materiaux (B)
Avenue Hadj Ahmed Cherkaoui
BP 753
Agdal, Rabat
p. (+212) 037 68 02 30
zaydi@enim.ac.ma
http://www.enim.ac.ma/formation/materiaux/

ONAREP - Naional Office of Petroleum Research

ONAREP - Naional Office of Petroleum Research (B)
PO Box 8030
Rabat 10050
p. +212 37 28-1616

ONHYM

National Office of Hydrocarbons and Mines
rh@onhym.com
http://www.onhym.com/

School of Mines of Marrakech

School of Mines (B,M,D)
Rue Machaar, El Harm Quartier Issil-B.P. 38
Marrakech
p. 212-30-97-79

Universite Abdelmalek Essaadi

Dept de Sciences de la Terre (B)
p. +212 539 97 93 16
presidence@uae.ma
http://www.uae.ma/portail/FR/

Faculte de Sciences de Tetouen (B)
BP 2121
Tetouan 93002
p. +212 (0) 39 97 24 23
vdap.fs@gmail.com
http://www.fst.ac.ma/stu.html

Universite Cadi Ayyad

Dept of Geology
Prince Moulay Abdellah
BP 515
Marakech
p. +212 (0)5 24 43 46 49
http://www.fssm.ucam.ac.ma/pages/geologie.php

Universite Hassan 1er - Settat

Dept de Geologie Apliquee (B)
BP 577
Settat 26000
p. 05.23.40.07.36
http://www.fsts.ac.ma/fsts/index/php?option=com_content&task=view&id=37&Itemid=12

Universite Hassan II

Dept of Geology (B)
9, Rue Tarik Bnou Zia
Anfa Casablanca
p. 0522 23 06 80
z.hilmi@fsac.ac.ma
http://www.fsac.ac.ma/depart/geo

Universite Hassan II - Mohammedia

Dept de Sciences de la Terre (B)
279 Cite Yassmina
Mohammedia, Casablanca
p. +212(33)314635
presidence@univh2m.ac.ma

Universite Ibn Tofail

Dept de Geologie (B,M,D) ● ☒
Abdelhakboua@Yahoo.com
Kenitra 14 000
p. +212 061984449
ammed.benammi@uit.ac.ma
http://www.univ-ibntofail.ac.ma/fre/departments.php?esp=4&rub=14&srub=36&srub_=96
Enrollment (2020): B: 11 (0) M: 35 (0) D: 20 (0)

Universite Ibn Zohr, Agadir

Dept de Sciences de la Terre (B)
BP 32/S
Agadir 80000
p. +212 (28) 22 71 25
contact@uiz.ac.ma
http://www.esta.ac.ma/

Universite Mohammed 1er (Oujda)

Ecole Nationale des Sciences Appliquees d'Al Hoceima (ENSAH) (B)
BP 724
Oujda 60000
p. +212(56)500612
presidence@ump.ma

http://webserver1.ump.ma/ecoles_facultes/ensah

Dept des Sciences de la Terre (B,M,D)
BP 717
60000 Oujda
p. (212)536500601/02
fso@fso.ump.ma
http://sciences1.univ-oujda.ac.ma

Universite Mohammed V

Dept of Geology
BP 554
Rue Michlifen Agdal
Rbat-Chellah
p. +212 05 37 77 54 71
http://www.fsr.ac.ma/ancien/index.php/departement/giologie.html

Universite Mohammed V (Agdal)

Dept de Genie Minerale (B)
Avenue Ibnsina
BP 765
Adgal Raba
p. (212 - 537) 77.26.47 - 77.19.0
contact@emi.ac.ma
http://www.emi.ac.ma

Universite Moulay Ismail, Meknes

Dept de Sciences de la Terre (B)
Marjane II BP 298
Meknes 5003
p. +212 5 35 53 78 96
doyen@fs-umi.ac.ma
http://ww.umi.ac.ma/

Universite Sidi Mohammed Ben Abdallah

Dept of Geology
Dhar El Mahraz
BP 42
Atlas-Fes
p. 06 61 35 04 81
dept_geo@fsdmfes.ac.ma
http://www.fsdmfes.ac.ma/Presentation/Departements/Geologie.php

University Mohammed I

Dept of Geology (B,M,D)
Faculté des Sciences Bd. Mohammed VI
BP 717
Oujda, Oujda-Angad 60000
p. (212)667772215
mbouabdellah2002@yahoo.fr
http://webserver1.ump.ma/

Mozambique

Universidade Eduardo Mondlane

Escola Superior de Ciencias Marinhas e Costeiras (B)
CP 128
Avenida 1 de Julho
Bairro Chuabo Dembe
Didad de Quelimane, Zambezia
p. +258 24900500/1
valera.dias@uem.mz
http://www.marine.uem.mz/

Laboratorio de Gemologia (B,M) ☒
Av. Julius Nyerere 3453, Departamento de Física
Campus Universitário
Maputo
p. (+258) 21 497003
uthui@zebra.uem.mz
http://www.fisica.uem.mz/index.php?option=com_content&view=article&id=51&Itemid=57
Programs: Geology

Associate Professor:
Akil Askharhodjaev, (D), URSS, 1985
Rogério J. Uthui, (D), Sweden, 1996

Universidade Lurio

Faculdade de Engenharia e Ciencias Naturais (B)
Av. 25 de Setembro
N. 958
Pemba, Cabo Delgado
p. (+258) 27 221238
fecn@unilurio.ac.mz
http://www.unilurio.ac.mz/faculdades_pmb_pt.htm

Namibia

Geological Survey of Namibia in Windhoek

Geological Survey of Namibia in Windhoek
Private Bag 13297
Windhoek
p. +264-61-2848111
gschneider@mme.gov.na
http://www.mme.gov.na/gsn/

University of Namibia

Dept of Geology (B,M,D) ●☒
P.O. Box 1727
Gordon Street,
Kronlein,
Keetmanshoop 9000
p. (+264 63) 2202038
ttjipura@unam.na
http://www.unam.na/faculties/science/departments.html
Programs: Geology (B); Applied Geology in Exploration (M); Hydrogeology (D)
Enrollment (2020): B: 47 (7) M: 10 (2) D: 4 (0)

Senior Lecturer:
Innocent Muchingami, (D), W Cape, 2014, Ye

Associate Professor:
Holgar Sommer , (D), Mainz, 2004, Gp
Akah Tses, (D), Nigeria, , Gec

Lecturer:
Gilbert Chongwain, (D), Pan Africa, 2017, Go
Josephine Hamutoko, (D), Namibia, 2017, Hw
Martin Harris, (M), Geosciences (Beijing), 2013
Albertina Nakwafila, (M), Stellenbosch, 2014
Ester Shalimba, (M), China Geosciences, 2014, GgpEz
Collen Uahengo, (D), China Geosciences, 2020, PgGoEg
Shoopala Uugulu, (M), France, 2012
Absai Vatuva, (D), China Geosciences, 2016, EgGtc

Nepal

Tribhuvan University

Central Dept of Geology (M,D) O☒
Kirtipur, Kathmandu
head@cdgl.tu.edu.np
http://www.cdgltu.edu.np
Programs: MSc Geology and MSc Engineering Geology
Enrollment (2022): M: 71 (59) D: 23 (0)

Netherlands

Delft University of Technology

Department of Geoscience & Engineering (B,M,D) O
Stevinweg 1
2628 Delft
p. +31 15 2781328
M.A.Hicks@tudelft.nl
http://www.tudelft.nl

Dept of Geoscience & Engineering (A,B,M,D) ●☒
Stevinweg 1
2628 CN Delft
The Netherlands
Delft 2628 CN
p. +31 15 27 81423
m.j.m.ammerlaan@tudelft.nl
https://www.tudelft.nl/citg/over-faculteit/afdelingen/geoscience-engineering/
f: https://www.facebook.com/AppliedEarthSciencesTUDelft/
Programs: Applied Geophysics, Environmental Engineering, Geo-Engineering, Geo-Energy Engineering, Geo-Resource Engineering
Certificates: -

National Geological Survey of the Netherlands

National Geological Survey of the Netherlands
Princetonlaan 6
3584 CB Utrecht
p. (+31)30-2564256
http://www.en.geologicalsurvey.nl/

Royal NIOZ

Jaap Sinninghe Damste (D) ☒
PO Box 59
Den Burg, Texel 1790 AB
p. +31222369550
Jaap.Damste@nioz.nl
https://www.nioz.nl/en/about/mmb
Programs: Organic Biogeochemistry; Marine Microbiology; Marine Microbial Ecology (w/ Utrecht Univ)

Physical Oceanography ☒
Landsdiep 4
't Horntje 1797 SZ
secretary@nioz.nl
http://www.nioz.nl

State University of Groningen

Faculty of Spatial Sciences (B,M) ☒
P.O. Box 800
Groningen 9700 AV
p. +31 50 363 8891
i.l.veen@rug.nl
http://www.rug.nl/frw
f: http://www.facebook.com/FRWRUG
t: @FRW_RUG
Programs: Geography, urban planning, demography, real estate, education, environment, infrastructure, sustainability
Enrollment (2014): B: 190 (0) M: 60 (0)

State University of Utrecht

Faculty of Geoscience (B,M,D)
P.O. Box 80.115
Utrecht 3508 TA
p. (+31) (0)30 253 2044
info@uu.nl
http://www.uu.nl/faculty/geosciences/EN/Contact/Pages/

Universiteit Utrecht

Faculty of Geosciences (B,M)
Budapestlaan 4b
Utrecht 3584 CD
p. +31 30 253 7890
mscinfo.geo@uu.nl
http://www.geo.uu.nl/

University of Amsterdam

Dept of Geology
Science Park 904
1098 XH Amsterdam
p. +31 (0)20525 8626
info-science@uva.nl

University of Twente

Faculty of Geo-Information Sciences and Earth Observation (ITC) (M,D) ●✓
P.O. Box 217
Enschede 7500 AE
p. +31 (0)53 487 44 44
info-itc@utwente.nl
https://www.itc.nl
f: https://www.facebook.com/ITC.UTwente
Programs: Master's Geo-information Science and Earth Observation with specialisations in Applied Remote Sensing; Natural Hazards and Disaster Risk Reduction; Geoinformatics; Land Administration; Natural Resources Management; Urban Planning and Management; Water Resources and Environmental Management

Master's Spatial Engineering
Certificates: See: https://www.itc.nl/studyfinder

University of Utrecht

Dept of Earth Sciences ☒
Heidelberglaan 2
P.O. Box 80115
Utrecht
info@uu.nl

Vrije Universiteit Amsterdam

Faculty of Beta Science, Dep. of Earth Sciences (B,M,D) ●☒
De Boelelaan 1085-1087
1081 HV Amsterdam 1081 HV
p. +31 (0)20 59 87310
f.bosse@vu.nl
https://vu.nl/nl/over-de-vu/faculteiten/faculteit-der-betawetenschappen/afdelingen/aardwetenschappen
Enrollment (2015): B: 248 (56) M: 67 (45)

Professor:
Gareth R. Davies, (D), GiCcGf
Han J. Dolman, (D), CbAmHg
Sander Houweling, (D), AscAp
Klaudia Kuiper, (D), Cc
Wouter Schellart, (D), Gt
Ronald R.T. van Balen, (D), GgmGt
Guido R. van der Werf, (D), Zr
Wim van Westrenen, (D), CpGz
Jan Wijbrans, (D), CcGtv

Associate Professor:
Fraukje M. Brouwer, (D), Utrecht, 2000, GptGc
Gerald M. Ganssen, (D), ObPe
Kees Kasse, (D), GmsSd
Janne Koornneef, (D), GvCa
Frank J.C. Peeters, (D), PmOg
Didier Roche, (D), AsZn
Sander Veraverbeke, (D), Zr
Jorien Vonk, (D), Hg
Pieter Vroon, (D), GvCcl

Assistant Professor:
Pepijn Bakker, PcAtZo
Kay J. Beets, Cg

Anouk Beniest, (D), ZyGcg
Wouter Berghuijs , (D), Hq
Kim Naudts, (D), HgPe
Maarten A. Prins, (D), GsOg
Duna Roda Boluda, Gs
Monica Sanchez Roman, (D), GdsGg
Marlies ter Voorde, (D), YgGt
Willem Toonen, GmsGn
Jeroen van der Lubbe, (D), Gs
Nathalie Van der Putten, (D), PecZy
Ype van der Velde, (D), Hw

Research Associate:
Antoon Meesters, (D), AmGq

Dr:
Bernd Andeweg, (D), 2002, GtcGg
Ron Kaandorp, (D), Gg
Els Ufkes, (D), PmZgGu

Netherlands Antilles

Royal NIOZ

Marine Geology
Landsdiep 4
Den Burg 1797 SZ
p. +31(0)222 369 300
Jens.Greinert@nioz.nl
http://www.nioz.nl

New Zealand

Institute of Geological and Nuclear Sciences Ltd (New Zealand)

Institute of Geological and Nuclear Sciences Ltd (New Zealand)
1 Fairway Drive, Avalon 5010. PO Box 30-368, 5040.
Lower Hutt
p. +64-4-570-1444
G.Alderwick@gns.cri.nz
http://www.gns.cri.nz/
f: https://www.facebook.com/gnsscience

Massey University

Soil and Earth Sciences Group
Private Bag
Palmerston North
p. 0800 MASSEY (0800 627 739)
C.W.N.Anderson@massey.ac.nz
http://www.massey.ac.nz/massey/learning/programme-course-paper/programme.cfm?major_code=2044&prog_id=92431

University of Auckland

School of Environment (B,M,D) ☒
Private Bag 92019
Auckland Mail Center
Auckland 1142
p. 64 9 3737599
p.kench@auckland.ac.nz
http://www.env.auckland.ac.nz
Programs: Geography; Earth Sciences; Environmental Science; Environmental Management; Geophysics; Geology

University of Canterbury

Dept of Geological Sciences (B,M,D) ●☒
Private Bag 4800
Christchurch 8140
p. (643) 3694384
geology@canterbury.ac.nz
http://www.geol.canterbury.ac.nz/title.html
Enrollment (2014): B: 28 (0) M: 0 (35) D: 0 (20)

Professor Hazard & Disaster Management:
Timothy R H Davies, (D)

Professor:
James W. Cole, (D), Wellington, 1966, Gv
Andy Nicol, (D), Canterbury (NZ), 1991, GctGo
Jarg R. Pettinga, (D), Auckland, 1981, Gtc

Senior Lecturer Hazard & Disaster Management:
Thomas Wilson, (D), Canterbury (NZ), Gv

Senior Lecturer:
Kari N. Bassett, (D), Minnesota, 1995, GstGd
Catherine Reid, (D), Pg

Associate Professor:
Ben Kennedy, (D), Gv

Lecturer:
Darren Gravley, (D), Gv
Samuel J. Hampton, (D), Canterbury (NZ), 2010, GvZe
Alex Nichols, (D)

University of Otago

Dept of Geology (B,M,D) O☒
P. O. Box 56
360 Leith Walk
Dunedin 9054
p. +64 3 479-7519
geology@otago.ac.nz
http://www.otago.ac.nz/geology/

Head of Department:
James D. L. White, (D), California (Santa Barbara), 1989, Gvs

Professor:
Dave Craw, (D), Otago (NZ), 1981, Eg
R. Ewan Fordyce, (D), Canterbury (NZ), 1981, Pv
David J. Prior, (D), Leeds (UK), 1988, Gct

Senior Lecturer:
Candace E. Martin, (D), Yale, 1990, ClcGz
Virginia G. Toy, (D), Otago (NZ), 2008, Gct

Dr:
Christopher M. Moy, (D), Gus

Associate Professor:
Andrew R. Gorman, (D), British Columbia, 2001, YrgYe
Daphne E. Lee, (D), Otago (NZ), 1980, Pib

Senior Lecturer:
J. Michael Palin, (D), Yale, 1992, CcGip

Lecturer:
Christina R. Riesselman, (D), Stanford, 2011, Ou
James M. Scott, (D), Otago (NZ), 2008, Gp
Steven AF Smith, (D)

Emeritus:
Douglas S. Coombs, (D), Gi
Alan F. Cooper, (D), Otago (NZ), 1970, GpiGt
Rick H. Sibson, (D), Imperial Coll (UK), 1977, Gc

University of Waikato

Dept of Earth Sciences
Private Bag 3105
Hamilton 3240
p. +64 7 838 4625
science@waikato.ac.nz
http://sci.waikato.ac.nz/study/subjects/earth-sciences

Victoria University of Wellington

School of Geography, Environment and Earth Sciences (A,B,M,D) O⊘
P. O. Box 600
Wellington 6140
p. +6444635337
geo-enquiries@vuw.ac.nz
http://www.wgtn.ac.nz/sgees
Programs: Earth Sciences, Geology, Development Studies, Environmental Studies, Human Geography, Physical Geography, Geophysics, Meteorology, Environmental Science, GIS, Climate Change and Science Policy
Enrollment (2012): B: 153 (107) M: 51 (22)

Emeritus:
Tim Little, (D), Stanford, 1988, Gct

Dept of Geology
P. O. Box 600
Wellington
p. +64 4 463 5337
geo-enquiries@vuw.ac.nz
http://www.victoria.ac.nz/sgees/study/postgraduate-study/geology/

Nicaragua

University of Managua

Centro de Investigaciones Geocientificas
Colonia Miguel Bonilla No. 165, P.O. Box A-131 Cc. Managua, Managua

Niger

Institut de Recherche pour le Developpement au Niger

Institut de Recherche pour le Developpement au Niger (B)
Avenue de Maradi
Niamey BP 11 416
p. (227) 20 75 38 27
gfar-secretariat@fao.org
http://www.ird.ne/

Universite Abdou Moumouni de Niamey

Dept de Geologie
B.P. 10662
Niamey
p. (227) 20 31 50 72
http://www.secheresse.info/article.php3?id_article=1664
Doyen:
Abdoulaye M. Alassane

Nigeria

Adekunle Ajasin University

Dept of Earth Sciences
(B)
Akungba Akoko, Ikare Akoko 23401
p. +234 (34) 246444
iosazuwa@yahoo.com
http://www.ajasin.edu.ng/academics

African University of Science & Technology

Petroleum Engineering (B,M)
PMB 681
Abuja F.C.T.
Km 10 Airport Road
Garki
p. +234 9 7800680
http://aust.edu.ng/

Ahmadu Bello University

Dept of Geology
Sokoto Road
Samaru-Zaria
Zaria 2222, Kaduna
p. +234 (69) 50 691
academicoffice@abu.edu.ng
http://www.abu.edu.ng/

American University of Nigeria

School of Arts and Sciences (B)
Lamido Zubairu Way
PMB 2250, Yola 23455
p. +234 (805) 5485 702
dsas@aun.edu.ng
http://www.abti-american.edu.ng

Caleb University

Dept of Surveying and Geoinformatics (B)
PMB 21238
Ikeja GPO, Lagos
p. +234-1-764-7312
info@calebuniversity.edu.ng
http://calebuniversity.edu.ng/courses.php?collUid=4

Dept of Environmental Protection and Management (B)
PMB 21238
Ikeja GPO, Lagos
p. +234-1-764-7312
info@calebuniversity.edu.ng
http://calebuniversity.edu.ng/courses.php?coll_id=4

Crawford University

Dept of Mathematics and Physical Sciences (B) ○☒
PMB 2001
Faith City, Ogun 112102
p. +2348023003436
info@crawforduniversity.edu.ng
http://www.crawforduniversity.edu.ng
f: https://www.facebook.com/CrawfordCRU/
t: @unicrawford
Programs: Geology and Mineral Sciences

Delta State University

Dept of Geology (B)
PMB 1
Abraka
p. +234 (54) 66009
keaweto@delsu.edu.ng
http://www.delsunigeria.net/
Director:
S. H. O. Egboh, (D)
Chair:
E. Adaikpoh, (D)

Enugu State University of Science and Technology

Dept of Geology and Mining (B)
MB 01660, Enugu
p. +234 (42) 451319
dean.fans@esut.edu.ng
http://www.esut.edu.ng/

Dept of Geography and Meteorology (B)
MB 01660, Enugu
p. +234 (42) 451319
info@esut.edu.ng
http://www.esut.edu.ng/

Federal Polytechnic, Bida

Dept of Survey and Geoinformatics (B)
http://www.fedpolybidaportal.com/

Dept of Quantitative Surveying (B)
http://www.fedpolybidaportal.com/

Federal Polytechnic, Offa

School of Environmental Studies (B)
http://www.fedpoffa.edu.ng/

Federal University of Technology, Akure

Earth & Mineral Sciences (B)
PMB 704
Akure
p. +234 (34) 243744
sems@futa.edu.ng
http://www.futa.edu.ng/sems/

Dept of Applied Geophysics (B,M,D)
PMB 704
Akure, Ondo
p. +234 803-595-9029
joamigun@futa.edu.ng
http://agp.futa.edu.ng/page.php?pageid=179

Applied Geology (B,M,D) ●☒
PMB 704
Akure, Ondo
p. +234 8036672708
agy@futa.edu.ng
http://www.futa.edu.ng/agy/
Programs: Bachelor of Technology, Master of Technology and Doctor of Philosophy degree programs in Applied Geology
Certificates: Post Graduate Diploma program in Applied Geology

Enrollment (2019): B: 96 (96) M: 8 (8) D: 1 (1)
Professor:
Saka Adelayo Opeloye, (D), Bauchi, 2003, GsPgGo
Professor:
Yinusa Ayodele Asiwaju-Bello, (D), Leeds (UK), 1991, HwNgGu
Idowu Bolofinde Odeyemi, (D), Ibadan, 1976, GgxZr
Chukwuemeka Tony Okonkwo, (D), Keele, 1991, GcxGg
Dr:
Anthony Odunyemi Ademeso, (D), Akure, 2011, NgrNt
Doctor:
Peter Sunday Ola, (D), Akure, 2002, GosGg
Mr:
Blessing Adeoti, (M), Akure, 2015, GcxGg
Adeshina Luqman Adisa, (M), Akure, 2012, CgeGc
Isaac Ominyi Ajigo, (M), Jos, 2013, GgCgGx
Sunday Olabisi Daramola, (M), Akure, 2012, GeNtHw
Oladimeji Rilwan Egbeyemi, (M), Ibadan, 2004, GgCeEg
Simeon O. Idowu, (M), Ago-Iwoye, 2007, NgHwGg
Emmanuel Eghietseme Igonor, (M), Ibadan, 2007, GgCeGi
Osazuwa Abifade Ogbahon, (M), Akure, 2011, PlGsPm
Franklin Oluwaseun Olabode, (M), Akure, 2016, HwNgGg
Dr Mrs:
Oluwaseyi Adunola Bamisaiye, (D), Pretoria, 2018, GcZrRn
Dr:
Mohammed Olatoye Adepoju, (D), Akure, 2017, CeEgGg
Stephen O. Ariyo, (D), Ibadan, 2014, YegYx
Timothy Iyobosa Asowata, (D), Ibadan, 2017, GeiGg
Olusiji Samuel Ayodele, (D), Ago-Iwoye, 2015, CeGcZr
Oladotun Afolabi Oluwajana, (D), Ibadan, 2017, GsPgGo
Joshua Oluwasanmi Owoseni, (D), Akure, 2019, HwNgRn

Federal University of Technology, Yola

Dept of Surveying & Geoinformatics (B)
PMB 2076
Yola
p. +234 (803) 6065646
info@mautech.edu.ng
http://www.futy.edu.ng/academic/dsurvey.htm

Dept of Soil Science (B)
PMB 2076
Yola
p. +234 805 3372 784
info@mautech.edu.ng
http://www.futy.edu.ng/academic/dsoil.htm

Dept of Geography (B)
PMB 2076
Yola
p. +234 (803) 6065646
info@mautech.edu.ng
http://www.futy.edu.ng/academic/dgeography.htm

Dept of Geology (B)
PMB 2076
Yola
p. +234 (803) 6065646
info@mautech.edu.ng
http://www.futy.edu.ng/academic/dgeology.htm

Gombe State University

Dept of Geography (B)
contact@gsu.edu.ng
http://www.gomsu.org/

Dept of Geology (B)
info@gsu.edu.ng
http://www.gomsu.org/

Ladoke Akintola University of Technology

Dept of Earth Sciences (B) ☒
PMB 4000
Ibadan , Oyo 4000
p. 08101043484
contact@lautech.edu.ng
http://www.lautech.edu.ng/Academics/undergraduates/FPAS
t: @Embrock4real
Programs: Geology
Certificates: GIS
Enrollment (2015): B: 13 (13)
Dr:
Oyelowo Gabriel Bayowa, (D), Obafemi Awolowo, 2013, GggGg
Abosede Olufunmi Adewoye, (D), Ladoke Akintola, 2014, GeeGe
Moruffdeen Adedapo Adabanija, (D), Ibadan, 2009, GggGg
Mr:
Olukayode Adegoke Afolabi, (M), Ibadan, 2000, GxxGx
Ismaila Abiodun Akinlabi, (M), Ibadan, 2011, GggGg
Mustapha Taiwo Jimoh, (M), Ibadan, 2006, GxxGx
Lanre Lateef Kolawole, (M), Newcastle, 1985, GeeGe

Lagos State Polytechnic

School of Environmental Studies (Quantity Surveying) (B)
info@mylaspotech.edu.ng
http://mylaspotech.net/

National Open University of Nigeria

School of Science & Technology (B)
PO Box 1866
Bauchi
p. +234 1-8188849, +234 1-4820720
sst@noun.edu.ng
http://www.nou.edu.ng/noun/

Nnamdi Azikiwe University

Dept of Geology/Meteorology and Env Management (B)
PMB 5025
Awka, Nnewi 5025
p. +234 (46) 550018
emmaojukwu@unizik.edu.ng
http://www.unizikeduportal.org/

Dept of Geological Sciences (B,M,D) ●☒
Awka Campus
PMB 5025
Awka, Anambra 5025
p. +234 (46) 550018
ne.ajaegwu@unizik.edu.ng
http://www.unizikeduportal.org/
Programs: Geology
Dr.:
Norbert E. Ajaegwu, (D), Nnamdi Azikiwe (Nigeria), 2014, GorGg

Obafemi Awolowo University

Dept of Geology (B,M,D) ●
Ile-Ife, Osun State 220002
p. 08128181062
geoife@oauife.edu.ng
http://gly.oauife.edu.ng/
Enrollment (2014): B: 213 (0) M: 24 (0) D: 12 (0)
Head:
A. A. Adepelumi, (D), YgsNe
Professor:
S. A. Adekola, (D), GdoCg
O. Afolabi, Yg
J. O. Ajayi, No
T. R. Ajayi, CgEgGe
O. A. Alao, (D), Yg
U. K. Benjamin, GoCg
S. L. Fadiya, (D), God
A. O. Ige, Cg
A. O. Ige, Cg
E. James-Aworeni, Hg
J. I. Nwachukwu, GoCg
O. O. Ocan, GtzGp
S. B. Ojo, Yg
V. O. Olarewaju, GozCg
A. O. Olorunfemi, Go
M. O. Olorunfemi, Yg
A. A. Oyawale, GeoCg
M. A. Rahaman, GpcGi
B. M. Salami, Hg
I. A. Tubosun
Associate Professor:
A Adetunji, (D), EgGi

Assistant Professor:
Dele E. Falebita, (D), Obafemi Awolowo, YeRnZg
Research Associate:
M. A. Olayiwola, GugPi
O. M. Oyebanjo

Dept of Soil Science and Land Resources Management (B,M,D) ○⊘
Department of Soil Science and Land Resources Management
Department of Soil Science and Land Resources Management
Faculty of Agriculture
Ile-Ife, Osun 220005
p. (803) 827-2331
midowu@oauife.edu.ng
http://www.oauife.edu.ng/faculties/agric/soil_sci.html
Programs: B.Agric., M.Sc. and Ph.D. D. (Soil Science)

Department of Geography (B,M,D) ○
Ile-Ife
p. +234 (70) 37772355
sinaayanlade@yahoo.co.uk
http://www.oauife.edu.ng/faculties/soc_sciences/departments/geo/home.htm
Programs: Geography and GIS
Certificates: GIS
Assistant Professor:
Ayansina Ayanlade, (D), AtZir

Rivers State University of Science & Technology

Inst of Geoscience & Space Technology (IGST)
P.M.B. 5080
Nkpolu - Oroworukwo
Port Harcourt, Rivers State
info@ust.edu.ng
http://www.ust.edu.ng/

University of Ado - Ekiti

Dept of Agricultural Sciences (B)
PMB 5363
Ado-Ekiti
p. +234 (30) 250026
http://www.unadportal.com/

Geology Dept (B)
PMB 5363
Ado-Ekiti
p. +234 (30) 250026
info@eksu.edu.ng

http://www.unadportal.com/

University of Benin

Dept of Petroleum Engineering (B)
PMB 115
Benin City, Edo State
p. 234 0802 345 5681
deanengineering@uniben.edu
http://www.uniben.edu/UniversityOfBenin-PetroleumEngineering.html

Dept of Geology (B,M,D) ●
P. M. B. 1155
Benin City, Edo State
p. 234 0802 345 5681
registrar@uniben.edu
http://www.uniben.edu/physicalscielcdfaculty.html
Professor:
Godwin Osarenkhoe Asuen, (D), EcGsg
Professor:
Christopher Nnaemeka Akujieze, (D), GgHwGe
Williams Ogbevire Emofurieta, (D), GzCeGe
Tony Uzozili Onyeobi, (D), NrgGq
Associate Professor:
Isaac Okpeseyi Imasuen, (D), W Ontario, CgGee
Senior Lecturer:
Ben Obelenwa Ezenabor, (M), HgZrGe
Franklin A. Lucas, (D), PmlGr
Lecturer II:
Aitalokhai Joel Edegbai, (M), GosPi
Efetobore Gladys Maju-Oyovwikowhe, (M), GszGo
Sikiru Adeoye Salami, (D), YgeGg
Lecturer I:
Ovie Odokuma-Alonge, (M), CgGip
Graduate Assistant:
Tomilola Andre-Obayanju, (B), NgHgGe
Assistant Lecturer:
Aiyevbekpen Helen Akenzua-Adamcyzk, (M), GeHgGg
Nosa Samuel Igbinigie, (M), GorGg
Alexander Ogbamikhumi, (M), YsgGg
Sunday Erapkpower Okunuwadje, (M), GsoGd
Edoseghe Edwin Osagiede, (M), GctGx

University of Calabar

Dept of Geology (B)
PMB 1115 Eta Agbo Road,
Calabar Municipal, Cross River State
p. (+234) 8173740083
general@unical.edu.ng
http://www.unical.edu.ng/pages/programs_courses/sciences.php?nav=departments
Head:
Nse U. Essien

University of Ibadan

Dept of Geology (B,M,D)
Ibadan, Oyo 200284
p. 02-8101100-8101104
o.okunlola@mail.ui.edu.ng
http://sci.ui.edu.ng/geowelcome
Professor:
I. M. Akaegbobi, Cg
A. A. Elueze, Eg
Ougbenga. Akindeji. Okunlola, (D), Ibadan, 2001, EgCgGe
A. I. Olayinka, Yg
Moshood N. Tijani, (D), Muenster (Germany), 1997, HwCgGe
Lecturer:
A. E. Abimbola, Cg
O. C. Adeigbe, (D), Ibadan, 2009, GosGe
G. O. Adeyemi, NgHg
O. A. Boboye, Go
A. T. Bolarinwa, EgGz
M. E. Nton, Go
M. A. Oladunjoye, Yg
O. O. Osinowo, Yg
I. A. Oyediran, Ng

Dept of Geology (B,M,D) ○⊘
Ibadan
Ibadan, Oyo 200284
p. +234-8033819066
ehinola01@yahoo.com
http://www.ui.edu.ng/?q=departmentofgeology
Programs: Geology
Certificates: Training Certificate
COMEG
Enrollment (2021): B: 208 (202) M: 40 (33) D: 42 (18)
Aret Adams Professorial Chair:
Olugbenga A. Ehinola, (D), Ibadan, 2002, GoeCo
Professor:
Olugbenga Ajayi Ehinola, (D), Ibadan, 2002, GoeGs
A. S. Olatunji, (D), Ibadan, 2006, CgGeCt

Dept of Petroleum Engineering (B,M,D) ☒
Oyo Road
Ibadan
p. +234-2-813-004-968
falodelias@gmail.com
http://www.ui.edu.ng/?q=petengine
Programs: Petroleum Engineering
Enrollment (2016): B: 223 (43) M: 17 (17) D: 10 (2)
Professor:
Olugbenga Falode, (D), NpZa

NIGERIA

University of Ilorin

Dept of Geology and Mineral Sciences (B)
PMB 1515
p. +234 (31) 221691
ccafs@cgiar.org
http://www.unilorin.edu.ng/unilorin/index.php/sciences-dept/geology-mineral-sciences

Head:
J.I. D. Adekeye, (D), GoCg

Professor:
S. O. Akande, (D), EgCg
O. Ogunsanwo, Ng

Lecturer:
A. Abdurrahman, (M), CeEg
A. D. Adedcyin, (M), Eg
O. A. Adekeye, (D), Pm
S.M. A. Adelana, (M), HgYgGe
D. A. Alao, NgGz
O. O. Ige, (M), Ng
Kehinde A. Olawuyi, (D), Fed Tech (Nigeria), 2015, YeGge
O. A. Omotoso, (M), Ge
W. O. Raji, (M), Ye

University of Jos

Dept of Geosciences and Mining (B)
PMB 2084
Jos 930001
p. +234 (73) 559 52
vc@unijos.edu.ng
http://www.unijos.edu.ng/

University of Lagos

Dept of Metalurgical and Materials Engineering (B)
Akoka
Yaba, Lagos
p. +234 (1) 493 2600 3
informationunit@unilag.edu.ng
http://www.unilag.edu.ng/index.php?page=home

Dept of Civil and Environmental Engineering (B)
Akoka
Yaba, Lagos
p. +234 (1) 493 2660 3
informationunit@unilag.edu.ng
http://www.unilag.edu.ng/index.php?page=about_departmentdetail&sno=11

Dept of Petroleum and Gas Engineering (B)
Akoka
Yaba, Lagos
p. +234 (1) 493 2660 3
informationunit@unilag.edu.ng
http://www.unilag.edu.ng/index.php?page=home

Dept Surveying and Geoinformatics (B)
Akoka
Yaba, Lagos
p. +234 (1) 493 2660 3
informationunit@unilag.edu.ng
http://www.unilag.edu.ng/index.php?page=about_departmentdetail&sno=16

University of Maiduguri

Dept of Geology
PMB 1069 FEA
Maiduguri
info@unimaid.edu.ng
http://www.unimaid.edu.ng/root/faculty_of_sciences/dept_geology.html

University of Maiduguri, Borno State

Dept of Geology (B,M,D)
Bama Road
PMB 1069
Maiduguri
p. +234 (76) 231730
info@unimaid.edu.ng
http://www.unimaid.edu.ng/home.php

Professor:
I. B. Goni

Assistant Professor:
Saidu Baba

Lecturer:
Sani Adamu
Kyari M. Ajar
Musa Malana Aji
Fati Bukar
Shettima Bukar
Jalo M. El-Nafaty
Millitus V. Joseph
Samaila C. Kali
Yakubu B. Mohd
Mohammed Poukar
Aishatu Sani
Manaja Mijinyawa Uba
Sidi M. Waru
Asabe Kuku Yahaya
Soloman N. Yusuf
A. Adam Zarma

University of Nigeria

Dept of Geology (B)
Nsukka
micah.osilike@unn.edu.ng
http://www.unn.edu.ng/physicalsciences/content/view/700/601/

University of Nigeria Nsukka

Dept of Geology (B)
anthony.okonta@unn.edu.ng
http://unn.edu.ng/department/geology

Lecturer:
Alloysuis Okwudili Anyiam
Luke I. Manah
Smart Chicka Obiora

Dept of Geoinformatics and Surveying (B)
johnson.urama@unn.edu.ng
http://www.unn.edu.ng/environmentalstudies/content/view/27/44/

University of Port Harcourt

Dept of Geography & Environmental Management (B)
East/ West Road
PMB 5323
Choba, Port Harcourt 500001
p. +234 (84) 230890-99
edulms@chimpgroup.com
http://uniport.edu.ng/

Dept of Geology (B)
East/ West Road
P.M.B. 5323
Choba, Port Harcourt 500001
p. +234 (0)84 817 941
uniport@uniport.edu.ng
http://www.uniport.edu.ng/faculties/science.html

Dept of Petroleum and Gas Engineering (B)
East/ West Road
PMB 5323
Choba, Port Harcourt 500001
p. +234 (84) 230890-99
boniface.oriji@uniport.edu.ng
http://uniport.edu.ng/

University of Uyo

Faculty of Science Education (B,M,D)
PMB 1017
Uyo
p. +234 (85) 200303
vc@uniuyo.edu.ng
http://www.uniuyo.edu.ng

Faculty of Social Science (B,M,D)
PMB 1017
Uyo
p. +234 (85) 200303
vc@uniuyo.edu.ng
http://www.uniuyo.edu.ng

Faculty of Engineering (B,M,D)
PMB 1017
Uyo
p. +234 (85) 200303
vc@uniuyo.edu.ng
http://www.uniuyo.edu.ng

Faculty of Agriculture (B,M,D)
PMB 1017
Uyo, Akwa Ibom State 520001
p. +234 (85) 200303
vc@uniuyo.edu.ng
http://www.uniuyo.edu.ng

Faculty of Environmental Studies (B,M,D)
PMB 1017
Uyo
p. +234 (85) 200303
vc@uniuyo.edu.ng
http://www.uniuyo.edu.ng

Western Delta University

Geology and Petroleum Studies (B)
PMB 10
Oghara
p. +234 (70) 35400531
info@wdu.edu.ngg
http://www.wduniversity.org/

Wukari Jubilee University

Dept of Geography and Environmental Conservation (B)
PMB 1019
Wukari, Taraba State
p. +234 (080) 80329138
wukari_jubilee@yahoo.com
http://wukarijubileeuniversity.org/

Norway

Norges Geologiske Undersokelse

Geological Survey of Norway ☒
Postboks 6315 Torgarden
7491 Trondheim
ngu@ngu.no
http://www.ngu.no/no/

Norwegian University of Life Sciences (NMBU)

Section of Soil and Water Sciences, Faculty of Environmental Sciences and Natural Resource Management (B,M,D) O☒
P.O. Box 5003
1432 AS
Ås 1432
p. +47 67 23 00 00
post-mina@nmbu.no
https://www.nmbu.no/en/faculty/mina
Programs: Environmental Sciences (B, M); Ecology and Natural Resource Management; Renewable Energy; Sustainable Tourism; Forestry

Professor:
Åsgeir Rosseland Almås, (D), Sc
Helen Kristine French, (D), HwYg
Tore Krogstad, (D), Sc
Jan Mulder, (D), Cb
Gunnhild Riise, (D), Hs
Trine Sogn, (D), Sb
Rohrlack Thomas, (D), Hs

Senior Scientist:
Vegard Martinsen, (D), Scf

Associate Professor:
Susanne Eich-Greatorex, (D), Sc
Mona Henriksen, (D), 2004, ZyGs
Jan Marten Huizenga, (D), GzcEd
Nils-Otto Kitterød, (D), Hg
Attila Nemes, (D), Sp
Danni Pearce, (D), GlmZr
Line Tau Strand, (D), Sfd

Norwegian University of Science and Technology

Dept of Geology and Mineral Resources Engineering
N-7034 Trondheim
mai.britt.mork@ntnu.no

Faculty of Engineering Science and Technology
S. P. Andersens vei 15A
N7034 Trondheim
p. 73594779
ingvald.strommen@ntnu.no

NTNU Norwegian University of Science and Technology

Dept of Geology and Mineral Resources Engineering (B,M,D)
Sem Sælands vei 1
7491 Trondheim
Trondheim
p. (+47) 73 59 48 00
iigb-info@ivt.ntnu.no
http://www.ntnu.no/igb
f: https://www.facebook.com/geologibergNTNU

University of Bergen

Dept of Earth Science (B,M,D) O☒
Allegaten 41
N-5007 Bergen
p. +47 55583600
post@geo.uib.no
http://www.geo.uib.no/
f: https://www.facebook.com/earthscienceuib
t: @uibgeo

University of Oslo

Deptartment of Geosciences (B,M,D) ☒
Post Box 1047, Blindern
N-0316 Oslo
p. +47 22856656
geosciences@geo.uio.no
http://www.mn.uio.no/geo/english/
f: https://www.facebook.com/uiogeo/
Programs: Two BSc programmes in Geosciences (in Norwegian), Masterprogramme in Geosciences (master's two years)

Mineralogical-Geological Museum
Sars gate 1
P.O. Box 1172
Blindern, Oslo 0318
p. +47 228 55050
postmottak@nhm.uio.no
http://www.nhm.uio.no/english/

University of Tromso

Geology Dept
N-9037 Tromso
matthias.forwick@uit.no
http://www.ibg.uit.no/geologi/geo_eng_end.html

Oman

Sultan Qaboos University

Dept of Earth Science (B,M,D) ●
123 Al-Khod
P.O. Box 36
Muscat
p. +968 2414 6832

alsayigh@squ.edu.om
http://www.squ.edu.om/earth-sci/tabid/11718/language/en-US/
Enrollment (2013): B: 85 (60) M: 5 (25) D: 2 (1)
Head:
Salah Al-Khirbash, (D)

Pakistan

COMSATS

Institute of Information Technology
Park Road, Chak Shahzad
Islamabad
p. +92-51-9247000-2
info@ciit.edu.pk
http://www.comsats.edu.pk/

Geological Survey of Pakistan

p. (+958) 9211032
qta@gsp.gov.pk
http://www.gsp.gov.pk/

Institute of Space Technology

Institute of Space Technology (B,M,D)
P.O. Box 2750
Islamabad 44000
p. 92.51.9075100
info@ist.edu.pk
http://www.ist.edu.pk/

National University of Sciences and Technology

Institute of Geographic Information System
RIMMS building, NUST H-12 Campus
Islamabad
p. +92-51-90854400
info@igis.nust.edu.pk
http://igis.nust.edu.pk/

Sind University Pakistan

Dept of Geology
University of Sindh, Allama I.I. Kazi Campus
Jamshoro
Sindh 76080
p. 92-22-9213172-9213181-90 Ext:
dean.science@usindh.edu.pk

University of Engineering and Technology

Geological Engineering (B,M,D) ●☒
Lahore, Punjab 54890
p. 0092-42-99029487
mzubairab1977@gmail.com
http://www.uet.edu.pk/faculties/facultiesinfo/geological/?RID=introduction
Programs: B.Sc. Geological Engineering, M.Sc. Geological Engineering, M.Sc. Geological Sciences, Ph.D. Geological Engineering, Ph.D. Geological Sciences
Enrollment (2018): B: 50 (50)
Chair:
Muhammad Zubair Abu Bakar, (D), 2012, NrmNg
Associate Professor:
Muhammad Farooq Ahmed, (D), Missouri (Rolla), 2013, NgxNt

University of Karachi

Dept of Geology
KARACHI-75270
p. 99261300-06 Ext: 2295
vc@uok.edu.pk
http://uok.edu.pk/faculties/geology/

University of Sargodha

Dept of Earth Science (B,M,D) ●☒
Department of Earth Sciences
University of Sargodha-
Sargodha-Pakistan.
Sargodha , Punjab 40100
p. +92-48-9230591
earthscience@uos.edu.pk
http://www.uos.edu.pk
Programs: BS Geology 4 Year
MS Geology 2 Year
BS Geography 4 Year
M.Sc. Geography 2 Year
Regular and permanent:
Dr. Syed Haroon Ali, (D), 2020, GsoGt

College of Agriculture
Sargodha
info@uos.edu.pk
http://www.uos.edu.pk

University of the Punjab

Department of Geography (A,B,M,D) ●☒
New Campus
Lahore 54590
shirazi.geog@pu.edu.pk
http://pu.edu.pk/home/department/46/Department-of-Geography
Programs: BS,MS and PhD in Geography,Remote Sensing and GIS
Enrollment (2021): B: 347 (0) M: 35 (0) D: 22 (0)

Panama

Universidad de Panama

Facultad de Ciencias Naturales, Exactas y Technologia
Urbanizacion El Cangrejo, Estafeta Universitaria
ciencias@ancon.up.ac.pa

Paraguay

Universidad Nacional se Asuncion

Dept de Geologia
University Campus Km. 11, Asuncion
facen@facen.una.py

Peru

Geological Survey of Peru

Instituto Geologico Minero y Metalurgico - INGEMMET
Av. Canada 1470 San Borja
Lima
p. 051-1-6189800
webmaster@ingemmet.gob.pe
http://www.ingemmet.gob.pe/

San Agustin's Natl University

Escuela Academico Profesional de Ingenieria Geologica y Geofisica (M) ☒
Av. Indraendencia #1015
Arequipa
p. (054) 244498
geologia@unsa.edu.pe
http://www.unsa.edu.pe

Univ Nacional de Ingeneiria

Escuela Academico Profesional de Ingenieria Geologica
Av. Tupac Amaru S/N
Rimac, Lima
esanchez@uingemmet.gob.pe
http://www.uni.edu.pe/sitio/academico/facultades/geologica/

Univ Nacional Mayor San Marcos

Escuela Academico Profesional de Ingenieria Geologica
Av. Venezuela S/N
Lima
j_jacay@yahoo.com
http://www.unmsm.edu.pe

Philippines

University of the Philippines

National Institute of Geological Sciences (B,M,D) ●☒
C.P. Garcia corner Velasquez Street
Diliman, Quezon City 1101
p. (+632) 9296046
admin@nigs.upd.edu.ph
https://nigs.science.upd.edu.ph
f: https://www.facebook.com/upnigs

Poland

AGH University of Science and Technology

Faculty of Geology, Geophysics, and Environmental Protection
A. Mickiewicza 30 Ave.
30-059 Krakow
p. +48 12 633 29 36
aglown@geol.agh.edu.pl
http://www.wggios.agh.edu.pl/

Dept of Mining and Geoengineering
Inst. of Drilling & Oil Exploration
Krakow
p. +48 12 617 21 15
wgig@agh.edu.pl
https://www.agh.edu.pl/en/wydzialy/faculty-of-mining-and-geo-engineering/

Faculty of Mining and Geoengineering
Al. Mickiewicza 30
PL-30059 Krakow
p. +48 12 617 21 15
gorn@agh.edu.pl

Jagiellonian University

Institute of Geological Sciences
Oleandry 2a Str.
30-063 Krakow
sekretariat.ing@uj.edu.pl

http://www.ing.uj.edu.pl

Ministry of Environmental Protection, Natural Resources and Forestry

Dept of Geology
st. Wawelska 52/54
Warsaw 00-922
p. +48 22 3692449
Departament.Geologii.i.Koncesji.Geologicznych@mos.gov.pl
http://www.mos.gov.pl/dg/dga1.htm

Panstwowy Instytut Geologiczny

Polish Geological Institute
Ul. Rakowiecka 4 ,
PL-00-975 Warszawa
sekretariat@pgi.gov.pl
http://www.pgi.gov.pl/

Polish Academy of Sciences

Institute of Geological Sciences
Al Zwirki i Wigury 93
02-089 Warsaw
p. (+48 22) 620 33 46, 656 60 63
ingpan@twarda.pan.pl

Silesian University of Technology

Faculty of Mining and Geology (B,M,D) ●
Wydzial Gornictwa i Geologii
Politechnika Slaska
ul.Akademicka 2
Gliwice 44-100
p. +48 32 2371283
rg@polsl.pl
http://www.polsl.pl/Wydzialy/RG/Strony/Witamy.aspx
Enrollment (2014): B: 2556 (615) M: 327 (283) D: 98 (5)

University of Wroclaw

Inst of Geological Sciences (B,M,D)
Pl. Maksa Borna 9
50-205 Wroclaw
p. (+4871) 321-10-76
sekretariat.ing@uwr.edu.pl
http://www.ing.uni.wroc.pl/home-page
Enrollment (2016): D: 14 (8)

Director:
Krystyna Choma-Moryl, (D), 1983, Ng

Professor:
Pawel Aleksandrowski, (D), Polish Acad of Sci, 1985, GctYe
Marek Awdankiewicz, (D), 1997, Gvi
Piotr Gunia, (D), 1983, Gaz
Jacek Gurwin, (D), 1997, HwyZf
Henryk Marszalek, (D), 1994, Hy
Jacek Puziewicz, (D), 1985, Gix
Jerzy Sobotka, (D), AGH (Poland), 1992, YeuEo
Andrzej Solecki, (D), 1987, Eg
Stanislaw Stasko, (D), 1986, Hg
Jacek Szczepanski, (D), 1999, Ggc

Adjunct Professor:
Anna Gorecka-Nowak, (D), 1992, Pg
Maciej Gorka, (D), 2007, AsCa
Jakub Kierczak, (D), 2007, GeSc
Antoni Muszer, (D), 1994, Eg
Anna Pietranik, (D), Wroclaw, 2006, CgGie
Marta Rauch, (D), 2002, Gc
Robert Tarka, (D), 1995, Hg
Jurand Wojewoda, (D), 1989, Gs

Portugal

Istituto Geologico e Mineiro (IGM)

Geological and Mining Institute of Portugal
Estrada da Portela
Zambujal
atendimento@ineti.pt
http://www.ineti.pt/

Oporto University

Dept of Geology
Faculdade de Ciencias do Porto
Praca de Gomes Teixeira
4099-002 PORTO
dg.sec@fc.up.pt
http://www.fc.up.pt/depts/geo/

Universidade da Madeira

Dept de Biologia (B)
Campus Universitario da Penteada
9000-390 Funchal
p. +251 291 705 380
mgouveia@uma.pt
http://www.uma.pt/Unidades/Biologia/

Universidade de Aveiro

Dept de Geociencias (B,M,D) ☒
Campo Universitario Santiago
Aveiro 3810-193
p. +351 234 370 357
eafsilva@ua.pt
http://www.ua.pt/geo
Programs: Geology BSc, Geological Engineering MSc, Geosciences PhD

Professor:
Eduardo Ferreira da Silva, (D), CeGbCl
Fernando Rocha, (D), Aveiro, 1994, EnGzb

Universidade de Coimbra

Dept de Ciencias da Terra (A,B,M,D) O☒
Rua Silvio Lima, Univ. Coimbra, Pólo II Coimbra
Coimbra 3030-790 Coimbra
p. +351-239 860 514
cgeo@ci.uc.pt

http://www.dct.uc.pt
Programs: Geology
Professor:
Rui Pena dos Reis, (D), GorGs
Associate Professor:
Alcides Castilho Pereira, (D), Coimbra, 1992, GebZr

Universidade de Lisboa

Dept of Geology (A,B,M,D) ● ☒
Ed. C6, Campo Grande
Campo Grande
Lisboa 1749-016
p. +35121 750 00 66 / + 351 21 75
dgeologia@fc.ul.pt
http://www.fc.ul.pt/en/dg?refer=1
f: https://www.facebook.com/Departamento-de-Geologia-FCUL-192772970763173
Programs: Geology
Certificates: Applied Geology
Petroleum Geosciences
Enrollment (2018): B: 314 (51) M: 87 (13) D: 30 (2)
Professor:
Maria da Conceição Pombo Freitas, (D), Lisbon, 1996, GusGe
Professor:
César Andrade, (D), Gu
Ana Azerêdo, (D), Gd
Maria da Conceiç Freitas, (D), GseGu
António Mateus, (D), Em
Associate Professor:
Mário Cachão, (D), PgeGg
Nuno Pimentel, (D), GrsGo
Assistant Professor:
Maria do Rosário Carvalho, (D), HwyZn
Telmo Bento dos Santos, (D), GptGi
Carla Kullberg, (D), Gt
Carlos Marques Silva, (D), Pg
Catrina Silva, (D), Hg
António Brum Silveira, (D), Zf

Universidade de Trás-os-Montes e Alto Douro

Departamento de Geologia (B,M,D) ⃠
Apartado 1013
Vila Real 5001-801
p. +351259350224
geologia@utad.pt
http://www.geologia.utad.pt
Enrollment (2020): B: 7 (0) M: 3 (0) D: 2 (0)
Professor:
Maria E.P. Gomes, (D), UTAD, 1996, GiCgGz
Associate Professor:
Ana M.P. Alencoão, (D), UTAD, 1998, HgGgm
Artur A.A. Sá, (D), UTAD, 2005, PgiPs
Assistant Professor:
Maria E.P.S. Abreu, (D), UTAD, 2012, GaaGa
João C.C.V. Baptista, (D), UTAD, 1998, GmmGa
Maria R.M. Costa, (D), UTAD, 2000, Hw
Paulo J.C. Favas, (D), UTAD, 2008, GeCtGg
José M.M. Lourenço, (D), UTAD, 2006, YeZin
Alcino S. Oliveira, (D), UTAD, 2002, Hww
Anabela R.R.C. Oliveira, (D), UTAD, 2011, Gss
Fernando A.L. Pacheco, (D), UTAD, 2001, Hq
Luís M.O. Sousa, (D), 2001, EnGgNg
Rui J.S. Teixeira, (D), UTAD, 2008
Nuno M.O.C.M. Vaz, (D), UTAD, 2010, PmlYg

Universidade do Minho

Dept de Ciencias da Terra (B,M,D)
Campus de Gualtar
Braga 4710-057
p. 351 253 604 300
sec@dct.uminho.pt
http://www.dct.uminho.pt/

Universidade Nova de Lisboa

Dept de Ciencias da Terra
Qta. da Torre
2829-516 Caparica
p. (351) 212 948 573
dct.secretariado@fct.unl.pt
http://www.dct.fct.unl.pt/

Qatar

University of Qatar

Dept of Geology (B)
PO Box 2713
Doha
Doha
hamadsaad@qu.edu.qa
http://www.angelfire.com/ms/GeoQU/
Professor:
Sobhi J. Nasir, (D), Germany, 1986, Gx
Associate Professor:
Hamad A. Al-Saad, (D), Egypt, 1996, PsGsu
Abdulali A. Sadiq, (D), Southampton (UK), 1995, Zr
Assistant Professor:
Latifa B. AL-Nouimy, (D), Egypt, 1994, Hgw
Sief A. Alhajari, (D), North Carolina, 1992, Hgw

Reunion

Universite de la Reunion

Faculte de Sciences et Technologie - Sciences de la Terre (B)
15, avenue Rene Cassin
CS 92003
Saint-Denis, Cedex 9 97744
p. (262) 0262 938697
jean-lambert.join@univ-reunion.fr
http://sciences.univ-reunion.fr/rubrique.php3?id_rubrique=72
Professor:
Jean-Lambert Join, (D), Hw
Associate Professor:
Fabrice R. Fontaine, (D), Yg

Laboratoire GeoSciences Reunion (B,M,D)
15 avenue Rene Cassin
BP 7151
Saint Denis, Messag Cedex 9 97744
vfamin@univ-reunion.fr
http://www.geosciencesreunion.fr/
Professor:
Laurent Michon, (D), GtvGg
Professor:
Jean-Lambert Join, HwGsHg
Associate Professor:
Vincent Famin, (D), GcCgGg
Fabrice R. Fontaine, (D), Ysg
Claude Smutek, NrZrg

Romania

Alexandru Ioan Cuza

Department of Geology (B,M,D) ☒
geology@uaic.ro
Bulevardul Carol I nr. 20A
Iasi 700505
geology@uaic.ro
http://geology.uaic.ro
Enrollment (2015): B: 59 (26) M: 40 (18) D: 11 (5)
Head:
Nicolae Buzgar, (D), A I Cuza (Iasi), 1998, GziGd
Professor:
Sorin Dorin Baciu, (D), A I Cuza (Iasi), 2001, GcPv
Dumitru Bulgariu, (D), Cga
Ovidiu Gabriel Iancu, (D), A I Cuza (Iasi), 1998, GpeCg
Crina Genoveva Miclaus, (D), A I Cuza (Iasi), 2001, Gsr
Dan Stumbea, (D), Bolyai Cluj, 1998, Gez
Associate Professor:
Andrei Ionut Apopei, (D), A I Cuza (Iasi), 2014, GziGg
Andrei Buzatu, (D), A I Cuza (Iasi), 2014, GziGg
Traian Gavriloaiei, (D), A I Cuza (Iasi), 1999, CaAsCg
Viorel Ionesi, (D), A I Cuza (Iasi), 2005, PmGgHg
Daniel Tabara, (D), A I Cuza (Iasi), 2000, Pl
Paul Tibuleac, (D), A I Cuza (Iasi), 1999, Pgs

Assistant Professor:
Iuliana Buliga, (D), A I Cuza (Iasi), 2014, Gep
Florentina Pascariu, (D), A I Cuza (Iasi), Nt
Anca Maria Seserman, (D), A I Cuza (Iasi), 2015, Gs
Lecturer:
Laviniu Apostoae, (D), A I Cuza (Iasi), 1999, Eg
Dan Astefanei, (D), A I Cuza (Iasi), 2012, Cga
Ciprian Chelariu, (D), A I Cuza (Iasi), 2013, Go
Claudia Cirimpei, (D), A I Cuza (Iasi), 2010, PgGr
Tony Cristian Dumitriu, (D), A I Cuza (Iasi), 2015, GcZir
Dan Bogdan Hanu, (D), Yge
Mitica Pintilei, (D), A I Cuza (Iasi), 2010, CgeCt
Bogdan Gabriel Ratoi, (D), A I Cuza (Iasi), 2013, PvGr
Smaranda Doina Sîrbu, (D), A I Cuza (Iasi), 2002, Ge
Oana Stan, (D), A I Cuza (Iasi), 2009, GedCb
Visiting Professor:
Mihai Brânzila, (D), A I Cuza (Iasi), 1998, GgrPg

Babes-Bolyai University

Department of Geology (B,M,D) ⃠
Str. M. Kogalniceanu nr.1
RO - 400084 Cluj-Napoca
Cluj-Napoca 400084
p. +40-264-405300
nicolae.har@ubbcluj.ro
http://bioge.ubbcluj.ro/pages/geologie/
Programs: Geology; Applied Geology
Enrollment (2021): B: 11 (18) M: 10 (5) D: 1 (1)
Professor:
Ioan Bucur, (D), Babes-Bolyai, 1991, PgmGs
Vlad Codrea, (D), Babes-Bolyai, 1995, PvGu
Sorin Filipescu, (D), Babes-Bolyai, 1996, GrPm
Associate Professor:
Ion Marcel Benea, 2006, GzaEn
Carme Mariana Chira, 2003, PemZe
Ferenc Lazar Forray, 2016, CglPc
Nicolae Har, 2007, GipNr
Emanoil Florin Sasaran, 2013, Gds
Calin Gbriel Tamas, 2016, EdGzNm
Ioan Tantau, 2014, PlePc
Assistant Professor:
Agnes Gal, 2011, GzEgGa
Lecturer:
Constantin Balica, 2012, GtCgc
Horea Stefan Bedelean, 2005, GcSdGe
Raluca Emilia Haitonic, 2016, PmGrPs
Boglarka Mercedes Kis, 2018, ScCqHg
Emilia Mosonyi, 2005, GpiGc
George Ples, 2016, PmsYg
Mirela Violetta Popa, 2005, GgPgZe
Lorand Silye, PmgGr
Tudor Liviu Tamas, 2005, GzZmn

Institul Geologic al Romaniei

Geological Institute of Romania ☒
1 Caransebes St, Sector 1
Bucuresti 012271
p. +40314933400
office@igr.ro
http://www.igr.ro/

University of Bucharest

Department of Geophysics - Faculty of Geology and Geophysics (B,M) ●☒
6 Traian Vuia Street
Sector 2
Bucharest, Bucharest 020956
p. +40-21-3125024
bogdan.niculescu@unibuc.ro
https://gg.unibuc.ro/organizare/departamentul-geofizica/
Programs: Geophysics; Geological Engineering
Certificates: N/A
Enrollment (2022): B: 20 (20) M: 12 (12)
Associate Professor:
Bogdan Mihai NICULESCU, (D), Bucharest, 2002, YgeYu

Rwanda

Universite Nationale du Rwanda

Dept of Geology (B)
Faculty of Science
PO Box 56
Butare
p. +250 (0) 252 530 122
pro.cst@ur.ac.rw
http://www.nur.ac.rw/

Dept of Geography (B)
PO Box 56
Butare
p. +250 (0) 252 530 122
info@nur.ac.rw
http://ww.nur.ac.rw/

Saudi Arabia

King Abdulaziz University

Dept of Geology
P. O. Box 1744
Building No 55
Jeddah 21441
jha25@hotmail.com
http://earthscience.kau.edu.sa/?Site_ID=145&Lng=EN

King Fahd University of Petroleum and Minerals

Geosciences Department (B,M,D) ●☒
P. O. Box 5070
KFUPM-CPG
Dhahran, Eastern Province 31261
p. +96638602620
c-es@kfupm.edu.sa
https://cpg.kfupm.edu.sa/academics/departments/department-of-geosciences/
Programs: BS Geology, BS Geophysics, MS Geology, MS Geophysics, MS Environmental Science, PhD Geology, PhD Geophysics, Professional Master's in Reservoir Characterization
Enrollment (2021): B: 72 (18) M: 37 (11) D: 19 (1)
Professor:
Abdullatif Al-Shuhail, (D), Texas A&M, 1998, Ygs
Michael A. Kaminski, (D), MIT/WHOI, 1988, GuPm
Panteleimon Soupios, (D), Aristotle, 2000, Yg
Chairman:
John Dean Humphrey, (D), Brown, 1987, GdoCs
Associate Professor:
Khalid Al-Ramadan, (D), Uppsala, 2006, GodGs
Husam Baalousha, (D), Aachen Tech, 2003, HwyGe
Sherif Hanafy, (D), Kiel, 2002, Yg
Mohammad Makkawi, (D), Colorado State, 1998, HwGoq
Bassam S. Tawabini, (D), King Fahd, 2002, GeHsCa
Scott Whattam, (D), Hong Kong, GxXc
Assistant Professor:
Abdullah Al-Shuhail, (D), Calgary, 2011, Ye
Ammar Elhusseiny, Stanford, Yg
Ismail Kaka, (D), Carleton, 2006, Ys
Ardiansyah Koeshidayatullah, (D), Manchester, 2018, GdCs
Umair Bin Waheed, (D), Kaust, 2015, Yg
Lecturer:
Ayman Al-Lehyani, (M), King Fahd, Yg
Mutasim Sami Osman, (M), King Fahd, 2014, GosGc

Saudi Geological Survey

Saudi Geological Survey
P.O. Box: 54141
21514
p. 966-2-619-5000
sgs@sgs.org.sa
http://www.sgs.org.sa/Arabic/Pages/

Senegal

Universite Cheikh Anta Diop

Dept de Geologie (B)
Faculte Des Sciences
BP 5005
Dakar-Fann
p. +221 869.27.66
fst@ucad.edu.sn
http://fst.ucad.sn/index.php?option=com_content&task=view&id=19&Itemid=35

Institut des Sciences de l' Environnement (ISE) (B) ☒
BP 5005
Dakar-Fann
p. +221 869.27.66
diengdiomaye@yahoo.fr

Sierra Leone

Njala University

Dept of Mining and Metallurgy (B)
Freetown
p. +232-22-228788, +232-22-226851
nuc@sierratel.sl
http://www.nu-online.com/

Dept of Soil Science (A,B,M,D)
Freetown
p. +232-22-228788, +232-22-226851
nuc@sierratel.sl
http://ww.nu-online.com/

Dept of Geography & Rural Development (B)
Freetown
p. +232-22-228788, +232-22-226851
nuc@sierratel.sl
http://www.nu-online.com/

Institute of Environmental Management and Quality Control (B)
Freetown
p. +232-22-228788, +232-22-226851
nuc@sierratel.sl
http://www.nu-online.com/

University of Sierra Leone

Dept of Geology
PO Box 87
Freetown
b.williams@slgeosciences.org
http://www.tusol.org/programmes

Slovakia

Comenius University in Bratislava

Department of Geology and Palaeontology (A,B,M,D) ○☒
Faculty of Natural Sciences
Ilkovičova 6
Mlynska dolina G
Bratislava 842 15
p. 00421260296529
kgp@fns.uniba.sk
http://geopaleo.fns.uniba.sk/
Programs: Geology (B); Palaeobiology (B); Dynamic Geology (M); Palaeontology (M,D); Tectonics (D); Sedimentology (D)
Enrollment (2016): B: 12 (0) M: 1 (0) D: 5 (0)

Professor:
Michal Kovac, (D), Gsg
Dusan Plasienka, (D), Gt
Daniela Rehakova, (D), Pms
Aubrecht Roman, (D), Gdr

Associate Professor:
Jozef Hok, (A), Ggt
Natalia H. Hudackova, (A), PimPg
Marianna Kovácová, (A), PlcPe
Martin Sabol, (D), Comenius (Slovakia), 2000, PvgPs
Rastislav Vojtko, (D), GctGm

Associate Scientist:
Matus Hyzny, Pi
Jan Schlogl, (D), Pig

Assistant Professor:
Peter Joniak, (D), Pv

Dionyz Stur Institute of Geology

Geology Dept
Mlynska Dolina 1
81704 Bratislava 11
p. 02 / 59375111
secretary@geology.sk
http://www.geology.sk/new/en

Geologicka Sluzba Slovenskej Republiky

Geological Survey of Slovak Republic
Mill Valley 1
817 04 Bratislava 11
secretary@geology.sk
http://www.sguds.sk

Kosice Technical University

Faculty of Mining, Ecology, Process Control and Geotechnology
Fakulta BERG
Park Komenskeho 12
040 01 Kosice
p. +421-55-602 1111
sekrd.fberg@tuke.sk
http://www.fberg.tuke.sk/bergweb/index.php?IdLang=0&Selection=4

Slovenia

Geoloski zavod Slovenije

Geological Survey of Slovenia (M,D)
Dimiceva ulica 14
Ljubljana SI - 1000
p. +386 1 2809702
www@geo-zs.si
http://www.geo-zs.si

University of Ljubljana

Faculty of natural science and engineering
Aškerčeva c. 12
Ljubljana 1000
p. 01/470-45-00
tanja.kocevar@ntf.uni-lj.si
http://www.uni-lj.si/academies_and_faculties/faculties/2013071111502957/

Oddelek za Geologijo
Askerceva 12
Ljubljana
spela.turic@ntf.uni-lj.si
http://www.ntf.uni-lj.si/og

Somalia

Burao University

Dept of Community Development (B)
p. +252 7126481
info@universityofburao.com
http://www.buraouniversity.com/rural_and_environmental_studies.htm

Mogadishu University

Somali Centre for Water and Environment (B)
p. (252)-5-932454/ 223433/ 658479
info@mu.edu.so
https://mu.edu.so/somali-center-for-water-environment/

Nugaal University

Institute of Geology (B,M) ●
Lasanod

p. +252 275 4063 / 7412619
nugaaluniversity@gmail.com
http://www.nugaaluniversity.com

South Africa

Cape Peninsula University of Technology

Dept of Environmental and Occupational Studies (B)
Cape Town
p. (+27) 021 959 6230
vanderwesthuizenh@cput.ac.za
http://www.cput.ac.za/

Geological Survey of South Africa

Council for Geoscience ●
Private Bag X112
Pretoria 0001
p. 0027128411911
info@geoscience.org.za
http://www.geoscience.org.za/
Executive Manager:
Fhatuwani L. Ramagwede, (M), 2011, EgGzEg

Nelson Mandela Metropolitan University

Dept of Geoscience (B,M,D)
PO Box 77000
Summerstrand, Port Elizabeth 6031
p. 041 504 2325
sheila.entress@nmmu.ac.za
http://geosci.nmmu.ac.za/
Administrative Assistant: Sheila Entress
Head:
Nigel Webb, (D)
Professor:
Vincent Kakembo, Sp
Associate Professor:
Moctar Doucoure, (D)
Daniel Mikes, (D)
Post-Doctoral Fellowship:
Bastien Linol, (D), NMMU, 2013, Gs
Lecturer:
Callum Anderson, (M), Port Elizabeth, Gzs
Wilma Britz, (D)
Gideon Brunsdon, (M)
Anton de Wit, (D)
Pakama Syongwana, (D)
Nicolas Tonnelier, (D)
Leizel Williams-Bruinders, (M), Zn
Emeritus:
Peter Booth, (D)
Anthony Christopher
Related Staff:
Paul Baldwin
Willie Deysel

Centre for African Conservation Ecology (M,D) ○
Summerstand Campus (South)
PO Box 77000
Port Elizabeth 6031
p. +27 41 504 2308
Graham.Kerley@nmmu.ac.za
http://ace.nmmu.ac.za/

Nelson Mandela University

Coastal and Marine Research Institute (A,B,M,D) ☒
Summerstand Campus (South)
PO Box 77000
Port Elizabeth 6031
p. +27 41 504 2852
cmr@mandela.ac.za
https://cmr.mandela.ac.za/
f: Institute for Coastal and Marine Research
t: CMR_MandelaUni

North-West University

School of Environmental and Health Sciences (B)
Private Bag x 1290
Potchefstroom 2520
p. (27) 18 299 2528
monica.mosala@nwu.ac.za
http://www.puk.ac.za/opencms/export/PUK/html/fakulteite/natur/geol/index_e.html

Rhodes University

Dept of Geography (B)
PO Box 94
Grahamstwon 6140
p. +27 (0)46 603 8111
registrar@ru.ac.za
http://oldwww.ru.ac.za/academic/departments/geography/

Institute for Water Research (M,D)
PO Box 94
Grahamstown 6140
p. +27 46 62224014 / 6222428 / 62
registrar@ru.ac.za
http://www.ru.ac.za/static/institutes/iwr//?request=institutes/iwr/
Professor:
Denis Hughes, (D), Wales, 1978, Hqs

Dept of Geology (B,M,D)
PO Box 94
Grahamstown , Eastern Cape Provinc 6140
p. +27 (0)46 603 8309
geolsec@ru.ac.za
http://www.ru.ac.za/geology/
Administrative Assistant: Ashley Goddard
Administrative Assistant: Vuyokazi Nkayi
Head:
Steve Prevec, Cg
Professor:
Annette Gotz, Gs
Peter Horvath, (D), Gp
Bantubonke Izwe Ntsaluba, Gi
Briony Proctor
Hari Tsikos, CgEg
Yon Yao, Eg
Associate Professor:
Steffen Buttner, (D), Frankfurt (Germany), 1997, Gcp
Associate Scientist:
Billy De Klerk, Pg
Robert Gess, Pg
Rose Prevec
Mike Skinner
Emeritus:
Roger Jacobs
Julian S. Marsh, (D), Cape Town, 1973, GiCgGv
Laboratory Director:
Gelu Costin, Gp
Cooperating Faculty:
John Hepple

Stellenbosch University

Dept of Process Engineering (Chemical and Mineral Processing) (B,M,D)
Private Bag X1
Matieland 7602
p. +27 21 808-4485
chemeng@sun.ac.za
http://www.chemeng.sun.ac.za/

Centre for Geographical Analysis ☒
Private Bag X1
Matieland 7602
p. (27) 21 808 3218
garth@sun.ac.za
http://www0.sun.ac.za/cga/
f: https://www.facebook.com/cga.sun/
Certificates: Introduction to Geographic Information Systems (GIS), Introduction to Earth Observation (EO), Introduction to Geospatial Object-based Image Analysis (GEOBIA)
Director:
Adriaan van Niekerk, (D), Stellenbosch, Zri
Manager:
Garth Stephenson, (M), Stellenbosch, 2010, Zri

Dept of Earth Sciences (B,M,D)
Private Bag X1
Matieland, Western Cape 7602
p. +27 (0)21 808 3219
lcon@sun.ac.za
http://www.sun.ac.za/earthSci/
Professor:
Alakendra N. Roychoudhury, (D), Georgia Tech, 1999, ClmOc

Tshwane University of Technology

Dept of Environmental, Water & Earth Sciences (B,M,D)
Private Bag x680
Pretoria 0001
p. 012 382 6232
gerberme@tut.ac.za
http://www.tut.ac.za/Students/facultiesdepartments/science/departments/environscience/Pages/
Administrative Assistant: Sarah Galebies
Administrative Assistant: Retha Gerber
Enrollment (2015): B: 59 (37) M: 1 (0)
Lecturer:
Chamunorwa Kambewa, (M), 1998, CgeZe
Thando Majodina, (M), GgzGe
Skhumbuzo Sibeko, (B), GggGg
Other:
Mlindelwa Lupankwa, (D), HywGg

Dept of Environmental Health (B)
Private Bag X680
Staatsartillerie Road
Pretoria West 0001
p. +27 (0)12 382 5911
vanrooyenps@tut.ac.za
http://www.tut.ac.za/Pages/

University of Cape Town

Dept of Oceanography (B)
Private Bag X3
Rhondebosch 7701
p. +27 (0) 21 650-3277
isabelle.ansorge@uct.ac.za
http://www.sea.uct.ac.za

Dept of Environmental and Geographical Science (B)
Shell Environmental & Geographical Science Building
South Lane, Upper Campus
Private Bad X3
Rondebosch 7701
p. (27) 21 - 6502873 / 4
michael.meadows@uct.ac.za
http://www.egs.uct.ac.za/

Department of Geological Sciences (B,M,D) ● ☒
Private Bag X3
Rondebosch 7701, Western Cape
p. +27 (0)21-650-2931 
head.geologicalsciences@uct.ac.za
http://www.geology.uct.ac.za/
Programs: Geology, Geochemistry
Administrative Assistant: Nathalie Barends, Elretha Roos
Enrollment (2021): B: 20 (16) M: 18 (7) D: 16 (5)
Head:
Chris Harris, GiCsGg
Prof.:
Phillip Janney, (D), California (San Diego), 1996, GiCaXc
Dr. :
Petrus J. Le Roux, (D), Cape Town, 2000, CaGia
Ms.:
Christel Tinguely, (M), Clermont-Ferrand, 2000
Associate Professor:
Emese M. Bordy, (D), Rhodes, 2001, GsPes
Johann Diener, (D), Melbourne, 2008, Gpc
Ms:
Kerryn Gray, (M), Cape Town, 2004
Associate Scientist:
Nicholas Laidler, (B), Cape Town
Fayrooza Rawoot

Dr.:
Alastair Sloan, (D), Cambridge, 2012, GcYgu
Rosalie Tostevin, (D), Univ Coll (London), 2014, CmlCb
Lecturer:
Miengah Abrahams, (D), Cape Town, 2019, PgGsPv
Geoffrey H. Howarth, (D), Rhodes, 2014, EgGiv
Diego Quiros Ugalde, (D), Cornell U., 2015, YsgEo

University of Fort Hare

Dept of Geology
Sobukwe Walk 8
Livingstone Hall
Alice, Eastern Cape 5700
p. +27 (0)40 602 2011
maliber@ufh.ac.za
http://www.ufh.ac.za/departments/geology/
Head:
Oswald Gwavava
Lecturer:
CJ Gunter
Vuyokazi Mazomba
Related Staff:
Luzuko Sigabi

Dept of Geographic Information Systems (B)
Private Bag X1314
Alice 5700
p. +27 (0)40 602 2011
wkoll@ufh.ac.za
http://www.ufh.ac.za/departments/gis/gishome.html

Dept of Geography, Land Use, and Environmental Sciences (B)
Private Bag X1314
Alice 5700
p. +27 (0)40 602 2011
wkoll@ufh.ac.za
http://www.ufh.ac.za/

University of Johannesburg

Dept of Geology (B)
PO Box 524
Kingsway & University (APK Campus)
Auckland Park 2006
p. +27 (11) 559-4701
mdekock@uj.ac.za
http://www.uj.ac.za/?alias=www.uj.ac.za/geology

Dept of Geography, Environmental Management & Energy Studies (B,M,D)
PO Box 524
Auckland Park, Gauteng 2006
p. +27 (0)11 559 2433
science@uj.ac.za
http://www.uj.ac.za/?alias=www.uj.ac.za/geography
Dr:
Isaac T. Rampedi, (D), South Africa, 2010, PgSdHg

Dept of Mining Engineering and Mine Surveying (B) ☒
PO Box 17011
Doornfontein 2028
p. +27 (0)11 559-6628
hgrobler@uj.ac.za
http://www.uj.ac.za/?alias=www.uj.ac.za/mining
Programs: Mine Surveying; Mineral Resource Management; Mining Engineering
Certificates: Gyroscope Surveying

Dept of Mine Surveying (B) ● ☒
P.O. Box 17011
Doornfontein
Johannesburg, Gauteng 2028
p. +27 (0)11 559-6186
hgrobler@uj.ac.za
http://www.uj.ac.za/?alias=www.uj.ac.za/minesurv
Programs: Bachelor of Mine Surveying
Bachelor of Mine Surveying (Honours)
Masters of Sustainable Mining
Certificates: Gyroscopic surveying techniques

Enrollment (2019): B: 10 (10)
Associate Professor:
Hendrik Christoffe Grobler, (D), Witwatersrand, 2015, ZnNm

University of KwaZulu-Natal

School of Civil Engineering Surveying & Construction (B)
Centenary Building
King George V Avenue
Durban 4041
p. (+27) 2603065
troisc@ukzn.ac.za
http://engineering.ukzn.ac.za/Home.aspx

Centre for Water Resources Research (B,M,D) ☒
Room 203
Rabie Saunders Building
Pietermaritzburg, KwaZulu-Natal 3201
p. +27-(0)33-260 5490
smithers@ukzn.ac.za
http://cwrr.ukzn.ac.za/
Programs: Hydrology; Water Resources; Agricultural Engineering
Enrollment (2017): B: 55 (0) M: 30 (0) D: 14 (0)

School of Environmental Sciences (B,M,D) ☒
Discipline of Geography
King Edward Avenue
Pietermaritzburg Campus
Pietermaritzburg, KwaZulu-Natal 3209
p. + 27 (0)33 260 5346
gijsbertsen@ukzn.ac.za
http://saees.ukzn.ac.za/Homepage.aspx
f: https://www.facebook.com/FriendsOfUkznAgriculture

Discipline of Geological Sciences, School of Agricultural, Earth and Environmental Sciences (B,M,D) ●☒
Private Bag X 54001
Durban, KwaZulu-Natal 4000
p. +27 (0)31 260 2516
mccoshp@ukzn.ac.za
http://www.geology.ukzn.ac.za/
Programs: Environmental and Engineering Geology (B); Geology and Ore Deposits (B)
Enrollment (2018): B: 195 (51) M: 14 (1) D: 6 (2)
Professor:
Tesfaye K. Birke, (D), France, 1999, GcYmGg
Emmanuel John Carranza, (D), Tech (Delft), 2002, EdCeZr
Associate Professor:
Andrew Green, (D), KwaZulu-Natal, 2009, GuYgGs
Senior Lecturer:
Molla Demlie, (D), Ruhr (Germany), 2007, HwGeg
Jeremy Woodard, (D), Turku (Finland), 2010, CcGiz
Developmental Lecturer:
Nonkululeko Dladla, (M), KwaZulu-Natal, 2013, GusGr
Palesa Leuta-Madondo, (M), W Cape, 2010, CgGig
Lecturer:
Warwick William Hastie, (D), KwaZulu-Natal, 2013, NrgGc
Egerton Hingston, (D), Leeds (UK), 2008, NtxNg
Lauren Hoyer, (D), KwaZulu-Natal, 2016, EgGc
Philani Mavimbela, (M), Pretoria, 2013, GpzEg
Saumitra Misra, (D), Calcutta, 1993, GiCgu

University of Limpopo

Dept of Soil Sciences (B)
Turfloop Campus
Private Bag X1106
Sovenga 0727
p. +27 (0)15 268 9111
paulina.mabapa@ul.ac.za
http://www.ul.ac.za/index.php?Entity=agri_soil_scie

Geography & Environmental Studies Dept (B)
Turfloop Campus
Private Bag X1106
Sovenga 0727
p. (+27) (0)15 268 3756
salphy.ramokolo@ul.ac.za
http://www.ul.ac.za/index.php?Entity=agri_geo_environ

School of Physical and Mineral Science (B)
Turfloop Campus
Private Bag X1106
Sovenga 0727
p. +27(0) 15 268 3492
mirriam.letsoalo@ul.ac.za
https://www.ul.ac.za/index.php?Entity=phy_geo
Administrative Assistant: M. D.. Ramusi
Head:
K. E. Rammutia
Professor:
John Dunlevey, (D), Stellenbosch, 1985, GzxEg
M. Khanyi
M. A. Letsoalo
M. A. Mahladisa
R. M. Makwela
T, Mobakazi
T. E. Mosuang
J. M. T. Mphahlele
T. T. Netshisaulu
M. Netsianda
P. Ntoahee
O. O. Nubi
M. Phala
M. J. Ramusi
L. Wilsenach
Other:
J. P. T. Crafford, (N), ZnNgZn

University of Pretoria

Centre of Environmental Studies (M,D)
Pretora 0002
p. +27 (0)12 420-3111
willemferguson@zoology.up.ac.za
http://www.up.ac.za/centre-environmental-studies

University of Pretoria Natural Hazards Centre, Department of Geology (B,M,D) ⦸
Department of Geology
University of Pretoria, Private Bag X20, Hatfield,
PRETORIA, 0028
Pretoria, Gauteng 0028
p. +27 (0)12 420 3613
andrzej.kijko@up.ac.za
http://www.up.ac.za/university-of-pretoria-natural-hazard-centre-africa
Programs: The Centre liaise with various academic departments at the University of Pretoria to serve a primary or co-supervisor for Masters and Doctoral students in the respective departments.
Enrollment (2021): M: 0 (1) D: 1 (0)
Professor:
Andrzej Kijko, (D), AGH (Poland), 1979, YsgNe
Dr.:
Ansie Smit, (D), Mathematical Statistics, 2019, Rn

Dept of Mining Engineering (B,M,D)
Pretoria 0002
p. +27 0 12 420-3111
ssc@up.ac.za
https://www.up.ac.za/mining-engineering

Department of Geography, Geoinformatics and Meteorology (B,M,D) ☒
Room 1-3
Geography Building
Cnr Lynnwood and University Roads
Hatfield, Pretoria 0083
p. +27 (0)12 420 3536
ggm.support@up.ac.za
https://www.up.ac.za/ggm
Programs: Geography, Environmental Science, Environment and Society, Geoinformatics, Meteorology
Certificates: GIS, GIS professional practice, Introductory remote sensing, strategic environmental management, Ethics of geospatial inforamtion
Administrative Assistant: Lunga Ngcongo
Administrative Assistant: Martha C. van Aardt

SOUTH AFRICA

Enrollment (2020): B: 266 (77) M: 65 (10) D: 32 (5)
Professor:
Serena Coetzee, (D), Pretoria, 2009, Zi
Associate Professor:
Emma Archer, (D), Clark, At
Greg Breetzke, (D), Pretoria
Liesl Dyson, (D), Pretoria, Am
Willem A. Landman, (D), Witwatersrand, Ams
Abel Ramoele, Zr
Lecturer:
Adedayo Adeleke, (D), Cape Town, YdZf
Nerhene Davis, (D), W Cape
JJ Gregory
Christel D. Hansen, (D), Rhodes, 2018, ZiGmZy
Natalie S. Haussmann, (D), Stellenbosch, Gm
Samy Katumba, (M), Zf
Michael J. Loubser, (M), Pretoria, GmZy
Thando Ndarana, (D), Johns Hopkins, Ams
Victoria J. Rautenbach, (D), Pretoria, Zif
Philemon L. Tsele, (D), Pretoria, Zfr
Barend van der Merwe, (M), Pretoria, Gm
Francis Wasswa Nsubuga, (D), Pretoria
Support Staff:
Popi Mahlangu
Erika Pretorius, (M), Pretoria, Zi
Senior Cartographer:
Lourens Snyman, (M), Pretoria, Zi

Dept of Geology (B,M,D) ●☒
Faculty of Natural and Agricultural Sciences
Hatfield Campus
Lynnwood Road
Pretoria, Gauteng 0002
p. +27 (0)12 420 2454
lucia.moyo@up.ac.za
https://www.up.ac.za/geology
Programs: Geology, Engineering Geology, Hydro Geology, Structural Geology,Physical Volcanology & Sedimentology, Igneous & Metamorphic Petrology, Economic Geology, Geostatistics, Rock mechanics, Earth History & Mineralogy

University of the Free State

Department of Geology (B,M,D) ●☒
PO Box 339
Bloemfontein, Free State Province 9300
p. +27 (0)51 401 2515
markramj1@ufs.ac.za
https://www.ufs.ac.za/natagri/departments-and-divisions/geology-home
Programs: Geology; Environmental Geology; Geochemistry; Mineral Resource Management
Enrollment (2020): B: 140 (7) M: 6 (10) D: 8 (1)
Head:
Bisrat Yibas, (D), Witwatersrand, 2000, GegEg
Prof:
Frederick Roelofse, (D), Witwatersrand, 2010, Gip
Ms:
Jarlen J. Beukes, (M), Free State, Gi
Justine Magson, (M), Free State, 2016, CgGi
Thendo Mapholi, (B), Free State, Ggp
Makhadi Rinae, (B), Free State, Gge
Mr:
Justin Nel, (B), Free State, Gc
Adriaan Odendaal, (B), Free State, 2010, GgsGr
Dr:
Martin D. Clark, Austria, ZriGc
Robert N. Hansen, (D), Stellenbosch, 2014, CgGe
Matthew Huber, (D), Vienna, EgXm
Hendrik Minnaar, (D), Free State, Gc
Ms:
Megan Purchase, (M), Free State, Gx
Mr:
Pelele Lehloenya, (M), Free State, Cg

Institute for Groundwater Studies (B,M,D)
PO Box 339
Internal Bus 56
Bloemfontein 9300
p. +27(0)51-4019111
vermeulend@ufs.ac.za
http://www.ufs.ac.za/igs

Centre for Environmental Management (B)
Po Box 338
Bloemfontein 9300
p. +27(0)51-4019111
avenantmf@ufs.ac.za

http://www.ufs.ac.za/faculties/index.php?FCode=04&DCode=106

University of the Western Cape

International Ocean Institute of Southern Africa (B)
Private Bag X17
Bellville 7535
p. +27 21 959 3088
ioi-sa@uwc.ac.za
http://www.ioisa.org.za/

Dept of Earth Science (B)
Modderdam Road
Private Bag X17
Bellville, Cape Town 7530
p. +27 (0)21 959 2223
wdavids@uwc.ac.za
http://www.uwc.ac.za/Faculties/NS/EarthScience/Pages/
Administrative Assistant: Caroline Barnard, Wasielah Davids, Chantal Johannes, Mandy Naidoo
Head:
Charles Okujeni, (D), Berlin
Chair:
Jacqueline Goldin, (D), UCT
Yongxin Xu
Professor:
Dominic Mazvimavi, (D), Wageningen (Neth), 2003
Jan M. van Bever Donker, (D), Cape Town, 1979, GcpGt
Associate Professor:
Ebernard Braune
Lecturer:
Marcelene Andrews
James Ayuk Ayuk, (M), W Cape
Lewis Jonkey, (M)
Thokozani Kanyerere
Mimonitu Opuwari, (D), W Cape
Henok Solomon, (M), W Cape
VLIR Coordinator:
Shauib Dustay
HIVE Manager:
Yafah Hoosain, (M), W Cape
Deputy Head:
Theo Scheepers, (M), Stellenbosch
Related Staff:
Janine Becorney
Shamiel Davids
Peter Meyer

University of the Witwatersrand

School of Geography, Archaeology and Environmental Studies (B,M,D) ☒
Private Bag 32050
Wits, Johannesburg
p. +27 11 717 6503
donna.koch@wits.ac.za
http://web.wits.ac.za/Academic/Science/Geography/Home.htm
Programs: Geography

School of Mining Engineering (B,M,D) ●
Private Bag 3
WITS, Johannesburg 2050
p. +27 11-717-7003
Zeenath.Adam@wits.ac.za
http://web.wits.ac.za/Academic/EBE/MiningEng/

School of Geosciences (B,M,D) ●
Faculty of Science
Private Bag 3

Wits, Johannesburg 2050
p. 27 11 717 6547
gillian.drennan@wits.ac.za
http://web.wits.ac.za/Academic/Science/GeoSciences/Home.htm

Professor:
Lewis D. Ashwal, (D), Princeton, 1979, GxCgGt
Roger L. Gibson, (D), Cambridge, 1990, GcpGg
Kim AA Hein, (D), Tasmania, 1995, GgEmGc
Judith A. Kinnaird, (D), St Andrews, 1987, EmGi

Associate Professor:
Paul A.M. Nex, (D), Univ Coll (Cork), 1997, Eg

Dr:
Michael QW Jones, (D), Witwatersrand, 1981, YhGtYg

Electron Microprobe Scientist:
Peter Horvath, (D), Eotvos Lorand, 2002, GpCpGz

Dr:
Grant M. Bybee, (D), Witwatersrand, 2013, Gi
Katie A. Smart, (D), Alberta, 2011, CsGx
Zubair A. Jinnah, (D), Witwatersrand, 2011, GsPg

Emeritus:
Carl R. Anhaeusser, (D), Witwatersrand, 1983, GgEgZg

University of Venda

Dept of Geography and Geo-Information Sciences (B)
University of Venda
Private Bag X5050
Thohoyandou, Limpopo 0950
p. +27 15 962 8593
nthaduleni.nethengwe@univen.ac.za
http://www.univen.ac.za/enviornmental_sciences/dep_geography_geo_sciences.html

Head:
Nthaduleni S. Nethengwe, (D), West Virginia, RnZiRw

Acting Head:
T. M. Nelwamondo, (D), UP, 2010, GeZuy

Lecturer:
E. Kori, (M), Univen, GmZgAt
M. J. Mokgoebo, (M), Univen
N. V. Mudau, (D), Univen, 2015, Rm
A. Muyoki, (D), Howard
M. Nembudani, (M), Stellenbosch

Related Staff:
K. H. Mathivha, (M), Univen

Mining and Environmental Geology (A,B,M,D) ☒
School of Environmental Science
Private Bag X5050
Thohoyandou , Limpopo Province 0950
p. +27 159628580
info@univen.ac.za
http://www.univen.ac.za/environmental_sciences/dep_mining_environmental.html
Programs: Geology and Earth Sciences

Enrollment (2019): B: 288 (46) M: 26 (6) D: 6 (1)

Dr.:
Milton Kataka, (D), Witwatersrand, 2003

University of Venda for Science & Technology

GIS Resource Centre (B,M,D) ⊘
GIS Resource Centre
Private Bag x5050
Thohoyandou, Limpopo 0950
p. +27 15 962 8044
farai.dondofema@univen.ac.za
http://www.univen.ac.za/environmental_sciences/gis_centre.html
t: @chinomukutu3
Programs: GIS, Geography, Remote sensing, Geology, Mining, Hydrology, Environmental Science, Ecology, Urban and Rural planning
Certificates: GIS, Ecology
Enrollment (2018): B: 46 (35) M: 13 (9) D: 4 (1)

Chief Technician:
Farai Dondofema, (M), Zimbabwe, 2007, ZrHgZi

Institute of Semi-Arid Environment and Disaster Management (B)
p. (+27) 015 962 8513
ndidzum@univen.ac.za
http://www.univen.ac.za/index.php?Entity=Institute%20of%20Semi-Arid%20Environment&Sch=3

Dept of Hydrology and Water Resources (B)
University of Venda
Private Bag X5050
Thohoyandou, Limpopo 0950
p. +27 015 962 8513
environmental@univen.ac.za
http://www.univen.ac.za/environmental_sciences/dep_hydrology_water.html

Head:
J. O. Odiyo, (D)

Lecturer:
J. R. Gumbo, (D)
P. M. Kundu, (D)
R. Makungo, (M), Venda, 2010, HzqHw
Tinyiko Rivers Nkuna, (M), Venda, 2012, HgwAs

University of Zululand

Dept of Geography and Environmental Studies (B)
Private Bag X1001
KwaDlangezwa 3886
p. +27 (035) 902 6282
kamwendog@unizulu.ac.za
http://www.uzulu.ac.za/scie_geo_env.php

Dept of Hydrology (B,M,D) O☒
Private Bag X1001
KwaDlangezwa 3886
p. +27 (035) 902 6282
SimonisJ@unizulu.ac.za
http://www.uzulu.ac.za/scie_hydro.php
Programs: Hydrology, Geohydrology, Geology
Certificates: GIS
Enrollment (2016): B: 332 (79) M: 15 (2) D: 2 (0)

Prof:
Jean Simonis, (D), UNIZULU, 2009, HyGgHg

Spain

Coruna University

University Institute of Geology (D)
Edificio Servicios Centrales de Investigación
Campus de Elviña s/n
La Coruna 15071
p. 0034 981167000
xeoloxia@udc.es
http://www.udc.es/iux
Enrollment (2015): D: 8 (80)

Director:
Juan Ramon Vidal Romaní, (D), Complutense (Madrid), 1983, GmlGc

Professor:
Antonio Paz Gonzalez, (D), Santiago de Compostela, 1982, Sd

Paleontologist:
Aurora Grandal d'Anglade, (D), Coruña, 1993, Pv

Associate Professor:
Elena Pilar de Uña Alvarez, (D), Santiago de Compostela, 1986, Zy
Cruz Iglesias, (D), Corunna, 2001, GaZmNr
Jorge Sanjurjo, (D), Corunna, 2005, GeCcGa
María Teresa Taboada Castro, (D), Santiago de Compostela, 1990, Sd

Associate Scientist:
Marcos Vaqueiro Rodriguez, (M), Vigo (Spain), GcmYd

Institute of Marine Sciences (CSIC)

Marine Geosciences Departmen ☒
Passeig Maritim de la Barceloneta
37-49
Barcelona 08003
p. (+34) 63 230 95 00
secredir@icm.csic.es
https://www.icm.csic.es/en/department/marine-geosciences

SPAIN

Instituto Geologico y Minero de Espana

Instituto Geologico y Minero de Espana
Rios Rosas, 23
28003 Madrid
igme@igme.es
http://www.igme.es

Univ Complutense de Madrid

Facultad de Ciencias Geologicas (B,M,D) O⃠
C/ Jose Antonio Novais 12
Cuidad Universitaria
Madrid E-28040
p. 34 91394 4837
secre.adm@geo.ucm.es
https://geologicas.ucm.es
t: @geologicasUCM
Programs: Geology (Bachelor)
Engineering Geology (Bachelor)
Environmental Geology (Master)
Engineering Geology (Master)
Advanced Paleontology (Master)
Oil and Ore deposits exploration (Master)
Certificates: Paleontology (on line)
Mineral deposits (on line)
Enrollment (2021): B: 135 (80) M: 62 (54) D: 28 (14)

Vicedean for Research:
Javier Martín Chivelet, (D), PcGs

Professor:
Ana María Alonso, (D), Gd
Eumenio Ancochea, (D), Gv
Ricardo Arenas, (D), Gp
Juan Luis Arsuaga, (D), Zn
César Casquet, (D), Gip
Lourdes Fernández, (D), Gz
Sixto Fernández, (D), Pg
Sol López, (D), Gz
Sergio Rodríguez, (D), PgGg
Carlos Villaseca, (D), Gi

Vicedean for quality:
María de la Luz García Lorenzo, (D), Cl

Vicedean for Master programs and international affairs:
María Luisa Canales, (D), PmgGg

Vicedean for Bachelor programs :
Eugenia Arribas, (D), Gd

Vicedean for academic issues:
Alfonso Muñoz, (D), Ygr

Dean:
Lorena Ortega, (D), 1993, EmgGg

Associate Professor:
Jacobo Abati, (D), Gp
Saturnino Alba, (D), Gm
José Antonio Alvarez, (D), Gc
Pilar Andonaegui, (D), Gg
Carmen Arias, (D), Gg
José Manuel Astilleros, (D), Gz
María Isabel Benito, (D), Gs
Manuel Bustillo, (D), Eg
Pedro Castiñeiras, (D), Gp
María José Comas, (D), Pi
Elena Crespo, (D), EnGz
Cristina de Ignacio, (D), Cg
Raúl de la Horra, (D), Gs
Lucia de Stefano, (D), Hw
Gerardo de Vicente, (D), Gt
Javier Fernández Suárez, (D), Cc
José María Fernández-Barrenechea, (D), Gz
María Antonia Fregenal, (D), Gsg
Fernando Gacía Joral, (D), Pi
Nuria García, (D), Pv
Alejandra García Frank, (D), Gg
Emilia García Romero, (D), Gz
Laura González, (D), Gr
José Luis Granja, (D), Yg
Manuel Hernández, (D), Pv
Concepción Herrero, (D), Pm
María Josefa Herrero, (D), Gs
María José Huertas, (D), Gi
Juan Miguel Insua, (D), NtGc
María del Pilar Llanes, (D), Yr
Francisco Javier Luque, (D), Gz
José Francisco Martín Duque, (D), Gm
Gemma Martínez, (D), Pg
José Jesús Martínez, (D), GcRn
Pedro Martínez, (D), Hw
Nieves Meléndez, (D), Ggs
Esperanza Montero, (D), Hw
Belén Muñoz, (D), Gr
David Orejana, (D), Gig
Cecilia Pérez-Soba, (D), Gi
Rubén Piña, (D), EmGz
Marta Rodríguez, (D), Ggs
Martín Jesús Rodríguez, (D), Ng
Ignacio Romeo, (D), GcXg
Carlos Rossi, (D), GdEo
Javier Ruiz, (D), Xg
Nuria Sánchez, (D), Gz
Sonia Sánchez, (D), Gp
Yolanda Sánchez, (D), Gs
Juan Ignacio Santisteban, (D), Gr
Esther Sanz, (D), Gd
Miguel Angel Sanz, (D), Zi
Paloma Sevilla, (D), Pi
Meaza Tsige, (D), Ng
David Uribelarrea, (D), Ga
María Josefa Varas, (D), Gd
Cristóbal Viedma, (D), Gz

Assistant Professor:
Laura Domingo, (D), PveCs
Julio Garrote, (D), Gm

Adjunct Professor:
Enrique Aracil, (B), Yu
Sonia Bautista, (M), Ng
María Druet, (D), Yr
Alejandro Faúndez, (B), Ng
Luis Ramón Fernández, (M), Nt
David Jiménez, (B), GcNt
Cristina Martín, (D), Gm
Svetlana Melentijevic, (D), Nx
Luis Eugenio Suárez, (B), Zn

Emeritus:
José Arribas, (D), Gd

Universidad Autonoma de Madrid

Departamento de Geologia y Geoquimica (M,D) O☒
Faculdad de ciencias
Campus de Cantoblanco, C/ Francisco Tomás y Valiente, 7
Módulo 06, 6ª planta
Madrid 28049
p. 91 497 48 00
directora.geologia@uam.es
http://www.uam.es/GyG
Programs: Earth Science; Biology; Chemistry
Certificates: GIS
Enrollment (2017): M: 9 (0) D: 8 (0)

Universidad de Alicante

Environment and Earth Sciences Dept (B,M,D)
Facultad de Ciencia
San Vicente de Raspeig
Alicante E-03080
p. (+34) 96 590 3552
dctma@ua.es
http://dctma.ua.es/en/environment-and-earth-sciences-department.html

Universidad de Granada

Dept de Geologia
Andalusian Institute of Geophysics
Campus Universitario de la Cartuja
18071 Granada
jaguirre@ugr.es
http://www.ugr.es/iag/iagpds.html

Universidad de Huelva

Geologia (B,M,D)
Huelva
p. 959219809
secgeo@uhu.es
http://www.uhu.es/dgeo/

Universidad de Jaen

Departamento de Geologia (B,M,D) ○☒
Edificio B-3
Campus Universitario
Jaen 23071
p. +34-953-212295
jmmolina@ujaen.es
http://geologia.ujaen.es/

Universidad de Las Palmas de Gran Canaria

Dept de Fisica (B)
C/Juan de Quesada, nº 30
, Las Palmas de Gran C 35001
p. (+34) 928 451 000/023
universidad@ulpgc.es
http://www.dfis.ulpgc.es/

Facultad de Ciencias del Mar (B,M,D)
Edificio de Ciencias Básicas
Campus Universitario de Tafira
Las Palmas de Gran Canaria, Las Palmas 35017
p. +34 928 452 900
sec_dec_fcm@ulpgc.es
http://www.fcm.ulpgc.es

Universidad de Murcia

Dept of Geography and Regional Planning

Facultad de Sciencia
Calle Santo Cristo 1
Murcia 30100
p. (868) 88-7446
jagb@um.es

Universidad de Oviedo

Dept de Geologia
Campus de Llamaquique
Jesus Arias de Velasco, s/n
33005 Oviedo E-33005
geodir@geol.uniovi.es

http://www.geol.uniovi.es/

Universidad de Pais Vascu

Dept de Geologia
Facultad de Sciencia
37008 Salamanca
p. 946015491
sec-centro.fct@ehu.es

Universidad de Palma de Mallorca

Dept de Geologia/Geofisica
Facultad de Sciencia
07012 Palma

Universidad de Sevilla

Applied Geology to Civil Engineering (D)
Departamento de Cristalografía y Mineralogía
calle Profesor García González 1
Seville 410012
p. +34954556318
igonza@us.es
http://www.departamento.us.es/dcmqa/
Professor:
Isabel González, (D), GzeSc

Universidad de Valencia

Dept de Geologia (M,D)
Faculty of Biological Sciences
Building A (2nd and 3rd floor)
C/ Dr. Moliner, 50
Burjassot- (Valencia) 46100
p. (+34) 96 354 46 02
dep.geologia@uv.es
http://www.uv.es/geologia

Universidad de Valladolid

Dept of Geography
Facultad de Faculty of Philosophy and Arts
Palacio Santa Cruz/Plaza Santa Cruz B
Valladolid
p. +34 983 423 005
belart@fyl.uva.es

Universidad de Zaragoza

Facultad de Ciencias
50.009 Zaragoza
seccienz@unizar.es
http://wzar.unizar.es/acad/fac/geolo/adepo.html

Museo de Ciencias Naturales (D) ○☒
Museo de Ciencias Naturales
Universidad de Zaragoza
Edificio Paraninfo. Plaza Basilio Paraiso
Zaragoza, Aragón 50005
p. +34 976762096
museonat@unizar.es
http://museonat.unizar.es/
f: https://www.facebook.com/museopaleounizar
t: @museonat
Professor:
José Ignacio Canudo, (D), Pv

Universitat Autonoma de Barcelona

Geologia (B,M,D) ●☒
Facultat de Ciencies, Edifici C
Campus UAB
Bellaterra, Catalunya 08193
p. +34 93 581 3022
d.geologia@uab.es
http://departaments.uab.cat/geologia/
Programs: B.S. Geology, B.S. Environmental Science, MSc. Reservoir Geology and Geophysics, MSc. Paleobiology and Fossil Record, MSc. Resources and Risks, PhD Geology
Enrollment (2018): B: 284 (46) M: 47 (47) D: 14 (7)
Chair:
Antoni Teixell Cácharo, (D), 1992, GtcYx
Professor:
María Luísa Arboleya Cimadevilla, (D), Gc
Joan Bach Plaza, (D), Gg
Esteve Cardellach López, (D), GzeCs
David Gómez Gras, (D), Gd
Juan Francesc Piniella Febrer, (D), Gz
Eduard Remacha Grau, (D), Univ Aut, 1983, Gso
Antonio Teixell, (D), 1992, Gtc
Associate Professor:
Lluís Casas Duocastella, (D), GzaYm
Mercè Corbella Cordomí, (D), GzCl
Elena Druguet Tantiña, (D), Gc
Joan Estalrich López, (D), HgGe
Gúmer Galán garcía, (D), GiCg
Rogelio Linares, (D), NgYx
Oriol Oms Llobet, (D), GrsYm
Joan Reche Estrada, (D), GpCp
Enric Vicens Batet, (D), Pg
Assistant Professor:
Carme Boix, (D), Pm
Isaac Corral, (D), Gx
Victor Fondevilla, (D), PgGr
Marc Furió, (D), Pv
Ona Margalef, (D), Ge
Didac Navarro, (D), Gz
Joan Poch Serra, (B), Auto (Barcelona), 2003, GrsZe

Felix Sacristan, (M), Ge
Eduard Saura, (D), Gc
Bernat Vila, (D), Pv
Lecturer:
Albert Griera Artigas, (D), GcCl
Mario Zarroca Hernández, (D), YgNg
Emeritus:
Francisco Martínez Fernández, (D), GpCp

Universitat de Barcelona

Estratigrafia, Paleontologia i Geociencies marines
c/Marti Franques, s/n
08028
Barcelona
p. +33 934021384
secretaria-geologia@ub.edu
http://www.ub.es/dpep/1welcom3.htm

Dept de Geoquimica, Petrologia I Prospeccio Geologica
Marti Franques, s/n
Barcelona
8028
dp.mineralogia@ub.edu
http://www.ub.es/geoquimi

Facultat de Ciències de la Terra (Earth Sciences Faculty) (B,M,D) ●☒
Martí Franqués s/n
Barcelona, Catalonia 08028
p. +34 934 021 335
deganat-geologia@ub.edu
http://www.ub.edu/geologia/en/
f: https://ca-es.facebook.com/geologiaub/
t: https://twitter.com/geologiaub
Programs: Geology degree
Marine Sciences degree
xxxx Master degree
Certificates: Gemmology

Universitat de les Illes Balears

Dept de Ciencies de la Terra (B,M,D)
Carretera de Valldemossa, km 7.5
Palma, Balearic Islands 07122
p. +34 971172362
dct@uib.es
http://www.uib.es/depart/dctweb/home.htm

University of the Basque Country UPV/EHU

Departamento de Geodinamica (B,M,D) ○☒
Facultad de Ciencia y Tecnología
Barrio Sarriena, s/n
Leioa, Vizcaya 48940 Leioa
p. + 34. 94 601 2563
julia.cuevas@ehu.eus
http://www.geodinamica.ehu.es/s0001-home1/es/
Programs: Geology
Environmental Sciences
Enrollment (2017): B: 45 (0) D: 2 (0)
Professor:
Julia Cuevas, (D), 1988, GctGg
Professor:
Benito Abalos, (D), Basque (Spain), 1990, GctGp
Iñaki Antigüedad, (D), Hw
Luis Eguiluz, (D), 1988, Ggt
Jose M. Tubia, (D), 1985, GctYg
Senior Scientist:
Aitor Aranguren, (D), GcYm
Nestor Vegas, (D), Pais Vasco UPV/EHU, 2002, GctYm
Associate Professor:
Jose Julian Esteban, (D), GtCcYg
Tomas Morales, (D), NgHg
Pablo Puelles, (D), Gct
Jesus Angel Uriarte, (D), NgHg
Assistant Professor:
Arturo Apraiz, (D), Gpt
Vicente Iribar, (D), HwGm
Luis Miguel Martinez Torres, (D), Ggt
Fernando Sarrionandia, (D), Basque (Spain), 2006, GigGv
Emeritus:
Hilario Llanos, (D), Hgy

Univesidad de Malaga

Dept de Geography
El Ejido
29071 Malaga
rlarrubia@uma.es

Sri Lanka

Sabaragamuwa University of Sri Lanka

Dept of Natural Resources
P.O Box 02
Belihuloya , Sabaragamuwa Provinc
p. 0094-45-2280293
head_nr@sab.ac.lk
http://www.sab.ac.lk

Sri Lanka Geological Survey & Mines Bureau

Sri Lanka Geological Survey & Mines Bureau
Senanayake Building, No 4, Galle Road,
Dehiwala
info@gsmb.gov.lk
http://www.gsmb.gov.lk/

University of Moratuwa

Dept of Earth Resources Engineering
Katubedda, Moratuwa
Western Province
Moratuwa
p. 0094-11-2650353
shiromi@earth.mrt.ac.lk
http://www.ere.mrt.ac.lk/

University of Peradeniya

Dept of Geology
Central Province
Peradeniya 22000
p. 0094-81 2394 200/201
geology@pdn.ac.lk
http://www.pdn.ac.lk/sci/geology/

Uwa Wellassa University of Sri Lanka

Dept Mineral Resources & Technology
2nd Mile Post Passara Road
Badulla , Uwa province
p. 0094-55-2226400
info@uwu.ac.lk
http://www.uwu.ac.lk/

Sudan

Cairo University

Dept of Geology
Khartoum Branch
PO Box 1055
Khartoum

El-Neelain University

Dept of Geology (B)
Gamhuria Street
Khartoum
p. +249 (183) 77-441
salihsas@gmail.com
https://www.neelain.edu.sd/sites/colleges/14/dept/79

Gama'at El Khartoum

Dept of Geology
PO Box 321
Khartoum
dc.itna@uofk.edu
http://staffpages.uofk.edu/department-of-geology/

International University of Africa

Faculty of Minerals and Petroleum (B)
PO Box 2469
Khartoum
p. (+249) 183 223211
tassas.iua@gmail.com
https://metals.iua.edu.sd/

Sudan University of Science and Technology

Surveying Engineering Dept (B)
Southern and Northern Campus
Khartoum
p. +249183468622
ceng@sustech.edu
http://engineering.sustech.edu/index.php/College_of_Engineering/Surveying_Engineering_Department/en/sections

College of Water & Environmental Engineering (B)
p. +24985200512 / 985200511 / 918
m.ginaya@sustech.edu
http://www.sustech.edu/faculty_en/index.php?coll_no=13&chk=cdf8953eb5124fa6f08f046043be8bf

College of Petroleum Engineering and Technology (B)
Southern Campus
Khartoum
zeinabkhaleel@sustech.edu
http://www.sustech.edu/faculty_en/index.php?coll_no=23&chk=1d4f14cee52598202e0fc32a12b2dfc2

University of Dongola

Dept of Geology (B)
PO Box 47
Northern Province, Dongola
p. +249-241-821-516, +249-241-821
http://www.uofd.edu.sd/index.php/ar/

University of Gezira

Dept of Geology
Faculty of Science and Technology
PO Box 20
Wad Medani
p. (002) 49511825724
sncs@uofg.edu.sd
http://snc.uofg.edu.sd/EN/Department.aspx?dep=6

University of Juba

Dept of Geology and Mining (B)
PO Box 321/1
Khartoum Center, Juba
p. +249 (83) 222125
info@juba.edu.sd
http://www.juba.edu.sd/

Dept of Environmental Studies (B)
PO Box 321/1
Khartoum Center, Juba
p. +249 (83) 222125
info@juba.edu.sd
http://www.juba.edu.sd/

University of Khartoum

Dept of Geology (B,M,D)
PO Box 321
Khartoum 11115
geology@uofk.edu
http://www.uofk.edu/
Head:
Ibahim Abdu Mohamed
Professor:
Saad Eldin Hamad Moha Ali
Abdelhalim Hassan Elnadi, Goz
Associate Professor:
Abdelwahab Yousif Abbas
Salah Bashir Abdalla, Ng
Osman Mahmoud Abdelatif
Samia Abdelrahman
Badr Eldin Khal Agmed
Fath Elrahman ali Birair
Omar Elbadri ali Elmaki
Abdelhafiz Gad Almula
Mohamed Zaid Awad
Insaf Sanhoory Babiker
Almed Sulaiman Daood
Abdalla Guma Farwa
Instructor:
Eltayeb Elasha Abdalia
Sami Osman Ibrahim
Amro Shaikh Idris Ahmed
Waleed Elmahdi Siddig
Saif Eldin Sir Elkhatim
Lecturer:
Amany Ali Badi
Related Staff:
Walaa Elnasir Ibrahim

Marine Research Laboratory in Suakin (B)
PO Box 321
Khartoum 11115
sbabdalla@hotmail.com
https://www.fkm.utm.my/marine/mtl/?Album:Visitor

University of Kordofan

Dept of Soil and Water Sciences (B)
p. 860008-611-00249
hadiaabdelatif@kordofan.edu.sd
http://science.kordofan.edu.sd/

University of Neelain

Dept of Geology (B)
science_dean@neelain.edu.sd
http://www.neelain.edu.sd/

University of the Red Sea

Dept of Geology (B)
PO Box 24
Red Sea, Port Sudan
p. +249-311-219-28
redseauniv44@hotmail.com

Wad El Magbout

Training Centre for Earth Science Technicians (B)

Suriname

Anton de Kom University of Suriname

Dept of Geology & Mining
Leysweg, P.O. Box 9212, Universiteits Complex, Gebouw VI

Environmental Sciences Dept (B) O
Universiteitscomplex Leysweg
P.O.B. 9212
Gebouw 17
Paramaribo
p. 597465558 (Ext. 308)
s.carilho@uvs.edu
http://adekus.uvs.edu/section/sectSection.php?secID=5304&lb=Onderwijs%20.%20Technologie%20.%20Milieu
f: https://www.facebook.com/AdeKUS.official/

Swaziland

University of Swaziland

Dept of Geography, Environmental Science and Planning (B,M,D)
Private Bag 4
Kwaluseni M201
p. (+268) 2517-0000
Saicos@Uniswa.Sz
http://www.uniswa.sz/academics/science/gep

Sweden

Chalmers University of Technology

Dept of Geology
S-412 96 Goteborg
Gothenburg
sofie.hallden@chalmers.se
http://geo.chalmers.se

Karlstad University

Department of Risk and Environmental Studies (B,M,D) ●
Karlstad University
Karlstad 65188
p. +46547001000
Magnus.Johansson@kau.se
https://www.kau.se/en/risk-and-environmental-studies
Programs: Bachelor: Risk and environmental studies
Master: Risk management in Society

Lund University

Dept of Geology (B,M,D) ● ☒
Solvegatan 12
223 62 Lund
p. (+46) 462221424
anders.schersten@geol.lu.se
https://www.geology.lu.se/
Programs: Geology with specialities in Quaternary geology, Palaeontology and Mineralogy & petrology

Stockholm University

Dept of Geological Sciences (B,M,D) ●
Svante Arrheniusväg 8
106 91 Stockholm
volker.bruchert@geo.su.se
http://www.geo.su.se

Sveriges geologiska Undersokning

Geological Survey of Sweden
Box 670
751 28 Uppsala
sgu@sgu.se
http://www.sgu.se/

Umea Universitet

Dept of Ecology and Environmental Science
Natural Sciences, Johan Bures Road 14, Umeå
S-901 87 Umea
p. +46 90 786 50 00
jolina.orrell@emg.umu.se
http://www.emg.umu.se

University of Gothenberg

Master of Earth Science
Box 100
S-405 30 Gothenburg
p. +46 31-786 0000

University of Stockholm

Dept of Geology and Geochemistry
106 91 Stockholm
volker.bruchert@geo.su.se
http://www.geo.su.se/

Uppsala Universitet

Dept of Earth Sciences (B,M,D) ●
Villavagen 16
UPPSALA SE-752 36
p. +46-18-4710000
PREFEKT@geo.uu.se
http://www.geo.uu.se/?pageid=1&lan=1

Switzerland

ETH Hoenggerberg

Geophysical Inst ETH
Sonneggstrasse 5
8092 Zurich
p. +41 44 633 26 05
christoph.baerlocher@erdw.ethz.ch
http://www.geophysics.ethz.ch/

ETH Zurich

Department of Earth Sciences (B,M,D) ☒
Sonneggstrasse 5
CH-8092 Zurich
department@erdw.ethz.ch
https://erdw.ethz.ch/en/
Programs: Earth Sciences, Climate Sciences, Applied Geophysics

Federal Office for the Environment (FOEN)

Federal Office for the Environment (FOEN)
info@bafu.admin.ch
Bern 3003
p. 0041313229311
info@bafu.admin.ch
http://www.bafu.admin.ch/
t: @bafuCH

Université de Lausanne

Dept of Geosciences and Environment (B,M,D) ☒
Décanat FGSE
Quartier Mouline
Géopolis
Lausanne, Vaud 1015
p. 0041216923500
doyen.gse@unil.ch
http://www.unil.ch/gse
Programs: Master of Science in Geology, Master of Science in Environmental Geosciences, Master of Science in Geography, Master of Science in Biogeosciences, Master in Tourism Studies, Master in foundations and practices of sustainability

University of Basel

Geol & Palaeontological Inst
Bernouillistrasse 32
CH-4056 Basale
sonja.zuercher@unibas.ch
http://www.unibas.ch/earth/GPI/paleo

Inst of Earth Sciences
Bernoullistrasse 30
Basel 4056
eberhard.parlow@unibas.ch
http://therion.minpet.unibas.ch/minpet

Dept of Earth Sciences
Bernoullistr.32
CH-4056 Basel
Joelle.Glanzmann@unibas.ch
http://duw.unibas.ch/

University of Fribourg

Div of Earth Sciences
Chemin du Musee 6
Perolles
CH-1700 Fribourg
nicole.bruegger@unifr.ch
http://www.unifr.ch/geology/

University of Geneva

Section of Earth Sciences and Environment
13, rue des Maraichers
Geneva CH-1205
p. 0041223796628
elisabeth.lagut@unige.ch

http://www.unige.ch/sciences/terre/

University of Lausanne

Faculty of Geosciences and Environment (B,M,D) ●⃠
Geopolis
Lausanne CH-1015
p. +41 21 692 35 00
doyen.gse@unil.ch
http://www.unil.ch/gse/home.html
Programs: geography, environmental sciences, geology, biogeosciences, tourism studies, foundations and practices of sustainability
Enrollment (2021): B: 4 (90) M: 7 (45) D: 0 (21)

University of Neuchatel

Centre for Hydrogeology and Geothermics (B,M,D)
Rue Emile-Argand 11
CH-2000 Neuchatel
secretariat.chyn@unine.ch
http://www.unine.ch/chyn

Universität Bern

Institut für Geologie (B,M,D)
Baltzerstrasse 1+3
Bern CH-3012
p. +41 (0) 31 631 87 61
info@geo.unibe.ch
http://www.geo.unibe.ch

Syria

Damascus University

Dept of Geology
Jammah Dimasq
Damscus

Taiwan

Academia Sinica

Institute of Earth Sciences (D) ☒
128, Section 2, Academia Road, Nangang
Taipei 11529
p. 886-2-2783-9910
jhwang@earth.sinica.edu.tw
http://www.earth.sinica.edu.tw/index_e.php
Distinguished Researcg Fellow:
Jeen-Hwa Wang, (D), SUNY, 1982, Ysn
Associate Research Fellow:
Wu-Cheng Chi, (D), California (Berkeley), GtYrs
Assistant Research Fellow:
Wen-che Yu, (D), SUNY (Stony Brook), 2007, Ys

Central Geological Survey (MOEA) of Taiwan

Central Geological Survey (MOEA) of Taiwan
District No. 2, Lane 109
Taipei 235
cgs@moeacgs.gov.tw
http://www.moeacgs.gov.tw/main.jsp

Chinese Culture University

Dept of Geology
Hanaoka Yangmingshan
Taipei 1111455
p. (02) 2861-0511 rpm 26105
crssge@staff.pccu.edu.tw

National Cheng-Kung University

Dept Earth Sciences
Tainan
wong56@mail.ncku.edu.tw

National Chung Cheng University

Inst of Seismology
168 University Rd
Min-Hsiung Chia-Yi
p. (886) -5--2720 x 9 ext: 61201?61209)
seismo@ccu.edu.tw
http://www.eq.ccu.edu.tw

National Taiwan University

Inst of Geology
245 Choushan Road, Taipei 106-17

Tanzania

ARDHI University

School of Geospatial Sciences and Technolgy (B,M,D)
P.O.Box 35176
Dar es Salaam
aru@aru.ac.tz
http://www.aru.ac.tz/page.php?id=63
Enrollment (2012): B: 120 (112) M: 10 (2) D: 6 (0)

Geological Survey of Tanzania

Geological Survey of Tanzania
P .O Box 903
Dodoma
madini.do@gst.go.tz
http://www.gst.go.tz/

University of Dar es Salaam

Dept of Geology (B,M,D) ●
PO Box 35052
Dar es Salaam
p. +255 22 2410013
geology@udsm.ac.tz
http://www.conas.udsm.ac.tz/geology/
Enrollment (2015): B: 65 (53) M: 17 (0)
Head:
Nelson Boniface, (D), Kiel, GcpGz
Professor:
Makenya A. Maboko, (D), Australian Nat, 1991, GpCgGx
Senior Lecturer:
Charles Z. Kaaya, (D), Cologne, 1993, GrsGu
Isaac Muneji Marobhe, (D), Helsinki Tech, 1990, YegYx
Associate Professor:
Shukrani Manya, (D), Dar Es Salaam, 2008, CgGzg
Hudson H. Nkotagu, (D), Tech (Berlin), 1994, HwgGe
Lecturer:
Kasanzu Charles, (D), 2014, GgRh
Emmanuel O. Kazimoto, (D), Kiel, 2014, EgGpCg
Elisante E. Mshiu, (D), Martin Luther (Germany), 2014, ZrCgZi
Elisante E. Mshiu, (D), Martin Luther (Germany), 2014, GgCeZr
Gabriel D. Mulibo, (D), Penn State, 2013, Ysg
Ferdinand W. Richard, (D), Uppsala, 1999, YsGt

Institute of Marine Sciences (B)
Mizingani Road
PO Box 35091
Zanzibar
p. 255-24-2232128/2230741
nyandwi@ims.udsm.ac.tz
http://www.ims.udsm.ac.tz

University of Dodoma

Dept of Geology and Petroleum Studies
P.O Box 259
Dodoma
p. +255 26 2310000
vc@udom.ac.tz
http://www.udom.ac.tz/

School of Mines and Petroleum Engineering
PO Box 259
Dodoma
p. +255 22 2410013
vc@udom.ac.tz
http://www.udom.ac.tz

Thailand

Asian Institute of Technology

Dept of Geology
P.O. Box 4
Klong Luang
Pathumthani 12120
Bangkok 10501
p. +66 (0) 2524 6057
supamas@ait.ac.th
http://www.set.ait.ac.th/page.php?fol=gte&page=gte

Dept of Mineral Resources

75/10 Rama 6 Road, Phayathai
Bangkok 10400
saraban@dmr.mail.go.th
http://www.dmr.go.th/

Togo

Universite de Lome

Dept de Geologie (B)
BP 1515
Lomé
p. (+228) 22-25-50-93

Tunisia

Birzeit University

Dept of Geology
Faculty of Science
P. O. Box 14
7021 Jarzouna Tunis
Bizerte
p. 216590717
fsb@fsb.rnu.tn
http://www.fsb.rnu.tn/index_fr/contact.html

Universite de Carthage

Dept de Geologie (B)
Jarzouna 7021
p. +21671841353
fsb@fsb.rnu.tn
http://www.fsb.rnu.tn/fsbindex.htm

Universite de Gabes

Dept de Sciences de la Terre (B)
Cite Riadh
Zerig, Gabes 6072
p. +216 75 394 800
mail@fsg.rnu.tn
http://www/fsg.rnu.tn/PRESENTATION.htm

Inst Superieur des Sciences et Techniques des Eaux de Gabes (B)
Cite Riadh
Zerig, Gabes 6072
p. +216 75 394 800
samir.kamal@isstegb.rnu.tn
http://www.isstegb.rnu.tn/francais

Universite de Sfax

Dept de Sciences de la Terre (B)
Route de l Aeroport km 0.5
, Sfax 3029
p. 216 74 276 400
contact@fss.rnu.tn
http://www.fss.rnu.tn

Dept de Genie Materiaux (B)
B.P:w.3038
, Sfax
p. (216) 74 274 088
enis@enis.rnu.tn
http://www.enis.rnu.tn/content/enis00003.htm

Dept de Genie Georessources et Environnement (B)
B.P:w.3038
, Sfax
p. (216) 74 274 088
enis@enis.rnu.tn
http://www.enis.rnu.tn/content/enis00003.htm

Universite de Tunis

Dept of Geology
Faculty of Science
1 Rue de Beja
Tunis 2092
p. 21671872600
http://www.fst.rnu.tn/fr

Turkey

Canakkale Onsekiz Mart University

Jeoloji Bolumu
Terzioglu Campus
Canakkale 17020
sztutkun@comu.edu.tr
http://jeoloji.comu.edu.tr/

Cukurova Universitesi

Jeoloji Muhendisligi Bolumu (B,M,D) ●
Faculty of Engineering and Architecture, Department of Geological Engineering
01330 Adana
Balcali
p. (+ 90 322) 338 67 15
parlak@cu.edu.tr
http://jeoloji.cu.edu.tr
Professor:
Osman Parlak, (D), Geneva, 1996, GipCc

Eskisehir Universitesi

Maden Muhendisligi Bolumu
Muhendislik Fakultesi
Yunus Emre Kampusu
Eskisehir

Firat University

Dept of Geological Engineering (A,B,M,D) ○☒
Fırat University, Department of Geological Engineering
23119 Elazıg Turkey
Elazig 23100
p. 424-2370000-5979
asasmaz@firat.edu.tr
http://jeo.muh.firat.edu.tr/tr/node/104
f: https://www.facebook.com/groups/firatjeoloji/
Programs: Geological Engineering
Enrollment (2018): A: 1 (1) B: 10 (10) M: 4 (4) D: 2 (2)
Chair:
Ahmet Sasmaz, (D), Firat, 2006, EgGeCe
Professor:
Ercan Aksoy, (A), Ggt
Melahat Beyarslan, (D), GipGx
Ahmet Feyzi Bingol, (A), GipGv
Bahattin Cetindag, (A), NgYv
Zulfu Gurocak, (A), Nrg
Leyla Kalender, (A), CgcCs
Ahmet Sagiroglu, (A), Egm
Associate Professor:
Bunyamin Akgul, (D), GzyGi
Dicle Bal Akkoca, (D), EnCo
Hasan Celik, (A), GscEo
Calibe Koc Tasgin, (A), GsdGg
Sevcan Kurum, (A), GivGx
Assistant Professor:
Ayse Didem Kilic, (D), GxpGy
Esra Ozel Yildirim, (D), Gxi
Ozlem Oztekin Okan, (D), HgwHs
Melek Ural, (D), GivCc
Research Assistant:
Mehmet Kokum, (D), Firat, 2017, GtcGg

Research Assistant:
Hatice Kara, (M), Cg
Nevin Ozturk, (D), Firat, 2017, Cg
Research Asistant:
Elif Akgun, (M), GgcGt
Onur Alkac, GgsGu
Gizem Arslan, (M), GxiGp
Yasemin Aslan, Nr
Serap Colak Erol, (D), 2014, GgcGt
Mehmet Ali Erturk, (D), 2016, GiCc
Mustafa Kanik, (D), 2015, NrSo
Sibel Kaygili, (D), 2016, GgPgs
Mahmut Palutoglu, (D), YgsGc
Mustafa Eren Rizeli, (M), Gxi
Abdullah Sar, (M), GviGp
Ismail Yildirim, ScGix

Hacettepe University

Geological Engineering Dept
Muhendislik Fakultesi
Beytepe Kampusu
tunay@hacettepe.edu.tr
http://www.jeo.hun.edu.tr/800x600.htm
Assistant Professor:
Turker Kurttas, (D), Hacettepe, 1997, HwgZr

Hydrogeological Engineering Programme (B,M,D) ●
Beytepe Campus
Ankara 06800
p. +90 (312) 297730
ekmekci@hacettepe.edu.tr
http://www.hacettepe.edu.tr/
Enrollment (2015): B: 235 (42) M: 12 (0) D: 11 (3)
Professor:
Serdar Bayari, (D), Hacettepe, 1991, Hw
Mehmet Ekmekci, (D), HwyCs
Sakir Simsek, (D), Istanbul, 1982, Hw
Galip Yüce, (D), Hw
Nur Naciye Özyurt, (D), Hacettepe, Hw
Assistant Professor:
Levent Tezcan, (D), Hacettepe, 1993, Hw

Istanbul Universitesi

Jeoloji Bolumu
Muhendislik Fakultesi
Beyazit
p. 0 (212) 473 70 70 ex. 17600
Muhendislik.dekan@iuc.edu.tr
http://muhendislik.istanbul.edu.tr/jeoloji/?p=6909

ITU

Jeoloji Bolumu
Maden Fakultesi
80394, Macka
Istanbul

Middle East Technical University

Inst of Marine Science
Div. of Marine Geology & Geophysics
P.O. Box 28
33731 Erdemli-Mersin
p. +90-324 521 3434
admin@ims.metu.edu.tr
http://www.ims.metu.edu.tr/

Dept of Geological Engineering (B,M,D)
Inonu Bulvari
ODTU
Ankara TR-06531
p. +90-312-2102682
gesin@metu.edu.tr
http://www.geoe.metu.edu.tr/

Mineral Research and Exploration General Directorate of Turkey

p. +90 312 201 11 51
mta@mta.gov.tr
http://www.mta.gov.tr/

Selcuk Universitesi

Faculty of Sciences
Muhendislik Fakultesi
42040 Konya
p. (+)903322412484
fen@selcuk.edu.tr

Uganda

Gulu University

Faculty of Agriculture and Environment (B,M,D)
P.O. Box 166
Gulu
p. +256782673491
d.ongeng@gu.ac.ug
http://www.gu.ac.ug

Dept of Geography (B) ●
P.O. Box 166
Gulu
p. +256772517488
charles.okumu52@gmail.com
http://www.gu.ac.ug/index.php?option=com_content&view=category&layout=blog&id=48&Itemid=66
Lecturer:
Expedito Nuwategeka, (D), Gulu, 2015, ZyyGv

Kabale University

Dept of Environmental Science and Natural Resources (B)
PO Box 317
Kabale
p. 256-4864-22803
akampabf@gmail.com
http://www.kabaleuniversity.ac.ag/

Makerere University

Dept of Geology and Petroleum Studies
P. O. Box 7062
Kampala 041
p. +256 41 532631-4
http://mak.ac.ug/
Instructor:
Kevin Aanyu, (M), Mak
Wycliff Kawule, (M), ITC
Robert Mamgbi, (M), CT-Prague
Lecturer:
Erasmus Barifajio, (D), Mak
John Mary Kiberu, (D), Tubingen
Agnes Alaba Kuterama, (M), ITC
Andrew Muwanga, (D), Braunschweig (Germany)
Immaculate Nakimera Ssemanda, (D), Mak

Dept of Geography, Geoinformatics & Climatic Sciences (B)
PO Box 7062
Kampala
p. +256-41-53126 1
fmugagga@gmail.com
http://www.geography.mak.ac.ug/

Institute of Environment and Natural Resources (B)
PO Box 7062
Kampala 041
p. 256 414 530134
http://muienr.mak.ac.ug/

Ndejje University

Faculty of Forest Science & Environmental Management (B)
PO Box 7088
Kampala, Ndejje Hill
p. +256-0392-730326
forest@ndejjeuniversity.ac.ug
http://www.ndejjeuinversity.ac.ug/academics.htm

Ukraine

National Mining University (in Ukraine)

State Mining University of Ukraine
19 Karl Marx Avenue
Dnepropetrovsk
320600
ues@uesukraine.com
http://www.apex.dp.ua/english/uageo/ukraine.html

United Arab Emirates

United Arab Emirates University

Dept of Geology (B,M,D) ●
P.O. Box 15551
College of Science, UAE University
Jamma Street
Al-Ain, Abu Dhabi 9713
p. +971-3-7136380
Ahmed.murad@uaeu.ac.ae

United Kingdom

Aberystwyth University

Dept of Geography & Earth Sciences (B,M,D) ☒
Llandinam Building
Penglais
Aberystwyth SY23 3DB
p. 01970 622606
gesstaff@aber.ac.uk
http://www.aber.ac.uk/en/iges

Professor:
Ron Fuge, (D), Wales, GbCgGe

Emeritus:
Michael M J Hambrey, (D), Manchester (UK), 1974, GlsGm

Anglia Ruskin University

Dept of Life Sciences
Cambridge Campus
East Road
Cambridge CB1 1PT
p. +44 845 271 3333
michael.cole@anglia.ac.uk
http://www.anglia-ruskin.ac.uk/ruskin/en/home/faculties/fst/departments/lifesciences.html

Bangor University

School of Ocean Sciences (B,M,D) ☒
Menai Bridge,
Isle of Anglesey
LL59 5AB
p. +44-1248-382854
oss082@bangor.ac.uk
http://www.sos.bangor.ac.uk/
Programs: Geological Oceanography (BSc and MSci)
Ocean and Geophysics (BSc)
Physical Geography and Oceanography (BSc)
Marine Environmental Science (BSc)
Ocean Science (BSc)
Physical Oceanography (MSci and MSc)
Applied Marine Geoscience (MSc)
Certificates: N/A

Senior Lecturer:
Dei Huws, (D), Wales, 1992, YruNt

Birkbeck College

Dept of Earth and Planetary Sciences
Department of Earth and Planetary Sciences
University of London - Birkbeck
Malet Street
Bloomsbury, London WC1E 7HX
p. +44 (0)20 7631 6665
p.gaunt@bbk.ac.uk
http://www.bbk.ac.uk/geology/
Administrative Assistant: Peter Gaunt

Head:
Gerald Roberts, (D), Durham (UK), 1991, Gct

Professor:
Charlie Bristow, (D), Leeds (UK), 1987, Gs
Andy Carter, Zg
Ian Crawford, Zn
Hilary Downes, Cg

Instructor:
Karen Hudson-Edwards, Gez

Lecturer:
Andy Beard
Simon Drake, Gv
Dominic Fortes, Zg
Steve Hirons, Cl
Philip Hopley, Pe
Phillip Pogge von Strandmann, (D), Csl
Vincent C. H. Tong, Yg
Charlie Underwood, Pg

British Antarctic Survey

British Antarctic Survey ☒
High Cross, Madingley Road
Cambridge CB3 0ET
p. +44 (0)1223 221400
trr@bas.ac.uk
https://www.bas.ac.uk/team/science-teams/geosciences/
t: @BAS_News

British Geological Survey

British Geological Survey
Environmental Science Centre
Nicker Hill
Keyworth
Nottingham NG12 5GG
p. 0115 936 3100
enquiries@bgs.ac.uk
http://www.bgs.ac.uk/

Cardiff University

School of Earth and Ocean Sciences (B,M,D)
Main Building
Park Place
Cardiff, Wales CF10 3YE
p. +44 (0)29 2087 4830
earth-ug@cf.ac.uk
http://www.cardiff.ac.uk/earth/

Head:
Ian R. Hall, (D), GuCsGs
R J. Parkes

Professor:
J A. Cartwright, (D), Ys
Dianne Edwards, (D), Pb
Dianne Edwards, (D), Pb
I Hall, Zn
A Harris
C Harris, Ge
A C. Kerr, (D), Durham (UK), 1994, Gxi
Bernard Elgey Leake, (D), Bristol, 1974, GipGz
R J. Lisle, Gc
C MacLeod, Gu
J A. Pearce, Cg
P N. Pearson, Pe
D Rickard, Cg
V P. Wright, Gs

Associate Professor:
Tiago Alves, (D), Manchester (UK), 2002, GouGt

Commander:
N Rodgers, (D), Am

Lecturer:
Rhoda Ballinger, (D), Zn
Stephen Barker, Pe
C M. Berry, (D), Pb
P J. Brabham, (D), Ye
L C. Cherns, Pg
Jose Constantine, (D), Hy
J H. Davies, Yg
I Fryett, Zn
TC Hales, (D), Gm
A Hemsley, Pl

T Jones, Ge
C Lear, Ou
Johan Lissenberg
Sergio Lourenco, (D), Sp
R Perkins, Zn
J Pike, Zn
H M. Prichard, Gx
H Sass
H D. Smith, Zn
S J. Wakefield, Cm
C F. Wooldridge, Zn
Y Yang, Hg

Durham University

Dept of Earth Sciences (B,M,D) ☒
Science Laboratories
South Road
Durham DH1 3LE
p. +44 0191 3342300
earth.sciences@durham.ac.uk
http://www.dur.ac.uk/earthsciences/
f: https://www.facebook.com/DurUniEarthSci/
t: @DurUniEarthSci
Programs: Geology
Geophysics with Geology
Geoscience
Environmental Geoscience
Earth Science

Professor:
Mark Allen, Gt
Andrew Aplin, Go
Kevin Burton, Cg
Jon Gluyas, Eo
Chris Greenwell, (D), CqEoCg
David Harper, Pg
Richard Hobbs, Ys
Robert E. Holdsworth, (D), Leeds (UK), Gc
Colin G. Macpherson, (D), London (UK), 1994, CgGiCs
Simon Mathias, Hy
Kenneth J. McCaffrey, (D), Durham (UK), Gc
Stefan Nielsen
Yaoling Niu, (D), Hawaii, 1992, Git
Christine Peirce, (D), Cambridge, Yg
Dave Selby, Cc
Peter Talling, Rn
Fred Worrall, (D), Cambridge, Cg

Associate Professor:
James Baldini, Pc
Richard James Brown, Gv
Nicola De Paola, Gc
Darren Grocke, Cs
Claire Horwell, Rn
Stuart Jones, Gs
Ed Llewellin, Gv
Julie Prytulak, Cg
Jeroen van Hunen, (D), 2001, Gt

Assistant Professor:
Martin Smith, Pe
Fabian Wadsworth, Gv
Richard Walters, Yd

Edinburgh University

School of Geosciences (B,M,D) ●⃠
Kings Buildings
West Mains Road
Edinburgh EH9 3JW
p. +44 (0) 131 651 7068
Susan.Orr@ed.ac.uk
http://www.geos.ed.ac.uk/
f: https://www.facebook.com/geosciences/
t: @geosciencesed
Programs: Geology; Geology and Physical Geography; Geophysics; Environmental GeoSciences; Geoenergy; GIS; Earth Observation; Geoinformation Management
Enrollment (2019): B: 1000 (1000) M: 450 (450) D: 250 (250)

Director:
Bruce M. Gittings, (M), Edinburgh, Zi

Head:
Alexander W. Tudhope, (D), Cs

Chair:
Chris Dibben, Zn
Mark D. A. Rounsevell, (D), Zu
Charles W. J. Withers, (D)

Professor:
Rob Bingham, (D), Zc
Tim Cresswell, (D), Zn
Andrew Curtis, Ys
Andrew J. Dugmore, (D), ZycGa
J. Godfrey Fitton, Gi
Simon L. Harley, Gt
R. Stuart Haszeldine, (D), Strathclyde, GoRmZo
Gabriele C. Hegerl, (D), As
Simon Kelley, (D), London (UK), 1985, CcsZg
Dick Kroon, Gg
Hayden Lorimer, (D), Zn
Ian G. Main, (D), Ys
John B. Moncrieff, As
Peter W. Neinow, (D), Cantab, Gl
Paul Palmer, As
Jamie R. Pearce, (D), Ge
Alastair H. F. Robertson, (D), Leicester (UK), 1975, GtgGs
Hugh D. Sinclair, (D), So
David Stevenson, (D), As
Simon F. B. Tett, Am
Kathryn A. Whaler, (D), Cambridge, 1981, YmRn
Mathew Williams, (D), E Anglia (UK), 1994, ZcrZu
Wyn Williams, Ym
Iain H. Woodhouse, (D), Zi
Rachel A. Woods, (D), Open, Gs
Anton M. Ziolkowski, (D), Cambridge, 1971, YesYg

Senior Scientist:
Nicola Cayzer, (D), Edinburgh, Yg
Richard Hinton, Cg

Associate Professor:
Stuart M. V. Gilfillan, (D), Manchester (UK), 2006, CgsCe

Assistant Professor:
Massimo A. Bollasina, (D), Maryland, 2010, Asp
Steve Brusatte, (D), Pv

Research Associate:
Andrew Bell, Gt
Ian B. Butler, (D), Cg
John Craven, Yg
Jan C. de Hoog, (D), Utrecht, 2001, GxzGv
Chris L. Hayward, Ga
Sian F. Henley, (D), Edinburgh, 2013, CmOcCs
Mark Naylor, (D), Edinburgh, Gt
Anthony J. Newton, (D), Edinburgh, Zn
Laetitia Pichevin, (D), Bordeaux, Cg

Instructor:
Richard L. H. Essery, As
Raja Ganeshram, Pg
Bryne T. Ngwenya, (D), Reed, Co
David Reay, (D), As
Tom Slater, (D), London (UK), Zn
Samantha Staddon

Lecturer:
Peter Alexander, (D), Zc
Mikael Attal, (D), Joseph Fourier, 2003, GmtZy
Clare Barnes, (D), Zn
Geoffrey Bromily, (D), Ge
Eliza Calder, Gv
Mark Chapman, (D), Yg
Murray Collins, (D), Zi
Gregory L. Cowie, Co
Kate Crowley, (D), Zc
Julie Cupples, Zn
Kyle Dexter
Ruth Doherty, (D), Edinburgh, As
Zhiqiang Feng, (D), Zn
Janet Fisher, (D), Zn
Florian Fusseis, Gc
Alfy Gathorne-Hardy, (D), Zc
Daniel Goldberg
Noel Gourmelen
Margaret C. Graham, Ge
Steven Hancock, (D), Zi

Kate V. Heal, Co
Sebastian Hennige, (D), Zc
Lea-Anne Henry, Zn
Rachel Hunt, Zn
Gail D. Jackson, PbGe
Simon Jung, Og
Hamish Kallin, (D), Zn
Aidan Keane, (D), Zc
Linda Kirstein, Gt
sukanya Krishnamurthy, (D), Zn
Eric Laurier, Zn
Fraser MacDonald
William A. Mackaness, Zi
Ondrej Masek, (D), Ng
Christopher I. McDermott, Hg
Marc J. Metzger, (D), Wageningen (Neth), Ge
Nina J. Morris
Simon N. Mudd, (D), Hg
Caroline Nichol
Eva Panagiotakopulu, Pe
Genevieve Patenaude, (D), Ge
Hugh C. Pumphrey, (D), Mississippi, AscYv
Kanchana Ruwanpura
Casey Ryan
Simon J. Shackley, Eg
Niamh K. Shortt, Eg
Sran P. Sohi, (D), London (UK), So
Neil Stuart, Zi
Daniel Swanton, (D), Durham (UK)
Alexander Thomas
Dan van der Horst, (D), Eg
Mark Wilkinson, (D)
Merwether Wilson, Ob

Teaching Fellow:
Thomas Challads, Hg

Other:
Andrew S. Hein, (D), Edinburgh
Tetsuya Komabayashi, (D), Gz
Isla Myers-Smith
Jan Penrose, (D), Toronto
Marisa Wilson, (D), Oxcon

Exeter University

Camborne School of Mines
College of Engineering, Mathematics and Physical Sciences
University of Exeter
Penryn Campus
Penryn, Cornwall TR10 9FE
p. (+44) 1392 661000
cornwall@exeter.ac.uk
http://emps.exeter.ac.uk/geology/contact/

Director:
John Coggan, (D), Newcastle, Nx
Charlie Moon, (D), Imperial Coll (UK), 1983, CeEmZi
Neill Wood, (B), Imperial Coll (UK), 1983, YuZen

Professor:
Hylke J. Glass, (D), NxmEm
Stephen Hasselbo, Gs
Kip Jeffrey, Nx
Bernd Lottermoser, Ca

Lecturer:
Jens C. Andersen, (D), Copenhagen, GiEmCa
Ian Bailey, (D), Univ Coll (UK), Zn
Robert Barley
Christopher Bryan, Nx
Patrick Foster, Nm
Sam Hughes, Gt
Gareth Kennedy, Nm
Kate L. Littler, (D), Univ Coll (UK), 2011, PeCl
John Macadam, Ge
Lewis Meyer, (D), Exeter (UK), 2001, Nmr
Kathryn Moore, (D), Bristol, 1999, Ge
Kim Moreton, (D), Exeter (UK), 2008, EnZur
Richard Pascoe, (D), Gz
Duncan Pirrie
Robin Shail, (D), Keele, 1992, GtcEm
Ross Stickland
David Watkins, Hg

Andrew Wetherelt, Nm
Paul Wheeler, Nx
Ben Williamson, (D), London (UK), 1991, EmGv

Geological Survey of Northern Ireland

Geological Survey of Northern Ireland ☒
Dundonald House
Upper Newtownards Road
Belfast BT4 3SB
gsni@economy-ni.gov.uk
https://www2.bgs.ac.uk/gsni/

Heriot-Watt University

Institute of Petroleum Engineering (M,D) O✓
Research Park, Riccarton
Heriot-Watt University
Edinburgh Campus
Edinburgh EH14 4AS
p. +44 (0) 131 451 3543
egis-staff@hw.ac.uk
https://www.hw.ac.uk/schools/energy-geoscience-infrastructure-society/research/ige.htm
f: https://www.facebook.com/hwu.egis
t: @HWUPetroleum
Programs: Petroleum Engineering
Applied Petroleum Geoscience
Reservoir Evaluation and Management
Subsurface Energy Systems
Mature Field Management
Enrollment (2019): M: 83 (0) D: 121 (0)

Director:
Andy Gardiner
Mahmoud Jamiolahmady, (D), Ins Pet Eng, 2001, Zn
Eric Mackay
Uisdean Nicholson
Asghar Shams, Yg
Rink van Dijke

Head:
Sebastian Geiger, (D), ETH (Switzerland), 2004, NpEoYh

Chair:
John Underhill

Professor:
Patrick Corbett, Np
Gary Couples, Np
David Davies, Ng
Colin MacBeth
Ken Sorbie
Dorrik Stow

Associate Professor:
Andreas Busch
Babak Jafarizadeh, (D), Eo
Jingsheng Ma, (D), Np
Kamaljit Singh, (D), Np

Assistant Professor:
Elli-Maria (Elma) Charalampidou
Morteza Haghighat Sefat

Research Associate:
Ross Anderson, Eo
Lorraine Boak
Matthew Booth
Rachel Brackenridge
Rod Burgass
Antonio Carvalho
Antonin Chapoy
Romain Chassagne
Ilya Fursov
Alexader Graham
Sally Hamilton, Gs
Oleg Ishkov
Rachel Jamieson
Marco Lorusso
Maria-Daphne Mangriotis, Ys
Pedro Martinez Garcia
Robin Shields, Cg
Mike Singleton
Oscar Vazquez
Francesca Watson
Jinhai Yang

Zhen Yin
Zhao Zhang
Zhao Zhang, (D)
Stephanie Zihms

Instructor:
Steve McDougall

Lecturer:
Vasily Demyanov, Zi
Florian Doster, Hg
Ahmed H. Elsheikh
Zeyun Jiang
Helen Lever, (D), James Cook, 2003, Go
Helen Lewis, (D), Gc
Khafiz Muradov, (D)
Gillian E. Pickup, (D), Np
Karl D. Stephen

Laboratory Director:
Jim Buckman
Mike Christie

Other:
Bahman Tohidi

Imperial College

Earth Science and Engineering
Imperial College London
South Kensington Campus
London SW7 2AZ
p. +44 (0)20 7589 5111
philip.allen@imperial.ac.uk

http://www3.imperial.ac.uk/earthscienceandengineering

Director:
Nigel Brandon, Ge

Head:
Johannes Cilliers, (D), Cape Town, 1994, Gz

Chair:
Alastair Fraser, (D), Glasgow, 1995, Go
Matthew Jackson, (D), Liverpool (UK), 1997
Howard Johnson, Go
Peter King, (D), Cambridge, 1982, Np
Ann H. Muggeridge, (D), Oxford, 1986, NpHwEo
Yanghua Wang, (D), Imperial Coll (UK), 1997, Yg

Professor:
Martin J. Blunt, (D), Cambridge, 1988, NpHw
John Cosgrove, (D), Imperial Coll, 1972, Gc
Sevket Durucan, (D), Nottingham (UK), 1981, NmEcNr
Sanjeev Gupta, Gs
Joanna Morgan, (D), Cambridge, 1988, Yg
Jane Plant, Ge
Mark Rehkamper, (D), Mainz, 1995, CsaCc
Mark Sephton, (D), Open, 1996, Xm
Velisa Vesovic, (D), Imperial Coll (UK), 1998, Np
Michael Warner, (D), York, 1979, Ys
Dominik Weiss, (D), Bern, 1998, ClaCq
Robert W. Zimmerman, (D), California (Berkeley), 1984, NrHwNp

Associate Professor:
Gerard Gorman, (D), Imperial Coll, 2005, Yg
Samuel Krevor, (D), Eo
Lidia Lonergan, (D), Oxford, 1991, GtcGs

Research Associate:
Rebecca Bell, (D), Southampton (UK), 2008, Gt
Branko Bijeljic, (D), Imperial Coll (UK), 2000, Np
Raphael Blumenfeld, (D), Cambridge, Zn
Rhordri Davies, (D), Cardiff, 2007, Yg
Zita Martins, (D), Leiden (Neth), 2007, Xm
Christopher Pain, Yg
Randall Perry, Zg

Instructor:
Jenny Collier, (D), Cambridge, 1989, Yr
Saskia Goes, (D), California (Santa Cruz), 1995, Yg
Gary Hampson, (D), Liverpool (UK), 1995, Gd
Christopher Jackson, (D), Manchester (UK), 2002, Gt
Anna Korre, Ng
Jian Guo Liu, (D), Imperial Coll (UK), 2002, Zr
Stephen Neethling, (D), UMIST, 1999, Gz
Emma Passmore
Matthew Piggot, Bath, 2001, Hg

Lecturer:
Ian Bastow, (D), Leeds (UK), 2005, Yg
Gareth Collins, (D), Imperial Coll, 2002, Xm
Matthew Genge, (D), Univ Coll (UK), 1993, Xm
Kathryn Hadler, (D), Manchester (UK), 2006, Gz
Cedric John, (D), Potsdam, 2003, Gs
John-Paul Latham, (D), Imperial Coll (UK), 2003
Philippa J. Mason, Zr
Adrian Muxworthy, (D), Oxford, 1998, Ym
Julie Prytulak, (D), Bristol, 2008, Gg
Mark Sutton, (D), Wales, 1996, Pg
Tina van de Flierdt, (D), ETH (Switzerland), 2003, Cs
Alex Whittaker, (D), Edinburgh, 2007, Gt

Visiting Professor:
Rosalind Coogon, Cs

Emeritus:
Alain Gringarten, (D), Stanford, 1971, Np
John Woods, (D), Imperial Coll (UK), 1964

Keele University

Geography, Geology and the Environment (B,M,D) ☒
William Smith Building,
Keele University
Keele, Staffordshire ST5 5BG
p. (+44) 01782 733615
gge@keele.ac.uk
http://www.keele.ac.uk/gge/
f: https://www.facebook.com/KeeleGeology
t: @KeeleGeology
Programs: BSc Geology (single & combined honours)
M. Geology
MSc Geoscience Research

Enrollment (2014): B: 149 (45) M: 29 (15) D: 10 (2)

Dr:
Stuart Egan, (D), Keele, Gtc
Jamie K. Pringle, (D), Heriot-Watt, YuGfa

Associate Professor:
Ralf Gertisser, (D), Freiburg (Germany), 2001, GivGz
Ian G. Stimpson, (D), Wales, 1987, YeuRc

Dr:
Glenda Jones, (D), Keele, NgYg
Stu Clarke, (D), Keele, GsoGd
Ralf Halama, (D), Tubingen, GpiCa
Guido Meinhold, (D), Mainz, GdCa
Michael Montenari, Tubingen, PmgPs
Steven L. Rogers, (D), Keele, PgGs

Kingston University

School of Geography, Geology and Environment
Penrhyn Road Centre
Penrhyn Road
Kingston upon Thames
Surrey KT1 2EE
p. +44 (0)20 8417 9000
g.haritos@kingston.ac.uk
http://www.kingston.ac.uk/geolsci/

Director:
Stuart Downward

Head:
Gavin Gillmore, (D), Univ Coll (UK), Gb

Professor:
Ian Jarvis, (D), Oxon, 1980, CgGsr
Peter Treloar, (D), Glasgow, 1978, Gc
Nigel Walford, (D), London (UK), 1981, Zi

Instructor:
Peter Hooda, (D), London (UK), 1992, Em
Annie Hughes, (D), Bristol, 1997, Zi
Mike Smith, (D), Sheffield, 1999, Zr

Lecturer:
Andy Adam-Bredford
Alistair Baird
Douglas Brown, Zi
Kerry Brown, (D), SUNY (Stony Brook), 2004, Ge
Norman Cheun, Ge
Tracey Coates
Hadrian Cook, (D), E Anglia (UK), 1986, Ge
Peter Garside, (D), Liverpool (UK)

Paul Grant
Ian Greatbatch, (D), 2008, Zi
Frances Harris, (D), Kingston, 1995
Mary Kelly
David Kidd, (D), St Andrews, 2005, Ziy
James Lambert-Smith, (D), Kingston, 2014, Gg
Andrew Miles, (D), Edinburgh, 2012, Gz
Stephanie Mills, (D), Witwatersrand, 2006, Gl
Pamela Murphy
Colin Ryall
Neil Thomas

Visiting Professor:
Rosalind Taylor

Emeritus:
Richard Moody
Andy Rankin

Other:
Andrew Swan

Leicester University

School of Geography, Geology and the Environment (B,M) ☒
University Road
Leicester LE1 7RH
p. +44 (0)116 252 3933
geology@le.ac.uk
http://www2.le.ac.uk/departments/geology
Programs: Geology, Applied & Environmental Geology, Geology with Palaeontology.
Geography with Geology.
Geography, Human Geography
Environmental Science
GIS
Satellite Data Science

Head:
Stewart Fishwick, Yg

Professor:
Mike Branney, Gv
Sarah Davies, (D), Gs
Sarah Gabbott, Pg
Gawen Jenkin
Mark Purnell, Pg
Mark Williams, Pg
Jens Zinke, (D)

Dr:
Cat Greenfield, (D), Yg

Lecturer:
Tiff Barry
Tom Harvey, Pg
Dave Holwell, Ge
Sarah Lee
Andrew Miles
Marc Reichow
Dan Smith, Ge

Liverpool John Moores University

School of Natural Sciences and Psychology (B,M,D)
Byrom Street
Liverpool L3 3AF
p. (+44) 0151 904 6300
j.r.kirby@ljmu.ac.uk
https://www.ljmu.ac.uk/research/centres-and-institutes/environment-research-group
f: https://www.facebook.com/groups/LJMU.geography.env.sci/
t: @ljmugeog
Enrollment (2016): B: 72 (0) M: 3 (0)

Professor:
Andy Tattersall

Dr:
Jason Kirby

Nottingham Trent University

Dept of Animal, Rural, and Environmental Sciences (B,M,D) ☒
Burton Street
Nottingham NG1 4BU
p. +44 (0)115 941 8418
dawn.scott@ntu.ac.uk
http://www.ntu.ac.uk

Open University

Dept of Environment, Earth & Ecosystems (B,M,D)
Faculty of Science
The Open University
Walton Hall
Milton Keynes MK7 6AA
p. +44 (0) 1908 652886
STEM-EEES-Enquiries@open.ac.uk
http://www.open.ac.uk/science/environment-earth-ecosystems/
f: https://www.facebook.com/pages/OU-Environment-Earth-Ecosystems/234968859948707
t: @OU_EEE

Head:
Arlene G. Hunter, (D), Open, 1993, ZeGiCg

Professor:
Fabrizio Ferrucci, Yg
David Gowing, (D), Lancaster (UK), 1991, Pb
Nigel B.W. Harris, (D), Cambridge, 1973, GtiGp
Walter Oechel, As

Associate Professor:
Mark A. Brandon, (D), Cambridge, 1995, OgGlOp
Angela L. Coe, (D), Oxford, GsrCg
Dave McGarvie, (D), Lancaster (UK), 1985, GviCt

Associate Scientist:
Peter Sheldon

Research Associate:
Marie-Laure Bagard
Gareth Davies, (D), Cranfield, Ze

Reader:
Neil Edwards, (D)

Instructor:
Stephen Blake, (D), Lancaster (UK), 1982, Gv

Lecturer:
Pallavi Anand, (D), Cambridge, 2002, ClPe
Tom Argles, (D), Oxon, Zi
Anthony Cohen, (D), Cambridge, 1988, Cs
Sarah Davies, (D), Sheffield, Ze
Miranda Dyson, (D), Witwatersrand, 1989, Pg
Tamsin Edwards, (D)
Richard Holliman, (D), Open, Ze
Philip Sexton, (D), Southampton (UK), 2000, Ge
Carlton Wood

Visiting Professor:
Stephen Self, (D)
Edward Youngs, (D), Cambridge, 1972, Sp

Emeritus:
Phil Potts
Robert Spicer, (D), Imperial Coll, 1975

Polytechnic Southwest

Dept of Earth Sciences
Drake Circus
Plymouth, Devon PL4 8AA
p. (+44) (0)1752 584584
science.environment@plymouth.ac.uk
http://www5.plymouth.ac.uk/schools/school-of-geography-earth-and-environmental-sciences/earth-sciences

Queen's University Belfast

School of Natural and Built Environment (Geography) (B,M,D) ●☒
University Road
Belfast BT7 1NN
p. + 44 (0) 28 90973186
p.warke@qub.ac.uk
http://www.qub.ac.uk/schools/NBE/
f: https://www.facebook.com/QUBGeography/
Programs: Geography (B); Geography with a Language (French or Spanish) (B)
Enrollment (2016): B: 320 (106) M: 6 (6) D: 40 (5)

Head:
Patricia Warke, (D), Queen's (Ireland), 1994, GmZym

Professor:
Keith Lilley, (D)
David Livingstone, (D)

Dr:
Andrew Newton, (D), 2017, GmlGs
Associate Professor:
Merav Amir, (D)
Oliver Dunnett, (D)
Paul Ell, (D)
Diarmd Finnegan, (D)
Nuala Johnson, (D)
M. Satish Kumar, (D), Jawaharlal Nehru (India), 1991, ZnnZn
Jennifer McKinley, (D)
Donal Mullan, (D)
Helen Roe, (D)
Alastair Ruffell, (D)
Ian Shuttleworth, (D)
Tristan Sturm, (D)
Lecturer:
Niall Majury, (D), Toronto, 1999

Royal Holloway University of London

Department of Earth Sciences (B,M,D) ●⊘
Department of Earth Sciences
Royal Holloway University of London
Egham Hill
Egham , Surrey TW20 0EX
p. +44 (0) 1784 443 581
info@es.rhul.ac.uk
https://www.royalholloway.ac.uk/earthsciences/home.aspx
f: https://www.facebook.com/RhulEarthSci
t: @RHULEarthSci
Programs: BSc/MSci Geology; Environmental Geology; Petroleum Geology; Geoscience; Digital Geoscience; Environmental Geoscience; Geosciences & Sustainable Energy; Earth, Climate & Environmental Change, MSc Environmental Diagnosis & Management; Energy Geosciences; Petroleum Geosciences
Head of Department:
Jurgen Adam, (D), Gc
Emeritus:
Robert Hall, (D), Univ Coll (UK), 1974, Gt
Dave Mattey, (D), Cs
Ken R. McClay, (D), Imperial Coll (UK), 1978, GctGo
Andrew Cunningham Scott, (D), Pb
Professor:
Margaret Collinson, (D), Pg
Howard Falcon-Lang, (D), Pg
Richard Ghail, (D), Gt
Agust Gudmundsson, (D), Gc
F. Javier Hernandez Molina, (D), Granada (Spain), 1993, GsuOu
Martin King, (D), As
Euan G. Nisbet, (D), Cambridge, 1974, ZgAcGi
Matthew Thirlwall, Cs
Dave Waltham, (D), Yg
Senior Lecturer:
Anirban Basu, (D), Illinois (Urbana), CglCs
Christina Manning, (D), Cg
Nicola Scarselli, (D), Go
Giulio Solferino, (D), Gg
Ian M. Watkinson, (D), 2009, GtcRn
Reader:
Kevin C. Clemitshaw, (D), E Anglia (UK), AsZg
Alex Dickson, (D), Cas
David Lowry, (D), CsZg
Associate Professor:
Domenico Chiarella, (D), Basilicata (Italy), 2011, Gs
Research Fellow:
Paula Koellemeijer, (D), Ys
Leturer:
Jonathan Paul, (D), Hw
Lecturer:
Queenie Chan, (D), XmCa
Rebecca Fisher, (D), As
Amy Gough, (D), Gs
Nathalie Grassineau, (D), Cs
Honorary Lecturer:
Dave Alderton, (D), Gz

Staffordshire Polytechnic

Dept of Applied Sciences
College Road, Stroke-on-Trent
Staffordshire ST4 2DE
p. (+44) 1782 294000
comms@staffs.ac.uk
http://www.staffs.ac.uk/academic_depts/sciences/subjects/environment/

Geography and the Environment
j.w.wheeler@staffs.ac.uk
http://www.staffs.ac.uk/sands/scis/geology/geology.html#

Swansea University Prifysgol Abertawe

Dept of Earth Sciences
Wallace Building
Swansea University
Singleton Park
Swansea, Wales SA2 8PP
p. +44 (0) 1792 205678 ext 8112
geography@swansea.ac.uk
http://www.swansea.ac.uk/geography/

Teesside University

School of Science and Engineering (B,M)
Tees Valley
Middlesborough TS1 3BA
p. 01642 738800
sse-admissions@tees.ac.uk
http://www.tees.ac.uk/schools/sse

Ulster University

School of Geography and Environmental Science (B,M,D)
Coleraine Campus
Cromore Road
Coleraine BT52 1SA
ges@ulster.ac.uk
https://www.ulster.ac.uk/faculties/life-and-health-sciences/geography-and-environmental-sciences
t: @UlsterUniGES
Programs: BSc programmes on campus in Geography, Marine Science and Environmental Science., PgDip/MSc programmes fully online and part-time on: Geographic Information Systems, Remote Sensing and Geographic Information Systems, Environmental Toxicology and Pollution Monitoring, Environmental Management, Environmental Management with Geographic Information Systems
Certificates: The modules related to the PgDip and MSc programmes are offered as stand-alone CPD options too.
Enrollment (2022): B: 38 (61) D: 22 (8)

Ulster University, Coleraine

School of Geography and Environmental Sciences (A,B,M,D) ○
Cromore Road
Coleraine BT521SA
p. 0044(0)2870124401
a.moore@ulster.ac.uk
https://www.ulster.ac.uk/faculties/life-and-health-sciences/schools/geography-and-environmental-sciences
t: @UlsterUniGES
Programs: MSc Geographic Information Systems; MSc Remote Sensing and Geographic Information Systems; MSc Environmental Management; MSc Environmental Management and Geographic Information Systems; MSc Environmental Management with Geographic Information Systems; MSc Environmental Pollution and Ecotoxicology;
Certificates: Geographic Information Systems; Remote Sensing and Geographic Information Systems; Environmental Management; Environmental Management and Geographic Information Systems; Environmental Management with Geographic Information Systems; Environmental Pollution and Ecotoxicology;
Enrollment (2019): B: 320 (100) M: 170 (25) D: 30 (8)
Professor:
Adrian Joseph Moore, (D), Ulster, 1993, ZifZn

University College London

Dept of Earth Sciences (B,M,D) ☒
Gower Street

, London WC1E 6BT
p. +44 (0)20 3108 6339
earthsci@ucl.ac.uk
http://www.ucl.ac.uk/es
Programs: Earth Sciences, Geology, Geophysics, Environmental Geoscience, Natural Sciences

Professor:
Dario Alfe, (D), Intl Sch for Adv Studies (Italy), 1997, ZmGyYh
Paul Bown, (D), Univ Coll (UK), 1986, Pm
John Brodholt, (D), Bristol, 1992, Yg
David Dobson, (D), Univ Coll (UK), 1995, Gz
Adrian P. Jones, (D), Durham (UK), 1980, GiCgGg
Chris Kilburn, (D), Univ Coll (UK), 1984, Yx
John M. McArthur, (D), Imperial Coll (UK), 1974, ClHwPs
Philip Meredith, (D), Imperial Coll (UK), 1983, Nr
Chris Rapley, (D), London (UK), 1976, ZgrGl
Graham Shields-Zhou, (D), Eidgenössische Tech Hochschule Zürich, 1997, Cg
Julienne Stroeve, (D), Colorado, 1996, GlZr
Juergen Thurow, (D), Eberhard-Karls, 1987, Gs
Paul Upchurch, (D), Cambridge, 1993, Pg
Lidunka Vocadlo, (D), Univ Coll (UK), 1993, Gy
Bridget S. Wade, (D), Edinburgh, PmeCl
Ian Wood, (D), Univ Coll (UK), 1977, Gz

Associate Professor:
Nicolas Dr Brantut, (D), Nr
Anna Ferreira, (D), Oxford, 2005, Ys
Philip Dr Mannion, (D), Pvi
Tom Mitchell, Gc
Alex Song, (D), Caltech, Ysg
Andrew Dr Thomson, (D), Cpg
Michel Dr Tsamados, (D), Zo
Pieter Vermeesch, (D), Stanford, 2005, Cc

Lecturer:
Maxim Dr Ballmer, (D), Yg
Matthew Dr Fox, (D), Gm
Martin Dr Homann, (D), Gs
Peter Dr Irvine, (D), ZcNx
Susan Dr Little, (D), Cg
Emma Dr Liu, (D), Gv
Katie Dr McFall, (D), Eg
Dominic Papineau, (D), Colorado, 2006, Cg

University College of Swansea

Dept of Geography
Singleton Park
Swansea SA2 8PP

University of Aberdeen

Dept of Geology and Petroleum Geology (B,M)
Meston Building
King's College
Aberdeen AB24 3UE
geology@abdn.ac.uk
http://www.abdn.ac.uk/geology/

Head:
David Jolley, (D), PlsGv

Chair:
Ian Alsop, (D)
Rob Butler
Adrian Hartley
John Howell, (D), Birmingham (UK), 1992, GsoGc
Andrew Hurst
Ben Kneller
David MacDonald
Randell A. Stephenson, (D), Dalhousie, 1981, YgGtZg

Associate Professor:
Clare Bond, (D), Gct

Research Associate:
Steven Andrews
Robert Daly
Jyldz Tabyldy Kyzy
Sam Spinks
Christian Vallejo

Instructor:
David Muirhead

Lecturer:
Stephen Bowden, (D), Newcastle upon Tyne, CoGoCb
David G. Cornwell, (D), Leicester (UK), 2008, YsGtYe
Dave Healy
Malcolm J. Hole
David Iacopini
Joyce Neilson
Colin North
Nick Schofield

University of Bedfordshire

Dept of Life Sciences
Park Square
Luton, Bedfordshire LU1 3JU
p. +44 1234 400400
international@beds.ac.uk
http://www.beds.ac.uk/howtoapply/departments/science

University of Birmingham

School of Geography, Earth & Environmental Sciences (B,M,D) ☒
School of Geography, Earth and Environmental Sciences
University of Birmingham
Edgbaston, Birmingham B15 2TT
p. +44 (0)121 414 5531
j.d.oldfield@bham.ac.uk
http://www.birmingham.ac.uk/schools/gees
Programs: Geology; Geology and Physical Geography; Palaeontology and Geology; Global Environmental Change & Sustainability

Director:
Rob MacKenzie

Head:
David Hannah, Hg

Chair:
Eugenia Valsami-Jones, (D), ClGeCa

Professor:
Stuart Harrad
Roy M. Harrison, (D), Birmingham (UK), 1989, As
Alexander Milner
Tim Reston
Jon Sadler
John H. Tellam, (D), Birmingham (UK), 1983, Hw

Associate Professor:
Rosie Day, (D), Univ Coll (UK), Zn
Lloyd Jenkins
Stephen M. Jones, (D), Cambridge, 2000, Zg
Jonathan Oldfield, (D), Zy

Research Associate:
Mohamed A. Abdallah
David Beddows
Ian Boomer
Simon J. Dixon, (D), Gms
Sophie Hadfield-Hill
Kieran Khamis
Zongbo Shi
Sebastian Watt

Deputy Head of School:
William Bloss

Instructor:
James Bendle
Lee Chapman
Jason Hilton
Dominique Moran
Ian Phillips
Jo Southworth

Lecturer:
Austin Barber
Nicholas Barrand
Rebecca Bartlett
Lesley Batty
Mike Beazley
Chris Bradley, (D), Leicester (UK), 1994, Hg
Julian Clark
Steven Emery
Sara Fregonese
Phil Jones
Nick Kettridge
Stephen Krause
Mark Ledger

Peter Lee
Iseult Lynch
Zena Lynch
Patricia Noxolo
Francis Pope
Jessica Pykett
Adam Ramadan
John Round
Ivan Sansom
Greg Sambrook Smith
Carl Stevenson
James Wheeley
Martin Widman

Birmingham Fellow & Academic Keeper, Lapworth Museum of Geology:
Richard Butler

Other:
Tom Dunkley Jones

University of Bristol

School of Earth Sciences (B,M,D) ☒
Wills Memorial Building
Queens Road
Bristol BS8 1RJ
p. +44 117 954 5400
m.j.walter@bristol.ac.uk
http://www.bristol.ac.uk/earthsciences/
f: https://www.facebook.com/School-of-Earth-Sciences-Bristol-University-146277648746662/
t: @UOBEarthscience
Programs: Geology; Environmental Geoscience; Geophysics; Palaeobiology
Enrollment (2013): B: 25 (9) M: 44 (5)

Professor:
Michael Walter , (D), 1991, GiCpg

University of Cambridge

Dept of Earth Sciences (B,M,D) ●☒
Downing Street
Cambridge , Cambridgeshire CB2 3EQ
p. +44 (0)1223 333400
satr@cam.ac.uk
http://www.esc.cam.ac.uk
f: https://www.facebook.com/pages/Department-of-Earth-Sciences-University-of-Cambridge/131259610400860
t: @EarthSciCam
Programs: Earth Sciences

Head:
Simon Redfern, (D), Cambridge, 1989, Gzy

Professor:
Michael Bickle, (D), Oxford, 1973, GtYgGo
Nicholas J. Butterfield, Pg
Michael Carpenter, Gy
David Hodell, Ge
Tim Holland, (D), Oxford, GptGz
Marian Holness, Gi
James A. Jackson, (D), Cambridge, 1980, YgGt
Simon Conway Morris, Pg
Keith Priestley, YgGt
Nick Rawlinson, Ygg
Ekhard Salje, Gy
Nicky White, YgGt
Robert White, YgGt
Eric W. Wolff, (D), Cambridge, 1992, GlPe
Andy Woods, YgGt

Lecturer:
David Al-Attar, YdGt
Alex Copley, YgGt
Sanne Cottaar, Ys
Neil Davies, Gs
Marie Edmonds, Gv
Ian Farnan, Gz
Sally Gibson, Gi
Richard Harrison, Gz
John Maclennan, Gi
Kenneth McNamara, Pg
Jerome Anthony Neufeld, GoYgGt
David Norman, Pg
John Rudge, YgGt
Luke C. Skinner, (D), Cambridge, 2005, GeCmOg
Ed Tipper, Cl
Alexandra Turchyn, Ge

Emeritus:
Nigel H. Woodcock, (D), Imperial Coll (UK), 1973, GcsGt

University of Derby

Geographical, Earth and Environmental Sciences (B,M)
Kedleston Road
Derby DE22 1GAB
p. +44 (0) 1332 591703
fehs@derby.ac.uk
http://www.derby.ac.uk/science/gees/
Enrollment (2013): B: 120 (30) M: 9 (10)

Head:
Hugh Rollinson, Cg

Professor:
Aradhana Mehra, Co

Instructor:
Jacob Adentunji, Ge

Lecturer:
Martin Whiteley, Go

University of East Anglia

School of Env Sciences
Norwich Research Park
Norwich NR4 7TJ
p. +44 (0)1603 592542
J.Alexander@uea.ac.uk
http://www.uea.ac.uk/environmental-sciences

Director:
Corinne Le Qu, (D), Zg

Professor:
Jan Alexander, (D), Leeds (UK), GseGd
Julian E. Andrews, (D), Leicester (UK), 1984, CsGsCl
Simon Clegg, (D), E Anglia (UK), 1986, AsOcCg
Brett Day, (D), UEA, 2004, Ge
Alastair Grant, (D), Wales, 1983, GeObCt
Karen Heywood, Op
Kevin Hiscock, (D), Birmingham (UK), 1987, Hwg
Tim Jickells, (D), Southampton (UK), Ob
Phillip D. Jones, (D), Newcastle upon Tyne, 1977, AsPeHg
Andy Jordan, Ge
Andrew A. Lovett, (D), Wales, 1990, Ziu
Coling Murrell, As
Timothy J. Osborn, (D), E Anglia (UK), 1996, At
Carlos Peres, Sf
Ian Renfrew, Am
Ian Renfrew, Am
Bill Sturges, As
Roland von Glasow, (D), Max Planck, Yg
Andrew Watkinson, Ge
Sir Robert Watson

Assistant Professor:
Gill Malin, (D), Liverpool (UK), 1983, ObcCb

Research Associate:
Amy Binner

Instructor:
Alex Baker, (D), Plymouth (UK), On
Jenni Barclay, (D), Bristol, Gv
Paul Burton, (D), Cambridge, Ys
Mark Chapman, (D), E Anglia (UK), Pm
Paul Dolman, Zu
Jan Kaiser, Cs
Adrian Matthews, (D), Reading (UK), Am
Claire Reeves, (D), UEA, Cg
Carol Robinson, Ob
Parvadha Suntharalingam, Oc
Cock von Oosterhout, (D), Leiden (Neth)
Rachel Warren, Eg

Lecturer:
Annela Anger-Kraavi, EgGe
Annela Anger-Kraavi, Eg
Victor Bense, Hg
Alan Bond, (D), Lancaster (UK), Ge
Jason Chilvers, (D), Univ Coll (UK), Ge
Pietro Cosentino, Yg
Stephen Dorling, Am

Aldina Franco, (D), UEA, Ge
Robert Hall, (D), Proudman Ocean Lab, On
Tom Hargreaves, Ge
Richard Herd, (D), Lancaster (UK), Gv
Martin Johnson, Cm
Manoj Joshi, (D), Oxford, Zi
Iain Lake, Ge
Irene Lorenzoni, Eg
Nikolai Pedentchouk, (D), Penn State, 2004, Cs
Jane Powell, Ge
Brian Reid, GeCg
Gill Seyfang, Og
Congxiao Shang, (D), Queen Mary (London), Hw
Peter Simmons, Eg
Trevor Tolhurst, Sb
Jenni Turner, (D), 2008, Gt
Naomi Vaughan, (D), UEA, Am
Charlie Wilson, (D), British Columbia, Ge
Xiaoming Zhai, (D), Dalhousie, Og

Emeritus:
Peter S. Liss, (D), OcCbAc

Professorial Fellow:
Kerry Turner, GeEg

University of Exeter

Camborne School of Mines
Redruth, Cornwall
TR15 3SE
C.Jeffrey@exeter.ac.uk
http://www.ex.ac.uk/CSM/

University of Glasgow

School of Geographical and Earth Sciences
The Gregory Building
East Quadrangle
University Avenue
Glasgow G12 8QQ
p. +44 (0) 141 330 5436
GES-General@glasgow.ac.uk
http://www.gla.ac.uk/schools/ges/

Head:
Maggie Cusack, Gz

Professor:
Paul Bishop, (D)
John Briggs, Zu
Roderick Brown, Cc
Gordon Curry, (D), Imperial Coll (UK), 1979, GegPg
Deborah Dixon
Jim Hansom, (D), Aberdeen, 1979, OnGmZy
Trevor Hoey, Gs
Martin Lee, Gz
Christopher Philo

Senior Scientist:
Nicolas Beaudoin

Associate Professor:
David Forrest, (D), Glasgow, 1996, Zi

Research Associate:
Enateri Alakpa
Susan Fitzer, Ob
Michael Gallagher
Lydia Hallis
Ulrich Kelka
Angela Last
Paula Lindgren, As
Alistair Mcgowan, Pg
Larissa Naylor
Alan W. Owen, (D), Glasgow, 1977, PgsPg

Instructor:
Anne Dunlop
Mhairi Harvey
Hayden Lorimer
Rhian Meara
Hester Parr
Vernon Phoenix
Daisy Rood

Lecturer:
Brian Bell, So
David Brown, Gvs

Seamus Coveney, Zi
Tim Dempster, Gz
Jane Drummond, Zi
Derek Fabel, Xg
David Featherstone
Nick Kamenos
Ozan Karaman
Daniel Koehn, Gt
Hannah Mathers, (D), GmlGg
Cheryl Mcgeachan
Simon Naylor
Cristina Persano

University of Leeds

School of Earth and Environment (B,M,D) ☒
Maths/Earth and Environment Building
The University of Leeds
Leeds LS2 9JT
p. +44 113 343 2846
enquiries@see.leeds.ac.uk
https://environment.leeds.ac.uk/see
f: https://www.facebook.com/SchoolofEarthandEnvironment/
t: @SEELeeds

Director:
Piers M. Forster, (D), Reading (UK), 1994, As

Head:
Simon Bottrell

Chair:
Christopher Collier
Simon Poulton, (D), Cls
Peter Taylor, Ge

Professor:
John Barrett, Ge
Alan Blyth, As
Ian Brooks, (D), UMIST, Zn
Ken Carslaw, (D), E Anglia (UK), 1994, As
Andy Challinor, (D), Leeds, 1999, Am
Martyn P. Chipperfield, (D), Cambridge, 1990, AscAp
Surage Dessai, (D), E Anglia (UK)
Andy Dougil, (D), Sheffield, So
Paul Field
Quentin Fisher, Np
Paul W.J. Glover, (D), E Anglia (UK), 1989, YxGoa
Andy Gouldson, Ge
Alan Haywood, Pe
Steve Hencher, Ng
Andy Hooper, Ydg
Greg Houseman, (D), Cambridge, Yg
Bill McCaffrey, Gs
Jurgen Neuberg, (D), Colorado, Gv
Jouni Paavola, Ge
Doug Parker, (D), Reading (UK), Am
Jeff Peakall, Gs
Andrew Shepherd
Lindsay C. Stinger, (D), Zn
Graham Stuart, Ys
Paul Wignall, PgGs
Tim Wright, Yd
Bruce W.D. Yardley, (D), Bristol, 1974, CgEgGg
William Young, (D), Ge

Associate Professor:
Doug Angus, Ys
Stephen Arnold, (D), Leeds (UK), As
Wolfgang Buermann, (D), Boston, 2012, Zu
Ian Burke, (D), Southampton (UK), Ge
Robert J. Chapman, (D), Leeds (UK), 1990, EmCe
Steven Dobbie, (D), Dalhousie, As
Luuk Flesken, (D), Wageningen (Neth), Ge
Phil Livermore, Ym
Christian Maerz, (D), Bremen, 2008, CmOuEm
Jim McQuaid, (D), Leeds (UK), 1999, As
Daniel Morgan, (D), Open (UK), 2003, Giv
Jon Mound, (D), Toronto, 2001, Gt
Noelle Odling, (D), Queens, Hg
Caroline Peacock, (D), Bristol, Sb
Sebastian Rost, (D), Universit, 2000, Ysg
Sally Russell, (D), Queensland, Ge
Dominick Spracklen, (D), Leeds (UK), 1999, Zn

Julia Steinberger, Zu
Jared West, Hg
Research Associate:
David Banks, Cg
Lauren Gregoire, (D), Bristol
Ruza Ivanovic, (D), Bristol, 2013, OgAsPe
John Marsham, (D), Edinburgh, 2003, As
Instructor:
Nick Dixon, (D), Leeds (UK), 2002, Zu
Tim Foxon, Ge
Chris Green
Dave Hodgson
Jacqueline Houghton
Damian Howells
Geoff Lloyd, (D), Birmingham (UK), 1984
Lecturer:
Emma Bramham, Ym
Roger A. Clark, (D), Leeds (UK), 1982, YseYs
Martin Dallimer, (D), Edinburgh, 2001, Ge
Monica Di Gregorio, (D), London Schl of Econ, Eg
Jen Dyer, (D), Leeds (UK), Ge
Alan Gadian
Fiona Gill, (D), PgCa
Clare E. Gordon, (B), Zi
Dabo Guan, Eg
Jason Harvey, (D), 2005, Cg
Mark Hildyard
George Holmes, Ge
Andrea Jackson, (D), Lancaster (UK), As
Julia Leventon, (D), Ge
Crispin Little, (D), Bristol, 1995, Pg
Piroska Lorinczi, (D), Leeds (UK), 2006, Np
Vernon Manville
Andrew McCair, (D), Cambridge, 1983, Gc
Lucie Middlemiss, (D), Leeds (UK), 2009, Zn
Nigel Mountney, Gs
Rob Newton, (D), Leeds (UK), Ge
Alice OWen, (D), Leeds (UK)
Douglas Paton, Gc
Richard Phillips, (D), Oxford, Gt
Claire Quinn, (D), King's Coll (UK), Ge
Andrew Ross, (D), Cambridge, Am
Susannah Sallu, Ga
Ivan Savov, (D), S Florida, 2004, Cg
Anne Tallontire
Mark Thomas, (D), Ng
Taija Torvela, (D), Abo Akademi, 2007, GctEg
James van Alstine, Ge
Xianyun Wen, (D), Sichuan (China), 1982, NoAsm
Visiting Professor:
Jane Francis, Pe
Tim Needham, Gc

University of Liverpool

Dept of Earth, Ocean, and Ecological Sciences
Jane Herdman Building
4 Brownlow Street
Liverpool L69 3GP
p. +44 (0)151 795 4642
faulkner@liverpool.ac.uk
http://www.liv.ac.uk/earth-ocean-and-ecological-sciences/
Head:
Andreas Rietbrock, YsGv
Professor:
Daniel Faulkner, Ys
Richard Holme, Ym
Nick Kusznir, Yd
Yan Lavallee, Gv
Rob Marrs, So
Jim Marshall, Cs
Jonathan Sharples, Ocp
Stan van der Berg, Em
John Wheeler, (D), GpcGz
Richard Worden, Gs
Research Associate:
Katherine Allen, Zi
James Ball
Helen Bloomfield, Ob
Stephen Crowley, Gs
Jonathan Lauderdale, Og
Claire Melet
Andreas Nilsson, Ym
Vassil M. Roussenov, (D), Sofia (Bulgaria), 1987, OpZo
Anu Thompson, Oc
Xiao Wang
Tsuyuko Yamanaka, Ob
Lecturer:
Charlotte Jeffrey Abt, Zn
Andy Biggin, Ym
Alan P. Boyle, (D), Univ Coll (UK), 1982, GzEmZe
Bryony Caswell, Ob
Silvio De Angelis, Ys
Rob Duller, Gvs
Liz Fisher, Co
Jonathan Green, Ob
Mimi Hill, Ym
Janine Kavanagh, Gv
Helen Kinvig, Gs
Harry Leach, (D), Leeds (UK), 1975, Og
Claire Mahaffey, Oc
Elisabeth Mariani, Nr
Kate Parr, Pe
John Piper, YmGt
Leonie Robinson, Ob
Isabelle Ryder, Gt
Matthew Spencer, Og
Neil Suttie
Alessandro Tagliabue, Oc
Jack Thomson, Ob
Visiting Professor:
Peter Kokelaar, (D), Wales, 1977, GstGv
Other:
Sara Henton De Angelis
Jackie Kendrick
Rachel Salaun, Oc
Felix von Aulock

School of Environmental Sciences (B,M,D) ☒
Jane Herdman Building
4 Brownlow Street
Liverpool, Merseyside L69 3GP
p. +44 151 794 5146
envsci@liv.ac.uk
http://www.liv.ac.uk/environmental-sciences

University of Manchester

School of Earth and Environmental Sciences
Williamson Building
Oxford Road
, Manchester M13 9PL
p. +44 (0) 161 306 9360
earth.support@manchester.ac.uk
Administrative Assistant: Steven Olivier
Director:
Stephen Boult, Hg
Chair:
Jonathan Redfern, (D), Bristol, 1990, GsoGl
Professor:
Mike Bowman, (D), Wales, Go
Andrew T. Chamberlain, (D), Liverpool (UK), 1987, GaSaPv
Thomas Choularton, (D), Manchester (UK), 1987, As
Hue Coe, (D), UMIST, 1993, As
Stephen Flint, (D), Leeds (UK), 1985, GrsGo
Martin Gallagher, (D), UMIDST, 1986, As
Jamie Gilmour, (D), Sheffield
Colin Hughes, (D), Open, 1992, Go
Francis Livens, Cg
Jonathan Lloyd, (D), Canterbury (NZ), 1993, Ge
Ian Lyon, (D), 1993, Xc
Gordon McFiggans, (D), UEA, 2000, As
Carl Percival, (D), Oxford, 1995, As
David Poyla, (D), Manchester (UK), 1987, Cm
Ernest H. Rutter, (D), Imperial Coll (UK), 1970, NrGct
David Schultz, Am
Kevin Taylor, Go
David Vaughan, (D), Oxford, 1971, GzClGe

Geraint Vaughan, (D), Oxford, 1982, As
Roy Wogelius, (D), Northwestern, 1990, Cg
Associate Professor:
Neil C. Mitchell, (D), Oxford, 1989, GumYg
Research Associate:
Alastair Booth, As
Gerard Capes, As
Deborah Chavrit, (D), Nantes (France), 2010, Gi
Patricia L. Clay, (D), Open (UK), 2010, CgaGg
Filipa Cox
Ian Crawford, As
Sarah Crowther, (D), Oxford, 2003, As
Christopher Dearden, As
Patrick Dowey
Helen Downie, Ge
Nicholas Edwards, Pg
Christopher Emersic, (D)
Torsten Henkel, Cs
Hazel Jones, As
Richard Kift, As
Kimberly Leather, As
Dantong Liu, As
Douglas Lowe, As
William Morgan, As
Miquel Poyatos-More
Laura Richards, Ge
Hugo Ricketts, As
Athanasios Rizoulis, Ge
Andrew Smedley, As
Robert Sparkes, (D), Cambridge, 2012
Jonathan Taylor, (D), Manchester (UK)
Karen Theis, (D), Manchester (UK), 2008, Zn
Paul Williams, As
Lecturer:
Grant Allen, (D), Leicester (UK), 2005, As
Simon Brocklehurst, (D), MIT, 2002, Gm
Kate Brodie, (D), Imperial Coll (UK), 1979, Gc
Rufus Brunt, (D), Leeds (UK), Gs
Michael Buckley, (D), York, 2008, Ga
Ray Burgess, (D), Open, Cs
Victoria Coker, (D), Manchester (UK), 2007, Gz
Paul Connolly, (D), Manchester (UK), 2006, As
Stephen Covery-Crump, (D), Univ Coll (UK), 1992
Giles Droop, (D), Oxford, 1979, Gp
Victoria Egerton, Pg
David Hodgetts, (D), Keele, 1995, Gsc
Cathy Hollis, (D), Aberdeen, 1995, Gs
Merren Jones, (D), Wyoming, 1997, Gst
Julian Mecklenburgh, Gc
John Nudds, (D), Dunhelm, 1975, Pg
Clare Robinson, (D), Lancaster (UK), 1990, Co
Stefan Schroeder, (D), Bern, 2000, Go
Bart van Dongen, Utrecht, 2003, Co
Emeritus:
Christopher Henderson
Peter Jonas
Grenville Turner, (D), Oxford, 1962, XcCc

University of Newcastle Upon Tyne

School of Civil Engineering and Geosciences
Newcastle University
Newcastle upon Tyne
NE1 7RU
p. +44 (0)191 208 6323
engineering@newcastle.ac.uk
http://www.ncl.ac.uk/ceg/
Professor of Soil Science / Head of School :
David Manning
Professor:
Margaret Carol Bell, Ge
Andras Bordossy, Zg
Peter J. Clarke, (D), Oxford, 1997, YdZri
Stephen Larter, Gg
Zhenhong Li, Yd
Phillip Moore, Yd
Research Associate:
David Alderson, Zi
Joana Baptista, Ng
Stephen Blenkinson, Zn
Bernard Bowler, Go
Aidan Burton
Allistair Ford, Gq
Kirill Palamartchouk, Yd
Instructor:
James Barhurst, Gs
Neil Gray, Pg
Helen Talbot, Co
Lecturer:
Jamie Amezaga, Ge
Stuart Barr, Zi
Colin Davie, (D), Glasgow, 2002, Ngr
Stuart Edwards, Gq
Gaetano Elia, Ng
David Fairbairn, Gq
Rachel Gaulton, Zr
Jean Hall, Ng
Geoffrey Parkin, Hg
Nigel Penna, Gq
Paul Quinn, Hg
Visiting Professor:
Rick Brassington, Hg
Related Staff:
Peter Cunningham, Hg

University of Nottingham

Dept of Mineral Resources
University Park
Nottingham HG7 2RD
wpadmin@nottingham.ac.uk

University of Oxford

Dept of Earth Sciences ☒
Department of Earth Sciences
South Parks Road
Oxford, Oxfordshire OX1 3AN
p. +44 1865 272000
reception@earth.ox.ac.uk
http://www.earth.ox.ac.uk/
t: @OxUniEarthSci
Head:
Alex Halliday, Cs
Chair:
Christopher Ballentine, Xg
Phillip England, Gt
Professor:
Martin Brasier, Pg
Joe Cartwright, Gs
Shamita Das, Gtv
Donalf G. Fraser, Cg
Gideon Henderson, Cg
Hugh C. Jenkyns, (D), Leicester (UK), 1970, GrsCs
Samar Khatiwala
Conall Mac Niocaill, (D), GtgYm
Tamsin A. Mather, (D), Gv
Barry Parsons, Yd
David M. Pyle, (D), Cambridge, 1990, GviRh
Ros Rickaby, Gz
Mike Searle, Gt
Anthony B. Watts, (D), Durham (UK), 1970, GutYv
Bernard Wood, Gz
John Woodhouse, Yg
Associate Professor:
Stuart Robinson, (D), Oxford, 2002, PeGr
Karin Sigloch, (D), Princeton, 2008, YgsGt
Lecturer:
Roger Benson, Pg
Heather Bouman, Cg
Matt Friedman, Pg
Lars Hansen, Gz
Helen Johnson, Og
Richard Katz, Cg
Graeme Lloyd, Gs
Don Porcelli, Cg
Richard Walker, Zr
Emeritus:
Dave Waters, (D), Oxford, 1976, GpzGc

University of Plymouth

Dept of Earth Sciences
Drake Circus
Plymouth, Devon PL4 8AA
p. +44 (0)1752 584584
science.technology@plymouth.ac.uk
http://www.plymouth.ac.uk/schools/sogees

Head:
Mark Anderson, (D), Wales, Gct

Professor:
Antony Morris, (D), Edinburgh, 1990, YmGtu
Greogry Price, (D), Reading (UK), 1994, Gsr

Associate Professor:
Paul D. Cole, (D), 1990, Gvg
Stephen Grimes, (D), Cardiff, 1998, Cs
Martin Stokes, (D), Plymouth (UK), 1997, Gm
Graeme Taylor, Yg
Matthew Watkinson, (D), Open, 1989, Gs

Lecturer:
Sarah Boulton, (D), Edinburgh, 2005, Zr
Arjan Dijkstra, (D), Utrecht, 2001, Gi
Meriel E.J. FitzPatrick, (D), Plymouth (UK), 1992, PlgGr
Luca Menegon, (D), Padua, 2006, Gct
Andrew Merritt, (D), Leeds (UK), 2010, Ng
Kevin Page, (D), Univ Coll (UK), 1988, Gr
Christopher Smart, (D), Southampton (UK), 1993, Pg
Colin Wilkins, (D), James Cook, 1991, Eg

Emeritus:
Malcolm B. Hart, (D), London (UK), 1993, PmGur

University of Portsmouth

School of Earth and Environmental Sciences
Burnaby Road
Portsmouth, Hampshire PO1 3QL
p. +44 (0)23 9284 2257
sees.enquiries@port.ac.uk
http://www.sci.port.ac.uk/departments/academic/sees

Head:
Rob Strachan, (D), Keele, 1982, Cc

Professor:
Andrew Gale, (D), King's Coll (UK), 1984, PgGrCl
Jim Smith, (D), Ge
Craig D. Storey, (D), Leicester (UK), 2002, GptCc

Associate Professor:
David Giles, (D), NzGlm

Research Associate:
Emilie Braund, (D), Graz, 2011, Gz
Fay Couceiro, (D), On
Penny Lancaster, (D), Bristol, 2011, Gz
Robert Loveridge
Darren Naish
David Ray, (D)
Alan Raybould, (D)
Steve Sweetman, (D), Pg
Mark Whitton, (D), Portsmouth, 2008, Pg

Instructor:
Chris Dewdney, (D), Birkbeck, 1983, Yg
Gary Fones, (D), C Lancashire, Cm

Lecturer:
John Allen, (D), Southampton (UK), 1996, Og
Hooshyar Assadullahi
Philip Benson, (D)
Michelle Bloor, (D), Hg
Dean Bullen, (D), Gt
Anthony Butcher, (D), PlmZe
James Darling, (D), Bristol, 2009, Cs
Mike Fowler, (D), Imperial Coll, Gi
David Franklin, (D), Ym
Martyn Gardiner, (D)
Andy Gibson, (D), Eg
Michelle Hale, (D), Flinders, Og
Nick Koor
David K. Loydell, (D), Aberystwyth, 1989, GrPg
Nicholas Minter, (D), Bristol, 2007, Gs
Derek Rust, (D), Gt
Camen Solana, (D), Gv
Richard Teeuw, (D), Stirring (UK), 1985, AsGmZi
Melvin M. Vopson, (D), C Lancashire, 2002, Yg
Nick Walton, Hw
John Whalley, (D), Gc
Malcolm Whitworth, (D)

Adjunct Professor:
David Martill, (D), PvgPe

University of Reading

Soil Research Centre (B,M,D)
Department of Geography and Environmental Science
School of Human and Environmental Sciences
Whiteknights
Reading RG6 6AB
p. +44 (0)118 378 8911
shes@reading.ac.uk
http://www.reading.ac.uk/soil-research-centre

Professor:
Chris Collins, (D), Sb

Associate Professor:
Stuart Black, (D), Lancaster (UK), Cc
Chris Collins, (D), Sc
McGoff Hazel, (D), Liverpool (UK), Pg
Steve Robinson, (D), Sf
Liz Shaw, (D), Sb

Dr:
Joanna Clark, (D), Leeds (UK), Sc

University of South Wales

Geology Section
Department of Science
Pontypridd

University of Southampton

School of Ocean and Earth Science (B,M,D) ● ☒
University of Southampton Waterfront Campus
National Oceanography Centre Southampton
European Way
Southampton, Hampshire SO14 3ZH
p. +44 (0)23 8059 2011
soes@soton.ac.uk
http://www.southampton.ac.uk/oes
f: https://www.facebook.com/UoSOceanography
t: @OceanEarthUoS
Programs: Geology with Physical Geography (B); Geophysical Sciences (B); Oceanography (B); Oceanography with Physical Geography (B); Marine Biology with Oceanography (B); Geology (B,M), Geology with Study Abroad (M); Geophysics (M); Geophysics with study abroad (M); Engineering/Physics/Maths/Geophysics (M); Biology and Marine Biology (M); Marine Biology (M); Marine Biology with study abroad (M); Oceanography (M); Oceanography with French (M); Oceanography with study abroad (M); Foundation Year in Science
Enrollment (2021): B: 45 (11) M: 57 (15) D: 184 (36)

Head of Marine Biogeochemistry Research Group:
Christopher Mark Moore, (D), Southampton (UK), 2002, Ob

Professor:
Stephen Roberts, (D), Open (UK), 1986, Cg

Head of Physical Oceanography Research Group:
Alberto Naveira Garabato, (D), Liverpool (UK), 1999, Op

Head of Palaeoceanography and Palaeoclimate Research Group:
Paul A. Wilson, (D), Cambridge, 1995, GsCg

Head of NOCS Graduate School and Inspire:
Martin R. Palmer, (D), Leeds (UK), 1984, Cg

Head of Marine Biology & Ecology Research Group:
Jorg Wiedenmann, (D), Ulm (Germany), 2000, Ob

Head of Geology and Geophysics Research Group:
Justin Dix, (D), St Andrews, 1994, GuYuNx

Faculty Director of Graduate School:
Timothy A. Minshull, (D), Cambridge, 1990, Yr

Director, GAU-Radioanalytical, Professorial Fellow:
Phillip Warwick, (D), Southampton (UK), 1999, CaGe

Deputy Head of School Research and Enterprise:
Rachael James, (D), Cambridge, 1995, Oc

Deputy Head of School International:
Andy Cundy, (D), Southampton (UK), 1994

Deputy Head of School Education:
Chris Hauton, (D), Southampton (UK), 1995, Ob

Professor:
Nicholas R. Bates, (D), Southampton (UK), 1995, Oc
Thomas Bibby, (D), Imperial Coll, 2003, Ob
Jonathan M. Bull, (D), Edinburgh, 1990, YrGct
Sybren Drijfhout, (D), Urtecht, 1992, Op
Gavin Foster, (D), Open, 2000, Csg
Tim Henstock, (D), Cambridge, 1994, YgrYe
Alan Kemp, (D), Edinburgh, 1985, Pe
Maeve Lohan, (D), Southampton (UK), 2003, Ob
Robert Marsh, (D), Southampton (UK), 2000, Op
John E.A Marshall, (D), Bristol, 1981, PlGro
Lisa McNeill, (D), Oregon State, 1998, GtYr
Duncan A. Purdie, (D), Wales, 1982, Ob
David Sims, (D), Plymouth (UK), 1994, Ob
Martin Solan, (D), National (Ireland), 2000, Ob
Damon A.H Teagle, (D), Cambridge, 1993, CgEmGp
Toby Tyrrell, (D), Edinburgh, 1993, OgPe
Head of Geochemistry Research Group:
Juerg M. Matter, (D), Swiss Fed Inst Tech, 2001, Ng
Associate Professor:
Jonathan Copley, (D), Southampton (UK), 1998, Ob
Thomas M. Gernon, (D), Bristol, 2007, ZgGvi
Ivan D. Haigh, (D), Southampton (UK), 2009, On
Antony Jensen, (D), Southampton (UK), 1982, Ob
Derek Keir, (D), Royal Holloway (UK), 2006, Gtv
Phyllis Lam, (D), Hawaii, 2004, Ob
Cathy Lucas, (D), Southampton (UK), 1993, Ob
Catherine A. Rychert, (D), Brown, 2007, Yg
Rex N. Taylor, (D), Southampton (UK), 1987, GvCa
Sven Thatje, (D), Bremen, 2003, Ob
Clive Trueman, (D), Bristol, 1997, Cg
Jessica H. Whiteside, (D), Columbia, 2006, Pe
Teaching and Research Fellow :
Hachem Kassem, (D), Southampton (UK), 2016, OnGsZi
SMMI Lecturer:
Steven Bohaty, (D), California (Santa Cruz), 2006, Og
Senior Tutor, Principle Teaching Fellow:
Simon R. Boxall, (D), Liverpool (UK), 1985, Op
Senior Research Fellow:
Kenneth Collins, (D), Southampton (UK), 1979, Ob
NERC Advanced Senior Research Fellow:
Tom H.G Ezard, (D), Imperial Coll, 2007
Lecturer:
Maria C. D'Angelo, Ob
Phillip Fenberg, (D), California (San Diego), 2008, Ob
Jasmin A. Godbold, (D), Aberdeen, 2009, Ob
Phillip A. Goodwin, (D), Liverpool (UK), 2007, Pe
Nicholas Harmon, (D), Brown, 2007, Yg
Anna Hickman, (D), Southampton (UK), 2007, Og
Kevin Oliver, (D), E Anglia (UK), 2003, Op
Ryan Reisinger
Marc Rius, (D), Barcelona, 2008, Znn
Martina Stiasny
Esther Sumner, (D), Bristol, 2009, Gs
Chuang Xuan, (D), Florida, 2010, Ym
Dean, Faculty of Environmental and Life Sciences:
Rachel A. Mills, (D), Cambridge, 1992, Cm
Visiting Professor:
Ian Townend, (B), ETH (Switzerland), 1975
Emeritus Fellow:
Lawrence Hawkins, (D), Southampton (UK), 1985, Ob
Visiting Fellow:
Mathis P. Hain, Princeton, 2013
Torben Stichel, (D), Christian-Albrechts (Germany), 2011
Teaching Fellow, Project Manager for Oman Drilling Project:
Jude A. Coggon, (D), Durham (UK), 2010, Cg
Senior Research Fellow:
Matthew Cooper, (D), Cambridge, 1999, Oc
Royal Society University Research Fellow:
Rosalind M. Coggon, (D), Southampton (UK), 2006, Ce
Samantha J. Gibbs, (D), Cambridge, 2002, Zg
Ben A. Ward, (D), Southampton (UK), 2009, Ob

University of St. Andrews

Dept of Earth and Environmental Science (B,M,D)
Irvine Building
North St.
Fife, Scotland KY16 9AL
p. (+44) (0)1334 463940
earthsci@st-andrews.ac.uk
http://earthsci.st-andrews.ac.uk/
f: https://www.facebook.com/EarthSciStA/
t: @EarthSciStA
Director:
Richard Bates, (D), Wales, Yx
Ruth Robinson, (D), Penn State, 1997, Gs
Head:
Tony Prave, (D), Penn State, 1986, GgrGs
Associate Professor:
Andrea Burke, (D), MIT/WHOI, 2011, ClPc
Assistant Professor:
Timothy D. Raub, (D), Yale, 2008, YmGgPc
Research Associate:
Nicky Allison, (D), Edinburgh, 1994, Oc
Mark Claire, (D), Washington, As
Catherine Cole, (D), Southampton (UK), 2013, Cm
Ruth Hindshaw, (D), ETH (Switzerland), 2011, Cs
Gareth Izon, (D), Cg
Coralie Mills
James Rae, (D), Bristol, Cg
Vincent Rinterknecht, (D), CcGma
Lecturer:
Adriam Finch, Gz
Timothy Hill, (D), Edinburgh, 2003, Ge
Michael Singer, (D), California (Santa Barbara), 2003, Hg
John Walden, (D), Wolverhampton Poly, 1990, Ym
Robert Wilson, (D), W Ontario, 2003, Zn
Aubrey Zerkle, (D), Penn State, 2006, Co

School of Geography & Geosciences
Division of Geology
St. Andrews
Fife KY16 9ST
Emeritus:
Colin K. Ballantyne, (D), Edinburgh, 1981, ZyGml

University of Sunderland

Faculty of Applied Science
Edinburgh Building
City Campus
Chester Rd.
Sunderland SR1 3SD
p. (+44) 191 515 2000
john.macintyre@sunderland.ac.uk
http://www.sunderland.ac.uk/faculties/apsc/ourdepartments/cet/

University of Wales

School of Ocean Sciences
Bangor University
Menai Bridge
Anglesey, Bangor LL59 5AB
p. (01248) 382851
oss011@bangor.ac.uk
http://www.bangor.ac.uk/oceansciences/
Director:
Colin Jago
Professor:
David Bowers
Alan Davies
Michel Kaiser
Hilary Kennedy
Chris Richardson
Tom Rippeth
James Scourse
John Simpson
David Thomas
Associate Professor:
Simon P. Neill, (D), Strathclyde, 2001, OnpOg
Instructor:
Jan Geert Hiddink
Stuart Jenkins
Lecturer:
Martin Austin
Jaco H. Baas
Paul Butler
Lui Gimenez

Mattias Green
Cara Hughes
Dei G. Huws, (D), Wales, 1992, YruNt
Suzanna Jackson
Lewis LeVay
Shelagh Malham
Irene Martins
Ian McCarthy
Gay Mitchelson-Jacob
Anna Pienkowski
Martin Skov
John Russel Turner
Katrien van Landeghem
Stephanie Wilson
Related Staff:
Timothy Whitton

Geography and Earth Sciences ☒
Llandinam Building
Penglais Campus
Aberystwyth SY23 3DB
p. +44(0) 1970 622 606
dgostaff@aber.ac.uk
http://www.aber.ac.uk/en/iges/
Director:
Neil Glasser
Head:
Rhys Jones
Professor:
Paul A. Brewer, (D), Gm
John Grattan
Bryn Hubbard
David Kay
Henry Lamb
Richard Lucas
Nick Pearce
Andrew D. Thomas, (D), Swansea, 1996, SbGmSf
Mark Whitehead
Michael Woods
Instructor:
Peter Merriman
Andrew Mitchell
Helen Roberts
Stephen Tooth
Dr:
Morgan Jones
Lecturer:
Peter Bunting
Marie Busfield
Rhys Dafydd Jones
Sarah Davies
Elizabeth Gagen
Hywel Griffiths
Andrew Hardy
Jesse Heley
Tom Holt
Gareth Hoskins
Tristram Irvine-Fynn
Cerys Jones
Sinéad O'Connor
Bill Perkings
Mitch Rose
Bethany Simmonds

Dept of Earth and Ocean Sciences
College of Cardiff
Main Building
Park Place
, Cardiff CF1 3YE
p. +44 (0)29 208 74830
earth-ug@cf.ac.uk
http://www.cardiff.ac.uk/earth/
Professor:
Thomas Blenkinspo
Ian R. Hall
Richard Lisle
Chris MacLeod, Og
Wolfgang Maier
Paul Pearson, PmCg
Instructor:
Stephen Barker
Huw Davies
Andrew Kerr, Og
Jenny Pike, Og
Lecturer:
Tiego Alves, Og
Liz Bagshaw
Rhoda Ballinger, On
Peter Brabham, Ng
David Buchs
Alan Channing, Gd
Jose Constantine
T. C. Hales, Gt
Alan Hemsley, Pl
Tim Jones
C. Johan Lissenberg
Iain MacDonald, Ca
Rupert Perkins
Phil Renforth
David Reynolds
Henrik Sass, Co
Simon Wakefield
Emeritus:
Hazel Prichard, (D), EgGze
Laboratory Director:
Caroline Lear, Og
Tutorial Fellow:
Ian Fryett
Nick Rodgers, Xm

Uruguay

Direccion Nacional de Mineria y Geologia de Uruguay.

Direccion Nacional de Mineria y Geologia de Uruguay
Hervidero 2861
Montevideo, Montevideo 11800
p. + 5982 2001951
infomiem@miem.gub.uy
http://www.dinamige.gub.uy/

Universidad de la Republica Montevideo

Dept de Geologia
Avenida 18 de Julio 1968, Montevideo

Universidad de la Republica Oriental del Uruguay (UDELAR)

Instituto de Geología y Paleontología
Avenida 18 de Julio 1968, Montevideo
CP11 400
Montevideo 10773
http://www.fcien.edu.uy/menu2/estructura2/ingepa2.html

Deptartamento de Geografia (B) ☒
Iguá 4225 Piso 14 Sur C.P: 11400
Montevideo, Montevideo 11400
p. (598-2) 525 15 52
geotecno@fcien.edu.uy
http://geografia.fcien.edu.uy/
Professor:
Virginia Fernández, (M), Girona, 2001, ZirZn
Assistant Professor:
Yuri Resnichenko, (M), la República, 2010, ZirZn
Related Staff:
Carlos Miguel, (B), Valladolid, 2004, ZirSf
Virginia Pedemonte, (B), Fac de Arquitectura, 2005, ZirZu

Dept de Suelos y Aguas
p. (598 2) 359 82 72
suelosyaguas@fagro.edu.uy
http://suelosyaguas.fagro.edu.uy/

Uzbekistan

Institute of Geology and Geophysics
http://www.ingeo.uz/

Institute of Mineral Resources
info@gpniimr.uz
http://www.gpniimr.uz/

National University of Uzbekistan
Faculty of Geology
Tashkent
University City 100095
p. +998712460224
geology@nuu.uz
http://nuu.uz/geolog

State Committee for Nature Protection
100 159, Tashkent, pl. Independence, 5
info@uznature.uz
http://www.uznature.uz

State Committee of the Republic of Uzbekistan on Geology and Mineral Resources
11, T.Shevchenko str., Tashkent, Republic of Uzbekistan 100060
p. +998 (71) 256-8653
geolcom@bcc.com.uz
http://www.uzgeolcom.uz

Tashkent State Technical University
Faculty of Geology
Uzbekistan, Tashkent, 10095 Universitetskaya street, 2
p. +998712464600
TFTU_info@mail.ru

Uzbekistan National Oil and Gas Company - Uzbekneftegaz (UNG)
100047, city Street, Tashkent.
Istiqbol, 21
p. +998 (71) 233-5757
kans@uzneftegaz.uz
http://www.uzneftegaz.uz

Venezuela

Universidad Central de Venezuela
Inst de Ciencias de la Tierra
Los Chaguaramos, Apdo. 3895, 1010-A Caracas
p. 0212 6366236
ictdir@ciens.ucv.ve
http://www.coordinv.ciens.ucv.ve/investigacion/genci/sitios/35

Escuela de Geologia, Minas y Geofisica
Ciudad Universitaria, 47028 Caracas 1041 A

Universidad de la Este, Cumana
Dept de Geologia y Minas
Apartado Postal 245, Cumana (Estado Sucre)

Universidad de Los Andes
Dept de Geologia y Minas
Avenida 3, Independencia, La Hechlcera, Merida
ocamacho@ula.ve
http://llama.adm.ula.ve/pingenieria/index.php?option=com_content&view=article&id=313&Itemid=215

Escuela de Ingenieria Geologica
Av. Tulio Febres C., Merida 5101

Universidad de Oriente, Nucleo Bolivar
Escuela de Ciencias de la Tierra
Av. Universidad, Campus Universitario La Sabanita, Ciudad Bolivar

Universidad Simon Bolivar
Coordinacion de Ingenieria Geofisica
Valle de Sarteneja, Baruta, Edo. Miranda 80659, Caracas 1080
p. 9063545
coord-geo@usb.ve
http://www.gc.usb.ve/geocoordweb

Vietnam

Hanoi University of Mining & Geology
Faculty of Geology (A,B,M,D) O
Duc Thang Ward - North Tu Liem Distr.
Hanoi
p. +84-4-38387567
diachat@humg.edu.vn
http://khoadiachat.edu.vn
Enrollment (2015): B: 2800 (420) M: 205 (50) D: 24 (6)
Dean of Faculty:
Lam Van Nguyen, (D), Hanoi Mining and Geology, 2002, HwGeHg
Deputy Dean of Faculty:
Thanh Xuan Ngo, (D), Okayama, 2009, GtcGi

Hanoi University of Science
Dept of Environmental Science
334 Nguyen Trai St, Dist Thanh Xuan
Hanoi
p. 04-8584615/8581419
hus@vnu.edu.vn
http://www.huse.edu.vn

Institute for Environment Science and Technology (INEST)
1 Dai Co Viet Road
Hanoi
p. (844) 38681686
inest@mail.hut.edu.vn
http://inest.hut.edu.vn/trang-chu

Ho Chi Minh City University of Technology
Geology and Petroleum Engineering
268 Ly Thuong Kiet Street
District 10
Ho Chi Minh City
geopet@hcmut.edu.vn
http://www.hcmut.edu.vn/en/

Ho Chi Minh University
Dept of Geology
227 Nguyen Van Cu St., Dist 5
Ho Chi Minh City, City S1
p. 84.08.8 355 271
nkhoang@hcmus.edu.vn
http://www.geology.hcmus.edu.vn/english

Vietnam Maritime University
Vietnam Maritime University
484 Lach Tray Road
Hai Phong City
p. 84.031.3735931
info@vimaru.edu.vn
http://www.vimaru.edu.vn

Water Resource University
Faculty of Water Resources Engineering & Environment
175 Tay Son Road, Dist Dong Da
Hanoi
p. 844 3852 2201
wru@wru.edu.vn
http://www.wru.edu.vn/en/web/

Zambia

Copperbelt University
Mining Dept (B)
Kitwe
deansot@cbu.ac.zm
http://www.cbu.edu.zm/technology

University of Zambia

Dept of Metallurgy and Mineral Processing (B)
PO Box 32379
Lusaka
p. +360-21-1-250871
registrar@unza.zm
http://www.unza.zm/index.php?option=com_content&task=view&id=479&Itemid=574

Zimbabwe

Africa University

Faculty of Agriculture and Natural Sciences (B)
Fairview Rd (Off- Nyanga Rd)
PO Box 1320
Old Mutare, Mutare
p. +2632060075
info@africau.edu
http://www.africau.edu/academic/

University of Zimbabwe

Dept of Geology (B,M,D)
Building B047
University of Zimbabwe
P.O. Box MP167
Mount Pleasant, Harare
p. 303211 Ext. 15032
gchipari@science.uz.ac.zw
http://www.uz.ac.zw/index.php/2013-07-09-08-51-40/the-department-of-geology/226-sci/dept-sci/geology-dpt/826-dr-lrm-nhamo
Enrollment (2012): B: 27 (17) M: 1 (0) D: 3 (1)

Chair:
LRM Nhamo

Lecturer:
Trendai Jnila, (D)
Isidro Rafael Vit Manuel, (D)
Maideyi Lydia Meck, (D)

Dept of Geography & Environmental Science (B,M,D)
PO Box MP167
Mount Pleasant, Harare
p. +263-04-303211
geography@arts.uz.ac.zw
http://www.uz.ac.zw/science/geography/

Institute of Mining Research (B)
PO Box MP167
Mount Pleasant, Harare
p. +263-4-336418
speka@science.uz.ac.zw
http://www.uz.ac.zw/

Dept of Mining Engineering (B)
PO Box MP167
Mount Pleasant, Harare
p. (263) 4 -3335 x ext: 17089)
kudzie@eng.uz.ac.zw
http://www.uz.ac.zw/index.php/mining-about

Dept of Geoinformatics and Surveying (B)
PO Box MP167
Mount Pleasant, Harare
p. +263 772 318 473
infor@eng.uz.ac.zw
http://www.uz.ac.zw/index.php/fac-of-eng/department-of-geoinformatics-and-surveying

THESES AND DISSERTATIONS, 2020

The following section documents all of the 2020 geoscience dissertations and theses from U.S. and Canadian institutions that were reported to GeoRef Information Services. If you have questions about the data or to make sure your institution's data is included in the future, please contact Monika Long at ml@americangeosciences.org.

Auburn University

Doctoral

Soltis, Nicholas Andrew, *Exploring and measuring the teaching and development of Earth systems thinking skills in undergraduate geoscience courses*

Master's

Barrett, Michael A., *Detrital zircon U-Pb age constraints on the sandstone provenance of the Upper Cretaceous Eutaw Formation from central Alabama to western Georgia*

Cain, Connor, *Natural and anthropogenic contaminants in groundwater and cancer cases in Fruithurst, Alabama*

Fischer, Alicia Brooke, *Field and laboratory investigations of groundwater arsenic sequestration in biogenic pyrite at an industrial site in Florida*

Naher, Jasmin, *Petrofacies analysis and detrital geochronology of Oligocene sediments from the Bengal Basin and southeast Shillong, northeast India*

Perritano, Stefan, *A speleothem precipitation record from Alabama spanning the last 12 thousand years; links between Northern Hemisphere high-latitude cooling and Southeastern US hydroclimate*

Ricketts, Sandor O., *Nature, age, and origin of the informal lower Cenozoic Red Bank Group, northern Belize*

Steltenpohl, Gregory, *A high-resolution paleoclimate record of precipitation variability from a stalagmite in northern Cuba spanning the late Holocene*

Baylor University

Doctoral

Ahmmed, Bulbul, *Numerical modeling, uncertainty analyses, and machine learning for decision support in the geosciences*

Baumgartner, Aly, *Filling the gaps in the floral record; implications for paleobotanical paleoclimate models*

Breen, Katherine H., *Machine learning and remote sensing in the physical sciences; applications in Earth surface and atmospheric modeling*

Flynn, Andrew G., *Early Paleocene fossil floras, paleoclimate, and magnetostratigraphy from the San Juan Basin, New Mexico, USA*

Jiang, Jiajun, *New approaches to characterize and simulate reservoirs by machine learning and model-coupling strategies*

Jin, Cong, *Late Triassic climate reconstructions derived from sandstones, petrified woods and paleosols of the Chinle Formation at Petrified Forest National Park, Arizona*

Sepulveda, Frank, *Edge computing with an "Internet of Things" based sensor array; an innovative approach to near real time seismic exploration and monitoring*

Master's

Casteel, Wynne Marcus, *Using microgravity and passive seismic methods jointly to explore the Brazos River alluvium aquifer*

Tew-Todd, Victoria Lee, *Late Quaternary stratigraphy and chronology by thermal-transfer optically stimulated luminescence dating of quartz grains for the Monahans eolian system, Winkler County, Texas*

Brigham Young University

Master's

Barton, Katelyn Jean, *Links between eruptive styles, magmatic evolution, and morphology of low-shield volcanoes; Snake River plain, Idaho*

Gunnell, Nathan Vaun, *A study of the anthropogenic impact in Farmington Bay through isotopic and elemental analysis*

Henze, Porter, *Implications of geochemistry and textures of titanite for the geologic histories of the Notch Peak Intrusion and Little Cottonwood Stock, Utah*

Parker, Sheena Marie, *Dynamics of internal phosphorus cycling in a highly eutropic, shallow, fresh water lake in Utah Lake State Park, Utah, USA*

Voorhees, Jacob Isaac, *Structure of collisional metamorphism, soft-sediment deformation, and low-angle normal faulting in the Beaver Dam Mountains*

Brock University

Master's

Pentesco, Justin, *Assessing environmental controls for candidate microbial ichnofossils in hydrothermally altered basaltic tuffs of Upsal Hogback, NV*

Brown University

Doctoral

Bramble, Michael Stephen, *Advancing thermal emission spectroscopy of ordinary chondrites and their parent bodies with laboratory experiments in a simulated asteroid environment*

Palumbo, Ashley, *Characterizing the climate and hydrologic cycle of early Mars; assessing the role of rainfall, groundwater, and transient atmospheric heating*

California Institute of Technology

Doctoral

Anderson, Dana Eklund, *Tracking volatile elements in protoplanetary disks and on planetary surfaces*

Bailey, Elizabeth, *Interior and orbital dynamics at the innermost and outermost reaches of planetary systems*

Chadwick, Austin John, *Mechanics of river avulsions on lowland river deltas*

Chadwick, Grayson Lee, *How to beat diffusion; explorations of energetics and spatial relationships in microbial ecosystems*

Finke, Cody Enslin, *An experimental and economic analysis of electrochemical technologies to reduce greenhouse gas emissions*

Lai, Voon Hui, *Seismic waveform modeling of natural hazards and sharp structural boundaries*

Leask, Ellen Kathleen, *Investigating the evolution of surface water on Mars through spectroscopy of secondary minerals*

Martin, Peter Eckels, *Detection and analysis of Martian low-temperature geochemistry*

Naviaux, John David, *Chemical and physical mechanisms of calcite dissolution in seawater*

Rooney, Brigitte Lee, *Modeling the impact of biomass combustion on atmospheric aerosol*

Stein, Nathaniel Thomas, *Investigation of past habitable environments through remote sensing of planetary surfaces*

Zhu, Yanzhe, *3D microfluidics for environmental pathogen detection and single-cell phenotype-to-genotype analysis*

California State University at Bakersfield

Master's

Mohammadi, Obeyd, *A late Pleistocene record of highstand shedding and dust flux from the inner sea of the Maldives; Site U1467, IODP 359*

California State University at Fresno

Master's

Larsen-Van Vleet, Amalie, *Transtensional strains tap magmas from the middle crust, and low-degree melts in crystal-liquid mush zones; evidence from the Sierra Crest-Little Walker volcanic center, Central California*

Platt, Bryant William, *Constraining internal eruption trigger mechanisms for flows at Brokeoff Volcano, Lassen volcanic center, California*

Polon, Casey Katherine, *65 km of dextral slip across Owens Valley, CA since 83 Ma; possible, but not necessary*

California State University at Long Beach

Master's

Boyd, Daniel, *Characterizing the geometry, kinematics, and potential connectivity at the southern termination of the Palos Verdes Fault*

Dombrowski, Anthony P., *Cycle counting procedures as basis for magnitude scaling effects in liquefaction risk models*

Farrell, John C., *Lateral variation of fine-grained, organic-rich sediments of the Miocene Monterey Formation, Belridge Field area, California*

Kelly, Clay, *Episodic deformation along a blind thrust in the past 125 kyr; the Santa Ynez River fault in the western Transverse Ranges of California*

Sedlak, Alexander, *Chemostratigraphy of the Miocene Monterey Formation in the San Joaquin Basin, California*

California State University at Los Angeles

Master's

Rodriguez, Jessica, *Site catchment analysis of paleoethnobotanical data from Chumash sites in the Santa Monica Mountains, California*

Carleton University

Doctoral

Davey, Sarah, *Testing the paleogeography of late Archean supercraton Superia using pre-breakup 2.51-1.98 Ga dike and sill provinces; with a focus on the relationship between the Karelia-Kola and Superior cratonic fragments*

Mann, Arjan, *A systematic and ecomorphological investigation of the early amniotes from Mazon Creek, Francis Creek Shale, Illinois, USA*

Nasser, Nawaf, *Lacustrine Arcellinida (testate lobose amoebae) as bioindicators of arsenic contamination; a new tool for environmental risk assessment*

Master's

Adams, Gabrielle Ruth, *Description of Calligenethlon watsoni based on computed tomography and resulting implications for the phylogenetic placement of embolomeres*

Dyer, Sabastien Coltrane, *The Early Jurassic metamorphic history of the Yukon-Tanana Terrane of northwestern British Columbia; insights from a new inverse garnet fractionation modelling technique*

Heldwein, Olivier Karl Anton, *Metamorphic evolution of the lower nappes of the Kalak Nappe Complex in eastern Finnmark (Norway)*

Khawja, Sara, *Tesserae on Venus; evidence of fluvial erosion and characterization of lineaments*

Little, Kathryn Jean, *Geology and geochemistry of the 25 Ma Underdown Caldera tuffs and the tuff of Clipper Gap, western Nevada volcanic field, north-central Nevada*

Robb, Samuel, *Distribution of platinum group elements in the New Afton alkalic Cu-Au porphyry deposit, Kamloops, British Columbia*

Stead, Cassandra Megan, *Using a heated column experiment to investigate the long-term effects of thermal recovery operations on the release, mobilization, and attenuation of arsenic in groundwater*

Wyenberg-Henzler, Taia Constance Alexis, *Ontogenetic niche shifts in megaherbivorous dinosaurs of Late Cretaceous North America and their ecological implications*

College of William and Mary

Doctoral

Coleman, Daniel J., *The role of suspended sediment in assessing coastal wetland vulnerability*

Backelor's

Laughlin, Evan, *Deposition and deformation of marbles at the Blue Ridge-Piedmont transition, central Virginia*

Volante, Paul, *The advantages of ^{241}Am as a dating tool in sedimentary systems*

Walter, Ryan, *Unmasking a terrane triple point in central Virginia*

Zach, Terri, *Structural analysis of the Shores Mélange at the Seven Islands, central Virginia Piedmont*

Colorado School of Mines

Doctoral

Adekunle, Olawale O., *Laboratory study of the effect of chemical osmosis on the elastic rock properties of Pierre Shale and implication for oil recovery*

Colorado School of Mines

Doctoral

Cardona, Sebastian, *Finding order in chaos; a multi scale study of mass-transport deposits*

Cardona, Sebastian, *Finding order in chaos; a multi-scale study of mass-transport deposits*

Colborne, Jacqueline, *A multi-scale approach to reservoir characterization of the Wolfcamp A, Delaware Basin, Texas*

Forrester, Mary Michael, *Understanding water fluxes through land, vegetation and air with coupled atmospheric and hydrologic modeling at extreme scales*

Hansford, Mark R., *The effects of climate on fluvial discharge and key controls of fluvial fans; a quantitative study*

Hernandez Uribe, David, *Constraining the tectonothermal evolution of subducted oceanic crust using thermodynamic phase equilibrium modeling*

Hou, Pengfei, *Facies trend and statistical characterization of deepwater systems during remnant ocean-early foreland transition; Lower Pennsylvanian of the Ouachita Mountains, USA*

Hu, Wei, *Developing a system to measure rheology of conditioned soil for soft-ground TBM tunneling*

Illig, Peter Edward, *The porphyry and skarn deposits of the Chillagoe mining district, northeast Queensland, Australia*

Li, Haipeng, *Environmental signal propagation, preservation, and identification in sediment routing systems*

Li, Zhen, *Water and heat transport in shallow subsurface soil and across the soil-air interface; simulation, experiments and parameterization*

Lu, Hui, *Uncertainty analysis in rock mass classification and its application to reliability evaluation in tunnel construction*

Pettinga, Luke A., *Submarine canyons, channels, levees, and lobes; an investigation of the linkages between depositional processes, stratigraphy, and sedimentology*

Pommer, Rania Eldam, *From pore-scale to watershed; parameterizing heterogeneous rock properties across scales to inform predictive environmental models*

Rey, David, *Evolving subsurface connectivity in Arctic and alpine ecosystems; the impact of permafrost and seasonally frozen ground*

Semmens, Stephen, *Geospatial and logistic regression assessment of factors contributing to backwards erosion piping*

Severson, Allison R., *Across- and along-strike structural and geochronological variations of the Nashoba-Putnam and Avalon terranes, eastern Massachusetts, Connecticut, and Rhode Island, southeastern New England Appalachians*

Master's

Al Ghaithi, Aun, *Deep learning methods for shear log predictions in the Volve Field Norwegian North Sea*

Arabnia, Omid, *The depositional timing of lobate landforms in the Park and Leon creek drainages with comparisons to West Salt Creek landslide, Grand Mesa area, Mesa County, Colorado*

Beetle-Moorcroft, Fern, *Exploring geologic controls on infiltration and groundwater recharge on an ephemeral river; a coupled geophysics and modeling approach*

Blake, Jenny A., *Geologic reservoir characterization of the Avalon Shale and 1st Bone Spring Formation in the southern Delaware Basin*

Bone, Courtney R., *Petroleum geology of the Turner Sandstone Member of the Carlile Shale, Weston County, Wyoming*

Bonner, Edward P. T., *Shortwave infrared spectral analysis of hydrothermal alteration in the main and southwest Tintic mining district and exploration implications of the Silver City porphyry Cu-Mo-Au prospect, Tintic mining district, Utah, USA*

da Silva, Atilas Meneses, *Rock physics modeling for quantitative seismic characterization of tight-gas sandstones in Neuquen Basin, Argentina*

Damasceno, Andrea, *4D quantitative interpretation of Jubarte Field (Brazil); an integrated approach*

Dwihusna, Nadima, *Seismic and well log based machine learning facies classification in the Panoma-Hugoton Field, Kansas and Raudhatain Field, north Kuwait*

Griffin, Alden A., *Mechanisms of sediment movement from topset to toeset in the Cenozoic clinoforming NW Australian margin, and implications for reservoir development, a seismic geomorphology study*

Gross, Evan C., *Three-dimensional eolian bounding surface architecture of the Entrada Sandstone, Utah; expanding our understanding of reservoir heterogeneity in wet eolian systems*

Lehman, Miranda Rae, *U-Th-Pb monazite constraints on the timing of Paleoproterozoic metamorphism in Big Thompson Canyon, Colorado Front Range*

Livingston, Kelsey, *The peralkaline Mount Rosa Granite; contrasting mineralogy and geochemistry observed in the Mount Rosa Granite, Pikes Peak Batholith, Colorado*

Marberry, Kristen, *Subsurface characterization and liquefaction potential assessment of the Spirit Lake debris blockage at Mount St. Helens, Washington using multi-method geophysics*

Noosri, Ratthapon, *Petroleum exploration in the mature Phitsanulok Basin, Thailand; success and failure analysis, and characterization of source rocks and crude oils*

Powell, Logan, *The Proterozoic geology of the northern half of the Mount Evans 7.5-minute quadrangle*

Radach, Kyle C., *Characterization, analysis, and remediation of the Cedar Pass landslide complex, Badlands National Park, South Dakota*

Sullivan, C. Bane, *PyVista; managing & visualizing geospatial data using an open-source framework*

Tello, Matthew, *Optimization of landslide susceptibility modeling; a Puerto Rico case study*

Wallace, Cory S., *A new mobility measure and scoring system for predicting long-runout landslides*

Colorado State University

Doctoral

Baker, Michael G., *Ambient and teleseismic elastic wavefields of the Ross Ice Shelf, Antarctica, and their application to crustal scale seismic imaging*

Seymour, Nikki M., *Magmatism, deformation, and mineralization along the intra-arc Atacama fault system, northern Chile*

Master's

Cole, Hank M., *Tidally induced seismicity at the grounded margins of the Ross Ice Shelf, Antarctica*

Gilkerson, Tyler, *Understanding carbon nanoparticle transport in saturated porous media; the influence of dissolved organic matter*

Grabowski, Julia, *Logjam attenuation of annual sediment waves in eolian-fluvial environments, North Park, Colorado*

Heath, David C., *Detection and relocation of earthquakes in the sparsely instrumented Mackenzie Mountains region, Yukon and Northwest Territories, Canada*

Spansel, Joel, *False Bakken interval-sediment patterns and depositional architecture at the facies boundary between siliciclastic mudstones and carbonates, Lodgepole Formation, Mississippian in the Williston Basin, ND*

Sturdivant, Matthew, *Hydrogeophysical investigation of unconfined aquifer drainage behavior using temporal microgravity and water level data*

Vento, Noah Francis Ryoichi, *Hypothesis-based machine learning for deep-water channel systems*

Dalhousie University

Doctoral

Aali, Masoud, *Analysis of the cyclic character of siliciclastic sediments at the New Jersey continental shelf using ultra-high resolution 3D seismic and well data*

Master's

Rotteveel, Lobke, *Trends, patterns, and drivers of freshwater aluminium concentrations; revisiting the conceptual model of freshwater acidification*

Tobey, Dawn E., *Water and wind; the fluvial and eolian forces behind the Pennsylvanian-Permian Halgaito Formation, Utah*

Backelor's

Berry, Bay, *Quantifying impacts of spatial resolution on pixel and object-based methods of image classification; a case study of identifying estuarine morphology in Cobequid Bay, Nova Scotia, Canada*

Rolfe, Olivia, *The strength evolution of a crustal scale shear zone*

Florida International University

Doctoral

Samajpati, Eshita, *Volcanic history of the Izu Bonin Mariana Arc prior to the first arc rift from volcaniclastic sediments of DSDP Site 296 and IODP Site 1438*

Florida State University

Doctoral

Ahmed, Nur, *Isotope analysis and groundwater modeling for advanced understanding of lake water and groundwater mixing through lake sinkholes in north Florida*

Almuntshry, Nawwaf Awad, *Structural evolution of the Mulberry recess eastern Blue Ridge-Talladega; northwest Georgia Appalachians*

Bowman, Chelsie Nicole, *Coupled evolution of the biosphere and global marine redox conditions in the Late Silurian*

Cruz, Jarrett W., *Utilizing Indian Ocean calcareous nannofossil biostratigraphy of Miocene sediments to investigate tectonic and climatic forcing during the evolution of the Himalayas*

Master's

Eke, James E., *Geology of the Mountville 7.5' Quadrangle, Georgia and preliminary investigation of the "Odessadale" granite gneiss of the Opelika Group*

Govert, Daniel John, *Investigating the role of marine redox conditions and orbital forcings in an Early Silurian mass extinction (Ireviken event); a deeper water perspective*

Illinois State University

Master's

Bosompemaa, Patience, *Nitrate transport in the unsaturated zone*

Konop, Preston J., *Impacts of land-use change on the Menominee River in Jo Daviess County, Illinois*

Iowa State University of Science and Technology

Doctoral

Thompson, Natalie Sarah, *A multi-proxy reconstruction of sea ice and primary productivity during Marine Isotope Stage 11 in the Bering Sea*

Lakehead University

Master's

Ramsay, Brittany, *Environmental control of seawater geochemistry in a Mesoarchean peritidal system, Woman Lake, Superior Province*

Marshall University

Backelor's

White, Olivia, *Quantitative analysis using energy-dispersive X-ray spectroscopy; standardization and the effects of carbon coat thickness*

Massachusetts Institute of Technology

Doctoral

Bramante, James F., *Wave-driven geomorphology of Pacific carbonate coastlines; from landscape to wavelength scale*

Chen, Christine Y., *U-Th dating of lacustrine carbonates*

Collinet, Max, *Experimental and analytical studies of partial melting in planetesimals and the Martian mantle*

Golos, Eva Marie, *Imaging and interpreting seismic heterogeneity in the North American lithosphere*

Marks, G. Serrato, *Investigating Mexican paleoclimate with precisely dated speleothems*

Moore, Kelsey Reed, *Cyanobacterial evolution and interactions with the Proterozoic world*

Rhim, Jeemin Hannah, *Experimental investigations of isotopologue fractionation during microbial methanogenesis*

Sutherland, Kevin M., *New insights into the marine oxygen cycle from manganese oxide minerals and reactive oxygen species*

Wallace, Elizabeth Jane, *High resolution sedimentary archives of past millennium hurricane activity in the Bahama Archipelago*

Master's

Dabrowski, Jessica Stephanie, *Radium isotopes and radon-222 as tracers of sediment-water interaction in Arctic coastal and lacustrine environments*

Safrit, Taylor K., *Centaur shapes and the origin of bilobate Jupiter-family comets*

Togaibekov, Anuar Zhanybekovich, *Monitoring of oil-production-induced subsidence and uplift*

Backelor's

Morgenstein, Kyle Jarad, *Exploring the silicification of microbes and understanding their role in the fossil record*

Rushlow, Matthew R., *Using machine learning, particle tracking, and grain shape modeling to characterize bedload sediment transport*

Tweedy, Ruth Rosegrant, *Tufa-based reconstructions of Huasco Basin lake levels*

McMaster University

Doctoral

McNeill-Jewer, Chelsi, *Development and application of μXRF-CS Cl as a proxy for Holocene drought and hurricane conditions in the Yucatan Aquifer, Mexico*

Shatilla, Nadine Joan, *New insights into catchment dynamics using novel approaches*

Thorne, Robin F., *Soil respiration dynamics in response to climate oscillations and shelterwood harvesting in a temperate pine forest*

Master's

Dereviankin, Mikhail, *Monitoring spatial distribution of solvent extractable organics in pit lake fluid fine tailings*

Elliott, Claire, *Mobilization of entrapped gases in quasi-saturated groundwater systems contaminated with biofuel additives*

Ma, Yueqian, *Analysis and modeling of soil CO_2 emissions within temperate coniferous and deciduous forests*

Miami University (Ohio)

Doctoral

Cymes, Brittany Allison, *Catalytic properties of novel microporous minerals*

Kugler, Alex, *Environmental interactions of phyllosilicates; microbial habitation, respiration, and organomodified photoreductive scaffolding*

Master's

Blakemore, Daniel, *Insights into the history of pyrite mineralization at the Round Mountain gold mine, Nevada; a detailed microanalytical study of the type 2 ore*

Emproto, Christopher Robert, *Column anion and trace element chemistry of apatite from crustal carbonatite deposits in the Grenville Province; implications for crustal carbonatite genesis*

Mississippi State University

Doctoral

Cartwright, John H., *Assessment of coastal watershed erosion potential using geographic information systems and expert input for decision support*

Cho, Youngwoo, *Integration of virtual learning in college geology education*

Master's

French, Joseph, *Optimization of marsh terracing as a wetland restoration technique; mitigation of cohesive sediment erosion by waves associated with frontal passage*

Intsiful, Adjoa Dwamena, *Glacier change assessment of the Columbia Icefield in the Canadian Rocky Mountains, Canada (1985-2018)*

Missouri University of Science and Technology

Doctoral

Abbas, Ahmed Khudhair, *An integrated wellbore stability study to mitigate expensive wellbore instability problems while drilling into Zubair shale/sand sequence, southern Iraq*

Al Saedi, Ahmed Qasim, *Exploring wellbore and reservoir temperature profiles in fluid circulation and production operations*

Al-Saedi, Hasan N., *Novel insights into low salinity water flooding enhanced oil recovery in sandstone reservoirs*

Alfuqara, Diya Ali Ahmad, *Geophysical assessment of the Lake Chesterfield site, Missouri*

Aljabbab, Abdulrahman Ahmed, *Geophysical characteristics of nickel and copper deposits in Wadi Al Khadra Prospect, southwestern of Saudi Arabia*

Almahfood, Mustafa M., *Experimental study on a novel EOR method; polymeric nanogel combined with surfactant and low salinity water flooding to enhance oil recovery*

Alzahrani, Hassan Abdullah, *Imaging in karst terrain using the electrical resistivity tomography (ERT) and multi-channel analysis of surface waves (MASW) methods*

Ding, Haifeng, *Factors impacting the transport and enhanced oil recovery potential of polymeric nanogel in sandstone*

Fakher, Sherif M., *Investigating the factors impacting the success of immiscible carbon dioxide injection in unconventional shale reservoirs; an experimental study*

Hernandez, Elisandra, *History of natural and anthropogenic activities in Lake Izabal, eastern Guatemala; using geochemical evidence to record recent contamination*

Ilesanmi, Olufeyisayo B., *Predictive geohazard mapping using LiDAR and satellite imagery in Missouri and Oklahoma, USA*

Mohamod, Majid S., *Development of tools for water management in the Hatra watershed (northwestern Iraq) using satellite technologies*

Wang, Dan, *A broad Iceland Plume associated with two phase transitions at the 660 km discontinuity; constraints from receiver functions*

Wang, Tuo, *Seismological investigations of lithospheric structure beneath the East African Rift system and the Kalahari Craton*

Zhang, Yandong, *Fabrication of micro-/nanofluidic models and their applications for enhanced oil recovery mechanism study*

Master's

Clinton, Justin James, *Determining variable depth to bedrock using ERT and MASW; a geophysical investigation in St. Louis, MO*

Lin, Baihua, *Microgel placement and plugging performance in sand filled fracture for conformance control*

Miller, William Joseph, *Micro-analysis of contaminants in suspended load and bank deposits from the Mineral Fork Creek Watershed in Potosi, Missouri, 2019-2020*

Quick, J. Daniel, *The role of tectonic inheritance, plate-reorganization, and magma flare-ups in the evolution of the Sevier Orogeny*

Sehly, Turki Essam A., *Evaluation of engineering properties of soil in Al Butayn, Mudarraj Al Qassim and Turubah-Hail Saudi Arabia*

Sun, Qiaoqi, *Strain partitioning across the Polochic-Motagua Fault System in Guatemala; insight from kinematic modeling*

Tunnell, Kaylin Agnata, *Mineral compositions of meteorite samples ALHA 77005, NWA 7397, NWA 5480, and Ausson; implications for geodynamic processes on their parent bodies and meteorite reclassification*

Waqar, Ali, *Data analytics of natural gas injection enhanced oil recovery for unconventional reservoirs*

Yin, Shize, *Evaluation of EOR potential of low salinity polymer in enhancing heavy oil recovery on the Alaska North Slope*

Montana State University

Master's

Howlett, Caden James, *Syn-orogenic magmatism, mid-crust exhumation, and placer gold deposition; the Anaconda metamorphic core complex of western Montana*

Montclair State University

Master's

Sandefur, Mathew L., *The behavior of the Baltic Sea Ice Stream during the deglaciation of the Baltic Ice Lake, recorded in the particle size and geochemistry within Bornholm Basin, IODP Site M0065*

Murray State University

Master's

Elliott, Zachariah Jordan, *Comparison of hyperspectral imaging and portable X-ray fluorescence in ceramic analysis and sourcing*

Stewart, Christopher Lance, *Paleofloods and landform development influence carbon storage in a humid-subtropical river valley*

Backelor's

Meyer, Matthew T., *Reconstructing late Postclassic exchange networks in the Cerro el Vigia area of the southern Gulf lowlands, Veracruz, Mexico*

White, Marie Nicole, *Fossil soil variability during the late Pleistocene and African Humid Period at Gona, Afar region, Ethiopia*

New Mexico State University at Las Cruces

Master's

Ridl, Shay Parker, *Provenance and sediment mixing trends of the Plio-Pleistocene ancestral Rio Grande fluvial system, Rio Grande Rift, New Mexico*

Northern Arizona University

Master's

Head, Eric, *Channel narrowing and floodplain inundation in Meander Canyon, Colorado River, Canyonlands National Park*

Morrison, Heather Iona, *Portable X-ray fluorescence compositional analysis and recording techniques of pictographs at Canyon de Chelly National Monument*

Northern Illinois University

Master's

Mastro, Joseph R., *Estimating paleotemperatures in the Amundsen Sea for the last 620,000 years through diatom morphometrics (IODP Expedition 379)*

Ohio State University

Backelor's

Brown, Alexander, *Characterization of surficial sediments from an island of Wax Lake delta, Louisiana (USA)*

Harris, Kira, *Geomorphic and sedimentological controls on hyporheic flow in an alpine river*

Harris, Mason, *XCT analysis of decreasing porosity in the Trenton Limestone of Williams County, Ohio*

Kleban, Lillian, *Dissolution kinetics of volcanic ash*

Kube, Johnathan, *Evidence of symbiotic zooxanthellae in a rugose coral, Siphonophrentis, from the Devonian of Ohio*

Lucas, Elijah, *Chemical weathering yields in relation to land use in the Cuyahoga and Maumee Watersheds*

Midraj, Nawal, *Abiotic methane geologic sites on Earth and the relation to methane on Mars*

Monagle, Caitlin, *The geochemistry and impacts of chemical weathering on the central Namib Desert gravel plains soils*

Piergallini, Brianna, *Analysis of acid-leacable barium, copper, iron, lead, and zinc concentrations in Taylor Valley, Antarctic stream sediments*

Schneider, Samuel, *Stability analysis of the ancient Sevier gravity slide, southwest Marysvale volcanic field, Utah*

Old Dominion University

Master's

Close, Brian James, *Constraining respired carbon storage in the eastern tropical Pacific over the last 25 thousand years using benthic foraminiferal boron/calcium ratios*

Watkins, Colton Steele, *Sodium-calcium ratios in the planktic foraminifera Trilobatus sacculifer as a proxy for sea surface salinity*

Oregon State University

Doctoral

Kuna, Vaclav M., *Investigation of slip and tectonics of the Blanco transform fault using high-resolution ocean bottom seismic data*

Manopkawee, Pichawut, *Spatial and temporal landscape responses to progressive tectonic forcing along California fault systems*

Ross, Ann Elizabeth Morey, *Investigating the paleoseismic potential of small, southern Cascadia lakes*

Pennsylvania State University at University Park

Doctoral

Ajayi, Seyi, *Examining the fidelity of carbon isotope records in the Paleocene-Eocene Thermal Maximum*

Jones, Heather L., *The recovery of calcareous nannoplankton following the Cretaceous-Paleogene (K-Pg) mass extinction event 66.0 million years ago*

Backelor's

Alsaad, Nouf, *Iceberg calving, meltwater plumes, and glacial retreat at Helheim Glacier, Greenland*

Alshimmari, Al Reem, *Analyzing the subsurface of the SSHCZO using seismic waves*

Alzhanova, Albina, *The feasibility of utilizing AMD in hydraulic fracturing in Pennsylvania*

Andersen, Eric D., *Oreodont morphological change in response to grassland expansion and its attribution to body size*

Aquino, Carl Fredrick, *Probabilistic inversion of structured expert judgment to inform a simple Antarctic ice sheet model*

Asri, Lukman Nurhakim Mohd, *Earthquake triggering along Vanuatu Islands*

Baharudin, Fatin Farihah, *Assessing tephra correlations from Ambare to the fossiliferous Asboli region in the Ledi-Geraru Project in Ethiopia*

Barr, Alyssa Marie, *Correlating 2.5 Ma tephras from Afar, Ethiopia*

Bin Mohamed Zaini, Muhammad Faris Zafir, *Constructing the structure of Thwaites Glacier's ice shelf by seismic imaging*

Binti Mohd Najib, Nuriffa Syahira, *Landslide induced tsunami hazard in narrow inlets in Southeast Asia*

Block, Elizabeth, *Mineralogical and textural analysis of pyrite nodules and central Pennsylvania host bentonite; a qualitative and quantitative assessment*

Carbaugh, Robin, *Investigating volcanic tephra for potential correlations to other feldspars in Afar, Ethiopia*

Crescenzo, Nicholas, *The Mercer Formation as a source of lithium in Clearfield County, Pennsylvania*

Dechdacho, Porraket, *Analyzing the weathering processes in the Susquehanna Shale Hills critical zone observatory using geochemistry and seismic modeling*

Elliott, Samuel, *Interpreting changes in bedform along flow beneath Kamb ice stream in West Antarctica through radar data*

Fulginiti, John, *Comparing biomineralized and organomineralized elemental sulfur*

Furfaro, Dion, *A comparison of gas and fluid production for the Utica/Pt. Pleasant and Marcellus Formation shales*

Gagnon, Cathy, *Hydrologic and landscape evolution near a Neolithic settlement in Dalmatia, Croatia*

Ju, John, *Analysis of groundwater overflow resulting from urbanization in Seoul*

Leddy, Matthew, *Late Quaternary paleoecological and paleosedimentological responses to paleoclimatic shifts at Bear Meadows Natural Area*

Loucks, Emily, *The role of sediment grain size and bedrock abundance in setting the morphology of headwater channels in McKittrick Canyon in the Guadalupe Mountains, NM/TX*

Marshall, Whitney, *Investigating the possibility of using thunderquakes by Penn State's fiber optic sensors to map the velocity structure underneath State College*

Ming, Cissy, *Rapid remobilization of intermediate arc magma; insights from textural analysis, mineralogy, and geochemistry of lava samples from Momotombo Volcano, Nicaragua*

Pippin, James, *Investigation of mid-lithospheric discontinuities beneath North America using P-wave receiver functions*

Raadnui, Sapol, *Basin thermal evolution of the Phitsanulok Basin, Thailand*

Reilly, Katherine, *Water quality analysis in the Marcellus Basin; constraining* *source contaminants via two-dimensional gas chromatography and strontium isotopes*

Rosland, Iman Nasuha, *Sedimentary & bulk organic geochemical characterization of newly discovered stratigraphy from Olduvai Gorge*

Sabattier, Alexandra, *The Bushveld Complex; evidence for heterogeneous upper crustal structure*

Swaminathan, Varsha, *Geological consequences of mineral precipitation and dissolution; implications of subsurface mineral carbon sequestration*

Tamburri, Christine, *Nitrogen cycling in a modern analogue of mid-Proterozoic euxinic environments; observations from Fayetteville Green Lake*

Tierney, James, *Analysis of trace metal contaminants in stormwater impacted sediment at the Penn State Duck Pond Basin*

Yusri, Muhamad Nazmi, *Quantifying the impact of coarse sediment input from landslides and tributaries on river morphology in the Lele River, Taiwan*

Zaidy, Amir Aiman, *Texture analysis of fibrous veins of celestine in Bellwood, PA and calcite in Bear Meadows, PA*

Portland State University

Doctoral

Cahoon, Emily Bogdan, *Distribution, geochronology, and petrogenesis of the Picture Gorge Basalt with special focus on petrological relationships to the main Columbia River Basalt Group*

Master's

Glenn, Bryce Allen, *Assessing airborne radar to map glacier elevations in alpine terrain including estimated glacier volume change*

Martin, Gregory David Linscott, *The Rocky Ledge Fault, Shasta County, NE California; development and morphology of a Quaternary oblique normal fault in basalt*

Princeton University

Doctoral

Harris, Rachel Lee, *Life on the fringe; surveying the ecophysiological tenacity of methanogens and anaerobic methanotrophs in the oligotrophic deep subsurface biosphere*

Kasbohm, Jennifer Jean, *Calibrating Archean and Miocene large igneous province emplacement and geologic timescales with high-precision U-Pb geochronology*

Kast, Emma Ruth, *Development and application of biogenic mineral-bound nitrogen isotope measurements to the million-year timescale*

Langer, Leah, *Impact of topography on coseismic modeling and earthquake static slip inversions*

Schlesinger, Danielle Rebecca, *Salting the Earth; biogeochemical cycling of chlorinated and brominated natural organic compounds in coastal ecosystems*

Simon, Joel David, *Recording earthquakes in the oceans for global seismic tomography by freely-drifting robots*

Sun, Xin, *New processes and microbes in the marine nitrogen cycle*

Rensselaer Polytechnic Institute

Doctoral

Pettitt, Elizabeth A., *New constraints on ancient atmospheric gas concentrations from fluid inclusions in surface materials*

Rice University

Doctoral

Blank, David G., *Investigating earthquake rupture using particle dynamics simulations*

Carlson, Brandee N., *The geomorphic evolution of abandoned deltaic lobes and distributary channels*

Chowdhury, Proteek, *Geochemical fractionation during melting of sulfide and sulfate bearing mantle lithologies*

Dong, Tian Yang, *Dynamics of delta building across different scales as informed by the Selenga River delta, Russia*

Moodie, Andrew J., *Assessing deltaic landscape management strategies based on studies from the Yellow River delta, China*

Seales, Johnny, *Uncertainty and hypothesis testing in planetary thermal history models*

Semple, Alana Gambini, *Earth's asthenosphere; connecting observations and mantle dynamics*

Suciu, Loredana G., *Anhydrosugars as tracers of fire air quality effects, carbon cycling and paleoclimate*

Wu, Chenliang, *Coupling non-uniform channel hydrodynamics and development of fluvial-deltaic stratigraphy, over scales ranging from bedforms to basins*

Zhang, Julin, *The applications of data science to petrology and geochemistry*

Master's

Baykut, Tanyel, *3D modern reef morphologies along the upper slopes of the Lighthouse Reef (Belize); potential archive of punctuated sea level fluctuations, using a global comparative basin analysis approach*

Cole, Trevor L., *The hydrochemical signatures of (in) congruent weathering in Iceland*

Sharton-Bierig, Eytan, *Widespread disequilibrium of phosphorus in olivine*

Backelor's

Nirenberg, Jared, *Glacially-controlled variations in the biological pump of the Ross Sea in the mid-to-late Pliocene*

Rutgers University

Master's

McLellan, Oliver John, *Late middle to late Miocene paleoceanography at Rockall Plateau Site 982*

Rutgers, The State University of New Jersey at New Brunswick

Doctoral

Hong, Isabel J., *Developing proxies to reconstruct the intensity and magnitude of prehistoric tropical cyclones and earthquakes*

Rutgers, The State University of New Jersey at Newark

Master's

Juarez, Karen M., *Calcite recrystallization at equilibrium as influenced by initial growth rate and pH*

Simon Fraser University

Doctoral

Chateau, Chloe, *Reconstructing depositional architecture and stratigraphy of coastal- to shallow-marine strata in a low-accommodation system; McMurray Formation, Alberta, Canada*

Le Moigne, Yannick, *Investigating Canada's deadliest volcanic eruption and mitigating future hazards*

Master's

Binner, Megan, *Implications of early halo type veins at the IKE copper-molybdenum-silver porphyry deposit, British Columbia, Canada; alteration studies as a guide to ore grade*

Dewit, Megan, *Applications of terrestrial LiDAR, infrared thermography, and photogrammetry for mapping volcanic rocks in southern BC*

Diaz, Nakari, *Sequence stratigraphy and facies analysis of the Viking Formation in Crossfield and adjacent areas, Alberta, Canada*

Ogloff, Anastasia, *The Murray dike swarm and its bearing on Cretaceous magmatism and tectonics in the Canadian Cordillera*

Rosales Ramirez, Teresa, *Modelling wastewater spills and mapping areas most vulnerable to groundwater quality deterioration in northeast British Columbia*

Walters, Miranda, *Integrated ichnological, sedimentological, and geochronological analysis of the Late Cretaceous upper Nanaimo Group, Saltspring Island, B.C., Canada*

Warwick, Rachel, *A comprehensive volcanic hazard assessment for Mount Meager volcanic complex, B.C.*

Southern Methodist University

Doctoral

DeGrandpre, K., *Characterization of magma storage and dynamics at Akutan, Semisopochnoi, and Okmok Volcanoes from analytical and numerical models of geodetic, seismic, and petrologic data*

Eshqi Molan, Yusuf, *Soil moisture contributions to InSAR phase and decorrelation*

Stanford University

Doctoral

Al-Ismail, Fatimah Mohammad, *Towards an extensible local magnitude scale*

Boag, Thomas Howard, *The role of ecophysiology and paleoenvironmental dynamics in the Ediacaran biostratigraphic record*

Brooks, Nina Rebecca, *Global health and the environment; studies on air pollution and reproductive health policies*

Burton, Zachary Florentino Murguia, *Characterizing oil, gas hydrate, and sedimentary systems via geochemistry, computational modeling, and global synthesis*

Chheda, Tanvi Dhiren, *Optimizing exploration decisions under geologic uncertainty in basin and petroleum system modeling*

Chu, Shanna, *Seismological constraints on the mechanisms of intermediate-depth earthquakes in Japan*

Hodgskiss, Malcolm Stewart William, *Characterising the Orosirian Earth system*

Kim, Hajin, *Markets, morality, and the environment*

Lund Snee, Jens-Erik, *State of stress in North America; seismicity, tectonics, and unconventional energy development*

Lund, Andrea, *Human-environment dynamics in the eco-epidemiology of schistosomiasis*

McCormack, Kevin Lynn, *Seismological and geomechanical controls on the properties of the brittle crust and economic reservoirs*

Michaelides, Roger John, *Quantifying permafrost processes and soil moisture with interferometric phase and closure phase*

Naughton, Hannah Rose, *Soil carbon cycling constrained by oxygen-dependent enzyme activity and microbial energetics*

Onda, Chikara, *Climate, jobs, and inequity; models of worker mobility and* *distribution under carbon pricing*

Sekar, Samantha, *Misrepresented; understanding the gap between US public opinion and policy on climate change*

Sheng, Yixiao, *Empirical analysis of complex wavefields*

Suwannasri, Krongrath, *Monitoring and modeling the elastic and transport properties of organic-rich marl during maturation through pyrolysis experiments*

Watson, Leighton Myles, *Geophysical applications of sound; volcano infrasound and marine seismic sources*

Witter, Molly Rose, *Timescales of magma genesis and the dynamics of processes that produce differentiated alkalic volcanism*

Womble, Philip James, *Water rights markets in the 21st century; transaction costs and optimal environmental water portfolios in Colorado, USA*

Wong, Ying Qi, *The dynamics of conduit flow in dome-forming volcanic eruptions; steady-state and time-dependent models applied to the 2004-2008 Mount St. Helens eruption*

Xiaowei Li, *Evaluating the effects of carbonate saturation state on carbonate platform geometry and porosity evolution; a case study from the Great Bank of Guizhou of south China*

Stephen F. Austin State University

Master's

Blizzard, Alex Parker, *Chemostratigraphy of carbonate gravity flows of the Wolfcamp Formation in Crockett County, Midland Basin, Texas*

Bridges, Jarrod R., *Delineating of the Utica Shale/Point Pleasant Formation play system to determine influence of the Precambrian basement in northeastern Ohio*

Case, Zachery T., *Stratigraphy, petrology, diagenesis, paleontology, and depositional environments of the Harrisburg Member of the Kaibab Formation in northern Arizona and southern Utah*

Dailey, Heather Jaclyn, *Characterization of surface karst using LiDAR and field traverses, Fort Hood military installation, Coryell County, Texas*

Galle, Travis Scott, *Petrology of the gradational intrusive contact between the Packsaddle Schist and the Town Mountain Granite, Enchanted Rock State Natural Area, Llano Uplift, central Texas*

Robison, Annie, *Delineating subsurface karst potential using electrical resistivity in the Shell Mountain province, Fort Hood military installation, Texas*

SUNY, Binghamton University

Master's

Braiotta, Franco, *Autotrophic and heterotrophic antimonate reduction by an arsenic reducing bacterium*

Novello, Joshua A., *Unlocking the secrets of the Wasatch and Green River Formations using lithostratigraphy, XRF chemostratigraphy, and sedimentary petrology in the Great Divide Basin, Wyoming, U.S.A.*

Rosenberg, Christiana D., *An expanded view of mid- to late Pliocene West Antarctic ice sheet dynamics*

Schneider, Jonathan B., *Geochemical impacts of drill cuttings on landfill leachate from unconventional energy development in the Appalachian Basin*

Souto, Chelsie, *Induced seismicity in northern Pennsylvania; directly monitoring seismicity from hydraulic fracturing*

SUNY, Stony Brook University

Master's

Braunthal, Ashley, *Studying paleo-redox conditions recorded in laminated sediments of Bowers Ridge, Bering Sea (IODP Expedition 323, Site U1342)*

Rueda, Fabian Ricardo Jaime, *Process steps and economics of mineral extraction from geothermal brines*

SUNY, University at Buffalo

Doctoral

Narkevic, David Ash, *Fusion of altimetry and digital elevation models for change detection at the Northeast Greenland ice stream*

Roberts, Carolyn E., *Hydrologic controls on ice dynamics inferred at Helheim Glacier, southeast Greenland, from high-resolution surface-elevation record*

Master's

Grogan, Amy Leigh, *Increase in northern Arctic moisture in winter precipitation on southwestern Greenland through the Holocene*

Rodgers, Samuel, *Deposition and fracture patterns within the Herkimer Quadrangle, NY; implications for the location and character of the Herkimer Fault*

Sbarra, Christopher, *On the maximum extent of the Greenland ice sheet in southwestern Greenland during the LGM*

Schiff, Nicholas Louis Gray, *Probable ice-rich deposits on north-facing slopes in Alba Patera, Mars*

Stock, Jeremy M., *Modeled patterns of crevassing induced by supraglacial lake drainage in western Greenland*

Syracuse University

Doctoral

DeMott, Laura Michelle, *Depositional constraints on lacustrine carbonates from the Winnemucca dry lake basin, Nevada, USA*

Judd, Emily J., *Seasonality in a changing climate; insights from the modern ocean with application to the Eocene Epoch*

Master's

Brigham, John Mark, *Mineral paragenesis of a partially serpentinized dunite in East Dover, Vermont*

Davis, Julianne M. S., *Evaluating the geomorphic channel response to beaver dam analogue installation using unoccupied aerial vehicles*

Texas A&M University

Doctoral

Prince, Kieron Caleb, *Sea-level versus bottom-current influenced diagenesis*

Backelor's

Sayprasith, Kimberly, *TEX_{86} and U37k applications to amplify information on western boundary currents during the mid- to late Miocene*

University of Alaska at Anchorage

Master's

Williams, Timothy Scott, *Petrophysical analysis and mudstone lithofacies classification of the Hue and HRZ Formations, North Slope, Alaska*

University of Alaska at Fairbanks

Master's

Gurav, Yojana Shivaji, *Application of design of experiments for well pattern optimization in Umiat oil field; a natural petroleum reserve of Alaska case study*

McClernan, Mark Timothy, *Analysis of the 2015 Sagavanirktok River flood; associated permafrost degradation using InSAR and change detection techniques*

Thompson, William D., *Geochemical, spatial, and temporal relationships of the intrusives and meta-intrusives of the Pogo Deposit, eastern Interior, AK*

University of Alberta

Doctoral

Barclay, Kristina, *Exploring the past, present, and future of predator-prey interactions between crabs and their gastropod prey*

Booker, Simone Donna, *Variations in climatic conditions from the Cayman Islands through stable isotope and element analysis from corals and sediment cores; a 500,000 year record*

Hardman, Matthew Fraser, *Improving the use of eclogitic garnet as a diamond indicator mineral, and constraining the origin of eclogites in the subcontinental lithospheric mantle*

Lypaczewski, Philip, *Hyperspectral imaging applications in the mining industry; refining alteration footprints*

Norris, Sophie L., *Dynamic response of the southwestern Laurentide ice sheet during the last deglaciation*

Playter, Tiffany L., *Pairing geochemistry and sedimentology; an evaluation of Alberta's earliest Triassic deposits and the proxies used to study them*

Regier, Margo Elaine, *Subduction and the genesis of mantle roots and diamonds*

Vaisblat, Noga, *Controls on reservoir quality in the Lower Triassic Montney Formation*

Master's

Finley, Theron, *Fault-hosted geothermal systems in southeastern British Columbia*

Harris, Brette S., *Chemostratigraphy and facies analysis of the Hare Indian Formation in the Mackenzie Mountains and central Mackenzie Valley, Northwest Territories, Canada*

Li, Yuhao, *Using a novel approach to precisely measure the sedimentation velocity of a (bio)chemical sediment; cyanobacteria-ferrihydrite aggregates*

Lord, Elaine Loretta, *Identifying and modeling evaporite facies using well logs; characterising the Devonian Elk Point Group salt giant*

MacPherson, Neil, *Solubility products of chloride and sulfate green rust during transformation to magnetite*

Mendonca, Steven Edward, *Measurement and interpretation of shell repair due to predation*

Seip, Adam John, *Lithium recovery from hydraulic fracturing flowback and produced water using a manganese-based sorbent*

Siva-Jothy, William Karim, *Studies of inclusions and their host diamonds from the Gahcho Kue#2. mine, Canada*

Smyth, Denelle, *Petrology, geochemistry, and geochronology of the Pikoo kimberlites, Saskatchewan*

Staniszewska, Katarzyna Joanna, *Legacy contaminant release from the cryosphere to downstream systems*

Veglio, Christian, *Behaviour of ore-forming elements in the sub-continental lithospheric mantle below the Slave Craton*

Zaremehrjardy, Majid, *Modelling snow dynamics and groundwater-surface water interactions in mountainous, foothill and plain regions in North Saskatchewan River basin, Canada*

Zhou, Haolin, *High-resolution geochemical analysis of controls on organic matter accumulation in the Middle-Upper Devonian Horn River Group*

University of British Columbia

Doctoral

Amini, Afshin, *Investigation of induced seismicity mechanisms and magnitude distributions under different stress regimes, geomechanical factors, and fluid injection parameters*

Lesage, Guillaume, *Distribution of district-scale hydrothermal alteration, vein orientations and white mica compositions in the Highland Valley Copper district, British Columbia, Canada; implications for the evolution of porphyry Cu-Mo systems*

McDowell, Graham, *Adaptation to glacio-hydrological change in high mountains*

McMahen, Katie, *Investigation of ecosystem legacy-based mine reclamation techniques; effects of soil transfers, forest connectivity and plant-soil feedbacks on soil food web recovery and plant establishment*

McMillan, Rhylan, *Micro-analytical geochemical and spectroscopic investigations of the original context and condition of archaeological biominerals and mineraloids*

Mitchell, Michael A., *Methodologies for the use of electrical and electromagnetic methods in complex, subterranean environments*

Ramos Musalem, Ana Karina, *Transport through submarine canyons*

Ryan, Amy Grace, *The timescales and consequences of solid-state sintering in volcanic systems*

St. Arnault, Melanie, *The assessment of mineralogical properties and hydrological or physicochemical controls on the drainage chemistry of mine waste rock*

Thompson, Katharine Jean, *Phototrophic iron oxidation and implications for biogeochemical cycling in the Archean Eon*

Xu, Cong, *Discrete element analysis of the mechanical response of bolted steel mesh supporting unstable rock on steep slopes*

Master's

Cano Roncagliolo, Alvaro, *The Large-Scale Mining (LSM) - informal Artisanal-Scale Mining (ASM) interface in Peru; exploring a key policy "bottleneck" through Qualitative Comparative Analysis (QCA)*

Cantor, Samuel Frank, *Stable isotopes as an exploration tool; tracking cryptic alteration surrounding the Iscaycruz Zn (Pb-Cu-Ag) skarn-CRD deposit, central Peru*

Chopra, Chitra, *Quantification and mapping of methane emissions using eddy covariance in a controlled subsurface synthetic natural gas release experiment*

Cone, Dylan, *Crystallization of megacrysts by carbonatitic metasomatism; evidence from the Muskox kimberlite, Nunavut, Canada*

Dehghanniri, Vahid, *An experimental study of volcanic tremor driven by magma wagging*

Eksir Monfared, Alireza, *Georgia sedimentary basin effects on the response of modern tall RC shear wall buildings to M9 Cascadia subduction zone earthquakes*

Ersay, Lutfi, *Multistage, magmatic and hydrothermal evolution of the Debert Lake rare earth element (REE) prospect, Nova Scotia, Canada*

Guajardo, Cristian Andres, *Numerical investigation on strength upscaling and its application to a back analysis of an open pit slope failure*

Guitart, Agustin, *The geology, alteration and timing of porphyry intrusions and breccias associated with the development of Los Helados porphyry copper-gold deposit, Chile*

Iulianella Phillips, Bianca P., *Microorganisms as sensors for concealed mineral deposits; application and development of microbiological mineral exploration in the Northwest Territories and British Columbia, Canada*

Kim, Rachel S. Y., *Evolution of Late Cretaceous Kasalka Group volcanics, Nechako Plateau, central British Columbia*

Li, Mu, *Temperature-pressure-composition model for melt viscosity in the Di-An-Ab system*

Lu, Xueya, *Characterization of ultramafic mine tailings reactivity for carbon capture and storage*

Manco, Julian David, *Geology, geochronology and geochemistry of the El Alacran Deposit, San Matias District, Cordoba-Colombia*

Misbahuddin, Mohammed, *Predicting the properties of rock using data-driven methods*

Niyazova, Sofya, *Felsic xenoliths assimilation in the Renard 65 kimberlite pipe, Quebec, Canada*

Nogare, Marianni de Aragao, *Measuring and modeling sediment dispersion in small streams*

Qiao, Tianyang, *Seismic performance of controlled outrigger rocking wall system with different types of energy dissipation devices*

Sandl, Elyse, *Investigating factors associated with fugitive gas migration in northeast British Columbia, Canada*

Soares, Julia Valentina, *Characterization of gas migration and surface emissions through a controlled release experiment at the Hudson's Hope Field research station, BC, Canada*

Turley, Michael, *Exploring the importance of sediment disconnectivity and connectivity in glacierized catchments, Tahoma Creek, WA*

Vincent, Lauren, *The importance of secondary processes on alluvial fan morphology, channel behaviour, and flood hazards*

Wang, Yinlue, *On the development of bed surface structures in response to variable flow regimes*

Wesolowski, Michelle, *Application of computed tomography for visualizing three-dimensional fabric and microstructure of Fraser River delta silt*

Zubrycky, Sophia, *Spatial impact trends on debris flow fans in southwestern British Columbia*

University of California at Davis

Doctoral

Pauloo, Richard Augustus, III, *Emerging consequences of regional-scale aquifer depletion; data-driven and numerical models of well failure, basin salinization, and contaminant transport*

University of California at San Diego

Master's

Shelton, Bronte, *Characterization of ten microbial isolates from two serpentinite seamounts, Asut Tesoru and Fantangisna, in the Mariana Forearc*

University of California at Santa Barbara

Doctoral

Adams, Jenna Victoria, *A thermodynamic and geochemical evaluation of magmatic processes beneath the Samoan Islands and the thermodynamics of crystalline garnet solutions*

de Jong, Menso Thomas, *Constraining hydraulic properties in oceanic crust near the Juan de Fuca spreading center*

Spratt, Rachel M., *Glacial-interglacial changes in sea level and seawater delta O-18 gradients*

Wrobel, Alex, *Exploring the tectonic and thermal evolution of the northern Snake Range metamorphic core complex; insights from field study and potassium feldspar >40`Ar/>39`Ar thermochronology*

Master's

Adamaitis, Chandler, *Chronology and recurrence of high-magnitude debris flows in the Santa Barbara and Montecito, California areas*

Condon, Scott James, *Joint study of the 1952 Kern County, California earthquake*

Everett, Christopher John, *A new species of Pinnarctidion from the Pysht Formation of Washington and phylogenetic analysis of basal pinnipedimorphs (Eutheria, Carnivora)*

Ma, Yue, *Southern African monsoon strengthening and weakening of Mozambique Channel through flow during Heinrich events*

Rose, Kaelynn, *Remote hydroacoustic and infrasonic detection and characterization of Anak Krakatau eruptive activity leading to, during, and following the December 2018 major flank collapse and tsunami*

Theilen, Brittany, *Changes in beach deposit characteristics on Joinville and Livingston Islands, Antarctica*

University of Chicago

Doctoral

Farahat, Navah X., *Numerical models for a better understanding of off-axis hydrothermal systems*

Lungmus, Jacqueline K., *Forelimb shape, disparity, and functional morphology in the deep evolutionary history of synapsida*

University of Cincinnati

Doctoral

Gangidine, Andrew, *Trace element concentrations in microbial fossils as a novel biosignature for life on ancient Earth and beyond*

Hannon, Jeffrey S., *Reconstructing the generation, evolution, and migration of arc magmatism using the whole-rock geochemistry of bentonites; a case study from the Cretaceous Idaho-Farallon Arc system*

Murry, Maisha M., *Dissolution and sequential extraction of select radioactive and stable elements in soil and lunar regolith simulants*

Parker, Wesley G., *Holocene climate change in the subtropical eastern North Atlantic; integrating high-resolution sclerochronology and shell midden archaeology in the Canary Islands, Spain*

Schlanser, Kristen M., *Advancing leaf carbon isotopes as a paleo proxy*

Master's

Bootes, Nathaniel R., *Late Cenozoic offshore record of exhumation and sediment routing in southeast Alaska from detrital zircon U/Pb and FT double-dating*

Ferry, Nicholas, *Role of a rigid bedrock substrate on emplacement of the Blue Diamond landslide, Basin and Range Province, eastern Spring Mountains, southern Nevada*

Garvey, Samuel T., *A new high-latitude Tylosaurus (Squamata, Mosasauridae) from Canada with unique dentition*

Sivapalan, Vinothan S., *Vegetation, hydroclimate, and fire dynamics from the late Illinois glaciation (130 ka) to the late Holocene*

University of Colorado at Boulder

Doctoral

Bergstrom, Anna J., *The effect of sediment on hydrological and biogeochemical connectivity of glaciers within the McMurdo dry valley ecosystem, Antarctica*

Boudinot, Frank Garrett, *Black dirt live again; black shale organic and stable isotope geochemistry reveals ecosystem responses to global environmental change*

Ginsberg Reitman, Nadine, *Strain across scales; exploring geologic and geomorphic records of past earthquakes*

Kochanski, Kelly, *The growth of snow bedforms*

Nothaft, Daniel Blake, *Subsurface microbial ecosystems and the origin of methane in serpentinites of the Samail Ophiolite, Oman*

Plescia, Steven, *Geometric considerations in seismic studies of Earth's upper crust*

Rongstad, Brigitta Lee, *Investigating past El Nino Southern Oscillation using Mg/Ca in individual planktic Foraminifera*

Simon, Rebekah Elizabeth, *Integrating the effects of diagenesis and catagenesis on pore network evolution in an unconventional chalk reservoir; Cretaceous Niobrara Formation, Colorado*

Yarce, Jefferson, *Microearthquake detection and seismic imaging using ocean bottom seismometers in the Hikurangi Margin, New Zealand*

Master's

Baumann, Sarah Pearl, *Map-reading skill development; historical perspectives and instructional implications*

Bell, Jackson, *Seismic activity in the northern Sangre de Cristo fault zone*

Zhang, Melody, *Insight into subsurface structure using Sp arrivals in the northern Hikurangi subduction zone, New Zealand*

University of Delaware

Master's

Kaiser, Emily Ann, *Orbital-scale controls on biogenic silica accumulation in the Indian sector of the Southern Ocean over the past 600,000 years*

University of Georgia

Doctoral

Brown, Garett Mattingly, *Beyond water depth; the environmental controls on marine benthic invertebrate assemblages in the modern and fossil record*

Callihan, Melanie Bergmann, *Growth of complex fracture systems in extensional tectonic settings on Earth and the Moon with applications to tectonic geomorphology, fault rock evolution, and dike emplacement*

Rotz, Rachel R., *Geomorphic-geologic indicators of zones of hydrologic flux in drylands on Earth and Mars*

Master's

Donaldson, Yonesha Yashelle, *Identification of inland freshwater lenses in central Florida using remote sensing and geostatistics*

Jordan, Barrett, *Geology of the Calhoun Critical Zone Observatory*

Larramendi, Gustavo, *Investigation of the Late Paleozoic collision between Gondwana and Laurentia and the resulting deformational response of the upper mantle beneath the southern Appalachians using global phase seismic events*

Lee, Sydney Paige, *Testing two paleogeographic interpretations of the Jurassic of western North America*

Medley, Melissa Jean, *Investigation of hydrology and climate of national wildlife refuges in Florida using water resource inventories and assessments*

Ramsey, Sierra Rhaye, *The petrogenesis of shergottites constrained by major and minor elemental variations in olivine*

University of Hawaii at Manoa

Doctoral

Costello, Emily S., *Impact impacts on the Moon, Mercury, and Europa*

Egan, Miles Jacob, *Character of signal and noise sources in dispersive and static Fourier transform remote-sensing Raman spectrometers*

Sandford, Macey W., *Innovative remote spectroscopic techniques for planetary exploration*

Master's

Avery, Jon, *Analysis of smartphone earthquake early warning networks in Chile and Costa Rica*

Beucler, Olliander, *Laser-induced breakdown spectroscopy (LIBS) for investigating surficial condensation of semi-volatiles on volcanic spatter*

Evans, Krista, *Precursory volcanic activity and cultural response to the late Bronze Age eruption of Santorini (Thera), Greece*

Foerder, Andrew Bounds, *Chemical alteration and soil provenance of polar desert sediments from the McMurdo Dry Valleys, Antarctica; an analog for alteration processes on Mars*

Genabe, Angelo, *Particle background and methods for cosmic dust collection at Mauna Loa Observatory*

McDonald, Kristian, *UAS surveys reveal high spatiotemporal variability in beach morphology including subcell sand exchange and accretion during swell events; Waikiki, Hawai'i*

Tavares, Kammie-Dominique, *Rate of beach loss greatest with near-term sea level rise*

Zinn, Malia, *Investigating volcanic eruptions at the seamount Tafu; chemical and geographic variation within the NELSC*

Backelor's

Ackerman, Katherine, *Observing the effects of the surf zone on local sea salt aerosol production*

Plazas, Marisol, *Mineral composition found in blue mud samples taken from the Mariana Forearc aboard IODP Expedition 366*

Suter, Ingrid, *Geoelectric data acquisition and modeling of dynamic hydrogeologic systems in Hawai'i*

University of Iowa

Master's

Trembath, Matthew J., *Structural characterization of the Porcupine Shear Zone in northeastern Alaska*

University of Louisiana at Lafayette

Doctoral

Yang, Xu, *Production data analysis and potential evaluation of Tuscaloosa Marine Shale oil wells*

Master's

Ajala, Samsideen, *Depositional environment and sequence stratigraphy of the Lower Tuscaloosa Sandstone in Washington Parish Louisiana*

Ariyibi, Olawale E., *High-resolution magnetostratigraphy, relative geomagnetic paleointensity and paleoenvironmental change from Eocene to Pleistocene sediments, IODP Expedition 369, Site U1514 (Mentelle Basin, Australia)*

Ariyibi, Olawale E., *High-resolution magnetostratigraphy, relative geomagnetic paleointensity and paleoenvironmental change from Eocene to Pleistocene sediments, IODP Expedition 369, Site U1514 (Mentelle Basin, Australia)*

Braun, Andrew Michael, *Geochemical analysis of lake sediments to highlight early human environmental impact, Poverty Point World Heritage Site, northeastern Louisiana*

Breaux, Abigail K., *Facies characterization of the Upper, Middle, and Lower Smackover Formation encountered in the Palmer #1 Caraland 26-9 well, Manila Embayment, southwest Alabama*

Cecil, Joshua, *Characterization of Upper Jurassic shoals across the Conecuh Embayment, Alabama*

Chapman, Joseph Cyle, *Formation of a large field of sub-circular depressions due to fluid migration in the Gippsland Basin (SE Australia)*

Hardt, Joshua, *Analysis of freshwater ostracods from Fossil Lake, Oregon, to determine local Pleistocene paleoclimate*

Harris, J. Mason, *Evaluating coastal wetland restoration in Louisiana using drones*

Hendrix, Gage Alexander, *Analysis of rock property trend for prediction of seismic response in southwestern Louisiana*

Howie, Christopher M., *Lithologic and structural analysis of the Chicot Aquifer confining unit and location of shallow sands in Acadia Parish, Louisiana*

Kramarov, Vladyslav, *Fracture toughness determination and strain analysis of anisotropic rocks using semicircular bend test and digital image correlation*

LaHaye, Olivia Ann, *Assessment of aquifer storage and recovery feasibility in replenishing coastal aquifers and addressing flood mitigation using numerical modeling, geospatial, and statistical techniques; application in southwest Louisiana*

Moore, Christopher D., *Streamline-based modeling of foam population in porous media*

Rask, James Allen, *A structural and lithologic characterization of the Chicot Aquifer surficial confining unit in Vermilion Parish, Louisiana*

Ruse, Cristina Mariana, *Characterization of elastic mechanical properties of Tuscaloosa Marine Shale from well logs using the VTI model*

Scates, Allison Robin, *Geomorphic and shallow subsurface expression of growth faults in Mississippi River delta Holocene sediment; Golden Meadow, Louisiana*

Smith, Brock Wyatt, *Alteration styles at the Trenton Canyon site, Battle Mountain, Nevada*

University of Maine

Doctoral

Burpee, Benjamin, *Assessing the factors that alter ecological responses of cold, oligotrophic lakes to nutrient subsidies*

Hathaway, Amber Emily, *Radiometric dating; a new technique*

Walters, Jesse B., *Sulfur and trace metal cycling at convergent margins*

Wei, Xinyuan, *A synthesis program; reducing uncertainties of the terrestrial biosphere carbon cycle at various spatio-temporal scales*

Zhao, Jianhen, *The future of Maine's forests under alternative socioeconomic, climate and conservation pathways*

Master's

Ayres, Stephanie L., *Three-dimensional analytical model of tidal flow in the Damariscotta River estuary, ME*

Gomez Arevalo, Efren D., *A groundwater flow model to aid in water resource management for the Carraipia Basin in the coastal semi-arid region of La Guajira State (Colombia)*

Lukasik, Simona, *Developing a long-term perspective on the sensitivity of lake trophic state changes in Maine*

Schultz, Kathleen M. T., *A numerical survey of multi-planet systems' inclina-*

tion excitation and survival under the influence of an oblate, tilted star

Spoth, Meghan M., *Variations in the Southern Hemisphere westerlies over the last 23,000 years from lake records in the Falkland Islands*

Tavora, Juliana, *An algorithm to estimate suspended particulate matter concentrations and associated uncertainties from remote sensing reflectance in coastal environments*

Walther, Tess Louise, *Reconstruction of Koettlitz Glacier, southern McMurdo Sound, Antarctica, during the last glacial maximum and termination*

University of Massachusetts at Amherst

Doctoral

Keisling, Benjamin Andrew, *Modeling the Pleistocene history of the Greenland Ice Sheet*

Lam, Adriane, *Neogene history of the Kuroshio current extension and planktic foraminifera evolutionary implications*

Mudiyanselage, Sampath Rathnayaka, *3-D shear wave velocity structures of the crust and upper mantle beneath Cascadia and New Zealand from full-wave ambient noise tomography*

Pless, Claire Rose, *Implications and significance of partial melting on the tectonic history of the Eastern Adirondack Mountains*

Master's

Geer, Phillip S., *Magnetite mineralization of the Hammondville Pluton; a syn-magmatic origin to poly-phase Kiruna type IOCG magnetite-apatite deposits in the Lyon Mountain Granite*

Hillenbrand, Ian, *Crustal evolution of the New England Appalachians; the rise and fall of a long-lived orogenic plateau*

Mistikawy, Justin, *Seeing through the Ottawan overprint, Adirondack Mtns., NY; integrating microstructural analysis, geothermobarometry, and in-situ monazite petrochronology*

Seidenstein, Julia, *Plio-Pleistocene paleoceanography of the Ross Sea, Antarctica based on Foraminifera from IODP Sites U1523, U1522, and U1521*

University of Memphis

Master's

Menietto, Krista F., *Investigating intra-site variability in the use of buildings during the Mississippian period at the Ames Town site (40FY7) in western Tennessee*

University of Nebraska at Lincoln

Master's

Hobbs, Kirby P., *Characterizing peridotite xenoliths from southern Vietnam; insight into the underlying lithospheric mantle*

University of Nevada at Reno

Doctoral

Hatch-Ibarra, Rachel L., *Characterization of recent earthquake sequences and swarms within the Walker Lane tectonic region*

Milliard, Justin B., *Genetic links between extensional tectonomagmatism and the formation of low sulfidation epithermal Au-Ag deposits; defining a mineral system*

Master's

O'Connor, Kyle, *Fate and transport of contaminants from acid mine drainage in the Perry Canyon Abandoned Mine Land, Washoe County, Nevada*

University of New Hampshire

Doctoral

Melendez Oyola, Melissa, *Effects of nearshore processes on carbonate chemistry dynamics and ocean acidification*

Master's

Bennett, Kathryn A., *Using stable isotopes to determine dominant methane production pathways of thaw ponds in a subarctic peatland*

Burkett, Emma M., *Geochemical and remote sensing reconstructions of timescales for magma assembly beneath Augustine Volcano*

Guimaraes, Ivan Bodra, *Obtaining a reference for calibrating broadband multibeam seabed backscatter*

Hampson, Patrick, *Glacial troughs eject wind-driven shelf circulation to the slope*

Macleod, Douglas R., *Characterization of a silty methane-hydrate reservoir in the Gulf of Mexico; analysis of full sediment grain-size distributions*

University of New Mexico

Doctoral

Clary, Wesley A., *Quaternary shelf-slope development in the northern Gulf of Alaska from combined geophysical and geomorphologic analysis*

University of North Carolina at Chapel Hill

Doctoral

Cao, Cheng, *Geochemical investigation of continental weathering and paleo-ocean redox evolution*

University of Notre Dame

Doctoral

Hartman, Michael A., *Engineering evaluation of nearshore bathymetric change using remotely sensed data*

Master's

Ozkaya, Omer Faruk, *Geochemical, isotopic and geochronological investigation of the Magnet Cove carbonatite complex, Arkansas, USA*

University of Oklahoma

Doctoral

Karam, Pierre, *Characterizing transfer zones with seismic and experimental models*

Kim, Yuji, *Machine learning applications for seismic processing and interpretation*

Kolawole, Folarin, *The roles of structural inheritance in regions of induced seismicity and active tectonics*

Lyu, Bin, *Full wave-equation based passive seismic imaging and multispectral seismic geometric attributes*

Patel, Swetal, *Geophysical characterization of resource plays*

Pennington, Colin, *Earthquake source and rupture characterization from microearthquakes to megathrust earthquakes*

Pfeifer, Lily S., *Permian loess and equatorial glaciation in eastern equatorial Pangea*

Qin, Yan, *Role of fluid injection, stress state, geological structure, and earthquake interaction on Oklahoma earthquakes*

Richmond, Dean, *Geological and paleontological investigations of the Upper Jurassic Morrison Formation at the basin margins of central Montana and the western panhandle of Oklahoma*

Suriamin, Fnu, *Integrated reservoir characterization of a mixed siliciclastic-carbonate system, Mississippian strata of northern and central Oklahoma*

Torres Parada, Emilio, *Woodford Shale enclosed mini-basin fill on the Hunton paleo shelf; a depositional model for unconventional resource shales*

Master's

Brown, Andrew, *High resolution sequence stratigraphy of the Leonardian Bone Spring Formation outcrop of Bone Canyon, Guadalupe Mountains, West Texas and its correlation to the subsurface*

Buist, Carl, *Sub-seismic reef characterization using machine learning and multi-attribute analysis*

Chenin, Julian, *Examining seismic amplitude responses of gaseous media using unsupervised machine learning*

Dingmore, Lindy, *Stratigraphic variability of mineralogy, rock types and porosity of the Wolfcamp Formation using multi-mineral petrophysical analysis and seismically constrained reservoir modeling, Northwest Shelf, Delaware Basin*

La Marca, Karelia, *Seismic attribute optimization with unsupervised machine learning techniques for deepwater seismic facies interpretation; users vs. machines*

Layden, Andrew, *High resolution XRF-Schmidt hammer based sequence and geomechanical stratigraphy of the Leonardian Bone Spring Formation, Shumard Canyon, Guadalupe Mountain National Park*

Lynch, Matthew, *Integrated petrophysical-seismic sequence stratigraphy of the mixed carbonate clastic sediment gravity flows of the Bone Spring Formation, New Mexico*

Maples, Brandon, *Evaluating cation exchange as a driver for manganese release during carbonate dissolution*

Mason, Daniel, *Raman spectroscopy and its application to ice-sediment-brine matrices in recurring slope lineae on Mars*

Ortega Romo, Angie, *Aspects of seismicity clustering, subsurface structure, and stress orientations*

Plowman, Aimee, *Paleoenvironment, depositional history, and process sedimentology of the upper Guadalupian Yates Formation; northwest shelf, New Mexico*

Putri, Tara, *The organic geochemical variability in depositional environmental processes within a stratigraphic sequence of the Woodford Shale in the Ardmore Basin, Oklahoma*

Rodriquez, Andrew, *Observation of Mars analogous basalt in near saturated brines*

Tamborello, Brittney Lynn, *Potential sources of oil from Hugoton Embayment reservoirs*

Webb, Nina Danielle Szabo, *Effects of mass wasting on fluvial sediments in Puerto Rico following Hurricane Maria*

Welch, Katie, *Cambrian trilobites from the Steptoean-Sunwaptan boundary interval (Jiangshanian), Nevada and Utah*

Williams, Madison, *Seismic-constrained lithology prediction and reservoir-quality modeling using pre-stack inversion and Bayesian classification, Mississippian Meramec strata, eastern Anadarko Basin, Oklahoma*

University of Oregon

Doctoral

Acosta, Marisa Dawn, *Titanium in quartz during growth and deformation under hydrothermal-magmatic conditions*

Bodmer, Miles, *Seismic imaging of the Cascadia subduction zone and Juan de Fuca Plate system mantle structure; a bottom up approach to subduction dynamics*

Cahalan, Ryan Cain, *Explosive subaqueous eruptions; the influence of volcanic jets on eruption dynamics and tephra dispersal in underwater eruptions*

Heath, Benjamin, *Interactions between tectonism and magmatism at Santorini; insights from an active source seismic experiment*

Lerner, Allan, *The depths and locations of magma reservoirs and their consequences for the behavior of sulfur and volcanic degassing*

Morriss, Matthew Connor, *The dynamic interplay between Miocene to post-Miocene magmatism, tectonism, and geomorphology in NE Oregon*

Zakharov, David, *Triple oxygen isotope in high-temperature hydrothermally altered rocks; a record of paleoclimate and ancient hydrosphere-rock interactions*

Master's

Flora, Holley May, *A genus-level phylogenetic analysis of Antilocapridae and implications for the evolution of headgear morphology and paleoecology*

Gardner, Kevin, *Stratigraphy, sedimentology, and tidal processes of the late Miocene to early Pliocene southern Bouse Formation, southeast Palo Verde Mountains, California*

Hunter, Pierce, *Quantifying the effects of lateral advection on shear margin thermal structure and meltwater production*

McQuillan, Maria, *Modelling fluid resonance in conduit-crack coupled systems; an application to englacial geometries*

Niday, William Bartholomew, *Origins of complex seismic anisotropy beneath the Wallowa Mountains, northeast Oregon*

Orland, Elijah, *Deep learning as a tool to forecast hydrologic response for landslide-prone hillslopes*

University of Rochester

Doctoral

Dyonisius, Michael N., *Using ancient ice from Taylor Glacier, Antarctica to study the past methane budget and in situ cosmogenic ^{14}C production rates in ice*

University of Texas at Rio Grande Valley

Master's

Youwakim, Joy Z., *Predicting patterns of heavy metal contamination in Rio Grande Valley agricultural soils*

University of Texas at San Antonio

Master's

Bernardo, Miguel, *The development of porosity in the Upper Cretaceous Austin Chalk Group of Texas*

Nunu, Rebecca R., *Using geochemical and statistical analyses to identify local and regional flow to a multi-outlet spring system; San Solomon Springs, Texas, USA*

University of Toronto

Doctoral

Tsang, Man Yin, *Modeling the effects of microbially-mediated sulfate reduction on porewater S-isotope ratios in marine sediments*

University of Tulsa

Master's

Bobka, Alana M., *Paleoceanographic implications from Silurian sedimentation and carbon isotope chemostratigraphy in the Arbuckle Mountains, Oklahoma*

Farr, C. Layne, *Permeability anisotropy and meandering fluvial facies architecture of the Bartlesville Sandstone, Nowata County, Oklahoma*

University of Victoria

Doctoral

Seyedali, Minasadat, *An investigation of low-temperature off-axis hydrothermal systems using lithium isotopes and trace element geochemistry*

University of Wisconsin at Madison

Master's

Hartman, Allison, *Microbially-mediated weathering of iron-bearing minerals in Central Sands soils*

University of Wyoming

Doctoral

Alexander, Jason S., *Scales, controls, and habitats of fluvial macroform sandbars*

Ashley, Thomas C., *Granular morphodynamics of fluvial bedforms*

Ayani, Mohit, *Probabilistic methods for inversion of marine controlled source electromagnetic data using rock physics modeling for reservoir characterization*

Brown, Tyler C., *Textures of gabbroic rocks formed at fast-spreading mid-ocean ridges*

Gure, Alero Jennifer, *Photoelectrochemical production and degradation, and deliquescence-driven transport, of perchlorate and other oxychlorines; implications for Mars*

Haupt, Ryan J., *Slow food movements; exploring the dietary ecology of extinct and extant sloths using stable isotope analysis of foods, tissues, and feces*

Liefert, David T., *Patterns and processes of lake response to Quaternary climate change in Rocky Mountain watersheds and across North America*

Overstreet, Brandon T., *Mapping riverscapes using high spatial resolution remote sensing data*

Master's

Aydinian, Karen, *Termination of Laramide arches at orogen margins; fracture and paleomagnetic evidence for partitioned strike-slip deformation in the northern Bighorn Arch and Pryor Mountains*

Caro, Derek A., *Geochemical evaluation of flowback and produced water and insight to fluid-rock interaction between alkaline fracturing fluid and Niobrara Formation chalk and marl, Denver-Julesburg Basin, Colorado, USA*

Cotton, Bradlee K., *Geomorphic analysis and flushing flow feasibility study of lower Shoshone River near Cody, Wyoming*

Da Prat, Fabio A., *Archean history of the northern Laramie Mountains*

Flaim, Andrew, *Causes, timing, and scales of subalpine ecosystem changes over the past six millennia; lessons from Libby Flats, Snowy Range, Medicine Bow Mountains, Wyoming*

Gess, Michelle R., *The dynamic accretion of fast-spread oceanic crust; insights from Pito Deep (Southeast Pacific)*

Granke, Sarah B., *Porosity production in granitic saprolite; the relative importance of mass loss and volumetric strain along a steep altitudinal gradient in climate and vegetation*

Mahan, Chase M., *A systematic survey and analysis of Wyoming archaeological assemblages of lithic raw materials*

Swaim, Nathan T., *Distribution and styles of hypogene alteration and mineralization in the Miami-Inspiration porphyry Cu-Mo deposit, Globe-Miami District, Arizona*

Thompson-Munson, Megan E., *Observations and implications of three-dimensional deformation in the Greenland ice sheet*

Utah State University

Master's

Scholpp, Jesse L., *Pre-eruptive evolution of Izu-Bonin boninite melts; mixing, cooling, and crystallization*

Vanderbilt University

Doctoral

Gibson, Brandt M., *New insights into ancient life; the Ediacara biota through the windows of taphonomy and fluid mechanics*

Alabama

Geological Survey of Alabama

Geological Survey of Alabama ☒ (2021)
420 Hackberry Lane
P.O. Box 869999
Tuscaloosa, AL 35486-6999
p. (205) 349-2852
ntew@gsa.state.al.us
http://www.gsa.state.al.us/

State Geologist:
Berry H (Nick) Tew , (D), Alabama, 1999, GroGs
Division Manager, Geologic Investigations Division:
Sandy M. Ebersole, (D), Alabama, 2009, Zi
Division Manager, Energy Investigations Program:
Denise J. Hills, (M), Delaware, 1998, EoZfYe
Division Manager, Ecosystems Investigations :
Stuart W. McGregor, (M), Tennessee Tech, 1987, Hs
Division Manager, Coastal Resources:
Stephen C. Jones, (M), Alabama, 1996, HyGeCg
Manager, Geologic Mapping:
Gene Daniel Irvin, (M), Alabama, 1994, Gg
Visiting Professor:
William A. Thomas, (D), Virginia Tech, 1960, GtcGr
Geologist & Paleontology Curator:
T. Lynn Harrell, Jr., (D), Alabama, 2016, PvGg
Geologist:
Guohai Jin, (M), Zhejiang, 1989, Np
Mac McKinney, (B), Alabama, 2007, Hws
Neil E. Moss, (M), Alabama, 1987, Hw
Marcella Redden, (M), Alabama, 2004, Gg
David Tidwell, (M), S Florida, 2005, Gm
Dane S. VanDervoort, (M), Auburn, 2016, GcpZi
Chemist:
Rick Wagner, (B), Texas (San Antonio), 1982, Cg
Biologist:
Rebecca A. Bearden, (B), Auburn, 2007, Hs
Chief of Staff, Asst. State Geologist:
Bennett L. Bearden, (D), McGeorge Law, 2011, HgEo

Alaska

Alaska Division of Geological & Geophysical Surveys

Department of Natural Resources (2017)
3354 College Road
Fairbanks, AK 99709-3707
p. (907) 451-5010
dggspubs@alaska.gov
http://www.dggs.alaska.gov
f: http://www.facebook.com/pages/Fairbanks-AK/Alaska-DGGS/346699054500
t: @akdggs

Director & State Geologist:
Steven S. Masterman, (M), Alaska (Fairbanks), 1990, EgNg
Petroleum Geologist I:
David Lepain, (D), Alaska, 1993, EoGos
GeoScientist I:
Melanie B. Werdon, (D), Alaska (Fairbanks), Eg
Geologist V:
Janet R. G. Schaefer, (M), Alaska (Fairbanks), Gv
De Anne S.P. Stevens, (M), Alaska (Fairbanks), Ng
Geologist IV:
Marwan A. Wartes, (M), Wisconsin, GsoGt
Gabriel J. Wolken, (D), Gm
Geohydrologist-Geologist IV:
Ronald P. Daanen, (D), Minnesota, 2004, HyGme
Division Operation Manager:
Kenneth R. Papp, (M), Alaska (Fairbanks), Gg

Alberta

Alberta Geological Survey

Alberta Geological Survey ☒ (2018)
402, Twin Atria Building
4999 - 98 Avenue
Edmonton, AB T6B 2X3
p. (780) 638-4491
AGS-Info@aer.ca
https://ags.aer.ca/

Arizona

University of Arizona

Arizona Geological Survey ☒ (2021)
1955 E 6th St.
PO Box 210184
Tucson, AZ 85721
p. (520) 621-2352
fmconway@arizona.edu
http://www.azgs.arizona.edu
f: https://www.facebook.com/AZ.Geological.Survey
t: @AZGeology

Director & State Geologist:
Philip A. Pearthree, (D), Arizona, Ge
Senior Research Scientist:
Michael F. Conway, (D), Michigan Tech, 1993, RcGvg
Chief, Environmental Geology:
Ann Youberg, (D), Arizona, Ge
Senior Scientist:
Andrew A. Zaffos, (D), Cincinnati, 2014, Zf
Research Scientist:
Victor H. Garcia, (D), Texas (El Paso)
Thompson A. Lisa, (M), N Arizona, 2005
Research Geologist:
Joseph P. Cook, (M), Arizona, 2006, Gm
Research Scientist III:
Rebecca Beers, (M), N Arizona, 2020
Research Scientist:
Laura Bookman, (M), Arizona
Senior Research Scientist:
Carson A. Richardson, (D), Arizona, 2019, Emg
Research Geologist:
Charles A. Ferguson, (D), Calgary, Gcv
Brian F. Gootee, (M), Arizona State, Gm
Brad Johnson, (D), Carleton, 1994, Gc
Jeri J. Young, (D), Arizona State, Ys

Arkansas

Arkansas Geological Survey

Arkansas Geological Survey (2015)
Vardelle Parham Geology Center
3815 West Roosevelt Road
Little Rock, AR 72204
p. (501) 296-1877
ags@arkansas.gov
http://www.geology.ar.gov
Administrative Assistant: Laure Hinze

Director & State Geologist:
Bekki C. White, (M), Centenary, 1993, GoEoGg
Geology Supervisor:
William D. Hanson, (M), Memphis State, 1991, EgnGs
Geologist Supervisor:
William L. Prior, (M), Memphis State, 1979, GggEc
Information Systems Analyst:
James K. Curry, (M), S Methodist, 1978, Zn
Geologist:
Sandra Chandler, Ze
Andrew Haner, Zi
David Johnston, Gg
Lea Nondorf, Cg
Senior Petroleum Geologist:
Peng Li, (D), Alabama, 2007, GoEoCo
M. Ed Ratchford, (D), Idaho, 1994, GocEo

Professional Geologist:
Richard S. Hutto, (B), Arkansas Tech, 1994, Gg
GIS Analyst:
Nathan H. Taylor, (B), 2007, ZiyGe
Geology Supervisor:
Scott Ausbrooks, (B), Arkansas, 2001, YsGeg
Angela Chandler, (M), Arkansas, 1996, GgrGs
Geologist:
Ty Johnson, (M), Arkansas, 2008, GgZiGr
Daniel S. Rains, (M), Arkansas, 2009, GgZi
Deputy Director & Asst State Geologist:
Mac B. Woodward, (B), Sacramento State, 1957, Gog

British Columbia

British Columbia Geological Survey

British Columbia Geological Survey ☒ (2021)
PO Box 9333 Stn Prov Govt
Victoria, BC V8W 9N3
Geological.Survey@gov.bc.ca
https://www2.gov.bc.ca/gov/content/industry/mineral-exploration-mining/british-columbia-geological-survey

California

California Geological Survey

California Geological Survey ☒ (2019)
801 K Street
Suite 1200
Sacramento, CA 95814
p. (916) 445-1825
cgshq@consrv.ca.gov
http://www.consrv.ca.gov/CGS/Pages
State Geologist:
John G. Parrish, (D)
Supervising Engineering Geologist:
John Clinkenbeared, (B), Eg
Supervising Engineering Geologist:
Timothy P. McCrink, (M), New Mexico Tech, 1982, Ng
William Short, (M), Ng
Senior Sesimologist:
Hamid Haddidi, (D), Ys
Senior Geologist:
Chris T. Higgins, (M), California (Davis), 1977, Gg
Senior Engineering Geologist (Specialist):
Jennifer Thornburg, (M), California (Santa Cruz), Ng
Senior Engineering Geologist:
David Branum, (B), Ng
Rui Chen, (D), Edmonton, Nr
Ron C. Churchill, (D), Minnesota, 1980, Cg
Tim Dawson, (M), Ng
Marc Delattre, (M), Ng
Fred Gius, (M), Gz
Will Harris, (B), Ng
Jeremy Lancaster, (M), Ng
Jeremy Lancaster, (B), Gg
Donald Lindsay, (M), Ng
David Longstreth, (M), Ng
Anne Rosinski, (M), Ng
Michael A. Silva, (B), California (Davis), 1978, Ng
Senior Engineer:
Moh J. Huang, (D), Caltech, 1983, Ne
Materials Engineer:
Badie Rowshandel, (D), Ne
Senior Scientist:
Badie Rowshandel, (D), Ys
Geologist:
Lawrence Busch, (A), Eg
Engineering Geologist:
Jackie Bott, (D), Gc
Patrick Brand, (M), Ng
Kevin Doherty, (M), Ng
Michael Fuller, (B), Ng
Joshua Goodwin, (M), Gg
Carlos Guiterrez, (B), Sacramento State, Ng
Wayne Haydon, (M), Ng
Cheryl Hayhurst, (B), Ng
Janis Hernandez, (M), Ng
Peter Holland, (M), Ng
Maxime Mareschal, (M), Ng
Michael McKinney, (M), Gg
Ante Mlinarevic, (M), Ng
Matt O'Neal, (B), Gi
Brian Olson, (M), Ng
John Oswald, (B), Ng
Florante Perez, Ng
Cindy L. Pridmore, (M), San Diego State, 1983, Ng
Pete Roffers, (M), Ng
Ron Rubin, (M), Ng
Gordon Seitz, (D), Ng
Eleanor Spangler, (M), Ng
Ellie Spangler, (M), Gc
Brian Swanson, (M), Ng
Mark Welgers, (M), Ng
Chase White, (M), Ng
Judy Zachariasen, (D), Gg
Civil Engineer:
Daniel Swensen, (D), Ne
Engineering Geologist:
Rick I. Wilson, (B), Fresno State, 1987, Ge
Associate Scientist:
Lijam Hagos, (D), Ys
Carla Rosa, (M), Ng

Colorado

Colorado Geological Survey

Colorado Geological Survey ☒ (2019)
1801 Moly Rd.
Golden, CO 80401
p. 303-384-2655
cgs_pubs@mines.edu
http://www.coloradogeologicalsurvey.org/
f: https://www.facebook.com/ColoradoGeologicalSurvey
State Geologist:
Karen Berry, (B), Colorado Mines, NgZu
Senior Mapping Geologist:
Matt Morgan, (M), Colorado Mines, 2006, GgmXm
Senior Hydrogeologist:
Peter Barkmann, (M), Montana, 1984, HwGcg
Senior Geothermal Geologist:
Paul Morgan, (D), Imperial Coll (UK), 2003, YhGtYg
Senior Engineering Geologist:
Jonathan Lovekin, (M), Colorado Mines, 2007, NgZuGs
Engineering Geologist:
Jill Carlson, (B), Wesleyan, 1987, NgZu
Senior Engineering Geologist:
Jonathan White, (B), E Illinois, 1983, NgZuGg
Hydrogeologist:
Lesley Sebol, (D), Waterloo, 2005, HwGeZg
GIS Hazard Analyst:
Francis Scot Fitzgerald, (M), Denver, 2011, ZirGg
Geologist:
Kassandra Lindsey, (M), Portland State, 2015, NgGmZi
Mike O'Keeffe, (M), New Mexico Tech, 1994, GgeEg
Engineering Geologist:
Kevin McCoy, (D), Colorado Mines, 2015, NgZiHw
Scientific & Technical Graphic Designer:
Larry Scott, (B), Univ Arts, 1985, Zy

Connecticut

Dept of Energy and Environmental Protection

Connecticut Geological Survey ☒ (2022)
Office of Information Management
79 Elm Street, 6th floor
Hartford, CT 06106-5127
p. (860) 424-3540
deep.ctgeosurvey@ct.gov
http://www.ct.gov/deep/geology
State Geologist:
Meghan M. Seremet, (M), Connecticut, 2007, Gg
Senior Research Associate:
Randolph P. Steinen, (D), Brown, 1973, Gsg
Research Associate:
Lindsey Belliveau, (M), Connecticut, 2016, Gmt

Teresa K. Gagnon, (M), Boston Coll, 1992, Gg
Resource Assistant:
James M. Bogart, (B), S Connecticut State, 2015, Ge

Delaware

University of Delaware

Delaware Geological Survey ☒ (2022)
257 Academy Street
Newark, DE 19716-7501
p. (302) 831-2833
DelGeoSurvey@udel.edu
http://www.dgs.udel.edu
Administrative Assistant: Denise T. Heldorfer, Laura K. Wisk
State Geologist:
David R. Wunsch, (D), Kentucky, 1992, ClHwNg
Senior Scientist:
Peter P. McLaughlin, (D), Louisiana State, 1989, HgPm
Hydrogeologist:
Thomas E. McKenna, (D), Texas, 1997, Hg
Hydrogeologist:
Changming He, (D), Nevada (Reno), 2004, GqHg
Associate Scientist:
Stefanie J. Baxter, (M), Delaware, 1994, On
Research Associate:
Jaime L. Tomlinson, (M), Delaware, 2006, Zn
Emeritus:
John H. Talley, (M), Franklin and Marshall, 1974, As
Scientist:
Kelvin W. Ramsey, (D), Delaware, 1988, As
GIS Specialist:
Lillian T. Wang, (M), Delaware, 2005, Zi

Florida

Florida Geological Survey

Florida Dept of Environmental Protection ☒ (2021)
Commonwealth Building
3000 Commonwealth Blvd, Suite 1
Tallahassee, FL 32303
p. (850) 617-0300
guy.means@floridadep.us
https://floridadep.gov/fgs
State Geologist:
Guy H. Means, (M), Florida State, 2009, PgGe

Georgia

Georgia Dept of Natural Resources

Georgia Environmental Protection Div (2016)
Environmental Protection Division
2 Martin Luther King Jr. Dr., Suite 1152
East Tower
Atlanta, GA 30334-9004
p. (404) 657-5947
askepd@gaepd.org
http://epd.georgia.gov/
Senior Scientist:
James Kennedy, (D), Texas A&M, 1981

Hawaii

Dept of Land & Natural Resources

Commission on Water Resource Management ☒ (2018)
Kalanimoku Building
P.O. Box 621
1151 Punchbowl Street, #227
Honolulu, HI 96809
p. (808) 587-0214
dlnr.cwrm@hawaii.gov
http://dlnr.hawaii.gov/cwrm

Idaho

University of Idaho

Idaho Geological Survey ☒ (2022)
875 Perimeter Dr. MS 3014
University of Idaho
Moscow, ID 83844-3014
p. (208) 885-7991
igs@uidaho.edu
http://www.idahogeology.org/
f: https://www.facebook.com/IDGeoSurvey/
t: @IDGeoSurvey
Director:
Claudio Berti, (D), Chieti (Italy), 2009, GgmGc
Research Geologist:
Virginia S. Gillerman, (D), California (Berkeley), 1982, Eg
Reed S. Lewis, (D), Oregon State, 1990, GgiEg
Senior Petroleum Geologist:
Mark Barton
Senior Geologist:
Russell Di Fiori, (D), Washington State
Dennis M. Feeney, (M), W Washington, 2008, GgiYg
Hydrogeologist:
Alexandra Clark, (M)
Hazard Geologist:
Zachary Lifton, (D), Georgia Tech
Geoscience Manager:
Nathan Hopkins, (D), Lehigh, 2016, GlmGg

Illinois

Illinois State Geological Survey

Energy and Minerals ☒ (2021)
615 E. Peabody Drive.
Champaign, IL 61820-6964
p. (217) 244-2430
sgwhit@illinois.edu
Director:
Steven Whittaker, (D), Goe
Principal Scientist:
Scott Elrick, (M)
Scott M. Frailey, (D), Missouri (Rolla), 1989, Np
Sallie Greenberg, (D), Illinois
Sherilyn Williams-Stroud, (D)
Donna Willlette, (D)
Senior Scientist:
Jared Freiburg, (M)
Scientist:
Joseph A. Devera, (M), S Illinois, 1985, Gr
Principal Scientist:
Hannes E. Leetaru, (D), Illinois, 1997, Eo
Associate Scientist:
Cheri A. Chenoweth, (B), Illinois, 1979, Ec
Assistant Scientist:
Christopher P. Korose, (B), Illinois, 1995, Ec
Vinodkumar A. Patel, (B), Inst of Tech, 1973, Zn
Assistant Scientist:
Kathleen M. Henry, (B), Illinois State, 1982, Gg

Illinois State Geological Survey ☒ (2017)
615 East Peabody Drive
Champaign, IL 61820-6964
p. 217-333-4747
info@isgs.illinois.edu
https://www.isgs.illinois.edu/
f: https://www.facebook.com/ILGeoSurvey
t: @ILGeoSurvey
Administrative Assistant: Tamra S. Montgomery
Director:
Richard C. Berg, (D), Illinois, 1979, GleGm
Senior Geologist:
Anne L. Ellison, (M), Ge
Geoscience Information Stewardship:
Mark Yacucci, (M), Illinois State
Associate Geologist:
Scott D. Elrick, (M), California (Riverside), 1999, EcGsPg
Senior Geologist:
Zakaria Lasemi, (D), Illinois, Eg
Senior Geochemist:
Samuel V. Panno, (M), S Illinois, 1978, ClGgHw
Head Geochemistry:
Randall Locke, (M), Cl
Chief Scientist:
Steven E. Brown, (M), Wisconsin, 1990, Glm

Associate Quaternary Geologist:
David A. Grimley, (D), Illinois, 1996, GmlSc
Senior Scientist:
Richard A. Cahill, (M), Maryland, 1974, CaGeCt
Associate Quaternary Geologist:
Andrew C. Phillips, (D), Illinois (Chicago), 1993, GsmZi
Adjunct Assistant Professor:
Ethan Theuerkauf, (D), North Carolina, 2016, GmuGs
Assistant Petroleum Geologist:
Nathan D. Webb, (M), Illinois (Urbana), 2009, EoGo
Illinois State Geologist:
E. Donald McKay III, (D), Illinois, 1977, GlsGr
Wetlands Geology Specialist:
Colleen Long, (M), North Carolina, 2012, Gg
Jessica Monson, (M), Gg
Wetlands Geologist and Head:
James J. Miner, (M), Gg
Wetlands Geologist:
Jessica R. Ackerman, Zn
Team Leader/Associate Geologist:
Dale R. Schmidt, (M), Ge
Team Leader:
D. Adomaitis, Ge
Senior Petroleum Geologist:
Hannes E. Leetaru, (D), Illinois, 1997, Gor
Senior Paleontologist:
Joseph A. Devera, (M), Pg
Senior Geophysicist:
Timothy H. Larson, (D)
Remote Sensing Data Manager:
Janet Carmarca
Program Manager, Illinois Height Modernization Project:
Sheena K. Beaverson, Zr
Principal Engineering Geologist:
R. A. Bauer, (M), Ng
Media and Information Technology Administrator:
Daniel Byers
Map Standards Coordinator:
Jennifer Carrell, Zg
Hydrogeologist and Assistant Section Head:
Yu-Feng Forrest Lin, (D), Wisconsin, 2002, Hg
Geospatial Applications Developer/GIS Specialist:
Melony Barrett, Zi
Geologic Specialist:
Zohreh Askari, (M), Azad, 1997, Gsc
Alan R. Myers, (B), EcGo
Jennifer M. Obrad, (M), EcGo
Enviromental Data Coordinator:
Clint Beccue, Ge
Associate Wetlands Geologist:
Steven Benton, Gg
Keith W. Carr, (M), Gg
Eric T. Plankell, (M), Hg
Geoff Pociask, (M), Hw
Associate Quaternary Geologist:
Olivier J. Caron, (D), Québec (Montréal), Ng
Associate Geologist:
Cheri Chenoweth, GoEc
Christopher P. Korose, (M), EcGo
Associate Geohydrologist:
Jason F. Thomason, (D), Hw
Associate Geochemist:
Shari E. Effert-Fanta, (M), Cas
Associate Engineering Geologist:
Andrew Anderson, Ng
Greg A. Kientop, (M), Ge
Associate Economic Geologist:
F. Brett Denny, (M), GzEg
Associate Director Advanced Energy Technology Initiative:
Sallie Greenberg, (D), Illinois (Urbana), 2013, ClZg
Assistant Wetlands Geologist:
Kathleen E. Bryant, (M), Gg
Assistant Section Head, Environmental Assessments:
Mark Collier, Ge
Assistant Section Head:
B. Brandon Curry, (D), Illinois (Urbana), 1995, Gel
Assistant Geologist:
James Damico, (M)
Craig R. Decker, (B)
Scott R. Ellis, (B), Ge
Bradley Ettlie, (B), Ge
Jared Freiburg, (M), GgzGx
James W. Geiger, (B), Ge
Nathan P. Grigsby, (B)
Matthew P. Spaeth, (M), Ge
Assistant Geochemist:
Peter M. Berger, (M), Illinois (Urbana), 2008, Cg
Other:
Hong Wang, Illinois (Urbana), 1996, Cg

University of Illinois

Illinois State Geological Survey/Prairie Research Institute ☒ (2018)
615 E. Peabody Dr.
Champaign, IL 61820-6964
p. (217) 333-4747
info@isgs.illinois.edu
http://www.isgs.edu
f: https://www.facebook.com/ILGeoSurvey/
t: @ILGeoSurvey
State Geologist:
Richard Berg, (D), Illinois (Urbana), 1979, GlRwZu
Energy & Minerals Group:
Steven Whittaker, (D), Saskatchewan
Principle Geologist:
B. Brandon Curry, Illinois, Gl
Principal Scientist:
Randall A. Locke II, (M), CgHw
Principal Geologist:
Scott D. Elrick, (M), California (Riverside), 1998, EcGgs
Assoociate Geologist:
Geoffrey Pociask, (M), Wisconsin, Hs
Head:
Anne L. Ellison, (M), Ge
Hannes Leetaru, (D), Illinois, Go
Jason Thomason, (D), Iowa State, Gl
Mark A. Yacucci, (M), Zi
Senior Scientist:
Yongqi Lu, (D), Zm
Chief Scientist:
Steven E. Brown, (M), Gl

Illinois State Water Survey, Prairie Research Institute ☒ (2022)
2204 Griffith Drive
Champaign, IL 61820
p. (217) 244-5459
info@isws.illinois.edu
http://www.isws.illinois.edu/
Programs: N/A
Certificates: N/A
Head, Groundwater Science Section:
Walton R. Kelly, (D), Virginia, 1993, Hw
Senior Scientist:
Steven D. Wilson, (M), Illinois, 1988, Hw
Associate Scientist:
Daniel Abrams, (D), Indiana, 2013, Hw
Daniel Hadley, (M), N Arizona, 2014, Hw
Devin Mannix, (M), S Illinois, 2013, Hw
Hydrogeologist:
Vlad Iordache, (M), Illinois, 2017, Hw
Research Associate:
Cecilia Cullen, (M), Illinois, Hw
Allan Jones, (D), Texas, 2017, Hw
Michael Krasowski, (M), Iowa, Hw

Indiana

Indiana University, Bloomington

Indiana Geological and Water Survey (2017)
611 North Walnut Grove Avenue
Bloomington, IN 47405
p. (812) 855-7636
igsinfo@indiana.edu
http://igws.indiana.edu/
State Geologist & Director:
Todd A. Thompson, (D), Indiana, 1987, GsmGu

Senior Scientist:
Maria Mastalerz, (D), Silesian Tech, 1988, Ec
Assitant Director for Research:
Lee J. Florea, (D), S Florida, 2006, HqGeCl
Associate Scientist:
Sally L. Letsinger, (D), Indiana, 2001, HqZir
Research Associate:
Patrick I. McLaughlin, (D), Cincinnati, 2006, GrClGs
Reservoir Geologist:
Cristian R. Medina, (M), Indiana, 2007, GoeHw
Research Geophysicist/Hydrologist:
Kevin M. Ellett, (M), California (Davis), 2002, YeHw
Research Geologist:
José Luis Antinao, (D), Dalhousie, 2009, CcGlr
Robert J. Autio, (M), Indiana, 1992, HgGle
Alyssa M. Bancroft, (D), Ohio State, 2014, PmGr
Tracy D. Branam, (M), Indiana, 1991, CasEm
Christopher Dintaman, (M), Indiana, 1997, Hw
Agnieszka Drobniak, (D), Sci Tech (Krakow), 2002, Ec
Jayson Eldridge, (B), California (Davis), 2012, EoGr
Nancy R. Hasenmueller, (M), Ohio State, 1969, Gre
Rebecca A. Meyer, Ec
Gary J. Motz, (M), Akron, 2010, PqgPi
Shawn C. Naylor, (M), Montana, 2006, Hw
John A. Rupp, (M), E Washington, 1980, Go
Quaternary Geologist:
Henry M. Loope, (D), Wisconsin, 2013, Gls

Iowa

Iowa Dept of Natural Resources

Iowa Geological and Water Survey (2015)
109 Trowbridge Hall
Iowa City, IA 52242-1319
p. (319) 335-1575
MaryPat.Heitman@dnr.iowa.gov
http://www.igsb.uiowa.edu
Administrative Assistant: Mary Pat. Heitman
Research Geologist:
Richard A. Langel, (M), Iowa, 1996, Gg
State Geologist:
Robert D. Libra, Gg
Section Supervisor, Geology and Groundwater Studies:
J. Michael Gannon, (M), Arizona, Gg
Section Supervisor, Geographic Information:
Chris Ensminger
Research Geologist:
Mary R. Howes, Gg
Lynette S. Seigley, (M), Iowa, Gg
Natural Resource Biologist:
Jacklyn Gautsch, (B), Wisconsin, Zn
GIS Technician:
Chris Kahle, (M), Kansas, Zi
Geologist 3, Research Geologist:
Paul Hiaibao Liu, (D), Nebraska, Gg
Robert M. McKay, (B), Tulane, Gg
Deborah J. Quade, (M), Iowa, 1992, GleGm
Robert Rowden, (M), Iowa, Gg
Keith Schilling, (M), Iowa State, Gg
Stephanie Tassier-Surine, (M), Massachusetts, Gg
Paul E. VanDorpe, (M), Wayne State, Gg
Geologist 3, Remote Sensing Analyst:
James D. Giglierano, (M), Purdue, GgZri
Pete Kollasch, (M), Iowa, GgZri
Geologist 3, GIS Analyst:
Kathryne Clark, (M), New Mexico, GgZi
Geologist 3, Geographic Information System Analyst:
Calvin Wolter, (B), Arizona, GgZi
Geologist 2, Research Geologist:
Michael Bounk, (M), Iowa, Gg
Chad Fields, (M), N Iowa, Gg
Geologist 3, NRGIS Library Manager and GIS Analyst:
Casey Kohrt, (B), Iowa State, GgZi

Kansas

University of Kansas

Kansas Geological Survey ☒ (2021)
1930 Constant Avenue
West Campus
Lawrence, KS 66047-3724
p. (785) 864-3965
http://www.kgs.ku.edu/
f: www.facebook.com/KansasGeologicalSurvey
t: @ksgeology
Director:
Rolfe D. Mandel, (D), Kansas, 1991, Ga
Section Chief, Senior Scientist:
Richard D. Miller, (D), Leoben, 2007, Ye
Research Project Director:
Kenneth A. Nelson, (B), Kansas, 1993, Zy
Manager, Wichita Well Sample Library:
Mike Dealy, (B), Fort Hays State, 1979, Gg
Manager, Geohydrology Support Services:
Blake Brownie Wilson, (M), Kansas State, 1993, Zi
Senior Scientific Fellow:
Evan K. Franseen, (D), Wisconsin, 1989, Gs
Section Chief, Senior Scientist:
Greg A. Ludvigson, (D), Iowa, 1988, GrCsPc
Senior Scientist:
James J. Butler, Jr., (D), Stanford, 1987, Hw
Associate Scientist:
Geoffrey C. Bohling, (D), Kansas, 1999, Hw
Gaisheng Liu, (D), Alabama, 2004, Hy
Kerry D. Newell, (D), Kansas, 1996, Go
Petroleum Engineer:
Yehven I. Holubnyak, (M), North Dakota, 2008, Np
Assistant Researcher Senior:
Edward Reboulet, (M), Boise State, 2003, Hy
Assistant Research Professor:
Julian Ivanov, (D), Kansas, 2002, Yg
Research Engineer Senior:
Brett Bennet, (B), Kansas, 1982, Ng
Outreach Manager:
Blair Schneider, Gg
DASC Manager:
Eileen Battles, (B), Kansas, 1993, Zy
Assistant Scientist:
Jon J. Smith, (D), Kansas, 2007, Gg

Kentucky

University of Kentucky

Kentucky Geological Survey ☒ (2020)
228 Mining & Minerals Resources Building
504 Rose Street
Lexington, KY 40506-0107
p. (859) 257-5500
KGSmail@uky.edu
http://www.uky.edu/KGS
f: www.uky.edu/KGS
State Geologist and Director:
William C. Haneberg, (D), Cincinnati, 1989, NgRnNp
Head, Water Resources Section:
Charles J. Taylor, (M), Kentucky, 1992, HwgCl
Head, Geoscience Information Management:
Doug C. Curl, (M), Tennessee, 1998, ZfGgc
Head, Geologic Mapping Section:
William M. Andrews, (D), Kentucky, 2004, GmZiEc
Head, Geologic Hazards Section:
Zhenming Wang, (D), Kentucky, 1998, Ys
Head, Energy & Minerals:
David C. Harris, (M), SUNY (Stony Brook), 1982, Go
Head, Administration:
Kathryn E. Ellis, (M), Kentucky, 2012, Zn

Louisiana

Louisiana State University

Basin Research Energy Section (2015)
Louisiana Geological Survey/LSU
208 Howe Russell Geoscience Complex

Baton Rouge, LA 70803-4101
p. (225) 578-8328
bharde1@lsu.edu
http://www.bri.lsu.edu
Office Coordinator: Cherri B. Webre
Associate Professor:
Ronald K. Zimmerman, (D), Louisiana State, 1966, Go
Assistant Professor:
Clayton F. Breland, (D), Tennessee, 1980, Ye
John B. Echols, (D), Louisiana State, 1966, Go
Research Associate:
Brian J. Harder, (B), Louisiana State, 1981, Eo
Bobby L. Jones, (B), Louisiana State, 1953, Go
Phillip W. Lemay, (B), Centenary, 1999, Gg
Michael B. Miller, (M), North Carolina, 1982, Go
Lloyd R. Milner, (B), Louisiana State, 1985, Gg
Patrick M. O'Neill, (B), Louisiana State, 1985, Sf
Computer Analyst:
Reed J. Bourgeois, (B), Louisiana State, 1983, Zn
Accountant Technician:
Carla Domingue, Zn

Louisiana Geological Survey (2015)
3079-Energy,
Coast and Environment Bldg.
Baton Rouge, LA 70803
p. (225) 578-5320
hammer@lsu.edu
http://www.lgs.lsu.edu
State Geologist and Professor:
Chacko J. John, (D), Delaware, 1977, GosGe
Assistant Professor:
Douglas A. Carlson, (D), Wisconsin (Milwaukee), 2001, Hw
Marty R. Horn, (D), Texas (Arlington), 1996, EoGgr
GIS Coordinator:
R Hampton Peele, (M), Louisiana State, 2000, ZirZg
Computer Analyst:
Reed J. Bourgeois, (B), Nicholls State, 1985, Zn
Research Associate:
Brian J. Harder, (B), Louisiana State, 1981, Eo
Paul V. Heinrich, (M), Illinois, 1982, GsaGm
Bobby Jones
Richard P. McCulloh, (M), Texas, 1977, Gg
Lloyd R. Milner, (B), Louisiana State, 1985, Gg
Patrick M. O'Neill, (B), Louisiana State, 1985, Sf
Robert L. Paulsell, (B), Louisiana State, 1987, ZyGg
Lisa G. Pond, (B), Louisiana State, 1987, Zg
Arren Schulingkamp, Gg
Cartographic Manager:
John I. Snead, (B), Louisiana State, 1978, Gm
Assistant Director:
John E. Johnston, III, (M), Texas, 1977, EoGe
Office Coordinator:
Melissa H. Esnault, (B), Louisiana Tech, Zn

Maine

Dept of Agriculture, Conservation, and Forestry

Maine Geological Survey ☒ (2021)
93 State House Station
Augusta, ME 04333-0093
p. (207) 287-2801
mgs@maine.gov
http://www.maine.gov/dacf/mgs/
Administrative Assistant: Tammara Roberts
State Geologist:
Robert G. Marvinney, (D), Syracuse, 1986, Gg
Senior Geologist:
Lindsay Spigel, (D), Wisconsin, 2006, Gm
Amber Whittaker, (M), New Mexico, 2006, GcZiCg
Physical Geologist:
Henry N. Berry IV, (D), Massachusetts, 1989, Gg
Marine Geologist:
Stephen M. Dickson, (D), Maine, 1999, OnGuOu
Peter Slovinsky, (M), South Carolina, 2001, Gu
Hydrogeologist:
Ryan Gordon, (D), Syracuse, 2014, HwGm
Daniel B. Locke, (B), Maine, 1982, Hw

Director, Earth Resources Information:
Christian Halsted, (M), Wisconsin, 2020, Zi

Manitoba

Manitoba Geological Survey

Manitoba Agriculture and Resource Development ☒ (2021)
360-1395 Ellice Avenue
Winnipeg, MB R3G 3P2
p. 1-800-223-5215
michelle.nicolas@gov.mb.ca
http://www.manitoba.ca/iem
Head:
Michelle P.B. Nicolas, (M), Manitoba, 1997, GrsGo
Senior Scientist:
Christian Bohm, (D), ETH (Switzerland), 1996, GgCgEg

Maryland

Maryland Department of Natural Resources

Maryland Geological Survey ☒ (2021)
2300 St. Paul Street
Baltimore, MD 21218-5210
p. (410) 554-5500
mgsinfo.dnr@maryland.gov
http://www.mgs.md.gov/
State Geologist:
Richard A. Ortt, Jr., (B), Johns Hopkins, 1991, NtGeYr
Geologist/Program Chief:
Stephen Van Ryswick, (B), Maryland, 2002, Ge
Senior Scientist:
David K. Brezinski, (D), Pittsburgh, 1984, Gr
Hydrogeologist:
Emelia Furlong, (D), Caltech, HwCqSp
Andrew W. Staley, (M), Wisconsin, 1992, Hw
Geologist:
Christopher B. Connallon, (M), Michigan State, 2015, GmZi
William D. Junkin, (M), California (Santa Barbara), 2018, GvrGg
Rebecca H. Kavage Adams, (M), Virginia Tech, 2002, GmZiGe
Heather A. Quinn, (M), Florida, 1988, GgHwg
Elizabeth R. Sylvia, (M), Towson, 2013, Ge
Education Specialist:
Dale W. Shelton, (B), Towson, 1986, ZeGg

Massachusetts

Massachusetts Geological Survey

Dept of Geosciences ☒ (2022)
Univ of Massachusetts (Amherst)
627 North Pleasant Street
Amherst, MA 01003
p. (413) 545-4814
mabes@umass.edu
State Geologist:
Stephen B. Mabee, (D), Massachusetts, 1992, HwGl

Minnesota

University of Minnesota

Minnesota Geological Survey ☒ (2021)
2609 West Territorial Road
Saint Paul, MN 55114-1009
p. 612-626-2969
mgs@umn.edu
https://cse.umn.edu/mgs
f: https://www.facebook.com/UMNMGS
Director:
Harvey Thorleifson, (D), Colorado, 1989, Gl
Lead Geologist:
Anthony Runkel, (D), Texas, 1988, Gr
Geologist:
Emily Bauer, (B), Wisconsin (River Falls), 1989, Zi
Amy Block, (M), Minnesota (Duluth), 2013, Gp
Angela Gowan, (M), Minnesota (Duluth), 1993, Gl
Rich Lively, (M), Michigan State, 1977, Cc
Gary Meyer, (M), North Dakota, 1979, Gl
Andrew Retzler, (M), Idaho State, 2013, GrsPi
Julia Steenberg, (M), Idaho State, 2008, Gr

Associate Director:
Barb Lusardi, (M), Maine, 1992, Gl
Geophysicist:
Aaron Hirsch, (D), Rhode Island, 2021, Yg
Geologist:
Angela Berthold, (M), Minnesota (Duluth), 2014, Gl
Maximiliano Bezada Diaz, (D), Venezuelan Inst Sci Inv, 1989, Gl
Jarrod Cicha, (M), Maine, 2015, Zi
Emma Collins, (M), Idaho State, 2021, Gl
Daniel Conrad, (M), N Arizona, 2018, Gl
Sarah Francis, (M), Western Washington, 2019, Zi
Jacqueline Hamilton, (M), Minnesota, 2007, ZiGe
Ethan Kurak, (M), Lehigh, 2021, Gl
Jordan Mayer, (M), Michigan Tech, 2016, Zi
Jennifer McDonald, (M), Toledo, 2015, Gl
Maurice Nguyen, (M), W Ontario, 2014, Gl
Eric Nowariak, (M), Missouri-Columbia, 2016, Gp
Allison Severson, (D), Colorado Mines, 2020, Gp
Technical Editor:
Lori Robinson, (M), Alaska (Fairbanks), 1999, Zi
Information Science:
Corey Betchwars, (M), Minnesota, 2014, Zi

Mississippi

Mississippi Office of Geology

Geospatial Resources Div (2015)
P.O. Box 2279
700 N. State St
Jackson, MS 39225-2279
p. (601) 961-5500
barbara_yassin@deq.state.ms.us
http://www.deq.state.ms.us
Geologist:
Steven D. Champlin, (B), Alabama, 1976, Go
GIS Analyst:
Barbara E. Yassin, (B), Illinois State, 1989, ZyiGm
Geologist:
Peter S. Hutchins, (B), Millsaps, 1990, Zn

Mississippi Dept of Environmental Quality (2016)
P.O. Box 2279
Jackson, MS 39225-2279
p. (601) 961-5500
mbograd@mdeq.ms.gov
http://www.deq.state.ms.us/
State Geologist:
Michael B. E. Bograd, (M), Mississippi, 2002, Gg

Missouri

Missouri Dept of Natural Resources

Dam & Reservoir Safety (2015)
PO Box 250
Buehler Bldg/111 Fairgrounds Rd
Rolla, MO 65402
p. (573) 368-2175
geology@dnr.mo.gov
Professional Staff:
Robert Clay, (M), Oklahoma State, 1977, Zn
Professional Staff:
Glenn D. Lloyd, (B), Zn
Other:
Paul Simon, (B), Missouri Tech
Ryan Stack, (B), Missouri Tech

Div of Geology and Land Survey (2015)
PO Box 250
111 Fairgrounds Road
Rolla, MO 65402-0250
p. (573) 368-2100
joe.gillman@dnr.mo.gov
http://www.dnr.state.mo.us/geology.htm
Administrative Assistant: Tami L.. Allison
State Geologist & Division Director:
Mimi R. Garstang, (B), SW Missouri State, Ge
Deputy Director & Assistant State Geologist:
James W. Duley, (B), C Missouri, 1975, Hy

Geological Survey Program ☒ (2019)
PO Box 250
111 Fairgrounds Rd
Rolla, MO 65402-0250
p. (573) 368-2100
geology@dnr.mo.gov
http://www.dnr.mo.gov/geology
State Geologist:
Joe Gillman, (B), Missouri State, 1992, Gg
Senior Scientist:
Pat Mulvany, (D), Missouri S&T, 1996
Unit Chief:
Justin Davis, (M), Missouri S&T, 2012
Jeremiah Jackson, (M), Missouri State, 2011, Gg
Brenna McDonald, (B), SE Missouri State, 1997, Ge
Cheryl M. Seeger, (D), Missouri S&T, 2003, Gig
Vicki Voigt, (B), Missouri S&T, 2010, Gg
Section Chief:
Larry Pierce, (M), Zn
Kyle Rollins, (B), Missouri State, 1986, ZyGg
Sherri Stoner, (B), Missouri State, 1991
Program Director:
Amber Steele, (M), Missouri, 2011, Sd
Geologist:
Peter Bachle, (B), Missouri S&T, 1997, Gg
Fletcher Bone, (B), C Missouri, 2007, Gg
David L. Bridges, (D), Missouri S&T, 2011, GgiPg
Kyle Brown, Mizzou, 2015, Gg
John H. Corley, (M), Missouri, 2014, Geg
Jeff Crews, (M), Missouri S&T, 2004, Hy
Trevor Ellis, (B), Missouri S&T, 2010, Gg
Kyle Ganz, (M), Missouri S&T, 2013
Airin J. Haselwander, (B), Missouri S&T, 2011, Gg
Terry Hawkins, (B), Brigham Young, 1988, Geg
Katelyn Kane-DeVries, (M), N Illinois, 2016, Gg
Lisa Lori, (M), Missouri S&T, 2016, Gp
Brad Mitchell, (M), Missouri S&T, 2010
Matt Parker, (B), Missouri S&T, 1993
Kirsten Schaefer, (B), Illinois State, 2016
Michael A. Siemens, (M), Wichita State, 1985, GgHgZi
Molly Starkey, (M), Missouri State, 2011
Chris Vierrether, (M), Missouri S&T, 1988, GgoGe
Geological Tech:
Cecil Boswell, Zn
Karen Loveland, Zn
Dan Nordwald, Zn
Patrick Scheel, Zn
Fred Shaw, Zn
Environmental Specialist:
Andrew Combs
Michelle Oglesby
Deputy Director:
Jerry L. Prewett, (B), Missouri State, 1992, GgHw

Water Resources Program ☒ (2022)
PO Box 250
111 Fairgrounds Rd
Rolla, MO 65401-0250
p. (573) 368-2175
mowaters@dnr.mo.gov
http://www.dnr.state.mo.us/mgs/
Geologist:
Scott Kaden, Hw
Associate Scientist:
Robert Bacon, Hs
Professional Staff:
Charles Du Charme, (B), Hs

Montana

Montana Tech of The University of Montana

Montana Bureau of Mines & Geology (A,B,M,D) ☒ (2018)
1300 West Park Street
Butte, MT 59701-8997
p. (406) 496-4180
jmetesh@mtech.edu
http://www.mbmg.mtech.edu
f: https://www.facebook.com/MontanaGeology/
Administrative Assistant: Margaret Delaney

State Geologist:
John J. Metesh, (D), Montana, 2003, HwCa
Assistant Director RET:
Marvin R. Miller, (M), Indiana, 1965, Hw
Research Division Chief:
Thomas W. Patton, (M), Montana Tech, 1987, Hw
Program Manager Geologist:
Catherine McDonald, (M), California (Davis), 1992, Grs
Assistant Research Geologist:
Jeff Lonn, (M), Montana, 1985, GgtGc
Senior Hydrogeologist:
Jon C. Reiten, (M), North Dakota, 1983, Hw
Geologist:
Susan M. Vuke, (M), Montana, 1982, GgrGs
Professor:
Colleen G. Elliott, (D), New Brunswick, 1988, GctGg
Senior Research Hydrogeologist:
John R. Wheaton, (M), Montana, 1987, HwEc
Program Manager Hydrogeologist:
Ginette Abdo, (M), Penn State, 1989, Hw
John I. La Fave, (M), Texas, 1987, Hw
Hydrogeologist:
Andrew L. Bobst, (M), SUNY (Binghamton), 2000, HwCgEo
Gary Icopini, (D), Michigan State, 2000, ClHwCb
Geologist:
Michael C. Stickney, (M), Montana, 1980, YsGtm
Publications Editor:
Susan A. Barth, (M), Montana Tech, 2009, Zn
Assistant Research Hydrogeologist:
Camela Carstarphen, (M), Oregon State, 1991, Hw
Assistant Research Geologist:
Phyllis Hargrave, (M), Montana Tech, 1990, Gg
Sr. Hydrogeologist:
Kirk B. Waren, (M), Wright State, 1988, HwsZe
Hydrogeologist:
Terence E. Duaime, (B), Montana Tech, 1978, Hws
Senior Research Geologist, Museum Curator:
Richard B. Berg, (D), Montana, 1964, EnGz
Seismic Analyst:
Deborah Smith
Research Assistant III:
Jaqueline R. Timmer, (B), Montana Tech, 1990, Ca
Hydrogeologist:
Daniel D. Blythe, (B), Montana State, 2006, Hws
GIS Specialist:
Ken L. Sandau, (A), Montana Tech, 1999, Zi
Paul R. Thale, (M), Montana State, 1994, Zi
Geologic Cartographer:
Susan M. Smith, (B), Montana State, 1970, Zn
Director of Sponsored Programs:
Joanne Lee
Chemist:
Ashley Huft, (B), Montana Tech, 2008, Ca
Associate Research Professor:
Steve F. McGrath, (M), Montana Tech, 1992, CaEgCe
Assistant Hydrogeologist:
Mary K. Sutherland, (M), Montana, 2009, Hws
Computer Software Eng/Applications:
Luke Buckley, (B), Montana Tech, 1995, Zn

Nebraska

Unversity of Nebraska, Lincoln

Conservation & Survey Division ☒ (2022)
Conservation & Survey Division, SNR
3310 Holdrege Street
616 Hardin Hall
Lincoln, NE 68583-0996
p. (402) 472-3471
rjoeckel3@unl.edu
http://snr.unl.edu/csd/
Director:
R. M. (Matt) Joeckel, (D), Iowa, 1993, GrsGm
Professor:
Paul Hanson, (D), Nebraska, 2005, Gm
Jozsef Szilagyi, (D), California (Davis), 1997, HqAs
Senior Scientist:
Mark E. Burbach, (D), Nebraska, 2004, Hw
Susan Olafsen-Lackey, (B), SD Mines, 1982, Hw
Associate Professor:
Troy E. Gilmore, (D), NC State, 2015, Hwg
Jesse T. Korus, (D), Nebraska, 2015, GsrHw
Associate Scientist:
Mohamed Khalil Aboushanab, (D), Justus Liebig, 2002, YgHw
Jeffrey Westrop, (D), Nebraska, 2021, HwClb
Assistant Professor:
Judith Turk, (D), California (Riverside), 2012, Sd
Emeritus:
Jerry Ayers, (D), Washington State, 1980, Hw
Marvin P. Carlson, (D), Nebraska, 1969, Gt
Robert F. Diffendal, Jr., (D), Nebraska, 1971, GrPiRh
Duane A. Eversoll, (M), Nebraska, 1977, Ng
James W. Goeke, (M), Colorado State, 1970, HwGmYe
Mark S. Kuzila, (D), Nebraska, 1988, Sd
James B. Swinehart, (M), Nebraska, 1979, GrsGe
Geological Research Specialist:
Michele M. Waszgis, (B), Nebraska, 2015, Gg
Survey Hydrogeologist:
Kathleen A. Cameron, (B), SUNY Binghamton, 1999, HwZfGg
Douglas R. Hallum, (B), Wisconsin (Eau Claire), 1996, HwGog
Survey Geologist:
Aaron R. Young, (M), Nebraska, 2008, HwGmZf

Nevada

University of Nevada, Reno

Nevada Bureau of Mines and Geology ☒ (2018)
Mail Stop 178
University of Nevada
Reno, NV 89557-0088
p. (775) 682-8766
jfaulds@unr.edu
http://www.nbmg.unr.edu/
f: https://www.facebook.com/Nevada-Bureau-of-Mines-and-Geology-106397989390636/
Administrative Assistant: Alex Nesbitt
State Geologist/Professor:
James E. Faulds, (D), New Mexico, 1989, Gc
Professor:
Geoffrey Blewitt, (D), Caltech, 1996, Yd
William C. Hammond, (D), Oregon, 2000, Yd
Christopher D. Henry, (D), Texas, 1975, GtvCc
Corne Kreemer, (D), SUNY (Stony Brook), 2001, YdRn
Geologic Mapping Specialist:
Seth Dee, (M), Oregon, 2006, Gg
Director, Center for Research in Economic Geology:
John Muntean, (D), Stanford, 1998, Eg
Associate Professor:
Bridget Ayling, (D), Australian Nat, 2006, CgYhGo
Craig M. dePolo, (D), Nevada (Reno), 1998, Ng
Assistant Professor:
Rich Koehler, (D), Nevada (Reno), 2009, GmtNg
Mike Ressel, (D), Nevada (Reno), 2005, EgGiCc
Andrew V. Zuza, (D), California (Los Angeles), 2016, Gct
Adjunct Professor:
Mark F. Coolbaugh, (D), Nevada (Reno), 2003, Eg
Emeritus:
John W. Bell, (M), Arizona State, 1974, NgGm
Stephen B. Castor, (D), Nevada (Reno), 1972, Eg
Larry J. Garside, (M), Nevada (Reno), 1968, Gg
Liang-Chi Hsu, (D), California (Los Angeles), 1966, Cp
Daphne D. LaPointe, (M), Montana, 1977, Gg
Paul J. Lechler, (D), Nevada (Reno), 1995, Cg
Jonathan G. Price, (D), California (Berkeley), 1977, Gg
Alan R. Ramelli, (M), Nevada (Reno), 1988, Ng
Lisa Shevenell, (D), Nevada, 1990, Cg
Joseph V. Tingley, (M), Nevada (Reno), 1963, Em
Geologic Information Specialist:
David A. Davis, (M), Nevada (Reno), 1990, Gg

New Hampshire

New Hampshire Geological Survey

New Hampshire Dept of Environmental Services ☒ (2022)
29 Hazen Drive
P.O.Box 95
Concord, NH 03302-0095

p. (603) 271-1975
Shane.J.Csiki@des.nh.gov
https://www.des.nh.gov/land/geology
State Geologist and Director:
Shane Csiki, (D), Illinois (Urbana), 2014, ZyGm
Program Geologist:
Joshua A. Keeley, (M), Idaho State, 2011, GgHyZi
Geoscience Program Specialist:
Michael W. Howley, (M), New Hampshire, 2008, GgYgHw
Geological Survey Outreach Coordinator:
Rebecca M. LeCain, (M), New Hampshire, 2010, Gr
Assistant Flood Hazards Geoscientist:
Brian K. Hauschild, (B), New Hampshire, 2018, HgsHw

New Jersey

New Jersey Geological and Water Survey

New Jersey Geological and Water Survey ☒ (2022)
PO Box 420, Mail Code:29-01
Trenton, NJ 08625-0420
p. (609) 292-1185
njgsweb@dep.nj.gov
http://www.njgeology.org
Administrative Assistant: Tenika Jacobs
State Geologist:
Jeffrey L. Hoffman, (M), Princeton, 1981, HqwRw
Bureau Chief:
David L. Pasicznyk, (B), Temple, 1979, Yg
Research Scientist 1:
Peter J. Sugarman, (D), Rutgers, 1995, Gr
Supervising Geologist:
Eric W. Roman, (M), Temple, 1999, Hw
Supervising Env Specialist:
Helen L. Rancan, (M), Stevens, 1995, Hs
Supervising Env Engineer:
Richard Shim-Chim, (B), Toronto, 1973, Hw
Research Scientist 2:
Donald H. Monteverde, (D), Rutgers, 2008, Grg
Research Scientist 1:
Scott D. Stanford, (D), Rutgers, 2001, Gml
GIS Specialist 1:
Zehdreh Allen-Lafayette, (M), Syracuse, 1994, Zi
Mark A. French, (B), Rutgers, 1989, Hq
Ted J. Pallis, (M), Montclair State, 1994, Zi
Ronald Pristas, (B), Penn State, 1990, Zi
Principal Environmental Specialist:
Raymond T. Bousenberry, (M), New Jersey Inst Tech, 2007, Ge
Steven E. Domber, (M), Wisconsin, 2000, Hw
Gregg M. Steidl, (B), Rutgers, 1995, Ge
Research Scientist 2:
John H. Dooley, (M), New Mexico Tech, 1983, CgGfe
GIS Specialist 2:
Michael W. Girard, (B), Bloomsburg, 1996, Zi
Manager:
Richard F. Dalton, (B), Rutgers, 1967, Gr
Investigator:
Walter Marzulli, Zy

New Mexico

New Mexico Institute of Mining and Technology

New Mexico Bureau of Geology & Mineral Resources ⦸ (2021)
801 Leroy Place
Socorro, NM 87801-4796
p. (575) 835-5302
connie.apache@nmt.edu
http://geoinfo.nmt.edu/
Director and State Geologist:
Nelia W. Dunbar, (D), New Mexico Tech, 1989, GivCa
Economic Geologist:
Alexander Gysi, (D), Iceland, 2011, EgCpGz
Geochemist-Microbeam Analyst:
Nels Iverson, (D), New Mexico Tech, 2017, GvCga
Adjunct Associate Research Professor of Geology:
Dana Ulmer-Scholle, (D), S Methodist, 1992, Gsd
Karst Hydrogeologist:
Lewis A. Land, (D), North Carolina (Chapel Hill), 1999, Hg
Sr. Principal Environmental Geologist Emeritus:
David W. Love, (D), New Mexico, 1980, Ge
Sr. Industrial Mineralogist Emeritus:
George S. Austin, (D), Iowa, 1971, Gz
Sr. Geologist, Head of Archives & Collections Emeritus:
Maureen Wilks, (D), New Mexico Tech, 1991, Gp
Sr. Field Geologist Emeritus:
Steven Cather, (D), Texas, 1986, Gg
Sr. Extractive Metallurgist Emeritus:
Ibrahim H. Gundiler, (D), New Mexico Tech, 1975, Nx
Senior Volcanologist, Geochronologist Emeritus:
William C. McIntosh, (D), New Mexico Tech, 1990, CcGv
Senior Mining Engineer Emeritus:
Robert W. Eveleth, (B), New Mexico Tech, 1969, Nm
Senior Field Geologist Emeritus:
Richard M. Chamberlin, (D), Colorado Mines, 1980, Gg
Senior Environmental Geologist Emeritus:
John W. Hawley, (D), Illinois (Urbana), 1962, Ge
Senior Chemist Emeritus:
Lynn A. Brandvold, (M), North Dakota State, 1964, Ca
Principal Senior Petroleum Geologist:
Ronald F. Broadhead, (M), Cincinnati, 1979, Go
Principal Senior Geophysicist Emeritus:
Marshall A. Reiter, (D), Virginia Tech, 1969, Yh
Principal Geologist Emeritus:
Paul W. Bauer, (D), New Mexico Tech, 1988, Gp
Industrial Mineralogist Emeritus:
James M. Barker, (M), California (Santa Barbara), 1972, En
Director and State Geologist Emeritus:
Charles E. Chapin, (D), Colorado Mines, 1966, Gt
L Greer Price, (M), Washington, 1974, Gg
Peter Scholle, (D), Princeton, 1970, Gd
Water Data Specialist:
Maria Magdalena Sandoval Donahue, (D), New Mexico, 2016, GmsGt
Subsurface Lab Associate:
Alyssa Baca, (B), New Mexico Tech, 2019, HwZi
Sr. Field Geologist:
Daniel Koning, (M), New Mexico, 1999, GsRnGm
Specialist GIS TechnicianCartographer:
Andi Knight, (M), New Mexico State, 2007, Zi
Senior Field Geologist:
Bruce D. Allen, (D), New Mexico, 1993, Gm
Senior Data/GIS specialist:
Richard Kelley, (M), S Methodist, 1982, ZiGg
Seismic analyst:
Amy Record, (M), Utah, 2020, Ys
Research Scientist II:
Jake Ross, (D), New Mexico Tech, 2014, Cc
Research Scientist:
Kristin Pearthree, (M), New Mexico, 2019, Gge
Publications Resource Specialist:
Michael Carroll
Publications Program Manager:
Barbara Horowitz, (B), Maryland Inst Art, 1976
Production Editor:
Belinda Harrison, (B), New Mexico Tech, 1998
PostDoctoral Fellow in Computational Thermodynamics:
Ruiguang Pan, (D), IUPU Indianapolis, 2021, CpGpEd
Post Doctoral Fellow in Experimental Geochemistry:
Hannah Juan Han, (D), North Dakota, 2021, CuqCa
Petroleum Information Coordinator:
Annabelle Lopez, Go
Petroleum Geologist:
Joseph Grigg, (M), New Mexico Tech, 2016, Go
Luke Martin, (M), N Arizona, 2005, GorGd
Network Systems Administrator:
David Kasefang, (M), Bowling Green, 1986
Museum Curator & X-Ray Diffraction Lab Manager:
Kelsey McNamara, (M), Montana State, 2010, Gz
Mechanic II:
Brian Wheeler
Manager, Publications Sales Office:
Kelly Luster, (B), New Mexico Tech, 2009
Manager, NMT Seismological Observatory:
Mairi Litherland, (D), Stanford, 2018, Ys
Manager, Education Outreach:
Cynthia Connolly, (M), New Mexico Tech, 2003, Ze
Lead Maintenance Carpenter:
Onesimo (Albert) Baca

Hydrogeologist/GIS Specialist:
Amanda Doherty, (M), Colorado State, 2019, HwCg
Hydrogeologist:
Marissa Fichera, (M), Colorado State, 2019, Hwq
Ethan Mamer, (M), SUNY (Buffalo), 2013, Hw
Cristopher Morton, (M), C Washington, 2018, Hw
Brad Talon Newton, (D), Queens, 2013, HwCgs
Hydrogeological Lab Technician:
Cathryn Pokorny, (B), Arizona, 2004
Hydrogeological Field Technician:
Scott Christenson, (M), Missouri, 1977, Hw
GIS Technician/Cartographer:
Amy Dunn, (B), Michigan Tech, 2010, Zi
GIS Specialist/Cartographer:
Justine Nicolette, (B), W Michigan, Zi
GIS Analyst I:
Mark Mansell, (B), New Mexico Tech, 2001, Zif
Geological Archives Coordinator:
Amy Trivitt-Kracke, (B), New Mexico Tech, 1999, ZfGg
Geochronology Research Scientist II:
Julia Ricci, (D), Paris-Saclay, 2014, CcGvi
Geochemist/Chemistry Lab Manager:
Bonnie A. Frey, (M), New Mexico Tech, 2002, Cg
Field Geologist – Surficial Mapper:
Kevin Hobbs, (D), New Mexico, 2016, GsrSd
Field Geologist II:
Matthew Zimmerer, (D), New Mexico Tech, 2011, Gvg
Field Geologist:
Snir Attia, (D), S California, 2020, GcgGi
Jacob Thacker, (D), New Mexico, 2020, GtcGr
Electron Microprobe Lab Manager/Geochemist:
Lynn Heizler, (B), Minnesota, 1983, Ca
Database Administrator:
Mark Leo-Russell, (B), New Mexico Tech
Coordinator/Map Production:
Phillip Miller, (B), New Mexico Tech, 2014, Zi
Computer Systems Administrator:
Christopher Armijo
Chemistry Lab Tech.:
Dustin Baca, Cag
Cartographer II:
Stephanie Chavez, (B), Southwest Visual Arts, 2010
Associate Director for Hydrogeology Programs:
Stacy Timmons, (M), Oregon State, 2002, Hwg
Associate Director for Finance:
Valentina Avramidi, (B), Rostov State (Russia), 1984
Aquifer Mapping Program Manager:
Laila Sturgis, (M), New Mexico Tech, 2003, Hw
Administrative Services Coordinator:
Connie Apache
Sr. Mineralogist/Economic Geologist/Museum Director:
Virgil W. Lueth, (D), Texas (El Paso), 1988, Gg
Sr. Geochronologist/Associate Director for Labs:
Matthew T. Heizler, (D), California (Los Angeles), 1993, Cc
Sr. Field Geologist:
Geoffrey Rawling, (D), New Mexico Tech, 2002, Gg
Senior Geophysicist, Field Geologist:
Shari Kelley, (D), S Methodist, 1984, Yg
Principal Sr. Coal Geologist Emeritus:
Gretchen K. Hoffman, (M), Arizona, 1979, Ec
Principal Sr. Economic Geologist:
Virginia McLemore, (D), Texas (El Paso), 1993, Em
Geologist, Webmaster, ABQ Office Manager:
Adam S. Read, (M), New Mexico, 1997, Gg
Associate Director for Map Program, Deputy Director:
J. Michael Timmons, (D), New Mexico, 2004, Gg

New York

New York State Geological Survey

New York State Geological Survey (2015)
3000 Cultural Education Center
Madison Avenue
Albany, NY 12230
p. (518) 473-6262
nysmdirector@nysed.gov
http://www.nysm.nysed.gov/research-collections/geology
Administrative Assistant: Donna Jornov
Curator of Sedimentary Geology:
Charles Ver Straeten, (D), Rochester, 1996, Grs
Curator of Geology:
Marian V. Lupulescu, (D), IASI (Romania), 1987, GzxEm
Senior Scientist:
Andrew Kozlowski, (D), W Michigan, 2002, Gl
Langhorne Smith, (D), Virginia Tech, 1996, Go
Museum Scientist 1:
Brian Bird, (D), GlZi
Project Geologist:
James Leone, (B), SUNY (Albany), 2006, EoGoCg
Brian Slater, (M), SUNY (Albany), 2007, GoEoGr
Education Specialist:
Kathleen Bonk, (B), SUNY (Albany), 2009, GoCg
Brandon L. Graham
State Paleontologist and Paleontology Curator:
Ed Landing, (D), Michigan, 1979, PscCc
Emeritus:
Michael A. Hawkins, (B), USNY Regents, 1972, Gz
Paleontology Collections Technician:
Frank Mannolini
Technician:
Michael Pascussi, (A), Schenectady Comm Coll, 1999, Eo
Related Staff:
Linda A. VanAller-Hernick, (B), St Rose, 1974, Pgb
Cooperating Faculty:
Barry Floyd, (M), Rensselaer, 1987, Zn

North Carolina

N.C. Department of Environmental Quality

North Carolina Geological Survey ☒ (2020)
1612 Mail Service Center
Raleigh, NC 27699-1612
p. 919-707-9210
kenneth.b.taylor@ncdenr.gov
http://portal.ncdenr.org/lr/geological_home
Administrative Assistant: Joyce Sanford
State Geologist:
Kenneth B. Taylor, (D), Saint Louis, 1991, YsGgEo
Senior Geologist:
Philip Bradley, (M), North Carolina State, 1996, Gcg
Senior Geologist :
Bart Cattanach, (M), North Carolina State, 1998, Gc
Senior Geologist:
Kathleen M. Farrell, (D), Louisiana State, 1989, Gs
Engineering Geologist:
Richard M. Wooten, (M), Georgia, 1980, RnGc
Senior Scientist:
Dwain Veach
Associate Scientist:
William Blocker, (M), Ohio State
Geologist I:
Nick Bozdog, (B), W Carolina, 2005, Gg
Kenny Gay, (M), E Carolina, 1980, GzsPi
Sierra J. Isard, (M), Iowa, 2014, Gcg

North Dakota

North Dakota Geological Survey

North Dakota Geological Survey ☒ (2021)
1016 E. Calgary Ave.
600 East Boulevard Avenue
Bismarck, ND 58505-0840
p. (701) 328-8000
emurphy@nd.gov
https://www.dmr.nd.gov/ndgs/
State Geologist:
Edward C. Murphy, (M), North Dakota, 1983, Zn
Paleontologist:
Jeff Person, Pg
Geologist:
Ned Kruger, Gg

Northern Territory

Northern Territory Government Minerals and

GEOLOGICAL SURVEYS

Energy

Northern Territory Government Minerals and Energy (2015)
48-50 Smith St
Paspalis Centrepoint Building
GPO Box 3000
Darwin, NT 0800
p. +61 8 8999 5511
ceo.ditt@nt.gov.au
http://www.minerals.nt.gov.au

Northwest Territories Geological Survey

Industry Tourism and Investment - Government of the Northwest Territories ☒ (2020)
P.O. Box 1320
Yellowknife, NT X1A 2L9
p. (867) 767-9211
ntgs@gov.nt.ca
http://www.nwtgeoscience.ca/

Nova Scotia

Geological Survey of Canada

Atlantic Div (2015)
Bedford Institute of Oceanography
1 Challenger Drive
P.O. Box 1006
Dartmouth, NS B2Y 4A2
p. (902) 426-4386
Pat.Dennis@NRCan-RNCan.gc.ca
http://gsc.nrcan.gc.ca/org/atlantic/index_e.php
Senior Scientist:
Kumiko Azetsu-Scott, (D), Dalhousie, 1992, Ocp

Nova Scotia Natural Resources

Nova Scotia Natural Resources (2015)
P.O. Box 698
Founders Square
Halifax, NS B3J 3M8
p. (902) 424-5935
http://www.gov.ns.ca/natr/

Ohio

Ohio Dept of Natural Resources

Div of Geological Survey ☒ (2018)
2045 Morse Road Bldg. C-1
Columbus, OH 43229
p. (614) 265-6576
geo.survey@dnr.state.oh.us
http://www.ohiodnr.com/geosurvey/
Programs: N/A
Certificates: N/A
Geologist:
Michael P. Solis, (M), Kentucky, 2010, GgcGt
Senior Scientist:
Frank Fugitt, (B), Ohio, GrHw
Geologist:
Daniel R. Blake, (M), Wright State, 2013, Ysg
Geologic Assistant:
Madge R. Fitak, (B), Mt Union, 1972, Gg

Oklahoma

University of Oklahoma

Oklahoma Geological Survey ☒ (2022)
100 East Boyd
Energy Center
Suite N-131
Norman, OK 73019-0628
p. (405) 325-3031
hayman@ou.edu
http://www.ogs.ou.edu/homepage.php
State Geologist:
Nicholas W. Hayman, (D), Washington, 2003, Gct
Geologist:
Thomas M. Stanley, (D), Kansas, 2000, PiGsc

Ontario

Ontario Geological Survey

Ontario Geological Survey ☒ (2021)
933 Ramsey Lake Road
Sudbury, ON P3E6B5
p. (705) 670-5614
mines.library.ndm@ontario.ca
http://www.mndm.gov.on.ca/en/mines-and-minerals
f: www.facebook.com/OGSgeology
t: @OGSgeology
Director:
Steve . Beneteau
Senior Manager, Earth Resources & Geoscience Mapping Section:
John Hechler, (M), Laurentian
Senior Manager, OGS Geoservices Section:
John Beals

Oregon

Oregon Dept of Geology and Mineral Industries

Coastal Field Office (2015)
PO Box 1033
Newport, OR 97365
p. (541) 574-6658
dogami-info@oregon.gov
Regional Geologist:
Rob Witter, (D), Oregon, 1999, Zg
Geologist:
George R. Priest, (D), Oregon State, 1980, Gg
Coastal Section Supervisor:
Jonathan C. Allan, (D), Canterbury (NZ), 1998, Gm

Oregon Dept of Geology and Mineral Industries ☒ (2022)
800 NE Oregon Street
Suite 965
Portland, OR 97232-2162
p. (971) 673-1555
dogami-info@dogami.oregon.gov
http://www.oregongeology.org
State Geologist:
Ruarri J. Day-Stirrat, (D), Great Britain, 2006, GodEd
Program Manager:
Sarah Lewis, (M)
Hydrogeologist:
Robert Brinkmann, (B), Colorado State, 1983, HwGoEm
Engineering Gelogist:
Bill Burns, (M), Portland State, 1999, Ng
Eastern Oregon Regional Geologist:
Jason McClaughry, (M), WSU, 2003, Gg
Coastal Geomorphologist:
Jonathan C. Allan, (D), Canterbury (NZ), 1998, Gm

Pennsylvania

DCNR- Pennsylvania Bureau of Geological Survey

Pennsylvania Geological Survey ☒ (2023)
3240 Schoolhouse Road
Middletown, PA 17057-3534
p. (717) 702-2017
http://www.dcnr.pa.gov/Geology/Pages
f: https://www.facebook.com/padcnr
Programs: none
Certificates: none
Administrative Assistant: Emily Parkovic
State Geologist and Bureau Director:
Gale C. Blackmer, (D), Penn State, 1992, Gc
Geoscience Manager:
Kristin M. Carter, (M), Lehigh, 1993, EoCqEg
Alfred C. Guiseppe, (M), Wyoming, 1997, ZiHwGg
Kyle S. Rybacki, (D), Penn State, 2015, ClEgXb
Senior Geoscientist:
Robin Anthony, (B), Case Western, Ego
Rose-Anna Behr, (M), New Mexico Tech, 1999, GcRnGr
Aaron D. Bierly, (B), Pittsburgh(Johnstown), GgZn
Helen L. Delano, (M), SUNY (Binghamton), 1979, GgmNg
Clifford H. Dodge, (M), Northwestern, 1976, GrEcRh

Craig M. Ebersole, (B), Juniata, 2009, ZiGgHg
Ellen R. Fehrs, (M), Colorado Mines, 2019, Zir
Adam J. Ianno, (D), Southern California, 2015, CaGxEg
Ty C. Johnson, (M), Arkansas, 2008, GlRnHg
Antonette K. Markowski, (M), S Illinois, 1990, Eoc
Victoria V. Neboga, (M), Kiev, 1985, GgHwNx
Christopher Oest, (M), Temple, 2015, GrdGg
Caron E. Pawlicki, (M), Pittsburgh, 1986, ZiGgc
Katherine W. Schmid, (M), Pittsburgh, 2005, GoCcXb
Theodore Tesler, (B), Lafayette, 1991, HgRwZr

Geoscientist Supervisor:
Michele L. Cooney, (M), Indiana (Penn), 2016, EgoGe
Kristen L. Hand, (B), Nicholls State, Ggr
Stephen G. Shank, (D), Penn State, 1993, GziGp
Simeon Suter, (B), Millersville, 1982, HwGg

Geospatial Specialist:
Morgen C. Weiant, (B), Lock Haven, 2017, Zi

Geoscientist:
Hailey Filippelli, (M), West Chester (Penn), 2021, GgrZi
Leonard J. Lentz, (M), North Carolina State, 1983, Ec
John C. Neubaum, (M), American Military, 2001, EgGg

Librarian:
Jody L. Smale, (M), Clarion, 2010, Zn

IT Generalist:
Mark A. Dornes
David Fletcher

IT Adminstrator:
Sandipkumar P. Patel, (B)

Clerk Typist:
Gail Fireman, (M), Pittsburgh, 1993, Zn
Jody L. Rebuck, (B), Messiah, Zn

Puerto Rico

Puerto Rico Bureau of Geology

Dept of Natural & Environmental Resources (2015)
Apartado 9066600
Puerta de Tierra Station
San Juan, PR 00906-6600
p. (787) 722-2526

Director:
Vanessa del S. Rodriguez, Zn

Quebec

Ministère de l'Énergie et des Ressources naturelles Québec

Ministère de l'Énergie et des Ressources naturelles Québec ☒ (2021)
5700 4eme avenue ouest
local A-409
Quebec, QC G1H 6R1
p. 1-866-248-6936
services.clientele@mern.gouv.qc.ca
http://www.mern.gouv.qc.ca/

Queensland

Queensland Environment and Resource Management

Queensland Environment and Resource Management (2015)
GPO Box 2454
Brisbane, QLD 4001
info@derm.qld.gov.au
http://www.derm.qld.gov.au/

Rhode Island

University of Rhode Island

Rhode Island Geological Survey (D) (2015)
9 East Alumni Ave.
314 Woodward Hall
University of Rhode Island
Kingston, RI 02881
p. 401.874.2191
rigsurv@etal.uri.edu
http://www.uri.edu/cels/geo/GEO_risurvey.html

State Geologist (Research Professor Emeritus):
Jon C. Boothroyd, (D), South Carolina, 1974, GslOn

Professor:
Thomas Boving, (D), Arizona, 1999, Hg
David Fastovsky, (D), Wisconsin, 1986, PvGsPs

Assistant Professor:
Dawn Cardace, (D), Washington (St Louis), 2006, Pg
Simon Engelhard, (D), Pennsylvania, 2010, Gg
Brian Savage, (D), Caltech, 2004, Ys

Lecturer:
Elizabeth Laliberte, (D), Rhode Island, 1997, Og

Research Associate:
Bryan A. Oakley, (D), Rhode Island, 2012, Gsl

Saskatchewan

Saskatchewan Ministry Energy and Resources

Saskatchewan Ministry Energy and Resources ☒ (2020)
1000 - 2103 - 11th Avenue
Regina, SK S4P 3Z8
p. (306) 787-9580
dmoadmin.er@gov.sk.ca
http://www.saskatchewan.ca/

South Carolina

South Carolina Dept of Natural Resources

Geological Survey ☒ (2022)
5 Geology Road
Columbia, SC 29212
p. 803.896.7931
howards@dnr.sc.gov
http://www.dnr.sc.gov/geology/

State Geologist:
C. Scott Howard, (D), Delaware

Geologist III:
William R. Doar, III, (D), South Carolina, 2014, GrdGp
Katherine E. Luciano, (M), Coll of Charleston, GusGm
Robert H. Morrow, (M), North Carolina (Wilmington), 2015, GcxGt

GIS Analyst:
Darby DeBruhl, (B), South Carolina, Zi
Jerry Krieger, (B), South Carolina, Zi

Geologist II:
Crystal Wilson, (M), Tennessee, Gct
Andy Wykel, (M), South Carolina, Gs

Geologist I:
Mason Payeur, (B), South Carolina, Gg
Zach Zelaya, (B), South Carolina, Gg

Program Manager-Drilling:
Joe Koch, Zg

South Dakota

South Dakota Dept of Environment and Natural Resources

Geological Survey Program ☒ (2017)
Akeley-Lawrence Science Center
University of South Dakota
414 East Clark Street
Vermillion, SD 57069-2390
p. (605) 677-5227
tim.cowman@usd.edu
http://www.sdgs.usd.edu/

State Geologist:
Derric L. Iles, (M), Iowa State, 1977, HwGg

Geologist:
Darren J. Johnson, (M), Washington, 2009, GgEo

Geologist:
Nicholas J. Krohe, (M), Idaho State, 2016, Gg

Tennessee

Tennessee Geological Survey

Dept of Environment & Conservation ☒ (2022)
William R. Snodgrass TN Tower
312 Rosa L. Parks Ave., 12th Floor
Nashville, TN 37243
p. (615) 532-1502

Ronald.Zurawski@tn.gov
https://www.tn.gov/environment/program-areas/tennessee-geological-survey.html
State Geologist:
Ronald P. Zurawski, (M), Vanderbilt, 1973, GgPzEg
Environmental Consultant:
Peter J. Lemiszki, (D), Tennessee, 1992, GcZiGg
Barry W. Miller, (M), Tennessee, 1989, EcZiGg
Environmental Scientist:
Vince Antonacci, (M), Ball State, 1987, GgZi
Ronald J. Clendening, (B), Tennessee Tech, 1986, GgcGm
Admin. Services Asst. 2 :
Carolyn A. Patton

Texas

University of Texas at Austin, Jackson School of Geosciences

Bureau of Economic Geology (2015)
University Station, Box X
Austin, TX 78713-8924
p. (512) 471-1534
begmail@beg.utexas.edu
http://www.beg.utexas.edu
Director:
Scott W. Tinker, (D), Colorado, 1996, Gor
Associate Director:
Jay P. Kipper, (M), Trinity (San Antonio), 1983, Zn
Eric C. Potter, (M), Oregon State, 1975, Eo
Senior Research Scientist:
Shirley P. Dutton, (D), Texas, 1986, Gd
Bob A. Hardage, (D), Oklahoma State, 1967, Yg
Susan D. Hovorka, (D), Texas, 1990, Gse
Michael R. Hudec, (D), Wyoming, 1990, GcEo
Martin P. A. Jackson, (D), Cape Town, 1976, Gc
Stephen E. Laubach, (D), Illinois, 1986, GcNrGz
Bob Loucks, (D), Texas, 1976, Go
Bridget R. Scanlon, (D), Kentucky, 1985, Hw
Research Scientist Associate V:
Tucker F. Hentz, (M), Kansas, 1982, GrsGo
Research Scientist:
William A. Ambrose, (M), Texas, 1983, Gs
Peter Eichhubl, (D), California (Santa Barbara), 1997, GcCgNr
Jean-Philippe Nicot, (D), Texas, 1998, HwCg
Senior Scientist:
Ian Duncan, (D), British Columbia, 1982, Gg
F. Jerry Lucia, (M), Minnesota, 1954, Go
Stephen C. Ruppel, (D), Tennessee, 1979, Gd
Senior Research Scientist:
Kitty L. Milliken
Research Scientist:
Sergey Fomel, (D), Stanford, 2002, Yx
Jeffrey G. Paine, (D), Washington, 1991, Gr
Julia Stowell Gale, (D), Exeter (UK), 1987, Gc
Hongliu Zeng, (D), Texas, 1994, Gs
Associate Professor:
Charles Kerans, (D), Carleton, 1982, Gd
Research Scientist:
Tim Dooley, (D), London (UK), 1994, Gc
Associate Scientist:
Katherine D. Romanak, (D), Texas, 1997, CgScGe
Research Scientist Associate IV:
Robert C. Reedy, (M), New Mexico Tech, 1996, Hgw
Research Scientist Associate:
John R. Andrews, (B), North Carolina, 1990, Zi
Research Associate:
Bruce Cutright
Qilong Fu, (D), Regina, 2005, Gso
H, Scott Hamlin, (D), Texas, Gr
Ursula Hammes, (D), Colorado, 1992, Gg
Farzam Javadpour, (D)
Timothy A. Meckel, (D)
Osareni C. Ogiesoba, (D)
Katherine D. Romanak, (D)
Diana Sava, (D), Stanford, 2004, Yg
Changbing Yang, (D)
Christopher K. Zahm, (D)
Beverly Blakeney DeJarnett, (M), Penn State, 1986, GrsGx
Edward W. Collins, (M), Stephen F. Austin, 1978, Ge
Micheal V. DeAngelo, (M), Texas (El Paso), 1988, Ye
Xavier Janson, (D), Miami, 2002, Gr
Research Scientist Associate:
Seay Nance, (D), Texas, 1988, Gg
Research Scientist Associate IV:
Robert M. Reed, (D), Texas, 1999, GxoGc
IT Manager:
Ron Russell, (M), Oklahoma, 1993, Zn
Associate Director:
Michael H. Young, Arizona, 1995, Ge
512-471-7135:
George Bush, (B), Texas Tech
Research Scientist Associate IV:
Thomas A. Tremblay, (M), Texas, 1992, Zy
Research Scientist Associate:
Caroline Breton, (B), Texas, 2001, Zi
Dallas B. Dunlap, (B), Texas, 1997, Ye
Tiffany Hepner, (M), S Florida, 2000, Og
Project Manager:
Rebecca C. Smyth, (M), Texas, 1995, Hg
Ramon H. Trevino, (M), Texas (Arlington), 1988, GsoGe
Related Staff:
Joseph S. Yeh, (B), Fu-Jen Catholic, 1977, Gm

Utah

Utah Geological Survey

Dept of Natural Resources ☒ (2022)
1594 West North Temple, Ste 3110
Box 146100
Salt Lake City, UT 84114-6100
p. (801) 537-3300
https://geology.utah.gov
f: https://www.facebook.com/UTGeologicalSurvey/
t: @utahgeological
Director/State Geologist:
Bill Keach, (M), Cornell, 1986, YegZg
Program Manager:
Steve D. Bowman, (D), Nevada (Reno), 2002, NgGe
Hugh Hurlow, (D), Washington, 1992, Hw
Stefan M. Kirby, (M), Utah State, 2005, GgcCq
Mark Milligan, (M), Utah, 1994, Ze
Marshall Robinson, (B), Brigham Young, 2007, GgZi
Michael Vanden Berg, (M), Utah, 2003, Eo
Librarian:
Suzanne Sawyer, (M), N Texas, 2012, Zn
Financial Manager:
Jodi T. Patterson, (M), Weber State, 2002, Zn
Deputy Director:
Michael D. Hylland, (M), Oregon State, 1990, Gg
Core Center Curator:
Katie Cummings, (M), East Carolina, 2014, Gg
Environmental Sciences Program (2015)
1594 West North Temple, Ste 3110
PO Box 146100
Salt Lake City, UT 84114-6100
p. (801) 537-3389
billkeach@utah.gov
http://geology.utah.gov
Manager:
Mike Lowe, (M), Utah State, 1985, Hw
Geologist:
Charles E. Bishop, (B), Utah, 1986, Hw
Hugh A. Hurlow, (D), Washington, 1982, Hw
Janae Wallace, (M), N Arizona, 1993, Hw
Senior Geologist:
James I. Kirkland, (D), Colorado, 1990, Pg
Senior Scientist:
David B. Madsen, (D), Missouri, 1973, Ga
Research Associate:
Martha C. Hayden, (B), Utah, 1978, Pv
Geotechnician:
Alison Corey, Zn
Geologic Hazards Program (2015)
1594 West North Temple
P O Box 146100
Salt Lake City, UT 84114-6100
p. (801) 537-3300

rickallis@utah.gov

http://geology.utah.gov/ghp/
Program Manager:
Steve D. Bowman, (D), Nevada (Reno), 2002, NgZn
Senior Scientist:
William R. Lund, (B), Idaho, 1970, Ng
Paleoseismologist:
Christopher B. DuRoss, (M), Utah, 2004, Ng
Landslide Geologist:
Gregg S. Beukelman, (M), Boise State, 1997, NgZi
Hazards Mapping Geologist:
Jessica Castleton, (B), Weber State, 2005, Ng
Adam McKean, (M), Brigham Young, 2011, Gmg
Hazards Geologist:
Tyler R. Knudsen, (M), Nevada, 2005, Ng
Gregory N. McDonald, (B), Utah, 1992, Ng
Hazard Mapping Geologist:
Ben A. Erickson, (M), Utah, 2011, Ng
Debris Flow/Landslide Geologist:
Richard E. Giraud, (M), Idaho, 1986, Ng
Geologist:
Mike Hylland, (M), Oregon State, 1990, PgYs
GIS Analyst:
Corey Unger, (B), Weber State, 2001, Zi

Geologic Mapping Program (2015)
1594 West North Temple, Ste 3110
PO Box 146100
Salt Lake City, UT 84114-6100
p. (801) 537-3355
grantwillis@utah.gov
http://geology.utah.gov
Senior Scientist:
Robert Biek, (M), N Illinois, 1988, Gg
Adjunct Associate Professor:
James I. Kirkland, (D), Colorado, 1990, PvsPi
Senior Geologist:
Donald Clark, (M), N Illinois, 1987, Gg
Jonathan K. King, (M), Wyoming, 1984, Gg
Douglas A. Sprinkel, (M), Utah State, 1977, GosGg
Manager:
Grant C. Willis, (M), Brigham Young, 1983, Gg
GIS Analyst:
Basia Matyjasik, (B), Warsaw, 1988, Gg
Senior GIS Analyst:
Kent D. Brown, Zn

Vermont

Agency of Natural Resources, Dept of Environmental Conservation

Vermont Geological Survey ☒ (2018)
1 National Life Drive
Main 2
Montpelier, VT 05620-3902
p. (802) 522-5210
jon.kim@vermont.gov
https://dec.vermont.gov/geological-survey
State Geologist:
Marjorie H. Gale, (M), Vermont, 1980, GgcGe
Geologist -Environmental Scientist:
Jonathan Kim, (D), Buffalo, 1995, GetCg

Victoria

Geological Survey of Victoria, Australia

Victoria - Department of Jobs, Precincts and Regions ☒ (2022)
GPO Box 4509
Melbourne, VIC 3001
p. +61 3 1300 366 356
customer.service@ecodev.vic.gov.au
http://www.earthresources.vic.gov.au/earth-resources

Victoria - Dept of Sustainability and Environment

Dept of Sustainability and Environment (2015)
8 Nicholson Street
East Melbourne, VIC 3002
p. +61 3 5332 5000
peter.walsh@parliament.vic.gov.au
http://www.dse.vic.gov.au/

Virginia

Geology and Mineral Resources Program

Virginia Dept of Energy ☒ (2022)
Fontaine Research Park
900 Natural Resources Drive
Suite 500
Charlottesville, VA 22903
p. (434) 951-6341
GeologyMineralResources@energy.virginia.gov
https://energy.virginia.gov/geology/geologymineralresources.shtml
State Geologist and Program Director:
Matthew J. Heller, (M), North Carolina State, 1996, GgRnGe
Manager, Economic Geology:
William L. Lassetter, (M), Nevada (Reno), 1996, EgHg

University of Virginia

Virginia State Climatology Office (2015)
291 McCormick Road
P.O. Box 400123
Charlottesville, VA 22904-4123
p. 434-924-0548
pjs2i@virginia.edu
http://climate.virginia.edu/home.htm
Director:
Patrick J. Michaels, (D), Wisconsin, 1979, As

Washington

Geological Survey of Western Australia

Geological Survey of Western Australia (2016)
Mineral House
100 Plain Street
East Perth , WA 6004
p. +61 8 9222 3333
media@dmirs.wa.gov.au
http://www.dmp.wa.gov.au/Geological-Survey/Geological-Survey-262.aspx

Washington Geological Survey

Washington Dept of Natural Resources ☒ (2021)
1111 Washington Street, SE, MS 47007
Olympia, WA 98504-7007
p. (360) 902-1450
geology@dnr.wa.gov
http://www.dnr.wa.gov/geology/
State Geologist:
Casey R. Hanell, (M), W Washington, 2011, Gg
Geology Librarian:
Stephanie Earls, (M), Washington, 2010, Gge

West Virginia

West Virginia Geological & Economic Survey

West Virginia Geological & Economic Survey (2016)
Mont Chateau Research Center
1 Mont Chateau Road
Morgantown, WV 26508-8079
p. (304) 594-2331
info@wvgs.wvnet.edu
http://www.wvgs.wvnet.edu
f: https://www.facebook.com/WVGeoSurvey
State Geologist:
Michael Ed. Hohn, (D), Indiana, 1976, Eg
Adjunct Professor:
Paula J. Hunt, (M), Purdue, 1988, GgHw

Wisconsin

University of Wisconsin, Madison, Division of

Extension

Wisconsin Geological and Natural History Survey ☒ (2022)
3817 Mineral Point Road
Madison, WI 53705-5100
p. (608) 262-1705
sushmita.lotlikar@wisc.edu
https://wgnhs.wisc.edu
f: https://www.facebook.com/WGNHS
t: @WGNHS

Interim Director and State Geologist:
Eric C. Carson, (D), Wisconsin, 2003, Gl
Hydrogeologist:
David J. Hart, (D), Wisconsin, 2000, HwYg
Geologist:
J. Elmo Rawling, (D), Wisconsin (Milwaukee), 2002, Gml
Professor:
John W. Attig, (D), Wisconsin, 1984, Gl
Thomas J. Evans, (D), Wisconsin, 1994, Eg
Director and State Geologist:
Kenneth R. Bradbury, (D), Wisconsin, 1982, Hw
James M. Robertson, (D), Michigan, 1972, Eg
Emeritus:
Bruce A. Brown, (D), Manitoba, 1984, GcEg
Ron G. Hennings, (M), Wisconsin, 1977
M. Carol L. McCartney, (D), Wisconsin, 1979, RcGee
Roger M. Peters, (B), Wisconsin, 1969, Gg
Web Developer:
David Sibley, (B), Wisconsin, 2009
Samples Coordinator:
Carsyn Ames, (M), Iowa State, 2018
Project Quaternary Geologist:
Kacie Stolzman, (B), Winona State, 2019, Gm
Project Geoscientist:
Sarah Bremmer, (M), Wisconsin, 2018, Ggt
William Fitzpatrick, (M), Arizona, 2020, EgGi
Lisa Haas, (M), Wisconsin, 2021
Project Geologist:
Stefanie Dodge, (B), Wisconsin (Milwaukee), Gg
Hydrogeologist:
Maureen Muldoon, (D), Wisconsin, 1999, GeHw
Mike J. Parsen, (M), Hw
GIS Specialist:
Steve Mauel
Matthew Rehwald, (M), Wisconsin, 2005, Zi
Caroline Rose, (M), Wisconsin, 2015, Zi
Geotechnician:
Peter M. Chase, (B), Wisconsin (Milwaukee), 1985, HwYgRw
Geoscience Program Coordinator:
Ian Orland, (D), Wisconsin, 2012, PcCls
Geologist:
Irene D. Lippelt, (B), Manitoba, 1978, Gg
Eric Stewart, (D), Texas A&M, 2015, GgcGi
Esther K. Stewart, (M), Idaho State, 2008, Gg
Geographic Information Systems:
Nick Rompa, (B), Wisconsin (Stevens Point), 2005
Assistant Director for Administration:
Sushmita Lotlikar, (M), Wisconsin, 2014
Assistant Director:
Peter Schoephoester, (M), Indiana, 2003, Zi
Archivist:
Bradford Gottschalk, (M), Wisconsin, 2007, Rh
Related Staff:
Amber Boudreau
Grace E. Graham, (B), Beloit, 2013, Hw

Wyoming

Wyoming State Geological Survey

Wyoming State Geological Survey ☒ (2022)
P.O. Box 1347
Laramie, WY 82073
p. (307) 766-2286
wsgs-info@wyo.gov
http://www.wsgs.wyo.gov/
f: https://www.facebook.com/Wyoming.Geological.Survey
t: https://twitter.com/wygeosurvey

Wyoming State Geologist:
Erin A. Campbell, (D), Wyoming, 1997, Gc
Geologist:
James P. Mauch, GmRnGg

Yukon

Yukon Geological Survey

Yukon Geological Survey (2015)
Whitehorse, YT
suzanne.roy@gov.yk.ca
http://www.geology.gov.yk.ca/

Agencies and Intl Organizations

United Nations Education, Scientific, and Cultural Organization

International Geoscience Programme (2015)
Division of Ecological and Earth Sciences
1 rue Miollis
Paris, Cedex 15 F-75732
p. +33 (0)1 45 68 10 00
k.vandenberghe@unesco.org
http://www.unesco.org/new/en/natural-sciences/environment/earth-sciences/international-geoscience-programme/
IGCP Chairperson:
Patricia Vickers-Rich, (D), Columbia, 1972
Team Leader, Hydrogeology:
Gil Mahe
Team Leader, Global Change and Evolution of Life:
Guy Narbonne
Team Leader, Geohazards:
Andrej Gosar
Team Leader, Geodynamic:
George Gibson
Team Leader, Earth Resources:
Robert Moritz
IGCP/SIDA Representative:
Vivi Vajda, (D), Lund (Sweden), 1998

Landmælingar Íslands

National Land Survey of Iceland ☒ (2021)
Stillholt 16-18 300 Akranes
Akranes 300
p. 4309000
lmi@lmi.is
http://www.lmi.is/

Banaras Hindu University

HARI SRIVASTAVA (B,M,D) ☒ (2022)
316/17
Ambrosia Appartment
Lanka
Varanasi, U.P. 221005
p. +91941535360
hbsrivastava@gmail.com
http://www.bhu.ac.in/Geology
Programs: Geology
Enrollment (2017): B: 150 (0) M: 56 (0) D: 50 (0)
Professor:
HARI BAHADUR SRIVASTAVA, (D), Banaras Hindu, 1980, GctGp

Geological Survey of India

Geological Survey of India O☒ (2021)
27, J.L.Nehru Road
Kolkata
Kolkata, West Bengal 700016
dg.gsi@gov.in
http://www.gsi.gov.in
f: https://www.facebook.com/gsipage/
t: @GeologyIndia
Certificates: Advanced GIS, Digital Cartography, Geological Mapping, Mineral Exploration, Geophysical Survey, etc

Ministry of Environmental Protection, Natural Resources and Forestry

Dept of Geology (2016)
st. Wawelska 52/54
Warsaw 00-922
p. +48 22 3692449
Departament.Geologii.i.Koncesji.Geologicznych@mos.gov.pl
http://www.mos.gov.pl/dg/dga1.htm

Sveriges geologiska Undersokning

Geological Survey of Sweden (2016)
Box 670
751 28 Uppsala
sgu@sgu.se
http://www.sgu.se/

Argonne National Laboratory

Environmental Research Div (2015)
9700 South Cass Avenue
Argonne, IL 60439
p. (630) 252-3879
negri@anl.gov
http://www.evs.anl.gov/
Senior Scientist:
Jeffrey S. Gaffney, (D), California (Riverside), 1975, As
Raymond M. Miller, (D), Illinois State, 1975, Sb
Marvin L. Wesely, (D), Wisconsin, 1970, As
Associate Scientist:
Jacqueline C. Burton, (D), Tennessee, 1978, Cg
Richard L. Coulter, (D), Penn State, 1976, As
Paul V. Doskey, (D), Wisconsin, 1982, As
Julie D. Jastrow, (D), Chicago, 1994, Sb
Lorraine M. LaFreniere, (D), Wisconsin, 1980, Zn
In Young Lee, (D), California (Los Angeles), 1975, As
Barry M. Lesht, (D), Chicago, 1977, Op
Nancy A. Marley, (D), Florida State, 1984, As
William T. Meyer, (D), Imperial Coll (UK), 1973, Cg
Robert A. Sedivy, (M), Georgia Tech, 1979, Hw
Jack D. Shannon, (D), Oklahoma, 1975, Am
Douglas L. Sisterson, (M), Wyoming, 1975, As
Mohamed Sultan, (D), Washington (St Louis), 1984, Cg
John L. Walker, (D), Imperial Coll (UK), 1964, Cg
Y Eugene Yan, (D), Ohio State, 1998, Gg
Research Associate:
Richard H. Becker, (M), Washington (St Louis), Zr
Clyde B. Dennis, (M), Florida, 1979, Zn
Richard L. Hart, (B), Illinois Inst Tech, 1970, As
Timothy J. Martin, (B), Beloit, 1974, As
Barney W. Nashold, (M), Illinois (Chicago Circle), 1976, Zn
Kent A. Orlandini, (B), Illinois, 1957, Cc
Candace M. Rose, (B), Benedictine Coll, 1985, Zn

Energy Systems Div (2015)
9700 South Cass Avenue
Bldg 362 E-340
Argonne, IL 60439-4815
p. (630) 252-3392
energy_systems@anl.gov
Administrative Assistant: Barbara Sullivan
Head:
Donald O. Johnson, (D), Illinois, 1972, Gr
Senior Scientist:
Lyle D. McGinnis, (D), Illinois, 1965, Yg
Scientist:
Kenneth L. Brubaker, (D), Wisconsin, 1972, As
Dorland E. Edgar, (D), Purdue, 1976, Gm
Steven F. Miller, (B), Knox, 1984, Gg
R. Eric Zimmerman, (D), Northwestern, 1972, Ng
Research Associate:
Paul C. Heigold, (D), Illinois, 1969, Yg
Theresa C. Scholtz, (B), N Illinois, 1985, Ge
Michael D. Thompson, (M), N Illinois, 1989, Yg
Related Staff:
John F. Schneider, (M), N Illinois, 1977, Ca
Linda M. Shem, (M), Northwestern, 1991, Ge
Patrick L. Wilkey, (M), Illinois, 1976, Ng

Chemical Technology Div (2015)
9700 South Cass Avenue
Building 205
Argonne, IL 60439
p. (630) 252-4383
ebunel@anl.gov

Senior Scientist:
Milton Blander, (D), Yale, 1953, Xc
Paul A. Fenter, (D), Pennsylvania, 1990, ClZm
Associate Scientist:
Allen J. Bakel, (D), Oklahoma, 1990, Co
Ronald P. Chiarello, (D), Northeastern, 1990, Zm
Donald G. Graczyk, (D), Wisconsin, 1975, Cs
Ben D. Holt, (D), Illinois Inst Tech, 1969, Ca
James J. Mazer, (D), Northwestern, 1987, Cl
David J. Wronkiewicz, (D), New Mexico Tech, 1989, Cg
Adjunct Professor:
Neil C. Sturchio, (D), Washington (St Louis), 1983, Cg
Related Staff:
Alice M. Essling, (B), St Xavier, 1956, Ca
Edmund A. Huff, (M), Chicago, 1957, Ca
Francis J. Markun, (B), Lewis, 1961, Cc
Florence P. Smith, (B), Southern, 1968, Ca

Geosciences & Information Technology Section (2015)
9700 South Cass Avenue
Argonne, IL 60439
p. (630) 252-6034
djmiller@anl.gov
Department Secretary: Sue Baumann
Manager, Geosciences & Information Tech:
Lisa A. Durham, (M), Purdue, 1989, Hw
Assistant System Engineer:
Cheong-yip R. Yuen, (D), Wisconsin (Milwaukee), 1986, Ge
Associate Scientist:
Robert L. Johnson, (D), Cornell, 1991, Zn
Research Associate:
John Ditmars, (D), Caltech, 1971, Zn
Jennifer Herbert, (B), N Illinois, 1997, Gg
Zhenhua Jiang, (D), Duke, 1992, Zn
David S. Miller, (D), Johns Hopkins, 1995, Ge
Terri L. Patton, (M), NE Illinois, 1989, Cl
John J. Quinn, (M), Minnesota, 1992, Hq
Geology Librarian:
Swati Wagh, Zn

Bureau of Land Management

Colorado State Office (2015)
2850 Youngfield Street
Lakewood, CO 80215
p. (303) 239-3600
hhankins@blm.gov

Headquarters Directorate (2015)
1849 C Street NW
Rm 5665
Washington, DC 20240
p. (202) 208-3801
director@blm.gov

National Operations Center (2015)
PO Box 25047
Denver, CO 80225
p. (303) 236-8857
lgraham@blm.gov

National Training Center (2015)
9828 North 31st Avenue
Phoenix, AZ 85051
p. (602) 906-5500
dwilkins@blm.gov

Alaska State Office (2015)
222 West Seventh Ave
#13
Anchorage, AK 99513
p. (907) 271-3212
dlassuy@blm.gov

Arizona State Office (2015)
One North Central Ave
Suite 800
Phoenix, AZ 85004
p. (602) 417-9500
egomez@blm.gov

Eastern States Office (2015)
7450 Boston Boulevard
Springfield, VA 22153
p. (703) 440-1600
es_general_web@blm.gov

New Mexico State Office (2015)
301 Dinosaur Trail
Santa Fe, NM 87502
p. (505) 954-2098
jjuen@blm.gov

Idaho State Office (2015)
1387 South Vinnell Way
Boise, ID 83709
p. (208) 373-4000
sellis@blm.gov

Montana State Office (2015)
5001 Southgate Drive
Billings, MT 59101
p. (406) 896-5012
kiszler@blm.gov

Nevada State Office (2015)
1340 Financial Blvd
Reno, NV 89502
p. (775) 861-6400
jswickard@blm.gov

Oregon State Office (2015)
333 SW 1st Avenue
Portland, OR 97204
p. (503) 808-6001
b2jackso@blm.gov

Utah State Office (2015)
440 West 200 South
Suite 500
Salt Lake City, UT 84101
p. (801) 539-4001
utsomail@blm.gov

Wyoming State Office (2015)
PO Box 1828
Cheyenne, WY 82003
p. (307) 775-6256
jcamargo@blm.gov

California State Office ☒ (2021)
2800 Cottage Way
Suite W-1623
Sacramento, CA 95825
p. (916) 978-4400
nml_ca_web_so@blm.gov
https://www.blm.gov/office/california-state-office
f: https://www.blm.gov/media/social-media

Department of Commerce

National Institute of Standards & Technology ☒ (2021)
100 Bureau Drive
Stop 1070
Gaithersburg, MD 20899-3460
p. (301) 975-2000
inquiry@nist.gov
https://www.nist.gov/
f: https://www.facebook.com/NIST
t: @NIST

Department of Interior Office of Surface Mining Reclamation and Enforcement

OSMRE Geospatial Information Services ☒ (2019)
OSMRE Attn: David Carter
1999 Broadway, Suite 3320
Denver, CO 80225

p. (303) 293-5019
dcarter@osmre.gov
https://www.tips.osmre.gov/geospatial.shtm
Branch Manager:
David Carter, (M), Denver, 1999, Zir
Physical Scientist:
Carrie A. Middleton, (M), Colorado Mines, 2008, ZrGfZi

Dept of Agriculture

Natural Resources Conservation Service (2016)
14th and Independence Ave., SW
Washington, DC 20250
p. (202) 720-7246
jason.weller@wdc.usda.gov
http://www.nrcs.usda.gov

Dept of Commerce

National Oceanic & Atmospheric Administration (2015)
Silver Spring Metro Center 3
1315 East-West Highway
Silver Spring, MD 20910-3282
p. (202) 482-3436
edward.horton@noaa.gov
http://www.noaa.gov

Dept of Defense

Office of Naval Research (2015)
Ballston Towers #1
800 North Quincy Street
Arlington, VA 22203
p. (703) 696-4767
http://www.onr.navy.mil/

U.S. Army Corps of Engineers (2015)
441 G Street, NW
Washington, DC 20314
p. (202) 761-0001
http://www.usace.army.mil/working.html

Air Force Center for Environmental Excellence (2015)
3207 North Road
Brooks Air Force Base, TX 78235-5363
p. (210) 536-2162

U.S. Army Chemical & Biological Defense Command (2015)
The Pentagon
Washington, DC 20310
p. (703) 695-0363
http://www.army.mil/csa/

U.S. Army Research Laboratory Command (2015)
The Pentagon
Washington, DC 20310
p. (703) 695-0363
http://www.army.mil/csa/

Defense Threat Reduction Agency (2015)
8725 John T. Kingman Road
MS 6201
Ft. Belvoir, VA 22060-6201
p. (703) 767-4883
dtra.publicaffairs@dtra.mil
http://www.dtra.mil

Space Command (2015)
150 Vandenberg Street
Peterson Air Force Base, CO 80914-4020
p. (719) 554-3001
http://www.af.mil/sites/afspc.shtml

Naval Intelligence Command (2015)
4251 Suitland Road
Washington, DC 20395-5720
p. (301) 669-4000
http://www.nmic.navy.mil/

Space & Naval Warfare Systems Command (2015)
4301 Pacific Highway
San Diego, CA 92110-3127
p. (619) 524-7053
http://www.spawar.navy.mil/

Naval Oceanography Command (2015)
U.S. Naval Observatory
34th Street & Massachusetts Avenue, NW
Washington, DC 20007
p. (202) 762-1020
http://www.oceanographer.navy.mil

U.S. Special Operations Command (Air Force) (2015)
100 Bartley Street
Hurlburt Field, FL 32544-5273
p. (850) 884-2323
http://www.af.mil/sites/afsoc.html

Air Weather Service (2015)
106 Peacekeeper Drive
Offutt Air Force Base, NE 68113-4039
p. (402) 294-5749

Dept of Energy

Office of Energy Projects (2015)
Federal Energy Regulatory Commission
888 First Street, NE
Washington, DC 20426
p. (202) 219-2700

Office of Resource Management (2015)
Forrestal Building
1000 Independence Avenue, SW
Washington, DC 20585-0620
p. (202) 586-3521

Office of Nuclear Energy, Science & Technology (2015)
Forrestal Building
1000 Independence Avenue, SW
Washington, DC 20585-0117
p. (202) 586-6630
http://www.ne.doe.gov/

Assistant Secretary for Energy Efficiency & Renewable Energy (2015)
Forrestal Building
1000 Independence Avenue, SW
Washington, DC 20585-0121
p. (202) 586-9220
http://www.eren.doe.gov/

Assistant Secretary for Fossil Energy (2015)
Forrestal Building
1000 Independence Avenue, SW
Washington, DC 20585-0301
p. (202) 586-6660
FECommunications@hq.doe.gov
http://www.fe.doe.gov/

Office of Oil & Gas (2015)
Forrestal Building
1000 Independence Avenue, SW
Washington, DC 20585-0640
p. (202) 586-6012

Assistant Secretary for Environment, Safety & Health (2015)
Forrestal Building
1000 Independence Avenue, SW
Washington, DC 20585-0119
p. (202) 586-6151
http://www.eh.doe.gov/

Energy Information Administration (2015)
Forrestal Building
1000 Independence Avenue, SW
Washington, DC 20585-0601
p. (202) 586-8800

howard.gruenspecht@eia.gov
http://www.eia.doe.gov/

Office of Coal, Nuclear, Electric & Alternate Fuels (2015)
COMSAT Building
950 L'Enfant Plaza, SW
Washington, DC 20024
p. (202) 287-7990

Coal & Power Systems (2015)
Forrestal Building
1000 Independence Avenue, SW
Washington, DC 20585-0320
p. (202) 586-1650

Dept of Health & Human Services

Agency for Toxic Substances & Disease Registry (2015)
1600 Clifton Road
Atlanta, GA 30333
p. (404) 639-7000
http://www.atsdr.cdc.gov/atsdrhome.html

Centers for Disease Control & Prevention (2015)
1600 Clifton Rd.
Atlanta, GA 30333
p. (404) 639-3535
http://www.cdc.gov

Dept of Labor

Mine Safety & Health Administration (2015)
Balston Tower #3
4015 Wilson Boulevard
Arlington, VA 22203
p. (703) 235-1385
http://www.msha.gov

Dept of State

Intl Boundary & Water Commission, US & Mexico (2015)
The Commons
4171 North Mesa, Suite C-310
El Paso, TX 79902-1441
p. 1-800-262-8857
john.merino@ibwc.state.gov

Dept of the Interior

National Park Service (2015)
1849 C. Street N.W.
Washington, DC 20240
p. (202) 208-4621
http://www.nps.gov

Office of the Secretary (2015)
1849 C. Street N.W.
Washington, DC 20240
p. (202) 208-3100
feedback@ios.doi.gov
http://www.doi.gov

Bureau of Reclamation (2015)
1849 C. Street N.W.
Washington, DC 20240
p. (202) 513-0501
http://www.usbr.gov

Bureau of Indian Affairs (2015)
1849 C. Street N.W.
Washington, DC 20240
p. (202) 208-7163
http://www.doi.gov/bureau-indian-affairs.html

Bureau of Land Management (2015)
Office of Public Affairs
1849 C Street, Room 406-LS
Washington, DC 20240
p. (202) 208-3801
BLM_Press@blm.gov
http://www.blm.gov

Office of Surface Mining, Reclamation & Enforcement (2015)
1951 Constitution Ave. N.W.
Washington, DC 20240
p. (202) 208-2565
osm-getinfo@osmre.gov
http://www.osmre.gov/

Dept of the Treasury

Internal Revenue Service (2015)
1111 Constitution Avenue, NW
Washington, DC 20224
p. (202) 622-9511
http://www.irs.ustreas,gov

Dept of Transportation

U.S. Coast Guard (2015)
2100 Second Street, SW
Washington, DC 20593
p. (202) 366-4000
http://www.uscg.mil

Federal Aviation Administration (2015)
800 Independence Avenue, SW
Washington, DC 20591
p. (202) 267-3484
http://www.faa.gov

Environmental Protection Agency

Assistant Administrator for Environmental Information
1200 Pennsylvania Avenue, NW
Washington, DC 20460
p. (202) 564-6665
Nelson.kimberly@Epa.gov

Office of Atmospheric Programs (2015)
501 Third Street, NW
Washington, DC 20001
p. (202) 564-9140
McCabe.janet@Epa.gov

Office of Wetlands, Oceans & Watersheds (2015)
Fairchild Building
499 South Capitol Street, SW
Washington, DC 20003
p. (202) 260-7166
best-wong.benita@epa.gov

Assistant Administrator for Water Programs ☒ (2017)
1200 Pennsylvania Avenue, NW
Washington, DC 20460
p. (202) 260-5700
Beauvais.joel@Epa.gov
https://www.epa.gov/aboutepa/about-office-water

Office of Wastewater Management (2015)
1200 Pennsylvania Avenue, NW
Washington, DC 20460
p. (202) 564-0748
ow-general@epa.gov

Office of Underground Storage Tanks (2015)
Crystal Gateway One
1235 Jefferson Davis Highway
Arlington, VA 22202
p. (703) 603-9900
hoskinson.carolyn@epa.gov

Office of Emergency & Remedial Response (Superfund/Oil Programs) (2015)
Crystal Gateway One
1235 Jefferson Davis Highway
Arlington, VA 22202
p. (703) 603-8960
oilinfo@epamail.epa.gov
http://www.epa.gov/superfund

Office of Solid Waste (2015)
1301 Constitutiion Ave. NW
Arlington, VA 22202
p. (202) 566-0200
aastanislaus@epa.gov
http://www2.epa.gov/aboutepa/about-office-solid-waste-and-emergency-response-oswer

Safety, Health & Environmental Management Div (2015)
1200 Pennsylvania Avenue, NW
Washington, DC 20460
p. (202) 564-1640
dfe@epa.gov

National Air & Radiation Environmental Laboratory (2015)
540 South Morris Avenue
Montgomery, AL 36115-2601
p. (334) 270-3404
griggs.john@epa.gov

Office of Research and Development ☒ (2020)
USEPA-ORD-8101R
1200 Pennsylvania Avenue, NW
Washington, DC 20460
p. 202-564-3512
frithsen.jeff@epa.gov
http://www.epa.gov/research

National Center for Environmental Research (2015)
Ronald Reagan Building
1300 Pennsylvania Avenue, NW
Washington, DC 20004
p. (202) 564-6825
johnson.jim@epa.gov

Office of Ground Water & Drinking Water (2015)
1200 Pennsylvania Avenue, NW
Washington, DC 20460
p. (202) 260-5400
grevatt.peter@epa.gov

Office of Science & Technology (2015)
1200 Pennsylvania Avenue, NW
Washington, DC 20460
p. (202) 260-5400
Southerland.elizabeth@Epa.gov

Office of Science Policy (2019)
1200 Pennsylvania Ave., NW
MC-8104R
Washington, DC 20460
p. (202) 564-6705
hauchman.fred@epa.gov

Federal Emergency Management Agency

Office of the Director (2015)
500 C Street, SW
Washington, DC 20472
p. (202) 646-4600
http://www.fema.gov/

Jet Propulsion Laboratory

Earth & Space Sciences Div (2015)
Geology & Planetology Section
Pasadena, CA 91109
p. (818) 354-3440
daniel.j.mccleese@jpl.nasa.gov
Department Administrative Manager: Murray Geller
Lead Scientist:
Bruce E. Banerdt, (D), S California, 1983, Xy
Section Manager:
Ronald G. Blom, (D), California (Santa Barbara), 1987, Zr
Section Member:
Michael J. Abrams, (M), Caltech, 1973, Zr
Ronald E. Alley, (M), Northwestern, 1972, Zr
Diana L. Blaney, (D), Hawaii, 1990, Xg
Bonnie J. Buratti, (D), Cornell, 1983, Zr
Robert E. Crippen, (D), California (Santa Barbara), 1989, Zr
Joy A. Crisp, (D), Princeton, 1984, Gv
Thomas C. Duxbury, (M), Purdue, 1966, Zn
Diane L. Evans, (D), Washington, 1981, Gm
Tom G. Farr, (D), Washington, 1981, Gm
Ken E. Herkenhoff, (D), Caltech, 1989, Xg
Simon J. Hook, (D), Durham (UK), 1989, Zr
Erik R. Ivins, (M), California (Los Angeles), 1976, Gt
Anne B. Kahle, (D), California (Los Angeles), 1975, Zr
Harold R. Lang, (D), Calgary, 1983, Gr
Kyle C. McDonald, (D), Michigan, 1991, Zn
Robert M. Nelson, (D), Pittsburgh, 1978, Xg
Eni G. Njoku, (D), MIT, 1976, Zr
Frank D. Palluconi, (M), Penn State, 1963, Zr
David C. Pieri, (D), Cornell, 1979, Gt
Jeffrey J. Plant, (D), Washington (St Louis), 1991, Zn
Jeffrey Plescia, (D), S California, 1985, Zn
Carol A. Raymond, (D), Columbia, 1989, Yr
Suzanne E. Smrekar, (D), S Methodist, 1990, Xy
Linda J. Spilker, (D), California (Los Angeles), 1992, XyZr
Ellen R. Stofan, (D), Brown, 1989, Xg
Glenn J. Veeder, (D), Caltech, 1974, Zn
Section Member:
Matthew P. Golombek, (D), Massachusetts, 1981, XgyGt
Senior Scientist:
Ronald S. Sanders, (D), Brown, 1970, Gr

Lawrence Livermore National Laboratory

Atmospheric, Earth and Energy (2015)
7000 East Avenue
L-203
Livermore, CA 94550
p. (925) 423-1848
antoun1@llnl.gov
https://www-pls.llnl.gov/?url=about_pls-atmospheric_earth_and_energy_division
Administrative Assistant: Laura Long
Head:
Kenneth J. Jackson, (D), California (Berkeley), 1983, Cl
Senior Scientist:
Roger D. Aines, (D), Caltech, 1984, Cg
Bill L. Bourcier, (D), Oregon State, 1983, Cg
Thomas A. Buscheck, (D), California (Berkeley), 1984, Ng
Steven Carle, (D), California (Davis), 1996, Hw
Charles R. Carrigan, (D), California (Los Angeles), 1977, Yg
Susan A. Carroll, (D), Northwestern, 1988, Cl
M Lee Davisson, (M), California (Davis), 1992, Cg
Quingyun Duan, (D), Arizona, 1991, Hg
Robert C. Finkel, (D), California (San Diego), 1974, Cg
Samuel (Julio) J. Friedmann, (D), MIT, Gg
Richard B. Knapp, (D), Arizona, 1978, Ng
Gayle A. Pawloski, (B), Cal State (Hayward), 1979, Gg
Abelardo L. Ramirez, (M), Purdue, 1979, Ng
Sarah K. Roberts, (M), Arizona State, 1990, Cl
Andrew F B Tompson, (D), Princeton, 1985, HwSpCc
Jeffrey L. Wagoner, (M), California (Riverside), 1977, Gs
Ananda M. Wijesinghe, (D), MIT, 1978, Nr
Sun Yunwei, (D), Israel Inst of Tech, 1995, Ng
Mavrik Zavarin, Yr

Los Alamos National Laboratory

Chemistry Div (B,M,D) (2015)
P.O. Box 1663
Los Alamos, NM 87545
p. (505) 667-4457
chemistry@lanl.gov
http://pearl1.lanl.gov/external/
Division Leader:
Alexander J. Gancarz, (D), Caltech, 1976, Cc
Staff:
Kent D. Abney, (D), Colorado State, 1987, Zn
Stephen F. Agnew, (D), Washington State, 1981, Zn
Moses Attrep, (D), Arkansas, 1965, Ze
Timothy M. Benjamin, (D), Caltech, 1979, Ca
Scott M. Bowen, (D), New Mexico, 1983, Ca
James R. Brainard, (D), Indiana, 1979, Zn
Jeffrey C. Bryan, (D), Washington, 1988, Zn
Carol J. Burns, (D), California (Berkeley), 1987, Zn
Timothy P. Burns, (D), Nebraska, 1986, Co

AGENCIES & INTL ORGANIZATIONS

Gilbert W. Butler, (D), California (Berkeley), 1967, Cc
Edwin P. Chamberlain, (D), Texas A&M, 1971, Cc
David L. Clark, (D), Indiana, 1986, Zn
Dean A. Cole, (D), Iowa, 1985, Co
David B. Curtis, (D), Oregon State, 1974, Cc
Paul R. Dixon, (D), Yale, 1989, Zn
Robert J. Donahoe, (D), North Carolina State, 1985, Zn
Stephen K. Doorn, (D), Northwestern, 1989, Zn
Clarence J. Duffy, (D), British Columbia, 1977, Cg
Deward W. Efurd, (D), Arkansas, 1975, Ct
Phillip G. Eller, (D), Ohio State, 1971, Zn
June T. Fabryka-Martin, (D), Arizona, 1988, Hg
Bryan L. Fearey, (D), Iowa State, 1986, Cc
David L. Finnegan, (D), Maryland, 1984, As
John R. Fitzpatrick, (M), New Mexico, 1983, Zn
Malcolm M. Fowler, (D), Washington (St Louis), 1972, As
Sammy R. Garcia, (B), New Mexico Highlands, 1972, Ca
Russell E. Gritzo, (B), New Mexico State, 1983, Zn
Richard C. Heaton, (D), Illinois, 1973, Zn
Sara B. Helmick, (B), Southwestern, 1955, Zn
David R. Janecky, (D), Minnesota, 1982, Cl
Kung King-Hsi, (D), Cornell, 1989, Zn
Scott Kinkead, (D), Idaho, 1983, Zn
Gregory J. Kubas, (D), Northwestern, 1970, As
Pat J. Langston-Unkefer, (D), Texas A&M, 1978, Co
Patrick A. Longmire, (D), New Mexico, 1991, Zn
Michael MacInnes, (M), Wisconsin, 1969, Zn
Allen S. Mason, (D), Miami, 1974, As
Charles M. Miller, (D), Stanford, 1980, Cc
Geoffrey G. Miller, (D), Rensselaer, 1984, Cc
Terrance L. Morgan, (M), Colorado State, 1984, Ng
David Morris, (D), North Carolina State, 1984, Cc
Eugene J. Mroz, (D), Maryland, 1976, As
Michael T. Murrell, (D), California (San Diego), 1980, Cc
Allen E. Ogard, (D), Chicago, 1957, Hw
Jose A. Olivares, (D), Iowa State, 1985, Cs
Hain Oona, (D), Arizona, 1979, Zn
Kevin C. Ott, (D), Caltech, 1983, Zn
Edward S. Patera, (D), Arizona State, 1982, Cg
Richard E. Perrin, (B), Denver, 1952, Cc
Eugene J. Peterson, (D), Arizona State, 1976, Ct
Dennis Phillips, (D), Hawaii, 1976, Co
Jane Poths, (D), Chicago, 1982, Cc
Pamela Z. Rogers, (D), California (Berkeley), 1981, Cl
Donald J. Rokop, (D), Lake Forest, 1985, Cc
Robert S. Rundberg, (D), CUNY, 1978, Gq
Nancy N. S. Sauer, (D), Iowa State, 1986, Zn
Norman C. Schroeder, (D), Iowa State, 1985, Cc
Louis A. Silks, (B), Suffolk, 1978, Zn
Paul H. Smith, (D), California (Berkeley), 1987, Zn
Zita V. Svitra, (B), Roosevelt, 1967, Zn
Basil I. Swanson, (D), Northwestern, 1969, Zn
C. Drew Tait, (D), North Carolina State, 1984, Cp
Wayne A. Taylor, (B), New Mexico, 1978, Cl
Kimberly W. Thomas, (D), California (Berkeley), 1978, Ct
Joseph L. Thompson, (D), Penn State, 1963, Ct
Ines Triay, (D), Miami, 1985, Hg
Clifford J. Unkefer, (D), Minnesota, 1981, Zn
David J. Vieira, (D), California (Berkeley), 1978, Zn
Jerry B. Wilhelmy, (D), California (Berkeley), 1969, Zn
Kurt Wolfsberg, (D), Washington (St Louis), 1959, Cc
William H. Woodruff, (D), Purdue, 1972, Zn
Mary Anne Yates, (D), Carnegie Mellon, 1976, Zn

Emeritus:
Ernest A. Bryant, (D), Washington (St Louis), 1956, Cg
Merle E. Bunker, (D), Indiana, 1950, Zn
William R. Daniels, (D), New Mexico, 1965, Ct
Donald L. Hull, (M), Iowa State, 1974, Zn

Earth & Environmental Sciences Div (2016)
P.O. Box 1663
Los Alamos, NM 87545
p. (505) 667-3644
wallacet@lanl.gov

Technical Staff:
James E. Bossert, (D), Colorado State, 1990, As

Director:
Terry C. Wallace, Jr., (D), Caltech, 1983, Ys

Technical Staff:
Kay H. Birdsell, (M), Colorado, 1985, Hqy
James W. Carey, (D), Harvard, 1990, CpYxNr
Lianjie Huang, (D), Paris, 1994, Yg
Kenneth H. Wohletz, (D), Arizona State, 1980, YxGvRm

Technical Staff:
Paul Aamodt, (M), Nevada, 1991, Zn
Douglas Alde, (D), Illinois, 1979, Zn
M. James Aldrich, (D), New Mexico, 1972, Gc
Fairley J. Barnes, (D), New Mexico State, 1986, Zn
Naomi M. Becker, (D), Wisconsin, 1991, Hg
James D. Blacic, (D), California (Los Angeles), 1971, Nr
Rainer Bleck, (D), Penn State, 1968, Am
Christopher R. Bradley, (D), MIT, 1993, Yg
Thomas L. Brake, (M), New Mexico State, 1985, Nr
David E. Broxton, (M), New Mexico, 1976, Gx
Wendee M. Brunish, (D), Illinois, 1981, Zn
Gilles Y. Bussod, (D), California (Los Angeles), 1988, Yx
Katherine Campbell, (D), New Mexico, 1979, Gq
Theodore C. Carney, (M), Colorado State, 1981, Ng
Gregory L. Cole, (D), Arizona, 1990, Zg
Steffanie Coonley, (M), New Mexico, 1984, Zn
Keeley R. Costigan, (D), Colorado State, 1992, As
William Cottingame, (D), Texas, 1984, As
James L. Craig, (B), Nevada, 1974, Gg
Bruce M. Crowe, (D), California (Santa Barbara), 1974, Gv
Zora V. Dash, (M), New Mexico, 1988, Zn
Deborah J. Daymon, (M), Idaho, 1994, Zn
Kalpak Dighe, (D), Clemson, 1994, Zn
John C. Dinsmoor, (B), Colorado Mines, 1989, Nm
Alison M. Dorries, (D), Harvard, 1986, Zn
Donald S. Dreesen, (B), New Mexico, 1968, Np
David V. Duchane, (D), Michigan, 1978, Zn
Michael H. Ebinger, (D), Purdue, 1988, So
C L Edwards, (D), New Mexico Tech, 1975, Yg
C. James Elliott, (D), Yale, Zn
Scott M. Elliott, (D), California, 1983, As
Perry D. Farley, (D), Oklahoma, 1981, Zn
George T. Farmer, (D), Cincinnati, 1968, Hw
David N. Fogel, (M), Cal State, 1997, Zn
Carl W. Gable, (D), Harvard, 1989, Gt
Edward S. Gaffney, (D), Caltech, 1973, Yg
Anthony F. Gallegos, (D), Colorado State, 1970, Pg
Jamie N. Gardner, (D), California (Davis), 1985, Gg
Fraser Goff, (D), California (Santa Cruz), 1977, Cg
Jeffrey C. Hansen, (B), Weber State, 1971, Zn
Charles D. Harrington, (D), Indiana, 1970, Gm
Hans E. Hartse, (D), New Mexico Tech, 1991, Ys
Ward L. Hawkins, (B), Nevada (Reno), 1977, Zg
Grant Heiken, (D), California (Santa Barbara), 1972, Gv
Donald D. Hickmott, (D), MIT, 1988, Gp
Steve T. Hildebrand, (D), Texas (Dallas), 1993, Ys
Emil F. Homuth, (D), Washington, 1974, Ye
Paul A. Johnson, (M), Arizona, 1984, Yg
Eric M. Jones, (D), Wisconsin, 1970, As
Hemendra N. Kalia, (D), Missouri Sch of Mines, 1970, Nm
Jim C. Kao, (D), Illinois, 1985, As
Danny Katzman, (M), New Mexico, 1991, HyGem
Elizabeth Keating, (D), Wisconsin, 1995, Gg
Sharad Kelkar, (M), Texas, 1979, Np
C. F. Keller, Jr., (D), Indiana, 1969, As
Richard G. Kovach, (M), Naval Postgrad Sch, 1979, Nm
Donathon J. Krier, (M), New Mexico, 1980, Gv
Thomas D. Kunkle, (D), Hawaii, 1978, Xy
Edward M. Kwicklis, (M), Colorado, 1987, Gg
Chung Chieng A. Lai, (D), Texas A&M, 1984, Am
Schon S. Levy, (M), Texas, 1975, Gi
Peter C. Lichtner, (D), Mainz, 1974, Zn
Rodman Linn, (D), New Mexico State, 1997, Ng
Lynn McDonald, (M), Cal State, Zn
Maureen A. McGraw, (D), California (Berkeley), 1996, Zn
Laurie A. McNair, (D), Carnegie Mellon, 1995, Zn
Wayne R. Meadows, (B), Cal State (Bakersfield), 1974, Yg
Theodore Mockler, (M), Carnegie Mellon, 1986, Zn
Orrin B. Myers, (D), Colorado State, 1992, Pg
Balu Nadiga, (D), Caltech, 1992, Zn
Brent D. Newman, (D), New Mexico Tech, 1996, Gg
John W. Nyhan, (D), Colorado State, 1972, So
Ronald D. Oliver, (B), Oregon Inst of Tech, 1972, Yg

Howard J. Patton, (D), MIT, 1978, Yg
Frank V. Perry, (D), California, 1988, Gv
William S. Phillips, (D), MIT, 1985, Yg
Eugene W. Pokorny, (B), Missouri, 1979, Nm
William M. Porch, (D), Washington, 1971, Yg
Allyn R. Pratt, (M), Boise State, 1977, Ge
George Randall, (D), SUNY (Binghamton), Yg
Steen Rasmussen, (D), Tech (Denmark), 1985, Yg
Jon M. Reisner, (D), Iowa State, As
Steven L. Reneau, (D), California (Berkeley), 1988, Gm
Douglas O. Revelle, (D), Michigan, 1974, As
Peter Roberts, (D), MIT, 1989, Ys
Bruce A. Robinson, (D), MIT, 1985, Zn
R. Roussel-Dupre, (D), Colorado, 1979, Zn
Thomas J. Shankland, (D), Harvard, 1966, Yx
Catherine H. Smith, (D), New Mexico State, 1995, Ca
Wendy E. Soll, (D), MIT, 1991, Hw
Everett P. Springer, (D), Utah State, 1983, Hq
Lee Steck, (D), California, Gg
Robert P. Swift, (D), Washington, 1969, Nr
E.M.D. Symbalisty, (D), Chicago, 1984, Zn
Steven R. Taylor, (D), MIT, 1980, Ys
James Tencate, (D), Texas, 1992, Zn
Bryan J. Travis, (D), Florida State, 1974, Hq
David T. Vaniman, (D), California (Santa Cruz), 1976, Gx
Richard G. Warren, (M), New Mexico, 1972, Gi
Douglas J. Weaver, (M), Nevada, 1995, Zn
Thomas A. Weaver, (D), Chicago, 1973, Yg
Rodney W. Whitaker, (D), Indiana, 1976, As
Earl M. Whitney, (D), Utah, Zn
Judith L. Winterkamp, (B), Texas, 1977, Zn
Giday WoldeGabriel, (D), Case Western, 1987, Gx
Andrew V. Wolfsberg, (D), Stanford, 1993, Hw
George A. Zyvoloski, (D), California (Santa Barbara), 1975, Ng

Project Leader:
Mark T. Peters, (D), Chicago, 1992, Yg

Technical Staff:
James N. Albright, (D), Chicago, 1969, Yg
Donald W. Brown, (M), California (Los Angeles), 1961, Nr
Leigh S. House, (D), Columbia, 1982, Ys
Larry Allan Jones, (B), Nebraska, 1972, Ng

Technical Staff:
W. Scott Baldridge, (D), Caltech, 1979, GlcGt

Other:
Christina "Tina" Behr-Andres, (D), Michigan Tech, 1992, Zn

National Aeronautics & Space Administration

Lyndon B. Johnson Space Center (2015)
Houston, TX 77058-3696
p. (281) 483-5309
ellen.ochoa-1@nasa.gov

NASA Headquarters (2015)
Washington, DC 20546-0001
p. (202) 358-2345
info-center@hq.nasa.gov
http://www.hq.nasa.gov/

Ames Research Center (2015)
Building BN200
Moffett Field, CA 94035-1000
p. (650) 604-5111
michael.s.mewhinney@nasa.gov

Chair:
Penelope Boston, (D), Colorado, 1985, Zg

Goddard Space Flight Center (2015)
Greenbelt Road
Greenbelt, MD 20771-0001
p. (301) 286-5121
stephanie.s.keene@nasa.gov

George C. Marshall Space Flight Center (2015)
Marshall Space Flight Center, AL 35812-0001
p. (256) 544-1910
patrick.e.scheuermann@nasa.gov

National Oceanic and Atmospheric Administration

National Centers for Environmental Information ☒ (2018)
David Skaggs Research Center
325 Broadway
Boulder, CO 80303
p. (303) 497-6826
ncei.info@noaa.gov
http://www.ngdc.noaa.gov/
f: http://www.facebook.com/NOAANCEIoceangeo
t: @NOAANCEIocngeo

National Science Foundation

Office of Polar Programs ☒ (2019)
2415 Eisenhower Avenue
Alexandria, VA 22314
p. (703) 292-8030
kfalkner@nsf.gov
https://www.nsf.gov/div/index.jsp?div=opp
f: https://www.facebook.com/NSFOPP/

Office of Polar Programs:
Kelly K. Falkner, (D)

Arctic Sciences Section:
Simon Stephenson, (M)

Antarctic Sciences Section:
Alexandra Isern, (D)

Antarctic Infrastructure and Logistics Section:
Stephanie A. Short, (M)

Polar Environment, Safety and Health Section:
Renee Crain

Technical Information Specialist, Office of Polar Programs:
David Friscic

Senior Advisor, Environment, Polar Environment, Safety and Health Section:
Polly A. Penhale, (D)

Senior Advisor, Business Analyst, Office of Polar Programs:
Scot Arnold, (D)

Science Assistant, Office of Polar Programs:
Beverly Walker

Science Assistant, Antarctic Sciences Section:
Wilson Sauthoff

Safety & Occupational Health Manager, Polar Environment, Safety and Health Section:
Gwendolyn Adams

Safety & Health Officer, Polar Environment, Safety and Health Section:
Jon Fentress

Program Support Contract Specialist, Arctic Sciences Section:
Randall V. Sisco

Program Support Contract Specialist, Antarctic Infrastructure & Logistics Section:
Russell McElyea

Program Specialist, Polar Environment, Safety and Health Section:
Karen Sloane

Program Specialist, Office of Polar Programs:
Erika N. Davis
Angela Lyons

Program Manager, Techology Development, Antarctic Infrastructure & Logistics Section:
Patrick D. Smith

Program Manager, Organisms & Ecosystems, Antarctic Sciences Section:
Karla Heidelberg, (D)

Program Manager, Operations, Antarctic Infrastructure & Logistics Section:
Margaret A. Knuth

Program Manager, Office of Polar Programs:
Erica Sahler

Program Manager, Ocean Projects, Antarctic Infrastructure & Logistics Section:
Timothy M. McGovern, (M), Hawaii, 2011

Program Manager, IT Support & USAP Info Security, Antarctic Infrastructure & Logistics Section:
Tim Howard

Program Manager, Facilities Engineering Projects, Antarctic Infrastructure & Logistics Section:
Ben D. Roth

Program Manager, Facilities Construction & Maintenance, Antarctic Infrastructure & Logistics Section:
Michael Gencarelli

Program Manager, Capital Planning, Antarctic Infrastructure & Logistics

Section:
Brian MacDonald
Program Manager, Aviation, Antarctic Infrastructure & Logistics Section:
Gary James
Program Manager, Arctic Research Support and Logistics, Arctic Sciences Section:
Frank Rack, (B)
Program Manager, Arctic Research Support and Logistics, Acrtic Sciences Section:
Jennifer Mercer, (D)
Program Manager, Arctic Research Support & Logistics, Arctic Sciences Section:
Patrick R. Haggerty
Program Manager, Antarctic Research Support, Antarctic Infrastructure & Logistics Section:
Jessie L. Crain
Program Manager, Antarctic Integrated System Sciences, Antarctic Sciences Section:
Jennifer Burns, (D)
Program Manager Polar Outreach, Office of Polar Programs:
Peter West, (B)
Program Director, Ocean & Atmospheric Sciences, Antarctic Sciences Section:
Peter Milne, (D)
Program Director, Glaciology, Antarctic Sciences Section:
Paul Cutler, (D)
Program Director, Earth Sciences, Antarctic Sciences Section:
Michael Jackson, (D)
Program Director, Astrophysics & Geospace, Antarctic Sciences Section:
Robert Moore, (D)
Program Director, Arctic System Sciences, Arctic Sciences Section:
Gregory J. Anderson, (D)
Program Director, Arctic Observing Networks, Arctic Sciences Section:
Roberto Delgado, (D)
Program Director, Arctic Natural Sciences, Arctic Sciences Section:
Xujing Davis, (D)
Cynthia Suchman, (D)
Program Director, Arctic Natural Resources, Arctic Sciences Section:
Marc Stieglitz, (D)
Program Director, Antarctic Research Facilities & Special Projects, Antarctic Sciences Section:
Vladimir O. Papitashvili, (D), Russian Acad of Sci, 1981, YmXpa
Program Coordination Specialist, Arctic Sciences Section:
Linda Izzard
Program Coordination Specialist, Antarctic Sciences Section:
Desiree Marshall
Program Coordination Specialist, Antarctic Infrastructure & Logistics Section:
Carlena Fooks
Program Analyst, Office of Polar Programs:
Kimiko S. Bowens-Knox
Management & Program Analyst, Office of Polar Programs:
Andrew Backe, (D)
IT Specialist:
Dan King
Environmental Policy Specialist, Polar Environment, Safety and Health Section:
Nature McGinn, (D)
Communications Specialist, Office of Polar Programs:
Terri Edillon
Chief Program Manager, Antarctic Infrastructure & Logistics Section:
Paul Sheppard
Associate Program Director, Arctic System Sciences, Arctic Sciences Section:
Colleen Strawhacker, (D)
Administrative Officer, Office of Polar Programs:
Pawnee Maiden
AAAS Sciences and Technology Fellow, Arctic Sciences Section:
Yekaterina Kontar, (D)

Ocean Sciences Division (2016)
4201 Wilson Boulevard
Arlington, VA 22230
p. (703) 292-8580
rwmurray@nsf.gov
http://www.nsf.gov/div/index.jsp?div=OCE
Division Director:
Richard W. Murray
Section Head for Integrative Programs Section:
Bauke Houtman
Program Director for Ship Operations Program:
Rose Dufour
Program Director for Physical Oceanography Program:
Eric C. Itsweire
Baris Mete Uz
Program Director for Oceanographic Instrumentation and Technical Service Programs:
James Holik
Program Director for Ocean Observatories Initiative:
Jean M. McGovern
Program Director for Ocean Education Programs:
Elizabeth L. Rom
Program Director for Ocean Drilling Programs:
James F. Allan
Program Director for Marine Geology and Geophysics Program:
Candace O. Major
Barbara Ransom
Program Director for Chemical Oceanography Program:
Donald Rice
Program Director for Biological Oceanography Program:
David L. Garrison
Program Director:
John Walter
Program Directof for Ocean Drilling Programs:
Thomas Janacek
Associate Program Director for Oceanographic Technology and Interdisciplinary Coordination Program:
Kandace S. Binkley

Earth Sciences Div ☒ (2017)
4201 Wilson Boulevard
Arlington, VA 22230
p. (703) 292-8550
cfrost@nsf.gov
http://www.nsf.gov/div/index.jsp?div=EAR
t: @NSF_EAR
Program Director for Hydrologic Sciences:
Thomas Torgersen
Program Director for Instrumentation and Facilities:
David Lambert
Section Head:
Gregory J. Anderson
Section Head :
Lina Patino, (D)
Section for Antarctic Sciences:
Scott G. Borg, (D), Arizona State, 1984, Git
Program Director for Tectonics Program:
David M. Fountain
Stephen S. Harlan
Program Director for Petrology and Geochemistry Program:
Sonia Esperanca
Jennifer Wade
Program Director for Integrated Earth Systems:
Leonard E. Johnson
Program Director for Instrumentation and Facilities:
Russell C. Kelz
Program Director for Geophysics Program:
Robin Reichlin
Program Director for Geobiology and low-T Geochemistry:
Jonathan Wynn
Program Director for Geobiology and Low Temp Geochemistry:
Enriqueta C. Barrera
Program Director for Earthscope:
Margaret Benoit

Office of the Director (2015)
4201 Wilson Boulevard
Arlington, VA 22230
p. (703) 292-5111
fcordova@nsf.gov
http://www.nsf.gov
f: https://www.facebook.com/US.NSF
t: @NSF

Atmospheric and Geospace Sciences Div (2015)
4201 Wilson Boulevard

Arlington, VA 22230
p. (703) 292-8520
pshepson@nsf.gov
http://www.nsf.gov/div/index.jsp?div=AGS
Division Director:
Paul B. Shepson
Section Head for NCAR/Facilities Section:
Stephan P. Nelson
Section Head for Geospace Section:
Richard A. Behnke
Section Head for Atmosphere Section:
David J. Verardo
Program Director for Solar Terrestrial Research Program:
Therese M. Jorgensen
Program Director for Physical and Dynamic Meteorology Program:
A. Gannet Hallar
Chungu Lu
Bradley F. Smull
Program Director for Paleoclimate Program:
Candace O. Major
Program Director for Magnetospheric Physics Program:
Raymond J. Walker
Program Director for Geospace Facilities:
Robert M. Robinson
Program Director for Education and Cross Disciplinary Activities Program:
Linda George
Program Director for Climate and Large Scale Dynamics Program & Carbon and Water in Earth Systems Program:
Eric T. DeWeaver
Program Director for Climate and Large Scale Dynamics Program:
Anjuli Bamzai
Program Director for Atmospheric Chemistry Program:
Peter Milne
Anne-Marie Schmoltner
Program Director for Aeronomy Program:
Anja Stromme
Program Coodinator for NCAR/Facilities Section:
Sarah L. Ruth
Facilities Program Manager for NCAR/Facilities Section:
Linnea M. Avallone
Assistant Program Director for Atmosphere Section:
Nicholas F. Anderson

Directorate for Geosciences ☒ (2018)
2415 Eisenhower Avenue
Room C8000
Alexandria, VA 22314
p. (703) 292-8500
sborg@nsf.gov
https://www.nsf.gov/dir/index.jsp?org=GEO
f: https://www.facebook.com/US.NSF
t: @NSF
Program Director for International Activities:
Maria Uhle

Nuclear Regulatory Commission

NRC Headquarters (2015)
One White Flint North Building
11555 Rockville Pike
Rockville, MD 20852
p. (301) 415-8200
http://www.nrc.gov/

Oak Ridge National Laboratory

Energy Div (2015)
P.O. Box 2008
Oak Ridge, TN 37831-6187
p. (615) 574-5510
robbkr@ornl.gov
Administrative Assistant: Teresa D. Ferguson
Director:
Robert B. Shelton, (D), S Illinois, 1970, Zn
Chair:
Donald W. Lee, (D), Michigan, 1977, Hw
Senior Scientist:
Richard H. Ketelle, (M), Tennessee, 1977, Ge
Russell Lee, (D), McMaster, 1978, Zy
William P. Staub, (D), Iowa State, 1969, Ng
Research Associate:
Arthur C. Curtis, (M), Colorado State, 1993, Ge
Robert O. Johnson, (D), Tennessee, 1984, Hg
Richard R. Lee, (M), Temple, 1982, Gr
John D. Tauxe, (D), Texas, 1994, Hw

Environmental Sciences Div (2015)
P.O. Box 2008
Mail Stop 6035
Oak Ridge, TN 37831
p. (865) 574-7374
pierceem@ornl.gov
http://www.esd.ornl.gov
Director:
Stephen G. Hildebrand, (D), Michigan, 1973, Zn
Professor:
Baohua Gu, (D), California (Berkeley), 1991, CbqSb
Senior Scientist:
Marshall Adams, (D), North Carolina, 1974, Ob
Jeff Amthor, (D), Yale, 1987, Sf
Tom Ashwood, (M), Murray State, 1975, Ge
Mark Bevelhimer, (D), Tennessee, 1990, Hs
Terence J. Blasing, (D), Wisconsin, 1975, As
Thomas Boden, (M), Miami, 1985, As
Craig Brandt, (M), Tennessee, 1988, Ge
Scott C. Brooks, (D), Virginia, 1994, Cl
Robert S. Burlage, (D), Tennessee, 1990, Sb
Meng-Dawn Cheng, (D), Illinois, 1986, As
Robert B. Cook, (D), Columbia, 1981, Cg
William E. Doll, (D), Wisconsin, 1983, Yg
Thomas O. Early, (D), Washington (St Louis), 1969, Cl
Philip M. Jardine, (D), Virginia Tech, 1985, Sc
Liyuan Liang, (D), Caltech, 1988, Cl
Steven E. Lindberg, (D), Florida State, 1979, Cl
John F. McCarthy, (D), Rhode Island, 1975, Co
Gerilynn R. Moline, (D), Wisconsin, 1992, Hw
Tony V. Palumbo, (D), North Carolina State, 1980, Sb
Tommy J. Phelps, (D), Wisconsin, 1985, Sb
Ellen D. Smith, (M), Wisconsin, 1979, Hg
Brian P. Spalding, (D), Cornell, 1976, Sc
Robert S. Turner, (D), Pennsylvania, 1983, Cl
David B. Watson, (M), New Mexico Tech, 1983, Hw
Olivia M. West, (D), MIT, 1991, Ng
Associate Scientist:
Tammy Beaty, Zi
Mary Anna Bogle, (M), Miami, 1975, Ge
Norman D. Farrow, (B), Oregon State, 1974, Hg
Deputy Director:
Gary K. Jacobs, (D), Penn State, 1981, Cg

Pacific Northwest National Laboratory

Applied Geology & Geochemistry (2015)
Environmental Technology Div
Wayne J. Martin, Manager
PO Box 999 MSIN K6-81
Richland, WA 99352
p. (509) 376-5952
Christopher.Brown@pnnl.gov
http://www.pnl.gov/agg
Administrative Assistant: Charissa J. Chou
Head:
George R. Holdren, (D), Johns Hopkins, 1977, Cl
Staff Scientist:
Christopher J. Murray, (D), Stanford, 1992, Gq
Senior Research Engineer:
Mark D. White, (D), Colorado State, 1986, Hy
Lab Fellow:
Bernard P. McGrail, (D), Columbia S, 1996, Cg
Senior Scientist:
Douglas B. Barnett, (M), E Washington, 1985, Em
Bruce N. Bjornstad, (M), E Washington, 1980, Gs
Kirk J. Cantrell, (D), Georgia Tech, 1989, Cl
Amy P. Gamerdinger, (D), Cornell, 1989, Cl
Tyler J. Gilmore, (M), Idaho, 1987, Hg
Floyd N. Hodges, (D), Texas, 1975, Cg
Duane G. Horton, (D), Illinois, 1983, Gz
Kenneth M. Krupka, (D), Penn State, 1984, Cl

George V. Last, (M), Washington State, 1997, Ge
Jonathan W. Lindberg, (M), Washington State, 1995, Ec
Shas V. Mattigod, (D), Washington State, 1976, Cl
Alan C. Rohay, (D), Washington, 1982, Ys
Herbert T. Schaef, (M), Texas Tech, 1991, Cg
R. Jeffrey Serne, (B), Washington, 1969, CgSoCl
Mark D. Sweeney, (B), C Washington, 1985, Yg
Bruce A. Williams, (B), Colorado Mines, 1980, Ng

Associate Scientist:
Yi-Ju Chien, (M), Stanford, 1998, Gq
Jonathan P. Icenhower, (D), Oklahoma, 1995, Cl
David C. Lanigan, (B), Michigan Tech, 1980, Hg
Virginia L. Legore, (B), Oregon State, 1974, Cg
Clark W. Lindenmeier, (M), E Washington, 1995, Cg
Paul F. Martin, (B), Pacific Lutheran, 1982, Cg
Kent E. Parker, (M), Washington State, 1995, Cg

Research Associate:
Alexandra B. Amonette, (M), CUNY, 1976, Cg
Deborah S. Burke, (B), E Washington, 1992, Sc
Elsa A. Camacho, (B), Heritage, 1996, Cg
Matthew J. O'Hara, (B), Montana, 1996, Cg
Robert D. Orr, (B), N State, 1994, Ca

Sandia National Laboratory

Geoscience and Environment (2015)
1515 Eubank SE
P.O. Box 5800
Albuquerque, NM 87185
knorri@sandia.gov
http://www.sandia.gov/

Research Scientist:
Thomas Dewers, (D), Indiana, 1990, Yx

Smithsonian Institution

Smithsonian Astrophysical Observatory ☒ (2017)
60 Garden Street
Cambridge, MA 02138
p. (617) 495-7100
calcock@cfa.harvard.edu
https://www.cfa.harvard.edu/sao
f: https://www.facebook.com/HarvardSmithsonianCenterFo-rAstrophysics

U.S. Geological Survey

Director, Office of Science Quality & Integrity (2015)
12201 Sunrise Valley Drive, MS 911
Reston, VA 20192
p. (703) 648-6601
athornhill@usgs.gov
http://www.usgs.gov

Regional Director, Southeast (2015)
1770 Corporate Drive, Suite 500
Norcross, GA 30093
p. (770) 409-7701
jdweaver@usgs.gov
http://www.usgs.gov

Regional Director, Northeast (2015)
12201 Sunrise Valley Drive, MS 953
Reston, VA 20192
p. (703) 648-6660
druss@usgs.gov
http://www.usgs.gov

Regional Director, Midwest (2015)
1451 Green Road
Ann Arbor, MI 48105
p. (737) 214-7207
lcarl@usgs.gov
http://www.usgs.gov

Regional Director, Northwest (2015)
909 1st Avenue
Seattle, WA 98104
p. (206) 220-4600
ddlynch@usgs.gov
http://www.usgs.gov

Regional Director, Southwest ☒ (2022)
2800 Cottage Way
Suite W-1619
Sacramento, CA 95825
p. 916 978-2161
gs-swr_info@usgs.gov
http://www.usgs.gov

Associate Director, Human Capital (2015)
12201 Sunrise Valley Drive, MS 201
Reston, VA 20192
p. (703) 648-7261
dwade@usgs.gov
http://www.usgs.gov

Associate Director, Office of Budget, Planning, & Integration
12201 Sunrise Valley Drive, MS 105
Reston, VA 20192
p. (703) 648-4443
cburzyk@usgs.gov
http://www.usgs.gov

Associate Director, Office of Communications & Publishing
12201 Sunrise Valley Drive, MS 119
Reston, VA 20192
p. (703) 648-5750
bwainman@usgs.gov
http://www.usgs.gov

Associate Director, Administration & Enterprise Information
12201 Sunrise Valley Drive, MS 201
Reston, VA 20192
p. (703) 648-7261
timothy_quinn@ios.doi.gov
http://www.usgs.gov

Associate Director, Core Science Systems (M) (2017)
12201 Sunrise Valley Drive, MD 108
Reston, VA 20192
p. (703) 648-5747
kgallagher@usgs.gov
http://www.usgs.gov

Natural Hazards Mission Area ☒ (2022)
12201 Sunrise Valley Drive, MS 111
Reston, VA 20192
p. (703) 648-6600
applegate@usgs.gov
http://www.usgs.gov/natural_hazards/

Regional Safety Manager ☒ (2018)
PO Box 596
Brightwood, OR 97011
p. (503) 622-4432
wsimonds@usgs.gov
http://www.usgs.gov

Associate Director, Climate and Land-Use Change (2015)
12201 Sunrise Valley Drive, MD 409
Reston, VA 20192
p. (703) 648-5215
sryker@usgs.gov
http://www.usgs.gov

Office of the Director (2015)
12201 Sunrise Valley Drive, MS 100
Reston, VA 20192
p. (703) 648-7411
abwade@usgs.gov
http://www.usgs.gov

Associate Director, Ecosystems (M) (2016)
12201 Sunrise Valley Drive, MS 300
Reston, VA 20192
p. (703) 648-4050
akinsinger@usgs.gov
http://www.usgs.gov

SPECIALTY CODES

Specialty codes are used to indicate the research or teaching specialties of faculty members listed in the directory. Bold numbers are total number of individuals for each major category. Numbers in parentheses are individual specialty totals.

GEOLOGY 6683

Gg General Geology (1357)
Ga Archaeological Geology (143)
Ge Environmental Geology (751)
Gm Geomorphology (652)
Gl Glaciology and Glacial Geology (341)
Gu Marine Geology (273)
Gz Mineralogy & Crystallography (565)
Gn Paleolimnology (70)
Go Petroleum Geology (304)
Gx General Petrology (226)
Gi Igneous Petrology (574)
Gp Metamorphic Petrology (312)
Gd Sedimentary Petrology (193)
Gs Sedimentology (901)
Gr Physical Stratigraphy (358)
Gc Structural Geology (794)
Gt Tectonics (780)
Gv Volcanology (390)
Gq Mathematical Geology (66)
Gy Mineral Physics (79)
Gb Medical Geology (20)
Gf Forensic Geology (12)

ECONOMIC GEOLOGY 609

Eg General Economic Geology (280)
Ec Coal (48)
Em Metallic Ore Deposits (160)
En Industrial Minerals (27)
Eo Oil and Gas (112)
Es Construction Materials (SSG) (2)
Ed Ore Deposits (Other) (17)

GEOCHEMISTRY 2357

Cg General Geochemistry (651)
Ca Analytical Geochemistry (197)
Cp Experimental Petrology/Phase Equilibria (106)
Ce Exploration Geochemistry (66)
Cc Geochronology & Radioisotopes (362)
Cl Low-temperature Geochemistry (395)
Cm Marine Geochemistry (203)
Co Organic Geochemistry (140)
Cs Stable Isotopes (373)
Ct Trace Element Distribution (81)
Cb Biogeochemistry (124)
Cu High-temperature Geochemistry (41)
Cq Aqueous Geochemistry (52)

GEOPHYSICS 1764

Yg General Geophysics (641)
Yx Experimental Geophysics (105)
Ye Exploration Geophysics (252)
Yd Geodesy (110)
Ym Geomagnetism & Paleomagnetism (157)
Yv Gravity (33)
Yh Heat Flow (28)
Ys Seismology (577)
Yr Marine Geophysics (163)
Yn Nonlinear Geophysics (6)
Yu Near-surface Geophysics (60)

PALEONTOLOGY 1625

Pg General Paleontology (535)
Ps Paleostratigraphy (118)
Pm Micropaleontology (182)
Pb Paleobotany (64)
Pl Palynology (64)
Pq Quantitative Paleontology (28)
Pv Vertebrate Paleontology (262)
Pi Invertebrate Paleontology (249)
Pe Paleoecology (340)
Po Geomicrobiology (30)
Pc Paleoclimatology (110)

HYDROLOGY 1497

Hg General Hydrology (356)
Hw Ground Water/Hydrogeology (724)
Hq Quantitative Hydrology (124)
Hs Surface Waters (175)
Hy Geohydrology (106)
Ht Materials Transport (12)

SOIL SCIENCE 828

Sp Soil Physics/Hydrology (106)
Sc Soil Chemistry/Mineralogy (188)
Sd Pedology/Classification/Morphology (91)
Sf Forest Soils/Rangelands/Wetlands (66)
Sb Soil Biology/Biochemistry (123)
Sa Paleopedology/Archeology (30)
So Other Soil Science (224)

ENGINEERING GEOLOGY 700

Ng General Engineering Geology (290)
Ne Earthquake Engineering (19)
Nx Mining Tech/Extractive Metallurgy (46)
Nm Mining Engineering (81)
Np Petroleum Engineering (77)
Nr Rock Mechanics (118)
No Ocean Engineering/Mining (29)
Nt Geotechnical Engineering (17)
Nx Geological Engineering (23)

OCEANOGRAPHY 1552

Og General Oceanography (222)
Ob Biological Oceanography (459)
Oc Chemical Oceanography (253)
Ou Geological Oceanography (152)
Op Physical Oceanography (409)
On Shore and Nearshore Processes (129)
Oo Paleooceanography (29)

ASTRONOMICAL SCI 447

Xs Astronomy (7)
Xc Cosmochemistry (71)
Xg Extraterrestrial Geology (177)
Xy Extraterrestrial Geophysics (95)
Xm Meteorites & Tektites (80)
Xa Astrophysics (50)
Xp Heliophysics (3)
Xb Astrobiology (18)

ATMOSPHERIC SCI 1220

As Atmospheric Sciences (908)
Am Meteorology (326)
Ac Atmospheric Chemistry (33)
Ap Atmospheric Physics (25)
At Climatology (74)

GEOSCIENCE & SOCIETY 212

Rn Natural Hazards (89)
Rm Manmade Hazards (12)
Rw Water Quality/Use (51)
Rc Geoscience Communication (23)
Rh History of Geoscience (45)

OTHER 3396

Zg General Earth Sciences (403)
Ze Earth Science Education (356)
Zy Physical Geography (352)
Zr Remote Sensing (493)
Zm Material Science (61)
Zu Land Use/Urban Geology (128)
Zi Geographic Information Systems (581)
Zf Geoinformatics (55)
Zc Global Change (115)
Za Nanogeoscience (12)
Zo Climate Modeling (44)
Zn Not Elsewhere Classified (1284)

Total 23,954

FACULTY SPECIALTY INDEX

GEOLOGY

General Geology

Abbott, Lon, Univ of Colorado
Abousleiman, Younane N., Univ of Oklahoma
Abushagur, Sulaiman, El Paso Comm Coll
Adabanija, Moruffdeen A., Ladoke Akintola Univ of Tech
Adinolfi, Bryan, Southern Connecticut State Univ
Agbogun, Henry, Fort Hays State Univ
Ahmed, Samir, Graduate School of the City Univ of New York
Ahola, John, Bristol Comm Coll
Ajigo, Isaac O., Federal Univ of Tech, Akure
Akgun, Elif , Firat Univ
Akinlabi, Ismaila A., Ladoke Akintola Univ of Tech
Aksoy, Ercan, Firat Univ
Akujieze, Christopher N., Univ of Benin
Albin, Ed, Georgia State Univ, Perimeter Coll, Online
Alexander, Dane, Western Michigan Univ
Alexander, Elaine, McLennan Comm Coll
Alfsen, Sheila, Portland State Univ
Alkac, Onur, Firat Univ
Allison, Alivia J., Emporia State Univ
Altermann, Wladyslaw, Ludwig-Maximilians-Universitaet Muenchen
Amaach, Noureddin, Coll of Staten Island/CUNY
Amato, James A., Univ of Wisconsin, Oshkosh
Anderson, Paula, Univ of Arkansas, Fayetteville
Andonaegui, Pilar, Univ Complutense de Madrid
Andrews, William, Univ of Kentucky
Andronicos, Chris, Purdue Univ
Anhaeusser, Carl R., Univ of the Witwatersrand
Antonacci, Vince, Tennessee Geological Survey
Aranda, Angela, California State Univ, Fullerton
Archibald, Donnelly, Saint Francis Xavier Univ
Arias, Carmen, Univ Complutense de Madrid
Ashe, Douglas, Tulsa Comm Coll
Athaide, Dileep, Univ of British Columbia
Athaide, Dileep J., Capilano Univ
Athey, Jennifer E., Alaska Div of Geological & Geophysical Surveys
Bach Plaza, Joan, Universitat Auto de Barcelona
Bachle, Peter, Missouri Dept of Nat Resources
Baer, Eric M., Highline Coll
Bagley, Evan, Univ of Southern Mississippi
Balakrishnan, Meena, Tarrant County Coll, Northeast Campus
Ballero, Deniz Z., Georgia State Univ, Perimeter Coll, Online
Balsley, Christopher, Southern Connecticut State Univ
Banda, Anna F., El Centro Coll - Dallas Comm Coll District
Bandosz, Teresa, Graduate School of the City Univ of New York
Barcena, Homar S., Graduate School of the City Univ of New York
Barcheck, Grace, Cornell Univ
Barclay, Julie L., SUNY, Cortland
Barminski, Robert, Hartnell Coll
Barnes, Brenda, El Paso Comm Coll
Bassett, Damon, Missouri State Univ
Baxter, P.G., James E., Harrisburg Area Comm Coll
Bayowa, Oyelowo G., Ladoke Akintola Univ of Tech
Beatty, Lynne, Johnson County Comm Coll
Beausoleil, Denis, Douglas Coll
Beltz, John F., Univ of Akron
Bender, Hallock J., Mt. San Antonio Coll
Bennett, Sara, Western Illinois Univ
Benson, Donna M., Mesa Comm Coll
Bentley, Callan, Northern Virginia Comm Coll - Annandale
Benton, Steven, Illinois State Geological Survey
Beranek, Luke, Memorial Univ of Newfoundland
Berry IV, Henry N., Dept of Agriculture, Conservation, and Forestry
Berti, Claudio, Univ of Idaho
Best, Myron G., Brigham Young Univ
Bice, David M., Pennsylvania State Univ, Univ Park
Biek, Robert, Utah Geological Survey
Bierly, Aaron D., DCNR- Pennsylvania Bureau of Geological Survey
Birner, Johannes, Ludwig-Maximilians-Universitaet Muenchen
Blaylock, Glenn W., Laredo Coll
Blome, Margaret, East Carolina Univ
Boehm, David A., SUNY, The Coll at Brockport
Bograd, Michael B., Mississippi Office of Geology
Bohm, Christian, Manitoba Geological Survey
Bone, Fletcher, Missouri Dept of Nat Resources
Booth, James, Graduate School of the City Univ of New York
Boroughs, Terry J., American River Coll
Bouker, Polly A., Georgia State Univ, Perimeter Coll, Newton Campus
Bounk, Michael, Iowa Dept of Nat Resources
Bour, William, Northern Virginia Comm Coll - Loudoun Campus
Bouse, Robin, El Camino Coll
Bowen, Esther E., Argonne Nat Lab
Bowles, Zack, Mesa Comm Coll
Bozdog, Nick, N.C. Department of Enval Quality
Bradshaw, Lanna K., Brookhaven Coll
Branciforte, Chloe, Mesa Comm Coll
Branco, Brett, Graduate School of the City Univ of New York
Brânzila, Mihai, Alexandru Ioan Cuza
Bremmer, Sarah, Univ of Wisconsin, Madison, Div of Extension
Bridges, David L., Missouri Dept of Nat Resources
Bries Korpik, Jill, Inver Hills Comm Coll
Brill, Jr., Richard C., Honolulu Comm Coll
Brock, Patrick W. G., Queens Coll (CUNY)
Brown, Alan, Middle Tennessee State Univ
Brown, Kyle, Missouri Dept of Nat Resources
Brown, Summer, Univ of Kentucky
Browning, James V., Rutgers, The State Univ of New Jersey
Browning, James M., Texas Tech Univ
Browning, Sharon, Baylor Univ
Broze, Elliot, Whitman Coll
Brueckner, Hannes K., Queens Coll (CUNY)
Brunengo, Matthew , Portland State Univ
Bruschke, Freddi Jo, California State Univ, Fullerton
Bryan, Sean, Colorado State Univ
Bryant, Kathleen E., Illinois State Geological Survey
Bryant, Marita, Bates Coll
Buchanan, Paul C., Kilgore Coll
Buchwald, Caryl E., Carleton Coll
Bucke, David P., Univ of Vermont
Bultman, John, Asheville-Buncombe Technical Comm Coll
Burd, Aurora, Antelope Valley Coll
Burns, Emily, Comm Coll of Rhode Island
Burris, John H., San Juan Coll
Burtis, Erik, Northern Virginia Comm Coll - Woodbridge
Butcher, William, Bowling Green State Univ
Butkos, Darryl J., Suffolk County Comm Coll, Ammerman Campus
Büchel, Georg, Friedrich-Schiller-Univ Jena
Caldwell, Andy, Front Range Comm Coll - Larimer
Caldwell, Marianne O., Hillsborough Comm Coll
Callahan, Caitlin N., Grand Valley State Univ
Cameron, Kevin, Simon Fraser Univ
Canalda, Sabrina, El Paso Comm Coll
Caputi, Samantha , Tarrant County Coll, Southeast Campus
Carpi, Anthony, Graduate School of the City Univ of New York
Carr, Keith W., Illinois State Geological Survey
Cather, Steven, New Mexico Tech
Cervato, Cinzia C., Iowa State Univ of Science & Tech
Chadima, Sarah A., South Dakota Dept of Env and Nat Resources
Chamberlin, Richard M., New Mexico Tech
Chandler, Angela, Arkansas Geological Survey
Charles, Kasanzu, Univ of Dar es Salaam
Chasteen, Hayden R., Tarrant County Coll, Northeast Campus
Cherukupalli, Nehru, Hofstra Univ
Cholnoky, Jennifer, Skidmore Coll
Christensen, Wesley P., South Dakota Dept of Env and Nat Resources
Ciciarelli, John A., Pennsylvania State Univ, Monaca
Clark, Donald, Utah Geological Survey
Clark, Kathryne, Iowa Dept of Nat Resources
Clark, Ken P., Univ of Puget Sound
Clayton, Rodney L., Tidewater Comm Coll
Clendening, Ronald J., Tennessee Geological Survey
Coffman, David, El Centro Coll - Dallas Comm Coll District
Coffman, Stephanie, El Centro Coll - Dallas Comm Coll District
Colak Erol, Serap, Firat Univ
Coleman, Alvin L., Cape Fear Comm Coll
Collins, Laura R., Middle Tennessee State Univ

FACULTY BY SPECIALTY

Colosimo, Amanda, Monroe Comm Coll
Conquy, Xenia, Broward Coll
Cooper, Jonathon L., Carleton Coll
Cornamusini, Gianluca, Univ of Siena
Cornell, Kristie, Univ of Louisiana at Lafayette
Cornell, Sean R., Shippensburg Univ
Cornell , William , Univ of Texas, El Paso
Couroux, Emile, El Paso Comm Coll
Coutand, Isabelle, Dalhousie Univ
Craig, James L., Los Alamos Nat Lab
Crain, John R., Coll of the Sequoias
Crawford, Vernon J., Southern Oregon Univ
Creveling, Jessica, Oregon State Univ
Criswell, James, Cape Fear Comm Coll
Cron, Mitch, Drexel Univ
Cronoble, James M., Metropolitan State Coll of Denver
Cummings, Katie, Utah Geological Survey
Cummings, Michael L., Portland State Univ
Curtis, Lynn A., Broward Coll
Cvetko Tešoviæ, Blanka, Univ of Zagreb
Damir, Buckoviæ, Univ of Zagreb
Davarpanah, Armita, Georgia State Univ
Davidson, Gavin, Bucknell Univ
Davis, David A., Univ of Nevada, Reno
Davis, Timothy, Temple Univ
Dawes, Ralph, Wenatchee Valley Coll
Day, Damon P., Mt. San Antonio Coll
Day, Mackenzie, Univ of California, Los Angeles
Deakin, Joann, Cochise Coll
Dealy, Mike, Univ of Kansas
Dee, Seth, Univ of Nevada, Reno
Deering, Chad, Michigan Technological Univ
Delano, Helen L., DCNR- Pennsylvania Bureau of Geological Survey
Devaney, Kathleen, El Paso Comm Coll
Devlahovich, Vincent A., Coll of the Canyons
DeVries-Zimmerman, Suzanne, Hope Coll
Dhavaskar, Poornima, Goa Univ
Di Fiori, Sara, El Camino Coll
DiFrancesco, Nicholas, SUNY, Buffalo
Dinklage, William, Santa Barbara City Coll
Dittmer, Eric, Southern Oregon Univ
Dodge, Stefanie, Univ of Wisconsin, Madison, Div of Extension
Doelling, Helmut H., Univ of Utah
Donnelly, Thomas W., Binghamton Univ
Dooley, Brett, Patrick Henry Comm Coll
Dougan, Bernie, Whatcom Comm Coll
Drobny, Gerry, Cosumnes River Coll
Dudley, Jutta S., SUNY, The Coll at Brockport
Duncan, Ian, Univ of Texas at Austin, Jackson Schl of Geosciences
Durazzo, Aldo, Indiana Univ Northwest
Dushman, Beth, American River Coll
Dyab, Ahmed I., Alexandria Univ
Earls, Stephanie, Washington Geological Survey
Easterday, Cary, Bellevue Coll
Eaton, Timothy, Graduate School of the City Univ of New York
Ebel, Denton, Graduate School of the City Univ of New York
Edson, Carol, Las Positas Coll
Egbeyemi, Oladimeji R., Federal Univ of Tech, Akure
Eguiluz, Luis, Univ of the Basque Country UPV/EHU
Eide, Elizabeth A., Nat Acad of Sci, Eng, and Med
Eisenhart, Ralph, York Coll of Pennsylvania
El Dakkak, Mohamed W., Alexandria Univ
Ellis, Trevor, Missouri Dept of Nat Resources
Enders, M. Stephen , Colorado School of Mines
Engelhard, Simon, Univ of Rhode Island
England, Richard, Leicester Univ
Erlandsson, Viktor, Univ Leoben
Escartin, Javier, Woods Hole Oceanographic Inst
Escuer, Joan, Universitat Auto de Barcelona
Evans, Lucas, Acadia Univ
Evans, Thomas J., Univ of Wisconsin, Extension
Ewas, Galal A., Alexandria Univ
Faatz, Renee M., Snow Coll
FABRE, Cecile, Universite de Lorraine - Faculte des Sciences et Technologies
Fagnan, Brian A., South Dakota Dept of Env and Nat Resources
Fahrenbach, Mark D., South Dakota Dept of Env and Nat Resources
Farquharson, Phil, MiraCosta Coll
Farrell, Michael, Cuyamaca Coll
Farris, David, East Carolina Univ
Feeney, Dennis M., Univ of Idaho
Fengler, Keegan , Central Washington Univ
Fernandez, Nicole, Cornell Univ
Fields, Chad, Iowa Dept of Nat Resources
Figg, Sean, Palomar Coll
Filippelli, Hailey, DCNR- Pennsylvania Bureau of Geological Survey
Filkorn, Harry, Los Angeles Pierce Coll
Finley, Mark, Heartland Comm Coll
Fitak, Madge R., Ohio Dept of Nat Resources
Fitchett, Rebekah, Northeastern Illinois Univ
Fitz, Thomas J., Northland Coll
Fleck, Michelle C., Utah State Univ Eastern
Flower, Hilary, Saint Petersburg Coll, Clearwater
Flowers Falls, Emily, Washington & Lee Univ
Flynn, John T., Graduate School of the City Univ of New York
Forno, Maria Gabriella, Università di Torino
Forsman, Nels F., Univ of North Dakota
Frederick, Bruce, Johnson County Comm Coll
Fredrick, Kyle, California Univ of Pennsylvania
Freed, Andrew M., Purdue Univ
Frei, Allan, Graduate School of the City Univ of New York
Frei, Michaela, Ludwig-Maximilians-Universitaet Muenchen
Freiburg, Jared, Illinois State Geological Survey
Friedmann, Samuel (Julio) J., Lawrence Livermore Nat Lab
Friedrich, Anke, Ludwig-Maximilians-Universitaet Muenchen
Frieman, Richard, SUNY, Oswego
Fromm, Jeanne M., Univ of South Dakota
Frost, Gina M., San Joaquin Delta Coll
Fuerst, Alice, Metropolitan Comm Coll-Kansas City
Gagnon, Teresa K., Dept of Energy and Enval Protection
Galal, Galal M., Alexandria Univ
Gale, Marjorie H., Agency of Nat Resources, Dept of Enval Conservation
Gannon, J. Michael, Iowa Dept of Nat Resources
García Frank, Alejandra, Univ Complutense de Madrid
Gardner, Eleanor E., Univ of Tennessee, Martin
Gardner, Jamie N., Los Alamos Nat Lab
Garrett, Maureen, Arizona Western Coll
Garside, Larry J., Univ of Nevada, Reno
Garwood, Phil, Cape Fear Comm Coll
Gawloski, Joan, Midland Coll
Gawu, Simon K., Kwame Nkrumah Univ of Science and Tech
Gentry, Amanda L., Weber State Univ
Gianotti, Franco, Università di Torino
Giegengack, Jr., Robert F., Univ of Pennsylvania
Giglierano, James D., Iowa Dept of Nat Resources
Giles, Antony, Midland Coll
Gillman, Joe, Missouri Dept of Nat Resources
Gishlick, Alan D., Bloomsburg Univ
Glynn, William G., SUNY, The Coll at Brockport
González-Juarez, Marco, Universidad Estatal de Sonora
Goodkin, Nathalie , Graduate School of the City Univ of New York
Goodwin, Joshua, California Geological Survey
Gorokhovich, Yuri, Graduate School of the City Univ of New York
Gould, Joseph C., Saint Petersburg Coll, Clearwater
Grace, Cathy A., Univ of Mississippi
Granger, Darryl E., Purdue Univ
Grassman, Jean, Graduate School of the City Univ of New York
Greenaway, Darren, Tarrant County Coll, Southeast Campus
Greene, Mott, Univ of Washington
Griffin, William R., Univ of Texas, Dallas
Griffing, Corinne, Douglas Coll
Gross, Amy, Univ of North Carolina, Pembroke
Grover, Jeffrey A., Cuesta Coll
Grupp, Steve, Everett Comm Coll
Guggino, Steve N., Georgia Southern Univ
Gušiæ, Ivan, Univ of Zagreb
Hagadorn, James W., Denver Museum of Nature & Science
Hakimian, Adina, Hofstra Univ
Hale, Leslie J., Smithsonian Inst / Nat Museum of Nat History
Haley, John C., Virginia Wesleyan Coll
Hall, Clarence A., Univ of California, Los Angeles
Hall, Tracy, Georgia Highlands Coll
Hamdan, Abeer, Phoenix Coll

Hamecher, Emily A., California State Univ, Fullerton
Hammes, Ursula, Univ of Texas at Austin, Jackson Schl of Geosciences
Hams, Jacquelyn E., Los Angeles Valley Coll
Hand, Kristen L., DCNR- Pennsylvania Bureau of Geological Survey
Hand, Linda M., Coll of San Mateo
Hanell, Casey R., Washington Geological Survey
Hanna, Ruth L., Las Positas Coll
Hansen, Cindy, Simon Fraser Univ
Hanson, Gilbert N., Stony Brook Univ
Harder, Vicki, Univ of Texas, El Paso
Hargrave, Phyllis, Montana Tech of The Univ of Montana
Harma, Roberta L., Moorpark Coll
Harone, Imad, Coll of Staten Island/CUNY
Harrell, Michael, Seattle Central Coll
Harris, Ann, Eastern Kentucky Univ
Harris, Michelle, Centralia Coll
Harrison, Linda , Western Michigan Univ
Hart, George, Louisiana State Univ
Hart, Roger M., Comm Coll of Rhode Island
Hartley, Susan, Lake Superior Coll
Haselwander, Airin J., Missouri Dept of Nat Resources
Hauer, Kendall L., Miami Univ
Hayes, Garry F., California State Univ, Stanislaus
Hayes, Garry F., Modesto Junior Coll
Heck, Frederick R., Ferris State Univ
Hegner, Ernst, Ludwig-Maximilians-Universitaet Muenchen
Heidlauf, Lisa, Wheaton Coll
Hein, Kim A., Univ of the Witwatersrand
Helba, Hossam EL-Din A., Alexandria Univ
Heller, Matthew J., Geology and Mineral Resources Program
Helmke, Vicky, West Chester Univ
Hendrey, George, Graduate School of the City Univ of New York
Henley, Barbara D., Mercer Univ
Henry, Kathleen M., Illinois State Geological Survey
Hepburn, J. Christopher, Boston Coll
Herbert, Jennifer, Argonne Nat Lab
Hernandez, Larry, MiraCosta Coll
Herring Mayo, Lisa L., Motlow State Comm Coll
Herzig, Chuck, El Camino Coll
Hetherington, Jean, Diablo Valley Coll
Hickcox, Charles W., Emory Univ
Higgins, Chris T., California Geological Survey
Hill, Catherine R., Arizona Western Coll
Hill, Chris, Grossmont Coll
Hinthorne, James R., Univ of Texas, Rio Grande Valley
Hinz, Nicholas, Univ of Nevada, Reno
Ho, Anita, Flathead Valley Comm Coll
Hobbs, Thomas M., Lonestar Coll - North Harris
Hochstaedter, Alfred, Monterey Peninsula Coll
Hodder, Donald R., SUNY, New Paltz
Hok, Jozef, Comenius Univ in Bratislava
Holail, Hanafy M., Alexandria Univ
Holmes, Stevie L., South Dakota Dept of Env and Nat Resources
Hood, Teresa A., Univ of Miami
Hood, William C., Colorado Mesa Univ
Hopkins, David M., Univ of Alaska, Fairbanks
Hoppe, Andreas, Technische Universitaet Darmstadt
Horn, John, Metropolitan Comm Coll-Kansas City
House, Martha, Pasadena City Coll
Howard, Katie, Olympic Coll
Howard, Kieren, Graduate School of the City Univ of New York
Howes, Mary R., Iowa Dept of Nat Resources
Howley, Michael W., New Hampshire Geological Survey
Hunt, Paula J., West Virginia Geological & Economic Survey
Hural, Kirsten, Dickinson Coll
Hutto, Richard S., Arkansas Geological Survey
Huxta, Stephen, Drexel Univ
Huycke, David, Yakima Valley Coll
Hylland, Michael D., Utah Geological Survey
Hylton, Alisa, Central Piedmont Comm Coll
Hynes, Joanna , California State Univ, Fullerton
Iaccheri, Maria Linda, Ludwig-Maximilians-Universitaet Muenchen
Ianno, Adam J., Juniata Coll
Igonor, Emmanuel E., Federal Univ of Tech, Akure
Ikwuazorm, Okia, Northern Virginia Comm Coll - Loudoun Campus
Irvin, Gene D., Geological Survey of Alabama
Isenor, Fenton M., Cape Breton Univ
Iwaz, Nicole, Cerritos Coll
Jackson, Hiram, Cosumnes River Coll
Jackson, Jeremiah, Missouri Dept of Nat Resources
Jacobs, Jon, Western Univ
Jacobson, William , Univ of Wisconsin, Green Bay
Jaecks, Glenn, American River Coll
Jans, Urs, Graduate School of the City Univ of New York
Janusz, Robert, San Antonio Comm Coll
Jellinek, Mark, Univ of British Columbia
Jensen, Ann R., South Dakota Dept of Env and Nat Resources
Jessee, Anna, Indiana Univ - Purdue Univ Indianapolis
Jiron, Rebecca, William & Mary
Johnson, Darren J., South Dakota Dept of Env and Nat Resources
Johnson, Edward, Coll of Staten Island/CUNY
Johnson, Kurt, Alaska Div of Geological & Geophysical Surveys
Johnson, Ty, Arkansas Geological Survey
Johnston, David, Arkansas Geological Survey
Johnston, Paul L., Emporia State Univ
Jones, Charles E., Univ of Pittsburgh
Jones, Gwyn, Bellevue Coll
Judith, Sausse, Ecole Nate Supérieure de Géologie (ENSG)
Judkins, Gabriel, Univ of Nevada, Las Vegas
Jurena, Dwight, Alamo Colls, San Antonio Coll
Kaandorp, Ron, Vrije Universiteit Amsterdam
Kaldor, Michael, Miami-Dade Coll (Wolfson Campus)
Kane-DeVries, Katelyn, Missouri Dept of Nat Resources
Kapp, Jessica, Univ of Arizona
Karwoski, Todd, Univ of Maryland
Kaye, John M., Mississippi State Univ
Kaygili, Sibel, Firat Univ
Keane, Christopher M., American Geosciences Inst
Keating, Elizabeth, Los Alamos Nat Lab
Keattch, Sharen , Eastern Washington Univ
Keeley, Joshua A., New Hampshire Geological Survey
Kelley, Neil P., Vanderbilt Univ
Kempe, Stephan, Technische Universitaet Darmstadt
Kerwin, Charles M., Keene State Coll
Kerwin, Michael W., Univ of Denver
Kessler, Carsten, Graduate School of the City Univ of New York
Ketcham, Richard A., Univ of Texas, Austin
Khalil Ebeid, Khalil I., Alexandria Univ
Khandaker, Nazul, Graduate School of the City Univ of New York
Kim, Keonho, Midland Coll
King, Jonathan K., Utah Geological Survey
Kirby, Stefan M., Utah Geological Survey
Kirkby, Kent C., Univ of Minnesota, Twin Cities
Klaus, James S., Univ of Miami
Klee, Thomas M., Hillsborough Comm Coll
Klosterman, Sue, Univ of Dayton
Knowlton, Kelly, Northwestern State Univ
Kodosky, Larry, Oakland Comm Coll
Koenig, Brian, Coll of the Desert
Kohrt, Casey, Iowa Dept of Nat Resources
Kolkas, Mosbah, Coll of Staten Island/CUNY
Kollasch, Pete, Iowa Dept of Nat Resources
Koutavas, Athanasios, Graduate School of the City Univ of New York
Kowallis, Bart J., Brigham Young Univ
Kraft, Kaatje, Whatcom Comm Coll
Kraft, Kaatje, Mesa Comm Coll
Krakauer, Nir, Graduate School of the City Univ of New York
Kramer, J. Curtis, Univ of the Pacific
Kramer, Kate, McHenry County Coll
Kramer, Walter V., Del Mar Coll
Kranz, Dwight S., Houston Comm Coll System
Krohe, Nicholas J., South Dakota Dept of Env and Nat Resources
Kroon, Dick, Edinburgh Univ
Kruger, Ned , North Dakota Geological Survey
Kuehner, Scott, Univ of Washington
Kuhlman, Robert, Montgomery County Comm Coll
Kuhnhenn, Gary L., Eastern Kentucky Univ
Kukoè, Duje, Univ of Zagreb
Kumpf, Amber C., Muskegon Comm Coll
Kuntz, Mark R., Elgin Comm Coll
Kutis, Michael, Ball State Univ
Kwicklis, Edward M., Los Alamos Nat Lab

Lacy, Tor, Cerritos Coll
Lambert-Smith, James, Kingston Univ
Lammerer, Bernd L., Ludwig-Maximilians-Universitaet Muenchen
Lancaster, Jeremy, California Geological Survey
Langel, Richard A., Iowa Dept of Nat Resources
Langille, Amanda, Memorial Univ of Newfoundland
LaPointe, Daphne D., Univ of Nevada, Reno
Larter, Stephen, Univ of Newcastle Upon Tyne
Lauziere, Kathleen, Universite du Quebec
Leite, Michael B., Chadron State Coll
Lemay, Phillip W., Louisiana State Univ
Leszczynski, Raymond F., Cayuga Comm Coll
Levy, Melissa H., American River Coll
Lewis, Chris, City Coll of San Francisco
Lewis, Mary, Contra Costa Coll
Lewis, Reed S., Univ of Idaho
Lewis, Tammy, Graduate School of the City Univ of New York
Liauw, Henri L., Broward Coll
Libra, Robert D., Iowa Dept of Nat Resources
Liebling, Richard, Hofstra Univ
Lippelt, Irene D., Univ of Wisconsin, Madison, Div of Extension
Liu, Paul Hiaibao, Iowa Dept of Nat Resources
Lombard, Armand J., Mesa Comm Coll
Long, Colleen, Illinois State Geological Survey
Lonn, Jeff, Montana Tech of The Univ of Montana
Lovett, Cole, Lake Michigan Coll
Lu, Yuehan, Univ of Alabama
Ludman, Allan, Graduate School of the City Univ of New York
Ludman, Allan, Queens Coll (CUNY)
Lueth, Virgil W., New Mexico Tech
Luna, Daniel, Boise State Univ
Lužar-Oberiter, Borna, Univ of Zagreb
Lyle, Mike, Tidewater Comm Coll
Lyle, Mitchell, Oregon State Univ
Lyman, John C., Bakersfield Coll
Mabee, Stephen B., Univ of Massachusetts, Amherst
Macdonald, Francis, Univ of California, Santa Barbara
Macias, Steve E., Olympic Coll
MacLachlan, James, Metropolitan State Coll of Denver
Magee, Robert, Virginia Wesleyan Coll
Mahaffee, Tina, Middle Georgia Coll
Mahmoud, Sara A., Alexandria Univ
Majodina, Thando, Tshwane Univ of Tech
Major, Penni, Lonestar Coll - North Harris
Major, Ruth H., Hudson Valley Comm Coll
Malone, Shawn, Univ of Wisconsin, Green Bay
Mandziuk, William S., Univ of Manitoba
Mansour, Ahmed S., Alexandria Univ
Mapholi, Thendo, Univ of the Free State
Marchand, Gerard, Mount Holyoke Coll
Marchisin, John, New Jersey City Univ
Marco, Shmuel, Tel Aviv Univ
Marco, Shmulik, Tel Aviv Univ
Marshall, Thomas R., South Dakota Dept of Env and Nat Resources
Martin, Gale D., Coll of Southern Nevada - West Charleston Campus
Martinez Torres, Luis Miguel, Univ of the Basque Country UPV/EHU
Martinuš, Maja, Univ of Zagreb
Martz, Todd, Univ of California, San Diego
Marvinney, Robert G., Dept of Agriculture, Conservation, and Forestry
Massey, Matthew, Univ of Kentucky
Mattox, Tari, Grand Rapids Comm Coll
Matyjasik, Basia, Utah Geological Survey
May, S J., Collin Coll - Preston Ridge Campus
Mayer, Larry, Woods Hole Oceanographic Inst
McCall, Rosemary, Coll of Staten Island/CUNY
McCauley, Steven, El Centro Coll - Dallas Comm Coll District
McClaughry, Jason, Oregon Dept of Geology and Mineral Industries
McCoy, Floyd W., Windward Comm Coll
McCulloh, Richard P., Louisiana State Univ
McCutchen, William T., Univ of Tennessee, Martin
McDermott, Frank, Univ Coll Dublin
McDowell, Robin J., Georgia State Univ, Perimeter Coll, Dunwoody Campus
McDowell, Ronald, Fairmont State Univ
McIlrath, Judy, Univ of South Florida
McIlroy, Duncan, Memorial Univ of Newfoundland
McKay, Robert M., Iowa Dept of Nat Resources
McKinney, Michael, California Geological Survey
McMonagle, Julie, Wilkes Univ
Meave, Edgardo, Universidad Nacional Auto de Mexico
Meister, Paul A., Illinois State Univ
Melcher, Frank, Univ Leoben
Meldahl, Keith H., MiraCosta Coll
Meléndez, Nieves, Univ Complutense de Madrid
Metz, Cheyl L., Blinn Coll
Metzler, Christopher V., MiraCosta Coll
Mey, Jacob, Graduate School of the City Univ of New York
Meyer, Jeffrey W., Santa Barbara City Coll
Michel, Suzanne, Cuyamaca Coll
Millan, Christina, Ohio State Univ
Millen, Timothy M., Elgin Comm Coll
Miller, Steven F., Argonne Nat Lab
Miller-Hicks, Bryan, Cuyamaca Coll
Milner, Lloyd R., Louisiana State Univ
Milner, Lloyd R., Louisiana State Univ
Miner, James J., Illinois State Geological Survey
Minium, Deborah, Bellevue Coll
Mitzman, Stephanie, Tarrant County Coll, Southeast Campus
Moll, Nancy E., Coll of the Desert
Monet, Julie, California State Univ, Chico
Monson, Jessica, Illinois State Geological Survey
Montayne, Simone, Alaska Div of Geological & Geophysical Surveys
Montwill, Gail F., Santiago Canyon Coll
Mora, Claudia, Univ of Texas, Austin
Morealli, Sarah A., Univ of Mary Washington
Morgan, Matt, Colorado Geological Survey
Morris, Billy, Georgia Highlands Coll
Moshary, Fred , Graduate School of the City Univ of New York
Mosteller, Joey D., Appalachian State Univ
Mshiu, Elisante E., Univ of Dar es Salaam
Muhlbauer, Jason , Univ of Tennessee, Chattanooga
Munn, Barbara J., Sacramento State Univ
Murphy, Cindy, Saint Francis Xavier Univ
Murphy, Dan, Asheville-Buncombe Technical Comm Coll
Muskatt, Herman, Utica Coll
Mutti, Laurel, SUNY, New Paltz
Nagorski, Sonia, Univ of Alaska, Southeast
Nagy-Shadman, Elizabeth, Pasadena City Coll
Nance, Seay, Univ of Texas at Austin, Jackson Schl of Geosciences
Neboga, Victoria V., DCNR- Pennsylvania Bureau of Geological Survey
Nelson, Jennifer, Indiana Univ - Purdue Univ Indianapolis
Newman, Brent D., Los Alamos Nat Lab
Newman, Jamie, American Museum of Nat History
Ni-Meister, Wenge, Graduate School of the City Univ of New York
Nielsen, Gregory B., Weber State Univ
Norouzi, Hamidreza, Graduate School of the City Univ of New York
NUDE, Prosper M., Univ of Ghana
Nugent, Barnes, Fairmont State Univ
O'Brien, Lawrence E., Orange County Comm Coll
O'Brien, Philip, Alvin Comm Coll
O'Keeffe, Mike, Colorado Geological Survey
Oakes-Miller, Hollie, Portland Comm Coll - Sylvania Campus
Odendaal, Adriaan, Univ of the Free State
Odeyemi, Idowu B., Federal Univ of Tech, Akure
Oldfield, Bruce K., Broome Comm Coll
Oldham, Richard L., American River Coll
Opperman, William, Broward Coll
Orr, William , Portland State Univ
Oswald, Peter J., Univ of Alaska, Anchorage
Pande, Srikant k., Pt. Ravishankar Shukla Univ
Pantoja-Alor, Jerjes, Universidad Nacional Auto de Mexico
Papp, Kenneth R., Alaska Div of Geological & Geophysical Surveys
Parish, Cynthia L., Lamar Univ
Parnella, Bill, Univ of Delaware
Parrick, Brittany, Ohio Dept of Nat Resources
Parsons, Michael, Dalhousie Univ
Paschert, Karin, Ludwig-Maximilians-Universitaet Muenchen
Patterson, Gary, Univ of Memphis
Patton, Jason A., Arkansas Tech Univ
Patton, Terri, Argonne Nat Lab
Patwardhan, Kaustubh, SUNY, New Paltz
Pawloski, Gayle A., Lawrence Livermore Nat Lab
Payeur, Mason, South Carolina Dept of Nat Resources

Pearthree, Kristin, New Mexico Tech
Pellowski, Christopher J., South Dakota School of Mines & Tech
Perez, Adriana, Univ of Texas, El Paso
Perez, Adriana, El Paso Comm Coll
Perkis, Bill, Gogebic Comm Coll
Peters, Roger M., Univ of Wisconsin, Madison, Div of Extension
Peterson, Joseph E., Elgin Comm Coll
Phillips, Michael, Illinois Valley Comm Coll
Piasecki, Michael, Graduate School of the City Univ of New York
Pierce, David, Lakeland Comm Coll
Pirie, Diane H., Florida InterNat Univ
Poole, T. Craig, Fresno City Coll
Popa, Mirela Violetta, Babes-Bolyai Univ
Posiloviæ, Hrvoje, Univ of Zagreb
Pound, Kate S., Saint Cloud State Univ
Powers, Elizabeth, California State Univ, Bakersfield
Prave, Tony, Univ of St. Andrews
Prewett, Jerry L., Missouri Dept of Nat Resources
Price, Jonathan G., Univ of Nevada, Reno
Price, L G., New Mexico Tech
Prichystal, Antonín, Masaryk Univ
Priesendorf, Carl, Metropolitan Comm Coll-Kansas City
Priest, George R., Oregon Dept of Geology and Mineral Industries
Priewisch, Alexandra, Fresno City Coll
Prior, William L., Arkansas Geological Survey
Prothero, Donald, Los Angeles Pierce Coll
Proudhon, Benoit, Institut Polytechnique LaSalle Beauvais (ex-IGAL)
Prytulak, Julie, Imperial Coll
Quick, Thomas J., Univ of Akron
Quinn, Heather A., Maryland Department of Nat Resources
Ragland, Deborah, Univ of New Mexico, Taos
Rahman, Ata U., Austin Comm Coll District
Rains, Daniel S., Arkansas Geological Survey
Ranhofer, Melissa, Furman Univ
Rantitsch, Gerd, Univ Leoben
Rashall, Jenny M., Stephen F. Austin State Univ
Rashed, Mohamed A., Alexandria Univ
Rath, Carolyn, California State Univ, Fullerton
Rathburn, Sara L., Colorado State Univ
Ratschbacher, Barbara C., Univ of California, Davis
Rawling, Geoffrey, New Mexico Tech
Ray, Waverly, Cuyamaca Coll
Raymer, Janet, Metropolitan Comm Coll-Kansas City
Read, Adam S., New Mexico Tech
Redden, Marcella, Geological Survey of Alabama
Reesman, Arthur L., Vanderbilt Univ
Reif, Samantha, Lincoln Land Comm Coll
Reinmann, Andrew , Graduate School of the City Univ of New York
Rennie, Colette, Saint Francis Xavier Univ
Repka, James, Saddleback Comm Coll
Ricchezza, Victor J., Georgia State Univ, Perimeter Coll, Clarkston Campus
Richards, Bill, North Idaho Coll
Riegel, Hannah B., Appalachian State Univ
Riker-Coleman, Kristin E., Univ of Wisconsin, Superior
Rinae, Makhadi, Univ of the Free State
Riordan, Jean, Alaska Div of Geological & Geophysical Surveys
Rivera, Mark, Univ of Alaska, Anchorage
Roberson, Philip, Tennessee Tech Univ
Roberson, Randal P., Austin Peay State Univ
Robinson, Marshall, Utah Geological Survey
Robinson, Sarah, United States Air Force Acad
Rocha, Guillermo, Brooklyn Coll (CUNY)
Roche, James E., Louisiana State Univ
Rodríguez, Marta, Univ Complutense de Madrid
Rodriguez-Castaneda, Jose L., Universidad Nacional Auto de Mexico
Rohbaugh, Robert, El Paso Comm Coll
Ross, Theodore W., Lawrence Univ
Rossbach, Thomas, Indiana Univ - Purdue Univ Indianapolis
Rothemund, Kirk, El Paso Comm Coll
Rounds, Steven W., Sacramento State Univ
Rowden, Robert, Iowa Dept of Nat Resources
Ruffel, Alice, El Centro Coll - Dallas Comm Coll District
Rumrill, Julie, Southern Connecticut State Univ
Ruppert, Kelly R., California State Univ, Fullerton
Rutberg, Randye, Graduate School of the City Univ of New York
Rygel, Michael C., Indiana Univ, Bloomington
Sacramentogrilo, Isabelle, San Diego State Univ
Sadowski, Andrew, Central Washington Univ
Salle, Bethan, El Centro Coll - Dallas Comm Coll District
Salmun, Haydee, Graduate School of the City Univ of New York
Samimi, Naser, Univ of Tabriz
Sassi, Raffaele, Università degli Studi di Padova
Sato, Yoko, Montclair State Univ
Schafer, Carl M., Macomb Comm Coll, Center Campus
Scheffler, Joanna, Mesa Comm Coll
Schenck, William S., Univ of Delaware
Schilling, Keith, Iowa Dept of Nat Resources
Schimmrich, Steven, SUNY, Ulster County Comm Coll
Schmidt, Bennetta, Lamar Univ
Schneider, Blair, Univ of Kansas
Schulingkamp, Arren, Louisiana State Univ
Schulte, Kimberly D., Georgia State Univ, Perimeter Coll, Online
Schultz, Jan, Santa Barbara City Coll
Sediek, Kadry N., Alexandria Univ
Seigley, Lynette S., Iowa Dept of Nat Resources
Seremet, Meghan M., Dept of Energy and Enval Protection
Severs, Matthew , Stockton Univ
Shaaban, Mohamad N., Alexandria Univ
Shafer, Erik, Portland State Univ
Shakun, Jeremy D., Boston Coll
Shalimba, Ester, Univ of Namibia
Shannon, Jeremy, Michigan Technological Univ
shedied, ahmad g., Fayoum Univ
Shields, Stephen, Angelo State Univ
Shiels, Christine, Douglas Coll
Shimizu, Melinda, Mesa Comm Coll
Shinn, Eugene, Univ of South Florida
Shorey, Christian V., Colorado School of Mines
Shroba, Cynthia S., Coll of Southern Nevada - West Charleston Campus
Sibeko, Skhumbuzo, Tshwane Univ of Tech
Sicard, Karri, Alaska Div of Geological & Geophysical Surveys
Siegelberg, Alan, Long Island Univ, Brooklyn Campus
Siemens, Michael A., Missouri Dept of Nat Resources
Sipola, Maija, Miami Univ
Skinner, Randall, Brigham Young Univ
Sloan, Heather , Graduate School of the City Univ of New York
Smaglik, Suzanne M., Central Wyoming Coll
Smith, Jason J., Broome Comm Coll
Smith, Jon J., Univ of Kansas
Smith, Mike, Front Range Comm Coll - Larimer
Smithson, Jayne, Contra Costa Coll
Snyder, Daniel, Middle Georgia Coll
Solferino, Giulio, Royal Holloway Univ of London
Solis, Michael P., Ohio Dept of Nat Resources
Sorenson, Adrienne, Central Washington Univ
Sori, Michael, Purdue Univ
Speed, Don, Phoenix Coll
Sperone, Felice, Wayne State Univ
Stakes, Debra, Cuesta Coll
Stanton, Kathryn, Sacramento City Coll
Stanton, Kelsay, Wenatchee Valley Coll
Steck, Lee, Los Alamos Nat Lab
Steffens, Katja, Ludwig-Maximilians-Universitaet Muenchen
Steinberg, Roger T., Del Mar Coll
Stewart, Eric, Univ of Wisconsin, Madison, Div of Extension
Stewart, Esther K., Univ of Wisconsin, Madison, Div of Extension
Straight, William, Northern Virginia Comm Coll - Loudoun Campus
Straub, Kyle M., Tulane Univ
Summers, Sara, Weber State Univ
Sumrall, Jonathan B., Fort Hays State Univ
Sundell, Ander, Coll of Western Idaho
Sundell, Kent A., Casper Coll
Sunderlin, David, Lafayette Coll
Suszek, Thomas J., Univ of Wisconsin, Oshkosh
Szczepanski, Jacek, Univ of Wroclaw
Söllner, Frank, Ludwig-Maximilians-Universitaet Muenchen
Tajik, Atieh, Georgia State Univ
Tamish, Mohamed M., Alexandria Univ
Tassier-Surine, Stephanie, Iowa Dept of Nat Resources
Tavarnelli, Enrico, Univ of Siena
Taylor, Carolyn, Mesa Comm Coll
Taylor, Penny M., Mount Holyoke Coll

Teklay, Mengist, Georgia State Univ, Perimeter Coll, Decatur Campus
Tetrault, Denis, Univ of Windsor
Thole, Jeffrey T., Macalester Coll
Thomas, Margaret A., Dept of Energy and Enval Protection
Thompson, Todd, Indiana Univ, Bloomington
Thorleifson, Harvey, Univ of Minnesota, Twin Cities
Thornberry-Ehrlich, Trista L., Colorado State Univ
Thul, David, Univ of Utah
Tibbits, Matthew, Chadron State Coll
Tibbits, Tawny, Chadron State Coll
Timmons, J. Michael, New Mexico Tech
Tolley, William, Southern Connecticut State Univ
Tomiæ, Vladimir, Univ of Zagreb
Toni, Rousine T., Alexandria Univ
Towery, Brooke L., Pensacola State Coll
Tucker, Eva, Pennsylvania State Univ, Erie
Turbeville, John, MiraCosta Coll
Turner, Wesley L., Stephen F. Austin State Univ
Turner, III, Henry, Univ of Arkansas, Fayetteville
Tvelia, Sean, Suffolk County Comm Coll, Ammerman Campus
Ugland, Richard, Southern Oregon Univ
Ullrich, Alexander D., Georgia State Univ, Perimeter Coll, Dunwoody Campus
Urquhart, Joanne, Bowdoin Coll
Valenti, Christine D., Georgia State Univ, Perimeter Coll, Online
Valenti, Christine, Montclair State Univ
van Balen, Ronald R., Vrije Universiteit Amsterdam
Van Brocklin, Matthew F., St. Lawrence Univ
VanDorpe, Paul E., Iowa Dept of Nat Resources
Vickery, Nancy, Univ of New England
Vidoviæ, Jelena, Univ of Zagreb
Viens, Rob, Bellevue Coll
Vierrether, Chris, Missouri Dept of Nat Resources
Vig, Pradeep K., Kaskaskia Coll
Voigt, Vicki, Missouri Dept of Nat Resources
Vuke, Susan M., Montana Tech of The Univ of Montana
Vuori, Saku, Geological Survey of Finland
Wacker, Michael, Florida InterNat Univ
Wadleigh, Hank, Cypress Coll
Waggoner, Karen, Midland Coll
Wakefield, Kelli, Mesa Comm Coll
Wang, Zhengrong, Graduate School of the City Univ of New York
Waszgis, Michele M., Univ of Nebraska, Lincoln
Waugh, John, Tidewater Comm Coll
Weatherford, Jonathan, Portland Comm Coll - Sylvania Campus
Webber, Jeffrey, Stockton Univ
Weber, Diane, Illinois Central Coll
Weber-Diefenbach, Klaus, Ludwig-Maximilians-Universitaet Muenchen
Weborg-Benson, Kimberly, SUNY, Fredonia
Weeden, Lori, Univ of Massachusetts, Lowell
Wein, Agatha, Cuyamaca Coll
Werhner, Matthew J., Hillsborough Comm Coll
Wernette, Shelly J., Texas State Univ
West, Robert, East Los Angeles Coll
Wheeler, Richard F., Austin Peay State Univ
White, Paul, Comm Coll of Rhode Island
Whitehead, James, Univ of New Brunswick
Whitehead, Peter W., James Cook Univ
Whitehill, Matthew, Lake Superior Coll
Whittington, Carla, Highline Coll
Wilbur, Bryan, Pasadena City Coll
Wilcots, Julia, Princeton Univ
Willahan, Duane, Gavilan Coll
Williams, Curtis J., Cypress Coll
Willis, Grant C., Utah Geological Survey
Willis, Marc, Fullerton Coll
Willsey, Shawn P., Coll of Southern Idaho
Wilson, Jeffrey A., Univ of Michigan
Wolfe, Ben, Metropolitan Comm Coll-Kansas City
Wolkersdorfer, Christian, Ludwig-Maximilians-Universitaet Muenchen
Wolter, Calvin, Iowa Dept of Nat Resources
Woodall, Debra W., Daytona State Coll
Worcester, Peter, Eastern Kentucky Univ
Wortmann, Ulrich B., Univ of Toronto
Wypych, Alicja, Alaska Div of Geological & Geophysical Surveys
Yalcin, Kaplan, Oregon State Univ
Yalcin, Rebecca, Oregon State Univ
Yan, Y E., Argonne Nat Lab
Yoshida, Glenn, Los Angeles Southwest Coll
Young, Mike, Dalhousie Univ
Zabel, Garrett E., Colorado Mountain Coll
Zabielski, Victor, Northern Virginia Comm Coll - Alexandria
Zachariasen, Judy, California Geological Survey
Zanetti, Kathleen, Univ of Nevada, Las Vegas
Zayac, John M., Los Angeles Pierce Coll
Zeitlhöfler, Matthias, Ludwig-Maximilians-Universitaet Muenchen
Zelaya, Zach, South Carolina Dept of Nat Resources
Zentner, Nick, Central Washington Univ
Zimmer, Brian W., Appalachian State Univ
Zurawski, Ronald P., Tennessee Geological Survey

Archaeological Geology

Abreu, Maria E., Universidade de Trás-os-Montes e Alto Douro
Adams, Kenneth, Univ of Nevada, Reno
Adovasio, James M., Mercyhurst Univ
Ballard, Robert D., Univ of Rhode Island
Bowers, Peter, Univ of Alaska, Fairbanks
Buckley, Michael, Univ of Manchester
Chamberlain, Andrew T., Univ of Manchester
D'Alpoim Guedes, Jade, Univ of California, San Diego
Degryse, Patrick, Katholieke Universiteit Leuven
Dunn, Richard K., Norwich Univ
Dye, David H., Univ of Memphis
Elkins, Nichole, Bowling Green State Univ
Fadem, Cynthia M., Earlham Coll
Farrand, William, Univ of Michigan
Ferguson, Terry A., Wofford Coll
Ferring, C. Reid, Univ of North Texas
Garrison, Ervan G., Univ of Georgia
Greenlee, Diana M., Univ of Louisiana, Monroe
Gundersen, James N., Wichita State Univ
Gunia, Piotr, Univ of Wroclaw
Haynes, Jr., C. Vance, Univ of Arizona
Hayward, Chris L., Edinburgh Univ
Holliday, Vance T., Univ of Arizona
Holliday, Vance, Univ of Arizona
Iglesias, Cruz, Coruna Univ
Jones, Jeri L., York Coll of Pennsylvania
Kelley, Alice R., Univ of Maine
Kvamme, Kenneth L., Boston Univ
Lee-Gorishti, Yolanda, Southern Connecticut State Univ
Madsen, David B., Utah Geological Survey
Mandel, Rolfe D., Univ of Kansas
Mandel, Rolfe, Univ of Kansas
Manning, Sturt W., Cornell Univ
Mason, Owen, Univ of Alaska, Fairbanks
Mickelson, Andrew M., Univ of Memphis
Milne, Brooke, Univ of Manitoba
Nagaoka, Lisa A., Univ of North Texas
Ortmann, Anthony L., Murray State Univ
Page, David, Desert Research Inst
Parish, Ryan M., Univ of Memphis
Peros, Matthew, McMaster Univ
Pike, Scott, Willamette Univ
Pope, Richard, Univ of Derby
Pournelle, Jennifer R., Univ of South Carolina
Powell, Wayne G., Brooklyn Coll (CUNY)
Reitz, Elizabeth J., Univ of Georgia
Rivera-Collazo, Isabel, Univ of California, San Diego
Rogers, Joe D., West Texas A&M Univ
Sallu, Susannah, Univ of Leeds
Schiffman, Robert A., Bakersfield Coll
Smith, Jen R., Washington Univ in St. Louis
Smith, Jennifer R., Washington Univ in St. Louis
Stafford, C. Russell, Indiana State Univ
Swyrtek, Sheila, Charles Stewart Mott Comm Coll
Thieme, Donald, Valdosta State Univ
Tomaso, Matthew S., Montclair State Univ
Towner, Ronald, Univ of Arizona
Uribelarrea, David, Univ Complutense de Madrid
Venter, Marcie, Murray State Univ
Vento, Frank, Mercyhurst Univ
Waters, Michael R., Texas A&M Univ

Whitney , Sandra , Univ of Georgia
Wilson, Lucy A., Univ of New Brunswick Saint John
Wilson, Michael C., Douglas Coll
Wolverton, Steve, Univ of North Texas
Wygal, Brian, Adelphi Univ

Environmental Geology

Abbasnejad, Ahmad, Shahid Bahonar Univ of Kerman
Aden, Douglas J., Ohio Dept of Nat Resources
Adentunji, Jacob, Univ of Derby
Adewoye, Abosede O., Ladoke Akintola Univ of Tech
Adomaitis, D., Illinois State Geological Survey
Akenzua-Adamcyzk, Aiyevbekpen H., Univ of Benin
Albee-Scott, Steven R., Jackson Coll
Amezaga, Jamie, Univ of Newcastle Upon Tyne
Apotsos, Alex, Williams Coll
Argyilan, Erin, Indiana Univ Northwest
Ashwood, Tom, Oak Ridge Nat Lab
Asowata, Timothy I., Federal Univ of Tech, Akure
Baker-Treloar, Elizabeth, Univ of San Diego
Bang, John, North Carolina Central Univ
Barber, Donald C., Bryn Mawr Coll
Barone, Jessica, Monroe Comm Coll
Barrett, John, Univ of Leeds
Bartholemew, Paul, Univ of New Haven
Bartolucci, Valerio, Broward Coll
Beccue, Clint, Illinois State Geological Survey
Bechtel, Timothy D., Franklin and Marshall Coll
Beebe, Alex, Univ of South Alabama
Bell, Margaret C., Univ of Newcastle Upon Tyne
Bernstein, Neil P., Univ of Iowa
Bhattacharya, Tripti, Syracuse Univ
Bilanovic, Dragoljub D., Bemidji State Univ
Binda, Pier L., Univ of Regina
Bircher, Harry, Youngstown State Univ
Bogart, James M., Dept of Energy and Enval Protection
Bogle, Mary Anna, Oak Ridge Nat Lab
Bond, Alan, Univ of East Anglia
Bousenberry, Raymond T., New Jersey Geological and Water Survey
Brake, Sandra S., Indiana State Univ
Brandon, Nigel, Imperial Coll
Brandt, Craig, Oak Ridge Nat Lab
Branfireun, Brian, Western Univ
Brodie, Gregory, Univ of Tennessee, Chattanooga
Bromily, Geoffrey, Edinburgh Univ
Brown, Huntting (Hunt), Wright State Univ
Brown, Kerry, Kingston Univ
Brown, Thomas W., Austin Comm Coll District
Bryan, Mark, Wayland Baptist Univ
Budkewitsch, Paul, Brock Univ
Buliga, Iuliana, Alexandru Ioan Cuza
Burke, Ian, Univ of Leeds
Cadet, Eddy, Utah Valley Univ
Callison, James, Utah Valley Univ
Carey, Jordan , Central Washington Univ
Cathcart, Eric, Univ of San Diego
Chappell, James R., Colorado State Univ
Chaput, Julien, Univ of Texas, El Paso
Chaudhuri, Sambhudas, Kansas State Univ
Chernoff, Barry, Wesleyan Univ
Cheun, Norman, Kingston Univ
Chhetri, Parveen, California State Univ, Dominguez Hills
Chilvers, Jason, Univ of East Anglia
Chipman, Melissa, Syracuse Univ
Cloutier, Danielle, Universite Laval
Clutter, Melissa, Fort Lewis Coll
Cobb, Kim, Brown Univ
Colegial Gutierrez, Juan D., Universidad Industrial de Santander
Collier, Mark, Illinois State Geological Survey
Collins, Edward W., Univ of Texas at Austin, Jackson Schl of Geosciences
Congleton, John D., Univ of West Georgia
Constantopoulos, James T., Eastern New Mexico Univ
Cook, Hadrian, Kingston Univ
Corley, John H., Missouri Dept of Nat Resources
Cowart, Richard, Coastal Bend Coll
Cox, Christena, Ohio State Univ
Coxon, Catherine, Trinity Coll
Cummins, R. Hays, Miami Univ
Curry, B. B., Illinois State Geological Survey
Curry, B. B., Northern Illinois Univ
Curry, Gordon, Univ of Glasgow
Curtis, Arthur C., Oak Ridge Nat Lab
Dallimer, Martin, Univ of Leeds
Daramola, Sunday O., Federal Univ of Tech, Akure
Davis, R. Laurence, Univ of New Haven
Davis, Steven J., Univ of California, Irvine
Day, Brett, Univ of East Anglia
Denniston, Rhawn F., Cornell Coll
deWet, Andrew P., Franklin and Marshall Coll
Di Lorenzo, Emanuele, Brown Univ
Dimmick, Charles W., Central Connecticut State Univ
Diochon, Amanda, Lakehead Univ
Downie, Helen, Univ of Manchester
Dubey, C S., Sri Sri Univ
Duncan, Ian J., Univ of Texas, Austin
Durrant, Jeffrey O., Brigham Young Univ
Dyer, Jen, Univ of Leeds
Edwards, Robin, Trinity Coll
Elliott, Susan J., McMaster Univ
Ellis, Scott R., Illinois State Geological Survey
Ellison, Anne L., Illinois State Geological Survey
Ellison, Anne L., Univ of Illinois
Ettlie, Bradley, Illinois State Geological Survey
Eyles, John D., McMaster Univ
Faulkner, Melinda S., Stephen F. Austin State Univ
Favas, Paulo J., Universidade de Trás-os-Montes e Alto Douro
Field, Cathryn K., Bowdoin Coll
Flesken, Luuk, Univ of Leeds
Foley, Duncan, Pacific Lutheran Univ
Foret, Jim, Univ of Louisiana at Lafayette
Foxon, Tim, Univ of Leeds
Franco, Aldina, Univ of East Anglia
Frappier, Amy, Skidmore Coll
Galicki, Stan, Millsaps Coll
Gallen, Sean, Colorado State Univ
Garrison, Trent, Northern Kentucky Univ
Garstang, Mimi R., Missouri Dept of Nat Resources
Geary, Lindsey, Utica Coll
Geiger, James W., Illinois State Geological Survey
Gerald, Carresse, North Carolina Central Univ
Geršlová, Eva, Masaryk Univ
Gidigasu, Solomon S., Kwame Nkrumah Univ of Science and Tech
Glenn, Craig R., Univ of Hawai'i, Manoa
Goetz, Heinrich, Collin Coll - Preston Ridge Campus
Gomes, Nuno N., Universidade Independente de Angola
Gomezdelcampo, Enrique, Bowling Green State Univ
Gouldson, Andy, Univ of Leeds
Graham, Margaret C., Edinburgh Univ
Grant, Alastair, Univ of East Anglia
Halfman, John D., Hobart & William Smith Colls
Hanes, Daniel M., Saint Louis Univ
Hargreaves, Tom, Univ of East Anglia
Harris, Ann G., Youngstown State Univ
Harris, C, Cardiff Univ
Harvey, Omar R., Texas Christian Univ
Havenith, Hans-Balder, Universite de Liege
Hawkins, Terry, Missouri Dept of Nat Resources
Hawley, John W., New Mexico Tech
Hawley, John W., New Mexico Tech
Helgers, Karen, SUNY, Ulster County Comm Coll
Hickey, James, Northwest Missouri State Univ
Hill, Nicole, Bentley Univ
Hill, Timothy, Univ of St. Andrews
Hodell, David, Univ of Cambridge
Holmes, George, Univ of Leeds
Holwell, Dave, Leicester Univ
Hoppie, Bryce W., Minnesota State Univ
Hubeny, J B., Salem State Univ
Hudson-Edwards, Karen, Birkbeck Coll
Humphreys, Robin, Coll of Charleston
Ibarra, Daniel, Brown Univ
Isaacson, Carl, Bemidji State Univ

Jackman, Toni K., Wichita State Univ
Jacobs, Alan M., Youngstown State Univ
Jamieson, Heather E., Queen's Univ
Jin, Lixin, Univ of Texas, El Paso
Jones, T, Cardiff Univ
Jordan, Andy, Univ of East Anglia
Jovanovic, Vladimir, Coll of Staten Island/CUNY
Kairies-Beatty, Candace L., Winona State Univ
Keller, John E., Coll of Southern Nevada - West Charleston Campus
Kertes, Randy, Rider Univ
Ketelle, Richard H., Oak Ridge Nat Lab
Kientop, Greg A., Illinois State Geological Survey
Kierczak, Jakub, Univ of Wroclaw
Kim, Jonathan, Agency of Nat Resources, Dept of Enval Conservation
Kirchner, James W., Univ of California, Berkeley
Kolawole, Lanre L., Ladoke Akintola Univ of Tech
Krekeler, Mark, Miami Univ
Kretzschmar, Thomas, Centro de Inv Científica y de Ed Superior de Ensenada
Lake, Iain, Univ of East Anglia
Last, George V., Pacific Northwest Nat Lab
Lawry-Berkins, Cynthia, Lonestar Coll - Montgomery
Lee, Arthur C., Roane State Comm Coll - Oak Ridge
Lee, Cindy M., Clemson Univ
Lee-Cullin, Joe, Albion Coll
Lene, Gene W., Saint Mary's Univ
Leventon, Julia, Univ of Leeds
Lloyd, Jonathan, Univ of Manchester
Love, David W., New Mexico Tech
Macadam, John, Exeter Univ
Maher, Damien, Southern Cross Univ
Manser, Nathan, Michigan Technological Univ
Margalef, Ona, Universitat Auto de Barcelona
Martin, Scott C., Youngstown State Univ
Masciocco, Luciano, Università di Torino
Matthews, Robert A., Univ of California, Davis
May, Michael, Western Kentucky Univ
Mayer, Margaret, Dine' Coll
McCullough, Jr., Edgar J., Univ of Arizona
McDonald, Brenna, Missouri Dept of Nat Resources
McGivern, Tiffany, Utica Coll
McKinney, Michael L., Univ of Tennessee, Knoxville
Metzger, Marc J., Edinburgh Univ
Meyer, Brian, Georgia State Univ
Meyer, W. Craig, Los Angeles Pierce Coll
Miller, David S., Argonne Nat Lab
Mitre-Salazar, Luis M., Universidad Nacional Auto de Mexico
Moncur, Michael, Univ of Saskatchewan
Monroy-Sanchez, Jaime I., Universidad Auto de Baja California Sur
Moore, Kathryn, Exeter Univ
Morse, Linda D., William & Mary
Mouat, David A., Desert Research Inst
Muldoon, Maureen, Univ of Wisconsin, Madison, Div of Extension
Nelwamondo, T. M., Univ of Venda
Newbold, K. B., McMaster Univ
Newton, Rob, Univ of Leeds
Null, E. Jan, San Francisco State Univ
O'Connor, Yuet-Ling, Pasadena City Coll
Omotoso, O. A., Univ of Ilorin
Orndorff, Richard L., Eastern Washington Univ
Oyawale, A. A., Obafemi Awolowo Univ
Paavola, Jouni, Univ of Leeds
Paltseva, Anna, Univ of Louisiana at Lafayette
Panko, Andrew W., Brock Univ
Patenaude, Genevieve, Edinburgh Univ
Pearce, Jamie R., Edinburgh Univ
Pearthree, Philip A., Univ of Arizona
Perault, David R., Univ of Lynchburg
Perdrial, Nicolas, Univ of Vermont
Pereira, Alcides C., Universidade de Coimbra
Perkons, Eriks, Towson Univ
Peterson, Jon W., Hope Coll
Pinho, Jennie, Lafayette Coll
Piotrowski, Alexander, Univ of Cambridge
Pisani-Gareau, Tara, Boston Coll
Plant, Jane, Imperial Coll
Pohopien, Kazimierz M., Mt. San Antonio Coll
Pope, Jeanette K., DePauw Univ
Pospelova, Vera, Univ of Minnesota, Twin Cities
Poudel, Durga, Univ of Louisiana at Lafayette
Powell, Jane, Univ of East Anglia
Pratt, Allyn R., Los Alamos Nat Lab
Quinn, Claire, Univ of Leeds
Radko, Nicholas, Georgia Southern Univ
Rainwater, Maira, California State Univ, Stanislaus
Rasmussen, Pat E., Univ of Ottawa
Reeves, Donald Matt, Univ of Alaska, Anchorage
Reid, Brian, Univ of East Anglia
Rezaie-Boroon, Mohammad H., California State Univ, Los Angeles
Rice, Thomas L., Cedarville Univ
Richards, Laura, Univ of Manchester
Rizoulis, Athanasios, Univ of Manchester
Roberts, Ray L., Hill Coll
Roberts, Sheila M., Univ of Montana Western
Roberts-Semple, Dawn, York Coll (CUNY)
Rocheford, MaryKathryn (Kat), Lake Superior State Univ
Rogers, William C., West Texas A&M Univ
Rucklidge, John C., Univ of Toronto
Russell, Sally, Univ of Leeds
Rutter, Nathaniel W., Univ of Alberta
Ryan, Anne-Marie, Dalhousie Univ
Sacristan, Felix, Universitat Auto de Barcelona
Sanjurjo, Jorge, Coruna Univ
Sarah, Willig B., Univ of Pennsylvania
Satkoski, Alina A., Austin Comm Coll District
Schleifer, Stanley, York Coll (CUNY)
Schmidt, Dale R., Illinois State Geological Survey
Schoenemann, Spruce W., Univ of Montana Western
Scholtz, Theresa C., Argonne Nat Lab
Scott, Robert B., Univ of Texas, Austin
Segall, Marylin, Univ of Utah
Sewall, Jacob, Kutztown Univ of Pennsylvania
Sexton, Philip, Open Univ
Sharma, Mukul, Dartmouth Coll
Shem, Linda M., Argonne Nat Lab
Sims, Douglas , Coll of Southern Nevada - West Charleston Campus
Sîrbu, Smaranda D., Alexandru Ioan Cuza
Skinner, Luke C., Univ of Cambridge
Slobodnik, Marek, Masaryk Univ
Smith, Dan, Leicester Univ
Smith, Jim, Univ of Portsmouth
Snider, Henry I., Eastern Connecticut State Univ
Snow, Daniel, Univ of Nebraska, Lincoln
Spaeth, Matthew P., Illinois State Geological Survey
Spahr, Paul, Ohio Dept of Nat Resources
Spooner, Ian S., Acadia Univ
Stahle, David W., Univ of Arkansas, Fayetteville
Stan, Oana, Alexandru Ioan Cuza
Stearman, Will, Queensland Univ of Tech
Steidl, Gregg M., New Jersey Geological and Water Survey
Stevens, Anthony, Montgomery County Comm Coll
Stumbea, Dan, Alexandru Ioan Cuza
Sturm, Diana, Univ of South Alabama
Sugden, John , Central Washington Univ
Sylvia, Elizabeth R., Maryland Department of Nat Resources
Tank, Ronald W., Lawrence Univ
Tawabini, Bassam S., King Fahd Univ of Petroleum and Minerals
Taylor, Eric, Kent State Univ at Stark
Taylor, Peter, Univ of Leeds
Thorpe, Mary S., Del Mar Coll
Toran, Marta, Appalachian State Univ
Triplett, Laura, Gustavus Adolphus Coll
Tses, Akah, Univ of Namibia
Turchyn, Alexandra, Univ of Cambridge
Turner, Derek, Douglas Coll
Turner, Kerry, Univ of East Anglia
Tuttle, Samuel, Syracuse Univ
Vaezi, Reza, Univ of Tabriz
van Alstine, James, Univ of Leeds
Van Horn, Stephen R., Muskingum Univ
Van Ryswick, Stephen, Maryland Department of Nat Resources
VanGundy, Robert D., Univ of Virginia's Coll at Wise
Vann, David R., Univ of Pennsylvania

Villa, Jorge, Univ of Louisiana at Lafayette
Walsh, Maud M., Louisiana State Univ
Watkinson, Andrew, Univ of East Anglia
Weber, Karen M., Virginia Polytechnic Inst & State Univ
Welling, Tim, Dutchess Comm Coll
Wen, Tao, Syracuse Univ
Weppner, Kerrie, Boise State Univ
Wilson, Charlie, Univ of East Anglia
Wilson, Rick I., California Geological Survey
Wunsch, David R., Univ of Delaware
Yiannakoulias, Niko, McMaster Univ
Yibas, Bisrat, Univ of the Free State
Youberg, Ann, Univ of Arizona
Young, Michael H., Univ of Texas at Austin, Jackson Schl of Geosciences
Young, William, Univ of Leeds
Yuen, Cheong-yip R., Argonne Nat Lab
Zavada, Michael, Univ of Texas Permian Basin

Geomorphology

Ros Fatimah M., Univ of Malaya
Aalto, Rolf E., Univ of Washington
Adams, Kenneth D., Desert Research Inst
Adams, Peter N., Univ of Florida
Alba, Saturnino, Univ Complutense de Madrid
Allan, Jonathan C., Oregon Dept of Geology and Mineral Industries
Allen, Bruce D., New Mexico Tech
Allen, Phillip, Frostburg State Univ
Anders, Alison M., Univ of Illinois, Urbana-Champaign
Anderson, Robert S., Univ of California, Santa Cruz
Anderson, Robert, Univ of Colorado
Andrews, William M., Univ of Kentucky
Antinao, JoseLuis, Desert Research Inst
Applegarth, Michael T., Shippensburg Univ
Armour, Jake, Univ of North Carolina, Charlotte
Aslan, Andres, Colorado Mesa Univ
Attal, Mikael, Edinburgh Univ
Bacon, Steven N., Desert Research Inst
Baker, Sophie, Desert Research Inst
Baker, Victor R., Univ of Arizona
Baptista, João C., Universidade de Trás-os-Montes e Alto Douro
Barendregt, Rene W., Univ of Lethbridge
Bauer, Bernard O., Univ of British Columbia, Okanagan Campus
Beget, James E., Univ of Alaska, Fairbanks
Belliveau, Lindsey, Dept of Energy and Enval Protection
Benedetti, Michael M., Univ of North Carolina, Wilmington
Bergeron, Normand, Universite du Quebec
Berta, Susan, Indiana State Univ
Bierman, Paul R., Univ of Vermont
Bird, Broxton, Indiana Univ - Purdue Univ Indianapolis
Blisniuk, Kimberly, San Jose State Univ
Booth, Adam M., Portland State Univ
Booth, Derek, Univ of Washington
Bradley, William C., Univ of Colorado
Brardinoni, Francesco, Alma Mater Studiorum Università di Bologna
Breilin, Olli, Geological Survey of Finland
Brewer, Paul A., Univ of Wales
Brock-Hon, Amy, Univ of Tennessee, Chattanooga
Brocklehurst, Simon, Univ of Manchester
Bull, William B., Univ of Arizona
Bullard, Thomas F., Desert Research Inst
Bullard, Tom, Univ of Nevada, Reno
Burbank, Doug, Univ of California, Santa Barbara
Burke, Raymond M., Humboldt State Univ
Buynevich, Ilya, Temple Univ
Campbell, Ian A., Univ of Alberta
Carson, Robert J., Whitman Coll
Carton, Alberto, Università degli Studi di Padova
Carver, Gary A., Humboldt State Univ
Castillon, David A., Missouri State Univ
Caudill, Michael, Marshall Univ
Chenoweth, M. S., Univ of Louisiana, Monroe
Chin, Anne, Texas A&M Univ
Clark, Jeffrey J., Lawrence Univ
Clark, Marin, Univ of Michigan
Coates, Donald R., Binghamton Univ
Colgan, Patrick M., Grand Valley State Univ
Collins, Brian, Univ of Washington
Collins, Charles W., Univ of Wisconsin, Platteville
Connallon, Christopher B., Maryland Department of Nat Resources
Constantine, Jose A., Williams Coll
Cook, Joseph P., Univ of Arizona
Cooke, M. J., Austin Comm Coll District
Cornwell, Kevin J., Sacramento State Univ
Cremonini, Stefano, Alma Mater Studiorum Università di Bologna
Crosby, Benjamin T., Idaho State Univ
Currey, Donald R., Univ of Utah
Dale, Janis, Univ of Regina
Daly, Julia F., Univ of Maine, Farmington
Daniels, J. Michael, Univ of Denver
Day, Stephanie S., North Dakota State Univ
de Waele, Jo, Alma Mater Studiorum Università di Bologna
Dethier, David P., Williams Coll
DiBiase, Roman , Pennsylvania State Univ, Univ Park
Dietrich, William E., Univ of California, Berkeley
Dixon, Jean, Montana State Univ
Dixon, John C., Univ of Arkansas, Fayetteville
Dixon, Simon J., Univ of Birmingham
Dogwiler, Toby J., Illinois State Univ
Dolliver, Holly A., Univ of Wisconsin, River Falls
Dort, Jr., Wakefield, Univ of Kansas
Dortch, Jason, Univ of Kentucky
Durbin, James, Univ of Southern Indiana
Duvall, Alison, Univ of Washington
Eaton, Lewis S., James Madison Univ
Edgar, Dorland E., Argonne Nat Lab
Ely, Lisa L., Central Washington Univ
England, John, Univ of Alberta
Evans, Diane L., Jet Propulsion Lab
Fagherazzi, Sergio, Boston Univ
Fantozzi, Pier Lorenzo, Univ of Siena
Farr, Tom G., Jet Propulsion Lab
Ferrier, Ken, Univ of Wisconsin, Madison
Finnegan, Noah J., Univ of California, Santa Cruz
Fontana, Alessandro, Università degli Studi di Padova
Ford, Derek C., McMaster Univ
Ford, Richard L., Weber State Univ
Fox, Matthew D., Univ Coll London
Fubelli, Giandomenico, Università di Torino
Furbish, David J., Vanderbilt Univ
Gabet, Emmanuel, San Jose State Univ
Garcia, Antonio F., California Polytechnic State Univ
Garcia, Marcelo H., Univ of Illinois, Urbana-Champaign
Gardner, Thomas W., Trinity Univ
Garrote, Julio, Univ Complutense de Madrid
Gartner, John, Norwich Univ
Gasparini, Nicole, Tulane Univ
Germanoski, Dru, Lafayette Coll
Ghoneim, Eman, Boston Univ
Giambastiani, Beatrice M., Alma Mater Studiorum Università di Bologna
Giardino, John R., Texas A&M Univ
Giardino, Marco, Università di Torino
Gill, Thomas E., Univ of Texas, El Paso
Gillam, Mary L., Fort Lewis Coll
Gillespie, Alan R., Univ of Washington
Gonzalez, Juan L., Univ of Texas, Rio Grande Valley
Gootee, Brian F., Univ of Arizona
Gordon, Steven J., United States Air Force Acad
Gosse, John, Univ of Kansas
Gran, Karen B., Univ of Minnesota Duluth
Greenberg, Harvey, Univ of Washington
Grimley, David A., Illinois State Geological Survey
Guccione, Margaret J., Univ of Arkansas, Fayetteville
Hales, TC, Cardiff Univ
Hallet, Bernard, Univ of Washington
Hancock, Gregory S., William & Mary
Hansen, Edward C., Hope Coll
Hanson, Lindley S., Salem State Univ
Hanson, Paul, Univ of Nebraska, Lincoln
Hanson, Paul, Univ of Nebraska, Lincoln
Harbor, David J., Washington & Lee Univ
Harbor, Jon M., Purdue Univ
Hardage, Sarah M., Univ of Texas, Rio Grande Valley

FACULTY BY SPECIALTY

Harrington, Charles D., Los Alamos Nat Lab
Harrison, Bruce I., New Mexico Tech
Harvey, Jon, Fort Lewis Coll
Hasbargen, Leslie E., SUNY, Oneonta
Haussmann, Natalie S., Univ of Pretoria
Heinzel, Chad E., Univ of Northern Iowa
Heitmuller, Franklin, Univ of Southern Mississippi
Hesp, Patrick A., Louisiana State Univ
Hesp, Patrick, Flinders Univ
Higgins, Charles G., Univ of California, Davis
Hopkins, Kenneth D., Univ of Northern Colorado
Howard, Alan D., Univ of Virginia
Hubbard, Trent, Alaska Div of Geological & Geophysical Surveys
Hungr, Oldrich, Univ of British Columbia
Hyatt, James A., Eastern Connecticut State Univ
Isacks, Bryan L., Cornell Univ
Jerolmack, Douglas, Univ of Pennsylvania
Jewell, Paul W., Univ of Utah
Johnson, Sarah E., Northern Kentucky Univ
Kasse, Kees, Vrije Universiteit Amsterdam
Kavage Adams, Rebecca H., Maryland Department of Nat Resources
Kehew, Alan E., Western Michigan Univ
Keller, Edward A., Univ of California, Santa Barbara
Kelsey, Harvey M., Humboldt State Univ
Kemmerly, Phillip R., Austin Peay State Univ
Kendrick, Katherine J., Univ of California, Riverside
Kirwan, Matthew L., William & Mary
Kiss, Timea, Univesity of Szeged
Kite, J. Steven, West Virginia Univ
Knott, Jeffrey R., California State Univ, Fullerton
Kochel, R. Craig, Bucknell Univ
Koehler, Rich, Univ of Nevada, Reno
Kori, E., Univ of Venda
Lamb, Michael P., California Inst of Tech
Lancaster, Nick, Univ of Nevada, Reno
Larsen, Isaac J., Univ of Massachusetts, Amherst
Larson, Erik B., Shawnee State Univ
Lasca, Norman P., Univ of Wisconsin, Milwaukee
Lehre, Andre K., Humboldt State Univ
Leverington, David W., Texas Tech Univ
Levine, Rebekah, Univ of Montana Western
Li, Gen, Univ of California, Santa Barbara
Li, Junran, Univ of Tulsa
Li, Yanan, Texas State Univ
Lifton, Nathaniel A., Purdue Univ
Lininger, Katherine, Univ of Colorado
Linneman, Scott R., Western Washington Univ
Lips, Elliott W., Univ of Utah
Long, Ann D., Univ of Illinois, Urbana-Champaign
Loring, Arthur P., York Coll (CUNY)
Loubser, Michael J., Univ of Pretoria
MacGregor, Kelly, Macalester Coll
Marchant, David R., Boston Univ
Marchetti, David W., Western Colorado Univ
Marshall, Jeffrey S., California State Polytechnic Univ, Pomona
Marshall, Jill A., Univ of Arkansas, Fayetteville
Martín, Cristina, Univ Complutense de Madrid
Martín Duque, José Francisco, Univ Complutense de Madrid
Mason, Joseph, Univ of Wisconsin, Madison
Mathers, Hannah, Univ of Glasgow
Mauch, James P., Wyoming State Geological Survey
McCoy, Scott W., Univ of Nevada, Reno
McDermott, Jeni, Univ of Saint Thomas
McDowell, Patricia, Univ of Oregon
McKean, Adam, Utah Geological Survey
McKenney, Rosemary, Pacific Lutheran Univ
McKenzie, Garry D., Ohio State Univ
Menking, Kirsten M., Vassar Coll
Merritts, Dorothy J., Franklin and Marshall Coll
Meyer, Grant A., Univ of New Mexico
Michaud, Yves, Universite du Quebec
Miller, Jerry R., Western Carolina Univ
Misner, Tamara, Edinboro Univ of Pennsylvania
Montgomery, David R., Univ of Washington
Moon, Seulgi, Univ of California, Los Angeles
Moore, Andrew, Earlham Coll
Moore, Laura J., Univ of North Carolina, Chapel Hill
Morgan, Daniel J., Vanderbilt Univ
Mount, Jeffrey F., Univ of California, Davis
Mozzi, Paolo, Università degli Studi di Padova
Murray, A. Bradshaw, Duke Univ
Musselman, Zachary A., Millsaps Coll
Muthukrishnan, Suresh, Furman Univ
Mylroie, John E., Mississippi State Univ
Namikas, Steven, Louisiana State Univ
Napieralski, Jacob A., Univ of Michigan, Dearborn
Nash, David, Univ of Cincinnati
Nelson, Robert S., Illinois State Univ
Neubeck, William S., Union Coll
Newton, Andrew, Queen's Univ Belfast
Newton, Robert M., Smith Coll
Nichols, Kyle K., Skidmore Coll
Nielsen, Dennis N., Winona State Univ
Okunade, Samuel, Central State Univ
Olsen, Paul E., Columbia Univ
ONeal, Michael, Univ of Delaware
Orme, Amalie, California State Univ, Northridge
Ortiz, Alejandra C., Colby Coll
Osborn, Gerald D., Univ of Calgary
Oskin, Michael, Univ of California, Davis
Ouimet, William, Univ of Connecticut
Oviatt, Charles G., Kansas State Univ
Owen, Lewis A., North Carolina State Univ
Owen, Lewis, Univ of Washington
Owen, Lewis, Univ of Cincinnati
Papangelakis, Elli, McMaster Univ
Paradise, Thomas R., Univ of Arkansas, Fayetteville
Parker, Gary, Univ of Illinois, Urbana-Champaign
Pasternack, Gregory B., Univ of California, Davis
Patton, Peter C., Wesleyan Univ
Pavlowsky, Robert T., Missouri State Univ
Pazzaglia, Frank J., Lehigh Univ
Pearson, Adam, SUNY Potsdam
Pederson, Joel L., Utah State Univ
Pelletier, Jon D., Univ of Arizona
Perron, Taylor, Massachusetts Inst of Tech
Persico, Lyman P., Whitman Coll
Pfeiffer, Allison, Western Washington Univ
Pierce, Jennifer L., Boise State Univ
Pieruccini, Pierluigi, Università di Torino
Pinter, Nicholas, Univ of California, Davis
Pitlick, John, Univ of Colorado
Pizzuto, James E., Univ of Delaware
Plug, Lawrence, Dalhousie Univ
Polk, Jason, Western Kentucky Univ
Pope, Gregory A., Montclair State Univ
Potter, Jr., Noel, Dickinson Coll
Putkonen, Jaakko, Univ of North Dakota
Rawling, J. E., Univ of Wisconsin, Madison, Div of Extension
Rayburn, John A., SUNY, New Paltz
Rech, Jason, Miami Univ
Reed, Denise J., Univ of New Orleans
Refsnider, Kurt, Prescott Coll
Reneau, Steven L., Los Alamos Nat Lab
Rhoads, Bruce, Univ of Illinois, Urbana-Champaign
Rhodes, Dallas D., Georgia Southern Univ
Rice-Snow, R. Scott, Ball State Univ
Rittenour, Tammy M., Utah State Univ
Ritter, John B., Wittenberg Univ
Robinson, Cordula, Boston Univ
Rockwell, Thomas K., San Diego State Univ
Roering, Joshua J., Univ of Oregon
Rogerson, Robert J., Univ of Lethbridge
Roth, Danica, Colorado School of Mines
Sandoval Donahue, Maria Magdalena, New Mexico Tech
Savina, Mary E., Carleton Coll
Sawyer, Carol F., Univ of South Alabama
Schaller, Mirjam, Univ of Michigan
Schiefer, Erik, Northern Arizona Univ
Schmidt, Amanda H., Oberlin Coll
Schmutz, Phillip P., Univ of West Florida
Scuderi, Louis A., Univ of New Mexico

Shepherd, Stephanie L., Auburn Univ
Shobe, Charles, West Virginia Univ
Shreve, Ronald L., Univ of California, Los Angeles
Shroder, Jr., John F., Univ of Nebraska, Omaha
Silvestri, Sonia, Alma Mater Studiorum Università di Bologna
Sipos, György, Univesity of Szeged
Sklar, Leonard, San Francisco State Univ
Smith, Richard, Univ of Arkansas, Fayetteville
Snead, John I., Louisiana State Univ
Snyder, Jeffrey A., Bowling Green State Univ
Snyder, Noah, Boston Coll
Souch, Catherine J., Indiana Univ, Indianapolis
Spigel, Lindsay, Dept of Agriculture, Conservation, and Forestry
Springer, Gregory S., Ohio Univ
Springston, George E., Norwich Univ
Stanford, Scott D., New Jersey Geological and Water Survey
Stewart, USA (ret), Alexander K., St. Lawrence Univ
Stokes, Martin, Univ of Plymouth
Stolzman, Kacie, Univ of Wisconsin, Madison, Div of Extension
Straffin, Eric, Edinboro Univ of Pennsylvania
Strasser, Jeffrey C., Augustana Coll
Stroup, Justin, SUNY, Oswego
Surian, Nicola, Università degli Studi di Padova
Swanson, Benjamin, Univ of Iowa
Sweeney, Mark R., Univ of South Dakota
Székely, Balázs, Eotvos Lorand Univ
Taylor, Stephen B., Western Oregon Univ
Tchakerian, Vatche P., Texas A&M Univ
Ten Brink, Norman W., Grand Valley State Univ
Theuerkauf, Ethan, Illinois State Geological Survey
Thomas, Paul, Western Washington Univ
Thorson, Robert M., Univ of Connecticut
Tidwell, David, Geological Survey of Alabama
Tomkin, Jonathan H., Univ of Illinois, Urbana-Champaign
Toonen, Willem, Vrije Universiteit Amsterdam
Toy, Terrence J., Univ of Denver
Tranel, Lisa M., Illinois State Univ
Trenhaile, Alan S., Univ of Windsor
Tucker, Gregory E., Univ of Colorado
Valentine, Leslie, Middle Tennessee State Univ
van der Merwe, Barend, Univ of Pretoria
van Dijk, Deanna, Calvin Coll
Vandeberg, Gregory S., Univ of North Dakota
Vidal Romaní, Juan Ramon, Coruna Univ
Vitek, John D., Oklahoma State Univ
Wakabayashi, John, Fresno State Univ
Walters, James C., Univ of Northern Iowa
Walther, Suzanne, Univ of San Diego
Ward, Dylan, Univ of Cincinnati
Warke, Patricia, Queen's Univ Belfast
Wegmann, Karl, North Carolina State Univ
Weirich, Frank H., Univ of Iowa
Werner, Bradley T., Univ of California, San Diego
West, Nicole, Central Michigan Univ
Whipple, Kelin, Arizona State Univ
Whisner, Jennifer B., Bloomsburg Univ
Whiting, Peter J., Case Western Reserve Univ
Whittecar, Jr., G. Richard, Old Dominion Univ
Wickert, Andrew D., Univ of Minnesota, Twin Cities
Wilcock, Peter W., Johns Hopkins Univ
Wilcox, Andrew, Univ of Montana
Williams, Harry F. L., Univ of North Texas
Williams, Kevin K., Buffalo State Coll
Wilson, Fred L., Angelo State Univ
Wilson, Greg C., Grand Valley State Univ
Wohl, Ellen E., Colorado State Univ
Wolken, Gabriel J., Alaska Div of Geological & Geophysical Surveys
Wood, Spencer H., Boise State Univ
Yanites, Brian, Indiana Univ, Bloomington
Yeh, Joseph S., Univ of Texas at Austin, Jackson Schl of Geosciences
Young, Richard A., SUNY, Geneseo

Glaciology and Glacial Geology

Aber, James S., Emporia State Univ
Aizen, Vladimir, Univ of Idaho
Alley, Richard B., Pennsylvania State Univ, Univ Park
Andrews, John T., Univ of Colorado
Angle, Michael P., Ohio Dept of Nat Resources
Armstrong, William H., Appalachian State Univ
Arnaud, Emmanuelle, Univ of Guelph
Attig, John W., Univ of Wisconsin, Madison, Div of Extension
Attig, John W., Univ of Wisconsin, Extension
Barclay, David J., SUNY, Cortland
Bartholomaus, Timothy, Univ of Idaho
Bassis, Jeremy, Univ of Michigan
Berg, Richard, Univ of Illinois
Berg, Richard C., Illinois State Geological Survey
Berthold, Angela, Univ of Minnesota
Bevis, Kenneth A., Hanover Coll
Bezada Diaz, Maximiliano, Univ of Minnesota
Bird, Brian, New York State Geological Survey
Birkel, Sean, Univ of Maine
Blake, Kevin, Eastern Michigan Univ
Blankenship, Donald D., Univ of Texas, Austin
Booth, Adam, Imperial Coll
Breckenridge, Andy, Univ of Wisconsin, Superior
Brigham-Grette, Julie, Univ of Massachusetts, Amherst
Briner, Jason P., SUNY, Buffalo
Broster, Bruce E., Univ of New Brunswick
Brown, Steven E., Illinois State Geological Survey
Brown, Steven E., Univ of Illinois
Brugger, Keith A., Univ of Minnesota, Morris
Campbell, Seth W., Univ of Maine
Carlson, Anders, Oregon State Univ
Carson, Eric C., Univ of Wisconsin, Madison, Div of Extension
Carson, Eric C., Univ of Wisconsin, Extension
Catania, Ginny A., Univ of Texas, Austin
Christianson, Knut, Univ of Washington
Clark, Douglas H., Western Washington Univ
Clark, Peter U., Oregon State Univ
Clayton, Lee, Univ of Wisconsin, Extension
Clebnik, Sherman M., Eastern Connecticut State Univ
Collins, Emma, Univ of Minnesota
Conrad, Daniel, Univ of Minnesota
Conway, Howard B., Univ of Washington
Cotter, James F., Univ of Minnesota, Morris
Creyts, Timothy T., Columbia Univ
Crossen, Kristine J., Univ of Alaska, Anchorage
Csatho, Beata M., SUNY, Buffalo
Culberg, Riley, Princeton Univ
Curry, B. B., Univ of Illinois
Das, Sarah B., Woods Hole Oceanographic Inst
Davis, P. Thompson, Bentley Univ
Denton, George H., Univ of Maine
Doughty, Alice, Bates Coll
Douglass, Daniel C., Northeastern Univ
Dupont, Todd, Univ of California, Irvine
Dupont, Todd, Miami Univ
Easterbrook, Don J., Western Washington Univ
Enderlin, Ellyn, Boise State Univ
Evenson, Edward B., Lehigh Univ
Eyles, Nicholas, Univ of Toronto
Fahnestock, Mark A., Univ of New Hampshire
Fisher, David A., Univ of Ottawa
Fisher, Timothy G., Univ of Toledo
Fleisher, P. Jay, SUNY, Oneonta
Flowers, Gwenn, Simon Fraser Univ
Fountain, Andrew G., Portland State Univ
Franzi, David A., SUNY, Plattsburgh
Froese, Duane, Univ of Alberta
Fudge, T.J., Univ of Washington
Goldstein, Barry, Univ of Puget Sound
Gowan, Angela, Univ of Minnesota
Hall, Brenda L., Univ of Maine
Ham, Nelson R., Saint Norbert Coll
Hambrey, Michael M., Aberystwyth Univ
Harper, Joel, Univ of Montana
Hawley, Bob, Univ of Washington
Hawley, Robert L., Dartmouth Coll
Hicock, Stephen R., Western Univ
Hock, Regine M., Univ of Alaska, Fairbanks
Holschuh, Nicholas D., Amherst Coll

Hooke, Roger L., Univ of Minnesota, Twin Cities
Hopkins, Nathan, Univ of Idaho
Howat, Ian M., Ohio State Univ
Hughes III, Richard O., Crafton Hills Coll
Hulbe, Christina L., Portland State Univ
Humphrey, Neil F., Univ of Wyoming
Hutchings, Jennifer, Oregon State Univ
Iverson, Neal R., Iowa State Univ of Science & Tech
Jiskoot, Hester, Univ of Lethbridge
Johnson, James B., Colorado Mesa Univ
Johnson, Ty C., DCNR- Pennsylvania Bureau of Geological Survey
Joughin, Ian, Univ of Washington
Kaplan, Michael, Columbia Univ
Karrow, Paul, Univ of Waterloo
Keisling, Benjamin, Univ of Texas, Austin
Kelly, Meredith, Dartmouth Coll
Kiver, Eugene P., Eastern Washington Univ
Koutnik, Michelle, Univ of Washington
Kozlowski, Andrew, New York State Geological Survey
Kurak, Ethan, Univ of Minnesota
La Frenierre, Jeff, Gustavus Adolphus Coll
Laabs, Benjamin J., North Dakota State Univ
Lamothe, Michel, Universite du Quebec a Montreal
Lampkin, Derrick, Univ of Maryland
Langway, Chester C., SUNY, Buffalo
Lea, Peter D., Bowdoin Coll
Lempe, Bernhard, Technical Univ of Munich
Leonard, Eric M., Colorado Coll
Levson, Victor M., Univ of Victoria
Levy, Joseph, Colgate Univ
Licciardi, Joseph M., Univ of New Hampshire
Licciardi, Joseph M., Univ of New Hampshire
Licht, Kathy J., Indiana Univ - Purdue Univ Indianapolis
Loope, Henry M., Indiana Univ, Bloomington
Lowell, Thomas V., Univ of Cincinnati
Lusardi, Barb, Univ of Minnesota
Malone, Andrew, Univ of Illinois Chicago
Marcott, Shaun, Univ of Wisconsin, Madison
Matsuoka, Kenichi, Univ of Washington
McCarthy, Daniel P., Brock Univ
McDonald, Jennifer, Univ of Minnesota
McGrath, Daniel, Colorado State Univ
McKay III, E. Donald, Illinois State Geological Survey
Menzies, John, Brock Univ
Meyer, Gary, Univ of Minnesota
Mickelson, David M., Univ of Wisconsin, Madison
Miller, Gifford H., Univ of Colorado
Mills, Stephanie, Kingston Univ
Mode, William N., Univ of Wisconsin, Oshkosh
Mooers, Howard D., Univ of Minnesota Duluth
Morgan, Alan V., Univ of Waterloo
Munroe, Jeffrey S., Middlebury Coll
Muto, Atsuhiro, Temple Univ
Nash, Thomas A., Ohio Dept of Nat Resources
Neinow, Peter W., Edinburgh Univ
Neumann, Tom, Univ of Washington
Nguyen, Maurice, Univ of Minnesota
Nowicki, Sophie, SUNY, Buffalo
Pair, Donald, Univ of Dayton
Parent, Michel, Universite du Quebec
Pearce, Danni, Norwegian Univ of Life Sciences (NMBU)
Pendleton, Simon, Plymouth State Univ
Pico, Tamara, Univ of California, Santa Cruz
Poinar, Kristin, SUNY, Buffalo
Prothro, Lindsay O., Texas A&M Univ, Corpus Christi
Putnam, Aaron E., Univ of Maine
Quade, Deborah J., Iowa Dept of Nat Resources
Radic, Valentina, Univ of British Columbia
Raymond, Charles F., Univ of Washington
Retelle, Michael J., Bates Coll
Rexius, James E., Schoolcraft Coll
Ridge, John C., Tufts Univ
Rignot, Eric, Univ of California, Irvine
Rodbell, Donald T., Union Coll
Roy, Martin, Universite du Quebec a Montreal
Rutford, Robert H., Univ of Texas, Dallas
Rysgaard, Soren, Univ of Manitoba
Schneider, Allan F., Univ of Wisconsin, Parkside
Schoof, Christian, Univ of British Columbia
Schroeder, Dustin M., Stanford Univ
Schulz, Layne D., South Dakota Dept of Env and Nat Resources
Sharp, Martin J., Univ of Alberta
Shaw, John, Univ of Alberta
Smith, Ben , Univ of Washington
Spencer, Matt, Lake Superior State Univ
Stansell, Nathan D., Northern Illinois Univ
Starr, Paepin K., Austin Comm Coll District
Stearns, Leigh, Univ of Kansas
Stroeve, Julienne, Univ Coll London
Swanger, Kate, Univ of Massachusetts, Lowell
Syverson, Kent M., Univ of Wisconsin, Eau Claire
Szymanski, Jason, Monroe Comm Coll
Taylor, Lawrence D., Albion Coll
Tedesco, Marco, Columbia Univ
Thackray, Glenn D., Idaho State Univ
Thomason, Jason, Univ of Illinois
Thompson, Lonnie G., Ohio State Univ
Thorleifson, Harvey, Univ of Minnesota
Todd, Claire E., Pacific Lutheran Univ
Totten, Stanley M., Hanover Coll
Tulaczyk, Slawek, Univ of California, Santa Cruz
Ullman, David J., Northland Coll
Waddington, Edwin D., Univ of Washington
Wake, Cameron P., Univ of New Hampshire
Ward, Brent C., Simon Fraser Univ
Werner, Alan, Mount Holyoke Coll
Westgate, John A., Univ of Toronto
Wilch, Thomas I., Albion Coll
Wiles, Gregory C., Coll of Wooster
Willis, Michael J., Virginia Polytechnic Inst & State Univ
Winebrenner, Dale P., Univ of Washington
Wolff, Eric W., Univ of Cambridge
Wright, Stephen F., Univ of Vermont
Young, Duncan, Univ of Texas, Austin
Young, Robert R., Univ of British Columbia, Okanagan Campus
Zoet, Lucas, Univ of Wisconsin, Madison

Marine Geology

Abrams, Lewis J., Univ of North Carolina, Wilmington
Aksu, Ali E., Memorial Univ of Newfoundland
Allen, Katherine A., Univ of Maine
Aluwihare, Lihini I., Univ of California, San Diego
Anderson, John B., Rice Univ
Andrade, César, Universidade de Lisboa
Attias, Eric, Univ of Texas, Austin
Austin, Jr., James A., Univ of Texas, Austin
Bacchus, Tania S., Northern Vermont Univ-Lyndon
Bangs, Nathan L., Univ of Texas, Austin
Barclay, Andrew, Univ of Washington
Barrie, Vaughn, Univ of Victoria
Becel, Anne, Columbia Univ
Belknap, Daniel F., Univ of Maine
Benoit-Bird, Kelly, Oregon State Univ
Bentley, Samuel J., Louisiana State Univ
Boston, Brian B., Columbia Univ
Brooks, Gregg R., Eckerd Coll
Brown, Kevin M., Univ of California, San Diego
Browne, Kathleen M., Rider Univ
Calmus, Thierry, Universidad Nacional Auto de Mexico
Carlin, Joe, California State Univ, Fullerton
Carson, Bobb, Lehigh Univ
Chase, Richard L., Univ of British Columbia
Cleary, William J., Univ of North Carolina, Wilmington
Coffey, Ruth, Univ of San Diego
Coffroth, Mary Alice, SUNY, Buffalo
Constantino, Renata R., Columbia Univ
Conway, Flaxen D., Oregon State Univ
Cook, Ann, Ohio State Univ
Crone, Timothy, Columbia Univ
Cruz-Orozco, Rodolfo, Universidad Auto de Baja California Sur
Curray, Joseph R., Univ of California, San Diego
Damuth, John E., Univ of Texas, Arlington

Daniell, James J., James Cook Univ
Das, Indrani, Columbia Univ
De Deckker, Patrick, Australian Nat Univ
Delaney, John R., Univ of Washington
Dix, Justin, Univ of Southampton
Dladla, Nonkululeko, Univ of KwaZulu-Natal
Dobson, David M., Guilford Coll
Donnelly, Jeffrey, Woods Hole Oceanographic Inst
Driscoll, Neal W., Univ of California, San Diego
Dugan, Brandon, Colorado School of Mines
Eberli, Gregor P., Univ of Miami
Edwards, Margo H., Univ of Hawai'i, Manoa
Fornari, Daniel J., Woods Hole Oceanographic Inst
Freeman-Lynde, Raymond, Univ of Georgia
Freitas, Maria da Conceição P., Universidade de Lisboa
Fryer, Patricia B., Univ of Hawai'i, Manoa
Fulthorpe, Craig S., Univ of Texas, Austin
Gee, Jeffrey S., Univ of California, San Diego
Georgen, Jennifer, Old Dominion Univ
Georgen, Jennifer, Florida State Univ
Georgiopopoulou, Aggeliki, Univ Coll Dublin
Gilerson, Alexander, Graduate School of the City Univ of New York
Giosan, Liviu, Woods Hole Oceanographic Inst
Glickson, Deborah, Nat Acad of Sci, Eng, and Med
Goes, Joaquim, Columbia Univ
Goff, John A., Univ of Texas, Austin
Goni, Miguel A., Oregon State Univ
Gonzalez-Fernandez, Antonio, Centro de Inv Científica y de Ed Sup de Ensenada
Goodwillie, Andrew, Columbia Univ
Grabowski, Jonathan, Northeastern Univ
Green, Andrew, Univ of KwaZulu-Natal
Greene, H. Gary, Moss Landing Marine Lab
Greenfield, Dianne I., Graduate School of the City Univ of New York
Griggs, Gary, Univ of California, Santa Cruz
Grossman, Eric, Western Washington Univ
Guerin, Gilles, Columbia Univ
Hall, Ian R., Cardiff Univ
Harte, Michael, Oregon State Univ
Hine, Albert C., Univ of South Florida
Humphris, Susan E., Woods Hole Oceanographic Inst
Jacobson, Gary L., Grossmont Coll
Jamieson, John, Memorial Univ of Newfoundland
Johnson, Harlan P., Univ of Washington
Johnson, Joel E., Univ of New Hampshire
Juraèiæ, Mladen, Univ of Zagreb
Kalangutkar, Niyati, Goa Univ
Kaminski, Michael A., King Fahd Univ of Petroleum and Minerals
Kastens, Kim A., Columbia Univ
Katuna, Michael P., Coll of Charleston
Kearney, Micheal S., Univ of Maryland
Keller, Randall A., Oregon State Univ
Kelley, Joseph T., Univ of Maine
Kidd, Karen, McMaster Univ
Kienast, Stephanie, Dalhousie Univ
Klasik, John A., California State Polytechnic Univ, Pomona
Klaus, Adam, Texas A&M Univ
Laine, Edward P., Bowdoin Coll
Lasker, Howard R., SUNY, Buffalo
Leonard, Lynn A., Univ of North Carolina, Wilmington
Liu, Jingpu P., North Carolina State Univ
Lonsdale, Peter F., Univ of California, San Diego
Luciano, Katherine E., South Carolina Dept of Nat Resources
MacLeod, C, Cardiff Univ
Malinverno, Alberto , Columbia Univ
Mallinson, David, East Carolina Univ
Manfrino, Carrie M., Kean Univ
Marinelli, Roberta, Oregon State Univ
Mayer, Larry A., Univ of New Hampshire
McCoy, Jr., Floyd W., Univ of Hawai'i, Manoa
Meadows, Guy A., Michigan Technological Univ
Mekik, Figen A., Grand Valley State Univ
Meylan, Maurice A., Univ of Southern Mississippi
Micallef, Aaron, Univ of Malta
Miller, Kenneth G., Rutgers, The State Univ of New Jersey
Miot da Silva, Graziela, Flinders Univ
Mitchell, Neil C., Univ of Manchester
Moberly, Ralph, Univ of Hawai'i, Manoa
Mountain, Gregory S., Rutgers, The State Univ of New Jersey
Moy, Christopher M., Univ of Otago
Neitzke Adamo, Lauren, Rutgers, The State Univ of New Jersey
Nicholson, David, Woods Hole Oceanographic Inst
Nitsche, Frank, Columbia Univ
Nittrouer, Charles A., Univ of Washington
Norris, Richard D., Univ of California, San Diego
Oehlert, Amanda M., Univ of Miami
Olayiwola, M. A., Obafemi Awolowo Univ
Patton, Jason R., Humboldt State Univ
Peterson, Larry C., Univ of Miami
Phipps, James B., Grays Harbor Coll
Pietro, Kathryn R., Woods Hole Oceanographic Inst
Piper, David J., Dalhousie Univ
Porter, David, Columbia Univ
Pratt, Thomas, Univ of Washington
Purkis, Sam J., Univ of Miami
Ramirez, Wilson R., Univ of Puerto Rico
Raphael, Constantine N., Eastern Michigan Univ
Rasmussen, Kenneth, Northern Virginia Comm Coll - Annandale
Rea, David K., Univ of Michigan
Reed, Donald L., San Jose State Univ
Richardson, Mary Jo, Texas A&M Univ
Richaud, Mathieu, Fresno State Univ
Riggs, Stanley R., East Carolina Univ
Rodolfo, Kelvin S., Univ of Illinois Chicago
Roland, Emily, Western Washington Univ
Romagnoli, Claudia, Alma Mater Studiorum Università di Bologna
Ross, David A., Woods Hole Oceanographic Inst
Ruddiman, William F., Univ of Virginia
Saffer, Demian, Univ of Texas, Austin
Scher, Howie, Univ of South Carolina
Schwartz, David, Cabrillo Coll
Scott, Steven D., Univ of Toronto
Skarke, Adam, Mississippi State Univ
Slagle, Angela, Columbia Univ
Slovinsky, Peter, Dept of Agriculture, Conservation, and Forestry
Tinto, Kirsteen, Columbia Univ
Tucholke, Brian E., Woods Hole Oceanographic Inst
Ward, Larry G., Univ of New Hampshire
Watts, Anthony B., Univ of Oxford
Wehmiller, John F., Univ of Delaware
Wheeler, Andy J., Univ Coll Cork
Wilcock, William S., Univ of Washington
Winguth, Cornelia, Univ of Texas, Arlington
Ziervogel, Kai, Univ of New Hampshire

Mineralogy & Crystallography

Abdu, Yassir, Univ of Manitoba
Ague, Jay J., Yale Univ
Akgul, Bunyamin, Firat Univ
Alderton, Dave, Royal Holloway Univ of London
Aldushin, Kirill, Ludwig-Maximilians-Universitaet Muenchen
Altaner, Stephen P., Univ of Illinois, Urbana-Champaign
Anderson, Callum, Nelson Mandela Metropolitan Univ
Antao, Sytle, Univ of Calgary
Apopei, Andrei I., Alexandru Ioan Cuza
Artioli, Gilberto, Università degli Studi di Padova
Astilleros, José Manuel, Univ Complutense de Madrid
Austin, George S., New Mexico Tech
Ball, Neil, Univ of Manitoba
Bassett, William A., Cornell Univ
Beane, Rachel J., Bowdoin Coll
Belley, Philippe, Memorial Univ of Newfoundland
Belluso, Elena, Università di Torino
Benea, Ion Marcel , Babes-Bolyai Univ
Benna, Piera, Università di Torino
Bermanec, Vladimir, Univ of Zagreb
Bish, David L., Indiana Univ, Bloomington
Bladh, Kenneth W., Wittenberg Univ
Bonhoure, Jessica, Institut Polytechnique LaSalle Beauvais (ex-IGAL)
Boyle, Alan P., Univ of Liverpool
Boysen, Hans, Ludwig-Maximilians-Universitaet Muenchen
Braund, Emilie, Univ of Portsmouth
Brearley, Adrian J., Univ of New Mexico

Brian, Lejeune, Lafayette Coll
Brown, Jr., Gordon E., Stanford Univ
Bruno, Emiliano, Università di Torino
Bruno, Marco, Università di Torino
Burisch, Mathias, Colorado School of Mines
Burnham, Charles W., Harvard Univ
Burns, Peter C., Univ of Notre Dame
Buzatu, Andrei, Alexandru Ioan Cuza
Buzgar, Nicolae, Alexandru Ioan Cuza
Cardellach López, Esteve, Universitat Auto de Barcelona
Carrigan, Charles W., Olivet Nazarene Univ
Cartier, Laurent E., Univ of Delaware
Casas Duocastella, Lluís, Universitat Auto de Barcelona
Catlos, Elizabeth J., Univ of Texas, Austin
Cavosie, Aaron, Curtin Univ
Cenpírek, Jan, Masaryk Univ
Chakhmouradian, Anton, Univ of Manitoba
Chen, Ning, Univ of Saskatchewan
Chiarenzelli, Jeffrey R., St. Lawrence Univ
Choi, Seon-Gyu, Korea Univ
Cilliers, Johannes, Imperial Coll
Clement, Stephen C., William & Mary
Cody, Robert, Iowa State Univ of Science & Tech
Cofer, Harland E., Georgia Southwestern State Univ
Coker, Victoria, Univ of Manchester
Cole, Kevin C., Grand Valley State Univ
Collins, Lorence G., California State Univ, Northridge
Cooper, Mark, Univ of Manitoba
Copjakova, Renata, Masaryk Univ
Corbella Cordomí, Mercè, Universitat Auto de Barcelona
Cordonnier, Benoit, Ludwig-Maximilians-Universitaet Muenchen
Costa, Emanuele, Università di Torino
Cox, Richard, Dalhousie Univ
Crandall, Jake R., Eastern Illinois Univ
Cusack, Maggie, Univ of Glasgow
Dalconi, Maria C., Università degli Studi di Padova
Darling, Robert S., SUNY, Cortland
De Campos, Christina, Ludwig-Maximilians-Universitaet Muenchen
De Nault, Kenneth J., Univ of Northern Iowa
De Pablo, Liberto, Universidad Nacional Auto de Mexico
Dempster, Tim, Univ of Glasgow
Deng, Youjun, Texas A&M Univ
Denny, F. B., Illinois State Geological Survey
Dobson, David, Univ Coll London
Dollase, Wayne A., Univ of California, Los Angeles
Dorfman, Alexander, Ludwig-Maximilians-Universitaet Muenchen
Dorfner, Thomas, Ludwig-Maximilians-Universitaet Muenchen
Downs, James W., Ohio State Univ
Duchesne, Josee, Universite Laval
Dunlevey, John, Univ of Limpopo
Dutrow, Barbara L., Louisiana State Univ
Ehm, Lars, Stony Brook Univ
Elsen, Jan, Katholieke Universiteit Leuven
Emofurieta, Williams O., Univ of Benin
Ercit, T. Scott, Carleton Univ
Erickson, Timmons, Curtin Univ
Ertel-Ingrisch, Werner, Ludwig-Maximilians-Universitaet Muenchen
Ewing, Rodney C., Stanford Univ
Ewing, Rodney E., Univ of New Mexico
Farfan, Gabriela, Smithsonian Inst / Nat Museum of Nat History
Farnan, Ian, Univ of Cambridge
Farthing, Dori J., SUNY, Geneseo
Farzaneh, Akbar, Univ of Tabriz
Feather, Russell, Smithsonian Inst / Nat Museum of Nat History
Feely, Martin, Nat Univ of Ireland Galway
Fehr, Karl Thomas, Ludwig-Maximilians-Universitaet Muenchen
Feinglos, Mark N., Duke Univ
Ferguson, Robert B., Univ of Manitoba
Fernández, Lourdes, Univ Complutense de Madrid
Fernández-Barrenechea, José María, Univ Complutense de Madrid
Ferraris, Giovanni, Università di Torino
Field, Stephen W., Tarleton State Univ
Finch, Adriam, Univ of St. Andrews
Flemming, Roberta L., Western Univ
Foit, Jr., Franklin F., Washington State Univ
Fransolet, Andre-Mathieu, Universite de Liege
Freed, Robert L., Trinity Univ
Frenette, Jean, Universite Laval
Frey, Friedrich, Ludwig-Maximilians-Universitaet Muenchen
Gadikota, Greeshma, Univ of Wisconsin, Madison
Gait, Robert I., Royal Ontario Museum
Gal, Agnes, Babes-Bolyai Univ
Gann, Delbert E., Millsaps Coll
Garapic, Gordana, SUNY, New Paltz
García Romero, Emilia, Univ Complutense de Madrid
García-Martínez, David, Universidad Estatal de Sonora
Garvie, Laurence, Arizona State Univ
Gasparotto, Giorgio, Alma Mater Studiorum Università di Bologna
Gaudio, Sarah J., SUNY, Geneseo
Gay, Kenny, N.C. Department of Enval Quality
Gibbs, Jerry, Virginia Polytechnic Inst & State Univ
Gieré, Reto, Univ of Pennsylvania
Gigler, Alexander, Ludwig-Maximilians-Universitaet Muenchen
Gille, Peter, Ludwig-Maximilians-Universitaet Muenchen
Gius, Fred, California Geological Survey
Giustetto, Roberto, Università di Torino
Gollmer, Steven, Cedarville Univ
González, Isabel , Universidad de Sevilla
Graham, Ian T., Univ of New South Wales
Grießl, Stefan, Ludwig-Maximilians-Universitaet Muenchen
Griffen, Dana T., Brigham Young Univ
Groat, Lee A., Univ of British Columbia
Grover, John E., Univ of Cincinnati
Guggenheim, Stephen J., Univ of Illinois Chicago
GUILLAUME, Damien, Université Jean Monnet, Saint-Etienne
Gunter, Mickey E., Univ of Idaho
Göttlich, Hagen, Ludwig-Maximilians-Universitaet Muenchen
Hadler, Kathryn, Imperial Coll
Hall, Anne M., Emory Univ
Hansen, Lars, Univ of Oxford
Harlow, George E., American Museum of Nat History
Harlow, George E., Graduate School of the City Univ of New York
Harrison, Richard, Univ of Cambridge
Hawkins, David, Wellesley Coll
Hawkins, Michael A., New York State Geological Survey
Hawthorne, Frank C., Univ of Manitoba
Hazen, Robert M., Carnegie Inst for Science
Heaney, Peter J., Pennsylvania State Univ, Univ Park
Heckel, Wolfgang, Ludwig-Maximilians-Universitaet Muenchen
Heimann, Adriana, East Carolina Univ
Henderson, Grant S., Univ of Toronto
Hennemeyer, Marc, Ludwig-Maximilians-Universitaet Muenchen
Hess, Kai-Uwe, Ludwig-Maximilians-Universitaet Muenchen
Heuss-Aßbichler, Soraya, Ludwig-Maximilians-Universitaet Muenchen
Hollabaugh, Curtis L., Univ of West Georgia
Hooper, Robert L., Univ of Wisconsin, Eau Claire
Horton, Duane G., Pacific Northwest Nat Lab
Hovis, Guy L., Lafayette Coll
Huff, Warren D., Univ of Cincinnati
Hughes, John M., Miami Univ
Hughes, John M., Univ of Vermont
Huizenga, Jan M., Norwegian Univ of Life Sciences (NMBU)
Hummer, Daniel R., Southern Illinois Univ Carbondale
Izawa, Matt, Western Univ
Joesten, Raymond, Univ of Connecticut
Johanesen, Katharine, Juniata Coll
Johnson, Neil E., Virginia Polytechnic Inst & State Univ
Jones, Rhian, Univ of New Mexico
Jordan, Guntram, Ludwig-Maximilians-Universitaet Muenchen
Kaiser, Jason, Southern Utah Univ
Kaiser-Bischoff, Ines, Ludwig-Maximilians-Universitaet Muenchen
Kamhi, Samuel R., Long Island Univ, Brooklyn Campus
Kampf, Anthony R., Los Angeles County Museum of Nat History
Kearns, Lance E., James Madison Univ
Keller, Dianne M., Colgate Univ
Kerestedjian, Thomas N., Bulgarian Acad of Sciences
Kim, Bojeong, Temple Univ
Kleebe, Hans-Joachim, Technische Universitaet Darmstadt
Komabayashi, Tetsuya, Edinburgh Univ
Krol, Michael A., Bridgewater State Univ
Kunzmann, Thomas, Ludwig-Maximilians-Universitaet Muenchen
Lackinger, Markus, Ludwig-Maximilians-Universitaet Muenchen

Lancaster, Penny, Univ of Portsmouth
Langenhorst, Falko H., Friedrich-Schiller-Univ Jena
Lapen, Thomas, Univ of Houston
Lascu, Ioan, Smithsonian Inst / Nat Museum of Nat History
Leavens, Peter B., Univ of Delaware
Lee, Martin, Univ of Glasgow
Leimer, H. Wayne, Tennessee Tech Univ
Leitz, Robert E., Metropolitan State Coll of Denver
Leung, Irene S., Lehman Coll (CUNY)
Levesque, Andre, Universite Laval
Levine, Jamie S., Appalachian State Univ
Lian, Jie, Univ of Michigan
López, Sol, Univ Complutense de Madrid
Losh, Steven, Minnesota State Univ
Losos, Zdenek, Masaryk Univ
Lueth, Virgil L., New Mexico Tech
Lupulescu, Marian V., New York State Geological Survey
Luque, Francisco Javier, Univ Complutense de Madrid
Lübbe, Maike, Ludwig-Maximilians-Universitaet Muenchen
Macris, Catherine, Indiana Univ - Purdue Univ Indianapolis
Majzlan, Juraj, Friedrich-Schiller-Univ Jena
Manu, Johnson, Univ of Ghana
Marshall, Craig, Univ of Kansas
Martin, Robert F., McGill Univ
Masch, Ludwig, Ludwig-Maximilians-Universitaet Muenchen
McCausland, Phil, Brock Univ
McCormack, John K., Univ of Nevada, Reno
McCormick, George R., Univ of Iowa
McCurry, Michael O., Idaho State Univ
McDonald, Andrew M., Laurentian Univ, Sudbury
McHenry, Lindsay J., Univ of Wisconsin, Milwaukee
McNamara, Kelsey, New Mexico Tech
McNamee, Brittani D., Univ of North Carolina, Asheville
Meisterernst, Götz, Ludwig-Maximilians-Universitaet Muenchen
Merkel, Timo Casjen, Ludwig-Maximilians-Universitaet Muenchen
Miles, Andrew, Kingston Univ
Miyagi, Lowell, Univ of Utah
Moritz, Wolfgang, Ludwig-Maximilians-Universitaet Muenchen
Myers, George, Temple Univ
Müller, Lena, Ludwig-Maximilians-Universitaet Muenchen
Müller-Sohnius, Dieter, Ludwig-Maximilians-Universitaet Muenchen
Navarro, Didac, Universitat Auto de Barcelona
Neethling, Stephen, Imperial Coll
Nestola, Fabrizio, Università degli Studi di Padova
Nord, Julia, George Mason Univ
Novak, Milan , Masaryk Univ
Nusbaum, Robert L., Coll of Charleston
Olson Hoal, Karin E., Cornell Univ
Paktunc, Dogan, Univ of Ottawa
Paquette, Jeanne, McGill Univ
Parise, John B., Stony Brook Univ
Park, Sohyun, Ludwig-Maximilians-Universitaet Muenchen
Pascoe, Richard, Exeter Univ
Pavese, Alessandro, Università di Torino
Peacor, Donald R., Univ of Michigan
Pentcheva, Rossitza, Ludwig-Maximilians-Universitaet Muenchen
Peterson, Ronald C., Queen's Univ
Phillips, Brian L., Stony Brook Univ
Phillips, William R., Brigham Young Univ
Piniella Febrer, Juan Francesc, Universitat Auto de Barcelona
Pohwat, Paul, Smithsonian Inst / Nat Museum of Nat History
Post, Jeffrey E., Smithsonian Inst / Nat Museum of Nat History
Prencipe, Mauro, Università di Torino
Price, Nancy A., SUNY, Plattsburgh
Putnis, Andrew, Curtin Univ
Radcliffe, Dennis, Hofstra Univ
Raudsepp, Mati, Univ of British Columbia
Redfern, Simon, Univ of Cambridge
Regan, Sean, Univ of Alaska, Fairbanks
Reuss, Robert L., Tufts Univ
Rickaby, Ros, Univ of Oxford
Riedel, Oliver, Ludwig-Maximilians-Universitaet Muenchen
Robinson, George W., St. Lawrence Univ
Rocholl, Alexander, Ludwig-Maximilians-Universitaet Muenchen
Rodriguez-Blanco, Juan Diego, Trinity Coll
Rojas-Soriano, Humberto, Universidad Auto de Baja California Sur
Rosenberg, Philip E., Washington State Univ
Ross, Nancy L., Virginia Polytechnic Inst & State Univ
Rossman, George R., California Inst of Tech
Rouse, Roland C., Univ of Michigan
Rubio-Sierra, Javier, Ludwig-Maximilians-Universitaet Muenchen
Rutstein, Martin S., SUNY, New Paltz
Salviulo, Gabriella, Università degli Studi di Padova
Sánchez, Nuria, Univ Complutense de Madrid
Schmahl, Wolfgang, Ludwig-Maximilians-Universitaet Muenchen
Schneider, Julius, Ludwig-Maximilians-Universitaet Muenchen
Schroeder, Paul A., Univ of Georgia
Scott Smith, Barbara H., Univ of British Columbia
Secco, Luciano, Università degli Studi di Padova
Sethi, Parvinder S., Radford Univ
Shank, Stephen G., DCNR- Pennsylvania Bureau of Geological Survey
Sharp, Thomas G., Arizona State Univ
Sharp, W. Edwin, Univ of South Carolina
Sheets, Julie, Ohio State Univ
Sherriff, Barbara L., Univ of Manitoba
Shim, Sang-Heon, Arizona State Univ
Shuller-Nickles, Lindsay C., Clemson Univ
Silvestri, Alberta, Università degli Studi di Padova
Simmons, William B., Univ of New Orleans
Simmons, William B., Univ of Michigan
Singer, David M., Kent State Univ
Skoda, Radek, Masaryk Univ
Smith, Michael S., Univ of North Carolina, Wilmington
Smyth, Joseph R., Univ of Colorado
Sokolova, Elena, Univ of Manitoba
Solorio-Munguia, Jose G., Universidad Nacional Auto de Mexico
Sparks, Carolyn R., Hope Coll
Spieler, Oliver, Ludwig-Maximilians-Universitaet Muenchen
Spilde, Michael N., Univ of New Mexico
Stark, Robert, Ludwig-Maximilians-Universitaet Muenchen
Stefano, Christopher J., Michigan Technological Univ
Stewart, Dion C., Georgia State Univ, Perimeter Coll, Alpharetta Campus
Strasser, Stefan, Ludwig-Maximilians-Universitaet Muenchen
Tait, Kimberly T., Royal Ontario Museum
Tamas, Tudor Liviu, Babes-Bolyai Univ
Tettenhorst, Rodney T., Ohio State Univ
Thibault, Yves, Western Univ
Tibljaš, Darko, Univ of Zagreb
Timms, Nick, Curtin Univ
Tomašiæ, Nenad, Univ of Zagreb
Trixler, Frank, Ludwig-Maximilians-Universitaet Muenchen
Uzochukwu, Godfrey A., North Carolina Agricultural & Tech State Univ
Valdrè, Giovanni, Alma Mater Studiorum Università di Bologna
Vassiliou, Andreas H., Rutgers, The State Univ of New Jersey, Newark
Vaughan, David, Univ of Manchester
Viedma, Cristóbal, Univ Complutense de Madrid
Viti, Cecilia, Univ of Siena
Walker, Jeffrey R., Vassar Coll
Walther, Ferdinand, Ludwig-Maximilians-Universitaet Muenchen
Warner, Richard D., Clemson Univ
Weinbruch, Stephan, Technische Universitaet Darmstadt
Wenk, Hans-Rudolf, Univ of California, Berkeley
Wicks, Frederick J., Royal Ontario Museum
Wicks, Frederick J., Univ of Toronto
Wilson, James R., Weber State Univ
Winchell, Robert E., California State Univ, Long Beach
Wise, Michael A., Smithsonian Inst / Nat Museum of Nat History
Wise, Michael A., Univ of New Orleans
Witter-Shelleman, Molly, Univ of Akron
Wood, Bernard, Univ of Oxford
Wood, Ian, Univ Coll London
Wylie, Ann G., Univ of Maryland
Xu, Huifang, Univ of Wisconsin, Madison
Yoshiasa, Akira, Kumamoto Univ
Yurtsever, Ayhan, Ludwig-Maximilians-Universitaet Muenchen
Zekri, Hamid, Univ of New South Wales
Zimmerman, Brian S., Edinboro Univ of Pennsylvania
Zink, Albert, Ludwig-Maximilians-Universitaet Muenchen
Zurevinski, Shannon, Lakehead Univ
Šæavnièar, Stjepan, Univ of Zagreb
Žigoveèki Gobac, Željka, Univ of Zagreb

Paleolimnology

Anderson, Roger Y., Univ of New Mexico
Axford, Yarrow L., Northwestern Univ
Brady, Kristina, Univ of Minnesota, Twin Cities
Brant, Lynn A., Univ of Northern Iowa
Brenner, Mark, Univ of Florida
Buchheim, Paul, Loma Linda Univ
Davies, Caroline P., Univ of Missouri, Kansas City
De Batist, Marc, Ghent Univ
Ferris, Grant F., Univ of Toronto
Gennari, Rocco, Università di Torino
Glaser, Paul H., Univ of Minnesota, Twin Cities
Heyvaert, Alan, Univ of Nevada, Reno
Kaplan, Samantha W., Univ of Wisconsin, Stevens Point
Kirby, Matthew E., California State Univ, Fullerton
Lisiecki, Lorraine, Univ of California, Santa Barbara
McGlue, Michael M., Univ of Kentucky
Noren, Anders, Univ of Minnesota, Twin Cities
Porter, Susannah, Univ of California, Santa Barbara
Rosen, Michael R., Univ of Nevada, Reno
Russell, James M., Brown Univ
Shapley, Mark, Univ of Minnesota, Twin Cities
Smith, Alison J., Kent State Univ
Steinman, Byron A., Univ of Minnesota Duluth
Stevens (Landon), Lora R., California State Univ, Long Beach
Stone, Alexander, Univ of Minnesota, Twin Cities
Theissen, Kevin, Univ of Saint Thomas
Waters, Matthew N., Auburn Univ

Petroleum Geology

Adeigbe, O. C., Univ of Ibadan
Adekeye, J.I. D., Univ of Ilorin
Aden, Leon, Univ of Iowa
Ajaegwu, Norbert E., Nnamdi Azikiwe Univ
Al-Ramadan, Khalid, King Fahd Univ of Petroleum and Minerals
Ali, Hendratta N., Fort Hays State Univ
Alves, Tiago, Cardiff Univ
Aplin, Andrew, Durham Univ
Arnold, Dan, Heriot-Watt Univ
Asquith, George B., Texas Tech Univ
Avary, Katherine L., West Virginia Univ
Bartlett, Wendy, Marietta Coll
Bartok, Peter, Univ of Houston
Bellian, Jerome, Miami Univ
Bend, Stephen L., Univ of Regina
Benjamin, U. K., Obafemi Awolowo Univ
Bible, Gary G., Univ of Tennessee, Knoxville
Bissell, Clinton R., Texas A&M Univ, Corpus Christi
Bloxson, Julie M., Stephen F. Austin State Univ
Boboye, O. A., Univ of Ibadan
Bonk, Kathleen, New York State Geological Survey
Bowersox, John R., Univ of Kentucky
Bowler, Bernard, Univ of Newcastle Upon Tyne
Bowman, Mike, Univ of Manchester
Broadhead, Ronald F., New Mexico Tech
Burton, Bradford R., Western Colorado Univ
Carr, Timothy R., Univ of Kansas
Carr, Timothy R., West Virginia Univ
Cassidy, Martin, Univ of Houston
Champlin, Steven D., Mississippi Office of Geology
Cheadle, Burns A., Western Univ
Chelariu, Ciprian, Alexandru Ioan Cuza
Chenoweth, Cheri, Illinois State Geological Survey
Chongwain, Gilbert, Univ of Namibia
Cochrane, Claudia, Western Univ
Cranganu, Constantin, Brooklyn Coll (CUNY)
Cunningham, Robert , Univ of Texas, Austin
Curtis, John B., Colorado School of Mines
Dallegge, Todd A., Univ of Northern Colorado
Day-Stirrat, Ruarri J., Oregon Dept of Geology and Mineral Industries
Denne, Richard A., Texas Christian Univ
Echols, John B., Louisiana State Univ
Edegbai, Aitalokhai J., Univ of Benin
Ehinola, Olugbenga A., Univ of Ibadan
Ehinola, Olugbenga A., Univ of Ibadan
Elnadi, Abdelhalim H., Univ of Khartoum
Fadiya, S. L., Obafemi Awolowo Univ
Fakhari, Mohammad D., Ohio Dept of Nat Resources
Finley, Robert J., Univ of Illinois, Urbana-Champaign
Folorunso, I. O., Univ of Ilorin
Fraser, Alastair, Imperial Coll
Giles, Katherine A., Univ of Texas, El Paso
Gillespie, Janice, California State Univ, Bakersfield
Gingras, Murray, Univ of Alberta
Graham, Stephan A., Stanford Univ
Grigg, Joseph, New Mexico Tech
Hanks, Catherine L., Univ of Alaska, Fairbanks
Harris, David C., Univ of Kentucky
Harun, Nina, Alaska Div of Geological & Geophysical Surveys
Haszeldine, R. S., Edinburgh Univ
Henk, Jr., Floyd, Texas Christian Univ
Hileman, Mary E., Oklahoma State Univ
Hofmann, Michael, Univ of Montana
Holford, Simon, Univ of Adelaide
Hu, Qinhong (Max), Univ of Texas, Arlington
Hudson, Samuel M., Brigham Young Univ
Hughes, Colin, Univ of Manchester
Igbinigie, Nosa S., Univ of Benin
Jiang , Shu, Univ of Utah
John, Chacko J., Louisiana State Univ
Johnson, Howard, Imperial Coll
Jones, Bobby L., Louisiana State Univ
Kaldi, John, Univ of Adelaide
Kavousi Ghahfarohki, Payam, West Virginia Univ
Keach, William, Univ of Utah
Leetaru, Hannes E., Univ of Illinois, Urbana-Champaign
Leetaru, Hannes E., Illinois State Geological Survey
Lever, Helen, Heriot-Watt Univ
Li, Peng, Arkansas Geological Survey
Lopez, Annabelle, New Mexico Tech
Loucks, Bob, Univ of Texas at Austin, Jackson Schl of Geosciences
Lucia, F. Jerry, Univ of Texas at Austin, Jackson Schl of Geosciences
Machel, Hans G., Univ of Alberta
Malinverno, Alberto, Columbia Univ
Manzocchi, Tom, Univ Coll Dublin
Martin, Luke, New Mexico Tech
Mazzullo, Salvatore J., Wichita State Univ
Mc Mahon, Michelle, Lonestar Coll - North Harris
McCabe, Peter, Univ of Adelaide
Meddaugh, W. S., Midwestern State Univ
Medina, Cristian R., Indiana Univ, Bloomington
Miller, Michael B., Louisiana State Univ
Minzoni, Marcello, Univ of Alabama
Neufeld, Jerome A., Univ of Cambridge
Newell, Kerry D., Univ of Kansas
Nixon, R. Paul, Brigham Young Univ
Nordeng, Stephan, Univ of North Dakota
Norrish, Winston, Central Washington Univ
North, F. K., Carleton Univ
Nton, M. E., Univ of Ibadan
Nunez-Velazco, Miriam, Universidad Auto de Baja California Sur
Nwachukwu, J. I., Obafemi Awolowo Univ
Ola, Peter S., Federal Univ of Tech, Akure
Olarewaju, V. O., Obafemi Awolowo Univ
Olorunfemi, A. O., Obafemi Awolowo Univ
Osman, Mutasim S., King Fahd Univ of Petroleum and Minerals
Panah, Assad I., Univ of Pittsburgh, Bradford
Pashin, Jack C., Mississippi State Univ
Pedersen, Per, Univ of Calgary
Pena dos Reis, Rui, Universidade de Coimbra
Peng, Sheng, Univ of Texas, Austin
Pertl, David, Amarillo Coll
Pigott, John D., Univ of Oklahoma
Pranter, Matthew J., Univ of Oklahoma
Puckette, James, Oklahoma State Univ
Ratchford, M. E., Arkansas Geological Survey
Resnic, Victor S., Lonestar Coll - North Harris
Richards-McClung, Bryony, Univ of Utah
Riess, Carolyn M., Austin Comm Coll District
Ronck, Catherine, Tarleton State Univ
Rupp, John A., Indiana Univ, Bloomington
Saleh, Saad, Univ of Texas, Austin

Scarselli, Nicola, Royal Holloway Univ of London
Schmid, Katherine W., DCNR- Pennsylvania Bureau of Geological Survey
Schneider, Robert V., Texas A&M Univ, Kingsville
Schroeder, Stefan, Univ of Manchester
Seyyed, Hosseini, Univ of Texas, Austin
Sharman, Glenn R., Univ of Arkansas, Fayetteville
Shumaker, Robert C., West Virginia Univ
Simpson, Altus, Cypress Coll
Slater, Brian, New York State Geological Survey
Smith, Greg, Curtin Univ
Smith, Langhorne, New York State Geological Survey
Song, Liaosha, California State Univ, Bakersfield
Sonnenberg, Stephen A., Colorado School of Mines
Sprinkel, Douglas A., Utah Geological Survey
Stewart, Gary, Oklahoma State Univ
Stright, Lisa, Colorado State Univ
Sullivan, Raymond, San Francisco State Univ
Sun, Yuefeng, Texas A&M Univ
Tan, Denis N., Univ of Malaya
Taylor, Kevin, Univ of Manchester
Temples, Tommy, Clemson Univ
Tinker, Scott W., Univ of Texas at Austin, Jackson Schl of Geosciences
Tinker, Scott W., Univ of Texas, Austin
Trentham, Robert C., Univ of Texas Permian Basin
Van Kooten, Gerald K., Calvin Coll
Vautier, Yannick, Institut Polytechnique LaSalle Beauvais (ex-IGAL)
Wach, Grant D., Dalhousie Univ
Wardlaw, Norman C., Univ of Calgary
Weimer, Paul, Univ of Colorado
White, Bekki C., Arkansas Geological Survey
Whiteley, Martin, Univ of Derby
Whittaker, Steven, Illinois State Geological Survey
Woodward, Mac B., Arkansas Geological Survey
Xie, Xiangyang, Texas Christian Univ
Zimmerman, Ronald K., Louisiana State Univ
Zonneveld, John-Paul, Univ of Alberta

General Petrology

Afolabi, Olukayode A., Ladoke Akintola Univ of Tech
Ague, Jay J., Yale Univ
Aird, Hannah M., California State Univ, Chico
Anderson, Alan J., Saint Francis Xavier Univ
Arslan, Gizem, Firat Univ
Ashwal, Lewis D., Univ of the Witwatersrand
Balen, Dražen, Univ of Zagreb
Barr, Sandra, Dalhousie Univ
Basu, Asish, Univ of Texas, Arlington
Bickel, Charles E., San Francisco State Univ
Biševac, Vanja, Univ of Zagreb
Blackburn, William H., Univ of Kentucky
Bonamici, Chloe, Univ of Wisconsin, Madison
Borghi, Alessandro, Università di Torino
Botcharnikov, Roman, Universität Mainz
Brady, John B., Smith Coll
Brandriss, Mark E., Smith Coll
Brown, Michael, Univ of Maryland
BrownLee, Sarah J., Wayne State Univ
Broxton, David E., Los Alamos Nat Lab
Cain, J. Allan, Univ of Rhode Island
Callegari, Ezio, Università di Torino
Carmichael, Sarah, Appalachian State Univ
Castelli, Daniele, Università di Torino
Chaumba, Jeff B., Univ of North Carolina, Pembroke
Chu, Xu, Univ of Toronto
Compagnoni, Roberto, Università di Torino
Compton, Robert R., Stanford Univ
Corral, Isaac, Universitat Auto de Barcelona
Crawford, Maria Luisa B., Bryn Mawr Coll
Dasgupta, Rajdeep, Rice Univ
Day, Howard W., Univ of California, Davis
de Hoog, Jan C., Edinburgh Univ
DeBari, Susan M., Western Washington Univ
Deininger, Robert W., Univ of Memphis
Dobrzhinetskaya, Larissa F., Univ of California, Riverside
Doolan, Barry L., Univ of Vermont
Eklund, Olav, Abo Akademi Univ
Elliot, David H., Ohio State Univ
Encarnacion, John, Saint Louis Univ
Erickson, Rolfe C., Sonoma State Univ
Ferrando, Simona, Università di Torino
Ferreiro Maehlmann, Rafael, Technische Universitaet Darmstadt
Finn, Gregory C., Brock Univ
Flynn, Natalie, Temple Univ
Frank, Charles O., Southern Illinois Univ Carbondale
Ganguly, Jiba, Univ of Arizona
Giamello, Marco, Univ of Siena
Goble, Ronald G., Univ of Nebraska, Lincoln
Gonzales, David A., Fort Lewis Coll
Groppo, Chiara Teresa, Università di Torino
Haileab, Bereket, Carleton Coll
Hanson, Richard E., Texas Christian Univ
Hatheway, Richard B., SUNY, Geneseo
Hayob, Jodie, Univ of Mary Washington
Hetherington, Callum J., Texas Tech Univ
Hollocher, Kurt T., Union Coll
Hutchison, David, Hartwick Coll
Ihinger, Phillip D., Univ of Wisconsin, Eau Claire
Jercinovic, Michael J., Univ of Massachusetts, Amherst
Jimoh, Mustapha T., Ladoke Akintola Univ of Tech
Johnson, Julie, Univ of Memphis
Jolliff, Bradley L., Washington Univ in St. Louis
Kerr, A C., Cardiff Univ
Keskinen, Mary J., Univ of Alaska, Fairbanks
Kilic, Ayse Didem, Firat Univ
Leichmann, Jaromir, Masaryk Univ
Lewy, Robert, Bakersfield Coll
Malcuit, Robert J., Denison Univ
Mallik, Ananya, Univ of Rhode Island
Manning, Craig E., Univ of California, Los Angeles
Maritan, Lara, Università degli Studi di Padova
Mathez, Edmond A., American Museum of Nat History
Mazza, Sarah, Smith Coll
Mazzoli, Claudio, Università degli Studi di Padova
McLeod, Claire, Miami Univ
Mirnejad, Hassan , Miami Univ
Moecher, David P., Univ of Kentucky
Murphy, J. Brendan, Dalhousie Univ
Mutti, Laurence J., Juniata Coll
Nasir, Sobhi J., Univ of Qatar
Nesse, William D., Univ of Northern Colorado
Nielsen, Peter A., Keene State Coll
Noblett, Jeffrey B., Colorado Coll
Owens, Brent E., William & Mary
Ozel Yildirim, Esra , Firat Univ
Pan, Yuanming, Univ of Saskatchewan
Peterson, Virginia L., Grand Valley State Univ
Plank, Terry A., Columbia Univ
Plymate, Thomas G., Missouri State Univ
Polley, Greg, Smithsonian Inst / Nat Museum of Nat History
Prichard, H M., Cardiff Univ
Ptacek, Anton D., San Diego State Univ
Purchase, Megan, Univ of the Free State
Purdom, William B., Southern Oregon Univ
Ranson, William A., Furman Univ
Reed, Robert M., Univ of Texas at Austin, Jackson Schl of Geosciences
Reyes-Cortes, Ignacio A., Universidad Auto de Chihuahua
Rizeli, Mustafa Eren, Firat Univ
Roadifer, Jack E., Colorado Mesa Univ
Rolfo, Franco, Università di Torino
Romanak, Katherine , Univ of Texas, Austin
Rougvie, James R., Beloit Coll
Saini-Eidukat, Bernhardt, North Dakota State Univ
Severs, Matthew R., Stockton Univ
Sharma, Shiv K., Univ of Hawai'i, Manoa
Shoemaker, Kurt A., Shawnee State Univ
Spandler, Carl, James Cook Univ
Stelling, Pete, Western Washington Univ
Sun, Chenguang, Univ of Texas, Austin
Thy, Peter, Univ of California, Davis
Torres-Agulera, Juan Manuel, Universidad Auto de San Luis Potosi
Van Buer, Nicholas J., California State Polytechnic Univ, Pomona
Vaniman, David T., Los Alamos Nat Lab

Vanko, David A., Towson Univ
Vieira Ribeiro, Bruno, Curtin Univ
Walker, David, Columbia Univ
Walowski, Kristina, Western Washington Univ
Warren, Jessica, Univ of Delaware
Weaver, Stephen G., Colorado Coll
Welsh, James L., Gustavus Adolphus Coll
Westerman, David S., Norwich Univ
Whattam, Scott, King Fahd Univ of Petroleum and Minerals
Whelan, Peter M., Univ of Minnesota, Morris
WoldeGabriel, Giday, Los Alamos Nat Lab
Wortel, Matthew J., Univ of Iowa
Yurkovich, Steven P., Western Carolina Univ

Igneous Petrology

Ackerson, Michael, Smithsonian Inst / Nat Museum of Nat History
Albee, Arden L., California Inst of Tech
Amedjoe, Godfrey C., Kwame Nkrumah Univ of Science and Tech
Amel, Nasir, Univ of Tabriz
Andersen, Jens C., Exeter Univ
Anderson, James Lawford, Boston Univ
Annesley, Irvine, Univ of Saskatchewan
Aranda, Jorge, Universidad Nacional Auto de Mexico
Atlas, Zachary, Univ of South Florida
Badger, Robert L., SUNY Potsdam
Bailey, David G., Hamilton Coll
Baldridge, W. Scott, Los Alamos Nat Lab
Barboni, Melanie, Arizona State Univ
Barnes, Calvin G., Texas Tech Univ
Barr, Sandra M., Acadia Univ
Barth, Andrew P., Indiana Univ - Purdue Univ Indianapolis
Barton, Michael, Ohio State Univ
Beard, James S., Virginia Polytechnic Inst & State Univ
Bedard, Jean H., Universite du Quebec
Bender, E. E., Orange Coast Coll
Bender, John F., Univ of North Carolina, Charlotte
Benimoff, Alan I., Coll of Staten Island/CUNY
Bergantz, George W., Univ of Washington
Beswick, Anthony E., Laurentian Univ, Sudbury
Beukes, Jarlen J., Univ of the Free State
Beyarslan, Melahat, Firat Univ
Bingol, Ahmet Feyzi, Firat Univ
Bizimis, Michael, Univ of South Carolina
Blackerby, Bruce A., Fresno State Univ
Bladh, Katherine L., Wittenberg Univ
Bloomer, Sherman H., Oregon State Univ
Bohrson, Wendy A., Central Washington Univ
Borg, Scott G., Nat Science Foundation
Boroughs, Scott, Washington State Univ
Boudreau, Alan E., Duke Univ
Brandon, Alan, Univ of Houston
Brandriss, Mark E., Williams Coll
Brehm, Sarah, St. Lawrence Univ
Brophy, James G., Indiana Univ, Bloomington
Brown, Cathleen, Smithsonian Inst / Nat Museum of Nat History
Brown, Kenneth L., DePauw Univ
Browne, Brandon L., Humboldt State Univ
Brueseke, Matthew E., Kansas State Univ
Bryan, Scott E., Queensland Univ of Tech
Bybee, Grant M., Univ of the Witwatersrand
Byerly, Gary R., Louisiana State Univ
Canil, Dante, Univ of Victoria
Carley, Tamara, Lafayette Coll
Carmichael, Ian S., Univ of California, Berkeley
Casquet, César, Univ Complutense de Madrid
Castillo, Paterno R., Univ of California, San Diego
Cathey, Henrietta, Queensland Univ of Tech
Cecil, M. R., California State Univ, Northridge
Cepeda, Joseph C., West Texas A&M Univ
Chadwick, John, Coll of Charleston
Chavrit, Deborah, Univ of Manchester
Cheney, John T., Amherst Coll
Chesner, Craig A., Eastern Illinois Univ
Chilson-Parks, Benjamin, Univ of Connecticut
Christiansen, Eric H., Brigham Young Univ
Claiborne, Lily L., Vanderbilt Univ
Clark, Russell G., Albion Coll
Clarke, D B., Dalhousie Univ
Clausen, Benjamin L., Loma Linda Univ
Clemens-Knott, Diane, California State Univ, Fullerton
Coleman, Robert G., Stanford Univ
Collins, Bill, Curtin Univ
Condit, Christopher D., Univ of Massachusetts, Amherst
Constantin, Marc, Universite Laval
Coogan, Laurence, Univ of Victoria
Coombs, Douglas S., Univ of Otago
Cooper, Kari M., Univ of California, Davis
Cooper, Roger W., Lamar Univ
Cornell, Winton, Univ of Tulsa
Cote, Denis, Universite du Quebec a Chicoutimi
Cottle, John, Univ of California, Santa Barbara
Cousens, Brian L., Carleton Univ
Crawford, Anthony J., Univ of Tasmania
Cribb, Warner, Middle Tennessee State Univ
Davies, Gareth R., Vrije Universiteit Amsterdam
Dean, Robert, Dickinson Coll
DELACOUR, Adélie, Université Jean Monnet, Saint-Etienne
Delgado-Argote, Luis A., Centro de Inv Científica y de Ed Superior de Ensenada
Dick, Henry J B., Woods Hole Oceanographic Inst
Dickson, Loretta D., Lock Haven Univ
Dijkstra, Arjan, Univ of Plymouth
Dingwell, Donald Bruce, Ludwig-Maximilians-Universitaet Muenchen
Dolejs, David, Charles Univ
Dorais, Michael J., Brigham Young Univ
Doucet, Luc, Curtin Univ
Downs, Robert T., Univ of Arizona
Dunbar, Nelia W., New Mexico Tech
Duncan, Megan S., Virginia Polytechnic Inst & State Univ
Dunn, John Todd, Univ of New Brunswick
Dygert, Nick, Univ of Tennessee, Knoxville
Dymek, Robert F., Washington Univ in St. Louis
Edwards, Benjamin R., Dickinson Coll
Elardo, Stephen, Univ of Florida
Elders, Wilfred A., Univ of California, Riverside
Elkins, Lynne J., Univ of Nebraska, Lincoln
Erturk, Mehmet Ali, Firat Univ
Fagan, Amy, Western Carolina Univ
Fernandez, Louis A., California State Univ, San Bernardino
First, Emily, Macalester Coll
Fitton, J. G., Edinburgh Univ
Fleming, Thomas H., Southern Connecticut State Univ
Flower, Martin F., Univ of Illinois Chicago
Fodor, Ronald V., North Carolina State Univ
Fournelle, John, Univ of Wisconsin, Madison
Fowler, Mike, Univ of Portsmouth
Fowler, Sarah, Katholieke Universiteit Leuven
Fox, Lydia K., Univ of the Pacific
Francis, Don, McGill Univ
Furman, Tanya, Pennsylvania State Univ, Univ Park
Gaetani, Glenn A., Woods Hole Oceanographic Inst
Galán garcía, Gúmer, Universitat Auto de Barcelona
Gamble, John, Univ Coll Cork
Ganguly, Sohini, Goa Univ
Garcia, Michael O., Univ of Hawai'i, Manoa
Garrison, Jennifer M., California State Univ, Los Angeles
Gaynor, Sean, Princeton Univ
Geist, Dennis J., Univ of Idaho
GERBE, Marie-Christine, Université Jean Monnet, Saint-Etienne
Gertisser, Ralf, Keele Univ
Ghani, Azman Abdul, Univ of Malaya
Gibson, Sally, Univ of Cambridge
Gittins, John, Univ of Toronto
Glazner, Allen F., Univ of North Carolina, Chapel Hill
Gomes, Maria E., Universidade de Trás-os-Montes e Alto Douro
Gonzalez, Maria M., Central Michigan Univ
Gorring, Matthew L., Montclair State Univ
Green, John C., Univ of Minnesota Duluth
Greenough, John D., Univ of British Columbia, Okanagan Campus
Grove, Timothy L., Massachusetts Inst of Tech
Grunder, Anita L., Oregon State Univ
Gualda, Guilherme, Vanderbilt Univ
Gust, David A., Queensland Univ of Tech

Haggerty, Stephen E., Florida InterNat Univ
Halsor, Sid P., Wilkes Univ
Hammer, Julia E., Univ of Hawai'i, Manoa
Hammersley, Lisa, Sacramento State Univ
Hammond, Paul E., Portland State Univ
Hanger, Brendan, Colorado School of Mines
Hannah, Judith L., Colorado State Univ
Hanson, Sarah L., Adrian Coll
Har, Nicolae, Babes-Bolyai Univ
Hari, Kosiyath R., Pt. Ravishankar Shukla Univ
Harris, Chris, Univ of Cape Town
Hart, William K., Miami Univ
Hebert, Rejean J., Universite Laval
Hermes, O D., Univ of Rhode Island
Hess, Paul C., Brown Univ
Hidalgo, Paulo J., Georgia State Univ
Higgins, Michael D., Universite du Quebec a Chicoutimi
Hirschmann, Marc M., Univ of Minnesota, Twin Cities
Hirt, William H., Coll of the Siskiyous
Holness, Marian, Univ of Cambridge
Hudgins, Thomas, Univ of Puerto Rico
Huertas, María José, Univ Complutense de Madrid
Husch, Jonathan M., Rider Univ
Huysken, Kristin, Indiana Univ Northwest
Jagoutz, Oliver, Massachusetts Inst of Tech
Jahangiri, Ahmad, Univ of Tabriz
Janney, Phillip, Univ of Cape Town
Jaye, Shelley, Northern Virginia Comm Coll - Annandale
Jean, Marlon , Univ of Maine, Farmington
Johnson, Kenneth, Univ of Houston Downtown
Johnson, Kevin T. M., Univ of Hawai'i, Manoa
Jones, Adrian P., Univ Coll London
Jones, Norris W., Univ of Wisconsin, Oshkosh
Jordan, Brennan T., Univ of South Dakota
Kay, Robert W., Cornell Univ
Kay, Suzanne M., Cornell Univ
Kelemen, Peter B., Woods Hole Oceanographic Inst
Kelemen, Peter B., Columbia Univ
Kempton, Pamela D., Kansas State Univ
Kent, Adam J., Oregon State Univ
Klein, Emily M., Duke Univ
Kopylova, Maya G., Univ of British Columbia
Koteas, G. Christopher, Norwich Univ
Kreiger, William (Bill), York Coll of Pennsylvania
Kurum, Sevcan, Firat Univ
Kyle, Philip R., New Mexico Tech
Kysar Mattietti, Giuseppina, George Mason Univ
Lackey, Jade Star, Pomona Coll
Landenberger, Bill, Univ of Newcastle
Lange, Rebecca A., Univ of Michigan
Le Roux, Veronique, Woods Hole Oceanographic Inst
Leake, Bernard E., Cardiff Univ
Leeman, William P., Rice Univ
Lehnert, Kerstin, Columbia Univ
Lesher, Charles E., Univ of California, Davis
Levy, Schon S., Los Alamos Nat Lab
Lidiak, Edward G., Univ of Pittsburgh
Lindline, Jennifer, New Mexico Highlands Univ
Luth, Robert W., Univ of Alberta
Lynn, Kendra J., Univ of Delaware
Lytwyn, Jennifer N., Univ of Houston
MacDonald, James H., Florida Gulf Coast Univ
Maclennan, John, Univ of Cambridge
Mahood, Gail A., Stanford Univ
Manduca, Cathryn A., Carleton Coll
Marsh, Julian S., Rhodes Univ
Mattioli, Glen, Univ of Texas, Arlington
Matty, David J., Weber State Univ
Mayborn, Kyle R., Western Illinois Univ
McCanta, Molly, Univ of Tennessee, Knoxville
McClellan, Elizabeth, Radford Univ
McMillan, Nancy J., New Mexico State Univ, Las Cruces
Medaris, Jr., Levi G., Univ of Wisconsin, Madison
Memeti, Valbone, California State Univ, Fullerton
Mertzman, Stanley A., Franklin and Marshall Coll
Michael, Peter J., Univ of Tulsa
Miller, Calvin F., Vanderbilt Univ
Miller, James D., Univ of Minnesota Duluth
Miller, Jonathan S., San Jose State Univ
Mills, James G., DePauw Univ
Misra, Saumitra, Univ of KwaZulu-Natal
Mitchell, Roger H., Lakehead Univ
Moayyed, Mohsen -., Univ of Tabriz
Morgan, Daniel, Univ of Leeds
Morse, Stearns A., Univ of Massachusetts, Amherst
MOYEN, Jean-François, Université Jean Monnet, Saint-Etienne
Mungall, James E., Carleton Univ
Mustart, David A., San Francisco State Univ
Myers, James D., Univ of Wyoming
Nabelek, Peter I., Univ of Missouri
Nash, Barbara P., Univ of Utah
Naslund, H. Richard, Binghamton Univ
Naumann, Terry R., Univ of Alaska, Anchorage
Neal, Clive R., Univ of Notre Dame
Nelson, Stephen A., Tulane Univ
Nelson, Wendy, Towson Univ
Nicholls, James W., Univ of Calgary
Nicholson, Kirsten N., Ball State Univ
Nicolaysen, Kirsten P., Whitman Coll
Nielsen, Roger L., Oregon State Univ
Niu, Yaoling, Durham Univ
Ntsaluba, Bantubonke I., Rhodes Univ
O'Neal, Matt, California Geological Survey
Orejana, David, Univ Complutense de Madrid
Ouimette, Mark A., Hardin-Simmons Univ
Page, Philippe, Universite du Quebec a Chicoutimi
Panter, Kurt S., Bowling Green State Univ
Parker, Don F., Baylor Univ
Parker, Don, Wayland Baptist Univ
Parlak, Osman, Cukurova Universitesi
Patino-Douce, Marta, Univ of Georgia
Peate, David W., Univ of Iowa
Pérez-Soba, Cecilia, Univ Complutense de Madrid
Perez-Venzor, Jose A., Universidad Auto de Baja California Sur
Perfit, Michael R., Univ of Florida
Petrinec, Zorica, Univ of Zagreb
Polat, Ali, Univ of Windsor
Potter, Katherine E., Utah State Univ
Price, Jonathan D., Midwestern State Univ
Puffer, John H., Rutgers, The State Univ of New Jersey, Newark
Puziewicz, Jacek, Univ of Wroclaw
Raia, Federica, City Coll (CUNY)
Ratajeski, Kent, Univ of Kentucky
Reagan, Mark K., Univ of Iowa
Reavy, John, Univ Coll Cork
Reid, Arch M., Univ of Houston
Reidel, Stephen P., Washington State Univ
Ren, Minghua , Univ of Nevada, Las Vegas
Reyes-Cortes, Manuel, Universidad Auto de Chihuahua
Roden, Michael F., Univ of Georgia
Roelofse, Frederick, Univ of the Free State
Rooney, Tyrone, Michigan State Univ
Ross, Martin E., Northeastern Univ
Ruprecht, Philipp P., Univ of Nevada, Reno
Sarrionandia, Fernando, Univ of the Basque Country UPV/EHU
Sas, Mai, Western Washington Univ
Schmidt, Mariek, Brock Univ
Schrader, Christian M., SUNY Potsdam
Schulze, Daniel J., Univ of Toronto
Schwab, Brandon E., Western Carolina Univ
Schwartz, Joshua J., California State Univ, Northridge
Scoates, James S., Univ of British Columbia
Seaman, Sheila J., Univ of Massachusetts, Amherst
Seeger, Cheryl M., Missouri Dept of Nat Resources
Sen, Gautam, Florida InterNat Univ
Shaw, Cliff S., Univ of New Brunswick
Shearer, Jr., Charles K., Univ of New Mexico
Shervais, John W., Utah State Univ
Shirey, Steven B., Carnegie Inst for Science
Shuster, Robert D., Univ of Nebraska, Omaha
Sigurdson, David, California State Univ, Dominguez Hills
Silva-Mora, Luis, Universidad Nacional Auto de Mexico

Sinton, John M., Univ of Hawai'i, Manoa
Sirbescu, Mona, Central Michigan Univ
Size, William B., Emory Univ
Sjoblom, Megan, Brigham Young Univ - Idaho
Smedes, Harry W., Southern Oregon Univ
Smith, Diane R., Trinity Univ
Smith, Douglas, Univ of Texas, Austin
Smith, Eugene I., Univ of Nevada, Las Vegas
Smith, Matthew C., Univ of Florida
Smith, Terence E., Univ of Windsor
Somarin, Alireza, Brandon Univ
Spera, Frank J., Univ of California, Santa Barbara
Springer, Robert K., Brandon Univ
Srogi, LeeAnn, West Chester Univ
Stachel, Thomas, Univ of Alberta
Stearns, Michael, Utah Valley Univ
Stern, Charles R., Univ of Colorado
Stewart, Michael A., Univ of Illinois, Urbana-Champaign
Stork, Allen L., Western Colorado Univ
Stormer, Jr., John C., Rice Univ
Streck, Martin J., Portland State Univ
Swanson , Sam , Univ of Georgia
Szymanski, David, Bentley Univ
Tacinelli, John C., Rochester Comm & Technical Coll
Taib, Nur Iskandar, Univ of Malaya
Templeton, Jeffrey H., Western Oregon Univ
Tepley, III, Frank J., Oregon State Univ
Tepper, Jeffrey H., Univ of Puget Sound
Thakurta, Joyashish, Western Michigan Univ
Thomas, Jay, Syracuse Univ
Till, Christy, Arizona State Univ
Turner, Stephen, Univ of Massachusetts, Amherst
Udry, Arya, Univ of Nevada, Las Vegas
Ural, Melek, Firat Univ
Urbanczyk, Kevin, Sul Ross State Univ
Van Tongeren, Jill A., Rutgers, The State Univ of New Jersey
Vance, Joseph A., Univ of Washington
Vance, Robert K., Georgia Southern Univ
Vander Auwera, Jacqueline, Universite de Liege
VanTongeren, Jill, Tufts Univ
Vetter, Scott K., Centenary Coll of Louisiana
Viegas, Anthony V., Goa Univ
Villaseca, Carlos, Univ Complutense de Madrid
von der Handt, Anette, Univ of Minnesota, Twin Cities
Wadsworth, William B., Whittier Coll
Walker, Barry , Portland State Univ
Walowski, Kristina J., Middlebury Coll
Walter , Michael , Univ of Bristol
Wanless, Dorsey, Boise State Univ
Warren, Richard G., Los Alamos Nat Lab
Webster, John R., Minot State Univ
Wehner, Peter J., Austin Comm Coll District
Weiblen, Paul W., Univ of Minnesota, Twin Cities
Weiland, Thomas J., Georgia Southwestern State Univ
Wendlandt, Richard F., Colorado School of Mines
Wenner, Jennifer M., Univ of Wisconsin, Oshkosh
White, Craig M., Boise State Univ
White, John C., Eastern Kentucky Univ
Whitney, James A., Univ of Georgia
Whittington, Alan, Univ of Missouri
Wiese, Katryn, City Coll of San Francisco
Wittke, James, Northern Arizona Univ
Wobus, Reinhard A., Williams Coll
Wolf, Michael B., Augustana Coll
Wolff, John A., Washington State Univ
Wolff, John A., Washington State Univ
Worcester, Peter A., Hanover Coll
Wulff, Andrew H., Whittier Coll
Yogodzinkski, Gene, Univ of South Carolina
Young, Davis A., Calvin Coll
Zamzow, Craig E., Clarion Univ

Metamorphic Petrology

Abati, Jacobo, Univ Complutense de Madrid
Allen, Gary C., Univ of New Orleans
Appleyard, Edward C., Univ of Waterloo
Apraiz, Arturo, Univ of the Basque Country UPV/EHU
Arenas, Ricardo, Univ Complutense de Madrid
Ashton, Kenneth E., Univ of Regina
Baldwin, Julia, Univ of Montana
Bartoli, Omar, Università degli Studi di Padova
Bauer, Paul W., New Mexico Tech
Bebout, Gray E., Lehigh Univ
Berg, Christopher A., Univ of West Georgia
Berg, Christopher A., Orange Coast Coll
Berman, Robert, Carleton Univ
Blake, David E., Univ of North Carolina, Wilmington
Block, Amy, Univ of Minnesota
Braga, Roberto, Alma Mater Studiorum Università di Bologna
Brouwer, Fraukje M., Vrije Universiteit Amsterdam
Brown, Edwin H., Western Washington Univ
Brown, Michael, Curtin Univ
Burnley, Pamela C., Univ of Nevada, Las Vegas
Burwash, Ronald A., Univ of Alberta
Caddick, Mark, Virginia Polytechnic Inst & State Univ
Cardace, Dawn, Univ of Rhode Island
Carillo-Chavez, Alejandro J., Universidad Auto de Baja California Sur
Carlson, William D., Univ of Texas, Austin
Castiñeiras, Pedro, Univ Complutense de Madrid
Castro, Adrian, Wellesley Coll
Cesare, Bernardo, Università degli Studi di Padova
Chacko, Thomas, Univ of Alberta
Clark, Chris, Curtin Univ
Clarke, Geoffrey L., Univ of Sydney
Cooper, Alan F., Univ of Otago
Costin, Gelu, Rhodes Univ
Cruz-Uribe, Alicia M., Univ of Maine
Daniel, Christopher G., Bucknell Univ
Davidson, Cameron, Carleton Coll
Dawson, Jane P., Iowa State Univ of Science & Tech
DeAngelis, Michael T., Univ of Arkansas at Little Rock
DesOrmeau, Joel, Univ of Nevada, Reno
Devore, George W., Florida State Univ
Diener, Johann, Univ of Cape Town
Dietsch, Craig, Univ of Cincinnati
Dipple, Gregory M., Univ of British Columbia
dos Santos, Telmo B., Universidade de Lisboa
Droop, Giles, Univ of Manchester
Duke, Edward F., South Dakota School of Mines & Tech
Dunn, Steven R., Mount Holyoke Coll
Dyck, Brendan, Univ of British Columbia, Okanagan Campus
Dymek, Robert F., Washington Univ in St. Louis
Eckert, James O., Yale Univ
El Shazly, Aley, Marshall Univ
Elias-Herrera, Mariano, Universidad Nacional Auto de Mexico
Evans, Bernard W., Univ of Washington
Evans, Katy, Curtin Univ
Faryad, Shah Wali, Charles Univ
Fawcett, J. J., Univ of Toronto
Fitzsimons, Ian, Curtin Univ
Fletcher, John, Centro de Inv Científica y de Ed Superior de Ensenada
Flores, Kennet, Univ of North Carolina, Chapel Hill
Flores, Kennet, Brooklyn Coll (CUNY)
Fornash, Katherine, Ohio Univ
Foster Jr., Charles T., Univ of Iowa
Frost, B. Ronald, Univ of Wyoming
Ghent, Edward D., Univ of Calgary
Giaramita, Mario J., California State Univ, Stanislaus
Gillis, Kathryn, Univ of Victoria
Goodge, John W., Univ of Minnesota Duluth
Gordon, Stacia M., Univ of Nevada, Reno
Grew, Edward S., Univ of Maine
Grew, Priscilla C., Univ of Nebraska, Lincoln
Grover, Timothy W., Castleton Univ
Grover, Timothy W., Univ of Northern Colorado
Guevara, Victor, Skidmore Coll
Guevara, Victor E., Amherst Coll
Hacker, Bradley R., Univ of California, Santa Barbara
Hajialioghli, Robab, Univ of Tabriz
Halama, Ralf, Keele Univ
Hallett, Benjamin W., Univ of Wisconsin, Oshkosh
Henry, Darrell J., Louisiana State Univ

Hepburn, J. Christopher, Boston Coll
Hernández Uribe, David, Univ of Illinois Chicago
Hickmott, Donald D., Los Alamos Nat Lab
Hoisch, Thomas D., Northern Arizona Univ
Holdaway, Michael J., Southern Methodist Univ
Holland, Tim, Univ of Cambridge
Hollister, Lincoln S., Princeton Univ
Horvath, Peter, Univ of the Witwatersrand
Horvath, Peter, Rhodes Univ
Huizenga, Jan Marten, James Cook Univ
Hunt, Emma, Furman Univ
Iancu, Ovidiu G., Alexandru Ioan Cuza
Indares, Aphrodite D., Memorial Univ of Newfoundland
James, Richard, Laurentian Univ, Sudbury
Johnson, Eric L., Hartwick Coll
Johnson, Tim, Curtin Univ
Jones, Jon W., Univ of Calgary
Kamber, Balz, Queensland Univ of Tech
Kays, Marvin A., Univ of Oregon
Kerrick, Derrill M., Pennsylvania State Univ, Univ Park
Kim, Hyeong Soo, Korea Univ
Laird, Jo, Univ of New Hampshire
Lamb, Will, Texas A&M Univ
LAURENT, Antonin, Université Jean Monnet, Saint-Etienne
Law, Eric W., Muskingum Univ
Leech, Mary L., San Francisco State Univ
Liou, Juhn G., Stanford Univ
Lori, Lisa, Missouri Dept of Nat Resources
Maboko, Makenya A., Univ of Dar es Salaam
Magloughlin, Jerry F., Colorado State Univ
Mahan, Kevin H., Univ of Colorado
Manecan, Teodosia, Hunter Coll (CUNY)
Manon, Matthew R., Union Coll
Martínez Fernández, Francisco, Universitat Auto de Barcelona
Mattinson, Chris, Central Washington Univ
Mavimbela, Philani, Univ of KwaZulu-Natal
Menold, Carrie A., Albion Coll
Metcalf, Rodney, Univ of Nevada, Las Vegas
Metzger, Ellen P., San Jose State Univ
Moazzen, Mohssen -., Univ of Tabriz
Mogk, David W., Montana State Univ
Morton, Douglas M., Univ of California, Riverside
Mosonyi, Emilia, Babes-Bolyai Univ
Mulcahy, Sean R., Western Washington Univ
Nishiyama, Tadao, Kumamoto Univ
Nowariak, Eric, Univ of Minnesota
Ortega-Gutierrez, Fernando, Universidad Nacional Auto de Mexico
Page, F Zeb, Oberlin Coll
Panish, Peter T., Univ of Massachusetts, Amherst
Parkinson, Christopher D., Univ of New Orleans
Pattison, David R., Univ of Calgary
Peck, William H., Colgate Univ
Penniston-Dorland, Sarah, Univ of Maryland
Perkins, III, Dexter, Univ of North Dakota
Pervunina, Aelita, Inst of Geology (Karelia, Russia)
Plummer, Charles C., Sacramento State Univ
Potel, Sébastien, Institut Polytechnique LaSalle Beauvais (ex-IGAL)
Raeside, Robert P., Acadia Univ
Rahaman, M. A., Obafemi Awolowo Univ
Reche Estrada, Joan, Universitat Auto de Barcelona
Rietmeijer, Frans J., Univ of New Mexico
Rivers, Toby C. J. S., Memorial Univ of Newfoundland
Roberts, Frank, Montgomery County Comm Coll
Roselle, Gregory T., Brigham Young Univ - Idaho
Rumble, III, Douglas, Carnegie Inst for Science
Sánchez, Sonia, Univ Complutense de Madrid
Sawyer, Edward W., Universite du Quebec a Chicoutimi
Schiffman, Peter, Univ of California, Davis
Scott, James M., Univ of Otago
Selverstone, Jane E., Univ of New Mexico
Severson, Allison, Univ of Minnesota
Sisson, Virginia, Univ of Houston
Sommer , Holgar, Univ of Namibia
Spear, Frank S., Rensselaer Polytechnic Inst
Spiess, Richard, Università degli Studi di Padova
Spray, John G., Univ of New Brunswick
Staples, Reid, Simon Fraser Univ
Staples, Reid, Douglas Coll
Stevens, Liane M., Stephen F. Austin State Univ
Stoddard, Edward F., North Carolina State Univ
Storey, Craig D., Univ of Portsmouth
Stout, James H., Univ of Minnesota, Twin Cities
Stowell, Harold H., Univ of Alabama
Swapp, Susan M., Univ of Wyoming
Thomson, Jennifer A., Eastern Washington Univ
Tinkham, Doug, Laurentian Univ, Sudbury
Toraman, Erkan, Salem State Univ
Turnock, Allan C., Univ of Manitoba
van Rooyen, Deanne, Acadia Univ
Vitale Brovarone, Alberto, Alma Mater Studiorum Università di Bologna
Walsh, Emily O., Cornell Coll
Walsh, Emily O., Univ of Iowa
Waters, Dave, Univ of Oxford
Weber, Bodo, Centro de Inv Científica y de Ed Superior de Ensenada
Wheeler, John, Univ of Liverpool
Whitney, Donna L., Univ of Minnesota, Twin Cities
Wilks, Maureen, New Mexico Tech
Wintsch, Robert P., Indiana Univ, Bloomington
Woodland, Bertram G., Field Museum of Nat History
Zhai, Xiaoming, Coll of Lake County

Sedimentary Petrology

Adekola, S. A., Obafemi Awolowo Univ
Al-Aasm, Ihsan S., Univ of Windsor
Alberstadt, Leonard P., Vanderbilt Univ
Alonso, Ana María, Univ Complutense de Madrid
Arribas, Eugenia, Univ Complutense de Madrid
Arribas, José, Univ Complutense de Madrid
Azerêdo, Ana , Universidade de Lisboa
Babek, Ondrej, Masaryk Univ
Basu, Abhijit, Indiana Univ, Bloomington
Bertog, Janet, Northern Kentucky Univ
Bertok, Carlo, Università di Torino
Boulvain, Frederic P., Universite de Liege
Brengman, Latisha A., Univ of Minnesota Duluth
Brenner, Robert, Univ of Iowa
Bruner, Katherine R., West Virginia Univ
Budd, David A., Univ of Colorado
Cavazza, William, Alma Mater Studiorum Università di Bologna
Chafetz, Henry S., Univ of Houston
Channing, Alan, Univ of Wales
Choh, Suk-Joo, Korea Univ
Corcoran, Patricia, Western Univ
Cousineau, Pierre, Universite du Quebec a Chicoutimi
dela Pierre, Francesco, Università di Torino
Dockal, James A., Univ of North Carolina, Wilmington
Donovan, R N., Texas Christian Univ
Dutton, Shirley P., Univ of Texas at Austin, Jackson Schl of Geosciences
Elmore, R. Douglas, Univ of Oklahoma
Fitzgerald, David, Saint Mary's Univ
Franseen, Evan K., Univ of Kansas
Galli-Olivier, Carlos A., Universidad Auto de Baja California Sur
Gómez Gras, David, Universitat Auto de Barcelona
Griffing, David H., Hartwick Coll
Grigsby, Jeffry D., Ball State Univ
Hampson, Gary , Imperial Coll
Horton, Jr., Robert A., California State Univ, Bakersfield
Humphrey, John D., King Fahd Univ of Petroleum and Minerals
James, Noel P., Queen's Univ
Kaczmarek, Stephen E., Western Michigan Univ
Kerans, Charles, Univ of Texas at Austin, Jackson Schl of Geosciences
Khan, Mohammad Wahdat Y., Pt. Ravishankar Shukla Univ
Khandaker, Nazrul I., York Coll (CUNY)
Kirkland, Brenda L., Mississippi State Univ
Koeshidayatullah, Ardiansyah, King Fahd Univ of Petroleum and Minerals
Laury, Robert L., Southern Methodist Univ
Lowe, Donald R., Stanford Univ
Lukoczki, Georgina, Univ of Kentucky
Lumsden, David N., Univ of Memphis
Mackey, Tyler, Univ of New Mexico
Martin, Arturo, Centro de Inv Científica y de Ed Superior de Ensenada
Martire, Luca, Università di Torino

McBride, Earle F., Univ of Texas, Austin
McKean , Rebecca, Saint Norbert Coll
McKnight, Brian K., Univ of Wisconsin, Oshkosh
Measures, Elizabeth A., Sul Ross State Univ
Meinhold, Guido, Keele Univ
Melim, Leslie A., Western Illinois Univ
Milliken, Kitty L., Univ of Texas, Austin
Montañez, Isabel P., Univ of California, Davis
Moore, Bruce R., Univ of Kentucky
Moshier, Stephen O., Wheaton Coll
Muntean, Thomas, Adrian Coll
Nothdurft, Luke, Queensland Univ of Tech
Pedone, Vicki A., California State Univ, Northridge
Perkins, Ronald D., Duke Univ
Portner, Ryan, San Jose State Univ
Richardson, Darlene S., Indiana Univ of Pennsylvania
Roman, Aubrecht, Comenius Univ in Bratislava
Rossi, Carlos, Univ Complutense de Madrid
Ruppel, Stephen C., Univ of Texas at Austin, Jackson Schl of Geosciences
Saja, David, Cleveland Museum of Nat History
Sanchez Roman, Monica, Vrije Universiteit Amsterdam
Sandberg, Philip A., Univ of Illinois, Urbana-Champaign
Sanz, Esther, Univ Complutense de Madrid
Sasaran, Emanoil Florin, Babes-Bolyai Univ
Savrda, Charles E., Auburn Univ
Schacht, Ulrike, Univ of Adelaide
Scholle, Peter, New Mexico Tech
Schreiber, B. Charlotte, Queens Coll (CUNY)
Schwab, Frederick L., Washington & Lee Univ
Siegrist, jr, Henry G., Univ of Guam
Stefani, Cristina, Università degli Studi di Padova
Sutterlin, Peter G., Wichita State Univ
Suttner, Lee J., Indiana Univ, Bloomington
Textoris, Daniel A., Univ of North Carolina, Chapel Hill
Thayer, Paul A., Univ of North Carolina, Wilmington
Uddin, Ashraf, Auburn Univ
Varas, María Josefa, Univ Complutense de Madrid
Vice, Mari A., Univ of Wisconsin, Platteville
Videtich, Patricia E., Grand Valley State Univ
Walker, Theodore R., Univ of Colorado
Young, Harvey R., Brandon Univ
Zeigler, E. Lynn, Georgia State Univ, Perimeter Coll, Clarkston Campus

Sedimentology

Abbott, Mark B., Univ of Pittsburgh
Abbott, Patrick L., San Diego State Univ
Adams, Robert W., SUNY, The Coll at Brockport
Alexander, Jan, Univ of East Anglia
Alexander, Jane L., Coll of Staten Island/CUNY
Algeo, Thomas J., Univ of Cincinnati
Ali, Dr. Syed Haroon, Univ of Sargodha
Allison, Mead A., Tulane Univ
Alvarez-Arellano, Alejandro, Universidad Auto de Baja California Sur
Ambrose, William, Univ of Texas, Austin
Ambrose, William A., Univ of Texas at Austin, Jackson Schl of Geosciences
Amorosi, Alessandro, Alma Mater Studiorum Università di Bologna
Amos, Kathryn, Univ of Adelaide
Anani, Chris Y., Univ of Ghana
Andersen, David W., San Jose State Univ
Anderson, Thomas B., Sonoma State Univ
Anderson-Folnagy, Heidi, Univ of Montana Western
Anfinson, Owen, Sonoma State Univ
Anglen, Jeff, Fresno State Univ
Arnott, R. William, Univ of Ottawa
Ashley, Gail M., Rutgers, The State Univ of New Jersey
Asiedu, Daniel K., Univ of Ghana
Askari, Zohreh, Illinois State Geological Survey
Austin, Steve A., Cedarville Univ
Autin, Whitney J., SUNY, The Coll at Brockport
Balog-Szabo, Anna, Virginia Western Comm Coll
Barbeau, Jr., David, Univ of South Carolina
Barham, Milo, Curtin Univ
Barhurst, James, Univ of Newcastle Upon Tyne
Bartley, Julie K., Gustavus Adolphus Coll
Bassett, Kari N., Univ of Canterbury
Beck, Catherine C., Hamilton Coll
Behl, Richard J., California State Univ, Long Beach
Bekker, Andrey, Univ of California, Riverside
Benito, María Isabel, Univ Complutense de Madrid
Bergmann, Kristin, Massachusetts Inst of Tech
Bershaw, John T., Portland State Univ
Best, James L., Univ of Illinois, Urbana-Champaign
Bestland, Erick, Flinders Univ
Bhattacharya, Janok, Univ of Houston
Birgenheier, Lauren, Univ of Utah
Bjornstad, Bruce N., Pacific Northwest Nat Lab
Blake, Bascombe (Mitch), West Virginia Univ
Blum, Michael D., Univ of Kansas
Bodenbender, Brian E., Hope Coll
Boothroyd, Jon C., Univ of Rhode Island
Bordy, Emese M., Univ of Cape Town
Borowski, Walter S., Eastern Kentucky Univ
Bourgeois, Joanne (Jody), Univ of Washington
Bowen, Brenda , Univ of Utah
Brand, Brittany, Boise State Univ
Brandon, Christine M., Bridgewater State Univ
Breda, Anna, Università degli Studi di Padova
Breyer, John, Texas Christian Univ
Bristow, Charlie, Birkbeck Coll
Broda, James E., Woods Hole Oceanographic Inst
Brunt, Rufus, Univ of Manchester
Burns, Diane M., Eastern Illinois Univ
Buyce, M. Raymond, Mercyhurst Univ
Campo, Bruno, Alma Mater Studiorum Università di Bologna
Capaldi, Tomas, Univ of Nevada, Las Vegas
Caputo, Mario V., San Diego State Univ
Carrapa, Barbara, Univ of Arizona
Carroll, Alan R., Univ of Wisconsin, Madison
Cartwright, Joe, Univ of Oxford
Caruthers, Andrew, Western Michigan Univ
Cassel, Elizabeth J., Univ of Idaho
Castle, James W., Clemson Univ
Catuneanu, Octavian, Univ of Alberta
Celik, Hasan, Firat Univ
Chan, Marjorie A., Univ of Utah
Cheel, Richard J., Brock Univ
Chiarella, Domenico, Royal Holloway Univ of London
Chow, Nancy, Univ of Manitoba
Chowns, Timothy M., Univ of West Georgia
Christie-Blick, Nicholas, Columbia Univ
Clarke, Stu, Keele Univ
Clift, Peter, Louisiana State Univ
Coch, Nicholas K., Queens Coll (CUNY)
Coe, Angela L., Open Univ
Cole, Rex D., Colorado Mesa Univ
Connely, Melissa, Casper Coll
Conrad, Susan H., Dutchess Comm Coll
Cope, Tim D., DePauw Univ
Corsetti, Frank A., Univ of Southern California
Covault, Jacob, Univ of Texas, Austin
Cowan, Clinton A., Carleton Coll
Cowan, Ellen A., Appalachian State Univ
Cox, Ronadh, Williams Coll
Crowley, Stephen, Univ of Liverpool
Cullen, James L., Salem State Univ
Cuneen, Jane, Curtin Univ
Currie, Brian S., Miami Univ
Dade, William B., Dartmouth Coll
Dalrymple, Robert W., Queen's Univ
Dansereau, Pauline, Universite Laval
Dashtgard, Shahin, Simon Fraser Univ
Davatzes, Alexandra K., Temple Univ
Davies, Neil, Univ of Cambridge
Davies, Sarah, Leicester Univ
de la Horra, Raúl, Univ Complutense de Madrid
de Wet, Carol B., Franklin and Marshall Coll
Dehler, Carol M., Utah State Univ
Deibert, Jack, Austin Peay State Univ
Demicco, Robert V., Binghamton Univ
Deocampo, Daniel M., Georgia State Univ
Desrochers, Andre, Univ of Ottawa
Diemer, John A., Univ of North Carolina, Charlotte

Dix, George R., Carleton Univ
Dominic, David F., Wright State Univ
Donaldson, J. Allan, Carleton Univ
Donovan, Art, Texas A&M Univ
Dravis, Jeffrey J., Rice Univ
Droxler, Andre W., Rice Univ
Dröllner, Max, Curtin Univ
Dumas, Simone, Univ of Ottawa
Dunagan, Stan P., Univ of Tennessee, Martin
Dutta, Prodip K., Indiana State Univ
Dutton, Shirley P., Univ of Texas, Austin
Dwyer, Gary S., Duke Univ
Edmonds, Douglas A., Indiana Univ, Bloomington
Edwards, Cole T., Appalachian State Univ
Elliott, Jr., William S., Univ of Southern Indiana
Elrick, Maya, Univ of New Mexico
Enos, Paul, Univ of Kansas
Eriksson, Kenneth A., Virginia Polytechnic Inst & State Univ
Ethridge, Frank G., Colorado State Univ
Evans, James E., Bowling Green State Univ
Ewing, Ryan, Texas A&M Univ
Eyles, Carolyn H., McMaster Univ
Fagel, Nathalie, Universite de Liege
Fan, Majie, Univ of Texas, Arlington
Farkas, Steven, Central Washington Univ
Farrell, Kathleen M., N.C. Department of Enval Quality
Fedo, Christopher, Univ of Tennessee, Knoxville
Fedorchuk, Nicholas D., Southern Connecticut State Univ
Fielding, Chris, Univ of Connecticut
Fielding, Christopher R., Univ of Nebraska, Lincoln
Fillmore, Robert P., Western Colorado Univ
Finkenbinder, Matthew S., Wilkes Univ
Finzel, Emily, Univ of Iowa
Fischer, Alfred G., Univ of Southern California
Fisher, William L., Univ of Texas, Austin
Flaig, Peter P., Univ of Texas, Austin
Fletcher, Charles H., Univ of Hawai'i, Manoa
Fouke, Bruce W., Univ of Illinois, Urbana-Champaign
Fox, James E., South Dakota School of Mines & Tech
Fralick, Philip W., Lakehead Univ
Frank, Tracy D., Univ of Nebraska, Lincoln
Franseen, Evan K., Univ of Kansas
Fregenal, María Antonia, Univ Complutense de Madrid
Freile, Deborah, New Jersey City Univ
Freitas, Maria d., Universidade de Lisboa
Fu, Qilong, Univ of Texas, Austin
Fu, Qilong, Univ of Texas at Austin, Jackson Schl of Geosciences
Gabel, Sharon, Lee Coll
Gaines, Robert R., Pomona Coll
Galli, Kenneth G., Boston Coll
Galloway, William E., Univ of Texas, Austin
Galloway, William E., Univ of Texas, Austin
Gani, M. Royhan, Western Kentucky Univ
Gannaway Dalton, Evelyn, Utah State Univ
Gaylord, David R., Washington State Univ
Gerhard, Lee C., Univ of Kansas
Ghinassi, Massimiliano, Università degli Studi di Padova
Gianniny, Gary, Fort Lewis Coll
Gibling, Martin R., Dalhousie Univ
Gilbert, Meagan, Univ of Saskatchewan
Gilleaudeau, Geoff, George Mason Univ
Gillespie, Robb, Western Michigan Univ
Gilley, Brett H., Univ of British Columbia
Glumac, Bosiljka, Smith Coll
Goldstein, Robert H., Univ of Kansas
Goodbred, Jr., Steven L., Vanderbilt Univ
Gore, Pamela J. W., Emory Univ
Gore, Pamela J. W., Georgia State Univ, Perimeter Coll, Clarkston Campus
Gorte, Mary C., Delta Coll
Gotz, Annette, Rhodes Univ
Goudge, Timothy, Univ of Texas, Austin
Gough, Amy, Royal Holloway Univ of London
Grammer, Michael, Oklahoma State Univ
Greb, Stephen F., Univ of Kentucky
Greene, Todd, California State Univ, Chico
Greer, Lisa, Washington & Lee Univ
Gregg, Jay M., Oklahoma State Univ
Gregor, C. B., Wright State Univ
Grimm, Kurt A., Univ of British Columbia
Grippo, Alessandro, Santa Monica Coll
Grothe, Pamela, Univ of Mary Washington
Grotzinger, John P., California Inst of Tech
Grove, Karen, San Francisco State Univ
Guo, Junhua, California State Univ, Bakersfield
Gupta, Sanjeev, Imperial Coll
Hajek, Elizabeth, Pennsylvania State Univ, Univ Park
Hall, Mary Jo, Rider Univ
Halverson, Galen, McGill Univ
Hamilton, Sally, Heriot-Watt Univ
Hampton, Brian A., New Mexico State Univ, Las Cruces
Harris, Clay D., Middle Tennessee State Univ
Harris, Mark T., Univ of Wisconsin, Milwaukee
Harris, Nicholas, Univ of Alberta
Harrison, III, William B., Western Michigan Univ
Hasselbo, Stephen, Exeter Univ
Hayes, Miles O., Univ of New Orleans
Haywick, Douglas W., Univ of South Alabama
Hazel, Jr., Joseph E., Northern Arizona Univ
Heinrich, Paul V., Louisiana State Univ
Henderson, Miles, Univ of Texas Permian Basin
Hendrix, Marc S., Univ of Montana
Hernandez Molina, F. J., Royal Holloway Univ of London
Herrero, María Josefa, Univ Complutense de Madrid
Hess-Tanguay, Lillian, Long Island Univ, C.W. Post Campus
Hesse, Reinhard, McGill Univ
Heubeck, Christoph, Friedrich-Schiller-Univ Jena
Hiatt, Eric E., Univ of Wisconsin, Oshkosh
Hickson, Thomas A., Univ of Saint Thomas
Higgins, Sean M., Columbia Univ
Hilbert-Wolf, Hannah L., James Cook Univ
Hill, Paul S., Dalhousie Univ
Hill, Philip R., Univ of Victoria
Hinderer, Matthias, Technische Universitaet Darmstadt
Hiscott, Richard N., Memorial Univ of Newfoundland
Hobbs, Kevin, New Mexico Tech
Hodgetts, David, Univ of Manchester
Hoey, Trevor, Univ of Glasgow
Holbrook, John M., Texas Christian Univ
Hollis, Cathy, Univ of Manchester
Holmes, Mary Anne, Univ of Nebraska, Lincoln
Homann, Martin D., Univ Coll London
Horton, Brian, Univ of Texas, Austin
Houston, William (Bill), Lake Superior State Univ
Hovorka, Susan D., Univ of Texas, Austin
Hovorka, Susan D., Univ of Texas at Austin, Jackson Schl of Geosciences
Howell, John, Univ of Aberdeen
Howell, Kimberly, Univ of Massachusetts, Lowell
Hubbard, Dennis K., Oberlin Coll
Hubbard, Stephen M., Univ of Calgary
Hurtgen, Matthew T., Northwestern Univ
Husinec, Antun, St. Lawrence Univ
Hussin, Azhar Hj, Univ of Malaya
Ielpi, Alessandro, Laurentian Univ, Sudbury
Ingersoll, Raymond V., Univ of California, Los Angeles
Isbell, John L., Univ of Wisconsin, Milwaukee
Jackson, Chester W., Georgia Southern Univ
Jackson, William T., Univ of South Alabama
Jaeger, John M., Univ of Florida
Janson, Xavier, Univ of Texas, Austin
Jiang, Ganqing, Univ of Nevada, Las Vegas
Jinnah, Zubair A., Univ of the Witwatersrand
John, Cedric, Imperial Coll
Johnson, Cari, Univ of Utah
Johnson, Joel P., Univ of Texas, Austin
Jones, David S., Amherst Coll
Jones, Lawrence S., Colorado Mesa Univ
Jones, Merren, Univ of Manchester
Jones, Stuart, Durham Univ
Jordan, William M., Millersville Univ
Kah, Linda, Univ of Tennessee, Knoxville
Kale, Makarand G., Savitribai Phule Pune Univ,
Kamola, Diane, Univ of Kansas

Kanfoush, Sharon L., Utica Coll
Kendall, Christopher G., Univ of South Carolina
Kent, Donald M., Univ of Regina
Kerans, Charles, Univ of Texas, Austin
Kerr, Dennis R., Univ of Tulsa
Kidwell, Susan M., Univ of Chicago
Kilibarda, Zoran, Indiana Univ Northwest
Kinvig, Helen, Univ of Liverpool
Koc Tasgin, Calibe, Firat Univ
Kocurek, Gary A., Univ of Texas, Austin
Kocurko, John, Midwestern State Univ
Kokelaar, Peter, Univ of Liverpool
Koning, Daniel, New Mexico Tech
Korus, Jesse, Univ of Nebraska, Lincoln
Korus, Jesse T., Univ of Nebraska, Lincoln
Kotha, Mahender, Goa Univ
Kovac, Michal, Comenius Univ in Bratislava
Kovaèiæ, Marijan, Univ of Zagreb
Krantz, David E., Univ of Toledo
Kraus, Mary J., Univ of Colorado
Krause, Federico F., Univ of Calgary
Krissek, Lawrence A., Ohio State Univ
Kurtanjek, Dražen, Univ of Zagreb
Langford, Richard P., Univ of Texas, El Paso
Larson, Toti, Univ of Texas, Austin
Last, William M., Univ of Manitoba
Lavoie, Denis, Universite du Quebec
Lawton, Timothy, Univ of Texas, Austin
Laya, Juan Carlos, Texas A&M Univ
Leary, Ryan, New Mexico Tech
Leckie, Dale A., Carleton Univ
LeFever, Richard D., Univ of North Dakota
Lehman, Thomas M., Texas Tech Univ
Leier, Andrew L., Univ of South Carolina
Leithold, Elana L., North Carolina State Univ
Leorri, Eduardo, East Carolina Univ
Levin, Naomi, Johns Hopkins Univ
Levy, Laura B., Humboldt State Univ
Licht, Alexis, Univ of Washington
Liddell, W. David, Utah State Univ
Link, Paul K., Idaho State Univ
Linn, Anne M., Nat Acad of Sci, Eng, and Med
Linol, Bastien, Nelson Mandela Metropolitan Univ
Liutkus-Pierce, Cynthia M., Appalachian State Univ
Lloyd, Graeme, Univ of Oxford
Lock, Brian E., Univ of Louisiana at Lafayette
Long, Darrel, Laurentian Univ, Sudbury
Loope, David B., Univ of Nebraska, Lincoln
Loucks, Robert G., Univ of Texas, Austin
Lowe, David, Memorial Univ of Newfoundland
Ludvigson, Gregory A., Univ of Kansas
Lynds, Ranie, Univ of Wyoming
Mac Eachern, James A., Simon Fraser Univ
MacCarthy, Ivor, Univ Coll Cork
Macdonald, R. Heather, William & Mary
MacInnes, Breanyn, Central Washington Univ
Macintyre, Ian G., Smithsonian Inst / Nat Museum of Nat History
MacLachlan, John, McMaster Univ
Mahoney, J. Brian, Univ of Wisconsin, Eau Claire
Maju-Oyovwikowhe, Efetobore G., Univ of Benin
Malpica-Cruz, Victor M., Universidad Nacional Auto de Mexico
Manker, John P., Georgia Southwestern State Univ
Mankiewicz, Carol, Beloit Coll
Manning-Berg, Ashley, Univ of Tennessee, Chattanooga
Manos, Constantine, SUNY, New Paltz
Marjanac, Tihomir, Univ of Zagreb
Marsaglia, Kathleen M., California State Univ, Northridge
Marshall, Madeline, Albion Coll
Martini, I. Peter, Univ of Guelph
Martini, Ivan, Univ of Siena
Martino, Ronald L., Marshall Univ
Matsuda, Hiroki, Kumamoto Univ
McCaffrey, Bill, Univ of Leeds
McCarthy, Paul, Univ of Alaska, Fairbanks
McCrone, Alistair W., Humboldt State Univ
McDowell, Ronald, West Virginia Univ
McElroy, Brandon, Univ of Wyoming
McNeill, Donald F., Univ of Miami
Mehrtens, Charlotte J., Univ of Vermont
Melas, Faye, Hunter Coll (CUNY)
Merrill, Robert D., Fresno State Univ
Meyer, Rudi, Univ of Calgary
Meyers, Jamie A., Winona State Univ
Miall, Andrew D., Univ of Toronto
Miclaus, Crina G., Alexandru Ioan Cuza
Middleton, Larry T., Northern Arizona Univ
Milligan, Timothy, Dalhousie Univ
Minter, Nicholas, Univ of Portsmouth
Mitchell, Simon F., Univ of the West Indies Mona Campus
Mohrig, David, Univ of Texas, Austin
Monecke, Katrin, Wellesley Coll
Moscardelli, Lorena, Univ of Texas, Austin
Moskalski, Susanne, Stockton Univ
Mount, Jeffrey F., Univ of California, Davis
Mountney, Nigel, Univ of Leeds
Mozley, Peter S., New Mexico Tech
Mrinjek, Ervin, Univ of Zagreb
Munoz, Samuel, Northeastern Univ
Murray, Bryan P., California State Polytechnic Univ, Pomona
Myers, Jeffrey A., Western Oregon Univ
Myrow, Paul M., Colorado Coll
Nalin, Ronald, Loma Linda Univ
Natalicchio, Marcello, Università di Torino
Neal, William J., Grand Valley State Univ
Nehyba, Slavomír, Masaryk Univ
Nelson, Bruce W., Univ of Virginia
Neser, Laura, Virginia Polytechnic Inst & State Univ
Neuweiler, Fritz, Universite Laval
Nick, Kevin, Loma Linda Univ
Noffke, Nora K., Old Dominion Univ
O'Brien, John M., New Jersey City Univ
Oakley, Bryan A., Univ of Rhode Island
Oaks, Jr., Robert Q., Utah State Univ
OConnell, Suzanne B., Wesleyan Univ
Okunuwadje, Sunday E., Univ of Benin
Oliveira, Anabela R., Universidade de Trás-os-Montes e Alto Douro
Oluwajana, Oladotun A., Federal Univ of Tech, Akure
Opeloye, Saka A., Federal Univ of Tech, Akure
Oppo, Davide, Univ of Louisiana at Lafayette
Ortiz, Joseph D., Kent State Univ
Paola, Christopher, Univ of Minnesota, Twin Cities
Pattison, Simon A., Brandon Univ
Peakall, Jeff, Univ of Leeds
Pearson, Eugene F., Univ of the Pacific
Peck, John A., Univ of Akron
Penman, Donald E., Utah State Univ
Perscio, Lyman, Mercyhurst Univ
Peters, Shanan, Univ of Wisconsin, Madison
Petersen, Morris S., Brigham Young Univ
Phillips, Andrew C., Illinois State Geological Survey
Pickering, Kevin P., Univ Coll London
Pietras, Jeff, Binghamton Univ
Pikelj, Kristina, Univ of Zagreb
Piper, David, Dalhousie Univ
Platt, Brian F., Univ of Mississippi
Plink-Bjorklund, Piret, Colorado School of Mines
Plint, A G., Western Univ
Potter, Paul E., Univ of Cincinnati
Potter-McIntyre, Sally, Southern Illinois Univ Carbondale
Powell, Ross D., Northern Illinois Univ
Pratson, Lincoln F., Duke Univ
Preto, Nereo, Università degli Studi di Padova
Price, Gregory, Univ of Plymouth
Prins, Maarten A., Vrije Universiteit Amsterdam
Prothro, Lindsay , Texas A&M Univ, Corpus Christi
Qing, Hairuo, Univ of Regina
Rainbird, R. R., Carleton Univ
Ramsey, Kelvin W., Univ of Delaware
Rankey, Gene, Univ of Kansas
Read, J. Fred, Virginia Polytechnic Inst & State Univ
Reams, Max W., Olivet Nazarene Univ
Redfern, Jonathan, Univ of Manchester

Redmond, Brian T., Wilkes Univ
Reece, Julia, Texas A&M Univ
Rees, Margaret, Univ of Nevada, Las Vegas
Reid, Pam, Univ of Miami
Reid, Steven K., Morehead State Univ
Remacha Grau, Eduard, Universitat Auto de Barcelona
Renaut, Robin W., Univ of Saskatchewan
Reynolds, Laura, Worcester State Univ
Ridgway, Kenneth D., Purdue Univ
Rigo, Manuel, Università degli Studi di Padova
Rigsby, Catherine A., East Carolina Univ
Ritts, Malinda, Univ of Idaho
Roberts, Eric M., James Cook Univ
Roberts, Harry H., Louisiana State Univ
Robinson, Ruth, Univ of St. Andrews
Roda Boluda, Duna, Vrije Universiteit Amsterdam
Rogers, Raymond R., Macalester Coll
Romans, Brian W., Virginia Polytechnic Inst & State Univ
Rossi, Veronica, Alma Mater Studiorum Università di Bologna
Runkel, Anthony, Univ of Minnesota, Twin Cities
Ruppel, Stephen C., Univ of Texas, Austin
Rygel, Michael C., SUNY Potsdam
Saltzman, Matthew R., Ohio State Univ
Sánchez, Yolanda, Univ Complutense de Madrid
Sawyer, J. F., South Dakota School of Mines & Tech
Saylor, Beverly Z., Case Western Reserve Univ
Scheingross, Joel S., Univ of Nevada, Reno
Schieber, Juergen, Indiana Univ, Bloomington
Schneiderman, Jill S., Vassar Coll
Scholz, Christopher A., Syracuse Univ
Schreiber, Charlotte, Univ of Washington
Schwimmer, Reed A., Rider Univ
Scott, Jenni, Univ of Saskatchewan
Seggie, Robert, Curtin Univ
Self, Robert P., Univ of Tennessee, Martin
Seserman, Anca M., Alexandru Ioan Cuza
Shane, Tyrrell, Nat Univ of Ireland Galway
Shanmugam, Ganapathy, Univ of Tennessee, Knoxville
Shaw, John B., Univ of Arkansas, Fayetteville
Shea, James H., Univ of Wisconsin, Parkside
Shimer, Grant, Southern Utah Univ
Siewers, Fredrick D., Western Kentucky Univ
Simms, Alexander, Univ of California, Santa Barbara
Simonson, Bruce M., Oberlin Coll
Simpson, Edward L., Kutztown Univ of Pennsylvania
Simpson, Frank -., Univ of Windsor
Singer, Jill K., Buffalo State Coll
Singler, Charles R., Youngstown State Univ
Sirocko, Frank, Universität Mainz
Slingerland, Rudy L., Pennsylvania State Univ, Univ Park
Sloss, Craig, Queensland Univ of Tech
Smith, Jon, Univ of Kansas
Smith, Larry N., Montana Tech of the Univ of Montana
Smith, Michael E., Northern Arizona Univ
Smith, Norman D., Univ of Nebraska, Lincoln
Smith, Shane V., Youngstown State Univ
Smith-Engle, Jennifer M., Texas A&M Univ, Corpus Christi
Soreghan, Gerilyn S., Univ of Oklahoma
Soreghan, Michael J., Univ of Oklahoma
Soster, Frederick M., DePauw Univ
Southard, John B., Massachusetts Inst of Tech
Staub, James R., Univ of Montana
Steel, Ronald J., Univ of Texas, Austin
Steinen, Randolph P., Dept of Energy and Enval Protection
Stoner, Joseph, Oregon State Univ
Strauss, Justin V., Dartmouth Coll
Stutenbecker, Laura, Universitaet Muenster
Sumner, Dawn Y., Univ of California, Davis
Sumner, Esther, Univ of Southampton
Surpless, Kathleen D., Trinity Univ
Swanson, Travis, Georgia Southern Univ
Sweet, Alisan C., Texas Tech Univ
Sweet, Dustin E., Texas Tech Univ
Swennen, Rudy, Katholieke Universiteit Leuven
Teller, James T., Univ of Manitoba
Thomas, Robert C., Univ of Montana Western
Thomka, James R., SUNY, Plattsburgh
Thompson, Todd A., Indiana Univ, Bloomington
Thornburg, Jesse, Temple Univ
Thurow, Juergen, Univ Coll London
Totten, Matthew W., Kansas State Univ
Totten, Rebecca, Univ of Alabama
Toullec, Renaud, Institut Polytechnique LaSalle Beauvais (ex-IGAL)
Trevino, Ramon H., Univ of Texas at Austin, Jackson Schl of Geosciences
Triplehorn, Donald M., Univ of Alaska, Fairbanks
Trofimovs, Jessica, Queensland Univ of Tech
Trop, Jeffrey M., Bucknell Univ
Törnqvist, Torbjörn E., Tulane Univ
Ulmer-Scholle, Dana, New Mexico Tech
Ulmer-Scholle, Dana S., New Mexico Tech
Underwood, Michael B., Univ of Missouri
Van Daele, Maarten, Ghent Univ
Van De Poll, Henk W., Univ of New Brunswick
van der Lubbe, Jeroen, Vrije Universiteit Amsterdam
Villegas, Monica B., Universidad Nacional de Rio Cuarto
Voice, Peter J., Western Michigan Univ
Wagoner, Jeffrey L., Lawrence Livermore Nat Lab
Walsh, Tim R., Wayland Baptist Univ
Walton, Anthony W., Univ of Kansas
Wang, Ping, Univ of South Florida
Wanless, Harold R., Univ of Miami
Warme, John E., Colorado School of Mines
Wartes, Marwan A., Alaska Div of Geological & Geophysical Surveys
Washburn, Robert H., Juniata Coll
Watkinson, Matthew, Univ of Plymouth
Weislogel, Amy L., West Virginia Univ
Wellner, Julia, Univ of Houston
Wells, Neil A., Kent State Univ
Whalen, Michael T., Univ of Alaska, Fairbanks
Whisonant, Robert C., Radford Univ
Whiteaker, Timothy, Univ of Texas, Austin
Wilkinson, Bruce, Syracuse Univ
Wilkinson, Bruce H., Univ of Michigan
Wilson, Carol A., Louisiana State Univ
Wilson, Kat, Boston Coll
Wilson, Paul A., Univ of Southampton
Winkelstern, Ian Z., Grand Valley State Univ
Wittkop, Chad, Minnesota State Univ
Wizevich, Michael, Central Connecticut State Univ
Wojewoda, Jurand, Univ of Wroclaw
Wolak, Jeannette, Tennessee Tech Univ
Wood, Lesli, Colorado School of Mines
Woodruff, Jonathan D., Univ of Massachusetts, Amherst
Woods, Adam D., California State Univ, Fullerton
Woods, Adam D., Santiago Canyon Coll
Woods, Rachel A., Edinburgh Univ
Worden, Richard, Univ of Liverpool
Wright, V P., Cardiff Univ
Wykel, Andy, South Carolina Dept of Nat Resources
Wynn, Thomas C., Lock Haven Univ
Yang, Wan, Wichita State Univ
Yang, Wan, Missouri Univ of Science and Tech
Yeager, Kevin M., Univ of Kentucky
Yellen, Brian, Univ of Massachusetts, Amherst
Young, Robert S., Western Carolina Univ
Zaleha, Michael J., Wittenberg Univ
Zattin, Massimiliano, Università degli Studi di Padova
Zeng, Hongliu, Univ of Texas at Austin, Jackson Schl of Geosciences

Physical Stratigraphy

Abdullah, Nuraiteng Tee, Univ of Malaya
Aitchison, Jonathan, Univ of Sydney
Alvarez, Walter, Univ of California, Berkeley
Anderson, Raymond R., Univ of Iowa
Archer, Allen W., Kansas State Univ
Aschoff, Jennifer, Univ of Alaska, Anchorage
Atchley, Stacy C., Baylor Univ
Baird, Gordon C., SUNY, Fredonia
Balgord, Elizabeth A., Weber State Univ
Bancroft, Alyssa M., Univ of Iowa
Bart, Philip J., Louisiana State Univ
Bartholomew, Alexander J., SUNY, New Paltz

Benison, Kathleen , West Virginia Univ
Bhattacharya, Janok, McMaster Univ
Blakeney DeJarnett, Beverly, Univ of Texas at Austin, Jackson Schl of Geosciences
Blakey, Ronald C., Northern Arizona Univ
Boone, Peter A., Austin Comm Coll District
Bowen, David W., Montana State Univ
Brezinski, David K., Maryland Department of Nat Resources
Busch, Richard M., West Chester Univ
Byers, Charles W., Univ of Wisconsin, Madison
Capozzi, Rossella, Alma Mater Studiorum Università di Bologna
Catacosinos, Paul A., Delta Coll
Chipping, David H., California Polytechnic State Univ
Clark, Joseph C., Indiana Univ of Pennsylvania
Colgan, Mitchell W., Coll of Charleston
Colombi, Carina, Miami Univ
Coniglio, Mario, Univ of Waterloo
Cramer, Bradley D., Univ of Iowa
d'Atri, Anna, Università di Torino
Dalton, Richard F., New Jersey Geological and Water Survey
Devera, Joseph A., Illinois State Geological Survey
Diecchio, Richard J., George Mason Univ
Diffendal, Jr., Robert F., Univ of Nebraska, Lincoln
Doar, III, William R., South Carolina Dept of Nat Resources
Dodge, Clifford H., DCNR- Pennsylvania Bureau of Geological Survey
Dorsey, Rebecca J., Univ of Oregon
Drzewiecki, Peter A., Eastern Connecticut State Univ
Ebert, James R., SUNY, Oneonta
Elliott, Monty A., Southern Oregon Univ
Evanoff, Emmett, Univ of Northern Colorado
Evans, Kevin R., Missouri State Univ
Feibel, Craig S., Rutgers, The State Univ of New Jersey
Ferrusquia, Ismael, Universidad Nacional Auto de Mexico
Fichter, Lynn S., James Madison Univ
Filipescu, Sorin, Babes-Bolyai Univ
Flemings, Peter B., Univ of Texas, Austin
Flint, Stephen, Univ of Manchester
Foreman, Brady Z., Western Washington Univ
Freeman, Rebecca, Univ of Kentucky
Fritz, William J., Coll of Staten Island/CUNY
Fugitt, Frank, Ohio Dept of Nat Resources
Gentile, Richard J., Univ of Missouri, Kansas City
González, Laura, Univ Complutense de Madrid
Gonzalez Leon, Carlos M., Universidad Nacional Auto de Mexico
Hagen, Cedric, Princeton Univ
Hamlin, H, Scott, Univ of Texas at Austin, Jackson Schl of Geosciences
Harris, W. Burleigh, Univ of North Carolina, Wilmington
Harvey, Cyril H., Guilford Coll
Hasenmueller, Nancy R., Indiana Univ, Bloomington
Heckel, Philip H., Univ of Iowa
Heermance, Richard V., California State Univ, Northridge
Helenes, Javier, Centro de Inv Científica y de Ed Superior de Ensenada
Hentz, Tucker F., Univ of Texas at Austin, Jackson Schl of Geosciences
Heron, Duncan, Duke Univ
Hinnov, Linda, George Mason Univ
Hippensteel, Scott P., Univ of North Carolina, Charlotte
Holland, Steven M., Univ of Georgia
Hyland, Ethan, North Carolina State Univ
Jacques Ayala, Cesar, Universidad Nacional Auto de Mexico
Janson, Xavier, Univ of Texas at Austin, Jackson Schl of Geosciences
Jenkyns, Hugh C., Univ of Oxford
Joeckel, R. M., Univ of Nebraska, Lincoln
Johnson, Donald O., Argonne Nat Lab
Johnson, Gary D., Dartmouth Coll
Johnson, William C., Univ of Kansas
Jordan, Robert R., Univ of Delaware
Jordan, Teresa E., Cornell Univ
Kaaya, Charles Z., Univ of Dar es Salaam
King, Norman R., Univ of Southern Indiana
King, Jr., David T., Auburn Univ
Kulp, Mark A., Univ of New Orleans
Lang, Harold R., Jet Propulsion Lab
Lash, Gary G., SUNY, Fredonia
LeCain, Rebecca M., New Hampshire Geological Survey
Lee, Richard R., Oak Ridge Nat Lab
Lincoln, Jonathan M., Montclair State Univ
Loydell, David K., Univ of Portsmouth
Ludvigson, Greg A., Univ of Kansas
Lundblad, Steven P., Univ of Hawai'i, Hilo
Manger, Walter, Univ of Arkansas, Fayetteville
Manspeizer, Warren, Rutgers, The State Univ of New Jersey, Newark
Martinez-Gutierrez, Genaro, Universidad Auto de Baja California Sur
Maytorena-Silva, Jesus F., Universidad Estatal de Sonora
McDonald, Catherine, Montana Tech of The Univ of Montana
McKee, James W., Univ of Wisconsin, Oshkosh
McLaughlin, Patrick I., Indiana Univ, Bloomington
McLaughlin, Patrick I., Univ of Wisconsin, Extension
McLaughlin, Jr., Peter P., Univ of Delaware
McLaurin, Brett T., Bloomsburg Univ
Meckel, Timothy A., Univ of Texas, Austin
Merriam, Daniel F., Wichita State Univ
Metz, Robert, Kean Univ
Monreal-Saavedra, Rogelio, Universidad Estatal de Sonora
Monteverde, Don, Rutgers, The State Univ of New Jersey
Monteverde, Donald H., New Jersey Geological and Water Survey
Morris, Thomas H., Brigham Young Univ
Muñoz, Belén, Univ Complutense de Madrid
Nicolas, Michelle P., Manitoba Geological Survey
Nielson, R. LaRell, Stephen F. Austin State Univ
Obrist Farner, Jonathan, Missouri Univ of Science and Tech
Oest, Christopher, DCNR- Pennsylvania Bureau of Geological Survey
Olariu, Cornel, Univ of Texas, Austin
Olariu, Mariana, Univ of Texas, Austin
Oms Llobet, Oriol, Universitat Auto de Barcelona
Owen, Donald E., Lamar Univ
Page, Kevin, Univ of Plymouth
Paine, Jeffrey G., Univ of Texas at Austin, Jackson Schl of Geosciences
Parcell, William C., Wichita State Univ
Pashin, Jack C., Oklahoma State Univ
Peterson, Gary L., San Diego State Univ
Peterson, Larry C., Univ of Miami
Pimentel, Nuno , Universidade de Lisboa
Poch Serra, Joan, Universitat Auto de Barcelona
Pope, Michael , Texas A&M Univ
Retzler, Andrew, Univ of Minnesota
Runkel, Anthony, Univ of Minnesota
Sadler, Peter M., Univ of California, Riverside
Sanders, Ronald S., Jet Propulsion Lab
Santisteban, Juan Ignacio, Univ Complutense de Madrid
Simmons, Lizanne V., Santiago Canyon Coll
Snyder, Walter S., Boise State Univ
Steenberg, Julia, Univ of Minnesota
Stieglitz, Ronald D., Univ of Wisconsin, Green Bay
Sugarman, Peter P., Rutgers, The State Univ of New Jersey
Sugarman, Peter J., New Jersey Geological and Water Survey
Swenson, John B., Univ of Minnesota Duluth
Swinehart, James B., Univ of Nebraska, Lincoln
Tew , Berry H., Geological Survey of Alabama
Tierney, Kate, Univ of Iowa
Turner, Elizabeth C., Laurentian Univ, Sudbury
Vail, Peter R., Rice Univ
Valenzuela-Renteria, Manuel, Universidad Estatal de Sonora
Van de Water, Peter, Fresno State Univ
Ver Straeten, Charles, New York State Geological Survey
Verlekar, Purushottam, Goa Univ
Vondra, Carl F., Iowa State Univ of Science & Tech
Wainman, Carmine, Univ of Adelaide
Watney, W. Lynn, Univ of Kansas
Wilson, Andy, Curtin Univ
Witzke, Brian J., Univ of Iowa
Young, Grant M., Western Univ
Zachry, Doy L., Univ of Arkansas, Fayetteville

Structural Geology

Abalos, Benito, Univ of the Basque Country UPV/EHU
Abdelsalam, Mohamed G., Oklahoma State Univ
Abolins, Mark J., Middle Tennessee State Univ
Adam, Jurgen, Royal Holloway Univ of London
Adams, John B., Univ of Washington
Adeoti, Blessing, Federal Univ of Tech, Akure
Akciz, Sinan, California State Univ, Fullerton
Alaniz Alvarez, Susana A., Universidad Nacional Auto de Mexico

Aldrich, M. James, Los Alamos Nat Lab
Aleksandrowski, Pawel, Univ of Wroclaw
Allard, Stephen T., Winona State Univ
Allen, Joseph L., Concord Univ
Allen, Richardson B., Univ of Utah
Allison, David T., Univ of South Alabama
Allmendinger, Richard W., Cornell Univ
Alsleben, Helge, Texas Christian Univ
Alvarez, José Antonio, Univ Complutense de Madrid
Amato, Jeffrey M., New Mexico State Univ, Las Cruces
Amenta, Roddy V., James Madison Univ
Amos, Colin B., Western Washington Univ
Amrouch, Khalid, Univ of Adelaide
Anastasio, David J., Lehigh Univ
Anderson, James L., Univ of Hawai'i, Hilo
Anderson, Mark, Univ of Plymouth
Anderson, Ryan, Idaho State Univ
Andrew, Joseph E., Youngstown State Univ
Antonellini, Marco, Alma Mater Studiorum Università di Bologna
Aranguren, Aitor, Univ of the Basque Country UPV/EHU
Arboleya Cimadevilla, María Luísa, Universitat Auto de Barcelona
Armstrong, Phillip A., California State Univ, Fullerton
Aronoff, Ruth F., Furman Univ
Arrowsmith, Ramon, Arizona State Univ
Attia, Snir, New Mexico Tech
Axen, Gary, New Mexico Tech
Aydin, Atilla, Stanford Univ
Babaie, Hassan A., Georgia State Univ
Baciu, Sorin D., Alexandru Ioan Cuza
Bailey, Christopher M., William & Mary
Baird, Graham, Univ of Northern Colorado
Balestro, Gianni, Università di Torino
Bamisaiye, Oluwaseyi A., Federal Univ of Tech, Akure
Bangs, Nathan L., Univ of Texas, Austin
Banoeng-Yakubo, Bruce K., Univ of Ghana
Barineau, Clinton I., Columbus State Univ
Barnes, Charles W., Northern Arizona Univ
Barquero-Molina, Miriam, Univ of Missouri
BASCOU, Jérôme, Université Jean Monnet, Saint-Etienne
Bauer, Paul W., New Mexico Tech
Bedelean, Horea Stefan, Babes-Bolyai Univ
Behr, Rose-Anna, DCNR- Pennsylvania Bureau of Geological Survey
Bernard, Rachael E., Amherst Coll
Berry, Ron F., Univ of Tasmania
Bethune, Kathryn, Univ of Regina
Betka, Paul, George Mason Univ
Beutel, Erin K., Coll of Charleston
Bhattacharyya, Prajukti, Univ of Wisconsin, Whitewater
Bidgoli, Tandis, Univ of Kansas
Bier, Sara E., SUNY Potsdam
Bilodeau, William L., California Lutheran Univ
Birke, Tesfaye K., Univ of KwaZulu-Natal
Bjorklund, Tom, Univ of Houston
Bjornerud, Marcia, Lawrence Univ
Blackmer, Gale C., DCNR- Pennsylvania Bureau of Geological Survey
Bobyarchick, Andy R., Univ of North Carolina, Charlotte
Bodhankar, Ninad, Pt. Ravishankar Shukla Univ
Bond, Clare, Univ of Aberdeen
Boniface, Nelson, Univ of Dar es Salaam
Bosbyshell, Howell, West Chester Univ
Bothner, Wallace A., Univ of New Hampshire
Bott, Jackie, California Geological Survey
Bradbury, Kelly K., Utah State Univ
Bradley, Michael, Eastern Michigan Univ
Bradley, Philip, N.C. Department of Enval Quality
Braid, James, Saint Francis Xavier Univ
Brandon, Mark T., Yale Univ
Brisbin, William C., Univ of Manitoba
Brodie, Kate, Univ of Manchester
Brown, Bruce A., Univ of Wisconsin, Madison, Div of Extension
Bunds, Michael, Utah Valley Univ
Burberry, Caroline M., Univ of Nebraska, Lincoln
Burchfiel, B. C., Massachusetts Inst of Tech
Burger, H. Robert, Smith Coll
Burks, Rachel J., Towson Univ
Buttner, Steffen, Rhodes Univ
Bykerk-Kauffman, Ann, California State Univ, Chico
Byrne, Timothy, Univ of Connecticut
Cadoppi, Paola, Università di Torino
Camilleri, Phyllis A., Austin Peay State Univ
Campbell, Erin A., Wyoming State Geological Survey
Campbell, Patricia A., Slippery Rock Univ
Carlson, Diane H., Sacramento State Univ
Carosi, Rodolfo, Università di Torino
Carpenter, Brett, Univ of Oklahoma
Casale, Gabriele M., Appalachian State Univ
Cashman, Susan M., Humboldt State Univ
Caskey, John, San Francisco State Univ
Cather, Steven M., New Mexico Tech
Cattanach, Bart, N.C. Department of Enval Quality
Çemen, Ibrahim, Univ of Alabama
Chamberlin, Richard, New Mexico Tech
Chapman, Alan, Macalester Coll
Chapman, James, Univ of Wyoming
Chester, Frederick M., Texas A&M Univ
Chester, Judith, Texas A&M Univ
Cladouhos, Trenton, Univ of Washington
Clayton, Robert W., Brigham Young Univ - Idaho
Cloos, Mark P., Univ of Texas, Austin
Coleman, Jr., Paul J., Univ of California, Los Angeles
Connelly, Jeffrey B., Univ of Arkansas at Little Rock
Connors, Christopher, Washington & Lee Univ
Conti, Paolo, Univ of Siena
Cooke, Michele L., Univ of Massachusetts, Amherst
Cooper-Boemmels, Jennifer, Southern Connecticut State Univ
Coppinger, Walter, Trinity Univ
Cosgrove, John, Imperial Coll
Cowan, Darrel S., Univ of Washington
Cox, Randel T., Univ of Memphis
Crespi, Jean M., Univ of Connecticut
Crider, Juliet G., Univ of Washington
Cronin, Vincent S., Baylor Univ
Crowley, Peter D., Amherst Coll
Cruikshank, Kenneth M., Portland State Univ
Cuevas, Julia, Univ of the Basque Country UPV/EHU
Culshaw, Nicholas, Dalhousie Univ
Czeck, Dyanna M., Univ of Wisconsin, Milwaukee
D'Allura, Jad A., Southern Oregon Univ
Daigneault, Real, Universite du Quebec a Chicoutimi
Darin, Mike, Univ of Nevada, Reno
Davatzes, Nicholas, Temple Univ
Davis, George H., Univ of Arizona
Davis, Peter B., Pacific Lutheran Univ
Dawers, Nancye H., Tulane Univ
De Paola, Nicola, Durham Univ
Dean, Sarah L., Hope Coll
DeGraff, James, Michigan Technological Univ
Di Toro, Giulio, Università degli Studi di Padova
DiPietro, Joseph A., Univ of Southern Indiana
Dixon, John M., Queen's Univ
Dooley, Tim, Univ of Texas at Austin, Jackson Schl of Geosciences
Douglas, Bruce, Indiana Univ, Bloomington
Draper, Grenville, Florida InterNat Univ
Druguet Tantiña, Elena, Universitat Auto de Barcelona
Duebendorfer, Ernest M., Northern Arizona Univ
Dumitriu, Tony C., Alexandru Ioan Cuza
Dunne, George C., California State Univ, Northridge
Dunne, William M., Univ of Tennessee, Knoxville
Dutch, Steven I., Univ of Wisconsin, Green Bay
Eichhubl, Peter, Univ of Texas, Austin
Eichhubl, Peter, Univ of Texas at Austin, Jackson Schl of Geosciences
Elders, Chris, Curtin Univ
Elliott, Colleen G., Montana Tech of The Univ of Montana
Erdmer, Philippe, Univ of Alberta
Erslev, Eric A., Colorado State Univ
Erslev, Eric A., Univ of Wyoming
Eusden, J. Dykstra, Bates Coll
Evans, James P., Utah State Univ
Evans, Mark, Central Connecticut State Univ
Famin, Vincent, Universite de la Reunion
Faulds, James E., Univ of Nevada, Reno
Faure, Stephane, Universite du Quebec a Montreal

Fayon, Annia K., Univ of Minnesota, Twin Cities
Ferguson, Charles A., Univ of Arizona
Festa, Andrea, Università di Torino
Fischer, Mark P., Northern Illinois Univ
Fisher, Donald M., Pennsylvania State Univ, Univ Park
Forte, Adam, Louisiana State Univ
Fraser, Alastair, Univ of Utah
French, Melodie E., Rice Univ
Fryxell, Joan E., California State Univ, San Bernardino
Fueten, Frank, Brock Univ
Fusseis, Florian, Edinburgh Univ
Gale, Julia F., Univ of Texas, Austin
Gani, Nahid, Western Kentucky Univ
Garihan, John M., Furman Univ
Gates, Alexander E., Rutgers, The State Univ of New Jersey, Newark
Gattiglio, Marco, Università di Torino
Gibson, H. Daniel (Dan), Simon Fraser Univ
Gibson, Roger L., Univ of the Witwatersrand
Gibson, Ronald C., Golden West Coll
Gillespie, Thomas, Coll of New Jersey
Gilotti, Jane A., Univ of Iowa
Giorgis, Scott D., SUNY, Geneseo
Girty, Gary H., San Diego State Univ
Gleason, Gayle C., SUNY, Cortland
Godin, Laurent, Queen's Univ
Gold, David (Duff) P., Pennsylvania State Univ, Univ Park
Goodwin, Laurel B., Univ of Wisconsin, Madison
Gottardi, Raphael, Univ of Louisiana at Lafayette
Goulet, Normand, Universite du Quebec a Montreal
Gray, Mary Beth, Bucknell Univ
Greene, David C., Denison Univ
Griera Artigas, Albert, Universitat Auto de Barcelona
Griffith, Ashley, Univ of Texas, Arlington
Griffith, William A., Ohio State Univ
Gross, Michael, Florida InterNat Univ
Groszos, Mark S., Valdosta State Univ
Gudmundsson, Agust, Royal Holloway Univ of London
Guitierrez-Alonso, Gabriel, Florida InterNat Univ
Gulick, Sean S., Univ of Texas, Austin
Guth, Peter L., United States Naval Acad
Hack, Alistair, Univ of Newcastle
Hacker, David B., Kent State Univ
Hall, Frank W., Indiana Univ of Pennsylvania
Hanley, Thomas B., Columbus State Univ
Hanmer, Simon, Carleton Univ
Harrap, Rob, Queen's Univ
Harris, Lyal B., Universite du Quebec
Harris, Ron, Brigham Young Univ
Harrison, Michael J., Tennessee Tech Univ
Hatcher Jr., Robert D., Univ of Tennessee, Knoxville
Hawkins, John F., Auburn Univ
Hayman, Nicholas W., Univ of Oklahoma
Heimsath, Arjun, Arizona State Univ
Helmstaedt, Herwart, Queen's Univ
Helper, Mark A., Univ of Texas, Austin
Hendrix, Thomas E., Grand Valley State Univ
Hennings, Peter, Univ of Texas, Austin
Hennings, Peter H., Univ of Wyoming
Hetherington, Eric D., Coll of the Sequoias
Hetzel, Ralf, Universitaet Muenster
Hibbard, James P., North Carolina State Univ
Hill, Mary Louise, Lakehead Univ
Hinz, Nick, Univ of Nevada, Reno
Hirth, Greg, Brown Univ
Hirth, Gregory, Woods Hole Oceanographic Inst
Hoffman, Paul F., Harvard Univ
Hogan, John, Missouri Univ of Science and Tech
Holdsworth, Robert E., Durham Univ
Holm, Daniel K., Kent State Univ
Holst, Timothy, Univ of Minnesota Duluth
Holyoke, Caleb, Univ of Akron
Hooks, Benjamin P., Univ of Tennessee, Martin
Horsman, Eric M., East Carolina Univ
Hozik, Michael J., Stockton Univ
Hudec, Michael R., Univ of Texas at Austin, Jackson Schl of Geosciences
Hudleston, Peter J., Univ of Minnesota, Twin Cities
Hughes, Kenneth S., Univ of Puerto Rico
Hull, Joseph M., Seattle Central Coll
Hynes, Andrew J., McGill Univ
Iaccarino, Salvatore, Università di Torino
Isard, Sierra J., N.C. Department of Enval Quality
Ismat, Zeshan, Franklin and Marshall Coll
Jackson, Martin P. A., Univ of Texas at Austin, Jackson Schl of Geosciences
Jackson, Richard A., Long Island Univ, Brooklyn Campus
Jacobi, Robert D., SUNY, Buffalo
Jessup, Micah, Univ of Tennessee, Knoxville
Jiang, Dazhi, Western Univ
Jiménez, David, Univ Complutense de Madrid
John, Barbara E., Univ of Wyoming
Johnson, Brad, Univ of Arizona
Johnson, Scott E., Univ of Maine
Johnston, Scott, California Polytechnic State Univ
Johnston, Stephen T., Univ of Alberta
Johnston, Stephen T., Univ of Victoria
Jones, Gustavo Tolson, Universidad Nacional Auto de Mexico
Kalakay, Thomas J., Rocky Mountain Coll
Karabinos, Paul, Williams Coll
Karimi, Bobak, Wilkes Univ
Kath, Randal L., Univ of West Georgia
Kelsch, Jesse, Sul Ross State Univ
Kelty, Thomas, California State Univ, Long Beach
Kennedy, Lori, Univ of British Columbia
Kirkpatrick, James, McGill Univ
Klepeis, Keith A., Univ of Vermont
Klimczak, Christian, Univ of Georgia
Knapp, James, Oklahoma State Univ
Kopf, Christopher F., Mansfield Univ
Kruckenberg, Seth, Boston Coll
Kruckenberg, Seth C., Boston Coll
Krugh, W C., California State Univ, Bakersfield
Kuiper, Yvette D., Colorado School of Mines
Kulander, Byron, Wright State Univ
Lafrance, Bruno, Laurentian Univ, Sudbury
Lageson, David R., Montana State Univ
Langille, Jackie M., Univ of North Carolina, Asheville
Lao, Daniel A., Oklahoma State Univ
Lapusta, Nadia, California Inst of Tech
Laubach, Stephen E., Univ of Texas, Austin
Laubach, Stephen E., Univ of Texas at Austin, Jackson Schl of Geosciences
Laurent-Charvet, Sébastien, Institut Polytechnique LaSalle Beauvais (ex-IGAL)
Law, Richard D., Virginia Polytechnic Inst & State Univ
Lee, Jeffrey, Central Washington Univ
Lemiszki, Peter J., Univ of Tennessee, Knoxville
Lemiszki, Peter J., Pellissippi State Comm Coll
Lemiszki, Peter J., Tennessee Geological Survey
Lenardic, Adrian, Rice Univ
Lennox, Paul G., Univ of New South Wales
Lewis, Helen, Heriot-Watt Univ
Lewis, Jon C., Indiana Univ of Pennsylvania
Lincoln, Beth Z., Albion Coll
Lisle, R J., Cardiff Univ
Little, Tim, Victoria Univ of Wellington
Livaccari, Richard F., Colorado Mesa Univ
Long , Sean P., Washington State Univ
Lowman, Julian, Univ of Toronto
Lowry, Wallace D., Virginia Polytechnic Inst & State Univ
Luther, Amy, Louisiana State Univ
MacDonald, William D., Binghamton Univ
Maher, Jr., Harmon D., Univ of Nebraska, Omaha
Mali, Heinrich, Univ Leoben
Malinconico, Lawrence L., Lafayette Coll
Malone, David H., Illinois State Univ
Mana, Sara, Salem State Univ
Markley, Michelle J., Mount Holyoke Coll
Marrett, Randall A., Univ of Texas, Austin
Marshak, Stephen, Univ of Illinois, Urbana-Champaign
Martin, Silvana, Università degli Studi di Padova
Martínez, José Jesús, Univ Complutense de Madrid
Martinez-Macias, Panfilo R., Universidad Auto de San Luis Potosi
Mattson, Peter H., Queens Coll (CUNY)
McCaffrey, Kenneth J., Durham Univ
McCair, Andrew, Univ of Leeds

McClay, Ken R., Royal Holloway Univ of London
McFadden, Rory, Gustavus Adolphus Coll
McGill, George E., Univ of Massachusetts, Amherst
McKay, Matthew P., Missouri State Univ
McNaught, Mark A., Mount Union Coll
McNulty, Brendan A., California State Univ, Dominguez Hills
McQuarrie, Nadine, Univ of Pittsburgh
Mecklenburgh, Julian, Univ of Manchester
Meere, Pat, Univ Coll Cork
Meigs, Andrew J., Oregon State Univ
Menegon, Luca, Univ of Plymouth
Merguerian, Charles M., Hofstra Univ
Metcalf, Kathryn, SUNY, Oneonta
Mezger, Jochen E., Univ of Alaska, Fairbanks
Mies, Jonathan W., Univ of Tennessee, Chattanooga
Miller, Elizabeth L., Stanford Univ
Miller, Marli G., Univ of Oregon
Miller, Robert B., San Jose State Univ
Milligan, G. Clinton, Dalhousie Univ
Minnaar, Hendrik, Univ of the Free State
Miranda, Elena A., California State Univ, Northridge
Mitchell, Tom, Univ Coll London
Mitra, Shankar, Univ of Oklahoma
Mookerjee, Matty, Sonoma State Univ
Morgan, Julia K., Rice Univ
Morgan, Julia K., Rice Univ
Morgan, Sven S., Iowa State Univ of Science & Tech
Morrow, Robert H., South Carolina Dept of Nat Resources
Mosher, Sharon, Univ of Texas, Austin
Muller, Otto H., Alfred Univ
Murphy, Michael, Univ of Houston
Needham, Tim, Univ of Leeds
Nel, Justin, Univ of the Free State
Nemcok, Michal, Univ of Utah
Neuhauser, Kenneth R., Fort Hays State Univ
Newman, Julie, Texas A&M Univ
Nicol, Andy, Univ of Canterbury
Northrup, Clyde J., Boise State Univ
Nourse, Jonathan A., California State Polytechnic Univ, Pomona
O'Hara, Kieran D., Univ of Kentucky
O'Meara, Stephanie A., Colorado State Univ
Okonkwo, Chukwuemeka T., Federal Univ of Tech, Akure
Onasch, Charles M., Bowling Green State Univ
Osagiede, Edoseghe E., Univ of Benin
Palmquist, John C., Lawrence Univ
Paterson, Scott R., Univ of Southern California
Paton, Douglas, Univ of Leeds
Patton, Regan L., Washington State Univ
Paulsen, Timothy S., Univ of Wisconsin, Oshkosh
Pawley, Alison, Univ of Manchester
Pearson, David M., Idaho State Univ
Pennacchioni, Giorgio, Università degli Studi di Padova
Perry, Bruce, California State Univ, Long Beach
Petrie, Elizabeth S., Western Colorado Univ
Phillips, Noah J., Lakehead Univ
Phipps, Stephen P., Univ of Pennsylvania
Pirouz, Mortaza X., Univ of Texas, Dallas
Platt, John P., Univ of Southern California
Pogue, Kevin R., Whitman Coll
Pollard, David D., Stanford Univ
Potter, Jr., Donald B., Sewanee: Univ of the South
Prior, David J., Univ of Otago
Pritchard, Chad, Eastern Washington Univ
Puelles, Pablo, Univ of the Basque Country UPV/EHU
Quintero, Odranoel, Universidad Nacional Auto de Mexico
Raffini, Sylvain, Universite du Quebec a Chicoutimi
Rafini, Silvain, Universite du Quebec a Chicoutimi
Raharimahefa, Tsilavo, Univ of Regina
Rauch, Marta, Univ of Wroclaw
Reber, Jacqueline, Iowa State Univ of Science & Tech
Ree, Jin-Han, Korea Univ
Reese, Joseph F., Edinboro Univ of Pennsylvania
Reinen, Linda A., Pomona Coll
Resor, Phillip G., Wesleyan Univ
Reynolds, Stephen J., Arizona State Univ
Ricketts, Jason, Univ of Texas, El Paso
Roberts, Gerald, Birkbeck Coll
Robin, Pierre-Yves F., Univ of Toronto
Robinson, Alexander, Univ of Houston
Robinson, Kevin, San Diego State Univ
Robinson, Peter, Univ of Massachusetts, Amherst
Rodgers, David W., Idaho State Univ
Rogers, Robert D., California State Univ, Stanislaus
Rojo-Garcia, Paulino, Universidad Auto de Baja California Sur
Romeo, Ignacio, Univ Complutense de Madrid
Rousell, Don H., Laurentian Univ, Sudbury
Rowe, Christie, McGill Univ
Roy, Denis W., Universite du Quebec a Chicoutimi
Rusmore, Margaret E., Occidental Coll
Saillet, Elodie, Institut Polytechnique LaSalle Beauvais (ex-IGAL)
Saja, David, Case Western Reserve Univ
Sak, Peter B., Dickinson Coll
Saleeby, Jason B., California Inst of Tech
Samaniego, Angel Nieto, Universidad Nacional Auto de Mexico
Sanchez, Veronica I., Texas A&M Univ, Kingsville
Sanislav, Ioan, James Cook Univ
Satterfield, Joseph I., Angelo State Univ
Saura, Eduard, Universitat Auto de Barcelona
Scharnberger, Charles K., Millersville Univ
Schlische, Roy W., Rutgers, The State Univ of New Jersey
Schmidt, Keegan L., Lewis-Clark State Coll
Schoonmaker, Adam, Utica Coll
Schrank, Christoph, Queensland Univ of Tech
Schwerdtner, Walfried M., Univ of Toronto
Sears, James W., Univ of Montana
Shami, Malek, York Coll (CUNY)
Shaw, Colin, Montana State Univ
Shuib, Mustaffa Kamal, Univ of Malaya
Sibson, Rick H., Univ of Otago
Siddoway, Christine S., Colorado Coll
Simony, Philip S., Univ of Calgary
Singleton, John, Colorado State Univ
Skehan, James W., Boston Coll
Skemer, Philip, Washington Univ in St. Louis
Sloan, Alastair, Univ of Cape Town
Snyder, Morgan E., Acadia Univ
Solar, Gary S., Buffalo State Coll
Sorkhabi, Rasoul, Univ of Utah
Soucy de la Roche, Renaud, Universite du Quebec
Spangler, Ellie, California Geological Survey
Spratt, Deborah A., Univ of Calgary
SRIVASTAVA, HARI B., Banaras Hindu Univ
Stauffer, Mel R., Univ of Saskatchewan
Stearns, David W., Univ of Oklahoma
Stewart, Kevin G., Univ of North Carolina, Chapel Hill
Stinson, Amy L., Santiago Canyon Coll
Stinson, Amy L., Irvine Valley Coll
Stowell Gale, Julia, Univ of Texas at Austin, Jackson Schl of Geosciences
Strayer, Luther M., California State Univ, East Bay
Sullivan, Walter, Colby Coll
Surpless, Benjamin E., Trinity Univ
Talbot, James L., Western Washington Univ
Tapp, J. B., Univ of Tulsa
Taylor, Lansing, Univ of Utah
Tewksbury, Barbara J., Hamilton Coll
Tewksbury-Christle, Carolyn, Fort Lewis Coll
Teyssier, Christian P., Univ of Minnesota, Twin Cities
Thiessen, Eric, Memorial Univ of Newfoundland
Thompson, Margaret D., Wellesley Coll
Tikoff, Basil, Univ of Wisconsin, Madison
Timmons, J M., New Mexico Tech
Tindall, Sarah E., Kutztown Univ of Pennsylvania
Titus, Sarah J., Carleton Coll
Toro, Jaime, West Virginia Univ
Torvela, Taija, Univ of Leeds
Toy, Virginia G., Univ of Otago
Treloar, Peter, Kingston Univ
Tremblay, Alain, Universite du Quebec a Montreal
Trudgill, Bruce D., Colorado School of Mines
Trullenque, Ghislain, Institut Polytechnique LaSalle Beauvais (ex-IGAL)
Tso, Jonathan L., Radford Univ
Tubía, Jose M., Univ of the Basque Country UPV/EHU

Tull, James F., Florida State Univ
Tullis, Jan A., Brown Univ
Twiss, Robert J., Univ of California, Davis
Ustaszewski, Kamil M., Friedrich-Schiller-Univ Jena
Van Arsdale, Roy B., Univ of Memphis
van Bever Donker, Jan M., Univ of the Western Cape
van der Pluijm, Ben, Univ of Michigan
VanDervoort, Dane S., Geological Survey of Alabama
Vaqueiro Rodriguez, Marcos, Coruna Univ
Vegas, Nestor, Univ of the Basque Country UPV/EHU
Vignaroli, Gianluca, Alma Mater Studiorum Università di Bologna
Viola, Giulio, Alma Mater Studiorum Università di Bologna
Vogl, Jim, Univ of Florida
Voicu, Gabriel-Constantin, Universite du Quebec a Montreal
Vojtko, Rastislav , Comenius Univ in Bratislava
Vollmer, Frederick W., SUNY, New Paltz
Waldien, Trevor S., South Dakota School of Mines & Tech
Waldron, John W., Univ of Alberta
Walker, J. Douglas, Univ of Kansas
Walsh, Talor B., Millersville Univ
Washington, Paul A., Marietta Coll
Waters-Tormey, Cheryl, Western Carolina Univ
Waterworth, Lauren H., Appalachian State Univ
Watkinson, A. John, Washington State Univ
Waugh, Richard A., Univ of Wisconsin, Platteville
Webber, Jeffrey R., Stockton Univ
Weber, John C., Grand Valley State Univ
Weil, Arlo B., Bryn Mawr Coll
Weldon, Ray J., Univ of Oregon
Wells, Michael L., Univ of Nevada, Las Vegas
West, David P., Middlebury Coll
Wetmore, Paul, Univ of South Florida
Whalley, John, Univ of Portsmouth
Whisner, S. Christopher, Bloomsburg Univ
White, Joseph C., Univ of New Brunswick
Whitmeyer, Steven J., James Madison Univ
Whittaker, Amber, Dept of Agriculture, Conservation, and Forestry
Wickham, John S., Univ of Texas, Arlington
Wilkerson, M. S., Univ of Illinois, Urbana-Champaign
Wilkerson, M. Scott, DePauw Univ
Williams, Michael L., Univ of Massachusetts, Amherst
Williams, Paul F., Univ of New Brunswick
Wilson, Crystal, South Carolina Dept of Nat Resources
Wilson, Terry J., Ohio State Univ
Winslow, Margaret A., City Coll (CUNY)
Withjack, Martha O., Rutgers, The State Univ of New Jersey
Wojtal, Steven F., Oberlin Coll
Woodcock, Nigel H., Univ of Cambridge
Woodward, Lee A., Univ of New Mexico
Woodwell, Grant R., Univ of Mary Washington
Workman-Ford, Kerry, Fresno State Univ
Yikilmaz, Burak, Univ of California, Davis
Yonkee, W. A., Weber State Univ
Yoshinobu, Aaron S., Texas Tech Univ
Young, David, Ohio State Univ
Young, Jeffrey, Univ of Manitoba
Zampieri, Dario, Università degli Studi di Padova
Zimmerman, Jr., Jay, Southern Illinois Univ Carbondale
Zuza, Andrew , Univ of Nevada, Reno
Zuza, Andrew V., Univ of Nevada, Reno

Tectonics

Abercrombie, Rachel, Boston Univ
Allen, Mark, Durham Univ
Anderson, Robert G., Univ of British Columbia
Andeweg, Bernd , Vrije Universiteit Amsterdam
Arkle, Jenny C., Augustana Coll
Ault, Alexis K., Utah State Univ
Avouac, Jean-Philippe, California Inst of Tech
Balica, Constantin, Babes-Bolyai Univ
Barth, Nicolas, Univ of California, Riverside
Bartholomew, Mervin J., Univ of Memphis
Baughman, Jaclyn S., Humboldt State Univ
Bell, Andrew, Edinburgh Univ
Bell, Rebecca, Imperial Coll
Bemis, Sean P., Virginia Polytechnic Inst & State Univ
Bickle, Michael, Univ of Cambridge
Bieler, David B., Centenary Coll of Louisiana
Bird, G. Peter, Univ of California, Los Angeles
Boggs, Katherine, Mount Royal Univ
Boland, Irene B., Winthrop Univ
Boutelier, David, Univ of Newcastle
Bradshaw, John, Univ of Canterbury
Bullen, Dean, Univ of Portsmouth
Burgette, Reed J., New Mexico State Univ, Las Cruces
Burgmann, Roland, Univ of California, Berkeley
Busby, Cathy J., Univ of California, Davis
Camacho, Alfredo, Univ of Manitoba
Cao, Wenrong, Univ of Nevada, Reno
Carlson, Marvin P., Univ of Nebraska, Lincoln
Carr, Sharon D., Carleton Univ
Casey, John F., Univ of Houston
Cawood, Peter A., Monash Univ
Centeno Garcia, Elena, Universidad Nacional Auto de Mexico
Chapin, Charles E., New Mexico Tech
Chi, Wu-Cheng, Academia Sinica
Choi, Eunseo, Univ of Memphis
Church, William R., Western Univ
Contreras, Juan, Centro de Inv Científica y de Ed Superior de Ensenada
Corrigan, David, Univ of Manitoba
Cottle, John M., Univ of Tennessee, Knoxville
Cowgill, Eric S., Univ of California, Davis
Cox, Randel T., Univ of Memphis
Cunningham, Dickson, Eastern Connecticut State Univ
Curry, Magdalena E., North Carolina State Univ
Dallmeyer, R. David, Univ of Georgia
Dalziel, Ian W. D., Univ of Texas, Austin
Das, Shamita, Univ of Oxford
Davis, Gregory A., Univ of Southern California
De Csema, Zoltan, Universidad Nacional Auto de Mexico
de Vicente, Gerardo, Univ Complutense de Madrid
DeCelles, Peter G., Univ of Arizona
Derakhshani, Reza, Shahid Bahonar Univ of Kerman
DeVecchio, Duane, Arizona State Univ
Dewey, John, Univ of California, Davis
Dickinson, William R., Univ of Arizona
Dilek, Yildirm, Miami Univ
Dinter, David A., Univ of Utah
Dolan, James F., Univ of Southern California
Dong, Yunpeng, Western Univ
Dooley, Tim P., Univ of Texas, Austin
Ducea, Mihai N., Univ of Arizona
Dumond, Gregory, Univ of Arkansas, Fayetteville
Eddy, Michael, Purdue Univ
Egan, Stuart, Keele Univ
Engebretson, David C., Western Washington Univ
England, Phillip, Univ of Oxford
Esteban, Jose Julian, Univ of the Basque Country UPV/EHU
Feigl, Kurt L., Univ of Wisconsin, Madison
Ferrari Pedraglio, Luca, Universidad Nacional Auto de Mexico
Fitzgerald, Paul G., Syracuse Univ
Foster, David A., Univ of Florida
Furlong, Kevin P., Pennsylvania State Univ, Univ Park
Gable, Carl W., Los Alamos Nat Lab
Gamaleldien, Hamed, Curtin Univ
Gans, Phillip B., Univ of California, Santa Barbara
Garver, John I., Union Coll
Gehrels, George E., Univ of Arizona
Geissman, John D., Univ of Texas, Dallas
Gerbi, Christopher C., Univ of Maine
German, Chris, Woods Hole Oceanographic Inst
Ghail, Richard, Royal Holloway Univ of London
Gifford, Jennifer N., Univ of Mississippi
Gomez, Francisco, Univ of Missouri
Gries, John C., Wichita State Univ
Grujic, Djordje, Dalhousie Univ
Guenthner, Willy, Univ of Illinois, Urbana-Champaign
Hales, T. C., Univ of Wales
Haley, J C., Miami Univ
Hall, Robert, Royal Holloway Univ of London
Hansen, Vicki L., Univ of Minnesota Duluth
Haq, Saad, Purdue Univ

FACULTY BY SPECIALTY

Harley, Simon L., Edinburgh Univ
Harms, Tekla A., Amherst Coll
Harris, Nigel B., Open Univ
Hemphill-Haley, Mark, Humboldt State Univ
Henderson, Robert, James Cook Univ
Henry, Christopher D., Univ of Nevada, Reno
Herman, Mathew, California State Univ, Bakersfield
Hughes, Sam, Exeter Univ
Huntington, Katharine W., Univ of Washington
Hurtado, Jose M., Univ of Texas, El Paso
Insel, Nadja, Northeastern Illinois Univ
Isacks, Bryan L., Cornell Univ
Ivins, Erik R., Jet Propulsion Lab
Jackson, Christopher, Imperial Coll
Jacobson, Carl E., Iowa State Univ of Science & Tech
Jadamec, Margarete, SUNY, Buffalo
Janecke, Susanne U., Utah State Univ
John, Hickman B., Univ of Kentucky
Judge, Shelley, Coll of Wooster
Kapp, Paul A., Univ of Arizona
Karlstrom, Karl E., Univ of New Mexico
Karson, Jeffrey, Syracuse Univ
Kato, Terence T., California State Univ, Chico
Keir, Derek, Univ of Southampton
Kelley, Shari A., New Mexico Tech
Keppie, J. Duncan, Universidad Nacional Auto de Mexico
Keppie, John D., Universidad Nacional Auto de Mexico
Kirby, Eric, Univ of North Carolina, Chapel Hill
Kirby, Eric, Oregon State Univ
Kirstein, Linda, Edinburgh Univ
Kleinspehn, Karen L., Univ of Minnesota, Twin Cities
Klemperer, Simon L., Stanford Univ
Knapp, Camelia, Oklahoma State Univ
Knuepfer, Peter L. K., Binghamton Univ
Koehn, Daniel, Univ of Glasgow
Kokum, Mehmet, Firat Univ
Koons, Peter O., Univ of Maine
Kullberg, Carla , Universidade de Lisboa
Lamb, Melissa A., Univ of Saint Thomas
Larson, Kyle, Univ of British Columbia, Okanagan Campus
Lavier, Luc L., Univ of Texas, Austin
Lin, Jian, Woods Hole Oceanographic Inst
Lonergan, Lidia, Imperial Coll
Loveless, Jack, Smith Coll
Lynner, Colton, Univ of Delaware
Mac Niocaill, Conall, Univ of Oxford
Malinconico, Mary Ann, Lafayette Coll
Mallard, Laura D., Appalachian State Univ
Malone, Shawn J., Ball State Univ
Mann, Paul, Univ of Texas, Austin
Marsellos, Antonios, Hofstra Univ
Martinez, Juventino, Universidad Nacional Auto de Mexico
Martinez-Reyes, Juventino, Universidad Nacional Auto de Mexico
May, Daniel J., Univ of New Haven
McClelland, William C., Univ of Iowa
McGill, Sally F., California State Univ, San Bernardino
McGrew, Allen J., Univ of Dayton
McNeill, Lisa, Univ of Southampton
McPherson, Robert C., Humboldt State Univ
McRivette, Michael, Albion Coll
Melichar, Rostislav, Masaryk Univ
Melosh, Henry J., Univ of Arizona
Mesbahi, Fatemeh, Univ of Tabriz
Metcalf, Kathryn, California State Univ, Fullerton
Michalak, Melanie, Humboldt State Univ
Michon, Laurent, Universite de la Reunion
Mickus, Kevin L., Missouri State Univ
Milan, Luke, Univ of New England
Molnar, Peter, Univ of Colorado
Moore, Gregory F., Univ of Hawai'i, Manoa
Moores, Eldridge M., Univ of California, Davis
Moser, Desmond, Western Univ
Moucha, Robert, Syracuse Univ
Mound, Jon, Univ of Leeds
Mueller, Karl J., Univ of Colorado
Murphy, J. Brendan, Saint Francis Xavier Univ
Möller, Andreas, Univ of Kansas
Nabelek, John L., Oregon State Univ
Nadin, Elisabeth S., Univ of Alaska, Fairbanks
Nakov, Radoslav A., Bulgarian Acad of Sciences
Nance, R. Damian, Ohio Univ
Nash, Greg, Univ of Utah
Naylor, Mark, Edinburgh Univ
Ngo, Thanh X., Hanoi Univ of Mining & Geology
Niemi, Nathan A., Univ of Michigan
Niemi, Tina M., Univ of Missouri, Kansas City
Ocan, O. O., Obafemi Awolowo Univ
Odlum , Margo , Univ of Nevada, Las Vegas
Onderdonk, Nate, California State Univ, Long Beach
Partin, Camille, Univ of Saskatchewan
Passchier, Cees W., Universität Mainz
Pavlis, Terry L., Univ of New Orleans
Peacock, Simon M., Univ of British Columbia
Perez, Nicholas, Texas A&M Univ
Peterman, Emily M., Bowdoin Coll
Pettinga, Jarg R., Univ of Canterbury
Phillips, Richard, Univ of Leeds
Pieri, David C., Jet Propulsion Lab
Plasienka, Dusan, Comenius Univ in Bratislava
Pluhar, Christopher J., Fresno State Univ
Price, Raymond A., Queen's Univ
Rahl, Jeffrey, Washington & Lee Univ
Regalla, Christine, Boston Univ
Reusch, Douglas N., Univ of Maine, Farmington
Richter, Frank M., Univ of Chicago
Robertson, Alastair H., Edinburgh Univ
Robinson, Delores, Univ of Alabama
Roeske, Sarah M., Univ of California, Davis
Rowan, Christopher J., Kent State Univ
Rowley, David B., Univ of Chicago
Royden, Leigh H., Massachusetts Inst of Tech
Rubin, Charles M., Central Washington Univ
Rust, Derek, Univ of Portsmouth
Ryder, Isabelle, Univ of Liverpool
Sager, William W., Univ of Houston
Sample, James C., Northern Arizona Univ
Schellart, Wouter , Vrije Universiteit Amsterdam
Schermer, Elizabeth R., Western Washington Univ
Schmidt, David , Univ of Washington
Schneider, David, Univ of Ottawa
Scotese, Christopher R., Univ of Texas, Arlington
Searle, Mike, Univ of Oxford
Shail, Robin, Exeter Univ
Shaw, John, Harvard Univ
Sinha, A. Krishna, Virginia Polytechnic Inst & State Univ
Sintubin, Manuel, Katholieke Universiteit Leuven
Sparks, David, Texas A&M Univ
Spotila, James A., Virginia Polytechnic Inst & State Univ
Srimal, Neptune, Florida InterNat Univ
Stanley, Jessica, Univ of Idaho
Steltenpohl, Mark G., Auburn Univ
Stern, Robert J., Univ of Texas, Dallas
Stock, Joann M., California Inst of Tech
Streepey Smith, Meg, Earlham Coll
Stump, Edmund, Arizona State Univ
Suppe, John, Univ of Houston
Suter-Cargnelutti, Max, Universidad Nacional Auto de Mexico
Takeuchi, Akira, Univ of Toyama
Taylor, Brian, Univ of Hawai'i, Manoa
Taylor, Frederick W., Univ of Texas, Austin
Taylor, Michael H., Univ of Kansas
Teixell, Antonio, Universitat Auto de Barcelona
Teixell Cácharo, Antoni, Universitat Auto de Barcelona
Thacker, Jacob, New Mexico Tech
Thigpen, Ryan, Univ of Kentucky
Thomas, William A., Geological Survey of Alabama
Thomas, William A., Univ of Kentucky
Thorkelson, Derek J., Simon Fraser Univ
Toke', Nathan, Utah Valley Univ
Turner, Jenni, Univ of East Anglia
Umhoefer, Paul J., Northern Arizona Univ
Unsworth, Martyn, Univ of Alberta

Valentino, David W., SUNY, Oswego
van Hunen, Jeroen, Durham Univ
van Kranendonk, Martin, Univ of New South Wales
Van Wijk, Jolante, New Mexico Tech
Wallace, Laura, Univ of Texas, Austin
Watkinson, Ian M., Royal Holloway Univ of London
Webb, Casey, Southern Utah Univ
Webb, Laura E., Univ of Vermont
Wernicke, Brian P., California Inst of Tech
Whittaker, Alex, Imperial Coll
Willis, Julie B., Brigham Young Univ - Idaho
Wong, Martin, Colgate Univ
Wright, James (Jim) E., Univ of Georgia
Wright, Vashan, Univ of California, San Diego
Wu, Jonny, Univ of Houston
Yeats, Robert S., Oregon State Univ
Yin, An, Univ of California, Los Angeles
Yule, J. Douglas, California State Univ, Northridge
Zamani, Behzad, Univ of Tabriz

Volcanology

Anthony F., Universite de la Reunion
Aguirre Diaz, Gerardo J., Universidad Nacional Auto de Mexico
Ancochea, Eumenio, Univ Complutense de Madrid
Anderson, Steven W., Univ of Northern Colorado
Anderson, Jr., Alfred T., Univ of Chicago
Andrews, Benjamin, Smithsonian Inst / Nat Museum of Nat History
Awdankiewicz, Marek, Univ of Wroclaw
Bailey, John E., Gustavus Adolphus Coll
Banik, Tenley, Illinois State Univ
Barclay, Jenni, Univ of East Anglia
Befus, Kenneth S., Baylor Univ
Bennis, Kadie, Smithsonian Inst / Nat Museum of Nat History
Blake, Stephen, Open Univ
Bohrson, Wendy, Colorado School of Mines
Boroughs, Scott, Washington State Univ
Branney, Mike, Leicester Univ
Brown, David, Univ of Glasgow
Brown, Richard J., Durham Univ
Brunstad, Keith A., SUNY, Oneonta
Burkett, Brett, Collin Coll - Central Park Campus
Burkett, Shannon, Collin Coll - Central Park Campus
Bursik, Marcus I., SUNY, Buffalo
Cabaniss, Haley, Eastern Kentucky Univ
Calder, Eliza, Edinburgh Univ
Cameron, Barry I., Univ of Wisconsin, Milwaukee
Cameron, Cheryl, Alaska Div of Geological & Geophysical Surveys
Camp, Victor E., San Diego State Univ
Canon-Tapia, Edgardo, Centro de Inv Científica y de Ed Superior de Ensenada
Carr, Michael J., Rutgers, The State Univ of New Jersey
Carrasco-Nunez, Gerardo, Universidad Nacional Auto de Mexico
Casares, Heather, Arizona Western Coll
Castro, Jonathan M., Universität Mainz
Chapin, Charles E., New Mexico Tech
Chapman, Marshall, Morehead State Univ
Charbonnier, Sylvain, Univ of South Florida
Chevrel, M. Oryaelle, SUNY, Buffalo
Cigolini, Corrado, Università di Torino
Clarke, Amanda, Arizona State Univ
Cole, James W., Univ of Canterbury
Cole, Paul D., Univ of Plymouth
Connor, Charles B., Univ of South Florida
Cook, Geoffrey, Univ of California, San Diego
Coppola, Diego, Università di Torino
Coulson, Ian, Univ of Regina
Crisp, Joy A., Jet Propulsion Lab
Crowe, Bruce M., Los Alamos Nat Lab
de Silva, Shanika, Oregon State Univ
de' Michieli Vitturi, Mattia, SUNY, Buffalo
Deardorff, Nicholas, Indiana Univ of Pennsylvania
Domagall, Abigail M., Black Hills State Univ
Drake, Simon, Birkbeck Coll
Dufek, Josef, Georgia Inst of Tech
Dufek, Josef, Univ of Oregon
Duffield, Wendell A., Northern Arizona Univ
Duller, Rob, Univ of Liverpool
Edmonds, Marie, Univ of Cambridge
Eggers, Albert A., Univ of Puget Sound
Fagents, Sarah, Univ of Hawai'i, Manoa
Fauria, Kristen, Vanderbilt Univ
Fischer, Tobias, Univ of New Mexico
Flood, Tim P., Saint Norbert Coll
Frey, Holli M., Union Coll
Gansecki, Cheryl A., Univ of Hawai'i, Hilo
Gardner, James E., Univ of Texas, Austin
Genareau, Kimberly, Univ of Alabama
Genna, Dominique, Universite du Quebec a Chicoutimi
Germa, Aurelie, Univ of South Florida
Giachetti, Thomas, Univ of Oregon
Giordano, Daniele, Università di Torino
Girard, Guillaume, Northern Illinois Univ
Godchaux, Martha M., Mount Holyoke Coll
Gonnermann, Helge, Rice Univ
Graettinger, Alison, Univ of Missouri, Kansas City
Grapenthin, Ronni, New Mexico Tech
Gravley, Darren, Univ of Canterbury
Green, Jack, California State Univ, Long Beach
Gregg, Tracy K. P., SUNY, Buffalo
Gutmann, James T., Wesleyan Univ
Hampton, Samuel J., Univ of Canterbury
Harpp, Karen, Colgate Univ
Hasenaka, Toshiaki, Kumamoto Univ
Hausback, Brian, Sacramento State Univ
Haynie, Kirstie, SUNY, Buffalo
Head, Elisabet M., Northeastern Illinois Univ
Heiken, Grant, Los Alamos Nat Lab
Heliker, Christina C., Univ of Hawai'i, Hilo
Herd, Richard, Univ of East Anglia
Holm, Richard F., Northern Arizona Univ
Hon, Ken, Univ of Hawai'i, Hilo
Houghton, Bruce F., Univ of Hawai'i, Manoa
Houle, Michel, Univ of Manitoba
Huber, Christian, Brown Univ
Hughes, Scott S., Idaho State Univ
Iverson, Nels, New Mexico Tech
Jackson, Marie D., Univ of Utah
Johnson, Emily R., New Mexico State Univ, Las Cruces
Junkin, William D., Maryland Department of Nat Resources
Kavanagh, Janine, Univ of Liverpool
Kennedy, Ben, Univ of Canterbury
Kieffer, Susan W., Univ of Illinois, Urbana-Champaign
Klemetti, Erik, Denison Univ
Knesel, Kurt, Trinity Univ
Kobs-Nawotniak, Shannon E., Idaho State Univ
Koleszar, Alison, Colgate Univ
Kolzenburg, Stephan, SUNY, Buffalo
Koornneef, Janne, Vrije Universiteit Amsterdam
Krier, Donathon J., Los Alamos Nat Lab
Lang, Nicholas, Mercyhurst Univ
Larsen, Jessica F., Univ of Alaska, Fairbanks
Lavallee, Yan, Univ of Liverpool
Le Mevel, Helene, Carnegie Inst for Science
Lee, Rachel J., SUNY, Oswego
Lehto, Heather L., Angelo State Univ
LeMasurier, Wesley E., Univ of Colorado, Denver
Lev, Einat , Columbia Univ
LeVeque, Randy, Univ of Washington
Leyrit, Hervé, Institut Polytechnique LaSalle Beauvais (ex-IGAL)
Liu, Emma D., Univ Coll London
Llewellin, Ed, Durham Univ
Lockwood, John P., Univ of Hawai'i, Hilo
Lucchi, Federico, Alma Mater Studiorum Università di Bologna
Magsino, Sammantha L., Nat Acad of Sci, Eng, and Med
Manga, Michael, Univ of California, Berkeley
Maria, Tony, Univ of Southern Indiana
Mather, Tamsin A., Univ of Oxford
Matoza, Robin, Univ of California, Santa Barbara
McGarvie, Dave, Open Univ
McPhie, Jocelyn, Univ of Tasmania
Michelfelder, Gary, Missouri State Univ
Mnich, Marissa, Sonoma State Univ
Moore, Joseph N., Univ of Utah

Moran, Seth, Univ of Washington
Morton, Ronald L., Univ of Minnesota Duluth
Neuberg, Jurgen, Univ of Leeds
O'Sullivan, Katie, California State Univ, Bakersfield
Ort, Michael H., Northern Arizona Univ
Ottavi-Pupier, Elsa, Institut Polytechnique LaSalle Beauvais (ex-IGAL)
Pallister, John S., Univ of Pittsburgh
Perry, Frank V., Los Alamos Nat Lab
Phipps Morgan, Jason, Cornell Univ
Pistone , Mattia , Univ of Georgia
Putirka, Keith D., Fresno State Univ
Pyle, David M., Univ of Oxford
Quick, James E., Southern Methodist Univ
Rader, Erika, Univ of Idaho
Reynolds, Robert W., Central Oregon Comm Coll
Rhodes, J. Michael, Univ of Massachusetts, Amherst
Riggs, Nancy, Northern Arizona Univ
Rodgers, Mel, Univ of South Florida
Rodriguez, Lizzette A., Univ of Puerto Rico
Roldan Quintana, Jaime, Universidad Nacional Auto de Mexico
Roman, Diana C., Carnegie Inst for Science
Rose, William I., Michigan Technological Univ
Ross, Pierre-Simon, Universite du Quebec
Rowland, Scott K., Univ of Hawai'i, Manoa
Ruiz Cubillo, Paulo, Universidad de Costa Rica
Russell, James K., Univ of British Columbia
Sanchez-Rubio, Gerardo, Universidad Nacional Auto de Mexico
Sar, Abdullah, Firat Univ
Schaefer, Janet R. G., Alaska Div of Geological & Geophysical Surveys
Sennert, Sally K., Smithsonian Inst / Nat Museum of Nat History
Shamloo, Hannah, Central Washington Univ
Shea, Thomas, Univ of Hawai'i, Manoa
Sheridan, Michael F., SUNY, Buffalo
Smith, Alan L., California State Univ, San Bernardino
Solana, Camen, Univ of Portsmouth
Soldati, Arianna, North Carolina State Univ
Sonder, Ingo, SUNY, Buffalo
Soule, S. Adam, Woods Hole Oceanographic Inst
Stix, John, McGill Univ
Straub, Susanne, Columbia Univ
Swanson, Donald A., Univ of Hawai'i, Manoa
Taylor, Rex N., Univ of Southampton
Teasdale, Rachel, California State Univ, Chico
Townsend, Meredith, Univ of Oregon
Tranne, Claudio, Alma Mater Studiorum Università di Bologna
Ukstins, Ingrid, Univ of Iowa
Valentine, Gregory, SUNY, Buffalo
Van Ry, Michael, Orange Coast Coll
Vanderkluysen, Loyc, Drexel Univ
Venzke, Edward, Smithsonian Inst / Nat Museum of Nat History
Vigouroux-Caillibot, Nathalie, Douglas Coll
Voight, Barry, Pennsylvania State Univ, Univ Park
Vroon, Pieter, Vrije Universiteit Amsterdam
Wadsworth, Fabian, Durham Univ
Wallace, Paul, Univ of Oregon
Webber, Karen L., Univ of New Orleans
Webber, Karen L., Univ of Michigan
White, James D., Univ of Otago
Williams, Stanley N., Arizona State Univ
Williams-Jones, Glyn, Simon Fraser Univ
Wilson, Thomas, Univ of Canterbury
Wulff, Andrew, Western Kentucky Univ
Zimmerer, Matthew, New Mexico Tech

Mathematical Geology

Barton, Christopher C., Wright State Univ
Bolton, Edward W., Yale Univ
Campbell, Katherine, Los Alamos Nat Lab
Cathles, Lawrence M., Cornell Univ
Chien, Yi-Ju, Pacific Northwest Nat Lab
Chou, Charissa J., Pacific Northwest Nat Lab
Cornejo-Luna, Efrain, Universidad Auto de Baja California Sur
Doveton, John H., Univ of Kansas
Edwards, Stuart, Univ of Newcastle Upon Tyne
Faccenda, Manuele, Università degli Studi di Padova
Fairbairn, David, Univ of Newcastle Upon Tyne
Ford, Allistair, Univ of Newcastle Upon Tyne
Ford, Arianne, James Cook Univ
Gordon, Terence M., Univ of Calgary
Harbaugh, John W., Stanford Univ
Harding, Chris, Iowa State Univ of Science & Tech
He, Changming, Univ of Delaware
Hesse, Marc A., Univ of Texas, Austin
Hunt, Allen G., Wright State Univ
Johnson, Glenn W., Univ of Utah
Journel, Andre G., Stanford Univ
Lee, Jejung, Univ of Missouri, Kansas City
Lutz, Timothy M., West Chester Univ
MAZZELLA, Antonio A., Universita di Cagliari
Murray, Christopher J., Pacific Northwest Nat Lab
Penna, Nigel, Univ of Newcastle Upon Tyne
Rogova, Galina L., SUNY, Buffalo
Rundberg, Robert S., Los Alamos Nat Lab
Sonder, Leslie J., Dartmouth Coll
Spiegelman, Marc W., Columbia Univ
Therrien, Pierre, Universite Laval
Tuttle, Samuel , Mount Holyoke Coll
Walton, Ian, Univ of Utah

Mineral Physics

Abramson, Evan H., Univ of Washington
Bass, Jay D., Univ of Illinois, Urbana-Champaign
Brown, J. Michael, Univ of Washington
Bukowinski, Mark S., Univ of California, Berkeley
Carpenter, Michael, Univ of Cambridge
Chen, Jiuhua, Florida InterNat Univ
Cohen, Ronald E., Carnegie Inst for Science
Cooper, Reid F., Brown Univ
Deng, Jie, Princeton Univ
Duffy, Thomas S., Princeton Univ
Farver, John R., Bowling Green State Univ
Fenter, Paul, Univ of Illinois Chicago
Fischer, Rebecca A., Harvard Univ
Geballe, Zachary, Carnegie Inst for Science
Goncharov, Alexander F., Carnegie Inst for Science
Heinz, Dion L., Univ of Chicago
Hugo, Richard, Portland State Univ
Jackson, Jennifer M., California Inst of Tech
Jacobsen, Steven D., Northwestern Univ
Jeanloz, Raymond, Univ of California, Berkeley
Karandikar, Amol, Carnegie Inst for Science
Karato, Shun-ichiro, Yale Univ
Kavner, Abby, Univ of California, Los Angeles
Knittle, Elise, Univ of California, Santa Cruz
Kronenberg, Andreas, Texas A&M Univ
Lieberman, Robert C., Stony Brook Univ
Lin, Jung-Fu, Univ of Texas, Austin
Militzer, Burkhard, Univ of California, Berkeley
Pommier, Anne, Carnegie Inst for Science
Salje, Ekhard, Univ of Cambridge
Saxena, Surendra K., Florida InterNat Univ
Secco, Richard A., Western Univ
Stracher, Glenn B., East Georgia State Coll
Struzhkin, Viktor V., Carnegie Inst for Science
Tschauner, Oliver, Univ of Nevada, Las Vegas
Tyburczy, James A., Arizona State Univ
Ulian, Gianfranco, Alma Mater Studiorum Università di Bologna
Vocadlo, Lidunka, Univ Coll London
Weidner, Donald J., Stony Brook Univ
Williams, Quentin, Univ of California, Santa Cruz
Zhang, Jin, Univ of New Mexico
Zhang, Jin, Texas A&M Univ

Medical Geology

Buck, Brenda J., Univ of Nevada, Las Vegas
Finkelman, Robert B., Univ of Texas, Dallas
Fuge, Ron, Aberystwyth Univ
Gillmore, Gavin, Kingston Univ
Hunt, Andrew, Univ of Texas, Arlington
Markowitz, Steven, Queens Coll (CUNY)
Morabia, Alfredo, Queens Coll (CUNY)
Pasteris, Jill D., Washington Univ in St. Louis
Williams, Lynda, Arizona State Univ

Forensic Geology
Williams, Thomas, Univ of Idaho

ECONOMIC GEOLOGY

General Economic Geology
Adedcyin, A. D., Univ of Ilorin
Adetunji, A, Obafemi Awolowo Univ
Akabzaa, Thomas M., Univ of Ghana
Akande, S. O., Univ of Ilorin
Alavi, Ghafour, Univ of Tabriz
Alexandre, Paul, Brandon Univ
Alexandrov, Eugene A., Queens Coll (CUNY)
Allard, Gilles O., Univ of Georgia
Anger-Kraavi, Annela, Univ of East Anglia
Ansdell, Kevin M., Univ of Saskatchewan
Anthony, Robin, DCNR- Pennsylvania Bureau of Geological Survey
Apostoae, Laviniu, Alexandru Ioan Cuza
Ashley, Paul, Univ of New England
Atkinson, Jr., William W., Univ of Colorado
Banerjee, Neil, Western Univ
Barnes, Sarah- J., Universite du Quebec a Chicoutimi
Barton, Mark D., Univ of Arizona
Beaudoin, Georges, Universite Laval
Bolarinwa, A. T., Univ of Ibadan
Bornhorst, Theodore J., Michigan Technological Univ
Bouzari, Farhad, Univ of British Columbia
Bradshaw, Peter M., Univ of British Columbia
Brimhall, George H., Univ of California, Berkeley
Burt, Donald M., Arizona State Univ
Busch, Lawrence, California Geological Survey
Bustillo, Manuel, Univ Complutense de Madrid
Calagari, Ali Asghar, Univ of Tabriz
Camprubi, Antoni, Universidad Nacional Auto de Mexico
Castor, Stephen B., Univ of Nevada, Reno
Chang, Zhaoshan, Colorado School of Mines
Cheney, Eric S., Univ of Washington
Cheng, Yanbo, James Cook Univ
Chi, Guoxiang, Univ of Regina
Clark, Kenneth F., Univ of Texas, El Paso
Clinkenbeared, John, California Geological Survey
Conly, Andrew G., Lakehead Univ
Cook, Robert B., Auburn Univ
Cooke, David R., Univ of Tasmania
Coolbaugh, Mark F., Univ of Nevada, Reno
Cooney, Michele L., DCNR- Pennsylvania Bureau of Geological Survey
Corbineau, Lucien, Institut Polytechnique LaSalle Beauvais (ex-IGAL)
Coron, Cynthia R., Southern Connecticut State Univ
Corral, Isaac, James Cook Univ
Coveney Jr, Raymond M., Univ of Missouri, Kansas City
Craig, James R., Virginia Polytechnic Inst & State Univ
Craw, Dave, Univ of Otago
Crawford, Thomas J., Univ of West Georgia
Dewaele, Stijn, Ghent Univ
Di Gregorio, Monica, Univ of Leeds
Donovan, John F., Winona State Univ
Duke, Norman A., Western Univ
Elueze, A. A., Univ of Ibadan
Evans, Thomas J., Univ of Wisconsin, Madison, Div of Extension
Fitzpatrick, William, Univ of Wisconsin, Madison, Div of Extension
Franco-Rubio, Miguel, Universidad Auto de Chihuahua
Garcia-Gutierrez, Luis, Universidad Auto de San Luis Potosi
Garofalo, Paolo S., Alma Mater Studiorum Università di Bologna
Gauthier, Michel, Universite du Quebec a Montreal
Gemmell, J B., Univ of Tasmania
Gibson, Andy, Univ of Portsmouth
Gillerman, Virginia S., Boise State Univ
Gillerman, Virginia S., Univ of Idaho
Guan, Dabo, Univ of Leeds
Guha, Jayanta, Universite du Quebec a Chicoutimi
Gysi, Alexander, New Mexico Tech
Hannington, Mark, Univ of Ottawa
Hanson, William D., Arkansas Geological Survey
Harris, DeVerle P., Univ of Arizona
Hayman, Patrick C., Queensland Univ of Tech
Hedenquist, Jeffrey, Univ of Ottawa
Hickey, Kenneth A., Univ of British Columbia
Hickson, Catherine J., Univ of British Columbia
Hohn, Michael E., West Virginia Geological & Economic Survey
Holley, Elizabeth, Colorado School of Mines
Hollings, Peter N., Lakehead Univ
Howarth, Geoffrey H., Univ of Cape Town
Hoyer, Lauren, Univ of KwaZulu-Natal
Huber, Matthew, Univ of the Free State
Ikonnikova, Svetlana, Univ of Texas, Austin
Jebrak, Michel, Universite du Quebec a Montreal
Jugo, Pedro J., Laurentian Univ, Sudbury
Kamilli, Robert J., Univ of Arizona
Karginoglu, Yusuf , Firat Univ
Kazimoto, Emmanuel O., Univ of Dar es Salaam
Keith, Jeffrey D., Brigham Young Univ
Kelly, William C., Univ of Michigan
Kesler, Stephen E., Univ of Michigan
Kontak, Daniel J., Laurentian Univ, Sudbury
La Berge, Gene L., Univ of Wisconsin, Oshkosh
Laki, Sam, Central State Univ
Large, Ross R., Univ of Tasmania
Lasemi, Zakaria, Illinois State Geological Survey
Lassetter, William L., Geology and Mineral Resources Program
Layton-Matthews, Daniel, Queen's Univ
Lehrberger, Gerhard, Technical Univ of Munich
Li, Chusi, Indiana Univ, Bloomington
Linnen, Robert, Univ of Waterloo
Locmelis, Marek, Missouri Univ of Science and Tech
Lorenzoni, Irene, Univ of East Anglia
Maher, Kierran, New Mexico Tech
Mantilla Figueroa, Luis C., Universidad Industrial de Santander
Marschik, Robert, Ludwig-Maximilians-Universitaet Muenchen
Marting, Andrew S., Univ of Nevada, Las Vegas
Masterman, Steven S., Alaska Div of Geological & Geophysical Surveys
Mattieu, Lucie, Universite du Quebec a Chicoutimi
mcClenaghan, Seán h., Trinity Coll
McFall, Katie D., Univ Coll London
McLemore, Virginia T., New Mexico Tech
Mensah, Emmanuel, Kwame Nkrumah Univ of Science and Tech
Misra, Kula C., Univ of Tennessee, Knoxville
Mohammad Reza, Hosseinzadeh, Univ of Tabriz
Moreno-Rodriguez, Veronica, Universidad Estatal de Sonora
Morton, Penelope C., Univ of Minnesota Duluth
Mumin, A. Hamid, Brandon Univ
Muntean, John, Univ of Nevada, Reno
Murphy, Sean C., Austin Comm Coll District
Muszer, Antoni, Univ of Wroclaw
Nasraoui, Mohamed, Institut Polytechnique LaSalle Beauvais (ex-IGAL)
Neubaum, John C., DCNR- Pennsylvania Bureau of Geological Survey
Nex, Paul A., Univ of the Witwatersrand
Nimis, Paolo, Università degli Studi di Padova
Okunlola, Ougbenga. A., Univ of Ibadan
Olivo, Gema, Queen's Univ
Orozco-Villasenor, Francisco Javier, Universidad Auto de San Luis Potosi
Paez, H. A., McMaster Univ
Paterson, Colin J., South Dakota School of Mines & Tech
Perrouty, Stéphane, Laurentian Univ, Sudbury
Pietrzak-Renaud, Natalie, Western Univ
Portnov, Alexey, Univ of Texas, Austin
Potra, Adriana, Univ of Arkansas, Fayetteville
Prichard, Hazel, Univ of Wales
Pride, Douglas E., Ohio State Univ
Ramagwede, Fhatuwani L., Geological Survey of South Africa
Rambaud, Fabienne M., Austin Comm Coll District
Ressel, Mike, Univ of Nevada, Reno
Ridley, John R., Colorado State Univ
Robertson, Daniel E., Monroe Comm Coll
Robertson, James M., Univ of Wisconsin, Extension
Robertson, James M., Univ of Wisconsin, Madison
Sagiroglu, Ahmet, Firat Univ
Sasmaz, Ahmet, Firat Univ
Seedorff, Eric, Univ of Arizona
Shackley, Simon J., Edinburgh Univ
Shaver, Stephen A., Sewanee: Univ of the South
Sherlock, Ross, Laurentian Univ, Sudbury
Shortt, Niamh K., Edinburgh Univ
Siahcheshm, Kamal, Univ of Tabriz

Simmons, Peter, Univ of East Anglia
Simon, Jowitt, Univ of Nevada, Las Vegas
Solecki, Andrzej, Univ of Wroclaw
Stucker, James D., Ohio Dept of Nat Resources
Talley, John H., Univ of Delaware
Thompson, Tommy, Colorado State Univ
Titley, Spencer R., Univ of Arizona
Tribe, Selina, Douglas Coll
Tritlla, Jordi, Universidad Nacional Auto de Mexico
Twelker, Evan, Alaska Div of Geological & Geophysical Surveys
Ulloa, Salvador, Universidad Nacional Auto de Mexico
van der Horst, Dan, Edinburgh Univ
van Hinsberg, Vincent, McGill Univ
Vatuva, Absai , Univ of Namibia
Warren, Rachel, Univ of East Anglia
Werdon, Melanie B., Alaska Div of Geological & Geophysical Surveys
Wilkins, Colin, Univ of Plymouth
Wolfgram, Diane, Montana Tech of the Univ of Montana
Yao, Yon, Rhodes Univ
Yates, Martin G., Univ of Maine
Yellich, John, Western Michigan Univ
Zentilli, Marcos, Dalhousie Univ

Coal

Abdullah, Wan Hasiah, Univ of Malaya
Asuen, Godwin O., Univ of Benin
Bailey, Judy, Univ of Newcastle
Bustin, R. Marc, Univ of British Columbia
Calder, John, Dalhousie Univ
Chenoweth, Cheri A., Illinois State Geological Survey
Drobniak, Agnieszka, Indiana Univ, Bloomington
Elrick, Scott D., Illinois State Geological Survey
Elrick, Scott D., Univ of Illinois
Flood, Peter, Univ of New England
Hoffman, Gretchen K., New Mexico Tech
Hower, James C., Univ of Kentucky
Huggett, William, Southern Illinois Univ Carbondale
Khawaja, Ikram U., Youngstown State Univ
Korose, Christopher P., Illinois State Geological Survey
Lentz, Leonard J., DCNR- Pennsylvania Bureau of Geological Survey
Lindberg, Jonathan W., Pacific Northwest Nat Lab
Mastalerz, Maria, Indiana Univ, Bloomington
Meyer, Rebecca A., Indiana Univ, Bloomington
Miller, Barry W., Tennessee Geological Survey
Mukhopadhyay, Prasanta, Dalhousie Univ
Myers, Alan R., Illinois State Geological Survey
Obrad, Jennifer M., Illinois State Geological Survey
Rimmer, Susan M., Southern Illinois Univ Carbondale
Wright, Chris, Ohio Dept of Nat Resources

Metallic Ore Deposits

Arribas, Antonio, Univ of Texas, El Paso
Barnett, Douglas B., Pacific Northwest Nat Lab
Bennett, Jason, Curtin Univ
Bissig, Thomas, Univ of British Columbia
Broughton, William A., Univ of Wisconsin, Platteville
Brown, Philip E., Univ of Wisconsin, Madison
Cedillo, Esteban, Universidad Nacional Auto de Mexico
Chang, Zhaoshan, James Cook Univ
Chapman, Robert J., Univ of Leeds
Chavez, Jr., William X., New Mexico Tech
Chen, Yuan, Univ of British Columbia, Okanagan Campus
Clark, Alan H., Queen's Univ
Cline, Jean S., Univ of Nevada, Las Vegas
Corona-Esquivel, Rodolfo, Universidad Nacional Auto de Mexico
Corriveau, Louise, Universite du Quebec
Crowe, Douglas E., Univ of Georgia
Dilles, John H., Oregon State Univ
Dube, Benoit, Universite du Quebec
Durazo-Tapia, Gustavo E., Universidad Estatal de Sonora
Fifarek, Richard, Southern Illinois Univ Carbondale
Friehauf, Kurt, Kutztown Univ of Pennsylvania
Gaboury, Damien, Universite du Quebec a Chicoutimi
Gibson, Harold, Carleton Univ
Gibson, Harold L., Laurentian Univ, Sudbury
Good, David, Western Univ
Graf, Jr., Joseph L., Southern Oregon Univ
Guilbert, John M., Univ of Arizona
Hagni, Richard D., Missouri Univ of Science and Tech
Hannington, Mark D., Carleton Univ
Hart, Craig J., Univ of British Columbia
Hooda, Peter, Kingston Univ
Keays, Reid R., Laurentian Univ, Sudbury
Kinnaird, Judith A., Univ of the Witwatersrand
Kissin, Stephen A., Lakehead Univ
Koestel, Mark, Mt. San Antonio Coll
Kyle, J. Richard, Univ of Texas, Austin
Lesher, Michael, Laurentian Univ, Sudbury
Linnen, Robert, Western Univ
Lodge, Robert, Univ of Wisconsin, Eau Claire
Macfarlane, Andrew W., Florida InterNat Univ
Mateus, António, Universidade de Lisboa
McAllister, Arnold L., Univ of New Brunswick
McLemore, Virginia, New Mexico Tech
Melchiorre, Erik, California State Univ, San Bernardino
Menuge, Julian F., Univ Coll Dublin
Monecke, Thomas, Colorado School of Mines
Morel-Kraepiel, Anne, Princeton Univ
Morton, Roger D., Univ of Alberta
Muchez, Philippe, Katholieke Universiteit Leuven
Muntean, John L., Univ of Nevada, Reno
Naldrett, Anthony J., Univ of Toronto
Newberry, Rainer J., Univ of Alaska, Fairbanks
Ortega, Lorena, Univ Complutense de Madrid
Petersen, Erich U., Univ of Utah
Piña, Rubén, Univ Complutense de Madrid
Powell, Wayne G., Graduate School of the City Univ of New York
Richardson, Carson A., Univ of Arizona
Romberger, Samuel B., Colorado School of Mines
Ruvalcaba-Ruiz, Delfino C., Universidad Auto de San Luis Potosi
Salaun, Pascal, Univ of Liverpool
Shelton, Kevin L., Univ of Missouri
Simmonds, John, Curtin Univ
Simmons, Stuart, Univ of Utah
Sinclair, Alastair J., Univ of British Columbia
Spry, Paul G., Iowa State Univ of Science & Tech
Thompson, John F., Cornell Univ
Thompson, John F. H., Univ of British Columbia
Tingley, Joseph V., Univ of Nevada, Reno
Tosdal, Richard, Univ of British Columbia
Uzunlar, Nuri, South Dakota School of Mines & Tech
van der Berg, Stan, Univ of Liverpool
Vasallo, Fernando, Universidad Nacional Auto de Mexico
Vassallo-Morales, Luis F., Universidad Nacional Auto de Mexico
Walker, Alex, Curtin Univ
Watkinson, David H., Carleton Univ
Wheeler, Greg, Sacramento State Univ
Williamson, Ben, Exeter Univ

Industrial Minerals

Bal Akkoca, Dicle, Firat Univ
Barker, James M., New Mexico Tech
Berg, Richard B., Montana Tech of The Univ of Montana
Calengas, Peter L., Western Illinois Univ
Crespo, Elena, Univ Complutense de Madrid
Moreton, Kim, Exeter Univ
Power, W. Robert, Georgia State Univ
Rocha, Fernando, Universidade de Aveiro
Simandl, George J., Univ of Victoria
Sousa, Luís M., Universidade de Trás-os-Montes e Alto Douro
Spooner, Edward T. C., Univ of Toronto
Van Straaten, H. Peter, Univ of Guelph

Oil and Gas

Anderson, Ross, Heriot-Watt Univ
Balázs, László, Eotvos Lorand Univ
Becker, Mauro, Texas A&M Univ
Begg, Stephen, Univ of Adelaide
Brown, Alan, West Virginia Univ
C., Jeffrey C., Youngstown State Univ
Carter, Kristin M., DCNR- Pennsylvania Bureau of Geological Survey
Clarkson, Christopher , Univ of Calgary
Cox, John, Mount Royal Univ
Deisher, Jeffrey , Ohio Dept of Nat Resources

Eldridge, Jayson, Indiana Univ, Bloomington
Gillis, Robert, Alaska Div of Geological & Geophysical Surveys
Gluyas, Jon, Durham Univ
Harder, Brian J., Louisiana State Univ
Herriott, Trystan, Alaska Div of Geological & Geophysical Surveys
Hills, Denise J., Geological Survey of Alabama
Hooks, Chris H., Geological Survey of Alabama
Horn, Marty R., Louisiana State Univ
Hosseini, Seyyed Abolfazi, Univ of Texas, Austin
Hulett, Sam, Ohio Dept of Nat Resources
Jafarizadeh, Babak, Heriot-Watt Univ
Johnston, III, John E., Louisiana State Univ
King, Carey, Univ of Texas, Austin
Krevor, Samuel, Imperial Coll
Leetaru, Hannes E., Illinois State Geological Survey
Leighton, Morris W., Univ of Illinois, Urbana-Champaign
Leone, James, New York State Geological Survey
Lepain, David, Alaska Div of Geological & Geophysical Surveys
Levey, Raymond A., Univ of Utah
Markowski, Antonette K., DCNR- Pennsylvania Bureau of Geological Survey
Nicholas, Chris, Trinity Coll
Oldershaw, Michael, Bakersfield Coll
Pascussi, Michael, New York State Geological Survey
Potter, Eric C., Univ of Texas at Austin, Jackson Schl of Geosciences
Shew, Roger D., Univ of North Carolina, Wilmington
Vanden Berg, Michael, Utah Geological Survey
Waid, Christopher, Ohio Dept of Nat Resources
Webb, Nathan D., Illinois State Geological Survey
Zeidouni, Mehdi, Univ of Texas, Austin

Construction Materials (SSG)

Cnudde, Veerle, Ghent Univ

Ore Deposits (Other)

Baig, Ayat, Brandon Univ
Carranza, Emmanuel John, Univ of KwaZulu-Natal
Haroldson, Erik L., Austin Peay State Univ
Mantei, Erwin J., Missouri State Univ
Runyon, Simone, Univ of Wyoming
Tamas, Calin Gbriel, Babes-Bolyai Univ

GEOCHEMISTRY

General Geochemistry

Abimbola, A. E., Univ of Ibadan
Adisa, Adeshina L., Federal Univ of Tech, Akure
Aines, Roger D., Lawrence Livermore Nat Lab
Ajayi, T. R., Obafemi Awolowo Univ
Akaegbobi, I. M., Univ of Ibadan
Allen, Douglas, Salem State Univ
Allen-King, Richelle, SUNY, Buffalo
Ameen, SM M., Curtin Univ
Amonette, Alexandra B., Pacific Northwest Nat Lab
Anderson, Robert F., Columbia Univ
April, Richard, Colgate Univ
Astefanei, Dan, Alexandru Ioan Cuza
Atekwana, Eliot A., Univ of Delaware
Ayers, John C., Vanderbilt Univ
Ayling, Bridget, Univ of Nevada, Reno
Babcock, R. S., Western Washington Univ
Banks, David, Univ of Leeds
Barnes, Hubert L., Pennsylvania State Univ, Univ Park
Barnes, Jaime D., Univ of Texas, Austin
Basu, Anirban, Royal Holloway Univ of London
Beets, Kay J., Vrije Universiteit Amsterdam
Beiersdorfer, Raymond E., Youngstown State Univ
Belmonte, Donato, Universita di Genova
Bender, Michael L., Princeton Univ
Berger, Peter M., Illinois State Geological Survey
Bergquist, Bridget, Univ of Toronto
Berlo, Kim, McGill Univ
Bernhard, Joan M., Woods Hole Oceanographic Inst
Bernier, Luc, McMaster Univ
Bird, Dennis K., Stanford Univ
Bissada, K. K., Rice Univ
Blake, Ruth E., Yale Univ
Blowes, David W., Univ of Waterloo
Bodnar, Robert J., Virginia Polytechnic Inst & State Univ
Boger, Phillip D., SUNY, Geneseo
Bohm, Christian, Univ of Manitoba
Bolge, Louise, Columbia Univ
Bouman, Heather, Univ of Oxford
Bourcier, Bill L., Lawrence Livermore Nat Lab
Brantley, Susan L., Pennsylvania State Univ, Univ Park
Brounce, Maryjo, Univ of California, Riverside
Bryant, Ernest A., Los Alamos Nat Lab
Bryce, Julia G., Univ of New Hampshire
Bulgariu, Dumitru, Alexandru Ioan Cuza
Burton, Jacqueline C., Argonne Nat Lab
Burton, Kevin, Durham Univ
Buseck, Peter R., Arizona State Univ
Butler, Ian B., Edinburgh Univ
Cai, Yue, Columbia Univ
Camacho, Elsa A., Pacific Northwest Nat Lab
Catling, David C., Univ of Washington
Centorbi, Tracey, Univ of Maryland
Chatterjee, Nilanjan, Massachusetts Inst of Tech
Chen, Chu-Yung, Univ of Illinois, Urbana-Champaign
Chen, Heng, Columbia Univ
Chillrud, Steven, Columbia Univ
Chorover, Jonathan D., Univ of Arizona
Churchill, Ron C., California Geological Survey
Claeys, Philippe, Vrije Univ Brussel
Class, Connie, Columbia Univ
Clay, Patricia L., Univ of Manchester
Coe, Douglas A., Montana Tech of the Univ of Montana
Coggon, Jude A., Univ of Southampton
Cole, David R., Ohio State Univ
Cook, Robert B., Oak Ridge Nat Lab
Cowman, Tim C., South Dakota Dept of Env and Nat Resources
Darrah, Thomas, Ohio State Univ
Davisson, M L., Lawrence Livermore Nat Lab
Dawson, M. Robert, Iowa State Univ of Science & Tech
Day, James, Univ of California, San Diego
de Ignacio, Cristina, Univ Complutense de Madrid
Deng, Baolin, New Mexico Tech
Derry, Louis A., Cornell Univ
Ding, Kang, Univ of Minnesota, Twin Cities
Ding, Shuo, Columbia Univ
Dong, Hailiang, Miami Univ
Dooley, John H., New Jersey Geological and Water Survey
Dostal, Jarda, Dalhousie Univ
Downes, Hilary, Birkbeck Coll
Drew, Douglas A., Montana Tech of the Univ of Montana
Duffy, Clarence J., Los Alamos Nat Lab
Dunbar, Nelia W., New Mexico Tech
Durham, William, Massachusetts Inst of Tech
Eby, G. Nelson, Univ of Massachusetts, Lowell
Eiler, John M., California Inst of Tech
Erhardt, Andrea M., Univ of Kentucky
Evans, Owen C., Stony Brook Univ
Fajkoviæ, Hana, Univ of Zagreb
Fall, Andras, Univ of Texas, Austin
Farley, Kenneth A., California Inst of Tech
Farmer, G. Lang, Univ of Colorado
Faure, Gunter, Ohio State Univ
Feigenson, Mark D., Rutgers, The State Univ of New Jersey
Fernandez, Diego, Univ of Utah
Finkel, Robert C., Lawrence Livermore Nat Lab
Finkelstein, David, Hobart & William Smith Colls
Flegal, Russell, Univ of California, Santa Cruz
Flowers, Rebecca M., Univ of Colorado
Forray, Ferenc Lazar, Babes-Bolyai Univ
Foustoukos, Dionysis, Carnegie Inst for Science
Frank, Tracy, Univ of Connecticut
Fraser, Donalf G., Univ of Oxford
Frey, Bonnie A., New Mexico Tech
Gambrell, Robert P., Louisiana State Univ
Gammons, Christopher H., Montana Tech of the Univ of Montana
Garbesi, Karina, California State Univ, East Bay
George, Graham, Univ of Saskatchewan
Ghiorso, Mark , Univ of Washington
Gilfillan, Stuart M., Edinburgh Univ

FACULTY BY SPECIALTY

Gili, Stefania, Princeton Univ
Gleason, James D., Univ of Michigan
Goff, Fraser, Univ of New Mexico
Goff, Fraser, Los Alamos Nat Lab
Goldstein, Steven, Columbia Univ
Gouldey, Jeremy C., Grand Valley State Univ
Goulet, Richard, Univ of Ottawa
Grant, Willard H., Emory Univ
Gurlea, Lawrence P., Youngstown State Univ
Gustin, Mae, Univ of Nevada, Reno
Gutierrez, Melida, Missouri State Univ
Halden, Norman M., Univ of Manitoba
Hall, Cynthia V., West Chester Univ
Hanchar, John, Memorial Univ of Newfoundland
Hansen, Robert N., Univ of the Free State
Hart, Stanley R., Woods Hole Oceanographic Inst
Harvey, Jason, Univ of Leeds
Hattori, Keiko, Univ of Ottawa
Hausrath, Elisabeth M., Univ of Nevada, Las Vegas
Hemming, Sidney, Columbia Univ
Henderson, Gideon, Univ of Oxford
Herndon, Elizabeth M., Univ of Tennessee, Knoxville
Hervig, Richard, Arizona State Univ
Hickey-Vargas, Rosemary, Florida InterNat Univ
Hinman, Nancy W., Univ of Montana
Hinton, Richard, Edinburgh Univ
Hodges, Floyd N., Pacific Northwest Nat Lab
Hudson, Robert J. M., Univ of Illinois, Urbana-Champaign
Hupp, Brittany, George Mason Univ
Ige, A. O., Obafemi Awolowo Univ
Imasuen, Isaac O., Univ of Benin
Irving, Tony, Univ of Washington
Izon, Gareth, Univ of St. Andrews
Jacobs, Gary K., Oak Ridge Nat Lab
Jacobson, Roger, Univ of Nevada, Reno
Jarvis, Ian, Kingston Univ
Jerde, Eric, Morehead State Univ
Jin, Qusheng, Univ of Oregon
John, Bratton F., Wayne State Univ
Johnson, Clark M., Univ of Wisconsin, Madison
Johnston, David T., Harvard Univ
Kalender, Leyla, Firat Univ
Kallemeyn, Gregory, Univ of California, Los Angeles
Kamber, Balz S., Trinity Coll
Kambewa, Chamunorwa , Tshwane Univ of Tech
Kamenov, George D., Univ of Florida
Kaplan, Isaac R., Univ of California, Los Angeles
Kara, Hatice, Firat Univ
Kaszuba, John P., Univ of Wyoming
Katz, Richard, Univ of Oxford
Kellman, Lisa M., Dalhousie Univ
Kenna, Timothy, Columbia Univ
Kim, Sang-Tae, McMaster Univ
Kiro Feldman, Yael, Columbia Univ
Kirschenfeld, Taylor, Univ of West Florida
Knudsen, Andrew, Lawrence Univ
Koeman-Shields, Elizabeth, Angelo State Univ
Koppers, Anthony, Oregon State Univ
Koretsky, Carla, Western Michigan Univ
Kotzer, Tom, Univ of Saskatchewan
Krawczynski, Michael J., Washington Univ in St. Louis
Ku, Teh-Lung, Univ of Southern California
Kuzyk, Zou Zou, Univ of Manitoba
Labidi, Jabrane, Univ of Maryland
Lamp, Jennifer L., Columbia Univ
Lange, Eric, Ball State Univ
Langmuir, Charles H., Harvard Univ
Layne, Graham, Memorial Univ of Newfoundland
Le Roex, Anton, Univ of Cape Town
Lechler, Paul J., Univ of Nevada, Reno
Lee, Cin-Ty A., Rice Univ
Legore, Virginia L., Pacific Northwest Nat Lab
Lehloenya, Pelele, Univ of the Free State
Lentz, David R., Univ of New Brunswick
Lerman, Abraham, Northwestern Univ
Leuta-Madondo, Palesa, Univ of KwaZulu-Natal
Léveillé, Richard, McGill Univ
Li, Lihan, Univ of Iowa
Lindenmeier, Clark W., Pacific Northwest Nat Lab
Little, Susan D., Univ Coll London
Liu, Xiaolei , Univ of Oklahoma
Liu, Xiaoming, Univ of North Carolina, Chapel Hill
Livens, Francis, Univ of Manchester
Locke II, Randall A., Univ of Illinois
Longerich, Henry, Memorial Univ of Newfoundland
Lopez, Dina L., Ohio Univ
Loyd, Sean, California State Univ, Fullerton
Lu, Zunli, Syracuse Univ
Luttge, Andreas, Rice Univ
Lyons, Timothy W., Univ of California, Riverside
Macdougall, J. Douglas, Univ of California, San Diego
Macpherson, Colin G., Durham Univ
Magson, Justine, Univ of the Free State
Mango, Helen N., Castleton Univ
Manning, Christina, Royal Holloway Univ of London
Manya, Shukrani, Univ of Dar es Salaam
Marquez, L. Lynn, Millersville Univ
Martin, Paul F., Pacific Northwest Nat Lab
Martin, Scot T., Harvard Univ
Marzoli, Andrea, Università degli Studi di Padova
McClelland, James W., Univ of Texas, Austin
McDonough, William F., Univ of Maryland
McGrail, Bernard P., Pacific Northwest Nat Lab
McIntosh, Jennifer C., Univ of Arizona
McKibben, Michael A., Univ of California, Riverside
McKinley, Galen, Columbia Univ
McLennan, Scott M., Stony Brook Univ
Meduniæ, Gordana, Univ of Zagreb
Meixner, Thomas, Univ of Arizona
Mensing, Teresa, Ohio State Univ
Meyer, William T., Argonne Nat Lab
Meyzen, Christine M., Università degli Studi di Padova
Michel, Jacqueline, Univ of New Orleans
Middleton, Jennifer L., Columbia Univ
Miller, Nari, California State Univ, Stanislaus
Moldovan, Brett, Univ of Saskatchewan
Morrill, Penny, Memorial Univ of Newfoundland
Mortlock, Richard , Rutgers, The State Univ of New Jersey
Moussallam, Yves, Columbia Univ
Mukasa, Samuel, Univ of Minnesota, Twin Cities
Mukhopadhyay, Sujoy, Univ of California, Davis
Murgulet, Valeriu, Texas A&M Univ, Corpus Christi
Murowchick, James B., Univ of Missouri, Kansas City
Murphy, David T., Queensland Univ of Tech
Naranjo, Ramon, Univ of Nevada, Reno
Niespolo, Elizabeth, Princeton Univ
Nondorf, Lea, Arkansas Geological Survey
Nyame, Frank K., Univ of Ghana
O'Driscoll, Nelson, Acadia Univ
O'Hara, Matthew J., Pacific Northwest Nat Lab
O'Neil, Jonathan, Univ of Ottawa
Odokuma-Alonge, Ovie, Univ of Benin
Odom, LeRoy A., Florida State Univ
Okulewicz, Steven C., Hofstra Univ
Olatunji, A. S., Univ of Ibadan
Olsen, Khris B., Pacific Northwest Nat Lab
Olson, Stephanie, Purdue Univ
Olszewski, Kathy, SUNY, Maritime Coll
Otamendi, Juan E., Universidad Nacional de Rio Cuarto
Ozturk, Nevin, Firat Univ
Palinkaš, Ladislav, Univ of Zagreb
Palmer, Martin R., Univ of Southampton
Pant, Hari, Lehman Coll (CUNY)
Papineau, Dominic, Univ Coll London
Parker, Kent E., Pacific Northwest Nat Lab
Parris, Thomas M., Univ of Kentucky
Patera, Edward S., Los Alamos Nat Lab
Pearce, J A., Cardiff Univ
Peng, Yongbo, Louisiana State Univ
Piccoli, Philip M., Univ of Maryland
Pichevin, Laetitia, Edinburgh Univ
Pickering, Ingrid J., Univ of Saskatchewan

Pierce, Nicholas, Alma Mater Studiorum Università di Bologna
Pietranik, Anna, Univ of Wroclaw
Pintilei, Mitica, Alexandru Ioan Cuza
Plank, Terry, Columbia Univ
Plante, Martin, Universite Laval
Porcelli, Don, Univ of Oxford
Pourmand, Ali, Univ of Miami
Pourret, Olivier, Institut Polytechnique LaSalle Beauvais (ex-IGAL)
Powell, Brian A., Clemson Univ
Pracny, Pavel, Masaryk Univ
Prevec, Steve, Rhodes Univ
Price, Douglas M., Youngstown State Univ
Prohiæ, Esad, Univ of Zagreb
Protano, Giuseppe, Univ of Siena
Prowse, Terry D., Univ of Waterloo
Prytulak, Julie, Durham Univ
Ptacek, Carol J., Univ of Waterloo
Radoman-Shaw, Brandon, Texas State Univ
Rae, James, Univ of St. Andrews
Reed, Mark H., Univ of Oregon
Reeves, Claire, Univ of East Anglia
Reiners, Peter, Univ of Arizona
Richardson, Steve, Univ of Cape Town
Rickard, D, Cardiff Univ
Riedinger, Natascha, Oklahoma State Univ
Roberts, Stephen, Univ of Southampton
Rohs, C. Renee, Northwest Missouri State Univ
Rollinson, Hugh, Univ of Derby
Romanak, Katherine D., Univ of Texas at Austin, Jackson Schl of Geosciences
Rose, Arthur W., Pennsylvania State Univ, Univ Park
Rose, Catherine V., Trinity Coll
Rudnick, Roberta L., Univ of Maryland
Ruiz, Joaquin, Univ of Arizona
Ruttenberg, Kathleen, Univ of Hawai'i, Manoa
Saal, Alberto E., Brown Univ
Sack, Richard, Univ of Washington
Sakyi, Patrick A., Univ of Ghana
Salters, Vincent J., Florida State Univ
Samson, Iain M., Univ of Windsor
Savov, Ivan, Univ of Leeds
Schaef, Herbert T., Pacific Northwest Nat Lab
Schaefer, Joerg, Columbia Univ
Schillling, Kathrin, Columbia Univ
Schlosser, Peter, Columbia Univ
Schrag, Daniel P., Harvard Univ
Schubert, Brian, Univ of Louisiana at Lafayette
Seitz, Jeffery C., California State Univ, East Bay
Senko, John M., Univ of Akron
Serne, R. Jeffrey, Pacific Northwest Nat Lab
Shahar, Anat, Carnegie Inst for Science
Shevenell, Lisa, Univ of Nevada, Reno
Shields, Robin, Heriot-Watt Univ
Shields-Zhou, Graham, Univ Coll London
Shock, Everett, Arizona State Univ
Sims, Kenneth W., Univ of Wyoming
Skippen, George B., Carleton Univ
Slater, Gregory F., McMaster Univ
Smethie, William, Columbia Univ
Speidel, David H., Queens Coll (CUNY)
Spencer, Ronald J., Univ of Calgary
Sprain, Courtney, Univ of Florida
Stack, Andrew, Georgia Inst of Tech
Stebbins, Jonathan F., Stanford Univ
Steele, Kenneth F., Univ of Arkansas, Fayetteville
Stevenson, Ross, Universite du Quebec a Montreal
Strmiæ Palinkaš, Sabina, Univ of Zagreb
Stubbins, Aron, Northeastern Univ
Sturchio, Neil C., Argonne Nat Lab
Sultan, Mohamed, Argonne Nat Lab
Susak, Nicholas J., Univ of New Brunswick
Sutton, Sally, Colorado State Univ
Taylor, Richard P., Carleton Univ
Teagle, Damon A., Univ of Southampton
Tefend, Karen S., Univ of West Georgia
Teng, Fangzhen, Univ of Washington
Thomas, Jim, Univ of Nevada, Reno
Tierney, Kate E., Denison Univ
Tomascak, Paul B., SUNY, Oswego
Trueman, Clive, Univ of Southampton
Tsikos, Hari, Rhodes Univ
Van Geen, Alexander, Columbia Univ
Van Orman, James A., Case Western Reserve Univ
Vann, Jamie, West Chester Univ
Varekamp, Johan C., Wesleyan Univ
Venturelli, Ryan, Colorado School of Mines
Vulava, Vijay M., Coll of Charleston
Wagner, Rick, Geological Survey of Alabama
Walker, John L., Argonne Nat Lab
Walter, Michael, Carnegie Inst for Science
Walther, John V., Southern Methodist Univ
Wang, Hong, Illinois State Geological Survey
Wang, Yang, Florida State Univ
Warburton, David L., Florida Atlantic Univ
Warren, Lesley A., McMaster Univ
Watkins, James, Univ of Oregon
White, William B., Pennsylvania State Univ, Univ Park
White, William M., Cornell Univ
Wigley, Rochelle, Univ of New Hampshire
Williams, Jeremy C., Kent State Univ
Wilson, Emily, Franklin and Marshall Coll
Winckler, Gisela, Columbia Univ
Wogelius, Roy, Univ of Manchester
Worrall, Fred, Durham Univ
Wronkiewicz, David J., Argonne Nat Lab
Wyman, Derek A., Univ of Saskatchewan
Yan, Beizhan, Columbia Univ
Yardley, Bruce W., Univ of Leeds
Young, Edward, Univ of California, Los Angeles
Young, Nicolas, Columbia Univ
Zanazzi, Alessandro, Utah Valley Univ
Zeman, Josef, Masaryk Univ
Zhang, Tongwei, Univ of Texas, Austin
Zheng, Yan, Queens Coll (CUNY)
Zhu, Chen, Indiana Univ, Bloomington
Zolotov, Mikhail Y., Arizona State Univ

Analytical Geochemistry

Aguilera-Ortiz, Irma, Universidad Nacional Auto de Mexico
Ash, Richard, Univ of Maryland
Baca, Dustin, New Mexico Tech
Barnes, Melanie A., Texas Tech Univ
Beauchamp, Marc, Western Univ
Bedard, Paul, Universite du Quebec a Chicoutimi
Benjamin, Timothy M., Los Alamos Nat Lab
Blamey, Nigel J., New Mexico Tech
Bowen, Scott M., Los Alamos Nat Lab
Branam, Tracy D., Indiana Univ, Bloomington
Brandvold, Lynn A., New Mexico Tech
Burton, Edward, Southern Cross Univ
Cahill, Richard A., Illinois State Geological Survey
Castañeda, Isla, Univ of Massachusetts, Amherst
Catalano, Jeff, Washington Univ in St. Louis
Chappaz, Anthony, Central Michigan Univ
Cheng, Zhongqi, Brooklyn Coll (CUNY)
Dewey, Janet, Univ of Wyoming
Dickson, Alex, Royal Holloway Univ of London
Dore, John, Montana State Univ
Effert-Fanta, Shari E., Illinois State Geological Survey
Elam, Tim, Univ of Washington
Essling, Alice M., Argonne Nat Lab
Evans, Noreen, Curtin Univ
Frew, Nelson M., Woods Hole Oceanographic Inst
Gabitov, Rinat, Mississippi State Univ
Gagnon, Joel E., Univ of Windsor
Garcia, Sammy R., Los Alamos Nat Lab
Gavriloaiei, Traian, Alexandru Ioan Cuza
Greenwood, Paul, Curtin Univ
Halbig, Joseph B., Univ of Hawai'i, Hilo
Hart, Brian R., Western Univ
Heizler, Lynn, New Mexico Tech
Hemming, N. G., Queens Coll (CUNY)
Hinthorne, James R., Central Washington Univ

Hoe, Teh Guan, Univ of Malaya
Holman, Alex, Curtin Univ
Holt, Ben D., Argonne Nat Lab
Hopper, Peter, Curtin Univ
Hu, Wan-Ping (Sunny), Queensland Univ of Tech
Huff, Edmund A., Argonne Nat Lab
Huft, Ashley, Montana Tech of The Univ of Montana
Ianno, Adam J., DCNR- Pennsylvania Bureau of Geological Survey
Jackson, Brian P., Dartmouth Coll
Jedrysek, Mariusz O., Univ of Wroclaw
Johnson, Carl G., Woods Hole Oceanographic Inst
Johnston, Scott, Southern Cross Univ
Jones, Graham, Southern Cross Univ
Kersten, Michael, Universität Mainz
Klinkhammer, Gary P., Oregon State Univ
Knaack, Charles, Washington State Univ
Knudstrup, Renee, Salem State Univ
Kuehn, Stephen C., Concord Univ
Kuentz, David C., Miami Univ
Landry, Peter B., Woods Hole Oceanographic Inst
Le Roux, Petrus J., Univ of Cape Town
Lottermoser, Bernd, Exeter Univ
MacDonald, Iain, Univ of Wales
Marshall, Dan D., Simon Fraser Univ
McGrath, Steve F., Montana Tech of The Univ of Montana
McMurtry, Gary M., Univ of Hawai'i, Manoa
Messo, Charles W., Univ of Dar es Salaam
Mock, R. Stephen, Univ of Montana Western
Moore, Lowell P., Virginia Polytechnic Inst & State Univ
Mora-Klepeis, Gabriela, Univ of Vermont
Mujumba, Jean K., Univ of Dar es Salaam
Navarre-Sitchler, Alexis, Colorado School of Mines
Neill, Owen K., Washington State Univ
Olesik, John W., Ohio State Univ
Olmsted, Wayne, Montana Tech of the Univ of Montana
Orouji, Maqsood, Univ of Tabriz
Orr, Robert D., Pacific Northwest Nat Lab
Papike, James J., Univ of New Mexico
Pietruszka, Aaron, Univ of Hawai'i, Manoa
Pike, Steven M., Woods Hole Oceanographic Inst
Pun, Aurora, Univ of New Mexico
Ragland, Paul C., Florida State Univ
Ranville, James, Colorado School of Mines
Rose, Timothy, Smithsonian Inst / Nat Museum of Nat History
Rouff, Ashaki, Queens Coll (CUNY)
Savard, Dany, Universite du Quebec a Chicoutimi
Schneider, John F., Argonne Nat Lab
Shen, Linhan, Princeton Univ
Shimizu, Nobumichi, Woods Hole Oceanographic Inst
Smith, Catherine H., Los Alamos Nat Lab
Smith, Florence P., Argonne Nat Lab
Snow, Jonathan, Univ of Houston
Student, James J., Central Michigan Univ
Thomas, Donald M., Univ of Hawai'i, Manoa
Thompson, Christopher J., Pacific Northwest Nat Lab
Timmer, Jaqueline R., Montana Tech of The Univ of Montana
Van Loon, Lisa, Western Univ
Villasenor-Cabral, Maria G., Universidad Nacional Auto de Mexico
Voelker, Bettina, Colorado School of Mines
Volborth, Alexis, Montana Tech of the Univ of Montana
Wang, Alian, Washington Univ in St. Louis
Warwick, Phillip, Univ of Southampton
Weis, Dominique A., Univ of British Columbia
Williams, Kim R., Colorado School of Mines
Yang, Panseok, Univ of Manitoba

Experimental Petrology/Phase Equilibria

Agee, Carl A., Univ of New Mexico
Anderson, G M., Univ of Toronto
Asimow, Paul D., California Inst of Tech
Boujibar, Asmaa, Western Washington Univ
Brenan, James M., Dalhousie Univ
Candela, Philip A., Univ of Maryland
Carey, James W., Los Alamos Nat Lab
Cottrell, Elizabeth, Univ of Maryland
Cottrell, Elizabeth, Smithsonian Inst / Nat Museum of Nat History
Davis, Fred A., Univ of Minnesota Duluth
Eggler, David H., Pennsylvania State Univ, Univ Park
Ernst, W. Gary, Stanford Univ
Fedortchouk, Yana, Dalhousie Univ
Fei, Yingwei, Univ of Maryland
Fei, Yingwei, Carnegie Inst for Science
Feineman, Maureen D., Pennsylvania State Univ, Univ Park
Frank, Mark R., Northern Illinois Univ
Gilbert, M. Charles, Univ of Oklahoma
Herzberg, Claude T., Rutgers, The State Univ of New Jersey
Hewitt, David A., Virginia Polytechnic Inst & State Univ
Hsu, Liang-Chi, Univ of Nevada, Reno
Jenkins, David M., Binghamton Univ
Johnston, A. Dana, Univ of Oregon
Kilinc, Attila I., Univ of Cincinnati
Koster Van Groos, August F., Univ of Illinois Chicago
Koziol, Andrea M., Univ of Dayton
Lambart, Sarah, Univ of Utah
Lazar, Codi, California State Univ, San Bernardino
Liang, Yan, Brown Univ
Lindsley, Donald H., Stony Brook Univ
London, David, Univ of Oklahoma
Lundstrom, Craig C., Univ of Illinois, Urbana-Champaign
McCauley, Marlene, Guilford Coll
Minarik, William G., McGill Univ
Mosenfelder, Jed, Univ of Minnesota, Twin Cities
Munoz, James L., Univ of Colorado
Mysen, Bjorn O., Carnegie Inst for Science
Nekvasil, Hanna, Stony Brook Univ
Newton, Robert C., Univ of California, Los Angeles
Nielsen, Roger L., South Dakota School of Mines & Tech
Pan, Ruiguang, New Mexico Tech
Parman, Stephen, Brown Univ
Presnall, Dean C., Univ of Texas, Dallas
Robert, Genevieve, Bates Coll
Rutherford, Malcolm J., Brown Univ
Seewald, Jeffrey S., Woods Hole Oceanographic Inst
Simmons, Craig, Colorado School of Mines
Steiner, Jeffrey, City Coll (CUNY)
Stolper, Edward M., California Inst of Tech
Tait, C. Drew, Los Alamos Nat Lab
Tattitch, Brian, Curtin Univ
Thomson, Andrew D., Univ Coll London
Trail, Dustin, Univ of Rochester
Ustunisik, Gokce K., South Dakota School of Mines & Tech
van Westrenen, Wim, Vrije Universiteit Amsterdam
Walker, David, Columbia Univ
Watson, E. Bruce, Rensselaer Polytechnic Inst
Windom, Kenneth E., Iowa State Univ of Science & Tech
Withers, Tony, Western Univ
Wyllie, Peter J., California Inst of Tech

Exploration Geochemistry

Abdurrahman, A., Univ of Ilorin
Adepoju, Mohammed O., Federal Univ of Tech, Akure
Aspandiar, Mehrooz, Curtin Univ
Ayodele, Olusiji S., Federal Univ of Tech, Akure
Borojeviæ Šoštariæ, Sibila, Univ of Zagreb
Closs, L. Graham, Colorado School of Mines
Coggon, Rosalind M., Univ of Southampton
Cohen, David R., Univ of New South Wales
da Silva, Eduardo F., Universidade de Aveiro
Dreschoff, Gisela, Univ of Kansas
Fletcher, William K., Univ of British Columbia
Gazel, Esteban, Cornell Univ
Goodell, Philip C., Univ of Texas, El Paso
Mathur, Ryan, Juniata Coll
McGoldrick, Peter J., Univ of Tasmania
Moon, Charlie, Exeter Univ
Pearson, Graham, Univ of Alberta
Price, Vaneaton, Clemson Univ
Stanley, Clifford R., Acadia Univ
White, William M., Cornell Univ
Whitehead, Robert E., Laurentian Univ, Sudbury
Williams-Jones, Anthony E., McGill Univ
Yanez, Carlos, Universidad Nacional Auto de Mexico

Geochronology & Radioisotopes

Allen, Charlotte M., Queensland Univ of Tech
Amidon, Will, Middlebury Coll
Antinao, José Luis , Indiana Univ, Bloomington
Apen, Francisco, Princeton Univ
Archibald, Doug A., Queen's Univ
Asmerom, Yemane, Univ of New Mexico
Baksi, Ajoy K., Louisiana State Univ
Baldwin, Suzanne L., Syracuse Univ
Baskaran, Mark, Wayne State Univ
Bauer, Annie, Univ of Wisconsin, Madison
Baxter, Ethan, Boston Coll
Beard, Brian L., Univ of Wisconsin, Madison
Bell, Keith, Carleton Univ
Bickford, M. E., Syracuse Univ
Black, Stuart, Univ of Reading
Blackburn, Terrence, Univ of California, Santa Cruz
Blenkinsop, John, Carleton Univ
Blusztajn, Jurek, Woods Hole Oceanographic Inst
Blythe, Ann, Occidental Coll
Brown, Roderick, Univ of Glasgow
Brown, Tom, Univ of Washington
Butler, Gilbert W., Los Alamos Nat Lab
Carlson, Richard W., Carnegie Inst for Science
Chamberlain, Edwin P., Los Alamos Nat Lab
Chamberlain, Kevin R., Univ of Wyoming
Cheng, Hai, Univ of Minnesota, Twin Cities
Chew, David, Trinity Coll
Clark, George S., Univ of Manitoba
Coleman, Drew S., Univ of North Carolina, Chapel Hill
Copeland, Peter, Univ of Houston
Cowart, James B., Florida State Univ
Creaser, Robert A., Univ of Alberta
Crowley, James, Boise State Univ
Crowley, Quentin G., Trinity Coll
Curtice, Joshua M., Woods Hole Oceanographic Inst
Curtis, David B., Los Alamos Nat Lab
Curtis, Garniss H., Univ of California, Berkeley
Davis, Donald W., Univ of Toronto
De Grave, Johan, Ghent Univ
DePaolo, Donald J., Univ of California, Berkeley
Dickin, Alan P., McMaster Univ
Dulai, Henrietta, Univ of Hawai'i, Manoa
Dunning, Gregory R., Memorial Univ of Newfoundland
Edwards, R. Lawrence, Univ of Minnesota, Twin Cities
Enkelmann, Eva, Univ of Calgary
Fearey, Bryan L., Los Alamos Nat Lab
Fenton, Cassandra, Colorado Mesa Univ
Fernández Suárez, Javier, Univ Complutense de Madrid
Foland, Kenneth A., Ohio State Univ
Forman, Stephen, Baylor Univ
Fosdick, Julie, Univ of Connecticut
Frost, Carol D., Univ of Wyoming
Frouin, Marine , Stony Brook Univ
Fullagar, Paul D., Univ of North Carolina, Chapel Hill
Gallup, Christina D., Univ of Minnesota Duluth
Gancarz, Alexander J., Los Alamos Nat Lab
Gaschnig, Richard M., Univ of Massachusetts, Lowell
Gaudette, Henri E., Univ of New Hampshire
Goehring, Brent M., Tulane Univ
Gosse, John C., Dalhousie Univ
Gromet, L. P., Brown Univ
Hall, Chris, Univ of Michigan
Hames, Willis E., Auburn Univ
Hamilton, Michael A., Univ of Toronto
Hanan, Barry B., San Diego State Univ
Hanes, John A., Queen's Univ
Harrison, T. Mark, Univ of California, Los Angeles
Hartnady, Michael, Curtin Univ
Hayatsu, Akio, Western Univ
Heaman, Larry M., Univ of Alberta
Heatherington, Ann L., Univ of Florida
Heizler, Matthew T., New Mexico Tech
Hodges, Kip V., Arizona State Univ
Hourigan, Jeremy, Univ of California, Santa Cruz
Hughen, Konrad A., Woods Hole Oceanographic Inst
Idleman, Bruce D., Lehigh Univ
Jacobsen, Stein B., Harvard Univ
Joannes-Boyau, Renaud, Southern Cross Univ
Jourdan, Fred, Curtin Univ
Jull, A. J. Timothy, Univ of Arizona
Karner, Daniel B., Sonoma State Univ
Kelley, Simon, Edinburgh Univ
Kimbrough, David L., San Diego State Univ
Kinny, Pete, Curtin Univ
Kirkland, Chris, Curtin Univ
Konrad, Kevin , Univ of Nevada, Las Vegas
Konter, Jasper G., Univ of Hawai'i, Manoa
Krummenacher, Daniel, San Diego State Univ
Kuiper, Klaudia, Vrije Universiteit Amsterdam
Kurz, Mark D., Woods Hole Oceanographic Inst
Lakatos, Stephen, York Coll (CUNY)
Lassiter, John, Univ of Texas, Austin
Lepper, Kenneth, North Dakota State Univ
Liebmann, Janne, Curtin Univ
Lively, Rich, Univ of Minnesota
Long, Leon E., Univ of Texas, Austin
Lopez, Margarita, Centro de Inv Científica y de Ed Superior de Ensenada
Maneiro, Kathryn A., Wheaton Coll
Manton, William I., Univ of Texas, Dallas
Markun, Francis J., Argonne Nat Lab
McIntosh, William C., New Mexico Tech
McIntosh, William C., New Mexico Tech
McLean, Noah, Univ of Kansas
Menninga, Clarence, Calvin Coll
Mertz, Dieter, Universität Mainz
Miller, Brent, Texas A&M Univ
Miller, Charles M., Los Alamos Nat Lab
Miller, Geoffrey G., Los Alamos Nat Lab
Min, Kyoungwon, Univ of Florida
Monteleone, Brian D., Woods Hole Oceanographic Inst
Moore, Willard S., Univ of South Carolina
Moran-Zenteno, Dante J., Universidad Nacional Auto de Mexico
Morris, David, Los Alamos Nat Lab
Mortensen, James K., Univ of British Columbia
Mueller, Paul A., Univ of Florida
Mukasa, Samuel B., Univ of Michigan
Murrell, Michael T., Los Alamos Nat Lab
Nana Yobo, Lucien, Texas A&M Univ
Nelson, Bruce K., Univ of Washington
Nemchin, Alexander, Curtin Univ
Olierook, Hugo, Curtin Univ
Omar, Gomaa I., Univ of Pennsylvania
Oretega Rivera, Amabel M., Universidad Nacional Auto de Mexico
Orlandini, Kent A., Argonne Nat Lab
Osmond, John K., Florida State Univ
Palin, J. Michael, Univ of Otago
Pardi, Richard R., William Paterson Univ
Perrin, Richard E., Los Alamos Nat Lab
Pidgeon, Bob, Curtin Univ
Placzek, Christa, James Cook Univ
Polyak, Victor J., Univ of New Mexico
Poths, Jane, Los Alamos Nat Lab
Puchtel, Igor, Univ of Maryland
Ramos, Frank C., New Mexico State Univ, Las Cruces
Rasbury, E. Troy, Stony Brook Univ
Renne, Paul, Univ of California, Berkeley
Rhodes, Edward J., Univ of California, Los Angeles
Ricci, Julia, New Mexico Tech
Rink, W. J., McMaster Univ
Rinterknecht, Vincent, Univ of St. Andrews
Rioux, Matt, Univ of California, Santa Barbara
Rokop, Donald J., Los Alamos Nat Lab
Rooney, Alan, Yale Univ
Ross, Jake, New Mexico Tech
Rubin, Kenneth H., Univ of Hawai'i, Manoa
Samson, Scott D., Syracuse Univ
Schindler, Kimberly, Univ of Kentucky
Schmitz, Mark D., Boise State Univ
Schneider, Robert J., Woods Hole Oceanographic Inst
Schoene, Blair, Princeton Univ
Schroeder, Norman C., Los Alamos Nat Lab

Seidemann, David E., Graduate School of the City Univ of New York
Seidemann, David E., Brooklyn Coll (CUNY)
Selby, Dave, Durham Univ
Shea, Erin, Univ of Alaska, Anchorage
Shumlyanskyy, Leonid, Curtin Univ
Silver, Leon T., California Inst of Tech
Singer, Bradley S., Univ of Wisconsin, Madison
Sole, Jesu, Universidad Nacional Auto de Mexico
Southon, John, Univ of California, Irvine
Spencer, Joel Q., Kansas State Univ
Srinivasan, Balakrishnan, Pondicherry Univ
Stevens-Goddard, Andrea L., Indiana Univ, Bloomington
Stewart, Brian W., Univ of Pittsburgh
Stockli, Daniel, Univ of Texas, Austin
Stone, John O., Univ of Washington
Strachan, Rob, Univ of Portsmouth
Stuiver, Minze, Univ of Washington
Swanson, Karen, William Paterson Univ
Swanson, Terry W., Univ of Washington
Swisher III, Carl C., Rutgers, The State Univ of New Jersey
Thibodeau, Alyson M., Dickinson Coll
Tomlinson, Emma L., Trinity Coll
Tremblay, Marissa, Purdue Univ
Turek, Andrew, Univ of Windsor
Turrin, Brent D., Rutgers, The State Univ of New Jersey
Van Schmus, W. Randall, Univ of Kansas
Vermeesch, Pieter, Univ Coll London
Vervoort, Jeffrey D., Washington State Univ
Walker, Richard J., Univ of Maryland
Walter, Robert C., Franklin and Marshall Coll
Wampler, J. Marion, Georgia State Univ
Widom, Elisabeth, Miami Univ
Wijbrans, Jan, Vrije Universiteit Amsterdam
Wilde, Simon, Curtin Univ
Wolfsberg, Kurt, Los Alamos Nat Lab
Woodard, Jeremy, Univ of KwaZulu-Natal
Yin, Qing-zhu, Univ of California, Davis
Zeitler, Peter K., Lehigh Univ
Zou, Haibo, Auburn Univ
Zreda, Marek, Univ of Arizona

Low-temperature Geochemistry

Achten, Christine, Universitaet Muenster
Agrawal, Abinash, Wright State Univ
Aja, Stephen U., Graduate School of the City Univ of New York
Aja, Stephen U., Brooklyn Coll (CUNY)
Alessi, Daniel S., Univ of Alberta
Alexander, Clark R., Georgia Inst of Tech
Amend, Jan, Univ of Southern California
Anand, Pallavi, Open Univ
Andersen, C. B., Clemson Univ
Andersen, C. Brannon, Furman Univ
Angino, Ernest E., Univ of Kansas
Azmy, Karem, Memorial Univ of Newfoundland
Baker, Leslie L., Univ of Idaho
Balogh-Brunstad, Zsuzsanna, Hartwick Coll
Banner, Jay L., Univ of Texas, Austin
Basak, Chandranath, Univ of Delaware
Bataille, Clement, Univ of Ottawa
Beckingham, Barbara, Coll of Charleston
Benner, Shawn, Boise State Univ
Bergslien, Elisa T., Buffalo State Coll
Bertine, Kathe K., San Diego State Univ
Bickmore, Barry R., Brigham Young Univ
Bird, MIchael I., James Cook Univ
Bischoff, William D., Wichita State Univ
Blanton, Jackson O., Georgia Inst of Tech
Bogner, Jean E., Univ of Illinois Chicago
Borrok, David M., Missouri Univ of Science and Tech
Bowser, Carl J., Univ of Wisconsin, Madison
Brabander, Daniel J., Wellesley Coll
Brand, Uwe, Brock Univ
Breecker, Daniel O., Univ of Texas, Austin
Brook, Edward J., Oregon State Univ
Brooks, Scott C., Oak Ridge Nat Lab
Burke, Andrea, Univ of St. Andrews
Cantrell, Kirk J., Pacific Northwest Nat Lab
Cao, Hongsheng, Wichita State Univ
Capo, Rosemary C., Univ of Pittsburgh
Carroll, Susan A., Lawrence Livermore Nat Lab
Casey, William H., Univ of California, Davis
Catalano, Jeffrey G., Washington Univ in St. Louis
Cerling, Thure E., Univ of Utah
Chermak, John A., Virginia Polytechnic Inst & State Univ
Chesworth, Ward, Univ of Guelph
Christina, Thomas D., Georgia Inst of Tech
Clark, Jordan F., Univ of California, Santa Barbara
Cody, Anita M., Iowa State Univ of Science & Tech
Crossey, Laura J., Univ of New Mexico
Curtin, Lorelei, Bucknell Univ
Datta, Saugata, Kansas State Univ
Derry, Louis A., Cornell Univ
Diebold, Frank E., Montana Tech of the Univ of Montana
Dinelli, Enrico, Alma Mater Studiorum Università di Bologna
Donahoe, Rona J., Univ of Alabama
Dostie, Philip, Bates Coll
Dove, Patricia M., Virginia Polytechnic Inst & State Univ
Drake, John C., Univ of Vermont
Druschel, Greg, Indiana Univ - Purdue Univ Indianapolis
Dworkin, Stephen I., Baylor Univ
Early, Thomas O., Oak Ridge Nat Lab
Eggleston, Carrick M., Univ of Wyoming
Elliott, W. Crawford, Georgia State Univ
Elswick, Erika R., Indiana Univ, Bloomington
Elwood Madden, Andrew S., Univ of Oklahoma
Elwood Madden, Megan E., Univ of Oklahoma
Engel, Annette S., Univ of Tennessee, Knoxville
Evans, Les J., Univ of Guelph
Fein, Jeremy B., Univ of Notre Dame
Fenter, Paul A., Argonne Nat Lab
Ferrell, Jr., Ray E., Louisiana State Univ
Fortner, Sarah K., Wittenberg Univ
Frape, Shaun K., Univ of Waterloo
Frisia, Silvia, Univ of Newcastle
Fryer, Brian J., Univ of Windsor
Gamerdinger, Amy P., Pacific Northwest Nat Lab
García Lorenzo, María de la Luz, Univ Complutense de Madrid
Gerke, Tammie, Miami Univ
Gilhooly, William, Indiana Univ - Purdue Univ Indianapolis
Graf, Donald L., Univ of Illinois, Urbana-Champaign
Grandstaff, David E., Temple Univ
Graney, Joseph R., Binghamton Univ
Graustein, William C., Yale Univ
Greenberg, Sallie, Illinois State Geological Survey
Haas, Johnson R., Western Michigan Univ
Hannigan, Robyn, Univ of Massachusetts, Boston
Hanor, Jeffrey S., Louisiana State Univ
Harris, Pat, Sam Houston State Univ
Harrison, Wendy J., Colorado School of Mines
Hasenmueller, Elizabeth, Saint Louis Univ
Herman, Janet S., Univ of Virginia
Higgins, John A., Princeton Univ
Hirons, Steve, Birkbeck Coll
Hishamunda, Valens, Princeton Univ
Hixon, Amy E., Univ of Notre Dame
Hluchy, Michele M., Alfred Univ
Hochella, Jr., Michael F., Virginia Polytechnic Inst & State Univ
Holdren, George R., Pacific Northwest Nat Lab
Hounslow, Arthur, Oklahoma State Univ
Hutcheon, Ian E., Univ of Calgary
Icenhower, Jonathan P., Pacific Northwest Nat Lab
Icopini, Gary, Montana Tech of The Univ of Montana
Ito, Emi, Univ of Minnesota, Twin Cities
Jacinthe, Pierre-Andre, Indiana Univ - Purdue Univ Indianapolis
Jackson, Kenneth J., Lawrence Livermore Nat Lab
Jacobson, Andrew D., Northwestern Univ
Jahnke, Richard, Georgia Inst of Tech
Janecky, David R., Los Alamos Nat Lab
Karlsson, Haraldur R., Texas Tech Univ
Kaste, James, William & Mary
Kelly, Jacque L., Georgia Southern Univ
Kettler, Richard M., Univ of Nebraska, Lincoln

Kirby, Carl S., Bucknell Univ
Knapp, Elizabeth P., Washington & Lee Univ
Kneeshaw, Tara A., Grand Valley State Univ
Krabbenhoft, David, Univ of Wisconsin, Madison
Kramer, James R., McMaster Univ
Krupka, Kenneth M., Pacific Northwest Nat Lab
Ku, Timothy C., Wesleyan Univ
Kubicki, James D., Univ of Texas, El Paso
Kump, Lee R., Pennsylvania State Univ, Univ Park
Kurtz, Andrew, Boston Univ
Langman, Jeffrey, Univ of Idaho
Larsen, Daniel, Univ of Memphis
Latimer, Jennifer C., Indiana State Univ
Lawrence, James, Univ of Houston
Leckie, James O., Stanford Univ
Li, Qingyun, Stony Brook Univ
Liang, Liyuan, Oak Ridge Nat Lab
Lindberg, Steven E., Oak Ridge Nat Lab
Lindsay, Matthew B., Univ of Saskatchewan
Locke, Randall, Illinois State Geological Survey
Long, David T., Michigan State Univ
Lowenstein, Tim K., Binghamton Univ
Macalady, Donald L., Colorado School of Mines
Martin, Candace E., Univ of Otago
Martin, Ellen E., Univ of Florida
Martin, Jonathan B., Univ of Florida
Martini, Anna M., Amherst Coll
Mattigod, Shas V., Pacific Northwest Nat Lab
Mayes, Melanie A., Univ of Tennessee, Knoxville
Maynard, J. Barry, Univ of Cincinnati
Mazer, James J., Argonne Nat Lab
McArthur, John M., Univ Coll London
McGee, David, Massachusetts Inst of Tech
McGehee, Thomas L., Texas A&M Univ, Kingsville
Merino, Enrique, Indiana Univ, Bloomington
Moore, Joel, Towson Univ
Morel, Francois M M., Princeton Univ
Munk, LeeAnn, Univ of Alaska, Anchorage
Murray, Kendra, Idaho State Univ
Nagy, Kathryn L., Univ of Illinois Chicago
Nelson, Marc A., Univ of Arkansas, Fayetteville
Nesbitt, H W., Western Univ
Neumann, Klaus, Ball State Univ
Newell, Dennis L., Utah State Univ
Nezat, Carmen A., Eastern Washington Univ
Niemitz, Jeffery W., Dickinson Coll
Noll, Mark R., SUNY, The Coll at Brockport
Norton, Stephen A., Univ of Maine
Nzengung, Valentine A., Univ of Georgia
Olsen, Amanda A., Univ of Maine
Oster, Jessica L., Vanderbilt Univ
Panno, Samuel V., Illinois State Geological Survey
Parnell, Roderic A., Northern Arizona Univ
Patton, Terri L., Argonne Nat Lab
Perdrial, Julia, Univ of Vermont
Peters, Stephen C., Lehigh Univ
Pingitore, Jr., Nicholas E., Univ of Texas, El Paso
Poulson, Simon, Univ of Nevada, Reno
Poulton, Simon, Univ of Leeds
Rademacher, Laura K., Univ of the Pacific
Raven, Morgan, Univ of California, Santa Barbara
Reardon, Eric J., Univ of Waterloo
Reeder, Richard J., Stony Brook Univ
Rhodes, Amy L., Smith Coll
Riebe, Clifford S., Univ of Wyoming
Roberts, Sarah K., Lawrence Livermore Nat Lab
Rogers, Pamela Z., Los Alamos Nat Lab
Romanek, Christopher, Furman Univ
Rouff, Ashaki, Rutgers, The State Univ of New Jersey, Newark
Roychoudhury, Alakendra N., Stellenbosch Univ
Rugenstein, Jeremy C., Colorado State Univ
Runnells, Donald D., Univ of Colorado
Rustad, James R., Univ of California, Davis
Ryan, Peter C., Middlebury Coll
Rybacki, Kyle S., DCNR- Pennsylvania Bureau of Geological Survey
Schlautman, Mark, Clemson Univ
Scholz, Denis, Universität Mainz
Schoonen, Martin A., Stony Brook Univ
Schroth, Andrew W., Univ of Vermont
Shiel, Alyssa E., Oregon State Univ
Sigman, Daniel M., Princeton Univ
Sjostrom, Derek, Rocky Mountain Coll
Skidmore, Mark L., Montana State Univ
Stillings, Lisa L., Univ of Nevada, Reno
Stoessell, Ronald K., Univ of New Orleans
Stone, Alan T., Johns Hopkins Univ
Sturchio, Neil C., Univ of Delaware
Suarez, Celina, Univ of Arkansas, Fayetteville
Sverjensky, Dimitri A., Johns Hopkins Univ
Swart, Peter K., Univ of Miami
Taillefert, Martial, Georgia Inst of Tech
Taylor, Wayne A., Los Alamos Nat Lab
Telmer, Kevin, Univ of Victoria
Templeton, Alexis, Univ of Colorado
Thurber, David L., Queens Coll (CUNY)
Tipper, Ed, Univ of Cambridge
Toner, Brandy M., Univ of Minnesota, Twin Cities
Turner, Robert S., Oak Ridge Nat Lab
Valsami-Jones, Eugenia, Univ of Birmingham
Van Der Flier-Keller, Eileen, Univ of Victoria
Varner, Ruth K., Univ of New Hampshire
Veizer, Jan, Univ of Ottawa
Velbel, Michael A., Michigan State Univ
Voglesonger, Kenneth M., Northeastern Illinois Univ
Walder, Ingar F., New Mexico Tech
Walker, Charles T., California State Univ, Long Beach
Wallace, Adam F., Univ of Delaware
Walter, Lynn M., Univ of Michigan
Wang, Jianwei, Louisiana State Univ
Wasylenki, Laura, Northern Arizona Univ
Wehmiller, John F., Univ of Delaware
Weiss, Dominik, Imperial Coll
Welch, Susan A., Ohio State Univ
Weldeghebriel, Mebrahtu, Princeton Univ
West, A. Joshua (Josh), Univ of Southern California
Whittemore, Donald O., Univ of Kansas
Widanagamage, Inoka, Univ of Mississippi
Wronkiewicz, David, Missouri Univ of Science and Tech
Wunsch, David R., Univ of Delaware
Xu, Jie, Univ of Texas, El Paso
Yuretich, Richard F., Univ of Massachusetts, Amherst
Zhaohui, Li, Univ of Wisconsin, Parkside

Marine Geochemistry

Adkins, Jess F., California Inst of Tech
Aller, Robert C., Stony Brook Univ
Barry, Peter, Woods Hole Oceanographic Inst
Berelson, William M., Univ of Southern California
Betzer, Peter R., Univ of South Florida
Biscaye, Pierre E., Columbia Univ
Bollinger, Marsha S., Winthrop Univ
Breier, John, Univ of Texas, Rio Grande Valley
Broecker, Wallace S., Columbia Univ
Byrne, Robert H., Univ of South Florida
Cahill, Kevin, Woods Hole Oceanographic Inst
Calvert, Stephen E., Univ of British Columbia
Cave, Rachel R., Nat Univ of Ireland Galway
Charette, Matthew A., Woods Hole Oceanographic Inst
Church, Thomas M., Univ of Delaware
Coble, Paula G., Univ of South Florida
Cole, Catherine, Univ of St. Andrews
Compton, John S., Univ of Cape Town
Cuomo, Carmela, Univ of New Haven
Dai, Minhan, Woods Hole Oceanographic Inst
Davis, Catherine, North Carolina State Univ
De Carlo, Eric H., Univ of Hawai'i, Manoa
Dickson, Andrew G., Univ of California, San Diego
Dunlea, Ann, Woods Hole Oceanographic Inst
Dutkiewicz, Stephanie, Massachusetts Inst of Tech
Fehrenbacher, Jennifer, Oregon State Univ
Feng, Huan E., Montclair State Univ
Filippelli, Gabriel M., Indiana Univ - Purdue Univ Indianapolis

Fones, Gary, Univ of Portsmouth
Francois, Roger, Univ of British Columbia
Fulweiler, Robinson, Boston Univ
Galy, Valier, Woods Hole Oceanographic Inst
Gerwick, William H., Univ of California, San Diego
Glazer, Brian, Univ of Hawai'i, Manoa
Godfrey, Linda, Rutgers, The State Univ of New Jersey
Gospodinova, Kalina D., Woods Hole Oceanographic Inst
Goudreau, Joanne, Woods Hole Oceanographic Inst
Graham, David W., Oregon State Univ
Grassian, Vicki, Univ of California, San Diego
Hammerschmidt, Chad, Wright State Univ
Hammond, Douglas E., Univ of Southern California
Hansel, Colleen, Woods Hole Oceanographic Inst
Hardison, Amber K., Univ of Texas, Austin
Hardison, Amber, William & Mary
Harrison, Peter, Southern Cross Univ
Heath, G. Ross, Univ of Washington
Henderson, Paul, Woods Hole Oceanographic Inst
Henley, Sian F., Edinburgh Univ
Hoenisch, Baerbel, Columbia Univ
Holycross, Megan, Cornell Univ
Horner, Tristan, Woods Hole Oceanographic Inst
Hu, Xinping, Texas A&M Univ, Corpus Christi
Huber, Julie, Woods Hole Oceanographic Inst
Ingall, Ellery D., Georgia Inst of Tech
Jayakumar, Amal, Princeton Univ
Johnson, Martin, Univ of East Anglia
Juranek, Lauren W., Oregon State Univ
Kastner, Miriam, Univ of California, San Diego
Kim, Hyewon (Heather), Woods Hole Oceanographic Inst
Klein, Frieder, Woods Hole Oceanographic Inst
Levas, Stephen J., Univ of Wisconsin, Whitewater
Li, Yuan-Hui, Univ of Hawai'i, Manoa
Long, Matthew, Woods Hole Oceanographic Inst
Luther, III, George W., Univ of Delaware
Mackenzie, Fred T., Univ of Hawai'i, Manoa
Maerz, Christian, Univ of Leeds
Malahoff, Alexander, Univ of Hawai'i, Manoa
Marconi, Dario, Princeton Univ
McCorkle, Daniel C., Woods Hole Oceanographic Inst
McDermott, Jill I., Lehigh Univ
McKay, Jennifer L., Oregon State Univ
McManus, Jerry, Columbia Univ
Mills, Rachel A., Univ of Southampton
Montoya, Joseph, Georgia Inst of Tech
Moore, Bradley S., Univ of California, San Diego
Mucci, Alfonso, McGill Univ
Murray, Richard, Boston Univ
Nuester, Jochen, California State Univ, Chico
Orians, Kristin J., Univ of British Columbia
Osburn, Chris, North Carolina State Univ
Owen, Robert M., Univ of Michigan
Patterson, Molly , Binghamton Univ
Pedersen, Thomas F., Univ of Victoria
Peucker-Ehrenbrink, Bernhard, Woods Hole Oceanographic Inst
Popp, Brian N., Univ of Hawai'i, Manoa
Poyla, David, Univ of Manchester
Ravizza, Gregory, Univ of Hawai'i, Manoa
Ries, Justin, Northeastern Univ
Rosenthal, Yair, Rutgers, The State Univ of New Jersey
Russell, Ann D., Univ of California, Davis
Rutberg, Randye L., Hunter Coll (CUNY)
Sansone, Francis J., Univ of Hawai'i, Manoa
Schoonmaker, Jane E., Univ of Hawai'i, Manoa
Sedwick, Peter, Old Dominion Univ
Sericano, Jose L., Texas A&M Univ
Seyfried, Jr., William E., Univ of Minnesota, Twin Cities
Sherrell, Robert M., Rutgers, The State Univ of New Jersey
Simpson, Jr., H. James, Columbia Univ
Smith, Joseph P., United States Naval Acad
Soares Cruz, Anna Paula ., California State Univ, Bakersfield
Spero, Howard J., Univ of California, Davis
Stillings, Lisa, Univ of Nevada, Reno
Sulanowska, Margaret, Woods Hole Oceanographic Inst
Sundby, Bjorn, McGill Univ
Sutherland, Bruce, Univ of Alberta
Takahashi, Taro, Columbia Univ
Tan, Chunyang, Univ of Minnesota, Twin Cities
Them, II, Theodore R., Coll of Charleston
Thompson, Geoffrey, Woods Hole Oceanographic Inst
Tivey, Margaret K., Woods Hole Oceanographic Inst
Torres, Marta E., Oregon State Univ
Tostevin, Rosalie, Univ of Cape Town
Tripati, Aradhna, Univ of California, Los Angeles
Ullman, William J., Univ of Delaware
Valentine, David, Univ of California, Santa Barbara
Wakefield, S J., Cardiff Univ
Wan, Xianhui S., Princeton Univ
Wang, Weihong, Utah Valley Univ
Wang, Z. Aleck, Woods Hole Oceanographic Inst
Wankel, Scott, Woods Hole Oceanographic Inst
Ward, Collin, Woods Hole Oceanographic Inst
Weiss, Ray F., Univ of California, San Diego
Winckler, Gisela, Columbia Univ
Windom, Herbert L., Georgia Inst of Tech
Wong, Chi S., Univ of British Columbia

Organic Geochemistry

Abdulla, Hussain, Texas A&M Univ, Corpus Christi
Amend, Jan P., Washington Univ in St. Louis
Anderson, Ken B., Southern Illinois Univ Carbondale
Aravena, Ramon, Univ of Waterloo
Bada, Jeffrey L., Univ of California, San Diego
Bakel, Allen J., Argonne Nat Lab
Banfield, Jillian, Univ of California, Berkeley
Barker, James F., Univ of Waterloo
Bianchi, Thomas S., Univ of Florida
Bissada, Adry, Univ of Houston
Blair, Neal E., Northwestern Univ
Bowden, Stephen, Univ of Aberdeen
Brassell, Simon C., Indiana Univ, Bloomington
Burns, Timothy P., Los Alamos Nat Lab
Chen, Bob, Univ of Massachusetts, Boston
Cody, George D., Carnegie Inst for Science
Cole, Dean A., Los Alamos Nat Lab
Cowie, Gregory L., Edinburgh Univ
Diefendorf, Aaron, Univ of Cincinnati
Elson, Amy, Curtin Univ
Engel, Michael H., Univ of Oklahoma
Eyre, Bradley, Southern Cross Univ
Farrington, John W., Woods Hole Oceanographic Inst
Feakins, Sarah, Univ of Southern California
Fernandez, Loretta, Northeastern Univ
Fisher, Liz, Univ of Liverpool
Francis, Robert D., California State Univ, Long Beach
Freeman, Katherine H., Pennsylvania State Univ, Univ Park
Fu, Qi, Univ of Houston
Grice, Kliti, Curtin Univ
Hanson, Andrew D., Univ of Nevada, Las Vegas
Hartnett, Hilairy E., Arizona State Univ
Harvey, H. Rodger, Old Dominion Univ
Heal, Kate V., Edinburgh Univ
Hernes, Peter J., Univ of California, Davis
Hockaday, William C., Baylor Univ
Huang, Yongsong, Brown Univ
Ioppolo-Armanios, Marisa, Curtin Univ
Kenig, Fabien, Univ of Illinois Chicago
Kotulova, Julia, Univ of Utah
Kruge, Michael A., Montclair State Univ
Langston-Unkefer, Pat J., Los Alamos Nat Lab
Love, Gordon D., Univ of California, Riverside
MacCarthy, Patrick, Colorado School of Mines
Macko, Stephen A., Univ of Virginia
McCarthy, John F., Oak Ridge Nat Lab
Mehra, Aradhana, Univ of Derby
Meyers, Philip A., Univ of Michigan
Mitra, Siddhartha, East Carolina Univ
Mitterer, Richard M., Univ of Texas, Dallas
Moslow, Thomas F., Univ of Calgary
Nelson, Robert K., Woods Hole Oceanographic Inst
Ngwenya, Bryne T., Edinburgh Univ

Nolan, Robert P., Graduate School of the City Univ of New York
Perdue, Edward Michael, Georgia Inst of Tech
Petsch, Steven, Univ of Massachusetts, Amherst
Phillips, Dennis, Los Alamos Nat Lab
Philp, R. Paul, Univ of Oklahoma
Pratt, Lisa M., Indiana Univ, Bloomington
Pruell, Richard J., Univ of Rhode Island
Reddy, Christopher M., Woods Hole Oceanographic Inst
Repeta, Daniel J., Woods Hole Oceanographic Inst
Robinson, Clare, Univ of Manchester
Rohrssen, Megan, Central Michigan Univ
Sass, Henrik, Univ of Wales
Schaefer, Bettina, Curtin Univ
Schwark, Lorenz, Curtin Univ
Sepúlveda, Julio C., Univ of Colorado
Sessions, Alex L., California Inst of Tech
Silliman, James, Texas A&M Univ, Corpus Christi
Spongberg, Alison L., Univ of Toledo
Steen, Andrew, Univ of Tennessee, Knoxville
Stierle, Donald, Montana Tech of the Univ of Montana
Summons, Roger, Massachusetts Inst of Tech
Talbot, Helen, Univ of Newcastle Upon Tyne
Thiel, Dr., Volker, Georg-August Univ of Goettingen
Thomas, Elizabeth K., SUNY, Buffalo
Tierney, Jessica, Univ of Arizona
Townsend-Small, Amy, Univ of Cincinnati
van Dongen, Bart, Univ of Manchester
Van Vleet, Edward S., Univ of South Florida
Venkatesan, M I., Univ of California, Los Angeles
Visscher, Pieter, Univ of Connecticut
Visscher, Pieter T., Univ of Connecticut
Voorhees, Kent J., Colorado School of Mines
Wagner, Sasha, Rensselaer Polytechnic Inst
Wakeham, Stuart G., Georgia Inst of Tech
Weisener, Christopher, Univ of Windsor
Werne, Josef, Univ of Pittsburgh
Whelan, Jean K., Woods Hole Oceanographic Inst
Whiteside, Jessica, Curtin Univ
Whiticar, Michael J., Univ of Victoria
Williams, Amy, Univ of Florida
Zerkle, Aubrey, Univ of St. Andrews
Zhang, Yige, Texas A&M Univ
Zhuang, Guangsheng, Louisiana State Univ
Zimmerman, Andrew, Univ of Florida
Ziolkowski, Lori A., Univ of South Carolina

Stable Isotopes

Akhtar, Alliya, Princeton Univ
Anderson, Thomas F., Univ of Illinois, Urbana-Champaign
Anderson, William T., Florida InterNat Univ
Andrews, Julian E., Univ of East Anglia
Atekwana, Eliot A., Univ of California, Davis
Atudorei, Nieu-Viorel, Univ of New Mexico
Bao, Huiming, Louisiana State Univ
Bedaso, Zelalem, Univ of Dayton
Berke, Melissa, Univ of Notre Dame
Bindeman, Ilya N., Univ of Oregon
Blum, Joel D., Univ of Michigan
Bowen, Gabriel, Univ of Utah
Bowman, John R., Univ of Utah
Bray, Colin, Univ of Toronto
Brunner, Benjamin, Univ of Texas, El Paso
Buczynska, Anna, Northern Illinois Univ
Budde, Gerrit, Brown Univ
Burgess, Ray, Univ of Manchester
Burns, Stephen J., Univ of Massachusetts, Amherst
Cadeau, Pierre, Princeton Univ
Campbell, Andrew R., New Mexico Tech
Cao, Xiaobin, Louisiana State Univ
Clayton, Robert N., Univ of Chicago
Cohen, Anthony, Open Univ
Conroy, Jessica, Univ of Illinois, Urbana-Champaign
Coogon, Rosalind, Imperial Coll
Criss, Robert E., Washington Univ in St. Louis
Crowley, Brooke E., Univ of Cincinnati
Darling, James, Univ of Portsmouth
Davidson, Garry J., Univ of Tasmania
Denniston, Rhawn F., Univ of Iowa
Dettman, David, Univ of Arizona
Deuser, Werner G., Woods Hole Oceanographic Inst
Dodd, Justin P., Northern Illinois Univ
Dorale, Jeffrey A., Univ of Iowa
Douglas, Peter, McGill Univ
Eastoe, Chris J., Univ of Arizona
Economos, Rita C., Southern Methodist Univ
Edwards, Thomas W. D., Univ of Waterloo
Elder, Kathryn L., Woods Hole Oceanographic Inst
Fantle, Matthew S., Pennsylvania State Univ, Univ Park
Farquhar, James, Univ of Maryland
Fayek, Mostafa, Univ of Manitoba
Feng, Xiahong, Dartmouth Coll
Fike, David A., Washington Univ in St. Louis
Fisher, David E., Univ of Miami
Foster, Gavin, Univ of Southampton
Fricke, Henry C., Colorado Coll
Gagnon, Alan R., Woods Hole Oceanographic Inst
Gao, Yongjun, Univ of Houston
Gazis, Carey A., Central Washington Univ
Gibson, John J., Univ of Waterloo
Gilbert, Kathleen W., Wellesley Coll
Gilg, Hans A., Technical Univ of Munich
Gill, Benjamin, Virginia Polytechnic Inst & State Univ
Gillikin, David P., Union Coll
Goldstein, Steven L., Columbia Univ
Gonzalez, Luis A., Univ of Kansas
Graczyk, Donald G., Argonne Nat Lab
Grassineau, Nathalie, Royal Holloway Univ of London
Gregory, Robert T., Southern Methodist Univ
Griffith, Elizabeth, Ohio State Univ
Griffith, Liz, Univ of Texas, Arlington
Grimes, Stephen, Univ of Plymouth
Grocke, Darren, Durham Univ
Grossman, Ethan L., Texas A&M Univ
Guo, Weifu, Woods Hole Oceanographic Inst
Haley, Brian, Oregon State Univ
Halliday, Alex, Univ of Oxford
Haynes, Shannon, Princeton Univ
Helie, Jean-François, Universite du Quebec a Montreal
Hemming, Sidney R., Columbia Univ
Henkel, Torsten, Univ of Manchester
Henkes, Gregory, Stony Brook Univ
Hillaire-Marcel, Claude, Universite du Quebec a Montreal
Hindshaw, Ruth, Univ of St. Andrews
Holk, Gregory J., California State Univ, Long Beach
Holmden, Chris, Univ of Saskatchewan
Horton, Travis, Univ of Canterbury
Howe, Stephen S., SUNY, Albany
Hren, Michael, Univ of Connecticut
Ingram, B. Lynn, Univ of California, Berkeley
Johnson, Beverly J., Bates Coll
Junium, Christopher, Syracuse Univ
Kaiser, Jan, Univ of East Anglia
Karhu, Juha A., Univ of Helsinki
Kaufman, Alan J., Univ of Maryland
Keigwin, Lloyd D., Woods Hole Oceanographic Inst
Kieffer, Bruno, Univ of British Columbia
Kienast, Markus, Dalhousie Univ
Kirtland Turner, Sandra, Univ of California, Riverside
Knauth, L. Paul, Arizona State Univ
Kohn, Matthew, Boise State Univ
Krantz, David E., Univ of Delaware
Kreutz, Karl J., Univ of Maine
Krishnamurthy, R. V., Western Michigan Univ
Lachniet, Matthew S., Univ of Nevada, Las Vegas
Larson, Peter B., Washington State Univ
LAVASTRE, Véronique, Université Jean Monnet, Saint-Etienne
Law, Kim, Western Univ
Leavitt, Steven W., Univ of Arizona
Lefticariu, Liliana, Southern Illinois Univ Carbondale
Li, Long, Univ of Alberta
Lini, Andrea, Univ of Vermont
Lohmann, Kyger C., Univ of Michigan

Longstaffe, Frederick J., Western Univ
Lowry, David, Royal Holloway Univ of London
Luzincourt, Marc R., Universite du Quebec
Marshall, Jim, Univ of Liverpool
Masterson, Andrew, Northwestern Univ
Matheney, Ronald K., Univ of North Dakota
Mattey, Dave, Royal Holloway Univ of London
Mayer, Bernhard, Univ of Calgary
Menicucci, Anthony, Univ of South Florida
Michalski, Greg, Purdue Univ
Mix, Alan C., Oregon State Univ
Mora, Claudia I., Univ of Tennessee, Knoxville
Muehlenbachs, Karlis, Univ of Alberta
Nelson, Stephen T., Brigham Young Univ
Nielsen, Sune G., Woods Hole Oceanographic Inst
O'Neil, James, Univ of Michigan
Ohmoto, Hiroshi, Pennsylvania State Univ, Univ Park
Oleynik, Sergey, Princeton Univ
Olivares, Jose A., Los Alamos Nat Lab
Padden, Maureen, McMaster Univ
Parai, Rita, Washington Univ in St. Louis
Partin, Jud, Univ of Texas, Austin
Passey, Benjamin, Univ of Michigan
Patterson, William P., Univ of Saskatchewan
Pedentchouk, Nikolai, Univ of East Anglia
Pinti, Daniele Luigi, Universite du Quebec a Montreal
Planavsky, Noah, Yale Univ
Pogge von Strandmann, Phillip, Birkbeck Coll
Poirier, Andre, Universite du Quebec a Montreal
Ponton, Camilo, Western Washington Univ
Popp, Brian N., Univ of Hawai'i, Manoa
Poulson, Simon R., Univ of Nevada, Reno
Ravelo, Christina, Univ of California, Santa Cruz
Rehkamper, Mark, Imperial Coll
Reid, Rachel E., Virginia Polytechnic Inst & State Univ
Reynard, Linda, Boise State Univ
Richards, Ian J., Southern Methodist Univ
Rye, Danny M., Yale Univ
Savin, Samuel M., Case Western Reserve Univ
Schaller, Morgan F., Rensselaer Polytechnic Inst
Schauble, Edwin A., Univ of California, Los Angeles
Schimmelmann, Arndt, Indiana Univ, Bloomington
Schwarcz, Henry P., McMaster Univ
Sedlacek, Alexa, Univ of Northern Iowa
Shahar, Anat, Univ of Maryland
Sharma, Shikha, West Virginia Univ
Sharp, Zachary D., Univ of New Mexico
Sherwood, Owen, Dalhousie Univ
Sherwood Lollar, Barbara, Univ of Toronto
Shieh, Yuch-Ning, Purdue Univ
Showers, William J., North Carolina State Univ
Shultis, Aaron, Univ of Kentucky
Smart, Katie A., Univ of the Witwatersrand
Smirnov, Anna, Universite du Quebec
Sowers, Todd, Pennsylvania State Univ, Univ Park
Steig, Eric J., Univ of Washington
Steig, Eric J., Univ of Washington
Strauss, Harald, Universitaet Muenster
Stute, Martin, Columbia Univ
Syrup, Krista A., Moraine Valley Comm Coll
Szynkiewicz, Anna, Univ of Tennessee, Knoxville
Taylor, Hugh P., California Inst of Tech
Theiling, Bethany P., Univ of Tulsa
Thirlwall, Matthew, Royal Holloway Univ of London
Tudhope, Alexander W., Edinburgh Univ
Valley, John W., Univ of Wisconsin, Madison
van de Flierdt, Tina, Imperial Coll
Wanamaker, Alan D., Iowa State Univ of Science & Tech
Webb, Elizabeth A., Western Univ
Welp, Lisa, Purdue Univ
Wenner, David B., Univ of Georgia
White, James W. C., Univ of Colorado
Widory, David, Universite du Quebec a Montreal
Wright, James D., Rutgers, The State Univ of New Jersey
Yapp, Crayton J., Southern Methodist Univ
Yeung, Laurence Y., Rice Univ
Young, Edward D., Univ of California, Los Angeles
Yun, Misuk, Univ of Manitoba
Zhang, Lin, Texas A&M Univ, Corpus Christi
Zhang, Ren, Baylor Univ
Zheng, Xinyuan, Univ of Minnesota, Twin Cities
Ziegler, Karen, Univ of New Mexico
Zierenberg, Robert A., Univ of California, Davis

Trace Element Distribution

Coale, Kenneth H., Moss Landing Marine Lab
Condie, Kent C., New Mexico Tech
Croot, Peter, Nat Univ of Ireland Galway
Cullers, Robert L., Kansas State Univ
Daniels, William R., Los Alamos Nat Lab
Efurd, Deward W., Los Alamos Nat Lab
Frey, Frederick A., Massachusetts Inst of Tech
Gabrielli, Paolo, Ohio State Univ
Hale, Beverley A., Univ of Guelph
Kar, Aditya, Fort Valley State Univ
Kyte, Frank T., Univ of California, Los Angeles
LaFleche, Marc R., Universite du Quebec
Marcantonio, Franco, Texas A&M Univ
Newton, Robert, Columbia Univ
Nicolescu, Stefan, Yale Univ
Peterson, Eugene J., Los Alamos Nat Lab
Rader, Shelby, Indiana Univ, Bloomington
Ritter, Charles J., Univ of Dayton
Ryan, Jeffrey G., Univ of South Florida
Schartup, Amina, Univ of California, San Diego
Seifert, Karl E., Iowa State Univ of Science & Tech
Simonetti, Tony, Univ of Notre Dame
Thomas, Kimberly W., Los Alamos Nat Lab
Thompson, Joseph L., Los Alamos Nat Lab
Von Reden, Karl F., Woods Hole Oceanographic Inst
Weaver, Barry L., Univ of Oklahoma
Wildeman, Thomas R., Colorado School of Mines

Biogeochemistry

Abou Najm, Majdi, Univ of California, Davis
Amon, Rainier M., Texas A&M Univ
Anbar, Ariel D., Arizona State Univ
Babbin, Andrew, Massachusetts Inst of Tech
Barker, Joel D., Ohio State Univ
Berns-Herrboldt, Erin, Univ of Wisconsin, Green Bay
Black, Erin, Univ of Rochester
Bradley, Alexander S., Washington Univ in St. Louis
Buick, Roger, Univ of Washington
Cassar, Nicolas, Duke Univ
Clayton, Sophie, Old Dominion Univ
Costa, Jr, Ozeas S., Ohio State Univ
Curtis, P. J., Univ of British Columbia, Okanagan Campus
Dolman, Han J., Vrije Universiteit Amsterdam
Elliott, Emily M., Univ of Pittsburgh
Fortin, Danielle, Univ of Ottawa
Fowle, David A., Univ of Kansas
Fulton, James, Baylor Univ
Gu, Baohua, Oak Ridge Nat Lab
Hain, Mathis, Univ of California, Santa Cruz
Hornibrook, Edward R., Univ of British Columbia, Okanagan Campus
Jaisi, Deb, Univ of Delaware
Jones, Daniel, New Mexico Tech
Kavanaugh, Maria, Oregon State Univ
Kellman, Lisa M., Saint Francis Xavier Univ
Kwon, Man Jae, Korea Univ
Lau, Kimberly, Univ of Wyoming
Lazcano, Cristina, Univ of California, Davis
Liang, Liyuan, Univ of Tennessee, Knoxville
Mahmoudi, Nagissa, McGill Univ
Marin-Spiotta, Erika, Univ of Wisconsin, Madison
Matisoff, Gerald, Case Western Reserve Univ
McCormick, Michael L., Hamilton Coll
Mellinger, David, Oregon State Univ
Mulder, Jan, Norwegian Univ of Life Sciences (NMBU)
Okie, Jordan, Arizona State Univ
Pearson, Ann, Harvard Univ
Phelps, Tommy J., Univ of Tennessee, Knoxville
Philben, Michael , Hope Coll

Reddy, K R., Univ of Florida
Reji, Linta, Princeton Univ
Richardson, Justin, Univ of Massachusetts, Amherst
Richey, Jeffrey E., Univ of Washington
Riding, Robert, Univ of Tennessee, Knoxville
Roberts, Jennifer A., Univ of Kansas
Sanford, Robert A., Univ of Illinois, Urbana-Champaign
Scarlett, Alan, Curtin Univ
Smith, Robert, Univ of Idaho
Steenwerth, Kerri, Univ of California, Davis
Tang, Weiyi, Princeton Univ
Throop, Heather, Arizona State Univ
Thurber, Andrew, Oregon State Univ
Torres, Mark, Rice Univ
Trembath-Reichert, Elizabeth, Arizona State Univ
Tufillaro, Nicholas, Oregon State Univ
Vineis, Joseph, Princeton Univ
Wang, Xingchen, Boston Coll
Weiss, Peter, Univ of California, Santa Cruz
Winbourne, Joy, Univ of Massachusetts, Lowell
Winnick, Matthew, Univ of Massachusetts, Amherst
Xu, Tingying, Oklahoma State Univ
Zhang, Xinning, Princeton Univ

High-temperature Geochemistry

Babechuk, Michael, Memorial Univ of Newfoundland
Baker, Don, McGill Univ
Chin, Emily J., Univ of Tennessee, Knoxville
Chin, Emily, Univ of California, San Diego
Han, Hannah Juan, New Mexico Tech
Jackson, Matt, Univ of California, Santa Barbara
Le Voyer, Marion, Smithsonian Inst / Nat Museum of Nat History
Luo, Haiyang, Princeton Univ
Muñoz-Saez, Carolina, Univ of Nevada, Reno
Pollock, Meagen, Coll of Wooster
Reid, Mary R., Northern Arizona Univ
Schwartz, Darin, Boise State Univ
Smithies, Hugh, Curtin Univ
Zajacz, Zoltan, Univ of Toronto
Zhang, Youxue, Univ of Michigan

Aqueous Geochemistry

Chin, Yu-Ping, Ohio State Univ
Clark, Elyse, Fitchburg State Univ
Cox, Alysia, Montana Tech of the Univ of Montana
Duan, Gan, Curtin Univ
Greenwell, Chris, Durham Univ
Hasan, M A., Univ of Dhaka
Hinkle, Margaret A., Washington & Lee Univ
Hon, Rudolph, Boston Coll
Kurz, Marie J., Drexel Univ
MacKay, Allison A., Ohio State Univ
Matiasek, Sandrine, California State Univ, Chico
O'Shea, Bethany, Univ of San Diego
Pasek, Matthew, Univ of South Florida
Pincus, Lauren N., Princeton Univ
Polizzotto, Matthew, Univ of Oregon
Ridley, Moira K., Texas Tech Univ
Saccocia, Peter J., Bridgewater State Univ
Wolff-Boenisch, Dom, Curtin Univ
Wood, Scott A., North Dakota State Univ
Yun, Seong-Taek, Korea Univ

GEOPHYSICS

General Geophysics

Adams, Aubreya, Colgate Univ
Adepelumi, A. A., Obafemi Awolowo Univ
Afolabi, O., Obafemi Awolowo Univ
Ahern, Judson L., Univ of Oklahoma
Ahmed, Mohamed, Texas A&M Univ, Corpus Christi
Ajo-Franklin, Jonathan, Rice Univ
Al-Lehyani, Ayman, King Fahd Univ of Petroleum and Minerals
Al-Shuhail, Abdullatif, King Fahd Univ of Petroleum and Minerals
Alao, O. A., Obafemi Awolowo Univ
Albarello, Dario, Univ of Siena
Albright, James N., Los Alamos Nat Lab
Anderson, Megan, Central Washington Univ
Angelopoulos, Vassilis, Univ of California, Los Angeles
Anyiam, Uzonna O., Hope Coll
Armah, Thomas K., Univ of Ghana
Audet, Pascal, Univ of Ottawa
Aurnou, Jonathan M., Univ of California, Los Angeles
Bachtadse, Valerian, Ludwig-Maximilians-Universitaet Muenchen
Bailey, Richard C., Univ of Toronto
Ballmer, Maxim D., Univ Coll London
Barbi, Greta, Ludwig-Maximilians-Universitaet Muenchen
Barklage, Mitchell, Northwestern Univ
Barsch, Robert, Ludwig-Maximilians-Universitaet Muenchen
Bastow, Ian, Imperial Coll
Beaumont, Christopher, Dalhousie Univ
Becker, Alex, Univ of California, Berkeley
Becker, Thorsten, Univ of Texas, Austin
Bedle, Heather, Univ of Oklahoma
Behn, Mark, Boston Coll
Bendick, Rebecca, Univ of Montana
Bercovici, David, Yale Univ
Beresnev, Igor A., Iowa State Univ of Science & Tech
Billen, Magali I., Univ of California, Davis
Billings, Stephen, Univ of British Columbia
Bina, Craig R., Northwestern Univ
Blankenship, Donald D., Univ of Texas, Austin
Bleibinhaus, Florian, Ludwig-Maximilians-Universitaet Muenchen
Bloxham, Jeremy, Harvard Univ
Boaga, Jacopo, Università degli Studi di Padova
Bokelmann, Goetz H., Univ of Vienna
Boschi, Lapo, Università degli Studi di Padova
Bradley, Christopher R., Los Alamos Nat Lab
Braun, Alexander, Queen's Univ
Briggs, Derek E., Yale Univ
Britzke, Gilbert, Ludwig-Maximilians-Universitaet Muenchen
Brodholt, John, Univ Coll London
Brooks, Debra A., Santiago Canyon Coll
Brown, Wesley A., Stephen F. Austin State Univ
Buck, IV, W. R., Columbia Univ
Bunge, Hans-Peter, Ludwig-Maximilians-Universitaet Muenchen
Bustin, Amanda, Univ of British Columbia
Butler, Karl, Univ of New Brunswick
Butler, Sam, Univ of Saskatchewan
Calvin, Wendy, Univ of Nevada, Reno
Carlson, Barry A., Delta Coll
Carpenter, Philip J., Northern Illinois Univ
Carrigan, Charles R., Lawrence Livermore Nat Lab
Cassiani, Giorgio, Università degli Studi di Padova
Cayzer, Nicola, Edinburgh Univ
Chapman, Mark, Edinburgh Univ
Chase, Clement G., Univ of Arizona
Cheadle, Michael J., Univ of Wyoming
Chen, Jingyi, Univ of Tulsa
Clark, H. C., Rice Univ
Clarke, Garry K. C., Univ of British Columbia
Coakley, Bernard, Univ of Alaska, Fairbanks
Colgan, William, York Univ
Colin, Farquharson, Memorial Univ of Newfoundland
Comas, Xavier, Florida Atlantic Univ
Cooper, Catherine , Washington State Univ
Copley, Alex, Univ of Cambridge
Cosentino, Pietro, Univ of East Anglia
Couture, Gilles, Universite du Quebec a Montreal
Craig, Mitchell S., California State Univ, East Bay
Craven, John, Edinburgh Univ
Daly, Eve, Nat Univ of Ireland Galway
Daniels, Jeffrey J., Ohio State Univ
Davies, Gregory, Princeton Univ
Davies, J H., Cardiff Univ
Davies, Rhodri, Imperial Coll
Davis, Daniel M., Stony Brook Univ
Davis, Tara, Univ Coll Cork
Delph, Jonathan, Purdue Univ
Dengler, Lorinda, Humboldt State Univ
Dewdney, Chris, Univ of Portsmouth
Dickman, Steven R., Binghamton Univ
Dieterich, James H., Univ of California, Riverside

Dmochowski, Jane, Univ of Pennsylvania
Doll, William E., Oak Ridge Nat Lab
Dorman, Leroy M., Univ of California, San Diego
Dunbar, John A., Baylor Univ
Dunning, Jeremy D., Indiana Univ, Bloomington
Durland, Theodore, Oregon State Univ
Dzunic, Aleksandar, Curtin Univ
Ebel, John E., Boston Coll
Eckert, Andreas, Missouri Univ of Science and Tech
Edwards, C L, Los Alamos Nat Lab
Ehlers, Todd A., Univ of Michigan
Ekstrom, Goran, Columbia Univ
Elhusseiny, Ammar , King Fahd Univ of Petroleum and Minerals
Ellwood, Brooks, Univ of Texas, Arlington
Emmer, Barbara, Ludwig-Maximilians-Universitaet Muenchen
Erickson, Brittany, Univ of Oregon
Fehler, Michael, Massachusetts Inst of Tech
Ferguson, John F., Univ of Texas, Dallas
Ferrucci, Fabrizio, Open Univ
Fialko, Yuri A., Univ of California, San Diego
Fichtner, Andreas, Ludwig-Maximilians-Universitaet Muenchen
Filina, Irina, Univ of Nebraska, Lincoln
Fishwick, Stewart, Leicester Univ
Flesch, Lucy M., Purdue Univ
Foley, Bradford, Pennsylvania State Univ, Univ Park
Fontaine, Fabrice R., Universite de la Reunion
Forte, Alessandro Marco, Universite du Quebec a Montreal
Fortier, Richard, Universite Laval
Gaffney, Edward S., Los Alamos Nat Lab
Gaidos, Eric J., Univ of Hawai'i, Manoa
Gallovic, Frantisek, Ludwig-Maximilians-Universitaet Muenchen
Galsa, Attila, Eotvos Lorand Univ
Gao, Stephen , Missouri Univ of Science and Tech
Gebrande, Helmut, Ludwig-Maximilians-Universitaet Muenchen
Ghazala, Hosni H., El Mansoura Univ
Goes, Saskia, Imperial Coll
Goforth, Thomas T., Baylor Univ
Goldsby, David, Univ of Pennsylvania
Goodliffe, Andrew, Univ of Alabama
Gorman, Gerard, Imperial Coll
Gosnold, William D., Univ of North Dakota
Graham, Gina R., Alaska Div of Geological & Geophysical Surveys
Grand, Stephen P., Univ of Texas, Austin
Granja, José Luis, Univ Complutense de Madrid
Greenfield, Cat, Leicester Univ
Greenfield, Roy J., Pennsylvania State Univ, Univ Park
Gregg, Patricia, Univ of Illinois, Urbana-Champaign
Gribenko, Alex, Univ of Utah
Grindlay, Nancy R., Univ of North Carolina, Wilmington
Haas, Christian, York Univ
Haber, Eldad, Univ of British Columbia
Hamzaoui, Cherif, Universite du Quebec a Montreal
Hanafy, Sherif , King Fahd Univ of Petroleum and Minerals
Hanu, Dan B., Alexandru Ioan Cuza
Hardage, Bob A., Univ of Texas at Austin, Jackson Schl of Geosciences
Harder, Steven H., Univ of Texas, El Paso
Harmon, Nicholas, Univ of Southampton
Harrington, Rebecca, McGill Univ
Harry, Dennis L., Colorado State Univ
Hayes, Jorden L., Dickinson Coll
Heigold, Paul C., Argonne Nat Lab
Hellio, Gabrielle, Univ of Southampton
Henson, Harvey, Southern Illinois Univ Carbondale
Henstock, Tim, Univ of Southampton
Henyey, Thomas L., Univ of Southern California
Higgs, Bettie M., Univ Coll Cork
Hirsch, Aaron, Univ of Minnesota
Hofmeister, Anne M., Washington Univ in St. Louis
Hornbach, Matthew, Univ of Texas, Austin
Horowitz, Franklin G., Cornell Univ
Horváth, Ferenc, Eotvos Lorand Univ
Houseman, Greg, Univ of Leeds
Huang, Chuan, Curtin Univ
Huang, Lianjie, Los Alamos Nat Lab
Huerta, Audrey, Central Washington Univ
Hussein, Musa, Univ of Texas, El Paso
Iaffaldano, Giampiero, Ludwig-Maximilians-Universitaet Muenchen
Igel, Heiner, Ludwig-Maximilians-Universitaet Muenchen
Ishii, Miaki, Harvard Univ
Ito, Garrett T., Univ of Hawai'i, Manoa
Ivanov, Julian, Univ of Kansas
Jackson, James A., Univ of Cambridge
Jacob, Robert W., Bucknell Univ
Jaiswal, Priyank, Oklahoma State Univ
James, Peter B., Baylor Univ
Jarvis, Gary T., York Univ
Jayawickreme, Dushmantha, Southern Connecticut State Univ
Jensen, Olivia G., McGill Univ
Jezek, Kenneth, Ohio State Univ
Jiang, Junle , Univ of Oklahoma
Johnson, Kaj, Indiana Univ, Bloomington
Johnson, Paul A., Los Alamos Nat Lab
Johnson, Verner C., Colorado Mesa Univ
Jomeiri, Rahim, Univ of Tabriz
Jurdy, Donna M., Northwestern Univ
Kafka, Alan L., Boston Coll
Kaplinski, Matthew A., Northern Arizona Univ
Kaus, Boris P., Universität Mainz
Kavner, Abby, Univ of California, Los Angeles
Keller, G. Randy, Univ of Oklahoma
Keller , George R., Univ of Texas, El Paso
Kelley, Shari, New Mexico Tech
Kellogg, James N., Univ of South Carolina
Kenyon, Patricia M., City Coll (CUNY)
Kenyon, Patricia M., Graduate School of the City Univ of New York
Khalil Aboushanab, Mohamed, Univ of Nebraska, Lincoln
King, Scott D., Virginia Polytechnic Inst & State Univ
Kingdon, Kevin, Univ of British Columbia
Kinsland, Gary L., Univ of Louisiana at Lafayette
Korenaga, Jun, Yale Univ
Kroeger, Glenn C., Trinity Univ
Kruger, Joseph M., Lamar Univ
Kruse, Sarah E., Univ of South Florida
Kukowski, Nina, Friedrich-Schiller-Univ Jena
Kuo, John T., Columbia Univ
Käser, Martin, Ludwig-Maximilians-Universitaet Muenchen
Lai, Ching-Yao, Princeton Univ
Lat, Che Noorliza, Univ of Malaya
Lathrop, Daniel, Univ of Maryland
Lavallee, Yan, Ludwig-Maximilians-Universitaet Muenchen
Lavier, Luc L., Univ of Texas, Austin
Leitch, Alison, Memorial Univ of Newfoundland
Lenardic, Adrian, Rice Univ
Leonard, Lucinda, Univ of Victoria
Lewis, Stephen D., Fresno State Univ
Li, Mingming, Arizona State Univ
Li, Yunyue E., Purdue Univ
Lin, Fan-Chi, Univ of Utah
Lin, Guoqing, Univ of Miami
Lithgow-Bertelloni, Carolina, Univ of California, Los Angeles
Liu, Lanbo, Univ of Connecticut
Liu, Lijun, Univ of Illinois, Urbana-Champaign
Liu, Mian, Univ of Missouri
Liu, Qinya, Univ of Toronto
Liu, Yajing, McGill Univ
Lohman, Rowena B., Cornell Univ
López-Pineda, Leobardo, Universidad Estatal de Sonora
Louie, John, Univ of Nevada, Reno
Lozos, Julian, California State Univ, Northridge
Luttrell, Karen, Louisiana State Univ
Lutz, Pascale, Institut Polytechnique LaSalle Beauvais (ex-IGAL)
Malcolm, Alison, Memorial Univ of Newfoundland
Malservisi, Rocco, Ludwig-Maximilians-Universitaet Muenchen
Mareschal, Jean-Claude, Universite du Quebec a Montreal
Martens, Hilary R., Univ of Montana
Mavko, Gerald M., Stanford Univ
Maxwell, Michael, Univ of British Columbia
McElwaine, Jim, Durham Univ
McGinnis, Lyle D., Argonne Nat Lab
Meade, Brendan, Harvard Univ
Meadows, Wayne R., Los Alamos Nat Lab
Meert, Joseph G., Univ of Florida

Merriam, James B., Univ of Saskatchewan
Merrill, Ronald T., Univ of Washington
Milne, Glenn A., Univ of Ottawa
Minchew, Brent, Massachusetts Inst of Tech
Mitrovica, Jerry X., Harvard Univ
Mohr, Marcus, Ludwig-Maximilians-Universitaet Muenchen
Montesi, Laurent G., Univ of Maryland
Morgan, F D., Massachusetts Inst of Tech
Morgan, Joanna, Imperial Coll
Morris, William A., McMaster Univ
Morrison, H. Frank, Univ of California, Berkeley
Morse, David L., Univ of Texas, Austin
Muñoz, Alfonso, Univ Complutense de Madrid
Murphy, Vincent, Boston Coll
NICULESCU, Bogdan M., Univ of Bucharest
Niu, Qifei, Boise State Univ
Nunn, Jeffrey A., Louisiana State Univ
Nur, Amos M., Stanford Univ
Nyblade, Andrew A., Pennsylvania State Univ, Univ Park
Nyquist, Jonathan, Temple Univ
O'Farrell, Keely A., Univ of Kentucky
Oeser, Jens, Ludwig-Maximilians-Universitaet Muenchen
Ojo, S. B., Obafemi Awolowo Univ
Oladunjoye, M. A., Univ of Ibadan
Olayinka, A. I., Univ of Ibadan
Oldenburg, Douglas W., Univ of British Columbia
Oliver, Ronald D., Los Alamos Nat Lab
Olorunfemi, M. O., Obafemi Awolowo Univ
Olson, Peter L., Johns Hopkins Univ
Osinowo, O. O., Univ of Ibadan
Paillet, Fred, Univ of Arkansas, Fayetteville
Pain, Christopher, Imperial Coll
Palmer, Donald F., Kent State Univ
Palutoglu, Mahmut, Firat Univ
Panero, Wendy R., Ohio State Univ
Parmentier, E. Marc, Brown Univ
Parsekian, Andrew D., Univ of Wyoming
Pasicznyk, David L., New Jersey Geological and Water Survey
Patton, Howard J., Los Alamos Nat Lab
Peavy, Samuel T., Georgia Southwestern State Univ
Peirce, Christine, Durham Univ
Pelton, John R., Boise State Univ
Peters, Mark T., Los Alamos Nat Lab
Petersen, Nikolai, Ludwig-Maximilians-Universitaet Muenchen
Pfuhl, Helen, Ludwig-Maximilians-Universitaet Muenchen
Phillips, William S., Los Alamos Nat Lab
Piazzoni, Antonio Sebastian, Ludwig-Maximilians-Universitaet Muenchen
Pimentel-Hernandez, Ramon, Universidad Auto de Baja California Sur
Plag, Hans-Peter, Old Dominion Univ
Plattner, Christina, Ludwig-Maximilians-Universitaet Muenchen
Pollack, Henry N., Univ of Michigan
Porch, William M., Los Alamos Nat Lab
Porter, Ryan, Northern Arizona Univ
Priestley, Keith, Univ of Cambridge
Pysklywec, Russell N., Univ of Toronto
Raine, Robin, Nat Univ of Ireland Galway
Ranalli, Giorgio, Carleton Univ
Randall, George, Los Alamos Nat Lab
Rasmussen, Steen, Los Alamos Nat Lab
Ravat, Dhananjay, Univ of Kentucky
Rawlinson, Nick, Univ of Cambridge
Rempel, Alan W., Univ of Oregon
Revetta, Frank A., SUNY Potsdam
Rice, James R., Harvard Univ
Richard, Robert, Univ of California, Los Angeles
Richards, Mark A., Univ of California, Berkeley
Richardson, Randall M., Univ of Arizona
Roach, Michael, Univ of Tasmania
Roberts, Mark L., Woods Hole Oceanographic Inst
Robinson, Judith, Rutgers, The State Univ of New Jersey, Newark
Rochester, Michael G., Memorial Univ of Newfoundland
Roecker, Steven W., Rensselaer Polytechnic Inst
Romanovsky, Vladimir, Univ of Alaska, Fairbanks
Rothman, Daniel H., Massachusetts Inst of Tech
Rubin, Allan M., Princeton Univ
Rudge, John, Univ of Cambridge
Rudolph, Maxwell L., Univ of California, Davis
Russell, Christopher T., Univ of California, Los Angeles
Russell, R. Doncaster, Univ of British Columbia
Russo, Raymond, Univ of Florida
Ryall, Patrick J., Dalhousie Univ
Rychert, Catherine A., Univ of Southampton
Sabra, Karim, Georgia Inst of Tech
Sahakian, Valerie J., Univ of Oregon
Salami, Sikiru A., Univ of Benin
Sammis, Charles G., Univ of Southern California
Sauck, William A., Western Michigan Univ
Sava, Diana, Univ of Texas at Austin, Jackson Schl of Geosciences
Sava, Diana C., Univ of Texas, Austin
Sawyer, Derek, Ohio State Univ
Schmandt, Brandon, Univ of New Mexico
Schriver, David, Univ of California, Los Angeles
Schubert, Gerald, Univ of California, Los Angeles
Schuberth, Bernhard, Ludwig-Maximilians-Universitaet Muenchen
Schultz, Adam, Oregon State Univ
Searls, Mindi L., Univ of Nebraska, Lincoln
Segall, Paul, Stanford Univ
Serzu, Mulu, Univ of Manitoba
Shams, Asghar, Heriot-Watt Univ
Shcherbakov, Robert, Western Univ
Sherrod, Laura, Kutztown Univ of Pennsylvania
Siegler, Matthew, Southern Methodist Univ
Sigloch, Karin, Univ of Oxford
Simila, Gerald W., California State Univ, Northridge
Simmons, M. G., Massachusetts Inst of Tech
Simms, Janet E., Mississippi State Univ
Sit, Stefany , Univ of Illinois Chicago
Sleep, Norman H., Stanford Univ
Smylie, Douglas E., York Univ
Solomatov, Viatcheslav S., Washington Univ in St. Louis
Sonett, Charles P., Univ of Arizona
Soupios, Panteleimon , King Fahd Univ of Petroleum and Minerals
Spanos, T.J.T (Tim), Univ of Alberta
Sparks, David, Texas A&M Univ
Stein, Carol A., Univ of Illinois Chicago
Stephenson, Randell A., Univ of Aberdeen
Stidham, Christiane W., Stony Brook Univ
Stixrude, Lars, Univ of California, Los Angeles
Stupazzini, Marco, Ludwig-Maximilians-Universitaet Muenchen
Suckale, Jenny, Stanford Univ
Swain, Chris, Curtin Univ
Sweeney, Mark D., Pacific Northwest Nat Lab
Szeliga, Walter, Central Washington Univ
Szeto, Anthony M. K., York Univ
Taib, Samsudin Hj., Univ of Malaya
Tatham, Robert H., Univ of Texas, Austin
Taylor, Graeme, Univ of Plymouth
ten Brink, Uri S., Woods Hole Oceanographic Inst
ter Voorde, Marlies, Vrije Universiteit Amsterdam
Thiruvathukal, John V., Montclair State Univ
Thompson, Michael D., Argonne Nat Lab
Timár, Gábor, Eotvos Lorand Univ
Tisato, Nicola, Univ of Texas, Austin
Tobin, Harold, Univ of Wisconsin, Madison
Tong, Vincent C., Birkbeck Coll
Tracy, Sally June, Carnegie Inst for Science
Tsai, Victor, California Inst of Tech
Tsoflias, George P., Univ of Kansas
Turcotte, Donald L., Univ of California, Davis
Utgard, Russell O., Ohio State Univ
Van Avendonk, Harm, Univ of Texas, Austin
van Keken, Peter J., Univ of Michigan
Velli, Marco, Univ of California, Los Angeles
Vernhes, Jean-David, Institut Polytechnique LaSalle Beauvais (ex-IGAL)
Versteeg, Roelof, Columbia Univ
Vincent, Robert K., Bowling Green State Univ
Viti, Marcello, Univ of Siena
von Glasow, Roland, Univ of East Anglia
Vopson, Melvin M., Univ of Portsmouth
Waheed, Umair B., King Fahd Univ of Petroleum and Minerals
Waltham, Dave, Royal Holloway Univ of London
Wang, Chi-Yuen, Univ of California, Berkeley

Wang, Herbert F., Univ of Wisconsin, Madison
Wang, Kelin, Univ of Victoria
Wang, Xiaoyu , Princeton Univ
Wang, Yanghua, Imperial Coll
Warren, Linda M., Saint Louis Univ
Wassermann, Joachim, Ludwig-Maximilians-Universitaet Muenchen
Weaver, Thomas A., Los Alamos Nat Lab
Weeraratne, Dayanthie, California State Univ, Northridge
Wettlaufer, John S., Yale Univ
White, Nicky, Univ of Cambridge
White, Robert, Univ of Cambridge
Whitman, Dean, Florida InterNat Univ
Williams, Ian S., Univ of Wisconsin, River Falls
Wilson, Clark R., Univ of Texas, Austin
Winberry, Paul, Central Washington Univ
Winklhofer, Michael, Ludwig-Maximilians-Universitaet Muenchen
Wisely, Beth , Casper Coll
Wishart, De Bonne N., Central State Univ
Woodhouse, John, Univ of Oxford
Woods, Andy, Univ of Cambridge
Worthington, Lindsay L., Univ of New Mexico
Yang, Xiaotao, Purdue Univ
Yelisetti, Subbarao, Texas A&M Univ, Kingsville
Zahm, Christopher K., Univ of Texas, Austin
Zaja, Annalisa, Università degli Studi di Padova
Zakharova, Natalia, Central Michigan Univ
Zarroca Hernández, Mario, Universitat Auto de Barcelona
Zaspel, Craig E., Univ of Montana Western
Zelt, Colin A., Rice Univ
Zha, Changsheng, Carnegie Inst for Science
Zhang, Chi, Univ of Kansas
Zoback, Mark D., Stanford Univ

Experimental Geophysics

Anderson, Jake , Boise State Univ
Anderson, Orson L., Univ of California, Los Angeles
Askari, Roohollah, Michigan Technological Univ
Bates, Richard, Univ of St. Andrews
Bevis, Michael G., Ohio State Univ
Bohlen, Thomas, Karlsruhe Inst of Tech
Bona, Andrej, Curtin Univ
Bridges, Frank, Univ of California, Santa Cruz
Bussod, Gilles Y., Los Alamos Nat Lab
Carson, Michael, Curtin Univ
Christensen, Nikolas I., Univ of Wisconsin, Madison
Comina, Cesare, Università di Torino
Dewers, Thomas, Sandia Nat Lab
Dillman, Amanda, Univ of Minnesota, Twin Cities
Dyaur, Nikolay, Univ of Houston
Evans, J. B., Massachusetts Inst of Tech
Fomel, Sergey, Univ of Texas at Austin, Jackson Schl of Geosciences
Fratta, Dante, Univ of Wisconsin, Madison
Glover, Paul W., Univ of Leeds
Glover, Paul W., Universite Laval
Glubokovskikh, Stanislav, Curtin Univ
Goebel, Thomas H., Univ of Memphis
Graham, Jr., Earl K., Pennsylvania State Univ, Univ Park
Gross, Gerardo W., New Mexico Tech
Hansen, Lars N., Univ of Minnesota, Twin Cities
Kaip, Galen M., Univ of Texas, El Paso
Karlstrom, Leif, Univ of Oregon
Kilburn, Chris, Univ Coll London
Kitajima, Hiroko, Texas A&M Univ
Kohlstedt, David L., Univ of Minnesota, Twin Cities
Lebedev, Maxim, Curtin Univ
Li, Baosheng, Stony Brook Univ
Manghnani, Murli H., Univ of Hawai'i, Manoa
Mao, Ho-kwang, Carnegie Inst for Science
Marone, Chris, Pennsylvania State Univ, Univ Park
Marshall, Hans-Peter, Boise State Univ
Marshall, Scott T., Appalachian State Univ
Mei, Shenghua, Univ of Minnesota, Twin Cities
Mikhaltsevitch, Vassily, Curtin Univ
Narod, Barry, Univ of British Columbia
Parks, George K., Univ of Washington
Pevzner, Roman, Curtin Univ
Saltiel, Seth, Univ of Nevada, Reno
Schmitt, Douglas R., Purdue Univ
Scholz, Christopher H., Columbia Univ
Shankland, Thomas J., Los Alamos Nat Lab
Shaw, George H., Union Coll
Shieh, Sean R., Western Univ
Tullis, Terry E., Brown Univ
Vanorio, Tiziana, Stanford Univ
Vaughan, Michael T., Stony Brook Univ
Velasco, Aaron A., Univ of Texas, El Paso
Vialle, Stéphanie, Curtin Univ
Watanabe, Tohru, Univ of Toyama
Wohletz, Kenneth H., Los Alamos Nat Lab
Wong, Teng-fong, Stony Brook Univ
Yong, Wenjun, Western Univ
Zebker, Howard A., Stanford Univ
Zimmerman, Mark, Univ of Minnesota, Twin Cities

Exploration Geophysics

Al-Shuhail, Abdullah, King Fahd Univ of Petroleum and Minerals
Alumbaugh, David L., Univ of California, Berkeley
Ariyo, Stephen O., Federal Univ of Tech, Akure
Armadillo, Egidio, Universita di Genova
B, Veeraiah, Osmania Univ
Backus, Milo M., Univ of Texas, Austin
Becker, Alex, Univ of California, Berkeley
Biondi, Biondo L., Stanford Univ
Black, Ross A., Univ of Kansas
Brabham, P J., Cardiff Univ
Brassea, Jesus M., Centro de Inv Científica y de Ed Superior de Ensenada
Breland, Clayton F., Louisiana State Univ
Brown, Larry D., Cornell Univ
Bunch, Mark, Univ of Adelaide
Carmichael, Robert S., Univ of Iowa
Carson, Michael, Curtin Univ
Castagna, John P., Univ of Houston
Chandler, Val W., Univ of Minnesota, Twin Cities
Claerbout, Jon F., Stanford Univ
Collet, Olivia, Curtin Univ
Cook, Frederick A., Univ of Calgary
Corbato, Charles E., Ohio State Univ
Coruh, Cahit, Virginia Polytechnic Inst & State Univ
DeAngelo, Micheal V., Univ of Texas at Austin, Jackson Schl of Geosciences
Donaldson, Paul R., Boise State Univ
Dubba, Vijay K., Osmania Univ
Dunlap, Dallas B., Univ of Texas at Austin, Jackson Schl of Geosciences
Dunn, Bernie, Western Univ
Ellett, Kevin M., Indiana Univ, Bloomington
Esparza, Francisco, Centro de Inv Científica y de Ed Superior de Ensenada
Espinosa, Juan M., Centro de Inv Científica y de Ed Superior de Ensenada
Falebita, Dele E., Obafemi Awolowo Univ
Ferguson, Ian J., Univ of Manitoba
Ferguson, Robert, Univ of Calgary
Flores, Carlos, Centro de Inv Científica y de Ed Superior de Ensenada
Fomel, Sergey B., Univ of Texas, Austin
Gajewski, Dirk J., Universitaet Hamburg
Galvin, Robert, Curtin Univ
Gao, Dengliang, West Virginia Univ
Gendzwill, Donald J., Univ of Saskatchewan
Giroux, Bernard, Universite du Quebec
Gloaguen, Erwan, Universite du Quebec
Gomez, Enrique, Centro de Inv Científica y de Ed Superior de Ensenada
Gonzalez-Escobar, Mario, Centro de Inv Científica y de Ed Sup de Ensenada
Grana, Dario, Univ of Wyoming
Gurevich, Boris, Curtin Univ
Han, De-hua, Univ of Houston
Hardage, Bob A., Univ of Texas, Austin
Harris, Brett, Curtin Univ
Harris, James B., Millsaps Coll
Hauser, Ernest C., Wright State Univ
Herrmann, Felix J., Georgia Inst of Tech
Hilterman, Fred, Univ of Houston
Hinze, William J., Purdue Univ
Hole, John A., Virginia Polytechnic Inst & State Univ
Holt, John W., Univ of Texas, Austin
Homuth, Emil F., Los Alamos Nat Lab

Howman, Dominic J., Curtin Univ
Hu, Hao, Univ of Houston
Innanen, Kristopher, Univ of Calgary
Jin, Ge, Colorado School of Mines
Jiracek, George R., San Diego State Univ
Johnson, Ansel G., Portland State Univ
Jones, Francis H., Univ of British Columbia
Keach, Bill, Utah Geological Survey
Keach II, R. William, Brigham Young Univ
Kepic, Anton W., Curtin Univ
Khalil, Mohamed, Montana Tech of the Univ of Montana
Lawton, Donald C., Univ of Calgary
Lee, Ki Ha, Univ of California, Berkeley
Levander, Alan R., Rice Univ
Li, Yaoguo, Colorado School of Mines
Link, Curtis A., Montana Tech of the Univ of Montana
Liu, Kelly, Missouri Univ of Science and Tech
Louie, John N., Univ of Nevada, Reno
Lourenço, José M., Universidade de Trás-os-Montes e Alto Douro
Madadi, Mahyar, Curtin Univ
Margrave, Gary F., Univ of Calgary
Marobhe, Isaac M., Univ of Dar es Salaam
Mathur, Ram Raj, Osmania Univ
McBride, John H., Brigham Young Univ
McPherson, Brian, Univ of Utah
Mikhaltsevitch, Vassily, Curtin Univ
Milkereit, Bernd, Univ of Toronto
Miller, Richard D., Univ of Kansas
Morrison, Huntly Frank, Univ of California, Berkeley
Muchingami, Innocent , Univ of Namibia
Nguyen, Hoang, Curtin Univ
Olawuyi, Kehinde A., Univ of Ilorin
Olsen, Kim B., San Diego State Univ
Perez, Marco A., Centro de Inv Científica y de Ed Superior de Ensenada
Pethick, Andrew, Curtin Univ
Potter, David, Univ of Alberta
Pratt, R. Gerhard, Western Univ
Pujol, Jose, Univ of Memphis
Puzyrev, Vladimir, Curtin Univ
Raef, Abdelmoneam E., Kansas State Univ
Raji, W. O., Univ of Ilorin
Rector, James W., Univ of California, Berkeley
Reshef, Moshe, Tel Aviv Univ
Richard, Benjamin H., Wright State Univ
Roche, Steven L., Univ of Tulsa
Romo, Jose M., Centro de Inv Científica y de Ed Superior de Ensenada
Sarwar, A. K. Mostofa, Univ of New Orleans
Sava, Paul, Colorado School of Mines
Sbar, Marc, Univ of Arizona
Scanlin, Michael A., Elizabethtown Coll
Schmidt, Britney E., Cornell Univ
Seifoullaev, Roustam K., Univ of Texas, Austin
Sen, Mrinal K., Univ of Texas, Austin
Sexton, John L., Southern Illinois Univ Carbondale
Shaw, Kenneth L., Univ of Utah
Slater, Lee S., Rutgers, The State Univ of New Jersey, Newark
Smith, Richard S., Laurentian Univ, Sudbury
Smithyman, Brenden, Western Univ
Snieder, Roel, Colorado School of Mines
Sobotka, Jerzy, Univ of Wroclaw
Speece, Marvin A., Montana Tech of the Univ of Montana
Spikes, Kyle T., Univ of Texas, Austin
Sprenke, Kenneth F., Univ of Idaho
Steer, David N., Univ of Akron
Sternberg, Ben K., Univ of Arizona
Stewart, Robert R., Univ of Houston
Stewart, Robert, Univ of Houston
Stimpson, Ian G., Keele Univ
Stoffa, Paul L., Univ of Texas, Austin
Sturmer, Daniel M., Univ of Cincinnati
Swidinsky, Andrei, Colorado School of Mines
Tertyshnikov, Konstantin, Curtin Univ
Thomsen, Leon, Univ of Houston
Tsvankin, Ilya D., Colorado School of Mines
Turpening, Roger M., Michigan Technological Univ
Unsworth, Martyn, Cornell Univ
Urosevic, Milovan, Curtin Univ
Vazquez, Rogelio, Centro de Inv Científica y de Ed Superior de Ensenada
Verma, Sumit, Univ of Texas Permian Basin
von Frese, Ralph R., Ohio State Univ
Wannamaker, Phillip E., Univ of Utah
Watts, Doyle, Wright State Univ
Weglein, Arthur B., Univ of Houston
Weiss, Chester J., Univ of New Mexico
Wilson, Thomas H., West Virginia Univ
Wolfe, Paul J., Wright State Univ
Zhang, Bo, Univ of Alabama
Zhang, Rui, Univ of Louisiana at Lafayette
Zhdanov, Michael S., Univ of Utah
Ziolkowski, Anton M., Edinburgh Univ
Ziramov, Sasha, Curtin Univ

Geodesy

Adeleke, Adedayo, Univ of Pretoria
Al-Attar, David, Univ of Cambridge
Allahvirdizadeh, Amir, Curtin Univ
Araya , María C., Universidad de Costa Rica
Bennett, Richard, Univ of Arizona
Blewitt, Geoffrey, Univ of Nevada, Reno
Caporali, Alessandro, Università degli Studi di Padova
Claessens, Sten, Curtin Univ
Clarke, Peter J., Univ of Newcastle Upon Tyne
Crossley, David J., Saint Louis Univ
Davis, James , Columbia Univ
Dixon, Tim, Univ of South Florida
El-Mowafy, Ahmed, Curtin Univ
Evans, Eileen L., California State Univ, Northridge
Feigl, Kurt, Univ of Wisconsin, Madison
Filmer, Mick, Curtin Univ
Flake, Rex , Central Washington Univ
Fotopoulos, Georgia, Queen's Univ
Funning, Gareth J., Univ of California, Riverside
Gonzalez, Jose J., Centro de Inv Científica y de Ed Superior de Ensenada
Grapenthin, Ronni, Univ of Alaska, Fairbanks
Hammond, William C., Univ of Nevada, Reno
Herring, Thomas A., Massachusetts Inst of Tech
Hiroji, Anand D., Univ of Southern Mississippi
Hooper, Andy, Univ of Leeds
James, Thomas S., Univ of Victoria
Jekeli, Christopher, Ohio State Univ
Johnston, Paul, Curtin Univ
Kirby, Jon, Curtin Univ
Kogan, Mikhail, Columbia Univ
Kreemer, Corne, Univ of Nevada, Reno
Kuhn, Michael, Curtin Univ
Kusznir, Nick, Univ of Liverpool
Kyriakopoulos, Christodoulos, Univ of Memphis
LaFemina, Peter C., Pennsylvania State Univ, Univ Park
Li, Zhenhong, Univ of Newcastle Upon Tyne
Lopez, Alberto, Univ of Puerto Rico
Lowry, Anthony R., Utah State Univ
Malservisi, Rocco, Univ of South Florida
Masterlark, Timothy L., South Dakota School of Mines & Tech
Maurer, Jeremy , Missouri Univ of Science and Tech
Melbourne, Timothy I., Central Washington Univ
Melgar, Diego, Univ of Oregon
Miller, M. Meghan, Central Washington Univ
Moore, Phillip, Univ of Newcastle Upon Tyne
Mueller, Ivan, Ohio State Univ
Murray, Mark, New Mexico Tech
Newman, Andrew V., Georgia Inst of Tech
Odera, Patroba A., Jomo Kenyatta Univ of Agriculture & Tech
Pagiatakis, Spiros, York Univ
Palamartchouk, Kirill, Univ of Newcastle Upon Tyne
Parsons, Barry, Univ of Oxford
Pratt-Sitaula, Beth, Central Washington Univ
Pritchard, Matthew E., Cornell Univ
Rapp, Richard H., Ohio State Univ
Rundle, John, Univ of California, Davis
Samsonov, Sergey V., Nat Resources Canada
Santillan, Marcelo, Central Washington Univ
Schaffrin, Burkhard A., Ohio State Univ

Shum, CK, Ohio State Univ
Smalley, Jr., Robert, Univ of Memphis
Smith-Konter, Bridget R., Univ of Hawai'i, Manoa
Spinler, Joshua C., Univ of Arkansas at Little Rock
Stamps, D. S., Virginia Polytechnic Inst & State Univ
Stegman, Dave, Univ of California, San Diego
Teunissen, Peter, Curtin Univ
Tiampo, Kristy F., Western Univ
van Dam, Tonie, Univ of Utah
Waithaka, Hunja, Jomo Kenyatta Univ of Agriculture & Tech
Walters, Richard, Durham Univ
Wdowinski, Shimon, Florida InterNat Univ
Wells, David E., Univ of Southern Mississippi
Wright, Tim, Univ of Leeds
Xu, Xiaohua (Eric), Univ of Texas, Austin
Yi, Yuchan, Ohio State Univ

Geomagnetism & Paleomagnetism

Aldridge, Keith D., York Univ
Backus, George E., Univ of California, San Diego
Banerjee, Subir K., Univ of Minnesota, Twin Cities
Beardmore, Josh, Curtin Univ
Biggin, Andy, Univ of Liverpool
Bilardello, Dario, Univ of Minnesota, Twin Cities
Bogue, Scott W., Occidental Coll
Booker, John, Univ of Washington
Bowles, Julie, Univ of Wisconsin, Milwaukee
Braginsky, Stanislav I., Univ of California, Los Angeles
Bramham, Emma , Univ of Leeds
Brown, Laurie L., Univ of Massachusetts, Amherst
Burmester, Russell F., Western Washington Univ
Channell, James E., Univ of Florida
Cioppa, Maria T., Univ of Windsor
Clough, Gene A., Bates Coll
Constable, Catherine G., Univ of California, San Diego
Cottrell, Rory D., Univ of Rochester
DeMets, D. Charles, Univ of Wisconsin, Madison
Dharmasoth, Manjula, Osmania Univ
Doh, Seong-Jae, Korea Univ
Driscoll, Peter E., Carnegie Inst for Science
Dulin, Shannon, Univ of Oklahoma
Ellwood, Brooks B., Louisiana State Univ
Evans, David A., Yale Univ
Evans, Michael E., Univ of Alberta
Everett, Mark, Texas A&M Univ
Feinberg, Joshua, Univ of Minnesota, Twin Cities
Ferre', Eric C., Univ of Louisiana at Lafayette
Franklin, David, Univ of Portsmouth
Fraser-Smith, Antony C., Stanford Univ
Fu, Roger, Harvard Univ
Garcia Lasanta, Cristina, Western Washington Univ
Geissman, John W., Univ of New Mexico
Geissman, John W., Univ of Michigan
Gilder, Stuart, Ludwig-Maximilians-Universitaet Muenchen
Gordon, Richard G., Rice Univ
Guerrero-Garcia, Jose C., Universidad Nacional Auto de Mexico
Hall, Stuart A., Univ of Houston
Halls, Henry C., Univ of Toronto
Harbert, William P., Univ of Pittsburgh
Hill, Mimi, Univ of Liverpool
Hodych, Joseph P., Memorial Univ of Newfoundland
Holme, Richard, Univ of Liverpool
Housen, Bernard A., Western Washington Univ
Jackson, Michael, Univ of Minnesota, Twin Cities
Jones, F. Walter, Univ of Alberta
Kean, Jr., William F., Univ of Wisconsin, Milwaukee
Kelso, Paul R., Lake Superior State Univ
Kent, Dennis V., Rutgers, The State Univ of New Jersey
Kirscher, Uwe, Curtin Univ
Kodama, Kenneth P., Lehigh Univ
Kravchinsky, Vadim, Univ of Alberta
Larson, Edwin E., Univ of Colorado
Lepre, Christopher H., Rutgers, The State Univ of New Jersey
Li, Zheng-Xiang, Curtin Univ
Lima, Eduardo A., Massachusetts Inst of Tech
Lippert, Peter C., Univ of Utah
Liu, Yebo, Curtin Univ
Livermore, Phil, Univ of Leeds
Lund, Steven P., Univ of Southern California
Maloof, Adam C., Princeton Univ
Marshall, Monte, San Diego State Univ
Márton, Péter, Eotvos Lorand Univ
McCausland, Phil J., Western Univ
Merrill, Ronald T., Univ of Washington
Morris, Antony, Univ of Plymouth
Moskowitz, Bruce M., Univ of Minnesota, Twin Cities
Muxworthy, Adrian, Imperial Coll
Negrini, Robert M., California State Univ, Bakersfield
Nilsson, Andreas, Univ of Liverpool
Noltimier, Hallan C., Ohio State Univ
Palmer, H. C., Western Univ
Panzik, Joeseph, Univ of South Florida
Papitashvili, Vladimir O., Nat Science Foundation
Pares, Josep M., Univ of Michigan
Park, Stephen K., Univ of California, Riverside
Parker, Robert L., Univ of California, San Diego
Paty, Carol M., Georgia Inst of Tech
Petronis, Michael S., New Mexico Highlands Univ
Piper, John, Univ of Liverpool
Pisarevskiy, Sergei, Curtin Univ
Plattner, Alain, Univ of Alabama
Rankin, David, Univ of Alberta
Raub, Timothy D., Univ of St. Andrews
Richter, Carl, Univ of Louisiana at Lafayette
Roberts, Paul H., Univ of California, Los Angeles
Shibuya, Hidetoshi, Kumamoto Univ
Smirnov, Aleksey V., Michigan Technological Univ
Srnka, Len, Univ of California, San Diego
Sternberg, Robert S., Franklin and Marshall Coll
Stone, David B., Univ of Alaska, Fairbanks
Symons, David T., Univ of Windsor
Tarduno, John A., Univ of Rochester
Tauxe, Lisa, Univ of California, San Diego
Tikoo, Sonia M., Rutgers, The State Univ of New Jersey
Valentine, Michael J., Univ of Puget Sound
Van der Voo, Rob, Univ of Michigan
Verosub, Kenneth L., Univ of California, Davis
Walden, John, Univ of St. Andrews
Weaver, John T., Univ of Victoria
Whaler, Kathryn A., Edinburgh Univ
Williams, Wyn, Edinburgh Univ
Xuan, Chuang, Univ of Southampton
Zanella, Elena, Università di Torino

Gravity

Eppelbaum, Lev, Tel Aviv Univ
Garcia, Juan, Centro de Inv Científica y de Ed Superior de Ensenada
Grannell, Roswitha B., California State Univ, Long Beach
Kratt, Chris, Univ of Nevada, Reno
Kusumoto, Shigekazu, Univ of Toyama
Steckler, Michael, Columbia Univ

Heat Flow

Beltrami, Hugo, Saint Francis Xavier Univ
Blackwell, David D., Southern Methodist Univ
Harris, Robert N., Oregon State Univ
Hsieh, Wen-Pin, Academia Sinica
Huang, Shaopeng, Univ of Michigan
Jones, Michael Q., Univ of the Witwatersrand
Lenkey, László, Eotvos Lorand Univ
Morgan, Paul, Colorado Geological Survey
Morgan, Paul, Northern Arizona Univ
Reiter, Marshall A., New Mexico Tech
Sclater, John G., Univ of California, San Diego

Seismology

Abers, Geoffrey A., Cornell Univ
Achauer, Ulrich, Universite de Strasbourg
Agnew, Duncan C., Univ of California, San Diego
Alexander, Shelton S., Pennsylvania State Univ, Univ Park
Allen, Clarence R., California Inst of Tech
Allen, Richard M., Univ of California, Berkeley
Ammon, Charles J., Pennsylvania State Univ, Univ Park

Ampuero, Jean-Paul, California Inst of Tech
Anandakrishnan, Sridhar, Pennsylvania State Univ, Univ Park
Anderson, John G., Univ of Nevada, Reno
Angus, Doug, Univ of Leeds
Archuleta, Ralph J., Univ of California, Santa Barbara
Arrowsmith, Stephen J., Southern Methodist Univ
Assatourians, Karen, Western Univ
Aster, Richard C., Colorado State Univ
Atkinson, Gail M., Western Univ
Ausbrooks, Scott, Arkansas Geological Survey
Austermann, Jacqueline, Columbia Univ
Bachmann, Etienne, Princeton Univ
Bagley, Brian, Univ of Minnesota, Twin Cities
Balch, Robert S., New Mexico Tech
Barazangi, Muawia, Cornell Univ
Barclay, Andrew, Columbia Univ
Barker, Jeffrey S., Binghamton Univ
Beaumont, Christopher, Dalhousie Univ
Beck, Susan L., Univ of Arizona
Beghein, Caroline, Univ of California, Los Angeles
Ben-Zion, Yehuda, Univ of Southern California
Benavides-Iglesias, Alfonso, Texas A&M Univ
Bennett, Scott, Univ of Washington
Benoit, Margaret, Coll of New Jersey
Beroza, Gregory C., Stanford Univ
Bezada, Max, Univ of Minnesota, Twin Cities
Biasi, Glenn P., Univ of Nevada, Reno
Bilek, Susan L., New Mexico Tech
Biundo, Marc, Univ of Washington
Blake, Daniel R., Ohio Dept of Nat Resources
Bodin, Paul A., Univ of Washington
Bollinger, G. A., Virginia Polytechnic Inst & State Univ
Bostock, Michael G., Univ of British Columbia
Bozdag, Ebru, Colorado School of Mines
Braile, Lawrence W., Purdue Univ
Braunmiller, Jochen, Univ of South Florida
Brodsky, Emily, Univ of California, Santa Cruz
Brudzinski, Michael, Miami Univ
Brumbaugh, David S., Northern Arizona Univ
Brune, James N., Univ of Nevada, Reno
Burdick, Scott, Wayne State Univ
Burton, Paul, Univ of East Anglia
Byrnes, Joseph, Univ of Minnesota, Twin Cities
Calvert, Andrew J., Simon Fraser Univ
Cann, Johnson R., Woods Hole Oceanographic Inst
Caplan-Auerbach, Jackie, Western Washington Univ
Cardimona, Steve, Mendocino Coll
Carpenter, N. Seth, Univ of Kentucky
Cartwright, J A., Cardiff Univ
Cassidy, John, Univ of Victoria
Castro, Raul, Centro de Inv Científica y de Ed Superior de Ensenada
Cattania, Camilla, Massachusetts Inst of Tech
Chapman, Martin C., Virginia Polytechnic Inst & State Univ
Chen, Po, Univ of Wyoming
Chen, Wang-Ping, Univ of Illinois, Urbana-Champaign
Chen, Xiaowei, Univ of Oklahoma
Chen, Xiaowei, Texas A&M Univ
Chesnokov, Evgeny, Univ of Houston
Chiu, Jer-Ming, Univ of Memphis
Christensen, Douglas, Univ of Alaska, Fairbanks
Cicerone, Robert D., Bridgewater State Univ
Cipar, John J., Boston Coll
Clark, Roger A., Univ of Leeds
Clayton, Robert W., California Inst of Tech
Clowes, Ronald M., Univ of British Columbia
Cochran, Elizabeth, Univ of California, Riverside
Collins, John A., Woods Hole Oceanographic Inst
Conder, James A., Southern Illinois Univ Carbondale
Cormier, Vernon F., Univ of Connecticut
Cornwell, David G., Univ of Aberdeen
Cottaar, Sanne , Univ of Cambridge
Cramer, Chris, Univ of Memphis
Creager, Kenneth C., Univ of Washington
Crosson, Robert S., Univ of Washington
Crotwell, Philip, Univ of South Carolina
Curtis, Andrew, Edinburgh Univ
D'Amico, Sebastiano, Univ of Malta
Dalton, Colleen, Brown Univ
Darbyshire, Fiona Ann, Universite du Quebec a Montreal
Davis, Paul M., Univ of California, Los Angeles
Day, Steven M., San Diego State Univ
De Angelis, Silvio, Univ of Liverpool
Denolle, Marine, Harvard Univ
dePolo, Diane, Univ of Nevada, Reno
DeShon, Heather R., Southern Methodist Univ
Dettmer, Jan, Univ of Calgary
Doll, William E., Univ of Tennessee, Knoxville
Dongmo Wamba, Mathurin, Princeton Univ
Dorman, James, Univ of Memphis
Douilly, Roby, Univ of California, Riverside
Dreger, Douglas S., Univ of California, Berkeley
Duan, Benchuan, Texas A&M Univ
Dueker, Kenneth G., Univ of Wyoming
Dunham, Eric M., Stanford Univ
Eaton, David, Univ of Calgary
Ebel, John E., Boston Coll
Eilon, Zach, Univ of California, Santa Barbara
Ekstrom, Goran, Columbia Univ
Ellis, Robert M., Univ of British Columbia
Ellsworth, William L., Stanford Univ
Fan, Wenyuan, Univ of California, San Diego
Farrell, Jamie, Univ of Utah
Faulkner, Daniel, Univ of Liverpool
Ferreira, Anna, Univ Coll London
Finley, William R., Univ of Louisiana at Lafayette
Fischer, Karen M., Brown Univ
Fontaine, Fabrice R., Universite de la Reunion
Forbriger, Thomas, Karlsruhe Inst of Tech
Ford, Heather, Univ of California, Riverside
Fortin, Will Frances J., Columbia Univ
Fox, Jeff, Ohio Dept of Nat Resources
Frankel, Arthur, Univ of Washington
Frazer, L. N., Univ of Hawai'i, Manoa
Frederiksen, Andrew, Univ of Manitoba
Frez, Jose, Centro de Inv Científica y de Ed Superior de Ensenada
Frohlich, Cliff, Univ of Texas, Austin
Frohlich, Clifford A., Univ of Texas, Austin
Frost, Daniel, Univ of South Carolina
Fujita, Kazuya, Michigan State Univ
Gabriel, Alice, Univ of California, San Diego
Galea, Pauline, Univ of Malta
Gao, Haiying, Univ of Massachusetts, Amherst
Garces, Milton A., Univ of Hawai'i, Manoa
Garnero, Edward, Arizona State Univ
Ghosh, Abhijit, Univ of California, Riverside
Gilbert, Hersh, Univ of Calgary
Gilpin, Bernard J., Golden West Coll
Glowacka, Ewa, Centro de Inv Científica y de Ed Superior de Ensenada
Gomberg, Joan, Univ of Washington
Grand, Stephen P., Univ of Texas, Austin
Gu, Jeff, Univ of Alberta
Gurevich, Boris, Curtin Univ
Gurnis, Michael C., California Inst of Tech
Gurrola, Harold, Texas Tech Univ
Haddidi, Hamid, California Geological Survey
Hager, Bradford H., Massachusetts Inst of Tech
Hagerty, Michael, Boston Coll
Hagos, Lijam, California Geological Survey
Hajnal, Zoltan, Univ of Saskatchewan
Hall, Jeremy, Memorial Univ of Newfoundland
Hamburger, Michael W., Indiana Univ, Bloomington
Hammer, Philip T., Univ of British Columbia
Hansen, Samantha E., Univ of Alabama
Harrington, Rebecca, Karlsruhe Inst of Tech
Harris, Jerry M., Stanford Univ
Hartog, Renate, Univ of Washington
Hartse, Hans E., Los Alamos Nat Lab
Hassanpour Sedghi, Mohammad, Univ of Tabriz
Hauksson, Egill, California Inst of Tech
Hawman, Robert B., Univ of Georgia
Hayward, Christopher T., Southern Methodist Univ
Hearn, Thomas M., New Mexico State Univ, Las Cruces

Helmberger, Donald V., California Inst of Tech
Herrmann, Robert B., Saint Louis Univ
Hildebrand, Steve T., Los Alamos Nat Lab
Hobbs, Richard, Durham Univ
Holt, William E., Stony Brook Univ
Holtzman, Benjamin, Columbia Univ
Honjas, Bill, Univ of Nevada, Reno
Horton, Stephen P., Univ of Memphis
House, Leigh S., Los Alamos Nat Lab
Houston, Heidi B., Univ of Southern California
Howell, Jr., Benjamin F., Pennsylvania State Univ, Univ Park
Huerfano, Victor, Univ of Puerto Rico
Humphreys, Eugene D., Univ of Oregon
Hyndman, Roy D., Univ of Victoria
Ismail, Ahmed, Oklahoma State Univ
Jackson, David D., Univ of California, Los Angeles
Jacob, Klaus H., Columbia Univ
Janiszewski, Helen, Univ of Hawai'i, Manoa
Jasbinsek, John J., California Polytechnic State Univ
Jaumé, Steven C., Coll of Charleston
Ji, Chen, Univ of California, Santa Barbara
Johnson, Jeffrey , Boise State Univ
Johnson, Lane R., Univ of California, Berkeley
Johnson, Roy A., Univ of Arizona
Johnston, Archibald C., Univ of Memphis
Jones, Alan, Binghamton Univ
Jones, Craig H., Univ of Colorado
Jordan, Thomas H., Univ of Southern California
Kafka, Alan L., Boston Coll
Kaka, Ismail, King Fahd Univ of Petroleum and Minerals
Kanamori, Hiroo, California Inst of Tech
Karplus, Marianne, Univ of Texas, El Paso
Kent, Graham, Univ of Nevada, Reno
Keranen, Kade M., Cornell Univ
Kijko, Andrzej, Univ of Pretoria
Kim, Seong Ryong, Korea Univ
Kim, Won-Young, Columbia Univ
Knopoff, Leon, Univ of California, Los Angeles
Koellemeijer, Paula, Royal Holloway Univ of London
Kovach, Robert L., Stanford Univ
Krebes, Edward S., Univ of Calgary
Langston, Charles A., Univ of Memphis
Laske, Gabi, Univ of California, San Diego
Lay, Thorne, Univ of California, Santa Cruz
Lees, Jonathan M., Univ of North Carolina, Chapel Hill
Lekic, Vedran, Univ of Maryland
Lenz, Brandi, Texas A&M Univ
Lerner-Lam, Arthur L., Columbia Univ
Levander, Alan, Rice Univ
Levin, Vadim, Rutgers, The State Univ of New Jersey
Li, Aibing, Univ of Houston
Li, Yong-Gang, Univ of Southern California
Liberty, Lee M., Boise State Univ
Linde, Alan T., Carnegie Inst for Science
Liner, Christopher, Univ of Arkansas, Fayetteville
Litherland, Mairi, New Mexico Tech
Liu, Qinya, Univ of Toronto
Lizarralde, Daniel, Woods Hole Oceanographic Inst
Lohman, Rowena B., Cornell Univ
Long, Maureen D., Yale Univ
Lorenzo, Juan M., Louisiana State Univ
Lough, Amanda, Drexel Univ
Louie, John, Univ of Nevada, Reno
Lumley, David, Univ of Texas, Dallas
Ma, Shuo, San Diego State Univ
Magnani, M. Beatrice, Southern Methodist Univ
Main, Ian G., Edinburgh Univ
Malin, Peter E., Duke Univ
Mallick, Subhashis, Univ of Wyoming
Malone, Stephen D., Univ of Washington
Mangriotis, Maria-Daphne , Heriot-Watt Univ
Marfurt, Kurt J., Univ of Oklahoma
Masters, T. Guy, Univ of California, San Diego
McCarthy, Christine, Columbia Univ
McClelland, Lori, Univ of Nevada, Reno
McGuire, Jeffrey J., Woods Hole Oceanographic Inst
McLaskey, Greg C., Cornell Univ
McMechan, George A., Univ of Texas, Dallas
McNutt, Steve, Univ of South Florida
Meltzer, Anne S., Lehigh Univ
Meng, Lingsen, Univ of California, Los Angeles
Menke, William H., Columbia Univ
Mereu, Robert F., Western Univ
Michel, Gero, Western Univ
Minster, J. Bernard H., Univ of California, San Diego
Molnar, Sheri, Western Univ
Morozov, Igor B., Univ of Saskatchewan
Moulis, Anastasia, Boston Coll
Mulibo, Gabriel D., Univ of Dar es Salaam
Munguia, Luis, Centro de Inv Científica y de Ed Superior de Ensenada
Nakamura, Yosio, Univ of Texas, Austin
Nava, Alejandro, Centro de Inv Científica y de Ed Superior de Ensenada
Nedimovic, Mladen, Dalhousie Univ
Neighbors, Corrie, Western New Mexico Univ
Nettles, Meredith, Columbia Univ
Ni, James F., New Mexico State Univ, Las Cruces
Nikulin, Alex, Binghamton Univ
Nissen-Meyer, Targe, Univ of Oxford
Niu, Fenglin, Rice Univ
Nowack, Robert L., Purdue Univ
Nyland, Edo, Univ of Alberta
Ogbamikhumi, Alexander, Univ of Benin
Oglesby, David D., Univ of California, Riverside
Okal, Emile A., Northwestern Univ
Okaya, David A., Univ of Southern California
Oliver, Adolph A., Chabot Coll
Olson, Ted L., Snow Coll
Olugboji, Tolulope, Univ of Rochester
Owens, Thomas J., Univ of South Carolina
Pancha, Aasha, Univ of Nevada, Reno
Pankow, Kristine L., Univ of Utah
Park, Jeffrey J., Yale Univ
Pavlis, Gary L., Indiana Univ, Bloomington
Pechmann, James C., Univ of Utah
Peng, Zhigang, Georgia Inst of Tech
Persaud, Patricia, Louisiana State Univ
Pevzner, Roman, Curtin Univ
Phinney, Robert A., Princeton Univ
Plank, Gabriel, Univ of Nevada, Reno
Polet, Jascha, California State Polytechnic Univ, Pomona
Powell, Christine A., Univ of Memphis
Pride, Steven, Univ of California, Berkeley
Pulliam, Jay, Baylor Univ
Pulliam, Robert J., Baylor Univ
Pulliam, Robert J., Univ of Texas, Austin
Quiros Ugalde, Diego, Univ of Cape Town
Reading, Anya, Univ of Tasmania
Record, Amy, New Mexico Tech
Rector, III, James W., Univ of California, Berkeley
Rennie, Tom, Univ of Nevada, Reno
Revenaugh, Justin, Univ of Minnesota, Twin Cities
Richard, Ferdinand W., Univ of Dar es Salaam
Richards, Paul G., Columbia Univ
Richardson, Eliza, Pennsylvania State Univ, Univ Park
Rieger, Duayne, Comm Coll of Rhode Island
Rietbrock, Andreas, Univ of Liverpool
Ritsema, Jeroen, Univ of Michigan
Ritter, Joachim R., Karlsruhe Inst of Tech
Roberts, Peter, Los Alamos Nat Lab
Rogers, Garry C., Univ of Victoria
Rohay, Alan C., Pacific Northwest Nat Lab
Romanowicz, Barbara A., Univ of California, Berkeley
Rost, Sebastian, Univ of Leeds
Rowe, Charlotte A., New Mexico Tech
Rowshandel, Badie, California Geological Survey
Ruff, Larry J., Univ of Michigan
Ruffman, Alan, Dalhousie Univ
Ruhl, Christine J., Univ of Tulsa
Sacchi, Mauricio D., Univ of Alberta
Sacks, I. Selwyn, Carnegie Inst for Science
Sahay, Pratap, Centro de Inv Científica y de Ed Superior de Ensenada
Sandvol, Eric A., Univ of Missouri

Savage, Brian K., Univ of Rhode Island
Savage, Brian, Univ of Rhode Island
Savage, Heather M., Columbia Univ
Savvaidis, Alexandros, Univ of Texas, Austin
Sawyer, Dale S., Rice Univ
Schaff, David, Columbia Univ
Schlue, John W., New Mexico Tech
Scholz, Christopher H., Columbia Univ
Schutt, Derek L., Colorado State Univ
Schwab, Fred, Univ of California, Los Angeles
Schwartz, Susan, Univ of California, Santa Cruz
Seeber, Leonardo, Columbia Univ
Serpa, Laura F., Univ of New Orleans
Shashkin, Pavel, Curtin Univ
Shaw, Bruce, Columbia Univ
Shearer, Peter M., Univ of California, San Diego
Sheehan, Anne F., Univ of Colorado
Shen, Weisen, Stony Brook Univ
Sherrod, Brian L., Univ of Washington
Shipley, Thomas H., Univ of Texas, Austin
Shragge, Jeffrey, Colorado School of Mines
Simon, Joel D., Princeton Univ
Simons, Frederik J., Princeton Univ
Simons, Mark, California Inst of Tech
Slater, David, Univ of Nevada, Reno
Slawinski, Michael A., Memorial Univ of Newfoundland
Smith, Ken D., Univ of Nevada, Reno
Smith, Stewart W., Univ of Washington
Smythe, William, Univ of California, Los Angeles
Snoke, J. Arthur, Virginia Polytechnic Inst & State Univ
Song, Alex, Univ Coll London
Song, Xiaodong, Univ of Illinois, Urbana-Champaign
Spence, George D., Univ of Victoria
Spetzler, Hartmut A., Univ of Colorado
Spiegelman, Marc, Columbia Univ
Stachnik, Joshua, Lehigh Univ
Steele, Bill, Univ of Washington
Steeples, Don W., Univ of Kansas
Stein, Seth, Northwestern Univ
Stephen, Ralph A., Woods Hole Oceanographic Inst
Stickney, Michael C., Montana Tech of The Univ of Montana
Street, Ronald L., Univ of Kentucky
Streig, Ashley, Portland State Univ
Stuart, Graham, Univ of Leeds
Stump, Brian W., Southern Methodist Univ
Sverdrup, Keith A., Univ of Wisconsin, Milwaukee
Sykes, Lynn R., Columbia Univ
Talwani, Pradeep, Univ of South Carolina
Tanimoto, Toshiro, Univ of California, Santa Barbara
Tape, Carl, Univ of Alaska, Fairbanks
Taylor, Kenneth B., N.C. Department of Enval Quality
Taylor, Steven R., Los Alamos Nat Lab
Teng, Ta-liang, Univ of Southern California
Tertyshnikov, Konstantin, Curtin Univ
Thomas, Amanda, Univ of Oregon
Thompson, zglenn, Univ of South Florida
Thorne, Michael S., Univ of Utah
Thurber, Clifford, Univ of Wisconsin, Madison
Tibuleac, Ileana, Univ of Nevada, Reno
Toksoz, M N., Massachusetts Inst of Tech
Toomey, Douglas R., Univ of Oregon
Tromp, Jeroen, Princeton Univ
Tsai, Victor, Brown Univ
Tura, Ali, Colorado School of Mines
Van Avendonk, Harm J., Univ of Texas, Austin
van der Hilst, Robert, Massachusetts Inst of Tech
van der Lee, Suzan, Northwestern Univ
van Keken, Peter E., Carnegie Inst for Science
Vanacore, Elizabeth A., Univ of Puerto Rico
Vidal, Antonio, Centro de Inv Científica y de Ed Superior de Ensenada
Vidale, John E., Univ of Southern California
Vlahovic, Gordana, North Carolina Central Univ
von Seggern, David H., Univ of Nevada, Reno
Wagner, Lara, Carnegie Inst for Science
Waite, Gregory P., Michigan Technological Univ
Waldhauser, Felix, Columbia Univ
Wallace, Jr., Terry C., Los Alamos Nat Lab
Wamba, Mathurin D., Princeton Univ
Wang, Jeen-Hwa, Academia Sinica
Wang, Zhenming, Univ of Kentucky
Ward, Kevin M., South Dakota School of Mines & Tech
Warner, Michael, Imperial Coll
Waszek, Lauren, New Mexico State Univ, Las Cruces
Wen, Lianxing, Stony Brook Univ
Wentzcovitch, Renata, Columbia Univ
Wesnousky, Steven, Univ of Nevada, Reno
Wesnousky, Steven G., Univ of Nevada, Reno
Wiens, Douglas A., Washington Univ in St. Louis
Williams, Erik, Univ of Nevada, Reno
Wirth, Erin, Univ of Washington
Withers, Mitchell M., Univ of Memphis
Wolf, Lorraine W., Auburn Univ
Wolny, Dave, Colorado Mesa Univ
Wu, Francis T., Binghamton Univ
Wust-Bloch, Gilles H., Tel Aviv Univ
Wysession, Michael E., Washington Univ in St. Louis
Young, Jeri J., Univ of Arizona
Yu, Wen-che, Academia Sinica
Zandt, George, Univ of Arizona
Zelt, Colin A., Rice Univ
Zeng, Hongliu, Univ of Texas, Austin
Zhan, Zhongwen, California Inst of Tech
Zhao, Zeyu, Univ of Texas, Austin
Zhou, Hua-Wei, Univ of Houston
Zhou, Ying, Virginia Polytechnic Inst & State Univ
Zhu, Hejun, Univ of Texas, Dallas
Zhu, Lupei, Saint Louis Univ
Ziramov, Sasha, Curtin Univ
Ziv, Alon, Tel Aviv Univ

Marine Geophysics

Archer, David, Univ of Chicago
Austin, Jr., James A., Univ of Texas, Austin
Becker, Keir, Univ of Miami
Behn, Mark D., Woods Hole Oceanographic Inst
Bell, Robin E., Columbia Univ
Ben-Avraham, Zvi, Tel Aviv Univ
Bohnenstiehl, DelWayne, North Carolina State Univ
Bokuniewicz, Henry J., Stony Brook Univ
Bonatti, Enrico, Columbia Univ
Bowin, Carl O., Woods Hole Oceanographic Inst
Buck, Roger W., Columbia Univ
Buizert, Christo, Oregon State Univ
Bull, Jonathan M., Univ of Southampton
Butterfield, David A., Univ of Washington
Canales Cisneros, Juan Pablo, Woods Hole Oceanographic Inst
Carbotte, Suzanne, Columbia Univ
Chamberlain, C. Page, Stanford Univ
Christeson, Gail L., Univ of Texas, Austin
Cochran, James R., Columbia Univ
Collier, Jenny, Imperial Coll
Commane, Roisin, Columbia Univ
Constable, Steven C., Univ of California, San Diego
Deans, Jeremy, Univ of Southern Mississippi
Delaney, John R., Univ of Washington
Druet, María, Univ Complutense de Madrid
Duncan, Alec, Curtin Univ
Dunn, Robert A., Univ of Hawai'i, Manoa
Edwards, Richard N., Univ of Toronto
Egbert, Gary D., Oregon State Univ
Eloranta, Edwin W., Univ of Wisconsin, Madison
Evans, Robert L., Woods Hole Oceanographic Inst
Forsyth, Donald W., Brown Univ
Fricker, Helen A., Univ of California, San Diego
Gavrilov, Alexander, Curtin Univ
Goff, John A., Univ of Texas, Austin
Goldberg, David S., Columbia Univ
Goldfinger, Chris, Oregon State Univ
Gorelick, Steven, Stanford Univ
Gorman, Andrew R., Univ of Otago
Gulick, Sean S., Univ of Texas, Austin
Han, Shuoshuo, Univ of Texas, Austin

Henderson, Naomi, Columbia Univ
Hildebrand, John A., Univ of California, San Diego
Hooft, Emilie E., Univ of Oregon
Hornbach, Matthew J., Southern Methodist Univ
Huang, Li, Western Univ
Huws, Dei G., Univ of Wales
Huws, Dei, Bangor Univ
Ji, Peng, Columbia Univ
Johnson, H. Paul, Univ of Washington
Karig, Daniel E., Cornell Univ
Kingslake, Jonathan, Columbia Univ
Krastel, Sebastian , Dalhousie Univ
Kummerow, Christian D., Colorado State Univ
Kurapov, Alexander, Oregon State Univ
Lawver, Lawrence A., Univ of Texas, Austin
Lewis, Brian T. R., Univ of Washington
Llanes, María del Pilar, Univ Complutense de Madrid
Louden, Keith E., Dalhousie Univ
MacAyeal, Douglas R., Univ of Chicago
Macelloni, Leonardo, Univ of Southern Mississippi
Mack, Seymour, Fresno State Univ
Madsen, John A., Univ of Delaware
Manley, Patricia L., Middlebury Coll
Martinez, Fernando, Univ of Hawai'i, Manoa
McClain, James S., Univ of California, Davis
McIntosh, Kirk D., Univ of Texas, Austin
McMahon, Darryl, Curtin Univ
Minshull, Timothy A., Univ of Southampton
Mittelstaedt, Eric L., Univ of Idaho
Muller, Dietmar, Univ of Sydney
Mutter, John C., Columbia Univ
Oakley, Adrienne, Kutztown Univ of Pennsylvania
Odom, Robert I., Univ of Washington
Orcutt, John A., Univ of California, San Diego
Parnum, Iain, Curtin Univ
Penrose, John, Curtin Univ
Purdy, G. Michael, Columbia Univ
Raymond, Carol A., Jet Propulsion Lab
Rodriguez Simental, Ana María, Universidad Nacional Auto de Mexico
Ryan, William B. F., Columbia Univ
Sandwell, David T., Univ of California, San Diego
Schouten, Hans, Woods Hole Oceanographic Inst
Sheridan, Robert E., Rutgers, The State Univ of New Jersey
Shipley, Thomas H., Univ of Texas, Austin
Shreve, Ronald L., Univ of Washington
Siwabessy, Justy, Curtin Univ
Sloan, Heather, Lehman Coll (CUNY)
Sohn, Robert A., Woods Hole Oceanographic Inst
Spitz, Yvette H., Oregon State Univ
Stephens, Jason H., Austin Comm Coll District
Swift, Stephen A., Woods Hole Oceanographic Inst
Talwani, Manik, Rice Univ
Thompson, David W. J., Colorado State Univ
Tivey, Maurice A., Woods Hole Oceanographic Inst
Tolstoy, Maria, Columbia Univ
Trehu, Anne M., Oregon State Univ
Van Rooij, David, Ghent Univ
Wattrus, Nigel J., Univ of Minnesota Duluth
Webb, Spahr C., Columbia Univ
Webb, Spahr, Columbia Univ
Wessel, Paul, Univ of Hawai'i, Manoa
Wetzel, Laura R., Eckerd Coll
White, Scott M., Univ of South Carolina
Wilkes, Daniel, Curtin Univ
Wooding, Frank B., Woods Hole Oceanographic Inst
Xu, Li, Woods Hole Oceanographic Inst
Zavarin, Mavrik, Lawrence Livermore Nat Lab
Zhu, Wenlu, Univ of Maryland

Near-surface Geophysics

Alam, Md Iftekhar, Univ of Tennessee, Knoxville
Aracil, Enrique, Univ Complutense de Madrid
Bradford, John, Colorado School of Mines
Clement, William P., Univ of Massachusetts, Amherst
Gemail, Khaled S., El Zagazig Univ
Herman, Rhett B., Radford Univ
Holbrook, W. S., Virginia Polytechnic Inst & State Univ
Irons, Trevor , Montana Tech of the Univ of Montana
Katumwehe, Andrew, Midwestern State Univ
Keating, Kristina M., Rutgers, The State Univ of New Jersey, Newark
Kidanu, Shishay, Univ of Alaska Fairbanks
Lamb, Andrew P., Univ of Arkansas, Fayetteville
Martin, Eileen, Colorado School of Mines
Neil, Foley T., Univ of Montana Western
Ntarlagiannis, Dimitrios, Rutgers, The State Univ of New Jersey, Newark
Pringle, Jamie K., Keele Univ
Wood, Neill, Exeter Univ
Woolery, Edward W., Univ of Kentucky

PALEONTOLOGY

General Paleontology

Abrahams, Miengah, Univ of Cape Town
Adrain, Tiffany S., Univ of Iowa
Allmon, Warren D., Cornell Univ
Allmon, Warren D., Paleontological Research Inst
Altenbach, Alexander, Ludwig-Maximilians-Universitaet Muenchen
Amler, Michael, Ludwig-Maximilians-Universitaet Muenchen
Anderson, Wayne I., Univ of Northern Iowa
Arens, Nan Crystal, Hobart & William Smith Colls
Arnone, John, Desert Research Inst
Arp, Dr., Gernot, Georg-August Univ of Goettingen
Ausich, William I., Ohio State Univ
Awramik, Stanley M., Univ of California, Santa Barbara
Babcock, Loren E., Ohio State Univ
Bailey, Jack B., Western Illinois Univ
Bailey, Jake, Univ of Minnesota, Twin Cities
Bajraktareviæ, Zlatan, Univ of Zagreb
Barbieri, Roberto, Alma Mater Studiorum Università di Bologna
Baumiller, Tomasz K., Univ of Michigan
Beatty, William L., Winona State Univ
Beck, Charles, Univ of Michigan
Belasky, Paul, Ohlone Coll
Bell, Phil R., Univ of New England
Benson, Roger, Univ of Oxford
Bergstrom, Stig M., Ohio State Univ
Berman, David S., Carnegie Museum of Nat History
Bill, Steven D., Keene State Coll
Bjork, Phillip R., South Dakota School of Mines & Tech
Blackwelder, Patricia L., Univ of Miami
Bonde, Aubrey, Univ of Nevada, Las Vegas
Bordeaux, Yvette, Univ of Pennsylvania
Bosak, Tanja, Massachusetts Inst of Tech
Branstrator, Jon W., Earlham Coll
Brasier, Martin, Univ of Oxford
Bray, Emily, Univ of Colorado
Brink, Alyson, Univ of Southern Mississippi
Britt, Brooks B., Brigham Young Univ
Brzobohaty, Rostislav, Masaryk Univ
Bucur, Ioan, Babes-Bolyai Univ
Budd, Ann F., Univ of Iowa
Busbey, Arthur B., Texas Christian Univ
Bush, Andrew M., Univ of Connecticut
Butterfield, Nicholas J., Univ of Cambridge
Cachão, Mário, Universidade de Lisboa
Caldwell, Roy, Univ of California, Berkeley
Cardace, Dawn, Univ of Rhode Island
Carew, James L., Coll of Charleston
Carlson, Sandra J., Univ of California, Davis
Carnevale, Giorgio, Università di Torino
Caron, Jean-Bernard, Royal Ontario Museum
Carter, Burchard D., Georgia Southwestern State Univ
Chamberlain, John A., Brooklyn Coll (CUNY)
Chamberlain, John A., Graduate School of the City Univ of New York
Cheetham, Alan H., Smithsonian Inst / Nat Museum of Nat History
Cherns, L C., Cardiff Univ
Chin, Karen, Univ of Colorado
Christie, Max L., Univ of Illinois, Urbana-Champaign
Cirimpei, Claudia, Alexandru Ioan Cuza
Cisne, John L., Cornell Univ
Clementz, Mark T., Univ of Wyoming
Clyde, William C., Univ of New Hampshire

Cohen, Phoebe A., Williams Coll
Collette, Joseph, Minot State Univ
Collins, Laurel S., Florida InterNat Univ
Collinson, James W., Ohio State Univ
Collinson, Margaret, Royal Holloway Univ of London
Conley, Catharine A., New Mexico Tech
Coorough Burke, Patricia J., Milwaukee Public Museum
Coulson, Alan B., Clemson Univ
Cowen, Richard, Univ of California, Davis
Cramer, Howard R., Emory Univ
Curran, H. Allen, Smith Coll
Darroch, Simon A., Vanderbilt Univ
De Klerk, Billy, Rhodes Univ
Deboo, Phili B., Univ of Memphis
Delfino, Massimo, Università di Torino
Derstler, Kraig L., Univ of New Orleans
Devera, Joseph A., Illinois State Geological Survey
Dietl, Gregory, Paleontological Research Inst
Dietz, Anne D., San Antonio Comm Coll
Diggins, Thomas P., Youngstown State Univ
Domack, Cynthia R., Hamilton Coll
Droser, Mary L., Univ of California, Riverside
Drumheller-Horton, Stephanie K., Univ of Tennessee, Knoxville
Duda, Jan-Peter, Georg-August Univ of Goettingen
Dyson, Miranda, Open Univ
Edwards, Nicholas, Univ of Manchester
Egerton, Victoria, Univ of Manchester
Erickson, J. Mark, St. Lawrence Univ
Falcon-Lang, Howard, Royal Holloway Univ of London
Fall, Leigh M., SUNY, Oneonta
Farmer, Jack D., Arizona State Univ
Felton, Richard M., Northwest Missouri State Univ
Fernández, Sixto, Univ Complutense de Madrid
Ferriz, Sergio Cevallos, Universidad Nacional Auto de Mexico
Fio, Karmen, Univ of Zagreb
Fisher, Frank, Univ of Colorado
Flessa, Karl W., Univ of Arizona
Flood, Beverly, Univ of Minnesota, Twin Cities
Fondevilla, Victor, Universitat Auto de Barcelona
Foote, Michael J., Univ of Chicago
Fornaciari, Eliana, Università degli Studi di Padova
Fournier, Gregory, Massachusetts Inst of Tech
Fowler, Gerald A., Univ of Wisconsin, Parkside
Fox, David L., Univ of Minnesota, Twin Cities
Fox, Nathaniel, South Dakota School of Mines & Tech
Fox-Dobbs, Kena L., Univ of Puget Sound
Frederick, Daniel L., Austin Peay State Univ
Freeman, Veronica, Marietta Coll
Friedman, Matt, Univ of Oxford
Fritsen, Christian H., Desert Research Inst
Gabbott, Sarah, Leicester Univ
Gagnaison, Cyril, Institut Polytechnique LaSalle Beauvais (ex-IGAL)
Gahn, Forest J., Brigham Young Univ - Idaho
Gale, Andrew, Univ of Portsmouth
Gallegos, Anthony F., Los Alamos Nat Lab
Ganeshram, Raja, Edinburgh Univ
Garb, Matt, Brooklyn Coll (CUNY)
Gatto, Roberto, Università degli Studi di Padova
Geary, Dana H., Univ of Wisconsin, Madison
Gess, Robert, Rhodes Univ
Gibson, Michael A., Univ of Tennessee, Martin
Gill, Fiona, Univ of Leeds
Glamoclija, Mihaela, Rutgers, The State Univ of New Jersey, Newark
Glasauer, Susan, Univ of Guelph
Gold, David, Univ of California, Davis
Goldman, Daniel, Univ of Dayton
Gorecka-Nowak, Anna, Univ of Wroclaw
Gorman-Lewis, Drew J., Univ of Washington
Gray, Lee M., Mount Union Coll
Gray, Neil, Univ of Newcastle Upon Tyne
Green, Keith E., Cypress Coll
Greenwood, David , Univ of Saskatchewan
Haggart, James, Univ of British Columbia
Haiar, Brooke, Univ of Lynchburg
Hansen, Thor A., Western Washington Univ
Harper, David, Durham Univ
Harris, Ann, Marshall Univ
Harvey, Tom, Leicester Univ
Hawkins, Andrew, Towson Univ
Hays, James D., Columbia Univ
Hazel, McGoff, Univ of Reading
Heim, Noel A., Tufts Univ
Henderson, Wayne G., California State Univ, Fullerton
Hendy, Ingrid, Univ of Michigan
Henßel, Katja, Ludwig-Maximilians-Universitaet Muenchen
Herrera-Gil, Luis A., Universidad Auto de Baja California Sur
Hickman, Carole S., Univ of California, Berkeley
Hintze, Lehi F., Brigham Young Univ
Hlusko, Leslea, Univ of California, Berkeley
Hopkins, Samantha, Univ of Oregon
House, Christopher H., Pennsylvania State Univ, Univ Park
Hughes, Nigel C., Univ of California, Riverside
Hurley, Sarah J., Columbia Univ
Hylland, Mike, Utah Geological Survey
Immel, Harald, Ludwig-Maximilians-Universitaet Muenchen
Ivanov, Martin, Masaryk Univ
Jackson, Jeremy B. C., Univ of California, San Diego
Johnson, Gerald H., William & Mary
Johnson, Kayleigh, South Dakota School of Mines & Tech
Johnston, Carl G., Youngstown State Univ
Jones, Douglas S., Univ of Florida
Jorge, Maria Luisa, Vanderbilt Univ
Kazmer, Miklos, Eotvos Lorand Univ
Kelley, Patricia H., Univ of North Carolina, Wilmington
Kendrick, David C., Hobart & William Smith Colls
Kern, J. Philip, San Diego State Univ
Khaleel, Tasneem, Montana State Univ, Billings
Kirkland, James I., Utah Geological Survey
Kirschvink, Joseph L., California Inst of Tech
Kleffner, Mark A., Ohio State Univ
Klemow, Kenneth M., Wilkes Univ
Koehl, Mimi A. R., Univ of California, Berkeley
Konhauser, Kurt, Univ of Alberta
Kowalke, Thorsten, Ludwig-Maximilians-Universitaet Muenchen
Kozdon, Reinhard, Columbia Univ
Kulp, Thomas, Binghamton Univ
Leinfelder, Reinhold, Ludwig-Maximilians-Universitaet Muenchen
Lenhart, Stephen W., Radford Univ
Li, Rong-Yu, Brandon Univ
Lieberman, Bruce S., Univ of Kansas
Lindberg, David R., Univ of California, Berkeley
Little, Crispin, Univ of Leeds
Lockwood, Rowan, William & Mary
Love, Renee, Univ of Idaho
Lozar, Francesca, Università di Torino
Macalady, Jennifer L., Pennsylvania State Univ, Univ Park
MacKenzie, Lindsay, Eastern Washington Univ
MacLeod, Kenneth A., Univ of Missouri
Macomber, Richard, Long Island Univ, Brooklyn Campus
Mandra, York T., San Francisco State Univ
Marenco, Pedro J., Bryn Mawr Coll
Marra, John, Columbia Univ
Martindale, Rowan, Univ of Texas, Austin
Martínez, Gemma, Univ Complutense de Madrid
McBeth, Joyce M., Univ of Saskatchewan
McCall, Peter L., Case Western Reserve Univ
McCollum, Linda B., Eastern Washington Univ
McGhee, Jr., George R., Rutgers, The State Univ of New Jersey
Mcgowan, Alistair, Univ of Glasgow
McHugh, Julia, Colorado Mesa Univ
McKenzie, Scott C., Mercyhurst Univ
McMenamin, Mark, Mount Holyoke Coll
McNamara, Kenneth, Univ of Cambridge
McRoberts, Christopher A., SUNY, Cortland
Mead, James I., Northern Arizona Univ
Means, Guy H., Florida Geological Survey
Melchin, Michael, Dalhousie Univ
Mezga, Aleksandar, Univ of Zagreb
Miller, Joshua H., Univ of Cincinnati
Miller, Randall F., Univ of New Brunswick
Mitchell, Charles E., SUNY, Buffalo
Mock, Thomas, Univ of East Anglia

Moe-Hoffman, Amy P., Mississippi State Univ
Monaco, Pat, Univ of Colorado
Monari, Stefano, Università degli Studi di Padova
Montgomery, Homer A., Univ of Texas, Dallas
Moro, Alan, Univ of Zagreb
Morris, Simon C., Univ of Cambridge
Munro, Ildi, Carleton Univ
Murray, Alison E., Desert Research Inst
Murray, John, Nat Univ of Ireland Galway
Muscente, Drew, Cornell Coll
Musil, Rudolf, Masaryk Univ
Myers, Orrin B., Los Alamos Nat Lab
Naumann, Malik, Ludwig-Maximilians-Universitaet Muenchen
Nealson, Kenneth, Univ of Southern California
Newman, Dianne K., California Inst of Tech
Newton, Cathryn R., Syracuse Univ
Norman, David, Univ of Cambridge
Nudds, John, Univ of Manchester
O'Mullan, Gregory, Queens Coll (CUNY)
Oboh-Ikuneobe, Francisca, Missouri Univ of Science and Tech
Olcott Marshall, Alison, Univ of Kansas
Ono, Shuhei, Massachusetts Inst of Tech
Orchard, Michael, Univ of British Columbia
Orphan, Victoria, California Inst of Tech
Orsi, William, Ludwig-Maximilians-Universitaet Muenchen
Owen, Alan W., Univ of Glasgow
Parsons-Hubbard, Karla M., Oberlin Coll
Patzkowsky, Mark E., Pennsylvania State Univ, Univ Park
Pavia, Giulio, Università di Torino
Peng, Lee Chai, Univ of Malaya
Perez-Huerta, Alberto, Univ of Alabama
Person, Jeff, North Dakota Geological Survey
Peter, Mark, Ohio Dept of Nat Resources
Peterson, Joseph E., Univ of Wisconsin, Oshkosh
Petuch, Edward J., Florida Atlantic Univ
Pezelj, Đurđica, Univ of Zagreb
Plotnick, Roy E., Univ of Illinois Chicago
Poropat, Steve, Curtin Univ
Porter, Susannah, Univ of California, Santa Barbara
Poteet, Mary F., Univ of Texas, Austin
Pratt, Brian R., Univ of Saskatchewan
Pruss, Sara B., Smith Coll
Puchalski, Stephaney, Highline Coll
Purnell, Mark, Leicester Univ
Rampedi, Isaac T., Univ of Johannesburg
Read, Michael T., Stephen F. Austin State Univ
Reichenbacher, Bettina, Ludwig-Maximilians-Universitaet Muenchen
Reid, Catherine, Univ of Canterbury
Reitner, Joachim, Georg-August Univ of Goettingen
Rhenberg, Elizabeth, Univ of Memphis
Risk, Michael J., McMaster Univ
Roden, Eric E., Univ of Wisconsin, Madison
Rodríguez, Sergio, Univ Complutense de Madrid
Rogers, Karyn L., Rensselaer Polytechnic Inst
Rogers, Kristina C., Univ of Minnesota, Twin Cities
Rogers, Raymond, Univ of Minnesota, Twin Cities
Rogers, Steven L., Keele Univ
Ross, Robert M., Cornell Univ
Ross, Robert M., Paleontological Research Inst
Runnegar, Bruce, Univ of California, Los Angeles
Rößner, Gertrud, Ludwig-Maximilians-Universitaet Muenchen
Sá, Artur A., Universidade de Trás-os-Montes e Alto Douro
Santelli, Cara M., Univ of Minnesota, Twin Cities
Sanudo, Sergio, Univ of Southern California
Schiffbauer, James, Univ of Missouri
Schmid, Dieter, Ludwig-Maximilians-Universitaet Muenchen
Schopf, J. William, Univ of California, Los Angeles
Schrenk, Matt, Michigan State Univ
Seibt, Ulrike, Univ of California, Los Angeles
Shapiro, Russell S., California State Univ, Chico
Shropshire, K. Lee, Univ of Northern Colorado
Silva, Carlos M., Universidade de Lisboa
Silva-Pineda, Alicia, Universidad Nacional Auto de Mexico
Slatkin, Montgomery, Univ of California, Berkeley
Slaughter, Richard, Univ of Wisconsin, Madison
Smart, Christopher, Univ of Plymouth
Smith, Dena M., Univ of Colorado
Smith, Paul L., Univ of British Columbia
Southam, Gordon, Western Univ
Spencer, Patrick K., Whitman Coll
Stafford, Emily, Western Carolina Univ
Steele, Michael A., Wilkes Univ
Steinker, Don C., Bowling Green State Univ
Steinker, Paula J., Bowling Green State Univ
Stetter, Karl O., Univ of California, Los Angeles
Stifel, Peter B., Univ of Maryland
Stigall, Alycia L., Ohio Univ
Superchi-Culver, Tonia, Univ of Colorado
Sutherland, Stuart, Univ of British Columbia
Sutton, Mark, Imperial Coll
Sweetman, Steve, Univ of Portsmouth
Tesmer, Irving, Buffalo State Coll
Thomas, Roger D. K., Franklin and Marshall Coll
Thompson, Jann W. M., Smithsonian Inst / Nat Museum of Nat History
Thompson, Joel B., Eckerd Coll
Thuy, Ben, Univ of Tennessee, Knoxville
Tibuleac, Paul, Alexandru Ioan Cuza
Tice, Mike, Texas A&M Univ
Toots, Heinrich, Long Island Univ, C.W. Post Campus
Toth, Natalie, Denver Museum of Nature & Science
Uahengo, Collen, Univ of Namibia
Underwood, Charlie, Birkbeck Coll
Upchurch, Paul, Univ Coll London
Valentine, James W., Univ of California, Berkeley
Van Alstine, James B., Univ of Minnesota, Morris
Van Iten, Heyo, Hanover Coll
VanAller-Hernick, Linda A., New York State Geological Survey
Vega Vera, Francisco J., Universidad Nacional Auto de Mexico
Vicens Batet, Enric, Universitat Auto de Barcelona
Wake, David B., Univ of California, Berkeley
Wake, Marvalee H., Univ of California, Berkeley
Wallace, Steve, Univ of Colorado
Ward, Peter D., Univ of Washington
Weber, Karrie A., Univ of Nebraska, Lincoln
Weber-Gobel, Reinhard, Universidad Nacional Auto de Mexico
Wehrmann, Laura, Ludwig-Maximilians-Universitaet Muenchen
West, Ronald R., Kansas State Univ
White, Lisa D., San Francisco State Univ
Whitmore, John H., Cedarville Univ
Whitton, Mark, Univ of Portsmouth
Wignall, Paul, Univ of Leeds
Wilf, Peter D., Pennsylvania State Univ, Univ Park
Williams, Mark, Leicester Univ
Woerheide, Gert, Ludwig-Maximilians-Universitaet Muenchen
Yacobucci, Peg M., Bowling Green State Univ
Yancey, Thomas E., Texas A&M Univ
Yee, Nathan, Rutgers, The State Univ of New Jersey
Zelenitsky, Darla, Univ of Calgary

Paleostratigraphy

Al-Saad, Hamad A., Univ of Qatar
Arden, Daniel D., Georgia Southwestern State Univ
Baarli, B. Gudveig, Williams Coll
Baird, Gordon C., SUNY, Fredonia
Barrick, James E., Texas Tech Univ
Bauer, Jeffrey A., Shawnee State Univ
Berggren, William A., Woods Hole Oceanographic Inst
Brett, Carlton E., Cincinnati Museum Center
Buatois, Luis, Univ of Saskatchewan
Caldwell, William G. E., Univ of Saskatchewan
Cohen, Andrew S., Univ of Arizona
Davila-Alcocer, Victor M., Universidad Nacional Auto de Mexico
Eaton, Jeffrey G., Weber State Univ
Ettensohn, Frank R., Univ of Kentucky
Finney, Stanley C., California State Univ, Long Beach
Grubb, Barbara, Mt. San Antonio Coll
Isaacson, Peter E., Univ of Idaho
Jeffery, David L., Marietta Coll
Johnson, David B., New Mexico Tech
Johnson, Markes E., Williams Coll
Jones, Brian, Univ of Alberta
Kalvoda, Jiri, Masaryk Univ

Kumpan, Tomas, Masaryk Univ
Kurtz, Vincent E., Missouri State Univ
Landing, Ed, New York State Geological Survey
Landman, Neil H., Graduate School of the City Univ of New York
Leatham, W. Britt, California State Univ, San Bernardino
Lewis, Gerald L., Pasadena City Coll
Liebe, Richard M., SUNY, The Coll at Brockport
Mangano, Gabriela, Univ of Saskatchewan
Martin, James E., South Dakota School of Mines & Tech
Maurrasse, Florentin J., Florida InterNat Univ
McLaughlin, Patrick, Indiana Univ, Bloomington
McWilliams, Robert G., Miami Univ
Miller, James F., Missouri State Univ
Miller, Richard H., San Diego State Univ
Murphy, Michael A., Univ of California, Riverside
Olson, Hillary C., Univ of Texas, Austin
Over, D. Jeffrey, SUNY, Geneseo
Ritter, Scott M., Brigham Young Univ
Rosario, Jose, California State Univ, East Bay
Sageman, Bradley B., Northwestern Univ
Shaw, Frederick C., Lehman Coll (CUNY)
Smith, Gerald R., Univ of Michigan
Stearn, Colin, McGill Univ
Stevens, Calvin H., San Jose State Univ
Taylor, John F., Indiana Univ of Pennsylvania
Thoms, Richard E., Portland State Univ
Titus, Robert C., Hartwick Coll
Tobin, Tom S., Univ of Alabama
Trexler, J. Peter, Juniata Coll
Van Niewenhuise, Donald, Univ of Houston
Vandenbroucke, Thijs, Ghent Univ
Zinsmeister, William J., Purdue Univ

Micropaleontology

Adekeye, O. A., Univ of Ilorin
Adojoh, Onema, Case Western Reserve Univ
Agnini, Claudia, Università degli Studi di Padova
Arnold, Anthony J., Florida State Univ
Aubry, Marie-Pierre, Rutgers, The State Univ of New Jersey
Bancroft, Alyssa M., Indiana Univ, Bloomington
Barnes, Christopher R., Univ of Victoria
Barrier, Pascal, Institut Polytechnique LaSalle Beauvais (ex-IGAL)
Beauchamp, Benoit, Univ of Calgary
Berggren, William A., Rutgers, The State Univ of New Jersey
Boix, Carme, Universitat Auto de Barcelona
Bollmann, Jörg, Univ of Toronto
Borrelli, Chiara, Univ of Rochester
Bown, Paul, Univ Coll London
Burke, Collette D., Wichita State Univ
Burkett, Ashley, Oklahoma State Univ
Buzas, Martin A., Smithsonian Inst / Nat Museum of Nat History
Caldwell, W. Glen E., Western Univ
Canales, María Luisa, Univ Complutense de Madrid
Carreno, Ana L., Universidad Nacional Auto de Mexico
Chapman, Mark, Univ of East Anglia
Chin, Shamar C., Univ of Iowa
Clark, David L., Univ of Wisconsin, Madison
Corliss, Bruce H., Duke Univ
Culver, Stephen J., East Carolina Univ
Curry, William B., Woods Hole Oceanographic Inst
Czaja, Andrew D., Univ of Cincinnati
Davydov, Vladimir, Boise State Univ
de Vernal, Anne, Universite du Quebec a Montreal
Eicher, Don L., Univ of Colorado
Eicher, Donald, Univ of Colorado
Finger, Ken, Univ of California, Berkeley
Fluegeman, Richard H., Ball State Univ
Foresi, Luca Maria, Univ of Siena
Fritts, Paul J., California State Univ, Long Beach
Giusberti, Luca, Università degli Studi di Padova
Goldstein, Susan T., Univ of Georgia
Gorog, Agnes, Eotvos Lorand Univ
Grimm, Kirsten I., Universität Mainz
Haitonic, Raluca Emilia, Babes-Bolyai Univ
Hart, Malcolm B., Univ of Plymouth
Harwood, David M., Univ of Nebraska, Lincoln
Hasegawa, Shiro, Kumamoto Univ
Hemphill-Haley, Eileen, Humboldt State Univ
Henderson, Charles M., Univ of Calgary
Herrero, Concepción, Univ Complutense de Madrid
Huber, Brian T., Smithsonian Inst / Nat Museum of Nat History
Hunt, Gene, Smithsonian Inst / Nat Museum of Nat History
Ingle, Jr., James C., Stanford Univ
Ionesi, Viorel, Alexandru Ioan Cuza
Ishman, Scott E., Southern Illinois Univ Carbondale
Kanungo, Sudeep, Univ of Utah
Katz, Miriam E., Rensselaer Polytechnic Inst
Keller, Gerta, Princeton Univ
Kelly, D. Clay, Univ of Wisconsin, Madison
Klapper, Gilbert, Northwestern Univ
Knox, Larry W., Tennessee Tech Univ
Leadbetter, Jared R., California Inst of Tech
Leckie, R. Mark, Univ of Massachusetts, Amherst
Lewandowski, Katherine, Eastern Illinois Univ
Lobegeier, Melissa, Middle Tennessee State Univ
Louwye, Stephen, Ghent Univ
Lucas, Franklin A., Univ of Benin
Lundin, Robert F., Arizona State Univ
Lupia, Richard, Univ of Oklahoma
Maddocks, Rosalie F., Univ of Houston
Martin, Ronald E., Univ of Delaware
McCarthy, Francine G., Brock Univ
McCartney, Kevin, Univ of Maine, Presque Isle
Mendelson, Carl V., Beloit Coll
Metcalfe, Ian, Univ of New England
Metzger, Ronald A., Southwestern Oregon Comm Coll
Montenari, Michael, Keele Univ
Nathan , Stephen, Eastern Connecticut State Univ
Nestell, Galina P., Univ of Texas, Arlington
Nestell, Merlynd K., Univ of Texas, Arlington
Noble, Paula J., Univ of Nevada, Reno
Olsson, Richard K., Rutgers, The State Univ of New Jersey
Oppo, Delia W., Woods Hole Oceanographic Inst
Orr, William N., Univ of Oregon
Patterson, R. Timothy, Carleton Univ
Pearson, Paul, Univ of Wales
Peeters, Frank J., Vrije Universiteit Amsterdam
Pessagno, Jr., Emile A., Univ of Texas, Dallas
Ples, George, Babes-Bolyai Univ
Puckett, Mark, Univ of Southern Mississippi
Pujana, Ignacio, Univ of Texas, Dallas
Rathburn, Anthony E., California State Univ, Bakersfield
Rehakova, Daniela, Comenius Univ in Bratislava
Reinhardt, Edward G., McMaster Univ
Roberts, Charles, Mt. San Antonio Coll
Robinson, Edward, Florida InterNat Univ
Rodrigues, Cyril G., Univ of Windsor
Ross, Charles A., Western Washington Univ
Rosscoe, Steven, Hardin-Simmons Univ
Sartipzadeh, Siavosh, Univ of Tabriz
Scherer, Reed P., Northern Illinois Univ
Schroder-Adams, Claudia, Carleton Univ
Scott, David B., Dalhousie Univ
Sen Gupta, Barun K., Louisiana State Univ
Siesser, William G., Vanderbilt Univ
Silye, Lorand , Babes-Bolyai Univ
Sloan, Doris, Univ of California, Berkeley
Sloan, Jon R., California State Univ, Northridge
Speijer, Robert, Katholieke Universiteit Leuven
Stott, Lowell D., Univ of Southern California
Thomas, Ellen, Wesleyan Univ
Thomas, Ellen, Yale Univ
Toth, Emoke, Eotvos Lorand Univ
Ufkes, Els , Vrije Universiteit Amsterdam
Vaiani, Stefano C., Alma Mater Studiorum Università di Bologna
Vaz, Nuno M., Universidade de Trás-os-Montes e Alto Douro
von Bitter, Peter H., Univ of Toronto
Wade, Bridget S., Univ Coll London
Walker, Sally E., Univ of Georgia
Watkins, David K., Univ of Nebraska, Lincoln
Webb, Peter N., Ohio State Univ
Wise, Jr, Sherwood W., Florida State Univ

Paleobotany
Basinger, James F., Univ of Saskatchewan
Beck, John H., Boston Coll
Berry, C M., Cardiff Univ
Bomfleur, Benjamin, Universitaet Muenster
Burnham, Robyn J., Univ of Michigan
Carrillo Martinez, Miguel, Universidad Nacional Auto de Mexico
Currano, Ellen, Univ of Wyoming
DiMichele, William A., Smithsonian Inst / Nat Museum of Nat History
Donoghue, Michael J., Yale Univ
Donovan, Michael, Cleveland Museum of Nat History
Doyle, James A., Univ of California, Davis
Edwards, Dianne, Cardiff Univ
Erwin, Diane, Univ of California, Berkeley
Gandolfo-Nixon, Maria A., Cornell Univ
Gowing, David, Open Univ
Griffin, Kevin L., Columbia Univ
Heath, Kathleen M., Indiana State Univ
Hu, Shusheng, Yale Univ
Jackson, Gail D., Edinburgh Univ
Knoll, Andrew H., Harvard Univ
Looy, Cynthia, Univ of California, Berkeley
Manchester, Steven R., Univ of Florida
Martinetto, Edoardo, Università di Torino
McCoy, Victoria, Univ of Wisconsin, Milwaukee
McElwain, Jenny, Field Museum of Nat History
Miller, Ian, Denver Museum of Nature & Science
Mustoe, George, Western Washington Univ
Peppe, Daniel J., Baylor Univ
Pfefferkorn, Hermann W., Univ of Pennsylvania
Punyasena, Surangi W., Univ of Illinois, Urbana-Champaign
Raymond, Anne, Texas A&M Univ
Reichgelt, Tammo, Univ of Connecticut
Retallack, Gregory J., Univ of Oregon
Rigby, John, Queensland Univ of Tech
Scott, Andrew C., Royal Holloway Univ of London
Stromberg, Caroline, Univ of Washington
Taggart, Ralph E., Michigan State Univ
Taylor, Edith, Univ of Kansas
Thomasson, Joseph R., Fort Hays State Univ
Tiffney, Bruce H., Univ of California, Santa Barbara
Wing, Scott L., Smithsonian Inst / Nat Museum of Nat History

Palynology
Anderson, Patricia M., Univ of Washington
Asselin, Esther, Nat Resources Canada
Atta-Peters, David, Univ of Ghana
Brenner, Gilbert J., SUNY, New Paltz
Brush, Grace S., Johns Hopkins Univ
Butcher, Anthony, Univ of Portsmouth
Calcote, Randy, Univ of Minnesota, Twin Cities
Darrell, James H., Georgia Southern Univ
Davis, Owen K., Univ of Arizona
Dolakova, Nela, Masaryk Univ
Eble, Cortland F., Univ of Kentucky
Farley, Martin B., Univ of North Carolina, Pembroke
FitzPatrick, Meriel E., Univ of Plymouth
Fowell, Sarah J., Univ of Alaska, Fairbanks
Habib, Daniel, Queens Coll (CUNY)
Head, Martin J., Brock Univ
Hebda, Richard J., Univ of Victoria
Hemsley, A, Cardiff Univ
Hemsley, Alan, Univ of Wales
Higgs, Ken, Univ Coll Cork
Ivory, Sarah, Pennsylvania State Univ, Univ Park
Jacobs, Bonnie F., Southern Methodist Univ
Jarvis, Ed, Univ Coll Cork
Javaux, Emmanuelle, Universite de Liege
Jolley, David, Univ of Aberdeen
Kovácová, Marianna, Comenius Univ in Bratislava
Mack, Charlotte, Curtin Univ
Maher, Jr., Louis J., Univ of Wisconsin, Madison
Marshall, John E., Univ of Southampton
Martinez Hernandez, Enrique, Universidad Nacional Auto de Mexico
McAndrews, John H., Univ of Toronto
Milne, Lynne, Curtin Univ
Mudie, Peta J., Dalhousie Univ
Norris, Geoffrey, Univ of Toronto
O'Keefe, Jennifer, Morehead State Univ
Ogbahon, Osazuwa A., Federal Univ of Tech, Akure
Rich, Fredrick J., Georgia Southern Univ
Rueger, Bruce F., Colby Coll
Smith, Jim, Univ Coll Cork
Stefanova, Ivanka, Univ of Minnesota, Twin Cities
Strother, Paul L., Boston Coll
Strother, Paul K., Boston Coll
Tabara, Daniel, Alexandru Ioan Cuza
Tantau, Ioan, Babes-Bolyai Univ
Warny, Sophie, Louisiana State Univ
Zobaa, Mohamed, Univ of Texas Permian Basin

Quantitative Paleontology
Bapst, David, Texas A&M Univ
Boyle, James, SUNY, Buffalo
Daley, Gwen M., Winthrop Univ
Huntley, John, Univ of Missouri
Marshall, Charles R., Univ of California, Berkeley
Miller, Arnold I., Cincinnati Museum Center
Motz, Gary J., Indiana Univ, Bloomington
Parker, William C., Florida State Univ
Sallan, Lauren, Univ of Pennsylvania

Vertebrate Paleontology
Adams, Thomas, San Antonio Comm Coll
Applegate-Pleasants, Shelton, Universidad Nacional Auto de Mexico
Archer, Michael, Univ of New South Wales
Baltensperger, David D., Texas A&M Univ
Bamforth, Emily, Univ of Saskatchewan
Barnosky, Anthony D., Univ of California, Berkeley
Bartels, William S., Albion Coll
Beard, K. Christopher, Carnegie Museum of Nat History
Behrensmeyer, Anna K., Smithsonian Inst / Nat Museum of Nat History
Bell, Christopher J., Univ of Texas, Austin
Bhullar, Bhart-Anjan, Yale Univ
Bloch, Jonathan, Univ of Florida
Bolt, John R., Field Museum of Nat History
Brand, Leonard R., Loma Linda Univ
Brochu, Christopher A., Univ of Iowa
Brown, Caleb M.,
Brusatte, Steve, Edinburgh Univ
Burger, Benjamin J., Utah State Univ
Caldwell, Michael W., Univ of Alberta
Campbell, Kenneth E., Los Angeles County Museum of Nat History
Canudo, José Ignacio, Universidad de Zaragoza
Carpenter, Kenneth, Univ of Colorado
Carrano, Matthew T., Smithsonian Inst / Nat Museum of Nat History
Carranza-Castaneda, Oscar, Universidad Nacional Auto de Mexico
Chatterjee, Sankar, Texas Tech Univ
Chiappe, Luis M., Los Angeles County Museum of Nat History
Cicimurri, Christian M., Clemson Univ
Cicimurri, Christian M., Clemson Univ
Cicimurri, David J., Clemson Univ
Clarke, Julia A., Univ of Texas, Austin
Clemens, William A., Univ of California, Berkeley
Codrea, Vlad, Babes-Bolyai Univ
Colleary, Caitlin, Cleveland Museum of Nat History
Cooke, H.B. S., Dalhousie Univ
Coombs, Margery C., Amherst Coll
Covert, Herbert, Univ of Colorado
Cumbaa, Steve L., Carleton Univ
Curry Rogers, Kristina A., Macalester Coll
Daeschler, Ted, Drexel Univ
Davis, Edward B., Univ of Oregon
Dawson, Mary, Univ of Colorado
Dawson, Mary R., Carnegie Museum of Nat History
Delson, Eric, Graduate School of the City Univ of New York
DeSantis, Larisa R., Vanderbilt Univ
Domingo, Laura, Univ Complutense de Madrid
Druckenmiller, Patrick S., Univ of Alaska, Fairbanks
Dundas, Robert G., Fresno State Univ
Dzanh, Trihn, Univ of Colorado
Eberle, Jaelyn J., Univ of Colorado
Elliott, David K., Northern Arizona Univ

Emry, Robert, Smithsonian Inst / Nat Museum of Nat History
Engelmann, George F., Univ of Nebraska, Omaha
Esperante, Raul, Loma Linda Univ
Evans, David C., Royal Ontario Museum
Fanti, Federico, Alma Mater Studiorum Università di Bologna
Fastovsky, David, Univ of Rhode Island
Fiorillo, Anthony R., Southern Methodist Univ
Flynn, John J., Columbia Univ
Flynn, John J., American Museum of Nat History
Fordyce, R. Ewan, Univ of Otago
Froehlich, David J., Austin Comm Coll District
Furió, Marc, Universitat Auto de Barcelona
Gaffney, Eugene S., American Museum of Nat History
Gallagher, William B., Rider Univ
García, Nuria, Univ Complutense de Madrid
Garcia, William, Univ of North Carolina, Charlotte
Gardner, James D.,
Gauthier, Jacques, Yale Univ
Gauthier, Jacques A., Yale Univ
Gillette, David D., Northern Arizona Univ
Gnidovec, Dale M., Ohio State Univ
Goodwin, Mark B., Univ of California, Berkeley
Gottfried, Michael D., Michigan State Univ
Graham, Russell W., Pennsylvania State Univ, Univ Park
Grandal d'Anglade, Aurora, Coruna Univ
Grande, Lance, Field Museum of Nat History
Gray, Robert S., Santa Barbara City Coll
Green, Harry W., Univ of California, Berkeley
Hammer, William R., Augustana Coll
Hand, Suzanne J., Univ of New South Wales
Hargrave, Jennifer E., Univ of Louisiana at Lafayette
Harrell, Jr., T. L., Geological Survey of Alabama
Harris, Judith A., Univ of Colorado
Hayden, Martha C., Utah Geological Survey
Heaton, Timothy H., Univ of South Dakota
Heckert, Andrew B., Appalachian State Univ
Henderson, Donald M.,
Henrici, Amy C., Carnegie Museum of Nat History
Hernández, Manuel, Univ Complutense de Madrid
Hitz, Ralph B., Tacoma Comm Coll
Holroyd, Pat, Univ of California, Berkeley
Holtz, Jr., Thomas R., Univ of Maryland
Holwerda, Femke M.,
Horner, John R., Montana State Univ
Hungerbuehler, Axel, Mesalands Comm Coll
Hunt, Robert M., Univ of Nebraska, Lincoln
Indeck, Jeff, Univ of Colorado
Jackson, Frankie, Montana State Univ
Jacobs, Louis L., Southern Methodist Univ
Jasinski, Steven E., State Museum of Pennsylvania
Johnson, Ned K., Univ of California, Berkeley
Joniak, Peter , Comenius Univ in Bratislava
Juntunen, Kristopher, Tarleton State Univ
Keenan, Sarah W., South Dakota School of Mines & Tech
Kirkland, James I., Utah Geological Survey
Koch, Paul L., Univ of California, Santa Cruz
Konishi, Takuya, Cincinnati Museum Center
Krishtalka, Leonard, Univ of Kansas
Lamanna, Matthew C., Univ of Pittsburgh
Lamanna, Matthew C., Carnegie Museum of Nat History
Lamb, James P., Univ of West Alabama
Lidicker, Jr., William Z., Univ of California, Berkeley
Lindsay, Everett H., Univ of Arizona
Loewen, Mark, Univ of Utah
Luo, Zhexi, Carnegie Museum of Nat History
Lyson, Tyler, Denver Museum of Nature & Science
MacFadden, Bruce J., Univ of Florida
Maisey, John G., American Museum of Nat History
Makovicky, Peter, Univ of Minnesota, Twin Cities
Makovicky, Peter J., Field Museum of Nat History
Mannion, Philip D., Univ Coll London
Marcot, Jonathan, Univ of Colorado
Martill, David, Univ of Portsmouth
Martin, James E., Univ of Louisiana at Lafayette
Massare, Judy A., SUNY, The Coll at Brockport
McDonald, Greg, Univ of Colorado
McFadden, Bruce J., Univ of Florida
McGee, Amanda R., Cleveland Museum of Nat History
McLeod, Samuel A., Los Angeles County Museum of Nat History
Meng, Jin, American Museum of Nat History
Meng, Jin, Graduate School of the City Univ of New York
Merck, Jr., John W., Univ of Maryland
Miller, Wade E., Brigham Young Univ
Miller-Camp, Jessica, Indiana Univ, Bloomington
Montellano-Ballesteros, Marisol, Universidad Nacional Auto de Mexico
Morales, Michael A., Emporia State Univ
Motani, Ryosuke, Univ of California, Davis
Naples, Virginia, Northern Illinois Univ
Nesbitt, Sterling J., Virginia Polytechnic Inst & State Univ
Njau, Jackson K., Indiana Univ, Bloomington
Norell, Mark A., American Museum of Nat History
Norell, Mark A., Columbia Univ
Novacek, Michael J., American Museum of Nat History
O'Leary, Maureen, Stony Brook Univ
Olsen, Paul E., Columbia Univ
Osi, Attila, Eotvos Lorand Univ
Padian, Kevin, Univ of California, Berkeley
Pagnac, Darrin C., South Dakota School of Mines & Tech
Parham, James, California State Univ, Fullerton
Patton, James L., Univ of California, Berkeley
Polly, P. David, Indiana Univ, Bloomington
Ratoi, Bogdan G., Alexandru Ioan Cuza
Ray, Clayton E., Smithsonian Inst / Nat Museum of Nat History
Rensberger, John M., Univ of Washington
Rieppel, Olivier C., Field Museum of Nat History
Robinson, Peter, Univ of Colorado
Rowe, Timothy B., Univ of Texas, Austin
Russell, Dale A., North Carolina State Univ
Sabol, Martin, Comenius Univ in Bratislava
Sagebiel, J. C., San Bernardino County Museum
Samonds, Karen, Northern Illinois Univ
Sankey, Julia, California State Univ, Stanislaus
Schiebout, Judith A., Louisiana State Univ
Schultz, Gerald E., West Texas A&M Univ
Schwartz, Hilde, Univ of California, Santa Cruz
Schwimmer, David R., Columbus State Univ
Scott, Craig S.,
Sears, Karen, Univ of Colorado
Secord, Ross, Univ of Nebraska, Lincoln
Semken, Jr., Holmes A., Univ of Iowa
Sertich, Joseph, Denver Museum of Nature & Science
Shaw, Christopher A., Los Angeles County Museum of Nat History
Shell, Ryan C., Cincinnati Museum Center
Shiller, Thomas, Sul Ross State Univ
Sidor, Christian A., Univ of Washington
Smith, Kathlyn M., Georgia Southern Univ
Stearley, Ralph F., Calvin Coll
Stocker, Michelle, Virginia Polytechnic Inst & State Univ
Storrs, Glenn W., Cincinnati Museum Center
Storrs, Glenn W., Univ of Cincinnati
Sumida, Stuart S., California State Univ, San Bernardino
Tabrum, Alan R., Carnegie Museum of Nat History
Tedford, Richard H., American Museum of Nat History
Therrien, Francois,
Tumarkin-Deratzian, Allison, Temple Univ
Tütken, Thomas, Universität Mainz
Uhen, Mark D., George Mason Univ
VanRegenmorter, John M., Grand Valley State Univ
Varricchio, David J., Montana State Univ
Vietti, Laura, Univ of Wyoming
Vila, Bernat, Universitat Auto de Barcelona
Voorhies, Michael R., Univ of Nebraska, Lincoln
Vrba, Elisabeth S., Yale Univ
Wilkins, Kenneth, Baylor Univ
Williamson, Thomas E., Univ of New Mexico
Wilson, Gregory, Univ of Washington
Wilson, Laura E., Fort Hays State Univ
Winkler, Alisa, Southern Methodist Univ
Winkler, Dale A., Southern Methodist Univ
Wood, Aaron R., Iowa State Univ of Science & Tech
Woodburne, Michael O., Univ of California, Riverside
Wuerthele, Norman, Carnegie Museum of Nat History

Wyss, Andre R., Univ of California, Santa Barbara
Yarborough, Vicki, Virginia Polytechnic Inst & State Univ
Zakrzewski, Richard J., Fort Hays State Univ

Invertebrate Paleontology

(Wittmer) Malinowski, Jacalyn M., SUNY, Geneseo
Adrain, Jonathan M., Univ of Iowa
Alencaster-Ybarra, Gloria, Universidad Nacional Auto de Mexico
Anderson, Laurie C., South Dakota School of Mines & Tech
Bailey, Richard H., Northeastern Univ
Batten, Roger L., American Museum of Nat History
Becker, Ralph Thomas, Universitaet Muenster
Bishop, Gale A., Georgia Southern Univ
Blake, Daniel B., Univ of Illinois, Urbana-Champaign
Bonem, Rena M., Baylor Univ
Bonuso, Nicole, California State Univ, Fullerton
Bork, Kennard B., Denison Univ
Brandt, Danita S., Michigan State Univ
Briggs, Derek E., Yale Univ
Broadhead, Thomas W., Univ of Tennessee, Knoxville
Brown, Lewis M., Lake Superior State Univ
Buitron-Sanchez, Blanca E., Universidad Nacional Auto de Mexico
Buss, Leo W., Yale Univ
Butts, Susan H., Yale Univ
Cairns, Stephen, Smithsonian Inst / Nat Museum of Nat History
Caldwell, Roy L., Univ of California, Berkeley
Camp, Mark J., Univ of Toledo
Caron, Jean-Bernard, Univ of Toronto
Chatelain, Edward E., Valdosta State Univ
Chatterton, Brian D., Univ of Alberta
Cintra-Buenrostro, Carlos, Univ of Texas, Rio Grande Valley
Clark, Murlene W., Univ of South Alabama
Comas, María José, Univ Complutense de Madrid
Copper, Paul, Laurentian Univ, Sudbury
Cuffey, Roger J., Pennsylvania State Univ, Univ Park
Dahl, Robyn, Western Washington Univ
Day, James E., Illinois State Univ
Deline, Brad, Univ of West Georgia
Dutro, Thomas, Smithsonian Inst / Nat Museum of Nat History
Eldredge, Niles, American Museum of Nat History
Elias, Robert J., Univ of Manitoba
Erwin, Douglas H., Smithsonian Inst / Nat Museum of Nat History
Evanoff, Emmett, Univ of Colorado
Feldmann, Rodney M., Kent State Univ
Finks, Robert M., Queens Coll (CUNY)
Finnegan, Seth, Univ of California, Berkeley
Fisher, Daniel C., Univ of Michigan
Fraiser, Margaret L., Univ of Wisconsin, Milwaukee
Full, Robert J., Univ of California, Berkeley
Gacía Joral, Fernando, Univ Complutense de Madrid
Gilmour, Ernest H., Eastern Washington Univ
Glass, Alexander, Duke Univ
Gonzalez-Arreola, Celestina, Universidad Nacional Auto de Mexico
Goodwin, David H., Denison Univ
Hageman, Steven J., Appalachian State Univ
Hanger, Rex, Univ of Wisconsin, Whitewater
Hannibal, Joseph T., Cleveland Museum of Nat History
Hannibal, Joseph T., Case Western Reserve Univ
Harnik, Paul, Colgate Univ
Harper, Jr., Charles W., Univ of Oklahoma
Hartman, Joseph H., Univ of North Dakota
Hegna, Thomas A., Western Illinois Univ
Hembree, Daniel, Ohio Univ
Hendricks, Jonathan R., Paleontological Research Inst
Hudackova, Natalia H., Comenius Univ in Bratislava
Hughes, Nigel C., Cincinnati Museum Center
Hunda, Brenda R., Univ of Cincinnati
Hunda, Brenda R., Cincinnati Museum Center
Hyzny, Matus, Comenius Univ in Bratislava
Jablonski, David, Univ of Chicago
James, Matthew J., Sonoma State Univ
Jin, Jisuo, Western Univ
Johns, Ronald A., Austin Comm Coll District
Johnston, Paul, Mount Royal Univ
Kammer, Thomas W., West Virginia Univ
Key, Jr., Marcus M., Dickinson Coll
Kowalewski, Michal, Univ of Florida
Kues, Barry S., Univ of New Mexico
Kukalova-Peck, Jarmila, Carleton Univ
Labandeira, Conrad C., Smithsonian Inst / Nat Museum of Nat History
Landman, Neil H., American Museum of Nat History
Lee, Daphne E., Univ of Otago
Leggitt, Leroy, Loma Linda Univ
Lehrmann, Daniel J., Trinity Univ
Leighton, Lindsey, Univ of Alberta
Lenz, Alfred C., Western Univ
Leonard-Pingel, Jill, Ohio State Univ
Lewis, Ronald D., Auburn Univ
Lidgard, Scott H., Field Museum of Nat History
Lindberg, David R., Univ of California, Berkeley
Lockley, Martin G., Univ of Colorado, Denver
LoDuca, Steven T., Eastern Michigan Univ
Marenco, Katherine N., Bryn Mawr Coll
Mason, Charles E., Morehead State Univ
McAfee, Gerald B., Odessa Coll
Melchin, Michael J., Saint Francis Xavier Univ
Merrill, Glen K., Univ of Houston Downtown
Meyer, David L., Univ of Cincinnati
Meyer, David L., Cincinnati Museum Center
Miller, Arnold I., Univ of Cincinnati
Miller, William C., Humboldt State Univ
Morris, Robert W., Wittenberg Univ
Nagel-Myers, Judith, St. Lawrence Univ
Narbonne, Guy M., Queen's Univ
Nitecki, Matthew H., Field Museum of Nat History
Oleinik, Anton, Florida Atlantic Univ
Paterson, John R., Univ of New England
Perrilliat-Montoya, Maria del Carmen, Universidad Nacional Auto de Mexico
Petsios, Elizabeth, Baylor Univ
Pietsch, Carlie, San Jose State Univ
Pojeta, John, Smithsonian Inst / Nat Museum of Nat History
Pope, John K., Miami Univ
Poty, Edouard, Universite de Liege
Powell, Matthew G., Juniata Coll
Prichonnet, Gilbert P., Universite du Quebec a Montreal
Ritterbush, Linda A., California Lutheran Univ
Robison, Richard A., Univ of Kansas
Rodland, David L., Muskingum Univ
Rohr, David M., Sul Ross State Univ
Rosenberg, Gary, Drexel Univ
Rowell, Albert J., Univ of Kansas
Rowland, Stephen, Univ of Nevada, Las Vegas
Sandy, Michael R., Univ of Dayton
Santos, Hernan, Univ of Puerto Rico
Savage, Norman M., Univ of Oregon
Schiappa, Tamra A., Slippery Rock Univ
Schlogl, Jan , Comenius Univ in Bratislava
Schmachtenbert, William, Virginia Polytechnic Inst & State Univ
Schmidt, David, Wright State Univ
Schwalbach, Cameron E., Cincinnati Museum Center
Schweitzer, Carrie E., Kent State Univ at Stark
Schweitzer, Carrie E., Kent State Univ
Segura-Vernis, Luis R., Universidad Auto de Baja California Sur
Selden, Paul A., Univ of Kansas
Sevilla, Paloma, Univ Complutense de Madrid
Sheehan, Peter M., Univ of Wisconsin, Milwaukee
Sheehan, Peter M., Milwaukee Public Museum
Shubak, Kenneth A., Univ of Wisconsin, Platteville
Soja, Constance M., Colgate Univ
Sorauf, James E., Binghamton Univ
Sprinkle, James T., Univ of Texas, Austin
Squires, Richard L., California State Univ, Northridge
Stanley, Thomas M., Univ of Oklahoma
Stanley, Jr., George D., Univ of Montana
Stephen, Daniel, Utah Valley Univ
Sumrall, Colin, Univ of Tennessee, Knoxville
Szente, Istvan, Eotvos Lorand Univ
Szlavecz, Katalin, Johns Hopkins Univ
Tapanila, Leif, Idaho State Univ
Tarhan, Lidya, Yale Univ
Tshudy, Dale, Edinboro Univ of Pennsylvania
Tsujita, Cameron J., Western Univ

Villasenor-Martinez, Ana B., Universidad Nacional Auto de Mexico
Visaggi, Christy, Georgia State Univ
Votaw, Robert, Indiana Univ Northwest
Wagner, Peter J., Univ of Nebraska, Lincoln
Wagner, Peter J., Field Museum of Nat History
Waller, Thomas R., Smithsonian Inst / Nat Museum of Nat History
Webber, Andrew J., Cincinnati Museum Center
Webster, Gary D., Washington State Univ
Webster, Mark, Univ of Chicago
Westrop, Stephen R., Univ of Oklahoma
Wilson, Mark A., Coll of Wooster
Wyse Jackson, Patrick N., Trinity Coll
Xiao, Shuhai, Virginia Polytechnic Inst & State Univ
Young, Graham A., Manitoba Museum
Young, Graham A., Univ of Manitoba
Zachos, Louis G., Univ of Mississippi
Zodrow, Erwin L., Cape Breton Univ

Paleoecology

Allabush, Kathleen, Univ of Utah
Almendinger, James E., Univ of Minnesota, Twin Cities
Andreu-Hayles, Laia, Columbia Univ
Arkle, Kelsey M., Augustana Coll
Ashworth, Allan C., North Dakota State Univ
Baker, Richard G., Univ of Iowa
Balescu, Sanda, Universite du Quebec a Montreal
Barker, Stephen, Cardiff Univ
Bartlein, Patrick, Univ of Oregon
Begin, Christian, Universite du Quebec
Belanger, Christina, Texas A&M Univ
Bennington, J Bret, Hofstra Univ
Billups, Katharina, Univ of Delaware
Blanchet, Jean-Pierre, Universite du Quebec a Montreal
Blyth, Alison, Curtin Univ
Bottjer, David J., Univ of Southern California
Boyd, Bill, Southern Cross Univ
Bralower, Timothy J., Pennsylvania State Univ, Univ Park
Brett, Carlton E., Univ of Cincinnati
Briskin, Madeleine, Univ of Cincinnati
Buckley, Brendon, Columbia Univ
Caissie, Beth E., Iowa State Univ of Science & Tech
Camill III, Philip, Bowdoin Coll
Capraro, Luca, Università degli Studi di Padova
Casey, Michelle , Towson Univ
Chapin, F. Stuart, Univ of California, Berkeley
Charles, Christopher D., Univ of California, San Diego
Chira, Carme Mariana, Babes-Bolyai Univ
Cobb, Kim M., Georgia Inst of Tech
Cohen, Anne L., Woods Hole Oceanographic Inst
Cole, Julia E., Univ of Arizona
Crowley, Brooke E., Cincinnati Museum Center
Curtin, Tara M., Hobart & William Smith Colls
Curtis, Jason H., Univ of Florida
D'Andrea, William J., Columbia Univ
de Menocal, Peter B., Columbia Univ
Dekens, Petra, San Francisco State Univ
deMenocal, Peter B., Columbia Univ
Dietl, Gregory P., Cornell Univ
Dornbos, Stephen Q., Univ of Wisconsin, Milwaukee
Duhamel, Solange, Columbia Univ
Edlund, Mark B., Univ of Minnesota, Twin Cities
Elick, Jennifer M., Susquehanna Univ
Endres, Anthony E., Univ of Waterloo
Engstrom, Daniel R., Univ of Minnesota, Twin Cities
Evans, Michael N., Univ of Maryland
Fariduddin, Mohammad, Northeastern Illinois Univ
Fawcett, Peter J., Univ of New Mexico
Francis, Jane, Univ of Leeds
Francus, Pierre, Universite du Quebec
Fritz, Sherilyn C., Univ of Nebraska, Lincoln
Gabler, Christopher, Univ of Texas, Rio Grande Valley
Goodwin, Phillip A., Univ of Southampton
Grigg, Laurie, Norwich Univ
Guber, Albert L., Pennsylvania State Univ, Univ Park
Halfar, Jochen, Univ of Toronto
Hasiotis, Stephen T., Univ of Kansas
Hays, James D., Columbia Univ
Haywood, Alan, Univ of Leeds
He, Helen, Univ of East Anglia
Herbert, Gregory, Univ of South Florida
Herbert, Timothy D., Brown Univ
Herrmann, Achim, Louisiana State Univ
Hill, Tessa M., Univ of California, Davis
Hopley, Philip, Birkbeck Coll
Hu, Feng-Sheng, Univ of Illinois, Urbana-Champaign
Hughes, Malcolm K., Univ of Arizona
Huybers, Peter, Harvard Univ
Ivany, Linda C., Syracuse Univ
Johnson, Claudia C., Indiana Univ, Bloomington
Johnson, Robert G., Univ of Minnesota, Twin Cities
Jones, Douglas S., Univ of Florida
Kemp, Alan, Univ of Southampton
Kennedy, Lisa M., Virginia Polytechnic Inst & State Univ
Knight, Tiffany, Washington Univ in St. Louis
Koutavas, Athanasios, Coll of Staten Island/CUNY
Kutzbach, John E., Univ of Wisconsin, Madison
Lea, David W., Univ of California, Santa Barbara
Lebold, Joe, West Virginia Univ
Linsley, Braddock, Columbia Univ
Littler, Kate L., Exeter Univ
Livingstone, Daniel A., Duke Univ
Lohmann, George P., Woods Hole Oceanographic Inst
Long, Colin, Univ of Wisconsin Oshkosh
Lozano Garcia, Maria S., Universidad Nacional Auto de Mexico
Mann, Daniel H., Univ of Alaska, Fairbanks
Marchal, Olivier, Woods Hole Oceanographic Inst
Martin, Anthony J., Emory Univ
Mayewski, Paul A., Univ of Maine
McIntyre, Andrew, Queens Coll (CUNY)
McManus, Jerry F., Columbia Univ
McManus, Jerry F., Woods Hole Oceanographic Inst
McWethy, David B., Montana State Univ
Medina Elizalde, Martin A., Auburn Univ
Meehan, Kimberly, SUNY, Buffalo
Miller, Joshua H., Cincinnati Museum Center
Miller, Keith B., Kansas State Univ
Miller, Molly F., Vanderbilt Univ
Mishler, Brent, Univ of California, Berkeley
Morgan, Ryan, Tarleton State Univ
Muscente, Drew, Univ of Iowa
Myers, Corinne E., Univ of New Mexico
Nichols, Jonathan E., Columbia Univ
Ortegren, Jason, Univ of West Florida
Panagiotakopulu, Eva, Edinburgh Univ
Parr, Kate, Univ of Liverpool
Pearson, P N., Cardiff Univ
Pekar, Stephen, Queens Coll (CUNY)
Persons, W. Scott, Coll of Charleston
Peteet, Dorothy M., Columbia Univ
Phelps, William, Riverside City Coll
Poli, Maria-Serena, Eastern Michigan Univ
Pospelova, Vera, Univ of Victoria
Poulsen, Christopher J., Univ of Michigan
Power, Mary E., Univ of California, Berkeley
Quinn, Terry, Univ of Texas, Austin
Raymo, Maureen, Columbia Univ
Rhode, David E., Desert Research Inst
Rindsberg, Andrew K., Univ of West Alabama
Robinson, Stuart, Univ of Oxford
Royer, Dana, Wesleyan Univ
Sahagian, Dork, Lehigh Univ
Salzer, Matthew W., Univ of Arizona
Sauer, Peter, Indiana Univ, Bloomington
Scarponi, Daniele, Alma Mater Studiorum Università di Bologna
Schaefer, Joerg, Columbia Univ
Schellenberg, Stephen A., San Diego State Univ
Schwert, Donald P., North Dakota State Univ
Schöne, Bernd R., Universität Mainz
Sessa, Jocelyn A., Drexel Univ
Severinghaus, Jeffrey P., Univ of California, San Diego
Shanahan, Timothy M., Univ of Texas, Austin
Sheffield, Sarah, Univ of South Florida

Shroat-Lewis, Rene A., Univ of Arkansas at Little Rock
Shuman, Bryan N., Univ of Wyoming
Sinha, Ashish, California State Univ, Dominguez Hills
Smerdon, Jason, Columbia Univ
Smiley, Tara M., Indiana Univ, Bloomington
Smith, Dena M., Univ of Colorado
Smith, Martin, Durham Univ
Speer, James, Indiana State Univ
Sremac, Jasenka, Univ of Zagreb
Stoykova, Kristalina C., Bulgarian Acad of Sciences
Surge, Donna M., Univ of North Carolina, Chapel Hill
Swetnam, Thomas W., Univ of Arizona
Tackett, Lydia S., North Dakota State Univ
Talyor, Frederick W., Univ of Texas, Austin
Taylor, Christopher, Curtin Univ
Thompson, William G., Woods Hole Oceanographic Inst
Trouet, Valerie, Univ of Arizona
Uno, Kevin, Columbia Univ
Van der Putten, Nathalie, Vrije Universiteit Amsterdam
Verardo, Stacey, George Mason Univ
Vermeij, Geerat J., Univ of California, Davis
Waddington, J. M., McMaster Univ
Walker, Ian R., Univ of British Columbia, Okanagan Campus
Warnock, Jonathan P., Indiana Univ of Pennsylvania
Webb, III, Thompson, Brown Univ
Weldeab, Syee, Univ of California, Santa Barbara
Westgate, James W., Lamar Univ
Whiteside, Jessica H., Univ of Southampton
Whitlock, Cathy, Montana State Univ
Williams, Christopher J., Franklin and Marshall Coll
Williams, John W., Univ of Wisconsin, Madison
Woodhouse, Connie, Univ of Arizona
Yanes, Yurena, Univ of Cincinnati
Yu, Zicheng, Lehigh Univ
Æosoviæ, Vlasta, Univ of Zagreb

Geomicrobiology

Ashbolt, Nicholas J., Southern Cross Univ
Boase, Katelyn, Curtin Univ
Chan, Clara S., Univ of Delaware
Coolen, Marco, Curtin Univ
Dhami, Navdeep, Curtin Univ
Flannery, David, Queensland Univ of Tech
Frantz, Carie M., Weber State Univ
Hazen, Terry C., Univ of Tennessee, Knoxville
Liang, Renxing, Princeton Univ
Lower, Steven K., Ohio State Univ
McDermott, Timothy, Montana State Univ
McKay, Luke, Montana State Univ
Meyer Dombard, DArcy, Univ of Illinois Chicago
Osburn, Magdalena R., Northwestern Univ
Pérez-Rodríguez, Ileana, Univ of Pennsylvania
Priscu, John, Montana State Univ
Ward, David M., Montana State Univ
Wilhelm, Steven W., Univ of Tennessee, Knoxville

Paleoclimatology

Andrus, C. Fred T., Univ of Alabama
Baker, Paul A., Duke Univ
Bakker, Pepijn, Vrije Universiteit Amsterdam
Baldini, James, Durham Univ
Bush, Rosemary, Northwestern Univ
Cleaveland, Malcolm K., Univ of Arkansas, Fayetteville
Cook, Tim, Univ of Massachusetts, Amherst
Cotton, Jennifer, California State Univ, Northridge
De Vleeschouwer, David, Universitaet Muenster
Doner, Lisa A., Plymouth State Univ
Feng, Ran, Univ of Connecticut
Gulbranson, Erik, Gustavus Adolphus Coll
Hoenisch, Baerbel, Columbia Univ
Johnson, Thomas C., Univ of Minnesota Duluth
Kaspari, Susan, Central Washington Univ
Keegan, Kaitlin M., Univ of Nevada, Reno
Koffman, Bess, Colby Coll
Konecky, Bronwen L., Washington Univ in St. Louis
Martín Chivelet, Javier, Univ Complutense de Madrid
Maxbauer, Dan, Carleton Coll
McKay, Nicholas, Northern Arizona Univ
Meyers, Stephen R., Univ of Wisconsin, Madison
Michel, Lauren, Tennessee Tech Univ
Neff, Peter, Univ of Minnesota, Twin Cities
Onac, Petroniu (Bogdan), Univ of South Florida
Orland, Ian, Univ of Wisconsin, Madison, Div of Extension
Osleger, David A., Univ of California, Davis
Quinton, Page, Indiana Univ, Bloomington
Quinton, Page C., SUNY Potsdam
Railsback, L. Bruce, Univ of Georgia
Saenger, Casey, Western Washington Univ
Salacup, Jeff, Univ of Massachusetts, Amherst
Shackelton, Sarah, Princeton Univ
Stone, Jeffery, Indiana State Univ
Sullivan, Donald G., Univ of Denver
Tabor, Clay, Univ of Connecticut
Touchan, Ramzi, Univ of Arizona
van Beynen, Phil, Univ of South Florida
Weiss, Anna M., Midwestern State Univ
Winsor, Kelsey, Northern Arizona Univ
Zachos, James, Univ of California, Santa Cruz
Zambito, James J., Beloit Coll

HYDROLOGY

General Hydrology

Adelana, S.M. A., Univ of Ilorin
AL-Nouimy, Latifa B., Univ of Qatar
Alencoão, Ana M., Universidade de Trás-os-Montes e Alto Douro
Alhajari, Sief A., Univ of Qatar
Allan, Craig J., Univ of North Carolina, Charlotte
Allen, Peter M., Baylor Univ
Alsdorf, Douglas E., Ohio State Univ
Arain, M. A., McMaster Univ
Autio, Robert J., Indiana Univ, Bloomington
Bain, Daniel J., Univ of Pittsburgh
Barnard, Holly R., Univ of Colorado
Baron, Dirk, California State Univ, Bakersfield
Bearden, Bennett L., Geological Survey of Alabama
Becker, Naomi M., Los Alamos Nat Lab
Bennett, Philip, Univ of Texas, Austin
Bense, Victor, Univ of East Anglia
Bergstrom, Anna, Boise State Univ
Bloor, Michelle, Univ of Portsmouth
Borsa, Adrian, Univ of California, San Diego
Boult, Stephen, Univ of Manchester
Boving, Thomas, Univ of Rhode Island
Bowman, Jean A., Texas A&M Univ
Boyle, Douglas, Univ of Nevada, Reno
Bradley, Chris, Univ of Birmingham
Brassington, Rick, Univ of Newcastle Upon Tyne
Brooks, Paul D., Univ of Utah
Brown, David L., California State Univ, Chico
Burkhart, Patrick A., Slippery Rock Univ
Byrne, James M., Univ of Lethbridge
Campana, Michael E., Oregon State Univ
Cardenas, Bayani, Univ of Texas, Austin
Carey, Sean, McMaster Univ
Carillo Chavez, Alejandro J., Universidad Nacional Auto de Mexico
Carman, Cary D., Angelo State Univ
Caylor, Kelly, Univ of California, Santa Barbara
Challads, Thomas, Edinburgh Univ
Cheng, Chu-Lin, Univ of Texas, Rio Grande Valley
Cianfrani, Christina, Hampshire Coll
Cirmo, Christopher P., SUNY, Cortland
Coulibaly, Paulin, McMaster Univ
Crawford, Nicholas, Western Kentucky Univ
Cunningham, Peter, Univ of Newcastle Upon Tyne
Dahlke, Helen E., Univ of California, Davis
Darby, Jeannie, Univ of California, Davis
Dee, Kato , Univ of Oklahoma
DeVries, Stephanie, Univ of Tennessee, Chattanooga
Doster, Florian, Heriot-Watt Univ
Dowling, Carolyn B., Ball State Univ
Duan, Quingyun, Lawrence Livermore Nat Lab
Duex, Timothy W., Univ of Louisiana at Lafayette

Dymond, Salli F., Univ of Minnesota Duluth
Eaton, Timothy, Queens Coll (CUNY)
El Kadiri, Racha, Middle Tennessee State Univ
Enright, Richard L., Bridgewater State Univ
Entekhabi, Dara, Massachusetts Inst of Tech
Eshleman, Keith N., Univ of Maryland
Estalrich López, Joan, Universitat Auto de Barcelona
Ezenabor, Ben O., Univ of Benin
Fabryka-Martin, June T., Los Alamos Nat Lab
Faramarzi, Monireh, Univ of Alberta
Farrow, Norman D., Oak Ridge Nat Lab
Fedele, Juan J., Saint Cloud State Univ
Florea, Lee, Indiana Univ, Bloomington
Foufoula-Georgiou, Efi, Univ of Minnesota, Twin Cities
Fox, Haydn A., Texas A&M Univ, Commerce
Frasson, Renato , Ohio State Univ
Gary, Marcus, Univ of Texas, Austin
Gilmore, Tyler J., Pacific Northwest Nat Lab
Godsey, Sarah E., Idaho State Univ
Green, Mark B., Case Western Reserve Univ
Grieneisen, Michael L., Univ of California, Davis
Groves, Christopher, Western Kentucky Univ
Guan, Huade, Flinders Univ
Gutierrez-Jurado, Hugo A., Univ of Texas, El Paso
Hamann, Hillary, Univ of Denver
Hannah, David, Univ of Birmingham
Hart, David J., Univ of Wisconsin, Madison
Harter, Thomas, Univ of California, Davis
Hasan, Mohamed Ali, Univ of Malaya
Hauschild, Brian K., New Hampshire Geological Survey
Hauswirth, Scott, California State Univ, Northridge
Heimann, William H., Fort Hays State Univ
Heitz, Leroy F., Univ of Guam
Heo, Joon, Univ of Texas Permian Basin
Hernes, Peter J., Univ of California, Davis
Hildebrandt, Anke, Friedrich-Schiller-Univ Jena
Hill, Mary C., Univ of Kansas
Hooyer, Thomas S., Univ of Wisconsin, Madison
Hornberger, George, Vanderbilt Univ
Hubbard, John E., SUNY, The Coll at Brockport
Huntington, Justin, Univ of Nevada, Reno
Ibaraki, Motomu, Ohio State Univ
Inamdar, Shreeram, Univ of Delaware
Jacobs, Katharine L., Univ of Arizona
James-Aworeni, E., Obafemi Awolowo Univ
Jefferson, Anne J., Kent State Univ
Jefferson, Anne, Univ of North Carolina, Charlotte
Johannesson, Karen H., Tulane Univ
Johnson, Robert L., Argonne Nat Lab
Johnson, Robert O., Oak Ridge Nat Lab
Kandel, Hari, Lake Superior State Univ
Kang, Peter, Univ of Minnesota, Twin Cities
Kasenow, Michael, Eastern Michigan Univ
Kesel, Richard H., Louisiana State Univ
Keyantash, John, California State Univ, Dominguez Hills
Kilroy, Kathyrn, Minot State Univ
Kinner, David A., Western Carolina Univ
Kitterød, Nils-Otto, Norwegian Univ of Life Sciences (NMBU)
Kuchovsky, Tomas, Masaryk Univ
Land, Lewis A., New Mexico Tech
Lane, Charles L., Southern Oregon Univ
Lanigan, David C., Pacific Northwest Nat Lab
Lavanchy, G. Thomas, Univ of Denver
Laycock, Arleigh H., Univ of Alberta
Lee, Michael D., California State Univ, East Bay
Levy, Morgan, Univ of California, San Diego
Lightbody, Anne F., Univ of New Hampshire
Lin, Yu-Feng F., Illinois State Geological Survey
Liu, Gaisheng, Univ of Kansas
Llanos, Hilario, Univ of the Basque Country UPV/EHU
Ma, Lin, Univ of Texas, El Paso
Marshall, Adrienne, Colorado School of Mines
Masteller, Claire C., Washington Univ in St. Louis
Matheny, Ashley, Univ of Texas, Austin
McColl, Kaighin, Harvard Univ
McDermott, Christopher I., Edinburgh Univ
McDowell, William H., Univ of New Hampshire
McElwee, Carl D., Univ of Kansas
McHale, Michael , SUNY, Buffalo
McKenna, Thomas E., Univ of Delaware
McLaughlin, Peter P., Univ of Delaware
McNamara, James P., Boise State Univ
Megdal, Sharon B., Univ of Arizona
Meindl, Christopher, Univ of South Florida
Molotch, Noah, Univ of Colorado
Monohan, Carrie, California State Univ, Chico
Moysey, Stephen, East Carolina Univ
Mudd, Simon N., Edinburgh Univ
Mueller, Erich , Southern Utah Univ
Naudts, Kim, Vrije Universiteit Amsterdam
Nicot, Jean-Philippe, Univ of Texas, Austin
Niu, Guo-Yue, Univ of Arizona
Nkuna, Tinyiko R., Univ of Venda for Science & Tech
Nolin, Anne , Univ of Nevada, Reno
O'Reilly, Catherine M., Illinois State Univ
Odling, Noelle, Univ of Leeds
Ortiz-Aguirre, Ramon, Universidad Auto de San Luis Potosi
Oztekin Okan, Ozlem, Firat Univ
Palmer, Arthur N., SUNY, Oneonta
Papuga, Shirley, Wayne State Univ
Parkin, Geoffrey, Univ of Newcastle Upon Tyne
Pavelsky, Tamlin M., Univ of North Carolina, Chapel Hill
Payn, Robert, Montana State Univ
Pettijohn, J. C., Univ of Illinois, Urbana-Champaign
Piccinini, Leonardo, Università degli Studi di Padova
Piggot, Matthew, Imperial Coll
Plankell, Eric T., Illinois State Geological Survey
Poole, Geoff, Montana State Univ
Prestegaard, Karen L., Univ of Maryland
Puente, Carlos E., Univ of California, Davis
Quinn, Paul, Univ of Newcastle Upon Tyne
Rakovan, Monica, Miami Univ
Reay, William G., William & Mary
Reedy, Robert C., Univ of Texas at Austin, Jackson Schl of Geosciences
Reidenbach, Matthew A., Univ of Virginia
Reinfelder, Ying Fan, Rutgers, The State Univ of New Jersey
Rempe, Daniella, Univ of Texas, Austin
Renshaw, Carl E., Dartmouth Coll
Rice, Karen C., Univ of Virginia
Richards, Paul L., SUNY, The Coll at Brockport
Ricka, Adam, Masaryk Univ
Root, Tara L., Florida Atlantic Univ
Saar, Martin O., Univ of Minnesota, Twin Cities
Salami, B. M., Obafemi Awolowo Univ
Sandoval, Samuel, Univ of California, Davis
Sandoval Solis, Samuel, Univ of California, Davis
Scanlon, Bridget R., Univ of Texas, Austin
Scanlon, Todd M., Univ of Virginia
Scanlon, Todd M., Univ of Virginia
Schenk, Xizhen (Jenny), Western Carolina Univ
Schilling, Keith E., Univ of Iowa
Scotton, Paolo, Università degli Studi di Padova
Screaton, Elizabeth J., Univ of Florida
Silva, Catrina, Universidade de Lisboa
Singer, Michael, Univ of St. Andrews
Slade, Jr., Raymond M., Austin Comm Coll District
Smith, Ellen D., Oak Ridge Nat Lab
Smith, James A., Princeton Univ
Smith, Michael, Geological Survey of Alabama
Smith, Ronald M., Pacific Northwest Nat Lab
Smyth, Rebecca C., Univ of Texas at Austin, Jackson Schl of Geosciences
Sorooshian, Soroosh, Univ of California, Irvine
Spane, Frank A., Pacific Northwest Nat Lab
Sritharan, Subramania I., Central State Univ
Stasko, Stanislaw, Univ of Wroclaw
Steenhuis, Tammo S., Cornell Univ
Sterling, Shannon, Dalhousie Univ
Tarka, Robert, Univ of Wroclaw
Tesler, Theodore, DCNR- Pennsylvania Bureau of Geological Survey
Thorne, Paul D., Pacific Northwest Nat Lab
Triay, Ines, Los Alamos Nat Lab
Turek, Marni, Univ of British Columbia, Okanagan Campus

Turner, Anne, Austin Comm Coll District
Tyler, Scott, Univ of Nevada, Reno
Van Stan, John, Georgia Southern Univ
Vermeul, Vincent R., Pacific Northwest Nat Lab
Vivoni, Enrique, Arizona State Univ
Vonk, Jorien, Vrije Universiteit Amsterdam
Wallender, Wes W., Univ of California, Davis
Watkins, David, Exeter Univ
West, Jared, Univ of Leeds
White, Jeffrey R., Indiana Univ, Bloomington
Witt, Emma, Stockton Univ
Wolf, Aaron T., Oregon State Univ
Woo, Ming-Ko, McMaster Univ
Wood, Eric F., Princeton Univ
Wymore, Adam, Univ of New Hampshire
Xu, Shangping, Univ of Wisconsin, Milwaukee
Yan, Eugene, Univ of Illinois Chicago
Yang, Y, Cardiff Univ
Ye, Ming, Florida State Univ
Yusoff, Ismail, Univ of Malaya
Zaccaria, Daniele, Univ of California, Davis
Zhang, Minghua , Univ of California, Davis
Ziemkiewicz, , Paul F., West Virginia Univ
Zume, Joseph T., Shippensburg Univ

Ground Water/Hydrogeology
Abdo, Ginette, Montana Tech of The Univ of Montana
Abrams, Daniel, Univ of Illinois
Ahmed, Kazi M., Univ of Dhaka
Al, Tom, Univ of Ottawa
Alexander, Scott, Univ of Minnesota, Twin Cities
Alexander, Jr., E. Calvin, Univ of Minnesota, Twin Cities
Ali, Bukari, Kwame Nkrumah Univ of Science and Tech
Ali, Genevieve, Univ of Manitoba
Allen, Diana M., Simon Fraser Univ
Amin, Isam E., Youngstown State Univ
Anderson, Mary P., Univ of Wisconsin, Madison
Anderson, William P., Appalachian State Univ
Andres, A. S., Univ of Delaware
Antigüedad, Iñaki, Univ of the Basque Country UPV/EHU
Appiah-Adjei, Emmanuel K., Kwame Nkrumah Univ of Science and Tech
Appold, Martin S., Univ of Missouri
Arthur, Jonathan D., Florida Geological Survey
Asante, Joseph, Tennessee Tech Univ
Asghari Moghadam, Asghar, Univ of Tabriz
Asiwaju-Bello, Yinusa A., Federal Univ of Tech, Akure
Atkinson, Timothy P., Univ Coll London
Ayers, Jerry, Univ of Nebraska, Lincoln
Baalousha, Husam, King Fahd Univ of Petroleum and Minerals
Baca, Alyssa, New Mexico Tech
Bagtzoglou, Ross, Columbia Univ
Bahr, Jean M., Univ of Wisconsin, Madison
Bair, E. Scott, Ohio State Univ
Bajjali, William, Univ of Wisconsin, Superior
Baker, Andy, Univ of New South Wales
Baldwin, Jr., A. Dwight, Miami Univ
Banks, Eddie, Flinders Univ
Barbecot, Florent, Universite du Quebec a Montreal
Barbee, Gary C., West Texas A&M Univ
Barkmann, Peter, Colorado Geological Survey
Barrash, Warren, Boise State Univ
Batelaan, Okke, Flinders Univ
Batelaan, Okke, Katholieke Universiteit Leuven
Bayari, Serdar, Hacettepe Univ
Becker, Matthew, California State Univ, Long Beach
Beckie, Roger D., Univ of British Columbia
Bennett, Steven W., Western Illinois Univ
Bentley, Laurence R., Univ of Calgary
Bergamaschi, Brian, Sacramento State Univ
Besancon, James, Wellesley Coll
Bethke, Craig M., Univ of Illinois, Urbana-Champaign
Bishop, Charles E., Utah Geological Survey
BlackEagle, Cory, Eastern Kentucky Univ
Blythe, Daniel D., Montana Tech of The Univ of Montana
Bobst, Andrew L., Montana Tech of The Univ of Montana
Bohling, Geoff, Univ of Kansas
Bohling, Geoffrey C., Univ of Kansas
Bolster, Diogo, Univ of Notre Dame
Bordeleau, Geneviève, Universite du Quebec
Boutt, David, Univ of Massachusetts, Amherst
Boving, Thomas B., Univ of Rhode Island
Bradbury, Kenneth R., Univ of Wisconsin, Madison
Bradbury, Kenneth R., Univ of Wisconsin, Madison, Div of Extension
Brahana, John V., Univ of Arkansas, Fayetteville
Brame, Scott E., Clemson Univ
Bratcher, Susan, Fresno State Univ
Breitmeyer, Ronald, Univ of Nevada, Reno
Brikowski, Tom H., Univ of Texas, Dallas
Brinkmann, Robert, Oregon Dept of Geology and Mineral Industries
Brookfield, Andrea, Univ of Kansas
Brown, Megan R., Northern Illinois Univ
Brusseau, Mark L., Univ of Arizona
Burbach, Mark E., Univ of Nebraska, Lincoln
Burbey, Thomas J., Virginia Polytechnic Inst & State Univ
Burgess, William P., Univ Coll London
Burnett, Joree, Tarleton State Univ
Butler, Jr., James J., Univ of Kansas
Callaghan, Kerry, Univ of Illinois Chicago
Callahan, Timothy J., Coll of Charleston
Cameron, Kathleen A., Univ of Nebraska, Lincoln
Capuano, Regina M., Univ of Houston
Cardiff, Michael A., Univ of Wisconsin, Madison
Cardiff, Mike, Univ of Wisconsin, Madison
Carey, Anne E., Ohio State Univ
Carle, Steven, Lawrence Livermore Nat Lab
Carling, Gregory T., Brigham Young Univ
Carlson, Catherine A., Eastern Connecticut State Univ
Carlson, Douglas A., Louisiana State Univ
Carstarphen, Camela, Montana Tech of The Univ of Montana
Carvalho, Maria d., Universidade de Lisboa
Casillas, Angelica, Universidad Nacional Auto de Mexico
Cassidy, Daniel P., Western Michigan Univ
Celia, Michael A., Princeton Univ
Cey, Edwin, Univ of Calgary
Chase, Peter M., Univ of Wisconsin, Madison, Div of Extension
Chavez-Aguirre, Rafael, Universidad Auto de Chihuahua
Cheng, Songlin, Wright State Univ
Cheng, Tao, Memorial Univ of Newfoundland
Cherkauer, Douglas S., Univ of Wisconsin, Milwaukee
Cherry, John A., Univ of Waterloo
Cherubini, Claudia, Institut Polytechnique LaSalle Beauvais (ex-IGAL)
Chesnaux, Romain, Universite du Quebec a Chicoutimi
Chowdhury, Shafiul H., SUNY, New Paltz
Christenson, Scott, New Mexico Tech
Clarey, Timothy L., Delta Coll
Clark, Ian D., Univ of Ottawa
Conant, Brewster, Univ of Waterloo
Cook, Edward R., Columbia Univ
Cook, Peter, Flinders Univ
Costa, Maria R., Universidade de Trás-os-Montes e Alto Douro
Crewdson, Robert, California State Univ, Bakersfield
Croxen III, Fred W., Arizona Western Coll
Cullen, Cecilia, Univ of Illinois
Davidson, Gregg R., Univ of Mississippi
Davis, Arden D., South Dakota School of Mines & Tech
Davis, J. Matthew, Univ of New Hampshire
de Stefano, Lucia, Univ Complutense de Madrid
Decker, Dave, Univ of Nevada, Reno
Del Arenal-Capetillo, Rodolfo, Universidad Nacional Auto de Mexico
Demlie, Molla, Univ of KwaZulu-Natal
Denizman, Can, Valdosta State Univ
Devlin, J. F., Univ of Kansas
Dhar, Ratan K., York Coll (CUNY)
Dimova, Natasha T., Univ of Alabama
Dintaman, Christopher, Indiana Univ, Bloomington
Doherty, Amanda, New Mexico Tech
Domber, Steven E., New Jersey Geological and Water Survey
Doran, Peter, Louisiana State Univ
Doss, Paul K., Univ of Southern Indiana
Douglas, Ellen, Univ of Massachusetts, Boston
Druhan, Jennifer, Univ of Illinois, Urbana-Champaign
Duaime, Terence E., Montana Tech of The Univ of Montana

Dunkle, Kallina M., Austin Peay State Univ
Durham, Lisa A., Argonne Nat Lab
Durham, Lisa, Argonne Nat Lab
Earman, Sam, Millersville Univ
Ekmekci, Mehmet, Hacettepe Univ
El-Kadi, Aly I., Univ of Hawai'i, Manoa
Engle, Mark, Univ of Texas, El Paso
Erickson, Melinda L., Univ of Minnesota, Twin Cities
Esling, Steven P., Southern Illinois Univ Carbondale
Evans, Sarah G., Appalachian State Univ
Fabbri, Paolo, Università degli Studi di Padova
Fairley, Jerry P., Univ of Idaho
Famiglietti, James, Univ of California, Irvine
Fandel, Chloé, Carleton Coll
Farmer, George T., Los Alamos Nat Lab
Feinstein, Daniel T., Univ of Wisconsin, Milwaukee
Ferre, Paul A., Univ of Arizona
Fichera, Marissa, New Mexico Tech
Fisher, Andrew T., Univ of California, Santa Cruz
Fitzpatrick, Stephan D., Georgia State Univ, Perimeter Coll, Clarkston Campus
Flowers, George C., Tulane Univ
Fogg, Graham E., Univ of California, Davis
Foglia, Laura, Univ of California, Davis
Fountain, John C., North Carolina State Univ
French, Helen K., Norwegian Univ of Life Sciences (NMBU)
Frisbee, Marty, Purdue Univ
Fryar, Alan E., Univ of Kentucky
Furlong, Emelia, Maryland Department of Nat Resources
Gakka, Udaya Laxm, Osmania Univ
Gardner, Payton, Univ of Montana
Gargini, Alessandro, Alma Mater Studiorum Università di Bologna
Garing, Charlotte , Univ of Georgia
Ge, Shemin, Univ of Colorado
Geary, Phil, Univ of Newcastle
Geidel, Gwendelyn, Univ of South Carolina
Gemperline, Johanna M., Maryland Department of Nat Resources
Gerla, Philip J., Univ of North Dakota
Gilmore, Troy, Univ of Nebraska, Lincoln
Gilmore, Troy E., Univ of Nebraska, Lincoln
Goeke, James W., Univ of Nebraska, Lincoln
Gordon, Ryan, Dept of Agriculture, Conservation, and Forestry
Gotkowitz, Madeline B., Univ of Wisconsin, Extension
Gouzie, Douglas R., Missouri State Univ
Graham, Grace E., Univ of Wisconsin, Madison, Div of Extension
Grasby, Stephen E., Univ of Calgary
Grismer, Mark E., Univ of California, Davis
Grundl, Timothy J., Univ of Wisconsin, Milwaukee
Gulley, Jason, Univ of South Florida
Guo, Bo, Univ of Arizona
Gurdak, Jason, San Francisco State Univ
Gurwin, Jacek, Univ of Wroclaw
Haacker, Erin, Univ of Nebraska, Lincoln
Habana, Nathan C., Univ of Guam
Hadley, Daniel, Univ of Illinois
Haggerty, Roy D., Oregon State Univ
Haile, Estifanos, Eastern Kentucky Univ
Halihan, Todd, Oklahoma State Univ
Hallum, Douglas R., Univ of Nebraska, Lincoln
Hampton, Duane R., Western Michigan Univ
Hamutoko, Josephine, Univ of Namibia
Harrington, Glenn, Flinders Univ
Harris, Brett D., Curtin Univ
Hart, David J., Univ of Wisconsin, Madison, Div of Extension
Hart, David J., Univ of Wisconsin, Extension
Harter, Thomas, Univ of California, Davis
Harter, Thomas L., Univ of California, Davis
Hatch, Christine, Univ of Massachusetts, Amherst
Hayashi, Masaki, Univ of Calgary
Hays, Phillip D., Univ of Arkansas, Fayetteville
Heiss, James, Univ of Massachusetts, Lowell
Helmke, Martin F., West Chester Univ
Hendrickx, Jan M., New Mexico Tech
Hendry, Jim, Univ of Saskatchewan
Henry, Eric J., Univ of North Carolina, Wilmington
Henry, Tiernan, Nat Univ of Ireland Galway
Herman, Ellen K., Bucknell Univ
Hermans, Thomas, Ghent Univ
Hershey, Ronald L., Univ of Nevada, Reno
Hibbs, Barry, California State Univ, Los Angeles
Hiscock, Kevin, Univ of East Anglia
Hoff, Jean L., Saint Cloud State Univ
Hoover, Karin A., California State Univ, Chico
Horner, Tim C., Sacramento State Univ
Howard, Kenneth W., Univ of Toronto
Howe III, Thomas R., Western Michigan Univ
Hu, Bill X., Florida State Univ
Hudak, Paul F., Univ of North Texas
Huizar-Alvarez, Rafael, Universidad Nacional Auto de Mexico
Hurlow, Hugh A., Utah Geological Survey
Hurlow, Hugh, Utah Geological Survey
Huysmans, Marijke, Katholieke Universiteit Leuven
Iles, Derric L., South Dakota Dept of Env and Nat Resources
Iordache, Vlad, Univ of Illinois
Iqbal, Mohammad Z., Univ of Northern Iowa
Iribar, Vicente , Univ of the Basque Country UPV/EHU
Irvine, Dylan, Flinders Univ
Jannik, Nancy O., Winona State Univ
Jarvis, W. T., Oregon State Univ
Jenson, John W., Univ of Guam
Johnson, Thomas M., Univ of Illinois, Urbana-Champaign
Join, Jean-Lambert, Universite de la Reunion
Jones, Allan, Univ of Illinois
Juster, Thomas C., Univ of South Florida
Kaden, Scott, Missouri Dept of Nat Resources
Keen, Kerry L., Univ of Wisconsin, River Falls
Keller, C. Kent, Washington State Univ
Kelly, Bryce, Univ of New South Wales
Kelly, Walton R., Univ of Illinois
Kent, Douglas, Oklahoma State Univ
Khalequzzaman, Md., Lock Haven Univ
Kim, Yongsang (Barry), Univ of Guam
Kirk, Scott, Ohio Dept of Nat Resources
Knappett, Peter, Texas A&M Univ
Knoll, Martin A., Sewanee: Univ of the South
Koonce, Jeremy, Univ of Nevada, Las Vegas
Krasowski, Michael, Univ of Illinois
Kreamer, David, Univ of Nevada, Reno
Kurttas, Turker, Hacettepe Univ
La Fave, John I., Montana Tech of The Univ of Montana
Lachhab, Ahmed, Susquehanna Univ
Lachmar, Thomas E., Utah State Univ
LaFreniere, Lorraine, Argonne Nat Lab
Lahiri, Chayan, Adams State Univ
Lambert, Carolyn D., Univ of Northern Colorado
Larocque, Marie, Universite du Quebec a Montreal
Laton, W. R., California State Univ, Fullerton
Lee, David R., Univ of Waterloo
Lee, Donald W., Oak Ridge Nat Lab
Lee, Eung Seok, Ohio Univ
Lee, Ming-Kuo, Auburn Univ
Lee, Soon Jae, Korea Univ
Lefebvre, Rene, Universite du Quebec
Lemke, Lawrence D., Central Michigan Univ
Lenczewski, Melissa E., Northern Illinois Univ
Leonhart, Leo S., Univ of Arizona
Leslie, Deborah, Univ of Memphis
Levy, Jonathan, Miami Univ
Li, Liangping, South Dakota School of Mines & Tech
Locke, Daniel B., Dept of Agriculture, Conservation, and Forestry
Loh, Yvonne A., Univ of Ghana
Loheide, Steven P., Univ of Wisconsin, Madison
Lopez-Ferreira, Cesar A., Universidad Auto de Baja California Sur
Lord, Mark L., Western Carolina Univ
Love, Andew, Flinders Univ
Lowe, Mike, Utah Geological Survey
Lowry, Christopher S., SUNY, Buffalo
Luczaj, John A., Univ of Wisconsin, Green Bay
Luhmann, Andrew J., Wheaton Coll
Luinstra, Brian, Curtin Univ
Lutz, Alexandra, Univ of Nevada, Reno
Lyons, William B., Ohio State Univ
Mabee, Stephen B., Massachusetts Geological Survey

Mace, Robert, Texas State Univ
Macpherson, Gwendolyn L., Univ of Kansas
Maharjan, Madan, Univ of North Carolina, Pembroke
Makkawi, Mohammad, King Fahd Univ of Petroleum and Minerals
Mamer, Ethan, New Mexico Tech
Manda, Alex K., East Carolina Univ
Mannix, Devin, Univ of Illinois
Marino, M. A., Univ of California, Davis
Martel, Richard, Universite du Quebec
Martínez, Pedro, Univ Complutense de Madrid
Mastropaolo, Carl, Drexel Univ
Matyjasik, Marek, Weber State Univ
May, James, Mississippi State Univ
Mayer, James R., Univ of West Georgia
Mayo, Alan L., Brigham Young Univ
McKenzie, Jeffrey M., McGill Univ
McKinney, Mac, Geological Survey of Alabama
Mendoza, Carl A., Univ of Alberta
Mendoza, Carl, Univ of Alberta
Metesh, John J., Montana Tech of The Univ of Montana
Meyer, Jessica R., Univ of Iowa
Michael, Holly A., Univ of Delaware
Michel, Fred A., Carleton Univ
Milewski, Adam, Univ of Georgia
Miller, Marvin R., Montana Tech of The Univ of Montana
Mitchell, Robert J., Western Washington Univ
Moline, Gerilynn R., Oak Ridge Nat Lab
Montero, Esperanza, Univ Complutense de Madrid
Montgomery, William W., New Jersey City Univ
Moortgat, Joachim, Ohio State Univ
Moran, Jean E., California State Univ, East Bay
Morton, Cristopher, New Mexico Tech
Moss, Neil E., Geological Survey of Alabama
Moysey, Stephen M., Clemson Univ
Muldoon, Maureen A., Univ of Wisconsin, Oshkosh
Murdoch, Lawrence C., Clemson Univ
Murgulet, Dorina, Texas A&M Univ, Corpus Christi
Murray, Kent S., Univ of Michigan, Dearborn
Nadiri, Ata Allah, Univ of Tabriz
Narasimhan, T. N., Univ of California, Berkeley
Naylor, Shawn C., Indiana Univ, Bloomington
Naymik, Thomas G., Ohio State Univ
Nedunuri, Krishna K., Central State Univ
Nelson, Craig, Ohio Dept of Nat Resources
Neuman, Shlomo P., Univ of Arizona
Newcomer, Darrell R., Pacific Northwest Nat Lab
Newton, Brad T., New Mexico Tech
Nguyen, Lam V., Hanoi Univ of Mining & Geology
Nichol, Craig F., Univ of British Columbia, Okanagan Campus
Nicot, Jean-Philippe, Univ of Texas at Austin, Jackson Schl of Geosciences
Nkotagu, Hudson H., Univ of Dar es Salaam
Noonan, Mathew T., South Dakota Dept of Env and Nat Resources
O'Brien, Arnold L., Univ of Massachusetts, Lowell
O'Reilly, Andrew M., Univ of Mississippi
Oberdorfer, June A., San Jose State Univ
Ogard, Allen E., Los Alamos Nat Lab
Olabode, Franklin O., Federal Univ of Tech, Akure
Olafsen-Lackey, Susan, Univ of Nebraska, Lincoln
Oliveira, Alcino S., Universidade de Trás-os-Montes e Alto Douro
Olyphant, Greg A., Indiana Univ, Bloomington
Ophori, Duke U., Montclair State Univ
Opper, Carl, Saint Petersburg Coll, Clearwater
Ortega, Adrian, Universidad Nacional Auto de Mexico
Osborn, Stephen G., California State Polytechnic Univ, Pomona
Owoseni, Joshua O., Federal Univ of Tech, Akure
Ozsvath, David L., Univ of Wisconsin, Stevens Point
Pan, Feifei, Univ of North Texas
Paniconi, Claudio, Universite du Quebec
Paradis, Charles, Univ of Wisconsin, Milwaukee
Parizek, Richard R., Pennsylvania State Univ, Univ Park
Parkin, Gary, Univ of Guelph
Parsen, Mike J., Univ of Wisconsin, Madison, Div of Extension
Patton, Thomas W., Montana Tech of The Univ of Montana
Paul, Jonathan, Royal Holloway Univ of London
Pederson, Darryll T., Univ of Nebraska, Lincoln
Peters, Catherine A., Princeton Univ
Peterson, Eric W., Illinois State Univ
Peterson, Holly, Guilford Coll
Pfannkuch, Hans O., Univ of Minnesota, Twin Cities
Phillips, Fred M., New Mexico Tech
Pleasants, Mark S., Ohio Dept of Nat Resources
Pociask, Geoff, Illinois State Geological Survey
Pohll, Greg, Univ of Nevada, Reno
Post, Vincent , Flinders Univ
Prestegaard, Karen L., Univ of Maryland
Price, Rene, Florida InterNat Univ
Puente-Muniz, Carlos Francisco, Universidad Auto de San Luis Potosi
Puleo, Jack, Univ of Delaware
Quinn, John, Argonne Nat Lab
Raab, James, Ohio Dept of Nat Resources
Rains, Mark C., Univ of South Florida
Rauch, Henry W., West Virginia Univ
Reeve, Andrew S., Univ of Maine
Reeves, Donald Matthew M., Western Michigan Univ
Reichard, James S., Georgia Southern Univ
Reiten, Jon C., Montana Tech of The Univ of Montana
Remenda, Victoria H., Queen's Univ
Remson, Irwin, Stanford Univ
Riemersma, Peter E., Grand Valley State Univ
Rios-Sanchez, Miriam, Bemidji State Univ
Ritzi, Jr., Robert W., Wright State Univ
Robbins, Gary A., Univ of Connecticut
Robertson, Wendy, Central Michigan Univ
Robertson, Will, Univ of Waterloo
Roman, Eric W., New Jersey Geological and Water Survey
Romanowicz, Edwin A., SUNY, Plattsburgh
Ronayne, Michael J., Colorado State Univ
Rose, Seth E., Georgia State Univ
Rostron, Benjamin J., Univ of Alberta
Rotz, Rachel, Florida Gulf Coast Univ
Rouleau, Alain, Universite du Quebec a Chicoutimi
Royo-Ochoa, Miguel, Universidad Auto de Chihuahua
Russo, Tess A., Pennsylvania State Univ, Univ Park
Russoniello, Christopher, West Virginia Univ
Ryan, Cathy, Univ of Calgary
Saffer, Demian M., Pennsylvania State Univ, Univ Park
Saint, Prem K., California State Univ, Fullerton
Salleolini, Massimo, Univ of Siena
Salvage, Karen M., Binghamton Univ
Samuelson, Alan C., Ball State Univ
Sanders, Laura L., Northeastern Illinois Univ
Sanford, William E., Colorado State Univ
Sasowsky, Ira D., Univ of Akron
Sawyer, Audrey , Ohio State Univ
Scanlon, Bridget R., Univ of Texas at Austin, Jackson Schl of Geosciences
Scheidt, Brian, Mineral Area Coll
Schincariol, Robert A., Western Univ
Schlosser, Peter, Columbia Univ
Schmitz, Darrel W., Mississippi State Univ
Schreiber, Madeline E., Virginia Polytechnic Inst & State Univ
Schueth, Christoph, Technische Universitaet Darmstadt
Schulze-Makuch, Dirk, Washington State Univ
Schwartz, Frank W., Ohio State Univ
Sebol, Lesley, Colorado Geological Survey
Sedivy, Robert A., Argonne Nat Lab
Sedivy, Robert, Argonne Nat Lab
Sendlein, Lyle V. A., Univ of Kentucky
Seyoum, Wondwosen M., Illinois State Univ
Shanafield, Margaret, Flinders Univ
Shang, Congxiao, Univ of East Anglia
Shaw, Glenn D., Montana Tech of the Univ of Montana
Sheldon, Amy L., SUNY, Geneseo
Shevenell, Lisa, Univ of Nevada, Reno
Shim-Chim, Richard, New Jersey Geological and Water Survey
Shimada, Jun, Kumamoto Univ
Sibray, Steven S., Univ of Nebraska, Lincoln
Siegel, Donald I., Syracuse Univ
Simmons, Craig, Flinders Univ
Simpkins, William W., Iowa State Univ of Science & Tech
Simsek, Sakir, Hacettepe Univ
Singha, Kamini, Colorado School of Mines
Skalbeck, John D., Univ of Wisconsin, Parkside

Smith, J. Leslie, Univ of British Columbia
Smith, James E., McMaster Univ
Smith, Karen, Argonne Nat Lab
Smith, Ryan , Missouri Univ of Science and Tech
Soll, Wendy E., Los Alamos Nat Lab
Solomon, Douglas K., Univ of Utah
Soltanian, Reza, Univ of Cincinnati
Spinelli, Glenn, New Mexico Tech
Springer, Abraham E., Northern Arizona Univ
Stafford, Kevin W., Stephen F. Austin State Univ
Staley, Andrew W., Maryland Department of Nat Resources
Stotler, Randy, Univ of Kansas
Strack, Otto D., Univ of Minnesota, Twin Cities
Sturgis, Laila, New Mexico Tech
Sudicky, Edward A., Univ of Waterloo
Suen, C. J., Fresno State Univ
Sukop, Michael C., Florida InterNat Univ
Sun, Hongbing, Rider Univ
Suter, Simeon, DCNR- Pennsylvania Bureau of Geological Survey
Sutherland, Mary K., Montana Tech of The Univ of Montana
Swanson, Susan K., Beloit Coll
Szilagyi, Jozsef, Univ of Nebraska, Lincoln
Tabidian, M. Ali, California State Univ, Northridge
Tauxe, John D., Oak Ridge Nat Lab
Taylor, Charles J., Univ of Kentucky
Tellam, John H., Univ of Birmingham
Tezcan, Levent, Hacettepe Univ
Therrien, Rene, Universite Laval
Thomason, Jason F., Illinois State Geological Survey
Tick, Geoffrey, Univ of Alabama
Tidwell, Vincent C., New Mexico Tech
Tijani, Moshood N., Univ of Ibadan
Tilton, Eric E., Univ of Colorado
Timmons, Stacy, New Mexico Tech
Tompson, Andrew F., Lawrence Livermore Nat Lab
Toran, Laura, Temple Univ
Tóth, József, Univ of Alberta
Totsche, Kai U., Friedrich-Schiller-Univ Jena
Tucker, Carla M., Lamar Univ
Tyler, Scott W., Univ of Nevada, Reno
Unger, Andre, Univ of Waterloo
Valerio, Mitch, Ohio Dept of Nat Resources
Van der Kamp, Garth, Univ of Waterloo
van der Velde, Ype, Vrije Universiteit Amsterdam
Veeger, Anne, Univ of Rhode Island
Veeger, Anne I., Univ of Rhode Island
Vengosh, Avner, Duke Univ
Vesper, Dorothy, West Virginia Univ
Wallace, Janae, Utah Geological Survey
Walraevens, Kristine, Ghent Univ
Walter, Julien, Universite du Quebec a Chicoutimi
Walton, Nick, Univ of Portsmouth
Wang, Dong, Fresno State Univ
Waren, Kirk B., Montana Tech of The Univ of Montana
Wassenaar, Len, Univ of Saskatchewan
Watson, David B., Oak Ridge Nat Lab
Weissmann, Gary S., Univ of New Mexico
Welby, Charles W., North Carolina State Univ
Werner, Adrian , Flinders Univ
Westrop, Jeffrey, Univ of Nebraska, Lincoln
Wheatcraft, Steve, Univ of Nevada, Reno
Wheaton, John R., Montana Tech of The Univ of Montana
Whiting, Duane L., Univ of Utah
Wicks, Carol M., Louisiana State Univ
Widdowson, Mark A., Virginia Polytechnic Inst & State Univ
Wilcox, Jeffrey D., Univ of North Carolina, Asheville
Wilson, Alicia M., Univ of South Carolina
Wilson, John L., New Mexico Tech
Wilson, Steven D., Univ of Illinois
Wolaver, Brad, Univ of Texas, Austin
Wolfsberg, Andrew V., Los Alamos Nat Lab
Woods, Juliette, Univ of South Australia
Yan, Eugene, Argonne Nat Lab
Yang, Changbing, Univ of Texas, Austin
Yang, Jianwen, Univ of Windsor
Yelderman, Jr., Joe C., Baylor Univ
YEO, MYEONG-HO (CHRIS), Univ of Guam
Yidana, Sandow M., Univ of Ghana
Young, Aaron R., Univ of Nebraska, Lincoln
Yüce, Galip, Hacettepe Univ
Zarnetske, Jay, Michigan State Univ
Zhan, Hongbin, Texas A&M Univ
Zhang, Pengfei, City Coll (CUNY)
Zhang, You-Kuan, Univ of Iowa
Zhu, Junfeng, Univ of Kentucky
Ziegler, Brady A., Trinity Univ
Zimmer, Margaret, Univ of California, Santa Cruz
Zlotnik, Vitaly A., Univ of Nebraska, Lincoln
Zouhri, Lahcen, Institut Polytechnique LaSalle Beauvais (ex-IGAL)
Zreda, Marek G., Univ of Arizona
Özyurt, Nur N., Hacettepe Univ

Quantitative Hydrology

Benson, David A., Colorado School of Mines
Berghuijs , Wouter, Vrije Universiteit Amsterdam
Birdsell, Kay H., Los Alamos Nat Lab
Bywater-Reyes, Sharon, Univ of Northern Colorado
Carroll, Rosemary, Univ of Nevada, Reno
Chebana, Fateh, Universite du Quebec
Condon, Laura, Univ of Arizona
Covington, Matthew, Univ of Arkansas, Fayetteville
Danko, George, Univ of Nevada, Reno
Dingman, S. Lawrence, Univ of New Hampshire
Donovan, Joseph J., West Virginia Univ
Dymond, Randel, Virginia Polytechnic Inst & State Univ
Evans, David G., Sacramento State Univ
Falta, Ronald W., Clemson Univ
Filippini, Maria, Alma Mater Studiorum Università di Bologna
Florea, Lee J., Indiana Univ, Bloomington
Flores, Alejandro, Boise State Univ
French, Mark A., New Jersey Geological and Water Survey
Frind, Emil O., Univ of Waterloo
Gillham, Robert W., Univ of Waterloo
Gupta, Hoshin V., Univ of Arizona
Harpold, Adrian, Univ of Nevada, Reno
Hermance, John F., Brown Univ
Hoffman, Jeffrey L., New Jersey Geological and Water Survey
Holt, Robert M., Univ of Mississippi
Hubbart, Jason A., West Virginia Univ
Hughes, Denis, Rhodes Univ
Kandiah, Ramanitharan, Central State Univ
Kiem, Anthony, Univ of Newcastle
Letsinger, Sally L., Indiana Univ, Bloomington
Li, Dan, Boston Univ
Looney, Brian, Clemson Univ
Maddock, III, Thomas, Univ of Arizona
Marston, Landon, Virginia Polytechnic Inst & State Univ
Martin-Hayden, James, Univ of Toledo
Meko, David M., Univ of Arizona
Michaud, Jene D., Univ of Hawai'i, Hilo
Nicholl, Michael J., Univ of Nevada, Las Vegas
Pacheco, Fernando A., Universidade de Trás-os-Montes e Alto Douro
Pangle, Luke, Georgia State Univ
Parashar, Rishi, Univ of Nevada, Reno
Park, Young-Jin, Univ of Waterloo
Perkins, Robert B., Portland State Univ
Pollyea, Ryan M., Virginia Polytechnic Inst & State Univ
Puente, Carlos E., Univ of California, Davis
Quinn, John J., Argonne Nat Lab
Ravi, Sujith, Temple Univ
Reinhold, Ann Marie, Montana State Univ
Restrepo, Jorge I., Florida Atlantic Univ
Rudolph, David L., Univ of Waterloo
Saksena, Suddarth, Virginia Polytechnic Inst & State Univ
Salvucci, Guido D., Boston Univ
Schumer, Rina, Univ of Nevada, Reno
Sharp, Jr., John M., Univ of Texas, Austin
Spellman, Patricia, Univ of South Florida
Springer, Everett P., Los Alamos Nat Lab
Stockton, Charles W., Univ of Arizona
Szilagyi, Jozsef, Univ of Nebraska, Lincoln
Telyakovskiy, Aleksey S., Univ of Nevada, Reno

Thorbjarnarson, Kathryn W., San Diego State Univ
Travis, Bryan J., Los Alamos Nat Lab
Verdon-Kidd, Danielle, Univ of Newcastle
Vitale, Sarah A., Univ of Wisconsin, Eau Claire
Wallis, Ilka, Flinders Univ
Wei, Xiaohua (Adam), Univ of British Columbia, Okanagan Campus
Whitman, Brian E., Wilkes Univ
Yeh, Tian-Chyi J., Univ of Arizona
Zhang, Yong, Univ of Alabama

Surface Waters

Acharya, Kumud, Univ of Nevada, Reno
Aitkenhead-Peterson, Jacqueline A., Texas A&M Univ
Allen, George H., Virginia Polytechnic Inst & State Univ
Aryal, Niroj, North Carolina Agricultural & Tech State Univ
Bacon, Robert, Missouri Dept of Nat Resources
Bauer, Carl J., Univ of Arizona
Bearden, Rebecca A., Geological Survey of Alabama
Benavides, Jude, Univ of Texas, Rio Grande Valley
Bevelhimer, Mark, Oak Ridge Nat Lab
Beyer, Patricia J., Bloomsburg Univ
Blanka-Végi, Viktória, Univesity of Szeged
Blanken, Peter D., Univ of Colorado
Blersch, Stacey S., Columbus State Univ
Blumentritt, Dylan, Winona State Univ
Cadol, Daniel, New Mexico Tech
Chandra, Sudeep, Univ of Nevada, Reno
Christian, Alan D., Univ of Massachusetts, Boston
D'Alpaos, Andrea, Università degli Studi di Padova
Demchak, Jennifer, Mansfield Univ
Dennett, Keith E., Univ of Nevada, Reno
Du Charme, Charles, Missouri Dept of Nat Resources
Duan, Shuiwang, Univ of Maryland
Durand, Michael T., Ohio State Univ
Franek, Benjamin, Bloomsburg Univ
Franz, Kristie, Iowa State Univ of Science & Tech
Goodrich, David C., Univ of Arizona
Grable, Judy, Valdosta State Univ
Grant, Stanley, Virginia Polytechnic Inst & State Univ
Gulliver, John S., Univ of Minnesota, Twin Cities
Hamlet, Alan, Univ of Notre Dame
Harris, Randa R., Univ of West Georgia
Hasan, Khaled W., Austin Comm Coll District
Helz, George, Univ of Maryland
Hester, Erich, Virginia Polytechnic Inst & State Univ
Julian, Jason, Texas State Univ
Kaiser, Kendra, Boise State Univ
Kindelman, Julie, Univ of Wisconsin, Parkside
Kisila, Ben O., Univ of Mary Washington
Kunza, Lisa, South Dakota School of Mines & Tech
Lancaster, Stephen, Oregon State Univ
Lansey, Kevin E., Univ of Arizona
Ledford, Sarah, Georgia State Univ
Lund, Jay R., Univ of California, Davis
Maneta, Marco, Univ of Montana
Mayfield, Michael W., Appalachian State Univ
McGregor, Stuart W., Geological Survey of Alabama
McIsaac, Gregory F., Univ of Illinois, Urbana-Champaign
Meitzen, Kimberly, Texas State Univ
Neukrug, Howard, Univ of Pennsylvania
Noe, Garry, Virginia Wesleyan Coll
Pociask, Geoffrey , Univ of Illinois
Pradhanang, Soni M., Univ of Rhode Island
Rancan, Helen L., New Jersey Geological and Water Survey
Reichelt-Brushett, Amanda, Southern Cross Univ
Richard, Gigi A., Colorado Mesa Univ
Ridenour, Gregory D., Austin Peay State Univ
Riise, Gunnhild, Norwegian Univ of Life Sciences (NMBU)
Royer, Todd, Indiana Univ, Bloomington
Smith, Brianne, Brooklyn Coll (CUNY)
Smith, Erik, Univ of South Carolina
Smith, Laurence C., Univ of California, Los Angeles
Smith, Sean, Univ of Maine
Splinter, Dale, Univ of Wisconsin, Whitewater
Stieglitz, Marc, Georgia Inst of Tech
Strom, Kyle, Virginia Polytechnic Inst & State Univ
Swanson, Sherman, Univ of Nevada, Reno
Thomas, Rohrlack, Norwegian Univ of Life Sciences (NMBU)
Troch, Peter A., Univ of Arizona
Valdes, Juan B., Univ of Arizona
Wampler, Peter J., Grand Valley State Univ
Ward, Adam, Indiana Univ, Bloomington
Wilderman, Candie, Dickinson Coll
Zollweg, James A., SUNY, The Coll at Brockport

Geohydrology

Attinger, Sabine, Friedrich-Schiller-Univ Jena
Baedke, Steven J., James Madison Univ
Beddows, Patricia A., Northwestern Univ
Burkart, Michael R., Iowa State Univ of Science & Tech
Castro, Maria Clara, Univ of Michigan
Chavez-Rodriguez, Adolfo, Universidad Auto de Chihuahua
Constantine, Jose, Cardiff Univ
Cooper, Clay A., Univ of Nevada, Reno
Crews, Jeff, Missouri Dept of Nat Resources
Daanen, Ronald P., Alaska Div of Geological & Geophysical Surveys
De Luca, Domenico Antonio, Università di Torino
Dowd, John F., Univ of Georgia
Drake, Lon D., Univ of Iowa
Duley, James W., Missouri Dept of Nat Resources
Fogg, Graham E., Univ of California, Davis
Gamage, Kusali R., Austin Comm Coll District
Gartner, Janette, Bentley Univ
Gierke, John S., Michigan Technological Univ
Haile, Estifanos, Univ of Tennessee, Knoxville
Hong, Sung-ho, Murray State Univ
Hyndman, David W., Michigan State Univ
Jones, Stephen C., Geological Survey of Alabama
Katzman, Danny, Los Alamos Nat Lab
Khanbilvardi, Reza M., Graduate School of the City Univ of New York
Knight, Rosemary J., Stanford Univ
Koch, Magaly, Boston Univ
Land, Lewis, New Mexico Tech
Lauer, Rachel, Univ of Calgary
Liu, Gaisheng, Univ of Kansas
Lupankwa, Mlindelwa , Tshwane Univ of Tech
Marszalek, Henryk, Univ of Wroclaw
Mathias, Simon, Durham Univ
McIntosh, Jennifer, Univ of Arizona
McKay, Larry D., Univ of Tennessee, Knoxville
McKenna, Thomas E., Univ of Delaware
Ng, Crystal, Univ of Minnesota, Twin Cities
Noyes, Joanne M., South Dakota Dept of Env and Nat Resources
Person, Mark A., New Mexico Tech
Raymond, Jasmin, Universite du Quebec
Rayne, Todd W., Hamilton Coll
Reboulet, Edward, Univ of Kansas
Rovey, Charles W., Missouri State Univ
Sargent, Kenneth A., Furman Univ
Simonis, Jean, Univ of Zululand
Sushama, Laxmi, Universite du Quebec a Montreal
Tamborski, Joseph, Old Dominion Univ
Tobin, Benjamin, Univ of Kentucky
Torres, Raymond, Univ of South Carolina
Wang, Lixin, Indiana Univ - Purdue Univ Indianapolis
White, Mark D., Pacific Northwest Nat Lab
Zhang, Ye, Univ of Wyoming

Materials Transport

Pham van Bang, Damien, Universite du Quebec
Troy, Marleen, Wilkes Univ
You, Kehua, Univ of Texas, Austin

SOIL SCIENCE

Soil Physics/Hydrology

Amoozegar, Aziz, North Carolina State Univ
Anderson, Stephen H., Univ of Missouri
Arnone, John J., Univ of Nevada, Reno
Baker, John M., Univ of Minnesota, Twin Cities
Barta, Károly, University of Szeged
Berli, Marcus, Univ of Nevada, Reno
Black, T. Andrew, Univ of British Columbia

Bland, William L., Univ of Wisconsin, Madison
Boast, Charles W., Univ of Illinois, Urbana-Champaign
Brown, Sandra, Univ of British Columbia
Casey, Francis, North Dakota State Univ
Cassel, Donald K., North Carolina State Univ
Cushman, John, Purdue Univ
Ellsworth, Timothy R., Univ of Illinois, Urbana-Champaign
Fermanich, Kevin J., Univ of Wisconsin, Green Bay
Feyereisen, Gary, Univ of Minnesota, Twin Cities
Flury, Markus, Washington State Univ
Fritton, Daniel D., Pennsylvania State Univ, Univ Park
Galbraith, John M., Virginia Polytechnic Inst & State Univ
Glade, Rachel, Univ of Rochester
Grattan, Stephen R., Univ of California, Davis
Groenevelt, Pieter H., Univ of Guelph
Gupta, Satish C., Univ of Minnesota, Twin Cities
Hook, James E., Univ of Georgia
Horton, Robert, Iowa State Univ of Science & Tech
Hutson, John , Flinders Univ
Jawitz, James W., Univ of Florida
Johnson, Mark S., Univ of British Columbia
Jones, Tim L., New Mexico State Univ, Las Cruces
Kakembo, Vincent, Nelson Mandela Metropolitan Univ
Kanemasu, Edward T., Univ of Georgia
Kang, James, Univ of Texas, Rio Grande Valley
Kirkham, Mary Beth, Kansas State Univ
Kluitenberg, Gerard J., Kansas State Univ
Knappenberger, Thorsten J., Auburn Univ
Kung, King-Jau S., Univ of Wisconsin, Madison
Lal, Rattan, Ohio State Univ
Lourenco, Sergio, Cardiff Univ
Lowery, Birl, Univ of Wisconsin, Madison
McCoy, Edward L., Ohio State Univ
McDonald, Eric, Univ of Nevada, Reno
McInnes, Kevin J., Texas A&M Univ
Molz, Fred, Clemson Univ
Momm, Henrique G., Middle Tennessee State Univ
Moncrief, John F., Univ of Minnesota, Twin Cities
Morgan, Cristine L., Texas A&M Univ
Mulla, David J., Univ of Minnesota, Twin Cities
Neely, Haly L., Texas A&M Univ
Nemes, Attila, Norwegian Univ of Life Sciences (NMBU)
Nkedi-Kizza, Peter, Univ of Florida
Nocco, Mallika, Univ of California, Davis
Norman, John M., Univ of Wisconsin, Madison
Papendick, Robert I., Washington State Univ
Perfect, Edmund, Univ of Tennessee, Knoxville
Persaud, Naraine, Virginia Polytechnic Inst & State Univ
Radcliffe, David E., Univ of Georgia
Ressler, Daniel E., Susquehanna Univ
Rogowski, Andrew S., Pennsylvania State Univ, Univ Park
Sammis, Theodore W., New Mexico State Univ, Las Cruces
Scott, David F., Univ of British Columbia, Okanagan Campus
Selim, Hussein M., Louisiana State Univ
Sharma, Vasudha, Univ of Minnesota, Twin Cities
Simmons, F. William, Univ of Illinois, Urbana-Champaign
Spokas, Kurt, Univ of Minnesota, Twin Cities
Stone, Loyd, Kansas State Univ
Thomas, John, Univ of Florida
Venterea, Rodney T., Univ of Minnesota, Twin Cities
Wang, Zhi (Luke), Fresno State Univ
Warrick, Arthur W., Univ of Arizona
Welch, Steve M., Kansas State Univ
Wierenga, Peter J., Univ of Arizona
Yeh, Jim, Univ of Arizona
Young, Michael H., Univ of Texas, Austin
Youngs, Edward, Open Univ

Soil Chemistry/Mineralogy

Adriano, Domy C., Virginia Polytechnic Inst & State Univ
Adriano, Domy C., Univ of Georgia
Ahmad, Abd Rashid, Univ of Malaya
Alley, Marcus M., Virginia Polytechnic Inst & State Univ
Almås, Åsgeir R., Norwegian Univ of Life Sciences (NMBU)
Appel, Christopher (Chip) S., California Polytechnic State Univ
Artiola, Janick F., Univ of Arizona
Baligar, V. C., Virginia Polytechnic Inst & State Univ
Barak, Phillip W., Univ of Wisconsin, Madison
Basta, Nicholas T., Ohio State Univ
Bertsch, Paul M., Univ of Georgia
Blanchar, Robert W., Univ of Missouri
Bleam, William F., Univ of Wisconsin, Madison
Bloom, Paul R., Univ of Minnesota, Twin Cities
Bomke, Arthur A., Univ of British Columbia
Bostick, Benjamin C., Columbia Univ
Burke, Deborah S., Pacific Northwest Nat Lab
Callanan, Jennifer R., William Paterson Univ
Chirenje, Tait, Stockton Univ
Chorover, Jon, Univ of Arizona
Chorover, Jonathan, Univ of Arizona
Clark, Joanna, Univ of Reading
Collins, Chris, Univ of Reading
Crouse, David A., North Carolina State Univ
Curry, Joan E., Univ of Arizona
Donohue, Stephen J., Virginia Polytechnic Inst & State Univ
Dowdy, Robert H., Univ of Minnesota, Twin Cities
Dudley, Lynn M., Florida State Univ
Eckert, Donald J., Ohio State Univ
Eich-Greatorex, Susanne, Norwegian Univ of Life Sciences (NMBU)
Eick, Matthew J., Virginia Polytechnic Inst & State Univ
Elzinga, Evert J., Rutgers, The State Univ of New Jersey, Newark
Engel, Richard E., Montana State Univ
Evangelou, V. P., Iowa State Univ of Science & Tech
Evanylo, Gregory K., Virginia Polytechnic Inst & State Univ
Fernandez, Fabian G., Univ of Minnesota, Twin Cities
Gamble, Audrey V., Auburn Univ
Gascho, Gary J., Univ of Georgia
Gaston, Lewis A., Louisiana State Univ
Ginder-Vogel, Matt, Univ of Wisconsin, Madison
Goos, Robert J., North Dakota State Univ
Grandy, Stuart, Univ of New Hampshire
Guertal, Elizabeth A., Auburn Univ
Han, Nizhou, Virginia Polytechnic Inst & State Univ
Harris, Jr., Willie G., Univ of Florida
Harsh, James B., Washington State Univ
Hassett, John J., Univ of Illinois, Urbana-Champaign
Havlin, John L., North Carolina State Univ
He, Zhenli, Univ of Florida
Heil, Dean, New Mexico State Univ, Las Cruces
Helmke, Philip A., Univ of Wisconsin, Madison
Hemzacek Laukant, Jean M., Northeastern Illinois Univ
Hesterberg, Dean L., North Carolina State Univ
Hettiarachchi, Ganga, Kansas State Univ
Horwath, William R., Univ of California, Davis
Howe, Julie, Texas A&M Univ
Huluka, Gobena E., Auburn Univ
Inskeep, William P., Montana State Univ
Jaouich, Alfred, Universite du Quebec a Montreal
Jardine, Philip M., Oak Ridge Nat Lab
Johnston, Cliff, Purdue Univ
Jones, Robert L., Univ of Illinois, Urbana-Champaign
Kabengi, Nadine, Georgia State Univ
Kaiser, Daniel E., Univ of Minnesota, Twin Cities
Kis, Boglarka Mercedes, Babes-Bolyai Univ
Kissel, David E., Univ of Georgia
Komarneni, Sridhar, Pennsylvania State Univ, Univ Park
Koskinen, William C., Univ of Minnesota, Twin Cities
Krause, Samantha, Texas State Univ
Krogstad, Tore, Norwegian Univ of Life Sciences (NMBU)
Kuo, Shiou, Washington State Univ
Laird, David A., Iowa State Univ of Science & Tech
Lamb, John A., Univ of Minnesota, Twin Cities
Lee, Young Jae, Korea Univ
Lewis, Katie, Texas A&M Univ
Li, Yuncong, Univ of Florida
Ma, Lena Q., Univ of Florida
Malzer, Gary L., Univ of Minnesota, Twin Cities
Marinos, Richard, SUNY, Buffalo
Marion, Giles, Desert Research Inst
Martinsen, Vegard, Norwegian Univ of Life Sciences (NMBU)
Massey, Michael S., California State Univ, East Bay
McCaslin, Bobby D., New Mexico State Univ, Las Cruces

McLaughlin, Richard A., North Carolina State Univ
Miao, Yuxin, Univ of Minnesota, Twin Cities
Miller, William P., Univ of Georgia
Moore, Duane M., Univ of New Mexico
Mowrer, Jake, Texas A&M Univ
Mullins, Gregory L., Virginia Polytechnic Inst & State Univ
Mulvaney, Richard L., Univ of Illinois, Urbana-Champaign
Mylavarapu, S R., Univ of Florida
Myneni, Satish C B., Princeton Univ
Nair, Vimala D., Univ of Florida
O'Connor, George A., Univ of Florida
Olson, Kenneth R., Univ of Illinois, Urbana-Champaign
Pagliari, Paulo H., Univ of Minnesota, Twin Cities
Parikh, Sanjai, Univ of California, Davis
Peck, Theodore R., Univ of Illinois, Urbana-Champaign
Peryea, Frank J., Washington State Univ
Pierzynski, Gary M., Kansas State Univ
Plank, Owen C., Univ of Georgia
Provin, Tony L., Texas A&M Univ
Quade, Jay, Univ of Arizona
Rehm, George W., Univ of Minnesota, Twin Cities
Reneau, Raymond B., Virginia Polytechnic Inst & State Univ
Robarge, Wayne P., North Carolina State Univ
Rosen, Carl J., Univ of Minnesota, Twin Cities
Russelle, Michael P., Univ of Minnesota, Twin Cities
Schmitt, Michael A., Univ of Minnesota, Twin Cities
Schwab, Paul, Texas A&M Univ
Segars, William P., Univ of Georgia
Seyfferth, Angelia L., Univ of Delaware
Shuman, Larry M., Univ of Georgia
Siebe Grabach, Christina D., Universidad Nacional Auto de Mexico
Sims, Albert L., Univ of Minnesota, Twin Cities
Smyth, Thomas J., North Carolina State Univ
Spalding, Brian P., Oak Ridge Nat Lab
Strawn, Daniel G., Univ of Idaho
Stucki, Joseph W., Univ of Illinois, Urbana-Champaign
Szulczewski, Melanie, Univ of Mary Washington
Thien, Steve J., Kansas State Univ
Thompson, Michael L., Iowa State Univ of Science & Tech
Toor, Gurpal, Univ of Florida
Trumbore, Susan E., Univ of California, Irvine
Werts, Scott P., Winthrop Univ
Wilson, Melissa L., Univ of Minnesota, Twin Cities
Wright, Alan, Univ of Florida
Yildirim, Ismail, Firat Univ
Zelazny, Lucian W., Virginia Polytechnic Inst & State Univ

Pedology/Classification/Morphology

Anderson, James L., Univ of Minnesota, Twin Cities
Baker, James C., Virginia Polytechnic Inst & State Univ
Bell, James C., Univ of Minnesota, Twin Cities
Bettis III, E. A., Univ of Iowa
Bigham, Jerry M., Ohio State Univ
Birkeland, Peter W., Univ of Colorado
Bourgault, Rebecca, Bloomsburg Univ
Burras, Lee, Iowa State Univ of Science & Tech
Buscaroli, Alessandro, Alma Mater Studiorum Università di Bologna
Calhoun, Frank G., Ohio State Univ
Ciolkosz, Edward J., Pennsylvania State Univ, Univ Park
Collins, Mary E., Univ of Florida
Cooper, Terence H., Univ of Minnesota, Twin Cities
Cornelis, Jean-Thomas, Univ of British Columbia
Daniels, Walter L., Virginia Polytechnic Inst & State Univ
Darmody, Robert G., Univ of Illinois, Urbana-Champaign
Daugherty, LeRoy A., New Mexico State Univ, Las Cruces
Dere, Ashlee L., Univ of Nebraska, Omaha
Evans, Barry M., Pennsylvania State Univ, Univ Park
Ewing, Stephanie, Montana State Univ
Farsang, Andrea, Univesity of Szeged
Frederick, Holly, Wilkes Univ
Heck, Richard, Univ of Guelph
Hoover, Michael T., North Carolina State Univ
Hopkins, David G., North Dakota State Univ
Jacobs, Peter, Univ of Wisconsin, Whitewater
Jelinski, Nicolas, Univ of Minnesota, Twin Cities
Johnson-Maynard, Jodi, Univ of Idaho
Kleiss, Harold J., North Carolina State Univ
Kuzila, Mark S., Univ of Nebraska, Lincoln
Lavkulich, Leslie M., Univ of British Columbia
Lindbo, David, North Carolina State Univ
Madison, Frederick W., Univ of Wisconsin, Madison
Mbila, Monday O., Alabama A&M Univ
McDaniel, Paul A., Univ of Idaho
McDonald, Eric, Desert Research Inst
McFadden, Leslie M., Univ of New Mexico
McGahan, Donald, Texas A&M Univ
McSweeney, Kevin, Univ of Wisconsin, Madison
Miles, Randall J., Univ of Missouri
Miller, Gerald A., Iowa State Univ of Science & Tech
Montagne, Cliff, Montana State Univ
Nater, Edward A., Univ of Minnesota, Twin Cities
Nielsen, Gerald A., Montana State Univ
Nordt, Lee C., Baylor Univ
Paz Gonzalez, Antonio, Coruna Univ
Petersen, Gary W., Pennsylvania State Univ, Univ Park
Post, Donald F., Univ of Arizona
Ransom, Michel D., Kansas State Univ
Ryder, Roy, Univ of South Alabama
Sandor, Jonathan A., Iowa State Univ of Science & Tech
Shaw, Joey N., Auburn Univ
Slater, Brian K., Ohio State Univ
Steele, Amber , Missouri Dept of Nat Resources
Taboada Castro, María T., Coruna Univ
Tabor, Neil J., Southern Methodist Univ
Thomas, Pamela J., Virginia Polytechnic Inst & State Univ
Turk, Judith, Univ of Nebraska, Lincoln
Turk, Judith, Univ of Nebraska, Lincoln
Tyler, E. Jerry, Univ of Wisconsin, Madison
Vepraskas, Michael J., North Carolina State Univ
Verburg, Paul, Univ of Nevada, Reno
West, Larry T., Univ of Georgia
Yoo, Kyungsoo, Univ of Minnesota, Twin Cities

Forest Soils/Rangelands/Wetlands

Amthor, Jeff, Oak Ridge Nat Lab
Andrus, Richard E., Binghamton Univ
Balster, Nick J., Univ of Wisconsin, Madison
Bauder, James W., Montana State Univ
Bockheim, James G., Univ of Wisconsin, Madison
Breiter, Jack, Univ of Illinois Chicago
Broome, Stephen W., North Carolina State Univ
Chambers, Jeanne, Univ of Nevada, Reno
Clark, Mark W., Univ of Florida
Comerford, Nicholas B., Univ of Florida
David, Mark B., Univ of Illinois, Urbana-Champaign
Dietze, Michael, Boston Univ
Ferris, Dawn, Ohio State Univ
Fick, Walter H., Kansas State Univ
Franz, Eldon H., Washington State Univ
Gunderson, Lance, Emory Univ
Hallock, Brent G., California Polytechnic State Univ
Hook, Paul B., Montana State Univ
Inglett, Patrick, Univ of Florida
Kleindl, William, Montana State Univ
Krzic, Maja, Univ of British Columbia
Laingen, Christopher R., Eastern Illinois Univ
Laney, Melissa, Indiana Univ, Bloomington
Meretsky, Vicky J., Indiana Univ, Bloomington
O'Neill, Patrick M., Louisiana State Univ
Osborne, Todd, Univ of Florida
Owensby, Clenton E., Kansas State Univ
Peres, Carlos, Univ of East Anglia
Posler, Gerry L., Kansas State Univ
Rains, Kai, Univ of South Florida
Riha, Susan, Cornell Univ
Robinson, Steve, Univ of Reading
Smalley, Glendon W., Sewanee: Univ of the South
Smith, Alistair, Univ of Idaho
Smith, C. Ken, Sewanee: Univ of the South
Smoak, Joseph (Donny), Univ of South Florida
Strand, Line T., Norwegian Univ of Life Sciences (NMBU)
Taskey, Ronald D., California Polytechnic State Univ

Torreano, Scott, Sewanee: Univ of the South
Wegner, John, Emory Univ

Soil Biology/Biochemistry

Allan, Deborah L., Univ of Minnesota, Twin Cities
Armstrong, Felicia P., Youngstown State Univ
Balser, Teri C., Univ of Wisconsin, Madison
Berry, Duane F., Virginia Polytechnic Inst & State Univ
Bezdicek, David F., Washington State Univ
Bird, Jeffrey, Graduate School of the City Univ of New York
Bird, Jeffrey, Queens Coll (CUNY)
Blum, Linda K., Univ of Virginia
Bollag, Jean-Marc, Pennsylvania State Univ, Univ Park
Burlage, Robert S., Oak Ridge Nat Lab
Cates, Anna, Univ of Minnesota, Twin Cities
Chanway, Christopher, Univ of British Columbia
Collins, Chris, Univ of Reading
Crozier, Carl, North Carolina State Univ
Davenport, Joan R., Washington State Univ
Dick, Richard P., Ohio State Univ
Dick, Warren A., Ohio State Univ
Dunfield, Kari, Univ of Guelph
Eldon, Jon, Indiana Univ, Bloomington
Feng, Yucheng, Auburn Univ
Frey, Serita, Univ of New Hampshire
Fultz, Lisa M., Louisiana State Univ
Geisseler, Daniel, Univ of California, Davis
Gentry, Terry, Texas A&M Univ
Gerba, Charles P., Univ of Arizona
Gilliam, James W., North Carolina State Univ
Graham, Jr., James H., Univ of Florida
Graves, Alexandria, North Carolina State Univ
Grayston, Sue, Univ of British Columbia
Groffman, Peter, Brooklyn Coll (CUNY)
Gutknecht, Jessica L., Univ of Minnesota, Twin Cities
Hagedorn, Charles, Virginia Polytechnic Inst & State Univ
Hagedorn, Jake, Univ of North Carolina, Asheville
Harris, Glendon H., Univ of Georgia
Harris, Robin F., Univ of Wisconsin, Madison
Hartel, Peter G., Univ of Georgia
Hickey, William J., Univ of Wisconsin, Madison
Hladik, Christine, Georgia Southern Univ
Hoyt, Greg D., North Carolina State Univ
Ishii, Satoshi, Univ of Minnesota, Twin Cities
Israel, Daniel W., North Carolina State Univ
Jaffe, Peter R., Princeton Univ
Jastrow, Julie D., Argonne Nat Lab
Kennedy, Ann C., Washington State Univ
Knudsen, Guy, Univ of Idaho
Kremer, Robert J., Univ of Missouri
Lindemann, William C., New Mexico State Univ, Las Cruces
Lybrand, Rebecca A., Univ of California, Davis
McNicol, Gavin, Univ of Illinois Chicago
Miller, Raymond M., Argonne Nat Lab
Molina, Jean-Alex E., Univ of Minnesota, Twin Cities
Morra, Matthew J., Univ of Idaho
Ogram, Andrew V., Univ of Florida
Oikawa, Patricia, California State Univ, East Bay
Osmond, Deanna L., North Carolina State Univ
Palumbo, Tony V., Oak Ridge Nat Lab
Pan, William L., Washington State Univ
Parker, David B., West Texas A&M Univ
Peacock, Caroline, Univ of Leeds
Pereira, Engil I., Univ of Texas, Rio Grande Valley
Phelps, Tommy J., Oak Ridge Nat Lab
Plante, Alain F., Univ of Pennsylvania
Prasad, Rishi, Auburn Univ
Prescott, Cindy, Univ of British Columbia
Ribbons, Relena R., Lawrence Univ
Rice, Chuck W., Kansas State Univ
Rodrigues, Jorge, Univ of California, Davis
Ruehr, Thomas A., California Polytechnic State Univ
Sadowsky, Michael J., Univ of Minnesota, Twin Cities
Scow, Kate M., Univ of California, Davis
Shaw, Liz, Univ of Reading
Shi, Wei, North Carolina State Univ
Silva, Lucas C., Univ of California, Davis
Simard, Suzanne, Univ of British Columbia
Smith, Jeffery L., Washington State Univ
Sogn, Trine, Norwegian Univ of Life Sciences (NMBU)
Somenhally, Anil, Texas A&M Univ
Stevens, Robert G., Washington State Univ
Thomas, Andrew D., Univ of Wales
Tolhurst, Trevor, Univ of East Anglia
Torrents, Alba, Univ of Maryland
Voroney, R. Paul, Univ of Guelph
Wagger, Michael G., North Carolina State Univ
Wenzel, Christopher, Eastern Wyoming Coll
Wilkie, Ann C., Univ of Florida

Paleopedology/Archeology

Beck, Colleen M., Desert Research Inst
Busacca, Alan J., Washington State Univ
Driese, Steven G., Univ of Tennessee, Knoxville
Driese, Steven G., Baylor Univ
Follmer, Leon R., Univ of Illinois, Urbana-Champaign
Herrmann, Edward W., Indiana Univ, Bloomington
Irmis, Randall B., Univ of Utah
Monger, H. C., New Mexico State Univ, Las Cruces
Terry, Dennis O., Temple Univ

Other Soil Science

Adamsen, Floyd, Univ of Arizona
Adee, Eric, Kansas State Univ
Bader, Nicholas E., Whitman Coll
Balser, Teri , Univ of Florida
Bell, Brian, Univ of Glasgow
Berg, Peter, Univ of Virginia
Bonczek, James, Univ of Florida
Budke, William, Ventura Coll
Bundy, Larry G., Univ of Wisconsin, Madison
Cabrera, Miguel L., Univ of Georgia
Cambardella, Cynthia, Iowa State Univ of Science & Tech
Ciampitti, Ignacio, Kansas State Univ
Claassen, Mark, Kansas State Univ
Claassen, Victor P., Univ of California, Davis
Coleman, Tommy L., Alabama A&M Univ
Contosta, Alexandra, Univ of New Hampshire
Cramer, Gary, Kansas State Univ
Cruse, Richard M., Iowa State Univ of Science & Tech
Dahlgren, Randy A., Univ of California, Davis
Daroub, Samira H., Univ of Florida
DeLaune, Paul, Texas A&M Univ
DeSutter, Tom, North Dakota State Univ
Dingus, Delmar D., California Polytechnic State Univ
Dormaar, John, Univ of Lethbridge
Dou, Fugen, Texas A&M Univ
Dougil, Andy, Univ of Leeds
Drescher, Andrew, Univ of Minnesota, Twin Cities
Duncan, Stewart, Kansas State Univ
Eberle, Bill, Kansas State Univ
Ebinger, Michael H., Los Alamos Nat Lab
Ehler, Stan, Kansas State Univ
Eppes, Martha C., Univ of North Carolina, Charlotte
Favorito, Jessica, Stockton Univ
Ferre, Paul (Ty), Univ of Arizona
Fjell, Dale, Kansas State Univ
Flores-Delgadillo, Maria L., Universidad Nacional Auto de Mexico
Flores-Roman, David, Universidad Nacional Auto de Mexico
Fritz, Allan, Kansas State Univ
Gale, Paula M., Univ of Tennessee, Martin
Gantzer, Clark J., Univ of Missouri
Geisseler, Daniel, Univ of California, Davis
Gerber, Stefan, Univ of Florida
Gordon, Barney, Kansas State Univ
Goss, Michael J., Univ of Guelph
Haag, Lucas, Kansas State Univ
Hall, III, John R., Virginia Polytechnic Inst & State Univ
Halverson, Larry, Iowa State Univ of Science & Tech
Hartshorn, Tony, Montana State Univ
Henning, Stanley J., Iowa State Univ of Science & Tech
Hernandez-Silva, Gilberto, Universidad Nacional Auto de Mexico
Hochmuth, George J., Univ of Florida

Holman, John, Kansas State Univ
Hopmans, Jan W., Univ of California, Davis
Horwath, William R., Univ of California, Davis
Houlton, Ben, Univ of California, Davis
Inglett, Kanika Sharma, Univ of Florida
Jackson, Louise E., Univ of California, Davis
Jacobsen, Jeffrey, Montana State Univ
Janssen, Keith, Kansas State Univ
Jasoni, Richard, Desert Research Inst
Johnson, Arthur H., Univ of Pennsylvania
Johnson, Jane M., Univ of Minnesota, Twin Cities
Jones, Clain, Montana State Univ
Jones, Julia A., Oregon State Univ
Jugulam, Mithila, Kansas State Univ
Karkanis, Pano G., Univ of Lethbridge
Karlen, Douglas L., Iowa State Univ of Science & Tech
Kaspar, Thomas A., Iowa State Univ of Science & Tech
Keeney, Dennis R., Iowa State Univ of Science & Tech
Kelling, Keith A., Univ of Wisconsin, Madison
Killorn, Randy J., Iowa State Univ of Science & Tech
Kitchen, Newell R., Univ of Missouri
Kolka, Randall K., Univ of Minnesota, Twin Cities
Kramer, Marc, Univ of Florida
Kusssow, Wayne R., Univ of Wisconsin, Madison
Laboski, Carrie A., Univ of Wisconsin, Madison
Landschoot, Peter J., Pennsylvania State Univ, Univ Park
Lauzon, John, Univ of Guelph
Lerch, Robert N., Univ of Missouri
Liang, George, Kansas State Univ
Liebens, Johan, Univ of West Florida
Loutkova, Klavdia Oleschko, Universidad Nacional Auto de Mexico
Loynachan, Thomas E., Iowa State Univ of Science & Tech
Mackowiak, Cheryl, Univ of Florida
Mallarino, Antonio W., Iowa State Univ of Science & Tech
Marrs, Rob, Univ of Liverpool
McBride, Raymond G., Univ of Guelph
Mengel, David, Kansas State Univ
Miller, Murray H., Univ of Guelph
Mills, Aaron L., Univ of Virginia
Min, Doo-Hong, Kansas State Univ
Mishra, Umakant, Argonne Nat Lab
Moorberg, Colby, Kansas State Univ
Moorman, Thomas B., Iowa State Univ of Science & Tech
Morgan, Kelly, Univ of Florida
Morris, Geoffrey, Kansas State Univ
Nelson, Nathan, Kansas State Univ
Norman, John M., Univ of Wisconsin, Madison
Nyhan, John W., Los Alamos Nat Lab
O'Geen, Toby, Univ of California, Davis
O'Geen, Toby A., Univ of California, Davis
O'Halloran, Ivan, Univ of Guelph
Obour, Augustine, Kansas State Univ
Obreza, Thomas A., Univ of Florida
Palacios-Mayorga, Sergio, Universidad Nacional Auto de Mexico
Parikh, Sanjai, Univ of California, Davis
Pedersen, Joel A., Univ of Wisconsin, Madison
Perumal, Ram, Kansas State Univ
Peterson, Dallas, Kansas State Univ
Polito, Thomas A., Iowa State Univ of Science & Tech
Powell, J. Mark, Univ of Wisconsin, Madison
Prasad, Vara , Kansas State Univ
Presley, DeAnn, Kansas State Univ
Qualls, Robert G., Univ of Nevada, Reno
Randall, Gyles W., Univ of Minnesota, Twin Cities
Rechcigl, John E., Univ of Florida
Reganold, John P., Washington State Univ
Regehr, David, Kansas State Univ
Rice, Pamela J., Univ of Minnesota, Twin Cities
Richards, James H., Univ of California, Davis
Richter, Daniel D., Duke Univ
Risk, Dave A., Saint Francis Xavier Univ
Rolston, Dennis E., Univ of California, Davis
Roozeboom, Kraig, Kansas State Univ
Roth, Gregory W., Pennsylvania State Univ, Univ Park
Ruark, Matthew D., Univ of Wisconsin, Madison
Ruiz-Diaz, Dorivar, Kansas State Univ
Santos, Eduardo, Kansas State Univ
Sartain, Jerry B., Univ of Florida
Sassenrath, Gretchen, Kansas State Univ
Schaap, Marcel, Univ of Arizona
Schapaugh, Bill T., Kansas State Univ
Schardt, Lawrence A., Pennsylvania State Univ, Univ Park
Schlegel, Alan, Kansas State Univ
Schumann, Arnold W., Univ of Florida
Scow, Kate, Univ of California, Davis
Sherwood, William C., James Madison Univ
Shoup, Doug, Kansas State Univ
Shroyer, Jim, Kansas State Univ
Shuford, James W., Alabama A&M Univ
Silveira, Maria, Univ of Florida
Silvertooth, Jeffrey C., Univ of Arizona
Sinclair, Hugh D., Edinburgh Univ
Sletten, Ronald S., Univ of Washington
Smith, Terry L., California Polytechnic State Univ
Smukler, Sean, Univ of British Columbia
Sohi, Sran P., Edinburgh Univ
Soldat, Douglas J., Univ of Wisconsin, Madison
Southard, Randal J., Univ of California, Davis
Stahlman, Phillip, Kansas State Univ
Strock, Jeffrey S., Univ of Minnesota, Twin Cities
Stubler, Craig, California Polytechnic State Univ
Szecsody, James E., Pacific Northwest Nat Lab
Tesso, Tesfaye, Kansas State Univ
Thompson, Curtis, Kansas State Univ
Tomlinson, Peter, Kansas State Univ
Tubana, Brenda S., Louisiana State Univ
Vanderlip, Richard, Kansas State Univ
Voss, Regis D., Iowa State Univ of Science & Tech
Walworth, James, Univ of Arizona
Wesley Wood, Charley, Univ of Florida
White, Jeffrey G., North Carolina State Univ
Whitney, D.A., Kansas State Univ
Wilson, P. Christopher, Univ of Florida
Woida, Kathleen, Univ of Iowa
Zhang, Guorong, Kansas State Univ

ENGINEERING GEOLOGY

General Engineering Geology

Abdalla, Salah B., Univ of Khartoum
Acevedo-Arroyo, Jose Refugio, Universidad Auto de San Luis Potosi
Adams, Herbert G., California State Univ, Northridge
Ademeso, Anthony O., Federal Univ of Tech, Akure
Adeyemi, G. O., Univ of Ibadan
Ahmed, Muhammad F., Univ of Engineering and Tech
Aimone-Martin, Cathrine T., New Mexico Tech
Alao, D. A., Univ of Ilorin
Anderson, Andrew, Illinois State Geological Survey
Andre-Obayanju, Tomilola, Univ of Benin
Axelbaum, Richard, Washington Univ in St. Louis
Aydin, Adnan, Univ of Mississippi
Baker, Paul, Curtin Univ
Baptista, Joana, Univ of Newcastle Upon Tyne
Bauer, R. A., Illinois State Geological Survey
Bautista, Sonia, Univ Complutense de Madrid
Bell, David, Univ of Canterbury
Bell, John W., Univ of Nevada, Reno
Bennet, Brett, Univ of Kansas
Berry, Karen, Colorado Geological Survey
Berti, Matteo, Alma Mater Studiorum Università di Bologna
Beukelman, Gregg S., Utah Geological Survey
Biswas, Pratim, Washington Univ in St. Louis
Bonetto, Sabrina Maria Rita, Università di Torino
Bowman, Steve D., Utah Geological Survey
Brabham, Peter, Univ of Wales
Brady, Roland H., Fresno State Univ
Brand, Patrick, California Geological Survey
Branum, David, California Geological Survey
Brugger, Matthias, Technical Univ of Munich
Brunkal, Holly, Western Colorado Univ
Buchanan, George, Montgomery County Comm Coll
Burns, Bill, Oregon Dept of Geology and Mineral Industries

Burns, Scott F., Portland State Univ
Buscheck, Thomas A., Lawrence Livermore Nat Lab
Carlson, Jill, Colorado Geological Survey
Carney, Theodore C., Los Alamos Nat Lab
Caron, Olivier J., Illinois State Geological Survey
Castleton, Jessica, Utah Geological Survey
Cato, Kerry, California State Univ, San Bernardino
Cetindag, Bahattin, Firat Univ
CHEGBELEH, Larry-Pax, Univ of Ghana
Choma-Moryl, Krystyna, Univ of Wroclaw
Crawford, Matthew, Univ of Kentucky
Davie, Colin, Univ of Newcastle Upon Tyne
Davies, David, Heriot-Watt Univ
Dawson, Tim, California Geological Survey
Delattre, Marc, California Geological Survey
dePolo, Craig M., Univ of Nevada, Reno
Disperati, Leonardo, Univ of Siena
Doherty, Kevin, California Geological Survey
DuRoss, Christopher B., Utah Geological Survey
Dusseault, Maurice B., Univ of Waterloo
Eberhardt, Erik, Univ of British Columbia
Edil, Tuncer B., Univ of Wisconsin, Madison
Elia, Gaetano, Univ of Newcastle Upon Tyne
Erickson, Ben A., Utah Geological Survey
Evans, Stephen G., Univ of Waterloo
Eversoll, Duane A., Univ of Nebraska, Lincoln
Faúndez, Alejandro, Univ Complutense de Madrid
Ferriz, Horacio, California State Univ, Stanislaus
Floris, Mario, Università degli Studi di Padova
Foster, John H., California State Univ, Fullerton
Fournier, Benoit, Universite Laval
Fratta, Dante, Univ of Wisconsin, Madison
Fuller, Michael, California Geological Survey
Galgaro, Antonio, Università degli Studi di Padova
Gamperl, Moritz, Technical Univ of Munich
Gautam, Tej P., Marietta Coll
Giraud, Richard E., Utah Geological Survey
Greene, Brian M., Youngstown State Univ
Guiterrez, Carlos, California Geological Survey
Hall, Jean, Univ of Newcastle Upon Tyne
Haneberg, William, New Mexico Tech
Haneberg, William C., Univ of Kentucky
Harris, Will, California Geological Survey
Hasan, Syed E., Univ of Missouri, Kansas City
Haydon, Wayne, California Geological Survey
Hayhurst, Cheryl, California Geological Survey
Hencher, Steve, Univ of Leeds
Henika, William S., Virginia Polytechnic Inst & State Univ
Hernandez, Janis, California Geological Survey
Ho, I-Hsuan, Univ of North Dakota
Holland, Peter, California Geological Survey
Horns, Daniel, Utah Valley Univ
Hudec, Peter P., Univ of Windsor
Ibrahim, Ahmad Tajuddin Hj, Univ of Malaya
Idowu, Simeon O., Federal Univ of Tech, Akure
Ige, O. O., Univ of Ilorin
Jamaluddin, Tajul Anuar, Univ of Malaya
Jo, Ho Young, Korea Univ
Johnson, William P., Univ of Utah
Jones, Glenda, Keele Univ
Jones, Larry Allan, Los Alamos Nat Lab
Joyce, James, Univ of Puerto Rico
Karanfil, Tanju, Clemson Univ
Kavanaugh, Jeffrey, Univ of Alberta
Knapp, Richard B., Lawrence Livermore Nat Lab
Knizek, Martin, Masaryk Univ
Knudsen, Tyler R., Utah Geological Survey
Korre, Anna, Imperial Coll
Kreylos, Oliver, Univ of California, Davis
Krohn, James P., Los Angeles Pierce Coll
Kudlac, John J., Point Park Univ
Lancaster, Jeremy, California Geological Survey
Likos, William J., Univ of Wisconsin, Madison
Linares, Rogelio, Universitat Auto de Barcelona
Lindsay, Donald, California Geological Survey
Lindsey, Kassandra, Colorado Geological Survey
Linn, Rodman, Los Alamos Nat Lab
Locat, Jacques E., Universite Laval
Lokau, Katja R., Technical Univ of Munich
Longstreth, David, California Geological Survey
Lovekin, Jonathan, Colorado Geological Survey
Lund, William R., Utah Geological Survey
Lupogo, Keneth, Univ of Dar es Salaam
Madrigal-Rubio, Rafael, Universidad Auto de Chihuahua
Mareschal, Maxime, California Geological Survey
Martel, Stephen J., Univ of Hawai'i, Manoa
Martin, Beth, Washington Univ in St. Louis
Masek, Ondrej, Edinburgh Univ
Matter, Juerg M., Univ of Southampton
McCoy, Kevin, Colorado Geological Survey
McCrink, Timothy P., California Geological Survey
McDonald, Gregory N., Utah Geological Survey
McMahon, Katherine, Univ of Wisconsin, Madison
Mena-Zambrano, Teodulo, Universidad Auto de Chihuahua
Menschik, Bettina, Technical Univ of Munich
Merifield, Paul M., Univ of California, Los Angeles
Merritt, Andrew, Univ of Plymouth
Mlinarevic, Ante, California Geological Survey
Moeglin, Thomas D., Missouri State Univ
Mogilevskaya, Sonia, Univ of Minnesota, Twin Cities
Moore, Jeffrey, Univ of Utah
Morales, Tomas, Univ of the Basque Country UPV/EHU
Morgan, Terrance L., Los Alamos Nat Lab
Nickmann, Marion, Technical Univ of Munich
Niemann, William L., Marshall Univ
Ogunsanwo, O., Univ of Ilorin
Olson, Brian , California Geological Survey
Origlia, Hector D., Universidad Nacional de Rio Cuarto
Oswald, John, California Geological Survey
Oyediran, I. A., Univ of Ibadan
Patrick, David M., Univ of Southern Mississippi
Perez, Florante, California Geological Survey
Perret, Didier H., Universite du Quebec
Pestrong, Raymond, San Francisco State Univ
Pipkin, Bernard W., Univ of Southern California
Poeter, Eileen P., Colorado School of Mines
Potten, Martin, Technical Univ of Munich
Potter, Kenneth, Univ of Wisconsin, Madison
Poulton, Mary, Univ of Arizona
Pridmore, Cindy L., California Geological Survey
Rahn, Perry H., South Dakota School of Mines & Tech
Raj, John K., Univ of Malaya
Ramelli, Alan R., Univ of Nevada, Reno
Ramirez, Abelardo L., Lawrence Livermore Nat Lab
Rockaway, John D., Northern Kentucky Univ
Rodríguez, Martín Jesús, Univ Complutense de Madrid
Roffers, Pete, California Geological Survey
Roggenthen, William M., South Dakota School of Mines & Tech
Rosa, Carla, California Geological Survey
Rosinski, Anne, California Geological Survey
Rubin, Ron, California Geological Survey
Santi, Paul M., Colorado School of Mines
Sass, Ingo, Technische Universitaet Darmstadt
Savigny, K. Wayne, Univ of British Columbia
Schiff, Sherry L., Univ of Waterloo
Seitz, Gordon, California Geological Survey
Sepulveda, Sergio, Simon Fraser Univ
Short, William, California Geological Survey
Silva, Michael A., California Geological Survey
Song, Lin-Ping, Univ of British Columbia
Sonnenwald, Margreta, Technical Univ of Munich
Spangler, Eleanor, California Geological Survey
Staub, William P., Oak Ridge Nat Lab
Stead, Douglas, Simon Fraser Univ
Steinberg, Laura, Boston Coll
Stevens, De Anne S., Alaska Div of Geological & Geophysical Surveys
Swanson, Brian, California Geological Survey
Tester, Jefferson W., Cornell Univ
Thomas, Mark, Univ of Leeds
Thornburg, Jennifer, California Geological Survey
Thuro, Kurosch, Technical Univ of Munich
Tinjum, James M., Univ of Wisconsin, Madison

Tsige, Meaza, Univ Complutense de Madrid
Uriarte, Jesus Angel, Univ of the Basque Country UPV/EHU
Urzua, Alfredo, Boston Coll
Villeneuve, Marlene C., Univ of Canterbury
Vorosmarty, Charles, Graduate School of the City Univ of New York
Walton, Gabriel, Colorado School of Mines
Wang, Dongmei, Univ of North Dakota
Watts, Chester F., Radford Univ
Weigers, Mark, California Geological Survey
West, Olivia M., Oak Ridge Nat Lab
West, Terry R., Purdue Univ
White, Chase, California Geological Survey
White, Jonathan, Colorado Geological Survey
White, Owen L., Univ of Waterloo
Wilkey, Patrick L., Argonne Nat Lab
Williams, Bruce A., Pacific Northwest Nat Lab
Williams, John W., San Jose State Univ
Wu, Chin, Univ of Wisconsin, Madison
Yarbrough, Lance D., Univ of Mississippi
Yunwei, Sun, Lawrence Livermore Nat Lab
Zhou, Wendy W., Colorado School of Mines
Zimmerman, R. Eric, Argonne Nat Lab
Zyvoloski, George A., Los Alamos Nat Lab

Earthquake Engineering

Acosta, Jose G., Centro de Inv Científica y de Ed Superior de Ensenada
Atkinson, Gail M., Carleton Univ
Heaton, Thomas H., California Inst of Tech
Huang, Moh J., California Geological Survey
Kutter, Bruce, Univ of California, Davis
Mendoza, Luis H., Centro de Inv Científica y de Ed Superior de Ensenada
Michaels, Paul, Boise State Univ
Oommen, Thomas, Michigan Technological Univ
Reyes, Alfonso, Centro de Inv Científica y de Ed Superior de Ensenada
Rowshandel, Badie, California Geological Survey
Swensen, Daniel, California Geological Survey

Mining Tech/Extractive Metallurgy

Aguirre-Moriel, Socorro I., Universidad Auto de Chihuahua
Anderson, Corby , Colorado School of Mines
Bryan, Christopher, Exeter Univ
Coggan, John, Exeter Univ
Demopoulos, George, McGill Univ
Doyle, Fiona M., Univ of California, Berkeley
Duby, Paul F., Columbia Univ
Evans, James W., Univ of California, Berkeley
Feunstenau, Maurice, Univ of Nevada, Reno
Fuerstenau, Douglas W., Univ of California, Berkeley
Glass, Hylke J., Exeter Univ
Griswold, George B., New Mexico Tech
Gundiler, Ibrahim H., New Mexico Tech
Guthrie, Roderick I., McGill Univ
Hasan, Mainul, McGill Univ
Hiskey, J. Brent, Univ of Arizona
Hurtado de La Ree, Maria B B., Universidad Estatal de Sonora
Jeffrey, Kip, Exeter Univ
Jonas, John J., McGill Univ
Kinabo, Crispin P., Univ of Dar es Salaam
Lee, Jaeheon, Univ of Arizona
Lee, Jaeheon, Colorado School of Mines
López, Maricela , Universidad Estatal de Sonora
Mendoza-Aguilar, Hector M., Universidad Auto de Chihuahua
Minor-Velazquez, Hector, Universidad Auto de Chihuahua
Murguía-Romero, Christian A., Universidad Estatal de Sonora
Murillo-Valenzuela, Hector O., Universidad Estatal de Sonora
Nesbitt, Carl, Univ of Nevada, Reno
Ruiz-Cisneros, David H., Universidad Auto de Chihuahua
Salazar- Avila, Arnulfo , Universidad Estatal de Sonora
Santos-Munguia, Paula C., Universidad Estatal de Sonora
Sastry, Kalanadh V. S., Univ of California, Berkeley
Seal, Thom, Univ of Nevada, Reno
Smith, Karl A., Univ of Minnesota, Twin Cities
Somasundaran, Ponisseril, Columbia Univ
Spiller, Erik, Colorado School of Mines
Themelis, Nickolas J., Columbia Univ
Trevizo-Cano, Luis M., Universidad Auto de Chihuahua
Wheeler, Paul, Exeter Univ
Yue, Stephen, McGill Univ

Mining Engineering

Bessinger, Stephen, Univ of Utah
Boshkov, Stefan H., Columbia Univ
Brune, Jürgen F., Colorado School of Mines
Calizaya, Felipe, Univ of Utah
Chatterjee, Snehamoy, Michigan Technological Univ
Dagdelen, Kadri, Colorado School of Mines
Danko, George, Univ of Nevada, Reno
Dimitrakopoulos, Roussos, McGill Univ
Dinsmoor, John C., Los Alamos Nat Lab
Donovan, James, Univ of Utah
Durucan, Sevket, Imperial Coll
Duzgun, Sebnem, Colorado School of Mines
Eveleth, Robert W., New Mexico Tech
Foster, Patrick, Exeter Univ
Hassani, Ferri, McGill Univ
Honkaer, Rick, Univ of Kentucky
Ibarra-Dessens, Gonzalo de Jesus, Universidad Estatal de Sonora
Kalia, Hemendra N., Los Alamos Nat Lab
Kaunda, Rennie, Colorado School of Mines
Kennedy, Gareth, Exeter Univ
Kohler, Jeffery L., Pennsylvania State Univ, Univ Park
Kovach, Richard G., Los Alamos Nat Lab
Liu, Shimin, Pennsylvania State Univ, Univ Park
Lusk, Braden , Univ of Kentucky
MANCA, Pierpaolo P., Universita di Cagliari
MASSACCI, Giorgio, Universita di Cagliari
Mayer, Ulrich, Univ of British Columbia
McCarter, Michael K., Univ of Utah
Meyer, Lewis, Exeter Univ
Miller, Hugh, Colorado School of Mines
Mitri, Hani, McGill Univ
Momayez, Moe, Univ of Arizona
Monteverde- Gutierrez, Gerardo, Universidad Estatal de Sonora
Mousset-Jones, Pierre, Univ of Nevada, Reno
Navarro, Ricardo, Universidad Nacional Auto de Mexico
Nelson, Michael G., Univ of Utah
Nieto Antunez, Antonio, Universidad Nacional Auto de Mexico
Novak, Thomas, Univ of Kentucky
Pokorny, Eugene W., Los Alamos Nat Lab
Silva-Castro, Jhon, Univ of Kentucky
Slistan- Grijalva, Angel, Universidad Estatal de Sonora
Sottile, Jr., Joseph, Univ of Kentucky
Squelch, Andrew P., Curtin Univ
Taylor, Danny L., Univ of Nevada, Reno
Tenorio, Victor O., Univ of Arizona
Vite-Picazo, Luis G., Universidad Estatal de Sonora
Wala, Andrew M., Univ of Kentucky
Wane, Malcolm T., Columbia Univ
Wedding, William C., Univ of Kentucky
Wetherelt, Andrew, Exeter Univ
Whyatt, Jeffrey, Univ of Utah
Yegulalp, Tuncel M., Columbia Univ
Zavodni, Zavis, Univ of Utah
Zhang, Jinhong, Univ of Arizona

Petroleum Engineering

Ahmed, Ramadan, Univ of Oklahoma
Alvarado, Vladimir, Univ of Wyoming
Badalyan, Alex, Univ of Adelaide
Bedrikovetski, Pavel, Univ of Adelaide
Bijeljic, Branko, Imperial Coll
Blasingame, Tom, Texas A&M Univ
Blunt, Martin J., Imperial Coll
Borazjani, Sara, Univ of Adelaide
Callard, Jeff, Univ of Oklahoma
Carageorgos, Themis, Univ of Adelaide
Civan, Faruk, Univ of Oklahoma
Cooper, George A., Univ of California, Berkeley
Corbett, Patrick, Heriot-Watt Univ
Couples, Gary, Heriot-Watt Univ
Devegowda, Deepak, Univ of Oklahoma
Dreesen, Donald S., Los Alamos Nat Lab
El-Monier, Ilham, Univ of Oklahoma
Fahes, Mashhad, Univ of Oklahoma

Falode, Olugbenga, Univ of Ibadan
Fisher, Quentin , Univ of Leeds
Frailey, Scott M., Illinois State Geological Survey
Geiger, Sebastian, Heriot-Watt Univ
Gonzalez-Perdomo, Maria, Univ of Adelaide
Gringarten, Alain, Imperial Coll
Haghighi, Manouchehr, Univ of Adelaide
Holubnyak, Yehven I., Univ of Kansas
Jamili, Ahmad, Univ of Oklahoma
Jin, Guohai, Geological Survey of Alabama
Kadkhodai Ilkhchi, Ali, Univ of Tabriz
Kelkar, Sharad, Los Alamos Nat Lab
King, Peter, Imperial Coll
Knapp, Roy M., Univ of Oklahoma
Lorinczi, Piroska, Univ of Leeds
Ma, Jingsheng, Heriot-Watt Univ
Misra, Siddharth, Univ of Oklahoma
Moghanloo, Rouzbeh, Univ of Oklahoma
Muggeridge, Ann H., Imperial Coll
Nikolinakou, Maria-Aikaterini, Univ of Texas, Austin
Patzek, Tad W., Univ of California, Berkeley
Pickup, Gillian E., Heriot-Watt Univ
Pournik, Maysam, Univ of Oklahoma
Rai, Chandra S., Univ of Oklahoma
Rose, Peter E., Univ of Utah
Sakhaee-Pour, Ahmad, Univ of Oklahoma
Salmachi, Alireza, Univ of Adelaide
Sayyafzadeh, Mohammad, Univ of Adelaide
Shah, Subhash N., Univ of Oklahoma
Sharma, Suresh, Univ of Oklahoma
Singh, Kamaljit, Heriot-Watt Univ
Sondergeld, Carl H., Univ of Oklahoma
Teodoriu, Catalin, Univ of Oklahoma
Vesovic, Velisa, Imperial Coll
Wu, Kan, Texas A&M Univ
Wu, Xingru, Univ of Oklahoma
Zaman, Musharraf, Univ of Oklahoma
Zeinijahromi, Abbas, Univ of Adelaide

Rock Mechanics

Abousleiman, Younane, Univ of Oklahoma
Abu Bakar, Muhammad Z., Univ of Engineering and Tech
Agioutantis, Zach, Univ of Kentucky
Archambault, Guy, Universite du Quebec a Chicoutimi
Aslan, Yasemin, Firat Univ
Barakat, Bassam, Institut Polytechnique LaSalle Beauvais (ex-IGAL)
Barzegari, Ghodrat, Univ of Tabriz
Blacic, James D., Los Alamos Nat Lab
Brahmi, Sadek, Institut Polytechnique LaSalle Beauvais (ex-IGAL)
Brake, Thomas L., Los Alamos Nat Lab
Brantut, Nicolas D., Univ Coll London
Brown, Donald W., Los Alamos Nat Lab
Chen, Rui, California Geological Survey
Cruden, David M., Univ of Alberta
Cundall, Peter A., Univ of Minnesota, Twin Cities
Daemen, Jaak, Univ of Nevada, Reno
Detournay, Emmanuel M., Univ of Minnesota, Twin Cities
Diederichs, Mark, Queen's Univ
Enderlin, Milton, Texas Christian Univ
Engelder, Terry, Pennsylvania State Univ, Univ Park
Ferrero, Anna Maria, Università di Torino
Germanovich, Leonid, Georgia Inst of Tech
Ghassemi, Ahmed, Univ of Oklahoma
Gordon, Robert B., Yale Univ
Gurocak, Zulfu, Firat Univ
Guzina, Bojan B., Univ of Minnesota, Twin Cities
Haimson, Bezalel C., Univ of Wisconsin, Madison
Hastie, Warwick W., Univ of KwaZulu-Natal
Jung, SJ, Univ of Idaho
Kanik, Mustafa, Firat Univ
Kemeny, John M., Univ of Arizona
Kitajima, Hiroko, Texas A&M Univ
Kronenberg, Andreas, Texas A&M Univ
Käsling, Heiko, Technical Univ of Munich
Labuz, Joseph F., Univ of Minnesota, Twin Cities
MacLaughlin, Mary M., Montana Tech of the Univ of Montana
Mariani, Elisabeth, Univ of Liverpool
Meredith, Philip, Univ Coll London
Mojtabai, Navid, New Mexico Tech
Nelson, Priscilla P., Colorado School of Mines
Onyeobi, Tony U., Univ of Benin
Oravecz, Kalman I., New Mexico Tech
Pariseau, William G., Univ of Utah
Pec, Matej, Massachusetts Inst of Tech
Perry , Kyle, Univ of Kentucky
Persson, Per-Anders, New Mexico Tech
Prasad, Manika, Colorado School of Mines
Roegiers, Jean-Claude, Univ of Oklahoma
Rostami, Jamal, Colorado School of Mines
Rutter, Ernest H., Univ of Manchester
Saeidi, Ali, Universite du Quebec a Chicoutimi
Shreedharan, Srisharan, Utah State Univ
Simoni, Alessandro, Alma Mater Studiorum Università di Bologna
Sinha, Krishna P., Univ of Utah
Smutek, Claude, Universite de la Reunion
Sone, Hiroki, Univ of Wisconsin, Madison
Stockinger, Georg M., Technical Univ of Munich
Swift, Robert P., Los Alamos Nat Lab
Unrug, Kot F., Univ of Kentucky
Vinciguerra, Sergio Carmelo, Università di Torino
Watters, Robert J., Univ of Nevada, Reno
Wijesinghe, Ananda M., Lawrence Livermore Nat Lab
Zimmerman, Robert W., Imperial Coll

Ocean Engineering/Mining

Ajayi, J. O., Obafemi Awolowo Univ
Cambazoglu, Mustafa K., Univ of Southern Mississippi
Carpenter, Roy, Univ of Washington
Gibson, Carl H., Univ of California, San Diego
Hodgkiss, Jr., William S., Univ of California, San Diego
Hoopes, John A., Univ of Wisconsin, Madison
Irish, Jennifer, Virginia Polytechnic Inst & State Univ
Nootz, Gero, Univ of Southern Mississippi
Wen, Xianyun, Univ of Leeds

Geotechnical Engineering

Fernández, Luis Ramón, Univ Complutense de Madrid
Hingston, Egerton, Univ of KwaZulu-Natal
Insua, Juan Miguel, Univ Complutense de Madrid
Ortt, Jr., Richard A., Maryland Department of Nat Resources
Pascariu, Florentina, Alexandru Ioan Cuza
Sanchez, Marcelo J., Texas A&M Univ

Geological Engineering

Darrow, Margaret M., Univ of Alaska Fairbanks
Gardner, Michael H., Univ of Nevada, Reno
Garnier Villareal, Maximiliano, Universidad de Costa Rica
Higgins, Jerry D., Colorado School of Mines
Katzenstein, Kurt W., South Dakota School of Mines & Tech
Melentijevic, Svetlana, Univ Complutense de Madrid
Stetler, Larry D., South Dakota School of Mines & Tech

OCEANOGRAPHY

General Oceanography

Abernathy, Ryan, Columbia Univ
Allen, John, Univ of Portsmouth
Alves, Tiego, Univ of Wales
Barros, Tony, Miami-Dade Coll (Wolfson Campus)
Becker, Janet M., Univ of Hawai'i, Manoa
Bohaty, Steven, Univ of Southampton
Brandes, Jay, Georgia Inst of Tech
Brandon, Mark A., Open Univ
Brown, Erik T., Univ of Minnesota Duluth
Buchanan, Donald G., San Bernardino Valley Coll
Calder, Brian, Univ of New Hampshire
Chamberlin, William S., Fullerton Coll
Chandler, Cyndy, Woods Hole Oceanographic Inst
Ciannelli, Lorenzo, Oregon State Univ
Cooney, Michael, Univ of Hawai'i, Manoa
Curewitz, Daniel, Syracuse Univ
DeJesus, Roman, Fullerton Coll
Denton, George, Univ of South Florida

Dijkstra, Semme, Univ of New Hampshire
Dong, Charles, El Camino Coll
Falkowski, Paul G., Univ of Hawai'i, Manoa
Field, Richard T., Univ of Delaware
Florie, Joshua, Tulsa Comm Coll
Fritsen, Christian H., Univ of Nevada, Reno
Froelich, Philip, Florida State Univ
Gerlach, Dana, Woods Hole Oceanographic Inst
Gomez, Natalya, McGill Univ
Gordon, Elizabeth S., Fitchburg State Univ
Grossman, Walter, San Bernardino Valley Coll
Hale, Michelle, Univ of Portsmouth
Heimbach, Patrick, Univ of Texas, Austin
Hepner, Tiffany, Univ of Texas at Austin, Jackson Schl of Geosciences
Herdendorf, Charles E., Ohio State Univ
Hickman, Anna, Univ of Southampton
Holland, Christina, Univ of Texas, Austin
Holliday, Joseph W., El Camino Coll
Horton, Radley, Columbia Univ
Hoyt, William H., Univ of Northern Colorado
Hughes-Clarke, John E., Univ of New Hampshire
Ivanovic, Ruza, Univ of Leeds
Johnson, Ashanti, Univ of Texas, Arlington
Johnson, Helen, Univ of Oxford
Judkins, Heather L., Saint Petersburg Coll, Clearwater
Jung, Simon, Edinburgh Univ
Keller, Klaus, Pennsylvania State Univ, Univ Park
Kerr, Andrew, Univ of Wales
Kim, Yong Hoon, West Chester Univ
King, Andrew, Univ of Illinois Chicago
Kustka, Adam B., Rutgers, The State Univ of New Jersey, Newark
Laliberte, Elizabeth, Univ of Rhode Island
Lauderdale, Jonathan, Univ of Liverpool
Leach, Harry, Univ of Liverpool
Lear, Caroline, Univ of Wales
Lee, Alyce, Concord Univ
Lee, Stephen C., Los Angeles Pierce Coll
Leinen, Margaret, Univ of California, San Diego
Lenz, Petra H., Univ of Hawai'i, Manoa
Lepore, Chiara, Columbia Univ
Lindo Atachati, David, Graduate School of the City Univ of New York
Liu, Zhanfei, Univ of Texas, Austin
Ma, Yanxia, Louisiana State Univ
MacLeod, Chris, Univ of Wales
Maxwell, Arthur E., Univ of Texas, Austin
McPhaden, Michael J., Univ of Washington
McWilliams, James, Univ of California, Los Angeles
Min, Dong-Ha, Univ of Texas, Austin
Montenegro, Alvaro, Ohio State Univ
Moore, Dennis, Univ of Hawai'i, Manoa
Mosher, David C., Univ of New Hampshire
Muir, William, San Bernardino Valley Coll
Myers, Paul, Univ of Alberta
Noyes, Jim, El Camino Coll
Ogston, Andrea, Univ of Washington
Peredo-Jaime, Jose I., Universidad Auto de Baja California Sur
Perry, Malcolm, Curtin Univ
Pike, Jenny, Univ of Wales
Pincus, Robert, Columbia Univ
Quan, Tracy, Oklahoma State Univ
Radulski, Robert, Southern Connecticut State Univ
Rappe, Michael S., Univ of Hawai'i, Manoa
Roesler, Collin, Bowdoin Coll
Salisbury, Joseph, Univ of New Hampshire
Schlegel, Mary A., Millersville Univ
Schneider, Niklas, Univ of Hawai'i, Manoa
Schroeder, William W., Dauphin Island Sea Lab
Scott, Robert B., Univ of Texas, Austin
Seyfang, Gill, Univ of East Anglia
Spencer, Matthew, Univ of Liverpool
Steger, John M., Miami Dade Coll (Kendall Campus)
Sullivan-Watts, Barbara K., Univ of Rhode Island
Swanson, R. L., SUNY, Stony Brook
Thompson, LuAnne, Univ of Washington
Thomson, Richard, Univ of Victoria
Thurman, Harold V., Mt. San Antonio Coll
Tyrrell, Toby, Univ of Southampton
Ulanski, Stanley L., James Madison Univ
Veeramony, Jayaram, Mississippi State Univ
Walker, Brett, Univ of Ottawa
Walker, Nan D., Louisiana State Univ
Wallace, William G., Graduate School of the City Univ of New York
Westervelt, Daniel, Columbia Univ
Wheatcroft, Rob, Oregon State Univ
Wiberg, Patricia L., Univ of Virginia
Wiesenburg, Denis A., Univ of Southern Mississippi
Williams, Ric, Univ of Liverpool
Wiltshire, John C., Univ of Hawai'i, Manoa
Wolcott, Ray, Cuyamaca Coll
Wolff, George, Univ of Liverpool
Wu, Yutian, Columbia Univ
Yeager, Lauren A., Univ of Texas, Austin
Yin, Jianjun, Univ of Arizona
Yon, Lisa, Palomar Coll
Zhai, Xiaoming, Univ of East Anglia

Biological Oceanography

Adams, Marshall, Oak Ridge Nat Lab
Adams, Michael S., Univ of Wisconsin, Madison
Algar, Christopher, Dalhousie Univ
Allam, Bassem, SUNY, Stony Brook
Allen, Eric E., Univ of California, San Diego
Aller, Josephine Y., SUNY, Stony Brook
Anderson, George C., Univ of Washington
Antoine, David, Curtin Univ
Armbrust, Virginia E., Univ of Washington
Atkinson, Marlin J., Univ of Hawai'i, Manoa
Audet, Celine, Universite du Quebec a Rimouski
Auster, Peter, Univ of Connecticut
Azam, Farooq, Univ of California, San Diego
Baltz, Donald M., Louisiana State Univ
Banse, Karl, Univ of Washington
Barber, Bruce, Univ of South Florida
Barber, Richard T., Duke Univ
Barlow, Jay P., Univ of California, San Diego
Baross, John A., Univ of Washington
Bartlett, Douglas H., Univ of California, San Diego
Batta-Lona, Paola, Univ of Connecticut
Baumann, Hannes, Univ of Connecticut
Baumann, Zofia, Univ of Connecticut
Baumgartner, Mark, Dalhousie Univ
Beck, Capri, Curtin Univ
Benfield, Mark C., Louisiana State Univ
Benner, Ron, Univ of South Carolina
Bibby, Thomas, Univ of Southampton
Bienfang, Paul K., Univ of Hawai'i, Manoa
Biggs, Douglas C., Texas A&M Univ
Blake, Norman J., Univ of South Florida
Bloomfield, Helen, Univ of Liverpool
Boar, Rosalind, Univ of East Anglia
Bochdansky, Alexander, Old Dominion Univ
Bollens, Stephen, Washington State Univ
Boudrias, Michel A., Univ of San Diego
Bowen, Jennifer, Univ of Massachusetts, Boston
Bowser, Paul, SUNY, Stony Brook
Breitbart, Mya, Univ of South Florida
Brethes, Jean-Claude, Universite du Quebec a Rimouski
Briggs, John C., Univ of South Florida
Bruno, Barbara, Univ of Hawai'i, Manoa
Bucklin, Ann, Univ of Connecticut
Burton, Ronald S., Univ of California, San Diego
Buskey, Edward J., Univ of Texas, Austin
Campbell, Lisa, Texas A&M Univ
Carlile, Amy L., Univ of New Haven
Carney, Robert S., Louisiana State Univ
Carpenter, Steve, Univ of Wisconsin, Madison
Carr, Mark, Univ of California, Santa Cruz
Case, James M., Wilkes Univ
Caswell, Bryony, Univ of Liverpool
Cerrato, Robert M., SUNY, Stony Brook
Checkley, David M., Univ of California, San Diego
Christian, James R., Univ of Victoria

Church, Matthew, Univ of Hawai'i, Manoa
Collie, Jeremy S., Univ of Rhode Island
Collier, Jackie, SUNY, Stony Brook
Collins, Kenneth, Univ of Southampton
Conover, David O., SUNY, Stony Brook
Copley, Jonathan, Univ of Southampton
Coull, Bruce, Univ of South Carolina
Cranford, Peter, Dalhousie Univ
Croll, Don, Univ of California, Santa Cruz
Cuba, Thomas, Univ of South Florida
Cuker, Benjamin E., Hampton Univ
Cullen, John J., Dalhousie Univ
D'Angelo, Maria C., Univ of Southampton
D'Elia, Christopher, Univ of South Florida
Daly, Kendra L., Univ of South Florida
Dam, Hans G., Univ of Connecticut
Dayton, Paul K., Univ of California, San Diego
Dean, John Mark, Univ of South Carolina
Demers, Serge, Universite du Quebec a Rimouski
Deming, Jody W., Univ of Washington
Denton, Gary R. W., Univ of Guam
DeWitt, Calvin B., Univ of Wisconsin, Madison
DiBacco, Claudio, Dalhousie Univ
Dieterle, Dwight A., Univ of South Florida
Dobbs, Fred C., Old Dominion Univ
Donaghay, Percy, Univ of Rhode Island
Doty, Thomas, Roger Williams Univ
Dower, John, Univ of Victoria
Drazen, Jeffrey C., Univ of Hawai'i, Manoa
Ducklow, Hugh, Columbia Univ
Duffy, Tara , Northeastern Univ
Dunn, Robert, Univ of South Carolina
Dunton, Kenneth H., Univ of Texas, Austin
Durbin, Edward G., Univ of Rhode Island
Dyhrman, Sonya, Columbia Univ
Edgcomb, Virginia, Woods Hole Oceanographic Inst
Eggleston, David B., North Carolina State Univ
Epp, Leonard G., Mount Union Coll
Erbe, Christine, Curtin Univ
Erdner, Deana L., Univ of Texas, Austin
Erisman, Brad, Univ of Texas, Austin
Esbaugh, Andrew J., Univ of Texas, Austin
Falkowski, Paul G., Rutgers, The State Univ of New Jersey
Fast, Mark, SUNY, Stony Brook
Felbeck, Horst, Univ of California, San Diego
Feller, Robert, Univ of South Carolina
Fenberg, Phillip, Univ of Southampton
Fennel, Katja, Dalhousie Univ
Fisher, Jr., Thomas R., Univ of Maryland
Fisk, Aaron, Univ of Windsor
Fitzer, Susan, Univ of Glasgow
Fletcher, Madilyn, Univ of South Carolina
Fortin, Samantha, Princeton Univ
Fournier, Robert O., Dalhousie Univ
Francis, Robert, Univ of Washington
Frank, Kenneth, Dalhousie Univ
Franks, Peter J. S., Univ of California, San Diego
Friedrichs, Marjorie, William & Mary
Frost, Bruce W., Univ of Washington
Fuiman, Lee A., Univ of Texas, Austin
Gaasterland, Terry, Univ of California, San Diego
Gallagher, Eugene, Univ of Massachusetts, Boston
Ganssen, Gerald M., Vrije Universiteit Amsterdam
Gardner, Wayne S., Univ of Texas, Austin
Gates, Ruth D., Univ of Hawai'i, Manoa
Gibson, Deidre M., Hampton Univ
Gifford, Dian J., Univ of Rhode Island
Gilbert, Patricia M., Univ of Maryland
Gobler, Christopher, SUNY, Stony Brook
Godbold, Jasmin A., Univ of Southampton
Goetze, Erica, Univ of Hawai'i, Manoa
Gomes, Helga, Columbia Univ
Gosselin, Michel, Universite du Quebec a Rimouski
Gould, Mark D., Roger Williams Univ
Graham, Linda K., Univ of Wisconsin, Madison
Graham, Michael, Moss Landing Marine Lab
Grant, Jonathan, Dalhousie Univ
Green, Jonathan, Univ of Liverpool
Greene, Charles H., Cornell Univ
Grigg, Richard W., Univ of Hawai'i, Manoa
Grottoli, Andrea G., Ohio State Univ
Grunbaum, Daniel, Univ of Washington
Hargraves, Paul E., Univ of Rhode Island
Hargreaves, Bruce R., Lehigh Univ
Harvey, James T., Moss Landing Marine Lab
Hastings, Philip A., Univ of California, San Diego
Hauton, Chris, Univ of Southampton
Hawkins, Lawrence, Univ of Southampton
Healey, Michael, Univ of British Columbia
Heil, Cynthia, Univ of South Florida
Hicks, David, Univ of Texas, Rio Grande Valley
Hill, Victoria, Old Dominion Univ
Holland, Nicholas D., Univ of California, San Diego
Holt, Gloria J., Univ of Texas, Austin
Hood, Raleigh, Univ of Maryland
Hopkins, Thomas L., Univ of South Florida
Hopkins, Thomas S., Dauphin Island Sea Lab
Hotchkiss, Sarah, Univ of Wisconsin, Madison
Humm, Harold J., Univ of South Florida
Hunsucker, Kelli, Florida Inst of Tech
Hunt, Brian, Univ of British Columbia
Jeffries, H. Perry, Univ of Rhode Island
Jensen, Antony, Univ of Southampton
Jickells, Tim, Univ of East Anglia
Johnson, Kevin, Florida Inst of Tech
Jones, William J., Univ of South Carolina
Jordan, Robert A., Hampton Univ
Juhl, Andrew, Columbia Univ
Jumars, Peter A., Univ of Washington
Juniper, Kim, Univ of Victoria
Kamykowski, Daniel, North Carolina State Univ
Kana, Todd M., Univ of Maryland
Karl, David M., Univ of Hawai'i, Manoa
Kaufmann, Ronald S., Univ of San Diego
Kelley, Christopher, Univ of Hawai'i, Manoa
Kelly, John, Univ of New Haven
Kemp, Paul, Univ of Hawai'i, Manoa
Kenney, Robert D., Univ of Rhode Island
Kimball, Matthew, Univ of South Carolina
Kitchell, James F., Univ of Wisconsin, Madison
Kooyman, Gerald L., Univ of California, San Diego
Kraemer, George P., SUNY, Purchase
Krieger-Brockett, Barbara B., Univ of Washington
Lam, Phyllis, Univ of Southampton
Landry, Michael R., Univ of California, San Diego
LaRock, Paul A., Louisiana State Univ
Lee, Carol, Univ of Wisconsin, Madison
Leichter, James J., Univ of California, San Diego
Lessard, Evelyn J., Univ of Washington
Letelier, Ricardo, Oregon State Univ
Levin, Lisa A., Univ of California, San Diego
Levinton, Jeffrey, SUNY, Stony Brook
Lewin, Joyce C., Univ of Washington
Lewis, Alan G., Univ of British Columbia
Lewis, Marlon R., Dalhousie Univ
Li, William K., Dalhousie Univ
Lilley, Marvin D., Univ of Washington
Lin, Senjie, Univ of Connecticut
Lohan, Maeve, Univ of Southampton
Longoria, Lorena, Univ of Texas, Rio Grande Valley
Lonsdale, Darcy J., SUNY, Stony Brook
Lopez, Glenn R., SUNY, Stony Brook
Lucas, Cathy, Univ of Southampton
Lyons, Dierdre, Univ of California, San Diego
MacIntyre, Hugh, Dalhousie Univ
Mackas, David L., Univ of Victoria
Madhusudhana, Shyam, Curtin Univ
Mahegan, Mairead, Woods Hole Oceanographic Inst
Maldonado, Maria, Univ of British Columbia
Malin, Gill, Univ of East Anglia
Marra, John, Brooklyn Coll (CUNY)
Martiny, Adam, Univ of California, Irvine

Matassa, Catherine, Univ of Connecticut
McCarthy, James T., Harvard Univ
McCauley, Rob, Curtin Univ
McElroy, Anne E., SUNY, Stony Brook
McGlathery, Karen, Univ of Virginia
McGlathery, Karen J., Univ of Virginia
McGowan, John A., Univ of California, San Diego
McManus, George B., Univ of Connecticut
Mendelssohn, Irving A., Louisiana State Univ
Metaxas, Anna, Dalhousie Univ
Meyer-Gutbrod, Erin, Univ of South Carolina
Meylan, Anne, Univ of South Florida
Miller, Douglas C., Univ of Delaware
Mills, Eric L., Dalhousie Univ
Milroy, Scott P., Univ of Southern Mississippi
Mobley, Curtis D., Univ of Washington
Mojica, Kristina, Univ of Southern Mississippi
Monger, Bruce, Cornell Univ
Montagna, Paul, Texas A&M Univ, Corpus Christi
Moore, Christopher M., Univ of Southampton
Moran, Dawn, Woods Hole Oceanographic Inst
Mueller, Erich M., Univ of South Alabama
Mulholland, Margaret, Old Dominion Univ
Munch, Stephan, SUNY, Stony Brook
Murray, Alison, Univ of Nevada, Reno
Murtugudde, Raghuram G., Univ of Maryland
Napora, Theodore A., Univ of Rhode Island
Napp, Jeffrey, Univ of Washington
Newton, Jan, Univ of Washington
Nixon, Scott W., Univ of Rhode Island
Nosal, Andrew, Univ of San Diego
Nye, Janet, SUNY, Stony Brook
Ohman, Mark D., Univ of California, San Diego
Oliver, John S., Moss Landing Marine Lab
Oliver, Matthew J., Univ of Delaware
Oschlies, Andreas, Dalhousie Univ
Oviatt, Candace, Univ of Rhode Island
Padilla, Dianna K., SUNY, Stony Brook
Pakhomov, Evgeny, Univ of British Columbia
Palenik, Brian, Univ of California, San Diego
Parida, Chandanlal, Curtin Univ
Paul, John H., Univ of South Florida
Peebles, Ernst B., Univ of South Florida
Pellerin, Jocelyne, Universite du Quebec a Rimouski
Perry, Mary Jane, Univ of Washington
Peterson, Bradley, SUNY, Stony Brook
Petrik, Colleen, Univ of California, San Diego
Pikitch, Ellen K., SUNY, Stony Brook
Pinckney, James, Univ of South Carolina
Plotkin, Pamela, Texas A&M Univ
Pollack, Jennifer, Texas A&M Univ, Corpus Christi
Potts, Donald C., Univ of California, Santa Cruz
Powers, Sean, Univ of South Alabama
Prairie, Jennifer, Univ of San Diego
Purdie, Duncan A., Univ of Southampton
Quattro, Joseph, Univ of South Carolina
Rabalais, Nancy N., Louisiana State Univ
Ragotzkie, Robert A., Univ of Wisconsin, Madison
Ray, G. Carleton, Univ of Virginia
Redalje, Donald G., Univ of Southern Mississippi
Resplandy, Laure, Princeton Univ
Richardson, Tammi, Univ of South Carolina
Robinson, Carol, Univ of East Anglia
Robinson, Leonie, Univ of Liverpool
Rocap, Gabrielle L., Univ of Washington
Roman, Charles T., Univ of Rhode Island
Roughgarden, Joan, Stanford Univ
Rouse, Gregory W., Univ of California, San Diego
Rowe, Gilbert T., Texas A&M Univ
Roy, Suzanne, Universite du Quebec a Rimouski
Ruttan, Lore, Emory Univ
Saila, Saul B., Univ of Rhode Island
Sambrotto, Raymond N., Columbia Univ
Schnetzer, Astrid, North Carolina State Univ
Schulze, Anja, Texas A&M Univ
Scott, Tim, Roger Williams Univ
Searcy, Steven, Univ of San Diego
Selph, Karen E., Univ of Hawai'i, Manoa
Shaw, Richard F., Louisiana State Univ
Shuman, Randy, Univ of Washington
Shumway, Sandra, Univ of Connecticut
Sibert, John R., Univ of Hawai'i, Manoa
Sieburth, John M., Univ of Rhode Island
Silver, Mary W., Univ of California, Santa Cruz
Sims, David, Univ of Southampton
Smalley, Gabriela W., Rider Univ
Smayda, Theodore J., Univ of Rhode Island
Smith, Craig R., Univ of Hawai'i, Manoa
Smith, David E., Univ of Virginia
Smith, David C., Univ of Rhode Island
Smith, Jennifer E., Univ of California, San Diego
Solan, Martin, Univ of Southampton
Sommer, Ulrich , Dalhousie Univ
Sousa, Wayne P., Univ of California, Berkeley
Specker, Jennifer, Univ of Rhode Island
Stanley, Emily, Univ of Wisconsin, Madison
Steppe, Cecily N., United States Naval Acad
Steward, Grieg F., Univ of Hawai'i, Manoa
Stickney, Robert R., Texas A&M Univ
Stoecker, Diane, Univ of Maryland
Stoner, Betsy, Bentley Univ
Strickland, Richard M., Univ of Washington
Subramaniam, Ajit, Columbia Univ
Sugihara, George, Univ of California, San Diego
Suttle, Curtis, Univ of British Columbia
Swain, Geoffrey, Florida Inst of Tech
Swift, Elijah V., Univ of Rhode Island
Sylvan, Jason B., Texas A&M Univ
Taggart, Christopher T., Dalhousie Univ
Talley, Drew, Univ of San Diego
Taylor, Frank J. R., Univ of British Columbia
Taylor, Gordon T., SUNY, Stony Brook
Thatje, Sven, Univ of Southampton
Thomas, Florence, Univ of Hawai'i, Manoa
Thomas, Peter, Univ of Texas, Austin
Thompson, Diane, Boston Univ
Thomsen, Laurenz A., Univ of Washington
Thomson, Jack, Univ of Liverpool
Thornton, Daniel C., Texas A&M Univ
Tlusty, Michael, Univ of Massachusetts, Boston
Toonen, Robert, Univ of Hawai'i, Manoa
Torres, Joseph J., Univ of South Florida
Tortell, Philippe, Univ of British Columbia
Treude, Tina, Univ of California, Los Angeles
Tunnicliffe, Verena, Univ of Victoria
Turner, Robert E., Louisiana State Univ
Tynan, Cynthia T., Univ of Washington
Urban-Rich, Juanita, Univ of Massachusetts, Boston
Vacquier, Victor D., Univ of California, San Diego
Vaillancourt, Robert, Millersville Univ
Varela, Diana, Univ of Victoria
Vargo, Gabriel A., Univ of South Florida
Vaudrey, Jamie, Univ of Connecticut
Villalard-Bohnsack, Martine, Roger Williams Univ
Villareal, Tracy A., Univ of Texas, Austin
Waldbusser, George, Oregon State Univ
Waldman, John, Graduate School of the City Univ of New York
Walsh, John J., Univ of South Florida
Ward, Ben A., Univ of Southampton
Ward, Bess B., Princeton Univ
Ward, J. Evan, Univ of Connecticut
Warren, Joseph, SUNY, Stony Brook
Weber, Thomas, Univ of Rochester
Wei, Chong, Curtin Univ
Wells, Randy, Univ of South Florida
Welschmeyer, Nicholas A., Moss Landing Marine Lab
Weng, Kevin, Univ of Hawai'i, Manoa
Wetz, Michael, Texas A&M Univ, Corpus Christi
Wiedenmann, Jorg, Univ of Southampton
Wilson, Merwether, Edinburgh Univ
Wishner, Karen, Univ of Rhode Island
Wolcott, Donna L., North Carolina State Univ

Wolcott, Thomas G., North Carolina State Univ
Woodin, Sarah A., Univ of South Carolina
Yamanaka, Tsuyuko, Univ of Liverpool
Yandle, Tracy, Emory Univ
Young, Richard E., Univ of Hawai'i, Manoa
Zakardjian, Bruno, Universite du Quebec a Rimouski
Zedler, Joy, Univ of Wisconsin, Madison
Zehr, Jonathan P., Univ of California, Santa Cruz
Zhang, Huan, Univ of Connecticut
Zimmerman, Richard, Old Dominion Univ

Chemical Oceanography

Allison, Nicky, Univ of St. Andrews
Andersen, Raymond J., Univ of British Columbia
Anderson, Robert F., Columbia Univ
Andersson, Andreas, Univ of California, San Diego
Andren, Anders W., Univ of Wisconsin, Madison
Andrews, Allen H., Univ of Hawai'i, Manoa
Annett, Amber, Univ of Southampton
Apprill, Amy, Woods Hole Oceanographic Inst
Armstrong, David E., Univ of Wisconsin, Madison
Armstrong, Robert A., SUNY, Stony Brook
Azetsu-Scott, Kumiko, Geological Survey of Canada
Bacon, Michael P., Woods Hole Oceanographic Inst
Balistrieri, Laurie, Univ of Washington
Barbeau, Katherine A., Univ of California, San Diego
Bates, Nicholas R., Univ of Southampton
Benitez-Nelson, Claudia, Univ of South Carolina
Benway, Heather, Woods Hole Oceanographic Inst
Bishop, James K., Univ of California, Berkeley
Boudreau, Bernard P., Dalhousie Univ
Bourbonnais, Annie, Univ of South Carolina
Boyle, Edward A., Massachusetts Inst of Tech
Brownawell, Bruce J., SUNY, Stony Brook
Bruland, Kenneth W., Univ of California, Santa Cruz
Buchwald, Carolyn, Dalhousie Univ
Buckley, Lawrence J., Univ of Rhode Island
Buesseler, Ken O., Woods Hole Oceanographic Inst
Bullister, John, Univ of Washington
Burdige, David J., Old Dominion Univ
Chan, Kwan M., California State Univ, Long Beach
Chapman, Piers, Texas A&M Univ
Chappell, P. D., Old Dominion Univ
Cherrier, Jennifer, Brooklyn Coll (CUNY)
Cochran, J. K., Stony Brook Univ
Cochran, J. Kirk, SUNY, Stony Brook
Coffin, Richard, Texas A&M Univ, Corpus Christi
Cooper, Matthew, Univ of Southampton
Cornwell, Jeffery C., Univ of Maryland
Crutzen, Paul, Univ of California, San Diego
Culberson, Charles H., Univ of Delaware
Cullen, Jay, Univ of Victoria
Cutter, Gregory A., Old Dominion Univ
deAngelis, Marie, SUNY, Maritime Coll
Delaney, Margaret L., Univ of California, Santa Cruz
DeMaster, David J., North Carolina State Univ
Devol, Allan H., Univ of Washington
Doney, Scott, Univ of Virginia
Druffel, Ellen R. M., Univ of California, Irvine
Drysdale, Jessica, Woods Hole Oceanographic Inst
Duce, Robert A., Texas A&M Univ
Eglinton, Tim, Woods Hole Oceanographic Inst
Emerson, Steven R., Univ of Washington
Emile-Geay, Julien, Univ of Southern California
Engel, Anga, SUNY, Stony Brook
Estapa, Margaret, Woods Hole Oceanographic Inst
Estapa, Margaret, Skidmore Coll
Fanning, Kent A., Univ of South Florida
Feely, Richard A., Univ of Washington
Fenical, William H., Univ of California, San Diego
Fisher, Nicholas S., SUNY, Stony Brook
Fitzgerald, William F., Univ of Connecticut
Fitzsimmons, Jessica N., Texas A&M Univ
Fogel, Marilyn L., Univ of Delaware
Fox, Austin, Florida Inst of Tech
Fox, Lewis, Broward Coll
Fredricks, Helen, Woods Hole Oceanographic Inst
Gagne, Jean-Pierre, Universite du Quebec a Rimouski
Gammon, Richard H., Univ of Washington
Garcia-Rubio, Luis H., Univ of South Florida
Gieskes, Joris M., Univ of California, San Diego
Glover, David M., Woods Hole Oceanographic Inst
Gold-Bouchot, Gerardo, Texas A&M Univ
Granger, Julie, Univ of Connecticut
Hales, Burke R., Oregon State Univ
Hamme, Roberta C., Univ of Victoria
Hanson, Jr., Alfred K., Univ of Rhode Island
Hayes, Christopher T., Univ of Southern Mississippi
Haymet, Anthony D., Univ of California, San Diego
Ho, David, Univ of Hawai'i, Manoa
Hollander, David J., Univ of South Florida
Huebert, Barry J., Univ of Hawai'i, Manoa
Ingalls, Anitra E., Univ of Washington
James, Rachael, Univ of Southampton
Johnson, Bruce D., Dalhousie Univ
Keil, Richard G., Univ of Washington
Kessler, John , Univ of Rochester
Kiene, Ronald P., Univ of South Alabama
Knap, Anthony, Texas A&M Univ
Kujawinski, Elizabeth B., Woods Hole Oceanographic Inst
LaVigne, Michèle, Bowdoin Coll
Lee, Cindy, SUNY, Stony Brook
Letscher, Robert T., Univ of New Hampshire
Lima, Ivan D., Woods Hole Oceanographic Inst
Liss, Peter S., Univ of East Anglia
Loder, Theodore C., Univ of New Hampshire
Longnecker, Krista, Woods Hole Oceanographic Inst
Lott, Dempsey E., Woods Hole Oceanographic Inst
Lucotte, Marc Michel, Universite du Quebec a Montreal
Lund, David, Univ of Connecticut
Lwiza, Kamazima M., SUNY, Stony Brook
Mahaffey, Claire, Univ of Liverpool
Manning, Cara, Univ of Connecticut
Marchitto, Thomas M., Univ of Colorado
Martin, William R., Woods Hole Oceanographic Inst
Mason, Robert P., Univ of Connecticut
Matsumoto, Katsumi, Univ of Minnesota, Twin Cities
McDuff, Russell E., Univ of Washington
McIlvin, Matt, Woods Hole Oceanographic Inst
McKinley, Galen, Univ of Wisconsin, Madison
McNichol, Ann P., Woods Hole Oceanographic Inst
Measures, Christopher, Univ of Hawai'i, Manoa
Merrill, John T., Univ of Rhode Island
Moffett, James W., Univ of Southern California
Moore, Robert M., Dalhousie Univ
Moran, S. Bradley, Univ of Rhode Island
Murray, James W., Univ of Washington
Najjar, Raymond G., Pennsylvania State Univ, Univ Park
Nuzzio, Donald B., Univ of Delaware
Palevsky, Hilary I., Boston Coll
Pellenbarg, Robert, Coll of the Desert
Pelletier, Emilien, Universite du Quebec a Rimouski
Pennock, Jonathan R., Dauphin Island Sea Lab
Pilson, Michael E., Univ of Rhode Island
Precedo, Laura, Broward Coll
Presley, Bobby J., Texas A&M Univ
Pyrtle, Ashanti J., Univ of South Florida
Quay, Paul D., Univ of Washington
Quinn, James G., Univ of Rhode Island
Rahn, Kenneth A., Univ of Rhode Island
Ravelo, Ana C., Univ of California, Santa Cruz
Reimer, Andreas, Georg-August Univ of Goettingen
Reimers, Clare, Oregon State Univ
Resing, Joseph A., Univ of Washington
Rheuban, Jennie, Woods Hole Oceanographic Inst
Roethel, Frank J., SUNY, Stony Brook
Russell, Joellen, Univ of Arizona
Sabine, Christopher L., Univ of Washington
Saito, Mak A., Woods Hole Oceanographic Inst
Salaun, Rachel, Univ of Liverpool
Santschi, Peter H., Texas A&M Univ
Sarmiento, Jorge L., Princeton Univ

Sayles, Frederick L., Woods Hole Oceanographic Inst
Schwartz, Matthew C., Univ of West Florida
Scranton, Mary I., SUNY, Stony Brook
Shamberger, Kathryn E., Texas A&M Univ
Sharp, Jonathan H., Univ of Delaware
Sharples, Jonathan, Univ of Liverpool
Shiller, Alan M., Univ of Southern Mississippi
Sholkovitz, Edward R., Woods Hole Oceanographic Inst
Simjouw, Jean-Paul, Univ of New Haven
Skoog, Annelie, Univ of Connecticut
Sonzogni, William C., Univ of Wisconsin, Madison
Stewart, Gillian, Graduate School of the City Univ of New York
Stewart, Gillian, Queens Coll (CUNY)
Sundby, Bjorn, Universite du Quebec a Rimouski
Suntharalingam, Parvadha, Univ of East Anglia
Swarr, Gretchen, Woods Hole Oceanographic Inst
Sylva, Sean, Woods Hole Oceanographic Inst
Tagliabue, Alessandro, Univ of Liverpool
Takahashi, Taro, Columbia Univ
Tanhua, Toste, Dalhousie Univ
Thomas, Helmuth, Dalhousie Univ
Thompson, Anu, Univ of Liverpool
Timmermans, Mary-Louise, Yale Univ
Tobias, Craig, Univ of Connecticut
Ullman, William J., Univ of Delaware
Van Mooy, Benjamin, Woods Hole Oceanographic Inst
Velinsky, David, Drexel Univ
Veron, Alain J., Univ of Delaware
Vlahos, Penny, Univ of Connecticut
Wade, Terry L., Texas A&M Univ
Wallace, Douglas, Dalhousie Univ
Wallace, Douglas W. R., SUNY, Stony Brook
Williams, Dave E., Univ of British Columbia
Windsor, Jr., John G., Florida Inst of Tech
Yao, Wensheng, Univ of South Florida
Yvon-Lewis, Shari A., Texas A&M Univ
Zafiriou, Oliver C., Woods Hole Oceanographic Inst
Zhu, Qingzhi, SUNY, Stony Brook

Geological Oceanography

Agardy, Rachael, Central Michigan Univ
Aiello, Ivano, Moss Landing Marine Lab
Alexander, Clark R., Georgia Southern Univ
Alt, Jeffrey C., Univ of Michigan
Arthur, Michael A., Pennsylvania State Univ, Univ Park
Bain, Olivier, Institut Polytechnique LaSalle Beauvais (ex-IGAL)
Baker, Edward T., Univ of Washington
Baldauf, Jack G., Texas A&M Univ
Berquist, Jr., Carl R., William & Mary
Billups, Katharina, Univ of Delaware
Black, David, SUNY, Stony Brook
Busch, William H., Univ of New Orleans
Bush, David M., Univ of West Georgia
Carey, Steven N., Univ of Rhode Island
Clemens, Steven C., Brown Univ
Coffin, Richard, Texas A&M Univ, Corpus Christi
Cook, Mea S., Williams Coll
Creager, Joe S., Univ of Washington
D'Hondt, Steven L., Univ of Rhode Island
Darby, Dennis A., Old Dominion Univ
Dellapenna, Timothy M., Texas A&M Univ
Diercks, Arne R., Univ of Southern Mississippi
Dudley, Walter, Univ of Hawai'i, Manoa
Emerick, Christina M., Univ of Washington
Farmer, Emma C., Hofstra Univ
Farrell, John, Univ of Rhode Island
Ferrini, Victoria, Columbia Univ
Finney, Bruce P., Univ of Alaska, Fairbanks
Flood, Roger D., SUNY, Stony Brook
Gagnon, Katie, Seattle Central Coll
Gardner, Wilford D., Texas A&M Univ
Hale, Rip, Old Dominion Univ
Hallock-Muller, Pamela, Univ of South Florida
Harris, Sara, Univ of British Columbia
Hawkins, James W., Univ of California, San Diego
He, Zhuoya, Rice Univ
Hein, Christopher J., William & Mary
Holcomb, Robin T., Univ of Washington
Holcombe, Troy L., Texas A&M Univ
Holmes, Mark L., Univ of Washington
Honjo, Susumu, Woods Hole Oceanographic Inst
Hovan, Steven A., Indiana Univ of Pennsylvania
Kelley, Deborah S., Univ of Washington
Kincaid, Christopher, Univ of Rhode Island
King, John, Univ of Rhode Island
Kominz, Michelle A., Western Michigan Univ
Kuehl, Steven A., William & Mary
Larson, Roger, Univ of Rhode Island
Lear, C, Cardiff Univ
Leinen, Margaret, Univ of Rhode Island
Leventer, Amy, Colgate Univ
Locker, Stanley D., Univ of South Florida
Longworth, Brett, Woods Hole Oceanographic Inst
Manganini, Steven J., Woods Hole Oceanographic Inst
Martinez-Noriega, Cesar, Universidad Auto de Baja California Sur
McHugh, Cecilia, Queens Coll (CUNY)
McHugh, Cecilia M., Graduate School of the City Univ of New York
McManus, Dean A., Univ of Washington
Moore, Theodore C., Univ of Michigan
Mosher, David C., Dalhousie Univ
Mottl, Michael J., Univ of Hawai'i, Manoa
Muza, Jay P., Broward Coll
Naar, David F., Univ of South Florida
Nittrouer, Charles A., Univ of Washington
Nowell, Arthur R. M., Univ of Washington
Ogston, Andrea S., Univ of Washington
Oostdam, Bernard L., Millersville Univ
Parsons, Jeffrey D., Univ of Washington
Pilkey, Jr., Orrin H., Duke Univ
Pinet, Paul, Colgate Univ
Prell, Warren L., Brown Univ
Quinn, Terrence, Univ of South Florida
Riesselman, Christina R., Univ of Otago
Sautter, Leslie R., Coll of Charleston
Schilling, Jean-Guy E., Univ of Rhode Island
Schmidt, Matthew, Old Dominion Univ
Shen, Yang, Univ of Rhode Island
Sigurdsson, Haraldur, Univ of Rhode Island
Slowey, Niall C., Texas A&M Univ
Smith, Stephen V., Univ of Hawai'i, Manoa
Smythe, Wendy F., Univ of Minnesota Duluth
Snoeckx, Hilde, Univ of West Florida
Sommerfield, Christopher K., Univ of Delaware
St. John, Kristen E., James Madison Univ
Stanley, Daniel J., Smithsonian Inst / Nat Museum of Nat History
Sternberg, Richard W., Univ of Washington
Thomas, Debbie, Texas A&M Univ
Trembanis, Art, Univ of Delaware
Tyce, Robert, Univ of Rhode Island
Uchupi, Elazar, Woods Hole Oceanographic Inst
Wallace, Davin, Univ of Southern Mississippi
Whitman, Jill M., Pacific Lutheran Univ
Wilcock, William S., Univ of Washington
Zarillo, Gary , Florida Inst of Tech
Zeebe, Richard E., Univ of Hawai'i, Manoa

Physical Oceanography

Aagaard, Knut, Univ of Washington
Abernathey, Ryan P., Columbia Univ
Alford, Matthew H., Univ of Washington
Allen, Susan E., Univ of British Columbia
Andrefouet, Serge, Univ of South Florida
Anis, Ayal, Texas A&M Univ
Armi, Laurence, Univ of California, San Diego
Barclay, David R., Dalhousie Univ
Baum, Steven K., Texas A&M Univ
Becker, Janet, Univ of California, San Diego
Bishop, Stuart P., North Carolina State Univ
Bogucki, Darek, Texas A&M Univ, Corpus Christi
Bohlen, Walter F., Univ of Connecticut
Boicourt, William, Univ of Maryland
Bordoni, Simona, California Inst of Tech

Bostater, Charles, Florida Inst of Tech
Bourke, Robert H., Naval Postgraduate School
Bowman, Malcolm J., SUNY, Stony Brook
Boxall, Simon R., Univ of Southampton
Boyd, John, Univ of Michigan
Bracco, Annalisa, Georgia Inst of Tech
Breaker, Laurence, Moss Landing Marine Lab
Brooks, David A., Texas A&M Univ
Buckingham, Michael J., Univ of California, San Diego
Buijsman, Maarten, Univ of Southern Mississippi
Bulusu, Subrahmanyam, Univ of South Carolina
Cane, Mark A., Columbia Univ
Cannon, Glenn A., Univ of Washington
Carder, Kendall L., Univ of South Florida
Carmack, Edward, Univ of British Columbia
Carter, Glenn, Univ of Hawai'i, Manoa
Carton, James, Univ of Maryland
Cessi, Paola, Univ of California, San Diego
Chang, Ping, Texas A&M Univ
Chao, Shenn-Yu, Univ of Maryland
Chapman, Ross N., Univ of Victoria
Chen, Dake, Columbia Univ
Chiu, Ching-Sang, Naval Postgraduate School
Chu, Peter C., Naval Postgraduate School
Cole, Rick, Univ of South Florida
Collins, Curtis A., Naval Postgraduate School
Cornillon, Peter, Univ of Rhode Island
Craig, Susanne , Dalhousie Univ
Crease, James, Univ of Delaware
Criminale, Jr., William O., Univ of Washington
Cronin, Meghan F., Univ of Washington
D'Asaro, Eric A., Univ of Washington
Davis, Leslie M., Austin Comm Coll District
de Szoeke, Simon P., Oregon State Univ
Denman, Kenneth L., Univ of Victoria
Dever, Edward P., Oregon State Univ
Dewey, Richard K., Univ of Victoria
Di Lorenzo, Emanuele, Georgia Inst of Tech
Dierssen, Heidi, Univ of Connecticut
DiMarco, Steven F., Texas A&M Univ
Donohue, Kathleen, Univ of Rhode Island
Donovan, Jeff C., Univ of South Florida
Dorman, Clive E., San Diego State Univ
Dosso, Stanley E., Univ of Victoria
Drago, Aldo, Univ of Malta
Drijfhout, Sybren, Univ of Southampton
Dushaw, Brian D., Univ of Washington
Edson, James, Univ of Connecticut
Ellis, Dale, Dalhousie Univ
English, David C., Univ of South Florida
Eriksen, Charles C., Univ of Washington
Ewart, Terry E., Univ of Washington
Ezer, Tal, Old Dominion Univ
Fedorov, Alexey V., Yale Univ
Ferrari, Raffaele, Massachusetts Inst of Tech
Firing, Eric, Univ of Hawai'i, Manoa
Flagg, Charles, SUNY, Stony Brook
Flament, Pierre J., Univ of Hawai'i, Manoa
Flierl, Glenn R., Massachusetts Inst of Tech
Flynn, Russell L., Cypress Coll
Follows, Michael, Massachusetts Inst of Tech
Foreman, Michael G., Univ of Victoria
Foreman, Michael G. G., Univ of British Columbia
Fox-Kemper, Baylor, Brown Univ
Fram, Jonathan, Oregon State Univ
Galea, Anthony, Univ of Malta
Galperin, Boris, Univ of South Florida
Garfield, Newell (Toby), San Francisco State Univ
Garrett, Christopher J. R., Univ of Victoria
Garwood, Roland W., Naval Postgraduate School
Gauci, Adam, Univ of Malta
Gerbi, Greg, Skidmore Coll
Giddings, Sarah, Univ of California, San Diego
Giese, Benjamin S., Texas A&M Univ
Gille, Sarah T., Univ of California, San Diego
Ginis, Isaac, Univ of Rhode Island
Gnanadesikan, Anand, Johns Hopkins Univ
Gong, Donglai, William & Mary
Gonzalez, Frank I., Univ of Washington
Gordon, Arnold L., Columbia Univ
Greatbatch, Richard , Dalhousie Univ
Greenberg, David, Dalhousie Univ
Gregg, Michael C., Univ of Washington
Guinasso, Norman L., Texas A&M Univ
Haine, Thomas W., Johns Hopkins Univ
Hara, Tetsu, Univ of Rhode Island
Harrison, Cheryl, Univ of Texas, Rio Grande Valley
Harrison, Don E., Univ of Washington
Hautala, Susan L., Univ of Washington
Hay, Alex E., Dalhousie Univ
He, Ruoying, North Carolina State Univ
Heavers, Richard, Roger Williams Univ
Hebert, David, Dalhousie Univ
Hebert, David L., Univ of Rhode Island
Hendershott, Myrl C., Univ of California, San Diego
Herbers, Thomas H., Naval Postgraduate School
Hermann, Albert, Univ of Washington
Hetland, Robert D., Texas A&M Univ
Heywood, Karen, Univ of East Anglia
Hickey, Barbara M., Univ of Washington
Hines, Paul, Dalhousie Univ
Hofmann, Eileen E., Old Dominion Univ
Howard, Matthew K., Texas A&M Univ
Howden, Stephan, Univ of Southern Mississippi
Howe, Bruce M., Univ of Washington
Hsieh, William W., Univ of British Columbia
Huang, Bohua, George Mason Univ
Jackson, George A., Texas A&M Univ
Jacobs, Stanley, Columbia Univ
Jochens, Ann E., Texas A&M Univ
Johnson, Gregory C., Univ of Washington
Justic, Dubravko, Louisiana State Univ
Kaplan, Alexey, Columbia Univ
Kawase, Mitsuhiro, Univ of Washington
Kelley, Dan, Dalhousie Univ
Kelly, Kathryn A., Univ of Washington
Kessler, William S., Univ of Washington
Klinck, John M., Old Dominion Univ
Klinger, Barry, George Mason Univ
Kloosterziel, Rudolf C., Univ of Hawai'i, Manoa
Klymak, Jody, Univ of Victoria
Knauss, John A., Univ of Rhode Island
Knowles, Charles E., North Carolina State Univ
Kosro, Michael, Oregon State Univ
Koutitonsky, Vladimir G., Universite du Quebec a Rimouski
Kumar, Ajoy, Millersville Univ
Kuperman, William A., Univ of California, San Diego
Lebow, Ruth Y., Los Angeles Pierce Coll
Lee, Craig M., Univ of Washington
Lerczak, Jim, Oregon State Univ
Lesht, Barry M., Argonne Nat Lab
Lesht, Barry, Univ of Illinois Chicago
Liu, Zheng-yu, Univ of Wisconsin, Madison
Liu, Zhengyu, Univ of Wisconsin, Madison
Lozier, M. Susan, Duke Univ
Lu, Youyu, Dalhousie Univ
Lucas, Andrew, Univ of California, San Diego
Lukas, Roger, Univ of Hawai'i, Manoa
Luther, Douglas S., Univ of Hawai'i, Manoa
Luther, Mark E., Univ of South Florida
MacCready, Parker, Univ of Washington
Magaard, Lorenz, Univ of Hawai'i, Manoa
Malanotte-Rizzoli, Paola M., Massachusetts Inst of Tech
Manley, Thomas O., Middlebury Coll
Marsh, Robert, Univ of Southampton
Marshall, John C., Massachusetts Inst of Tech
Martin, Seelye, Univ of Washington
Martinson, Douglas G., Columbia Univ
Martinson, Douglas G., Columbia Univ
Maslowski, Wieslaw, Naval Postgraduate School
Matano, Ricardo P., Oregon State Univ
Maul, George A., Florida Inst of Tech

Mazzini, Pierro, William & Mary
McCreary, Julian P., Univ of Hawai'i, Manoa
McManus, Margaret Anne, Univ of Hawai'i, Manoa
Mellor, George, Princeton Univ
Merrifield, Mark A., Univ of Hawai'i, Manoa
Meyer, Florian, Univ of California, San Diego
Mitchum, Gary T., Univ of South Florida
Mofjeld, Harold O., Univ of Washington
Moore, Dennis W., Univ of Washington
Moore, Jefferson K., Univ of California, Irvine
Morison, James H., Univ of Washington
Moum, James N., Oregon State Univ
Muller, Andrew C., United States Naval Acad
Muller, Peter, Univ of Hawai'i, Manoa
Myers, Paul, Univ of Alberta
Mysak, Lawrence A., McGill Univ
Nash, Jonathan, Oregon State Univ
Naveira Garabato, Alberto, Univ of Southampton
Nechaev, Dmitri, Univ of Southern Mississippi
Nowlin, Jr., Worth D., Texas A&M Univ
Nystuen, Jeffrey A., Univ of Washington
O'Donnell, James, Univ of Connecticut
Oliver, Eric, Dalhousie Univ
Oliver, Kevin, Univ of Southampton
Oltman-Shay, Joan M., Univ of Washington
Omta, Anne W., Case Western Reserve Univ
Orsi, Alejandro H., Texas A&M Univ
Paduan, Jeffrey D., Naval Postgraduate School
Pawlowicz, Richard A., Univ of British Columbia
Perrie, William, Dalhousie Univ
Perry, David J., Chabot Coll
Philander, S. George H., Princeton Univ
Pieters, Roger, Univ of British Columbia
Pinkel, Robert, Univ of California, San Diego
Pond, Stephen G., Univ of British Columbia
Potter, Henry, Texas A&M Univ
Powell, Brian, Univ of Hawai'i, Manoa
Powell, Thomas M., Univ of California, Berkeley
Primeau, Francois, Univ of California, Irvine
Pringle, James M., Univ of New Hampshire
Purkey, Sarah, Univ of California, San Diego
Qiu, Bo, Univ of Hawai'i, Manoa
Radko, Timour, Naval Postgraduate School
Rhines, Peter B., Univ of Washington
Richards, Kelvin J., Univ of Hawai'i, Manoa
Riser, Stephen C., Univ of Washington
Rocha, Cesar, Univ of Connecticut
Roden, Gunnar I., Univ of Washington
Roemmich, Dean H., Univ of California, San Diego
Romero, Leonel, Univ of Connecticut
Ross, Tetjana, Dalhousie Univ
Rossby, Hans T., Univ of Rhode Island
Rothrock, David A., Univ of Washington
Rothstein, Lewis, Univ of Rhode Island
Rouse, Jr., Lawrence J., Louisiana State Univ
Roussenov, Vassil M., Univ of Liverpool
Ruddick, Barry R., Dalhousie Univ
Rudnick, Daniel L., Univ of California, San Diego
Salmon, Richard L., Univ of California, San Diego
Salmun, Haydee, Hunter Coll (CUNY)
Samelson, Roger, Oregon State Univ
Sanford, Lawrence P., Univ of Maryland
Sanford, Thomas B., Univ of Washington
Schopf, Paul S., George Mason Univ
Send, Uwe, Univ of California, San Diego
Shaw, Ping-Tung, North Carolina State Univ
Shearman, Kipp, Oregon State Univ
Shen, Jian, William & Mary
Sheng, Jinyu, Dalhousie Univ
Smith, Peter C., Dalhousie Univ
Smith, Robert L., Oregon State Univ
Smyth, William D., Oregon State Univ
Southam, John R., Univ of Miami
Spindel, Robert C., Univ of Washington
Stanton, Timothy P., Naval Postgraduate School
Stewart, Robert H., Texas A&M Univ
Stoessel, Achim, Texas A&M Univ
Stoessel, Marion, Texas A&M Univ
Stramski, Dariusz, Univ of California, San Diego
Straub, David, McGill Univ
Sutherland, Dave, Univ of Oregon
Swaters, Gordon E., Univ of Alberta
Talley, Lynne D., Univ of California, San Diego
Thompson, Andrew F., California Inst of Tech
Thompson, Keith R., Dalhousie Univ
Thompson, LuAnne, Univ of Washington
Thomson, Richard E., Univ of British Columbia
Thurnherr, Andreas M., Columbia Univ
Thurnherr, Andreas, Columbia Univ
Timmermann, Axel, Univ of Hawai'i, Manoa
Tokmakian, Robin T., Naval Postgraduate School
Tremblay, Bruno, McGill Univ
Tziperman, Eli, Harvard Univ
Villas Boas, Bia, Colorado School of Mines
Von Schwind, Joseph J., Naval Postgraduate School
Wang, Harry, William & Mary
Wang, Zhankun, Texas A&M Univ
Warner, Mark J., Univ of Washington
Watts, D. Randolph, Univ of Rhode Island
Weaver, Andrew J., Univ of Victoria
Webster, Ferris, Univ of Delaware
Weisberg, Robert H., Univ of South Florida
Wells, Neil C., Univ of Southampton
White, Martin, Nat Univ of Ireland Galway
Whitney, Michael, Univ of Connecticut
Wiederwohl, Chrissy, Texas A&M Univ
Wiggert, Jerry, Univ of Southern Mississippi
Williams, Kevin L., Univ of Washington
Wilson, Robert E., SUNY, Stony Brook
Wimbush, Mark, Univ of Rhode Island
Winguth, Arne M., Univ of Texas, Arlington
Wolfe, Christopher, SUNY, Stony Brook
Woodgate, Rebecca A., Univ of Washington
Wu, Jin, Univ of Delaware
Wunsch, Carl, Harvard Univ
Wunsch, Carl I., Massachusetts Inst of Tech
Yankovsky, Alexander E., Univ of South Carolina
Young, William R., Univ of California, San Diego
Yuan, Xiaojun, Columbia Univ
Zappa, Christopher, Columbia Univ
Zappa, Christopher J., Columbia Univ
Zhang, Rong, Princeton Univ
Zhang, Xiaodong, Univ of Southern Mississippi
Zhang, Y. J., William & Mary

Shore and Nearshore Processes

Ashton, Andrew, Woods Hole Oceanographic Inst
Baker, Alex, Univ of East Anglia
Ballinger, Rhoda, Univ of Wales
Barth, Jack A., Oregon State Univ
Baxter, Stefanie J., Univ of Delaware
Blomberg, Brittany N., Austin Comm Coll District
Boardman, Mark R., Miami Univ
Bowen, Anthony J., Dalhousie Univ
Branco, Brett, Brooklyn Coll (CUNY)
Briggs, Tiffany Roberts M., Florida Atlantic Univ
Buonaiuto, Frank, Hunter Coll (CUNY)
Burbanck, George P., Hampton Univ
Camann, Eleanor J., Red Rocks Comm Coll
Couceiro, Fay, Univ of Portsmouth
Dickson, Stephen M., Dept of Agriculture, Conservation, and Forestry
Dunbar, Robert B., Stanford Univ
Dupre, William R., Univ of Houston
Dura, Christina, Virginia Polytechnic Inst & State Univ
Duxbury, Alyn C., Univ of Washington
Engelhart, Simon E., Univ of Rhode Island
Farnsworth, Katie, Indiana Univ of Pennsylvania
Farrell, Stewart C., Stockton Univ
Fenster, Michael S., Randolph-Macon Coll
FitzGerald, Duncan M., Boston Univ
Friedrichs, Carl T., William & Mary
Gibeaut, James, Texas A&M Univ, Corpus Christi

Giese, Graham S., Woods Hole Oceanographic Inst
Guza, Robert T., Univ of California, San Diego
Haigh, Ivan D., Univ of Southampton
Hall, Robert, Univ of East Anglia
Haller, Merrick C., Oregon State Univ
Hansom, Jim, Univ of Glasgow
Harris, Courtney K., William & Mary
Harris, M. Scott, Coll of Charleston
Heise, Elizabeth, Univ of Texas, Rio Grande Valley
Helmuth, Brian, Northeastern Univ
Jackson, Chester M., Georgia Southern Univ
Kassem, Hachem, Univ of Southampton
Kemp, Andrew, Tufts Univ
Kennedy, Andrew, Univ of Notre Dame
Kineke, Gail C., Boston Coll
Koppelman, Lee K., SUNY, Stony Brook
Lippmann, Thomas C., Univ of New Hampshire
Loftis, Jon D., William & Mary
MacMahan, Jamie, Naval Postgraduate School
McBride, Randolph, George Mason Univ
McManus, Margaret A., Univ of California, Santa Cruz
Neill, Simon P., Univ of Wales
Oakley, Bryan, Eastern Connecticut State Univ
Oxner-Jones, D. M., Ohio Dept of Nat Resources
Ozkan-Haller, Tuba, Oregon State Univ
Paine, Jeffrey G., Univ of Texas, Austin
Rosen, Peter S., Northeastern Univ
Ruggiero, Peter, Oregon State Univ
Seibel, Erwin, San Francisco State Univ
Shroyer, Emily L., Oregon State Univ
Siedlecki , Samantha, Univ of Connecticut
Strosnider, William, Univ of South Carolina
Thornton, Edward B., Naval Postgraduate School
Vander Zanden, Jake, Univ of Wisconsin, Madison
Voulgaris, George, Univ of South Carolina
Westerink, Joannes J., Univ of Notre Dame
Wilson, Greg, Oregon State Univ
Winant, Clinton D., Univ of California, San Diego
Wu, Chin, Univ of Wisconsin, Madison

Paleooceanography
Came, Rosemarie E., Univ of New Hampshire
Huang, Huai-Hsuan, Princeton Univ
Ivanochko, Tara, Univ of British Columbia
Lawrence, Kira, Lafayette Coll
Lowery, Chris, Univ of Texas, Austin
Lynch-Stieglitz, Jean, Georgia Inst of Tech
Tappa, Eric, Univ of South Carolina
Tems, Caitlin E., Weber State Univ

ASTRONOMICAL SCIENCES

Astronomy
Chambers, John, Carnegie Inst for Science
Gaidos, Eric J., Univ of Hawai'i, Manoa
Malhotra, Renu, Univ of Arizona
Sheppard, Scott S., Carnegie Inst for Science
Teague, Richard , Massachusetts Inst of Tech
Wood, David A., San Antonio Comm Coll

Cosmochemistry
Agee, Carl B., Univ of New Mexico
Alexander, Conel M., Carnegie Inst for Science
Bermingham, Katherine R., Univ of Maryland
Blake, Geoffrey A., California Inst of Tech
Bland, Phil, Curtin Univ
Blander, Milton, Argonne Nat Lab
Bodenheimer, Peter, Univ of California, Santa Cruz
Bose, Maitrayee, Arizona State Univ
Bouvier, Audrey, Western Univ
Boynton, William V., Univ of Arizona
Brown, Robert H., Univ of Arizona
Burnett, Donald S., California Inst of Tech
Ciesla, Fred, Univ of Chicago
Dauphas, Nicolas, Univ of Chicago
Davis, Andrew, Univ of Chicago
Drake, Michael J., Univ of Arizona
Fegley, Bruce, Washington Univ in St. Louis
Fegley, M. Bruce, Washington Univ in St. Louis
Greenwood, James P., Wesleyan Univ
Gross, Juliane, Rutgers, The State Univ of New Jersey
Herd, Christopher D., Univ of Alberta
Humayun, Munir, Florida State Univ
Huntress, Jr., Wesley T., Carnegie Inst for Science
Jordan, Jim L., Lamar Univ
Korycansky, Don, Univ of California, Santa Cruz
Kring, David, Univ of Arizona
Lauretta, Dante, Univ of Arizona
Lewis, John S., Univ of Arizona
Lin, Douglas, Univ of California, Santa Cruz
Lodders-Fegley, Katharina, Washington Univ in St. Louis
Lunine, Jonathan I., Cornell Univ
Lunine, Jonathan I., Univ of Arizona
Lyon, Ian, Univ of Manchester
MacPherson, Glenn J., Smithsonian Inst / Nat Museum of Nat History
McKeegan, Kevin D., Univ of California, Los Angeles
McSween, Jr., Harry Y., Univ of Tennessee, Knoxville
Mojzsis, Stephen J., Univ of Colorado
Smith, William H., Washington Univ in St. Louis
Swindle, Timothy, Univ of Arizona
Telus, Myriam, Univ of California, Santa Cruz
Turner, Grenville, Univ of Manchester
Vogt, Steven, Univ of California, Santa Cruz
Wadhwa, Meenakshi, Arizona State Univ
Wang, Kun, Washington Univ in St. Louis

Extraterrestrial Geology
Adams, Mark, Lamar Univ
Adcock, Christopher, Univ of Nevada, Las Vegas
Anderson, Jennifer L., Winona State Univ
Arvidson, Raymond E., Washington Univ in St. Louis
Baker, Victor R., Univ of Arizona
Ballentine, Christopher, Univ of Oxford
Batygin, Konstantin, California Inst of Tech
Bell, James, Arizona State Univ
Benedix, Gretchen, Curtin Univ
Blaney, Diana L., Jet Propulsion Lab
Bleamaster III, Leslie F., Trinity Univ
Bourassa, Matthew, Western Univ
Bramson, Ali, Purdue Univ
Byrne, Paul, Washington Univ in St. Louis
Calvin, Wendy M., Univ of Nevada, Reno
Christensen, Philip R., Arizona State Univ
Coles, Kenneth S., Indiana Univ of Pennsylvania
Craddock, Robert A., Smithsonian Inst, Nat Air & Space Museum
Cull-Hearth, Selby, Bryn Mawr Coll
Delaney, Jeremy S., Rutgers, The State Univ of New Jersey
Dietz, Richard D., Univ of Northern Colorado
Ehlmann, Bethany L., California Inst of Tech
Elachi, Charles, California Inst of Tech
Elkins-Tanton, Linda, Arizona State Univ
Fabel, Derek, Univ of Glasgow
Fishman, George , Bentley Univ
Gao, Peter, Carnegie Inst for Science
Gilmore, Martha S., Wesleyan Univ
Glotch, Timothy, Stony Brook Univ
Golombek, Matthew P., Jet Propulsion Lab
Grant, John A., Smithsonian Inst, Nat Air & Space Museum
Grieve, Richard, Western Univ
Grieve, Richard A. F., Univ of New Brunswick
Grosfils, Eric B., Pomona Coll
Hardgrove, Craig, Arizona State Univ
Hardgrove, Craig J., Univ of Tennessee, Knoxville
Hayes, Alexander G., Cornell Univ
Head, III, James W., Brown Univ
Herkenhoff, Ken E., Jet Propulsion Lab
Horgan, Briony, Purdue Univ
Howell, Robert R., Univ of Wyoming
Hurowitz, Joel A., Stony Brook Univ
Hynek, Brian M., Univ of Colorado
Irwin, Rossman P., Smithsonian Inst, Nat Air & Space Museum
Jacobsen, Abram, Univ of Washington
Jakosky, Bruce M., Univ of Colorado

Jurena, Dwight J., San Antonio Comm Coll
Karuntillake, Suniti, Louisiana State Univ
Kennedy, Trudi, Curtin Univ
Klimczak, Christian, Univ of Tennessee, Knoxville
Knutson, Heather, California Inst of Tech
Korotev, Randy L., Washington Univ in St. Louis
Kraal, Erin, Kutztown Univ of Pennsylvania
Lagain, Anthony, Curtin Univ
Landa, Keith, SUNY, Purchase
Leake, Martha A., Valdosta State Univ
Lorenz, Ralph, Univ of Arizona
Mack, John E., Buffalo State Coll
Martin, Emily S., Smithsonian Inst, Nat Air & Space Museum
Massironi, Matteo, Università degli Studi di Padova
Maxwell, Ted A., Smithsonian Inst, Nat Air & Space Museum
McEwen, Alfred S., Univ of Arizona
McEwen, Alfred, Univ of Arizona
McGowan, Eileen, Univ of Massachusetts, Amherst
McKinnon, William B., Washington Univ in St. Louis
Milam, Keith A., Ohio Univ
Minton, David, Purdue Univ
Moersch, Jeffery E., Univ of Tennessee, Knoxville
Mouginis-Mark, Peter J., Univ of Hawai'i, Manoa
Neish, Catherine, Western Univ
Nelson, Robert M., Jet Propulsion Lab
Newsom, Horton E., Univ of New Mexico
Osinski, Gordon R., Western Univ
Parsons, Reid A., Fitchburg State Univ
Payré, Valerie, Univ of Iowa
Piatex, Jennifer L., Central Connecticut State Univ
Prockter, Louise M., Univ of Tennessee, Knoxville
Radebaugh, Jani, Brigham Young Univ
Rice, Melissa S., Western Washington Univ
Robinson, Mark, Arizona State Univ
Rogers, A. D., Stony Brook Univ
Ruff, Steven, Arizona State Univ
Ruiz, Javier , Univ Complutense de Madrid
Runyon, Cassandra R., Coll of Charleston
Schultz, Peter H., Brown Univ
Shepard, Michael K., Bloomsburg Univ
Siebach, Kirsten L., Rice Univ
Stofan, Ellen R., Jet Propulsion Lab
Strom, Robert G., Univ of Arizona
Teske, Johanna K., Carnegie Inst for Science
Thompson, Michelle, Purdue Univ
Thomson, Bradley J., Univ of Tennessee, Knoxville
Tornabene, Livio, Western Univ
Towner, Martin, Curtin Univ
Turtle, Elizabeth, Univ of Arizona
Viviano-Beck, Christina E., Univ of Tennessee, Knoxville
Warner, Nicholas H., SUNY, Geneseo
Watters, Thomas R., Smithsonian Inst, Nat Air & Space Museum
Williams, David, Arizona State Univ
Williams, Kevin K., Buffalo State Coll
Wilson Purdy, Sharon A., Smithsonian Inst, Nat Air & Space Museum
Wordsworth, Robin, Harvard Univ
Zimbelman, James R., Smithsonian Inst, Nat Air & Space Museum

Extraterrestrial Geophysics

Angelopoulos, Vassilis, Univ of California, Los Angeles
Arkani-Hamed, Jafar, McGill Univ
Asphaug, Erik I., Univ of Arizona
Banerdt, Bruce E., Jet Propulsion Lab
Benacquista, Matt, Montana State Univ, Billings
Bord, Don, Univ of Michigan, Dearborn
Brown, Michael E., California Inst of Tech
Brown, Peter, Western Univ
Cao, Hao, Univ of California, Los Angeles
Dombard, Andrew, Case Western Reserve Univ
Dombard, Andrew J., Univ of Illinois Chicago
Fenrich, Francis, Univ of Alberta
Ferencz, Orsolya, Eotvos Lorand Univ
Garrick-Bethell, Ian, Univ of California, Santa Cruz
Giacalone, Joe, Univ of Arizona
Goldreich, Peter M., California Inst of Tech
Greenberg, Richard J., Univ of Arizona
Grima, Cyril , Univ of Texas, Austin
Hapke, Bruce W., Univ of Pittsburgh
Harnett, Erika M., Univ of Washington
Hauck, II, Steven A., Case Western Reserve Univ
Hedgepath, Joshua, Curtin Univ
Holzworth, Robert H., Univ of Washington
Hood, Lonnie L., Univ of Arizona
Hubbard, William B., Univ of Arizona
Jewitt, David, Univ of California, Los Angeles
Jokipii, Jack R., Univ of Arizona
Kivelson, Margaret G., Univ of California, Los Angeles
Kunkle, Thomas D., Los Alamos Nat Lab
Larsen, Kristine, Central Connecticut State Univ
Line, Michael, Arizona State Univ
Lucey, Paul G., Univ of Hawai'i, Manoa
Margot, Jean-Luc, Univ of California, Los Angeles
Matzke, David, Univ of Michigan, Dearborn
McCarthy, Michael P., Univ of Washington
McGovern, Patrick J., Rice Univ
McPherron, Robert L., Univ of California, Los Angeles
Miljkovic, Katarina, Curtin Univ
Muhleman, Duane O., California Inst of Tech
Nakajima, Miki, Univ of Rochester
Newman, William I., Univ of California, Los Angeles
Nimmo, Francis, Univ of California, Santa Cruz
O'Rourke, Joseph, Arizona State Univ
Paige, David A., Univ of California, Los Angeles
Petrovay, Kristof, Eotvos Lorand Univ
Rankin, Robert, Univ of Alberta
Robinson, R. Craig, Central Connecticut State Univ
Rostoker, Gordon, Univ of Alberta
Russell, Christopher T., Univ of California, Los Angeles
Samson, John C., Univ of Alberta
Schlichting, Hilke, Univ of California, Los Angeles
Schmerr, Nicholas C., Univ of Maryland
Seager, Sara, Massachusetts Inst of Tech
Smrekar, Suzanne E., Jet Propulsion Lab
Soderlund, Krista, Univ of Texas, Austin
Solomon, Sean , Columbia Univ
Spilker, Linda J., Jet Propulsion Lab
Steele, Andrew, Carnegie Inst for Science
Stevenson, David J., California Inst of Tech
Sydora, Richard D., Univ of Alberta
Urquhart, Mary, Univ of Texas, Dallas
Walker, Raymond J., Univ of California, Los Angeles
Winglee, Robert M., Univ of Washington
Zuber, Maria T., Massachusetts Inst of Tech

Meteorites & Tektites

Alpert, Sam, American Museum of Nat History
Binzel, Richard P., Massachusetts Inst of Tech
Cartwright, Julia, Univ of Alabama
Chan, Queenie, Royal Holloway Univ of London
Collins, Gareth, Imperial Coll
Corrigan, Catherine, Smithsonian Inst / Nat Museum of Nat History
Dunn, Tasha L., Colby Coll
Ebel, Denton S., American Museum of Nat History
Falquero, Kelsey, Smithsonian Inst / Nat Museum of Nat History
Forman, Lucy, Curtin Univ
French, Bevan, Smithsonian Inst / Nat Museum of Nat History
Genge, Matthew, Imperial Coll
Glass, Billy P., Univ of Delaware
Hartman, Ron N., Mt. San Antonio Coll
Harvey, Ralph P., Case Western Reserve Univ
Heymann, Dieter, Rice Univ
Hildebrand, Alan R., Univ of Calgary
Hoskin, Julie, Smithsonian Inst / Nat Museum of Nat History
Howie, Robert, Curtin Univ
Hutson, Melinda, Portland State Univ
Hutson, Melinda, Portland Comm Coll - Sylvania Campus
Jurewicz, Amy , Arizona State Univ
Karner, James M., Univ of Utah
Krot, Alexander N., Univ of Hawai'i, Manoa
Martins, Zita, Imperial Coll
McCoy, Timothy J., Smithsonian Inst / Nat Museum of Nat History
Park, Jisun, Graduate School of the City Univ of New York

Rodgers, Nick, Univ of Wales
Rubin, Alan E., Univ of California, Los Angeles
Ruzicka, Alexander (Alex) M., Portland State Univ
Sansom, Ellie, Curtin Univ
Schrader, Devin, Arizona State Univ
Sephton, Mark, Imperial Coll
Simon, Steven B., Univ of New Mexico
Sipiera, Paul P., Mercer Univ
Swindle, Timothy D., Univ of Arizona
Tait, Kim, Western Univ
Taylor, G. Jeffrey, Univ of Hawai'i, Manoa
Tera, Fouad, Carnegie Inst for Science
Wadhwa, Meenakshi, Field Museum of Nat History
Warren, Paul, Univ of California, Los Angeles
Wasson, John T., Univ of California, Los Angeles

Astrophysics
Borthakur, Sanchayeeta, Arizona State Univ
Boss, Alan, Carnegie Inst for Science
Bowman, Judd, Arizona State Univ
Bryan, Sean A., Arizona State Univ
Butler, Nathaniel, Arizona State Univ
Butler, Paul, Carnegie Inst for Science
Cruzen, Shawn, Columbus State Univ
Desch, Steven, Arizona State Univ
Devillepoix, Hadrien, Curtin Univ
Gagne, Marc R., West Chester Univ
Groppi, Christopher, Arizona State Univ
Hawthorne Allen, Alice M., Concord Univ
Horst, Sarah, Johns Hopkins Univ
Jacobs, Daniel, Arizona State Univ
Kulkarni, Shrinivas R., California Inst of Tech
MacIssac, Daniel E., Buffalo State Coll
Mauskopf, Philip, Arizona State Univ
Morgan, Siobahn M., Univ of Northern Iowa
Noble, Allison, Arizona State Univ
Paglione, Timothy, York Coll (CUNY)
Patience, Jennifer, Arizona State Univ
Puckett, Andrew, Columbus State Univ
Scannapieco, Evan, Arizona State Univ
Schwarz, Karen, West Chester Univ
Scowen, Paul, Arizona State Univ
Shkolnik, Evgenya, Arizona State Univ
Simon, Molly N., Arizona State Univ
Starrfield, Sumner, Arizona State Univ
Timmes, Francis, Arizona State Univ
Van Engelen, Alexander, Arizona State Univ
Weinberger, Alycia J., Carnegie Inst for Science
Williams, Rosa, Columbus State Univ
Windhorst, Rogier, Arizona State Univ
Young, Patrick, Arizona State Univ
Ziemba, Tim, Univ of Washington

Heliophysics
Bossert, Katrina, Arizona State Univ
Moldwin, Mark B., Univ of Michigan

Astrobiology
Cavalazzi, Barbara, Alma Mater Studiorum Università di Bologna
Rasmussen, Cornelia, Univ of Texas, Austin
Walker, Sara, Arizona State Univ

ATMOSPHERIC SCIENCES

Atmospheric Sciences
Ackerman, Thomas P., Univ of Washington
Adames-Corraliza, Angel F., Univ of Michigan
Adegoke, Jimmy, Univ of Missouri, Kansas City
Agee, Ernest M., Purdue Univ
Aiyyer, Anantha, North Carolina State Univ
Aldrich, Eric A., Univ of Missouri
Alexander, Becky, Univ of Washington
Allan, Andrea, Oregon State Univ
Allard, Jason, Valdosta State Univ
Allen, Grant, Univ of Manchester
Allen, Robert J., Univ of California, Riverside
Alpert, Pinhas, Tel Aviv Univ
Anastasio, Cort, Univ of California, Davis
Ancell, Brian C., Texas Tech Univ
Anderson, Bruce, Boston Univ
Anderson, James G., Harvard Univ
Anderson, Mark R., Univ of Nebraska, Lincoln
Antonescu, Adrian, Univ of Manchester
Aquilina, Noel , Univ of Malta
Arakawa, Akio, Univ of California, Los Angeles
Ariya, Parisa A., McGill Univ
Arnfield, A. John, Ohio State Univ
Arnold, David L., Frostburg State Univ
Arnold, Stephen, Univ of Leeds
Arritt, Raymond W., Iowa State Univ of Science & Tech
Arya, Satyapal S., North Carolina State Univ
Austin, Philip, Univ of British Columbia
Back, Larissa E., Univ of Wisconsin, Madison
Baker, Marcia B., Univ of Washington
Baldwin, Michael, Purdue Univ
Balmforth, Neil, Univ of British Columbia
Barlow, Mathew, Univ of Massachusetts, Lowell
Barnes, Elizabeth A., Colorado State Univ
Barnes, Jeffrey R., Oregon State Univ
Barrett, Bradford S., United States Naval Acad
Barrett, Kevin M., Tarrant County Coll, Northeast Campus
Bartello, Peter, McGill Univ
Barthelmie, Rebecca J., Cornell Univ
Bates, Timothy S., Univ of Washington
Bathke, Deborah J., Univ of Nebraska, Lincoln
Battisti, David S., Univ of Washington
Beard, Kenneth V., Univ of Illinois, Urbana-Champaign
Behrangi, Ali, Univ of Arizona
Bell, Michael M., Colorado State Univ
Bennartz, Ralf, Vanderbilt Univ
Bergin, Michael H., Georgia Inst of Tech
Betterton, Eric, Univ of Arizona
Bierly, Gregory, Indiana State Univ
Bitz, Cecilia M., Univ of Washington
Black, Benjamin, Graduate School of the City Univ of New York
Black, Robert X., Georgia Inst of Tech
Blanchard, Edward, Univ of Washington
Blasing, Terence J., Oak Ridge Nat Lab
Blechman, Jerome B., SUNY, Oneonta
Blyth, Alan, Univ of Leeds
Boag, Thomas, Princeton Univ
Boden, Thomas, Oak Ridge Nat Lab
Boering, Kristie, Univ of California, Berkeley
Bollasina, Massimo A., Edinburgh Univ
Bond, Nicholas A., Univ of Washington
Booth, Alastair, Univ of Manchester
Bosart, Lance F., SUNY, Albany
Bossert, James E., Los Alamos Nat Lab
Bou-Zeid, Elie, Princeton Univ
Boybeyi, Zafer, George Mason Univ
Bradley, Raymond S., Univ of Massachusetts, Amherst
Bretherton, Christopher S., Univ of Washington
Bromwich, David H., Ohio State Univ
Brooks, Sarah D., Texas A&M Univ
Brown, Robert A., Univ of Washington
Brubaker, Kenneth L., Argonne Nat Lab
Bruning, Eric C., Texas Tech Univ
Budikova, Dagmar, Illinois State Univ
Burls, Natalie, George Mason Univ
Bush, Andrew B., Univ of Alberta
Bush, Andrew B. G., Univ of Alberta
Businger, Joost A., Univ of Washington
Calhoun, Joseph, Millersville Univ
Camargo, Suzana, Columbia Univ
Cannon, Alex, Univ of British Columbia
Capehart, William J., South Dakota School of Mines & Tech
Capes, Gerard, Univ of Manchester
Carslaw, Ken, Univ of Leeds
Case Hanks, Anne T., Univ of Louisiana, Monroe
Castro, Christopher, Univ of Arizona
Castro, Mark S., Univ of Maryland
Cess, Robert D., SUNY, Stony Brook
Chameides, William L., Duke Univ

Chang, Edmund K., SUNY, Stony Brook
Chang, Hai-ru, Georgia Inst of Tech
Chang, Young-Soo, Argonne Nat Lab
Charlevoix, Donna J., Univ of Illinois, Urbana-Champaign
Charlson, Bob, Univ of Washington
Charlson, Robert J., Univ of Washington
Chavas, Daniel , Purdue Univ
Chen, Shuyi S., Univ of Washington
Chen, Tsing-Chang, Iowa State Univ of Science & Tech
Chen, Yi-Leng, Univ of Hawai'i at Manoa
Chen, Yongsheng, York Univ
Cheng, Meng-Dawn, Oak Ridge Nat Lab
Chipperfield, Martyn P., Univ of Leeds
Chiu, Christine, Colorado State Univ
Chiu, Long S., George Mason Univ
Choate, Jessica (Jessie), Univ of Illinois, Urbana-Champaign
Choularton, Thomas, Univ of Manchester
Chu, Pao-Shin, Univ of Hawai'i at Manoa
Chuang, Patrick Y., Univ of California, Santa Cruz
Clabo, Darren R., South Dakota School of Mines & Tech
Claire, Mark, Univ of St. Andrews
Clark, Richard D., Millersville Univ
Clarke, Antony D., Univ of Hawai'i, Manoa
Clegg, Simon, Univ of East Anglia
Clemitshaw, Kevin C., Royal Holloway Univ of London
Cobb, Steven R., Texas Tech Univ
Coe, Hue, Univ of Manchester
Colle, Brian, SUNY, Stony Brook
Collett, Jr., Jeffrey L., Colorado State Univ
Collins, Jennifer, Univ of South Florida
Collins, William D., Univ of California, Berkeley
Collis, Scott, Argonne Nat Lab
Colman, Bradley R., Univ of Washington
Colucci, Stephen J., Cornell Univ
Connolly, Paul, Univ of Manchester
Cook, Kerry H., Univ of Texas, Austin
Corcoran, William, Missouri State Univ
Costigan, Keeley R., Los Alamos Nat Lab
Côté, Jean, Universite du Quebec a Montreal
Cottingame, William, Los Alamos Nat Lab
Coulter, Richard L., Argonne Nat Lab
Covert, David S., Univ of Washington
Cowan, Nicolas B., McGill Univ
Crawford, Ian, Univ of Manchester
Crawford, James, Georgia Inst of Tech
Cronin, Timothy W., Massachusetts Inst of Tech
Crowther, Sarah, Univ of Manchester
Curry, Judith A., Georgia Inst of Tech
Cziczo, Daniel, Purdue Univ
Davis, Robert, Univ of Virginia
Dawson, Daniel, Purdue Univ
de Foy, Benjamin, Saint Louis Univ
Dearden, Christopher, Univ of Manchester
DeCaria, Alex J., Millersville Univ
DeConto, Robert, Univ of Massachusetts, Amherst
DeGaetano, Arthur, Cornell Univ
Delcambre, Sharon, Portland Comm Coll - Sylvania Campus
DelSole, Timothy, George Mason Univ
Dempsey, David P., San Francisco State Univ
Deng, Yi, Georgia Inst of Tech
Denning, A. Scott, Colorado State Univ
Derome, Jacques F., McGill Univ
Desai, Ankur R., Univ of Wisconsin, Madison
Dessler, Alex, Texas A&M Univ
Dessler, Andrew E., Texas A&M Univ
Detwiler, Andrew G., South Dakota School of Mines & Tech
Deutsch, Curtis, Princeton Univ
DeWekker, Stephan F., Univ of Virginia
Didlake, Jr., Anthony, Pennsylvania State Univ, Univ Park
Dirmeyer, Paul A., George Mason Univ
Dixon, Richard, Texas State Univ
Dobbie, Steven, Univ of Leeds
Dobler, Scott, Western Kentucky Univ
Doherty, Ruth, Edinburgh Univ
Dominguez, Francina, Univ of Illinois, Urbana-Champaign
Dong, Xiquan, Univ of Arizona
Doskey, Paul V., Argonne Nat Lab
Dugas, Bernard, Universite du Quebec a Montreal
Dunn, Allison L., Worcester State Univ
Durkee, Josh, Western Kentucky Univ
Durran, Dale R., Univ of Washington
Edgell, Dennis J., Univ of North Carolina, Pembroke
Elliott, Scott M., Los Alamos Nat Lab
Ellul, Raymond , Univ of Malta
Epifanio, Craig, Texas A&M Univ
Essery, Richard L., Edinburgh Univ
Esslinger, Kelly L., Arizona Western Coll
Fabry, Frederic, McGill Univ
Faloona, Ian C., Univ of California, Davis
Fan, Xingang, Western Kentucky Univ
Faraji, Maedeh, Austin Comm Coll District
Feng, Song, Univ of Arkansas, Fayetteville
Feng, Yan, Argonne Nat Lab
Fernando, Joe, Univ of Notre Dame
Ferruzza, David, Elizabethtown Coll
Fink, Uwe, Univ of Arizona
Finnegan, David L., Los Alamos Nat Lab
Fiore, Arlene, Columbia Univ
Fiore, Arlene M., Columbia Univ
Fischer, Emily, Colorado State Univ
Fisher, Rebecca, Royal Holloway Univ of London
Fishman, Jack, Saint Louis Univ
Fitzjarrald, David R., SUNY, Albany
Flato, Gregory M., Univ of Victoria
Fletcher, Roy J., Univ of Lethbridge
Forster, Piers M., Univ of Leeds
Fovell, Robert, Univ of California, Los Angeles
Fowler, Malcolm M., Los Alamos Nat Lab
Fox, Neil I., Univ of Missouri
Frame, Jeffrey, Univ of Illinois, Urbana-Champaign
Frederick, John E., Univ of Chicago
French, Adam, South Dakota School of Mines & Tech
Frierson, Dargan M., Univ of Washington
Frolking, Stephen E., Univ of New Hampshire
Fthenakis, Vasilis M., Columbia Univ
Fu, Qiang, Univ of Washington
Fueglistaler, Stephan A., Princeton Univ
Fuentes, Jose D., Pennsylvania State Univ, Univ Park
Fuhrmann, Christopher, Mississippi State Univ
Fung, Inez, Univ of California, Berkeley
Fyfe, John C., Univ of Victoria
Gachon, Philippe, Universite du Quebec a Montreal
Gaffney, Jeffrey S., Argonne Nat Lab
Gallagher, Martin, Univ of Manchester
Galloway, James N., Univ of Virginia
Gallus, William A., Iowa State Univ of Science & Tech
Gao, Yuan, Rutgers, The State Univ of New Jersey, Newark
Garcia, Oswaldo, San Francisco State Univ
Garrett, Timothy J., Univ of Utah
Gedzelman, Stanley, City Coll (CUNY)
Geller, Marvin A., SUNY, Stony Brook
Gervais, Melissa, Pennsylvania State Univ, Univ Park
Ghate, Virendra P., Argonne Nat Lab
Ghil, Michael, Univ of California, Los Angeles
Giannini, Alessandra, Columbia Univ
Gill, Swarndeep S., California Univ of Pennsylvania
Gluhovsky, Alexander, Purdue Univ
Goddard, Lisa M., Columbia Univ
Godfrey, Christopher, Univ of North Carolina, Asheville
Goldblatt, Colin, Univ of Victoria
Gonzalez, Alex, Iowa State Univ of Science & Tech
Goodman, Paul, Univ of Arizona
Goodrich, Greg, Western Kentucky Univ
Gordon, Mark, York Univ
Gordon, Rob J., Univ of Guelph
Gorka, Maciej, Univ of Wroclaw
Graves, Charles E., Saint Louis Univ
Grenfell, Thomas C., Univ of Washington
Griffis, Timothy J., Univ of Minnesota, Twin Cities
Grise, Kevin M., Univ of Virginia
Grotjahn, Richard , Univ of California, Davis
Guinan, Patrick E., Univ of Missouri

Gutzler, David J., Univ of New Mexico
Hage, Keith D., Univ of Alberta
Hakim, Gregory J., Univ of Washington
Hallar, Anna G., Univ of Utah
Hameed, Sultan, SUNY, Stony Brook
Hanrahan, Janel, Northern Vermont Univ-Lyndon
Hansen, William, Pace Univ, New York Campus
Hanson, Howard, Florida Atlantic Univ
Harnik, Nili, Tel Aviv Univ
Harrison, Halstead, Univ of Washington
Harrison, Roy M., Univ of Birmingham
Harshvardhan, Purdue Univ
Hart, Richard L., Argonne Nat Lab
Hartmann, Dennis L., Univ of Washington
Hastenrath, Stefan L., Univ of Wisconsin, Madison
Hastings, Meredith, Brown Univ
Hawkins, Timothy W., Shippensburg Univ
Hayes, Michael J., Univ of Nebraska, Lincoln
Heald, Colette, Massachusetts Inst of Tech
Hegerl, Gabriele C., Edinburgh Univ
Hegg, Dean A., Univ of Washington
Heikes, Brian G., Univ of Rhode Island
Helsdon, John H., South Dakota School of Mines & Tech
Hence, Deanna A., Univ of Illinois, Urbana-Champaign
Henderson, Gina R., United States Naval Acad
Hennon, Chris, Univ of North Carolina, Asheville
Hickmon, Nicki, Argonne Nat Lab
Hindman, Edward E., City Coll (CUNY)
Hirschboeck, Katherine K., Univ of Arizona
Hitchcock, Peter, Cornell Univ
Hitchman, Matthew H., Univ of Wisconsin, Madison
Hjelmfelt, Mark R., South Dakota School of Mines & Tech
Holberg, Jay B., Univ of Arizona
Holgood, Jay S., Ohio State Univ
Holloway, Tracey, Univ of Wisconsin, Madison
Holzer, Mark, Univ of British Columbia
Horel, John, Univ of Utah
Horton, Daniel E., Northwestern Univ
Houston, Adam L., Univ of Nebraska, Lincoln
Houweling, Sander, Vrije Universiteit Amsterdam
Houze, Robert A., Univ of Washington
Howard, Evan, Princeton Univ
Hoyos, Carlos, Georgia Inst of Tech
Hu, Qi S., Univ of Nebraska, Lincoln
Hu, Qi (Steve), Univ of Nebraska, Lincoln
Huang, Alex, Univ of North Carolina, Asheville
Huber, Matthew, Purdue Univ
Huey, Gregory L., Georgia Inst of Tech
Hugli, Wilbur G., Univ of West Florida
Hunten, Donald M., Univ of Arizona
Hurrell, James, Colorado State Univ
Hutyra, Lucy, Boston Univ
Idone, Vincent P., SUNY, Albany
Igel, Adele, Univ of California, Davis
Igel, Matthew, Univ of California, Davis
Ingersoll, Andrew P., California Inst of Tech
Jablonowski, Christiane, Univ of Michigan
Jackson, Andrea, Univ of Leeds
Jacob, Daniel J., Harvard Univ
Jaegle, Lyatt, Univ of Washington
Jaffe, Daniel A., Univ of Washington
Jain, Atul K., Univ of Illinois, Urbana-Champaign
Jenkins, Mary Ann, York Univ
Jewett, Brian, Univ of Illinois, Urbana-Champaign
Johnson, Alexandria, Purdue Univ
Johnson, Brandon, Purdue Univ
Johnson, Graham, Queensland Univ of Tech
Jones, Eric M., Los Alamos Nat Lab
Jones, Hazel, Univ of Manchester
Jones, Phillip D., Univ of East Anglia
Kaiser, Jason, Northern Vermont Univ-Lyndon
Kang, Song-Lak, Texas Tech Univ
Kang, Wanying, Massachusetts Inst of Tech
Kao, Jim C., Los Alamos Nat Lab
Karamperidou, Christina, Univ of Hawai'i at Manoa
Karmalkar, Ambarish, Univ of Massachusetts, Amherst
Kasting, James F., Pennsylvania State Univ, Univ Park
Kauffman, Chad, California Univ of Pennsylvania
Keables, Michael J., Univ of Denver
Keeling, Ralph F., Univ of California, San Diego
Keene, William, Univ of Virginia
Keesee, Robert G., SUNY, Albany
Keller, Linda M., Univ of Wisconsin, Madison
Keller, Jr., C. F., Los Alamos Nat Lab
Kennel, Charles F., Univ of California, San Diego
Keyser, Daniel, SUNY, Albany
Khairoutdinov, Marat, SUNY, Stony Brook
Kidder, Steven, Graduate School of the City Univ of New York
Kieu, Chanh Q., Indiana Univ, Bloomington
Kift, Richard, Univ of Manchester
Kim, Daehyun, Univ of Washington
Kim, Hyemi, Georgia Inst of Tech
Kim, Saewung, Univ of California, Irvine
King, Kenneth M., Univ of Guelph
King, Martin, Royal Holloway Univ of London
Kinter, Jim, George Mason Univ
Kirkpatrick, Cody, Indiana Univ, Bloomington
Kirlin, R. Lynn, Univ of British Columbia
Kirshbaum, Daniel, McGill Univ
Kishcha, Pavel, Tel Aviv Univ
Klaassen, Gary P., York Univ
Klees, Alicia, Univ of Illinois, Urbana-Champaign
Kliche, Donna V., South Dakota School of Mines & Tech
Kluver, Daria, Central Michigan Univ
Knapp, Warren, Cornell Univ
Knight, David, SUNY, Albany
Knopf, Daniel A., SUNY, Stony Brook
Koehler, Thomas , United States Air Force Acad
Konigsberg, Alvin S., SUNY, New Paltz
Kopacz, Dawn, Univ of Nebraska, Lincoln
Korty, Robert, Texas A&M Univ
Kossin, James, Univ of Wisconsin, Madison
Kota, Jozsef, Univ of Arizona
Kotamarthi, Rao, Argonne Nat Lab
Kreidenweis, Sonia M., Colorado State Univ
Krueger, Steven, Univ of Utah
Kuang, Zhiming, Harvard Univ
Kubas, Gregory J., Los Alamos Nat Lab
Kucharik, Chris, Univ of Wisconsin, Madison
Kursinski, Robert, Univ of Arizona
Kushnir, Yochanan, Columbia Univ
L'Ecuyer, Tristan S., Univ of Wisconsin, Madison
Lackmann, Gary M., North Carolina State Univ
Laird, Neil, Hobart & William Smith Colls
Laprise, Rene, Universite du Quebec a Montreal
Lasher-Trapp, Sonia, Univ of Illinois, Urbana-Champaign
Lazarus, Steven M., Florida Inst of Tech
Leather, Kimberly, Univ of Manchester
Lee, In Young, Argonne Nat Lab
Lee, Meehye, Korea Univ
Leibensperger, Eric, SUNY, Plattsburgh
Leighton, Henry G., McGill Univ
Leppert, Ken, Univ of Louisiana, Monroe
Lew, Jeffrey, Univ of California, Los Angeles
Li, Laifang, Pennsylvania State Univ, Univ Park
Li, Tim, Univ of Hawai'i at Manoa
Li, Xiangshan, Univ of Houston
Lichtenberger, János, Eotvos Lorand Univ
Light, Bonnie, Univ of Washington
Lin, Hai, McGill Univ
Lin, Jialin, Ohio State Univ
Lin, John C., Univ of Utah
Lin, Jonathan, Cornell Univ
Lin, Wuyin, SUNY, Stony Brook
Lindgren, Paula, Univ of Glasgow
Liou, Kuo Nan, Univ of California, Los Angeles
Liu, Chuntao, Texas A&M Univ, Corpus Christi
Liu, Dantong, Univ of Manchester
Liu, Jiping, Georgia Inst of Tech
Liu, Ping, SUNY, Stony Brook
Lombardo, Kelly A., Pennsylvania State Univ, Univ Park
Lora, Juan, Yale Univ

Loriot, George, Northern Vermont Univ-Lyndon
Lowe, Douglas, Univ of Manchester
Lozowski, Edward P., Univ of Alberta
Lundquist, Jessica D., Univ of Washington
Luo, Chao, Georgia Inst of Tech
Lupo, Anthony R., Univ of Missouri
Lutsko, Nicholas, Univ of California, San Diego
Lynch, Amanda H., Brown Univ
Lyons, Lawrence, Univ of California, Los Angeles
Lyons, Walter A., Univ of Northern Colorado
Maasch, Kirk A., Univ of Maine
Mace, Gerald, Univ of Utah
Magee, Nathan, Coll of New Jersey
Magnusdottir, Gudrun, Univ of California, Irvine
Mahowald, Natalie M., Cornell Univ
Mak, John E., SUNY, Stony Brook
Mak, Mankin, Univ of Illinois, Urbana-Champaign
Malcolm, Elizabeth, Virginia Wesleyan Coll
Maloney, Eric, Colorado State Univ
Manabe, Syukuro, Princeton Univ
Mann, Michael E., Pennsylvania State Univ, Univ Park
Mann, Michael E., Pennsylvania State Univ, Univ Park
Manning, Andrew, Univ of East Anglia
Marchand, Roger T., Univ of Washington
Margolskee, Andrew, Princeton Univ
Mark, Bryan G., Ohio State Univ
Market, Patrick S., Univ of Missouri
Marley, Nancy A., Argonne Nat Lab
Marsham, John, Univ of Leeds
Martin, Randal S., New Mexico Tech
Martin, Timothy J., Argonne Nat Lab
Martin, Walter, Univ of North Carolina, Charlotte
Mason, Allen S., Los Alamos Nat Lab
Mass, Clifford F., Univ of Washington
Mauzerall, Denise L., Princeton Univ
Mawalagedara, Rachindra, Iowa State Univ of Science & Tech
Maykut, Gary A., Univ of Washington
Mayor, Shane D., California State Univ, Chico
McCollor, Doug, Univ of British Columbia
McDonald, Kyle, Graduate School of the City Univ of New York
McElroy, Michael B., Harvard Univ
McElroy, Tom, York Univ
McFarlane, Norman, Univ of Victoria
McFiggans, Gordon, Univ of Manchester
McGregor, Kent M., Univ of North Texas
McMurdie, Lynn A., Univ of Washington
McQuaid, Jim, Univ of Leeds
McWilliams, James C., Univ of California, Los Angeles
Mechoso, Carlos R., Univ of California, Los Angeles
Meng, Yue (Olivia), Princeton Univ
Merlis, Timothy, McGill Univ
Meskhidze, Nicholas, North Carolina State Univ
Metz, Nicholas, Hobart & William Smith Colls
Meyer, Steven J., Univ of Wisconsin, Green Bay
Michaels, Patrick J., Univ of Virginia
Miller, Doug, Univ of North Carolina, Asheville
Miller, Ronald L., Columbia Univ
Millet, Dylan B., Univ of Minnesota, Twin Cities
Mitchell, Jonathan, Univ of California, Los Angeles
Modzelewski, Henryk, Univ of British Columbia
Molina, Mario J., Univ of California, San Diego
Molinari, John E., SUNY, Albany
Monahan, Edward C., Univ of Connecticut
Moncrieff, John B., Edinburgh Univ
Montenegro, Alvaro, Ohio State Univ
Monteverdi, John P., San Francisco State Univ
Moody, Jennie L., Univ of Virginia
Morawska, Lidia, Queensland Univ of Tech
Morgan, William, Univ of Manchester
Mosley-Thompson, Ellen E., Ohio State Univ
Mote, Philip, Oregon State Univ
Mote, Philip W., Univ of Washington
Moyer, Elisabeth, Univ of Chicago
Moyer, Kerry A., Edinboro Univ of Pennsylvania
Mroz, Eugene J., Los Alamos Nat Lab
Mullen, Steven L., Univ of Arizona
Murphy, Todd, Univ of Louisiana, Monroe
Murrell, Coling, Univ of East Anglia
Murthy, Prahlad N., Wilkes Univ
Nakamura, Noboru, Univ of Chicago
Nappo, Carmen, Georgia Inst of Tech
Nathan, Terrence R., Univ of California, Davis
Nathan, Terry, Univ of California, Davis
Navarra, Antonio, Alma Mater Studiorum Università di Bologna
Neelin, J. David, Univ of California, Los Angeles
Nenes, Athanasios, Georgia Inst of Tech
Nesbitt, Stephen W., Univ of Illinois, Urbana-Champaign
Newman, Steven B., Central Connecticut State Univ
Nielsen-Gammon, John, Texas A&M Univ
Niyogi, Dev, Univ of Texas, Austin
Nogueira, Ricardo, Georgia State Univ
Norris, Joel R., Univ of California, San Diego
North, Gerald, Texas A&M Univ
Nugent, Alison D., Univ of Hawai'i at Manoa
O'Brien, Travis A., Univ of California, Davis
O'Brien, Travis, Indiana Univ, Bloomington
O'Gorman, Paul, Massachusetts Inst of Tech
Obrist, Daniel, Univ of Massachusetts, Lowell
Oechel, Walter, Open Univ
Ojala, Carl F., Eastern Michigan Univ
Okumura, Yuko, Univ of Texas, Austin
Oppenheimer, Michael, Princeton Univ
Orlanski, Isidoro, Princeton Univ
Orlove, Benjamin S., Columbia Univ
Ortiz, Luis, George Mason Univ
Orville, Richard E., Texas A&M Univ
Osterberg, Erich C., Dartmouth Coll
Overland, James E., Univ of Washington
Overpeck, Jonathan T., Univ of Arizona
Palmer, Paul, Edinburgh Univ
Pan, Ying, Pennsylvania State Univ, Univ Park
Pan, Zaitao, Saint Louis Univ
Pani, Eric A., Univ of Louisiana, Monroe
Paulson, Suzanne, Univ of California, Los Angeles
Paw U, Kyaw Tha, Univ of California, Davis
Pendergrass, Angeline G., Cornell Univ
Percell, Peter, Univ of Houston
Percival, Carl, Univ of Manchester
Perez, Richard R., SUNY, Albany
Perry, Kevin D., Univ of Utah
Persad, Geeta, Univ of Texas, Austin
Peterson, Richard E., Texas Tech Univ
Petters, Markus, North Carolina State Univ
Petty, Grant W., Univ of Wisconsin, Madison
Pierce, Jeff, Colorado State Univ
Plumb, Raymond A., Massachusetts Inst of Tech
Polzin, Dierk T., Univ of Wisconsin, Madison
Powell, Mark L., Los Angeles Pierce Coll
Prabhakar, Gouri, Purdue Univ
Prather, Kimberly A., Univ of California, San Diego
Prather, Michael J., Univ of California, Irvine
Preston, Aaron, Northern Vermont Univ-Lyndon
Preston, Thomas C., McGill Univ
Previdi, Michael, Columbia Univ
Price, Colin G., Tel Aviv Univ
Prinn, Ronald G., Massachusetts Inst of Tech
Proistosescu, Cristian (Cristi), Univ of Illinois, Urbana-Champaign
Pryor, Sara C., Cornell Univ
Pu, Zhaoxia, Univ of Utah
Pumphrey, Hugh C., Edinburgh Univ
Pusede, Sally, Univ of Virginia
Radke, Lawrence F., Univ of Washington
Raman, Sethu S., North Carolina State Univ
Ramanathan, V., Univ of California, San Diego
Ramsey, Kelvin W., Univ of Delaware
Randall, David A., Colorado State Univ
Rappenglueck, Bernhard, Univ of Houston
Rasmussen, Kristen L., Colorado State Univ
Rauber, Robert M., Univ of Illinois, Urbana-Champaign
Ravishankara, A.R., Colorado State Univ
Rawlins, Michael, Univ of Massachusetts, Amherst
Ray, Pallav, Florida Inst of Tech

Rayner, John N., Ohio State Univ
Raysoni, Amit , Univ of Texas, Rio Grande Valley
Reay, David, Edinburgh Univ
Reichler, Thomas, Univ of Utah
Reisner, Jon M., Los Alamos Nat Lab
Reiss, Nathan, Pace Univ, New York Campus
Reuter, Gerhard W., Univ of Alberta
Revelle, Douglas O., Los Alamos Nat Lab
Rey-Sanchez, Camilo, North Carolina State Univ
Richter, David, Univ of Notre Dame
Ricketts, Hugo, Univ of Manchester
Riemer, Nicole, Univ of Illinois, Urbana-Champaign
Riordan, Allen J., North Carolina State Univ
Ristovski, Zoran, Queensland Univ of Tech
Ritsche, Michael, Argonne Nat Lab
Robertson, Andrew W., Columbia Univ
Robinson, Walter, North Carolina State Univ
Robinson, Walter A., Univ of Illinois, Urbana-Champaign
Roche, Didier, Vrije Universiteit Amsterdam
Roe, Gerard H., Univ of Washington
Rogers, Jeffrey C., Ohio State Univ
Ross, Robert S., Millersville Univ
Rowe, Clinton M., Univ of Nebraska, Lincoln
Rupp, David, Oregon State Univ
Russell, Armistead G., Georgia Inst of Tech
Russell, Lynn M., Univ of California, San Diego
Rutledge, Steven A., Colorado State Univ
Saltzman, Eric S., Univ of California, Irvine
Samson, Perry, Univ of Michigan
Sandel, Bill R., Univ of Arizona
Sarachik, Edward, Univ of Washington
Sarachik, Edward S., Univ of Washington
Saravanan, R., Texas A&M Univ
Schade, Gunnar, Texas A&M Univ
Schlosser, C. Adam, Massachusetts Inst of Tech
Schmittner, Andreas, Oregon State Univ
Schneider, Edwin K., George Mason Univ
Schneider, Tapio, California Inst of Tech
Schroeder, John L., Texas Tech Univ
Schumacher, Courtney, Texas A&M Univ
Schumacher, Russ, Colorado State Univ
Seeley, Mark W., Univ of Minnesota, Twin Cities
Selin, Noelle, Massachusetts Inst of Tech
Semazzi, Fred H. M., North Carolina State Univ
Serreze, Mark, Univ of Colorado
Shafer, Jason, Northern Vermont Univ-Lyndon
Shell, Karen M., Oregon State Univ
Shellito, Lucinda, Univ of Northern Colorado
Shen, Samuel S., Univ of Alberta
Shen, Xinhua, Univ of Northern Iowa
Shepherd, Gordon G., York Univ
Shepherd, Mark A., Austin Comm Coll District
Shepson, Paul B., Purdue Univ
Sherman-Morris, Kathleen M., Mississippi State Univ
Shinoda, Toshiaki, Texas A&M Univ, Corpus Christi
Shirley, Terry, Univ of North Carolina, Charlotte
Shukla, Jagadish, George Mason Univ
Shulski, Martha D., Univ of Nebraska, Lincoln
Sikora, Todd D., Millersville Univ
Simpson, Robert M., Univ of Tennessee, Martin
Sisterson, Doug, Argonne Nat Lab
Sisterson, Douglas L., Argonne Nat Lab
Siuta, David, Northern Vermont Univ-Lyndon
Smedley, Andrew, Univ of Manchester
Smith, Paul L., South Dakota School of Mines & Tech
Smith, Phillip J., Purdue Univ
Smith, Ronald B., Yale Univ
Snow, Julie A., Slippery Rock Univ
Snyder, Peter K., Univ of Minnesota, Twin Cities
Snyder, Richard L., Univ of California, Davis
Sobel, Adam H., Columbia Univ
Sobel, Adam, Columbia Univ
Sokolik, Irina, Georgia Inst of Tech
Solomon, Susan, Massachusetts Inst of Tech
Somerville, Richard C., Univ of California, San Diego
Soule, Peter T., Appalachian State Univ
Srivastava, Ramesh C., Univ of Chicago
Sriver, Ryan, Univ of Illinois, Urbana-Champaign
St. John, James C., Georgia Inst of Tech
Stan, Cristiana, George Mason Univ
Staten, Paul W., Indiana Univ, Bloomington
Steenburgh, Jim, Univ of Utah
Stevens, Philip S., Indiana Univ, Bloomington
Stevenson, David, Edinburgh Univ
Steyn, Douw G., Univ of British Columbia
Stickel, Robert, Georgia Inst of Tech
Stoelinga, Mark T., Univ of Washington
Stone, Peter H., Massachusetts Inst of Tech
Straub, Derek J., Susquehanna Univ
Straus, David M., George Mason Univ
Strobel, Darrell F., Johns Hopkins Univ
Strong, Courtenay, Univ of Utah
Stull, Roland B., Univ of British Columbia
Sturges, Bill, Univ of East Anglia
Stutz, Jochen P., Univ of California, Los Angeles
Sun, Wen-Yih, Purdue Univ
Suvocarev, Kosana, Univ of California, Davis
Suyker, Andrew E., Univ of Nebraska, Lincoln
Swann, Abigail L., Univ of Washington
Swanson, Brian, Univ of Washington
Szunyogh, Istvan, Texas A&M Univ
Takle, Eugene S., Iowa State Univ of Science & Tech
Talley, John H., Univ of Delaware
Tanamachi, Robin, Purdue Univ
Tatarskii, Viatcheslav, Georgia Inst of Tech
Taylor, Elwynn, Iowa State Univ of Science & Tech
Taylor, Peter A., York Univ
Teeuw, Richard, Univ of Portsmouth
Thorncroft, Christopher D., SUNY, Albany
Thorne, Richard M., Univ of California, Los Angeles
Thornton, Joel A., Univ of Washington
Thurtell, George W., Univ of Guelph
Tillman, James E., Univ of Washington
Ting, Mingfang, Columbia Univ
Tomasko, Martin G., Univ of Arizona
Tongue, Jeffrey, SUNY, Stony Brook
Trapp, Robert J., Univ of Illinois, Urbana-Champaign
Tsikalas, Stephen, Eastern Washington Univ
Tung, Ka-Kit, Univ of Washington
Tung, Wen-wen, Purdue Univ
Turco, Richard, Univ of California, Los Angeles
Twine, Tracy E., Univ of Minnesota, Twin Cities
Tzortziou, Maria, Graduate School of the City Univ of New York
Ullrich, Paul, Univ of California, Davis
Van Den Broeke, Matthew S., Univ of Nebraska, Lincoln
van den Heever, Sue, Colorado State Univ
van Leeuwen, Peter Jan, Colorado State Univ
Vaughan, Geraint, Univ of Manchester
Veres, Michael, SUNY, Oswego
Vimont, Daniel J., Univ of Wisconsin, Madison
Vogelmann, Andrew, SUNY, Stony Brook
von Salzen, Knut, Univ of Victoria
von Salzen, Knut, Univ of British Columbia
Vukovich, George, York Univ
Wagner, Timothy J., Creighton Univ
Walcek, Christopher J., SUNY, Albany
Waliser, Duane E., SUNY, Stony Brook
Wallace, John M., Univ of Washington
Walter-Shea, Elizabeth A., Univ of Nebraska, Lincoln
Wang, Hsiang-Jui, Georgia Inst of Tech
Wang, Jian, SUNY, Stony Brook
Wang, Lei, Purdue Univ
Wang, Pao-Kuan, Univ of Wisconsin, Madison
Wang, Wei-Chyung, SUNY, Albany
Wang, Yongji, Princeton Univ
Wang, Yuan, Purdue Univ
Wang, Yuhang, Georgia Inst of Tech
Wang, Yuqing, Univ of Hawai'i at Manoa
Wang, Yuxuan, Univ of Houston
Wang, Zhuo, Univ of Illinois, Urbana-Champaign
Warland, Jon, Univ of Guelph
Warren, Stephen G., Univ of Washington

Waugh, Darryn W., Johns Hopkins Univ
Wax, Charles L., Mississippi State Univ
Weare, B. C., Univ of California, Davis
Weaver, Justin E., Texas Tech Univ
Weber, Rodney J., Georgia Inst of Tech
Webster, Peter J., Georgia Inst of Tech
Weinbeck, Robert S., SUNY, The Coll at Brockport
Weiss, Christopher C., Texas Tech Univ
Wennberg, Paul O., California Inst of Tech
Wesely, Marvin L., Argonne Nat Lab
Wettstein, Justin, Oregon State Univ
Whitaker, Rodney W., Los Alamos Nat Lab
Whiteway, James, York Univ
Wilhelmson, Robert B., Univ of Illinois, Urbana-Champaign
Wilks, Daniel, Cornell Univ
Williams, John M., SUNY, The Coll at Brockport
Williams, Paul, Univ of Manchester
Willoughby, Hugh E., Florida InterNat Univ
Wilson, John D., Univ of Alberta
Wine, Paul H., Georgia Inst of Tech
Wofsy, Steven C., Harvard Univ
Wood, Kim, Mississippi State Univ
Wood, Robert, Univ of Washington
Wu, Shiliang, Michigan Technological Univ
Wu, Xiaoqing, Iowa State Univ of Science & Tech
Wuebbles, Donald J., Univ of Illinois, Urbana-Champaign
Wysocki, Mark, Cornell Univ
Xi, Xin, Michigan Technological Univ
Xie, Lian, North Carolina State Univ
Yalda, Sepideh, Millersville Univ
Yang, Da, Univ of California, Davis
Yang, Ping, Texas A&M Univ
Yang, Wenchang, Princeton Univ
Yarger, Douglas N., Iowa State Univ of Science & Tech
Yau, Man Kong, McGill Univ
Yi, Chuixiang, Queens Coll (CUNY)
Young, John A., Univ of Wisconsin, Madison
Yu, Jin-Yi, Univ of California, Irvine
Yung, Yuk L., California Inst of Tech
Yuter, Sandra, North Carolina State Univ
Zaitchik, Benjamin, Johns Hopkins Univ
Zarzycki, Colin, Pennsylvania State Univ, Univ Park
Zawadzki, Isztar, McGill Univ
Zehnder, Joseph A., Creighton Univ
Zender, Charles, Univ of California, Irvine
Zeng, Ning, Univ of Maryland
Zhang, Chidong, Univ of Washington
Zhang, Henian, Georgia Inst of Tech
Zhang, Minghua, SUNY, Stony Brook
Zhang, Renyi, Texas A&M Univ
Zhang, Xi, Univ of California, Santa Cruz
Zhu, Ping, Florida InterNat Univ
Zhuang, Qianlai, Purdue Univ
Zipser, Edward, Univ of Utah
Zuend, Andreas, McGill Univ

Meteorology

Adams, Manda S., Univ of North Carolina, Charlotte
Allen, John, Central Michigan Univ
Ault, Toby R., Cornell Univ
Babb, David M., Pennsylvania State Univ, Univ Park
Bahrmann, Chad, Pennsylvania State Univ, Univ Park
Bannon, Peter R., Pennsylvania State Univ, Univ Park
Bartholy, Judit, Eotvos Lorand Univ
Baxter, Martin, Central Michigan Univ
Bernhardt, Jase, Hofstra Univ
Biasutti, Michela, Columbia Univ
Billings, Brian J., Saint Cloud State Univ
Blackwell, Keith G., Univ of South Alabama
Bleck, Rainer, Los Alamos Nat Lab
Bohren, Craig F., Pennsylvania State Univ, Univ Park
Bowman, Kenneth P., Texas A&M Univ
Brown, Michael E., Mississippi State Univ
Brown, Paul W., Univ of Arizona
Brune, William H., Pennsylvania State Univ, Univ Park
Brunet, Gilbert, McGill Univ
Businger, Joost, Univ of Washington
Businger, Steven, Univ of Hawai'i at Manoa
Cahir, John J., Pennsylvania State Univ, Univ Park
Carlson, Richard E., Iowa State Univ of Science & Tech
Carlson, Toby N., Pennsylvania State Univ, Univ Park
Carroll, David, Virginia Polytechnic Inst & State Univ
Challinor, Andy, Univ of Leeds
Chang, Chih-Pei, Naval Postgraduate School
Chen, Hway-Jen, Naval Postgraduate School
Chen, Jing, McMaster Univ
Clark, John H., Pennsylvania State Univ, Univ Park
Clothiaux, Eugene E., Pennsylvania State Univ, Univ Park
Colby, Frank P., Univ of Massachusetts, Lowell
Cook, David R., Argonne Nat Lab
Corona, Thomas J., Metropolitan State Coll of Denver
Crafts, Christine, SUNY, The Coll at Brockport
Creasey, Robert L., Naval Postgraduate School
Croft, Paul J., Kean Univ
Croft, Paul J., Mississippi State Univ
Cullen, Heidi, Georgia Inst of Tech
Czarnetzki, Alan C., Univ of Northern Iowa
Dahl, Johannes M., Texas Tech Univ
Davidson, Kenneth L., Naval Postgraduate School
Davis, Kenneth J., Pennsylvania State Univ, Univ Park
Dixon, P. Grady, Fort Hays State Univ
Dorling, Stephen, Univ of East Anglia
Duncan Tabb, Neva, Saint Petersburg Coll, Clearwater
Durkee, Philip A., Naval Postgraduate School
Dutton, John A., Pennsylvania State Univ, Univ Park
Dyer, Jamie, Mississippi State Univ
Dyson, Liesl, Univ of Pretoria
Eastin, Matt, Univ of North Carolina, Charlotte
Elsberry, Russell L., Naval Postgraduate School
Emanuel, Kerry A., Massachusetts Inst of Tech
Evans, Jenni L., Pennsylvania State Univ, Univ Park
Farrell, Brian F., Harvard Univ
Ferger, Marisa, Pennsylvania State Univ, Univ Park
Fillion, Luc, McGill Univ
Finley, Jason P., Los Angeles Pierce Coll
Fitzpatrick, Patrick J., Mississippi State Univ
Flory, David M., Iowa State Univ of Science & Tech
Flynn, Wendilyn, Univ of Northern Colorado
Frank, William M., Pennsylvania State Univ, Univ Park
Fraser, Alistair B., Pennsylvania State Univ, Univ Park
Frederickson, Paul A., Naval Postgraduate School
Fritsch, J. Michael, Pennsylvania State Univ, Univ Park
Fullmer, James W., Southern Connecticut State Univ
Galewsky, Joseph, Univ of New Mexico
Garza, Jessica, Angelo State Univ
Gauthier, Pierre, Universite du Quebec a Montreal
Gillespie, Terry J., Univ of Guelph
Girard, Eric, Universite du Quebec a Montreal
Godek, Melissa, SUNY, Oneonta
Greybush, Steven J., Pennsylvania State Univ, Univ Park
Griswold, Jennifer S., Univ of Hawai'i at Manoa
Guest, Peter S., Naval Postgraduate School
Gutowski, William J., Iowa State Univ of Science & Tech
Gyakum, John R., McGill Univ
Hacker, Joshua P., Naval Postgraduate School
Hamill, Paul, McHenry County Coll
Haney, Christa M., Mississippi State Univ
Haney, Robert, Naval Postgraduate School
Hansen, Anthony R., Saint Cloud State Univ
Harlim, John, Pennsylvania State Univ, Univ Park
Harr, Patrick A., Naval Postgraduate School
Harrington, Jerry Y., Pennsylvania State Univ, Univ Park
Hartmann, Heike, Slippery Rock Univ
Heyniger, William C., Kean Univ
Hilliker, Joby, West Chester Univ
Hoffman, Eric G., Plymouth State Univ
Hosler, Charles L., Pennsylvania State Univ, Univ Park
Illari, Lodovica, Massachusetts Inst of Tech
Jenkins, Gregory S., Pennsylvania State Univ, Univ Park
Jessup, Stephen, Columbus State Univ
Jin, Fei Fei, Univ of Hawai'i at Manoa
Jordan, Mary S., Naval Postgraduate School

Karmosky, Christopher, SUNY, Oneonta
Kaster, Mark A., Wilkes Univ
Keeler, Jason, Central Michigan Univ
Kimball, Sytske K., Univ of South Alabama
Kubesh, Rodney, Saint Cloud State Univ
Kumjian, Matthew R., Pennsylvania State Univ, Univ Park
Lai, Chung Chieng A., Los Alamos Nat Lab
Lamb, Dennis, Pennsylvania State Univ, Univ Park
Lander, Mark A., Univ of Guam
Landman, Willem A., Univ of Pretoria
Lawson, Merlin P., Univ of Nebraska, Lincoln
Lee, Sukyoung, Pennsylvania State Univ, Univ Park
Leonard, Meredith L., Los Angeles Valley Coll
Lerach, David G., Univ of Northern Colorado
Lindzen, Richard S., Massachusetts Inst of Tech
Liu, Yongqiang, Georgia Inst of Tech
Lyons, Steven, Angelo State Univ
Mandia, Scott A., Suffolk County Comm Coll, Ammerman Campus
Manzi, Anthony, SUNY, Maritime Coll
Martin, Jonathan E., Univ of Wisconsin, Madison
Matthews, Adrian, Univ of East Anglia
McGauley, Michael G., Miami Dade Coll (Kendall Campus)
Meesters, Antoon, Vrije Universiteit Amsterdam
Mendenhall, Larry, Mt. San Antonio Coll
Mercer, Andrew, Mississippi State Univ
Montgomery, Michael T., Naval Postgraduate School
Moore, Richard W., Naval Postgraduate School
Morgan, Michael C., Univ of Wisconsin, Madison
Morschauser, Lindsey, Mississippi State Univ
Mower, Richard, Central Michigan Univ
Murphree, James Thomas, Naval Postgraduate School
Murray, Andrew, Univ of South Alabama
Ndarana, Thando, Univ of Pretoria
Nese, Jon, Pennsylvania State Univ, Univ Park
Nielsen, Kurt E., Naval Postgraduate School
Nietfeld, Daniel, Creighton Univ
Noone, David, Oregon State Univ
Nordstrom, Greg, Mississippi State Univ
North, Gerald R., Texas A&M Univ
Nowotarski, Christopher J., Texas A&M Univ
Nuss, Wendell A., Naval Postgraduate School
O'Neill, Larry, Oregon State Univ
Panetta, Richard L., Texas A&M Univ
Park, Myung-Sook, Naval Postgraduate School
Parker, Doug, Univ of Leeds
Parker, Matthew, North Carolina State Univ
Pasken, Robert W., Saint Louis Univ
Penny, Andrew, Naval Postgraduate School
Person, Arthur, Pennsylvania State Univ, Univ Park
Polvani, Lorenzo M., Columbia Univ
Priest, Eric, Coll of Lake County
Renard, Robert J., Naval Postgraduate School
Renfrew, Ian, Univ of East Anglia
Reuter, Gerhard, Univ of Alberta
Richardson, Yvette P., Pennsylvania State Univ, Univ Park
Ritchie, Harold C., Dalhousie Univ
Ritz, Richard, Creighton Univ
Rochette, Scott M., SUNY, The Coll at Brockport
Rockwood, Anthony A., Metropolitan State Coll of Denver
Rodgers, N, Cardiff Univ
Rodriguez-Plasencia, Oscar, Universidad Auto de Baja California Sur
Ross, Andrew, Univ of Leeds
Russell, William H., Los Angeles Pierce Coll
Ryan, William F., Pennsylvania State Univ, Univ Park
Sanabia, Elizabeth R., United States Naval Acad
Scarnato, Barbara V., Naval Postgraduate School
Schaffer, David L., Wichita State Univ
Schrage, Jon M., Creighton Univ
Schultz, David, Univ of Manchester
Seager, Richard, Columbia Univ
Seaman, Nelson L., Pennsylvania State Univ, Univ Park
Shannon, Jack D., Argonne Nat Lab
Shirer, Hampton N., Pennsylvania State Univ, Univ Park
Skubis, Steven T., SUNY, Oswego
Skyllingstad, Eric, Oregon State Univ
Stamm, Alfred J., SUNY, Oswego
Stauffer, David R., Pennsylvania State Univ, Univ Park
Steiger, Scott, SUNY, Oswego
Stensrud, David J., Pennsylvania State Univ, Univ Park
Straub, Katherine H., Susquehanna Univ
Syrett, William, Pennsylvania State Univ, Univ Park
Terwey, Wes, Univ of South Alabama
Tett, Simon F., Edinburgh Univ
Thomson, Dennis W., Pennsylvania State Univ, Univ Park
Todd, Steve, Portland Comm Coll - Sylvania Campus
Torlaschi, Enrico, Universite du Quebec a Montreal
Tripoli, Gregory J., Univ of Wisconsin, Madison
Vaughan, Naomi, Univ of East Anglia
Verlinde, Johannes, Pennsylvania State Univ, Univ Park
Vernon, James Y., Los Angeles Pierce Coll
Wagner Riddle, Claudia, Univ of Guelph
Wallace, Tim, Mississippi State Univ
Wang, Bin, Univ of Hawai'i at Manoa
Wang, Qing, Naval Postgraduate School
Wash, Carlyle H., Naval Postgraduate School
Weisman, Robert A., Saint Cloud State Univ
Williams, Aaron, Univ of South Alabama
Williams, Forrest, Naval Postgraduate School
Williams, Roger T., Naval Postgraduate School
Wilson, John D., Univ of Alberta
Wyngaard, John C., Pennsylvania State Univ, Univ Park
Wysong, Jr., James F., Hillsborough Comm Coll
Yang, Zong-Liang, Univ of Texas, Austin
Yoh, Shing, Kean Univ
Young, George S., Pennsylvania State Univ, Univ Park
Zhao, Jingxia, Univ of Hawai'i at Manoa
Zois, Constantine S., Kean Univ
Zunkel, Paul, Emporia State Univ

Atmospheric Chemistry

Anastasio, Cort, Univ of California, Davis
Aneja, Viney P., North Carolina State Univ
Dibb, Jack E., Univ of New Hampshire
Elshorbany, Yasin, Univ of South Florida
Faloona, Ian, Univ of California, Davis
Francisco, Joseph, Univ of Pennsylvania
Hess, Peter G., Cornell Univ
Miljevic, Branka, Queensland Univ of Tech
Murray, Lee, Univ of Rochester
Petrenko, Vasilii, Univ of Rochester
Raff, Jonathan, Indiana Univ, Bloomington
Schwab, James J., SUNY, Albany
Silva, Sam, Univ of Southern California
Talbot, Robert, Univ of Houston
Tong, Daniel, George Mason Univ

Atmospheric Physics

Atreya, Sushil, Univ of Michigan
Chen, Shu-Hua, Univ of California, Davis
Kok, Jasper F., Univ of Tennessee, Knoxville
Malakar, Nabin, Worcester State Univ
Micallef, Alfred, Univ of Malta
Ming, Yi, Boston Coll
Torri , Giuseppe, Univ of Hawai'i at Manoa
Warren, Stephen G., Univ of Washington

Climatology

Archer, Emma, Univ of Pretoria
Ayanlade, Ayansina, Obafemi Awolowo Univ
Berkelhammer, Max, Univ of Illinois Chicago
Bradley, Alice, Williams Coll
Burakowski, Elizabeth, Univ of New Hampshire
Burnette, Dorian, Univ of Memphis
Daoust, Mario, Missouri State Univ
DeCosmo, Janice, Univ of Washington
Diem, Jeremy E., Georgia State Univ
Druckenbrod, Daniel L., Rider Univ
Feldl, Nicole, Univ of California, Santa Cruz
Frei, Allan, Hunter Coll (CUNY)
Frye, John , Univ of Wisconsin, Whitewater
Hass, Alisa L., Middle Tennessee State Univ
Kelsey, Eric, Plymouth State Univ
Kravitz, Ben, Indiana Univ, Bloomington

Lehner, Flavio, Cornell Univ
Monier, Erwan, Univ of California, Davis
Osborn, Timothy J., Univ of East Anglia
Payne, Ashley E., Univ of Michigan
Pidwirny, Michael, Univ of British Columbia, Okanagan Campus
Roop, Heidi, Univ of Minnesota, Twin Cities
Saunders, Ian , Univ of British Columbia, Okanagan Campus
Schultze, Steven R., Univ of South Alabama
Skinner, Christopher , Univ of Massachusetts, Lowell
Stine, Alexander, San Francisco State Univ
Suriano, Zac, Univ of Nebraska, Omaha
Tian, Di, Auburn Univ
Tudhope, Alexander, Univ of Washington
Vega, Anthony J., Clarion Univ
Yow, Donald M., Eastern Kentucky Univ
Zarzar, Christopher, North Carolina Central Univ

GEOSCIENCE & SOCIETY

Natural Hazards
Atwater, Brian, Univ of Washington
Blanchard, R. Denise, Texas State Univ
Blodgett, Robert H., Austin Comm Coll District
Clague, John J., Simon Fraser Univ
Cullen, Cheila, Graduate School of the City Univ of New York
Danielson, Stentor B., Slippery Rock Univ
De Scally, Fes, Univ of British Columbia, Okanagan Campus
Dewan, Ashraf, Curtin Univ
Goda, Katsu, Western Univ
Hagelgans, Duane E., Millersville Univ
Hagelman III, Ronald, Texas State Univ
Haney, Jennifer, Bloomsburg Univ
Hill, Arleen A., Univ of Memphis
Horwell, Claire, Durham Univ
Johanson, Erik, Florida Atlantic Univ
Koehler, Rich D., Univ of Nevada, Reno
Nethengwe, Nthaduleni S., Univ of Venda
Pilarczyk, Jessica, Simon Fraser Univ
Smit, Ansie, Univ of Pretoria
Talling, Peter, Durham Univ
Tiefenbacher, John, Texas State Univ
Vidale, Jon , Univ of Washington
Weiss, Robert, Virginia Polytechnic Inst & State Univ
Wooten, Richard M., N.C. Department of Enval Quality

Manmade Hazards
Keil, Chris, Wheaton Coll

Water Quality/Use
Alsharif, Kamal, Univ of South Florida
Berger, Karen, Univ of Rochester
Bouwer, Edward, Johns Hopkins Univ
Bradford, Joel, Utah Valley Univ
Crootof, Arica, Univ of Montana Western
Frankel, Tyler E., Univ of Mary Washington
Hardy, Scott, Case Western Reserve Univ
Keller, Troy, Columbus State Univ
Korths, Patrick, SUNY, Plattsburgh
Kuwabara, James, City Coll of San Francisco
Martin, Zackary E., Austin Comm Coll District
Muller, Marc, Univ of Notre Dame
Rippy, Megan, Virginia Polytechnic Inst & State Univ
Sigler, Adam, Montana State Univ
Steinschneider, Scott, Cornell Univ
Zhang, Ning, Central State Univ

Geoscience Communication
Boss, Stephen K., Univ of Arkansas, Fayetteville
Conway, Michael F., Univ of Arizona
McCartney, M. Carol L., Univ of Wisconsin, Madison, Div of Extension

History of Geoscience
Case, Stephen, Olivet Nazarene Univ
Dezzani, Raymond, Univ of Idaho
Drobnyk, John W., Southern Connecticut State Univ
Gottschalk, Bradford, Univ of Wisconsin, Madison, Div of Extension
Graber, Anna, Univ of Minnesota, Twin Cities
Kohlstedt, Sally G., Univ of Minnesota, Twin Cities
Oreskes, Naomi, Harvard Univ
Rosenberg, Gary D., Milwaukee Public Museum
Strick, James E., Franklin and Marshall Coll

OTHER

General Earth Sciences
Admire, Amanda R., Humboldt State Univ
Agutu, Nathan O., Jomo Kenyatta Univ of Agriculture & Tech
Antipova, Anzhelika, Univ of Memphis
Balch, Jennifer K., Univ of Colorado
Bank, Carl-Georg, Univ of Toronto
Beach Davis, Janet, Heartland Comm Coll
Berry, Leonard, Florida Atlantic Univ
Boger, Rebecca, Graduate School of the City Univ of New York
Boitt, Mark, Jomo Kenyatta Univ of Agriculture & Tech
Booher, Gary, El Camino Coll
Boorstein, Margaret F., Long Island Univ, C.W. Post Campus
Bordossy, Andras, Univ of Newcastle Upon Tyne
Boston, Penelope, Nat Aeronautics & Space Administration
Boyce, Joseph I., McMaster Univ
Bradley, Jeffrey, Northwest Missouri State Univ
Branlund, Joy, Southwestern Illinois Coll - Sam Wolf Granite City Campus
Bries-Korpik, Jill, Century Coll
Britton, Gloria C., Cuyahoga Comm Coll - Western Campus
Brod, Joseph D., Hillsborough Comm Coll
Burns, Danny, Coastal Bend Coll
Busa, Mark, Middlesex Comm Coll
Carlson, Galen R., California State Univ, Fullerton
Carlson, Heath, Eastern Connecticut State Univ
Carrell, Jennifer, Illinois State Geological Survey
Carter, Andy, Birkbeck Coll
Champion, Kyle M., Hillsborough Comm Coll
Chapa, Laura E., Austin Comm Coll District
Chege, George W., Jomo Kenyatta Univ of Agriculture & Tech
Chen, jiquan, Michigan State Univ
Cheng, Zhongqi (Joshua), Graduate School of the City Univ of New York
Cherrier, Jennifer, Graduate School of the City Univ of New York
Childers, Daniel, Delaware County Comm Coll
Cihacek, Larry J., North Dakota State Univ
Clark, Michael, Mount Royal Univ
Cleveland, Natasha, Frederick Comm Coll
Cole, Gregory L., Los Alamos Nat Lab
Condreay, Denise, Central Comm Coll
Cordero, David I., Lower Columbia Coll
Cornebise, Michael W., Eastern Illinois Univ
Cotterill, Carol J., Columbia Univ
Craig, Cameron D., Eastern Illinois Univ
Cranganu, Constantin, Graduate School of the City Univ of New York
Davis, James A., Eastern Illinois Univ
de Wit, Cary, Univ of Alaska, Fairbanks
Deka, Jennifer, Cuyahoga Comm Coll - Western Campus
DeLima, Lynn-Ann, Eastern Connecticut State Univ
Dennison, Robert, Heartland Comm Coll
Dod, Bruce D., Mercer Univ
Dolgoff, Anatole, Pace Univ, New York Campus
Douglas, Arthur V., Creighton Univ
Douthat, Jr., James R., Hillsborough Comm Coll
Dutton, Jessica, Suffolk County Comm Coll, Ammerman Campus
Dyhrman, Sonya, Columbia Univ
Eisenberger, Peter, Columbia Univ
Erski, Theodore, McHenry County Coll
Ezerskis, John L., Cuyahoga Comm Coll - Western Campus
Falk, Lisa, North Carolina State Univ
Fatherree, James W., Hillsborough Comm Coll
Fielding, Lynn, El Camino Coll
Fields, Nancy, El Centro Coll - Dallas Comm Coll District
Flanagan, Michael, Suffolk County Comm Coll, Ammerman Campus
Flores, Kennet, Graduate School of the City Univ of New York
Fondran, Carol, Cuyahoga Comm Coll - Western Campus
Fortes, Dominic, Birkbeck Coll
Foss, Donald J., Coll of Marin
Fritz, William, Graduate School of the City Univ of New York
Gernon, Thomas M., Univ of Southampton
Gibbs, Samantha J., Univ of Southampton

Goetz, Andrew R., Univ of Denver
Goulden, Michael L., Univ of California, Irvine
Groffman, Peter , Graduate School of the City Univ of New York
Harrington, Philip, Suffolk County Comm Coll, Ammerman Campus
Haverluk, Terry W., United States Air Force Acad
Hawkins, Ward L., Los Alamos Nat Lab
Hayes, Van E., Hillsborough Comm Coll
Hendrey, George, Queens Coll (CUNY)
Hendricks, Jonathan R., Cornell Univ
Hillier, John, Grays Harbor Coll
Hobbs, Chasidy, Univ of West Florida
Hoke, Gregory, Syracuse Univ
Hoppe, Kathryn A., Green River Comm Coll
Hughes, Noah C., Modesto Junior Coll
Hurd, David, Edinboro Univ of Pennsylvania
Ibrahim, Mohamed, Hunter Coll (CUNY)
Ide, Kayo, Univ of California, Los Angeles
Imwati, Andrew, Jomo Kenyatta Univ of Agriculture & Tech
Inglis, Michael, Suffolk County Comm Coll, Ammerman Campus
Jackson, Allyson K., SUNY, Purchase
Johnson, Jean M., Dalton State Coll
Johnson, Kathleen, Univ of California, Irvine
Jones, Stephen M., Univ of Birmingham
Kane, Mustapha, Florida Gateway Coll
Keefover-Ring, Ken, Univ of Wisconsin, Madison
Keeling, David J., Western Kentucky Univ
Kellner, Patricia, Los Angeles Harbor Coll
Kelly, Maria, Edmonds Coll
Khade, Vishnu R., Eastern Connecticut State Univ
Koch, Joe, South Carolina Dept of Nat Resources
Lane, Joseph M., Cuyahoga Comm Coll - Western Campus
Lawlor, Kiera, Bridgewater State Univ
Le Qu, Corinne, Univ of East Anglia
Leighty, Robert S., Mesa Comm Coll
Lerch, Derek, Feather River Coll
Levasseur, Emile , Eastern Connecticut State Univ
Lew, Alan A., Northern Arizona Univ
Locke, James L., Coll of Marin
Lofthouse, Stephen T., Pace Univ, New York Campus
Lomaga, Margaret, Suffolk County Comm Coll, Ammerman Campus
Longpre, Marc-Antoine, Graduate School of the City Univ of New York
Luo, Johnny, Graduate School of the City Univ of New York
Mack, John, Los Angeles Harbor Coll
Mackenzie, Kristen, Denver Museum of Nature & Science
MacNeil, James E., Hillsborough Comm Coll
Marchisin, John, Pace Univ, New York Campus
Marlowe, Brian W., Hillsborough Comm Coll
Marra, John, Graduate School of the City Univ of New York
Martek, Lynnette F., Univ of South Carolina, Lancaster
Mason, Sherri A., SUNY, Fredonia
McConnaughhay, Mark, Dutchess Comm Coll
McDougall, Jim, Tacoma Comm Coll
McGinn, Chris, North Carolina Central Univ
McNicol, Barbara, Mount Royal Univ
Meador, Cindy D., West Texas A&M Univ
Medina, Francisco, Universidad Nacional Auto de Mexico
Meisner, Caroline B., Great Basin Coll
Mignery, Jill, Miami Univ
Mitra, Chandana, Auburn Univ
Moreland, Amy L., Austin Comm Coll District
Moreno, Rafael, Metropolitan State Coll of Denver
Morton, Bruce, Eastern Connecticut State Univ
Motyka, James, Eastern Connecticut State Univ
Munasinghe, Tissa, Los Angeles Harbor Coll
Nagel, Athena, Mississippi State Univ
Narey, Martha A., Univ of Denver
Neumann, Patricia, El Camino Coll
Neves, Douglas, El Camino Coll
Newlon, Charles F. J., William Jewell Coll
Nisbet, Euan G., Royal Holloway Univ of London
North, Leslie, Western Kentucky Univ
Norton-Krane, Abby N., Cuyahoga Comm Coll - Western Campus
O'Mullan, Gregory, Graduate School of the City Univ of New York
Olney, Jessica L., Hillsborough Comm Coll
Olney, Matthew P., Hillsborough Comm Coll
Osborn, Joe, Century Coll
Oxford, Jeremiah, Western Oregon Univ
Oymayan, Avo, Florida Gateway Coll
Pant, Hari, Graduate School of the City Univ of New York
Pappas, Matthew, Suffolk County Comm Coll, Ammerman Campus
Parrot, Lael, Univ of British Columbia, Okanagan Campus
Pedley, Kate, Univ of Canterbury
Pekar, Stephen , Graduate School of the City Univ of New York
Peprah, Ebenezer, El Camino Coll
Perry, Baker, Appalachian State Univ
Perry, Randall, Imperial Coll
Pettengill, Gordon H., Massachusetts Inst of Tech
Phillippi, Nathan E., Univ of North Carolina, Pembroke
Pittman, Jason, Folsom Lake Coll
Pond, Lisa G., Louisiana State Univ
Potess, Marla, Hardin-Simmons Univ
Pringle, Patrick T., Centralia Coll
Rapley, Chris, Univ Coll London
Renfrew, Melanie, Los Angeles Harbor Coll
Ritter, Paul, Heartland Comm Coll
Rock, Jessie L., North Dakota State Univ
Rock, Tommy, Princeton Univ
Rosa, Lynn C., West Texas A&M Univ
Roth, Leonard T., Hillsborough Comm Coll
Ryan, Susan, California Univ of Pennsylvania
Sasowsky, Kathryn, Cuyahoga Comm Coll - Western Campus
Saunders, Charles, Randolph-Macon Coll
Saunders, Ralph, California State Univ, Dominguez Hills
Schreier, Hans D., Univ of British Columbia
Schulze, Karl, Waubonsee Comm Coll
Schumacher, Matthew, Saint Francis Xavier Univ
Shaw, Katy, Green River Comm Coll
Shugart, Jr., Herman H., Univ of Virginia
Smilnak, Roberta A., Metropolitan State Coll of Denver
Smith, Grant, Western Oregon Univ
Smith, H. Dixon, Metropolitan State Coll of Denver
Smith, Steven J., Univ of Northern Iowa
Smith, Thomas M., Univ of Virginia
Snyder, Jennifer L., Delaware County Comm Coll
Soash, Norman E., Hillsborough Comm Coll
Spooner, Alecia, Everett Comm Coll
Springer, Kathleen B., San Bernardino County Museum
Stermer, Ed, Illinois Central Coll
Stone, Jim, Aims Comm Coll
Stucky de Quay, Gaia, Massachusetts Inst of Tech
Sumrall, Jeanne L., Fort Hays State Univ
Tarr, Alexander, Worcester State Univ
Thomas, Ray G., Univ of Florida
Thurston, Phillips C., Laurentian Univ, Sudbury
Tidwell, Allan, Chipola Coll
Tinsley, Mark, Central Virginia Comm Coll
Tongdee, Poetchanaporn, Hillsborough Comm Coll
Travers, Steven, Heartland Comm Coll
Valencia, Victor A., Washington State Univ
Valentine, Michael, Tacoma Comm Coll
Van Cleave, Kevan A., Hillsborough Comm Coll
van de Gevel, Saskia, Appalachian State Univ
Veblen, Thomas T., Univ of Colorado
Voorhees, David H., Waubonsee Comm Coll
Vorwald, Brian, Suffolk County Comm Coll, Ammerman Campus
Waddington, David C., Douglas Coll
Walter, Nathan A., Univ of West Georgia
Wang, Enru, Univ of North Dakota
Warger, Jane, Chaffey Coll
Webb, Craig A., Mt. San Antonio Coll
West, Keith, Univ of Wisconsin, Green Bay
White, Susan, Los Angeles Harbor Coll
Williams, Mark W., Univ of Colorado
Wills, William V., Hillsborough Comm Coll
Winterbottom, Wesley, Eastern Connecticut State Univ
Witter, Rob, Oregon Dept of Geology and Mineral Industries
Woods, Karen M., Lamar Univ
XU, Yingfeng, Univ of Louisiana at Lafayette
Yacucci, Mark, Heartland Comm Coll
Yi, Chuixiang, Graduate School of the City Univ of New York
Zaleha, Robert, Cuyahoga Comm Coll - Western Campus
Zhang, Pengfei, Graduate School of the City Univ of New York

Earth Science Education

Albach, Suzanne M., Washtenaw Comm Coll
Armstrong, Anne, Worcester State Univ
Atchison, Christopher L., Univ of Cincinnati
Attrep, Moses, Los Alamos Nat Lab
Aylward, Linda, Illinois Central Coll
Baker, Brett, Univ of Texas, Austin
Barnett, Michael, Boston Coll
Bednarski, Marsha, Central Connecticut State Univ
Bednarz, Robert S., Texas A&M Univ
Bednarz, Sarah W., Texas A&M Univ
Benner, Jake, Univ of Tennessee, Knoxville
Bergwerff, Kenneth A., Calvin Coll
Bernard-Kingsley, Noell, Univ of Washington
Black, Alice (Jill), Missouri State Univ
Blanford, William , Graduate School of the City Univ of New York
Block, Karin, Graduate School of the City Univ of New York
Bowman, Cassandra, Arizona State Univ
Buonaiuto Jr., Frank S., Graduate School of the City Univ of New York
Burns, Sandra, Central Connecticut State Univ
Cadieux, Sarah, Rensselaer Polytechnic Inst
Carden-Jessen, Melanie E., Missouri State Univ
Carrick, Tina L., El Paso Comm Coll
Cercone, Karen Rose, Indiana Univ of Pennsylvania
Chandler, Sandra, Arkansas Geological Survey
Charbonneau, Luc, American Geosciences Inst
Clark, Scott K., Univ of Wisconsin, Eau Claire
Clary, Renee M., Mississippi State Univ
Clepper, Marta L., SUNY, Oneonta
Coburn, Daniel, Southern Connecticut State Univ
Coleman, Catherine, Arizona State Univ
Connolly, Cynthia, New Mexico Tech
Cooney, Timothy M., Univ of Northern Iowa
Crowder, Margaret, Western Kentucky Univ
d'Alessio, Matthew, California State Univ, Northridge
Dakin, Susan, Univ of Lethbridge
Dalman, Michael, Blinn Coll
Davies, Gareth, Open Univ
Davies, Sarah, Open Univ
De Kock, Walter E., Univ of Northern Iowa
de Ritter, Samantha, Flinders Univ
Dehne, Kevin T., Delta Coll
DeKraker, Dan, Cerritos Coll
Denton-Hedrick, Meredith Y., Austin Comm Coll District
DePriest, Thomas A., Univ of Tennessee, Martin
Dieckmann, Melissa S., Eastern Kentucky Univ
Douglass, David N., Pasadena City Coll
Egger, Anne E., Central Washington Univ
Elkins, Joe T., Univ of Northern Colorado
Emerson, Cheryl R., Illinois Central Coll
Ensign, Todd I., Fairmont State Univ
Falkena, Lee, Univ of Illinois Chicago
Faust, Megan, Portland State Univ
Ferguson, Julie , Univ of California, Irvine
Flanagan, Timothy, Berkshire Comm Coll
Forcino, Frank, Western Carolina Univ
Frankic, Anamarija, Univ of Massachusetts, Boston
Geraghty Ward, Emily, Rocky Mountain Coll
Gibbons, Doug, Univ of Washington
Glesener, Gary, Virginia Polytechnic Inst & State Univ
Godsey, Holly, Univ of Utah
Goodell, Laurel P., Princeton Univ
Gosselin, David C., Univ of Nebraska, Lincoln
Gray, Kyle R., Univ of Northern Iowa
Hammond, Amie C., Austin Comm Coll District
Harrell, Michael, Univ of Washington
Heid, Kelly L., Grand Valley State Univ
Hemler, Deb, Fairmont State Univ
Herman, Theodore C., West Valley Coll
Holliman, Richard, Open Univ
Hollister, Ryan, California State Univ, Stanislaus
Hubenthal, Michael, Binghamton Univ
Hunter, Arlene G., Open Univ
Isava, Virginia, California State Univ, Fullerton
Jahanyar, Alireza, Univ of Tabriz
Jo, Injeong, Texas State Univ
Johnston, Thomas, Univ of Lethbridge
Kornreich Wolf, Susan, Augustana Coll
Kortz, Karen M., Comm Coll of Rhode Island
Krause, Lois B., Clemson Univ
Kusnick, Judith E., Sacramento State Univ
LaDue, Nicole D., Northern Illinois Univ
Libarkin, Julie C., Michigan State Univ
Longoria, Jose F., Florida InterNat Univ
Ludwikoski, David J., Comm Coll of Baltimore County, Catonsville
MacKinnon, Stuart J., Univ of British Columbia, Okanagan Campus
MacLachlan, Ian R., Univ of Lethbridge
Maltese, Adam V., Indiana Univ, Bloomington
Martin-Vermilyea, Laurie, Montgomery County Comm Coll
Mattox, Stephen R., Grand Valley State Univ
Maxwell, James "Sandy", West Chester Univ
McCollough, Cherie, Texas A&M Univ, Corpus Christi
McConnell, David A., North Carolina State Univ
McEvoy, Jamie , Montana State Univ
McGeary, Susan, Univ of Delaware
McMullin, David W., Acadia Univ
McNeal, Karen S., Auburn Univ
Messina, Paula, San Jose State Univ
Miller, Heather R., Austin Comm Coll District
Milligan, Mark, Utah Geological Survey
Miyares, Ines, Hunter Coll (CUNY)
Morgan, Emory, Hampton Univ
Morton, Allan E., Central Arizona Coll
Mulkey, Sean, Illinois Central Coll
Muniz-Solari, Oswaldo, Texas State Univ
Munski, Douglas C., Univ of North Dakota
Murray, Daniel P., Univ of Rhode Island
Newcomer, April, Virginia Polytechnic Inst & State Univ
O'Neal, Claire J., Univ of Delaware
Oxley, Meghan, Univ of Washington
Papcun, George, New Jersey City Univ
Pepple, Christopher, Bowling Green State Univ
Petit, Martin A., Illinois Central Coll
Phillips, Paul, Fort Hays State Univ
Preston, William L., California Polytechnic State Univ
Pyle, Eric J., James Madison Univ
Rakonczai, János, Univesity of Szeged
Reano, Darryl, Arizona State Univ
Refenes, James L., Concordia Univ
Revels, Josh, Fairmont State Univ
Roemmele, Christopher, West Chester Univ
Russ, Jean M., Coll of Southern Maryland
Ryker, Katherine, Univ of South Carolina
Schaffer, Linda J., SUNY, The Coll at Brockport
Schmitt, Danielle M., Princeton Univ
Scott, Vernon, Oklahoma State Univ
Semken, Steven, Arizona State Univ
Shelton, Dale W., Maryland Department of Nat Resources
Shepardson, Daniel P., Purdue Univ
Skinner, Lisa, Northern Arizona Univ
Slattery, William, Wright State Univ
Smay, Jessica J., San Jose City Coll
Spurr, Aaron, Univ of Northern Iowa
Straw, Byron, Univ of Northern Colorado
Sundareshwar, P. V., South Dakota School of Mines & Tech
Sutton, Connie J., Indiana Univ of Pennsylvania
Swyrtek, Sheila M., Mott Comm Coll
Taylor, Ta-Shana A., Univ of Miami
Teed, Rebecca, Wright State Univ
Thompson, Kenneth W., Emporia State Univ
Tonon, Marco Davide, Università di Torino
van der Flier-Keller, Eileen, Simon Fraser Univ
Villalobos, Joshua, El Paso Comm Coll
Viskupic, Karen, Boise State Univ
Wagner, John R., Clemson Univ
White, Lisa D., Univ of California, Berkeley
Wood, Howard, Texas A&M Univ, Corpus Christi
Wright, Carrie L., Univ of Southern Indiana
Yuan, Yihong, Texas State Univ
Zafian, Holly, Austin Comm Coll District
Zipper, Carl E., Virginia Polytechnic Inst & State Univ
Zlotkin, Howard, New Jersey City Univ

Zwick, Thomas T., Montana State Univ, Billings

Physical Geography

Aay, Henry, Calvin Coll
Abend, Martin, New Jersey City Univ
Aber, Jeremy, Middle Tennessee State Univ
Akiwumi, Fenda, Univ of South Florida
Albright, Thomas , Univ of Nevada, Reno
Allen, Kelly E., Tulsa Comm Coll
Andresen, Christian, Univ of Wisconsin, Madison
Applegate, Toby, Univ of Massachusetts, Amherst
Atkins, Dane, Texas State Univ
Aubert, John E., American River Coll
Balascio, Nicholas, William & Mary
Ballantyne, Colin K., Univ of St. Andrews
Barnett, Roger T., Univ of the Pacific
Barton, James H., Thiel Coll
Bascom, Johnathan, Calvin Coll
Bass, David, Flinders Univ
Bates, Mary, Coll of the Canyons
Battles, Eileen, Univ of Kansas
Bein, F. L., Indiana Univ, Indianapolis
Bekker, Matthew F., Brigham Young Univ
Beniest, Anouk, Vrije Universiteit Amsterdam
Berry, Kate, Univ of Nevada, Reno
Black, Kathryn, Montclair State Univ
Blackburn, William, Western Kentucky Univ
Boester, Michael, Monroe Comm Coll
Brothen, Jerry, El Camino Coll
Brothers, Timothy S., Indiana Univ, Indianapolis
Butzow, Dean G., Lincoln Land Comm Coll
Cairns, David M., Texas A&M Univ
Camille, Michael A., Univ of Louisiana, Monroe
Campbell, Glenn A., Eastern Kentucky Univ
Carter, Greg, Univ of Southern Mississippi
Chacko, Elizabeth, George Washington Univ
Chaddock, Lisa, Cuyamaca Coll
Chatterjee, Meera, Univ of Akron
Clarke, Beverley, Flinders Univ
Clennan, Patrick D., Coll of Southern Nevada - West Charleston Campus
Coburn, Craig, Univ of Lethbridge
Comrie, Andrew, Univ of Arizona
Coveney, Eamonn, Fort Hays State Univ
Croft, Gary M., San Bernardino Valley Coll
Cross, John A., Univ of Wisconsin Oshkosh
Csank, Adam, Univ of Nevada, Reno
Csiki, Shane, New Hampshire Geological Survey
Curry, Janel M., Calvin Coll
Dascher, Erin, Eastern Washington Univ
Davidson, Fiona M., Univ of Arkansas, Fayetteville
Dawsey, Cyrus B., Auburn Univ
de Uña Alvarez, Elena P., Coruna Univ
DeWeese, Georgina G., Univ of West Georgia
Dexter, Leland R., Northern Arizona Univ
Dixon, Richard W., Texas State Univ
Dubsky, Scott , United States Air Force Acad
Dugmore, Andrew J., Edinburgh Univ
Duncan, Ian, City Coll of San Francisco
Ebiner, Matt, El Camino Coll
Eisenhart, Karen, Edinboro Univ of Pennsylvania
Engstrom, Vanessa, San Bernardino Valley Coll
Feeney, Thomas P., Shippensburg Univ
Fonstad, Mark, Univ of Oregon
Frolking, Tod A., Denison Univ
Gamble, Douglas W., Univ of North Carolina, Wilmington
Gaubatz, Piper, Univ of Massachusetts, Amherst
Gavin, Dan, Univ of Oregon
Girhard, T S., San Antonio Comm Coll
Goudge, Theodore L., Northwest Missouri State Univ
Graham, Barbara, Coll of Southern Nevada - West Charleston Campus
Greene, Don M., Baylor Univ
Gripshover, Margaret , Western Kentucky Univ
Hall, Luke D., Ventura Coll
Hampson, Arthur, Univ of Utah
Harley, Grant , Univ of Idaho
Harrison, Robert S., Long Island Univ, C.W. Post Campus
Hart, Evan A., Tennessee Tech Univ
Harty, John P., Johnson County Comm Coll
Hay, Iain, Flinders Univ
Hazen, Helen, Univ of Denver
Hedquist, Brent, Texas A&M Univ, Kingsville
Heibel, Todd, San Bernardino Valley Coll
Heil, Melissa, Illinois State Univ
Henriksen, Mona, Norwegian Univ of Life Sciences (NMBU)
Hepner, George F., Univ of Utah
Hess, Darrel E., City Coll of San Francisco
Heywood, Neil C., Univ of Wisconsin, Stevens Point
Hindle, Tobin, Florida Atlantic Univ
Holland, Edward C., Univ of Arkansas, Fayetteville
Holt, David, Univ of Southern Mississippi
Houston, Serin D., Mount Holyoke Coll
Ivey, James, Univ of South Florida
Jeon, Sung Bae, Eastern Kentucky Univ
Jones, Alice, Eastern Kentucky Univ
Jordan, Karen J., Univ of South Alabama
Karpilo, Jr, Ronald J., Colorado State Univ
Kavanagh, Trudy A., Univ of British Columbia, Okanagan Campus
Kay, Jonathan, William & Mary
Kebbede, Girma, Mount Holyoke Coll
Kennedy, Linda, Mansfield Univ
Key, Doug, Palomar Coll
Khan, Belayet H., Eastern Illinois Univ
Kiage, Lawrence M., Georgia State Univ
Kilcoyne, John R., Metropolitan State Coll of Denver
Kimber, Clarissa T., Texas A&M Univ
Kung, Hsiang-Te, Univ of Memphis
Kuzera, Kristopher , Univ of Denver
Ladányi, Zsuzsanna, Univesity of Szeged
Laity, Julie E., California State Univ, Northridge
Lambert, Dean P., San Antonio Comm Coll
Lambert, Dean, Alamo Colls, San Antonio Coll
League, Larry D., Dickinson State Univ
Lee, Jeffrey A., Texas Tech Univ
Lee, Russell, Oak Ridge Nat Lab
Little, Jonathan, Monroe Comm Coll
Liu, Jing, Santa Monica Coll
LoVetere, Crystal, Cerritos Coll
Lynn, Resler M., Virginia Polytechnic Inst & State Univ
MacQuarrie, Pamela, Mount Royal Univ
Maher, John, Johnson County Comm Coll
Maio, Chris, Univ of Alaska, Fairbanks
Mann, Dan, Univ of Alaska, Fairbanks
Manos, Leah D., Northwest Missouri State Univ
Markwith, Scott H., Florida Atlantic Univ
Marzulli, Walter, New Jersey Geological and Water Survey
Mbatu, Richard, Univ of South Florida
McAfee, Stephanie , Univ of Nevada, Reno
McCoy, William D., Univ of Massachusetts, Amherst
McGwire, Kenneth C., Desert Research Inst
Mensing, Scott A., Univ of Nevada, Reno
Mezősi, Gábor, Univesity of Szeged
Minnich, Richard A., Univ of California, Riverside
Mizak, Connie, Univ of South Florida
Montoya, Judith, Northern Arizona Univ
Morris, Pete, Santa Monica Coll
Motta, Luigi, Università di Torino
Motta, Michele, Università di Torino
Nelson, Kenneth A., Univ of Kansas
Ngoy, Kikombo, Kean Univ
Nicholas, Joseph W., Mary Washington Coll
Nuwategeka, Expedito, Gulu Univ
Ogbuchiekwe, Edmund Jekwu, San Bernardino Valley Coll
Oldfield, Jonathan, Univ of Birmingham
Olson, Kimberly, American River Coll
Paulsell, Robert L., Louisiana State Univ
Peake, Jeffrey S., Univ of Nebraska, Omaha
Pennington, Deana, Univ of Texas, El Paso
Pesses, Michael, Antelope Valley Coll
Phillips, Nathan, Boston Univ
Pierce, Heather, Monroe Comm Coll
Pinnt, Todd, Arizona Western Coll
Ponette-Gonzalez, Alexandra, Univ of North Texas

Precht, Francis L., Frostburg State Univ
Price, Marie D., George Washington Univ
Privette, David, Central Piedmont Comm Coll
Ptak, Thomas, Texas State Univ
Rahman, Abu, Antelope Valley Coll
Reese, Andy, Univ of Southern Mississippi
Rettig, Andrew, Univ of Dayton
Riley, James D., Eastern Illinois Univ
Robinson, Michael A., Santa Barbara City Coll
Rodgers, John C., Mississippi State Univ
Rollins, Kyle, Missouri Dept of Nat Resources
Rose, Leanna S., Univ of West Georgia
Ross, Thomas E., Univ of North Carolina, Pembroke
Rudnicki, Ryan E., San Antonio Comm Coll
Rudnicki, Ryan E., Alamo Colls, San Antonio Coll
Saladyga, Thomas, Concord Univ
Sandlin, Stephen H., San Bernardino Valley Coll
Santelmann, Mary V., Oregon State Univ
Schafer, Tom, Fort Hays State Univ
Schmidt, Lisa, San Bernardino Valley Coll
Scott, Larry, Colorado Geological Survey
Scott, Thomas A., Univ of California, Riverside
Sharkey, Debra, Cosumnes River Coll
Smith, C. K., San Antonio Comm Coll
Smith, Charles K., Alamo Colls, San Antonio Coll
Smith, Steven C., American River Coll
Spinney, Jamie, Southern Utah Univ
Stampone, Mary D., Univ of New Hampshire
Stern, Herschel I., MiraCosta Coll
Stevens, Stan, Univ of Massachusetts, Amherst
Stocks, Lee, Mansfield Univ
Swarts, Stanley W., Northern Arizona Univ
Thomsen, Charles E., American River Coll
Townsend, Christi, Texas State Univ
Tremblay, Thomas A., Univ of Texas at Austin, Jackson Schl of Geosciences
van Leeuwen, Willem, Univ of Arizona
Veverka, Laura, Metropolitan Comm Coll-Kansas City
Vogel, Eve, Univ of Massachusetts, Amherst
Walter, Thomas, Hunter Coll (CUNY)
Watson, James R., Oregon State Univ
Webb, Craig, Eastern Kentucky Univ
Weisberg, Michael, Graduate School of the City Univ of New York
White, Justin, Utah Valley Univ
Wilkie, Richard W., Univ of Massachusetts, Amherst
Wilkins, David E., Boise State Univ
Wilson, Jeffey S., Indiana Univ, Indianapolis
Wu, Shuang-Ye, Univ of Dayton
Wyckoff, John W., Univ of Colorado, Denver
Yassin, Barbara E., Mississippi Office of Geology
Zhou, Yuyu, Iowa State Univ of Science & Tech

Remote Sensing

Aanstoos, James V., Iowa State Univ of Science & Tech
Abdalati, Waleed, Univ of Colorado
Abrams, Michael J., Jet Propulsion Lab
Ackerman, Steven A., Univ of Wisconsin, Madison
Aiken, Carlos L., Univ of Texas, Dallas
Ali, K. Adem, Coll of Charleston
Alley, Ronald E., Jet Propulsion Lab
Asner, Gregory, Arizona State Univ
Barros, Jose Antonio, Florida InterNat Univ
Beahan, Lyn, Curtin Univ
Beaverson, Sheena K., Illinois State Geological Survey
Becker, Richard H., Univ of Toledo
Becker, Richard H., Argonne Nat Lab
Berlin, Graydon L., Northern Arizona Univ
Bernier, Monique, Universite du Quebec
Blom, Ronald G., Jet Propulsion Lab
Boulton, Sarah, Univ of Plymouth
Broadfoot, Lyle A., Univ of Arizona
Buchanan, Brian, Eastern Washington Univ
Buratti, Bonnie J., Jet Propulsion Lab
Cablk, Mary, Desert Research Inst
Calvin, Wendy, Univ of Nevada, Reno
Campbell, Bruce A., Smithsonian Inst, Nat Air & Space Museum
Campbell, James B., Virginia Polytechnic Inst & State Univ
Cantelli, Luigi, Alma Mater Studiorum Università di Bologna
Carn, Simon, Michigan Technological Univ
Cetin, Haluk, Murray State Univ
Chen, Shu-Hua, Univ of California, Davis
Chen, Xianfeng, Slippery Rock Univ
Chopping, Mark J., Montclair State Univ
Clark, Martin D., Univ of the Free State
Crippen, Robert E., Jet Propulsion Lab
Currit, Nathan, Texas State Univ
Dai, Chunli, Ohio State Univ
Dana, Gayle L., Univ of Nevada, Reno
Das, Jnaneshwar, Arizona State Univ
Dash, Padmanava, Mississippi State Univ
Datt, Bisun, Curtin Univ
Delahunty, Tina, Bloomsburg Univ
Di Girolamo, Larry, Univ of Illinois, Urbana-Champaign
Dodge, Rebecca L., Midwestern State Univ
Dondofema, Farai, Univ of Venda for Science & Tech
Duchesne, Rocio, Univ of Wisconsin, Whitewater
Easson, Gregory L., Univ of Mississippi
El-Baz, Farouk, Boston Univ
Engstrom, Ryan, George Washington Univ
Fenstermaker, Lynn, Desert Research Inst
Ferris, Michael H., California State Univ, Dominguez Hills
Fildes, Stephen , Flinders Univ
Flynn, Luke P., Univ of Hawai'i, Manoa
Forster, Richard R., Univ of Utah
Frankenberg, Christian, California Inst of Tech
Frazier, Ryan J., Weber State Univ
Friedl, Mark A., Boston Univ
Friedl, Mark, Boston Univ
FROGER, Jean-Luc, Université Jean Monnet, Saint-Etienne
Frost, Eric G., San Diego State Univ
Gaitan-Moran, Javier, Universidad Auto de Baja California Sur
Gammack-Clark, James, Florida Atlantic Univ
Gamon, John, Univ of Alberta
Gaulton, Rachel, Univ of Newcastle Upon Tyne
Gaya, Charles , Jomo Kenyatta Univ of Agriculture & Tech
Ghent, Rebecca, Univ of Toronto
Gibson, Glen, United States Air Force Acad
Gilbes, Fernando, Univ of Puerto Rico
Gimmestad, Gary G., Georgia Inst of Tech
Giraldo, Mario A., California State Univ, Northridge
Glenn, Nancy, Boise State Univ
Goetz, Alexander, Univ of Colorado
Hacker, Jorg, Flinders Univ
Hanna, Stephen P., Mary Washington Coll
Hardin, Perry J., Brigham Young Univ
Haritashya, Umesh, Univ of Dayton
Hay, Rodrick, California State Univ, Dominguez Hills
Heidinger, Andrew, Univ of Wisconsin, Madison
Hinojosa, Alejandro, Centro de Inv Científica y de Ed Superior de Ensenada
Hobbs, Richard D., Amarillo Coll
Homayouni, Saied, Universite du Quebec
Hook, Simon J., Jet Propulsion Lab
Hossain, A K M Azad, Univ of Tennessee, Chattanooga
Hu, Baoxin, York Univ
Hu, Zhiyong, Univ of West Florida
Huang, Yi, McGill Univ
Humes, Karen, Univ of Idaho
Hung, Ming-Chih, Northwest Missouri State Univ
Hutchinson, Charles F., Univ of Arizona
Jaworski, Eugene, Eastern Michigan Univ
Jensen, Jennifer, Texas State Univ
Ji, Wei, Univ of Missouri, Kansas City
Jin, Yufang, Univ of California, Davis
Johnson, Elias, Missouri State Univ
Jones, Eriita, Curtin Univ
Kahle, Anne B., Jet Propulsion Lab
Kasar, Sheila , Clarion Univ
Kenduiywo, Benson K., Jomo Kenyatta Univ of Agriculture & Tech
Kennedy, Robert, Oregon State Univ
Kern, Anikó, Eotvos Lorand Univ
Kessler, Fritz, Frostburg State Univ
Key, Jeffrey R., Univ of Wisconsin, Madison
Khan, Shuhab D., Univ of Houston

Klein, Andrew G., Texas A&M Univ
Klemas, Victor, Univ of Delaware
Kollias, Pavlos, McGill Univ
Kovács, Ferenc, Univesity of Szeged
Kudela, Raphael M., Univ of California, Santa Cruz
Lakhan, V. Chris, Univ of Windsor
Larson, Eric J., Univ of Wisconsin, Stevens Point
Lawrence, Rick L., Montana State Univ
Lee, Keenan, Colorado School of Mines
Lee, Zhongping, Univ of Massachusetts, Boston
Li, Juan, Curtin Univ
Li, Lin, Indiana Univ - Purdue Univ Indianapolis
Liu, Jian G., Imperial Coll
Lu, Zhong, Southern Methodist Univ
Lulla, Kamlesh, Univ of Delaware
Lynch, Merv, Curtin Univ
Lyon, Ronald J. P., Stanford Univ
Malhotra, Rakesh, North Carolina Central Univ
Manning, Aspen , Southern Utah Univ
Mason, Philippa J., Imperial Coll
Mbuh, Mbongowo J., Univ of North Dakota
Mccoy, Roger M., Univ of Utah
McDade, Ian C., York Univ
McGrath, Andrew, Flinders Univ
McGwire, Kenneth, Univ of Nevada, Reno
Meyer, Franz J., Univ of Alaska, Fairbanks
Miao, Xin, Missouri State Univ
Middleton, Carrie A., Department of Interior Office of Surface Mining Reclamation and Enforcement
Milan-Navarro, Joel, Universidad Auto de San Luis Potosi
Miller, John, York Univ
Millette, Thomas L., Mount Holyoke Coll
Minor, Timothy B., Desert Research Inst
Molnár, Gábor, Eotvos Lorand Univ
Molnia, Bruce F., Duke Univ
Moon, Wooil M., Univ of Manitoba
Mshiu, Elisante E., Univ of Dar es Salaam
Muller-Karger, Frank E., Univ of South Florida
Mushkin, Amit, Univ of Washington
Mustard, John F., Brown Univ
Myneni, Ranga , Boston Univ
Nduati, Eunice W., Jomo Kenyatta Univ of Agriculture & Tech
Newland, Franz, York Univ
Ngigi, Thomas G., Jomo Kenyatta Univ of Agriculture & Tech
Ni-Meister, Wenge, Hunter Coll (CUNY)
Njoku, Eni G., Jet Propulsion Lab
Noh, Myoung-Jong, Ohio State Univ
O'Banion, Matthew S., United States Military Acad
Palluconi, Frank D., Jet Propulsion Lab
Palmer, Richard, Curtin Univ
Parrish, Christopher E., Univ of New Hampshire
Pathirana, Sumith, Southern Cross Univ
Peddle, Derek R., Univ of Lethbridge
Peltzer, Gilles, Univ of California, Los Angeles
Petrie, Gregg, Univ of Washington
Pieters, Carle M., Brown Univ
Platnick, Steven, Univ of Wisconsin, Madison
Powell, Rebecca , Univ of Denver
Powell, Scott, Montana State Univ
Price, Kevin, Kansas State Univ
Pu, Ruiliang, Univ of South Florida
Qin, Rongjun, Ohio State Univ
Quisenberry, Dan R., Mercer Univ
Rahman, Abdullah F., Univ of Texas, Rio Grande Valley
Ramage, Joan, Lehigh Univ
Ramoele, Abel, Univ of Pretoria
Ramsey, Michael S., Univ of Pittsburgh
Ramspott, Matthew E., Frostburg State Univ
Rapp, Anita, Texas A&M Univ
Reed, Wallace E., Univ of Virginia
Ridd, Merrill K., Univ of Utah
Rivard, Benoit, Univ of Alberta
Roberts, Charles E., Florida Atlantic Univ
Rundquist, Bradley C., Univ of North Dakota
Sadiq, Abdulali A., Univ of Qatar
Sahr, John D., Univ of Washington
Salvini, Riccardo, Univ of Siena
Schaaf, Crystal, Boston Univ
Schaaf, Crystal, Univ of Massachusetts, Boston
Schenk, Anton, SUNY, Buffalo
Seirer, Jami J., Fort Hays State Univ
Shirzaei, Manoochehr, Virginia Polytechnic Inst & State Univ
Siegfried, Matthew, Colorado School of Mines
Small, Christopher, Columbia Univ
Smith, Mike, Kingston Univ
Smith, Peter, Univ of Arizona
Stephenson, Garth, Stellenbosch Univ
Strahler, Alan, Boston Univ
Styers, Diane M., Western Carolina Univ
Sui, Daniel Z., Texas A&M Univ
Sultan, Mohamed, Western Michigan Univ
Szatmári, József, Univesity of Szeged
Thomas, Jeremy, Univ of Washington
Torres Parisian, Cathleen, Univ of Minnesota, Twin Cities
Toth, Charles K., Ohio State Univ
Tullis, Jason A., Univ of Arkansas, Fayetteville
Ustin, Susan L., Univ of California, Davis
van der Werf, Guido R., Vrije Universiteit Amsterdam
van Niekerk, Adriaan, Stellenbosch Univ
Vandemark, Douglas C., Univ of New Hampshire
Velicogna, Isabella, Univ of California, Irvine
Ventura, Stephen J., Univ of Wisconsin, Madison
Veraverbeke, Sander, Vrije Universiteit Amsterdam
Viertel, David C., Eastern Illinois Univ
Walker, Richard, Univ of Oxford
Wang, Lei, Ohio State Univ
Warner, Timothy A., West Virginia Univ
Wasomi, Charles B., Jomo Kenyatta Univ of Agriculture & Tech
Watson, Kelly, Eastern Kentucky Univ
Wei, Xiaofang, Central State Univ
Weng, Qihao, Indiana State Univ
Werth, Susanna, Virginia Polytechnic Inst & State Univ
White, Joseph D., Baylor Univ
Woodcock, Curtis, Boston Univ
Woodcock, Curtis E., Boston Univ
Wright, William C., United States Military Acad
Yan, Xiao-Hai, Univ of Delaware
Yang, Zhiming, North Carolina Central Univ
Yool, Stephen R., Univ of Arizona
Yu, Qian, Univ of Massachusetts, Amherst
Zhang, Caiyun, Florida Atlantic Univ
Zhou, Xiaobing, Montana Tech of the Univ of Montana

Material Science

Alfe, Dario, Univ Coll London
Calaway, Wallis F., Argonne Nat Lab
Chiarello, Ronald P., Argonne Nat Lab
Chrzan, Daryl, Univ of California, Berkeley
Cook, Jane, Pennsylvania State Univ, Univ Park
deFontaine, Didier, Univ of California, Berkeley
DeJonghe, Lutgard, Univ of California, Berkeley
Devine, Thomas M., Univ of California, Berkeley
Eberhart, Mark E., Colorado School of Mines
Ferrari, Mauro, Univ of California, Berkeley
Glaeser, Andreas, Univ of California, Berkeley
Gronsky, Ronald, Univ of California, Berkeley
Gruen, Dieter M., Argonne Nat Lab
Haller, Eugene, Univ of California, Berkeley
Hayden, Geoffrey W., Mercer Univ
Hojjatie, Barry, Valdosta State Univ
Lu, Yongqi, Univ of Illinois
Morris, Jr., J. W., Univ of California, Berkeley
Nakote, Heinz, New Mexico State Univ, Las Cruces
Navrotsky, Alexandra, Univ of California, Davis
Pellin, Michael J., Argonne Nat Lab
Pylypenko, Svitlana, Colorado School of Mines
Ritchie, Robert O., Univ of California, Berkeley
Rovella, Natalia, Alma Mater Studiorum Università di Bologna
Sands, Timothy, Univ of California, Berkeley
Strobel, Timothy A., Carnegie Inst for Science
Valiunas, Jonas K., Lakehead Univ
Voller, Vaughan R., Univ of Minnesota, Twin Cities

Weber, Eicke, Univ of California, Berkeley

Land Use/Urban Geology

Agrawal, Sandeep , Univ of Alberta
Bassett, Scott, Univ of Nevada, Reno
Bedore, Melanie, McMaster Univ
Bereitschaft, Bradley, Univ of Nebraska, Omaha
Birchall, Jeff, Univ of Alberta
Bodenman, John E., Bloomsburg Univ
Bonine, Michael E., Univ of Arizona
Bremer, Keith A., Fort Hays State Univ
Briggs, John, Univ of Glasgow
Buermann, Wolfgang, Univ of Leeds
Clark, Jessie , Univ of Nevada, Reno
Day, Rick L., Pennsylvania State Univ, Univ Park
Dixon, Nick, Univ of Leeds
Dolman, Paul, Univ of East Anglia
Emmi, Philip C., Univ of Utah
Ford, Andrew, Washington State Univ
Fürst, Christine, Martin-Luther-Universitaet Halle-Wittenberg
Godfrey, Brian J., Vassar Coll
Gosnell, Hannah, Oregon State Univ
Haggerty, Julia H., Montana State Univ
Harris, Richard S., McMaster Univ
Hines, Mary E., Univ of North Carolina, Wilmington
Hintz, Rashauna, Univ of Arkansas, Fayetteville
Holl, Karen D., Univ of California, Santa Cruz
Hungerford, Hilary, Utah Valley Univ
James, Valentine U., Clarion Univ
Jantz, Claire A., Shippensburg Univ
Kaushal, Sujay, Univ of Maryland
Kisekka, Isaya, Univ of California, Davis
Kpienbaareh, Daniel, Illinois State Univ
Larson, David J., California State Univ, East Bay
Lee, Chung M., Univ of Utah
Liao, Haifeng (Felix), Univ of Idaho
Marston, Sallie A., Univ of Arizona
Mercier, Michael , McMaster Univ
Millington, Andrew, Flinders Univ
Moscovici, Daniel, Stockton Univ
Nevins, Joseph, Vassar Coll
Odogba, Ismaila, Univ of Wisconsin, Stevens Point
Ormerod, Kerri Jean , Univ of Nevada, Reno
Otterstrom, Samuel M., Brigham Young Univ
Platt, Rutherford H., Univ of Massachusetts, Amherst
Pomeroy, George M., Shippensburg Univ
Price, Benjamin, Texas State Univ
Ptak, Tom, Univ of Idaho
Ray, Rosalie, Texas State Univ
Riebesell, John, Univ of Michigan, Dearborn
Rounsevell, Mark D., Edinburgh Univ
Shirgaokar, Manish, Univ of Alberta
Singh, Amrita, Univ of Alberta
Skillen, James, Calvin Coll
Smith, Betty E., Eastern Illinois Univ
Solecki, William, Hunter Coll (CUNY)
Solecki, William, Montclair State Univ
Steinberger, Julia, Univ of Leeds
Sullivan, Jack B., Univ of Maryland
Szilassi, Péter, Univesity of Szeged
Thomas, Valerie, Georgia Inst of Tech
Tilt, Jenna, Oregon State Univ
Troost, Kathy G., Univ of Washington
Van Den Hoek, Jamon, Oregon State Univ
Wickham, Thomas, California Univ of Pennsylvania
Wollheim, Wil, Univ of New Hampshire
Zhou, Yu, Vassar Coll

Geographic Information Systems

Adhikari, Sanchayeeta, California State Univ, Northridge
Ahearn, Sean C., Hunter Coll (CUNY)
Albrecht, Jochen, Hunter Coll (CUNY)
Alderson, David, Univ of Newcastle Upon Tyne
Aldrich, Stephen, Indiana State Univ
Algeo, Catherine, Western Kentucky Univ
Allen, Katherine, Univ of Liverpool
Allen-Lafayette, Zehdreh, New Jersey Geological and Water Survey
Alvarez, Laura, Univ of Texas, El Paso
Ambinakudige, Shrinidhi, Mississippi State Univ
Amer, Reda, Tulane Univ
Andrews, John R., Univ of Texas at Austin, Jackson Schl of Geosciences
Argles, Tom, Open Univ
Artigas, Francisco, Rutgers, The State Univ of New Jersey, Newark
Ayad, Yasser M., Clarion Univ
Badillo, Elvis A., Indiana State Univ
Badruddin, Abu Z., Cayuga Comm Coll
Bailey, Keiron D., Univ of Arizona
Barr, Stuart, Univ of Newcastle Upon Tyne
Barrett, Melony, Illinois State Geological Survey
Bauer, Emily, Univ of Minnesota
Beaty, Tammy, Oak Ridge Nat Lab
Becker, Lorene Y., Oregon State Univ
Benger, Simon, Flinders Univ
Benson, Jane L., Murray State Univ
Betchwars, Corey, Univ of Minnesota
Bloechle, Amber, Univ of West Florida
Boger, Rebecca, Brooklyn Coll (CUNY)
Boroushaki, Soheil, California State Univ, Northridge
Bottenberg, H. Carrie, Idaho State Univ
Bourbonnais, Mathieu L., Univ of British Columbia, Okanagan Campus
Bowlick, Forrest J., Univ of Massachusetts, Amherst
Breton, Caroline, Univ of Texas at Austin, Jackson Schl of Geosciences
Brown, Douglas, Kingston Univ
Bruns, Dale A., Wilkes Univ
Brunskill, Jeffrey C., Bloomsburg Univ
Burton, Christopher G., Auburn Univ
Busby, Michael R., Murray State Univ
Cao, Guofeng, Texas Tech Univ
Carstensen, Laurence W., Virginia Polytechnic Inst & State Univ
Carter, David, Dept of Interior Off of Surface Mining Reclama and Enforcement
Cary, Kevin, Western Kentucky Univ
Cheng, Qiuming, York Univ
Cheung, Wing H., Palomar Coll
Chokmani, Karem, Universite du Quebec
Chow, Edwin, Texas State Univ
Christopherson, Gary, Univ of Arizona
Cicha, Jarrod, Univ of Minnesota
Cloutis, Ed, Western Univ
Coetzee, Serena, Univ of Pretoria
Colby, Jeff, Appalachian State Univ
Collins, Murray, Edinburgh Univ
Compas, Eric, Univ of Wisconsin, Whitewater
Congalton, Russell, Univ of New Hampshire
Cothren, Jackson D., Univ of Arkansas, Fayetteville
Coulibaly, Mamadou, Univ of Wisconsin Oshkosh
Cova, Thomas J., Univ of Utah
Coveney, Seamus, Univ of Glasgow
Crook, Steven, Palomar Coll
Cunningham, Mary A., Vassar Coll
Curri, Neil, Vassar Coll
Curry, Susan, Univ of Florida
Dai, Dajun, Georgia State Univ
Davis, Trevor J., Univ of Utah
Deakin, Ann K., SUNY, Fredonia
Deal, Richard, Edinboro Univ of Pennsylvania
DeBruhl, Darby, South Carolina Dept of Nat Resources
Delparte, Donna M., Idaho State Univ
Demyanov, Vasily, Heriot-Watt Univ
Diver, Kim, Wesleyan Univ
Dixon, Barnali, Univ of South Florida
Dogwiler, Toby, Missouri State Univ
Dong, Pinliang, Univ of North Texas
Downs, Joni, Univ of South Florida
Drake, Luke, California State Univ, Northridge
Drummond, Jane, Univ of Glasgow
Drzyzga, Scott A., Shippensburg Univ
Dunn, Amy, New Mexico Tech
Earle, Robert, American River Coll
Ebersole, Craig M., DCNR- Pennsylvania Bureau of Geological Survey
Eichenbaum, Jack, Hunter Coll (CUNY)
Fan, Chao, Univ of Idaho
Fan, Weihong, Stockton Univ
Feeney, Alison E., Shippensburg Univ

Fehrs, Ellen R., DCNR- Pennsylvania Bureau of Geological Survey
Feng, Jia , Univ of Nevada, Reno
Fernández, Virginia, Universidad de la Rep Oriental del Uruguay (UDELAR)
Fitzgerald, Francis S., Colorado Geological Survey
Forrest, David, Univ of Glasgow
Fox, Nicholas, Montana State Univ
Fox, Tim, Itasca Comm Coll
Francis, Sarah, Univ of Minnesota
Garren, Sandra J., Hofstra Univ
Gathany, Mark, Cedarville Univ
Gatrell, Jay, Eastern Illinois Univ
Gauthier, Donald J., Los Angeles Valley Coll
Gentry, Christopher, Austin Peay State Univ
Giordano, Alberto, Texas State Univ
Girard, Michael W., New Jersey Geological and Water Survey
Gittings, Bruce M., Edinburgh Univ
Gopal, Sucharita, Boston Univ
Gordon, Clare E., Univ of Leeds
Gorsevski, Peter, Bowling Green State Univ
Grala, Katarzynz, Mississippi State Univ
Graniero, Phil A., Univ of Windsor
Greatbatch, Ian, Kingston Univ
Greene, Chris, Dalhousie Univ
Grunwald, Sabine., Univ of Florida
Guiseppe, Alfred C., DCNR- Pennsylvania Bureau of Geological Survey
Haddock, Gregory D., Northwest Missouri State Univ
Halls, Joanne N., Univ of North Carolina, Wilmington
Halsted, Christian, Dept of Agriculture, Conservation, and Forestry
Hamilton, Jacqueline, Univ of Minnesota
Han, Su Y., Texas State Univ
Hancock, Gregory, Univ of Newcastle
Hancock, Steven, Edinburgh Univ
Haner, Andrew, Arkansas Geological Survey
Hansen, Christel D., Univ of Pretoria
Hansen, William J., Worcester State Univ
Haugerud, Ralph, Univ of Washington
He, Hong S., Univ of Missouri
Hedberg, Russell, Shippensburg Univ
Hernandez, Michael W., Weber State Univ
Hick, Steven R., Univ of Denver
Hill, Malcolm D., Northeastern Univ
Hintz, John G., Bloomsburg Univ
Hong, Jessie, Univ of West Georgia
Hotz, Helenmary, Univ of Massachusetts, Boston
Howard, Hugh H., American River Coll
Huang, Jane, Fitchburg State Univ
Huffman, F. T., Eastern Kentucky Univ
Hughes, Annie, Kingston Univ
Humagain, Kamal, SUNY Potsdam
Jennings, Nathan, American River Coll
Jensen, Ryan R., Brigham Young Univ
Jeu, Amy, Hunter Coll (CUNY)
Jialing, Wang, Slippery Rock Univ
Jin, He (Hannah), Univ of South Florida
Joshi, Manoj, Univ of East Anglia
Kahle, Chris, Iowa Dept of Nat Resources
Kambesis, Patricia, Western Kentucky Univ
Kar, Bandana, Univ of Southern Mississippi
Karnes, Victoria M., South Dakota School of Mines & Tech
Kelleher, Cole, Univ of Minnesota, Twin Cities
Kelley, Richard, New Mexico Tech
Kelley, Scott , Univ of Nevada, Reno
Kidd, David, Kingston Univ
Kienzle, Stefan, Univ of Lethbridge
Kimerling, A. Jon, Oregon State Univ
Kirimi, Fridah K., Jomo Kenyatta Univ of Agriculture & Tech
Klancher, Jacki, Central Wyoming Coll
Knight, Andi, New Mexico Tech
Kobara, Shinichi, Texas A&M Univ
Kohler, Nicholas, Univ of Oregon
Kostelnick, John, Illinois State Univ
Krieger, Jerry, South Carolina Dept of Nat Resources
Krizek, Jeffrey, San Bernardino Valley Coll
Kronenfeld, Barry J., Eastern Illinois Univ
Kwon, Youngsang, Univ of Memphis
Laffan, Shawn, Univ of New South Wales
lalap-Ayça, Seda, Univ of Massachusetts, Amherst
Landry, Shawn, Univ of South Florida
LaRue, Michelle, Univ of Minnesota, Twin Cities
Law, Zada, Middle Tennessee State Univ
Le, Yanfen, Northwest Missouri State Univ
Lee, Wook, Edinboro Univ of Pennsylvania
Lethbridge, Mark, Flinders Univ
Levine, Norman S., Coll of Charleston
Li, Jing, Univ of Denver
Li, Ping-Chi, Tennessee Tech Univ
Liang, Lu, Univ of North Texas
Lin, Meimei, Georgia Southern Univ
Lisichenko, Richard, Fort Hays State Univ
Liu, Weibo, Florida Atlantic Univ
Livingston, Jack, Slippery Rock Univ
Lobben, Amy, Univ of Oregon
Logsdon, Miles G., Univ of Washington
Longbrake, David, Univ of Denver
Lovett, Andrew A., Univ of East Anglia
Lu, Tianjun, California State Univ, Dominguez Hills
Lu, Yongmei, Texas State Univ
Lukinbeal, Christopher, Univ of Arizona
Luo, Jun, Missouri State Univ
Lupo, Tom, American River Coll
Maantay, Juliana, Lehman Coll (CUNY)
Maas, Regan, California State Univ, Northridge
Machovina, Brett , United States Air Force Acad
Mackaness, William A., Edinburgh Univ
Magondu, Moffat G., Jomo Kenyatta Univ of Agriculture & Tech
Mansell, Mark, New Mexico Tech
Marcano, Eugenio J., Mount Holyoke Coll
Marin, Liliana, Baylor Univ
Marsan, Yvonne, Univ of North Carolina, Wilmington
Marzen, Luke J., Auburn Univ
Maxwell, Dave , Southern Utah Univ
May, Cynthia, Northland Coll
Mayer, Jordan, Univ of Minnesota
Mazariegos, Ruben, Univ of Texas, Rio Grande Valley
Meentemeyer, Ross, Univ of North Carolina, Charlotte
Meng, Qingmin, Mississippi State Univ
Metcalf, Meredith, Eastern Connecticut State Univ
Micander, Rachel, Univ of Nevada, Reno
Miguel, Carlos, Universidad de la Republica Oriental del Uruguay (UDELAR)
Miller, Harvey J., Ohio State Univ
Miller, Phillip, New Mexico Tech
Mitasova, Helena, North Carolina State Univ
Moore, Adrian J., Ulster Univ, Coleraine
Moreno, Hernan, Univ of Texas, El Paso
Morgan, John D., Univ of West Florida
Morgan, Tamie, Texas Christian Univ
Morris, John A., Mississippi State Univ
Mucsi, László, Univesity of Szeged
Mueller, Thomas, California Univ of Pennsylvania
Mulligan, Kevin R., Texas Tech Univ
Mulrooney, Timothy, North Carolina Central Univ
Mutua, Felix N., Jomo Kenyatta Univ of Agriculture & Tech
Mwangi, Nancy , Jomo Kenyatta Univ of Agriculture & Tech
Mwaniki, Mercy W., Jomo Kenyatta Univ of Agriculture & Tech
Nagihara, Seiichi, Texas Tech Univ
Nemon, Amy, Western Kentucky Univ
Ngo, Anh, Curtin Univ
Nicolette, Justine, New Mexico Tech
Nimako, Solomon Nana Kwaku, San Bernardino Valley Coll
Obermeyer, Nancy J., Indiana State Univ
Oduor, Peter, North Dakota State Univ
Olson, Jeff, Univ of Wisconsin, Whitewater
Ozdenerol, Esra, Univ of Memphis
Palladino, Steve D., Ventura Coll
Pallis, Ted J., New Jersey Geological and Water Survey
Palmer, Evan, United States Air Force Acad
Pavlovskaya, Marianna, Hunter Coll (CUNY)
Pawlicki, Caron E., DCNR- Pennsylvania Bureau of Geological Survey
Pedemonte, Virginia, Univ de la Republica Oriental del Uruguay (UDELAR)
Peele, R Hampton, Louisiana State Univ
Percy, David, Portland State Univ
Phillips, Zachary, Saint Louis Univ

Porter, Claire, Univ of Minnesota, Twin Cities
Porter, John H., Univ of Virginia
Portillo, Danny, United States Air Force Acad
Pretorius, Erika, Univ of Pretoria
Price, Maribeth H., South Dakota School of Mines & Tech
Price, Peter E., Lonestar Coll - North Harris
Pristas, Ronald, New Jersey Geological and Water Survey
Proctor, Cameron, Univ of Windsor
Proctor, Elizabeth, City Coll of San Francisco
Qi, Feng, Kean Univ
Qiang, Yi, Univ of South Florida
Qiu, Xiaomin, Missouri State Univ
Raber, George, Univ of Southern Mississippi
Rautenbach, Victoria J., Univ of Pretoria
Reader, Steven , Univ of South Florida
Rehwald, Matthew, Univ of Wisconsin, Madison, Div of Extension
Resnichenko, Yuri, Univ de la Republica Oriental del Uruguay (UDELAR)
Rice, Keith W., Univ of Wisconsin, Stevens Point
Robinson, Lori, Univ of Minnesota
Robinson, Michael, Georgia Southern Univ
Roof, Steven, Hampshire Coll
Rose, Caroline, Univ of Wisconsin, Madison, Div of Extension
Rowley, Rex J., Illinois State Univ
Saalfeld, Alan J., Ohio State Univ
Sadd, James L., Occidental Coll
Sagan, Vasit, Saint Louis Univ
Sanchez-Azofeifa, G. Arturo, Univ of Alberta
Sandau, Ken L., Montana Tech of The Univ of Montana
Sanz, Miguel Angel, Univ Complutense de Madrid
Schoephoester, Peter, Univ of Wisconsin, Madison, Div of Extension
Scott, Darren M., McMaster Univ
Seong, Jeong C., Univ of West Georgia
Sleezer, Richard O., Emporia State Univ
Snyman, Lourens, Univ of Pretoria
Stachowiak, Lauren , Eastern Washington Univ
Starek, Michael , Texas A&M Univ, Corpus Christi
Stuart, Neil, Edinburgh Univ
Su, Haibin, Texas A&M Univ, Kingsville
Sun, Jeff, Casper Coll
Sun, Yifei, California State Univ, Northridge
Szpakowski, David, Western Oregon Univ
Tadesse, Tsegaye, Univ of Nebraska, Lincoln
Tao, Ran, Univ of South Florida
Taylor, Nathan H., Arkansas Geological Survey
Taylor, Ryan W., SUNY, Purchase
Templeton, Alan R., Washington Univ in St. Louis
Thale, Paul R., Montana Tech of The Univ of Montana
Thayn, Jonathan B., Illinois State Univ
Tinkler, Dorothy, Treasure Valley Comm Coll
Tiwari, Chetan, Univ of North Texas
Tobak, Zalán, Univesity of Szeged
Tong, Daoqin, Univ of Arizona
Tran, Linda C., Lonestar Coll - North Harris
Trotter, Lewis, Curtin Univ
Tu, Wei, Georgia Southern Univ
Unger, Corey, Utah Geological Survey
van Leeuwen, Boudewijn, Univesity of Szeged
VanHorn, Jason E., Calvin Coll
Veisze, Paul M., American River Coll
Vincent, Paul C., Valdosta State Univ
Vlcan, Jennifer, Univ of Nevada, Reno
Walford, Nigel, Kingston Univ
Walls, Charles C., Dalhousie Univ
Walters, Steven, Univ of Washington
Wang, Lillian T., Univ of Delaware
Warren, Stacy, Eastern Washington Univ
Weber, Keith, Idaho State Univ
Weiant, Morgen C., DCNR- Pennsylvania Bureau of Geological Survey
Weiss, Alfred W., Waubonsee Comm Coll
Wen, Yuming, Univ of Guam
Whiteaker, Timothy L., Univ of Texas, Austin
Williamson, Douglas, Hunter Coll (CUNY)
Wilson, Blake B., Univ of Kansas
Wilson, John R., Lafayette Coll
Wilson, Roy R., Eastern Connecticut State Univ
Woodhouse, Iain H., Edinburgh Univ
Xia, Jianhong (Cecilia), Curtin Univ
Xie, Zhixiao, Florida Atlantic Univ
Xu, Wei, Univ of Lethbridge
Yacucci, Mark A., Univ of Illinois
Yan, Jun, Western Kentucky Univ
Ye, Gordon, City Coll of San Francisco
Yin, Zhi-Yong, Univ of San Diego
Zhan, F Benjamin, Texas State Univ
Zhang, Guiming, Univ of Denver
Zhang, Qiaofeng (Robin), Murray State Univ
Zhao, Bo, Oregon State Univ

Geoinformatics

Arnold, Lesley, Curtin Univ
Awange, Joseph, Curtin Univ
Belton, Dave, Curtin Univ
Bhaskaran, Sunil, Graduate School of the City Univ of New York
Bulatov, Dimitri, Curtin Univ
Connor, Laura, Univ of South Florida
Curl, Doug C., Univ of Kentucky
Helmholz, Petra, Curtin Univ
Ivanova, Ivana, Curtin Univ
Katumba, Samy, Univ of Pretoria
Lichti, Derek, Curtin Univ
Limp, William (., Univ of Arkansas, Fayetteville
Mansur, Adam, Smithsonian Inst / Nat Museum of Nat History
Marr, Paul G., Shippensburg Univ
O'Dean, Emily, Univ of Nevada, Reno
Pham, Hoa Thi, Curtin Univ
Robinson, Todd, Curtin Univ
Silveira, António B., Universidade de Lisboa
Sui, Daniel, Univ of Arkansas, Fayetteville
Trivitt-Kracke, Amy, New Mexico Tech
Tsele, Philemon L., Univ of Pretoria
West, Geoff, Curtin Univ
Yilmaz, Alper, Ohio State Univ
Zaffos, Andrew A., Univ of Arizona

Global Change

Alexander, Peter, Edinburgh Univ
Bingham, Rob, Edinburgh Univ
Boschmann, E. Eric, Univ of Denver
Brookshire, Jack, Montana State Univ
Crimmins, Michael A., Univ of Arizona
Crowley, Kate, Edinburgh Univ
Doran, Peter T., Univ of Illinois Chicago
Ducey, Mark, Univ of New Hampshire
El Masri, Bassil, Murray State Univ
Ernakovich, Jessica, Univ of New Hampshire
Felzer, Benjamin S., Lehigh Univ
Gathorne-Hardy, Alfy, Edinburgh Univ
Giefer, Madeline M., Austin Peay State Univ
Giesche, Alena, Colby Coll
Gilligan, Jonathan M., Vanderbilt Univ
Goode, Ryan , Cerritos Coll
Hennige, Sebastian, Edinburgh Univ
Hernandez, Rebecca, Univ of California, Davis
Hicke, Jeff, Univ of Idaho
Hicke, Jeffrey, Univ of Idaho
Irvine, Peter D., Univ Coll London
Kaufman, Darrell S., Northern Arizona Univ
Kopp, Robert E., Rutgers, The State Univ of New Jersey
Lawrence, Deborah, Univ of Virginia
Marks-Block, Tony, California State Univ, East Bay
Mathenge, Christine, Austin Peay State Univ
Novick, Kimberly, Indiana Univ, Bloomington
Nyantakyi-Frimpong, Hanson, Univ of Denver
Oches, Rick, Bentley Univ
Ollinger, Scott, Univ of New Hampshire
Penn, Justin, Princeton Univ
Rood, Richard B., Univ of Michigan
Rooney-Varga, Juliette, Univ of Massachusetts, Lowell
Sutton, Paul C., Univ of Denver
Taylor, Matthew, Univ of Denver
Tierney, Sean, Univ of Denver
Trigoso, Erika, Univ of Denver
Watson, Elizabeth, Drexel Univ

Wexler, Anthony S., Univ of California, Davis
Williams, Mathew, Edinburgh Univ
Wood, Gillen D., Univ of Illinois, Urbana-Champaign

Nanogeoscience
Fougerouse, Denis, Curtin Univ
Michel, F. M., Virginia Polytechnic Inst & State Univ
Reddy, Steve, Curtin Univ
Tacchetto, Tommaso, Curtin Univ

Climate Modeling
Abatzoglou, John, Univ of Idaho
Akinsanola, Akintomide, Univ of Illinois Chicago
Cheng, Linyin, Univ of Arkansas, Fayetteville
Coats, Sloan, Univ of Hawai'i, Manoa
Dee, Sylvia G., Rice Univ
Galbraith, Eric, McGill Univ
Gutowski, William J., Iowa State Univ of Science & Tech
Hegewisch, Katherine, Univ of Idaho
Jackson, Charles, Univ of Texas, Austin
Lora, Juan, Yale Univ
Marinov, Irina, Univ of Pennsylvania
Ricke, Katherine, Univ of California, San Diego
Sun, Fengpeng, Univ of Missouri, Kansas City
Tian, Hanqin, Boston Coll
Tsamados, Michel D., Univ Coll London
Vecchi, Gabriel, Princeton Univ

Not Elsewhere Classified
(Deep) McGregor, Mary, Montgomery Coll
Aamodt, Paul, Los Alamos Nat Lab
Abney, Kent D., Los Alamos Nat Lab
Abt, Charlotte J., Univ of Liverpool
Ackerman, Jessica R., Illinois State Geological Survey
Acosta, Patricia E., Williams Coll
Adams, Debi, West Texas A&M Univ
Agnew, Stephen F., Los Alamos Nat Lab
Aguilera, Juan Manuel Torres, Universidad Auto de San Luis Potosi
Ahmed, Waquar, Univ of North Texas
Alavi, Hedy, Johns Hopkins Univ
Alberts, Heike C., Univ of Wisconsin Oshkosh
Albrecht, Lorraine, Univ of Waterloo
Albright, Katia, Univ of Nevada, Reno
Alde, Douglas, Los Alamos Nat Lab
Algin, Barbara, Columbia Univ
Ali-Bray, Julie, Mt. San Antonio Coll
Allen, Karen, Furman Univ
Allison, Tami L., Missouri Dept of Nat Resources
Ambruster, W. Scott, Univ of Alaska, Fairbanks
Amelung, Falk, Univ of Miami
Amor, John, Univ of British Columbia
Anand, Madhur, Univ of Guelph
Antar, Ali A., Central Connecticut State Univ
Anthony, Nina, Furman Univ
Appel, Heidi, Univ of Toledo
Arabas, Karen, Willamette Univ
Argenbright, Robert T., Univ of North Carolina, Wilmington
Armstrong, Andrew, Univ of New Hampshire
Arsuaga, Juan Luis, Univ Complutense de Madrid
Arundale, Wendy H., Univ of Alaska, Fairbanks
Arvidson, Raymond E., Washington Univ in St. Louis
Ashour-Abdalla, Maha, Univ of California, Los Angeles
Atisha-Castillo, Hector David, Universidad Auto de San Luis Potosi
Au, Whitlow W L., Univ of Hawai'i, Manoa
Aufrecht, Walter E., Univ of Lethbridge
Ayers, Joseph , Northeastern Univ
Bailey, Ian, Exeter Univ
Baldizon, Ileana, Miami-Dade Coll (Wolfson Campus)
Ball, Elizabeth, Univ of Nevada, Reno
Ball, William P., Johns Hopkins Univ
Ballinger, Rhoda, Cardiff Univ
Bannon, Ann, Dalhousie Univ
Barmore, Garrett, Univ of Nevada, Reno
Barnbaum, Cecilia S., Valdosta State Univ
Barnes, Clare, Edinburgh Univ
Barnes, Fairley J., Los Alamos Nat Lab
Barnes, Jessica, Univ of South Carolina
Barnes, Philip, Univ of South Carolina
Barnes, Randal J., Univ of Minnesota, Twin Cities
Barra, Monica, Univ of South Carolina
Barron, Elizabeth, Univ of Wisconsin Oshkosh
Barron, George, Univ of Guelph
Barth, Susan A., Montana Tech of The Univ of Montana
Bartl, Simona, Moss Landing Marine Lab
Baskin, Perry A., Valdosta State Univ
Bass, Jerry, Univ of Southern Mississippi
Basso, Bruno, Michigan State Univ
Baumann, Sue, Argonne Nat Lab
Beatty, Merrill Ann, Univ of New Brunswick
Beck, Alan, Western Univ
Becker, Laurence, Oregon State Univ
Becker, Udo, Univ of Michigan
Bederman, Sanford H., Georgia State Univ
Behr-Andres, Christina "Tina", Los Alamos Nat Lab
Bellew, Angela, Univ of Iowa
Berchem, Jean, Univ of California, Los Angeles
Betancourt, Julio, Univ of Arizona
Bezdecny, Kris, California State Univ, Los Angeles
Bienvenu, Nadean S., Univ of Louisiana at Lafayette
Billiot, Fereshteh, Texas A&M Univ, Corpus Christi
Black, Annjeannette , Comm Coll of Baltimore County, Catonsville
Blankenship, Robert, Washington Univ in St. Louis
Blatner , Keith , Washington State Univ
Blenkinson, Stephen, Univ of Newcastle Upon Tyne
Blumenfeld, Raphael, Imperial Coll
Boelman, Natalie, Columbia Univ
Boggs, Carol, Univ of South Carolina
Boisvert, Eric, Universite du Quebec
Boland, Greg J., Univ of Guelph
Boland, John J., Johns Hopkins Univ
Booth, Robert K., Lehigh Univ
Bosman, Martin, Univ of South Florida
Bossenbroek, Jonathon , Univ of Toledo
Boswell, Cecil, Missouri Dept of Nat Resources
Bourgeois, Reed J., Louisiana State Univ
Bowen, Dawn S., Mary Washington Coll
Bowen, Jennifer, Northeastern Univ
Bowen, Robert, Univ of Massachusetts, Boston
Bowersox, Joe, Willamette Univ
Brainard, James R., Los Alamos Nat Lab
Bratman, Eve, Franklin and Marshall Coll
Braught, Patricia, Dickinson Coll
Brewer, Margene, Calvin Coll
Bridgeman, Thomas, Univ of Toledo
Brooks, Ian, Univ of Leeds
Brown, Bradford D., Univ of Idaho
Brown, Kent D., Utah Geological Survey
Brown, Rob, Appalachian State Univ
Brunetto, Eileen, Middlebury Coll
Brunish, Wendee M., Los Alamos Nat Lab
Bryan, Jeffrey C., Los Alamos Nat Lab
Bryant, Anita M., Univ of West Georgia
Buchan, Stuart, Curtin Univ
Buck, Daniel, Univ of Oregon
Buck, Gregory, Texas A&M Univ, Corpus Christi
Buckley, Luke, Montana Tech of The Univ of Montana
Bullamore, Henry W., Frostburg State Univ
Bunker, Merle E., Los Alamos Nat Lab
Burk, Sue, Univ of Maryland
Burns, Carol J., Los Alamos Nat Lab
Burns, John W., Mt. San Antonio Coll
Busan, Daniel-Cristian, Curtin Univ
Busse, Friedrich H., Univ of California, Los Angeles
Butts, Thomas R., Univ of Texas, Dallas
Cahill, Abigail, Williams Coll
Cahill, Michael J., SUNY, Stony Brook
Cahoy, Kerri, Massachusetts Inst of Tech
Cailliet, Gregor M., Moss Landing Marine Lab
Calegari, Pat, Pace Univ, New York Campus
Cammerata, Kirk , Texas A&M Univ, Corpus Christi
Campana, Michael E., Univ of New Mexico
Cannon, Kevin, Colorado School of Mines
Cantarero, Debra A., Pasadena City Coll

OTHER: **NOT ELSEWHERE CLASSIFIED**

Capoccia, Mary, Ohio State Univ
Carey, Kristine M., Lakehead Univ
Carlin, Scott, Long Island Univ, C.W. Post Campus
Carr, David E., Univ of Virginia
Carroll, Matt, Washington State Univ
Carroll, Tracy, Missouri State Univ
Catalano, Valerie, Roger Williams Univ
Catau, John C., Missouri State Univ
Cattolico, Rose Ann, Univ of Washington
Caupp, Craig L., Frostburg State Univ
Cavello, Seth, Univ of South Florida
Chametzky, Peter, Univ of South Carolina
Chan, Selene, Univ of British Columbia
Chaney, Phil L., Auburn Univ
Chase, Anne, Univ of Arizona
Chase, Jon M., Washington Univ in St. Louis
Chatterjee, Ipsita, Univ of North Texas
Cheek, William H., Missouri State Univ
Chen, Kai Loon, Johns Hopkins Univ
Cheney, Donald, Northeastern Univ
Chief, Karletta, Univ of Arizona
Childs, Geoff, Washington Univ in St. Louis
Choquette, Marc, Universite Laval
Chouinard, Vera, McMaster Univ
Christopher, Micol, Mt. San Antonio Coll
Church, Warren, Columbus State Univ
Cich, Anna, Weber State Univ
Clark, David L., Los Alamos Nat Lab
Clark, James S., Duke Univ
Clark, Robert O., Northern Arizona Univ
Clark, Shannon, Univ of New Mexico
Clay, Robert, Missouri Dept of Nat Resources
Cochran, David, Univ of Southern Mississippi
Coffin, Richard B., Univ of Delaware
Cohen, Alice, Acadia Univ
Cohen, Joel E., Columbia Univ
Cohen, Matt, Furman Univ
Cohen, Shaul, Univ of Oregon
Colby, Bonnie C., Univ of Arizona
Colby, Sarah, Northern Arizona Univ
Coleman, Gary D., Univ of Maryland
Collins, Damian, Univ of Alberta
Collins, Lisa J., Northwestern Univ
Colwell, Frederick (Rick), Oregon State Univ
Confer, John, California Univ of Pennsylvania
Conkle, Jeremy, Texas A&M Univ, Corpus Christi
Cook, Jacob, Curtin Univ
Cook, Stanton, Univ of Oregon
Cook, Steve, Oregon State Univ
Coonley, Steffanie, Los Alamos Nat Lab
Copeman, Louise A., Oregon State Univ
Corcoran, Deborah, Missouri State Univ
Corey, Alison, Utah Geological Survey
Costea, Lidia, Northeastern Illinois Univ
Costello, Margaret, California State Univ, Long Beach
Couch, Libby, Denver Museum of Nature & Science
Cox, Shelah, Temple Univ
Crafford, J. P., Univ of Limpopo
Crail, Todd, Univ of Toledo
Crawford, Ian, Birkbeck Coll
Crepeau, Richard J., Appalachian State Univ
Cresswell, Tim, Edinburgh Univ
Cromar, Nancy, Flinders Univ
Crumbly, Isaac J., Fort Valley State Univ
Cupak, Martin, Curtin Univ
Cupples, Julie, Edinburgh Univ
Curry, Barbara, Univ of Texas, Dallas
Curry, James K., Arkansas Geological Survey
D'Arrigo, Rosanne, Columbia Univ
Damschen, Ellen, Washington Univ in St. Louis
Dash, Zora V., Los Alamos Nat Lab
Dasvarma, Gour, Flinders Univ
Daugherty, Carolyn M., Northern Arizona Univ
Davey, Patricia M., Brown Univ
Davis, James A., Brigham Young Univ
Dawson, Sarah, Franklin and Marshall Coll
Day, Rosie, Univ of Birmingham
Daymon, Deborah J., Los Alamos Nat Lab
De Beus, Barbara, Montclair State Univ
De Santo, Eilzabeth, Franklin and Marshall Coll
de Wit, Julien, Massachusetts Inst of Tech
Dean, Jeffrey S., Univ of Arizona
Deaton, Tami, Univ of North Texas
DeGraff, Jerry, Fresno State Univ
Delawder, Sandra, James Madison Univ
Demanet, Laurent, Massachusetts Inst of Tech
Deming, Jody, Univ of Manitoba
Dennis, Clyde B., Argonne Nat Lab
Derrick, Sharon, Texas A&M Univ, Corpus Christi
Deschaine, Sylvia, Bates Coll
Dessler, Alexander, Univ of Arizona
Detrich, William, Northeastern Univ
Dibben, Chris, Edinburgh Univ
Dickerson, Richard E., Univ of California, Los Angeles
Dickhoff, Willem H., Washington Univ in St. Louis
Dighe, Kalpak, Los Alamos Nat Lab
Dill, Carilee, Bucknell Univ
Dingman, Steve, San Antonio Comm Coll
Ditmars, John, Argonne Nat Lab
Divine, Aaron, Northern Arizona Univ
Dixon, Clifton, Univ of Southern Mississippi
Dixon, Michael, Univ of Guelph
Dixon, Paul R., Los Alamos Nat Lab
Domingue, Carla, Louisiana State Univ
Donahoe, Robert J., Los Alamos Nat Lab
Doorn, Stephen K., Los Alamos Nat Lab
Doro, Kennedy, Univ of Toledo
Dorries, Alison M., Los Alamos Nat Lab
Downey, Fergus, Curtin Univ
Driever, Steven L., Univ of Missouri, Kansas City
Driscoll, John R., SUNY, Cortland
Duchane, David V., Los Alamos Nat Lab
Dudukovic, Mike, Washington Univ in St. Louis
Duff, John, Univ of Massachusetts, Boston
Duffey, Patricia, Fort Hays State Univ
Dunton, Karen K., Illinois State Univ
Duxbury, Thomas C., Jet Propulsion Lab
Dvorzak, Marie, Univ of Wisconsin, Madison
Dwyer, Daryl F., Univ of Toledo
Earls, Sandy, Oklahoma State Univ
Ebert, David, Moss Landing Marine Lab
Eby, Stephanie, Northeastern Univ
Edwards, John , Flinders Univ
Eggers, Delores M., Univ of North Carolina, Asheville
Eisenberger, Peter M., Columbia Univ
Eller, Phillip G., Los Alamos Nat Lab
Ellins, Katherine K., Univ of Texas, Austin
Elliott, C. James, Los Alamos Nat Lab
Ellis, Hugh, Johns Hopkins Univ
Ellis, Kathryn E., Univ of Kentucky
Emmett, Chad, Brigham Young Univ
Epstein, Howard E., Univ of Virginia
Esnault, Melissa H., Louisiana State Univ
Esser, Corinne, Univ of California, Davis
Evans, Claude, Washington Univ in St. Louis
Evans, Krista M., Missouri State Univ
Fadiman, Maria, Florida Atlantic Univ
Fallowfield, Howard, Flinders Univ
Fares, Mary A., Valdosta State Univ
Farley, Perry D., Los Alamos Nat Lab
Farrell, Mark O., Point Park Univ
Favor, Michael, Univ of Michigan, Dearborn
Felix, Joseph D., Texas A&M Univ, Corpus Christi
Feng, Zhiqiang, Edinburgh Univ
Ferguson, Teresa D., Oak Ridge Nat Lab
Ferreira, Maryanne F., Woods Hole Oceanographic Inst
Ferren, Richard L., Berkshire Comm Coll
Fifer, Fred L., Univ of Texas, Dallas
Fireman, Gail, DCNR- Pennsylvania Bureau of Geological Survey
Fish, Jennifer M., Univ of California, Santa Cruz
Fisher, Janet, Edinburgh Univ
Fitzpatrick, John R., Los Alamos Nat Lab

Fitzsimmons, Kevin, Univ of Arizona
Flaherty, Frank A., Valdosta State Univ
Fletcher, Austin, Univ of Guelph
Flores-Garcia, Mari C., California State Univ, Northridge
Floyd, Barry, New York State Geological Survey
Fogel, David N., Los Alamos Nat Lab
Foley, John P., Montana Tech of The Univ of Montana
Foster, Michael, Moss Landing Marine Lab
Foti, Pamela, Northern Arizona Univ
Fox, Laurel R., Univ of California, Santa Cruz
Freiermuth, Sue, Univ of Wisconsin, River Falls
Frisk, DeAnn M., Iowa State Univ of Science & Tech
Fry, Matthew, Univ of North Texas
Fryett, I, Cardiff Univ
Fuente, David E., Univ of South Carolina
Fuentes, Edna L., Univ of Illinois Chicago
Gagosian, Carol, Wellesley Coll
Gallick, Roberta T., Point Park Univ
Ganey-Curry, Patricia E., Univ of Texas, Austin
Gao, Mengsheng, Univ of Florida
Gao, Oliver H., Cornell Univ
Garcia, Jan, Occidental Coll
Gartner, John F., Univ of Waterloo
Garvin, Theresa D., Univ of Alberta
Garza, Darlene, Univ of Southern California
Gautsch, Jacklyn, Iowa Dept of Nat Resources
Gehrels, Tom, Univ of Arizona
Geller, Jonathan, Moss Landing Marine Lab
Geller, Murray, Jet Propulsion Lab
Gerhardt, Hannes, Univ of West Georgia
Gerry, Janelle, Univ of Nebraska, Lincoln
Ghattas, Omar, Univ of Texas, Austin
Ghosheh, Baher A., Edinboro Univ of Pennsylvania
Giammar, Daniel, Washington Univ in St. Louis
Gibbs, Gaynell, Louisiana State Univ
Gibson, Deana, Missouri State Univ
Gifford-Gonzalez, Dianne, Univ of California, Santa Cruz
Gilbert, Co'Quesie, Columbia Univ
Gillespie, Thomas, Emory Univ
Gillette, David P., Univ of North Carolina, Asheville
Glenn, Ed, Univ of Arizona
Gong, Hongmian, Hunter Coll (CUNY)
Good, Daniel B., Georgia Southern Univ
Goodin, Ruth, Univ of Miami
Goodwin, Paul, Univ of Guelph
Gordon, Andrew, Univ of Guelph
Gouhier, Tarik, Northeastern Univ
Grace, Shannon M., Austin Comm Coll District
Grande, Anthony, Hunter Coll (CUNY)
Granshaw, Frank D., Portland State Univ
Grant, Jacqueline , Southern Utah Univ
Gravely, Cynthia Rae, Clemson Univ
Gray, Sarah, Univ of San Diego
Green, Brittany, Kansas State Univ
Griffin, Kevin, Columbia Univ
Griffith, Caitlin, Univ of Arizona
Gritzo, Russell E., Los Alamos Nat Lab
Grobler, Hendrik C., Univ of Johannesburg
Grossman, Lawrence S., Virginia Polytechnic Inst & State Univ
Grushow, Alex, Rider Univ
Guerra, Oralia, Austin Comm Coll District
Guerrieri, Mary, Univ of Arizona
Guikema, Seth, Johns Hopkins Univ
Guzman, Ernesto, Univ of Guelph
Habash, Marc, Univ of Guelph
Habron, Geoffrey, Furman Univ
Haddad, Brent, Univ of California, Santa Cruz
Haff, Peter K., Duke Univ
Hafner, James A., Univ of Massachusetts, Amherst
Haggart, Renee, Univ of British Columbia
Halfman, Barbara, Hobart & William Smith Colls
Hall, Christopher C., Univ of Guelph
Hall, I, Cardiff Univ
Hall, Jude, Denison Univ
Hall, Robert, Univ of Guelph
Hall, Ron, Univ of Lethbridge
Halladay, Sylvia R., Ohio Dept of Nat Resources
Hallett, Rebecca, Univ of Guelph
Hamilton, George, Berkshire Comm Coll
Hammersley, Charles, Northern Arizona Univ
Hanke, Steve H., Johns Hopkins Univ
Hankins, Katherine B., Georgia State Univ
Hanna, Kevin, Univ of British Columbia, Okanagan Campus
Hansen, Jeffrey C., Los Alamos Nat Lab
Hardy, R. Dean, Univ of South Carolina
Hardy, Shaun J., Carnegie Inst for Science
Harris, Virginia M., Wesleyan Univ
Harrison, Conor, Univ of South Carolina
Hartig, Ben, Curtin Univ
Hawley, Rebecca D., Northern Arizona Univ
Haynes, Kyle J., Univ of Virginia
Haynes, Kyle J., Univ of Virginia
Heaton, Jill, Univ of Nevada, Reno
Heaton, Richard C., Los Alamos Nat Lab
Heatwole, Charles A., Hunter Coll (CUNY)
Heckathorn, Scott, Univ of Toledo
Heenan, Cleo J., South Dakota School of Mines & Tech
Hegarty, Patricia, Univ Coll Cork
Heimpel, Moritz, Univ of Alberta
Heitman, Mary P., Iowa Dept of Nat Resources
Helmick, Sara B., Los Alamos Nat Lab
Hendrikx, Jordy, Montana State Univ
Henry, Lea-Anne, Edinburgh Univ
Herr, Larry, Univ of Lethbridge
Hicks, Andrea, Univ of Wisconsin, Madison
Higinbotham, Pamela, California Univ of Pennsylvania
Hildebrand, Stephen G., Oak Ridge Nat Lab
Hill, Christopher, Massachusetts Inst of Tech
Hill, Julie, Univ of Nevada, Reno
Hilpert, Markus, Johns Hopkins Univ
Hilts, Stewart G., Univ of Guelph
Hinderaker, Pam, Iowa State Univ of Science & Tech
Hintz, William, Univ of Toledo
Hirsch, Eric, Franklin and Marshall Coll
Hiser, Susan, Marietta Coll
Hobbs, Benjamin F., Johns Hopkins Univ
Hobbs, John D., Montana Tech of the Univ of Montana
Hockey, Thomas A., Univ of Northern Iowa
Hodges, Jackie, Fort Valley State Univ
Holbrook, Amanda, Morehead State Univ
Holdren, John P., Harvard Univ
Hollenbeck, Diane, Metropolitan State Coll of Denver
Holstein, Thomas J., Roger Williams Univ
Holt, Caecilia, Kutztown Univ of Pennsylvania
Homan, Amy L., Pennsylvania State Univ, Univ Park
Hommel, Demian, Oregon State Univ
Hrepic, Zdeslav, Columbus State Univ
Hsiang, Tom, Univ of Guelph
Hsiao, Theodore C., Univ of California, Davis
Huang, Norden E., Univ of Delaware
Huckabey, Marsha, Univ of Missouri
Huffman, Debra E., Univ of South Florida
Huffman, Robert L., Mercer Univ
Hughes, Randall, Northeastern Univ
Hull, Donald L., Los Alamos Nat Lab
Hunt, Kathy, Univ of Maryland
Hunt, Rachel, Edinburgh Univ
Hunt, Shelley, Univ of Guelph
Huntoon, Laura, Univ of Arizona
Hutchins, Peter S., Mississippi Office of Geology
Hysell, David L., Cornell Univ
Hysell, David, Cornell Univ
Iantria, Linnea, Missouri State Univ
Iaquinta, Connie, Univ of Windsor
Immonen, Wilma, Montana Tech of the Univ of Montana
Ioannides, Dimitri, Missouri State Univ
Ivy, Russell L., Florida Atlantic Univ
Jackson, Robert B., Duke Univ
Jacobs, Tenika, New Jersey Geological and Water Survey
Jafari, Mohammad , Columbus State Univ
Jamiolahmady, Mahmoud, Heriot-Watt Univ
Janetos, Tony, Boston Univ

Jarcho, Kari A., Univ of Minnesota, Twin Cities
Jensen, Scott W., South Dakota Dept of Env and Nat Resources
Jiang, Zhenhua, Argonne Nat Lab
Joanna, Lucero, New Mexico Tech
Johns, Rebecca, Univ of South Florida
Johnson, Daniel L., Univ of Lethbridge
Johnson, Emily P., Boston Univ
Johnson, Jeanne L., Louisiana State Univ
Johnson, Robert L., Argonne Nat Lab
Jones, Gwilym, Northeastern Univ
Jones, Minnie O., Univ of Illinois Chicago
Jones, III, John P., Univ of Arizona
Jordan-Hernandez, Rafael, Universidad Estatal de Sonora
Jornov, Donna, New York State Geological Survey
Joseph, Miranda, Univ of Arizona
Juanes, Ruben, Massachusetts Inst of Tech
Jun, Young-Shin, Washington Univ in St. Louis
Jurmanovich, Barb, Delta Coll
Juszczyk , Carmen , Univ of Colorado
Jutla, Rajinder S., Missouri State Univ
Kaczynski, Kristen M., California State Univ, Chico
Kallin, Hamish, Edinburgh Univ
Karl, Tami S., Florida State Univ
Kastrisios, Christos , Univ of New Hampshire
Kay, Richard F., Duke Univ
Keala, Lori, Pomona Coll
Keaton, Jeffrey R., Univ of Utah
Kelly, Kimberly, Montgomery Coll
Kelly, Sherrie, St. Lawrence Univ
Kennedy, Christina B., Northern Arizona Univ
Kennel, Charles F., Univ of California, Los Angeles
Kevan, Peter, Univ of Guelph
Keyser, Jan, Univ of South Alabama
Khurana, Krishan, Univ of California, Los Angeles
Kidder, T.R., Washington Univ in St. Louis
Kidwell, Jen, Stanford Univ
Kifer, Lauri A., SUNY, The Coll at Brockport
Kim, Stacy, Moss Landing Marine Lab
Kimberly, Shaw, Columbus State Univ
Kimbro, David, Northeastern Univ
King-Hsi, Kung, Los Alamos Nat Lab
Kinkead, Scott, Los Alamos Nat Lab
Kipper, Jay P., Univ of Texas at Austin, Jackson Schl of Geosciences
Kneas, David, Univ of South Carolina
Komoto, Cary, Normandale Comm Coll
Kontuly, Thomas M., Univ of Utah
Koralek, Susan, Southern Oregon Univ
Kosinski, Leszek A., Univ of Alberta
Kostov, Svilen, Georgia Southwestern State Univ
Kressler, Sharon J., Univ of Minnesota, Twin Cities
Krieble, Kelly, Moravian Coll
Krishnamurthy, sukanya, Edinburgh Univ
Kruse, Jennifer, Gustavus Adolphus Coll
Kumar, M. Satish, Queen's Univ Belfast
Kuntz, Kara, Fort Hays State Univ
LaBella, Joel, Wesleyan Univ
LaDochy, Steve, California State Univ, Los Angeles
LaFreniere, Lorraine M., Argonne Nat Lab
Lagowski, Alison A., SUNY, Buffalo
Lam, Anita, Univ of British Columbia
Lane, Mark, Palomar Coll
Larkin, Patrick, Texas A&M Univ, Corpus Christi
Larock, B. E., Univ of California, Davis
Larson, Harold P., Univ of Arizona
Laughlin, Andrew, Univ of North Carolina, Asheville
Laurier, Eric, Edinburgh Univ
Leal-Bejarano, Arturo, Universidad Auto de Chihuahua
Lebofsky, Larry A., Univ of Arizona
Ledbetter, Cynthia E., Univ of Texas, Dallas
Lee, Alexis, Univ of North Carolina, Wilmington
Lee, Hung, Univ of Guelph
Lehane, Mary, Univ Coll Cork
Leland, John, Univ of Nevada, Reno
Lemmond, Peter C., Woods Hole Oceanographic Inst
Lener, Edward, Virginia Polytechnic Inst & State Univ
Lerdau, Manuel, Univ of Virginia
Lemurier, Nathalie, Institut Polytechnique LaSalle Beauvais (ex-IGAL)
Leveque, Connie, Univ of Maine, Presque Isle
Li, Wenhong, Duke Univ
Lichtner, Peter C., Los Alamos Nat Lab
Lin, Shoufa, Univ of Waterloo
Lin, Xiaomao, Kansas State Univ
Lindley, Stacy, Sacramento State Univ
Linky, Edward, Hunter Coll (CUNY)
Lipeles, Maxine I., Washington Univ in St. Louis
Lippert, Sarah, Univ of Chicago
Lloyd, Glenn D., Missouri Dept of Nat Resources
Loeb, Valerie, Moss Landing Marine Lab
Long, Laura, Lawrence Livermore Nat Lab
Long, Lisa, Ohio Dept of Nat Resources
Longmire, Patrick A., Los Alamos Nat Lab
Lorimer, Hayden, Edinburgh Univ
Lotterhos, Kathleen, Northeastern Univ
Loveland, Karen, Missouri Dept of Nat Resources
Lovseth, John, Principia Coll
Lowry, William R., Washington Univ in St. Louis
Loxsom, Fred, Eastern Connecticut State Univ
Lucas, Beth, Virginia Polytechnic Inst & State Univ
Lujan-Lopez, Arturo, Universidad Auto de Chihuahua
Lumpkin, Thomas A., Washington State Univ
Lynch, Casey, Univ of Nevada, Reno
Maciha, Mark J., Northern Arizona Univ
MacInnes, Michael, Los Alamos Nat Lab
Magee, Elizabeth, Northeastern Univ
Mahler, Robert L., Univ of Idaho
Maier, Raina M., Univ of Arizona
Maksoudian, Michell, California Polytechnic State Univ
Malagon, Teresa Soledad Medina, Universidad Nacional Auto de Mexico
Malega, Ron, Missouri State Univ
Mangel, Mark S., Univ of California, Santa Cruz
Marks, Dennis W., Valdosta State Univ
Marshall, Stephen, Univ of Guelph
Martinez-Macias, Panfilo R., Universidad Auto de San Luis Potosi
Masiello, Caroline A., Rice Univ
Maulud, Mat Ruzlin, Univ of Malaya
May, Diane M., Missouri State Univ
May, Fred E., Univ of Utah
Mayefsky, Cheryl, American Museum of Nat History
Mayer, Christine M., Univ of Toledo
Maynard, David, California State Univ, San Bernardino
McArthur, Russell, San Francisco State Univ
McConnell, Joseph, Univ of Nevada, Reno
McCoy, Sue, Cosumnes River Coll
McDermott, Thomas M., SUNY, The Coll at Brockport
McDonald, Kyle C., Jet Propulsion Lab
McDonald, Lynn, Los Alamos Nat Lab
McFadden, Jennifer, Elizabethtown Coll
McGee, Tara, Univ of Alberta
McGraw, Maureen A., Los Alamos Nat Lab
McKenzie, Phyllis, Smithsonian Inst / Nat Museum of Nat History
McKenzie, Ross, Univ of Lethbridge
McLaughlin, Richard, Texas A&M Univ, Corpus Christi
McLeod, Clara, Washington Univ in St. Louis
McMillan, Robert S., Univ of Arizona
McNair, Laurie A., Los Alamos Nat Lab
Medlin, Peggy, Tennessee Tech Univ
Mehes, Roxane J., Laurentian Univ, Sudbury
Mendoza, Blanca, Universidad Nacional Auto de Mexico
Meyer, Christopher, Smithsonian Inst / Nat Museum of Nat History
Meyer, Judith, Missouri State Univ
Meyerson, Rohana, Lafayette Coll
Michel, Shirley, Univ of Toledo
Middlemiss, Lucie, Univ of Leeds
Miller, Judy, Monroe Comm Coll
Miller, Mark, Univ of Southern Mississippi
Miller, Ted R., South Dakota Dept of Env and Nat Resources
Milligan, Richard, Georgia State Univ
Mills, Suzanne, McMaster Univ
Milner, Jennifer, Augustana Coll
Minkowski, Valeska, Furman Univ
Mitchneck, Beth A., Univ of Arizona
Mittleider, Stacy, Chadron State Coll

Mockler, Theodore, Los Alamos Nat Lab
Moe-Hoffman, Amy P., Mississippi State Univ
Mollner, Daniel, Gustavus Adolphus Coll
Momen, Nasim, Boston Univ
Monahan, Adam, Univ of Victoria
Montgomery, Laurie A., SUNY, Geneseo
Montgomery, Tamra S., Illinois State Geological Survey
Moodie, T. Bryant, Univ of Alberta
Moody, Eva, Tarleton State Univ
Mooney, Phillip, Sonoma State Univ
Moore, Clyde, Louisiana State Univ
Moorhead, Daryl L., Univ of Toledo
Moran-Taylor, Michelle, Univ of Denver
Morehouse, Barbara, Univ of Arizona
Morgan, Gary, Hampton Univ
Morin, Paul, Univ of Minnesota, Twin Cities
Morris, Donald P., Lehigh Univ
Morrison, Shaunna, Carnegie Inst for Science
Moss, Patti, Whitman Coll
Mozzachiodi, Riccardo, Texas A&M Univ, Corpus Christi
Mueller, Amy, Northeastern Univ
Murphy, Alexander, Univ of Oregon
Murphy, Edward C., North Dakota Geological Survey
Myers, Clifford D., Berkshire Comm Coll
Myers, Tammy, Shippensburg Univ
Nadiga, Balu, Los Alamos Nat Lab
Nakatsuka, James, Univ of California, Los Angeles
Narinder, Kaushik, Univ of Guelph
Nashold, Barney W., Argonne Nat Lab
Nesbitt, Alex, Univ of Nevada, Reno
Nevins, Susan K., SUNY, Cortland
Newman, Jonathan, Univ of Guelph
Newton, Anthony J., Edinburgh Univ
Newton, Seth A., Geological Survey of Alabama
Nichols, Terry E., Univ of Arkansas, Fayetteville
Nielsen, Mary, Univ of South Dakota
Niemiller, Matthew L., Univ of Tennessee, Knoxville
njoh, Ambe, Univ of South Florida
Noll, Michael G., Valdosta State Univ
Nordwald, Dan, Missouri Dept of Nat Resources
Norman, Catherine, Johns Hopkins Univ
Nussear, Kenneth, Univ of Nevada, Reno
O'Callaghan, Mick, Univ Coll Cork
O'Day, Sandra, Central Connecticut State Univ
O'Melia, Charles R., Johns Hopkins Univ
O'Neil, Jennifer, Univ of Nevada, Reno
Occhiuzzi, Tony, Mesa Comm Coll
Odland, Sarah K., Columbia Univ
Oglesby, Elizabeth, Univ of Arizona
Okafor, Florence A., Alabama A&M Univ
Olson, Derek, Naval Postgraduate School
Olson, James R., South Dakota Dept of Env and Nat Resources
Oona, Hain, Los Alamos Nat Lab
Oppong, Joseph R., Univ of North Texas
Ormsby, Alison, Univ of North Carolina, Asheville
Orrock, John, Washington Univ in St. Louis
Osborn, Dan P., Univ Coll London
Osburn, Glenn (Bob), Washington Univ in St. Louis
Otis, Gard, Univ of Guelph
Ott, Kevin C., Los Alamos Nat Lab
Owens, Tamera, Muskegon Comm Coll
Oza, Rupal, Hunter Coll (CUNY)
Pace, Michael L., Univ of Virginia
Pacia, Christina, Kean Univ
Palace, Michael W., Univ of New Hampshire
Palmer, Christina, California State Univ, San Bernardino
Palmer, Clare, Washington Univ in St. Louis
Papaleo, Silvanna, Univ of Toronto
Parendes, Laurie A., Edinboro Univ of Pennsylvania
Parker, Jim, Univ of Houston
Parker, Joan, Moss Landing Marine Lab
Parker, Stephen R., Montana Tech of the Univ of Montana
Parkovic, Emily, DCNR- Pennsylvania Bureau of Geological Survey
Parrish, Pia, San Diego State Univ
Pasteris, Jill D., Washington Univ in St. Louis
Patel, Vinodkumar A., Illinois State Geological Survey
Patino-Douce, Alberto E., Univ of Georgia
Patronas, Dennis, Dauphin Island Sea Lab
Patterson, Jodi T., Utah Geological Survey
Patterson, Mark, Northeastern Univ
Penney, Paulette, Baylor Univ
Pepper, Ian L., Univ of Arizona
Peri, Francesco, Univ of Massachusetts, Boston
Perkins, R, Cardiff Univ
Petersen, Bruce, Washington Univ in St. Louis
Peterson, Peter A., Iowa State Univ of Science & Tech
Pickering, Dan, Carnegie Museum of Nat History
Pickett, Nicki, Broward Coll
Pierce, Larry, Missouri Dept of Nat Resources
Pike, J, Cardiff Univ
Plane, David A., Univ of Arizona
Plant, Jeffrey J., Jet Propulsion Lab
Plescia, Jeffrey, Jet Propulsion Lab
Pociask, Walter, Wayne State Univ
Podolak, Morris, Tel Aviv Univ
Pollak, Robert, Washington Univ in St. Louis
Polovina, Jeffrey J., Univ of Hawai'i, Manoa
Polsky, Colin, Florida Atlantic Univ
Potter, Amy E., Georgia Southern Univ
Powers, Roger W., Univ of Alaska, Fairbanks
Poynton, Helen, Univ of Massachusetts, Boston
Prabhu, Anirudh, Carnegie Inst for Science
Prewitt, Charles, Univ of Arizona
Puig, Claire, Smithsonian Inst / Nat Museum of Nat History
Pulkkinen, Tuija, Univ of Michigan
Pundsack, Jonathan, Univ of Minnesota, Twin Cities
Purtle, Jennifer M., Univ of Arkansas, Fayetteville
Qian, Song, Univ of Toledo
Quinn, Courtney, Furman Univ
Quintero, Sylvia, Univ of Arizona
Radil, Steven, Univ of Idaho
Rallis, Donald N., Mary Washington Coll
Rasmussen, Tab, Washington Univ in St. Louis
Rauber, Carolyn, Univ of Minnesota, Twin Cities
Ravela, Sai, Massachusetts Inst of Tech
Reaven, Sheldon, SUNY, Stony Brook
Rebuck, Jody L., DCNR- Pennsylvania Bureau of Geological Survey
Refsnider-Streby, Jeanine, Univ of Toledo
Rennie, Tim, Univ of Guelph
Reusch, David B., New Mexico Tech
Reyhanoglu, Mahmut, Columbus State Univ
Rhine, Michelle, Eastern Illinois Univ
Rice, Murray, Univ of North Texas
Ricker, Alison, Oberlin Coll
Ridgwell, Andy, Univ of California, Riverside
Ridky, Alice M., Colby Coll
Rieke, George H., Univ of Arizona
Riley, James, Univ of Arizona
Ritter, Leonard, Univ of Guelph
Rius, Marc, Univ of Southampton
Robas, Sheryl A., Princeton Univ
Roberts, A. Lynn, Johns Hopkins Univ
Robinson, Bruce A., Los Alamos Nat Lab
Robinson, David, Open Univ
Robinson, William, Univ of Massachusetts, Boston
Rodriguez, Vanessa del S., Puerto Rico Bureau of Geology
Roe, Carol, William & Mary
Roemer, Elizabeth, Univ of Arizona
Rogers, Jefferson S., Univ of Tennessee, Martin
Rogers, William J., West Texas A&M Univ
Roinstad, Lori L., South Dakota Dept of Env and Nat Resources
Rojas, Adena, Johns Hopkins Univ
Rollinson, Paul A., Missouri State Univ
Roman-Alcalá, Antonio, California State Univ, East Bay
Rooney, Neil, Univ of Guelph
Rose, Candace M., Argonne Nat Lab
Rose, Dan, Univ Coll Cork
Rosengaus, Rebeca, Northeastern Univ
Ross, Kirstin, Flinders Univ
Ross, Paula, New Mexico State Univ, Las Cruces
Rossell, Irene M., Univ of North Carolina, Asheville
Rostron, Ben, Univ of Alberta

Rouhani, Farhang, Mary Washington Coll
Rouse, Jesse, Univ of North Carolina, Pembroke
Roussel-Dupre, R., Los Alamos Nat Lab
Rozmus, Wojciech, Univ of Alberta
Rumstay, Kenneth S., Valdosta State Univ
Rupert, Denise, Lock Haven Univ
Russell, Ron, Univ of Texas at Austin, Jackson Schl of Geosciences
Russell, Terry P., Univ of Victoria
Russell, Theresa J., Univ of Arkansas, Fayetteville
Russo, Mary Rose, Princeton Univ
Ruzicka, Jaromir, Univ of Hawai'i, Manoa
Saikia, Udoy, Flinders Univ
Saito, Heather, Univ of Hawai'i, Manoa
Saku, James C., Frostburg State Univ
Sammarco, Paul W., Louisiana State Univ
Sanchez, Charles, Univ of Arizona
Sandhu, Harpinder, Flinders Univ
Santander, Erma, Univ of Arizona
Sapigao, Gladys, Queens Coll (CUNY)
Sarin, Manmohan, Univ of Delaware
Sarmiento, Jorge L., Princeton Univ
Sauer, Nancy N. S., Los Alamos Nat Lab
Sawyer, Suzanne, Utah Geological Survey
Scapelli, Krista, Texas Christian Univ
Scarpino, Samuel, Northeastern Univ
Schaal, Barbara, Washington Univ in St. Louis
Scheel, Patrick, Missouri Dept of Nat Resources
Scheidemen, Kathy J., Univ of California, Santa Barbara
Schlesinger, William H., Duke Univ
Schlumpberger, Debbie, Saint Cloud State Univ
Schmidt, Jonathan, Univ of Guelph
Schoenberger, Erica, Johns Hopkins Univ
Schroeder, Kathleen, Appalachian State Univ
Schwenke, Eszter, Univ of New Brunswick
Schwob, Stephanie L., Southern Methodist Univ
Scott, Kathy, Univ of British Columbia
Scott-Dupree, Cynthia, Univ of Guelph
Scyphers, Steven, Northeastern Univ
Seifu, Abiye, Columbus State Univ
Seymour, Dorie, Montana State Univ
Shade, Janet, Univ of Pittsburgh, Bradford
Shah, Ashru, North Carolina State Univ
Sharma, Govind, Alabama A&M Univ
Shaw, Fred, Missouri Dept of Nat Resources
Shchukin, Eugene D., Johns Hopkins Univ
Shelton, Robert B., Oak Ridge Nat Lab
Shields, Nancy, Virginia Polytechnic Inst & State Univ
Showman, Adam, Univ of Arizona
Shumway, Matthew J., Brigham Young Univ
Sibley, Paul, Univ of Guelph
Sigler, William V., Univ of Toledo
Silks, Louis A., Los Alamos Nat Lab
Simonetti, Stephanie, Univ of Notre Dame
Singh, Hanumant, Northeastern Univ
Singh, Harbans, Montclair State Univ
Sissom, David , West Texas A&M Univ
Sitwell, O.F. George, Univ of Alberta
Skeel, Loreene, Univ of Montana
Slater, Tom, Edinburgh Univ
Slavetskas, Carol, Binghamton Univ
Smale, Jody L., DCNR- Pennsylvania Bureau of Geological Survey
Smith, Derald, Univ of Lethbridge
Smith, Donald R., Univ of California, Santa Cruz
Smith, H D., Cardiff Univ
Smith, Jim, Flinders Univ
Smith, K. L., Wichita State Univ
Smith, Langdon, Slippery Rock Univ
Smith, Nicole, Colorado School of Mines
Smith, Paul H., Los Alamos Nat Lab
Smith, Peter J., Univ of Alberta
Smith, Susan M., Montana Tech of The Univ of Montana
Snipes, Brittany, Univ of Florida
Solomon, Keith, Univ of Guelph
Somers, Jr., Arnold E., Valdosta State Univ
Southwell, Benjamin, Lake Superior State Univ
Spaeth, Bridget, Bowdoin Coll
Spanbauer, Trisha, Univ of Toledo
Sperazza, Michael, Stony Brook Univ
Spracklen, Dominick, Univ of Leeds
Stadnyk, Leona, Mount Royal Univ
Standridge, Debbie, Georgia Southwestern State Univ
Starr, Richard, Moss Landing Marine Lab
Steadman, Todd A., Clemson Univ
Stechmann, Samuel, Univ of Wisconsin, Madison
Steller, Diana, Moss Landing Marine Lab
Stephenson, Gerry, Univ of Guelph
Sterling, Tracy M., Montana State Univ
Sternberg, Rolf, Montclair State Univ
Stevens, Joan M., California Polytechnic State Univ
Stierle, Andrea, Montana Tech of the Univ of Montana
Stimer, Debra, Kent State Univ at Stark
Stinger, Lindsay C., Univ of Leeds
Stone, Glenn D., Washington Univ in St. Louis
Strangeway, Robert J., Univ of California, Los Angeles
Streby, Henry, Univ of Toledo
Strong, Ellen, Smithsonian Inst / Nat Museum of Nat History
Stunz, Greg, Texas A&M Univ, Corpus Christi
Su, Xiaobo, Univ of Oregon
Suárez, Luis Eugenio, Univ Complutense de Madrid
Subulwa, Angela G., Univ of Wisconsin Oshkosh
Sullivan, Barbara, Argonne Nat Lab
Summers, Robert, Univ of Alberta
Sussman, Robert W., Washington Univ in St. Louis
Sutherland, Bruce R., Univ of Alberta
Svitra, Zita V., Los Alamos Nat Lab
Swanson, Basil I., Los Alamos Nat Lab
Swetnam, Thomas W., Univ of Arizona
Swindall, Diane, Univ of California, Davis
Symbalisty, E.M.D., Los Alamos Nat Lab
Sywensky, Ann, Moravian Coll
Taney, R. Marieke, Northern Arizona Univ
Tanji, K. K., Univ of California, Davis
Tavener, Kristi, Lakehead Univ
Tay, Chelsea, Curtin Univ
Taylor, Michael, Flinders Univ
Taylor, Robert W., Montclair State Univ
Tencate, James, Los Alamos Nat Lab
Teplitski, Max, Univ of Florida
Ter-Simonian, Vardui, Univ of Southern California
Thacker, Heather, Shawnee State Univ
Theis, Karen, Univ of Manchester
Thériault, Julie Mireille, Universite du Quebec a Montreal
Thompson, Glennis, Univ of California, Berkeley
Thomson, Cary, Univ of British Columbia
Thomson, Cynthia, Columbia Univ
Tingey, David G., Brigham Young Univ
Tippett, John, Univ of Mary Washington
Tissot, Philippe , Texas A&M Univ, Corpus Christi
Todd, Brenda, Univ of Nebraska, Omaha
Tomlinson, Jaime L., Univ of Delaware
Townshend, Ivan, Univ of Lethbridge
Tracy, Matthew, United States Air Force Acad
Trevors, Jack, Univ of Guelph
Triplehorn, Judy, Univ of Alaska, Fairbanks
Trost, G K., Univ of Arkansas, Fayetteville
Trout, Jennifer, Western Michigan Univ
Trussell, Geoffrey, Northeastern Univ
Tuller, Markus, Univ of Arizona
Turner, Jay, Washington Univ in St. Louis
Tyning, Thomas F., Berkshire Comm Coll
Unkefer, Clifford J., Los Alamos Nat Lab
Unsworth, Martyn, Univ of Alberta
Urquhart, Alvin, Univ of Oregon
van den Akker, Ben, Flinders Univ
Van Eerd, Laura, Univ of Guelph
Van Otten, George A., Northern Arizona Univ
Van Roosendaal, Susan, Univ of Utah
Varadi, Ferenc D., Univ of California, Los Angeles
Varady, Robert G., Univ of Arizona
Veeder, Glenn J., Jet Propulsion Lab
Vieira, David J., Los Alamos Nat Lab
Vollmer, Steve, Northeastern Univ

Wada, Ikuko, Univ of Minnesota, Twin Cities
Wagh, Swati, Argonne Nat Lab
Walker, Mark, Univ of Nevada, Reno
Walker, Peter, Univ of Oregon
Walker, Raymond J., Univ of California, Los Angeles
Walraven, Brenda, Mercer Univ
Walrod, Amanda G., Univ of Arkansas, Fayetteville
Walsh, Christopher, Univ of Maryland
Walsh, Ellen c., Lawrence Univ
Ward, Landon, Univ of North Carolina, Asheville
Ward, Tristan, Curtin Univ
Wardell, Rob, Smithsonian Inst / Nat Museum of Nat History
Warkentin, Alicia, Univ of British Columbia
Waterstone, Marvin, Univ of Arizona
Watson, Alan, Univ of Guelph
Wauthier, Christelle, Pennsylvania State Univ, Univ Park
Weaver, Douglas J., Los Alamos Nat Lab
Webre, Cherri B., Louisiana State Univ
Weinstein, Charles E., Berkshire Comm Coll
Weintraub, Michael N., Univ of Toledo
Wells, Steve G., Univ of Nevada, Reno
Welton, Leicha, Univ of Alaska, Fairbanks
Wenz, Helmut C., Univ of Tennessee, Martin
West, Alan, Univ of Virginia's Coll at Wise
Wetzel, Dan L., Univ of Alaska, Fairbanks
Whiley, Harriet, Flinders Univ
White, George W., Frostburg State Univ
White, Paul , Univ of Nevada, Reno
Whitney, Earl M., Los Alamos Nat Lab
Wiederspahn, Mark, Univ of Texas, Austin
Wilder, Margaret, Univ of Arizona
Wilhelmy, Jerry B., Los Alamos Nat Lab
Wille, Frank, California State Polytechnic Univ, Pomona
William, Nancy, Kansas State Univ
Williams, Allison M., McMaster Univ
Williams-Bruinders, Leizel, Nelson Mandela Metropolitan Univ
Willson, Lee, Rice Univ
Wilmut, Michael, Univ of Victoria
Wilson, Robert, Univ of St. Andrews
Wilson, Sarah B., Washington & Lee Univ
Wilson, Thomas B., Univ of Arizona
Wilton, Robert D., McMaster Univ
Winston, Barbara, Washington Univ in St. Louis
Winterkamp, Judith L., Los Alamos Nat Lab
Winton, Alison, Texas Tech Univ
Wisdom, Jack, Massachusetts Inst of Tech
Wise, Dana, Southern Illinois Univ Carbondale
Withers, Kim, Texas A&M Univ, Corpus Christi
Wixman, Ronald, Univ of Oregon
Wolfe, Karen M., Middle Tennessee State Univ
Wollan, Jacinda, North Dakota State Univ
Wong, Cindy, Buffalo State Coll
Wood, Carolyn F., Dauphin Island Sea Lab
Woodley, Teresa, Univ of British Columbia
Woodring, Hannah, Univ of Southern Indiana
Woodruff, William H., Los Alamos Nat Lab
Wooldridge, C F., Cardiff Univ
Wright, Kathyrn, Western Michigan Univ
Wu, David T., Colorado School of Mines
Wyckoff, William K., Montana State Univ
Yarbrough, Robert A., Georgia Southern Univ
Yates, Mary Anne, Los Alamos Nat Lab
Yelle, Roger, Univ of Arizona
Yoskowitz, David , Texas A&M Univ, Corpus Christi
Young, Priscilla E., South Dakota Dept of Env and Nat Resources
Yutzy, Gale, Frostburg State Univ
Zajac, Roman N., Univ of New Haven
Zaniewski, Kazimierz J., Univ of Wisconsin Oshkosh
Zeiger, Elaine, Field Museum of Nat History
Zelizer, Nora, Princeton Univ
Zhang, Max, Cornell Univ
Zimba, Paul, Texas A&M Univ, Corpus Christi
Zink, Paul, Colorado School of Mines
Zotti, Maria, Flinders Univ
Zurick, David, Eastern Kentucky Univ
Zwiefelhofer, Luke, Winona State Univ

FACULTY INDEX

A

Aagaard, Knut, (206) 543-8942 aagaard@apl.washington.edu, Univ of Washington; Op

Aalto, Rolf E., +44 (0) 1392-26-3344 aalto@uw.edu, Univ of Washington; Gms

Aamodt, Paul, (505) 665-1331 plaamod@lanl.gov, Los Alamos; Zn

Aanstoos, James V., 515-294-1032 aanstoos@iastate.edu, Iowa State Univ of Sci & Tech; ZrAs

Aay, Henry, (616) 526-7033 aay@calvin.edu, Calvin Coll; Zyu

Abalos, Benito, benito.abalos@ehu.eus, Univ of the Basque Country UPV/EHU; GctGp

Abati, Jacobo, abati@ucm.es, Univ Compl de Madrid; Gp

Abatzoglou, John, 208-885-6239 jabatzoglou@uidaho.edu, Univ of Idaho; ZoyZc

Abbasnejad, Ahmad, abbasnejadahmad@yahoo.com, Shahid Bahonar Univ of Kerman ; GemHg

Abbott, Lon, (303) 492-6172 lon.abbott@colorado.edu, Univ of Colorado; GgtGm

Abbott, Mark B., (412) 624-1408 mabbott1@pitt.edu, Univ of Pittsburgh; Gs

Abbott, Patrick L., (619) 594-5591 pabbott@mail.sdsu.edu, San Diego State Univ; Gs

Abdalati, Waleed, waleed.abdalati@colorado.edu, Univ of Colorado; Zr

Abdalla, Salah B., geology@uofk.edu, Univ of Khartoum; Ng

Abdelsalam, Mohamed G., 405-744-6358 mohamed.abdel_salam@okstate.edu, Oklahoma State Univ; GctYg

Abdo, Ginette, 406-496-4152 gabdo@mtech.edu, Montana Tech of The Univ of Montana; Hw

Abdu, Yassir, (204) 474-7356 abdu@cc.umanitoba.ca, Univ of Manitoba; Gz

Abdulla, Hussain, 361-825-6050 hussain.abdulla@tamucc.edu, Texas A&M Univ, Corpus Christi; Co

Abdullah, Nuraiteng Tee, 03-79674229 naiteng@um.edu.my, Univ of Malaya; Gr

Abdullah, Wan Hasiah, 03-79674232 wanhasia@um.edu.my, Univ of Malaya; Ec

Abdurrahman, A., aabdrrahman@unilorin.edu.ng, Univ of Ilorin; CeEg

Abend, Martin, (201) 200-3161 mabend@njcu.edu, New Jersey City Univ; Zy

Aber, James S., jaber@emporia.edu, Emporia State Univ; Gl

Aber, Jeremy, jeremy.aber@mtsu.edu, Middle Tennessee State Univ; Zyi

Abercrombie, Rachel, rea@bu.edu, Boston Univ; Gt

Abernathey, Ryan P., 845-365-8185 rpa@ldeo.columbia.edu, Columbia Univ; Op

Abernathy, Ryan, rpa@ldeo.columbia.edu, Columbia Univ; Og

Abers, Geoffrey A., 607-255-3879 abers@cornell.edu, Cornell Univ; YsGt

Abimbola, A. E., af.abimbola@mail.ui.edu.ng, Univ of Ibadan; Cg

Abney, Kent D., (505) 665-3894 Los Alamos; Zn

Abolins, Mark J., (615)904-8372 mark.abolins@mtsu.edu, Middle Tennessee State Univ; Gc

Abou Najm, Majdi, mabounajm@ucdavis.edu, Univ of California, Davis; CbSb

Abousleiman, Younane, 405-325-2900 yabousle@ou.edu, Univ of Oklahoma; Nr

Abrahams, Miengah, miengah.abrahams@uct.ac.za, Univ of Cape Town; PgGsPv

Abrams, Daniel, (217) 244-1520 dbabrams@illinois.edu, Univ of Illinois; Hw

Abrams, Lewis J., 910-962-2350 abramsl@uncw.edu, Univ of North Carolina, Wilmington; Gu

Abrams, Michael J., (818) 354-0937 Jet Propulsion Lab; Zr

Abramson, Evan H., 206-616-4388 evana@uw.edu, Univ of Washington; Gy

Abreu, Maria E., msabreu@utad.pt, Unive de Trás-os-Montes e Alto Douro; GaaGa

Abt, Charlotte J., +440151 794 5178, chj@liverpool.ac.uk, Univ of Liverpool; Zn

Abu Bakar, Muhammad Z., 0092-42-99029487 mzubairab1977@gmail.com, Univ of Eng and Tech; NrmNg

Abushagur, Sulaiman, 915-831-2539 sulaiman@epcc.edu, El Paso Comm Coll; Gg

Acharya, Kumud, (702) 862-5371 kumud.acharya@dri.edu, Univ of Nevada, Reno; HsGeHq

Achauer, Ulrich, ulrich.achauer@unistra.fr, Univ de Strasbourg; Ysg

Achten, Christine, +49-251-83-33941 achten@uni-muenster.de, Univ Muenster; CloEg

Ackerman, Jessica R., 217-333-4258 jracker@illinois.edu, Illinois State Geo Survey; Zn

Ackerman, Steven A., (608)263-3647 stevea@ssec.wisc.edu, Univ of Wisconsin, Madison; Zr

Ackerman, Thomas P., (206) 221-2767 ackerman@atmos.washington.edu, Univ of Washington; As

Ackerson, Michael, (202) 633-1876 ackersonm@si.edu, Smithsonian Inst / Natl Mus of Natural History; Gip

Acosta, Jose G., jchang@cicese.mx, Centro de Investigación Científica y de Educación Superior de Ensenada; Ne

Acosta, Patricia E., (413) 597-2221 patricia.e.acosta@williams.edu, Williams Coll; Zn

Adabanija, Moruffdeen A., +2348037786592 maadabanija@lautech.edu.ng, Ladoke Akintola Univ of Tech; GggGg

Adam, Jurgen, +44 1784 414258 jurgen.adam@rhul.ac.uk, Royal Holloway Univ of London; Gc

Adames-Corraliza, Angel F., afadames@umich.edu, Univ of Michigan; Ast

Adams, Aubreya, 315-228-7202 aadams@colgate.edu, Colgate Univ; Yg

Adams, Debi, 806-651-2570 dadams@wtamu.edu, West Texas A&M Univ; Zn

Adams, Herbert G., (818) 677-2575 herb.adams@csun.edu, California State Univ, Northridge; Ng

Adams, John B., adams@ess.washington.edu, Univ of Washington; GcXgZr

Adams, Kenneth, 775.673.7345 ken.adams@dri.edu, Univ of Nevada, Reno; Gam

Adams, Manda S., 704-687-5984 manda.adams@uncc.edu, Univ of North Carolina, Charlotte; Am

Adams, Mark, capt.mark.adams@gmail.com, Lamar Univ; Xg

Adams, Marshall, (865) 574-7335 sma@ornl.gov, Oak Ridge; Ob

Adams, Michael S., (608) 263-5994 Univ of Wisconsin, Madison; Ob

Adams, Peter N., (352) 846-0825 adamsp@ufl.edu, Univ of Florida; Gm

Adams, Thomas, 210-486-0045 tadams67@alamo.edu, San Antonio Comm Coll; PvGg

Adamsen, Floyd, (602) 437-1702 f.j.adamsen@gmail.com, Univ of Arizona; So

Adcock, Christopher, (702) 774-1429 christopher.adcock@unlv.edu, Univ of Nevada, Las Vegas; Xg

Adedcyin, A. D., deloadedoyin@yahoo.com, Univ of Ilorin; Eg

Adee, Eric, (785) 354-7236 eadee@ksu.edu, Kansas State Univ; So

Adegoke, Jimmy, (816) 235-2978 adegokej@umkc.edu, Univ of Missouri, Kansas City; AsZyr

Adeigbe, O. C., oc.adeigbe@mail.ui.edu.ng, Univ of Ibadan; GosGe

Adekeye, J.I. D., adekeye2001@yahoo.com, Univ of Ilorin; GoCg

Adekeye, O. A., geologistolabisi@yahoo.com, Univ of Ilorin; Pm

Adekola, S. A., 234-708-928-5170 adekoladsolo@gmail.com, Obafemi Awolowo Univ; GdoCg

Adelana, S.M. A., adelana@gmx.net, Univ of Ilorin; HgYgGe

Adeleke, Adedayo, +27 (0) 12 420 3536 adedayo.adeleke@up.ac.za, Univ of Pretoria; YdZf

Ademeso, Anthony O., +2348034738470 aoademeso@futa.edu.ng, Federal Univ of Tech, Akure; NgrNt

Aden, Douglas J., (614) 265-6579 douglas.aden@dnr.ohio.gov, Ohio Dept of Nat Res; GeRnGl

Aden, Leon, leon-aden@uiowa.edu, Univ of Iowa; Go

Adentunji, Jacob, j.adetunji@derby.ac.uk, Univ of Derby; Ge

Adeoti, Blessing, +2347065810638 badeoti@futa.edu.ng, Federal Univ of Tech, Akure; GcxGg

Adepelumi, A. A., 234-8128181062 aadepelu@oauife.edu.ng, Obafemi Awolowo Univ; YgsNe

Adepoju, Mohammed O., +2348034722855 moadepoju@futa.edu.ng, Federal Univ of Tech, Akure; CeEgGg

Adetunji, A, 234-8137433622 Obafemi Awolowo Univ; EgGi

Adewoye, Abosede O., +2347032045714, +2347036602812 aoadewoye55@lautech.edu.ng, Ladoke Akintola Univ of Tech; GeeGe

Adeyemi, G. O., go.adeyemi@mail.ui.edu.ng, Univ of Ibadan; NgHg

Adhikari, Sanchayeeta, 818-677-5630 sadhikari@csun.edu, California State Univ, Northridge; Zi

Adinolfi, Bryan, adinolfib1@southernct.edu, Southern Connecticut State Univ; GgZg

Adisa, Adeshina L., +2348029458886 aladisa@futa.edu.ng, Federal Univ of Tech, Akure; CgeGc

Adkins, Jess F., (626) 395-8550 jess@gps.caltech.edu,CalTech; Cm

Admire, Amanda R., 707-826-3111 ara11@humboldt.edu, Humboldt State Univ; Zg

Adojoh, Onema, onema.adojoh@case.edu, Case Western Reserve Univ; Pm

Adomaitis, D., 217-244-8872 dadomait@illinois.edu, Illinois State Geo Survey; Ge

Adovasio, James M., 814-824-2581 adovasio@mercyhurst.edu, Mercyhurst

Univ; GafGs
Adrain, Jonathan M., (319) 335-1539 jonathan-adrain@uiowa.edu, Univ of Iowa; Pi
Adrain, Tiffany S., (319) 335-1822 tiffany-adrain@uiowa.edu, Univ of Iowa; Pg
Adriano, Domy C., (803) 725-2752 Univ of Georgia; Sc
Afolabi, O., 234-803-512-4838 toltobi@yahoo.com, Obafemi Awolowo Univ; Yg
Afolabi, Olukayode A., +2348052158722 oaafolabi@lautech.edu.ng, Ladoke Akintola Univ of Tech; GxxGx
Agardy, Rachael, 989-774-7596 agard1rv@cmich.edu, Central Michigan Univ; OuGrs
Agbogun, Henry, 785-628-4574 hmagobogun@fhsu.edu, Fort Hays State Univ; GgYg
Agee, Carl A., (505) 277-1644 agee@unm.edu, Univ of New Mexico; Cp
Agee, Ernest M., eagee@purdue.edu, Purdue Univ; As
Agioutantis, Zach, 859-257-953 zach.agioutantis@uky.edu, Univ of Kentucky; NrmEc
Agnew, Duncan C., (858) 534-2590 dagnew@ucsd.edu, Univ of California, San Diego; Ys
Agnew, Stephen F., (505) 665-1764 Los Alamos; Zn
Agnini, Claudia, 39-049-827918 claudia.agnini@unipd.it, Univ degli Studi di Padova; PmeGu
Agrawal, Abinash, (937) 775-3455 abinash.agrawal@wright.edu, Wright State Univ; ClHwPg
Agrawal, Sandeep , 780-492-1230 sagrawal@ualberta.ca, Univ of Alberta; Zu
Ague, Jay J., 203-432-3171 jay.ague@yale.edu, Yale Univ; Gzx
Agutu, Nathan O., nagutu@jkuat.ac.ke, Jomo Kenyatta Univ of Agriculture & Tech; ZgiZr
Ahearn, Sean C., (212) 772-5327 sahearn@hunter.cuny.edu, Hunter Coll (CUNY); Zi
Ahern, Judson L., (405) 325-4480 jahern@ou.edu, Univ of Oklahoma; Yg
Ahmad, Abd Rashid, 03-79674156 abrashid@um.edu.my, Univ of Malaya; Sc
Ahmed, Kazi M., +880 1711846840 kmahmed@du.ac.bd, Univ of Dhaka ; HwGgHg
Ahmed, Mohamed, 361-825-3278 mohamed.ahmed@tamucc.edu, Texas A&M Univ, Corpus Christi; YgZri
Ahmed, Muhammad F., mfanr5@mst.edu, Univ of Eng and Tech; NgxNt
Ahmed, Ramadan, r.ahmed@ou.edu, Univ of Oklahoma; Np
Ahmed, Samir, 212-650-7250 ahmed@ccny.cuny.edu, Graduate Sch of the City Univ of New York; Gg
Ahmed, Waquar, 940-565-2721 waquar.ahmed@unt.edu, Univ of North Texas; Zn
Ahola, John, john.ahola@bristolcc.edu, Bristol Comm Coll; Gg
Aiello, Ivano, (831) 771-4400 iaiello@mlml.calstate.edu, Moss Landing Marine Lab; OuGmu
Aines, Roger D., (925) 423-7184 aines1@llnl.gov, Lawrence Livermore; Cg
Aird, Hannah M., 530-898-6369 haird@csuchico.edu, California State Univ, Chico; GxEg
Aitchison, Jonathan, geoscience.headofschool@sydney.edu.au, Univ of Sydney; GrtPm
Aitkenhead-Peterson, Jacqueline A., (979) 845-3682 jacqui_a-p@tamu.edu, Texas A&M Univ; HsSoGf
Aiyyer, Anantha, 919-515-7973 aaiyyer@ncsu.edu, North Carolina State Univ; As
Aizen, Vladimir, 208-885-5888 aizen@uidaho.edu, Univ of Idaho; GlHg
Aja, Stephen U., (718) 951-5405 Graduate Sch of the City Univ of New York; Cl
Ajaegwu, Norbert E., (+234) 8023649729 ne.ajaegwu@unizik.edu.ng, Nnamdi Azikiwe Univ; GorGg
Ajayi, J. O., 234-803-401-4357 owoajayi@oauife.edu.ng, Obafemi Awolowo Univ; No
Ajayi, T. R., 234-803-725-8924 traajayi@oauife.edu.ng, Obafemi Awolowo Univ; CgEgGe
Ajigo, Isaac O., +2348032107038 ioajigo@futa.edu.ng, Federal Univ of Tech, Akure; GgCgGx
Ajo-Franklin, Jonathan, 713.348.3962 ja62@rice.edu, Rice Univ; Yg
Akabzaa, Thomas M., 233-246325685 akabzaa@ug.edu.gh, Univ of Ghana; EgGeNm
Akaegbobi, I. M., izumike2002@yahoo.com, Univ of Ibadan; Cg
Akande, S. O., samoakande2004@yahoo.com, Univ of Ilorin; EgCg
Akciz, Sinan, (657) 278-4371 sakciz@fullerton.edu, California State Univ, Fullerton; GcRn
Akenzua-Adamcyzk, Aiyevbekpen H., aiyevbekpen.akenzua@uniben.edu, Univ of Benin; GeHgGg
Akgul, Bunyamin, 00904242370000-5988 bakgul@firat.edu.tr, Firat Univ; GzyGi
Akgun, Elif, 00904242370000-5972 efiratligil@firat.edu.tr, Firat Univ; GgcGt
Akhtar, Alliya, 609-258-4101 aakhtar@princeton.edu, Princeton Univ; CsOoPc
Akinlabi, Ismaila A., +2348050225113 iaakinlabi@lautech.edu.ng, Ladoke Akintola Univ of Tech; GggGg
Akinsanola, Akintomide, aakinsan@uic.edu, Univ of Illinois Chicago; Zo
Akiwumi, Fenda, fakiwumi@usf.edu, Univ of South Florida; Zy
Aksoy, Ercan, 00904242370000-5974 Firat Univ; Ggt
Aksu, Ali E., (709) 737-8385 aaksu@mun.ca, Memorial Univ of Newfoundland; Gu
Akujieze, Christopher N., krisjiaku@yahoo.com, Univ of Benin; GgHwGe
Al, Tom, tom.al@uottawa.ca, Univ of Ottawa; HwCgZg
Al-Aasm, Ihsan S., 519-253-3000 x2494 alaasm@uwindsor.ca, Univ of Windsor; GdClGo
Al-Attar, David, +44 (0) 1223 348935 da380@cam.ac.uk, Univ of Cambridge; YdGt
Al-Lehyani, Ayman, +96638601661 allehyani@kfupm.edu.sa, King Fahd Univ of Petroleum and Minerals; Yg
Al-Ramadan, Khalid, +96638607175 ramadank@kfupm.edu.sa, King Fahd Univ of Petroleum and Minerals; GodGs
Al-Saad, Hamad A., hamadsaad@qu.edu.qa, Univ of Qatar; PsGsu
Al-Shuhail, Abdullah, +966138602538 shuhail@kfupm.edu.sa, King Fahd Univ of Petroleum and Minerals; Ye
Al-Shuhail, Abdullatif, +96638603584 ashuhail@kfupm.edu.sa, King Fahd Univ of Petroleum and Minerals; Ygs
Alam, Md Iftekhar, 865-974-3665 malam11@utk.edu, Univ of Tennessee, Knoxville; Yu
Alao, D. A., lomalao@yahoo.com, Univ of Ilorin; NgGz
Alao, O. A., 234-805-466-7314 olade77@yahoo.com, Obafemi Awolowo Univ; Yg
Alavi, Hedy, (410) 516-7091 alavi@jhu.edu, Johns Hopkins Univ; Zn
Alba, Saturnino, sdeal@baucm.es, Univ Compl de Madrid; Gm
Albach, Suzanne M., 734-677-5111 salbach@wccnet.edu, Washtenaw Comm Coll; ZegGe
Albarello, Dario, dario.albarello@unisi.it, Univ of Siena; YgsYx
Albee, Arden L., 626.395.6260 aalbee@caltech.edu,CalTech; GizXg
Albee-Scott, Steven R., (517) 796-8526 albeescsteven@jccmi.edu, Jackson Coll; GePeSf
Alberstadt, Leonard P., 615-322-2160 leonard.p.alberstadt@vanderbilt.edu, Vanderbilt Univ; Gd
Alberts, Heike C., (920) 424-4105 alberts@uwosh.edu, Univ of Wisconsin Oshkosh; Zn
Albin, Ed, ealbin@gsu.edu, Georgia State Univ, Perimeter Coll, Online; GgXg
Albrecht, Jochen, (212) 772-5221 jochen@hunter.cuny.edu, Hunter Coll (CUNY); ZiyZu
Albrecht, Lorraine, klalbrec@uwaterloo.ca, Univ of Waterloo; Zn
Albright, James N., (505) 667-4318 j_albright@lanl.gov, Los Alamos; Yg
Albright, Katia, (775) 682-8370 kalbright@unr.edu, Univ of Nevada, Reno; Zn
Albright, Thomas , (775) 784-6673 talbright@unr.edu, Univ of Nevada, Reno; Zy
Alde, Douglas, (505) 667-0488 dxa@lanl.gov, Los Alamos; Zn
Alderson, David, +44 (0) 191 208 7121 david.alderson@ncl.ac.uk, Univ of Newcastle Upon Tyne; Zi
Alderton, Dave, +44 1784 443585 d.alderton@rhul.ac.uk, Royal Holloway Univ of London; Gz
Aldrich, Eric A., 573-882-6301 aldriche@missouri.edu, Univ of Missouri; As
Aldrich, M. James, (505) 667-1495 jaldrich@lanl.gov, Los Alamos; Gc
Aldridge, Keith D., (416) 399-0124 keith@yorku.ca, York Univ; YmdYh
Aldushin, Kirill, 089/2180 4337 kirill.aldushin@lrz.uni-muenchen.de, Ludwig-Maximilians-Univ Muenchen; Gz
Aleksandrowski, Pawel, pawel.aleksandrowski@uwr.edu.pl, Univ of Wroclaw; GctYe
Alencoão, Ana M., alencoao@utad.pt, Unive de Trás-os-Montes e Alto Douro; HgGgm
Alessi, Daniel S., (780) 492-8019 alessi@ualberta.ca, Univ of Alberta; ClGeHw
Alexander, Becky, (206) 543-0164 beckya@atmos.washington.edu, Univ of Washington; AsCsPe
Alexander, Clark R., 912-598-2329 clark.alexander@skio.usg.edu, Georgia Southern Univ; OuGsOn
Alexander, Conel M., (202) 478-8478 calexander@carnegiescience.edu, Carnegie Inst for Sci; Xc
Alexander, Dane, (269) 387-5486 dane.alexander@wmich.edu, Western Michigan Univ; GgZnn
Alexander, Elaine, (254) 299-8442 ealexander@mclennan.edu, McLennan Comm Coll; Gg

Alexander, Jan, +44 (0)1603 59 3759 j.alexander@uea.ac.uk, Univ of East Anglia; GseGd
Alexander, Jane L., 718-982-3013 jane.alexander@csi.cuny.edu, Coll of Staten Island/CUNY; GsClt
Alexander, Peter, peter.alexander@ed.ac.uk, Edinburgh Univ; Zc
Alexander, Scott, 612-626-4164 alexa107@umn.edu, Univ of Minnesota, Twin Cities; Hw
Alexander, Shelton S., (814) 863-7246 shel@geosc.psu.edu, Pennsylvania State Univ, Univ Park; Ys
Alexander, Jr., E. Calvin, (612) 624-3517 alexa001@umn.edu, Univ of Minnesota, Twin Cities; HwCcl
Alexandre, Paul, 204-727-9693 alexandrep@brandonu.ca, Brandon Univ; EgCgGz
Alfe, Dario, +44 202 7679 32361 d.alfe@ucl.ac.uk, Univ Coll London; ZmGyYh
Alford, Matthew H., 206-221-3257 malford@apl.washington.edu, Univ of Washington; Op
Alfsen, Sheila, sheila.alfsen@pdx.edu, Portland State Univ; GgPgOu
Algar, Christopher, (902) 494-7192 calgar@dal.ca, Dalhousie Univ; Obc
Algeo, Catherine, (270) 745-5922 katie.algeo@wku.edu, Western Kentucky Univ; Zi
Algeo, Thomas J., (513) 556-4195 thomas.algeo@uc.edu, Univ of Cincinnati; Gs
Algin, Barbara, (212) 854-2905 ba110@columbia.edu, Columbia Univ; Zn
Ali, Bukari, +233 20 330 7976 bukariali@yahoo.co.uk, Kwame Nkrumah Univ of Sci and Tech; HwNgYg
Ali, Dr. Syed Haroon, +923213221343 haroon.ali@uos.edu.pk, Univ of Sargodha; GsoGt
Ali, Genevieve, 204-474-7266 genevieve.ali@umanitoba.ca, Univ of Manitoba; Hw
Ali, Hendratta N., (785) 628-4608 hnali@fhsu.edu, Fort Hays State Univ; GogYs
Ali, K. Adem, (843) 953-0877 alika@cofc.edu, Coll of Charleston; ZrHwNg
Alkac, Onur, 00904242370000-5970 oalkac@firat.edu.tr, Firat Univ; GgsGu
Allabush, Kathleen, 801-581-7062 k.ritterbush@utah.edu, Univ of Utah; Pe
Allam, Bassem, (632) 632-8745 bassem.allam@stonybrook.edu, SUNY, Stony Brook; Ob
Allan, Andrea, (541) 737-3427 aallan@coas.oregonstate.edu, Oregon State Univ; As
Allan, Craig J., 704-687-5999 cjallan@email.uncc.edu, Univ of North Carolina, Charlotte; Hg
Allan, Deborah L., (612) 625-3158 dallan@soils.umn.edu, Univ of Minnesota, Twin Cities; Sb
Allan, Jonathan C., (541) 574-6658 Oregon Dept of Geology and Mineral Ind; Gm
Allard, Gilles O., (706) 542-2420 goallar@uga.edu, Univ of Georgia; Eg
Allard, Jason, (229) 333-5752 jmallard@valdosta.edu, Valdosta State Univ; AsZg
Allard, Stephen T., 507-457-2739 sallard@winona.edu, Winona State Univ; GctGx
Allen, Bruce D., 575-366-2531 bruce.allen@nmt.edu, New Mexico Tech; Gm
Allen, Charlotte M., +61 7 3138 0177 cm.allen@qut.edu.au, Queensland Univ of Tech; CcaGi
Allen, Clarence R., (626) 395-6904 allen@gps.caltech.edu,CalTech; Ys
Allen, Diana M., (778) 782-3967 dallen@sfu.ca, Simon Fraser Univ; Hw
Allen, Douglas, dallen@salemstate.edu, Salem State Univ; CgHsCt
Allen, Eric E., (858) 534-2570 eallen@ucsd.edu, Univ of California, San Diego; Ob
Allen, Gary C., 828-747-7818 gallen@uno.edu, Univ of New Orleans; Gp
Allen, George H., 540-231-6521 geoallen@vt.edu, Virginia Tech; HsZrGm
Allen, Grant, +44 0161 306-6851 grant.allen@manchester.ac.uk, Univ of Manchester; As
Allen, John, 989-774-1923 allen4jt@cmich.edu, Central Michigan Univ; Ams
Allen, Joseph L., 304-384-5238 allenj@concord.edu, Concord Univ; Gct
Allen, Karen, (864)294-2504 karen.allen@furman.edu, Furman Univ; Zn
Allen, Katherine, +44 0151 795 4646 k.a.allen@liverpool.ac.uk, Univ of Liverpool; Zi
Allen, Kelly E., 918-595-7085 kely.e.allen@tulsacc.edu, Tulsa Comm Coll; ZyiZe
Allen, Mark, +44 191 33 42344 m.b.allen@durham.ac.uk, Durham Univ; Gt
Allen, Peter M., (254)710-2189 peter_allen@baylor.edu, Baylor Univ; HgNg
Allen, Phillip, 301/687-4891 ppallen@frostburg.edu, Frostburg State Univ; GmePe
Allen, Richard M., (510) 642-1275 rallen@seismo.berkeley.edu, Univ of California, Berkeley; Ys
Allen, Richardson B., (801) 581-7574 pallen@egi.utah.edu, Univ of Utah; GctZi
Allen, Robert J., (951) 827-4870 robert.allen@ucr.edu, Univ of California, Riverside; As
Allen, Susan E., (604) 822-2828 sallen@eos.ubc.ca, Univ of British Columbia; Op
Allen-King, Richelle, (716) 645-4287 richelle@buffalo.edu, SUNY, Buffalo; Cg
Allen-Lafayette, Zehdreh, (609) 292-2576 zehdreh.allen-lafayette@dep.nj.gov, New Jersey Geological and Water Survey; Zi
Aller, Josephine Y., (631) 632-8655 josephine.aller@stonybrook.edu, SUNY, Stony Brook; ObAs
Aller, Robert C., (631) 632-8746 robert.aller@stonybrook.edu, SUNY, Stony Brook; Cm
Alley, Marcus M., (540) 231-9777 malley@vt.edu, Virginia Tech; Sc
Alley, Richard B., (814) 863-1700 rba6@psu.edu, Pennsylvania State Univ, Univ Park; Gl
Alley, Ronald E., (818) 354-0751 ron@lithos.jpl.nasa.gov, Jet Propulsion Lab; Zr
Allison, Alivia J., 620-341-5984 aalliso2@emporia.edu, Emporia State Univ; GgaGm
Allison, David T., 251-460-6381 dallison@southalabama.edu, Univ of South Alabama; Gc
Allison, Mead A., (504) 862-3270 meadallison@tulane.edu, Tulane Univ; Gs
Allison, Nicky, +44 01334 463952 na9@st-andrews.ac.uk, Univ of St. Andrews; Oc
Allison, Tami L., tami.allison@dnr.mo.gov, Missouri Dept of Nat Res; Zn
Allmendinger, Richard W., rwa1@cornell.edu, Cornell Univ; Gc
Allmon, Warren D., (607) 273-6623 (Ext. 114) wda1@cornell.edu, Paleontological Research Inst; Pg
Almendinger, James E., 651-433-5953 dinger@smm.org, Univ of Minnesota, Twin Cities; Pe
Almås, Åsgeir R., asgeir.almas@nmbu.no, Norwegian Univ of Life Scis (NMBU); Sc
Alonso, Ana María, alonsoza@ucm.es, Univ Compl de Madrid; Gd
Alpert, Pinhas, pinhas@post.tau.ac.il, Tel Aviv Univ; AsZrAm
Alpert, Sam, 212-769-5383 salpert@amnh.org, American Mus of Natural History; Xm
Alsdorf, Douglas E., 614 247-6908 alsdorf.1@osu.edu, Ohio State Univ; HgYg
Alsharif, Kamal, kalshari@usf.edu, Univ of South Florida; Rw
Alsleben, Helge, (817) 257-5455 h.alsleben@tcu.edu, Texas Christian Univ; GctGg
Alt, Jeffrey C., (734) 764-8380 jalt@umich.edu, Univ of Michigan; Ou
Altaner, Stephen P., (217) 244-1244 altaner@illinois.edu, Univ of Illinois, Urbana-Champaign; Gz
Altenbach, Alexander, 089/2180 6598 a.altenbach@lrz.uni-muenchen.de, Ludwig-Maximilians-Univ Muenchen; Pg
Altermann, Wladyslaw, 089/2180 6552 wlady.altermann@iaag.geo.uni-muenchen.de, Ludwig-Maximilians-Univ Muenchen; Gg
Alumbaugh, David L., (510) 215-4227 dalumbaugh@berkeley.edu, Univ of California, Berkeley; Ye
Aluwihare, Lihini I., (858) 822-4886 laluwihare@ucsd.edu, Univ of California, San Diego; Gu
Alvarado, Vladimir, 307-766-6464 valvarad@uwyo.edu, Univ of Wyoming; Np
Alvarez, José Antonio, jaalvare@ucm.es, Univ Compl de Madrid; Gc
Alvarez, Laura, alvarez@utep.edu, Univ of Texas, El Paso; ZiGs
Alvarez, Walter, 510-642-2602 platetec@berkeley.edu, Univ of California, Berkeley; GrRh
Alves, Tiago, alvest@cf.ac.uk, Cardiff Univ; GouGt
Alves, Tiego, +44(0)29 208 76754 alvest@cf.ac.uk, Univ of Wales; Og
Amato, James A., (920) 424-2268 amatoj@uwosh.edu, Univ of Wisconsin, Oshkosh; GgZi
Amato, Jeffrey M., (575) 646-3017 amato@nmsu.edu, New Mexico State Univ, Las Cruces; GcCc
Ambinakudige, Shrinidhi, (662) 268-1032 x210 shrinidhi@geosci.msstate.edu, Mississippi State Univ; ZiyZr
Ambrose, William, 512-471-0258 william.ambrose@beg.utexas.edu, Univ of Texas, Austin; GsrEo
Ambruster, W. Scott, (907) 474-7161 ffwsa@aurora.alaska.edu, Univ of Alaska, Fairbanks; Zn
Amedjoe, Godfrey C., +233 24 5961 073 chiri.amedjoe@gmail.com, Kwame Nkrumah Univ of Sci and Tech; GidGc
Amel, Nasir, +98 (411) 339 2696 amel@tabrizu.ac.ir, Univ of Tabriz; GivGg
Amelung, Falk, 305 421-4949 famelung@rsmas.miami.edu, Univ of Miami; Zn
Amend, Jan, (213) 740-0652 janamend@usc.edu, Univ of Southern California; ClmCo
Amenta, Roddy V., (540) 568-6674 amentarv@jmu.edu, James Madison Univ; Gc

FACULTY INDEX – A

Amer, Reda, 504-862-3220 ramer1@tulane.edu, Tulane Univ; ZirGe
Amezaga, Jamie, +44 (0) 191 208 4876 jaime.amezaga@ncl.ac.uk, Univ of Newcastle Upon Tyne; Ge
Amidon, Will, 802-443-5980 wamidon@middlebury.edu, Middlebury Coll; CcGmt
Amin, Isam E., (330) 941-2293 ieamin@ysu.edu, Youngstown State Univ; Hw
Amler, Michael, 089/2180 6602 amler@lrz.uni-muenchen.de, Ludwig-Maximilians-Univ Muenchen; Pg
Ammon, Charles J., (814) 865-2310 cammon@geosc.psu.edu, Pennsylvania State Univ, Univ Park; Ys
Amon, Rainier M., 409 740 4719 amonr@tamug.edu, Texas A&M Univ; CbOcb
Amonette, Alexandra B., (509) 376-5019 Pacific Northwest; Cg
Amoozegar, Aziz, (919) 515-3967 North Carolina State Univ; Sp
Amor, John, (604) 822-6933 jamor@eos.ubc.ca, Univ of British Columbia; Zn
Amos, Colin B., 360-650-3587 colin.amos@wwu.edu, Western Washington Univ; Gc
Amos, Kathryn, kathryn.amos@adelaide.edu.au, Univ of Adelaide; Gs
Ampuero, Jean-Paul, 626.395.6958 ampuero@gps.caltech.edu,CalTech; Ys
Amrouch, Khalid, khalid.amrouch@adelaide.edu.au, Univ of Adelaide; Gc
Amthor, Jeff, (865) 576-2773 Oak Ridge; Sf
Anand, Madhur, (519) 824-4120 Ext.56254 manand@uoguelph.ca, Univ of Guelph; Zn
Anand, Pallavi, +44 (0) 1908 652225 x 52225 pallavi.anand@open.ac.uk, Open Univ; ClPe
Anandakrishnan, Sridhar, (814) 863-6742 sak@essc.psu.edu, Pennsylvania State Univ, Univ Park; Ys
Anani, Chris Y., cyanani@ug.edu.gh, Univ of Ghana; Gsd
Anastasio, Cort, 530-754-6095 canastasio@ucdavis.edu, Univ of California, Davis; Acs
Anastasio, David J., (610) 758-5117 dja2@lehigh.edu, Lehigh Univ; Gca
Anbar, Ariel D., (480) 965-0767 anbar@asu.edu, Arizona State Univ; CbsXb
Ancell, Brian C., 806-834-3143 brian.ancell@ttu.edu, Texas Tech Univ; As
Ancochea, Eumenio, anco@ucm.es, Univ Compl de Madrid; Gv
Anders, Alison M., 217-244-3917 amanders@illinois.edu, Univ of Illinois, Urbana-Champaign; Gm
Andersen, C. B., brannon.andersen@furman.edu, Clemson Univ; Cl
Andersen, C. Brannon, (864) 294-3366 brannon.andersen@furman.edu, Furman Univ; ClGe
Andersen, David W., (408) 924-5014 david.andersen@sjsu.edu, San Jose State Univ; GsCl
Andersen, Jens C., +44 01326 371836 j.c.andersen@exeter.ac.uk, Exeter Univ; GiEmCa
Andersen, Raymond J., (604) 822-4511 randersn@eos.ubc.ca, Univ of British Columbia; Oc
Anderson, Alan J., (902) 867-2309 aanderso@stfx.ca, Saint Francis Xavier Univ; Gx
Anderson, Andrew, 217-244-0995 acandrsn@illinois.edu, Illinois State Geo Survey; Ng
Anderson, Bruce, brucea@bu.edu, Boston Univ; AsOp
Anderson, Callum, 27 41 504 2811 callum.anderson@nmmu.ac.za, Nelson Mandela Metropolitan Univ; Gzs
Anderson, Corby , 303-273-3580 cganders@mines.edu, Colorado Sch of Mines; Nx
Anderson, G M., (416) 978-2062 Univ of Toronto; Cp
Anderson, Jake , jacobanderson152@boisestate.edu , Boise State Univ; YxGv
Anderson, James G., (617) 495-5922 anderson@huarp.harvard.edu, Harvard Univ; As
Anderson, James Lawford, lawford@bu.edu, Boston Univ; Gi
Anderson, Jennifer L., 507-457-2457 jlanderson@winona.edu, Winona State Univ; XgYgZe
Anderson, John B., (713) 348-4652 johna@rice.edu, Rice Univ; Gu
Anderson, Ken B., 618-453-7389 kanderson@geo.siu.edu, Southern Illinois Univ Carbondale; Co
Anderson, Laurie C., (605) 394-1290 laurie.anderson@sdsmt.edu, South Dakota Sch of Mines & Tech; PiePq
Anderson, Mark, +44 1752 584768 m.anderson@plymouth.ac.uk, Univ of Plymouth; Gct
Anderson, Mary P., (914) 365-8335 andy@geology.wisc.edu, Univ of Wisconsin, Madison; Hw
Anderson, Orson L., (310) 825-2386 Univ of California, Los Angeles; Yx
Anderson, Patricia M., pata@uw.edu, Univ of Washington; Ple
Anderson, Paula, 479-575-3355 pea001@uark.edu, Univ of Arkansas, Fayetteville; Gg
Anderson, Raymond R., (319) 335-1589 raymond.anderson@dnr.iowa.gov, Univ of Iowa; Gr
Anderson, Robert, 303-735-4684 robert.s.anderson@colorado.edu, Univ of Colorado; Gm
Anderson, Ross, +44 (0) 131 451 3798 r.anderson@hw.ac.uk, Heriot-Watt Univ; Eo
Anderson, Ryan, (208) 282-5024 ryananderson@isu.edu, Idaho State Univ; GcCaZe
Anderson, Stephen H., 573-882-6303 andersons@missouri.edu, Univ of Missouri; Sp
Anderson, Steven W., (970) 351-2973 steven.anderson@unco.edu, Univ of Northern Colorado; GvZeXg
Anderson, Thomas B., (775) 747-1438 tom.anderson@sonoma.edu, Sonoma State Univ; GsrGd
Anderson, Wayne I., (319) 273-3125 Univ of Northern Iowa; Pg
Anderson, William P., (828) 262-7540 andersonwp@appstate.edu, Appalachian State Univ; HwqHs
Anderson, Jr., Alfred T., (773) 702-8138 Univ of Chicago; Gv
Anderson-Folnagy, Heidi, (406) 683-7134 heidi.anderson@umwestern.edu, Univ of Montana Western; Gs
Andersson, Andreas, (858) 822-2486 aandersson@ucsd.edu, Univ of California, San Diego; Oc
Andeweg, Bernd , +31 20 5987339 bernd.andeweg@vu.nl, Vrije Univit Amsterdam; GtcGg
Andonaegui, Pilar, andonaeg@ucm.es, Univ Compl de Madrid; Gg
Andrade, César, candrade@fc.ul.pt, Unive de Lisboa; Gu
Andre-Obayanju, Tomilola, tomilola.obayanju@uniben.edu, Univ of Benin; NgHgGe
Andres, A. S., (302) 831-2833 asandres@udel.edu, Univ of Delaware; HwZi
Andresen, Christian, candresen@wisc.edu, Univ of Wisconsin, Madison; Zy
Andreu-Hayles, Laia, lah@ldeo.columbia.edu, Columbia Univ; Pe
Andrews, Allen H., allen.andrews@noaa.gov, Univ of Hawai'i, Manoa; Oc
Andrews, Benjamin, (202) 633-1818 andrewsb@si.edu, Smithsonian Inst / Natl Mus of Natural History; GvCpGi
Andrews, John R., 512-471-1534 john.andrews@beg.utexas.edu, Univ of Texas at Austin, Jackson Sch of Geosciences; Zi
Andrews, Julian E., +44 (0)1603 59 2536 j.andrews@uea.ac.uk, Univ of East Anglia; CsGsCl
Andrews, William, 859-323-0506 wandrews@uky.edu, Univ of Kentucky; Gg
Andronicos, Chris, candroni@purdue.edu, Purdue Univ; Gg
Andrus, C. Fred T., (205) 348-5177 fandrus@ua.edu, Univ of Alabama; PcGaCs
Andrus, Richard E., (607) 777-2453 Binghamton Univ; Sf
Aneja, Viney P., 91951557808 viney_aneja@ncsu.edu, North Carolina State Univ; Ac
Anfinson, Owen, anfinson@sonoma.edu, Sonoma State Univ; GsdGg
Angelopoulos, Vassilis, (310) 794-7090 vassilis@ucla.edu, Univ of California, Los Angeles; XyAsYm
Anger-Kraavi, Annela, +44 (0)1603 59 2633 a.anger-kraavi@uea.ac.uk, Univ of East Anglia; Eg
Angino, Ernest E., (785) 843-7503 Univ of Kansas; Cl
Angle, Michael P., 614 265 6602 mike.angle@dnr.state.oh.us, Ohio Dept of Nat Res; GlgHw
Angus, Doug, +4401133431326 d.angus@leeds.ac.uk, Univ of Leeds; Ys
Anhaeusser, Carl R., 011-7176581 carl.anhaeusser@wits.ac.za, Univ of the Witwatersrand; GgEgZg
Anis, Ayal, (409) 740-4987 anisa@tamug.tamu.edu, Texas A&M Univ; Opn
Annett, Amber, +44 023 80596041 a.l.annett@soton.ac.uk, Univ of Southampton; Oc
Ansdell, Kevin M., (306) 966-5698 kevin.ansdell@usask.ca, Univ of Saskatchewan; EgCgGt
Antao, Sytle, 403 220-3083 antao@ucalgary.ca, Univ of Calgary; Gz
Antar, Ali A., (860) 832-2931 antar@ccsu.edu, Central Connecticut State Univ; Zn
Anthony, Nina, 864-294-2052 nina.anthony@furman.edu, Furman Univ; Zn
Anthony, Robin, 412-442-4295 robanthony@pa.gov, DCNR- Pennsylvania Bur of Geo Survey; Ego
Antigüedad, Iñaki, inaki.antiguedad@ehu.eus, Univ of the Basque Country UPV/EHU; Hw
Antinao, José Luis , (812) 855-1366 jantinao@indiana.edu, Indiana Univ, Bloomington; CcGlr
Antinao, JoseLuis, 775-673-7450 joseluis.antinao@dri.edu, Desert Research Inst; Gm
Antipova, Anzhelika, 901-678-2178 antipova@memphis.edu, Univ of Memphis; Zg
Antonacci, Vince, (615) 532-1507 vince.antonacci@tn.gov, Tennessee Geo Survey; GgZi
Antonescu, Adrian, +44 0161 306-3911 bogdan.antonescu@manchester.ac.uk, Univ of Manchester; As
Anyiam, Uzonna O., (616) 395-7112 anyiam@hope.edu, Hope Coll; Yge

Apen, Francisco, 609-258-4101 feapen@princeton.edu, Princeton Univ; CcGpt
Aplin, Andrew, +44 191 33 42332 Durham Univ; Go
Apopei, Andrei I., 0040747760718 andrei.apopei@uaic.ro, Alexandru Ioan Cuza; GziGg
Apostoae, Laviniu, 00400232201463 laviniu@uaic.ro, Alexandru Ioan Cuza; Eg
Apotsos, Alex, 413-597-5082 alex.apotsos@williams.edu, Williams Coll; Ge
Appel, Christopher (Chip) S., (805) 756-1691 cappel@calpoly.edu, California Polytechnic State Univ; Sc
Appel, Heidi, (419) 530-2812 heidi.appel@utoledo.edu, Univ of Toledo; Zn
Appiah-Adjei, Emmanuel K., +233 20 7934 556 ekappiah-adjei.soe@knust.edu.gh, Kwame Nkrumah Univ of Sci and Tech; HwgGg
Applegarth, Michael T., (717) 477-1712 mtappl@ship.edu, Shippensburg Univ; GmZrSo
Applegate, Toby, 413-545-1535 tapplega@geo.umass.edu, Univ of Massachusetts, Amherst; Zy
Appold, Martin S., (573) 882-0701 appoldm@missouri.edu, Univ of Missouri; HwEm
Apprill, Amy, 508-289-2649 aapprill@whoi.edu, Woods Hole Oceanographic Inst; Oc
Apraiz, Arturo, arturo.apraiz@ehu.eus, Univ of the Basque Country UPV/EHU; Gpt
April, Richard, (315) 228-7212 rapril@colgate.edu, Colgate Univ; CgGze
Aquilina, Noel , noel.aquilina@um.edu.mt, Univ of Malta; Asm
Arabas, Karen, 503-370-6666 karabas@willamette.edu, Willamette Univ; Zn
Aracil, Enrique, earacil@ucm.es, Univ Compl de Madrid; Yu
Arain, M. A., (905) 525-9140 (Ext. 27941)) arainm@mcmaster.ca, McMaster Univ; Hg
Arakawa, Akio, (310) 825-9874 aar@atmos.ucla.edu, Univ of California, Los Angeles; As
Aranda, Angela, (657) 278-3551 angperez@fullerton.edu, California State Univ, Fullerton; Gg
Aranguren, Aitor, aitor.aranguren@ehu.eus, Univ of the Basque Country UPV/EHU; GcYm
Araya , María C., mariacristina.araya@ucr.ac.cr, Univ de Costa Rica; YdGtYs
Arboleya Cimadevilla, María Luísa, ++935811951 marialuisa.arboleya@uab.cat, Univ Auto de Barcelona; Gc
Archambault, Guy, garchambault@uqac.ca, Univ du Quebec a Chicoutimi; Nr
Archer, Allen W., (785) 532-2244 aarcher@ksu.edu, Kansas State Univ; Gr
Archer, David, (773) 702-0823 Univ of Chicago; Yr
Archer, Emma, +27 (0) 12 420 2881 emma.archer@up.ac.za, Univ of Pretoria; At
Archer, Michael, m.archer@unsw.edu.au, Univ of New South Wales; Pv
Archibald, Donnelly, darchiba@stfx.ca, Saint Francis Xavier Univ; Gg
Archibald, Doug A., (613) 545-6594 Queen's Univ; Cc
Archuleta, Ralph J., (805) 893-8441 archuleta@geol.ucsb.edu, Univ of California, Santa Barbara; Ys
Arenas, Ricardo, rarenas@ucm.es, Univ Compl de Madrid; Gp
Arens, Nan Crystal, 315-781-3930 arens@hws.edu, Hobart & William Smith Colls; Pg
Argenbright, Robert T., (910) 962-3498 argenbrightr@uncw.edu, Univ of North Carolina, Wilmington; Zn
Argles, Tom, tom.argles@open.ac.uk, Open Univ; Zi
Argyilan, Erin, 219-980-7124 eargyila@iun.edu, Indiana Univ Northwest; GeHwAm
Arias, Carmen, cariasf@ucm.es, Univ Compl de Madrid; Gg
Ariya, Parisa A., (514) 398-3615 parisa.ariya@mcgill.ca, McGill Univ; As
Ariyo, Stephen O., +234-8038047475 Federal Univ of Tech, Akure; YegYx
Arkani-Hamed, Jafar, 514-398-6767 jafar@physics.utoronto.ca, McGill Univ; Xy
Arkle, Jenny C., 309-794-7844 jennyarkle@augustana.edu, Augustana Coll; GtmCc
Arkle, Kelsey M., 309-794-7487 kelseyarkle@augustana.edu, Augustana Coll; PeGsg
Armadillo, Egidio, +390103538085 egidio@dipteris.unige.it, Universita di Genova; Ye
Armah, Thomas K., tekarmah@ug.edu.gh, Univ of Ghana; YgeYe
Armbrust, Virginia E., (206) 616-1783 armbrust@ocean.washington.edu, Univ of Washington; Ob
Armi, Laurence, (858) 534-6843 larmi@ucsd.edu, Univ of California, San Diego; Op
Armour, Jake, 704-687-5968 jarmour@uncc.edu, Univ of North Carolina, Charlotte; GmlGg
Armstrong, Andrew, 603-862-4559 andya@ccom.unh.edu, Univ of New Hampshire; Zn
Armstrong, Anne, 508-929-8606 aarmstrong3@worcester.edu, Worcester State Univ; Zec
Armstrong, David E., (608) 262-2470 Univ of Wisconsin, Madison; Oc
Armstrong, Felicia P., 330-941-1385 fparmstrong@ysu.edu, Youngstown State Univ; Sb
Armstrong, Phillip A., (657) 278-3169 parmstrong@fullerton.edu, California State Univ, Fullerton; GcYg
Armstrong, Robert A., (609) 258-5260 robert.armstrong@stonybrook.edu, SUNY, Stony Brook; Ocb
Armstrong, William H., armstrongwh@appstate.edu, Appalachian State Univ; Glm
Arnaud, Emmanuelle, 519-824-4120 x58087 earnaud@uoguelph.ca, Univ of Guelph; Gls
Arnold, Anthony J., (850) 644-4228 arnold@gly.fsu.edu, Florida State Univ; Pm
Arnold, Dan, +44 (0) 131 451 8298 d.arnold@hw.ac.uk, Heriot-Watt Univ; Go
Arnold, David L., 301/687-4053 dlarnold@frostburg.edu, Frostburg State Univ; As
Arnold, Stephen, +44(0) 113 34 37245 s.arnold@leeds.ac.uk, Univ of Leeds; As
Arnone, John, (775) 673-7445 jarnone@dri.edu, Desert Research Inst; Pg
Arnott, R. William, warnott@uottawa.ca, Univ of Ottawa; Gsd
Aronoff, Ruth F., (864)294-3363 ruth.aronoff@furman.edu, Furman Univ; Gc
Arp, Dr., Gernot, +49 (0)551 397986 garp@gwdg.de, Georg-August Univ of Goettingen; PggPg
Arribas, Antonio, aarribas@utep.edu, Univ of Texas, El Paso; EmCe
Arribas, Eugenia, earribas@ucm.es, Univ Compl de Madrid; Gd
Arribas, José, arribas@ucm.es, Univ Compl de Madrid; Gd
Arritt, Raymond W., (515) 294-9870 rwarritt@iastate.edu, Iowa State Univ of Sci & Tech; As
Arrowsmith, Ramon, (480) 965-3541 ramon.arrowsmith@asu.edu, Arizona State Univ; GcmGt
Arrowsmith, Stephen J., sarrowsmith@smu.edu, Southern Methodist Univ; Ys
Arslan, Gizem, 00904242370000-5972 gturus@firat.edu.tr, Firat Univ; GxiGp
Arsuaga, Juan Luis, azara@ucm.es, Univ Compl de Madrid; Zn
Arthur, Jonathan D., (850) 617-0320 jonathan.arthur@floridadep.gov, Florida Geo Survey; HwCgGe
Arthur, Michael A., (814) 863-6054 maa6@psu.edu, Pennsylvania State Univ, Univ Park; OuClGs
Artigas, Francisco, 201-460-2801 francisco.artigas@rutgers.edu, Rutgers, The State Univ of New Jersey, Newark; Zi
Artiola, Janick F., (520) 621-3516 jartiola@email.arizona.edu, Univ of Arizona; ScHsCa
Artioli, Gilberto, 39-049-8279162 gilberto.artioli@unipd.it, Univ degli Studi di Padova; GzyZm
Arundale, Wendy H., (907) 474-7039 Univ of Alaska, Fairbanks; Zn
Arvidson, Raymond E., (314) 935-5679 arvidson@wunder.wustl.edu, Washington Univ in St. Louis; Xg
Arya, Satyapal S., (919) 515-7002 pal_arya@ncsu.edu, North Carolina State Univ; As
Aryal, Niroj, (336) 285-3832 naryal@ncat.edu, North Carolina Agricultural & Tech State Univ; HsgZi
Asante, Joseph, (931) 372-3576 jasante@tntech.edu, Tennessee Tech Univ; HwZrGe
Aschoff, Jennifer, (907) 786-1442 jaschoff@uaa.alaska.edu, Univ of Alaska, Anchorage; GrdGo
Asghari Moghadam, Asghar, +98 (411) 339 2703 moghadam@tabrizu.ac.ir, Univ of Tabriz; HwqHy
Ash, Richard, (301) 405-7504 rdash@umd.edu, Univ of Maryland; Ca
Ashbolt, Nicholas J., nick.ashbolt@scu.edu.au, Southern Cross Univ; PoCb
Ashe, Douglas, douglas.ashe@tulsacc.edu, Tulsa Comm Coll; GgZe
Ashley, Gail M., (848) 445-2221 gmashley@eps.rutgers.edu, Rutgers, The State Univ of New Jersey; GsmSa
Ashley, Paul, +61-2-67732348 pashley@une.edu.au, Univ of New England; EgGeEm
Ashour-Abdalla, Maha, (310) 825-8881 mabdalla@igpp.ucla.edu, Univ of California, Los Angeles; Zn
Ashton, Andrew, 508-289-3751 aashton@whoi.edu, Woods Hole Oceanographic Inst; On
Ashwal, Lewis D., +27 11 717 6652 lewis.ashwal@wits.ac.za, Univ of the Witwatersrand; GxCgGt
Ashwood, Tom, (865) 574-7542 Oak Ridge; Ge
Ashworth, Allan C., 701-231-7919 allan.ashworth@ndsu.edu, North Dakota State Univ; Pe
Asiedu, Daniel K., dasiedu@ug.edu.gh, Univ of Ghana; GsCgGd
Asimow, Paul D., (626) 395-4133 asimow@gps.caltech.edu,CalTech; CpGzy

Asiwaju-Bello, Yinusa A., +2348036672708 ayoasiwajubello@futa.edu.ng, Federal Univ of Tech, Akure; HwNgGu
Askari, Roohollah, (906) 487-2029 raskari@mtu.edu, Michigan Technological Univ; YxxNp
Askari, Zohreh, (217) 300-1819 askari@illinois.edu, Illinois State Geo Survey; Gsc
Aslan, Andres, (970) 248-1614 aaslan@coloradomesa.edu, Colorado Mesa Univ; Gm
Aslan, Yasemin, 00904242370000-5959 yaslan@hotmail.com, Firat Univ; Nr
Asmerom, Yemane, (505) 277-4204 asmerom@unm.edu, Univ of New Mexico; CcPeCg
Asner, Gregory, gregasner@asu.edu, Arizona State Univ; ZrGmSb
Asowata, Timothy I., +2348065348350 tiasowata@futa.edu.ng, Federal Univ of Tech, Akure; GeiGg
Asphaug, Erik I., (831) 566-7877 asphaug@lpl.arizona.edu, Univ of Arizona; XygXa
Asquith, George B., 806-834-0497 george.asquith@ttu.edu, Texas Tech Univ; Go
Assatourians, Karen, 519-661-2111, ext. 84715 kassatou@uwo.ca, Western Univ; Ys
Asselin, Esther, (418) 654-2612 esther.asselin@canada.ca, Nat Res Canada; PlGuRh
Astefanei, Dan, 004023201455 astdan@uaic.ro, Alexandru Ioan Cuza; Cga
Aster, Richard C., (970) 491-7606 rick.aster@colostate.edu, Colorado State Univ; YsGv
Astilleros, José Manuel, jmastill@ucm.es, Univ Compl de Madrid; Gz
Asuen, Godwin O., godwin.asuen@uniben.edu, Univ of Benin; EcGsg
Atchison, Christopher L., (513) 556-3613 christopher.atchison@uc.edu, Univ of Cincinnati; Ze
Atchley, Stacy C., (254) 710-2196 stacy_atchley@baylor.edu, Baylor Univ; GroGo
Atekwana, Eliot A., 302-8312569 eatekwan@udel.edu, Univ of Delaware; CgsCq
Athaide, Dileep, (604) 984-1771 dathaide@eos.ubc.ca, Univ of British Columbia; Gg
Athey, Jennifer E., (907) 451-5028 jennifer.athey@alaska.gov, Alaska Division of Geological & Geophysical Surveys; Gg
Atkinson, Gail M., 519-661-2111 x.84207 gatkins6@uwo.ca, Western Univ; YssYs
Atkinson, Marlin J., 235-2224 mja@hawaii.edu, Univ of Hawai'i, Manoa; Ob
Atkinson, Timothy P., t.atkinson@ucl.ac.uk, Univ Coll London; Hw
Atkinson, Jr., William W., (303) 492-6103 william.atkinson@colorado.edu, Univ of Colorado; EgmCg
Atlas, Zachary, zatlas@usf.edu, Univ of South Florida; Gi
Atreya, Sushil, (734) 936-0489 atreya@umich.edu, Univ of Michigan; ApZy
Atta-Peters, David, datta@ug.edu.gh, Univ of Ghana; PlsPl
Attal, Mikael, +44 (0) 131 650 8533 mikael.attal@ed.ac.uk, Edinburgh Univ; GmtZy
Attia, Snir, 575-835-5290 snir.attia@nmt.edu, New Mexico Tech; GcgGi
Attias, Eric, (+1) 808-829-6559 attias@ig.utexas.edu, Univ of Texas, Austin; GuYr
Attig, John W., john.w.attig@wisc.edu, Univ of Wisconsin, Madison, Division of Extension; Gl
Attinger, Sabine, 0049(0)3641/948651 sabine.attinger@ufz.de, Friedrich-Schiller-Univ Jena; Hy
Attrep, Moses, (505) 667-0088 Los Alamos; Ze
Atudorei, Nieu-Viorel, atudorei@unm.edu, Univ of New Mexico; Cs
Atwater, Brian, 206-553-2927 atwater@uw.edu, Univ of Washington; RnGr
Au, Whitlow W L., 808-247-5026 wau@hawaii.edu, Univ of Hawai'i, Manoa; Zn
Aubert, John E., (916) 484-8637 aubertj@arc.losrios.edu, American River Coll; Zy
Aubry, Marie-Pierre, (732) 445-0822 aubry@eps.rutgers.edu, Rutgers, The State Univ of New Jersey; Pms
Audet, Celine, (418) 723-1986 (Ext. 1744) celine_audet@uqar.qc.ca, Univ du Quebec a Rimouski; Ob
Audet, Pascal, pascal.audet@uottawa.ca, Univ of Ottawa; Ygs
Aufrecht, Walter E., (403) 329-2485 aufrecht@uleth.ca, Univ of Lethbridge; Zn
Ault, Alexis K., aault@email.arizona.edu, Utah State Univ; GtCc
Ault, Toby R., 607-255-1509 tra38@cornell.edu, Cornell Univ; Am
Aurnou, Jonathan M., (310) 825-2054 aurnou@epss.ucla.edu, Univ of California, Los Angeles; YgmYx
Ausbrooks, Scott, 501 683-0119 scott.ausbrooks@arkansas.gov, Arkansas Geo Survey; YsGeg
Ausich, William I., (614) 292-3353 ausich.1@osu.edu, Ohio State Univ; PgGs
Auster, Peter, 860-405-9118 peter.auster@uconn.edu, Univ of Connecticut; Ob
Austermann, Jacqueline, jackya@ldeo.columbia.edu, Columbia Univ; Ys
Austin, George S., gaustin@sdc.org, New Mexico Tech; Gz
Austin, Philip, (604) 822-2175 paustin@eos.ubc.ca, Univ of British Columbia; As
Austin, Jr., James A., (512) 471-0450 jamie@ig.utexas.edu, Univ of Texas, Austin; Gut
Autin, Whitney J., 585-395-5738 dirtguy@esc.brockport.edu, SUNY, The Coll at Brockport; Gs
Autio, Robert J., (812) 856-9104 rjautio@indiana.edu, Indiana Univ, Bloomington; HgGle
Avary, Katherine L., 304 594 2331 avarygeo@gmail.com, West Virginia Univ; Go
Avouac, Jean-Philippe, 626.395.2350 avouac@gps.caltech.edu,CalTech; GtYs
Awdankiewicz, Marek, marek.awdankiewicz@uwr.edu.pl, Univ of Wroclaw; Gvi
Awramik, Stanley M., (805) 893-3830 awramik@geol.ucsb.edu, Univ of California, Santa Barbara; Pg
Axelbaum, Richard, rla@me.wustl.edu, Washington Univ in St. Louis; Ng
Axen, Gary, 575.835.5178 gaxen@ees.nmt.edu, New Mexico Tech; Gct
Axford, Yarrow L., 847.467.2268 yarrow@earth.northwestern.edu, Northwestern Univ; GnPmi
Ayad, Yasser M., yayad@clarion.edu, Clarion Univ; Zi
Ayanlade, Ayansina, (803) 777-2355 sinaayanlade@yahoo.co.uk, Obafemi Awolowo Univ; AtZir
Aydin, Adnan, (662) 915-1342 aaydin@olemiss.edu, Univ of Mississippi; NgrYe
Aydin, Atilla, (650) 725-8708 aydin@pangea.stanford.edu, Stanford Univ; Gc
Ayers, John C., 615-322-2158 john.c.ayers@vanderbilt.edu, Vanderbilt Univ; CgGx
Ayers, Joseph , 7815817370 x309 lobster@neu.edu, Northeastern Univ; Zn
Ayling, Bridget, (775) 682-8768 bayling@unr.edu, Univ of Nevada, Reno; CgYhGo
Aylward, Linda, (309) 694-5256 Illinois Central Coll; Ze
Ayodele, Olusiji S., +2347060566760 osayodele@futa.edu.ng, Federal Univ of Tech, Akure; CeGcZr
Azam, Farooq, (858) 534-6850 fazam@ucsd.edu, Univ of California, San Diego; Ob
Azerêdo, Ana , acazeredo@fc.ul.pt, Unive de Lisboa; Gd
Azetsu-Scott, Kumiko, (902) 426-8572 kumiko.azetsu-scott@mar.dfo-mpo.gc.ca, Geo Survey of Canada; Ocp
Azmy, Karem, 709 864 6731 kazmy@mun.ca, Memorial Univ of Newfoundland; Cl

B

Baalousha, Husam, husam.baalousha@kfupm.edu.sa, King Fahd Univ of Petroleum and Minerals; HwyGe
Baarli, B. Gudveig, (413) 597-2329 gudveig.baarli@williams.edu, Williams Coll; PssPi
Babaie, Hassan A., (404) 413-5766 hbabaie@gsu.edu, Georgia State Univ; GcZf
Babb, David M., (814) 863-3918 dmb16@psu.edu, Pennsylvania State Univ, Univ Park; Am
Babbin, Andrew, (617) 253-2181 babbin@mit.edu, Massachusetts Inst of Tech; Cb
Babcock, Loren E., (614) 292-2103 babcock.5@osu.edu, Ohio State Univ; Pg
Babcock, R. S., babcock@wwu.edu, Western Washington Univ; Cg
Babechuk, Michael, 709 864 6095 mbabechuk@mun.ca, Memorial Univ of Newfoundland; Cu
Babek, Ondrej, +420 549 49 3163 Masaryk Univ; Gds
Baca, Alyssa, 575-835-5627 alyssa.baca@nmt.edu, New Mexico Tech; HwZi
Baca, Dustin, 575-835-5416 dustin.baca@nmt.edu, New Mexico Tech; Cag
Bacchus, Tania S., (802) 635-1329 tania.bacchus@jsc.edu, Northern Vermont Univ-Lyndon; GuAm
Bach Plaza, Joan, ++935811272 joan.bach@uab.cat, Univ Auto de Barcelona; Gg
Bachle, Peter, (573) 368-2472 peter.bachle@dnr.mo.gov, Missouri Dept of Nat Res; Gg
Bachmann, Etienne, 609-258-4101 etienneb@princeton.edu, Princeton Univ; Ys
Bachtadse, Valerian, 089/2180 4237 valerian@geophysik.uni-muenchen.de, Ludwig-Maximilians-Univ Muenchen; Yg
Baciu, Sorin D., 0040740404546 dsbaciu@gmail.com, Alexandru Ioan Cuza; GcPv
Back, Larissa E., 608-262-0776 lback@wisc.edu, Univ of Wisconsin,

Madison; As
Backus, George E., (858) 534-2468 gebackus@ucsd.edu, Univ of California, San Diego; YmsOp
Bacon, Robert, (573) 526-0807 Missouri Dept of Nat Res; Hs
Bacon, Steven N., 775-673-7473 steve.bacon@dri.edu, Desert Research Inst; Gm
Bada, Jeffrey L., (858) 534-4258 jbada@ucsd.edu, Univ of California, San Diego; Co
Badalyan, Alex, alex.badalyan@adelaide.edu.au, Univ of Adelaide; Np
Bader, Nicholas E., (509) 527-5113 baderne@whitman.edu, Whitman Coll; SoHgZi
Badger, Robert L., badgerrl@potsdam.edu, SUNY Potsdam; GipGc
Badruddin, Abu Z., 315 2948610 ext.231 badruddin@cayuga-cc.edu, Cayuga Comm Coll; Zi
Baedke, Steven J., (540) 568-6156 baedkesj@jmu.edu, James Madison Univ; Hy
Baer, Eric M., (206) 592-3513 ebaer@highline.edu, Highline Coll; GgvZg
Bagley, Brian, 612-624-8539 bagl0025@umn.edu, Univ of Minnesota, Twin Cities; YsZn
Bagley, Evan, 601-266-4729 evan.bagley@usm.edu, Univ of Southern Mississippi; GgEo
Bagtzoglou, Ross, (212) 854-3154 Columbia Univ; Hw
Bahr, Jean M., (608) 262-2396 jmbahr@geology.wisc.edu, Univ of Wisconsin, Madison; Hw
Bahrmann, Chad, 814-865-9500 cbahrmann@psu.edu, Pennsylvania State Univ, Univ Park; Am
Baig, Ayat, 204-227-7499 baiga@branedonu.ca, Brandon Univ; EdmEg
Bailey, Christopher M., (757) 221-2445 cmbail@wm.edu, William & Mary; Gc
Bailey, David G., (315) 859-4142 dbailey@hamilton.edu, Hamilton Coll; GizGa
Bailey, Ian, +4401326 259322 i.bailey@exeter.ac.uk, Exeter Univ; Zn
Bailey, Jack B., (309) 298-1481 jb-bailey@wiu.edu, Western Illinois Univ; Pg
Bailey, Jake, (612) 624-1603 baileyj@umn.edu, Univ of Minnesota, Twin Cities; Pg
Bailey, John E., (507) 933-6307 bailey@gustavus.edu, Gustavus Adolphus Coll; GvmZr
Bailey, Judy, 61 02 4921 5415 judy.bailey@newcastle.edu.au, Univ of Newcastle; Ec
Bailey, Keiron D., kbailey@email.arizona.edu, Univ of Arizona; Zi
Bailey, Richard C., (416) 978-3231 bailey@physics.utoronto.ca, Univ of Toronto; Yg
Bain, Daniel J., (412) 624-8780 dbain@pitt.edu, Univ of Pittsburgh; HgGmZu
Bain, Olivier, +33(0)3 44069304 olivier.bain@lasalle-beauvais.fr, Institut Polytechnique LaSalle Beauvais (ex-IGAL); OuGgZi
Bair, E. Scott, 614-592-3900 bair.1@osu.edu, Ohio State Univ; HwGe
Baird, Gordon C., (716) 673-3840 baird@fredonia.edu, SUNY, Fredonia; PseGs
Baird, Graham, 970-351-2830 graham.baird@unco.edu, Univ of Northern Colorado; GcgGz
Bajjali, William, (715) 394 8056 wbajjali@uwsuper.edu, Univ of Wisconsin, Superior; HwCsg
Bajraktareviæ, Zlatan, +38514606098 zbajrak@geol.pmf.hr, Univ of Zagreb; PgvPm
Bakel, Allen J., (630) 252-5486 Argonne; Co
Baker, Alex, +44 (0)1603 59 1529 alex.baker@uea.ac.uk, Univ of East Anglia; On
Baker, Andy, a.baker@unsw.edu.au, Univ of New South Wales; HwCs
Baker, Brett, 361-749-6774 brett_baker@utexas.edu, Univ of Texas, Austin; Ze
Baker, Don, 514-398-7485 don.baker@mcgill.ca, McGill Univ; CuGvCp
Baker, Edward T., (206) 526-6251 baker@pmel.noaa.gov, Univ of Washington; Ou
Baker, James C., (540) 231-9785 suebrown@vt.edu, Virginia Tech; Sd
Baker, John M., (612) 625-4249 jbaker@soils.umn.edu, Univ of Minnesota, Twin Cities; Sp
Baker, Leslie L., (208) 885-7625 lbaker@uidaho.edu, Univ of Idaho; ClXgGe
Baker, Marcia B., mbbaker@u.washington.edu, Univ of Washington; As
Baker, Paul A., (919) 684-6450 pbaker@duke.edu, Duke Univ; PcCgGg
Baker, Richard G., (319) 335-1827 dick-baker@uiowa.edu, Univ of Iowa; PebPc
Baker, Sophie, 775-673-7434 sophie.baker@dri.edu, Desert Research Inst; Gm
Baker, Victor R., (520) 621-7875 baker@email.arizona.edu, Univ of Arizona; XgGmRh
Baker-Treloar, Elizabeth, (619) 260-6822 ebaker@sandiego.edu, Univ of San Diego; Geu
Bakker, Pepijn, +31 20 59 83258 p.bakker@vu.nl, Vrije Univit Amsterdam; PcAtZo
Baksi, Ajoy K., (225) 388-3422 abaksi@geol.lsu.edu, Louisiana State Univ; Cc
Bal Akkoca, Dicle, 00904242370000-5982 dbal@firat.edu.tr, Firat Univ; EnCo
Balakrishnan, Meena, 817-515-6360 meena.balakrishnan@tccd.edu, Tarrant County Coll, Northeast Campus; GggGg
Balascio, Nicholas, 757-221-2880 nbalascio@wm.edu, William & Mary; ZyGnl
Balázs, László, balazslaszlo@windowslive.com, Eotvos Lorand Univ; EoYeg
Balch, Jennifer K., jennifer.balch@colorado.edu, Univ of Colorado; Zg
Balch, Robert S., 575-835-5305 balch@prrc.nmt.edu, New Mexico Tech; Yse
Baldauf, Jack G., (979) 845-3651 jbaldauf@tamu.edu, Texas A&M Univ; Ou
Baldini, James, +44 191 33 42334 james.baldini@durham.ac.uk, Durham Univ; Pc
Baldizon, Ileana, (305) 237-3658 ileana.baldizo@mdc.edu, Miami-Dade Coll (Wolfson Campus); Zn
Baldridge, W. Scott, (505) 667-4338 sbaldridge@lanl.gov, Los Alamos; GicGt
Baldwin, Julia, 406-243-5778 jbaldwin@mso.umt.edu, Univ of Montana; Gp
Baldwin, Michael, baldwin@purdue.edu, Purdue Univ; As
Baldwin, Suzanne L., (315) 443-2672 sbaldwin@syr.edu, Syracuse Univ; Cc
Baldwin, Jr., A. Dwight, dbbaldwin@comcast.net, Miami Univ; Hw
Balen, Dražen, +38514605967 drbalen@geol.pmf.hr, Univ of Zagreb; GxpGi
Balescu, Sanda, 514-987-3000 #3126 sanda.balescu@club-internet.fr, Univ du Quebec a Montreal; Pe
Balestro, Gianni, gianni.balestro@unito.it, Univ di Torino; Gc
Balgord, Elizabeth A., (801) 626-6225 balgord@email.arizona.edu, Weber State Univ; Grt
Balica, Constantin, constantin.balica@ubbcluj.ro, Babes-Bolyai Univ; GtCgc
Balistrieri, Laurie, (206) 543-8966 balistri@ocean.washington.edu, Univ of Washington; Oc
Ball, Elizabeth, (775) 682-7351 ball@mines.unr.edu, Univ of Nevada, Reno; Zn
Ball, Neil, (204) 474-8075 na_ball@umanitoba.ca, Univ of Manitoba; Gz
Ball, William P., (410) 516-5434 bball@jhu.edu, Johns Hopkins Univ; Zn
Ballantyne, Colin K., ckb@st-andrews.ac.uk, Univ of St. Andrews; ZyGml
Ballard, Robert D., (401) 874-6115 bballard@gso.uri.edu, Univ of Rhode Island; Ga
Ballentine, Christopher, +44 (1865) 272000 (72938) chrisb@earth.ox.ac.uk, Univ of Oxford; Xg
Ballero, Deniz Z., 678-891-3750 dballero@gsu.edu, Georgia State Univ, Perimeter Coll, Online; GgPm
Ballinger, Rhoda, +44(0)29 208 76671 ballingerrc@cf.ac.uk, Univ of Wales; On
Ballmer, Maxim D., m.ballmer@ucl.ac.uk, Univ Coll London; Yg
Balmforth, Neil, (604) 822-9835 njb@math.ubc.ca, Univ of British Columbia; As
Balog-Szabo, Anna, 540-857-7222 abalog-szabo@virginiawestern.edu, Virginia Western Comm Coll; Gs
Balogh-Brunstad, Zsuzsanna, (607) 431-4734 balogh_brunz@hartwick.edu, Hartwick Coll; ClHgSo
Balser, Teri , 352-294-3157 tcbalser@ufl.edu, Univ of Florida; Soo
Balser, Teri C., (608) 262-0132 tcbalser@wisc.edu, Univ of Wisconsin, Madison; Sb
Balsley, Christopher, 203-392-6647 balsleyc1@southernct.edu, Southern Connecticut State Univ; Gg
Balster, Nick J., (608) 263-5719 njbalster@wisc.edu, Univ of Wisconsin, Madison; Sf
Baltensperger, David D., (979) 845-3041 dbaltensperger@tamu.edu, Texas A&M Univ; PvCbSb
Baltz, Donald M., (225) 388-6512 Louisiana State Univ; Ob
Bamisaiye, Oluwaseyi A., +2347031533939 oabamisaye@futa.edu.ng, Federal Univ of Tech, Akure; GcZrRn
Bancroft, Alyssa M., (812) 856-5313 ambancro@indiana.edu, Indiana Univ, Bloomington; PmGr
Banda, Anna F., (903) 802-0186 aperry@dcccd.edu, El Centro Coll - Dallas Comm Coll District; GgeGv
Bandosz, Teresa, 212-650-6017 tbandosz@ccny.cuny.edu, Graduate Sch of the City Univ of New York; Gge
Banerdt, Bruce E., (818) 354-5413 Jet Propulsion Lab; Xy
Banerjee, Neil, (519) 661-2111 x.83727 nbanerj3@uwo.ca, Western Univ; EgCgGx
Banerjee, Subir K., (612) 624-5722 banerjee@tc.umn.edu, Univ of Minnesota, Twin Cities; Ym
Banfield, Jillian, 510-642-9488 jill@seismo.berkeley.edu, Univ of California,

Berkeley; CoGz
Bang, John, (919) 530-6569 jjbang@nccu.edu, North Carolina Central Univ; GeZa
Bangs, Nathan L., (512) 471-0424 nathan@ig.utexas.edu, Univ of Texas, Austin; Gu
Banik, Tenley, 309-438-8922 tjbanik@ilstu.edu, Illinois State Univ; GviCc
Bank, Carl-Georg, (416) 978-4381 bank@geology.utoronto.ca, Univ of Toronto; Zg
Banks, David, +44(0) 113 34 35244 d.banks@see.leeds.ac.uk, Univ of Leeds; Cg
Banks, Eddie, eddie.banks@flinders.edu.au, Flinders Univ; Hw
Banner, Jay L., (512) 471-5016 banner@mail.utexas.edu, Univ of Texas, Austin; Cl
Bannon, Ann, ann.bannon@dal.ca, Dalhousie Univ; Zn
Bannon, Peter R., (814) 863-1309 bannon@ems.psu.edu, Pennsylvania State Univ, Univ Park; Am
Banoeng-Yakubo, Bruce K., "kofi bruce babkbanoeng-yakubo@ug.edu.gh, Univ of Ghana; GcHw
Banse, Karl, (206) 543-5079 banse@ocean.washington.edu, Univ of Washington; Ob
Bao, Huiming, (225) 578-3419 bao@lsu.edu, Louisiana State Univ; CsAcCu
Baptista, Joana, +44 (0) 191 208 5899 joana.baptista@ncl.ac.uk, Univ of Newcastle Upon Tyne; Ng
Baptista, João C., jbaptist@utad.pt, Unive de Trás-os-Montes e Alto Douro; GmmGa
Barak, Phillip W., (608) 890-0689 phillip.barak@wisc.edu, Univ of Wisconsin, Madison; ScGzRw
Barakat, Bassam, +33(0)3 44068973 bassam.barakat@lasalle-beauvais.fr, Institut Polytechnique LaSalle Beauvais (ex-IGAL); NrgGq
Barazangi, Muawia, (607) 255-6411 mb44@cornell.edu, Cornell Univ; Ys
Barbeau, Katherine A., (858) 822-4339 kbarbeau@ucsd.edu, Univ of California, San Diego; Oc
Barbeau, Jr., David, dbarbeau@geol.sc.edu, Univ of South Carolina; Gs
Barbecot, Florent, 514-987-3000 #7786 barbecot.florent@uqam.ca, Univ du Quebec a Montreal; Hw
Barbee, Gary C., (806) 651-2294 gbarbee@wtamu.edu, West Texas A&M Univ; HwSoZi
Barber, Donald C., (610) 526-5110 dbarber@brynmawr.edu, Bryn Mawr Coll; GesOn
Barber, Richard T., (919) 728-2111 rbarber@duke.edu, Duke Univ; Ob
Barbi, Greta, 089/2180 4234 barbi@gophysik.uni-muenchen.de, Ludwig-Maximilians-Univ Muenchen; Yg
Barboni, Melanie, (480) 965-0612 melanie.barboni@asu.edu, Arizona State Univ; Gi
Barcena, Homar S., 718-368-5758 homar.barcena@kingsborough.edu, Graduate Sch of the City Univ of New York; Gg
Barcheck, Grace, grace.barcheck@cornell.edu, Cornell Univ; Gg
Barclay, Andrew, 206-543-8956 barclay@ldeo.columbia.edu, Univ of Washington; Gu
Barclay, David J., (607) 753-2921 david.barclay@cortland.edu, SUNY, Cortland; Glm
Barclay, Jenni, +44 (0)1603 59 3887 j.barclay@uea.ac.uk, Univ of East Anglia; Gv
Barclay, Julie L., julie.barclay@cortland.edu, SUNY, Cortland; Ggm
Barendregt, Rene W., (403) 329-2530 barendregt@uleth.ca, Univ of Lethbridge; Gm
Barhurst, James, +44 (0) 191 208 5431 james.bathurst@ncl.ac.uk, Univ of Newcastle Upon Tyne; Gs
Barineau, Clinton I., 706-507-8092 barineau_clinton@columbusstate.edu, Columbus State Univ; GctGg
Barker, James M., 575-835-5322 james.barker@nmt.edu, New Mexico Tech; En
Barker, Jeffrey S., (607) 777-2522 jbarker@binghamton.edu, Binghamton Univ; Ys
Barker, Joel D., (740) 725-6097 barker.246@osu.edu, Ohio State Univ; CblPe
Barker, Stephen, barkers3@cardiff.ac.uk, Cardiff Univ; Pe
Barklage, Mitchell, mitch@earth.northwestern.edu, Northwestern Univ; YgeYs
Barkmann, Peter, 303-384-2642 barkmann@mines.edu, Colorado Geo Survey; HwGcg
Barlow, Jay P., (858) 546-7178 jbarlow@ucsd.edu, Univ of California, San Diego; Ob
Barlow, Mathew, (978) 934-3908 mathew_barlow@uml.edu, Univ of Massachusetts, Lowell; As
Barminski, Robert, 831-770-7056 rbarminski@aol.com, Hartnell Coll; GgOg
Barmore, Garrett, 775-784-4528 gbarmore@unr.edu, Univ of Nevada, Reno; Zn
Barnard, Holly R., holly.barnard@colorado.edu, Univ of Colorado; Hg
Barnbaum, Cecilia S., (229) 249-2645 cbarnbau@valdosta.edu, Valdosta State Univ; Zn
Barnes, Calvin G., 806-834-7389 cal.barnes@ttu.edu, Texas Tech Univ; Gi
Barnes, Charles W., (928) 774-6079 chuck.barnes@nau.edu, Northern Arizona Univ; GcXg
Barnes, Christopher R., (250) 721-8847 crbarnes@uvic.ca, Univ of Victoria; PmOgPe
Barnes, Clare, c.barnes@ed.ac.uk, Edinburgh Univ; Zn
Barnes, Elizabeth A., eabarnes@atmos.colostate.edu, Colorado State Univ; As
Barnes, Fairley J., (505) 667-4933 fyb@lanl.gov, Los Alamos; Zn
Barnes, Hubert L., (814) 865-7573 barnes@psu.edu, Pennsylvania State Univ, Univ Park; CgEmCe
Barnes, Jaime D., jdbarnes@jsg.utexas.edu, Univ of Texas, Austin; CgsCc
Barnes, Jeffrey R., (541) 737-5685 barnes@coas.oregonstate.edu, Oregon State Univ; As
Barnes, Jessica, 803-777-9945 jebarnes@mailbox.sc.edu, Univ of South Carolina; Zn
Barnes, Melanie A., (806) 834-7965 melanie.barnes@ttu.edu, Texas Tech Univ; CaGie
Barnes, Philip, pbarnes@environ.sc.edu, Univ of South Carolina; Zn
Barnes, Randal J., (612) 625-5828 Univ of Minnesota, Twin Cities; Zn
Barnes, Sarah- J., 418 545 5011 sjbarnes@uqac.ca, Univ du Quebec a Chicoutimi; EgGiCt
Barnett, Douglas B., (509) 376-3416 brent.barnett@pnl.gov, Pacific Northwest; Em
Barnett, Michael, 617-552-8300 barnetge@bc.edu, Boston Coll; Ze
Barnett, Roger T., rbarnett@pacific.edu, Univ of the Pacific; Zy
Barnosky, Anthony D., barnosky@berkeley.edu, Univ of California, Berkeley; Pv
Barone, Jessica, 585-292-2448 jbarone@monroecc.edu, Monroe Comm Coll; Ge
Baross, John A., (206) 543-0833 Univ of Washington; Ob
Barquero-Molina, Miriam, (573) 882-9557 barqueromolinam@missouri.edu, Univ of Missouri; Gct
Barr, Sandra M., (902) 585-1340 sandra.barr@acadiau.ca, Acadia Univ; GitGg
Barr, Stuart, +44 (0) 191 208 6449 stuart.barr@ncl.ac.uk, Univ of Newcastle Upon Tyne; Zi
Barra, Monica, (803) 576-8340 mbarra@seie.sc.edu, Univ of South Carolina; Zn
Barrash, Warren, wbarrash@boisestate.edu, Boise State Univ; HwYxGe
Barrett, Bradford S., 410-293-6567 bbarrett@usna.edu, United States Naval Acad; As
Barrett, John, +44(0) 113 34 32394 j.r.barrett@leeds.ac.uk, Univ of Leeds; Ge
Barrett, Kevin M., (817) 515-6352 kevin.barrett@tccd.edu, Tarrant County Coll, Northeast Campus; AsZgr
Barrett, Melony, 217-333-7917 mebarret@illinois.edu, Illinois State Geo Survey; Zi
Barrick, James E., (806) 834-2717 jim.barrick@ttu.edu, Texas Tech Univ; Psm
Barrie, Vaughn, (250) 363-6424 Univ of Victoria; Gu
Barrier, Pascal, +33(0)3 44068975 pascal.barrier@lasalle-beauvais.fr, Institut Polytechnique LaSalle Beauvais (ex-IGAL); PmGsPs
Barron, Elizabeth, (920) 424-4105 barrone@uwosh.edu, Univ of Wisconsin Oshkosh; Zn
Barron, George, gbarron@uoguelph.ca, Univ of Guelph; Zn
Barros, Tony, (305) 237-3754 tbarros@mdc.edu, Miami-Dade Coll (Wolfson Campus); Og
Barry, Peter, pbarry@whoi.edu, Woods Hole Oceanographic Inst; Cm
Barsch, Robert, 089/2180 4201 barsch@geophysik.uni-muenchen.de, Ludwig-Maximilians-Univ Muenchen; Yg
Bart, Philip J., (225) 388-3109 pbart@geol.lsu.edu, Louisiana State Univ; Gr
Barta, Károly, +3662544156 barta@geo.u-szeged.hu, Univesity of Szeged; SpZy
Bartello, Peter, (514) 398-8075 peter.bartello@mcgill.ca, McGill Univ; As
Bartels, William S., (517) 629-0313 wbartels@albion.edu, Albion Coll; Pv
Barth, Andrew P., (317) 274-1243 ibsz100@iupui.edu, Indiana Univ - Purdue Univ Indianapolis; Gi
Barth, Jack A., (541) 737-1607 barth@coas.oregonstate.edu, Oregon State Univ; On
Barth, Nicolas, (951) 827-3138 nic.barth@ucr.edu, Univ of California, Riverside; Gt
Barth, Susan A., (406) 496-4687 sbarth@mtech.edu, Montana Tech of The Univ of Montana; Zn
Barthelmie, Rebecca J., rb737@cornell.edu, Cornell Univ; Ass
Bartholemew, Paul, pbartholomew@newhaven.edupbartholomew@newhaven.edu, Univ of New Haven; GeZiGz

Bartholomaus, Timothy, (208) 885-6217 tbartholomaus@uidaho.edu, Univ of Idaho; Gl
Bartholomew, Alexander J., 845-257-3765 barthola@newpaltz.edu, SUNY, New Paltz; GrPi
Bartholomew, Mervin J., (901) 678-1613 jbrthlml@memphis.edu, Univ of Memphis; Gtc
Bartholy, Judit, +36 20 3722945 bartholy@caesar.elte.hu, Eotvos Lorand Univ; Am
Bartl, Simona, 831-771-4400 sbartl@mlml.calstate.edu, Moss Landing Marine Lab; Zn
Bartlein, Patrick, (541) 346-4967 bartlein@uoregon.edu, Univ of Oregon; Pe
Bartlett, Douglas H., (858) 534-5233 dbartlett@ucsd.edu, Univ of California, San Diego; Ob
Bartlett, Wendy, 740-376-4782 bartletw@marietta.edu, Marietta Coll; Goe
Bartley, Julie K., 507-933-7307 jbartley@gustavus.edu, Gustavus Adolphus Coll; GsPgo
Bartok, Peter, 713-444-7751 peter@bartokinc.com, Univ of Houston; GoYeNr
Bartolucci, Valerio, (954) 201-6678 vbartolu@broward.edu, Broward Coll; Ge
Barton, Christopher C., (727) 215-5538 chris.barton@wright.edu, Wright State Univ; GqYgHq
Barton, James H., (412) 589-2821 Thiel Coll; Zy
Barton, Mark D., (520) 621-8529 mdbarton@email.arizona.edu, Univ of Arizona; Eg
Barton, Michael, (614) 292-3132 barton.2@osu.edu, Ohio State Univ; GiCpt
Barzegari, Ghodrat, gbarzegari@gmail.com, Univ of Tabriz; Nr
Basak, Chandranath, 302-831-8230 cbasak, Univ of Delaware; ClOcCs
Bascom, Johnathan, (616) 526-7053 jbascom@calvin.edu, Calvin Coll; Zy
BASCOU, Jérôme, jerome.bascou@univ-st-etienne.fr, Université Jean Monnet, Saint-Etienne; Gcy
Basinger, James F., (306) 966-5684 jim.basinger@usask.ca, Univ of Saskatchewan; Pb
Baskaran, Mark, (313) 577-3262 baskaran@wayne.edu, Wayne State Univ; Ccl
Baskin, Perry A., (229) 259-5052 pbaskin@valdosta.edu, Valdosta State Univ; Zn
Bass, David, david.bass@flinders.edu.au, Flinders Univ; Zy
Bass, Jay D., (217) 333-1018 jaybass@illinois.edu, Univ of Illinois, Urbana-Champaign; Gy
Bass, Jerry, 601 266-4732 joby@usm.edu, Univ of Southern Mississippi; Zn
Bassett, Damon, (417) 836-4897 dbassett@missouristate.edu, Missouri State Univ; Gg
Bassett, Kari N., (03) 369-4495 kari.bassett@canterbury.ac.nz, Univ of Canterbury; GstGd
Bassett, Scott, (775) 784-1434 sbassett@unr.edu, Univ of Nevada, Reno; Zu
Bassett, William A., wab7@cornell.edu, Cornell Univ; Gz
Bassis, Jeremy, (734) 615-3606 jbassis@umich.edu, Univ of Michigan; GlOgZg
Basta, Nicholas T., 614-292-6282 basta.4@osu.edu, Ohio State Univ; Sc
Bastow, Ian, +44 20 759 42974 i.bastow@imperial.ac.uk, Imperial Coll; Yg
Basu, Abhijit, 8128556654/5581 basu@indiana.edu, Indiana Univ, Bloomington; Gd
Basu, Anirban, +44 1784 414083 anirban.basu@rhul.ac.uk, Royal Holloway Univ of London; CglCs
Basu, Asish, (817) 272-2987 abasu@uta.edu, Univ of Texas, Arlington; GxCcs
Bataille, Clement, clement.bataille@uottawa.ca, Univ of Ottawa; ClGeCs
Batelaan, Okke, okke.batelaan@ees.kuleuven.be, Katholieke Univit Leuven; Hw
Bates, Mary, (661) 362-5054 mary.bates@canyons.edu, Coll of the Canyons; Zyg
Bates, Nicholas R., n.r.bates@soton.ac.uk, Univ of Southampton; Oc
Bates, Richard, +44 01334 463997 crb@st-andrews.ac.uk, Univ of St. Andrews; Yx
Bates, Timothy S., (206) 526-6248 tim.bates@noaa.gov, Univ of Washington; As
Bathke, Deborah J., dbathke2@unl.edu, Univ of Nebraska, Lincoln; AsZe
Batta-Lona, Paola, paola.batta_lona@uconn.edu, Univ of Connecticut; Ob
Battisti, David S., (206) 543-2019 battisti@uw.edu, Univ of Washington; As
Battles, Eileen, (785) 864-2129 battles@kgs.ku.edu, Univ of Kansas; Zy
Batygin, Konstantin, 626-395-2920 kbatygin@gps.caltech.edu,CalTech; Xg
Bauder, James W., (406) 994-5685 jbauder@montana.edu, Montana State Univ; Sf
Bauer, Annie, annie.bauer@wisc.edu, Univ of Wisconsin, Madison; Cca
Bauer, Bernard O., bernard.bauer@ubc.ca, Univ of British Columbia, Okanagan Campus; GmOnZy
Bauer, Carl J., cjbauer@email.arizona.edu, Univ of Arizona; Hs
Bauer, Emily, (612) 626-0909 bauer010@umn.edu, Univ of Minnesota; Zi
Bauer, Jeffrey A., jbauer@shawnee.edu, Shawnee State Univ; Ps
Bauer, Paul W., (575) 835-5106 bauer@nmt.edu, New Mexico Tech; Gc
Bauer, R. A., 217-244-2394 rabauer@illinois.edu, Illinois State Geo Survey; Ng
Baughman, Jaclyn S., 707-826-3959 jsb119@humboldt.edu, Humboldt State Univ; GtCgGc
Baum, Steven K., (979) 845-0793 sbaum@ocean.tamu.edu, Texas A&M Univ; Op
Baumann, Hannes, 860-405-9297 hannes.baumann@uconn.edu, Univ of Connecticut; Ob
Baumann, Zofia, (860) 405-9281 zofia.baumann@uconn.edu, Univ of Connecticut; ObcCm
Baumgartner, Mark, mbaumgartner@whoi.edu, Dalhousie Univ; Ob
Baumiller, Tomasz K., (734) 764-7543 tomaszb@umich.edu, Univ of Michigan; Pg
Bautista, Sonia, sbautist@ucm.es, Univ Compl de Madrid; Ng
Baxter, Ethan, 617-552-3640 ethan.baxter@bc.edu, Boston Coll; Ccg
Baxter, Martin, 989-774-2055 baxte1ma@cmich.edu, Central Michigan Univ; Ams
Baxter, Stefanie J., 302-831-1576 steff@udel.edu, Univ of Delaware; On
Baxter, P.G., James E., 717-780-2377 jebaxter@hacc.edu, Harrisburg Area Comm Coll; GgHwGm
Bayari, Serdar, +90 (312) 2977740 serdar@hacettepe.edu.tr, Hacettepe Univ; Hw
Bayowa, Oyelowo G., +2348030820291 obayowa@lautech.edu.ng, Ladoke Akintola Univ of Tech; GggGg
Beach Davis, Janet, (309) 268-8513 janet.beach-davis@heartland.edu, Heartland Comm Coll; Zg
Beane, Rachel J., (207) 725-3160 rbeane@bowdoin.edu, Bowdoin Coll; GzxZe
Beard, Brian L., 608-262-1806 beardb@geology.wisc.edu, Univ of Wisconsin, Madison; Cc
Beard, James S., (276) 634-4170 jim.beard@vmnh.virginia.gov, Virginia Tech; GiClg
Beard, K. Christopher, (412) 622-5782 beardc@carnegiemnh.org, Carnegie Mus of Natural History; Pv
Beard, Kenneth V., (217) 333-1676 k-beard@uiuc.edu, Univ of Illinois, Urbana-Champaign; As
Bearden, Bennett L., 205-247-3683 bbearden@gsa.state.al.us, Geo Survey of Alabama; HgEo
Bearden, Rebecca A., 205-247-3623 rbearden@gsa.state.al.us, Geo Survey of Alabama; Hs
Beatty, Lynne, 9134698500 x3785 lbeatty@jccc.edu, Johnson County Comm Coll; GgZy
Beatty, Merrill Ann, (506) 453-4803 mbeatty@unb.ca, Univ of New Brunswick; Zn
Beatty, William L., 507-474-5789 wbeatty@winona.edu, Winona State Univ; PgiGs
Beaty, Tammy, (865) 574-0119 taw@ornl.gov, Oak Ridge; Zi
Beauchamp, Marc, 519-661-2111 ext 88104 mbeauch6@uwo.ca, Western Univ; Ca
Beaudoin, Georges, (418) 656-3141 beaudoin@ggl.ulaval.ca, Univ Laval; EgGzCa
Beaumont, Christopher, (902) 494-3779 chris.beaumont@dal.ca, Dalhousie Univ; Yg
Beausoleil, Denis, 604-777-6019 beausoleild@douglascollege.ca, Douglas Coll; GgZg
Beaverson, Sheena K., 217-244-9306 sbeavers@illinois.edu, Illinois State Geo Survey; Zr
Bebout, Gray E., (610) 758-5831 geb0@lehigh.edu, Lehigh Univ; Gp
Beccue, Clint, 217-265-5161 cbeccue@illinois.edu, Illinois State Geo Survey; Ge
Becel, Anne, annebcl@ldeo.columbia.edu, Columbia Univ; Gu
Bechtel, Timothy D., 717 358--4750 timothy.bechtel@fandm.edu, Franklin and Marshall Coll; GeYg
Beck, Catherine C., 315-859-4847 ccbeck@hamilton.edu, Hamilton Coll; GsnGr
Beck, Charles, (734) 763-5089 chbeck@umich.edu, Univ of Michigan; Pg
Beck, Colleen M., (702) 795-8077 colleen@dri.edu, Desert Research Inst; Sa
Beck, John H., 617-552-8300 john.beck@bc.edu, Boston Coll; Pb
Beck, Susan L., (520) 621-8628 slbeck@email.arizona.edu, Univ of Arizona; YsGt
Becker, Alex, (510) 643-9181 Univ of California, Berkeley; Yg
Becker, Janet, jmbecker@ucsd.edu, Univ of California, San Diego; Op
Becker, Keir, (305) 421-4661 kbecker@rsmas.miami.edu, Univ of Miami; YrGu
Becker, Laurence, (541) 737-9504 beckerla@geo.oregonstate.edu, Oregon State Univ; Zn

Becker, Lorene Y., 541-737-6993 beckelo@geo.oregonstate.edu, Oregon State Univ; Zi
Becker, Matthew, 562-985-8983 mbecker3@csulb.edu, California State Univ, Long Beach; Hwg
Becker, Naomi M., (505) 667-2165 nmb@lanl.gov, Los Alamos; Hg
Becker, Ralph Thomas, +49-251-83-33951 rbecker@uni-muenster.de, Univ Muenster; PisPm
Becker, Richard H., (630) 252-7595 Argonne; Zr
Becker, Thorsten, twb@ig.utexas.edu, Univ of Texas, Austin; Yg
Becker, Udo, (734) 615-6894 ubecker@umich.edu, Univ of Michigan; Zn
Beckie, Roger D., (604) 822-6462 rbeckie@eos.ubc.ca, Univ of British Columbia; HwCl
Beckingham, Barbara, (843) 953-0483 beckinghamba@cofc.edu, Coll of Charleston; ClHwGe
Bedard, Jean H., (418) 654-2671 jbedard@nrcan.gc.ca, Univ du Quebec; Gi
Bedard, Paul, 4185455011, 2276 pbedard@uqac.ca, Univ du Quebec a Chicoutimi; CaGzEg
Bedaso, Zelalem, (937) 229-2393 zbedaso1@udayton.edu, Univ of Dayton; CslGs
Beddows, Patricia A., (847) 491-7460 patricia@earth.northwestern.edu, Northwestern Univ; HyGma
Bedelean, Horea Stefan, horea.bedelean@ubbcluj.ro, Babes-Bolyai Univ; GcSdGe
Bedle, Heather, hbedle@ou.edu, Univ of Oklahoma; Yg
Bednarski, Marsha, (860) 832-2943 bednarskim@ccsu.edu, Central Connecticut State Univ; Ze
Bednarz, Robert S., r-bednarz@tamu.edu, Texas A&M Univ; Zeu
Bednarz, Sarah W., (979) 845-1579 s-bednarz@tamu.edu, Texas A&M Univ; Ze
Bedore, Melanie, (905) 525-9140 (Ext. 23525)) bedorem@mcmaster.ca, McMaster Univ; Zu
Bedrikovetski, Pavel, pavel.bedrikovetski@adelaide.edu.au, Univ of Adelaide; Np
Beebe, Alex, dbeebe@southalabama.edu, Univ of South Alabama; Ge
Beets, Kay J., +31 20 59 87357 c.j.beets@vu.nl, Vrije Univit Amsterdam; Cg
Befus, Kenneth S., (254)710-2192 kenneth_befus@baylor.edu, Baylor Univ; GviGv
Beget, James E., ffjeb1@uaf.edu, Univ of Alaska, Fairbanks; Gm
Begg, Stephen, steve.begg@adelaide.edu.au, Univ of Adelaide; Eo
Beghein, Caroline, (310) 825-0742 cbeghein@ucla.edu, Univ of California, Los Angeles; Ysg
Begin, Christian, (418) 654-2648 christian.begin@canada.ca, Univ du Quebec; PeGmSo
Behl, Richard J., (562) 985-5850 behl@csulb.edu, California State Univ, Long Beach; GsPeGd
Behn, Mark, 617 552-2180 mark.behn@bc.edu, Boston Coll; YgrGt
Behr, Rose-Anna, 717-702-2035 rosbehr@pa.gov, DCNR- Pennsylvania Bur of Geo Survey; GcRnGr
Behr-Andres, Christina "Tina", (505) 667-3644 behr-andres@lanl.gov, Los Alamos; Zn
Behrangi, Ali, (520) 626-6639 behrangi@email.arizona.edu, Univ of Arizona; As
Behrensmeyer, Anna K., (202) 633-1307 Smithsonian Inst / Natl Mus of Natural History; Pv
Beiersdorfer, Raymond E., (330) 941-1753 rebeiersdorfer@ysu.edu, Youngstown State Univ; Cg
Bein, F. L., (317) 274-1100 Indiana Univ, Indianapolis; Zy
Bekker, Andrey, (951) 827-4611 andrey.bekker@ucr.edu, Univ of California, Riverside; GsrCg
Bekker, Matthew F., (801) 422-1961 matthew_bekker@byu.edu, Brigham Young Univ; Zy
Belasky, Paul, 510-676-0945 pbelasky@ohlone.edu, Ohlone Coll; PgqGs
Belknap, Daniel F., (207) 581-2159 belknap@maine.edu, Univ of Maine; GusGa
Bell, Andrew, +44 (0) 131 650 4918 a.bell@ed.ac.uk, Edinburgh Univ; Gt
Bell, Brian, +44 01413306898 brian.bell@glasgow.ac.uk, Univ of Glasgow; So
Bell, Christopher J., (512) 471-7301 cjbell@mail.utexas.edu, Univ of Texas, Austin; Pv
Bell, David, (03) 3642-717 david.bell@canterbury.ac.nz, Univ of Canterbury; Ng
Bell, James, (480) 965-1044 jim.bell@asu.edu, Arizona State Univ; Xg
Bell, John W., (775) 784-1939 jbell@unr.edu, Univ of Nevada, Reno; NgGm
Bell, Keith, keith_bell@carleton.ca, Carleton Univ; Cc
Bell, Margaret C., +44 (0) 191 208 7936 margaret.bell@ncl.ac.uk, Univ of Newcastle Upon Tyne; Ge
Bell, Michael M., mmbell@colostate.edu, Colorado State Univ; As
Bell, Phil R., pbell23@une.edu.au, Univ of New England; PgvPg
Bell, Rebecca, +44 20 759 40903 rebecca.bell@imperial.ac.uk, Imperial Coll; Gt
Bell, Robin E., robinb@ldeo.columbia.edu, Columbia Univ; Yr
Bellew, Angela, (319) 335-0150 angela-bellew@uiowa.edu, Univ of Iowa; Zn
Belley, Philippe, 709 864-4385 pbelley@mun.ca, Memorial Univ of Newfoundland; Gz
Bellian, Jerome, jerry.bellian@whiting.com, Miami Univ; Go
Belliveau, Lindsey, 860-424-3581 lindsey.belliveau@ct.gov, Dept of Energy and Env Protection; Gmt
Belluso, Elena, elena.belluso@unito.it, Univ di Torino; Gz
Belmonte, Donato, +39 010 353 8136 Universita di Genova; Cg
Beltrami, Hugo, 902-867-2326 hbeltram@stfx.ca, Saint Francis Xavier Univ; YhAs
Beltz, John F., 330 972-6687 jfb4@uakron.edu, Univ of Akron; GgRh
Bemis, Sean P., 540-231-6521 sbemis@vt.edu, Virginia Tech; Gt
Ben-Avraham, Zvi, zviba@tau.ac.il, Tel Aviv Univ; YrGt
Ben-Zion, Yehuda, (213) 740-6734 ybz@earth.usc.edu, Univ of Southern California; Ys
Benacquista, Matt, (406) 657-2341 Montana State Univ, Billings; Xy
Benavides, Jude, 956 883-5938 jude.benavides@utrgv.edu, Univ of Texas, Rio Grande Valley; Hsq
Benavides-Iglesias, Alfonso, a.benavides@geos.tamu.edu, Texas A&M Univ; Ys
Bender, E. E., (714) 432-5681 ebender@occ.cccd.edu, Orange Coast Coll; GitGz
Bender, John F., 704-687- 5956 jfbender@email.uncc.edu, Univ of North Carolina, Charlotte; GiCtOu
Bender, Michael L., 609-258-5807 bender@princeton.edu, Princeton Univ; Cg
Bendick, Rebecca, 406-243-5774 bendick@mso.umt.edu, Univ of Montana; Yg
Benea, Ion Marcel , ion.benea@ubbcluj.ro, Babes-Bolyai Univ; GzaEn
Benedetti, Michael M., (910) 962-7650 benedettim@uncw.edu, Univ of North Carolina, Wilmington; GmSaZy
Benfield, Mark C., (225) 388-6372 mbenfie@lsu.edu, Louisiana State Univ; Ob
Benger, Simon, simon.benger@flinders.edu.au, Flinders Univ; Ziy
Beniest, Anouk, +31 20 59 81509 a.beniest@vu.nl, Vrije Univit Amsterdam; ZyGcg
Benimoff, Alan I., 718-982-2835 alan.benimoff@csi.cuny.edu, Coll of Staten Island/CUNY; GizGp
Benison, Kathleen , 304-293-5603 West Virginia Univ; GrsCm
Benitez-Nelson, Claudia, 803-777-0525 benitezn@mailbox.sc.edu, Univ of South Carolina; Oc
Benito, María Isabel, mibenito@ucm.es, Univ Compl de Madrid; Gs
Benjamin, Timothy M., (505) 667-5154 Los Alamos; Ca
Benjamin, U. K., 234-8038312732 uzochukwu.benjamin@oauife.edu.ng, Obafemi Awolowo Univ; GoCg
Benna, Piera, piera.benna@unito.it, Univ di Torino; Gz
Bennartz, Ralf, ralf.bennartz@vanderbilt.edu, Vanderbilt Univ; AsZr
Benner, Jake, 865-974-6013 jbenner@utk.edu, Univ of Tennessee, Knoxville; ZePg
Benner, Ron, benner@mailbox.sc.edu, Univ of South Carolina; Ob
Benner, Shawn, shawnbenner@boisestate.edu, Boise State Univ; Clt
Bennet, Brett, 785-864-2117 bb@kgs.ku.edu, Univ of Kansas; Ng
Bennett, Philip, (512) 471-3587 pbennett@mail.utexas.edu, Univ of Texas, Austin; Hg
Bennett, Richard, (520) 621-2324 rab@geo.arizona.edu, Univ of Arizona; YdGt
Bennett, Sara, 309/298-1905 sc-bennett@wiu.edu, Western Illinois Univ; Gg
Bennett, Scott, scottekb@uw.edu, Univ of Washington; YsGtm
Bennett, Steven W., (309) 298-1256 sw-bennett1@wiu.edu, Western Illinois Univ; Hw
Bennington, J Bret, 516 463-5568 geojbb@hofstra.edu, Hofstra Univ; PeGs
Bennis, Kadie, bennisk@si.edu, Smithsonian Inst / Natl Mus of Natural History; Gv
Benoit-Bird, Kelly, 541-737-2063 kbenoit@coas.oregonstate.edu, Oregon State Univ; Gu
Bense, Victor, +44 (0)1603 59 1297 v.bense@uea.ac.uk, Univ of East Anglia; Hg
Benson, David A., 303-273-3806 dbenson@mines.edu, Colorado Sch of Mines; Hq
Benson, Donna M., (480) 461-7247 donnabenson@mesacc.edu, Mesa Comm Coll; Gg
Benson, Jane L., (270) 809-3106 jbenson1@murraystate.edu, Murray State Univ; Zi
Benson, Roger, +44 (1865) 272000 roger.benson@earth.ox.ac.uk, Univ of

Oxford; Pg
Bentley, Callan, (703) 323-3276 cbentley@nvcc.edu, Northern Virginia Comm Coll - Annandale; Gg
Bentley, Laurence R., (403) 220-4512 Univ of Calgary; Hw
Bentley, Samuel J., 225-578-5735 sjb@lsu.edu, Louisiana State Univ; Gu
Benton, Steven, 217-244-0082 s-benton@illinois.edu, Illinois State Geo Survey; Gg
Benway, Heather, (508) 289-2838 hbenway@whoi.edu, Woods Hole Oceanographic Inst; OcCmOo
Beranek, Luke, 709 864 4588 lberanek@mun.ca, Memorial Univ of Newfoundland; GgsCc
Berchem, Jean, (310) 206-6484 jberchem@igpp.ucla.edu, Univ of California, Los Angeles; Zn
Bercovici, David, (203) 432-3168 david.bercovici@yale.edu, Yale Univ; Yg
Bereitschaft, Bradley, 402-554-2674 bbereitschaft@unomaha.edu, Univ of Nebraska, Omaha; Zuy
Berelson, William M., (213) 740-5828 berelson@usc.edu, Univ of Southern California; Cm
Beresnev, Igor A., (515) 294-7529 beresnev@iastate.edu, Iowa State Univ of Sci & Tech; YgsYe
Berg, Christopher A., (714)432-0202 x21229 cberg5@occ.cccd.edu, Orange Coast Coll; GpZeGt
Berg, Peter, (804) 924-1318 pb8n@virginia.edu, Univ of Virginia; So
Berg, Richard, 217-244-2776 rberg@illinois.edu, Univ of Illinois; GlRwZu
Bergantz, George W., 206-685-4972 bergantz@uw.edu, Univ of Washington; Gi
Berger, Karen, (585) 275-7611 karen.berger@rochester.edu, Univ of Rochester; Rw
Berger, Peter M., 217-333-7078 pmberger@illinois.edu, Illinois State Geo Survey; Cg
Bergeron, Normand, (418) 654-3703 normand.bergeron@ete.inrs.ca, Univ du Quebec; Gm
Berggren, William A., (508) 289-2593 wberggren@whoi.edu, Woods Hole Oceanographic Inst; Ps
Berghuijs , Wouter, +31 20 59 87968 w.r.berghuijs@vu.nl, Vrije Univit Amsterdam; Hq
Bergin, Michael H., (404) 894-9723 mike.bergin@ce.gatech.edu, Georgia Inst of Tech; As
Bergmann, Kristin, (617) 253-9852 kdberg@mit.edu, Massachusetts Inst of Tech; GsCl
Bergquist, Bridget, bergquist@es.utoronto.ca, Univ of Toronto; Cgb
Bergslien, Elisa T., 716-878-3793 bergslet@buffalostate.edu, Buffalo State Coll; ClHw
Bergstrom, Anna, annabergstrom@boisestate.edu , Boise State Univ; HgGlCb
Bergstrom, Stig M., (614) 292-4473 bergstrom.1@osu.edu, Ohio State Univ; Pg
Bergwerff, Kenneth A., (616) 526-6371 kbergwer@calvin.edu, Calvin Coll; Ze
Berke, Melissa, melissa.berke.1@nd.edu, Univ of Notre Dame; Cs
Berkelhammer, Max, 312-413-8271 berkelha@uic.edu, Univ of Illinois Chicago; At
Berli, Marcus, 702.862.5452 markus.berli@dri.edu, Univ of Nevada, Reno; Sp
Berlin, Graydon L., lenn.berlin@nau.edu, Northern Arizona Univ; Zr
Berlo, Kim, 514-398-5884 kim.berlo@mcgill.ca, McGill Univ; CgGv
Berman, David S., (412) 622-3248 bermand@carnegiemnh.org, Carnegie Mus of Natural History; PggPv
Bermanec, Vladimir, +38514605972 vberman@public.carnet.hr, Univ of Zagreb; GzEnGi
Bermingham, Katherine R., (301) 405-2707 kberming@umd.edu, Univ of Maryland; XcCcg
Bernard, Rachael E., rbernard@amherst.edu, Amherst Coll; Gc
Bernard-Kingsley, Noell, 206-616-8511 noelleon@uw.edu, Univ of Washington; Ze
Bernhard, Joan M., 508-289-3480 jbernhard@whoi.edu, Woods Hole Oceanographic Inst; Cg
Bernhardt, Jase, 516-463-5731 jase.e.bernhardt@hofstra.edu, Hofstra Univ; Ams
Bernier, Luc, (905) 525-9140 (Ext. 26364) berniejm@mcmaster.ca, McMaster Univ; Cg
Bernier, Monique, 418 654-2585 monique.bernier@ete.inrs.ca, Univ du Quebec; Zr
Berns-Herrboldt, Erin, (920) 465-2446 bernse@uwgb.edu, Univ of Wisconsin, Green Bay; CbSp
Bernstein, Neil P., neil-bernstein@uiowa.edu, Univ of Iowa; Ge
Beroza, Gregory C., (650) 723-4958 beroza@stanford.edu, Stanford Univ; Ys
Berquist, Jr., Carl R., (757) 221-2448 crberq@wm.edu, William & Mary; Ou
Berry, C M., chris.berry@earth.cf.ac.uk, Cardiff Univ; Pb
Berry, Duane F., (540) 231-9792 duberry@vt.edu, Virginia Tech; Sb
Berry, Karen, 303-384-2640 kaberry@mines.edu, Colorado Geo Survey; NgZu
Berry, Kate, (775) 784-6344 kberry@unr.edu, Univ of Nevada, Reno; Zy
Berry, Leonard, 561 297-2935 berry@fau.edu, Florida Atlantic Univ; Zg
Berry, Ron F., 61 3 6226 2456 ron.berry@utas.edu.au, Univ of Tasmania; Gc
Berry IV, Henry N., 207-287-7179 henry.n.berry@maine.gov, Dept of Agriculture, Conservation, and Forestry; Gg
Bershaw, John T., (503) 725-3378 bershaw@pdx.edu, Portland State Univ; GsCsHy
Berta, Susan, 812-237-2260 sberta@indstate.edu, Indiana State Univ; GmZr
Berthold, Angela, (612) 626-6744 berth084@d.umn.edu, Univ of Minnesota; Gl
Berti, Claudio, (208) 885-7479 cberti@uidaho.edu, Univ of Idaho; GgmGc
Bertine, Kathe K., (619) 594-6369 kbertine@geology.sdsu.edu, San Diego State Univ; Cl
Bertog, Janet, 859-572-1523 bertogj@nku.edu, Northern Kentucky Univ; Gd
Bertok, Carlo, carlo.bertok@unito.it, Univ di Torino; Gd
Bertsch, Paul M., (803) 725-2752 Univ of Georgia; Sc
Besancon, James, (781) 283-3030 jbesancon@wellesley.edu, Wellesley Coll; Hww
Best, James L., 217-244-1839 jimbest@illinois.edu, Univ of Illinois, Urbana-Champaign; Gs
Best, Myron G., best_myron_g@byu.edu, Brigham Young Univ; Gg
Bestland, Erick, erick.bestland@flinders.edu.au, Flinders Univ; GsSa
Beswick, Anthony E., (705) 675-1151 tonybeswick@xplornet.ca, Laurentian Univ, Sudbury; Gi
Betancourt, Julio, jlbetanc@usgs.gov, Univ of Arizona; Zn
Betchwars, Corey, (612) 625-3507 betch038@umn.edu, Univ of Minnesota; Zi
Betka, Paul, pbetka@gmu.edu, George Mason Univ; GctGg
Betterton, Eric, better@email.arizona.edu, Univ of Arizona; As
Bettis III, E. A., art-bettis@uiowa.edu, Univ of Iowa; SdGmSa
Betzer, Peter R., (727) 553-1130 pbetzer@marine.usf.edu, Univ of South Florida; Cm
Beukelman, Gregg S., (801) 903-8138 greggbeukelman@msn.com, Utah Geo Survey; Ng
Beukes, Jarlen J., (051) 401-2318 beukesj@ufs.ac.za, Univ of the Free State; Gi
Beutel, Erin K., beutele@cofc.edu, Coll of Charleston; Gct
Bevelhimer, Mark, (865) 576-0266 mb2@ornl.gov, Oak Ridge; Hs
Bevis, Kenneth A., (812) 866-7307 bevis@hanover.edu, Hanover Coll; GlmGe
Bevis, Michael G., 614 247 5071 bevis.6@osu.edu, Ohio State Univ; Yx
Beyarslan, Melahat, 00904242370000-5985 mbeyarslan@firat.edu.tr, Firat Univ; GipGx
Beyer, Patricia J., (570) 389-4570 pbeyer@bloomu.edu, Bloomsburg Univ; HsZyGm
Bezada, Max, 614-624-3280 mbezada@umn.edu, Univ of Minnesota, Twin Cities; Ys
Bezada Diaz, Maximiliano, 612-625-3507 mbezadad@umn.edu, Univ of Minnesota; Gl
Bezdecny, Kris, (323) 343-2400 kbezdec@calstatela.edu, California State Univ, Los Angeles; ZniZn
Bezdicek, David F., (509) 335-3644 bezdicek@wsu.edu, Washington State Univ; Sb
Bhaskaran, Sunil, 718-289-5566 sunil.bhaskaran@bcc.cuny.edu, Graduate Sch of the City Univ of New York; ZfrZi
Bhattacharya, Janok, bhattaj@mcmaster.ca, Univ of Houston; Gsr
Bhattacharya, Tripti, 315-443-1348 trbhatta@syr.edu, Syracuse Univ; Ge
Bhattacharyya, Prajukti, (262) 472-5257 bhattacj@uww.edu, Univ of Wisconsin, Whitewater; Gce
Bhullar, Bhart-Anjan, 402-689-5998 Yale Univ; Pv
Bianchi, Thomas S., 352-392-6138 tbianchi@ufl.edu, Univ of Florida; Co
Biasi, Glenn P., (775) 784-4576 glenn@seismo.unr.edu, Univ of Nevada, Reno; Ys
Biasutti, Michela, biasutti@ldeo.columbia.edu, Columbia Univ; Am
Bibby, Thomas, +44 (0)23 8059 6446 tsb@noc.soton.ac.uk, Univ of Southampton; Ob
Bible, Gary G., garybible11@gmail.com, Univ of Tennessee, Knoxville; Go
Bice, David M., 814-865-4477 dbice@geosc.psu.edu, Pennsylvania State Univ, Univ Park; Gg
Bickel, Charles E., (415) 338-1963 bickel@sfsu.edu, San Francisco State Univ; Gx
Bickford, M. E., (315) 443-2672 mebickfo@syr.edu, Syracuse Univ; CcGiz
Bickle, Michael, +44 (0) 1223 333484 mb72@esc.cam.ac.uk, Univ of Cambridge; GtYgGo

Bickmore, Barry R., (801) 422-4680 barry_bickmore@byu.edu, Brigham Young Univ; ClGz
Bidgoli, Tandis, (785) 864-3315 bidgoli@kgs.ku.edu, Univ of Kansas; GcoCc
Biek, Robert, (801) 537-3356 bobbiek@utah.gov, Utah Geo Survey; Gg
Bieler, David B., (318) 869-5234 dbieler@centenary.edu, Centenary Coll of Louisiana; GtrYe
Bienfang, Paul K., 808-956-7402 bienfang@soest.hawaii.edu, Univ of Hawai'i, Manoa; Ob
Bienvenu, Nadean S., (337) 482-6647 nadean@louisiana.edu, Univ of Louisiana at Lafayette; Zn
Bier, Sara E., (315) 267-3482 bierse@potsdam.edu, SUNY Potsdam; GctGg
Bierly, Aaron D., 717-702-2034 aabierly@pa.gov, DCNR- Pennsylvania Bur of Geo Survey; GgZn
Bierly, Gregory, 812-237-3225 gregory.bierly@indstate.edu, Indiana State Univ; Asm
Bierman, Paul R., (802) 656-4411 pbierman@uvm.edu, Univ of Vermont; GmeCc
Biggin, Andy, +440151 794 3460 a.biggin@liverpool.ac.uk, Univ of Liverpool; Ym
Biggs, Douglas C., (979) 845-3423 d-biggs@tamu.edu, Texas A&M Univ; Ob
Bigham, Jerry M., (614) 292-9066 bigham.1@osu.edu, Ohio State Univ; Sd
Bijeljic, Branko, +44 20 759 46420 b.bijeljic@imperial.ac.uk, Imperial Coll; Np
Bilanovic, Dragoljub D., 218-755-2801 dbilanovic@bemidjistate.edu, Bemidji State Univ; GePgCa
Bilardello, Dario, (612) 624-5049 dario@umn.edu, Univ of Minnesota, Twin Cities; Ym
Bilek, Susan L., 575.835.6510 sbilek@nmt.edu, New Mexico Tech; Ys
Bill, Steven D., sbill@keene.edu, Keene State Coll; Pg
Billen, Magali I., (530) 752-4169 mibillen@ucdavis.edu, Univ of California, Davis; Yg
Billings, Brian J., 320-308-3298 bjbillings@stcloudstate.edu, Saint Cloud State Univ; Am
Billings, Stephen, sbillings@skyri.com, Univ of British Columbia; Yg
Billiot, Fereshteh, 361-825-6067 fereshteh.billiot@tamucc.edu, Texas A&M Univ, Corpus Christi; Zn
Billups, Katharina, 302-645-4249 kbillups@udel.edu, Univ of Delaware; Pe
Bilodeau, William L., (805) 493-3264 bilodeau@callutheran.edu, California Lutheran Univ; GcsGg
Bina, Craig R., (847) 491-5097 craig@earth.northwestern.edu, Northwestern Univ; YgZmGy
Bindeman, Ilya N., 541-346-3817 bindeman@uoregon.edu, Univ of Oregon; CsGvCl
Bingham, Rob, r.bingham@ed.ac.uk, Edinburgh Univ; Zc
Bingol, Ahmet Feyzi, 00904242370000-5973 fbingol@firat.edu.tr, Firat Univ; GipGv
Binzel, Richard P., (617) 253-6486 rpb@mit.edu, Massachusetts Inst of Tech; Xm
Biondi, Biondo L., (650) 723-9831 biondo@sep.stanford.edu, Stanford Univ; Ye
Birchall, Jeff, (780) 248-5758 jeff.birchall@ualberta.ca, Univ of Alberta; ZuyZg
Bircher, Harry, 330-549-9051 harrybircher@buckeyecivildesign.com, Youngstown State Univ; Ge
Bird, Brian, bbird@mail.nysed.gov, New York State Geo Survey; GlZi
Bird, Broxton, bwbird@iupui.edu, Indiana Univ - Purdue Univ Indianapolis; Gmn
Bird, Dennis K., (650) 723-1664 bird@pangea.stanford.edu, Stanford Univ; Cg
Bird, G. Peter, (310) 825-1126 pbird@epss.ucla.edu, Univ of California, Los Angeles; GtYsRn
Bird, Jeffrey, 718-997-3332 jeffrey.bird@qc.cuny.edu, Graduate Sch of the City Univ of New York; Sbf
Bird, MIchael I., 61742421137 michael.bird@jcu.edu.au, James Cook Univ; ClsSa
Birdsell, Kay H., (505) 665-0260 khb@lanl.gov, Los Alamos; Hqy
Birgenheier, Lauren, (801) 585-3158 lauren.birgenheier@utah.edu, Univ of Utah; Gso
Birke, Tesfaye K., (031) 260-8781 birket@ukzn.ac.za, Univ of KwaZulu-Natal; GcYmGg
Birkel, Sean, sean.birkel@maine.edu, Univ of Maine; Gl
Birkeland, Peter W., (303) 492-6985 birkelap@stripe.colorado.edu, Univ of Colorado; Sd
Birner, Johannes, 089/2180 6565 johannes.birner@web.de, Ludwig-Maximilians-Univ Muenchen; Gg
Biscaye, Pierre E., biscaye@ldeo.columbia.edu, Columbia Univ; Cm
Bischoff, William D., (316) 978-6659 Wichita State Univ; Cl
Bish, David L., (812) 855-2039 bish@indiana.edu, Indiana Univ, Bloomington; Gz
Bishop, Charles E., (801) 537-3361 nrugs.cbishop@state.ut.us, Utah Geo Survey; Hw
Bishop, Gale A., (912) 681-5361 Georgia Southern Univ; Pi
Bishop, James K., (510) 495-2457 bishop@atmos.berkeley.edu, Univ of California, Berkeley; Oc
Bishop, Stuart P., 919-515-7894 spbishop@ncsu.edu, North Carolina State Univ; OpGq
Bissada, Adry, 713-743-4026 kbissada@uh.edu, Univ of Houston; Co
Bissell, Clinton R., randy.bissell@tamucc.edu, Texas A&M Univ, Corpus Christi; GosGr
Bissig, Thomas, 604-822-5503 tbissig@eos.ubc.ca, Univ of British Columbia; EmCg
Biswas, Pratim, pratim.biswas@seas.wustl.edu, Washington Univ in St. Louis; Ng
Bitz, Cecilia M., (206) 543-1339 bitz@atmos.washington.edu, Univ of Washington; As
Biundo, Marc, mbiundo@uw.edu, Univ of Washington; Ys
Bizimis, Michael, 803-777-5565 mbizimis@geol.sc.edu, Univ of South Carolina; GiCg
Biševac, Vanja, +38514605969 vabisevac@geol.pmf.hr, Univ of Zagreb; GxpGz
Bjorklund, Tom, 281-589-6846 tbjorklund@uh.edu, Univ of Houston; Gco
Bjornerud, Marcia, (920) 832-7015 marcia.bjornerud@lawrence.edu, Lawrence Univ; GctRc
Bjornstad, Bruce N., (509) 373-6948 bruce.bjornstad@pnl.gov, Pacific Northwest; Gs
Blacic, James D., (505) 667-6815 jblacic@lanl.gov, Los Alamos; Nr
Black, Alice (Jill), (417) 836-5300 ablack@missouristate.edu, Missouri State Univ; ZeAsHw
Black, Annjeannette , 443-840-4560 ablack@ccbcmd.edu, Comm Coll of Baltimore County, Catonsville; Zn
Black, Benjamin, 212-650-7027 bblack@ccny.cuny.edu, Graduate Sch of the City Univ of New York; As
Black, David, (631) 632-8676 david.black@stonybrook.edu, SUNY, Stony Brook; Ou
Black, Erin, (585) 275-3440 e.black@rochester.edu, Univ of Rochester; Cbt
Black, Kathryn, (973) 655-4448 Montclair State Univ; Zy
Black, Robert X., (404) 894-1756 rob.black@eas.gatech.edu, Georgia Inst of Tech; As
Black, Ross A., (785) 864-2740 black@ku.edu, Univ of Kansas; Ye
Black, Stuart, s.black@rdg.ac.uk, Univ of Reading; Cc
Black, T. Andrew, (604) 822-2730 Univ of British Columbia; Sp
Blackburn, Terrence, terryb@ucsc.edu, Univ of California, Santa Cruz; Cc
Blackburn, William, (270) 745-8849 will.blackburn@wku.edu, Western Kentucky Univ; Zy
BlackEagle, Cory, cory.blackeagle@eku.edu, Eastern Kentucky Univ; HwZg
Blackmer, Gale C., 717-702-2032 gblackmer@pa.gov, DCNR- Pennsylvania Bur of Geo Survey; Gc
Blackwelder, Patricia L., 305 421-4677 pblackwelder@rsmas.miami.edu, Univ of Miami; Pg
Blackwell, Keith G., 251-460-6302 kblackwell@southalabama.edu, Univ of South Alabama; Am
Bladh, Katherine L., klbladh@wittenberg.edu, Wittenberg Univ; Gi
Bladh, Kenneth W., kbladh@wittenberg.edu, Wittenberg Univ; GzEm
Blair, Neal E., (847) 491-8790 n-blair@northwestern.edu, Northwestern Univ; CoSbOc
Blake, Bascombe (Mitch), (304) 594-2331 blake@geosrv.wvnet.edu, West Virginia Univ; GsEcPb
Blake, Daniel B., (217) 333-3833 dblake@illinois.edu, Univ of Illinois, Urbana-Champaign; Pi
Blake, David E., (910) 962-3387 blaked@uncw.edu, Univ of North Carolina, Wilmington; Gp
Blake, Geoffrey A., (626) 395-6296 gab@gps.caltech.edu,CalTech; Xc
Blake, Norman J., (727) 553-1521 nblake@marine.usf.edu, Univ of South Florida; Ob
Blake, Ruth E., (203) 432-3191 ruth.blake@yale.edu, Yale Univ; Cg
Blake, Stephen, stephen.blake@open.ac.uk, Open Univ; Gv
Blakeney DeJarnett, Beverly, 713-896-6740 bev.dejarnett@beg.utexas.edu, Univ of Texas at Austin, Jackson Sch of Geosciences; GrsGx
Blakey, Ronald C., (928) 523-2740 ronald.blakey@nau.edu, Northern Arizona Univ; Gr
Blamey, Nigel J., 905-688-3365 nblamey@brocku.ca, New Mexico Tech; CaeCl
Blanchard, Edward, (206) 543-5219 ed@atmos.washington.edu, Univ of Washington; As
Blanchard, R. Denise, (512) 245-2170 Texas State Univ; RnmRc
Blanchet, Jean-Pierre, 514-987-3000 #3316 blanchet.jean-pierre@uqam.ca,

Univ du Quebec a Montreal; Pe
Bland, William L., (608) 262-0221 wlbland@wisc.edu, Univ of Wisconsin, Madison; Sp
Blander, Milton, (630) 252-4548 Argonne; Xc
Blaney, Diana L., (818) 354-5419 Jet Propulsion Lab; Xg
Blanford, William , 718-997-3303 william.blanford!qc.cuny.edu, Graduate Sch of the City Univ of New York; Ze
Blanka-Végi, Viktória, +3662544156 blankav@geo.u-szeged.hu, Univesity of Szeged; HsZyRn
Blanken, Peter D., blanken@colorado.edu, Univ of Colorado; HsAsm
Blankenship, Donald D., 512-471-0489 blank@ig.utexas.edu, Univ of Texas, Austin; Gl
Blankenship, Robert, blankship@wustl.edu, Washington Univ in St. Louis; Zn
Blasing, Terence J., (865) 574-7368 blasingtj@ornl.gov, Oak Ridge; As
Blasingame, Tom, (979) 845-2292 t-blasingame@spindletop.tamu.edu, Texas A&M Univ; Np
Blatner , Keith , 509-335-4499 blatner@wsu.edu, Washington State Univ; Zn
Blaylock, Glenn W., 956-764-5715 glenn.blaylock@laredo.edu, Laredo Coll; Ggv
Bleam, William F., (608) 262-9956 wfbleam@wisc.edu, Univ of Wisconsin, Madison; Sc
Bleamaster III, Leslie F., (210) 999-7740 lbleamas@trinity.edu, Trinity Univ; Xg
Blechman, Jerome B., (607) 436-3322 jerome.blechman@oneonta.edu, SUNY, Oneonta; As
Bleck, Rainer, (505) 665-9150 bleck@lanl.gov, Los Alamos; Am
Bleibinhaus, Florian, 089/2180 4202 bleibi@geophysik.uni-muenchen.de, Ludwig-Maximilians-Univ Muenchen; Yg
Blenkinson, Stephen, +44 (0)191 208 7933 stephen.blenkinsop@ncl.ac.uk, Univ of Newcastle Upon Tyne; Zn
Blenkinsop, John, jblenkin@ccs.carleton.ca, Carleton Univ; Cc
Blersch, Stacey S., 706-507-8068 blersch_stacey@columbusstate.edu, Columbus State Univ; HsgRw
Blewitt, Geoffrey, (775) 682-8778 gblewitt@unr.edu, Univ of Nevada, Reno; Yd
Blisniuk, Kimberly, 408 924-5045 kimberly.blisniuk@sjsu.edu, San Jose State Univ; Gmt
Bloch, Jonathan, jbloch@flmnh.ufl.edu, Univ of Florida; Pv
Block, Amy, (612) 626-4434 rada0042@d.umn.edu, Univ of Minnesota; Gp
Block, Karin, 212-650-8543 kblock@ccny.cuny.edu, Graduate Sch of the City Univ of New York; Ze
Blodgett, Robert H., (512) 964-3351 rblodget@austincc.edu, Austin Comm Coll District; RnGsZe
Bloechle, Amber, (850) 857-6121 abloechle@uwf.edu, Univ of West Florida; Zi
Blom, Ronald G., (818) 354-4681 Jet Propulsion Lab; Zr
Blomberg, Brittany N., brittany.blomberg@austincc.edu, Austin Comm Coll District; OnbRn
Blome, Margaret, 252 328 6360 blome19@ecu.edu, East Carolina Univ; Ggn
Bloom, Paul R., (612) 625-4711 pbloom@soils.umn.edu, Univ of Minnesota, Twin Cities; Sc
Bloomer, Sherman H., (541) 737-4811 sherman.bloomer@oregonstate.edu, Oregon State Univ; Gi
Bloomfield, Helen, +440151 795 4652 h.j.bloomfield@liverpool.ac.uk, Univ of Liverpool; Ob
Bloor, Michelle, +44 023 92 842295 michelle.bloor@port.ac.uk, Univ of Portsmouth; Hg
Bloxham, Jeremy, (617) 495-9517 jeremy_bloxham@harvard.edu, Harvard Univ; Yg
Bloxson, Julie M., 936-468-3701 bloxsonjm@sfasu.edu, Stephen F. Austin State Univ; Go
Blum, Joel D., (734) 764-1435 jdblum@umich.edu, Univ of Michigan; CsaCl
Blum, Linda K., (434) 924-0560 lkb2e@virginia.edu, Univ of Virginia; Sb
Blum, Michael D., 785-864-4974 Univ of Kansas; GsrGm
Blumenfeld, Raphael, r.blumenfeld@imperial.ac.uk, Imperial Coll; Zn
Blumentritt, Dylan, 507.457.5234 dblumentritt@winona.edu, Winona State Univ; HsGm
Blunt, Martin J., 442075946500 m.blunt@imperial.ac.uk, Imperial Coll; NpHw
Blusztajn, Jurek, (508) 289-2692 jblusztajn@whoi.edu, Woods Hole Oceanographic Inst; CcGi
Blyth, Alan, +44(0) 113 34 31632 a.m.blyth@leeds.ac.uk, Univ of Leeds; As
Blythe, Ann, 323 259-2553 ablythe@oxy.edu, Occidental Coll; Cc
Blythe, Daniel D., 406.496.4379 dblythe@mtech.edu, Montana Tech of The Univ of Montana; Hws
Boag, Thomas, 609-258-4101 tb9285@princeton.edu, Princeton Univ; As
Boaga, Jacopo, jacopo.boaga@unipd.it, Univ degli Studi di Padova; Yg
Boar, Rosalind, +44 (0)1603 59 3103 r.boar@uea.ac.uk, Univ of East Anglia; Ob
Boardman, Mark R., (513) 529-3230 boardman@miamioh.edu, Miami Univ; On
Boast, Charles W., (217) 333-1278 Univ of Illinois, Urbana-Champaign; Sp
Boboye, O. A., oa.boboye@mail.ui.edu.ng, Univ of Ibadan; Go
Bobst, Andrew L., 406.496.4409 abobst@mtech.edu, Montana Tech of The Univ of Montana; HwCgEo
Bobyarchick, Andy R., 704-687-5998 arbobyar@email.uncc.edu, Univ of North Carolina, Charlotte; Gc
Bochdansky, Alexander, 757-6834933 abochdan@odu.edu, Old Dominion Univ; Ob
Bockheim, James G., (608) 263-5903 bockheim@wisc.edu, Univ of Wisconsin, Madison; Sf
Boden, Thomas, (865) 241-4842 tab@ornl.gov, Oak Ridge; As
Bodenbender, Brian E., (616) 395-7541 bodenbender@hope.edu, Hope Coll; GsPiGe
Bodenheimer, Peter, (831) 459-2064 peter@ucolick.org, Univ of California, Santa Cruz; Xc
Bodenman, John E., (570) 389-4697 boden@bloomu.edu, Bloomsburg Univ; Zu
Bodin, Paul A., 206-616-7315 bodin@uw.edu, Univ of Washington; Ys
Bodnar, Robert J., (540) 231-7455 rjb@vt.edu, Virginia Tech; CgEmGv
Boehm, David A., dirtboy@rochester.rr.com, SUNY, The Coll at Brockport; Gg
Boelman, Natalie, nboelman@ldeo.columbia.edu, Columbia Univ; Zn
Boering, Kristie, (510) 642-3472 boering@cchem.berkeley.edu, Univ of California, Berkeley; As
Boester, Michael, mboester@monroecc.edu, Monroe Comm Coll; Zy
Bogart, James M., james.bogart@ct.gov, Dept of Energy and Env Protection; Ge
Boger, Rebecca, 718-951-5000 x2159 rboger@brooklyn.cuny.edu, Brooklyn Coll (CUNY); ZiHs
Boggs, Carol, boggscl@mailbox.sc.edu, Univ of South Carolina; Zn
Boggs, Katherine, (403) 440-6645 kboggs@mtroyal.ca, Mount Royal Univ; Gt
Bogle, Mary Anna, (865) 574-7824 amb@ornl.gov, Oak Ridge; Ge
Bogner, Jean E., jbogner@uic.edu, Univ of Illinois Chicago; ClSbHy
Bograd, Michael B., (601) 961-5528 mbograd@mdeq.ms.gov, Mississippi Office of Geology; Gg
Bogucki, Darek, 361-825-825-2836 darek.bogucki@tamucc.edu, Texas A&M Univ, Corpus Christi; Op
Bogue, Scott W., (323) 259-2563 bogue@oxy.edu, Occidental Coll; Ym
Bohaty, Steven, +44 (0)23 8059 3040 s.bohaty@noc.soton.ac.uk, Univ of Southampton; Og
Bohlen, Thomas, +49-721-6084416 thomas.bohlen@kit.edu, Karlsruhe Inst of Tech; Yx
Bohlen, Walter F., 860-405-9176 walter.bohlen@uconn.edu, Univ of Connecticut; Op
Bohling, Geoff, 785-864-2093 geoff@kgs.ku.edu, Univ of Kansas; Hw
Bohling, Geoffrey C., (785) 864-2093 geoff@kgs.ku.edu, Univ of Kansas; Hw
Bohm, Christian, 204-945-6549 christian.bohm@gov.mb.ca, Univ of Manitoba; Cg
Bohnenstiehl, DelWayne, (919) 515-7449 drbohnen@ncsu.edu, North Carolina State Univ; YrsGt
Bohren, Craig F., (814) 865-2951 bohren@ems.psu.edu, Pennsylvania State Univ, Univ Park; Am
Bohrson, Wendy, (303) 273-3066 bohrson@mines.edu, Colorado Sch of Mines; Gv
Boicourt, William, (410) 221-8426 boicourt@hpl.umces.edu, Univ of Maryland; Op
Boisvert, Eric, (418) 654-3705 eboisvert@nrcan.gc.ca, Univ du Quebec; Zn
Boitt, Mark, mboitt@jkuat.ac.ke, Jomo Kenyatta Univ of Agriculture & Tech; ZgrZi
Boix, Carme, carme.boix@uab.cat, Univ Auto de Barcelona; Pm
Bokelmann, Goetz H., goetz.bokelmann@univie.ac.at, Univ of Vienna; Ygs
Bokuniewicz, Henry J., (631) 632-8674 henry.bokuniewicz@stonybrook.edu, SUNY, Stony Brook; Yr
Boland, Greg J., (519) 824-4120 x52755 gboland@uoguelph.ca, Univ of Guelph; Zn
Boland, Irene B., 803-323-4949 bolandi@winthrop.edu, Winthrop Univ; GtgZe
Boland, John J., (410) 516-7103 jboland@jhu.edu, Johns Hopkins Univ; Zn
Bolarinwa, A. T., at.bolarinwa@mail.ui.edu.ng, Univ of Ibadan; EgGz
Bolge, Louise, bolge@ldeo.columbia.edu, Columbia Univ; Cg
Bollag, Jean-Marc, (814) 863-0843 jmbollag@psu.edu, Pennsylvania State Univ, Univ Park; Sb

Bollasina, Massimo A., +44 (0) 131 651 3464 massimo.bollasina@ed.ac.uk, Edinburgh Univ; Asp
Bollens, Stephen, 360-546-9116 sbollens@vancouver.wsu.edu, Washington State Univ; Ob
Bollinger, G. A., (540) 231-6521 Virginia Tech; Ys
Bollinger, Marsha S., 803-323-4944 bollingerm@winthrop.edu, Winthrop Univ; CmOcCc
Bollmann, Jörg, (416) 978-2061 bollmann@geology.utoronto.ca, Univ of Toronto; Pm
Bolster, Diogo, diogo.bolster.5@nd.edu, Univ of Notre Dame; Hw
Bolt, John R., (312) 665-7629 jbolt@fieldmuseum.org, Field Mus of Natural History; Pv
Bolton, Edward W., (203) 432-3149 edward.bolton@yale.edu, Yale Univ; GqHqCq
Bomfleur, Benjamin, +49-251-83-23966 bbomfleur@uni-muenster.de, Univ Muenster; PblGg
Bomke, Arthur A., (604) 822-6534 Univ of British Columbia; Sc
Bona, Andrej, +61 8 9266 7194 a.bona@curtin.edu.au, Curtin Univ; Yx
Bonamici, Chloe, bonamici@wisc.edu, Univ of Wisconsin, Madison; GxCa
Bonatti, Enrico, bonatti@ldeo.columbia.edu, Columbia Univ; Yr
Bonczek, James, 352-294-3112 bonczek@ufl.edu, Univ of Florida; SoHsg
Bond, Alan, +44 (0)1603 59 3402 alan.bond@uea.ac.uk, Univ of East Anglia; Ge
Bond, Clare, +44 (0)1224 273492 clare.bond@abdn.ac.uk, Univ of Aberdeen; Gct
Bond, Nicholas A., (206) 526-6459 nicholas.bond@noaa.gov, Univ of Washington; As
Bonde, Aubrey, (702) 895-1753 aubrey.bonde@unlv.edu, Univ of Nevada, Las Vegas; Pg
Bone, Fletcher, (573) 368-2183 fletcher.bone@dnr.mo.gov, Missouri Dept of Nat Res; Gg
Bonem, Rena M., (254) 710-2187 rena_bonem@baylor.edu, Baylor Univ; PieOb
Bonetto, Sabrina Maria Rita, sabrina.bonetto@unito.it, Univ di Torino; Ng
Bonhoure, Jessica, +33(0)3 44068994 jessica.bonhoure@lasalle-beauvais.fr, Institut Polytechnique LaSalle Beauvais (ex-IGAL); GziCc
Boniface, Nelson, +255222410013 nelson.boniface@udsm.ac.tz, Univ of Dar es Salaam; GcpGz
Bonine, Michael E., kebonine@u.arizona.edu, Univ of Arizona; Zu
Bonk, Kathleen, 518-473-9988 kbonk@mail.nysed.gov, New York State Geo Survey; GoCg
Bonuso, Nicole, (657) 278-8451 nbonuso@fullerton.edu, California State Univ, Fullerton; Pie
Booher, Gary, (310) 660-3593 gbooher@elcamino.edu, El Camino Coll; Zg
Booker, John, 206-543-0489 jrbooker@uw.edu, Univ of Washington; YmGct
Boone, Peter A., pboone@austincc.edu, Austin Comm Coll District; Gro
Boorstein, Margaret F., (516) 299-3058 maboorst@liu.edu, Long Island Univ, C.W. Post Campus; Zg
Booth, Adam, +44 20 759 46528 a.booth@imperial.ac.uk, Imperial Coll; Gl
Booth, Alastair, murray.booth@manchester.ac.uk, Univ of Manchester; As
Booth, Derek, 206-914-5031 dbooth@uw.edu, Univ of Washington; Gm
Booth, James, 212-650-6471 jbooth@ccny.cuny.edu, Graduate Sch of the City Univ of New York; Gge
Booth, Robert K., rkb205@lehigh.edu, Lehigh Univ; ZnnPc
Boothroyd, Jon C., (401) 874-2265 jon_boothroyd@uri.edu, Univ of Rhode Island; GslOn
Borazjani, Sara, sara.borazjani@adelaide.edu.au, Univ of Adelaide; Np
Bord, Don, (313) 593-5483 dbord@umich.edu, Univ of Michigan, Dearborn; Xy
Bordeaux, Yvette, (215) 898-9191 bordeaux@sas.upenn.edu, Univ of Pennsylvania; Pg
Bordeleau, Geneviève, 418 654-2543 genevieve.bordeleau@ete.inrs.ca, Univ du Quebec; Hw
Bordoni, Simona, 626.395.2672 bordoni@gps.caltech.edu,CalTech; Op
Bordossy, Andras, +44 (0) 191 208 6319 andras.bardossy@ncl.ac.uk, Univ of Newcastle Upon Tyne; Zg
Bordy, Emese M., 021-650-2901 emese.bordy@uct.ac.za, Univ of Cape Town; GsPes
Borg, Scott G., 703-292-8500 sborg@nsf.gov, Natl Sci Foundation; Git
Borghi, Alessandro, alessandro.borghi@unito.it, Univ di Torino; Gx
Bork, Kennard B., (928) 554-4942 bork@denison.edu, Denison Univ; PiRhGs
Bornhorst, Theodore J., 906-487-2721 tjb@mtu.edu, Michigan Technological Univ; EgCgGx
Borojeviæ Šoštariæ, Sibila, +38514605961 sborojsost@geol.pmf.hr, Univ of Zagreb; CegCc
Boroughs, Scott, geoentoptics@wsu.edu, Washington State Univ; Gv
Boroughs, Terry J., (916) 331-8596 borougt@arc.losrios.edu, American River Coll; GgXgCg
Boroushaki, Soheil, 818-677-4715 soheil.boroushaki@csun.edu, California State Univ, Northridge; Zi
Borowski, Walter S., (859) 622-1277 w.borowski@eku.edu, Eastern Kentucky Univ; GsrOu
Borrelli, Chiara, (585) 275-7884 cborrelli@ur.rochester.edu, Univ of Rochester; PmcCb
Borrok, David M., 573-341-6784 borrokd@mst.edu, Missouri Univ of Sci and Tech; ClqCs
Borsa, Adrian, (858) 534-6895 aborsa@ucsd.edu, Univ of California, San Diego; Hg
Borthakur, Sanchayeeta, (480) 965-3171 sanchayeeta.borthakur@asu.edu, Arizona State Univ; Xa
Bosak, Tanja, 617-324-3959 tbosak@mit.edu, Massachusetts Inst of Tech; Pg
Bosart, Lance F., (518) 442-4564 bosart@atmos.albany.edu, SUNY, Albany; As
Bosbyshell, Howell, (610) 436-2805 hbosbyshell@wcupa.edu, West Chester Univ; GcpGt
Boschmann, E. Eric, 303-871-4387 eric.boschmann@du.edu, Univ of Denver; Zc
Bose, Maitrayee, (480) 965-4244 maitrayee.bose@asu.edu, Arizona State Univ; Xc
Bosman, Martin, bosman@usf.edu, Univ of South Florida; Zn
Boss, Alan, aboss@carnegiescience.edu, Carnegie Inst for Sci; Xa
Boss, Stephen K., (479) 575-7134 sboss@uark.edu, Univ of Arkansas, Fayetteville; RcYgGg
Bossenbroek, Jonathon , 419-530-4595 jonathon.bossenbroek@utoledo.edu, Univ of Toledo; Zn
Bossert, James E., (505) 667-6268 bossert@lanl.gov, Los Alamos; As
Bossert, Katrina, katrina.bossert@asu.edu, Arizona State Univ; Xp
Bostater, Charles, (321) 674-8096 bostater@fit.edu, Florida Inst of Tech; Op
Bostick, Benjamin C., bostick@ldeo.columbia.edu, Columbia Univ; Sc
Bostock, Michael G., (604) 822-2082 bostock@eos.ubc.ca, Univ of British Columbia; YsGt
Boston, Brian B., bbb2138@columbia.edu, Columbia Univ; Gu
Boston, Penelope, 604 224 1141 penelope.j.boston@nasa.gov, Natl Aeronautics & Space Administration; Zg
Boswell, Cecil, 573-368-2146 cecil.boswell@dnr.mo.gov, Missouri Dept of Nat Res; Zn
Bothner, Wallace A., (603) 862-1718 wally.bothner@unh.edu, Univ of New Hampshire; Gc
Bottenberg, H. Carrie, 208-282-3538 bottcarr@isu.edu, Idaho State Univ; ZiGtc
Bottjer, David J., (213) 740-6100 dbottjer@usc.edu, Univ of Southern California; Pe
Bou-Zeid, Elie, (609) 258-5429 ebouzeid@princeton.edu, Princeton Univ; As
Boudreau, Alan E., (919) 684-5646 boudreau@duke.edu, Duke Univ; Gi
Boudreau, Bernard P., (902) 494-8895 bernie.boudreau@dal.ca, Dalhousie Univ; OcCml
Boudrias, Michel A., (619) 260-4794 boum@sandiego.edu, Univ of San Diego; Ob
Boujibar, Asmaa, (360) 650-2439 boujiba@wwu.edu, Western Washington Univ; CpXcy
Bouker, Polly A., (770) 278-1320 pbouker@gsu.edu, Georgia State Univ, Perimeter Coll, Newton Campus; Gg
Boult, Stephen, +44 0161 275-3867 s.boult@manchester.ac.uk, Univ of Manchester; Hg
Boulton, Sarah, +44 1752 584762 sarah.boulton@plymouth.ac.uk, Univ of Plymouth; Zr
Boulvain, Frederic P., 32 4 366 22 52 fboulvain@ulg.ac.be, Univ de Liege; GdgGs
Bouman, Heather, +44 (1865) 272019 heather.bouman@earth.ox.ac.uk, Univ of Oxford; Cg
Bounk, Michael, michael.bounk@dnr.iowa.gov, Iowa Dept of Nat Res; Gg
Bour, William, (703) 450-2612 wbour@nvcc.edu, Northern Virginia Comm Coll - Loudoun Campus; Gge
Bourassa, Matthew, 519-661-2111 ext 80370 mbouras@uwo.ca, Western Univ; Xg
Bourbonnais, Annie, (803) 777-7042 abourbonnais@seoe.sc.edu, Univ of South Carolina; Oc
Bourbonnais, Mathieu L., 250-807-8186 mathieu.bourbonnais@ubc.ca, Univ of British Columbia, Okanagan Campus; ZifRn
Bourcier, Bill L., (925) 423-3745 bourcier1@llnl.gov, Lawrence Livermore; Cg
Bourgault, Rebecca, rbourgault@bloomu.edu, Bloomsburg Univ; SdfSb
Bourgeois, Joanne (Jody), (206) 685-2443 jbourgeo@uw.edu, Univ of Washington; GsrRh
Bourgeois, Reed J., (225) 388-8879 reed@lgs.bri.lsu.edu, Louisiana State Univ; Zn

Bourke, Robert H., (831) 656-2673 rbourke@nps.edu, Naval Postgrad Sch; Op
Bouse, Robin, (310) 660-3593 rbouse@elcamino.edu, El Camino Coll; Gg
Bousenberry, Raymond T., (609) 984-6587 raymond.bousenberry@dep.nj.gov, New Jersey Geological and Water Survey; Ge
Boutt, David, (413) 545-2724 dboutt@geo.umass.edu, Univ of Massachusetts, Amherst; HwqHy
Bouvier, Audrey, 519-661-2111 x.88516 audrey.bouvier@uwo.ca, Western Univ; Xc
Bouwer, Edward, (410) 516-7437 bouwer@jhu.edu, Johns Hopkins Univ; RwHws
Bouzari, Farhad, (604) 822-1874 fbouzari@eos.ubc.ca, Univ of British Columbia; Eg
Boving, Thomas, (401) 874-7053 boving@uri.edu, Univ of Rhode Island; Hg
Bowden, Stephen, +44 (0)1224 273467 s.a.bowden@abdn.ac.uk, Univ of Aberdeen; CoGoCb
Bowen, Anthony J., (902) 494-7082 tony.bowen@dal.ca, Dalhousie Univ; On
Bowen, Brenda , (801) 585-5326 brenda.bowen@utah.edu, Univ of Utah; Gse
Bowen, David W., dbowen@montana.edu, Montana State Univ; GroGs
Bowen, Dawn S., (540) 654-1491 dbowen@mwc.edu, Mary Washington Coll; Zn
Bowen, Esther E., (630) 252-7553 ebowen@anl.gov, Argonne; Gg
Bowen, Gabriel, (801) 585-7925 gabe.bowen@utah.edu, Univ of Utah; Cs
Bowen, Jennifer, 7815817370 x346 je.bowen@northeastern.edu, Northeastern Univ; Zn
Bowen, Robert, bob.bowen@umb.edu, Univ of Massachusetts, Boston; ZnnZn
Bowen, Scott M., (505) 667-4313 Los Alamos; Ca
Bowersox, Joe, 503-370-6220 jbowerso@willamette.edu, Willamette Univ; Zn
Bowersox, John R., (859) 257-1147 j.r.bowersox@uky.edu, Univ of Kentucky; God
Bowin, Carl O., (508) 289-2572 cbowin@whoi.edu, Woods Hole Oceanographic Inst; YrGtg
Bowler, Bernard, +44 (0) 191 208 5931 bernard.bowler@ncl.ac.uk, Univ of Newcastle Upon Tyne; Go
Bowles, Julie, 414-229-6110 bowlesj@uwm.edu, Univ of Wisconsin, Milwaukee; Ym
Bowles, Zack, zbowles@mesacc.edu, Mesa Comm Coll; Gg
Bowlick, Forrest J., (413)-577-3816 fbowlick@umass.edu, Univ of Massachusetts, Amherst; ZieZg
Bowman, Cassandra, c.bowman@asu.edu, Arizona State Univ; Ze
Bowman, Jean A., (409) 862-6544 jbowman@tamu.edu, Texas A&M Univ; Hg
Bowman, John R., (801) 581-7250 john.bowman@utah.edu, Univ of Utah; Cs
Bowman, Judd, (480) 965-8880 judd.bowman@asu.edu, Arizona State Univ; Xa
Bowman, Kenneth P., (979) 845-4060 k-bowman@tamu.edu, Texas A&M Univ; Am
Bowman, Malcolm J., (631) 632-8669 malcolm.bowman@stonybrook.edu, SUNY, Stony Brook; Op
Bowman, Mike, michael.bowman@manchester.ac.uk, Univ of Manchester; Go
Bowman, Steve D., (801) 537-3304 stevebowman@utah.gov, Utah Geo Survey; NgGe
Bown, Paul, +44 020 7679 32431 p.bown@ucl.ac.uk, Univ Coll London; Pm
Bowser, Carl J., (608) 262-8955 bowser@geology.wisc.edu, Univ of Wisconsin, Madison; Cl
Bowser, Paul, prb4@cornell.edu, SUNY, Stony Brook; Ob
Boxall, Simon R., +44 (0)23 80592744 simon.boxall@soton.ac.uk, Univ of Southampton; Op
Boybeyi, Zafer, (703) 993-1560 zboybeyi@gmu.edu, George Mason Univ; As
Boyce, Joseph I., (905) 525-9140 (Ext. 24188) boycej@mcmaster.ca, McMaster Univ; Zg
Boyd, Bill, william.boyd@scu.edu.au, Southern Cross Univ; Pe
Boyd, John, (734) 764-3338 jpboyd@umich.edu, Univ of Michigan; OpAs
Boyle, Alan P., +44 0151 794 5154 apboyle@liverpool.ac.uk, Univ of Liverpool; GzEmZe
Boyle, Douglas, 775-784-6995 geography@unr.edu, Univ of Nevada, Reno; Hg
Boyle, Edward A., (617) 253-3388 eaboyle@mit.edu, Massachusetts Inst of Tech; Oc
Boyle, James, (716) 645-4293 jamesboy@buffalo.edu, SUNY, Buffalo; Pq
Boynton, William V., (520) 621-6941 wboynton@lpl.arizona.edu, Univ of Arizona; XcZr
Boysen, Hans, 089/2180 4333 boysen@lmu.de, Ludwig-Maximilians-Univ Muenchen; Gz
Bozdag, Ebru, 303-273-3578 bozdag@mines.edu, Colorado Sch of Mines; Ys
Bozdog, Nick, 828-296-4635 nick.bozdog@ncdenr.gov , N.C. Dept of Env Quality; Gg
Brabander, Daniel J., 781-283-3056 dbraband@wellesley.edu, Wellesley Coll; Cl
Brabham, P J., brabham@cf.ac.uk, Cardiff Univ; Ye
Brabham, Peter, +44(0)29 208 74334 brabham@cardiff.ac.uk, Univ of Wales; Ng
Bracco, Annalisa, (404) 894-1749 annalisa@eas.gatech.edu, Georgia Inst of Tech; OpZnn
Bradbury, Kelly K., kellykbradbury@gmail.com, Utah State Univ; Gc
Bradbury, Kenneth R., (608) 263-7921 ken.bradbury@uwex.edu, Univ of Wisconsin, Madison; Hw
Bradford, Joel, (801) 863-7263 bradfojo@uvu.edu, Utah Valley Univ; RwGe
Bradford, John, 303-273-3938 jbradford@mines.edu, Colorado Sch of Mines; Yue
Bradley, Alexander S., (314) 935-6333 abradley@levee.wustl.edu, Washington Univ in St. Louis; CboGg
Bradley, Alice, (413) 597-4649 alice.c.bradley@williams.edu, Williams Coll; AtOn
Bradley, Chris, +44 (0)121 414 8097 c.bradley@bham.ac.uk, Univ of Birmingham; Hg
Bradley, Christopher R., (505) 665-6713 cbradley@lanl.gov, Los Alamos; Yg
Bradley, Jeffrey, (660) 562-1818 jbradle@nwmissouri.edu, Northwest Missouri State Univ; Zg
Bradley, Michael, (313) 487-0218 Eastern Michigan Univ; Gc
Bradley, Philip, 919-707-9241 pbradley@ncdenr.gov, N.C. Dept of Env Quality; Gcg
Bradley, Raymond S., (413) 545-2120 rbradley@geo.umass.edu, Univ of Massachusetts, Amherst; As
Bradshaw, John, (03) 3667-001 x7487 john.bradshaw@canterbury.ac.nz, Univ of Canterbury; Gt
Bradshaw, Lanna K., 972/860-4713 lannabradshaw@dcccd.edu, Brookhaven Coll; GgZg
Bradshaw, Peter M., (604) 681-8600 pbradshaw@firstpointminerals.com, Univ of British Columbia; EgCeGe
Brady, John B., (413) 585-3953 jbrady@smith.edu, Smith Coll; GxzZe
Brady, Kristina, 612-626-7889 brad0311@umn.edu, Univ of Minnesota, Twin Cities; Gn
Brady, Roland H., (559) 278-2391 roland_brady@csufresno.edu, Fresno State Univ; NgZuGa
Braginsky, Stanislav I., (310) 794-1331 jbragin@ucla.edu, Univ of California, Los Angeles; Ym
Brahana, John V., 479-575-3355 brahana@uark.edu, Univ of Arkansas, Fayetteville; HwGe
Brahmi, Sadek, +33(0)3 44068978 sadek.brahmi@lasalle-beauvais.fr, Institut Polytechnique LaSalle Beauvais (ex-IGAL); Nr
Braid, James, jbraid@stfx.ca, Saint Francis Xavier Univ; Gc
Braile, Lawrence W., (765) 494-5979 braile@purdue.edu, Purdue Univ; Ys
Brainard, James R., (505) 667-0150 Los Alamos; Zn
Brake, Sandra S., (812) 237-2270 sandra.brake@indstate.edu, Indiana State Univ; GeEm
Brake, Thomas L., (702) 794-7828 brake_thomas_l@lanl.gov, Los Alamos; Nr
Bralower, Timothy J., 814-863-1240 bralower@psu.edu, Pennsylvania State Univ, Univ Park; Pe
Brame, Scott E., (864) 656-7167 scott@clemson.edu, Clemson Univ; Hw
Bramham, Emma , +44(0) 113 34 35595 e.k.bramham@leeds.ac.uk, Univ of Leeds; Ym
Bramson, Ali, 765-494-0279 bramsona@purdue.edu, Purdue Univ; Xg
Branam, Tracy D., (812) 855-8390 tbranam@indiana.edu, Indiana Univ, Bloomington; CasEm
Branciforte, Chloe, cbranciforte@mesacc.edu, Mesa Comm Coll; Gg
Branco, Brett, 718-951-5000, ext. 6441 bbranco@brooklyn.cuny.edu, Graduate Sch of the City Univ of New York; Gg
Brand, Brittany, brittanybrand@boisestate.edu, Boise State Univ; GsvRn
Brand, Leonard R., (909) 558-4530 lbrand@llu.edu, Loma Linda Univ; Pv
Brand, Uwe, (905) 688-5550 x3529 uwe.brand@brocku.ca, Brock Univ; ClsAs
Brandes, Jay, (912) 598-2361 jay.brandes@skio.usg.edu, Georgia Inst of Tech; Og
Brandon, Alan, 713-743-2359 abrandon@uh.edu, Univ of Houston; GiCac
Brandon, Christine M., (508) 531-2125 christine.brandon@bridgew.edu, Bridgewater State Univ; Gs
Brandon, Mark A., 01908 655172 mark.brandon@open.ac.uk, Open Univ; OgGlOp

Brandon, Nigel, +44 20 759 45704 n.brandon@imperial.ac.uk, Imperial Coll; Ge
Brandriss, Mark E., mbrandri@smith.edu, Williams Coll; Gi
Brandt, Craig, (865) 574-1921 fcb@ornl.gov, Oak Ridge; Ge
Brandt, Danita S., (517) 355-6595 brandt@msu.edu, Michigan State Univ; PiGr
Brandvold, Lynn A., 575-835-5517 lynn.brandvold@nmt.edu, New Mexico Tech; Ca
Branfireun, Brian, bbranfir@uwo.ca, Western Univ; Ge
Branlund, Joy, (618) 797-7451 joy.branlund@swic.edu, Southwestern Illinois Coll - Sam Wolf Granite City Campus; Zg
Branney, Mike, +440116 252 3647 mjb26@le.ac.uk, Leicester Univ; Gv
Branstrator, Jon W., (765) 983-1339 jonb@earlham.edu, Earlham Coll; Pg
Brant, Lynn A., (319) 273-6160 lynn.brant@uni.edu, Univ of Northern Iowa; Gn
Brantley, Susan L., (814) 863-1739 brantley@eesi.psu.edu, Pennsylvania State Univ, Univ Park; Cg
Brantut, Nicolas D., n.brantut@ucl.ac.uk, Univ Coll London; Nr
Brânzila, Mihai, 004023201495 mib@uaic.ro, Alexandru Ioan Cuza; GgrPg
Brasier, Martin, +44 (1865) 272074 martin.brasier@earth.ox.ac.uk, Univ of Oxford; Pg
Brassea, Jesus M., jbrassea@cicese.mx, Centro de Investigación Científica y de Educación Superior de Ensenada; Ye
Brassell, Simon C., (812) 855-3786 simon@indiana.edu, Indiana Univ, Bloomington; Cob
Brassington, Rick, +44 (0) 192 576 6754 rick.brassington@ncl.ac.uk, Univ of Newcastle Upon Tyne; Hg
Bratman, Eve, eve.bratman@fandm.edu, Franklin and Marshall Coll; Zn
Braught, Patricia, (717) 245-1355 braught@dickinson.edu, Dickinson Coll; Zn
Braun, Alexander, (613) 533-6621 braun@queensu.ca, Queen's Univ; YgdYv
Braund, Emilie, +44 023 9284 2257 emilie.bruand@port.ac.uk, Univ of Portsmouth; Gz
Braunmiller, Jochen, jbraunmiller@usf.edu, Univ of South Florida; Ys
Bray, Colin, (416) 978-6516 cjbray@quartz.geology.utoronto.ca, Univ of Toronto; Cs
Breaker, Laurence, 831-771-4400 lbreaker@mlml.calstate.edu, Moss Landing Marine Lab; Op
Brearley, Adrian J., (505) 277-4163 brearley@unm.edu, Univ of New Mexico; Gz
Breckenridge, Andy, abrecken@uwsuper.edu, Univ of Wisconsin, Superior; GlnGs
Breda, Anna, anna.breda@unipd.it, Univ degli Studi di Padova; GsrGg
Breecker, Daniel O., 512-471-6166 breecker@jsg.utexas.edu, Univ of Texas, Austin; ClsSc
Brehm, Sarah, (315) 229-5142 sbrehm@stlawu.edu, St. Lawrence Univ; Gi
Breier, John, 956 882-6625 john.breier@utrgv.edu, Univ of Texas, Rio Grande Valley; Cma
Breilin, Olli, +358 (0)29 503 5200 olli.breilin@gtk.fi, Geo Survey of Finland; GmlGs
Breitbart, Mya, (727) 553-3520 mya@marine.usf.edu, Univ of South Florida; Ob
Breiter, Jack, breiter3@uic.edu, Univ of Illinois Chicago; Sf
Breitmeyer, Ronald, 775.682.6049 rbreitmeyer@unr.edu, Univ of Nevada, Reno; HwGe
Breland, Clayton F., (225) 388-8300 clayton@lgs.bri.lsu.edu, Louisiana State Univ; Ye
Bremer, Keith A., (785) 628-4644 kabremer@fhsu.edu, Fort Hays State Univ; ZuiZn
Bremmer, Sarah, (608) 265-5323 sarah.bremmer@wisc.edu, Univ of Wisconsin, Madison, Division of Extension; Ggt
Brenan, James M., 902-494-2355 jbrenan@dal.ca, Dalhousie Univ; CptEm
Brengman, Latisha A., (218) 726-7586 lbrengma@d.umn.edu, Univ of Minnesota Duluth; Gd
Brenner, Mark, (352) 392-9617 brenner@ufl.edu, Univ of Florida; Gn
Brenner, Robert, robert-brenner@uiowa.edu, Univ of Iowa; Gds
Bretherton, Christopher S., (206) 685-7414 breth@washington.edu, Univ of Washington; As
Brethes, Jean-Claude, (418) 724-1779 cjmichau@globetrotter.net, Univ du Quebec a Rimouski; Ob
Breton, Caroline, 512-471-0322 cari.breton@beg.utexas.edu, Univ of Texas at Austin, Jackson Sch of Geosciences; Zi
Brett, Carlton E., (513) 556-4556 carlton.brett@uc.edu, Univ of Cincinnati; Pe
Brewer, Margene, (616) 526-8415 mkb25@calvin.edu, Calvin Coll; ZnGgZn
Brewer, Paul A., +44 (0)1970 622 586 pqb@aber.ac.uk, Univ of Wales; Gm
Breyer, John, j.breyer@tcu.edu, Texas Christian Univ; GsdEo
Brezinski, David K., (410) 554-5526 david.brezinski@maryland.gov, Maryland Dept of Nat Res; Gr
Brian, Lejeune, 401-441-8699 lejeuneb@lafayette.edu, Lafayette Coll; GzXmZm
Bridgeman, Thomas, 419-530-8373 thomas.bridgeman@utoledo.edu, Univ of Toledo; Zn
Bridges, David L., (573) 368-2656 david.bridges@dnr.mo.gov, Missouri Dept of Nat Res; GgiPg
Bridges, Frank, (831) 459-2893 bridges@cats.ucsc.edu, Univ of California, Santa Cruz; Yx
Bries Korpik, Jill, (651) 450-3726 jkorpik@inverhills.mnscu.edu, Inver Hills Comm Coll; Gg
Bries-Korpik, Jill, (651) 779-3434 jill.brieskorpik@century.edu, Century Coll; Zg
Briggs, Derek E., (203) 432-8590 derek.briggs@yale.edu, Yale Univ; Yg
Briggs, John, +44 0141 330 8744 john.briggs@glasgow.ac.uk, Univ of Glasgow; Zu
Briggs, Tiffany Roberts M., (561) 297-4669 briggst@fau.edu, Florida Atlantic Univ; On
Brigham-Grette, Julie, (413) 577-2270 juliebg@geo.umass.edu, Univ of Massachusetts, Amherst; GlPcGu
Brikowski, Tom H., (972) 883-6242 brikowi@utdallas.edu, Univ of Texas, Dallas; Hw
Brill, Jr., Richard C., (808) 845-9488 brill@hawaii.edu, Honolulu Comm Coll; Gg
Brimhall, George H., 510.642.5868 brimhall@eps.berkeley.edu, Univ of California, Berkeley; Eg
Briner, Jason P., 716-645-4326 jbriner@buffalo.edu, SUNY, Buffalo; Gl
Brink, Alyson, 601-266-4729 alyson.brink@usm.edu, Univ of Southern Mississippi; PgvGg
Brinkmann, Robert, (541) 967-2039 Oregon Dept of Geology and Mineral Ind; HwGoEm
Brisbin, William C., (204) 474-9454 w_brisbin@umanitoba.ca, Univ of Manitoba; Gc
Briskin, Madeleine, madeleine.briskin@uc.edu, Univ of Cincinnati; Pe
Bristow, Charlie, +44 020 3073 8025 c.bristow@ucl.ac.uk, Birkbeck Coll; Gs
Britt, Brooks B., 801-422-7316 brooks_britt@byu.edu, Brigham Young Univ; Pg
Britton, Gloria C., (216) 987-5228 gloria.britton@tri-c.edu, Cuyahoga Comm Coll - Western Campus; ZgGmZn
Britzke, Gilbert, 089/2180 4209 gilbert.brietzke@geophysik.uni-muenchen.de, Ludwig-Maximilians-Univ Muenchen; Yg
Broadfoot, Lyle A., (520) 621-4303 broadfoot@vega.lpl.arizona.edu, Univ of Arizona; Zr
Broadhead, Ronald F., 575-835-5202 ron.broadhead@nmt.edu, New Mexico Tech; Go
Broadhead, Thomas W., (865) 974-1151 twbroadhead@utk.edu, Univ of Tennessee, Knoxville; Pi
Brochu, Christopher A., (319) 353-1808 chris-brochu@uiowa.edu, Univ of Iowa; Pv
Brock, Patrick W. G., (718) 997-3328 patrick.brock@qc.cuny.edu, Queens Coll (CUNY); Gg
Brock-Hon, Amy, (423) 425-4409 amy-brock-hon@utc.edu, Univ of Tennessee, Chattanooga; GmSc
Brocklehurst, Simon, +44 0161 275-3037 simon.h.brocklehurst@manchester.ac.uk, Univ of Manchester; Gm
Brod, Joseph D., 813-744-8018 jbrod@hccfl.edu, Hillsborough Comm Coll; Zg
Broda, James E., (508) 289-2466 jbroda@whoi.edu, Woods Hole Oceanographic Inst; Gs
Brodholt, John, +44 020 7679 32622 j.brodholt@ucl.ac.uk, Univ Coll London; Yg
Brodie, Gregory, (423) 425-5915 gregory-brodie@utc.edu, Univ of Tennessee, Chattanooga; Ge
Brodie, Kate, +44 0161 275-3948 kate.brodie@manchester.ac.uk, Univ of Manchester; Gc
Brodsky, Emily, (831) 459-1854 brodsky@ucsc.edu, Univ of California, Santa Cruz; YsgHw
Broecker, Wallace S., (845) 365-8413 broecker@ldeo.columbia.edu, Columbia Univ; CmOcPe
Bromily, Geoffrey, +44 (0) 131 650 8519 geoffrey.bromiley@ed.ac.uk, Edinburgh Univ; Ge
Bromwich, David H., (614) 688-5314 bromwich.1@osu.edu, Ohio State Univ; As
Brook, Edward J., (541) 737-8197 brooke@geo.oregonstate.edu, Oregon State Univ; ClPeGl
Brookfield, Andrea, 785-864-2199 andrea@ku.edu, Univ of Kansas; Hw
Brooks, David A., (979) 845-5527 dbrooks@ocean.tamu.edu, Texas A&M Univ; Op
Brooks, Debra A., (714) 564-4788 Santiago Canyon Coll; Yg
Brooks, Gregg R., (727) 864-8992 brooksgr@eckerd.edu, Eckerd Coll; Gu

Brooks, Ian, +44(0) 113 34 36743 i.m.brooks@leeds.ac.uk, Univ of Leeds; Zn
Brooks, Paul D., (801) 585-2858 paul.brooks@utah.edu, Univ of Utah; HgCbZg
Brooks, Sarah D., (979) 845-5632 sbrooks@tamu.edu, Texas A&M Univ; AsZc
Brooks, Scott C., (865) 574-6398 3sb@ornl.gov, Oak Ridge; Cl
Brookshire, Jack, jbrookshire@montana.ede, Montana State Univ; Zc
Broome, Stephen W., (919) 515-2643 North Carolina State Univ; Sf
Brophy, James G., (812) 855-6417 brophy@indiana.edu, Indiana Univ, Bloomington; Gi
Broster, Bruce E., (506) 453-4804 broster@unb.ca, Univ of New Brunswick; Gl
Brothen, Jerry, (310) 660-3593 jbrothen@elcamino.edu, El Camino Coll; Zy
Brothers, Timothy S., (317) 274-1101 tbrother@iupui.edu, Indiana Univ, Indianapolis; Zy
Brounce, Maryjo, (951) 827-3151 maryjo.brounce@ucr.edu, Univ of California, Riverside; Cg
Brouwer, Fraukje M., +31-20-598 7335 fraukje.brouwer@vu.nl, Vrije Univit Amsterdam; GptGc
Brown, Alan, abrown11@houston.oilfield.slb.com, West Virginia Univ; EoYe
Brown, Bradford D., (208) 722-6701 Univ of Idaho; Zn
Brown, Bruce A., (608) 263-3201 babrown1@wisc.edu, Univ of Wisconsin, Madison, Division of Extension; GcEg
Brown, Cathleen, (202) 633-1788 brownc@si.edu, Smithsonian Inst / Natl Mus of Natural History; GiCgGz
Brown, David, +44 01413307410 david.brown@glasgow.ac.uk, Univ of Glasgow; Gvs
Brown, Donald W., (505) 667-1926 dwb@lanl.gov, Los Alamos; Nr
Brown, Douglas, +44 208 417 2245 doug.brown@kingston.ac.uk, Kingston Univ; Zi
Brown, Edwin H., ehbrown@wwu.edu, Western Washington Univ; Gp
Brown, Erik T., (218) 726-8891 etbrown@d.umn.edu, Univ of Minnesota Duluth; Og
Brown, Huntting (Hunt), 937 775-4996 hunt.brown@wright.edu, Wright State Univ; Ge
Brown, J. Michael, 206-616-6058 jmbrown5@uw.edu, Univ of Washington; Gy
Brown, Kenneth L., (765) 658-6767 kennethbrown@depauw.edu, DePauw Univ; GizGe
Brown, Kent D., 801-537-3350 kentbrown@utah.gov, Utah Geo Survey; Zn
Brown, Kerry, k.brown@kingston.ac.uk, Kingston Univ; Ge
Brown, Kevin M., (858) 534-5368 kmbrown@ucsd.edu, Univ of California, San Diego; Gu
Brown, Kyle, 573-368-2168 kyle.brown@dnr.mo.gov, Missouri Dept of Nat Res; Gg
Brown, Larry D., (607) 255-6346 ldb7@cornell.edu, Cornell Univ; Ye
Brown, Laurie L., (413) 545-0245 lbrown@geo.umass.edu, Univ of Massachusetts, Amherst; Ymg
Brown, Lewis M., (906) 635-2155 lbrown@lssu.edu, Lake Superior State Univ; PiZe
Brown, Megan R., 815753-1778 mbrown18@niu.edu, Northern Illinois Univ; HwRw
Brown, Michael, (301) 405-4080 mbrown@umd.edu, Univ of Maryland; GxpGt
Brown, Paul W., (520) 621-1319 pbrown@ag.arizona.edu, Univ of Arizona; Am
Brown, Peter, 5196612111 ext.86458 pbrown@uwo.ca, Western Univ; Xym
Brown, Philip E., (608) 262-5954 pebrown@wisc.edu, Univ of Wisconsin, Madison; EmGzp
Brown, Richard J., +44 (0) 191 33 42303 richard.brown3@durham.ac.uk, Durham Univ; Gv
Brown, Rob, (828) 262-7222 brownrn@appstate.edu, Appalachian State Univ; Zn
Brown, Robert A., rabrown@atmos.washington.edu, Univ of Washington; As
Brown, Roderick, +44 01413305460 roderick.brown@glasgow.ac.uk, Univ of Glasgow; Cc
Brown, Sandra, 604-822-5965 sandra.brown@ubc.ca, Univ of British Columbia; Sp
Brown, Steven E., 217-333-5143 steebrow@illinois.edu, Univ of Illinois; Gl
Brown, Summer, 859-218-0873 summer.brown@uky.edu, Univ of Kentucky; GgZei
Brown, Thomas W., tbrown1@austincc.edu, Austin Comm Coll District; Ges
Brown, Tom, (925) 423-8507 tabrown@llnl.gov, Univ of Washington; CcAt
Brown, Wesley A., (936) 468-2422 brownwa1@sfasu.edu, Stephen F. Austin State Univ; YgsGt
Brown, Jr., Gordon E., (650) 723-9168 gordon@pangea.stanford.edu, Stanford Univ; Gz
Brownawell, Bruce J., (631) 632-8658 bruce.brownawell@stonybrook.edu, SUNY, Stony Brook; Oc
Browne, Brandon L., (707) 826-3950 brandon.browne@humboldt.edu, Humboldt State Univ; GivZe
Browne, Kathleen M., (609) 896-5408 browne@rider.edu, Rider Univ; GuZe
Browning, James M., 806-834-7782 james.browning@ttu.edu, Texas Tech Univ; Ggx
Browning, Sharon, (254)710-2159 sharon_browning@baylor.edu, Baylor Univ; Gg
BrownLee, Sarah J., (313) 577-6223 sarah.brownlee@wayne.edu, Wayne State Univ; GxcGy
Broxton, David E., 505-667-2492 broxton@lanl.gov, Los Alamos; Gx
Broze, Elliot, brozeea@whitman.edu, Whitman Coll; Gg
Brubaker, Kenneth L., (630) 252-7630 Argonne; As
Brudzinski, Michael, 513-529-9758 brudzimr@miamioh.edu, Miami Univ; Ys
Brueckner, Hannes K., (845) 521-4637 hannes@ldeo.columbia.edu, Queens Coll (CUNY); GgtCg
Brueseke, Matthew E., (785) 532-6724 brueseke@ksu.edu, Kansas State Univ; GitGv
Brugger, Keith A., (320) 589-6310 bruggeka@mrs.umn.edu, Univ of Minnesota, Morris; GllGm
Brugger, Matthias, +49 89 289 - 25886 m.brugger@tum.de, Technical Univ of Munich; NgGg
Bruland, Kenneth W., (831) 459-4587 bruland@cats.ucsc.edu, Univ of California, Santa Cruz; OcCtm
Brumbaugh, David S., (928) 523-7191 david.brumbaugh@nau.edu, Northern Arizona Univ; Ys
Brune, James N., (775) 784-4974 brune@seismo.unr.edu, Univ of Nevada, Reno; Ys
Brune, Jürgen F., jbrune@mines.edu, Colorado Sch of Mines; Nm
Brune, William H., (814) 865-3286 whb2@psu.edu, Pennsylvania State Univ, Univ Park; Am
Brunengo, Matthew , 503-725-3391 mbruneng@pdx.edu, Portland State Univ; GgNgGm
Bruner, Katherine R., (304) 293-5603 bruner@geo.wvu.edu, West Virginia Univ; Gd
Brunet, Gilbert, 514 421-4617 gilbert.brunet@canada.ca, McGill Univ; Am
Brunetto, Eileen, (802) 443-5970 efahey@middlebury.edu, Middlebury Coll; Zn
Bruning, Eric C., 806-834-3120 eric.bruning@ttu.edu, Texas Tech Univ; As
Brunish, Wendee M., (505) 667-5724 wb@lanl.gov, Los Alamos; Zn
Brunkal, Holly, 970.943.2180 hbrunkal@western.edu, Western Colorado Univ; NgGm
Brunner, Benjamin, 915-747-5507 bbrunner@utep.edu, Univ of Texas, El Paso; Cs
Bruno, Barbara, (808) 956-0901 barb@hawaii.edu, Univ of Hawai'i, Manoa; Ob
Bruno, Emiliano, emiliano.bruno@unito.it, Univ di Torino; Gz
Bruno, Marco, marco.bruno@unito.it, Univ di Torino; Gz
Bruns, Dale A., 570-408-4603 dale.bruns@wilkes.edu, Wilkes Univ; ZirSf
Brunskill, Jeffrey C., 570-389-4355 jbrunski@bloomu.edu, Bloomsburg Univ; ZiAmZy
Brunstad, Keith A., (607) 436-3066 keith.brunstad@oneonta.edu, SUNY, Oneonta; GviEg
Brunt, Rufus, +44 0161 306-6816 rufus.brunt@manchester.ac.uk, Univ of Manchester; Gs
Brusatte, Steve, stephen.brusatte@ed.ac.uk, Edinburgh Univ; Pv
Bruschke, Freddi Jo, 657-278-3551 fbruschke@fullerton.edu, California State Univ, Fullerton; Gg
Brush, Grace S., (410) 516-7107 gbrush@jhu.edu, Johns Hopkins Univ; PleGn
Brusseau, Mark L., (520) 621-3244 brusseau@ag.arizona.edu, Univ of Arizona; Hw
Bryan, Christopher, +44 01326 259482 c.g.bryan@exeter.ac.uk, Exeter Univ; Nx
Bryan, Scott E., +61 7 3138 4827 scott.bryan@qut.edu.au, Queensland Univ of Tech; GitGs
Bryan, Sean, sean.bryan@colostate.edu, Colorado State Univ; GgOoPc
Bryant, Anita M., (678) 839-4051 abryant@westga.edu, Univ of West Georgia; Zn
Bryant, Ernest A., (505) 667-2422 Los Alamos; Cg
Bryant, Kathleen E., 217-244-9045 kebryant@illinois.edu, Illinois State Geo Survey; Gg
Bryant, Marita, (207) 786-6452 mbryant@bates.edu, Bates Coll; Gg
Bryce, Julia G., (603) 862-3139 julie.bryce@unh.edu, Univ of New Hampshire; CgcCb
Brzobohaty, Rostislav, +420 549 49 3326 rosta@sci.muni.cz, Masaryk Univ; PgsPe
Buatois, Luis, luis.buatois@usask.ca, Univ of Saskatchewan; Ps

Buchanan, Brian, bbuchanan1@ewu.edu, Eastern Washington Univ; ZrGaZi
Buchanan, Donald G., 9093844399 ext. 5467 dbuchanan@sbccd.cc.ca.us, San Bernardino Valley Coll; Og
Buchanan, George, gbuchana@mc3.edu, Montgomery County Comm Coll; Ng
Buchanan, Paul C., (903) 983-8253 pbuchanan@kilgore.edu, Kilgore Coll; GgiXm
Buchheim, Paul, (909) 558-4530 pbuchheim@llu.edu, Loma Linda Univ; GnsPe
Buchwald, Carolyn, (902) 494-3666 cbuchwald@dal.ca, Dalhousie Univ; Oc
Buck, Brenda J., (702) 895-1694 buckb@unlv.nevada.edu, Univ of Nevada, Las Vegas; GbSd
Buck, Daniel, (541) 346-2353 danielb@uoregon.edu, Univ of Oregon; Zn
Buck, Gregory, 361-825-3717 gregory.buck@tamucc.edu, Texas A&M Univ, Corpus Christi; Zn
Buck, Roger W., buck@ldeo.columbia.edu, Columbia Univ; Yr
Buck, IV, W. R., (845) 365-8592 buck@ldeo.columbia.edu, Columbia Univ; YgGt
Bucke, David P., (802) 899-3584 david.bucke@uvm.edu, Univ of Vermont; GgdGs
Buckingham, Michael J., (858) 534-7977 mbuckingham@ucsd.edu, Univ of California, San Diego; Op
Buckley, Brendon, bmb@ldeo.columbia.edu, Columbia Univ; Pe
Buckley, Lawrence J., (401) 874-6671 lbuckley@gso.uri.edu, Univ of Rhode Island; Oc
Buckley, Luke, (406) 496-4677 lbuckley@mtech.edu, Montana Tech of The Univ of Montana; Zn
Buckley, Michael, +44(0)161 306 5175 m.buckley@manchester.ac.uk, Univ of Manchester; Ga
Bucklin, Ann, 860-405-9260 ann.bucklin@uconn.edu, Univ of Connecticut; Ob
Bucur, Ioan, +40-264-405371 ioan.bucur@ubbcluj.ro, Babes-Bolyai Univ; PgmGs
Buczynska, Anna, 815-753-7945 abuczynska@niu.edu, Northern Illinois Univ; Csa
Budd, Ann F., ann-budd@uiowa.edu, Univ of Iowa; Pgi
Budd, David A., (303) 492-3988 david.budd@colorado.edu, Univ of Colorado; GdsEo
Budikova, Dagmar, (309) 438-2546 dbudiko@ilstu.edu, Illinois State Univ; AsZyi
Buermann, Wolfgang, +44(0) 113 34 34958 w.buermann@leeds.ac.uk, Univ of Leeds; Zu
Buesseler, Ken O., (508) 289-2309 kbuesseler@whoi.edu, Woods Hole Oceanographic Inst; Oc
Buick, Roger, 206-543-1913 buick@uw.edu, Univ of Washington; CbXbPm
Buijsman, Maarten, (228) 688-2385 maarten.buijsman@usm.edu, Univ of Southern Mississippi; Op
Buizert, Christo, 541-737-1572 buizertc@science.oregonstate.edu, Oregon State Univ; Yr
Bukowinski, Mark S., (510) 642-0977 Univ of California, Berkeley; Gy
Bulgariu, Dumitru, 0040232201656 dbulgariu@yahoo.com, Alexandru Ioan Cuza; Cga
Buliga, Iuliana, 0040744297849 iuliana.buliga@uaic.ro, Alexandru Ioan Cuza; Gep
Bull, Jonathan M., +44 (0)23 8059 3078 bull@soton.ac.uk, Univ of Southampton; YrGct
Bull, William B., (520) 621-6024 Univ of Arizona; Gm
Bullamore, Henry W., (301) 687-4413 hbullamore@frostburg.edu, Frostburg State Univ; Zn
Bullard, Thomas F., (775) 673-7420 tbullard@dri.edu, Desert Research Inst; Gm
Bullard, Tom, (775) 673-7420 tbullard@dri.edu, Univ of Nevada, Reno; Gm
Bullen, Dean, +44 023 92 842289 dean.bullen@port.ac.uk, Univ of Portsmouth; Gt
Bullister, John, (206) 526-6741 bullister@pmel.noaa.gov, Univ of Washington; Oc
Bultman, John, (828) 254-1921 ext 319 jbultman@abtech.edu, Asheville-Buncombe Technical Comm Coll; Gg
Bulusu, Subrahmanyam, 803-777-2572 sbulusu@geol.sc.edu, Univ of South Carolina; Op
Bunch, Mark, mark.bunch@adelaide.edu.au, Univ of Adelaide; Ye
Bunds, Michael, (801) 863-6306 michael.bunds@uvu.edu, Utah Valley Univ; GctRn
Bundy, Larry G., (608) 263-2889 lgbundy@wisc.edu, Univ of Wisconsin, Madison; So
Bunge, Hans-Peter, 089/2180 4225 bunge@lmu.de, Ludwig-Maximilians-Univ Muenchen; Yg
Buonaiuto, Frank, 212-650-3092 frank.buonaiuto@hunter.cuny.edu, Hunter Coll (CUNY); On
Buonaiuto Jr., Frank S., 212-650-3092 jbuonaiu@hunter.cuny.edu, Graduate Sch of the City Univ of New York; Ze
Burakowski, Elizabeth, elizabeth.burakowski@unh.edu, Univ of New Hampshire; At
Buratti, Bonnie J., (818) 354-7427 Jet Propulsion Lab; Zr
Burbach, Mark E., mburback1@unl.edu, Univ of Nebraska, Lincoln; Hw
Burbanck, George P., (757) 727-5783 george.burbanck@hamptonu.edu, Hampton Univ; On
Burbank, Doug, (805) 893-7858 burbank@ucsb.edu, Univ of California, Santa Barbara; GmtGs
Burberry, Caroline M., (402) 472-7157 cburberry2@unl.edu, Univ of Nebraska, Lincoln; Gct
Burbey, Thomas J., (540) 231-6696 tjburbey@vt.edu, Virginia Tech; HwyZr
Burchfiel, B. C., bcburch@mit.edu, Massachusetts Inst of Tech; Gc
Burd, Aurora, aburd@avc.edu, Antelope Valley Coll; GgZgYm
Burdick, Scott, (313) 577-6412 sburdick@wayne.edu, Wayne State Univ; Ysg
Burdige, David J., (757) 683-4930 dburdige@odu.edu, Old Dominion Univ; OcCbo
Burger, Benjamin J., 435.722.1778 benjamin.burger@usu.edu, Utah State Univ; PvGs
Burger, H. Robert, rburger@smith.edu, Smith Coll; GcYeZi
Burgess, Ray, +44 0161 275-3958 ray.burgess@manchester.ac.uk, Univ of Manchester; Cs
Burgess, William P., william.burgess@ucl.ac.uk, Univ Coll London; Hw
Burgette, Reed J., 575-646-3782 burgette@nmsu.edu, New Mexico State Univ, Las Cruces; Gtc
Burgmann, Roland, (510) 643-9545 burgmann@seismo.berkeley.edu, Univ of California, Berkeley; GtYd
Burisch, Mathias, 303-384-2456 burisch@mines.edu, Colorado Sch of Mines; Gz
Burk, Sue, 301-405-6244 Univ of Maryland; Zn
Burkart, Michael R., mburkart@iastate.edu, Iowa State Univ of Sci & Tech; Hy
Burke, Andrea, +44 01334 463910 ab276@st-andrews.ac.uk, Univ of St. Andrews; ClPc
Burke, Collette D., (316) 978-3140 Wichita State Univ; Pm
Burke, Deborah S., (509) 372-2483 deborah.burke@pnl.gov, Pacific Northwest; Sc
Burke, Ian, +44(0) 113 34 37532 lab 33965 i.t.burke@leeds.ac.uk, Univ of Leeds; Ge
Burke, Raymond M., (707) 826-4292 rmb2@humboldt.edu, Humboldt State Univ; Gm
Burkett, Ashley, 405-744-6358 ashley.burkett@okstate.edu, Oklahoma State Univ; Pme
Burkett, Brett, (976) 548-6510 bburkett@collin.edu, Collin Coll - Central Park Campus; Gvg
Burkett, Shannon, (972) 548-6611 sburkett@collin.edu, Collin Coll - Central Park Campus; Gv
Burkhart, Patrick A., (724) 738-2502 patrick.burkhart@sru.edu, Slippery Rock Univ; HgGl
Burks, Rachel J., (410) 704-3005 rburks@towson.edu, Towson Univ; Gc
Burlage, Robert S., (865) 574-7321 Oak Ridge; Sb
Burls, Natalie, 703-993-5756 nburls@gmu.edu, George Mason Univ; AsOp
Burmester, Russell F., russ.burmester@wwu.edu, Western Washington Univ; Ym
Burnett, Donald S., (626) 395-6117 burnett@gps.caltech.edu,CalTech; Xc
Burnett, Joree, 254-968-1868 jburnett@tarleton.edu, Tarleton State Univ; HwGe
Burnette, Dorian, 901-678-4452 djbrntte@memphis.edu, Univ of Memphis; Atm
Burnham, Charles W., burnham_c@fortlewis.edu, Harvard Univ; GzyGe
Burnham, Robyn J., (734) 764-0489 rburnham@umich.edu, Univ of Michigan; Pb
Burnley, Pamela C., 702-895-2536 pamela.burnley@unlv.edu, Univ of Nevada, Las Vegas; GpyGg
Burns, Bill, (971) 673-1555 Oregon Dept of Geology and Mineral Ind; Ng
Burns, Carol J., (505) 665-1765 Los Alamos; Zn
Burns, Danny, 361-354-2405 deburns@coastalbend.edu, Coastal Bend Coll; Zgn
Burns, Diane M., (217) 581-2626 dmburns@eiu.edu, Eastern Illinois Univ; GsrGe
Burns, Emily, eburns@ccri.edu, Comm Coll of Rhode Island; GgOgZi
Burns, Peter C., (574) 631-5380 peter.burns.50@nd.edu, Univ of Notre Dame; Gz
Burns, Sandra, (860) 832-2934 burns@ccsu.edu, Central Connecticut State Univ; Ze
Burns, Scott F., (503) 725-3389 burnss@pdx.edu, Portland State Univ; Ng
Burns, Stephen J., (413) 545-0142 sburns@geo.umass.edu, Univ of

Massachusetts, Amherst; Cs
Burns, Timothy P., (505) 667-4600 Los Alamos; Co
Burras, Lee, (515) 294-0559 lburras@iastate.edu, Iowa State Univ of Sci & Tech; Sd
Burris, John H., (505) 566-3325 burrisj@sanjuancollege.edu, San Juan Coll; GgRhGz
Bursik, Marcus I., (716) 645-4265 mib@buffalo.edu, SUNY, Buffalo; Gv
Burt, Donald M., (480) 965-6180 donald.burt@asu.edu, Arizona State Univ; EgCgGz
Burtis, Erik, (703) 878-5614 eburtis@nvcc.edu, Northern Virginia Comm Coll - Woodbridge; GgcGi
Burton, Bradford R., 970.943.2252 bburton@western.edu, Western Colorado Univ; GocGt
Burton, Christopher G., 334-844-3418 cgb0038@abuurn.edu, Auburn Univ; Zi
Burton, Edward, ed.burton@scu.edu.au, Southern Cross Univ; CaHy
Burton, Jacqueline C., (630) 252-8795 jcburton@anl.gov, Argonne; Cg
Burton, Kevin, +44 (0) 191 33 44298 kevin.burton@durham.ac.uk, Durham Univ; Cg
Burton, Paul, +44 (0)1603 59 2982 p.burton@uea.ac.uk, Univ of East Anglia; Ys
Burton, Ronald S., (858) 822-5784 rburton@ucsd.edu, Univ of California, San Diego; Ob
Burwash, Ronald A., (780) 492-3085 ronald.burwash@telus.net, Univ of Alberta; Gp
Busa, Mark, 860-343-5779 mbusa@mxcc.commnet.edu, Middlesex Comm Coll; ZgEgYg
Busacca, Alan J., (509) 335-1859 busacca@wsu.edu, Washington State Univ; Sa
Busbey, Arthur B., (817) 257-7301 a.busbey@tcu.edu, Texas Christian Univ; PgvGr
Busby, Cathy J., cjbusby@ucdavis.edu, Univ of California, Davis; Gt
Busby, Michael R., (270) 809-3370 mbusby@murraystate.edu, Murray State Univ; Zi
Busch, Richard M., (610) 436-2716 rbusch@wcupa.edu, West Chester Univ; GrPgZe
Busch, William H., (503) 392-3341 whbusch@gmail.com, Univ of New Orleans; Ou
Buscheck, Thomas A., (925) 423-9390 buscheck1@llnl.gov, Lawrence Livermore; Ng
Buseck, Peter R., (480) 965-3945 pbuseck@asu.edu, Arizona State Univ; Cg
Bush, Andrew B., (780) 492-0351 andrew.bush@ualberta.ca, Univ of Alberta; AsOpPe
Bush, Andrew B. G., (780) 492-0351 andrew.bush@ualberta.ca, Univ of Alberta; As
Bush, David M., (678) 839-4057 dbush@westga.edu, Univ of West Georgia; Ou
Bush, Rosemary, (847) 491-8916 rosemary.bush@northwestern.edu, Northwestern Univ; Pc
Businger, Joost, 206-543-4250 Univ of Washington; Am
Businger, Steven, businger@hawaii.edu, Univ of Hawai'i at Manoa; Am
Buskey, Edward J., (361) 749-3102 ed.buskey@utexas.edu, Univ of Texas, Austin; Ob
Buss, Leo W., 203-432-3869 leo.buss@yale.edu, Yale Univ; Pi
Busse, Friedrich H., (310) 825-7698 Univ of California, Los Angeles; Zn
Bussod, Gilles Y., (505) 667-7220 gbussod@lanl.gov, Los Alamos; Yx
Bustillo, Manuel, bustillo@ucm.es, Univ Compl de Madrid; Eg
Bustin, Amanda, abustin@eos.ubc.ca, Univ of British Columbia; YgNg
Bustin, R. Marc, (604) 822-6179 mbustin@eos.ubc.ca, Univ of British Columbia; Ec
Butcher, Anthony, +44 023 92 842486 anthony.butcher@port.ac.uk, Univ of Portsmouth; PlmZe
Butkos, Darryl J., (631) 451-4354 butkosd@sunysuffolk.edu, Suffolk County Comm Coll, Ammerman Campus; GgHw
Butler, Gilbert W., (505) 667-6005 Los Alamos; Cc
Butler, Ian B., +44 (0) 131 650 5885 ian.butler@ed.ac.uk, Edinburgh Univ; Cg
Butler, Karl, (506) 453-4804 Univ of New Brunswick; Yg
Butler, Nathaniel, (480) 965-8207 nathaniel.butler@asu.edu, Arizona State Univ; Xa
Butler, Paul, pbutler@carnegiescience.edu, Carnegie Inst for Sci; Xa
Butler, Sam, (306) 966-5702 sam.butler@usask.ca, Univ of Saskatchewan; Yg
Butler, Jr., James J., (785) 864-2116 jbutler@kgs.ku.edu, Univ of Kansas; Hw
Butterfield, David A., (206) 526-6722 dab3@u.washington.edu, Univ of Washington; Yr
Butterfield, Nicholas J., +44 (0) 1223 333379 njb1005@esc.cam.ac.uk, Univ of Cambridge; Pg
Buttner, Steffen, +27 (0)46-603-8775 s.buettner@ru.ac.za, Rhodes Univ; Gcp
Butts, Susan H., (203) 432-3037 susan.butts@yale.edu, Yale Univ; PiGsPe
Butzow, Dean G., 217-786-4923 dean.butzow@llcc.edu, Lincoln Land Comm Coll; Zy
Buyce, M. Raymond, rbuyce@mercyhurst.edu, Mercyhurst Univ; GsaOn
Buynevich, Ilya, (215) 204-3635 coast@temple.edu, Temple Univ; GmPeOn
Buzas, Martin A., (202) 633-1313 Smithsonian Inst / Natl Mus of Natural History; Pm
Buzatu, Andrei, 0040232201463 andrei.buzatu@uaic.ro, Alexandru Ioan Cuza; GziGg
Buzgar, Nicolae, 0040232201462 nicolae.buzgar@uaic.ro, Alexandru Ioan Cuza; GziGd
Bybee, Grant M., 011 717 6633 grant.bybee@wits.ac.za, Univ of the Witwatersrand; Gi
Büchel, Georg, +49(0)3641 948640 georg.buechel@uni-jena.de, Friedrich-Schiller-Univ Jena; Gg
Byerly, Gary R., (225) 578-5318 glbyer@lsu.edu, Louisiana State Univ; Gi
Byers, Charles W., cwbyers@wisc.edu, Univ of Wisconsin, Madison; Grs
Bykerk-Kauffman, Ann, 530-898-6305 abykerk-kauffman@csuchico.edu, California State Univ, Chico; GcZe
Byrne, James M., (403) 329-2002 byrne@uleth.ca, Univ of Lethbridge; Hg
Byrne, Paul, paul.byrne@wustl.edu, Washington Univ in St. Louis; Xg
Byrne, Robert H., (727) 553-1508 byrne@marine.usf.edu, Univ of South Florida; Cm
Byrne, Timothy, (860) 486-3142 tim.byrne@uconn.edu, Univ of Connecticut; Gc
Byrnes, Joseph, jsbyrnes@umn.edu, Univ of Minnesota, Twin Cities; Ys
Bywater-Reyes, Sharon, 970-351-1086 sharon.bywaterreyes@unco.edu, Univ of Northern Colorado; HqsHt

C

C., Jeffrey C., (330) 941-3612 jcdick@ysu.edu, Youngstown State Univ; EoNgHy
Cabaniss, Haley, (859) 622-6061 haley.cabaniss@eku.edu, Eastern Kentucky Univ; GvuYr
Cablk, Mary, (775) 673-7371 mcablk@dri.edu, Desert Research Inst; Zr
Cabrera, Miguel L., (706) 542-1242 Univ of Georgia; So
Cachão, Mário, mcachao@fc.ul.pt, Unive de Lisboa; PgeGg
Caddick, Mark, 540-231-9919 caddick@vt.edu, Virginia Tech; GpCa
Cadeau, Pierre, 609-258-4101 pc2862@princeton.edu, Princeton Univ; Cs
Cadet, Eddy, (801) 863-8881 cadeted@uvu.edu, Utah Valley Univ; Ge
Cadieux, Sarah, (518) 276-6474 cadies@rpi.edu, Rensselaer Polytechnic Inst; ZeCb
Cadol, Daniel, 575-835-5645 dcadol@ees.nmt.edu, New Mexico Tech; HsGe
Cadoppi, Paola, paola.cadoppi@unito.it, Univ di Torino; Gc
Cahill, Abigail, abigail.cahill@williams.edu, Williams Coll; Zn
Cahill, Kevin, (508) 289-2925 kcahill@whoi.edu, Woods Hole Oceanographic Inst; Cm
Cahill, Michael J., (631) 588-8778 mjc@germanocahill.com, SUNY, Stony Brook; Zn
Cahill, Richard A., 217-244-2532 racahill49@gmail.com, Illinois State Geo Survey; CaGeCt
Cahir, John J., (814) 863-8358 cahir@ems.psu.edu, Pennsylvania State Univ, Univ Park; Am
Cahoy, Kerri, 617-324-6005 kcahoy@mit.edu, Massachusetts Inst of Tech; Zn
Cai, Yue, cai@ldeo.columbia.edu, Columbia Univ; Cg
Cairns, David M., (979) 845-2783 cairns@geog.tamu.edu, Texas A&M Univ; Zy
Cairns, Stephen, cairnss@si.edu, Smithsonian Inst / Natl Mus of Natural History; Pi
Caissie, Beth E., 515-294-7528 bethc@iastate.edu, Iowa State Univ of Sci & Tech; Pe
Calagari, Ali Asghar, +98 (411) 339 2699 calagari@tabrizu.ac.ir, Univ of Tabriz; EgmCg
Calaway, Wallis F., (630) 972-3586 Argonne; Zm
Calcote, Randy, calco001@umn.edu, Univ of Minnesota, Twin Cities; Ple
Calder, Brian, 603-862-0526 brc@ccom.unh.edu, Univ of New Hampshire; Og
Calder, Eliza, +44 (0) 131 650 4910 ecalder@staffmail.ed.ac.uk, Edinburgh Univ; Gv
Calder, John, (902) 424-2778 Dalhousie Univ; Ec
Caldwell, Andy, 970-204-8228 andrew.caldwell@frontrange.edu, Front Range Comm Coll - Larimer; Gg
Caldwell, Marianne O., mcaldwell@hccfl.edu, Hillsborough Comm Coll; Ggs
Caldwell, Michael W., 780-492-3458 mw.caldwell@ualberta.ca, Univ of Alberta; Pv
Caldwell, Roy, (510) 642-1391 rlcaldwell@berkeley.edu, Univ of California,

Berkeley; Pg
Caldwell, W. Glen E., 519-661-3187 gcaldwel@uwo.ca, Western Univ; Pm
Caldwell, William G. E., (519) 661-3857 Univ of Saskatchewan; Ps
Calegari, Pat, (212) 346-1502 Pace Univ, New York Campus; Zn
Calengas, Peter L., (309) 298-1151 pl-calengas@wiu.edu, Western Illinois Univ; En
Calhoun, Frank G., (330) 263-3722 scalhoun@coas.oregonstate.edu, Ohio State Univ; Sd
Calhoun, Joseph, 717-872-3289 jcalhoun@hearst.com, Millersville Univ; As
Calizaya, Felipe, (801) 581-5422 felipe.calizaya@utah.edu, Univ of Utah; Nm
Callaghan, Kerry, 312-996-0455 kerryc@uic.edu, Univ of Illinois Chicago; Hw
Callahan, Caitlin N., 616-331-3601 callahac@gvsu.edu, Grand Valley State Univ; GgZg
Callahan, Timothy J., (843) 953-5589 callahant@cofc.edu, Coll of Charleston; Hwq
Callanan, Jennifer R., (973) 720-3979 callananj@wpunj.edu, William Paterson Univ; ScGmSz
Callard, Jeff, callard@ou.edu, Univ of Oklahoma; Np
Callegari, Ezio, ezio.callegari@unito.it, Univ di Torino; Gx
Callison, James, (801) 863-8679 jcallison@uvu.edu, Utah Valley Univ; GeSfo
Calvert, Andrew J., (778) 782-5511 acalvert@sfu.ca, Simon Fraser Univ; Ys
Calvert, Stephen E., (604) 822-5210 calvert@eos.ubc.ca, Univ of British Columbia; Cm
Calvin, Wendy, (775) 784-1785 wcalvin@unr.edu, Univ of Nevada, Reno; ZrYgXg
Camacho, Alfredo, (204) 474-7413 camacho@cc.umanitoba.ca, Univ of Manitoba; Gt
Camacho, Elsa A., (509) 376-5473 elsa.camacho@pnl.gov, Pacific Northwest; Cg
Camann, Eleanor J., 303-914-6290 eleanor.camann@rrcc.edu, Red Rocks Comm Coll; OnZeGs
Camargo, Suzana, suzana@ldeo.columbia.edu, Columbia Univ; As
Cambardella, Cynthia, (515) 294-2921 cindy.cambardella@ars.usda.gov, Iowa State Univ of Sci & Tech; So
Cambazoglu, Mustafa K., 228.688.3009 kemal.cambazoglu@usm.edu, Univ of Southern Mississippi; No
Came, Rosemarie E., 603-862-1720 rosemarie.came@unh.edu, Univ of New Hampshire; OoZnn
Cameron, Barry I., 414-229-3136 bcameron@uwm.edu, Univ of Wisconsin, Milwaukee; Gvi
Cameron, Cheryl, (907) 451-5012 cheryl.cameron@alaska.gov, Alaska Division of Geological & Geophysical Surveys; Gv
Cameron, Kathleen A., (402) 419-4798 kcameron2@unl.edu, Univ of Nebraska, Lincoln; HwZfGg
Cameron, Kevin, (778) 782-4703 kjc@sfu.ca, Simon Fraser Univ; Ggz
Camill III, Philip, 207-721-5149 pcamill@bowdoin.edu, Bowdoin Coll; PebSb
Camille, Michael A., (318) 342-1750 Univ of Louisiana, Monroe; Zy
Camilleri, Phyllis A., (931) 221-7317 camillerip@apsu.edu, Austin Peay State Univ; Gct
Cammerata, Kirk , 361-825-2468 kirk.cammerata@tamucc.edu, Texas A&M Univ, Corpus Christi; Zn
Camp, Mark J., (419) 530-2398 mark.camp@utoledo.edu, Univ of Toledo; PiRhEs
Camp, Victor E., (619) 594-7170 vcamp@mail.sdsu.edu, San Diego State Univ; Gv
Campana, Michael E., (541) 737-2413 michael.campana@oregonstate.edu, Oregon State Univ; Hg
Campbell, Andrew R., (575) 835-5327 campbell@nmt.edu, New Mexico Tech; CsEmGz
Campbell, Bruce A., 202-633-2472 campbellb@si.edu, Smithsonian Inst, Natl Air & Space Mus; Zr
Campbell, Erin A., 307-766-2286 erin.campbell@wyo.gov, Wyoming State Geo Survey; Gc
Campbell, Glenn A., (859) 622-6474 glenn.campbell@eku.edu, Eastern Kentucky Univ; Zy
Campbell, Ian A., ian.campbell@ualberta.ca, Univ of Alberta; Gm
Campbell, James B., (540) 231-5841 Virginia Tech; Zr
Campbell, Katherine, (505) 667-2799 ksc@lanl.gov, Los Alamos; Gq
Campbell, Kenneth E., (213) 763-3425 kcampbell@nhm.org, Los Angeles County Mus of Natural History; PvGgr
Campbell, Lisa, (979) 845-5706 lisacampbell@tamu.edu, Texas A&M Univ; Ob
Campbell, Patricia A., (724) 738-4405 patricia.campbell@sru.edu, Slippery Rock Univ; Gc
Campbell, Seth W., seth.campbell@maine.edu, Univ of Maine; Gl
Canalda, Sabrina, 915-831- 2617 mren@epcc.edu, El Paso Comm Coll; Gg
Canales, María Luisa, 34 913944819 mcanales@ucm.es, Univ Compl de Madrid; PmgGg
Canales Cisneros, Juan Pablo, (508) 289-2893 jpcanales@whoi.edu, Woods Hole Oceanographic Inst; Yr
Candela, Philip A., (301) 405-2783 candela@geol.umd.edu, Univ of Maryland; CpgEg
Cane, Mark A., mcane@ldeo.columbia.edu, Columbia Univ; Op
Canil, Dante, (250) 472-4180 dcanil@uvic.ca, Univ of Victoria; Gi
Cannon, Alex, acannon@eoas.ubc.ca, Univ of British Columbia; As
Cannon, Kevin, 303-273-3819 cannon@mines.edu, Colorado Sch of Mines; ZnnZr
Canon-Tapia, Edgardo, ecanon@cicese.mx, Centro de Investigación Científica y de Educación Superior de Ensenada; GvYmGt
Cantarero, Debra A., (626) 585-7138 dacantarero@pasadena.edu, Pasadena City Coll; Zn
Cantrell, Kirk J., (509) 376-2136 kirk.cantrell@pnl.gov, Pacific Northwest; Cl
Canudo, José Ignacio, jicanudo@unizar.es, Univ de Zaragoza; Pv
Cao, Guofeng, 806-834-8920 guofeng.cao@ttu.edu, Texas Tech Univ; Zi
Cao, Hao, 310-825-1986 hcao@epss.ucla.edu, Univ of California, Los Angeles; Xy
Cao, Hongsheng, 316-978-3140 Wichita State Univ; Cl
Cao, Wenrong, 775-784-1770 wenrongc@unr.edu, Univ of Nevada, Reno; Gtc
Cao, Xiaobin, xcao@lsu.edu, Louisiana State Univ; Cs
Capaldi, Tomas, 702-895-1774 tomas.capaldi@unlv.edu, Univ of Nevada, Las Vegas; Gst
Capehart, William J., (605) 394-2291 william.capehart@sdsmt.edu, South Dakota Sch of Mines & Tech; AsZoHq
Capes, Gerard, gerard.capes@manchester.ac.uk, Univ of Manchester; As
Caplan-Auerbach, Jackie, (360) 650-4153 caplanj@wwu.edu, Western Washington Univ; Ys
Capo, Rosemary C., (412) 624-8873 rcapo@pitt.edu, Univ of Pittsburgh; Cl
Capoccia, Mary, (614) 292-8522 capoccia.6@osu.edu, Ohio State Univ; Zn
Caporali, Alessandro, 39-049-8279122 alessandro.caporali@unipd.it, Univ degli Studi di Padova; YdsZr
Capraro, Luca, 39-049-8279182 luca.capraro@unipd.it, Univ degli Studi di Padova; PelGg
Capuano, Regina M., 713-743-2957 capuano@uh.edu, Univ of Houston; HwCqs
Caputi, Samantha , samantha.caputi@tccd.edu, Tarrant County Coll, Southeast Campus; Gg
Caputo, Mario V., (909) 214-7742 mvcaputo@earthlink.net, San Diego State Univ; GsdOn
Carageorgos, Themis, themis.carageorgos@adelaide.edu.au, Univ of Adelaide; Np
Carbotte, Suzanne, carbotte@ldeo.columbia.edu, Columbia Univ; Yr
Cardace, Dawn, 401-874-9384 cardace@uri.edu, Univ of Rhode Island; GpPg
Cardellach López, Esteve, ++34935813091 esteve.cardellach@uab.cat, Univ Auto de Barcelona; GzeCs
Carden-Jessen, Melanie E., (417) 836-3231 mcardenjessen@missouristate.edu, Missouri State Univ; Ze
Cardenas, Bayani, cardenas@jsg.utexas.edu, Univ of Texas, Austin; Hg
Cardiff, Michael A., (608) 262-8960 cardiff@wisc.edu, Univ of Wisconsin, Madison; HwYg
Cardiff, Mike, cardiff@wisc.edu, Univ of Wisconsin, Madison; Hw
Cardimona, Steve, (707) 468-3219 scardimo@mendocino.edu, Mendocino Coll; Ys
Carew, James L., (843) 953-5592 carewj@cofc.edu, Coll of Charleston; PgGde
Carey, Anne E., (614) 292-2375 carey.145@osu.edu, Ohio State Univ; HwCg
Carey, James W., (505) 667-5540 bcarey@lanl.gov, Los Alamos; CpYxNr
Carey, Jordan , jordan.carey@cwu.edu, Central Washington Univ; Geg
Carey, Kristine M., (807) 343-8010 Ext. 8461 kristine.carey@lakeheadu.ca, Lakehead Univ; Zn
Carey, Sean, 905-525-9140 (Ext. 20134) careysk@mcmaster.ca, McMaster Univ; Hg
Carey, Steven N., (401) 874-6209 scarey@gso.uri.edu, Univ of Rhode Island; Ou
Carle, Steven, (925) 423-5039 carle1@llnl.gov, Lawrence Livermore; Hw
Carley, Tamara, 610-330-5199 carleyt@lafayette.edu, Lafayette Coll; GiCgc
Carlile, Amy L., 203-479-4257 acarlile@newhaven.edu, Univ of New Haven; ObZn
Carlin, Joe, (657) 278-3054 jcarlin@fullerton.edu, California State Univ, Fullerton; GuOuGs
Carlin, Scott, 516-299-2318 scott.carlin@liu.edu, Long Island Univ, C.W. Post Campus; Zn

Carling, Gregory T., (801) 422-2622 greg.carling@byu.edu, Brigham Young Univ; HwCts
Carlson, Anders, 541-737-3625 acarlson@coas.oregonstate.edu, Oregon State Univ; GlCt
Carlson, Barry A., bacarlso@delta.edu, Delta Coll; Yg
Carlson, Catherine A., 860-465-5218 carlsonc@easternct.edu, Eastern Connecticut State Univ; Hwg
Carlson, Diane H., carlsondh@csus.edu, Sacramento State Univ; Gc
Carlson, Douglas A., 225/578-3671 dcarlson@lsu.edu, Louisiana State Univ; Hw
Carlson, Galen R., 657-278-3882 gcarlson@exchange.fullerton.edu, California State Univ, Fullerton; Zg
Carlson, Heath, carlsonh@easternct.edu, Eastern Connecticut State Univ; Zg
Carlson, Jill, 303-384-2643 carlson@mines.edu, Colorado Geo Survey; NgZu
Carlson, Marvin P., (402) 472-3471 rockdrmpc@hotmail.com, Univ of Nebraska, Lincoln; Gt
Carlson, Richard E., (515) 294-9868 richard@iastate.edu, Iowa State Univ of Sci & Tech; Am
Carlson, Sandra J., (530) 752-2834 sjcarlson@ucdavis.edu, Univ of California, Davis; Pg
Carlson, Toby N., (814) 863-1582 tnc@psu.edu, Pennsylvania State Univ, Univ Park; Am
Carlson, William D., (512) 471-4770 wcarlson@jsg.utexas.edu, Univ of Texas, Austin; GpzCp
Carmack, Edward, (250) 363-6585 carmack@dfo-mpo.gc.ca, Univ of British Columbia; Op
Carman, Cary D., 325-944-4600 cdcarman@usgs.gov, Angelo State Univ; Hg
Carmichael, Ian S., (510) 642-2577 ian@eps.berkeley.edu, Univ of California, Berkeley; Gi
Carmichael, Robert S., (319) 337-6499 robert-carmichael@uiowa.edu, Univ of Iowa; YexGg
Carmichael, Sarah, (828) 262-8471 carmichaelsk@appstate.edu, Appalachian State Univ; GxCgGd
Carn, Simon, 906/487-1756 scarn@mtu.edu, Michigan Technological Univ; ZrGv
Carnevale, Giorgio, giorgio.carnevale@unito.it, Univ di Torino; Pg
Carney, Robert S., (225) 388-6511 Louisiana State Univ; Ob
Carney, Theodore C., (505) 667-3415 tedc@lanl.gov, Los Alamos; Ng
Caron, Jean-Bernard, jcaron@rom.on.ca, Royal Ontario Mus; PgiPg
Caron, Olivier J., 217-300-0198 caron@illinois.edu, Illinois State Geo Survey; Ng
Carosi, Rodolfo, rodolfo.carosi@unito.it, Univ di Torino; Gc
Carpenter, Brett, brett.carpenter@ou.edu, Univ of Oklahoma; Gc
Carpenter, Michael, +44 (0) 1223 333483 mc43@esc.cam.ac.uk, Univ of Cambridge; Gy
Carpenter, N. Seth, (859) 323-0563 seth.carpenter@uky.edu, Univ of Kentucky; Ys
Carpenter, Philip J., (815) 753-1523 pjcarpenter@niu.edu, Northern Illinois Univ; YgHwNg
Carpenter, Roy, (206) 543-8535 rcarp@u.washington.edu, Univ of Washington; No
Carpi, Anthony, 212-237-8195 acarpi@jjay.cuny.edu, Graduate Sch of the City Univ of New York; Gg
Carr, David E., 540-837-1758 blandy@virginia.edu, Univ of Virginia; Zn
Carr, Keith W., 217-265-0267 kw-carr@illinois.edu, Illinois State Geo Survey; Gg
Carr, Mark, (831) 459-3958 carr@biology.ucsc.edu, Univ of California, Santa Cruz; Ob
Carr, Michael J., carr@rutgers.edu, Rutgers, The State Univ of New Jersey; Gv
Carr, Sharon D., scarr@ccs.carleton.ca, Carleton Univ; Gt
Carr, Timothy R., (785) 864-2135 tcarr@kgs.ku.edu, Univ of Kansas; Go
Carrano, Matthew T., (202) 633-1506 Smithsonian Inst / Natl Mus of Natural History; Pv
Carranza, Emmanuel John, (031) 260-2803 carranzae@ukzn.ac.za, Univ of KwaZulu-Natal; EdCeZr
Carrapa, Barbara, 520-621-4910 bcarrapa@email.arizona.edu, Univ of Arizona; Gs
Carrell, Jennifer, 217-244-2764 jcarrell@illinois.edu, Illinois State Geo Survey; Zg
Carrick, Tina L., 915-831-8905 tcarrick@epcc.edu, El Paso Comm Coll; ZeGgYg
Carrigan, Charles R., (925) 422-3941 carrigan1@llnl.gov, Lawrence Livermore; Yg
Carroll, Alan R., (608) 262-2368 acarroll@geology.wisc.edu, Univ of Wisconsin, Madison; Gs
Carroll, David, (540) 231-5469 carrolld@vt.edu, Virginia Tech; Am
Carroll, Matt, 509-335-2235 carroll@wsu.edu, Washington State Univ; Zn
Carroll, Rosemary, rosemary.carroll@dri.edu, Univ of Nevada, Reno; Hq
Carroll, Susan A., (925) 423-7552 Lawrence Livermore; Cl
Carroll, Tracy, (417) 836-5800 tracycarroll@missouristate.edu, Missouri State Univ; Zn
Carslaw, Ken, +44(0) 113 34 31597 k.s.carslaw@leeds.ac.uk, Univ of Leeds; As
Carson, Bobb, bc00@lehigh.edu, Lehigh Univ; GusGt
Carson, Eric C., 608-890-1998 eccarson@wisc.edu, Univ of Wisconsin, Extension; Gl
Carson, Michael, +618 9266-4973 michael.carson@curtin.edu.au, Curtin Univ; Yx
Carson, Robert J., (509) 527-5224 carsonrj@whitman.edu, Whitman Coll; GmeGl
Carstarphen, Camela, 406-496-4633 ccarstarphen@mtech.edu, Montana Tech of The Univ of Montana; Hw
Carstensen, Laurence W., (540) 231-2600 Virginia Tech; Zi
Carter, Andy, +44 020 7679 2418 a.carter@ucl.ac.uk, Birkbeck Coll; Zg
Carter, Burchard D., (229) 931-2325 burchard.carter@gsw.edu, Georgia Southwestern State Univ; Pg
Carter, David, (303) 293-5014 dcarter@osmre.gov, Dept of Interior Office of Surface Mining Reclamation and Enforcement; Zir
Carter, Glenn, (808) 956-9267 gscarter@hawaii.edu, Univ of Hawai'i, Manoa; Op
Carter, Greg, (228) 214-3305 greg.carter@usm.edu, Univ of Southern Mississippi; Zyi
Carter, Kristin M., 412.442.4234 krcarter@pa.gov, DCNR- Pennsylvania Bur of Geo Survey; EoCqEg
Cartier, Laurent E., laurent.cartier@unil.ch, Univ of Delaware; Gz
Carton, Alberto, +390498279141 alberto.carton@unipd.it, Univ degli Studi di Padova; Gm
Carton, James, (301) 405-5365 Univ of Maryland; Op
Cartwright, J A., cartwrightja@cf.ac.uk, Cardiff Univ; Ys
Cartwright, Joe, +44 (1865) 272000 joe.cartwright@earth.ox.ac.uk, Univ of Oxford; Gs
Cartwright, Julia, jacartwright@ua.edu, Univ of Alabama; XmgCc
Caruthers, Andrew, (269) 387-8633 andrew.caruthers@wmich.edu, Western Michigan Univ; GsCsGr
Carvalho, Maria d., mdrcarvalho@fc.ul.pt, Unive de Lisboa; HwyZn
Carver, Gary A., (907) 487-4551 Humboldt State Univ; Gmt
Cary, Kevin, (270) 745-2981 kevin.cary@wku.edu, Western Kentucky Univ; Zi
Casale, Gabriele M., casalegm@appstate.edu, Appalachian State Univ; Gct
Casares, Heather, heather.casares@azwestern.edu, Arizona Western Coll; Gvi
Casas Duocastella, Lluís, ++34935868365 lluis.casas@uab.cat, Univ Auto de Barcelona; GzaYm
Case, James M., james.case@wilkes.edu, Wilkes Univ; Ob
Case, Stephen, (815) 939-5681 scase@olivet.edu, Olivet Nazarene Univ; RhXg
Case Hanks, Anne T., (318) 342-1822 casehanks@ulm.edu, Univ of Louisiana, Monroe; AsZe
Casey, Francis, francis.casey@ndsu.edu, North Dakota State Univ; SpHq
Casey, John F., 713-743-2824 jfcasey@uh.edu, Univ of Houston; GtiGu
Casey, Michelle , (410) 704-2744 mcasey@towson.edu, Towson Univ; Pe
Casey, William H., (916) 752-3211 Univ of California, Davis; Cl
Cashman, Susan M., (707) 826-3114 smc1@humboldt.edu, Humboldt State Univ; Gc
Caskey, John, (415) 405-0353 caskey@sfsu.edu, San Francisco State Univ; Gc
Casquet, César, casquet@ucm.es, Univ Compl de Madrid; Gip
Cassar, Nicolas, 919 681-8865 nicolas.cassar@duke.edu, Duke Univ; CbOcb
Cassel, Donald K., (919) 515-1457 North Carolina State Univ; Sp
Cassel, Elizabeth J., (208) 885-4289 ecassel@uidaho.edu, Univ of Idaho; Gs
Cassiani, Giorgio, +390498279189 giorgio.cassiani@unipd.it, Univ degli Studi di Padova; YgHs
Cassidy, Daniel P., (269) 387-5324 daniel.cassidy@wmich.edu, Western Michigan Univ; Hw
Cassidy, John, cassidy@pgc.nrcan.gc.ca, Univ of Victoria; Ys
Cassidy, Martin, 713-616-5853 mcassidy@uh.edu, Univ of Houston; GoCo
Castagna, John P., 713-743-8699 jpcastagna@uh.edu, Univ of Houston; Ye
Castañeda, Isla, 413 577-1124 isla@geo.umass.edu, Univ of Massachusetts, Amherst; Ca
Castelli, Daniele, daniele.castelli@unito.it, Univ di Torino; Gx
Castillo, Paterno R., (858) 534-0383 pcastillo@ucsd.edu, Univ of California, San Diego; Giu
Castillon, David A., (417) 836-5801 davidcastillon@missouristate.edu, Missouri State Univ; Gm
Castiñeiras, Pedro, castigar@ucm.es, Univ Compl de Madrid; Gp

Castle, James W., (864) 656-5015 jcastle@clemson.edu, Clemson Univ; GseHw
Castleton, Jessica, (801) 537-3381 jessicacastleton@utah.gov, Utah Geo Survey; Ng
Castor, Stephen B., 775-682-8766 scastor@unr.edu, Univ of Nevada, Reno; Eg
Castro, Adrian, ac114@wellesley.edu, Wellesley Coll; Gp
Castro, Christopher, (520) 626-5617 clcastro@email.arizona.edu, Univ of Arizona; As
Castro, Maria Clara, 734-615-3812 mccastro@umich.edu, Univ of Michigan; Hy
Castro, Mark S., (301) 689-3115 castro@al.umces.edu, Univ of Maryland; As
Castro, Raul, raul@cicese.mx, Centro de Investigación Científica y de Educación Superior de Ensenada; Ys
Caswell, Bryony, +440151 795 4390 b.a.caswell@liverpool.ac.uk, Univ of Liverpool; Ob
Catalano, Jeff, 314-935-6015 catalano@levee.wustl.edu, Washington Univ in St. Louis; Ca
Catalano, Jeffrey G., (314) 935-6015 catalano@levee.wustl.edu, Washington Univ in St. Louis; Cl
Catania, Ginny A., (512) 471-0403 gcatania@utig.ig.utexas.edu, Univ of Texas, Austin; Gl
Catau, John C., (417) 836-5801 johncatau@missouristate.edu, Missouri State Univ; Zn
Cates, Anna, 612-625-3135 catesa@umn.edu, Univ of Minnesota, Twin Cities; Sb
Cathcart, Eric, (619) 260-4099 cathcart@sandiego.edu, Univ of San Diego; Geg
Cather, Steven, 575-835-5153 steven.cather@nmt.edu, New Mexico Tech; Gg
Cathey, Henrietta, +61 7 3138 0416 henrietta.cathey@qut.edu.au, Queensland Univ of Tech; Giv
Cathles, Lawrence M., (607) 255-2844 lmc19@cornell.edu, Cornell Univ; GqYgEg
Catling, David C., (206) 543-8653 dcatling@uw.edu, Univ of Washington; CgAsXb
Catlos, Elizabeth J., 512-471-4762 ejcatlos@gmail.com, Univ of Texas, Austin; GzCg
Cato, Kerry, (909) 537-5409 kerry.cato@csusb.edu, California State Univ, San Bernardino; Ng
Cattanach, Bart, 828-296-4634 bart.cattanach@ncdenr.gov, N.C. Dept of Env Quality; Gc
Cattania, Camilla, camcat@mit.edu, Massachusetts Inst of Tech; Ys
Cattolico, Rose Ann, (206) 543-9363 racat@u.washington.edu, Univ of Washington; Zn
Catuneanu, Octavian, (780) 492-6569 octavian.catuneanu@ualberta.ca, Univ of Alberta; Gs
Caudill, Michael, (304) 696-6721 Marshall Univ; GmNgHw
Caupp, Craig L., (301) 687-4755 ccaupp@frostburg.edu, Frostburg State Univ; ZnHs
Cave, Rachel R., +353 (0)91 492 351 rachel.cave@nuigalway.ie, Natl Univ of Ireland Galway; Cm
Cavello, Seth, cavello@usf.edu, Univ of South Florida; Zn
Cawood, Peter A., (+44) (0)1334 463911 peter.cawood@monash.edu, Monash Univ; GtcGg
Caylor, Kelly, caylor@ucsb.edu, Univ of California, Santa Barbara; HgZcy
Cayzer, Nicola, +44 (0) 131 650 8527 nicola.cayzer@ed.ac.uk, Edinburgh Univ; Yg
Cecil, M. R., (818) 677-7009 robinson.cecil@csun.edu, California State Univ, Northridge; GiCcGz
Celia, Michael A., (609) 258-5425 Princeton Univ; Hw
Celik, Hasan, 00904242370000-5980 hasancelik@firat.edu.tr, Firat Univ; GscEo
Çemen, Ibrahim, 205-348-8019 icemen@ua.edu, Univ of Alabama; Gct
Cenpírek, Jan, +420 549 49 6450 jcemp@sci.muni.cz, Masaryk Univ; Gz
Centorbi, Tracey, (301) 405-6965 tlcento@umd.edu, Univ of Maryland; Cg
Cepeda, Joseph C., 806-651-2272 jcepeda@mail.wtamu.edu, West Texas A&M Univ; GiHw
Cercone, Karen Rose, (724) 357-5623 kcercone@iup.edu, Indiana Univ of Pennsylvania; Zeg
Cerling, Thure E., (801) 581-5558 thure.cerling@utah.edu, Univ of Utah; ClsZc
Cerrato, Robert M., (631) 632-8666 robert.cerrato@stonybrook.edu, SUNY, Stony Brook; Ob
Cervato, Cinzia C., (515) 294-7583 cinzia@iastate.edu, Iowa State Univ of Sci & Tech; GgZe
Cesare, Bernardo, +390498279148 bernardo.cesare@unipd.it, Univ degli Studi di Padova; Gp
Cessi, Paola, (858) 534-0622 pcessi@ucsd.edu, Univ of California, San Diego; Op
Cetin, Haluk, (270) 809-2085 hcetin@murraystate.edu, Murray State Univ; ZrGe
Cetindag, Bahattin, 00904242370000-5957 bcetindag@firat.edu.tr, Firat Univ; NgYv
Cey, Edwin, (403) 220-8393 ecey@ucalgary.ca, Univ of Calgary; Hw
Chacko, Elizabeth, (202) 994-6185 echack@gwu.edu, George Washington Univ; Zy
Chacko, Thomas, (780) 492-5395 tom.chacko@ualberta.ca, Univ of Alberta; Gp
Chaddock, Lisa, 6196604000 x3033 lisa.chaddock@gcccd.edu, Cuyamaca Coll; Zy
Chadima, Sarah A., 605-677-6166 sarah.chadima@usd.edu, South Dakota Dept of Environment and Nat Res; Gg
Chadwick, John, (843) 953-5950 chadwickj@cofc.edu, Coll of Charleston; GivZr
Chafetz, Henry S., (713) 743-3227 hchafetz@uh.edu, Univ of Houston; Gds
Chakhmouradian, Anton, (204) 474-7278 anton.chakhmouradian@umanitoba.ca, Univ of Manitoba; GziCg
Challads, Thomas, +44 (0) 131 650 8543 tchallan@staffmail.ed.ac.uk, Edinburgh Univ; Hg
Challinor, Andy, +44(0) 113 34 33194 a.j.challinor@leeds.ac.uk, Univ of Leeds; Am
Chamberlain, Andrew T., +44 (0)161 306 4176 andrew.chamberlain@manchester.ac.uk, Univ of Manchester; GaSaPv
Chamberlain, Edwin P., (718) 951-5416 Los Alamos; Cc
Chamberlain, John A., (718) 951-5926 johnc@brooklyn.cuny.edu, Graduate Sch of the City Univ of New York; Pg
Chamberlain, Kevin R., (307) 766-2914 kchamber@uwyo.edu, Univ of Wyoming; CcGt
Chamberlin, Richard, (575) 835-5310 richard@nmbg.nmt.edu, New Mexico Tech; GcvGg
Chamberlin, William S., (714) 992-7443 schamberlin@fullcoll.edu, Fullerton Coll; OgbZe
Chambers, Jeanne, 775-784-5329 chambers@unr.edu, Univ of Nevada, Reno; SfHg
Chambers, John, jchambers@carnegiescience.edu, Carnegie Inst for Sci; Xs
Chameides, William L., 613-8004 bill.chameides@duke.edu, Duke Univ; AsGe
Chametzky, Peter, 803-777-8833 chametzk@mailbox.sc.edu, Univ of South Carolina; Zn
Champion, Kyle M., 813-253-7326 Hillsborough Comm Coll; Zg
Champlin, Steven D., (601) 961-5506 stephen_champlin@deq.state.ms.us, Mississippi Office of Geology; Go
Chan, Clara S., 302-831-1819 cschan@udel.edu, Univ of Delaware; PoObGz
Chan, Kwan M., (562) 985-4817 kmchan@csulb.edu, California State Univ, Long Beach; Oc
Chan, Marjorie A., (801) 581-6551 marjorie.chan@utah.edu, Univ of Utah; Gs
Chan, Queenie, +44 1784 443896 queenie.chan@rhul.ac.uk, Royal Holloway Univ of London; XmCa
Chan, Selene, (604) 822-2034 schan@eos.ubc.ca, Univ of British Columbia; Zn
Chandler, Angela, (501) 683-0111 angela.chandler@arkansas.gov, Arkansas Geo Survey; GgrGs
Chandler, Cyndy, (508) 289-2765 cchandler@whoi.edu, Woods Hole Oceanographic Inst; Og
Chandler, Sandra, (501) 683-0125 sandra.chandler@arkansas.gov, Arkansas Geo Survey; Ze
Chandler, Val W., 612-627-4780 (Ext. 203) chand004@umn.edu, Univ of Minnesota, Twin Cities; Ye
Chandra, Sudeep, 775.784.6221 sudeep@unr.edu, Univ of Nevada, Reno; Hs
Chaney, Phil L., (334) 844-3420 chanepl@auburn.edu, Auburn Univ; ZnnZn
Chang, Chih-Pei, 831-656-2840 cpchang@nps.edu, Naval Postgrad Sch; Am
Chang, Edmund K., (631) 632-6170 kar.chang@stonybrook.edu, SUNY, Stony Brook; As
Chang, Hai-ru, hrc@eas.gatech.edu, Georgia Inst of Tech; As
Chang, Ping, (979) 845-8196 pchang@ocean.tamu.edu, Texas A&M Univ; Op
Chang, Young-Soo, (630) 252-4076 changy@anl.gov, Argonne; As
Chang, Zhaoshan, 303-384-2127 chang@mines.edu, Colorado Sch of Mines; Eg
Channell, James E., 352-392-3658 jetc@ufl.edu, Univ of Florida; YmGt
Channing, Alan, +44(0)29 208 76213 channinga@cardiff.ac.uk, Univ of Wales; Gd
Chanway, Christopher, 604–822–3716 christopher.chanway@ubc.ca, Univ of British Columbia; Sb
Chao, Shenn-Yu, (410) 221-8427 chao@hpl.umces.edu, Univ of Maryland;

Op
Chapa, Laura E., (512) 820-9079 laura.chapa@austincc.edu, Austin Comm Coll District; ZgeZr
Chapin, Charles E., cechapin2@comcast.net, New Mexico Tech; Gt
Chapin, F. Stuart, (510) 642-1003 Univ of California, Berkeley; Pe
Chapman, Alan, chapman@macalester.edu, Macalester Coll; GctCc
Chapman, James, jay.chapman@uwyo.edu, Univ of Wyoming; Gct
Chapman, Mark, +44 (0)1603 59 3114 mark.chapman@uea.ac.uk, Univ of East Anglia; Pm
Chapman, Marshall, (606) 783-5397 m.chapman@moreheadstate.edu, Morehead State Univ; Gv
Chapman, Martin C., (540) 231-5036 mcc@vt.edu, Virginia Tech; Ys
Chapman, Piers, (979) 845-9399 piers.chapman@tamu.edu, Texas A&M Univ; OcCm
Chapman, Robert J., +44(0) 113 34 33190 r.j.chapman@leeds.ac.uk, Univ of Leeds; EmCe
Chapman, Ross N., (250) 472-4340 chapman@uvic.ca, Univ of Victoria; Op
Chappaz, Anthony, 989-774-4388 chapp1a@cmich.edu, Central Michigan Univ; Cat
Chappell, James R., (970) 491-5147 jim_chappell@partner.nps.gov, Colorado State Univ; Ge
Chappell, P. D., 757-683-4937 pdchappe@odu.edu, Old Dominion Univ; Oc
Chaput, Julien, 915-747-7003 jachaput@utep.edu, Univ of Texas, El Paso; GeYes
Charbonnier, Sylvain, sylvain@usf.edu, Univ of South Florida; Gv
Charette, Matthew A., (508) 289-3205 mcharette@whoi.edu, Woods Hole Oceanographic Inst; Cm
Charles, Christopher D., (858) 534-5911 ccharles@ucsd.edu, Univ of California, San Diego; Pe
Charles, Kasanzu, kcharls16@yahoo.com, Univ of Dar es Salaam; GgRh
Charlevoix, Donna J., (217) 244-9575 charlevo@atmos.uiuc.edu, Univ of Illinois, Urbana-Champaign; As
Charlson, Bob, 206-543-2537 Univ of Washington; As
Charlson, Robert J., charlson@chem.washington.edu, Univ of Washington; As
Chase, Anne, (520) 621-6004 achase@email.arizona.edu, Univ of Arizona; Zn
Chase, Clement G., (520) 621-2417 cgchase@email.arizona.edu, Univ of Arizona; Yg
Chase, Jon M., jchase@biology2.wustl.edu, Washington Univ in St. Louis; Zn
Chase, Peter M., (608) 265.6003 pete.chase@wisc.edu, Univ of Wisconsin, Madison, Division of Extension; HwYgRw
Chase, Richard L., (604) 822-3086 rchase@eos.ubc.ca, Univ of British Columbia; Gu
Chasteen, Hayden R., 817-515-6694 hayden.chasteen@tccd.edu, Tarrant County Coll, Northeast Campus; GgeGg
Chatelain, Edward E., (229) 333-5758 echatela@valdosta.edu, Valdosta State Univ; PiGlPv
Chatterjee, Ipsita, 940-565-2372 ipsita.chatterjee@unt.edu, Univ of North Texas; Zn
Chatterjee, Meera, 330 972-2394 meera@uakron.edu, Univ of Akron; Zy
Chatterjee, Nilanjan, (617) 253-1995 nchat@mit.edu, Massachusetts Inst of Tech; Cg
Chatterjee, Sankar, 806-834-4590 sankar.chatterjee@ttu.edu, Texas Tech Univ; PvGct
Chatterjee, Snehamoy, 906-487-2516 schatte1@mtu.edu, Michigan Technological Univ; NmGqNx
Chatterton, Brian D., (780) 492-3085 brian.chatterton@ualberta.ca, Univ of Alberta; Pi
Chaudhuri, Sambhudas, (785) 532-2246 ksuncsc@ksu.edu, Kansas State Univ; GeoCc
Chaumba, Jeff B., (910) 522-5787 jeff.chaumba@uncp.edu, Univ of North Carolina, Pembroke; GxEgGz
Chavas, Daniel , 765-496-6661 dchavas@purdue.edu, Purdue Univ; As
Chavrit, Deborah, +44 0161 275-0760 deborah.chavrit@manchester.ac.uk, Univ of Manchester; Gi
Cheadle, Burns A., 519-661-2111 x.89009 bcheadle@uwo.ca, Western Univ; Go
Cheadle, Michael J., (307) 766-3206 cheadle@uwyo.edu, Univ of Wyoming; Yg
Chebana, Fateh, fateh.chebana@ete.inrs.ca, Univ du Quebec; Hq
Checkley, David M., (858) 534-4228 dcheckley@ucsd.edu, Univ of California, San Diego; Ob
Cheek, William H., (417) 836-5801 billcheek@missouristate.edu, Missouri State Univ; Zn
Cheel, Richard J., (905) 688-5550 (Ext. 3512) rcheel@brocku.ca, Brock Univ; Gs
CHEGBELEH, Larry-Pax, lpchegbeleh@ug.edu.gh, Univ of Ghana; Ng
Chege, George W., gchege@jkuat.ac.ke, Jomo Kenyatta Univ of Agriculture & Tech; ZgrZi
Chelariu, Ciprian, 0040232201486 ciprian.chelariu@uaic.ro, Alexandru Ioan Cuza; Go
Chen, Bob, bob.chen@umb.edu, Univ of Massachusetts, Boston; CoOc
Chen, Chu-Yung, cychen7@illinois.edu, Univ of Illinois, Urbana-Champaign; Cg
Chen, Dake, dchen@ldeo.columbia.edu, Columbia Univ; Op
Chen, Heng, hc3158@columbia.edu, Columbia Univ; Cg
Chen, Hway-Jen, 831-656-3788 hjchen@nps.edu, Naval Postgrad Sch; Am
Chen, Jing, chenj@geog.utoronto.ca, McMaster Univ; AmZry
Chen, Jingyi, 918-631-3014 jingyi-chen@utulsa.edu, Univ of Tulsa; Ygs
Chen, jiquan, (419) 932-1517 jqchen@msu.edu, Michigan State Univ; Zg
Chen, Jiuhua, 305-348-3030 chenj@fiu.edu, Florida InterNatl Univ; Gyz
Chen, Ning, ning.chen@lightsource.ca, Univ of Saskatchewan; Gz
Chen, Po, 307-766-3086 pchen@uwyo.edu, Univ of Wyoming; YseYg
Chen, Shu-Hua, 530-752-1822 shachen@ucdavis.edu, Univ of California, Davis; ZrAsm
Chen, Shuyi S., (206) 685-1736 shuyic@uw.edu, Univ of Washington; As
Chen, Tsing-Chang, (515) 294-9874 tmchen@iastate.edu, Iowa State Univ of Sci & Tech; As
Chen, Wang-Ping, wpchen@illinois.edu, Univ of Illinois, Urbana-Champaign; YsGt
Chen, Xianfeng, (724) 738-2385 xianfeng.chen@sru.edu, Slippery Rock Univ; Zr
Chen, xiaowei, (979)845-2400 xiaowei.chen@exchange.tamu.edu, Texas A&M Univ; YsuGt
Chen, Xiaowei, 508-289-3820 xiaowei.fengr@gmail.com, Univ of Oklahoma; YsGt
Chen, Yi-Leng, (808) 956-2570 yileng@hawaii.edu, Univ of Hawai'i at Manoa; Asm
Chen, Yongsheng, (416)736-2100 #40124 yochen@yorku.ca, York Univ; As
Chen, Yuan, 250-807-9594 yuan.chen@ubc.ca, Univ of British Columbia, Okanagan Campus; EmCeGz
Cheney, Donald, d.cheney@neu.edu, Northeastern Univ; Zn
Cheney, Eric S., 206-543-1163 vaalbara@uw.edu, Univ of Washington; Eg
Cheney, John T., (413) 542-2311 jtcheney@amherst.edu, Amherst Coll; Gi
Cheng, Chu-Lin, 956 665-2464 chulin.cheng@utrgv.edu, Univ of Texas, Rio Grande Valley; HgwHq
Cheng, Hai, 86-29-83395119 cheng021@xjtu.edu.cn, Univ of Minnesota, Twin Cities; CcPe
Cheng, Linyin, 479-575-5569 lc032@uark.edu, Univ of Arkansas, Fayetteville; Zo
Cheng, Meng-Dawn, (423) 241-5918 ucn@ornl.gov, Oak Ridge; As
Cheng, Qiuming, (416)736-2100 #22842 qiuming@yorku.ca, York Univ; Zi
Cheng, Songlin, 937 775-3455 songlin.cheng@wright.edu, Wright State Univ; Hw
Cheng, Tao, 709 864 8924 tcheng@mun.ca, Memorial Univ of Newfoundland; Hw
Cheng, Yanbo, 07 47816808 yanbo.cheng1@jcu.edu.au, James Cook Univ; EgGgEm
Cheng, Zhongqi, 7189515000 x2647 zcheng@brooklyn.cuny.edu, Brooklyn Coll (CUNY); CaScGe
Cheng, Zhongqi (Joshua), 718-951-5000, ext. 2647 zcheng@brooklyn.cuny.edu, Graduate Sch of the City Univ of New York; Zg
Chenoweth, Cheri, 217-244-4610 cchenowe@illinois.edu, Illinois State Geo Survey; GoEc
Chenoweth, M. S., (318) 342-1887 chenoweth@ulm.edu, Univ of Louisiana, Monroe; GmZir
Cherkauer, Douglas S., (262) 628-3672 aquadoc@uwm.edu, Univ of Wisconsin, Milwaukee; HwGe
Chermak, John A., 540-239-4504 jchermak@vt.edu, Virginia Tech; ClGeEd
Chernoff, Barry, 860 685-2452 bchernoff@wesleyan.edu, Wesleyan Univ; Ge
Cherns, L C., cherns@cf.ac.uk, Cardiff Univ; Pg
Cherrier, Jennifer, 718-951-5000, ext. 2927 jennifer.cherrier18@brooklyn.cuny.edu, Graduate Sch of the City Univ of New York; Zg
Cherubini, Claudia, +33(0)3 44068977 claudia.cherubini@lasalle-beauvais.fr, Institut Polytechnique LaSalle Beauvais (ex-IGAL); HwGe
Cherukupalli, Nehru, 516 463-6545 geonec@hofstra.edu, Hofstra Univ; Gg
Chesnaux, Romain, 418-545-5011 ext: 5426 rchesnaux@uqac.ca, Univ du Quebec a Chicoutimi; Hw
Chesner, Craig A., (217) 581-2626 cachesner@eiu.edu, Eastern Illinois Univ; GivGz
Chesnokov, Evgeny, 713-743-2579 emchesnokov@uh.edu, Univ of Houston; YseYx
Chester, Frederick, (979) 845-3296 chesterf@tamu.edu, Texas A&M Univ; GcYxGt

Chester, Judith, (979) 845-1380 chesterj@geo.tamu.edu, Texas A&M Univ; GcNr
Chesworth, Ward, (519) 824-4120 (Ext. 52457) wcheswor@uoguelph.ca, Univ of Guelph; ClGe
Cheun, Norman, +44 020 8417 2811 k.w.cheung@kingston.ac.uk, Kingston Univ; Ge
Cheung, Wing H., 7607441150 ext. 3652 wcheung@palomar.edu, Palomar Coll; ZiRnZr
Chevrel, M. Oryaelle, mchevrel@buffalo.edu, SUNY, Buffalo; Gv
Chew, David, + 353 1 8963481 chewd@tcd.ie, Trinity Coll; CcGgt
Chhetri, Parveen, (310) 243-3377 pchhetri@csudh.edu, California State Univ, Dominguez Hills; Geg
Chi, Wu-Cheng, 886-2-2783-9910 ext 510 wchi@sinica.edu.tw, Academia Sinica; GtYrs
Chiappe, Luis M., (213) 863-3323 lchiappe@nhm.org, Los Angeles County Mus of Natural History; Pv
Chiarella, Domenico, +44 1784 443890 domenico.chiarella@rhul.ac.uk, Royal Holloway Univ of London; Gs
Chiarello, Ronald P., (630) 252-9327 Argonne; Zm
Chiarenzelli, Jeffrey R., 315-229-5202 jchiarenzelli@stlawu.edu, St. Lawrence Univ; GzCg
Chief, Karletta, (520) 626-5598 kchief@email.arizona.edu, Univ of Arizona; Zn
Chien, Yi-Ju, (509) 373-4822 yi-hu.chien@pnl.gov, Pacific Northwest; Gq
Childers, Daniel, (610) 359-5242 dchilder@dccc.edu, Delaware County Comm Coll; ZgOuZi
Childs, Geoff, gchilds@wustl.edu, Washington Univ in St. Louis; Zn
Chillrud, Steven, chilli@ldeo.columbia.edu, Columbia Univ; Cg
Chilson-Parks, Benjamin, benjami.chilson-parks@uconn.edu, Univ of Connecticut; Gi
Chilvers, Jason, +44 (0)1603 59 3130 jason.chilvers@uea.ac.uk, Univ of East Anglia; Ge
Chin, Anne, (409) 845-7141 Texas A&M Univ; Gm
Chin, Emily, e8chin@ucsd.edu, Univ of California, San Diego; CuGx
Chin, Emily J., e8chin@ucsd.edu, Univ of Tennessee, Knoxville; Cu
Chin, Karen, 303-735-3074 karen.chin@colorado.edu, Univ of Colorado; Pg
Chin, Shamar C., (319) 467-3148 shamar-chin@uiowa.edu, Univ of Iowa; PmOoPe
Chin, Yu-Ping, (614) 292-6953 chin.15@osu.edu, Ohio State Univ; CqHw
Chipman, Melissa, 315-443-2489 mlchipma@syr.edu, Syracuse Univ; GelGn
Chipperfield, Martyn P., +44(0) 113 34 36459 m.chipperfield@leeds.ac.uk, Univ of Leeds; AscAp
Chipping, David H., (805) 528-0914 dchippin@calpoly.edu, California Polytechnic State Univ; Grt
Chira, Carme Mariana, carmen.chira@ubbcluj.ro, Babes-Bolyai Univ; PemZe
Chirenje, Tait, 609-652-4588 tait.chirenje@stockton.edu, Stockton Univ; Sc
Chiu, Ching-Sang, (831) 656-3239 chiu@nps.edu, Naval Postgrad Sch; Op
Chiu, Christine, christine.chiu@colostate.edu, Colorado State Univ; As
Chiu, Jer-Ming, (901) 678-2007 jerchiu@memphis.edu, Univ of Memphis; Ys
Chiu, Long S., (703) 993-1984 lchiu@gmu.edu, George Mason Univ; AsZrAm
Choate, Jessica (Jessie), choate2@illinois.edu, Univ of Illinois, Urbana-Champaign; Asm
Choh, Suk-Joo, 82-2-3290-3180 sjchoh@korea.ac.kr, Korea Univ; GdsPe
Choi, Eunseo, (901) 678-4923 echoi2@memphis.edu, Univ of Memphis; Gtq
Choi, Seon-Gyu, seongyu@korea.ac.kr, Korea Univ; GzEmg
Chokmani, Karem, karem.chokmani@ete.inrs.ca, Univ du Quebec; Zi
Cholnoky, Jennifer, (518) 580-8127 jcholnok@skidmore.edu, Skidmore Coll; Gg
Choma-Moryl, Krystyna, krystyna.choma-moryl@uwr.edu.pl, Univ of Wroclaw; Ng
Chongwain, Gilbert, gchongwai@unam.na, Univ of Namibia; Go
Chopping, Mark J., 973-655-4448 choppingm@mail.montclair.edu, Montclair State Univ; ZriZe
Choquette, Marc, choquette@ggl.ulaval.ca, Univ Laval; Zn
Chorover, Jon, 520-626-5635 chorover@cals.arizona.edu, Univ of Arizona; Sc
Chorover, Jonathan, (520) 621-7228 chorover@cals.arizona.edu, Univ of Arizona; ScClo
Chou, Charissa J., (509) 372-3804 charissa.chou@pnl.gov, Pacific Northwest; Gq
Chouinard, Vera, (905) 525-9140 (Ext. 23518) chouinar@mcmaster.ca, McMaster Univ; Zn
Choularton, Thomas, +44 0161 306-3950 choularton@manchester.ac.uk, Univ of Manchester; As
Chow, Edwin, (512) 245-2170 Texas State Univ; ZiRn
Chow, Nancy, (204) 474-6451 n_chow@umanitoba.ca, Univ of Manitoba; Gs
Chowdhury, Shafiul H., chowdhus@newpaltz.edu, SUNY, New Paltz; HwGe
Chowns, Timothy M., (678) 839-4052 tchowns@westga.edu, Univ of West Georgia; GsrGg
Christensen, Douglas, doug@giseis.alaska.edu, Univ of Alaska, Fairbanks; Ys
Christensen, Nikolas I., (608) 265-4469 chris@geology.wisc.edu, Univ of Wisconsin, Madison; Yx
Christensen, Philip R., (480) 965-7105 phil.christensen@asu.edu, Arizona State Univ; Xg
Christensen, Wesley P., 605-677-6149 wes.christensen@usd.edu, South Dakota Dept of Environment and Nat Res; Gg
Christenson, Scott, scott.christenson@nmt.edu, New Mexico Tech; Hw
Christeson, Gail L., 512-471-0463 gail@ig.utexas.edu, Univ of Texas, Austin; Yr
Christian, Alan D., 617-287-6639 alan.christian@umb.edu, Univ of Massachusetts, Boston; HsGmCs
Christian, James R., jim.christian@canada.ca, Univ of Victoria; Obc
Christiansen, Eric H., 801422 2113 eric_christiansen@byu.edu, Brigham Young Univ; GiXgGv
Christianson, Knut, knut@uw.edu, Univ of Washington; GlZr
Christie, Max L., 217-333-3540 mlc7@illinois.edu, Univ of Illinois, Urbana-Champaign; PgGr
Christie-Blick, Nicholas, ncb@ldeo.columbia.edu, Columbia Univ; Gs
Christopherson, Gary, garych@casa.arizona.edu, Univ of Arizona; Zi
Chrzan, Daryl, (510) 643-1624 dcchrzan@berkeley.edu, Univ of California, Berkeley; Zm
Chu, Pao-Shin, chu@hawaii.edu, Univ of Hawai'i at Manoa; As
Chu, Peter C., (831) 656-3688 pcchu@nps.edu, Naval Postgrad Sch; OpZr
Chuang, Patrick Y., (831) 459-1501 pchuang@pmc.ucsc.edu, Univ of California, Santa Cruz; As
Church, Matthew, (808) 956-8779 mjchurch@hawaii.edu, Univ of Hawai'i, Manoa; Ob
Church, Thomas M., (302) 831-2558 tchurch@udel.edu, Univ of Delaware; Cm
Church, Warren, 706-507-8093 church_warren@columbusstate.edu, Columbus State Univ; ZnPcZc
Church, William R., (519) 661-3192 wrchurch@uwo.ca, Western Univ; Gt
Churchill, Ron C., (916) 327-0745 California Geo Survey; Cg
Ciampitti, Ignacio, (785) 532-6940 ciampitti@ksu.edu, Kansas State Univ; So
Cianfrani, Christina, ccns@hampshire.edu, Hampshire Coll; Hg
Ciannelli, Lorenzo, 541-737-3142 lciannelli@coas.oregonstate.edu, Oregon State Univ; Og
Cicerone, Robert D., (508) 531-2713 rcicerone@bridgew.edu, Bridgewater State Univ; YsgXy
Cich, Anna, (801) 626-7139 annacich@weber.edu, Weber State Univ; Zn
Cicha, Jarrod, 612-626-4468 cich0060@morris.umn.edu, Univ of Minnesota; Zi
Ciciarelli, John A., (412) 773-3867 jac7@psu.edu, Pennsylvania State Univ, Monaca; Gg
Cicimurri, Christian M., (864)656-4602 cmcici@clemson.edu, Clemson Univ; Pv
Cicimurri, David J., (864) 656-4601 dcheech@clemson.edu, Clemson Univ; Pv
Ciesla, Fred, (773) 702-8169 Univ of Chicago; Xcg
Cigolini, Corrado, corrado.cigolini@unito.it, Univ di Torino; Gv
Cihacek, Larry J., (701) 231-8572 North Dakota State Univ; Zg
Cilliers, Johannes, +44 20 759 47360 j.j.cilliers@imperial.ac.uk, Imperial Coll; Gz
Cintra-Buenrostro, Carlos, 956 882-5746 carlos.cntra@utrgv.edu, Univ of Texas, Rio Grande Valley; PiCsPe
Ciolkosz, Edward J., (814) 865-1530 f8i@psu.edu, Pennsylvania State Univ, Univ Park; Sd
Cioppa, Maria T., 519-253-3000 x2502 mcioppa@uwindsor.ca, Univ of Windsor; Ym
Cipar, John J., 617-552-8300 cipar@bc.edu, Boston Coll; Ys
Cirimpei, Claudia, (023) 220-1496 claudia.cirimpei@yahoo.com, Alexandru Ioan Cuza; PgGr
Cirmo, Christopher P., (607) 753-2924 cirmoc@cortland.edu, SUNY, Cortland; Hg
Cisne, John L., john.cisne@cornell.edu, Cornell Univ; PgsPe
Civan, Faruk, (405) 325-6778 fcivan@ou.edu, Univ of Oklahoma; Np
Claassen, Mark, (785) 532-6101 mclaasse@ksu.edu, Kansas State Univ; So
Claassen, Victor P., vpclaassen@ucdavis.edu, Univ of California, Davis; So
Clabo, Darren R., 605-394-1996 darren.clabo@sdsmt.edu, South Dakota Sch

FACULTY INDEX – C

of Mines & Tech; AssAs
Cladouhos, Trenton, 206 729-2400 ttcladouhos@gmail.com, Univ of Washington; Gcg
Claerbout, Jon F., (650) 723-3717 Stanford Univ; Ye
Claeys, Philippe, (322) 629-3391 phclaeys@vub.be, Vrije Univ Brussel; CgaXc
Clague, John J., (778) 782-4657 Simon Fraser Univ; Rn
Claiborne, Lily L., (615) 343-4515 lily.claiborne@vanderbilt.edu, Vanderbilt Univ; GivCg
Claire, Mark, +44 01334 463688 mc229@st-andrews.ac.uk, Univ of St. Andrews; As
Clarey, Timothy L., (989) 686-9252 tlclarey@delta.edu, Delta Coll; HwGcPv
Clark, Alan H., (613) 533-6187 Queen's Univ; Em
Clark, David L., (505) 665-0005 Los Alamos; Zn
Clark, Donald, (801) 537-3344 donclark@utah.gov, Utah Geo Survey; Gg
Clark, Douglas H., (360) 650-7939 doug.clark@wwu.edu, Western Washington Univ; Gl
Clark, George S., (204) 474-7343 gs_clark@umanitoba.ca, Univ of Manitoba; Cc
Clark, H. C., (713) 527-4887 hcclark@owlnet.rice.edu, Rice Univ; YgGeYu
Clark, Ian D., idclark@uottawa.ca, Univ of Ottawa; HwGel
Clark, James S., (919) 660-7402 jimclark@duke.edu, Duke Univ; Zn
Clark, Jeffrey J., (920) 832-6733 Lawrence Univ; Gm
Clark, Jessie , (775) 784-6345 jessieclark@unr.edu, Univ of Nevada, Reno; Zu
Clark, Joanna, j.m.clark@reading.ac.uk, Univ of Reading; Sc
Clark, John H., 814-863-1581 clark@ems.psu.edu, Pennsylvania State Univ, Univ Park; Am
Clark, Jordan F., (805) 893-7838 jfclark@geol.ucsb.edu, Univ of California, Santa Barbara; ClHg
Clark, Joseph C., (724) 357 5622 Indiana Univ of Pennsylvania; Gr
Clark, Kathryne, kathryne.clark@dnr.iowa.gov, Iowa Dept of Nat Res; GgZi
Clark, Ken P., (253) 879-3138 kclark@pugetsound.edu, Univ of Puget Sound; GgcGi
Clark, Kenneth F., 915-581-8371 clark@utep.edu, Univ of Texas, El Paso; Eg
Clark, Marin, 734-615-0484 marinkc@umich.edu, Univ of Michigan; Gm
Clark, Mark W., (352) 392-1803 (Ext. 316) clarkmw@ufl.edu, Univ of Florida; Sf
Clark, Martin D., 051 401 3892 Univ of the Free State; ZriGc
Clark, Michael, (403) 440-8944 mdclark@mtroyal.ca, Mount Royal Univ; Zg
Clark, Murlene W., 251-460-6381 mclark@southalabama.edu, Univ of South Alabama; PiGrg
Clark, Peter U., (541) 737-1247 clarkp@geo.oregonstate.edu, Oregon State Univ; Gl
Clark, Richard D., 717-872-3930 richard.clark@millersville.edu, Millersville Univ; As
Clark, Robert O., (928) 523-1321 robert.clark@nau.edu, Northern Arizona Univ; Zn
Clark, Roger A., +44(0) 113 34 35221 r.a.clark@leeds.ac.uk, Univ of Leeds; YseYs
Clark, Russell G., (517) 629-0312 rgclark@albion.edu, Albion Coll; GiZi
Clark, Scott K., (715) 836-2958 clarksco@uwec.edu, Univ of Wisconsin, Eau Claire; Ze
Clark, Shannon, (505) 277-1644 skclark@unm.edu, Univ of New Mexico; Zn
Clarke, Amanda, 480-965-6590 amanda.clarke@asu.edu, Arizona State Univ; Gv
Clarke, Antony D., (808) 956-6215 tclarke@soest.hawaii.edu, Univ of Hawai'i, Manoa; As
Clarke, Beverley, beverley.clarke@flinders.edu.au, Flinders Univ; Zy
Clarke, Garry K. C., (604) 822-3602 gclarke@eos.ubc.ca, Univ of British Columbia; Yg
Clarke, Geoffrey L., +61293512919 geoffrey.clarke@usyd.edu.au, Univ of Sydney; Gp
Clarke, Julia A., (512) 232-7563 julia_clarke@jsg.utexas.edu, Univ of Texas, Austin; PvgPq
Clarke, Peter J., +44 (0) 191 208 6351 peter.clarke@ncl.ac.uk, Univ of Newcastle Upon Tyne; YdZri
Clarke, Stu, (+44) 01782 733171 s.m.clarke@keele.ac.uk, Keele Univ; GsoGd
Clarkson, Christopher , 403 220-6445 clarksoc@ucalgary.ca, Univ of Calgary; Eo
Clary, Renee M., (662) 268-1032 x215 rclary@geosci.msstate.edu, Mississippi State Univ; ZePgGe
Class, Connie, class@ldeo.columbia.edu, Columbia Univ; Cg
Clausen, Benjamin L., (909) 558-4548 bclausen@llu.edu, Loma Linda Univ; GiCtYg
Clay, Patricia L., +44 0161 275-0407 patricia.clay@manchester.ac.uk, Univ of Manchester; CgaGg
Clay, Robert, 573/368-2177 bob.clay@dnr.mo.gov, Missouri Dept of Nat Res; Zn
Clayton, Lee, (608) 263-6839 lclayton@wisc.edu, Univ of Wisconsin, Extension; Gl
Clayton, Robert N., (773) 702-7777 Univ of Chicago; Cs
Clayton, Rodney L., 757-822-7089 rclayton@tcc.edu, Tidewater Comm Coll; GgOg
Clayton, Sophie, (757) 683-3867 sclayton@odu.edu, Old Dominion Univ; Cb
Cleary, William J., 910-962-2320 clearyw@uncw.edu, Univ of North Carolina, Wilmington; Gu
Cleaveland, Malcolm K., (479) 575-3355 mcleavel@uark.edu, Univ of Arkansas, Fayetteville; PcZc
Clebnik, Sherman M., (860) 465-4323 clebniks@easternct.edu, Eastern Connecticut State Univ; Gl
Clegg, Simon, +44 (0)1603 59 3185 s.clegg@uea.ac.uk, Univ of East Anglia; AsOcCg
Clemens, Steven C., (401) 863-1964 steven_clemens@brown.edu, Brown Univ; Ou
Clemens, William A., (510) 642-6675 Univ of California, Berkeley; Pv
Clemens-Knott, Diane, (657) 278-2369 dclemensknott@fullerton.edu, California State Univ, Fullerton; Gig
Clement, William P., (413) 545-5910 wclement@geo.umass.edu, Univ of Massachusetts, Amherst; Yue
Clementz, Mark T., (307) 766-6048 mclement1@uwyo.edu, Univ of Wyoming; Pg
Clemitshaw, Kevin C., +44 1784 414026 k.clemitshaw@rhul.ac.uk, Royal Holloway Univ of London; AsZg
Clendening, Ronald J., (615) 532-1504 ron.clendening@tn.gov, Tennessee Geo Survey; GgcGm
Clennan, Patrick D., 702-651-7501 patrick.clennan@csn.edu, Coll of Southern Nevada - West Charleston Campus; ZyGe
Clepper, Marta L., 607-436-3736 marta.clepper@oneonta.edu, SUNY, Oneonta; ZeGgs
Cleveland, Natasha, 301-846-2563 ncleveland@frederick.edu, Frederick Comm Coll; Zge
Clift, Peter, 225-578-2153 pclift@lsu.edu, Louisiana State Univ; GsuGo
Cline, Jean S., (702) 575-9968 jean.cline@unlv.edu, Univ of Nevada, Las Vegas; EmCe
Cloos, Mark P., (512) 471-4170 cloos@jsg.utexas.edu, Univ of Texas, Austin; Gcx
Closs, L. Graham, (303) 273-3856 lcloss@mines.edu, Colorado Sch of Mines; Ce
Clothiaux, Eugene E., 814-865-2915 eec3@psu.edu, Pennsylvania State Univ, Univ Park; Am
Clough, Gene A., (207) 786-6396 gclough@bates.edu, Bates Coll; Ym
Cloutier, Danielle, (418) 656-7679 danielle.cloutier@ggl.ulaval.ca, Univ Laval; GeOg
Cloutis, Ed, 204-786-9386 e.cloutis@uwinnipeg.ca, Western Univ; Zi
Clowes, Ronald M., (604) 822-4138 clowes@eos.ubc.ca, Univ of British Columbia; Ys
Clutter, Melissa, 970-247-6170 mjclutter@fortlewis.edu, Fort Lewis Coll; Ge
Clyde, William C., 603-862-3148 will.clyde@unh.edu, Univ of New Hampshire; PgYmGs
Cnudde, Veerle, veerle.cnudde@ugent.be, Ghent Univ; EsHt
Coakley, Bernard, 907-474-5385 bernard.coakley@gi.alaska.edu, Univ of Alaska, Fairbanks; Yg
Coale, Kenneth H., (831) 771-4400 coale@mlml.calstate.edu, Moss Landing Marine Lab; CtcOc
Coats, Sloan, scoats@hawaii.edu, Univ of Hawai'i, Manoa; Zo
Cobb, Kim, kim_cobb@brown.edu, Brown Univ; GeOp
Cobb, Steven R., 806-834-1395 steve.cobb@ttu.edu, Texas Tech Univ; As
Coble, Paula G., (727) 553-1631 pcoble@marine.usf.edu, Univ of South Florida; Cm
Coburn, Craig, (403) 317-2818 craig.coburn@uleth.ca, Univ of Lethbridge; Zy
Coburn, Daniel, 203-392-5835 coburnd1@southernct.edu, Southern Connecticut State Univ; Ze
Coch, Nicholas K., (718) 997-3326 nicholas.coch@qc.cuny.edu, Queens Coll (CUNY); GseOu
Cochran, David, 601-266-6014 david.cochran@usm.edu, Univ of Southern Mississippi; Zn
Cochran, Elizabeth, cochran@ucr.edu, Univ of California, Riverside; Ys
Cochran, J. K., (631) 632-8746 kcochran@notes.cc.sunysb.edu, Stony Brook Univ; Oc
Cochran, J. Kirk, (631) 632-8733 kirk.cochran@stonybrook.edu, SUNY,

Stony Brook; Oc
Cochran, James R., jrc@ldeo.columbia.edu, Columbia Univ; Yr
Cochrane, Claudia, ccochra3@uwo.ca, Western Univ; Go
Codrea, Vlad, +40-264-405300 ext 5205 vlad.codrea@ubbcluj.ro, Babes-Bolyai Univ; PvGu
Cody, Anita M., amcody@iastate.edu, Iowa State Univ of Sci & Tech; Cl
Cody, George D., 202-478-8980 gcody@carnegiescience.edu, Carnegie Inst for Sci; Co
Cody, Robert, rdcody@iastate.edu, Iowa State Univ of Sci & Tech; Gz
Coe, Angela L., angela.coe@open.ac.uk, Open Univ; GsrCg
Coe, Douglas A., (406) 496-4207 dcoe@mtech.edu, Montana Tech of the Univ of Montana; Cg
Coe, Hue, +44 0161 306-9362 hugh.coe@manchester.ac.uk, Univ of Manchester; As
Coetzee, Serena, +27 (0)12 420 3823 serena.coetzee@up.ac.za, Univ of Pretoria; Zi
Coffey, Ruth, rcoffey@sandiego.edu, Univ of San Diego; Gug
Coffin, Richard, 361-825-2814 richard.coffin@tamucc.edu, Texas A&M Univ, Corpus Christi; Ouc
Coffman, David, david.coffman@dcccd.edu, El Centro Coll - Dallas Comm Coll District; Gg
Coffman, Stephanie, svcoffman@dcccd.edu, El Centro Coll - Dallas Comm Coll District; Gg
Coffroth, Mary Alice, 716-645-4871 coffroth@buffalo.edu, SUNY, Buffalo; Gu
Coggan, John, +44 01326 371824 j.coggan@exeter.ac.uk, Exeter Univ; Nx
Coggon, Jude A., +44 (0)23 80596539 jude.coggon@soton.ac.uk, Univ of Southampton; Cg
Coggon, Rosalind M., 023 80596539 r.m.coggon@soton.ac.uk, Univ of Southampton; Ce
Cohen, Alice, 902-585-1126 alice.cohen@acadiau.ca, Acadia Univ; ZnnZn
Cohen, Andrew S., (520) 621-4691 cohen@email.arizona.edu, Univ of Arizona; Ps
Cohen, Anne L., (508) 289-2958 acohen@whoi.edu, Woods Hole Oceanographic Inst; PecOo
Cohen, Anthony, anthony.cohen@open.ac.uk, Open Univ; Cs
Cohen, David R., 612 9385 8084 d.cohen@unsw.edu.au, Univ of New South Wales; Cet
Cohen, Joel E., (212) 327-8883 cohen@rockvax.rockefeller.edu, Columbia Univ; Zn
Cohen, Matt, (864) 294 - 2505 matthew.cohen@furman.edu, Furman Univ; Zn
Cohen, Phoebe A., 413-597-2358 phoebe.a.cohen@williams.edu, Williams Coll; Pgg
Cohen, Ronald E., 202-478-8937 rcohen@carnegiescience.edu, Carnegie Inst for Sci; Gy
Cohen, Shaul, (541) 346-4500 scohen@uoregon.edu, Univ of Oregon; Zn
Coker, Victoria, +44 0161 275-3803 vicky.coker@manchester.ac.uk, Univ of Manchester; Gz
Colak Erol, Serap, 00904242370000-5995 serapcolak@firat.edu.tr, Firat Univ; GgcGt
Colby, Bonnie C., (520) 621-4775 bcolby@email.arizona.edu, Univ of Arizona; Zn
Colby, Frank P., (978) 934-3906 frank_colby@uml.edu, Univ of Massachusetts, Lowell; Am
Colby, Jeff, (828) 262-7126 colbyj@appstate.edu, Appalachian State Univ; Zi
Colby, Sarah, (928) 523-4561 Northern Arizona Univ; Zn
Cole, Catherine, +44 01334 464018 csc5@st-andrews.ac.uk, Univ of St. Andrews; Cm
Cole, David R., (614) 688-7407 cole.618@osu.edu, Ohio State Univ; CgsZa
Cole, Dean A., (505) 665-0832 Los Alamos; Co
Cole, Gregory L., (505) 667-1858 gcole@lanl.gov, Los Alamos; Zg
Cole, James W., (03) 3642-766 jim.cole@canterbury.ac.nz, Univ of Canterbury; Gv
Cole, Julia E., (520) 626-2341 jecole@email.arizona.edu, Univ of Arizona; PeCsAs
Cole, Kevin C., 616-331-3791 colek@gvsu.edu, Grand Valley State Univ; Gz
Cole, Paul D., +44 1752 585985 paul.cole@plymouth.ac.uk, Univ of Plymouth; Gvg
Cole, Rex D., (970) 248-1599 rcole@coloradomesa.edu, Colorado Mesa Univ; GsrGo
Cole, Rick, (727) 553-1522 rickcole@rdsea.com, Univ of South Florida; Op
Colegial Gutierrez, Juan D., (316) 237-2685 colegial@uis.edu.co, Univ Industrial de Santander; GeNgZr
Coleman, Alvin L., (910) 362-7365 acoleman@cfcc.edu, Cape Fear Comm Coll; Gg
Coleman, Catherine, cady-coleman@asu.edu, Arizona State Univ; Ze
Coleman, Drew S., 919-962-0705 dcoleman@email.unc.edu, Univ of North Carolina, Chapel Hill; CcGit
Coleman, Robert G., (650) 723-9205 coleman@pangea.stanford.edu, Stanford Univ; Gi
Coleman, Tommy L., (205) 851-5462 tcoleman@aamu.edu, Alabama A&M Univ; So
Coleman, Jr., Paul J., (310) 825-1776 Univ of California, Los Angeles; Gc
Coles, Kenneth S., (724) 357-5626 kcoles@iup.edu, Indiana Univ of Pennsylvania; XgZeYg
Colgan, Mitchell W., colganm@cofc.edu, Coll of Charleston; GrsZc
Colgan, Patrick M., (616) 313-3201 colganp@gvsu.edu, Grand Valley State Univ; GmlPe
Colgan, William, (416)736-2100 #77703 colgan@yorku.ca, York Univ; Yg
Colin, Farquharson, 709 864 6890 cgfarquh@mun.ca, Memorial Univ of Newfoundland; Yg
Colle, Brian, (631) 632-3174 brian.colle@stonybrook.edu, SUNY, Stony Brook; As
Colleary, Caitlin, (216) 231-4600 ccolleary@cmnh.org, Cleveland Mus of Natural History; Pv
Collett, Jr., Jeffrey L., collett@atmos.colostate.edu, Colorado State Univ; As
Collette, Joseph, 701-858-4142 joseph.collette@minotstateu.edu, Minot State Univ; PgGsr
Collie, Jeremy S., (401) 874-6859 jcollie@gso.uri.edu, Univ of Rhode Island; Ob
Collier, Jackie, (631) 632-8696 jackie.collier@stonybrook.edu, SUNY, Stony Brook; Ob
Collier, Jenny, +44 20 759 46547 r.coggon@imperial.ac.uk, Imperial Coll; Yr
Collier, Mark, 217-300-1171 mcollier@illinois.edu, Illinois State Geo Survey; Ge
Collins, Brian, 206-685-1910 bcollins@uw.edu, Univ of Washington; Gm
Collins, Chris, c.d.collins@reading.ac.uk, Univ of Reading; Sb
Collins, Curtis A., (831) 656-3271 collins@nps.edu, Naval Postgrad Sch; Op
Collins, Damian, 780-492-3197 damian.collins@ualberta.ca, Univ of Alberta; Zn
Collins, Edward W., (512) 471-6247 eddie.collins@beg.utexas.edu, Univ of Texas at Austin, Jackson Sch of Geosciences; Ge
Collins, Emma, 612-626-7429 coll1341@umn.edu, Univ of Minnesota; Gl
Collins, Gareth, +44 20 759 41518 g.collins@imperial.ac.uk, Imperial Coll; Xm
Collins, Jennifer, collinsjm@usf.edu, Univ of South Florida; As
Collins, John A., (508) 289-2733 jcollins@whoi.edu, Woods Hole Oceanographic Inst; Ys
Collins, Kenneth, +44 (0)23 8059 6010 kjc@noc.soton.ac.uk, Univ of Southampton; Ob
Collins, Laura R., (615) 494-8635 laura.collins@mtsu.edu, Middle Tennessee State Univ; Gg
Collins, Laurel S., (305) 348-1732 collinsl@fiu.edu, Florida InterNatl Univ; Pg
Collins, Lisa J., (847) 491-3238 earth@northwestern.edu, Northwestern Univ; ZnnZn
Collins, Lorence G., lorencec@sysmatrix.net, California State Univ, Northridge; Gzx
Collins, Mary E., (352) 392-1951 mec@ufl.edu, Univ of Florida; Sd
Collins, Murray, murray.collins@ed.ac.uk, Edinburgh Univ; Zi
Collins, William D., (510) 495-2407 wdcollins@berkeley.edu, Univ of California, Berkeley; As
Collinson, James W., collinson.1@osu.edu, Ohio State Univ; PgGsg
Collinson, Margaret, +44 1784 443607 m.collinson@rhul.ac.uk, Royal Holloway Univ of London; Pg
Collis, Scott, 630-252-0550 scollis@anl.gov, Argonne; As
Colman, Bradley R., brad.colman@noaa.gov, Univ of Washington; As
Colosimo, Amanda, (585) 292-2421 acolosimo@monroecc.edu, Monroe Comm Coll; Gg
Colucci, Stephen J., (607) 255-1752 sjc25@cornell.edu, Cornell Univ; As
Colwell, Frederick (Rick), (541) 737-5220 rcolwell@coas.oregonstate.edu, Oregon State Univ; Zn
Comas, María José, mjcomas@ucm.es, Univ Compl de Madrid; Pi
Comas, Xavier, 561 297-3256 xcomas@fau.edu, Florida Atlantic Univ; Yg
Comerford, Nicholas B., (850) 875-7100 nbc@ufl.edu, Univ of Florida; Sf
Comina, Cesare, cesare.comina@unito.it, Univ di Torino; Yx
Commane, Roisin, rcommane@ldeo.columbia.edu, Columbia Univ; Yr
Compagnoni, Roberto, roberto.compagnoni@unito.it, Univ di Torino; Gx
Compas, Eric, compase@uww.edu, Univ of Wisconsin, Whitewater; ZiRw
Compton, John S., john.compton@uct.ac.za, Univ of Cape Town; CmGa
Comrie, Andrew, (520) 621-1585 comrie@email.arizona.edu, Univ of Arizona; Zy
Conder, James A., (618) 453-7352 conder@geo.siu.edu, Southern Illinois Univ Carbondale; Ysr
Condie, Kent C., (575) 835-5531 kcondie@nmt.edu, New Mexico Tech; Ct
Condit, Christopher D., (413) 545-0272 ccondit@geo.umass.edu, Univ of

Massachusetts, Amherst; Gi
Condon, Laura, (520) 621-6601 lecondon@email.arizona.edu, Univ of Arizona; Hq
Condreay, Denise, 402-562-1216 dcondreay@cccneb.edu, Central Comm Coll; Zg
Confer, John, 724-938-4211 confer@calu.edu, California Univ of Pennsylvania; Zn
Congalton, Russell, russ.congalton@unh.edu, Univ of New Hampshire; Zi
Congleton, John D., (678) 839-4066 jconglet@westga.edu, Univ of West Georgia; GeZiHs
Conkle, Jeremy, 361-825-2682 jeremy.conkle@tamucc.edu, Texas A&M Univ, Corpus Christi; Zn
Conley, Catharine A., 202-358-3912 cassie.conley@nasa.gov, New Mexico Tech; Pg
Conly, Andrew G., (807) 343-8010 Ext. 8463 andrew.conly@lakeheadu.ca, Lakehead Univ; EgCgGz
Connallon, Christopher B., (410) 554 5545 christopher.connallon@maryland.gov, Maryland Dept of Nat Res; GmZi
Connelly, Jeffrey B., (501) 569-3546 jbconnelly@ualr.edu, Univ of Arkansas at Little Rock; GcNg
Connely, Melissa, 307-268-2017 mconnely@caspercollege.edu, Casper Coll; GsPev
Connolly, Cynthia, 575-835-5264 cynthia.connolly@nmt.edu, New Mexico Tech; Ze
Connolly, Paul, paul.connolly@manchester.ac.uk, Univ of Manchester; As
Connor, Charles B., cconnor@cas.usf.edu, Univ of South Florida; Gv
Connor, Laura, lconnor@usf.edu, Univ of South Florida; Zf
Connors, Christopher, (540) 458-8514 connorsc@wlu.edu, Washington & Lee Univ; GcYeGo
Conover, David O., david.conover@stonybrook.edu, SUNY, Stony Brook; Ob
Conquy, Xenia, xconquy@broward.edu, Broward Coll; Gg
Conrad, Daniel, 612-626-4819 dconrad@umn.edu, Univ of Minnesota; Gl
Conrad, Susan H., 845-431- 8534 conrad@sunydutchess.edu, Dutchess Comm Coll; GsmHs
Conroy, Jessica, 217-244-4855 jconro@illinois.edu, Univ of Illinois, Urbana-Champaign; Cs
Constable, Catherine G., (858) 534-3183 cconstable@ucsd.edu, Univ of California, San Diego; Ymg
Constable, Steven C., (858) 534-2409 sconstable@ucsd.edu, Univ of California, San Diego; Yr
Constantin, Marc, (418) 656-3139 marc.constantin@ggl.ulaval.ca, Univ Laval; Gi
Constantine, Jose, constantineja@cf.ac.uk, Cardiff Univ; Hy
Constantino, Renata R., rrc2136@columbia.edu, Columbia Univ; Gu
Constantopoulos, James T., (575) 562-2651 jim.constantopoulos@enmu.edu, Eastern New Mexico Univ; GezGx
Conti, Paolo, paolo.conti@unisi.it, Univ of Siena; GctGg
Contosta, Alexandra, alix.contosta@unh.edu, Univ of New Hampshire; So
Contreras, Juan, juanc@cicese.mx, Centro de Investigación Científica y de Educación Superior de Ensenada; Gt
Conway, Flaxen D., (541) 737-1339 fconway@coas.oregonstate.edu, Oregon State Univ; Gu
Conway, Howard B., (206) 685-8085 hcon@uw.edu, Univ of Washington; Gl
Conway, Michael F., 520.621.2352 fmconway@email.arizona.edu, Univ of Arizona; RcGvg
Coogan, Laurence, lacoogan@uvic.ca, Univ of Victoria; Gi
Coogon, Rosalind, +44 20 759 46547 r.coggon@imperial.ac.uk, Imperial Coll; Cs
Cook, Ann, (614) 247-6085 cook.1129@osu.edu, Ohio State Univ; GuYg
Cook, David R., (630) 252-5840 drcook@anl.gov, Argonne; AmsSo
Cook, Edward R., drdendro@ldeo.columbia.edu, Columbia Univ; Hw
Cook, Frederick A., (250) 537-8892 fcook@ucalgary.ca, Univ of Calgary; YeGtYs
Cook, Geoffrey, gwcook@ucsd.edu, Univ of California, San Diego; Gv
Cook, Hadrian, +44 020 8417 67756 h.cook@kingston.ac.uk, Kingston Univ; Ge
Cook, Jane, 814-863-8554 jbc6075@psu.edu, Pennsylvania State Univ, Univ Park; ZmeZn
Cook, Joseph P., 520-621-2470 joecook@email.arizona.edu, Univ of Arizona; Gm
Cook, Kerry H., 512-232-7931 kc@jsg.utexas.edu, Univ of Texas, Austin; AsPe
Cook, Mea S., 413-597-4541 mea.s.cook@williams.edu, Williams Coll; OuGe
Cook, Peter, peter.cook@flinders.edu.au, Flinders Univ; Hw
Cook, Robert B., cookrb@ornl.gov, Oak Ridge; Cg
Cook, Steve, (541) 737-0962 cooks@geo.oregonstate.edu, Oregon State Univ; Zn
Cook, Tim, 413-545-1831 tcook@geo.umass.edu, Univ of Massachusetts, Amherst; Pc
Cooke, David R., 61 3 6226 7605 d.cooke@utas.edu.au, Univ of Tasmania; Eg
Cooke, M. J., mary.cooke@austincc.edu, Austin Comm Coll District; GmCl
Cooke, Michele L., (413) 547-3142 cooke@geo.umass.edu, Univ of Massachusetts, Amherst; Gc
Coolbaugh, Mark F., mfc@unr.edu, Univ of Nevada, Reno; Eg
Coombs, Douglas S., +64 3 479-7505 doug.coombs@otago.ac.nz, Univ of Otago; Gi
Coombs, Margery C., mcoombs@amherst.edu, Amherst Coll; Pv
Cooney, Michael, (808) 956-7337 mcooney@hawaii.edu, Univ of Hawai'i, Manoa; Og
Cooney, Michele L., 412-442-4230 miccooney@pa.gov, DCNR- Pennsylvania Bur of Geo Survey; EgoGe
Cooney, Timothy M., (319) 273-2918 timothy.cooney@uni.edu, Univ of Northern Iowa; Ze
Coonley, Steffanie, (505) 665-2330 scoonley@lanl.gov, Los Alamos; Zn
Cooper, Alan F., +64 3 479-7515 alan.cooper@stonebow.otago.ac.nz, Univ of Otago; GpiGt
Cooper, Catherine , 509-335-1501 cmcooper@wsu.edu, Washington State Univ; Yg
Cooper, Clay A., (775) 673-7372 clay.cooper@dri.edu, Univ of Nevada, Reno; HySpHq
Cooper, Jonathon L., (507) 222-4401 jlcooper@carleton.edu, Carleton Coll; Gg
Cooper, Kari M., (530) 754-8826 kmcooper@ucdavis.edu, Univ of California, Davis; Gi
Cooper, Mark, (204) 474-8075 mark_cooper@umanitoba.ca, Univ of Manitoba; Gz
Cooper, Matthew, +44 (0)23 80592062 matthew.cooper@noc.soton.ac.uk, Univ of Southampton; Oc
Cooper, Reid F., (401) 863-2160 reid_cooper@brown.edu, Brown Univ; Gy
Cooper, Roger W., (409) 880-8239 roger.cooper@lamar.edu, Lamar Univ; Gi
Cooper, Terence H., (612) 625-7747 tcooper@soils.umn.edu, Univ of Minnesota, Twin Cities; Sd
Cooper-Boemmels, Jennifer, 203-392-5842 cooperj1@southernct.edu, Southern Connecticut State Univ; GcgOg
Coorough Burke, Patricia J., (414) 278-6155 coorough@mpm.edu, Milwaukee Public Mus; PgiPe
Cope, Tim D., (765) 658-6443 tcope@depauw.edu, DePauw Univ; GsZiGr
Copeland, Peter, 713-743-3649 copeland@uh.edu, Univ of Houston; CcGt
Copeman, Louise A., 541-737-6840 Oregon State Univ; Zn
Copjakova, Renata, +549 49 3073 copjakova@sci.muni.cz, Masaryk Univ; Gz
Copley, Alex, +44 (0) 1223 748937 acc41@cam.ac.uk, Univ of Cambridge; YgGt
Copley, Jonathan, +44 (0)23 8059 6621 jtc@noc.soton.ac.uk, Univ of Southampton; Ob
Copper, Paul, pcopper@laurentian.ca, Laurentian Univ, Sudbury; PieZc
Coppinger, Walter, (406) 287-2288 wcopping@trinity.edu, Trinity Univ; Gcg
Coppola, Diego, diego.coppola@unito.it, Univ di Torino; Gv
Corbato, Charles E., 614-451-6537 ccorbato@gmail.com, Ohio State Univ; Ye
Corbella Cordomí, Mercè, ++34935811973 merce.corbella@uab.cat, Univ Auto de Barcelona; GzCl
Corbett, Patrick, +44 (0) 131 451 3171 p.corbett@hw.ac.uk, Heriot-Watt Univ; Np
Corbineau, Lucien, +33(0)3 44068962 lucien.corbineau@lasalle-beauvais.fr, Institut Polytechnique LaSalle Beauvais (ex-IGAL); EgGgNm
Corcoran, Deborah, 417-836-6889 debcorcoran@missouristate.edu, Missouri State Univ; Zn
Corcoran, Patricia, 519-661-2111, ext. 86836 pcorcor@uwo.ca, Western Univ; Gd
Corcoran, William, (417) 836-5801 williamcorcoran@missouristate.edu, Missouri State Univ; As
Cordero, David I., 360-442-2883 dcordero@lowercolumbia.edu, Lower Columbia Coll; Zg
Cordonnier, Benoit, 089/2180 4271 cordonnier@min.uni-muenchen.de, Ludwig-Maximilians-Univ Muenchen; Gz
Corey, Alison, (801) 537-3122 nrugs.acorey@state.ut.us, Utah Geo Survey; Zn
Corley, John H., (573) 368-2132 john.corley@dnr.mo.gov, Missouri Dept of Nat Res; Geg
Corliss, Bruce H., (919) 684-2951 bruce.corliss@duke.edu, Duke Univ; Pm
Cormier, Vernon F., (860) 486-3547 vernon.cormier@uconn.edu, Univ of Connecticut; YsGt
Cornamusini, Gianluca, gianluca.cornamusini@unisi.it, Univ of Siena; GgsGr

Cornebise, Michael W., (217) 581-7486 mwcornebise@eiu.edu, Eastern Illinois Univ; Zg
Cornelis, Jean-Thomas, 604-827-4813 jt.cornelis@ubc.ca, Univ of British Columbia; Sd
Cornell, Kristie, 337-482-1455 kcornell@louisiana.edu, Univ of Louisiana at Lafayette; Gg
Cornell, Sean R., 717-477-1310 srcornell@ship.edu, Shippensburg Univ; GgsOn
Cornell , William , cornell@utep.edu, Univ of Texas, El Paso; Gg
Cornell, Winton, (918) 631-3248 winton-cornell@utulsa.edu, Univ of Tulsa; GivCa
Cornillon, Peter, (401) 874-6283 pcornillon@gso.uri.edu, Univ of Rhode Island; Op
Cornwell, David G., +44 (0)1224 273448 d.cornwell@abdn.ac.uk, Univ of Aberdeen; YsGtYe
Cornwell, Jeffery C., (410) 221-8445 cornwell@hpl.umces.edu, Univ of Maryland; Oc
Cornwell, Kevin J., (916) 278-6667 cornwell@csus.edu, Sacramento State Univ; GmeHs
Coron, Cynthia R., (203) 392-5840 coroncl@southernct.edu, Southern Connecticut State Univ; Eg
Corona, Thomas J., (303) 556-8525 Metropolitan State Coll of Denver; Am
Corral, Isaac, isaac.corral@uab.cat, Univ Auto de Barcelona; Gx
Corrigan, Catherine, (202) 633-1855 corriganc@si.edu, Smithsonian Inst / Natl Mus of Natural History; XmcXg
Corriveau, Louise, (418) 654-2672 louise.corriveau@canada.ca, Univ du Quebec; EmGpg
Corsetti, Frank A., (213) 740-6123 fcorsett@earth.usc.edu, Univ of Southern California; Gs
Coruh, Cahit, (386) 447-1289 coruh@vt.edu, Virginia Tech; YesGt
Cosentino, Pietro, p.cosentino@uea.ac.uk, Univ of East Anglia; Yg
Cosgrove, John, +44 20 759 46466 j.cosgrove@imperial.ac.uk, Imperial Coll; Gc
Costa, Emanuele, emanuele.costa@unito.it, Univ di Torino; Gz
Costa, Maria R., rpereira@utad.pt, Unive de Trás-os-Montes e Alto Douro; Hw
Costa, Jr, Ozeas S., (419) 755-4128 costa.47@osu.edu, Ohio State Univ; CblZu
Costea, Lidia, (773) 442-6050 l-costea@neiu.edu, Northeastern Illinois Univ; ZnnZn
Costello, Margaret, (562) 985-4809 mcostel2@csulb.edu, California State Univ, Long Beach; Zn
Costigan, Keeley R., (505) 665-4788 krc@lanl.gov, Los Alamos; As
Costin, Gelu, +27 (0)46-603-8316 g.costin@ru.ac.za, Rhodes Univ; Gp
Cote, Denis, (418) 545-5011 dcote@uqac.ca, Univ du Quebec a Chicoutimi; GiEgZi
Côté, Jean, 514 987-3000 #2351 cote.jean@uqam.ca, Univ du Quebec a Montreal; As
Cothren, Jackson D., (479) 575-6790 jcothre@uark.edu, Univ of Arkansas, Fayetteville; Zir
Cottaar, Sanne , sc845@cam.ac.uk, Univ of Cambridge; Ys
Cotter, James F., (320) 589-6312 cotterjf@mrs.umn.edu, Univ of Minnesota, Morris; Gl
Cotterill, Carol J., cjc2291@columbia.edu, Columbia Univ; Zg
Cottingame, William, (505) 667-8339 wcottingame@lanl.gov, Los Alamos; As
Cottle, John, (805) 893-3471 cottle@geol.ucsb.edu, Univ of California, Santa Barbara; Gi
Cotton, Jennifer, (818) 677-7978 jennifer.cotton@csun.edu, California State Univ, Northridge; Pc
Cottrell, Elizabeth, cottrell@si.edu, Univ of Maryland; CpgGx
Cottrell, Rory D., (585) 275-6025 rory.cottrell@rochester.edu, Univ of Rochester; Ym
Couceiro, Fay, +44 023 92 842294 fay.couceiro@port.ac.uk, Univ of Portsmouth; On
Couch, Libby, (303) 370-6367 libby.couch@dmns.org, Denver Mus of Nature & Sci; Zn
Coulibaly, Mamadou, (920) 424-4105 coulibalym@uwosh.edu, Univ of Wisconsin Oshkosh; ZiyHg
Coulibaly, Paulin, (905) 525-9140 (Ext. 23354) couliba@mcmaster.ca, McMaster Univ; Hg
Coull, Bruce, bccoull@sc.edu, Univ of South Carolina; Ob
Coulson, Alan B., (864) 656-1897 acoulso@clemson.edu, Clemson Univ; PgGgCs
Coulter, Richard L., (630) 252-5833 rl_coulter@anl.gov, Argonne; As
Couples, Gary, +44 (0) 131 451 3123 g.couples@hw.ac.uk, Heriot-Watt Univ; Np
Cousens, Brian L., 001-613-520-2600 x4436 brian.cousens@carleton.ca, Carleton Univ; GiCgc
Cousineau, Pierre, (418) 545-5011 pcousine@uqac.ca, Univ du Quebec a Chicoutimi; Gd
Couture, Gilles, 514-987-3000 #8905 couture.gilles@uqam.ca, Univ du Quebec a Montreal; Yg
Cova, Thomas J., (801) 581-7930 tom.cova@geog.utah.edu, Univ of Utah; Zi
Covault, Jacob, 512-475-9506 jake.covault@beg.utexas.edu, Univ of Texas, Austin; Gso
Coveney, Eamonn, ewcoveney@fhsu.edu, Fort Hays State Univ; Zyn
Coveney, Seamus, +44 0141 330 7750 seamus.coveney@glasgow.ac.uk, Univ of Glasgow; Zi
Coveney Jr, Raymond M., (816) 235-1334 coveneyr@umkc.edu, Univ of Missouri, Kansas City; EgmCe
Covert, David S., (206) 685-7461 dcovert@atmos.washington.edu, Univ of Washington; As
Covington, Matthew, 479-575-3876 mcoving@uark.edu, Univ of Arkansas, Fayetteville; HqGmg
Cowan, Clinton A., (507) 222-7021 ccowan@carleton.edu, Carleton Coll; GsPe
Cowan, Darrel S., (206) 543-4033 darrel@uw.edu, Univ of Washington; Gct
Cowan, Ellen A., (828) 262-2260 cowanea@appstate.edu, Appalachian State Univ; GsmGu
Cowan, Nicolas B., 514-398-1967 nicolas.cowan@mcgill.ca, McGill Univ; AsXyg
Cowart, James B., (850) 644-5784 jcowart@mailer.fsu.edu, Florida State Univ; Cc
Cowart, Richard, (361) 644-3049 recowart@coastalbend.edu, Coastal Bend Coll; GeEo
Cowen, Richard, richard@blueaokfarm.com, Univ of California, Davis; Pg
Cowgill, Eric S., (530) 754-6574 escowgill@ucdavis.edu, Univ of California, Davis; Gtc
Cowie, Gregory L., +44 (0) 131 650 8502 dr.greg.cowie@ed.ac.uk, Edinburgh Univ; Co
Cowman, Tim C., 605-677-6151 tim.cowman@usd.edu, South Dakota Dept of Environment and Nat Res; CgGg
Cox, Alysia, (406) 496-4185 acox@mtech.edu, Montana Tech of the Univ of Montana; CqlCm
Cox, Christena, 614-292-0138 cox.1@osu.edu, Ohio State Univ; GeoGc
Cox, John, (403) 440-6160 jcox@mtroyal.ca, Mount Royal Univ; EoGsHw
Cox, Randel T., (901) 678-4361 randycox@memphis.edu, Univ of Memphis; GcmGt
Cox, Richard, 902 494 3362 richard.cox@dal.ca, Dalhousie Univ; GzEdGg
Cox, Ronadh, (413) 597-2297 ronadh.cox@williams.edu, Williams Coll; GsXg
Cox, Shelah, 215 204-8227 scox@temple.edu, Temple Univ; Zn
Coxon, Catherine, + 353 1 8962235 cecoxon@tcd.ie, Trinity Coll; GeOu
Craddock, Robert A., 202-633-2473 craddockb@si.edu, Smithsonian Inst, Natl Air & Space Mus; XgGm
Crafford, J. P., +27 (0)15 268 2217 krappie.crafford@ul.ac.za, Univ of Limpopo; ZnNgZn
Craig, Cameron D., (217) 581-2626 cdcraig@eiu.edu, Eastern Illinois Univ; ZgAm
Craig, James L., (505) 665-7996 jlcraig@lanl.gov, Los Alamos; Gg
Craig, Mitchell S., (510) 885-3425 mitchell.craig@csueastbay.edu, California State Univ, East Bay; Yg
Craig, Susanne , (902) 494-4381 susanne.craig@dal.ca, Dalhousie Univ; Op
Crail, Todd, 419-530-4583 todd.crail@utoledo.edu, Univ of Toledo; Zn
Crain, John R., (209) 730-3812 Coll of the Sequoias; Gg
Cramer, Bradley D., (319) 335-0704 bradley-cramer@uiowa.edu, Univ of Iowa; GrPsGs
Cramer, Chris, (901) 678-2007 ccramer@memphis.edu, Univ of Memphis; Ys
Cramer, Gary, (620) 662-9021 gcramer@ksu.edu, Kansas State Univ; So
Crandall, Jake R., (217) 581-6245 jrcrandall@eiu.edu, Eastern Illinois Univ; GzXgEd
Cranford, Peter, (902) 426-3277 cranfordp@mar.dfo-mpo.gc.ca, Dalhousie Univ; Ob
Cranganu, Constantin, 7189515000 x2878 cranganu@brooklyn.cuny.edu, Brooklyn Coll (CUNY); GoYhHw
Craven, John, +44 (0) 131 650 7887 john.craven@ed.ac.uk, Edinburgh Univ; Yg
Craw, Dave, +64 3 479-7529 dave.craw@otago.ac.nz, Univ of Otago; Eg
Crawford, Anthony J., tony.crawford@utas.edu.au, Univ of Tasmania; Git
Crawford, Ian, +44 020 3073 8026 i.crawford@bbk.ac.uk, Birkbeck Coll; Zn
Crawford, James, 757-864-7231 j.h.crawford@larc.nasa.gov, Georgia Inst of Tech; As
Crawford, Maria Luisa B., (610) 526-5111 mcrawfor@brynmawr.edu, Bryn Mawr Coll; GxzGp
Crawford, Matthew, (859) 323-0510 mcrawford@uky.edu, Univ of

Kentucky; Ng
Crawford, Nicholas, (270) 745-5889 nicholas.crawford@wku.edu, Western Kentucky Univ; Hg
Crawford, Thomas J., (678) 839-4062 Univ of West Georgia; Eg
Crawford, Vernon J., (541) 552-6479 crawford@sou.edu, Southern Oregon Univ; Gg
Creager, Kenneth C., 206-685-2803 kcc@uw.edu, Univ of Washington; Ys
Crease, James, (302) 645-4240 Univ of Delaware; Op
Creaser, Robert A., (780) 492-2942 robert.creaser@ualberta.ca, Univ of Alberta; Cc
Creasey, Robert L., 831-656-3178 creasey@nps.edu, Naval Postgrad Sch; Am
Crepeau, Richard J., (828) 262-7052 crepeaurj@appstate.edu, Appalachian State Univ; Zn
Crespi, Jean M., (860) 486-0601 crespi@geol.uconn.edu, Univ of Connecticut; Gc
Crespo, Elena, ecrespo@ucm.es, Univ Compl de Madrid; EnGz
Cresswell, Tim, tim.cresswell@ed.ac.uk, Edinburgh Univ; Zn
Creveling, Jessica, 541-737-2112 crevelij@oregonstate.edu, Oregon State Univ; GgsGr
Crewdson, Robert, rcrewdson1@csub.edu, California State Univ, Bakersfield; HwYgCg
Crews, Jeff, (573) 368-2356 jeff.crews@dnr.mo.gov, Missouri Dept of Nat Res; Hy
Creyts, Timothy T., tcreyts@ldeo.columbia.edu, Columbia Univ; Gl
Cribb, Warner, (615) 898-2379 warner.cribb@mtsu.edu, Middle Tennessee State Univ; GipGz
Crider, Juliet G., 206-543-8715 criderj@uw.edu, Univ of Washington; GcmNr
Criminale, Jr., William O., (206) 543-9506 Univ of Washington; Op
Crimmins, Michael A., (520) 626-4244 crimmins@email.arizona.edu, Univ of Arizona; Zc
Crippen, Robert E., (818) 354-2475 robert.e.crippen@jpl.nasa.gov, Jet Propulsion Lab; Zr
Crisp, Joy A., (818) 354-9036 Jet Propulsion Lab; Gv
Criss, Robert E., (314) 935-7441 criss@levee.wustl.edu, Washington Univ in St. Louis; Cs
Criswell, James, (910) 392-7536 jcriswell@cfcc.edu, Cape Fear Comm Coll; Gg
Croft, Gary M., gcroft@valleycollege.edu, San Bernardino Valley Coll; Zy
Croft, Paul J., (908) 737-3737 pcroft@kean.edu, Kean Univ; AmsZe
Croll, Don, (831) 459-3610 dcroll@cats.ucsc.edu, Univ of California, Santa Cruz; Ob
Cromar, Nancy, nancy.cromar@flinders.edu.au, Flinders Univ; Zn
Cron, Mitch, mitch.cron@gmail.com, Drexel Univ; Gge
Crone, Timothy, tjc@ldeo.columbia.edu, Columbia Univ; Gu
Cronin, Meghan F., (206) 526-6449 meghan.f.cronin@noaa.gov, Univ of Washington; Op
Cronin, Timothy W., twcronin@mit.edu, Massachusetts Inst of Tech; As
Cronin, Vincent S., (254) 710-2174 vince_cronin@baylor.edu, Baylor Univ; GctNg
Cronoble, James M., (303) 556-3070 Metropolitan State Coll of Denver; Gg
Croot, Peter, + 353 (0)91 492 194 peter.croot@nuigalway.ie, Natl Univ of Ireland Galway; Ct
Crootof, Arica, 406-683-7075 arica.crootof@umwestern.edu, Univ of Montana Western; RwcZi
Crosby, Benjamin T., (208) 282-2949 crosby@isu.edu, Idaho State Univ; GmHsRn
Cross, John A., (920) 424-4105 cross@uwosh.edu, Univ of Wisconsin Oshkosh; ZyAst
Crossen, Kristine J., (907) 786-6838 kjcrossen@uaa.alaska.edu, Univ of Alaska, Anchorage; Gl
Crossey, Laura J., (505) 277-4204 lcrossey@unm.edu, Univ of New Mexico; Cl
Crossley, David J., (314) 977-3153 david.crossley@slu.edu, Saint Louis Univ; YdsXy
Crosson, Robert S., crosson@uw.edu, Univ of Washington; YsGct
Crotwell, Philip, 803-777-0955 crotwell@geol.sc.edu, Univ of South Carolina; Ys
Crouse, David A., (919) 515-7302 North Carolina State Univ; Sc
Crowder, Margaret, (270) 745-5973 margaret.crowder@wku.edu, Western Kentucky Univ; Ze
Crowe, Bruce M., (702) 794-7206 bmc@lanl.gov, Los Alamos; Gv
Crowe, Douglas E., 706-542-2382 crowe@gly.uga.edu, Univ of Georgia; Em
Crowley, Brooke E., (513) 556-7181 brooke.crowley@uc.edu, Univ of Cincinnati; CsPev
Crowley, James, jimcrowley@boisestate.edu, Boise State Univ; Ccg
Crowley, Kate, kate.crowley@ed.ac.uk, Edinburgh Univ; Zc
Crowley, Peter D., (413) 542-2715 pdcrowley@amherst.edu, Amherst Coll; Gc
Crowley, Quentin G., + 353 1 8962403 crowleyq@tcd.ie, Trinity Coll; Cc
Crowley, Stephen, +44 0151 794 5163 sfcrow@liverpool.ac.uk, Univ of Liverpool; Gs
Crowther, Sarah, +44 0161 275-0407 sarah.crowther@manchester.ac.uk, Univ of Manchester; As
Croxen III, Fred W., 928-344-7586 fred.croxen@azwestern.edu, Arizona Western Coll; HwPvZi
Crozier, Carl, (919) 793-4428 North Carolina State Univ; Sb
Cruden, David M., (780) 492-3085 dave.cruden@ualberta.ca, Univ of Alberta; Nr
Cruikshank, Kenneth M., (503) 725-3383 cruikshankk@pdx.edu, Portland State Univ; Gc
Crumbly, Isaac J., (912) 825-6454 Fort Valley State Univ; Zn
Cruse, Richard M., (515) 294-7850 rmc@iastate.edu, Iowa State Univ of Sci & Tech; So
Crutzen, Paul, pcrutzen@ucsd.edu, Univ of California, San Diego; Oc
Cruz-Uribe, Alicia M., (207) 581-4494 alicia.cruzuribe@maine.edu, Univ of Maine; GpCpt
Cruzen, Shawn, 706-649-1785 cruzen_shawn@columbusstate.edu, Columbus State Univ; Xa
Csank, Adam, 775-784-6663 acsank@unr.edu, Univ of Nevada, Reno; Zy
Csatho, Beata M., (716) 645-4325 bcsatho@buffalo.edu, SUNY, Buffalo; GlZrYg
Csiki, Shane, (603) 271-1975 shane.j.csiki@des.nh.gov, New Hampshire Geo Survey; ZyGm
Cuevas, Julia, julia.cuevas@ehu.eus, Univ of the Basque Country UPV/EHU; GctGg
Cuffey, Roger J., (814) 865-1293 rcuffey@psu.edu, Pennsylvania State Univ, Univ Park; PiePv
Cuker, Benjamin E., (757) 727-5783 benjamin.cuker@hamptonu.edu, Hampton Univ; Ob
Culberg, Riley, 609-258-4101 rtculberg@princeton.edu, Princeton Univ; GlYu
Cull-Hearth, Selby, scull@brynmawr.edu, Bryn Mawr Coll; XgGzx
Cullen, Cecilia, ccullen3@illinois.edu, Univ of Illinois; Hw
Cullen, Cheila, 718-289-5558 cheila.cullen@bcc.cuny.edu, Graduate Sch of the City Univ of New York; Rn
Cullen, Heidi, hcullen@weather.com, Georgia Inst of Tech; Am
Cullen, James L., jcullen@salemstate.edu, Salem State Univ; GsPmOu
Cullen, Jay, 250-721-6120 jcullen@uvic.ca, Univ of Victoria; Oc
Cullen, John J., (902) 494-6667 john.cullen@dal.ca, Dalhousie Univ; Ob
Cullers, Robert L., (785) 532-2240 rcullers@ksu.edu, Kansas State Univ; Ct
Culshaw, Nicholas, (902) 494-3501 Dalhousie Univ; Gc
Culver, Stephen J., (252) 328-6360 culvers@ecu.edu, East Carolina Univ; PmGu
Cummings, Katie, (801) 537-3359 kcummings@utah.gov, Utah Geo Survey; Gg
Cummings, Michael L., (503) 725-3395 cummingsm@pdx.edu, Portland State Univ; GgHg
Cummins, R. Hays, (513) 529-1338 cumminrh@miamioh.edu, Miami Univ; Ge
Cundall, Peter A., (612) 626-0369 pacundall@aol.com, Univ of Minnesota, Twin Cities; Nr
Cunningham, Dickson, 860-465-4321 cunninghamw@easternct.edu, Eastern Connecticut State Univ; GtcGx
Cunningham, Mary A., 845-437-5547 macunningham@vassar.edu, Vassar Coll; ZiyZu
Cunningham, Peter, +44 (0) 191 208 7932 peter.cunningham@ncl.ac.uk, Univ of Newcastle Upon Tyne; Hg
Cunningham, Robert , 325-248-5392 robert.cunningham@austin.utexas.edu, Univ of Texas, Austin; Go
Cuomo, Carmela, ccuomo@newhaven.edu, Univ of New Haven; CmmGu
Cupples, Julie, +44 (0) 131 651 4315 julie.cupples@ed.ac.uk, Edinburgh Univ; Zn
Curewitz, Daniel, 315-443-2672 dcurewit@syr.edu, Syracuse Univ; OgGct
Curl, Doug C., 859-323-0519 doug@uky.edu, Univ of Kentucky; ZfGgc
Curran, H. Allen, (413) 585-3943 acurran@smith.edu, Smith Coll; Pg
Currano, Ellen, (307) 766-2819 ecurrano@uwyo.edu, Univ of Wyoming; Pbe
Curray, Joseph R., (858) 459-2130 jcurray@ucsd.edu, Univ of California, San Diego; Gu
Currey, Donald R., (801) 581-8218 Univ of Utah; Gm
Curri, Neil, 845-437-7708 necurri@vassar.edu, Vassar Coll; Zi
Currie, Brian S., (513) 529-7578 curriebs@miamioh.edu, Miami Univ; GstGc
Currit, Nathan, (512) 245-2170 Texas State Univ; ZruZc
Curry, B. B., (217) 244-5787 b-curry@illinois.edu, Northern Illinois Univ;

GelPs
Curry, Gordon, +44 01413305444 gordon.curry@glasgow.ac.uk, Univ of Glasgow; GegPg
Curry, James K., (501) 682-6992 james.curry@arkansas.gov, Arkansas Geo Survey; Zn
Curry, Janel M., (616) 526-6869 Calvin Coll; Zy
Curry, Joan E., (520) 626-5081 curry@ag.arizona.edu, Univ of Arizona; Sc
Curry, Judith A., (404) 894-3948 judith.curry@eas.gatech.edu, Georgia Inst of Tech; AsZrn
Curry, Magdalena E., mcurry2@ncsu.edu, North Carolina State Univ; Gt
Curry, Susan, (352) 294-3147 scurry@ufl.edu, Univ of Florida; ZiSo
Curry, William B., (508) 289-2591 wcurry@whoi.edu, Woods Hole Oceanographic Inst; Pm
Curry Rogers, Kristina A., 651-696-6288 rogersk@macalester.edu, Macalester Coll; PveGg
Curtice, Joshua M., (508) 289-2618 jcurtice@whoi.edu, Woods Hole Oceanographic Inst; Cc
Curtin, Lorelei, (570) 577-1381 lc058@bucknell.edu, Bucknell Univ; Cl
Curtin, Tara M., (315) 781-3928 curtin@hws.edu, Hobart & William Smith Colls; PeGs
Curtis, Andrew, +44 (0) 131 650 8515 andrew.curtis@ed.ac.uk, Edinburgh Univ; Ys
Curtis, Arthur C., (615) 576-7642 Oak Ridge; Ge
Curtis, David B., (505) 667-4845 Los Alamos; Cc
Curtis, Garniss H., (510) 845-0333 Univ of California, Berkeley; Cc
Curtis, Jason H., (352) 392-2296 jcurtis@che.ufl.edu, Univ of Florida; Pe
Curtis, John B., jbcurtis@mines.edu, Colorado Sch of Mines; GoCo
Curtis, Lynn A., (954) 201-6942 lcurtis@broward.edu, Broward Coll; GgPg
Curtis, P. J., 250-807-9521 jeff.curtis@ubc.ca, Univ of British Columbia, Okanagan Campus; CbHsRw
Cusack, Maggie, +44 01413305491 maggie.cusack@glasgow.ac.uk, Univ of Glasgow; Gz
Cushman, John, jcushman@purdue.edu, Purdue Univ; Sp
Cutter, Gregory A., (757) 683-4929 gcutter@odu.edu, Old Dominion Univ; OcCtHs
Cvetko Tešoviæ, Blanka, +38514606107 blankacvetko@yahoo.com, Univ of Zagreb; GgPsm
Czaja, Andrew D., 513-556-3574 andrew.czaja@uc.edu, Univ of Cincinnati; PmCl
Czarnetzki, Alan C., (319) 273-2152 alan.czarnetzki@uni.edu, Univ of Northern Iowa; Am
Czeck, Dyanna M., 414-229-3948 dyanna@uwm.edu, Univ of Wisconsin, Milwaukee; Gc
Cziczo, Daniel, 765-494-3258 djcziczo@purdue.edu, Purdue Univ; As

D

d'Alessio, Matthew, (818) 677-3647 matthew.dalessio@csun.edu, California State Univ, Northridge; ZeGt
D'Allura, Jad A., (541) 552-6480 dallura@sou.edu, Southern Oregon Univ; Gc
D'Alpaos, Andrea, +390498279117 andrea.dalpaos@unipd.it, Univ degli Studi di Padova; Hs
D'Alpoim Guedes, Jade, jguedes@ucsd.edu, Univ of California, San Diego; Ga
D'Amico, Sebastiano, +356 2340 3101 sebdamico@gmail.com, Univ of Malta; YsgZg
D'Andrea, William J., dandrea@ldeo.columbia.edu, Columbia Univ; Pe
D'Angelo, Maria C., c.dangelo@soton.ac.uk, Univ of Southampton; Ob
D'Arrigo, Rosanne, druidrd@ldeo.columbia.edu, Columbia Univ; Zn
D'Asaro, Eric A., (206) 545-2982 dasaro@apl.washington.edu, Univ of Washington; Op
d'Atri, Anna, anna.datri@unito.it, Univ di Torino; Gr
D'Hondt, Steven L., (401) 792-6808 dhondt@gso.uri.edu, Univ of Rhode Island; Ou
da Silva, Eduardo F., eafsilva@ua.pt, Unive de Aveiro; CeGbCl
Daanen, Ronald P., 907-451-5965 ronald.daanen@alaska.gov, Alaska Division of Geological & Geophysical Surveys; HyGme
Dade, William B., 603-646-0286 william.b.dade@dartmouth.edu, Dartmouth Coll; GsHgEo
Daemen, Jaak, (775) 7844309 daemen@mines.unr.edu, Univ of Nevada, Reno; Nr
Daeschler, Ted, 215 299 1133 ebd29@drexel.edu, Drexel Univ; PvGgPg
Dagdelen, Kadri, (303) 273-3711 kdagdele@mines.edu, Colorado Sch of Mines; Nm
Dahl, Johannes M., 806-834-6197 johannes.dahl@ttu.edu, Texas Tech Univ; Ams
Dahl, Robyn, (360) 650-7207 robyn.dahl@wwu.edu, Western Washington Univ; PiZe
Dahlgren, Randy A., (530) 752-2814 radahlgren@ucdavis.edu, Univ of California, Davis; So
Dahlke, Helen E., (530) 302-5358 hdahlke@ucdavis.edu, Univ of California, Davis; HgsZi
Dai, Chunli, (614) 292-2269 dai.56@osu.edu, Ohio State Univ; ZrYg
Dai, Dajun, 404 413 5750 geoddd@langate.gsu.edu, Georgia State Univ; Zi
Daigneault, Real, (418) 545-5011 rdaignea@uqac.ca, Univ du Quebec a Chicoutimi; GcEgZi
Dakin, Susan, (403) 329-2279 susan.dakin@uleth.ca, Univ of Lethbridge; Ze
Dalconi, Maria C., +390498279163 mariachiara.dalconi@unipd.it, Univ degli Studi di Padova; Gz
Dale, Janis, (306) 585-4840 Univ of Regina; Gml
Daley, Gwen M., 803-323-4973 daleyg@winthrop.edu, Winthrop Univ; PqgGs
Dallegge, Todd A., 970-351-1086 todd.dallegge@unco.edu, Univ of Northern Colorado; GosGr
Dallimer, Martin, +44(0) 113 34 35279 m.dallimer@leeds.ac.uk, Univ of Leeds; Ge
Dallmeyer, R. David, (706) 542-7448 dallmeyr@uga.edu, Univ of Georgia; Gt
Dalman, Michael, (979) 830-4206 mdalman@blinn.edu, Blinn Coll; Ze
Dalrymple, Robert W., (613) 533-6186 dalrympl@queensu.ca, Queen's Univ; Gs
Dalton, Colleen, 401-863-5875 colleen_dalton@brown.edu, Brown Univ; Ys
Dalton, Richard F., (609) 292-2576 richard.dalton@dep.nj.gov, New Jersey Geological and Water Survey; Gr
Daly, Eve, +353 (0)91 492 183 eve.daly@nuigalway.ie, Natl Univ of Ireland Galway; Ygu
Daly, Julia F., (207) 778-7403 dalyj@maine.edu, Univ of Maine, Farmington; GmHsGu
Daly, Kendra L., (727) 553-1041 kdaly@marine.usf.edu, Univ of South Florida; Ob
Dalziel, Ian W. D., (512) 471-0431 ian@ig.utexas.edu, Univ of Texas, Austin; Gt
Dam, Hans G., (860) 405-9098 hans.dam@uconn.edu, Univ of Connecticut; Ob
Damir, Buckoviæ, +38514606106 buckovic@geol.pmf.hr, Univ of Zagreb; GgPms
Damschen, Ellen, damschen@wustl.edu, Washington Univ in St. Louis; Zn
Damuth, John E., (972) 623-7368 damuth@uta.edu, Univ of Texas, Arlington; GusYr
Dana, Gayle L., (775) 674-7538 gdana@dri.edu, Univ of Nevada, Reno; Zr
Daniel, Christopher G., (570) 577-1133 cdaniel@bucknell.edu, Bucknell Univ; Gp
Daniell, James J., james.daniell@jcu.edu.au, James Cook Univ; GumYr
Daniels, J. Michael, (303) 871-7531 j.michael.daniels@du.edu, Univ of Denver; GmSp
Daniels, Jeffrey J., 614-292-6704 daniels.9@osu.edu, Ohio State Univ; Yg
Daniels, Walter L., (540) 231-7175 wdaniels@vt.edu, Virginia Tech; SdGeCg
Daniels, William R., (505) 667-4546 Los Alamos; Ct
Danielson, Stentor B., 724.738.2564 stentor.danielson@sru.edu, Slippery Rock Univ; Rn
Danko, George, (775) 784-4284 danko@mines.unr.edu, Univ of Nevada, Reno; Hq
Dansereau, Pauline, dansero@ggl.ulaval.ca, Univ Laval; Gs
Daoust, Mario, 417-836-5301 mariodaoust@missouristate.edu, Missouri State Univ; AtZy
Daramola, Sunday O., +2348060256588 sunday.daramola@gmail.com, Federal Univ of Tech, Akure; GeNtHw
Darby, Dennis A., (757) 683-4701 ddarby@odu.edu, Old Dominion Univ; Ou
Darby, Jeannie, (916) 752-5670 Univ of California, Davis; Hg
Darbyshire, Fiona Ann, 514-987-3000 #5054 darbyshire.fiona_ann@uqam.ca, Univ du Quebec a Montreal; Ys
Darin, Mike, (775) 682-8751 mdarin@unr.edu, Univ of Nevada, Reno; Gct
Darling, James, +44 023 92 842247 james.darling@port.ac.uk, Univ of Portsmouth; Cs
Darling, Robert S., (607) 753-2923 darlingr@cortland.edu, SUNY, Cortland; GzpCp
Darmody, Robert G., (217) 333-9489 Univ of Illinois, Urbana-Champaign; Sd
Daroub, Samira H., (561) 993-1500 sdaroub@ufl.edu, Univ of Florida; So
Darrah, Thomas, (614) 688-2132 darrah.24@osu.edu, Ohio State Univ; Cg
Darrell, James H., (912) 478-5361 jdarrell@georgiasouthern.edu, Georgia Southern Univ; Pl
Darroch, Simon A., simon.a.darroch@vanderbilt.edu, Vanderbilt Univ; PgiPe
Darrow, Margaret M., (907) 474-7303 mmdarrow@alaska.edu, Univ of Alaska Fairbanks; NxZnn
Das, Indrani, indrani@ldeo.columbia.edu, Columbia Univ; Gu

Das, Jnaneshwar, (480) 727-4726 djnan@asu.edu, Arizona State Univ; Zr
Das, Sarah B., 508-289-2464 sdas@whoi.edu, Woods Hole Oceanographic Inst; GlAtZg
Dascher, Erin, edascher@ewu.edu, Eastern Washington Univ; ZySf
Dasgupta, Rajdeep, 713.348.2664 rajdeep.dasgupta@rice.edu, Rice Univ; GxCg
Dash, Padmanava, 662-325-3915 pd175@msstate.edu, Mississippi State Univ; ZrHg
Dash, Zora V., (505) 667-1923 zvd@lanl.gov, Los Alamos; Zn
Dashtgard, Shahin, 778-782-5492 sdashtga@sfu.ca, Simon Fraser Univ; Gs
Dasvarma, Gour, gour.dasvarma@flinders.edu.au, Flinders Univ; Zn
Datta, Saugata, (785) 532-2241 sdatta@ksu.edu, Kansas State Univ; ClHwGe
Daugherty, Carolyn M., carolyn.daugherty@nau.edu, Northern Arizona Univ; Zn
Daugherty, LeRoy A., (505) 646-3406 New Mexico State Univ, Las Cruces; Sd
Dauphas, Nicolas, (773) 702-2930 Univ of Chicago; Xc
Davatzes, Alexandra K., (215) 204-3907 alix@temple.edu, Temple Univ; GsXgZe
Davatzes, Nicholas, (215) 204-2837 davatzes@temple.edu, Temple Univ; GcNrZn
Davenport, Joan R., (509) 786-9384 jdavenp@tricity.wsu.edu, Washington State Univ; Sb
Davey, Patricia M., 863-2449 patricia_davey@brown.edu, Brown Univ; Zn
David, Mark B., (217) 333-4308 dmnicol@illinois.edu, Univ of Illinois, Urbana-Champaign; Sf
Davidson, Cameron, (507) 222-7144 cdavidso@carleton.edu, Carleton Coll; GptCc
Davidson, Fiona M., (479) 575-3879 fdavidso@comp.uark.edu, Univ of Arkansas, Fayetteville; Zy
Davidson, Garry J., 61 3 6226 2815 garry.davidson@utas.edu.au, Univ of Tasmania; Cs
Davidson, Gavin, (570) 577-1367 gd015@bucknell.edu, Bucknell Univ; Gg
Davidson, Gregg R., (662) 915-5824 davidson@olemiss.edu, Univ of Mississippi; Hw
Davidson, Kenneth L., 831-656-2309 kldavids@nps.edu, Naval Postgrad Sch; Am
Davie, Colin, +44 (0) 191 208 6458 colin.davie@ncl.ac.uk, Univ of Newcastle Upon Tyne; Ngr
Davies, Caroline P., daviesc@umkc.edu, Univ of Missouri, Kansas City; GnZce
Davies, David, +44 (0) 131 451 3569 d.davies@hw.ac.uk, Heriot-Watt Univ; Ng
Davies, Gareth, +44(0)7780864555 gareth.davies@open.ac.uk, Open Univ; Ze
Davies, Gregory, 609-258-4101 gd3@princeton.edu, Princeton Univ; Yg
Davies, J H., huw@earth.cf.ac.uk, Cardiff Univ; Yg
Davies, Neil, +44 (0) 1223 333453 nsd27@cam.ac.uk, Univ of Cambridge; Gs
Davies, Rhordri, +44 20 759 45722 rhodri.davies@imperial.ac.uk, Imperial Coll; Yg
Davies, Sarah, sarah.davies@open.ac.uk, Open Univ; Ze
Davis, Andrew, (773) 702-8164 Univ of Chicago; Xc
Davis, Arden D., (605) 394-2527 arden.davis@sdsmt.edu, South Dakota Sch of Mines & Tech; HwNx
Davis, Catherine, catherinedavis@ncsu.edu, North Carolina State Univ; Cm
Davis, Daniel M., 631-632-8217 daniel.davis@sunysb.edu, Stony Brook Univ; Yg
Davis, David A., (775) 682-8767 ddavis@unr.edu, Univ of Nevada, Reno; Gg
Davis, Donald W., (416) 946-0365 dond@geology.utoronto.ca, Univ of Toronto; Cc
Davis, Edward B., 541-346-3461 edavis@uoregon.edu, Univ of Oregon; PvePq
Davis, Fred A., (218) 726-8331 fdavis@d.umn.edu, Univ of Minnesota Duluth; CpGi
Davis, George H., (520) 621-1856 gdavis@u.arizona.edu, Univ of Arizona; Gc
Davis, Gregory A., (213) 740-6726 gdavis@usc.edu, Univ of Southern California; GtcGe
Davis, J. Matthew, (603) 862-1718 matt.davis@unh.edu, Univ of New Hampshire; HwYh
Davis, James , jdavis@ldeo.columbia.edu, Columbia Univ; Yd
Davis, James A., (217) 581-5528 jadavis2@eiu.edu, Eastern Illinois Univ; Zg
Davis, Kenneth J., (814) 863-8601 kjd10@psu.edu, Pennsylvania State Univ, Univ Park; Am
Davis, Leslie M., ldavis1@austincc.edu, Austin Comm Coll District; Op
Davis, Owen K., (520) 621-7953 odavis@email.arizona.edu, Univ of Arizona; Pl
Davis, P. Thompson, (781) 891-3479 pdavis@bentley.edu, Bentley Univ; GlmZc
Davis, Paul M., (310) 825-1343 pdavis@epss.ucla.edu, Univ of California, Los Angeles; YsmYg
Davis, Peter B., 253-538-5770 davispb@plu.edu, Pacific Lutheran Univ; GcpGz
Davis, R. Laurence, (203) 932-7108 rldavis@newhaven.edu, Univ of New Haven; GemHg
Davis, Robert, (804) 924-0579 red3u@virginia.edu, Univ of Virginia; AsmZy
Davis, Steven J., (949) 824-1821 sjdavis@uci.edu, Univ of California, Irvine; GeCsZu
Davis, Tara, +353 21 4903696 t.davis@ucc.ie, Univ Coll Cork; Yg
Davis, Trevor J., (801) 587-9019 Univ of Utah; Zi
Davisson, M L., (925) 423-5993 davisson2@llnl.gov, Lawrence Livermore; Cg
Davydov, Vladimir, vdavydov@boisestate.edu , Boise State Univ; PmGr
Dawers, Nancye H., (504) 865-5198 ndawers@tulane.edu, Tulane Univ; GcoGm
Dawes, Ralph, 509-682-6754 rdawes@wvc.edu, Wenatchee Valley Coll; GgeAm
Dawsey, Cyrus B., (334) 844-3418 dawsecb@auburn.edu, Auburn Univ; Zy
Dawson, Daniel, 765-494-5999 dandawson@purdue.edu, Purdue Univ; As
Dawson, Jane P., (515) 294-6302 jpdawson@iastate.edu, Iowa State Univ of Sci & Tech; Gp
Dawson, Mary R., (412) 622-3246 dawsonm@carnegiemnh.org, Carnegie Mus of Natural History; Pv
Dawson, Sarah, (717)358-5870 sarah.dawson@fandm.edu, Franklin and Marshall Coll; Zn
Day, Brett, +44 (0)1603 59 1413 brett.day@uea.ac.uk, Univ of East Anglia; Ge
Day, Howard W., (530) 752-2882 hwday@ucdavis.edu, Univ of California, Davis; Gx
Day, James, (858) 534-5431 jmdday@ucsd.edu, Univ of California, San Diego; Cg
Day, Mackenzie, daym@epss.ucla.edu, Univ of California, Los Angeles; Ggs
Day, Rick L., (814) 863-1615 r4d@psu.edu, Pennsylvania State Univ, Univ Park; Zu
Day, Rosie, +44 (0)121 41 48096 r.j.day@bham.ac.uk, Univ of Birmingham; Zn
Day, Stephanie S., 701-231-8837 stephanie.day@ndsu.edu, North Dakota State Univ; GmZiu
Day, Steven M., (619) 594-2663 sday@mail.sdsu.edu, San Diego State Univ; Ys
Day-Stirrat, Ruarri J., (971) 610-8968 Oregon Dept of Geology and Mineral Ind; GodEd
Daymon, Deborah J., (505) 667-9021 deba@lanl.gov, Los Alamos; Zn
Dayton, Paul K., (858) 534-6740 pdayton@ucsd.edu, Univ of California, San Diego; Ob
De Angelis, Silvio, +44-151-794-5161 s.de-angelis@liverpool.ac.uk, Univ of Liverpool; Ys
De Batist, Marc, marc.debatist@ugent.be, Ghent Univ; GnuGs
De Beus, Barbara, (973) 655-4448 debeusb@mail.montclair.edu, Montclair State Univ; Zn
De Campos, Christina, 089/2180 4264 campos@min.uni-muenchen.de, Ludwig-Maximilians-Univ Muenchen; Gz
De Carlo, Eric H., (808) 956-5924 edecarlo@soest.hawaii.edu, Univ of Hawai'i, Manoa; Cm
De Deckker, Patrick, +61-2-6125 2070 patrick.dedeckker@anu.edu.au, Australian Natl Univ; GuPmGn
de Foy, Benjamin, (314) 977-3122 benjamin.defoy@slu.edu, Saint Louis Univ; As
De Grave, Johan, johan.degrave@ugent.be, Ghent Univ; CcZgGx
de Hoog, Jan C., +44 (0) 131 650 8525 ceesjan.dehoog@ed.ac.uk, Edinburgh Univ; GxzGv
de Ignacio, Cristina, cris@ucm.es, Univ Compl de Madrid; Cg
De Klerk, Billy, +27 (0)46-622-2312 b.deklerk@ru.ac.za, Rhodes Univ; Pg
De Kock, Walter E., (319) 266-6577 Univ of Northern Iowa; Ze
de la Horra, Raúl, rhorraba@ucm.es, Univ Compl de Madrid; Gs
De Luca, Domenico Antonio, domenico.deluca@unito.it, Univ di Torino; Hy
de Menocal, Peter B., (845) 365-8483 peter@ldeo.columbia.edu, Columbia Univ; PeOuCg
De Paola, Nicola, nicola.de-paola@durham.ac.uk, Durham Univ; Gc
De Santo, Eilzabeth, (717) 358-4555 edesanto@fandm.edu, Franklin and Marshall Coll; Zn
De Scally, Fes, fes.descally@ubc.ca, Univ of British Columbia, Okanagan Campus; RnZyRm
de Silva, Shanika, (541) 737-1212 desilvsh@geo.oregonstate.edu, Oregon State Univ; Gv

de Stefano, Lucia, luciads@ucm.es, Univ Compl de Madrid; Hw
de Szoeke, Simon P., 541-737-8391 sdeszoek@coas.oregonstate.edu, Oregon State Univ; OpAs
de Uña Alvarez, Elena P., 00 34 988387137 edeuna@uvigo.es, Coruna Univ; Zy
de Vernal, Anne, 514-987-3000 #8599 devernal.anne@uqam.ca, Univ du Quebec a Montreal; Pm
de Vicente, Gerardo, gdv@ucm.es, Univ Compl de Madrid; Gt
De Vleeschouwer, David, ddevlees@uni-muenster.de, Univ Muenster; PcGg
de Wet, Carol B., 717-358-4388 cdewet@fandm.edu, Franklin and Marshall Coll; GsdGn
de Wit, Cary, (907) 474-7141 cwdewit@alaska.edu, Univ of Alaska, Fairbanks; Zg
de Wit, Julien, (617) 258-0209 jdewit@mit.edu, Massachusetts Inst of Tech; Zn
de' Michieli Vitturi, Mattia, mattiade@buffalo.edu, SUNY, Buffalo; Gv
Deakin, Ann K., (716) 673-3303 deakin@fredonia.edu, SUNY, Fredonia; ZiGgZr
Deakin, Joann, 520-249-9042 deakinj@cochise.edu, Cochise Coll; Ggg
Deal, Richard, (814) 732-1733 rdeal@edinboro.edu, Edinboro Univ of Pennsylvania; Zi
Dealy, Mike, (316) 943-2343 mdealy@kgs.ku.edu, Univ of Kansas; Gg
Dean, Jeffrey S., (520) 621-2320 jdean@ltrr.arizona.edu, Univ of Arizona; Zn
Dean, John Mark, jmdean36@gmail.com, Univ of South Carolina; Ob
Dean, Robert, 717-245-1109 deanr@dickinson.edu, Dickinson Coll; Gi
Dean, Sarah L., 616 395 7306 deansl@hope.edu, Hope Coll; Gc
deAngelis, Marie, 718-409-7380 mdeangelis@sunymaritime.edu, SUNY, Maritime Coll; Oc
DeAngelis, Michael T., 569-3542 mtdeangelis@ualr.edu, Univ of Arkansas at Little Rock; GpCp
DeAngelo, Micheal V., (512) 232-3373 michael.deangelo@beg.utexas.edu, Univ of Texas at Austin, Jackson Sch of Geosciences; Ye
Deans, Jeremy, 601-266-4729 jeremy.deans@usm.edu, Univ of Southern Mississippi; YrGc
Dearden, Christopher, +44 0161 306-3911 christopher.dearden@manchester.ac.uk, Univ of Manchester; As
Deardorff, Nicholas, 724-357-5625 n.deardorff@iup.edu, Indiana Univ of Pennsylvania; Gv
Deaton, Tami, (940) 565-2091 tami.deaton@unt.edu, Univ of North Texas; Zn
DeBari, Susan M., (360) 650-3588 debari@wwu.edu, Western Washington Univ; GxZe
Deboo, Phili B., (901) 678-4424 pdeboo@memphis.edu, Univ of Memphis; Pg
DeBruhl, Darby, debruhld@dnr.sc.gov, South Carolina Dept of Nat Res; Zi
DeCaria, Alex J., (717) 871-4739 alex.decaria@millersville.edu, Millersville Univ; As
DeCelles, Peter G., (520) 621-4910 decelles@email.arizona.edu, Univ of Arizona; GtsGc
Decker, Dave, 775.673.7353 dave.decker@dri.edu, Univ of Nevada, Reno; Hw
DeConto, Robert, (413) 545-3426 deconto@geo.umass.edu, Univ of Massachusetts, Amherst; As
DeCosmo, Janice, 206-221-6178 jdecosmo@uw.edu, Univ of Washington; AtZe
Dee, Kato , kdee@ou.edu, Univ of Oklahoma; HgCg
Dee, Seth, (775) 682-7704 sdee@unr.edu, Univ of Nevada, Reno; Gg
Dee, Sylvia G., 713.348.4889 sylvia.dee@rice.edu, Rice Univ; Zo
Deering, Chad, 906/487-1187 cddeerin@mtu.edu, Michigan Technological Univ; Ggi
DeGaetano, Arthur, (607) 255-0385 atd2@cornell.edu, Cornell Univ; As
DeGraff, James, jmdegraf@mtu.edu, Michigan Technological Univ; Gc
Degryse, Patrick, patrick.degryse@ees.kuleuven.be, Katholieke Univit Leuven; GaCeGf
Dehler, Carol M., (435) 797-0764 carol.dehler@usu.edu, Utah State Univ; Gs
Dehne, Kevin T., (989) 686-9326 ktdehne@delta.edu, Delta Coll; Ze
Deibert, Jack, (931) 221-6318 deibertj@apsu.edu, Austin Peay State Univ; GsrGo
Deininger, Robert W., (901) 678-2177 Univ of Memphis; Gx
Deisher, Jeffrey , 740 2012765 jeffrey.deischer@dnr.state.oh.us, Ohio Dept of Nat Res; Eo
DeJesus, Roman, 714-992-7462 rdejesus@fullcoll.edu, Fullerton Coll; Og
Deka, Jennifer, (216) 987-5827 jennifer.deka@tri-c.edu, Cuyahoga Comm Coll - Western Campus; Zg
Dekens, Petra, (415) 338-6015 dekens@sfsu.edu, San Francisco State Univ; Pe
DeKraker, Dan, (562) 860-2451 x2668 ddekraker@cerritos.edu, Cerritos Coll; ZeOpGg
dela Pierre, Francesco, francescodelapierre@unito.it, Univ di Torino; Gd
DELACOUR, Adélie, adelie.delacour@univ-st-etienne.fr, Université Jean Monnet, Saint-Etienne; GiCsOu
Delahunty, Tina, (570) 389-5181 tdelahun@bloomu.edu, Bloomsburg Univ; Zri
Delaney, Jeremy S., jsd@eps.rutgers.edu, Rutgers, The State Univ of New Jersey; XgmGx
Delaney, John R., 206-543-4830 jdelaney@ocean.washington.edu, Univ of Washington; GuYr
Delaney, Margaret L., (831) 459-4736 delaney@cats.ucsc.edu, Univ of California, Santa Cruz; Oc
Delano, Helen L., 717.702.2031 hdelano@pa.gov, DCNR- Pennsylvania Bur of Geo Survey; GgmNg
DeLaune, Paul, 940-552-9941x207 pbdelaune@ag.tamu.edu, Texas A&M Univ; So
Delawder, Sandra, delawdsa@jmu.edu, James Madison Univ; Zn
Delcambre, Sharon, sharon.delcambre@pcc.edu, Portland Comm Coll - Sylvania Campus; AsOg
Delfino, Massimo, massimo.delfino@unito.it, Univ di Torino; Pg
Delgado-Argote, Luis A., ldelgado@cicese.mx, Centro de Investigación Científica y de Educación Superior de Ensenada; GitGu
DeLima, Lynn-Ann, delimal@easternct.edu, Eastern Connecticut State Univ; Zg
Deline, Brad, (678) 839-4061 bdeline@westga.edu, Univ of West Georgia; Pi
Dellapenna, Timothy M., (409) 740-4952 dellapet@tamug.tamu.edu, Texas A&M Univ; Ou
Delparte, Donna M., 208-282-4419 delparte@isu.edu, Idaho State Univ; Zir
Delph, Jonathan, 765-494-5979 jdelph@purdue.edu, Purdue Univ; Yg
DelSole, Timothy, (703) 993-5715 tdelsole@gmu.edu, George Mason Univ; As
Delson, Eric, (718) 960-8405 eric.delson@lehman.cuny.edu, Graduate Sch of the City Univ of New York; PvqPs
Demanet, Laurent, (617) 253-2614 laurent@math.mit.edu, Massachusetts Inst of Tech; Zn
DeMaster, David J., (919) 515-7026 david_demaster@ncsu.edu, North Carolina State Univ; OcCbc
Demchak, Jennifer, 570-662-4613 jdemchak@mansfield.edu, Mansfield Univ; Hs
deMenocal, Peter B., peter@ldeo.columbia.edu, Columbia Univ; Pe
Demers, Serge, (418) 723-1986 (Ext. 1483) Univ du Quebec a Rimouski; Ob
DeMets, D. Charles, (608) 262-8598 chuck@geology.wisc.edu, Univ of Wisconsin, Madison; Ym
Demicco, Robert V., (607) 777-2604 demicco@binghamton.edu, Binghamton Univ; Gs
Deming, Jody, jdeming@u.washington.edu, Univ of Manitoba; Zn
Demlie, Molla, (031) 260-7380 demliem@ukzn.ac.za, Univ of KwaZulu-Natal; HwGeg
Demopoulos, George, (514) 398-4755 McGill Univ; Nx
Dempsey, David P., (415) 338-7716 ddempsey@norte.sfsu.edu, San Francisco State Univ; As
Dempster, Tim, +4401413305445 tim.dempster@glasgow.ac.uk, Univ of Glasgow; Gz
Demyanov, Vasily, +44 (0) 131 451 8298 v.demyanov@hw.ac.uk, Heriot-Watt Univ; Zi
Deng, Baolin, (505) 835-5505 New Mexico Tech; Cg
Deng, Jie, 609-258-4101 jie.deng@princeton.edu, Princeton Univ; Gy
Deng, Yi, 404-385-1821 yi.deng@eas.gatech.edu, Georgia Inst of Tech; AsZoAm
Deng, Youjun, 979-862-8476 yjd@tamu.edu, Texas A&M Univ; GzSc
Dengler, Lorinda, (707) 826-3115 lad1@humboldt.edu, Humboldt State Univ; Yg
Denizman, Can, (229) 333-5752 Valdosta State Univ; Hw
Denman, Kenneth L., (250) 363-8230 denmank@ec.gc.ca, Univ of Victoria; Op
Denne, Richard A., 817-257-4423 r.denne@tcu.edu, Texas Christian Univ; GoPms
Dennett, Keith E., (775) 784-4056 kdennett@unr.edu, Univ of Nevada, Reno; Hs
Denning, A. Scott, (970) 491-8359 scott.denning@colostate.edu, Colorado State Univ; AsZg
Dennis, Clyde B., (630) 252-5999 cbdennis@anl.gov, Argonne; Zn
Dennison, Robert, (309) 268-8646 robert.dennison@heartland.edu, Heartland Comm Coll; Zgn
Denniston, Rhawn F., (319) 895-4306 rdenniston@cornellcollege.edu, Cornell Coll; GeCs
Denny, F. B., 618-985-3394 x240 fdenny@illinois.edu, Illinois State Geo Survey; GzEg
Denolle, Marine, (617) 496-2708 mdenolle@fas.harvard.edu, Harvard Univ; Ys

Denton, Gary R. W., (671) 735-2690 gdenton@triton.uog.edu, Univ of Guam; Ob
Denton, George H., (207) 581-2193 debbies@maine.edu, Univ of Maine; Gl
Denton-Hedrick, Meredith Y., mdentonh@austincc.edu, Austin Comm Coll District; ZeYeEo
Deocampo, Daniel M., 404413-5750 deocampo@gsu.edu, Georgia State Univ; Gsn
DePaolo, Donald J., (510) 642-7686 djdepaolo@lbl.gov, Univ of California, Berkeley; Cc
dePolo, Craig M., (775) 682-8770 eq_dude@sbcglobal.net, Univ of Nevada, Reno; Ng
dePolo, Diane, (775) 784-4976 diane@seismo.unr.edu, Univ of Nevada, Reno; Ys
DePriest, Thomas A., (731) 881-7441 tdepriest@utm.edu, Univ of Tennessee, Martin; Ze
Derakhshani, Reza, derakhshani@uk.ac.ir, Shahid Bahonar Univ of Kerman ; GtmGc
Dere, Ashlee L., 402 554 3317 adere@unomaha.edu, Univ of Nebraska, Omaha; SdGmCl
Derome, Jacques F., (514) 398-6079 jacques.derome@mcgill.ca, McGill Univ; As
Derrick, Sharon, 361-825-3637 sharon.derrick@tamucc.edu, Texas A&M Univ, Corpus Christi; Zn
Derry, Louis A., lad9@cornell.edu, Cornell Univ; Cg
Derstler, Kraig L., (504) 280-6799 kderstle@uno.edu, Univ of New Orleans; PgiPv
Desai, Ankur R., (608) 265-9201 desai@aos.wisc.edu, Univ of Wisconsin, Madison; Asm
DeSantis, Larisa R., 615-343-7831 larisa.desantis@vanderbilt.edu, Vanderbilt Univ; Pve
Desch, Steven, (480) 965-7742 steve.desch@asu.edu, Arizona State Univ; Xa
Deschaine, Sylvia, (207) 786-6490 sdescha2@bates.edu, Bates Coll; Zn
DeShon, Heather R., 214-768-2916 hdeshon@smu.edu, Southern Methodist Univ; Ysr
DesOrmeau, Joel, 775-784-6054 jdesormeau@unr.edu, Univ of Nevada, Reno; Gp
Desrochers, Andre, adesro@uottawa.ca, Univ of Ottawa; Gs
Dessler, Alexander, (520) 621-4589 dessler@vega.lpl.arizona.edu, Univ of Arizona; Zn
Dessler, Andrew E., 979-862-1427 adessler@tamu.edu, Texas A&M Univ; As
DeSutter, Tom, 701-231-8690 thomas.desutter@ndsu.edu, North Dakota State Univ; So
Dethier, David P., (413) 597-2078 david.p.dethier@williams.edu, Williams Coll; Gm
Detournay, Emmanuel M., (612) 625-5522 detou001@tc.umn.edu, Univ of Minnesota, Twin Cities; NrpNm
Detrich, William, (617) 373-4495 iceman@neu.edu, Northeastern Univ; Zn
Dettman, David, (520) 621-4618 dettman@email.arizona.edu, Univ of Arizona; Cs
Dettmer, Jan, 403 220 4606 jan.dettmer@ucalgary.ca, Univ of Calgary; Ys
Detwiler, Andrew G., (605) 394-1995 andrew.detwiler@sdsmt.edu, South Dakota Sch of Mines & Tech; AspAm
Deuser, Werner G., (508) 289-2551 wdeuser@whoi.edu, Woods Hole Oceanographic Inst; CsGsu
Deutsch, Curtis, 609-258-4101 cdeutsch@princeton.edu, Princeton Univ; As
Devaney, Kathleen, 915-831-5161 kdevaney@epcc.edu, El Paso Comm Coll; Gg
DeVecchio, Duane, (480) 727-2636 duane.devecchio@asu.edu, Arizona State Univ; GtmGc
Devegowda, Deepak, deepak.devegowda@ou.edu, Univ of Oklahoma; Np
Dever, Edward P., 541-737-2749 edever@coas.oregonstate.edu, Oregon State Univ; Op
Devera, Joseph A., 618-985-3394 devera@isgs.uiuc.edu, Illinois State Geo Survey; Gr
Devlahovich, Vincent A., (661) 362-3658 vincent.devlahovich@canyons.edu, Coll of the Canyons; GgZye
Devlin, J. F., 785-864-4994 jfdevlin@ku.edu, Univ of Kansas; Hw
Devol, Allan H., (206) 543-1292 devol@u.washington.edu, Univ of Washington; Oc
DeVries-Zimmerman, Suzanne, (616) 395-7297 zimmerman@hope.edu, Hope Coll; Gge
Dewaele, Stijn, +32 9 264 4618 stijndg.dewaele@ugent.be, Ghent Univ; EgGzCg
Dewdney, Chris, +44 023 92 842417 chris.dewdney@port.ac.uk, Univ of Portsmouth; Yg
DeWeese, Georgina G., (678) 839-4065 gdeweese@westga.edu, Univ of West Georgia; Zy
DeWekker, Stephan F., (434) 924-3324 sfd3d@virginia.edu, Univ of Virginia; As
Dewers, Thomas, 505-845-0631 tdewers@sandia.gov, Sandia; Yx
deWet, Andrew P., (717) 291-3815 andy.dewet@fandm.edu, Franklin and Marshall Coll; Ge
Dewey, Janet, (307) 223-2265 jdewey2@uwyo.edu, Univ of Wyoming; CaGg
Dewey, John, jfdewey@ucdavis.edu, Univ of California, Davis; Gt
Dewey, Richard K., (250) 472-4009 rdewey@uvic.ca, Univ of Victoria; Opg
Dexter, Leland R., lee.dexter@nau.edu, Northern Arizona Univ; Zy
Dezzani, Raymond, 208-885-7360 dezzani@uidaho.edu, Univ of Idaho; RhZnn
Dhar, Ratan K., 718-262-2889 rdhar@york.cuny.edu, York Coll (CUNY); Hw
Dharmasoth, Manjula, +914027097116 manju.geophysics@gmail.com, Osmania Univ; YmeYg
Di Fiori, Sara, (310) 660-3593 sdifiori@elcamino.edu, El Camino Coll; GgOg
Di Girolamo, Larry, (217) 333-3080 gdi@illinois.edu, Univ of Illinois, Urbana-Champaign; Zr
Di Gregorio, Monica, +44(0) 113 34 31592 m.digregorio@leeds.ac.uk, Univ of Leeds; Eg
Di Lorenzo, Emanuele, (404) 894-3994 edl@gatech.edu, Georgia Inst of Tech; OpZnn
Di Toro, Giulio, +390498279105 giulio.ditoro@unipd.it, Univ degli Studi di Padova; Gc
DiBacco, Claudio, (902) 426-9778 claudio.dibacco@dfo-mpo.gc.ca, Dalhousie Univ; Ob
Dibb, Jack E., 603-862-3063 jack.dibb@unh.edu, Univ of New Hampshire; AcClAs
Dibben, Chris, +44 (0) 131 650 2552 chris.dibben@ed.ac.uk, Edinburgh Univ; Zn
DiBiase, Roman , 814-865-7388 rdibiase@psu.edu, Pennsylvania State Univ, Univ Park; Gmt
Dick, Henry J B., (508) 289-2590 hdick@whoi.edu, Woods Hole Oceanographic Inst; GitGc
Dick, Richard P., 614-247-7605 dick.78@osu.edu, Ohio State Univ; Sb
Dick, Warren A., (330) 263-3877 dick.5@osu.edu, Ohio State Univ; SbCbHs
Dickerson, Richard E., (310) 825-5864 Univ of California, Los Angeles; Zn
Dickhoff, Willem H., (314) 935-4169 wimd@howdy.wustl.edu, Washington Univ in St. Louis; Zn
Dickin, Alan P., (905) 525-9140 (Ext. 24365) dickin@mcmaster.ca, McMaster Univ; Cc
Dickinson, William R., (520) 299-5220 wrdickin@dakotacom.net, Univ of Arizona; Gt
Dickman, Steven R., (607) 777-2857 dickman@binghamton.edu, Binghamton Univ; Yg
Dickson, Alex, +44 1784 443834 alex.dickson@rhul.ac.uk, Royal Holloway Univ of London; Cas
Dickson, Andrew G., (858) 822-2990 adickson@ucsd.edu, Univ of California, San Diego; Cm
Dickson, Loretta D., (570) 484-2068 ldickson@lockhaven.edu, Lock Haven Univ; GizGe
Dickson, Stephen M., (207) 287-7174 stephen.m.dickson@maine.gov, Dept of Agriculture, Conservation, and Forestry; OnGuOu
Didlake, Jr., Anthony, 814-863-4636 acd171@psu.edu, Pennsylvania State Univ, Univ Park; As
Diebold, Frank E., (406) 496-4215 Montana Tech of the Univ of Montana; Cl
Diecchio, Richard J., (703) 993-5394 rdiecchi@gmu.edu, George Mason Univ; Gr
Dieckmann, Melissa S., (859) 622-1274 melissa.dieckmann@eku.edu, Eastern Kentucky Univ; ZeCl
Diederichs, Mark, 613-533-6504 diederim@queensu.ca, Queen's Univ; Nr
Diefendorf, Aaron, (513) 556-3787 aaron.diefendorf@uc.edu, Univ of Cincinnati; Co
Diem, Jeremy E., (404) 413-5770 jdiem@gsu.edu, Georgia State Univ; At
Diemer, John A., 704-687-5994 jadiemer@email.uncc.edu, Univ of North Carolina, Charlotte; Gs
Diener, Johann, 021-650-2925 johann.diener@uct.ac.za, Univ of Cape Town; Gpc
Diercks, Arne R., 228-688-3177 arne.diercks@usm.edu, Univ of Southern Mississippi; Ou
Dierssen, Heidi, 860-405-9239 heidi.dierssen@uconn.edu, Univ of Connecticut; Op
Dieterich, James H., (951) 827-2976 james.dieterich@ucr.edu, Univ of California, Riverside; Yg
Dieterle, Dwight A., (727) 553-1114 ddieterle@marine.usf.edu, Univ of South Florida; Ob
Dietl, Gregory, 6072736623 x117 gpd3@cornell.edu, Paleontological Research Inst; Pg

Dietrich, William E., (510) 642-2633 bill@geomorph.berkeley.edu, Univ of California, Berkeley; Gm
Dietsch, Craig, (513) 556-2547 craig.dietsch@uc.edu, Univ of Cincinnati; Gp
Dietz, Anne D., 210-486-0470 adietz@alamo.edu, San Antonio Comm Coll; PgZg
Dietz, Richard D., (970) 351-2950 richard.dietz@unco.edu, Univ of Northern Colorado; Xg
Dietze, Michael, dietze@bu.edu, Boston Univ; SfZr
Diffendal, Jr., Robert F., 402-472-7546 rfd@unl.edu, Univ of Nebraska, Lincoln; GrPiRh
DiFrancesco, Nicholas, 716-645-3489 ndifranc@buffalo.edu, SUNY, Buffalo; Gg
Diggins, Thomas P., 330-941-3605 tpdiggins@ysu.edu, Youngstown State Univ; Pg
Dighe, Kalpak, (505) 665-1701 kdighe@lanl.gov, Los Alamos; Zn
Dijkstra, Arjan, +44 1752 584774 arjan.dijkstra@plymouth.ac.uk, Univ of Plymouth; Gi
Dijkstra, Semme, 603-862-0525 semmed@ccom.unh.edu, Univ of New Hampshire; Og
Dilek, Yildirm, (513) 529-2212 dileky@miamioh.edu, Miami Univ; Gt
Dilles, John H., (541) 737-1245 dillesj@geo.oregonstate.edu, Oregon State Univ; Em
Dillman, Amanda, dillm004@umn.edu, Univ of Minnesota, Twin Cities; Yx
DiMarco, Steven F., (979) 862-4168 sdimarco@tamu.edu, Texas A&M Univ; OpnOg
DiMichele, William A., (202) 633-1319 Smithsonian Inst / Natl Mus of Natural History; Pb
Dimmick, Charles W., (860) 832-2936 dimmick@ccsu.edu, Central Connecticut State Univ; Ge
Dimova, Natasha T., 205-348-0256 ntdimova@ua.edu, Univ of Alabama; HwCcm
Ding, Kang, 612-626-1860 mlcd@umn.edu, Univ of Minnesota, Twin Cities; Cg
Dingman, S. Lawrence, (603) 862-1718 Univ of New Hampshire; Hq
Dingus, Delmar D., (805) 756-2753 ddingus@calpoly.edu, California Polytechnic State Univ; So
Dingwell, Donald Bruce, 089/21804136 dingwell@min.uni-muenchen.de, Ludwig-Maximilians-Univ Muenchen; Giv
Dinklage, William, 805-730-4114 wsdinklage@sbcc.edu, Santa Barbara City Coll; Gge
Dinsmoor, John C., (702) 295-6189 john_dinsmoor@lanl.gov, Los Alamos; Nm
Dintaman, Christopher, (812) 856-5654 cdintama@indiana.edu, Indiana Univ, Bloomington; Hw
Dinter, David A., (801) 581-7937 david.dinter@utah.edu, Univ of Utah; GtcGu
Diochon, Amanda, (807) 343-8010 Ext. 8444 adiochon@lakeheadu.ca, Lakehead Univ; Ge
DiPietro, Joseph A., (812) 465-7041 dipietro@usi.edu, Univ of Southern Indiana; GcpGm
Dipple, Gregory M., (604) 827-0653 gdipple@eoas.ubc.ca, Univ of British Columbia; Gp
Dirmeyer, Paul A., (703) 993-5363 pdirmeye@gmu.edu, George Mason Univ; As
Disperati, Leonardo, leonardo.disperati@unisi.it, Univ of Siena; NgtZr
Ditmars, John, (630) 252-5953 Argonne; Zn
Dittmer, Eric, (541) 552-6496 dittmer@sou.edu, Southern Oregon Univ; Gg
Diver, Kim, 860-685-2610 kdiver@wesleyan.edu, Wesleyan Univ; ZiyZg
Divine, Aaron, (928) 523-7835 aaron.divine@nau.edu, Northern Arizona Univ; ZnnZn
Dix, George R., (613) 520-2600 george.dix@carleton.ca, Carleton Univ; Gsd
Dix, Justin, +44 (0)23 8059 3057 j.k.dix@soton.ac.uk, Univ of Southampton; GuYuNx
Dixon, Barnali, bdixon@usf.edu, Univ of South Florida; Zi
Dixon, Clifton, 601-266-4731 c.dixon@usm.edu, Univ of Southern Mississippi; Zn
Dixon, Jean, (406) 994-3342 jean.dixon@montana.edu, Montana State Univ; GmScZg
Dixon, John C., (479) 422-2115 jcdixon@uark.edu, Univ of Arkansas, Fayetteville; GmXgSo
Dixon, Michael, (519) 824-4120 Ext.52555 dixon@ces.uoguelph.ca, Univ of Guelph; Zn
Dixon, Nick, +44(0) 113 34 34931 fuensd@leeds.ac.uk, Univ of Leeds; Zu
Dixon, P. Grady, (785) 628-4536 pgdixon@fhsu.edu, Fort Hays State Univ; AmZyAs
Dixon, Richard, (512) 245-2170 Texas State Univ; AsOpRn
Dixon, Simon J., s.j.dixon@bham.ac.uk, Univ of Birmingham; Gms
Dixon, Tim, thd@usf.edu, Univ of South Florida; Yd
Dladla, Nonkululeko, (031) 260-2801 dladlan2@ukzn.ac.za, Univ of KwaZulu-Natal; GusGr
Dmochowski, Jane, 215-5735388 janeed@sas.upenn.edu, Univ of Pennsylvania; Yg
Doar, III, William R., (803) 609-7065 doarw@dnr.sc.gov, South Carolina Dept of Nat Res; GrdGp
Dobbie, Steven, +44(0) 113 34 36725 j.s.e.dobbie@leeds.ac.uk, Univ of Leeds; As
Dobbs, Fred C., (757) 683-4301 fdobbs@odu.edu, Old Dominion Univ; Ob
Dobler, Scott, (270) 745-7078 scott.dobler@wku.edu, Western Kentucky Univ; As
Dobrzhinetskaya, Larissa F., (951) 827-2028 larissa@ucrac1.ucr.edu, Univ of California, Riverside; Gx
Dobson, David, +44 020 7679 32398 d.dobson@ucl.ac.uk, Univ Coll London; Gz
Dockal, James A., (563) 845-1034 dockal@uncw.edu, Univ of North Carolina, Wilmington; GdcEm
Dodd, Justin P., 815-753-7949 jdodd@niu.edu, Northern Illinois Univ; Cs
Dodge, Clifford H., 717.702.2036 cdodge@pa.gov, DCNR- Pennsylvania Bur of Geo Survey; GrEcRh
Dodge, Rebecca L., rebecca.dodge@msutexas.edu, Midwestern State Univ; Zre
Dodge, Stefanie, (414) 324-5676 stefanie.dodge@wisc.edu, Univ of Wisconsin, Madison, Division of Extension; Gg
Dogwiler, Toby, tdogwiler@missouristate.edu, Missouri State Univ; ZiGmZy
Doh, Seong-Jae, sjdoh@korea.ac.kr, Korea Univ; YmgYe
Doherty, Amanda, 575-835-5038 amanda.doherty@nmt.edu, New Mexico Tech; HwCg
Doherty, Ruth, +44 (0) 131 650 6759 ruth.doherty@ed.ac.uk, Edinburgh Univ; As
Dolakova, Nela, +420 549 49 3542 nela@sci.muni.cz, Masaryk Univ; Plg
Dolan, James F., (213) 740-8599 dolan@usc.edu, Univ of Southern California; Gt
Dolejs, David, +420-221951525 Charles Univ; GiCpg
Dolgoff, Anatole, (212) 346-1502 Pace Univ, New York Campus; Zg
Doll, William E., d8e@ornl.gov, Oak Ridge; Yg
Dollase, Wayne A., (310) 825-3823 dollase@ucla.edu, Univ of California, Los Angeles; Gz
Dolliver, Holly A., holly.dolliver@uwrf.edu, Univ of Wisconsin, River Falls; GmSd
Dolman, Han J., +31 20 5987358 han.dolman@vu.nl, Vrije Univit Amsterdam; CbAmHg
Dolman, Paul, +44 (0)1603 59 3175 p.dolman@uea.ac.uk, Univ of East Anglia; Zu
Domack, Cynthia R., (315) 859-4710 cdomack@hamilton.edu, Hamilton Coll; Pg
Domagall, Abigail M., (605) 642-6506 abigail.domagall@bhsu.edu, Black Hills State Univ; GveZe
Dombard, Andrew J., (312) 996-9206 adombard@uic.edu, Univ of Illinois Chicago; Xy
Domber, Steven E., (609) 984-6587 steven.domber@dep.nj.gov, New Jersey Geological and Water Survey; Hw
Domingo, Laura, ldomingo@ucm.es, Univ Compl de Madrid; PveCs
Domingue, Carla, (225) 388-8407 carla@lgs.bri.lsu.edu, Louisiana State Univ; Zn
Dominguez, Francina, 217-265-5483 francina@illinois.edu, Univ of Illinois, Urbana-Champaign; AsHg
Dominic, David F., 937 775-2201 david.dominic@wright.edu, Wright State Univ; GsrRc
Donaghay, Percy, (401) 874-6944 donaghay@gso.uri.edu, Univ of Rhode Island; Ob
Donahoe, Robert J., (505) 667-7603 Los Alamos; Zn
Donahoe, Rona J., 205-348-1879 rdonahoe@geo.ua.edu, Univ of Alabama; Cl
Donaldson, Paul R., pdonalds@boisestate.edu, Boise State Univ; Ye
Dondofema, Farai, +27 15 962 80044 farai.dondofema@univen.ac.za, Univ of Venda for Sci & Tech; ZrHgZi
Doner, Lisa A., 603-535-2245 ladoner@plymouth.edu, Plymouth State Univ; PcGes
Doney, Scott, 434-924-0570 scd5c@virginiae.edu, Univ of Virginia; Oc
Dong, Charles, (310) 660-3593 El Camino Coll; Og
Dong, Hailiang, (513) 529-2517 dongh@miamioh.edu, Miami Univ; Cg
Dong, Pinliang, (940) 565-2377 pinliang.dong@unt.edu, Univ of North Texas; Zir
Dong, Xiquan, (520) 621-4652 xdong@email.arizona.edu, Univ of Arizona; As
Dong, Yunpeng, ydong265@uwo.ca, Western Univ; GtcCg
Dongmo Wamba, Mathurin, 609-258-4101 mw1685@princeton.edu, Princeton Univ; YsnYg
Donnelly, Jeffrey, 508-289-2994 jdonnelly@whoi.edu, Woods Hole

Oceanographic Inst; Gu
Donoghue, Michael J., 203-432-1935 michael.donoghue@yale.edu, Yale Univ; Pbg
Donohue, Kathleen, (401) 874-6615 kdonohue@gso.uri.edu, Univ of Rhode Island; Op
Donohue, Stephen J., (540) 231-9740 donohue@vt.edu, Virginia Tech; Sc
Donovan, James, (801) 585-3029 james.donovan@utah.edu, Univ of Utah; NmrZr
Donovan, Jeff C., (727) 553-1116 jdonovan@marine.usf.edu, Univ of South Florida; Op
Donovan, Joseph J., (304) 293-5603 donovan@geo.wvu.edu, West Virginia Univ; Hq
Donovan, Michael, 216-231-4600 mdonovan@cmnh.org, Cleveland Mus of Natural History; Pb
Donovan, R N., (817) 257-7214 r.donovan@tcu.edu, Texas Christian Univ; GdcGt
Doolan, Barry L., (802) 656-0248 bdoolan@uvm.edu, Univ of Vermont; Gx
Dooley, Brett, bdooley@ph.vccs.edu, Patrick Henry Comm Coll; Gg
Dooley, John H., (609) 292-2576 john.dooley@dep.state.nj.us, New Jersey Geological and Water Survey; CgGfe
Dooley, Tim, 512-471-8261 tim.dooley@beg.utexas.edu, Univ of Texas at Austin, Jackson Sch of Geosciences; Gc
Dorais, Michael J., (801) 422-1347 dorais@byu.edu, Brigham Young Univ; Gi
Dorale, Jeffrey A., (319) 335-0822 jeffrey-dorale@uiowa.edu, Univ of Iowa; CsPc
Doran, Peter, pdoran@lsu.edu, Louisiana State Univ; Hws
Dore, John, (406) 600-5932 jdore@montana.edu, Montana State Univ; Ca
Dorfman, Alexander, 089/2180 4275 Ludwig-Maximilians-Univ Muenchen; Gz
Dorfner, Thomas, 089/2180 4278 dorfner@min.uni-muenchen.de, Ludwig-Maximilians-Univ Muenchen; Gz
Dorling, Stephen, +44 (0)1603 59 2533 s.dorling@uea.ac.uk, Univ of East Anglia; Am
Dormaar, John, (403) 327-4561 Univ of Lethbridge; So
Dorman, Clive E., (619) 594-5707 cdorman@mail.sdsu.edu, San Diego State Univ; OpAmt
Dorman, James, (901) 678-2007 dorman@comcast.net, Univ of Memphis; Yse
Dorman, Leroy M., (858) 534-2406 ldorman@ucsd.edu, Univ of California, San Diego; Yg
Dornbos, Stephen Q., sdornbos@uwm.edu, Univ of Wisconsin, Milwaukee; PeiPo
Doro, Kennedy, (419) 530-2811 kennedy.doro@utoledo.edu, Univ of Toledo; Zn
Dorries, Alison M., (505) 665-6952 adorries@lanl.gov, Los Alamos; Zn
Dorsey, Rebecca J., (541) 346-4431 rdorsey@uoregon.edu, Univ of Oregon; Gr
Dort, Jr., Wakefield, (785) 864-4974 Univ of Kansas; Gm
Dortch, Jason, (513) 667-6128 jason.m.dortch@uky.edu, Univ of Kentucky; GmZr
dos Santos, Telmo B., tmsantos@fc.ul.pt, Unive de Lisboa; GptGi
Doskey, Paul V., (630) 252-7662 pvdoskey@anl.gov, Argonne; As
Doss, Paul K., (812) 465-7132 pdoss@usi.edu, Univ of Southern Indiana; HwGeZu
Dosso, Stanley E., (250) 472-4341 sdosso@uvic.ca, Univ of Victoria; Op
Dostal, Jarda, (902) 420-5747 Dalhousie Univ; Cg
Doster, Florian, +44 (0)131 451 3171 f.doster@hw.ac.uk, Heriot-Watt Univ; Hg
Dostie, Philip, (207) 786-6485 pdostie@bates.edu, Bates Coll; Cl
Doty, Thomas, (401) 254-3066 Roger Williams Univ; Ob
Dou, Fugen, (409) 752-2741 ext. 2223 f-dou@aesrg.tamu.edu, Texas A&M Univ; So
Dougan, Bernie, (360) 383-3877 bdougan@whatcom.edu, Whatcom Comm Coll; Gg
Doughty, Alice, 207-786-6113 adoughty@bates.edu, Bates Coll; Gl
Dougil, Andy, +44(0) 113 34 36782 a.j.dougill@leeds.ac.uk, Univ of Leeds; So
Douglas, Arthur V., (402) 280-2464 sonora@creighton.edu, Creighton Univ; Zg
Douglas, Bruce, 812-855-3848 douglasb@indiana.edu, Indiana Univ, Bloomington; Gcg
Douglas, Ellen, ellen.douglas@umb.edu, Univ of Massachusetts, Boston; HwqHy
Douglas, Peter, 514-398-6772 peter.douglas@mcgill.ca, McGill Univ; Cs
Douglass, Daniel C., (617) 373-4381 d.douglass@northeastern.edu, Northeastern Univ; GlSdPc
Douglass, David N., (626) 585-7036 dndouglass@pasadena.edu, Pasadena City Coll; Ze
Douilly, Roby, (951) 827-3180 roby.douilly@ucr.edu, Univ of California, Riverside; YsGtYg
Dove, Patricia M., (540) 231-2444 pdove@vt.edu, Virginia Tech; Cl
Doveton, John H., 785-864-3965 doveton@kgs.ku.edu, Univ of Kansas; Gq
Dowd, John F., (706) 542-2383 jdowd@uga.edu, Univ of Georgia; Hy
Dowdy, Robert H., (612) 625-7058 bdowdy@soils.umn.edu, Univ of Minnesota, Twin Cities; Sc
Dower, John, 250-721-6120 dower@uvic.ca, Univ of Victoria; Ob
Dowling, Carolyn B., 765-285-8274 cbdowling@bsu.edu, Ball State Univ; HgGeCg
Downes, Hilary, +44 020 3073 8027 h.downes@ucl.ac.uk, Birkbeck Coll; Cg
Downie, Helen, helen.downie@manchester.ac.uk, Univ of Manchester; Ge
Downs, James W., downs.1@osu.edu, Ohio State Univ; Gz
Downs, Joni, downs@usf.edu, Univ of South Florida; Zi
Downs, Robert T., (520) 626-8092 rdowns@email.arizona.edu, Univ of Arizona; Gi
Doyle, James A., (530) 752-7591 jadoyle@ucdavis.edu, Univ of California, Davis; Pb
Drago, Aldo, aldo.drago@um.edu.mt, Univ of Malta; Opg
Drake, John C., (802) 656-0244 jdrake@uvm.edu, Univ of Vermont; Cl
Drake, Lon D., (319) 335-1826 Univ of Iowa; Hy
Drake, Luke, 818-677-3508 luke.drake@csun.edu, California State Univ, Northridge; Zi
Drake, Michael J., (520) 621-6962 drake@lpl.arizona.edu, Univ of Arizona; Xc
Drake, Simon, +440203 073 8024 drakesimon1@gmail.com, Birkbeck Coll; Gv
Draper, Grenville, (305) 348-3087 draper@fiu.edu, Florida InterNatl Univ; GctRh
Drazen, Jeffrey C., (808) 956-6567 jdrazen@hawaii.edu, Univ of Hawai'i, Manoa; Ob
Dreesen, Donald S., (505) 667-1913 dreesen@lanl.gov, Los Alamos; Np
Dreger, Douglas S., (510) 643-1719 dreger@seismo.berkeley.edu, Univ of California, Berkeley; Ys
Drescher, Andrew, (612) 625-2374 Univ of Minnesota, Twin Cities; So
Dreschoff, Gisela, (785) 864-4517 Univ of Kansas; Ce
Drew, Douglas A., (404) 496-4202 ddrew@mtech.edu, Montana Tech of the Univ of Montana; Cg
Driese, Steven G., 254-710-2361 steven_driese@baylor.edu, Univ of Tennessee, Knoxville; SaClGs
Driever, Steven L., (816) 235-2971 drievers@umkc.edu, Univ of Missouri, Kansas City; ZnnZn
Drijfhout, Sybren, +44 (0)23 8059 6202 s.s.drijfhout@soton.ac.uk, Univ of Southampton; Op
Driscoll, John R., (607) 753-2926 driscollj@cortland.edu, SUNY, Cortland; Zn
Driscoll, Neal W., (858) 822-5026 ndriscoll@ucsd.edu, Univ of California, San Diego; Gu
Driscoll, Peter E., pdriscoll@carnegiescience.edu, Carnegie Inst for Sci; Ym
Drobniak, Agnieszka, (812) 855-2687 agdrobni@indiana.edu, Indiana Univ, Bloomington; Ec
Drobny, Gerry, (916) 691-7204 Cosumnes River Coll; Gg
Droop, Giles, +44 0161 275-3809 giles.droop@manchester.ac.uk, Univ of Manchester; Gp
Droser, Mary L., 951-827-3797 mary.droser@ucr.edu, Univ of California, Riverside; Pg
Droxler, Andre W., andre@rice.edu, Rice Univ; Gs
Druckenbrod, Daniel L., 609-895-5422 ddruckenbrod@rider.edu, Rider Univ; AtZu
Druckenmiller, Patrick S., 907-474-6954 ffpsd@uaf.edu, Univ of Alaska, Fairbanks; Pv
Druet, María, mdruetve@ucm.es, Univ Compl de Madrid; Yr
Druffel, Ellen R. M., (949) 824-2116 edruffel@uci.edu, Univ of California, Irvine; Oc
Druguet Tantiña, Elena, ++935813730 elena.druguet@uab.cat, Univ Auto de Barcelona; Gc
Druhan, Jennifer, (217) 300-0142 jdruhan@illinois.edu, Univ of Illinois, Urbana-Champaign; HwCs
Drumheller-Horton, Stephanie K., 865-974-7759 sdrumhel@utk.edu, Univ of Tennessee, Knoxville; Pg
Drummond, Jane, +4401413304208 jane.drummond@glasgow.ac.uk, Univ of Glasgow; Zi
Druschel, Greg, gdrusche@iupui.edu, Indiana Univ - Purdue Univ Indianapolis; Cl
Drysdale, Jessica, (508) 289-3739 jdrysdale@whoi.edu, Woods Hole Oceanographic Inst; Oc
Drzewiecki, Peter A., (860) 465-4322 drzewieckip@easternct.edu, Eastern Connecticut State Univ; GrsGd
Drzyzga, Scott A., (717) 477-1307 sadrzy@ship.edu, Shippensburg Univ;

ZiuGl
Duaime, Terence E., (406) 496-4157 tduaime@mtech.edu, Montana Tech of The Univ of Montana; Hws
Duan, Benchuan, (979) 845-3297 bduan@tamu.edu, Texas A&M Univ; YsgGq
Duan, Benchun, (979) 845-3297 bduan@tamu.edu, Texas A&M Univ; YsGtYg
Duan, Quingyun, (925) 422-7704 duan2@llnl.gov, Lawrence Livermore; Hg
Duan, Shuiwang, (301) 405-2407 sduan@umd.edu, Univ of Maryland; HsZu
Dubba, Vijay K., +919550627034 ddvkumar7@gmail.com, Osmania Univ; YevYs
Dube, Benoit, (418) 654-2669 bdube@nrcan.gc.ca, Univ du Quebec; Em
Dubey, C S., (981) 107-4867 csdubey@gmail.com, Sri Sri Univ; GexGt
Dubsky, Scott , 719-333-3080 scott.dubsky@usafa.edu, United States Air Force Acad; ZyGm
Duby, Paul F., (212) 854-2928 pfd1@columbia.edu, Columbia Univ; Nx
Duce, Robert A., (979) 845-5756 rduce@ocean.tamu.edu, Texas A&M Univ; Oc
Ducea, Mihai N., (520) 621-5171 ducea@email.arizona.edu, Univ of Arizona; Gt
Ducey, Mark, 603-862-4429 mark.ducey@unh.edu, Univ of New Hampshire; Zc
Duchane, David V., (505) 667-9893 duchane@lanl.gov, Los Alamos; Zn
Duchesne, Josee, (418) 656-2177 duchesne@ggl.ulaval.ca, Univ Laval; Gz
Duchesne, Rocio, duchesnr@uww.edu, Univ of Wisconsin, Whitewater; Zri
Ducklow, Hugh, 845-365-8167 hducklow@ldeo.columbia.edu, Columbia Univ; ObCb
Duda, Jan-Peter, +49 551 397960 jan-peter.duda@geo.uni-goettingen.de, Georg-August Univ of Goettingen; PgCoGs
Dudley, Lynn M., 850-644-4214 dudley@gly.fsu.edu, Florida State Univ; Sc
Dudley, Walter, 808-933-3905 dudley@hawaii.edu, Univ of Hawai'i, Manoa; OuRn
Dudukovic, Mike, dudu@che.wustl.edu, Washington Univ in St. Louis; Zn
Duebendorfer, Ernest M., (928) 523-7510 ernie.d@nau.edu, Northern Arizona Univ; Gct
Dueker, Kenneth G., (307) 766-2657 dueker@uwyo.edu, Univ of Wyoming; Ys
Duex, Timothy W., (337) 482-6222 twd7693@louisiana.edu, Univ of Louisiana at Lafayette; HgGex
Dufek, Josef, 541-346-4788 jdufek@uoregon.edu, Univ of Oregon; Gv
Duff, John, john.duff@umb.edu, Univ of Massachusetts, Boston; ZnnZn
Duffey, Patricia, 785-628-5389 pduffey@fhsu.edu, Fort Hays State Univ; Zn
Duffield, Wendell A., (928) 523-4852 Northern Arizona Univ; Gv
Duffy, Clarence J., (505) 667-5154 Los Alamos; Cg
Duffy, Tara , (617) 373-6323 t.duffy@neu.edu, Northeastern Univ; Ob
Duffy, Thomas S., 609-258-6769 duffy@princeton.edu, Princeton Univ; GyYgZm
Dugan, Brandon, 303-273-3512 dugan@mines.edu, Colorado Sch of Mines; GuYrHw
Dugas, Bernard, 514 987-3000 #5319 dugas.bernard@uqam.ca, Univ du Quebec a Montreal; As
Dugmore, Andrew J., +44 (0) 131 650 8156 andrew.dugmore@ed.ac.uk, Edinburgh Univ; ZycGa
Duhamel, Solange, sduhamel@ldeo.columbia.edu, Columbia Univ; Pe
Duke, Edward F., (605) 394-2388 edward.duke@sdsmt.edu, South Dakota Sch of Mines & Tech; GpiZr
Duke, Norman A., (519) 661-3199 nduke@uwo.ca, Western Univ; Eg
Dulai, Henrietta, (808) 956-0720 hdulaiov@hawaii.edu, Univ of Hawai'i, Manoa; CcHw
Duley, James W., 573-368-2105 bill.duley@dnr.mo.state, Missouri Dept of Nat Res; Hy
Dulin, Shannon, (405) 325-5488 sdulin@ou.edu, Univ of Oklahoma; YmGs
Duller, Rob, +44-151-794-5179 robert.duller@liverpool.ac.uk, Univ of Liverpool; Gvs
Dumas, Simone, dumas@uottawa.ca, Univ of Ottawa; GsOg
Dumitriu, Tony C., 0040756584797 dtonycristian@yahoo.com, Alexandru Ioan Cuza; GcZir
Dumond, Gregory, 479-575-3411 gdumond@uark.edu, Univ of Arkansas, Fayetteville; GtcGg
Dunagan, Stan P., (731) 881-7437 sdunagan@utm.edu, Univ of Tennessee, Martin; GsSa
Dunbar, John A., (254) 710-2191 john_dunbar@baylor.edu, Baylor Univ; YgHg
Dunbar, Nelia W., 575-835-5783 nelia.dunbar@nmt.edu, New Mexico Tech; GivCa
Dunbar, Robert B., (650) 725-6830 dunbar@pangea.stanford.edu, Stanford Univ; On
Duncan, Ian, 512-471-5117 ian.duncan@beg.utexas.edu, Univ of Texas at Austin, Jackson Sch of Geosciences; Gg
Duncan, Megan S., (540) 231-5980 msduncan19@vt.edu, Virginia Tech; GiCp
Duncan, Stewart, 785532277 sduncan@ksu.edu, Kansas State Univ; So
Duncan Tabb, Neva, 721-791-2758 duncantabb.neva@spcollege.edu, Saint Petersburg Coll, Clearwater; Am
Dundas, Robert G., (559) 278-6984 rdundas@csufresno.edu, Fresno State Univ; PvGr
Dunfield, Kari, (519) 824-4120 x58088 dunfield@uoguelph.ca, Univ of Guelph; SbZn
Dunkle, Kallina M., (931) 221-7451 dunklek@apsu.edu, Austin Peay State Univ; HwGle
Dunlap, Dallas B., (512) 471-4858 dallas.dunlap@beg.utexas.edu, Univ of Texas at Austin, Jackson Sch of Geosciences; Ye
Dunlea, Ann, 508-289-3423 adunlea@whoi.edu, Woods Hole Oceanographic Inst; Cm
Dunlevey, John, +27 (0)15 268 3483 john.dunlevey@ul.ac.za, Univ of Limpopo; GzxEg
Dunn, Allison L., (508) 929-8641 adunn@worcester.edu, Worcester State Univ; AsZyHg
Dunn, Amy, 575-835-5501 amy.dunn@nmt.edu, New Mexico Tech; Zi
Dunn, Bernie, 519-850-2362 bdunn@uwo.ca, Western Univ; Ye
Dunn, John Todd, (506) 453-4804 Univ of New Brunswick; Gi
Dunn, Richard K., (802) 485-2304 rdunn@norwich.edu, Norwich Univ; GalGs
Dunn, Robert, 843-904-9026 robert@baruch.sc.edu, Univ of South Carolina; Ob
Dunn, Steven R., 413-538-2531 sdunn@mtholyoke.edu, Mount Holyoke Coll; Gp
Dunn, Tasha L., (207) 859-5808 tldunn@colby.edu, Colby Coll; XmGiz
Dunne, George C., (818) 677-2511 george.dunne@csun.edu, California State Univ, Northridge; Gc
Dunne, William M., (423) 974-4161 wdunne@utk.edu, Univ of Tennessee, Knoxville; Gc
Dunning, Gregory R., (709) 864-8481 gdunning@mun.ca, Memorial Univ of Newfoundland; CcGit
Dunning, Jeremy D., (812) 856-4448 dunning@indiana.edu, Indiana Univ, Bloomington; Yg
Dunton, Karen K., (309) 438-7640 kkedunt@ilstu.edu, Illinois State Univ; Zn
Dunton, Kenneth H., (361) 749-6744 ken.dunton@utexas.edu, Univ of Texas, Austin; Ob
Dupont, Todd, (513) 529-9734 dupontt@miamioh.edu, Miami Univ; Gl
Dupre, William R., 713-743-4987 wdupre@uh.edu, Univ of Houston; On
Dura, Christina, (540) 231-3785 tinadura@vt.edu, Virginia Tech; OnRnPm
Durand, Michael T., (614)247-4835 durand.8@osu.edu, Ohio State Univ; HsZr
Durazo-Tapia, Gustavo E., gustavo.durazo@ues.mx, Univ Estatal de Sonora; Em
Durazzo, Aldo, aldurazz@iun.edu, Indiana Univ Northwest; Ggi
Durbin, Edward G., (401) 874-6850 edurbin@gso.uri.edu, Univ of Rhode Island; Ob
Durbin, James, 812-465-1208 jdurbin@usi.edu, Univ of Southern Indiana; Gm
Durham, Lisa, 630-252-3170 ladurham@anl.gov, Argonne; Hw
Durham, William, 617-253-5810 wbdurham@mit.edu, Massachusetts Inst of Tech; Cg
Durkee, Josh, joshua.durkee@wku.edu, Western Kentucky Univ; AsZy
Durkee, Philip A., 831-656-2517 durkee@nps.edu, Naval Postgrad Sch; AmZr
Durland, Theodore, 541-737-5058 tdurland@coas.oregonstate.edu, Oregon State Univ; Yg
DuRoss, Christopher B., (801) 537-3348 christopherduross@utah.gov, Utah Geo Survey; Ng
Durran, Dale R., (206) 543-7440 drdee@uw.edu, Univ of Washington; As
Durrant, Jeffrey O., (801) 422-4116 jodurrant@byu.edu, Brigham Young Univ; Ge
Durucan, Sevket, +44 20 759 47354 s.durucan@imperial.ac.uk, Imperial Coll; NmEcNr
Dushaw, Brian D., (206) 685-4198 dushaw@apl.washington.edu, Univ of Washington; Op
Dutch, Steven I., (920) 465-2371 dutchs@uwgb.edu, Univ of Wisconsin, Green Bay; GcgGe
Dutkiewicz, Stephanie, 617-253-8039 stephd@mit.edu, Massachusetts Inst of Tech; Cm
Dutro, Thomas, (202) 633-1322 Smithsonian Inst / Natl Mus of Natural History; Pi
Dutrow, Barbara L., (225) 388-2525 dutrow@geol.lsu.edu, Louisiana State Univ; Gz

Dutta, Prodip K., (812) 237-2268 prodip.dutta@indstate.edu, Indiana State Univ; Gs
Dutton, Jessica, duttonj@sunysuffolk.edu, Suffolk County Comm Coll, Ammerman Campus; Zg
Dutton, John A., 814-865-1534 dutton@ems.psu.edu, Pennsylvania State Univ, Univ Park; Am
Dutton, Shirley P., 512-471-0329 shirley.dutton@beg.utexas.edu, Univ of Texas, Austin; Gs
Duvall, Alison, 206-211-8311 aduvall@uw.edu, Univ of Washington; GmcGt
Duxbury, Thomas C., (818) 354-4301 Jet Propulsion Lab; Zn
Duzgun, Sebnem, 303-384-2357 duzgun@mines.edu, Colorado Sch of Mines; Nm
Dvorzak, Marie, (608) 262-8956 mdvorzak@geology.wisc.edu, Univ of Wisconsin, Madison; Zn
Dworkin, Stephen I., (254) 710-2186 steve_dworkin@baylor.edu, Baylor Univ; ClGd
Dwyer, Daryl F., 419-5302661 daryl.dwyer@utoledo.edu, Univ of Toledo; Zn
Dwyer, Gary S., (919) 681-8164 gsd3@duke.edu, Duke Univ; Gs
Dyab, Ahmed I., 002-03-3921595 seawolf_5000@yahoo.com, Alexandria Univ; Gg
Dyaur, Nikolay, 713-743-6539 ndyaur@uh.edu, Univ of Houston; Yxs
Dyck, Brendan, (250) 807-8934 brendan.dyck@ubc.ca, Univ of British Columbia, Okanagan Campus; Gpx
Dye, David H., 901-678-3330 daviddye@memphis.edu, Univ of Memphis; Ga
Dyer, Jamie, (662) 268-1032 Ext 220 jamie.dyer@msstate.edu, Mississippi State Univ; Am
Dyer, Jen, +44(0) 113 34 39086 j.dyer@leeds.ac.uk, Univ of Leeds; Ge
Dygert, Nick, 865-974-6022 ndygert1@utk.edu, Univ of Tennessee, Knoxville; Giz
Dyhrman, Sonya, 845-365-8165 sdyhrman@ldeo.columbia.edu, Columbia Univ; Ob
Dymek, Robert F., (314) 935-5344 bob_d@levee.wustl.edu, Washington Univ in St. Louis; GpiGa
Dymond, Randel, dymond@vt.edu, Virginia Tech; Hq
Dymond, Salli F., (218) 726-6582 dymon003@umn.edu, Univ of Minnesota Duluth; Hg
Dyson, Liesl, +27 (0)12 420 2469 liesl.dyson@up.ac.za, Univ of Pretoria; Am
Dyson, Miranda, +44 (0)1908 653398 x 53398 miranda.dyson@open.ac.uk, Open Univ; Pg
Dzunic, Aleksandar, +61 8 9266 2297 a.dzunic@curtin.edu.au, Curtin Univ; YgsYe

E

Earls, Sandy, 405-744-6358 sandy.earls@okstate.edu, Oklahoma State Univ; Zn
Earls, Stephanie, (360) 902-1473 stephanie.earls@dnr.wa.gov, Washington Geo Survey; Gge
Early, Thomas O., eot@ornl.gov, Oak Ridge; Cl
Earman, Sam, (717) 871-4336 searman@millersville.edu, Millersville Univ; HwCsl
Easson, Gregory L., (662) 915-5995 geasson@olemiss.edu, Univ of Mississippi; Zr
Easterbrook, Don J., (360) 319-5389 don.easterbrook@wwu.edu, Western Washington Univ; GlmGe
Easterday, Cary, cary.easterday@bellevuecollege.edu, Bellevue Coll; GgPg
Eastin, Matt, 704-687-5914 mdeastin@uncc.edu, Univ of North Carolina, Charlotte; Am
Eastoe, Chris J., (520) 791-7430 eastoe@email.arizona.edu, Univ of Arizona; CsGg
Eaton, David, 403 220-4233 eatond@ucalgary.ca, Univ of Calgary; Ys
Eaton, Jeffrey G., jeaton@weber.edu, Weber State Univ; PsGs
Eaton, Lewis S., (540) 568-3339 eatonls@jmu.edu, James Madison Univ; Gm
Eaton, Timothy, 718-997-3327 timothy.eaton@qc.cuny.edu, Queens Coll (CUNY); HgwGg
Ebel, Denton, 212-769-5381 debel@amnh.org, Graduate Sch of the City Univ of New York; Gg
Ebel, John E., (617) 552-3399 ebel@bc.edu, Boston Coll; Ys
Eberhardt, Erik, (604) 827-5573 eeberhardt@eos.ubc.ca, Univ of British Columbia; Ng
Eberhart, Mark E., (303) 273-3726 meberhar@mines.edu, Colorado Sch of Mines; Zm
Eberle, Bill, (785) 532-6101 weberle@ksu.edu, Kansas State Univ; SoZu
Eberle, Jaelyn J., 303-492-8069 jaelyn.eberle@colorado.edu, Univ of Colorado; Pv
Eberli, Gregor P., (305) 421-4678 geberli@rsmas.miami.edu, Univ of Miami; GusYx
Ebersole, Craig M., 717-702-2023 craebersol@pa.gov, DCNR- Pennsylvania Bur of Geo Survey; ZiGgHg
Ebert, David, 831-771-4400 debert@mlml.calstate.edu, Moss Landing Marine Lab; Zn
Ebert, James R., (607) 436-3065 james.ebert@oneonta.edu, SUNY, Oneonta; GrsZe
Ebiner, Matt, (310) 660-3593 mebiner@elcamino.edu, El Camino Coll; Zy
Ebinger, Michael H., (505) 667-3147 mhe@lanl.gov, Los Alamos; So
Eble, Cortland F., 8592575500 x 149 eble@uky.edu, Univ of Kentucky; Pl
Eby, G. Nelson, (978) 934-3907 nelson_eby@uml.edu, Univ of Massachusetts, Lowell; CgGiCt
Eby, Stephanie, (617) 373-4380 s.eby@neu.edu, Northeastern Univ; Zn
Echols, John B., (225) 388-8573 john@lgs.bri.lsu.edu, Louisiana State Univ; Go
Eckert, Andreas, 573-341-4876 eckertan@mst.edu, Missouri Univ of Sci and Tech; Yg
Eckert, Donald J., (614) 292-9048 eckert.1@osu.edu, Ohio State Univ; Sc
Eckert, James O., (203) 432-3181 xwn6w3i02@sneakemail.com, Yale Univ; Gp
Economos, Rita C., 214-768-3313 reconomos@smu.edu, Southern Methodist Univ; CsGc
Eddy, Michael, mpeddy@purdue.edu, Purdue Univ; Gt
Edegbai, Aitalokhai J., aitalokhai.edegbai@uniben.edu, Univ of Benin; GosPi
Edgar, Dorland E., (630) 252-7596 Argonne; Gm
Edgcomb, Virginia, 508-289-3734 vedgcomb@whoi.edu, Woods Hole Oceanographic Inst; ObZc
Edgell, Dennis J., (910) 521-6479 dennis.edgell@uncp.edu, Univ of North Carolina, Pembroke; As
Edil, Tuncer B., (608) 262-3225 edil@engr.wisc.edu, Univ of Wisconsin, Madison; Ng
Edlund, Mark B., mbedlund@smm.org, Univ of Minnesota, Twin Cities; Pe
Edmonds, Douglas A., (812) 855-4512 edmondsd@indiana.edu, Indiana Univ, Bloomington; Gsm
Edmonds, Marie, +44 (0) 1223 333463 medm06@esc.cam.ac.uk, Univ of Cambridge; Gv
Edson, Carol, 925-424-1336 cedson@laspositascollege.edu, Las Positas Coll; Gg
Edson, James, james.edson@uconn.edu, Univ of Connecticut; Op
Edwards, Benjamin R., (717) 254-8934 edwardsb@dickinson.edu, Dickinson Coll; GivCg
Edwards, C L, (505) 667-8464 cledwards@lanl.gov, Los Alamos; Yg
Edwards, Cole T., edwardsct4@appstate.edu, Appalachian State Univ; GsCs
Edwards, Dianne, +44(0)29 208 74264 edwardsd2@cardiff.ac.uk, Cardiff Univ; Pb
Edwards, John , john.edwards@flinders.edu.au, Flinders Univ; Zn
Edwards, Margo H., (808) 956-5232 margo@soest.hawaii.edu, Univ of Hawai'i, Manoa; Gu
Edwards, Nicholas, +44 0161 275-3810 nicholas.edwards@manchester.ac.uk, Univ of Manchester; Pg
Edwards, R. Lawrence, 612-626-0207 edwar001@umn.edu, Univ of Minnesota, Twin Cities; Cc
Edwards, Richard N., (416) 978-2267 edwards@core.physics.utoronto.ca, Univ of Toronto; Yr
Edwards, Robin, robin.edwards@tcd.ie, Trinity Coll; Ge
Edwards, Stuart, +44 (0) 191 208 3986 stuart.edwards@ncl.ac.uk, Univ of Newcastle Upon Tyne; Gq
Effert-Fanta, Shari E., 217-244-2192 sfanta@illinois.edu, Illinois State Geo Survey; Cas
Efurd, Deward W., (505) 667-2437 Los Alamos; Ct
Egan, Stuart, (+44) 01782 733174 s.s.egan@keele.ac.uk, Keele Univ; Gtc
Egbert, Gary D., (541) 737-2947 egbert@coas.oregonstate.edu, Oregon State Univ; Yr
Egbeyemi, Oladimeji R., +2348167781509 oregbeyemi@futa.edu.ng, Federal Univ of Tech, Akure; GgCeEg
Egerton, Victoria, victoria.egerton@manchester.ac.uk, Univ of Manchester; Pg
Egger, Anne E., annegger@geology.cwu.edu, Central Washington Univ; ZeGtRn
Eggers, Delores M., (828) 251-6654 eggers@unca.edu, Univ of North Carolina, Asheville; Zn
Eggler, David H., (814) 571-1960 dhe1@psu.eu, Pennsylvania State Univ, Univ Park; CpGiv
Eggleston, Carrick M., (307) 766-6769 carrick@uwyo.edu, Univ of Wyoming; ClqCb
Eggleston, David B., (919) 515-7840 eggleston@ncsu.edu, North Carolina State Univ; Ob
Eguiluz, Luis, luis.eguiluz@ehu.eus, Univ of the Basque Country UPV/EHU; Ggt

Ehinola, Olugbenga A., +2348033819066 oa.ehinola@mail.ui.edu.ng, Univ of Ibadan; GoeCo
Ehler, Stan, (785) 532-6101 swehler@ksu.edu, Kansas State Univ; So
Ehlers, Todd A., (734) 763-5112 tehlers@umich.edu, Univ of Michigan; Yg
Ehlmann, Bethany L., 626.395.6720 ehlmann@caltech.edu,CalTech; XgZrGz
Ehm, Lars, 631-632-8058 lars.ehm@stonybrook.edu, Stony Brook Univ; Gz
Eich-Greatorex, Susanne, susanne.eich@nmbu.no, Norwegian Univ of Life Scis (NMBU); Sc
Eichenbaum, Jack, (718) 961-8406 jaconet@aol.com, Hunter Coll (CUNY); Zie
Eicher, Don L., (303) 492-6187 Univ of Colorado; Pm
Eichhubl, Peter, (512) 475-8829 peter.eichhubl@beg.utexas.edu, Univ of Texas at Austin, Jackson Sch of Geosciences; GcCgNr
Eick, Matthew J., (540) 231-8943 eick@vt.edu, Virginia Tech; Sc
Eide, Elizabeth A., (202) 334-2392 eeide@nas.edu, Natl Academies of Scis, Eng, and Medicine; GgoGt
Eiler, John M., (626) 395-6239 eiler@gps.caltech.edu,CalTech; Cg
Eilon, Zach, eilon@geol.ucsb.edu, Univ of California, Santa Barbara; Ys
Eisenberger, Peter, petere@ldeo.columbia.edu, Columbia Univ; Zg
Eisenhart, Karen, (814) 732-2398 keisenhart@edinboro.edu, Edinboro Univ of Pennsylvania; Zy
Eisenhart, Ralph, (717) 846-7788 York Coll of Pennsylvania; Gg
Eklund, Olav, +358442956429 olav.eklund@abo.fi, Abo Akademi Univ; GxeGz
Ekmekci, Mehmet, +90 312 2977730 ekmekci@hacettepe.edu.tr, Hacettepe Univ; HwyCs
Ekstrom, Goran, ekstrom@ldeo.columbia.edu, Columbia Univ; Yg
El Dakkak, Mohamed W., 002-03-3921595 waguihm@yahoo.com, Alexandria Univ; GgsGd
El Kadiri, Racha, 615-484-7641 racha.elkadiri@mtsu.edu, Middle Tennessee State Univ; HgGge
El Masri, Bassil, (270) 809-3110 belmasri@murraystate.edu, Murray State Univ; Zc
El Shazly, Aley, (304) 696-6756 elshazly@marshall.edu, Marshall Univ; GpiCg
El-Baz, Farouk, (617) 353-5081 farouk@bu.edu, Boston Univ; Zr
El-Kadi, Aly I., (808) 956-6331 elkadi@hawaii.edu, Univ of Hawai'i, Manoa; Hw
El-Monier, Ilham, ilham.el-monier@ou.edu, Univ of Oklahoma; Np
Elachi, Charles, 626.395.8858 charles.elachi@jpl.nasa.gov,CalTech; Xg
Elam, Tim, 206-685-3092 wtelam@apl.uw.edu, Univ of Washington; Cag
Elardo, Stephen, 352-392-2634 selardo@ufl.edu, Univ of Florida; GiXgCp
Elder, Kathryn L., (508) 289-2513 kelder@whoi.edu, Woods Hole Oceanographic Inst; Cs
Elders, Wilfred A., wilfred.elders@ucr.edu, Univ of California, Riverside; Gi
Eldredge, Niles, (212) 769-5723 American Mus of Natural History; Pi
Eldridge, Jayson, (812) 856-2382 jayeldri@indiana.edu, Indiana Univ, Bloomington; EoGr
Elhusseiny, Ammar , +966138602625 ammar.elhusseiny@kfupm.edu.sa, King Fahd Univ of Petroleum and Minerals; Yg
Elia, Gaetano, +44 (0) 191 208 7934 Univ of Newcastle Upon Tyne; Ng
Elias, Robert J., (204) 474-8862 eliasrj@ms.umanitoba.ca, Univ of Manitoba; Pi
Elick, Jennifer M., 570-372-4214 elick@susqu.edu, Susquehanna Univ; Pe
Elkins, Joe T., 970-351-3060 joe.elkins@unco.edu, Univ of Northern Colorado; ZeGgCl
Elkins, Lynne J., (402) 472-7563 lelkins@unl.edu, Univ of Nebraska, Lincoln; GivCt
Elkins-Tanton, Linda, (480) 727-2451 ltelkins@asu.edu, Arizona State Univ; Xg
Eller, Phillip G., (505) 667-5830 Los Alamos; Zn
Ellett, Kevin M., 812-856-3671 kmellett@indiana.edu, Indiana Univ, Bloomington; YeHw
Ellins, Katherine K., (512) 232-3251 kellins@ig.utexas.edu, Univ of Texas, Austin; Zn
Elliot, David H., (614) 292-5076 elliot.1@osu.edu, Ohio State Univ; GxtGv
Elliott, C. James, (505) 667-9517 cje@lanl.gov, Los Alamos; Zn
Elliott, Colleen G., (406) 496-4143 celliott@mtech.edu, Montana Tech of The Univ of Montana; GctGg
Elliott, David K., (928) 523-7188 david.elliott@nau.edu, Northern Arizona Univ; Pv
Elliott, Emily M., (412) 624-8780 eelliott@pitt.edu, Univ of Pittsburgh; Cbs
Elliott, Monty A., (541) 552-6482 Southern Oregon Univ; Gr
Elliott, Scott M., (505) 667-0949 sme@lanl.gov, Los Alamos; As
Elliott, Susan J., (519) 888-4567 (Ext. 33923) elliotts@uwaterloo.ca, McMaster Univ; Ge
Elliott, W. Crawford, (404) 413-5756 wcelliott@gsu.edu, Georgia State Univ; Clc
Elliott, Jr., William S., (812) 228-5053 wselliott@usi.edu, Univ of Southern Indiana; GsrCl
Ellis, Dale, daledellis@gmail.com, Dalhousie Univ; Op
Ellis, Hugh, (901) 678-2007 hugh.ellis@jhu.edu, Johns Hopkins Univ; Zn
Ellis, Kathryn E., (859) 323-0504 kathryn.ellis@uky.edu, Univ of Kentucky; Zn
Ellis, Robert M., (604) 822-6574 rellis@eos.ubc.ca, Univ of British Columbia; Ys
Ellis, Scott R., 217-265-5105 ellis3@illinois.edu, Illinois State Geo Survey; Ge
Ellis, Trevor, 573-368-2153 trevor.ellis@dnr.mo.gov, Missouri Dept of Nat Res; Gg
Ellison, Anne L., 217-244-2502 aerdmann@illinois.edu, Univ of Illinois; Ge
Ellsworth, Timothy R., (217) 333-2055 Univ of Illinois, Urbana-Champaign; Sp
Ellsworth, William L., (650) 723-9390 wellsworth@stanford.edu, Stanford Univ; Ys
Ellwood, Brooks B., (225) 578-3416 ellwood@lsu.edu, Louisiana State Univ; Ym
Elmore, R. Douglas, delmore@ou.edu, Univ of Oklahoma; GdYm
Elnadi, Abdelhalim H., geology@uofk.edu, Univ of Khartoum; Goz
Eloranta, Edwin W., (608)262-7327 eloranta@lidar.ssec.wisc.edu, Univ of Wisconsin, Madison; Yr
Elrick, Maya, (505) 277-5077 dolomite@unm.edu, Univ of New Mexico; Gs
Elrick, Scott D., (217) 333-3222 elrick@illinois.edu, Univ of Illinois; EcGgs
Elsberry, Russell L., 831-656-2373 elsberry@nps.edu, Naval Postgrad Sch; Am
Elsen, Jan, jan.elsen@ees.kuleuven.be, Katholieke Univit Leuven; Gz
Elshorbany, Yasin, elshorbany@usf.edu, Univ of South Florida; Ac
Elswick, Erika R., (812) 855-2493 eelswick@indiana.edu, Indiana Univ, Bloomington; Cl
Elueze, A. A., aa.elueze@mail.ui.edu.ng, Univ of Ibadan; Eg
Elwood Madden, Andrew S., (405) 325-3253 amadden@ou.edu, Univ of Oklahoma; ClZaCb
Elwood Madden, Megan E., (405) 325-3253 melwood@ou.edu, Univ of Oklahoma; ClXgCg
Ely, Lisa L., (509) 963-2177 lisa.ely@cwu.edu, Central Washington Univ; GmHs
Elzinga, Evert J., 973-353-5238 elzinga@newark.rutgers.edu, Rutgers, The State Univ of New Jersey, Newark; Sc
Emanuel, Kerry A., (617) 253-2462 emanuel@mit.edu, Massachusetts Inst of Tech; AmRn
Emerick, Christina M., (206) 543-2491 tina@ocean.washington.edu, Univ of Washington; Ou
Emerson, Cheryl R., (309) 694-5373 Illinois Central Coll; Ze
Emerson, Steven R., (206) 543-0428 emerson@u.washington.edu, Univ of Washington; OcCmOg
Emile-Geay, Julien, 213-740-2945 julieneg@usc.edu, Univ of Southern California; Oc
Emmer, Barbara, 089/2180 4231 Ludwig-Maximilians-Univ Muenchen; Yg
Emmett, Chad, (801) 422-7886 chad_emmett@byu.edu, Brigham Young Univ; Zn
Emmi, Philip C., (801) 581-5562 pcemmi@geog.utah.edu, Univ of Utah; Zu
Emofurieta, Williams O., wemofu@uniben.edu, Univ of Benin; GzCeGe
Emry, Robert, (202) 633-1323 Smithsonian Inst / Natl Mus of Natural History; Pv
Encarnacion, John, (314) 977-3119 john.encarnacion@slu.edu, Saint Louis Univ; Gx
Enderlin, Ellyn, ellynenderlin@boisestate.edu , Boise State Univ; GlYg
Enderlin, Milton, 817 257 5318 m.enderlin@tcu.edu, Texas Christian Univ; Nr
Enders, M. Stephen , menders@mines.edu, Colorado Sch of Mines; Gg
Endres, Anthony E., (519) 888-4567 (Ext. 3552) Univ of Waterloo; Pe
Engebretson, David C., david.engebretson@wwu.edu, Western Washington Univ; Gt
Engel, Anga, mpeschke@geomar.de, SUNY, Stony Brook; Oc
Engel, Annette S., 865-974-0402 aengel1@utk.edu, Univ of Tennessee, Knoxville; Cl
Engel, Michael H., (405) 325-4435 ab1635@ou.edu, Univ of Oklahoma; Co
Engel, Richard E., (406) 994-5295 rengel@montana.edu, Montana State Univ; Sc
Engelder, Terry, (814) 865-3620 engelder@geosc.psu.edu, Pennsylvania State Univ, Univ Park; Nr
Engelhard, Simon, (401) 874-2187 engelhart@uri.edu, Univ of Rhode Island; Gg
Engelhart, Simon E., 401-874-2187 engelhart@uri.edu, Univ of Rhode Island; OnGmYs

Engelmann, George F., 402-554-4804 gengelmann@unomaha.edu, Univ of Nebraska, Omaha; PvGdr
England, John, (780) 492-5673 john.england@ualberta.ca, Univ of Alberta; Gml
England, Phillip, +44 (1865) 272000 philip@earth.ox.ac.uk, Univ of Oxford; Gt
England, Richard, rwe5@le.ac.uk, Leicester Univ; Gg
Engle, Mark, maengle@utep.edu, Univ of Texas, El Paso; HwGe
English, David C., (727) 553-1503 denglish@marine.usf.edu, Univ of South Florida; Op
Engstrom, Daniel R., (612) 433-5953 (Ext. 18) dre@umn.edu, Univ of Minnesota, Twin Cities; Pe
Engstrom, Ryan, rengstro@gwu.edu, George Washington Univ; Zr
Engstrom, Vanessa, vengstrom@valleycollege.edu, San Bernardino Valley Coll; Zy
Enkelmann, Eva, eva.enkelmann@ucalgary.ca, Univ of Calgary; CcGmt
Enos, Paul, (785) 864-9714 enos@ku.edu, Univ of Kansas; GsdGu
Enright, Richard L., (508) 531-1390 enright@bridgew.edu, Bridgewater State Univ; HgZrEg
Ensign, Todd I., 304-367-8438 tensign@fairmontstate.edu, Fairmont State Univ; Ze
Entekhabi, Dara, (617) 253-9698 darae@mit.edu, Massachusetts Inst of Tech; Hg
Epifanio, Craig, cepi@tamu.edu, Texas A&M Univ; As
Eppelbaum, Lev, +97236405086 levap@post.tau.ac.il, Tel Aviv Univ; YvmGt
Eppes, Martha C., 704-687-5993 meppes@uncc.edu, Univ of North Carolina, Charlotte; So
Epstein, Howard E., (434) 924-4308 hee2b@virginia.edu, Univ of Virginia; Zn
Erdmer, Philippe, (403) 492-2676 philippe.erdmer@ualberta.ca, Univ of Alberta; Gc
Erdner, Deana L., (361) 749-6719 derdner@utexas.edu, Univ of Texas, Austin; Ob
Erhardt, Andrea M., 859-257-6931 andrea.erhardt@uky.edu, Univ of Kentucky; CgsCm
Erickson, Ben A., (801) 537-3379 benerickson@utah.gov, Utah Geo Survey; Ng
Erickson, Brittany, 541-346-1350 bae@cs.uoregon.edu, Univ of Oregon; Ygs
Erickson, J. Mark, (315) 379-5198 St. Lawrence Univ; Pgi
Erickson, Melinda L., 763-783-3231 mericksо@usgs.gov, Univ of Minnesota, Twin Cities; Hw
Erickson, Rolfe C., (707) 664-2334 rolfe.erickson@sonoma.edu, Sonoma State Univ; Gx
Eriksen, Charles C., (206) 543-6528 eriksen@u.washington.edu, Univ of Washington; Op
Eriksson, Kenneth A., (540) 231-4680 kaeson@vt.edu, Virginia Tech; Gs
Erisman, Brad, 361-749-6833 berisman@utexas.edu, Univ of Texas, Austin; Ob
Ernst, W. Gary, (650) 723-2750 ernst@pangea.stanford.edu, Stanford Univ; Cp
Erski, Theodore, (815) 455-8992 tederski@gmail.com, McHenry County Coll; ZgEg
Erslev, Eric A., (970) 231-2654 eric.erslev@colostate.edu, Colorado State Univ; GctGg
Ertel-Ingrisch, Werner, 089/2180 4275 ertel@min.uni-muenchen.de, Ludwig-Maximilians-Univ Muenchen; Gz
Erturk, Mehmet Ali, 00904242370000-5981 maerturk@firat.edu.tr, Firat Univ; GiCc
Erwin, Diane, (510) 642-3921 dmerwin@berkeley.edu, Univ of California, Berkeley; Pb
Erwin, Douglas H., (202) 633-1324 Smithsonian Inst / Natl Mus of Natural History; Pi
Esbaugh, Andrew J., (361) 749-6835 a.esbaugh@austin.utexas.edu, Univ of Texas, Austin; Ob
Escuer, Joan, joan.escuer@uab.cat, Univ Auto de Barcelona; Gg
Eshleman, Keith N., (301) 689-7170 eshleman@al.umces.edu, Univ of Maryland; Hg
Esling, Steven P., (618) 453-7376 esling@siu.edu, Southern Illinois Univ Carbondale; HwGl
Esnault, Melissa H., (225) 578-5320 mesnau1@lsu.edu, Louisiana State Univ; Zn
Esparza, Francisco, fesparz@cicese.mx, Centro de Investigación Científica y de Educación Superior de Ensenada; Ye
Esperante, Raul, resperante@llu.edu, Loma Linda Univ; Pv
Espinosa, Juan M., jespinos@cicese.mx, Centro de Investigación Científica y de Educación Superior de Ensenada; Ye
Esser, Corinne, (530) 752-3668 caesser@ucdavis.edu, Univ of California, Davis; Zn
Essery, Richard L., +44 (0) 131 651 9093 richard.essery@ed.ac.uk, Edinburgh Univ; As
Essling, Alice M., (630) 252-3493 Argonne; Ca
Esslinger, Kelly L., kelly.esslinger@azwestern.edu, Arizona Western Coll; AsOpGg
Estalrich López, Joan, ++935811270 joan.estalrich@uab.cat, Univ Auto de Barcelona; HgGe
Estapa, Margaret, 518-589-5477 mestapa@skidmore.edu, Skidmore Coll; Oc
Esteban, Jose Julian, jj.esteban@ehu.es, Univ of the Basque Country UPV/EHU; GtCcYg
Ethridge, Frank G., (970) 491-6195 fredpet@cnr.colostate.edu, Colorado State Univ; Gs
Ettensohn, Frank R., (859) 257-1401 f.ettensohn@uky.edu, Univ of Kentucky; PsGdPi
Ettlie, Bradley, 217-265-6543 ettlie76@illinois.edu, Illinois State Geo Survey; Ge
Eusden, J. Dykstra, (207) 786-6015 deusden@bates.edu, Bates Coll; Gc
Evangelou, V. P., (515) 294-9237 Iowa State Univ of Sci & Tech; Sc
Evanoff, Emmett, 970-351-2647 emmett.evanoff@unco.edu, Univ of Northern Colorado; GrPgGs
Evans, Barry M., (814) 863-3531 bmel@psu.edu, Pennsylvania State Univ, Univ Park; Sd
Evans, Bernard W., 206-543-1163 bwevans@uw.edu, Univ of Washington; GpzGi
Evans, Claude, (314) 935-6684 Washington Univ in St. Louis; Zn
Evans, David A., (203) 432-3127 david.evans@yale.edu, Yale Univ; YmGt
Evans, Diane L., (818) 354-2418 Jet Propulsion Lab; Gm
Evans, Eileen L., (818) 677-5026 eileen.evans@csun.edu, California State Univ, Northridge; YdGtYg
Evans, J. B., (617) 253-2856 brievans@mit.edu, Massachusetts Inst of Tech; Yx
Evans, James E., (419) 372-2414 evansje@bgsu.edu, Bowling Green State Univ; GsHsGe
Evans, Jenni L., (814) 865-3240 jle7@psu.edu, Pennsylvania State Univ, Univ Park; Am
Evans, Kevin R., (417) 836-5590 kevinevans@missouristate.edu, Missouri State Univ; GrsGt
Evans, Krista M., (417) 836-5688 kristaevans@missouristate.edu, Missouri State Univ; Zn
Evans, Les J., (519) 824-4120 (Ext. 53017) levans@lrs.uoguelph.ca, Univ of Guelph; Cl
Evans, Lucas, (902) 585-1513 lucas.evans@acadiau.ca, Acadia Univ; Gg
Evans, Mark, evansmaa@ccsu.edu, Central Connecticut State Univ; GciGz
Evans, Michael E., (780) 492-5517 evans@phys.ualberta.ca, Univ of Alberta; Ym
Evans, Owen C., (631) 632-8061 owen.evans@sunysb.edu, Stony Brook Univ; Cg
Evans, Robert L., (508) 289-2673 revans@whoi.edu, Woods Hole Oceanographic Inst; Yrr
Evans, Sarah G., evanssg@appstate.edu, Appalachian State Univ; Hwq
Evans, Thomas J., (608) 263-4125 tevans@wisc.edu, Univ of Wisconsin, Extension; Gg
Evanylo, Gregory K., (540) 231-9739 gevanylo@vt.edu, Virginia Tech; Sc
Eveleth, Robert W., 575-835-5325 bob.eveleth@nmt.edu, New Mexico Tech; Nm
Evenson, Edward B., (610) 758-3659 ebe0@lehigh.edu, Lehigh Univ; Gl
Everett, Mark, (979) 862-2129 everett@geo.tamu.edu, Texas A&M Univ; Ym
Eversoll, Duane A., (402) 472-7524 deversoll2@unl.edu, Univ of Nebraska, Lincoln; Ng
Ewart, Terry E., (206) 543-1327 ewart@apl.washington.edu, Univ of Washington; Op
Ewas, Galal A., 002-03-3921595 goueiss@gmail.com, Alexandria Univ; GgNr
Ewing, Rodney C., (650) 497-6203 rewing1@stanford.edu, Stanford Univ; GzeCl
Ewing, Stephanie, stephanie.ewing@montana.edu, Montana State Univ; Sd
Eyles, Carolyn H., (905) 525-9140 (Ext. 24077) eylesc@mcmaster.ca, McMaster Univ; GslGm
Eyles, John D., (905) 525-9140 (Ext. 23152) eyles@mcmaster.ca, McMaster Univ; Ge
Eyles, Nicholas, 416-287-7195 eyles@utsc.utoronto.ca, Univ of Toronto; Gl
Eyre, Bradley, bradley.eyre@scu.edu.au, Southern Cross Univ; CoGeCs
Ezenabor, Ben O., benedict.ezenabor@uniben.edu, Univ of Benin; HgZrGe
Ezer, Tal, (757)-683-5631 tezer@odu.edu, Old Dominion Univ; Op
Ezerskis, John L., (216) 268-0493 john.ezerskis@tri-c.edu, Cuyahoga Comm Coll - Western Campus; ZgGcHw

F

Faatz, Renee M., (435) 283-7519 renee.faatz@snow.edu, Snow Coll; Gg

Fabbri, Paolo, +390498279124 paolo.fabbri@unipd.it, Univ degli Studi di Padova; Hw
Fabel, Derek, +4401413305473 derek.fabel@glasgow.ac.uk, Univ of Glasgow; Xg
FABRE, Cecile, cecile.fabre@univ-lorraine.fr, Univ de Lorraine - Faculte des Scis et Technologies; Gg
Fabry, Frederic, (514) 398-7733 frederic.fabry@staff.mcgill.ca, McGill Univ; As
Fabryka-Martin, June T., (505) 665-2300 Los Alamos; Hg
Faccenda, Manuele, +390498279159 manuele.faccenda@unipd.it, Univ degli Studi di Padova; Gq
Fadem, Cynthia M., (765) 983-1231 fademcy@earlham.edu, Earlham Coll; GaSoCs
Fadiman, Maria, 561 297-3314 mfadiman@fau.edu, Florida Atlantic Univ; Zn
Fadiya, S. L., 234-803-332-0230 fadiyalawrence@yahoo.co.uk, Obafemi Awolowo Univ; God
Fagan, Amy, 828 227 3820 alfagan@wcu.edu, Western Carolina Univ; GiXg
Fagel, Nathalie, +32 4 3662209 nathalie.fagel@ulg.ac.be, Univ de Liege; GsuGe
Fagents, Sarah, 808-956-3163 fagents@higp.hawaii.edu, Univ of Hawai'i, Manoa; Gv
Fagherazzi, Sergio, (617) 353-2092 sergio@bu.edu, Boston Univ; GmOnGs
Fagnan, Brian A., 605-394-6652 brian.fagnan@state.sd.us, South Dakota Dept of Environment and Nat Res; Gg
Fahes, Mashhad, mashhad.fahes@ou.edu, Univ of Oklahoma; Np
Fahnestock, Mark A., 603-862-0322 mark.fahnestock@unh.edu, Univ of New Hampshire; Gl
Fahrenbach, Mark D., 605-394-6830 mark.fahrenbach@state.sd.us, South Dakota Dept of Environment and Nat Res; Gg
Fairbairn, David, +44 (0) 191 208 6353 david.fairbairn@ncl.ac.uk, Univ of Newcastle Upon Tyne; Gq
Fairley, Jerry P., (208) 885-9259 jfairley@uidaho.edu, Univ of Idaho; HwGv
Fajkoviæ, Hana, +38514605969 hanaf@geol.pmf.hr, Univ of Zagreb; CgGeCm
Fakhari, Mohammad D., 614 6584 mohammad.fakhari@dnr.state.oh.us, Ohio Dept of Nat Res; GocNx
Falcon-Lang, Howard, +44 1784 414039 howard.falcon-lang@rhul.ac.uk, Royal Holloway Univ of London; Pg
Falebita, Dele E., 234-703-298-7836 delefale@oauife.edu.ng, Obafemi Awolowo Univ; YeRnZg
Falk, Lisa, 919-515-8458 esfalk@ncsu.edu, North Carolina State Univ; Zg
Falkena, Lee, 312-996-4499 lfalkena@uic.edu, Univ of Illinois Chicago; ZcGig
Falkowski, Paul G., falco@imcs.rutgers.edu, Univ of Hawai'i, Manoa; Og
Fall, Andras, (512) 471-8334 andras.fall@beg.utexas.edu, Univ of Texas, Austin; CgGcEg
Fall, Leigh M., 607 436-2615 leigh.fall@oneonta.edu, SUNY, Oneonta; Pgq
Fallowfield, Howard, howard.fallowfield@flinders.edu.au, Flinders Univ; Zn
Falode, Olugbenga, 08130049657 falodelias@gmail.com, Univ of Ibadan; NpZa
Faloona, Ian, 530-752-2044 icfaloona@ucdavis.edu, Univ of California, Davis; AcpAm
Falquero, Kelsey, falquerok@si.edu, Smithsonian Inst / Natl Mus of Natural History; Xm
Falta, Ronald W., (864) 656-0125 faltar@clemson.edu, Clemson Univ; Hq
Famiglietti, James, (949)824-9434 jfamigli@uci.edu, Univ of California, Irvine; Hw
Fan, Chao, 208-885-0949 cfan@uidaho.edu, Univ of Idaho; Zir
Fan, Majie, (817) 272-2987 mfan@uta.edu, Univ of Texas, Arlington; GstCs
Fan, Weihong, weihong.fan@stockton.edu, Stockton Univ; Zi
Fan, Wenyuan, wenyuanfan@ucsd.edu, Univ of California, San Diego; Ys
Fan, Xingang, xingang.fan@wku.edu, Western Kentucky Univ; As
Fandel, Chloé, 507 222 5800 cfandel@carleton.edu, Carleton Coll; Hw
Fanning, Kent A., (727) 553-1594 kaf@marine.usf.edu, Univ of South Florida; OcGu
Fantle, Matthew S., 814-863-9968 mfantle@geosc.psu.edu, Pennsylvania State Univ, Univ Park; Cs
Fantozzi, Pier Lorenzo, piero.fantozzi@unisi.it, Univ of Siena; Gme
Faraji, Maedeh, maedeh.faraji@austincc.edu, Austin Comm Coll District; AsZge
Faramarzi, Monireh, monireh.faramarzi@ualberta.ca, Univ of Alberta; HgwHq
Fares, Mary A., (229) 333-5755 mfares@valdosta.edu, Valdosta State Univ; Zn
Farfan, Gabriela, (202) 633-1820 farfang@si.edu, Smithsonian Inst / Natl Mus of Natural History; GzCbGu
Fariduddin, Mohammad, 773/442-6059 m-fariduddin@neiu.edu, Northeastern Illinois Univ; Pe
Farley, Kenneth A., (626) 395-6005 farley@gps.caltech.edu,CalTech; Cg
Farley, Martin B., 910-521-6478 martin.farley@uncp.edu, Univ of North Carolina, Pembroke; PlZe
Farley, Perry D., (505) 667-2415 dfarley@lanl.gov, Los Alamos; Zn
Farmer, Emma C., 516 463-5568 geoecf@hofstra.edu, Hofstra Univ; Ou
Farmer, G. Lang, 303-492-6534 farmer@colorado.edu, Univ of Colorado; Cg
Farmer, George T., (505) 665-0225 gfarmer@lanl.gov, Los Alamos; Hw
Farmer, Jack D., (480) 965-6748 jfarmer@asu.edu, Arizona State Univ; Pg
Farnan, Ian, +44 (0) 1223 333431 ifarnan@esc.cam.ac.uk, Univ of Cambridge; Gz
Farnsworth, Katie, 724-357-3406 katie.farnsworth@iup.edu, Indiana Univ of Pennsylvania; Ong
Farquhar, James, (301) 405-5034 jfarquha@umd.edu, Univ of Maryland; Cs
Farquharson, Phil, pfarquharson@miracosta.edu, MiraCosta Coll; Gg
Farr, Tom G., (818) 354-9057 tom.farr@jpl.nasa.gov, Jet Propulsion Lab; Gm
Farrand, William, (734) 764-1435 wfarrand@umich.edu, Univ of Michigan; Ga
Farrell, Brian F., (617) 495-2998 farrell@seas.harvard.edu, Harvard Univ; Am
Farrell, Jamie, 801-581-7856 jamie.farrell@utah.edu, Univ of Utah; Ys
Farrell, John, (401) 874-6561 jfarrell@gso.uri.edu, Univ of Rhode Island; Ou
Farrell, Kathleen M., (919) 733-7353 kathleen.farrell@ncdenr.gov, N.C. Dept of Env Quality; Gs
Farrell, Mark O., (412) 392-3879 Point Park Univ; Zn
Farrell, Michael, michaelfarrell@myefolio.com, Cuyamaca Coll; Gg
Farrell, Stewart C., (609) 652-4245 Stockton Univ; On
Farrington, John W., 5082741926 (cell Phone) jfarrington@whoi.edu, Woods Hole Oceanographic Inst; ComOc
Farris, David, 252 328 6773 farrisd19@ecu.edu, East Carolina Univ; Ggc
Farrow, Norman D., ndf@ornl.gov, Oak Ridge; Hg
Farsang, Andrea, +3662544156 farsang@geo.u-szeged.hu, Univesity of Szeged; SdZe
Farthing, Dori J., 585-245-5298 farthing@geneseo.edu, SUNY, Geneseo; Gz
Farver, John R., (419) 372-7203 jfarver@bgnet.bgsu.edu, Bowling Green State Univ; Gy
Faryad, Shah Wali, +420221951521 faryad@natur.cuni.cz, Charles Univ; GpzGi
Fastovsky, David, (401) 874-2185 defastov@uri.edu, Univ of Rhode Island; PvGsPs
Fatherree, James W., 813-253-7906 jfatherree@hccfl.edu, Hillsborough Comm Coll; Zg
Faulds, James, (775) 682-8751 jfaulds@unr.edu, Univ of Nevada, Reno; Gc
Faulkner, Daniel, +440151 794 5169 faulkner@liverpool.ac.uk, Univ of Liverpool; Ys
Faulkner, Melinda S., (936) 468-2236 mgshaw@sfasu.edu, Stephen F. Austin State Univ; GeClEg
Faúndez, Alejandro, afaundez@ucm.es, Univ Compl de Madrid; Ng
Faure, Gunter, faure.1@osu.edu, Ohio State Univ; CgHw
Faure, Stephane, 514-987-3000 #2369 faure.stephane@uqam.ca, Univ du Quebec a Montreal; Gc
Fauria, Kristen, (615) 322-2976 kristen.fauria@vanderbilt.edu, Vanderbilt Univ; GvqYu
Faust, Megan, megfaust@pdx.edu, Portland State Univ; ZeGr
Favas, Paulo J., 351259350220 pjcf@utad.pt, Unive de Trás-os-Montes e Alto Douro; GeCtGg
Favor, Michael, (313) 593-5235 Univ of Michigan, Dearborn; Zn
Favorito, Jessica, jessica.favorito@stockton.edu, Stockton Univ; So
Fawcett, J. J., (416) 978-3027 fawcett@quartz.geology.utoronto.ca, Univ of Toronto; Gp
Fawcett, Peter J., (505) 277-3867 fawcett@unm.edu, Univ of New Mexico; Pe
Fayek, Mostafa, (204) 474-7982 fayek@cc.umanitoba.ca, Univ of Manitoba; Cs
Fayon, Annia K., 612-626-9805 fayon001@umn.edu, Univ of Minnesota, Twin Cities; Gc
Feakins, Sarah, (213) 740-7168 feakins@usc.edu, Univ of Southern California; CoGuCs
Fearey, Bryan L., (505) 665-2423 Los Alamos; Cc
Feather, Russell, (202) 633-1793 featherr@si.edu, Smithsonian Inst / Natl Mus of Natural History; Gz
Fedele, Juan J., 320-308-1049 jjfedele@stcloudstate.edu, Saint Cloud State Univ; Hg
Fedo, Christopher, (865) 974-6002 cfedo@utk.edu, Univ of Tennessee, Knoxville; GsXgCg
Fedorchuk, Nicholas D., (203) 392-2586 fedorchukn1@southernct.edu, Southern Connecticut State Univ; GslGg
Fedorov, Alexey V., (203) 432-3153 alexey.fedorov@yale.edu, Yale Univ; Op

Fedortchouk, Yana, 902 494 8432 yana@dal.ca, Dalhousie Univ; CpEdGi
Feely, Martin, +353 (0)91 492 129 martin.feely@nuigalway.ie, Natl Univ of Ireland Galway; GzxEg
Feely, Richard A., (206) 526-6214 feely@pmel.noaa.gov, Univ of Washington; Oc
Feeney, Alison E., (717) 477-1319 aefeen@ship.edu, Shippensburg Univ; ZinZn
Feeney, Dennis M., (208) 885-5203 dmfeeney@uidaho.edu, Univ of Idaho; GgiYg
Feeney, Thomas P., (717) 477-1297 tpfeen@ship.edu, Shippensburg Univ; ZyGeHw
Fegley, Bruce, (314) 935-4852 bfegley@levee.wustl.edu, Washington Univ in St. Louis; Xc
Fegley, M. Bruce, (314) 935-4852 bfegley@levee.wustl.edu, Washington Univ in St. Louis; Xc
Fehler, Michael, 617-253-3589 fehler@mit.edu, Massachusetts Inst of Tech; Yg
Fehr, Karl Thomas, 089/2180 4256 fehr@min.uni-muenchen.de, Ludwig-Maximilians-Univ Muenchen; Gz
Fehrenbacher, Jennifer, 541-737-6285 fehrenje@coas.oregonstate.edu, Oregon State Univ; Cm
Fehrs, Ellen R., efehrs@pa.gov, DCNR- Pennsylvania Bur of Geo Survey; Zir
Fei, Yingwei, fei@gl.ciw.edu, Univ of Maryland; CpGxCg
Feibel, Craig S., (848) 445-2721 feibel@eps.rutgers.edu, Rutgers, The State Univ of New Jersey; GrsGa
Feigenson, Mark D., (848) 445-3149 feigy@eps.rutgers.edu, Rutgers, The State Univ of New Jersey; CgGiv
Feigl, Kurt, 608-262-0176 feigl@wisc.edu, Univ of Wisconsin, Madison; Yd
Fein, Jeremy B., (574) 631-6101 fein.1@nd.edu, Univ of Notre Dame; Cl
Feinberg, Joshua, (612) 624-8429 feinberg@umn.edu, Univ of Minnesota, Twin Cities; YmGze
Feineman, Maureen D., 814) 863-6649 mdf12@psu.edu, Pennsylvania State Univ, Univ Park; Cp
Feinglos, Mark N., (919) 668-1367 mark.feinglos@duke.edu, Duke Univ; Gz
Feinstein, Daniel T., (414) 962-2582 dtfeinst@usgs.gov, Univ of Wisconsin, Milwaukee; Hw
Felbeck, Horst, (858) 534-6647 hfelbeck@ucsd.edu, Univ of California, San Diego; Ob
Feldl, Nicole, 831-459-3693 nfeldl@ucsc.edu, Univ of California, Santa Cruz; At
Feldmann, Rodney M., (330) 672-2506 rfeldman@kent.edu, Kent State Univ; Pi
Felix, Joseph D., 361-825-4180 joseph.felix@tamucc.edu, Texas A&M Univ, Corpus Christi; Zn
Feller, Robert, feller@biol.sc.edu, Univ of South Carolina; Ob
Felton, Richard M., (660) 562-1569 rfelton@nwmissouri.edu, Northwest Missouri State Univ; Pg
Felzer, Benjamin S., (610) 758-3536 bsf208@lehigh.edu, Lehigh Univ; ZcoZn
Fenberg, Phillip, +44 (0)23 80592729 p.b.fenberg@soton.ac.uk, Univ of Southampton; Ob
Feng, Huan E., (973) 655-7549 fengh@mail.montclair.edu, Montclair State Univ; Cm
Feng, Jia , (775)682-7251 jiaf@unr.edu, Univ of Nevada, Reno; Zi
Feng, Ran, ran.feng@uconn.edu, Univ of Connecticut; PcAt
Feng, Song, (479) 575-4748 songfeng@uark.edu, Univ of Arkansas, Fayetteville; AsZyAm
Feng, Xiahong, 603-646-1712 xiahong.feng@dartmouth.edu, Dartmouth Coll; Cs
Feng, Yan, 630-252-2550 yfeng@anl.gov, Argonne; As
Feng, Yucheng, (334) 844-3967 yfeng@auburn.edu, Auburn Univ; SbCb
Feng, Zhiqiang, zhiqiang.feng@ed.ac.uk, Edinburgh Univ; Zn
Fengler, Keegan , 509.963.2719 keegan@geology.cwu.edu, Central Washington Univ; Gg
Fenical, William H., (858) 534-2133 wfenical@ucsd.edu, Univ of California, San Diego; Oc
Fennel, Katja, (902) 494-4526 katja.fennel@dal.ca, Dalhousie Univ; Ob
Fenrich, Francis, (780) 492-2149 frances@space.ualberta.ca, Univ of Alberta; Xy
Fenster, Michael S., (804)752-3745 mfenster@rmc.edu, Randolph-Macon Coll; OnGue
Fenstermaker, Lynn, (702) 862-5412 lynn@dri.edu, Desert Research Inst; Zr
Fenter, Paul, (630) 252-7053 fenter@anl.gov, Univ of Illinois Chicago; Gy
Fenton, Cassandra, 970-248-1077 cfenton@coloradomesa.edu, Colorado Mesa Univ; CcGmCl
Ferencz, Orsolya, orsi@sas.elte.hu, Eotvos Lorand Univ; XyZr
Ferger, Marisa, (814) 863-4229 mferger@psu.edu, Pennsylvania State Univ, Univ Park; Am
Ferguson, Charles A., 520.621.2470 caf@email.arizona.edu, Univ of Arizona; Gcv
Ferguson, Ian J., (204) 474-9154 ij_ferguson@umanitoba.ca, Univ of Manitoba; YemYu
Ferguson, John F., ferguson@utdallas.edu, Univ of Texas, Dallas; Yg
Ferguson, Julie , (949)824-9411 julie.ferguson@uci.edu, Univ of California, Irvine; Ze
Ferguson, Robert, 403 220-3269 rjfergus@ucalgary.ca, Univ of Calgary; Ye
Ferguson, Terry A., (864) 597-4527 fergusonta@wofford.edu, Wofford Coll; Ga
Fermanich, Kevin J., (920) 465-2240 fermanik@uwgb.edu, Univ of Wisconsin, Green Bay; SpRwHt
Fernandez, Diego, (801) 587-9366 diego.fernandez@utah.edu, Univ of Utah; Cg
Fernandez, Fabian G., 612-625-7460 fabiangf@umn.edu, Univ of Minnesota, Twin Cities; ScbSo
Fernandez, Loretta, 617.373.5461 l.fernandez@neu.edu, Northeastern Univ; Co
Fernandez, Louis A., (909) 537-5024 California State Univ, San Bernardino; Gi
Fernández, Lourdes, lfdiaz@ucm.es, Univ Compl de Madrid; Gz
Fernández, Luis Ramón, lrfernandez@ucm.es, Univ Compl de Madrid; Nt
Fernandez, Nicole, n.fernandez@cornell.edu, Cornell Univ; Gg
Fernández, Sixto, sixto@ucm.es, Univ Compl de Madrid; Pg
Fernández, Virginia, (598) 2525 1552 vivi@fcien.edu.uy, Univ de la Republica Oriental del Uruguay (UDELAR); ZirZn
Fernández Suárez, Javier, jfsuarez@ucm.es, Univ Compl de Madrid; Cc
Fernández-Barrenechea, José María, barrene@ucm.es, Univ Compl de Madrid; Gz
Fernando, Joe, harindra.j.fernando.10@nd.edu, Univ of Notre Dame; As
Ferrando, Simona, simona.ferrando@unito.it, Univ di Torino; Gx
Ferrari, Raffaele, (617) 253-1291 raffaele@mit.edu, Massachusetts Inst of Tech; OpAsOb
Ferraris, Giovanni, giovanni.ferraris@unito.it, Univ di Torino; Gz
Ferre, Paul A., (520) 621-2952 ty@hwr.arizona.edu, Univ of Arizona; Hw
Ferre, Paul (Ty), (520) 621-6082 ty@hwr.arizona.edu, Univ of Arizona; So
Ferre', Eric C., 337-482-2228 eric.ferre@louisiana.edu, Univ of Louisiana at Lafayette; YmGc
Ferreira, Anna, +44 020 7679 37704 a.ferreira@ucl.ac.uk, Univ Coll London; Ys
Ferreira, Maryanne F., (508) 289-2266 mferreira@whoi.edu, Woods Hole Oceanographic Inst; Zn
Ferrell, Jr., Ray E., (225) 388-5306 rferrell@lsu.edu, Louisiana State Univ; Cl
Ferren, Richard L., (413) 236-4553 rferren@berkshirecc.edu, Berkshire Comm Coll; Zn
Ferrero, Anna Maria, anna.ferrero@unito.it, Univ di Torino; Nr
Ferrier, Ken, kferrier@wisc.edu, Univ of Wisconsin, Madison; GmZo
Ferring, C. Reid, (940) 565-2993 reid.ferring@unt.edu, Univ of North Texas; Ga
Ferrini, Victoria, ferrini@ldeo.columbia.edu, Columbia Univ; Ou
Ferris, Dawn, 419-755-3909 dferris@sciencesocieties.org, Ohio State Univ; Sf
Ferris, Grant F., 416-978-0526 ferris@quartz.geology.utoronto.ca, Univ of Toronto; Gn
Ferris, Michael H., 310 243-3405 mferris@csudh.edu, California State Univ, Dominguez Hills; Zr
Ferriz, Horacio, 209 667-3874 hferriz@csustan.edu, California State Univ, Stanislaus; NgHgYe
Ferrucci, Fabrizio, fabrizio.ferrucci@open.ac.uk, Open Univ; Yg
Ferruzza, David, (not) li-sted ferruzzad@etown.edu, Elizabethtown Coll; AsZn
Festa, Andrea, andrea.festa@unito.it, Univ di Torino; Gc
Feunstenau, Maurice, mcf@unr.edu, Univ of Nevada, Reno; Nx
Feyereisen, Gary, 612-625-0968 gfeyer@umn.edu, Univ of Minnesota, Twin Cities; Sp
Fialko, Yuri A., (858) 822-5028 yfialko@ucsd.edu, Univ of California, San Diego; YgdGt
Fichera, Marissa, 575-835-5308 marissa.fichera@nmt.edu, New Mexico Tech; Hwq
Fichter, Lynn S., (540) 568-6531 fichtels@jmu.edu, James Madison Univ; Gr
Fichtner, Andreas, 089/2180 4230 andreas.fichtner@geophysik.uni-muenchen.de, Ludwig-Maximilians-Univ Muenchen; Yg
Fick, Walter H., (785) 532-7223 whfick@ksu.edu, Kansas State Univ; Sf
Field, Cathryn K., cfield@bowdoin.edu, Bowdoin Coll; Ge
Field, Richard T., 302-831-2695 Univ of Delaware; Og
Field, Stephen W., (254) 968-9887 field@tarleton.edu, Tarleton State Univ; Gz
Fielding, Chris, christopher.fielding@uconn.edu, Univ of Connecticut; Gs

Fielding, Christopher R., (402) 472-9801 cfielding2@unl.edu, Univ of Nebraska, Lincoln; Gs
Fielding, Lynn, (310) 660-3593 lfielding@elcamino.edu, El Camino Coll; Zg
Fields, Chad, chad.fields@dnr.iowa.gov, Iowa Dept of Nat Res; Gg
Fields, Nancy, 214-860-2429 nfields@dcccd.edu, El Centro Coll - Dallas Comm Coll District; Zg
Fifarek, Richard, fifarek@geo.siu.edu, Southern Illinois Univ Carbondale; Em
Figg, Sean, (760)744-1150 ext 2513 sfigg@palomar.edu, Palomar Coll; Gg
Fike, David A., (314) 935-6607 dfike@levee.wustl.edu, Washington Univ in St. Louis; CsmPg
Fildes, Stephen , stephen.fildes@flinders.edu.au, Flinders Univ; Zri
Filina, Irina, (402) 472-2077 ifilina2@unl.edu, Univ of Nebraska, Lincoln; YgGtYe
Filipescu, Sorin, +40-264-405300 ext 5206 sorin.filipescu@ubbcluj.ro, Babes-Bolyai Univ; GrPm
Filippelli, Gabriel M., (317) 274-3795 gfilippe@iupui.edu, Indiana Univ - Purdue Univ Indianapolis; CmGb
Filippelli, Hailey, 717-702-2028 hfilippell@pa.gov, DCNR- Pennsylvania Bur of Geo Survey; GgrZi
Filkorn, Harry, filkornh@piercecollege.edu, Los Angeles Pierce Coll; GgPi
Fillion, Luc, 514-421-4770 luc.fillion@ec.gc.ca, McGill Univ; Ams
Fillmore, Robert P., (970) 943-2650 rfillmore@western.edu, Western Colorado Univ; Gs
Finch, Adriam, +44 (0) 1334 462384 aafl@st-and.ac.uk, Univ of St. Andrews; Gz
Finger, Ken, 510-6432559 kfinger@berkeley.edu, Univ of California, Berkeley; PmZnn
Fink, Uwe, (520) 2736 uwefink@lpl.arizona.edu, Univ of Arizona; As
Finkel, Robert C., (925) 422-2044 finkel1@llnl.gov, Lawrence Livermore; Cg
Finkelman, Robert B., (972) 883-2459 bobf@utdallas.edu, Univ of Texas, Dallas; GbEcCa
Finkelstein, David, (315) 781-4443 finkelstein@hws.edu, Hobart & William Smith Colls; CgbCs
Finkenbinder, Matthew S., (570) 408-3871 matthew.finkenbinder@wilkes.edu, Wilkes Univ; GsmGl
Finks, Robert M., (718) 997-3305 Queens Coll (CUNY); Pi
Finley, Jason P., finleyjp@piercecollege.edu, Los Angeles Pierce Coll; Am
Finley, Mark, (309) 268-8642 mark.finley@heartland.edu, Heartland Comm Coll; Gg
Finley, Robert J., 217-244-8389 finley@isgs.uiuc.edu, Univ of Illinois, Urbana-Champaign; Go
Finley, William R., 337-482-6468 wfinley@rocmando.com, Univ of Louisiana at Lafayette; YsZi
Finn, Gregory C., (905) 688-5550 x3528 greg.finn@brocku.ca, Brock Univ; Gx
Finnegan, David L., (505) 667-6548 Los Alamos; As
Finnegan, Noah J., 831-459-5110 nfinnegan@pmc.ucsc.edu, Univ of California, Santa Cruz; Gm
Finnegan, Seth, sethf@berkeley.edu, Univ of California, Berkeley; Pig
Finney, Bruce P., (907) 474-7724 Univ of Alaska, Fairbanks; Ou
Finney, Stanley C., (562) 985-8637 scfinney@csulb.edu, California State Univ, Long Beach; Psi
Finzel, Emily, (319) 335-0405 emily-finzel@uiowa.edu, Univ of Iowa; GstEo
Fio, Karmen, +38514606088 karmen.fio@gmail.com, Univ of Zagreb; PgGg
Fiore, Arlene, amfiore@ldeo.columbia.edu, Columbia Univ; As
Fiorillo, Anthony R., (214) 421-3466 Southern Methodist Univ; Pv
Fireman, Gail, 412-442-4236 gfireman@pa.gov, DCNR- Pennsylvania Bur of Geo Survey; Zn
Firing, Eric, (808) 956-7894 efiring@hawaii.edu, Univ of Hawai'i, Manoa; Op
Fischer, Alfred G., (213) 740-5821 fisher@usc.edu, Univ of Southern California; Gs
Fischer, Emily, evf@rams.colostate.edu, Colorado State Univ; As
Fischer, Karen M., (401) 863-1360 karen_fischer@brown.edu, Brown Univ; Ys
Fischer, Mark P., (815) 753-0523 mfischer@niu.edu, Northern Illinois Univ; GcoHy
Fischer, Rebecca A., 617-384-6992 rebeccafischer@fas.harvard.edu, Harvard Univ; Gy
Fischer, Tobias, (505) 277-0683 fischer@unm.edu, Univ of New Mexico; Gv
Fish, Jennifer M., 831-459-1235 jmsfish@ucsc.edu, Univ of California, Santa Cruz; Zn
Fisher, Andrew T., (831) 459-5598 afisher@ucsc.edu, Univ of California, Santa Cruz; Hw
Fisher, Daniel C., (734) 764-0488 dcfisher@umich.edu, Univ of Michigan; Pi
Fisher, David A., 613-996-7623 fisher@nrcan-rncan.gc.ca, Univ of Ottawa; Gl
Fisher, Donald M., (814) 865-3206 fisher@geosc.psu.edu, Pennsylvania State Univ, Univ Park; GctZm
Fisher, Janet, janet.fisher@ed.ac.uk, Edinburgh Univ; Zn
Fisher, Liz, +440151 795 4390 e.h.fisher@liverpool.ac.uk, Univ of Liverpool; Co
Fisher, Nicholas S., (631) 632-8649 nicholas.fisher@stonybrook.edu, SUNY, Stony Brook; Oc
Fisher, Quentin , +44(0) 113 34 31920 q.j.fisher@leeds.ac.uk, Univ of Leeds; Np
Fisher, Rebecca, +44 1784 443628 r.e.fisher@rhul.ac.uk, Royal Holloway Univ of London; As
Fisher, Timothy G., (419) 530-2009 timothy.fisher@utoledo.edu, Univ of Toledo; GlmGn
Fisher, William L., (512) 471-5600 wfisher@mail.utexas.edu, Univ of Texas, Austin; GsrGo
Fisher, Jr., Thomas R., (410) 221-8432 fisher@hpl.umces.edu, Univ of Maryland; ObZiCl
Fishman, George , (781) 891-2721 gfishman@bentley.edu, Bentley Univ; XgbXc
Fishman, Jack, (314) 977-3132 jack.fishman@slu.edu, Saint Louis Univ; As
Fishwick, Stewart, +44 0116 252 3810 hos-geoggeolenv@leicester.ac.uk, Leicester Univ; Yg
Fisk, Aaron, 519-253-3000 x4740 afisk@uwindsor.ca, Univ of Windsor; ObCs
Fitak, Madge R., 614-265-6585 madge.fitak@dnr.state.oh.us, Ohio Dept of Nat Res; Gg
Fitchett, Rebekah, 773-442-6052 r-fitchett@neiu.edu, Northeastern Illinois Univ; GgCg
Fitton, J. G., +44 (0) 131 650 8529 godfrey.fitton@ed.ac.uk, Edinburgh Univ; Gi
Fitz, Thomas J., (715) 682-1852 tfitz@northland.edu, Northland Coll; GgxGc
Fitzer, Susan, +44 01413305442 susan.fitzer@glasgow.ac.uk, Univ of Glasgow; Ob
Fitzgerald, David, (210) 436-3235 Saint Mary's Univ; Gd
FitzGerald, Duncan M., (617) 353-2530 dunc@bu.edu, Boston Univ; OnGsu
Fitzgerald, Francis S., 303-384-2644 ffitzger@mines.edu, Colorado Geo Survey; ZirGg
Fitzgerald, Paul G., (315) 443-2672 pgfitzge@syr.edu, Syracuse Univ; GtCc
Fitzgerald, William F., 860-405-9158 william.fitzgerald@uconn.edu, Univ of Connecticut; Oc
Fitzjarrald, David R., (518) 437-8735 dfitzjarrald@albany.edu, SUNY, Albany, Asm
Fitzpatrick, John R., (505) 667-4761 Los Alamos; Zn
FitzPatrick, Meriel E., +44 1752 584769 m.e.fitzpatrick@plymouth.ac.uk, Univ of Plymouth; PlgGr
Fitzpatrick, Patrick J., (228) 688-1157 fitz@gri.msstate.edu, Mississippi State Univ; Am
Fitzpatrick, Stephan D., (678) 891-3773 sfitzpatrick1@gsu.edu, Georgia State Univ, Perimeter Coll, Clarkston Campus; HwCgSo
Fitzpatrick, William, (608) 263-0972 wfitzpatric3@wisc.edu, Univ of Wisconsin, Madison, Division of Extension; EgGi
Fitzsimmons, Jessica N., 979-845-5137 jessfitz@tamu.edu, Texas A&M Univ; OcCta
Fitzsimmons, Kevin, (520) 626-3324 kevfitz@ag.arizona.edu, Univ of Arizona; Zn
Fjell, Dale, (785) 532-5833 dfjell@ksu.edu, Kansas State Univ; So
Flagg, Charles, (631) 632-3184 charles.flagg@stonybrook.edu, SUNY, Stony Brook; Op
Flaherty, Frank A., (229) 333-5665 flaherty@valdosta.edu, Valdosta State Univ; Zn
Flaig, Peter P., peter.flaig@beg.utexas.edu, Univ of Texas, Austin; GsrEo
Flake, Rex , 509.963.1114 rex@geology.cwu.edu, Central Washington Univ; Yd
Flament, Pierre J., (808) 956-6663 pflament@hawaii.edu, Univ of Hawai'i, Manoa; Op
Flanagan, Michael, flanagam@sunysuffolk.edu, Suffolk County Comm Coll, Ammerman Campus; Zg
Flanagan, Timothy, (413) 236-4503 tflanaga@berkshirecc.edu, Berkshire Comm Coll; Zei
Flannery, David, +61 7 3138-1615 david.flannery@qut.edu.au, Queensland Univ of Tech; PoCaXg
Flato, Gregory M., (250) 363-8223 greg.flato@ec.gc.ca, Univ of Victoria; As
Fleck, Michelle C., (435) 613-5232 Utah State Univ Eastern; GgZy
Flegal, Russell, (831) 459-2093 rflegal@es.ucsc.edu, Univ of California, Santa Cruz; Cg
Fleisher, P. Jay, (607) 286-7541 fleishpj@oneonta.edu, SUNY, Oneonta; GlmGe
Fleming, Thomas H., 203-392-5837 flemingt1@southernct.edu, Southern

Connecticut State Univ; GiCgc
Flemings, Peter B., 512-471-6156 flemings@ig.utexas.edu, Univ of Texas, Austin; Gr
Flemming, Roberta L., (519) 661-3143 rflemmin@uwo.ca, Western Univ; GzXmZm
Flesch, Lucy M., (765) 494-0263 lmflesch@purdue.edu, Purdue Univ; Yg
Flesken, Luuk, +31(0) 317 485467 l.fleskens@leeds.ac.uk, Univ of Leeds; Ge
Flessa, Karl W., (520) 621-7336 kflessa@email.arizona.edu, Univ of Arizona; PgeGe
Fletcher, Austin, rfletche@uoguelph.ca, Univ of Guelph; Zn
Fletcher, Charles H., (808) 956-2582 fletcher@soest.hawaii.edu, Univ of Hawai'i, Manoa; GsOnZu
Fletcher, John, jfletche@cicese.mx, Centro de Investigación Científica y de Educación Superior de Ensenada; Gpt
Fletcher, Madilyn, fletcher@sc.edu, Univ of South Carolina; Ob
Fletcher, William K., (604) 822-2392 wfletcher@eos.ubc.ca, Univ of British Columbia; Ce
Flierl, Glenn R., (617) 253-4692 glenn@lake.mit.edu, Massachusetts Inst of Tech; Op
Flint, Stephen, +44 0161 306-6971 stephen.flint@manchester.ac.uk, Univ of Manchester; GrsGo
Flood, Beverly, 612-624-1603 floo0017@umn.edu, Univ of Minnesota, Twin Cities; Pg
Flood, Peter, +61-2-67732329 pflood@une.edu.au, Univ of New England; EcGt
Flood, Roger D., (631) 632-6971 roger.flood@stonybrook.edu, SUNY, Stony Brook; Ou
Flood, Tim P., (920) 403-1356 tim.flood@snc.edu, Saint Norbert Coll; GvZe
Florea, Lee, (812) 855-1376 lflorea@indiana.edu, Indiana Univ, Bloomington; Hg
Flores, Alejandro, lejoflores@boisestate.edu, Boise State Univ; HqAs
Flores, Carlos, cflores@cicese.mx, Centro de Investigación Científica y de Educación Superior de Ensenada; Ye
Flores, Kennet, 718-951-5000 x3253 keflores@brooklyn.cuny.edu, Brooklyn Coll (CUNY); Gpc
Flores-Garcia, Mari C., (818) 677-3541 mari.flores@csun.edu, California State Univ, Northridge; Zn
Florie, Joshua, joshua.florie@tulsacc.edu, Tulsa Comm Coll; OgGg
Floris, Mario, +390498279121 mario.floris@unipd.it, Univ degli Studi di Padova; NgZr
Flory, David M., (515) 294-0264 flory@iastate.edu, Iowa State Univ of Sci & Tech; AmsZn
Flower, Hilary, flower.hilary@spcollege.edu, Saint Petersburg Coll, Clearwater; Gg
Flower, Martin F., +1 312 391-7632 flower@uic.edu, Univ of Illinois Chicago; GiCuGt
Flowers, George C., (504) 862-3192 flowers@tulane.edu, Tulane Univ; Hw
Flowers, Gwenn, (778)782-6638 gflowers@sfu.ca, Simon Fraser Univ; Gl
Flowers, Rebecca M., (303) 492-5135 rebecca.flowers@colorado.edu, Univ of Colorado; CgcGt
Flowers Falls, Emily, (540) 458-8868 fallse@wlu.edu, Washington & Lee Univ; Gg
Floyd, Barry, 518-473-9988 esogis@mail.nysed.gov, New York State Geo Survey; Zn
Fluegeman, Richard H., (765) 285-8267 rfluegem@bsu.edu, Ball State Univ; PmGro
Flynn, John J., (212) 769-5806 American Mus of Natural History; Pv
Flynn, Luke P., (808) 956-3154 flynn@higp.hawaii.edu, Univ of Hawai'i, Manoa; Zr
Flynn, Wendilyn, 970-351-1071 wendilyn.flynn@unco.edu, Univ of Northern Colorado; Ams
Fodor, Ronald V., (919) 515-7177 ron_fodor@ncsu.edu, North Carolina State Univ; Gi
Fogel, David N., (505) 665-3305 fogel@lanl.gov, Los Alamos; Zn
Fogg, Graham E., (916) 752-6810 Univ of California, Davis; Hy
Foglia, Laura, lfoglia@ucdavis.edu, Univ of California, Davis; HwZc
Foit, Jr., Franklin F., (509) 335-3093 Washington State Univ; Gz
Foland, Kenneth A., foland.1@osu.edu, Ohio State Univ; CcgCs
Foley, Bradford, (814) 863-3591 bjf5382@psu.edu, Pennsylvania State Univ, Univ Park; YgXyGt
Foley, Duncan, (253) 535-7568 foleyd@plu.edu, Pacific Lutheran Univ; GeHyZe
Foley, John P., (406) 496-4414 jfoley@mtech.edu, Montana Tech of The Univ of Montana; Zn
Follmer, Leon R., (217) 359-2090 lfollmer@illinois.edu, Univ of Illinois, Urbana-Champaign; SaGlm
Follows, Michael, (617) 253-5939 mick@mit.edu, Massachusetts Inst of Tech; Op
Folorunso, I. O., foloisgood@yahoo.com, Univ of Ilorin; GoCg
Fomel, Sergey, 512-475-9573 sergey.fomel@beg.utexas.edu, Univ of Texas at Austin, Jackson Sch of Geosciences; Yx
Fondevilla, Victor, victor.fondevilla@uab.cat, Univ Auto de Barcelona; PgGr
Fondran, Carol, (216) 987-5227 carol.fondran@tri-c.edu, Cuyahoga Comm Coll - Western Campus; Zg
Fones, Gary, +44 023 92 842252 gary.fones@port.ac.uk, Univ of Portsmouth; Cm
Fonstad, Mark, fonstad@uoregon.edu, Univ of Oregon; Zy
Fontaine, Fabrice R., +262262938207 fabrice.fontaine@univ-reunion.fr, Univ de la Reunion; Yg
Fontana, Alessandro, +390498279118 alessandro.fontana@unipd.it, Univ degli Studi di Padova; Gm
Foote, Michael J., (773) 702-4320 mfoote@midway.uchicago.edu, Univ of Chicago; Pg
Forbriger, Thomas, +49-721-6084593 thomas.forbriger@kit.edu, Karlsruhe Inst of Tech; Ys
Forcino, Frank, (828) 227-2888 flforcino@wcu.edu, Western Carolina Univ; ZePgGr
Ford, Allistair, +44 (0) 191 208 7121 alistair.ford@ncl.ac.uk, Univ of Newcastle Upon Tyne; Gq
Ford, Andrew, 509-335-7846 forda@wsu.edu, Washington State Univ; Zu
Ford, Arianne, arianne.ford@jcu.edu.au, James Cook Univ; GqZiEg
Ford, Derek C., (905) 525-9140 x20132 dford@mcmaster.ca, McMaster Univ; GmHwCc
Ford, Heather, (951) 827-3194 heather.ford@ucr.edu, Univ of California, Riverside; Ys
Ford, Richard L., (801) 626-6942 rford@weber.edu, Weber State Univ; GmZeEo
Fordyce, R. Ewan, +64 3 479-7510/7535 ewan.fordyce@otago.ac.nz, Univ of Otago; Pv
Foreman, Brady Z., (360) 650-2546 brady.foreman@wwu.edu, Western Washington Univ; Grs
Foreman, Michael G., 250-363-6306 mike.foreman@dfo-mpo.gc.ca, Univ of Victoria; OpZo
Foreman, Michael G. G., (250) 363-6306 foremanm@pac.dfo-mpo.gc.ca, Univ of British Columbia; Op
Foresi, Luca Maria, luca.foresi@unisi.it, Univ of Siena; PmGs
Foret, Jim, 337-482-6064 jaf3663@louisiana.edu, Univ of Louisiana at Lafayette; Ge
Forman, Stephen, (254)710-2495 steve_forman@baylor.edu, Baylor Univ; CcPc
Fornaciari, Eliana, +390498279185 eliana.fornaciari@unipd.it, Univ degli Studi di Padova; Pgm
Fornari, Daniel J., (508) 289-2857 dfornari@whoi.edu, Woods Hole Oceanographic Inst; Gu
Fornash, Katherine, 740-593-1104 kffornash@ohio.edu, Ohio Univ; GpCga
Forno, Maria Gabriella, gabriella.forno@unito.it, Univ di Torino; Gg
Forray, Ferenc Lazar, ferenc.forray@ubbcluj.ro, Babes-Bolyai Univ; CglPc
Forrest, David, +4401413305401 david.forrest@glasgow.ac.uk, Univ of Glasgow; Zi
Forsman, Nels F., (701) 777-4349 nels.forsman@engr.und.edu, Univ of North Dakota; GgXgRh
Forster, Piers M., +44(0) 113 34 36476 p.m.forster@leeds.ac.uk, Univ of Leeds; As
Forster, Richard R., (801) 581-3611 rick.forster@geog.utah.edu, Univ of Utah; Zr
Forsyth, Donald W., (401) 863-1699 donald_forsyth@brown.edu, Brown Univ; YrsGt
Forte, Adam, aforte8@lsu.edu, Louisiana State Univ; Gcm
Forte, Alessandro Marco, 514-987-3000 #5607 forte.alessandro@uqam.ca, Univ du Quebec a Montreal; Yg
Fortes, Dominic, andrew.fortes@ucl.ac.uk, Birkbeck Coll; Zg
Fortier, Richard, (418) 656-2746 richard.fortier@ggl.ulaval.ca, Univ Laval; Yg
Fortin, Danielle, dfortin@uottawa.ca, Univ of Ottawa; CblGe
Fortin, Samantha, 609-258-4101 sf3033@princeton.edu, Princeton Univ; ObCbOc
Fortin, Will Frances J., wff2106@columbia.edu, Columbia Univ; Ys
Fortner, Sarah K., (937) 327-7328 sfortner@wittenberg.edu, Wittenberg Univ; ClZeHg
Fosdick, Julie, julie.fosdick@uconn.edu, Univ of Connecticut; CcGt
Foss, Donald J., (415) 485-9523 jcprice@metro.net, Coll of Marin; Zg
Foster, David A., (352) 392-2231 dafoster@ufl.edu, Univ of Florida; GtCc
Foster, Gavin, +44 (0)23 8059 3786 gavin.foster@noc.soton.ac.uk, Univ of Southampton; Csg
Foster, John H., 657-278-3882 jfoster@exchange.fullerton.edu, California State Univ, Fullerton; Ng
Foster, Patrick, +44 01326 371828 p.j.foster@exeter.ac.uk, Exeter Univ; Nm

Foster Jr., Charles T., (319) 335-1801 tom-foster@uiowa.edu, Univ of Iowa; GptGg
Foti, Pamela, (928) 523-6196 pam.foti@nau.edu, Northern Arizona Univ; Zn
Fotopoulos, Georgia, (613) 533-6639 georgia.fotopoulos@queensu.ca, Queen's Univ; Yd
Foufoula-Georgiou, Efi, (612) 626-0369 Univ of Minnesota, Twin Cities; Hg
Fouke, Bruce W., (217) 244-5431 fouke@illinois.edu, Univ of Illinois, Urbana-Champaign; GsdGb
Fountain, Andrew G., (503) 725-3386 andrew@pdx.edu, Portland State Univ; Gl
Fountain, John C., 919-515-3717 fountain@ncsu.edu, North Carolina State Univ; Hw
Fournelle, John, (608) 438-7480 johnf@geology.wisc.edu, Univ of Wisconsin, Madison; GiZn
Fournier, Benoit, (418) 656-3930 benoit.fournier@ggl.ulaval.ca, Univ Laval; NgZmGg
Fournier, Gregory, 617-324-6164 g4nier@mit.edu, Massachusetts Inst of Tech; Pg
Fournier, Robert O., (902) 494-3666 robert.fournier@dal.ca, Dalhousie Univ; Ob
Foustoukos, Dionysis, dfoustoukos@carnegiescience.edu, Carnegie Inst for Sci; Cg
Fovell, Robert, (310) 206-9956 fovell@atmos.ucla.edu, Univ of California, Los Angeles; As
Fowell, Sarah J., (907) 474-7810 sjfowell@alaska.edu, Univ of Alaska, Fairbanks; Pl
Fowle, David A., (785) 864-1955 fowle@ku.edu, Univ of Kansas; Cb
Fowler, Gerald A., fowler@uwp.edu, Univ of Wisconsin, Parkside; PgGg
Fowler, Malcolm M., (505) 667-5439 Los Alamos; As
Fowler, Mike, +44 023 92 842293 mike.fowler@port.ac.uk, Univ of Portsmouth; Gi
Fowler, Sarah, sarah.fowler@ees.kuleuven.be, Katholieke Univit Leuven; Gip
Fox, Austin, (321) 674-7463 afox@fit.edu, Florida Inst of Tech; Oc
Fox, David L., 612-624-6361 dlfox@umn.edu, Univ of Minnesota, Twin Cities; Pg
Fox, James E., (605) 394-2461 james.fox@sdsmt.edu, South Dakota Sch of Mines & Tech; Gs
Fox, Jeff, 740 2012762 jeff.fox@dnr.state.oh.us, Ohio Dept of Nat Res; YsmZe
Fox, Laurel R., (831) 459-2533 fox@biology.ucsc.edu, Univ of California, Santa Cruz; Zn
Fox, Lewis, 954-201-6674 lfox@broward.edu, Broward Coll; Oc
Fox, Lydia K., (209) 946-2481 lkfox@pacific.edu, Univ of the Pacific; GizZe
Fox, Matthew D., m.fox@ucl.ac.uk, Univ Coll London; Gm
Fox, Nathaniel, 605-394-2487 nathaniel.fox@sdsmt.edu, South Dakota Sch of Mines & Tech; Pgv
Fox, Neil I., 573-882-2144 foxn@missouri.edu, Univ of Missouri; As
Fox, Nicholas, nicholas.fox3@montana.edu, Montana State Univ; Zi
Fox, Tim, 218-322-2364 timothy.fox@itascacc.edu, Itasca Comm Coll; Zi
Fox-Dobbs, Kena L., 253-879-2458 kena@pugetsound.edu, Univ of Puget Sound; PgCsGe
Fox-Kemper, Baylor, 401-863-3979 baylor@brown.edu, Brown Univ; OpYnAt
Foxon, Tim, +44(0) 113 34 37910 t.j.foxon@leeds.ac.uk, Univ of Leeds; Ge
Frailey, Scott M., 217-244-2430 sfrailey@illinois.edu, Illinois State Geo Survey; Np
Fraiser, Margaret L., mfraiser@uwm.edu, Univ of Wisconsin, Milwaukee; PiePg
Fralick, Philip W., (807) 343-8010 Ext. 8288 philip.fralick@lakeheadu.ca, Lakehead Univ; Gs
Fram, Jonathan, 541-737-3966 jfram@coas.oregonstate.edu, Oregon State Univ; Op
Frame, Jeffrey, 217-244-9575 frame@illinois.edu, Univ of Illinois, Urbana-Champaign; Asm
Francis, Don, donald.francis@mcgill.ca, McGill Univ; Gi
Francis, Jane, j.e.francis@leeds.ac.uk, Univ of Leeds; Pe
Francis, Robert, (206) 543-7345 Univ of Washington; Ob
Francis, Sarah, 612-626-6901 franc944@umn.edu, Univ of Minnesota; Zi
Francisco, Joseph, frjoseph@sas.upenn.edu, Univ of Pennsylvania; Ac
Franco, Aldina, +44 (0)1603 59 2721 a.franco@uea.ac.uk, Univ of East Anglia; Ge
Francois, Roger, rfrancois@eos.ubc.ca, Univ of British Columbia; Cm
Francus, Pierre, pierre.francus@ete.inrs.ca, Univ du Quebec; PeGse
Franek, Benjamin, (570) 389-4567 bfranek@bloomu.edu, Bloomsburg Univ; HsZy
Frank, Charles O., (618) 453-7365 frank@geo.siu.edu, Southern Illinois Univ Carbondale; Gx
Frank, Kenneth, (902) 426-3498 frankk@mar.dfo-mpo.gc.ca, Dalhousie Univ; Ob
Frank, Mark R., (815) 753-8395 mfrank@niu.edu, Northern Illinois Univ; Cp
Frank, Tracy, tracy.frank@uconn.edu, Univ of Connecticut; CgGs
Frank, William M., frank@ems.psu.edu, Pennsylvania State Univ, Univ Park; Am
Frankel, Arthur, 206-553-0626 afrankel@uw.edu, Univ of Washington; Ys
Frankel, Tyler E., tfrankel@umw.edu, Univ of Mary Washington; Rw
Frankenberg, Christian, 626-395-6143 cfranken@caltech.edu,CalTech; Zr
Frankic, Anamarija, anamarija.frankic@umb.edu, Univ of Massachusetts, Boston; Ze
Franklin, David, +44 023 92 843540 david.franklin@port.ac.uk, Univ of Portsmouth; Ym
Franks, Peter J. S., (858) 534-7528 pfranks@ucsd.edu, Univ of California, San Diego; Ob
Franseen, Evan K., (785) 864-2072 evanf@kgs.ku.edu, Univ of Kansas; Gs
Fransolet, Andre-Mathieu, 32 4 366 22 06 amfransolet@ulg.ac.be, Univ de Liege; Gz
Frantz, Carie M., (801) 626-6181 cariefrantz@weber.edu, Weber State Univ; PoCbZn
Franz, Eldon H., (206) 300-9259 franz@wsu.edu, Washington State Univ; Sf
Franz, Kristie, (515) 294-7454 kfranz@iastate.edu, Iowa State Univ of Sci & Tech; Hs
Franzi, David A., (518) 564-4033 franzida@plattsburgh.edu, SUNY, Plattsburgh; Gl
Frappier, Amy, 518-580-8371 afrappie@skidmore.edu, Skidmore Coll; GePe
Fraser, Alastair, 44-20-7594-6530 afraser@egi.utah.edu, Univ of Utah; Gco
Fraser, Alistair B., abf1@psu.edu, Pennsylvania State Univ, Univ Park; Am
Fraser, Donalf G., +44 (1865) 272033 don@earth.ox.ac.uk, Univ of Oxford; Cg
Fraser-Smith, Antony C., (650) 723-3684 Stanford Univ; Ym
Frasson, Renato , frasson.1@osu.edu, Ohio State Univ; Hg
Fratta, Dante, 608-265-5644 fratta@wisc.edu, Univ of Wisconsin, Madison; Yx
Frazer, L. N., (808) 956-3724 neil@soest.hawaii.edu, Univ of Hawai'i, Manoa; Ys
Frazier, Ryan J., (801) 626-7435 ryanfrazier@weber.edu, Weber State Univ; ZriZc
Frederick, Bruce, bfreder6@jccc.edu, Johnson County Comm Coll; Gg
Frederick, Daniel L., (931) 221-7455 frederickd@apsu.edu, Austin Peay State Univ; PgGsr
Frederick, Holly, 570-408-4880 holly.frederick@wilkes.cdi, Wilkes Univ; Sdb
Frederick, John E., (773) 702-3237 Univ of Chicago; As
Frederickson, Paul A., 831-595-5212 pafreder@nps.edu, Naval Postgrad Sch; Am
Frederiksen, Andrew, (204) 474-9460 andrew_frederiksen@umanitoba.ca, Univ of Manitoba; YsGt
Fredrick, Kyle, (724) 938-4180 fredrick@calu.edu, California Univ of Pennsylvania; GgHgGm
Fredricks, Helen, 508-289-3678 hfredricks@whoi.edu, Woods Hole Oceanographic Inst; Oc
Freed, Andrew M., freed@purdue.edu, Purdue Univ; Gg
Freed, Robert L., (210) 999-7092 bfreed@trinity.edu, Trinity Univ; Gz
Freeman, Katherine H., (814) 863-8177 khf4@psu.edu, Pennsylvania State Univ, Univ Park; Cos
Freeman, Rebecca, (859) 257-3758 rebecca.freeman@uky.edu, Univ of Kentucky; GrPi
Freeman, Veronica, 740-376-4779 freemanv@marietta.edu, Marietta Coll; Pg
Freeman-Lynde, Raymond, rfreeman@uga.edu, Univ of Georgia; Gu
Fregenal, María Antonia, mariana@ucm.es, Univ Compl de Madrid; Gsg
Frei, Allan, 212-772-5322 afrei@hunter.cuny.edu, Graduate Sch of the City Univ of New York; Gg
Frei, Michaela, 089/2180 6590 michaela.frei@iaag.geo.uni-muenchen.de, Ludwig-Maximilians-Univ Muenchen; Gg
Freiburg, Jared, 217-244-2495 freiburg@illinois.edu, Illinois State Geo Survey; GgzGx
Freiermuth, Sue, (715) 425-3345 susan.m.freiermuth@uwrf.edu, Univ of Wisconsin, River Falls; Zn
Freile, Deborah, (201) 200-3161 dfreile@njcu.edu, New Jersey City Univ; GseGg
Freitas, Maria d., cfreitas@fc.ul.pt, Unive de Lisboa; GseGu
Freitas, Maria da Conceição P., +351217500352 cfreitas@fc.ul.pt, Unive de Lisboa; GusGe
French, Adam, (605) 394-1649 adam.french@sdsmt.edu, South Dakota Sch of Mines & Tech; Asm
French, Bevan, (202) 633-1326 Smithsonian Inst / Natl Mus of Natural History; Xm

French, Helen K., helen.french@nmbu.no, Norwegian Univ of Life Scis (NMBU); HwYg
French, Mark A., (609) 984-6587 mark.french@dep.state.nj.us, New Jersey Geological and Water Survey; Hq
French, Melodie E., 713.348.5088 mefrench@rice.edu, Rice Univ; Gc
Frenette, Jean, 418-656-8123 frenette@ggl.ulaval.ca, Univ Laval; Gz
Frew, Nelson M., (508) 289-2489 nfrew@whoi.edu, Woods Hole Oceanographic Inst; Ca
Frey, Bonnie A., 575-835-5160 bonnie.frey@nmt.edu, New Mexico Tech; Cg
Frey, Frederick A., (617) 253-2818 fafrey@mit.edu, Massachusetts Inst of Tech; CtGvi
Frey, Friedrich, 089/21804332 f.frey@lmu.de, Ludwig-Maximilians-Univ Muenchen; Gz
Frey, Holli M., 518-388-6418 freyh@union.edu, Union Coll; GvCag
Frey, Serita, serita.frey@unh.edu, Univ of New Hampshire; Sb
Frez, Jose, jofrez@cicese.mx, Centro de Investigación Científica y de Educación Superior de Ensenada; Ys
Fricke, Henry C., (719) 389-6514 hfricke@coloradocollege.edu, Colorado Coll; Cs
Fricker, Helen A., (858) 534-6145 hafricker@ucsd.edu, Univ of California, San Diego; Yr
Friedl, Mark, friell@bu.edu, Boston Univ; Zr
Friedman, Matt, +44 (1865) 272035 matt.friedman@earth.ox.ac.uk, Univ of Oxford; Pg
Friedmann, Samuel (Julio) J., (925) 423-0585 friedmann2@llnl.gov, Lawrence Livermore; Gg
Friedrich, Anke, friedrich@iaag.geo.uni-muenchen.de, Ludwig-Maximilians-Univ Muenchen; GgtCc
Friedrichs, Carl T., (804) 684-7303 cfried@vims.edu, William & Mary; On
Friedrichs, Marjorie, (804)684-7695 marjy@vims.edu, William & Mary; Obp
Friehauf, Kurt, (610) 683-4446 friehauf@kutztown.edu, Kutztown Univ of Pennsylvania; EmCgGx
Frieman, Richard, richard.frieman@oswego.edu, SUNY, Oswego; Gg
Frierson, Dargan M., (206) 685-7364 dargan@atmos.washington.edu, Univ of Washington; As
Frind, Emil O., (519) 888-4567 (Ext. 3959) Univ of Waterloo; Hq
Frisbee, Marty, (765) 494-8678 mdfrisbee@purdue.edu, Purdue Univ; Hws
Frisia, Silvia, +61 2 4921 5402 silvia.frisia@newcastle.edu.au, Univ of Newcastle; ClPeGd
Frisk, DeAnn M., (515) 294-4477 dfrisk@iastate.edu, Iowa State Univ of Sci & Tech; Zn
Fritsch, J. Michael, (814) 863-1842 fritsch@ems.psu.edu, Pennsylvania State Univ, Univ Park; Am
Fritsen, Christian H., (775) 673-7487 cfritsen@dri.edu, Univ of Nevada, Reno; Og
Fritton, Daniel D., (814) 865-1143 ddf@psu.edu, Pennsylvania State Univ, Univ Park; Sp
Fritz, Allan, (785) 532-7245 akf@ksu.edu, Kansas State Univ; So
Fritz, Sherilyn C., (402) 472-6431 sfritz2@unl.edu, Univ of Nebraska, Lincoln; Pem
Fritz, William, 718-982-2400 william.fritz@csi.cuny.edu, Graduate Sch of the City Univ of New York; Zg
Froehlich, David J., (512) 223-4894 eohippus@austincc.edu, Austin Comm Coll District; PvZn
Froelich, Philip, 850-644-4331 froelich@ocean.fsu.edu, Florida State Univ; Og
Froese, Duane, (780) 492-1968 duane.froese@ualberta.ca, Univ of Alberta; GlmPe
FROGER, Jean-Luc, jean.luc.froger@univ-st-etienne.fr, Université Jean Monnet, Saint-Etienne; Zr
Frohlich, Cliff, (512) 471-0460 cliff@ig.utexas.edu, Univ of Texas, Austin; Ys
Frohlich, Clifford A., 512-471-0460 cliff@ig.utexas.edu, Univ of Texas, Austin; YsGt
Frolking, Stephen E., (603) 862-0244 steve.frolking@unh.edu, Univ of New Hampshire; As
Frolking, Tod A., (740) 587-6222 frolking@denison.edu, Denison Univ; Zy
Fromm, Jeanne M., (605) 582-3416 jeanne.fromm@usd.edu, Univ of South Dakota; GgHs
Frost, B. Ronald, (307) 399-5585 rfrost@uwyo.edu, Univ of Wyoming; GpiEd
Frost, Bruce W., 206-543-7186 frost@ocean.washington.edu, Univ of Washington; Ob
Frost, Carol D., frost@uwyo.edu, Univ of Wyoming; CcGit
Frost, Daniel, 803-777-4530 dfrost@seoe.sc.edu, Univ of South Carolina; Ys
Frost, Eric G., (619) 594-5003 eric.frost@sdsu.edu, San Diego State Univ; Zr
Frost, Gina M., 209 954 5380 gfrost@deltacollege.edu, San Joaquin Delta Coll; Gg
Frouin, Marine , 631-632-1429 marine.frouin@stonybrook.edu, Stony Brook Univ; Cc
Fry, Matthew, 940-369-7576 mfry@unt.edu, Univ of North Texas; Zn
Fryar, Alan E., (859) 257-4392 alan.fryar@uky.edu, Univ of Kentucky; HwGeCl
Frye, John , fryej@uww.edu, Univ of Wisconsin, Whitewater; AtZr
Fryer, Brian J., 519-253-3000 x3750 bfryer@uwindsor.ca, Univ of Windsor; Cla
Fryer, Patricia B., (808) 956-3146 pfryer@soest.hawaii.edu, Univ of Hawai'i, Manoa; GuiGt
Fryett, I, fryetti@cf.ac.uk, Cardiff Univ; Zn
Fryxell, Joan E., (909) 537-5311 jfryxell@csusb.edu, California State Univ, San Bernardino; Gct
Fthenakis, Vasilis M., (516) 282-2830 Columbia Univ; As
Fu, Qi, (713) 743-3660 qfu3@uh.edu, Univ of Houston; CosCa
Fu, Qiang, (206) 685-2070 qfu@atmos.washington.edu, Univ of Washington; As
Fu, Qilong, (512) 232-9372 qilong.fu@beg.utexas.edu, Univ of Texas at Austin, Jackson Sch of Geosciences; Gso
Fu, Roger, 617-384-6991 rogerfu@fas.harvard.edu, Harvard Univ; YmXym
Fubelli, Giandomenico, giandomenico.fubelli@unito.it, Univ di Torino; Gm
Fudge, T.J., 206-543-0162 tjfudge@uw.edu, Univ of Washington; Gl
Fueglistaler, Stephan A., 609-258-8238 stf@princeton.edu, Princeton Univ; As
Fuente, David E., 803 777-2757 fuente@seoe.sc.edu, Univ of South Carolina; Zn
Fuentes, Edna L., (312) 996-6123 eriver15@uic.edu, Univ of Illinois Chicago; Zn
Fuentes, Jose D., jdfuentes@psu.edu, Pennsylvania State Univ, Univ Park; As
Fueten, Frank, (905) 688-5550 (Ext. 3856) ffueten@brocku.ca, Brock Univ; Gc
Fuge, Ron, +44 (0)1970 622 642 rrf@aber.ac.uk, Aberystwyth Univ; GbCgGe
Fugitt, Frank, (614) 265-6759 frank.fugitt@dnr.state.oh.us, Ohio Dept of Nat Res; GrHw
Fuhrmann, Christopher, (662) 268-1032 Ext 219 cmf396@msstate.edu, Mississippi State Univ; As
Fuiman, Lee A., (361) 749-6775 lee.fuiman@utexas.edu, Univ of Texas, Austin; Ob
Fujita, Kazuya, 517-355-0142 fujita@msu.edu, Michigan State Univ; Ys
Full, Robert J., (510) 642-9896 rjfull@garnet.berkeley.edu, Univ of California, Berkeley; Pi
Fullagar, Paul D., fullagar@unc.edu, Univ of North Carolina, Chapel Hill; CcGaZe
Fullmer, James W., fullmerj1@southernct.edu, Southern Connecticut State Univ; AmXa
Fulthorpe, Craig S., (512) 471-0459 craig@ig.utexas.edu, Univ of Texas, Austin; Gu
Fulton, James, (254)710-2184 james_fulton@baylor.edu, Baylor Univ; Cbo
Fultz, Lisa M., 225-578-1344 lfutz@agcenter.lsu.edu, Louisiana State Univ; Sb
Fulweiler, Robinson, 617-358-5466 rwf@bu.edu, Boston Univ; Cm
Fung, Inez, (510) 643-9367 inez@atmos.berkeley.edu, Univ of California, Berkeley; As
Funning, Gareth J., gareth.funning@ucr.edu, Univ of California, Riverside; YdgYs
Furbish, David J., 615-322-2137 david.j.furbish@vanderbilt.edu, Vanderbilt Univ; GmHg
Furió, Marc, marc.furio@uab.cat, Univ Auto de Barcelona; Pv
Furlong, Emelia, emeliaa.furlong@maryland.gov, Maryland Dept of Nat Res; HwCqSp
Furlong, Kevin P., (814) 863-0567 kevin@geodyn.psu.edu, Pennsylvania State Univ, Univ Park; Gt
Furman, Tanya, (814) 865-5782 furman@psu.edu, Pennsylvania State Univ, Univ Park; GiCu
Fusseis, Florian, +44 (0) 131 650 6755 ffusseis@staffmail.ed.ac.uk, Edinburgh Univ; Gc
Fyfe, John C., (250) 363-8236 john.fyfe@ec.gc.ca, Univ of Victoria; As
Fürst, Christine, +345 5526017 christine.fuerst@geo.uni-halle.de, Martin-Luther-Univ Halle-Wittenberg; ZucZi

G

Gaasterland, Terry, (858) 822-4600 tgaasterland@ucsd.edu, Univ of California, San Diego; Ob
Gabbott, Sarah, +440116 252 3636 sg21@le.ac.uk, Leicester Univ; Pg
Gabel, Sharon, 281-425-6335 sgabel@lee.edu, Lee Coll; GseGg
Gabet, Emmanuel, 408 924-5035 manny.gabet@sjsu.edu, San Jose State Univ; Gm

Gabitov, Rinat, (662) 268-1032 Ext 218 rg850@msstate.edu, Mississippi State Univ; Cas
Gable, Carl W., (505) 665-3533 gable@lanl.gov, Los Alamos; Gt
Gabler, Christopher, 956 882 7656 christopher.gabler@utrgv.edu, Univ of Texas, Rio Grande Valley; PeSf
Gaboury, Damien, 418-545-5011 dgaboury@uqac.ca, Univ du Quebec a Chicoutimi; EmCe
Gabriel, Alice, algabriel@ucsd.edu, Univ of California, San Diego; Ys
Gabrielli, Paolo, 614 292 6664 gabrielli.1@osu.edu, Ohio State Univ; CtPeGl
Gachon, Philippe, 514-987-3000 #2601 gachon.philippe@uqam.ca, Univ du Quebec a Montreal; As
Gacía Joral, Fernando, fgjoral@ucm.es, Univ Compl de Madrid; Pi
Gadikota, Greeshma, gadikota@wisc.edu, Univ of Wisconsin, Madison; GzCqu
Gaetani, Glenn A., (508) 289-3724 ggaetani@whoi.edu, Woods Hole Oceanographic Inst; Gi
Gaffney, Edward S., (505) 665-6387 gaffney@lanl.gov, Los Alamos; Yg
Gaffney, Eugene S., (212) 769-5801 American Mus of Natural History; Pv
Gaffney, Jeffrey S., (630) 252-5178 gaffney@anl.gov, Argonne; As
Gagnaison, Cyril, +33(0)3 44068997 cyril.gagnaison@lasalle-beauvais.fr, Institut Polytechnique LaSalle Beauvais (ex-IGAL); PgGsa
Gagne, Jean-Pierre, 418-723-1986 ext 1870 jean-pierre_gagne@uqar.ca, Univ du Quebec a Rimouski; OcCmo
Gagne, Marc R., (610) 436-3014 mgagne@wcupa.edu, West Chester Univ; Xa
Gagnon, Alan R., (508) 289-2961 agagnon@whoi.edu, Woods Hole Oceanographic Inst; Cs
Gagnon, Joel E., 519-253-3000 x2496 jgagnon@uwindsor.ca, Univ of Windsor; CaqEg
Gagnon, Katie, 206-516-3161 kgagnon@sccd.ctc.edu, Seattle Central Coll; OuZg
Gagnon, Teresa K., (860) 424-3680 teresa.gagnon@ct.gov, Dept of Energy and Env Protection; Gg
Gagosian, Carol, 781-283-3151 cgagosian@wellesley.edu, Wellesley Coll; Zn
Gahn, Forest J., (208) 496-7677 gahnf@byui.edu, Brigham Young Univ - Idaho; PggGs
Gaidos, Eric J., (808) 956-7897 gaidos@hawaii.edu, Univ of Hawai'i, Manoa; Yg
Gaines, Robert R., (909) 621-8674 robert.gaines@pomona.edu, Pomona Coll; GsClPi
Gajewski, Dirk J., +4940428382975 dirk.gajewski@uni-hamburg.de, Univ Hamburg; YesYg
Gakka, Udaya Laxm, +914027097116 udayalaxmi.g@gmail.com, Osmania Univ; HwYge
Galal, Galal M., 002-03-3921595 galalgalal2004@yahoo.com, Alexandria Univ; GgPeGe
Galán garcía, Gúmer, ++935812835 gumer.galan@uab.cat, Univ Auto de Barcelona; GiCg
Galbraith, Eric, (514) 398-3677 eric.galbraith@mcgill.ca, McGill Univ; Zo
Galbraith, John M., (540) 231-9784 ttcf@vt.edu, Virginia Tech; Sp
Gale, Andrew, +44 023 92 846127 andy.gale@port.ac.uk, Univ of Portsmouth; PgGrCl
Gale, Julia F., (512) 232-7957 julia.gale@beg.utexas.edu, Univ of Texas, Austin; GctGo
Gale, Marjorie H., (802) 522-5210 marjorie.gale@vermont.gov, Agency of Nat Res, Dept of Env Conservation; GgcGe
Gale, Paula M., (731) 881-7326 pgale@utm.edu, Univ of Tennessee, Martin; SodSc
Galea, Anthony, anthony.galea@um.edu.mt, Univ of Malta; Opg
Galea, Pauline, +356 2340 3034 pauline.galea@um.edu.mt, Univ of Malta; YsgZg
Galewsky, Joseph, 505-277-4204 galewsky@unm.edu, Univ of New Mexico; Am
Galgaro, Antonio, +390498279123 antonio.galgaro@unipd.it, Univ degli Studi di Padova; Ng
Galicki, Stan, (601) 974-1340 galics@millsaps.edu, Millsaps Coll; Ged
Gallagher, Eugene, eugene.gallagher@umb.edu, Univ of Massachusetts, Boston; Ob
Gallagher, Martin, +44 0161 306-3937 martin.gallagher@manchester.ac.uk, Univ of Manchester; As
Gallagher, William B., 609-896-5000 ext. 7784 wgallagher@rider.edu, Rider Univ; PvgGr
Gallegos, Anthony F., (505) 665-0862 agallegos@lanl.gov, Los Alamos; Pg
Gallen, Sean, sean.gallen@colostate.edu, Colorado State Univ; GemGt
Galli, Kenneth G., (617) 552-4504 kenneth.galli@bc.edu, Boston Coll; GsdGg
Gallovic, Frantisek, 089/2180 4209 frantisek.gallovic@geophysik.uni-muenchen.de, Ludwig-Maximilians-Univ Muenchen; Yg
Galloway, James N., (804) 924-1303 jng@virginia.edu, Univ of Virginia; As
Galloway, William E., (512) 471-5673 galloway@austin.utexas.edu, Univ of Texas, Austin; GsoGr
Gallup, Christina D., (218) 726-8984 cgallup@d.umn.edu, Univ of Minnesota Duluth; Cct
Gallus, William A., (515) 294-2270 wgallus@iastate.edu, Iowa State Univ of Sci & Tech; As
Galperin, Boris, (727) 553-1101 bgalperin@marine.usf.edu, Univ of South Florida; Op
Galsa, Attila, gali@pangea.elte.hu, Eotvos Lorand Univ; YgHwYe
Galvin, Robert, r.galvin@curtin.edu.au, Curtin Univ; Ye
Galy, Valier, (508) 289-2340 vgaly@whoi.edu, Woods Hole Oceanographic Inst; Cm
Gamage, Kusali R., kgamage@austincc.edu, Austin Comm Coll District; HyOuPm
Gamble, Audrey V., (334) 844-4100 avg0001@auburn.edu, Auburn Univ; Sc
Gamble, Douglas W., (910) 962-3778 gambled@uncw.edu, Univ of North Carolina, Wilmington; Zy
Gamble, John, +353 21 4903955 j.gamble@ucc.ie, Univ Coll Cork; Gi
Gambrell, Robert P., (225) 388-6426 Louisiana State Univ; Cg
Gamerdinger, Amy P., (509) 373-3077 amy.gamerdinger@pnl.gov, Pacific Northwest; Cl
Gammack-Clark, James, 561 297-0314 jgammack@fau.edu, Florida Atlantic Univ; Zri
Gammon, Richard H., (206) 543-1609 gammon@macmail.chem.washington.edu, Univ of Washington; Oc
Gammons, Christopher H., (406) 496-4763 cgammons@mtech.edu, Montana Tech of the Univ of Montana; CgEmCs
Gamon, John, 780-492-0345 gamon@ualberta.ca, Univ of Alberta; Zr
Gamperl, Moritz, +49 89 289 25885 moritz.gamperl@tum.de, Technical Univ of Munich; NgrYu
Gancarz, Alexander J., (505) 667-2606 Los Alamos; Cc
Gandolfo-Nixon, Maria A., 607-255-3273 mag4@cornell.edu, Cornell Univ; Pb
Ganeshram, Raja, +44 (0) 131 650 7364 r.ganeshram@ed.ac.uk, Edinburgh Univ; Pg
Ganey-Curry, Patricia E., (512) 471-0408 patty@ig.utexas.edu, Univ of Texas, Austin; Zn
Ganguly, Jiba, 520-621-6006 ganguly@email.arizona.edu, Univ of Arizona; Gx
Gani, M. Royhan, (270) 745-5977 royhan.gani@wku.edu, Western Kentucky Univ; GsrGg
Gani, Nahid, nahid.gani@wku.edu, Western Kentucky Univ; Gc
Gann, Delbert E., 601-974-1341 gannde@millsaps.edu, Millsaps Coll; Gz
Gannaway Dalton, Evelyn, evey.gannawaydalton@usu.edu, Utah State Univ; GstZe
Gannon, J. Michael, mike.gannon@dnr.iowa.gov, Iowa Dept of Nat Res; Gg
Gans, Phillip B., (805) 893-2642 gans@geol.ucsb.edu, Univ of California, Santa Barbara; Gt
Gansecki, Cheryl A., (808) 932-7549 gansecki@hawaii.edu, Univ of Hawai'i, Hilo; GviZe
Ganssen, Gerald M., +31 20 59 87369 g.m.ganssen@vu.nl, Vrije Univit Amsterdam; ObPe
Gantzer, Clark J., 573-882-0611 Univ of Missouri; So
Gao, Dengliang, 304-293-3310 dengliang.gao@mail.wvu.edu, West Virginia Univ; Ye
Gao, Haiying, 413 577-1250 haiyinggao@geo.umass.edu, Univ of Massachusetts, Amherst; Ys
Gao, Mengsheng, 352-392-1951 msgao@ufl.edu, Univ of Florida; Zn
Gao, Oliver H., hg55@cornell.edu, Cornell Univ; Zn
Gao, Peter, pgao@carnegiescience.edu, Carnegie Inst for Sci; Xg
Gao, Stephen , 573-341-6676 sgao@mst.edu, Missouri Univ of Sci and Tech; Yg
Gao, Yongjun, 713-743-4382 yongjungao@uh.edu, Univ of Houston; CstCa
Gao, Yuan, 973-353-1139 yuangaoh@newark.rutgers.edu, Rutgers, The State Univ of New Jersey, Newark; As
Garb, Matt, mgarb@brooklyn.cuny.edu, Brooklyn Coll (CUNY); Pg
Garbesi, Karina, 510-885-3172 karina.garbesi@csueastbay.edu, California State Univ, East Bay; Cg
Garces, Milton A., (808) 325-1558 milton@higp.hawaii.edu, Univ of Hawai'i, Manoa; Ys
Garcia, Antonio F., (805) 756-2430 afgarcia@calpoly.edu, California Polytechnic State Univ; Gmg
Garcia, Jan, (323) 259-2823 jangarcia@oxy.edu, Occidental Coll; Zn
Garcia, Juan, jgarcia@cicese.mx, Centro de Investigación Científica y de Educación Superior de Ensenada; Yv
Garcia, Marcelo H., (217) 356-9482 mhgarcia@illinois.edu, Univ of Illinois, Urbana-Champaign; Gmu
Garcia, Michael O., (808) 956-6641 garcia@soest.hawaii.edu, Univ of

Hawai'i, Manoa; Gi
García, Nuria, nugarcia@ucm.es, Univ Compl de Madrid; Pv
Garcia, Oswaldo, (415) 338-2061 ogarcia@sfsu.edu, San Francisco State Univ; As
Garcia, Sammy R., (505) 667-4151 Los Alamos; Ca
Garcia, William, 704-687-5982 wjgarcia@uncc.edu, Univ of North Carolina, Charlotte; Pv
García Frank, Alejandra, agfrank@ucm.es, Univ Compl de Madrid; Gg
Garcia Lasanta, Cristina, (360) 650-3835 cristina.garcialasanta@wwu.edu, Western Washington Univ; YmGtc
García Lorenzo, María de la Luz, mglorenzo@ucm.es, Univ Compl de Madrid; Cl
García Romero, Emilia, mromero@ucm.es, Univ Compl de Madrid; Gz
García-Martínez, David, david.garcia@ues.mx, Univ Estatal de Sonora; GzEmNx
Garcia-Rubio, Luis H., (727) 553-1246 garcia@marine.usf.edu, Univ of South Florida; Oc
Gardner, Eleanor E., 731-881-7444 egardne3@utm.edu, Univ of Tennessee, Martin; GgPg
Gardner, James E., (512) 471-0953 gardner@mail.utexas.edu, Univ of Texas, Austin; Gv
Gardner, Jamie N., (505) 667-1799 jgardner@lanl.gov, Los Alamos; Gg
Gardner, Michael H., 775-327-5060 mhgardner@unr.edu, Univ of Nevada, Reno; NxRn
Gardner, Payton, (406) 243-2458 payton.gardner@umontana.edu, Univ of Montana; Hw
Gardner, Thomas W., tgardner@trinity.edu, Trinity Univ; GmtHg
Gardner, Wayne S., (361) 749-6823 wayne.gardner@utexas.edu, Univ of Texas, Austin; Ob
Gardner, Wilford D., (979) 845-7211 wgardner@ocean.tamu.edu, Texas A&M Univ; OugZr
Garfield, Newell (Toby), (415) 338-3713 garfield@sfsu.edu, San Francisco State Univ; Op
Garihan, John M., (864) 294-2052 jack.garihan@furman.edu, Furman Univ; Gct
Garing, Charlotte , charlotte.garing@uga.edu, Univ of Georgia; Hw
Garnero, Edward, (480) 965-7653 garnero@asu.edu, Arizona State Univ; Ys
Garnier Villareal, Maximiliano, maximiliano.garniervillarreal@ucr.ac.cr, Univ de Costa Rica; NxYgGg
Garren, Sandra J., 516-463-5564 sandra.j.garren@hofstra.edu, Hofstra Univ; Zi
Garrett, Christopher J. R., (250) 721-7702 garrett@uvphys.phys.uvic.ca, Univ of Victoria; Op
Garrett, Maureen, maureen.garrett@azwestern.edu, Arizona Western Coll; GgeOg
Garrett, Timothy J., 801-581-5768 tim.garrett@utah.edu, Univ of Utah; As
Garrick-Bethell, Ian, 831-459-1277 igarrick@ucsc.edu, Univ of California, Santa Cruz; Xy
Garrison, Ervan G., (706) 542-1097 egarriso@uga.edu, Univ of Georgia; GamGn
Garrison, Jennifer M., (323) 343-2412 jgarris@calstatela.edu, California State Univ, Los Angeles; GiCcGv
Garrison, Trent, (859) 572-6683 garrisont@nku.edu, Northern Kentucky Univ; GeHwEc
Garrote, Julio, juliog@ucm.es, Univ Compl de Madrid; Gm
Garside, Larry J., 775-784-6693 lgarside@unr.edu, Univ of Nevada, Reno; Gg
Garstang, Mimi R., 573-368-2101 mimi.garstang@dnr.mo.gov, Missouri Dept of Nat Res; Ge
Gartner, Janette, (781) 891-2901 jgartner@bentley.edu, Bentley Univ; Hy
Gartner, John, jgartner@norwich.edu, Norwich Univ; GmHy
Garver, John I., (518) 388-6770 garverj@union.edu, Union Coll; GtCcGr
Garvie, Laurence, (480) 965-3361 lgarvie@asu.edu, Arizona State Univ; GzXc
Garvin, Theresa D., (780) 492-4593 theresa.garvin@ualberta.ca, Univ of Alberta; Zn
Garwood, Phil, 910 392 7111 pgarwood@cfcc.edu, Cape Fear Comm Coll; Gg
Garwood, Roland W., garwood@nps.edu, Naval Postgrad Sch; Op
Gary, Marcus, doc.mog@gmail.com, Univ of Texas, Austin; Hgs
Garza, Darlene, (213) 740-6109 garzadar@usc.edu, Univ of Southern California; Zn
Garza, Jessica, 325-486-6987 jessica.garza@angelo.edu, Angelo State Univ; AmGg
Gaschnig, Richard M., (978) 934-3706 richard_gaschnig@uml.edu, Univ of Massachusetts, Lowell; CctCa
Gascho, Gary J., (912) 386-3329 Univ of Georgia; Sc
Gasparini, Nicole, 504-862-3197 ngaspari@tulane.edu, Tulane Univ; GmHs
Gaston, Lewis A., (504) 388-1323 lagaston@agcenter.lsu.edu, Louisiana State Univ; Sc
Gates, Alexander E., 973-353-5034 agates@rutgers.edu, Rutgers, The State Univ of New Jersey, Newark; Gc
Gates, Ruth D., (808) 236-7420 rgates@hawaii.edu, Univ of Hawai'i, Manoa; Ob
Gathany, Mark, 937-766-3823 mgathany@cedarville.edu, Cedarville Univ; ZiuSb
Gathorne-Hardy, Alfy, a.gathorne-hardy@ed.ac.uk, Edinburgh Univ; Zc
Gatrell, Jay, (217) 581-2121 dgatrell@eiu.edu, Eastern Illinois Univ; Zi
Gattiglio, Marco, marco.gattiglio@unito.it, Univ di Torino; Gc
Gatto, Roberto, +390498279172 roberto.gatto@unipd.it, Univ degli Studi di Padova; Pg
Gaubatz, Piper, (413) 545-0768 gaubatz@geo.umass.edu, Univ of Massachusetts, Amherst; Zy
Gauci, Adam, adam.gauci@um.edu.mt, Univ of Malta; Op
Gaudette, Henri E., (603) 862-1718 Univ of New Hampshire; Cc
Gaudio, Sarah J., (585) 245-5016 gaudio@geneseo.edu, SUNY, Geneseo; GzgGg
Gaulton, Rachel, +44 (0) 191 208 6577 rachel.gaulton@ncl.ac.uk, Univ of Newcastle Upon Tyne; Zr
Gautam, Tej P., 740-376-4371 tej.gautam@marietta.edu, Marietta Coll; NgGeZi
Gauthier, Donald J., (818) 778-5514 gauthidj@lavc.edu, Los Angeles Valley Coll; ZiyAm
Gauthier, Jacques, (203) 432-3150 jacques.gauthier@yale.edu, Yale Univ; Pv
Gauthier, Michel, 514-987-3000 #4560 gauthier.michel@uqam.ca, Univ du Quebec a Montreal; Eg
Gauthier, Pierre, 514-987-3000 #3304 gauthier.pierre@uqam.ca, Univ du Quebec a Montreal; Am
Gautsch, Jacklyn, 319-335-1761 jackie.gautsch@dnr.iowa.gov, Iowa Dept of Nat Res; Zn
Gavin, Dan, (541) 346-5787 dgavin@uoregon.edu, Univ of Oregon; ZyPle
Gavriloaiei, Traian, 0040232201462 tgavrilo@uaic.ro, Alexandru Ioan Cuza; CaAsCg
Gawloski, Joan, 432-685-4630 jgawloski@midland.edu, Midland Coll; GgzGe
Gawu, Simon K., +233 244 067804 skygawu@yahoo.com, Kwame Nkrumah Univ of Sci and Tech; GgEgZn
Gay, Kenny, 9197337353 x28 kenny.gay@ncdenr.gov, N.C. Dept of Env Quality; GzsPi
Gaya, Charles , cogaya@jkuat.ac.ke, Jomo Kenyatta Univ of Agriculture & Tech; ZrgZi
Gaylord, David R., (509) 335-8127 gaylordd@wsu.edu, Washington State Univ; Gs
Gaynor, Sean, 609-258-4101 sean.gaynor@princeton.edu, Princeton Univ; GiEmCc
Gazel, Esteban, egazel@cornell.edu, Cornell Univ; CeGiv
Gazis, Carey A., 509.963-2820 cgazis@geology.cwu.edu, Central Washington Univ; CsHw
Ge, Shemin, (303) 492-8323 ges@colorado.edu, Univ of Colorado; HwEoYg
Geary, Dana H., dana@geology.wisc.edu, Univ of Wisconsin, Madison; Pg
Geary, Lindsey, (315)792-3134 Utica Coll; Geg
Geary, Phil, 61 02 4921 6726 phil.geary@newcastle.edu.au, Univ of Newcastle; HwSoHw
Geballe, Zachary, zgeballe@carnegiescience.edu, Carnegie Inst for Sci; Gy
Gebrande, Helmut, 089/2180 4325 gebrande@geophysik.uni-muenchen.de, Ludwig-Maximilians-Univ Muenchen; YgsEo
Gedzelman, Stanley, (212) 650-6470 City Coll (CUNY); As
Gee, Jeffrey S., (858) 534-4707 jsgee@ucsd.edu, Univ of California, San Diego; Gu
Gehrels, George E., (520) 349-4702 ggehrels@email.arizona.edu, Univ of Arizona; Gt
Gehrels, Tom, (520) 621-6970 tgehrels@lpl.arizona.edu, Univ of Arizona; Zn
Geidel, Gwendelyn, (803) 777-7171 geidel@environ.sc.edu, Univ of South Carolina; HwGeCg
Geiger, James W., 217-265-8989 jgeiger@illinois.edu, Illinois State Geo Survey; Ge
Geiger, Sebastian, s.geiger@hw.ac.uk, Heriot-Watt Univ; NpEoYh
Geisseler, Daniel, 530-754-9637 djgeisseler@ucdavis.edu, Univ of California, Davis; Sb
Geissman, John D., 972-883-2403 geissman@utdallas.edu, Univ of Texas, Dallas; GtYm
Geist, Dennis J., (208) 885-6491 dgeist@uidaho.edu, Univ of Idaho; Giv
Geller, Jonathan, 831-771-4400 geller@mlml.calstate.edu, Moss Landing Marine Lab; Zn
Geller, Marvin A., (631) 632-8701 marvin.geller@stonybrook.edu, SUNY, Stony Brook; As
Gemail, Khaled S., 01145031722 khaledgemail@zu.edu.eg, El Zagazig

Univ; YuRcYe
Gemmell, J B., 61 3 6226 2893 bruce.gemmell@utas.edu.au, Univ of Tasmania; Eg
Gemperline, Johanna M., (410) 554-5552 johanna.gemperline@maryland.gov, Maryland Dept of Nat Res; Hw
Genareau, Kimberly, 205-348-1878 kdg@ua.edu, Univ of Alabama; Gvi
Gendzwill, Donald J., don.gendzwill@usask.ca, Univ of Saskatchewan; Ye
Genge, Matthew, +44 20 759 46499 m.genge@imperial.ac.uk, Imperial Coll; Xm
Genna, Dominique, gennadomi@hotmail.com, Univ du Quebec a Chicoutimi; GvEn
Gennari, Rocco, rocco.gennari@unito.it, Univ di Torino; Gn
Gentile, Richard J., (816) 235-2974 gentiler@umkc.edu, Univ of Missouri, Kansas City; GrsNx
Gentry, Amanda L., (303) 319-0695 amandagentry@weber.edu, Weber State Univ; GgsGg
Gentry, Christopher, 931-221-7478 gentryc@apsu.edu, Austin Peay State Univ; Ziy
Gentry, Terry, 979-845-5323 tjgentry@tamu.edu, Texas A&M Univ; Sb
George, Graham, 306-966-5722 g.george@usask.ca, Univ of Saskatchewan; Cg
Georgen, Jennifer, 850-645-4987 georgen@gly.fsu.edu, Florida State Univ; Gu
Georgiopopoulou, Aggeliki, (+353) 1 716 2062 aggie.georg@ucd.ie, Univ Coll Dublin; GusOg
Geraghty Ward, Emily, emily.ward@rocky.edu, Rocky Mountain Coll; ZeGc
Gerald, Carresse, (919) 530-7117 cgerald6@nccu.edu, North Carolina Central Univ; GeZg
Gerba, Charles P., (520) 621-6906 gerba@ag.arizona.edu, Univ of Arizona; Sb
GERBE, Marie-Christine, 33-477485123 gerbe@univ-st-etienne.fr, Université Jean Monnet, Saint-Etienne; GiCsg
Gerber, Stefan, 352294-3174 sgerber@ufl.edu, Univ of Florida; So
Gerbi, Christopher C., 207 581-2153 Univ of Maine; Gt
Gerbi, Greg, 518-580-5127 ggerbi@skidmore.edu, Skidmore Coll; OpYg
Gerhard, Lee C., 78538643965 leegtn37@gmail.com, Univ of Kansas; GsoGd
Gerhardt, Hannes, (678) 839-4064 hgerhard@westga.edu, Univ of West Georgia; Zn
Gerke, Tammie, 513-727-3268 gerketl@miamioh.edu, Miami Univ; ClGea
Gerla, Philip J., (701) 777-3305 phil.gerla@engr.und.edu, Univ of North Dakota; Hw
Gerlach, Dana, 508-289-3438 dgerlach@whoi.edu, Woods Hole Oceanographic Inst; Og
Germa, Aurelie, agerma@usf.edu, Univ of South Florida; Gv
German, Chris, 508-289-2853 cgerman@whoi.edu, Woods Hole Oceanographic Inst; Gt
Germanoski, Dru, (610) 330-5196 germanod@lafayette.edu, Lafayette Coll; Gm
Germanovich, Leonid, (404) 894-2284 leonid@ce.gatech.edu, Georgia Inst of Tech; NrYg
Gernon, Thomas M., +44 (0)23 8059 2670 thomas.gernon@noc.soton.ac.uk, Univ of Southampton; ZgGvi
Gerry, Janelle, 402-472-2663 Univ of Nebraska, Lincoln; Zn
Gertisser, Ralf, (+44) 01782 733181 r.gertisser@keele.ac.uk, Keele Univ; GivGz
Gervais, Melissa, mmg62@psu.edu, Pennsylvania State Univ, Univ Park; As
Gerwick, William H., (858) 534-0578 wgerwick@ucsd.edu, Univ of California, San Diego; Cm
Geršlová, Eva, +420 549 49 4027 gerslova@sci.muni.cz, Masaryk Univ; Ge
Gess, Robert, +27 (0)82-7595848 robg@imaginet.co.za, Rhodes Univ; Pg
Ghail, Richard, +44 1784 276766 richard.ghail@rhul.ac.uk, Royal Holloway Univ of London; Gt
Ghani, Azman Abdul, 03-79674234 azmangeo@um.edu.my, Univ of Malaya; Gi
Ghassemi, Ahmed, ahmad.ghassemi@ou.edu, Univ of Oklahoma; Nr
Ghate, Virendra P., (630) 252-1609 vghate@anl.gov, Argonne; AsmZr
Ghattas, Omar, (512) 232-4304 omar@ices.utexas.edu, Univ of Texas, Austin; ZnYg
Ghazala, Hosni H., (109) 688-7904 ghazala@mans.edu.eg, El Mansoura Univ; Ygx
Ghent, Edward D., (403) 220-5847 ghent@geo.ucalgary.ca, Univ of Calgary; Gp
Ghent, Rebecca, (416) 978-0597 ghentr@geology.utoronto.ca, Univ of Toronto; Zr
Ghil, Michael, (310) 206-2285 ghil@atmos.ucla.edu, Univ of California, Los Angeles; As
Ghinassi, Massimiliano, +390498279181 massimiliano.ghinassi@unipd.it, Univ degli Studi di Padova; Gs
Ghiorso, Mark , 206-550-1850 ghiorso@uw.edu, Univ of Washington; CgGx
Ghoneim, Eman, ghoneime@uncw.edu, Boston Univ; Gm
Ghosh, Abhijit, (951) 827-4493 aghosh@ucr.edu, Univ of California, Riverside; YsGt
Ghosheh, Baher A., (814) 732-2207 ghosheh@edinboro.edu, Edinboro Univ of Pennsylvania; Zn
Giacalone, Joe, 520-626-8365 giacalone@lpl.arizona.edu, Univ of Arizona; Xy
Giachetti, Thomas, tgiachet@uoregon.edu, Univ of Oregon; Gvu
Giamello, Marco, marco.giamello@unisi.it, Univ of Siena; GxCgZm
Giammar, Daniel, degiammar@seas.wustl.edu, Washington Univ in St. Louis; Zn
Giannini, Alessandra, (845) 680-4473 alesall@iri.columbia.edu, Columbia Univ; AsOpZc
Gianniny, Gary, (970) 247-7254 gianniny_g@fortlewis.edu, Fort Lewis Coll; Gs
Gianotti, Franco, franco.gianotti@unito.it, Univ di Torino; Gg
Giaramita, Mario J., (209) 667-3558 mgiaramita@csustan.edu, California State Univ, Stanislaus; Gpz
Giardino, John R., (409) 867-9067 rickg@tamu.edu, Texas A&M Univ; Gm
Giardino, Marco, marco.giardino@unito.it, Univ di Torino; Gm
Gibbons, Doug, 206-685-8180 dgibbons@uw.edu, Univ of Washington; ZeYs
Gibbs, Gaynell, ocean@lsu.edu, Louisiana State Univ; Zn
Gibbs, Samantha J., +44 (0)23 80592003 samantha.gibbs@noc.soton.ac.uk, Univ of Southampton; Zg
Gibeaut, James, 361-825-2060 james.gibeaut@tamucc.edu, Texas A&M Univ, Corpus Christi; OnZi
Gibling, Martin R., mgibling@is.dal.ca, Dalhousie Univ; Gs
Gibson, Andy, +44 023 92 842654 andy.gibson@port.ac.uk, Univ of Portsmouth; Eg
Gibson, Carl H., (858) 534-3184 cgibson@ucsd.edu, Univ of California, San Diego; No
Gibson, Deana, (417) 836-5801 deanagibson@missouristate.edu, Missouri State Univ; Zn
Gibson, Deidre M., 757.727.5883 deidre.gibson@hamptonu.edu, Hampton Univ; Ob
Gibson, Glen, 719-333-3080 glen.gibson@usafa.edu, United States Air Force Acad; Zri
Gibson, H. Daniel (Dan), (778) 782-7057 hdgibson@sfu.ca, Simon Fraser Univ; GcCcGp
Gibson, Harold L., 7056751151 x2371 hgibson@laurentian.ca, Laurentian Univ, Sudbury; EmGv
Gibson, Michael A., (731) 881-7435 mgibson@utm.edu, Univ of Tennessee, Martin; PgiZe
Gibson, Roger L., 011 717 6553 roger.gibson@wits.ac.za, Univ of the Witwatersrand; GcpGg
Gibson, Ronald C., (714) 895-8194 rgibson@gwc.cccd.edu, Golden West Coll; Gc
Gibson, Sally, +44 (0) 1223 333401 sally@esc.cam.ac.uk, Univ of Cambridge; Gi
Giddings, Sarah, sgiddings@ucsd.edu, Univ of California, San Diego; Op
Gidigasu, Solomon S., +233 27 7807 707 ssrgidigasu.soe@knust.edu.gh, Kwame Nkrumah Univ of Sci and Tech; GeZrNr
Giefer, Madeline M., 931-221-7473 gieferm@apsu.edu, Austin Peay State Univ; ZcrZi
Giegengack, Jr., Robert F., (215) 898-5191 gieg@sas.upenn.edu, Univ of Pennsylvania; Gg
Gieré, Reto, giere@sas.upenn.edu, Univ of Pennsylvania; GzCgGe
Gierke, John S., (906) 487-2535 jsgierke@mtu.edu, Michigan Technological Univ; Hy
Giesche, Alena, 207-859-5868 agiesche@colby.edu, Colby Coll; ZcGa
Giese, Benjamin S., (979) 845-2306 bgiese@ocean.tamu.edu, Texas A&M Univ; Op
Giese, Graham S., (508) 289-2297 ggiese@whoi.edu, Woods Hole Oceanographic Inst; On
Gieskes, Joris M., (858) 534-4257 jgieskes@ucsd.edu, Univ of California, San Diego; Oc
Gifford, Dian J., (401) 874-6690 dgifford@uri.edu, Univ of Rhode Island; Ob
Gifford, Jennifer N., 662-915-2079 jngiffor@olemiss.edu, Univ of Mississippi; GtxCg
Gifford-Gonzalez, Dianne, (831) 459-2633 Univ of California, Santa Cruz; Zn
Gigler, Alexander, 089/2180 4185 Ludwig-Maximilians-Univ Muenchen; Gz
Giglierano, James D., james.giglierano@dnr.iowa.gov, Iowa Dept of Nat Res; GgZri
Gilbert, Co'Quesie, (619) 534-2470 cag65@columbia.edu, Columbia Univ; Zn
Gilbert, Hersh, 403 220-6446 hersh.gilbert@ucalgary.ca, Univ of Calgary; Ys

Gilbert, Kathleen W., 781-283-3086 kgilbert@wellesley.edu, Wellesley Coll; Cs
Gilbert, M. Charles, (405) 325-3253 mcgilbert@ou.edu, Univ of Oklahoma; Cp
Gilbert, Patricia M., (410) 221-8422 Univ of Maryland; Ob
Gilbes, Fernando, 787 2653845 fernando.gilbes@upr.edu, Univ of Puerto Rico; ZrGuZi
Gilder, Stuart, 089/2180 4239 stuart.gilder@geophysik.uni-muenchen.de, Ludwig-Maximilians-Univ Muenchen; Ym
Gilerson, Alexander, gilerson@ccny.cuny.edu, Graduate Sch of the City Univ of New York; Gu
Giles, Antony, (432) 685-5580 agiles@midland.edu, Midland Coll; GgvGi
Giles, David, +44 023 92 842248 david.giles@port.ac.uk, Univ of Portsmouth; NzGlm
Giles, Katherine A., 915-747-7075 kagiles@utep.edu, Univ of Texas, El Paso; Go
Gilfillan, Stuart M., +44 (0) 131 651 3462 stuart.gilfillan@ed.ac.uk, Edinburgh Univ; CgsCe
Gilg, Hans A., +49 89 289 25855 agilg@tum.de, Technical Univ of Munich; CsEgn
Gilhooly, William, wgilhool@iupui.edu, Indiana Univ - Purdue Univ Indianapolis; ClOc
Gili, Stefania, 609-258-4101 sgili@princeton.edu, Princeton Univ; CgGgCo
Gill, Benjamin, 540-231-7485 bcgill@vt.edu, Virginia Tech; CsPcGs
Gill, Fiona, +44(0) 113 34 35190 f.gill@leeds.ac.uk, Univ of Leeds; PgCa
Gill, Swarndeep S., (724) 938-1677 gill@calu.edu, California Univ of Pennsylvania; As
Gill, Thomas E., (915) 747-5168 tegill@utep.edu, Univ of Texas, El Paso; GmAsCl
Gillam, Mary L., gillam@rmi.net, Fort Lewis Coll; Gm
Gille, Peter, 089/2180 4355 peter.gille@lrz.uni-muenchen.de, Ludwig-Maximilians-Univ Muenchen; GzZm
Gille, Sarah T., (858) 822-4425 sgille@ucsd.edu, Univ of California, San Diego; Op
Gilleaudeau, Geoff, (703) 993-3289 ggilleau@gmu.edu, George Mason Univ; Gs
Gillerman, Virginia S., (208) 426-4002 vgillerm@boisestate.edu, Boise State Univ; Eg
Gillespie, Alan R., arg3@uw.edu, Univ of Washington; GmlZr
Gillespie, Janice, jgillespie@csub.edu, California State Univ, Bakersfield; GorZi
Gillespie, Robb, (269) 387-5364 robb.gillespie@wmich.edu, Western Michigan Univ; GsoGm
Gillespie, Terry J., (519) 824-4120 (Ext. 54276) tgillesp@uoguelph.ca, Univ of Guelph; Am
Gillespie, Thomas, (609) 771-2569 Coll of New Jersey; Gc
Gillette, David D., david.gillette@nau.edu, Northern Arizona Univ; Pv
Gilley, Brett H., bgilley@eoas.ubc.ca, Univ of British Columbia; GsZeRn
Gilliam, James W., (919) 515-2040 North Carolina State Univ; Sb
Gilligan, Jonathan M., (615) 322-2420 jonathan.gilligan@vanderbilt.edu, Vanderbilt Univ; ZcRwZu
Gillikin, David P., (518) 388-6679 gillikid@union.edu, Union Coll; CsmCl
Gillis, Kathryn, 250-721-6120 kgillis@uvic.ca, Univ of Victoria; Gp
Gillis, Robert, (907) 451-5024 robert.gillis@alaska.gov, Alaska Division of Geological & Geophysical Surveys; Eo
Gillman, Joe, (573) 368-2101 joe.gillman@dnr.mo.gov, Missouri Dept of Nat Res; Gg
Gillmore, Gavin, +44 020 8417 2518 g.gillmore@kingston.ac.uk, Kingston Univ; Gb
Gilmore, Martha S., (860) 685-3129 mgilmore@wesleyan.edu, Wesleyan Univ; XgGmZr
Gilmore, Troy E., (402) 470-1741 gilmore@unl.edu, Univ of Nebraska, Lincoln; Hwg
Gilmore, Tyler J., (509) 376-2370 tyler.gilmore@pnl.gov, Pacific Northwest; Hg
Gilmour, Ernest H., (509) 359-7480 egilmour@ewu.edu, Eastern Washington Univ; Pi
Gilotti, Jane A., (319) 335-1097 jane-gilotti@uiowa.edu, Univ of Iowa; GctGp
Gilpin, Bernard J., (714) 895-8233 bgilpin@gwc.cccd.edu, Golden West Coll; Ys
Gimmestad, Gary G., (404) 493-1331 gary.gimmestad@gmail.com, Georgia Inst of Tech; ZrAs
Ginder-Vogel, Matt, 608-262-0768 mgindervogel@wisc.edu, Univ of Wisconsin, Madison; Sc
Gingras, Murray, 780-492-1963 mgringras@ualberta.ca, Univ of Alberta; Go
Ginis, Isaac, (401) 874-6484 iginis@gso.uri.edu, Univ of Rhode Island; Op
Giordano, Alberto, (512) 245-2170 Texas State Univ; Zi
Giordano, Daniele, daniele.giordano@unito.it, Univ di Torino; Gv
Giorgis, Scott D., (585) 245-5293 giorgis@geneseo.edu, SUNY, Geneseo; GctYm
Giosan, Liviu, 508-289-2257 lgiosan@whoi.edu, Woods Hole Oceanographic Inst; GusPc
Giraldo, Mario A., (818) 677-4431 mario.giraldo@csun.edu, California State Univ, Northridge; ZriZg
Girard, Eric, 514-987-3000 #3325 girard.eric@uqam.ca, Univ du Quebec a Montreal; Am
Girard, Guillaume, 815-753-1778 ggirard@niu.edu, Northern Illinois Univ; GviCu
Girard, Michael W., (609) 292-2576 mike.girard@dep.state.nj.us, New Jersey Geological and Water Survey; Zi
Giraud, Richard E., (801) 537-3351 richardgiraud@utah.gov, Utah Geo Survey; Ng
Girhard, T S., 486-0045 tgirhard@alamo.edu, San Antonio Comm Coll; ZyAm
Giroux, Bernard, bernard.giroux@ete.inrs.ca, Univ du Quebec; Yeg
Girty, Gary H., (619) 594-2552 ggirty@mail.sdsu.edu, San Diego State Univ; Gc
Gishlick, Alan D., agishlic@bloomu.edu, Bloomsburg Univ; GgPgOg
Gittings, Bruce M., +44 (0) 131 650 2558 bruce@ed.ac.uk, Edinburgh Univ; Zi
Gittins, John, (416) 483-9345 j.gittins@utoronto.ca, Univ of Toronto; GipGz
Giusberti, Luca, +390498279183 Univ degli Studi di Padova; Pm
Giustetto, Roberto, roberto.giustetto@unito.it, Univ di Torino; Gz
Glade, Rachel, (585) 273-4059 rachel.glade@rochester.edu, Univ of Rochester; SpHt
Glamoclija, Mihaela, (973) 353-2509 m.glamoclija@rutgers.edu, Rutgers, The State Univ of New Jersey, Newark; PgXgGe
Glasauer, Susan, (519) 824-4120 xt52453 glasauer@uoguelph.ca, Univ of Guelph; Pg
Glaser, Paul H., 612-624-8395 glase001@umn.edu, Univ of Minnesota, Twin Cities; Gn
Glass, Alexander, (919) 684-6167 alex.glass@duke.edu, Duke Univ; PiGgZe
Glass, Billy P., (302) 449-2464 bglass@udel.edu, Univ of Delaware; XmcGz
Glass, Hylke J., +44 01326 371823 h.j.glass@exeter.ac.uk, Exeter Univ; NxmEm
Glazer, Brian, 808-956-6658 glazer@hawaii.edu, Univ of Hawai'i, Manoa; Cm
Glazner, Allen F., (919) 962-0689 afg@unc.edu, Univ of North Carolina, Chapel Hill; GitZf
Gleason, Gayle C., (607) 753-2816 gleasong@cortland.edu, SUNY, Cortland; Gct
Gleason, James D., 734-764-9523 jdgleaso@umich.edu, Univ of Michigan; Cg
Glenn, Craig R., (808) 956-2200 glenn@soest.hawaii.edu, Univ of Hawai'i, Manoa; GeCmGs
Glenn, Ed, (520) 626-2664 eglenn@ag.arizona.edu, Univ of Arizona; Zn
Glenn, Nancy, (208) 426-2933 nancyglenn@boisestate.edu, Boise State Univ; ZrNgZi
Glesener, Gary, 540-231-6521 gbgmedl@vt.edu, Virginia Tech; Ze
Glickson, Deborah, (202) 334-2024 dglickson@nas.edu, Natl Academies of Scis, Eng, and Medicine; GutGg
Gloaguen, Erwan, erwan.gloaguen@ete.inrs.ca, Univ du Quebec; Ye
Glotch, Timothy, (631) 632-1168 timothy.glotch@stonybrook.edu, Stony Brook Univ; XgGz
Glover, David M., (508) 289-2656 dglover@whoi.edu, Woods Hole Oceanographic Inst; Oc
Glover, Paul W., 418-656-5180 paul.glover@ggl.ulaval.ca, Univ Laval; Yx
Glowacka, Ewa, glowacka@cicese.mx, Centro de Investigación Científica y de Educación Superior de Ensenada; Ys
Glubokovskikh, Stanislav, +618 9266-7190 stanislav.glubokovskikh@curtin.edu.au, Curtin Univ; Yxe
Gluhovsky, Alexander, aglu@purdue.edu, Purdue Univ; As
Glumac, Bosiljka, (413) 585-3680 bglumac@smith.edu, Smith Coll; Gs
Gluyas, Jon, +44 (0) 191 33 42302 j.g.gluyas@durham.ac.uk, Durham Univ; Eo
Glynn, William G., wglynn@brockport.edu, SUNY, The Coll at Brockport; Gg
Gnanadesikan, Anand, (410) 516-0722 gnanades@jhu.edu, Johns Hopkins Univ; OpcAs
Gnidovec, Dale M., (614) 292-6896 gnidovec.1@osu.edu, Ohio State Univ; PvGr
Gobler, Christopher, (631) 632-5043 christopher.gobler@stonybrook.edu, SUNY, Stony Brook; Ob
Goda, Katsu, kgoda2@uwo.ca, Western Univ; Rn
Godbold, Jasmin A., +44 (0)23 80593639 j.a.goldbold@soton.ac.uk, Univ of Southampton; Ob
Godchaux, Martha M., (208) 882-9062 Mount Holyoke Coll; Gv

Goddard, Lisa M., (845) 680-4430 goddard@iri.columbia.edu, Columbia Univ; AstZc
Godek, Melissa, (607) 436-3375 melissa.godek@oneonta.edu, SUNY, Oneonta; AmsZg
Godfrey, Brian J., 845-437-5544 godfrey@vassar.edu, Vassar Coll; Zu
Godfrey, Christopher, 828-232-5160 cgodfrey@unca.edu, Univ of North Carolina, Asheville; AsZnn
Godfrey, Linda, godfrey@marine.rutgers.edu, Rutgers, The State Univ of New Jersey; CmlCq
Godin, Laurent, (613) 533-3223 godinl@queensu.ca, Queen's Univ; Gct
Godsey, Holly, (801) 587-7865 holly.godsey@utah.edu, Univ of Utah; Ze
Godsey, Sarah E., 208-282-3170 godsey@isu.edu, Idaho State Univ; HgsHq
Goebel, Thomas H., (901) 678-4885 thgoebel@memphis.edu, Univ of Memphis; Yxs
Goehring, Brent M., (504) 862-3196 bgoehrin@tulane.edu, Tulane Univ; CcGlm
Goeke, James W., (308) 530-4437 jgoeke@unl.edu, Univ of Nebraska, Lincoln; HwGmYe
Goes, Joaquim, jig@ldeo.columbia.edu, Columbia Univ; Gu
Goes, Saskia, +44 20 759 46434 s.goes@imperial.ac.uk, Imperial Coll; Yg
Goetz, Alexander, 303-492-5086 goetz@cses.colorado.edu, Univ of Colorado; Zr
Goetz, Andrew R., (303) 871-2674 agoetz@du.edu, Univ of Denver; Zgu
Goetz, Heinrich, (972) 377-1079 hgoetz@collin.edu, Collin Coll - Preston Ridge Campus; Ge
Goetze, Erica, (808) 956-7156 egoetze@hawaii.edu, Univ of Hawai'i, Manoa; Ob
Goff, Fraser, (505) 667-8060 fraser@lanl.gov, Los Alamos; Cg
Goff, John A., (512) 471-0476 goff@ig.utexas.edu, Univ of Texas, Austin; Yr
Goforth, Thomas T., (254) 710-2183 tom_goforth@baylor.edu, Baylor Univ; Yg
Golabi, Mohammad H., 671-735-2143 mgolabi@triton.uog.edu, Univ of Guam; SzpSc
Gold, David, 530 754-5361 dgold@ucdavis.edu, Univ of California, Davis; Pg
Gold, David (Duff) P., (814) 865-9993 gold@ems.psu.edu, Pennsylvania State Univ, Univ Park; GcgGx
Gold-Bouchot, Gerardo, (979) 845-9826 ggold@tamu.edu, Texas A&M Univ; OcCm
Goldberg, David S., goldberg@ldeo.columbia.edu, Columbia Univ; Yr
Goldblatt, Colin, 250-721-6120 czg@uvic.ca, Univ of Victoria; As
Goldfinger, Chris, (541) 737-9622 gold@coas.oregonstate.edu, Oregon State Univ; Yr
Goldman, Daniel, (937) 229-5637 dgoldman1@udayton.edu, Univ of Dayton; PgqGr
Goldreich, Peter M., (626) 395-6193 pmg@gps.caltech.edu,CalTech; Xy
Goldsby, David, 215-746-0090 dgoldsby@sas.upenn.edu, Univ of Pennsylvania; Yg
Goldstein, Barry, (253) 879-3822 goldstein@pugetsound.edu, Univ of Puget Sound; GlmGs
Goldstein, Robert H., (785) 864-2738 gold@ku.edu, Univ of Kansas; Gs
Goldstein, Steven, steveg@ldeo.columbia.edu, Columbia Univ; Cg
Goldstein, Susan T., (706) 542-2397 sgoldst@uga.edu, Univ of Georgia; Pm
Gollmer, Steven, 937-766-7764 gollmers@cedarville.edu, Cedarville Univ; GzAmOg
Golombek, Matthew P., (818) 393-7948 mgolombek@jpl.nasa.gov, Jet Propulsion Lab; XgyGt
Gomberg, Joan, 206-616-5581 gomberg@usgs.gov, Univ of Washington; YsGt
Gomes, Helga, helga@ldeo.columbia.edu, Columbia Univ; Ob
Gomes, Maria E., 351259350261 mgomes@utad.pt, Unive de Trás-os-Montes e Alto Douro; GiCgGz
Gomes, Nuno N., +244924987900 ngomes999@gmail.com, Unive Independente de Angola; GesGu
Gomez, Enrique, egomez@cicese.mx, Centro de Investigación Científica y de Educación Superior de Ensenada; Ye
Gomez, Francisco, (573) 882-9744 fgomez@missouri.edu, Univ of Missouri; Gt
Gomez, Natalya, 514-398-4885 natalya.gomez@mcgill.ca, McGill Univ; Og
Gómez Gras, David, ++935813093 david.gomez@uab.cat, Univ Auto de Barcelona; Gd
Gomezdelcampo, Enrique, 419 372 2886 egomezd@bgsu.edu, Bowling Green State Univ; Ge
Goncharov, Alexander F., 202-478-8947 agoncharov@carnegiescience.edu, Carnegie Inst for Sci; GyYxh
Gong, Donglai, 804-684-7529 gong@vims.edu, William & Mary; Op
Gong, Hongmian, (212) 772-4658 gong@hunter.cuny.edu, Hunter Coll (CUNY); Zn
Goni, Miguel A., (541) 737-0578 mgoni@coas.oregonstate.edu, Oregon State Univ; Gu
Gonnermann, Helge, (713) 348-6263 helge.m.gonnermann@rice.edu, Rice Univ; Gv
Gonzales, David A., (970) 247-7378 gonzales_d@fortlewis.edu, Fort Lewis Coll; Gx
Gonzalez, Alex, (515) 294-8729 agon@iastate.edu, Iowa State Univ of Sci & Tech; As
Gonzalez, Frank I., 206-290-0903 figonzal@uw.edu, Univ of Washington; OpRn
González, Isabel , +34954556317 igonza@us.es, Univ de Sevilla; GzeSc
Gonzalez, Jose J., javier@cicese.mx, Centro de Investigación Científica y de Educación Superior de Ensenada; Yd
Gonzalez, Juan L., (956) 665-3523 juan.l.gonzalez@utrgv.edu, Univ of Texas, Rio Grande Valley; GmsGa
González, Laura, lgacebron@ucm.es, Univ Compl de Madrid; Gr
Gonzalez, Luis A., 785-864-2743 lgonzlez@ku.edu, Univ of Kansas; CsGd
Gonzalez, Maria M., 989-774-2295 gonza1mm@cmich.edu, Central Michigan Univ; Gig
Gonzalez Leon, Carlos M., cmgleon@servidor.unam.mx, Univ Nacional Auto de Mexico; Gr
Gonzalez-Escobar, Mario, mgonzale@cicese.mx, Centro de Investigación Científica y de Educación Superior de Ensenada; YesYs
Gonzalez-Fernandez, Antonio, mindundi@cicese.mx, Centro de Investigación Científica y de Educación Superior de Ensenada; Gu
Gonzalez-Perdomo, Maria, maria.gonzalezperdomo@adelaide.edu.au, Univ of Adelaide; Np
Good, Daniel B., (912) 478-5361 dangood@georgiasouthern.edu, Georgia Southern Univ; Zn
Good, David, dgood3@uwo.ca, Western Univ; Em
Goodbred, Jr., Steven L., (615) 343-6424 steven.goodbred@vanderbilt.edu, Vanderbilt Univ; GsOn
Goode, Ryan , rgoode@cerritos.edu, Cerritos Coll; Zc
Goodell, Laurel P., 609-258-1043 laurel@princeton.edu, Princeton Univ; ZeGgc
Goodell, Philip C., 915-747-5593 goodell@utep.edu, Univ of Texas, El Paso; Ce
Goodge, John W., (218) 726-7491 jgoodge@d.umn.edu, Univ of Minnesota Duluth; Gp
Goodin, Ruth, 305-284-4253 ruthgoodin@miami.edu, Univ of Miami; Zn
Goodkin, Nathalie , 212-769-5379 ngoodkin@amnh.org, Graduate Sch of the City Univ of New York; Gg
Goodliffe, Andrew, amg@ua.edu, Univ of Alabama; YgeGt
Goodman, Paul, 520-621-8484 pgoodman@email.arizona.edu, Univ of Arizona; As
Goodrich, David C., (520) 670-6380 goodrich@tucson.ars.ag.gov, Univ of Arizona; Hs
Goodrich, Greg, gregory.goodrich@wku.edu, Western Kentucky Univ; Ast
Goodwin, David H., (740) 587-5621 goodwind@denison.edu, Denison Univ; PiCs
Goodwin, Laurel B., (608) 265-4234 laurel@geology.wisc.edu, Univ of Wisconsin, Madison; Gc
Goodwin, Mark B., (505) 835-5178 mark@berkeley.edu, Univ of California, Berkeley; Pv
Goodwin, Paul, (519) 824-4120 Ext.52754 pgoodwin@uoguelph.ca, Univ of Guelph; Zn
Goodwin, Phillip A., +44 (0)23 80596161 p.a.goodwin@soton.ac.uk, Univ of Southampton; Pe
Goos, Robert J., (701) 231-8581 North Dakota State Univ; Sc
Gootee, Brian F., 602.708.8846 bgootee@email.arizona.edu, Univ of Arizona; Gm
Gopal, Sucharita, suchi@bu.edu, Boston Univ; Zi
Gordon, Andrew, (519) 824-4120 ext 52415 agordon@uoguelph.ca, Univ of Guelph; Zn
Gordon, Arnold L., agordon@ldeo.columbia.edu, Columbia Univ; Op
Gordon, Barney, (785) 532-6101 bgordon@ksu.edu, Kansas State Univ; So
Gordon, Clare E., +44(0) 113 34 35210 c.e.gordon@leeds.ac.uk, Univ of Leeds; Zi
Gordon, Mark, (416)736-2100 mgordon@yorku.ca, York Univ; As
Gordon, Richard G., (713) 348-5279 rgg@rice.edu, Rice Univ; YmGtYd
Gordon, Rob J., (519) 824-4120 Ext.52285 rjgordon@uoguelph.ca, Univ of Guelph; As
Gordon, Robert B., (203) 432-3125 robert.gordon@yale.edu, Yale Univ; Nr
Gordon, Ryan, 207-287-7178 ryan.gordon@maine.gov, Dept of Agriculture, Conservation, and Forestry; HwGm
Gordon, Stacia M., 775-784-6476 staciag@unr.edu, Univ of Nevada, Reno; Gpz
Gordon, Steven J., (719) 333-3067 usafa.dfeg@usafa.edu, United States Air Force Acad; Gm
Gordon, Terence M., (403) 220-8301 tmg@geo.ucalgary.ca, Univ of Calgary;

Gq
Gore, Pamela J. W., 678-891-3754 pgore@gsu.edu, Georgia State Univ, Perimeter Coll, Clarkston Campus; GsZeGn
Gorecka-Nowak, Anna, anna.gorecka-nowak@uwr.edu.pl, Univ of Wroclaw; Pg
Gorelick, Steven, (650) 725-2950 gorelick@geo.stanford.edu, Stanford Univ; Yr
Gorka, Maciej, maciej.gorka@uwr.edu.pl, Univ of Wroclaw; AsCa
Gorman, Andrew R., +64 3 479-7516 andrew.gorman@otago.ac.nz, Univ of Otago; YrgYe
Gorman, Gerard, +44 20 759 49985 g.gorman@imperial.ac.uk, Imperial Coll; Yg
Gorman-Lewis, Drew J., 206-543-3541 dgormanl@uw.edu, Univ of Washington; PgCl
Gorog, Agnes, gorog@ludens.elte.hu, Eotvos Lorand Univ; Pmi
Gorokhovich, Yuri, 718-960-1981 yuri.gorokhovich@lehman.cuny.edu, Graduate Sch of the City Univ of New York; Gg
Gorring, Matthew L., (973) 655-5409 gorringm@mail.montclair.edu, Montclair State Univ; Gi
Gorsevski, Peter, (419) 372-7201 peterg@bgsu.edu, Bowling Green State Univ; ZirGm
Gosnell, Hannah, (541) 737-1222 gosnell@colorado.edu, Oregon State Univ; Zu
Gosnold, William D., (701) 777-2631 will.gosnold@engr.und.edu, Univ of North Dakota; YghCt
Gospodinova, Kalina D., 508-289-3212 kgospodinova@whoi.edu, Woods Hole Oceanographic Inst; Cm
Goss, Michael J., (519) 824-4120 (Ext. 2491) mgoss@uoguelph.ca, Univ of Guelph; So
Gosse, John, 902-494-6632 jcgosse@is.dal.ca, Univ of Kansas; Gm
Gosselin, David C., 402-472-8919 dgosselin2@unl.edu, Univ of Nebraska, Lincoln; Ze
Gosselin, Michel, 4187241761 x1284 michel_gosselin@uqar.ca, Univ du Quebec a Rimouski; Ob
Gotkowitz, Madeline B., (608) 262-1580 mbgotkow@wisc.edu, Univ of Wisconsin, Extension; Hw
Gottardi, Raphael, 337-482-6177 rxg0121@louisiana.edu, Univ of Louisiana at Lafayette; Gcp
Gottfried, Michael D., 517-432-5480 gottfrie@msu.edu, Michigan State Univ; Pv
Gottschalk, Bradford, (608) 263-7389 bradford.gottschalk@wisc.edu, Univ of Wisconsin, Madison, Division of Extension; Rh
Gotz, Annette, +27 (0)46-603-8313 a.gotz@ru.ac.za, Rhodes Univ; Gs
Goudge, Theodore L., (660) 562-1798 tgoudge@nwmissouri.edu, Northwest Missouri State Univ; Zy
Goudge, Timothy, tgoudge@jsg.utexas.edu, Univ of Texas, Austin; Gs
Goudreau, Joanne, (508) 289-2560 jgoudreau@whoi.edu, Woods Hole Oceanographic Inst; Cm
Gough, Amy, +44 1784 44 amy.gough@rhul.ac.uk, Royal Holloway Univ of London; Gs
Gouhier, Tarik, 7815817370 x302 t.gouhier@neu.edu, Northeastern Univ; Zn
Gould, Joseph C., gould.joe@spcollege.edu, Saint Petersburg Coll, Clearwater; Gg
Gould, Mark D., (401) 254-3087 Roger Williams Univ; Ob
Goulden, Michael L., (949) 824-1983 mgoulden@uci.edu, Univ of California, Irvine; Zg
Gouldey, Jeremy C., 616-331-2995 gouldjer@gvsu.edu, Grand Valley State Univ; CgPeCg
Gouldson, Andy, +44(0) 113 34 36417 a.gouldson@leeds.ac.uk, Univ of Leeds; Ge
Goulet, Normand, (514) 987-3375 r27254@er.uqam.ca, Univ du Quebec a Montreal; Gc
Goulet, Richard, 613-943-9922 richard.goulet@cnsc-ccsn.gc.ca, Univ of Ottawa; CgPgGe
Gouzie, Douglas R., (417) 836-5228 douglasgouzie@missouristate.edu, Missouri State Univ; HwGeCl
Gowan, Angela, (612) 626-6451 gowa0001@umn.edu, Univ of Minnesota; Gl
Gowing, David, +44 (0)1908 659468 x 59468 david.gowing@open.ac.uk, Open Univ; Pb
Graber, Anna, agraber@umn.edu, Univ of Minnesota, Twin Cities; Rh
Grable, Judy, (229) 333-5752 Valdosta State Univ; Hs
Grabowski, Jonathan, 7815817370 x337 j.grabowski@neu.edu, Northeastern Univ; GuEg
Grace, Cathy A., (662)915-1799 cag@olemiss.edu, Univ of Mississippi; Gg
Grace, Shannon M., (512) 223-4891 sgrace@austincc.edu, Austin Comm Coll District; Zn
Graczyk, Donald G., (630) 252-3489 Argonne; Cs
Graettinger, Alison, 816-235-6701 graettingera@umkc.edu, Univ of Missouri, Kansas City; Gv
Graf, Jr., Joseph L., (541) 552-6861 graf@sou.edu, Southern Oregon Univ; Em
Graham, Barbara, 702-651-4173 barbara.graham@csn.edu, Coll of Southern Nevada - West Charleston Campus; ZyAm
Graham, David W., (541) 737-4140 dgraham@coas.oregonstate.edu, Oregon State Univ; Cm
Graham, Gina R., 907-451-5031 gina.graham@alaska.gov, Alaska Division of Geological & Geophysical Surveys; Yg
Graham, Grace E., (608) 263-4125 grace.graham@wisc.edu, Univ of Wisconsin, Madison, Division of Extension; Hw
Graham, Ian T., i.graham@unsw.edu.au, Univ of New South Wales; Gzx
Graham, Margaret C., +44 (0) 131 650 4767 margaret.graham@ed.ac.uk, Edinburgh Univ; Ge
Graham, Michael, (831) 771-4400 mgraham@mlml.calstate.edu, Moss Landing Marine Lab; Ob
Graham, Russell W., (814) 865-6336 graham@ems.psu.edu, Pennsylvania State Univ, Univ Park; Pv
Graham, Stephan A., (650) 723-0507 graham@pangea.stanford.edu, Stanford Univ; Go
Graham, Jr., Earl K., (814) 865-2273 graham@ems.psu.edu, Pennsylvania State Univ, Univ Park; Yx
Graham, Jr., James H., (863) 956-1151 jhgraham@ufl.edu, Univ of Florida; Sb
Grala, Katarzynz, (662) 268-1032 Ext 222 kg160@msstate.edu, Mississippi State Univ; Zi
Grammer, Michael, 405-744-6358 michael.grammer@okstate.edu, Oklahoma State Univ; GsEoGr
Gran, Karen B., (218) 726-7406 kgran@d.umn.edu, Univ of Minnesota Duluth; Gms
Grana, Dario, 307-766-3449 dgrana@uwyo.edu, Univ of Wyoming; Ye
Grand, Stephen P., (512) 471-3005 steveg@maestro.geo.utexas.edu, Univ of Texas, Austin; Ys
Grandal d'Anglade, Aurora, 00 34 981 167000 xeaurora@udc.es, Coruna Univ; Pv
Grande, Anthony, 212-772-5265 agrande@hunter.cuny.edu, Hunter Coll (CUNY); Zn
Grande, Lance, (312) 665-7632 lgrande@fieldmuseum.org, Field Mus of Natural History; Pv
Grandstaff, David E., (215) 204-8228 grand@temple.edu, Temple Univ; Cl
Grandy, Stuart, stuart.grandy@unh.edu, Univ of New Hampshire; Sc
Graney, Joseph R., (607) 777-6347 jgraney@binghamton.edu, Binghamton Univ; ClGeHs
Granger, Darryl E., (765) 494-0043 dgranger@purdue.edu, Purdue Univ; Gg
Granger, Julie, 860-405-9094 julie.granger@uconn.edu, Univ of Connecticut; Oc
Graniero, Phil A., 519-253-3000 x2485 graniero@uwindsor.ca, Univ of Windsor; ZinZy
Granja, José Luis, jlgranja@ucm.es, Univ Compl de Madrid; Yg
Grannell, Roswitha B., (562) 985-4927 grannell@csulb.edu, California State Univ, Long Beach; Yv
Granshaw, Frank D., (503) 725-3391 fgransha@pdx.edu, Portland State Univ; ZnGl
Grant, Alastair, +44 (0)1603 59 2537 a.grant@uea.ac.uk, Univ of East Anglia; GeObCt
Grant, Jacqueline , jacquelinegrant@suu.edu, Southern Utah Univ; Zn
Grant, John A., 202-633-2474 Smithsonian Inst, Natl Air & Space Mus; XgGm
Grant, Jonathan, (902) 494-2021 jon.grant@dal.ca, Dalhousie Univ; Ob
Grant, Stanley, 703-361-5606 stanleyg@vt.edu, Virginia Tech; HsRw
Grapenthin, Ronni, 575-835-5924 rg@nmt.edu, New Mexico Tech; Gv
Grassian, Vicki, (858) 534-2499 vhgrassian@ucsd.edu, Univ of California, San Diego; Cm
Grassineau, Nathalie, +44 1784 443810 nathalie.grassineau@rhul.ac.uk, Royal Holloway Univ of London; Cs
Grassman, Jean, 718-951-5000 ext 2752 Graduate Sch of the City Univ of New York; Gg
Grattan, Stephen R., 530-752-4618 srgrattan@ucdavis.edu, Univ of California, Davis; Sp
Graustein, William C., (203) 287-2853 william.graustein@yale.edu, Yale Univ; Cl
Gravely, Cynthia Rae, (864) 656-3438 gravelc@clemson.edu, Clemson Univ; Zn
Graves, Alexandria, 919-513-0635 alexandria_graves@ncsu.edu, North Carolina State Univ; Sb
Graves, Charles E., (314) 977-3121 charles.graves@slu.edu, Saint Louis Univ; As
Gravley, Darren, +64 3 3667001 Ext 45683 darren.gravley@canterbury.ac.nz, Univ of Canterbury; Gv

Gray, Kyle R., 319-273-2809 kyle.gray@uni.edu, Univ of Northern Iowa; ZeGg
Gray, Lee M., (330) 823-3605 graylm@mountunion.edu, Mount Union Coll; PgGg
Gray, Mary Beth, (570) 577-1146 mbgray@bucknell.edu, Bucknell Univ; Gc
Gray, Neil, +44 (0) 191 208 4887 neil.gray@ncl.ac.uk, Univ of Newcastle Upon Tyne; Pg
Gray, Robert S., (805) 965-0581 gray@sbcc.net, Santa Barbara City Coll; PvGzx
Gray, Sarah, (619) 260-4098 sgray@sandiego.edu, Univ of San Diego; Zn
Grayston, Sue, 604–822–5928 sue.grayston@ubc.ca, Univ of British Columbia; Sb
Greatbatch, Ian, +44 020 8417 2879 i.greatbatch@kingston.ac.uk, Kingston Univ; Zi
Greatbatch, Richard , rgreatbatch@geomar.de, Dalhousie Univ; Op
Greb, Stephen F., 8592575500 x136 greb@uky.edu, Univ of Kentucky; GsEcGr
Green, Andrew, (031) 260-2516 greena1@ukzn.ac.za, Univ of KwaZulu-Natal; GuYgGs
Green, Brittany, (785) 532-6101 bdgreen@ksu.edu, Kansas State Univ; Zn
Green, Harry W., (510) 642-3059 Univ of California, Berkeley; Pv
Green, Jack, (562) 985-4198 jgreen3@csulb.edu, California State Univ, Long Beach; Gv
Green, John C., (218) 726-7208 jgreen@d.umn.edu, Univ of Minnesota Duluth; GieGv
Green, Jonathan, +44 0151 795 4385 jonathan.green@liverpool.ac.uk, Univ of Liverpool; Ob
Green, Mark B., mbg78@case.edu, Case Western Reserve Univ; Hg
Greenaway, Darren, 817-515-3606 darren.greenaway@tccd.edu, Tarrant County Coll, Southeast Campus; Gg
Greenberg, David, (902) 426-2431 greenbergd@mar.dfo-mpo.gc.ca, Dalhousie Univ; Op
Greenberg, Harvey, (206) 685-7981 hgreen@uw.edu, Univ of Washington; Gml
Greenberg, Richard J., (520) 621-6940 greenberg@lpl.arizona.edu, Univ of Arizona; Xy
Greenberg, Sallie, (217) 244-4068 sallieg@illinois.edu, Illinois State Geo Survey; ClZg
Greene, Brian M., 412-395-7323 brian.greene@usace.army.mil, Youngstown State Univ; Ng
Greene, Charles H., chg2@cornell.edu, Cornell Univ; Ob
Greene, Chris, 902-494-7018 csgreene@dal.ca, Dalhousie Univ; Zi
Greene, David C., (740) 587-6476 greened@denison.edu, Denison Univ; GctGe
Greene, Don M., (254) 710-2193 don_greene@baylor.edu, Baylor Univ; ZyAm
Greene, Mott, 206-713-0917 mgreene@uw.edu, Univ of Washington; GgAmOg
Greene, Todd, 530-898-5546 tjgreene@csuchico.edu, California State Univ, Chico; Gsr
Greenfield, Cat, cg331@le.ac.uk, Leicester Univ; Yg
Greenfield, Dianne I., 212-413-3154 dianne.greenfield@asrc.cuny.edu, Graduate Sch of the City Univ of New York; Gu
Greenfield, Roy J., (814) 237-1810 rkg4@psu.edu, Pennsylvania State Univ, Univ Park; YgsYu
Greenlee, Diana M., (318) 926-3314 greenlee@ulm.edu, Univ of Louisiana, Monroe; GaCo
Greenough, John D., 250-807-9520 john.greenough@ubc.ca, Univ of British Columbia, Okanagan Campus; GiCtGa
Greenwell, Chris, +44 (0) 191 33 42324 chris.greenwell@durham.ac.uk, Durham Univ; CqEoCg
Greenwood, James P., 860 685-2545 jgreenwood@wesleyan.edu, Wesleyan Univ; Xc
Greer, Lisa, 540-458-8870 greerl@wlu.edu, Washington & Lee Univ; GsPe
Gregg, Jay M., 405-744-6358 jay.gregg@okstate.edu, Oklahoma State Univ; Gs
Gregg, Michael C., (206) 543-1353 gregg@apl.washington.edu, Univ of Washington; Op
Gregg, Patricia, 217-333-3540 pgregg@illinois.edu, Univ of Illinois, Urbana-Champaign; YgCg
Gregg, Tracy K. P., (716) 645-4328 tgregg@buffalo.edu, SUNY, Buffalo; Gv
Gregor, C. B., 937 775-3455 Wright State Univ; Gs
Gregory, Robert T., (214) 768-3075 bgregory@smu.edu, Southern Methodist Univ; Cs
Grenfell, Thomas C., tcg@atmos.washington.edu, Univ of Washington; As
Grew, Edward S., (207) 581-2169 esgrew@maine.edu, Univ of Maine; Gpz
Grew, Priscilla C., (402) 472-2095 pgrew1@unl.edu, Univ of Nebraska, Lincoln; GpeZe
Greybush, Steven J., 814-867-4926 sjg213@psu.edu, Pennsylvania State Univ, Univ Park; Am
Gribenko, Alex, (801) 585-6484 alex.gribenko@utah.edu, Univ of Utah; Yg
Grieneisen, Michael L., mgrien@ucdavis.edu, Univ of California, Davis; Hg
Griera Artigas, Albert, ++935811035 albert.griera@uab.cat, Univ Auto de Barcelona; GcCl
Gries, John C., (316) 978-3140 Wichita State Univ; Gt
Grieve, Richard, rgrieve2@uwo.ca, Western Univ; Xg
Grieve, Richard A. F., (613) 995-5372 geology@unb.ca, Univ of New Brunswick; Xg
Grießl, Stefan, 089/2180 4188 stefan.griessl@uni.muenchen.de, Ludwig-Maximilians-Univ Muenchen; Gz
Griffen, Dana T., 801-422-2305 dana_griffen@byu.edu, Brigham Young Univ; Gz
Griffin, Kevin, griff@ldeo.columbia.edu, Columbia Univ; Zn
Griffin, William R., (972) 883-2430 griffin@utdallas.edu, Univ of Texas, Dallas; Gg
Griffing, Corinne, 604-527-5217 griffingc@douglascollege.ca, Douglas Coll; GgeGm
Griffing, David H., 607-431-4629 griffingd@hartwick.edu, Hartwick Coll; GduGm
Griffis, Timothy J., (612) 625-3117 tgriffis@soils.umn.edu, Univ of Minnesota, Twin Cities; As
Griffith, Ashley, (817) 272-2987 wagriff@uta.edu, Univ of Texas, Arlington; Gct
Griffith, Caitlin, (520) 626-3806 griffith@lpl.arizona.edu, Univ of Arizona; Zn
Griffith, Elizabeth, griffith.906@osu.edu, Ohio State Univ; CsOoCb
Griffith, Liz, (817) 272-2987 lgriff@uta.edu, Univ of Texas, Arlington; CslCm
Griffith, William A., griffith.233@osu.edu, Ohio State Univ; GcNr
Grigg, Joseph, 575-835-5336 joseph.grigg, New Mexico Tech; Go
Grigg, Laurie, 802 485 3323 lgrigg@norwich.edu, Norwich Univ; PeGnZi
Grigg, Richard W., (808) 956-7186 rgrigg@soest.hawaii.edu, Univ of Hawai'i, Manoa; Ob
Griggs, Gary, griggs@ucsc.edu, Univ of California, Santa Cruz; GuOnp
Grigsby, Jeffry D., (765) 285-8270 jgrigsby@bsu.edu, Ball State Univ; GdClGo
Grima, Cyril , 512-471-0377 cyril.grima@utexas.edu, Univ of Texas, Austin; XyGlYg
Grimes, Stephen, +44 1752 584759 stephen.grimes@plymouth.ac.uk, Univ of Plymouth; Cs
Grimley, David A., 217-244-7324 dgrimley@illinois.edu, Illinois State Geo Survey; GmlSc
Grimm, Kurt A., (604) 822-9258 kgrimm@eos.ubc.ca, Univ of British Columbia; Gs
Grindlay, Nancy R., (910) 962-2352 grindlayn@uncw.edu, Univ of North Carolina, Wilmington; Yg
Gringarten, Alain, +44 20 759 47440 a.gringarten@imperial.ac.uk, Imperial Coll; Np
Grippo, Alessandro, grippo_alessandro@smc.edu, Santa Monica Coll; Gs
Gripshover, Margaret , margaret.gripshover@wku.edu, Western Kentucky Univ; Zy
Grise, Kevin M., (434) 924-0433 kmg3r@virginia.edu, Univ of Virginia; As
Grismer, Mark E., (530) 752-3245 megrismer@ucdavis.edu, Univ of California, Davis; Hw
Griswold, George B., (505) 299-6192 New Mexico Tech; Nx
Griswold, Jennifer S., smalljen@hawaii.edu, Univ of Hawai'i at Manoa; AmZr
Gritzo, Russell E., (505) 667-0481 Los Alamos; Zn
Groat, Lee A., (604) 822-1289 groat@mail.ubc.ca, Univ of British Columbia; GzEg
Grobler, Hendrik C., +27115596838 hgrobler@uj.ac.za, Univ of Johannesburg; ZnNm
Grocke, Darren, +44 (0) 191 33 42282 Durham Univ; Cs
Groenevelt, Pieter H., (519) 824-4120 (Ext. 53585) pgroenev@lrs.uoguelph.ca, Univ of Guelph; Sp
Groffman, Peter , 719-951-5000, ext. 5416 peter.groffman@brooklyn.cuny.edu, Graduate Sch of the City Univ of New York; Zg
Groffman, Peter, 212-413-3143 peter.groffman@brooklyn.cuny.edu, Brooklyn Coll (CUNY); SbCl
Gromet, L. P., 401-863-1920 peter_gromet@brown.edu, Brown Univ; Cc
Groppi, Christopher, (480) 965-6436 cgroppi@asu.edu, Arizona State Univ; XaZrn
Groppo, Chiara Teresa, chiara.groppo@unito.it, Univ di Torino; Gx
Grosfils, Eric B., (909) 621-8673 egrosfils@pomona.edu, Pomona Coll; XgGvq
Gross, Amy, 910-521-6588 amy.gross@uncp.edu, Univ of North Carolina, Pembroke; Gg
Gross, Gerardo W., grossgw@nmt.edu, New Mexico Tech; Yx

Gross, Juliane, (848) 445-3619 jgross@eps.rutgers.edu, Rutgers, The State Univ of New Jersey; Xcm
Gross, Michael, (305) 348-3932 grossm@fiu.edu, Florida InterNatl Univ; Gc
Grossman, Eric, (360) 650-4697 eric.grossman@wwu.edu, Western Washington Univ; GusGm
Grossman, Ethan L., (979) 845-0637 e-grossman@tamu.edu, Texas A&M Univ; Csl
Grossman, Lawrence S., (540) 231-5116 Virginia Tech; Zn
Grossman, Walter, wgrossman@yahoo.com, San Bernardino Valley Coll; Og
Groszos, Mark S., (229) 333-5664 msgroszo@valdosta.edu, Valdosta State Univ; GctEg
Grothe, Pamela, pgrothe@umw.edu, Univ of Mary Washington; GsOgZc
Grotjahn, Richard , 530-752-2246 grotjahn@ucdavis.edu, Univ of California, Davis; As
Grottoli, Andrea G., 614 292 5782 grottoli.1@osu.edu, Ohio State Univ; ObCbOo
Grotzinger, John P., 626.395.6785 grotz@gps.caltech.edu,CalTech; GsPeXg
Grove, Karen, (415) 338-2617 kgrove@sfsu.edu, San Francisco State Univ; Gs
Grove, Timothy L., (617) 253-2878 tlgrove@mit.edu, Massachusetts Inst of Tech; Gi
Grover, Jeffrey A., 8055463100 x2759 jgrover@cuesta.edu, Cuesta Coll; Gg
Grover, John E., john.grover@uc.edu, Univ of Cincinnati; Gz
Grover, Timothy W., (802) 468-1289 tim.grover@castleton.edu, Castleton Univ; GptGc
Groves, Christopher, (270) 745-5974 chris.groves@wku.edu, Western Kentucky Univ; Hg
Gruen, Dieter M., (630) 972-3513 Argonne; Zm
Grujic, Djordje, (902) 494-2208 dgrujic@is.dal.ca, Dalhousie Univ; Gt
Grunbaum, Daniel, (206) 221-6594 grunbaum@ocean.washington.edu, Univ of Washington; Ob
Grunder, Anita L., (541) 737-5189 grundera@geo.oregonstate.edu, Oregon State Univ; Gi
Grundl, Timothy J., 414-229-4765 grundl@uwm.edu, Univ of Wisconsin, Milwaukee; HwCl
Grunwald, Sabine -., (352) 294-3145 sabgru@ufl.edu, Univ of Florida; ZiSoZr
Grupp, Steve, 425-388-9450 sgrupp@everettcc.edu, Everett Comm Coll; GgOg
Grushow, Alex, grushow@rider.edu, Rider Univ; Zn
Gu, Baohua, 865-574-7286 gub1@ornl.gov, Oak Ridge; CbqSb
Gu, Jeff, (780) 492-2292 jgu@phys.ualberta.ca, Univ of Alberta; Ys
Gualda, Guilherme, 615-322-2976 g.gualda@vanderbilt.edu, Vanderbilt Univ; GivGz
Guan, Dabo, +44(0) 113 34 37432 d.guan@leeds.ac.uk, Univ of Leeds; Eg
Guan, Huade, huade.guan@flinders.edu.au, Flinders Univ; HgsHg
Guber, Albert L., guber@ems.psu.edu, Pennsylvania State Univ, Univ Park; Pe
Guccione, Margaret J., (479) 575-3354 guccione@comp.uark.edu, Univ of Arkansas, Fayetteville; GmaGg
Gudmundsson, Agust, +44 1784 276345 agust.gudmundsson@rhul.ac.uk, Royal Holloway Univ of London; Gc
Guenthner, Willy, wrg@illinois.edu, Univ of Illinois, Urbana-Champaign; GtzCg
Guerin, Gilles, guerin@ldeo.columbia.edu, Columbia Univ; Gu
Guerra, Oralia, (512) 223-7483 oguerra1@austincc.edu, Austin Comm Coll District; Zn
Guerrieri, Mary, (520) 621-2828 mary@lpl.arizona.edu, Univ of Arizona; Zn
Guertal, Elizabeth A., eguertal@acesag.auburn.edu, Auburn Univ; Sc
Guest, Peter S., 831-656-2451 pguest@nps.edu, Naval Postgrad Sch; Am
Guevara, Victor, vguevara@skidmore.edu, Skidmore Coll; Gp
Guggenheim, Stephen J., (312) 996-3263 xtal@uic.edu, Univ of Illinois Chicago; Gz
Guggino, Steve N., (912) 478-5361 sguggino@georgiasouthern.edu, Georgia Southern Univ; GgCg
Guha, Jayanta, 418 545 5222 jguha@uqac.c, Univ du Quebec a Chicoutimi; Eg
Guilbert, John M., (520) 621-6024 j.guilbert@comcast.net, Univ of Arizona; Em
GUILLAUME, Damien, damien.guillaume@univ-st-etienne.fr, Université Jean Monnet, Saint-Etienne; Gz
Guinan, Patrick E., 573-882-5909 guinanp@missouri.edu, Univ of Missouri; As
Guinasso, Norman L., (979) 862-2323 norman@geos.tamu.edu, Texas A&M Univ; Opc
Guiseppe, Alfred C., 717.702.2024 aguiseppe@pa.gov, DCNR- Pennsylvania Bur of Geo Survey; ZiHwGg
Guitierrez-Alonso, Gabriel, gabi@gugu.usal.es, Florida InterNatl Univ; Gc
Gulbranson, Erik, (507) 933-6475 erikgulbranson@gustavus.edu, Gustavus Adolphus Coll; PcGsSc
Gulick, Sean S., 512-471-0483 sean@ig.utexas.edu, Univ of Texas, Austin; Gc
Gulley, Jason, jdgulley@usf.edu, Univ of South Florida; Hw
Gulliver, John S., (612) 625-4080 gulli003@tc.umn.edu, Univ of Minnesota, Twin Cities; Hs
Gundersen, James N., (316) 978-3140 Wichita State Univ; Ga
Gunderson, Lance, 404 727 8108 lgunder@emory.edu, Emory Univ; Sf
Gundiler, Ibrahim H., 575-835-5730 ibrahim.gundiler@nmt.edu, New Mexico Tech; Nx
Gunia, Piotr, piotr.gunia@uwr.edu.pl, Univ of Wroclaw; Gaz
Gunter, Mickey E., (208) 885-6015 mgunter@uidaho.edu, Univ of Idaho; GzEnGb
Guo, Bo, (520) 626-9971 boguo@email.arizona.edu, Univ of Arizona; Hw
Guo, Junhua, jguo1@csub.edu, California State Univ, Bakersfield; Gs
Guo, Weifu, (508) 289-3380 wguo@whoi.edu, Woods Hole Oceanographic Inst; CslPe
Gupta, Hoshin V., (520) 275-5534 hoshin.gupta@hwr.arizona.edu, Univ of Arizona; HqZgAs
Gupta, Sanjeev, +44 20 759 46527 s.gupta@imperial.ac.uk, Imperial Coll; Gs
Gupta, Satish C., (612) 625-1241 sgupta@soils.umn.edu, Univ of Minnesota, Twin Cities; Sp
Gurdak, Jason, (415) 338-6869 jgurdak@sfsu.edu, San Francisco State Univ; HwCqAt
Gurevich, Boris, +61 8 9266-7359 b.gurevich@curtin.edu.au, Curtin Univ; Ye
Gurlea, Lawrence P., lisobar@aol.com, Youngstown State Univ; Cg
Gurnis, Michael C., (626) 395-6979 gurnis@gps.caltech.edu,CalTech; Ys
Gurocak, Zulfu, 00904242370000-5991 zgurocak@firat.edu.tr, Firat Univ; Nrg
Gurrola, Harold, 806-834-8625 harold.gurrola@ttu.edu, Texas Tech Univ; Ys
Gurwin, Jacek, jacek.gurwin@uwr.edu.pl, Univ of Wroclaw; HwyZf
Gust, David A., 61 7 3138 2217 d.gust@qut.edu.au, Queensland Univ of Tech; GiCpGt
Gustin, Mae, 775.784.4203 mgustin@cabnr.unr.edu, Univ of Nevada, Reno; Cg
Guth, Peter L., (410) 293-6560 pguth@usna.edu, United States Naval Acad; GcZiy
Guthrie, Roderick I., (514) 398-4755 McGill Univ; Nx
Gutierrez, Melida, (417) 836-5967 mgutierrez@missouristate.edu, Missouri State Univ; CgHy
Gutierrez-Jurado, Hugo A., 915-747-5159 hagutierrez@utep.edu, Univ of Texas, El Paso; HgsHw
Gutknecht, Jessica L., 612-626-8435 jgutknec@umn.edu, Univ of Minnesota, Twin Cities; Sb
Gutmann, James T., (860) 685-2258 jgutmann@wesleyan.edu, Wesleyan Univ; GviGz
Gutowski, William J., (515) 294-5632 gutowski@iastate.edu, Iowa State Univ of Sci & Tech; Am
Gutzler, David J., (505) 277-3328 gutzler@unm.edu, Univ of New Mexico; As
Guza, Robert T., (858) 534-0585 rguza@ucsd.edu, Univ of California, San Diego; On
Guzina, Bojan B., (612) 626-0789 guzina@wave.ce.umn.edu, Univ of Minnesota, Twin Cities; Nr
Guzman, Ernesto, (519) 824-4120 Ext.53609 eguzman@uoguelph.ca, Univ of Guelph; Zn
Gušiæ, Ivan, +38514606102 ivangusic@yahoo.com, Univ of Zagreb; GgPsm
Gyakum, John R., (514) 398-6076 john.gyakum@mcgill.ca, McGill Univ; Am
Gysi, Alexander, 575-835-5754 alexander.gysi@nmt.edu, New Mexico Tech; EgCpGz
Göttlich, Hagen, 089/2180 5615 goettlich@ennab.de, Ludwig-Maximilians-Univ Muenchen; Gz

H

Haag, Lucas, (785) 462-6281 lhaag@ksu.edu, Kansas State Univ; So
Haas, Christian, (416)736-2100 #77705 haasc@yorku.ca, York Univ; Yg
Haas, Johnson R., (269) 387-2878 johnson.haas@wmich.edu, Western Michigan Univ; Cl
Habana, Nathan C., (671) 735-2693 nhabana@triton.uog.edu, Univ of Guam; Hw
Habash, Marc, (519) 824-4120 Ext.52748 mhabash@uoguelph.ca, Univ of Guelph; Zn
Haber, Eldad, (604) 822-4525 haber@eos.ubc.ca, Univ of British Columbia; YgZn
Habib, Daniel, (718) 997-3333 daniel.habib@qc.cuny.edu, Queens Coll (CUNY); Pl

Habron, Geoffrey, 864-294-3413 geoffrey.habron@furman.edu, Furman Univ; Zn
Hacker, Bradley R., (805) 893-7952 hacker@geol.ucsb.edu, Univ of California, Santa Barbara; Gp
Hacker, David B., 330-672-8831 dhacker@kent.edu, Kent State Univ; GcHw
Hacker, Jorg, jorg.hacker@flinders.edu.au, Flinders Univ; ZrAs
Hacker, Joshua P., 831-656-2722 jphacker@nps.edu, Naval Postgrad Sch; Am
Haddad, Brent, (831) 459-4149 bhaddad@cats.ucsc.edu, Univ of California, Santa Cruz; Zn
Haddock, Gregory D., (660) 562-1719 haddock@nwmissouri.edu, Northwest Missouri State Univ; Zi
Hadler, Kathryn, +44 20 759 47198 k.hadler@imperial.ac.uk, Imperial Coll; Gz
Hadley, Daniel, drhadley@illinois.edu, Univ of Illinois; Hw
Haff, Peter K., (919) 684-5902 haff@duke.edu, Duke Univ; Zn
Hafner, James A., (413) 545-0778 hafner@geo.umass.edu, Univ of Massachusetts, Amherst; Zn
Hagadorn, James W., (303) 370-6058 james.hagadorn@dmns.org, Denver Mus of Nature & Sci; GgsPg
Hagedorn, Charles, (540) 231-4895 chagedor@vt.edu, Virginia Tech; Sb
Hagedorn, Jake, 828-232-5183 jghagedo@unca.edu, Univ of North Carolina, Asheville; SbZun
Hagelgans, Duane E., 717-871-7536 duane.hagelgans@millersville.edu, Millersville Univ; Rnm
Hagelman III, Ronald, 51422452170 Texas State Univ; Rnm
Hageman, Steven J., (828) 262-6609 hagemansj@appstate.edu, Appalachian State Univ; PieGg
Hagen, Cedric, 609-258-4101 ch0934@princeton.edu, Princeton Univ; GrsPc
Hager, Bradford H., (617) 253-0126 bhhager@mit.edu, Massachusetts Inst of Tech; Ys
Hagerty, Michael, 617-552-8300 hagertmb@bc.edu, Boston Coll; Ys
Haggart, James, jim.haggart@canada.ca, Univ of British Columbia; PgGr
Haggart, Renee, (604) 822-2789 rhaggart@eos.ubc.ca, Univ of British Columbia; Zn
Haggerty, Julia H., (406) 994-6904 julia.haggerty@montana.edu, Montana State Univ; Zu
Haggerty, Roy D., (541) 737-0663 roy.haggerty@oregonstate.edu, Oregon State Univ; Hw
Haggerty, Stephen E., (305) 348-7338 haggerty@fiu.edu, Florida InterNatl Univ; GiEgGz
Haghighi, Manouchehr, manouchehr.haghighi@adelaide.edu.au, Univ of Adelaide; Np
Haiar, Brooke, haiar@lynchburg.edu, Univ of Lynchburg; PgGeg
Haigh, Ivan D., +44 (023) 80596501 i.d.haigh@soton.ac.uk, Univ of Southampton; On
Haile, Estifanos, 865-974-8340 ehaile1@utk.edu, Univ of Tennessee, Knoxville; Hy
Haileab, Bereket, (507) 222-5746 bhaileab@carleton.edu, Carleton Coll; GxzGi
Haimson, Bezalel C., (608) 262-2563 bhaimson@wisc.edu, Univ of Wisconsin, Madison; Nr
Hain, Mathis, mhain@ucsc.edu, Univ of California, Santa Cruz; Cb
Haine, Thomas W., (410) 516-7048 thomas.haine@jhu.edu, Johns Hopkins Univ; OpgAs
Haitonic, Raluca Emilia, raluca.haitonic@ubbcluj.ro, Babes-Bolyai Univ; PmGrPs
Hajek, Elizabeth, hajek@psu.edu, Pennsylvania State Univ, Univ Park; Gs
Hajnal, Zoltan, (306) 966-5694 zoltan.hajnal@usask.ca, Univ of Saskatchewan; YsGtc
Hakim, Gregory J., (206) 685-2439 hakim@atmos.washington.edu, Univ of Washington; As
Hakimian, Adina, 516-463-6545 adina.i.hakimian@hofstra.edu, Hofstra Univ; Gg
Halama, Ralf, +44 (0) 1782 7 34960 r.halama@keele.ac.uk, Keele Univ; GpiCa
Halbig, Joseph B., halbig@wazoo.com, Univ of Hawai'i, Hilo; Ca
Halden, Norman M., (204) 474-6910 nm_halden@umanitoba.ca, Univ of Manitoba; Cg
Hale, Beverley A., (519) 824-4120 (Ext. 53434) bhale@uoguelph.ca, Univ of Guelph; Ct
Hale, Leslie J., (202) 633-1796 halel@si.edu, Smithsonian Inst / Natl Mus of Natural History; GgxZf
Hale, Michelle, +44 023 92 842290 michelle.hale@port.ac.uk, Univ of Portsmouth; Og
Hale, Rip, (757) 683-3959 rphale@odu.edu, Old Dominion Univ; Ou
Hales, Burke R., (541) 737-8121 bhales@coas.oregonstate.edu, Oregon State Univ; Oc
Hales, T. C., +44(0)29 208 74329 halest@cardiff.ac.uk, Univ of Wales; Gt
Hales, TC, halest@cf.ac.uk, Cardiff Univ; Gm
Haley, Brian, 541-737-2649 bhaley@coas.oregonstate.edu, Oregon State Univ; Cs
Haley, J C., jachaley@vwc.edu, Miami Univ; Gts
Haley, John C., 757-455-3407 jchaley@vwc.edu, Virginia Wesleyan Coll; GgeZi
Halfar, Jochen, (905) 828-5419 jochen.halfar@utoronto.ca, Univ of Toronto; Pe
Halfman, John D., (315) 781-3918 halfman@hws.edu, Hobart & William Smith Colls; GeHsGn
Halihan, Todd, 405-744-6358 todd.halihan@okstate.edu, Oklahoma State Univ; HwYuRc
Hall, Anne M., (404) 727-2863 ahall04@emory.edu, Emory Univ; Gz
Hall, Brenda L., (207) 581-2191 brendah@maine.edu, Univ of Maine; Gl
Hall, Chris, (734) 764-6391 cmhall@umich.edu, Univ of Michigan; Cc
Hall, Christopher C., (519) 824-4120 Ext.52740 jchall@uoguelph.ca, Univ of Guelph; Zn
Hall, Clarence A., (310) 825-1010 hall@epss.ucla.edu, Univ of California, Los Angeles; GgtGe
Hall, Cynthia V., (610) 436-1003 chall@wcupa.edu, West Chester Univ; Cg
Hall, I, hall@cf.ac.uk, Cardiff Univ; Zn
Hall, Ian R., +44(0)29 208 75612 hall@cardiff.ac.uk, Cardiff Univ; GuCsGs
Hall, Jean, +44 (0) 191 208 8783 jean.hall@ncl.ac.uk, Univ of Newcastle Upon Tyne; Ng
Hall, Jeremy, (709) 864-7569 jeremyh@mun.ca, Memorial Univ of Newfoundland; YsGtYr
Hall, Jude, (740) 587-6217 hall@denison.edu, Denison Univ; Zn
Hall, Luke D., (805) 642-3211 lhall@vcccd.net, Ventura Coll; Zy
Hall, Robert, rhall@uoguelph.ca, Univ of Guelph; Zn
Hall, Stuart A., 713-743-3416 sahgeo@uh.edu, Univ of Houston; Ym
Hall, Tracy, 678-872-8415 thall@highlands.edu, Georgia Highlands Coll; Gg
Hall, III, John R., (504) 231-6305 jrhall3@vt.edu, Virginia Tech; So
Halladay, Sylvia R., 614 265 6624 sylvia.halladay@dnr.state.oh.us, Ohio Dept of Nat Res; Zn
Hallar, Anna G., (801) 587-7238 gannet.hallar@utah.edu, Univ of Utah; AspAc
Haller, Merrick C., (541) 737-9141 merrick.haller@oregonstate.edu, Oregon State Univ; OnZrOp
Hallet, Bernard, 206-685-2409 hallet@uw.edu, Univ of Washington; Gml
Hallett, Benjamin W., 920-424-0868 hallettb@uwosh.edu, Univ of Wisconsin, Oshkosh; Gpx
Hallett, Rebecca, (519) 824-4120 Ext.54488 rhallett@uoguelph.ca, Univ of Guelph; Zn
Halliday, Alex, +44 (1865) 272969 alex.halliday@earth.ox.ac.uk, Univ of Oxford; Cs
Hallock, Brent G., (805)756-2436 bhallock@calpoly.edu, California Polytechnic State Univ; Sf
Hallock-Muller, Pamela, (727) 553-1567 pmuller@usf.edu, Univ of South Florida; OuGsPm
Halls, Henry C., 905-828-5363 hlhalls@utm.utoronto.ca, Univ of Toronto; Ym
Halls, Joanne N., (910) 962-7614 hallsj@uncw.edu, Univ of North Carolina, Wilmington; Zi
Hallum, Douglas R., (308) 696-6743 dhallum2@unl.edu, Univ of Nebraska, Lincoln; HwGog
Halsor, Sid P., (570) 408-4611 sid.halsor@wilkes.edu, Wilkes Univ; Giv
Halsted, Christian, (207) 287-7175 christian.h.halsted@maine.gov, Dept of Agriculture, Conservation, and Forestry; Zi
Halverson, Galen, 514-398-4894 galen.halverson@mcgill.ca, McGill Univ; GsrCs
Halverson, Larry, (515) 294-0495 larryh@iastate.edu, Iowa State Univ of Sci & Tech; So
Ham, Nelson R., (920) 403-3977 nelson.ham@snc.edu, Saint Norbert Coll; Gl
Hamann, Hillary, (303) 871-3977 hillary.hamann@du.edu, Univ of Denver; HgZy
Hambrey, Michael M., +44 (0)1970 621 860 mjh@aber.ac.uk, Aberystwyth Univ; GlsGm
Hamburger, Michael W., (812) 855-2934 hamburg@indiana.edu, Indiana Univ, Bloomington; YsGtv
Hamdan, Abeer, abeer.hamdan@phoenixcollege.edu, Phoenix Coll; Gg
Hamecher, Emily A., 626-660-8314 ehamecher@fullerton.edu, California State Univ, Fullerton; GgzGi
Hameed, Sultan, (631) 632-8319 sultan.hameed@stonybrook.edu, SUNY, Stony Brook; As
Hames, Willis E., (334) 844-4881 hameswe@auburn.edu, Auburn Univ; CcGpt
Hamill, Paul, (815) 455-8698 phamill@mchenry.edu, McHenry County Coll;

AmZg
Hamilton, George, (413) 499-4660 bhamilton@berkshirecc.edu, Berkshire Comm Coll; Zn
Hamilton, Jacqueline, (612) 626-8292 stub0035@umn.edu, Univ of Minnesota; ZiGe
Hamilton, Michael A., (416) 946-7424 mahamilton@es.utoronto.ca, Univ of Toronto; CcGiCu
Hamilton, Sally, +44 (0)131 451 3198 s.hamilton@hw.ac.uk, Heriot-Watt Univ; Gs
Hamlet, Alan, alan.hamlet.1@nd.edu, Univ of Notre Dame; Hs
Hamlin, H, Scott, 512-4759527 scott.hamlin@beg.utexas.edu, Univ of Texas at Austin, Jackson Sch of Geosciences; Gr
Hamme, Roberta C., (250) 472-4014 rhamme@uvic.ca, Univ of Victoria; Oc
Hammer, Julia E., 808-956-5996 jhammer@hawaii.edu, Univ of Hawai'i, Manoa; GiCpGv
Hammer, Philip T., (604) 822-5703 phammer@eos.ubc.ca, Univ of British Columbia; Ys
Hammer, William R., (309) 794-7487 williamhammer@augustana.edu, Augustana Coll; Pv
Hammerschmidt, Chad, 937 775-3457 chad.hammerschmidt@wright.edu, Wright State Univ; CmOcCb
Hammersley, Charles, (928) 523-6655 charles.hammersley@nau.edu, Northern Arizona Univ; Zn
Hammersley, Lisa, 916-278-7200 hammersley@csus.edu, Sacramento State Univ; Gig
Hammes, Ursula, 512-471-1891 ursula.hammes@beg.utexas.edu, Univ of Texas at Austin, Jackson Sch of Geosciences; Gg
Hammond, Amie C., ahammond@austincc.edu, Austin Comm Coll District; ZeGg
Hammond, Douglas E., (213) 740-5837 dhammond@usc.edu, Univ of Southern California; Cm
Hammond, Paul E., (503) 725-3387 hammondp@pdx.edu, Portland State Univ; Gi
Hammond, William C., (775) 784-6436 whammond@unr.edu, Univ of Nevada, Reno; Yd
Hampson, Arthur, (801) 585-5698 spike.hampson@geog.utah.edu, Univ of Utah; Zy
Hampson, Gary , +44 20 759 46475 g.j.hampson@imperial.ac.uk, Imperial Coll; Gd
Hampton, Brian A., 575-646-2997 bhampton@nmsu.edu, New Mexico State Univ, Las Cruces; Gst
Hampton, Duane R., (269) 387-5496 duane.hampton@wmich.edu, Western Michigan Univ; HwSpHg
Hampton, Samuel J., +64 3 3667001 Ext 6770 samuel.hampton@canterbury.ac.nz, Univ of Canterbury; GvZe
Hams, Jacquelyn E., (818) 778-5566 hamsje@lavc.edu, Los Angeles Valley Coll; GgOuZe
Hamutoko, Josephine, jhamutoko@unam.na, Univ of Namibia; Hw
Hamzaoui, Cherif, 514-987-3000 #6837 hamzaoui.cherif@uqam.ca, Univ du Quebec a Montreal; Yg
Han, De-hua, 713-743-9293 dhan@uh.edu, Univ of Houston; Ye
Han, Hannah Juan, 575-835-5452 hannah.juan.han@nmt.edu, New Mexico Tech; CuqCa
Han, Nizhou, (540) 231-2403 nhan@vt.edu, Virginia Tech; Sc
Han, Shuoshuo, 512-471-0415 han@ig.utexas.edu, Univ of Texas, Austin; YrGt
Hanafy, Sherif , +9661368603765 sherif.mahmoud@kfupm.edu.sa, King Fahd Univ of Petroleum and Minerals; Yg
Hanan, Barry B., (619) 594-6710 bhanan@mail.sdsu.edu, San Diego State Univ; Cc
Hanchar, John, 709 864 6785 jhanchar@mun.ca, Memorial Univ of Newfoundland; Cg
Hancock, Gregory, 61 02 4921 5090 greg.hancock@newcastle.edu.au, Univ of Newcastle; ZiGmm
Hancock, Steven, steven.hancock@ed.ac.uk, Edinburgh Univ; Zi
Hand, Kristen L., 717–702–2046 khand@pa.gov, DCNR- Pennsylvania Bur of Geo Survey; Ggr
Hand, Linda M., 650 574-6633 hand@smccd.edu, Coll of San Mateo; GgPgOg
Hand, Suzanne J., s.hand@unsw.edu.au, Univ of New South Wales; Pv
Haneberg, William, (505) 255-8005 New Mexico Tech; Ng
Hanell, Casey R., (360) 902-1439 casey.hanell@dnr.wa.gov, Washington Geo Survey; Gg
Haner, Andrew, (501) 683-0153 andrew.haner@arkansas.gov, Arkansas Geo Survey; Zi
Hanes, Daniel M., (314) 977-3703 dan.hanes@slu.edu, Saint Louis Univ; GesGm
Hanes, John A., (613) 533-6188 hanes@queensu.ca, Queen's Univ; Cc
Haney, Christa M., 662-268-1032 Ext 224 meloche@geosci.msstate.edu, Mississippi State Univ; Am
Haney, Jennifer, (570) 389-5305 jhaney@bloomu.edu, Bloomsburg Univ; Rn
Haney, Robert, rlhaney@nps.edu, Naval Postgrad Sch; Am
Hanger, Brendan, 303-273-3828 hanger@mines.edu, Colorado Sch of Mines; GiCg
Hanger, Rex, hangerr@uww.edu, Univ of Wisconsin, Whitewater; PigGs
Hanke, Steve H., (410) 516-7183 hanke@jhu.edu, Johns Hopkins Univ; Zn
Hankins, Katherine B., 404 413-5775 khankins@gsu.edu, Georgia State Univ; Zn
Hanks, Catherine L., 907-474-5562 chanks@gi.alaska.edu, Univ of Alaska, Fairbanks; Go
Hanna, Kevin, 250-807-9265 kevin.hanna@ubc.ca, Univ of British Columbia, Okanagan Campus; ZnRcZu
Hanna, Ruth L., 925-424-1319 rhanna@laspositascollege.edu, Las Positas Coll; Gg
Hanna, Stephen P., (540) 654-1490 shanna@umw.edu, Mary Washington Coll; Zr
Hannah, David, +44 (0)121 41 46925 d.m.hannah@bham.ac.uk, Univ of Birmingham; Hg
Hannah, Judith L., (970) 491-1329 jhannah@warnercnr.colostate.edu, Colorado State Univ; GiCc
Hannibal, Joseph T., (216) 231-4600 jhanniba@cmnh.org, Case Western Reserve Univ; Pi
Hannigan, Robyn, robyn.hannigan@umb.edu, Univ of Massachusetts, Boston; ClmCt
Hannington, Mark, mark.hannington@uottawa.ca, Univ of Ottawa; EgGuEm
Hanor, Jeffrey S., (225) 388-3418 hanor@geol.lsu.edu, Louisiana State Univ; Cl
Hanrahan, Janel, 802-626-6370 janel.hanrahan@northernvermont.edu, Northern Vermont Univ-Lyndon; AsZoRc
Hansel, Colleen, (508) 289-3738 chansel@whoi.edu, Woods Hole Oceanographic Inst; Cm
Hansen, Anthony R., 320-308-2009 arhansen@stcloudstate.edu, Saint Cloud State Univ; Am
Hansen, Christel D., +27 (0)12 420 4255 christel.hansen@up.ac.z, Univ of Pretoria; ZiGmZy
Hansen, Cindy, 778-782-8518 cdhansen@sfu.ca, Simon Fraser Univ; Gg
Hansen, Edward C., hansen@hope.edu, Hope Coll; GmpGc
Hansen, Jeffrey C., (505) 667-5043 jchansen@lanl.gov, Los Alamos; Zn
Hansen, Lars, +44 (1865) 272000 lars.hansen@earth.ox.ac.uk, Univ of Oxford; Gz
Hansen, Robert N., +27 (0)51 401 2712 hansenr@ufs.ac.za, Univ of the Free State; CgGe
Hansen, Samantha E., (205) 348-7089 shansen@ua.edu, Univ of Alabama; YsGt
Hansen, Thor A., thor.hansen@wwu.edu, Western Washington Univ; Pg
Hansen, Vicki L., (218) 726-8628 vhansen@d.umn.edu, Univ of Minnesota Duluth; GtXgGc
Hansen, William, (212) 346-1502 Pace Univ, New York Campus; As
Hansom, Jim, +4401413305406 jim.hansom@glasgow.ac.uk, Univ of Glasgow; OnGmZy
Hanson, Andrew D., (702) 895-1092 Univ of Nevada, Las Vegas; Co
Hanson, Gilbert N., (631) 632-8210 gilbert.hanson@stonybrook.edu, Stony Brook Univ; GgZe
Hanson, Howard, 561 297 2460 hphanson@fau.edu, Florida Atlantic Univ; AsOg
Hanson, Lindley S., lhanson@salemstate.edu, Salem State Univ; GmgOn
Hanson, Paul, (402) 472-7762 phanson2@unl.edu, Univ of Nebraska, Lincoln; Gm
Hanson, Richard E., (817) 257-7996 r.hanson@tcu.edu, Texas Christian Univ; GxvGt
Hanson, Sarah L., 517-264-3944 slhanson@adrian.edu, Adrian Coll; Giz
Hanson, William D., 501 683-0115 doug.hanson@arkansas.gov, Arkansas Geo Survey; EgnGs
Hanson, Jr., Alfred K., (401) 874-6899 akhanson@gso.uri.edu, Univ of Rhode Island; Oc
Hanu, Dan B., 0040232201464 bogdan.hanu@gmail.com, Alexandru Ioan Cuza; Yge
Hapke, Bruce W., (412) 624-8876 hapke@pitt.edu, Univ of Pittsburgh; Xy
Haq, Saad, (765) 496-7206 haq@purdue.edu, Purdue Univ; Gt
Har, Nicolae, +40745511022 nicolae.har@ubbcluj.ro, Babes-Bolyai Univ; GipNr
Hara, Tetsu, (401) 874-6509 thara@uri.edu, Univ of Rhode Island; Op
Harbaugh, John W., (650) 723-3365 harbaugh@pangea.stanford.edu, Stanford Univ; Gq
Harbert, William P., (412) 624-8874 harbert@pitt.edu, Univ of Pittsburgh; Ym
Harbor, David J., 540 458 8871 harbord@wlu.edu, Washington & Lee Univ; Gm

Harbor, Jon M., jharbor@purdue.edu, Purdue Univ; GmlGe
Hardage, Bob A., (512) 471-0300 bob.hardage@beg.utexas.edu, Univ of Texas at Austin, Jackson Sch of Geosciences; Yg
Hardage, Sarah M., 956 665 3023 sarah.fearnleyhardage@utrgv.edu, Univ of Texas, Rio Grande Valley; GmZiGg
Harder, Brian J., (225) 578-8533 bharde1@lsu.edu, Louisiana State Univ; Eo
Harder, Steven H., (915) 747-5746 harder@utep.edu, Univ of Texas, El Paso; Yg
Harder, Vicki, 915-747-5501 vmharder@utep.edu, Univ of Texas, El Paso; Gg
Hardgrove, Craig, (480) 727-2170 craig.hardgrove@asu.edu, Arizona State Univ; Xg
Hardin, Perry J., (801) 378-6062 perry_hardin@byu.edu, Brigham Young Univ; Zr
Harding, Chris, (515) 294-7521 charding@iastate.edu, Iowa State Univ of Sci & Tech; Gq
Hardison, Amber, akhardison@vims.edu, William & Mary; Cm
Hardy, R. Dean, 803-777-7593 dhardy@seoe.sc.edu, Univ of South Carolina; Zn
Hardy, Shaun J., (202) 478-7960 shardy@carnegiescience.edu, Carnegie Inst for Sci; Zn
Hargrave, Jennifer E., 337-482-0678 jhargrave@louisiana.edu, Univ of Louisiana at Lafayette; PvGrZe
Hargrave, Phyllis, 406-496-4606 phargrave@mtech.edu, Montana Tech of The Univ of Montana; Gg
Hargraves, Paul E., (401) 874-6241 pharg@gso.uri.edu, Univ of Rhode Island; Ob
Hargreaves, Bruce R., (610) 758-3683 brh0@lehigh.edu, Lehigh Univ; Ob
Hargreaves, Tom, +44 (0)1603 59 3116 tom.hargreaves@uea.ac.uk, Univ of East Anglia; Ge
Hari, Kosiyath R., krharigeology@gmail.com, Pt. Ravishankar Shukla Univ; GiiCp
Haritashya, Umesh, 937-229-2939 umesh.haritashya@notes.udayton.edu, Univ of Dayton; ZrGm
Harley, Grant , 208-885-0950 gharley@uidaho.edu, Univ of Idaho; ZyAtRn
Harley, Grant, 208-885-0950 gharley@uidaho.edu, Univ of Idaho; ZynZn
Harley, Simon L., +44 (0) 131 650 4839 simon.harley@ed.ac.uk, Edinburgh Univ; Gt
Harlim, John, jharlim@psu.edu, Pennsylvania State Univ, Univ Park; Am
Harlow, George E., 212-769-5378 gharlow@amnh.org, American Mus of Natural History; Gz
Harmon, Nicholas, +44 (0)23 80594783 n.harmon@noc.soton.ac.uk, Univ of Southampton; Yg
Harms, Tekla A., (413) 542-2711 taharms@amherst.edu, Amherst Coll; Gt
Harnett, Erika M., 206-543-0212 eharnett@uw.edu, Univ of Washington; XyGev
Harnik, Nili, 03-640-6359 harnik@tau.ac.il, Tel Aviv Univ; As
Harnik, Paul, (315) 228-6802 pharnik@colgate.edu, Colgate Univ; PieOb
Haroldson, Erik L., 931-221-7449 haroldsone@apsu.edu, Austin Peay State Univ; EdGzCg
Harper, David, +44 (0) 191 33 47143 david.harper@durham.ac.uk, Durham Univ; Pg
Harper, Joel, (406) 243-2341 joel.harper@umontana.edu, Univ of Montana; GlZc
Harper, Jr., Charles W., (405) 325-7725 charper@gcn.ou.edu, Univ of Oklahoma; Pi
Harpold, Adrian, (775) 784-6759 aharpold@cabnr.unr.edu, Univ of Nevada, Reno; HqsZr
Harpp, Karen, (315) 228-7211 kharpp@colgate.edu, Colgate Univ; GvCgGi
Harr, Patrick A., 831-656-3787 paharr@nps.edu, Naval Postgrad Sch; Am
Harrap, Rob, 613-533-2553 harrap@queensu.ca, Queen's Univ; Gc
Harrell, Michael, 206-543-0367 mdh666@uw.edu, Univ of Washington; Ze
Harrell, Jr., T. L., 205-247-3559 tlharrell@gsa.sate.al.us, Geo Survey of Alabama; PvGg
Harrington, Charles D., (505) 667-0078 charrington@lanl.gov, Los Alamos; Gm
Harrington, Glenn, glenn.harrington@flinders.edu.au, Flinders Univ; Hw
Harrington, Jerry Y., harring@meteo.psu.edu, Pennsylvania State Univ, Univ Park; Am
Harrington, Philip, harrinp@sunysuffolk.edu, Suffolk County Comm Coll, Ammerman Campus; Zg
Harrington, Rebecca, +49-721-6084625 rebecca.harrington@kit.edu, Karlsruhe Inst of Tech; Ys
Harris, Ann, ann.harris@eku.edu, Eastern Kentucky Univ; GgZy
Harris, Brett D., +618 9266-3089 b.harris@curtin.edu.au, Curtin Univ; HwYve
Harris, C, harrisc@cf.ac.uk, Cardiff Univ; Ge
Harris, Chris, 021-650-4886 chris.harris@uct.ac.za, Univ of Cape Town; GiCsGg
Harris, Clay D., 615-904-8019 clay.harris@mtsu.edu, Middle Tennessee State Univ; Gse
Harris, Courtney K., (804) 684-7194 ckharris@vims.edu, William & Mary; On
Harris, David C., (859) 323-0545 dcharris@uky.edu, Univ of Kentucky; Go
Harris, DeVerle P., (520) 621-6024 dharris@geo.arizona.edu, Univ of Arizona; Eg
Harris, Glendon H., (706) 542-2968 Univ of Georgia; Sb
Harris, James B., 601-974-1343 harrijb@millsaps.edu, Millsaps Coll; Ye
Harris, Jerry M., (650) 723-0496 Stanford Univ; Ys
Harris, Lyal B., lyal.harris@ete.inrs.ca, Univ du Quebec; GcYg
Harris, M. Scott, (843) 953-0864 harriss@cofc.edu, Coll of Charleston; OnGam
Harris, Mark T., (414) 229-5483 mtharris@uwm.edu, Univ of Wisconsin, Milwaukee; GsrGd
Harris, Michelle, (360) 623-8472 michelle.harris@centralia.edu, Centralia Coll; Gg
Harris, Nicholas, (780) 492-0356 nharris@ualberta.ca, Univ of Alberta; Gso
Harris, Nigel B., n.b.w.harris@open.ac.uk, Open Univ; GtiGp
Harris, Pat, (936) 294-1452 pat-harris@shsu.edu, Sam Houston State Univ; ClGdSc
Harris, Randa R., (678) 839-4056 rharris@westga.edu, Univ of West Georgia; Hs
Harris, Richard S., (905) 525-9140 (Ext. 27216) harrisr@mcmaster.ca, McMaster Univ; Zu
Harris, Robert N., (541) 737-4370 rharris@coas.oregonstate.edu, Oregon State Univ; Yhr
Harris, Robin F., (608) 263-5691 rfharris@wisc.edu, Univ of Wisconsin, Madison; Sb
Harris, Ron, (801) 422-9264 rharris@byu.edu, Brigham Young Univ; Gc
Harris, Sara, (604) 822-5674 sara@eos.ubc.ca, Univ of British Columbia; Ou
Harris, Virginia M., (860) 685-2244 vharris@wesleyan.edu, Wesleyan Univ; Zn
Harris, W. Burleigh, (910) 962-3492 harrisw@uncw.edu, Univ of North Carolina, Wilmington; GrdGs
Harris, Jr., Willie G., (352) 294-3110 apatite@ufl.edu, Univ of Florida; Scd
Harrison, Bruce I., (575) 835-5864 bruce@nmt.edu, New Mexico Tech; GmSof
Harrison, Cheryl, 956 882-8456 cheryl.harrison@utrgv.edu, Univ of Texas, Rio Grande Valley; Op
Harrison, Conor, 803-576-6010 cmharris@mailbox.sc.edu, Univ of South Carolina; Zn
Harrison, Don E., (206) 526-6225 harrison@pmel.noaa.gov, Univ of Washington; Op
Harrison, Halstead, harrison@atmos.washington.edu, Univ of Washington; As
Harrison, Linda , (269) 387-8642 linda.harrison@wmich.edu, Western Michigan Univ; GgPmZn
Harrison, Michael J., (931) 372-3751 mharrison@tntech.edu, Tennessee Tech Univ; Gc
Harrison, Peter, peter.harrison@scu.edu.au, Southern Cross Univ; Cm
Harrison, Richard, +44 (0) 1223 333380 rjh40@esc.cam.ac.uk, Univ of Cambridge; Gz
Harrison, Robert S., (516) 299-2318 Long Island Univ, C.W. Post Campus; Zy
Harrison, Roy M., +44 (0)121 41 43494 r.m.harrison@bham.ac.uk, Univ of Birmingham; As
Harrison, T. Mark, (310) 825-7970 tmh@argon.ess.ucla.edu, Univ of California, Los Angeles; CcGt
Harrison, Wendy J., (303) 273-3821 wharriso@mines.edu, Colorado Sch of Mines; Cl
Harrison, III, William B., (269) 387-8691 william.harrison_iii@wmich.edu, Western Michigan Univ; GsrPg
Harry, Dennis L., (970) 491-2714 dharry@warnercnr.colostate.edu, Colorado State Univ; Yg
Harsh, James B., (509) 335-3650 harsh@wsu.edu, Washington State Univ; Sc
Harshvardhan, , harsh@purdue.edu, Purdue Univ; As
Hart, Brian R., bhart@uwo.ca, Western Univ; Ca
Hart, Craig J., (604) 822-5149 chart@eoas.ubc.ca, Univ of British Columbia; EmGiCc
Hart, David J., (608) 262-2307 david.hart@wisc.edu, Univ of Wisconsin, Madison, Division of Extension; HwYg
Hart, Evan A., (931) 372-3121 ehart@tntech.edu, Tennessee Tech Univ; Zy
Hart, Malcolm B., +441752584761 mhart@plymouth.ac.uk, Univ of Plymouth; PmGur
Hart, Richard L., (630) 252-5839 rlhart@anl.gov, Argonne; As
Hart, Roger M., rmhart1@ccri.edu, Comm Coll of Rhode Island; Gg
Hart, Stanley R., (520) 625-4543 shart@whoi.edu, Woods Hole

Oceanographic Inst; CgGvCp
Hart, William K., (513) 529-3217 hartwk@miamioh.edu, Miami Univ; GivCc
Harte, Michael, mharte@coas.oregonstate.edu, Oregon State Univ; Gu
Hartel, Peter G., (706) 542-0898 Univ of Georgia; Sb
Harter, Thomas, 530-752-2709 thharter@ucdavis.edu, Univ of California, Davis; Hw
Hartley, Susan, (218)279-2661 s.hartley@lsc.edu, Lake Superior Coll; Gg
Hartman, Joseph H., (701) 777-5055 joseph.hartman@engr.und.edu, Univ of North Dakota; Pi
Hartmann, Dennis L., (206) 543-7460 dennis@atmos.washington.edu, Univ of Washington; As
Hartmann, Heike, 724.738.4376 heike.hartmann@sru.edu, Slippery Rock Univ; Amt
Hartnett, Hilairy E., (480) 965-5593 h.hartnett@asu.edu, Arizona State Univ; ComOc
Hartog, Renate, 206-685-7079 jrhartog@uw.edu, Univ of Washington; Ys
Hartse, Hans E., (505) 664-8495 Los Alamos; Ys
Hartshorn, Tony, anthony.hartshorn@montana.edu, Montana State Univ; So
Harty, John P., jharty1@jccc.edu, Johnson County Comm Coll; Zy
Harun, Nina, 907-451-5085 nina.harun@alaska.gov, Alaska Division of Geological & Geophysical Surveys; Go
Harvey, Cyril H., (336) 316-2238 charvey@guilford.edu, Guilford Coll; Gr
Harvey, H. Rodger, 757-683-5657 rharvey@odu.edu, Old Dominion Univ; Com
Harvey, James T., harvey@mlml.calstate.edu, Moss Landing Marine Lab; Ob
Harvey, Jason, +44(0) 113 34 34033 feejh@leeds.ac.uk, Univ of Leeds; Cg
Harvey, Jon, 970-247-7129 jeharvey@fortlewis.edu, Fort Lewis Coll; Gm
Harvey, Omar R., 817-257-4272 omar.harvey@tcu.edu, Texas Christian Univ; GeHwg
Harvey, Ralph P., (216) 368-0198 rph@case.edu, Case Western Reserve Univ; Xmg
Harvey, Tom, +440116 252 3644 thph2@le.ac.uk, Leicester Univ; Pg
Harwood, David M., (402) 472-2648 dharwood1@unl.edu, Univ of Nebraska, Lincoln; PmGul
Hasan, Khaled W., khaled.hasan@austincc.edu, Austin Comm Coll District; HsZri
Hasan, M A., mahasan@du.ac.bd, Univ of Dhaka ; CqHwGg
Hasan, Mainul, (514) 398-4755 McGill Univ; Nx
Hasan, Mohamed Ali, 03-79674203/4141 alihasan@um.edu.my, Univ of Malaya; Hg
Hasan, Syed E., (816) 235-2976 hasans@umkc.edu, Univ of Missouri, Kansas City; Ng
Hasbargen, Leslie E., (607) 436-2741 leslie.hasbargen@oneonta.edu, SUNY, Oneonta; GmHgZr
Haselwander, Airin J., (573) 368-2196 airin.haselwander@dnr.mo.gov, Missouri Dept of Nat Res; Gg
Hasenmueller, Elizabeth, (314) 977-7518 elizabeth.hasenmueller@slu.edu, Saint Louis Univ; ClHs
Hasenmueller, Nancy R., (812) 855-1360 hasenmue@indiana.edu, Indiana Univ, Bloomington; Gre
Hasiotis, Stephen T., (785) 864-4941 hasiotis@ku.edu, Univ of Kansas; Pe
Hass, Alisa L., (615) 898-2730 alisa.hass@mtsu.edu, Middle Tennessee State Univ; AtZy
Hassani, Ferri, (514) 398-4755 McGill Univ; Nm
Hassanpour Sedghi, Mohammad, +98 (411) 339 2697 Univ of Tabriz; YsNeYg
Hasselbo, Stephen, +44 01326 253651 s.p.hesselbo@exeter.ac.uk, Exeter Univ; Gs
Hassett, John J., (217) 333-9472 Univ of Illinois, Urbana-Champaign; Sc
Hastenrath, Stefan L., (608)262-3659 slhasten@wisc.edu, Univ of Wisconsin, Madison; As
Hastie, Warwick W., 031-2602519 hastiew@ukzn.ac.za, Univ of KwaZulu-Natal; NrgGc
Hastings, Meredith, 401-863-3658 meredith_hastings@brown.edu, Brown Univ; As
Hastings, Philip A., (858) 822-2913 phastings@ucsd.edu, Univ of California, San Diego; Ob
Haszeldine, R. S., +44 (0) 131 650 8549 stuart.haszeldine@ed.ac.uk, Edinburgh Univ; GoRmZo
Hatch, Christine, 413 577-2245 chatch@geo.umass.edu, Univ of Massachusetts, Amherst; HwgSp
Hatcher Jr., Robert D., 865-974-6565 bobmap@utk.edu, Univ of Tennessee, Knoxville; Gct
Hatheway, Richard B., hatheway@geneseo.edu, SUNY, Geneseo; GxzGp
Hattori, Keiko, khattori@uottawa.ca, Univ of Ottawa; CgEmGz
Hauck, II, Steven A., (216) 368-3675 hauck@case.edu, Case Western Reserve Univ; XyYgXg
Hauer, Kendall L., (513) 529-3220 hauerkl@miamioh.edu, Miami Univ; GgZeCg
Haugerud, Ralph, 206-713-7453 haugerud@uw.edu, Univ of Washington; Zi
Hauksson, Egill, (626) 395-6954 hauksson@gps.caltech.edu,CalTech; YsGtYg
Hausback, Brian, (916) 278-6521 hausback@csus.edu, Sacramento State Univ; Gv
Hauschild, Brian K., brian.k.hauschild@des.nh.gov, New Hampshire Geo Survey; HgsHw
Hauser, Ernest C., 937 775-3443 ernest.hauser@wright.edu, Wright State Univ; YeGtYu
Hausrath, Elisabeth M., 702-895-1134 elisabeth.hausrath@unlv.edu, Univ of Nevada, Las Vegas; CgScCl
Haussmann, Natalie S., +27 (0)12 420 4049 natalie.haussmann@up.ac.za, Univ of Pretoria; Gm
Hauswirth, Scott, (818) 677-4880 scott.hauswirth@csun.edu, California State Univ, Northridge; HgCg
Hautala, Susan L., (206) 543-0596 susanh@ocean.washington.edu, Univ of Washington; Op
Hauton, Chris, +44 (0)23 80595784 ch10@noc.soton.ac.uk, Univ of Southampton; Ob
Havenith, Hans-Balder, 32 4 3662035 hb.havenith@ulg.ac.be, Univ de Liege; Ge
Haverluk, Terry W., (719) 333-8746 usafa.dfeg@usafa.edu, United States Air Force Acad; Zg
Havlin, John L., (919) 515-2655 havlin@ncsu.edu, North Carolina State Univ; Sc
Hawkins, Andrew, 410-704-3187 dhawkins@towson.edu, Towson Univ; Pg
Hawkins, David, 781-283-3554 dhawkins@wellesley.edu, Wellesley Coll; Gzi
Hawkins, James W., (858) 534-2161 jhawkins@ucsd.edu, Univ of California, San Diego; OuGip
Hawkins, John F., (334) 844-4894 jfh0005@auburn.edu, Auburn Univ; GcZeGt
Hawkins, Lawrence, +44 (0)23 80593426 leh@noc.soton.ac.uk, Univ of Southampton; Ob
Hawkins, Michael A., (518) 486-2011 michael.hawkins@nysed.gov, New York State Geo Survey; Gz
Hawkins, Terry, (573) 368-2164 terry.hawkins@dnr.mo.gov, Missouri Dept of Nat Res; Geg
Hawkins, Timothy W., (717) 477-1662 twhawk@ship.edu, Shippensburg Univ; AsmHg
Hawkins, Ward L., (505) 667-5835 whawkins@lanl.gov, Los Alamos; Zg
Hawley, Bob, robert.l.hawley@dartmouth.edu, Univ of Washington; Gl
Hawley, John W., (505) 255-4847 hgeomatters@gmail.com, New Mexico Tech; GeHwGr
Hawley, Rebecca D., (928) 523-1251 d.hawley@nau.edu, Northern Arizona Univ; Zn
Hawley, Robert L., 603-646-2373 robert.hawley@dartmouth.edu, Dartmouth Coll; Gl
Hawman, Robert B., (706) 542-2398 rob@seismo.gly.uga.edu, Univ of Georgia; YseYu
Hawthorne, Frank C., (204) 474-8861 frank_hawthorne@umanitoba.ca, Univ of Manitoba; Gz
Hawthorne Allen, Alice M., 304-384-6273 amhallen@concord.edu, Concord Univ; Xa
Hay, Alex E., (902) 494-6657 alex.hay@dal.ca, Dalhousie Univ; OpnGu
Hay, Iain, iain.hay@flinders.edu.au, Flinders Univ; Zyy
Hay, Rodrick, (310) 243-2547 rhay@csudh.edu, California State Univ, Dominguez Hills; Zr
Hayashi, Masaki, (403) 220-2794 hayashi@ucalgary.ca, Univ of Calgary; HwSp
Hayden, Geoffrey W., (912) 752-2597 Mercer Univ; Zm
Hayden, Martha C., (801) 537-3311 nrugs.mhayden@state.ut.us, Utah Geo Survey; Pv
Hayes, Alexander G., (607) 255-1712 agh4@cornell.edu, Cornell Univ; XgZrGs
Hayes, Christopher T., (228) 688-3469 christopher.t.hayes@usm.edu, Univ of Southern Mississippi; Oc
Hayes, Garry F., ghayes@csustan.edu, California State Univ, Stanislaus; Gg
Hayes, Jorden L., (717) 245-8303 hayesjo@dickinson.edu, Dickinson Coll; Ygs
Hayes, Michael J., (402) 472-4271 mhayes2@unl.edu, Univ of Nebraska, Lincoln; As
Hayes, Miles O., 504-280-6325 mhayes@uno.edu, Univ of New Orleans; Gs
Hayes, Van E., 813-253-7685 vhayes@hccfl.edu, Hillsborough Comm Coll; Zg
Hayman, Nicholas W., 405-325-7968 hayman@ou.edu, Univ of Oklahoma; Gct

Hayman, Patrick C., +61 7 3138 5259 patrick.hayman@qut.edu.au, Queensland Univ of Tech; EgGvi
Haymet, Anthony D., (858) 534-2827 thaymet@ucsd.edu, Univ of California, San Diego; OcZgn
Haynes, Kyle J., (540) 837-1758 kjh8w@virginia.edu, Univ of Virginia; Zn
Haynes, Shannon, 609-258-4101 sjhaynes@princeton.edu, Princeton Univ; CsOoGg
Haynes, Jr., C. Vance, (520) 621-6307 Univ of Arizona; Ga
Haynie, Kirstie, klhaynie@buffalo.edu, SUNY, Buffalo; Gv
Hayob, Jodie, (540) 654-1425 jhayob@umw.edu, Univ of Mary Washington; GxzGg
Hays, James D., (845) 365-8403 jimhays@ldeo.columbia.edu, Columbia Univ; PemOu
Hays, Phillip D., (479) 575-7343 pdhays@uark.edu, Univ of Arkansas, Fayetteville; HwCsGg
Hayward, Chris L., +44 (0) 131 650 5827 chris.hayward@ed.ac.uk, Edinburgh Univ; Ga
Hayward, Christopher T., hayward@smu.edu, Southern Methodist Univ; Ys
Haywick, Douglas W., (334) 460-6381 dhaywick@jaguar1.usouthal.edu, Univ of South Alabama; Gs
Haywood, Alan, +44(0) 113 34 38657 earamh@leeds.ac.uk, Univ of Leeds; Pe
Hazel, McGoff, h.j.mcgoff@rdg.ac.uk, Univ of Reading; Pg
Hazel, Jr., Joseph E., (928) 523-9145 joseph.hazel@nau.edu, Northern Arizona Univ; Gs
Hazen, Helen, 303-871-2672 helen.hazen@du.edu, Univ of Denver; Zy
Hazen, Robert M., (202) 478-8962 rhazen@carnegiescience.edu, Carnegie Inst for Sci; Gz
Hazen, Terry C., 865-974-7709 tchazen@utk.edu, Univ of Tennessee, Knoxville; Po
He, Changming, (302) 831-4917 hchm@udel.edu, Univ of Delaware; GqHg
He, Helen, +44 (0)1603 59 2091 yi.he@uea.ac.uk, Univ of East Anglia; Pe
He, Hong S., 15738827717 heh@missouri.edu, Univ of Missouri; ZicZi
He, Ruoying, 919-513-0943 ruoying_he@ncsu.edu, North Carolina State Univ; Op
He, Zhenli, 772-468-3922 zhe@ufl.edu, Univ of Florida; Sc
He, Zhuoya, 713.348.4886 zhuoya.he@rice.edu, Rice Univ; Ou
Head, Elisabet M., 773-442-6055 e-head@neiu.edu, Northeastern Illinois Univ; GvZrCg
Head, Martin J., (905) 688-5550 (Ext. 5216) mjhead@brocku.ca, Brock Univ; Pl
Head, III, James W., (401) 863-2526 james_head_iii@brown.edu, Brown Univ; XgGvt
Heal, Kate V., +44 (0) 131 650 5420 k.heal@ed.ac.uk, Edinburgh Univ; Co
Heald, Colette, (617) 324-5666 heald@mit.edu, Massachusetts Inst of Tech; As
Healey, Michael, (604) 822-4705 mhealey@eos.ubc.ca, Univ of British Columbia; Ob
Heaman, Larry M., (780) 492-2778 larry.heaman@ualberta.ca, Univ of Alberta; Cc
Heaney, Peter J., (814) 865-6821 pjheaney@psu.edu, Pennsylvania State Univ, Univ Park; GzClGe
Hearn, Thomas M., (505) 646-5076 thearn@nmsu.edu, New Mexico State Univ, Las Cruces; Ys
Heath, G. Ross, (206) 543-3153 rheath@u.washington.edu, Univ of Washington; Cm
Heath, Kathleen M., (812) 237-3004 kheath@indstate.edu, Indiana State Univ; Pb
Heatherington, Ann L., (352) 392-6220 aheath@ufl.edu, Univ of Florida; CcGti
Heaton, Jill, (775) 784-8056 jheaton@unr.edu, Univ of Nevada, Reno; ZniZy
Heaton, Richard C., (505) 667-1141 Los Alamos; Zn
Heaton, Thomas H., (626) 395-6897 heaton@gps.caltech.edu,CalTech; Ne
Heaton, Timothy H., (605) 677-6122 timothy.heaton@usd.edu, Univ of South Dakota; PvOg
Heatwole, Charles A., (212) 772-5323 cah@geo.hunter.cuny.edu, Hunter Coll (CUNY); Zn
Heavers, Richard, (401) 254-3095 Roger Williams Univ; Op
Hebda, Richard J., (250) 387-5493 Univ of Victoria; Pl
Hebert, David, (902) 426-1216 david.hebert@dfo-mpo.gc.ca, Dalhousie Univ; Op
Hebert, Rejean J., 418-656-3137 herbert@ggl.ulaval.ca, Univ Laval; Gi
Heck, Frederick R., (231) 591-2588 heckf@ferris.edu, Ferris State Univ; GgZg
Heck, Richard, (519) 824-4120 (Ext. 52450) rheck@uoguelph.ca, Univ of Guelph; Sd
Heckathorn, Scott, 419-530-4328 scott.heckathorn@utoledo.edu, Univ of Toledo; Zn
Heckel, Philip H., (319) 335-1804 philip-heckel@uiowa.edu, Univ of Iowa; GrPsGd
Heckel, Wolfgang, 089/2180 4331 heckl@lmu.de, Ludwig-Maximilians-Univ Muenchen; Gz
Heckert, Andrew B., (828) 262-7609 heckertab@appstate.edu, Appalachian State Univ; PvgZe
Hedberg, Russell, 717-477-1515 rchedberg@ship.edu, Shippensburg Univ; ZiSoZc
Hedenquist, Jeffrey, jhedenquist@gmail.com, Univ of Ottawa; Eg
Hedquist, Brent, 361-593-3586 brent.hedquist@tamuk.edu, Texas A&M Univ, Kingsville; ZyAtZi
Heenan, Cleo J., (605) 394-2461 cleo.heenan@sdsmt.edu, South Dakota Sch of Mines & Tech; Zn
Heermance, Richard V., (818) 677-4357 richard.heermance@csun.edu, California State Univ, Northridge; GrCc
Hegarty, Patricia, +353 21 4902533 p.hegarty@ucc.ie, Univ Coll Cork; Zn
Hegerl, Gabriele C., +44 (0) 131 651 9092 gabi.hegerl@ed.ac.uk, Edinburgh Univ; As
Hegewisch, Katherine, 208-885-6622 khegewisch@uidaho.edu, Univ of Idaho; Zo
Hegg, Dean A., (206) 543-1984 deanhegg@atmos.washington.edu, Univ of Washington; As
Hegna, Thomas A., (309) 298-1366 ta-hegna@wiu.edu, Western Illinois Univ; PigPg
Hegner, Ernst, 089/2180 4274 hegner@lmu.de, Ludwig-Maximilians-Univ Muenchen; Gg
Heibel, Todd, 909-384-8638 theibel@sbccd.cc.ca.us, San Bernardino Valley Coll; Zyi
Heid, Kelly L., 616-331-3783 heidke@gvsu.edu, Grand Valley State Univ; Ze
Heidinger, Andrew, 608-263-6757 heidinger@ssec.wisc.edu, Univ of Wisconsin, Madison; Zr
Heidlauf, Lisa, lisa.heidlauf@wheaton.edu, Wheaton Coll; Gg
Heigold, Paul C., (630) 252-7861 Argonne; Yg
Heiken, Grant, (505) 667-8477 heiken@lanl.gov, Los Alamos; Gv
Heikes, Brian G., (401) 874-6810 zagar@notos.gso.uri.edu, Univ of Rhode Island; As
Heil, Dean, (505) 646-3405 New Mexico State Univ, Las Cruces; Sc
Heil, Melissa, 13094387846 mkheil@ilstu.edu, Illinois State Univ; Zy
Heim, Noel A., 617-627-2207 noel.heim@tufts.edu, Tufts Univ; PgiPs
Heimann, Adriana, 252 328 5206 heimanna@ecu.edu, East Carolina Univ; GzpGi
Heimann, William H., (785) 625-5663 whheimann@fhsu.edu, Fort Hays State Univ; HgwGm
Heimbach, Patrick, heimbach@utexas.edu, Univ of Texas, Austin; Og
Heimpel, Moritz, (780) 492-3519 mheimpel@phys.ualberta.ca, Univ of Alberta; Zn
Heimsath, Arjun, 480-965-5585 arjun.heimsath@asu.edu, Arizona State Univ; Gc
Hein, Christopher J., 804-684-7533 hein@vims.edu, William & Mary; Ou
Hein, Kim A., 011 717 6623 kim.ncube0hein@wits.ac.za, Univ of the Witwatersrand; GgEmGc
Heinrich, Paul V., 225-578-4398 heinric@lsu.edu, Louisiana State Univ; GsaGm
Heinz, Dion L., (773) 702-3046 heinz@uchicago.edu, Univ of Chicago; GyzYx
Heinzel, Chad E., (319) 273-6168 chad.heinzel@uni.edu, Univ of Northern Iowa; GmaSa
Heise, Elizabeth, 956 882-6769 elizabeth.heise@utrgv.edu, Univ of Texas, Rio Grande Valley; OnZeGg
Heiss, James, (978) 934-6304 james_heiss@uml.edu, Univ of Massachusetts, Lowell; Hw
Heitman, Mary P., marypat.heitman@dnr.iowa.gov, Iowa Dept of Nat Res; Zn
Heitmuller, Franklin, 601-266-5423 franklin.heitmuller@usm.edu, Univ of Southern Mississippi; GmsGe
Heitz, Leroy F., lheitz@triton.uog.edu, Univ of Guam; Hg
Heizler, Lynn, 575-835-5166 lynn.heizler@nmt.edu, New Mexico Tech; Ca
Heizler, Matthew T., (575) 835-5244 matt@nmt.edu, New Mexico Tech; Cc
Helba, Hossam EL-Din A., 002-03-3921595 kahralran2009@yahoo.com, Alexandria Univ; GgZm
Helenes, Javier, jhelenes@cicese.mx, Centro de Investigación Científica y de Educación Superior de Ensenada; Gr
Helgers, Karen, (845) 687-5224 helgersk@sunyulster.edu, SUNY, Ulster County Comm Coll; GeZgy
Helie, Jean-François, 514-987-3000 #2413 helie.jean-francois@uqam.ca, Univ du Quebec a Montreal; Cs
Heliker, Christina C., (808) 967-8807 cheliker@usgs.gov, Univ of Hawai'i, Hilo; Gv
Heller, Matthew J., (434) 951-6351 matt.heller@energy.virginia.gov,

Geology and Mineral Resources Program; GgRnGe
Hellio, Gabrielle, +44 023 80599258 g.c.hellio@soton.ac.uk, Univ of Southampton; Yg
Helmberger, Donald V., (626) 395-6998 helm@gps.caltech.edu,CalTech; Ys
Helmick, Sara B., (505) 667-9583 Los Alamos; Zn
Helmke, Martin F., 610-436-3565 mhelmke@wcupa.edu, West Chester Univ; HwSdZg
Helmke, Philip A., (608) 263-4947 pahelmke@wisc.edu, Univ of Wisconsin, Madison; Sc
Helmke, Vicky, (610) 738-0308 vhelmke@wcupa.edu, West Chester Univ; Gg
Helmstaedt, Herwart, (613) 533-6175 Queen's Univ; Gc
Helmuth, Brian, 7815817370 x307 b.helmuth@northeastern.edu, Northeastern Univ; OnZcRc
Helper, Mark A., (512) 471-1009 helper@jsg.utexas.edu, Univ of Texas, Austin; GctXg
Helsdon, John H., (605) 394-2291 john.helsdon@sdsmt.edu, South Dakota Sch of Mines & Tech; As
Hembree, Daniel, (740) 597-1495 hembree@ohio.edu, Ohio Univ; Pi
Hemler, Deb, (304) 367-4393 dhemler@fairmontstate.edu, Fairmont State Univ; Ze
Hemming, N. G., 718-997-3335 Queens Coll (CUNY); CaOcGg
Hemming, Sidney, sidney@ldeo.columbia.edu, Columbia Univ; Cg
Hemphill-Haley, Eileen, emh21@humboldt.edu, Humboldt State Univ; Pm
Hemphill-Haley, Mark, (707) 826-3933 mah54@humboldt.edu, Humboldt State Univ; Gt
Hemsley, A, hemsleyar@cf.ac.uk, Cardiff Univ; Pl
Hemsley, Alan, +44(0)29 208 75367 hemsleyar@cardiff.ac.uk, Univ of Wales; Pl
Hemzacek Laukant, Jean M., 773/442-6056 j-hemzacek@neiu.edu, Northeastern Illinois Univ; Sc
Hence, Deanna A., 217-244-6014 dhence@illinois.edu, Univ of Illinois, Urbana-Champaign; AsmAp
Hencher, Steve, s.r.hencher@leeds.ac.uk, Univ of Leeds; Ng
Hendershott, Myrl C., (858) 534-5705 mhendershott@ucsd.edu, Univ of California, San Diego; Op
Henderson, Charles M., (403) 220-6170 henderson@geo.ucalgary.ca, Univ of Calgary; Pm
Henderson, Gideon, gideon.henderson@earth.ox.ac.uk, Univ of Oxford; Cg
Henderson, Gina R., 410-293-6555 ghenders@usna.edu, United States Naval Acad; As
Henderson, Grant S., (416) 978-6041 henders@geology.utoronto.ca, Univ of Toronto; Gz
Henderson, Miles, (432) 552-2245 henderson_m@utpb.edu, Univ of Texas Permian Basin; GsdGg
Henderson, Naomi, naomi@ldeo.columbia.edu, Columbia Univ; Yr
Henderson, Paul, (508) 289-3466 phenderson@whoi.edu, Woods Hole Oceanographic Inst; Cm
Henderson, Robert, 747814746 bob.henderson@jcu.edu.au, James Cook Univ; GtPiGg
Henderson, Wayne G., (657) 278-3561 whenderson@fullerton.edu, California State Univ, Fullerton; PgGg
Hendrey, George, 718-997-3325 george.hendrey@qc.cuny.edu, Queens Coll (CUNY); Zg
Hendricks, Jonathan R., 6072736623 x120 jrh42@cornell.edu, Paleontological Research Inst; Pig
Hendricks, Jonathan R., 607-273-6623 x120 jrh42@cornell.edu, Cornell Univ; Zg
Hendrickx, Jan M., (575) 835-5892 hendrick@nmt.edu, New Mexico Tech; Hw
Hendrikx, Jordy, (406) 994-6918 jordy.hendrikx@montana.edu, Montana State Univ; ZnHsAs
Hendrix, Marc S., 406-243-5867 marc.hendrix@umontana.edu, Univ of Montana; Gs
Hendrix, Thomas E., 616-331-3728 Grand Valley State Univ; Gc
Hendry, Jim, (306) 966-5720 jim.hendry@usask.ca, Univ of Saskatchewan; HwClHy
Hendy, Ingrid, 734615-6892 ihendy@umich.edu, Univ of Michigan; Pg
Henika, William S., (540) 231-4298 bhenika@vt.edu, Virginia Tech; Ng
Henkel, Torsten, +44 0161 275-3810 torsten.henkel@manchester.ac.uk, Univ of Manchester; Cs
Henkes, Gregory, 631-632-2905 gregory.henkes@stonybrook.edu, Stony Brook Univ; Cs
Henley, Sian F., +44 (0) 131 650 7010 s.f.henley@ed.ac.uk, Edinburgh Univ; CmOcCs
Hennemeyer, Marc, 089/2180 4340 marc@hennemeyer.de, Ludwig-Maximilians-Univ Muenchen; Gz
Hennige, Sebastian, s.hennige@ed.ac.uk, Edinburgh Univ; Zc
Henning, Stanley J., (515) 294-7846 sjhennin@iastate.edu, Iowa State Univ of Sci & Tech; So
Hennings, Peter, 512-471-0156 peter.hennings@beg.utexas.edu, Univ of Texas, Austin; GcoYs
Hennon, Chris, (828) 232-5159 chennon@unca.edu, Univ of North Carolina, Asheville; AstZn
Henrici, Amy C., (412) 622-3265 henricia@carnegiemnh.org, Carnegie Mus of Natural History; Pv
Henriksen, Mona, mona.henriksen@nmbu.no, Norwegian Univ of Life Scis (NMBU); ZyGs
Henry, Christopher D., (775) 682-8753) chenry@unr.edu, Univ of Nevada, Reno; GtvCc
Henry, Darrell J., (225) 388-2693 dhenry@geol.lsu.edu, Louisiana State Univ; Gp
Henry, Eric J., 910-962-7622 henrye@uncw.edu, Univ of North Carolina, Wilmington; Hw
Henry, Kathleen M., (217) 244-8994 kmhenry@illinois.edu, Illinois State Geo Survey; Gg
Henry, Lea-Anne, l.henry@ed.ac.uk, Edinburgh Univ; Zn
Henry, Tiernan, +353 (0)91 495 096 tiernan.henry@nuigalway.ie, Natl Univ of Ireland Galway; HwyEg
Henson, Harvey, 618-536-6666 henson@geo.siu.edu, Southern Illinois Univ Carbondale; Yg
Henstock, Tim, +44 (0)23 80596491 then@noc.soton.ac.uk, Univ of Southampton; YgrYe
Hentz, Tucker F., (512) 471-7281 tucker.hentz@beg.utexas.edu, Univ of Texas at Austin, Jackson Sch of Geosciences; GrsGo
Henyey, Thomas L., (213) 740-5832 henyey@usc.edu, Univ of Southern California; Yg
Henßel, Katja, 089/2180 6624 k.henssel@lrz.uni-muenchen.de, Ludwig-Maximilians-Univ Muenchen; Pg
Heo, Joon, heo_j@utpb.edu, Univ of Texas Permian Basin; Hg
Hepburn, J. Christopher, (617) 552-3642 john.hepburn@bc.edu, Boston Coll; Gpt
Hepner, George F., (801) 581-6021 george.hepner@geog.utah.edu, Univ of Utah; Zy
Hepner, Tiffany, (512) 475-9572 tiffany.hepner@beg.utexas.edu, Univ of Texas at Austin, Jackson Sch of Geosciences; Og
Herbers, Thomas H., (831) 656-2917 herbers@nps.edu, Naval Postgrad Sch; Op
Herbert, Gregory, gherbert@usf.edu, Univ of South Florida; Pe
Herbert, Jennifer, (630) 252-0493 Argonne; Gg
Herbert, Timothy D., (401) 863-1207 timothy_herbert@brown.edu, Brown Univ; Pe
Herd, Christopher D., (780) 492-5798 herd@ualberta.ca, Univ of Alberta; XcGiCp
Herd, Richard, +44 (0)1603 59 3667 r.herd@uea.ac.uk, Univ of East Anglia; Gv
Herdendorf, Charles E., 440-934-1514 herdendorf.1@osu.edu, Ohio State Univ; OgGuOn
Herkenhoff, Ken E., (818) 354-3539 ken.e.herkenhoff@jpl.nasa.gov, Jet Propulsion Lab; Xg
Herman, Ellen K., (570) 577-3088 ekh008@bucknell.edu, Bucknell Univ; Hw
Herman, Janet S., (434) 924-0553 jsh5w@virginia.edu, Univ of Virginia; ClHw
Herman, Mathew, mherman2@csub.edu, California State Univ, Bakersfield; GtYsRn
Herman, Rhett B., 540 831-5441 rherman@radford.edu, Radford Univ; YuXaGa
Hermance, John F., (508) 252-3116 john_hermance@brown.edu, Brown Univ; HqZrHw
Hermann, Albert, 206-526-6495 albert.j.hermann@noaa.gov, Univ of Washington; Opb
Hermans, Thomas, thomas.hermans@ugent.be, Ghent Univ; HwYeu
Hermes, O D., dhermes@uri.edu, Univ of Rhode Island; Gi
Hernandez, Larry, (760) 757-2121, x6329 lhernandez@miracosta.edu, MiraCosta Coll; Gg
Hernández, Manuel, hdezfdez@ucm.es, Univ Compl de Madrid; Pv
Hernandez, Michael W., (801) 626-8186 mhernandez@weber.edu, Weber State Univ; Zir
Hernandez, Rebecca, 530-752-5471 rrhernandez@ucdavis.edu, Univ of California, Davis; Zcn
Hernandez Molina, F. J., javier.hernandez-molina@rhul.ac.uk, Royal Holloway Univ of London; GsuOu
Hernández Uribe, David, davidhu@uic.edu, Univ of Illinois Chicago; Gpt
Herndon, Elizabeth M., 865-974-2366 eherndo4@utk.edu, Univ of Tennessee, Knoxville; Cg
Hernes, Peter J., 530-754-4327 pjhernes@ucdavis.edu, Univ of California, Davis; CoHg

Heron, Duncan, (919) 684-5321 duncan.heron@duke.edu, Duke Univ; Grg
Herrero, Concepción, cherrero@ucm.es, Univ Compl de Madrid; Pm
Herrero, María Josefa, mjherrero@ucm.es, Univ Compl de Madrid; Gs
Herring, Thomas A., (617) 253-5941 tah@mit.edu, Massachusetts Inst of Tech; Yd
Herring Mayo, Lisa L., 931-393-2136 lmayo@mscc.edu, Motlow State Comm Coll; GgZge
Herriott, Trystan, (907) 451-5011 trystan.herriott@alaska.gov, Alaska Division of Geological & Geophysical Surveys; Eo
Herrmann, Achim, 225-578-3016 aherrmann@lsu.edu, Louisiana State Univ; PeGd
Herrmann, Edward W., (812) 856-0587 edherrma@indiana.edu, Indiana Univ, Bloomington; Sa
Herrmann, Felix J., +1 (404) 385-7069 felix.herrmann@gatech.edu, Georgia Inst of Tech; YesZn
Herrmann, Robert B., (314) 977-3120 robert.herrmann@slu.edu, Saint Louis Univ; Ys
Hershey, Ronald L., (775) 673-7393 ron.hershey@dri.edu, Univ of Nevada, Reno; HwCls
Hervig, Richard, (480) 965-8427 hervig@asu.edu, Arizona State Univ; Cg
Herzberg, Claude T., (848) 445-3154 herzberg@eps.rutgers.edu, Rutgers, The State Univ of New Jersey; CpGi
Herzig, Chuck, (310) 660-3593 cherzig@elcamino.edu, El Camino Coll; GgOg
Hesp, Patrick, patrick.hesp@flinders.edu.au, Flinders Univ; GmZy
Hess, Darrel E., (415) 239-3104 dhess@ccsf.edu, City Coll of San Francisco; Zy
Hess, Kai-Uwe, 089/2180 4275 hess@min.uni-muenchen.de, Ludwig-Maximilians-Univ Muenchen; Gz
Hess, Paul C., (401) 863-1929 paul_hess@brown.edu, Brown Univ; Gi
Hess, Peter G., 607-255-2495 peter.hess@cornell.edu, Cornell Univ; Acs
Hess-Tanguay, Lillian, (516) 299-2318 lhess@liu.edu, Long Island Univ, C.W. Post Campus; Gs
Hesse, Marc A., (512) 471-0768 mhesse@jsg.utexas.edu, Univ of Texas, Austin; GqoYg
Hesse, Reinhard, 514-398-3627 reinhard.hesse@mcgill.ca, McGill Univ; Gs
Hester, Erich, (540) 231-9758 ehester@vt.edu, Virginia Tech; Hsw
Hesterberg, Dean L., (919) 515-2636 North Carolina State Univ; Sc
Hetherington, Callum J., (806) 834-3110 callum.hetherington@ttu.edu, Texas Tech Univ; GxzCg
Hetherington, Eric D., (209) 730-3812 Coll of the Sequoias; Gc
Hetherington, Jean, (925) 685-1230 (Ext. 462) jhetheri@dvc.edu, Diablo Valley Coll; Gg
Hetland, Robert D., (979) 458-0096 rhetland@ocean.tamu.edu, Texas A&M Univ; Op
Hettiarachchi, Ganga, (785) 532-7209 ganga@ksu.edu, Kansas State Univ; Sc
Hetzel, Ralf, +49-251-83-33908 rahetzel@uni-muenster.de, Univ Muenster; Gcm
Heubeck, Christoph, 0049(0)3641/948620 christoph.heubeck@uni-jena.de, Friedrich-Schiller-Univ Jena; GsdPg
Heuss-Aßbichler, Soraya, 089/21804252 soraya@min.uni-muenchen.de, Ludwig-Maximilians-Univ Muenchen; Gz
Hewitt, David A., (540) 231-6521 dhewitt@vt.edu, Virginia Tech; Cp
Heymann, Dieter, (713) 348-4890 dieter@owlnet.rice.edu, Rice Univ; Xm
Heyniger, William C., (908) 737-3660 wheynige@kean.edu, Kean Univ; Am
Heyvaert, Alan, 775.673.7322 alan.heyvaert@dri.edu, Univ of Nevada, Reno; GnHs
Heywood, Karen, +44 (0)1603 59 2555 k.heywood@uea.ac.uk, Univ of East Anglia; Op
Heywood, Neil C., (715) 346-4452 nheywood@uwsp.edu, Univ of Wisconsin, Stevens Point; Zy
Hiatt, Eric E., (920) 424-7001 hiatt@uwosh.edu, Univ of Wisconsin, Oshkosh; GsOcGu
Hibbard, James P., (919) 515-7242 jphibbar@ncsu.edu, North Carolina State Univ; GctZn
Hibbs, Barry, (323) 343-2414 bhibbs@calstatela.edu, California State Univ, Los Angeles; Hw
Hick, Steven R., 303-871-2535 shick@du.edu, Univ of Denver; Zi
Hickcox, Charles W., (404) 727-0118 geocwh@emory.edu, Emory Univ; Gg
Hicke, Jeff, 208-885-6240 jhicke@uidaho.edu, Univ of Idaho; ZcyZr
Hicke, Jeffrey, 208-885-6240 jhicke@uidaho.edu, Univ of Idaho; ZcyZn
Hickey, Barbara M., (206) 543-4737 bhickey@u.washington.edu, Univ of Washington; Op
Hickey, James, (660) 562-1817 jhickey@nwmissouri.edu, Northwest Missouri State Univ; Ge
Hickey, Kenneth A., (778) 384-7074 khickey@eos.ubc.ca, Univ of British Columbia; EgGcg
Hickey, William J., (608) 262-9018 wjhickey@wisc.edu, Univ of Wisconsin, Madison; Sb
Hickey-Vargas, Rosemary, (305) 348-3471 hickey@fiu.edu, Florida InterNatl Univ; Cg
Hickman, Anna, +44 (0)23 80592132 a.hickman@noc.soton.ac.uk, Univ of Southampton; Og
Hickman, Carole S., (510) 642-3429 Univ of California, Berkeley; Pg
Hickmon, Nicki, 630-252-7662 nhickmon@anl.gov, Argonne; As
Hickmott, Donald D., (505) 667-8753 dhickmott@lanl.gov, Los Alamos; Gp
Hicks, Andrea, hicks5@wisc.edu, Univ of Wisconsin, Madison; Zn
Hicks, David, 956 882 5055 david.hicks@utrgv.edu, Univ of Texas, Rio Grande Valley; Ob
Hickson, Catherine J., (604) 761-5573 ttgeo@telus.net, Univ of British Columbia; EgGvZe
Hickson, Thomas A., (651) 962-5241 tahickson@stthomas.edu, Univ of Saint Thomas; GsmGe
Hicock, Stephen R., shicock@uwo.ca, Western Univ; Gl
Hidalgo, Paulo J., (404) 413-5789 phidalgo@gsu.edu, Georgia State Univ; GivGz
Higgins, Charles G., paulah88@hotmail.com , Univ of California, Davis; Gm
Higgins, Chris T., (916) 322-9997 California Geo Survey; Gg
Higgins, Jerry D., (303) 273-3817 jhiggins@mines.edu, Colorado Sch of Mines; NxgNt
Higgins, John A., 609-258-2756 jahiggin@princeton.edu, Princeton Univ; ClPcCs
Higgins, Michael D., 4185455011 x 5052 mhiggins@uqac.ca, Univ du Quebec a Chicoutimi; Gi
Higgins, Sean M., sean@ldeo.columbia.edu, Columbia Univ; Gs
Higgs, Bettie M., +353 860476789 b.higgs@ucc.ie, Univ Coll Cork; YgGqRh
Higgs, Ken, +353 21 4902290 k.higgs@ucc.ie, Univ Coll Cork; Pl
Higinbotham, Pamela, 724-938-4180 higinbotham@calu.edu, California Univ of Pennsylvania; Zn
Hilbert-Wolf, Hannah L., hannah.hilbertwolf@my.jcu.edu.au, James Cook Univ; GsCc
Hildebrand, Alan R., 403-220-2291 ahildebr@ucalgary.ca, Univ of Calgary; XmgXc
Hildebrand, John A., (858) 534-4069 jhildebrand@ucsd.edu, Univ of California, San Diego; Yr
Hildebrand, Stephen G., hildebrandsg@ornl.gov, Oak Ridge; Zn
Hildebrand, Steve T., (505) 667-4318 hildebrand@lanl.gov, Los Alamos; Ys
Hildebrandt, Anke, 349-3641-948651 hildebrandt.a@uni-jena.de, Friedrich-Schiller-Univ Jena; HgSpf
Hileman, Mary E., (405) 744-4341 mary.hileman@okstate.edu, Oklahoma State Univ; GorGd
Hill, Arleen A., 901-678-2589 aahill@memphis.edu, Univ of Memphis; Rn
Hill, Catherine R., 928344-7719 catherine.hill@azwestern.edu, Arizona Western Coll; GgOgHw
Hill, Chris, chris.hill@gcccd.edu, Grossmont Coll; GgOg
Hill, Christopher, 617-253-6430 cnh@mit.edu, Massachusetts Inst of Tech; Zn
Hill, Julie, (775) 784-6987 juliehill@unr.edu, Univ of Nevada, Reno; Zn
Hill, Malcolm D., (617) 373-4377 m.hill@neu.edu, Northeastern Univ; ZiGze
Hill, Mary C., 785-864-2728 mchill@ku.edu, Univ of Kansas; HgqHw
Hill, Mary Louise, (807) 343-8010 Ext. 8319 mary.louise.hill@lakeheadu.ca, Lakehead Univ; Gc
Hill, Mimi, +44 0151 794 3462 m.hill@liverpool.ac.uk, Univ of Liverpool; Ym
Hill, Nicole, nhill@bentley.edu, Bentley Univ; Ge
Hill, Paul S., (902) 494-2266 paul.hill@dal.ca, Dalhousie Univ; Gs
Hill, Philip R., philip.hill@canada.ca, Univ of Victoria; Gsu
Hill, Tessa M., (530) 752-0179 tmhill@ucdavis.edu, Univ of California, Davis; Pe
Hill, Timothy, +44 01334 464013 tch2@st-andrews.ac.uk, Univ of St. Andrews; Ge
Hill, Victoria, (757) 683-4911 vhill@odu.edu, Old Dominion Univ; Ob
Hillaire-Marcel, Claude, 514-987-3000 #3376 hillaire-marcel.claude@uqam.ca, Univ du Quebec a Montreal; Cs
Hillier, John, jhillier@ghc.edu, Grays Harbor Coll; ZgCa
Hilliker, Joby, (610) 436-2213 jhilliker@wcupa.edu, West Chester Univ; Am
Hills, Denise J., (205) 247-3694 dhills@gsa.state.al.us, Geo Survey of Alabama; EoZfYe
Hilterman, Fred, 713-850-7600 x3318 fhilterman@uh.edu, Univ of Houston; Ye
Hilts, Stewart G., (519) 824-4120 (Ext. 52448) shilts@uoguelph.ca, Univ of Guelph; Zn
Hindle, Tobin, 561 297-2846 thindle@fau.edu, Florida Atlantic Univ; ZyiZc
Hindman, Edward E., (212) 650-6469 City Coll (CUNY); As

Hindshaw, Ruth, +44 01334 463936 rh71@st-andrews.ac.uk, Univ of St. Andrews; Cs
Hine, Albert C., (727) 553-1161 hine@usf.edu, Univ of South Florida; Gus
Hines, Mary E., (910) 962-3012 hinese@uncw.edu, Univ of North Carolina, Wilmington; Zu
Hines, Paul, phines50@gmail.com, Dalhousie Univ; Op
Hingston, Egerton, (031) 260-2805 hingstone@ukzn.ac.za, Univ of KwaZulu-Natal; NtxNg
Hinkle, Margaret A., (540) 458-8271 hinklem@wlu.edu, Washington & Lee Univ; Cqb
Hinman, Nancy W., (406) 243-5277 nancy.hinman@umontana.edu, Univ of Montana; Cgl
Hinnov, Linda, 703-993-3082 lhinnov@gmu.edu, George Mason Univ; GrYg
Hinojosa, Alejandro, alhinc@cicese.mx, Centro de Investigación Científica y de Educación Superior de Ensenada; Zr
Hinthorne, James R., james.hinthorne@utrgv.edu, Univ of Texas, Rio Grande Valley; GgzCa
Hinton, Richard, +44 (0) 131 650 8548 richard.hinton@ed.ac.uk, Edinburgh Univ; Cg
Hintz, John G., (570)389-4140 jhintz@bloomu.edu, Bloomsburg Univ; Zin
Hintz, Rashauna, 479-575-3355 rmicken@uark.edu, Univ of Arkansas, Fayetteville; Zu
Hintz, William, (419) 530-5079 william.hintz@utoledo.edu, Univ of Toledo; Zn
Hintze, Lehi F., 801 422-6361 lehi.hintze@gmail.com, Brigham Young Univ; Pg
Hinz, Nicholas, (775)784-1446 nhinz@unr.edu, Univ of Nevada, Reno; Gg
Hinz, Nick, 775-784-1446 nhinz@unr.edu, Univ of Nevada, Reno; Gcg
Hinze, William J., wjh730@comcast.net, Purdue Univ; YevGt
Hippensteel, Scott P., 704-687-5992 shippens@email.uncc.edu, Univ of North Carolina, Charlotte; Gr
Hiroji, Anand D., (228) 688-1510 anand.hiroji@usm.edu, Univ of Southern Mississippi; Yd
Hirons, Steve, +44 020 3073 8028 s.hirons@ucl.ac.uk, Birkbeck Coll; Cl
Hirsch, Aaron, 612-626-4028 ahirsch@umn.edu, Univ of Minnesota; Yg
Hirsch, Eric, eric.hirsch@fandm.edu, Franklin and Marshall Coll; Zn
Hirschboeck, Katherine K., (520) 621-6466 katie@ltrr.arizona.edu, Univ of Arizona; AsHsZc
Hirschmann, Marc M., 612-625-6698 hirsc022@umn.edu, Univ of Minnesota, Twin Cities; Gi
Hirt, William H., (530) 938-5255 hirt@siskiyous.edu, Coll of the Siskiyous; Gi
Hirth, Greg, (401) 863-7063 greg_hirth@brown.edu, Brown Univ; GcyYg
Hirth, Gregory, 508-289-2776 greg_hirth@brown.edu, Woods Hole Oceanographic Inst; Gc
Hiscock, Kevin, +44 (0)1603 59 3104 k.hiscock@uea.ac.uk, Univ of East Anglia; Hwg
Hiscott, Richard N., (709) 737-8394 rhiscott@mun.ca, Memorial Univ of Newfoundland; Gs
Hiser, Susan, 740-376-4775 sch030@marietta.edu, Marietta Coll; Zn
Hishamunda, Valens, 609-258-4101 vh2421@princeton.edu, Princeton Univ; ClsCq
Hiskey, J. Brent, (520) 621-6185 jbh@engr.arizona.edu, Univ of Arizona; Nx
Hitchcock, Peter, 607 255 3398 aph28@cornell.edu, Cornell Univ; As
Hitchman, Matthew H., (608)262-4653 matt@aos.wisc.edu, Univ of Wisconsin, Madison; As
Hitz, Ralph B., (253) 566-5299 rhitz@tacomacc.edu, Tacoma Comm Coll; PvSaZi
Hixon, Amy E., (574) 631-1872 ahixon@nd.edu, Univ of Notre Dame; ClqCc
Hjelmfelt, Mark R., (605) 394-2291 mark.hjelmfelt@sdsmt.edu, South Dakota Sch of Mines & Tech; As
Hladik, Christine, (912) 478-0338 chladik@georgiasouthern.edu, Georgia Southern Univ; SbZr
Hluchy, Michele M., (607) 871-2838 Alfred Univ; Cl
Hlusko, Leslea, (510) 643-8838 hlusko@berkeley.edu, Univ of California, Berkeley; PgZnn
Ho, Anita, (406) 756-3873 aho@fvcc.edu, Flathead Valley Comm Coll; Gg
Ho, David, (808) 956-3311 ho@hawaii.edu, Univ of Hawai'i, Manoa; OcCb
Ho, I-Hsuan, (701) 777-6156 ihsuan.ho@engr.und.edu, Univ of North Dakota; Ng
Hobbs, Benjamin F., (410) 516-4681 bhobbs@jhu.edu, Johns Hopkins Univ; Zn
Hobbs, Chasidy, (850) 474-2735 chobbs@uwf.edu, Univ of West Florida; ZgeZy
Hobbs, John D., dhobbs@mtech.edu, Montana Tech of the Univ of Montana; Zn
Hobbs, Kevin, 575-835-6213 kevin.hobbs@nmt.edu, New Mexico Tech; GsrSd
Hobbs, Richard, +44 (0) 191 33 44295 r.w.hobbs@durham.ac.uk, Durham Univ; Ys
Hobbs, Thomas M., 281-618-5796 tom.hobbs@nhmccd.edu, Lonestar Coll - North Harris; Gg
Hochella, Jr., Michael F., (540) 231-6227 hochella@vt.edu, Virginia Tech; ClGze
Hochmuth, George J., 352-392-1803 318 hoch@ufl.edu, Univ of Florida; So
Hochstaedter, Alfred, (831) 646-4149 ahochstaedter@mpc.edu, Monterey Peninsula Coll; GgOgZg
Hock, Regine M., 907-474-7691 regine.hock@gi.alaska.edu, Univ of Alaska, Fairbanks; Gl
Hockaday, William C., (254)7102639 william_hockaday@baylor.edu, Baylor Univ; CoaSb
Hockey, Thomas A., (319) 273-2065 thomas.hockey@uni.edu, Univ of Northern Iowa; Zn
Hodder, Donald R., (914) 257-3757 SUNY, New Paltz; Gg
Hodell, David, +44 (0) 1223 330270 dhod07@esc.cam.ac.uk, Univ of Cambridge; Ge
Hodges, Floyd N., (509) 376-4627 floyd.hodges@pnl.gov, Pacific Northwest; Cg
Hodges, Kip V., 480-965-5331 kvhodges@asu.edu, Arizona State Univ; CcGtXg
Hodgetts, David, david.hodgetts@manchester.ac.uk, Univ of Manchester; Gsc
Hodgkiss, Jr., William S., (858) 534-1798 whodgkiss@ucsd.edu, Univ of California, San Diego; No
Hodych, Joseph P., (709) 864-7567 jhodych@mun.ca, Memorial Univ of Newfoundland; Ym
Hoe, Teh Guan, 03-79674231 tehgh@um.edu.my, Univ of Malaya; Ca
Hoenisch, Baerbel, (845) 365-8828 hoenisch@ldeo.columbia.edu, Columbia Univ; PcClm
Hoey, Trevor, +4401413307736 trevor.hoey@glasgow.ac.uk, Univ of Glasgow; Gs
Hoff, Jean L., (320) 308-5914 jhoff@stcloudstate.edu, Saint Cloud State Univ; HwZeGg
Hoffman, Eric G., 603-535-2321 ehoffman@plymouth.edu, Plymouth State Univ; Am
Hoffman, Gretchen K., 575-835-6105 gretchen.hoffman@nmt.edu, New Mexico Tech; Ec
Hoffman, Jeffrey L., (609) 292-1185 jeffrey.l.hoffman@dep.nj.gov, New Jersey Geological and Water Survey; HqwRw
Hoffman, Paul F., (617) 496-6380 paulfhoffman@gmail.com, Harvard Univ; Gc
Hofmann, Eileen E., (757) 683-5334 ehofmann@odu.edu, Old Dominion Univ; Op
Hofmann, Michael, (406) 243-5855 michael.hofmann@umontana.edu, Univ of Montana; GosGr
Hofmeister, Anne M., (314) 935-7440 hofmeist@levee.wustl.edu, Washington Univ in St. Louis; YghXa
Hogan, John, 573-341-4618 jhogan@mst.edu, Missouri Univ of Sci and Tech; Gc
Hohn, Michael E., 304 594-2331 hohn@geosrv.wvnet.edu, West Virginia Geological & Economic Survey; Eg
Hoisch, Thomas D., (928) 523-1904 thomas.hoisch@nau.edu, Northern Arizona Univ; Gp
Hojjatie, Barry, (229) 333-5753 bhojjati@valdosta.edu, Valdosta State Univ; Zm
Hok, Jozef, jozef.hok@uniba.sk, Comenius Univ in Bratislava; Ggt
Hoke, Gregory, (315) 443-1903 gdhoke@syr.edu, Syracuse Univ; ZgGmt
Holail, Hanafy M., 002-03-3921595 hanafyholail@hotmail.com, Alexandria Univ; Ggs
Holberg, Jay B., (520) 621-4571 holberg@vega.lpl.arizona.edu, Univ of Arizona; As
Holbrook, Amanda, (606) 783-2381 a.holbrook@moreheadstate.edu, Morehead State Univ; Zn
Holbrook, John M., 817-257-6275 john.holbrook@tcu.edu, Texas Christian Univ; Gsr
Holbrook, W. S., 540-231-6521 wstevenh@vt.edu, Virginia Tech; Yur
Holcomb, Robin T., (206) 543-5274 rholcomb@ocean.washington.edu, Univ of Washington; Ou
Holcombe, Troy L., 979 845-3528 or 979 690-7398 tholcombe@ocean.tamu.edu, Texas A&M Univ; OuGcg
Holdaway, Michael J., (214) 692-2750 Southern Methodist Univ; Gp
Holdren, George R., (509) 376-2242 rich.holdren@pnl.gov, Pacific Northwest; Cl
Holdren, John P., 617-495-1498 john_holdren@hks.harvard.edu, Harvard Univ; Zn
Holdsworth, Robert E., 0191 3742529 r.e.holdsworth@durham.ac.uk,

Durham Univ; Gc
Hole, John A., (540) 231-3858 hole@vt.edu, Virginia Tech; Ye
Holford, Simon, simon.holford@adelaide.edu.au, Univ of Adelaide; Go
Holgood, Jay S., (614) 292-3999 hobgood.1@osu.edu, Ohio State Univ; As
Holk, Gregory J., 562-985-5006 gholk@csulb.edu, California State Univ, Long Beach; CsGxEm
Holl, Karen D., (831) 459-3668 kholl@cats.ucsc.edu, Univ of California, Santa Cruz; Zu
Hollabaugh, Curtis L., (678) 839-4050 chollaba@westga.edu, Univ of West Georgia; GzCg
Holland, Edward C., 479-575-6635 echollan@uark.edu, Univ of Arkansas, Fayetteville; Zy
Holland, Nicholas D., (858) 534-2085 nholland@ucsd.edu, Univ of California, San Diego; Ob
Holland, Steven M., (706) 542-0424 stratum@gly.uga.edu, Univ of Georgia; Gr
Holland, Tim, +44 (0) 1223 333466 tjbh@esc.cam.ac.uk, Univ of Cambridge; GptGz
Hollander, David J., (727) 553-1019 davidh@usf.edu, Univ of South Florida; Oc
Holley, Elizabeth, 303-273-3409 eholley@mines.edu, Colorado Sch of Mines; Eg
Holliday, Joseph W., 3106603593 xt. 3371 jholliday@elcamino.edu, El Camino Coll; Og
Holliday, Vance, vthollid@email.arizona.edu, Univ of Arizona; Ga
Holliman, Richard, +44(0) 1908 654 646 x 54646 richard.holliman@open.ac.uk, Open Univ; Ze
Hollings, Peter N., (807) 343-8010 Ext. 8329 peter.hollings@lakeheadu.ca, Lakehead Univ; Eg
Hollis, Cathy, +44 0161 306-6583 cathy.hollis@manchester.ac.uk, Univ of Manchester; Gs
Hollister, Lincoln S., 609-258-4106 linc@princeton.edu, Princeton Univ; GptGz
Hollister, Ryan, rhollister@csustan.edu, California State Univ, Stanislaus; ZeGg
Hollocher, Kurt T., (518) 388-6518 hollochk@union.edu, Union Coll; GxCga
Holloway, Tracey, taholloway@wisc.edu, Univ of Wisconsin, Madison; As
Holm, Daniel K., (330) 672-4094 dholm@kent.edu, Kent State Univ; Gc
Holm, Richard F., richard.holm@nau.edu, Northern Arizona Univ; Gv
Holman, John, (620) 276-8286 jholman@ksu.edu, Kansas State Univ; So
Holmden, Chris, (306) 966-5697 chris.holmden@usask.ca, Univ of Saskatchewan; Csc
Holme, Richard, +44 0151 794 5254 r.t.holme@liverpool.ac.uk, Univ of Liverpool; Ym
Holmes, George, +44(0) 113 34 31163 g.holmes@leeds.ac.uk, Univ of Leeds; Ge
Holmes, Mark L., 425 487-2639 mholmes@uw.edu, Univ of Washington; OuGl
Holmes, Mary Anne, 402-588-2492 mholmes2@unl.edu, Univ of Nebraska, Lincoln; GsOuZn
Holmes, Stevie L., 605-677-6147 stevie.holmes@usd.edu, South Dakota Dept of Environment and Nat Res; Gg
Holness, Marian, +44 (0) 1223 333434 marian@esc.cam.ac.uk, Univ of Cambridge; Gi
Holroyd, Pat, (510) 642-3733 pholroyd@berkeley.edu, Univ of California, Berkeley; Pv
Holschuh, Nicholas D., 413-542-2285 nholschuh@amherst.edu, Amherst Coll; GlYg
Holst, Timothy, tholst@d.umn.edu, Univ of Minnesota Duluth; GcYnGt
Holstein, Thomas J., (401) 254-3097 Roger Williams Univ; Zn
Holt, Ben D., (630) 252-4347 Argonne; Ca
Holt, Caecilia, 610 683-4445 holt@kutztown.edu, Kutztown Univ of Pennsylvania; Zn
Holt, David, 228-214-3255 david.h.holt@usm.edu, Univ of Southern Mississippi; Zyi
Holt, Gloria J., (361) 749-6716 joanholt@utexas.edu, Univ of Texas, Austin; Ob
Holt, John W., 512-471-0487 jack@ig.utexas.edu, Univ of Texas, Austin; Ye
Holt, Robert M., (662) 915-6687 rmholt@olemiss.edu, Univ of Mississippi; Hq
Holt, William E., 631-632-8215 william.holt@stonybrook.edu, Stony Brook Univ; Ys
Holtz, Jr., Thomas R., (301) 405-4084 tholtz@umd.edu, Univ of Maryland; PvgPg
Holtzman, Benjamin, benh@ldeo.columbia.edu, Columbia Univ; Ys
Holubnyak, Yehven I., (785) 864-2070 eugene@kgs.ku.edu, Univ of Kansas; Np
Holwell, Dave, +440116 252 3804 dah29@le.ac.uk, Leicester Univ; Ge
Holycross, Megan, holycross@cornell.edu, Cornell Univ; CmGv
Holyoke, Caleb, (330) 972-7635 cholyoke@uakron.edu, Univ of Akron; GcNrYx
Holzer, Mark, (604) 822-0531 mholzer@langara.bc.ca, Univ of British Columbia; As
Holzworth, Robert H., (206) 685-7410 bobholz@uw.edu, Univ of Washington; XyAsZg
Homan, Amy L., (814) 865-2622 aqs3@psu.edu, Pennsylvania State Univ, Univ Park; Zn
Homann, Martin D., m.homann@ucl.ac.uk, Univ Coll London; Gs
Homayouni, Saied, 418 654-2687 saeid.homayouni@ete.inrs.ca, Univ du Quebec; Zrf
Hommel, Demian, 541-737-5070 hommeld@geo.oregonstate.edu, Oregon State Univ; Zn
Homuth, Emil F., (702) 794-7351 fred_homuth@notes.ymp.gov, Los Alamos; Ye
Hon, Ken, (808) 974-7302 kenhon@hawaii.edu, Univ of Hawai'i, Hilo; Gvi
Hon, Rudolph, (617) 552-3656 hon@bc.edu, Boston Coll; Cq
Hong, Jessie, (678) 839-5466 jhong@westga.edu, Univ of West Georgia; ZieZy
Hong, Sung-ho, 270-809-6319 shong4@murraystate.edu, Murray State Univ; HyGe
Honjas, Bill, (775) 784-6613 bhonjas@optimsoftware.com , Univ of Nevada, Reno; Ys
Honjo, Susumu, (508) 289-2589 shonjo@whoi.edu, Woods Hole Oceanographic Inst; Ou
Honkaer, Rick, 859-257-1108 rick.honaker@uky.edu, Univ of Kentucky; Nm
Hood, Lonnie L., (520) 621-6936 lon@lpl.arizona.edu, Univ of Arizona; XyAs
Hood, Raleigh, (305) 361-4668 rhood@umces.edu, Univ of Maryland; Ob
Hood, Teresa A., 305-284-8647 t.hood@miami.edu, Univ of Miami; Gg
Hood, William C., (970) 241-8020 Colorado Mesa Univ; Gg
Hooda, Peter, +44 020 8417 2155 p.hooda@kingston.ac.uk, Kingston Univ; Em
Hooft, Emilie E., (541) 346-4762 emilie@uoregon.edu, Univ of Oregon; Yr
Hook, James E., (912) 386-3182 Univ of Georgia; Sp
Hook, Paul B., paul@intermountainaquatics.com, Montana State Univ; Sf
Hook, Simon J., (818) 354-0974 simon@lithos.jpl.nasa.gov, Jet Propulsion Lab; Zr
Hooke, Roger L., 207-581-2203 rhooke@verizon.net, Univ of Minnesota, Twin Cities; Gl
Hooks, Benjamin P., (731) 881-7430 bhooks@utm.edu, Univ of Tennessee, Martin; GciNr
Hooks, Chris H., (205) 247-3721 chooks@gsa.state.al.us, Geo Survey of Alabama; EoGxo
Hooper, Andy, +44(0) 113 34 37723 a.hooper@leeds.ac.uk, Univ of Leeds; Ydg
Hooper, Robert L., (715) 836-4932 hooperrl@uwec.edu, Univ of Wisconsin, Eau Claire; Gz
Hoover, Karin A., 530-898-6269 khoover@csuchico.edu, California State Univ, Chico; Hw
Hoover, Michael T., (919) 515-7305 North Carolina State Univ; Sd
Hooyer, Thomas S., 608-263-4175 hooyer@uwm.edu, Univ of Wisconsin, Madison; Hg
Hopkins, David G., (701) 231-8948 North Dakota State Univ; Sd
Hopkins, Kenneth D., (970) 351-2853 kenneth.hopkins@unco.edu, Univ of Northern Colorado; Gml
Hopkins, Nathan, (208) 885-1508 nhopkins@uidaho.edu, Univ of Idaho; GlmGg
Hopkins, Samantha, 541-346-5976 shopkins@uoregon.edu, Univ of Oregon; Pg
Hopkins, Thomas S., (205) 348-1791 Dauphin Island Sea Lab; Ob
Hopley, Philip, +44 020 3073 8029 p.hopley@ucl.ac.uk, Birkbeck Coll; Pe
Hopmans, Jan W., (916) 752-3060 Univ of California, Davis; So
Hoppe, Kathryn A., 2538339111 x 4323 khoppe@greenriver.edu, Green River Comm Coll; ZgOgPg
Hoppie, Bryce W., (507) 389-2315 bryce.hoppie@mnsu.edu, Minnesota State Univ; GeHw
Horel, John, 801-581-7091 john.horel@utah.edu, Univ of Utah; AsZf
Horgan, Briony, (765) 496-2290 briony@purdue.edu, Purdue Univ; XgZr
Horn, John, (816) 604-3132 john.horn@mcckc.edu, Metropolitan Comm Coll-Kansas City; GgZy
Horn, Marty R., 225-578-2681 mhorn@lsu.edu, Louisiana State Univ; EoGgr
Hornbach, Matthew J., 214-768-2389 mhornbach@smu.edu, Southern Methodist Univ; YrhYe
Hornberger, George, (615) 343-1144 george.m.hornberger@vanderbilt.edu, Vanderbilt Univ; HgwHs
Horner, John R., (406) 994-3982 jhorner@montana.edu, Montana State Univ; Pv

Horner, Tim C., (916) 278-5635 hornertc@csus.edu, Sacramento State Univ; HwGs
Horner, Tristan, (508) 289-3825 thorner@whoi.edu, Woods Hole Oceanographic Inst; Cm
Hornibrook, Edward R., 250-807-8059 ed.hornibrook@ubc.ca, Univ of British Columbia, Okanagan Campus; CbsSc
Horns, Daniel, (801) 863-8582 hornsda@uvu.edu, Utah Valley Univ; NgGeRn
Horowitz, Franklin G., +16072754955 frank.horowitz@cornell.edu, Cornell Univ; YgZnYe
Horsman, Eric M., 252 3285265 horsmane@ecu.edu, East Carolina Univ; GcYgGt
Horst, Sarah, 410-516-5286 sarah.horst@jhu.edu, Johns Hopkins Univ; XaAcXb
Horton, Brian, 512-471-5172 horton@mail.utexas.edu, Univ of Texas, Austin; Gs
Horton, Daniel E., 847-467-6185 danethan@earth.northwestern.edu, Northwestern Univ; AsPeZn
Horton, Duane G., (509) 376-6868 duane.horton@pnl.gov, Pacific Northwest; Gz
Horton, Radley, hortonr@ldeo.columbia.edu, Columbia Univ; Og
Horton, Robert, (515) 294-7843 rhorton@iastate.edu, Iowa State Univ of Sci & Tech; Sp
Horton, Stephen P., (901) 678-2007 shorton@memphis.edu, Univ of Memphis; Ys
Horton, Travis, +64 3 3667001 Ext 7734 travis.horton@canterbury.ac.nz, Univ of Canterbury; Cs
Horton, Jr., Robert A., 661-654-3059 rhorton@csub.edu, California State Univ, Bakersfield; GdsGo
Horváth, Ferenc, frankh@ludens.elte.hu, Eotvos Lorand Univ; YgRhGc
Horvath, Peter, +27 (0)46-603-8312 p.horvath@ru.ac.za, Rhodes Univ; Gp
Horwath, William R., 530-754-6029 wrhorwath@ucdavis.edu, Univ of California, Davis; ScCo
Horwath, William R., (530) 754-6029 wrhorwath@ucdavis.edu, Univ of California, Davis; Sob
Horwell, Claire, +44 (0) 191 33 42253 claire.horwell@durham.ac.uk, Durham Univ; Rn
Hoskin, Julie, hoskinj@si.edu, Smithsonian Inst / Natl Mus of Natural History; Xm
Hosler, Charles L., (814) 863-8358 hosler@ems.psu.edu, Pennsylvania State Univ, Univ Park; Am
Hossain, A K M Azad, 423-425-4404 azad-hossain@utc.edu, Univ of Tennessee, Chattanooga; ZriZg
Hosseini, Seyyed Abolfazi, 512-471-1534 seyyed.hosseini@beg.utexas.edu, Univ of Texas, Austin; Eo
Hotz, Helenmary, helenmary.hotz@umb.edu, Univ of Massachusetts, Boston; Zi
Houghton, Bruce F., (808) 956-2561 bhought@soest.hawaii.edu, Univ of Hawai'i, Manoa; Gv
Houlton, Ben, bzhoulton@ucdavis.edu, Univ of California, Davis; So
Hounslow, Arthur, 405-372-2328 Oklahoma State Univ; Cl
Hourigan, Jeremy, 831-459-2879 hourigan@ucsc.edu, Univ of California, Santa Cruz; Cc
House, Christopher H., (814) 865-8802 chouse@geosc.psu.edu, Pennsylvania State Univ, Univ Park; Pg
House, Leigh S., house@lanl.gov, Los Alamos; Ys
House, Martha, mahouse@pasadena.edu, Pasadena City Coll; Gg
Houseman, Greg, +44(0) 113 34 35206 g.a.houseman@leeds.ac.uk, Univ of Leeds; Yg
Housen, Bernard A., (360) 650-3581 bernieh@wwu.edu, Western Washington Univ; Ym
Houston, Adam L., ahouston2@unl.edu, Univ of Nebraska, Lincoln; Asm
Houston, Heidi B., 310 691 9152 houstonh@usc.edu, Univ of Southern California; YsGt
Houston, Serin D., 413-538-2055 shouston@mtholyoke.edu, Mount Holyoke Coll; Zyn
Houston, William (Bill), 906-635-2155 whouston1@lssu.edu, Lake Superior State Univ; Gsg
Houweling, Sander, +31 20 59 83687 s.houweling@vu.nl, Vrije Univit Amsterdam; AscAp
Houze, Robert A., (206) 543-6922 houze@atmos.washington.edu, Univ of Washington; As
Hovan, Steven A., (724) 357-2379 hovan@iup.edu, Indiana Univ of Pennsylvania; Ou
Hovis, Guy L., 610-330-5193 hovisguy@lafayette.edu, Lafayette Coll; GzxCg
Hovorka, Susan D., 512-471-4863 susan.hovorka@beg.utexas.edu, Univ of Texas, Austin; Gs
Howard, Alan D., (804) 924-0563 ah6p@virginia.edu, Univ of Virginia; GmXg
Howard, Evan, 609-258-4101 evanhoward@princeton.edu, Princeton Univ; As
Howard, Hugh H., 916 484-8805 howardh@arc.losrios.edu, American River Coll; Zi
Howard, Katie, 360-475-7700 khoward@llion.org, Olympic Coll; Gg
Howard, Kenneth W., 416-287-7233 Univ of Toronto; Hw
Howard, Kieren, 917-974-6168 khoward@amnh.org, Graduate Sch of the City Univ of New York; Ggz
Howard, Matthew K., (979) 862-4169 mkhoward@tamu.edu, Texas A&M Univ; Op
Howarth, Geoffrey H., geoffrey.howarth@uct.ac.za, Univ of Cape Town; EgGiv
Howat, Ian M., (614) 292-4839 howat.4@osu.edu, Ohio State Univ; GlZr
Howden, Stephan, (228) 688-5284 stephan.howden@usm.edu, Univ of Southern Mississippi; Op
Howe, Bruce M., (206) 543-9141 billhowe@cs.washington.edu, Univ of Washington; Op
Howe, Julie, jhowe@tamu.edu, Texas A&M Univ; Sc
Howe, Stephen S., (518) 442-5053 showe@albany.edu, SUNY, Albany; Cs
Howe III, Thomas R., (269) 387-5492 thomas.r.howe@wmich.edu, Western Michigan Univ; HwGem
Howell, John, +44 (0)1224 272606 john.howell@abdn.ac.uk, Univ of Aberdeen; GsoGc
Howell, Kimberly, 978-934-3917 kimberly_howell1@uml.edu, Univ of Massachusetts, Lowell; Gso
Howell, Robert R., (307) 766-6296 rhowell@uwyo.edu, Univ of Wyoming; XgZr
Howell, Jr., Benjamin F., (814) 863-0886 howellbf@aol.com, Pennsylvania State Univ, Univ Park; Ys
Hower, James C., (859) 257-0261 james.hower@uky.edu, Univ of Kentucky; Ec
Howes, Mary R., mary.howes@dnr.iowa.gov, Iowa Dept of Nat Res; Gg
Howley, Michael W., (603) 271-7332 michael.w.howley@des.nh.gov, New Hampshire Geo Survey; GgYgHw
Howman, Dominic J., +61 8 9266-2329 d.howman@curtin.edu.au, Curtin Univ; Ye
Hoyer, Lauren, 031-2601613 hoyerl@ukzn.ac.za, Univ of KwaZulu-Natal; EgGc
Hoyos, Carlos, carlos.hoyos@eas.gatech.edu, Georgia Inst of Tech; As
Hoyt, Greg D., (919) 684-3562 North Carolina State Univ; Sb
Hoyt, William H., (970) 351-2487 william.hoyt@unco.edu, Univ of Northern Colorado; OgGsZe
Hozik, Michael J., 609 652-4277 hozikm@stockton.edu, Stockton Univ; GcYmg
Hren, Michael, 860-486-9511 michael.hren@uconn.edu, Univ of Connecticut; CslCo
Hrepic, Zdeslav, 706-569-3052 hrepic_zdeslav@columbusstate.edu, Columbus State Univ; Zn
Hsiang, Tom, (519) 824-4120 Ext.52753 thsiang@uoguelph.ca, Univ of Guelph; Zn
Hsieh, Wen-Pin, 886-227839910-509 wphsieh@earth.sinica.edu.tw, Academia Sinica; YhGy
Hsieh, William W., (250) 388-0508 whsieh@eos.ubc.ca, Univ of British Columbia; OpAs
Hsu, Liang-Chi, (775) 682-8746 lihsu@unr.edu, Univ of Nevada, Reno; Cp
Hu, Baoxin, (416)736-2100 #20557 baoxin@yorku.ca, York Univ; Zr
Hu, Bill X., 850-644-3943 hu@gly.fsu.edu, Florida State Univ; Hw
Hu, Feng-Sheng, (217) 244-2982 fhu@illinois.edu, Univ of Illinois, Urbana-Champaign; Pe
Hu, Hao, hhu5@central.uh.edu, Univ of Houston; Ye
Hu, Qi S., 402-472-6642 qhu2@unl.edu, Univ of Nebraska, Lincoln; As
Hu, Qi (Steve), qhu2@unl.edu, Univ of Nebraska, Lincoln; As
Hu, Qinhong (Max), (817) 272-5398 maxhu@uta.edu, Univ of Texas, Arlington; GoHwNg
Hu, Shusheng, 203-432-3790 shusheng.hu@yale.edu, Yale Univ; PblGs
Hu, Wan-Ping (Sunny), +61 7 3138 7314 sunny.hu@qut.edu.au, Queensland Univ of Tech; Ca
Hu, Xinping, 361-825-3395 xinping.hu@tamucc.edu, Texas A&M Univ, Corpus Christi; Cms
Hu, Zhiyong, 850-474-3494 zhu@uwf.edu, Univ of West Florida; Zr
Huang, Alex, (828) 232-5157 Univ of North Carolina, Asheville; As
Huang, Bohua, (703) 993-6084 bhuang@gmu.edu, George Mason Univ; Op
Huang, Huai-Hsuan, 609-258-4101 hh6077@princeton.edu, Princeton Univ; OoPm
Huang, Li, (519) 661-3188 lhuang3@uwo.ca, Western Univ; Yr
Huang, Lianjie, (505) 665-1108 ljh@lanl.gov, Los Alamos; Yg
Huang, Moh J., (916) 322-9304 California Geo Survey; Ne

Huang, Norden E., (301) 286-8879 Univ of Delaware; Zn
Huang, Shaopeng, (734) 763-3169 shaopeng@umich.edu, Univ of Michigan; YhZcGt
Huang, Yi, 514-398-8217 yi.huang@mcgill.ca, McGill Univ; ZrYgAs
Huang, Yongsong, yongsong_huang@brown.edu, Brown Univ; Co
Hubbard, Dennis K., 440-775-8346 dennis.hubbard@oberlin.edu, Oberlin Coll; Gsu
Hubbard, Stephen M., 403 220-6236 shubbard@ucalgary.ca, Univ of Calgary; Gs
Hubbard, Trent, (907) 451-5009 trent.hubbard@alaska.gov, Alaska Division of Geological & Geophysical Surveys; GmNgZi
Hubbard, William B., (520) 621-6942 hubbard@lpl.arizona.edu, Univ of Arizona; XyYvAs
Hubbart, Jason A., (304) 293-2472 jason.hubbart@mail.wvu.edu, West Virginia Univ; HqgHs
Hubenthal, Michael, (607) 777-4612 hubenth@iris.edu, Binghamton Univ; ZeYg
Hubeny, J B., bhubeny@salemstate.edu, Salem State Univ; GeOnCs
Huber, Brian T., (202) 633-1328 Smithsonian Inst / Natl Mus of Natural History; Pm
Huber, Christian, christian_huber@brown.edu, Brown Univ; Gv
Huber, Julie, 508-289-2556 jhuber@whoi.edu, Woods Hole Oceanographic Inst; Cm
Huber, Matthew, huberm@purdue.edu, Purdue Univ; As
Huckabey, Marsha, (573) 882-2040 huckabeym@missouri.edu, Univ of Missouri; Zn
Hudackova, Natalia H., natalia.hudackova@uniba.sk, Comenius Univ in Bratislava; PimPg
Hudak, Paul F., (940) 565-4312 paul.hudak@unt.edu, Univ of North Texas; Hw
Hudec, Michael R., (512) 471-1428 michael.hudec@beg.utexas.edu, Univ of Texas at Austin, Jackson Sch of Geosciences; GcEo
Hudec, Peter P., 519-253-3000 x2491 hudec@uwindsor.ca, Univ of Windsor; NgEmGe
Hudgins, Thomas, (787) 265-3845 thomas.hudgins@upr.edu, Univ of Puerto Rico; GiCpGv
Hudleston, Peter J., 612-625-0046 hudle001@umn.edu, Univ of Minnesota, Twin Cities; Gc
Hudson, Robert J. M., (217) 333-7641 rjhudson@uiuc.edu, Univ of Illinois, Urbana-Champaign; Cg
Hudson, Samuel M., sam.hudson@byu.edu, Brigham Young Univ; GosGr
Hudson-Edwards, Karen, +44 0203 073 8030 k.hudson-edwards@bbk.ac.uk, Birkbeck Coll; Gez
Huebert, Barry J., (808) 258-4673 huebert@hawaii.edu, Univ of Hawai'i, Manoa; OcCg
Huerfano, Victor, (787) 265-3845 victor.huerfano@upr.edu, Univ of Puerto Rico; YsgGt
Huerta, Audrey, 509.963.2718 huerta@geology.cwu.edu, Central Washington Univ; Ygd
Huertas, María José, huertas@ucm.es, Univ Compl de Madrid; Gi
Huey, Gregory L., (404) 894-5541 greg.huey@eas.gatech.edu, Georgia Inst of Tech; As
Huff, Edmund A., (630) 252-3633 Argonne; Ca
Huff, Warren D., (513) 556-3731 warren.huff@uc.edu, Univ of Cincinnati; GzvGr
Huffman, Debra E., (727) 553-3930 debrah@usf.edu, Univ of South Florida; Zn
Huffman, F. T., (859) 622-6968 tyler.huffman@eku.edu, Eastern Kentucky Univ; Zi
Huffman, Robert L., (912) 752-2704 Mercer Univ; Zn
Huft, Ashley, (406) 496-4789 ahuft@mtech.edu, Montana Tech of The Univ of Montana; Ca
Huggett, William, 618-453-7392 huggett@geo.siu.edu, Southern Illinois Univ Carbondale; Ec
Hughen, Konrad A., (508) 289-3353 khughen@whoi.edu, Woods Hole Oceanographic Inst; Cc
Hughes, Annie, +44 020 8417 2603 ku08925@kingston.ac.uk, Kingston Univ; Zi
Hughes, Colin, colin.hughes@manchester.ac.uk, Univ of Manchester; Go
Hughes, Denis, +27 46 6224014 d.hughes@ru.ac.za, Rhodes Univ; Hqs
Hughes, John M., (802) 656-9443 jmhughes@uvm.edu, Univ of Vermont; Gz
Hughes, Kenneth S., (787) 265-3845 kenneth.hughes@upr.edu, Univ of Puerto Rico; Gcm
Hughes, Malcolm K., (520) 621-6470 mhughes@ltrr.arizona.edu, Univ of Arizona; Pe
Hughes, Nigel C., (951) 827-3098 nigel.hughes@ucr.edu, Univ of California, Riverside; Pg
Hughes, Noah C., (209) 575-6172 hughesn@yosemite.edu, Modesto Junior Coll; ZgAs
Hughes, Randall, 7815817370 x314 rhughes@neu.edu, Northeastern Univ; Zn
Hughes, Sam, +44 07727 096492 s.p.hughes@exeter.ac.uk, Exeter Univ; Gt
Hughes, Scott S., (530) 533-1933 hughscot@isu.edu, Idaho State Univ; GvXgGi
Hughes III, Richard O., 909-389-3237 rihughes@craftonhills.edu, Crafton Hills Coll; GlZe
Hughes-Clarke, John E., 603-862-5505 jhc@ccom.unh.edu, Univ of New Hampshire; Og
Hugli, Wilbur G., 850-474-3470 whugli@uwf.edu, Univ of West Florida; As
Hugo, Richard, 503-725-3356 hugo@pdx.edu, Portland State Univ; Gy
Huizenga, Jan M., jan.marten.huizenga@nmbu.no, Norwegian Univ of Life Scis (NMBU); GzcEd
Huizenga, Jan Marten, (074) 781-4597 jan.huizenga@jcu.edu.au, James Cook Univ; GpCgGc
Hulbe, Christina L., chulbe@pdx.edu, Portland State Univ; Gl
Hulett, Sam, 614 265 6573 sam.hulett@dnr.state.oh.us, Ohio Dept of Nat Res; Eo
Hull, Donald L., (505) 667-4151 Los Alamos; Zn
Hull, Joseph M., jhull@sccd.ctc.edu, Seattle Central Coll; Gc
Huluka, Gobena E., (334) 844-4100 hulukgo@auburn.edu, Auburn Univ; Sc
Humagain, Kamal, (315) 267-3356 humagak@potsdam.edu, SUNY Potsdam; Zir
Humayun, Munir, 850-644-5860 humayun@magnet.fsu.edu, Florida State Univ; XcCgGi
Humes, Karen, 208-885-6506 khumes@uidaho.edu, Univ of Idaho; ZriZn
Hummer, Daniel R., 618-453-7386 daniel.hummer@siu.edu, Southern Illinois Univ Carbondale; GzyCu
Humphrey, John D., +966 13 860 7982 humphrey@kfupm.edu.sa, King Fahd Univ of Petroleum and Minerals; GdoCs
Humphrey, Neil F., (307) 314-2332 neil@uwyo.edu, Univ of Wyoming; GlmYh
Humphreys, Eugene D., (541) 852-0091 genehumphreys@gmail.com, Univ of Oregon; YsGt
Humphreys, Robin, (843) 953-7424 humphreysr@cofc.edu, Coll of Charleston; Ge
Humphris, Susan E., (508) 289-3451 shumphris@whoi.edu, Woods Hole Oceanographic Inst; GuCm
Hunda, Brenda R., (513) 455-7160 bhunda@cincymuseum.org, Cincinnati Mus Center; Pi
Hung, Ming-Chih, (660) 562-1797 mhung@nwmissouri.edu, Northwest Missouri State Univ; Zr
Hungerbuehler, Axel, (575) 461-3466 axelh@mesalands.edu, Mesalands Comm Coll; PvgGg
Hungerford, Hilary, (801) 863-7160 hilary.hungerford@uvu.edu, Utah Valley Univ; ZuRw
Hungr, Oldrich, (604) 822-8471 ohungr@eos.ubc.ca, Univ of British Columbia; Gm
Hunsucker, Kelli, (321) 674-8437 khunsucker@fit.edu, Florida Inst of Tech; Ob
Hunt, Allen G., (937) 775-3116 allen.hunt@wright.edu, Wright State Univ; GqSpCl
Hunt, Andrew, 817 272 2987 hunt@uta.edu, Univ of Texas, Arlington; GbCg
Hunt, Brian, (604) 822-9135 bhunt@eos.ubc.ca, Univ of British Columbia; Ob
Hunt, Emma, (864) 294-3364 emma.hunt@furman.edu, Furman Univ; GpzEg
Hunt, Gene, (202) 633-1331 Smithsonian Inst / Natl Mus of Natural History; Pm
Hunt, Kathy, bhunt@ipst.umd.edu, Univ of Maryland; Zn
Hunt, Paula J., (304) 594-2331 phunt@geosrv.wvnet.edu, West Virginia Geological & Economic Survey; GgHw
Hunt, Rachel, rachel.hunt@ed.ac.uk, Edinburgh Univ; Zn
Hunt, Robert M., (402) 472-4604 rhunt2@unl.edu, Univ of Nebraska, Lincoln; Pv
Hunt, Shelley, (519) 824-420 Ext.53065 shunt@uoguelph.ca, Univ of Guelph; Zn
Hunten, Donald M., (520) 621-4002 dhunten@lpl.arizona.edu, Univ of Arizona; As
Hunter, Arlene G., +44 (0) 1908 655400 x55400 a.g.hunter@open.ac.uk, Open Univ; ZeGiCg
Huntington, Justin, 775.673.7670 justin.huntington@dri.edu, Univ of Nevada, Reno; HgZr
Huntington, Katharine W., 206-543-1750 kate1@uw.edu, Univ of Washington; Gt
Huntley, John, (573) 884-8083 Univ of Missouri; PqgPe
Huntoon, Laura, huntoon@email.arizona.edu, Univ of Arizona; Zn
Huntress, Jr., Wesley T., (202) 478-8910 whuntress@carnegiescience.edu,

Carnegie Inst for Sci; Xc
Hupp, Brittany, bhupp@gmu.edu, George Mason Univ; Cg
Hural, Kirsten, (717) 245-1632 Dickinson Coll; Gg
Hurd, David, (814) 732-2493 dhurd@edinboro.edu, Edinboro Univ of Pennsylvania; Zg
Hurlow, Hugh, (801) 537-3385 hughhurlow@utah.gov, Utah Geo Survey; Hw
Hurowitz, Joel A., 631-632-5355 joel.hurowitz@stonybrook.edu, Stony Brook Univ; Xg
Hurrell, James, james.hurrell@colostate.edu, Colorado State Univ; As
Hurtado, Jose M., (915) 747-5669 jhurtado@utep.edu, Univ of Texas, El Paso; Gt
Hurtgen, Matthew T., (847) 491-7539 matt@earth.northwestern.edu, Northwestern Univ; GsPgGr
Husch, Jonathan M., husch@rider.edu, Rider Univ; GieXg
Husinec, Antun, (315) 229-5248 ahusinec@stlawu.edu, St. Lawrence Univ; GsCsGo
Hussein, Musa, (915) 747-5424 mjhussein@utep.edu, Univ of Texas, El Paso; YgeYm
Hussin, Azhar Hj, 03-79674203 azharhh@um.edu.my, Univ of Malaya; Gs
Hutcheon, Ian E., (403) 220-6744 ian@earth.geo.ucalgary.ca, Univ of Calgary; Cl
Hutchings, Jennifer, (541) 737-4453 jhutchings@coas.oregonstate.edu, Oregon State Univ; Gl
Hutchins, Peter S., (601) 961-5505 peter_hutchins@deq.state.ms.us, Mississippi Office of Geology; Zn
Hutchinson, Charles F., chuck@ag.arizona.edu, Univ of Arizona; Zr
Hutchison, David, 607-431-4730 hutchisond@hartwick.edu, Hartwick Coll; GxgGi
Hutson, John , john.hutson@flinders.edu.au, Flinders Univ; Spc
Hutson, Melinda, 971-722-4146 mhutson@pcc.edu, Portland Comm Coll - Sylvania Campus; Xm
Hutto, Richard S., (501) 683-0151 richard.hutto@arkansas.gov, Arkansas Geo Survey; Gg
Hutyra, Lucy, lrhutyra@bu.edu, Boston Univ; AsZr
Huws, Dei, d.g.huws@bangor.ac.uk, Bangor Univ; YruNt
Huybers, Peter, 617-495-8391 phuybers@fas.harvard.edu, Harvard Univ; Pe
Huycke, David, 509-574-4817 dhuycke@yvcc.edu, Yakima Valley Coll; Gg
Huysken, Kristin, (219) 980-6739 khuysken@iun.edu, Indiana Univ Northwest; GizGc
Huysmans, Marijke, marijke.huysmans@ees.kuleuven.be, Katholieke Univit Leuven; Hw
Hyatt, James A., (860) 465-5789 hyattj@easternct.edu, Eastern Connecticut State Univ; Gml
Hyland, Ethan, 207-999-9047 ehyland@unity.ncsu.edu, North Carolina State Univ; GrCsPe
Hylland, Michael D., (801) 537-3382 mikehylland@utah.gov, Utah Geo Survey; Gg
Hylland, Mike, 801.537.3382 mikehylland@utah.gov, Utah Geo Survey; PgYs
Hylton, Alisa, (704)330-6297 alisa.hylton@cpcc.edu, Central Piedmont Comm Coll; Gg
Hyndman, David W., 517-353-4442 hyndman@msu.edu, Michigan State Univ; Hy
Hyndman, Roy D., (250) 363-6428 Univ of Victoria; Ys
Hynek, Brian M., 303-735-4312 brian.hynek@colorado.edu, Univ of Colorado; Xg
Hynes, Andrew J., andrew.hynes@mcgill.ca, McGill Univ; Gc
Hynes, Joanna , (657) 278-5464 jfantozzi@fullerton.edu, California State Univ, Fullerton; Gg
Hysell, David, dlh37@cornell.edu, Cornell Univ; Znr
Hyzny, Matus, hyzny@fns.uniba.sk, Comenius Univ in Bratislava; Pi

I

Iaccarino, Salvatore, salvatore.iaccarino@unito.it, Univ di Torino; Gc
Iaccheri, Maria Linda, 089/2180 4274 Ludwig-Maximilians-Univ Muenchen; Gg
Iaffaldano, Giampiero, 089/2180 4220 giampiero@geophysik.uni-muenchen.de, Ludwig-Maximilians-Univ Muenchen; Yg
Iancu, Ovidiu G., 0040232201455 ogiancu@uaic.ro, Alexandru Ioan Cuza; GpeCg
Ianno, Adam J., 717-702-2033 aianno@pa.gov, DCNR- Pennsylvania Bur of Geo Survey; CaGxEg
Iantria, Linnea, (417) 836-4486 liantria@missouristate.edu, Missouri State Univ; Zn
Iaquinta, Connie, 519-253-3000 x2528 connie@uwindsor.ca, Univ of Windsor; Zn
Ibaraki, Motomu, (614) 292-7528 ibaraki.1@osu.edu, Ohio State Univ; Hg
Ibarra, Daniel, daniel_ibarra@brown.edu, Brown Univ; GeCm
Ibrahim, Ahmad Tajuddin Hj, 03-79674230 ahmadt@um.edu.my, Univ of Malaya; Ng
Ibrahim, Mohamed, 212-772-5267 mibrahim@hunter.cuny.edu, Hunter Coll (CUNY); ZgRwZc
Icenhower, Jonathan P., (509) 372-0078 jon.icenhower@pnl.gov, Pacific Northwest; Cl
Icopini, Gary, 406-496-4841 gicopini@mtech.edu, Montana Tech of The Univ of Montana; ClHwCb
Ide, Kayo, (310) 206-6484 Univ of California, Los Angeles; Zg
Idleman, Bruce D., bdi2@lehigh.edu, Lehigh Univ; Cc
Idone, Vincent P., (518) 442-4577 vpi@atmos.albany.edu, SUNY, Albany; As
Idowu, Simeon O., +234-8060227871 Federal Univ of Tech, Akure; NgHwGg
Ielpi, Alessandro, (705) 675-1151 aielpi@laurentian.ca, Laurentian Univ, Sudbury; GsmGr
Igbinigie, Nosa S., nosa.igbinigie@uniben.edu, Univ of Benin; GorGg
Ige, A. O., aige@oauife.edu.ng, Obafemi Awolowo Univ; Cg
Ige, O. O., vickeyeg2002@yahoo.com, Univ of Ilorin; Ng
Igel, Adele, (530)752-1018 aigel@ucdavis.edu, Univ of California, Davis; Asp
Igel, Heiner, 089/2180 4204 heiner.igel@lmu.de, Ludwig-Maximilians-Univ Muenchen; Yg
Igel, Matthew, (530) 752-6280 migel@ucdavis.edu, Univ of California, Davis; Asm
Iglesias, Cruz, 0034981167000 cruzi@udc.es, Coruna Univ; GaZmNr
Igonor, Emmanuel E., +2348066943346 eeigonor@futa.edu.ng, Federal Univ of Tech, Akure; GgCeGi
Ihinger, Phillip D., (715) 836-2158 ihinger@uwec.edu, Univ of Wisconsin, Eau Claire; Gx
Ikonnikova, Svetlana, 512-232-9464 s.ikonnikova@mail.utexas.edu, Univ of Texas, Austin; Ego
Iles, Derric L., (605) 677-6148 derric.iles@usd.edu, South Dakota Dept of Environment and Nat Res; HwGg
Illari, Lodovica, (617) 253-2286 illari@mit.edu, Massachusetts Inst of Tech; Am
Imasuen, Isaac O., okpeseyi.imasuen@uniben.edu, Univ of Benin; CgGee
Immel, Harald, 089/2180 6615 h.immel@lrz.uni-muenchen.de, Ludwig-Maximilians-Univ Muenchen; Pg
Immonen, Wilma, (406) 496-4182 wimmonen@mtech.edu, Montana Tech of the Univ of Montana; Zn
Inamdar, Shreeram, 302-831-8877 inamdar@udel.edu, Univ of Delaware; Hg
Indares, Aphrodite D., (709) 737-2456 aindares@mun.ca, Memorial Univ of Newfoundland; Gp
Ingall, Ellery D., (404) 894-3883 ellery.ingall@eas.gatech.edu, Georgia Inst of Tech; CmlOc
Ingalls, Anitra E., 206-221-6748 aingalls@u.washington.edu, Univ of Washington; Oc
Ingersoll, Andrew P., 626 395-6167 api@gps.caltech.edu,CalTech; As
Ingersoll, Raymond V., (310) 825-8634 ringer@epss.ucla.edu, Univ of California, Los Angeles; Gst
Ingle, Jr., James C., (650) 723-3366 ingle@pangea.stanford.edu, Stanford Univ; Pm
Inglett, Kanika Sharma, 352-294-3164 Univ of Florida; So
Inglett, Patrick, 352-392-1803 pinglett@ufl.edu, Univ of Florida; Sf
Inglis, Michael, 631-451-4120 ingism@sunysuffolk.edu, Suffolk County Comm Coll, Ammerman Campus; Zg
Ingram, B. Lynn, (510) 643-1474 ingram@eps.berkeley.edu, Univ of California, Berkeley; Cs
Innanen, Kristopher, 403 210-6837 k.innanen@ucalgary.ca, Univ of Calgary; Ye
Insel, Nadja, 773-442-6058 n-insel@neiu.edu, Northeastern Illinois Univ; Gt
Inskeep, William P., (406) 994-5077 binskeep@montana.edu, Montana State Univ; Sc
Insua, Juan Miguel, insuarev@ucm.es, Univ Compl de Madrid; NtGc
Ioannides, Dimitri, 417-836-5801 dioannides@missouristate.edu, Missouri State Univ; Zn
Ionesi, Viorel, vioion@uaic.ro, Alexandru Ioan Cuza; PmGgHg
Iordache, Vlad, 217 300 8779 iordach1@illinois.edu, Univ of Illinois; Hw
Iqbal, Mohammad Z., (319) 273-2998 m.iqbal@uni.edu, Univ of Northern Iowa; Hw
Iribar, Vicente , vicente.iribar@ehu.eus, Univ of the Basque Country UPV/EHU; HwGm
Irish, Jennifer, (540) 231-2298 jirish@vt.edu, Virginia Tech; NoOg
Irmis, Randall B., (801) 581-6555 irmis@umnh.utah.edu, Univ of Utah; Sa
Irons, Trevor , 406-496-4144 tirons@mtech.edu, Montana Tech of the Univ of Montana; Yue
Irvin, Gene D., 205/247-3542 dirvin@gsa.state.al.us, Geo Survey of Alabama; Gg

Irvine, Dylan, dylan.irvine@flinders.edu.au, Flinders Univ; Hw
Irvine, Peter D., p.irvine@ucl.ac.uk, Univ Coll London; ZcNx
Irving, Tony, irvingaj@uw.edu, Univ of Washington; CgXmGi
Irwin, Rossman P., 202-633-3632 irwinr@si.edu, Smithsonian Inst, Natl Air & Space Mus; XgGm
Isaacson, Carl, 218-755-4104 cisaacson@bemidjistate.edu, Bemidji State Univ; Ge
Isaacson, Peter E., (208) 885-7969 isaacson@uidaho.edu, Univ of Idaho; Psi
Isacks, Bryan L., 607-316-8266 bli1@cornell.edu, Cornell Univ; GtYsGm
Isard, Sierra J., 828-296-4629 sierra.isard@ncdenr.gov, N.C. Dept of Env Quality; Gcg
Isava, Virginia, 657-278-5616 visava@fullerton.edu, California State Univ, Fullerton; ZeRc
Isbell, John L., 414-229-2877 jisbell@uwm.edu, Univ of Wisconsin, Milwaukee; Gs
Isenor, Fenton M., fenton_isenor@cbu.ca, Cape Breton Univ; GgNgGm
Ishii, Miaki, 617-384-8066 ishii@eps.harvard.edu, Harvard Univ; Yg
Ishii, Satoshi, 612-624-7902 ishi0040@umn.edu, Univ of Minnesota, Twin Cities; SbPg
Ishman, Scott E., (618) 453-7377 sishman@siu.edu, Southern Illinois Univ Carbondale; Pmc
Ismail, Ahmed, 405-744-6358 ahmed.ismail@okstate.edu, Oklahoma State Univ; Ysu
Ismat, Zeshan, 717-358-4485 zeshan.ismat@fandm.edu, Franklin and Marshall Coll; Gc
Israel, Daniel W., (919) 515-2388 North Carolina State Univ; Sb
Ito, Emi, 612-624-7881 eito@umn.edu, Univ of Minnesota, Twin Cities; ClGnCs
Ito, Garrett T., (808) 956-9717 gito@hawaii.edu, Univ of Hawai'i, Manoa; YgrGt
Ivanochko, Tara, (604) 827-3179 tivanochko@eos.ubc.ca, Univ of British Columbia; OoZce
Ivanov, Julian, (785) 864-2089 jivanov@kgs.ku.edu, Univ of Kansas; Yg
Ivanov, Martin, +420 549 49 4600 mivanov@sci.muni.cz, Masaryk Univ; PgiPe
Ivanovic, Ruza, +44(0) 113 34 34945 r.ivanovic@leeds.ac.uk, Univ of Leeds; OgAsPe
Ivany, Linda C., (315) 443-3626 lcivany@syr.edu, Syracuse Univ; PegCs
Iverson, Neal R., (515) 294-8048 niverson@iastate.edu, Iowa State Univ of Sci & Tech; Gl
Iverson, Nels, 575-835-5319 nels.iverson@nmt.edu, New Mexico Tech; GvCga
Ivey, James, jivey@usf.edu, Univ of South Florida; Zy
Ivins, Erik R., (818) 354-4785 Jet Propulsion Lab; Gt
Ivory, Sarah, sarah_ivory@psu.edu, Pennsylvania State Univ, Univ Park; Pl
Ivy, Russell L., ivy@fau.edu, Florida Atlantic Univ; Zn
Iwaz, Nicole, niwaz@cerritos.edu, Cerritos Coll; GgeOg
Izawa, Matt, mizawa2@uwo.ca, Western Univ; Gz
Izon, Gareth, +44 01334 463936 gji3@st-andrews.ac.uk, Univ of St. Andrews; Cg

J

Jablonowski, Christiane, (734) 763-6238 cjablono@umich.edu, Univ of Michigan; AsmZo
Jablonski, David, (773) 702-8163 djablons@midway.uchicago.edu, Univ of Chicago; Pi
Jacinthe, Pierre-Andre, pjacinth@iupui.edu, Indiana Univ - Purdue Univ Indianapolis; Cl
Jackson, Allyson K., (914) 251-6646 allyson.jackson@purchase.edu, SUNY, Purchase; ZgcZe
Jackson, Andrea, +44(0) 113 34 36728 a.v.jackson@leeds.ac.uk, Univ of Leeds; As
Jackson, Brian P., (603) 646-1272 brian.jackson@dartmouth.edu, Dartmouth Coll; CaGeCt
Jackson, Charles, (512) 471-0401 charles@ig.utexas.edu, Univ of Texas, Austin; ZoOoAt
Jackson, Chester M., (912) 598-2328 cjackson@georgiasouthern.edu, Georgia Southern Univ; OnGs
Jackson, Christopher, +44 20 759 47450v c.jackson@imperial.ac.uk, Imperial Coll; Gt
Jackson, David D., (310) 825-0421 djackson@ucla.edu, Univ of California, Los Angeles; YsdYg
Jackson, Frankie, (406) 994-6642 frankiej@montana.edu, Montana State Univ; Pv
Jackson, Gail D., +44 (0) 131 650 5436 g.jackson@ed.ac.uk, Edinburgh Univ; PbGe
Jackson, George A., (979) 845-0405 gjackson@ocean.tamu.edu, Texas A&M Univ; Op
Jackson, Hiram, (916) 691-7605 Cosumnes River Coll; Gg
Jackson, James A., +44 (0) 1223 337197 jaj2@cam.ac.uk, Univ of Cambridge; YgGt
Jackson, Jennifer M., (626) 395-6780 jackson@gps.caltech.edu,CalTech; Gy
Jackson, Jeremiah, 573-368-2182 jeremiah.jackson@dnr.mo.gov, Missouri Dept of Nat Res; Gg
Jackson, Jeremy B. C., (858) 518-7613 jbjackson@ucsd.edu, Univ of California, San Diego; PgePi
Jackson, Kenneth J., (925) 422-6053 jackson8@llnl.gov, Lawrence Livermore; Cl
Jackson, Louise E., 530-754-9116 lejackson@ucdavis.edu, Univ of California, Davis; So
Jackson, Marie D., 801-581-7062 m.d.jackson@utah.edu, Univ of Utah; Gv
Jackson, Martin P. A., (512) 475-9548 martin.jackson@beg.utexas.edu, Univ of Texas at Austin, Jackson Sch of Geosciences; Gc
Jackson, Matt, jackson@geol.ucsb.edu, Univ of California, Santa Barbara; Cu
Jackson, Michael, 612-624-5274 jacks057@umn.edu, Univ of Minnesota, Twin Cities; Ym
Jackson, Richard A., (718) 460-1476 Long Island Univ, Brooklyn Campus; Gc
Jackson, Robert B., 919 660 7408 jackson@duke.edu, Duke Univ; Zn
Jackson, William T., wjackson@southalabama.edu, Univ of South Alabama; GsCc
Jacob, Daniel J., (617) 495-1794 djj@io.harvard.edu, Harvard Univ; As
Jacob, Klaus H., jacob@ldeo.columbia.edu, Columbia Univ; Ys
Jacob, Robert W., (570) 577-1791 rwj003@bucknell.edu, Bucknell Univ; Yg
Jacobi, Robert D., (716) 645-3489 rdjacobi@buffalo.edu, SUNY, Buffalo; GctGs
Jacobs, Alan M., 330-941-2933 amjacobs@ysu.edu, Youngstown State Univ; Ge
Jacobs, Bonnie F., bjacobs@smu.edu, Southern Methodist Univ; Pl
Jacobs, Daniel, (480) 727-2227 daniel.c.jacobs@asu.edu, Arizona State Univ; Xa
Jacobs, Gary K., jacobsgk@gmail.com, Oak Ridge; Cg
Jacobs, Jon, 519-661-2111 ext 86752 jjacobs@uwo.ca, Western Univ; Gg
Jacobs, Katharine L., (520) 626-0684 jacobsk@email.arizona.edu, Univ of Arizona; Hg
Jacobs, Louis L., (214) 768-2773 Southern Methodist Univ; Pv
Jacobs, Peter, jacobsp@uww.edu, Univ of Wisconsin, Whitewater; SdaGm
Jacobs, Stanley, sjacobs@ldeo.columbia.edu, Columbia Univ; Op
Jacobs, Tenika, (609) 984-6587 tenika.jacobs@dep.state.nj.us, New Jersey Geological and Water Survey; Zn
Jacobsen, Abram, 360-734-5019 abramj@uw.edu, Univ of Washington; Xg
Jacobsen, Stein B., (617) 495-5233 jacobsen@neodymium.harvard.edu, Harvard Univ; Cc
Jacobsen, Steven D., 847-467-1825 s-jacobsen@northwestern.edu, Northwestern Univ; GyzZm
Jacobson, Andrew D., (847) 491-3132 adj@earth.northwestern.edu, Northwestern Univ; ClPeCa
Jacobson, Carl E., cejac@iastate.edu, Iowa State Univ of Sci & Tech; GtcGp
Jacobson, Gary L., gary.jacobson@gcccd.edu, Grossmont Coll; GucEm
Jacobson, Roger, (775) 673-7364 roger@dri.edu, Univ of Nevada, Reno; Cg
Jacobson, William , jacobsow@uwgb.edu, Univ of Wisconsin, Green Bay; GgZg
Jadamec, Margarete, 716-645-4262 mjadamec@buffalo.edu, SUNY, Buffalo; GtYgZf
Jaecks, Glenn, jaecksg@arc.losrios.edu, American River Coll; GgOgPg
Jaeger, John M., (352) 846-1381 jmjaeger@ufl.edu, Univ of Florida; Gs
Jaegle, Lyatt, (206) 685-2679 jaegle@atmos.washington.edu, Univ of Washington; As
Jafari, Mohammad , jafari_mohammad@columbusstate.edu, Columbus State Univ; Zn
Jafarizadeh, Babak, b.jafarizadeh@hw.ac.uk, Heriot-Watt Univ; Eo
Jaffe, Daniel A., (425) 352-5357 djaffe@u.washington.edu, Univ of Washington; As
Jaffe, Peter R., (609) 258-4653 jaffe@princeton.edu, Princeton Univ; SbHgw
Jagoutz, Oliver, (617) 324-5514 jagoutz@mit.edu, Massachusetts Inst of Tech; GitGc
Jahangiri, Ahmad, +98 (411) 339 2695 jahangiri@tabriu.ac.ir, Univ of Tabriz; GivGg
Jain, Atul K., (217) 333-2128 jain1@illinois.edu, Univ of Illinois, Urbana-Champaign; As
Jaisi, Deb, 302-831-1376 jaisi@udel.edu, Univ of Delaware; Cb
Jaiswal, Priyank, 405-744-6358 priyank.jaiswal@okstate.edu, Oklahoma State Univ; Yg
Jakosky, Bruce M., 303-492-8004 bruce.jakosky@colorado.edu, Univ of Colorado; Xg
Jamaluddin, Tajul Anuar, 03-79674152 taj@um.edu.my, Univ of Malaya; Ng

James, Matthew J., (707) 664-2301 james@sonoma.edu, Sonoma State Univ; PiRhPg
James, Noel P., (613) 533-6170 jamesn@queensu.ca, Queen's Univ; Gd
James, Peter B., (254)710-2351 p_james@baylor.edu, Baylor Univ; Ygg
James, Rachael, +44 (0)23 80599005 r.h.james@soton.ac.uk, Univ of Southampton; Oc
James, Richard, 7056751151 x2271 rjames@laurentian.ca, Laurentian Univ, Sudbury; GpiEm
James, Thomas S., (250) 363-6403 thomas.james@canada.ca, Univ of Victoria; YdvYg
James, Valentine U., (814) 393-1938 vjames@clarion.edu, Clarion Univ; ZuRwZn
James-Aworeni, E., 234-705-326-6216 dawnjames@gmail.com, Obafemi Awolowo Univ; Hg
Jamieson, Heather E., (613) 545-6181 Queen's Univ; Ge
Jamieson, John, 709 864 2273 jjamieson@mun.ca, Memorial Univ of Newfoundland; Gu
Jamili, Ahmad, ahmad.jamili@ou.edu, Univ of Oklahoma; Np
Jamiolahmady, Mahmoud, +44 (0) 131 451 3122 m.jamiolahmady@pet.hw.ac.uk, Heriot-Watt Univ; Zn
Janecke, Susanne U., (435) 797-3877 susanne.janecke@usu.edu, Utah State Univ; Gt
Janecky, David R., (505) 667-7603 Los Alamos; Cl
Janetos, Tony, ajanetos@bu.edu, Boston Univ; Zn
Janiszewski, Helen, hajanisz@hawaii.edu, Univ of Hawai'i, Manoa; Ys
Janney, Phillip, +27-21-650-2929 phil.janney@uct.ac.za, Univ of Cape Town; GiCaXc
Jans, Urs, 212-650-8369 ujans@ccny.cuny.edu, Graduate Sch of the City Univ of New York; Gg
Janson, Xavier, 512-475-9524 xavier.janson@beg.utexas.edu, Univ of Texas, Austin; Gs
Janssen, Keith, (785) 532-6101 kjanssen@ksu.edu, Kansas State Univ; So
Jantz, Claire A., (717) 477-1399 cajant@ship.edu, Shippensburg Univ; ZuiZc
Janusz, Robert, 210-486-0045 San Antonio Comm Coll; Gg
Jaouich, Alfred, 514-987-3000 #3378 jaouich.alfred@uqam.ca, Univ du Quebec a Montreal; Sc
Jarcho, Kari A., (612) 625-5251 kjarcho@umn.edu, Univ of Minnesota, Twin Cities; Zn
Jardine, Philip M., ipj@ornl.gov, Oak Ridge; Sc
Jarvis, Ed, +353 21 4902698 e.jarvis@ucc.ie, Univ Coll Cork; Pl
Jarvis, Gary T., (416)736-2100 #77710 jarvis@yorku.ca, York Univ; Yg
Jarvis, Ian, +44 020 8417 2526 i.jarvis@kingston.ac.uk, Kingston Univ; CgGsr
Jarvis, W. T., 541-737-4032 todd.jarvis@oregonstate.edu, Oregon State Univ; Hw
Jasbinsek, John J., (805) 756-2013 jjasbins@calpoly.edu, California Polytechnic State Univ; YsuHw
Jasinski, Steven E., (717) 783-9897 c-sjasinsk@pa.gov, State Mus of Pennsylvania; PvgPg
Jasoni, Richard, 775-673-7472 richard.jasoni@dri.edu, Desert Research Inst; So
Jastrow, Julie D., (630) 252-3226 jdjastrow@anl.gov, Argonne; Sb
Jaumé, Steven C., (843) 953-1802 jaumes@cofc.edu, Coll of Charleston; YsRnYg
Javaux, Emmanuelle, 32 4 366 54 22 ej.javaux@ulg.ac.be, Univ de Liege; Pl
Jawitz, James W., 352-392-1951 (203) jawitz@ufl.edu, Univ of Florida; SpHsw
Jayakumar, Amal, 609-258-6294 ajayakum@princeton.edu, Princeton Univ; Cm
Jayawickreme, Dushmantha, (203) 392-7286 jayawickred1@southernct.edu, Southern Connecticut State Univ; YgSpNg
Jaye, Shelley, (703) 425-5180 sjaye@nvcc.edu, Northern Virginia Comm Coll - Annandale; Giy
Jean, Marlon , 207-778-7402 marlon.jean@maine.edu, Univ of Maine, Farmington; GizGp
Jeanloz, Raymond, (510) 642-2639 jeanloz@uclink.berkeley.edu, Univ of California, Berkeley; Gy
Jebrak, Michel, 514-987-3000 #3986 jebrak.michel@uqam.ca, Univ du Quebec a Montreal; Eg
Jedrysek, Mariusz O., mariusz.jedrysek@uwr.edu.pl, Univ of Wroclaw; CaGe
Jefferson, Anne, 704-687-5977 ajefferson@uncc.edu, Univ of North Carolina, Charlotte; HgGm
Jeffery, David L., 740-376-4844 jefferyd@marietta.edu, Marietta Coll; PsGo
Jeffrey, Kip, +4401326259442 c.jeffrey@exeter.ac.uk, Exeter Univ; Nx
Jeffries, H. Perry, (401) 874-6222 jeffries@uri.edu, Univ of Rhode Island; Ob
Jekeli, Christopher, 614 292 7117 jekeli.1@osu.edu, Ohio State Univ; Ydv
Jelinski, Nicolas, 612-626-9936 jeli0026@umn.edu, Univ of Minnesota, Twin Cities; SdCs
Jellinek, Mark, mjellinek@eos.ubc.ca, Univ of British Columbia; Gg
Jenkins, David M., (607) 777-2736 dmjenks@binghamton.edu, Binghamton Univ; CpGzp
Jenkins, Gregory S., 814-865-0479 gsj1@psu.edu, Pennsylvania State Univ, Univ Park; Am
Jenkins, Mary Ann, (416)736-2100 #22992 maj@yorku.ca, York Univ; As
Jenkyns, Hugh C., +44 (1865) 272023 hugh.jenkyns@earth.ox.ac.uk, Univ of Oxford; GrsCs
Jensen, Ann R., 605-677-6159 ann.jensen@usd.edu, South Dakota Dept of Environment and Nat Res; Gg
Jensen, Antony, +44 (0)23 80593428 acj@noc.soton.ac.uk, Univ of Southampton; Ob
Jensen, Jennifer, (512) 245-2170 Texas State Univ; Zru
Jensen, Olivia G., 514-398-3587 olivia.jensen@mcgill.ca, McGill Univ; Yg
Jensen, Ryan R., (801) 422-5386 rjensen@byu.edu, Brigham Young Univ; Zii
Jensen, Scott W., 605-677-6869 scott.jensen@usd.edu, South Dakota Dept of Environment and Nat Res; Zn
Jenson, John W., (671) 735-2689 jjenson@triton.uog.edu, Univ of Guam; HwGel
Jeon, Sung Bae, sungbae.jeon@eku.edu, Eastern Kentucky Univ; Zy
Jercinovic, Michael J., (413) 545-2431 mjj@geo.umass.edu, Univ of Massachusetts, Amherst; Gx
Jerde, Eric, (606) 783-5406 e.jerde@moreheadstate.edu, Morehead State Univ; Cg
Jerolmack, Douglas, 215-746-2823 sediment@sas.upenn.edu, Univ of Pennsylvania; GmYxHq
Jessee, Anna, manowick@iu.edu, Indiana Univ - Purdue Univ Indianapolis; GgYg
Jessup, Micah, (865) 974-2366 mjessup@utk.edu, Univ of Tennessee, Knoxville; Gc
Jessup, Stephen, 706-569-4571 jessup_stephen@columbusstate.edu, Columbus State Univ; Am
Jeu, Amy, (212) 772-4019 ajeu@hunter.cuny.edu, Hunter Coll (CUNY); Zi
Jewell, Paul W., (801) 581-6636 paul.jewell@utah.edu, Univ of Utah; GmHg
Jewett, Brian, 217-333-3957 bjewett@illinois.edu, Univ of Illinois, Urbana-Champaign; As
Jewitt, David, (310) 825-3880 jewitt@epss.ucla.edu, Univ of California, Los Angeles; XyZnn
Jezek, Kenneth, 614-292-7973 jezek.1@osu.edu, Ohio State Univ; YgPcGl
Ji, Chen, (805) 893-2923 ji@geol.ucsb.edu, Univ of California, Santa Barbara; Ys
Ji, Peng, pengji@ldeo.columbia.edu, Columbia Univ; Yr
Ji, Wei, (816) 235-1334 jiwei@umkc.edu, Univ of Missouri, Kansas City; ZriZy
Jialing, Wang, 724.738.2383 jialing.wang@sru.edu, Slippery Rock Univ; Zi
Jiang, Dazhi, 519-661-3192 djiang3@uwo.ca, Western Univ; GcgGg
Jiang, Ganqing, (702) 895-2708 ganqing.jiang@unlv.edu, Univ of Nevada, Las Vegas; GsClGr
Jiang, Junle , jiang@ou.edu, Univ of Oklahoma; Yg
Jiang , Shu, 801-585-9816 sjiang@egi.utah.edu, Univ of Utah; GoYsNp
Jiang, Zhenhua, (410) 436-6864 Argonne; Zn
Jickells, Tim, +44 (0)1603 59 3117 t.jickells@uea.ac.uk, Univ of East Anglia; Ob
Jiménez, David, davidj03@ucm.es, Univ Compl de Madrid; GcNt
Jimoh, Mustapha T., +2348052472942 mtjimoh@lautech.edu.ng, Ladoke Akintola Univ of Tech; GxxGx
Jin, Fei Fei, 808-956-4645 jff@hawaii.edu, Univ of Hawai'i at Manoa; AmsZo
Jin, Ge, 303-273-3455 gjin@mines.edu, Colorado Sch of Mines; Ye
Jin, Guohai, 205-247-3560 gjin@gsa.state.al.us, Geo Survey of Alabama; Np
Jin, He (Hannah), hjin1@usf.edu, Univ of South Florida; Zi
Jin, Jisuo, (519) 661-4061 jjin@uwo.ca, Western Univ; Pi
Jin, Lixin, (915) 747-5559 ljin2@utep.edu, Univ of Texas, El Paso; Ge
Jin, Qusheng, 541-346-4999 qjin@uoregon.edu, Univ of Oregon; Cg
Jin, Yufang, (530) 601 9805 jin.yufang@gmail.com, Univ of California, Davis; Zr
Jinnah, Zubair A., (011) 717-6554 zubair.jinnah@wits.ac.za, Univ of the Witwatersrand; GsPg
Jiracek, George R., (619) 594-5160 gjiracek@mail.sdsu.edu, San Diego State Univ; Ye
Jiron, Rebecca, 757-221-2484 rljiron@wm.edu, William & Mary; GgsZi
Jiskoot, Hester, (403) 329-2739 hester.jiskoot@uleth.ca, Univ of Lethbridge; GlmZy
Jo, Ho Young, 82-2-3290-3179 hyjo@korea.ac.kr, Korea Univ; NgCaGe
Jo, Injeong, (512) 245-2170 Texas State Univ; Ze
Joannes-Boyau, Renaud, renaud.joannes-boyau@scu.edu.au, Southern Cross Univ; CcaCs

Jochens, Ann E., (979) 845-6714 ajochens@ocean.tamu.edu, Texas A&M Univ; Op
Joeckel, R. M., 402-472-7520 rjoeckel3@unl.edu, Univ of Nebraska, Lincoln; GrsGm
Joeckel, R. M., (402) 472-7520 rjoeckel3@unl.edu, Univ of Nebraska, Lincoln; GrsGm
Joeckel, R.M., (402)472-7520 rjoeckel3@unl.edu, Univ of Nebraska, Lincoln; Grs
Joesten, Raymond, raymond.joesten@uconn.edu, Univ of Connecticut; GzZmGp
Johanesen, Katharine, (814) 641-3601 johanesen@juniata.edu, Juniata Coll; GzxGc
Johannesson, Karen H., 504-862-3193 kjohanne@tulane.edu, Tulane Univ; HgCg
Johanson, Erik, (561) 297-4153 ejohanson@fau.edu, Florida Atlantic Univ; RnZy
John, Barbara E., (307) 223-1951 bjohn@uwyo.edu, Univ of Wyoming; Gci
John, Bratton F., (313) 577-2506 jbratton@limno.com, Wayne State Univ; Cg
John, Cedric, +44 20 759 46461 cedric.john@imperial.ac.uk, Imperial Coll; Gs
John, Chacko J., (225) 578-8681 cjohn@lsu.edu, Louisiana State Univ; GosGe
John, Hickman B., 859-323-0541 jhickman@uky.edu, Univ of Kentucky; Gto
Johns, Rebecca, rjohns@usf.edu, Univ of South Florida; Zn
Johns, Ronald A., (512) 223-7491 rjohns@austincc.edu, Austin Comm Coll District; Pi
Johnson, Alexandria, avjohns@purdue.edu, Purdue Univ; As
Johnson, Ansel G., (503) 725-3381 Portland State Univ; Ye
Johnson, Arthur H., ahj@sas.upenn.edu, Univ of Pennsylvania; So
Johnson, Ashanti, (817) 272-2987 ashanti@uta.edu, Univ of Texas, Arlington; OgGe
Johnson, Beverly J., 207 786-6062 bjohnso3@bates.edu, Bates Coll; CsbCl
Johnson, Brad, 520-621-2470 bradjohnson@email.arizona.edu, Univ of Arizona; Gc
Johnson, Brandon, bcjohnson@purdue.edu, Purdue Univ; As
Johnson, Bruce D., (902) 494-3259 bruce.johnson@dal.ca, Dalhousie Univ; Oc
Johnson, Cari, (801) 585-3782 cari.johnson@utah.edu, Univ of Utah; Gs
Johnson, Carl G., (508) 289-2304 cjohnson@whoi.edu, Woods Hole Oceanographic Inst; Ca
Johnson, Clark M., (828) 265-8680 clarkj@geology.wisc.edu, Univ of Wisconsin, Madison; CgcCs
Johnson, Claudia C., 812-855-0646 claudia@indiana.edu, Indiana Univ, Bloomington; Peg
Johnson, Daniel L., (403) 327-4561 Univ of Lethbridge; Zn
Johnson, Darren J., 605-677-6162 darren.j.johnson@usd.edu, South Dakota Dept of Environment and Nat Res; GgEo
Johnson, David B., (575) 835-5635 djohnson@nmt.edu, New Mexico Tech; Ps
Johnson, Donald O., (630) 252-3392 Argonne; Gr
Johnson, Elias, (417) 836-5801 Missouri State Univ; Zr
Johnson, Emily P., (617) 353-9709 Boston Univ; Zn
Johnson, Eric L., 1-607-4314658 johnsone@hartwick.edu, Hartwick Coll; GpcGi
Johnson, Gary D., (603) 636-2371 gary.d.johnson@dartmouth.edu, Dartmouth Coll; GrdYm
Johnson, Glenn W., (801) 581-6151 gjohnson@egi.utah.edu, Univ of Utah; Gq
Johnson, Graham, +61 7 3138 4371 g.johnson@qut.edu.au, Queensland Univ of Tech; Asp
Johnson, Gregory C., (206) 526-6806 Univ of Washington; Op
Johnson, H. Paul, (206) 543-8474 johnson@ocean.washington.edu, Univ of Washington; Yr
Johnson, Harlan P., 206-543-8474 johnson@ocean.washington.edu, Univ of Washington; Gu
Johnson, Helen, +44 01865 272142 helen.johnson@earth.ox.ac.uk, Univ of Oxford; Og
Johnson, Howard, +44 20 759 46461 h.d.johnson@imperial.ac.uk, Imperial Coll; Go
Johnson, Jane M., (320) 589-3411 Univ of Minnesota, Twin Cities; So
Johnson, Jean M., (706) 272-2666 jmjohnson@daltonstate.edu, Dalton State Coll; Zg
Johnson, Jeanne L., (225) 354-9562 jeannej@lsu.edu, Louisiana State Univ; Zn
Johnson, Jeffrey , jeffreybjohnson@boisestate.edu , Boise State Univ; YsGv
Johnson, Joel E., 603-862-1718 joel.johnson@unh.edu, Univ of New Hampshire; Gus
Johnson, Julie, 901-678-4217 jjhnsn79@memphis.edu, Univ of Memphis; Gx
Johnson, Kaj, 812-855-3612 kajjohns@indiana.edu, Indiana Univ, Bloomington; Yg
Johnson, Kathleen, 949-824-6174 kathleen.johnson@uci.edu, Univ of California, Irvine; Zg
Johnson, Kayleigh, kayleigh.johnson@sdsmt.edu, South Dakota Sch of Mines & Tech; Pgv
Johnson, Kenneth, (713) 222-5375 johnsonk@uhd.edu, Univ of Houston Downtown; GiCac
Johnson, Kevin, 321-674-7186 johnson@fit.edu, Florida Inst of Tech; Ob
Johnson, Kevin T. M., 808-956-3444 kjohnso2@hawaii.edu, Univ of Hawai'i, Manoa; Gi
Johnson, Kurt, (907) 696-0079 kurt.johnson@alaska.gov, Alaska Division of Geological & Geophysical Surveys; Gg
Johnson, Lane R., lrjohnson@lbl.gov, Univ of California, Berkeley; Ys
Johnson, Mark S., (778) 999-2102 mark.johnson.2102@gmail.com, Univ of British Columbia; SpHsq
Johnson, Markes E., 413-597-2329 markes.e.johnson@williams.edu, Williams Coll; PseGm
Johnson, Martin, +44 (0)1603 59 1299 martin.johnson@uea.ac.uk, Univ of East Anglia; Cm
Johnson, Ned K., (510) 642-3059 Univ of California, Berkeley; Pv
Johnson, Neil E., 540-231-1785 johnsonne@vt.edu, Virginia Tech; GzEg
Johnson, Paul A., (507) 933-7442 Los Alamos; Yg
Johnson, Robert G., 612-626-0853 johns088@umn.edu, Univ of Minnesota, Twin Cities; Pe
Johnson, Roy A., (520) 621-4890 johnson6@email.arizona.edu, Univ of Arizona; Ys
Johnson, Sarah E., 859-572-6907 johnsonsa@nku.edu, Northern Kentucky Univ; GmNgZi
Johnson, Scott E., (207) 581-2142 johnsons@maine.edu, Univ of Maine; Gc
Johnson, Thomas C., tcj@d.umn.edu, Univ of Minnesota Duluth; PcGn
Johnson, Ty, (501) 683-0153 ty.johnson@arkansas.gov, Arkansas Geo Survey; GgZiGr
Johnson, Verner C., (970) 248-1672 vjohnson@coloradomesa.edu, Colorado Mesa Univ; Yg
Johnson, William C., 785-864-5548 wcj@ku.edu, Univ of Kansas; Gm
Johnson, William P., (801) 581-5033 william.johnson@utah.edu, Univ of Utah; Ng
Johnston, A. Dana, (541) 346-5588 adjohn@uoregon.edu, Univ of Oregon; Cp
Johnston, Archibald C., (901) 678-2007 ajohnstn@memphis.edu, Univ of Memphis; Ys
Johnston, Carl G., 330-941-7151 cgjohnston@ysu.edu, Youngstown State Univ; Pg
Johnston, Cliff, cliffjohnston@purdue.edu, Purdue Univ; Sc
Johnston, David, (501) 683-0126 david.johnston@arkansas.gov, Arkansas Geo Survey; Gg
Johnston, Paul, (403) 440-6174 pajohnston@mtroyal.ca, Mount Royal Univ; Pi
Johnston, Scott, 805-756-1650 scjohnst@calpoly.edu, California Polytechnic State Univ; GcpCc
Johnston, Stephen T., (780) 492-5249 stjohnst@ualberta.ca, Univ of Alberta; GctGg
Johnston, Thomas, (403) 329-2534 johnston@uleth.ca, Univ of Lethbridge; Ze
Johnston, III, John E., (225) 578-8657 hammer@lsu.edu, Louisiana State Univ; EoGe
Join, Jean-Lambert, +262262938697 jean-lambert.join@univ-reunion.fr, Univ de la Reunion; Hw
Jokipii, Jack R., (520) 621-4256 jokipii@lpl.arizona.edu, Univ of Arizona; Xy
Jolley, David, +44 (0)1224 273450 d.jolley@abdn.ac.uk, Univ of Aberdeen; PlsGv
Jolliff, Bradley L., (314) 935-5622 blj@levee.wustl.edu, Washington Univ in St. Louis; GxzGi
Jomeiri, Rahim, +98 (411) 339 2704 Univ of Tabriz; Yg
Jonas, John J., (514) 398-4755 McGill Univ; Nx
Jones, Adrian P., +44 020 7679 32415 adrian.jones@ucl.ac.uk, Univ Coll London; GiCgGg
Jones, Alan, (607) 777-2518 Binghamton Univ; Ys
Jones, Alice, (859) 622-1424 alice.jones@eku.edu, Eastern Kentucky Univ; Zyn
Jones, Allan, alljones@illinois.edu, Univ of Illinois; Hw
Jones, Bobby L., (225) 388-8328 Louisiana State Univ; Go
Jones, Brian, (780) 492-3074 brian.jones@ualberta.ca, Univ of Alberta; Ps
Jones, Charles E., (412) 624-6347 cejones@pitt.edu, Univ of Pittsburgh; GgsPg

Jones, Clain, clainj@montana.edu, Montana State Univ; So
Jones, Craig H., (303) 492-6994 craig.jones@colorado.edu, Univ of Colorado; YsGtYg
Jones, Daniel, daniel.s.jones@nmt.edu, New Mexico Tech; Cb
Jones, David S., (413) 542-2714 djones@amherst.edu, Amherst Coll; GsClGr
Jones, Douglas S., (352) 273-1901 dsjones@flmnh.ufl.edu, Univ of Florida; Pg
Jones, Eric M., (505) 667-6386 honais@lanl.gov, Los Alamos; As
Jones, F. Walter, (780) 492-0667 wjones@phys.ualberta.ca, Univ of Alberta; Ym
Jones, Francis H., (604) 822-2138 fjones@eoas.ubc.ca, Univ of British Columbia; YeZe
Jones, Glenda, +44 1782 734309 g.m.jones@keele.ac.uk, Keele Univ; NgYg
Jones, Gwilym, g.jones@neu.edu, Northeastern Univ; Zn
Jones, Gwyn, gwjones@bellevuecollege.edu, Bellevue Coll; Gg
Jones, Hazel, hazel.jones@manchester.ac.uk, Univ of Manchester; As
Jones, Jon W., (403) 220-5024 Univ of Calgary; Gp
Jones, Julia A., (541) 737-1224 jonesj@geo.oregonstate.edu, Oregon State Univ; So
Jones, Larry Allan, (505) 667-0142 ljones@lanl.gov, Los Alamos; Ng
Jones, Lawrence S., (970) 248-1708 lajones@coloradomesa.edu, Colorado Mesa Univ; GsmGr
Jones, Merren, merren.a.jones@manchester.ac.uk, Univ of Manchester; Gst
Jones, Michael Q., (011) 717-6628 michael.jones@wits.ac.za, Univ of the Witwatersrand; YhGtYg
Jones, Minnie O., (312) 996-3154 mojones@uic.edu, Univ of Illinois Chicago; Zn
Jones, Norris W., (920) 424-4460 jonesnw@uwosh.edu, Univ of Wisconsin, Oshkosh; Gi
Jones, Phillip D., +44 (0)1603 59 2090 p.jones@uea.ac.uk, Univ of East Anglia; AsPeHg
Jones, Rhian, 505-277-4204 rjones@unm.edu, Univ of New Mexico; Gz
Jones, Robert L., (217) 333-9490 Univ of Illinois, Urbana-Champaign; Sc
Jones, Stephen C., (205) 247-3601 sjones@gsa.state.al.us, Geo Survey of Alabama; HyGeCg
Jones, Stuart, +44 (0) 191 33 42319 stuart.jones@durham.ac.uk, Durham Univ; Gs
Jones, T, jonestp@cf.ac.uk, Cardiff Univ; Ge
Jones, Tim L., (505) 646-3405 New Mexico State Univ, Las Cruces; Sp
Jones, William J., 803-777-7171 jjones@seoe.sc.edu, Univ of South Carolina; Ob
Jones, III, John P., jpjones@email.arizona.edu, Univ of Arizona; Zn
Joniak, Peter , joniak@fns.uniba.sk, Comenius Univ in Bratislava; Pv
Jordan, Andy, +44 (0)1603 59 2552 a.jordan@uea.ac.uk, Univ of East Anglia; Ge
Jordan, Brennan T., (605) 677-6143 brennan.jordan@usd.edu, Univ of South Dakota; GiAm
Jordan, Guntram, 089/2180 4353 guntram.jodan@lrz.uni-muenchen.de, Ludwig-Maximilians-Univ Muenchen; Gz
Jordan, Jim L., (409) 880-8211 jim.jordan@lamar.edu, Lamar Univ; XcgXm
Jordan, Karen J., 251-460-6381 kjordan@southalabama.edu, Univ of South Alabama; Zyr
Jordan, Mary S., 831-656-7571 jordan@nps.edu, Naval Postgrad Sch; Am
Jordan, Robert A., (757) 727-5783 robert.jordan@hamptonu.edu, Hampton Univ; Ob
Jordan, Teresa E., (607) 255-3596 tej1@cornell.edu, Cornell Univ; GrtZg
Jordan, Thomas H., (213) 821-1237 tjordan@usc.edu, Univ of Southern California; Ys
Jorge, Maria Luisa, (615) 322-2160 malu.jorge@vanderbilt.edu, Vanderbilt Univ; Pg
Jornov, Donna, djornov@mail.nysed.gov, New York State Geo Survey; Zn
Joseph, Miranda, mirandaj@email.arizona.edu, Univ of Arizona; Zn
Joshi, Manoj, +44 (0)1603 59 3647 m.joshi@uea.ac.uk, Univ of East Anglia; Zi
Joughin, Ian, 206-221-3177 ian@apl.washington.edu, Univ of Washington; GlZrAt
Journel, Andre G., (650) 723-1594 journel@pangea.stanford.edu, Stanford Univ; Gq
Joyce, James, (787) 265-3845 james.joyce@upr.edu, Univ of Puerto Rico; NgGcp
Juanes, Ruben, 617-253-7191 juanes@mit.edu, Massachusetts Inst of Tech; Zn
Judge, Shelley, (330) 263-2297 sjudge@wooster.edu, Coll of Wooster; GtsGc
Judith, Sausse, +33 3 72 74 46 00 judith.sausse@univ-lorraine.fr, Ecole Natle Supérieure de Géologie (ENSG); GgeGq
Judkins, Gabriel, (702) 895-4302 gabriel.judkins@unlv.edu, Univ of Nevada, Las Vegas; Gg
Judkins, Heather L., judkins.heather@spcollege.edu, Saint Petersburg Coll, Clearwater; Og
Jugo, Pedro J., 7056751151 x2106 pjugo@laurentian.ca, Laurentian Univ, Sudbury; EgCpGi
Jugulam, Mithila, (785) 532-2755 mithila@ksu.edu, Kansas State Univ; So
Juhl, Andrew, andyjuhl@ldeo.columbia.edu, Columbia Univ; Ob
Julian, Jason, (512) 245-2170 Texas State Univ; HsRwZc
Jull, A. J. Timothy, (520) 621-6816 jull@u.arizona.edu, Univ of Arizona; CclCg
Jun, Young-Shin, ysjun@seas.wustl.edu, Washington Univ in St. Louis; Zn
Jung, Simon, +44 (0) 131 650 4837 simon.jung@ed.ac.uk, Edinburgh Univ; Og
Jung, SJ, (208) 885-6481 sjung@uidaho.edu, Univ of Idaho; NrxNr
Juniper, Kim, 250-721-6120 kjuniper@uvic.ca, Univ of Victoria; Ob
Junium, Christopher, 315-443-8969 ckjunium@syr.edu, Syracuse Univ; CsPeCo
Junkin, William D., (410) 554-5543 william.junkin@maryland.gov, Maryland Dept of Nat Res; GvrGg
Juntunen, Kristopher, juntunen@tarleton.edu, Tarleton State Univ; Pv
Juraèiæ, Mladen, +38514606099 mjuracic@geol.pmf.hr, Univ of Zagreb; GueCm
Juranek, Lauren W., 541-737-2368 ljuranek@coas.oregonstate.edu, Oregon State Univ; Cm
Jurdy, Donna M., (847) 491-7163 donna@earth.northwestern.edu, Northwestern Univ; YgXyg
Jurena, Dwight, djurena@alamo.edu, Alamo Colls, San Antonio Coll; GgZg
Jurewicz, Amy , (480) 773-1796 amy.jurewicz@asu.edu, Arizona State Univ; Xm
Jurmanovich, Barb, 989-686-9445 Delta Coll; Zn
Juster, Thomas C., juster@usf.edu, Univ of South Florida; Hw
Juszczyk , Carmen , (303) 492-2330 carmenj@colorado.edu, Univ of Colorado; Zn
Jutla, Rajinder S., (417) 836-5298 rajinderjutla@missouristate.edu, Missouri State Univ; Zn

K

Kaandorp, Ron, +31 20 59 87375 ron.kaandorp@falw.vu.nl, Vrije Univit Amsterdam; Gg
Kabengi, Nadine, 404-413-5207 kabengi@gsu.edu, Georgia State Univ; ScClGe
Kaczmarek, Stephen E., (269) 387-5479 stephen.kaczmarek@wmich.edu, Western Michigan Univ; GdCgGz
Kaczynski, Kristen M., 530-898-6340 kkaczynski@csuchico.edu, California State Univ, Chico; ZnHg
Kaden, Scott, 573-368-2194 nrkades@mail.dnr.state.mo.us, Missouri Dept of Nat Res; Hw
Kadkhodai Ilkhchi, Ali, +98 (411) 339 2724 Univ of Tabriz; NpEo
Kafka, Alan L., (617) 552-3650 kafka@bc.edu, Boston Coll; Yg
Kah, Linda, (865) 974-2366 lckah@utk.edu, Univ of Tennessee, Knoxville; Gs
Kahle, Anne B., (818) 354-7265 anne@aster.jpl.nasa.gov, Jet Propulsion Lab; Zr
Kahle, Chris, chris.kahle@dnr.iowa.gov, Iowa Dept of Nat Res; Zi
Kaip, Galen M., (915) 747-6817 gkaip@utep.edu, Univ of Texas, El Paso; Yx
Kairies-Beatty, Candace L., (507) 474-5789 ckairiesbeatty@winona.edu, Winona State Univ; GeClRw
Kaiser, Daniel E., 612-624-3482 dekaiser@umn.edu, Univ of Minnesota, Twin Cities; Sc
Kaiser, Jan, +44 (0)1603 59 3393 j.kaiser@uea.ac.uk, Univ of East Anglia; Cs
Kaiser, Jason, 802-626-6685 jason.kaiser@northernvermont.edu, Northern Vermont Univ-Lyndon; AsZn
Kaiser, Kendra, kendrakaiser@boisestate.edu, Boise State Univ; Hsq
Kaiser-Bischoff, Ines, 089/2180 4314 kaiser-bischoff@lmu.de, Ludwig-Maximilians-Univ Muenchen; Gz
Kaka, Ismail, +96638603879 skaka@kfupm.edu.sa, King Fahd Univ of Petroleum and Minerals; Ys
Kakembo, Vincent, 27 41 504 4516 vincent.kakembo@nmmu.ac.za, Nelson Mandela Metropolitan Univ; Sp
Kalakay, Thomas J., 406-657-1101 kalakayt@rocky.edu, Rocky Mountain Coll; GcpGi
Kalangutkar, Niyati, (866) 960-9198 Goa Univ; GugGe
Kaldi, John, john.kaldi@adelaide.edu.au, Univ of Adelaide; Go
Kaldor, Michael, (305) 237-3025 michael.kaldor@mdc.edu, Miami-Dade Coll (Wolfson Campus); Gg
Kale, Makarand G., +919922050970 mgkale@unipune.ac.in, Savitribai Phule Pune Univ,; Gso
Kalender, Leyla, 00904242370000-5984 leylakalender@firat.edu.tr, Firat

Univ; CgcCs
Kalia, Hemendra N., (702) 295-5767 hemendra_kalia@notes.ymp.gov, Los Alamos; Nm
Kallemeyn, Gregory, (310) 825-3202 Univ of California, Los Angeles; Cg
Kallin, Hamish, h.kallin@ed.ac.uk, Edinburgh Univ; Zn
Kalvoda, Jiri, +420 549 49 4756 dino@sci.muni.cz, Masaryk Univ; PsePm
Kamber, Balz, +61 7 3138 2261 balz.kamber@qut.edu.au, Queensland Univ of Tech; GpCgGz
Kambesis, Patricia, (270) 745-5984 pat.kambesis@wku.edu, Western Kentucky Univ; ZigHw
Kambewa, Chamunorwa , 078 321 6710 kambewac@tut.ac.za, Tshwane Univ of Tech; CgeZe
Kamenov, George D., (352) 846-3599 kamenov@ufl.edu, Univ of Florida; CgaEm
Kamhi, Samuel R., (718) 460-1476 Long Island Univ, Brooklyn Campus; Gz
Kamilli, Robert J., (520) 670-5576 bkamilli@usgs.gov, Univ of Arizona; EgmCg
Kaminski, Michael A., +966138602344 kaminski@kfupm.edu.sa, King Fahd Univ of Petroleum and Minerals; GuPm
Kammer, Thomas W., (304) 293-5603 tkammer@wvu.edu, West Virginia Univ; Pi
Kamola, Diane, (785) 864-2724 dlkamola@ku.edu, Univ of Kansas; Gs
Kampf, Anthony R., (213) 763-3328 akampf@nhm.org, Los Angeles County Mus of Natural History; Gz
Kamykowski, Daniel, (919) 859-3428 dkamyko@ncsu.edu, North Carolina State Univ; ObpOc
Kana, Todd M., (410) 221-8481 kana@hpl.umces.edu, Univ of Maryland; Ob
Kanamori, Hiroo, (626) 395-6914 hiroo@gps.caltech.edu,CalTech; Ysg
Kandel, Hari, 906-635-2618 hkandel@lssu.edu, Lake Superior State Univ; HgZi
Kandiah, Ramanitharan, (937) 376-6260 rkandiah@centralstate.edu, Central State Univ; HqSoZi
Kane, Mustapha, 386.754. 4452 mustapha.kane@fgc.edu, Florida Gateway Coll; Zg
Kane-DeVries, Katelyn, 573-368-2148 katelyn.kane-devries@dnr.mo.gov, Missouri Dept of Nat Res; Gg
Kanemasu, Edward T., (706) 542-2151 Univ of Georgia; Sp
Kanfoush, Sharon L., (315) 792-3134 skanfoush@utica.edu, Utica Coll; GsOuGn
Kang, James, 956 665 3526 jihoon.kang@utrgv.edu, Univ of Texas, Rio Grande Valley; SpcZu
Kang, Peter, pkkang@umn.edu, Univ of Minnesota, Twin Cities; Hg
Kang, Song-Lak, (806) 834-1139 song-lak.kang@ttu.edu, Texas Tech Univ; AsHg
Kang, Wanying, wanying@mit.edu, Massachusetts Inst of Tech; AsZo
Kanik, Mustafa, 00904242370000-5964 Firat Univ; NrSo
Kanungo, Sudeep, 801-585-7852 skanungo@egi.utah.edu, Univ of Utah; Pms
Kao, Jim C., (505) 667-9226 Los Alamos; As
Kaplan, Alexey, alexeyk@ldeo.columbia.edu, Columbia Univ; Op
Kaplan, Isaac R., (310) 825-5076 Univ of California, Los Angeles; Cg
Kaplan, Michael, mkaplan@ldeo.columbia.edu, Columbia Univ; Gl
Kaplan, Samantha W., (715) 346-4149 skaplan@uwsp.edu, Univ of Wisconsin, Stevens Point; GnPlGd
Kaplinski, Matthew A., (928) 523-9145 Northern Arizona Univ; Yg
Kapp, Jessica, 520-626-5701 jkapp@email.arizona.edu, Univ of Arizona; Gg
Kapp, Paul A., (520) 626-8763 pkapp@email.arizona.edu, Univ of Arizona; GtcGm
Kar, Aditya, (912) 825-6844 Fort Valley State Univ; Ct
Kar, Bandana, 601-266-5786 bandana.kar@usm.edu, Univ of Southern Mississippi; Zi
Kara, Hatice, 00904242370000-5965 haticekara@firat.edu.tr, Firat Univ; Cg
Karabinos, Paul, (413) 597-2079 paul.m.karabinos@williams.edu, Williams Coll; Gc
Karamperidou, Christina, 808-956-2565 ckaramp@hawaii.edu, Univ of Hawai'i at Manoa; AsZoc
Karandikar, Amol, akaran@carnegiescience.edu, Carnegie Inst for Sci; Gy
Karanfil, Tanju, (864) 656-1005 tkaranf@clemson.edu, Clemson Univ; NgZnn
Karato, Shun-ichiro, (203) 432-3147 shun-ichiro.karato@yale.edu, Yale Univ; Gy
Karginoglu, Yusuf, 00904242370000-5989 Firat Univ; Eg
Karhu, Juha A., 358-294150834 juha.karhu@helsinki.fi, Univ of Helsinki; Csg
Karig, Daniel E., dek9@cornell.edu, Cornell Univ; YrGl
Karimi, Bobak, 570-408-4698 bobak.karimi@wilkes.edu, Wilkes Univ; GctYg
Karkanis, Pano G., (403) 317-0156 Res. (403) 330-9445 Cel. karkanis@telusplanet.net, Univ of Lethbridge; So
Karl, David M., (808) 956-8964 dkarl@soest.hawaii.edu, Univ of Hawai'i, Manoa; Ob
Karl, Tami S., (850) 644-5861 karl@gly.fsu.edu, Florida State Univ; Zn
Karlen, Douglas L., (515) 294-3336 doug.karlen@ars.usda.gov, Iowa State Univ of Sci & Tech; So
Karlsson, Haraldur R., 806-834-7978 hal.karlsson@ttu.edu, Texas Tech Univ; Cl
Karlstrom, Karl E., (505) 277-4346 kek1@unm.edu, Univ of New Mexico; Gt
Karlstrom, Leif, (541) 346-4323 leif@uoregon.edu, Univ of Oregon; Yx
Karmalkar, Ambarish, akarmalkar@geo.umass.edu, Univ of Massachusetts, Amherst; As
Karmosky, Christopher, christopher.karmosky@oneonta.edu, SUNY, Oneonta; AmsZy
Karner, Daniel B., (707) 664-2854 karner@sonoma.edu, Sonoma State Univ; Cc
Karner, James M., 801-585-5976 j.karner@utah.edu, Univ of Utah; Xm
Karnes, Victoria M., (605) 394-2461 victoria.karnes@sdsmt.edu, South Dakota Sch of Mines & Tech; ZifZr
Karpilo, Jr, Ronald J., (970) 225-3500 ron_karpilo@partner.nps.gov, Colorado State Univ; Zy
Karplus, Marianne, (915) 747-5413 mkarplus@utep.edu, Univ of Texas, El Paso; Ysg
Karson, Jeffrey, 315-443-7976 jakarson@syr.edu, Syracuse Univ; GtcGv
Karuntillake, Suniti, 225-578-2337 sunitiw@lsu.edu, Louisiana State Univ; Xg
Karwoski, Todd, (301) 405-0084 karwoski@geol.umd.edu, Univ of Maryland; Gg
Kasar, Sheila , 814-393-2718 skazar@clarion.edu, Clarion Univ; ZreGe
Kaspar, Thomas A., (515) 294-8873 tom.kaspar@ars.usda.gov, Iowa State Univ of Sci & Tech; So
Kaspari, Susan, kaspari@geology.cwu.edu, Central Washington Univ; PcAtGl
Kasse, Kees, +31 20 59 87381 c.kasse@vu.nl, Vrije Univit Amsterdam; GmsSd
Kassem, Hachem, +44 (0)23 8059 6205 hachem.kassem@soton.ac.uk, Univ of Southampton; OnGsZi
Kaste, James, 757-221-2591 jmkaste@wm.edu, William & Mary; ClcSc
Kastens, Kim A., kak2@columbia.edu, Columbia Univ; Gu
Kaster, Mark A., 570-408-5046 mark.kaster@wilkes.edu, Wilkes Univ; Am
Kasting, James F., (814) 865-3207 jfk4@psu.edu, Pennsylvania State Univ, Univ Park; As
Kastner, Miriam, (858) 534-2065 mkastner@ucsd.edu, Univ of California, San Diego; CmlCm
Kastrisios, Christos , (603) 862-0618 mailto:christos.kastrisios@unh.edu, Univ of New Hampshire; Zn
Kaszuba, John P., 307-766-3392 jkaszub1@uwyo.edu, Univ of Wyoming; CgGze
Kath, Randal L., (678) 839-4063 rkath@westga.edu, Univ of West Georgia; GcNgHg
Kato, Terence T., 530-898-5262 tkato@csuchico.edu, California State Univ, Chico; GtcGp
Katumba, Samy, samy.katumba@up.ac.za, Univ of Pretoria; Zf
Katumwehe, Andrew, (940) 397-4031 andrew.katumwehe@msutexas.edu, Midwestern State Univ; YuvYe
Katuna, Michael P., katunam@cofc.edu, Coll of Charleston; Gus
Katz, Miriam E., (518) 276-8521 katzm@rpi.edu, Rensselaer Polytechnic Inst; PmeGu
Katz, Richard, +44-1865-282122 richard.katz@earth.ox.ac.uk, Univ of Oxford; Cg
Katzenstein, Kurt W., (605) 394-2346 kurt.katzenstein@sdsmt.edu, South Dakota Sch of Mines & Tech; NxtZr
Katzman, Danny, (505) 667-0599 katzman@lanl.gov, Los Alamos; HyGem
Kauffman, Chad, 724 938-5760 kauffman@cup.edu, California Univ of Pennsylvania; As
Kaufman, Alan J., (301) 405-0395 kaufman@geol.umd.edu, Univ of Maryland; Csg
Kaufman, Darrell S., (928) 523-7192 darrell.kaufman@nau.edu, Northern Arizona Univ; Zc
Kaufmann, Ronald S., (619) 260-5904 kaufmann@sandiego.edu, Univ of San Diego; Ob
Kaunda, Rennie, 303-273-3772 rkaunda@mines.edu, Colorado Sch of Mines; Nm
Kaushal, Sujay, (301) 405-0454 skaushal@umd.edu, Univ of Maryland; ZuGeHs
Kavage Adams, Rebecca H., (410) 554-5553 rebecca.adams@maryland.gov, Maryland Dept of Nat Res; GmZiGe

Kavanagh, Janine, +44-151-794-5150 janine.kavanagh@liverpool.ac.uk, Univ of Liverpool; Gv
Kavanagh, Trudy A., 250-807-8564 trudy.kavanagh.ubc.ca, Univ of British Columbia, Okanagan Campus; ZyRcZg
Kavanaugh, Jeffrey, 780-492-1740 jeff.kavanaugh@ualberta.ca, Univ of Alberta; Ng
Kavanaugh, Maria, mkavanau@coas.oregonstate.edu, Oregon State Univ; Cb
Kavner, Abby, (310) 206-3675 akavner@ucla.edu, Univ of California, Los Angeles; GyYg
Kavousi Ghahfarohki, Payam, 304-293-5603 pkavousi@mail.wvu.edu, West Virginia Univ; Gog
Kawase, Mitsuhiro, (206) 543-0766 kawase@ocean.washington.edu, Univ of Washington; Op
Kay, Jonathan, 757-221-2106 jpkay@wm.edu, William & Mary; ZyiGc
Kay, Richard F., (919) 684-2143 richard.kay@duke.edu, Duke Univ; Zn
Kay, Robert W., (607) 255-3461 rwk6@cornell.edu, Cornell Univ; Giz
Kay, Suzanne M., (607) 227-5143 smk16@cornell.edu, Cornell Univ; GiCuGz
Kaye, John M., (662) 325-3915 Mississippi State Univ; Gg
Kaygili, Sibel, 00904242370000-5962 skaygili@firat.edu.tr, Firat Univ; GgPgs
Kays, Marvin A., (541) 346-4578 makays@oregon.uoregon.edu, Univ of Oregon; GpcGt
Kazimoto, Emmanuel O., +255222410013 ekazimoto@udsm.ac.tz, Univ of Dar es Salaam; EgGpCg
Kazmer, Miklos, +36-1-372-2500 ext. 8627 mkazmer@gmail.com, Eotvos Lorand Univ; PggRh
Keables, Michael J., (303) 871-2653 michael.keables@du.edu, Univ of Denver; AstAm
Keach, Bill, (801) 537-3301 billkeach@utah.gov, Utah Geo Survey; YegZg
Keach, William, 801-585-1717 bkeach@egi.utah.edu, Univ of Utah; GoYes
Keach II, R. William, 801-857-7728 bkeach@byu.edu, Brigham Young Univ; YeEoGg
Keala, Lori, (909) 621-8675 lkeala@pomona.edu, Pomona Coll; Zn
Kean, Jr., William F., 414-229-5231 wkean@uwm.edu, Univ of Wisconsin, Milwaukee; Ym
Keane, Christopher M., (703) 379-2480 keane@americangeosciences.org, American Geosciences Inst; GgmZf
Kearney, Micheal S., (301) 405-4057 kearneym@umd.edu, Univ of Maryland; Gu
Kearns, Lance E., 540-568-6130 kearnsle@jmu.edu, James Madison Univ; Gz
Keating, Elizabeth, (505) 665-6714 ekeating@lanl.gov, Los Alamos; Gg
Keating, Kristina M., 973-353-1263 kmkeat@newark.rutgers.edu, Rutgers, The State Univ of New Jersey, Newark; Yug
Keaton, Jeffrey R., (801) 581-8218 Univ of Utah; Zn
Keattch, Sharen , 509-359-7358 skeattch@ewu.edu, Eastern Washington Univ; Gg
Keays, Reid R., (705) 675-1151 Laurentian Univ, Sudbury; Em
Kebbede, Girma, 413-538-2004 gkebbede@mtholyoke.edu, Mount Holyoke Coll; Zy
Keegan, Kaitlin M., 775-682-6049 kkeegan@unr.edu, Univ of Nevada, Reno; PcNg
Keeler, Jason, 989-774-3179 keele1j@cmich.edu, Central Michigan Univ; Am
Keeley, Joshua A., 603-271-2875 joshua.a.keeley@des.nh.gov, New Hampshire Geo Survey; GgHyZi
Keeling, David J., (270) 745-4555 david.keeling@wku.edu, Western Kentucky Univ; Zg
Keeling, Ralph F., (858) 534-7582 rkeeling@ucsd.edu, Univ of California, San Diego; As
Keen, Kerry L., (715) 425-3729 kerry.l.keen@uwrf.edu, Univ of Wisconsin, River Falls; HwGse
Keenan, Sarah W., (605) 394-2461 sarah.keenan@sdsmt.edu, South Dakota Sch of Mines & Tech; PvClb
Keene, William, (804) 924-0586 wck@virginia.edu, Univ of Virginia; As
Keeney, Dennis R., (515) 294-8066 drkeeney@iastate.edu, Iowa State Univ of Sci & Tech; So
Keesee, Robert G., (518) 442-4566 rgk@atmos.albany.edu, SUNY, Albany; As
Kehew, Alan E., (269) 387-5495 alan.kehew@wmich.edu, Western Michigan Univ; GmlHw
Keigwin, Lloyd D., (508) 289-2784 lkeigwin@whoi.edu, Woods Hole Oceanographic Inst; CsGul
Keil, Chris, (630) 752-7271 chris.keil@wheaton.edu, Wheaton Coll; Rm
Keil, Richard G., (206) 616-1947 rickkeil@ocean.washington.edu, Univ of Washington; Oc
Keir, Derek, +44 (023) 8059 6614 d.keir@soton.ac.uk, Univ of Southampton; Gtv
Keisling, Benjamin, keisling@ig.utexas.edu, Univ of Texas, Austin; GlZc
Keith, Jeffrey D., 801-422-2189 jeff_keith@byu.edu, Brigham Young Univ; Eg
Kelemen, Peter, peterk@ldeo.columbia.edu, Columbia Univ; Gi
Kelkar, Sharad, (505) 667-4639 kelkar@vega.lanl.gov, Los Alamos; Np
Kelleher, Cole, 612-626-0505 Univ of Minnesota, Twin Cities; Zi
Keller, C. Kent, (509) 335-3040 ckkeller@wsu.edu, Washington State Univ; Hw
Keller, Dianne M., (315) 228-7893 dkeller@colgate.edu, Colgate Univ; Gz
Keller, Edward A., (805) 893-4207 keller@geol.ucsb.edu, Univ of California, Santa Barbara; Gm
Keller, G. Randy, (405) 255-0608 grkeller@ou.edu, Univ of Oklahoma; YgGtEo
Keller , George R., keller@utep.edu, Univ of Texas, El Paso; Yg
Keller, Gerta, 609-258-4117 gkeller@princeton.edu, Princeton Univ; Pm
Keller, John E., 702-651-5887 john.keller@csn.edu, Coll of Southern Nevada - West Charleston Campus; GeHw
Keller, Klaus, (814) 865-6718 kkeller@geosc.psu.edu, Pennsylvania State Univ, Univ Park; Og
Keller, Linda M., (608) 265-2209 lmkeller@wisc.edu, Univ of Wisconsin, Madison; As
Keller, Randall A., 541-737-7648 kellerr@geo.oregonstate.edu, Oregon State Univ; Gu
Keller, Troy, 706-507-8099 keller_troy@columbusstate.edu, Columbus State Univ; RwZc
Keller, Jr., C. F., (505) 667-0920 cfk@lanl.gov, Los Alamos; As
Kelley, Alice R., (207) 581-2056 akelley@maine.edu, Univ of Maine; GalGm
Kelley, Christopher, (808) 956-7437 ckelley@hawaii.edu, Univ of Hawai'i, Manoa; Ob
Kelley, Dan, (902) 494-1694 dan.kelley@dal.ca, Dalhousie Univ; Op
Kelley, Deborah S., (206) 685-9556 dskelley@uw.edu, Univ of Washington; OuGvp
Kelley, Joseph T., (207) 581-2162 jtkelley@maine.edu, Univ of Maine; Gua
Kelley, Neil P., neil.p.kelley@vanderbilt.edu, Vanderbilt Univ; GgOgPg
Kelley, Patricia H., (910) 448-0526 kelleyp@uncw.edu, Univ of North Carolina, Wilmington; Pgi
Kelley, Richard, 505-412-2967 richard.kelley@nmt.edu, New Mexico Tech; ZiGg
Kelley, Scott , 775-784-6705 scottkelley@unr.edu, Univ of Nevada, Reno; Zi
Kelley, Shari, 575-835-5306 shari.kelley@nmt.edu, New Mexico Tech; Yg
Kelley, Simon, +44 (0)131 650 2537 simon.kelley@ed.ac.uk, Edinburgh Univ; CcsZg
Kelling, Keith A., (608) 263-2795 kkelling@wisc.edu, Univ of Wisconsin, Madison; So
Kellman, Lisa M., (902) 867-5086 lkellman@stfx.ca, Saint Francis Xavier Univ; CbSoCs
Kellner, Patricia, pakellner@socal.rr.com, Los Angeles Harbor Coll; Zg
Kellogg, James N., (803) 777-4501 kellogg@geol.sc.edu, Univ of South Carolina; Yg
Kelly, Bryce, bryce.kelly@unsw.edu.au, Univ of New South Wales; HwAcGe
Kelly, D. Clay, (608) 262-1698 ckelly@geology.wisc.edu, Univ of Wisconsin, Madison; Pm
Kelly, Jacque L., (912) 478-8677 jkelly@georgiasouthern.edu, Georgia Southern Univ; Cl
Kelly, John, 203-479-4822 jkelly@newhaven.edu, Univ of New Haven; ObZn
Kelly, Kathryn A., (206) 543-9810 kkelly@apl.washington.edu, Univ of Washington; Op
Kelly, Kimberly, (240) 567-5227 kimberly.kelly@montgomerycollege.edu, Montgomery Coll; Zn
Kelly, Maria, (425) 640-1918 mkelly@edmonds.edu, Edmonds Coll; Zg
Kelly, Meredith, 603-646-9647 meredith.kelly@dartmouth.edu, Dartmouth Coll; Gl
Kelly, Sherrie, 315-229-5851 skelly@stlawu.edu, St. Lawrence Univ; Zn
Kelly, Walton R., (217) 333-3729 wkelly@illinois.edu, Univ of Illinois; Hw
Kelly, William C., (734) 764-1435 billkell@umich.edu, Univ of Michigan; Eg
Kelsch, Jesse, (432) 837-8657 jkelsch@sulross.edu, Sul Ross State Univ; Gct
Kelsey, Eric, ekelsey2@plymouth.edu, Plymouth State Univ; AtsGl
Kelsey, Harvey M., (707) 826-3991 hmk1@humboldt.edu, Humboldt State Univ; Gmt
Kelso, Paul R., (906) 635-2158 pkelso@lssu.edu, Lake Superior State Univ; YmGct
Kelty, Thomas, (562) 985-4589 t.kelty@csulb.edu, California State Univ, Long Beach; Gc
Kemeny, John M., (520) 621-4448 kemeny@email.arizona.edu, Univ of Arizona; Nr

Kemmerly, Phillip R., (931) 221-7471 kemmerlyp@apsu.edu, Austin Peay State Univ; GmeHg
Kemp, Alan, +44 (0)23 80592788 aesk@noc.soton.ac.uk, Univ of Southampton; Pe
Kemp, Andrew, (617) 627-0869 andrew.kemp@tufts.edu, Tufts Univ; OnPmOg
Kemp, Paul, (808) 956-6220 paulkemp@hawaii.edu, Univ of Hawai'i, Manoa; Ob
Kempton, Pamela D., (785) 532-6724 pkempton@ksu.edu, Kansas State Univ; GiCtc
Kendall, Christopher G., kendall@geol.sc.edu, Univ of South Carolina; Gs
Kendrick, David C., 315-781-3929 kendrick@hws.edu, Hobart & William Smith Colls; PgGg
Kendrick, Katherine J., (951) 276-4418 kendrick@gps.caltech.edu, Univ of California, Riverside; Gm
Kenduiywo, Benson K., bkenduiywo@jkuat.ac.ke, Jomo Kenyatta Univ of Agriculture & Tech; ZriYd
Kenig, Fabien, (312) 996-3020 fkenig@uic.edu, Univ of Illinois Chicago; CoXbGe
Kenna, Timothy, tkenna@ldeo.columbia.edu, Columbia Univ; Cg
Kennedy, Andrew, andrew.b.kennedy.117@nd.edu, Univ of Notre Dame; On
Kennedy, Ann C., (509) 335-1554 akennedy@wsu.edu, Washington State Univ; Sb
Kennedy, Ben, +64 3 3667001 Ext 7775 ben.kennedy@canterbury.ac.nz, Univ of Canterbury; Gv
Kennedy, Christina B., tina.kennedy@nau.edu, Northern Arizona Univ; Zn
Kennedy, Gareth, +44 01326 371876 g.a.kennedy@exeter.ac.uk, Exeter Univ; Nm
Kennedy, Linda, 570-662-4609 lkennedy@mansfield.edu, Mansfield Univ; Zy
Kennedy, Lisa M., (540) 231-1422 kennedy1@vt.edu, Virginia Tech; Pe
Kennedy, Lori, (604) 822-1811 lkennedy@eos.ubc.ca, Univ of British Columbia; Gc
Kennedy, Robert, (541) 737-6332 rkennedy@coas.oregonstate.edu, Oregon State Univ; Zri
Kennel, Charles F., (310) 825-4018 Univ of California, Los Angeles; Zn
Kenney, Robert D., (401) 874-6664 rkenney@uri.edu, Univ of Rhode Island; Ob
Kent, Adam J., (541) 737-1205 adam.kent@geo.oregonstate.edu, Oregon State Univ; GiCga
Kent, Dennis V., 848-445-7049 dvk@rutgers.edu, Rutgers, The State Univ of New Jersey; YmZcGr
Kent, Douglas, 405-377-0166 Oklahoma State Univ; Hw
Kent, Graham, (775) 784-4977 gkent@seismo.unr.edu, Univ of Nevada, Reno; Ys
Kenyon, Patricia M., (212) 650-6472 City Coll (CUNY); Yg
Kepic, Anton W., +61 8 9266-7503 a.kepic@curtin.edu.au, Curtin Univ; YexYg
Keppie, J. Duncan, 56224303 duncan@servidor.unam.mx, Univ Nacional Auto de Mexico; Gt
Keranen, Kade M., kmk299@cornell.edu, Cornell Univ; Ys
Keranen, Katie M., kmk299@cornell.edu, Cornell Univ; Ys
Kerans, Charles, (512) 471-1368 charles.kerans@beg.utexas.edu, Univ of Texas at Austin, Jackson Sch of Geosciences; Gd
Kerestedjian, Thomas N., +359 2 979 2244 thomas@geology.bas.bg, Bulgarian Acad of Scis; GzeEg
Kern, Anikó, anikoc@nimbus.elte.hu, Eotvos Lorand Univ; ZrAms
Kern, J. Philip, jphilipkern@gmail.com, San Diego State Univ; Pg
Kerr, A C., kerra@cf.ac.uk, Cardiff Univ; Gxi
Kerr, Andrew, +44(0)29 208 74578 kerra@cardiff.ac.uk, Univ of Wales; Og
Kerr, Dennis R., 918 631 3020 dennis-kerr@utulsa.edu, Univ of Tulsa; GsrGt
Kerrick, Derrill M., (814) 865-7574 kerrick@geosc.psu.edu, Pennsylvania State Univ, Univ Park; Gp
Kertes, Randy, rkertes@rider.edu, Rider Univ; Ge
Kerwin, Charles M., 603.358.2405 ckerwin@keene.edu, Keene State Coll; Gg
Kerwin, Michael W., 303-871-3998 mkerwin@du.edu, Univ of Denver; GgeZc
Kesel, Richard H., (225) 578-5880 gakesel@lsu.edu, Louisiana State Univ; Hg
Keskinen, Mary J., (907) 474-7769 mjkeskinen@alaska.edu, Univ of Alaska, Fairbanks; Gxz
Kesler, Stephen E., (734) 763-5057 skesler@umich.edu, Univ of Michigan; Eg
Kessler, Carsten, hello@carsten.io, Graduate Sch of the City Univ of New York; Gg
Kessler, Fritz, (301) 687-4266 fkessler@frostburg.edu, Frostburg State Univ; Zr
Kessler, John , 585-273-4572 john.kessler@rochester.edu, Univ of Rochester; Oc
Kessler, William S., (206) 526-6221 kessler@pmel.noaa.gov, Univ of Washington; Op
Ketcham, Richard A., (512) 471-6942 ketcham@jsg.utexas.edu, Univ of Texas, Austin; GgCc
Kettler, Richard M., (402) 472-0882 rkettler1@unl.edu, Univ of Nebraska, Lincoln; Cl
Kevan, Peter, pkevan@uoguelph.ca, Univ of Guelph; Zn
Key, Doug, 7607441150 ext.2515 dkey@palomar.edu, Palomar Coll; Zy
Key, Jeffrey R., (608) 263-2605 jkey@ssec.wisc.edu, Univ of Wisconsin, Madison; ZrAs
Key, Jr., Marcus M., (717) 245-1448 key@dickinson.edu, Dickinson Coll; PiGsa
Keyantash, John, (310) 243-2363 jkeyantash@csudh.edu, California State Univ, Dominguez Hills; HgAsZe
Keyser, Daniel, (518) 442-4559 dkeyser@albany.edu, SUNY, Albany; As
Khade, Vishnu R., khadev@easternct.edu, Eastern Connecticut State Univ; Zg
Khairoutdinov, Marat, (631) 632-6339 marat.khairoutdinov@stonybrook.edu, SUNY, Stony Brook; As
Khalequzzaman, Md., (570) 484-2075 mkhalequ@lockhaven.edu, Lock Haven Univ; HwOnZi
Khalil, Mohamed, 406-496-4716 mkhalil@mtech.edu, Montana Tech of the Univ of Montana; YeuGe
Khalil Aboushanab, Mohamed, (308) 632-1230 maboushanab2@unl.edu, Univ of Nebraska, Lincoln; YgHw
Khalil Ebeid, Khalil I., 01227431425 kebeid@yahoo.com, Alexandria Univ; GgEmm
Khan, Belayet H., 217-581-2626 bhkhan@eiu.edu, Eastern Illinois Univ; Zy
Khan, Mohammad Wahdat Y., (982) 719-7331 mwykhan@rediffmail.com, Pt. Ravishankar Shukla Univ; GdEgCl
Khan, Shuhab D., (713) 743-5404 sdkhan@uh.edu, Univ of Houston; ZrGtYg
Khanbilvardi, Reza M., 212-650-8009 khanbilvardi@ccny.cuny.edu, Graduate Sch of the City Univ of New York; Hy
Khandaker, Nazrul I., (718) 262-2079 nkhandaker@york.cuny.edu, York Coll (CUNY); GdeZe
Khandaker, Nazul, 781-262-2079 nkhandaker@york.cuny.edu, Graduate Sch of the City Univ of New York; Gg
Khawaja, Ikram U., john@cis.ysu.edu, Youngstown State Univ; Ec
Khurana, Krishan, (310) 825-8240 Univ of California, Los Angeles; Zn
Kiage, Lawrence M., (404) 413-5777 geolkk@langate.gsu.edu, Georgia State Univ; ZyPlZr
Kidanu, Shishay, (907) 474-5988 stkdanu@alaska.edu, Univ of Alaska Fairbanks; Yu
Kidd, David, +44 020 8417 62541 david.kidd@kingston.ac.uk, Kingston Univ; Ziy
Kidd, Karen, 905-525-9140 (Ext. 23550) karenkidd@mcmaster.ca, McMaster Univ; Gu
Kidder, Steven, 212-650-8431 skidder@ccny.cuny.edu, Graduate Sch of the City Univ of New York; As
Kidder, T.R., trkidder@wustl.edu, Washington Univ in St. Louis; Zn
Kidwell, Jen, (650) 723-0891 jparis@stanford.edu, Stanford Univ; Zn
Kidwell, Susan M., (773) 702-3008 skidwell@uchicago.edu, Univ of Chicago; GsPe
Kieffer, Bruno, bkieffer@eos.ubc.ca, Univ of British Columbia; Cs
Kieffer, Susan W., (217) 244-6206 skieffer@illinois.edu, Univ of Illinois, Urbana-Champaign; Gv
Kienast, Markus, (902) 494-8338 markus.kienast@dal.ca, Dalhousie Univ; Cs
Kienast, Stephanie, (902) 494-2203 stephanie.kienast@dal.ca, Dalhousie Univ; Gu
Kiene, Ronald P., (251) 861-7526 rkiene@jaguar1.usouthal.edu, Univ of South Alabama; Oc
Kientop, Greg A., 217-265-6581 gkientop@illinois.edu, Illinois State Geo Survey; Ge
Kienzle, Stefan, stefan.kienzle@uleth.ca, Univ of Lethbridge; Zi
Kierczak, Jakub, jakub.kierczak@uwr.edu.pl, Univ of Wroclaw; GeSc
Kieu, Chanh Q., (812) 856-5704 ckieu@indiana.edu, Indiana Univ, Bloomington; As
Kifer, Lauri A., (585) 395-2636 lmulley@brockport.edu, SUNY, The Coll at Brockport; Zn
Kift, Richard, +44 0161 306-8770 richard.kift@manchester.ac.uk, Univ of Manchester; As
Kijko, Andrzej, +27 12 420 3613 andrzej.kijko@up.ac.za, Univ of Pretoria; YsgNe
Kilburn, Chris, +44 020 7679 37194 c.kilburn@ucl.ac.uk, Univ Coll London; Yx
Kilcoyne, John R., (303) 556-4258 Metropolitan State Coll of Denver; Zy

Kilibarda, Zoran, (219) 980-6753 zkilibar@iun.edu, Indiana Univ Northwest; GsmGd
Kilic, Ayse Didem, 00904242370000-5969 adkilic@firat.edu.tr, Firat Univ; GxpGy
Kilinc, Attila I., (513) 556-5967 attila.kilinc@uc.edu, Univ of Cincinnati; CpGvCu
Killorn, Randy J., (515) 294-1923 rkillorn@iastate.edu, Iowa State Univ of Sci & Tech; So
Kilroy, Kathyrn, 701-858-3114 kathryn.kilroy@minotstateu.edu, Minot State Univ; Hg
Kim, Bojeong, 215-204-3304 bkim@temple.edu, Temple Univ; GzZa
Kim, Daehyun, (206) 221-8935 daehyun@atmos.washington.edu, Univ of Washington; As
Kim, Hyemi, (404) 894-1738 hyemi.kim@eas.gatech.edu, Georgia Inst of Tech; As
Kim, Hyeong Soo, haskim2@korea.ac.kr, Korea Univ; GpxGt
Kim, Hyewon (Heather), hkim@whoi.edu, Woods Hole Oceanographic Inst; Cm
Kim, Jonathan, (802) 522-5401 jon.kim@vermont.gov, Agency of Nat Res, Dept of Env Conservation; GetCg
Kim, Keonho, 432-685-4739 kkim@midland.edu, Midland Coll; GgAmPi
Kim, Saewung, (949)824-4531 saewungk@uci.edu, Univ of California, Irvine; As
Kim, Sang-Tae, (905) 525-9140 (Ext. 26494) sangtae@mcmaster.ca, McMaster Univ; Cg
Kim, Seong Ryong, seongryongkim@korea.ac.kr, Korea Univ; Ys
Kim, Stacy, 831-771-4400 skim@mlml.calstate.edu, Moss Landing Marine Lab; Zn
Kim, Won-Young, wykim@ldeo.columbia.edu, Columbia Univ; Ys
Kim, Yong Hoon, (610) 436-2203 ykim@wcupa.edu, West Chester Univ; OgZr
Kim, Yongsang (Barry), (671) 735-1223 kimys@triton.uog.edu, Univ of Guam; Hw
Kimball, Matthew, matt@belle.baruch.sc.edu, Univ of South Carolina; Ob
Kimball, Sytske K., skimball@southalabama.edu, Univ of South Alabama; Am
Kimber, Clarissa T., (409) 845-7141 Texas A&M Univ; Zy
Kimberly, Shaw, 706-507-8344 Columbus State Univ; Zn
Kimbro, David, 7815817370 x310 d.kimbro@neu.edu, Northeastern Univ; Zn
Kimbrough, David L., (619) 594-1385 dkimbrough@mail.sdsu.edu, San Diego State Univ; Cc
Kimerling, A. Jon, kimerlia@geo.oregonstate.edu, Oregon State Univ; Zir
Kinabo, Crispin P., kinabo_2003@yahoo.co.uk, Univ of Dar es Salaam; NxGe
Kincaid, Christopher, (401) 874-6571 kincaid@gso.uri.edu, Univ of Rhode Island; Ou
Kineke, Gail C., (617) 552-3655 gail.kineke.1@bc.edu, Boston Coll; On
King, Andrew, alking@uic.edu, Univ of Illinois Chicago; Og
King, Carey, 512-471-5468 cking@jsg.utexas.edu, Univ of Texas, Austin; Eog
King, John, (401) 874-6594 jking@gso.uri.edu, Univ of Rhode Island; Ou
King, Jonathan K., (801) 537-3354 jonking@utah.gov, Utah Geo Survey; Gg
King, Kenneth M., (519) 824-4120 (Ext. 52453) kenmking@rogers.com, Univ of Guelph; As
King, Martin, +44 1784 414038 m.king@rhul.ac.uk, Royal Holloway Univ of London; As
King, Norman R., (812) 464-1794 nking@usi.edu, Univ of Southern Indiana; Gr
King, Peter, +44 20 759 47362 peter.king@imperial.ac.uk, Imperial Coll; Np
King, Scott D., 540-231-6521 sdk@vt.edu, Virginia Tech; Yg
King, Jr., David T., (334) 844-4882 kingdat@auburn.edu, Auburn Univ; GrZn
Kingdon, Kevin, kevin.kingdon@skyresearch.com, Univ of British Columbia; Yg
Kingslake, Jonathan, jonnyk@ldeo.columbia.edu, Columbia Univ; Yr
Kinkead, Scott, (505) 665-1760 Los Alamos; Zn
Kinnaird, Judith A., 27117176583 judith.kinnaird@wits.ac.za, Univ of the Witwatersrand; EmGi
Kinner, David A., (828) 227-3821 dkinner@wcu.edu, Western Carolina Univ; HgZe
Kinsland, Gary L., (337) 288-6421 glkinsland@louisiana.edu, Univ of Louisiana at Lafayette; YgeGt
Kinter, Jim, (703) 993-5700 ikinter@gmu.edu, George Mason Univ; As
Kinvig, Helen, +440151 795 4657 h.kinvig@liverpool.ac.uk, Univ of Liverpool; Gs
Kipper, Jay P., (512) 475-9505 jay.kipper@beg.utexas.edu, Univ of Texas at Austin, Jackson Sch of Geosciences; Zn
Kirby, Carl S., kirby@bucknell.edu, Bucknell Univ; Cl
Kirby, Eric, (919) 962-8035 ekirby@unc.edu, Univ of North Carolina, Chapel Hill; GtmGc
Kirby, Matthew E., (657) 278-2158 mkirby@fullerton.edu, California State Univ, Fullerton; GnPeGg
Kirby, Stefan M., (801) 537-3349 stefankirby@utah.gov, Utah Geo Survey; GgcCq
Kirchner, James W., (510) 643-8559 kirchner@seismo.berkeley.edu, Univ of California, Berkeley; Ge
Kirimi, Fridah K., fkirimi@jkuat.ac.ke, Jomo Kenyatta Univ of Agriculture & Tech; ZiiZr
Kirk, Scott, 614 265 6742 scott.kirk@dnr.state.oh.us, Ohio Dept of Nat Res; Hw
Kirkby, Kent C., (612) 624-1392 kirkby@tc.umn.edu, Univ of Minnesota, Twin Cities; Gg
Kirkham, Mary Beth, (785) 532-0422 mbk@ksu.edu, Kansas State Univ; Sp
Kirkland, Brenda L., 662-268-1032 Ext 228 blk39@msstate.edu, Mississippi State Univ; GdEoPo
Kirkland, James I., (801) 537-3307 nrugs.jkirklan@state.ut.us, Utah Geo Survey; Pg
Kirkpatrick, Cody, (812) 855-3481 codykirk@indiana.edu, Indiana Univ, Bloomington; As
Kirkpatrick, James, 514-398-7442 james.kirkpatrick@mcgill.ca, McGill Univ; Gct
Kirlin, R. Lynn, (250) 721-8681 lkirlin@comcast.net, Univ of British Columbia; As
Kiro Feldman, Yael, ykiro@ldeo.columbia.edu, Columbia Univ; Cg
Kirschenfeld, Taylor, 850-474-2746 Univ of West Florida; Cg
Kirschvink, Joseph L., (626) 395-6136 kirschvink@caltech.edu,CalTech; Pg
Kirshbaum, Daniel, 514-398-3347 daniel.kirshbaum@mcgill.ca, McGill Univ; Asm
Kirstein, Linda, +44 (0) 131 650 4838 linda.kirstein@ed.ac.uk, Edinburgh Univ; Gt
Kirtland Turner, Sandra, (951) 827-3191 sandra.kirtlandturner@ucr.edu, Univ of California, Riverside; Cs
Kirwan, Matthew L., 804-684-7054 kirwan@vims.edu, William & Mary; Gm
Kis, Boglarka Mercedes, boglarka.kis@ubbcluj.ro, Babes-Bolyai Univ; ScCqHg
Kisekka, Isaya, ikisekka@ucdavis.edu, Univ of California, Davis; ZuRw
Kishcha, Pavel, 972-3-6407411 pavelk@post.tau.ac.il, Tel Aviv Univ; As
Kisila, Ben O., (540) 654-1107 bkisila@umw.edu, Univ of Mary Washington; HsSfZu
Kiss, Timea, +3662544156 kisstimi@gmail.com, Univesity of Szeged; GmZyHs
Kissel, David E., (706) 542-0900 Univ of Georgia; Sc
Kissin, Stephen A., (807) 343-8010 Ext. 8220 stephen.kissin@lakeheadu.ca, Lakehead Univ; Em
Kitajima, Hiroko, (979) 458-2717 kitaji@tamu.edu, Texas A&M Univ; YxSpGc
Kitchen, Newell R., 573-882-1138 kitchenn@missouri.edu, Univ of Missouri; So
Kite, J. Steven, jkite@wvu.edu, West Virginia Univ; Gm
Kitterød, Nils-Otto, nils-otto.kitterod@nmbu.no, Norwegian Univ of Life Scis (NMBU); Hg
Kivelson, Margaret G., (310) 825-3435 mkivelson@igpp.ucla.edu, Univ of California, Los Angeles; Xy
Kiver, Eugene P., (509) 359-7959 eugene.kiver@ewu.edu, Eastern Washington Univ; Gl
Klaassen, Gary P., (416)736-2100 #77727 gklaass@yorku.ca, York Univ; As
Klancher, Jacki, 307-855-2205 jklanche@cwc.edu, Central Wyoming Coll; Zi
Klapper, Gilbert, (847) 732-1859 g-klapper@northwestern.edu, Northwestern Univ; Pm
Klasik, John A., 909-869-3453 jaklasik@cpp.edu, California State Polytechnic Univ, Pomona; Gu
Klaus, Adam, (979) 845-3055 aklaus@odpemail.tamu.edu, Texas A&M Univ; Gu
Klaus, James S., 305-284-3426 j.klaus@miami.edu, Univ of Miami; Gg
Klee, Thomas M., (813) 253-7259 tklee@hccfl.edu, Hillsborough Comm Coll; Gg
Klees, Alicia, aklees@illinois.edu, Univ of Illinois, Urbana-Champaign; As
Kleffner, Mark A., (419) 995-8208 kleffner.1@osu.edu, Ohio State Univ; PgsGr
Klein, Andrew G., (979) 845-7179 klein@geog.tamu.edu, Texas A&M Univ; ZriGl
Klein, Emily M., (919) 684-5965 ek4@duke.edu, Duke Univ; Gi
Klein, Frieder, fklein@whoi.edu, Woods Hole Oceanographic Inst; Cm
Kleindl, William, william.kleindl@montana.edu, Montana State Univ; Sf
Kleinspehn, Karen L., 612-624-0537 klein004@umn.edu, Univ of

Minnesota, Twin Cities; Gt
Kleiss, Harold J., (919) 515-2643 North Carolina State Univ; Sd
Klemas, Victor, (302) 831-8256 klemas@udel.edu, Univ of Delaware; Zr
Klemetti, Erik, 740-587-5788 klemettie@denison.edu, Denison Univ; GvxGg
Klemow, Kenneth M., (570) 408-4758 kenneth.klemow@wilkes.edu, Wilkes Univ; Pg
Klemperer, Simon L., (650) 723-8214 sklemp@stanford.edu, Stanford Univ; GtYes
Klepeis, Keith A., (802) 656-0247 kklepeis@uvm.edu, Univ of Vermont; Gct
Kliche, Donna V., 605-394-1957 donna.kliche@sdsmt.edu, South Dakota Sch of Mines & Tech; AssAs
Klimczak, Christian, klimczak@uga.edu, Univ of Tennessee, Knoxville; Xg
Klinck, John M., (757) 683-6005 klinck@ccpo.odu.edu, Old Dominion Univ; Op
Klinger, Barry, (703) 993-9227 bklinger@gmu.edu, George Mason Univ; Op
Klinkhammer, Gary P., (541) 737-5209 gklinkhammer@coas.oregonstate.edu, Oregon State Univ; CaHsRw
Kloosterziel, Rudolf C., 808-956-7668 rudolf@soest.hawaii.edu, Univ of Hawai'i, Manoa; Op
Klosterman, Sue, (937) 229-2661 sue.klosterman@notes.udayton.edu, Univ of Dayton; Gg
Kluitenberg, Gerard J., (785) 532-7215 gjk@ksu.edu, Kansas State Univ; Sp
Kluver, Daria, 989-774-1249 kluve1db@cmich.edu, Central Michigan Univ; Asm
Klymak, Jody, 250-721-6120 jklymak@uvic.ca, Univ of Victoria; Op
Knaack, Charles, (509) 335-6742 knaack@wsu.edu, Washington State Univ; Ca
Knap, Anthony, 979-862-2323 ext 111 tknap@tamu.edu, Texas A&M Univ; Oc
Knapp, Camelia, 405-744-6358 camelia.knapp@okstate.edu, Oklahoma State Univ; GtYgs
Knapp, Elizabeth P., (540) 458-8867 knappe@wlu.edu, Washington & Lee Univ; Cl
Knapp, James, 405-744-6358 james.knapp@okstate.edu, Oklahoma State Univ; GcYgs
Knapp, Richard B., (925) 423-3328 knapp4@llnl.gov, Lawrence Livermore; Ng
Knapp, Roy M., 405-325-6829 knapp@ou.edu, Univ of Oklahoma; Np
Knapp, Warren, (607) 255-3034 wwk2@cornell.edu, Cornell Univ; As
Knappenberger, Thorsten J., (334) 844-4100 tjk0013@auburn.edu, Auburn Univ; Sp
Knauss, John A., (401) 874-6141 jknauss@gso.uri.edu, Univ of Rhode Island; Op
Knauth, L. Paul, knauth@asu.edu, Arizona State Univ; Cs
Kneas, David, 803-777-1308 kneas@mailbox.sc.edu, Univ of South Carolina; Zn
Kneeshaw, Tara A., 616-331-8996 kneeshta@gvsu.edu, Grand Valley State Univ; ClGe
Knesel, Kurt, 210-999-7606 kknesel@trinity.edu, Trinity Univ; Gvi
Knight, Andi, 575-835-5594 andi.knight@nmt.edu, New Mexico Tech; Zi
Knight, David, (518) 442-4204 knight@atmos.albany.edu, SUNY, Albany; As
Knight, Rosemary J., (650) 723-4746 Stanford Univ; Hy
Knight, Tiffany, tknight@biology2.wustl.edu, Washington Univ in St. Louis; Pe
Knittle, Elise, (831) 459-4949 eknittle@pmc.ucsc.edu, Univ of California, Santa Cruz; Gy
Knizek, Martin, +420 549 49 6298 kniza@sci.muni.cz, Masaryk Univ; NgmNr
Knoll, Andrew H., (617) 495-9306 aknoll@harvard.edu, Harvard Univ; Pb
Knoll, Martin A., (931) 598-1713 mknoll@sewanee.edu, Sewanee: Univ of the South; Hw
Knopf, Daniel A., (631) 632-3092 daniel.knopf@stonybrook.edu, SUNY, Stony Brook; As
Knopoff, Leon, (310) 825-1885 knopoff@physics.ucla.edu, Univ of California, Los Angeles; Ys
Knott, Jeffrey R., (657) 278-5547 jknott@fullerton.edu, California State Univ, Fullerton; Gmg
Knowles, Charles E., (919) 515-3700 ernie_knowles@ncsu.edu, North Carolina State Univ; Op
Knox, Larry W., (931) 372-3523 lknox@tntech.edu, Tennessee Tech Univ; Pmg
Knudsen, Andrew, (920) 832-6731 knudsena@lawrence.edu, Lawrence Univ; CgGez
Knudsen, Guy, gknudsen@uidaho.edu, Univ of Idaho; Sb
Knudsen, Tyler R., (435) 865-9036 tylerknudsen@utah.gov, Utah Geo Survey; Ng
Knudstrup, Renee, rknudstrup@salemstate.edu, Salem State Univ; Ca
Knuepfer, Peter L. K., (607) 777-2389 Binghamton Univ; Gt
Knutson, Heather, 626.395.4268 hknutson@caltech.edu,CalTech; Xg
Kobara, Shinichi, 979 845 4089 shinichi@tamu.edu, Texas A&M Univ; Zi
Kobs-Nawotniak, Shannon E., (208) 282-3365 kobsshan@isu.edu, Idaho State Univ; Gv
Koc Tasgin, Calibe, 00904242370000-5976 calibekoc@firat.edu.tr, Firat Univ; GsdGg
Koch, Joe, (803) 896-4167 kochj@dnr.sc.gov, South Carolina Dept of Nat Res; Zg
Koch, Magaly, mkoch@bu.edu, Boston Univ; HyZri
Koch, Paul L., (831) 459-5861 pkoch@pmc.ucsc.edu, Univ of California, Santa Cruz; Pv
Kochel, R. Craig, kochel@bucknell.edu, Bucknell Univ; Gm
Kocurek, Gary A., (512) 471-5855 garyk@mail.utexas.edu, Univ of Texas, Austin; Gs
Kocurko, John, 940-397-4250 Midwestern State Univ; Gs
Kodama, Kenneth P., (610) 758-3663 kpk0@lehigh.edu, Lehigh Univ; Ym
Kodosky, Larry, 248-232-4538 lgkodosk@oaklandcc.edu, Oakland Comm Coll; GgvEg
Koehl, Mimi A. R., (510) 642-8103 Univ of California, Berkeley; Pg
Koehler, Rich, (775) 682-8763 rkoehler@unr.edu, Univ of Nevada, Reno; GmtNg
Koehler, Thomas , (719) 333-8712 thomas.koehler@usafa.edu, United States Air Force Acad; As
Koehn, Daniel, daniel.koehn@glasgow.ac.uk, Univ of Glasgow; Gt
Koellemeijer, Paula, paula.koelemeijer@rhul.ac.uk, Royal Holloway Univ of London; Ys
Koeman-Shields, Elizabeth, 325-486-6767 elizabeth.koeman-shields@angelo.edu, Angelo State Univ; CgXg
Koenig, Brian, bkoenig@collegeofthedesert.edu, Coll of the Desert; Gge
Koeshidayatullah, Ardiansyah, a.koeshidayatullah, King Fahd Univ of Petroleum and Minerals; GdCs
Koffman, Bess, bkoffman@colby.edu, Colby Coll; PcClZc
Kogan, Mikhail, (845) 365-8882 kogan@ldeo.columbia.edu, Columbia Univ; Yd
Kohler, Jeffery L., 814-865-9834 jk9@psu.edu, Pennsylvania State Univ, Univ Park; Nm
Kohler, Nicholas, (541) 346-4160 nicholas@uoregon.edu, Univ of Oregon; Zi
Kohlstedt, David L., (612) 626-1544 dlkohl@umn.edu, Univ of Minnesota, Twin Cities; YxGzi
Kohlstedt, Sally G., 612-624-9368 sgk@umn.edu, Univ of Minnesota, Twin Cities; Rh
Kohn, Matthew, mattkohn@boisestate.edu, Boise State Univ; CsGpPv
Kohrt, Casey, casey.kohrt@dnr.iowa.gov, Iowa Dept of Nat Res; GgZi
Kok, Jasper F., jfkok@ucla.edu, Univ of Tennessee, Knoxville; Ap
Kokelaar, Peter, p.kokelaar@liverpool.ac.uk, Univ of Liverpool; GstGv
Kokum, Mehmet, 00904242370000-5963 mkokum@firat.edu.tr, Firat Univ; GtcGg
Kolawole, Lanre L., +2348032277598 llkolawole@lautech.edu.ng, Ladoke Akintola Univ of Tech; GeeGe
Koleszar, Alison, (315) 228-7835 akoleszar@colgate.edu , Colgate Univ; GviCu
Kolka, Randall K., (218) 326-7100 Univ of Minnesota, Twin Cities; So
Kolkas, Mosbah, mossbah.kolkas@csi.cuny.edu, Coll of Staten Island/CUNY; Gg
Kollasch, Pete, pete.kollasch@dnr.iowa.gov, Iowa Dept of Nat Res; GgZri
Kollias, Pavlos, 514-398-1500 pavlos.kollias@stonybrook.edu, McGill Univ; ZrAs
Kolzenburg, Stephan, (716) 645-3489 stephank@buffalo.edu, SUNY, Buffalo; Gv
Komabayashi, Tetsuya, +44 (0) 131 650 8518 tetsuya.komabayashi@ed.ac.uk, Edinburgh Univ; Gz
Komarneni, Sridhar, (814) 865-1542 komarneni@psu.edu, Pennsylvania State Univ, Univ Park; Sc
Kominz, Michelle A., (269) 387-5340 michelle.kominz@wmich.edu, Western Michigan Univ; OuYgGu
Komoto, Cary, 952-358-9007 cary.komoto@normandale.edu, Normandale Comm Coll; ZnnZn
Konecky, Bronwen L., (415) 420-8083 bronwen.konecky@colorado.edu, Washington Univ in St. Louis; PcGn
Konhauser, Kurt, 780-492-2571 kurtk@ualberta.ca, Univ of Alberta; Pg
Koning, Daniel, 575-835-6950 dan.koning@nmt.edu, New Mexico Tech; GsRnGm
Konishi, Takuya, 513-556-9726 konishta@ucmail.uc.edu, Cincinnati Mus Center; Pv
Konrad, Kevin , 702-895-3696 kevin.konrad@unlv.eu, Univ of Nevada, Las Vegas; Cc
Kontak, Daniel J., 7056751151 x2352 dkontak@laurentian.ca, Laurentian

Univ, Sudbury; Eg
Konter, Jasper G., 808-956-8705 jkonter@hawaii.edu, Univ of Hawai'i, Manoa; CcGiCt
Kontuly, Thomas M., (801) 581-8218 thomas.kontuly@geog.utah.edu, Univ of Utah; Zn
Koonce, Jeremy, (702) 895-1092 Univ of Nevada, Las Vegas; Hw
Koons, Peter O., (207)-581-2158 peter.koons@maine.edu, Univ of Maine; Gt
Koornneef, Janne, +31 20 59 81824 j.m.koornneef@vu.nl, Vrije Univit Amsterdam; GvCa
Kooyman, Gerald L., (858) 534-2091 gkooyman@ucsd.edu, Univ of California, San Diego; Ob
Kopf, Christopher F., (570) 662-4615 ckopf@mansfield.edu, Mansfield Univ; Gcp
Kopp, Robert E., (732) 200-2705 robert.kopp@rutgers.edu, Rutgers, The State Univ of New Jersey; ZcOoZg
Koppers, Anthony, 541-737-5425 akoppers@coas.oregonstate.edu, Oregon State Univ; CgGv
Kopylova, Maya G., (604) 822-0865 mkopylova@eos.ubc.ca, Univ of British Columbia; Gi
Koralek, Susan, koralek@sou.edu, Southern Oregon Univ; Zn
Korenaga, Jun, (203) 432-7381 jun.korenaga@yale.edu, Yale Univ; YgsCg
Koretsky, Carla, (269) 387-4372 carla.koretsky@wmich.edu, Western Michigan Univ; CglHs
Kori, E., +27 15 962 8565 edmore.kori@univen.ac.za, Univ of Venda; GmZgAt
Kornreich Wolf, Susan, (309) 794-7369 susanwolf@augustana.edu, Augustana Coll; Ze
Korose, Christopher P., 217-333-7256 korose@illinois.edu, Illinois State Geo Survey; Ec
Korotev, Randy L., (314) 935-5637 korotev@wustl.edu, Washington Univ in St. Louis; XgCtXm
Korre, Anna, +44 20 759 47372 a.korre@imperial.ac.uk, Imperial Coll; Ng
Korths, Patrick, 518-564-2028 pkort001@plattsburgh.edu, SUNY, Plattsburgh; RwGg
Korty, Robert, 979-847-9090 korty@tamu.edu, Texas A&M Univ; As
Kortz, Karen M., kkortz@ccri.edu, Comm Coll of Rhode Island; ZeGg
Korus, Jesse, (402)472-7561 jkorus3@unl.edu, Univ of Nebraska, Lincoln; Gs
Korycansky, Don, (831) 459-5843 Univ of California, Santa Cruz; Xc
Koskinen, William C., (612) 625-4276 koskinen@soils.umn.edu, Univ of Minnesota, Twin Cities; Sc
Kosro, Michael, (541) 737-3079 kosro@coas.oregonstate.edu, Oregon State Univ; OpZrOn
Kossin, James, 608-265-5356 kossin@ssec.wisc.edu, Univ of Wisconsin, Madison; As
Kostelnick, John, 309-438-7679 jckoste@ilstu.edu, Illinois State Univ; Ziy
Koster Van Groos, August F., (312) 996-8678 kvg@uic.edu, Univ of Illinois Chicago; Cp
Kostov, Svilen, 229-931-2321 skostov@gsw.edu, Georgia Southwestern State Univ; Zn
Kota, Jozsef, (520) 621-4396 kota@lpl.arizona.edu, Univ of Arizona; As
Kotamarthi, Rao, 630-252-7164 vrkotamarthi@anl.gov, Argonne; As
Koteas, G. Christopher, 802 485 3321 gkoteas@norwich.edu, Norwich Univ; GitYh
Kotha, Mahender, +91-832-6519329 mkotha@unigoa.ac.in, Goa Univ; GsoZi
Kotulova, Julia, jkotulova@egi.utah.edu, Univ of Utah; CoGo
Koutavas, Athanasios, 718-932-2524 koutavas@mail.csi.cuny.edu, Graduate Sch of the City Univ of New York; Gg
Koutitonsky, Vladimir G., (418) 724-1986 (Ext. 1763) vgk@uqar.qc.ca, Univ du Quebec a Rimouski; Op
Koutnik, Michelle, 206-221-5041 mkoutnik@uw.edu, Univ of Washington; Gll
Kovac, Michal, +421260296555 kovacm@uniba.sk, Comenius Univ in Bratislava; Gsg
Kovach, Richard G., (702) 295-6180 rkovach@lanl.gov, Los Alamos; Nm
Kovach, Robert L., (650) 723-4827 Stanford Univ; Ys
Kovácová, Marianna, marianna.kovacova@uniba.sk, Comenius Univ in Bratislava; PlcPe
Kovács, Ferenc, +3662544156 kovacsf@geo.u-szeged.hu, Univesity of Szeged; ZriZy
Kovaèiæ, Marijan, +38514605963 mkovacic@geol.pmf.hr, Univ of Zagreb; GsdGx
Kowalewski, Michal, kowalewski@flmnh.ufl.edu, Univ of Florida; Pi
Kowalke, Thorsten, 089/2180 6733 t.kowalke@lrz.uni-muenchen.de, Ludwig-Maximilians-Univ Muenchen; Pg
Kowallis, Bart J., (801) 422-2467 bkowallis@byu.edu, Brigham Young Univ; GgCcGz
Kozdon, Reinhard, rkozdon@ldeo.columbia.edu, Columbia Univ; PgGg
Koziol, Andrea M., (937) 229-2954 andrea.koziol@notes.udayton.edu, Univ of Dayton; Cp
Kozlowski, Andrew, 518 486-2012 akozlows@mail.nysed.gov, New York State Geo Survey; Gl
Kpienbaareh, Daniel, 309 438-7084 dlkpien@ilstu.edu, Illinois State Univ; ZuyZr
Kraal, Erin, 484-646-5859 kraal@kutztown.edu, Kutztown Univ of Pennsylvania; XgGm
Krabbenhoft, David, (608) 821-3843 dpkrabbe@usgs.gov, Univ of Wisconsin, Madison; ClGeHw
Kraemer, George P., 914-251-6640 george.kraemer@purchase.edu, SUNY, Purchase; ObnCm
Kraft, Kaatje, (360) 383-3539 kkraft@whatcom.edu, Whatcom Comm Coll; Gg
Krakauer, Nir, 212-650-8009 nir@sci.ccny.cuny.edu, Graduate Sch of the City Univ of New York; Gg
Kramer, J. Curtis, (209) 946-2482 ckramer@pacific.edu, Univ of the Pacific; Gg
Kramer, James R., kramer@mcmaster.ca, McMaster Univ; Cl
Kramer, Kate, (815) 479-7877 kkramer@mchenry.edu, McHenry County Coll; GgZg
Kramer, Marc, 352-294-3165 mgkramer@ufl.edu, Univ of Florida; So
Kramer, Walter V., (361) 698-1385 wkramer@delmar.edu, Del Mar Coll; GgoGx
Krantz, David E., (419) 530-2662 david.krantz@utoledo.edu, Univ of Toledo; Gs
Kranz, Dwight S., (713) 718-5641 dwight.kranz@hccs.edu, Houston Comm Coll System; Gg
Krasowski, Michael, (217) 244-3166 krasows2@illinois.edu, Univ of Illinois; Hw
Krastel, Sebastian , skrastel@geophysik.uni-kiel.de, Dalhousie Univ; Yr
Kratt, Chris, (775) 784-4986 krattc@unr.edu, Univ of Nevada, Reno; YvZrYe
Kraus, Mary J., 303-492-7251 mary.kraus@colorado.edu, Univ of Colorado; GsSa
Krause, Federico F., (403) 220-5845 fkrause@ucalgary.ca, Univ of Calgary; GsdGo
Krause, Lois B., (864) 656-7653 Clemson Univ; Ze
Krause, Samantha, (512) 245-2170 Texas State Univ; ScHyZc
Kravchinsky, Vadim, (780) 492-5591 vkrav@phys.ualberta.ca, Univ of Alberta; Ym
Kravitz, Ben, (812) 855-4334 bkravitz@iu.edu, Indiana Univ, Bloomington; At
Krawczynski, Michael J., (314) 935-6328 mikekraw@levee.wustl.edu, Washington Univ in St. Louis; Cg
Kreamer, David, 702.895.3553 dave.kreamer@unlv.edu, Univ of Nevada, Reno; HwGn
Krebes, Edward S., (403) 220-5028 Univ of Calgary; Ys
Kreemer, Corne, 775 682-8780 kreemer@unr.edu, Univ of Nevada, Reno; YdRn
Kreidenweis, Sonia M., sonia@atmos.colostate.edu, Colorado State Univ; As
Kreiger, William (Bill), (717) 815-1379 wkreiger@ycp.edu, York Coll of Pennsylvania; GisZg
Krekeler, Mark, 513-785-3106 krekelmp@miamioh.edu, Miami Univ; Ge
Kremer, Robert J., 573-882-6408 Univ of Missouri; Sb
Kressler, Sharon J., 612-625-5068 kress004@umn.edu, Univ of Minnesota, Twin Cities; Zn
Kretzschmar, Thomas, tkretzsc@cicese.mx, Centro de Investigación Científica y de Educación Superior de Ensenada; Ge
Kreutz, Karl J., (207)-581-3011 karl.kreutz@maine.edu, Univ of Maine; Cs
Krevor, Samuel, s.krevor@imperial.ac.uk, Imperial Coll; Eo
Kreylos, Oliver, okreylos@ucdavis.edu, Univ of California, Davis; Ng
Krieble, Kelly, (610) 861-1437 krieblek@moravian.edu, Moravian Coll; ZnnZn
Krieger, Jerry, (803) 896-7931 kriegerj@dnr.sc.gov, South Carolina Dept of Nat Res; Zi
Krieger-Brockett, Barbara B., (206) 543-2216 krieger@cheme.washington.edu, Univ of Washington; Ob
Krier, Donathon J., (505) 665-7834 krier@lanl.gov, Los Alamos; Gv
Kring, David, (281) 486-2119 kring@lpi.usra.edu, Univ of Arizona; XcGiXg
Krishnamurthy, R. V., (269) 387-5501 r.v.krishnamurthy@wmich.edu, Western Michigan Univ; Cs
Krishnamurthy, sukanya, sukanya.krishnamurthy@ed.ac.uk, Edinburgh Univ; Zn
Krishtalka, Leonard, (785) 864-4540 krishtalka@ku.edu, Univ of Kansas; Pv
Krissek, Lawrence A., (614) 292-1924 krissek.1@osu.edu, Ohio State Univ; Gsu
Kroeger, Glenn C., (210) 999-7607 gkroeger@trinity.edu, Trinity Univ; Yg
Krogstad, Tore, tore.krogstad@nmbu.no, Norwegian Univ of Life Scis (NMBU); Sc

Krohe, Nicholas J., 605-677-3923 nick.krohe@usd.edu, South Dakota Dept of Environment and Nat Res; Gg
Krohn, James P., krohnjp@piercecollege.edu, Los Angeles Pierce Coll; NgGg
Krol, Michael A., (508) 531-2121 mkrol@bridgew.edu, Bridgewater State Univ; GzxGt
Kronenberg, Andreas, (979) 845-2400 a-kronenberg@tamu.edu, Texas A&M Univ; NrGyYx
Kronenfeld, Barry J., (217) 581-7014 bjkronenfeld@eiu.edu, Eastern Illinois Univ; Zi
Kroon, Dick, +44 (0) 131 651 7089 d.kroon@ed.ac.uk, Edinburgh Univ; Gg
Krot, Alexander N., 808-956-3900 sasha@higp.hawaii.edu, Univ of Hawai'i, Manoa; Xm
Kruckenberg, Seth, 617-552-8300 seth.kruckenberg@bc.edu, Boston Coll; Gc
Krueger, Steven, 801-581-3903 steve.krueger@utah.edu, Univ of Utah; As
Kruge, Michael A., 973-655-7668 krugem@mail.montclair.edu, Montclair State Univ; Co
Kruger, Joseph M., (409) 880-8233 joseph.kruger@lamar.edu, Lamar Univ; YgGgZi
Kruger, Ned , ekadrmas@nd.gov, North Dakota Geo Survey; Gg
Krugh, W C., 661-654-3126 wkrugh@csub.edu, California State Univ, Bakersfield; GcmCc
Krupka, Kenneth M., (509) 376-4412 ken.krupka@pnl.gov, Pacific Northwest; Cl
Kruse, Jennifer, (507) 933-7333 jkruse@gustavus.edu, Gustavus Adolphus Coll; Zn
Kruse, Sarah E., skruse@usf.edu, Univ of South Florida; Yg
Krzic, Maja, 604-822-0252 maja.krzic@ubc.ca, Univ of British Columbia; Sf
Ku, Teh-Lung, (213) 740-5826 rku@usc.edu, Univ of Southern California; CgcGn
Ku, Timothy C., (860)685-2265 tcku@wesleyan.edu, Wesleyan Univ; Cl
Kuang, Zhiming, 617-495-2354 kuang@eps.harvard.edu, Harvard Univ; As
Kubas, Gregory J., (505) 667-5846 Los Alamos; As
Kubesh, Rodney, 320-308-4217 rjkubesh@stcloudstate.edu, Saint Cloud State Univ; Am
Kubicki, James D., (915) 747-6552 jdkubicki@utep.edu, Univ of Texas, El Paso; ClGe
Kucharik, Chris, 608-890-3021 kucharik@wisc.edu, Univ of Wisconsin, Madison; As
Kuchovsky, Tomas, +420 549 49 5452 tomas@sci.muni.cz, Masaryk Univ; Hgy
Kudela, Raphael M., (831) 459-3290 kudela@cats.ucsc.edu, Univ of California, Santa Cruz; Zr
Kudlac, John J., (412) 392-3423 Point Park Univ; Ng
Kuehl, Steven A., (804) 684-7118 kuehl@vims.edu, William & Mary; Ou
Kuehn, Stephen C., 304-384-6322 sckuehn@concord.edu, Concord Univ; CaGv
Kuehner, Scott, 206-543-8393 kuehner@uw.edu, Univ of Washington; Ggz
Kuentz, David C., (513) 529-5992 kuentzdc@miamioh.edu, Miami Univ; Ca
Kues, Barry S., (505) 277-3626 bkues@unm.edu, Univ of New Mexico; Pi
Kuhlman, Robert, rkuhlman@mc3.edu, Montgomery County Comm Coll; Gg
Kuhnhenn, Gary L., (859) 622-8140 gary.kuhnhenn@eku.edu, Eastern Kentucky Univ; GgZe
Kuiper, Klaudia, +31 20 59 83559 k.f.kuiper@vu.nl, Vrije Univit Amsterdam; Cc
Kuiper, Yvette D., 303-273-3105 ykuiper@mines.edu, Colorado Sch of Mines; Gc
Kujawinski, Elizabeth B., (508) 289-3493 ekujawinski@whoi.edu, Woods Hole Oceanographic Inst; Oc
Kukoè, Duje, +38514606111 duje.kukoc@geol.pmf.hr, Univ of Zagreb; Gg
Kukowski, Nina, 0049(0)3641/948680 nina.kukowski@uni-jena.de, Friedrich-Schiller-Univ Jena; Ygx
Kulander, Byron, byron.kulander@wright.edu, Wright State Univ; Gc
Kulkarni, Shrinivas R., (626) 395-4010 srk@astro.caltech.edu,CalTech; Xa
Kullberg, Carla , ckullberg@fc.ul.pt, Unive de Lisboa; Gt
Kulp, Mark A., (504) 280-1170 mkulp@uno.edu, Univ of New Orleans; GrmGs
Kulp, Thomas, tkulp@binghamton.edu, Binghamton Univ; Pg
Kumar, Ajoy, (717) 871-2432 ajoy.kumar@millersville.edu, Millersville Univ; OpZr
Kumar, M. Satish, +442890973479 s.kumar@qub.ac.uk, Queen's Univ Belfast; ZnnZn
Kumjian, Matthew R., 814-863-1581 kumjian@psu.edu, Pennsylvania State Univ, Univ Park; Am
Kummerow, Christian D., kummerow@atmos.colostate.edu, Colorado State Univ; Yr
Kump, Lee R., (814) 863-1274 lkump@psu.edu, Pennsylvania State Univ, Univ Park; ClPeCm
Kumpan, Tomas, +420 549 49 3314 kumpan@sci.muni.cz, Masaryk Univ; Ps
Kumpf, Amber C., (231) 777-0289 amber.kumpf@muskegoncc.edu, Muskegon Comm Coll; GgYrOu
Kung, Hsiang-Te, (901) 678-4538 hkung@memphis.edu, Univ of Memphis; Zy
Kung, King-Jau S., (608) 262-6530 kskung@wisc.edu, Univ of Wisconsin, Madison; Sp
Kunkle, Thomas D., (505) 667-1259 Los Alamos; Xy
Kuntz, Kara, (785) 628-5804 kkuntz@fhsu.edu, Fort Hays State Univ; Zn
Kuntz, Mark R., (847) 697-1000 mkuntz@elgin.edu, Elgin Comm Coll; Gg
Kunza, Lisa, (605) 394-2449 lisa.kunza@sdsmt.edu, South Dakota Sch of Mines & Tech; HsZe
Kunzmann, Thomas, 089/2180 4292 kunzmann@min.uni-muenchen.de, Ludwig-Maximilians-Univ Muenchen; Gz
Kuo, Shiou, (206) 840-4573 skuo@wsu.edu, Washington State Univ; Sc
Kuperman, William A., (858) 534-3158 wkuperman@ucsd.edu, Univ of California, San Diego; Op
Kurak, Ethan, 612-626-6744 kurak006@umn.edu, Univ of Minnesota; Gl
Kurapov, Alexander, 541-737-2865 kurapov@coas.oregonstate.edu, Oregon State Univ; Yr
Kursinski, Robert, (520) 626-3338 kursinsk@atmo.arizona.edu, Univ of Arizona; As
Kurtanjek, Dražen, +38514605965 dkurtan@inet.hr, Univ of Zagreb; GsdGg
Kurttas, Turker, +90-312-2977760 kurttast@gmail.com, Hacettepe Univ; HwgZr
Kurtz, Andrew, kurtz@bu.edu, Boston Univ; Clg
Kurtz, Vincent E., (417) 836-5801 Missouri State Univ; Ps
Kurum, Sevcan, 00904242370000-5992 skurum@firat.edu.tr, Firat Univ; GivGx
Kurz, Marie J., mk3483@drexel.edu, Drexel Univ; CqbHw
Kurz, Mark D., (508) 289-2888 mkurz@whoi.edu, Woods Hole Oceanographic Inst; Cc
Kushnir, Yochanan, kushnir@ldeo.columbia.edu, Columbia Univ; As
Kusnick, Judith E., (916) 278-4692 Sacramento State Univ; Ze
Kusssow, Wayne R., (608) 263-3631 wrkussow@wisc.edu, Univ of Wisconsin, Madison; So
Kustka, Adam B., 973-353-5509 kustka@newark.rutgers.edu, Rutgers, The State Univ of New Jersey, Newark; OgCmHs
Kusumoto, Shigekazu, 81-76-445-6653 kusu@sci.u-toyama.ac.jp, Univ of Toyama; YvdGt
Kusznir, Nick, +44-151-794-5182 n.kusznir@liverpool.ac.uk, Univ of Liverpool; Yd
Kutis, Michael, 765-285-2487 mkutis@bsu.edu, Ball State Univ; Gg
Kuwabara, James, 415-452-7776 kuwabara@usgs.gov, City Coll of San Francisco; RwOgCg
Kuzera, Kristopher , 303-871-3378 kristopher.kuzera@du.edu, Univ of Denver; ZyHg
Kuzila, Mark S., (402) 472-7537 mkuzila@unl.edu, Univ of Nebraska, Lincoln; Sd
Kuzyk, Zou Zou, 204-272-1535 umkuzyk@cc.umanitoba.ca, Univ of Manitoba; Cg
Kvamme, Kenneth L., (617) 353-3415 Boston Univ; Ga
Kwicklis, Edward M., (505) 665-7408 kwicklis@lanl.gov, Los Alamos; Gg
Kwon, Man Jae, manjaekwon@korea.ac.kr, Korea Univ; CbSbGe
Kwon, Youngsang, 901-678-2979 ykwon@memphis.edu, Univ of Memphis; Zi
Kyle, J. Richard, (512) 471-4351 rkyle@jsg.utexas.edu, Univ of Texas, Austin; EmnCe
Kyle, Philip R., (575) 835-5995 kyle@nmt.edu, New Mexico Tech; Giv
Kyriakopoulos, Christodoulos, (901) 678-4830 ckyrkpls@memphis.edu, Univ of Memphis; Yd
Kysar Mattietti, Giuseppina, (703) 993-9269 gkysar@gmu.edu, George Mason Univ; GiZeGa
Kyte, Frank T., (310) 825-2015 kyte@igpp.ucla.edu, Univ of California, Los Angeles; Ct
Käser, Martin, 089/2180 4138 martin.kaeser@geophysik.uni-muenchen.de, Ludwig-Maximilians-Univ Muenchen; Yg
Käsling, Heiko, +49 89 289 25831 heiko.kaesling@tum.de, Technical Univ of Munich; NrmNg

L

L'Ecuyer, Tristan S., 608-890-2107 tlecuyer@wisc.edu, Univ of Wisconsin, Madison; As
La Berge, Gene L., (920) 424-4460 laberge@uwosh.edu, Univ of Wisconsin, Oshkosh; Eg
La Fave, John I., 406-496-4306 jlafave@mtech.edu, Montana Tech of The Univ of Montana; Hw

La Frenierre, Jeff, (507) 933-6332 jlafreni@gustavus.edu, Gustavus Adolphus Coll; GlHg
Laabs, Benjamin J., (701) 231-6197 benjamin.laabs@ndsu.edu, North Dakota State Univ; GlPcGm
Labandeira, Conrad C., (202) 633-1336 Smithsonian Inst / Natl Mus of Natural History; Pi
LaBella, Joel, (860)685-2242 jlabella@wesleyan.edu, Wesleyan Univ; Zn
Laboski, Carrie A., (608) 263-2795 laboski@wisc.edu, Univ of Wisconsin, Madison; So
Labuz, Joseph F., (612) 625-9060 jlabuz@umn.edu, Univ of Minnesota, Twin Cities; Nr
Lachhab, Ahmed, 570-374-4215 lachhab@susqu.edu, Susquehanna Univ; Hw
Lachmar, Thomas E., (435) 797-1247 tom.lachmar@gmail.com, Utah State Univ; Hw
Lachniet, Matthew S., 702-895-4388 matthew.lachniet@unlv.edu, Univ of Nevada, Las Vegas; Css
Lackey, Jade Star, (909) 621-8677 jadestar.lackey@pomona.edu, Pomona Coll; GipCs
Lackinger, Markus, 089/2180 Ludwig-Maximilians-Univ Muenchen; Gz
Lackmann, Gary M., 919-515-1439 gary@ncsu.edu, North Carolina State Univ; As
Lacy, Tor, (657) 464-0409 torlacy@yahoo.com, Cerritos Coll; GgcGc
Ladányi, Zsuzsanna, +3662544156 ladanyi@geo.u-szeged.hu, Univesity of Szeged; ZyeRn
LaDochy, Steve, (323) 343-3222 sladoch@calstatela.edu, California State Univ, Los Angeles; Zn
LaDue, Nicole D., (815) 753-7935 nladue@niu.edu, Northern Illinois Univ; Ze
LaFemina, Peter C., 814-865-7326 pfemina@geosc.psu.edu, Pennsylvania State Univ, Univ Park; Yd
Laffan, Shawn, shawn.laffan@unsw.edu.au, Univ of New South Wales; Ziy
LaFleche, Marc R., (418) 654-2670 marc.richer-lafleche@ete.inrs.ca, Univ du Quebec; Ct
Lafrance, Bruno, 7056751151 x2264 blafrance@laurentian.ca, Laurentian Univ, Sudbury; GcEmGt
LaFreniere, Lorraine, (630) 252-7969 lafreniere@anl.gov, Argonne; HwGoPs
Lageson, David R., (406) 994-6913 lageson@montana.edu, Montana State Univ; GctGs
Lagowski, Alison A., (716) 645-4856 aal@buffalo.edu, SUNY, Buffalo; Zn
Lahiri, Chayan, (719) 587-7357 chayanlahiri@adams.edu, Adams State Univ; HwZr
Lai, Ching-Yao, 609-258-4101 cylai@princeton.edu, Princeton Univ; YgZoGl
Lai, Chung Chieng A., (505) 665-6635 cal@lanl.gov, Los Alamos; Am
Laine, Edward P., edlaine@bowdoin.edu, Bowdoin Coll; Gu
Laingen, Christopher R., (217) 581-2626 crlaingen@eiu.edu, Eastern Illinois Univ; SfZy
Laird, David A., (515) 294-1581 dalaird@iastate.edu, Iowa State Univ of Sci & Tech; Sc
Laird, Jo, (603) 862-1718 jl@cisunix.unh.edu, Univ of New Hampshire; Gp
Laird, Neil, 315-781-3603 laird@hws.edu, Hobart & William Smith Colls; As
Laity, Julie E., (818) 677-3532 julie.laity@csun.edu, California State Univ, Northridge; Zy
Lakatos, Stephen, (718) 262-2589 York Coll (CUNY); Cc
Lake, Iain, +44 (0)1603 59 3744 i.lake@uea.ac.uk, Univ of East Anglia; Ge
Lakhan, V. Chris, 519-253-3000 x2183 lakan@uwindsor.ca, Univ of Windsor; ZriOn
Laki, Sam, (937) 376-6272 slaki@centralstate.edu, Central State Univ; EgSoHs
Lal, Rattan, 614-292-9069 lal.1@osu.edu, Ohio State Univ; Sp
lalap-Ayça, Seda, ssalapayca@umass.edu, Univ of Massachusetts, Amherst; Zi
Laliberte, Elizabeth, (401) 874-5512 elalib@mail.uri.edu, Univ of Rhode Island; Og
Lam, Anita, 604-822-2736 alam@eos.ubc.ca, Univ of British Columbia; Zn
Lam, Phyllis, +44 (0)23 8059 8388 p.lam@southampton.ac.uk, Univ of Southampton; Ob
Lamanna, Matthew C., (412) 624-8780 Univ of Pittsburgh; Pv
Lamb, Andrew P., 479-575-8685 aplamb@uark.edu, Univ of Arkansas, Fayetteville; YumYv
Lamb, Dennis, (814) 883-0174 lno@psu.edu, Pennsylvania State Univ, Univ Park; Am
Lamb, James P., 205 652 3725 jlamb@uwa.edu, Univ of West Alabama; PvePg
Lamb, John A., (612) 625-1772 jlamb@soils.umn.edu, Univ of Minnesota, Twin Cities; Sc
Lamb, Melissa A., (651) 962-5242 malamb@stthomas.edu, Univ of Saint Thomas; GtcGs
Lamb, Michael P., 626.395.3612 mpl@gps.caltch.edu,CalTech; Gm
Lamb, Will, (979) 845-3075 lamb@geo.tamu.edu, Texas A&M Univ; GpCpGz
Lambart, Sarah, 801-581-7062 sarah.lambart@utah.edu, Univ of Utah; Cpg
Lambert, Carolyn D., 970-351-2647 carolyn.lambert@unco.edu, Univ of Northern Colorado; Hw
Lambert, Dean, dlambert@alamo.edu, Alamo Colls, San Antonio Coll; Zy
Lambert-Smith, James, j.s.lambert-smith@kingston.ac.uk, Kingston Univ; Gg
Lammerer, Bernd L., lammerer@iaag.geo.uni-muenchen.de, Ludwig-Maximilians-Univ Muenchen; GgtGc
Lamothe, Michel, 514-987-3000 #3361 lamothe.michel@uqam.ca, Univ du Quebec a Montreal; Gl
Lamp, Jennifer L., jll2214@columbia.edu, Columbia Univ; Cg
Lampkin, Derrick, 301 405 7797 dlampkin@umd.edu, Univ of Maryland; GlAsZr
Lancaster, Nick, (775) 673-7304 nick@dri.edu, Univ of Nevada, Reno; Gm
Lancaster, Penny, +44 023 9284 2272 penny.lancaster@port.ac.uk, Univ of Portsmouth; Gz
Lancaster, Stephen, 541-737-9258 lancasts@geo.oregonstate.edu, Oregon State Univ; Hs
Land, Lewis, 575-887-5508 lland@nmbg.nmt.edu, New Mexico Tech; Hy
Landa, Keith, (914) 251-6450 keith.landa@purchase.edu, SUNY, Purchase; XgbGg
Landenberger, Bill, 61 02 4921 6366 bill.landenberger@newcastle.edu.au, Univ of Newcastle; GiCg
Lander, Mark A., (671) 735-2685 mlander@triton.uog.edu, Univ of Guam; Am
Landing, Ed, (518) 473-8071 elanding@mail.nysed.gov, New York State Geo Survey; PscCc
Landman, Neil H., (212) 769-5712 American Mus of Natural History; Pi
Landman, Willem A., +27 (0) 12 420 3713 willem.landman@up.ac.za, Univ of Pretoria; Ams
Landry, Michael R., (858) 534-4702 mlandry@ucsd.edu, Univ of California, San Diego; Ob
Landry, Peter B., (508) 289-3443 plandry@whoi.edu, Woods Hole Oceanographic Inst; Ca
Landry, Shawn, landry@usf.edu, Univ of South Florida; Zi
Landschoot, Peter J., (814) 863-1017 pcl11@psu.edu, Pennsylvania State Univ, Univ Park; So
Lane, Charles L., (541) 552-6114 lane@sou.edu, Southern Oregon Univ; Hg
Lane, Joseph M., (216) 987-5227 joseph.lane@tri-c.edu, Cuyahoga Comm Coll - Western Campus; Zg
Lane, Mark, 7607441150 xt. 2951 mlane@palomar.edu, Palomar Coll; Zn
Laney, Melissa, (812) 855-4556 Indiana Univ, Bloomington; SfZg
Lang, Harold R., (304) 293-5603 harold@lithos.jpl.nasa.gov, Jet Propulsion Lab; Gr
Lang, Nicholas, 814-824-3646 nlang@mercyhurst.edu, Mercyhurst Univ; GvXgGc
Lange, Eric, (765) 285-8272 eslange@bsu.edu, Ball State Univ; Cg
Lange, Rebecca A., (734) 764-7421 becky@umich.edu, Univ of Michigan; Gi
Langel, Richard A., 319-335-4102 richard.langel@dnr.iowa.gov, Iowa Dept of Nat Res; Gg
Langenhorst, Falko H., 0049(0)3641/948730 falko.langenhorst@uni-jena.de, Friedrich-Schiller-Univ Jena; Gz
Langford, Richard P., (915) 747-5968 langford@utep.edu, Univ of Texas, El Paso; Gs
Langille, Amanda, (709) 737-8349 alangille@mun.ca, Memorial Univ of Newfoundland; Gg
Langille, Jackie M., (828) 251-6453 jlangill@unca.edu, Univ of North Carolina, Asheville; GctGe
Langman, Jeffrey, (208) 885-0310 jlangman@uidaho.edu, Univ of Idaho; ClHws
Langmuir, Charles H., (617) 384-9948 langmuir@eps.harvard.edu, Harvard Univ; Cg
Langston, Charles A., (901) 678-4869 clangstn@memphis.edu, Univ of Memphis; Ys
Langston-Unkefer, Pat J., (505) 665-2556 Los Alamos; Co
Langway, Chester C., (716)645-3489 langway@buffalo.edu, SUNY, Buffalo; Gl
Lanigan, David C., (509) 376-9308 david.lanigan@pnl.gov, Pacific Northwest; Hg
Lansey, Kevin E., (520) 621-2512 lansey@engr.arizona.edu, Univ of Arizona; Hs
Lao, Daniel A., (405) 744-6358 daniel.lao_davila@okstate.edu, Oklahoma State Univ; GctGg
Lapen, Thomas, 713-743-6122 tjlapen@uh.edu, Univ of Houston; Gz

LaPointe, Daphne D., (775) 682-8772 dlapoint@unr.edu, Univ of Nevada, Reno; Gg
Laprise, Rene, 514-987-3000 #3302 laprise.rene@uqam.ca, Univ du Quebec a Montreal; As
Lapusta, Nadia, (626) 395-2277 lapusta@caltech.edu,CalTech; GcYs
Large, Ross R., 61 3 6226 2819 ross.large@utas.edu.au, Univ of Tasmania; Eg
Larkin, Patrick, 361-825-3258 patrick.larkin@tamucc.edu, Texas A&M Univ, Corpus Christi; Zn
LaRock, Paul A., (225) 388-6307 Louisiana State Univ; Ob
Larocque, Marie, 514-987-3000 #1515 larocque.marie@uqam.ca, Univ du Quebec a Montreal; Hw
Larsen, Daniel, (901) 678-4358 dlarsen@memphis.edu, Univ of Memphis; ClGsHw
Larsen, Isaac J., 413-545-0538 ilarsen@geo.umass.edu, Univ of Massachusetts, Amherst; Gm
Larsen, Jessica F., 907-474-7992 jflarsen@alaska.edu, Univ of Alaska, Fairbanks; Gv
Larsen, Kristine, (860) 832-2938 larsen@ccsu.edu, Central Connecticut State Univ; Xy
Larson, David J., 510-885-3132 david.larson@csueastbay.edu, California State Univ, East Bay; Zu
Larson, Edwin E., 303-492-6172 Univ of Colorado; Ym
Larson, Eric J., 715-346-4098 eric.larsen@uwsp.edu, Univ of Wisconsin, Stevens Point; Zr
Larson, Erik B., (740) 351-3144 elarson@shawnee.edu, Shawnee State Univ; GmsHg
Larson, Harold P., (520) 621-6943 hplarson@u.arizona.edu, Univ of Arizona; Zn
Larson, Kyle, 250-807-7564 kyle.larson@ubc.ca, Univ of British Columbia, Okanagan Campus; GtpCc
Larson, Peter B., (509) 335-3095 plarson@wsu.edu, Washington State Univ; Cs
Larson, Roger, (401) 874-6165 rlar@uri.edu, Univ of Rhode Island; Ou
Larson, Toti, 512-471-1856 toti.larson@beg.utexas.edu, Univ of Texas, Austin; GsoEo
Larter, Stephen, +44 (0) 191 208 5956 alex.leathard@ncl.ac.uk, Univ of Newcastle Upon Tyne; Gg
LaRue, Michelle, larue010@umn.edu, Univ of Minnesota, Twin Cities; Zi
Lasca, Norman P., (414) 229-4602 nplasca@uwm.edu, Univ of Wisconsin, Milwaukee; GmlSd
Lascu, Ioan, (202) 633-1815 lascui@si.edu, Smithsonian Inst / Natl Mus of Natural History; GzYm
Lasemi, Zakaria, (217) 244-6944 zlasemi@illinois.edu, Illinois State Geo Survey; Eg
Lash, Gary G., (716) 673-3842 lash@fredonia.edu, SUNY, Fredonia; Gr
Lasher-Trapp, Sonia, 217-244-4250 slasher@illinois.edu, Univ of Illinois, Urbana-Champaign; As
Laske, Gabi, (858) 534-8774 glaske@ucsd.edu, Univ of California, San Diego; Ys
Lasker, Howard R., 716-645-4870 hlasker@buffalo.edu, SUNY, Buffalo; Gu
Lassetter, William L., (434) 951-6361 william.lassetter@dmme.virginia.gov, Geology and Mineral Resources Program; EgHg
Lassiter, John, (512) 471-4002 lassiter1@mail.utexas.edu, Univ of Texas, Austin; Cc
Last, George V., (509) 376-3961 george.last@pnl.gov, Pacific Northwest; Ge
Last, William M., (204) 474-8361 wm_last@umanitoba.ca, Univ of Manitoba; GsnGo
Lat, Che Noorliza, 03-79674157 noorliza@um.edu.my, Univ of Malaya; Yg
Lathrop, Daniel, (301) 405-1594 lathrop@umd.edu, Univ of Maryland; Yg
Latimer, Jennifer C., (812) 237-2254 jen.latimer@indstate.edu, Indiana State Univ; ClGeb
Laton, W. R., (657) 278-7514 wlaton@fullerton.edu, California State Univ, Fullerton; HwGeHy
Lau, Kimberly, kimberly.lau@uwyo.edu, Univ of Wyoming; Cb
Laubach, Stephen E., (512) 471-6303 steve.laubach@beg.utexas.edu, Univ of Texas at Austin, Jackson Sch of Geosciences; GcNrGz
Lauderdale, Jonathan, j.lauderdale@liverpool.ac.uk, Univ of Liverpool; Og
Lauer, Rachel, 403 220-7923 rachel.lauer@ucalgary.ca, Univ of Calgary; Hy
Laughlin, Andrew, 828-250-3993 alaughli@unca.edu, Univ of North Carolina, Asheville; Zn
LAURENT, Antonin, antonin.laurent@univ-st-etienne.fr, Université Jean Monnet, Saint-Etienne; Gp
Laurent-Charvet, Sébastien, +33(0)3 44068995 sebastien.laurent-charvet@lasalle-beauvais.fr, Institut Polytechnique LaSalle Beauvais (ex-IGAL); GctZe
Lauretta, Dante, (520) 626-1138 lauretta@lpl.arizona.edu, Univ of Arizona; XcmXg
Laurier, Eric, +44 (0) 131 651 4303 eric.laurier@ed.ac.uk, Edinburgh Univ; Zn
Lauziere, Kathleen, (418) 654-2658 klauzier@nrcan.gc.ca, Univ du Quebec; Gg
Lauzon, John, (519) 824-4120 (Ext. 52459) lauzonj@uoguelph.ca, Univ of Guelph; So
Lavallee, Yan, +440151 794 5183 yan.lavallee@liverpool.ac.uk, Univ of Liverpool; Gv
Lavanchy, G. Thomas, 303-871-7521 thomas.lavanchy@du.edu, Univ of Denver; Hg
Lavier, Luc L., 512-471-0455 luc@ig.utexas.edu, Univ of Texas, Austin; Gt
LaVigne, Michèle, 207-798-4283 mlavign@bowdoin.edu, Bowdoin Coll; OcPe
Lavkulich, Leslie M., (604) 822-3477 Univ of British Columbia; Sd
Lavoie, Denis, (418) 654-2571 delavoie@nrcan.gc.ca, Univ du Quebec; Gs
Law, Eric W., (740) 826-8242 ericlaw@muskingum.edu, Muskingum Univ; Gp
Law, Kim, 519-661-2111 ext 83881 krlaw@uwo.ca, Western Univ; Cs
Law, Richard D., (540) 231-6685 rdlaw@vt.edu, Virginia Tech; Gc
Law, Zada, (615) 494-8805 zada.law@mtsu.edu, Middle Tennessee State Univ; Zin
Lawlor, Kiera, (508) 531 1390 k1lawlor@bridgew.edu, Bridgewater State Univ; Zg
Lawrence, Deborah, (434) 924-0581 dl3c@virginia.edu, Univ of Virginia; ZcoZg
Lawrence, James, jimrslawrence@gmail.com, Univ of Houston; Cl
Lawrence, Kira, (610) 330-5194 lawrenck@lafayette.edu, Lafayette Coll; OoPc
Lawrence, Rick L., (406) 994-5409 rickl@montana.edu, Montana State Univ; Zr
Lawry-Berkins, Cynthia, 936-273-7407 cynthia.lawryberkins@lonestar.edu, Lonestar Coll - Montgomery; Ges
Lawson, Merlin P., (402) 202-5392 mlawson1@unl.edu, Univ of Nebraska, Lincoln; AmtZr
Lawton, Donald C., (403) 220-5718 lawton@ucalgary.ca, Univ of Calgary; Ye
Lawton, Timothy, 512-471-6247 tim.lawton@beg.utexas.edu, Univ of Texas, Austin; Gst
Lawver, Lawrence A., 512-471-0433 lawver@ig.utexas.edu, Univ of Texas, Austin; Yr
Lay, Thorne, (831) 459-3164 tlay@pmc.ucsc.edu, Univ of California, Santa Cruz; Ys
Layne, Graham, 709 864 3766 gdlayne@mun.ca, Memorial Univ of Newfoundland; Cg
Layton-Matthews, Daniel, (613) 533-6338 dlayton@queensu.ca, Queen's Univ; EgCte
Lazar, Codi, 909-537-5586 clazar@csusb.edu, California State Univ, San Bernardino; Cp
Lazarus, Steven M., 321-674-2160 slazarus@fit.edu, Florida Inst of Tech; As
Lazcano, Cristina, clazcano@ucdavis.edu, Univ of California, Davis; CbSb
Le, Yanfen, 660-562-1525 le@nwmissouri.edu, Northwest Missouri State Univ; Zi
Le Mevel, Helene, hlemevel@carnegiescience.edu, Carnegie Inst for Sci; Gv
Le Qu, Corinne, +44 (0)1603 59 2840 c.lequere@uea.ac.uk, Univ of East Anglia; Zg
Le Roex, Anton, anton.leroex@uct.ac.za, Univ of Cape Town; Cg
Le Roux, Petrus J., 021-650-4139 petrus.leroux@uct.ac.za, Univ of Cape Town; CaGia
Le Roux, Veronique, (508) 289-3549 vleroux@whoi.edu, Woods Hole Oceanographic Inst; Gi
Le Voyer, Marion, levoyerm@si.edu, Smithsonian Inst / Natl Mus of Natural History; Cu
Lea, David W., (805) 893-8665 lea@geol.ucsb.edu, Univ of California, Santa Barbara; PeCm
Lea, Peter D., (207) 725-3439 plea@bowdoin.edu, Bowdoin Coll; Gl
Leach, Harry, +44 0151 794 4097 leach@liverpool.ac.uk, Univ of Liverpool; Og
Leadbetter, Jared R., 626.395.4182 jleadbetter@caltech.edu,CalTech; Pm
Leake, Bernard E., 00442920876421 leakeb@cf.ac.uk, Cardiff Univ; GipGz
Leake, Martha A., (229) 333-5756 mleake@valdosta.edu, Valdosta State Univ; Xg
Lear, C, carrie@earth.cf.ac.uk, Cardiff Univ; Ou
Lear, Caroline, +44(0)29 208 79004 learc@cardiff.ac.uk, Univ of Wales; Og
Leary, Ryan, ryan.leary@nmt.edu, New Mexico Tech; Gs
Leatham, W. Britt, (909) 537-5322 bleatham@csusb.edu, California State Univ, San Bernardino; PsmOg
Leather, Kimberly, kimberley.leather@manchester.ac.uk, Univ of Manchester; As
Leavens, Peter B., 302-831-8106 pbl@udel.edu, Univ of Delaware; Gz
Leavitt, Steven W., (520) 621-6468 sleavitt@ltrr.arizona.edu, Univ of

Arizona; Csc
Lebedev, Maxim, +61 8 9266 3519 m.lebedev@curtin.edu.au, Curtin Univ; YxNr
Lebofsky, Larry A., (520) 621-6947 lebofsky@lpl.arizona.edu, Univ of Arizona; ZngXm
Lebold, Joe, (304) 293-0749 joe.lebold@mail.wvu.edu, West Virginia Univ; PeGrg
LeCain, Rebecca M., (603) 271-1976 rebecca.m.lecain@des.nh.gov, New Hampshire Geo Survey; Gr
Lechler, Paul J., (775) 682-8773 plechler@unr.edu, Univ of Nevada, Reno; Cg
Leckie, James O., leckie@cive.stanford.edu, Stanford Univ; Cl
Leckie, R. Mark, (413) 545-1948 mleckie@geo.umass.edu, Univ of Massachusetts, Amherst; PmGru
Ledford, Sarah, 404-413-5780 sledford@gsu.edu, Georgia State Univ; HsRw
Lee, Alexis, (910) 962-3736 leea@uncw.edu, Univ of North Carolina, Wilmington; Zn
Lee, Alyce, 304-256-0270 alee@concord.edu, Concord Univ; Og
Lee, Arthur C., (865) 220-9145 leea@roanestate.edu, Roane State Comm Coll - Oak Ridge; Ges
Lee, Chung M., (801) 581-8218 chunglee@geog.utah.edu, Univ of Utah; Zu
Lee, Cin-Ty A., 713.348.5084 ctlee@rice.edu, Rice Univ; CgGiv
Lee, Cindy, (632) 220-2101 cindy.lee@stonybrook.edu, SUNY, Stony Brook; OcCo
Lee, Craig M., (206) 685-7656 craig@apl.washington.edu, Univ of Washington; Op
Lee, Daphne E., +64 3 479-7525 daphne.lee@otago.ac.nz, Univ of Otago; Pib
Lee, Eung Seok, (740) 593-1101 leee1@ohio.edu, Ohio Univ; Hw
Lee, Hung, (519) 824-4120 Ext.53828 hlee@uoguelph.ca, Univ of Guelph; Zn
Lee, In Young, (630) 252-8724 Argonne; As
Lee, Jaeheon, 520-626-4967 jaeheon@email.arizona.edu, Univ of Arizona; Nx
Lee, Jeffrey, jeff@geology.cwu.edu, Central Washington Univ; Gct
Lee, Jejung, leej@umkc.edu, Univ of Missouri, Kansas City; GqHw
Lee, Keenan, (303) 273-3808 klee@mines.edu, Colorado Sch of Mines; Zr
Lee, Martin, +4401413302634 martin.lee@glasgow.ac.uk, Univ of Glasgow; Gz
Lee, Meehye, 82-2-3290-3178 meehye@korea.ac.kr, Korea Univ; AsOgCm
Lee, Michael D., 510-885-3155 michael.lee@csueastbay.edu, California State Univ, East Bay; Hg
Lee, Ming-Kuo, (334) 844-4898 leeming@auburn.edu, Auburn Univ; HwGe
Lee, Rachel J., (315) 312-5506 rachel.lee@oswego.edu, SUNY, Oswego; GvZr
Lee, Soon Jae, soonjam@korea.ac.kr, Korea Univ; HwyCq
Lee, Stephen C., leesc@piercecollege.edu, Los Angeles Pierce Coll; Og
Lee, Sukyoung, (814) 863-1587 slg9@psu.edu, Pennsylvania State Univ, Univ Park; Am
Lee, Wook, (814) 732-2291 wlee@edinboro.edu, Edinboro Univ of Pennsylvania; Zi
Lee, Young Jae, 82-2-3290-3181 youngjlee@korea.ac.kr, Korea Univ; ScGze
Lee, Zhongping, zhongping.lee@umb.edu, Univ of Massachusetts, Boston; ZrOg
Lee-Cullin, Joe, jleecullin@albion.edu, Albion Coll; Ge
Lee-Gorishti, Yolanda, 203-392-6647 leegorishty1@southernct.edu, Southern Connecticut State Univ; GaZe
Leech, Mary L., (415) 338-1144 leech@sfsu.edu, San Francisco State Univ; GptGz
Leeman, William P., (713) 348-4892 leeman@rice.edu, Rice Univ; Gi
Lees, Jonathan M., (919) 962-0695 jonathan.lees@unc.edu, Univ of North Carolina, Chapel Hill; YsGv
Leetaru, Hannes, 217 333-5058 hleetaru@illinois.edu, Univ of Illinois; Go
Lefebvre, Rene, (418) 654-2651 rene.lefebvre@ete.inrs.ca, Univ du Quebec; Hw
LeFever, Richard D., (701) 777-3014 richard.lefever@engr.und.edu, Univ of North Dakota; GsrGo
Lefticariu, Liliana, (618) 453-7373 lefticariu@geo.siu.edu, Southern Illinois Univ Carbondale; CsbGe
Leggitt, Leroy, lleggitt@llu.edu, Loma Linda Univ; Pi
Legore, Virginia L., (509) 376-5019 virginia.legore@pnl.gov, Pacific Northwest; Cg
Lehane, Mary, +353 21 4902764 m.lehane@ucc.ie, Univ Coll Cork; Zn
Lehloenya, Pelele, (051) 401-3783 lehloenyapb@ufs.ac.za, Univ of the Free State; Cg
Lehman, Thomas M., 806-834-3148 tom.lehman@ttu.edu, Texas Tech Univ; Gs
Lehner, Flavio, flavio.lehner@cornell.edu, Cornell Univ; Atm
Lehnert, Kerstin, lehnert@ldeo.columbia.edu, Columbia Univ; Gi
Lehrberger, Gerhard, +49 89 289 25832 lehrberger@tum.de, Technical Univ of Munich; EgGxg
Lehre, Andre K., (707) 616-8573 akl1@humboldt.edu, Humboldt State Univ; GmHs
Lehrmann, Daniel J., 210-999-7654 dlehrmann@trinity.edu, Trinity Univ; PiGs
Lehto, Heather L., (325) 486-6990 heather.lehto@angelo.edu, Angelo State Univ; GvYsZe
Leibensperger, Eric, (518) 564-4104 eleib003@plattsburgh.edu, SUNY, Plattsburgh; AsOg
Leichmann, Jaromir, +420 549 49 5559 leichman@sci.muni.cz, Masaryk Univ; GxpGi
Leichter, James J., (858) 822-5330 jleichter@ucsd.edu, Univ of California, San Diego; Ob
Leier, Andrew L., 803-777-9941 aleier@geol.sc.edu, Univ of South Carolina; Gst
Leighton, Henry G., (514) 398-3766 henry.leighton@mcgill.ca, McGill Univ; As
Leighton, Lindsey, 780-492-3983 lleighto@ualberta.ca, Univ of Alberta; Pi
Leighty, Robert S., (480) 461-7021 rleighty@mesacc.edu, Mesa Comm Coll; Zg
Leimer, H. Wayne, (931) 372-3522 hwleimer@tntech.edu, Tennessee Tech Univ; Gz
Leinen, Margaret, (401) 874-6222 mleinen@nsf.gov, Univ of Rhode Island; Ou
Leinfelder, Reinhold, 089/2180 6629 r.leinfelder@lrz.uni-muenchen.de, Ludwig-Maximilians-Univ Muenchen; Pg
Leitch, Alison, (709) 737-3306 aleitch@mun.ca, Memorial Univ of Newfoundland; Yg
Leite, Michael B., (308) 432-6377 mleite@csc.edu, Chadron State Coll; Gg
Leithold, Elana L., (919) 515-7282 lonnie_leithold@ncsu.edu, North Carolina State Univ; Gs
Leitz, Robert E., (303) 556-3072 Metropolitan State Coll of Denver; Gz
Lekic, Vedran, (301) 405-4086 ved@umd.edu, Univ of Maryland; Ysg
Leland, John, 775-784-6670 jleland@unr.edu, Univ of Nevada, Reno; Zn
Lemay, Phillip W., (225) 388-6922 philip@lgs.bri.lsu.edu, Louisiana State Univ; Gg
Lemiszki, Peter J., (865) 594-5596 peter.lemiszki@tn.gov, Pellissippi State Comm Coll; GcgZi
Lemke, Lawrence D., (989) 774-1144 l.d.lemke@cmich.edu, Central Michigan Univ; HwGer
Lemmond, Peter C., (508) 289-2457 plemmond@whoi.edu, Woods Hole Oceanographic Inst; Zn
Lempe, Bernhard, +49 89289 25862 lempe@tum.de, Technical Univ of Munich; GlgNg
Lenardic, Adrian, (713) 348-4883 ajns@rice.edu, Rice Univ; Yg
Lenczewski, Melissa E., (815) 753-7937 lenczewski@niu.edu, Northern Illinois Univ; HwGeCo
Lene, Gene W., (210) 436-3011 glene@stmarytx.edu, Saint Mary's Univ; Ge
Lener, Edward, (540) 231-9249 lener@vt.edu, Virginia Tech; Zn
Lenhart, Stephen W., slenhart@radford.edu, Radford Univ; Pgg
Lenkey, László, lenkey@pangea.elte.hu, Eotvos Lorand Univ; YhHwGa
Lennox, Paul G., + 61 2 9385 8096 p.lennox@unsw.edu.au, Univ of New South Wales; GcgZe
Lentz, David R., (506) 547-2070 Univ of New Brunswick; Cg
Lentz, Leonard J., 717–702–2040 lelentz@pa.gov, DCNR- Pennsylvania Bur of Geo Survey; Ec
Lenz, Alfred C., (519) 661-3195 aclenz@uwo.ca, Western Univ; PiePi
Lenz, Petra H., 808-956-8003 petra@pbrc.hawaii.edu, Univ of Hawai'i, Manoa; Og
Leonard, Eric M., (719) 389-6513 eleonard@coloradocollege.edu, Colorado Coll; Gl
Leonard, Lynn A., 910-962-2338 lynnl@uncw.edu, Univ of North Carolina, Wilmington; Gu
Leonard, Meredith L., 818.778.5595 leonarml@lavc.edu, Los Angeles Valley Coll; AmZei
Leonard-Pingel, Jill, leonard-pingel.1@osu.edu, Ohio State Univ; Pi
Leone, James, 518-473-9988 jleone@mail.nysed.gov, New York State Geo Survey; EoGoCg
Leonhart, Leo S., (520) 881-7300 Univ of Arizona; Hw
Leorri, Eduardo, 252 737 2529 leorrie@ecu.edu, East Carolina Univ; GsPm
Lepain, David, 907-451-5085 david.lepain@alaska.gov, Alaska Division of Geological & Geophysical Surveys; EoGos
Lepper, Kenneth, (701) 231-6746 ken.lepper@ndsu.edu, North Dakota State Univ; CcGml
Leppert, Ken, (318) 342-1918 Univ of Louisiana, Monroe; As
Lepre, Christopher H., 848-445-3410 clepre@eps.rutgers.edu, Rutgers, The State Univ of New Jersey; YmSaPv

Lerach, David G., (970) 351-2853 david.lerach@unco.edu, Univ of Northern Colorado; Ams
Lerch, Derek, dlerch@frc.edu, Feather River Coll; Zg
Lerch, Robert N., 573-882-1402 Univ of Missouri; So
Lerczak, Jim, 541-737-6128 jlerczak@coas.oregonstate.edu, Oregon State Univ; Op
Lerdau, Manuel, (434) 924-3325 mtl5g@virginia.edu, Univ of Virginia; Zn
Lerman, Abraham, 847-491-7385 alerman@northwestern.edu, Northwestern Univ; CgGnCm
Lermurier, Nathalie, +33(0)3 44062540 nathalie.lermurier@lasalle-beauvais.fr, Institut Polytechnique LaSalle Beauvais (ex-IGAL); Zn
Lerner-Lam, Arthur L., lerner@ldeo.columbia.edu, Columbia Univ; Ys
Lesher, Charles E., (530) 752-9779 celesher@ucdavis.edu, Univ of California, Davis; Gi
Lesher, Michael, 7056751151 x2276 mlesher@laurentian.ca, Laurentian Univ, Sudbury; Em
Lesht, Barry, 312-413-3176 blesht@uic.edu, Univ of Illinois Chicago; Op
Leslie, Deborah, dlleslie@memphis.edu, Univ of Memphis; HwCl
Lessard, Evelyn J., (206) 543-8795 elessard@u.washington.edu, Univ of Washington; Ob
Leszczynski, Raymond F., 3152948613 ext. 2313 ray@cayuga-cc.edu, Cayuga Comm Coll; GglGm
Letelier, Ricardo, (541) 737-3890 letelier@coas.oregonstate.edu, Oregon State Univ; ObZr
Lethbridge, Mark, mark.lethbridge@flinders.edu.au, Flinders Univ; Ziy
Letscher, Robert T., 603-862-1008 robert.letscher@unh.edu, Univ of New Hampshire; OcCbZo
Letsinger, Sally L., (812) 855-1356 sletsing@indiana.edu, Indiana Univ, Bloomington; HqZir
Leung, Irene S., (718) 960-8572 Lehman Coll (CUNY); Gz
Leuta-Madondo, Palesa, (031) 260-2517 leutap@ukzn.ac.za, Univ of KwaZulu-Natal; CgGig
Lev, Einat , einatlev@ldeo.columbia.edu, Columbia Univ; Gv
Levander, Alan, (713) 348-6064 alan@rice.edu, Rice Univ; Ys
Levas, Stephen J., 262-472-6200 levass@uww.edu, Univ of Wisconsin, Whitewater; CmGe
Levasseur, Emile , levasseure@easternct.edu, Eastern Connecticut State Univ; Zg
Léveillé, Richard, richard.leveille@mcgill.ca, McGill Univ; CgXb
Leventer, Amy, (315) 228-7213 aleventer@colgate.edu, Colgate Univ; Ou
Leventon, Julia, +44(0) 113 34 31635 j.leventon@leeds.ac.uk, Univ of Leeds; Ge
LeVeque, Randy, 206.685.3037 rjl@uw.edu, Univ of Washington; GvYg
Lever, Helen, +44 (0) 131 451 4057 h.lever@hw.ac.uk, Heriot-Watt Univ; Go
Leverington, David W., 806-834-5310 david.leverington@ttu.edu, Texas Tech Univ; GmZrXg
Levesque, Andre, levesque.andre@ggl.ulaval.ca, Univ Laval; Gz
Levey, Raymond A., (801) 585-3826 rlevey@egi.utah.edu, Univ of Utah; Eo
Levin, Lisa A., (858) 534-3579 llevin@ucsd.edu, Univ of California, San Diego; Ob
Levin, Naomi, 410-516-4317 nlevin3@jhu.edu, Johns Hopkins Univ; Gs
Levin, Vadim, 848-445-5415 vlevin@eps.rutgers.edu, Rutgers, The State Univ of New Jersey; Ysg
Levine, Jamie S., levinejs@appstate.edu, Appalachian State Univ; Gzp
Levine, Norman S., (843) 953-5308 levinen@cofc.edu, Coll of Charleston; ZiGeNg
Levine, Rebekah, 406-683-7134 rebekah.levine@umwestern.edu, Univ of Montana Western; GmHsSd
Levinton, Jeffrey, (631) 632-8602 jeffrey.levinton@stonybrook.edu, SUNY, Stony Brook; Ob
Levson, Victor M., (250) 952-0391 vic.levson@gems9.gov.bc.ca, Univ of Victoria; Gl
Levy, Jonathan, (513) 529-1947 levyj@miamioh.edu, Miami Univ; Hw
Levy, Joseph, (315) 228-7834 jlevy@colgate.edu, Colgate Univ; GlsGm
Levy, Laura B., 707-826-3165 laura.levy@humboldt.edu, Humboldt State Univ; GslPc
Levy, Melissa H., (916) 484-8684 levym@arc.losrios.edu, American River Coll; GgZgy
Levy, Morgan, mclevy@ucsd.edu, Univ of California, San Diego; Hg
Levy, Schon S., (505) 667-9504 sslevy@lanl.gov, Los Alamos; Gi
Lew, Alan A., (928) 523-6567 alan.lew@nau.edu, Northern Arizona Univ; Zgn
Lew, Jeffrey, (310) 825-3023 lew@atmos.ucla.edu, Univ of California, Los Angeles; As
Lewandowski, Katherine, (217) 581-7270 kjlewandowski@eiu.edu, Eastern Illinois Univ; PmcZe
Lewis, Alan G., (604) 822-3626 alewis@eos.ubc.ca, Univ of British Columbia; Ob
Lewis, Brian T. R., (206) 543-7419 blewis@u.washington.edu, Univ of Washington; Yr
Lewis, Chris, cjlewis@ccsf.edu, City Coll of San Francisco; GgEmSo
Lewis, Gerald L., (626) 585-7137 gllewis@pasadena.edu, Pasadena City Coll; Ps
Lewis, Helen, +44 (0) 131 451 3691 h.lewis@hw.ac.uk, Heriot-Watt Univ; Gc
Lewis, John S., (360) 873-8781 jsl@u.arizona.edu, Univ of Arizona; Xcm
Lewis, Jon C., (724) 357-5624 jclewis@iup.edu, Indiana Univ of Pennsylvania; Gc
Lewis, Katie, 806-746-6101 krothlisberger@tamu.edu, Texas A&M Univ; Sco
Lewis, Marlon R., (902) 494-3513 marlon.lewis@dal.ca, Dalhousie Univ; Ob
Lewis, Mary, 5102357800 x.4284 mlewis@contracosta.edu, Contra Costa Coll; Gg
Lewis, Reed S., (208) 885-7472 reedl@uidaho.edu, Univ of Idaho; GgiEg
Lewis, Ronald D., (334) 844-4886 lewisrd@auburn.edu, Auburn Univ; Pi
Lewis, Stephen D., (559) 278-6956 slewis@csufresno.edu, Fresno State Univ; Yg
Lewis, Tammy, 718-951-5000 tlewis@brooklyn.cuny.edu, Graduate Sch of the City Univ of New York; Gg
Leyrit, Hervé, +33(0)3 44068998 herve.leyrit@lasalle-beauvais.fr, Institut Polytechnique LaSalle Beauvais (ex-IGAL); Gv
Li, Aibing, 713-743-2878 ali2@uh.edu, Univ of Houston; Ys
Li, Baosheng, 631-632-9642 baosheng.li@stonybrook.edu, Stony Brook Univ; Yx
Li, Chusi, (812) 855-1558 cli@indiana.edu, Indiana Univ, Bloomington; EgGiEm
Li, Dan, lidan@bu.edu, Boston Univ; HqAs
Li, Gen, ligen@ucsb.edu, Univ of California, Santa Barbara; Gms
Li, Jing, 303-871-4687 jing.li145@du.edu, Univ of Denver; Zi
Li, Junran, 918-631-3013 junran-li@utulsa.edu, Univ of Tulsa; Gme
Li, Laifang, lfl5240@psu.edu, Pennsylvania State Univ, Univ Park; AsZo
Li, Liangping, (605) 394-2461 liangping.li@sdsmt.edu, South Dakota Sch of Mines & Tech; HwqHt
Li, Lihan, lihan-li@uiowa.edu, Univ of Iowa; Cg
Li, Lin, ll3@iupui.edu, Indiana Univ - Purdue Univ Indianapolis; ZrXgZg
Li, Long, 780-492-9288 long4@ualberta.ca, Univ of Alberta; CsGpv
Li, Mingming, (480) 965-1733 mingming.li@asu.edu, Arizona State Univ; Yg
Li, Peng, (501) 683-0118 peng.li@arkansas.gov, Arkansas Geo Survey; GoEoCo
Li, Ping-Chi, (931) 372-3752 pli@tntech.edu, Tennessee Tech Univ; Zi
Li, Qingyun, 631-632-8336 qingyun.li@stonybrook.edu, Stony Brook Univ; Cl
Li, Rong-Yu, 204-727-9684 lir@brandonu.ca, Brandon Univ; PgmGg
Li, Tim, 808-956-9427 timli@hawaii.edu, Univ of Hawai'i at Manoa; Asm
Li, Wenhong, 919 684-5015 wl66@duke.edu, Duke Univ; Zn
Li, William K., (902) 426-6349 lib@mar.dfo-mpo.gc.ca, Dalhousie Univ; Ob
Li, Xiangshan, 713-743-0742 xli10@uh.edu, Univ of Houston; Asm
Li, Yanan, (512) 245-2170 Texas State Univ; GmCcPc
Li, Yaoguo, 303-273-3510 ygli@mines.edu, Colorado Sch of Mines; Yev
Li, Yong-Gang, (213) 740-3556 hylirenko@marshall.usc.edu, Univ of Southern California; Ys
Li, Yuan-Hui, 808-956-6297 yhli@soest.hawaii.edu, Univ of Hawai'i, Manoa; Cm
Li, Yuncong, (305) 246-7000 yunli@ufl.edu, Univ of Florida; Sco
Li, Yunyue E., li4017@purdue.edu, Purdue Univ; Yg
Li, Zhenhong, +44 (0) 191 208 5704 zhenhong.li@ncl.ac.uk, Univ of Newcastle Upon Tyne; Yd
Lian, Jie, 734-647-5704 jlian@umich.edu, Univ of Michigan; Gz
Liang, George, (785) 532-6101 gliang@ksu.edu, Kansas State Univ; So
Liang, Liyuan, 865-241-3933 2ll@ornl.gov, Oak Ridge; Cl
Liang, Lu, 940.369.5198 lu.liang@unt.edu, Univ of North Texas; Zi
Liang, Renxing, 609-258-4101 Princeton Univ; Po
Liang, Yan, (401) 863-9477 yan_liang@brown.edu, Brown Univ; Cp
Liao, Haifeng (Felix), (208) 885-6452 hliao@uidaho.edu, Univ of Idaho; ZugZi
Liauw, Henri L., hliauwap@broward.edu, Broward Coll; Gg
Libarkin, Julie C., (517) 355-8369 libarkin@msu.edu, Michigan State Univ; Zen
Liberty, Lee M., (208) 426-1166 lliberty@boisestate.edu, Boise State Univ; Ys
Libra, Robert D., robert.libra@dnr.iowa.gov, Iowa Dept of Nat Res; Gg
Licciardi, Joseph M., 603-862-1718 joe.licciardi@unh.edu, Univ of New Hampshire; Gl
Licht, Alexis, licht@uw.edu, Univ of Washington; Gsn
Licht, Kathy J., (317) 278-1343 klicht@iupui.edu, Indiana Univ - Purdue

Univ Indianapolis; Gl
Lichtenberger, János, lityi@sas.elte.hu, Eotvos Lorand Univ; AsZrXy
Lichtner, Peter C., (505) 667-3420 lichtner@lanl.gov, Los Alamos; Zn
Liddell, W. David, (435) 797-1261 dave.liddell@usu.edu, Utah State Univ; GsPg
Lidgard, Scott H., (312) 665-7625 slidgard@fieldmuseum.org, Field Mus of Natural History; Pi
Lidiak, Edward G., (412) 624-8871 egl@pitt.edu, Univ of Pittsburgh; Gi
Lidicker, Jr., William Z., (510) 642-3059 Univ of California, Berkeley; Pv
Liebe, Richard M., (585) 395-5100 (Ext. 7524) rliebe@weather.brockport.edu, SUNY, The Coll at Brockport; Ps
Liebens, Johan, 850-474-2065 liebens@uwf.edu, Univ of West Florida; So
Lieberman, Bruce S., (785) 864-2741 blieber@ku.edu, Univ of Kansas; Pgi
Lieberman, Robert C., (631) 632-8214 robert.liebermann@stonybrook.edu, Stony Brook Univ; GyYsx
Liebling, Richard, 516 463-6545 georsl@hofstra.edu, Hofstra Univ; Gg
Lifton, Nathaniel A., (765) 494-0754 nlifton@purdue.edu, Purdue Univ; GmCc
Light, Bonnie, (206) 543-9824 bonnie@apl.washington.edu, Univ of Washington; As
Lightbody, Anne F., 603-862--0711 anne.lightbody@unh.edu, Univ of New Hampshire; HgsZn
Likos, William J., 608-890-2662 likos@wisc.edu, Univ of Wisconsin, Madison; NgSp
Lilley, Marvin D., (206) 543-0859 lilley@ocean.washington.edu, Univ of Washington; Ob
Lima, Eduardo A., 617-324-2829 limaea@mit.edu, Massachusetts Inst of Tech; Ym
Lima, Ivan D., ilima@whoi.edu, Woods Hole Oceanographic Inst; Oc
Limp, William (., (479) 575-7909 flimp@uark.edu, Univ of Arkansas, Fayetteville; Zfi
Lin, Douglas, (831) 459-2732 lin@lick.ucsc.edu, Univ of California, Santa Cruz; Xc
Lin, Fan-Chi, (801) 581-4373 fanchi.lin@utah.edu, Univ of Utah; Ygs
Lin, Guoqing, glin@rsmas.miami.edu, Univ of Miami; Yg
Lin, Hai, hai.lin@ec.gc.ca, McGill Univ; As
Lin, Jialin, (614) 292-6634 lin.789@osu.edu, Ohio State Univ; As
Lin, Jian, (508) 289-2576 jlin@whoi.edu, Woods Hole Oceanographic Inst; Gt
Lin, John C., 801-581-7530 john.lin@utah.edu, Univ of Utah; As
Lin, Jonathan, jonathanlin@cornell.edu, Cornell Univ; As
Lin, Jung-Fu, 512-471-8054 afu@jsg.utexas.edu, Univ of Texas, Austin; GyYm
Lin, Meimei, (912) 344-2974 meimeilin@georgiasouthern.edu, Georgia Southern Univ; Zi
Lin, Senjie, (860) 405-9168 senjie.lin@uconn.edu, Univ of Connecticut; Ob
Lin, Shoufa, (519) 888-4567 (Ext. 6557) shoufa@uwaterloo.ca, Univ of Waterloo; Zn
Lin, Wuyin, (631) 632-3141 wuyin.lin@stonybrook.edu, SUNY, Stony Brook; As
Lin, Xiaomao, (785) 532-6816 xlin@ksu.edu, Kansas State Univ; Zn
Lin, Yu-Feng F., 217-333-0235 yflin@illinois.edu, Illinois State Geo Survey; Hg
Linares, Rogelio, ++935811259 rogelio.linares@uab.cat, Univ Auto de Barcelona; NgYx
Lincoln, Beth Z., (517) 629-0331 blincoln@albion.edu, Albion Coll; Gc
Lincoln, Jonathan M., (973) 655-7273 lincolnj@mail.montclair.edu, Montclair State Univ; Gr
Lindberg, David R., (510) 642-3926 Univ of California, Berkeley; Pi
Lindberg, Jonathan W., (509) 376-5005 jon.lindberg@pnl.gov, Pacific Northwest; Ec
Lindberg, Steven E., partnerships@ornl.gov, Oak Ridge; Cl
Lindbo, David, (919) 793-4428 North Carolina State Univ; Sd
Linde, Alan T., alinde@carnegiescience.edu, Carnegie Inst for Sci; Ys
Lindemann, William C., (505) 646-3405 New Mexico State Univ, Las Cruces; Sb
Lindenmeier, Clark W., (509) 376-8419 clark.lindenmeier@pnl.gov, Pacific Northwest; Cg
Lindgren, Paula, 441413305442 paula.lindgren@glasgow.ac.uk, Univ of Glasgow; As
Lindley, Stacy, (916) 278-6337 stacy.lindley@csus.edu, Sacramento State Univ; Zn
Lindline, Jennifer, (505) 426-2046 lindlinej@nmhu.edu, New Mexico Highlands Univ; GizZe
Lindo Atachati, David, 718-982-2919 david.lindo@csi.cuny.edu, Graduate Sch of the City Univ of New York; Og
Lindsay, Everett H., (520) 621-6024 ehlind@geo.arizona.edu, Univ of Arizona; Pv
Lindsay, Matthew B., (306) 966-5693 matt.lindsay@usask.ca, Univ of Saskatchewan; ClGg
Lindsey, Kassandra, 303-384-2660 kolindsey@mines.edu, Colorado Geo Survey; NgGmZi
Lindsley, Donald H., (631) 632-8195 donald.lindsley@stonybrook.edu, Stony Brook Univ; Cp
Lindzen, Richard S., (617) 253-2432 rlindzen@mit.edu, Massachusetts Inst of Tech; Am
Line, Michael, (480) 965-3971 michael.line@asu.edu, Arizona State Univ; Xy
Liner, Christopher, 479-575-5667 liner@uark.edu, Univ of Arkansas, Fayetteville; YseGg
Lini, Andrea, (802) 656-0245 andrea.lini@uvm.edu, Univ of Vermont; CsGn
Lininger, Katherine, katherine.lininger@colorado.edu, Univ of Colorado; Gm
Link, Curtis A., (406) 496-4165 clink@mtech.edu, Montana Tech of the Univ of Montana; Ye
Link, Paul K., (208) 282-3365 linkpaul@isu.edu, Idaho State Univ; GsgGr
Linky, Edward, (212) 772-5265 linky.edward@epamail.epa.gov, Hunter Coll (CUNY); Zn
Linn, Anne M., (202) 334-2744 alinn@nas.edu, Natl Academies of Scis, Eng, and Medicine; Gs
Linn, Rodman, (505) 665-6254 rrl@lanl.gov, Los Alamos; Ng
Linneman, Scott R., (360) 650-3446 scott.linneman@wwu.edu, Western Washington Univ; GmZe
Linnen, Robert, 519-661-2111 x89207 rlinnen@uwo.ca, Western Univ; EmCe
Linol, Bastien, bastien.aeon@gmail.com, Nelson Mandela Metropolitan Univ; Gs
Linsley, Braddock, blinsley@ldeo.columbia.edu, Columbia Univ; Pe
Liou, Juhn G., (650) 723-2716 liou@pangea.stanford.edu, Stanford Univ; Gp
Liou, Kuo Nan, (415) 723-2716 knliou@atmos.ucla.edu, Univ of California, Los Angeles; As
Lipeles, Maxine I., (314) 935-5482 milipeles@seas.wustl.edu, Washington Univ in St. Louis; Zn
Lippelt, Irene D., (608) 262-7430 irene.lippelt@wisc.edu, Univ of Wisconsin, Madison, Division of Extension; Gg
Lippert, Peter C., (801) 581-4599 pete.lippert@utah.edu, Univ of Utah; YmGtPe
Lippmann, Thomas C., 603-862-4450 lippmann@ccom.unh.edu, Univ of New Hampshire; Onp
Lips, Elliott W., (801) 581-8218 elliott.lips@geog.utah.edu, Univ of Utah; Gm
Lisichenko, Richard, 785-628-4159 rlisiche@fhsu.edu, Fort Hays State Univ; Zie
Lisiecki, Lorraine, (805) 893-4437 lisiecki@geol.ucsb.edu, Univ of California, Santa Barbara; Gn
Lisle, R J., lisle@cf.ac.uk, Cardiff Univ; Gc
Liss, Peter S., +44 (0)1603 59 2563 p.liss@uea.ac.uk, Univ of East Anglia; OcCbAc
Litherland, Mairi, 575-835-5921 mairi.litherland@nmt.edu, New Mexico Tech; Ys
Lithgow-Bertelloni, Carolina, 310-267-4719 clb@epss.ucla.edu, Univ of California, Los Angeles; Yg
Little, Crispin, +44(0) 113 34 36621 earctsl@leeds.ac.uk, Univ of Leeds; Pg
Little, Jonathan, jlittle@monroecc.edu, Monroe Comm Coll; ZyGlZi
Little, Susan D., susan.little@ucl.ac.uk, Univ Coll London; Cg
Little, Tim, +64 4 463 6198 tim.little@vuw.ac.nz, Victoria Univ of Wellington; Gct
Littler, Kate L., 01326255725 k.littler@exeter.ac.uk, Exeter Univ; PeCl
Liu, Chuntao, 361-825-3845 chuntao.liu@tamucc.edu, Texas A&M Univ, Corpus Christi; As
Liu, Dantong, dantong.liu@manchester.ac.uk, Univ of Manchester; As
Liu, Emma D., emma.liu@ucl.ac.uk, Univ Coll London; Gv
Liu, Gaisheng, 785-864-2115 gliu@kgs.ku.edu, Univ of Kansas; Hy
Liu, Jian G., +44 20 759 46418 j.g.liu@imperial.ac.uk, Imperial Coll; Zr
Liu, Jing, liu_jing@smc.edu, Santa Monica Coll; ZyiZr
Liu, Jingpu P., 919-515-3711 jpliu@ncsu.edu, North Carolina State Univ; Gu
Liu, Jiping, jiping.liu@eas.gatech.edu, Georgia Inst of Tech; As
Liu, Kelly, 573-341-6724 liukh@mst.edu, Missouri Univ of Sci and Tech; Ye
Liu, Lanbo, (860) 486-1388 lanbo.liu@uconn.edu, Univ of Connecticut; Yg
Liu, Lijun, (217) 300-0378 ljliu@illinois.edu, Univ of Illinois, Urbana-Champaign; YgGtm
Liu, Mian, (573) 882-3784 lium@missouri.edu, Univ of Missouri; Yg
Liu, Paul Hiaibao, paul.liu@dnr.iowa.gov, Iowa Dept of Nat Res; Gg
Liu, Ping, (631) 632-3195 ping.liu@stonybrook.edu, SUNY, Stony Brook; As
Liu, Qinya, 416-978-5434 Univ of Toronto; Ygx
Liu, Shimin, (814) 863-4491 szl3@psu.edu, Pennsylvania State Univ, Univ

Park; Nm
Liu, Weibo, 561 297-4965 liuw@fau.edu, Florida Atlantic Univ; Zi
Liu, Xiaolei , xlliu@ou.edu, Univ of Oklahoma; Cg
Liu, Xiaoming, (919) 962-0675 xiaomliu@unc.edu, Univ of North Carolina, Chapel Hill; Cgl
Liu, Yajing, 514-398-4085 yajing.liu@mcgill.ca, McGill Univ; Ygs
Liu, Yongqiang, (706) 559-4240 yliu@fs.fed.us, Georgia Inst of Tech; Am
Liu, Zhanfei, 361-749-6772 zhanfei.liu@utexas.edu, Univ of Texas, Austin; Og
Liu, Zhengyu, zliu3@wisc.edu, Univ of Wisconsin, Madison; Op
Liutkus-Pierce, Cynthia M., 828-262-6933 liutkuscm@appstate.edu, Appalachian State Univ; GsCsPs
Livaccari, Richard F., (970) 248-1081 rlivacca@coloradomesa.edu, Colorado Mesa Univ; Gc
Lively, Rich, (612) 626-3103 lively@umn.edu, Univ of Minnesota; Cc
Livens, Francis, +44 0161 275-4647 francis.livens@manchester.ac.uk, Univ of Manchester; Cg
Livermore, Phil, +44(0) 113 34 30379 p.w.livermore@leeds.ac.uk, Univ of Leeds; Ym
Livingston, Jack, 724.738.2897 jack.livingston@sru.edu, Slippery Rock Univ; ZiSo
Livingstone, Daniel A., (919) 684-3264 livingst@duke.edu, Duke Univ; Pe
Lizarralde, Daniel, 508-289-2942 dlizarralde@whoi.edu, Woods Hole Oceanographic Inst; Ys
Llanes, María del Pilar, pllanes@ucm.es, Univ Compl de Madrid; Yr
Llanos, Hilario, hilario.llanos@ehu.eus, Univ of the Basque Country UPV/EHU; Hgy
Llewellin, Ed, +44 (0) 191 33 42336 ed.llewellin@durham.ac.uk, Durham Univ; Gv
Lloyd, Glenn D., 573/368-2176 glenn.lloyd@dnr.mo.gov, Missouri Dept of Nat Res; Zn
Lloyd, Graeme, +44 (1865) 272056 graeme.lloyd@earth.ox.ac.uk, Univ of Oxford; Gs
Lloyd, Jonathan, +44 0161 275-7155 jon.lloyd@manchester.ac.uk, Univ of Manchester; Ge
Lobben, Amy, (541) 346-4566 lobben@uoregon.edu, Univ of Oregon; ZinZn
Lobegeier, Melissa, 615-898-2403 melissa.lobegeier@mtsu.edu, Middle Tennessee State Univ; Pmg
Locat, Jacques E., (418) 656-2179 locat@ggl.ulaval.ca, Univ Laval; Ng
Lock, Brian E., (337) 482-6823 belock@louisiana.edu, Univ of Louisiana at Lafayette; Gs
Locke, Daniel B., (207) 287-7171 daniel.b.locke@maine.gov, Dept of Agriculture, Conservation, and Forestry; Hw
Locke, James L., (415) 485-9526 jim@marin.edu, Coll of Marin; Zg
Locke, Randall, (217) 333-3866 rlocke@illinois.edu, Illinois State Geo Survey; Cl
Locke II, Randall A., 217-333-3866 rlocke@illinois.edu, Univ of Illinois; CgHw
Locker, Stanley D., (727) 553-1502 stan@marine.usf.edu, Univ of South Florida; Ou
Lockley, Martin G., (303) 556-4884 Univ of Colorado, Denver; Pi
Lockwood, John P., (808) 967-8579 jplockwood@volcanologist.com, Univ of Hawai'i, Hilo; Gvg
Lockwood, Rowan, (757) 221-2878 rxlock@wm.edu, William & Mary; PgePi
Locmelis, Marek, 573-341-4759 locmelism@mst.edu, Missouri Univ of Sci and Tech; Eg
Lodders-Fegley, Katharina, (314) 935-4851 lodders@levee.wustl.edu, Washington Univ in St. Louis; Xc
Lodge, Robert, 715-836-4361 lodgerw@uwec.edu, Univ of Wisconsin, Eau Claire; EmCuGv
Loeb, Valerie, 831-771-4400 loeb@mlml.calstate.edu, Moss Landing Marine Lab; Zn
Lofthouse, Stephen T., (212) 346-1760 slofthouse@fsmail.pace.edu, Pace Univ, New York Campus; Zg
Loftis, Jon D., (804)684-7876 jdloftis@vims.edu, William & Mary; OnpZi
Logsdon, Miles G., (206) 543-5334 mlog@u.washington.edu, Univ of Washington; Zi
Lohan, Maeve, +44 (0)23 80596449 m.lohan@soton.ac.uk, Univ of Southampton; Ob
Loheide, Steven P., (608) 265-5277 loheide@wisc.edu, Univ of Wisconsin, Madison; HwSp
Lohman, Rowena B., rbl62@cornell.edu, Cornell Univ; YsZrGt
Lohmann, George P., (508) 289-2840 glohmann@whoi.edu, Woods Hole Oceanographic Inst; Pe
Lohmann, Kyger C., (734) 763-2298 kacey@umich.edu, Univ of Michigan; CsGsCl
Lokau, Katja R., +49 89 289 25857 katja.lokau@tum.de, Technical Univ of Munich; Ng
Lomaga, Margaret, lomagam@sunysuffolk.edu, Suffolk County Comm Coll, Ammerman Campus; Zg
Lombard, Armand J., ajlombard@mesacc.edu, Mesa Comm Coll; Gg
Lombardo, Kelly A., lombardo@psu.edu, Pennsylvania State Univ, Univ Park; As
London, David, (405) 325-3253 dlondon@ou.edu, Univ of Oklahoma; CpGiz
Lonergan, Lidia, +44 20 759 46465 l.lonergan@imperial.ac.uk, Imperial Coll; GtcGs
Long, Ann D., (217) 244-6172 annlong@illinois.edu, Univ of Illinois, Urbana-Champaign; Gm
Long, Colin, (920) 424-4105 longco@uwosh.edu, Univ of Wisconsin Oshkosh; PeZyc
Long, Colleen, longcm@illinois.edu, Illinois State Geo Survey; Gg
Long, Darrel, dlong@laurentian.ca, Laurentian Univ, Sudbury; GsaGl
Long, David T., (517) 353-9618 long@msu.edu, Michigan State Univ; ClGeb
Long, Leon E., (512) 459-7838 leonlong@jsg.utexas.edu, Univ of Texas, Austin; Cc
Long, Lisa, 614 265 6590 lisa.long@dnr.state.oh.usm, Ohio Dept of Nat Res; Zn
Long, Matthew, 508-289-2798 mlong@whoi.edu, Woods Hole Oceanographic Inst; Cm
Long, Maureen D., (203) 432-5031 maureen.long@yale.edu, Yale Univ; Ysg
Long , Sean P., 509-335-8868 sean.p.long@wsu.edu, Washington State Univ; Gc
Longbrake, David, david.longbrake@du.edu, Univ of Denver; Zi
Longerich, Henry, (709) 737-8380 henry@mun.ca, Memorial Univ of Newfoundland; Cg
Longnecker, Krista, (508) 289-2824 klongnecker@whoil.edu, Woods Hole Oceanographic Inst; Oc
Longoria, Jose F., (786) 210-0590 longoria@fiu.edu, Florida InterNatl Univ; ZeGoe
Longoria, Lorena, 956 665-5124 lorena.longoria@utrgv.edu, Univ of Texas, Rio Grande Valley; Ob
Longpre, Marc-Antoine, 718-997-3259 mlongpre@qc.cuny.edu, Graduate Sch of the City Univ of New York; Zg
Longstaffe, Frederick J., (519) 661-3177 flongsta@uwo.ca, Western Univ; Cs
Longworth, Brett, 508-289-3559 blongworth@whoi.edu, Woods Hole Oceanographic Inst; Ou
Lonn, Jeff, (406) 496-4177 jlonn@mtech.edu, Montana Tech of The Univ of Montana; GgtGc
Lonsdale, Darcy J., (631) 632-8712 darcy.lonsdale@stonybrook.edu, SUNY, Stony Brook; Ob
Lonsdale, Peter F., (858) 534-2855 plonsdale@ucsd.edu, Univ of California, San Diego; Gu
Looney, Brian, 803-725-3692 brian02.looney@srnl.doe.gov, Clemson Univ; Hq
Loope, David B., (402) 472-2647 dloope1@unl.edu, Univ of Nebraska, Lincoln; Gsg
Loope, Henry M., (812) 856-3117 hloope@indiana.edu, Indiana Univ, Bloomington; Gls
Looy, Cynthia, 510-642-1607 looy@berkeley.edu, Univ of California, Berkeley; Pb
Lopez, Alberto, (787) 265-3845 alberto.lopez3@upr.edu, Univ of Puerto Rico; YdGtYs
Lopez, Annabelle, 575-835-5139 annabelle.lopez@nmt.edu, New Mexico Tech; Go
Lopez, Dina L., (740) 593-9435 lopezd@ohio.edu, Ohio Univ; CgHwGv
Lopez, Glenn R., (631) 632-8660 glenn.lopez@stonybrook.edu, SUNY, Stony Brook; Ob
Lopez, Margarita, marlopez@cicese.mx, Centro de Investigación Científica y de Educación Superior de Ensenada; Cc
López, Maricela , (662) 370-6262 maricela.lopez@gmail.com, Univ Estatal de Sonora; NxGg
López, Sol, antares@ucm.es, Univ Compl de Madrid; Gz
Lora, Juan, 203-432-2627 juan.lora@yale.edu, Yale Univ; ZoAs
Lord, Mark L., (828) 227-2271 mlord@wcu.edu, Western Carolina Univ; HwGmZe
Lorenz, Ralph, (520) 621-5585 rlorenz@lpl.arizona.edu, Univ of Arizona; Xg
Lorenzo, Juan M., (225) 388-2497 juan@geol.lsu.edu, Louisiana State Univ; Ys
Lorenzoni, Irene, +44 (0)1603 59 3173 i.lorenzoni@uea.ac.uk, Univ of East Anglia; Eg
Lori, Lisa, 573-368-2152 lisa.lori@dnr.mo.gov, Missouri Dept of Nat Res; Gp
Lorimer, Hayden, hayden.lorimer@ed.ac.uk, Edinburgh Univ; Zn
Lorinczi, Piroska, +44(0) 113 34 39245 earpl@leeds.ac.uk, Univ of Leeds;

Np
Loring, Arthur P., (718) 262-2079 York Coll (CUNY); Gm
Loriot, George, 802-626-6378 george.loriot@northernvermont.edu, Northern Vermont Univ-Lyndon; AsZn
Losh, Steven, (507) 389-6323 steven.losh@mnsu.edu, Minnesota State Univ; GzxGo
Losos, Zdenek, ++420 549 49 5623 losos@sci.muni.cz, Masaryk Univ; Gzy
Lott, Dempsey E., (508) 289-2929 dlott@whoi.edu, Woods Hole Oceanographic Inst; Oc
Lotterhos, Kathleen, (781) 581-7370 k.lotterhos@neu.edu, Northeastern Univ; Zn
Lottermoser, Bernd, +44 01326 255973 b.lottermoser@exeter.ac.uk, Exeter Univ; Ca
Loubser, Michael J., michael.loubser@up.ac.za, Univ of Pretoria; GmZy
Loucks, Bob, (512) 471-0366 bob.loucks@beg.utexas.edu, Univ of Texas at Austin, Jackson Sch of Geosciences; Go
Loucks, Robert G., (512) 471-0366 bob.loucks@beg.utexas.edu, Univ of Texas, Austin; Gso
Louden, Keith E., (902) 494-3452 keith.louden@dal.ca, Dalhousie Univ; Yr
Lough, Amanda, (215) 895 6456 amanda.c.lough@drexel.edu, Drexel Univ; Ysg
Louie, John, (775) 784-4219 louie@unr.edu, Univ of Nevada, Reno; Yse
Lourenço, José M., 00351259350280 martinho@utad.pt, Unive de Trás-os-Montes e Alto Douro; YeZin
Lourenco, Sergio, lourencosd@cf.ac.uk, Cardiff Univ; Sp
Louwye, Stephen, stephen.louwye@ugent.be, Ghent Univ; PmZgOo
Love, Andew, andrew.love@flinders.edu.au, Flinders Univ; Hw
Love, David W., (575) 835-5146 dave@nmbg.nmt.edu, New Mexico Tech; Ge
Love, Gordon D., glove@ucr.edu, Univ of California, Riverside; Cog
Love, Renee, (208) 885-4079 rlove@uidaho.edu, Univ of Idaho; PgEoGo
Lovekin, Jonathan, 303-384-2654 jlovekin@mines.edu, Colorado Geo Survey; NgZuGs
Loveland, Karen, 573-368-2142 karen.loveland@dnr.mo.gov, Missouri Dept of Nat Res; Zn
Loveless, Jack, (413) 585-2657 jloveles@smith.edu, Smith Coll; Gtc
LoVetere, Crystal, clovetere@cerritos.edu, Cerritos Coll; Zy
Lovett, Andrew A., +44 (0)1603 59 3126 a.lovett@uea.ac.uk, Univ of East Anglia; Ziu
Lovett, Cole, (269) 927-8744 lovett@lakemichigancollege.edu, Lake Michigan Coll; Gg
Lovseth, John, 618.374.5291 john.lovseth@principia.edu, Principia Coll; Zni
Lowe, David, 709 864-8398 dlowe@mun.ca, Memorial Univ of Newfoundland; Gs
Lowe, Donald R., (650) 725-3040 lowe@pangea.stanford.edu, Stanford Univ; Gd
Lowe, Douglas, douglas.lowe@manchester.ac.uk, Univ of Manchester; As
Lowe, Mike, (801) 537-3389 nrugs.mlowe@state.ut.us, Utah Geo Survey; Hw
Lowell, Thomas V., (513) 556-4165 thomas.lowell@uc.edu, Univ of Cincinnati; Gl
Lowenstein, Tim K., (607) 777-4254 lowenst@binghamton.edu, Binghamton Univ; Cl
Lower, Steven K., 614 292-1571 lower.9@osu.edu, Ohio State Univ; PoZa
Lowery, Birl, (608) 262-2752 blowery@wisc.edu, Univ of Wisconsin, Madison; Sp
Lowery, Chris, cmlowery@utexas.edu, Univ of Texas, Austin; OoPm
Lowry, Anthony R., (435) 797-7096 tony.lowry@usu.edu, Utah State Univ; YdsGt
Lowry, Christopher S., 716-645-4266 cslowry@buffalo.edu, SUNY, Buffalo; Hw
Lowry, David, +44 1784 443105 d.lowry@rhul.ac.uk, Royal Holloway Univ of London; CsZg
Lowry, Wallace D., (540) 231-6521 Virginia Tech; Gc
Lowry, William R., (314) 935-5821 Washington Univ in St. Louis; Zn
Loxsom, Fred, loxsomf@easternct.edu, Eastern Connecticut State Univ; Zn
Loyd, Sean, 657-278-4537 sloyd@fullerton.edu, California State Univ, Fullerton; CgGe
Loydell, David K., +44 023 92 842698 david.loydell@port.ac.uk, Univ of Portsmouth; GrPg
Loynachan, Thomas E., (515) 290-7154 teloynac@iastate.edu, Iowa State Univ of Sci & Tech; So
Lozar, Francesca, francesca.lozar@unito.it, Univ di Torino; Pg
Lozier, M. Susan, (919) 681-8199 s.lozier@duke.edu, Duke Univ; Op
Lozos, Julian, (818) 677-7977 julian.lozos@csun.edu, California State Univ, Northridge; YgGt
Lozowski, Edward P., (403) 492-0348 Univ of Alberta; As
Lu, Tianjun, (310) 243-3815 tilu@csudh.edu, California State Univ, Dominguez Hills; Zi
Lu, Yongmei, (512) 245-2170 Texas State Univ; Zi
Lu, Yongqi, 217 244-4985 yongqilu@illinois.edu, Univ of Illinois; Zm
Lu, Youyu, (902) 426-7780 youyu.lu@ec.gc.ca, Dalhousie Univ; Op
Lu, Yuehan, 205-348-1882 yuehanlu@as.ua.edu, Univ of Alabama; GgOuGe
Lu, Zhong, 214-768-0101 zhonglu@smu.edu, Southern Methodist Univ; Zr
Lu, Zunli, 315-443-0281 zunlilu@syr.edu, Syracuse Univ; Cg
Lucas, Andrew, ajlucas@ucsd.edu, Univ of California, San Diego; Op
Lucas, Beth, (540) 231-4595 blucas06@vt.edu, Virginia Tech; Zn
Lucas, Cathy, +44 (0)23 80596617 cathy.lucas@noc.soton.ac.uk, Univ of Southampton; Ob
Lucas, Franklin A., frank.lucas@uniben.edu, Univ of Benin; PmlGr
Lucey, Paul G., (808) 956-3137 lucey@higp.hawaii.edu, Univ of Hawai'i, Manoa; Xy
Lucia, F. Jerry, (512) 471-7367 jerry.lucia@beg.utexas.edu, Univ of Texas at Austin, Jackson Sch of Geosciences; Go
Luciano, Katherine E., (843) 953-6843 lucianok@dnr.sc.gov, South Carolina Dept of Nat Res; GusGm
Lucotte, Marc Michel, 514-987-3000 #3767 lucotte.marc_michel@uqam.ca, Univ du Quebec a Montreal; Oc
Luczaj, John A., (920) 465-5139 luczajj@uwgb.edu, Univ of Wisconsin, Green Bay; HwGsCl
Ludman, Allan, (718) 997-3271 allan.ludman@gc.cuny.edu, Graduate Sch of the City Univ of New York; Gg
Ludvigson, Greg A., (785) 864-2734 gludvigson@ku.edu, Univ of Kansas; GrCsPc
Ludvigson, Gregory A., 785-864-2734 gludvigson@kgs.ku.edu, Univ of Kansas; Gs
Ludwikoski, David J., 443-840-4216 dludwikoski@ccbcmd.edu, Comm Coll of Baltimore County, Catonsville; ZegZn
Lueth, Virgil L., 575-835-5140 vwlueth@nmt.edu, New Mexico Tech; GzEg
Luhmann, Andrew J., 630 752 7476 andrew.luhmann@wheaton.edu, Wheaton Coll; HwGmCl
Lukas, Roger, (808) 956-7896 rlukas@soest.hawaii.edu, Univ of Hawai'i, Manoa; Op
Lukinbeal, Christopher, clukinbe@email.arizona.edu, Univ of Arizona; Zi
Lukoczki, Georgina, 859 323 0235 gina.lukoczki@uky.edu, Univ of Kentucky; GdsCg
Lulla, Kamlesh, kamlesh.p.lulla@jsc.nasa.gov, Univ of Delaware; Zr
Lumley, David, (972) 883-2454 david.lumley@utdallas.edu, Univ of Texas, Dallas; YseYu
Lumpkin, Thomas A., (509) 335-2726 christie.lumpkin@wsu.edu, Washington State Univ; Zn
Lumsden, David N., (901) 678-2774 dlumsden@memphis.edu, Univ of Memphis; Gd
Luna, Daniel, dluna@boisestate.edu, Boise State Univ; GgPv
Lund, David, 860-405-9331 david.lund@uconn.edu, Univ of Connecticut; Oc
Lund, Jay R., (916) 752-5671 Univ of California, Davis; Hs
Lund, Steven P., (213) 740-5835 slund@usc.edu, Univ of Southern California; Ym
Lund, William R., (435) 865-9034 billlund@utah.gov, Utah Geo Survey; Ng
Lundblad, Steven P., 808-932-7548 slundbla@hawaii.edu, Univ of Hawai'i, Hilo; GraCz
Lundin, Robert F., robert.lundin@asu.edu, Arizona State Univ; Pm
Lundquist, Jessica D., (206) 685-7594 jdlund@uw.edu, Univ of Washington; As
Lundstrom, Craig C., (217) 244-6293 lundstro@illinois.edu, Univ of Illinois, Urbana-Champaign; Cpg
Lunine, Jonathan I., (520) 621-2789 jlunine@lpl.arizona.edu, Univ of Arizona; Xc
Luo, Chao, chao.luo@eas.gatech.edu, Georgia Inst of Tech; As
Luo, Haiyang, 609-258-4101 haiyang.luo@princeton.edu, Princeton Univ; CusGy
Luo, Johnny, 212-650-8936 luo@sci.ccny.cuny.edu, Graduate Sch of the City Univ of New York; Zg
Luo, Jun, (417) 836-4273 junluo@missouristate.edu, Missouri State Univ; Zi
Luo, Zhexi, (412) 622-6578 calleryb@carnegiemnh.org, Carnegie Mus of Natural History; Pv
Lupankwa, Mlindelwa , 012 382 6213 lupankwam@tut.ac.za, Tshwane Univ of Tech; HywGg
Lupia, Richard, (405) 325-7229 rlupia@ou.edu, Univ of Oklahoma; Pm
Lupo, Anthony R., (573) 884-1638 lupoa@missouri.edu, Univ of Missouri; As
Lupulescu, Marian V., (518) 474-1432 marian.lupulescu@nysed.gov, New York State Geo Survey; GzxEm
Luque, Francisco Javier, jluque@ucm.es, Univ Compl de Madrid; Gz
Lusardi, Barb, (612) 626-4791 lusar001@umn.edu, Univ of Minnesota; Gl

Lusk, Braden , 859-257-1105 braden.lusk@uky.edu, Univ of Kentucky; Nm
Luth, Robert W., (780) 492-2740 robert.luth@ualberta.ca, Univ of Alberta; Gi
Luther, Amy, (225) 578-2337 aluther@lsu.edu, Louisiana State Univ; GcZe
Luther, Douglas S., (808) 956-5875 dluther@soest.hawaii.edu, Univ of Hawai'i, Manoa; Op
Luther, Mark E., (727) 553-1528 mluther@marine.usf.edu, Univ of South Florida; Op
Luther, III, George W., (302) 645-4208 luther@udel.edu, Univ of Delaware; Cm
Lutsko, Nicholas, nlutsko@ucsd.edu, Univ of California, San Diego; As
Luttge, Andreas, (713) 348-6304 aluttge@rice.edu, Rice Univ; Cg
Luttrell, Karen, (225) 578-5620 kluttrell@lsu.edu, Louisiana State Univ; Yg
Lutz, Alexandra, 775.673.7418 alexandra.lutz@dri.edu, Univ of Nevada, Reno; Hw
Lutz, Pascale, +33(0)3 44068990 pascale.lutz@lasalle-beauvais.fr, Institut Polytechnique LaSalle Beauvais (ex-IGAL); YgxYd
Lutz, Timothy M., (610) 436-3498 tlutz@wcupa.edu, West Chester Univ; Gq
Luzincourt, Marc R., (418) 654-3715 mluzinco@nrcan.gc.ca, Univ du Quebec; Cs
Lužar-Oberiter, Borna, +38514606115 bluzar@geol.pmf.hr, Univ of Zagreb; Gg
Lwiza, Kamazima M., (631) 632-7309 kamazima.lwiza@stonybrook.edu, SUNY, Stony Brook; Oc
Lübbe, Maike, 089/2180 4337 meike.luebbe@lrz.uni-muenchen.de, Ludwig-Maximilians-Univ Muenchen; Gz
Lybrand, Rebecca A., ralybrand@ucdavis.edu, Univ of California, Davis; SbCb
Lyle, Mike, 757-822-7189 tclylem@tcc.edu, Tidewater Comm Coll; Gg
Lyle, Mitchell, 541-737-3427 mlyle@coas.oregonstate.edu, Oregon State Univ; GgYg
Lynch, Amanda H., amanda_lynch@brown.edu, Brown Univ; AsGe
Lynch, Casey, (775) 682-6192 caseylynch@unr.edu, Univ of Nevada, Reno; Zn
Lynch-Stieglitz, Jean, (404) 894-3944 jean@eas.gatech.edu, Georgia Inst of Tech; OoPcCs
Lynds, Ranie, (307) 766-2286 ranie.lynds@wyo.gov, Univ of Wyoming; Gsr
Lynn, Kendra J., klynn@usgs.gov, Univ of Delaware; Gi
Lynn, Resler M., (540) 231-5790 resler@vt.edu, Virginia Tech; Zy
Lynner, Colton, 302-831-4558 clynner@udel.edu, Univ of Delaware; GtYsGc
Lyon, Ian, ian.lyon@manchester.ac.uk, Univ of Manchester; Xc
Lyon, Ronald J. P., (650) 725-8077 lyon@pangea.stanford.edu, Stanford Univ; Zr
Lyons, Dierdre, d1lyons@ucsd.edu, Univ of California, San Diego; Ob
Lyons, Lawrence, (310) 206-7876 larry@atmos.ucla.edu, Univ of California, Los Angeles; As
Lyons, Steven, steve.lyons@noaa.gov, Angelo State Univ; Am
Lyons, Timothy W., (951) 827-3106 timothy.lyons@ucr.edu, Univ of California, Riverside; CgOcCl
Lyons, William B., (614) 688-3241 lyons.142@osu.edu, Ohio State Univ; HwCgZg
Lyson, Tyler, 303-370-6328 tyler.lyson@dmns.org, Denver Mus of Nature & Sci; Pv
Lytwyn, Jennifer N., 713-743-3296 jlytwyn@uh.edu, Univ of Houston; Gi

M

Ma, Jingsheng, +44 (0)131 451 8296 jingsheng.ma@pet.hw.ac.uk, Heriot-Watt Univ; Np
Ma, Lena Q., (352) 392-9063 lqma@ufl.edu, Univ of Florida; Sc
Ma, Lin, 915-747-5218 lma@utep.edu, Univ of Texas, El Paso; HgCg
Ma, Shuo, 619-594-3091 sma@mail.sdsu.edu, San Diego State Univ; Ys
Ma, Yanxia, yma@lsu.edu, Louisiana State Univ; Og
Maantay, Juliana, 718-960-8574 maantay@aol.com, Lehman Coll (CUNY); Zi
Maas, Regan, 818-677-3515 regan.maas@csun.edu, California State Univ, Northridge; Zi
Maasch, Kirk A., (207) 581-2197 kirk@iceage.umeqs.maine.edu, Univ of Maine; As
Mabee, Stephen B., (413) 545-4814 mabes@umass.edu, Massachusetts Geo Survey; HwGl
Maboko, Makenya A., +255222410013 mmaboko@uccmail.co.tz, Univ of Dar es Salaam; GpCgGx
Mac Eachern, James A., 778-782-5388 jmaceach@sfu.ca, Simon Fraser Univ; Gs
Mac Niocaill, Conall, +44 (0)1865 282135 conallm@earth.ox.ac.uk, Univ of Oxford; GtgYm
Macadam, John, j.d.macadam@exeter.ac.uk, Exeter Univ; Ge
Macalady, Donald L., dmacalad@mines.edu, Colorado Sch of Mines; Cl
Macalady, Jennifer L., (814) 865-6330 jlm80@psu.edu, Pennsylvania State Univ, Univ Park; PgClSb
MacAyeal, Douglas R., (773) 702-8027 drm7@midway.uchicago.edu, Univ of Chicago; Yr
MacCarthy, Ivor, +353 21 4902152 i.maccarthy@ucc.ie, Univ Coll Cork; Gs
MacCarthy, Patrick, pmaccart@mines.edu, Colorado Sch of Mines; Co
MacCready, Parker, (206) 685-9588 parker@ocean.washington.edu, Univ of Washington; Op
Macdonald, Francis, francism@geol.ucsb.edu, Univ of California, Santa Barbara; Gg
MacDonald, Iain, +44(0)29 208 77302 mcdonaldi1@cardiff.ac.uk, Univ of Wales; Ca
MacDonald, James H., (239) 590-7429 jmacdona@fgcu.edu, Florida Gulf Coast Univ; GizGt
Macdonald, R. Heather, (757) 221-2443 rhmacd@wm.edu, William & Mary; GsZe
MacDonald, William D., (607) 777-2863 Binghamton Univ; Gc
Macdougall, J. Douglas, jdmacdougall@ucsd.edu, Univ of California, San Diego; Cg
Mace, Gerald, 801-585-9489 jay.mace@utah.edu, Univ of Utah; As
Mace, Robert, (512) 245-2170 Texas State Univ; HwgHy
Macelloni, Leonardo, (228) 688-7099 leonardo.macelloni@usm.edu, Univ of Southern Mississippi; Yr
MacFadden, Bruce J., (352) 273-1937 bmacfadd@flmnh.ufl.edu, Univ of Florida; Pv
Macfarlane, Andrew W., (305) 348-3980 macfarla@fiu.edu, Florida InterNatl Univ; Em
MacGregor, Kelly, (651) 696-6441 macgregor@macalester.edu, Macalester Coll; GmlGg
Machel, Hans G., (780) 492-5659 hans.machel@ualberta.ca, Univ of Alberta; Go
Machovina, Brett , 719-333-3080 brett.machovina@usafa.edu, United States Air Force Acad; Zir
Macias, Steve E., (360) 475-7711 smacias@olympic.edu, Olympic Coll; Gg
Maciha, Mark J., (928) 523-8242 mark.maciha@nau.edu, Northern Arizona Univ; ZnnZn
MacInnes, Breanyn, 509-963-2827 macinnes@geology.cwu.edu, Central Washington Univ; GseGu
MacInnes, Michael, (505) 665-2154 Los Alamos; Zn
MacIntyre, Hugh, (902) 494-2932 hugh.macintyre@dal.ca, Dalhousie Univ; Ob
Macintyre, Ian G., (202) 633-1339 Smithsonian Inst / Natl Mus of Natural History; Gs
MacIssac, Daniel E., (716) 878-6731 Buffalo State Coll; XaZe
Mack, John E., (716) 878-5006 mackje@buffalostate.edu, Buffalo State Coll; Xg
Mackaness, William A., +44 (0) 131 650 8163 william.mackaness@ed.ac.uk, Edinburgh Univ; Zi
Mackas, David L., (250) 363-6442 mackas@ios.bc.ca, Univ of Victoria; Ob
MacKay, Allison A., 614-247-7652 mackay.49@osu.edu, Ohio State Univ; Cq
Mackenzie, Fred T., (808) 395-7907 fredm@soest.hawaii.edu, Univ of Hawai'i, Manoa; CmGes
Mackenzie, Kristen, 303-370-8372 kristen.mackenzie@dmns.org, Denver Mus of Nature & Sci; Zg
Mackey, Tyler, tjmackey@unm.edu, Univ of New Mexico; GdPo
MacKinnon, Stuart J., 250-807-8405 stuart.mackinnon@ubc.ca, Univ of British Columbia, Okanagan Campus; ZeRcZg
Macko, Stephen A., (434) 924-6849 sam8f@virginia.edu, Univ of Virginia; CosOc
Mackowiak, Cheryl, 850-875-7126 echo13@ufl.edu, Univ of Florida; So
MacLachlan, Ian R., (403) 329-2076 maclachlan@uleth.ca, Univ of Lethbridge; Ze
MacLachlan, John, (905) 525-9140 (Ext. 24195) maclacjc@mcmaster.ca, McMaster Univ; Gs
MacLaughlin, Mary M., (406) 496-4655 mmaclaughlin@mtech.edu, Montana Tech of the Univ of Montana; NrgNt
Maclennan, John, +44 (0) 1223 761602 jmac05@esc.cam.ac.uk, Univ of Cambridge; Gi
MacLeod, C, macleod@cf.ac.uk, Cardiff Univ; Gu
MacLeod, Chris, +44(0)29 208 74332 macleod@cardiff.ac.uk, Univ of Wales; Og
MacLeod, Kenneth A., (573) 884-3118 macleodk@missouri.edu, Univ of Missouri; Pg
MacMahan, Jamie, jhmacmah@nps.edu, Naval Postgrad Sch; Onp
Macomber, Richard, (718) 460-1476 Long Island Univ, Brooklyn Campus; Pg
Macpherson, Colin G., 0191 3342283 colin.macpherson@durham.ac.uk, Durham Univ; CgGiCs

MacPherson, Glenn J., (202) 633-1803 macphers@si.edu, Smithsonian Inst / Natl Mus of Natural History; XcGi
Macpherson, Gwendolyn L., (785) 864-2742 glmac@ku.edu, Univ of Kansas; Hw
MacQuarrie, Pamela, (403) 440-6176 pmacquarrie@mtroyal.ca, Mount Royal Univ; Zy
Macris, Catherine, camacris@iupui.edu, Indiana Univ - Purdue Univ Indianapolis; GzCu
Madadi, Mahyar, +61 8 9266-2324 mahyar.madadi@curtin.edu.au, Curtin Univ; Ye
Maddock, III, Thomas, (520) 621-7120 tm3@email.arizona.edu, Univ of Arizona; Hq
Maddocks, Rosalie F., 713-743-2772 rmaddocks@uh.edu, Univ of Houston; Pm
Madison, Frederick W., (608) 263-4004 fredmad@wisc.edu, Univ of Wisconsin, Madison; Sd
Madsen, David B., (801) 537-3314 nrugs.dmadsen@state.ut.us, Utah Geo Survey; Ga
Madsen, John A., (302) 831-1608 jmadsen@udel.edu, Univ of Delaware; YrOuZe
Maerz, Christian, +44 (0) 1133431504 c.maerz@leeds.ac.uk, Univ of Leeds; CmOuEm
Magaard, Lorenz, 808-956-7509 lorenz@hawaii.edu, Univ of Hawai'i, Manoa; Op
Magee, Elizabeth, 7815817370 x316 e.magee@neu.edu, Northeastern Univ; Zn
Magee, Robert, rmagee@vwc.edu, Virginia Wesleyan Coll; Gge
Magloughlin, Jerry F., (970) 491-1812 jerrym@cnr.colostate.edu, Colorado State Univ; Gpz
Magnani, M. Beatrice, 214-768-1751 mmagnani@smu.edu, Southern Methodist Univ; YsGtYe
Magnusdottir, Gudrun, (949) 824-3520 gudrun@uci.edu, Univ of California, Irvine; As
Magondu, Moffat G., mmagondu@jkuat.ac.ke, Jomo Kenyatta Univ of Agriculture & Tech; ZirZg
Magsino, Sammantha L., (202) 334-2744 smagsino@nas.edu, Natl Academies of Scis, Eng, and Medicine; GvNg
Magson, Justine, +27 (0)51 401 2373 markramj1@ufs.ac.za, Univ of the Free State; CgGi
Mahaffee, Tina, 478-934-3400 tmahaffee@mgc.edu, Middle Georgia Coll; Gg
Mahaffey, Claire, +44 0151 794 4090 claire.mahaffey@liverpool.ac.uk, Univ of Liverpool; Oc
Mahan, Kevin H., kevin.mahan@colorado.edu, Univ of Colorado; Gpc
Maharjan, Madan, 910-775-4589 madan.maharjan@uncp.edu, Univ of North Carolina, Pembroke; Hw
Mahegan, Mairead, 508-289-3931 mmaheigan@whoi.edu, Woods Hole Oceanographic Inst; Ob
Maher, Damien, damien.maher@scu.edu.au, Southern Cross Univ; GeCs
Maher, John, 9134698500 x5953 jmaher1@jccc.edu, Johnson County Comm Coll; Zy
Maher, Kierran, 575-835-6354 kmaher@nmt.edu, New Mexico Tech; Eg
Maher, Jr., Harmon D., 402-554-4807 harmon_maher@unomaha.edu, Univ of Nebraska, Omaha; GcsGt
Maher, Jr., Louis J., (608) 262-9595 maher@geology.wisc.edu, Univ of Wisconsin, Madison; Pl
Mahler, Robert L., (208) 885-7025 Univ of Idaho; Zn
Mahmoud, Sara A., 002-03-3921595 geo_soso2006@yahoo.com, Alexandria Univ; Gg
Mahmoudi, Nagissa, 514-398-2722 nagissa.mahmoudi@mcgill.ca, McGill Univ; Cb
Mahoney, J. Brian, (715) 836-4952 mahonej@uwec.edu, Univ of Wisconsin, Eau Claire; GstEg
Mahood, Gail A., (650) 723-1429 gail@pangea.stanford.edu, Stanford Univ; Gi
Mahowald, Natalie M., 607-255-5166 nmm63@cornell.edu, Cornell Univ; As
Maier, Raina M., (520) 621-7231 rmaier@ag.arizona.edu, Univ of Arizona; Zn
Main, Ian G., +44 (0) 131 650 4911 ian.main@ed.ac.uk, Edinburgh Univ; Ys
Maio, Chris, (907) 474-5651 cvmaio@alaska.edu, Univ of Alaska, Fairbanks; Zy
Maisey, John G., (212) 769-5811 American Mus of Natural History; Pv
Majodina, Thando, (012) 382-6283 majodinato@tut.ac.za, Tshwane Univ of Tech; GgzGe
Major, Penni, 281-765-7865 penny.westerfeld@nhmccd.edu, Lonestar Coll - North Harris; Gg
Major, Ruth H., 518-629-7131 Hudson Valley Comm Coll; GgRh
Maju-Oyovwikowhe, Efetobore G., efetobore.maju-oyovwikowhe@uniben.edu, Univ of Benin; GszGo
Majzlan, Juraj, +49(0)3641 948700 juraj.majzlan@uni-jena.de, Friedrich-Schiller-Univ Jena; Gz
Mak, John E., (631) 632-8673 john.mak@stonybrook.edu, SUNY, Stony Brook; AsCas
Mak, Mankin, (217) 333-8071 mak@atmos.uiuc.edu, Univ of Illinois, Urbana-Champaign; As
Makkawi, Mohammad, +966138602621 makkawi@kfupm.edu.sa, King Fahd Univ of Petroleum and Minerals; HwGoq
Makovicky, Peter, pmakovic@umn.edu, Univ of Minnesota, Twin Cities; Pv
Maksoudian, Michell, mmaksoud@calpoly.edu, California Polytechnic State Univ; Zn
Makungo, R., +27 15 962 8577 rachel.makungo@univen.ac.za, Univ of Venda for Sci & Tech; HzqHw
Malahoff, Alexander, (808) 956-6802 malahoff@hawaii.edu, Univ of Hawai'i, Manoa; Cm
Malakar, Nabin, nmalakar@worcester.edu, Worcester State Univ; Ap
Malanotte-Rizzoli, Paola M., (617) 253-2451 rizzoli@mit.edu, Massachusetts Inst of Tech; Op
Malcolm, Alison, (709) 864-2728 amalcolm@mun.ca, Memorial Univ of Newfoundland; Yg
Malcolm, Elizabeth, 757-233-8751 emalcolm@vwc.edu, Virginia Wesleyan Coll; AsCgOg
Malcuit, Robert J., malcuit@denison.edu, Denison Univ; Gx
Malega, Ron, 417-836-4556 rmalega@missouristate.edu, Missouri State Univ; Zn
Malhotra, Rakesh, rmalhotra@nccu.edu, North Carolina Central Univ; Zri
Malhotra, Renu, (520) 626-5899 renu@lpl.arizona.edu, Univ of Arizona; Xs
Malin, Gill, +44 (0)1603 59 2531 g.malin@uea.ac.uk, Univ of East Anglia; ObcCb
Malin, Peter E., (919) 681-8889 p.malin@auckland.ac.nz, Duke Univ; Ys
Malinconico, Lawrence L., (610) 330-5195 malincol@lafayette.edu, Lafayette Coll; GctYu
Malinconico, Mary Ann, 610-330-5195 lovem@lafayette.edu, Lafayette Coll; Gt
Malinverno, Alberto , (845) 365-8577 alberto@ldeo.columbia.edu, Columbia Univ; Gug
Malinverno, Alberto, alberto@ldeo.columbia.edu, Columbia Univ; Go
Mallard, Laura D., mallardl@appstate.edu, Appalachian State Univ; GtZe
Mallarino, Antonio W., (515) 294-6200 apmallar@iastate.edu, Iowa State Univ of Sci & Tech; So
Mallick, Subhashis, 307-766-2884 smallick@uwyo.edu, Univ of Wyoming; YseYg
Mallik, Ananya, 401)-874-2671 ananya_mallik@uri.edu, Univ of Rhode Island; Gx
Mallinson, David, (252) 328-1344 mallinsond@ecu.edu, East Carolina Univ; Gu
Malone, Andrew, 312-413-1868 amalone@uic.edu, Univ of Illinois Chicago; GlAtZc
Malone, David H., (309) 438-7649 dhmalon@ilstu.edu, Illinois State Univ; GcsEg
Malone, Shawn, (920) 465-2349 malones@uwgb.edu, Univ of Wisconsin, Green Bay; GgpGz
Malone, Stephen D., 206-685-3811 smalone@uw.edu, Univ of Washington; YsGt
Maloney, Eric, emaloney@atmos.colostate.edu, Colorado State Univ; As
Maloof, Adam C., 609-258-4101 maloof@princeton.edu, Princeton Univ; Ym
Malservisi, Rocco, 089/2180 4202 malservisi@rsmas.miami.edu, Ludwig-Maximilians-Univ Muenchen; Yg
Maltese, Adam V., (812) 856-8059 amaltese@indiana.edu, Indiana Univ, Bloomington; Ze
Malzer, Gary L., (612) 625-6728 gmalzer@soils.umn.edu, Univ of Minnesota, Twin Cities; Sc
Mamer, Ethan, 575-835-5118 ethan.mamer@nmt.edu, New Mexico Tech; Hw
Mana, Sara, smana@salemstate.edu, Salem State Univ; GciGv
Manabe, Syukuro, (609) 258-2790 manabe@princeton.edu, Princeton Univ; AstPc
MANCA, Pierpaolo P., 070-675-5529 ppmanca@unica.it, Universita di Cagliari; Nm
Manchester, Steven R., (352) 392-1721 steven@flmnh.ufl.edu, Univ of Florida; PbGs
Manda, Alex K., (252) 328-9403 mandaa@ecu.edu, East Carolina Univ; HwgHq
Mandel, Rolfe, (785) 864-2171 mandel@ku.edu, Univ of Kansas; Gam
Mandia, Scott A., 631-451-4104 mandias@sunysuffolk.edu, Suffolk County Comm Coll, Ammerman Campus; Amt
Mandra, York T., (415) 338-2061 ytmandra@sfsu.edu, San Francisco State

Univ; Pg
Manduca, Cathryn A., (507) 222-7096 cmanduca@carleton.edu, Carleton Coll; Gi
Mandziuk, William S., (204) 474-7826 mandziu0@cc.umanitoba.ca, Univ of Manitoba; Gg
Manecan, Teodosia, (212) 772-5265 tmanecan@hunter.cuny.edu, Hunter Coll (CUNY); Gp
Maneiro, Kathryn A., 630 752 7228 kathryn.maneiro@wheaton.edu, Wheaton Coll; CcGic
Maneta, Marco, 406-243-2454 marco.maneta@mso.umt.edu, Univ of Montana; Hsg
Manfrino, Carrie M., (908) 737-3697 cmanfrin@kean.edu, Kean Univ; GuOb
Manga, Michael, (510) 643-8532 manga@seismo.berkeley.edu, Univ of California, Berkeley; GvXyHw
Manganini, Steven J., (508) 289-2778 smanganini@whoi.edu, Woods Hole Oceanographic Inst; Ou
Mangano, Gabriela, 306-966-5730 gabriela.mangano@usask.ca, Univ of Saskatchewan; PsGs
Mangel, Mark S., (831) 459-5785 msmangel@cats.ucsc.edu, Univ of California, Santa Cruz; Zn
Manger, Walter, 479-575-3355 wmanger@uark.edu, Univ of Arkansas, Fayetteville; Gr
Manghnani, Murli H., (808) 956-7825 murli@soest.hawaii.edu, Univ of Hawai'i, Manoa; Yx
Mango, Helen N., (802) 468-1478 helen.mango@castleton.edu, Castleton Univ; CgGe
Mangriotis, Maria-Daphne , +44 (0) 131 451 3565 m.mangriotis@hw.ac.uk, Heriot-Watt Univ; Ys
Mankiewicz, Carol, mankiewi@beloit.edu, Beloit Coll; Gs
Manley, Patricia L., (802) 443-5430 patmanley@middlebury.edu, Middlebury Coll; YrGu
Manley, Thomas O., (802) 443-3114 tmanley@middlebury.edu, Middlebury Coll; Op
Mann, Dan, dhmann@alaska.edu, Univ of Alaska, Fairbanks; Zy
Mann, Daniel H., (907) 474-5872 Univ of Alaska, Fairbanks; Pe
Mann, Michael E., (814) 863-4075 mann@meteo.psu.edu, Pennsylvania State Univ, Univ Park; As
Mann, Paul, (512) 471-0452 paulm@ig.utexas.edu, Univ of Texas, Austin; Gt
Manning, Andrew, a.manning@uea.ac.uk, Univ of East Anglia; As
Manning, Aspen , aspenmanning@suu.edu, Southern Utah Univ; Zry
Manning, Cara, cara.manning@uconn.edu, Univ of Connecticut; Oc
Manning, Christina, +44 1784 443835 c.manning@es.rhul.ac.uk, Royal Holloway Univ of London; Cg
Manning, Craig E., (310) 206-3290 manning@epss.ucla.edu, Univ of California, Los Angeles; GxCp
Manning, Sturt W., 607-255-8650 sm456@cornell.edu, Cornell Univ; Ga
Mannion, Philip D., p.mannion@ucl.ac.uk, Univ Coll London; Pvi
Mannix, Devin, (217) 244-2419 mannix@illinois.edu, Univ of Illinois; Hw
Manon, Matthew R., (518) 388-8015 manonm@union.edu, Union Coll; Gp
Manos, Leah D., (660) 562-1385 lmanos@nwmissouri.edu, Northwest Missouri State Univ; Zy
Mansell, Mark, 575-835-5562 mark.mansell@nmt.edu, New Mexico Tech; Zif
Manser, Nathan, 906/487-2894 ndmanser@mtu.edu, Michigan Technological Univ; GeNm
Mansour, Ahmed S., 002-03-3921595 ah_sadek@hotmail.com, Alexandria Univ; Ggs
Manspeizer, Warren, (973) 353-5100 mansp@andromeda.rutgers.edu, Rutgers, The State Univ of New Jersey, Newark; Gr
Mansur, Adam, (202) 633-1828 mansura@si.edu, Smithsonian Inst / Natl Mus of Natural History; Zf
Mantei, Erwin J., emantei@missouristate.edu, Missouri State Univ; EdGez
Mantilla Figueroa, Luis C., lcmantil@uis.edu.co, Univ Industrial de Santander; EgCec
Manton, William I., (972) 883-2441 manton@utdallas.edu, Univ of Texas, Dallas; Cc
Manya, Shukrani, shukrani73@udsm.ac.tz, Univ of Dar es Salaam; CgGzg
Manzi, Anthony, 718-409-7371 amanzi@sunymaritime.edu, SUNY, Maritime Coll; Am
Manzocchi, Tom, (+353) 1 716 2605 tom.manzocchi@ucd.ie, Univ Coll Dublin; Goc
Mao, Ho-kwang, hmao@carnegiescience.edu, Carnegie Inst for Sci; Yx
Mapholi, Thendo, +27 (0)51 401 2724 mapholit@ufs.ac.za, Univ of the Free State; Ggp
Marcano, Eugenio J., (413) 538-3237 emarcano@mtholyoke.edu, Mount Holyoke Coll; ZiSoZn
Marcantonio, Franco, 979-845-9240 marcantonio@geo.tamu.edu, Texas A&M Univ; Ct
Marchal, Olivier, (508) 289-3374 omarchal@whoi.edu, Woods Hole Oceanographic Inst; Pe
Marchand, Gerard, (413) 538-3173 gmarchan@mtholyoke.edu, Mount Holyoke Coll; Gga
Marchand, Roger T., (206) 685-3757 rojmarch@u.washington.edu, Univ of Washington; Ass
Marchant, David R., (617) 353-3236 marchant@bu.edu, Boston Univ; Gm
Marchetti, David W., (970) 943-2367 dmarchetti@western.edu, Western Colorado Univ; GmClc
Marchisin, John, (212) 346-1502 Pace Univ, New York Campus; Zg
Marchitto, Thomas M., 303-492-7739 tom.marchitto@colorado.edu, Univ of Colorado; Oc
Marco, Shmuel, 972-3-6405689 shmulik.marco@gmail.com, Tel Aviv Univ; Gg
Marco, Shmulik, shmulikm@tau.ac.il, Tel Aviv Univ; Ggc
Marconi, Dario, 609-258-4101 dmarconi@princeton.edu, Princeton Univ; Cms
Marcott, Shaun, smarcott@wisc.edu, Univ of Wisconsin, Madison; GlPe
Marenco, Katherine N., (610) 526-5627 kmarenco@brynmawr.edu, Bryn Mawr Coll; PiePg
Marenco, Pedro J., (610) 526-7580 pmarenco@brynmawr.edu, Bryn Mawr Coll; PgCsPg
Mareschal, Jean-Claude, 514-987-3000 #6864 mareschal.jean-claude@uqam.ca, Univ du Quebec a Montreal; Yg
Marfurt, Kurt J., (405) 325-5669 kmarfurt@ou.edu, Univ of Oklahoma; Ys
Margalef, Ona, ona.margalef@uab.cat, Univ Auto de Barcelona; Ge
Margolskee, Andrew, 609-258-4101 am9753@princeton.edu, Princeton Univ; As
Margot, Jean-Luc, (310) 206-8345 jlm@epss.ucla.edu, Univ of California, Los Angeles; XyYdv
Margrave, Gary F., (403) 220-4606 Univ of Calgary; Ye
Maria, Tony, 812-461-5326 ahmaria@usi.edu, Univ of Southern Indiana; Gv
Mariani, Elisabeth, +440151 794 5180 mariani@liverpool.ac.uk, Univ of Liverpool; Nr
Marin, Liliana, (254)710-2648 liliana_marin@baylor.edu, Baylor Univ; ZiGgCc
Marinelli, Roberta, 541-737-5195 marinelr@oregonstate.edu, Oregon State Univ; GuOcGe
Marinos, Richard, (716) 645-3489 rmarinos@buffalo.edu, SUNY, Buffalo; Sc
Marinov, Irina, 215-898-1014 imarinov@sas.upenn.edu, Univ of Pennsylvania; ZoOg
Marion, Giles, (775) 673-7349 gmarion@dri.edu, Desert Research Inst; Sc
Maritan, Lara, +390498279143 lara.maritan@unipd.it, Univ degli Studi di Padova; Gx
Marjanac, Tihomir, +38514606109 marjanac@geol.pmf.hr, Univ of Zagreb; GsgYg
Mark, Bryan G., (614) 247-6180 mark.9@osu.edu, Ohio State Univ; As
Market, Patrick S., 573-882-1496 marketp@missouri.edu, Univ of Missouri; As
Markley, Michelle J., (413) 538-2814 mmarkley@mtholyoke.edu, Mount Holyoke Coll; GcgGt
Markowitz, Steven, 718-6704180 Queens Coll (CUNY); GbZn
Markowski, Antonette K., 717-702-2038 amarkowski@pa.gov, DCNR-Pennsylvania Bur of Geo Survey; Eoc
Marks, Dennis W., 229-244-5159 Valdosta State Univ; Zn
Marks-Block, Tony, (510) 885-3164 tony.marksblock@csueastbay.edu, California State Univ, East Bay; Zc
Markun, Francis J., (630) 252-7521 Argonne; Cc
Markwith, Scott H., 561 297-2102 smarkwit@fau.edu, Florida Atlantic Univ; Zy
Marley, Nancy A., (630) 252-5014 namarley@anl.gov, Argonne; As
Marlowe, Brian W., 813-253-7685 bmarlowe2@hccfl.edu, Hillsborough Comm Coll; Zg
Marobhe, Isaac M., 255222410013 marobhe@udsm.ac.tz, Univ of Dar es Salaam; YegYx
Marone, Chris, (814) 865-7964 cjm38@psu.edu, Pennsylvania State Univ, Univ Park; Yx
Marquez, L. Lynn, (717) 871-4339 lynn.marquez@millersville.edu, Millersville Univ; Cg
Marr, Paul G., (717) 477-1656 pgmarr@ship.edu, Shippensburg Univ; ZfyZi
Marra, John, (718) 951-5000 x2594 jfm7780@brooklyn.cuny.edu, Brooklyn Coll (CUNY); Ob
Marrett, Randall A., (512) 471-2113 marrett@mail.utexas.edu, Univ of Texas, Austin; Gc
Marrs, Rob, +44 0151 795 5172 calluna@liverpool.ac.uk, Univ of Liverpool; So
Marsaglia, Kathleen M., (818) 677-6309 kathie.marsaglia@csun.edu, California State Univ, Northridge; Gs

FACULTY INDEX – M

Marsan, Yvonne, 910-962-7288 marsany@uncw.edu, Univ of North Carolina, Wilmington; Zi
Marschik, Robert, 089/2180 6527 marschik@lmu.de, Ludwig-Maximilians-Univ Muenchen; EgmCg
Marsellos, Antonios, (516) 463-5567 antonios.e.marsellos@hofstra.edu, Hofstra Univ; GtqZf
Marsh, Julian S., +27 (0)46-603-7394 goonie.marsh@ru.ac.za, Rhodes Univ; GiCgGv
Marsh, Robert, +44 (0)23 8059 6214 rma@noc.soton.ac.uk, Univ of Southampton; Op
Marshak, Stephen, (217) 333-7705 smarshak@illinois.edu, Univ of Illinois, Urbana-Champaign; Gct
Marshall, Adrienne, 303-273-3886 adriennemarshall@mines.edu, Colorado Sch of Mines; HgAt
Marshall, Charles R., 510- 642-1821 crmarshall@berkeley.edu, Univ of California, Berkeley; PqgPi
Marshall, Craig, (785) 864-6029 cpmarshall@ku.edu, Univ of Kansas; GzZmXb
Marshall, Dan D., 778-782-5474 Simon Fraser Univ; Ca
Marshall, Hans-Peter, hpmarshall@boisestate.edu, Boise State Univ; YxGl
Marshall, Jeffrey S., 909-869-3461 marshall@csupomona.edu, California State Polytechnic Univ, Pomona; GmtHs
Marshall, Jill A., 479-575-2420 jillm@uark.edu, Univ of Arkansas, Fayetteville; Gm
Marshall, Jim, +44-151-794-5177 isotopes@liverpool.ac.uk, Univ of Liverpool; Cs
Marshall, John C., (617) 253-9615 jmarsh@mit.edu, Massachusetts Inst of Tech; Op
Marshall, Madeline, mmarshall@albion.edu, Albion Coll; GsPi
Marshall, Monte, mmarshall@mail.sdsu.edu, San Diego State Univ; YmGtc
Marshall, Scott T., 828-265-8680 marshallst@appstate.edu, Appalachian State Univ; YxGc
Marshall, Stephen, (519) 824-4120 Ext.52720 samarsha@uoguelph.ca, Univ of Guelph; Zn
Marshall, Thomas R., 605-677-6164 tom.marshall@usd.edu, South Dakota Dept of Environment and Nat Res; GgEo
Marsham, John, +44(0) 113 34 36422 j.marsham@leeds.ac.uk, Univ of Leeds; As
Marston, Landon, (540) 231-9023 lmarston@vt.edu, Virginia Tech; HqRw
Marston, Sallie A., marston@email.arizona.edu, Univ of Arizona; Zu
Marszalek, Henryk, henryk.marszalek@uwr.edu.pl, Univ of Wroclaw; Hy
Martek, Lynnette F., martek@mailbox.sc.edu, Univ of South Carolina, Lancaster; ZgAm
Martel, Richard, (418) 654-2683 rene.martel@ete.inrs.ca, Univ du Quebec; Hw
Martel, Stephen J., (808) 956-7797 martel@soest.hawaii.edu, Univ of Hawai'i, Manoa; Ng
Martens, Hilary R., (406) 243-6855 hilary.martens@umontana.edu, Univ of Montana; YgsYd
Martill, David, +44 023 92 842256 david.martill@port.ac.uk, Univ of Portsmouth; PvgPe
Martin, Anthony J., (740) 368-3621 geoam@learnlink.emory.edu, Emory Univ; Pe
Martin, Arturo, amartin@cicese.mx, Centro de Investigación Científica y de Educación Superior de Ensenada; Gd
Martin, Beth, 314-935-4136 martin@wustl.edu, Washington Univ in St. Louis; Ng
Martin, Candace E., +64 3 479-7526 candace.martin@otago.ac.nz, Univ of Otago; ClcGz
Martín, Cristina, crismartin@ucm.es, Univ Compl de Madrid; Gm
Martin, Eileen, eileenrmartin@mines.edu, Colorado Sch of Mines; YuZu
Martin, Ellen E., (352) 392-2141 eemartin@ufl.edu, Univ of Florida; Cl
Martin, Emily S., (202) 633-2254 martines@si.edu, Smithsonian Inst, Natl Air & Space Mus; XgGt
Martin, Gale D., 702-651- 3141 gale.martin@csn.edu, Coll of Southern Nevada - West Charleston Campus; Gg
Martin, James E., (337) 482-6468 geology@louisiana.edu, Univ of Louisiana at Lafayette; PvGg
Martin, Jonathan B., (352) 392-6219 jbmartin@ufl.edu, Univ of Florida; ClHg
Martin, Luke, 575-835-5283 luke.martin@nmt.edu, New Mexico Tech; GorGd
Martin, Paul F., (509) 376-2519 paul.martin@pnl.gov, Pacific Northwest; Cg
Martin, Randal S., (505) 835-5469 New Mexico Tech; As
Martin, Robert F., 514-324-2579 robert.martin@mcgill.ca, McGill Univ; GziEm
Martin, Ronald E., (302) 831-2569 daddy@udel.edu, Univ of Delaware; PmePg
Martin, Scot T., 617-495-7620 smartin@seas.harvard.edu, Harvard Univ; Cg
Martin, Scott C., 330-941-3026 scmartin@ysu.edu, Youngstown State Univ; Ge
Martin, Seelye, (206) 543-6438 seelye@ocean.washington.edu, Univ of Washington; Op
Martin, Silvana, +390498279104 silvana.martin@unipd.it, Univ degli Studi di Padova; Gc
Martin, Timothy J., (630) 252-8708 tjmartin@anl.gov, Argonne; As
Martin, Walter, 704-687-5954 wemartin@uncc.edu, Univ of North Carolina, Charlotte; As
Martin, William R., (508) 289-2836 wmartin@whoi.edu, Woods Hole Oceanographic Inst; Oc
Martin, Zackary E., zmartin@austincc.edu, Austin Comm Coll District; RwZn
Martín Chivelet, Javier, j.m.chivelet@ucm.es, Univ Compl de Madrid; PcGs
Martín Duque, José Francisco, josefco@ucm.es, Univ Compl de Madrid; Gm
Martin-Hayden, James, (419) 530-2634 jhayden@geology.utoledo.edu, Univ of Toledo; Hq
Martindale, Rowan, +1 512 475 6439 martindale@jsg.utexas.edu, Univ of Texas, Austin; PgGs
Martinetto, Edoardo, edoardo.martinetto@unito.it, Univ di Torino; Pb
Martinez, Fernando, (808) 956-6882 martinez@soest.hawaii.edu, Univ of Hawai'i, Manoa; Yr
Martínez, Gemma, gemmamar@ucm.es, Univ Compl de Madrid; Pg
Martínez, José Jesús, jmdiaz@ucm.es, Univ Compl de Madrid; GcRn
Martínez, Pedro, permatin@ucm.es, Univ Compl de Madrid; Hw
Martínez Fernández, Francisco, ++935811513 francisco.martinez@uab.cat, Univ Auto de Barcelona; GpCp
Martinez Torres, Luis Miguel, luismiguel.martinez@ehu.eus, Univ of the Basque Country UPV/EHU; Ggt
Marting, Andrew S., 702-895-4612 andrew.martin@unlv.edu, Univ of Nevada, Las Vegas; EgCg
Martini, Anna M., (413) 542-2067 ammartini@amherst.edu, Amherst Coll; Cl
Martini, I. Peter, (519) 824-4120 x52488 pmartini@uoguelph.ca, Univ of Guelph; Gsl
Martini, Ivan, ivan.martini@unisi.it, Univ of Siena; GsaGg
Martino, Ronald L., (304) 696-2715 martinor@marshall.edu, Marshall Univ; GsrPe
Martins, Zita, +44 20 759 49982 z.martins@imperial.ac.uk, Imperial Coll; Xm
Martinsen, Vegard, vegard.martinsen@nmbu.no, Norwegian Univ of Life Scis (NMBU); Scf
Martinson, Douglas G., dgm@ldeo.columbia.edu, Columbia Univ; Op
Martinuš, Maja, +38514606089 majamarti@geol.pmf.hr, Univ of Zagreb; Gg
Martiny, Adam, 949-824-9713 amartiny@uci.edu, Univ of California, Irvine; Ob
Martire, Luca, luca.martire@unito.it, Univ di Torino; Gd
Márton, Péter, martonp@ludens.elte.hu, Eotvos Lorand Univ; YmgYs
Martz, Todd, (858) 534-7466 trmartz@ucsd.edu, Univ of California, San Diego; Gg
Marvinney, Robert G., (207) 287-2804 robert.g.marvinney@maine.gov, Dept of Agriculture, Conservation, and Forestry; Gg
Marzen, Luke J., (334) 844-3462 marzelj@auburn.edu, Auburn Univ; Zir
Marzoli, Andrea, +390498279154 andrea.marzoli@unipd.it, Univ degli Studi di Padova; Cg
Marzulli, Walter, (609) 292-2576 walt.marzulli@dep.state.nj.us, New Jersey Geological and Water Survey; Zy
Masch, Ludwig, 089/2180 4273 masch@min.uni-muenchen.de, Ludwig-Maximilians-Univ Muenchen; Gz
Masciocco, Luciano, luciano.masciocco@unito.it, Univ di Torino; Ge
Masek, Ondrej, +44 (0) 131 650 5095 ondrej.masek@ed.ac.uk, Edinburgh Univ; Ng
Masiello, Caroline A., 713.348.5234 masiello@rice.edu, Rice Univ; Zn
Maslowski, Wieslaw, (831) 656-3162 maslowsk@nps.edu, Naval Postgrad Sch; Op
Mason, Allen S., (505) 667-4140 Los Alamos; As
Mason, Charles E., (606) 783-2166 c.mason@moreheadstate.edu, Morehead State Univ; PiGrs
Mason, Owen, (907) 474-6293 Univ of Alaska, Fairbanks; Ga
Mason, Philippa J., +44 20 759 46528 p.j.mason@imperial.ac.uk, Imperial Coll; Zr
Mason, Robert P., (860) 405-9129 robert.mason@uconn.edu, Univ of Connecticut; OcCtm
Mason, Sherri A., 716.673.3292 mason@fredonia.edu, SUNY, Fredonia; ZgGe
Mass, Clifford F., cliff@atmos.washington.edu, Univ of Washington; As
MASSACCI, Giorgio, 070-675-5530 massacci@unica.it, Universita di Cagliari; Nm
Massare, Judy A., (585) 395-2419 jmassare@brockport.edu, SUNY, The Coll

at Brockport; Pv
Massey, Matthew, 859.323.0517 matthew.massey@uky.edu, Univ of Kentucky; Ggc
Massey, Michael S., mike.massey@csueastbay.edu, California State Univ, East Bay; ScGeCg
Massironi, Matteo, +390498279116 matteo.massironi@unipd.it, Univ degli Studi di Padova; XgZr
Mastalerz, Maria, (812) 855-5412 mmastale@indiana.edu, Indiana Univ, Bloomington; Ec
Masteller, Claire C., (314) 935-3026 cmasteller@wustl.edu, Washington Univ in St. Louis; Hg
Masterlark, Timothy L., (605) 394-2461 timothy.masterlark@sdsmt.edu, South Dakota Sch of Mines & Tech; YdhNr
Masterman, Steven S., (907) 451-5007 steve.masterman@alaska.gov, Alaska Division of Geological & Geophysical Surveys; EgNg
Masters, T. Guy, (858) 534-4122 tmasters@ucsd.edu, Univ of California, San Diego; YsGy
Masterson, Andrew, andym@earth.northwestern.edu, Northwestern Univ; CslCa
Matano, Ricardo P., (541) 737-2212 rmatano@coas.oregonstate.edu, Oregon State Univ; OpAs
Matassa, Catherine, catherine.matassa@uconn.edu, Univ of Connecticut; Ob
Mateus, António, amateus@fc.ul.pt, Unive de Lisboa; Em
Matheney, Ronald K., (701) 777-4569 ronald.matheney@engr.und.edu, Univ of North Dakota; Cs
Mathenge, Christine, 931-221-6434 mathengec@apsu.ecu, Austin Peay State Univ; Zcy
Matheny, Ashley, ashley.matheny@jsg.utexas.edu, Univ of Texas, Austin; Hg
Mather, Tamsin A., +44 (0)1865 282125 tamsin.mather@earth.ox.ac.uk, Univ of Oxford; Gv
Mathers, Hannah, hannah.mathers@glasgow.ac.uk, Univ of Glasgow; GmlGg
Mathez, Edmond A., 212-769-5459 mathez@amnh.org, American Mus of Natural History; Gx
Mathias, Simon, s.a.mathias@durham.ac.uk, Durham Univ; Hy
Mathur, Ram Raj, +919347303173 ramrajmathur@gmail.com, Osmania Univ; YeuYm
Mathur, Ryan, 814-641-3725 mathur@juniata.edu, Juniata Coll; CeYgHw
Matiasek, Sandrine, 530-898-5096 smatiasek@csuchico.edu, California State Univ, Chico; CqRw
Matisoff, Gerald, (216) 368-3690 gerald.matisoff@case.edu, Case Western Reserve Univ; Cbc
Matoza, Robin, matoza@geol.ucsb.edu, Univ of California, Santa Barbara; Gv
Matsumoto, Katsumi, 612-624-0275 katsumi@umn.edu, Univ of Minnesota, Twin Cities; Oc
Matsuoka, Kenichi, +47 77 75 06 45 matsuoka@npolar.no, Univ of Washington; Gl
Matter, Juerg M., +44 (0)23 80593042 j.matter@southampton.ac.uk, Univ of Southampton; Ng
Mattey, Dave, +44 1784 443587 d.mattey@rhul.ac.uk, Royal Holloway Univ of London; Cs
Matthews, Adrian, +44 (0)1603 59 3733 a.j.matthews@uea.ac.uk, Univ of East Anglia; Am
Matthews, Robert A., (916) 752-0179 Univ of California, Davis; Ge
Mattieu, Lucie, 418-545-5011 x2538 lucie.mathieu1@uqac.ca, Univ du Quebec a Chicoutimi; EgmCe
Mattigod, Shas V., (509) 376-4311 shas.mattigod@pnl.gov, Pacific Northwest; Cl
Mattinson, Chris, 509.963.1628 mattinson@geology.cwu.edu, Central Washington Univ; GpCcGt
Mattioli, Glen, (817) 272-2987 mattioli@uta.edu, Univ of Texas, Arlington; GitYd
Mattox, Stephen R., (616) 331-3734 mattoxs@gvsu.edu, Grand Valley State Univ; ZeGv
Mattox, Tari, 616-234-2119 tmattox@grcc.edu, Grand Rapids Comm Coll; GgvCg
Mattson, Peter H., (718) 997-3335 Queens Coll (CUNY); Gc
Matty, David J., (801) 626-7195 dmatty@weber.edu, Weber State Univ; Gig
Matyjasik, Basia, (801) 537-3122 basiamatyjasik@utah.gov, Utah Geo Survey; Gg
Matyjasik, Marek, (801) 626-7726 mmatyjasik@weber.edu, Weber State Univ; HwGe
Matzke, David, (313) 593-5036 dmatzke@umich.edu, Univ of Michigan, Dearborn; Xy
Mauch, James P., 307-766-2286 james.mauch@wyo.gov, Wyoming State Geo Survey; GmRnGg
Maul, George A., 321-674-8096 gmaul@fit.edu, Florida Inst of Tech; Op
Maulud, Mat Ruzlin, 03-79674144 matruzlin@um.edu.my, Univ of Malaya; Zn
Maurer, Jeremy , 573-341-4852 jlmd9g@mst.edu , Missouri Univ of Sci and Tech; YdRn
Maurrasse, Florentin J., (305) 348-2350 maurrass@fiu.edu, Florida InterNatl Univ; PsGsCm
Mauskopf, Philip, (480) 965-3267 philip.mauskopf@asu.edu, Arizona State Univ; Xa
Mauzerall, Denise L., (609) 258-2498 mauzeral@princeton.edu, Princeton Univ; As
Mavimbela, Philani, (031) 260-7081 mavimbelap@ukzn.ac.za, Univ of KwaZulu-Natal; GpzEg
Mavko, Gerald M., (650) 723-9438 Stanford Univ; Yg
Mawalagedara, Rachindra, 515-294-8633 rmawala@iastate.edu, Iowa State Univ of Sci & Tech; As
Maxbauer, Dan, 507-222-5787 dmaxbauer@carleton.edu, Carleton Coll; PcYm
Maxwell, Arthur E., 512-471-0411 art@ig.utexas.edu, Univ of Texas, Austin; Og
Maxwell, Dave , maxwell@suu.edu, Southern Utah Univ; Zi
Maxwell, James “Sandy”, (610) 436-2240 jmaxwell@wcupa.edu, West Chester Univ; Ze
Maxwell, Michael, (604) 296-4218 michael_maxwell@golder.com, Univ of British Columbia; Yg
May, Cynthia, (715) 682-1499 cmay@northland.edu, Northland Coll; ZiyZn
May, Daniel J., (203) 932-7262 dmay@newhaven.edu, Univ of New Haven; GtxGe
May, Diane M., dianemay@missouristate.edu, Missouri State Univ; Zn
May, Fred E., (801) 581-8218 pscem.fmay@state.ut.us, Univ of Utah; Zn
May, James, (662) 325-5806 Mississippi State Univ; Hw
May, Michael, (502) 745-4555 michael.may@wku.edu, Western Kentucky Univ; Ge
May, S J., (972) 377-1635 smay@collin.edu, Collin Coll - Preston Ridge Campus; GgoGc
Mayborn, Kyle R., (309) 298-1577 kr-mayborn@wiu.edu, Western Illinois Univ; Gic
Mayefsky, Cheryl, 212-769-5390 cmayefsky@amnh.org, American Mus of Natural History; Zn
Mayer, Bernhard, (403) 220-5389 bmayer@ucalgary.ca, Univ of Calgary; CslGe
Mayer, Christine M., (419) 530-8377 christine.mayer@utoledo.edu, Univ of Toledo; Zn
Mayer, James R., (678) 839-4055 jmayer@westga.edu, Univ of West Georgia; Hw
Mayer, Jordan, 612-626-2672 jmayer@umn.edu, Univ of Minnesota; Zi
Mayer, Larry A., 603-862-2615 larry.mayer@unh.edu, Univ of New Hampshire; Gu
Mayer, Margaret, 928-724-6722 Dine' Coll; GeZn
Mayer, Ulrich, (604) 822-1539 umayer@eos.ubc.ca, Univ of British Columbia; Nm
Mayes, Melanie A., 865-574-7336 mayesma@ornl.gov, Univ of Tennessee, Knoxville; ClGeHw
Mayewski, Paul A., (207) 581-3019 paul.mayewski@maine.edu, Univ of Maine; Pe
Mayfield, Michael W., (828) 262-7058 mayfldmw@appstate.edu, Appalachian State Univ; Hs
Maynard, David, (909) 537-4321 dmaynard@csusb.edu, California State Univ, San Bernardino; ZnnZn
Maynard, J. Barry, maynarjb@uc.edu, Univ of Cincinnati; Cl
Mayo, Alan L., 801-422-2338 alan_mayo@byu.edu, Brigham Young Univ; Hw
Mayor, Shane D., 530-898-6337 sdmayor@csuchico.edu, California State Univ, Chico; AsZrAp
Mazariegos, Ruben, 956-665-2154 ruben.mazariegos@utrgv.edu, Univ of Texas, Rio Grande Valley; ZiYe
Mazer, James J., (630) 252-7362 Argonne; Cl
Mazza, Sarah, (413) 558-4396 smazza@smith.edu, Smith Coll; Gx
MAZZELLA, Antonio A., 070-675-5542 mazzella@unica.it, Universita di Cagliari; Gq
Mazzini, Pierro, pmazzini@vims.edu, William & Mary; Op
Mazzoli, Claudio, +390498279144 claudio.mazzoli@unipd.it, Univ degli Studi di Padova; Gx
Mazzullo, Salvatore J., (316) 978-3140 Wichita State Univ; Go
Mbatu, Richard, mbatu@usf.edu, Univ of South Florida; Zy
Mbila, Monday O., 256-372-4185 monday.mbila@aamu.edu, Alabama A&M Univ; Sd
Mbuh, Mbongowo J., 701-777-4587 mbongowo.mbuh@und.edu, Univ of North Dakota; ZrHs
Mc Mahon, Michelle, 281-765-7865 mmckubota@aol.com, Lonestar Coll - North Harris; Go

McAfee, Gerald B., (432) 335-6558 bmcafee@odessa.edu, Odessa Coll; Pi
McAfee, Stephanie , (775) 784-6999 smcafee@unr.edu, Univ of Nevada, Reno; Zy
McAllister, Arnold L., (506) 453-4804 Univ of New Brunswick; Em
McAndrews, John H., (416) 586-5609 Univ of Toronto; Pl
McArthur, John M., +44 020 7679 2376 j.mcarthur@ucl.ac.uk, Univ Coll London; ClHwPs
McArthur, Russell, (415) 338-1755 San Francisco State Univ; Zn
McBeth, Joyce M., joyce.mcbeth@usask.ca, Univ of Saskatchewan; PgGeSb
McBride, Earle F., (512) 471-1905 efmcbride@mail.utexas.edu, Univ of Texas, Austin; Gd
McBride, John H., (801) 422-5219 john_mcbride@byu.edu, Brigham Young Univ; YeGoe
McBride, Randolph, (703) 993-1642 rmcbride@gmu.edu, George Mason Univ; On
McBride, Raymond G., (519) 824-4120 (Ext. 52492) rmcbride@uoguelph.ca, Univ of Guelph; So
McCabe, Peter, peter.mccabe@adelaide.edu.au, Univ of Adelaide; Go
McCaffrey, Bill, +44(0) 113 34 36625 w.d.mccaffrey@leeds.ac.uk, Univ of Leeds; Gs
McCaffrey, Kenneth J., 0191 3742523 k.j.mccaffrey@durham.ac.uk, Durham Univ; Gc
McCair, Andrew, +44(0) 113 34 35219 a.m.mccaig@leeds.ac.uk, Univ of Leeds; Gc
McCall, Peter L., (216) 368-3676 plm4@case.edu, Case Western Reserve Univ; Pg
McCanta, Molly, 865-974-6003 mmccanta@utk.edu, Univ of Tennessee, Knoxville; GiXg
McCarter, Michael K., (801) 581-8603 k.mccarter@utah.edu, Univ of Utah; Nm
McCarthy, Christine, mccarthy@ldeo.columbia.edu, Columbia Univ; Ys
McCarthy, Daniel P., (905) 688-5550 (Ext. 3864) dmccarthy2@brocku.ca, Brock Univ; Gl
McCarthy, Francine G., (905) 688-5550 (Ext. 4286) francine@brocku.ca, Brock Univ; Pm
McCarthy, James T., 617-495-2330 jmccarthy@oeb.harvard.edu, Harvard Univ; Ob
McCarthy, Michael P., 206-685-2543 mccarthy@uw.edu, Univ of Washington; XyYx
McCarthy, Paul, (907) 474-6894 pjmccarthy@alaska.edu, Univ of Alaska, Fairbanks; GsSaPc
McCartney, Kevin, (207) 768-9482 kevin.mccartney@umpi.edu, Univ of Maine, Presque Isle; Pm
McCartney, M. Carol L., carol.mccartney@wisc.edu, Univ of Wisconsin, Madison, Division of Extension; RcGee
McCaslin, Bobby D., (505) 646-3405 New Mexico State Univ, Las Cruces; Sc
McCauley, Marlene, (336) 316-2236 mmccauley@guilford.edu, Guilford Coll; Cp
McCauley, Steven, 214-860-2398 smccauley@dcccd.edu, El Centro Coll - Dallas Comm Coll District; GgAm
McCausland, Phil, pmccausl@brocku.ca, Brock Univ; Gz
McClain, James S., (916) 752-7093 Univ of California, Davis; Yr
McClaughry, Jason, (541) 523-3133 Oregon Dept of Geology and Mineral Ind; Gg
McClay, Ken R., +44 1784 443618 k.mcclay@rhul.ac.uk, Royal Holloway Univ of London; GctGo
McClellan, Elizabeth, emcclellan@radford.edu, Radford Univ; GipGd
McClelland, James W., (361) 749-6756 jimm@utexas.edu, Univ of Texas, Austin; CgHsCs
McClelland, Lori, (775)784-4977 mcclella@unr.edu, Univ of Nevada, Reno; Ys
McClelland, William C., 319-335-1827 bill-mcclelland@uiowa.edu, Univ of Iowa; GtCcGc
mcClenaghan, Seán h., + 353 1 8961585 mcclens@tcd.ie, Trinity Coll; Eg
McColl, Kaighin, kmccoll@seas.harvard.edu, Harvard Univ; Hg
McCollor, Doug, doug.mccollor@bchydro.com, Univ of British Columbia; Asm
McCollough, Cherie, 361-825-3166 cherie.mccollough@tamucc.edu, Texas A&M Univ, Corpus Christi; Ze
McCollum, Linda B., (509) 359-7473 lmccollum@ewu.edu, Eastern Washington Univ; Pg
McConnaughhay, Mark, 845-431-8536 mcconn@sunydutchess.edu, Dutchess Comm Coll; ZgAmEo
McConnell, David A., (919) 515-0381 damcconn@ncsu.edu, North Carolina State Univ; ZeRcGc
McConnell, Joseph, (775) 673-7348 jmcconn@dri.edu, Univ of Nevada, Reno; Zn
McCorkle, Daniel C., (508) 289-2949 dmccorkle@whoi.edu, Woods Hole Oceanographic Inst; Cm
McCormack, John K., (775) 784-4518 mccormac@mines.unr.edu, Univ of Nevada, Reno; GzEm
McCormick, George R., 419-798-4812 erie-@msn.com, Univ of Iowa; GzEmCa
McCormick, Michael L., 315-859-4832 mmccormi@hamilton.edu, Hamilton Coll; Cb
McCoy, Edward L., (330) 263-3884 mccoy.13@osu.edu, Ohio State Univ; Sp
McCoy, Floyd W., 236-9115 fmccoy@hawaii.edu, Windward Comm Coll; GgaGu
McCoy, Kevin, 303-384-2632 kemccoy@mines.edu, Colorado Geo Survey; NgZiHw
Mccoy, Roger M., (801) 581-8218 Univ of Utah; Zr
McCoy, Scott W., 775-682-7205 scottmccoy@unr.edu, Univ of Nevada, Reno; GmNg
McCoy, Timothy J., (202) 633-2206 mccoyt@si.edu, Smithsonian Inst / Natl Mus of Natural History; Xmg
McCoy, Victoria, mccoyv@uwm.edu, Univ of Wisconsin, Milwaukee; Pb
McCoy, William D., (413) 545-1535 wdmccoy@geo.umass.edu, Univ of Massachusetts, Amherst; Zy
McCoy, Jr., Floyd W., (808) 235-7318 Univ of Hawai'i, Manoa; Gu
McCreary, Julian P., (808) 956-2216 jay@soest.hawaii.edu, Univ of Hawai'i, Manoa; Op
McCrink, Timothy P., (916) 324-2549 California Geo Survey; Ng
McCulloh, Richard P., (225) 578-5327 mccullo@lsu.edu, Louisiana State Univ; Gg
McCullough, Jr., Edgar J., (520) 621-6024 Univ of Arizona; Ge
McCurry, Michael O., mccumich@isu.edu, Idaho State Univ; Gzv
McDade, Ian C., (416)736-2100 #22859 mcdade@yorku.ca, York Univ; Zr
McDaniel, Paul A., (208) 885-7012 Univ of Idaho; Sd
McDermott, Christopher I., +44 (0) 131 650 5931 cmcdermo@staffmail.ed.ac.uk, Edinburgh Univ; Hg
McDermott, Frank, (+353) 1 716 2327 frank.mcdermott@ucd.ie, Univ Coll Dublin; GgCgZg
McDermott, Jeni, jamcdermott@stthomas.edu, Univ of Saint Thomas; GmHw
McDermott, Jill I., (610) 758-3683 jill.mcdermott@lehigh.edu, Lehigh Univ; Cm
McDermott, Thomas M., (585) 395-5718 SUNY, The Coll at Brockport; Zn
McDermott, Timothy, timmcder@montana.edu, Montana State Univ; Po
McDonald, Andrew M., 7056751151 x2266 amcdonald@laurentian.ca, Laurentian Univ, Sudbury; Gz
McDonald, Brenna, (573) 368-2163 brenna.mcdonald@dnr.mo.gov, Missouri Dept of Nat Res; Ge
McDonald, Catherine, (406) 496-4883 kmcdonald@mtech.edu, Montana Tech of The Univ of Montana; Grs
McDonald, Eric, (775) 673-7302 emcdonald@dri.edu, Desert Research Inst; Sd
McDonald, Gregory N., (801) 537-3383 gregmcdonald@utah.gov, Utah Geo Survey; Ng
McDonald, Jennifer, (612) 626-4067 jmhorton@umn.edu, Univ of Minnesota; Gl
McDonald, Kyle, 212-650-6452 kmcdonald@ccny.cuny.edu, Graduate Sch of the City Univ of New York; As
McDonald, Lynn, (505) 667-1582 lmcdonald@lanl.gov, Los Alamos; Zn
McDonough, William F., (301) 405-5561 mcdonoug@geol.umd.edu, Univ of Maryland; Cg
McDougall, Jim, 253-566-5060 jmcdougall@tacomacc.edu, Tacoma Comm Coll; Zg
McDowell, Patricia, (541) 346-4567 pmcd@uoregon.edu, Univ of Oregon; Gm
McDowell, Robin J., 770-274-5466 rmcdowell@gsu.edu, Georgia State Univ, Perimeter Coll, Dunwoody Campus; Gge
McDowell, Ronald, (304) 594-2331 mcdowell@geosrv.wvnet.edu, West Virginia Univ; GsPiCe
McDowell, William H., (603) 862-2249 bill.mcdowell@unh.edu, Univ of New Hampshire; HgRw
McDuff, Russell E., (360) 453-7877 mcduff@uw.edu, Univ of Washington; OcGu
McElroy, Anne E., (631) 632-8488 anne.mcelroy@stonybrook.edu, SUNY, Stony Brook; ObCb
McElroy, Brandon, (307) 766-3601 bmcelroy@uwyo.edu, Univ of Wyoming; GsmGr
McElroy, Michael B., (617) 495-9261 mbm@io.harvard.edu, Harvard Univ; As
McElroy, Tom, (416) 736-2100 x22113 tmcelroy@yorku.ca, York Univ; As
McElwain, Jenny, (312) 665-7635 jennifer.mcelwain@gmail.com, Field Mus of Natural History; Pb

McElwaine, Jim, james.mcelwaine@durham.ac.uk, Durham Univ; Yg
McElwee, Carl D., (785) 843-4164 cmcelwee@ku.edu, Univ of Kansas; HgYg
McEvoy, Jamie , (406) 994-4069 jamie.mcevoy@montana.edu, Montana State Univ; Ze
McEwen, Alfred, (520) 621-4573 mcewen@lpl.arizona.edu, Univ of Arizona; Xg
McFadden, Bruce J., (352) 846-2000 bmacfadd@flmnh.ufl.edu, Univ of Florida; Pv
McFadden, Jennifer, 717-361-1392 mcfaddenj@etown.edu, Elizabethtown Coll; ZnnZn
McFadden, Leslie M., (505) 277-6121 lmcfadnm@unm.edu, Univ of New Mexico; Sd
McFadden, Rory, 507-933-7335 rmcfadde@gustavus.edu, Gustavus Adolphus Coll; GcxZe
McFall, Katie D., k.mcfall@ucl.ac.uk, Univ Coll London; Eg
McFarlane, Norman, (250) 363-8227 norm.mcfarlane@ec.gc.ca, Univ of Victoria; As
McFiggans, Gordon, +44 0161 306-3954 gordon.b.mcfiggans@manchester.ac.uk, Univ of Manchester; As
McGahan, Donald, (254) 968-9701 mcgahan@tarleton.edu, Texas A&M Univ; Sdc
McGarvie, Dave, +44 (0) 131 549 7140 x 71140 dave.mcgarvie@open.ac.uk, Open Univ; GviCt
McGauley, Michael G., 305-237-2687 mmcgaule@mdc.edu, Miami Dade Coll (Kendall Campus); AmOp
McGeary, Susan, (302) 831-8174 smcgeary@udel.edu, Univ of Delaware; Ze
McGee, Amanda R., 216-231-4600 amcgee@cmnh.org, Cleveland Mus of Natural History; Pvg
McGee, David, (617) 324-3545 davidmcg@mit.edu, Massachusetts Inst of Tech; Cl
McGee, Tara, 780-492-3042 tmcgee@ualberta.ca, Univ of Alberta; Zn
McGehee, Thomas L., (512) 595-3590 kftlm00@tamuk.edu, Texas A&M Univ, Kingsville; Cl
McGhee, Jr., George R., 848-445-8523 mcghee@eps.rutgers.edu, Rutgers, The State Univ of New Jersey; PgqPi
McGill, George E., (413) 545-0140 Univ of Massachusetts, Amherst; Gc
McGill, Sally F., (909) 537-5347 smcgill@csusb.edu, California State Univ, San Bernardino; GtYdGm
McGinn, Chris, (919) 530-6269 cmcginn@nccu.edu, North Carolina Central Univ; Zgi
McGinnis, Lyle D., (630) 252-8722 Argonne; Yg
McGivern, Tiffany, (315)792-3134 Utica Coll; Geg
McGlathery, Karen, (804) 924-0558 kjm4k@virginia.edu, Univ of Virginia; Ob
McGlue, Michael M., 859-257-1952 michael.mcglue@uky.edu, Univ of Kentucky; GnrGm
McGoldrick, Peter J., 61 3 6226 7209 Univ of Tasmania; Ce
Mcgowan, Alistair, +4401413305449 alistair.mcgowan@glasgow.ac.uk, Univ of Glasgow; Pg
McGowan, Eileen, 413-586-8305 emcgowan@geo.umass.edu, Univ of Massachusetts, Amherst; Xg
McGowan, John A., (858) 534-2074 jmcgowan@ucsd.edu, Univ of California, San Diego; Ob
McGrail, Bernard P., (509) 376-9193 pete.mcgrail@pnl.gov, Pacific Northwest; Cg
McGrath, Andrew, andrew.mcgrath@flinders.edu.au, Flinders Univ; ZrAs
McGrath, Daniel, daniel.mcgrath@colostate.edu, Colorado State Univ; GlZrYu
McGrath, Steve F., 406-496-4767 smcgrath@mtech.edu, Montana Tech of The Univ of Montana; CaEgCe
McGraw, Maureen A., (505) 665-8128 mcgraw@lanl.gov, Los Alamos; Zn
McGregor, Kent M., (940) 565-2380 kent.mcgregor@unt.edu, Univ of North Texas; AsZg
McGregor, Stuart W., 205-247-3629 smcgregor@gsa.state.al.us, Geo Survey of Alabama; Hs
McGrew, Allen J., (937) 229-5638 amcgrew1@udayton.edu, Univ of Dayton; GtcGp
McGuire, Jeffrey J., (508) 289-3290 jmcguire@whoi.edu, Woods Hole Oceanographic Inst; Ys
McGwire, Kenneth, 775.673.7324 ken.mcgwire@dri.edu, Univ of Nevada, Reno; Zr
McHale, Michael , mrmchale@buffalo.edu, SUNY, Buffalo; Hg
McHenry, Lindsay J., (414) 229-3951 lmchenry@uwm.edu, Univ of Wisconsin, Milwaukee; GzaXg
McHugh, Cecilia, (718) 997-3330 cecilia.mchugh@qc.cuny.edu, Queens Coll (CUNY); Ou
McHugh, Julia, 970-248-1993 jumchugh@coloradomesa.edu, Colorado Mesa Univ; Pgv
McIlrath, Judy, jmcilrath@usf.edu, Univ of South Florida; Gg
McIlroy, Duncan, (709) 864-6722 dmcilroy@mun.ca, Memorial Univ of Newfoundland; Gg
McIlvin, Matt, 508-289-2884 mmcilvin@whoi.edu, Woods Hole Oceanographic Inst; Oc
McInnes, Kevin J., (979) 845-5986 k-mcinnes@tamu.edu, Texas A&M Univ; Sp
McIntosh, Jennifer, 520-626-2282 mcintosh@hwr.arizona.edu, Univ of Arizona; Hy
McIntosh, Kirk D., (512) 471-0480 kirk@ig.utexas.edu, Univ of Texas, Austin; Yr
McIntosh, William C., (575) 835-5324 mcintosh@nmt.edu, New Mexico Tech; Cc
McIntyre, Andrew, (718) 997-3329 Queens Coll (CUNY); Pe
McIsaac, Gregory F., (217) 333-9411 Univ of Illinois, Urbana-Champaign; Hs
McKay, Jennifer L., 541-737-4054 mckay@coas.oregonstate.edu, Oregon State Univ; Cm
McKay, Larry D., 865-974-5332 lmckay@utk.edu, Univ of Tennessee, Knoxville; HyGe
McKay, Luke, luke.mckay@montana.edu, Montana State Univ; Po
McKay, Matthew P., (417) 836-5318 matthewmckay@missouristate.edu, Missouri State Univ; Gct
McKay, Nicholas, 928-523-1918 nicholas.mckay@nau.edu, Northern Arizona Univ; PcZo
McKay, Robert M., robert.mckay@dnr.iowa.gov, Iowa Dept of Nat Res; Gg
McKay III, E. Donald, (217) 356-7338 emckay@illinois.edu, Illinois State Geo Survey; GlsGr
McKean, Adam, (801) 537-3386 adammckean@utah.gov, Utah Geo Survey; Gmg
McKean , Rebecca, 920-403-3227 rebecca.mckean@snc.edu, Saint Norbert Coll; GdPgRh
McKee, James W., (414) 424-4460 mckee@athenet.net, Univ of Wisconsin, Oshkosh; Gr
McKeegan, Kevin D., (310) 825-3580 mckeegan@epss.ucla.edu, Univ of California, Los Angeles; XcCsc
McKenna, Thomas E., 302-831-8257 mckennat@udel.edu, Univ of Delaware; Hg
McKenney, Rosemary, 253-535-8726 mckennra@plu.edu, Pacific Lutheran Univ; Gm
McKenzie, Garry D., mckenzie.4@osu.edu, Ohio State Univ; GmeZe
McKenzie, Jeffrey M., 514-398-6767 jeffrey.mckenzie@mcgill.ca, McGill Univ; Hw
McKenzie, Phyllis, (202) 633-1860 mckenzie@si.edu, Smithsonian Inst / Natl Mus of Natural History; Zn
McKenzie, Ross, ross.mckenzie@agric.gov.ab.ca, Univ of Lethbridge; Zn
McKenzie, Scott C., (814) 824-2382 smckenzie@mercyhurst.edu, Mercyhurst Univ; PgXmZe
McKibben, Michael A., (951) 827-3444 michael.mckibben@ucr.edu, Univ of California, Riverside; Cg
McKinley, Galen, mckinley@ldeo.columbia.edu, Univ of Wisconsin, Madison; OcpOg
McKinney, Mac, 205-247-3549 mmckinney@gsa.state.al.us, Geo Survey of Alabama; Hws
McKinney, Michael L., (865) 974-2366 mmckinne@utk.edu, Univ of Tennessee, Knoxville; GePg
McKinnon, William B., (314) 935-5604 mckinnon@wustl.edu, Washington Univ in St. Louis; Xg
McKnight, Brian K., (920) 233-4595 briankmcknight@yahoo.com, Univ of Wisconsin, Oshkosh; Gdu
McLaskey, Greg C., gcm8@cornell.edu, Cornell Univ; YsNr
McLaughlin, Patrick, (812) 855-1350 pimclaug@iu.edu, Indiana Univ, Bloomington; Ps
McLaughlin, Peter P., 302-831-8263 ppmclau@udel.edu, Univ of Delaware; HgPm
McLaughlin, Richard, 361-825-2010 richard.mclaughlin@tamucc.edu, Texas A&M Univ, Corpus Christi; Zn
McLaughlin, Jr., Peter P., 302-831-2833 ppmclau@udel.edu, Univ of Delaware; Gr
McLaurin, Brett T., 570-389-4142 bmclauri@bloomu.edu, Bloomsburg Univ; GrsGd
McLean, Noah, noahmc@ku.edu, Univ of Kansas; CcGq
McLemore, Virginia, 575-835-5521 virginia.mclemore@nmt.edu, New Mexico Tech; Em
McLennan, Scott M., (631) 632-8194 scott.mclennan@stonybrook.edu, Stony Brook Univ; Cg
McLeod, Claire, 513-529-9662 mcleodcl@miamioh.edu, Miami Univ; Gx
McLeod, Clara, cpmcleod@wustl.edu, Washington Univ in St. Louis; Zn
McLeod, Samuel A., (213) 763-3325 smcleod@ref.usc.edu, Los Angeles

County Mus of Natural History; Pv
McManus, Dean A., (206) 543-0587 mcmanus@u.washington.edu, Univ of Washington; Ou
McManus, George B., 860-405-9164 george.mcmanus@uconn.edu, Univ of Connecticut; Ob
McManus, Jerry, jmcmanus@ldeo.columbia.edu, Columbia Univ; Cm
McManus, Margaret A., (831) 459-4736 Univ of California, Santa Cruz; On
McManus, Margaret Anne, (808) 956-8623 mamc@hawaii.edu, Univ of Hawai'i, Manoa; Op
McMechan, George A., (972) 883-2419 mcmec@utdallas.edu, Univ of Texas, Dallas; Ys
McMenamin, Mark, 413-538-2280 mmcmenam@mtholyoke.edu, Mount Holyoke Coll; PgGst
McMillan, Nancy J., 575-646-5000 nmcmilla@nmsu.edu, New Mexico State Univ, Las Cruces; Gi
McMillan, Robert S., (520) 621-6968 bob@lpl.arizona.edu, Univ of Arizona; ZnnZn
McMonagle, Julie, (570) 408-4604 julie.mcmonagle@wilkes.edu, Wilkes Univ; Gge
McMullin, David W., (902) 585-1276 david.mcmullin@acadiau.ca, Acadia Univ; Ze
McMurdie, Lynn A., (206) 685-9405 mcmurdie@atmos.washington.edu, Univ of Washington; As
McMurtry, Gary M., (808) 956-6858 garym@soest.hawaii.edu, Univ of Hawai'i, Manoa; CaGvu
McNair, Laurie A., (505) 665-3328 mcnair@lanl.gov, Los Alamos; Zn
McNamara, James P., jmcnamar@boisestate.edu, Boise State Univ; Hg
McNamara, Kelsey, 575-835-5418 kelsey.mcnamara@nmt.edu, New Mexico Tech; Gz
McNamara, Kenneth, +44 (0) 1223 333410 kmcn07@esc.cam.ac.uk, Univ of Cambridge; Pg
McNamee, Brittani D., (828) 350-4554 bmcnamee@unca.edu, Univ of North Carolina, Asheville; GzEgGe
McNaught, Mark A., (330) 829-8226 mcnaugma@mountunion.edu, Mount Union Coll; Gcg
McNeal, Karen S., (334) 844-4282 ksm0041@auburn.edu, Auburn Univ; ZeGeCm
McNeill, Donald F., (305) 284-3360 dmcneill@rsmas.miami.edu, Univ of Miami; GsrGu
McNeill, Lisa, +44 (0)23 80593640 lcmn@noc.soton.ac.uk, Univ of Southampton; GtYr
McNichol, Ann P., (508) 289-3394 amcnichol@whoi.edu, Woods Hole Oceanographic Inst; Oc
McNicol, Barbara, (403) 440-6175 bmcnicol@mtroyal.ca, Mount Royal Univ; Zg
McNicol, Gavin, gmcnicol@uic.edu, Univ of Illinois Chicago; Sbf
McNulty, Brendan A., (310) 243-3412 bmcnulty@csudh.edu, California State Univ, Dominguez Hills; Gc
McNutt, Steve, smcnutt@usf.edu, Univ of South Florida; Ys
McPhaden, Michael J., (206) 526-6783 mcphaden@pmel.noaa.gov, Univ of Washington; Og
McPherron, Robert L., (310) 825-1882 rmcpherr@igpp.ucla.edu, Univ of California, Los Angeles; Xy
McPherson, Brian, 801-581-5634 bmcpherson@egi.utah.edu, Univ of Utah; Ye
McPherson, Robert C., (707) 826-5828 rm4@humboldt.edu, Humboldt State Univ; Gt
McPhie, Jocelyn, 61 3 6226 2892 jocelyn.mcphie@utas.edu.au, Univ of Tasmania; Gv
McQuaid, Jim, +44(0) 113 34 36724 j.b.mcquaid@leeds.ac.uk, Univ of Leeds; As
McQuarrie, Nadine, (412) 624-8870 nmcq@pitt.edu, Univ of Pittsburgh; Gct
McRivette, Michael, 517-629-0276 mmcrivette@albion.edu, Albion Coll; GtZir
McRoberts, Christopher A., (607) 753-2925 mcroberts@cortland.edu, SUNY, Cortland; Pg
McSween, Jr., Harry Y., (865) 974-9805 mcsween@utk.edu, Univ of Tennessee, Knoxville; XcGi
McSweeney, Kevin, (608) 262-0331 kmcsween@wisc.edu, Univ of Wisconsin, Madison; SdaSf
McWethy, David B., (406) 994-6915 dmcwethy@montana.edu, Montana State Univ; PeSb
McWilliams, James, (310) 206-2829 jcm@atmos.ucla.edu, Univ of California, Los Angeles; OgAs
McWilliams, Robert G., mcwillrg@miamioh.edu, Miami Univ; Ps
Mead, James I., mead@etsu.edu, Northern Arizona Univ; Pg
Meade, Brendan, 617-495-8921 meade@fas.harvard.edu, Harvard Univ; Yg
Meador, Cindy D., 806-651-2582 cmeador@wtamu.edu, West Texas A&M Univ; Zg
Meadows, Guy A., 906/487-1106 gmeadows@mtu.edu, Michigan Technological Univ; Gu
Meadows, Wayne R., (505) 665-0291 wmeadows@lanl.gov, Los Alamos; Yg
Means, Guy H., (850) 617-0312 guy.means@floridadep.gov, Florida Geo Survey; PgGe
Measures, Christopher, (808) 956-8693 chrism@soest.hawaii.edu, Univ of Hawai'i, Manoa; Oc
Measures, Elizabeth A., (432) 837-8117 measures@sulross.edu, Sul Ross State Univ; Gdq
Mechoso, Carlos R., (310) 825-3057 mechoso@atmos.ucla.edu, Univ of California, Los Angeles; As
Meckel, Timothy A., 512-471-4306 tip.meckel@beg.utexas.edu, Univ of Texas, Austin; Gr
Mecklenburgh, Julian, +440161 275-3821 julian.mecklenburgh@manchester.ac.uk, Univ of Manchester; Gc
Medaris, Jr., Levi G., (608) 833-4258 medaris@geology.wisc.edu, Univ of Wisconsin, Madison; GipSa
Meddaugh, W. S., (940) 397-4469 scott.meddaugh@msutexas.edu, Midwestern State Univ; GoEgCg
Medina, Cristian R., 855-9992 crmedina@indiana.edu, Indiana Univ, Bloomington; GoeHw
Medina Elizalde, Martin A., (334) 844-4966 mam0199@auburn.edu, Auburn Univ; PeHyCl
Medlin, Peggy, (931) 372-3121 pmedlin@tntech.edu, Tennessee Tech Univ; Zn
Meduniæ, Gordana, +38514605909 gpavlovi@inet.hr, Univ of Zagreb; CgScGe
Meehan, Kimberly, (716)645-4293 kcmeehan@buffalo.edu, SUNY, Buffalo; Pe
Meentemeyer, Ross, 704-687-5944 rkmeente@email.uncc.edu, Univ of North Carolina, Charlotte; Zi
Meere, Pat, +353 21 4903056 p.meere@ucc.ie, Univ Coll Cork; Gc
Meert, Joseph G., (352) 392-2231 jmeert@ufl.edu, Univ of Florida; YgGt
Meesters, Antoon, +31 20 59 87362 a.g.c.a.meesters@vu.nl, Vrije Univit Amsterdam; AmGq
Megdal, Sharon B., (520) 792-9591 smegdal@ag.arizona.edu, Univ of Arizona; Hg
Mehes, Roxane J., (705) 673-6575 rmehes@laurentian.ca, Laurentian Univ, Sudbury; Zn
Mehra, Aradhana, +44 01332 591133 a.mehra@derby.ac.uk, Univ of Derby; Co
Mehrtens, Charlotte J., (802) 656-0243 cmehrten@uvm.edu, Univ of Vermont; Gs
Mei, Shenghua, 612-626-0572 meixx002@umn.edu, Univ of Minnesota, Twin Cities; Yx
Meigs, Andrew J., (541) 737-1214 meigsa@geo.oregonstate.edu, Oregon State Univ; Gc
Meindl, Christopher, cmeindl@usf.edu, Univ of South Florida; Hg
Meinhold, Guido, +44 1782 733617 g.meinhold@keele.ac.uk, Keele Univ; GdCa
Meisner, Caroline B., (775) 340-7341 caroline.meisner@gbcnv.edu, Great Basin Coll; ZgSo
Meister, Paul A., (309) 438-7479 pameist@ilstu.edu, Illinois State Univ; GgHgZn
Meisterernst, Götz, 089/2180 4356 goetz.meisterernst@lrz.uni-muenchen.de, Ludwig-Maximilians-Univ Muenchen; Gz
Meitzen, Kimberly, (512) 245-2170 Texas State Univ; HsGmZy
Meixner, Thomas, 520-626-1532 tmeixner@email.arizona.edu, Univ of Arizona; CgRw
Mekik, Figen A., (616) 331-3020 mekikf@gvsu.edu, Grand Valley State Univ; GuCm
Meko, David M., (520) 621-3457 dmeko@ltrr.arizona.edu, Univ of Arizona; Hq
Melas, Faye, (212) 772-5265 ffmelas@yahoo.com, Hunter Coll (CUNY); Gs
Melbourne, Timothy I., (509) 963-2799 tim@geology.cwu.edu, Central Washington Univ; Yd
Melcher, Frank, frank.melcher@unileoben.ac.at, Univ Leoben; GgEgm
Melchin, Michael, (902) 867-5177 Dalhousie Univ; Pg
Melchiorre, Erik, (909) 537-7754 emelch@csusb.edu, California State Univ, San Bernardino; EmHwCl
Meldahl, Keith H., (760) 757-2121, x6412 kmeldahl@miracosta.edu, MiraCosta Coll; Gg
Meléndez, Nieves, nievesml@ucm.es, Univ Compl de Madrid; Ggs
Melentijevic, Svetlana, svmelent@ucm.es, Univ Compl de Madrid; Nx
Melgar, Diego, (541) 346-3488 dmelgar@uoregon.edu, Univ of Oregon; Yd
Melichar, Rostislav, +420 549 49 5812 melda@sci.muni.cz, Masaryk Univ; GtcGr
Melim, Leslie A., (309) 298-1377 la-melim@wiu.edu, Western Illinois Univ; Gd

Mellinger, David, 541-867-0372 david.mellinger@oregonstate.edu, Oregon State Univ; Cb
Mellor, George, (609) 258-6570 glmellor@princeton.edu, Princeton Univ; Op
Melosh, Henry J., (520) 621-2806 jmelosh@lpl.arizona.edu, Univ of Arizona; Gt
Meltzer, Anne S., (610) 758-3673 asm3@lehigh.edu, Lehigh Univ; Ys
Memeti, Valbone, (657) 278-2036 vmemeti@fullerton.edu, California State Univ, Fullerton; GivCa
Mendelson, Carl V., mendelsn@beloit.edu, Beloit Coll; PmGs
Mendelssohn, Irving A., (225) 388-6425 Louisiana State Univ; Ob
Mendoza, Blanca, blanca@ciencias.unam.mx, Univ Nacional Auto de Mexico; Zn
Mendoza, Carl, (780) 492-2664 carl.mendoza@ualberta.ca, Univ of Alberta; Hw
Mendoza, Luis H., lmendoza@cicese.mx, Centro de Investigación Científica y de Educación Superior de Ensenada; Ne
Menegon, Luca, +44 1752 584931 luca.menegon@plymouth.ac.uk, Univ of Plymouth; Gct
Meng, Jin, (212) 496-3337 American Mus of Natural History; Pv
Meng, Lingsen, 310-825-1885 meng@epss.ucla.edu, Univ of California, Los Angeles; Ys
Meng, Qingmin, 662-268-1032 Ext 240 Mississippi State Univ; Zig
Meng, Yue (Olivia), 609-258-4101 om3193@princeton.edu, Princeton Univ; As
Mengel, David, (785) 532-2166 dmengel@ksu.edu, Kansas State Univ; So
Menicucci, Anthony, amenicucci@usf.edu, Univ of South Florida; Cs
Menke, William H., (845) 365-8438 menke@ldeo.columbia.edu, Columbia Univ; YsGvq
Menking, Kirsten M., 845-437-5545 kimenking@vassar.edu, Vassar Coll; GmPeGc
Menninga, Clarence, menn@calvin.edu, Calvin Coll; CcGg
Menold, Carrie A., 517-629-0312 cmenold@albion.edu, Albion Coll; GpCsGt
Mensah, Emmanuel, +233 24 4186 193 emmamensah6@yahoo.com, Kwame Nkrumah Univ of Sci and Tech; EgGg
Menschik, Bettina, +49 89 289 25822 sellmeier@tum.de, Technical Univ of Munich; NgrNx
Mensing, Scott A., (775) 784-6346 smensing@unr.edu, Univ of Nevada, Reno; Zy
Mensing, Teresa, (740) 725-6234 mensing.1@osu.edu, Ohio State Univ; Cg
Menuge, Julian F., (+353) 1 716 2141 j.f.menuge@ucd.ie, Univ Coll Dublin; EmCcs
Menzies, John, (905) 688-5550 x3865 jmenzies@brocku.ca, Brock Univ; GlmGs
Mercer, Andrew, 662-268-1032 Ext 231 mercer@gri.msstate.edu, Mississippi State Univ; Ams
Mercier, Michael , (905) 525-9140 (Ext. 27597) mercieme@mcmaster.ca, McMaster Univ; Zu
Merck, Jr., John W., (301) 405-2808 jmerck@umd.edu, Univ of Maryland; Pv
Meredith, Philip, +44 020 7679 37824 p.meredith@ucl.ac.uk, Univ Coll London; Nr
Meretsky, Vicky J., (812) 855-5971 meretsky@indiana.edu, Indiana Univ, Bloomington; Sf
Mereu, Robert F., (519) 661-3605 a424@uwo.ca, Western Univ; Ys
Merguerian, Charles M., 516 463-5567 geocmm@hofstra.edu, Hofstra Univ; Gc
Merifield, Paul M., (310) 794-5019 pmerifie@ucla.edu, Univ of California, Los Angeles; Ng
Merino, Enrique, (812) 855-5088 merino@indiana.edu, Indiana Univ, Bloomington; Cl
Merkel, Timo Casjen, 089/2180 4337 casjen.merkel@lrz.uni-muenchen.de, Ludwig-Maximilians-Univ Muenchen; Gz
Merlis, Timothy, 514-398-3140 timothy.merlis@mcgill.ca, McGill Univ; Asm
Merriam, James B., (306) 966-5716 jim.merriam@usask.ca, Univ of Saskatchewan; Yg
Merrifield, Mark A., (808) 956-6161 markm@soest.hawaii.edu, Univ of Hawai'i, Manoa; Op
Merrill, Glen K., (713) 221-8168 merrillg@uhd.edu, Univ of Houston Downtown; Pi
Merrill, John T., (401) 874-6715 jmerrill@boreas.gso.uri.edu, Univ of Rhode Island; Oc
Merrill, Ronald T., (206) 543-6686 Univ of Washington; Ym
Merritt, Andrew, +44 1752 584702 andrew.merritt@plymouth.ac.uk, Univ of Plymouth; Ng
Merritts, Dorothy J., (717) 291-4398 dorothy.merritts@fandm.edu, Franklin and Marshall Coll; Gm
Mertzman, Stanley A., (717) 291-3818 stan.mertzman@fandm.edu, Franklin and Marshall Coll; GizZm
Meskhidze, Nicholas, 919-515-7243 nicholas_meskhidze@ncsu.edu, North Carolina State Univ; As
Messina, Paula, (408) 924-5027 paula.messina@sjsu.edu, San Jose State Univ; ZeGm
Metaxas, Anna, (902) 494-3021 metaxas@dal.ca, Dalhousie Univ; Ob
Metcalf, Kathryn, 657-278-3942 kametcalf@fullerton.edu, California State Univ, Fullerton; Gt
Metcalf, Meredith, (860) 465-4370 metcalfm@easternct.edu, Eastern Connecticut State Univ; ZirHw
Metcalfe, Ian, +61-2-67726297 imetcal2@une.edu.au, Univ of New England; PmGtg
Metesh, John J., (406) 496-4159 jmetesh@mtech.edu, Montana Tech of The Univ of Montana; HwCa
Metz, Cheyl L., 979 209-7461 cl.metz@blinn.edu, Blinn Coll; GgOg
Metz, Nicholas, (315) 781-3615 metz@hws.edu, Hobart & William Smith Colls; As
Metz, Robert, (908) 737-3687 rmetz@kean.edu, Kean Univ; Gr
Metzger, Ellen P., (408) 924-5048 ellen.metzger@sjsu.edu, San Jose State Univ; GpZe
Metzger, Marc J., +44 (0) 131 651 4446 mmetzger@staffmail.ed.ac.uk, Edinburgh Univ; Ge
Metzger, Ronald A., (541) 888-7216 rmetzger@socc.edu, Southwestern Oregon Comm Coll; PmZePs
Metzler, Christopher V., (760) 944-4449, x7738 cmetzler@miracosta.edu, MiraCosta Coll; Gg
Mey, Jacob, 718-368-5770 jacob.mey@kbcc.cuny.edu, Graduate Sch of the City Univ of New York; Gg
Meyer, Brian, bmeyer2@gsu.edu, Georgia State Univ; GesHw
Meyer, Christopher, meyerc@si.edu, Smithsonian Inst / Natl Mus of Natural History; ZnPi
Meyer, David L., (513) 556-4530 david.meyer@uc.edu, Univ of Cincinnati; Pi
Meyer, Florian, flmeyer@ucsd.edu, Univ of California, San Diego; Op
Meyer, Franz J., 907-474-7767 fmeyer@gi.alaska.edu, Univ of Alaska, Fairbanks; Zr
Meyer, Gary, (612) 626-1741 meyer015@umn.edu, Univ of Minnesota; Gl
Meyer, Grant A., (505) 277-4204 gmeyer@unm.edu, Univ of New Mexico; Gm
Meyer, Jeffrey W., (805) 965-0531 (Ext. 4270) meyerj@sbcc.edu, Santa Barbara City Coll; GgzGx
Meyer, Jessica R., (319) 335-1831 jessica-meyer@uiowa.edu, Univ of Iowa; Hw
Meyer, Judith, (417) 836-5604 judithmeyer@missouristate.edu, Missouri State Univ; Zn
Meyer, Lewis, +44 01326 253766 l.h.i.meyer@exeter.ac.uk, Exeter Univ; Nmr
Meyer, Rebecca A., (812) 855-2687 reameyer@indiana.edu, Indiana Univ, Bloomington; Ec
Meyer, Rudi, (403) 210-7848 rmeyer@ucalgary.ca, Univ of Calgary; Gs
Meyer, Steven J., (920) 465-5022 meyers@uwgb.edu, Univ of Wisconsin, Green Bay; AsmZg
Meyer, W. Craig, (818) 710-4241 meyerwc@piercecollege.edu, Los Angeles Pierce Coll; GeuPm
Meyer, William T., (630) 969-6586 meyer@iastate.edu, Argonne; Cg
Meyer Dombard, DArcy, 312-996-2423 drmd@uic.edu, Univ of Illinois Chicago; PoXb
Meyer-Gutbrod, Erin, 803-777-3552 emgutbrod@seoe.sc.edu, Univ of South Carolina; Ob
Meyers, Jamie A., jmeyers@winona.edu, Winona State Univ; Gsd
Meyers, Philip A., (734) 764-0597 pameyers@umich.edu, Univ of Michigan; CoGnu
Meyers, Stephen R., 608-890-2574 smeyers@geology.wisc.edu, Univ of Wisconsin, Madison; PcOoGs
Meyerson, Rohana, 610-330-5193 meyersor@lafayette.edu, Lafayette Coll; Zn
Meylan, Anne, (727) 896-8626 Univ of South Florida; Ob
Meylan, Maurice A., (601) 266-4527 mmeylan@otr.usm.edu, Univ of Southern Mississippi; Gu
Meyzen, Christine M., +390498279153 christine.meyzen@unipd.it, Univ degli Studi di Padova; Cg
Mezősi, Gábor, +3662544156 mezosi@geo.u-szeged.hu, Univesity of Szeged; ZyGmRn
Mezga, Aleksandar, +38514606116 amezga@geol.pmf.hr, Univ of Zagreb; Pg
Mezger, Jochen E., (907) 474-7809 jemezger@alaska.edu, Univ of Alaska, Fairbanks; GcpCc
Miall, Andrew D., (416) 978-8841 miall@es.utoronto.ca, Univ of Toronto;

FACULTY INDEX – M

Gso
Miao, Xin, (417) 836-5173 xinmiao@missouristate.edu, Missouri State Univ; Zri
Miao, Yuxin, 612-625-8101 ymiao@umn.edu, Univ of Minnesota, Twin Cities; ScZf
Micallef, Aaron, +356 23403662 aaron.micallef@um.edu.mt, Univ of Malta; GumOu
Micander, Rachel, 775-682-6351 rmicander@unr.edu, Univ of Nevada, Reno; Zie
Michael, Holly A., (302) 831-4197 hmichael@udel.edu, Univ of Delaware; Hwq
Michael, Peter J., (918) 631-3017 pjm@utulsa.edu, Univ of Tulsa; GivGz
Michaels, Patrick J., (804) 924-0549 pmichaels@cato.org, Univ of Virginia; As
Michaels, Paul, paulmichaels@boisestate.edu , Boise State Univ; Ne
Michalak, Melanie, 707-826-4978 mjm367@humboldt.edu, Humboldt State Univ; GtCcGc
Michalski, Greg, (765) 494-3704 gmichals@purdue.edu, Purdue Univ; CsGeAs
Michaud, Jene D., 808-974-7411 jene@hawaii.edu, Univ of Hawai'i, Hilo; HqwGm
Michaud, Yves, (418) 654-2647 Univ du Quebec; Gm
Michel, F. M., 540-231-3299 mfrede2@vt.edu, Virginia Tech; ZaGz
Michel, Fred A., fmichel@ccs.carleton.ca, Carleton Univ; Hw
Michel, Gero, gmichel3@uwo.ca, Western Univ; Ys
Michel, Jacqueline, 504-280-6325 jmichel@uno.edu, Univ of New Orleans; Cg
Michel, Lauren, (931) 372-3188 lmichel@tntech.edu, Tennessee Tech Univ; PcSaCl
Michel, Shirley, 419-530-5058 shirley.michel@utoledo.edu, Univ of Toledo; Zn
Michel, Suzanne, 6196447454 x3028 suzanne.michel@gcccd.edu, Cuyamaca Coll; Gg
Michelfelder, Gary, (417) 836-3171 garymichelfelder@missouristate.edu, Missouri State Univ; GviCg
Mickelson, Andrew M., 901-678-4505 amicklsn@memphis.edu, Univ of Memphis; Ga
Mickelson, David M., (608) 262-7863 mickelson@geology.wisc.edu, Univ of Wisconsin, Madison; Glm
Mickus, Kevin L., (417) 836-6375 kevinmickus@missouristate.edu, Missouri State Univ; GtYv
Miclaus, Crina G., 00402324095 crina_miclaus@yahoo.co.uk, Alexandru Ioan Cuza; Gsr
Middlemiss, Lucie, +44(0) 113 34 35246 l.k.middlemiss@leeds.ac.uk, Univ of Leeds; Zn
Middleton, Carrie A., (303) 293-5019 cmiddleton@osmre.gov, Dept of Interior Office of Surface Mining Reclamation and Enforcement; ZrGfZi
Middleton, Larry T., (928) 523-2429 larry.middleton@nau.edu, Northern Arizona Univ; Gs
Mies, Jonathan W., (423) 425-4606 jonathan-mies@utc.edu, Univ of Tennessee, Chattanooga; GctHg
Mignery, Jill, 513-529-03225 henryjm@miamioh.edu, Miami Univ; Zge
Miguel, Carlos, (598) 25251552 cmiguel@fcien.edu.uy, Univ de la Republica Oriental del Uruguay (UDELAR); ZirSf
Mikhaltsevitch, Vassily, +61 8 9266-4976 v.mikhaltsevitch@curtin.edu.au, Curtin Univ; Ye
Milam, Keith A., (740) 593-1106 milamk@ohio.edu, Ohio Univ; Xg
Milan, Luke, +61267732019 lmilan@une.edu.au, Univ of New England; GtcGp
Miles, Andrew, andrew.miles@kingston.ac.uk, Kingston Univ; Gz
Miles, Randall J., 573-882-6607 Univ of Missouri; Sd
Milewski, Adam, (706) 542-2652 milewski@uga.edu, Univ of Georgia; Hw
Militzer, Burkhard, militzer@seismo.berkeley.edu, Univ of California, Berkeley; Gy
Miljevic, Branka, +61 7 3138 3827 b.miljevic@qut.edu.au, Queensland Univ of Tech; Acs
Milkereit, Bernd, (416) 978-2466 bm@physics.utoronto.ca, Univ of Toronto; Ye
Millan, Christina, 614-292-0863 millan.2@osu.edu, Ohio State Univ; Ggc
Millen, Timothy M., 847-697-1000 tmillen@elgin.edu, Elgin Comm Coll; Gg
Miller, Arnold I., (513) 556-4022 arnold.miller@uc.edu, Univ of Cincinnati; Pi
Miller, Barry W., (865) 594-5599 barry.miller@tn.gov, Tennessee Geo Survey; EcZiGg
Miller, Brent, (979) 458-3671 bvmiller@geo.tamu.edu, Texas A&M Univ; Cc
Miller, Calvin F., 615-322-2232 calvin.miller@vanderbilt.edu, Vanderbilt Univ; Gi
Miller, Charles M., (505) 667-8415 Los Alamos; Cc
Miller, David S., (630) 252-7191 Argonne; Ge
Miller, Doug, 828-232-5158 dmiller@unca.edu, Univ of North Carolina, Asheville; AsZnn
Miller, Douglas C., (302) 645-4277 dmiller@udel.edu, Univ of Delaware; Ob
Miller, Elizabeth L., (650) 723-1149 miller@pangea.stanford.edu, Stanford Univ; Gc
Miller, Geoffrey G., (506) 643-2361 Los Alamos; Cc
Miller, Gerald A., (515) 291-3442 soil@iastate.edu, Iowa State Univ of Sci & Tech; SdoGm
Miller, Gifford H., 303-492-6962 gmiller@colorado.edu, Univ of Colorado; GlCc
Miller, Harvey J., (801) 585-3972 miller.81@osu.edu, Ohio State Univ; Zi
Miller, Heather R., heather.miller@austincc.edu, Austin Comm Coll District; Zeg
Miller, Hugh, (303) 273-3558 hbmiller@mines.edu, Colorado Sch of Mines; Nmg
Miller, Ian, 303-370-8351 ian.miller@dmns.org, Denver Mus of Nature & Sci; PbGg
Miller, James D., mille066@d.umn.edu, Univ of Minnesota Duluth; GiEd
Miller, Jerry R., (828) 227-2269 jmiller@wcu.edu, Western Carolina Univ; GmeGf
Miller, John, (416)736-5245 jrmiller@yorku.ca, York Univ; Zr
Miller, Jonathan S., (408) 924-5015 jonathan.miller@sjsu.edu, San Jose State Univ; GiCc
Miller, Joshua H., (513) 556-6704 joshua.miller@uc.edu, Univ of Cincinnati; PgvPe
Miller, Judy, jmiller264@monroecc.edu, Monroe Comm Coll; Zn
Miller, Keith B., (785) 532-2250 kbmill@ksu.edu, Kansas State Univ; Pe
Miller, Kenneth G., (848) 445-3622 kgm@eps.rutgers.edu, Rutgers, The State Univ of New Jersey; GurPm
Miller, Mark, 601-266-4729 m.m.miller@usm.edu, Univ of Southern Mississippi; Zn
Miller, Marli G., (541) 346-4410 millerm@uoregon.edu, Univ of Oregon; Gc
Miller, Marvin R., (406) 496-4155 mmiller@mtech.edu, Montana Tech of The Univ of Montana; Hw
Miller, Michael B., (225) 388-3412 byron@lgs.bri.lsu.edu, Louisiana State Univ; Go
Miller, Molly F., 615-322-3528 molly.miller@vanderbilt.edu, Vanderbilt Univ; PeGs
Miller, Murray H., (519) 824-4120 (Ext. 53758) jmmiller7@sympatico.ca, Univ of Guelph; So
Miller, Nari, nmiller9@csustan.edu, California State Univ, Stanislaus; CgYg
Miller, Phillip, 575-835-5487 phil.miller@nmt.edu, New Mexico Tech; Zi
Miller, Randall F., (506) 643-2361 Univ of New Brunswick; Pg
Miller, Raymond M., (630) 252-3395 rmmiller@anl.gov, Argonne; Sb
Miller, Richard D., (785) 864-3965 rmiller@kgs.ku.edu, Univ of Kansas; Ye
Miller, Robert B., (408) 924-5025 robert.b.miller@sjsu.edu, San Jose State Univ; Gct
Miller, Ronald L., (212) 678-5577 rmiller@giss.nasa.gov, Columbia Univ; As
Miller, Steven F., (506) 643-2361 Argonne; Gg
Miller, Ted R., 605-677-6867 ted.miller@usd.edu, South Dakota Dept of Environment and Nat Res; Zn
Miller, Wade E., 801 422-2321 wem@geology.byu.edu, Brigham Young Univ; Pv
Miller, William C., (707) 826-3110 wm1@humboldt.edu, Humboldt State Univ; Pi
Miller-Camp, Jessica, jessmc@iu.edu, Indiana Univ, Bloomington; Pv
Miller-Hicks, Bryan, bryan.miller-hicks@gcccd.edu, Cuyamaca Coll; Gg
Millet, Dylan B., 612-626-3259 dbm@umn.edu, Univ of Minnesota, Twin Cities; As
Millette, Thomas L., (413) 538-2813 tmillett@mtholyoke.edu, Mount Holyoke Coll; ZriGm
Milligan, Mark, (801) 537-3326 markmilligan@utah.gov, Utah Geo Survey; Ze
Milligan, Richard, 404 413-5778 rmilligan@gsu.edu, Georgia State Univ; Zn
Milligan, Timothy, (902) 426-3273 milligant@dfo-mpo.gc.ca, Dalhousie Univ; Gs
Milliken, Kitty L., (512) 471-6082 kittym@mail.utexas.edu, Univ of Texas, Austin; Gd
Millington, Andrew, andrew.millington@flinders.edu.au, Flinders Univ; ZuyZr
Mills, Aaron L., (804) 924-0564 alm7d@virginia.edu, Univ of Virginia; So
Mills, Eric L., (902) 471-2016 e.mills@dal.ca, Dalhousie Univ; ObZnn
Mills, James G., (765) 658-4669 jmills@depauw.edu, DePauw Univ; Gi
Mills, Rachel A., +44 (0)23 80592678 rachel.mills@soton.ac.uk, Univ of Southampton; Cm
Mills, Stephanie, +44 020 8417 2950 s.mills@kingston.ac.uk, Kingston

Univ; Gl
Mills, Suzanne, (905) 525-9140 smills@mcmaster.ca, McMaster Univ; Zn
Milne, Glenn A., gamilne@uottawa.ca, Univ of Ottawa; YgAsZg
Milner, Jennifer, (309) 794-7318 jennifermilner@augustana.edu, Augustana Coll; Zn
Milner, Lloyd R., (225) 578-3410 lmilne1@lsu.edu, Louisiana State Univ; Gg
Milroy, Scott P., (228) 688-7128 scott.milroy@usm.edu, Univ of Southern Mississippi; Ob
Min, Dong-Ha, 512-475-9290 dongha@austin.utexas.edu, Univ of Texas, Austin; Og
Min, Doo-Hong, dmin@ksu.edu, Kansas State Univ; So
Min, Kyoungwon, (352) 392-2720 kmin@ufl.edu, Univ of Florida; Cc
Minarik, William G., 514-398-2596 william.minarik@mcgill.ca, McGill Univ; Cp
Minchew, Brent, (617) 324-3704 minchew@mit.edu, Massachusetts Inst of Tech; Yg
Miner, James J., 217-244-5786 miner@illinois.edu, Illinois State Geo Survey; Gg
Ming, Yi, 617-552-3763 yi.ming@bc.edu, Boston Coll; ApZo
Minium, Deborah, (425) 564-5120 deborah.minium@bellevuecollege.edu, Bellevue Coll; Gg
Minkowski, Valeska, (864) 294-3366 valeska.minkowski@furman.edu, Furman Univ; Zn
Minnaar, Hendrik, +27 (0)51 401 2372 minnaarh@ufs.ac.za, Univ of the Free State; Gc
Minnich, Richard A., (951) 827-5515 richard.minnich@ucr.edu, Univ of California, Riverside; Zy
Minor, Timothy B., (775) 673-7477 tminor@dri.edu, Desert Research Inst; Zr
Minshull, Timothy A., +44 (0)23 80596569 tmin@noc.soton.ac.uk, Univ of Southampton; Yr
Minster, J. Bernard H., (858) 945-0693 jbminster@ucsd.edu, Univ of California, San Diego; Ys
Minter, Nicholas, +44 023 92 842288 nic.minter@port.ac.uk, Univ of Portsmouth; Gs
Minton, David, (765) 494-3292 daminton@purdue.edu, Purdue Univ; Xg
Minzoni, Marcello, (205)348-0768 marcello.minzoni@ua.edu, Univ of Alabama; GosEo
Miot da Silva, Graziela, graziela.miotdasilva@flinders.edu.au, Flinders Univ; Gug
Miranda, Elena A., 818-677-4671 elena.miranda@csun.edu, California State Univ, Northridge; Gc
Mirnejad, Hassan , 513-529-3216 mirnejh@miamioh.edu, Miami Univ; GxCg
Mishler, Brent, (510) 642-6810 bmishler@berkeley.edu, Univ of California, Berkeley; Pe
Mishra, Umakant, 630-252-1108 umishra@anl.gov, Argonne; So
Misner, Tamara, 814-732-1352 tmisner@edinboro.edu, Edinboro Univ of Pennsylvania; GmHs
Misra, Kula C., (865) 974-2366 kmisra@utk.edu, Univ of Tennessee, Knoxville; Eg
Misra, Saumitra, (031) 260-2800 misras@ukzn.ac.za, Univ of KwaZulu-Natal; GiCgu
Misra, Siddharth, 405-325-6787 misra@ou.edu, Univ of Oklahoma; Np
Mitasova, Helena, 919-513-1327 helena_mitasova@ncsu.edu, North Carolina State Univ; Zi
Mitchell, Charles E., (716) 645-4290 cem@buffalo.edu, SUNY, Buffalo; Pg
Mitchell, Jonathan, 310-825-2970 mitch@epss.ucla.edu, Univ of California, Los Angeles; AsXg
Mitchell, Neil C., neil.mitchell@manchester.ac.uk, Univ of Manchester; GumYg
Mitchell, Robert J., (360) 650-3591 robert.mitchell@wwu.edu, Western Washington Univ; Hw
Mitchell, Roger H., (807) 343-8010 Ext. 8287 roger.mitchell@lakeheadu.ca, Lakehead Univ; Gi
Mitchell, Simon F., 876-927-2728 geoggeol@uwimona.edu.jm, Univ of the West Indies Mona Campus; GsgGs
Mitchell, Tom, +44 (0)20 7679 7361 tom.mitchell@ucl.ac.uk, Univ Coll London; Gc
Mitchneck, Beth A., bethm@email.arizona.edu, Univ of Arizona; Zn
Mitchum, Gary T., (727) 553-3941 gmitchum@marine.usf.edu, Univ of South Florida; Op
Mitra, Chandana, (334) 844-4229 czm0033@auburn.edu, Auburn Univ; ZgAsZi
Mitra, Shankar, (405) 325-4462 smitra@ou.edu, Univ of Oklahoma; Gc
Mitra, Siddhartha, 252 328 6611 mitras@ecu.edu, East Carolina Univ; Co
Mitri, Hani, (514) 398-4755 McGill Univ; Nm
Mitrovica, Jerry X., 617-496-2732 jxm@eps.harvard.edu, Harvard Univ; Yg
Mittelstaedt, Eric L., (208) 885-2045 emittelstaedt@uidaho.edu, Univ of Idaho; Yr
Mitterer, Richard M., mitterer@utdallas.edu, Univ of Texas, Dallas; ColOc
Mitzman, Stephanie, stephanie.mitzman@tccd.edu, Tarrant County Coll, Southeast Campus; Gg
Mix, Alan C., 541-737-5212 amix@coas.oregonstate.edu, Oregon State Univ; Cs
Miyagi, Lowell, (801) 581-6619 lowell.miyagi@utah.edu, Univ of Utah; Gz
Miyares, Ines, (212) 772-5443 imiyares@hunter.cuny.edu, Hunter Coll (CUNY); Zen
Mizak, Connie, mizak@usf.edu, Univ of South Florida; Zy
Mnich, Marissa, mnichm@sonoma.edu, Sonoma State Univ; GvpGg
Moayyed, Mohsen -., +98 (413) 339 2616 moayyed@tabrizu.ac.ir, Univ of Tabriz; GigEg
Moazzen, Mohssen -., +98 (411) 339 2679 moazzen@tabrizu.ac.ir, Univ of Tabriz; GpCgGz
Moberly, Ralph, (808) 956-8765 ralph@soest.hawaii.edu, Univ of Hawai'i, Manoa; GutGs
Mobley, Curtis D., (206) 230-8166 Univ of Washington; Ob
Mock, R. Stephen, (406) 683-7261 steve.mock@umwestern.edu, Univ of Montana Western; Ca
Mock, Thomas, +44 (0)1603 59 2566 t.mock@uea.ac.uk, Univ of East Anglia; Pg
Mockler, Theodore, (505) 667-4318 mockler@lanl.gov, Los Alamos; Zn
Mode, William N., (920) 424-7004 mode@uwosh.edu, Univ of Wisconsin, Oshkosh; Gl
Modzelewski, Henryk, (604) 822-3591 hmodzelewski@eos.ubc.ca, Univ of British Columbia; As
Moe-Hoffman, Amy P., apm105@msstate.edu, Mississippi State Univ; PgZe
Moecher, David P., (859) 257-6939 moker@uky.edu, Univ of Kentucky; GxtCg
Moeglin, Thomas D., (417) 836-5800 Missouri State Univ; Ng
Moersch, Jeffery E., (865) 974-0371 jmoersch@utk.edu, Univ of Tennessee, Knoxville; XgZr
Moffett, James W., 213-740-5626 jmoffett@usc.edu, Univ of Southern California; OcCba
Mofjeld, Harold O., (206) 526-6819 mofjeld@pmel.noaa.gov, Univ of Washington; Op
Moghanloo, Rouzbeh, rouzbeh.gm@ou.edu, Univ of Oklahoma; Np
Mogilevskaya, Sonia, 612-625-4810 mogil003@umn.edu, Univ of Minnesota, Twin Cities; Ng
Mogk, David W., (406) 600-4071 mogk@montana.edu, Montana State Univ; GpZeGz
Mohammad Reza, Hosseinzadeh, +98 (411) 339 2697 Univ of Tabriz; Egm
Mohr, Marcus, 089/2180 4230 marcus.mohr@geophysik.uni-muenchen.de, Ludwig-Maximilians-Univ Muenchen; Yg
Mohrig, David, (512) 471-2282 mohrig@jsg.utexas.edu, Univ of Texas, Austin; GsmGr
Mojica, Kristina, (228) 688-3003 kristina.mojica@usm.edu, Univ of Southern Mississippi; Ob
Mojzsis, Stephen J., (303) 492-5014 stephen.mojzsis@colorado.edu, Univ of Colorado; XcCcGp
Moldwin, Mark B., (734) 647-3370 mmoldwin@umich.edu, Univ of Michigan; XpYz
Molina, Jean-Alex E., (651) 647-9865 jamolina@umn.edu, Univ of Minnesota, Twin Cities; SbCgSo
Molina, Mario J., (858) 534-1696 mjmolina@ucsd.edu, Univ of California, San Diego; As
Molinari, John E., (518) 442-4562 molinari@atmos.albany.edu, SUNY, Albany; As
Moll, Nancy E., 760-776-7272 nmoll@collegeofthedesert.edu, Coll of the Desert; Gge
Mollner, Daniel, (507) 933-7569 Gustavus Adolphus Coll; Zn
Molnár, Gábor, molnar@sas.elte.hu, Eotvos Lorand Univ; ZrYgZi
Molnar, Peter, 303-492-4936 peter.molnar@colorado.edu, Univ of Colorado; Gt
Molnar, Sheri, 519-661-2111, ext.87031 smolnar8@uwo.ca, Western Univ; Ys
Molnia, Bruce F., (703) 648-4120 bmolnia@usgs.gov, Duke Univ; ZrGlu
Molotch, Noah, noah.molotch@colorado.edu, Univ of Colorado; Hg
Momayez, Moe, (520) 621-6580 moe.momayez@arizona.edu, Univ of Arizona; NmYxNr
Momen, Nasim, (617) 353-5679 Boston Univ; Zn
Momm, Henrique G., (615) 904-8372 henrique.momm@mtsu.edu, Middle Tennessee State Univ; SpZfRw
Monahan, Adam, 250-721-6120 monahana@uvic.ca, Univ of Victoria; Zn
Monahan, Edward C., (860) 405-9110 edward.monahan@uconn.edu, Univ of Connecticut; AsOp
Monari, Stefano, +390498279171 stefano.monari@unipd.it, Univ degli Studi di Padova; Pg

Moncrief, John F., (612) 625-2771 moncr001@umn.ed, Univ of Minnesota, Twin Cities; Sp

Moncrieff, John B., +44 (0) 131 650 5402 j.moncrieff@ed.ac.uk, Edinburgh Univ; As

Monecke, Katrin, kmonecke@wellesley.edu, Wellesley Coll; Gsn

Monecke, Thomas, 303-273-3841 tmonecke@mines.edu, Colorado Sch of Mines; Em

Monet, Julie, 530-898-3460 jmonet@csuchico.edu, California State Univ, Chico; GgNg

Monger, Bruce, bcm3@cornell.edu, Cornell Univ; Ob

Monger, H. C., (505) 646-3405 New Mexico State Univ, Las Cruces; Sa

Monier, Erwan, 530-754-1837 emonier@ucdavis.edu, Univ of California, Davis; Atm

Monohan, Carrie, cmonohan@csuchico.edu, California State Univ, Chico; Hg

Monson, Jessica, 217-265-6895 jlbm@illinois.edu, Illinois State Geo Survey; Gg

Montagna, Paul, 361-825-2040 paul.montagna@tamucc.edu, Texas A&M Univ, Corpus Christi; ObCmHs

Montagne, Cliff, (406) 599-7755 montagne@montana.edu, Montana State Univ; Sd

Montañez, Isabel P., (530) 754-7823 ipmontanez@ucdavis.edu, Univ of California, Davis; Gd

Montayne, Simone, (907) 451-5036 simone.montayne@alaska.gov, Alaska Division of Geological & Geophysical Surveys; Gg

Monteleone, Brian D., (508) 289-2405 bmonteleone@whoi.edu, Woods Hole Oceanographic Inst; Cc

Montenari, Michael, (+44) 01782 733162 m.montenari@keele.ac.uk, Keele Univ; PmgPs

Montenegro, Alvaro, (614) 688-5451 montenegro.8@osu.edu, Ohio State Univ; OgAsZy

Montero, Esperanza, emontero@ucm.es, Univ Compl de Madrid; Hw

Montesi, Laurent G., (301) 405-7534 montesi@umd.edu, Univ of Maryland; YgXyGt

Monteverde, Don, dmonte@eps.rutgers.edu, Rutgers, The State Univ of New Jersey; GruGg

Monteverde, Donald H., (609) 292-2576 don.monteverde@dep.state.nj.us, New Jersey Geological and Water Survey; Grg

Monteverdi, John P., (415) 338-7728 montever@sfsu.edu, San Francisco State Univ; As

Montgomery, David R., 206-685-2560 bigdirt@uw.edu, Univ of Washington; Gm

Montgomery, Homer A., 972-883-2496 mont@utdallas.edu, Univ of Texas, Dallas; Pg

Montgomery, Laurie A., (585) 245-5291 montgome@geneseo.edu, SUNY, Geneseo; Zn

Montgomery, Michael T., (831) 656-2296 mtmontgo@nps.edu, Naval Postgrad Sch; AmsZg

Montgomery, Tamra S., 217-333-5105 tmntgmry@illinois.edu, Illinois State Geo Survey; Zn

Montgomery, William W., (201) 200-3161 wmontgomery@njcu.edu, New Jersey City Univ; Hw

Montoya, Joseph, (404) 385-0479 Georgia Inst of Tech; Cm

Montoya, Judith, (928) 523-8523 judith.montoya@nau.edu, Northern Arizona Univ; Zy

Montwill, Gail F., (909) 946-1796 Santiago Canyon Coll; Gg

Moodie, T. Bryant, (780) 492-5742 bryant.moodie@ualberta.ca, Univ of Alberta; Zn

Moody, Eva, (254) 968-9143 moody@tarleton.edu, Tarleton State Univ; Zn

Moody, Jennie L., (804) 924-0592 jlm8h@virginia.edu, Univ of Virginia; As

Mooers, Howard D., (218) 726-7239 hmooers@d.umn.edu, Univ of Minnesota Duluth; GleHw

Mookerjee, Matty, (707) 664-2002 matty.mookerjee@sonoma.edu, Sonoma State Univ; GcYgZi

Moon, Charlie, +44 01326 371822 c.j.moon@exeter.ac.uk, Exeter Univ; CeEmZi

Moon, Seulgi, 310-206-5760 sgmoon@g.ucla.edu, Univ of California, Los Angeles; Gm

Moon, Wooil M., (204) 474-9833 wmoon@cc.umanitoba.ca, Univ of Manitoba; ZrYnOp

Mooney, Phillip, (707) 664-2328 mooneyp@sonoma.edu, Sonoma State Univ; ZnGc

Moorberg, Colby, (785) 532-7207 Kansas State Univ; So

Moore, Adrian J., +442870123947 a.moore@ulster.ac.uk, Ulster Univ, Coleraine; ZifZn

Moore, Andrew, (765) 983-1672 moorean@earlham.edu, Earlham Coll; Gm

Moore, Bradley S., (858) 822-6650 bsmoore@ucsd.edu, Univ of California, San Diego; Cm

Moore, Christopher M., +44 (0)23 80594801 cmm297@noc.soton.ac.uk, Univ of Southampton; Ob

Moore, Dennis, dmoore@pmel.noaa.gov, Univ of Hawai'i, Manoa; Og

Moore, Duane M., 505-277-4204 dewey33@unm.edu, Univ of New Mexico; Sc

Moore, Gregory F., (808) 956-6854 gmoore@hawaii.edu, Univ of Hawai'i, Manoa; Gt

Moore, Jefferson K., 949-824-5391 jkmoore@uci.edu, Univ of California, Irvine; Op

Moore, Jeffrey, (801) 585-0491 jeff.moore@utah.edu, Univ of Utah; NgrGm

Moore, Joel, (410) 704-4245 moore@towson.edu, Towson Univ; ClsSc

Moore, Joseph N., (801) 585-6931 jmoore@egi.utah.edu, Univ of Utah; Gv

Moore, Kathryn, +44 (0)1326 255693 k.moore@exeter.ac.uk, Exeter Univ; Ge

Moore, Laura J., (919) 962-5960 moorelj@email.unc.edu, Univ of North Carolina, Chapel Hill; Gme

Moore, Lowell P., 540-231-6521 moorelr@vt.edu, Virginia Tech; Ca

Moore, Phillip, +44 (0) 191 208 5040 philip.moore@ncl.ac.uk, Univ of Newcastle Upon Tyne; Yd

Moore, Richard W., 831-656-1041 rwmoor1@nps.edu, Naval Postgrad Sch; Am

Moore, Robert M., (902) 494-3871 robert.moore@dal.ca, Dalhousie Univ; Oc

Moore, Theodore C., (734) 763-0202 tedmoore@umich.edu, Univ of Michigan; OuPmc

Moore, Willard S., moore@geol.sc.edu, Univ of South Carolina; Cc

Moores, Eldridge M., (916) 752-0352 Univ of California, Davis; Gt

Moorhead, Daryl L., (419) 530-2017 daryl.moorhead@utoledo.edu, Univ of Toledo; Zn

Moorman, Thomas B., (515) 294-2308 tom.moorman@ars.usda.gov, Iowa State Univ of Sci & Tech; So

Moortgat, Joachim, (614) 688-2140 moortgat.1@osu.edu, Ohio State Univ; HwNp

Mora, Claudia, claudia.mora@jsg.utexas.edu, Univ of Texas, Austin; Gg

Mora-Klepeis, Gabriela, 802-656-0246 gmora@uvm.edu, Univ of Vermont; CacGi

Morabia, Alfredo, 718-670-4180 Queens Coll (CUNY); GbZn

Morales, Michael A., (620) 794-0191 mmorales@emporia.edu, Emporia State Univ; PvGrg

Morales, Tomas, tomas.morales@ehu.eus, Univ of the Basque Country UPV/EHU; NgHg

Moran, Dawn, (508) 289-4918 dmoran@whoi.edu, Woods Hole Oceanographic Inst; Ob

Moran, Jean E., 510-885-2491 jean.moran@csueastbay.edu, California State Univ, East Bay; HwCl

Moran, S. Bradley, (401) 874-6530 moran@gso.uri.edu, Univ of Rhode Island; Oc

Moran, Seth, 360-993-8934 smoran@usgs.gov, Univ of Washington; Gv

Moran-Taylor, Michelle, 303-871-2513 michelle.moran-taylor@du.edu, Univ of Denver; Zn

Moran-Zenteno, Dante J., dante@tonatiuh.igeofcu.unam.mx, Univ Nacional Auto de Mexico; Cc

Morawska, Lidia, +61 7 3138 2616 l.morawska@qut.edu.au, Queensland Univ of Tech; Asc

Morealli, Sarah A., 540 654-1402 smoreall@umw.edu, Univ of Mary Washington; Gg

Morehouse, Barbara, morehoub@email.arizona.edu, Univ of Arizona; Zn

Morel, Francois M M., 609-258-2416 morel@princeton.edu, Princeton Univ; Cl

Morel-Kraepiel, Anne, 609-258-7415 kraepiel@princeton.edu, Princeton Univ; Em

Moreland, Amy L., amy.moreland@austincc.edu, Austin Comm Coll District; Zge

Moreno, Hernan, moreno@utep.edu, Univ of Texas, El Paso; Zir

Moreno, Rafael, (303) 556-8477 Metropolitan State Coll of Denver; Zg

Moreton, Kim, (+44) (0) 7854825368 k.moreton@exeter.ac.uk, Exeter Univ; EnZur

Morgan, Cristine L., (979) 845-3603 cmorgan@tamu.edu, Texas A&M Univ; Spd

Morgan, Daniel, +44(0) 113 34 35202 d.j.morgan@leeds.ac.uk, Univ of Leeds; Giv

Morgan, Emory, (757) 727-5783 emory.morgan@hampton.edu, Hampton Univ; Ze

Morgan, F D., (617) 253-7857 fdmorgan@mit.edu, Massachusetts Inst of Tech; Yg

Morgan, Gary, 757.727.5783 gary.morgan@hamptonu.edu, Hampton Univ; ZnnZn

Morgan, Joanna, +44 20 759 46423 j.v.morgan@imperial.ac.uk, Imperial Coll; Yg

Morgan, John D., 850-474-2224 jmorgan3@uwf.edu, Univ of West Florida;

Zi
Morgan, Julia K., (713) 348-6330 morganj@rice.edu, Rice Univ; GctGv
Morgan, Kelly, 239-658-3400 conserv@ufl.edu, Univ of Florida; So
Morgan, Matt, 303-384-2632 mmorgan@mines.edu, Colorado Geo Survey; GgmXm
Morgan, Michael C., (608)265-8159 morgan@aurora.aos.wisc.edu, Univ of Wisconsin, Madison; Am
Morgan, Paul, 303-384-2648 morgan@mines.edu, Colorado Geo Survey; YhGtYg
Morgan, Ryan, 254-968-9894 rmorgan@tarleton.edu, Tarleton State Univ; Pei
Morgan, Siobahn M., (319) 273-2389 siobahn.morgan@uni.edu, Univ of Northern Iowa; Xa
Morgan, Sven S., 515-294-1837 smorgan@iastate.edu, Iowa State Univ of Sci & Tech; Gc
Morgan, Tamie, (817) 257-7743 tamie.morgan@tcu.edu, Texas Christian Univ; Zi
Morgan, Terrance L., (505) 667-0837 Los Alamos; Ng
Morgan, William, +44 0161 306-6586 will.morgan@manchester.ac.uk, Univ of Manchester; As
Morin, Paul, 612-626-0505 lpaul@umn.edu, Univ of Minnesota, Twin Cities; Zn
Morison, James H., (206) 543-1394 morison@apl.washington.edu, Univ of Washington; Op
Moritz, Wolfgang, 089/2180 4336 w.moritz@lrz.uni-muenchen.de, Ludwig-Maximilians-Univ Muenchen; Gz
Moro, Alan, +38514606093 amoro@geol.pmf.hr, Univ of Zagreb; PgsGg
Morozov, Igor B., (306) 966-2761 igor.morozov@usask.ca, Univ of Saskatchewan; YsGy
Morra, Matthew J., (208) 885-6315 Univ of Idaho; Sb
Morrill, Penny, (709) 864-6729 pmorrill@mun.ca, Memorial Univ of Newfoundland; Cg
Morris, Antony, +44 1752 584766 a.morris@plymouth.ac.uk, Univ of Plymouth; YmGtu
Morris, Billy, 706-368-7528 bmorris@highlands.edu, Georgia Highlands Coll; Gg
Morris, David, (314) 935-6926 Los Alamos; Cc
Morris, Donald P., (610) 758-5175 dpm2@lehigh.edu, Lehigh Univ; Zn
Morris, Geoffrey, (785) 532-3397 gpmorris@ksu.edu, Kansas State Univ; So
Morris, John A., 662-268-1032 Ext 235 jam16@msstate.edu, Mississippi State Univ; ZirAm
Morris, Robert W., rmorris@wittenberg.edu, Wittenberg Univ; PiGg
Morris, Simon C., +44 (0) 1223 333414 sc113@esc.cam.ac.uk, Univ of Cambridge; Pg
Morris, Thomas H., (801) 422-3761 tom_morris@byu.edu, Brigham Young Univ; Gr
Morris, William A., (905) 525-9140 (Ext. 20116) morriswa@mcmaster.ca, McMaster Univ; Yg
Morrison, Shaunna, smorrison@carnegiescience.edu, Carnegie Inst for Sci; Zn
Morrow, Robert H., 803.896.1214 morrowr@dnr.sc.gov, South Carolina Dept of Nat Res; GcxGt
Morschauser, Lindsey, 662-268-1032 Ext 243 lcm193@msstate.edu, Mississippi State Univ; Am
Morse, David L., (512) 232-3241 Univ of Texas, Austin; Yg
Morse, Linda D., (757) 221-2444 ldmors@wm.edu, William & Mary; GeZg
Morse, Stearns A., (413) 545-0175 tm@geo.umass.edu, Univ of Massachusetts, Amherst; Gi
Mortensen, James K., (604) 822-6208 jmortensen@eos.ubc.ca, Univ of British Columbia; Cc
Mortlock, Richard , (848) 445-3423 rmortloc@eps.rutgers.edu, Rutgers, The State Univ of New Jersey; CgmCs
Morton, Allan E., (602) 426-4351 Central Arizona Coll; Ze
Morton, Bruce, mortonb@easternct.edu, Eastern Connecticut State Univ; Zg
Morton, Cristopher, 575-835-6669 cristopher.morton@nmt.edu, New Mexico Tech; Hw
Morton, Douglas M., (951) 276-6397 scamp@ucrac1.ucr.edu, Univ of California, Riverside; Gp
Morton, Penelope C., pmorton@d.umn.edu, Univ of Minnesota Duluth; EgGpCg
Morton, Roger D., (780) 492-3265 Univ of Alberta; Em
Morton, Ronald L., rmorton@d.umn.edu, Univ of Minnesota Duluth; GvEg
Moscardelli, Lorena, 512-471-0258 lorena.moscardelli@beg.utexas.edu, Univ of Texas, Austin; Gso
Moscovici, Daniel, (215) 688-2910 daniel.moscovici@stockton.edu, Stockton Univ; Zu
Mosenfelder, Jed, jmosenfe@umn.edu, Univ of Minnesota, Twin Cities; Cp
Moser, Desmond, 519-661-4214 dmoser22@uwo.ca, Western Univ; GtCcXg
Moshary, Fred , 212-684-7251 moshary@ccny.cuny.edu, Graduate Sch of the City Univ of New York; Gg
Mosher, David C., 603-862-5493 dmosher@ccom.unh.edu, Univ of New Hampshire; Og
Mosher, Sharon, (512) 471-4135 mosher@mail.utexas.edu, Univ of Texas, Austin; Gc
Moshier, Stephen O., (630) 752-5856 stephen.moshier@wheaton.edu, Wheaton Coll; Gd
Moskalski, Susanne, susanne.moskalski@stockton.edu, Stockton Univ; GsOn
Moskowitz, Bruce M., 612-624-1547 bmosk@umn.edu, Univ of Minnesota, Twin Cities; Ym
Mosley-Thompson, Ellen E., (614) 292-2580 thompson.4@osu.edu, Ohio State Univ; As
Moslow, Thomas F., (403) 269-6911 Univ of Calgary; Co
Mosonyi, Emilia, emilia.mosony@ubbcluj.ro, Babes-Bolyai Univ; GpiGc
Moss, Neil E., 205-247-3557 nmoss@gsa.state.al.us, Geo Survey of Alabama; Hw
Moss, Patti, 509-527-5225 mosspm@whitman.edu, Whitman Coll; Zn
Mosteller, Joey D., mostellerjd@appstate.edu, Appalachian State Univ; Gg
Motani, Ryosuke, 530-754-6284 rmotani@ucdavis.edu, Univ of California, Davis; Pv
Mote, Philip, 541-737-5694 pmote@coas.oregonstate.edu, Oregon State Univ; AsZg
Motta, Luigi, luigi.motta@unito.it, Univ di Torino; Zy
Motta, Michele, michele.motta@unito.it, Univ di Torino; Zy
Mottl, Michael J., (808) 956-7006 mmottl@soest.hawaii.edu, Univ of Hawai'i, Manoa; OuCgm
Motyka, James, motykaj@easternct.edu, Eastern Connecticut State Univ; Zg
Motz, Gary J., (812) 856-3500 garymotz@indiana.edu, Indiana Univ, Bloomington; PqgPi
Mouat, David A., (775) 673-7402 dmouat@dri.edu, Desert Research Inst; Ge
Moucha, Robert, 315-443-6239 rmoucha@syr.edu, Syracuse Univ; GtYeg
Mouginis-Mark, Peter J., (808) 956-3147 Univ of Hawai'i, Manoa; Xg
Moulis, Anastasia, 617-552-8300 anastasia.macherides@bc.edu, Boston Coll; Ys
Moum, James N., 541-737-2553 jmoum@coas.oregonstate.edu, Oregon State Univ; Op
Mound, Jon, +44(0) 113 34 35216 earjem@leeds.ac.uk, Univ of Leeds; Gt
Mount, Jeffrey F., (916) 752-7092 Univ of California, Davis; Gs
Mountain, Gregory S., 848-445-0817 gmtn@eps.rutgers.edu, Rutgers, The State Univ of New Jersey; GuYrGr
Mountney, Nigel, +44(0) 113 34 35249 n.p.mountney@leeds.ac.uk, Univ of Leeds; Gs
Mousset-Jones, Pierre, (775) 784-6959 mousset@mines.unr.edu, Univ of Nevada, Reno; Nm
Mower, Richard, 989-774-3821 mower1rn@cmich.edu, Central Michigan Univ; Am
Mowrer, Jake, 979-845-5366 jake.mowrer@tamu.edu, Texas A&M Univ; Sc
Moy, Christopher M., 64-3-479-5279 chris.moy@otago.ac.nz, Univ of Otago; Gus
MOYEN, Jean-François, (+33)477481510 jean.francois.moyen@univ-st-etienne.fr, Université Jean Monnet, Saint-Etienne; GiCtGp
Moyer, Elisabeth, (773) 834-2992 Univ of Chicago; As
Moyer, Kerry A., (814) 732-2454 kmoyer@edinboro.edu, Edinboro Univ of Pennsylvania; As
Moysey, Stephen, (864) 353-1517 moyseys18@ecu.edu, East Carolina Univ; Hg
Mozley, Peter S., (575) 835-5311 mozley@nmt.edu, New Mexico Tech; Gs
Mozzachiodi, Riccardo, 361-825-3634 riccardo.mozzachiodi@tamucc.edu, Texas A&M Univ, Corpus Christi; Zn
Mozzi, Paolo, +390498279190 paolo.mozzi@unipd.it, Univ degli Studi di Padova; Gm
Mrinjek, Ervin, +38514606057 ervin.mrinjek@zg.t-com.hr, Univ of Zagreb; GscGg
Mroz, Eugene J., (505) 667-7758 Los Alamos; As
Mshiu, Elisante E., *255768384430 mshiutz@gmail.com, Univ of Dar es Salaam; ZrCgZi
Mucci, Alfonso, 514-398-4892 alfonso.mucci@mcgill.ca, McGill Univ; CmlOc
Muchez, Philippe, 32 16 327584 philippe.muchez@kuleuven.be, Katholieke Univit Leuven; EmCgGd
Muchingami, Innocent , imuchingami@unam.na, Univ of Namibia; Ye
Mucsi, László, +3662544156 mucsi@geo.u-szeged.hu, Univesity of Szeged; ZirZy
Mudd, Simon N., +44 (0) 131 650 2535 simon.m.mudd@ed.ac.uk, Edinburgh Univ; Hg
Muehlenbachs, Karlis, (780) 492-2827 karlis.muehlenbachs@ualberta.ca, Univ of Alberta; Cs
Mueller, Amy, 617.373.8131 a.mueller@northeastern.edu, Northeastern

Univ; Zn

Mueller, Erich , (435) 865-8326 erichmueller@suu.edu, Southern Utah Univ; HgGm

Mueller, Erich M., (334) 460-7136 em256@cornell.edu, Univ of South Alabama; Ob

Mueller, Ivan, 860-726-2069 mueller.3@osu.edu, Ohio State Univ; YddYv

Mueller, Karl J., (303) 492-7336 karl.mueller@colorado.edu, Univ of Colorado; GtcGm

Mueller, Paul A., (352) 392-2231 pamueller@ufl.edu, Univ of Florida; CcGit

Mueller, Thomas, (724) 938-4255 mueller@calu.edu, California Univ of Pennsylvania; Zi

Muggeridge, Ann H., +44 20 759 47379 a.muggeridge@imperial.ac.uk, Imperial Coll; NpHwEo

Muhlbauer, Jason , jason-muhlbauer@utc.edu, Univ of Tennessee, Chattanooga; Gg

Muhleman, Duane O., (626) 395-6186 dom@gps.caltech.edu,CalTech; Xy

Muir, William, 909-387-1603 wmuir@sbccd.cc.ca.us, San Bernardino Valley Coll; Og

Mukasa, Samuel, mukasa@umn.edu, Univ of Minnesota, Twin Cities; Cg

Mukhopadhyay, Sujoy, (530) 752-4711 sujoy@ucdavis.edu, Univ of California, Davis; Cg

Mulcahy, Sean R., (360) 650-3645 sean.mulcahy@wwu.edu, Western Washington Univ; GpzGp

Mulder, Jan, jan.mulder@nmbu.no, Norwegian Univ of Life Scis (NMBU); Cb

Muldoon, Maureen, (608) 265-8181 maureen.a.muldoon@wisc.edu, Univ of Wisconsin, Madison, Division of Extension; GeHw

Mulholland, Margaret, (757) 683-3972 mmulholl@odu.edu, Old Dominion Univ; Ob

Mulibo, Gabriel D., gmbelwa@yahoo.com, Univ of Dar es Salaam; Ysg

Mulla, David J., (612) 625-6721 mulla003@umn.edu, Univ of Minnesota, Twin Cities; Sp

Mullen, Steven L., (520) 621-6842 mullen@email.arizona.edu, Univ of Arizona; As

Muller, Andrew C., (410) 293-6569 amuller@usna.edu, United States Naval Acad; OpnOu

Muller, Dietmar, (029) 351-3244 d.muller@usyd.edu.au, Univ of Sydney; Yrg

Muller, Marc, mmuller1@nd.edu, Univ of Notre Dame; Rw

Muller, Otto H., (607) 871-2208 Alfred Univ; Gc

Muller, Peter, (808) 956-8081 pmuller@soest.hawaii.edu, Univ of Hawai'i, Manoa; Op

Muller-Karger, Frank E., (727) 553-3335 Univ of South Florida; Zr

Mulligan, Kevin R., 806-834-0391 kevin.mulligan@ttu.edu, Texas Tech Univ; ZiGmSo

Mullins, Gregory L., (540) 231-4383 gmullins@vt.edu, Virginia Tech; Sc

Mulrooney, Timothy, (919) 530-6269 tmulroon@nccu.edu, North Carolina Central Univ; Zir

Mulvaney, Richard L., (217) 333-9467 Univ of Illinois, Urbana-Champaign; Sc

Mumin, A. Hamid, 204-727-9685 mumin@brandonu.ca, Brandon Univ; EgGzt

Munasinghe, Tissa, munasit@lahc.edu, Los Angeles Harbor Coll; Zg

Munch, Stephan, steve.munch@noaa.gov, SUNY, Stony Brook; Ob

Mungall, James E., 613 558 9337 jamesmungall@cunet.carleton.ca, Carleton Univ; GiEgCg

Munguia, Luis, lmunguia@cicese.mx, Centro de Investigación Científica y de Educación Superior de Ensenada; Ys

Muniz-Solari, Oswaldo, (512) 245-2170 Texas State Univ; Ze

Munk, LeeAnn, 907 786-6895 lamunk@uaa.alaska.edu, Univ of Alaska, Anchorage; Cle

Munn, Barbara J., (916) 278-6811 Sacramento State Univ; Ggp

Muñoz, Alfonso, amunoz@ucm.es, Univ Compl de Madrid; Ygr

Muñoz, Belén, mbmunoz@ucm.es, Univ Compl de Madrid; Gr

Munoz, Samuel, s.munoz@northeastern.edu, Northeastern Univ; GsPcRn

Muñoz-Saez, Carolina, (775) 682-7844 cmunozsaez@unr.edu, Univ of Nevada, Reno; CuEdYe

Munro, Ildi, imunro@ccs.carleton.ca, Carleton Univ; Pg

Munroe, Jeffrey S., (802) 443-3446 jmunroe@middlebury.edu, Middlebury Coll; Gln

Munski, Douglas C., (701) 777-4591 douglas.munski@und.edu, Univ of North Dakota; ZeRhZu

Muntean, John, (775) 682-8748 munteanj@unr.edu, Univ of Nevada, Reno; Eg

Muntean, Thomas, 517-264-3943 tmuntean@adrian.edu, Adrian Coll; Gde

Murdoch, Lawrence C., (864) 656-2597 lmurdoc@clemson.edu, Clemson Univ; Hw

Murgulet, Dorina, (361) 825-2309 dorina.murgulet@tamucc.edu, Texas A&M Univ, Corpus Christi; HwsGe

Murgulet, Valeriu, 361-825-6023 valeriu.murgulet@tamucc.edu, Texas A&M Univ, Corpus Christi; Cg

Murowchick, James B., (816) 235-2979 murowchickj@umkc.edu, Univ of Missouri, Kansas City; CgGz

Murphree, James Thomas, 831-656-2723 murphree@nps.edu, Naval Postgrad Sch; Am

Murphy, Alexander, (541) 346-4571 abmurphy@uoregon.edu, Univ of Oregon; Zn

Murphy, Cindy, (902) 867-2299 cmurphy@stfx.ca, Saint Francis Xavier Univ; Gg

Murphy, David T., 61 07 31382329 david.murphy@qut.edu.au, Queensland Univ of Tech; CgEgCc

Murphy, Edward C., (701) 328-8002 emurphy@nd.gov, North Dakota Geo Survey; Zn

Murphy, J. Brendan, (902) 867-2481 Dalhousie Univ; Gx

Murphy, Michael, 713-743-3564 mmurphy@central.uh.edu, Univ of Houston; Gct

Murphy, Sean C., sean.murphy@austincc.edu, Austin Comm Coll District; EgGpRc

Murphy, Todd, (318) 342-3428 murphy@ulm.edu, Univ of Louisiana, Monroe; As

Murphy, Vincent, 617-552-8300 Boston Coll; Yg

Murray, A. Bradshaw, (919) 684-5847 abmurray@duke.edu, Duke Univ; Gm

Murray, Alison, 775.673.7361 alison.murray@dri.edu, Univ of Nevada, Reno; ObCg

Murray, Andrew, 251-460-7325 amurray@southalabama.edu, Univ of South Alabama; Am

Murray, Bryan P., 909-869-3459 bpmurray@cpp.edu, California State Polytechnic Univ, Pomona; Gsg

Murray, Christopher J., (509) 376-5848 chris.murray@pnl.gov, Pacific Northwest; Gq

Murray, Daniel P., (401) 874-2265 dpmurray@uri.edu, Univ of Rhode Island; ZeGtp

Murray, James W., (206) 543-4730 jmurray@u.washington.edu, Univ of Washington; Oc

Murray, John, + 353 (0)91 495 095 john.murray@nuigalway.ie, Natl Univ of Ireland Galway; PgGs

Murray, Kendra, (208) 282-2949 kendramurray@isu.edu, Idaho State Univ; ClGtm

Murray, Kent S., (313) 436-9129 kmurray@umich.edu, Univ of Michigan, Dearborn; Hw

Murray, Lee, 585-275-2077 lee.murray@rochester.edu, Univ of Rochester; Ac

Murray, Mark, 575.835.6930 murray@ees.nmt.edu, New Mexico Tech; Yds

Murray, Richard, (617) 353-6532 rickm@bu.edu, Boston Univ; Cm

Murrell, Coling, +44 (0)1603 59 2959 j.c.murrell@uea.ac.uk, Univ of East Anglia; As

Murrell, Michael T., (505) 667-0967 Los Alamos; Cc

Murthy, Prahlad N., 570-408-4617 prahlad.murthy@wilkes.edu, Wilkes Univ; AsZnn

Murtugudde, Raghuram G., (301) 314-2622 ragu@umd.edu, Univ of Maryland; ObZr

Muscente, Drew, (319) 895-4309 dmuscente@cornellcollege.edu, Univ of Iowa; Peq

Mushkin, Amit, 972(02)5314254 mushkin@uw.edu, Univ of Washington; ZrCcGm

Musil, Rudolf, +420 549 49 5997 rudolf@sci.muni.cz, Masaryk Univ; PgvGe

Muskatt, Herman, 792-3028 ewelch@utica.edu, Utica Coll; GgrPg

Musselman, Zachary A., (601) 974-1344 musseza@millsaps.edu, Millsaps Coll; Gm

Mustard, John F., (401) 863-1264 john_mustard@brown.edu, Brown Univ; Zr

Mustart, David A., (415) 338-7729 mustart@sfsu.edu, San Francisco State Univ; GizGv

Mustoe, George, george.mustoe@wwu.edu, Western Washington Univ; PbCaPo

Muszer, Antoni, antoni.muszer@uwr.edu.pl, Univ of Wroclaw; Eg

Muthukrishnan, Suresh, (864) 294-3361 suresh.muthukrishnan@furman.edu, Furman Univ; GmZir

Muto, Atsuhiro, 215-204-3699 amuto@temple.edu, Temple Univ; GlYg

Mutter, John C., (646) 269-6942 jcm7@columbia.edu, Columbia Univ; YrGtYs

Mutti, Laurel, muttil@newpaltz.edu, SUNY, New Paltz; Gg

Mutti, Laurence J., (814) 641-3067 mutti@juniata.edu, Juniata Coll; GxzSc

Mutua, Felix N., fnmutua@jkuat.ac.ke, Jomo Kenyatta Univ of Agriculture & Tech; ZigZr

Muxworthy, Adrian, +44 20 759 46442 adrian.muxworthy@imperial.ac.uk, Imperial Coll; Ym

Muza, Jay P., (954) 736-8231 jmuza@broward.edu, Broward Coll; OuPmOo
Mwangi, Nancy , nwmwangi@jkuat.ac.ke, Jomo Kenyatta Univ of Agriculture & Tech; ZiiZr
Mwaniki, Mercy W., mmwaniki@jkuat.ac.ke, Jomo Kenyatta Univ of Agriculture & Tech; ZiiZr
Myers, Alan R., 217-300-2570 amyers76@illinois.edu, Illinois State Geo Survey; EcGo
Myers, Clifford D., (413) 236-4601 cmyers@berkshirecc.edu, Berkshire Comm Coll; Zn
Myers, Corinne E., 505.277.4204 cemyers@unm.edu, Univ of New Mexico; PeqPi
Myers, James D., (307) 766-2203 magma@uwyo.edu, Univ of Wyoming; Gi
Myers, Jeffrey A., 503-838-8365 myersj@wou.edu, Western Oregon Univ; Gs
Myers, Orrin B., (505) 665-3742 obm@lanl.gov, Los Alamos; Pg
Myers, Paul, (780) 492-6706 pmyers@ualberta.ca, Univ of Alberta; Op
Myers, Tammy, (717) 477-1685 tlmyers@ship.edu, Shippensburg Univ; Zn
Mylavarapu, S R., 352-392-1951 ext 202 raom@ufl.edu, Univ of Florida; Sc
Müller, Lena, 089/2180 4340 lena-mueller@lrz.uni-muenchen.de, Ludwig-Maximilians-Univ Muenchen; Gz
Müller-Sohnius, Dieter, 089/2180 4259 mueso@min.uni-muenchen.de, Ludwig-Maximilians-Univ Muenchen; Gz
Mylroie, John E., 662-268-1032 Ext 237 mylroie@geosci.msstate.edu, Mississippi State Univ; GmHyGs
Myneni, Satish C B., 609-258-5848 smyneni@princeton.edu, Princeton Univ; Sc
Myrow, Paul M., (719) 389-6790 pmyrow@coloradocollege.edu, Colorado Coll; Gsr
Mysak, Lawrence A., (514) 398-3768 lawrence.mysak@mcgill.ca, McGill Univ; Op
Mysen, Bjorn O., (202) 478-8975 bmysen@carnegiescience.edu, Carnegie Inst for Sci; Cp
Möller, Andreas, 785864358 amoller@ku.edu, Univ of Kansas; Gt

N

Naar, David F., (727) 553-1637 dnaar@usf.edu, Univ of South Florida; Ou
Nabelek, John L., 541-737-2757 nabelek@coas.oregonstate.edu, Oregon State Univ; GtYs
Nabelek, Peter I., (573) 884-6463 nabelekp@missouri.edu, Univ of Missouri; GipCu
Nadiga, Balu, (505) 667-9466 balu@lanl.gov, Los Alamos; Zn
Nadin, Elisabeth S., (907) 474-5181 enadin@alaska.edu, Univ of Alaska, Fairbanks; Gtc
Nadiri, Ata Allah, +989143003725 nadiri@tabrizu.ac.ir, Univ of Tabriz; HwGqe
Nagaoka, Lisa A., (940) 565-2510 lisa.nagaoka@unt.edu, Univ of North Texas; Ga
Nagel, Athena, (662) 268-1032 ext 238 amo58@msstate.edu, Mississippi State Univ; ZgiZr
Nagel-Myers, Judith, 315-229-5239 jnagel@stlawu.edu, St. Lawrence Univ; PisPe
Nagihara, Seiichi, 806-834-4481 seiichi.nagihara@ttu.edu, Texas Tech Univ; Zi
Nagorski, Sonia, (907) 796-6580 sanagorski@alaska.edu, Univ of Alaska, Southeast; GgZcCt
Nagy, Kathryn L., 312-355-3276 klnagy@uic.edu, Univ of Illinois Chicago; ClbGe
Nagy-Shadman, Elizabeth, 626 585-3369 eanagy-shadman@pasadena.edu, Pasadena City Coll; Gg
Nair, Vimala D., (352) 392-1803 Ext. 324 vdn@ufl.edu, Univ of Florida; Sc
Najjar, Raymond G., (814) 863-1586 rgn1@psu.edu, Pennsylvania State Univ, Univ Park; OcpAt
Nakajima, Miki, (585) 276-6617 mnakajima@rochester.edu, Univ of Rochester; Xyg
Nakamura, Noboru, (773) 702-3802 nnn@bethel.uchicago.edu, Univ of Chicago; As
Nakamura, Yosio, (512) 471-0428 yosio@ig.utexas.edu, Univ of Texas, Austin; Ys
Nakote, Heinz, (575) 646-7627 hnakotte@nmsu.edu, New Mexico State Univ, Las Cruces; Zm
Nakov, Radoslav A., +359-2-979 3472 radnac@geology.bas.bg, Bulgarian Acad of Scis; GtEgm
Naldrett, Anthony J., +441243531823 ajnaldrett@yahoo.com, Univ of Toronto; EmGiCp
Nalin, Ronald, (909) 558-7151 rnalin@llu.edu, Loma Linda Univ; Gsr
Namikas, Steven, (225) 578-6142 snamik1@lsu.edu, Louisiana State Univ; Gms
Nance, R. Damian, (740) 593-1107 nance@ohio.edu, Ohio Univ; Gtc
Nance, Seay, 512-471-6285 seay.nance@beg.utexas.edu, Univ of Texas at Austin, Jackson Sch of Geosciences; Gg
Napieralski, Jacob A., 313.593.5157 jnapiera@umich.edu, Univ of Michigan, Dearborn; GmlZi
Naples, Virginia, 815-753-7820 vlnaples@niu.edu, Northern Illinois Univ; Pv
Napora, Theodore A., (401) 874-6222 Univ of Rhode Island; Ob
Napp, Jeffrey, (206) 526-4148 jeff.napp@noaa.gov, Univ of Washington; Ob
Naranjo, Ramon, 775-887=7659 rnaranjo@usgs.gov, Univ of Nevada, Reno; CgHw
Narbonne, Guy M., (613) 533-6168 narbonne@queensu.ca, Queen's Univ; PigGs
Narey, Martha A., mnarey@du.edu, Univ of Denver; Zg
Narinder, Kaushik, nkaushik@uoguelph.ca, Univ of Guelph; Zn
Narod, Barry, (604) 822-2267 narod@eos.ubc.ca, Univ of British Columbia; Yx
Nash, Barbara P., (801) 581-8587 barb.nash@utah.edu, Univ of Utah; Gi
Nash, David, david.nash@uc.edu, Univ of Cincinnati; Gm
Nash, Greg, 801-585-9986 gnash@egi.utah.edu, Univ of Utah; Gt
Nash, Jonathan, 541-737-4573 nash@coas.oregonstate.edu, Oregon State Univ; Op
Nash, Thomas A., 614 265 6582 thomas.nash@dnr.state.oh.us, Ohio Dept of Nat Res; Gl
Nashold, Barney W., (630) 252-7698 bwnashold@anl.gov, Argonne; Zn
Nasir, Sobhi J., 00974-4852207 nasir54@qu.edu.qa, Univ of Qatar; Gx
Naslund, H. Richard, (607) 777-4313 Binghamton Univ; Gi
Nasraoui, Mohamed, +33 (0)344063813 mohmed.nasraoui@lasalle-beauvais.fr, Institut Polytechnique LaSalle Beauvais (ex-IGAL); EgGzEm
Natalicchio, Marcello, marcello.natalicchio@unito.it, Univ di Torino; Gs
Nater, Edward A., (612) 625-1725 enater@umn.edu, Univ of Minnesota, Twin Cities; Sd
Nathan , Stephen, 860-465-5579 nathans@easternct.edu, Eastern Connecticut State Univ; PmEo
Nathan, Terrence R., (530) 752-1609 trnathan@ucdavis.edu, Univ of California, Davis; Asm
Nathan, Terry, 530-752-1609 trnathan@ucdavis.edu, Univ of California, Davis; AsZc
Naudts, Kim, +31 20 59 87354 k.naudts@vu.nl, Vrije Univit Amsterdam; HgPe
Naumann, Malik, 089/2180 6714 Ludwig-Maximilians-Univ Muenchen; Pg
Naumann, Terry R., trnaumann@uaa.alaska.edu, Univ of Alaska, Anchorage; Gi
Nava, Alejandro, fnava@cicese.mx, Centro de Investigación Científica y de Educación Superior de Ensenada; Ys
Navarre-Sitchler, Alexis, 303-384-2219 asitchle@mines.edu, Colorado Sch of Mines; Ca
Navarro, Didac, didac.navarro@uab.cat, Univ Auto de Barcelona; Gz
Naveira Garabato, Alberto, +44 (0)23 80592680 acng@noc.soton.ac.uk, Univ of Southampton; Op
Navrotsky, Alexandra, (530) 752-3292 anavrotsky@ucdavis.edu, Univ of California, Davis; Zm
Naylor, Mark, +44 (0) 131 650 4918 mark.naylor@ed.ac.uk, Edinburgh Univ; Gt
Naylor, Shawn C., 812-855-2504 snaylor@indiana.edu, Indiana Univ, Bloomington; Hw
Naymik, Thomas G., 614-433-9280 tnaymik@geosyntec.com, Ohio State Univ; Hw
Ndarana, Thando, +27 (0) 12 420 5164 thando.ndarana@up.ac.za, Univ of Pretoria; Ams
Nduati, Eunice W., enduati@jkuat.ac.ke, Jomo Kenyatta Univ of Agriculture & Tech; ZrrZi
Neal, Clive R., (574) 631-8328 neal.1@nd.edu, Univ of Notre Dame; Gi
Neal, William J., (616) 331-3381 nealw@gvsu.edu, Grand Valley State Univ; GsdGr
Nealson, Kenneth, (213) 821-2271 knealson@usc.edu, Univ of Southern California; Pg
Neboga, Victoria V., 717–702–2026 vneboga@pa.gov, DCNR- Pennsylvania Bur of Geo Survey; GgHwNx
Nechaev, Dmitri, (228) 688-2573 dmitri.nechaev@usm.edu, Univ of Southern Mississippi; Op
Nedimovic, Mladen, 1 902 494 4524 mladen@dal.ca, Dalhousie Univ; Ysg
Nedunuri, Krishna K., (937) 376-6455 knedunuri@centralstate.edu, Central State Univ; HwCaSo
Needham, Tim, +44(0) 113 34 39104 d.t.needham@leeds.ac.uk, Univ of Leeds; Gc
Neelin, J. David, (310) 206-3734 neelin@atmos.ucla.edu, Univ of California, Los Angeles; As
Neely, Haly L., 979-845-3041 hneely@tamu.edu, Texas A&M Univ; Spd
Neethling, Stephen, +44 20 759 49341 s.neethling@imperial.ac.uk, Imperial Coll; Gz
Neff, Peter, pneff@umn.edu, Univ of Minnesota, Twin Cities; Pc

Negrini, Robert M., (661) 654-2185 rnegrini@csub.edu, California State Univ, Bakersfield; YmgGn
Nehyba, Slavomír, +420 549 49 6067 slavek@sci.muni.cz, Masaryk Univ; GsdGg
Neighbors, Corrie, (505) 538-6352 neighborsc@wnmu.edu, Western New Mexico Univ; Ysr
Neil, Foley T., (406) 683-7366 neil.foley@umwestern.edu, Univ of Montana Western; YuGl
Neill, Owen K., (509) 335-6770 owen.neill@wsu.edu, Washington State Univ; Ca
Neill, Simon P., 01248 383938 s.p.neill@bangor.ac.uk, Univ of Wales; OnpOg
Neinow, Peter W., +44 (0) 131 650 9139 peter.nienow@ed.ac.uk, Edinburgh Univ; Gl
Neish, Catherine, 519-661-2111, ext.83188 cneish@uwo.ca, Western Univ; Xg
Neitzke Adamo, Lauren, (848) 932-7243 lauren.adamo@rutgers.edu, Rutgers, The State Univ of New Jersey; GuPmZe
Nekvasil, Hanna, 631-632-8201 hanna.nekvasil@stonybrook.edu, Stony Brook Univ; Cp
Nel, Justin, (051) 401-3402 nelwj@ufs.ac.za, Univ of the Free State; Gc
Nelson, Bruce K., (206) 543-4434 bnelson@uw.edu, Univ of Washington; CcGie
Nelson, Craig, 614 265 6603 craig.nelson@dnr.state.oh.us, Ohio Dept of Nat Res; Hw
Nelson, Jennifer, jsembach@iupui.edu, Indiana Univ - Purdue Univ Indianapolis; Gg
Nelson, Kenneth A., (785) 864-2164 nelson@kgs.ku.edu, Univ of Kansas; Zy
Nelson, Marc A., (501) 575-3964 manelson@comp.uark.edu, Univ of Arkansas, Fayetteville; Cl
Nelson, Michael G., (801) 585-3064 nelsonelson@aol.com, Univ of Utah; Nm
Nelson, Nathan, (785) 532-5115 nonelson@ksu.edu, Kansas State Univ; So
Nelson, Priscilla P., 303-273-3700 pnelson@mines.edu, Colorado Sch of Mines; NrgNm
Nelson, Robert K., (508) 289-2656 rnelson@whoi.edu, Woods Hole Oceanographic Inst; Co
Nelson, Stephen A., (504) 862-3194 snelson@tulane.edu, Tulane Univ; GivRn
Nelson, Wendy, (410) 704-3133 wrnelson@towson.edu, Towson Univ; GiCc
Nelwamondo, T. M., +27 15 962 8582 tshililo.nelwamondo@univen.ac.za, Univ of Venda; GeZuy
Nemcok, Michal, +421-2-5463-0337 mnemcok@egi.utah.edu, Univ of Utah; GctGo
Nemes, Attila, attila.nemes@nmbu.no, Norwegian Univ of Life Scis (NMBU); Sp
Nemon, Amy, (270) 745-6952 amy.nemon@wku.edu, Western Kentucky Univ; Zig
Nenes, Athanasios, (404) 894-9225 nenes@eas.gatech.edu, Georgia Inst of Tech; As
Nesbitt, Alex, (775) 682-8762 anesbitt@unr.edu, Univ of Nevada, Reno; Zn
Nesbitt, Carl, 775-784-8287 carln@unr.edu, Univ of Nevada, Reno; Nx
Nesbitt, H W., (519) 661-3194 hwn@uwo.ca, Western Univ; Cl
Nesbitt, Stephen W., (217) 244-3740 snesbitt@illinois.edu, Univ of Illinois, Urbana-Champaign; As
Nesbitt, Sterling J., 540-231-6330 sjn2104@vt.edu, Virginia Tech; Pv
Nese, Jon, j2n@psu.edu, Pennsylvania State Univ, Univ Park; Am
Neser, Laura, (757) 641-0628 neser@vt.edu, Virginia Tech; Gsc
Nesse, William D., (970) 351-2830 w.d.nesse@gmail.com, Univ of Northern Colorado; Gxz
Nestell, Galina P., (817) 272-2987 gnestell@uta.edu, Univ of Texas, Arlington; PmsPi
Nestell, Merlynd K., (817) 272-2987 nestell@uta.edu, Univ of Texas, Arlington; PmiPs
Nestola, Fabrizio, +390498279160 fabrizio.nestola@unipd.it, Univ degli Studi di Padova; Gz
Nethengwe, Nthaduleni S., +27 15 962 8593 nthaduleni.nethengwe2@univen.ac.za, Univ of Venda; RnZiRw
Nettles, Meredith, nettles@ldeo.columbia.edu, Columbia Univ; YsGl
Neubaum, John C., 717-702-2039 jneubaum@pa.gov, DCNR- Pennsylvania Bur of Geo Survey; EgGg
Neubeck, William S., (518) 388-6519 neubeckw@union.edu, Union Coll; Gm
Neuberg, Jurgen, +44(0) 113 34 36769 j.neuberg@leeds.ac.uk, Univ of Leeds; Gv
Neufeld, Jerome A., +44 (0) 1223 765709 jn271@cam.ac.uk, Univ of Cambridge; GoYgGt
Neuhauser, Kenneth R., (785) 628-5349 kneuhaus@fhsu.edu, Fort Hays State Univ; GceGs
Neukrug, Howard, hneukrug@upenn.edu, Univ of Pennsylvania; Hs
Neuman, Shlomo P., (520) 621-7114 neuman@email.arizona.edu, Univ of Arizona; HwqGq
Neumann, Klaus, 765-285-8262 kneumann@bsu.edu, Ball State Univ; Cl
Neumann, Patricia, (310) 660-3593 pneumann@elcamino.edu, El Camino Coll; Zg
Neumann, Tom, 301-614-5923 thomas.neumann@nasa.gov, Univ of Washington; Gl
Neuweiler, Fritz, (418) 656-7479 fritz.neuweiler@ggl.ulaval.ca, Univ Laval; GsdPg
Neves, Douglas, (310) 660-3593 dneves@elcamino.edu, El Camino Coll; Zg
Nevins, Joseph, 845-437-7823 jonevins@vassar.edu, Vassar Coll; Zu
Nevins, Susan K., (607) 753-2815 geologydept@cortland.edu, SUNY, Cortland; Zn
Newberry, Rainer J., ffrn@uaf.edu, Univ of Alaska, Fairbanks; Em
Newbold, K. B., (905) 525-9140 (Ext. 27948) newbold@mcmaster.ca, McMaster Univ; Ge
Newcomer, April, gbgmedl@vt.edu apriln@vt.edu, Virginia Tech; Ze
Newcomer, Darrell R., (509) 376-1054 darrell.newcomer@pnl.gov, Pacific Northwest; Hw
Newell, Dennis L., dennis.newell@usu.edu, Utah State Univ; Cls
Newell, Kerry D., (785) 864-2183 dnewell@kgs.ku.edu, Univ of Kansas; Go
Newland, Franz, (416)736-2100 #22095 franz.newland@lassonde.yorku.ca, York Univ; Zr
Newman, Andrew V., (404) 894-3976 anewman@gatech.edu, Georgia Inst of Tech; YdsYg
Newman, Brent D., (505) 667-3021 bnewman@lanl.gov, Los Alamos; Gg
Newman, Dianne K., (636) 395-3543 dkn@caltech.edu,CalTech; Pg
Newman, Jamie, 212-769-5386 jamien@amnh.org, American Mus of Natural History; Gg
Newman, Jonathan, (519) 824-4120 Ext.52147 jnewma01@uoguelph.ca, Univ of Guelph; Zn
Newman, Julie, (979)845-2240 newman@geo.tamu.edu, Texas A&M Univ; Gct
Newman, Steven B., (860) 832-2940 ccsuwxman@gmail.com, Central Connecticut State Univ; As
Newman, William I., (310) 825-3912 win@ucla.edu, Univ of California, Los Angeles; Xy
Newsom, Horton E., (505) 277-0375 newsom@unm.edu, Univ of New Mexico; XgcXm
Newton, Anthony J., +44 (0) 131 650 2546 anthony.newton@ed.ac.uk, Edinburgh Univ; Zn
Newton, Brad T., 575-835-6668 talon.newton@nmt.edu, New Mexico Tech; HwCgs
Newton, Cathryn R., 315-443-2672 crnewton@syr.edu, Syracuse Univ; Pg
Newton, Jan, (206) 616-3641 newton@ocean.washington.edu, Univ of Washington; Ob
Newton, Rob, +44(0) 113 34 31631 d.oneill@leeds.ac.uk, Univ of Leeds; Ge
Newton, Robert, bnewton@ldeo.columbia.edu, Columbia Univ; Ct
Newton, Seth A., 205-247-3708 snewton@gsa.state.al.us, Geo Survey of Alabama; ZnGg
Nex, Paul A., 011 717 6563 paul.nex@wits.ac.za, Univ of the Witwatersrand; Eg
Nezat, Carmen A., (509) 359-7959 cnezat@ewu.edu, Eastern Washington Univ; ClGeSc
Ng, Crystal, 612-624-9243 gcng@umn.edu, Univ of Minnesota, Twin Cities; Hy
Ngigi, Thomas G., tgngigi@jkuat.ac.ke, Jomo Kenyatta Univ of Agriculture & Tech; ZriZr
Ngo, Thanh X., +84-9782 24246 ngoxuanthanh@humg.edu.vn, Hanoi Univ of Mining & Geology; GtcGi
Ngoy, Kikombo, (908) 737-3718 kngoy@kean.edu, Kean Univ; Zy
Nguyen, Maurice, 612-626-8739 nguyenm@umn.edu, Univ of Minnesota; Gl
Ngwenya, Bryne T., +44 (0) 131 650 8507 bryne.ngwenya@ed.ac.uk, Edinburgh Univ; Co
Ni, James F., (505) 646-1920 jni@nmsu.edu, New Mexico State Univ, Las Cruces; YsGtYg
Ni-Meister, Wenge, (212) 772-5321 wenge.ni-meister@hunter.cuny.edu, Hunter Coll (CUNY); Zr
Nichol, Craig F., 250-807-8087 craig.nichol@ubc.ca, Univ of British Columbia, Okanagan Campus; HwSpZe
Nicholas, Chris, +353 1 8962176 nicholyj@tcd.ie, Trinity Coll; Eo
Nicholas, Joseph W., (540) 654-1470 jnicola@mwc.edu, Mary Washington Coll; Zy
Nicholl, Michael J., 702-895-4616 michael.nicholl@unlv.edu, Univ of Nevada, Las Vegas; Hq
Nicholls, James W., (403) 220-7127 Univ of Calgary; Gi

Nichols, Jonathan E., jnichols@ldeo.columbia.edu, Columbia Univ; Pe
Nichols, Kyle K., 518-580-5194 knichols@skidmore.edu, Skidmore Coll; Gm
Nichols, Terry E., (501) 575-7317 tenichol@comp.uark.edu, Univ of Arkansas, Fayetteville; Zn
Nicholson, David, (508) 289-3547 dnicholson@whoi.edu, Woods Hole Oceanographic Inst; Gu
Nicholson, Kirsten N., 765-285-8268 knichols@bsu.edu, Ball State Univ; Gi
Nick, Kevin, (909) 558-4530 knick@llu.edu, Loma Linda Univ; GsbGr
Nickmann, Marion, +49 89 289 25853 nickmann@tum.de, Technical Univ of Munich; Ng
Nicol, Andy, andy.nicol@canterbury.ac.nz, Univ of Canterbury; GctGo
Nicolas, Michelle P., (204) 945-6571 michelle.nicolas@gov.mb.ca, Manitoba Geo Survey; GrsGo
Nicolaysen, Kirsten P., 509-527-4934 nicolakp@whitman.edu, Whitman Coll; GiCgGa
Nicolescu, Stefan, 203-432-3141 stefan.nicolescu@yale.edu, Yale Univ; CtGpz
Nicolette, Justine, 575-835-5613 justine.nicolette@nmt.edu, New Mexico Tech; Zi
Nicot, Jean-Philippe, (512) 471-6246 jp.nicot@beg.utexas.edu, Univ of Texas at Austin, Jackson Sch of Geosciences; HwCg
NICULESCU, Bogdan M., +40-21-3125024 bogdan.niculescu@g.unibuc.ro, Univ of Bucharest; YgeYu
Nielsen, Gerald A., (406) 994-5075 nielsenmontana@aol.com, Montana State Univ; Sd
Nielsen, Gregory B., (801) 626-6394 gnielsen@weber.edu, Weber State Univ; Gg
Nielsen, Kurt E., 831-656-2295 nielsen@nps.edu, Naval Postgrad Sch; Am
Nielsen, Mary, 605-677-5649 mary.nielsen@usd.edu, Univ of South Dakota; Zn
Nielsen, Peter A., pnielsen@keene.edu, Keene State Coll; GxzGc
Nielsen, Roger L., (541) 737-1235 nielsenr@geo.oregonstate.edu, Oregon State Univ; GizGv
Nielsen, Sune G., (508) 289-2837 snielsen@whoi.edu, Woods Hole Oceanographic Inst; Cs
Nielsen-Gammon, John, (979) 862-2248 n-g@tamu.edu, Texas A&M Univ; As
Nielson, R. LaRell, (936) 468-2248 rnielson@sfasu.edu, Stephen F. Austin State Univ; Grs
Niemann, William L., (304) 696-6721 niemann@marshall.edu, Marshall Univ; Ng
Niemi, Nathan A., 734-764-6377 naniemi@umich.edu, Univ of Michigan; Gtc
Niemi, Tina M., (816) 235-5342 niemit@umkc.edu, Univ of Missouri, Kansas City; Gt
Niemiller, Matthew L., matthew.niemiller@uah.edu, Univ of Tennessee, Knoxville; Zn
Niemitz, Jeffery W., niemitz@dickinson.edu, Dickinson Coll; ClOuGn
Niespolo, Elizabeth, 609-258-4101 niespolo@princeton.edu, Princeton Univ; CgGg
Nikolinakou, Maria-Aikaterini, 512-475-6613 mariakat@austin.utexas.edu, Univ of Texas, Austin; Np
Nikulin, Alex, (607) 777-2518 anikulin@binghamton.edu, Binghamton Univ; Ys
Nilsson, Andreas, andreas.nilsson@liverpool.ac.uk, Univ of Liverpool; Ym
Nimako, Solomon Nana Kwaku, solomon.nimako@gmail.com, San Bernardino Valley Coll; Zi
Nimis, Paolo, +390498279161 paolo.nimis@unipd.it, Univ degli Studi di Padova; Eg
Nimmo, Francis, 831-459-1783 fnimmo@ucsc.edu, Univ of California, Santa Cruz; Xy
Nisbet, Euan G., +44 1784 443809 e.nisbet@rhul.ac.uk, Royal Holloway Univ of London; ZgAcGi
Nissen-Meyer, Targe, +44 (1865) 282149 tarje.nissen-meyer@earth.ox.ac.uk, Univ of Oxford; Ys
Nitecki, Matthew H., (312) 665-7093 mnitecki@fmnh.org, Field Mus of Natural History; Pi
Nitsche, Frank, fnitsche@ldeo.columbia.edu, Columbia Univ; Gu
Nittrouer, Charles A., (206) 543-5099 nittroue@ocean.washington.edu, Univ of Washington; Ou
Niu, Fenglin, 713.348.4122 niu@rice.edu, Rice Univ; YsGt
Niu, Guo-Yue, 520-626-4820 niug@email.arizona.edu, Univ of Arizona; Hg
Niu, Qifei, qifeiniu@boisestate.edu , Boise State Univ; Ygu
Niu, Yaoling, +44 (0) 191 33 42311 yaoling.niu@durham.ac.uk, Durham Univ; Gi
Nixon, R. Paul, 801-422-4657 paul_nixon@byu.edu, Brigham Young Univ; Go
Nixon, Scott W., (401) 874-6803 Univ of Rhode Island; Ob
Niyogi, Dev, dev.niyogi@jsg.utexas.edu, Univ of Texas, Austin; As
Njau, Jackson K., (812) 856-3170 jknjau@indiana.edu, Indiana Univ, Bloomington; Pv
njoh, Ambe, njoh@usf.edu, Univ of South Florida; Zn
Njoku, Eni G., (818) 354-3693 eni.g.njoku@jpl.nasa.gov, Jet Propulsion Lab; Zr
Nkedi-Kizza, Peter, (352) 392-1951 kizza@ufl.edu, Univ of Florida; Sp
Nkotagu, Hudson H., nkotaguh@yahoo.com, Univ of Dar es Salaam; HwgGe
Nkuna, Tinyiko R., +27 15 962 8017 tinyiko.nkuna@univen.ac.za, Univ of Venda for Sci & Tech; HgwAs
Noble, Allison, 480-965-2552 allison.noble@asu.edu, Arizona State Univ; Xa
Noble, Paula J., 775-784-6211 noblepj@unr.edu, Univ of Nevada, Reno; PmGnPs
Noblett, Jeffrey B., (719) 389-6516 jnoblett@coloradocollege.edu, Colorado Coll; Gxv
Nocco, Mallika, manocco@ucdavis.edu, Univ of California, Davis; Sp
Noe, Garry, (757) 455-3284 gnoe@vwc.edu, Virginia Wesleyan Coll; HsZin
Noffke, Nora K., (757) 683-3313 nnoffke@odu.edu, Old Dominion Univ; GsPoGu
Nogueira, Ricardo, 404-413-5791 mogueira@gsu.edu, Georgia State Univ; Asm
Noh, Myoung-Jong, (614) 292-5445 noh.56@osu.edu, Ohio State Univ; Zr
Nolan, Robert P., (718) 951-4242 rnolan@gc.cuny.edu, Graduate Sch of the City Univ of New York; Co
Nolin, Anne , (775) 682-8148 anolin@unr.edu, Univ of Nevada, Reno; Hg
Noll, Mark R., (585) 395-5717 mnoll@esc.brockport.edu, SUNY, The Coll at Brockport; Cl
Noll, Michael G., (229) 333-7143 mgnoll@valdosta.edu, Valdosta State Univ; Zn
Noltimier, Hallan C., (614) 292-9796 noltimier.2@osu.edu, Ohio State Univ; YmgGt
Nondorf, Lea, (501) 683-0110 lea.nondorf@arkansas.gov, Arkansas Geo Survey; Cg
Noonan, Mathew T., 605-677-6152 matthew.noonan@usd.edu, South Dakota Dept of Environment and Nat Res; Hw
Noone, David, (541) 737-3629 dcn@coas.oregonstate.edu, Oregon State Univ; AmZg
Nootz, Gero, 228.688.3091 gero.nootz@usm.edu, Univ of Southern Mississippi; No
Nord, Julia, (703) 993-3395 jnord@gmu.edu, George Mason Univ; Gzg
Nordeng, Stephan, (701) 777-3455 stephan.nordeng@engr.und.edu, Univ of North Dakota; Go
Nordstrom, Greg, 662-268-1032 Ext 248 gjn2@msstate.edu, Mississippi State Univ; Am
Nordt, Lee C., (254) 710-4288 lee_nordt@baylor.edu, Baylor Univ; SdGa
Nordwald, Dan, 573-368-2451 dan.nordwald@dnr.mo.gov, Missouri Dept of Nat Res; Zn
Norell, Mark A., (212) 769-5804 American Mus of Natural History; Pv
Noren, Anders, 612-626-3298 noren021@umn.edu, Univ of Minnesota, Twin Cities; Gn
Norman, David, +44 (0) 1223 333426 dn102@esc.cam.ac.uk, Univ of Cambridge; Pg
Norman, John M., (608) 262-2633 jmnorman@wisc.edu, Univ of Wisconsin, Madison; Spo
Norouzi, Hamidreza, 718-260-5410 hnorouzi@city.tech.cuny.edu, Graduate Sch of the City Univ of New York; Gg
Norris, Geoffrey, (416) 978-4851 norris@quartz.geology.utoronto.ca, Univ of Toronto; Pl
Norris, Joel R., (858) 822-4420 jnorris@ucsd.edu, Univ of California, San Diego; As
Norris, Richard D., (858) 822-1868 rnorris@ucsd.edu, Univ of California, San Diego; Gu
North, Gerald, g-north@tamu.edu, Texas A&M Univ; As
North, Leslie, leslie.north@wku.edu, Western Kentucky Univ; Zg
Northrup, Clyde J., (208) 426-1581 cjnorth@boisestate.edu, Boise State Univ; Gc
Norton, Stephen A., (207) 581-2156 norton@maine.edu, Univ of Maine; Cl
Norton-Krane, Abby N., (216) 987-5227 abby.norton-krane@tri-c.edu, Cuyahoga Comm Coll - Western Campus; Zg
Nosal, Andrew, 6192604600 x2438 anosal@sandiego.edu, Univ of San Diego; Ob
Nothdurft, Luke, +61 7 3138 1531 l.nothdurft@qut.edu.au, Queensland Univ of Tech; GdCmGg
Nourse, Jonathan A., 909-869-3460 janourse@cpp.edu, California State Polytechnic Univ, Pomona; GctEm
Novacek, Michael J., (212) 769-5805 American Mus of Natural History; Pv
Novak, Milan , +420 549 49 6188 mnovak@sci.muni.cz, Masaryk Univ;

GzpGi
Novak, Thomas, 859-257-3818 thomas.novak@uky.edu, Univ of Kentucky; Nm
Nowack, Robert L., (765) 494-5978 nowack@purdue.edu, Purdue Univ; Ys
Nowariak, Eric, 612-626-3379 nowar005@d.umn.edu, Univ of Minnesota; Gp
Nowell, Arthur R. M., (206) 543-6605 nowell@cofs.washington.edu, Univ of Washington; Ou
Nowicki, Sophie, (716) 645-3489 sophien@buffalo.edu, SUNY, Buffalo; Gl
Nowlin, Jr., Worth D., (979) 846-6747 wnowlin@tamu.edu, Texas A&M Univ; OpZn
Nowotarski, Christopher J., (979) 845-3305 cjnowotarski@tamu.edu, Texas A&M Univ; Ams
Noyes, Jim, (310) 660-3593 tnoyes@elcamino.edu, El Camino Coll; Og
Noyes, Joanne M., (605) 394-6972 joanne.noyes@state.sd.us, South Dakota Dept of Environment and Nat Res; HyGg
Ntarlagiannis, Dimitrios, 973-353-5189 dimntar@newark.rutgers.edu, Rutgers, The State Univ of New Jersey, Newark; Yu
Nton, M. E., aa.elueze@mail.ui.edu.ng, Univ of Ibadan; Go
Ntsaluba, Bantubonke I., +27 (0)46-603-8309 b.ntsaluba@ru.ac.za, Rhodes Univ; Gi
Nudds, John, +44 0161 275-7861 john.nudds@manchester.ac.uk, Univ of Manchester; Pg
NUDE, Prosper M., +233-244-116879 pmnude@ug.edu.gh, Univ of Ghana; GgpGx
Nuester, Jochen, jnuester@csuchico.edu, California State Univ, Chico; Cm
Nugent, Alison D., 808-956-2878 anugent@hawaii.edu, Univ of Hawai'i at Manoa; AsZnn
Nugent, Barnes, barnes@geosrv.wvnet.edu, Fairmont State Univ; Gg
Null, E. Jan, (415) 338-2061 San Francisco State Univ; Ge
Nunn, Jeffrey A., (225) 388-6657 jeff@geol.lsu.edu, Louisiana State Univ; Yg
Nur, Amos M., (650) 723-9526 Stanford Univ; Yg
Nusbaum, Robert L., nusbaumr@cofc.edu, Coll of Charleston; Gz
Nuss, Wendell A., 831-656-2308 nuss@nps.edu, Naval Postgrad Sch; Am
Nussear, Kenneth , (775) 784-6612 knussear@unr.edu, Univ of Nevada, Reno; Zn
Nuwategeka, Expedito, +256 782 889 985 nuwategeka@gmail.com, Gulu Univ; ZyyGv
Nwachukwu, J. I., 234-803-725-8122 jnwachuk@oauife.edu.ng, Obafemi Awolowo Univ; GoCg
Nyantakyi-Frimpong, Hanson, 303-871-4175 hnyanta2@du.edu, Univ of Denver; Zc
Nyblade, Andrew A., (814) 863-8341 andy@geosc.psu.edu, Pennsylvania State Univ, Univ Park; Yg
Nye, Janet, (631) 632-3187 janet.nye@stonybrook.edu, SUNY, Stony Brook; Ob
Nyhan, John W., (505) 667-3163 Los Alamos; So
Nyland, Edo, (780) 492-5502 edo@phys.ualberta.ca, Univ of Alberta; Ys
Nyquist, Jonathan, (215) 204-7484 nyq@temple.edu, Temple Univ; Ygu
Nystuen, Jeffrey A., (206) 543-1343 nystuen@apl.washington.edu, Univ of Washington; Op
Nzengung, Valentine A., (706) 202-4296 vnzengun@uga.edu, Univ of Georgia; ClGeCq

O

O'Banion, Matthew S., 845-938-2326 matthew.obanion@westpoint.edu, United States Military Acad; ZrGg
O'Brien, Arnold L., (978) 934-3902 arnold_obrien@uml.edu, Univ of Massachusetts, Lowell; Hw
O'Brien, John M., (201) 200-3161 jobrien@njcu.edu, New Jersey City Univ; Gs
O'Brien, Lawrence E., (914) 341-4570 lobrien@sunyorange.edu, Orange County Comm Coll; Gg
O'Brien, Philip, (281) 756-5670 pobrien@alvincollege.edu, Alvin Comm Coll; Gg
O'Brien, Travis, (812) 269-2051 obrienta@iu.edu, Indiana Univ, Bloomington; As
O'Callaghan, Mick, +353 21 4902662 mick.ocallaghan@ucc.ie, Univ Coll Cork; Zn
O'Connor, George A., (352) 392-1803 (Ext. 329) gao@ufl.edu, Univ of Florida; Sc
O'Connor, Yuet-Ling, yloconnor@pasadena.edu, Pasadena City Coll; Ge
O'Dean, Emily, 775-682-9780 eodean@unr.edu, Univ of Nevada, Reno; Zf
O'Donnell, James, 860-405-9171 james.odonnell@uconn.edu, Univ of Connecticut; Op
O'Driscoll, Nelson, 902-585-1679 nelson.odriscoll@acadiau.ca, Acadia Univ; CgGgZn
O'Farrell, Keely A., 859-323-4876 k.ofarrell@uky.edu, Univ of Kentucky; Ygd
O'Geen, Toby, 530-752-2155 atogeen@ucdavis.edu, Univ of California, Davis; So
O'Gorman, Paul, (617) 452-3382 pog@mit.edu, Massachusetts Inst of Tech; As
O'Halloran, Ivan, (519) 674-1635 iohallo@ridgetownc.uoguelph.ca, Univ of Guelph; So
O'Hara, Kieran D., (859) 269-5161 geokoh@uky.edu, Univ of Kentucky; GcCg
O'Hara, Matthew J., (509) 373-1671 matt.ohara@pnl.gov, Pacific Northwest; Cg
O'Keefe, Jennifer, (606) 783-2349 j.okeefe@moreheadstate.edu, Morehead State Univ; PlEcZe
O'Keeffe, Mike, 303-384-2637 okeeffe@mines.edu, Colorado Geo Survey; GgeEg
O'Leary, Maureen, 631-444-3730 maureen.oleary@sunysb.edu, Stony Brook Univ; Pv
O'Meara, Stephanie A., (970) 491-6655 stephanie_o'meara@partner.nps.gov, Colorado State Univ; Gc
O'Melia, Charles R., (410) 516-7102 omelia@jhu.edu, Johns Hopkins Univ; Zn
O'Mullan, Gregory, 718-997-3452 gomullan@qc.cuny.edu, Queens Coll (CUNY); PgOb
O'Neal, Claire J., 302-831-8909 cjoneal@udel.edu, Univ of Delaware; Ze
O'Neil, James, (734) 764-1435 jro@umich.edu, Univ of Michigan; Cs
O'Neil, Jennifer, 775-682-8747 joneil@unr.edu, Univ of Nevada, Reno; Zn
O'Neil, Jonathan, jonathan.oneil@uottawa.ca, Univ of Ottawa; Cgc
O'Neill, Larry, (541) 737-2064 loneill@coas.oregonstate.edu, Oregon State Univ; AmOg
O'Neill, Patrick M., (504) 388-2681 pat@lgs.bri.lsu.edu, Louisiana State Univ; Sf
O'Reilly, Andrew M., 662-915-2483 aoreilly@olemiss.edu, Univ of Mississippi; Hwq
O'Reilly, Catherine M., 309-438-3493 cmoreil@ilstu.edu, Illinois State Univ; HgCgGg
O'Rourke, Joseph, jgorourk@asu.edu, Arizona State Univ; Xy
O'Shea, Bethany, (619) 260-4243 bethoshea@sandiego.edu, Univ of San Diego; CqGeCt
O'Sullivan, Katie, (661) 654-3991 kosullivan@csub.edu, California State Univ, Bakersfield; GviXg
Oakes-Miller, Hollie, hollie.oakesmiller@pcc.edu, Portland Comm Coll - Sylvania Campus; Gg
Oakley, Adrienne, 484-646-4334 oakley@kutztown.edu, Kutztown Univ of Pennsylvania; YrOun
Oakley, Bryan, (860) 465-0418 oakleyb@easternct.edu, Eastern Connecticut State Univ; OnGml
Oaks, Jr., Robert Q., (435) 752-0867 boboaks@comcast.net, Utah State Univ; GsHwYv
Oberdorfer, June A., (408) 924-5026 june.oberdorfer@sjsu.edu, San Jose State Univ; HwGeHy
Obermeyer, Nancy J., 812-237-4351 njobie28@kiva.net, Indiana State Univ; Zi
Oboh-Ikuneobe, Francisca, 573-341-6946 ikuenobe@mst.edu, Missouri Univ of Sci and Tech; Pg
Obour, Augustine, (785) 625-3425 aobour@ksu.edu, Kansas State Univ; So
Obrad, Jennifer M., 217-333-8741 jobrad@illinois.edu, Illinois State Geo Survey; EcGo
Obreza, Thomas A., 352-392-1951 ext 243 obreza@ufl.edu, Univ of Florida; So
Obrist, Daniel, (978) 934-3900 daniel_obrist@uml.edu, Univ of Massachusetts, Lowell; As
Obrist Farner, Jonathan, 573-341-7879 obristj@mst.edu, Missouri Univ of Sci and Tech; Gr
Ocan, O. O., 234-803-705-8796 Obafemi Awolowo Univ; GtzGp
Occhiuzzi, Tony, 4807528888 x 81369 tocchiuzzi.cds@tuhsd.k12.az.us, Mesa Comm Coll; Zn
Oches, Rick, (781) 891-2937 roches@bentley.edu, Bentley Univ; ZcePe
OConnell, Suzanne B., (860) 685-2262 soconnell@wesleyan.edu, Wesleyan Univ; GsuGe
Odendaal, Adriaan, +27 (0)51 401 9928 odendaalai@ufs.ac.za, Univ of the Free State; GgsGr
Odera, Patroba A., podera@jkuat.ac.ke, Jomo Kenyatta Univ of Agriculture & Tech; YdZge
Odeyemi, Idowu B., +2348033436110 ibodeyemi@futa.edu.ng, Federal Univ of Tech, Akure; GgxZr
Odland, Sarah K., 845-365-8633 odland@ldeo.columbia.edu, Columbia Univ; Zn
Odling, Noelle, +44(0) 113 34 32806 n.e.odling@leeds.ac.uk, Univ of Leeds; Hg

Odlum , Margo , 702-895-3092 margaret.odlum@unlv.edu, Univ of Nevada, Las Vegas; Gt
Odogba, Ismaila, 715-346-4451 iodogba@uwsp.edu, Univ of Wisconsin, Stevens Point; Zui
Odokuma-Alonge, Ovie, ovie.odokuma-alonge@uniben.edu, Univ of Benin; CgGip
Odom, LeRoy A., (850) 644-6706 odom@magnet.fsu.edu, Florida State Univ; Cg
Odom, Robert I., 206-685-3788 odom@apl.washington.edu, Univ of Washington; YrGu
Oduor, Peter, (701) 231-7145 peter.oduor@ndsu.edu, North Dakota State Univ; ZiRwZr
Oechel, Walter, walter.oechel@open.ac.uk, Open Univ; As
Oehlert, Amanda M., 305-421-4405 aoehlert@rsmas.miami.edu, Univ of Miami; Gu
Oeser, Jens, 089/21804208 jens.oeser@lmu.de, Ludwig-Maximilians-Univ Muenchen; Yg
Oest, Christopher, (717) 702-2017 coest@pa.gov, DCNR- Pennsylvania Bur of Geo Survey; GrdGg
Ogard, Allen E., (505) 667-6344 Los Alamos; Hw
Ogbahon, Osazuwa A., 08064253825 oaogbahon@futa.edu.ng, Federal Univ of Tech, Akure; PlGsPm
Ogbamikhumi, Alexander, alexander.ogbamikhumi@uniben.edu, Univ of Benin; YsgGg
Ogbuchiekwe, Edmund Jekwu, ogbuchiekwe@charter.net, San Bernardino Valley Coll; Zy
Oglesby, David D., (951) 827-2036 david.oglesby@ucr.edu, Univ of California, Riverside; Ys
Oglesby, Elizabeth, eoglesby@email.arizona.edu, Univ of Arizona; Zn
Ogram, Andrew V., 3523921951 (Ext 211) aogram@ufl.edu, Univ of Florida; Sb
Ogston, Andrea, ogston@uw.edu, Univ of Washington; Og
Ogunsanwo, O., femiogunsanwo2003@yahoo.com, Univ of Ilorin; Ng
Ohman, Mark D., (858) 534-2754 mohman@ucsd.edu, Univ of California, San Diego; Ob
Ohmoto, Hiroshi, (814) 865-4074 ohmoto@geosc.psu.edu, Pennsylvania State Univ, Univ Park; Cs
Oikawa, Patricia, patty.oikawa@csueastbay.edu, California State Univ, East Bay; SbCoSf
Ojala, Carl F., (313) 487-0218 Eastern Michigan Univ; As
Ojo, S. B., 234-803-703-7226 sbojo@oauife.edu.ng, Obafemi Awolowo Univ; Yg
Okafor, Florence A., (256) 372-4926 florence.okafor@aamu.edu, Alabama A&M Univ; ZnEn
Okal, Emile A., (847) 491-3194 emile@earth.northwestern.edu, Northwestern Univ; YsgYr
Okaya, David A., (213) 740-7452 okaya@earth.usc.edu, Univ of Southern California; Ys
Okie, Jordan, jordan.okie@asu.edu, Arizona State Univ; Cb
Okonkwo, Chukwuemeka T., +2348035778090 ctokonkwo@futa.edu.ng, Federal Univ of Tech, Akure; GcxGg
Okulewicz, Steven C., (516) 463-6545 geosco@hofstra.edu, Hofstra Univ; CgGue
Okumura, Yuko, 512-471-0383 yukoo@ig.utexas.edu, Univ of Texas, Austin; AsZoOp
Okunade, Samuel, (937) 376-6455 Central State Univ; Gm
Okunlola, Ougbenga. A., +2348033814070 o.okunlola@mail.ui.edu.ng, Univ of Ibadan; EgCgGe
Okunuwadje, Sunday E., sunday.okunuwadje@uniben.edu, Univ of Benin; GsoGd
Ola, Peter S., +2348035927594 psola@futa.edu.ng, Federal Univ of Tech, Akure; GosGg
Olabode, Franklin O., +2348063139107 ofolabode@futa.edu.ng, Federal Univ of Tech, Akure; HwNgGg
Oladunjoye, M. A., ma.oladunjoye@mail.ui.edu.ng, Univ of Ibadan; Yg
Olafsen-Lackey, Susan, (402) 371-6512 slackey@unl.edu, Univ of Nebraska, Lincoln; Hw
Olarewaju, V. O., 234-803-403-2030 volarewa@oauife.edu.ng, Obafemi Awolowo Univ; GozCg
Olariu, Cornel, 512-471-1519 cornelo@jsg.utexas.edu, Univ of Texas, Austin; Gr
Olariu, Mariana, iulialet@yahoo.com, Univ of Texas, Austin; GruGs
Olatunji, A. S., +2348055475865 as.olatunji@mail.ui.edu.ng, Univ of Ibadan; CgGeCt
Olawuyi, Kehinde A., +2348107331222 gideonola2001@yahoo.com, Univ of Ilorin; YeGge
Olayinka, A. I., aa.elueze@mail.ui.edu.ng, Univ of Ibadan; Yg
Olayiwola, M. A., 234-806-448-7416 mayiwola@oauife.edu.ng, Obafemi Awolowo Univ; GugPi
Olcott Marshall, Alison, (785) 864-1917 olcott@ku.edu, Univ of Kansas; PgCoPo
Oldenburg, Douglas W., (604) 822-5406 doldenburg@eos.ubc.ca, Univ of British Columbia; Yg
Oldfield, Jonathan, +44 (0)121 414 2943 j.d.oldfield@bham.ac.uk, Univ of Birmingham; Zy
Oldham, Richard L., 568-3100 (12147) oldhamr@arc.losrios.edu, American River Coll; Gg
Oleinik, Anton, (561) 297-3297 aoleinik@fau.edu, Florida Atlantic Univ; Pi
Olesik, John W., (614) 292-6954 olesik.2@osu.edu, Ohio State Univ; Ca
Oleynik, Sergey, 609-258-2390 soleynik@princeton.edu, Princeton Univ; Cs
Olivares, Jose A., (505) 665-2643 Los Alamos; Cs
Oliveira, Alcino S., soliveir@utad.pt, Unive de Trás-os-Montes e Alto Douro; Hww
Oliveira, Anabela R., anarreis@utad.pt, Unive de Trás-os-Montes e Alto Douro; Gss
Oliver, Adolph A., (415) 786-6865 Chabot Coll; Ys
Oliver, Kevin, +44 (0)23 80596490 k.oliver@noc.soton.ac.uk, Univ of Southampton; Op
Oliver, Matthew J., 302-645-4079 moliver@udel.edu, Univ of Delaware; ObZrOg
Oliver, Ronald D., (702) 794-7095 ron_oliver@lanl.gov, Los Alamos; Yg
Olivo, Gema, 613 533-6998 olivo@queensu.ca, Queen's Univ; Eg
Ollinger, Scott, scott.ollinger@unh.edu, Univ of New Hampshire; Zc
Olmsted, Wayne, (406) 496-4151 Montana Tech of the Univ of Montana; Ca
Olney, Jessica L., 813-253-7647 jolney2@hccfl.edu, Hillsborough Comm Coll; ZgGg
Olney, Matthew P., molney@hccfl.edu, Hillsborough Comm Coll; ZgGg
Olorunfemi, A. O., 234-803-392-1712 akinrewa@oauife.edu.ng, Obafemi Awolowo Univ; Go
Olorunfemi, M. O., 234-903-719-2169 mlorunfe@oauife.edu.ng, Obafemi Awolowo Univ; Yg
Olsen, Amanda A., 207 581-2194 amanda.olsen@umit.maine.edu, Univ of Maine; Cl
Olsen, Khris B., (509) 376-4114 kb.olsen@pnl.gov, Pacific Northwest; Cg
Olsen, Kim B., 619-594-2649 kbolsen@mail.sdsu.edu, San Diego State Univ; Ye
Olsen, Paul E., kolsen@ldeo.columbia.edu, Columbia Univ; Gm
Olson, Derek, dolson@nps.edu, Naval Postgrad Sch; Zn
Olson, Hillary C., (512) 653-8356 holson@austin.utexas.edu, Univ of Texas, Austin; Ps
Olson, James R., 605-677-6866 jim.olson@usd.edu, South Dakota Dept of Environment and Nat Res; Zn
Olson, Jeff, olsonjl@uww.edu, Univ of Wisconsin, Whitewater; Ziu
Olson, Kenneth R., (217) 333-9639 Univ of Illinois, Urbana-Champaign; Sc
Olson, Kimberly, olsonk@arc.losrios.edu, American River Coll; Zye
Olson, Peter L., (410) 516-7707 olson@jhu.edu, Johns Hopkins Univ; Yg
Olson, Stephanie, 765-496-0810 stephanieolson@purdue.edu, Purdue Univ; CgGg
Olson, Ted L., 435 283-7533 ted.olson@snow.edu, Snow Coll; Ys
Olson Hoal, Karin E., keo52@cornell.edu, Cornell Univ; Gz
Olsson, Richard K., 848-445-3043 olsson@eps.rutgers.edu, Rutgers, The State Univ of New Jersey; PmGr
Olszewski, Kathy, (718) 409-7366 kolszewski@sunymaritime.edu, SUNY, Maritime Coll; Cg
Oltman-Shay, Joan M., (425) 644-9660 Univ of Washington; Op
Olugboji, Tolulope, (585) 276-6609 tolulope.olugboji@rochester.edu, Univ of Rochester; Ys
Oluwajana, Oladotun A., +234-8037708544 Federal Univ of Tech, Akure; GsPgGo
Olyphant, Greg A., (812) 855-5154 olyphant@indiana.edu, Indiana Univ, Bloomington; Hw
Omar, Gomaa I., (215) 898-6908 gomar@sas.upenn.edu, Univ of Pennsylvania; Cc
Omotoso, O. A., deletoso2002@yahoo.com, Univ of Ilorin; Ge
Oms Llobet, Oriol, ++34935811218 joseporiol.oms@uab.cat, Univ Auto de Barcelona; GrsYm
Omta, Anne W., awo13@case.edu, Case Western Reserve Univ; Op
Onac, Petroniu (Bogdan), bonac@usf.edu, Univ of South Florida; Pc
Onasch, Charles M., (419) 372-7197 conasch@bgnet.bgsu.edu, Bowling Green State Univ; Gc
Onderdonk, Nate, (562) 985-2654 nate.onderdonk@csulb.edu, California State Univ, Long Beach; Gtm
ONeal, Michael, 302-831-8273 michael@udel.edu, Univ of Delaware; Gm
Ono, Shuhei, (617) 253-0474 sono@mit.edu, Massachusetts Inst of Tech; PgCs
Onyeobi, Tony U., tony.onyeobi@uniben.edu, Univ of Benin; NrgGq
Oommen, Thomas, (906) 487-2045 toommen@mtu.edu, Michigan

Technological Univ; NeZiNg
Oona, Hain, (505) 667-5685 Los Alamos; Zn
Oostdam, Bernard L., 717-871-9928 boostdam@hotmail.com, Millersville Univ; Ou
Opeloye, Saka A., +2348060981972 saopeloye@futa.edu.ng, Federal Univ of Tech, Akure; GsPgGo
Ophori, Duke U., (973) 655-7558 ophorid@mail.montclair.edu, Montclair State Univ; Hw
Oppenheimer, Michael, 609-258-2338 omichael@princeton.edu, Princeton Univ; AsGe
Opper, Carl, 727-791-2536 opper.carl@spcollege.edu, Saint Petersburg Coll, Clearwater; Hw
Opperman, William, wopperma@broward.edu, Broward Coll; Gg
Oppo, Davide, 337-482-6676 davide.oppo@louisiana.edu, Univ of Louisiana at Lafayette; Gs
Oppo, Delia W., 508-289-2681 doppo@whoi.edu, Woods Hole Oceanographic Inst; Pm
Oppong, Joseph R., (940) 565-2181 oppong@unt.edu, Univ of North Texas; Zn
Orchard, Michael, 604-669-0909 morchard@nrcan.gc.ca, Univ of British Columbia; PgGg
Orcutt, John A., (858) 534-2887 jorcutt@ucsd.edu, Univ of California, San Diego; YrsYd
Orejana, David, dorejana@ucm.es, Univ Compl de Madrid; Gig
Oreskes, Naomi, 617-495-3480 oreskes@fas.harvard.edu, Harvard Univ; Rh
Orians, Kristin J., (604) 822-6571 korians@eoas.ubc.ca, Univ of British Columbia; Cm
Origlia, Hector D., +5435846761998 doriglia@exa.unrc.edu.ar, Univ Nacional de Rio Cuarto; NgZn
Orland, Ian, (608) 890-3328 ian.orland@wisc.edu, Univ of Wisconsin, Madison, Division of Extension; PcCls
Orlandini, Kent A., (630) 252-4236 Argonne; Cc
Orlanski, Isidoro, (609) 258-1319 isidoro.orlanski@noaa.gov, Princeton Univ; AsZoc
Orlove, Benjamin S., bsorlove@ucdavis.edu, Columbia Univ; As
Orme, Amalie, (818) 677-3532 amalie.orme@csun.edu, California State Univ, Northridge; Gm
Ormerod, Kerri Jean , (775) 784-6347 kormerod@unr.edu, Univ of Nevada, Reno; Zu
Ormsby, Alison, 828-232-5182 aormsby@unca.edu, Univ of North Carolina, Asheville; Zn
Orndorff, Richard L., (509) 359-2855 rorndorff@ewu.edu, Eastern Washington Univ; Ge
Orphan, Victoria, (626) 395-1786 vorphan@gps.caltech.edu,CalTech; Pg
Orr, Robert D., (209) 376-2481 robert.orr@pnl.gov, Pacific Northwest; Ca
Orr, William , orrw@pdx.edu, Portland State Univ; GgOg
Orr, William N., (541) 346-4577 worr@darkwing.uoregon.edu, Univ of Oregon; Pm
Orrock, John, orrock@wustl.edu, Washington Univ in St. Louis; Zn
Orsi, Alejandro H., (979) 845-4014 aorsi@ocean.tamu.edu, Texas A&M Univ; Op
Ort, Michael H., (928) 523-9363 michael.ort@nau.edu, Northern Arizona Univ; Gv
Ortega, Lorena, 34 913944837 lortega@ucm.es, Univ Compl de Madrid; EmgGg
Ortegren, Jason, 850-474-2491 jortegren@uwf.edu, Univ of West Florida; PeZu
Ortiz, Alejandra C., acortiz@colby.edu, Colby Coll; Gm
Ortiz, Joseph D., 330-672-2225 jortiz@kent.edu, Kent State Univ; Gs
Ortiz, Luis, lortizur@gmu.edu, George Mason Univ; As
Ortmann, Anthony L., 270-809-6755 aortmann@murraystate.edu, Murray State Univ; GaYg
Ortt, Jr., Richard A., (410) 554-5503 richard.ortt@maryland.gov, Maryland Dept of Nat Res; NtGeYr
Orville, Richard E., (979) 845-7671 rorville@tamu.edu, Texas A&M Univ; As
Osagiede, Edoseghe E., edoseghe.osagiede@uniben.edu, Univ of Benin; GctGx
Osborn, Dan P., dan.osborn1@ucl.ac.uk, Univ Coll London; Zn
Osborn, Gerald D., (403) 220-6448 Univ of Calgary; Gm
Osborn, Joe, (651) 779-3434 joe.osborn@century.edu, Century Coll; Zg
Osborn, Stephen G., (909)869-3009 sgosborn@csupomona.edu, California State Polytechnic Univ, Pomona; HwClSp
Osborn, Timothy J., +44 (0)1603 59 2089 t.osborn@uea.ac.uk, Univ of East Anglia; At
Osborne, Todd, 352-392-1803 osbornet@ufl.edu, Univ of Florida; Sf
Osburn, Chris, 919-515-0382 chris_osburn@ncsu.edu, North Carolina State Univ; Cms
Osburn, Glenn (Bob), (314) 935-7361 osburn@wustl.edu, Washington Univ in St. Louis; Zn
Osburn, Magdalena R., 847-491-4254 maggie@earth.northwestern.edu, Northwestern Univ; PoCbs
Oschlies, Andreas, aoschlies@geomar.de, Dalhousie Univ; Ob
Osi, Attila, hungaros@freemail.hu, Eotvos Lorand Univ; PviPe
Osinski, Gordon R., 519-661-2111 ext.84208 gosinbski@uwo.ca, Western Univ; XgcGp
Oskin, Michael, (530) 752-3993 meoskin@ucdavis.edu, Univ of California, Davis; GmtGc
Osleger, David A., (530) 754-7824 daosleger@ucdavis.edu, Univ of California, Davis; Pc
Osman, Mutasim S., +966138603184 mutasimsami@kfupm.edu.sa, King Fahd Univ of Petroleum and Minerals; GosGc
Osmond, Deanna L., (919) 515-7303 deanna_osmond@ncsu.edu, North Carolina State Univ; Sb
Osmond, John K., (850) 644-5860 osmond@gly.fsu.edu, Florida State Univ; Cc
Oster, Jessica L., 615-322-1461 jessica.l.oster@vanderbilt.edu, Vanderbilt Univ; ClsPc
Osterberg, Erich C., erich.c.osterberg@dartmouth.edu, Dartmouth Coll; AsCl
Oswald, Peter J., pjoswald@uaa.alaska.edu, Univ of Alaska, Anchorage; Gg
Otamendi, Juan E., +543584676198 jotamendi@exa.unrc.edu.ar, Univ Nacional de Rio Cuarto; CgaCp
Otis, Gard, (519) 814-4120 Ext.52478 gotis@uoguelph.ca, Univ of Guelph; Zn
Ott, Kevin C., (505) 667-4600 Los Alamos; Zn
Ottavi-Pupier, Elsa, +33 (0)3 44069358 elsa.ottavi-pupier@lasalle-beauvais.fr, Institut Polytechnique LaSalle Beauvais (ex-IGAL); GvzGi
Otterstrom, Samuel M., (801) 422-7751 samuel_otterstrom@byu.edu, Brigham Young Univ; Zu
Ouimet, William, 860-486-3772 william.ouimet@uconn.edu, Univ of Connecticut; GmZy
Ouimette, Mark A., (325) 670-1383 ouimette@hsutx.edu, Hardin-Simmons Univ; GieGt
Over, D. Jeffrey, (585) 245-5294 over@geneseo.edu, SUNY, Geneseo; Ps
Overland, James E., james.e.overland@noaa.gov, Univ of Washington; As
Overpeck, Jonathan T., 520-626-4345 jto@email.arizona.edu, Univ of Arizona; As
Oviatt, Candace, (401) 874-6132 coviatt@gso.uri.edu, Univ of Rhode Island; Ob
Oviatt, Charles G., (505) 304-6515 joviatt@ksu.edu, Kansas State Univ; Gm
Owen, Alan W., +4401413305461 alan.owen@glasgow.ac.uk, Univ of Glasgow; PgsPg
Owen, Donald E., (409) 880-8234 donald.owen@lamar.edu, Lamar Univ; Gr
Owen, Lewis, 513-410-2339 lewis.owen@uc.edu, Univ of Washington; GmCcGt
Owen, Robert M., (734) 763-4593 rowen@umich.edu, Univ of Michigan; CmGu
Owens, Brent E., (757) 221-1813 beowen@wm.edu, William & Mary; Gx
Owens, Tamera, (231) 777-0289 tamera.owens@muskegoncc.edu, Muskegon Comm Coll; Zn
Owens, Thomas J., 803-777-4530 owens@seis.sc.edu, Univ of South Carolina; Ys
Owensby, Clenton E., (785) 532-7232 owensby@ksu.edu, Kansas State Univ; Sf
Owoseni, Joshua O., +2348060719411 joowoseni@futa.edu.ng, Federal Univ of Tech, Akure; HwNgRn
Oxford, Jeremiah, 503-838-8680 oxfordj@wou.edu, Western Oregon Univ; Zg
Oxley, Meghan, 206-543-8904 what@uw.edu, Univ of Washington; Ze
Oxner-Jones, D. M., 740 201 2761 dalton.jones@dnr.state.oh.us, Ohio Dept of Nat Res; OnGgZg
Oyawale, A. A., 234-803-331-2150 aoyawale@oauife.edu.ng, Obafemi Awolowo Univ; GeoCg
Oyediran, I. A., ia.oyediran@mail.ui.edu.ng, Univ of Ibadan; Ng
Oymayan, Avo, avo.oymayan@fgc.edu, Florida Gateway Coll; Zg
Oza, Rupal, 212-650-3035 rupal.oza@hunter.cuny.edu, Hunter Coll (CUNY); Zn
Ozdenerol, Esra, 901-678-2787 eozdenrl@memphis.edu, Univ of Memphis; Zi
Ozel Yildirim, Esra , 00904242370000-5953 eozel@firat.edu.tr, Firat Univ; Gxi
Ozkan-Haller, Tuba, 541-737-9170 ozkan@coas.oregonstate.edu, Oregon State Univ; On
Ozsvath, David L., (715) 346-2287 dozsvath@uwsp.edu, Univ of Wisconsin, Stevens Point; Hw
Oztekin Okan, Ozlem, 00904242370000-5983 ooztekin@firat.edu.tr, Firat Univ; HgwHs
Ozturk, Nevin, 00904242370000-5956 nevinozturk@firat.edu.tr, Firat Univ;

Cg

P

Paavola, Jouni, +44(0) 113 34 36787 j.paavola@leeds.ac.uk, Univ of Leeds; Ge
Pace, Michael L., (434) 924-6541 mlp5fy@virginia.edu, Univ of Virginia; Zn
Pacheco, Fernando A., fpacheco@utad.pt, Unive de Trás-os-Montes e Alto Douro; Hq
Pacia, Christina, (908) 737-3738 cpacia@kean.edu, Kean Univ; Zn
Padden, Maureen, (905) 525-9140 (Ext. 20118) paddenm@mcmaster.ca, McMaster Univ; CsPc
Padian, Kevin, (510) 642-7434 kpadian@berkeley.edu, Univ of California, Berkeley; Pv
Padilla, Dianna K., (631) 632-7434 dianna.padilla@stonybrook.edu, SUNY, Stony Brook; ObnZn
Paduan, Jeffrey D., (831) 656-3350 paduan@nps.edu, Naval Postgrad Sch; Op
Paez, H. A., 905-525-9140 (Ext. 26099) paezha@mcmaster.ca, McMaster Univ; Eg
Page, David, 775-673-7110 dave.page@dri.edu, Desert Research Inst; Ga
Page, F Zeb, (440) 775-6701 zeb.page@oberlin.edu, Oberlin Coll; GpCsg
Page, Kevin, +44 1752 584750 kevin.page@plymouth.ac.uk, Univ of Plymouth; Gr
Page, Philippe, philippe_page@uqac.ca, Univ du Quebec a Chicoutimi; GiCgEm
Pagiatakis, Spiros, (416)736-2100 #77757 spiros@yorku.ca, York Univ; Yd
Pagliari, Paulo H., 507-752-5065 pagli005@umn.edu, Univ of Minnesota, Twin Cities; Sc
Paglione, Timothy, 718-262-2654 tpaglione@york.cuny.edu, York Coll (CUNY); Xa
Pagnac, Darrin C., 605-394-2469 darrin.pagnac@sdsmt.edu, South Dakota Sch of Mines & Tech; PvGrZe
Paige, David A., (310) 825-4268 dap@epss.ucla.edu, Univ of California, Los Angeles; Xy
Paillet, Fred, 479-575-3355 fredp@cox.net, Univ of Arkansas, Fayetteville; YgHw
Pain, Christopher, +44 20 759 49322 c.pain@imperial.ac.uk, Imperial Coll; Yg
Paine, Jeffrey G., 512-471-1260 jeff.paine@beg.utexas.edu, Univ of Texas, Austin; On
Pair, Donald, (937) 229-2936 don.pair@notes.udayton.edu, Univ of Dayton; Gl
Pakhomov, Evgeny, (604) 827-5564 epakhomov@eos.ubc.ca, Univ of British Columbia; Ob
Paktunc, Dogan, 613-947-7061 dogan.packtunc@nrcan-rncan.gc.ca, Univ of Ottawa; Gz
Palace, Michael W., 603-862-4193 michael.palace@unh.edu, Univ of New Hampshire; Zn
Palamartchouk, Kirill, +44 191 208 6421 kirill.palamartckirill.palamartchouk@ncl.ac.uk, Univ of Newcastle Upon Tyne; Yd
Palenik, Brian, (858) 534-7505 bpalenik@ucsd.edu, Univ of California, San Diego; Ob
Palevsky, Hilary I., (617) 552-4936 hilary.palevsky@bc.edu, Boston Coll; Oc
Palin, J. Michael, +64 3 479-9083 michael.palin@otago.ac.nz, Univ of Otago; CcGip
Palinkaš, Ladislav, +38514605971 lpalinka@geol.pmf.hr, Univ of Zagreb; CgeCp
Palladino, Steve D., (805) 654-6400 (Ext. 1365) Ventura Coll; Zi
Pallis, Ted J., (609) 984-6587 ted.pallis@dep.nj.gov, New Jersey Geological and Water Survey; Zi
Pallister, John S., (360) 993-8964 jpallist@usgs.gov, Univ of Pittsburgh; GviGg
Palluconi, Frank D., (818) 354-8362 frank.d.palluconi@jpl.nasa.gov, Jet Propulsion Lab; Zr
Palmer, Arthur N., (607) 436-3064 arthur.palmer@oneonta.edu, SUNY, Oneonta; HgqGm
Palmer, Christina, (909) 537-5336 cpalmer@csusb.edu, California State Univ, San Bernardino; Zn
Palmer, Clare, cpalmer@wustl.edu, Washington Univ in St. Louis; Zn
Palmer, Donald F., (330) 672-2680 dpalmer@kent.edu, Kent State Univ; Yg
Palmer, Evan, 719-333-3080 ronald.palmer@usafa.edu, United States Air Force Acad; Zig
Palmer, H. C., (519) 661-2111 x86749 cpalmer@uwo.ca, Western Univ; YmGv
Palmer, Martin R., +44 (0)23 80596607 pmrp@noc.soton.ac.uk, Univ of Southampton; Cg
Palmer, Paul, +44 (0) 131 650 7724 paul.palmer@ed.ac.uk, Edinburgh Univ; As
Palmquist, John C., (414) 832-6732 palmquij@lawrence.edu, Lawrence Univ; GcpGg
Paltseva, Anna, anna.paltseva@louisiana.edu, Univ of Louisiana at Lafayette; Ge
Palumbo, Tony V., (423) 576-8002 avp@ornl.gov, Oak Ridge; Sb
Palutoglu, Mahmut, 00904242370000-5954 mpalutoglu@firat.edu.tr, Firat Univ; YgsGc
Pan, Feifei, 940-369-5109 feifei/pan@unt.edu, Univ of North Texas; Hw
Pan, Ruiguang, 575-835-5452 ruiguang.pan@nmt.edu, New Mexico Tech; CpGpEd
Pan, William L., (509) 335-3611 wlpan@wsu.edu, Washington State Univ; Sb
Pan, Ying, yyp5053@psu.edu, Pennsylvania State Univ, Univ Park; As
Pan, Yuanming, (706) 966-5699 yuanming.pan@usask.ca, Univ of Saskatchewan; Gx
Pan, Zaitao, (314) 977-3114 zaito.pan@eas.slu.edu, Saint Louis Univ; As
Panagiotakopulu, Eva, +44 (0) 131 650 2531 eva.p@ed.ac.uk, Edinburgh Univ; Pe
Panah, Assad I., (814) 362-7569 aap@pitt.edu, Univ of Pittsburgh, Bradford; GorGc
Pancha, Aasha, (775) 784-4254 pancha@seismo.unr.edu, Univ of Nevada, Reno; Ys
Panero, Wendy R., 614 292 6290 panero.1@osu.edu, Ohio State Univ; Yg
Panetta, Richard L., (979) 845-1386 r-panetta@tamu.edu, Texas A&M Univ; Am
Pangle, Luke, lpangle@gsu.edu, Georgia State Univ; HqCs
Pani, Eric A., (318) 342-1878 Univ of Louisiana, Monroe; As
Paniconi, Claudio, claudio.paniconi@ete.inrs.ca, Univ du Quebec; Hw
Panish, Peter T., (413) 545-2593 panish@geo.umass.edu, Univ of Massachusetts, Amherst; Gp
Pankow, Kristine L., 801-585-6484 pankow@seis.utah.edu, Univ of Utah; Ys
Panno, Samuel V., (217) 244-2456 s-panno@illinois.edu, Illinois State Geo Survey; ClGgHw
Pant, Hari, 718-960-5859 hari.pant@hunter.cuny.edu, Graduate Sch of the City Univ of New York; Zg
Panter, Kurt S., (419) 372-7337 kpanter@bgnet.bgsu.edu, Bowling Green State Univ; Giv
Panzik, Joeseph, jepanzik@usf.edu, Univ of South Florida; Ym
Paola, Christopher, 612-624-8025 cpaola@umn.edu, Univ of Minnesota, Twin Cities; Gs
Papaleo, Silvanna, spapaleo@zircon.geology.utoronto.ca, Univ of Toronto; Zn
Papangelakis, Elli, 905-525-9140 ext. 27594 papangee@mcmaster.ca, McMaster Univ; Gm
Papcun, George, 12012993161 gpapcun@njcu.edu, New Jersey City Univ; Ze
Papendick, Robert I., (509) 335-1552 papendick@wsu.edu, Washington State Univ; Sp
Papike, James J., (505) 277-1633 jpapike@unm.edu, Univ of New Mexico; Ca
Papineau, Dominic, d.papineau@ucl.ac.uk, Univ Coll London; Cg
Papitashvili, Vladimir O., 703-292-8033 vpapita@nsf.gov, Natl Sci Foundation; YmXpa
Papp, Kenneth R., (907) 696-0079 kenneth.papp@alaska.gov, Alaska Division of Geological & Geophysical Surveys; Gg
Pappas, Matthew, 631-451-4301 pappasm@sunysuffolk.edu, Suffolk County Comm Coll, Ammerman Campus; Zg
Papuga, Shirley, (313) 577-9436 shirley.papuga@wayne.edu, Wayne State Univ; Hgs
Paquette, Jeanne, 514-398-4402 jeanne.paquette@mcgill.ca, McGill Univ; Gz
Paradis, Charles, paradisc@uwm.edu, Univ of Wisconsin, Milwaukee; Hw
Paradise, Thomas R., (479) 575-3159 paradise@uark.edu, Univ of Arkansas, Fayetteville; GmZyRn
Parai, Rita, (314) 935-3974 rparai@levee.wustl.edu, Washington Univ in St. Louis; Cs
Parashar, Rishi, 775.673-7496 rishi.parashar@dri.edu, Univ of Nevada, Reno; Hq
Parcell, William C., (316)978- 3140 Wichita State Univ; Gr
Pardi, Richard R., (201) 595-2695 William Paterson Univ; Cc
Parendes, Laurie A., (814) 732-2840 lparendes@edinboro.edu, Edinboro Univ of Pennsylvania; Zn
Parent, Michel, (418) 654-2657 Univ du Quebec; Gl
Pares, Josep M., (734) 615-0472 jmpares@umich.edu, Univ of Michigan; Ym
Parham, James, (657) 278-2043 jparham@fullerton.edu, California State Univ, Fullerton; PvgGg
Parikh, Sanjai, 530-752-1265 sjparikh@ucdavis.edu, Univ of California, Davis; ScbCb
Parise, John B., (631) 632-8196 john.parise@stonybrook.edu, Stony Brook

Univ; Gz
Pariseau, William G., (801) 581-5164 w.pariseau@utah.edu, Univ of Utah; Nr
Parish, Cynthia L., cparish@gt.rr.com, Lamar Univ; Gg
Parish, Ryan M., 901-678-2606 rmparish@memphis.edu, Univ of Memphis; Ga
Parizek, Richard R., (814) 865-3012 parizek@ems.psu.edu, Pennsylvania State Univ, Univ Park; Hw
Park, Jeffrey J., (203) 432-3172 jeffrey.park@yale.edu, Yale Univ; YsZgc
Park, Jisun, 201-575-5939 jisun.park@kbcc.cuny.edu, Graduate Sch of the City Univ of New York; Xm
Park, Myung-Sook, 831-656-2858 mpark@nps.edu, Naval Postgrad Sch; Am
Park, Sohyun, 089/2180 4333 soohyun.park@physik.uni-muenchen.de, Ludwig-Maximilians-Univ Muenchen; Gz
Park, Stephen K., (951) 827-4501 magneto@ucrmt.ucr.edu, Univ of California, Riverside; Ym
Parker, David B., 806-651-4099 dparker@wtamu.edu, West Texas A&M Univ; Sbf
Parker, Don, 806-291-1121 don.parker@wbu.edu, Wayland Baptist Univ; GivCg
Parker, Doug, +44(0) 113 34 36739 d.j.parker@leeds.ac.uk, Univ of Leeds; Am
Parker, Gary, 217-244-6161 parkerg@illinois.edu, Univ of Illinois, Urbana-Champaign; Gm
Parker, Jim, 713-743-6750 jlparker9@uh.edu, Univ of Houston; Zn
Parker, Joan, 831-771-4400 parker@mlml.calstate.edu, Moss Landing Marine Lab; Zn
Parker, Kent E., (509) 373-6337 kent.parker@pnl.gov, Pacific Northwest; Cg
Parker, Matthew, 919-513-4367 mdparker@ncsu.edu, North Carolina State Univ; Am
Parker, Robert L., (858) 534-2475 rlparker@ucsd.edu, Univ of California, San Diego; Ym
Parker, Stephen R., (406) 496-4185 sparker@mtech.edu, Montana Tech of the Univ of Montana; Zn
Parker, William C., (850) 644-1568 parker@gly.fsu.edu, Florida State Univ; PqeGq
Parkin, Gary, (519) 824-4120 (Ext. 52452) gparkin@uoguelph.ca, Univ of Guelph; HwSp
Parkin, Geoffrey, +44 (0) 191 208 6146 geoff.parkin@ncl.ac.uk, Univ of Newcastle Upon Tyne; Hg
Parkinson, Christopher D., 504-280-6792 cparkins@uno.edu, Univ of New Orleans; Gp
Parkovic, Emily, 717-702-2063 eparkovic@pa.gov, DCNR- Pennsylvania Bur of Geo Survey; Zn
Parks, George K., (510) 643-5512 parks@ssl.berkeley.edu, Univ of Washington; YxXy
Parlak, Osman, 03223387081 parlak@cu.edu.tr, Cukurova Univsi; GipCc
Parman, Stephen, (401) 863-3352 stephen_parman@brown.edu, Brown Univ; CpgCt
Parmentier, E. Marc, (401) 863-1700 em_parmentier@brown.edu, Brown Univ; Yg
Parnell, Roderic A., (928) 523-3329 roderic.parnell@nau.edu, Northern Arizona Univ; Cl
Parnella, Bill, 302-831-3156 parnella@udel.edu, Univ of Delaware; Gg
Parr, Kate, +44 0151 795 4640 kate.parr@liverpool.ac.uk, Univ of Liverpool; Pe
Parrick, Brittany, 614 265 6581 brittany.parrick@dnr.state.oh.us, Ohio Dept of Nat Res; Gg
Parris, Thomas M., 859-257-1147 mparris@uky.edu, Univ of Kentucky; Cg
Parrish, Christopher E., 603-862-3438 cparrish@ccom.unh.edu, Univ of New Hampshire; Zr
Parrish, Pia, (619) 594-5587 pparrish@mail.sdsu.edu, San Diego State Univ; Zn
Parrot, Lael, 250-807-8122 lael.parrot@ubc.ca, Univ of British Columbia, Okanagan Campus; ZgiZf
Parsekian, Andrew D., (307) 766-3603 aparseki@uwyo.edu, Univ of Wyoming; Yg
Parsen, Mike J., 608629419 mike.parsen@wisc.edu, Univ of Wisconsin, Madison, Division of Extension; Hw
Parsons, Barry, +44 (1865) 272017 barry.parsons@earth.ox.ac.uk, Univ of Oxford; Yd
Parsons, Jeffrey D., (206) 221-6627 parsons@ocean.washington.edu, Univ of Washington; Ou
Parsons-Hubbard, Karla M., (440) 775-8353 karla.hubbard@oberlin.edu, Oberlin Coll; PgGu
Partin, Camille, camille.partin@usask.ca, Univ of Saskatchewan; Gtc
Partin, Jud, 512-471-5172 jpartin@ig.utexas.edu, Univ of Texas, Austin; CsAt
Pascariu, Florentina, 0040232201496 florentinavieru@yahoo.com, Alexandru Ioan Cuza; Nt
Paschert, Karin, 089/2180 6573 karin.paschert@lmu.de, Ludwig-Maximilians-Univ Muenchen; Gg
Pascoe, Richard, +4401326 371838 r.d.pascoe@exeter.ac.uk, Exeter Univ; Gz
Pascussi, Michael, (518) 486-4820 mpascucc@mail.nysed.gov, New York State Geo Survey; Eo
Pasek, Matthew, mpasek@usf.edu, Univ of South Florida; Cq
Pashin, Jack C., 405-744-6358 jack.pashin@okstate.edu, Oklahoma State Univ; GrcGo
Pasicznyk, David L., (609) 984-6587 dave.pasicznyk@dep.state.nj.us, New Jersey Geological and Water Survey; Yg
Pasken, Robert W., (314) 977-3125 robert.pasken@slu.edu, Saint Louis Univ; Am
Passey, Benjamin, (410) 340-9473 passey@umich.edu, Univ of Michigan; CsaCl
Pasteris, Jill D., (314) 935-5434 pasteris@levee.wustl.edu, Washington Univ in St. Louis; GbzGe
Pasternack, Gregory B., (530) 754-9243 gpast@ucdavis.edu, Univ of California, Davis; GmHg
Patel, Vinodkumar A., 217-244-0639 patel@isgs.uiuc.edu vapatel@illinois.edu, Illinois State Geo Survey; Zn
Patenaude, Genevieve, +44 (0) 131 651 4472 genevieve.patenaude@ed.ac.uk, Edinburgh Univ; Ge
Patera, Edward S., (505) 667-4457 Los Alamos; Cg
Paterson, Colin J., (605) 394-2461 colin.paterson@sdsmt.edu, South Dakota Sch of Mines & Tech; Eg
Paterson, John R., +61-2-67732101 jpater20@une.edu.au, Univ of New England; PigPs
Paterson, Scott R., (213) 740-6103 paterson@usc.edu, Univ of Southern California; Gc
Patience, Jennifer, (480) 727-8554 jennifer.patience@asu.edu, Arizona State Univ; Xa
Patino-Douce, Alberto E., (706) 542-2394 alpatino@uga.edu, Univ of Georgia; ZnnZn
Patino-Douce, Marta, (706) 542-2399 mapatino@uga.edu, Univ of Georgia; Gi
Paton, Douglas, +44(0) 113 34 35238 d.a.paton@leeds.ac.uk, Univ of Leeds; Gc
Patrick, David M., (601) 266-4530 Univ of Southern Mississippi; Ng
Patronas, Dennis, dpatronas@disl.org, Dauphin Island Sea Lab; Zn
Patterson, Gary, (901) 678-5264 glpttrsn@memphis.edu, Univ of Memphis; Gg
Patterson, Jodi T., (801) 537-3310 jpatters@utah.gov, Utah Geo Survey; Zn
Patterson, Mark, (781) 581-7370 m.patterson@neu.edu, Northeastern Univ; Zn
Patterson, Molly , (607) 777-2831 patterso@binghamton.edu, Binghamton Univ; CmGl
Patterson, R. Timothy, tpatters@ccs.carleton.ca, Carleton Univ; Pm
Patterson, William P., (306) 966-5691 bill.patterson@usask.ca, Univ of Saskatchewan; CsPeGn
Pattison, David R., (403) 220-3263 pattison@geo.ucalgary.ca, Univ of Calgary; Gp
Pattison, Simon A., 204-727-7468 pattison@brandonu.ca, Brandon Univ; GsoGd
Patton, Howard J., (505) 667-1003 patton@lanl.gov, Los Alamos; Yg
Patton, James L., (510) 642-3059 Univ of California, Berkeley; Pv
Patton, Jason A., 479-968-0676 jpatton@atu.edu, Arkansas Tech Univ; GgeGc
Patton, Peter C., (860) 685-2268 ppatton@wesleyan.edu, Wesleyan Univ; Gm
Patton, Regan L., (509) 534-3670 rpatton@wsu.edu, Washington State Univ; Gc
Patton, Terri, 630-252-3294 tlpatton@anl.gov, Argonne; Gg
Patton, Thomas W., 406-496-4153 tpatton@mtech.edu, Montana Tech of The Univ of Montana; Hw
Patwardhan, Kaustubh, 845-257-3738 patwardk@newpaltz.edu, SUNY, New Paltz; Ggi
Paty, Carol M., (404) 894-2860 carol.paty@eas.gatech.edu, Georgia Inst of Tech; YmgZn
Patzkowsky, Mark E., (814) 863-1959 brachio@geosc.psu.edu, Pennsylvania State Univ, Univ Park; Pg
Paul, John H., (727) 553-1168 jpaul@marine.usf.edu, Univ of South Florida; Ob
Paul, Jonathan, +44 1784 44 jonathan.paul@rhul.ac.uk, Royal Holloway Univ of London; Hw
Paulsell, Robert L., (225) 578-8655 rpaulsell@lsu.edu, Louisiana State Univ; ZyGg
Paulsen, Timothy S., (920) 424-7002 paulsen@uwosh.edu, Univ of

Wisconsin, Oshkosh; Gc
Paulson, Suzanne, (310) 206-4442 paulson@atmos.ucla.edu, Univ of California, Los Angeles; As
Pavelsky, Tamlin M., (919) 962-4239 pavelsky@unc.edu, Univ of North Carolina, Chapel Hill; Hg
Pavese, Alessandro, alessandro.pavese@unito.it, Univ di Torino; Gz
Pavia, Giulio, giulio.pavia@unito.it, Univ di Torino; Pg
Pavlis, Gary L., (812) 855-5141 pavlis@indiana.edu, Indiana Univ, Bloomington; YsGtYe
Pavlis, Terry L., (504) 280-6797 tpavlis@uno.edu, Univ of New Orleans; Gt
Pavlovskaya, Marianna, (212) 772-5320 mpavlov@hunter.cuny.edu, Hunter Coll (CUNY); Zi
Pavlowsky, Robert T., (417) 836-8473 robertpavlowsky@missouristate.edu, Missouri State Univ; Gm
Paw U, Kyaw Tha, 530-752-8172 ktpawu@ucdavis.edu, Univ of California, Davis; Asm
Pawley, Alison, +44 0161 275-3944 alison.pawley@manchester.ac.uk, Univ of Manchester; Gc
Pawlicki, Caron E., (717) 702-2042 cpawlicki@pa.gov, DCNR- Pennsylvania Bur of Geo Survey; ZiGgc
Pawloski, Gayle A., (925) 423-0437 pawloski1@llnl.gov, Lawrence Livermore; Gg
Pawlowicz, Richard A., (604) 822-1356 rpawlowicz@eos.ubc.ca, Univ of British Columbia; Op
Payeur, Mason, payeurm@dnr.sc.gov, South Carolina Dept of Nat Res; Gg
Payn, Robert, rpayn@montana.edu, Montana State Univ; Hg
Payne, Ashley E., 734-647-1991 aepayne@umich.edu, Univ of Michigan; At
Payré, Valerie, 319467-1546 valerie-payre@uiowa.edu, Univ of Iowa; XgGiz
Paz Gonzalez, Antonio, 34 981167000 tucho@udc.es, Coruna Univ; Sd
Pazzaglia, Frank J., (610) 758-3667 fjp3@lehigh.edu, Lehigh Univ; Gm
Peacock, Caroline, +44(0) 113 34 37877 c.l.peacock@leeds.ac.uk, Univ of Leeds; Sb
Peacock, Simon M., simon.peacock@ubc.ca, Univ of British Columbia; GtpYs
Peacor, Donald R., (734) 764-1452 drpeacor@umich.edu, Univ of Michigan; Gz
Peakall, Jeff, +44(0) 113 34 35205 j.peakall@leeds.ac.uk, Univ of Leeds; Gs
Peake, Jeffrey S., jpeake@unomaha.edu, Univ of Nebraska, Omaha; ZyAtRm
Pearce, Danni, danni.pearce@nmbu.no, Norwegian Univ of Life Scis (NMBU); GlmZr
Pearce, J A., pearceja@cf.ac.uk, Cardiff Univ; Cg
Pearce, Jamie R., +44 (0) 131 650 2294 jamie.pearce@ed.ac.uk, Edinburgh Univ; Ge
Pearson, Adam, 315 267-2951 pearsoaj@potsdam.edu, SUNY Potsdam; Gm
Pearson, Ann, (617) 384-8392 pearson@eps.harvard.edu, Harvard Univ; Cb
Pearson, David M., 208-282-3486 peardavi@isu.edu, Idaho State Univ; Gct
Pearson, Eugene F., (209) 946-2926 epearson@pacific.edu, Univ of the Pacific; GsPgZg
Pearson, Graham, (780) 492-4156 gdpearson@ualberta.ca, Univ of Alberta; Cec
Pearson, P N., pearsonp@cf.ac.uk, Cardiff Univ; Pe
Pearson, Paul, +44(0)29 208 74579 pearsonp@cardiff.ac.uk, Univ of Wales; PmCg
Pearthree, Kristin, 575-835-5320 kristin.pearthree@nmt.edu, New Mexico Tech; Gge
Pearthree, Philip A., 520-621-2470 pearthre@email.arizona.edu, Univ of Arizona; Ge
Peate, David W., (319) 335-0567 david-peate@uiowa.edu, Univ of Iowa; GiCa
Peavy, Samuel T., 229-931-2330 speavy@canes.gsw.edu, Georgia Southwestern State Univ; Yg
Pec, Matej, 617-324-7279 mpec@mit.edu, Massachusetts Inst of Tech; Nr
Pechmann, James C., (801) 581-3858 pechmann@seis.utah.edu, Univ of Utah; YsGt
Peck, John A., (330) 972-7659 jpeck@uakron.edu, Univ of Akron; GseGn
Peck, Theodore R., (217) 333-9486 Univ of Illinois, Urbana-Champaign; Sc
Peck, William H., (315) 228-7698 wpeck@colgate.edu, Colgate Univ; GpiCs
Peddle, Derek R., derek.peddle@uleth.ca, Univ of Lethbridge; Zr
Pedemonte, Virginia, (598) 25251552 vpedemonte@fcien.edu.uy, Univ de la Republica Oriental del Uruguay (UDELAR); ZirZu
Pedentchouk, Nikolai, +44 (0)1603 59 3395 n.pedentchouk@uea.ac.uk, Univ of East Anglia; Cs
Pedersen, Joel A., (608) 263-4971 joelpedersen@wisc.edu, Univ of Wisconsin, Madison; So
Pedersen, Per, 403 220-8454 pkpeders@ucalgary.ca, Univ of Calgary; Go
Pedersen, Thomas F., tfp@uvic.ca, Univ of Victoria; CmOcCs
Pederson, Darryll T., (402) 472-7563 dpederson2@unl.edu, Univ of Nebraska, Lincoln; Hw
Pederson, Joel L., (435) 797-7097 joel.pederson@usu.edu, Utah State Univ; Gm
Pedley, Kate, +64 3 3667001 Ext 3892 kate.pedley@canterbury.ac.nz, Univ of Canterbury; Zg
Pedone, Vicki A., (818) 677-3541 vicki.pedone@csun.edu, California State Univ, Northridge; GdCs
Peebles, Ernst B., (727) 553-3983 epeebles@mail.usf.edu, Univ of South Florida; ObCs
Peele, R Hampton, (225) 328-6651 hampton@lsu.edu, Louisiana State Univ; ZirZg
Peeters, Frank J., +31 20 5987419 f.j.c.peeters@vu.nl, Vrije Univit Amsterdam; PmOg
Peirce, Christine, 0191 3742515 christine.peirce@durham.ac.uk, Durham Univ; Yg
Pekar, Stephen , 718-997-3305 stephen.pekar@qc.cuny.edu, Graduate Sch of the City Univ of New York; Zg
Pekar, Stephen, (718) 997-3305 stephen.pekar@qc.cuny.edu, Queens Coll (CUNY); PeGrOu
Pellenbarg, Robert, rpellenbarg@collegeofthedesert.edu, Coll of the Desert; OcGu
Pellerin, Jocelyne, (418) 724-1704 jocelyne_pellerin@uqar.uquebec.ca, Univ du Quebec a Rimouski; Ob
Pelletier, Emilien, (418) 723-1986 (Ext. 1764) emilien_pelletier@uqar.ca, Univ du Quebec a Rimouski; Oc
Pelletier, Jon D., (520) 626-2126 jdpellet@email.arizona.edu, Univ of Arizona; Gm
Pellin, Michael J., (630) 972-3510 Argonne; Zm
Pellowski, Christopher J., (605) 394-2465 christopher.pellowski@sdsmt.edu, South Dakota Sch of Mines & Tech; Gg
Pelton, John R., (208) 426-3640 jpelton@boisestate.edu, Boise State Univ; YgeYs
Peltzer, Gilles, (310) 206-2156 peltzer@epss.ucla.edu, Univ of California, Los Angeles; ZrGt
Pena dos Reis, Rui, 965042654 penareis@dct.uc.pt, Unive de Coimbra; GorGs
Pendergrass, Angeline G., 607-255-3587 a.pendergrass@cornell.edu, Cornell Univ; As
Pendleton, Simon, simon.pendleton@plymouth.edu, Plymouth State Univ; GlCcGe
Peng, Lee Chai, 03-79674233 leecp@um.edu.my, Univ of Malaya; Pg
Peng, Sheng, 512-232-7619 sheng.peng@beg.utexas.edu, Univ of Texas, Austin; GoHt
Peng, Yongbo, 225-578-3413 ypeng@lsu.edu, Louisiana State Univ; Cg
Peng, Zhigang, 404-894-0231 zhigang.peng@eas.gatech.edu, Georgia Inst of Tech; YsgZn
Penman, Donald E., donald.penman@usu.edu, Utah State Univ; GsCbPc
Penn, Justin, 609-258-4101 jpenn@princeton.edu, Princeton Univ; ZcObCb
Penna, Nigel, +44 (0) 191 208 8747 nigel.penna@ncl.ac.uk, Univ of Newcastle Upon Tyne; Gq
Pennacchioni, Giorgio, +390498279106 giorgio.pennacchioni@unipd.it, Univ degli Studi di Padova; Gc
Penney, Paulette, 254710-2361 paulette_penney@baylor.edu, Baylor Univ; Zn
Pennington, Deana, 915-747-5867 ddpennington@utep.edu, Univ of Texas, El Paso; ZyfZe
Penniston-Dorland, Sarah, (301) 405-4087 sarahpd@geol.umd.edu, Univ of Maryland; Gp
Pennock, Jonathan R., (334) 861-7531 jonathan.pennock@unh.edu, Dauphin Island Sea Lab; Oc
Penny, Andrew, 831-656-3101 abpenny@nps.edu, Naval Postgrad Sch; Am
Pentcheva, Rossitza, 089/2180 4352 pentcheva@lrz.uni-muenchen.de, Ludwig-Maximilians-Univ Muenchen; Gz
Peppe, Daniel J., (254) 710-2629 daniel_peppe@baylor.edu, Baylor Univ; PbcYm
Pepper, Ian L., (520) 621-7234 ipepper@ag.arizona.edu, Univ of Arizona; Zn
Peprah, Ebenezer, (310) 660-3593 epeprah@elcamino.edu, El Camino Coll; Zg
Perault, David R., (434) 544-8370 perault@lynchburg.edu, Univ of Lynchburg; Ge
Percell, Peter, 713-743-6724 ppercell@uh.edu, Univ of Houston; As
Percival, Carl, c.percival@manchester.ac.uk, Univ of Manchester; As
Percy, David, (503) 725-3373 percy@pdx.edu, Portland State Univ; Zi
Perdrial, Julia, (802) 656-0665 julia.perdrial@uvm.edu, Univ of Vermont; ClSc
Perdrial, Nicolas, nicolas.perdrial@uvm.edu, Univ of Vermont; GezCl
Perdue, Edward Michael, michael.perdue@eas.gatech.edu, Georgia Inst of Tech; Co
Pereira, Alcides C., +351-239 860 563 apereira@dct.uc.pt, Unive de

Coimbra; GebZr

Pereira, Engil I., 956 665 2220 engil.pereira@utrgv.edu, Univ of Texas, Rio Grande Valley; SbZcSo

Peres, Carlos, +44 (0)1603 59 2549 c.peres@uea.ac.uk, Univ of East Anglia; Sf

Perez, Adriana, 915-747-5501 aperez13@utep.edu, Univ of Texas, El Paso; Gg

Perez, Florante, (213) 620-5026 California Geo Survey; Ng

Perez, Marco A., mperez@cicese.mx, Centro de Investigación Científica y de Educación Superior de Ensenada; Ye

Perez, Richard R., (518) 437-8751 perez@asrc.cestm.albany.edu, SUNY, Albany; As

Perez-Huerta, Alberto, (205) 348-8382 aphuerta@ua.edu, Univ of Alabama; PgClGz

Pérez-Rodríguez, Ileana, ileperez@sas.upenn.edu, Univ of Pennsylvania; Po

Pérez-Soba, Cecilia, pesoa@ucm.es, Univ Compl de Madrid; Gi

Perfect, Edmund, (865) 974-6017 eperfect@utk.edu, Univ of Tennessee, Knoxville; SpHwGq

Perfit, Michael R., (352) 392-2128 mperfit@ufl.edu, Univ of Florida; Giu

Peri, Francesco, francesco.peri@umb.edu, Univ of Massachusetts, Boston; Zn

Perkins, R, perkinsr@cf.ac.uk, Cardiff Univ; Zn

Perkins, Robert B., rperkins@pdx.edu, Portland State Univ; Hq

Perkins, Ronald D., (919) 684-3376 rperkins@duke.edu, Duke Univ; Gdo

Perkins, III, Dexter, (701) 777-2991 dexter.perkins@engr.und.edu, Univ of North Dakota; Gp

Perkis, Bill, bill.perkis@gogebic.edu, Gogebic Comm Coll; Gg

Perkons, Eriks, (410) 704-3187 eperkons@towson.edu, Towson Univ; Ge

Peros, Matthew, 819-822-9600 xt. 2783 mperos@ubishops.ca, McMaster Univ; GaPg

Perret, Didier H., (418) 654-2686 didier.perret@canada.ca, Univ du Quebec; NgeZu

Perrie, William, (902) 426-3985 william.perrie@dfo-mpo.gc.ca, Dalhousie Univ; OpAm

Perrin, Richard E., (505) 667-4755 Los Alamos; Cc

Perron, Taylor, (617) 253-5735 perron@mit.edu, Massachusetts Inst of Tech; Gm

Perrouty, Stéphane, (705) 675-1151 Ext. 2212 sperrouty@laurentian.ca, Laurentian Univ, Sudbury; EgGcYg

Perry, Baker, (828) 262-7597 perrylb@appstate.edu, Appalachian State Univ; Zg

Perry, David J., (415) 786-6887 Chabot Coll; Op

Perry, Frank V., (505) 667-1033 fperry@lanl.gov, Los Alamos; Gv

Perry, Kevin D., (801) 581-6138 kevin.perry@utah.edu, Univ of Utah; As

Perry , Kyle, 859-257-0133 kyle.perry@uky.edu, Univ of Kentucky; Nr

Perry, Randall, +44 20 759 46425 r.perry@imperial.ac.uk, Imperial Coll; Zg

Persad, Geeta, geeta.persad@jsg.utexas.edu, Univ of Texas, Austin; As

Persaud, Naraine, (540) 231-3817 npers@vt.edu, Virginia Tech; Sp

Persaud, Patricia, 225-578-5676 ppersaud@lsu.edu, Louisiana State Univ; Ys

Perscio, Lyman, 814-824-2076 lpersico@mercyhurst.edu, Mercyhurst Univ; GsSpHg

Persico, Lyman P., 509-527-5157 persiclp@whitman.edu, Whitman Coll; Gm

Person, Arthur, (814) 863-8568 aap1@psu.edu, Pennsylvania State Univ, Univ Park; Am

Person, Jeff, jjperson@nd.gov, North Dakota Geo Survey; Pg

Person, Mark A., (575) 835-6506 mperson@ees.nmt.edu, New Mexico Tech; HyYhg

Persons, W. Scott, personss@cofc.edu, Coll of Charleston; PegGe

Pertl, David, djpertl@actx.edu, Amarillo Coll; Go

Perumal, Ram, (785) 625-3425 perumal@ksu.edu, Kansas State Univ; So

Pervunina, Aelita, +78142766039 aelita@krc.karelia.ru, Inst of Geology (Karelia, Russia); GpvEm

Peryea, Frank J., (509) 663-8181 Washington State Univ; Sc

Pessagno, Jr., Emile A., (972) 883-2444 pessagno@utdallas.edu, Univ of Texas, Dallas; Pm

Pesses, Michael, mpesses@avc.edu, Antelope Valley Coll; Zyi

Pestrong, Raymond, (415) 338-2080 rayp@sfsu.edu, San Francisco State Univ; Ng

Peteet, Dorothy M., (845) 365-8420 peteet@ldeo.columbia.edu, Columbia Univ; PelGn

Peter, Mark, 614 265 6529 mark.peter@dnr.state.oh.us, Ohio Dept of Nat Res; PgZe

Peterman, Emily M., 207-725-3846 epeterma@bowdoin.edu, Bowdoin Coll; Gtp

Peters, Catherine A., (609) 258-5645 cap@phoenix.princeton.edu, Princeton Univ; Hw

Peters, Mark T., (202) 586-9279 mark_peters@notes.ymp.gov, Los Alamos; Yg

Peters, Roger M., roger.peters@wisc.edu, Univ of Wisconsin, Madison, Division of Extension; Gg

Peters, Shanan, 608-262-5987 peters@geology.wisc.edu, Univ of Wisconsin, Madison; Gs

Peters, Stephen C., 610/758-3660 scp2@lehigh.edu, Lehigh Univ; Cl

Petersen, Bruce, (314) 935-5643 petersen@wustl.edu, Washington Univ in St. Louis; Zn

Petersen, Erich U., (801) 581-7238 erich.petersen@utah.edu, Univ of Utah; Em

Petersen, Gary W., (814) 865-1540 gwp2@psu.edu, Pennsylvania State Univ, Univ Park; Sd

Petersen, Nikolai, 0049/89/2180-4233 petersen@geophysik.uni-muenchen.de, Ludwig-Maximilians-Univ Muenchen; Ygm

Peterson, Bradley, bradley.peterson@stonybrook.edu, SUNY, Stony Brook; Ob

Peterson, Dallas, (785) 532-0405 Kansas State Univ; So

Peterson, Eric W., (309) 438-7865 ewpeter@ilstu.edu, Illinois State Univ; HwsHt

Peterson, Eugene J., (505) 667-5182 Los Alamos; Ct

Peterson, Gary L., (619) 594-5594 gpeterson@geology.sdsu.edu, San Diego State Univ; Gr

Peterson, Holly, 336-316-2263 petersonhe@guilford.edu, Guilford Coll; HwGe

Peterson, Jon W., peterson@hope.edu, Hope Coll; Ge

Peterson, Joseph E., (847) 697-1000 jpeterson@elgin.edu, Elgin Comm Coll; Gg

Peterson, Larry C., 305-284-6821 l.peterson@miami.edu, Univ of Miami; Gr

Peterson, Peter A., (515) 294-9652 pap@iastate.edu, Iowa State Univ of Sci & Tech; Zn

Peterson, Richard E., 806-834-3418 richard.peterson@ttu.edu, Texas Tech Univ; As

Peterson, Ronald C., (613) 533-6180 peterson@queensu.ca, Queen's Univ; Gz

Peterson, Virginia L., (616) 331-2811 petersvi@gvsu.edu, Grand Valley State Univ; GxcGt

Pethick, Andrew, +618 9266-2297 andrew.pethick@curtin.edu.au, Curtin Univ; YevYg

Petit, Martin A., (309) 694-5327 Illinois Central Coll; Ze

Petrenko, Vasilii, 585-276-6094 vasilii.petrenko@rochester.edu, Univ of Rochester; Act

Petrie, Elizabeth S., (970) 943-2117 epetrie@western.edu, Western Colorado Univ; GcoYe

Petrie, Gregg, 206-552-6385 gregg.petrie@ieee.org, Univ of Washington; Zr

Petrik, Colleen, cpetrik@ucsd.edu, Univ of California, San Diego; Ob

Petrinec, Zorica, +38514605969 zoricap@geol.pmf.hr, Univ of Zagreb; GipGz

Petronis, Michael S., (505) 454-3513 mspetro@nmhu.edu, New Mexico Highlands Univ; YmGcv

Petrovay, Kristof, +36 20 3722500/6621 k.petrovay@astro.elte.hu, Eotvos Lorand Univ; Xy

Petsch, Steven, (413) 545-4413 spetsch@geo.umass.edu, Univ of Massachusetts, Amherst; Col

Petsios, Elizabeth, (254)710-2388 elizabeth_petsios@baylor.edu, Baylor Univ; Pie

Pettengill, Gordon H., (617) 253-4281 ghp@space.mit.edu, Massachusetts Inst of Tech; Zg

Petters, Markus, 919-515-7144 markus_petters@ncsu.edu, North Carolina State Univ; As

Pettijohn, J. C., 217-333-3540 jcpettij@illinois.edu, Univ of Illinois, Urbana-Champaign; Hg

Pettinga, Jarg R., (03) 3667-001 x7716 jarg.pettinga@canterbury.ac.nz, Univ of Canterbury; Gtc

Petty, Grant W., (608)263-3265 gpetty@aos.wisc.edu, Univ of Wisconsin, Madison; AsZrAm

Petuch, Edward J., (561) 297-2398 epetuch@fau.edu, Florida Atlantic Univ; Pg

Peucker-Ehrenbrink, Bernhard, (508) 289-2518 bpeucker@whoi.edu, Woods Hole Oceanographic Inst; CmlHs

Pevzner, Roman, +618 9266-9805 r.pevzner@curtin.edu.au, Curtin Univ; YxxYz

Pezelj, Đurđica, +38514606117 durpezelj@yahoo.com, Univ of Zagreb; PgbPm

Pfannkuch, Hans O., (612) 624-1620 h2olafpf@umn.edu, Univ of Minnesota, Twin Cities; HwGeRh

Pfefferkorn, Hermann W., (215) 898-5156 hpfeffer@sas.upenn.edu, Univ of Pennsylvania; PbePs

Pfeiffer, Allison, (360) 650-3654 allison.pfeiffer@wwu.edu, Western Washington Univ; Gm

Pfuhl, Helen, 089/2180 4230 helen.pfuhl@geophysik.uni-muenchen.de, Ludwig-Maximilians-Univ Muenchen; Yg
Pham van Bang, Damien, 418 654-2590 damien.pham_van_bang@ete.inrs.ca, Univ du Quebec; HtOn
Phelps, Tommy J., (423) 574-7290 partnership@ornl.gov, Oak Ridge; Sb
Phelps, William, 951-222-8350 william.phelps@rcc.edu, Riverside City Coll; PeOpPg
Philander, S. George H., 609-258-5683 gphlder@princeton.edu, Princeton Univ; Op
Philben, Michael , 616 395 6710 philben@hope.edu, Hope Coll; CboSb
Phillippi, Nathan E., 910-521-6588 nathan.phillippi@uncp.edu, Univ of North Carolina, Pembroke; Zgi
Phillips, Andrew C., (217) 333-2513 aphillps@illinois.edu, Illinois State Geo Survey; GsmZi
Phillips, Brian L., (631) 632-6853 brian.phillips@stonybrook.edu, Stony Brook Univ; GzCl
Phillips, Dennis, (505) 667-4253 Los Alamos; Co
Phillips, Fred M., (575) 835-5540 phillips@nmt.edu, New Mexico Tech; Hw
Phillips, Michael, (815) 224-0394 mike_phillips@ivcc.edu, Illinois Valley Comm Coll; GgeGm
Phillips, Noah J., 8073438010 ext 8328 nphilli2@lakeheadu.ca, Lakehead Univ; GcNrZm
Phillips, Paul, (785) 628-5969 pphillip@fhsu.edu, Fort Hays State Univ; Zen
Phillips, Richard, +44(0) 113 34 31728 r.j.phillips@leeds.ac.uk, Univ of Leeds; Gt
Phillips, William R., (801) 378-4545 Brigham Young Univ; Gz
Phillips, Zachary, zachary.phillips@slu.edu, Saint Louis Univ; Zi
Philp, R. Paul, (405) 325-4469 pphilp@ou.edu, Univ of Oklahoma; Co
Phinney, Robert A., 609-258-4118 rphinney@princeton.edu, Princeton Univ; Ys
Phipps, Stephen P., (215) 898-4602 sphipps@sas.upenn.edu, Univ of Pennsylvania; Gc
Phipps Morgan, Jason, jp369@cornell.edu, Cornell Univ; GvCmPe
Piasecki, Michael, 212-650-8000 mpiasecki@ccny.cuny.edu, Graduate Sch of the City Univ of New York; Gg
Piazzoni, Antonio Sebastian, 089/2180 4230 antonio.piazzoni@geophysik.uni-muenchen.de, Ludwig-Maximilians-Univ Muenchen; Yg
Piccinini, Leonardo, +390498279124 leonardo.piccinini@unipd.it, Univ degli Studi di Padova; Hg
Piccoli, Philip M., (301) 405-6966 piccoli@umd.edu, Univ of Maryland; Cg
Pichevin, Laetitia, +44 (0) 131 650 5980 laetitia.pichevin@ed.ac.uk, Edinburgh Univ; Cg
Pickering, Ingrid J., (306) 966-5706 ingrid.pickering@usask.ca, Univ of Saskatchewan; CgtZn
Pickering, Kevin P., +44 020 7679 31325 kt.pickering@ucl.ac.uk, Univ Coll London; Gs
Pickett, Nicki, 954-201-6677 npickett@broward.edu, Broward Coll; Zn
Pickup, Gillian E., +44 (0) 131 451 3168 g.pickup@hw.ac.uk, Heriot-Watt Univ; Np
Pico, Tamara, tpico@ucsc.edu, Univ of California, Santa Cruz; Gl
Pidwirny, Michael, 250-807-8758 michael.pidwirny, Univ of British Columbia, Okanagan Campus; AtZyc
Pierce, David, 440-525-7341 dpierce@lakelandcc.edu, Lakeland Comm Coll; GgHsAm
Pierce, Heather, (585) 292-2426 hpierce@monroecc.edu, Monroe Comm Coll; Zyi
Pierce, Jeff, jeffrey.pierce@colostate.edu, Colorado State Univ; As
Pierce, Jennifer L., (208) 426-5380 jenpierce@boisestate.edu, Boise State Univ; GmRcZc
Pierce, Larry, (573) 368-2191 larry.pierce@dnr.mo.gov, Missouri Dept of Nat Res; Zn
Pieri, David C., (818) 354-6299 Jet Propulsion Lab; Gt
Pieruccini, Pierluigi, pierluigi.pieruccini@unito.it, Univ di Torino; Gm
Pierzynski, Gary M., 785-532-6101 gmp@ksu.edu, Kansas State Univ; Sc
Pieters, Carle M., (401) 863-2417 carle_pieters@brown.edu, Brown Univ; Zr
Pieters, Roger, (604) 822-4297 rpieters@eos.ubc.ca, Univ of British Columbia; Op
Pietranik, Anna, anna.pietranik@uwr.edu.pl, Univ of Wroclaw; CgGie
Pietras, Jeff, (607) 777-3348 jpietras@binghamton.edu, Binghamton Univ; GsEo
Pietro, Kathryn R., (508) 289-3862 kpietro@whoi.edu, Woods Hole Oceanographic Inst; Gu
Pietruszka, Aaron, apietrus@hawaii.edu, Univ of Hawai'i, Manoa; Ca
Pietrzak-Renaud, Natalie, 519-473-3766 npietrz@uwo.ca, Western Univ; Eg
Pietsch, Carlie, 408 924-5279 carlie.pietsch@sjsu.edu, San Jose State Univ; PiePc
Pigott, John D., (405) 325-4498 jpigott@ou.edu, Univ of Oklahoma; GoYeGd
Pike, J, pikej@cf.ac.uk, Cardiff Univ; Zn
Pike, Jenny, +44(0)29 208 75181 pikej@cardiff.ac.uk, Univ of Wales; Og
Pike, Scott, (503) 370-6587 spike@willamette.edu, Willamette Univ; Gag
Pike, Steven M., (508) 289-2350 spike@whoi.edu, Woods Hole Oceanographic Inst; Ca
Pikelj, Kristina, +38514606113 kpikelj@geol.pmf.hr, Univ of Zagreb; Gs
Pikitch, Ellen K., (631) 632-9599 ellen.pikitch@stonybrook.edu, SUNY, Stony Brook; Ob
Pilarczyk, Jessica, 778-782-7288 jpilarcz@sfu.ca, Simon Fraser Univ; Rn
Pilkey, Jr., Orrin H., (919) 684-4238 opilkey@geo.duke.edu, Duke Univ; Ou
Pilson, Michael E., (401) 874-6104 pilson@gso.uri.edu, Univ of Rhode Island; Oc
Pimentel, Nuno , pimentel@fc.ul.pt, Unive de Lisboa; GrsGo
Piña, Rubén, rpinagar@ucm.es, Univ Compl de Madrid; EmGz
Pinckney, James, 803 777 7133 pinckney@sc.edu, Univ of South Carolina; Obn
Pincus, Lauren N., 609-258-4101 lpincus@princeton.edu, Princeton Univ; CqlCg
Pinet, Paul, (315) 228-7656 ppinet@colgate.edu, Colgate Univ; Ou
Pingitore, Jr., Nicholas E., 915-747-5754 npingitore@utep.edu, Univ of Texas, El Paso; Cl
Pinho, Jennie, 610-330-5215 pinhoj@lafayette.edu, Lafayette Coll; GeHwZu
Piniella Febrer, Juan Francesc, ++34935813088 juan.piniella@uab.cat, Univ Auto de Barcelona; Gz
Pinkel, Robert, (858) 534-2056 rpinkel@ucsd.edu, Univ of California, San Diego; Op
Pinnt, Todd, todd.pinnt@azwestern.edu, Arizona Western Coll; ZyiZg
Pinter, Nicholas, (530) 754-1041 npinter@ucdavis.edu, Univ of California, Davis; Gm
Pinti, Daniele Luigi, 514-987-3000 #2572 pinti.daniele@uqam.ca, Univ du Quebec a Montreal; Cs
Pintilei, Mitica, 0040232401494 mpintilei@gmail.com, Alexandru Ioan Cuza; CgeCt
Piotrowski, Alexander, +44 (0) 1223 333473 apio04@esc.cam.ac.uk, Univ of Cambridge; Ge
Piper, David J., (902) 426-6580 david.piper@canada.ca, Dalhousie Univ; GusGo
Piper, John, +44 0151 794 3461 sg04@liverpool.ac.uk, Univ of Liverpool; YmGt
Pipkin, Bernard W., (213) 740-6106 pipkin@usc.edu, Univ of Southern California; Ng
Pirie, Diane H., (305) 348-2876 Florida InterNatl Univ; Gg
Pirouz, Mortaza X., (972) 883-2406 mxp180004@utdallas.edu, Univ of Texas, Dallas; GcZiGc
Pisani-Gareau, Tara, (617) 552-0843 pisanoga@bc.edu, Boston Coll; Ge
Pistone , Mattia , mattia.pistone@uga.edu, Univ of Georgia; Gv
Pitlick, John, pitlick@colorado.edu, Univ of Colorado; Gm
Pittman, Jason, (916) 608-6668 pittmaj@flc.losrios.edu, Folsom Lake Coll; ZgiZi
Pizzuto, James E., (302) 831-2710 pizzuto@udel.edu, Univ of Delaware; Gm
Placzek, Christa, 0747 814 756 christa.placzek@jcu.edu.au, James Cook Univ; CclGe
Plag, Hans-Peter, (757) 683-5335 hpplag@odu.edu, Old Dominion Univ; YgZc
Planavsky, Noah, 203-432-9043 noah.planavsky@yale.edu, Yale Univ; CsbCm
Plane, David A., plane@email.arizona.edu, Univ of Arizona; Zn
Plank, Gabriel, (775) 784-7039 gabe@seismo.unr.edu, Univ of Nevada, Reno; Ys
Plank, Owen C., (706) 542-9072 Univ of Georgia; Sc
Plank, Terry, tplank@ldeo.columbia.edu, Columbia Univ; Cg
Plankell, Eric T., 217-265-8029 eplankel@illinois.edu, Illinois State Geo Survey; Hg
Plant, Jane, +44 20 759 47416 jane.plant@imperial.ac.uk, Imperial Coll; Ge
Plant, Jeffrey J., (818) 393-3799 plant@jpl.nasa.gov, Jet Propulsion Lab; Zn
Plante, Alain F., (215) 898-9269 aplante@sas.upenn.edu, Univ of Pennsylvania; SboCo
Plante, Martin, (418) 656-8121 martin.plante@ggl.ulaval.ca, Univ Laval; Cg
Plasienka, Dusan, 0042160296529 plasienka@fns.uniba.sk, Comenius Univ in Bratislava; Gt
Platnick, Steven, steven.platnick@nasa.gov, Univ of Wisconsin, Madison; Zr
Platt, Brian F., 662-915-5440 bfplatt@olemiss.edu, Univ of Mississippi; GsPe
Platt, John P., (213) 821-1194 jplatt@usc.edu, Univ of Southern California; GctGp
Platt, Rutherford H., (413) 545-2499 rplatt@geo.umass.edu, Univ of Massachusetts, Amherst; Zu
Plattner, Alain, 205-348-0387 amplattner@ua.edu, Univ of Alabama;

YmuYg
Plattner, Christina, 089/21804220 christina.plattner@geophysik.uni-muenchen.de, Ludwig-Maximilians-Univ Muenchen; Yg
Pleasants, Mark S., 614 265 6748 mark.pleasants@dnr.state.oh.us, Ohio Dept of Nat Res; HwSp
Ples, George, george.ples@ubbcluj.ro, Babes-Bolyai Univ; PmsYg
Plescia, Jeffrey, (202) 358-0295 Jet Propulsion Lab; Zn
Plink-Bjorklund, Piret, (303) 384-2042 pplink@mines.edu, Colorado Sch of Mines; Gs
Plint, A G., (519) 661-3179 gplint@uwo.ca, Western Univ; Gs
Plotkin, Pamela, 979-845-3902 plotkin@tamu.edu, Texas A&M Univ; Ob
Plotnick, Roy E., (312) 996-2111 plotnick@uic.edu, Univ of Illinois Chicago; PgiGq
Plug, Lawrence, 902-494-1200 lplug@is.dal.ca, Dalhousie Univ; Gm
Pluhar, Christopher J., (559) 278-1128 cpluhar@csufresno.edu, Fresno State Univ; GtYmNg
Plumb, Raymond A., (617) 253-6281 plumb@mit.edu, Massachusetts Inst of Tech; As
Plummer, Charles C., plummercc@csus.edu, Sacramento State Univ; Gp
Plymate, Thomas G., tomplymate@missouristate.edu, Missouri State Univ; Gx
Poch Serra, Joan, ++935811085 joan.poch@uab.cat, Univ Auto de Barcelona; GrsZe
Pociask, Geoff, 217-265-8212 pociask@illinois.edu, Illinois State Geo Survey; Hw
Pociask, Geoffrey , 217 265-8029 pociask@illinois.edu, Univ of Illinois; Hs
Pociask, Walter, (313) 577-2506 wpociask@gmail.com, Wayne State Univ; Zn
Podolak, Morris, (052) 838-0976 morris@post.tau.ac.il, Tel Aviv Univ; ZnnZn
Poeter, Eileen P., epoeter@mines.edu, Colorado Sch of Mines; Ng
Pogge von Strandmann, Phillip, +44 020 7679 3637 p.strandmann@ucl.ac.uk, Birkbeck Coll; Csl
Pogue, Kevin R., (509) 527-5955 pogue@whitman.edu, Whitman Coll; Gc
Pohll, Greg, 775-682-6349 greg.pohll@dri.edu, Univ of Nevada, Reno; Hwq
Pohwat, Paul, (202) 633-1812 pohwatp@si.edu, Smithsonian Inst / Natl Mus of Natural History; Gz
Poinar, Kristin, (716)645-4286 kpoinar@buffalo.edu, SUNY, Buffalo; Gl
Poirier, Andre, 514-987-3000 #1718 Univ du Quebec a Montreal; Cs
Pojeta, John, (202) 633-1347 Smithsonian Inst / Natl Mus of Natural History; Pi
Pokorny, Eugene W., (702) 295-7496 eugene_pokorny@lanl.gov, Los Alamos; Nm
Polat, Ali, 519-253-3000 x2495 polat@uwindsor.ca, Univ of Windsor; GixCc
Polet, Jascha, 909-869-3459 jpolet@cpp.edu, California State Polytechnic Univ, Pomona; Ysg
Polito, Thomas A., (515) 294-0513 tpolito@iastate.edu, Iowa State Univ of Sci & Tech; So
Polizzotto, Matthew, (541) 346-5217 mpolozzo@uoregon.edu, Univ of Oregon; CqSc
Polk, Jason, jason.polk@wku.edu, Western Kentucky Univ; Gm
Pollack, Henry N., (734) 763-0084 hpollack@umich.edu, Univ of Michigan; Yg
Pollack, Jennifer, 361-825-2041 jennifer.pollack@tamucc.edu, Texas A&M Univ, Corpus Christi; Ob
Pollak, Robert, pollak@wustl.edu, Washington Univ in St. Louis; Zn
Pollard, David D., (650) 723-4679 dpollard@pangea.stanford.edu, Stanford Univ; Gc
Polley, Greg, polleygj@si.edu, Smithsonian Inst / Natl Mus of Natural History; Gxz
Pollock, Meagen, (330) 263-2202 mpollock@wooster.edu, Coll of Wooster; CuGiv
Polly, P. David, (812) 855-7994 pdpolly@indiana.edu, Indiana Univ, Bloomington; PvgPq
Pollyea, Ryan M., 540-231-7929 rpollyea@vt.edu, Virginia Tech; Hq
Polovina, Jeffrey J., (808) 983-5390 jeffrey.polovina@noaa.gov, Univ of Hawai'i, Manoa; Zn
Polsky, Colin, (954) 236-1088 cpolsky@fau.edu, Florida Atlantic Univ; Znc
Polvani, Lorenzo M., (212) 854-7331 lmp3@columbia.edu, Columbia Univ; AmsGq
Polyak, Victor J., (505) 277-4204 polyak@unm.edu, Univ of New Mexico; CcGzm
Polzin, Dierk T., dtpolzin@wisc.edu, Univ of Wisconsin, Madison; As
Pomeroy, George M., (717) 477-1776 gmpome@ship.edu, Shippensburg Univ; ZuyZn
Pommier, Anne, apommier@carnegiescience.edu, Carnegie Inst for Sci; Gy
Pond, Lisa G., (225) 578-0401 lgpond@lsu.edu, Louisiana State Univ; Zg
Pond, Stephen G., (604) 822-2205 spond@eos.ubc.ca, Univ of British Columbia; Op
Ponette-Gonzalez, Alexandra, 940-565-4012 alexandra.ponette@unt.edu, Univ of North Texas; Zyc
Ponton, Camilo, (360) 650-3648 camilo.ponton@wwu.edu, Western Washington Univ; CsoPc
Poole, Geoff, gpoole@montana.edu, Montana State Univ; Hg
Poole, T. Craig, (559) 442-4600 craig.poole@fresnocitycollege.edu, Fresno City Coll; Gg
Popa, Mirela Violetta, mirela.popa@ubbcluj.ro, Babes-Bolyai Univ; GgPgZe
Pope, Gregory A., (973) 655-7385 popeg@mail.montclair.edu, Montclair State Univ; GmaZy
Pope, Jeanette K., (765) 658-4105 jpope@depauw.edu, DePauw Univ; Ge
Pope, Michael , (979) 845-4376 mcpope@geo.tamu.edu, Texas A&M Univ; GrCc
Pope, Richard, +44 (0)1332 591751 r.j.pope@derby.ac.uk, Univ of Derby; Gam
Popp, Brian N., (808) 956-6206 popp@soest.hawaii.edu, Univ of Hawai'i, Manoa; Cm
Porcelli, Don, +44 (1865) 282121 don.porcelli@earth.ox.ac.uk, Univ of Oxford; Cg
Porch, William M., (505) 667-0971 wporch@lanl.gov, Los Alamos; Yg
Porter, Claire, 612-626-0505 porte254@umn.edu, Univ of Minnesota, Twin Cities; Zi
Porter, John H., (804) 924-8999 jhp7e@virginia.edu, Univ of Virginia; Zif
Porter, Ryan, 928-523-2429 ryan.porter@nau.edu, Northern Arizona Univ; YgsGt
Porter, Susannah, (805) 893-8954 porter@geol.ucsb.edu, Univ of California, Santa Barbara; Gn
Portillo, Danny, (719) 333-3080 United States Air Force Acad; Zi
Portner, Ryan, ryan.portner@sjsu.edu, San Jose State Univ; Gdv
Portnov, Alexey, alexey.portnov@austin.utexas.edu, Univ of Texas, Austin; EgGx
Posiloviæ, Hrvoje, +38514606087 posilovic@geol.pmf.hr, Univ of Zagreb; Gg
Posler, Gerry L., (785) 532-6101 gposler@ksu.edu, Kansas State Univ; Sf
Pospelova, Vera, vpospe@umn.edu, Univ of Minnesota, Twin Cities; GePl
Post, Donald F., (520) 621-1262 postdf@ag.arizona.edu, Univ of Arizona; Sd
Post, Jeffrey E., (202) 633-1814 postj@si.edu, Smithsonian Inst / Natl Mus of Natural History; Gz
Post, Vincent , vincent.post@flinders.edu.au, Flinders Univ; Hw
Poteet, Mary F., 512-471-5209 mpoteet@jsg.utexas.edu, Univ of Texas, Austin; Pg
Potel, Sébastien, +33(0)3 44068999 sebastien.potel@lasalle-beauvais.fr, Institut Polytechnique LaSalle Beauvais (ex-IGAL); GpzGi
Potess, Marla, (325) 670-1395 marla.potess@hsutx.edu, Hardin-Simmons Univ; Zgu
Poths, Jane, (505) 667-1506 Los Alamos; Cc
Potra, Adriana, 479-575-6419 potra@uark.edu, Univ of Arkansas, Fayetteville; EgCcGg
Potten, Martin, +49 89 289 - 25885 martin.potten@tum.de, Technical Univ of Munich; NgrNx
Potter, Amy E., (912) 344-3612 amypotter@georgiasouthern.edu, Georgia Southern Univ; Zn
Potter, David, (780) 2481518 dkpotter@ualberta.ca, Univ of Alberta; Ye
Potter, Eric C., (512) 471-7090 eric.potter@beg.utexas.edu, Univ of Texas at Austin, Jackson Sch of Geosciences; Eo
Potter, Henry, hpotter@tamu.edu, Texas A&M Univ; Op
Potter, Katherine E., katie.potter@usu.edu, Utah State Univ; GiZeEg
Potter, Paul E., (513) 556-3732 Univ of Cincinnati; Gs
Potter, Jr., Donald B., (931) 598-1479 bpotter@sewanee.edu, Sewanee: Univ of the South; Gc
Potter, Jr., Noel, pottern@dickinson.edu, Dickinson Coll; Gmc
Potter-McIntyre, Sally, (618) 453-7372 pottermcintyre@siu.edu, Southern Illinois Univ Carbondale; Gs
Potts, Donald C., (831) 459-4417 potts@biology.ucsc.edu, Univ of California, Santa Cruz; Ob
Poty, Edouard, 32 4 366 52 83 e.poty@ulg.ac.be, Univ de Liege; PiGsPe
Poudel, Durga, 337-482-6163 ddpoudel@louisiana.edu, Univ of Louisiana at Lafayette; Ge
Poulsen, Christopher J., (734) 615-2236 poulsen@umich.edu, Univ of Michigan; Pe
Poulson, Simon, 775.784.1833 poulson@mines.unr.edu, Univ of Nevada, Reno; Cls
Poulton, Mary, 520-621-8391 mpoulton@email.arizona.edu, Univ of Arizona; Ng
Poulton, Simon, +44(0) 113 34 35237 s.poulton@leeds.ac.uk, Univ of Leeds; Cls
Pound, Kate S., 320-308-2014 kspound@stcloudstate.edu, Saint Cloud State Univ; GgZeGd

Pourmand, Ali, apourmand@rsmas.miami.edu, Univ of Miami; Cga
Pournelle, Jennifer R., jpournelle@environ.sc.edu, Univ of South Carolina; GaZuSa
Pournik, Maysam, maysam.pournik@ou.edu, Univ of Oklahoma; Np
Pourret, Olivier, +33(0)3 44068979 olivier.pourret@lasalle-beauvais.fr, Institut Polytechnique LaSalle Beauvais (ex-IGAL); CgaGe
Powell, Brian, (808) 956-6724 powellb@hawaii.edu, Univ of Hawai'i, Manoa; Op
Powell, Christine A., (901) 678-8455 capowell@memphis.edu, Univ of Memphis; Ys
Powell, J. Mark, (608) 890-0070 jmpowel2@wisc.edu, Univ of Wisconsin, Madison; So
Powell, Jane, +44 (0)1603 59 2822 j.c.powell@uea.ac.uk, Univ of East Anglia; Ge
Powell, Matthew G., (814) 641-3602 powell@juniata.edu, Juniata Coll; PiqPe
Powell, Rebecca , 303-871-2667 rebecca.l.powell@du.edu, Univ of Denver; Zr
Powell, Ross D., (815) 753-7952 rpowell@niu.edu, Northern Illinois Univ; GslZc
Powell, Scott, spowell@montana.edu, Montana State Univ; Zr
Powell, Thomas M., (510) 642-7455 zackp@socrates.berkeley.edu, Univ of California, Berkeley; Op
Powell, Wayne G., (718) 951-5761 wpowell@brooklyn.cuny.edu, Graduate Sch of the City Univ of New York; EmGaCs
Power, Mary E., (510) 643-7776 Univ of California, Berkeley; Pe
Powers, Elizabeth, epowers@csub.edu, California State Univ, Bakersfield; Gg
Powers, Roger W., (907) 474-6188 Univ of Alaska, Fairbanks; Zn
Powers, Sean, spowers@southalabama.edu, Univ of South Alabama; Ob
Poyla, David, +44 0161 275-3818 david.polya@manchester.ac.uk, Univ of Manchester; Cm
Poynton, Helen, helen.poynton@umb.edu, Univ of Massachusetts, Boston; Zn
Prabhakar, Gouri, gourip@purdue.edu, Purdue Univ; As
Prabhu, Anirudh, aprabhu@carnegiescience.edu, Carnegie Inst for Sci; Zn
Pracny, Pavel, +420 549 49 8523 pracny@sci.muni.cz, Masaryk Univ; Cg
Pradhanang, Soni M., (401) 874-5980 spradhanang@uri.edu, Univ of Rhode Island; HsqHq
Prairie, Jennifer, (619) 260-8820 jcprairie@sandiego.edu, Univ of San Diego; Ob
Pranter, Matthew J., (405) 325-3253 matthew.pranter@ou.edu, Univ of Oklahoma; Go
Prasad, Manika, 303-273-3457 mprasad@mines.edu, Colorado Sch of Mines; NrYx
Prasad, Rishi, (334) 844-4100 rzp0050@auburn.edu, Auburn Univ; Sb
Prasad, Vara , (785) 532-3746 vara@ksu.edu, Kansas State Univ; So
Prather, Kimberly A., (858) 822-5312 kprather@ucsd.edu, Univ of California, San Diego; AsCmHg
Prather, Michael J., (949) 824-5838 mprather@uci.edu, Univ of California, Irvine; As
Pratson, Lincoln F., (919) 681-8077 lincoln.pratson@duke.edu, Duke Univ; Gs
Pratt, Allyn R., (505) 667-4308 pratt_allyn_r@lanl.gov, Los Alamos; Ge
Pratt, Brian R., (306) 966-5725 brian.pratt@usask.ca, Univ of Saskatchewan; PgGsr
Pratt, Lisa M., (812) 855-9203 prattl@indiana.edu, Indiana Univ, Bloomington; Co
Pratt, R. Gerhard, 519-661-2111 ext. 86690 gpratt2@uwo.ca, Western Univ; Yeg
Pratt-Sitaula, Beth, beth psitaula@geology.cwu.edu, Central Washington Univ; YdZe
Prave, Tony, +44 (0)1334462381 ap13@st-andrews.ac.uk, Univ of St. Andrews; GgrGs
Precedo, Laura, 954-201-6674 lprecedo@broward.edu, Broward Coll; Oc
Precht, Francis L., (301) 687-4440 fprecht@frostburg.edu, Frostburg State Univ; Zy
Prell, Warren L., (401) 863-3221 warren_prell@brown.edu, Brown Univ; Ou
Prencipe, Mauro, mauro.prencipe@unito.it, Univ di Torino; Gz
Prescott, Cindy, 604–822–4701 cindy.prescott@ubc.ca, Univ of British Columbia; Sb
Presley, Bobby J., (979) 845-5136 bpresley@ocean.tamu.edu, Texas A&M Univ; Oc
Presley, DeAnn, 785532128 deann@ksu.edu, Kansas State Univ; So
Presnall, Dean C., (972) 883-2444 presnall@utdallas.edu, Univ of Texas, Dallas; Cp
Prestegaard, Karen L., (301) 405-6982 kpresto@geol.umd.edu, Univ of Maryland; Hgw
Preston, Aaron, 802-626-6496 aaron.preston@northernvermont.edu, Northern Vermont Univ-Lyndon; AsZrAc
Preston, Thomas C., 514-398-3766 thomas.preston@mcgill.ca, McGill Univ; AsZnn
Preston, William L., (805) 756-2210 wpreston@calpoly.edu, California Polytechnic State Univ; Ze
Preto, Nereo, +390498279174 nereo.preto@unipd.it, Univ degli Studi di Padova; Gs
Pretorius, Erika, +27 (0)12 420 3737 erika.pretorius@up.ac.za, Univ of Pretoria; Zi
Prevec, Steve, +27 (0)46-603-8309 s.prevec@ru.ac.za, Rhodes Univ; Cg
Previdi, Michael, mprevidi@ldeo.columbia.edu, Columbia Univ; As
Prewett, Jerry L., (573) 368-2101 jerry.prewett@dnr.mo.gov, Missouri Dept of Nat Res; GgHw
Price, Colin G., +97236406029 cprice@flash.tau.ac.il, Tel Aviv Univ; As
Price, Douglas M., 330-941-3019 dmprice@ysu.edu, Youngstown State Univ; Cg
Price, Greogry, +44 1752 584771 g.price@plymouth.ac.uk, Univ of Plymouth; Gsr
Price, Jonathan D., (940) 397-4288 jonathan.price@msutexas.edu, Midwestern State Univ; GiCuGz
Price, Kevin, (785) 532-6101 kpprice@ksu.edu, Kansas State Univ; Zr
Price, L G., lloydgreer.price@nmt.edu, New Mexico Tech; Gg
Price, Maribeth H., (605) 394-2461 maribeth.price@sdsmt.edu, South Dakota Sch of Mines & Tech; Zir
Price, Marie D., (202) 994-6187 George Washington Univ; Zy
Price, Nancy A., (518)564-4032 npric002@plattsburgh.edu, SUNY, Plattsburgh; GzCgGc
Price, Peter E., 281-765-7764 peter.e.price@nhmccd.edu, Lonestar Coll - North Harris; Zi
Price, Raymond A., (613) 533-6542 pricera@queensu.ca, Queen's Univ; GtcGe
Price, Rene, (305) 348-3119 pricer@fiu.edu, Florida InterNatl Univ; HwClRw
Prichard, H M., prichard@cf.ac.uk, Cardiff Univ; Gx
Prichard, Hazel, +44(0)29 208 74323 prichard@cardiff.ac.uk, Univ of Wales; EgGze
Prichonnet, Gilbert P., 514-987-3000 #3383 prichonnet.gilbert@uqam.ca, Univ du Quebec a Montreal; Pi
Prichystal, Antonín, +420 549 49 6699 prichy@sci.muni.cz, Masaryk Univ; GgaGv
Pride, Douglas E., 614-329-0941 pride.1@osu.edu, Ohio State Univ; Eg
Pride, Steven, (510) 495-2823 srpride@lbl.gov, Univ of California, Berkeley; Ys
Pridmore, Cindy L., (916) 925-6902 California Geo Survey; Ng
Priesendorf, Carl, (816) 604-2549 carl.priesendorf@mcckc.edu, Metropolitan Comm Coll-Kansas City; GgZyGe
Priest, Eric, (847) 543-2585 epriest@clcillinois.edu, Coll of Lake County; AmZe
Priest, George R., (541) 574-6642 george.priest@dogami.state.or.us, Oregon Dept of Geology and Mineral Ind; Gg
Priestley, Keith, +44 (0) 1223 337195 Univ of Cambridge; YgGt
Priewisch, Alexandra, 505-504-3675 alexandra.priewisch@fresnocitycollege.edu, Fresno City Coll; Gg
Primeau, Francois, 949-824-9435 fprimeau@uci.edu, Univ of California, Irvine; Op
Pringle, James M., (603) 862-5000 jpringle@cisunix.unh.edu, Univ of New Hampshire; Op
Pringle, Jamie K., (+44) 01782 733163 j.k.pringle@keele.ac.uk, Keele Univ; YuGfa
Pringle, Patrick T., 360-623-8584 ppringle@centralia.edu, Centralia Coll; ZgGvZn
Prinn, Ronald G., 617-253-2452 rprinn@mit.edu, Massachusetts Inst of Tech; As
Prins, Maarten A., +31 20 59 83635 m.a.prins@vu.nl, Vrije Univit Amsterdam; GsOg
Prior, David J., +64 3 479-5279 david.prior@otago.ac.nz, Univ of Otago; Gct
Prior, William L., 501 683-0117 bill.prior@arkansas.gov, Arkansas Geo Survey; GggEc
Priscu, John, jpriscu@montana.edu, Montana State Univ; Po
Pristas, Ronald, (609) 984-6587 ron.pristas@dep.state.nj.us, New Jersey Geological and Water Survey; Zi
Pritchard, Chad, 509-359-7026 cpritchard@ewu.edu, Eastern Washington Univ; GcHg
Pritchard, Matthew E., (607) 255-4870 mp337@cornell.edu, Cornell Univ; YdGvl
Privette, David, (704)330-6750 david.privette@cpcc.edu, Central Piedmont Comm Coll; Zy
Prockter, Louise M., prockter@lpi.usra.edu, Univ of Tennessee, Knoxville; Xg

Proctor, Cameron, 519-253-3000 x2183 cameron.proctor@uwindsor.ca, Univ of Windsor; Zi
Prohiæ, Esad, +38514605963 eprohic@geol.pmf.hr, Univ of Zagreb; CglGe
Proistosescu, Cristian (Cristi), cristi@illinois.edu, Univ of Illinois, Urbana-Champaign; AsGgAp
Protano, Giuseppe, giuseppe.protano@unisi.it, Univ of Siena; CgaSc
Prothro, Lindsay , 361-825-4062 lindsay.prothro@tamucc.edu, Texas A&M Univ, Corpus Christi; GsuGl
Prothro, Lindsay O., 361-825-4062 lindsay.prothro@tamucc.edu, Texas A&M Univ, Corpus Christi; GluGs
Proudhon, Benoit, +33(0)3 44068996 benoit.proudhon@lasalle-beauvais.fr, Institut Polytechnique LaSalle Beauvais (ex-IGAL); GgcGt
Provin, Tony L., (979) 862-4955 t-provin@tamu.edu, Texas A&M Univ; Sc
Pruell, Richard J., (401) 782-3000 Univ of Rhode Island; Co
Pruss, Sara B., (413) 585-3948 spruss@smith.edu, Smith Coll; Pg
Pryor, Sara C., 607-255-3376 sp2279@cornell.edu, Cornell Univ; AsZn
Prytulak, Julie, +44 20 759 46474 j.prytulak@imperial.ac.uk, Imperial Coll; Gg
Ptacek, Anton D., ptacek2@juno.com, San Diego State Univ; Gx
Ptak, Tom, 208-885-6238 tptak@uidaho.edu, Univ of Idaho; Zu
Pu, Ruiliang, rpu@usf.edu, Univ of South Florida; Zr
Pu, Zhaoxia, 801-585-3864 zhaoxia.pu@utah.edu, Univ of Utah; As
Puchalski, Stephaney, spuchalski@highline.edu, Highline Coll; PgGg
Puchtel, Igor, (301) 405-4054 ipuchtel@umd.edu, Univ of Maryland; CcgGi
Puckett, Andrew, 706-507-8098 puckett_andrew@columbusstate.edu, Columbus State Univ; Xa
Puckett, Mark, 601-266-4729 mark.puckett@usm.edu, Univ of Southern Mississippi; PmiGg
Puckette, James, 405-744-6358 jim.puckette@okstate.edu, Oklahoma State Univ; Go
Puelles, Pablo, pablo.puelles@ehu.eus, Univ of the Basque Country UPV/EHU; Gct
Puente, Carlos E., (916) 752-0689 Univ of California, Davis; Hg
Puffer, John H., (973) 353-5100 jpuffer@andromeda.rutgers.edu, Rutgers, The State Univ of New Jersey, Newark; GivGe
Puig, Claire, puigc@si.edu, Smithsonian Inst / Natl Mus of Natural History; Zn
Pujana, Ignacio, (972) 883-2461 pujana@utdallas.edu, Univ of Texas, Dallas; Pm
Pujol, Jose, (901) 678-4827 jpujol@memphis.edu, Univ of Memphis; Ye
Puleo, Jack, 302-831-2440 jpuleo@udel.edu, Univ of Delaware; Hw
Pulkkinen, Tuija, 734-615-3583 tuija@umich.edu, Univ of Michigan; Zn
Pulliam, Jay, (254) 710-2183 jay_pulliam@baylor.edu, Baylor Univ; YsgYe
Pulliam, Robert J., (512) 471-6156 jay@ig.utexas.edu, Univ of Texas, Austin; Ys
Pumphrey, Hugh C., +44 (0) 131 650 6026 hugh.pumphrey@ed.ac.uk, Edinburgh Univ; AscYv
Pun, Aurora, (505) 277-5629 apun@unm.edu, Univ of New Mexico; Ca
Pundsack, Jonathan, 612-626-0505 pundsack@umn.edu, Univ of Minnesota, Twin Cities; Zn
Punyasena, Surangi W., (217) 244-8049 punyasena@life.illinois.edu, Univ of Illinois, Urbana-Champaign; PblPe
Purchase, Megan, +27 (0)51 401 7158 purchasemd@ufs.ac.za, Univ of the Free State; Gx
Purdie, Duncan A., +44 (0)23 80592263 duncan.purdie@noc.soton.ac.uk, Univ of Southampton; Ob
Purdom, William B., (541) 552-6494 purdom@sou.edu, Southern Oregon Univ; Gx
Purdy, G. Michael, (845) 365-8348 mpurdy@ldeo.columbia.edu, Columbia Univ; Yr
Purkey, Sarah, spurkey@ucsd.edu, Univ of California, San Diego; Op
Purkis, Sam J., 305-421-4351 spurkis@rsmas.miami.edu, Univ of Miami; Gu
Purnell, Mark, +440116 252 3645 map2@le.ac.uk, Leicester Univ; Pg
Purtle, Jennifer M., (501) 575-7317 jms14@comp.uark.edu, Univ of Arkansas, Fayetteville; Zn
Pusede, Sally, (434) 924-4544 sep6a@virginia.edu, Univ of Virginia; As
Putirka, Keith D., (559) 278-4524 kputirka@csufresno.edu, Fresno State Univ; GviGp
Putkonen, Jaakko, (701) 777-3213 jaakko.putkonen@engr.und.edu, Univ of North Dakota; Gml
Putnam, Aaron E., 207-581-2186 aaron.putnam@maine.edu, Univ of Maine; Gl
Puziewicz, Jacek, jacek.puziewicz@uwr.edu.pl, Univ of Wroclaw; Gix
Pyle, David M., david.pyle@earth.ox.ac.uk, Univ of Oxford; GviRh
Pyle, Eric J., 540-568-7115 pyleej@jmu.edu, James Madison Univ; Ze
Pylypenko, Svitlana, (303) 384-2140 spylypen@mines.edu, Colorado Sch of Mines; Zmn
Pyrtle, Ashanti J., apyrtle@seas.marine.usf.edu, Univ of South Florida; Oc
Pysklywec, Russell N., (416) 978-4852 russ@geology.utoronto.ca, Univ of Toronto; Yg

Q

Qi, Feng, (908) 737-3702 fqi@kean.edu, Kean Univ; ZiyZg
Qian, Song, (419) 530-4230 song.qian@utoledo.edu, Univ of Toledo; Zn
Qiang, Yi, qiangy@usf.edu, Univ of South Florida; Zi
Qin, Rongjun, qin.324@osu.edu, Ohio State Univ; Zrf
Qiu, Bo, (808) 956-4098 bo@soest.hawaii.edu, Univ of Hawai'i, Manoa; Op
Qiu, Xiaomin, (417) 836-3219 qiu@missouristate.edu, Missouri State Univ; Zi
Quade, Deborah J., -319-337-9785 deborah.quade@dnr.iowa.gov, Iowa Dept of Nat Res; GleGm
Quade, Jay, 520-626-3223 quadej@email.arizona.edu, Univ of Arizona; Sc
Qualls, Robert G., (775) 327-5014 qualls@unr.edu, Univ of Nevada, Reno; SoHsSf
Quan, Tracy, 405-744-6358 tracy.quan@okstate.edu, Oklahoma State Univ; Og
Quattro, Joseph, quattro@biol.sc.edu, Univ of South Carolina; Ob
Quay, Paul D., (206) 545-8061 pdquay@u.washington.edu, Univ of Washington; Oc
Quick, Thomas J., (330) 972-6935 tquick@uakron.edu, Univ of Akron; GgCaYe
Quinn, Claire, +44(0) 113 34 38700 c.h.quinn@leeds.ac.uk, Univ of Leeds; Ge
Quinn, Courtney, (864)294-3655 courtney.quinn@furman.edu, Furman Univ; Zn
Quinn, Heather A., (410) 554-5522 heather.quinn@maryland.gov, Maryland Dept of Nat Res; GgHwg
Quinn, James G., (401) 874-6219 jgquinn@gso.uri.edu, Univ of Rhode Island; Oc
Quinn, John, (630) 252-5357 quinnj@anl.gov, Argonne; Hw
Quinn, Paul, +44 (0) 191 208 5773 p.f.quinn@ncl.ac.uk, Univ of Newcastle Upon Tyne; Hg
Quinn, Terry, 512-471-0377 quinn@ig.utexas.edu, Univ of Texas, Austin; Pe
Quintero, Sylvia, (520) 621-6025 squinter@email.arizona.edu, Univ of Arizona; Zn
Quinton, Page, 315-267-2815 quintopc@potsdam.edu, Indiana Univ, Bloomington; Pc
Quiros Ugalde, Diego, diego.quirosugalde@uct.ac.za, Univ of Cape Town; YsgEo

R

Raab, James, 614 265 6747 jim.raab@dnr.state.oh.us, Ohio Dept of Nat Res; Hw
Rabalais, Nancy N., (225) 851-2800 Louisiana State Univ; Ob
Raber, George, 601-266-5807 george.raber@usm.edu, Univ of Southern Mississippi; Zi
Radcliffe, David E., (706) 542-0897 Univ of Georgia; Sp
Radcliffe, Dennis, 516-463-6545 dennis.radcliffe@hofstra.edu, Hofstra Univ; Gzx
Radebaugh, Jani, 801 422-9127 jani.radebaugh@byu.edu, Brigham Young Univ; XgGvm
Rademacher, Laura K., (209) 946-7351 lrademacher@pacific.edu, Univ of the Pacific; ClHySo
Rader, Erika, (208) 885-1142 erader@uidaho.edu, Univ of Idaho; GviXg
Rader, Shelby, (812) 855-7508 shtrader@iu.edu, Indiana Univ, Bloomington; Ct
Radic, Valentina, 604-827-1446 vradic@eos.ubc.ca, Univ of British Columbia; GlAsZg
Radil, Steven, 208-885-5058 sradil@uidaho.edu, Univ of Idaho; Zn
Radke, Lawrence F., radke@ucar.edu, Univ of Washington; As
Radko, Nicholas, (912) 478-5361 nradko@georgiasouthern.edu, Georgia Southern Univ; Geg
Radko, Timour, (831) 656-3318 tradko@nps.edu, Naval Postgrad Sch; Op
Rae, James, +44 01334 464948 jwbr@st-andrews.ac.uk, Univ of St. Andrews; Cg
Raef, Abdelmoneam E., (785) 532-2240 abraef@ksu.edu, Kansas State Univ; YesZr
Raeside, Robert P., (902) 585-1323 rob.raeside@acadiau.ca, Acadia Univ; GptGc
Rafini, Silvain, 514-987-3000 poste 2369 silvain.rafini@gmail.com, Univ du Quebec a Chicoutimi; GcHwEg
Ragland, Paul C., (850) 644-5018 ragland@magnet.fsu.edu, Florida State Univ; Ca
Ragotzkie, Robert A., ragotzkie@wisc.edu, Univ of Wisconsin, Madison; Ob
Rahaman, M. A., 234-800-337-7441 mrahaman@oauife.edu.ng, Obafemi Awolowo Univ; GpcGi

Raharimahefa, Tsilavo, 306-585-4987 tsilavo.raharimahefa@uregina.ca, Univ of Regina; Gc
Rahl, Jeffrey, 540-458-8101 rahlj@wlu.edu, Washington & Lee Univ; Gt
Rahman, Abdullah F., 956-882-4074 abdullah.rahman@utrgv.edu, Univ of Texas, Rio Grande Valley; Zr
Rahman, Ata U., rata@austincc.edu, Austin Comm Coll District; GgdGe
Rahn, Kenneth A., (401) 874-6713 krahn@uri.edu, Univ of Rhode Island; Oc
Rahn, Perry H., (605) 394-2527 perry.rahn@sdsmt.edu, South Dakota Sch of Mines & Tech; Ng
Rai, Chandra S., (405) 325-6866 crai@ou.edu, Univ of Oklahoma; Np
Railsback, L. Bruce, (706) 542-3453 rlsbk@gly.uga.edu, Univ of Georgia; PcClGd
Raine, Robin, +353 (0)91 492271 robin.raine@nuigalway.ie, Natl Univ of Ireland Galway; Yg
Rains, Daniel S., 479-601-7521 daniel.rains@arkansas.gov, Arkansas Geo Survey; GgZi
Rains, Kai, krains@usf.edu, Univ of South Florida; Sf
Rains, Mark C., mrains@usf.edu, Univ of South Florida; Hw
Rainwater, Maira, mrainwater@csustan.edu, California State Univ, Stanislaus; GeHw
Raj, John K., 03-79674225 jkraj@um.edu.my, Univ of Malaya; Ng
Raji, W. O., lanreraji24@unilorin.edu.ng, Univ of Ilorin; Ye
Rakonczai, János, +3662544156 rjanos@earth.geo.u-szeged.hu, Univesity of Szeged; ZeHwZy
Rakovan, Monica, rakovamt@miamioh.edu, Miami Univ; Hg
Rallis, Donald N., (540) 654-1492 drallis@mwcgw.mwc.edu, Mary Washington Coll; Zn
Ramage, Joan, (610) 758-6410 jmr204@lehigh.edu, Lehigh Univ; ZrGlm
Ramagwede, Fhatuwani L., 0027128411911 info@geoscience.org.za, Geo Survey of South Africa ; EgGzEg
Raman, Sethu S., (919) 515-7144 sethu_raman@ncsu.edu, North Carolina State Univ; As
Ramanathan, V., (858) 534-0219 vramanathan@ucsd.edu, Univ of California, San Diego; As
Rambaud, Fabienne M., frambaud@austincc.edu, Austin Comm Coll District; Eg
Ramelli, Alan R., (775) 784-4151 ramelli@unr.edu, Univ of Nevada, Reno; Ng
Ramirez, Abelardo L., (925) 422-6919 ramirez3@llnl.gov, Lawrence Livermore; Ng
Ramirez, Wilson R., (787) 265-3845 wilson.ramirez1@upr.edu, Univ of Puerto Rico; GudOu
Ramos, Frank C., 575-646-2511 framos@nmsu.edu, New Mexico State Univ, Las Cruces; CcuGi
Rampedi, Isaac T., +27115592429 isaacr@uj.ac.za, Univ of Johannesburg; PgSdHg
Ramsey, Kelvin W., (302) 831-3586 kwramsey@udel.edu, Univ of Delaware; As
Ramsey, Michael S., (412) 624-8772 ramsey@ivis.eps.pitt.edu, Univ of Pittsburgh; Zr
Ramspott, Matthew E., 301/687-4412 meramspott@frostburg.edu, Frostburg State Univ; Zr
Ranalli, Giorgio, granalli@ccs.carleton.ca, Carleton Univ; Yg
Rancan, Helen L., 609-984-6587 helen.rancan@dep.state.nj.us, New Jersey Geological and Water Survey; Hs
Randall, David A., randall@atmos.colostate.edu, Colorado State Univ; As
Randall, George, (505) 667-9483 Los Alamos; Yg
Randall, Gyles W., (507) 835-3620 Univ of Minnesota, Twin Cities; So
Ranhofer, Melissa, (864)294-3647 melissa.ranhofer@furman.edu, Furman Univ; Gg
Rankey, Gene, (785) 864-6028 grankey@ku.edu, Univ of Kansas; GsrGu
Rankin, Robert, (780) 492-5082 rankin@phys.ualberta.ca, Univ of Alberta; Xy
Ransom, Michel D., (785) 532-7203 mdransom@ksu.edu, Kansas State Univ; SdcGm
Ranson, William A., (864) 294-3364 bill.ranson@furman.edu, Furman Univ; GxzGp
Ranville, James, (303) 273-3004 jranvill@mines.edu, Colorado Sch of Mines; Ca
Rapley, Chris, +44 020 310 86320 christopher.rapley@ucl.ac.uk, Univ Coll London; ZgrGl
Rapp, Anita, arapp@tamu.edu, Texas A&M Univ; Zr
Rapp, Richard H., rapp.1@osu.edu, Ohio State Univ; Yd
Rappe, Michael S., (808) 236-7464 rappe@hawaii.edu, Univ of Hawai'i, Manoa; Og
Rappenglueck, Bernhard, 713-743-2469 brappenglueck@uh.edu, Univ of Houston; As
Rasbury, E. Troy, 631-632-1488 troy.rasbury@stonybrook.edu, Stony Brook Univ; Cc
Rashall, Jenny M., 936-468-2340 jenny.rashall@sfasu.edu, Stephen F. Austin State Univ; GgPg
Rashed, Mohamed A., 002-03-3921595 rashedmohamed@yahoo.com, Alexandria Univ; GgSo
Rasmussen, Cornelia, crasmussen@utexas.edu, Univ of Texas, Austin; XbPgCc
Rasmussen, Craig, (520) 621-7223 crasmuss@cals.arizona.edu, Univ of Arizona; SzGmCl
Rasmussen, Kenneth, (703) 323-2139 krasmussen@nvcc.edu, Northern Virginia Comm Coll - Annandale; Gus
Rasmussen, Kristen L., kristenr@rams.colostate.edu, Colorado State Univ; As
Rasmussen, Pat E., (613) 868-8609 pat.rasmussen@canada.gc.ca, Univ of Ottawa; GeAsGb
Rasmussen, Steen, (505) 665-0052 steen@lanl.gov, Los Alamos; Yg
Rasmussen, Tab, (314) 935-4844 dtrasmus@wustl.edu, Washington Univ in St. Louis; Zn
Ratajeski, Kent, (859) 257-4444 kent.ratajeski@email.uky.edu, Univ of Kentucky; Gi
Ratchford, M. E., (501) 683-0118 ed.ratchford@arkansas.gov, Arkansas Geo Survey; GocEo
Rath, Carolyn, (657) 278-7096 crath@fullerton.edu, California State Univ, Fullerton; Gg
Rathburn, Anthony E., (661) 654-3281 arathburn@csub.edu, California State Univ, Bakersfield; PmOoCb
Rathburn, Sara L., (970) 491-6956 sara.rathburn@colostate.edu, Colorado State Univ; Ggm
Ratoi, Bogdan G., 0040742955966 bog21rat@gmail.com, Alexandru Ioan Cuza; PvGr
Ratschbacher, Barbara C., bratschbacher@ucdavis.edu, Univ of California, Davis; GgCg
Raub, Timothy D., +44 01334 464012 timraub@st-andrews.ac.uk, Univ of St. Andrews; YmGgPc
Rauber, Carolyn, 612-625-0317 cbishoff@umn.edu, Univ of Minnesota, Twin Cities; Zn
Rauber, Robert M., (217) 333-2835 r-rauber@illinois.edu, Univ of Illinois, Urbana-Champaign; As
Rauch, Henry W., (304) 293-2187 hrauch@wvu.edu, West Virginia Univ; HwGe
Rauch, Marta, marta.rauch@uwr.edu.pl, Univ of Wroclaw; Gc
Raudsepp, Mati, (604) 822-6396 mraudsepp@eoas.ubc.ca, Univ of British Columbia; GzeZm
Rautenbach, Victoria J., +27 (0) 12 420 3489 victoria.rautenbach@up.ac.za, Univ of Pretoria; Zif
Ravat, Dhananjay, (859) 257-4726 dhananjay.ravat@uky.edu, Univ of Kentucky; YgmYv
Ravela, Sai, 617-253-0997 ravela@mit.edu, Massachusetts Inst of Tech; Zn
Ravelo, Ana C., 831-459-3722 acr@pmc.ucsc.edu, Univ of California, Santa Cruz; OcGe
Ravelo, Christina, (831) 459-3722 acr@cats.ucsc.edu, Univ of California, Santa Cruz; Cs
Raven, Morgan, raven@geol.ucsb.edu, Univ of California, Santa Barbara; Cl
Ravi, Sujith, 215-204-7122 sravi@temple.edu, Temple Univ; HqZuRw
Ravishankara, A.R., a.r.ravishankara@colostate.edu, Colorado State Univ; AsZg
Ravizza, Gregory, 808-956-2916 gravizza@soest.hawaii.edu, Univ of Hawai'i, Manoa; Cm
Rawling, Geoffrey, 575-366-2535 geoffrey.rawling@nmt.edu, New Mexico Tech; Gg
Rawling, J. E., (608) 263-6839 elmo.rawling@wisc.edu, Univ of Wisconsin, Madison, Division of Extension; Gml
Rawlins, Michael, 413 545-0659 rawlins@geo.umass.edi, Univ of Massachusetts, Amherst; As
Rawlinson, Nick, nr441@cam.ac.uk, Univ of Cambridge; Yg
Ray, G. Carleton, (804) 924-0551 cr@virginia.edu, Univ of Virginia; Ob
Ray, Pallav, 321-674-7191 pray@fit.edu, Florida Inst of Tech; AstZo
Ray, Waverly, 6196447454 x3018 waverly.ray@gcccd.edu, Cuyamaca Coll; Gg
Rayburn, John A., 845-257-3767 rayburnj@newpaltz.edu, SUNY, New Paltz; Gml
Raymo, Maureen, raymo@ldeo.columbia.edu, Columbia Univ; Pe
Raymond, Anne, (979) 845-0644 raymond@geo.tamu.edu, Texas A&M Univ; Pb
Raymond, Carol A., (818) 354-8690 Jet Propulsion Lab; Yr
Raymond, Charles F., 206-685-9697 cfr2@uw.edu, Univ of Washington; Gl
Raymond, Jasmin, 418 654-2559 jasmin.raymond@ete.inrs.ca, Univ du Quebec; Hy
Rayne, Todd W., (315) 859-4698 trayne@hamilton.edu, Hamilton Coll; Hy
Raysoni, Amit , 956-882-8835 amit.raysoni@utrgv.edu, Univ of Texas, Rio

Grande Valley; Asc
Rea, David K., (734) 936-0521 davidrea@umich.edu, Univ of Michigan; Gu
Read, Adam S., 575-366-2533 adamread@gis.nmt.edu, New Mexico Tech; Gg
Read, J. Fred, (540) 231-5124 jread@vt.edu, Virginia Tech; Gs
Read, Michael T., 936-468-3701 michael.read@sfasu.edu, Stephen F. Austin State Univ; PgGg
Reader, Steven , sreader@usf.edu, Univ of South Florida; Zi
Reading, Anya, 61 3 6226 2477 anya.reading@utas.edu.au, Univ of Tasmania; Ysg
Reagan, Mark K., (319) 335-1802 mark-reagan@uiowa.edu, Univ of Iowa; Gi
Reams, Max W., (815) 939-5394 mreams@olivet.edu, Olivet Nazarene Univ; GsmPg
Reano, Darryl, darryl.reano@asu.edu, Arizona State Univ; Ze
Reaven, Sheldon, (631) 632-8765 sheldon.reaven@stonybrook.edu, SUNY, Stony Brook; Zn
Reavy, John, +353 21 4904574 j.reavy@ucc.ie, Univ Coll Cork; GicGt
Reay, David, +44 (0) 131 650 7723 david.reay@ed.ac.uk, Edinburgh Univ; As
Reay, William G., 804-684-7119 wreay@vims.edu, William & Mary; Hg
Reber, Jacqueline, (515) 294-7513 jreber@iastate.edu, Iowa State Univ of Sci & Tech; Gct
Reboulet, Edward, 785-864-2173 reboulet@kgs.ku.edu, Univ of Kansas; Hy
Rebuck, Jody L., 717-702-2073 ra-nr-topo_maindesk@pa.gov, DCNR-Pennsylvania Bur of Geo Survey; Zn
Rech, Jason, 513-529-1935 rechja@miamioh.edu, Miami Univ; Gm
Rechcigl, John E., 813-633-4111 rechcigl@ufl.edu, Univ of Florida; So
Reche Estrada, Joan, ++935811781 joan.reche@uab.cat, Univ Auto de Barcelona; GpCp
Record, Amy, amy.record@nmt.edu, New Mexico Tech; Ys
Rector, James W., (510) 643-7820 Univ of California, Berkeley; Ye
Redalje, Donald G., (228) 688-3177 donald.redalje@usm.edu, Univ of Southern Mississippi; Ob
Redden, Marcella, 205-247-3654 mmcintyre@gsa.state.al.us, Geo Survey of Alabama; Gg
Reddy, Christopher M., (508) 289-2316 creddy@whoi.edu, Woods Hole Oceanographic Inst; Co
Reddy, K R., (352) 295-3154 krr@ufl.edu, Univ of Florida; CbSbc
Redfern, Jonathan, +440161 275-3773 jonathan.redfern@manchester.ac.uk, Univ of Manchester; GsoGl
Redfern, Simon, +44 (0) 1223 333475 satr@cam.ac.uk, Univ of Cambridge; Gzy
Redmond, Brian T., (570) 696-2906 brian.redmond@wilkes.edu, Wilkes Univ; GsHwZg
Ree, Jin-Han, 82-2-3290-3175 reejh@korea.ac.kr, Korea Univ; Gct
Reed, Denise J., djreed@uno.edu, Univ of New Orleans; GmOn
Reed, Donald L., (408) 924-5036 donald.reed@sjsu.edu, San Jose State Univ; GuYg
Reed, Mark H., (541) 346-5587 mhreed@uoregon.edu, Univ of Oregon; CgEm
Reed, Robert M., (512) 471-0356 rob.reed@beg.utexas.edu, Univ of Texas at Austin, Jackson Sch of Geosciences; GxoGc
Reed, Wallace E., wer@virginia.edu, Univ of Virginia; Zr
Reeder, Richard J., (631) 632-8208 rjreeder@stonybrook.edu, Stony Brook Univ; ClGz
Reedy, Robert C., (512) 471-7244 bob.reedy@beg.utexas.edu, Univ of Texas at Austin, Jackson Sch of Geosciences; Hgw
Reese, Andy, 601-266-4729 andy.reese@usm.edu, Univ of Southern Mississippi; Zy
Reese, Joseph F., (814) 732-2529 jreese@edinboro.edu, Edinboro Univ of Pennsylvania; GctZe
Reesman, Arthur L., 615-322-2976 Vanderbilt Univ; Gg
Reeve, Andrew S., (207) 581-2353 asreeve@maine.edu, Univ of Maine; Hw
Reeves, Claire, +44 (0)1603 59 3625 c.reeves@uea.ac.uk, Univ of East Anglia; Cg
Reeves, Donald Matt, (907) 786-1372 dmreeves2@uaa.alaska.edu, Univ of Alaska, Anchorage; GeHgZn
Reeves, Donald Matthew M., (269) 387-5493 matt.reeves@wmich.edu, Western Michigan Univ; HwqNr
Refenes, James L., (734) 995-7594 james.refenes@cuaa.edu, Concordia Univ; Ze
Refsnider, Kurt, (928) 350-2256 kurt.refsnider@prescott.edu, Prescott Coll; GmlPe
Refsnider-Streby, Jeanine, (419) 530-2664 jeanine.refsnider@utoledo.edu, Univ of Toledo; Zn
Regalla, Christine, cregalla@bu.edu, Boston Univ; GtcGm
Regan, Sean, sregan5@alaska.edu, Univ of Alaska, Fairbanks; Gzx
Reganold, John P., (509) 335-8856 reganold@wsu.edu, Washington State Univ; Soo
Regehr, David, (785) 532-6101 dregehr@ksu.edu, Kansas State Univ; So
Rehakova, Daniela, +421260296700 rehakova@fns.uniba.sk, Comenius Univ in Bratislava; Pms
Rehkamper, Mark, +44 20 759 46391 markrehk@imperial.ac.uk, Imperial Coll; CsaCc
Rehm, George W., (612) 625-6210 grehm@soils.umn.edu, Univ of Minnesota, Twin Cities; Sc
Rehwald, Matthew, (608) 263-1357 matthew.rehwald@wisc.edu, Univ of Wisconsin, Madison, Division of Extension; Zi
Reichard, James S., (912) 478-5546 jreich@georgiasouthern.edu, Georgia Southern Univ; HwGe
Reichelt-Brushett, Amanda, amanda.reichelt-brushett@scu.edu.au, Southern Cross Univ; Hs
Reichenbacher, Bettina, 089/2180 6603 b.reichenbacher@lrz.uni-muenchen.de, Ludwig-Maximilians-Univ Muenchen; PgvPe
Reichgelt, Tammo, tammo.reichgelt@uconn.edu, Univ of Connecticut; Pbc
Reichler, Thomas, 801-585-0040 thomas.reichler@utah.edu, Univ of Utah; As
Reid, Arch M., (919) 386-1711 archreid@gmail.com, Univ of Houston; GizXm
Reid, Brian, +44 (0)1603 59 2357 b.reid@uea.ac.uk, Univ of East Anglia; GeCg
Reid, Catherine, 3667001 Ext 7764 catherine.reid@canterbury.ac.nz, Univ of Canterbury; Pg
Reid, Mary R., (928) 523-7200 mary.reid@nau.edu, Northern Arizona Univ; CucGi
Reid, Pam, 305 421-4606 preid@rsmas.miami.edu, Univ of Miami; Gs
Reid, Rachel E., gbgmedl@vt.edu rebreid@vt.edu, Virginia Tech; CsPcCb
Reid, Steven K., 606-783-5293 s.reid@morehead-st.edu, Morehead State Univ; GsoGe
Reidel, Stephen P., (509) 376-9932 sreidel@wsu.edu, Washington State Univ; GivGt
Reidenbach, Matthew A., (434) 243-4937 mar5jj@virginia.edu, Univ of Virginia; Hg
Reif, Samantha, 217-786-2764 samantha.reif@llcc.edu, Lincoln Land Comm Coll; Gg
Reimer, Andreas, +49 (0)551 392164 areimer@gwdg.de, Georg-August Univ of Goettingen; OcCmm
Reimers, Clare, 541-737-0220 creimers@coas.oregonstate.edu, Oregon State Univ; Oc
Reinen, Linda A., (909) 621-8672 lreinen@pomona.edu, Pomona Coll; Gct
Reiners, Peter, 520-626-2236 reiners@email.arizona.edu, Univ of Arizona; Cg
Reinfelder, Ying Fan, 848-445-2044 yingfan@eps.rutgers.edu, Rutgers, The State Univ of New Jersey; HgqGe
Reinhardt, Edward G., (905) 525-9140 (Ext. 27594) ereinhar@mcmaster.ca, McMaster Univ; PmGam
Reinhold, Ann Marie, (406) 994-5093 reinhold@montana.edu, Montana State Univ; Hq
Reinmann, Andrew , 212-413-3152 andrew.reinmann@asrc.cuny.edu, Graduate Sch of the City Univ of New York; Gg
Reisner, Jon M., (505) 665-1889 reisner@lanl.gov, Los Alamos; As
Reiss, Nathan, (212) 346-1502 Pace Univ, New York Campus; As
Reiten, Jon C., 406-657-2630 jreiten@mtech.edu, Montana Tech of The Univ of Montana; Hw
Reiter, Marshall A., 575-835-5306 marshall.reiter@nmt.edu, New Mexico Tech; Yh
Reitner, Joachim, +49 (0)551 397950 jreitne@gwdg.de, Georg-August Univ of Goettingen; PggPg
Reitz, Elizabeth J., (706) 542-1464 ereitz@arches.uga.edu, Univ of Georgia; Ga
Reji, Linta, 609-258-4101 lreji@princeton.edu, Princeton Univ; CbPoOb
Remacha Grau, Eduard, ++935811603 eduard.remacha@uab.cat, Univ Auto de Barcelona; Gso
Remenda, Victoria H., (613) 533-6594 remendav@queensu.ca, Queen's Univ; Hw
Rempe, Daniella, rempe@jsg.utexas.edu, Univ of Texas, Austin; Hg
Rempel, Alan W., (541) 346-6316 rempel@uoregon.edu, Univ of Oregon; YgNrGl
Remson, Irwin, (650) 723-9191 Stanford Univ; Hw
Ren, Minghua , 702-774-1465 minghua.ren@unlv.edu, Univ of Nevada, Las Vegas; Gig
Renard, Robert J., 831-375-2354 bobandotty@aol.com, Naval Postgrad Sch; Am
Renaut, Robin W., (306) 966-5705 robin.renaut@usask.ca, Univ of Saskatchewan; Gsn
Reneau, Raymond B., (540) 231-9779 reneau@vt.edu, Virginia Tech; Sc

Reneau, Steven L., (505) 665-3151 sreneau@lanl.gov, Los Alamos; Gm
Renfrew, Ian, +44 (0)1603 59 2557 i.renfrew@uea.ac.uk, Univ of East Anglia; Am
Renfrew, Melanie, renfremp@lahc.edu, Los Angeles Harbor Coll; Zg
Renne, Paul, (510) 644-1350 prenne@bgc.org, Univ of California, Berkeley; Cc
Rennie, Colette, (902) 867-2299 crennie@stfx.ca, Saint Francis Xavier Univ; Gg
Rennie, Tim, (613) 258-8336 Ext. 61286 trennie@kemptvillec.uoguelph.ca, Univ of Guelph; Zn
Rennie, Tom, (775) 682-6088 tom@seismo.unr.edu, Univ of Nevada, Reno; Ys
Rensberger, John M., rensb@uw.edu, Univ of Washington; Pvg
Renshaw, Carl E., (603) 646-3365 carl.renshaw@dartmouth.edu, Dartmouth Coll; HgGc
Repeta, Daniel J., (508) 289-2635 drepeta@whoi.edu, Woods Hole Oceanographic Inst; Co
Repka, James, (949) 582-4694 jrepka@saddleback.edu, Saddleback Comm Coll; GgmZe
Reshef, Moshe, 972-36406880 mosher@post.tau.ac.il, Tel Aviv Univ; Yeg
Resing, Joseph A., (206) 526-6184 resing@pmel.noaa.gov, Univ of Washington; Oc
Resnic, Victor S., 281-618-5800 rvictor@mcleodusa.net, Lonestar Coll - North Harris; Go
Resnichenko, Yuri, (598) 2525 1552 yresni@fcien.edu.uy, Univ de la Republica Oriental del Uruguay (UDELAR); ZirZn
Resor, Phillip G., 860 6853139 presor@wesleyan.edu, Wesleyan Univ; Gct
Resplandy, Laure, 609-258-4101 laurer@princeton.edu, Princeton Univ; Ob
Ressel, Mike, (775) 682-7844 mressel@unr.edu, Univ of Nevada, Reno; EgGiCc
Ressler, Daniel E., 570-372-4216 resslerd@susqu.edu, Susquehanna Univ; Sp
Restrepo, Jorge I., (561) 297-2795 restrepo@fau.edu, Florida Atlantic Univ; Hqw
Retallack, Gregory J., (541) 346-4558 gregr@uoregon.edu, Univ of Oregon; PbSoGa
Retelle, Michael J., (207) 786-6155 mretelle@bates.edu, Bates Coll; Gl
Rettig, Andrew, 937-229-2261 arettig1@udayton.edu, Univ of Dayton; Zy
Retzler, Andrew, (612) 626-3895 aretzler@umn.edu, Univ of Minnesota; GrsPi
Reusch, David B., 575-835-5404 dreusch@ees.nmt.edu, New Mexico Tech; Zn
Reusch, Douglas N., 207-778-7463 info@reuschlaw.de, Univ of Maine, Farmington; GtCgOu
Reuss, Robert L., 617-627-3494 bert.reuss@tufts.edu, Tufts Univ; Gzi
Reuter, Gerhard, (780) 492-0358 gerhard.reuter@ualberta.ca, Univ of Alberta; Am
Revelle, Douglas O., (505) 667-1256 revelle@lanl.gov, Los Alamos; As
Revenaugh, Justin, 612-624-7553 justinr@umn.edu, Univ of Minnesota, Twin Cities; Ys
Revetta, Frank A., (315) 267-3441 revettfa@potsdam.edu, SUNY Potsdam; YgGtg
Rexius, James E., 734-462-4400 jrexius@schoolcraft.edu, Schcraft Coll; Gl
Rey-Sanchez, Camilo, areysan@ncsu.edu, North Carolina State Univ; As
Reyes, Alfonso, reyeszca@cicese.mx, Centro de Investigación Científica y de Educación Superior de Ensenada; Ne
Reyhanoglu, Mahmut, 706-507-8154 reyhanoglu_mahmut@columbusstate.edu, Columbus State Univ; Zn
Reynard, Linda, lindareynard@boisestate.edu , Boise State Univ; Cs
Reynolds, Laura, 508-929-8574 lreynolds2@worcester.edu, Worcester State Univ; Gs
Reynolds, Robert W., (541) 383-7557 Central Oregon Comm Coll; Gv
Reynolds, Stephen J., (480) 965-9049 sreynolds@asu.edu, Arizona State Univ; GctZe
Rezaie-Boroon, Mohammad H., (323) 343-2406 mrezaie@calstatela.edu, California State Univ, Los Angeles; GeCg
Rhenberg, Elizabeth, (901) 678-2177 erhenbrg@memphis.edu, Univ of Memphis; Pg
Rheuban, Jennie, (508) 289-3782 jrheuban@whoi.edu, Woods Hole Oceanographic Inst; Oc
Rhine, Michelle, 217-581-2626 mdrhine@eiu.edu, Eastern Illinois Univ; Zn
Rhines, Peter B., (206) 543-0593 rhines@atmos.washington.edu, Univ of Washington; Op
Rhoads, Bruce, brhoads@illinois.edu, Univ of Illinois, Urbana-Champaign; Gm
Rhode, David E., (775) 673-7310 dave@dri.edu, Desert Research Inst; Pe
Rhodes, Amy L., (413) 585-3947 arhodes@smith.edu, Smith Coll; ClGe
Rhodes, Dallas D., (912) 478-5361 Georgia Southern Univ; Gm
Rhodes, Edward J., (310) 825-3880 erhodes@epss.ucla.edu, Univ of California, Los Angeles; CcGma
Rhodes, J. Michael, (413) 545-2841 jmrhodes@geo.umass.edu, Univ of Massachusetts, Amherst; GviCa
Ribbons, Relena R., relena.r.ribbons@lawrence.edu, Lawrence Univ; SbGe
Ricchezza, Victor J., 678-567-6904 vricchezza@gsu.edu, Georgia State Univ, Perimeter Coll, Clarkston Campus; Gg
Ricci, Julia, 575-835-5217 julia.ricci@nmt.edu, New Mexico Tech; CcGvi
Rice, Chuck W., (785) 532-7217 cwrice@ksu.edu, Kansas State Univ; Sb
Rice, James R., (617) 495-3445 rice@esag.harvard.edu, Harvard Univ; Ygu
Rice, Karen C., (434) 243-3429 kcr4y@virginia.edu, Univ of Virginia; HgCg
Rice, Keith W., 715-346-4454 krice@uwsp.edu, Univ of Wisconsin, Stevens Point; Zir
Rice, Melissa S., (360) 650-3592 melissa.rice@wwu.edu, Western Washington Univ; XgGsm
Rice, Murray, 940-565-3861 murray.rice@unt.edu, Univ of North Texas; Zn
Rice, Pamela J., (612) 625-1909 price@soils.umn.edu, Univ of Minnesota, Twin Cities; So
Rice, Thomas L., (937) 766-6140 trice@cedarville.edu, Cedarville Univ; GemEo
Rice-Snow, R. Scott, 765-285-8269 ricesnow@bsu.edu, Ball State Univ; Gm
Rich, Fredrick J., (912) 478-0849 frich@georgiasouthern.edu, Georgia Southern Univ; Pl
Richard, Benjamin H., benjamin.richard@wright.edu, Wright State Univ; Ye
Richard, Ferdinand W., rf@udsm.ac.tz, Univ of Dar es Salaam; YsGt
Richard, Gigi A., 970-248-1689 grichard@coloradomesa.edu, Colorado Mesa Univ; Hsg
Richard, Robert, (310) 825-6663 rrichard@igpp.ucla.edu, Univ of California, Los Angeles; Yg
Richards, Bill, (208) 769-3477 bill_richards@nic.edu, North Idaho Coll; GgzZe
Richards, Ian J., irichard@smu.edu, Southern Methodist Univ; Cs
Richards, James H., (530) 752-0170 jhrichards@ucdavis.edu, Univ of California, Davis; So
Richards, Kelvin J., 808-956-5399 rkelvin@hawaii.edu, Univ of Hawai'i, Manoa; Op
Richards, Laura, +44 0161 306-0361 laura.richards@manchester.ac.uk, Univ of Manchester; Ge
Richards, Mark A., mark_richards@berkeley.edu, Univ of California, Berkeley; Yg
Richards, Paul G., richards@ldeo.columbia.edu, Columbia Univ; Ys
Richards-McClung, Bryony, 801-585-0599 bmcclung@egi.utah.edu, Univ of Utah; Go
Richardson, Carson A., carichardson@email.arizona.edu, Univ of Arizona; Emg
Richardson, Eliza, (814) 863-2507 eliza@psu.edu, Pennsylvania State Univ, Univ Park; YsZe
Richardson, Justin, 413-545-4840 jbrichardson@umass.edu, Univ of Massachusetts, Amherst; CbSc
Richardson, Mary Jo, (979) 845-7966 mrichardson@ocean.tamu.edu, Texas A&M Univ; Gu
Richardson, Randall M., (520) 621-4950 rmr@email.arizona.edu, Univ of Arizona; Yg
Richardson, Steve, steve.richardson@uct.ac.za, Univ of Cape Town; Cg
Richardson, Tammi, richardson@biol.sc.edu, Univ of South Carolina; Ob
Richardson, Yvette P., yrichardson@psu.edu, Pennsylvania State Univ, Univ Park; Ams
Richaud, Mathieu, (559) 278-4557 mathieu@csufresno.edu, Fresno State Univ; GusCs
Richey, Jeffrey E., (206) 543-7339 jrichey@u.washington.edu, Univ of Washington; CbHs
Richter, Carl, (337) 482-5353 richter@louisiana.edu, Univ of Louisiana at Lafayette; Ym
Richter, Daniel D., 919-613-8031 drichter@duke.edu, Duke Univ; So
Richter, David, david.richter.26@nd.edu, Univ of Notre Dame; As
Richter, Frank M., (773) 702-8118 richter@geosci.uchicago.edu, Univ of Chicago; Gt
Ricka, Adam, +420 549 49 6605 ricka@sci.muni.cz, Masaryk Univ; Hgy
Rickaby, Ros, +44 (1865) 272034 rosalind.rickaby@earth.ox.ac.uk, Univ of Oxford; Gz
Rickard, D, rickard@cf.ac.uk, Cardiff Univ; Cg
Ricke, Katherine, kricke@ucsd.edu, Univ of California, San Diego; Zo
Ricketts, Hugo, +44 0161 306-3911 h.ricketts@manchester.ac.uk, Univ of Manchester; As
Ricketts, Jason, (915) 747-5599 jricketts@utep.edu, Univ of Texas, El Paso; GctGi
Ridd, Merrill K., (801) 581-7939 merrill.ridd@geog.utah.edu, Univ of Utah; Zr
Ridenour, Gregory D., (931) 221-7454 ridenourg@apsu.edu, Austin Peay State Univ; HsOgZn

Ridge, John C., (617) 627-3494 jack.ridge@tufts.edu, Tufts Univ; GlnGg
Ridgway, Kenneth D., (765) 494-3269 ridge@purdue.edu, Purdue Univ; Gs
Ridgwell, Andy, (951) 827-3186 andy@seao2.org, Univ of California, Riverside; Zn
Riding, Robert, 865-974-2366 rriding@utk.edu, Univ of Tennessee, Knoxville; Cb
Ridky, Alice M., (207) 859-5800 amridky@colby.edu, Colby Coll; Zn
Ridley, John R., (970) 491-5943 jridley@colostate.edu, Colorado State Univ; Eg
Ridley, Moira K., (806) 834-0627 moira.ridley@ttu.edu, Texas Tech Univ; CqlZm
Riebe, Clifford S., 307-766-3965 criebe@uwyo.edu, Univ of Wyoming; ClGme
Riebesell, John, (313) 593-5132 jriebese@umich.edu, Univ of Michigan, Dearborn; Zu
Riedel, Oliver, 089/2180 4335 oliver.riedl@lrz.uni-muenchen.de, Ludwig-Maximilians-Univ Muenchen; Gz
Riedinger, Natascha, 405-744-6358 natascha.riedinger@okstate.edu, Oklahoma State Univ; CgGu
Riegel, Hannah B., riefelhb@appstate.edu, Appalachian State Univ; Ggc
Rieger, Duayne, drieger@ccri.edu, Comm Coll of Rhode Island; Ys
Rieke, George H., (520) 621-2832 grieke@as.arizona.edu, Univ of Arizona; Zn
Riemer, Nicole, 217-244-2844 nriemer@illinois.edu, Univ of Illinois, Urbana-Champaign; As
Riemersma, Peter E., 616-331-3553 riemersp@gvsu.edu, Grand Valley State Univ; Hw
Rieppel, Olivier C., (312) 665-7630 orieppel@fieldmuseum.org, Field Mus of Natural History; Pv
Ries, Justin, 7815817370 x342 j.ries@neu.edu, Northeastern Univ; Cm
Riess, Carolyn M., (512) 468-1832 criess@austincc.edu, Austin Comm Coll District; GoeGg
Riesselman, Christina R., +64 3 479-7505 christina.riesselman@otago.ac.nz, Univ of Otago; Ou
Rietbrock, Andreas, +44-151-794-5181 a.rietbrock@liverpool.ac.uk, Univ of Liverpool; YsGv
Rietmeijer, Frans J., (505) 277-5733 fransjmr@unm.edu, Univ of New Mexico; Gp
Rigby, John, 61 7 3138 1638 j.rigby@qut.edu.au, Queensland Univ of Tech; Pb
Riggs, Nancy, (928) 523-9362 nancy.riggs@nau.edu, Northern Arizona Univ; Gv
Riggs, Stanley R., (252) 328-6015 riggss@ecu.edu, East Carolina Univ; GusGe
Rignot, Eric, (949) 824-3739 erignot@uci.edu, Univ of California, Irvine; GlOpZr
Rigo, Manuel, +390498279175 manuel.rigo@unipd.it, Univ degli Studi di Padova; Gs
Rigsby, Catherine A., (252) 328-4297 rigsbyc@ecu.edu, East Carolina Univ; Gs
Riha, Susan, (607) 255-1729 sjr4@cornell.edu, Cornell Univ; Sf
Riise, Gunnhild, gunnhild.riise@nmbu.no, Norwegian Univ of Life Scis (NMBU); Hs
Riker-Coleman, Kristin E., 715-394-8410 krikerco@uwsuper.edu, Univ of Wisconsin, Superior; Gg
Riley, James, (520) 626-6681 jjriley@ag.arizona.edu, Univ of Arizona; Zn
Rimmer, Susan M., (618) 453-7369 srimmer@siu.edu, Southern Illinois Univ Carbondale; EcCgGo
Rinae, Makhadi, +27 (0)51 401 9008 makhadi@ufs.ac.za, Univ of the Free State; Gge
Rindsberg, Andrew K., 205 652 3416 arindsberg@uwa.edu, Univ of West Alabama; PeGePi
Rink, W. J., 850 229-1443 rinkwj@mcmaster.ca, McMaster Univ; CcGam
Rinterknecht, Vincent, +33 (0)1 45 07 55 81 vincent.rinterknecht@lgp.cnrs.fr, Univ of St. Andrews; CcGma
Riordan, Allen J., (919) 515-7973 al_riordan@ncsu.edu, North Carolina State Univ; As
Riordan, Jean, (907) 696-0079 jean.riordan@alaska.gov, Alaska Division of Geological & Geophysical Surveys; Gg
Rios-Sanchez, Miriam, 218-755-2563 mriossanchez@bemidjistate.edu, Bemidji State Univ; HwZrGg
Rioux, Matt, rioux@geol.ucsb.edu, Univ of California, Santa Barbara; Cc
Rippy, Megan, 703-361-5606 mrippy@vt.edu, Virginia Tech; Rw
Riser, Stephen C., (206) 543-1187 riser@uw.edu, Univ of Washington; Op
Risk, Dave A., drisk@stfx.ca, Saint Francis Xavier Univ; SoZr
Risk, Michael J., riskmj@mcmaster.ca, McMaster Univ; Pg
Ristovski, Zoran, +61 7 3138 1129 z.ristovski@qut.edu.au, Queensland Univ of Tech; Asp
Ritchie, Harold C., (902) 494-5192 hritchie@phys.ocean.dal.ca, Dalhousie Univ; Am
Ritsche, Michael, 630-252-1554 mtritsche@anl.gov, Argonne; As
Ritsema, Jeroen, (734) 615-6405 jritsema@umich.edu, Univ of Michigan; Ysg
Rittenour, Tammy M., (435) 213-5756 tammy.rittenour@usu.edu, Utah State Univ; GmCcGa
Ritter, Charles J., (937) 229-2953 Univ of Dayton; Ct
Ritter, Joachim R., +49-721-60844539 joachim.ritter@kit.edu, Karlsruhe Inst of Tech; YsGtv
Ritter, John B., (937) 327-7332 jritter@wittenberg.edu, Wittenberg Univ; Gm
Ritter, Leonard, (519) 824-4120 Ext.52980 lritter@uoguelph.ca, Univ of Guelph; Zn
Ritter, Paul, (309) 268-8640 paul.ritter@heartland.edu, Heartland Comm Coll; Zg
Ritter, Scott M., 801-4224239 scott_ritter@byu.edu, Brigham Young Univ; Ps
Ritterbush, Linda A., ritterbu@callutheran.edu, California Lutheran Univ; Pi
Ritts, Malinda, (208) 885-1179 mritts@uidaho.edu, Univ of Idaho; GsCg
Ritz, Richard, (402) 280-2461 richard.ritz@afwa.af.mil, Creighton Univ; Am
Ritzi, Jr., Robert W., 937 775-3455 robert.ritzi@wright.edu, Wright State Univ; Hw
Rius, Marc, +44 (0)23 8059 3275 m.rius@soton.ac.uk, Univ of Southampton; Znn
Rivard, Benoit, (780) 492-0345 benoit.rivard@ualberta.ca, Univ of Alberta; Zr
Rivera, Mark, 907-786-1235 marivera@uaa.alaska.edu, Univ of Alaska, Anchorage; Gg
Rivera-Collazo, Isabel, iriveracollazo@ucsd.edu, Univ of California, San Diego; Ga
Rivers, Toby C. J. S., (709) 737-8392 trivers@mun.ca, Memorial Univ of Newfoundland; Gp
Rizeli, Mustafa Eren, 00904242370000-5961 merizeli@firat.edu.tr, Firat Univ; Gxi
Rizoulis, Athanasios, +44 0161 275-0311 a.rizoulis@manchester.ac.uk, Univ of Manchester; Ge
Roach, Michael, 61 3 6226 2474 Univ of Tasmania; Yg
Robarge, Wayne P., (919) 515-1454 North Carolina State Univ; Sc
Robas, Sheryl A., 609-258-6144 srobas@princeton.edu, Princeton Univ; Zn
Robbins, Gary A., (860) 486-2448 gary.robbins@uconn.edu, Univ of Connecticut; Hw
Roberson, Philip, (931) 372-3522 rproberson@tntech.edu, Tennessee Tech Univ; GgPg
Roberson, Randal P., (931) 221-1004 robersonr@apsu.edu, Austin Peay State Univ; Gg
Robert, Genevieve, (207) 786-6105 grobert@bates.edu, Bates Coll; CpGiz
Roberts, A. Lynn, (410) 516-4387 lroberts@jhu.edu, Johns Hopkins Univ; Zn
Roberts, Charles E., (561) 297-3254 croberts@fau.edu, Florida Atlantic Univ; Zri
Roberts, Eric M., +61747816947 eric.roberts@jcu.edu.au, James Cook Univ; GsPvCc
Roberts, Frank, (504) 388-2964 Montgomery County Comm Coll; Gp
Roberts, Gerald, +44 020 3073 8033 gerald.roberts@ucl.ac.uk, Birkbeck Coll; Gct
Roberts, Harry H., (225) 388-2964 harry@antares.esl.lsu.edu, Louisiana State Univ; Gs
Roberts, Jennifer A., (785) 864-4997 jenrob@ku.edu, Univ of Kansas; Cbl
Roberts, Mark L., (508) 289-3654 mroberts@whoi.edu, Woods Hole Oceanographic Inst; Yg
Roberts, Paul H., (310) 206-2707 roberts@math.ucla.edu, Univ of California, Los Angeles; Ym
Roberts, Peter, (505) 667-1199 proberts@lanl.gov, Los Alamos; Ys
Roberts, Ray L., 817295-7392 rroberts@hillcollege.edu, Hill Coll; Ge
Roberts, Sarah K., (925) 423-4112 roberts28@llnl.gov, Lawrence Livermore; Cl
Roberts, Sheila M., (406) 683-7017 sheila.roberts@umwestern.edu, Univ of Montana Western; GeSo
Roberts, Stephen, +44 ()023 80593246 steve.roberts@noc.soton.ac.uk, Univ of Southampton; Cg
Roberts-Semple, Dawn, 718-262-2775 drobertssemple@york.cuny.edu, York Coll (CUNY); Gem
Robertson, Alastair H., +44 (0) 131 650 8546 alastair.robertson@ed.ac.uk, Edinburgh Univ; GtgGs
Robertson, Andrew W., 845-680-4491 awr@iri.columbia.edu, Columbia Univ; As
Robertson, Daniel E., 585-292-2422 drobertson@monroecc.edu, Monroe Comm Coll; Eg
Robertson, James M., (608) 263-7384 jmrober1@facstaff.wisc.edu, Univ of Wisconsin, Madison; Eg
Robertson, Wendy, 989-774-7517 wendy.robertson@cmich.edu, Central

Michigan Univ; HwCg
Robin, Pierre-Yves F., 905 828-5419 Univ of Toronto; Gc
Robinson, Alexander, 713-743-2547 acrobinson@uh.edu, Univ of Houston; Gc
Robinson, Bruce A., (505) 667-1910 robinson@lanl.gov, Los Alamos; Zn
Robinson, Carol, +44 (0)1603 59 3174 carol.robinson@uea.ac.uk, Univ of East Anglia; Ob
Robinson, Clare, +44 0161 275-3296 a.rizoulis@manchester.ac.uk, Univ of Manchester; Co
Robinson, Cordula, cordula@crsa.bu.edu, Boston Univ; Gm
Robinson, David, +44 (0)1908 653493 x 53493 david.robinson@open.ac.uk, Open Univ; Zn
Robinson, Delores, dmr@ua.edu, Univ of Alabama; Gtc
Robinson, Edward, (305) 348-3572 draper@fiu.edu, Florida InterNatl Univ; Pm
Robinson, George W., grobinson@stlawu.edu, St. Lawrence Univ; Gz
Robinson, Judith, 973-353-1976 judy.robinson@rutgers.edu, Rutgers, The State Univ of New Jersey, Newark; Yg
Robinson, Kevin, (619) 594-1386 rockrobinson@gmail.com, San Diego State Univ; Gc
Robinson, Leonie, +44 0151 795 4387 leonie.robinson@liverpool.ac.uk, Univ of Liverpool; Ob
Robinson, Lori, (612) 626-7429 robin126@umn.edu, Univ of Minnesota; Zi
Robinson, Mark, (480) 727-9691 mark.s.robinson@asu.edu, Arizona State Univ; Xg
Robinson, Marshall, (801) 803-3898 marshallrobinson@utah.gov, Utah Geo Survey; GgZi
Robinson, Michael, (912) 598-3310 mike.robinson@skio.usg.edu, Georgia Southern Univ; ZiOn
Robinson, Peter, (303) 492-5108 peter.robinson@colorado.edu, Univ of Colorado; Pv
Robinson, R. Craig, (860) 832-2950 Central Connecticut State Univ; Xy
Robinson, Ruth, +44 01334 463996 rajr@st-andrews.ac.uk, Univ of St. Andrews; Gs
Robinson, Sarah, (719) 333-9287 sarah.robinson@usafa.edu, United States Air Force Acad; GgZi
Robinson, Steve, j.s.robinson@reading.ac.uk, Univ of Reading; Sf
Robinson, Stuart, +44 (1865) 272058 stuartr@earth.ox.ac.uk, Univ of Oxford; PeGr
Robinson, Walter, 919-515-7002 walter_robinson@ncsu.edu, North Carolina State Univ; As
Robinson, William, william.robinson@umb.edu, Univ of Massachusetts, Boston; Zn
Robison, Richard A., rrobisn@ku.edu, Univ of Kansas; PiGr
Rocha, Cesar, 860-405-9153 cesar.rocha@uconn.edu, Univ of Connecticut; Op
Rocha, Fernando, tavares.rocha@ua.pt, Unive de Aveiro; EnGzb
Rocha, Guillermo, 7189515000 x2887 grocha@brooklyn.cuny.edu, Brooklyn Coll (CUNY); GgeCg
Roche, Didier, +31 20 59 83077 didier.roche@vu.nl, Vrije Univit Amsterdam; AsZn
Roche, James E., (225) 388-2707 jroche@geol.lsu.edu, Louisiana State Univ; Gg
Roche, Steven L., 918-631-3307 sroche@utulsa.edu, Univ of Tulsa; YesYg
Rocheford, MaryKathryn (Kat), 906-635-2140 mrocheford@lssu.edu, Lake Superior State Univ; GeZiGa
Rochester, Michael G., (709) 737-7565 mrochest@mun.ca, Memorial Univ of Newfoundland; Yg
Rochette, Scott M., (585) 395-2603 srochett@brockport.edu, SUNY, The Coll at Brockport; Am
Rocholl, Alexander, 089/2180 4293 rocholl@min.uni-muenchen.de, Ludwig-Maximilians-Univ Muenchen; Gz
Rock, Jessie L., 701-231-7951 jessie.rock@ndsu.edu, North Dakota State Univ; ZgPcg
Rock, Tommy, 609-258-4101 tr6717@princeton.edu, Princeton Univ; ZgGg
Rockaway, John D., (859) 572-5412 rockawayj@nku.edu, Northern Kentucky Univ; Ng
Rockwell, Thomas K., (619) 594-4441 trockwell@mail.sdsu.edu, San Diego State Univ; Gm
Rockwood, Anthony A., (303) 556-8399 Metropolitan State Coll of Denver; Am
Roda Boluda, Duna, +31 2059 82857 d.c.rodaboluda@vu.nl, Vrije Univit Amsterdam; Gs
Rodbell, Donald T., (518) 388-6034 rodbelld@union.edu, Union Coll; GlmGe
Roden, Eric E., (608) 260-0724 eroden@geology.wisc.edu, Univ of Wisconsin, Madison; PgSbCl
Roden, Gunnar I., (206) 543-5627 giroden@u.washington.edu, Univ of Washington; Op
Roden, Michael F., (706) 542-2416 mroden@uga.edu, Univ of Georgia; Gi
Rodgers, David W., rodgdavi@isu.edu, Idaho State Univ; Gct
Rodgers, John C., (662) 325-0732 jcr100@msstate.edu, Mississippi State Univ; Zy
Rodgers, Mel, melrodgers@usf.edu, Univ of South Florida; Gv
Rodgers, N, rodgersn@cf.ac.uk, Cardiff Univ; Am
Rodgers, Nick, +44(0)29 208 79064 rodgersn@cardiff.ac.uk, Univ of Wales; Xm
Rodland, David L., (740) 826-8425 drodland@muskingum.edu, Muskingum Univ; PieGr
Rodolfo, Kelvin S., krodolfo@uic.edu, Univ of Illinois Chicago; GueGs
Rodrigues, Cyril G., 519-253-3000 ext. 2499 cgr@uwindsor.ca, Univ of Windsor; Pm
Rodrigues, Jorge, 530-341-4355 jmrodrigues@ucdavis.edu, Univ of California, Davis; SbPo
Rodriguez, Lizzette A., (787) 265-3845 lizzette.rodriguez1@upr.edu, Univ of Puerto Rico; GviZr
Rodríguez, Marta, martarm@ucm.es, Univ Compl de Madrid; Ggs
Rodríguez, Martín Jesús, martinjr@ucm.es, Univ Compl de Madrid; Ng
Rodríguez, Sergio, sergrodr@ucm.es, Univ Compl de Madrid; PgGg
Rodriguez, Vanessa del S., (787) 722-2526 Puerto Rico Bur of Geology; Zn
Rodriguez-Blanco, Juan Diego, +353 1 8961691 j.d.rodriguez-blanco@tcd.ie, Trinity Coll; GzClZm
Roe, Carol, 757-221-2440 crroex@wm.edu, William & Mary; Zn
Roe, Gerard H., (206) 543-4980 gerard@ess.washington.edu, Univ of Washington; As
Roecker, Steven W., (518) 276-6773 roecks@rpi.edu, Rensselaer Polytechnic Inst; Yg
Roegiers, Jean-Claude, (405) 255-5459 jroegiers@ou.edu, Univ of Oklahoma; Nr
Roelofse, Frederick, +27 (0)51 401 9001 roelofsef@ufs.ac.za, Univ of the Free State; Gip
Roemer, Elizabeth, (520) 621-2897 eroemer@pirlmail.lpl.arizona.edu, Univ of Arizona; Zn
Roemmele, Christopher, (610) 436-2108 croemmele@wcupa.edu, West Chester Univ; ZeGgZg
Roemmich, Dean H., (858) 534-2307 droemmich@ucsd.edu, Univ of California, San Diego; Op
Roering, Joshua J., (541) 346-5574 jroering@uoregon.edu, Univ of Oregon; Gm
Roeske, Sarah M., (530) 752-4933 smroeske@ucdavis.edu, Univ of California, Davis; GtcGp
Roesler, Collin, 207-725-3842 croesler@bowdoin.edu, Bowdoin Coll; OgpZr
Roethel, Frank J., (631) 632-8732 frank.roethel@stonybrook.edu, SUNY, Stony Brook; Oc
Rogers, A. D., 631-632-1509 deanne.rogers@stonybrook.edu, Stony Brook Univ; XgZr
Rogers, Garry C., (250) 363-6450 Univ of Victoria; Ys
Rogers, Jefferson S., (731) 881-7442 jrogers@utm.edu, Univ of Tennessee, Martin; Zn
Rogers, Jeffrey C., (614) 292-0148 rogers.21@osu.edu, Ohio State Univ; As
Rogers, Joe D., 806-651-2570 West Texas A&M Univ; Ga
Rogers, Karyn L., (518) 276-2372 rogerk5@rpi.edu, Rensselaer Polytechnic Inst; PgCl
Rogers, Pamela Z., (505) 667-1765 Los Alamos; Cl
Rogers, Raymond R., (651) 696-6434 rogers@macalester.edu, Macalester Coll; GsrPv
Rogers, Robert D., (209) 667-3466 rrogers1@csustan.edu, California State Univ, Stanislaus; GcmGt
Rogers, Steven L., (+44) 01782 733752 s.l.rogers@keele.ac.uk, Keele Univ; PgGs
Rogers, William J., 806-651-2581 West Texas A&M Univ; Zn
Rogerson, Robert J., (403) 329-5117 rogerson@uleth.ca, Univ of Lethbridge; Gm
Roggenthen, William M., (605) 394-2461 william.roggenthen@sdsmt.edu, South Dakota Sch of Mines & Tech; NgYg
Rogova, Galina L., (716)645-3489 rogova@buffalo.edu, SUNY, Buffalo; Gq
Rogowski, Andrew S., (814) 863-8758 asr@psu.edu, Pennsylvania State Univ, Univ Park; Sp
Rohay, Alan C., (509) 376-6925 alan.rohay@pnl.gov, Pacific Northwest; Ys
Rohbaugh, Robert, (915) 831-7144 rrohrba1@epcc.edu, El Paso Comm Coll; Ggt
Rohr, David M., (432) 837-8167 drohr@sulross.edu, Sul Ross State Univ; PiGd
Rohrssen, Megan, rohrs1m@cmich.edu, Central Michigan Univ; Co
Rohs, C. Renee, (660) 562-1201 rrohs@nwmissouri.edu, Northwest Missouri State Univ; Cg
Roinstad, Lori L., 605-677-6154 lori.roinstad@usd.edu, South Dakota Dept of Environment and Nat Res; Zn

Rokop, Donald J., (505) 667-4299 Los Alamos; Cc
Roland, Emily, 360-650-4078 emily.roland@wwu.edu, Western Washington Univ; GuYrOn
Rolfo, Franco, franco.rolfo@unito.it, Univ di Torino; Gx
Rollins, Kyle, (573) 368-2171 kyle.rollins@dnr.mo.gov, Missouri Dept of Nat Res; ZyGg
Rollinson, Hugh, +44 01332 591786 h.rollinson@derby.ac.uk, Univ of Derby; Cg
Rollinson, Paul A., paulrollinson@missouristate.edu, Missouri State Univ; Zn
Rolston, Dennis E., (916) 752-2113 Univ of California, Davis; So
Roman, Aubrecht, aubrecht@fns.uniba.sk, Comenius Univ in Bratislava; Gdr
Roman, Charles T., (401) 874-6885 croman@gso.uri.edu, Univ of Rhode Island; Ob
Roman, Diana C., droman@carnegiescience.edu, Carnegie Inst for Sci; Gv
Roman, Eric W., (609) 984-6587 eric.roman@dep.state.nj.us, New Jersey Geological and Water Survey; Hw
Roman-Alcalá, Antonio, (510) 885-3160 antonio.romanalcala@csueastbay.edu, California State Univ, East Bay; Zn
Romanak, Katherine , 512-471-6136 katherine.romanak@beg.utexas.edu, Univ of Texas, Austin; GxHt
Romanak, Katherine D., (512) 471-6136 katherine.romanak@beg.utexas.edu, Univ of Texas at Austin, Jackson Sch of Geosciences; CgScGe
Romanovsky, Vladimir, (907) 474-7459 ffver@uaf.edu, Univ of Alaska, Fairbanks; Yg
Romanowicz, Barbara A., 510-643-5690 barbara@seismo.berkeley.edu, Univ of California, Berkeley; Ys
Romanowicz, Edwin A., (518) 564-2152 romanoea@plattsburgh.edu, SUNY, Plattsburgh; HwGcYu
Romans, Brian W., 540-231-2234 romans@vt.edu, Virginia Tech; GsrEo
Romberger, Samuel B., sromberg@mines.edu, Colorado Sch of Mines; Em
Romeo, Ignacio, iromerobr@ucm.es, Univ Compl de Madrid; GcXg
Romero, Leonel, 860-405-9153 leonel.romero@uconn.edu, Univ of Connecticut; Op
Romo, Jose M., jromo@cicese.mx, Centro de Investigación Científica y de Educación Superior de Ensenada; Ye
Ronayne, Michael J., (970) 491-0666 michael.ronayne@colostate.edu, Colorado State Univ; Hwq
Ronck, Catherine, (254) 968-1862 ronck@tarleton.edu, Tarleton State Univ; God
Rood, Richard B., (734) 647-3530 rbrood@umich.edu, Univ of Michigan; ZcAsm
Rooney, Alan, 203-432-3761 alan.rooney@yale.edu, Yale Univ; Cc
Rooney, Neil, (519) 824-4120 Ext.52573 nrooney@uoguelph.ca, Univ of Guelph; Zn
Rooney, Tyrone, 517-432-5522 rooneyt@msu.edu, Michigan State Univ; Gi
Rooney-Varga, Juliette, (978) 934-4715 juliette_rooneyvarga@uml.edu, Univ of Massachusetts, Lowell; ZcRc
Roop, Heidi, hroop@umn.edu, Univ of Minnesota, Twin Cities; At
Root, Tara L., 561 297-3253 troot@fau.edu, Florida Atlantic Univ; Hg
Roozeboom, Kraig, (785) 532-3781 kraig@ksu.edu, Kansas State Univ; So
Rosa, Lynn C., lrosa@wtamu.edu, West Texas A&M Univ; Zge
Rosario, Jose, jose.rosario@csueastbay.edu, California State Univ, East Bay; PsSaGr
Rose, Arthur W., (814) 238-2838 awr1@psu.edu, Pennsylvania State Univ, Univ Park; Cge
Rose, Candace M., (630) 252-3499 cmrose@anl.gov, Argonne; Zn
Rose, Caroline, (608) 263-5495 caroline.rose@wisc.edu, Univ of Wisconsin, Madison, Division of Extension; Zi
Rose, Catherine V., +353 1 8961165 crose@tcd.ie, Trinity Coll; Cg
Rose, Dan, +353 21 4902189 Univ Coll Cork; Zn
Rose, Leanna S., (678) 839-4067 srose@westga.edu, Univ of West Georgia; ZyAsm
Rose, Peter E., (801) 585-7785 prose@egi.utah.edu, Univ of Utah; Np
Rose, Seth E., 404413-5750 geoser@langate.gsu.edu, Georgia State Univ; Hw
Rose, Timothy, (202) 633-1398 roset@si.edu, Smithsonian Inst / Natl Mus of Natural History; CaGv
Rose, William I., (906) 487-2367 raman@mtu.edu, Michigan Technological Univ; Gv
Roselle, Gregory T., (208) 496-7683 roselleg@byui.edu, Brigham Young Univ - Idaho; GpCaHg
Rosen, Carl J., (612) 625-8114 crosen@umn.edu, Univ of Minnesota, Twin Cities; Sc
Rosen, Michael R., 775.887.7683 mrosen@usgs.gov, Univ of Nevada, Reno; GnClHg
Rosen, Peter S., p.rosen@neu.edu, Northeastern Univ; OnZu
Rosenberg, Gary, 215.299.1033 gr347@drexel.edu, Drexel Univ; Pi
Rosenberg, Philip E., (509) 335-4368 rosenberg@wsu.edu, Washington State Univ; GzCg
Rosengaus, Rebeca, (617) 373-7032 r.rosengaus@neu.edu, Northeastern Univ; Zn
Rosenthal, Yair, 848-932-6555X227 rosentha@marine.rutgers.edu, Rutgers, The State Univ of New Jersey; CmlOc
Ross, Andrew, +44(0) 113 34 37590 a.n.ross@leeds.ac.uk, Univ of Leeds; Am
Ross, Charles A., charles.ross@wwu.edu, Western Washington Univ; Pm
Ross, David A., (508) 289-2578 dross@whoi.edu, Woods Hole Oceanographic Inst; GuOg
Ross, Jake, 575-835-5081 jake.ross@nmt.edu, New Mexico Tech; Cc
Ross, Kirstin, kirstin.ross@flinders.edu.au, Flinders Univ; Zn
Ross, Martin E., (617) 373-3263 m.ross@neu.edu, Northeastern Univ; Gie
Ross, Nancy L., (540) 231-6356 nross@vt.edu, Virginia Tech; Gz
Ross, Pierre-Simon, pierre-simon.ross@ete.inrs.ca, Univ du Quebec; GvEg
Ross, Robert M., 6072736623 x118 rmr16@cornell.edu, Paleontological Research Inst; PgZePc
Ross, Tetjana, (902) 494-1327 tetjana.ross@dal.ca, Dalhousie Univ; Op
Ross, Thomas E., (910) 521-6218 tom.ross@uncp.edu, Univ of North Carolina, Pembroke; Zy
Rossbach, Thomas, trossbac@iupui.edu, Indiana Univ - Purdue Univ Indianapolis; GgPg
Rossby, Hans T., (401) 874-6521 trossby@gso.uri.edu, Univ of Rhode Island; Op
Rosscoe, Steven, (325) 670-1387 srosscoe@hsutx.edu, Hardin-Simmons Univ; PmGsPs
Rossell, Irene M., (828) 232-5185 irossell@unca.edu, Univ of North Carolina, Asheville; Zn
Rossi, Carlos, crossi@ucm.es, Univ Compl de Madrid; GdEo
Rossman, George R., (626) 395-6471 grr@gps.caltech.edu,CalTech; GzCa
Rost, Sebastian, +44(0) 113 34 35212 s.rost@leeds.ac.uk, Univ of Leeds; Ysg
Rostami, Jamal, 303-273-3041 rostami@mines.edu, Colorado Sch of Mines; Nr
Rostoker, Gordon, (780) 492-5286 rostoker@space.ualberta.ca, Univ of Alberta; Xy
Rostron, Ben, (780) 492-2178 ben.rostron@ualberta.ca, Univ of Alberta; Zn
Rostron, Benjamin J., (780) 492-2178 ben.rostron@ualberta.ca, Univ of Alberta; HwGoCg
Roth, Danica, 303-273-3802 droth@mines.edu, Colorado Sch of Mines; GmYs
Roth, Gregory W., (814) 863-1018 gwr@psu.edu, Pennsylvania State Univ, Univ Park; So
Roth, Leonard T., troth@hccfl.edu, Hillsborough Comm Coll; Zg
Rothman, Daniel H., (617) 253-7861 dhr@mit.edu, Massachusetts Inst of Tech; Yg
Rothrock, David A., (206) 545-2262 rothrock@apl.washington.edu, Univ of Washington; Op
Rothstein, Lewis, (401) 874-6517 lrothstein@gso.uri.edu, Univ of Rhode Island; Op
Rotz, Rachel, 239-745-4684 rrotz@fgcu.edu, Florida Gulf Coast Univ; HwGg
Rouff, Ashaki, 718-997-3073 ashaki.rouff@qc.cuny.edu, Queens Coll (CUNY); Cac
Roughgarden, Joan, (650) 723-3648 Stanford Univ; Ob
Rougvie, James R., (608) 363-2268 rougviej@beloit.edu, Beloit Coll; GxpGz
Rouhani, Farhang, (540) 654-1895 frouhani@mwc.edu, Mary Washington Coll; Zn
Rouleau, Alain, 4185455011x5213 arouleau@uqac.ca, Univ du Quebec a Chicoutimi; Hw
Rounds, Steven W., (916) 278-7828 rounds@csus.edu, Sacramento State Univ; Gg
Rounsevell, Mark D., +44 (0) 131 651 4468 mark.rounsevell@ed.ac.uk, Edinburgh Univ; Zu
Rouse, Gregory W., (858) 534-7943 grouse@ucsd.edu, Univ of California, San Diego; Ob
Rouse, Jesse, 910-521-6387 jesse.rouse@uncp.edu, Univ of North Carolina, Pembroke; Zni
Rouse, Roland C., (734) 763-0952 rouserc@umich.edu, Univ of Michigan; Gz
Rouse, Jr., Lawrence J., (225) 388-2953 Louisiana State Univ; Op
Rousell, Don H., 7056751151 x2265 drousell@laurentian.ca, Laurentian Univ, Sudbury; Gc
Roussel-Dupre, R., (505) 667-9228 rroussel-dupre@lanl.gov, Los Alamos; Zn
Roussenov, Vassil M., +44 0151 794 4099 v.roussenov@liverpool.ac.uk, Univ of Liverpool; OpZo
Rovey, Charles W., (417) 836-6890 charlesrovey@missouristate.edu, Missouri State Univ; HyGle
Rowan, Christopher J., 330-672-7428 crowan5@kent.edu, Kent State Univ;

FACULTY INDEX – R

Gt
Rowden, Robert, robert.rowden@dnr.iowa.gov, Iowa Dept of Nat Res; Gg
Rowe, Charlotte A., 505-665-6404 char@lanl.gov, New Mexico Tech; Ys
Rowe, Christie, 514-398-2769 christie.rowe@mcgill.ca, McGill Univ; GctEm
Rowe, Clinton M., (402) 472-1946 crowe1@unl.edu, Univ of Nebraska, Lincoln; As
Rowe, Gilbert T., (409) 740-4458 roweg@tamug.edu, Texas A&M Univ; Ob
Rowe, Timothy B., (512) 471-1725 rowe@mail.utexas.edu, Univ of Texas, Austin; Pv
Rowell, Albert J., (785) 864-2747 arowell@ku.edu, Univ of Kansas; Pi
Rowland, Scott K., (808) 956-3150 scott@hawaii.edu, Univ of Hawai'i, Manoa; GvZn
Rowland, Stephen, (702) 895-3625 steve.rowland@unlv.edu, Univ of Nevada, Las Vegas; Pi
Rowley, David B., (773) 702-8146 rowley@geosci.uchicago.edu, Univ of Chicago; GtgGc
Rowley, Rex J., 309-438-7832 rjrowle@ilstu.edu, Illinois State Univ; Zi
Roy, Denis W., (418) 545-5011 dwroy@uqac.ca, Univ du Quebec a Chicoutimi; Gc
Roy, Martin, 514-987-3000 #7619 roy.matin@uqam.ca, Univ du Quebec a Montreal; Gl
Roy, Suzanne, (418) 723-1986 x1748 suzanne_roy@uqar.ca, Univ du Quebec a Rimouski; Ob
Roychoudhury, Alakendra N., +27 21 808 3124 roy@sun.ac.za, Stellenbosch Univ; ClmOc
Royden, Leigh H., (617) 253-1292 lhroyden@mit.edu, Massachusetts Inst of Tech; Gt
Royer, Dana, (860) 685-2836 droyer@wesleyan.edu, Wesleyan Univ; PegPb
Royer, Todd, 812-855-0563 troyer@indiana.edu, Indiana Univ, Bloomington; HsZg
Rozmus, Wojciech, (780) 492-8486 rozmus@phys.ualberta.ca, Univ of Alberta; Zn
Ruark, Matthew D., (608) 263-2889 mdruark@wisc.edu, Univ of Wisconsin, Madison; So
Rubin, Alan E., (310) 825-3202 rubin@igpp.ucla.edu, Univ of California, Los Angeles; Xm
Rubin, Allan M., 609-258-1506 arubin@princeton.edu, Princeton Univ; Yg
Rubin, Kenneth H., (808) 956-8973 krubin@hawaii.edu, Univ of Hawai'i, Manoa; CcGvCg
Rubio-Sierra, Javier, 089/2180 4317 rubio@lrz.uni-muenchen.de, Ludwig-Maximilians-Univ Muenchen; Gz
Rucklidge, John C., (416) 978-2061 jcr@quartz.geology.utoronto.ca, Univ of Toronto; Ge
Ruddick, Barry R., (902) 494-2505 barry.ruddick@dal.ca, Dalhousie Univ; Op
Ruddiman, William F., 540-348-1963 wfr5c@virginia.edu, Univ of Virginia; Gu
Rudge, John, +44 (0) 1223 765545 jfr23@cam.ac.uk, Univ of Cambridge; YgGt
Rudnick, Daniel L., (858) 534-7669 drudnick@ucsd.edu, Univ of California, San Diego; Op
Rudnick, Roberta L., rudnick@ucsb.edu, Univ of Maryland; CgsCt
Rudnicki, Ryan E., rrudnicki@alamo.edu, Alamo Colls, San Antonio Coll; Zy
Rudolph, Maxwell L., (530) 752-3669 maxrudolph@ucdavis.edu, Univ of California, Davis; Yg
Rueger, Bruce F., (207) 859-5806 bfrueger@colby.edu, Colby Coll; PlGg
Ruehr, Thomas A., (805) 756-2552 truehr@calpoly.edu, California Polytechnic State Univ; Sb
Ruff, Larry J., (734) 763-9301 ruff@umich.edu, Univ of Michigan; Ys
Ruff, Steven, (480) 965-6089 steve.ruff@asu.edu, Arizona State Univ; Xg
Ruffel, Alice, (214) 507-9014 aruffel@dcccd.edu, El Centro Coll - Dallas Comm Coll District; Gg
Ruffman, Alan, (902) 422-6482 Dalhousie Univ; Ys
Rugenstein, Jeremy C., jeremy.rugenstein@colostate.edu, Colorado State Univ; ClsPc
Ruggiero, Peter, (541) 737-1239 ruggierp@science.oregonstate.edu, Oregon State Univ; On
Ruhl, Christine J., 918-631-3018 cruhl@utulsa.edu, Univ of Tulsa; YsGtc
Ruiz, Javier , jaruiz@ucm.es, Univ Compl de Madrid; Xg
Ruiz, Joaquin, (520) 621-4090 jruiz@email.arizona.edu, Univ of Arizona; CgcEz
Ruiz Cubillo, Paulo, paulo.ruizcubillo@ucr.ac.cr, Univ de Costa Rica; GvmRn
Ruiz-Diaz, Dorivar, (785) 532-6183 ruizdiaz@ksu.edu, Kansas State Univ; So
Rumble, III, Douglas, 202-478-8990 drumble@carnegiescience.edu, Carnegie Inst for Sci; Gp
Rumrill, Julie, 203-392-5842 rumrillj1@southernct.edu, Southern Connecticut State Univ; GgeGl
Rumstay, Kenneth S., (229) 333-5754 krumstay@valdosta.edu, Valdosta State Univ; Znn
Rundberg, Robert S., (505) 667-4559 Los Alamos; Gq
Rundle, John, jbrundle@ucdavis.edu, Univ of California, Davis; Yd
Rundquist, Bradley C., (701) 777-4246 bradley.rundquist@und.edu, Univ of North Dakota; ZriZy
Runkel, Anthony, 612-627-4780 (Ext. 222) runke001@umn.edu, Univ of Minnesota, Twin Cities; Gs
Runnegar, Bruce, (310) 206-1738 Univ of California, Los Angeles; Pg
Runyon, Cassandra R., (843) 953-8279 runyonc@cofc.edu, Coll of Charleston; XgGmZe
Runyon, Simone, srunyon@uwyo.edu, Univ of Wyoming; EdGip
Rupert, Denise, (570) 484-2048 drupert@lockhaven.edu, Lock Haven Univ; Zn
Rupp, David, 541-737-5222 drupp@coas.oregonstate.edu, Oregon State Univ; AsOp
Rupp, John A., (812) 855-1323 rupp@indiana.edu, Indiana Univ, Bloomington; Go
Ruppel, Stephen C., (512) 471-2965 stephen.ruppel@beg.utexas.edu, Univ of Texas at Austin, Jackson Sch of Geosciences; Gd
Ruppert, Kelly R., (657) 278-3561 kruppert@fullerton.edu, California State Univ, Fullerton; Gg
Ruprecht, Philipp P., 775-682-6084 pruprecht@unr.edu, Univ of Nevada, Reno; Giv
Rusmore, Margaret E., 323 259 2565 rusmore@oxy.edu, Occidental Coll; Gc
Russ, Jean M., (301) 934-7814 jruss@csmd.edu, Coll of Southern Maryland; Zey
Russell, Ann D., (530) 752-3311 adrussell@ucdavis.edu, Univ of California, Davis; Cm
Russell, Armistead G., (404) 894-3079 ted.russell@ce.gatech.edu, Georgia Inst of Tech; As
Russell, Christopher T., (310) 825-3188 ctrussell@igpp.ucla.edu, Univ of California, Los Angeles; Yg
Russell, Dale A., (919) 515-1339 dale_russell@ncsu.edu, North Carolina State Univ; Pv
Russell, James K., (604) 822-2703 krussell@eos.ubc.ca, Univ of British Columbia; GviCg
Russell, Joellen, 520-626-2194 jrussell@email.arizona.edu, Univ of Arizona; Oc
Russell, Lynn M., (858) 534 4852 lmrussell@ucsd.edu, Univ of California, San Diego; As
Russell, R. Doncaster, (604) 822-2551 drussell@eos.ubc.ca, Univ of British Columbia; Yg
Russell, Ron, (512) 471-8831 ron.russell@beg.utexas.edu, Univ of Texas at Austin, Jackson Sch of Geosciences; Zn
Russell, Sally, +44(0) 113 34 35279 s.russell@leeds.ac.uk, Univ of Leeds; Ge
Russell, Terry P., (250) 721-6184 trussell@uvic.ca, Univ of Victoria; Zn
Russell, Theresa J., (501) 575-4403 trussell@comp.uark.edu, Univ of Arkansas, Fayetteville; Zn
Russelle, Michael P., (612) 625-8145 russelle@soils.umn.edu, Univ of Minnesota, Twin Cities; Sc
Russo, Mary Rose, 609 258-4101 mrusso@princeton.edu, Princeton Univ; Zn
Russo, Raymond, 352-392-6766 rrusso@ufl.edu, Univ of Florida; Yg
Russo, Tess A., 814-865-7389 russo@psu.edu, Pennsylvania State Univ, Univ Park; Hw
Russoniello, Christopher, chris.russoniello@mail.wvu.edu, West Virginia Univ; Hw
Rust, Derek, +44 023 92 842298 derek.rust@port.ac.uk, Univ of Portsmouth; Gt
Rustad, James R., jrrustad@ucdavis.edu, Univ of California, Davis; Cl
Rutberg, Randye, 212-772-5326 randye.rutberg@hunter.cuny.edu, Graduate Sch of the City Univ of New York; Gg
Rutford, Robert H., (972) 883-6470 rutford@utdallas.edu, Univ of Texas, Dallas; Gl
Rutherford, Malcolm J., (401) 863-1927 malcolm_rutherford@brown.edu, Brown Univ; Cp
Rutledge, Steven A., rutledge@atmos.colostate.edu, Colorado State Univ; As
Ruttan, Lore, 404 7274217 lruttan@emory.edu, Emory Univ; Ob
Ruttenberg, Kathleen, 808-956-9371 kcr@soest.hawaii.edu, Univ of Hawai'i, Manoa; Cg
Rutter, Ernest H., +44 0161 275-3945 e.rutter@manchester.ac.uk, Univ of Manchester; NrGct
Rutter, Nathaniel W., (780) 492-3085 nat.rutter@ualberta.ca, Univ of Alberta; Ge

Ruzicka, Alexander (Alex) M., (503) 725-3372 ruzickaa@pdx.edu, Portland State Univ; Xm
Ruzicka, Jaromir, jaromirr@hawaii.edu, Univ of Hawai'i, Manoa; Zn
Ryall, Patrick J., (902) 494-3465 pryall@is.dal.ca, Dalhousie Univ; Yg
Ryan, Anne-Marie, 902 494 3184 amryan@dal.ca, Dalhousie Univ; GeZeCg
Ryan, Cathy, (403) 220-2793 ryan@geo.ucalgary.ca, Univ of Calgary; Hw
Ryan, Jeffrey G., ryan@usf.edu, Univ of South Florida; Ct
Ryan, Peter C., (802) 443-2557 pryan@middlebury.edu, Middlebury Coll; Cl
Ryan, Susan, (724) 938-4531 ryan@calu.edu, California Univ of Pennsylvania; Zg
Ryan, William F., (814) 865-0478 wfr1@psu.edu, Pennsylvania State Univ, Univ Park; Am
Ryan, William B. F., billr@ldeo.columbia.edu, Columbia Univ; Yr
Rybacki, Kyle S., 717-702-2058 krybacki@pa.gov, DCNR- Pennsylvania Bur of Geo Survey; ClEgXb
Rychert, Catherine A., +44 (0)23 80598663 c.rychert@soton.ac.uk, Univ of Southampton; Yg
Ryder, Isabelle, +44-151-794-5143 i.ryder@liverpool.ac.uk, Univ of Liverpool; Gt
Ryder, Roy, rryder@southalabama.edu, Univ of South Alabama; SdZry
Rye, Danny M., (203) 432-3174 danny.rye@yale.edu, Yale Univ; Cs
Rygel, Michael C., (315) 267-3401 rygelmc@potsdam.edu, SUNY Potsdam; GsrZi
Ryker, Katherine, kryker@seoe.sc.edu, Univ of South Carolina; Ze
Rysgaard, Soren, 204-272-1611 rysgaard@umanitoba.ca, Univ of Manitoba; Gl
Rößner, Gertrud, 089/2180 6612 g.roessner@lrz.uni-muenchen.de, Ludwig-Maximilians-Univ Muenchen; Pg

S

Sá, Artur A., asa@utad.pt, Unive de Trás-os-Montes e Alto Douro; PgiPs
Saal, Alberto E., (401) 863-7238 alberto_saal@brown.edu, Brown Univ; CgGi
Saalfeld, Alan J., 614 292 6665 saalfeld.1@osu.edu, Ohio State Univ; Zi
Saar, Martin O., 612-625-7332 saar@umn.edu, Univ of Minnesota, Twin Cities; Hg
Sabine, Christopher L., (206) 526-4809 chris.sabine@noaa.gov, Univ of Washington; Oc
Sabol, Martin, +421-2-60296229 sabol@fns.uniba.sk, Comenius Univ in Bratislava; PvgPs
Sabra, Karim, (404) 385-6193 Georgia Inst of Tech; Yg
Sacchi, Mauricio D., (780) 492-1060 sacchi@phys.ualberta.ca, Univ of Alberta; Ys
Saccocia, Peter J., 508-531-2124 psaccocia@bridgew.edu, Bridgewater State Univ; CqsCm
Sack, Richard, 425-880-4418 rosack@uw.edu, Univ of Washington; CgZn
Sacks, I. Selwyn, ssacks@carnegiescience.edu, Carnegie Inst for Sci; Ys
Sacramentogrilo, Isabelle, 619-594-5607 isacramentogrilo@mail.sdsu.edu, San Diego State Univ; Gg
Sacristan, Felix, felix.sacristan@uab.cat, Univ Auto de Barcelona; Ge
Sadd, James L., 323 259 2518 jsadd@oxy.edu, Occidental Coll; ZiGeZr
Sadiq, Abdulali A., sadiqa@qu.edu.qa, Univ of Qatar; Zr
Sadler, Peter M., (951) 827-5616 peter.sadler@ucr.edu, Univ of California, Riverside; GrPs
Sadowski, Andrew, andrew.sadowski@dnr.wa.gov, Central Washington Univ; Gg
Sadowsky, Michael J., (612) 624-2706 sadowsky@soils.umn.edu, Univ of Minnesota, Twin Cities; Sb
Saeidi, Ali, (418) 545-5011 x2561 asaeidi@uqac.ca, Univ du Quebec a Chicoutimi; Nrm
Saenger, Casey, (360) 650-3597 casey.saenger@wwu.edu, Western Washington Univ; Pc
Saffer, Demian, demian@ig.utexas.edu, Univ of Texas, Austin; Gu
Sagan, Vasit, (314) 977-5156 vasit.sagan@slu.edu, Saint Louis Univ; Zi
Sagebiel, J. C., 909 307-2669 237 jsagebiel@sbcm.sbcounty.gov, San Bernardino County Mus; Pv
Sageman, Bradley B., (847) 467-2257 brad@earth.northwestern.edu, Northwestern Univ; PsCbPc
Sager, William W., (713) 743-3108 wwsager@uh.edu, Univ of Houston; GtYrm
Sagiroglu, Ahmet, 00904242370000-5990 sagiroglu@firat.edu.tr, Firat Univ; Egm
Sahagian, Dork, 610-758-6379 dork.sahagian@lehigh.edu, Lehigh Univ; PeGvr
Sahakian, Valerie J., 541-346-3401 vjs@uoregon.edu, Univ of Oregon; Yg
Sahay, Pratap, pratap@cicese.mx, Centro de Investigación Científica y de Educación Superior de Ensenada; Ys
Sahr, John D., (206) 616-7175 jdsahr@ee.washington.edu, Univ of Washington; ZrXym
Saikia, Udoy, udoy.saikia@flinders.edu.au, Flinders Univ; Zn
Saila, Saul B., (401) 874-6485 saila@gso.uri.edu, Univ of Rhode Island; Ob
Saillet, Elodie, +33(0)3 44067563 elodie.saillet@lasalle-beauvais.fr, Institut Polytechnique LaSalle Beauvais (ex-IGAL); GctGo
Saini-Eidukat, Bernhardt, (701) 231-8785 bernhardt.sainieiduk@ndsu.edu, North Dakota State Univ; GxCqEg
Saint, Prem K., psaint@fullerton.edu, California State Univ, Fullerton; Hw
Saito, Heather, (808) 956-8763 hsaito2@hawaii.edu, Univ of Hawai'i, Manoa; Zn
Saito, Mak A., 508-289-3696 msaito@whoi.edu, Woods Hole Oceanographic Inst; Oc
Saja, David, dbs11@cwru.edu, Cleveland Mus of Natural History; GdcZe
Sak, Peter B., (717) 245-1423 sakp@dickinson.edu, Dickinson Coll; Gcm
Sakhaee-Pour, Ahmad, 405-325-3306 sakhaee@ou.edu, Univ of Oklahoma; Np
Saksena, Suddarth, (540) 231-3478 ssaksena@vt.edu, Virginia Tech; HqZiRn
Saku, James C., (301) 687-4724 jsaku@frostburg.edu, Frostburg State Univ; Zn
Salacup, Jeff, 413-545-1913 jsalacup@geo.umass.edu, Univ of Massachusetts, Amherst; PcCb
Saladyga, Thomas, 304-384-6040 saladygat@concord.edu, Concord Univ; Zyi
Salami, B. M., 234-803-321-9685 salamibm@oauife.edu.ng, Obafemi Awolowo Univ; Hg
Salami, Sikiru A., sikiru.salami@uniben.edu, Univ of Benin; YgeGg
Salaun, Pascal, +44-151-794-4101 pascal.salaun@liverpool.ac.uk, Univ of Liverpool; Em
Salaun, Rachel, +440151 795 4649 rachel.jeffreys@liverpool.ac.uk, Univ of Liverpool; Oc
Saleeby, Jason B., (626) 395-6141 jason@gps.caltech.edu,CalTech; Gc
Saleh, Saad, 512-475-9508 saad.saleh@beg.utexas.edu, Univ of Texas, Austin; Go
Salisbury, Joseph, 603-862-0849 joe.salisbury@unh.edu, Univ of New Hampshire; Og
Salje, Ekhard, +44 (0) 1223 768321 ekhard@esc.cam.ac.uk, Univ of Cambridge; Gy
Sallan, Lauren, 215-898-5650 lsallan@sas.upenn.edu, Univ of Pennsylvania; PqvPe
Salle, Bethan, (214) 578-1914 bsalle@dcccd.edu, El Centro Coll - Dallas Comm Coll District; GgeZg
Salleolini, Massimo, massimo.salleolini@unisi.it, Univ of Siena; HwyHs
Sallu, Susannah, +44(0) 113 34 31641 s.sallu@leeds.ac.uk, Univ of Leeds; Ga
Salmachi, Alireza, alireza.salmachi@adelaide.edu.au, Univ of Adelaide; Np
Salmon, Richard L., (858) 534-2090 rsalmon@ucsd.edu, Univ of California, San Diego; Op
Salmun, Haydee, (212) 772-4159 hsalmun@hunter.cuny.edu, Hunter Coll (CUNY); OpZgAs
Salters, Vincent J., (850) 644-1934 salters@magnet.fsu.edu, Florida State Univ; Cg
Saltiel, Seth, ssaltiel@unr.edu, Univ of Nevada, Reno; Yxs
Saltzman, Eric S., 949-824-3936 esaltzma@uci.edu, Univ of California, Irvine; As
Saltzman, Matthew R., 614 292-0481 saltzman.11@osu.edu, Ohio State Univ; GsClg
Salvini, Riccardo, riccardo.salvini@unisi.it, Univ of Siena; ZruNg
Salviulo, Gabriella, +390498279157 gabriella.salviulo@unipd.it, Univ degli Studi di Padova; Gz
Salvucci, Guido D., (617) 353-8344 gdsalvuc@bu.edu, Boston Univ; Hq
Salzer, Matthew W., (520) 621-2946 msalzer@ltrr.arizona.edu, Univ of Arizona; PeGav
Sambrotto, Raymond N., sambrott@ldeo.columbia.edu, Columbia Univ; Ob
Samelson, Roger, 541-737-4752 rsamelson@coas.oregonstate.edu, Oregon State Univ; Op
Sammarco, Paul W., (225) 851-2800 Louisiana State Univ; Zn
Sammis, Charles G., (213) 740-5836 sammis@usc.edu, Univ of Southern California; Yg
Sammis, Theodore W., (505) 646-3405 New Mexico State Univ, Las Cruces; Sp
Samonds, Karen, 815-753-3201 ksamonds@niu.edu, Northern Illinois Univ; Pv
Sample, James C., (928) 523-0881 james.sample@nau.edu, Northern Arizona Univ; GtClOu
Samson, Iain M., 519-253-3000 ext. 2489 ims@uwindsor.ca, Univ of Windsor; CgEmGx
Samson, John C., (780) 492-3616 samson@space.ualberta.ca, Univ of Alberta; Xy
Samson, Perry, (734) 276-0815 samson@umich.edu, Univ of Michigan; AsZeAm

Samson, Scott D., 315-443-2672 sdsamson@syr.edu, Syracuse Univ; Cc
Samsonov, Sergey V., (613) 656-1534 sergey.samsonov@canada.ca, Nat Res Canada; YdZrYg
Samuelson, Alan C., (765) 285-8270 Ball State Univ; Hw
Sanabia, Elizabeth R., 410-293-6556 sanabia@usna.edu, United States Naval Acad; Am
Sanchez, Charles, (520) 782-3836 sanchez@ag.arizona.edu, Univ of Arizona; Zn
Sanchez, Marcelo J., (979) 862-6604 msanchez@civil.tamu.edu, Texas A&M Univ; NtrNx
Sánchez, Nuria, nsanchez@ucm.es, Univ Compl de Madrid; Gz
Sánchez, Sonia, s.sanchez@ucm.es, Univ Compl de Madrid; Gp
Sanchez, Veronica I., (361) 593-3590 veronica.sanchez@tamuk.edu, Texas A&M Univ, Kingsville; GctGm
Sánchez, Yolanda, yol@ucm.es, Univ Compl de Madrid; Gs
Sanchez Roman, Monica, m.sanchezroman@vu.nl, Vrije Univit Amsterdam; GdsGg
Sanchez-Azofeifa, G. Arturo, (780) 492-1822 arturo.sanchez@ualberta.ca, Univ of Alberta; ZirZg
Sandau, Ken L., 406-496-4151 ksandau@mtech.edu, Montana Tech of The Univ of Montana; Zi
Sandel, Bill R., (520) 621-4305 sandel@vega.lpl.arizona.edu, Univ of Arizona; As
Sanders, Laura L., (773) 442-6051 l-sanders@neiu.edu, Northeastern Illinois Univ; HwGe
Sanders, Ronald S., (818) 354-2867 sanders@jpl.nasa.gov, Jet Propulsion Lab; Gr
Sandhu, Harpinder, harpinder.sandhu@flinders.edu.au, Flinders Univ; Zn
Sandlin, Stephen H., 909-389-8644 ssandlin@sbccd.cc.ca.us, San Bernardino Valley Coll; Zy
Sandor, Jonathan A., (515) 294-2209 jasandor@iastate.edu, Iowa State Univ of Sci & Tech; Sd
Sandoval, Samuel, (530) 750-9722 samsandoval@ucdavis.edu, Univ of California, Davis; Hg
Sandoval Donahue, Maria Magdalena, dr. maria magdalena sandoval donahue — water data specialis, New Mexico Tech; GmsGt
Sandoval Solis, Samuel, 530-750-9722 samsandoval@ucdavis.edu, Univ of California, Davis; Hg
Sandvol, Eric A., (573) 884-9616 sandvole@missouri.edu, Univ of Missouri; Ys
Sandwell, David T., (858) 534-7109 dsandwell@ucsd.edu, Univ of California, San Diego; Yr
Sandy, Michael R., (937) 229-3436 Univ of Dayton; Pi
Sanford, Lawrence P., (410) 221-8429 lsanford@hpl.umces.edu, Univ of Maryland; Op
Sanford, Robert A., 217-244-7250 rsanford@illinois.edu, Univ of Illinois, Urbana-Champaign; CbSbPo
Sanford, Thomas B., (206) 543-1365 sanford@apl.washington.edu, Univ of Washington; Op
Sanford, William E., (970) 491-5929 bills@cnr.colostate.edu, Colorado State Univ; Hw
Sanislav, Ioan, ioan.sanislav@jcu.edu.au, James Cook Univ; GctGp
Sanjurjo, Jorge, 0034981167000 jsanjurjo@udc.es, Coruna Univ; GeCcGa
Sankey, Julia, (209) 667-3466 jsankey@csustan.edu, California State Univ, Stanislaus; Pve
Sansone, Francis J., 808-956-2912 sansone@hawaii.edu, Univ of Hawai'i, Manoa; Cm
Santander, Erma, (520) 621-7120 ermasan@arizona.edu, Univ of Arizona; Zn
Santelli, Cara M., santelli@umn.edu, Univ of Minnesota, Twin Cities; Pg
Santelmann, Mary V., (541) 737-1215 santelmm@geo.oregonstate.edu, Oregon State Univ; Zy
Santi, Paul M., 303-273-3108 psanti@mines.edu, Colorado Sch of Mines; Ng
Santillan, Marcelo, (509) 963-1107 marcelo@geology.cwu.edu, Central Washington Univ; Yd
Santisteban, Juan Ignacio, juancho@ucm.es, Univ Compl de Madrid; Gr
Santos, Eduardo, (785) 532-6932 esantos@ksu.edu, Kansas State Univ; So
Santos, Hernan, (787) 265-3845 hernan.santos@upr.edu, Univ of Puerto Rico; PiGdr
Santschi, Peter H., (409) 740-4476 santschip@tamug.edu, Texas A&M Univ; Oc
Sanudo, Sergio, 213-821-1302 sanudo@usc.edu, Univ of Southern California; Pg
Sanz, Esther, mesanz@ucm.es, Univ Compl de Madrid; Gd
Sanz, Miguel Angel, massinfo@ucm.es, Univ Compl de Madrid; Zi
Sapigao, Gladys, 718-997-3300 gladys.sapigao@qc.cuny.edu, Queens Coll (CUNY); ZnnZn
Sar, Abdullah, 00904242370000-5961 asasr@firat.edu.tr, Firat Univ; GviGp
Sarachik, Edward, 206-543-6720 sarachik@atmos.washington.edu, Univ of Washington; As
Sarah, Willig B., (215) 898-5724 sawillig@verizon.net, Univ of Pennsylvania; Ge
Saravanan, R., 979-845-0175 r.saravanan@tamu.edu, Texas A&M Univ; As
Sargent, Kenneth A., (864) 294-3362 ken.sargent@furman.edu, Furman Univ; Hy
Sarmiento, Jorge L., 609-258-6585 jls@princeton.edu, Princeton Univ; Oc
Sarrionandia, Fernando, fernando.sarrionandia@ehu.eus, Univ of the Basque Country UPV/EHU; GigGv
Sartain, Jerry B., (352) 392-1803 (Ext. 330) sartain@ufl.edu, Univ of Florida; So
Sarwar, A. K. Mostofa, (504) 280-6717 asarwar@uno.edu, Univ of New Orleans; Ye
Sas, Mai, 360-650-7363 mai.sas@wwu.edu, Western Washington Univ; GiCaGv
Sasaran, Emanoil Florin, emanoil.sasaran@ubbcluj.ro, Babes-Bolyai Univ; Gds
Sasmaz, Ahmet, +905323406642 asasmaz@firat.edu.tr, Firat Univ; EgGeCe
Sasowsky, Ira D., (330) 972-5389 ids@uakron.edu, Univ of Akron; HwGmo
Sasowsky, Kathryn, (216) 987-5227 kathryn.sasowsky@tri-c.edu, Cuyahoga Comm Coll - Western Campus; Zg
Sass, Henrik, +44(0)29 208 76001 sassh@cardiff.ac.uk, Univ of Wales; Co
Sassenrath, Gretchen, (620) 421-4826 gsassenrath@ksu.edu, Kansas State Univ; So
Sassi, Raffaele, +390498279149 raffaele.sassi@unipd.it, Univ degli Studi di Padova; Ggx
Satkoski, Alina A., 518-265-7183 alina.satkoski@austincc.edu, Austin Comm Coll District; GemGg
Sato, Yoko, (973) 655-4448 satoyo@mail.montclair.edu, Montclair State Univ; Gg
Satterfield, Joseph I., 325-486-6766 joseph.satterfield@angelo.edu, Angelo State Univ; Gc
Sauck, William A., (269) 567-8282 bill.sauck@wmich.edu, Western Michigan Univ; YgeYv
Sauer, Nancy N. S., (505) 665-3759 Los Alamos; Zn
Sauer, Peter, 812-855-6591 pesauer@indiana.edu, Indiana Univ, Bloomington; Pe
Saunders, Charles, charlessaunders@rmc.edu, Randolph-Macon Coll; ZgGrPg
Saunders, Ian , 250-807-8817 ian.saunders@ubc.ca, Univ of British Columbia, Okanagan Campus; AtGlZy
Saunders, Ralph, (310) 243-3284 rsaunders@csudh.edu, California State Univ, Dominguez Hills; Zg
Saura, Eduard, eduard.saura@uab.cat, Univ Auto de Barcelona; Gc
Sautter, Leslie R., sautterl@cofc.edu, Coll of Charleston; OuGuOn
Sava, Diana, 720-981-0522 diana.sava@beg.utexas.edu, Univ of Texas at Austin, Jackson Sch of Geosciences; Yg
Sava, Paul, 303-384-2362 psava@mines.edu, Colorado Sch of Mines; Ye
Savage, Brian, (401) 874-5392 savage@uri.edu, Univ of Rhode Island; Ys
Savage, Heather M., hsavage@ldeo.columbia.edu, Columbia Univ; Ys
Savage, Norman M., (541) 346-4585 nmsavage@uoregon.edu, Univ of Oregon; Pi
Savard, Dany, 418-545-5011 ddsavard@uqac.ca, Univ du Quebec a Chicoutimi; Ca
Savigny, K. Wayne, (604) 684-5900 Univ of British Columbia; Ng
Savin, Samuel M., (216) 368-6592 sms7@case.edu, Case Western Reserve Univ; Cs
Savina, Mary E., 507-222-4404 msavina@carleton.edu, Carleton Coll; Gm
Savov, Ivan, +44(0) 113 34 35199 earis@leeds.ac.uk, Univ of Leeds; Cg
Savrda, Charles E., (334) 844-4887 savrdce@auburn.edu, Auburn Univ; GdPeGr
Savvaidis, Alexandros, 512-475-9549 alexandros.savvaidis@beg.utexas.edu, Univ of Texas, Austin; Ys
Sawyer, Audrey , 614-292-8383 sawyer.143@osu.edu, Ohio State Univ; HwSpHs
Sawyer, Carol F., 251-460-6169 sawyer@southalabama.edu, Univ of South Alabama; GmZye
Sawyer, Dale S., (713) 348-5106 dale@rice.edu, Rice Univ; Ys
Sawyer, Derek, 614-292-7243 sawyer.144@osu.edu, Ohio State Univ; YgGoRn
Sawyer, Edward W., (418) 545-5011 x5636 edward-w_sawyer@uqac.ca, Univ du Quebec a Chicoutimi; GptGi
Sawyer, J. F., (605) 394-2462 foster.sawyer@sdsmt.edu, South Dakota Sch of Mines & Tech; GsoHw
Sawyer, Suzanne, (801) 537-3333 ssawyer@utah.gov, Utah Geo Survey; Zn
Saxena, Surendra K., (305) 348-3030 saxenas@fiu.edu, Florida InterNatl Univ; Gy
Sayles, Frederick L., (508) 289-2561 fsayles@whoi.edu, Woods Hole Oceanographic Inst; Oc

Saylor, Beverly Z., 216 368 3763 bzs@case.edu, Case Western Reserve Univ; Gs
Sayyafzadeh, Mohammad, mohammad.sayyafzadeh@adelaide.edu.au, Univ of Adelaide; Np
Sbar, Marc, msbar@email.arizona.edu, Univ of Arizona; Yes
Scanlin, Michael A., 717-361-1323 jte2@psu.edu, Elizabethtown Coll; Ye
Scanlon, Bridget R., (512) 471-8241 bridget.scanlon@beg.utexas.edu, Univ of Texas at Austin, Jackson Sch of Geosciences; Hw
Scanlon, Todd M., (434) 924-3382 tms2v@virginia.edu, Univ of Virginia; Hg
Scannapieco, Evan, (480) 727-6788 evan.scannapieco@asu.edu, Arizona State Univ; Xa
Scapelli, Krista, (817) 257-7270 k.l.scarpelli@tcu.edu, Texas Christian Univ; Zn
Scarnato, Barbara V., bvscarna@nps.edu, Naval Postgrad Sch; Am
Scarpino, Samuel, 617.373.2482 s.scarpino@northeastern.edu, Northeastern Univ; Zn
Scarselli, Nicola, +44 1784 443597 nicola.scarselli@rhul.ac.uk, Royal Holloway Univ of London; Go
Schaaf, Crystal, crystal.schaaf@umb.edu, Univ of Massachusetts, Boston; ZrAm
Schaal, Barbara, (314) 935-6822 schaal@wustl.edu, Washington Univ in St. Louis; Zn
Schaap, Marcel, (520) 626-4532 mschapp@cals.arizona.edu, Univ of Arizona; So
Schacht, Ulrike, ulrike.schacht@adelaide.edu.au, Univ of Adelaide; Gd
Schade, Gunnar, (979) 845-7671 schade@ariel.met.tamu.edu, Texas A&M Univ; As
Schaef, Herbert T., (509) 373-9949 todd.schaef@pnl.gov, Pacific Northwest; Cg
Schaefer, Janet R. G., (907) 451-5005 janet.schaefer@alaska.gov, Alaska Division of Geological & Geophysical Surveys; Gv
Schaefer, Joerg, schaefer@ldeo.columbia.edu, Columbia Univ; Pe
Schafer, Carl M., (586) 286-2154 schaferc@macomb.edu, Macomb Comm Coll, Center Campus; Gg
Schafer, Tom, 785-628-5969 tschafer@fhsu.edu, Fort Hays State Univ; ZyiRn
Schaff, David, dschaff@ldeo.columbia.edu, Columbia Univ; Ys
Schaffer, Linda J., schafm4@aol.com, SUNY, The Coll at Brockport; Ze
Schaffrin, Burkhard A., 614 292-0502 schaffrin.1@osu.edu, Ohio State Univ; Yd
Schaller, Mirjam, 734-615-4286 mirjam@umich.edu, Univ of Michigan; Gm
Schaller, Morgan F., (518) 276-3358 schall@rpi.edu, Rensselaer Polytechnic Inst; CsPe
Schapaugh, Bill T., (785) 532-7242 wts@ksu.edu, Kansas State Univ; So
Schardt, Lawrence A., (814) 863-7655 las233@psu.edu, Pennsylvania State Univ, Univ Park; SoHw
Scharnberger, Charles K., charles.scharnberger@millersville.edu, Millersville Univ; Gc
Schartup, Amina, aschartup@ucsd.edu, Univ of California, San Diego; Ct
Schauble, Edwin A., (310) 206-9292 schauble@ucla.edu, Univ of California, Los Angeles; Csg
Scheel, Patrick, 573-368-2243 patrick.scheel@dnr.mo.gov, Missouri Dept of Nat Res; Zn
Scheffler, Joanna, joanna.scheffler@gmail.com, Mesa Comm Coll; Gg
Scheidemen, Kathy J., (805) 893-7615 kathys@icess.ucsb.edu, Univ of California, Santa Barbara; Zn
Scheidt, Brian, 573-518-2314 bscheidt@mineralarea.edu, Mineral Area Coll; Hw
Scheingross, Joel S., (775) 682-9839 jscheingross@unr.edu, Univ of Nevada, Reno; Gsm
Schellart, Wouter , +31 20 59 8610 w.p.schellart@vu.nl, Vrije Univit Amsterdam; Gt
Schellenberg, Stephen A., 61959421039 saschellenberg@mail.sdsu.edu, San Diego State Univ; Pe
Schenck, William S., (302) 831-2833 rockman@udel.edu, Univ of Delaware; GgpGc
Schenk, Anton, (716)645-3489 afshenko@buffalo.edu, SUNY, Buffalo; Zr
Schenk, Xizhen (Jenny), xschenk@wcu.edu, Western Carolina Univ; HgRwZi
Scher, Howie, 803-777-2410 hscher@geol.sc.edu, Univ of South Carolina; GuCl
Scherer, Reed P., (815) 753-7951 reed@niu.edu, Northern Illinois Univ; Pm
Schermer, Elizabeth R., (360) 650-3658 schermer@geol.wwu.edu, Western Washington Univ; Gt
Schiappa, Tamra A., (724) 738-2829 tamra.schiappa@sru.edu, Slippery Rock Univ; PiGrZe
Schieber, Juergen, (812) 856-4740 jschiebe@indiana.edu, Indiana Univ, Bloomington; Gs
Schiebout, Judith A., (225) 578-2717 schiebout@geol.lsu.edu, Louisiana State Univ; Pv
Schiefer, Erik, (928) 523-6535 erik.schiefer@nau.edu, Northern Arizona Univ; GmZyi
Schiffbauer, James, (573) 882-9501 schiffbauerj@missouri.edu, Univ of Missouri; Pgi
Schiffman, Peter, (530) 752-3669 pschiffman@ucdavis.edu, Univ of California, Davis; Gp
Schilling, Jean-Guy E., (401) 423 1417 jgchilling@yahoo.com, Univ of Rhode Island; OuCuc
Schilling, Keith, keith.schilling@dnr.iowa.gov, Iowa Dept of Nat Res; Gg
Schimmelmann, Arndt, (812) 855-7645 aschimme@indiana.edu, Indiana Univ, Bloomington; CsGsPe
Schimmrich, Steven, (845) 687-7683 schimmrs@sunyulster.edu, SUNY, Ulster County Comm Coll; GgZgn
Schincariol, Robert A., (519) 661-3732 schincar@uwo.ca, Western Univ; HwsNg
Schindler, Kimberly, 859-257-5434 kimberly.schindler@uky.edu, Univ of Kentucky; Cc
Schlautman, Mark, (864) 656-4059 mschlau@clemson.edu, Clemson Univ; ClHgSc
Schlegel, Alan, (620) 376-4761 schlegel@ksu.edu, Kansas State Univ; So
Schleifer, Stanley, 718-262-2726 sschleifer@york.cuny.edu, York Coll (CUNY); Ge
Schlesinger, William H., (919) 613-8004 schlesin@duke.edu, Duke Univ; Zn
Schlichting, Hilke, (626) 316-3629 hilke@ucla.edu, Univ of California, Los Angeles; Xy
Schlische, Roy W., 848-445-3445 schlisch@eps.rutgers.edu, Rutgers, The State Univ of New Jersey; GctEo
Schlogl, Jan , schlogl@fns.uniba.sk, Comenius Univ in Bratislava; Pig
Schlosser, C. Adam, 617-253-3983 casch@mit.edu, Massachusetts Inst of Tech; As
Schlosser, Peter, (845) 365-8707 schlosser@ldeo.columbia.edu, Columbia Univ; Hw
Schlue, John W., jwschlue@yahoo.com, New Mexico Tech; YsgZg
Schlumpberger, Debbie, 320-308-3260 dmschlumpberger@stcloudstate.edu, Saint Cloud State Univ; Zn
Schmachtenbert, William, wschmachtenberg@gmail.com, Virginia Tech; Pi
Schmahl, Wolfgang, 089/2180 4311 wolfgang.schmahl@lrz.uni-muenchen.de, Ludwig-Maximilians-Univ Muenchen; Gz
Schmandt, Brandon, 505.277.4204 bschmandt@unm.edu, Univ of New Mexico; Yg
Schmerr, Nicholas C., 301-405-4385 nschmerr@umd.edu, Univ of Maryland; XyYs
Schmid, Dieter, 089/2180 6635 d.schmid@lrz.uni-muenchen.de, Ludwig-Maximilians-Univ Muenchen; Pg
Schmid, Katherine W., 412-442-4232 kschmid@pa.gov, DCNR- Pennsylvania Bur of Geo Survey; GoCcXb
Schmidt, Amanda H., 440-775-8351 amanda.schmidt@oberlin.edu, Oberlin Coll; Gm
Schmidt, Bennetta, bennetta.schmidt@lamar.edu, Lamar Univ; Gg
Schmidt, Britney E., britneys@cornell.edu, Cornell Univ; Ye
Schmidt, Dale R., 217-300-1169 schmidt2@illinois.edu, Illinois State Geo Survey; Ge
Schmidt, David , 206-685-3799 dasc@uw.edu, Univ of Washington; Gt
Schmidt, David, 937 775-3539 david.schmidt@wright.edu, Wright State Univ; PiGd
Schmidt, Jonathan, (519) 824-4120 Ext.53966 jonschm@uoguelph.ca, Univ of Guelph; Zn
Schmidt, Keegan L., (208) 790-2283 klschmidt@lcsc.edu, Lewis-Clark State Coll; Gc
Schmidt, Lisa, lschmidt@sbccd.cc.ca.us, San Bernardino Valley Coll; Zy
Schmidt, Mariek, mschmidt2@brocku.ca, Brock Univ; GivXg
Schmidt, Matthew, 757-683-4285 mwschmid@odu.edu, Old Dominion Univ; Ou
Schmitt, Danielle M., 609-258-7015 dschmitt@princeton.edu, Princeton Univ; Ze
Schmitt, Douglas R., 765-496-0507 schmitt@purdue.edu, Purdue Univ; YxNrYe
Schmitt, Michael A., (612) 625-7017 mschmitt@soils.umn.edu, Univ of Minnesota, Twin Cities; Sc
Schmittner, Andreas, 541-737-9952 aschmittner@coas.oregonstate.edu, Oregon State Univ; AsYr
Schmitz, Darrel W., 662-268-1032 Ext 241 schmitz@geosci.msstate.edu, Mississippi State Univ; Hw
Schmitz, Mark D., (208) 426-5907 markschmitz@boisestate.edu, Boise State Univ; CcaCg
Schmutz, Phillip P., 850-474-3418 pschmutz@uwf.edu, Univ of West Florida; Gm
Schneider, Blair, 785-864-2063 blair.schneider@ku.edu, Univ of Kansas; Gg

Schneider, David, david.schneider@uottawa.ca, Univ of Ottawa; GtCcGc
Schneider, Edwin K., (703) 993-5364 eschnei1@gmu.edu, George Mason Univ; As
Schneider, John F., (630) 252-8923 Argonne; Ca
Schneider, Julius, 089/2180 4354 julius.schneider@lrz.uni-muenchen.de, Ludwig-Maximilians-Univ Muenchen; Gz
Schneider, Niklas, (808) 956-8383 nschneid@hawaii.edu, Univ of Hawai'i, Manoa; Og
Schneider, Robert J., 508-289-2756 rschneider@whoi.edu, Woods Hole Oceanographic Inst; Cc
Schneider, Tapio, (626) 395-6143 tapio@caltech.edu,CalTech; AsmZg
Schneiderman, Jill S., 845-437-5542 schneiderman@vassar.edu, Vassar Coll; Gs
Schnetzer, Astrid, (919) 515-7837 aschnet@ncsu.edu, North Carolina State Univ; Ob
Schoenberger, Erica, (410) 516-6158 ericas@jhu.edu, Johns Hopkins Univ; Zn
Schoene, Blair, 609-258-5747 bschoene@princeton.edu, Princeton Univ; Cc
Schoenemann, Spruce W., 406 683-7624 spruce.schoenemann@umwestern.edu, Univ of Montana Western; GeCgs
Schoephoester, Peter, 608/262.2320 peter.schoephoester@wisc.edu, Univ of Wisconsin, Madison, Division of Extension; Zi
Scholle, Peter, peter.scholle@nmt.edu, New Mexico Tech; Gd
Scholtz, Theresa C., (630) 252-6499 Argonne; Ge
Scholz, Christopher A., 315-443-2672 cascholz@syr.edu, Syracuse Univ; Gs
Schoof, Christian, (604) 822-3063 cschoof@eos.ubc.ca, Univ of British Columbia; Gl
Schoonen, Martin A., (631) 632-8007 martin.schoonen@stonybrook.edu, Stony Brook Univ; Cl
Schoonmaker, Adam, 792-2577 adschoonmaker@utica.edu, Utica Coll; GczGx
Schoonmaker, Jane E., 808-956-9935 jane@soest.hawaii.edu, Univ of Hawai'i, Manoa; Cm
Schopf, J. William, (310) 825-1170 schopf@epss.ucla.edu, Univ of California, Los Angeles; Pg
Schopf, Paul S., (703) 993-5394 pschopf@gmu.edu, George Mason Univ; OpAs
Schouten, Hans, (508) 289-2574 hschouten@whoi.edu, Woods Hole Oceanographic Inst; Yr
Schrader, Christian M., 315-267-2285 schradcm@potsdam.edu, SUNY Potsdam; GiEmGv
Schrader, Devin, (480) 965-0720 devin.schrader@asu.edu, Arizona State Univ; Xm
Schrag, Daniel P., (617) 495-7676 schrag@eps.harvard.edu, Harvard Univ; Cg
Schrage, Jon M., (402) 280-5759 schragej@gmail.com, Creighton Univ; Am
Schrank, Christoph, +61 7 3138 1583 christoph.schrank@qut.edu.au, Queensland Univ of Tech; GcqGt
Schreiber, B. Charlotte, (718) 997-3300 Queens Coll (CUNY); Gd
Schreiber, Charlotte, 206-297-1454 geologo1@uw.edu, Univ of Washington; Gsd
Schreiber, Madeline E., (540) 231-3377 mschreib@vt.edu, Virginia Tech; Hw
Schreier, Hans D., (604) 822-4401 Univ of British Columbia; Zg
Schrenk, Matt, (517) 884-7966 schrenkm@msu.edu, Michigan State Univ; Pg
Schriver, David, (310) 825-6663 dave@igpp.ucla.edu, Univ of California, Los Angeles; Yg
Schroder-Adams, Claudia, csadams@ccs.carleton.ca, Carleton Univ; Pm
Schroeder, Dustin M., (650) 725-7861 dustin.m.schroeder@stanford.edu, Stanford Univ; Gl
Schroeder, John L., 806-834-5678 john.schroeder@ttu.edu, Texas Tech Univ; As
Schroeder, Kathleen, (828) 262-7055 schroederk@appstate.edu, Appalachian State Univ; Zn
Schroeder, Norman C., (505) 667-0967 Los Alamos; Cc
Schroeder, Paul A., (706) 542-2384 schroe@uga.edu, Univ of Georgia; GzEnCl
Schroeder, Stefan, +44 0161 306-6870 stefan.schroeder@manchester.ac.uk, Univ of Manchester; Go
Schroeder, William W., (334) 861-7528 wschroeder@disl.org, Dauphin Island Sea Lab; Og
Schroth, Andrew W., (802)656 3481 aschroth@uvm.edu, Univ of Vermont; ClqCb
Schubert, Brian, 337-482-6967 bas9777@louisiana.edu, Univ of Louisiana at Lafayette; Cg
Schubert, Gerald, (310) 825-4577 Univ of California, Los Angeles; Yg
Schuberth, Bernhard, 089/2180 4220 bernhard@geophysik.uni-muenchen.de, Ludwig-Maximilians-Univ Muenchen; Yg
Schulingkamp, Arren, 225-578-3412 warrenii@lsu.edu, Louisiana State Univ; Gg
Schulte, Kimberly D., kschulte1@gsu.edu, Georgia State Univ, Perimeter Coll, Online; Gg
Schultz, Adam, 541-737-9832 adam.schultz@oregonstate.edu, Oregon State Univ; YgmRn
Schultz, David, +44 0161 306-3909 david.schultz@manchester.ac.uk, Univ of Manchester; Am
Schultz, Gerald E., 806-651-2580 gschultz@wtamu.edu, West Texas A&M Univ; PvGzZg
Schultz, Jan, (805) 965-0581 (Ext. 2313) schultz@sbcc.edu, Santa Barbara City Coll; GgePg
Schultz, Peter H., (401) 863-2417 peter_schultz@brown.edu, Brown Univ; Xg
Schultze, Steven R., schultze@southalabama.edu, Univ of South Alabama; At
Schulz, Layne D., 605-67-76161 layne.schulz@usd.edu, South Dakota Dept of Environment and Nat Res; Glg
Schulze, Anja, 409 740 4540 schulzea@tamug.edu, Texas A&M Univ; Ob
Schulze, Daniel J., 905-8283970 dschulze@utm.utoronto.ca, Univ of Toronto; Gi
Schulze, Karl, (630) 466-2652 kschulze@waubonsee.edu, Waubonsee Comm Coll; ZgAms
Schulze-Makuch, Dirk, 509-335-1180 dirksm@wsu.edu, Washington State Univ; Hw
Schumacher, Courtney, (979) 845-5522 cschu@tamu.edu, Texas A&M Univ; As
Schumacher, Matthew, mschumac@stfx.ca, Saint Francis Xavier Univ; Zg
Schumacher, Russ, russ.schumacher@colostate.edu, Colorado State Univ; As
Schumann, Arnold W., (863) 956-1151 schumaw@ufl.edu, Univ of Florida; So
Schumer, Rina, (775) 673-7414 rina.schumer@dri.edu, Univ of Nevada, Reno; HqwGm
Schutt, Derek L., (970) 491-5786 schutt@warnercnr.colostate.edu, Colorado State Univ; Ysg
Schwab, Brandon E., (828) 227-7495 beschwab@wcu.edu, Western Carolina Univ; GizGv
Schwab, Fred, (310) 825-3123 schwab@igpp.ucla.edu, Univ of California, Los Angeles; Ys
Schwab, Frederick L., (540) 458-5830 schwabf@wlu.edu, Washington & Lee Univ; Gdg
Schwab, James J., (518) 437-8754 jschwab@albany.edu, SUNY, Albany; Acs
Schwab, Paul, 979-845-3663 pschwab@tamu.edu, Texas A&M Univ; Sc
Schwalbach, Cameron E., 513-287-7000 cschwalbach@cincymuseum.org, Cincinnati Mus Center; Pi
Schwarcz, Henry P., (905) 525-9140 (Ext. 24186) schwarcz@mcmaster.ca, McMaster Univ; Cs
Schwartz, Darin, darinschwartz@boisestate.edu , Boise State Univ; CuGi
Schwartz, David, (408) 479-6495 daschwar@cabrillo.edu, Cabrillo Coll; Gu
Schwartz, Frank W., (614) 292-6196 schwartz.11@osu.edu, Ohio State Univ; Hw
Schwartz, Hilde, (831) 459-5429 hschwartz@pmc.ucsc.edu, Univ of California, Santa Cruz; Pv
Schwartz, Joshua J., (818) 677-5813 joshua.schwartz@csun.edu, California State Univ, Northridge; GiCc
Schwartz, Matthew C., 850-474-3469 mschwartz@uwf.edu, Univ of West Florida; OcCl
Schwartz, Susan, 831-459-3133 syschwar@ucsc.edu, Univ of California, Santa Cruz; Ys
Schwarz, Karen, (610)436-2788 kschwarz@wcupa.edu, West Chester Univ; XaZe
Schweitzer, Carrie E., 330-672-2505 cschweit@kent.edu, Kent State Univ; Pi
Schwerdtner, Walfried M., (416) 978-5080 fried@quartz.geology.utoronto.ca, Univ of Toronto; GctGq
Schwert, Donald P., (701) 231-7496 donald.schwert@ndsu.edu, North Dakota State Univ; PeGe
Schwimmer, David R., 706 569-3028 schwimmer_david@columbusstate.edu, Columbus State Univ; Pv
Schwimmer, Reed A., 609-896-5346 rschwimmer@rider.edu, Rider Univ; GsOgn
Schwob, Stephanie L., 214-768-2770 sschwob@smu.edu, Southern Methodist Univ; Zn
Sclater, John G., (858) 534-3051 jsclater@ucsd.edu, Univ of California, San Diego; Yh
Scoates, James S., (604) 822-3667 jscoates@eoas.ubc.ca, Univ of British Columbia; GizGv
Scotese, Christopher R., (817) 272-2987 cscotese@exchange.uta.edu, Univ of Texas, Arlington; Gt
Scott, Andrew C., +44 1784 443608 a.scott@rhul.ac.uk, Royal Holloway Univ of London; Pb

Scott, Darren M., (905) 525-9140 (Ext. 24953) scottdm@mcmaster.ca, McMaster Univ; Zi
Scott, David B., (902) 494-3604 Dalhousie Univ; Pm
Scott, James M., +64 3 479-7520 james.scott@otago.ac.nz, Univ of Otago; Gp
Scott, Kathy, (604) 822-5606 kscott@eos.ubc.ca, Univ of British Columbia; Zn
Scott, Larry, 303-384-2631 lmscott@mines.edu, Colorado Geo Survey; Zy
Scott, Robert B., 512-471-0375 rscott@ig.utexas.edu, Univ of Texas, Austin; Og
Scott, Steven D., (416) 978-5424 scottsd@es.utoronto.ca, Univ of Toronto; GuEmCm
Scott, Thomas A., (951) 827-5115 thomas.scott@ucr.edu, Univ of California, Riverside; Zy
Scott, Tim, (401) 254-3108 Roger Williams Univ; Ob
Scott, Vernon, 405-744-6358 vscott@okstate.edu, Oklahoma State Univ; Ze
Scott Smith, Barbara H., (604) 984-9609 bhssmith@allstream.net, Univ of British Columbia; Gz
Scott-Dupree, Cynthia, (519) 824-4120 Ext.52477 cscottdu@uoguelph.ca, Univ of Guelph; Zn
Scotton, Paolo, +390498279120 paolo.scotton@unipd.it, Univ degli Studi di Padova; Hg
Scow, Kate, (530) 752-4632 kmscow@ucdavis.edu, Univ of California, Davis; So
Scowen, Paul, (480) 965-0938 paul.scowen@asu.edu, Arizona State Univ; Xa
Scranton, Mary I., (631) 632-8735 mary.scranton@stonybrook.edu, SUNY, Stony Brook; Oc
Screaton, Elizabeth J., (352) 3924612 screaton@ufl.edu, Univ of Florida; Hg
Scuderi, Louis A., (505) 277-4204 tree@unm.edu, Univ of New Mexico; GmPcGs
Scyphers, Steven, 781.581.7370 s.scyphers@northeastern.edu, Northeastern Univ; Zn
Seager, Richard, seager@ldeo.columbia.edu, Columbia Univ; Am
Seager, Sara, 617-253-6775 seager@mit.edu, Massachusetts Inst of Tech; Xy
Seal, Thom, 775-784-1813 tseal@unr.edu, Univ of Nevada, Reno; Nx
Seaman, Nelson L., (814) 863-1583 Pennsylvania State Univ, Univ Park; Am
Seaman, Sheila J., (413) 545-2822 sjs@geo.umass.edu, Univ of Massachusetts, Amherst; Gi
Searcy, Steven, (619) 260-2793 ssearcy@sandiego.edu, Univ of San Diego; Ob
Searle, Mike, +44 (1865) 272022 mike.searle@earth.ox.ac.uk, Univ of Oxford; Gt
Searls, Mindi L., 402-472-6934 Univ of Nebraska, Lincoln; YgGg
Sears, James W., (406) 243-5251 james.sears@umontana.edu, Univ of Montana; GctEg
Sebol, Lesley, (303) 384-2633 lsebol@mines.edu, Colorado Geo Survey; HwGeZg
Secco, Luciano, +390498279158 luciano.secco@unipd.it, Univ degli Studi di Padova; Gz
Secco, Richard A., (519) 661-4079 secco@uwo.ca, Western Univ; Gy
Secord, Ross, (402) 472-2663 rsecord2@unl.edu, Univ of Nebraska, Lincoln; PvCs
Sediek, Kadry N., 002-03-3921595 kknsed@yahoo.com, Alexandria Univ; GgsGd
Sedivy, Robert, 402-465-9021 rasedivy@anl.gov, Argonne; Hw
Sedlacek, Alexa, 319-273-3072 alexa.sedlacek@uni.edu, Univ of Northern Iowa; CsGr
Sedwick, Peter, (757) 683-4936 psedwick@odu.edu, Old Dominion Univ; CmOc
Seeber, Leonardo, nano@ldeo.columbia.edu, Columbia Univ; Ys
Seedorff, Eric, (520) 626-3921 seedorff@email.arizona.edu, Univ of Arizona; Eg
Seeger, Cheryl M., (573) 368-2184 cheryl.seeger@dnr.mo.gov, Missouri Dept of Nat Res; Gig
Seeley, Mark W., (612) 625-4724 mseeley@umn.edu, Univ of Minnesota, Twin Cities; As
Seewald, Jeffrey S., (508) 289-2966 jseewald@whoi.edu, Woods Hole Oceanographic Inst; Cp
Segall, Marylin, 801-585-5730 mpsegall@egi.utah.edu, Univ of Utah; GeOu
Segall, Paul, (650) 725-7241 Stanford Univ; Yg
Segars, William P., (706) 542-9072 Univ of Georgia; Sc
Seibel, Erwin, (415) 338-2061 San Francisco State Univ; On
Seibt, Ulrike, (310) 206-4442 useibt@ucla.edu, Univ of California, Los Angeles; PgCg
Seidemann, David E., 7189515000 x2882 dseidemann@earthlink.net, Brooklyn Coll (CUNY); Cc
Seifert, Karl E., (515) 294-5265 kseifert@iastate.edu, Iowa State Univ of Sci & Tech; Ct
Seifoullaev, Roustam K., (512) 232-3223 roustam@utig.ig.utexas.edu, Univ of Texas, Austin; Ye
Seifu, Abiye, 706-568-2187 seifu_abiye@columbusstate.edu, Columbus State Univ; Zn
Seigley, Lynette S., 319-335-1598 lynette.seigley@dnr.iowa.gov, Iowa Dept of Nat Res; Gg
Seirer, Jami J., jjseirer2@fhsu.edu, Fort Hays State Univ; ZriZn
Seitz, Jeffery C., (510) 885-3438 jeff.seitz@csueastbay.edu, California State Univ, East Bay; CgZeGx
Selby, Dave, +44 (0) 191 33 42294 david.selby@durham.ac.uk, Durham Univ; Cc
Selden, Paul A., selden@ku.edu, Univ of Kansas; Pig
Selim, Hussein M., (504) 388-2110 Louisiana State Univ; Sp
Selin, Noelle, (617) 324-2592 selin@mit.edu, Massachusetts Inst of Tech; As
Selph, Karen E., (808) 956-7941 selph@hawaii.edu, Univ of Hawai'i, Manoa; Ob
Selverstone, Jane E., selver@unm.edu, Univ of New Mexico; Gpt
Semazzi, Fred H. M., (919) 515-1434 fred_semazzi@ncsu.edu, North Carolina State Univ; As
Semken, Steven, (480) 965-7965 semken@asu.edu, Arizona State Univ; ZeGgZg
Semken, Jr., Holmes A., (319) 335-1830 holmes-semken@uiowa.edu, Univ of Iowa; Pv
Sen, Gautam, (305) 348-2299 seng@fiu.edu, Florida InterNatl Univ; Gi
Sen, Mrinal K., (512) 471-0466 Univ of Texas, Austin; Ye
Sen Gupta, Barun K., (225) 388-5984 barun@geol.lsu.edu, Louisiana State Univ; Pm
Send, Uwe, (858) 822-6710 usend@ucsd.edu, Univ of California, San Diego; Op
Senko, John M., 330 972-8047 senko@uakron.edu, Univ of Akron; Cg
Sennert, Sally K., (202) 633-1805 kuhns@si.edu, Smithsonian Inst / Natl Mus of Natural History; GvZr
Seong, Jeong C., (678) 839-4069 jseong@westga.edu, Univ of West Georgia; Zir
Sephton, Mark, +44 20 759 46542 m.a.sephton@imperial.ac.uk, Imperial Coll; Xm
Sepúlveda, Julio C., jsepulveda@colorado.edu, Univ of Colorado; Co
Sepulveda, Sergio, 778-782-5389 ssepulve@sfu.ca, Simon Fraser Univ; NgRn
Seremet, Meghan M., meghan.seremet@ct.gov, Dept of Energy and Env Protection; Gg
Sericano, Jose L., (979) 862-2323 jose@gerg.tamu.edu, Texas A&M Univ; Cm
Serne, R. Jeffrey, (360) 539-6162 jeff.serne@pnnl.gov, Pacific Northwest; CgSoCl
Serpa, Laura F., (504) 280-6801 lserpa@uno.edu, Univ of New Orleans; Ys
Serreze, Mark, mark.serreze@colorado.edu, Univ of Colorado; As
Sertich, Joseph, 303-370-6331 joe.sertich@dmns.org, Denver Mus of Nature & Sci; Pv
Seserman, Anca M., 0040232201456 anca.seserman@yahoo.com, Alexandru Ioan Cuza; Gs
Sessa, Jocelyn A., (215) 299-1149 jsessa@drexel.edu, Drexel Univ; PeCsPq
Sessions, Alex L., 626.395.6445 als@gps.caltech.edu,CalTech; CosCb
Sethi, Parvinder S., (540) 831-5619 psethi@radford.edu, Radford Univ; Gz
Severinghaus, Jeffrey P., (858) 822-2483 jseveringhaus@ucsd.edu, Univ of California, San Diego; Pe
Severs, Matthew , matthew.severs@stockton.edu, Stockton Univ; Gg
Severs, Matthew R., 609-626-6857 matthew.severs@stockton.edu, Stockton Univ; GxCgEg
Severson, Allison, 612-626-2969 sever270@d.umn.edu, Univ of Minnesota; Gp
Sevilla, Paloma, psevilla@ucm.es, Univ Compl de Madrid; Pi
Sewall, Jacob, 484-646-5864 sewall@kutztown.edu, Kutztown Univ of Pennsylvania; GeAs
Sexton, John L., (618) 453-7374 sexton@geo.siu.edu, Southern Illinois Univ Carbondale; Ye
Sexton, Philip, philip.sexton@open.ac.uk, Open Univ; Ge
Seyfang, Gill, +44 (0)1603 59 2956 g.seyfang@uea.ac.uk, Univ of East Anglia; Og
Seyfferth, Angelia L., 302-831-4865 angelias@udel.edu, Univ of Delaware; Sc
Seyfried, Jr., William E., 612-624-1333 wes@umn.edu, Univ of Minnesota, Twin Cities; Cm
Seymour, Dorie, (406) 994-5718 dorie.seymour@montana.edu, Montana State Univ; Zn
Seyoum, Wondwosen M., (309) 438-2833 wmseyou@ilstu.edu, Illinois State Univ; HwqZr
Seyyed, Hosseini, 512-471-2360 seyyed.hosseini@beg.utexas.edu, Univ of Texas, Austin; Go

Shaaban, Mohamad N., 002-03-3921595 moshaaban@yahoo.com, Alexandria Univ; GgsGd
Shackelton, Sarah, 609-258-4101 ss77@princeton.edu, Princeton Univ; Pc
Shackley, Simon J., +44 (0) 131 650 7862 simon.shackley@ed.ac.uk, Edinburgh Univ; Eg
Shade, Janet, 814-362-7560 jas144@pitt.edu, Univ of Pittsburgh, Bradford; Zn
Shafer, Erik, shafere@pdx.edu, Portland State Univ; Gg
Shafer, Jason, 802-626-6225 jason.shafer@northernvermont.edu, Northern Vermont Univ-Lyndon; As
Shah, Subhash N., (405) 325-6871 subhash@ou.edu, Univ of Oklahoma; Np
Shahar, Anat, (202) 478-8929 ashahar@carnegiescience.edu, Carnegie Inst for Sci; CgYxCs
Shail, Robin, +44 01326 371826 r.k.shail@exeter.ac.uk, Exeter Univ; GtcEm
Shakun, Jeremy D., 617-552-1625 jeremy.shakun@bc.edu, Boston Coll; Gg
Shalimba, Ester, +264-61-2063839 eshalimba@unam.na, Univ of Namibia; GgpEz
Shamberger, Kathryn E., (979) 845-5752 katie.shamberger@tamu.edu, Texas A&M Univ; OcZc
Shami, Malek, (718) 262-2654 mshami2@york.cuny.edu, York Coll (CUNY); Gcg
Shamloo, Hannah, hannah.shamloo@cwu.edu, Central Washington Univ; Gvi
Shams, Asghar, +44 (0)131 451 3904 a.shams@hw.ac.uk, Heriot-Watt Univ; Yg
Shanafield, Margaret, margaret.shanafield@flinders.edu.au, Flinders Univ; Hw
Shanahan, Timothy M., 512-232-7051 tshanahan@jsg.utexas.edu, Univ of Texas, Austin; PeGsCg
Shane, Tyrrell, +353 (0)91 494387 shane.tyrrell@nuigalway.ie, Natl Univ of Ireland Galway; Gs
Shang, Congxiao, +44 (0)1603 59 3123 c.shang@uea.ac.uk, Univ of East Anglia; Hw
Shank, Stephen G., 717-702-2021 stshank@pa.gov, DCNR- Pennsylvania Bur of Geo Survey; GziGp
Shankland, Thomas J., (505) 667-4907 shanklan@lanl.gov, Los Alamos; Yx
Shanmugam, Ganapathy, shanshanmugam@aol.com, Univ of Tennessee, Knoxville; Gs
Shannon, Jack D., (630) 252-5807 jack_shannon@anl.gov, Argonne; Am
Shannon, Jeremy, (906) 487-3573 jmshanno@mtu.edu, Michigan Technological Univ; Gg
Shapiro, Russell S., (530) 898-4300 rsshapiro@csuchico.edu, California State Univ, Chico; PgGdXb
Shapley, Mark, shap0029@umn.edu, Univ of Minnesota, Twin Cities; Gn
Sharkey, Debra, (916) 691-7210 Cosumnes River Coll; Zy
Sharma, Govind, (205) 851-5462 aamaxs01@aamu.edu, Alabama A&M Univ; Zn
Sharma, Mukul, 603-646-0024 mukul.sharma@dartmouth.edu, Dartmouth Coll; Ge
Sharma, Shikha, 304-293-5603 shikha.sharma@mail.wvu.edu, West Virginia Univ; Csa
Sharma, Shiv K., (808) 956-8476 Univ of Hawai'i, Manoa; Gx
Sharma, Suresh, ssharma@ou.edu, Univ of Oklahoma; Np
Sharma, Vasudha, 612-626-4986 vasudha@umn.edu, Univ of Minnesota, Twin Cities; SpHg
Sharman, Glenn R., 479-575-4471 gsharman@uark.edu, Univ of Arkansas, Fayetteville; Go
Sharp, Jonathan H., (302) 645-4259 jsharp@udel.edu, Univ of Delaware; Oc
Sharp, Martin J., (780) 492-4156 martin.sharp@ualberta.ca, Univ of Alberta; Gl
Sharp, Thomas G., 480-965-3071 tom.sharp@asu.edu, Arizona State Univ; Gz
Sharp, Zachary D., (505) 277-4204 zsharp@unm.edu, Univ of New Mexico; Cs
Sharp, Jr., John M., (512) 471-3317 jmsharp@jsg.utexas.edu, Univ of Texas, Austin; HqwGe
Sharples, Jonathan, +44 0151 794 4093 jonathan.sharples@liverpool.ac.uk, Univ of Liverpool; Ocp
Shaver, Stephen A., (931) 598-1116 sshaver@sewanee.edu, Sewanee: Univ of the South; Eg
Shaw, Bruce, shaw@ldeo.columbia.edu, Columbia Univ; Ys
Shaw, Christopher A., (323) 857-6317 chrissha@bcf.usc.edu, Los Angeles County Mus of Natural History; Pv
Shaw, Cliff S., 506-447-3195 cshaw@unb.ca, Univ of New Brunswick; Giv
Shaw, Colin, 406 994 6760 colin.shaw1@montana.edu, Montana State Univ; GcEmGp
Shaw, Fred, 573-368-2479 fred.shaw@dnr.mo.gov, Missouri Dept of Nat Res; Zn
Shaw, Frederick C., (212) 642-2202 Lehman Coll (CUNY); Ps
Shaw, George H., (518) 388-6310 shawg@union.edu, Union Coll; Yx
Shaw, Glenn D., 406-496-4809 gshaw@mtech.edu, Montana Tech of the Univ of Montana; HwCsHs
Shaw, Joey N., (334) 844-3957 shawjo1@auburn.edu, Auburn Univ; Sd
Shaw, John, (780) 492-3573 john.shaw@ualberta.ca, Univ of Alberta; Gl
Shaw, Katy, 2538339111 x 4325 kshaw@greenriver.edu, Green River Comm Coll; ZgOgGg
Shaw, Kenneth L., (212) 642-2202 kshaw@egi.utah.edu, Univ of Utah; Ye
Shaw, Liz, e.j.shaw@reading.ac.uk, Univ of Reading; Sb
Shaw, Ping-Tung, (919) 515-7276 ping-tung_shaw@ncsu.edu, North Carolina State Univ; Op
Shaw, Richard F., (225) 388-6734 Louisiana State Univ; Ob
Shcherbakov, Robert, 519-661-2111 ext.84212 rshcherb@uwo.ca, Western Univ; YgsYd
Shchukin, Eugene D., (410) 516-5079 shchukin@jhuvms.hcf.jhu.edu, Johns Hopkins Univ; Zn
Shea, Erin, 907-786-6846 eshea2@uaa.alaska.edu, Univ of Alaska, Anchorage; CcGcCg
Shea, Thomas, tshea@hawaii.edu, Univ of Hawai'i, Manoa; Gv
Shearer, Peter M., (858) 534-2260 pshearer@ucsd.edu, Univ of California, San Diego; Ysg
Shearer, Jr., Charles K., (505) 277-9159 cshearer@unm.edu, Univ of New Mexico; Gi
Shearman, Kipp, 541-737-1866 shearman@coas.oregonstate.edu, Oregon State Univ; Opn
shedied, ahmad g., 01005856505 ags00@fayoum.edu.eg, Fayoum Univ; GgHww
Sheehan, Anne F., (303) 492-4597 anne.sheehan@colorado.edu, Univ of Colorado; YszYe
Sheehan, Peter M., (414) 278-2741 sheehan@mpm.edu, Milwaukee Public Mus; Pi
Sheets, Julie, (614) 406-3298 sheets.2@osu.edu, Ohio State Univ; Gz
Sheffield, Sarah, ssheffield2@usf.edu, Univ of South Florida; Pe
Sheldon, Amy L., (585) 245-5988 sheldon@geneseo.edu, SUNY, Geneseo; Hw
Shell, Karen M., (541) 737-0980 kshell@coas.oregonstate.edu, Oregon State Univ; As
Shell, Ryan C., 414-297-3600 ryanshell501@gmail.com, Cincinnati Mus Center; Pv
Shellito, Lucinda, (970) 351-2491 lucinda.shellito@unco.edu, Univ of Northern Colorado; AsPeAm
Shelton, Dale W., (410) 554-5505 dale.shelton@maryland.gov, Maryland Dept of Nat Res; ZeGg
Shelton, Kevin L., (573) 882-1004 sheltonkl@missouri.edu, Univ of Missouri; EmCs
Shem, Linda M., (630) 252-3857 Argonne; Ge
Shen, Jian, (804) 684-7359 shen@vims.edu, William & Mary; Op
Shen, Linhan, 609-258-4101 linhans@princeton.edu, Princeton Univ; Cas
Shen, Samuel S., (780) 492-0216 shen@ualberta.ca, Univ of Alberta; As
Shen, Weisen, 631-632-8212 weisen.shen@stonybrook.edu, Stony Brook Univ; Ys
Shen, Xinhua, (319) 273-2536 xinhua.shen@uni.edu, Univ of Northern Iowa; Asm
Shen, Yang, (401) 874-6848 yshen@gso.uri.edu, Univ of Rhode Island; Ou
Sheng, Jinyu, (902) 494-2718 jinyu.sheng@dal.ca, Dalhousie Univ; Op
Shepard, Michael K., (570) 389-4568 mshepard@bloomu.edu, Bloomsburg Univ; XgYgZr
Shepardson, Daniel P., (765) 494-5284 dshep@purdue.edu, Purdue Univ; Ze
Shepherd, Gordon G., 4167362100 x33221 gordon@yorku.ca, York Univ; AsZrAp
Shepherd, Mark A., (512) 574-4227 mark.shepherd@austincc.edu, Austin Comm Coll District; AsObZn
Shepherd, Stephanie L., 334-844-4926 sls0070@auburn.edu, Auburn Univ; GmZi
Sheppard, Scott S., ssheppard@carnegiescience.edu, Carnegie Inst for Sci; Xs
Shepson, Paul B., pshepson@purdue.edu, Purdue Univ; As
Sheridan, Michael F., mfs@buffalo.edu, SUNY, Buffalo; Gv
Sheridan, Robert E., (848) 445-2015 rsheridan@eps.rutgers.edu, Rutgers, The State Univ of New Jersey; YrGu
Sherlock, Ross, (705) 675-1151 Ext. 2379 rsherlock@laurentian.ca, Laurentian Univ, Sudbury; Eg
Sherman-Morris, Kathleen M., (662) 268-1032 x242 kms5@geosci.msstate.edu, Mississippi State Univ; As
Sherrell, Robert M., (848) 932-6555 (Ext. 252) sherrell@ahab.rutgers.edu, Rutgers, The State Univ of New Jersey; CmOu
Sherriff, Barbara L., (204) 474-9786 bl_sherriff@umanitoba.ca, Univ of Manitoba; Gz
Sherrod, Brian L., 253-653-8358 bsherrod@uw.edu, Univ of Washington;

YsGt
Sherrod, Laura, (484) 646-4113 sherrod@kutztown.edu, Kutztown Univ of Pennsylvania; YgHwGa
Shervais, John W., (435) 797-1274 john.shervais@usu.edu, Utah State Univ; Gi
Sherwood, Owen, 902 494 3604 owen.sherwood@dal.ca, Dalhousie Univ; CsOoGo
Sherwood, William C., (540) 568-6473 sherwowc@jmu.edu, James Madison Univ; So
Sherwood Lollar, Barbara, 416-978-0770 bslollar@chem.utoronto.ca, Univ of Toronto; Cs
Shevenell, Lisa, (775) 784-1779 lisaas@unr.edu, Univ of Nevada, Reno; Hw
Shew, Roger D., 910-962-7676 shewr@uncw.edu, Univ of North Carolina, Wilmington; Eo
Shi, Wei, 919-513-4641 wei_shi@ncsu.edu, North Carolina State Univ; Sb
Shieh, Sean R., 5196612111 x82467 sshieh@uwo.ca, Western Univ; YxGzZm
Shieh, Yuch-Ning, ynshieh@purdue.edu, Purdue Univ; CsGpv
Shiel, Alyssa E., (541) 737-5209 ashiel@coas.oregonstate.edu, Oregon State Univ; ClsCt
Shields, Robin, +44 (0) 131 451 8215 robin.shields@hw.ac.uk, Heriot-Watt Univ; Cg
Shields, Stephen, 325-942-2242 stephen.shields@angelo.edu, Angelo State Univ; Ggm
Shields-Zhou, Graham, +44 020 7679 7821 g.shields@ucl.ac.uk, Univ Coll London; Cg
Shiels, Christine, 604-527-5217 shielsc@douglascollege.ca, Douglas Coll; GgxZg
Shiller, Alan M., (228) 688-1178 alan.shiller@usm.edu, Univ of Southern Mississippi; OcCtHs
Shiller, Thomas, (432) 837-8238 thomas.shiller@sulross.edu, Sul Ross State Univ; PvGrd
Shim, Sang-Heon, (480) 727-2876 shdshim@asu.edu, Arizona State Univ; Gz
Shim-Chim, Richard, (609) 984-6587 rich.shim-chim@dep.state.nj.us, New Jersey Geological and Water Survey; Hw
Shimer, Grant, 435 865 8429 grantshimer@suu.edu, Southern Utah Univ; Gs
Shimizu, Melinda, mel2066254@maricopa.edu, Mesa Comm Coll; Gg
Shimizu, Nobumichi, (508) 289-2963 Woods Hole Oceanographic Inst; Ca
Shinn, Eugene, (727) 893-3100 Univ of South Florida; Gg
Shinoda, Toshiaki, 361-825-3636 toshiaki.shinoda@tamucc.edu, Texas A&M Univ, Corpus Christi; As
Shipley, Thomas H., (512) 471-0430 tom@ig.utexas.edu, Univ of Texas, Austin; Yr
Shirer, Hampton N., (814) 863-1992 hns@psu.edu, Pennsylvania State Univ, Univ Park; Am
Shirey, Steven B., sshirey@carnegiescience.edu, Carnegie Inst for Sci; Gi
Shirgaokar, Manish, 780.492.2802 shirgaokar@ualberta.ca, Univ of Alberta; Zu
Shirley, Terry, 704-687-5925 trshirle@uncc.edu, Univ of North Carolina, Charlotte; Asm
Shirzaei, Manoochehr, 540-231-6521 shirzaei@vt.edu, Virginia Tech; ZrYd
Shkolnik, Evgenya, shkolnik@asu.edu, Arizona State Univ; Xa
Shobe, Charles, charles.shobe@mail.wvu.edu, West Virginia Univ; Gm
Shock, Everett, 480-965-0631 everett.shock@asu.edu, Arizona State Univ; Cg
Shoemaker, Kurt A., (740) 351-3395 kshoemaker@shawnee.edu, Shawnee State Univ; GxzGm
Sholkovitz, Edward R., (508) 289-2346 esholkovitz@whoi.edu, Woods Hole Oceanographic Inst; Oc
Shorey, Christian V., (303) 273-3556 cshorey@mines.edu, Colorado Sch of Mines; Gg
Shortt, Niamh K., +44 (0) 131 651 7130 niamh.shortt@ed.ac.uk, Edinburgh Univ; Eg
Shoup, Doug, (620) 421-1530 dshoup@ksu.edu, Kansas State Univ; So
Showers, William J., (919) 515-7143 w_showers@ncsu.edu, North Carolina State Univ; Cs
Showman, Adam, (520) 621-4021 showman@lpl.arizona.edu, Univ of Arizona; Zn
Shragge, Jeffrey, 303-273-3552 jshragge@mines.edu, Colorado Sch of Mines; Ys
Shreedharan, Srisharan, srisharan.shreedharan@usu.edu, Utah State Univ; NrYsGq
Shreve, Ronald L., (310) 825-5273 shreve@ess.ucla.edu, Univ of California, Los Angeles; Gm
Shroat-Lewis, Rene A., 501-683-7743 rashroatlew@ualr.edu, Univ of Arkansas at Little Rock; PeZe
Shroba, Cynthia S., 702-651- 7427 cindy.shroba@csn.edu, Coll of Southern Nevada - West Charleston Campus; Gg
Shroder, Jr., John F., (402) 554-2770 jshroder@unomaha.edu, Univ of Nebraska, Omaha; GmlZy
Shropshire, K. Lee, (970) 351-2285 leeshrop@att.net, Univ of Northern Colorado; Pg
Shroyer, Emily L., 541-737-1298 eshroyer@coas.oregonstate.edu, Oregon State Univ; On
Shroyer, Jim, (785) 532-6101 jshroyer@ksu.edu, Kansas State Univ; So
Shuford, James W., (205) 851-5462 jshufard@aamu.edu, Alabama A&M Univ; So
Shugart, Jr., Herman H., (804) 924-7642 hhs@virginia.edu, Univ of Virginia; Zg
Shuib, Mustaffa Kamal, 03-79674227 mustaffk@um.edu.my, Univ of Malaya; Gc
Shukla, Jagadish, (703) 993-1983 jshukla@gmu.edu, George Mason Univ; As
Shuller-Nickles, Lindsay C., 864656-1448 lshulle@clemson.edu, Clemson Univ; Gz
Shulski, Martha D., mshulski3@unl.edu, Univ of Nebraska, Lincoln; As
Shultis, Aaron, 859.257.6222 aaron.shultis@uky.edu, Univ of Kentucky; Csg
Shum, CK, 614 292 7118 ckshum@osu.edu, Ohio State Univ; YdHg
Shumaker, Robert C., (304) 293-5603 rshumaker@wvu.edu, West Virginia Univ; Go
Shuman, Bryan N., 307-766-6442 bshuman@uwyo.edu, Univ of Wyoming; PeGne
Shuman, Larry M., (404) 228-7276 Univ of Georgia; Sc
Shuman, Randy, (206) 296-8243 rshuman531@gmail.com, Univ of Washington; Ob
Shumway, Matthew J., (801) 422-2707 jms7@byu.edu, Brigham Young Univ; Zn
Shumway, Sandra, (860) 405-9282 sandra.shumway@uconn.edu, Univ of Connecticut; Ob
Shuster, Robert D., (402) 554-2457 rshuster@unomaha.edu, Univ of Nebraska, Omaha; GiaZe
Siahcheshm, Kamal, +98 (411) 339 2699 Univ of Tabriz; Egm
Sibeko, Skhumbuzo, 012 382 6271 sibekosg@tut.ac.za, Tshwane Univ of Tech; GggGg
Sibert, John R., sibert@hawaii.edu, Univ of Hawai'i, Manoa; Ob
Sibley, Paul, (519) 824-4120 Ext.52707 psibley@uoguelph.ca, Univ of Guelph; Zn
Sibray, Steven S., (720) 428-9331 sibray1@unl.edu, Univ of Nebraska, Lincoln; HwEog
Sibson, Rick H., +64 3 479-7506 rick.sibson@otago.ac.nz, Univ of Otago; Gc
Sicard, Karri, 907-451-5040 karri.sicard@alaska.gov, Alaska Division of Geological & Geophysical Surveys; Gg
Siddoway, Christine S., (719) 389-6717 csiddoway@coloradocollege.edu, Colorado Coll; GctGg
Sidor, Christian A., 206-221-3285 casidor@uw.edu, Univ of Washington; Pv
Siebach, Kirsten L., 713.348.6751 ksiebach@rice.edu, Rice Univ; Xg
Siedlecki , Samantha, samantha.siedlecki@uconn.edu, Univ of Connecticut; On
Siegel, Donald I., 315-443-2672 disiegel@syr.edu, Syracuse Univ; Hw
Siegfried, Matthew, 303-384-2004 siegfried@mines.edu, Colorado Sch of Mines; ZrGl
Siegler, Matthew, 214-768-2745 msiegler@smu.edu, Southern Methodist Univ; YgXy
Siegrist, jr, Henry G., 410 827 8095 g.a.siegrist@gmail.com, Univ of Guam; GdzHy
Siemens, Michael A., (573) 368-2134 mike.siemens@dnr.mo.gov, Missouri Dept of Nat Res; GgHgZi
Siesser, William G., 615-298-5659 william.g.siesser@vanderbilt.edu, Vanderbilt Univ; PmGsu
Siewers, Fredrick D., (270) 745-5988 fred.siewers@wku.edu, Western Kentucky Univ; GsPiZc
Sigler, Adam, asigler@montana.edu, Montana State Univ; Rw
Sigler, William V., (419) 530-2897 von.sigler@utoledo.edu, Univ of Toledo; Zn
Sigloch, Karin, +44 (1865) 272000 karin.sigloch@earth.ox.ac.uk, Univ of Oxford; YgsGt
Sigman, Daniel M., 609-258-2194 sigman@princeton.edu, Princeton Univ; ClsPc
Sigurdsson, Haraldur, (401) 874-6596 haraldur@gso.uri.edu, Univ of Rhode Island; Ou
Sikora, Todd D., 717-872-3292 todd.sikora@millersville.edu, Millersville Univ; As
Silliman, James, (361) 825-3718 james.silliman@tamucc.edu, Texas A&M Univ, Corpus Christi; Co
Silva, Carlos M., cmsilva@fc.ul.pt, Unive de Lisboa; Pg
Silva, Catrina, csilva@fc.ul.pt, Unive de Lisboa; Hg

Silva, Lucas C., lcsilva@ucdavis.edu, Univ of California, Davis; Sb
Silva, Michael A., (916) 324-0768 California Geo Survey; Ng
Silva, Sam, samsilva@usc.edu, Univ of Southern California; Ac
Silva-Castro, Jhon, 859-257-1173 john.silva@uky.edu, Univ of Kentucky; Nm
Silveira, António B., antonio.brum@fc.ul.pt, Unive de Lisboa; Zf
Silveira, Maria, 8637351314 x209 mlas@ufl.edu, Univ of Florida; So
Silver, Leon T., (626) 395-6490 lsilver@gps.caltech.edu,CalTech; Cc
Silver, Mary W., (831) 459-2908 msilver@cats.ucsc.edu, Univ of California, Santa Cruz; Ob
Silvertooth, Jeffrey C., (520) 621-7145 silver@ag.arizona.edu, Univ of Arizona; So
Silvestri, Alberta, +390498279142 alberta.silvestri@unipd.it, Univ degli Studi di Padova; Gz
Silye, Lorand , lorand.silye@ubbcluj.ro, Babes-Bolyai Univ; PmgGr
Simandl, George J., (250) 952-0413 gsimandle@galaxy.gov.bc.ca, Univ of Victoria; En
Simard, Suzanne, 604–822–1955 suzanne.simard@ubc.ca, Univ of British Columbia; Sb
Simila, Gerald W., (818) 677-3543 gerry.simila@csun.edu, California State Univ, Northridge; Yg
Simjouw, Jean-Paul, 203-932-1253 jsimjouw@newhaven.edu, Univ of New Haven; OcCma
Simmons, Craig, craig.simmons@flinders.edu.au, Flinders Univ; Hw
Simmons, F. William, (217) 333-4424 Univ of Illinois, Urbana-Champaign; Sp
Simmons, Lizanne V., (949) 589-0562 Santiago Canyon Coll; Gr
Simmons, M. G., gene.simmons29@outlook.com, Massachusetts Inst of Tech; Yg
Simmons, Peter, +44 (0)1603 59 3122 p.simmons@uea.ac.uk, Univ of East Anglia; Eg
Simmons, Stuart, 801-581-4122 ssimmons@egi.utah.edu, Univ of Utah; EmCqHg
Simmons, William B., 7342866325 , 5042806325 wsimmons@uno.edu, Univ of Michigan; Gz
Simms, Alexander, (805) 893-7292 asimms@geol.ucsb.edu, Univ of California, Santa Barbara; GsuGm
Simms, Janet E., (601) 634-3493 Mississippi State Univ; Yg
Simon, Joel D., 609-258-4101 jdsimon@princeton.edu, Princeton Univ; YsgYr
Simon, Jowitt, 702-895-2447 simon.jowitt@unlv.edu, Univ of Nevada, Las Vegas; Eg
Simon, Molly N., molly.n.simon@asu.edu, Arizona State Univ; Xa
Simon, Steven B., sbs8@unm.edu, Univ of New Mexico; XmcGi
Simonetti, Stephanie, stefanie.simonetti.4@nd.edu, Univ of Notre Dame; Zn
Simonetti, Tony, simonetti.tony@gmail.com, Univ of Notre Dame; Ct
Simonis, Jean, +27729773994 simonismtz@yahoo.com, Univ of Zululand; HyGgHg
Simons, Frederik J., 609-258-2598 fjsimons@princeton.edu, Princeton Univ; YsdYm
Simons, Mark, (626) 395-6984 simons@caltech.edu,CalTech; Ys
Simonson, Bruce M., bruce.simonson@oberlin.edu, Oberlin Coll; GsXmHw
Simony, Philip S., (403) 220-6679 Univ of Calgary; Gc
Simpkins, William W., (515) 294-7814 bsimp@iastate.edu, Iowa State Univ of Sci & Tech; HwGlCs
Simpson, Edward L., (610) 683-4447 simpson@kutztown.edu, Kutztown Univ of Pennsylvania; Gs
Simpson, Frank -., 519-253-3000 x 2487 franks@uwindsor.ca, Univ of Windsor; GsrGe
Simpson, Robert M., (731) 881-7439 msimpson@utm.edu, Univ of Tennessee, Martin; AsmZi
Simpson, Jr., H. James, (703) 779-2043 simpsonj@ldeo.columbia.edu, Columbia Univ; Cm
Sims, Albert L., (218) 281-8619 simsx008@umn.edu, Univ of Minnesota, Twin Cities; Sc
Sims, David, d.w.sims@soton.ac.uk, Univ of Southampton; Ob
Sims, Douglas , 702-651-4840 douglas.sims@csn.edu, Coll of Southern Nevada - West Charleston Campus; GeSp
Sims, Kenneth W., 307-766-3386 ksims7@uwyo.edu, Univ of Wyoming; CgGvCa
Simsek, Sakir, +90 (312) 2977770 ssimsek@hacettepe.edu.tr, Hacettepe Univ; Hw
Sinclair, Alastair J., (604) 822-3086 asinclair@eos.ubc.ca, Univ of British Columbia; EmGg
Sinclair, Hugh D., +44 (0) 131 650 2518 hugh.sinclair@ed.ac.uk, Edinburgh Univ; So
Singer, Bradley S., (608) 262-6366 bsinger@geology.wisc.edu, Univ of Wisconsin, Madison; Cc
Singer, David M., 330-672-3006 dsinger4@kent.edu, Kent State Univ; Gze
Singer, Jill K., (716) 878-4724 singerjk@buffalostate.edu, Buffalo State Coll; GsOp
Singer, Michael, +44 01334 462874 michael.singer@st-andrews.ac.uk, Univ of St. Andrews; Hg
Singh, Amrita, amrita4@ualberta.ca, Univ of Alberta; Zu
Singh, Hanumant, 617.373.7286 ha.singh@neu.edu, Northeastern Univ; Zn
Singh, Harbans, (973) 655-7383 singhh@mail.montclair.edu, Montclair State Univ; Zn
Singh, Kamaljit, k.singh@hw.ac.uk, Heriot-Watt Univ; Np
Singha, Kamini, 303-273-3822 ksingha@mines.edu, Colorado Sch of Mines; Hw
Singler, Charles R., 330-941-3611 crsingler@ysu.edu, Youngstown State Univ; Gs
Singleton, John, (970) 491-0740 john.singleton@colostate.edu, Colorado State Univ; Gct
Sinha, A. Krishna, (540) 231-5580 pitlab@vt.edu, Virginia Tech; Gt
Sinha, Ashish, (310) 243-3166 asinha@csudh.edu, California State Univ, Dominguez Hills; PeCs
Sinton, John M., (808) 956-7751 sinton@hawaii.edu, Univ of Hawai'i, Manoa; Gi
Sintubin, Manuel, manuel.sintubin@ees.kuleuven.be, Katholieke Univit Leuven; GtcGa
Sipola, Maija, 513-529-3230 sipolame@miamioh.edu, Miami Univ; Gg
Sipos, György, +3662544156 gysipos@geo.u-szeged.hu, University of Szeged; GmCcHs
Sirbescu, Mona, 989-774-4497 sirbe1mc@cmich.edu, Central Michigan Univ; GiCup
Sîrbu, Smaranda D., doina.sirbu@uaic.ro, Alexandru Ioan Cuza; Ge
Sissom, David , dsissom@wtamu.edu, West Texas A&M Univ; Zn
Sisson, Virginia, 713-743-7634 vbsisson@uh.edu, Univ of Houston; Gpt
Sisterson, Doug, 630-252-5836 dlsisterson@anl.gov, Argonne; As
Sisterson, Douglas L., (630) 252-5836 dsisterson@anl.gov, Argonne; As
Sit, Stefany , 312-413-1868 ssit@uic.edu, Univ of Illinois Chicago; YgsZe
Sitwell, O.F. George, sitwell@telus.net, Univ of Alberta; Zn
Siuta, David, 802-626-6238 david.siuta@northernvermont.edu, Northern Vermont Univ-Lyndon; AsZn
Size, William B., (404) 727-0203 wsize@emory.edu, Emory Univ; GivGa
Sjoblom, Megan, (208) 486-7678 pickardm@byui.edu, Brigham Young Univ - Idaho; GiZeGe
Sjostrom, Derek, (406) 238-7387 derek.sjostrom@rocky.edu, Rocky Mountain Coll; ClGsHs
Skalbeck, John D., 262-595-2490 skalbeck@uwp.edu, Univ of Wisconsin, Parkside; HwYgGg
Skarke, Adam, (662) 268-1032 ext 258 adam.skarke@msstate.edu, Mississippi State Univ; GuYrOn
Skehan, James W., 617-552-8312 james.skehan@bc.edu, Boston Coll; Gc
Skemer, Philip, 314-935-3584 pskemer@levee.wustl.edu, Washington Univ in St. Louis; GcCp
Skidmore, Mark L., (406) 994-7251 skidmore@montana.edu, Montana State Univ; ClPgGl
Skillen, James, 616-526-7546 jrs39@calvin.edu, Calvin Coll; Zu
Skinner, Christopher , (978) 934-3901 christopher_skinner@uml.edu, Univ of Massachusetts, Lowell; At
Skinner, Lisa, 928-523-5814 lisa.skinner@nau.edu, Northern Arizona Univ; ZeGvZi
Skinner, Luke C., +44 (0) 1223 764912 luke00@esc.cam.ac.uk, Univ of Cambridge; GeCmOg
Skinner, Randall, 801-422-6083 randy_skinner@byu.edu, Brigham Young Univ; Gg
Skippen, George B., gskippen@ccs.carleton.ca, Carleton Univ; Cg
Sklar, Leonard, 415-338-1204 leonard@sfsu.edu, San Francisco State Univ; Gm
Skoda, Radek, +420 549 49 7392 rskoda@sci.muni.cz, Masaryk Univ; GzZmGi
Skoog, Annelie, 860-405-9220 annelie.skoog@uconn.edu, Univ of Connecticut; Oc
Skubis, Steven T., (315) 312-2799 steven.skubis@oswego.edu, SUNY, Oswego; Am
Skyllingstad, Eric, 541-737-5697 skylling@coas.oregonstate.edu, Oregon State Univ; Am
Slade, Jr., Raymond M., rslade@austincc.edu, Austin Comm Coll District; HgsHq
Slagle, Angela, aslagle@ldeo.columbia.edu, Columbia Univ; Gu
Slater, Brian, 518-473-9988 bslater@mail.nysed.gov, New York State Geo Survey; GoEoGr
Slater, David, (775) 784-4893 dslater@seismo.unr.edu, Univ of Nevada, Reno; Ys
Slater, Gregory F., (905) 525-9140 (Ext. 26388) gslater@mcmaster.ca, McMaster Univ; Cg

Slater, Lee S., 973-353-5109 lslater@newark.rutgers.edu, Rutgers, The State Univ of New Jersey, Newark; YeHw
Slater, Tom, +44 (0) 131 650 9506 tom.slater@ed.ac.uk, Edinburgh Univ; Zn
Slatkin, Montgomery, (510) 642-6300 Univ of California, Berkeley; Pg
Slattery, William, 937 775-3455 william.slattery@wright.edu, Wright State Univ; ZeGr
Slaughter, Richard, 608/262-2399 rich@geology.wisc.edu, Univ of Wisconsin, Madison; Pg
Slavetskas, Carol, cslavets@binghamton.edu, Binghamton Univ; Zn
Slawinski, Michael A., (709) 737-7541 mslawins@mun.ca, Memorial Univ of Newfoundland; Ys
Sleep, Norman H., (650) 723-0882 Stanford Univ; Yg
Sleezer, Richard O., rsleezer@emporia.edu, Emporia State Univ; ZiSdHs
Sletten, Ronald S., 206-543-0571 sletten@uw.edu, Univ of Washington; Sob
Slingerland, Rudy L., (814) 865-6892 sling@geosc.psu.edu, Pennsylvania State Univ, Univ Park; Gs
Sloan, Alastair, +27-21-650-2924 alastair.sloan@uct.ac.za, Univ of Cape Town; GcYgu
Sloan, Doris, (510) 527-5710 dsloan@berkeley.edu, Univ of California, Berkeley; Pme
Sloan, Heather , 718-960-8008 heather.sloan@lehman.cuny.edu, Graduate Sch of the City Univ of New York; Gg
Sloan, Heather, 718-960-8008 heather.sloan@lehman.cuny.edu, Lehman Coll (CUNY); Yr
Sloan, Jon R., (818) 677-4880 jon.sloan@csun.edu, California State Univ, Northridge; Pm
Slobodnik, Marek, +420 549 49 7055 marek@sci.muni.cz, Masaryk Univ; GeEg
Sloss, Craig, 610731382610 c.sloss@qut.edu.au, Queensland Univ of Tech; GsrGm
Slovinsky, Peter, (207) 287-7173 peter.a.slovinsky@maine.gov, Dept of Agriculture, Conservation, and Forestry; Gu
Slowey, Niall C., (979) 845-8478 nslowey@ocean.tamu.edu, Texas A&M Univ; Ou
Smaglik, Suzanne M., 307-855-2146 ssmaglik@cwc.edu, Central Wyoming Coll; GgZeCg
Smale, Jody L., 717-702-2020 jsmale@pa.gov, DCNR- Pennsylvania Bur of Geo Survey; Zn
Small, Christopher, (845) 365-8354 small@ldeo.columbia.edu, Columbia Univ; ZrYr
Smalley, Gabriela W., 609-896-5097 gsmalley@rider.edu, Rider Univ; ObcOp
Smalley, Glendon W., (615) 598-5714 Sewanee: Univ of the South; Sf
Smalley, Jr., Robert, (901) 678-2007 rsmalley@memphis.edu, Univ of Memphis; Yd
Smart, Christopher, +44 1752 584764 c.smart@plymouth.ac.uk, Univ of Plymouth; Pg
Smart, Katie A., 011 717 6549 katie.smart2@wits.ac.za, Univ of the Witwatersrand; CsGx
Smay, Jessica J., 4082982181 x3933 jessica.smay@sjcc.edu, San Jose City Coll; ZeGm
Smayda, Theodore J., (401) 874-6171 tsmayda@gso.uri.edu, Univ of Rhode Island; Ob
Smedes, Harry W., (541) 552-6479 Southern Oregon Univ; Gi
Smedley, Andrew, +44 0161 306-8770 andrew.smedley@manchester.ac.uk, Univ of Manchester; As
Smerdon, Jason, jsmerdon@ldeo.columbia.edu, Columbia Univ; Pe
Smethie, William, bsmeth@ldeo.columbia.edu, Columbia Univ; Cg
Smiley, Tara M., (812) 855-3728 tmsmiley@iu.edu, Indiana Univ, Bloomington; Pe
Smilnak, Roberta A., (303) 556-3144 Metropolitan State Coll of Denver; Zg
Smirnov, Aleksey V., (906) 487-2365 asmirnov@mtu.edu, Michigan Technological Univ; YmgGt
Smirnov, Anna, (418) 654-3711 asmirnov@nrcan.gc.ca, Univ du Quebec; Cs
Smit, Ansie, +27124202282 ansie.smit@up.ac.za, Univ of Pretoria; Rn
Smith, Alan L., (909) 537-5409 alsmith@csusb.edu, California State Univ, San Bernardino; GviGz
Smith, Alison J., (330) 672-3709 alisonjs@kent.edu, Kent State Univ; GnPce
Smith, Alistair, (208) 885-1109 alistair@uidaho.edu, Univ of Idaho; SfZyAm
Smith, Ben , 206-616-9176 bsmith@apl.washington.edu, Univ of Washington; GlZr
Smith, Betty E., (217) 581-2626 besmith@eiu.edu, Eastern Illinois Univ; Zu
Smith, Brianne, 718-951-5000 x2689 brianne.smith43@brooklyn.cuny.edu, Brooklyn Coll (CUNY); Hs
Smith, C. K., 210-486-0062 csmith55@alamo.edu, San Antonio Comm Coll; ZyzZz
Smith, C. Ken, (931) 598-3219 ksmith@sewanee.edu, Sewanee: Univ of the South; Sf
Smith, Catherine H., (505) 667-0113 chsmith@lanl.gov, Los Alamos; Ca
Smith, Charles K., csmith55@alamo.edu, Alamo Colls, San Antonio Coll; Zy
Smith, Craig R., (808) 956-7776 csmith@soest.hawaii.edu, Univ of Hawai'i, Manoa; Ob
Smith, Dan, +440116 252 5355 djs40@le.ac.uk, Leicester Univ; Ge
Smith, David C., (401) 874-6172 dcsmith@gsosun1.gso.uri.edu, Univ of Rhode Island; Ob
Smith, Dena M., 303-735-2011 dena.smith@colorado.edu, Univ of Colorado; Pg
Smith, Diane R., (210) 999-7656 dsmith@trinity.edu, Trinity Univ; Gi
Smith, Donald R., (831) 459-5041 dsmith@scipp.ucsc.edu, Univ of California, Santa Cruz; Zn
Smith, Douglas, (512) 471-4261 doug@jsg.utexas.edu, Univ of Texas, Austin; GipGz
Smith, Erik, erik@belle.baruch.sc.edu, Univ of South Carolina; HsOb
Smith, Eugene I., (702) 895-3971 gsmith@ccmail.nevada.edu, Univ of Nevada, Las Vegas; Gi
Smith, Florence P., (630) 252-7980 Argonne; Ca
Smith, Gerald R., (734) 764-0491 grsmith@umich.edu, Univ of Michigan; Ps
Smith, Grant, 503-838-8862 smithg@wou.edu, Western Oregon Univ; Zg
Smith, H D., smithhd@cf.ac.uk, Cardiff Univ; Zn
Smith, J. Leslie, (604) 822-4108 lsmith@eos.ubc.ca, Univ of British Columbia; Hw
Smith, James A., (609) 258-4615 jsmith@princeton.edu, Princeton Univ; Hg
Smith, Jason J., 607-778-5116 smithjj@sunybroome.edu, Broome Comm Coll; Ggs
Smith, Jeffery L., (509) 335-7648 Washington State Univ; Sb
Smith, Jen R., jensmith@wustl.edu, Washington Univ in St. Louis; Ga
Smith, Jennifer E., (858) 246-0803 smithj@ucsd.edu, Univ of California, San Diego; Obn
Smith, Jim, j.smith@flinders.edu.au, Flinders Univ; Zn
Smith, Jon, 785-864-2179 jjsmith@kgs.ku.edu, Univ of Kansas; GsPe
Smith, Joseph P., 410-293-6568 jpsmith@usna.edu, United States Naval Acad; CmOg
Smith, K. L., (702) 895-3971 Wichita State Univ; Zn
Smith, Karen, 630-252-0136 smithk@anl.gov, Argonne; Hw
Smith, Karl A., (805) 756-2262 ksmith@tc.umn.edu, Univ of Minnesota, Twin Cities; Nx
Smith, Kathlyn M., (912) 478-5398 ksmith@georgiasouthern.edu, Georgia Southern Univ; Pv
Smith, Ken D., (775) 784-4218 ken@seismo.unr.edu, Univ of Nevada, Reno; YsGt
Smith, Langdon, 724.738.2389 langdon.smith@sru.edu, Slippery Rock Univ; Zn
Smith, Langhorne, 518-473-6262 lsmith@remove@this.mail.nysed.gov, New York State Geo Survey; Go
Smith, Larry N., (406) 496-4859 lsmith@mtech.edu, Montana Tech of the Univ of Montana; GsmEo
Smith, Laurence C., (310) 825-3154 lsmith@geog.ucla.edu, Univ of California, Los Angeles; Hs
Smith, Martin, martin.smith@durham.ac.uk, Durham Univ; Pe
Smith, Matthew C., (352) 392-2106 mcsmith@ufl.edu, Univ of Florida; Gi
Smith, Michael, 205-247-3724 msmith@gsa.state.al.us, Geo Survey of Alabama; Hg
Smith, Mike, +44020 8417 2500 michael.smith@kingston.ac.uk, Kingston Univ; Zr
Smith, Nicole, 303-273-3634 nmsmith@mines.edu, Colorado Sch of Mines; Zn
Smith, Norman D., (402) 472-5362 nsmith3@unl.edu, Univ of Nebraska, Lincoln; Gsm
Smith, Paul H., (505) 667-7494 Los Alamos; Zn
Smith, Peter, (520) 621-2725 psmith@lpl.arizona.edu, Univ of Arizona; Zr
Smith, Phillip J., pjsmith@purdue.edu, Purdue Univ; As
Smith, Richard, 479-575-3355 Univ of Arkansas, Fayetteville; Gm
Smith, Robert, (208) 885-2560 smithbob@uidaho.edu, Univ of Idaho; Cb
Smith, Ronald B., (203) 432-3129 ronald.smith@yale.edu, Yale Univ; As
Smith, Ryan , 573-341-4466 smithryang@mst.edu, Missouri Univ of Sci and Tech; HwYg
Smith, Sean, 207-581-2198 sean.m.smith@maine.edu, Univ of Maine; Hs
Smith, Shane V., 330-941-1752 svsmith@ysu.edu, Youngstown State Univ; Gs
Smith, Stephen V., svsmith@hawaii.edu, Univ of Hawai'i, Manoa; Ou
Smith, Steven C., 916-435-9180 smithsc@arc.losrios.edu, American River Coll; ZynZn
Smith, Stewart W., stews@uw.edu, Univ of Washington; Ys
Smith, Susan M., 406-496-4173 ssmith@mtech.edu, Montana Tech of The Univ of Montana; Zn
Smith, Terence E., tsmith@uwindsor.ca, Univ of Windsor; Gi
Smith, Terry L., (805) 756-2262 tsmith@calpoly.edu, California Polytechnic

State Univ; So
Smith, Thomas M., (804) 924-3107 tms9a@virginia.edu, Univ of Virginia; Zg
Smith, William H., (314) 935-5638 whsmith@dasi.wustl.edu, Washington Univ in St. Louis; Xc
Smith-Engle, Jennifer M., (361) 825-2436 jennifer.smith-engle@tamucc.edu, Texas A&M Univ, Corpus Christi; Gs
Smith-Konter, Bridget R., 808-956-3618 brkonter@hawaii.edu, Univ of Hawai'i, Manoa; YdXy
Smithson, Jayne, 5102357800 x.4284 jsmithson@contracosta.edu, Contra Costa Coll; Gg
Smithyman, Brenden, bsmithym@uwo.ca, Western Univ; Ye
Smoak, Joseph (Donny), smoak@usf.edu, Univ of South Florida; Sf
Smrekar, Suzanne E., (818) 354-4192 Jet Propulsion Lab; Xy
Smukler, Sean, 604–822–2795 sean.smukler@ubc.ca, Univ of British Columbia; So
Smylie, Douglas E., (416) 736-2100 #66438 doug@core.yorku.ca, York Univ; Yg
Smyth, Joseph R., 303-492-5521 joseph.smyth@colorado.edu, Univ of Colorado; Gz
Smyth, Rebecca C., (512) 471-0232 rebecca.smyth@beg.utexas.edu, Univ of Texas at Austin, Jackson Sch of Geosciences; Hg
Smyth, Thomas J., (919) 515-2838 North Carolina State Univ; Sc
Smyth, William D., (541) 737-3029 smyth@coas.oregonstate.edu, Oregon State Univ; Op
Smythe, Wendy F., (218) 726-7899 wsmythe@d.umn.edu, Univ of Minnesota Duluth; Ou
Smythe, William, (310) 825-2434 Univ of California, Los Angeles; Ys
Snead, John I., (225) 578-3454 snead@lsu.edu, Louisiana State Univ; Gm
Snider, Henry I., (860) 228-9815 snider@easternct.edu, Eastern Connecticut State Univ; Ge
Snieder, Roel, 303-273-3456 rsnieder@mines.edu, Colorado Sch of Mines; Ye
Snipes, Brittany, bsnipes@flmnh.ufl.edu, Univ of Florida; Zn
Snoeckx, Hilde, (850) 474-3377 snoeckx@uwf.edu, Univ of West Florida; OuZg
Snoke, J. Arthur, (540) 231-6028 snoke@vt.edu, Virginia Tech; Ys
Snow, Daniel, (402)472-7539 dsnow1@unl.edu, Univ of Nebraska, Lincoln; Ge
Snow, Jonathan, 713-743-3298 jesnow@uh.edu, Univ of Houston; Ca
Snow, Julie A., (724) 738-2503 julie.snow@sru.edu, Slippery Rock Univ; As
Snyder, Daniel, 478-274-7806 dsnyder@mgc.edu, Middle Georgia Coll; Gg
Snyder, Jeffrey A., (419) 372-0533 jasnyd@bgnet.bgsu.edu, Bowling Green State Univ; Gm
Snyder, Jennifer L., (610) 359-5291 jsnyder2@dccc.edu, Delaware County Comm Coll; Zg
Snyder, Morgan E., morgan.snyder@acadiau.ca, Acadia Univ; Gc
Snyder, Noah, 617-552-0839 snyderno@bc.edu, Boston Coll; GmHg
Snyder, Peter K., 612-625-8207 pksnyder@umn.edu, Univ of Minnesota, Twin Cities; As
Snyder, Richard L., rlsnyder@ucdavis.edu, Univ of California, Davis; As
Snyder, Walter S., wsnyder@boisestate.edu, Boise State Univ; GrcGt
Snyman, Lourens, +27 (0) 12 420 2862 lourens.snyman@up.ac.za, Univ of Pretoria; Zi
Soares Cruz, Anna Paula ., acruz@udel.edu, California State Univ, Bakersfield; CmOoPc
Sobel, Adam, ahs129@columbia.edu, Columbia Univ; As
Sobotka, Jerzy, jerzy.sobotka@uwr.edu.pl, Univ of Wroclaw; YeuEo
Soderlund, Krista, 512-471-0449 krista@ig.utexas.edu, Univ of Texas, Austin; XyYm
Sogn, Trine, trine.sogn@nmbu.no, Norwegian Univ of Life Scis (NMBU); Sb
Sohi, Sran P., +44 (0) 131 651 4471 saran.sohi@ed.ac.uk, Edinburgh Univ; So
Sohn, Robert A., (508) 289-3616 rsohn@whoi.edu, Woods Hole Oceanographic Inst; Yr
Soja, Constance M., (315) 228-7200 csoja@colgate.edu, Colgate Univ; PiGs
Sokolik, Irina, (404) 894-6180 isokolik@eas.gatech.edu, Georgia Inst of Tech; AsZr
Sokolova, Elena, (204) 747-8252 elena_sokolova@umanitoba.ca, Univ of Manitoba; Gz
Solan, Martin, +44 (0)23 80593755 m.solan@soton.ac.uk, Univ of Southampton; Ob
Solana, Camen, +44 023 92 842394 carmen.solana@port.ac.uk, Univ of Portsmouth; Gv
Solar, Gary S., (716) 878-4900 solargs@buffalostate.edu, Buffalo State Coll; GcpGt
Soldat, Douglas J., (608) 263-3631 djsoldat@wisc.edu, Univ of Wisconsin, Madison; So
Soldati, Arianna, a.soldati@ncsu.edu, North Carolina State Univ; Gv
Solecki, Andrzej, andrzej.solecki@uwr.edu.pl, Univ of Wroclaw; Eg
Solecki, William, (973) 655-5129 soleckiw@mail.montclair.edu, Montclair State Univ; Zu
Solferino, Giulio, +44 1784 443585 giulio.solferino@rhul.ac.uk, Royal Holloway Univ of London; Gg
Solis, Michael P., 614 265 6597 michael.solis@dnr.state.oh.us, Ohio Dept of Nat Res; GgcGt
Soll, Wendy E., (505) 665-6930 weasel@lanl.gov, Los Alamos; Hw
Solomatov, Viatcheslav S., (314) 935-7882 slava@wustl.edu, Washington Univ in St. Louis; Yg
Solomon, Douglas K., (801) 581-7231 kip.solomon@utah.edu, Univ of Utah; Hw
Solomon, Keith, (519) 824-4120 Ext 58792 ksolomon@uoguelph.ca, Univ of Guelph; Zn
Solomon, Sean , 845-365-8714 solomon@ldeo.columbia.edu, Columbia Univ; Xy
Solomon, Susan, solos@mit.edu, Massachusetts Inst of Tech; As
Soltanian, Reza, mohamadreza.soltanian@uc.edu, Univ of Cincinnati; Hw
Somarin, Alireza, 204-727-9680 somarina@brandonu.ca, Brandon Univ; Gig
Somasundaran, Ponisseril, (212) 854-2926 ps24@columbia.edu, Columbia Univ; Nx
Somenhally, Anil, asomenahally@tamu.edu, Texas A&M Univ; Sb
Somers, Jr., Arnold E., (229) 333-5664 gsomers@valdosta.edu, Valdosta State Univ; Zn
Somerville, Richard C., (858) 534-4644 rsomerville@ucsd.edu, Univ of California, San Diego; As
Sommer, Ulrich , usommer@geomar.de, Dalhousie Univ; Ob
Sommerfield, Christopher K., (302) 645-4255 cs@udel.edu, Univ of Delaware; OuGsCc
Sonder, Ingo, (716)645-6366 ingomark@buffalo.edu, SUNY, Buffalo; Gv
Sonder, Leslie J., 603-646-2372 leslie.sonder@dartmouth.edu, Dartmouth Coll; Gq
Sondergeld, Carl H., (405) 325-6870 csondergeld@ou.edu, Univ of Oklahoma; NpYex
Sone, Hiroki, (608) 890-0531 hsone@wisc.edu, Univ of Wisconsin, Madison; NrYxGt
Sonett, Charles P., (520) 621-6935 sonett@dakotacom.net, Univ of Arizona; Yg
Song, Alex, +44 07472526049 alex.song@ucl.ac.uk, Univ Coll London; Ysg
Song, Liaosha, lsong1@csub.edu, California State Univ, Bakersfield; GoYeEo
Song, Xiaodong, xsong@illinois.edu, Univ of Illinois, Urbana-Champaign; Ys
Sonnenberg, Stephen A., 303-384-2182 ssonnenb@mines.edu, Colorado Sch of Mines; Go
Sonnenwald, Margreta, ‭+49 89 28925822‬ margreta.sonnenwald@tum.de, Technical Univ of Munich; Ng
Soreghan, Gerilyn S., (405) 325-4482 lsoreg@ou.edu, Univ of Oklahoma; GsPcZg
Soreghan, Michael J., (405) 325-3393 msoreg@ou.edu, Univ of Oklahoma; Gs
Sorenson, Adrienne, adrienne.sorenson@cwu.edu, Central Washington Univ; Gg
Sori, Michael, 765-496-5235 msori@purdue.edu, Purdue Univ; Gg
Sorkhabi, Rasoul, 801-581-9070 rsorkhabi@egi.utah.edu, Univ of Utah; GctGc
Sorooshian, Soroosh, (949) 824-8825 soroosh@uci.edu, Univ of California, Irvine; Hg
Soster, Frederick M., (765) 658-4670 fsoster@depauw.edu, DePauw Univ; Gs
Sottile, Jr., Joseph, 859-257-4616 joe.sottile@uky.edu, Univ of Kentucky; Nm
Souch, Catherine J., (317) 274-1103 Indiana Univ, Indianapolis; Gm
Soucy de la Roche, Renaud, 418-654-2609 renaud.soucy_la_roche@ete.inrs.ca, Univ du Quebec; Gc
Soule, Peter T., (828) 262-7056 soulept@appstate.edu, Appalachian State Univ; As
Soule, S. Adam, (508) 289-3213 ssoule@whoi.edu, Woods Hole Oceanographic Inst; Gvu
Soupios, Panteleimon , +966138602689 panteleimon.soupios@kfupm.edu.sa, King Fahd Univ of Petroleum and Minerals; Yg
Sousa, Luís M., lsousa@utad.pt, Unive de Trás-os-Montes e Alto Douro; EnGgNg
Sousa, Wayne P., (510) 642-2435 Univ of California, Berkeley; Ob
Southam, Gordon, (519) 661-3197 gsoutham@uwo.ca, Western Univ; Pg
Southam, John R., (305) 284-1898 jsoutham@miami.edu, Univ of Miami; Op
Southard, John B., (617) 253-3397 southard@mit.edu, Massachusetts Inst

of Tech; Gs
Southard, Randal J., (530) 752-2199 rjsouthard@ucdavis.edu, Univ of California, Davis; So
Southon, John, (949) 824-3674 jsouthon@uci.edu, Univ of California, Irvine; CcOcPe
Southwell, Benjamin, (906) 635-2076 bsouthwell@lssu.edu, Lake Superior State Univ; Zn
Sowers, Todd, (814) 863-8093 sowers@geosc.psu.edu, Pennsylvania State Univ, Univ Park; Cs
Spaeth, Bridget, (207) 725-3628 bspaeth@bowdoin.edu, Bowdoin Coll; Zn
Spaeth, Matthew P., 217-265-6578 spaeth@illinois.edu, Illinois State Geo Survey; Ge
Spahr, Paul, 614 265 6577 paul.spahr@dnr.state.oh.us, Ohio Dept of Nat Res; GeHwZi
Spalding, Brian P., bps@ornl.gov, Oak Ridge; Sc
Spanbauer, Trisha, (419) 530-2823 trisha.spanbauer@utoledo.edu, Univ of Toledo; Zn
Spandler, Carl, carl.spandler@jcu.edu.au, James Cook Univ; GxEmCg
Spane, Frank A., (509) 371-7081 frank.spane@pnl.gov, Pacific Northwest; Hgw
Spanos, T.J.T (Tim), (780) 435-5245 tim@phys.ualberta.ca, Univ of Alberta; YgsEo
Sparks, Carolyn R., sparks@hope.edu, Hope Coll; GzHg
Sparks, David, (979) 458-1051 sparks@geo.tamu.edu, Texas A&M Univ; YgGqZe
Spear, Frank S., (518) 276-6103 spearf@rpi.edu, Rensselaer Polytechnic Inst; GpCpGc
Specker, Jennifer, (401) 874-6858 jspecker@gsosun1.gso.uri.edu, Univ of Rhode Island; Ob
Speece, Marvin A., (406) 496-4188 mspeece@mtech.edu, Montana Tech of the Univ of Montana; YeuYs
Speed, Don, (602) 285-7244 d.speed@phoenixcollege.edu, Phoenix Coll; Gg
Speer, James, (812) 237-2257 jim.speer@indstate.edu, Indiana State Univ; PeZyGe
Speidel, David H., (718) 261-9524 david.speidel@qc.cuny.edu, Queens Coll (CUNY); Cgp
Speijer, Robert, robert.speijer@kuleuven.be, Katholieke Univit Leuven; PmcOo
Spellman, Patricia, pdspellm@usf.edu, Univ of South Florida; Hq
Spence, George D., (250) 721-6187 gspence@uvic.ca, Univ of Victoria; Ys
Spencer, Joel Q., (785) 532-2249 joelspen@ksu.edu, Kansas State Univ; CcGs
Spencer, Matt, (906) 635-2085 mspencer1@lssu.edu, Lake Superior State Univ; Gl
Spencer, Matthew, +44 0151 795 4399 m.spencer@liverpool.ac.uk, Univ of Liverpool; Og
Spencer, Patrick K., (509) 527-5222 spencerp@whitman.edu, Whitman Coll; Pg
Spencer, Ronald J., (403) 220-6447 spencer@geo.ucalgary.ca, Univ of Calgary; Cg
Spera, Frank J., (805) 893-4880 spera@geol.ucsb.edu, Univ of California, Santa Barbara; Gi
Sperazza, Michael, 631-632-1687 michael.sperazza@stonybrook.edu, Stony Brook Univ; Zn
Spero, Howard J., (530) 752-3307 hjspero@ucdavis.edu, Univ of California, Davis; Cm
Sperone, Felice, (313) 577-2506 fgsperone@wayne.edu, Wayne State Univ; GgZr
Spetzler, Hartmut A., 303-492-6715 spetzler@colorado.edu, Univ of Colorado; Ys
Spiegelman, Marc, mspieg@ldeo.columbia.edu, Columbia Univ; Ys
Spieler, Oliver, 089/2180 4221 spieler@min.uni-muenchen.de, Ludwig-Maximilians-Univ Muenchen; Gz
Spiess, Richard, +390498279150 richard.spiess@unipd.it, Univ degli Studi di Padova; Gp
Spigel, Lindsay, (207) 287-7177 lindsay.spigel@maine.gov, Dept of Agriculture, Conservation, and Forestry; Gm
Spikes, Kyle T., 512-471-7674 kyle.spikes@jsg.utexas.edu, Univ of Texas, Austin; Ye
Spilde, Michael N., (505) 277-5430 mspilde@unm.edu, Univ of New Mexico; Gz
Spilker, Linda J., (818) 354-1647 linda.j.spilker@jpl.nasa.gov, Jet Propulsion Lab; XyZr
Spiller, Erik, 303-882-2682 espiller@mines.edu, Colorado Sch of Mines; Nx
Spindel, Robert C., (206) 543-1310 spindel@apl.washington.edu, Univ of Washington; Op
Spinelli, Glenn, 575.835.6512 spinelli@nmt.edu, New Mexico Tech; Hw
Spinler, Joshua C., (501) 569-3544 jxspinler@ualr.edu, Univ of Arkansas at Little Rock; YdGt
Spinney, Jamie, (435) 865-8244 jamiespinney@suu.edu, Southern Utah Univ; Zyu
Spitz, Yvette H., 541-737-3227 yspitz@coas.oregonstate.edu, Oregon State Univ; Yr
Splinter, Dale, splinted@uww.edu, Univ of Wisconsin, Whitewater; Hs
Spokas, Kurt, 612-626-2834 kurt.spokas@ars.usda.gov, Univ of Minnesota, Twin Cities; Spo
Spongberg, Alison L., (419) 530-4091 alison.spongberg@utoledo.edu, Univ of Toledo; Co
Spooner, Alecia, 425-388-9003 aspooner@everettcc.edu, Everett Comm Coll; Zg
Spooner, Edward T. C., (416) 978-3280 etcs@geology.utoronto.ca, Univ of Toronto; En
Spooner, Ian S., (902) 585-1312 ian.spooner@acadiau.ca, Acadia Univ; Ge
Spotila, James A., (540) 231-2109 spotila@vt.edu, Virginia Tech; Gtm
Spracklen, Dominick, +44(0) 113 34 37488 d.v.spracklen@leeds.ac.uk, Univ of Leeds; Zn
Sprain, Courtney, 352-294-6319 csprain@ufl.edu, Univ of Florida; CgYg
Spratt, Deborah A., (403) 220-6446 spratt@geo.ucalgary.ca, Univ of Calgary; Gc
Spray, John G., (506) 999-3544 jgs@unb.ca, Univ of New Brunswick; GpXgGz
Sprenke, Kenneth F., (208) 885-5791 ksprenke@uidaho.edu, Univ of Idaho; Ye
Springer, Abraham E., (928) 523-7198 abe.springer@nau.edu, Northern Arizona Univ; Hw
Springer, Everett P., (505) 667-0569 everetts@lanl.gov, Los Alamos; Hq
Springer, Gregory S., (740) 593-9431 springeg@ohio.edu, Ohio Univ; GmPcGa
Springer, Kathleen B., (909) 307-2669 242 kspringer@sbcm.sbcounty.gov, San Bernardino County Mus; Zg
Springer, Robert K., springer@brandonu.ca, Brandon Univ; Gi
Springston, George E., (802) 485-2734 gsprings@norwich.edu, Norwich Univ; GmZiYg
Sprinkel, Douglas A., (801) 391-1977 douglassprinkel@utah.gov, Utah Geo Survey; GosGg
Sprinkle, James T., (512) 471-4264 echino@jsg.utexas.edu, Univ of Texas, Austin; PigGr
Spry, Paul G., (515) 294-9637 pgspry@iastate.edu, Iowa State Univ of Sci & Tech; Em
Spurr, Aaron, (319) 273-3789 aaron.spurr@uni.edu, Univ of Northern Iowa; Zeg
Squelch, Andrew P., +61 8 9266 2324 a.squelch@curtin.edu.au, Curtin Univ; NmYeZe
Squires, Richard L., (818) 677-2514 richard.squires@csun.edu, California State Univ, Northridge; Pi
Sremac, Jasenka, +38514606108 jsremac@yahoo.com, Univ of Zagreb; PesPb
Srimal, Neptune, (305) 919-5969 srimal@fiu.edu, Florida InterNatl Univ; Gt
Srinivasan, Balakrishnan, +914132655008 sbala.esc@pondiuni.edu.in, Pondicherry Univ; CcGiz
Sritharan, Subramania I., (937) 376-6275 sri@centralstate.edu, Central State Univ; Hg
SRIVASTAVA, HARI B., +919415353606 hbsrivastava@gmail.com, Banaras Hindu Univ; GctGp
Srivastava, Ramesh C., (312) 502-7139 srivast@geosci.uchicago.edu, Univ of Chicago; AsmAm
Sriver, Ryan, 217-300-0364 rsriver@illinois.edu, Univ of Illinois, Urbana-Champaign; As
Srnka, Len, (858) 822-1510 lsrnka@ucsd.edu, Univ of California, San Diego; Ym
Srogi, LeeAnn, (610) 436-2721 esrogi@wcupa.edu, West Chester Univ; Gi
St. John, James C., (404) 894-1754 jim.stjohn@eas.gatech.edu, Georgia Inst of Tech; As
St. John, Kristen E., 540-568-6675 stjohnke@jmu.edu, James Madison Univ; Ou
Stachel, Thomas, (780) 492-0865 thomas.stachel@ualberta.ca, Univ of Alberta; GiCs
Stachnik, Joshua, 610-758-2581 jcs612@lehigh.edu, Lehigh Univ; Ys
Stack, Andrew, (404) 894-3895 stackag@ornl.gov, Georgia Inst of Tech; Cg
Stadnyk, Leona, (403) 440-6165 lstadnyk@mtroyal.ca, Mount Royal Univ; Zn
Stafford, C. Russell, (812) 237-3989 russell.stafford@indstate.edu, Indiana State Univ; Gam
Stafford, Emily, (828) 227-7367 esstafford@wcu.edu, Western Carolina Univ; PgGg
Stafford, Kevin W., (936) 468-2429 staffordk@sfasu.edu, Stephen F. Austin State Univ; Hw
Stahle, David W., (479) 575-3703 dstahle@uark.edu, Univ of Arkansas,

Fayetteville; GeZon
Stahlman, Phillip, (785) 625-3425 stahlman@ksu.edu, Kansas State Univ; So
Stakes, Debra, (805) 546-3100 dstakes@cuesta.edu, Cuesta Coll; Gg
Staley, Andrew W., (410) 260-8818 andrew.staley@maryland.gov, Maryland Dept of Nat Res; Hw
Stamm, Alfred J., (315) 312-2806 stamm@oswego.edu, SUNY, Oswego; Am
Stampone, Mary D., 603-862-3136 mary.stampone@unh.edu, Univ of New Hampshire; Zy
Stamps, D. S., 540-231-3651 dstamps@vt.edu, Virginia Tech; YdGtRn
Stan, Cristiana, (703) 993-5391 cstan@gmu.edu, George Mason Univ; As
Stan, Oana, 0040232201467 cristina.stan@uaic.ro, Alexandru Ioan Cuza; GedCb
Standridge, Debbie, deborah.standridge@gsw.edu, Georgia Southwestern State Univ; Zn
Stanford, Scott D., (609) 292-2576 scott.stanford@dep.nj.gov, New Jersey Geological and Water Survey; Gml
Stanley, Clifford R., (902) 585-1344 cliff.stanley@acadiau.ca, Acadia Univ; Cea
Stanley, Daniel J., (202) 633-1354 Smithsonian Inst / Natl Mus of Natural History; Ou
Stanley, George R., (210) 486-0045 gstanley@alamo.edu, San Antonio Comm Coll; XzZgRm
Stanley, Jessica, (208) 885-4704 jessicastanley@uidaho.edu, Univ of Idaho; Gt
Stanley, Thomas M., (405) 325-7281 tmstanley@ou.edu, Univ of Oklahoma; PiGsc
Stanley, Jr., George D., (406) 243-5693 george.stanley@umontana.edu, Univ of Montana; Pi
Stansell, Nathan D., 815-753-1943 nstansell@niu.edu, Northern Illinois Univ; Gl
Stanton, Kathryn, 916-558-2343 stantok@scc.losrios.edu, Sacramento City Coll; GgPg
Stanton, Kelsay, kstanton@wvc.edu, Wenatchee Valley Coll; GgZe
Stanton, Timothy P., (831) 656-3144 stanton@nps.edu, Naval Postgrad Sch; Op
Staples, Reid, rdstaple@sfu.ca, Simon Fraser Univ; GpcCg
Starek, Michael , 361-825-3978 michael.starek@tamucc.edu, Texas A&M Univ, Corpus Christi; Zi
Stark, Robert, 089/2180 4329 stark@lrz.uni-muenchen.de, Ludwig-Maximilians-Univ Muenchen; Gz
Starr, Paepin K., paepin.starr@austincc.edu, Austin Comm Coll District; GlZrRc
Starr, Richard, 831-771-4400 starr@mlml.calstate.edu, Moss Landing Marine Lab; Zn
Starrfield, Sumner, (480) 965-7569 sumner.starrfield@asu.edu, Arizona State Univ; XaZnn
Stasko, Stanislaw, stanislaw.stasko@uwr.edu.pl, Univ of Wroclaw; Hg
Staten, Paul W., (812) 856-5135 pwstaten@indiana.edu, Indiana Univ, Bloomington; As
Staub, James R., (406) 243-4953 james.staub@umontana.edu, Univ of Montana; GsrEo
Stauffer, David R., (814) 863-3932 stauffer@essc.psu.edu, Pennsylvania State Univ, Univ Park; Am
Stauffer, Mel R., (306) 966-5708 mel.stauffer@usask.ca, Univ of Saskatchewan; Gc
Stead, Douglas, 778-782-6670 Simon Fraser Univ; Ng
Steadman, Todd A., 864-656-2536 tsteadm@clemson.edu, Clemson Univ; ZnnZn
Stearley, Ralph F., (616) 526-6370 rstearle@calvin.edu, Calvin Coll; PvOgRh
Stearman, Will, +61 7 3138 4165 w.stearman@qut.edu.au, Queensland Univ of Tech; Geg
Stearns, David W., (405) 325-3253 mtdstearns@aol.com, Univ of Oklahoma; Gc
Stearns, Leigh, (785) 864-4202 stearns@ku.edu, Univ of Kansas; GlZrf
Stearns, Michael, (801) 863-8498 mstearns@uvu.edu, Utah Valley Univ; Gip
Stebbins, Jonathan F., (650) 723-1140 stebbins@pangea.stanford.edu, Stanford Univ; Cg
Stechmann, Samuel, 608-263-4351 stechmann@wisc.edu, Univ of Wisconsin, Madison; ZnAs
Steck, Lee, (505) 665-3528 Los Alamos; Gg
Steckler, Michael, steckler@ldeo.columbia.edu, Columbia Univ; Yv
Steel, Ronald J., (512) 471-0954 rsteel@mail.utexas.edu, Univ of Texas, Austin; Gs
Steele, Amber , 573-368-7152 amber.steele@dnr.mo.gov, Missouri Dept of Nat Res; Sd
Steele, Andrew, 202-478-8974 asteele@carnegiescience.edu, Carnegie Inst for Sci; Xy
Steele, Bill, 206-685-5880 wsteele@uw.edu, Univ of Washington; Ys
Steele, Kenneth F., (501) 575-4403 ksteele@comp.uark.edu, Univ of Arkansas, Fayetteville; Cg
Steele, Michael A., (570) 408-4763 michael.steele@wilkes.edu, Wilkes Univ; Pg
Steen, Andrew, 865-974-0821 asteen1@utk.edu, Univ of Tennessee, Knoxville; CoOb
Steenberg, Julia, (612) 626-1830 and01006@umn.edu, Univ of Minnesota; Gr
Steenburgh, Jim, 801-581-8727 jim.steenburgh@utah.edu, Univ of Utah; As
Steenhuis, Tammo S., (607) 255-2489 tss1@cornell.edu, Cornell Univ; HgSo
Steenwerth, Kerri, ksteenwerth@ucdavis.edu, Univ of California, Davis; Cb
Steeples, Don W., (785) 737-3399 don@ku.edu, Univ of Kansas; YseGe
Steer, David N., 330 972-2099 steer@uakron.edu, Univ of Akron; Yes
Stefani, Cristina, +390498279195 cristina.stefani@unipd.it, Univ degli Studi di Padova; Gd
Stefano, Christopher J., 906-487-3028 cjstefano@mtu.edu, Michigan Technological Univ; GzxCg
Stefanova, Ivanka, 651-646-0665 stefa014@umn.edu, Univ of Minnesota, Twin Cities; Pl
Steffens, Katja, 089/2180 6533 katja.steffens@lmu.de, Ludwig-Maximilians-Univ Muenchen; Gg
Steger, John M., 305-237-2609 jsteger@mdc.edu, Miami Dade Coll (Kendall Campus); OgAmZg
Stegman, Dave, (858) 822-0767 dstegman@ucsd.edu, Univ of California, San Diego; Yd
Steidl, Gregg M., (609) 984-6587 gregg.steidl@dep.state.nj.us, New Jersey Geological and Water Survey; Ge
Steig, Eric J., (206) 685-3715 steig@uw.edu, Univ of Washington; CsGlAs
Steiger, Scott, (315) 312-2802 steiger@oswego.edu, SUNY, Oswego; Am
Stein, Carol A., (312) 996-9349 cstein@uic.edu, Univ of Illinois Chicago; Yg
Stein, Seth, (847) 491-5265 seth@earth.northwestern.edu, Northwestern Univ; YsdYg
Steinberg, Laura, 617-552-3503 laura.steinberg@bc.edu, Boston Coll; Ng
Steinberg, Roger T., (361) 698-1665 rsteinb@delmar.edu, Del Mar Coll; GgoPg
Steinberger, Julia, +44 (0)785 607 9625 j.k.steinberger@leeds.ac.uk, Univ of Leeds; Zu
Steinen, Randolph P., (860) 933-2590 randolph.steinen@ct.gov, Dept of Energy and Env Protection; Gsg
Steiner, Jeffrey, (212) 650-6465 City Coll (CUNY); Cp
Steinker, Don C., (419) 372-7200 Bowling Green State Univ; Pg
Steinman, Byron A., (218) 726-7435 bsteinma@d.umn.edu, Univ of Minnesota Duluth; Gn
Steinschneider, Scott, 607-255-2155 ss3378@cornell.edu, Cornell Univ; Rw
Steller, Diana, 831-771-4400 dsteller@mlml.calstate.edu, Moss Landing Marine Lab; Zn
Stelling, Pete, pete.stelling@wwu.edu, Western Washington Univ; GxEgGv
Steltenpohl, Mark G., (334) 844-4893 steltmg@auburn.edu, Auburn Univ; Gtc
Stensrud, David J., (814) 863-7714 djs78@psu.edu, Pennsylvania State Univ, Univ Park; Am
Stephen, Daniel, (801) 863-8584 daniel.stephen@uvu.edu, Utah Valley Univ; PiGsPe
Stephen, Ralph A., (508) 289-2583 rstephen@whoi.edu, Woods Hole Oceanographic Inst; Ys
Stephens, Jason H., (512) 484-0874 jason.stephens@austincc.edu, Austin Comm Coll District; YrGuYe
Stephenson, Garth, (27) 21 808 4935 garth@sun.ac.za, Stellenbosch Univ; Zri
Stephenson, Gerry, gerry.stephenson@rogers.com, Univ of Guelph; Zn
Stephenson, Randell A., +44 (0)1224 274817 r.stephenson@abdn.ac.uk, Univ of Aberdeen; YgGtZg
Steppe, Cecily N., 410-293-6552 natunewi@usna.edu, United States Naval Acad; Obp
Sterling, Shannon, 902 494 7741 shannon.sterling@dal.ca, Dalhousie Univ; HgCbHs
Sterling, Tracy M., (406) 994-4605 tracy.sterling@montana.edu, Montana State Univ; Zn
Stermer, Ed, estermer@icc.cc.il.us, Illinois Central Coll; Zg
Stern, Charles R., 303-492-7170 charles.stern@colorado.edu, Univ of Colorado; Gi
Stern, Herschel I., hstern@miracosta.edu, MiraCosta Coll; Zy
Stern, Robert J., (972) 883-2442 rjstern@utdallas.edu, Univ of Texas, Dallas; Gt
Sternberg, Ben K., (520) 621-2439 bkslasi@u.arizona.edu, Univ of Arizona; Ye
Sternberg, Richard W., (206) 543-6487 rws@ocean.washington.edu, Univ of Washington; Ou
Sternberg, Robert S., (717) 291-4133 rob.sternberg@fandm.edu, Franklin

and Marshall Coll; YmGa
Sternberg, Rolf, (973) 655-7386 Montclair State Univ; Zn
Stetler, Larry D., (605) 394-2461 larry.stetler@sdsmt.edu, South Dakota Sch of Mines & Tech; NxGsm
Stevens, Calvin H., (408) 924-5029 calvin.stevens@sjsu.edu, San Jose State Univ; PsGrs
Stevens, De Anne S., (907) 451-5014 deanne.stevens@alaska.gov, Alaska Division of Geological & Geophysical Surveys; Ng
Stevens, Joan M., (805) 756-2261 jstevens@calpoly.edu, California Polytechnic State Univ; Zn
Stevens, Liane M., (936) 468-2024 stevenslm@sfasu.edu, Stephen F. Austin State Univ; Gpc
Stevens, Philip S., (812) 855-0732 pstevens@indiana.edu, Indiana Univ, Bloomington; As
Stevens, Robert G., (509) 786-9231 stevensr@wsu.edu, Washington State Univ; Sb
Stevens, Stan, 413-545-0773 sstevens@geo.umass.edu, Univ of Massachusetts, Amherst; Zy
Stevens (Landon), Lora R., 562-985-4817 lsteven2@csulb.edu, California State Univ, Long Beach; Gn
Stevens-Goddard, Andrea L., alsg@iu.edu, Indiana Univ, Bloomington; Cc
Stevenson, David, +44 (0) 131 650 6750 david.s.stevenson@ed.ac.uk, Edinburgh Univ; As
Stevenson, Ross, 514-987-3000 #7835 stevenson.ross@uqam.ca, Univ du Quebec a Montreal; Cg
Steward, Grieg F., 808-956-6775 grieg@hawaii.edu, Univ of Hawai'i, Manoa; Ob
Stewart, Brian W., (412) 624-8883 bstewart@pitt.edu, Univ of Pittsburgh; CclCq
Stewart, Dion C., 678-240-6227 dstewart29@gsu.edu, Georgia State Univ, Perimeter Coll, Alpharetta Campus; Gzi
Stewart, Eric, 2629403 eric.stewart@wisc.edu, Univ of Wisconsin, Madison, Division of Extension; GgcGi
Stewart, Esther K., (608) 263-3201 esther.stewart@wisc.edu, Univ of Wisconsin, Madison, Division of Extension; Gg
Stewart, Gary, 405-372-6063 Oklahoma State Univ; Go
Stewart, Gillian, 718-997-3104 gillian.stewart@qc.cuny.edu, Graduate Sch of the City Univ of New York; Oc
Stewart, Kevin G., (919) 962-0683 kgstewar@email.unc.edu, Univ of North Carolina, Chapel Hill; Gc
Stewart, Michael A., (217) 244-5025 stewart1@illinois.edu, Univ of Illinois, Urbana-Champaign; Gi
Stewart, Robert, 713-743-3081 rrstewart@uh.edu, Univ of Houston; Ye
Stewart, USA (ret), Alexander K., (315) 229-5087 astewart@stlawu.edu, St. Lawrence Univ; GmlHw
Steyn, Douw G., (604) 364-1266 dsteyn@eos.ubc.ca, Univ of British Columbia; Asm
Stickel, Robert, (404) 385-4413 robert.stickel@eas.gatech.edu, Georgia Inst of Tech; As
Stickney, Michael C., (406) 496-4332 mstickney@mtech.edu, Montana Tech of The Univ of Montana; YsGtm
Stickney, Robert R., (979) 845-3854 stickney@tamu.edu, Texas A&M Univ; Ob
Stidham, Christiane W., (631) 632-8059 christiane.stidham@stonybrook.edu, Stony Brook Univ; YgGe
Stieglitz, Marc, (404) 385-6530 marc.stieglitz@ce.gatech.edu, Georgia Inst of Tech; Hsg
Stieglitz, Ronald D., (920) 465-2371 stieglir@uwgb.edu, Univ of Wisconsin, Green Bay; GrlHy
Stierle, Andrea, (406) 496-4117 andrea.stierle@umontana.edu, Montana Tech of the Univ of Montana; Zn
Stierle, Donald, (406) 496-4117 donald.stierle@umontana.edu, Montana Tech of the Univ of Montana; Co
Stigall, Alycia L., (740) 593-0393 stigall@ohio.edu, Ohio Univ; Pg
Stillings, Lisa, (775) 784-5803 Univ of Nevada, Reno; Cm
Stimer, Debra, 330-244-3511 Kent State Univ at Stark; Zn
Stimpson, Ian G., (+44) 01782 733182 i.g.stimpson@keele.ac.uk, Keele Univ; YeuRc
Stine, Alexander, 415-338-1209 stine@sfsu.edu, San Francisco State Univ; At
Stinger, Lindsay C., +44(0) 113 34 37530 l.stringer@leeds.ac.uk, Univ of Leeds; Zn
Stinson, Amy L., 949-451-5622 astinson@ivc.edu, Irvine Valley Coll; Gc
Stix, John, 514-398-5391 john.stix@mcgill.ca, McGill Univ; Gv
Stixrude, Lars, lstixrude@epss.ucla.edu, Univ of California, Los Angeles; YgXy
Stock, Joann M., (626) 395-6938 jstock@gps.caltech.edu,CalTech; GtcYr
Stocker, Michelle, 540-231-8417 stockerm@vt.edu, Virginia Tech; Pv
Stockinger, Georg M., +49 89 289 25885 georg.stockinger@tum.de, Technical Univ of Munich; NrgNx
Stockli, Daniel, 512-475-6037 stockli@jsg.utexas.edu, Univ of Texas, Austin; CcGct
Stocks, Lee, (570) 662-4612 lstocks@mansfield.edu, Mansfield Univ; ZyGgm
Stockton, Charles W., (520) 621-7680 stockton@ltrr.arizona.edu, Univ of Arizona; Hq
Stoddard, Edward F., (919) 515-7939 skip_stoddard@ncsu.edu, North Carolina State Univ; Gp
Stoecker, Diane, (410) 221-8407 stoecker@hpl.umces.edu, Univ of Maryland; Ob
Stoelinga, Mark T., (206) 708-8588 mstoelinga@3tier.com, Univ of Washington; As
Stoessel, Achim, (979) 862-4170 astoessel@ocean.tamu.edu, Texas A&M Univ; OpAmGl
Stoessel, Marion, 979 845 7662 mstoessel@ocean.tamu.edu, Texas A&M Univ; Op
Stoessell, Ronald K., (504) 280-6795 rstoesse@uno.edu, Univ of New Orleans; Cl
Stofan, Ellen R., (818) 354-2076 ellen.r.stofan@jpl.nasa.gov, Jet Propulsion Lab; Xg
Stoffa, Paul L., (512) 471-6405 pauls@ig.utexas.edu, Univ of Texas, Austin; Ye
Stokes, Martin, +44 1752 584772 m.stokes@plymouth.ac.uk, Univ of Plymouth; Gm
Stolper, Edward M., (626) 395-6504 ems@gps.caltech.edu,CalTech; Cp
Stolzman, Kacie, (608) 800-6735 kstolzman@wisc.edu, Univ of Wisconsin, Madison, Division of Extension; Gm
Stone, Alan T., (410) 516-8476 astone@jhu.edu, Johns Hopkins Univ; ClSc
Stone, Alexander, alexm@umn.edu, Univ of Minnesota, Twin Cities; Gn
Stone, Glenn D., (314) 935-5239 stone@artsci.wustl.edu, Washington Univ in St. Louis; Zn
Stone, Jim, 970-339-6664 jim.stone@aims.edu, Aims Comm Coll; Zg
Stone, John O., 206-221-6332 stn@uw.edu, Univ of Washington; Cca
Stone, Loyd, (785) 532-6101 stoner@ksu.edu, Kansas State Univ; Sp
Stone, Peter H., phstone@mit.edu, Massachusetts Inst of Tech; As
Stoner, Betsy, (781) 891-7106 Bentley Univ; ObPe
Stoner, Joseph, 541-737-9002 jstoner@coas.oregonstate.edu, Oregon State Univ; Gsr
Storey, Craig D., +44 023 92 842245 craig.storey@port.ac.uk, Univ of Portsmouth; GptCc
Stork, Allen L., (970) 943-3044 astork@western.edu, Western Colorado Univ; Giv
Stormer, Jr., John C., (904) 456-7884 Rice Univ; Gi
Storrs, Glenn W., (513) 345-8500 Univ of Cincinnati; Pv
Stotler, Randy, (785) 864-6048 rstotler@ku.edu, Univ of Kansas; Hw
Stott, Lowell D., (213) 740-5120 stott@usc.edu, Univ of Southern California; Pm
Stout, James H., (612) 624-4344 jstout@umn.edu, Univ of Minnesota, Twin Cities; GpzGt
Stowell, Harold H., (205) 348-5098 hstowell@ua.edu, Univ of Alabama; GpCcGt
Stowell Gale, Julia, 512-232-7957 julia.gale@beg.utexas.edu, Univ of Texas at Austin, Jackson Sch of Geosciences; Gc
Stoykova, Kristalina C., +35929792213 stoykova@geology.bas.bg, Bulgarian Acad of Scis; PemPi
Strachan, Rob, +44 023 92 842279 rob.strachan@port.ac.uk, Univ of Portsmouth; Cc
Stracher, Glenn B., (478) 289-2073 stracher@ega.edu, East Georgia State Coll; GycEc
Strack, Otto D., (612) 625-3009 strac001@tc.umn.edu, Univ of Minnesota, Twin Cities; Hw
Straffin, Eric, (814) 732-1574 estraffin@edinboro.edu, Edinboro Univ of Pennsylvania; GmsSo
Strahler, Alan, (617) 353-5984 Boston Univ; Zr
Straight, William, (703) 948-7750 wstraight@nvcc.edu, Northern Virginia Comm Coll - Loudoun Campus; GgPv
Stramski, Dariusz, (858) 534-3353 dstramski@ucsd.edu, Univ of California, San Diego; Op
Strand, Line T., line.strand@nmbu.no, Norwegian Univ of Life Scis (NMBU); Sfd
Strangeway, Robert J., (310) 206-6247 strange@igpp.ucla.edu, Univ of California, Los Angeles; Zn
Strasser, Jeffrey C., (309) 794-7218 jeffreystrasser@augustana.edu, Augustana Coll; GmlGe
Strasser, Stefan, 089/2180 4340 stefan.strasser@vr-web.de, Ludwig-Maximilians-Univ Muenchen; Gz
Straub, David, (514) 398-8995 david.straub@mcgill.ca, McGill Univ; Op
Straub, Derek J., 570-372-4767 straubd@susqu.edu, Susquehanna Univ; As

Straub, Katherine H., (570) 372-4318 straubk@susqu.edu, Susquehanna Univ; Am
Straub, Kyle M., 504-862-3273 kmstraub@tulane.edu, Tulane Univ; GgCg
Straub, Susanne, smstraub@ldeo.columbia.edu, Columbia Univ; Gv
Straus, David M., (703) 993-5719 dstraus@gmu.edu, George Mason Univ; As
Strauss, Harald, +49-251-83-33932 hstrauss@uni-muenster.de, Univ Muenster; CsGg
Strauss, Justin V., (603) 646-6954 justin.v.strauss@dartmouth.edu, Dartmouth Coll; GsCgGt
Straw, Byron, (970) 351-2470 byron.straw@unco.edu, Univ of Northern Colorado; ZeGgl
Strawn, Daniel G., (208) 885-2713 dgstrawn@uidaho.edu, Univ of Idaho; Sc
Strayer, Luther M., (510) 885-3083 luther.strayer@csueastbay.edu, California State Univ, East Bay; Gc
Streby, Henry, (419) 530-8451 henry.streby@utoledo.edu, Univ of Toledo; Zn
Streck, Martin J., (503) 725-3379 streckm@pdx.edu, Portland State Univ; GivCt
Streepey Smith, Meg, (765) 973-2168 streeme@earlham.edu, Earlham Coll; Gt
Street, Ronald L., bhrstreet@yahoo.com, Univ of Kentucky; Ys
Streig, Ashley, 503-725-3371 streig@pdx.edu, Portland State Univ; YsGt
Strick, James E., (717) 291-3856 james.strick@fandm.edu, Franklin and Marshall Coll; RhGe
Strickland, Richard M., (206) 543-3131 strix@ocean.washington.edu, Univ of Washington; ObgZg
Stright, Lisa, (970) 491-4296 lisa.stright@colostate.edu, Colorado State Univ; GoNpYe
Strmiæ Palinkaš, Sabina, +38514605961 sabina.strmic@inet.hr, Univ of Zagreb; CgeCc
Strobel, Darrell F., (410) 516-7829 strobel@jhu.edu, Johns Hopkins Univ; AsZnn
Strobel, Timothy A., 202-478-8943 tstrobel@carnegiescience.edu, Carnegie Inst for Sci; Zm
Strock, Jeffrey S., (507) 752-7372 stroc001@umn.edu, Univ of Minnesota, Twin Cities; So
Stroeve, Julienne, j.stroeve@ucl.ac.uk, Univ Coll London; GlZr
Strom, Kyle, (540) 231-0979 strom@vt.edu, Virginia Tech; HsGs
Strom, Robert G., 520- 621-2720 rstrom@lpl.arizona.edu, Univ of Arizona; Xg
Stromberg, Caroline, 206-543-1687 caestrom@uw.edu, Univ of Washington; Pbe
Strong, Courtenay, 801-585-0049 court.strong@utah.edu, Univ of Utah; As
Strong, Ellen, stronge@si.edu, Smithsonian Inst / Natl Mus of Natural History; ZnPi
Strosnider, William, 843-904-9031 bill@baruch.sc.edu, Univ of South Carolina; On
Strother, Paul K., (617) 552-8395 strother@bc.edu, Boston Coll; PlbPg
Stroup, Justin, justin.stroup@oswego.edu, SUNY, Oswego; GmHwGe
Stuart, Graham, +44(0) 113 34 35217 g.w.stuart@leeds.ac.uk, Univ of Leeds; Ys
Stuart, Neil, +44 (0) 131 650 2549 n.stuart@ed.ac.uk, Edinburgh Univ; Zi
Stubbins, Aron, 617.373.5872 a.stubbins@northeastern.edu, Northeastern Univ; CgbZc
Stubler, Craig, (805) 756-2188 cstubler@calpoly.edu, California Polytechnic State Univ; So
Stucker, James D., (614) 265-6601 james.stucker@dnr.state.oh.us, Ohio Dept of Nat Res; EgCgGl
Stucki, Joseph W., (217) 333-9636 Univ of Illinois, Urbana-Champaign; Sc
Stucky de Quay, Gaia, gaias@mit.edu, Massachusetts Inst of Tech; ZgGms
Student, James J., 989-774-2295 stude1jj@cmich.edu, Central Michigan Univ; CaGiz
Stuiver, Minze, minze@uw.edu, Univ of Washington; Cc
Stull, Roland B., 604-822-5901 rstull@eos.ubc.ca, Univ of British Columbia; As
Stumbea, Dan, 0040232201464 dan.stumbea@uaic.ro, Alexandru Ioan Cuza; Gez
Stump, Brian W., (214) 768-1223 stump@smu.edu, Southern Methodist Univ; Ys
Stump, Edmund, ed.stump@asu.edu, Arizona State Univ; Gt
Stunz, Greg, 361-825-3254 greg.stunz@tamucc.edu, Texas A&M Univ, Corpus Christi; Zn
Stupazzini, Marco, 089/2180 4143 stupa@geophysik.uni-muenchen.de, Ludwig-Maximilians-Univ Muenchen; Yg
Sturchio, Neil C., (630) 252-3986 Argonne; Cg
Sturges, Bill, +44 (0)1603 59 2018 w.sturges@uea.ac.uk, Univ of East Anglia; As
Sturgis, Laila, 575-835-5327 laila.sturgis@nmt.edu, New Mexico Tech; Hw
Sturm, Diana, 251-460-6381 dsturm@southalabama.edu, Univ of South Alabama; GeZge
Sturmer, Daniel M., (513) 556-3718 daniel.sturmer@uc.edu, Univ of Cincinnati; YeGdo
Stute, Martin, (845) 365-8704 martins@ldeo.columbia.edu, Columbia Univ; CsHgGe
Stutenbecker, Laura, +49-251-83-33935 laura.stutenbecker@uni-muenster.de, Univ Muenster; GsgCc
Stutz, Jochen P., (310) 825-1217 jochen@atmos.ucla.edu, Univ of California, Los Angeles; As
Styers, Diane M., 828-227-3819 dmstyers@wcu.edu, Western Carolina Univ; Zrn
Su, Haibin, 361-593-3590 haibin.su@tamuk.edu, Texas A&M Univ, Kingsville; Zir
Su, Xiaobo, (541) 346-4568 xiaobo@uoregon.edu, Univ of Oregon; Zn
Suarez, Celina, (479) 575-4866 casuarez@uark.edu, Univ of Arkansas, Fayetteville; ClPqGg
Suárez, Luis Eugenio, luisesua@ucm.es, Univ Compl de Madrid; Zn
Subramaniam, Ajit, ajit@ldeo.columbia.edu, Columbia Univ; ObZr
Subulwa, Angela G., (920) 424-4105 subulwaa@uwosh.edu, Univ of Wisconsin Oshkosh; Zn
Suen, C. J., (559) 278-8656 john_suen@csufresno.edu, Fresno State Univ; HwCsGe
Sugarman, Peter J., (609) 292-2576 pete.sugarman@dep.state.nj.us, New Jersey Geological and Water Survey; Gr
Sugden, John , john.sugden@cwu.edu, Central Washington Univ; Geg
Sugihara, George, (858) 534-5582 gsugihara@ucsd.edu, Univ of California, San Diego; Ob
Sui, Daniel, 479-575-2470 Univ of Arkansas, Fayetteville; Zf
Sukop, Michael C., (305) 348-3117 sukopm@fiu.edu, Florida InterNatl Univ; HwqSp
Sulanowska, Margaret, (508) 289-2306 msulanowska@whoi.edu, Woods Hole Oceanographic Inst; Cm
Sullivan, Donald G., donald.sullivan@du.edu, Univ of Denver; Pc
Sullivan, Jack B., (301) 405-0106 jsull@umd.edu, Univ of Maryland; Zu
Sullivan, Raymond, (415) 338-2061 sullivan@sfsu.edu, San Francisco State Univ; Go
Sullivan, Walter, (207) 859-5800 wasulliv@colby.edu, Colby Coll; Gct
Sullivan-Watts, Barbara K., (401) 874-6659 bsull@gso.uri.edu, Univ of Rhode Island; Og
Sultan, Mohamed, (630) 252-1929 Argonne; Cg
Sumida, Stuart S., (909) 537-5346 California State Univ, San Bernardino; Pv
Summers, Robert, 780-492-0342 robert.summers@ualberta.ca, Univ of Alberta; Zn
Summers, Sara, (801) 626-7222 sarasummers@weber.edu, Weber State Univ; Gg
Summons, Roger, (617) 452-2791 rsummons@mit.edu, Massachusetts Inst of Tech; CobCs
Sumner, Dawn Y., (530) 752-5353 dysumner@ucdavis.edu, Univ of California, Davis; Gs
Sumner, Esther, +44 (0)23 80592067 e.j.sumner@soton.ac.uk, Univ of Southampton; Gs
Sumrall, Colin, (865) 974-2366 csumrall@utk.edu, Univ of Tennessee, Knoxville; Pi
Sumrall, Jeanne L., jlsumrall@fhsu.edu, Fort Hays State Univ; Zgn
Sumrall, Jonathan B., 785-628-5348 jbsumrall@fhsu.edu, Fort Hays State Univ; GgxZn
Sun, Chenguang, csun@jsg.utexas.edu, Univ of Texas, Austin; Gx
Sun, Fengpeng, 816-235-2973 sunf@umkc.edu, Univ of Missouri, Kansas City; ZoiZc
Sun, Hongbing, 609-895-5185 hsun@rider.edu, Rider Univ; HwCgSc
Sun, Jeff, 307 268 3560 jsun@caspercollege.edu, Casper Coll; Zi
Sun, Wen-Yih, wysun@purdue.edu, Purdue Univ; AsZoAm
Sun, Yifei, (818) 677-3532 yifei.sun@csun.edu, California State Univ, Northridge; Zi
Sun, Yuefeng, (979) 845-0635 sun@geos.tamu.edu, Texas A&M Univ; GoYe
Sundareshwar, P. V., 605-394-2492 psundareshwar@usaid.gov, South Dakota Sch of Mines & Tech; ZeCgSb
Sundby, Bjorn, (418) 723-1986 (Ext. 1767) b.sundby@uquebec.ca, Univ du Quebec a Rimouski; Oc
Sundell, Ander, 208.562.3354 andersundell@cwidaho.cc, Coll of Western Idaho; Ggc
Sundell, Kent A., (307) 268-2498 ksundell@caspercollege.edu, Casper Coll; GgvPv
Sunderlin, David, 610-330-5198 sunderld@lafayette.edu, Lafayette Coll; GgPg
Suntharalingam, Parvadha, +44 (0)1603 59 1423 p.suntharalingam@uea.ac.uk, Univ of East Anglia; Oc
Superchi-Culver, Tonia, 303-492-5211 toni.culver@colorado.edu, Univ of

Colorado; Pg
Suppe, John, jsuppe@uh.edu, Univ of Houston; Gtc
Surge, Donna M., 919-843-1994 donna64@unc.edu, Univ of North Carolina, Chapel Hill; Pe
Surian, Nicola, +390498279125 nicola.surian@unipd.it, Univ degli Studi di Padova; Gm
Suriano, Zac, 402-554-2662 zsuriano@unomaha.edu, Univ of Nebraska, Omaha; Atm
Surpless, Benjamin E., (210) 999-7110 bsurples@trinity.edu, Trinity Univ; GctGi
Surpless, Kathleen D., (210) 999-7365 ksurples@trinity.edu, Trinity Univ; Gst
Susak, Nicholas J., (506) 453-4803 nsusak@unb.ca, Univ of New Brunswick; CglEm
Sushama, Laxmi, 514-987-3000 #2414 sushama.laxmi@uqam.ca, Univ du Quebec a Montreal; Hy
Sussman, Robert W., (314) 935-5264 rwsussma@artsci.wustl.edu, Washington Univ in St. Louis; Zn
Suszek, Thomas J., suszek@uwosh.edu, Univ of Wisconsin, Oshkosh; Gg
Suter, Simeon, 717-702-2047 ssuter@pa.gov, DCNR- Pennsylvania Bur of Geo Survey; HwGg
Sutherland, Bruce, 780-492-0573 bruce.sutherland@ualberta.ca, Univ of Alberta; Cm
Sutherland, Dave, 541-346-8753 dsuth@uoregon.edu, Univ of Oregon; Op
Sutherland, Mary K., (406) 496-4410 msutherland@mtech.edu, Montana Tech of The Univ of Montana; Hws
Sutherland, Stuart, (604) 822-0176 ssutherland@eos.ubc.ca, Univ of British Columbia; Pg
Suttle, Curtis, (604) 822-8610 csuttle@eos.ubc.ca, Univ of British Columbia; Ob
Suttner, Lee J., (812) 855-4957 suttner@indiana.edu, Indiana Univ, Bloomington; Gd
Sutton, Mark, +44 20 759 47487 m.sutton@imperial.ac.uk, Imperial Coll; Pg
Sutton, Paul C., (303) 871-2399 psutton@du.edu, Univ of Denver; Zc
Sutton, Sally, sally.sutton@colostate.edu, Colorado State Univ; CgGd
Suvocarev, Kosana, 530-754-1929 ksuvocarev@ucdavis.edu, Univ of California, Davis; Asm
Suyker, Andrew E., asuyker@unl.edu, Univ of Nebraska, Lincoln; As
Sverdrup, Keith A., (414) 229-4017 sverdrup@uwm.edu, Univ of Wisconsin, Milwaukee; YsGt
Sverjensky, Dimitri A., (410) 516-8568 sver@jhu.edu, Johns Hopkins Univ; Cl
Svitra, Zita V., (505) 667-7616 Los Alamos; Zn
Swain, Geoffrey, (321) 674-8096 swain@fit.edu, Florida Inst of Tech; Ob
Swanger, Kate, (978) 934-2664 kate_swanger@uml.edu, Univ of Massachusetts, Lowell; Gl
Swann, Abigail L., (206) 616-0486 aswann@atmos.washington.edu, Univ of Washington; AsZn
Swanson, Basil I., (505) 667-5814 Los Alamos; Zn
Swanson, Benjamin, benjamin-swanson@uiowa.edu, Univ of Iowa; Gme
Swanson, Brian, 250-537-4330 brians@uw.edu, Univ of Washington; Asp
Swanson, Donald A., (808) 967-8863 donswan@usgs.gov, Univ of Hawai'i, Manoa; Gv
Swanson, Karen, (201) 595-2589 William Paterson Univ; Cc
Swanson, R. L., (631) 632-8704 larry.swanson@stonybrook.edu, SUNY, Stony Brook; Og
Swanson , Sam , sswanson@uga.edu, Univ of Georgia; Gi
Swanson, Sherman, (775) 784-4057 sswanson@agntl.ag.unr.edu, Univ of Nevada, Reno; Hs
Swanson, Susan K., (608) 363-2132 swansons@beloit.edu, Beloit Coll; HwGmZi
Swanson, Terry W., tswanson@uw.edu, Univ of Washington; CcGe
Swanson, Travis, (912) 478-0337 tswanson@georgiasouthern.edu, Georgia Southern Univ; Gsm
Swapp, Susan M., (307) 766-2513 swapp@uwyo.edu, Univ of Wyoming; Gp
Swarr, Gretchen, (508) 289-2558 gswarr@whoi.edu, Woods Hole Oceanographic Inst; Oc
Swart, Peter K., (305) 421-4103 pswart@rsmas.miami.edu, Univ of Miami; ClPsCs
Swaters, Gordon E., (780) 492-7159 gordon.swaters@ualberta.ca, Univ of Alberta; Op
Sweeney, Mark D., (509) 373-0703 mark.sweeney@pnl.gov, Pacific Northwest; Yg
Sweet, Alisan C., 806-834-2398 alisan.sweet@ttu.edu, Texas Tech Univ; Gs
Sweet, Dustin E., (806) 834-8390 dustin.sweet@ttu.edu, Texas Tech Univ; GsdGr
Sweetman, Steve, +44 023 9284 2257 steve.sweetman@port.ac.uk, Univ of Portsmouth; Pg
Swennen, Rudy, rudy.swennen@ees.kuleuven.be, Katholieke Univit Leuven; Gso
Swenson, John B., (218) 726-6844 jswenso2@d.umn.edu, Univ of Minnesota Duluth; GrsGm
Swetnam, Thomas W., tswetnam@email.arizona.edu, Univ of Arizona; Zn
Swidinsky, Andrei, andrei.swidinsky@utoronto.ca, Colorado Sch of Mines; Ye
Swift, Elijah V., (401) 874-6146 lige@gso.uri.edu, Univ of Rhode Island; Ob
Swift, Robert P., (505) 665-7871 bswift@lanl.gov, Los Alamos; Nr
Swift, Stephen A., (508) 289-2626 sswift@whoi.edu, Woods Hole Oceanographic Inst; Yr
Swindle, Timothy, 520-621-4128 tswindle@lpl.arizona.edu, Univ of Arizona; Xc
Swinehart, James B., (314) 968-6340 sdworm@fastmail.fm, Univ of Nebraska, Lincoln; GrsGe
Swisher III, Carl C., 848-445-5363 cswish@eps.rutgers.edu, Rutgers, The State Univ of New Jersey; Cc
Swyrtek, Sheila, (810)232-9312 sheila.swyrtek@mcc.edu, Charles Stewart Mott Comm Coll; Ga
Sydora, Richard D., (780) 492-3624 rsydora@phys.ualberta.ca, Univ of Alberta; Xy
Sykes, Lynn R., (845) 359-7428 sykes@ldeo.columbia.edu, Columbia Univ; YsGtZn
Sylva, Sean, 508-289-3546 ssylva@whoi.edu, Woods Hole Oceanographic Inst; Oc
Sylvan, Jason B., (979) 845-5105 jasonsylvan@tamu.edu, Texas A&M Univ; ObXb
Sylvia, Elizabeth R., 410 554-5542 elizabeth.sylvia@maryland.gov, Maryland Dept of Nat Res; Ge
Symbalisty, E.M.D., (505) 667-9670 esymbalisty@lanl.gov, Los Alamos; Zn
Symons, David T., 519-253-3000 x2493 dsymons@uwindsor.ca, Univ of Windsor; YmGtEm
Syrett, William, (814) 865-6172 wjs1@psu.edu, Pennsylvania State Univ, Univ Park; Am
Syrup, Krista A., (708) 974-5615 syrup@morainevalley.edu, Moraine Valley Comm Coll; Cs
Syverson, Kent M., (715) 836-3676 syverskm@uwec.edu, Univ of Wisconsin, Eau Claire; GlEnGe
Szatmári, József, +3662544156 szatmari@geo.u-szeged.hu, Univesity of Szeged; ZriZy
Szczepanski, Jacek, jacek.szczepanski@uwr.edu.pl, Univ of Wroclaw; Ggc
Szecsody, James E., (509) 372-6080 jim.szecsody@pnl.gov, Pacific Northwest; So
Székely, Balázs, balazs.szekely@ttk.elte.hu, Eotvos Lorand Univ; GmZri
Szeliga, Walter, 509-963-2705 walter@geology.cwu.edu, Central Washington Univ; YgsYd
Szente, Istvan, szente@ludens.elte.hu, Eotvos Lorand Univ; Pig
Szeto, Anthony M. K., (416)736-2100 #77703 szeto@yorku.ca, York Univ; Yg
Szilagyi, Jozsef, (402)472-9667 jszilagyi1@unl.edu, Univ of Nebraska, Lincoln; Hw
Szilassi, Péter, +3662544156 toto@geo.u-szeged.hu, Univesity of Szeged; Zuy
Szlavecz, Katalin, (410) 516-8947 szlavecz@jhu.edu, Johns Hopkins Univ; Pi
Szpakowski, David, szpakowskid@mail.wou.edu, Western Oregon Univ; ZirRn
Szulczewski, Melanie, (540) 654-2345 mszulcze@umw.edu, Univ of Mary Washington; ScClZg
Szunyogh, Istvan, (979) 458-0553 szunyogh@tamu.edu, Texas A&M Univ; As
Szymanski, David, (781) 891-2901 dszymanski@bentley.edu, Bentley Univ; GifCg
Szymanski, Jason, 585-292-2423 jszymanski@monroecc.edu, Monroe Comm Coll; GlPe
Szynkiewicz, Anna, 865-974-6006 aszynkie@utk.edu, Univ of Tennessee, Knoxville; CslXg
Söllner, Frank, 089/2180 6519 fank.soellner@iaag.geo.uni-muenchen.de, Ludwig-Maximilians-Univ Muenchen; Gg

T

Tabara, Daniel, 0040756105973 dan.tabara@yahoo.com, Alexandru Ioan Cuza; Pl
Tabidian, M. Ali, (818) 677-2536 ali.tabidian@csun.edu, California State Univ, Northridge; Hw
Taboada Castro, María T., 00 34 981 167000 teresat@udc.es, Coruna Univ; Sd
Tabor, Clay, clay.tabor@uconn.edu, Univ of Connecticut; PcAt
Tabor, Neil J., ntabor@smu.edu, Southern Methodist Univ; SdGs
Tabrum, Alan R., (412) 622-3265 tabruma@carnegiemnh.org, Carnegie Mus

of Natural History; Pv
Tacinelli, John C., (507) 285-7501 john.tacinelli@rctc.edu, Rochester Comm & Technical Coll; Gig
Tackett, Lydia S., 701-231-6164 lydia.tackett@ndsu.edu, North Dakota State Univ; PeGsPi
Tadesse, Tsegaye, ttadesse2@unl.edu, Univ of Nebraska, Lincoln; ZirAs
Taggart, Christopher T., (902) 494-7144 chris.taggart@dal.ca, Dalhousie Univ; Ob
Taggart, Ralph E., 517-353-5175 taggart@msu.edu, Michigan State Univ; Pb
Tagliabue, Alessandro, +44 0151 794 4651 a.tagliabue@liverpool.ac.uk, Univ of Liverpool; Oc
Taib, Samsudin Hj., 03-79674235 samsudin@um.edu.my, Univ of Malaya; Yg
Taillefert, Martial, (404) 894-6043 mtaillef@eas.gatech.edu, Georgia Inst of Tech; ClmCa
Tait, C. Drew, (505) 667-7603 Los Alamos; Cp
Tait, Kim, ktait@rom.on.ca, Western Univ; Xm
Tait, Kimberly T., (416) 586-5820 ktait@rom.on.ca, Royal Ontario Mus; GzyXm
Tajik, Atieh, 404 413 5790 geoatt@langate.gsu.edu, Georgia State Univ; Gg
Takahashi, Taro, taka@ldeo.columbia.edu, Columbia Univ; Cm
Takeuchi, Akira, 81-76-445-6654 takeuchi@sci.u-toyama.ac.jp, Univ of Toyama; GtcYe
Takle, Eugene S., (515) 294-9871 gstakle@iastate.edu, Iowa State Univ of Sci & Tech; Asm
Talbot, Helen, +44 (0) 191 208 6426 helen.talbot@ncl.ac.uk, Univ of Newcastle Upon Tyne; Co
Talbot, James L., (360) 733-4282 talbot@wwu.edu, Western Washington Univ; Gc
Talbot, Robert, 713-743-2725 rtalbot@uh.edu, Univ of Houston; ActAm
Talley, Drew, (619) 260-6810 dtalley@sandiego.edu, Univ of San Diego; ObZe
Talley, John H., 302-831-2833 waterman@udel.edu, Univ of Delaware; As
Talley, Lynne D., (858) 534-6610 ltalley@ucsd.edu, Univ of California, San Diego; OpZc
Talling, Peter, peter.j.talling@durham.ac.uk, Durham Univ; Rn
Talwani, Manik, (713) 348-6067 manik@rice.edu, Rice Univ; Yr
Talwani, Pradeep, talwani@geol.sc.edu, Univ of South Carolina; Ys
Talyor, Frederick W., fred@ig.utexas.edu, Univ of Texas, Austin; Pe
Tamas, Calin Gbriel, calin.tamas@ubbcluj.ro, Babes-Bolyai Univ; EdGzNm
Tamas, Tudor Liviu, tudor.tamas@ubbcluj.ro, Babes-Bolyai Univ; GzZmn
Tamborski, Joseph, (757) 683-6298 jtambors@odu.edu, Old Dominion Univ; Hy
Tamish, Mohamed M., 002-03-3921595 mtamish@hotmail.com, Alexandria Univ; GgsCg
Tan, Chunyang, tanc@umn.edu, Univ of Minnesota, Twin Cities; Cm
Tan, Denis N., denistan@um.edu.my, Univ of Malaya; Go
Tanamachi, Robin, 765-496-2866 rtanamachi@purdue.edu, Purdue Univ; Asm
Taney, R. Marieke, (928) 523-2384 marieke.taney@nau.edu, Northern Arizona Univ; ZnnZn
Tang, Weiyi, 609-258-4101 weiyit@princeton.edu, Princeton Univ; CbOcb
Tanhua, Toste, ttanhua@geomar.de, Dalhousie Univ; Oc
Tanimoto, Toshiro, (805) 893-8375 toshiro@geol.ucsb.edu, Univ of California, Santa Barbara; Ys
Tantau, Ioan, ioan.tantau@ubbcluj.ro, Babes-Bolyai Univ; PlePc
Tao, Ran, rtao@usf.edu, Univ of South Florida; Zi
Tapanila, Leif, 208-282-5417 tapaleif@isu.edu, Idaho State Univ; Pig
Tape, Carl, 907-474-5456 carltape@gi.alaska.edu, Univ of Alaska, Fairbanks; Ys
Tapp, J. B., jbt@utulsa.edu, Univ of Tulsa; GceZi
Tappa, Eric, 803-777-7473 tappa@geol.sc.edu, Univ of South Carolina; Oo
Tarduno, John A., john.tarduno@rochester.edu, Univ of Rochester; YmGtXm
Tarhan, Lidya, 203-432-8100 lidya.tarhan@yale.edu, Yale Univ; PiGsPe
Tarka, Robert, robert.tarka@uwr.edu.pl, Univ of Wroclaw; Hg
Tarr, Alexander, 508 929-8474 alexander.tarr@worcester.edu, Worcester State Univ; Zg
Taskey, Ronald D., (805) 756-1160 rtaskey@calpoly.edu, California Polytechnic State Univ; Sf
Tassier-Surine, Stephanie, stephanie.surine@dnr.iowa.gov, Iowa Dept of Nat Res; Gg
Tatarskii, Viatcheslav, (404) 894-9224 vvt@eas.gatech.edu, Georgia Inst of Tech; As
Tatham, Robert H., (512) 471-9129 tatham@mail.utexas.edu, Univ of Texas, Austin; Yg
Tauxe, John D., (423) 574-5348 Oak Ridge; Hw
Tauxe, Lisa, (858) 534-6084 ltauxe@ucsd.edu, Univ of California, San Diego; Ym
Tavarnelli, Enrico, 0577 233984 enrico.tavarnelli@unisi.it, Univ of Siena; GgcGt
Tavener, Kristi, (807) 343-8010 Ext. 8677 ktavener@lakeheadu.ca, Lakehead Univ; Zn
Tawabini, Bassam S., +96638607643 bassamst@kfupm.edu.sa, King Fahd Univ of Petroleum and Minerals; GeHsCa
Taylor, Brian, (808) 956-6182 taylorb@hawaii.edu, Univ of Hawai'i, Manoa; Gtu
Taylor, Carolyn, carolyn.taylor@mcmail.maricopa.edu, Mesa Comm Coll; Gg
Taylor, Charles J., 859-323-0523 charles.taylor@uky.edu, Univ of Kentucky; HwgCl
Taylor, Danny L., (775) 784-6922 dtaylor@mines.unr.edu, Univ of Nevada, Reno; Nm
Taylor, Edith, 785-864-3621 etaylor@ku.edu, Univ of Kansas; Pb
Taylor, Elwynn, (808) 956-3899 setaylor@iastate.edu, Iowa State Univ of Sci & Tech; As
Taylor, Frank J. R., (604) 822-4587 mtaylor@eos.ubc.ca, Univ of British Columbia; Ob
Taylor, Frederick W., (512) 471-0453 fred@ig.utexas.edu, Univ of Texas, Austin; GtmGe
Taylor, G. Jeffrey, (808) 956-3899 Univ of Hawai'i, Manoa; Xm
Taylor, Gordon T., (631) 632-8688 gordon.taylor@stonybrook.edu, SUNY, Stony Brook; Obc
Taylor, Graeme, +44 1752 584770 g.taylor@plymouth.ac.uk, Univ of Plymouth; Yg
Taylor, Hugh P., (626) 395-6116 hptaylor@gps.caltech.edu,CalTech; Cs
Taylor, John F., (724) 357-4469 jftaylor@iup.edu, Indiana Univ of Pennsylvania; PsiGs
Taylor, Kenneth B., (919) 707-9211 kenneth.b.taylor@ncdenr.gov, N.C. Dept of Env Quality; YsGgEo
Taylor, Kevin, +44 0161 275-8557 kevin.taylor@manchester.ac.uk, Univ of Manchester; Go
Taylor, Lansing, 801-581-8430 ltaylor@egi.utah.edu, Univ of Utah; Gc
Taylor, Lawrence D., (517) 629-0308 ltaylor@albion.edu, Albion Coll; Gl
Taylor, Matthew, 303-871-2656 mtaylor7@du.edu, Univ of Denver; Zc
Taylor, Michael, michael.taylor@flinders.edu.au, Flinders Univ; Zn
Taylor, Nathan H., 501 683-1085 nathan.taylor@arkansas.gov, Arkansas Geo Survey; ZiyGe
Taylor, Penny M., 413-538-3236 pmtaylor@mtholyoke.edu, Mount Holyoke Coll; Gg
Taylor, Peter, +44(0) 113 34 37169 p.g.taylor@leeds.ac.uk, Univ of Leeds; Ge
Taylor, Rex N., +44 (0)23 80592007 rex@noc.soton.ac.uk, Univ of Southampton; GvCa
Taylor, Richard P., richard_taylor@carleton.ca, Carleton Univ; Cg
Taylor, Robert W., (973) 655-4129 Montclair State Univ; Zn
Taylor, Ryan W., 914 251 6652 ryan.taylor@purchase.edu, SUNY, Purchase; ZiyZg
Taylor, Stephen B., 503-838-8398 taylors@wou.edu, Western Oregon Univ; Gm
Taylor, Steven R., (505) 667-1007 taylor@lanl.gov, Los Alamos; Ys
Taylor, Ta-Shana A., (305) 284-4254 t.taylor2@miami.edu, Univ of Miami; ZePvGe
Taylor, Wayne A., (505) 667-4253 Los Alamos; Cl
Tchakerian, Vatche P., (409) 845-7997 Texas A&M Univ; Gm
Teagle, Damon A., +44 (0)23 80592723 damon.teagle@southampton.ac.uk, Univ of Southampton; CgEmGp
Teague, Richard , rteague@mit.edu, Massachusetts Inst of Tech; XsZn
Teasdale, Rachel, 530-898-5547 rteasdale@csuchico.edu, California State Univ, Chico; Gv
Tedford, Richard H., (212) 769-5809 American Mus of Natural History; Pv
Teed, Rebecca, 937 775-3446 rebecca.teed@wright.edu, Wright State Univ; ZePeGn
Teeuw, Richard, +44 023 92 842267 richard.teeuw@port.ac.uk, Univ of Portsmouth; AsGmZi
Tefend, Karen S., ktefend@westga.edu, Univ of West Georgia; CgGe
Teixell, Antonio, antonio.teixell@uab.cat, Univ Auto de Barcelona; Gtc
Teixell Cácharo, Antoni, ++935811163 antonio.teixell@uab.cat, Univ Auto de Barcelona; GtcYx
Teklay, Mengist, mteklay@gsu.edu, Georgia State Univ, Perimeter Coll, Decatur Campus; Gg
Tellam, John H., +44 (0)121 41 46138 j.h.tellam@bham.ac.uk, Univ of Birmingham; Hw
Teller, James T., (204) 474-9270 tellerjt@ms.umanitoba.ca, Univ of Manitoba; Gs
Telmer, Kevin, ktelmer@uvic.ca, Univ of Victoria; Cl
Telus, Myriam, mtelus@ucsc.edu, Univ of California, Santa Cruz; Xc
Telyakovskiy, Aleksey S., 775.784.1364 alekseyt@unr.edu, Univ of Nevada,

Reno; Hqy

Temples, Tommy, (803) 348-0472 ttemples@sc.rr.com, Clemson Univ; GoYg

Templeton, Alan R., (314) 935-6868 temple_a@wustl.edu, Washington Univ in St. Louis; ZicZr

Templeton, Alexis, 303-735-6069 alexis.templeton@colorado.edu, Univ of Colorado; ClZaCq

Templeton, Jeffrey H., (503) 838-8858 templej@wou.edu, Western Oregon Univ; GivZe

Tems, Caitlin E., (801) 626-7421 caitlintems@weber.edu, Weber State Univ; OoCs

Ten Brink, Norman W., tenbrinn@gvsu.edu, Grand Valley State Univ; Gm

Tencate, James, (505) 665-6667 tencate@lanl.gov, Los Alamos; Zn

Teng, Fangzhen, 206-543-7615 fteng@uw.edu, Univ of Washington; Cg

Teng, Ta-liang, (213) 740-5838 lteng@usc.edu, Univ of Southern California; YsGtYe

Tenorio, Victor O., (520) 621-3858 vtenorio@email.arizona.edu, Univ of Arizona; Nmx

Teodoriu, Catalin, 405-325-6872 cteodoriu@ou.edu, Univ of Oklahoma; Np

Tepley, III, Frank J., 541-737-2064 ftepley@coas.oregonstate.edu, Oregon State Univ; GiCs

Teplitski, Max, 352-392-1951 maxtep@ufl.edu, Univ of Florida; Zn

Tepper, Jeffrey H., (253) 879-3820 jtepper@pugetsound.edu, Univ of Puget Sound; GiCgGv

ter Voorde, Marlies, +31 20 59 87343 m.ter.voorde@vu.nl, Vrije Univit Amsterdam; YgGt

Ter-Simonian, Vardui, (213) 740-6106 tersimon@usc.edu, Univ of Southern California; Zn

Tera, Fouad, ftera@carnegiescience.edu, Carnegie Inst for Sci; Xm

Terry, Dennis O., (215) 204-8226 doterry@temple.edu, Temple Univ; SaPcGr

Tertyshnikov, Konstantin, +61 8 9266 2297 konstantin.tertyshnikov@curtin.edu.au, Curtin Univ; Ye

Terwey, Wes, terwey@southalabama.edu, Univ of South Alabama; Am

Teske, Johanna K., jteske@carnegiescience.edu, Carnegie Inst for Sci; Xg

Tesler, Theodore, 717-702-2017 thtesler@pa.gov, DCNR- Pennsylvania Bur of Geo Survey; HgRwZr

Tesso, Tesfaye, (785) 532-7238 ttesso@ksu.edu, Kansas State Univ; So

Tester, Jefferson W., 607-254-7211 jwt54@cornell.edu, Cornell Univ; NgCg

Tetrault, Denis, 519-253-3000 x3409 deniskt@uwindsor.ca, Univ of Windsor; Gg

Tett, Simon F., +44 (0) 131 650 5341 simon.tett@ed.ac.uk, Edinburgh Univ; Am

Tettenhorst, Rodney T., 614 247-4246 tettenhorst.2@osu.edu, Ohio State Univ; Gz

Tew , Berry H., 205-247-3679 ntew@gsa.state.al.us, Geo Survey of Alabama; GroGs

Tewksbury, Barbara J., (315) 859-4713 btewksbu@hamilton.edu, Hamilton Coll; Gc

Tewksbury-Christle, Carolyn, cmtewksburychristl@fortlewis.edu, Fort Lewis Coll; Gc

Textoris, Daniel A., (919) 962-0690 dtextori@email.unc.edu, Univ of North Carolina, Chapel Hill; Gd

Teyssier, Christian P., (612) 624-6801 teyssier@umn.edu, Univ of Minnesota, Twin Cities; GctGg

Tezcan, Levent, +90 (312) 2977750 tezcan@hacettepe.edu.tr, Hacettepe Univ; Hw

Thacker, Heather, (740) 351-3456 hthacker@shawnee.edu, Shawnee State Univ; Zn

Thacker, Jacob, 575-835-5151 jacob.thacker@nmt.edu, New Mexico Tech; GtcGr

Thackray, Glenn D., (208) 282-3565 thacglen@isu.edu, Idaho State Univ; Glm

Thakurta, Joyashish, (269) 387-3667 joyashish.thakurta@wmich.edu, Western Michigan Univ; GiEg

Thale, Paul R., 406-496-4653 pthale@mtech.edu, Montana Tech of The Univ of Montana; Zi

Thatje, Sven, +44 (0)23 80592009 svth@noc.soton.ac.uk, Univ of Southampton; Ob

Thayer, Paul A., (910) 520-8719 thayer@uncw.edu, Univ of North Carolina, Wilmington; GdoHg

Thayn, Jonathan B., 309-438-8112 jthayn@ilstu.edu, Illinois State Univ; ZirZf

Theiling, Bethany P., bethany-theiling@utulsa.edu, Univ of Tulsa; Cs

Theis, Karen, +44 0161 275-0407 karen.theis@manchester.ac.uk, Univ of Manchester; Zn

Theissen, Kevin, 651-962-5243 kmtheissen@stthomas.edu, Univ of Saint Thomas; GnOg

Them, II, Theodore R., themtr@cofc.edu, Coll of Charleston; CmOcZc

Themelis, Nickolas J., (212) 854-2138 njt1@columbia.edu, Columbia Univ; Nx

Thériault, Julie Mireille, 514 987-3000 #4276 theriault.julie@uqam.ca, Univ du Quebec a Montreal; Zn

Therrien, Pierre, therrien@ggl.ulaval.ca, Univ Laval; Gq

Therrien, Rene, (418) 656-5400 rene.therrien@ggl.ulaval.ca, Univ Laval; HwGe

Theuerkauf, Ethan, ejtheu@illinois.edu, Illinois State Geo Survey; GmuGs

Thibault, Yves, 613-992-1376 yves.thibault@nrcan.gc.ca, Western Univ; Gz

Thibodeau, Alyson M., (717) 245-8337 thibodea@dickinson.edu, Dickinson Coll; CcGaCs

Thiel, Dr., Volker, +49 (0)551 3914395 vthiel@gwdg.de, Georg-August Univ of Goettingen; CooCo

Thieme, Donald, (229) 333-5752 dmthieme@valdosta.edu, Valdosta State Univ; GaSdGm

Thien, Steve J., (785) 532-7207 sjthien@ksu.edu, Kansas State Univ; Sc

Thiessen, Eric, 709 864-2582 ethiessen@mun.ca, Memorial Univ of Newfoundland; Gc

Thigpen, Ryan, 859-2181532 ryan.thigpen@uky.edu, Univ of Kentucky; Gtc

Thirlwall, Matthew, +44 1784 443609 m.thirlwall@rhul.ac.uk, Royal Holloway Univ of London; Cs

Thiruvathukal, John V., (973) 655-4417 Montclair State Univ; Yg

Thole, Jeffrey T., (651) 696-6426 thole@macalester.edu, Macalester Coll; GgCaHw

Thomas, Amanda, amthomas@uoregon.edu, Univ of Oregon; Ys

Thomas, Andrew D., +44 (0)1970 622 781 ant23@aber.ac.uk, Univ of Wales; SbGmSf

Thomas, Debbie, (979) 845-3651 dthomas@tamu.edu, Texas A&M Univ; Ou

Thomas, Donald M., (808) 956-6482 dthomas@soest.hawaii.edu, Univ of Hawai'i, Manoa; CaHwRn

Thomas, Elizabeth K., (716)645-4329 ekthomas@buffalo.edu, SUNY, Buffalo; Cos

Thomas, Ellen, (860) 685-2238 ellen.thomas@yale.edu, Yale Univ; Pm

Thomas, Florence, (808) 236-7418 fithomas@hawaii.edu, Univ of Hawai'i, Manoa; Ob

Thomas, Helmuth, (902) 494-7177 helmuth.thomas@dal.ca, Dalhousie Univ; Oc

Thomas, Jay, (315)443-7631 jthom102@syr.edu, Syracuse Univ; Gi

Thomas, Jeremy, 206-947-2678 jnt@uw.edu, Univ of Washington; ZrXa

Thomas, Jim, (775) 887-7648 tom_j_smith@usgs.gov, Univ of Nevada, Reno; Cg

Thomas, John, 352-392-1951 ext 216 thomas@ufl.edu, Univ of Florida; Sp

Thomas, Kimberly W., (505) 667-4379 Los Alamos; Ct

Thomas, Margaret A., (860) 424-3583 margaret.thomas@ct.gov, Dept of Energy and Env Protection; Gg

Thomas, Mark, +44(0) 113 34 35233 m.e.thomas@leeds.ac.uk, Univ of Leeds; Ng

Thomas, Paul, (360) 650-7796 paul.thomas@wwu.edu, Western Washington Univ; Gm

Thomas, Peter, 361749-6768 peter.thomas@utexas.edu, Univ of Texas, Austin; Ob

Thomas, Ray G., (352) 392-7984 rgthomas@ufl.edu, Univ of Florida; Zg

Thomas, Robert C., (406) 683-7615 rob.thomas@umwestern.edu, Univ of Montana Western; GseZe

Thomas, Roger D. K., (717) 358-4135 roger.thomas@fandm.edu, Franklin and Marshall Coll; PgRh

Thomas, Rohrlack, thomas.rohrlack@nmbu.no, Norwegian Univ of Life Scis (NMBU); Hs

Thomas, Valerie, (404) 385-7254 valerie.thomas@isye.gatech.edu, Georgia Inst of Tech; Zu

Thomas, William A., 205-247-3547 geowat@uky.edu, Univ of Kentucky; Gt

Thomason, Jason, 217 244-2508 jthomaso@illinois.edu, Univ of Illinois; Gl

Thomasson, Joseph R., (785) 628-5665 Fort Hays State Univ; Pb

Thomka, James R., (518)564-2227 jthom059@plattsburgh.edu, SUNY, Plattsburgh; GsPg

Thompson, Andrew F., 626.395.8345 andrewt@caltech.edu,CalTech; OpgZc

Thompson, Anu, +440151 794 4095 anu@liverpool.ac.uk, Univ of Liverpool; Oc

Thompson, Christopher J., (509) 376-6602 chris.thompson@pnl.gov, Pacific Northwest; Ca

Thompson, Curtis, (785) 532-5776 cthompso@ksu.edu, Kansas State Univ; So

Thompson, David W. J., davet@atmos.colostate.edu, Colorado State Univ; Yr

Thompson, Geoffrey, (508) 289-2397 gthompson@whoi.edu, Woods Hole Oceanographic Inst; Cm

Thompson, Glennis, (510) 642-7025 Univ of California, Berkeley; Zn

Thompson, Jann W. M., (202) 633-1357 Smithsonian Inst / Natl Mus of

Natural History; Pg
Thompson, Joel B., (727) 864-8991 thompsjb@eckerd.edu, Eckerd Coll; Pg
Thompson, John F., jft66@cornell.edu, Cornell Univ; EmgNx
Thompson, John F. H., (604) 687-1117 Univ of British Columbia; Em
Thompson, Joseph L., (505) 667-4559 Los Alamos; Ct
Thompson, Keith R., (902) 494-3491 keith.thompson@dal.ca, Dalhousie Univ; Op
Thompson, Kenneth W., 620-341-5985 kthompso@emporia.edu, Emporia State Univ; Ze
Thompson, Lonnie G., (614) 292-6652 thompson.3@osu.edu, Ohio State Univ; GlZgPc
Thompson, LuAnne, (206) 543-9965 luanne@ocean.washington.edu, Univ of Washington; Og
Thompson, Margaret D., (781) 283-3029 mthompson@wellesley.edu, Wellesley Coll; Gc
Thompson, Michael D., (630) 252-9269 Argonne; Yg
Thompson, Michelle, 765-494-8677 mthompson@purdue.edu, Purdue Univ; Xg
Thompson, Todd, (812) 855-7428 tthomps@indiana.edu, Indiana Univ, Bloomington; Gg
Thompson, Tommy, thompsontommyt@aol.com, Colorado State Univ; Egm
Thompson, William G., (508) 289-2630 wthompson@whoi.edu, Woods Hole Oceanographic Inst; PeCc
Thompson, zglenn, thompsong@usf.edu, Univ of South Florida; Ys
Thoms, Richard E., (503) 725-3379 dick@pdx.edu, Portland State Univ; Ps
Thomsen, Charles E., (916) 484-8184 thomsec@arc.losrios.edu, American River Coll; Zy
Thomsen, Laurenz A., l.thomsen@iu-bremen.de, Univ of Washington; Ob
Thomsen, Leon, 713-743-3386 lathomsen@uh.edu, Univ of Houston; Ye
Thomson, Andrew D., a.r.thomson@ucl.ac.uk, Univ Coll London; Cpg
Thomson, Bradley J., 865-974-2699 bthomso1@utk.edu, Univ of Tennessee, Knoxville; Xg
Thomson, Cary, 604-822-0653 cthomson@eos.ubc.ca, Univ of British Columbia; Zn
Thomson, Cynthia, 212-854-9896 cthomson@iri.columbia.edu, Columbia Univ; Zn
Thomson, Dennis W., dwt2@psu.edu, Pennsylvania State Univ, Univ Park; Am
Thomson, Jack, +44 0151 794 3594 jack.thomson@liverpool.ac.uk, Univ of Liverpool; Ob
Thomson, Jennifer A., (509) 359-7478 jthomson@ewu.edu, Eastern Washington Univ; Gpz
Thomson, Richard E., (250) 363-6555 thomsonr@pac.dfo-mpo.gc.ca, Univ of British Columbia; Op
Thorbjarnarson, Kathryn W., (619) 594-5586 kthorbjarnarson@mail.sdsu.edu, San Diego State Univ; Hq
Thorkelson, Derek J., 778-782-5390 dthorkel@sfu.ca, Simon Fraser Univ; Gt
Thorleifson, Harvey, (612) 626-2150 thorleif@umn.edu, Univ of Minnesota; Gl
Thornberry-Ehrlich, Trista L., (406) 420-2391 trista.thornberry@colostate.edu, Colorado State Univ; GgcGx
Thorncroft, Christopher D., (518) 442-4555 SUNY, Albany; As
Thorne, Michael S., (801) 585-9792 michael.thorne@utah.edu, Univ of Utah; YsxYv
Thorne, Paul D., (509) 372-4482 paul.thorne@pnl.gov, Pacific Northwest; HgyYg
Thorne, Richard M., (310) 614-6630 rmt@atmos.ucla.edu, Univ of California, Los Angeles; As
Thornton, Daniel C., 979 845 4092 dthornton@ocean.tamu.edu, Texas A&M Univ; Ob
Thornton, Edward B., (831) 656-2847 thornton@nps.edu, Naval Postgrad Sch; On
Thornton, Joel A., (206) 543-4010 thornton@atmos.washington.edu, Univ of Washington; As
Thorson, Robert M., (860) 486-1396 robert.thorson@uconn.edu, Univ of Connecticut; Gm
Throop, Heather, (480) 727-2617 heather.throop@asu.edu, Arizona State Univ; Cb
Thul, David, 801-585-7013 dthul@egi.utah.edu, Univ of Utah; Gg
Thurber, Andrew, 541-737-4500 athurber@coas.oregonstate.edu, Oregon State Univ; Cb
Thurber, Clifford, (608) 262-6027 thurber@geology.wisc.edu, Univ of Wisconsin, Madison; Ysx
Thurber, David L., (718) 997-3325 dlthumper@aol.com, Queens Coll (CUNY); Cl
Thurnherr, Andreas, ant@ldeo.columbia.edu, Columbia Univ; Op
Thuro, Kurosch, +49 89 28925850 thuro@tum.de, Technical Univ of Munich; NgrGg
Thurow, Juergen, +44 020 7679 32416 j.thurow@ucl.ac.uk, Univ Coll London; Gs
Thurston, Phillips C., 7056751151 x2372 pthurston@laurentian.ca, Laurentian Univ, Sudbury; ZgCgGr
Thurtell, George W., (519) 824-4120 (Ext. 52453) gthurtel@uoguelph.ca, Univ of Guelph; As
Thuy, Ben, ben.thuy@mnhn.lu, Univ of Tennessee, Knoxville; Pg
Thy, Peter, pthy@ucdavis.edu, Univ of California, Davis; Gx
Tiampo, Kristy F., ktiampo@uwo.ca, Western Univ; Yd
Tian, Di, (334) 844-4100 dzt0025@auburn.edu, Auburn Univ; At
Tian, Hanqin, hanqin.tian@bc.edu, Boston Coll; Zo
Tibbits, Matthew, mtibbits@csc.edu, Chadron State Coll; Gg
Tibbits, Tawny, ttibbits@csc.edu, Chadron State Coll; Ggz
Tibljaš, Darko, +38514605970 dtibljas@geol.pmf.hr, Univ of Zagreb; GzScGx
Tibuleac, Ileana, (775) 784-6256 ileana@seismo.unr.edu, Univ of Nevada, Reno; Ys
Tibuleac, Paul, paul.tibuleac@uaic.ro, Alexandru Ioan Cuza; Pgs
Tice, Mike, 979-845-3138 tice@geo.tamu.edu, Texas A&M Univ; Pg
Tick, Geoffrey, gtick@ua.edu, Univ of Alabama; Hw
Tidwell, Allan, (850) 526-2761 tidwella@chipola.edu, Chipola Coll; Zg
Tidwell, David, 205-247-3698 dtidwell@gsa.state.al.us, Geo Survey of Alabama; Gm
Tidwell, Vincent C., (505) 844-6025 vctidwe@sandia.gov, New Mexico Tech; Hwq
Tiefenbacher, John, (512) 245-2170 Texas State Univ; Rnm
Tierney, Jessica, 520-621-5377 jesst@email.arizona.edu, Univ of Arizona; Co
Tierney, Kate, (319) 335-0670 kate-tierney@uiowa.edu, Univ of Iowa; GrCsGg
Tierney, Sean, 303-871-3972 sean.tierney@du.edu, Univ of Denver; Zc
Tiffney, Bruce H., (805) 893-2959 bruce.tiffney@ccs.ucsb.edu, Univ of California, Santa Barbara; Pb
Tijani, Moshood N., +2348023252339 mn.tijani@mail.ui.edu.ng, Univ of Ibadan; HwCgGe
Tikoff, Basil, (608) 262-4678 basil@geology.wisc.edu, Univ of Wisconsin, Madison; Gc
Tikoo, Sonia M., (848) 445-3444 sonia.tikoo@rutgers.edu, Rutgers, The State Univ of New Jersey; YmXy
Till, Christy, (480) 727-2828 christy.till@asu.edu, Arizona State Univ; Gi
Tillman, James E., mars@atmos.washington.edu, Univ of Washington; As
Tilt, Jenna, 541-737-1232 tiltj@oregonstate.edu, Oregon State Univ; Zu
Tilton, Eric E., 303-735-5033 eric.small@colorado.edu, Univ of Colorado; Hw
Timár, Gábor, +36 1 3722700/1762 timar@caesar.elte.hu, Eotvos Lorand Univ; YgdZi
Timmer, Jaqueline R., 406-496-4842 jtimmer@mtech.edu, Montana Tech of The Univ of Montana; Ca
Timmermann, Axel, (808) 956-2720 axel@hawaii.edu, Univ of Hawai'i, Manoa; Op
Timmermans, Mary-Louise, 203 432-3167 mary-louise.timmermans@yale.edu, Yale Univ; Oc
Timmes, Francis, (480) 965-4274 francis.timmes@asu.edu, Arizona State Univ; Xa
Timmons, J M., 575-835-5237 mtimmons@nmbg.nmt.edu, New Mexico Tech; GctGs
Timmons, J. Michael, 575-835-5237 mike.timmons@nmt.edu, New Mexico Tech; Gg
Timmons, Stacy, 575-835-6951 stacy.timmons@nmt.edu, New Mexico Tech; Hwg
Tindall, Sarah E., (610) 683-4446 tindall@kutztown.edu, Kutztown Univ of Pennsylvania; Gc
Ting, Mingfang, (845) 365-8374 ting@ldeo.columbia.edu, Columbia Univ; As
Tingey, David G., 801-422-7752 david_tingey@byu.edu, Brigham Young Univ; Zn
Tingley, Joseph V., jtingley@unr.edu, Univ of Nevada, Reno; Em
Tinjum, James M., 608/262-0785 tinjum@epd.engr.wisc.edu, Univ of Wisconsin, Madison; Ng
Tinker, Scott W., (512) 471-1534 scott.tinker@beg.utexas.edu, Univ of Texas, Austin; Go
Tinkham, Doug, 7056751151 x2270 dtinkham@laurentian.ca, Laurentian Univ, Sudbury; Gp
Tinkler, Dorothy, 541-881-5967 dtinkler@tvcc.cc, Treasure Valley Comm Coll; Zi
Tinsley, Mark, (434) 832-7708 tinsleym@cvcc.vccs.edu, Central Virginia Comm Coll; Zg
Tinto, Kirsteen, tinto@ldeo.columbia.edu, Columbia Univ; Gu
Tipper, Ed, ett20@cam.ac.uk, Univ of Cambridge; Cl

Tippett, John, jtippett@umw.edu, Univ of Mary Washington; Zn
Tisato, Nicola, nicola.tisato@jsg.utexas.edu, Univ of Texas, Austin; Yg
Tissot, Philippe , 361-825-3776 philippe.tissot@tamucc.edu, Texas A&M Univ, Corpus Christi; Zn
Titley, Spencer R., (520) 621-6018 stitley@email.arizona.edu, Univ of Arizona; Eg
Titus, Robert C., (607) 431-4733 titusr@hartwick.edu, Hartwick Coll; Ps
Titus, Sarah J., (507) 222-4419 stitus@carleton.edu, Carleton Coll; Gc
Tivey, Margaret K., (508) 289-3362 mktivey@whoi.edu, Woods Hole Oceanographic Inst; Cm
Tivey, Maurice A., (508) 289-2265 mtivey@whoi.edu, Woods Hole Oceanographic Inst; Yrm
Tiwari, Chetan, 940-369-8103 chetan.tiwari@unt.edu, Univ of North Texas; Zi
Tlusty, Michael, mtlusty@neaq.org, Univ of Massachusetts, Boston; Ob
Tobak, Zalán, +3662544156 tobak@geo.u-szeged.hu, Univesity of Szeged; ZirZy
Tobias, Craig, 860-405-9140 craig.tobias@uconn.edu, Univ of Connecticut; Oc
Tobin, Benjamin, (859) 323-0526 benjamin.tobin@uky.edu, Univ of Kentucky; Hy
Tobin, Harold, 608-265-5796 htobin@geology.wisc.edu, Univ of Wisconsin, Madison; Yg
Tobin, Tom S., (205) 348-1878 ttobin@ua.edu, Univ of Alabama; PsCsPe
Todd, Brenda, (402) 554-2662 btodd@unomaha.edu, Univ of Nebraska, Omaha; Zn
Todd, Claire E., (253) 535-5163 toddce@plu.edu, Pacific Lutheran Univ; Glm
Todd, Steve, stephen.todd15@pcc.edu, Portland Comm Coll - Sylvania Campus; Am
Toke', Nathan, (801) 863-8117 nathan.toke@uvu.edu, Utah Valley Univ; GtmZi
Tokmakian, Robin T., (831) 656-3255 rtt@nps.edu, Naval Postgrad Sch; Op
Toksoz, M N., (617) 253-7852 toksoz@mit.edu, Massachusetts Inst of Tech; Ys
Tolhurst, Trevor, +44 (0)1603 59 3124 t.tolhurst@uea.ac.uk, Univ of East Anglia; Sb
Tolstoy, Maria, tolstoy@ldeo.columbia.edu, Columbia Univ; Yr
Tomascak, Paul B., (315) 312-2285 tomascak@oswego.edu, SUNY, Oswego; CgGzi
Tomasko, Martin G., (520) 621-6969 mtomasko@lpl.arizona.edu, Univ of Arizona; As
Tomaso, Matthew S., (973) 655-7990 thomasw@mail.montclair.edu, Montclair State Univ; Ga
Tomašiæ, Nenad, +38514605968 ntomasic@geol.pmf.hr, Univ of Zagreb; GziCg
Tomiæ, Vladimir, +38514606094 vtomic@geol.pmf.hr, Univ of Zagreb; GgPgZi
Tomkin, Jonathan H., 217-244-2928 tomkin@illinois.edu, Univ of Illinois, Urbana-Champaign; Gm
Tomlinson, Emma L., + 353 1 8963856 tomlinse@tcd.ie, Trinity Coll; Cc
Tomlinson, Jaime L., 302-831-2649 jaimet@udel.edu, Univ of Delaware; Zn
Tomlinson, Peter, (785) 532-3198 ptomlin@ksu.edu, Kansas State Univ; So
Tompson, Andrew F., (925) 422-6348 tompson1@llnl.gov, Lawrence Livermore; HwSpCc
Toner, Brandy M., (612) 624-1362 toner@umn.edu, Univ of Minnesota, Twin Cities; ClmSc
Tong, Daniel, 703-993-1210 qtong@gmu.edu, George Mason Univ; Ac
Tong, Daoqin, daoqin@email.arizona.edu, Univ of Arizona; Zi
Tong, Vincent C., +44020 3073 8035 vincent.tong@ucl.ac.uk, Birkbeck Coll; Yg
Tongdee, Poetchanaporn, ptongdee@hccfl.edu, Hillsborough Comm Coll; Zg
Toni, Rousine T., 002-03-3921595 dr_rosine@hotmail.com, Alexandria Univ; GgPg
Tonon, Marco Davide, marco.tonon@unito.it, Univ di Torino; Ze
Toomey, Douglas R., (541) 346-5576 drt@uoregon.edu, Univ of Oregon; YsGt
Toonen, Robert, (808) 236-7425 toonen@hawaii.edu, Univ of Hawai'i, Manoa; Ob
Toonen, Willem, w.h.j.toonen@vu.nl, Vrije Univit Amsterdam; GmsGn
Toor, Gurpal, 813-6334152 gstoor@ufl.edu, Univ of Florida; Sc
Toraman, Erkan, 978-542-6493 etoraman@salemstate.edu, Salem State Univ; GpiGt
Toran, Laura, (215) 204-2352 ltoran@temple.edu, Temple Univ; HwZu
Toran, Marta, toranml@appstate.edu, Appalachian State Univ; Ge
Torlaschi, Enrico, 514-987-3000 #6848 torlaschi.enrico@uqam.ca, Univ du Quebec a Montreal; Am
Tornabene, Livio, ltornabe@uwo.ca, Western Univ; XgZr
Toro, Jaime, 3042935603 ext. 4327 jtoro@wvu.edu, West Virginia Univ; GctGo
Torreano, Scott, (615) 598-1271 storrean@sewanee.edu, Sewanee: Univ of the South; Sf
Torrents, Alba, (301) 405-1979 alba@eng.umd.edu, Univ of Maryland; Sb
Torres, Joseph J., (727) 553-1169 jtorres@marine.usf.edu, Univ of South Florida; Ob
Torres, Mark, 713.348.2371 mt61@rice.edu, Rice Univ; Cb
Torres, Marta E., 541-737-2902 mtorres@coas.oregonstate.edu, Oregon State Univ; Cm
Torres, Raymond, (803) 777-4506 torres@geol.sc.edu, Univ of South Carolina; Hy
Torres Parisian, Cathleen, ctorresp@umn.edu, Univ of Minnesota, Twin Cities; Zr
Torri , Giuseppe, 808-956-2564 gtorri@hawaii.edu, Univ of Hawai'i at Manoa; ApsAm
Tortell, Philippe, (604) 822-4728 ptortell@eos.ubc.ca, Univ of British Columbia; Ob
Torvela, Taija, +44(0) 113 34 36620 t.m.torvela@leeds.ac.uk, Univ of Leeds; GctEg
Tosdal, Richard, rtosdal@eos.ubc.ca, Univ of British Columbia; Em
Tostevin, Rosalie, +27-21-650-2140 Univ of Cape Town; CmlCb
Toth, Charles K., 614-292-7681 toth.2@osu.edu, Ohio State Univ; ZrYdZi
Toth, Emoke, tothemoke.pal@gmail.com, Eotvos Lorand Univ; PmsPb
Tóth, József, (780) 492-1115 joe.toth@ualberta.ca, Univ of Alberta; Hw
Toth, Natalie, natalie.toth@dmns.org, Denver Mus of Nature & Sci; Pg
Totsche, Kai U., +49(0)3641 948650 kai.totsche@uni-jena.de, Friedrich-Schiller-Univ Jena; HwClSo
Totten, Matthew W., (785) 532-2227 mtotten@ksu.edu, Kansas State Univ; Gso
Totten, Rebecca, (205)348-6050 rebecca.minzoni@ua.edu, Univ of Alabama; GsPme
Totten, Stanley M., (812) 866-7245 totten@hanover.edu, Hanover Coll; Gl
Touchan, Ramzi, (520) 621-2992 rtouchan@ltrr.arizona.edu, Univ of Arizona; PcHsAs
Toullec, Renaud, +33(0)3 44069333 renaud.toullec@lasalle-beauvais.fr, Institut Polytechnique LaSalle Beauvais (ex-IGAL); GsoEo
Towery, Brooke L., (850) 484-2056 btowery@pensacolastate.edu, Pensacola State Coll; Gg
Towner, Ronald, (520) 621-6465 rht@email.arizona.edu, Univ of Arizona; Gae
Townsend, Christi, (512) 245-2170 Texas State Univ; Zy
Townsend, Meredith, 541-346-4573 mtownse4@uoregon.edu, Univ of Oregon; Gv
Townsend-Small, Amy, (513) 556-3762 amy.townsend-small@uc.edu, Univ of Cincinnati; Co
Toy, Virginia G., 64 3 479 7506 virginia@geology.co.nz, Univ of Otago; Gct
Tracy, Matthew, matthew.tracy@usafa.edu, United States Air Force Acad; Zn
Tracy, Sally June, sjtracy@carnegiescience.edu, Carnegie Inst for Sci; Yg
Trail, Dustin, (585) 276-7182 dustin.trail@rochester.edu, Univ of Rochester; Cp
Tran, Linda C., 281-765-7865 linda.c.tran@nhmccd.edu, Lonestar Coll - North Harris; Zi
Tranel, Lisa M., (309) 438-7966 ltranel@ilstu.edu, Illinois State Univ; GmtZi
Trapp, Robert J., 217-300-0967 jtrapp@illinois.edu, Univ of Illinois, Urbana-Champaign; As
Travers, Steven, (309) 268-8640 steve.travers@heartland.edu, Heartland Comm Coll; Zg
Travis, Bryan J., (505) 667-1254 bjtravis@lanl.gov, Los Alamos; Hq
Trehu, Anne M., 541-737-2655 trehu@coas.oregonstate.edu, Oregon State Univ; YrsGt
Treloar, Peter, +44020 8417 2525 p.treloar@kingston.ac.uk, Kingston Univ; Gc
Trembanis, Art, 302-831-2498 art@udel.edu, Univ of Delaware; Ou
Trembath-Reichert, Elizabeth, 480-727-2631 e.t.r@asu.edu, Arizona State Univ; Cb
Tremblay, Alain, 514-987-3000 #1397 tremblay.alain@uqam.ca, Univ du Quebec a Montreal; Gct
Tremblay, Bruno, (514) 398-4369 bruno.tremblay@mcgill.ca, McGill Univ; Op
Tremblay, Marissa, tremblam@purdue.edu, Purdue Univ; Cc
Tremblay, Thomas A., (512) 475-9537 tom.tremblay@beg.utexas.edu, Univ of Texas at Austin, Jackson Sch of Geosciences; Zy
Trenhaile, Alan S., 519-253-3000 x2184 tren@uwindsor.ca, Univ of Windsor; Gm
Trentham, Robert C., 432 552-2432 trentham_r@utpb.edu, Univ of Texas Permian Basin; Go
Treude, Tina, (310) 267-5213 ttreude@g.ucla.edu, Univ of California, Los Angeles; ObPgCm

Trevino, Ramon H., (512) 471-3362 ramon.trevino@beg.utexas.edu, Univ of Texas at Austin, Jackson Sch of Geosciences; GsoGe
Trevors, Jack, (519) 824-4120 Ext.53367 jtrevors@uoguelph.ca, Univ of Guelph; Zn
Triay, Ines, (505) 665-1755 Los Alamos; Hg
Tribe, Selina, 604-527-5471 tribes@douglascollege.ca, Douglas Coll; EgGge
Trigoso, Erika, 303-871-3936 erika.trigoso@du.edu, Univ of Denver; Zc
Tripati, Aradhna, ripple@epss.ucla.edu, Univ of California, Los Angeles; CmPeCs
Triplett, Laura, (507) 933-7442 triplett@gustavus.edu, Gustavus Adolphus Coll; GemCl
Tripoli, Gregory J., (608)262-3700 tripoli@aos.wisc.edu, Univ of Wisconsin, Madison; Am
Trivitt-Kracke, Amy, 575-835-5362 amy.trivitt@nmt.edu, New Mexico Tech; ZfGg
Trixler, Frank, 089/2179 509 trixler@lrz.uni-muenchen.de, Ludwig-Maximilians-Univ Muenchen; GzZmXc
Troch, Peter A., (520) 626-1277 patroch@hwr.arizona.edu, Univ of Arizona; Hs
Trofimovs, Jessica, +61 7 3138 2766 jessica.trofimovs@qut.edu.au, Queensland Univ of Tech; GsvGu
Tromp, Jeroen, 609-258-4128 jtromp@princeton.edu, Princeton Univ; Yss
Troost, Kathy G., 206-221-1770 ktroost@uw.edu, Univ of Washington; Zun
Trop, Jeffrey M., (570) 577-3027 jtrop@bucknell.edu, Bucknell Univ; Gs
Trost, G K., (501) 575-7317 garrett@dicksonstreet.com, Univ of Arkansas, Fayetteville; Zn
Trouet, Valerie, (520) 626-8004 trouet@email.arizona.edu, Univ of Arizona; PeZyAs
Trout, Jennifer, (269) 387-8633 jennifer.l.trout@wmich.edu, Western Michigan Univ; ZnnZn
Troy, Marleen, (570) 408-4615 marleen.troy@wilkes.edu, Wilkes Univ; Ht
Trudgill, Bruce D., (303) 273-3883 btrudgil@mines.edu, Colorado Sch of Mines; Gc
Trueman, Clive, +44 (0)23 80596571 trueman@noc.soton.ac.uk, Univ of Southampton; Cg
Trullenque, Ghislain, +33(0)3 44069327 ghislain.trullenque@laslle-beauvais.fr, Institut Polytechnique LaSalle Beauvais (ex-IGAL); GcNrGg
Trumbore, Susan E., (949) 824-6142 setrumbo@uci.edu, Univ of California, Irvine; Sc
Trussell, Geoffrey, 7815817370 x300 g.trussell@neu.edu, Northeastern Univ; Zn
Tsai, Victor, 626.395.6993 tsai@caltech.edu,CalTech; Yg
Tsamados, Michel D., m.tsamados@ucl.ac.uk, Univ Coll London; Zo
Tsele, Philemon L., +27 (0)12 420 4939 philemon.tsele@up.ac.za, Univ of Pretoria; Zfr
Tshudy, Dale, (814) 732-2453 dtshudy@edinboro.edu, Edinboro Univ of Pennsylvania; Pi
Tsige, Meaza, meaza@ucm.es, Univ Compl de Madrid; Ng
Tsikos, Hari, +27 (0)46-603-8511 h.tsikos@ru.ac.za, Rhodes Univ; CgEg
Tso, Jonathan L., (540) 831-5638 jtso@radford.edu, Radford Univ; Gcp
Tsoflias, George P., (785) 864-4584 tsoflias@ku.edu, Univ of Kansas; YgeGe
Tsujita, Cameron J., (519) 661-2111 (Ext. 86740) ctsujita@uwo.ca, Western Univ; Pi
Tsvankin, Ilya D., 303-273-3060 ilya@mines.edu, Colorado Sch of Mines; Ye
Tu, Wei, (912) 478-5233 wtu@georgiasouthern.edu, Georgia Southern Univ; Zi
Tubana, Brenda S., 225-578-9420 btubana@agcenter.lsu.edu, Louisiana State Univ; So
Tubía, Jose M., (+94) 601-5392 jm.tubia@ehu.eus, Univ of the Basque Country UPV/EHU; GctYg
Tucholke, Brian E., (508) 289-2494 btucholke@whoi.edu, Woods Hole Oceanographic Inst; Gut
Tucker, Carla M., (409) 880-8236 ctucker@lamar.edu, Lamar Univ; Hw
Tucker, Gregory E., (303) 492-6985 gregory.tucker@colorado.edu, Univ of Colorado; GmtGs
Tudhope, Alexander, +44 (0) 131 650 8508 sandy.tudhope@ed.ac.uk, Univ of Washington; AtPc
Tufillaro, Nicholas, 541-737-2064 nbt@coas.oregonstate.edu, Oregon State Univ; Cb
Tulaczyk, Slawek, (831) 459-5207 stulaczy@ucsc.edu, Univ of California, Santa Cruz; GlmNg
Tull, James F., (850) 644-1448 jtull@fsu.edu, Florida State Univ; Gct
Tuller, Markus, (520) 621-7225 mtuller@cals.arizona.edu, Univ of Arizona; Zn
Tullis, Jan A., (401) 863-1921 jan_tullis@brown.edu, Brown Univ; Gcy
Tullis, Jason A., (479) 575-8784 jatullis@uark.edu, Univ of Arkansas, Fayetteville; ZriZf
Tullis, Terry E., (401) 863-3829 terry_tullis@brown.edu, Brown Univ; Yx
Tumarkin-Deratzian, Allison, altd@temple.edu, Temple Univ; Pv
Tung, Ka-Kit, tung@amath.washington.edu, Univ of Washington; As
Tung, Wen-wen, (765) 494-0272 wwtung@purdue.edu, Purdue Univ; As
Tunnicliffe, Verena, (250) 721-7135 verenat@uvic.ca, Univ of Victoria; Ob
Tura, Ali, 303-273-3454 alitura@mines.edu, Colorado Sch of Mines; YsNr
Turbeville, John, (760) 944-4449, x6413 jturbeville@miracosta.edu, MiraCosta Coll; GgOg
Turchyn, Alexandra, +44 (0) 1223 333479 atur07@esc.cam.ac.uk, Univ of Cambridge; Ge
Turco, Richard, (310) 825-6936 turco@atmos.ucla.edu, Univ of California, Los Angeles; As
Turcotte, Donald L., (530) 752-6808 dlturcotte@ucdavis.edu, Univ of California, Davis; Yg
Turek, Andrew, 519-969-6710 turek@uwindsor.ca, Univ of Windsor; Cc
Turek, Marni, 250-807-8284 marni.turek@ubc.ca, Univ of British Columbia, Okanagan Campus; Hg
Turk, Judith, (402) 472-8024 Univ of Nebraska, Lincoln; Sd
Turner, Anne, aturner@austincc.edu, Austin Comm Coll District; Hg
Turner, Derek, 604-777-6568 turnerd1@douglascollege.ca, Douglas Coll; GemGl
Turner, Elizabeth C., 7056751151 x2267 eturner@laurentian.ca, Laurentian Univ, Sudbury; GrdEm
Turner, Grenville, +44 0161 275-0401 grenville.turner@manchester.ac.uk, Univ of Manchester; XcCc
Turner, Jay, (314) 935-5480 jrturner@wustl.edu, Washington Univ in St. Louis; Zn
Turner, Jenni, +44 (0)1603 59 3109 jenni.turner@uea.ac.uk, Univ of East Anglia; Gt
Turner, Kerry, +44 (0)1603 59 2551 r.k.turner@uea.ac.uk, Univ of East Anglia; GeEg
Turner, Robert E., (225) 388-6454 Louisiana State Univ; Ob
Turner, Stephen, sjturner@umass.edu, Univ of Massachusetts, Amherst; Giv
Turner, Wesley L., (936) 468-1049 turnerwl@sfasu.edu, Stephen F. Austin State Univ; Gg
Turner, III, Henry, (479) 575-7295 hturner@uark.edu, Univ of Arkansas, Fayetteville; GgtGc
Turnock, Allan C., (204) 474-6911 ac_turnock@umanitoba.ca, Univ of Manitoba; Gp
Turpening, Roger M., (906) 487-1784 roger@mtu.edu, Michigan Technological Univ; Ye
Turrin, Brent D., 848-445-3177 bturrin@eps.rutgers.edu, Rutgers, The State Univ of New Jersey; CcGvYm
Turtle, Elizabeth, (520) 621-8284 turtle@lpl.arizona.edu, Univ of Arizona; Xg
Tuttle, Samuel, 315-443-2672 setuttle@syr.edu, Syracuse Univ; Ge
Tuttle, Samuel , stuttle@mtholyoke.edu, Mount Holyoke Coll; Gq
Tvelia, Sean, (631) 451-4303 tvelias@sunysuffolk.edu, Suffolk County Comm Coll, Ammerman Campus; Ggl
Twelker, Evan, 907-451-5086 evan.twelker@alaska.gov, Alaska Division of Geological & Geophysical Surveys; Eg
Twine, Tracy E., 612-625-7278 twine@umn.edu, Univ of Minnesota, Twin Cities; As
Twiss, Robert J., (530) 752-1860 rjtwiss@ucdavis.edu, Univ of California, Davis; Gc
Tyburczy, James A., (480) 965-2637 jim.tyburczy@asu.edu, Arizona State Univ; GyzYx
Tyce, Robert, (401) 874-6879 tyce@oce.uri.edu, Univ of Rhode Island; Ou
Tyler, E. Jerry, (608) 262-0853 ejtyler@wisc.edu, Univ of Wisconsin, Madison; Sd
Tyler, Scott, (775) 784-6250 styler@unr.edu, Univ of Nevada, Reno; Hg
Tynan, Cynthia T., (206) 860-6793 ctynan@coastalstudies.org, Univ of Washington; Ob
Tyning, Thomas F., (413) 236-4502 ttyning@berkshirecc.edu, Berkshire Comm Coll; Zn
Tyrrell, Toby, +44 (0)23 80596110 toby.tyrrell@soton.ac.uk, Univ of Southampton; OgPe
Tziperman, Eli, (617) 384-8381 eli@eps.harvard.edu, Harvard Univ; Op
Tzortziou, Maria, 212-650-5769 mtzortziou@ccny.cuny.edu, Graduate Sch of the City Univ of New York; As
Törnqvist, Torbjörn E., 504-314-2221 tor@tulane.edu, Tulane Univ; Gs

U

Uahengo, Collen, +8613021208022 cuahengo@unam.na, Univ of Namibia; PgGoEg
Uchupi, Elazar, (508) 289-2830 Woods Hole Oceanographic Inst; Ou
Uddin, Ashraf, (334) 844-4885 uddinas@auburn.edu, Auburn Univ; GdoGt
Udry, Arya, (702) 895-1239 arya.udry@unlv.edu, Univ of Nevada, Las Vegas; GiXmGg
Ufkes, Els , +31 20 5989953 els.ufkes@falw.vu.nl, Vrije Univit Amsterdam;

PmZgGu

Ugland, Richard, 541-552.6479 lane@sou.edu, Southern Oregon Univ; Gg

Uhen, Mark D., (703) 993-5264 muhen@gmu.edu, George Mason Univ; Pv

Ukstins, Ingrid, ingrid-peate@uiowa.edu, Univ of Iowa; GviXg

Ulanski, Stanley L., (540) 568-6130 ulanskl@jmu.edu, James Madison Univ; Og

Ullman, David J., 715.682.1312 dullman@northland.edu, Northland Coll; GlPeGm

Ullman, William J., (302) 645-4302 ullman@udel.edu, Univ of Delaware; Cm

Ullrich, Alexander D., 770-274-5083 aullrich@gsu.edu, Georgia State Univ, Perimeter Coll, Dunwoody Campus; GgZnGu

Ullrich, Paul, 530-400-9817 paullrich@ucdavis.edu, Univ of California, Davis; As

Ulmer-Scholle, Dana, dana.ulmer-scholle@nmt.edu, New Mexico Tech; Gsd

Umhoefer, Paul J., (928) 523-6464 paul.umhoefer@nau.edu, Northern Arizona Univ; Gt

Underwood, Charlie, +44 020 3073 8036 Birkbeck Coll; Pg

Underwood, Michael B., (573) 882-4685 underwoodm@missouri.edu, Univ of Missouri; GsuGt

Unger, Corey, (801) 538-4810 coreyunger@utah.gov, Utah Geo Survey; Zi

Unkefer, Clifford J., (505) 665-2560 Los Alamos; Zn

Uno, Kevin, kevinuno@ldeo.columbia.edu, Columbia Univ; Pe

Unrug, Kot F., (606) 257-1883 Univ of Kentucky; Nr

Unsworth, Martyn, (780) 492-3041 martyn.unsworth@ualberta.ca, Univ of Alberta; GtvYe

Upchurch, Paul, +44 020 7679 37947 p.upchurch@ucl.ac.uk, Univ Coll London; Pg

Ural, Melek, 00904242370000-5960 melekural@firat.edu.tr, Firat Univ; GivCc

Urban-Rich, Juanita, juanita.urban-rich@umb.edu, Univ of Massachusetts, Boston; Ob

Urbanczyk, Kevin, (432) 837-8110 kevinu@sulross.edu, Sul Ross State Univ; GiZi

Uriarte, Jesus Angel, jesusangel.uriarte@ehu.eus, Univ of the Basque Country UPV/EHU; NgHg

Uribelarrea, David, uriben@ucm.es, Univ Compl de Madrid; Ga

Urosevic, Milovan, +61 8 9266-2296 m.urosevic@curtin.edu.au, Curtin Univ; YerYx

Urquhart, Joanne, jurquhar@bowdoin.edu, Bowdoin Coll; Gge

Urquhart, Mary, 972-883-2496 urquhart@utdallas.ede, Univ of Texas, Dallas; Xy

Urzua, Alfredo, 617-552-8339 alfredo.urzua@bc.edu, Boston Coll; Ng

Ustaszewski, Kamil M., 0049(0)3641/948623 kamil.u@uni-jena.de, Friedrich-Schiller-Univ Jena; Gct

Ustin, Susan L., (530) 752-0621 slustin@ucdavis.edu, Univ of California, Davis; Zr

Ustunisik, Gokce K., (605) 394-2461 gokce.ustunisik@sdsmt.edu, South Dakota Sch of Mines & Tech; Cp

Utgard, Russell O., 614 247-4246 utgard.1@osu.edu, Ohio State Univ; YgEnZe

Uzochukwu, Godfrey A., (336) 285-4866 uzo@ncat.edu, North Carolina Agricultural & Tech State Univ; Gz

Uzunlar, Nuri, (605) 394-2494 nuri.uzunlar@sdsmt.edu, South Dakota Sch of Mines & Tech; EmCeGc

V

Vacquier, Victor D., (858) 534-4803 vvacquier@ucsd.edu, Univ of California, San Diego; Ob

Vaezi, Reza, +98 (411) 339 2721 vaezi@tabrizu.ac.ir, Univ of Tabriz; GeHgs

Vail, Peter R., (713) 348-4888 vail@rice.edu, Rice Univ; Gr

Vaillancourt, Robert, 717-872-3294 robert.vaillancourt@millersville.edu, Millersville Univ; Obc

Valdes, Juan B., (520) 621-2266 jvaldes@email.arizona.edu, Univ of Arizona; Hs

Valencia, Victor A., (520) 300-1605 victor.valencia@wsu.edu, Washington State Univ; ZgCcEg

Valenti, Christine, (973) 655-4448 Montclair State Univ; Gg

Valentine, David, valentine@geol.ucsb.edu, Univ of California, Santa Barbara; Cm

Valentine, Gregory, (716) 645-4295 gav4@buffalo.edu, SUNY, Buffalo; Gv

Valentine, James W., (510) 643-5791 Univ of California, Berkeley; Pg

Valentine, Michael, 253-566-5060 mvalentine@tacomacc.edu, Tacoma Comm Coll; Zg

Valentino, David W., (315) 312-2798 david.valentino@oswego.edu, SUNY, Oswego; GtcYu

Valerio, Mitch, 614 265 6616 mitch.valerio@dnr.state.oh.us, Ohio Dept of Nat Res; Hw

Valiunas, Jonas K., (807) 343-8677 jvaliuna@lakeheadu.ca, Lakehead Univ; Zm

Valley, John W., (608) 263-5659 valley@geology.wisc.edu, Univ of Wisconsin, Madison; CsGxz

Valsami-Jones, Eugenia, +44 (0)121 414 5537 e.valsamijones@bham.ac.uk, Univ of Birmingham; ClGeCa

van Alstine, James, +44(0) 113 34 37531 j.vanalstine@leeds.ac.uk, Univ of Leeds; Ge

Van Alstine, James B., (320) 589-6313 vanalstj@mrs.umn.edu, Univ of Minnesota, Morris; Pg

Van Arsdale, Roy B., (901) 678-4356 rvanrsdl@memphis.edu, Univ of Memphis; Gcm

Van Avendonk, Harm, 512-471-0429 harm@ig.utexas.edu, Univ of Texas, Austin; Yg

van Balen, Ronald R., +31 20 59 87324 r.t.van.balen@vu.nl, Vrije Univit Amsterdam; GgmGt

van Bever Donker, Jan M., +27 021 959 3263 jvanbeverdonker@uwc.ac.za, Univ of the Western Cape; GcpGt

van Beynen, Phil, vanbeyne@usf.edu, Univ of South Florida; Pc

Van Brocklin, Matthew F., (315) 229-5197 mvanbrocklin@stlawu.edu, St. Lawrence Univ; GgZgn

Van Buer, Nicholas J., 909-869-3457 njvanbuer@cpp.edu, California State Polytechnic Univ, Pomona; GxtGz

Van Daele, Maarten, +3292644637 maarten.vandaele@ugent.be, Ghent Univ; GsRnZc

van de Flierdt, Tina, +44 20 759 41290 tina.vandeflierdt@imperial.ac.uk, Imperial Coll; Cs

van de Gevel, Saskia, (828) 262-7028 gevelsv@appstate.edu, Appalachian State Univ; Zg

Van De Poll, Henk W., (506) 453-4804 Univ of New Brunswick; Gs

Van de Water, Peter, 559-278-2912 pvandewater@csufresno.edu, Fresno State Univ; GrPbGf

van den Akker, Ben, ben.vandenakker@flinders.edu.au, Flinders Univ; Zn

Van Den Broeke, Matthew S., (402) 472-2418 mvandenbroeke2@unl.edu, Univ of Nebraska, Lincoln; AsZr

van den Heever, Sue, sue@atmos.colostate.edu, Colorado State Univ; As

Van Den Hoek, Jamon, 541-737-1229 vandenhj@oregonstate.edu, Oregon State Univ; Zur

van der Berg, Stan, +440151 794 4096 vandenberg@liverpool.ac.uk, Univ of Liverpool; Em

van der Flier-Keller, Eileen, evanderf@sfu.ca, Simon Fraser Univ; Ze

Van Der Flier-Keller, Eileen, (250) 472-4019 fkeller@uvic.ca, Univ of Victoria; Cl

van der Hilst, Robert, (617) 253-3382 hilst@mit.edu, Massachusetts Inst of Tech; Ys

van der Horst, Dan, +44 (0) 131 651 4467 dan.vanderhorst@ed.ac.uk, Edinburgh Univ; Eg

van der Lee, Suzan, (847) 491-8183 suzan@earth.northwestern.edu, Northwestern Univ; YsxYg

van der Lubbe, Jeroen, +31 20 59 87366 h.j.l.vander.lubbe@vu.nl, Vrije Univit Amsterdam; Gs

van der Merwe, Barend, +27 (0)12 420 3699 barend.vandermerwe@up.ac.za, Univ of Pretoria; Gm

van der Pluijm, Ben, (734) 763-0373 vdpluijm@umich.edu, Univ of Michigan; GcZce

Van der Putten, Nathalie, +31 20 59 83257 n.n.l.vanderputten@vu.nl, Vrije Univit Amsterdam; PecZy

van der Velde, Ype, +31 20 59 87402 y.vander.velde@vu.nl, Vrije Univit Amsterdam; Hw

Van der Voo, Rob, (734) 764-8322 voo@umich.edu, Univ of Michigan; YmGtc

van der Werf, Guido R., +31 20 59 85687 guido.vander.werf@vu.nl, Vrije Univit Amsterdam; Zr

van Dijk, Deanna, (616) 526-6510 dvandijk@calvin.edu, Calvin Coll; GmOnZy

van Dongen, Bart, +44 0161 306-7460 bart.vandongen@manchester.ac.uk, Univ of Manchester; Co

Van Eerd, Laura, (519) 674-1644 lvaneerd@ridgetownc.uoguelph.ca, Univ of Guelph; Zn

Van Engelen, Alexander, alexander.van.engelen@asu.edu, Arizona State Univ; Xa

Van Geen, Alexander, avangeen@ldeo.columbia.edu, Columbia Univ; Cg

van Hinsberg, Vincent, 514-398-8112 vincent.vanhinsberg@mcgill.ca, McGill Univ; EgCg

Van Horn, Stephen R., (740) 826-8306 svanhorn@muskingum.edu, Muskingum Univ; GeZiEo

van Hunen, Jeroen, +44 (0) 191 33 42293 jeroen.van-hunen@durham.ac.uk, Durham Univ; Gt

Van Iten, Heyo, (812) 866-7303 vaniten@hanover.edu, Hanover Coll; PgHgRn

FACULTY INDEX – V

van Keken, Peter E., pvankeken@carnegiescience.edu, Carnegie Inst for Sci; Ys
Van Kooten, Gerald K., (616) 526-6374 gvankooten1@gmail.com, Calvin Coll; GoCePi
van Leeuwen, Boudewijn, +3662544156 leeuwen@geo.u-szeged.hu, Univesity of Szeged; ZirZy
van Leeuwen, Peter Jan, peter.vanleeuwen@colostate.edu, Colorado State Univ; As
van Leeuwen, Willem, leeuw@email.arizona.edu, Univ of Arizona; Zy
Van Loon, Lisa, 306-657-3585 lisa.vanloon@lightsource.ca, Western Univ; Ca
Van Mooy, Benjamin, (508) 289-2740 bvanmooy@whoi.edu, Woods Hole Oceanographic Inst; Oc
van Niekerk, Adriaan, avn@sun.ac.za, Stellenbosch Univ; Zri
Van Niewenhuise, Donald, 713-743-3423 donvann@uh.edu, Univ of Houston; PsGoPe
Van Orman, James A., jav12@case.edu, Case Western Reserve Univ; CgGyCp
Van Rooij, David, david.vanrooij@ugent.be, Ghent Univ; YrGu
Van Roosendaal, Susan, (801) 581-8218 Univ of Utah; Zn
van Rooyen, Deanne, 902-585-1163 deanne.vanrooyen@acadiau.ca, Acadia Univ; Gpt
Van Ry, Michael, mvanry@occ.cccd.edu, Orange Coast Coll; Gv
Van Ryswick, Stephen, 410-554-5544 stephen.vanryswick@maryland.gov, Maryland Dept of Nat Res; Ge
Van Schmus, W. Randall, (785) 864-2727 rvschmus@ku.edu, Univ of Kansas; Cc
Van Stan, John, (912) 478-8040 jvanstan@georgiasouthern.edu, Georgia Southern Univ; HgClSb
Van Straaten, H. Peter, (519) 824-4120 x52454 pvanstra@uoguelph.ca, Univ of Guelph; EnScCb
Van Tongeren, Jill A., (848) 445-5363 jvantongeren@eps.rutgers.edu, Rutgers, The State Univ of New Jersey; GiCgGt
Van Vleet, Edward S., (727) 553-1165 vanvleet@usf.edu, Univ of South Florida; Co
van Westrenen, Wim, +31 20 59 87279 w.van.westrenen@vu.nl, Vrije Univit Amsterdam; CpGz
Van Wijk, Jolante, (575) 835-6661 jolante.vanwijk@nmt.edu, New Mexico Tech; GtoYe
Vanacore, Elizabeth A., (787) 265-3845 elizabeth.vanacore@upr.edu, Univ of Puerto Rico; YsuYx
VanAller-Hernick, Linda A., (518) 486-3699 lhernick@mail.nysed.gov, New York State Geo Survey; Pgb
Vance, Robert K., (912) 478-5353 rkvance@georgiasouthern.edu, Georgia Southern Univ; GiEgGg
Vandeberg, Gregory S., 701-777-4588 gregory.vandeberg@und.edu, Univ of North Dakota; GmZiGl
Vandemark, Douglas C., 603-862-0195 doug.vandemark@unh.edu, Univ of New Hampshire; Zr
Vanden Berg, Michael, (801) 538-5419 michaelvandenberg@utah.gov, Utah Geo Survey; Eo
Vandenbroucke, Thijs, thijs.vandenbroucke@ugent.be, Ghent Univ; PsmPc
Vander Auwera, Jacqueline, 32 4 366 22 53 jvdauwera@ulg.ac.be, Univ de Liege; Giv
Vanderkluysen, Loyc, (215) 571-4673 loyc@drexel.edu, Drexel Univ; GviGz
Vanderlip, Richard, (785) 532-6101 vanderrl@ksu.edu, Kansas State Univ; So
VanDervoort, Dane S., (205) 247-3626 dvandervoort@gsa.state.al.us, Geo Survey of Alabama; GcpZi
VanDorpe, Paul E., paul.vandorpe@dnr.iowa.gov, Iowa Dept of Nat Res; Gg
VanGundy, Robert D., (276) 376-4656 rdv4v@uvawise.edu, Univ of Virginia's Coll at Wise; GeHgZg
VanHorn, Jason E., (616) 526-7623 jev35@calvin.edu, Calvin Coll; ZirRh
Vaniman, David T., (505) 667-1863 vaniman@lanl.gov, Los Alamos; Gx
Vanko, David A., (410) 704-2121 dvanko@towson.edu, Towson Univ; Gxz
Vann, David R., (215) 898-4906 drvann@sas.upenn.edu, Univ of Pennsylvania; GeSfXm
Vann, Jamie, (610) 738-0598 jvann@wcupa.edu, West Chester Univ; CgGg
VanRegenmorter, John M., 616-331-3164 vanregjo@gvsu.edu, Grand Valley State Univ; PvGsr
VanTongeren, Jill, jill.vantongeren@tufts.edu, Tufts Univ; GiCu
Vaqueiro Rodriguez, Marcos, 0034637756150 mvaqueiro@frioya.es, Coruna Univ; GcmYd
Varadi, Ferenc D., (310) 206-6484 Univ of California, Los Angeles; Zn
Varady, Robert G., (520) 626-4393 rvarady@email.arizona.edu, Univ of Arizona; Zn
Varas, María Josefa, mjvaras@ucm.es, Univ Compl de Madrid; Gd
Varekamp, Johan C., (860) 685-2248 jvarekamp@wesleyan.edu, Wesleyan Univ; CgGzv
Varela, Diana, 250-721-6120 dvarela@uvic.ca, Univ of Victoria; Ob
Vargo, Gabriel A., (727) 864-2683 vargo@mail.usf.edu, Univ of South Florida; Ob
Varner, Ruth K., (603) 862-0853 ruth.varner@unh.edu, Univ of New Hampshire; Cl
Varricchio, David J., (406) 994-6907 djv@montana.edu, Montana State Univ; Pv
Vassiliou, Andreas H., 973-353-5100 ahvass@andromeda.rutgers.edu, Rutgers, The State Univ of New Jersey, Newark; Gz
Vatuva, Absai , avatuva@unam.na, Univ of Namibia; EgGtc
Vaudrey, Jamie, 860-405-9149 jamie.vaudrey@uconn.edu, Univ of Connecticut; Ob
Vaughan, David, +44 0161 275-3935 david.vaughan@manchester.ac.uk, Univ of Manchester; GzClGe
Vaughan, Geraint, +44 0161 306-3931 geraint.vaughan@manchester.ac.uk, Univ of Manchester; As
Vaughan, Michael T., 631 632-8030 michael.vaughan@stonybrook.edu, Stony Brook Univ; Yx
Vaughan, Naomi, +44 (0)1603 59 3904 n.vaughan@uea.ac.uk, Univ of East Anglia; Am
Vautier, Yannick, +33(0)3 44068991 yannick.vautier@lasalle-beauvais.fr, Institut Polytechnique LaSalle Beauvais (ex-IGAL); GodCo
Vaz, Nuno M., nunovaz@utad.pt, Unive de Trás-os-Montes e Alto Douro; PmlYg
Vazquez, Rogelio, rvazquez@cicese.mx, Centro de Investigación Científica y de Educación Superior de Ensenada; Ye
Veblen, Thomas T., veblen@colorado.edu, Univ of Colorado; Zg
Vecchi, Gabriel, 609-258-4101 gvecchi@princeton.edu, Princeton Univ; ZoOp
Veeder, Glenn J., (818) 354-7388 Jet Propulsion Lab; Zn
Veeger, Anne, (401) 874-4184 veeger@uri.edu, Univ of Rhode Island; Hw
Vega, Anthony J., (814) 226-2023 avega@clarion.edu, Clarion Univ; AtmOp
Vegas, Nestor, 0034946015374 nestor.vegas@ehu.eus, Univ of the Basque Country UPV/EHU; GctYm
Veizer, Jan, veizer@uottawa.ca, Univ of Ottawa; Cl
Velasco, Aaron A., (915) 747-5101 aavelasco@utep.edu, Univ of Texas, El Paso; Yx
Velbel, Michael A., (517) 355-4626 velbel@msu.edu, Michigan State Univ; ClGdSc
Velicogna, Isabella, 949-824-2685 isabella.velicogna@gmail.com, Univ of California, Irvine; Zr
Velinsky, David, 215 299 1147 velinsky@drexel.edu, Drexel Univ; OcCms
Velli, Marco, 310-206-4892 mvelli@ucla.edu, Univ of California, Los Angeles; Yg
Vengosh, Avner, 919 684 5847 vengosh@duke.edu, Duke Univ; Hw
Venkatesan, M I., (310) 206-2561 Univ of California, Los Angeles; Co
Venter, Marcie, (270) 809-3457 mventer@murraystate.edu, Murray State Univ; Ga
Venterea, Rodney T., (612) 624-7842 venterea@soils.umn.edu, Univ of Minnesota, Twin Cities; Sp
Vento, Frank, 814-824-2581 fvento@mercyhurst.edu, Mercyhurst Univ; GaSoGm
Ventura, Stephen J., (608) 262-6416 sventura@wisc.edu, Univ of Wisconsin, Madison; ZrSo
Venturelli, Ryan, 303-273-3800 venturelli@mines.edu, Colorado Sch of Mines; CgcZo
Venzke, Edward, (202) 633-1822 venzkee@si.edu, Smithsonian Inst / Natl Mus of Natural History; Gv
Vepraskas, Michael J., (919) 515-1458 North Carolina State Univ; Sd
Ver Straeten, Charles, (518) 486-2004 cverstrae@mail.nysed.gov, New York State Geo Survey; Grs
Verardo, Stacey, (703) 993-1045 sverardo@gmu.edu, George Mason Univ; Pei
Veraverbeke, Sander, +31 20 59 82975 s.n.n.veraverbeke@vu.nl, Vrije Univit Amsterdam; Zr
Verburg, Paul, 775-784-4511 pverburg@cabnr.unr.edu, Univ of Nevada, Reno; SdcSb
Veres, Michael, michael.veres@oswego.edu, SUNY, Oswego; As
Verlinde, Johannes, (814) 863-9711 jxv7@psu.edu, Pennsylvania State Univ, Univ Park; Am
Verma, Sumit, 432-552-2242 verma_s@utpb.edu, Univ of Texas Permian Basin; YeGo
Vermeesch, Pieter, +44 02076792418 p.vermeesch@ucl.ac.uk, Univ Coll London; Cc
Vermeij, Geerat J., (530) 752-2234 gjvermeij@ucdavis.edu, Univ of California, Davis; Pe
Vermeul, Vincent R., (509) 376-8316 vince.vermeul@pnl.gov, Pacific Northwest; Hg
Vernhes, Jean-David, +33(0)3 44062547 jean-david.vernhes@

lasalle-beauvais.fr, Institut Polytechnique LaSalle Beauvais (ex-IGAL); YgNgr
Verosub, Kenneth L., (530) 752-6911 klverosub@ucdavis.edu, Univ of California, Davis; Ym
Versteeg, Roelof, versteeg@ldeo.columbia.edu, Columbia Univ; Yg
Vervoort, Jeffrey D., (509) 335-5597 vervoort@wsu.edu, Washington State Univ; CcaGt
Vesovic, Velisa, +44 20 759 47352 v.vesovic@imperial.ac.uk, Imperial Coll; Np
Vesper, Dorothy, 304 293 5603 dorothy.vesper@mail.wvu.edu, West Virginia Univ; Hw
Vetter, Scott K., (318) 869-5055 svetter@centenary.edu, Centenary Coll of Louisiana; Gi
Vialle, Stephanie, stephanie.vialle@curtin.edu.au, Curtin Univ; YxCqNp
Vice, Mari A., (608) 342-1055 vice@uwplatt.edu, Univ of Wisconsin, Platteville; Gd
Vicens Batet, Enric, ++935811783 enric.vicens@uab.cat, Univ Auto de Barcelona; Pg
Vickery, Nancy, nvicker2@une.edu.au, Univ of New England; GgeYg
Vidal, Antonio, vidalv@cicese.mx, Centro de Investigación Científica y de Educación Superior de Ensenada; Ys
Vidal Romaní, Juan Ramon, 00 34 981 167000 xemoncho@udc.es, Coruna Univ; GmlGc
Vidale, John E., (206) 543-6790 vidale@uw.edu, Univ of Southern California; YsGtYg
Vidale, Jon , 310-210-2131 vidale@uw.edu, Univ of Washington; RnYs
Videtich, Patricia E., 616-331-3887 videticp@gvsu.edu, Grand Valley State Univ; Gd
Vidoviæ, Jelena, +38514606090 jelena.vidovic@geol.pmf.hr, Univ of Zagreb; Gg
Viedma, Cristóbal, viedma@ucm.es, Univ Compl de Madrid; Gz
Viegas, Anthony V., +91-8669609198 anthonyviegas@yahoo.com, Goa Univ; GiEmGx
Vieira, David J., (505) 667-7231 Los Alamos; Zn
Viens, Rob, (425) 564-3158 rob.viens@bellevuecollege.edu, Bellevue Coll; GgZg
Vierrether, Chris, 573-368-2370 chris.vierrether@dnr.mo.gov, Missouri Dept of Nat Res; GgoGe
Viertel, David C., 217-581-6244 dviertel@eiu.edu, Eastern Illinois Univ; Zru
Vietti, Laura, (307) 314-2024 lvietti@uwyo.edu, Univ of Wyoming; Pv
Vig, Pradeep K., 618.545.3373 pvig@kaskaskia.edu, Kaskaskia Coll; Gg
Vigouroux-Caillibot, Nathalie, 604-527-5860 vigourouxcaillibotn@douglascollege.ca, Douglas Coll; GviCg
Vila, Bernat, bernat.vila@uab.cat, Univ Auto de Barcelona; Pv
Vilcaez, Javier, 405744-6358 vilcaez@okstate.edu, Oklahoma State Univ; XzNpCl
Villa, Jorge, jorge.villa@louisiana.edu, Univ of Louisiana at Lafayette; Ge
Villalard-Bohnsack, Martine, (401) 254-3243 Roger Williams Univ; Ob
Villalobos, Joshua, (915) 831-7001 jvillal6@epcc.edu, El Paso Comm Coll; ZeYgu
Villareal, Tracy A., (361) 749-6732 tracyv@austin.utexas.edu, Univ of Texas, Austin; Ob
Villas Boas, Bia, villasboas@mines.edu, Colorado Sch of Mines; OpZrHs
Villaseca, Carlos, granito@ucm.es, Univ Compl de Madrid; Gi
Villegas, Monica B., 543584676198 mvillegas@exa.unrc.edu.ar, Univ Nacional de Rio Cuarto; Gsc
Villeneuve, Marlene C., +64 3 3667001 Ext 45682 marlene.villeneuve@canterbury.ac.nz, Univ of Canterbury; Ng
Vimont, Daniel J., (608)262-2828 dvimont@wisc.edu, Univ of Wisconsin, Madison; As
Vincent, Paul C., 229.333.5752 pvincent@valdosta.edu, Valdosta State Univ; Zig
Vincent, Robert K., (419) 372-0160 rvincen@bgnet.bgsu.edu, Bowling Green State Univ; Yg
Vinciguerra, Sergio Carmelo, sergiocarmelo.vinciguerra@unito.it, Univ di Torino; Nr
Vineis, Joseph, 609-258-4101 jv2474@princeton.edu, Princeton Univ; Cb
Visaggi, Christy, cvissagi@gsu.edu, Georgia State Univ; PiZe
Viskupic, Karen, (208) 426-3658 karenviskupic@boisestate.edu, Boise State Univ; ZeGc
Visscher, Pieter, (860) 486-4434 pieter.visscher@uconn.edu, Univ of Connecticut; Co
Vitale, Sarah A., 715-836-4300 vitalesa@uwec.edu, Univ of Wisconsin, Eau Claire; Hq
Vitek, John D., 405-744-6358 jvitek@okstate.edu, Oklahoma State Univ; Gm
Viti, Cecilia, cecilia.viti@unisi.it, Univ of Siena; Gz
Viti, Marcello, marcello.viti@unisi.it, Univ of Siena; YgsYh
Viviano-Beck, Christina E., christina.beck@jhuapl.edu, Univ of Tennessee, Knoxville; Xg
Vivoni, Enrique, (480) 727-3575 vivoni@asu.edu, Arizona State Univ; Hg
Vlahos, Penny, (860) 405-9269 penny.vlahos@uconn.edu, Univ of Connecticut; OcGeu
Vlahovic, Gordana, (919) 530-5172 gvlahovic@nccu.edu, North Carolina Central Univ; YsZeRn
Vlcan, Jennifer, 775-682-8759 mauldin@unr.edu, Univ of Nevada, Reno; Zi
Vocadlo, Lidunka, +44 020 7679 37919 l.vocadlo@ucl.ac.uk, Univ Coll London; Gy
Voelker, Bettina, (303) 273-3152 voelker@mines.edu, Colorado Sch of Mines; Ca
Vogel, Eve, 413-545-0778 evevogel@geo.umass.edu, Univ of Massachusetts, Amherst; Zy
Vogl, Jim, (352) 392-6987 jvogl@ufl.edu, Univ of Florida; Gct
Voglesonger, Kenneth M., 773-442-6053 k-voglesonger@neiu.edu, Northeastern Illinois Univ; ClGe
Vogt, Steven, (831) 459-2151 vogt@ucolick.org, Univ of California, Santa Cruz; Xc
Voice, Peter J., (269) 387-5446 peter.voice@wmich.edu, Western Michigan Univ; GsZn
Voicu, Gabriel-Constantin, 514-987-3000 #1648 gvoicu@videotron.ca, Univ du Quebec a Montreal; GcCg
Voight, Barry, (814) 238-4431 voight@ems.psu.edu, Pennsylvania State Univ, Univ Park; GveGg
Voigt, Vicki, 573-368-2128 vicki.voigt@dnr.mo.gov, Missouri Dept of Nat Res; Gg
Vojtko, Rastislav , vojtko@fns.uniba.sk, Comenius Univ in Bratislava; GctGm
Volborth, Alexis, (406) 496-4134 Montana Tech of the Univ of Montana; Ca
Voller, Vaughan R., (612) 625-0764 volle001@tc.umn.edu, Univ of Minnesota, Twin Cities; Zm
Vollmer, Frederick W., 845-257-3760 vollmerf@newpaltz.edu, SUNY, New Paltz; GctGx
Vollmer, Steve, 7815817370 x312 s.vollmer@neu.edu, Northeastern Univ; Zn
von Bitter, Peter H., 416-586-5591 peterv@rom.on.ca, Univ of Toronto; Pm
von der Handt, Anette, 612-624-7370 avdhandt@umn.edu, Univ of Minnesota, Twin Cities; GiCt
von Frese, Ralph R., (614) 292-5635 von-frese.3@osu.edu, Ohio State Univ; Ye
von Glasow, Roland, +44 (0)1603 59 3204 r.von-glasow@uea.ac.uk, Univ of East Anglia; Yg
Von Reden, Karl F., (508) 289-3384 kvonreden@whoi.edu, Woods Hole Oceanographic Inst; Ct
von Salzen, Knut, (250) 363-8287 knut.vonsalzen@canada.ca, Univ of British Columbia; As
von Seggern, David H., (775) 303 8461 vonseg@seismo.unr.edu, Univ of Nevada, Reno; YsgEo
Vondra, Carl F., (515) 294-5867 cfvondra@iastate.edu, Iowa State Univ of Sci & Tech; Gr
Vonk, Jorien, +31 20 59 87336 j.e.vonk@vu.nl, Vrije Univit Amsterdam; Hg
Voorhees, David H., (630) 466-2783 dvoorhees@waubonsee.edu, Waubonsee Comm Coll; ZgGg
Voorhees, Kent J., kvoorhee@mines.edu, Colorado Sch of Mines; Co
Vopson, Melvin M., +44 023 92 842246 melvin.vopson@port.ac.uk, Univ of Portsmouth; Yg
Voroney, R. Paul, (519) 824-4120 (Ext. 53057) pvoroney@uoguelph.ca, Univ of Guelph; Sb
Vorosmarty, Charles, 212-650-7042 cvorosmarty@ccny.cuny.edu, Graduate Sch of the City Univ of New York; Ng
Vorwald, Brian, vorwalb@sunysuffolk.edu, Suffolk County Comm Coll, Ammerman Campus; Zg
Voss, Regis D., (515) 294-1923 Iowa State Univ of Sci & Tech; So
Votaw, Robert, rvotaw@iun.edu, Indiana Univ Northwest; PiGrs
Voulgaris, George, (803) 777-2549 gvoulgaris@geol.sc.edu, Univ of South Carolina; OnpOg
Vrba, Elisabeth S., 203-432-5008 elisabeth.vrba@yale.edu, Yale Univ; Pvg
Vroon, Pieter, +31 20 59 87404 p.z.vroon@vu.nl, Vrije Univit Amsterdam; GvCcl
Vuke, Susan M., (406) 496-4326 svuke@mtech.edu, Montana Tech of The Univ of Montana; GgrGs
Vukovich, George, (416)736-2100 #30090 vukovich@yorku.ca, York Univ; As
Vulava, Vijay M., vulavav@cofc.edu, Coll of Charleston; CgHwSc
Vuori, Saku, +358 29 503 2459 saku.vuori@gtk.fi, Geo Survey of Finland; Gg

W

Wach, Grant D., (902) 494-8019 grant.wach@dal.ca, Dalhousie Univ; GorGs

Wada, Ikuko, (612) 301-9535 iwada@umn.edu, Univ of Minnesota, Twin Cities; ZnYgGt
Waddington, David C., (604) 527-5230 waddingtond@douglascollege.ca, Douglas Coll; ZgGuEd
Waddington, Edwin D., 206-543-4585 edw@uw.edu, Univ of Washington; Gl
Waddington, J. M., (905) 525-9140 (Ext. 23217) wadding@mcmaster.ca, McMaster Univ; Pe
Wade, Bridget S., +44 020 3108 6359 b.wade@ucl.ac.uk, Univ Coll London; PmeCl
Wade, Terry L., (979) 862-2325 t-wade@geos.tamu.edu, Texas A&M Univ; OcCao
Wadhwa, Meenakshi, (480) 965-0796 meenakshi.wadhwa@asu.edu, Arizona State Univ; Xc
Wadsworth, Fabian, fabian.b.wadsworth@durham.ac.uk, Durham Univ; Gv
Wadsworth, William B., (562) 907-4200 Whittier Coll; Gi
Wagger, Michael G., (919) 515-4269 North Carolina State Univ; Sb
Waggoner, Karen, 432-685-5540 kwaggoner@midland.edu, Midland Coll; GgRhGc
Wagner, John R., (864) 656-5024 jrwgnr@clemson.edu, Clemson Univ; ZeGm
Wagner, Lara, lwagner@carnegiescience.edu, Carnegie Inst for Sci; Ys
Wagner, Peter J., 312-665-7634 pwagner@fieldmuseum.org, Field Mus of Natural History; Pi
Wagner, Rick, 205-247-3622 rwagner@gsa.state.al.us, Geo Survey of Alabama; Cg
Wagner, Sasha, (518) 276-3075 wagnes3@rpi.edu, Rensselaer Polytechnic Inst; Cob
Wagner, Timothy J., (402) 280-2239 timothywagner@creighton.edu, Creighton Univ; As
Wagner Riddle, Claudia, (519) 824-4120 (Ext. 2787) cwagnerr@uoguelph.ca, Univ of Guelph; Am
Wagoner, Jeffrey L., (925) 422-1374 wagoner1@llnl.gov, Lawrence Livermore; Gs
Waheed, Umair B., +966138602297 umair.waheed@kfupm.edu.sa, King Fahd Univ of Petroleum and Minerals; Yg
Waid, Christopher, 614 265 6627 christopher.waid@dnr.state.oh.us, Ohio Dept of Nat Res; Eo
Wainman, Carmine, carmine.wainman@adelaide.edu.au, Univ of Adelaide; Grs
Waite, Gregory P., (906) 487-3554 gpwaite@mtu.edu, Michigan Technological Univ; YsGv
Waithaka, Hunja, hunja@eng.jkuat.ac.ke, Jomo Kenyatta Univ of Agriculture & Tech; YdZgi
Wakabayashi, John, 559-278-6459 jwakabayashi@csufresno.edu, Fresno State Univ; GmcGt
Wake, Cameron P., (603) 862-2329 cameron.wake@unh.edu, Univ of New Hampshire; Gl
Wake, David B., (510) 642-3059 wakelab@uclink.berkeley.edu, Univ of California, Berkeley; Pg
Wake, Marvalee H., (510) 642-4743 Univ of California, Berkeley; Pg
Wakefield, Kelli, kelliwakefield@hotmail.com, Mesa Comm Coll; Gg
Wakefield, S J., wakefield@cf.ac.uk, Cardiff Univ; Cm
Wala, Andrew M., 859-257-2959 Univ of Kentucky; Nm
Walcek, Christopher J., (518) 437-8720 SUNY, Albany; As
Waldbusser, George, 541-737-8964 waldbuss@coas.oregonstate.edu, Oregon State Univ; Ob
Walden, John, +44 01334 463688 jw9@st-andrews.ac.uk, Univ of St. Andrews; Ym
Waldhauser, Felix, (845) 365-8538 felixw@ldeo.columbia.edu, Columbia Univ; YsgGt
Waldien, Trevor S., (605) 394-2461 trevor.waldien@sdsmt.edu, South Dakota Sch of Mines & Tech; GctGp
Waldman, John, 718-997-3603 john.waldman@qc.cuny.edu, Graduate Sch of the City Univ of New York; ObHsZc
Waldron, John W., john.waldron@ualberta.ca, Univ of Alberta; Gc
Walford, Nigel, +44020 8417 2512 nwalford@kingston.ac.uk, Kingston Univ; Zi
Walker, Barry , bawalker@pdx.edu, Portland State Univ; GiCgGv
Walker, Brett, brett.walker@uottawa.ca, Univ of Ottawa; OgcCc
Walker, Charles T., (562) 985-4818 California State Univ, Long Beach; Cl
Walker, David, dwalker@ldeo.columbia.edu, Columbia Univ; Gx
Walker, Ian R., 250-807-9559 ian.walker@ubc.ca, Univ of British Columbia, Okanagan Campus; PeGnHs
Walker, J. Douglas, (785) 864-7711 jdwalker@ku.edu, Univ of Kansas; Gc
Walker, Jeffrey R., 845-437-5546 jewalker@vassar.edu, Vassar Coll; GzScGv
Walker, John L., (630) 252-6803 jlwalker@anl.gov, Argonne; Cg
Walker, Mark, (775) 784-1938 mwalker@cabnr.unr.edu, Univ of Nevada, Reno; Zn
Walker, Nan D., (225) 388-5331 Louisiana State Univ; Og
Walker, Peter, (541) 346-4541 pwalker@oregon.uoregon.edu, Univ of Oregon; Zn
Walker, Raymond J., (310) 825-7685 rwalker@igpp.ucla.edu, Univ of California, Los Angeles; Zn
Walker, Richard, +44 (1865) 282115 richard.walker@earth.ox.ac.uk, Univ of Oxford; Zr
Walker, Sally E., (706) 542-2396 swalker@gly.uga.edu, Univ of Georgia; Pm
Walker, Sara, (480) 727-2394 sara.i.walker@asu.edu, Arizona State Univ; Xb
Wallace, Adam F., (302) 831-1950 afw@udel.edu, Univ of Delaware; ClGzZm
Wallace, Davin, (228) 688-3060 davin.wallace@usm.edu, Univ of Southern Mississippi; Ou
Wallace, Douglas, (902) 494-4132 douglas.wallace@dal.ca, Dalhousie Univ; Oc
Wallace, Douglas W. R., (631) 344-2945 wallace@notes.cc.sunysb.edu, SUNY, Stony Brook; Oc
Wallace, Janae, (801) 537-3387 nrugs.jwallace@state.ut.us, Utah Geo Survey; Hw
Wallace, John M., (206) 543-7390 wallace@atmos.washington.edu, Univ of Washington; As
Wallace, Laura, lwallace@ig.utexas.edu, Univ of Texas, Austin; GtYdZr
Wallace, Paul, (541) 346-5985 pwallace@uoregon.edu, Univ of Oregon; GviCg
Wallace, Tim, 662-268-1032 Ext 244 tjw5@msstate.edu, Mississippi State Univ; Am
Wallace, William G., (718) 982-3876 william.wallace@csi.cuny.edu, Graduate Sch of the City Univ of New York; Og
Wallace, Jr., Terry C., (505) 667-3644 wallacet@lanl.gov, Los Alamos; Ys
Wallender, Wes W., 530.752.0688 wwwallender@ucdavis.edu, Univ of California, Davis; Hg
Waller, Thomas R., (202) 357-2127 Smithsonian Inst / Natl Mus of Natural History; Pi
Wallis, Ilka, ilka.wallis@flinders.edu.au, Flinders Univ; HqCg
Walowski, Kristina, (360) 650-4095 walowski@wwu.edu, Western Washington Univ; GxvCg
Walraevens, Kristine, (329) 264-4648 kristine.walraevens@ugent.be, Ghent Univ; HwCqYu
Walrod, Amanda G., (501) 575-7317 awalrod@comp.uark.edu, Univ of Arkansas, Fayetteville; Zn
Walsh, Christopher, (301) 405-4351 cswalsh@umd.edu, Univ of Maryland; Zn
Walsh, Ellen c., (920) 832-6739 ellen.c.walsh@lawrence.edu, Lawrence Univ; Zn
Walsh, Emily O., (319) 895-4302 ewalsh@cornellcollege.edu, Univ of Iowa; GptCc
Walsh, John J., (727) 553-1164 jwalsh@marine.usf.edu, Univ of South Florida; Ob
Walsh, Maud M., (225) 578-1211 evwals@lsu.edu, Louisiana State Univ; GeRcZe
Walsh, Talor B., 717-871-4270 talor.walsh@millersville.edu, Millersville Univ; GcmGe
Walsh, Tim R., 806-291-1123 Wayland Baptist Univ; GsPmGo
Walter, Julien, 41854550112680 julien_walter@uqac.ca, Univ du Quebec a Chicoutimi; HwqGm
Walter, Lynn M., (734) 763-4590 lmwalter@umich.edu, Univ of Michigan; Cl
Walter , Michael , 0117 9515007 m.j.walter@bristol.ac.uk, Univ of Bristol; GiCpg
Walter, Michael, mwalter@carnegiescience.edu, Carnegie Inst for Sci; Cg
Walter, Nathan A., (678) 839-4070 awalter@westga.edu, Univ of West Georgia; Zg
Walter, Robert C., 717 358-7198 robert.walter@fandm.edu, Franklin and Marshall Coll; Cc
Walter, Thomas, (212) 772-5457 twalter@hunter.cuny.edu, Hunter Coll (CUNY); ZyeAm
Walter-Shea, Elizabeth A., (402) 472-1553 ewalter-shea1@unl.edu, Univ of Nebraska, Lincoln; As
Walters, Richard, richard.walters@durham.ac.uk, Durham Univ; Yd
Walters, Steven, swalt826@uw.edu, Univ of Washington; Zir
Waltham, Dave, +44 1784 443617 d.waltham@rhul.ac.uk, Royal Holloway Univ of London; Yg
Walther, Ferdinand, 089/2180 4346 macferdi@gmx.org, Ludwig-Maximilians-Univ Muenchen; Gz
Walther, John V., (214) 768-3174 Southern Methodist Univ; Cg
Walther, Suzanne, (619) 260-4787 swalther@sandiego.edu, Univ of San Diego; GmZi
Walton, Anthony W., (785) 864-2726 twalton@ku.edu, Univ of Kansas; GsEoGv

Walton, Gabriel, (303) 273-2235 gwalton@mines.edu, Colorado Sch of Mines; NgrYg
Walton, Ian, 801-581-8497 iwalton@egi.utah.edu, Univ of Utah; Gq
Walton, Nick, +44 023 92 842263 nick.walton@port.ac.uk, Univ of Portsmouth; Hw
Walworth, James, (520) 626-3364 walworth@ag.arizona.edu, Univ of Arizona; Soc
Wamba, Mathurin D., 609-258-4101 mw1685@princeton.edu, Princeton Univ; Ys
Wampler, J. Marion, kayargon@earthlink.net, Georgia State Univ; Cc
Wampler, Peter J., 616-331-2834 wamplerp@gvsu.edu, Grand Valley State Univ; HsZiGm
Wan, Xianhui S., 609-258-4101 Princeton Univ; Cm
Wanamaker, Alan D., (515) 294-5142 adw@iastate.edu, Iowa State Univ of Sci & Tech; CsPeOg
Wang, Alian, (314) 935-5671 alianw@levee.wustl.edu, Washington Univ in St. Louis; Ca
Wang, Bin, wangbin@hawaii.edu, Univ of Hawai'i at Manoa; Am
Wang, Chi-Yuen, (510) 642-2288 chiyuen@seismo.berkeley.edu, Univ of California, Berkeley; Yg
Wang, Dongmei, (701) 777-6143 dongmei.wang@engr.und.edu, Univ of North Dakota; NgGo
Wang, Enru, enru.wang@und.edu, Univ of North Dakota; Zgi
Wang, Harry, (804) 684-7215 wang@vims.edu, William & Mary; Op
Wang, Herbert F., (608) 262-5932 hfwang@wisc.edu, Univ of Wisconsin, Madison; YgHwYx
Wang, Hong, 217-244-7692 hongwang@illinois.edu, Illinois State Geo Survey; Cg
Wang, Hsiang-Jui, (404) 894-3748 raywang@eas.gatech.edu, Georgia Inst of Tech; As
Wang, Jeen-Hwa, 886-2-27839910-326 jhwang@earth.sinica.edu.tw, Academia Sinica; Ysn
Wang, Jianwei, (225) 578-5532 jianwei@lsu.edu, Louisiana State Univ; ClZmGy
Wang, Jim, (225) 578-1360 jjwang@agcenter.lsu.edu, Louisiana State Univ; Szc
Wang, Kun, (314) 935-3855 kunwang@levee.wustl.edu, Washington Univ in St. Louis; XcCsXm
Wang, Lei, 765-494-3177 leiwang@purdue.edu, Purdue Univ; As
Wang, Lillian T., 302-831-1096 lillian@udel.edu, Univ of Delaware; Zi
Wang, Lixin, lxwang@iupui.edu, Indiana Univ - Purdue Univ Indianapolis; HyCl
Wang, Pao-Kuan, (608)263-6479 pao@windy.aos.wisc.edu, Univ of Wisconsin, Madison; As
Wang, Ping, pwang@usf.edu, Univ of South Florida; Gs
Wang, Qing, 831-656-7716 qwang@nps.edu, Naval Postgrad Sch; Am
Wang, Wei-Chyung, (518) 437-8708 wcwang@albany.edu, SUNY, Albany; As
Wang, Weihong, (801) 863-7607 weihong.wang@uvu.edu, Utah Valley Univ; CmGeZi
Wang, Xiaoyu , 609-258-4101 xw1796@princeton.edu, Princeton Univ; Yg
Wang, Xingchen, (617) 552-3327 xingchen.wang@bc.edu, Boston Coll; Cb
Wang, Yang, (850) 644-1121 ywang@magnet.fsu.edu, Florida State Univ; CgsPc
Wang, Yanghua, +44 20 759 41171 yanghua.wang@imperial.ac.uk, Imperial Coll; Yg
Wang, Yongji, 609-258-4101 yw1705@princeton.edu, Princeton Univ; As
Wang, Yuan, (765) 494-6581 yuanwang@purdue.edu, Purdue Univ; As
Wang, Yuhang, (404) 894-3995 yuhang.wang@eas.gatech.edu, Georgia Inst of Tech; As
Wang, Yuqing, yuqing@hawaii.edu, Univ of Hawai'i at Manoa; As
Wang, Yuxuan, 713-743-9049 ywang140@uh.edu, Univ of Houston; As
Wang, Z. Aleck, (508) 289-3676 zawang@whoi.edu, Woods Hole Oceanographic Inst; Cm
Wang, Zhankun, 979 458 3464 zhankunwang@tamu.edu, Texas A&M Univ; Op
Wang, Zhengrong, 212-650-5191 zwang1@ccny.cuny.edu, Graduate Sch of the City Univ of New York; Gge
Wang, Zhenming, 859-323-0564 zmwang@uky.edu, Univ of Kentucky; Ys
Wang, Zhi (Luke), (559) 278-4427 zwang@csufresno.edu, Fresno State Univ; SpHwZi
Wang, Zhuo, 217-244-4270 zhuowang@illinois.edu, Univ of Illinois, Urbana-Champaign; As
Wankel, Scott, (508) 289-3944 sdwankel@whoi.edu, Woods Hole Oceanographic Inst; Cm
Wanless, Dorsey, dwanless@boisestate.edu, Boise State Univ; Gi
Wanless, Harold R., (305) 284-2697 hwanless@miami.edu, Univ of Miami; GseOn
Wannamaker, Phillip E., (801) 581-3547 pewanna@egi.utah.edu, Univ of Utah; Ye
Warburton, David L., (561) 297-3312 warburto@fau.edu, Florida Atlantic Univ; CgGe
Ward, Adam, (812) 856-4820 adamward@indiana.edu, Indiana Univ, Bloomington; Hs
Ward, Ben A., +44 023 80596041 b.a.ward@soton.ac.uk, Univ of Southampton; Ob
Ward, Bess B., 609-258-5150 bbw@princeton.edu, Princeton Univ; ObCb
Ward, Brent C., (604) 291-4229 bcward@sfu.ca, Simon Fraser Univ; GlmGe
Ward, Collin, 508-289-2931 cward@whoi.edu, Woods Hole Oceanographic Inst; Cm
Ward, David M., (406) 994-3401 umbdw@montana.edu, Montana State Univ; Po
Ward, Dylan, (513) 556-2174 dylan.ward@uc.edu, Univ of Cincinnati; GmqZi
Ward, J. Evan, 860-405-9073 evan.ward@uconn.edu, Univ of Connecticut; Ob
Ward, Kevin M., (605) 394-2461 kevin.ward@sdsmt.edu, South Dakota Sch of Mines & Tech; YsGtYg
Ward, Landon, 828-350-4558 lward@unca.edu, Univ of North Carolina, Asheville; ZnnZn
Ward, Larry G., (603) 862-5132 larry.ward@unh.edu, Univ of New Hampshire; Gu
Ward, Peter D., 206-543-2962 swift@ocean.washington.edu, Univ of Washington; Pg
Wardell, Rob, (202) 633-1817 wardellr@si.edu, Smithsonian Inst / Natl Mus of Natural History; Zn
Wardlaw, Norman C., (403) 220-6429 nwardlaw@geo.ucalgary.ca, Univ of Calgary; Go
Waren, Kirk B., 406-496-4866 kwaren@mtech.edu, Montana Tech of The Univ of Montana; HwsZe
Warger, Jane, (909) 652-6485 jane.warger@chaffey.edu, Chaffey Coll; ZgGg
Warke, Patricia, p.warke@qub.ac.uk, Queen's Univ Belfast; GmZym
Warkentin, Alicia, 604-822-3146 awarkentin@eoas.ubc.ca, Univ of British Columbia; Zn
Warland, Jon, (519) 824-4120 (Ext. 6374) jwarland@uoguelph.ca, Univ of Guelph; As
Warme, John E., jwarme@mines.edu, Colorado Sch of Mines; GsPeGu
Warner, Mark J., (206) 543-0765 warner@u.washington.edu, Univ of Washington; Op
Warner, Michael, +44 20 759 46535 m.warner@imperial.ac.uk, Imperial Coll; Ys
Warner, Nicholas H., 585-245-5291 warner@geneseo.edu, SUNY, Geneseo; XgGrHg
Warner, Richard D., (864) 656-5023 wrichar@ces.clemson.edu, Clemson Univ; Gz
Warner, Timothy A., (304) 293-4725 tim.warner@mail.wvu.edu, West Virginia Univ; Zr
Warnock, Jonathan P., (724) 357-5627 jwarnock@iup.edu, Indiana Univ of Pennsylvania; PemPv
Warny, Sophie, (225) 578-5089 swarny@lsu.edu, Louisiana State Univ; Pl
Warren, Jessica, 302-831-2569 warrenj@udel.edu, Univ of Delaware; GxCpGt
Warren, Joseph, (631) 632-5045 joe.warren@stonybrook.edu, SUNY, Stony Brook; Ob
Warren, Lesley A., (905) 525-9140 (Ext. 27347) warrenl@mcmaster.ca, McMaster Univ; CgHg
Warren, Linda M., (314) 977-3197 linda.warren@slu.edu, Saint Louis Univ; YgsZn
Warren, Paul, (310) 825-2015 Univ of California, Los Angeles; Xm
Warren, Rachel, +44 (0)1603 59 3912 r.warren@uea.ac.uk, Univ of East Anglia; Eg
Warren, Richard G., (505) 667-7063 rgw@lanl.gov, Los Alamos; Gi
Warren, Stephen G., (206) 543-7230 sgw@atmos.washington.edu, Univ of Washington; ApGl
Warrick, Arthur W., (520) 621-1516 aww@ag.arizona.edu, Univ of Arizona; Sp
Wartes, Marwan A., (907) 451-5056 marwan.wartes@alaska.gov, Alaska Division of Geological & Geophysical Surveys; GsoGt
Warwick, Phillip, +44 (0)23 80592780 phil.warwick@noc.soton.ac.uk, Univ of Southampton; CaGe
Wash, Carlyle H., 831-656-7776 wash@nps.edu, Naval Postgrad Sch; Am
Washington, Paul A., pw005@marietta.edu, Marietta Coll; Gcs
Wasomi, Charles B., cwasomi@jkuat.ac.ke, Jomo Kenyatta Univ of Agriculture & Tech; ZrrZi
Wassenaar, Len, (306) 239-2270 Univ of Saskatchewan; Hw
Wassermann, Joachim, 08141/5346762 jowa@geophysik.uni-muenchen.de, Ludwig-Maximilians-Univ Muenchen; Yg
Wasson, John T., (310) 825-1986 jtwasson@ucla.edu, Univ of California,

Los Angeles; Xm
Wasylenki, Laura, laura.wasylenki@nau.edu, Northern Arizona Univ; ClOoCs
Waszek, Lauren, lwaszek@nmsu.edu, New Mexico State Univ, Las Cruces; Ys
Waszgis, Michele M., (402) 472-0693 mwaszgis@unl.edu, Univ of Nebraska, Lincoln; Gg
Watanabe, Tohru, 81-76-445-6650 twatnabe@sci.u-toyama.ac.jp, Univ of Toyama; YxsGv
Waters, Dave, +44 (1865) 282457 dave.waters@earth.ox.ac.uk, Univ of Oxford; GpzGc
Waters, Matthew N., (334) 844-4100 mwaters@auburn.edu, Auburn Univ; Gn
Waters, Michael R., (979) 845-5246 mwaters@tamu.edu, Texas A&M Univ; Ga
Waters-Tormey, Cheryl, (828) 227-3696 cherylwt@wcu.edu, Western Carolina Univ; Gct
Waterstone, Marvin, marvinw@email.arizona.edu, Univ of Arizona; Zn
Waterworth, Lauren H., waterworthlh@appstate.edu, Appalachian State Univ; Gcg
Watkins, David, d.c.watkins@exeter.ac.uk, Exeter Univ; Hg
Watkins, James, watkins4@uoregon.edu, Univ of Oregon; Cg
Watkinson, A. John, (509) 335-2470 watkinso@mail.wsu.edu, Washington State Univ; Gc
Watkinson, Andrew, +44 (0)1603 59 2267 a.watkinson@uea.ac.uk, Univ of East Anglia; Ge
Watkinson, David H., dwatkson@ccs.carleton.ca, Carleton Univ; Em
Watkinson, Ian M., +44 1784 414046 ian.watkinson@rhul.ac.uk, Royal Holloway Univ of London; GtcRn
Watkinson, Matthew, +44 1752 584765 m.p.watkinson@plymouth.ac.uk, Univ of Plymouth; Gs
Watney, W. Lynn, (785) 864-2184 lwatney@kgs.ku.edu, Univ of Kansas; Gr
Watson, Alan, awatson@uoguelph.ca, Univ of Guelph; Zn
Watson, David B., (865) 241-4749 v6i@ornl.gov, Oak Ridge; Hw
Watson, E. Bruce, (518) 276-8838 watsoe@rpi.edu, Rensselaer Polytechnic Inst; CpcCt
Watson, Elizabeth, 215.299.1109 elizabeth.b.watson@drexel.edu, Drexel Univ; Zc
Watson, James R., 541-737-2519 jrwatson@coas.oregonstate.edu, Oregon State Univ; Zy
Watson, Kelly, 859-622-1419 kelly.watson@eku.edu, Eastern Kentucky Univ; ZriZn
Watters, Robert J., (775) 784-6069 watters@mines.unr.edu, Univ of Nevada, Reno; Nrg
Watters, Thomas R., 202-633-2483 watterst@si.edu, Smithsonian Inst, Natl Air & Space Mus; Xg
Wattrus, Nigel J., (218) 726-7154 nwattrus@d.umn.edu, Univ of Minnesota Duluth; Yr
Watts, Anthony B., +44 (1865) 272032 tony@earth.ox.ac.uk, Univ of Oxford; GutYv
Watts, Chester F., (540) 831-5637 cwatts@radford.edu, Radford Univ; NgrHw
Watts, D. Randolph, (401) 874-6507 rwatts@gso.uri.edu, Univ of Rhode Island; Op
Watts, Doyle, 937 775-3455 doyle.watts@wright.edu, Wright State Univ; YeZr
Waugh, Darryn W., (410) 516-8344 waugh@jhu.edu, Johns Hopkins Univ; As
Waugh, John, 757-822-7436 tcwaugj@tcc.edu, Tidewater Comm Coll; Gg
Waugh, Richard A., (608) 342-1386 waugh@uwplatt.edu, Univ of Wisconsin, Platteville; Gc
Wauthier, Christelle, 814-865-6711 cuw25@psu.edu, Pennsylvania State Univ, Univ Park; Zn
Wax, Charles L., (662) 325-3915 wax@geosci.msstate.edu, Mississippi State Univ; As
Wdowinski, Shimon, (305) 348-6826 shimon.wdowinski@fiu.edu, Florida InterNatl Univ; YdGt
Weatherford, Jonathan, jonathan.weatherford@pcc.edu, Portland Comm Coll - Sylvania Campus; Gg
Weaver, Andrew J., (250) 472-4001 weaver@ocean.seos.uvic.ca, Univ of Victoria; Op
Weaver, Barry L., (405) 325-4492 bweaver@ou.edu, Univ of Oklahoma; Ct
Weaver, Douglas J., (702) 295-5916 douglas_weaver@lanl.gov, Los Alamos; Zn
Weaver, John T., (250) 721-6155 weaver@phys.uvic.ca, Univ of Victoria; Ym
Weaver, Justin E., 806-834-4610 justin.e.weaver@ttu.edu, Texas Tech Univ; As
Weaver, Stephen G., (719) 389-6954 sweaver@coloradocollege.edu, Colorado Coll; Gx
Weaver, Thomas A., (505) 667-8464 tweaver@lanl.gov, Los Alamos; Yg
Webb, Casey, caseywebb@suu.edu, Southern Utah Univ; GtZg
Webb, Craig, craig.webb@eku.edu, Eastern Kentucky Univ; Zyn
Webb, Elizabeth A., 5196612111 x80208 ewebb5@uwo.ca, Western Univ; CslSa
Webb, Laura E., (802) 656-8136 lewebb@uvm.edu, Univ of Vermont; GtCcGc
Webb, Nathan D., (217) 244-2426 ndwebb2@illinois.edu, Illinois State Geo Survey; EoGo
Webb, Peter N., (614) 292-7285 webb.3@osu.edu, Ohio State Univ; PmRhGr
Webb, Spahr, scw@ldeo.columbia.edu, Columbia Univ; Yr
Webb, III, Thompson, (401) 863-3128 thompson_webb_iii@brown.edu, Brown Univ; PeIAs
Webber, Andrew J., (513) 260-5863 webberaw@ucmail.uc.edu, Cincinnati Mus Center; Pi
Webber, Jeffrey, jeffrey.webber@stockton.edu, Stockton Univ; Gg
Webber, Karen L., (504) 280-6791 kwebber@uno.edu, Univ of New Orleans; Gv
Weber, Bodo, bweber@cicese.mx, Centro de Investigación Científica y de Educación Superior de Ensenada; Gp
Weber, John C., 616-331-3191 weberj@gvsu.edu, Grand Valley State Univ; GctYd
Weber, Karrie A., (402) 472-2739 kweber@unl.edu, Univ of Nebraska, Lincoln; PgCl
Weber, Keith, (208) 282-2757 webekeit@isu.edu, Idaho State Univ; Zi
Weber, Rodney J., (404) 894-1750 rweber@eas.gatech.edu, Georgia Inst of Tech; As
Weber, Thomas, 585-275-2103 t.weber@rochester.edu, Univ of Rochester; Obu
Weber-Diefenbach, Klaus, 089/2180 6549 klaus.diefenbach@iaag.geo.uni-muenchen.de, Ludwig-Maximilians-Univ Muenchen; Gg
Weborg-Benson, Kimberly, (716) 673-3293 kim.weborg-benson@fredonia.edu, SUNY, Fredonia; GgAsPg
Webre, Cherri B., (225) 388-8328 cherri@lgs.bri.lsu.edu, Louisiana State Univ; Zn
Webster, Ferris, (302) 645-4266 Univ of Delaware; Op
Webster, Gary D., (509) 335-4369 webster@wsu.edu, Washington State Univ; Pis
Webster, John R., (701) 858-3873 john.webster@minotstateu.edu, Minot State Univ; Giz
Webster, Peter J., (404) 894-1748 pjw@eas.gatech.edu, Georgia Inst of Tech; AsOp
Wedding, William C., 859-257-1883 chad.wedding@uky.edu, Univ of Kentucky; Nm
Weeden, Lori, (978) 934-3344 lori_weeden@uml.edu, Univ of Massachusetts, Lowell; GgeGs
Weeraratne, Dayanthie, (818) 677-2046 dsw@csun.edu, California State Univ, Northridge; Yg
Weglein, Arthur B., (713) 743-3848 aweglein@uh.edu, Univ of Houston; Ye
Wegmann, Karl, 919-515-0380 karl_wegmann@ncsu.edu, North Carolina State Univ; Gm
Wegner, John, 404 727 4206 jwegner@emory.edu, Emory Univ; Sf
Wehmiller, John F., (302) 831-2926 jwehm@udel.edu, Univ of Delaware; Cl
Wehner, Peter J., (512) 223-0201 pwehner@austincc.edu, Austin Comm Coll District; GieXs
Wehrmann, Laura, 089/2180 Ludwig-Maximilians-Univ Muenchen; Pg
Wei, Xiaofang, 937 376 6193 xwei@centralstate.edu, Central State Univ; ZrOgZe
Wei, Xiaohua (Adam), 250-807-8750 adam.wei@ubc.ca, Univ of British Columbia, Okanagan Campus; HqgSf
Weiant, Morgen C., 717-702-2027 DCNR- Pennsylvania Bur of Geo Survey; Zi
Weiblen, Paul W., 612-625-3477 pweib@umn.edu, Univ of Minnesota, Twin Cities; Gi
Weidner, Donald J., (631) 632-8211 donald.weidner@sunysb.edu, Stony Brook Univ; Gy
Weil, Arlo B., (610) 526-5113 aweil@brynmawr.edu, Bryn Mawr Coll; GctYm
Weiland, Thomas J., tjw@canes.gsw.edu, Georgia Southwestern State Univ; Gi
Weimer, Paul, (303) 492-3809 paul.weimer@colorado.edu, Univ of Colorado; Gor
Weinbeck, Robert S., weinbeck@ametsoc.org, SUNY, The Coll at Brockport; As
Weinberger, Alycia J., aweinberger@carnegiescience.edu, Carnegie Inst for Sci; Xa
Weinstein, Charles E., (413) 236-4556 cweinste@berkshirecc.edu, Berkshire

Comm Coll; Zn
Weintraub, Michael N., 419-530-2585 michael.weintraub@utoledo.edu, Univ of Toledo; Zn
Weirich, Frank H., (319) 335-0156 frank-weirich@uiowa.edu, Univ of Iowa; Gm
Weis, Dominique A., (604) 822-1697 dweis@eos.ubc.ca, Univ of British Columbia; CaGeCu
Weisberg, Michael, 718-997-3366 michael.weisberg@kbcc.cuny.edu, Graduate Sch of the City Univ of New York; Zy
Weisberg, Robert H., (727) 553-1568 rweisberg@marine.usf.edu, Univ of South Florida; Op
Weisener, Christopher, 519-253-3000 x3753 weisener@uwindsor.ca, Univ of Windsor; CoGePg
Weislogel, Amy L., 304-293-5603 amy.weislogel@mail.wvu.edu, West Virginia Univ; Gs
Weisman, Robert A., 320-308-3247 raweisman@stcloudstate.edu, Saint Cloud State Univ; Am
Weiss, Alfred W., 630.466.2720 aweiss@waubonsee.edu, Waubonsee Comm Coll; Ziy
Weiss, Anna M., (940) 397-4448 anna.weiss@msutexas.edu, Midwestern State Univ; PcgGs
Weiss, Chester J., 540-231-6521 cjweiss@unm.edu, Univ of New Mexico; Ye
Weiss, Christopher C., 806-834-4712 chris.weiss@ttu.edu, Texas Tech Univ; AssAs
Weiss, Dominik, +44 20 759 46383 d.weiss@imperial.ac.uk, Imperial Coll; ClaCq
Weiss, Peter, 831-459-1616 pweiss@ucsc.edu, Univ of California, Santa Cruz; Cb
Weiss, Ray F., (858) 534-2598 rfweiss@ucsd.edu, Univ of California, San Diego; Cm
Weiss, Robert, 540-231-2334 weiszr@vt.edu, Virginia Tech; RnOn
Weissmann, Gary S., 505-277-4204 weissman@unm.edu, Univ of New Mexico; HwGsr
Welby, Charles W., (919) 515-7158 cwwelby@unity.ncsu.edu, North Carolina State Univ; Hw
Welch, Steve M., (785) 532-7236 welchsm@ksu.edu, Kansas State Univ; Sp
Welch, Susan A., 614-292-9059 welch.318@osu.edu, Ohio State Univ; ClGeCs
Weldeab, Syee, 805-893-4903 weldeab@geol.ucsb.edu, Univ of California, Santa Barbara; PeCm
Weldeghebriel, Mebrahtu, 609-258-4101 mweldeg@princeton.edu, Princeton Univ; Cl
Weldon, Ray J., (541) 346-4584 ray@uoregon.edu, Univ of Oregon; Gc
Welling, Tim, 845-431-8536 welling@sunydutchess.edu, Dutchess Comm Coll; GeSo
Wellner, Julia, 713-743-2887 jwellner@uh.edu, Univ of Houston; GslOu
Wells, David E., (228) 688-3389 dew@unb.ca, Univ of Southern Mississippi; YdZi
Wells, Michael L., (702) 895-0828 michael.wells@unlv.edu, Univ of Nevada, Las Vegas; Gc
Wells, Neil A., (330) 672-2951 nwells@kent.edu, Kent State Univ; Gs
Wells, Randy, (941) 388-4441 Univ of South Florida; Ob
Wells, Steve G., (775) 673-7470 sgwells@dri.edu, Univ of Nevada, Reno; Zn
Welp, Lisa, (765) 496-6896 lwelp@purdue.edu, Purdue Univ; Cs
Welsh, James L., 507-933-7333 welsh@gac.edu, Gustavus Adolphus Coll; Gxc
Wen, Lianxing, (631) 632-1726 lianxing.wen@stonybrook.edu, Stony Brook Univ; Ys
Wen, Tao, 315-443-2672 twen08@syr.edu, Syracuse Univ; Ge
Wen, Xianyun, +44(0) 113 34 31425 x.wen@leeds.ac.uk, Univ of Leeds; NoAsm
Wen, Yuming, 16717352685 ywen@uguam.uog.edu, Univ of Guam; Zir
Wendlandt, Richard F., (303) 273-3809 rwendlan@mines.edu, Colorado Sch of Mines; GizCp
Weng, Kevin, 808-956-4109 kevin.weng@hawaii.edu, Univ of Hawai'i, Manoa; Ob
Weng, Qihao, 812-237-2255 qweng@indstate.edu, Indiana State Univ; Zru
Wenk, Hans-Rudolf, (510) 642-7431 wenk@berkeley.edu, Univ of California, Berkeley; Gz
Wennberg, Paul O., (626) 395-2447 wennberg@gps.caltech.edu,CalTech; As
Wenner, David B., (706) 542-2393 dwenner@uga.edu, Univ of Georgia; Cs
Wenner, Jennifer M., (920) 424-7003 wenner@uwosh.edu, Univ of Wisconsin, Oshkosh; Gi
Wentzcovitch, Renata, rmw2150@columbia.edu, Columbia Univ; Ys
Wenzel, Christopher, (307) 532-8293 chris.wenzel@ewc.wy.edu, Eastern Wyoming Coll; SbfHg
Weppner, Kerrie, kerrieweppner@boisestate.edu, Boise State Univ; GeZc
Werdon, Melanie B., (907) 451-5082 melanie.werdon@alaska.gov, Alaska Division of Geological & Geophysical Surveys; Eg
Werhner, Matthew J., (813) 253-7263 mwerhner@hccfl.edu, Hillsborough Comm Coll; Gg
Werne, Josef, (412) 624-8775 jwerne@pitt.edu, Univ of Pittsburgh; CosGn
Werner, Adrian , adrian.werner@flinders.edu.au, Flinders Univ; Hw
Werner, Alan, (413) 538-2134 awerner@mtholyoke.edu, Mount Holyoke Coll; GleGm
Werner, Bradley T., (858) 534-0583 bwerner@ucsd.edu, Univ of California, San Diego; Gm
Wernette, Shelly J., 512-245-0377 swernette@txstate.edu, Texas State Univ; GgPge
Wernicke, Brian P., (626) 395-6192 brian@gps.caltech.edu,CalTech; Gtc
Werth, Susanna, 540-231-6521 swerth@vt.edu, Virginia Tech; ZrHgYv
Werts, Scott P., 803-323-4930 wertss@winthrop.edu, Winthrop Univ; ScbPg
Wesely, Marvin L., (630) 252-5827 mlwesely@anl.gov, Argonne; As
Wesley Wood, Charley, 850-9837126 woodwes@ufl.edu, Univ of Florida; So
Wesnousky, Steve, (775) 784-6067 stevew@seismo.unr.edu, Univ of Nevada, Reno; Ys
Wesnousky, Steven, (775) 784-6067 stevew@seismo.unr.edu, Univ of Nevada, Reno; Ys
Wessel, Paul, (808) 956-4778 pwessel@hawaii.edu, Univ of Hawai'i, Manoa; YrGt
West, A. Joshua (Josh), (213) 740-6736 joshwest@usc.edu, Univ of Southern California; ClGmHs
West, Alan, 276-328-0203 maw4v@uvawise.edu, Univ of Virginia's Coll at Wise; Zn
West, David P., (802) 443-3476 dwest@middlebury.edu, Middlebury Coll; Gc
West, Jared, +44(0) 113 34 35253 l.j.west@leeds.ac.uk, Univ of Leeds; Hg
West, Keith, (715) 504-3337 westk@uwgb.edu, Univ of Wisconsin, Green Bay; Zgn
West, Larry T., (706) 542-0906 Univ of Georgia; Sd
West, Nicole, 989-774-4475 west2n@cmich.edu, Central Michigan Univ; Gm
West, Olivia M., (615) 576-0505 qm5@ornl.gov, Oak Ridge; Ng
West, Robert, 323-260-8115 westrb@elac.edu, East Los Angeles Coll; GgmSd
West, Ronald R., (785) 532-2248 rrwest@ksu.edu, Kansas State Univ; PgeGs
West, Terry R., (765) 494-3296 trwest@purdue.edu, Purdue Univ; Ng
Westerink, Joannes J., (574) 631-6475 Univ of Notre Dame; On
Westerman, David S., (802) 485-2337 westy@norwich.edu, Norwich Univ; Gxi
Westervelt, Daniel, danielmw@ldeo.columbia.edu, Columbia Univ; Og
Westgate, James W., (409) 880-8237 james.westgate@lamar.edu, Lamar Univ; Pev
Westgate, John A., (416) 287-7235 Univ of Toronto; Gl
Westrop, Jeffrey, (402) 472-9422 jwestrop@unl.edu, Univ of Nebraska, Lincoln; HwClb
Westrop, Stephen R., (405) 325-0542 swestrop@ou.edu, Univ of Oklahoma; Pi
Wetherelt, Andrew, +44 01326 371827 a.wetherelt@exeter.ac.uk, Exeter Univ; Nm
Wetmore, Paul, wetmore@usf.edu, Univ of South Florida; Gc
Wettlaufer, John S., (203) 432-0892 john.wettlaufer@yale.edu, Yale Univ; Yg
Wettstein, Justin, (541) 737-5177 justinw@coas.oregonstate.edu, Oregon State Univ; As
Wetz, Michael, 361-825-2132 michael.wetz@tamucc.edu, Texas A&M Univ, Corpus Christi; Ob
Wetzel, Dan L., (907) 488-3746 Univ of Alaska, Fairbanks; Zn
Wetzel, Laura R., (727) 864-8484 wetzellr@eckerd.edu, Eckerd Coll; YrOuGg
Wexler, Anthony S., 530-754-6558 aswexler@ucdavis.edu, Univ of California, Davis; ZcAsZn
Whalen, Michael T., 907-474-5302 ffmtw@uaf.edu, Univ of Alaska, Fairbanks; Gs
Whaler, Kathryn A., +44 (0) 131 650 4904 kathy.whaler@ed.ac.uk, Edinburgh Univ; YmRn
Whalley, John, +44 023 9284 2257 john.whalley@port.ac.uk, Univ of Portsmouth; Gc
Whattam, Scott, +9661368602993 scott.whattam@kfupm.edu.sa, King Fahd Univ of Petroleum and Minerals; GxXc
Wheatcraft, Steve, (775) 784-1973 steve@wheatcraftandassociates.com, Univ of Nevada, Reno; Hw
Wheatcroft, Rob, 541-737-3891 raw@coas.oregonstate.edu, Oregon State Univ; OgGg
Wheaton, John R., (406) 272-1603 jwheaton@mtech.edu, Montana Tech of The Univ of Montana; HwEc
Wheeler, Andy J., +353 21 4903951 a.wheeler@ucc.ie, Univ Coll Cork;

GusGg
Wheeler, Greg, 916-278-3919 wheelergr@csus.edu, Sacramento State Univ; Em
Wheeler, John, +440151 794 5172 johnwh@liverpool.ac.uk, Univ of Liverpool; GpcGz
Wheeler, Paul, +44 01326 255778 p.wheeler@exeter.ac.uk, Exeter Univ; Nx
Wheeler, Richard F., (931) 221-1004 wheelerr@apsu.edu, Austin Peay State Univ; Ggo
Whelan, Jean K., (508) 289-2819 jwhelan@whoi.edu, Woods Hole Oceanographic Inst; Co
Whelan, Peter M., (320) 589-6315 whelan@mrs.umn.edu, Univ of Minnesota, Morris; Gx
Whiley, Harriet, harriet.whiley@flinders.edu.au, Flinders Univ; Zn
Whipple, Kelin, 480-965-9508 kxw@asu.edu, Arizona State Univ; Gmt
Whisner, Jennifer B., (570) 389-4139 jwhisner@bloomu.edu, Bloomsburg Univ; GmHw
Whisner, S. Christopher, (570) 389-4746 swhisner@bloomu.edu, Bloomsburg Univ; GciGz
Whisonant, Robert C., (540) 831-5224 rwhisona@radford.edu, Radford Univ; GsaGg
Whitaker, Rodney W., (505) 667-7672 rww@lanl.gov, Los Alamos; As
White, Bekki C., 501 296-1877 bekki.white@arkansas.gov, Arkansas Geo Survey; GoEoGg
White, Craig M., cwhite@boisestate.edu, Boise State Univ; Giv
White, George W., (301) 687-4264 gwhite@frostburg.edu, Frostburg State Univ; Zn
White, James D., +64 3 479-9009 james.white@otago.ac.nz, Univ of Otago; Gvs
White, James W. C., 303-492-5494 jwhite@colorado.edu, Univ of Colorado; CsAt
White, Jeffrey G., (919) 515-2389 jeffrey_white@ncsu.edu, North Carolina State Univ; SoZri
White, John C., (859) 622-1276 john.white@eku.edu, Eastern Kentucky Univ; GiCgt
White, Jonathan, (303) 384-2650 jwhite@mines.edu, Colorado Geo Survey; NgZuGg
White, Joseph C., (506) 453-4804 clancy@unb.ca, Univ of New Brunswick; Gc
White, Justin, 801-863-6864 justin.white@uvu.edu, Utah Valley Univ; ZyiZu
White, Lisa D., 510-664-4966 ldwhite@berkeley.edu, Univ of California, Berkeley; ZePim
White, Mark D., (509) 372-6070 mark.white@pnl.gov, Pacific Northwest; Hy
White, Martin, +353 (0)91 493 214 martin.white@nuigalway.ie, Natl Univ of Ireland Galway; Op
White, Nicky, +44 (0) 1223 337063 njw10@cam.ac.uk, Univ of Cambridge; YgGt
White, Paul , (775) 682-6193 paulwhite@unr.edu, Univ of Nevada, Reno; Zn
White, Paul, pdwhite1@ccri.edu, Comm Coll of Rhode Island; Gg
White, Robert, +44 (0) 1223 337187 rsw1@cam.ac.uk, Univ of Cambridge; YgGt
White, Scott M., (803) 777-6304 swhite@geol.sc.edu, Univ of South Carolina; YrZr
White, Susan, just19@earthlink.net, Los Angeles Harbor Coll; Zg
White, William B., (814) 667-2709 wbw2@psu.edu, Pennsylvania State Univ, Univ Park; CgGzy
Whiteaker, Timothy L., 512-471-0570 twhit@mail.utexas.edu, Univ of Texas, Austin; Zi
Whitehead, James, (506) 453-4593 Univ of New Brunswick; Gg
Whitehead, Peter W., +61 742321200 peter.whitehead@jcu.edu.au, James Cook Univ; Ggv
Whitehead, Robert E., rwhitehead@laurentian.ca, Laurentian Univ, Sudbury; Ce
Whitehill, Matthew, 218-733-5981 m.whitehill@lsc.edu, Lake Superior Coll; Gg
Whiteley, Martin, +44 01332 593752 m.whiteley@derby.ac.uk, Univ of Derby; Go
Whiteside, Jessica H., +44 (0)23 80593199 j.whiteside@soton.ac.uk, Univ of Southampton; Pe
Whiteway, James, (416)736-2100 #22310 whiteway@yorku.ca, York Univ; As
Whiticar, Michael J., (250) 721-6514 whiticar@uvic.ca, Univ of Victoria; CosGo
Whiting, Peter J., (216) 368-3989 peter.whiting@case.edu, Case Western Reserve Univ; Gm
Whitlock, Cathy, (406) 994-6910 whitlock@montana.edu, Montana State Univ; Peg
Whitman, Brian E., (570) 408-4882 brian.whitman@wilkes.edu, Wilkes Univ; HqwSb
Whitman, Dean, (305) 348-3089 whitmand@fiu.edu, Florida InterNatl Univ; YgGet
Whitman, Jill M., (253) 535-8720 whitmaj@plu.edu, Pacific Lutheran Univ; OuGuZe
Whitmeyer, Steven J., whitmesj@jmu.edu, James Madison Univ; GctZi
Whitmore, John H., (937) 766-7947 johnwhitmore@cedarville.edu, Cedarville Univ; PgGsg
Whitney, D.A., (785) 532-6101 whitney@ksu.edu, Kansas State Univ; So
Whitney, Donna L., (612) 626-7582 dwhitney@umn.edu, Univ of Minnesota, Twin Cities; Gpt
Whitney, Earl M., (505) 667-3595 whitney@lanl.gov, Los Alamos; Zn
Whitney, James A., (706) 548-6894 jamesawhitney@hotmail.com, Univ of Georgia; GivCg
Whitney, Michael, 860-405-9157 michael.whitney@uconn.edu, Univ of Connecticut; Op
Whitney , Sandra , swhitney@uga.edu, Univ of Georgia; Ga
Whittaker, Alex, +44 20 759 47491 a.whittaker@imperial.ac.uk, Imperial Coll; Gt
Whittaker, Amber, 207-287-2803 amber.whittaker@maine.gov, Dept of Agriculture, Conservation, and Forestry; GcZiCg
Whittaker, Steven, (217) 244-3114 sgwhit@illinois.edu, Illinois State Geo Survey; Goe
Whittecar, Jr., G. Richard, (757) 683-5197 rwhittec@odu.edu, Old Dominion Univ; GmHwZe
Whittemore, Donald O., 785-864-3965 donwhitt@kgs.ku.edu, Univ of Kansas; Cl
Whittington, Alan, (573) 884-7625 whittingtona@missouri.edu, Univ of Missouri; GivCp
Whittington, Carla, (206) 878-3710 cwhittin@highline.edu, Highline Coll; GgvZg
Whitton, Mark, +44 023 9284 2257 mark.witton@port.ac.uk, Univ of Portsmouth; Pg
Wiberg, Patricia L., (434) 924-7546 pw3c@virginia.edu, Univ of Virginia; Og
Wickert, Andrew D., (612) 625-6878 awickert@umn.edu, Univ of Minnesota, Twin Cities; Gml
Wickham, John S., (817) 272-2987 wickham@uta.edu, Univ of Texas, Arlington; Gco
Wickham, Thomas, (724) 938-4180 wickham@calu.edu, California Univ of Pennsylvania; Zu
Wicks, Carol M., (225) 578-2692 cwicks@lsu.edu, Louisiana State Univ; HwClHy
Wicks, Frederick J., fredw@rom.on.ca, Royal Ontario Mus; Gz
Widanagamage, Inoka, 662-915-2154 ihwidana@olemiss.edu, Univ of Mississippi; ClGep
Widdowson, Mark A., (540) 231-7153 mwiddows@vt.edu, Virginia Tech; HwSpHq
Widom, Elisabeth, (513) 529-5048 widome@miamioh.edu, Miami Univ; CcGve
Widory, David, 514-987-3000 #1968 widory.david@uqam.ca, Univ du Quebec a Montreal; Cs
Wiedenmann, Jorg, +44 (0)23 80596497 joerg.wiedenmann@noc.soton.ac.uk, Univ of Southampton; Ob
Wiederspahn, Mark, (512) 471-0406 markw@ig.utexas.edu, Univ of Texas, Austin; Zn
Wiederwohl, Chrissy, 979-845-7191 chrissyw@tamu.edu, Texas A&M Univ; Op
Wiens, Douglas A., (314) 935-6517 doug@wustl.edu, Washington Univ in St. Louis; YsrYg
Wierenga, Peter J., (520) 621-1899 pjw@gmail.com, Univ of Arizona; Sp
Wiese, Katryn, (415) 452-5061 katryn.wiese@mail.ccsf.edu, City Coll of San Francisco; GiOg
Wiesenburg, Denis A., (601) 266-4937 denis.wiesenburg@usm.edu, Univ of Southern Mississippi; Og
Wiggert, Jerry, (228) 688-3177 jerry.wiggert@usm.edu, Univ of Southern Mississippi; Op
Wigley, Rochelle, 603-862-1135 rochelle@ccom.unh.edu, Univ of New Hampshire; CgGu
Wignall, Paul, +44(0) 113 34 35247 p.b.wignall@leeds.ac.uk, Univ of Leeds; PgGs
Wijbrans, Jan, +31 20 59 87296 j.r.wijbrans@vu.nl, Vrije Univit Amsterdam; CcGtv
Wijesinghe, Ananda M., (925) 423-0605 Lawrence Livermore; Nr
Wilbur, Bryan, 626 585-3118 bcwilbur@pasadena.edu, Pasadena City Coll; Gg
Wilch, Thomas I., (517) 629-0759 twilch@albion.edu, Albion Coll; GlvGm
Wilcock, Peter W., (410) 516-5421 Johns Hopkins Univ; Gm
Wilcock, William S., (206) 543-6043 wilcock@u.washington.edu, Univ of

Washington; Ou
Wilcots, Julia, 609-258-4101 jwilcots@princeton.edu, Princeton Univ; GgPc
Wilcox, Andrew, 406-243-4761 andrew.wilcox@mso.umt.edu, Univ of Montana; Gm
Wilcox, Jeffrey D., (828) 232-5184 jwilcox@unca.edu, Univ of North Carolina, Asheville; HwClGg
Wildeman, Thomas R., twildema@mines.edu, Colorado Sch of Mines; Ct
Wilder, Margaret, mwilder@email.arizona.edu, Univ of Arizona; Zn
Wilderman, Candie, (717) 245-1573 wilderma@dickinson.edu, Dickinson Coll; Hs
Wiles, Gregory C., (330) 263-2298 gwiles@wooster.edu, Coll of Wooster; Gl
Wilf, Peter D., (814) 865-6721 pwilf@geosc.psu.edu, Pennsylvania State Univ, Univ Park; Pg
Wilhelm, Steven W., wilhelm@utk.edu, Univ of Tennessee, Knoxville; Po
Wilhelmson, Robert B., (217) 333-8651 bw@ncsa.uiuc.edu, Univ of Illinois, Urbana-Champaign; As
Wilhelmy, Jerry B., (505) 665-3188 Los Alamos; Zn
Wilkerson, M. S., (765) 658-4666 mswilke@depauw.edu, Univ of Illinois, Urbana-Champaign; Gc
Wilkerson, M. Scott, (765) 658-4666 mswilke@depauw.edu, DePauw Univ; GctZi
Wilkey, Patrick L., (630) 252-6258 Argonne; Ng
Wilkie, Ann C., (352) 392-8699 acwilkie@ifas.ufl.edu, Univ of Florida; Sb
Wilkie, Richard W., (413) 253-5752 rwilkie@geo.umass.edu, Univ of Massachusetts, Amherst; ZynZn
Wilkins, Colin, +44 1752 584773 c.wilkins@plymouth.ac.uk, Univ of Plymouth; Eg
Wilkins, David E., dwilkins@boisestate.edu, Boise State Univ; ZyGm
Wilkinson, Bruce, 315-443-3869 eustasy@syr.edu, Syracuse Univ; Gs
Wilks, Daniel, (607) 255-1750 dsw5@cornell.edu, Cornell Univ; As
Wilks, Maureen, 575-835-5322 maureen.wilks@nmt.edu, New Mexico Tech; Gp
Willahan, Duane, (408) 848-4702 Gavilan Coll; Gg
Wille, Frank, 909-869-3648 frwille@cpp.edu, California State Polytechnic Univ, Pomona; Zn
William, Nancy, (785) 532-7257 nkw@ksu.edu, Kansas State Univ; Zn
Williams, Aaron, (251) 460-6915 bwilliams@southalabama.edu, Univ of South Alabama; Am
Williams, Allison M., (905) 525-9140 (Ext. 24334) awill@mcmaster.ca, McMaster Univ; Zn
Williams, Amy, 352-273-1287 amywilliams1@ufl.edu, Univ of Florida; Co
Williams, Bruce A., (509) 372-3799 bruce.williams@pnl.gov, Pacific Northwest; Ng
Williams, Christopher J., (717) 291-3814 chris.williams@fandm.edu, Franklin and Marshall Coll; PeSb
Williams, Curtis J., (714) 484-7000 (Ext. 48181) Cypress Coll; Gg
Williams, Dave E., williams.david@interchange.ubc.ca, Univ of British Columbia; Oc
Williams, David, (480) 965-7045 david.williams@asu.edu, Arizona State Univ; Xg
Williams, Erik, (775) 784-1396 eswilliams@unr.edu, Univ of Nevada, Reno; Ys
Williams, Forrest, 831-656-3274 fwillia@comcast.net, Naval Postgrad Sch; Am
Williams, Harry F. L., (940) 565-3317 harryf.williams@unt.edu, Univ of North Texas; Gm
Williams, Ian S., (709) 737-8395 ian.williams@uwrf.edu, Univ of Wisconsin, River Falls; Yg
Williams, Jeremy C., (330) 672-1459 jwill243@kent.edu, Kent State Univ; Cg
Williams, John W., (408) 924-5050 john.williams@sjsu.edu, San Jose State Univ; Ng
Williams, Kevin K., 716-878-4911 williakk@buffalostate.edu, Buffalo State Coll; Xg
Williams, Kim R., (303) 273-3245 krwillia@mines.edu, Colorado Sch of Mines; Ca
Williams, Lynda, (480) 965-0829 lynda.williams@asu.edu, Arizona State Univ; Gb
Williams, Mark, +44 0116 252 3642 mri@le.ac.uk, Leicester Univ; Pg
Williams, Mathew, +44 (0) 131 650 7776 mat.williams@ed.ac.uk, Edinburgh Univ; ZcrZu
Williams, Michael L., (413) 545-0745 mlw@geo.umass.edu, Univ of Massachusetts, Amherst; Gc
Williams, Paul, +44 0161 306-3905 paul.i.williams@manchester.ac.uk, Univ of Manchester; As
Williams, Quentin, (831) 459-3132 qwilliams@pmc.ucsc.edu, Univ of California, Santa Cruz; Gy
Williams, Ric, +44-151-794-5136 ric@liverpool.ac.uk, Univ of Liverpool; Og
Williams, Roger T., rtwillia@nps.edu, Naval Postgrad Sch; Am
Williams, Rosa, 706-649-1474 williams_rosa@columbusstate.edu, Columbus State Univ; Xa
Williams, Stanley N., (480) 965-1438 stanley.williams@asu.edu, Arizona State Univ; Gv
Williams, Thomas, (208) 885-6656 tomw@uidaho.edu, Univ of Idaho; GfzEm
Williams, Wyn, +44 (0) 131 650 4909 wyn.williams@ed.ac.uk, Edinburgh Univ; Ym
Williams-Bruinders, Leizel, +27 (0)41 504 4361 leizel.williams@nmmu.ac.za, Nelson Mandela Metropolitan Univ; Zn
Williams-Jones, Anthony E., 514-398-1676 anthony.williams-jones@mcgill.ca, McGill Univ; Ce
Williams-Jones, Glyn, (778) 782-3306 glynwj@sfu.ca, Simon Fraser Univ; GvCeYe
Williamson, Ben, +44 01326 371856 b.j.williamson@exeter.ac.uk, Exeter Univ; EmGv
Williamson, Douglas, 212-772-5265 douglas.williamson@hunter.cuny.edu, Hunter Coll (CUNY); Zi
Willis, Grant C., (801) 537-3355 grantwillis@utah.gov, Utah Geo Survey; Gg
Willis, Julie B., 208-496-7681 willisj@byui.edu, Brigham Young Univ - Idaho; GtcZi
Willis, Marc, 714-992-7446 mwillis@fullcoll.edu, Fullerton Coll; Gg
Willis, Michael J., 540-231-6521 mikewillis@vt.edu, Virginia Tech; GlZr
Willoughby, Hugh E., 305-348-0243 hugh.willoughby@fiu.edu, Florida InterNatl Univ; As
Wills, William V., 813-253-7809 wwills@hccfl.edu, Hillsborough Comm Coll; Zg
Willsey, Shawn P., (208) 732-6421 swillsey@csi.edu, Coll of Southern Idaho; GgcGv
Willson, Lee, 713.348.6219 Rice Univ; Zn
Wilmut, Michael, (250) 472-4343 Univ of Victoria; Zn
Wilson, Alicia M., (803) 777-1240 awilson@geol.sc.edu, Univ of South Carolina; Hw
Wilson, Blake B., (785) 864-2118 bwilson@kgs.ku.edu, Univ of Kansas; Zi
Wilson, Carol A., carolw@lsu.edu, Louisiana State Univ; GsOnCc
Wilson, Charlie, +44 (0)1603 59 1386 charlie.wilson@uea.ac.uk, Univ of East Anglia; Ge
Wilson, Clark R., (512) 471-5008 crwilson@jsg.utexas.edu, Univ of Texas, Austin; Yg
Wilson, Crystal, wilsonc@dnr.sc.gov, South Carolina Dept of Nat Res; Gct
Wilson, Emily, 717-358-3821 emily.wilson@fandm.edu, Franklin and Marshall Coll; CgGi
Wilson, Fred L., (325) 486-6984 fwilson@angelo.edu, Angelo State Univ; Gm
Wilson, Greg, 541-737-4015 wilsongr@coas.oregonstate.edu, Oregon State Univ; Onp
Wilson, Gregory, 206-543-8917 gpwilson@uw.edu, Univ of Washington; Pvg
Wilson, James R., jwilson@weber.edu, Weber State Univ; Gz
Wilson, Jeffey S., (317) 274-1128 jeswilso@iupui.edu, Indiana Univ, Indianapolis; Zy
Wilson, Jeffrey A., (734) 647-7461 wilsonja@umich.edu, Univ of Michigan; Gg
Wilson, John D., (780) 492-0353 Univ of Alberta; Am
Wilson, Kat, kathleen.wilson@bc.edu, Boston Coll; GsmOn
Wilson, Laura E., (785) 639-6192 lewilson6@fhsu.edu, Fort Hays State Univ; PvePg
Wilson, Lucy A., (506) 648-5607 lwilson@unbsj.ca, Univ of New Brunswick Saint John; Ga
Wilson, Mark A., (330) 263-2247 mwilson@wooster.edu, Coll of Wooster; Pi
Wilson, Melissa L., 612-625-4276 mlw@umn.edu, Univ of Minnesota, Twin Cities; Sco
Wilson, Merwether, +44 (0) 131 650 8636 meriwether.wilson@ed.ac.uk, Edinburgh Univ; Ob
Wilson, Michael C., wilsonmi@douglascollege.ca, Douglas Coll; GaPgGs
Wilson, P. Christopher, 772-468-3922 ext 119 pcwilson@ufl.edu, Univ of Florida; So
Wilson, Paul A., +44 (0)23 80596164 paul.wilson@noc.soton.ac.uk, Univ of Southampton; GsCg
Wilson, Rick I., (709) 737-8386 California Geo Survey; Ge
Wilson, Robert, +44 01334 463914 rjsw@st-andrews.ac.uk, Univ of St. Andrews; Zn
Wilson, Roy R., (860) 465-4370 wilsonr@easternct.edu, Eastern Connecticut State Univ; Zi
Wilson, Sarah B., (540) 458-8800 wilsons@wlu.edu, Washington & Lee Univ; Zn
Wilson, Steven D., (217) 333-0956 sdwilson@illinois.edu, Univ of Illinois; Hw

Wilson, Terry J., 614-292-0723 wilson.43@osu.edu, Ohio State Univ; GctGl
Wilson, Thomas, +64 3 3667001 Ext 45511 thomas.wilson@canterbury.ac.nz, Univ of Canterbury; Gv
Wilson Purdy, Sharon A., 202.633.2489 purdys@si.edu, Smithsonian Inst, Natl Air & Space Mus; XgGm
Wilton, Robert D., (905) 525-9140 (Ext. 24536) wiltonr@mcmaster.ca, McMaster Univ; Zn
Wiltshire, John C., (808) 956-6042 johnw@soest.hawaii.edu, Univ of Hawai'i, Manoa; Og
Wimbush, Mark, (401) 874-6515 m.wimbush@gso.uri.edu, Univ of Rhode Island; Op
Winant, Clinton D., cwinant@ucsd.edu, Univ of California, San Diego; On
Winberry, Paul, winberry@geology.cwu.edu, Central Washington Univ; Yg
Winbourne, Joy, (978) 934-3964 joy_winbourne@uml.edu, Univ of Massachusetts, Lowell; CbZc
Winchell, Robert E., (562) 985-4920 California State Univ, Long Beach; Gz
Winckler, Gisela, winckler@ldeo.columbia.edu, Columbia Univ; Cg
Windhorst, Rogier, (480) 965-7143 windhorst@asu.edu, Arizona State Univ; Xa
Windom, Kenneth E., kewindom@iastate.edu, Iowa State Univ of Sci & Tech; Cp
Windsor, Jr., John G., 321-674-8096 jwindsor@fit.edu, Florida Inst of Tech; Oc
Wine, Paul H., (404) 894-3425 pw7@prism.gatech.edu, Georgia Inst of Tech; As
Winebrenner, Dale P., 206-543-1393 dpw@apl.washington.edu, Univ of Washington; GlZr
Wing, Scott L., (202) 357-2649 Smithsonian Inst / Natl Mus of Natural History; Pb
Winglee, Robert M., 206-685-8160 winglee@uw.edu, Univ of Washington; XyYx
Winguth, Arne M., 817 272 2987 awinguth@uta.edu, Univ of Texas, Arlington; OpAs
Winguth, Cornelia, 817 272 2987 cwinguth@uta.edu, Univ of Texas, Arlington; GugZm
Winkelstern, Ian Z., 616-331-9219 winkelsi@gvsu.edu, Grand Valley State Univ; GsOo
Winkler, Dale A., (214) 768-2750 dwinkler@smu.edu, Southern Methodist Univ; Pv
Winklhofer, Michael, 089/2180 4143 michael@geophysik.uni-muenchen.de, Ludwig-Maximilians-Univ Muenchen; Yg
Winnick, Matthew, 413-545-1715 mwinnick@geo.umass.edu, Univ of Massachusetts, Amherst; Cbq
Winsor, Kelsey, kelsey.winsor@nau.edu, Northern Arizona Univ; PcGm
Winston, Barbara, (314) 935-7047 Washington Univ in St. Louis; Zn
Winterbottom, Wesley, winterbottomw@easternct.edu, Eastern Connecticut State Univ; Zg
Winterkamp, Judith L., (505) 667-1264 judyw@lanl.gov, Los Alamos; Zn
Winton, Alison, (806) 834-0497 alison.winton@ttu.edu, Texas Tech Univ; ZnnZn
Wintsch, Robert P., (812) 855-4018 wintsch@indiana.edu, Indiana Univ, Bloomington; GpcGt
Wirth, Erin, ewirth@uw.edu, Univ of Washington; YsRn
Wisdom, Jack, (617) 253-7730 wisdom@mit.edu, Massachusetts Inst of Tech; Zn
Wise, Dana, (618) 453-4362 dwise@siu.edu, Southern Illinois Univ Carbondale; Zn
Wise, Michael A., (202) 633-1826 wisem@si.edu, Smithsonian Inst / Natl Mus of Natural History; Gz
Wise, Jr, Sherwood W., (850) 644-6265 swise@fsu.edu, Florida State Univ; PmGu
Wisely, Beth , 307-268 2233 bwisely@caspercollege.edu, Casper Coll; YgHwGc
Wishart, De Bonne N., dwishart@centralstate.edu, Central State Univ; YgCa
Wishner, Karen, (401) 874-6402 kwishner@gso.uri.edu, Univ of Rhode Island; Ob
Withers, Kim, 361-825-5907 kim.withers@tamucc.edu, Texas A&M Univ, Corpus Christi; Zn
Withers, Mitchell M., (901) 678-2007 mwithers@memphis.edu, Univ of Memphis; Ys
Withers, Tony, 519-661-2111 x.88627 tony.withers@uwo.ca, Western Univ; Cp
Withjack, Martha O., (848) 445-3142 drmeow3@eps.rutgers.edu, Rutgers, The State Univ of New Jersey; GctEo
Witt, Emma, emma.witt@stockton.edu, Stockton Univ; Hg
Witter, Rob, 541-574-7969 rwitter@usgs.gov, Oregon Dept of Geology and Mineral Ind; Zg
Witter-Shelleman, Molly, 330-972-8046 mwittershelleman@uakron.edu, Univ of Akron; Gz
Wittke, James, 9285239565/9044 james.wittke@nau.edu, Northern Arizona Univ; Gi
Wittkop, Chad, (507) 389-6929 chad.wittkop@mnsu.edu, Minnesota State Univ; GsCgGl
Witzke, Brian J., brian-witzke@uiowa.edu, Univ of Iowa; GrPsg
Wixman, Ronald, (541) 346-4568 rwixman@uoregon.edu, Univ of Oregon; Zn
Wizevich, Michael, wizevichmic@ccsu.edu, Central Connecticut State Univ; GsdGm
Wobus, Reinhard A., (413) 597-2470 reinhard.a.wobus@williams.edu, Williams Coll; GivGz
Woerheide, Gert, geobiologie@geo.lmu.de, Ludwig-Maximilians-Univ Muenchen; Pgi
Wofsy, Steven C., (617) 495-4566 scw@io.harvard.edu, Harvard Univ; As
Wogelius, Roy, (+44)-(0)161-275 3841 roy.wogelius@manchester.ac.uk, Univ of Manchester; Cg
Wohl, Ellen E., (970) 491-5298 ellen.wohl@colostate.edu, Colorado State Univ; GmHs
Wohletz, Kenneth H., (505) 667-9202 wohletz@lanl.gov, Los Alamos; YxGvRm
Woida, Kathleen, kathleen-woida@uiowa.edu, Univ of Iowa; So
Wojewoda, Jurand, jurand.wojewoda@uwr.edu.pl, Univ of Wroclaw; Gs
Wojtal, Steven F., (440) 775-8352 steven.wojtal@oberlin.edu, Oberlin Coll; Gc
Wolak, Jeannette, (931) 372-3695 jluna@tntech.edu, Tennessee Tech Univ; Gsr
Wolaver, Brad, (512) 471-1368 brad.wolaver@beg.utexas.edu, Univ of Texas, Austin; HwGe
Wolcott, Donna L., (919) 515-7866 donna_wolcott@ncsu.edu, North Carolina State Univ; Ob
Wolcott, Ray, 6196447454 x3099 rwolcott@palomar.edu, Cuyamaca Coll; Og
Wolcott, Thomas G., (919) 515-7866 tom_wolcott@ncsu.edu, North Carolina State Univ; Ob
WoldeGabriel, Giday, 505-667-8749 wgiday@lanl.gov, Los Alamos; Gx
Wolf, Aaron T., (541) 737-2722 wolfa@geo.oregonstate.edu, Oregon State Univ; Hg
Wolf, Lorraine W., (334) 844-4878 wolflor@auburn.edu, Auburn Univ; Ysg
Wolf, Michael B., (309) 794-7304 michaelwolf@augustana.edu, Augustana Coll; Gi
Wolfe, Ben, (816) 604-6622 ben.wolfe@mcckc.edu, Metropolitan Comm Coll-Kansas City; GgZyg
Wolfe, Christopher, christopher.wolfe@stonybrook.edu, SUNY, Stony Brook; Op
Wolfe, Karen M., (615) 898-2726 karen.wolfe@mtsu.edu, Middle Tennessee State Univ; Zn
Wolfe, Paul J., 937 775-2201 paul.wolfe@wright.edu, Wright State Univ; Ye
Wolff, Eric W., +44 (0) 1223 333486 ew428@cam.ac.uk, Univ of Cambridge; GlPe
Wolff, George, +44-151-794-4094 wolff@liverpool.ac.uk, Univ of Liverpool; Og
Wolff, John A., (509) 335-2825 jawolff@wsu.edu, Washington State Univ; Giv
Wolfgram, Diane, (406) 496-4353 dwolfgram@mtech.edu, Montana Tech of the Univ of Montana; EgoNm
Wolfsberg, Andrew V., (505) 667-3599 awolf@lanl.gov, Los Alamos; Hw
Wolfsberg, Kurt, (505) 667-4464 Los Alamos; Cc
Wolken, Gabriel J., (907) 451-5018 gabriel.wolken@alaska.gov, Alaska Division of Geological & Geophysical Surveys; Gm
Wollan, Jacinda, mandy.looser@ndsu.edu, North Dakota State Univ; Zn
Wollheim, Wil, wil.wollheim@unh.edu, Univ of New Hampshire; Zu
Wolny, Dave, (970) 248-1154 dwolmy@mesastate.edu, Colorado Mesa Univ; Ys
Wolter, Calvin, calvin.wolter@dnr.iowa.gov, Iowa Dept of Nat Res; GgZi
Wolverton, Steve, 940-565-4987 steven.wolverton@unt.edu, Univ of North Texas; Ga
Wong, Chi S., (250) 363-6407 wongcs@pac.dfo-mpo.gc.ca, Univ of British Columbia; Cm
Wong, Cindy, (716) 878-6731 solargs@buffalostate.edu, Buffalo State Coll; Zn
Wong, Martin, (315) 228-7203 mswong@colgate.edu, Colgate Univ; Gtc
Wong, Teng-fong, (631) 632-8212 teng-fong.wong@stonybrook.edu, Stony Brook Univ; Yx
Woo, Ming-Ko, (905) 525-9140 woo@mcmaster.ca, McMaster Univ; HgZy
Wood, Aaron R., (515) 294-8862 awood@iastate.edu, Iowa State Univ of Sci & Tech; PvGrs
Wood, Bernard, +44 (1865) 272014 bernie.wood@earth.ox.ac.uk, Univ of Oxford; Gz
Wood, David A., 210-486-0063 dwood30@alamo.edu, San Antonio Comm

Coll; XsZnn
Wood, Eric F., (609) 258-4675 efwood@princeton.edu, Princeton Univ; HgZrHq
Wood, Gillen D., gdwood@illinois.edu, Univ of Illinois, Urbana-Champaign; Zc
Wood, Howard, (361) 825-3335 tony.wood@tamucc.edu, Texas A&M Univ, Corpus Christi; ZeHsZn
Wood, Ian, +44 020 7679 32405 ian.wood@ucl.ac.uk, Univ Coll London; Gz
Wood, Kim, 662-268-1032 kimberly.wood@msstate.edu, Mississippi State Univ; AsmZr
Wood, Lesli, 303-273-3801 lwood@mines.edu, Colorado Sch of Mines; GsoGm
Wood, Neill, +44 01326 255163 n.a.wood@exeter.ac.uk, Exeter Univ; YuZen
Wood, Robert, (206) 543-1203 robwood@atmos.washington.edu, Univ of Washington; As
Wood, Scott A., 701-231-8552 scott.wood@ndsu.edu, North Dakota State Univ; Cq
Wood, Spencer H., swood@boisestate.edu, Boise State Univ; Gmm
Woodall, Debra W., (386) 506-3765 woodald@daytonastate.edu, Daytona State Coll; GgOg
Woodard, Jeremy, (031) 260-2804 woodardj@ukzn.ac.za, Univ of KwaZulu-Natal; CcGiz
Woodburne, Michael O., (909) 787-5028 michael.woodburne@ucr.edu, Univ of California, Riverside; Pv
Woodcock, Curtis, curtis@bu.edu, Boston Univ; Zr
Woodcock, Nigel H., +44 (0) 1223 333430 nhw1@esc.cam.ac.uk, Univ of Cambridge; GcsGt
Woodgate, Rebecca A., (206) 221-3268 woodgate@apl.washington.edu, Univ of Washington; Op
Woodhouse, Connie, 520-626-0235 conniew1@email.arizona.edu, Univ of Arizona; Pe
Woodhouse, Iain H., +44 (0) 131 650 2527 i.h.woodhouse@ed.ac.uk, Edinburgh Univ; Zi
Woodhouse, John, +44 (1865) 272021 john.woodhouse@earth.ox.ac.uk, Univ of Oxford; Yg
Woodin, Sarah A., (803) 782-9727 swoodin@gmail.com, Univ of South Carolina; Ob
Wooding, Frank B., (508) 289-3334 Woods Hole Oceanographic Inst; Yr
Woodland, Bertram G., (312) 665-7648 Field Mus of Natural History; Gp
Woodley, Teresa, (604) 822-3146 twoodley@eos.ubc.ca, Univ of British Columbia; Zn
Woodring, Hannah, (812) 464-1701 hmwoodring@usi.edu, Univ of Southern Indiana; Zn
Woodruff, Jonathan D., 413-577-3831 woodruff@geo.umass.edu, Univ of Massachusetts, Amherst; Gs
Woodruff, William H., (505) 665-2557 Los Alamos; Zn
Woods, Adam D., (657) 278-2921 awoods@fullerton.edu, California State Univ, Fullerton; GsPe
Woods, Andy, +44 (0) 1223 765702 andy@bpi.cam.ac.uk, Univ of Cambridge; YgGt
Woods, Juliette, juliette.woods@flinders.edu.au, Univ of South Australia; Hwq
Woods, Karen M., (409) 880-2251 karen.woods@lamar.edu, Lamar Univ; Zg
Woods, Rachel A., +44 (0) 131 650 6014 rachel.wood@ed.ac.uk, Edinburgh Univ; Gs
Woodward, Lee A., (505) 277-5309 Univ of New Mexico; Gc
Woodward, Mac B., 501 683-0113 mac.woodward@arkansas.gov, Arkansas Geo Survey; Gog
Woodwell, Grant R., (540) 654-1427 gwoodwel@mwc.edu, Univ of Mary Washington; Gc
Wooldridge, C F., wooldridge@cf.ac.uk, Cardiff Univ; Zn
Woolery, Edward W., (859) 257-3016 woolery@uky.edu, Univ of Kentucky; YusNg
Wooten, Richard M., 828-296-4632 rick.wooten@ncdenr.gov, N.C. Dept of Env Quality; RnGc
Worcester, Peter, peter.worcester@eku.ed, Eastern Kentucky Univ; GgOuZe
Worden, Richard, 0151 794 5184 r.worden@liverpool.ac.uk, Univ of Liverpool; Gs
Wordsworth, Robin, rwordsworth@seas.harvard.edu, Harvard Univ; Xg
Worrall, Fred, 0191 3742525 fred.worrall@durham.ac.uk, Durham Univ; Cg
Wortel, Matthew J., (319) 335-3992 matthew-wortel@uiowa.edu, Univ of Iowa; Gx
Worthington, Lindsay L., 505.277.4204 lworthington@unm.edu, Univ of New Mexico; Yg
Wortmann, Ulrich B., 416-978-2084 Univ of Toronto; Gg
Wright, Alan, 561-992-1555 alwr@ufl.edu, Univ of Florida; Sc
Wright, Carrie L., (812) 465-1145 clwright@usieagles.org, Univ of Southern Indiana; Ze
Wright, Chris, 614 265 6632 chris.wright@dnr.state.oh.us, Ohio Dept of Nat Res; Ec
Wright, James D., (848) 445-5722 jdwright@eps.rutgers.edu, Rutgers, The State Univ of New Jersey; CsOuPm
Wright, James (Jim) E., (706) 542-4394 jwright@gly.uga.edu, Univ of Georgia; Gt
Wright, Kathyrn, (269) 387-5486 kathyrn.wright@wmich.edu, Western Michigan Univ; Zn
Wright, Stephen F., (802) 656-4479 swright@uvm.edu, Univ of Vermont; Glc
Wright, Tim, +44(0) 113 34 35258 t.j.wright@leeds.ac.uk, Univ of Leeds; Yd
Wright, V P., wrightvp@cf.ac.uk, Cardiff Univ; Gs
Wright, Vashan, vwright@ucsd.edu, Univ of California, San Diego; Gt
Wright, William C., 845-939-2063 william.wright@westpoint.edu, United States Military Acad; ZriYd
Wronkiewicz, David, 573-341-4679 wronk@mst.edu, Missouri Univ of Sci and Tech; Cl
Wu, Chin, (608) 263-3078 chinwu@engr.wisc.edu, Univ of Wisconsin, Madison; Onp
Wu, David T., (303) 273-2066 dwu@mines.edu, Colorado Sch of Mines; Zn
Wu, Francis T., (607) 777-2512 Binghamton Univ; Ys
Wu, Jonny, (713) 743-9624 jwu40@central.uh.edu, Univ of Houston; GtcGo
Wu, Kan, (979)862-7654 kan.wu@tamu.edu, Texas A&M Univ; NpYeGo
Wu, Shiliang, (906) 487-2590 slwu@mtu.edu, Michigan Technological Univ; AsCg
Wu, Shuang-Ye, (937) 229-1720 shuang-ye.wu@notes.udayton.edu, Univ of Dayton; Zy
Wu, Xiaoqing, (515) 294-9872 wuxq@iastate.edu, Iowa State Univ of Sci & Tech; As
Wu, Xingru, xingru.wu@ou.edu, Univ of Oklahoma; Np
Wu, Yutian, yutianwu@ldeo.columbia.edu, Columbia Univ; Og
Wuebbles, Donald J., (217) 244-1568 wuebbles@illinois.edu, Univ of Illinois, Urbana-Champaign; As
Wuerthele, Norman, (412) 622-3265 wuerthelen@carnegiemnh.org, Carnegie Mus of Natural History; Pv
Wulff, Andrew, (270) 745-5976 andrew.wulff@wku.edu, Western Kentucky Univ; Gv
Wunsch, Carl, (617) 496-2732 cwunsch@fas.harvard.edu, Harvard Univ; Op
Wunsch, David R., (302) 831-8258 dwunsch@udel.edu, Univ of Delaware; ClHwNg
Wust-Bloch, Gilles H., (03-) 640-5475 Tel Aviv Univ; Ys
Wyckoff, John W., (303) 556-2590 john.wyckoff@ucdenver.edu, Univ of Colorado, Denver; Zy
Wyckoff, William K., (406) 994-6914 bwyckoff@montana.edu, Montana State Univ; Zn
Wygal, Brian, (516) 877-4170 bwygal@adelphi.edu, Adelphi Univ; GaRmGe
Wykel, Andy, 803.896.7703 wykelc@dnr.sc.gov, South Carolina Dept of Nat Res; Gs
Wylie, Ann G., (301) 405-4079 awylie@umd.edu, Univ of Maryland; Gz
Wyllie, Peter J., (626) 395-6461 wyllie@gps.caltech.edu,CalTech; Cp
Wyman, Derek A., (618) 938-0117 Univ of Saskatchewan; Cg
Wymore, Adam, adam.wymore@unh.edu, Univ of New Hampshire; Hg
Wyngaard, John C., wyngaard@ems.psu.edu, Pennsylvania State Univ, Univ Park; Am
Wynn, Thomas C., (570) 484-2081 twynn@lockhaven.edu, Lock Haven Univ; GsPiEo
Wypych, Alicja, 907-451-5010 alicja.wypych@alaska.gov, Alaska Division of Geological & Geophysical Surveys; Gg
Wyse Jackson, Patrick N., + 353-1-8961477 wysjcknp@tcd.ie, Trinity Coll; PiRh
Wysession, Michael E., (314) 935-5625 michael@wucore.wustl.edu, Washington Univ in St. Louis; Ys
Wysocki, Mark, (607) 255-2568 mww3@cornell.edu, Cornell Univ; As
Wysong, Jr., James F., (813) 253-7805 jwysong@hccfl.edu, Hillsborough Comm Coll; Am
Wyss, Andre R., (805) 893-8628 wyss@geol.ucsb.edu, Univ of California, Santa Barbara; Pv

X

Xi, Xin, 906-487-2029 xinxi@mtu.edu, Michigan Technological Univ; AsZor
Xiao, Shuhai, 540-231-1366 xiao@vt.edu, Virginia Tech; Pi
Xie, Lian, (919) 515-1435 xie@ncsu.edu, North Carolina State Univ; As
Xie, Xiangyang, (817) 257-4395 x.xie@tcu.edu, Texas Christian Univ; GoEog
Xie, Zhixiao, (561) 297-2852 xie@fau.edu, Florida Atlantic Univ; ZirZy
Xu, Huifang, 608/265-5587 hfxu@geology.wisc.edu, Univ of Wisconsin, Madison; Gz

Xu, Jie, 915-747-7556 jxu2@utep.edu, Univ of Texas, El Paso; ClPoGz
Xu, Li, (508)289-3673 lxu@whoi.edu, Woods Hole Oceanographic Inst; Yr
Xu, Shangping, 414-229-6148 xus@uwm.edu, Univ of Wisconsin, Milwaukee; Hg
Xu, Tingying, 405-744-6358 tingying.xu@okstate.edu, Oklahoma State Univ; Cbq
Xu, Wei, wei.xu@uleth.ca, Univ of Lethbridge; Zi
Xu, Xiaohua (Eric), xiaohua.xu@austin.utexas.edu, Univ of Texas, Austin; Yd
XU, Yingfeng, 337-482-1113 yingfeng@louisiana.edu, Univ of Louisiana at Lafayette; ZgGe
Xuan, Chuang, +44 (0)23 80596401 c.xuan@soton.ac.uk, Univ of Southampton; Ym

Y

Yacobucci, Peg M., (419) 372-7982 mmyacob@bgsu.edu, Bowling Green State Univ; Pgi
Yacucci, Mark, (309) 268-8640 mark.yacucci@heartland.edu, Heartland Comm Coll; Zg
Yalcin, Kaplan, (541) 737-1230 yalcink@geo.oregonstate.edu, Oregon State Univ; GgeZe
Yalcin, Rebecca, yalcinr@onid.orst.edu, Oregon State Univ; Gg
Yalda, Sepideh, 717-872-3293 sepi.yalda@millersville.edu, Millersville Univ; As
Yamanaka, Tsuyuko, +44 0151 795 5291 t.yamanaka@liverpool.ac.uk, Univ of Liverpool; Ob
Yan, Beizhan, yanbz@ldeo.columbia.edu, Columbia Univ; Cg
Yan, Eugene, eyan1@uic.edu, Univ of Illinois Chicago; Hg
Yan, Jun, (270) 745-8952 jun.yan@wku.edu, Western Kentucky Univ; Zi
Yan, Xiao-Hai, (302) 831-3694 Univ of Delaware; Zr
Yan, Y E., (630) 252-6322 eyan@anl.gov, Argonne; Gg
Yancey, Thomas E., (979) 845-0643 yancey@geo.tamu.edu, Texas A&M Univ; Pg
Yandle, Tracy, 404 727 5652 tyandle@emory.edu, Emory Univ; Ob
Yanes, Yurena, yurena.yanes@uc.edu, Univ of Cincinnati; PeCsPg
Yang, Changbing, 512-471-4364 changbing.yang@beg.utexas.edu, Univ of Texas, Austin; Hw
Yang, Da, dayang@ucdavis.edu, Univ of California, Davis; Ast
Yang, Jianwen, 519-253-3000 x2181 jianweny@uwindsor.ca, Univ of Windsor; HwYg
Yang, Panseok, (204) 474-6910 panseok.yang@umanitoba.ca, Univ of Manitoba; CaGp
Yang, Ping, (979) 845-7679 pyang@tamu.edu, Texas A&M Univ; As
Yang, Wan, 316 -978-3140 Wichita State Univ; Gs
Yang, Wenchang, 609-258-4101 wenchang@princeton.edu, Princeton Univ; As
Yang, Xiaotao, 765-494-8681 xtyang@purdue.edu, Purdue Univ; Yg
Yang, Y, yangy6@cf.ac.uk, Cardiff Univ; Hg
Yang, Zhiming, (919) 530-5296 zyang@nccu.edu, North Carolina Central Univ; Zry
Yang, Zong-Liang, (512) 471-3824 liang@mail.utexas.edu, Univ of Texas, Austin; AmsHq
Yanites, Brian, (812) 855-6109 byanites@indiana.edu, Indiana Univ, Bloomington; Gm
Yankovsky, Alexander E., (803) 777-3550 ayankovsky@geol.sc.edu, Univ of South Carolina; OpnAm
Yao, Wensheng, (727) 553-3922 Univ of South Florida; Oc
Yao, Yon, +27 (0)46-603-7393 y.yao@ru.ac.za, Rhodes Univ; Eg
Yapp, Crayton J., (214) 768-3897 cjyapp@smu.edu, Southern Methodist Univ; Csl
Yarborough, Vicki, 540-231-6521 vickiy3@vt.edu, Virginia Tech; Pv
Yarbrough, Lance D., 662-915-7499 ldyarbro@olemiss.edu, Univ of Mississippi; Ng
Yarbrough, Robert A., 912-478-0846 ryarbrough@georgiasouthern.edu, Georgia Southern Univ; Zn
Yardley, Bruce W., b.w.d.yardley@leeds.ac.uk, Univ of Leeds; CgEgGg
Yarger, Douglas N., (515) 294-9872 doug@iastate.edu, Iowa State Univ of Sci & Tech; As
Yassin, Barbara E., (601) 961-5571 barbara_yassin@mdeq.ms.gov, Mississippi Office of Geology; ZyiGm
Yates, Martin G., (207) 581-2154 yates@maine.edu, Univ of Maine; EgGzx
Yates, Mary Anne, (505) 667-7090 Los Alamos; Zn
Yau, Man Kong, (514) 398-3719 peter.yau@mcgill.ca, McGill Univ; As
Ye, Ming, mingye@scs.fsu.edu, Florida State Univ; Hg
Yeager, Kevin M., (859) 257-5431 kevin.yeager@uky.edu, Univ of Kentucky; GsCmc
Yeager, Lauren A., 361-749-6776 lyeager@utexas.edu, Univ of Texas, Austin; Og
Yeats, Robert S., 541-740-3806 yeatsr@geo.oregonstate.edu, Oregon State Univ; GtYsNg
Yee, Nathan, 848-932-5714 nyee@envsci.rutgers.edu, Rutgers, The State Univ of New Jersey; PgCo
Yegulalp, Tuncel M., (212) 854-2984 yegulalp@columbia.edu, Columbia Univ; Nm
Yeh, Jim, (520) 621-5943 yeh@hwr.arizona.edu, Univ of Arizona; Sp
Yeh, Joseph S., (512) 471-3323 joseph.yeh@beg.utexas.edu, Univ of Texas at Austin, Jackson Sch of Geosciences; Gm
Yeh, Tian-Chyi J., (520) 621-5943 ybiem@arizona.edu, Univ of Arizona; Hq
Yelderman, Jr., Joe C., (254) 710-2185 joe_yelderman@baylor.edu, Baylor Univ; HwRwHw
Yelisetti, Subbarao, 361-593-4894 subbarao.yelisetti@tamuk.edu, Texas A&M Univ, Kingsville; Ygs
Yelle, Roger, (520) 621-6243 yelle@lpl.arizona.edu, Univ of Arizona; Zn
Yellen, Brian, byellen@geo.umass.edu, Univ of Massachusetts, Amherst; Gs
Yellich, John, (269) 387-8649 john.a.yellich@wmich.edu, Western Michigan Univ; EgGeHw
YEO, MYEONG-HO (CHRIS), (671)735-2693 yeom@triton.uog.edu, Univ of Guam; Hwy
Yeung, Laurence Y., (713) 348-6304 ly19@rice.edu, Rice Univ; CsAsCl
Yi, Chuixiang, 718-997-3366 chuixiang.yi@qc.cuny.edu, Graduate Sch of the City Univ of New York; Zg
Yi, Yuchan, 614 292 6005 yi.3@osu.edu, Ohio State Univ; Ydv
Yiannakoulias, Niko, (905) 525-9140 (Ext. 20117) yiannan@mcmaster.ca, McMaster Univ; Ge
Yibas, Bisrat, +270514013080 yibasbabsob@ufs.ac.za, Univ of the Free State; GegEg
Yikilmaz, Burak, mbyikilmaz@ucdavis.edu, Univ of California, Davis; Gc
Yildirim, Ismail, 00904242370000-5970 iyildirim@firat.edu.tr, Firat Univ; ScGix
Yilmaz, Alper, 614-247-4323 yilmaz.15@osu.edu, Ohio State Univ; ZfrZu
Yin, An, (310) 825-8752 yin@epss.ucla.edu, Univ of California, Los Angeles; Gt
Yin, Jianjun, 520-626-7453 yin@email.arizona.edu, Univ of Arizona; Og
Yin, Qing-zhu, 530-752-0934 qyin@ucdavis.edu, Univ of California, Davis; Cc
Yin, Zhi-Yong, (619) 260-8864 zyin@sandiego.edu, Univ of San Diego; Zir
Yogodzinkski, Gene, 803-777-9524 gyogodzin@geol.sc.edu, Univ of South Carolina; GiCuc
Yoh, Shing, (908) 737-3692 syoh@kean.edu, Kean Univ; Am
Yon, Lisa, 7607441150 x2369 lyon@palomar.edu, Palomar Coll; OgZg
Yong, Wenjun, 519-661-2111 ext 86628 wyong4@uwo.ca, Western Univ; Yx
Yonkee, W. A., (801) 626-7419 ayonkee@weber.edu, Weber State Univ; Gct
Yoo, Kyungsoo, 612-624-7784 kyoo@umn.edu, Univ of Minnesota, Twin Cities; Sd
Yool, Stephen R., yools@email.arizona.edu, Univ of Arizona; Zri
Yoshida, Glenn, (323) 241-5296 yoshidgy@lasc.edu, Los Angeles Southwest Coll; Gg
Yoshinobu, Aaron S., (806) 834-7715 aaron.yoshinobu@ttu.edu, Texas Tech Univ; GctGu
Yoskowitz, David , 361-825-2966 david.yoskowitz@tamucc.edu, Texas A&M Univ, Corpus Christi; Zn
You, Kehua, 512-471-0462 kehua@ig.utexas.edu, Univ of Texas, Austin; HtEo
Youberg, Ann, ayouberg@email.arizona.edu, Univ of Arizona; Ge
Young, Aaron R., (402) 472-8339 ayoung3@unl.edu, Univ of Nebraska, Lincoln; HwGmZf
Young, David, young.2536@osu.edu, Ohio State Univ; GcpCc
Young, Davis A., dotndave@comcast.net, Calvin Coll; GiRhGz
Young, Duncan, 512-471-0485 duncan@ig.utexas.edu, Univ of Texas, Austin; GlXy
Young, Edward D., (310) 267-4930 eyoung@epss.ucla.edu, Univ of California, Los Angeles; CsXc
Young, George S., (814) 863-4228 young@ems.psu.edu, Pennsylvania State Univ, Univ Park; Am
Young, Graham A., (204) 988-0648 gyoung@cc.umanitoba.ca, Univ of Manitoba; Pi
Young, Grant M., (519) 473-5692 gyoung@uwo.ca, Western Univ; GrCgGt
Young, Harvey R., 204-727-9798 young@brandonu.ca, Brandon Univ; Gd
Young, Jeffrey, (204) 474-8863 jeff.young@umanitoba.ca, Univ of Manitoba; GcZe
Young, Jeri J., 602.708.8558 jeribenhorin@email.arizona.edu, Univ of Arizona; Ys
Young, John A., (608)263-2374 jayoung@wisc.edu, Univ of Wisconsin, Madison; AstOp
Young, Michael H., (512) 475-8830 michael.young@beg.utexas.edu, Univ of Texas, Austin; SpHwGe
Young, Mike, 902 494 2364 mdyoung@dal.ca, Dalhousie Univ; GgcGp

Young, Nicolas, nicolasy@ldeo.columbia.edu, Columbia Univ; Cg
Young, Patrick, (480) 727-6581 patrick.young.1@asu.edu, Arizona State Univ; Xa
Young, Priscilla E., 605-677-6144 priscilla.young@usd.edu, South Dakota Dept of Environment and Nat Res; Zn
Young, Richard A., (585) 245-5296 young@geneseo.edu, SUNY, Geneseo; GmXgZr
Young, Robert R., 250-807-9523 robert.young@ubc.ca, Univ of British Columbia, Okanagan Campus; GlmZy
Young, William, +44(0) 113 34 31640 c.w.young@leeds.ac.uk, Univ of Leeds; Ge
Youngs, Edward, edward.youngs@open.ac.uk, Open Univ; Sp
Yow, Donald M., (859) 622-1420 don.yow@eku.edu, Eastern Kentucky Univ; AtZye
Yu, Jin-Yi, (949) 824-3878 jyyu@uci.edu, Univ of California, Irvine; AsOpZc
Yu, Qian, (413) 545-2095 qyu@geo.umass.edu, Univ of Massachusetts, Amherst; Zri
Yu, Wen-che, fgsyw@earth.sinica.edu.tw, Academia Sinica; Ys
Yu, Zicheng, 610-758-6751 ziy2@lehigh.edu, Lehigh Univ; Pe
Yuan, Xiaojun, xyuan@ldeo.columbia.edu, Columbia Univ; Op
Yuan, Yihong, (512) 245-2170 Texas State Univ; Ze
Yue, Stephen, (514) 398-4755 McGill Univ; Nx
Yuen, Cheong-yip R., (630) 252-4869 yuenr@anl.gov, Argonne; Ge
Yule, J. Douglas, (818) 677-6238 j.d.yule@csun.edu, California State Univ, Northridge; Gt
Yun, Misuk, (204) 474-8870 yun@cc.umanitoba.ca, Univ of Manitoba; Cs
Yun, Seong-Taek, 82-2-3290-3176 styun@korea.ac.kr, Korea Univ; CqHyCl
Yung, Yuk L., (626) 395-6940 yly@gps.caltech.edu,CalTech; As
Yunwei, Sun, (925)422-1587 sun4@llnl.gov, Lawrence Livermore; Ng
Yuretich, Richard F., (413) 545-0538 yuretich@geo.umass.edu, Univ of Massachusetts, Amherst; Cl
Yurkovich, Steven P., (828) 227-7367 yurkovich@wcu.edu, Western Carolina Univ; Gx
Yurtsever, Ayhan, 089/2180 4346 ayhan.yurtsever@ltz.uni-uemchen.de, Ludwig-Maximilians-Univ Muenchen; Gz
Yusoff, Ismail, 03-79674153 ismaily70@um.edu.my, Univ of Malaya; Hg
Yuter, Sandra, 919-513-7963 sandra_yuter@ncsu.edu, North Carolina State Univ; As
Yutzy, Gale, (301) 687-4369 gyutzy@frostburg.edu, Frostburg State Univ; Zn
Yvon-Lewis, Shari A., 979 458 1816 syvonlewis@ocean.tamu.edu, Texas A&M Univ; Oc

Z

Zabielski, Victor, (703) 845-6507 vzabielski@nvcc.edu, Northern Virginia Comm Coll - Alexandria; GgOg
Zaccaria, Daniele, 530 219-7502 dzaccaria@ucdavis.edu, Univ of California, Davis; HgRw
Zachos, James, 831-459-4644 jzachos@ucsc.edu, Univ of California, Santa Cruz; Pc
Zachos, Louis G., (662) 915-8827 lgzachos@olemiss.edu, Univ of Mississippi; PiGeNx
Zachry, Doy L., (479) 575-2785 dzachry@uark.edu, Univ of Arkansas, Fayetteville; Gr
Zaffos, Andrew A., 520.621.3635 azafazaffos@email.arizona.edu, Univ of Arizona; Zf
Zafian, Holly, holly.zafian@austincc.edu, Austin Comm Coll District; Ze
Zafiriou, Oliver C., (508) 289-2342 ozafiriou@whoi.edu, Woods Hole Oceanographic Inst; Oc
Zahm, Christopher K., 512-471-3159 chris.zahm@beg.utexas.edu, Univ of Texas, Austin; Yg
Zaitchik, Benjamin, 410-516-7135 bzaitch1@jhu.edu, Johns Hopkins Univ; As
Zaja, Annalisa, +390498279193 annalisa.zaja@unipd.it, Univ degli Studi di Padova; Yg
Zajac, Roman N., (203) 932-7114 rzajac@newhaven.edu, Univ of New Haven; Zni
Zajacz, Zoltan, 416-946-0278 zajacz@es.utoronto.ca, Univ of Toronto; Cu
Zakardjian, Bruno, (418) 723-1986 (Ext. 1570) bruno_zakardjian@uqar.qc.ca, Univ du Quebec a Rimouski; Ob
Zakharova, Natalia, 989-774-4496 zakha1n@cmich.edu, Central Michigan Univ; YgGe
Zakrzewski, Richard J., (785) 628-5389 rzakrzew@fhsu.edu, Fort Hays State Univ; PvsPe
Zaleha, Michael J., (937) 327-7331 mzaleha@wittenberg.edu, Wittenberg Univ; Gs
Zaleha, Robert, (216) 987-5278 robert.zaleha@tri-c.edu, Cuyahoga Comm Coll - Western Campus; Zg
Zaman, Musharraf, 405-325-2626 zaman@ou.edu, Univ of Oklahoma; Np
Zamani, Behzad, +98 (411) 339 2700 zamani@tabrizu.ac.ir, Univ of Tabriz; GtYs
Zambito, James J., (608) 363-2223 zambitoj@beloit.edu, Beloit Coll; PcGsPg
Zampieri, Dario, +390498279179 dario.zampieri@unipd.it, Univ degli Studi di Padova; Gc
Zamzow, Craig E., czamzow@clarion.edu, Clarion Univ; Gi
Zanazzi, Alessandro, (801) 863-5395 alessandro.zanazzi@uvu.edu, Utah Valley Univ; CgsPe
Zandt, George, (520) 621-2273 gzandt@email.arizona.edu, Univ of Arizona; Ys
Zanella, Elena, elena.zanella@unito.it, Univ di Torino; Ym
Zanetti, Kathleen, (702) 895-4789 Univ of Nevada, Las Vegas; Gg
Zaniewski, Kazimierz J., (920) 424-4105 zaniewsk@uwosh.edu, Univ of Wisconsin Oshkosh; Zn
Zappa, Christopher, zappa@ldeo.columbia.edu, Columbia Univ; Op
Zarillo, Gary , 321-674-7378 zarillo@fit.edu, Florida Inst of Tech; OupOn
Zarnetske, Jay, (517) 353-3249 jpz@msu.edu, Michigan State Univ; Hwg
Zarroca Hernández, Mario, ++935812033 mario.zarroca.hernandez@uab.cat, Univ Auto de Barcelona; YgNg
Zarzycki, Colin, czarzycki@psu.edu, Pennsylvania State Univ, Univ Park; As
Zaspel, Craig E., (406) 683-7366 craig.zaspel@umwestern.edu, Univ of Montana Western; Yg
Zattin, Massimiliano, +390498279186 massimiliano.zattin@unipd.it, Univ degli Studi di Padova; Gs
Zavada, Michael, (423) 946-1084 zavadam01@gmail.com, Univ of Texas Permian Basin; Ge
Zawadzki, Isztar, (514) 398-1034 isztar.zawadzki@mcgill.ca, McGill Univ; AsZr
Zayac, John M., (818) 710-2218 zayacjm@piercecollege.edu, Los Angeles Pierce Coll; Ggv
Zebker, Howard A., (650) 723-8067 zebker@stanford.edu, Stanford Univ; Yx
Zeebe, Richard E., (808) 956-6473 zeebe@soest.hawaii.edu, Univ of Hawai'i, Manoa; Ou
Zehnder, Joseph A., 402-280-2448 zehnder@creighton.edu, Creighton Univ; As
Zehr, Jonathan P., (831) 459-4009 zehrj@cats.ucsc.edu, Univ of California, Santa Cruz; Ob
Zeidouni, Mehdi, mehdi.zeidouni@beg.utexas.edu, Univ of Texas, Austin; Eo
Zeiger, Elaine, (312) 665-7627 Field Mus of Natural History; Zn
Zeigler, E. Lynn, 678-891-3767 ezeigler1@gsu.edu, Georgia State Univ, Perimeter Coll, Clarkston Campus; Gdg
Zeinijahromi, Abbas, abbas.zeinijahromi@adelaide.edu.au, Univ of Adelaide; Np
Zeitler, Peter K., (610) 758-3671 pkz0@lehigh.edu, Lehigh Univ; Cc
Zeitlhöfler, Matthias, 089/2180 6570 matthias@zeitlhoefler.de, Ludwig-Maximilians-Univ Muenchen; Gg
Zekri, Hamid, h.zekri@unsw.edu.au, Univ of New South Wales; Gz
Zelaya, Zach, zelayar@dnr.sc.gov, South Carolina Dept of Nat Res; Gg
Zelazny, Lucian W., (540) 231-9781 Virginia Tech; Sc
Zelenitsky, Darla, 403 220-8016 dkzeleni@ucalgary.ca, Univ of Calgary; Pg
Zelizer, Nora, 609 258-5809 nzelizer@princeton.edu, Princeton Univ; Zn
Zelt, Colin A., (713) 348-4757 czelt@rice.edu, Rice Univ; Ys
Zeman, Josef, +420 549 49 8295 jzeman@sci.muni.cz, Masaryk Univ; CgpCe
Zender, Charles, 949-824-2987 zender@uci.edu, Univ of California, Irvine; As
Zeng, Hongliu, 512-475-6382 hongliu.zeng@beg.utexas.edu, Univ of Texas, Austin; YsGs
Zeng, Ning, (301) 405-5377 zeng@umd.edu, Univ of Maryland; AsCgGl
Zentilli, Marcos, (902) 494-3873 Dalhousie Univ; Eg
Zentner, Nick, (509) 963-2828 nick@geology.cwu.edu, Central Washington Univ; Gg
Zerkle, Aubrey, +44 01334 464949 az29@st-andrews.ac.uk, Univ of St. Andrews; Co
Zhai, Xiaoming, 847 543-2504 Coll of Lake County; Gp
Zhan, F Benjamin, (512) 245-2170 Texas State Univ; Zi
Zhan, Hongbin, (979) 862-7961 zhan@geos.tamu.edu, Texas A&M Univ; Hw
Zhan, Zhongwen, 626-395-6906 zwzhan@caltech.edu,CalTech; Ys
Zhang, Bo, 205-348-4544 bzhang33@ua.edu, Univ of Alabama; YeGoEo
Zhang, Caiyun, 561-297-2648 czhang3@fau.edu, Florida Atlantic Univ; Zr
Zhang, Chi, 785-864-4974 chizhang@ku.edu, Univ of Kansas; YgxYu
Zhang, Chidong, (206) 526-6239 chidong.zhang@noaa.gov, Univ of Washington; AsmOp
Zhang, Guiming, 303-871-7908 guiming.zhang@du.edu, Univ of Denver; Zi

Zhang, Guorong, (785) 625-3425 gzhang@ksu.edu, Kansas State Univ; So
Zhang, Henian, (404) 894-1738 henian.zhang@eas.gatech.edu, Georgia Inst of Tech; As
Zhang, Huan, 860-405-9237 huan.zhang@uconn.edu, Univ of Connecticut; Ob
Zhang, Jin , 505-277-1607 Univ of New Mexico; Gyp
Zhang, Jin, (979)845-4959 jinzhang@tamu.edu, Texas A&M Univ; GytYx
Zhang, Jinhong, (520) 626-9656 jhzhang@email.arizona.edu, Univ of Arizona; Nm
Zhang, Lin, 361-825-2436 lin.zhang@tamucc.edu, Texas A&M Univ, Corpus Christi; Cs
Zhang, Max, 607-254-5402 kz33@cornell.edu, Cornell Univ; Zn
Zhang, Minghua , 530-752-4953 mhzhang@ucdavis.edu, Univ of California, Davis; Hg
Zhang, Minghua, 530-752-4953 mhzhang@ucdavis.edu, Univ of California, Davis; HgEg
Zhang, Ning, 937 376 6043 nzhang@centralstate.edu, Central State Univ; RwEo
Zhang, Pengfei, 212-650-5609 pzhang@ccny.cuny.edu, Graduate Sch of the City Univ of New York; Zg
Zhang, Qiaofeng (Robin), (270) 809-6760 qzhang@murraystate.edu, Murray State Univ; ZirZu
Zhang, Ren, (254)7102496 ren_zhang@baylor.edu, Baylor Univ; Cs
Zhang, Renyi, (979) 845-7671 renyi-zhang@geos.tamu.edu, Texas A&M Univ; As
Zhang, Rong, (609) 987-5061 rong.zhang@noaa.gov, Princeton Univ; Op
Zhang, Rui, 337-482-6920 Univ of Louisiana at Lafayette; Ye
Zhang, Tongwei, 512-232-1496 tongwei.zhang@beg.utexas.edu, Univ of Texas, Austin; Cgs
Zhang, Xi, (831) 502-8126 xiz@ucsc.edu, Univ of California, Santa Cruz; AsXy
Zhang, Xiaodong, (228) 688-3178 xiaodong.zhang@usm.edu, Univ of Southern Mississippi; Op
Zhang, Xinning, 609-258-7438 xinningz@princeton.edu, Princeton Univ; Cb
Zhang, Y. J., 804-684-7466 yjzhang@vims.edu, William & Mary; Opn
Zhang, Ye, 307-223-2292 yzhang9@uwyo.edu, Univ of Wyoming; HyGeYu
Zhang, Yige, yigezhang@fas.harvard.edu, Texas A&M Univ; Co
Zhang, Yong, (205) 348-3317 yzhang264@ua.edu, Univ of Alabama; Hqw
Zhang, You-Kuan, you-kuan-zhang@uiowa.edu, Univ of Iowa; Hw
Zhang, Youxue, (734) 763-0947 youxue@umich.edu, Univ of Michigan; Cu
Zhao, Bo, 541-737-3497 bzhao@coas.oregonstate.edu, Oregon State Univ; Zi
Zhao, Jingxia, 808-956-3736 jingxiaz@hawaii.edu, Univ of Hawai'i at Manoa; AmmAc
Zhao, Zeyu, zeyu@utexas.edu, Univ of Texas, Austin; Ys
Zhaohui, Li, (262) 595-2487 li@uwp.edu, Univ of Wisconsin, Parkside; ClGzHw
Zhdanov, Michael S., (801) 581-7750 michael.zhdanov@utah.edu, Univ of Utah; Ye
Zheng, Xinyuan, zhengxy@umn.edu, Univ of Minnesota, Twin Cities; CsOc
Zheng, Yan, (718) 997-3329 Queens Coll (CUNY); CgHwCm
Zhou, Hua-Wei, (713) 743-3440 hzhou@uh.edu, Univ of Houston; YseYg
Zhou, Wendy W., (303) 384-2181 wzhou@mines.edu, Colorado Sch of Mines; NgZir
Zhou, Xiaobing, (406) 496-4350 xzhou@mtech.edu, Montana Tech of the Univ of Montana; Zr
Zhou, Ying, 540-231-6521 yingz@vt.edu, Virginia Tech; Ys
Zhou, Yu, 845-437-5543 yuzhou@vassar.edu, Vassar Coll; Zun
Zhou, Yuyu, (515) 294-2842 yuyuzhou@iastate.edu, Iowa State Univ of Sci & Tech; Zyi
Zhu, Chen, (812) 856-1884 chenzhu@indiana.edu, Indiana Univ, Bloomington; CgHg
Zhu, Hejun, (972) 883-2406 hejun.zhu@utdallas.edu, Univ of Texas, Dallas; Ysx
Zhu, Junfeng, 859-257-1147 junfeng.zhu@uky.edu, Univ of Kentucky; Hw
Zhu, Lupei, (314) 977-3118 lupei.zhu@slu.edu, Saint Louis Univ; YsGt
Zhu, Ping, 305-348-7096 zhup@fiu.edu, Florida InterNatl Univ; As
Zhu, Qingzhi, (631) 632-8747 qing.zhu@stonybrook.edu, SUNY, Stony Brook; Oc
Zhu, Wenlu, (301) 405-1831 wzhu@umd.edu, Univ of Maryland; YrNrHg
Zhuang, Guangsheng, gzhuang@lsu.edu, Louisiana State Univ; CocGs
Zhuang, Qianlai, (765) 494-9610 qzhuang@purdue.edu, Purdue Univ; As
Ziegler, Brady A., bziegler@trinity.edu, Trinity Univ; HwCqGe
Ziegler, Karen, (505) 277-1437 kziegler@unm.edu, Univ of New Mexico; CsXc
Ziemba, Tim, 206-685-7437 ziemba@uw.edu, Univ of Washington; Xa
Ziemkiewicz, , Paul F., paul.ziemkiewicz@mail.wvu.edu, West Virginia Univ; Hg
Zierenberg, Robert A., (530) 752-1863 razierenberg@ucdavis.edu, Univ of California, Davis; Cs
Ziervogel, Kai, kai.ziervogel@unh.edu, Univ of New Hampshire; Gu
Zimba, Paul, 361-825-2768 paul.zimba@tamucc.edu, Texas A&M Univ, Corpus Christi; Zn
Zimbelman, James R., 202-633-2471 Smithsonian Inst, Natl Air & Space Mus; Xg
Zimmer, Brian W., 828-262-7517 zimmerbw@appstate.edu, Appalachian State Univ; Gg
Zimmer, Margaret, (831) 502-8714 margaret.zimmer@ucsc.edu, Univ of California, Santa Cruz; Hw
Zimmerer, Matthew, 575-835-5114 matthew.zimmerer@nmt.edu, New Mexico Tech; Gvg
Zimmerman, Andrew, 352-392-0070 azimmer@ufl.edu, Univ of Florida; ComSb
Zimmerman, Brian S., (814) 732-2529 bzimmerman@edinboro.edu, Edinboro Univ of Pennsylvania; Gzx
Zimmerman, Mark, 612-626-0572 zimme030@umn.edu, Univ of Minnesota, Twin Cities; Yx
Zimmerman, R. Eric, (630) 252-6816 Argonne; Ng
Zimmerman, Richard, (757) 683-5991 rzimmerm@odu.edu, Old Dominion Univ; Ob
Zimmerman, Robert W., +44 20 7594 7412 r.w.zimmerman@imperial.ac.uk, Imperial Coll; NrHwNp
Zimmerman, Ronald K., (225) 388-8302 ron@lgs.bri.lsu.edu, Louisiana State Univ; Go
Zimmerman, Jr., Jay, (618) 453-3351 zimmerman@geo.siu.edu, Southern Illinois Univ Carbondale; Gc
Zink, Albert, 089/2180 4340 Ludwig-Maximilians-Univ Muenchen; Gz
Zink, Paul, 303-271-3480 pzink@mines.edu, Colorado Sch of Mines; Zn
Zinsmeister, William J., wjzins@purdue.edu, Purdue Univ; Ps
Ziolkowski, Anton M., +44 (0) 131 650 8511 anton.ziolkowski@ed.ac.uk, Edinburgh Univ; YesYg
Ziolkowski, Lori A., 803-777-0035 lziolkowski@geol.dc.edu, Univ of South Carolina; CoOcCs
Zipper, Carl E., (540) 231-9782 czip@vt.edu, Virginia Tech; Ze
Zipser, Edward, 801-585-0467 ed.zipser@utah.edu, Univ of Utah; As
Ziramov, Sasha, +61 8 9266-4973 sasha.ziramov@curtin.edu.au, Curtin Univ; Ye
Zlotkin, Howard, 12012003161 hzlotkin@njcu.edu, New Jersey City Univ; Ze
Zlotnik, Vitaly A., (402) 472-2495 vzlotnik1@unl.edu, Univ of Nebraska, Lincoln; HwyHq
Zobaa, Mohamed, 432-552-2246 zobaa_m@utpb.edu, Univ of Texas Permian Basin; PlEoCo
Zoback, Mark D., (650) 725-9295 zoback@stanford.edu, Stanford Univ; Yg
Zodrow, Erwin L., erwin_zodrow@uccb.ca, Cape Breton Univ; Pi
Zoet, Lucas, lzoet@wisc.edu, Univ of Wisconsin, Madison; GllGm
Zois, Constantine S., (908) 737-3693 czois@kean.edu, Kean Univ; Am
Zollweg, James A., (585) 395-2352 jzollweg@brockport.edu, SUNY, The Coll at Brockport; Hs
Zolotov, Mikhail Y., (480) 965-4739 zolotov@asu.edu, Arizona State Univ; Cg
Zonneveld, John-Paul, 780-492-3287 zonneveld@ualberta.ca, Univ of Alberta; Go
Zotti, Maria, maria.zotti@flinders.edu.au, Flinders Univ; Zn
Zou, Haibo, (334) 844-4315 haibo.zou@auburn.edu, Auburn Univ; CcGiv
Zouhri, Lahcen, +33(0)3 44068976 lahcen.zouhri@lasalle-beauvais.fr, Institut Polytechnique LaSalle Beauvais (ex-IGAL); HwgGe
Zreda, Marek, 520-621-4072 marek@hwr.arizona.edu, Univ of Arizona; Cc
Zuber, Maria T., 617-253-6397 zuber@mit.edu, Massachusetts Inst of Tech; Xy
Zuend, Andreas, (514) 398-7420 andreas.zuend@mcgill.ca, McGill Univ; Asm
Zume, Joseph T., (717) 477-1548 jtzume@ship.edu, Shippensburg Univ; HgYug
Zunkel, Paul, pzunkel@emporia.edu, Emporia State Univ; AmZig
Zurawski, Ronald P., (615) 532-1502 ronald.zurawski@tn.gov, Tennessee Geo Survey; GgPzEg
Zurevinski, Shannon, (807) 343-8010 Ext. 8015 shannon.zurevinski@lakeheadu.ca, Lakehead Univ; Gz
Zurick, David, (859) 622-1427 david.zurick@eku.edu, Eastern Kentucky Univ; Zn
Zuza, Andrew , (775) 784-1446 azuza@unr.edu, Univ of Nevada, Reno; Gc
Zuza, Andrew V., (775) 682-8752 azuza@unr.edu, Univ of Nevada, Reno; Gct
Zwiefelhofer, Luke, 507-457-2778 lzwiefelhofer@winona.edu, Winona State Univ; Zn
Zyvoloski, George A., (505) 667-1581 Los Alamos; Ng

www.ingramcontent.com/pod-product-compliance
Lightning Source LLC
LaVergne TN
LVHW082015200926
848481LV00010B/2864
* 9 7 8 0 9 2 2 1 5 2 6 7 4 *